Indira's
Objective Agriculture

2nd Edition

(Multiple choice questions (MCQ) for competitive exams of Agriculture)

(ICAR-ARS/NET, JRF, SRF, IAS, IFS, PCS, SLETS, CSIR/UGC-NET, UPSC-CS, FCI & Ph.D.)

About the Author

Dr. R.L. Arya born on 3 March 1964 at Village Jagdish Pur (Bhekhepur), Auraiya - Etawah (UP), completed his **B.Sc.** (Ag & A.H.) in 1984 from Chandra Shekhar Azad University of Agriculture and Technology, Kanpur, **M.Sc.** and Ph.D. in Agronomy from Indian Agricultural Research Institute, Pusa, New Delhi in 1986 and 1989, respectively. He served as a scientist at CRIDA, Hyderabad, IGFRI, Jhansi, IIPR, Kanpur, CTRI, Rajahmundry, CTRI Research Station, Jeelugumilli (West Godavari dist., A.P.) and CTRI Research Station, Dinhata (Cooch Behar dist, W.B.). He has good number of publication to his credit about 250 research papers/ popular articles/papers presented in seminar symposia etc/book chapters in various national and international journals/megazines. Dr. Arya has vast experience (more than 20 years) in the field of dryland/rainfed agriculture, nutrient and water management, cropping system in pulses and pulse based cropping system, nutrient management and cropping system research in Jati and Motihari tobacco in West Bengal. He has also released a variety "Torsa" – an early maturing type of Motihari tobacco. He has authored several books such as Principles of Crop Production and Crops (2 Volume), Modern techniques of tobacco cultivation, Pulse production, Irrigation management, Weed management, Tobacco production in India, Indira Krishi Pratiyogita Darpan, Indira Competition Explorer, Indira Agricultural Competition Explorer, Indira's Objective in Agricultural Biotechnology, Indira's Objective in Agricultural Extension, Indira's Objective in Agronomy and Indira's Objective in Agriculture, Fundamentals of Agronomy and Soil and Plant Nutrition. He was also selected for Dr. Rajendra Prasad Puruskar by Indian Council of Agricultural Research, New Delhi and received best book award by Indian Society of Weed Management from NRC Weed Science, Jabalpur.

Dr. Sonam Arya has completed her graduation and post graduation in Biotechnology from College of Biotechnology in Chhatrapati Sahu Ji Maharaj University, Kanpur and Babasaheb Bhem Rao Ambedkar University (Central university), Lucknow, respectively. She obtained M.Phill degree in Life science from CSJM, Kanpur. She obtained Ph.D. degree in Agricultural Biotechnology in Department of Agricultural Biotechnology, Sardar Ballabh Bhai Patel University of Agriculture and Technology, Modipuram, Meerut. She has published many research papers/popular articles in national and international journal of repute. She has also authored several books entitled Indira Agricultural competition Explorer, Indira Competition Explorer, Indira's Objective in Agricultural Biotechnology and Indira's Objective in Agriculture, Home Science and Fundaments of Horticulture. She is awarded Rajiv Gandhi Fellowship from UGC for doing Ph.D.

Dr. Renu Arya has completed her graduation and post graduation in Home science with specialization in extension education and communication [illegible] ture and Technology, Kanpur. She was also [illegible] agement from Chandra Shekhar Azad University of Agricul- education and communication management in Pu[illegible] doctoral progra[illegible] [illegible] working as a Subject Matter Specialist in Home Scien[illegible] [illegible]ltural University, [illegible] Narendra Deo University of Agriculture and Technology, Kri[illegible]hi Vigyan Kendra, Be[illegible] research papers/popular articles in national and international journa[illegible] Faizabad. She has publi[illegible] books entitled Indira Krishi Pratiyogita Darpan in Hindi and Indira Competi[illegible]te. She has also authored [illegible] in Agricultural Extension and Indira's Objective in Agriculture, Home Science and F[illegible] Explorer, Indira's Objective [illegible]aments of Horticul- ture. She was awarded Rajiv Gandhi Fellowship from UGC for doing Ph.D.

Indira's

Objective Agriculture

2nd Edition

(Multiple choice questions (MCQ) for competitive exams of Agriculture)

(ICAR-ARS/NET, JRF, SRF, IAS, IFS, PCS, SLETS, CSIR/UGC-NET, UPSC-CS, FCI & Ph.D.)

Dr. R.L. Arya
(M.Sc. & Ph.D. in Agronomy, IARI, New Delhi)
Ex-Senior Scientist (Agronomy) And District
Consultant, National Food Security Mission, Jalaun (UP)

Renu Arya
Subject Matter Specialist,
Krishi Vigyan Kendra, Behraich (UP)

Sonam Arya
Lecturer (Biotech)
College of Applied Education
and Health Science, Marrut (UP)

SCIENTIFIC
PUBLISHERS

Published by:

SCIENTIFIC PUBLISHERS (INDIA)
5A, New Pali Road, P.O. Box 91
Jodhpur 342 001, India
E-mail: info@scientificpub.com
Website: www.scientificpub.com

ISBN : 978-93-90495-59-7
eISBN : 978-93-90495-68-9

Printed in India

Preface

Agriculture is a most important source for economic assistance and all round progress of our country, where more than 70 per cent of Indian population is dependent on agriculture.. India, with only 2.3 per cent land area, 4 per cent fresh water ad 1 per cent forest cover has to support about 17 per cent world's human population (over 1.18 billion) and 15 per cent livestock (479 million). Science based agriculture adopted for the past four decades in the country has brought appreciable change in the food production scenario of the country. Over the years, a six fold increase in food grain production has transformed food scarcity of early fifties into food sufficiency with ever highest production figure of 335.88 million tones. During 2019-20, the production of total food was rose upto 297 million tones but there is still need to increase the total food grain production for ever increasing human population in the country. The main issues of agricultural growth and development in India which needs scientific redressal are: food self–sufficiency versus food self–reliance; low productivity and declining rate of growth of factor productivity; low and unstable, yields especially in rainfed areas and increasing ecological and environmental insecurity leading to resource wars with stagnating the food production.

The competitive examinations for Agriculture have become a routine procedure of recruitment and admissions for higher positions and education. Now a days a large number of questions appear in the examination. These questions call for quick answering for success within a specified period of time. A sincere effort has been made by the authors to present them in most easy, short and understandable language for the benefit of scientists, students, farmers and those who are interested in the field of the agriculture.

"Indira's Objective Agriculture" for competitive exams in agriculture discipline contain 22 chapters covering all related discipline.The chapters included such as: General agriculture, Agricultural climatology, Genetics and plant breeding, Agricultural biotechnology, Plant physiology, Plant biochemistry, Agricultural microbiology, Seed science, Agronomy, Soil science, Entomology, Plant pathology, Horticulture, Agricultural extension, Agricultural economics, Animal husbandry and dairying, Agricultural statistics, Research methodology, Agroforestry, Fish culture/pisicluture and appendix have been given due importance and whole syllabus was covered as per ICAR syllabus and guidlines. Each chapters contains multiple choice questions and total about thousand objective questions with multiple choice have been framed and arranged sequentially for the easy understanding of the students. Recent information and development in the field of agriculture have been incorporated in the book. Thus this book is based on the syllabus of student of agricultural stream, it may be useful not only to students but also teachers, researchers, extension workers and development officers for reference and easy answering of many complicated questions.

The chapters are chosen in view to cover the course contents of competitive examinations like IAS, IFS, ARS, PCS, Banking services, states and national levels of diffirent competition in agricultural subjects. The entire book is prepared in most simple, clear and talking language so that the contents could be easily undertand by the readers. Hence this book can serve as a single plateform for preparation of different competitive examinations in agriculture.

Finally, I have belief and hope that this book will fulfill the requirement of students of agricultural stream for appearing in different competitive examinations.

Being first edition, there could be some shortcomings for which valuable and informative suggestions will be gratefully appreciated, so that these shortcoming may be incorporated in forthcoming edition for the benefit of readers.

Authors shall be grateful to the various books, periodicals, magazines, newsletter, news papers, journals etc., rendered help for bringing out of this books. Authors are also hightly grateful to M/S Scientific Publishers, Jodhpur and their staff members for printing in the form of book. The publisher shall also be grateful for bringing to their notice of any suggestions for further improvement.

Authors

R. L. Arya

Renu Arya

Sonam Arya

CONTENT

Chapter 1

GENERAL AGRICULTURE

1. The practice of raising plants and animals is called:
(A) Agriculture (B) Horticulture
(C) Agronomy (D) Silviculture

2. Agriculture word derived from:
(A) Latin (B) French
(C) Greek (D) English

3. Agriculture denotes:
(A) Agri + cultura (B) Ager + cultura
(C) Agri + culture (D) Agris + cultura

4. Agriculture is the branch of:
(A) Social science (B) Basic science
(C) Applied science (D) None of these

5. Agriculture originated about:
(A) 2300-3000 years back
(B) 20000-30000 years back
(C) 7000-13000 years back
(D) 90000-100000 years back

6. Kung tribe is:
(A) Primitive agriculturist
(B) Hunter-gatherer
(C) Advanced agriculturist
(D) Completely dependent on horticulture

7. In various parts of the world, agriculture:
(A) Came from Asia Minor
(B) Has spread from Yellow River Valley
(C) Originated independently
(D) Has developed to various degree

8. Fire helped in the cultivation of:
(A) Fruit plants (B) Fibre Plants
(C) Drug yielding plants(D) Grain yielding plants

9. Early human civilization was based on:
(A) Development of fibre crops (B) Fire
(C) Agriculture
(D) Fire and Agriculture

10. A nonagricultural society living today is:
(A) Kung tribe (B) Urban Society
(C) Hintza tribe (D) Kuntizha tribe

11. Early man preferred plants:
(A) Grain yielding (B) Fruit yielding
(C) Multipurpose (D) Fibre yielding

12. Cabbage originated in:
(A) Mexico and Central America
(B) Mediterranean
(C) Peruvian Andes
(D) South west Asia

13. Maize evolved in:
(A) Brazil (B) Mexico and Central America
(C) USA (D) Asia minor and Afganistan

14. The centre of origin of Almond and Apple is:
(A) South West Asia
(B) South East Asia
(C) China
(D) Asia minor and Afganistan

15, The origin home of rice is:
(A) South West Asia (B) South East Asia
(B) Mediterranean (D) China

16. Peruvian Andes are believed to be centre of origin of:
(A) Tomato
(B) Potato
(C) Tomato and Potato
(D) Tomato, Potato and Chillies

17. The origin of sunflower is believed to be in:
(A) Mexico and Central America
(B) Brazil
(C) USA
(D) Peruvian Andes

18. South East Asia is thought to be centre for the origin of:
(A) Rice
(B) Rice and Sugarcane
(C) Rice, Sugarcane and Mango
(D) Rice, Sugarcane, Mango and Banana

19. The famous cultivated plant which developed in China is:
(A) Tea (B) Coffee (C) Cotton (D) Orange

20. Production centre of Hevea Rubber is:
(A) Malayasia (B) Brazil
(C) Peruvian Andes (D) South West Asia

21. The centre of origin of Wheat is:
(A) South West Asia
(B) South East Asia
(C) Asia minor and Afganistan
(D) Mexico

22. Which is not a new world crop:
(A) Tapioca (B) Peanut
(C) Cashewnut (D) Poppy

23. Which one belong to old world:
(A) Squash (B) Pumpkin
(C) Water Melon (D) Guava

24. Select a crop which belongs to new world:
(A) Jackfruit (B) Papaya
(C) Spinach (D) Pear

25. Which crop belong to new world:
(A) Red Pepper (B) Black pepper
(C) Onion (D) Garlic

26. Which is the crop which originated in Peru and Brazil but is now mainly grown in India is:
(A) Groundnut (B) Potato
(C) Maize (D) None of these

27. The crop which originated in Peru but is most extensively grown in East Europe is:
(A) Coffee (B) Potato
(C) Rubber (D) Maize

28. Name the crop which had its birthplace in Tropical America but the centre of production is now Midwest USA:
(A) Pineapple (B) Oil palm
(C) Cocoa (D) Maize

29. Christopher Columbus landed in island of Bahmas (Carribean sea) in:
(A) 1692 (B) 1592 (C) 1492 (D) 1792

30. Tropical forests are being lost at the annual rate of:
(A) 11 million ha/yr (B) 110 million ha/yr
(C) 1 million ha/yr (D) 1100 million ha/yr

31. Post Vavilov megadiversity centre is:
(A) Australia (B) Siberia
(C) Andamana (D) Tibet

32. The number of centres of origin of crop plants recognized by Vavilove is:
(A) 8 (B) 11 (C) 7 (D) 21

33. Ethopia is the origin home of:
(A) Coffee (B) Maize (C) Rice (D) Cabbage

34. Potato famine of Ireland occurred in:
(A) 1645 (B) 1745 (C) 1845 (D) 1854

35. Potato famine of Ireland was caused by the attack of fungus
(A) *Phytophthora palmivora*
(B) *Plasmopora viticola*
(C) *Puccinia graminis*
(D) *Phytophthora infestans*

36. Natural resistance to late blight of potato is present in:
(A) *Solanum tuberosum* (B) *Solanum acaule*
(C) *S. demissum* (D) *S stoloniferum*

37. Gene bank are part of:
(A) Ex situ conservation
(B) In situ conservation
(C) Both ex situ and in situ conservation
(D) Tribal diet

38. *In situ* conservation are carried out through:
(A) Natural parks (B) Wildlife santuaries
(C) Biosphere reserves (D) All of these

39. *Ex situ* conservation is carried out through:
(A) Funneling the threatened species into trade
(B) Offsite collection
(C) Gene banks
(D) All of these

40. Gene banks comprise:
(A) Seed banks, orchards, tissue culture and cryopreservation
(B) Seed banks, tissue culture and cryopreservation
(C) Seed banks and cryopreservation
(D) Tissue culture and cryopreservation

41. Orthodox seeds can be preserved for long periods at a temperature of:
(A) -196 °C (B) -10 to -20 °C
(C) 0 – 4° C (D) 10 to 15 °C

42. Which one will be scarched for obtaining maximum genetic diversity:
(A) Recent varieties (B) Old varieties
(C) Wild relatives (D) All of these

43. Paraguay Tea is got from:
(A) Leaves of *Ilex*
(B) Seeds of *Ilex*
(C) Leaves of *Paullinia*
(D) Paste of crushed kernels of *Paullinia*

44. Arabian Tea is:
(A) Leaves and buds of *Paullinia*
(B) Seed of Ilex
(C) Leaves and buds of *Catha edulis*
(D) Seeds of Catha

45. Cryopreservation is carried out at:
(A) -10 to -20 °C (B) -50 to -60 °C
(C) -196 °C (D) -100 ° C

46. Co-varieties of sugarcane obtained red rot resistance from:
(A) *Saccharum munja* (B) *S. spontaneum*
(C) *S. arundinacearum* (D) *S. edula*

47. A calcitrant seed will be killed in:
(A) Low temperature (B) Dry air
(C) Further dehydration (D) All of these

48. A gift from new world to old world is:
(A) Barley (B) Rice (C) Wheat (D) Maize

49. Which plant was introduced from old to new world:
(A) Potato (B) Rice (C) Corn (D) Sweet potato

50. Dwarf wheat varieties brought into India from Mexico were:
(A) Sonora-64 and Sonalika
(B) Sharbati Sonora and Pusa Lerma
(C) Sonalika and Sharbati Sonora
(D) Sonora-64 and Lerma Rojo-64

51. Vavilov centres of origin of crop plants are located in:
(A) Tropical plains
(B) Mountains of tropical areas
(C) Mountains of temperate areas
(D) Both (B) and (C)

52. An under utilized oil crop is:
(A) *Brassica campestris* (B) *Cocos nucifera*
(C) *Helianthus annus* (D) *Azadiracta indica*

53. Liquid wax is got from:
(A) Jajoba (B) Winged Bean
(C) Guayule (D) Leucaena

54. Wheat varieties collected by Vavilov were:
(A) 26000 (B) 2600 (C) 4600 (D) 260

55. Hevea Rubber belongs to:
(A) Malayasia (B) Peruvial Andes
(C) Brazil (D) South east Asia

56. Germplasm is contained in:
(A) Somatic cells (B) Growing points
(C) Reproductive cells (D) Both (A) and (C)

57. Which one will be best for germplasm collection:
(A) Britain (B) France
(C) Portugal (D) Peru

58. Dwarf wheat were evolved by:
(A) Swaminathan (B) Vavilov
(C) Khush (D) Borlaug

59. In a gene bank, genetic material is stored as:
(A) Genes
(B) Seeds
(C) Vegetative material/tissue culture
(D) Both (B) and (C)

60. Genetic erosion is due to:
(A) Deforestation
(B) Shifting cultivation
(C) Adoption of genetically uniform varieties
(D) All of these

61. Exchange of germplasm is carried mostly through shoot tips because it is:
(A) Cheap (B) Small and handy
(C) Virus free (D) Genetically stable

62. Tomato and potato belong to:
(A) North America (B) South America
(C) Asia minor (D) Ethiopia

63. A plant introduced from new world to old world is:
(A) Rice (B) Wheat
(C) Sugarcane (D) Potato

64. National Botanical Research Institute is located at:
(A) Dehradun (B) Delhi
(C) Lucknow (D) Mumbai

65. In India, the largest herbarium is located at:
(A) Dehradun (B) Delhi
(C) Lucknow (D) Calcutta

66. Which of the following countries is the centre of origin of rice:
(A) India (B) China
(C) Indo-Malayan region (D) Russia

67. Central Sugarcane breeding Research Institute is situated at:
(A) Coimbatore (B) Lucknow
(C) Delhi (D) Bhopal

68. A new world spice that has become an essential part of Indian cuisine is:
(A) Red pepper (B) Black Pepper
(C) Cardamom (D) Ginger

69. A social forest tree is:
(A) Teak (B) Sandal wood
(C) Subabul (D) Rosewood

70. A plant species threatened with extinction can be preserved in:
(A) Herbarium (B) National park
(C) Botanical Garden (D) Gene Bank

71. A living fossil is:
(A) Gnetum (B) Ginkgo
(C) Riccia (D) Pinus

72. Jojoba is:
(A) Crotolaria (B) Leucaena
(C) Simmondsia (D) Parthenium

73. A place where endangered genetic materials of plants are kept is:
(A) Gene library (B) Genebank
(C) Gene pool (D) Herbarium

74. Wild varieties of plants must be conserved to:
(A) Feeding wild animals
(B) Maintain ecosystem
(C) Future evolution
(D) Incorporate useful traits in future crop varieties

75. The energy required daily by hunter-gatherer humans was:
(A) 2000-4000 kcal (B) 6000-10000 kcal
(C) 10000-70000 kcal (D) 10000-12000 kcal

76. Per capita daily energy consumption in nineteenth century industrialized society was:
(A) 6000 kcal (B) 10000 kcal
(C) 30000 kcal (D) 70000 kcal

77. Per capita daily energy consumption of USA is:
(A) 10000 kcal (B) 30000 kcal
(C) 70000 kcal (D) 250000 kcal

78. Which is the most important crop in the world:
(A) Maize (B) Sorghum (C) Barley (D) Wheat

79. The great popularity of wheat is on accounts of:
(A) Well balanced carbohydrates and proteins
(B) Inability for complete mechanized production
(C) Short storage period of grain and flour storage
(D) Limited to narrow range of climate

80. According to Vavilov, the common bread wheat originated in :
(A) The Caucasus-Turkey-Iraq area
(B) Temperate zone of America
(C) Naturally fertile regions of France
(D) Cool areas of China

81. *Acacia senegal* and *Ahlagi camelorum* are the xerophytes belonging to:
(A) Psammophytes (B) Psychrophytes
(C) Lithophytes (D) Halophytes

82. *Ceina parviflora* is a xerophyte that has the succulent:
(A) Stems (B) Leaves
(C) Roots (D) Fruits

83. "The dominant species with maximum canopy coverage of two or more species" represents a community character known as:
(A) Society (B) Location
(C) Consociation (D) Association

84. In —— the buds (life forms) are invariably buried in the soil or the substratum as the bulbs and rhizomes:
(A) Chamaephytes (B) Crypto/geophytes
(C) Hemicryptophytes (D) Therophytes

85. Collective farming system has originated from:
(A) Bulgaria (B) Poland
(C) Soviet Union (D) France

86. Commercial plantation farming comprises an:
(A) Import-oriented system
(B) Export-oriented system
(C) Indigenous use-oriented system
(D) None of these

87. Agricultural growth is very soon likely to be hampered by scarcity of:
(A) Natural resources
(B) Man-made resources
(C) International cooperation
(D) Local coordination

88. A set of shared, similar characteristics (as applied to logically deduced and symbolically expressed models and observations) may be called as:
(A) Negative isomorphism (B) Positive isomorphism
(C) Neutral isomorphism (D) None of these

89. Rubber has probably the —— range of tolerance for temperature:
(A) Narrow west (B) Widest
(C) Medium (D) None of these

90. The requirement and tolerance of a species may be considered as:
(A) Supplementary (B) Complementary
(C) Antagonistic (D) None of these

91. The cactus from North America and the jack rabbit from Europe were introduced into:
(A) Holland (B) Poland
(C) Austria (D) Australia

92. When competition effect is small for both the plant species so that the interspecific effects are —— the intraspecific effects, both species may be able to live together:
(A) More than (B) Equal to
(C) Less than (D) None of these

93. The self-crowding effect represent the —— competition:
(A) Interspecific (B) Intraspecific
(C) Intergeneric (D) Intrageneric

94. The interplant competition drastically affects the —— of the plant:
(A) Size and Shape (B) Number and Size
(C) Number and Shape (D) None of these

95. Banana (*Musa sapientum*) plant is an herbaceous —— which may grow to a height of about 30 feet:
(A) Annual (B) Biennial
(C) Perennial (D) None of these

96. —— has been the world's leading producer of cane sugar:
(A) India (B) Indonesia
(C) Hawaii (D) Cuba

97. Cotton has been grown in India as a crop of fibre for perhaps —— years and was known in Peru as early as 1000 BC:
(A) 3000 (B) 4000 (C) 5000 (D) 6000

98. When one batch species react with the environment to facilitate the entry of a new set of species, such a model is known as:
(A) Inhibition model (B) Tolerance model
(C) Facilitation model (D) Elucidation model

99. The centre of origin of wheat is:
(A) South western Asia (B) South-eastern Asia
(C) India (D) China

100. High yielding dwarf varieties of wheat were developed by:
(A) Dr. B.P. Pal (B) Dr. Norman E. Borlaug
(C) Dr. E.W. Burton (D) Dr. M.S. Swaminathan

101. The maximum area under wheat is in:
(A) USSR (east while) (B) India
(C) USA (D) China

102. The production of wheat is the highest in:
(A) USSR (eastwhile) (B) India
(C) China (D) USA

103. The *Triticum aestivum* is a wheat:
(A) Diploid (B) Tetraploid
(C) Hexaploid (D) Monoploid

104. Maize protein is deficient in:
(A) Betadine (B) Tryptophane and Lysine
(C) Lysine (D) Tryptophane

105. The country ranking first in respect of area under maize (corn) is:
(A) India (B) China (C) Brazil (D) USA

106. The largest maize producing country in the world is:
(A) USA (B) Mexico (C) India (D) Canada

107. The dent corn is also known as:
(A) *Zea mays identata* (B) *Zea mays indurata*
(C) *Zea mays tunicata* (D) *Zea mays everta*

108. All India Co-ordinated Maize Improvement Project was started from:
(A) 1957 (B) 1955 (C) 1965 (D) 1967

109. The place of origin of urdbean is:
(A) India (B) Burma (C) Bangladesh (D) Ceylon

110. The scientific name of early maturing bold and black seeded urdbean varieties is:
(A) *Vigna mungo var viridis*
(B) *Vigna mungo var niger*
(C) *Vigna mungo var macropode*
(D) *Vigna mungo var micropode*

111. Soybean besides good quality protein also contains about—— per cent of oil:
(A) 10 (B) 20 (C) 30 (D) 40

112. The protein content of soybean is about:
(A) 40 % (B) 20% (C) 30% (D) 10%

113. The grain legume which is also grown as an oilseed is:
(A) Chickpea (B) Urdbean
(C) Lentil (D) Soybean

114. The ———— is considered to be a progenitor of soybean:
(A) *Glycine max* (B) *Glycine sinensis*
(C) *Glycine ussuriensis* (D) *Glycine usitattisimum*

115. Among the oilseed crops, groundnuts ranks ——— in India:
(A) First (B) Second
(C) Third (D) Fourth

116. Groundnut seed normally contains per cent protein and oil:
(A) 25 and 45 (B) 22 and 43
(C) 45 and 26 (D) 43 and 22

117. Quadricarpellatum capsule of sesamum contains how many carpels:
(A) Two (B) Three
(C) Four (D) Six

118. Sesamum belongs to the family of:
(A) Leguminoceae (B) Papilionaceae
(C) Pedaliaceae (D) Chinipideaceae

119. In south India sesamum is grown in:
(A) Rabi season only (B) Kharif season only
(C) Rabi and kharif season (D) Summer season only

120. Castor belongs to family:
(A) Compositeae (B) Euphorbiaceae
(C) Leguminaceae (D) Gramineae

121. The largest producer of rapeseed and mustard in the world is:
(A) China (B) Japan
(C) India (D) Canada

122. Rapeseed and mustard belongs to the family:
(A) Leguminaceae (B) Compositeae
(C) Euphorbiaceae (D) Crucifereae

123. The 2n chromosome number in Rugosa is:
(A) 20 (B) 24 (C) 32 (D) 36

124. The fruit of rapeseed and mustard is known as:
(A) Pod (B) Silique (C) Grain (D) Caryopsis

125. The scientific name of erect type of ground nut is:
(A) Arachis hypogeae subspecies *fastigiata*
(B) Arachis hypogeae subspecies *procumbens*
(C) Arachis hypogeae subspecies *oleifera*
(D) Arachis hypogeae subspecies *gigantean*

126. The gynophore of groundnut is commonly referred to as the:
(A) Perinth (B) Peg (C) Ovate (D) Monadelph

127. Linseed can be used as a crop:
(A) Oilseed (B) Fibre
(C) Oilseed and fibre (D) None of these

128. The protein content in linseed cake has been estimated at:
(A) 12% (B) 24% (C) 36% (D) 48%

129. Linseed crop belongs to the family:
(A) Compositeae (B) Euphorbiaceae
(C) Leguminoceae (D) Linaceae

130. The fruit of linseed is known as:
(A) Pod (B) Canum
(C) Seedball (D) Dutum

131. Sunflower as an oilseed crop was introduced into India in:
(A) 1965 (B) 1969 (C) 1972 (D) 1985

132. Sunflower is a :
(A) Self pollinated crop
(B) Cross-pollinated crop
(C) Self and cross pollinated crop
(D) Parthenocarpous

133. Cotton belongs to the family of:
(A) Malvaceae (B) Compositeae
(C) Crucifereae (D) Lineaceae

134. The number of known species in genus Gossypium have been estimated at:
(A) 10 (B) 15 (C) 20 (D) 25

135. Jute belongs to the family of:
(A) Malvaceae (B) Compositeae
(C) Lineaceae (D) Tiliaceae

136. Tobacco has been classified as a crop:
(A) Cash (B) Catch (C) Inter (D) Relay

137. *Nicotiana rustica* has been derived as an amphidiploid from a cross between:
(A) *Nicotiana tabacum* x *Nicotiana paniculata*
(B) *Nicotiana paniculata* x *Nicotiana sylvestris*
(C) *Nicotiana undulata* x *Nicotiana sylvestris*
(D) *Nicotiana undulata* x *Nicotiana paniculata*

138. Tobacco belongs to the family:
(A) Compositeae (B) Solanaceae
(C) Crucifereae (D) Leguminaceae

139. Nicotine content of the leaves of *Nicotiana rustica* ranges between:
(A) 2.0 – 5.5 (B) 3.0-4.5
(C) 3.5-8.0 (D) 5.0 – 10.5

140. Nicotine content of the leaves of *Nicotiana tabacum* ranges between:
(A) 2.0 – 5.5 (B) 3.0-4.5
(C) 3.5-8.0 (D) 5.0 – 10.5

141. Out of the following — is commonly cultivated potato:
(A) *Solanum demissum* (B) *Solanum stenotonum*
(C) *Solanum tuberosum* (D) *Slanum andigenum*

142. The family of sugarbeet is a:
(A) Chenopodiaceae (B) Crucifereae
(C) Euphorbiaceae (D) Compositeae

143. A tuber represents:
(A) Enlarged underground root
(B) Enlarged underground stem
(C) Enlarged underground nodule
(D) None of these

144. Potato belongs to the family:
(A) Solanaceae (B) Compositeae
(C) Leguminaceae (D) Euphorbiaceae

145. Department of Revenue, agriculture and Commerce was set up in India on:
(A) April 17, 1871 (B) April 27, 1871
(C) April 27, 1872 (D) April 17, 1872

146. Agriculture colleges were opened in India first of all in year:
(A) 1905 (B) 1900 (C) 1907 (D) 1911

147. The Department of Revenue, Agriculture and Commerce was set up during the reign of:
(A) Lord Curzon (B) Lord Mayo
(C) Lord Wellington (D) None of these

148. Department of agriculture in provinces were set up in:
(A) 1871 (B) 1881 (C) 1891 (D) 1911

149. First Inspector General of Agriculture was appointed in year:
(A) 1901 (B) 1902 (C) 1900 (D) 1907

150. Imperial Agricultural Research Institute was set up in Pusa, Bihar in year:
(A) 1905 (B) 1901 (C) 1871 (D) 1872

151. IARI was shifted to New Delhi in year:
(A) 1925 (B) 1926 (C) 1936 (D) 1947

152. Imperial Council of Agricultural Research was set up in a year of:
(A) 1901 (B) 1926 (C) 1929 (D) 1936

153. Who was the first president of ICAR:
(A) Dr. B.P. Pal
(B) Khan Bahadur Sir Mod. Habibullah
(C) Dr. M.S. Swaminathan (D) Dr. Panjab Singh

154. Indian Council of Agricultural Research came into existence in:
(A) 1929 (B) 1936 (C) 1946 (D) 1947

155. Total net sown area of India is:
(A) 120 mha (B) 142 mha
(C) 180 mha (D) 95 mha

156. Total cropped area of India is:
(A) 120 mha (B) 142 mha
(C) 180 mha (D) 95 mha

157. The headquarter of National Bureau of Soil Survey and Land Use Planning is situated at:
(A) Nagpur (B) Pune
(C) New Delhi (D) Hyderabad

158. The headquarter of National Bureau of Plant Genetic Resources is situated at:
(A) Nagpur (B) Pune
(C) New Delhi (D) Hyderabad

159. Match the name of ICAR institute and their headquarters:

Institutes	Headquarter
1. IGFRI	i. Jhansi
2. CAZRI	ii. Cuttack
3. CRRI	iii. Jodhpur
4. IISR	iv. Lucknow

Which of the match is correct?

(A)	1. – i	2. – ii	3. – iii	4. - iv
(B)	1. – i	2. – ii	3. – iii	4. – iv
(C)	1. – i	2. – iii	3. – ii	4. – iv
(D)	1. – i	2. – iv	3. – iii	4. – ii

160. IIHR stands for:
(A) Indian Institute of Horticultural Research
(B) Indian Inland Horticulture Research
(C) Institute of Inland Horticulture Research
(D) None of these

161. Which one of the following is matched correctly
(A) CIAE – Bhopal (B) CSWCTI – Jodhpur
(C) CTCRI – Dehradun (D) NAARM – New Delhi

162. CRIDA is situated at:
(A) Hyderabad (B) New Delhi
(C) Bangalore (D) Lucknow

163. The ICAR institutes in 'A' and its headquarters in 'B' are given below:

A	B
1. CTRI	i.Trivendrum
2. CTCRI	ii. Rajahmundry
3. IASRI	iii. Nagpur
4. CICR	iv. New Delhi

Which of the match is correct?

(A)	1. – i	2. – ii	3. – iii	4. - iv
(B)	1. – i	2. – iv	3. – iii	4. – ii
(C)	1. – ii	2. – i	3. – iv	4. – iii
(D)	1. – iii	2. – iv	3. – i	4. – ii

164. CIMMYT is working mainly for:
(A) Wheat (B) Maize and Wheat
(C) Rice (D) Cotton

165. The name of International Research Institutes and its head quarters are given below:

Institutes	Head quarters
1. ICRISAT	Hyderabad
2. IRRI	Philippines
3.ICARDA	Syria
4. IBPGR	Rome

Which of the above pairs is matched correctly?
(A) 1 (B) 3
(C) 4 (D) All of these

166. Zein is known as:
(A) Rice protein (B) Wheat protein
(C) Maize protein (D) Millet protein

167. Maize protein is deficient in:
(A) Tryptophane (B) Lysine
(C) Both (A) and (B) (D) None of these

168. Origin of maize is:
(A) Mexico (B) India
(C) Africa (D) China

169. Pulses are rich source of:
(A) Fat (B) Carbohydrate
(C) Protein (D) Vitamin

170. Who introduce tobacco cultivation in India?
(A) ITC (B) Portugeese merchants
(C) Chinese (D) Persians

171. Central Tobacco Research Institute is situated in:
(A) UP (B) MP
(C) AP (D) Karnataka

172. Removal of terminal bud in tobacco is called:
(A) Topping (B) Desuckering
(C) Nipping (D) Curing

173. Jute (*Corchorus spp*) belongs to the family:
(A) Malvaceae (B) Tiliaceae
(C) Pedaliaceae (D) Solanaceae

174. Jute Agricultural Research Institute is situated in:
(A) Karnataka (B) Andhra Pradesh
(C) Tamilnadu (D) WB

175. The process by which the fibre in the bark get loosened and separated from the woody stalk is called as:
(A) Retting (B) Extraction
(C) Both (A) and (B) (D) None of these

176. Wheat protein is called as:
(A) Zein (B) Gluten
(C) Dhurin (D) Ricin

177. Which of the following states has the highest area of groundnut?
(A) Maharastra (B) Karnataka
(C) Punjab (D) Gujarat

178. The rice variety containing 'Dee-geo-woo-gen' gene is:
(A) Indrasan (B) Basmoti
(C) Tilak (D) IR 8

179. The origin place of soybean is:
(A) India (B) China (C) America (D) Brazil

180. Soybean is a:
(A) Day – neutral plant (B) Long day plant
(C) Short day plant (D) None of these

181. Origin of berseem crop is:
(A) USSR (B) Egypt
(C) India (D) China

182. *Phalaris minor* is also known as:
(A) Barmuda grass (B) Canery grass
(C) Water Hyacinth (D) Kasni

183. How much sugar can be prepared from 1000 quintals of sugar cane?
(A) 125 q (B) 100 q (C) 80 q (D) 50 q

184. The crop in which maximum pesticides are used?
(A) Tomato (B) Potato (C) Sugarcane (D) Cotton

185. Central Rice Research Institute is situated at:
(A) Cuttuck (B) Karnal
(C) Mumbai (D) Manila

186. One acre is equivalent to:
(A) 1.8 ha (B) 0.652 ha
(C) 0.404 ha (D) 0.038 ha

187. Origin of urd bean is:
(A) Burma (B) Pakistan
(C) India (D) China

188. Yellow revolution indicates the production of:
(A) Oilseeds (B) Sugar crops
(C) Fruits production (D) Pulses production

189. First Director General of ICAR was:
(A) Dr. M.S. Swaminathan (B) Dr. N.S. Randhawa
(C) Dr. G.S. Randhawa (D) Dr. B.P. Pal

190. Total food grain production in India in 2008-09 is:
(A) 207 mt (B) 275 mt
(C) 300 mt (D) 317 mt

191. To develop the mars (planet) on the earth's environment is called:
(A) Terra farming (B) Aeroponics
(C) Organic farming (D) Conservation agriculture

192. The form of agriculture aimed ar meeting the needs of present generation without engangering the resources of future generation is called:
(A) Sustainable agriculture
(B) Conservation agriculture
(C) Precision agriculture
(D) Organic agriculture

193. The technique where plants are grown without soil in the aerated moist atmosphere with the help of very small quantity of water is known as:
(A) Aeroponics (B) Hydroponics
(C) Slash and burn (D) None of these

194. First State Agricultural University of India is:
(A) Gujarat Agricultural University
(B) Punjab Agricultural University
(C) G. B. Pant University of Agricultural & Technology
(D) Chaudhary Charan Singh Hisar Agricultural University

195. Inflorescence of sugarcane is known as:
(A) Arrow (B) Tillers (C) Silk (D) Tassel

196. The highest productivity of wheat is in:
(A) Haryana (B) Punjab (C) UP (D) Bihar

197. The highest production of wheat is in:
(A) Haryana (B) Punjab (C) UP (D) Bihar

198. The principle male sterile line introduced for hybrid pearlmillet production in India was:
(A) Tift-23 A (B) BJ-104
(C) MHB -118 (D) CO-2

199. Essential oils are used in:
(A) Cosmetics and Toiletries (B) Food beverage
(C) Paint industry (D) All of these

200. First Face Pack was developed in:
(A) France (B) Rome
(C) Greece (D) None of these

201. Oil content in fresh herbage in menthol mint is:
(A) 0.1-0.5 % (B) 0.5-0.8 %
(C) 1.0-1.5 % (D) 1.5 – 2.0 %

202. Major constituent of *Occimum cannum* is:
(A) Methyl chavical (B) Thymal
(C) Linalool (D) Engenol

203. Origin of cow pea crop is:
(A) India (B) Pakistan
(C) China (D) Japan

204. Sunflower crop belongs to the family of:
(A) Cruciferеae (B) Compositeae
(C) Chenopodiaceae (D) Liliaceae

205. What is the optimum pulse requirement per capita per day?
(A) 20 g (B) 40 g (C) 80 g (D) 100 g

206. Origin of French bean is:
(A) America (B) India
(C) China (D) Pakistan

207. Protein content in rajmash is:
(A) 10.9 % (B) 22.9 % (C) 25.9 % (D) 32.9 %

208. Carbohydrate content in rajmash is:
(A) 50.6 % (B) 60.6 % (C) 70.6 % (D) 80.6 %

209. Protein content in pea is:
(A) 10 % (B) 22 % (C) 25 % (D) 32 %

210. Carbohydrate content in potato is:
(A) 40.3 % (B) 70.3 % (C) 81.3 % (D) 91.3 %

211. India occupies position in vegetable production in world:
(A) First (B) Second (C) Third (D) Fourth

212. India occupies position in production of onion in world:
(A) First (B) Second (C) Third (D) Fourth

213. Total cattle population in India during 2006-07 is:
(A) 181 million (B) 196 million
(C) 201 million (D) 210 million

214. Total buffalo population in India during 2006-07 is:
(A) 80 million (B) 90 million
(C) 95 million (D) 99 million

215. Total goat population in India during 2006-07 is:
(A) 100 million (B) 115 million
(C) 125 million (D) 135 million

216. Total sheep population in India during 2006-07 is:
(A) 35 million (B) 51 million
(C) 58 million (D) 63 million

217. Total camel population in India during 2006-07 is:
(A) 0.63 million (B) 0.75 million
(C) 0.76 million (D) 0.63 million

218. Total milk production in India during 2006-07 is:
(A) 77.7 million tonnes (B) 80.6 million tonnes
(C) 85.8 million tonnes (D) 100.9 million tonnes

219. The contribution of livestock to total agricultural output is:
(A) 20 % (B) 25 % (C) 30 % (D) 35 %

220. The contribution of livestock sector to total GDP is:
(A) 1 % (B) 3 % (C) 6 % (D) 8 %

221. India's share in total milk production of world is:
(A) 10 % (B) 13 % (C) 15 % (D) 21 %

222. The average productivity of Indian cow is:
(A) 810 kg/lactation (B) 987 kg/lactation
(C) 1200 kg/lactation (D) 1500 kg/lactation

223. Dairy sector provides employment to the total work force of:
(A) 5 % (B) 10 % (C) 15 % (D) 20 %

224. India's position in total cattle population is:
(A) First (B) Second
(C) Third (D) Fourth

225. Central Sheep and Wool Research Institute is situated at:
(A) Avikanagar (Raj) (B) Bangalore (KN)
(C) Bikaner (Raj) (D) Pusa (New Delhi)

226. Which of the following state has highest sheep production in India:
(A) Rajasthan (B) Gujarat
(C) Madhya Pradesh (D) Uttar Pradesh

227. India's share in goat population in the world is:
(A) 15% (B) 20% (C) 25% (D) 30%

228. Which of the following state is leading in goat and sheep population?
(A) Madhya Pradesh (B) Rajasthan
(C) J & K (D) Gujarat

229. The meat contribution by goat to total meat production in India is:
(A) 20 % (B) 30 % (C) 37 % (D) 48 %

230. The highest percentage of total goat production is found in the region:
(A) Western and semi-arid region
(B) South humid tropical region
(C) Eastern region
(D) North temperate region

231. The contribution of goat in total milk production of India is only:
(A) 1.7% (B) 3.2% (C) 5.7% (D) 6.1%

232. The mean rate of goat slaughtering is:
(A) 25% (B) 30% (C) 36% (D) 41%

233. Which of the state is leading in poultry production?
(A) UP (B) AP
(C) Karnataka (D) Maharastra

234. The total egg production in India is:
(A) 20.2 million (B) 35.1 million
(C) 50.7 million (D) 40.3 million

235. The highest poultry production in India is:
(A) UP (B) AP (C) MP (D) WB

236. Which of the country has the highest population of poultry birds?
(A) USA (B) UK (C) China (D) India

237. Per capita availability of animal protein in Indian diet is:
(A) 10 g/day (B) 15 g/day
(C) 18 g/day (D) 20 g/day

238. The protein per cent in egg yolk is:
(A) 5% (B) 11% (C) 16% (D) 18%

239. The highest producing state of duck is:
(A) UP (B) WB (C) Bihar (D) Assam

240. Turkey production is highest in the world is:
(A) USA (B) India (C) Brazil (D) China

241. Per capita availability of milk in India is:
(A) 200 g/day (B) 226 g/day
(C) 230 g/day (D) 240 g/day

242. India's rank in world egg production is:
(A) Ist (B) IInd (C) IIIrd (D) IVth

243. Total wool production in India during 2006-07is:
(A) 40 million kg (B) 50 million kg
(C) 45.1 million kg (D) 60 million kg

244. Total fish production in India during 2006-07 is:
(A) 68.69 million tones (B) 70.12 million tones
(C) 75.63 million tones (D) 80.06 million tones

245. Highest inland fish production during 2006-07 is
(A) 40.25 million tones (B) 38.45 million tones
(C) 42.68 million tones (D) 45.69 million tones

246. Highest marine fish production during 2006-07 is:
(A) 21.36 million tones (B) 25.36 million tones
(C) 30.24 million tones (D) 32.69 million tones

247. The number of milk recording centres at national level are:
(A) 30 (B) 50 (C) 94 (D) 115

248. The number of Central Cattle Breeding Centres in India are:
(A) 3 (B) 5 (C) 7 (D) 9

249. Central Cattle Breeding Centre is situated at:
(A) Suratgarh (Rajasthan) (B) Andeshnagar (UP)
(C) Alamadi (TN) (D) All of these

250. Central Frozen Semen Production and Training Institute is located at:
(A) Hisar (B) Suratgarh
(C) Hessarghatta (D) Bangalore

251. Central Sheep Breeding Farm is located at:
(A) Hisar (B) Suratgarh
(C) Hesarghatta (D) Bangalore

252. How many pig breeding farms have been established in India?
(A) 101 (B) 158 (C) 260 (D) 275

253. Central Fodder Seed Production Farm is located at:
(A) Jhansi (B) Hessarghatta
(C) Hyderbad (D) Hisar

254. India's share in world buffalo population is:
(A) 20 % (B) 40 % (C) 57 % (D) 75 %

255. India's share in world cattle population is:
(A) 10% (B) 12% (C) 15% (D) 18%

256. Central Institute of Fisheries, Nautical and Engineering Training is located at:
(A) Kochi (B) Mumbai
(C) Vishakhapattanam (D) Kolkata

257. Central Institute of Coastal Engineering for Fisheries is located at:
(A) Kochi (B) Mumbai
(C) Bangalore (D) Kolkata

258. The number of major fishing harbours in India is:
(A) 2 (B) 4 (C) 6 (D) 8

259. Total honey production in India is:
(A) 7500 tonnes (B) 10200 tonnes
(C) 13000 tonnes (D) 15000 tonnes

260. Which country has highest milk production is:
(A) India (B) Denmark
(C) Switzerland (D) USA

261. India's share in world bovine population is:
(A) 17% (B) 19% (C) 21% (D) 23%

262. Buffalo constitute the total bovine population of:
(A) 35% (B) 40% (C) 45% (D) 47%

263. What is the contribution of dairy sector to the employment:
(A) 5% of total work force (B) 8% of total work force
(C) 9% of total work force (D) 12% of total work force

264. Milk production in India during 2006-07 is:
(A) 72 m tonnes (B) 77 m tonnes
(C) 100.9 m tonnes (D) 100 m tones

265. Largest milk colony of Asia is:
(A) Aere milk colony (B) Allahabad dairy farm
(C) Rajasthan dairy farm (D) Kanpur diary farm

266. First Dairy farm in India was opened in year of:
(A) 1870 (B) 1877 (C) 1889 (D) 1890

267. First Dairy farm in India was opened at:
(A) Anand (B) Allahabad
(C) Karnal (D) New Delhi

268. Contribution of dairy sector to the total agricultural output is nearly:
(A) 21 % (B) 25 % (C) 29 % (D) 31 %

269. Contribution of dairy sector to the GDP is around:
(A) 4 % (B) 6 % (C) 8 % (D) 10 %

270. Before 1951, India's milk production was:
(A) Less than 20 mt (B) 20-30 mt
(C) 30-40 mt (D) More than 40 mt

271. India's share in world milk production is:
(A) 10% (B) 13% (C) 15% (D) 21%

272. Indian Dairy Corporation was setup mainly for:
(A) Success of Operation of Flood Program
(B) Implementation of Operation Flood Program
(C) Trade exchange related to Operation Flood Program
(D) All of these

273. The plan investment in Animal Husbandry and Dairying is about:
(A) 4% (B) 5% (C) 6% (D) 8%

274. India's share in Asian bovine population is:
(A) 27% (B) 38% (C) 51% (D) 60%

275. According to ICAR, minimum requirement of milk/day/head is:
(A) 200 g (B) 280 g (C) 290 g (D) 300 g

276. National Agricultural Insurance Scheme launched in India in year:
(A) 1972 (B) 1985 (C) 1999 (D) 2000

277. Comprehensive Crop Insurance Scheme was launched in year:
(A) 1975 (B) 1985 (C) 1995 (D) 2001

278. Central Institute of Agricultural Engineering is situated at:
(A) Kanpur (B) Bhopal
(C) Ludhiana (D) Nagpur

279. The intermediaries came into Indian Agriculture in which year:
(A) 1793 (B) 1783 (C) 1773 (D) 1763

280. The first bill for abolition of intermediaries or zamidari introduced in:
(A) 1946, UP (B) 1948, Bihar
(C) 1971, TN (D) 1967, Gujarat

281. Central Farm Machinery Training and Testing Institute is situated at:
(A) Budni, MP (B) Ludhiana, Punjab
(C) Kanpur, UP (D) Jabalpur, MP

282. Zamidary system was introduced in year:
(A) 1792 (B) 1793 (C) 1794 (D) 1795

283. Rayatwari system was introduced by:
(A) Thomas Munro and Captain Read
(B) Lord Powel
(C) Lord Curzon
(D Lord Cornwalish

284. Land Reform Implementation Committee was set up in 1963 under the chairmanship is:
(A) G.L. Nanda (B) Prof. Danlewala
(C) Arvind Ghosh (D) M.K. Gandhi

285. The minimum wages Act (Labour) was enacted during:
(A) 1932 (B) 1936 (C) 1948 (D) 1952

286. The Employees Provident Fund and Family Pension Act was enforced in:
(A) 1952 (B) 1962 (C) 1972 (D) 1982

287. The payment of Bonus Act was implemented in:
(A) 1948 (B) 1958 (C) 1965 (D) 1968

288. The payment of Gratuity Act was implemented in:
(A) 1952 (B) 1962 (C) 1972 (D) 1982

289. The Industrial Disputes Act was enacted in:
(A) 1937 (B) 1947 (C) 1957 (D) 1967

290. The Trade Union Act was enacted in:
(A) 1926 (B) 1936 (C) 1946 (D) 1956

291. The Bonded Labour System (Abolition) ordinance came into force in:
(A) 1971 (B) 1973 (C) 1975 (D) 1978

292. The Livestock Importation Act was implemented in:
(A) 1862 (B) 1872 (C) 1898 (D) 1903

293. The Agriculture Produce (grading) and Marketing Act was passed in:
(A) 1927 (B) 1937 (C) 1947 (D) 1957

294. The Cattle Trespass Act was implemented in:
(A) 1781 (B) 1881 (C) 1981 (D) 1991

295. The prevention of Cruelty to animals Act, 1950 was implemented in:
(A) 1857 (B) 1891 (C) 1907 (D) 1950

296. Dangerous Machines (Regulation) Act was passed in:
(A) 1971 (B) 1981 (C) 1983 (D) 1993

297. The Land Improvement Act was initiated in:
(A) 1869 (B) 1870 (C) 1871 (D) 1872

298. The Land Improvement Loan Act was started in:
(A) 1871 (B) 1876 (C) 1883 (D) 1891

299. The Agriculturists Loans Act was implemented in:
(A) 1881 (B) 1882 (C) 1883 (D) 1884

300. The National Co-operative Development Corporation Act was started in:
(A) 1947 (B) 1959 (C) 1962 (D) 1971

301. The Agricultural Produce (Development and Warehousing) Corporation Act was implemented in:
(A) 1947 (B) 1956 (C) 1962 (D) 1971

302. The National Co-operative Development Corporation was set up in:
(A) 1952 (B) 1962 (C) 1963 (D) 1964

303. The Agricultural Refinance and Development Corporation Act came into existence in:
(A) 1961 (B) 1962 (C) 1963 (D) 1964

304. The Warehouse Corporation Act was implemented in:
(A) 1960 (B) 1961 (C) 1962 (D) 1963

305. The Regional Rural Banks Act was enforced in India in:
(A) 1971 (B) 1974 (C) 1976 (D) 1981

306. The Multi-Unit Co-operative Societies Act was implemented in:
(A) 1940 (B) 1941 (C) 1942 (D) 1943

307. The All India Rural Credit Survey Committee was constituted by RBI in:
(A) 1931 (B) 1941 (C) 1951 (D) 1961

308. National Development Council Resolution was passed in
(A) 1949 (B) 1959 (C) 1969 (D) 1979

309. The New Co-operative Policy Resolution was adopted in:
(A) 1975 (B) 1976 (C) 1977 (D) 1978

310. National Cooperative Union of India was set up in:
(A) 1951 (B) 1953 (C) 1957 (D) 1961

311. Agricultural Finance Corporation was set up in:
(A) April 10, 1965 (B) April 10, 1966
(C) April 10, 1968 (D) April 10, 1969

312. Agricultural Refinance and Development Corporation (ARDC) was setup in:
(A) 1961 (B) 1962 (C) 1963 (D) 1964

313. Tribal Area Development Project was launched in:
(A) 1971-72 (B) 1973-74
(C) 1976-77 (D) 1978-79

314. State Trading Corporation Ltd was constituted in:
(A) 1956 (B) 1957 (C) 1958 (D) 1959

315. Indian Standard Institute was established in:
(A) 1945 (B) 1946 (C) 1947 (D) 1948

316. Marketing and Designing centre is situated at:
(A) Pune (B) Faridabad
(C) Nagpur (D) New Delhi

317. Central Agmark Laboratory is situated at:
(A) New Delhi (B) Nagpur
(C) Bangalore (D) Kanpur

318. Indian Standard Institute is situated at:
(A) Kolkata (B) New Delhi
(C) Madras (D) Hyderabad

319. Agricultural Produce (Grading and Marketing) Act was established in:
(A) 1932 (B) 1937 (C) 1939 (D) 1940

320. Central Marketing department was opened in:
(A) 1932 (B) 1035 (C) 1937 (D) 1941

321. The per capita availability of land has been decreased from 0.5 ha in 1950-51 to 0.14 ha in 2000 and will further reduced by 2020:
(A) 0.08 ha (B) 1.08 ha
(C) 1.18 ha (D) 1.28 ha

322. First Agricultural Marketing Advisor was appointed on:
(A) January 1, 1935 (B) January 26, 1935
(C) October 2, 1935 (D) August 15, 1935

323. Agricultural Price Commission was constitutes in:
(A) 1955 (B) 1965 (C) 1975 (D) 1985

324. Essential Commodities Act was passed on:
(A) 1954 (B) 1955 (C) 1956 (D) 1957

325. Cold Storage Order came into existence in:
(A) 1962 (B) 1963 (C) 1964 (D) 1965

326. Meat Food Products order was passed on:
(A) 1971 (B) 1972 (C) 1973 (D) 1974

327. Export (Quality Control and Inspection) Act was passed on:
(A) 1965 (B) 1968 (C) 1969 (D) 1971

328. Head office of Directorate of Marketing and Inspection is situated at:
(A) New Delhi (B) Faridabad
(C) Gurgaon (D) Ghaziabad

329. Food Corporation of India (FCI) was came into existence on:
(A) 1965 (B) 1966 (C) 1967 (D) 1968

330. A expert committee to review the Cooperative Marketng was set up in 1964 under the chairmanship of:
(A) Dr. M.L. Dantewala (B) R.N. Mirdha
(C) Raj Gopalacharya (D) R.L. Mehta

331. The Agricultural Finance sub committee was constituted under the chairmanship of:
(A) V.L. Mehta (B) D.R. Gadgil
(C) Sardar Ballabh Bhai Patel (D) J.L. Nehru

332. Annual loss due to pests, diseases and weeds annually is estimated nearly:
(A) >50000 crores (B) <50000 crores
(C) > 1.0 crores (D) < 1.0 crores

333. India is using biopesticides presently:
(A) 1% of total pesticide (B) 5 % to total pesticide
(C) 10 % of total pesticide (D) 15 % of total pesticide

334. Average biopesticide use globally is:
(A) 5 % of total pesticide (B) 10 % of total pesticide
(C) 12 % of total pesticide (D) 15 % of total pesticide

335. Total pesticide consumption in India during 2006-07 is :
(A) 35.65 million tonnes
(B) 37.95 million tonnes
(C) 39.26 million tonnes
(D) 42.56 million tonnes

336. Destructive Insects and Pest Act was launched in India in:
(A) 1907 (B) 1914 (C) 1921 (D) 1946

337. Directorate of Plant Protection Quarantine and Storage was started in:
(A) 1907 (B) 1914 (C) 1921 (D) 1946

338. Which of the airport does not have plant quarantine station?
(A) Lucknow (B) Kolkata
(C) New Delhi (D) Madras

339. Headquarter of Central Silk Board is:
(A) Lucknow (B) Bangalore
(C) Hyderabad (D) Behrampur

340. Central Sericulture Research Institute and Technology is situated at:
(A) New Delhi (B) Kolkata
(C) Behrampur (D) Madras

341. Headquarter of Tobacco Board is:
(A) Rajahmundry (B) Guntur
(C) Vijaywada (D) Vishakhapatannam

342. What is origin place of cauliflower:
(A) Canada (B) Italy
(C) India (D) China

343. What is origin place of cabbage:
(A) West Europe (B) West Africa
(C) West China (D) East India

344. What is origin place of bottle guard:
(A) Canada (B) USA
(C) India (D) China

345. What is origin place of pumpkin:
(A) Canada (B) America
(C) India (D) China

346. What is origin place of carrot:
(A) Canada (B) USA
(C) Asia (D) China

347. What is origin place of Sweet potato:
(A) North America (B) South America
(C) Australia (D) China

348. What is origin place of chilli:
(A) Brazil (B) Chile
(C) Gutemala (D) China

349. APEDA refers to:
(A) Agricultural and Processed Food Products Export Development Authority
(B) Agricultural Production and Export Development Authority
(C) Agricultural Production and Export Development Agency
(D) Asian Policy and Export Development Association

350. India is leading in production of:
(A) Turmeric (B) Black pepper
(C) Ginger (D) All of these

351. Total production of vegetables in India during 2006-07 is:
(A) 93.92 mt (B) 96.54 mt
(C) 102.11 mt (D) 120.71 mt

352. What is the contribution of India in world vegetable production:
(A) 10 % (B) 15 % (C) 20 % (D) 23 %

353. Which of the following vegetable is produced maximum in India is:
(A) Cauliflower (B) Cabbage
(C) Tomato (D) Brinjal

354. Whch of the following spices is most important in India is:
(A) Chilli (B) Turmeric
(C) Garlic (D) All of these

355. India's share in chilli production is:
(A) 27 % (B) 33.7 % (C) 45 % (D) 51.2 %

356. Minimum requirement of fruit per head per day for balanced diet is:
(A) 60 gm (B) 80 gm
(C) 85 gm (D) 120 gm

357. CFTRI is situated at:
(A) Mysore (B) Kolkata
(C) New Delhi (D) Bangalore

358. First fruit Product Research Laboratory was opened in India at:
(A) Kodour (B) Pune
(C) Lucknow (D) Bhopal

359. The origin place of banana is:
(A) South-East Asia (B) South-West Asia
(C) South Africa (D) South America

360. The origin place of loquat is:
(A) China (B) India
(C) South Africa (D) South America

361. The origin place of jackfruit is:
(A) China (B) Mexico
(C) India (D) South America

362. The origin place of grape is:
(A) Asia (B) China
(C) South Africa (D) South America

363. The origin place of peach is:
(A) Europe (B) China
(C) South Africa (D) South America

364. The origin place of pear is:
(A) Iran (B) Asia
(C) Africa (D) America

365. The origin place of gladiolus is:
(A) China (B) Egypt
(C) South Africa (D) South America

366. India's place in fruit production in world is:
(A) First (B) Second
(C) Third (D) Fourth

367. The fruit production in India is:
(A) 26 mt (B) 40 mt
(C) >45 mt (D) 80 mt

368. The area under fruit production in India is:
(A) 1.5 m ha (B) 2.5 m ha
(C) 3.8 m ha (D) 8.5 m ha

369. What is the total production of horticultural crops in India during 2001-02 is:
(A) 152.5 mt (B) 155.65 mt
(C) 171.0 mt (D) 185.13 mt

370. What is the share of India in total world fruit production?
(A) 5 % (B) 10 % (C) 15 % (D) 20 %

371. Total honey production in India is:
(A) 2.2 thousand tones (B) 8.7 thousand tones
(C) 10.2 thousand tones (D) 12.5 thousand tones

372. The area under herbal garden in India is:
(A) 200 ha (B) 226 ha
(C) 310 ha (D) 510 ha

373. The fruits are arranged in ascending order in respect of its area. Which of the following sequence is correct:
(A)Mango>Citrus>Banana>Apple
(B)Mango>Apple>Banana>Citrus
(C) Guava>Mango>Banana>Citrus
(D) Apple>Mango>Citrus>Litchi

374. What is the share of India in total mango production of world?
(A) 23% (B) 42% (C) 54% (D) 60%

375. Presently the total production of cut flowers in India during 2008-09 is:
(A) 550 million (B) 780 million
(C) 1952 million (D) 910 million

376. Which of the following fruit has its highest productivity in India?
(A) Grape (B) Mango
(C) Banana (D) Guava

377. Oasis theory of agricultural development has been prepared by:
(A) Raphael Pumpelly in 1908
(B) Robert Braidwood in 1948
(C) Bryan Hayden
(D) Carl Sauer

378. Hilly Flanks hypothesis suggests that agriculture began in the hilly flanks of the Taurus and Zagros mountains has been proposed by:
(A) Raphael Pumpelly in 1908
(B) Robert Braidwood in 1948
(C) Bryan Hayden
(D) Carl Sauer

379. The Feasting model suggests that agriculture was driven by ostentatious displays of power such as throwing feasts to exert dominance has been proposed by:
(A) Raphael Pumpelly in 1908
(B) Robert Braidwood in 1948
(C) Bryan Hayden
(D) Carl Sauer

380. Demographic theories describes an increasingly sedentary population, expanding upto the carrying capacity of the local environment and requiring more food than can be gathered has been proposed by:
(A) Raphael Pumpelly in 1908
(B) Robert Braidwood in 1948
(C) Bryan Hayden
(D) Carl Sauer

381. The evolutionary/intentionally theory suggests that agriculture is a co-evolutionary adaptation of plants and hmans has been proposed by:
(A) Davis Rindos (B) Daniel Quinn
(C) Bryan Hayden (D) Carl Sauer

382. The domestication theory states that first hman stayed in particular areas, giving up their nomadic ways than developed agriculture and animal domestication put forth by:
(A) Davis Rindos (B) Daniel Quinn
(C) Bryan Hayden (D) Carl Sauer

383. India stands —————— in the world, in geographical areas:
(A) Fifth (B) Sixth
(C) Seventh (D) Eight

384. India stands ———— in arable lands:
(A) Fifth (B) Fourth
(C) Third (D) Second

385. A multipurpose plant of India is
(A) Walnut (B) Almond
(C) Apple (D) Coconut

386. A grain producing plant where spathe covering helped in domestication
(A) Wheat (B) Maize
(C) Rice (D) Barley

387. The main source of irrigation in India is:
(A) Canals (B) Tubewells
(C) Wells (D) Tanks

388. The total food grain production in India during 2007-08 was:
(A) 202 million tones (B) 220 million tones
(C) 225 million tones (D) 230 million tones

389. Agriculture became state subject in year:
(A) 1910 (B) 1919 (C) 1947 (D) 1952

390. Indian Agricultural Services ceased to function in the year:
(A) 1919 (B) 1921 (C) 1924 (D) 1947

391. ICAR was born in:
(A) 1908 (B) 1910 (C) 1929 (D) 1949

392. What was the name of ICAR in initial years of its establishment?
(A) Indian Council of Agricultural Research
(B) Imperial Council of Agricultural Research
(C) Imperial Centre of agricultural Research
(D) Indian Centre of Agricultural Research

393. The establishment of ICAR is based on the recommendation of which commission?
(A) Royal Commission on Agriculture
(B) Famine Commission
(C) Grow More Food Commission
(D) Balwant Rai Mehta Commission

394. ICAR was renamed Indian Council of Agricultural Research in which year?
(A) 1929 (B) 1946 (C) 1949 (D) 1952

395. Who was the first Director General of ICAR and Secretary to Ministry of Agr culture?
(A) Dr. B.P. Pal (B) Dr. M.S. Swaminathan
(C) Dr. O.P. Gautam (D) Dr. R.S. Paroda

396. When did ARS service start?
(A) 1947 (B) 1952 (C) 1973 (D) 1982

397. What is the main mandate of ICAR?
(A) To plan aid, undertake, promote and coordinate agriculture research in the country
(B) To plan, aid undertake, promote and coordinate agricultural research in the country
(C) To plan, aid undertake, promote and coordinate agricultural education and research in the country
(D) To plan, aid undertake, promote and coordinate agricultural research, education and extension in the country

398. How many Deputy Director Generals are working in ICAR?
(A) 05 (B) 06 (C) 07 (D) 08

399. Who works as President of ICAR?
(A) The Prime Minister of India
(B) The Union Agriculture Minister
(C) The State Minister of Agriculture
(D) Director General, ICAR

400. The growth of ICAR started with the establishment of which of the following institutes, being the first one?
(A) Imperial Bacteriological Laboratory, Pune
(B) Agricultural Research Institute, Pusa, Bihar
(C) Indian Agricultural Research Institute, New Delhi
(D) Agriculture College, Kanpur

401. How many National Institutes are working in the ICAR system?
(A) 04 (B) 05 (C) 06 (D) 07

402. Which of the following was the first institute/university to be establishment only land Grant College System in 1958?
(A) Indian Agricultural Research Institute, New Delhi
(B) G.B. Pant University of Agriculture & Technology, Pantnagar
(C) College of Agriculture, Coimbatore
(D) College of Agriculture, Kanpur

403. NAARM (National Academy of Agricultural Research Management) Hyderabad was established to train scientists in which year?
(A) 1949 (B) 1973 (C) 1976 (D) 1986

404. AICRP stands for?
(A) All India Crop Research Programme
(B) All India Coordinated Research Project
(C) All India Cropping System Research Project
(D) None of these

405. MANAGE, Hyderabad is an institution working in the area of:
(A) Agricultural research management
(B) Agricultural extension management
(C) Agricultural education Management
(D) Agricultural research, education and extension management

406. What for ASRB stands?
(A) Agricultural Services Recruitment Board
(B) Agricultural Scientist Recruitment Board
(C) Agricultural Science Research Bureau
(D) Agricultural Scientist Research Bureau

407. How many Trainers Training Centres (TTCs) are in India?
(A) 05 (B) 08 (C) 13 (D) 21

408. IVLP (Institution Village Linkage Programme) centres are functioning at how many places?
(A) 70 (B) 79 (C) 83 (D) 89

409. How many Zonal Research Stations have been remandated to function as KVKs also?
(A) 50 (B) 53 (C) 57 (D) 59

410. How many central institutes of ICAR are working in agricultural science?
(A) 10 (B) 36 (C) 73 (D) 89

411. How many Bureaus have been established in India?
(A) 01 (B) 02 (C) 05 (D) 11

412. National Bureau of Plant Genetic Resources is established at which place?
(A) New Delhi (B) Modipuram
(C) Coimbatore (D) Bangalore

413. National Bureau of Soil Survey & Land Use Planning is established at which place?
(A) Jorhat (B) Nagpur
(C) New Delhi (D) Kochin

414. National Research Centre for Women in Agriculture is established at which place?
(A) Calcutta (B) Bhubaneshwar
(C) Jorhat (D) New Delhi

415. National Centre of Agricultural Economics & Policy Research is established at which place?
(A) New Delhi (B) Hyderabad
(C) Bangalore (D) Madras

416. How many agricultural universities are functional in India?
(A) 25 (B) 30 (C) 29 (D) 38

417. How many central agricultural universities are established in India?
(A) 01 (B) 02 (C) 03 (D) 04

418. How many ICAR institutes have been upgraded to the status of Deemed universities?
(A) 02 (B) 04 (C) 05 (D) 12

419. The demand of food grain production can be mmet only:
(A) Vertical (B) Horizontal
(C) Both (A) and (B) (D) None of these

420. ICRISAT stands for:
(A) International Crops Research Institute for the Semi Arid Tropics
(B) International Centre for Research in Sugarcane & Tobacco
(C) Indian Crop Research Institute for Semi Arid Tropics

421. Agriculture is the subject of which of the following?
(A) State Govt. (B) Central Govt.
(C) State & Central Govt. (D) None of these

422. ICAR/SAUs conduct what kind of extension?
(A) Public extension (B) Frontline extension
(C) Private extension (D) None of these

423. What is the main mandate of ICAR/SAUs in the area of agricultural extension?
(A) Technology transfer
(B) Technology assessment on refinement
(C) Technology assessment, refinement
(D) None of these

424. Extension Education Institutes (EEIs) impart training in the area of:
(A) Subject matter
(B) Training & Communication methodology
(C) Subject matter, training and communication methodology
(D) None of these

425. The most serious economic problem of India are:
(A) Poverty and Unemployment
(B) Unemployment and stagnation
(C) Underdevelopment and Poverty
(D) Corruption and Unemployment

426. The percentage of people below poverty line in India is:
(A) 25.1 (B) 27.8 (C) 20.3 (D) 21.45

427. By eradicating poverty, who wished to "wipe out every tear from every eye":
(A) Jawahar Lal Nehru
(B) Mahatma Gandhi
(C) Rabindra Nath Tagore
(D) Vinoba Bhave

428. The poverty lines in rural area on the basis of an average daily intake:
(A) 2100 calories (B) 2300 calories
(C) 2400 calories (D) 3400 calories

429. The 'Garibhi hatao' slogan was coined by:
(A) Jawahar Lal Nehru (B) Mahatma Gandhi
(C) Rabindra Nath Tagore (D) Indira Gandhi

430. The proportion of Indian population living in the villages:
(A) One-third (B) Half
(C) Two-third (D) Three-fourth

431. Who said 'If the villages perish, India will perish:
(A) Jawahar Lal Nehru (B) Mahatma Gandhi
(C) Rabindra Nath Tagore(D) Indira Gandhi

432. Rural development can be defined as a strategy designated to improve:
(A) Economic and social life of rural poor
(B) Condition of all section of people
(C) Economic condition of rural society
(D) Physical infrastructure

433. The basic guide, philosopher and friend in rural development:
(A) BDO (B) VDO
(C) Pradhan (D) Pramukh

434. Rural development programmes are related to:
(A) Self employment (B) Employment generation
(C) Poverty alleviation(D) All of these

435. The body responsible for rural development at base level:
(A) Panchayat (B) Block
(C) District (D) State

436. The basic operational unit for rural development in India is:
(A) Village (B) District
(C) Tehsil (D) Block

437. Who coined the term Green revolution:
(A) Willianm Gadd (B) M.S. Swaminathan
(C) Norman E. Borlaoug (D) C. Subramanian

438. Agricultural Scientists Recruitment Board (ASRB) was set up based on the recommendations of the committee headed by:
(A) Mohan Singh Mehta (B) M S Swaminathan
(C) Gajendra Gadkar (D) B P Sinha

439. Agricultural Scientists Recruitment Board (ASRB) was set up in the year:
(A) 1973 (B) 1975 (C) 1977 (D) 1979

440. Agriculture and livestock was published by:
(A) Ministry of Agriculture and Cooperation
(B) Ministry of Rural Development
(C) Indian Council of Agricultural Research
(D) Ministry of Animal Husbandry

441. The concept of agro-climatic zones were introduced during:
(A) NATP (B) NARP (C) NREP (D) NAIP

442. The concept of agro-eco system analysis was developed during:
(A) Late 60s (B) Late 70s
(C) Late 80s (D) Late 90s

443. The concept of agro-eco system analysis was developed in:
(A) USA (B) Philippines
(C) Thailand (D) Israel

444. The concept of agro-eco system was basically originated at University of Chiang Mai by:
(A) Tonnis Fermetty (B) Adam Groves
(C) Levis Carol (D) Gordon Conway

445. The first farm magazine in India is:
(A) Kheti (B) Phool Phul
(C) Krishi Chayanika (D) Yojana

446. The general micro-unit of an agro-ecosystem is:
(A) District (B) Village
(C) Farmer (D) Household

447. The headquarter of Food and Agricultural Organization are at:
(A) New York (B) New Delhi
(C) Geneva (D) Rome

448. The ideal types of social norms are:
(A) Traditional (B) Modern
(C) Both of these (D) None of these

449. The second most important hygiene factor according to Heirzberg is:
(A) Policy and administration
(B) Achievement
(C) Salary
(D) Working conditions

450. The impact of green revolution was largely noted in:
(A) Rice only (B) Wheat only
(C) Rice and Wheat (D) Rice, Wheat and Pulses

451. The National Steering Committee (NSC) is chaired by:
(A) Director General, ICAR
(B) Union Minister of Agriculture
(C) Minister of State for Agriculture
(D) Principle Secretary Agriculture

452. The president of ICAR is:
(A) Prime Minister (B) Agriculture Minister
(C) Director General (D) Finance Minister

453. Scientific name of Flint corn is:
(A) *Zea mays indurata* (B) *Zea mays tunicata*
(C) *Zea mays saccharata* (D) *Zea mays ceratina*

454. Scientific name of Pod corn is:
(A) *Zea mays indurata* (B) *Zea mays tunicata*
(C) *Zea mays saccharata* (D) *Zea mays ceratina*

455. Scientific name of Sweet corn is:
(A) *Zea mays indurata* (B) *Zea mays tunicata*
(C) *Zea mays saccharata* (D) *Zea mays ceratina*

456. Scientific name of Waxy corn is:
(A) *Zea mays indurata* (B) *Zea mays tunicata*
(C) *Zea mays saccharata* (D) *Zea mays ceratina*

457. Scientific name of Pop corn is:
(A) *Zea mays indurata* (B) *Zea mays tunicata*
(C) *Zea mays everta* (D) *Zea mays ceratina*

458. Scientific name of Soft or flour corn is:
(A) *Zea mays indurata* (B) *Zea mays amylacea*
(C) *Zea mays saccharata* (D) *Zea mays ceratina*

459. In America which crop is known as yellow gold:
(A) Maize (B) Rice
(C) Soybean (D) Wheat

460. Scientific name of green varieties of black gram is:
(A) *Vigna mungo var viridis*
(B) *Vigna mungi var niger*
(C) *Vigna mungo var macropode*
(D) *Vigna mungi var micropode*

461. Scientific name of bunch type of ground nut is:
(A) *Arachis hypogea sp fastigiata*
(B) *Arachis hypodea sp procumbens*
(C) *Arachis hypogea sp viridis*
(D) *Arachis hypogea sp niger*

462. Scientific name of spreading type of ground nut is:
(A) *Arachis hypogea sp fastigua*
(B) *Arachis hypodea sp procumbens*
(C) *Arachis hypogea sp viridis*
(D) *Arachis hypogea sp niger*

463. Scientific name of desi cotton is:
(A) *Gossypium herbacium* (B) *Gossypium arboreum*
(C) Both (A) and (B) (D) None of these

464. Scientific name of American cotton is:
(A) *Gossypium barbedensii* (B) *Gossypium hirsutum*
(C) Both (A) and (B) (D) None of these

465. Diploid species of cotton are:
(A) Desi (B) American
(C) Both (A) and (B) (D) None of these

466. Tetraploid species of cotton are:
(A) Desi (B) American
(C) Both A and B (D) None of these

467. Poor man's timber is:
(A) Mango wood (B) Bamboo
(C) Acacia wood (D) Eucalyptus wood

468. Tendu/Kendu leaves used as wrappers for bidis are got from:
(A) *Diospyros ebenum*
(B) *Dipterocarpus indicus*
(C) *Dipterocarpus tuberculatus*
(D) *Diospyros melonxylon*

469. Conservation is:
(A) Protection of natural resources
(B) Management of natural resources
(C) Proper use of natural resources
(D) All of these

470. MAB is:
(A) Man and Botany
(B) Man and Biosphere
(C) Man and Biotic community
(D) Man, Antibiotics and Bacteria

471. Silent valley has been preserved as it has:
(A) Recreational value (B) Valuable timber plants
(C) Natural forest (D) Rare plant and animals

472. Silent valley having rare plants and animals is located in:
(A) Kerala (B) Karnataka
(C) Jammu and Kashimir (D) Andhra Pradesh

473. Which one is connected with conservation of forests?
(A) Gir (B) Silent valley
(C) Kaziranga (D) Bharatpur

474. New approach to conservation is establishment of:
(A) Reserve forests (B) Santuries
(C) Biosphere reserves (D) National parks

475. Social forestry is:
(A) Management of forests by cooperative societies
(B) Growing and management of useful plants on common lands
(C) Growing one type of trees
(D) Growing different types of plants together

476. Decrease in species diversity in tropical countries is mainly due to:
(A) Urbanisation (B) Pollution
(C) Deforestation (D) Soil erosion

477. Niger is a:
(A) Pulse crop (B) Fibre crop
(C) Cereal crop (D) Oilseed crop

478. Tobacco is extensively grown in:
(A) Andhra Pradesh and Gujarat
(B) Jummu and Kashmir
(C) M.P. and Maharastra
(D) North-Eastern states

479. Cotton is cultivated on commercial lines in:
(A) U.P. and Bihar
(B) A.P., Karnataka and Maharastra
(C) Gujarat and Rajasthan
(D) None of these

480. Sunhemp is largely grown for fibre purpose in:
(A) Eastern U.P. (B) Eastern Bihar
(C) Eastern M.P. (D) Rajasthan

481. Emergence of flag leaf indicates emergence of:
(A) Panicle (B) Inflorescence
(C) Ear (D) Tillers

482. In wheat crop, crown roots emerge:
(A) 2 weeks after sowing (B) 3 weeks after sowing
(C) 4 weeks after sowing (D) 5 weeks after sowing

483. In view of area and production of opium the first place is occupied by:
(A) M.P. (B) U.P.
(C) Bihar (D) Rajasthan

484. Food grain production in India during 2010-11 increased in comparison to 2009-10 to the tune of :
(A) 22 million tones (B) 26 million tones
(C) 30 million tones (D) 35 million tones

485. Four major mustard producing states in India are:
(A) Rajasthan, U.P., M.P. and Haryana
(B) Rajasthan, M.P., U.P. and Haryana
(C) Rajasthan, Haryana, U.P. and M.P.
(D) Rajasthan, Punjab, U.P. and Haryana

486. Adulteration of mustard oil with *Argemone mexicana* seed oil caused in the recent past:
(A) Tuberculosis (B) Diarrhoea
(C) Dropsy (D) None of these

487. Wheat is a:
(A) Monoecious plant (B) Diocious plant
(C) Hermaphrodite plant (D) None of these

488. First of all coloured cotton developed in India in the state of:
(A) Madhya Pradesh (B) Maharastra
(C) Gujarat (D) Andhra Pradesh

489 Botanical name of kodon is:
(A) *Echinochloa frumentacea* L.
(B) *Eleusine coracana* Gaertn
(C) *Paspalum scrobiculatum*
(D) None of these

490. First hybrid variety of sorghum was evolved in the year:
(A) 1946 (B) 1956 (C) 1964 (D) 1974

491. First hybrid variety of bajra was evolved in the year:
(A) 1946 (B) 1956 (C) 1965 (D) 1974

492. According to Manchurian classification soybean has been divided into:
(A) Two groups (B) Three groups
(C) Four groups (D) Only one group

493. It is believed that soybean originated in:
(A) Eastern Asia or China (B) Western Asia
(C) Africa (D) None of these

494. Jute is divided into:
(A) Two groups (B) Three groups
(C) One group (D) None of these

495. Jute belongs to the family:
(A) Compositae (B) Linaceae
(C) Tiliaceae (D) None of these

496. In alkaline soil which crop can be grown successfully?
(A) Potato (B) Paddy
(C) Banana (D) Cauliflower

497. Pegging of groundnut refers to:
(A) Flower initiation
(B) Initiation of secondary branches
(C) Entering of fertilized ovary in soil
(D) Grain development

498. The edible part of cabbage is:
(A) Root (B) Flower (C) Leaf (D) Fruit

499. Krishak Bharti periodical is published from:
(A) New Delhi (B) Lucknow
(C) Pantnagar (D) Hisar

500. Directorate of Knowledge Management in Agriculture is situated at:
(A) Pusa, New Delhi (B) Pusa, Samasti pur
(C) Hyderabad, AP (D) Hisar, Haryana

501. Fakhurruddin Ali Ahmad award related to:
(A) Best research work on Tribal people upliftment
(B) Team work in agricultural development
(C) Best technical book in hindi
(D) None of these

502. Hari Om Trust Award related to:
(A) Best research work on Tribal people upliftment
(B) Team work in agricultural development
(C) Best technical book in hindi
(D) None of these

503. Dr. Rajendra Prasad award related to:
(A) Best research work on Tribal people upliftment
(B) Team work in agricultural development
(C) Best technical book in hindi
(D) None of these

504. Evergreen revolution is related to:
(A) Potato
(B) Food grain production
(C) Milk production
(D) Increasing production in all together

505. Rainbow revolution is related to:
(A) Potato
(B) Food grain production
(C) Milk production
(D) Increasing production in all together

506. Green revolution is related to:
(A) Potato
(B) Food grain production
(C) Milk production
(D) Increasing production in all together

507. Brown /Black revolution is related to:
(A) Potato (B) Food grain production
(C) Milk production (D) Non conventional energy

508. Round revolution is related to:
(A) Potato
(B) Food grain production
(C) Milk production
(D) Increasing production in all together

509. White revolution is related to:
(A) Potato
(B) Food grain production
(C) Milk production
(D) Increasing production in all together

510. Blue revolution is related to:
(A) Potato (B) Food grain production
(C) Milk production (D) Fish production

511. Grey revolution is related to:
(A) Fertilizer
(B) Food grain production
(C) Milk production
(D) Increasing production in all together

512. Red revolution is related to:
(A) Potato
(B) Meat/tomato production
(C) Milk production
(D) Increasing production in all together

513. Golden revolution is related to:
(A) Potato
(B) Food grain production
(C) Horticultural crop
(D) Increasing production in all together

514. Pink revolution is related to:
(A) Potato (B) Food grain production
(C) Milk production (D) Prawan production

515. Silver revolution is related to:
(A) Chicken/eggs production
(B) Food grain production
(C) Milk production
(D) Increasing production in all together

516. Yellow revolution is related to:
(A) Potato
(B) Oilseed production
(C) Milk production
(D) Increasing production in all together

517. The agricultural Price Commission was set up to evolve balanced and integrated price structure in the year:
(A) 1962 (B) 1963 (C) 1964 (D) 1965

518. In Uttar Pradesh Krishi Utpadan Mandi Adhiniyam was passed in:
(A) 1962 (B) 1963 (C) 1964 (D) 1965

519. International financing agency to support National Agricultural Technology Project in India is:
(A) FAO (B) World Bank
(C) Asian Bank (D) UNDP

520. Which year is declared as International year of rice:
(A) 2004 (B) 2005 (C) 2006 (D) 2007

521. Which year is declared as International year of parthenium/neem patented year:
(A) 2004 (B) 2005 (C) 2006 (D) 2007

522. Which year is declared as International year of Desert and desertification:
(A) 2004 (B) 2005 (C) 2006 (D) 2007

523. Which year is declared as International year of water:
(A) 2004 (B) 2005 (C) 2006 (D) 2007

524. Which year is declared as International year of potato:
(A) 2008 (B) 2005 (C) 2006 (D) 2007

525. Which year is declared as International year of Astronomy:
(A) 2008 (B) 2009 (C) 2010 (D) 2011

526. Which year is declared as International year of biodiversity:
(A) 2008 (B) 2009 (C) 2010 (D) 2011

527. Which year is declared as International year of forest:
(A) 2008 (B) 2009 (C) 2010 (D) 2011

528. Who says "Everything else can wait, but not agriculture"?
(A) Shri. J.L. Nehru (B) Dr. A.P.J. Abul Kalam
(C) Dr. H. Rembold (D) Mahatma Gandhi

529. Who says "Agriculture is the culture of the times as/or all culture"?
(A) Shri. J.L. Nehru (B) Dr. A.P.J. Abul Kalam
(C) Dr. H. Rembold (D) Mahatma Gandhi

530. Who says "We have to have the second green revolution for ever increasing population of the country"?
(A) Shri. J.L. Nehru (B) Dr. A.P.J. Abul Kalam
(C) Dr. H. Rembold (D) Mahatma Gandhi

531. Who says "India is sitting on gold mines will all its Neem trees"?
(A) Shri. J.L. Nehru (B) Dr. A.P.J. Abul Kalam
(C) Dr. H. Rembold (D) Mahatma Gandhi

532. Total food grain production in India during 2011- 12 was:
(A) 235.00 mt (B) 250.55 mt
(C) 257.44 mt (D) 267.00 mt

533. Total pulses production in India during 2011- 12 was:
(A) 17.2 mt (B) 18.2 mt
(C) 19.2 mt (D) 20.2 mt

534. Total Oilseeds production in India during 2011- 12 was:
(A) 23.5 mt (B) 30.0 mt
(C) 35.1 mt (D) 35.5 mt

535. Total wheat production in India during 2011- 12 was:
(A) 93.90 mt (B) 95.54 mt
(C) 95.98 mt (D) 96.31 mt

536. Total rice production in India during 2011- 12 was:
(A) 85.00 mt (B) 90.54 mt
(C) 95.98 mt (D) 104.3 mt

537. Food grain production in India during 2011- 12 increased in comparison to 2010-11 was:
(A) 12.95 mt (B) 13.95 mt
(C) 14.95 mt (D) 15.95 mt

538. Total pulses production in India during 2010-11was:
(A) 17.24 mt (B) 18.24 mt
(C) 19.24 mt (D) 20.24 mt

539. Total oilseed production in India during 2010-11was:
(A) 23.5 mt (B) 25.0 mt
(C) 32.5 mt (D) 35.5 mt

540. Wheat production in India during 2011- 12 increased in comparison to 2010-11was:
(A) 4.03 mt (B) 5.03 mt
(C) 6.03 mt (D) 7.03 mt

541. Rice production in India during 2011- 12 increased in comparison to 2010-11was:
(A) 8.34 mt (B) 9.34 mt
(C) 10.34 mt (D) 11.34 mt

542. Area under total food grain in India during 2011- 12 was:
(A) 115.05 mha (B) 125.03 mha
(C) 128.04 mha (D) 130.28 mha

543. Area under total pulses in India during 2011- 12 was:
(A) 23.28 mha (B) 26.40 mha
(C) 24.78 mha (D) 25.85 mha

544. Area under total oilseeds in India during 2011- 12 was:
(A) 27.56 mha (B) 25.96 mha
(C) 27.22 mha (D) 26.44 mha

545. Area under wheat in India during 2011- 12 was:
(A) 29.90 mha (B) 29.07 mha
(C) 28.46 mha (D) 27.75 mha

546. Area under rice in India during 2011- 12 was:
(A) 35.85 mha (B) 40.50 mha
(C) 43.97 mha (D) 45.85 mha

547. The Central AGMARK lab is located at:
(A) Lucknow (B) Nagpur
(C) Bhopal (D) Hyderabad

548. Cultivation of duram wheat is primarily confined to:
(A) Central India (B) Southern India
(C) Central & Southern India (D) UP hills

549. Tick the correct agro-climatic zone of UP which has highest average population density per km^2:
(A) Bundelkhand zone (B) Central plains zone
(C) Hills zone (D) North-eastern plains zone

550. National Food Security Mission started in which five year plan:
(A) 9th (B) 10th (C) 11th (D) 12th

551. Bharat Ratna Dr. C. Subramanian Award is related to:
(A) Best agriculture teacher
(B) Best technical book
(C) Tribal people upliftment
(D) Best research in agriculture

552. In India, during 1960 milk production was 2 crore metric tones, it was rose in 2010-11 to:
(A) 10.2 crore metric tones (B) 11.2 crore metric tones
(C) 12.2 crore metric tones (D) 13.2 crore metric tones

553. Father of White Revolution in India is:
(A) Dr. M.S. Swaminathan (B) Dr. Vergeese Kurien
(C) Dr. R.S. Paroda (D) Dr. B.P. Pal

554. National Dairy Development Board was established in:
(A) 1965 (B) 1966 (C) 1967 (D)1968

555. Operation of Flood was launched in:
(A) 1969 (B) 1970 (C) 1971 (D)1972

556. Gujarat Cooperative Milk Marketing Federation (GCMMF) was established in:
(A) 1971 (B) 1972 (C) 1973 (D)1074

557. India Rank in milk production during 2010-11 was:
(A) First (B) Second (C) Third (D) Fourth

558. Marketing of AMUL is to be done by:
(A) AMUL it self (B) GCMMF
(C) Government of India (D) Gujarat State

559. Who is known as milk man of India and Amul man is?
(A) Dr. M.S. Swaminathan (B) Dr. Vergeese Kurien
(C) Dr. R.S. Paroda (D) Dr. B.P. Pal

560. Growth rate of agricultural development in 9th five year plan was:
(A) 2.5% (B) 2.4% (C) 3.3% (D)4.0%

561. Growth rate of agricultural development in10th five year plan was:
(A) 2.5% (B) 2.4% (C) 3.3% (D) 4.0%

562. Growth rate of agricultural development in11th five year plan was:
(A) 2.5% (B) 2.4% (C) 3.3% (D) 4.0%

563. Growth rate of agricultural development in 12th five year plan will be:
(A) 2.5% (B) 2.4% (C) 3.3% (D) 4.0%

564. The wheat production during 1965-66 was 12.57 mt but now rose during 2011-12 to:
(A) 90 mt (B) 92 mt (C) 94 mt (D) 96 mt

565. The increase in production of pulses by 2020 AD to be:
(A) 29.0 mt (B) 30.3 mt
(C) 31.0 mt (D) 32.0 mt

566. The increase in production of pulses by 2030 AD to be:
(A) 29.0 mt (B) 30.3 mt
(C) 31.0 mt (D) 32.0 mt

567. Guarana is:
(A) Fodder plant (B) Oil plant
(C) Beverage plant (D) Food plant

568. Basic input of agriculture is:
(A) Irrigation (B) Fertilizer
(C) Soil (D) Seed

569. Which one of the following chemical is used for reclamation of alkali soil:
(A) Lime (B) Gypsum
(C) Sodium hydride (D) Bleaching chloride

570. Central Upland Rice Research Station located at:
(A) Cuttak (B) Phillipines
(C) New Delhi (D) Masodha

571. National Cooperative Development Corporation established in:
(A) 1962 (B) 1963 (C) 1964 (D) 1965

572. Who is the chairman of planning commission:
(A) President (B) Prime Minister
(C) Agriculture Minister (D) D.G. of ICAR

573. Name of agriculture scientist who won Nobel prize:
(A) N.E. Borlaug (B) M.S. Swaminathan
(C) B.P. Pal (D) P.J. Kurien

574. Green revolution mainly related to which crop:
(A) Wheat (B) Potato
(C) Pulses (D) Sugarcane

575. IARI established in Pusa Bihar:
(A) 1902 (B) 1904 (C) 1905 (D) 1906

576. IARI Pusa Bihar established to New Delhi on:
(A) 1935 (B) 1936 (C) 1937 (D) 1938

577. IARI got status of deemed university on:
(A) 1958 (B) 1959 (C) 1960 (D) 1961

578. IARI established during:
(A) Lord Curzon (B) Lord Macnaro
(C) Lord Carbvalis (D) None of these

579. Who was first secretary of ICAR:
(A) Mohd. Habibulla (B) S.A. Haidery
(C) B.P.Pal (D) M.S. Swaminathan

580. Who was first chairman of ICAR:
(A) Mohd. Habibulla (B) S.A. Haidery
(C) B.P.Pal (D) M.S. Swaminathan

581. When ICAR declared an autonomous body?
(A) 1964 (B) 1965 (C) 1966 (D) 1967

582. How any crops covered in National Food Security Mission are:
(A) Rice (B) Wheat
(C) Pulses (D) All of these

583. Who was first Indian director of IARI:
(A) B. Vishwanath (B) S. A. Haidery
(C) B. P. Pal (D) M. S. Swaminathan

584. Agricultural department was established on:
(A) 1880 (B) 1881 (C) 1882 (D) 1883

585. First irrigation commission was established on:
(A) 1898 (B) 1899 (C) 1900 (D) 1901

586. Central Agriculture Commission was established on:
(A) 1969 (B) 1970 (C) 1971 (D) 1972

587. Indian National Sugarcane Committee was established on:
(A) 1944 (B) 1945 (C) 1946 (D) 1947

588. First five year plan was started on:
(A) 1947-1951 (B) 1948-1952
(C) 1951-1956 (D) 1950-1955

589. NABARD established on:
(A) 1981 (B) 1982 (C) 1983 (D) 1984

590. Food Corporation of India was established on:
(A) 1963 (B) 1964 (C) 1965 (D) 1966

591. ICAR foundation day celebrated on:
(A) 16 July (B) 16 August
(C) 16 September (D) 16 October

592. National Agriculture Insurance was started on:
(A) 1998-1999 (B) 1999-2000
(C) 2000-2001 (D) 2001-2002

593. Now-a-days wheat cultivation started in:
(A) West Bengal (B) Andhra Pradesh
(C) Maharastra (D)North-eastern states

594. Maximum water used for irrigation in which river:
(A) Yamuna (B) Ganga
(C) Narmada (D) Ghaghara

595. Hand Book of Agriculture published from:
(A) New Delhi (B) Lucknow
(C) Bhopal (D) Jaipur

596. Rothamsted Agricultural Research Station started by:
(A) Larves and Gilbert (B) Phiphs
(C) Bourlag (D) Swaminathan

597. Maximum consumption of edible oils in India:
(A) Mustard (B) Coconut
(C) Groundnut (D) Sesame

598. Maximum agriculture culturable area in which country:
(A) America (B) Australia
(C) Canada (D) India

599. Queen of food grain is:
(A) Wheat (B) Rice
(C) Maize (D) Potato

600. Which crop is declared as a gold in America:
(A) Wheat (B) Rice
(C) Sorghum (D) Maize

601. Weight of a cotton bale is:
(A) 150 kg (B) 160 kg
(C) 170 kg (D) 180 kg

602. Queen of spices is:
(A) Ilaichi (B) Turmeric
(C) Chilli (D) Coriander

603. King of fruit is:
(A) Banana (B) Papaya
(C) Guava (D) Mango

604. Pseudo cereal is:
(A) Duram wheat (B) Buck wheat
(C) Macroni wheat (D) None of these

605. Maximum wheat production in:
(A) Punjab (B) Haryana
(C) Uttar Pradesh (D) Madhya Pradesh

606. Maximum potato production in:
(A) Punjab (B) Haryana
(C) Uttar Pradesh (D) Madhya Pradesh

607. Maximum rice production in:
(A) Tamilnadu (B) West Bengal
(C) Uttar Pradesh (D) Orissa

608. Maximum sugarcane production in:
(A) Punjab (B) Haryana
(C) Uttar Pradesh (D) Madhya Pradesh

609. Which state is most affected by water erosion:
(A) West Bengal (B) Uttar Pradesh
(C) Rajasthan (D) Madhya Pradesh

610. Which state is most affected by wind erosion:
(A) West Bengal (B) Uttar Pradesh
(C) Rajasthan (D) Madhya Pradesh

611. Which state has maximum area under alkali soils in India:
(A) West Bengal (B) Uttar Pradesh
(C) Gujarat (D) Bihar

612. Which state has maximum area under acidic soils in India:
(A) West Bengal (B) Uttar Pradesh
(C) Gujarat (D) Bihar

613. Maximum fertilizer consumption per hectare is:
(A) Punjab (B) Pondicherry
(C) Haryana (D) Uttar Pradesh

614. Maximum per capita availability of land (Approximately 57 ha) is:
(A) Canada (B) Australia
(C) America (D) England

615. Maximum area under non reporting land is:
(A) Arunachal Pradesh (B) Meghalaya
(C) Tripura (D) J & K

616. Area under problematic land is:
(A) 187 mha (B) 189 mha
(C) 195 mha (D) 198 mha

617. Per capita availability of pulses during 2011-12 is:
(A) 27 g/day (B) 28 g/day
(C) 29 g/day (D) 30 g/day

618. Maximum cigarette tobacco production is:
(A) Andhra Pradesh (B) Gujarat
(C) Bihar (D) Uttar Pradesh

619. Maximum Bidi tobacco production is:
(A) Andhra Pradesh (B) Gujarat
(C) Bihar (D) Uttar Pradesh

620. Maximum chewing tobacco production is:
(A) Andhra Pradesh (B) Gujarat
(C) Bihar (D) Uttar Pradesh

621. Highest productivity of wheat is:
(A) Uttar Pradesh (B) Punjab
(D) Haryana (D) Rajasthan

622. Highest total production of wheat in India is:
(A) Uttar Pradesh (B) Madhya Pradesh
(C) Rajasthan (D) Gujarat

623. Highest total production of pulses in India is:
(A) Uttar Pradesh (B) Madhya Pradesh
(C) Rajasthan (D) Gujarat

624. Highest total production of mustard in India is:
(A) Uttar Pradesh (B) Madhya Pradesh
(C) Rajasthan (D) Gujarat

625. Highest total production of groundnut in India is:
(A) Uttar Pradesh (B) Madhya Pradesh
(C) Rajasthan (D) Gujarat

626. Highest total production of jute in India is:
(A) West Bengal (B) Madhya Pradesh
(C) Bihar (D) Assam

627. Highest total production of tea in India is:
(A) West Bengal (B) Madhya Pradesh
(C) Bihar (D) Assam

628. Highest total production of coffee in India is:
(A) Karnataka (B) Andhra Pradesh
(C) Tamilnadu (D) Assam

629. Highest total production of rubber in India is:
(A) West Bengal (B) Madhya Pradesh
(C) Bihar (D) Kerala

630. Highest total production of onion in India is:
(A) Maharastra (B) Madhya Pradesh
(C) Bihar (D) Gujarat

631. Maximum production of sugarcane in the world is:
(A) Brazil (B) China
(C) India (D) France

632. Maximum production of cotton in the world is:
(A) Brazil (B) China
(C) India (D) France

633. Maximum production of tobacco in the world is:
(A) Brazil (B) China
(C) India (D) France

634. Maximum production of tea in the world is:
(A) Brazil (B) China
(C) India (D) France

635. Which is more suitable for germplasm collection:
(A) Centre of production (B) Homelands
(C) Tropical areas (D) Temperate areas

636. Maximum production of coffee in the world is:
(A) Brazil (B) China
(C) India (D) France

637. Maximum production of wheat in the world is:
(A) Brazil (B) China
(C) India (D) France

638. Maximum production of rice in the world is:
(A) Brazil (B) China
(C) India (D) France

639. Maximum production of groundnut in the world is:
(A) Brazil (B) China
(C) India (D) France

640. Maximum production of mustard in the world is:
(A) Brazil (B) China
(C) India (D) France

641. Maximum production of onion in the world is:
(A) Brazil (B) China
(C) India (D) France

642. Maximum production of fruits in the world is:
(A) Brazil (B) China
(C) India (D) France

643. Maximum production of potato in the world is:
(A) Brazil (B) China
(C) India (D) France

644. Maximum production of fishes in the world is:
(A) Brazil (B) China
(C) India (D) France

645. Maximum area under macroni wheat in India is:
(A) Punjab (B) West Bengal
(C) Punjab and West Bengal (D) Uttar Pradesh

646. Bowl of rice is known as:
(A) West Bengal (B) Kerala
(C) Tamilnadu (D) Karnataka

647. Bowl of spices is known as:
(A) West Bengal (B) Kerala
(C) Tamilnadu (D) Karnataka

648. City of gardens is known as:
(A) Bangalore (B) J & K
(C) New Delhi (D) Hyderabad

649. Minimum area under forest is:
(A) Rajasthan (B) Madhya Pradesh
(C) West Bengal (D) Uttar Pradesh

650. Maximum area under forest is:
(A) Rajasthan (B) Madhya Pradesh
(C) West Bengal (D) Uttar Pradesh

651. Per capita availability of milk in India is:
(A) 184 ml/day (B) 194 ml/day
(C) 204 ml/day (D) 214 ml/day

652. Sustainable agriculture is also known as:
(A) Perma culture (B) Permanent agriculture
(C) Both (A) and (B) (D) None of these

653. Pea cultivation is not recommended in:
(A) Normal soils (B) Acidic soils
(C) Alluvial soils (D) Alkali soils

654. Raring of sheep was started on:
(A) 8700 AD (B) 7700 AD
(C) 7500 AD (D) 6000 AD

655. Raring of goat was started on:
(A) 8700 AD (B) 7700 AD
(C) 7500 AD (D) 6000 AD

656. Raring of wheat and barley was started on:
(A) 8700 AD (B) 7700 AD
(C) 7500 AD (D) 6000 AD

657. Raring of cow and pig was started on:
(A) 8700 AD (B) 7700 AD
(C) 7500 AD (D) 6000 AD

658. Raring of silkworm was started on:
(A) 1700 AD (B) 2700 AD
(C) 3500 AD (D) 4000 AD

659. Cultivation of maize was started on:
(A) 4400 AD (B) 2900 AD
(C) 2300 AD (D) 2200 AD

660. Knowledge gain for ploughing and irrigation was on:
(A) 4400 AD (B) 2900 AD
(C) 2300 AD (D) 2200 AD

661. Cultivation of arhar, pine apple, mustard and cotton was started on:
(A) 4400 AD (B) 2900 AD
(C) 2300 AD (D) 2200 AD

662. Raring of poultry, buffalo and elephant was started on:
(A) 4400 AD (B) 2900 AD
(C) 2300 AD (D) 2200 AD

663. Cultivation of ragi and mandua was started on:
(A) 1800 AD (B) 1725 AD
(C) 1500 AD (D) 1400 AD

664. Cultivation of sorghum was started on:
(A) 1800 AD (B) 1725 AD
(C) 1500 AD (D) 1400 AD

665. Cultivation of sugarcane was started on:
(A) 1800 AD (B) 1725 AD
(C) 1500 AD (D) 1400 AD

666. Cultivation of potato was started on:
(A) 3500 AD (B) 4400 AD
(C) 4500 AD (D) 7500 AD

667. Which type of cultivation practices mainly responsible for agricultural development in India:
(A) Precision farming
(B) Shifting cultivation
(C) Mutistoryed cultivation
(D) Organic farming

668. Importance of animals and raring of animals was known as era of:
(A) Pastoral age (B) Neolithic age
(C) Bronze age (D) None of these

669. Cotton is commercially grown in:
(A) Andhra Pradesh (B) Karnataka
(C) Maharastra (D) All of these

670. Sorghum is known as camel due to:
(A) Important crop of millet
(B) Drought tolerant
(C) Important feed of camel
(D) Important crop of dry areas

671. Cultivation of duram wheat mainly in:
(A) Central and southern India
(B) North India
(C) North-eastern India
(D) Dry region of India

672. Spot satellite services was started by:
(A) UK (B) USA
(C) India (D) China

673. Which country is origin of tea but totally stopped for its cultivation:
(A) China (B) India
(C) Srilanka (D) Bhuan

674. Area and production of posta maximum in which state:
(A) Bihar (B) West Bengal
(C) Uttar Pradesh (D) Haryana

675. Panchgavya is an organic product has the potential to play the role of promoting:
(A) Growth (B) Productivity
(C) Growth and productivity (D) None of these

676. Panchgavya consist of:
(A) Cow dung and cow urine
(B) Cow milk and cow curd
(C) Cow ghee
(D) All of these

677. Sorghum is considered as the king of millets also known as:
(A) Camel crop (B) King of cereals
(C) Queen of cereals (D) Important crop of kharif

678. Total area under fodder production is:
(A) 7.3 mha (B) 8.3 mha
(C) 9.3 mha (D) 10.3 mha

679. Requirement of concentrate as feed for dairy in India is:
(A) 570 mt (B) 923 mt
(C) 1025 mt (D) 1135 mt

680. Requirement of green fodder as feed for dairy in India is:
(A) 570 mt (B) 923 mt
(C) 1025 mt (D) 1135 mt

681. Requirement of dry fodder as feed for dairy in India is:
(A) 570 mt (B) 923 mt
(C) 1025 mt (D) 1135 mt

682. Availability of concentrate as feed for dairy in India is:
(A) 45 mt (B) 390 mt
(C) 443 mt (D) 501 mt

683. Availability of green fodder as feed for dairy in India is:
(A) 45 mt (B) 390 mt
(C) 443 mt (D) 501 mt

684. Availability of dry fodder as feed for dairy in India is:
(A) 45 mt (B) 390 mt
(C) 443 mt (D) 501 mt

685. IRRI is the abbreviation for:
(A) International Raw Rice Institute
(B) Institute of Rice Research in India
(C) Indian Rice Research Institute
(D) International Rice Research Institute

686. The typical odour of garlic is due to:
(A) Sinigrin (B) Ficin
(C) Allicin (D) Alliin

687. Which crop is called as king of cereal crops?
(A) Rice (B) Wheat
(C) Oat (D) Barley

688. The term 'Tropical oils' refers to:
(A) Palm oil and coconut oil
(B) Olive oil and almond oil
(C) Corn oil and soyabean oil
(D) Sunflower oil and safflower oil

689. 'Touch me not' plant exhibits a reflex action that is due to:
(A) Special knee joint
(B) Sudden temperature control
(C) Exceptional nervous conduction
(D) Powerful water pressure alteration

690. "Tension wood" is the name given to that wood which reacts to:
(A) Drought (B) Low oxygen
(C) Drooping (D) Disease

691. Button lac is manufactured from:
(A) Molten lac (B) Crushed lac
(C) Shel lac (D) Seed lac

692. Essential oils are used in:
(A) Cosmetics and Toiletries
(B) Food beverage
(C) Paint industry
(D) All of these

693. First Face Pack was developed in:
(A) France (B) Rome
(C) Greece (D) None of these

694. Oil content in fresh herbage in menthol mint is:
(A) 0.1-0.5 % (B) 0.5-0.8 %
(C) 1.0-1.5 % (D) 1.5 – 2.0 %

695. Major constituent of *Occimum cannum* is:
(A) Methyl chavical (B) Thymal
(C) Linalool (D) Engenol

696. Productivity of rice is highest in:
(A) Desi rice (B) Hybrid rice
(C) Basmati rice (D) Composite rice

697. The Bengal famine of 1942 was due to failure of paddy crop caused by:
(A) *Helminthosporium oryzae*
(B) *Xanthomonas compestris pv oryzae*
(C) *Pyricularia oryzae*
(D) *Rhizoctonia solani*

698. Highest number of sugar mills in India is located in:
(A) Maharasthra (B) Tamilnadu
(C) Bihar (D) UP

699. Sustainable agriculture can be achieved by:
(A) More resources
(B) More subsidies
(C) More technological support
(D) Better prices of the produce

700. The cause of famous 'Irish famine' was:
(A) Early blight disease of potato
(B) Late blight disease of potato
(C) Blast disease of paddy
(D) Rust disease of wheat

701. Contribution of agriculture sector to GDP in India is:
(A) 10% (B) 20% (C) 30% (D) 40%

702. High yielding variety programme was started in the year :
(A) 1965 (B) 1966 (C) 1967 (D) 1968

703. N.E. Borlaug was awarded Nobel Prize in:
(A) Peace (B) Economic
(C) Agriculture (D) Physics

704. NATP as sponsored by:
(A) Planning commission (B) Government of India
(C) ICAR (D) World Bank

705. IVLP stands for:
(A) Institution Village Linkage Programme
(B) Integrated Village Lab to Land Programme
(C) India Village Linkage Programme
(D) Indian Village Linkage Plan

706. Dwarfing gene of wheat is:
(A) Pal -1 (B) Atomita – 1
(C) Norin – 10 (D) Pelita -2

707. Jojoba yields wax for cosmetics from its:
(A) Stem (B) Leaves
(C) Seeds (D) Fruit pulp

708. IR 36 was developed through breeding of:
(A) Six rice varieties and *Oryza nivara*
(B) 13 rice varieties and *Oryza nivara*
(C) *Oryza sativa* and *O. indica*
(D) *Oryza indica* and *O. nivara*

709. Which one is beverage plant:
(A) *Ilex paraguriensis* (B) *Caltha edulis*
(C) *Poullinia cupana* (D) All of these

710. The three top most major crops of the world in the order of total production are:
(A) Wheat>Rice > Maize (B) Rice>Wheat>Maize
(C) Wheat>Maize>Rice (D) Rice>Maize>Wheat

711. Potential food plant is:
(A) Winged bean (*Psophocarpus tetragonolobus*)
(B) Jajoba (*Simmondsia chinensis*)
(C) Guayule (*Parthenium argentatum*)
(D) *Leucaena leu co cephala*

712. A potential source of rubber is:
(A) Guayule (*Parthenium argentatum*)
(B) *Leucaena leucocephala*
(C) *Cucurbita foetidissima*
(D) *Citrullus colocynthis*

713. Cola, the common beverage, is obtained from:
(A) Seeds (B) Flowers
(C) Leaves (D) Stems

714. Maximum genetic diversity of crop plants occurs where agriculture is:
(A) Run on commercial lines (B) Advanced
(C) Rainfed (D) Primitive

715. Rastriya Krishi Vikas Yojna (RKVJ) launched on:
(A) 2007 (B) 2008
(C) 2010 (D) 2009

716. Which among the following is another name for vitamin B12?
(A) Niacin (B) Pyridoxal phosphate
(C) Cobalamine (D) Riboflavin

717. Who is the present chairman of Protection of Plant varieties and farmers Right Authority (PPV&FRA) :
(A) Dr. K.V. Prabhu (B) Dr. P.L. Gautam
(C) Dr. R.R. Hanchinal (D) Dr. S. Nagarajan

718. The National Bureau of Agricultural Important Insects was established by ICAR in .... was earlier known as ....
(A) Bangalore PDBC
(B) New Delhi, National Pusa Collection
(C) Ranchi, Indian Lac Research Institute
(D) New Delhi, NCIPM

719. The cause of the great Bengal famine was :
(A) Blast of Rice (B) Brown spot of rice
(C) Rust of Wheat (D) Karnal Bunt of Wheat

720. Which of the following microorganism causes fatal poisoning in canned fruits and vegetables?
(A) Aspergillus flavus
(B) Penicillium digitatum
(C) Clostridium botulinum
(D) Rhizoctonia solani

721. The most important sucking pests of cotton and rice are respectively
(A) Nilaparvata lugens and Aphis gossypii
(B) Aphis gossypii and Thrips oryzae
(C) Amrasca biguttula biguttula and Scirtothrips dorsalis
(D) Thrips gossyppei and Orseolia oryzae

722. A virus free clone from a virus infected plant can be obtained by :
(A) Cotyledonary leaf culture
(B) Axenic culture
(C) Stem culture
(D) Meristem tip culture

723. Which of the following is rot an objective of the National Food Security Mission?
(A) Sustainable increase in production of rice, wheat and pulses
(B) Restoring soil fertility and productivity at individual farm level
(C) Promoting use of biopesticides and organic fertilizer
(D) Creation of employement opportunities

724. Agmarknet, a portal for the dissemination of agricultural marketing information, is a joint endeavour of :
(A) DMI and NIC
(B) DMI and Ministry of Agriculture
(C) NIC and Ministry of Agriculture
(D) DMI and Directorate of Economics and statistics

725. Farmers first concept was proposed by :
(A) Paul leagens (B) Neils Rolling
(C) Robert Chamber (D) Indira Gandhi

726. Dog ridge is :
(A) Salt tolerant resistance of mango
(B) Salt tolerant resistance of guava
(C) Salt tolerant resistance of grape
(D) Salt tolerant resistance of citrus

727. Which of the following fertilizers is not produced in India?
(A) DAP (B) Urea
(C) Muriate of Potato (D) TSP

728. Pungency in mustard oil is due to :
(A) Glucoside (B) Glucosinolates
(C) Prussic acid (D) Glucosinic acid

729. Growing degree day concept is useful in forecasting?
(A) Rainfall (B) Drought
(C) Crop harvest dates (D) Frost

730. Which of the following is called "Green Gold"?
(A) Green gram (B) Sisal
(C) Bamboo (D) None of the above

731. Eurca sativa is a :
(A) Pulse crop (B) Tuber crop
(C) Oilseed crop (D) Vegetable crop

732. Per hectare consumption of $N + P_2O_5 + K_2O$ in India is about :
(A) 90 kg/ha (B) 102 kg/ha
(C) 117 kg/ha (D) 147 kg/ha

733. Percent area treated with fertilizers in India is maximum :
(A) Paddy (B) Wheat
(C) Sugarcane (D) Cotton

734. In which region, Hopkins (1988) bioclimate law is mainly applicable?
(A) Temperate (B) Arid
(C) Tropical (D) Semi-arid

735. Which of the following crops have been approved for comercial cultivation :
(A) Bt cotton and Bt brinjal
(B) Bt cotton and Golden rice
(C) Bt maize and Bt cotton
(D) Bt cotton only

736. Cotton belongs to the family :
(A) Cruciferae (B) Anacardiaceae
(C) Malvaceae (D) Solanaceae

737. According to the approach paper to the 12th Five Year plan is :
(A) Inclusive growth
(B) Sustainable growth
(C) Faster, more inclusive and sustainable growth
(D) Inclusive and sustainable growth

738. To address the problems of sustainable and holistic development of rainfed areas including appropriate farming and livelihood system approaches, the Government of India has set up the :
(A) National Rainfed Area Authority
(B) National Watershed Development Project for Rainfed Areas
(C) National Mission on Rainfed Areas
(D) Command Area Development and Water Management Authority

739. Which of the following sub-schemes are not covered under the Rastriya Krishi Vikas Yojna?
(A) Extending the Green Revolution to eastern India
(B) Development of 6000 pulses and oilseed villages in identified watersheds
(C) National Mission on Saffron
(D) National Mission on Bamboo

740. The species of rice (Oryza) other than O. sativa that is cultivated is :
(A) O. rufipygon (B) O. longisteminata
(C) O. glaberrima (D) O. nivera

741. Carthamus tinctorius is a :
(A) Fodder crop (B) Tuber crop
(C) Oilseed crop (D) Vegetable crops

742. Hydrocyanic acid in sorghum is synthesized in :
(A) Leaves (B) Root
(C) Stem (D) Flower premordia

743. Norman E. Barlaug got Nobel Prize in the field of :
(A) Agriculture (B) Medicine
(C) Peace (D) Economics

744. Protein Content of millet rice is about :
(A) 4-5% (B) 6-7%
(C) 9-10% (D) 12-13%

745. Boehmenia nivea is botanical name of
(A) Mesta (B) Indigo
(C) Ramie (D) Buckwheat

746. Pigeonpea is categorized as :
(A) Self pollinated (B) Cross pollinated
(C) Often self pollinated (D) Often cross pollinated

747. The UPOV (Union for the Protection of New Plant Varieties) head quarters is located in :
(A) Geneva (B) Berne (C) Rome (D) Paris

748. `Double digging' is a method of :
(A) Bio intensive agriculture
(B) Deforestation
(C) Aforestation
(D) Water conservation

749. Which one of the following is not the goal of sustainable agriculture in India :
(A) Maintaining productive top soil
(B) Mechanical farming
(C) Reduce the use of chemical fertilizer and pesticide
(D) Keep farms economically viable

750. Rice was first domesticated approximately 5000 years age in :
(A) Southeast Asia (B) South west Asia
(C) South Asia (D) Asia Pacific

751. Two major races of rice are :
(A) European and Tropical
(B) Indica and Japonicum
(C) Asiatic and American
(D) Temperate and Tropical

752. The specific heat of water is :
(A) 0.5 Cal/g (B) 1 Cal/g
(C) 2 Cal/g (D) 10 Cal/g

753. Non-edible plant suitable for bio-diesel :
(A) Castor (B) Rapseed
(C) Coconut (D) Jatropa

754. Plants capable of growing in rock crevices are called:
(A) Lithophytes (B) Calciphytes
(C) Chosmophytes (D) None of these

755. Total geographical area of India is :
(A) 328.9 ha (B) 328.9 mha
(C) 328.9 sq.km (D) 328.9 m sq.km

756. Net cultivated area in India is:
(A) 328 mha (B) 180 mha
(C) 150 mha (D) 143 mha

757. Contribution of agriculture in GDP is :
(A) 60% (B) 40% (C) 30% (D) 23%

758. Most of the wheat varieties are :
(A) Quantitative long day plants
(B) Qualitative long day plants
(C) Day neutral
(D) Short day plants

759. Mat nursery is related to :
(A) Onion (B) Tobacco
(C) Chillies (D) Rice

760. In Indo-gangetic plains, rice-wheat cropping system covers about:
(A) 8 mha (B) 10 mha
(C) 15 mha (D) 5 mha

761. The area under Bt-cotton in India is about:
(A) 6.4 mha (B) 5.4 mha
(C) 4.4 mha (D) 3.4 mha

762. The first maize hybrid in India was:
(A) Vikram (B) Kisan
(C) Vijay (D) Ganga-1

763. Which state contributes maximum pulse production?
(A) U.P. (B) M.P.
(C) Punjab (D) Maharashtra

764. The chemical responsible for lathyrism in Mammals is:
(A) BOAA (B) HCN
(C) 2,4-DB (D) NAN

765. Plants adopted to saline soils are called as :
(A) Lithophytes (B) Halophytes
(C) Chasmophytes (D) None of these

766. Which degradation plays a major role in the persistence and behaviour of herbicides in soil?
(A) Thermal (B) Physical
(C) Chemical (D) Microbial

767. In India, area under rice is about :
(A) 25 mha (B) 28 mha
(C) 45 mha (D) 57 mha

768. Ground water table is measured by:
(A) Pressure plate (B) Neutron Probe
(C) Tensiometer (D) Piezometer

769. Most widely cultivated mustard type in India is:
(A) Black mustard (B) Yellow/brown sarson
(C) Toria (D) Indian mustard

770. In India maximum area under wheat is occupied by the species
(A) Triticum durum (B) Triticum aestivum
(C) Triticum vulgare (D) Triticum dicoccum

771. Plants that grows on extremely dry soils are placed under:
(A) Hydrophytes (B) Xerophytes
(C) Thalophytes (D) Hydroponies

772. Study of soil in relation to higher plants is called as:
(A) Geophysics (B) Pedology
(C) Edaphology (D) Physiology

773. Which food is desginated as "boneless meat"?
(A) Potato (B) Banana
(C) Soybean (D) Topioca

774. Which of the following method involves analysis of plant tissue for nutrient management in sugarcane?
(A) STCR (B) DRIS
(C) Crop logging (D) Crop lodging

775. Sugarcane growers of a region sell their whole produce to a single sugar factory. Such situation is:
(A) Monopsomy (B) Monopolistic
(C) Moropoly (D) Oligopoly

776. Bushening operation is proformed in:
(A) Wheat (B) Rice
(C) Maize (D) Pearlmillet

777. General practice of cropping under dryland farming areas of arid and semi-arid regions of India is referred as:
(A) Relay cropping (B) Inter cropping
(C) Alternate cropping (D) Multiple cropping

778. The most common method of irrigation is India is:
(A) Sprinkler (B) Furrow
(C) Border strip (D) Check basin

779. The ratio between crop ET and pan ET, where the crop is raised on a large scale under optimum growing condition is called:
(A) CU (B) Crop coefficient
(C) Critical value (D) Crop efficiency

780. By making use of non-commercial portion of the crop for the protection or soil improvement activities is called as:
(A) Crop Production Technology
(B) Crop Technology
(C) Crop residue management
(D) Crop lodging

781. A green manure cum fibre cum minor fodder crops is:
(A) Mesta (B) Sisal
(C) Linseed (D) Sunhemp

782. Growing of five different crops with different canopy levels in coconut based FS is called:
(A) Mixed cropping
(B) Mixed farming
(C) Multi storied cropping
(D) Home stead farming

783. Basic requirement of zero tillage is:
(A) Raw surface is cleared of mulch
(B) Weed management by herbicide
(C) Seed placed in proper depth with good germination
(D) All of the above

784. Which portion of sugarcane stalk should normally be utilized for seed sowing?
(A) Top 1/4th portion
(B) Top 1/3th portion
(C) Bottom 1/4th portion
(D) Bottom 1/3rd to 1/2 portion

785. The practice of flashing out germinable weed seeds before crop sowing is called:
(A) Weed eradication (B) Soil solarization
(C) Stale seed bed (D) Weed preservation

786. The individual crop which is a part of cropping system are called as:
(A) Mixed crop (B) Component crop
(C) Intercrops (D) Alley crops

787. Repetitive growing of same crop over several years is called:
(A) Single cropping (B) Pure cropping
(C) Sole cropping (D) Monoculture

788. Collection and storage of water for subsequent use in dry period is called:
(A) Water logging (B) Water harvesting
(C) Watershed (D) Crop logging

789. The product formed after dehydration of green plant material so as to decrease the moisture content to 15% or less is known as:
(A) Kutti (B) Bhusa
(C) Hay (D) Silage

ANSWERSHEET: 1.GENERAL AGRICULTURE

1. (A); 2. (A); 3. (B); 4. (C); 5. (C); 6. (B); 7. (C); 8. (D); 9. (D); 10. (A); 11. (B); 12. (B); 13. (B); 14. (D); 15. (B); 16. (D); 17. (C); 18. (D); 19. (A); 20. (A); 21. (A); 22. (D); 23. (C); 24. (B); 25. (A); 26. (A); 27. (B); 28. (D); 29. (C); 30. (A); 31. (A); 32. (B); 33. (A); 34. (C); 35. (D); 36. (C); 37. (A); 38. (D); 39. (D); 40. (A); 41. (B); 42. (D); 43. (A); 44. (A); 45. (C); 46. (B); 47. (D); 48. (D); 49. (B); 50. (D); 51. (B); 52. (D); 53. (A); 54. (A); 55. (C); 56. (C); 57. (D); 58. (D); 59. (D); 60. (D); 61. (C); 62. (B); 63. (D); 64. (C); 65. (D); 66. (A); 67. (A); 68. (A); 69. (C); 70. (C); 71. (B); 72. (C); 73. (B); 74. (D); 75. (A); 76. (D); 77. (D); 78. (D); 79. (A); 80. (A); 81. (A); 82. (C); 83. (D); 84. (B); 85. (C); 86. (B); 87. (A); 88. (B); 89. (A); 90. (B); 91. (D); 92. (C); 93. (B); 94. (A); 95. (C); 96. (D); 97. (B); 98. (C); 99. (A); 100. (B); 101. (A); 102. (A); 103. (C); 104. (B); 105. (D); 106. (A); 107. (A); 108. (A); 109. (A); 110. (B); 111. (B); 112. (A); 113. (D); 114. (C); 115. (A); 116. (A); 117. (C); 118. (C); 119. (C); 120. (B); 121. (D); 122. (D); 123. (D); 124. (B); 125. (A); 126. (B); 127. (C); 128. (C); 129. (D); 130. (C); 131. (B); 132. (B); 133. (A); 134. (C); 135. (D); 136. (A); 137. (D); 138. (B); 139. (C); 140. (A); 141. (C); 142. (A); 143. (B); 144. (A); 145. (B); 146. (B); 147. (B); 148. (A); 149. (A); 150. (A); 151. (C); 152. (C); 153. (B); 154. (C); 155. (B); 156. (C); 157. (A); 158. (C); 159. (C); 160. (A); 161. (A); 162. (A); 163. (C); 164. (B); 165. (D); 166. (C); 167. (C); 168. (A); 169. (C); 170. (B); 171. (C); 172. (A); 173. (B); 174. (D); 175. (A); 176. (B); 177. (D); 178. (D); 179. (B); 180. (C); 181. (B); 182. (B); 183. (D); 184. (D); 185. (A); 186. (C); 187. (C); 188. (A); 189. (D); 190. (D); 191. (A); 192. (A); 193. (A); 194. (C); 195. (A); 196. (B); 197. (C); 198. (A); 199. (D); 200. (B); 201. (B); 202. (C); 203. (A); 204. (A); 205. (C); 206. (A); 207. (B); 208. (B); 209. (B); 210. (B); 211. (B); 212. (B); 213. (A); 214. (D); 215. (C); 216. (D); 217. (A); 218. (D); 219. (C); 220. (C); 221. (B); 222. (B); 223. (A); 224. (A); 225. (A); 226. (A); 227. (B); 228. (B); 229. (C); 230. (A); 231. (B); 232. (D); 233. (B); 234. (C); 235. (B); 236. (A); 237. (A); 238. (C); 239. (B); 240. (A); 241. (B); 242. (C); 243. (C); 244. (A); 245. (B); 246. (C); 247. (C); 248. (C); 249. (D); 250. (C); 251. (A); 252. (B); 253. (B); 254. (C); 255. (C); 256. (A); 257. (C); 258. (C); 259. (B); 260. (A); 261. (B); 262. (B); 263. (A); 264. (C); 265. (A); 266. (C); 267. (B); 268. (B); 269. (B); 270. (A); 271. (B); 272. (D); 273. (B); 274. (C); 275. (B); 276. (C); 277. (B); 278. (B); 279. (A); 280. (A); 281. (A); 282. (B); 283. (A); 284. (A); 285. (C); 286. (C); 287. (C); 288. (C); 289. (B); 290. (A); 291. (C); 292. (C); 293. (B); 294. (A); 295. (D); 296. (C); 297. (C); 298. (C); 299. (D); 300. (C); 301. (B); 302. (C); 303. (C); 304. (C); 305. (C); 306. (C); 307. (C); 308. (B); 309. (C); 310. (C); 311. (C); 312. (C); 313. (A); 314. (A); 315. (C); 316. (B); 317. (B); 318. (B); 319. (B); 320. (B); 321. (A); 322. (A); 323. (B); 324. (B); 325. (C); 326. (C); 327. (B); 328. (B); 329. (A); 330. (A); 331. (B); 332. (A); 333. (A); 334. (C); 335. (B); 336. (B); 337. (D); 338. (A); 339. (B); 340. (C); 341. (B); 342. (B); 343. (A); 344. (C); 345. (B); 346. (C); 347. (B); 348. (A); 349. (A); 350. (D); 351. (B); 352. (B); 353. (A); 354. (A); 355. (B); 356. (C); 357. (A); 358. (A); 359. (A); 360. (A); 361. (C); 362. (A); 363. (B); 364. (A); 365. (C); 366. (B); 367. (C); 368. (C); 369. (B); 370. (B); 371 (C); 372. (B); 373. (A); 374. (C); 375. (C); 376. (A); 377. (A); 378. (B); 379. (C); 380. (D); 381. (A); 382. (B); 383. (C); 384. (D); 385. (D); 386. (B); 387. (A); 388. (D); 389. (B); 390. (C); 391. (C); 392. (B); 393. (A); 394 (B); 395. (B); 396. (C); 397. (C); 398. (D); 399. (B); 400. (A); 401. (B); 402. (A); 403. (C); 404. (B); 405. (B); 406. (B); 407. (B); 408. (A); 409. (B); 410. (D); 411. (C); 412. (A); 413. (B); 414. (B); 415. (A); 416. (D); 417. (A); 418. (B); 419. (A); 420. (A); 421. (A); 422. (B); 423. (C); 424. (B); 425. (A); 426. (B); 427. (B); 428. (C); 429. (D); 430. (D); 431. (B); 432. (A); 433. (B); 434. (D); 435. (A); 436. (D); 437. (A); 438. (C); 439. (A); 440. (C); 441. (B); 442. (B); 443. (C); 444. (D); 445. (A); 446. (B); 447. (D); 448. (C); 449. (C); 450. (C); 451. (A); 452. (B); 453. (A); 454. (B); 455. (C); 456. (D); 457. (C); 458. (B); 459. (A); 460. (A); 461. (A); 462. (B); 463. (C); 464. (C); 465. (A); 466. (B); 467. (B); 468. (D); 469. (D); 470. (B); 471. (D); 472. (A); 473. (B); 474. (C); 475. (B); 476. (C); 477. (D); 478. (A); 479. (B); 480. (A); 481. (C); 482. (B); 483. (C);

484. (B); 485. (A); 486. (C); 487. (A); 488. (B); 489. (C);
490. (C); 491. (C); 492. (B); 493. (A); 494. (A); 495. (C);
496. (B); 497. (C); 498. (C); 499. (C); 500. (A); 501. (A);
502. (B); 503. (C); 504. (D); 505. (D); 506. (B); 507. (D);
508 (A); 509. (C); 510. (D); 511. (A); 512. (B); 513. (C);
514. (D); 515. (A); 516. (B); 517. (D); 518. (C); 519. (B);
520. (A); 521. (B); 522. (C); 523. (D); 524. (A); 525. (B);
526. (C); 527. (D); 528. (A); 529. (A); 530. (B); 531. (C);
532. (C); 533 (A); 534. (B); 535. (A); 536. (D); 537. (A);
538. (B); 539. (C); 540. (D); 541. (A); 542. (B); 543. (C);
544. (D); 545. (A); 546. (C); 547. (B); 548. (C); 549. (B);
550. (C); 551. (A); 552. (D); 553. (B); 554. (A); 555. (B);
556. (C); 557. (A); 558. (B); 559. (B); 560. (A); 561. (B);
562. (C); 563. (D); 564. (C); 565. (B); 566. (D); 567. (C);
568. (D); 569. (B); 570. (D); 571. (B); 572. (B); 573. (A);
574. (A); 575. (C); 576. (B); 577. (B); 578. (A); 579. (B);
580. (A); 581. (C); 582. (D); 583. (A); 584. (B); 585. (D);
586. (B); 587. (A); 588. (C); 589. (B); 590. (C); 591. (A);
592. (B); 593. (D); 594. (B); 595. (A); 596. (A); 597. (C);
598. (B); 599. (C); 600. (D); 601. (C); 602. (A); 603. (D);
604. (B); 605. (C); 606. (C); 607. (B); 608. (C); 609. (A);
610. (C); 611. (C); 612. (A); 613. (B); 614. (B); 615. (D);
616. (A) 617. (C); 618. (A); 619. (B); 620. (C); 621. (B);
622. (A); 623. (B); 624. (C); 625. (D); 626. (A); 627. (A);
628. (A); 629. (D); 630. (A); 631. (A); 632. (B); 633. (B);
634. (B); 635. (B) 636. (A) 637. (B) 638. (B) 639. (B)
640. (B); 641. (B); 642. (B); 643. (B); 644. (B); 645. (C);
646. (A); 647. (B); 648. (A) 649. (A) 650. (B); 651. (B);
652. (C); 653. (D); 654. (A); 655. (B); 656. (C); 657. (D);
658. (B); 659. (A); 660. (B); 661. (C); 662. (D); 663. (A);
664. (B) 665. (C); 666. (A); 667. (B) 668. (A); 669. (D);
670. (B) 671. (A) 672. (B); 673. (C) 674. (A); 675. (C)
676. (D); 677. (A); 678. (B); 679. (B); 680. (C); 681. (A);
682. (A); 683. (B); 684. (C); 685. (D); 686. (C); 687. (B);
688. (A); 689. (D); 690. (C); 691. (A); 692. (D); 693. (B);
694. (B); 695. (C); 696. (B); 697. (A); 698. (D); 699. (C);
700. (B); 701. (C); 702. (B); 703. (C); 704. (D); 705. (A);
706. (C); 707. (C); 708. (B); 709. (D); 710. (C); 711. (A);
712. (A) 713. (A); 714. (D); 715. (A). 716. (C); 717. (A);
706. (C); 707. (C); 708. (B); 709. (D); 710. (C); 711. (A);
712. (A) 713. (A); 714. (D); 715. (A). 716. (C); 717. (A);
718. (A); 719. (B); 720. (C); 721. (C); 722. (D); 723. (C);
724. (A) 725. (C); 726. (C); 727. (C). 728. (B); 729. (C);
730. (C) 731. (C); 732. (A); 733. (A). 734. (A); 735. (D);
736. (C); 737. (C); 738. (A); 739. (D); 740. (C); 741. (C);
742. (A) 743. (C); 744. (B); 745. (C). 746. (D); 747. (A);
748. (A); 749. (B); 750. (A); 751. (B); 752. (B); 753. (D);
754. (A) 755. (B); 756. (D); 757. (D). 758. (B); 759. (D);
760. (B); 761. (A); 762. (D); 763. (B); 764. (A); 765. (B);
766. (D) 767. (C); 768. (D); 769. (D). 770. (B); 771. (B);
772. (C); 773. (C); 774. (C); 775. (A); 776. (B); 777. (B);
778. (D) 779. (B); 780. (C); 781. (D). 782. (C); 783. (D);
784. (B); 785. (C); 786. (B); 787. (D); 788. (B); 789. (C);

Chapter 2

AGRICULTURAL CLIMATOLOGY

1. Climate word originated from:
(A) Greek (B) Latin
(C) English (D) Parsian

2. Climate word derived from:
(A) Klima+ Logos (B) Klima+Logo
(C) Clima + Logo (D) Clima+Logos

3. The basic meaning of klima is:
(A) Angle incidence of sun light
(B) Changes in temperature
(C) Incidence of humidity
(D) All of these

4. The science that seeks to describe and explain that nature of climate is called:
(A) Climatology (B) Meteorology
(C) Both of these (D) None of these

5. The study of physical status of climate is called:
(A) Physical climatology (B) Statistical climatology
(C) Descriptive climatology (D) Applied climatology

6. To study the physical atmospheric factors viz., temperature, relative humidity, clouds, wind velocity and direction is called:
(A) Physical climatology (B) Statistical climatology
(C) Descriptive climatology (D) Applied climatology

7. To study the different factors of climate such as temperature and rainfall, are divided into state level is called:
(A) Physical climatology (B) Statistical climatology
(C) Descriptive climatology (D) Applied climatology

8. To study the relationship with other branched of climatology is called:
(A) Physical climatology (B) Statistical climatology
(C) Descriptive climatology (D) Applied climatology

9. The weather information collected from observatory are depicted in tables is called:
(A) Weather chart (B) Weather map
(C) Both of these (D) None of these

10. The weather information are depicted through map is called:
(A) Weather chart (B) Weather map
(C) Both of these (D) None of these

11. World Meteorological Organisation (WMO) was established at:
(A) Geneva (Switzerland) (B) Spain
(C) USA (D) Canada

12. World Meteorological Organisation (WMO) was established in:
(A) 1950 (B) 1951 (C) 1952 (D) 1953

13. International Meteorological Organisation was established in:
(A) 1877 (B) 1878 (C) 1879 (D) 1880

14. The science of climatological condition which have direct relation or relevance to agriculture is called:
(A) Agro-meteorology (B) Agro-climatology
(C) Meteorology (D) Climatology

15. The composite or generalization of the variety of day to day weather conditions is called:
(A) Climate (B) Weather
(C) Season (D) All of these

16. How many years of weather is optimum for climate:
(A) 30 years (B) 31 years
(C) 32 years (D) 33 years

17. The observations recorded from open area of 2 meter height of soil surface is called:
(A) Macro climate (B) Micro climate
(C) Topo climate (D) Local climate

18. When weather data collected from sloppy land is called:
(A) Macro climate (B) Micro climate
(C) Topo climate (D) Local climate

19. When weather data collected form standing crop is called:
(A) Macro climate (B) Micro climate
(C) Topo climate (D) Local climate

20. The data recorded in macro climate is called:
(A) Synoptic Meteorological observation
(B) Asynoptic Meteorological observation
(C) Both of these
(D) None of these

21. Upto 15 km above the earth is called:
(A) Troposphere (B) Mesosphere
(C) Stratosphere (D) None of these

22. 50 km above the earth is called:
(A) Troposphere (B) Mesosphere
(C) Stratosphere (D) None of these

23. In atmosphere mainly which gas is determined to atmospheric activities is:
(A) O2 (B) Ozone
(C) Both of these (D) None of these

24. In atmosphere gasses found in:
(A) Mixed status (B) Without mixed status
(C) Alone (D) None of these

25. Four gasses- nitrogen, oxygen, argon and CO2 accounts:
(A) 97 % (B) 98 % (C) 99 % (D) 100 %

26. Nitrogen alone constituents nearly:
(A) One-third by volume (B) Two-third by volume
(C) Three-fourth by volume (D) Four-fifth by volume

27. Oxygen alone constituents nearly:
(A) One-third by volume (B) Two-third by volume
(C) One-fifth by volume (D) Four-fifth by volume

28. The blanket of gases, suspended liquids and solids the entirely envelops the earth is called:
(A) Atmosphere (B) Climate
(C) Clouds (D) Gasses

29. The powerful oxidizing agents and therefore a threat to life and property when concentrations become excessive is:
(A) Ozone (B) Nitrogen
(C) Oxygen (D) Carbon di oxide

30. The lower portion of the atmosphere, extending upto about 8 km at the poles and 16 km at the equator is called:
(A) Troposphere (B) Stratosphere
(C) Ionosphere (D) Exosphere

31. The upper portion of the toposphere, extending upto 50-60 km is called:
(A) Troposphere (B) Stratosphere
(C) Ionosphere (D) Exosphere

32. The upper portion of the stratosphere, extending upto 60-500 km is called:
(A) Troposphere (B) Stratosphere
(C) Ionosphere (D) Exosphere

33. The upper portion of the atmosphere, extending upto 500-1000 km is called:
(A) Troposphere (B) Stratosphere
(C) Ionosphere (D) Exosphere

34. Ozone gas mainly found in which region:
(A) Troposphere (B) Stratosphere
(C) Ionosphere (D) Exosphere

35. High air pressure and its affected area is called:
(A) Anticyclone (B) Cyclone
(C) Depression (D) Rainfall

36. Low air pressure and its affected area is called:
(A) Anticyclone (B) Cyclone and Depression
(C) Fog (D) Rainfall

37. Wavelengths longer than —— m/μ are not visible to the eye, and are called infrared:
(A) 450 (B) 550 (C) 650 (D) 750

38. It has been observed that in a number of outbreeding populations of rye grass, *Lolium perenne* could be brought about at low temperature (0-3°C) or ——:
(A) Long day (B) Short day
(C) Intermediate day (D) None of these

39. In ———, long days promote bulb formation whereas, short days promote flowering and seed production:
(A) Sweet potato (B) Artichoke
(C) Cassava (D) Onion

40. The ——— variety Maryland Mommoth in the latitude of Maryland (USA) produces only leaves and needs shorter days for seed production:
(A) Tomato (B) Carrots
(C) Radish (D) Tobacco

41. Beyond the lower atmosphere, extending out into space, is the stratosphere, in which temperature remains relatively constant from:
(A) -15 to -25 (B) -45 to -75
(C) -80 to -95 (D) -5 to -10

42. One of the characteristics of the stratosphere is the occurrence of a well-marked layer having an increased concentration of ——— at about 20 miles (32 km) above sea level:
(A) Carbon dioxide (B) Carbon monoxide
(C) Ozone (D) Hydrogen

43. In the ozone layer, the ozone does not exceed ——— ppm at 20 miles (32 km) above sea level (stratosphere):
(A) 5 (B) 10 (C) 15 (D) 20

44. The gaseous mixing in the atmosphere is, in general, similar at sea level and at height upto ——km.
(A) 20 (B) 40 (C) 60 (D) 80

45. Ozonated hexane led to distruption of —— in plants:
(A) Membrane permeability (B) Protein synthesis
(C) Lipid synthesis (D) Stem elongation

46. Animals and nongreen plants continuously use oxygen and liberate —— into the air:
(A) Sulfur dioxide (B) Carbon monoxide
(C) Carbon dioxide (D) Nitrous oxide

47. The rate of photosynthesis under the conditions of high light intensity may be controlled by:
(A) Light (B) Carbon dioxide
(C) Water (D) None of these

48. North of the equator, surface winds are known as:
(A) Northeast trade winds (B) Southeast winds
(C) Westerlies (D) Easterlies

49. In the humid tropics and subtropics the process of weathering (soil formation) is known as:
(A) Podzolization (B) Gleization
(C) Latosolization (D) None of these

50. The range of coffee production is limited approximately by the ——°F of isotherms for the coldest month:
(A) 25 (B) 35 (C) 45 (D) 55

51. The diversity pertaining to a plant composition in several plant communities in one range of environment is known as:
(A) Alpha index of diversity
(B) Beta index of diversity
(C) Gamma index of diversity
(D) None of these

52. Among the types of natural succession, the one which begins in water is known as:
(A) Xerarch (xerosere) (B) Halosere
(C) Hydrarch (hydrosere) (D) Psammosere

53. In the earth's atmosphere at —— cal/cm²/annum of solar energy is received:
(A) 15×10^6 (B) 15×10^8 (C) 15×10^{10} (D) 15×10^{12}

54. Which of the following crops utilizes even upto 10-12 per cent of the total solar energy:
(A) Sorghum (B) Rice
(C) Maize (D) Sugarcane

55. Fixation of each gramme of carbon from carbon dioxide to organic compounds during photosynthesis traps ——— of solar energy:
(A) 124 k cal (B) 134 k cal
(C) 144 k cal (D) 154 k cal

56. C_3 plants show a light saturation upto 40000 to 50000 lux, while C_4 plants upto full sun in tropics upto:
(A) 60000-70000 (B) 70000-80000
(C) 90000-100000 (D) 115000-120000

57. ——— pollutants are very harmful for human health and therefore such polluted water should essentially be purified before their supply for drinking purposes:
(A) Biological (B) Physical
(C) Mechanical (D) Metallic

58. Ionising ——— are very important and harmful pollutants:
(A) Reflection (B) Deflection
(C) Radiation (D) Absorptions

59. The sound beyond ———dB are regarded as pollutant as it harms hearing ability:
(A) 60 (B) 80 (C) 100 (D) 120

60. If all the moisture present in the air at any time to converted into precipitation, it would be equal to 2.5 cm or about ———of water supply by precipitation (average residence time):
(A) 5 days (B) 10 days
(C) 15 days (D) 20 days

61. About 2.5 cm rainfall sufficient to produce over 10 tonnes of stream flow water on a watershed area of about ——— cm^2:
(A) 3000 (B) 4000 (C) 5000 (D) 6000

62. Fresh water bodies like rivers and streams are known as:
(A) Palustrine (B) Lotic or riverine
(C) Lentic or lacustrice (D) None of these

63. One of the basis of temperature variation in the vertical column of water, or thermal stratification, lake ecosystem have been divided into the following zones wherein ——— comes in direct contact of sun and wind:
(A) Thermocline (B) Hypolimnion
(C) Epilimnion (D) None of these

64. Mercury level of ——— µg/g of fresh weight of fish is considered unsafe for human consumption:
(A) 0.5-1.0 (B) 1.6-2.0
(C) 2.1-2.5 (D) 2.6-3.0

65. Fish absorbs and may accumulate zinc in muscles amounting approximately upto how many time more than that in sea water:
(A) 3700 (B) 4700 (C) 5700 (D) 6700

66. Permissible limit of iron in drinking water is:
(A) 1.0 ppm (B) 1.5 ppm
(C) 2.0 ppm (D) 2.5 ppm

67. Potassium in natural waters never assume toxic proportions and remain less than ——— mg/litre:
(A) 10 (B) 15 (C) 20 (D) 25

68. When with chloride ion in drinking water, the dominant cation is potassium or magnesium, the salty taste is not felt even upto ——— mg/litre:
(A) 500 (B) 750 (C) 1000 (D) 1250

69. The characteristic pungent odorous taste of onion is due to :
(A) Magnesium (B) Calcium
(C) Potassium (D) Sulphur

70. Many factory waters contain how much mg/litre of nitrate nitrogen:
(A) 20 (B) 30 (C) 40 (D) 50

71. Chloride added to drinking water supply when encountered with ——— gets neutralized and chloroamines are formed:
(A) Nitrous oxide (B) Carbon dioxide
(C) Methane (D) Ammonia

72. The nitrate content say in the range of 30 to 40 mg/litre lead to poor oynation of blood haemoglobin and blood turns:
(A) Blackish (B) Whitish
(C) Bluish (D) Yellowish

73. In a study in USA, the earthworm inhabiting a foliarly fed DDT-tree, contained upto 80 ppm of DDT even after how many months:
(A) 5 (B) 7.5 (C) 10 (D) 10.5

74. The coliform bacteria count in drinking water is regarded safe if it is:
(A) 5 or less (B) 10 or less
(C) 15 or less (D) 20 or less

75. The PSI value range ——— has been classified as hazardous (air) causing premature on set of certain diseases and aggravation of symptoms:
(A) 200-299 (B) 300-399
(C) 100-199 (D) 50-100

76. Methane gas once released in the atmosphere stays for how many years:
(A) 3-4 (B) 4-5 (C) 5-7 (D) 8-10

77. The earth during the last one century has warmed up by——— °C as estimated:
(A) 0.3-0.7 (B) 0.8-1.0
(C) 1.2-1.3 (D) 1.4-1.5

78. The earth temperature during the past 20000 years has risen by — °C:
(A) 2-3 (B) 4-5 (C) 6-7 (D) 8-9

79. Ultraviolet radiation are of similar wavelengths than of visible light and those ranging from 320-400 nm are called:
(A) UV-C (B) UV-B (C) UV-A (D) UV-D

80. CFCs methane and nitrogen oxide gases have the capacity to react with ozone to convert into:
(A) Nitrogen (B) Hydrogen
(C) Oxygen (D) None of these

81. The estimates reveal that the maximum NO_2 emission in European countries is in ——— in the range of 2Tg/annum (Tetagramme = 10^{12} grammes):
(A) France (B) Germany
(C) England (D) None of these

82. The threshold concentration upto which no visible injury symptoms appear on plants due to NO_2 is 1 mg/m^2 on:
(A) Daily average (24 hrs) (B) Weekly average
(C) Monthly average (D) Annual average

83. The suggested threshold level of fluoride pollution in atmosphere is how much quantity (ppm)
(A) 10-40 (B) 50-80
(C) 90-100 (D) 110-115

84. Uranium contribute the radioactivity of about – pci/litre of ocean water:
(A) 2 (B) 3 (C) 4 (D) 5

85. The Department of Environment in independence India was formed in November:
(A) 1960 (B) 1970 (C) 1980 (D) 1990

86. Serious efforts on Environment protection in India began with the establishment of NCEPC by the Department of Science and Technology in February:
(A) 1952 (B) 1962 (C) 1972 (D) 1982

87. Environment Protection Act 1986 came into force in India on ——— 1968:
(A) October 19 (B) November 19
(C) September 19 (D) August 19

88. In December, 1989 standards were laid for the noise limits in day and night time in various area and silence zones upto ——— around educational institutions, hospitals and courts in India:
(A) 50 m (B) 100 m (C) 150 m (D) 1200 m

89. The optimum temperature range for ideal germination of wheat varied from:
(A) 25-30 °C (B) 20-25 °C
(C) 15-20 °C (D) 10-15 °C

90. The optimum rainfall range for growing maize is:
(A) 360-500 mm (B) 560-700 mm
(C) 300-400 mm (D) 460-600 mm

91. The arhar (pigeonpea) is commonly grows well in:
(A) Tropical and subtropical climate
(B) Temperate climate
(C) Humid and temperate climate
(D) Sub-humid and temperate climate

92. For urdbean, the suitable climate is:
(A) Hot and humid (B) Cool and humid
(C) Cool and dry (D) Hot and dry

93. High temperature above 41 °C at flowering of castor results in:
(A) Rotting of pollens and poor fertilization
(B) Blasting of flowers and poor seed set
(C) Delayed pollination and poor fertilization
(D) Reduced pollination and rotting of seed

94. The crop of —— is more liable to suffer from frost and cold:
(A) Toria (B) Pahari rai
(C) Sarson (D) Banarasi rai

95. Linseed is a which season crop:
(A) Warm (B) Cool (C) Hot (D) Dry

96. Safflower is a which season crop:
(A) Dry (B) Summer (C) Arid (D) Cool

97. The sunflower crop requires cool climate from:
(A) Germination to seedling stage
(B) Seedling to flowering stage
(C) Flowering to maturity stage
(D) Seedling to maturity stage

98. If temperature is high during maturation of sunflower, the linoleic acid:
(A) Remain unaffected
(B) Increases with slow rate
(C) Decreases
(D) Increases with rapid rate

99. Tobacco requires temperature for uniformly rapid germination:
(A) 17-22 °C (B) 19-25 °C
(C) 27-32 °C (D) 29-42 °C

100. The soil temperature for satisfactory growth of potato tuber should be:
(A) 7-9 °C (B) 11-13 °C
(C) 17-19 °C (D) 20-23 °C

101. Optimum temperature for germination of pigeonpea crop is:
(A) 15-20°C (B) 20-25°C
(C) 30-35°C (D) 40-45°C

102. Temperature required for good sugarcane germination is:
(A) 20-25° C (B) 25-32° C
(C) 35-41° C (D) 38-45° C

103. Optimum temperature for vegetative growth of gram is:
(A) 10-15° C (B) 15-20° C
(C) 20-25° C (D) 25-30° C

104. Troposphere of the atmosphere extending upto a height from ground to:
(A) 8-18 km (B) 50 km
(C) 40 km (D) >50 km

105. Fog, dew, mist, rain etc occur in the layer of:
(A) Thermosphere (B) Mesosphere
(C) Stratosphere (D) Troposphere

106. Troposphere is the densest part of atmosphere and contains about:
(A) 55% of atmospheric mass
(B) 75% of atmospheric mass
(C) 85% of atmospheric mass
(D) 95% of atmospheric mass

107. The layer separated the stratosphere and mesosphere is called:
(A) Tropopause (B) Stratopause
(C) Mesopause (D) None of these

108. Inosphore is present in:
(A) Troposphere (B) Stratosphere
(C) Mesosphere (D) Thermosphere

109. Heat can flow between substance or within substance by:
(A) Conduction (B) Convection
(C) Radiation (D) All of these

110. The reason (s) for the decline in temperature with increase altitude is:
(A) Air is heated by the earth
(B) Water vapour content of air decreases with increase in altitude
(C) Expansion of air at higher elevation
(D) All of these

111. Vertical temperature decrease is called lapse rate and the normal lapse rate is:
(A) 4.5° C/km (B) 6.5° C/km
(C) 8.5° C/km (D) None of these

112. Heat transfer into the soil is mainly by:
(A) Conduction (B) Convection
(C) Radiation (D) None of these

113. The rate of heat flux into soil depends on:
(A) Temperature gradient (B) Thermal conductivity
(C) Both (A) and (B) (D) None of these

114. Thermal conductivity of soil decrease with:
(A) Increase porosity
(B) Increase organic matter
(C) Decrease soil moisture
(D) All of these

115. What will be the effect of tillage operation on thermal conductivity of soil?
(A) Remain unchanged (B) Decrease
(C) Increase
(D) First increase then decrease

116. Minimum temperature for the germination of cotton seed is:
(A) 10°C (B) 16°C (C) 20°C (D) 26°C

117. Optimum temperature for retting of jute is:
(A) 20°C (B) 24°C (C) 30°C (D) 34°C

118. The optimum temperature for germination of soybean is:
(A) 10°C (B) 22°C (C) 17°C (D) 35°C

119. South-west monsoon contributes how much per cent of rainfall?
(A) 50 % (B) 75 % (C) 80 % (D) 95 %

120. Average rainfall in India is around:
(A) 2120 mm (B) 1120 mm
(C) 850 mm (D) 500 mm

121. Annual rainfall required for sugarcane crop is:
(A) 50-100 mm (B) 50-100 cm
(C) 100-125 cm (D) >125 cm

122. Annual rainfall required for cotton crop is:
(A) 25-50 cm (B) 50-75 cm
(C) 75-115 cm (D) >115 cm

123. Annual rainfall required for gram crop is:
(A) 40-50.cmm (B) 60-90 cm
(C) 100-110 cm (D) >110 cm

124. Annual rainfall required for sugarbeet crop is:
(A) 50 cm (B) 75 cm
(C) 100 cm (D) 125 cm

125. Which of the following disorder in mango is caused by low temperature?
(A) Black heart (B) Crotch injury
(C) Collor injury (D) All of these

126. How much rainfall is required for the cultivation of banana?
(A) 100-120 cm (B) !25-150 cm
(C) 170-200 cm (D) 300-350 cm

127. The biological amplication of DDT in the various trophic levels is known as:
(A) Green house effect (B) Biomagnification
(C) Entraphication (D) Pollution

128. Environment can be improved by:
(A) Wind erosion (B) Water erosion
(C) Conservation (D) None of these

129. Smog is produced due to:
(A) Air and metal dust (B) Water and nitrogen oxide
(C) Smoke and fog (D) None of these

130. The photochemical smog is formed by:
(A) Nitrogen oxides (B) Hydrocarbons
(C) Nitrogen oxides and hydrocarbons
(D) Solar radiations upon nitrogen oxides and hydro carbons

131. O3 promotes:
(A) Oxidation of biochemicals
(B) Destruction of chlorenchyma
(C) Both (A) and (B)
(D) Growth of plants

132. Fertiliers, pesticides and insecticides may cause:
(A) Air pollution (B) Water pollution
(C) Both (A) and (B) (D) None of these

133. Photochemical smog brings about:
(A) Defoliation
(B) Bleaching and blaing of foliage
(C) Floral bud shedding
(D) Discolouration and curling of petals

134. Water soacked appearance of leaves due to sulphur dioxide pollution is caused by:
(A) Destruction of chlorophyll
(B) Damage to membrane
(C) Both (A) and (B)
(D) Corrosive action over metals

135. Addition of phosphates and nitrate fertilizers into water leads to:
(A) Increased growth decompositors
(B) Reduced algal growth
(C) Increased algal growth
(D) Nutrient enrichment/eutrophication

136. Major reason for soil erosion in India is:
(A) Tilling of land
(B) Deforestation
(C) Overgrazing and deforestation
(D) Over grazing

137. Domestic water is:
(A) Effluent (B) Biodgradable
(C) Nonbiodegradable (D) Air pollutant

138. Inadequate drainage in a soil will lead to:
(A) Floods (B) Soil erosion
(C) Aridity (D) Salination

139. Jhuming cultivation generally practiced in India in:
(A) Hilly areas (B) Madhya Pradesh
(C) North-East states (D) Malabar area

140. Ecological balance of a mining area is restricted through:
(A) Conversion into agricultural area
(B) Revegetation of area
(C) Prevention of soil erosion
(D) Protection against grazing

141. Ultra-violet light causes:
(A) Formation of pyrimidines
(B) Sticky metaphases
(C) Photodynamic action
(D) Destruction of hydrogen bonds between complementary DNA strands

142. Sulphur dioxide affects:
(A) Cell wall (B) Plasmodesmata
(C) All membrane system (D) Nucleus

143. Ultra-violet radiation are lethal because of inactivation of:
(A) Proteins, nucleic acids and pigments
(B) Minerals, air and water
(C) Carbohydrates, fats and vitamins
(D) Water, carbon dioxide and oxygen

144. SO_2 and NO_2 produce pollution by increasing:
(A) Alkalinity (B) Acidity
(C) Neutrality (D) Buffer action

145. Spraying of DDT produces pollution of:
(A) Air (B) Air and water
(C) Air and soil (D) Air, soil and water

146. Most abundant water pollutant is:
(A) Detergents (B) Pesticides
(C) Industrial wastes (D) Ammonia

147. Ozone depletion in stratosphere shall result in:
(A) Forest fires
(B) Increased incidence of skin cancer
(C) Global warming
(D) None of these

148. DDT is:
(A) Green house gas (B) Degradable pollutant
(C) Nondegradable pollutant (D) None of these

149. Smog occurs in places of:
(A) Excess SO_2 (B) Low temperature
(C) High temperature (D) Excess NH_3

150. Acid rain is due to increase in atmospheric concentration of:
(A) Ozone and dust (B) CO_2 and CO
(C) SO_3 and CO (D) SO_2 and NO_2

151. Chief source of water and soil pollution is:
(A) Mining (B) Thermal power plant
(C) Agroindustry (D) All of these

152. Which causes water pollution?
(A) Smoke/flyash (B) Automobile exhausts
(C) Aeroplanes (D) 2,4-D and pesticides

153. Drawback of DDT as pesticide is:
(A) It becomes ineffective after some time
(B) It is less effective than others
(C) It is not easily/rapidly degraded in nature
(D) Its high cost

154. Which one is found in higher concentration in the atmospheric?
(A) Ozone
(B) Nitrous oxides
(C) Carbon dioxide, carbon monoxide and sulphur dioxide
(D) All of these

155. Carbon dioxide is a major pollutant in:
(A) Air (B) Soil
(C) Water (D) Noise

156. Major source of methane in India is:
(A) Rice fields (B) Sugarcane fields
(C) Wheat fields (D) Fruit orchards

157. Phosphate pollution is caused by:
(A) Weathering of phosphate rocks only
(B) Agricultural fertilizers
(C) Phosphate rocks and sewage
(D) Sewage and agricultural fertilizers

158. Which pigment protects plants from UV damage?
(A) Chlorophyll (B) Xanthophyll
(C) Carotenoids (D) Phycocyanin

159. Indescriminate use of fertilizers causes:
(A) Air pollution (B) Water pollution
(C) Land pollution (D) All of these

160. SO_2 pollution causes:
(A) Plasmolysis
(B) Destruction of cell wall
(C) Destruction of Golgi bodies
(D) Destruction of chlorophyll

161. Depletion of ozone layer is due to:
(A) Oxides of nitrogen (B) Oxides of carbon
(C) Oxides of sulphur (D) None of these

162. Chipko movement originated in the district:
(A) Allahabad (B) Chamoli
(C) Hardwar (D) Dehradun

163. Chipko movement originated in Gopeshwar in:
(A) 1953 (B) 1963 (C) 1973 (D) 1083

164. A movement similar to Chipko which is spread in south is:
(A) Uliso (B) Appiko
(C) Belesu (D) Balasu

165. Chipko movement originated in:
(A) Silent valley of Kerala
(B) Kangra valley of Himanchal Pradesh
(C) Tehri Garwal of Uttarakhand
(D) Panchmari of MP

166. Extinction of numerous wild animals is due to:
(A) Hunting (B) Deforestation
(C) Overcrowding (D) Scarcity of food

167. Chipko movement is connected with:
(A) Conservation of natural resources
(B) Plant/forest conservation
(C) Plant Breeding
(D) Project tiger

168. Forest control drought through:
(A) Lot of water plants
(B) Increasing rainfall
(C) Retention of water and prevention of erosion
(D) Functioning as water shed

169. Overgrazing causes:
(A) Negative pollution (B) Positive pollution
(C) Soil erosion (D) Reduction in crop yields

170. Deforestation brings about:
(A) Increased sunlight (B) Increased grazing area
(C) Weed control (D) Soil erosion

171. Deforestation will decrease:
(A) Soil erosion (B) Land slides
(C) Soil fertility (D) Rainfall

172. Deforestation is the major causal agent of:
(A) Depletion of natural resources
(B) Environmental pollution
(C) Desertification of habitat
(D) Genetic erosion

173. Deforestation does not lead to:
(A) Quick nutrient cycling
(B) Soil erosion
(C) Alteration of local weather conditions
(D) Destruction of natural habitat of wild animals

174. Forest destruction results in:
(A) Loss of wild life (B) Floods and drought
(C) Soil erosion (D) All of these

175. Afforestation should be with:
(A) Exotic species (B) Indigenous species
(C) Bamboos (D) Eucalyptus

176. If we remove half of the forest cover of earth, the crises that will occur:
(A) Many species would become extinct
(B) Population, pollution and ecological imbalance will rise
(C) Energy crises will commence
(D) The remaining forest will correct the imbalance

177. Decreased rainfall in northern areas is due to:
(A) Change in sun
(B) Global phenomenon
(C) Deforestation in catchment areas
(D) Deforestation in evaporating areas near equator

178. Increased incidence of floods in the plains of northern India are due to:
(A) Increased deforestation in catchment areas
(B) Increase in incidence of rainfall
(C) Silting of dams
(D) More area under cultivation

179. Soil erosion can be prevented by:
(A) Restricted human activity
(B) Good plant cover
(C) Checking movement of animals
(D) Wind screen alone

180. Adherence to two-children norms shall help in:
(A) Improving food web (B) Checking pollution
(C) Stabilising ecosystem (D) Fertility of soil

181. Mist is horizontal visibility of:
(A) 100 m (B) 200 m (C) 500 m (D) 1 km

182. Fog is formed by:
(A) Smoke
(B) Condensation of water vapours
(C) Condensation of water vapours over dust particles
(D) Both (A) and (C)

183. Succession on sand dune is:
(A) Lithosere (B) Psychrosere
(C) Psammosere (D) Halosere

184. The amount of biogenic nutrient present in the abiotic environment is called:
(A) Standing state (B) Standing crop
(C) Detritus (D) Reserve nutrient

185. Phosphorus has:
(A) Imperfect nutrient cycle (B) Perfect nutrient cycle
(C) Gasseous cycle (D) Mixed cycle

186. Carbosphere comprises:
(A) Atmosphere (B) Lithosphere
(C) Hydrosphere (D) All of these

187. Adiabatic expansion resulting in cooling is called:
(A) Coriolis effect (B) Joule-Kelvin effect
(C) Le Chatelier Principle (D) Principle of Parsimony

188. Cyclic/frontal rainfall generally occurs:
(A) All over the world
(B) Tropical areas
(C) Temperate areas
(D) Boundary of warm tropical air and cold polar air

189. Which area has four seasons:
(A) Tropics (B) Monsoonic
(C) Temperate (D) Arctic

190. The land of wind mills is:
(A) UK (B) USA
(C) Netherlands (D) Himachal Pradesh

191. The shining and thunder in clouds during rainy season:
(A) Start simulataneosly
(B) Occur thunder first and then shining starts
(C) Occur shining first and then thunder starts
(D) Occur mostly no thunder and no shining

192. Daily weather forecast is obtained in the country from:
(A) Pune (B) New Delhi
(C) Mumbai (D) Kolkata

193. The layer of atmosphere having uniform temperature is:
(A) Stratosphere (B) Exosphere
(C) Thermosphere (D) Troposphere

194. The physical state of the atmosphere at a given place and time is referred to as:
(A) Weather (B) Climate
(C) Both of these (D) None of these

195. The first meteorological department of India was started at Shimla in:
(A) 1863 (B) 1864 (C) 1865 (D) 1866

196. Layer of ozone is present in:
(A) Stratosphere (B) Thermosphere
(C) Troposphere (D) None of these

197. The atmosphere stratum nearest to earth is:
(A) Thermosphere (B) Stratosphere
(C) Troposphere (D) None of these

198. Part of earth covered by water is about:
(A) 73% (B) 50% (C) 92% (D) 87%

199. Floods can be checked by:
(A) Making dams (B) Deforestation
(C) Reforestation (D) None of these

200. Percentage of fresh water in the hydrosphere is:
(A) 3% (B) 30% (C) 13% (D) 33%

201. Percentage of fresh water found in lakes, ponds, swamp, streams and rivers is:
(A) 22.4% (B) 77.2% (C) 7.7% (D) 0.36%

202. Of the total fresh water, ground water constitutes:
(A) 90% (B) 10% (C) 22.4% (D) 44.2%

203. Of the total water evaporated from oceans, precipitation returns:
(A) 100% (B) 90% (C) 75% (D) 50%

204. Number of agro climatic zones in Uttar Pradesh are:
(A) 6 (B) 7 (C) 8 (D) 9

205. Planning Commission agro-climatic region the whole country has been divided into:
(A) 15 (B) 20 (C) 25 (D) 30

206. The concept of agro-climatic zones were introduced during:
(A) NATP (B) NARP (C) NREP (D) NAIP

207. The number of agro-climatic zones under NARP:
(A) 127 (B) 146 (C) 152 (D) 164

208. The number of agro-ecosystems identified under NATP:
(A) 6 (B) 9 (C) 14 (D) 21

209. Movement of air occurs in the atmosphere upto height of:
(A) 100km (B) 50 km
(C) 25 km (D) 10 km

210. Peak concentration of ozone in the stratosphere (ozonosphere) lies at a height of:
(A) 10 km (B) 25 km
(C) 40 km (D) 50 km

211. Planning Commission advice that the agro-climatic zones in which plan:
(A) Fifth Five Year Plan (B) Sixth Five Year Plan
(C) Seventh Five Year Plan (D) Eight Five Year Plan

212. The estimates postulate that the carbon dioxide content of atmosphere has increased from 280 parts per million to —— during last three decades
(A) 300 ppm (B) 350 ppm
(C) 400 ppm (D) None of these

213. Nitrate levels in drinking water above ——— mg per litre are considered a human health hazard:
(A) 5 (B) 10 (C) 15 (D) 20

214. The discovery of oxygen was done by which scientist who proved the keystone of other discoveries that went far toward unlocking the mystery of plant life:
(A) John Mayow (B) John Woodward
(C) Benjamin Franklin (D) Priestly

215. There was not only one "Principle of vegetation" as reflected by agricultural writings during 17th and 18th centuries, was observed by:
(A) William Michael (B) Guy Baird
(C) Francis Home (D) Jone Cummings

216. More than which per cent of earth's surface is covered by water in the form of seas, lakes and rivers:
(A) 40 (B) 50 (C) 60 (D) 70

217. Depending upon the temperature factor, the upper warmer region of lacks is called:
(A) Epilimnion (B) Hypolimnion
(C) Thermocline (D) None of these

218. Secondary consequences of increasing elevation, both agronomically and economically significant are:
(A) Decreased temperature
(B) Decreased precipitation
(C) Decreased wind velocity
(D) Increased soil fertility

219. "Diversion of water from mountain streams by diversion dams to canals of different length and breadth and ditches of varying size, and finally directed to individual fields" is an important practice in:
(A) China (B) Egypt
(C) Philippines (D) Japan

220. The success or failure of the cropping season is determined by the intensity of:
(A) Climatic factors (B) Political factors
(C) Social factors (D) None of these

221. The ideal conditions of temperature for crop production are between:
(A) 12-18 °C (B) 18-24 °C
(C) 24-30 °C (D) 30-36 °C

222. The direct effect of strong winds on plants is of which nature:
(A) Chemical (B) Biological
(C) Mechanical (D) None of these

223. Altitude, because of it's —— effects, induces a zonation of vegetation similar to that found through a wide range of latitude:
(A) Humidity (B) Radiation
(C) Temperature (D) None of these

224. Rice, maize and sugarcane are replaced by temperate zone species at elevations of possibly:
(A) 1000-3000' (B) 4000-6000'
(C) 7000-9000' (D) None of these

225. At 0 °C water is in equilibrium with a water vapour pressure of——— mm of mercury:
(A) 2.58 (B) 3.58 (C) 4.58 (D) 5.58

226. The atmosphere over one square foot of earth's surface at the Equator will contain approximately ——— pounds of water vapour:
(A) 20 (B) 30 (C) 40 (D) 50

227. It has been found that peas will remain in good harvest condition twice as long at ———°F than at 80 °F:
(A) 32 (B) 42 (C) 52 (D) 62

228. The changes that accompany hardening of plants include:
(A) Increased sugar concentration
(B) Decreased soluble proteins
(C) Decreased osmotic pressure
(D) Decreased bound water

229. Cacao, a tropical plant has been found to produce the maximum growth in approximately—— of full sunlight:
(A) 5% (B) 10% (C) 15% (D) 25%

230. When white light is passed through a prism, it is dispersed into different wavelengths of various colours and the orange colour appears at:
(A) 435-490 (B) 574-595
(C) 400-435 (D) 595-626

231. It has been demonstrated that the ——— light present in the white light prevents maximal growth rate:
(A) Blue (B) Red
(C) Green (D) Violet

232. The unit of atmospheric pressure is:
(A) Millibar (B) Bar
(C) Pascal (D) Gram

233. What per cent of isolation is absorbed in the atmosphere by gasses, clouds and suspended solids?
(A) 19 % (B) 17 % (C) 15 % (D)13 %

234. Oxygen and ozone at high levels absorb most of which radiation:
(A) Ultra violet radiation (B) Gama radiation
(C) Blue radiation (D) All of these

235. The ratio of the amount of radiation reflected to the amount received on a surface is termed as:
(A) Albedo (B) Radiation
(C) Reflection (D) All of these

236. The ability to admit most of the insolation yet retard losses by reradiation from the earth's surface commonly is known as:
(A) Greenhouse effect (B) Atmosphere effect
(C) Both (A) and (B) (D) None of these

237. The highest per cent of albedo is found:
(A) Desert (B) Tropical forest
(C) Deciduous forest (D) Coniferous forest

238. The instrument which measure sunshine and radiation is:
(A) Total radiation pyranometer
(B) Net radiometer
(C) Foster switch sunshine duration sensor
(D) All of these

239. The ratio of the mass of water vapour actually in the air to a unit of mass of air, including the water vapour is called:
(A) Specific humidity (B) Relative humidity
(C) Both (A) and (B) (D) None of these

240. The ratio of the amount of water vapour actually in a volume occupied by air to the amount the space could contain at saturation is called:
(A) Specific humidity (B) Relative humidity
(C) Both (A) and (B) (D) None of these

241. Early morning hours when temperatures are low the relative humidity will be:
(A) More (B) Less
(C) Not change (D) Unknown

242. The visible aggregates of water droplets, ice particles, or a mixture of both alongwith varying amounts of dust particles is called:
(A) Fog (B) Frost
(C) Clouds (D) All of these

243. The clouds are a cirriform layer or patch of small white flakes or tiny globules arranged in distinct group or lines is called:
(A) Cirrocumulus (B) Cirrostratus
(C) Altocumulus (D) Altostratus

244. The clouds is a thin white veil of cirrus is called:
(A) Cirrocumulus (B) Cirrostratus
(C) Altocumulus (D) Altostratus

245. The middle clouds predominate at height ranging from 2 to 8 km above ground levels is called:
(A) Altocumulus (B) Altostratus
(C) Both (A) and (B) (D) None of these

246. The banner clouds above peaks are of which type clouds is called:
(A) Cirrocumulus (B) Cirrostratus
(C) Altocumulus (D) Altostratus

247. When the cloud layer lowers to become some what thicker and darker and falling rains or snow obscures its base it is called:
(A) Cirrocumulus (B) Cirrostratus
(C) Altocumulus (D) Nimbostratus

248. The clouds are broken into fragments by the wind it is called:
(A) Cirrocumulus (B) Cirrostratus
(C) Fractostratus (D) Altostratus

249. The clouds are dense, dome shaped and have flat bases is called:
(A) Cumulus (B) Cirrostratus
(C) Altocumulus (D) Altostratus

250. The heavy showers of rain, snow or hail, often accompanied by lighting and thunder occurs by:
(A) Cirrocumulus (B) Cumulonimbus
(C) Altocumulus (D) Altostratus

251. More than 60 per cent rains occurs by:
(A) South-west monsoon (B) North-East monsoon
(C) Both (A) and (B) (D) None of these

252. Less than 30 per cent rains occurs by:
(A) South-west monsoon (B) North-East monsoon
(C) Both (A) and (B) (D) None of these

253. North-East monsoon also called:
(A) Post monsoon period (B) Post rainy season
(C) Both (A) and (B) (D) None of these

254. A counter clock wise rotation of wind in northern hemisphere is:
(A) Cyclone (B) Anticyclone
(C) Rainfall (D) Fog

255. A counter clock wise rotation of wind in southern hemisphere is:
(A) Cyclone (B) Anticyclone
(C) Rainfall (D) Fog

256. Wind speed measured by:
(A) Windvane (B) Anemometer
(C) Both (A) and (B) (D) None of these

257. The barometric pressure is highest at its centre and decreased out ward is indication of:
(A) Cyclone (B) Anticyclone
(C) Rainfall (D) Fog

258. The common pressure that initiates thunder storms development are:
(A) Heating and convection in moist air over warm land surfaces
(B) Passage of cold, moist air over warm water
(C) Radiational cooling at upper levels
(D) Advection of cold air aloft
(E) All of these

259. The height of a thunderstorms is related to:
(A) Latitude (B) Season
(C) Both (A) and (B) (D) None of these

260. Thunderstorms acused by surface heating over land are most common in summer in the:
(A) Afternoon (B) Early morning
(C) Both (A) and (B) (D) None of these

261. Most violent storms of the lower troposphere is:
(A) Tornado (B) Watersprout
(C) Both (A) and (B) (D) None of these

262. The moisture index is more in:
(A) Perhumid (B) Humid
(C) Semiarid (D) Arid

263. The moisture index is less in:
(A) Perhumid (B) Humid
(C) Semiarid (D) Arid

264. Human health and comfort suggest another possible approach to defining:
(A) Climatic types (B) Relative humidity
(C) Temperature (D) None of these

265. The summation of all atmospheric events an infinite time span does not changes is called:
(A) Climate (B) Weather
(C) Both (A) and (B) (D) None of these

266. The periods of atmospheric changes on the order of millions of years is:
(A) Paleoclimates (B) Taiga climate
(C) Polar climate (D) Ocean climate

267. The mean monthly temperatures are below 0^0C (snow, ice or barren rocks covers such areas) and vegetation is entirely lacking is called:
(A) Paleoclimates (B) Taiga climate
(C) Polar climate (D) Ocean climate

268. The key element in the water budget in which is also the link between moisture and energy exchange is:
(A) Evaporation (B) Transpiration
(C) Evapo-transpiration (D) None of these

269. Evapotranspiration can be measured by:
(A) Lysimeter (B) Evapotranspirometer
(C) Both (A) and (B) (D) None of these

270. The frost occurs on clear nights with a temperature inversion and usually results in formation of ice crystals on bold objects is called:
(A) Radiation frost (B) Advection frost
(C) Black frost (D) None of these

271. Advection frost is also called:
(A) Radiation frost (B) Advection frost
(C) Black frost (D) None of these

272. The frost which results when the temperature at the surface in an air mass is below freezing is called:
(A) Radiation frost (B) Advection frost
(C) Black frost (D) None of these

273. The deficit that results when soil moisture is insufficient to meet the demands of potential evapotranspiration is:
(A) Drought (B) Rainfed
(C) Dryland (D) None of these

274. Drought may be classified as:
(A) Permanent drought may be associated with arid climate
(B) Seasonal drought, which occurs in climate with distinct amount periods of dry weather
(C) Drought due to variability of precipitation
(D) All of these

275. Combating the drought effects are:
(A) Growing of less water requiring crops
(B) Decreasing the water need of crops
(C) Increasing the water supply
(D) All of these

276. When the annual rainfall is less than the normal over an area for a prolonged period is called:
(A) Meteorological drought (B) Hydrological drought
(C) Agricultural drought (D) None of these

277. When a prolonged drought results in drying of the available water sources like streams, tanks, reservoir, well etc. is called:
(A) Meteorological drought (B) Hydrological drought
(C) Agricultural drought (D) None of these

278. Shortage of supply of water to meet the demand for crop maturity is called:
(A) Meteorological drought (B) Hydrological drought
(C) Agricultural drought (D) None of these

279. During snowfall, the trees:
(A) Do not respire
(B) Do not photosynthesis
(C) Show minimum life function
(D) Show maximum transpiration

280. Suitable temperature for tuberization in potato is:
(A) 12-14°C (B) 18-20°C
(C) 22-24°C (D) 28-30°C

281. The pollutants which are already present in nature, but are released in substantial amounts by man are known as:
(A) Quantitative pollutants (B) Qualitative pollutants
(C) Degradable pollutants (D) Primary pollutants

282. The biological amplification of DDT in the various trophic levels is known as:
(A) Green house effect (B) Biomagnification
(C) Entrophication (D) Pollution

283. Environment can be improved by:
(A) Wind erosion (B) Excessive tree fall
(C) Conservation (D) None of these

284. Smog is produced due to :
(A) Air and metal dust
(B) Water and nitrogen oxide
(C) Smoke and fog
(D) None of these

285. The photochemical smog is formed by:
(A) Nitrogen oxides
(B) Hydrocarbons
(C) Nitrogen oxides and hydrocarbons
(D) Solar radiations upon nitrogen oxides and hydrocarbons

286. Photochemical smog brings about:
(A) Defoliation
(B) Bleaching and blazing of foliage
(C) Floral bud shedding
(D) Discolouration and curling of petals

287. In metropolitan cities the automobiles cause air pollution:
(A) 80% (B) 60% (C) 50% (D) 100%

288. Huge quantity of sewage is dumped in a river. The BOD will:
(A) Increase (B) Decrease
(C) Slightly increase (D) Remain unchanged

289. Jet aeroplane releases pollutants in air is called:
(A) Smog
(B) Photo-chemical oxidants
(C) Aerosols
(D) Algae and bacteria

290. Fertilizer-pesticides and insecticides may cause:
(A) Air pollution (B) Water pollution
(C) Both (A) and (B) (D) None of these

291. Pollutants directly emanating from human activities are:
(A) Quantitative pollutants (B) Qualitative pollutants
(C) Air pollutants (D) Water pollutants

292. Atmospheric pollutants formed by human activities constitute of the total:
(A) 95% (B) 75% (C) 25% (D) 0.05%

293. Most atmospheric pollutants do not rise above:
(A) 600 m (B) 1000 m (C) 2000 m (D) 200 m

294. Aerosol is a system of colloidal particles having a size of:
(A) 5 μm (B) 1 – 2 μm
(C) Less than 1 μm (D) More than 10 μm

295. A secondary pollutants is:
(A) O_3 (B) CO (C) CH_4 (D) Pb

296. A colourless, nonirritant, highly toxic gas that impairs respiration is:
(A) SO_2 (B) NO_2 (C) CO_2 (D) CO

297. Lead pollution in Indian cities is mainly due to:
(A) Lead pencils (B) Batteries
(C) Automobiles (D) Electronic gadgets

298. Pollutants likely to deplete ozone layer is:
(A) CO (B) Nitrogen oxides
(C) Chloroflurocarbons (D) Both (B) and (C)

299. Acid rain is caused by:
(A) Excess production of coal gas
(B) Excess release of CO_2 due to increasing combustion and respiration
(C) Excess release of SO_2 and NO_2 from burning fossil fuels
(D) Excess production of gaseous hydrocarbons

300. SO_2 pollution is indicated by destruction of:
(A) Climbers (B) Lichens
(C) Mosses (D) Grasses

301. Freom is:
(A) Ferrous sulphide pollutant
(B) Chloroflurocarbon
(C) Iron containing polycyclic hydrocarbons
(D) Metal fluoride

302. Maximum atmospheric NO_2 pollution is found in:
(A) Baroda (B) Chennai
(C) Nagpur (D) Mumbai

303. O_3 promotes:
(A) Oxidation of biochemicals
(B) Destruction of chlorenchyma
(C) Both (A) and (B)
(D) Growth of plants

304. Water soaked appearance of leaves due to sulphur dioxide pollution is caused by:
(A) Destruction of chlorophyll
(B) Damage to membranes
(C) Both (A) and (B)
(D) Corrosive action over metals

305. In addition to SO_2, metals and textiles are damaged by:
(A) Nitrogen oxides (B) Fluorides
(C) Carbon monoxide (D) Aldehydes

306. Photochemical reaction of nitrogen oxides with unsaturation hydrocarbons produces:
(A) Methyl isocyanate (B) Phosgene
(C) Benzpyrene (D) Peroxy-acyl nitrates

307. Level of atmospheric CO_2 has increased from:
(A) 290 to 340 ppp (B) 250 to 300 ppm
(C) 300 to 350 ppm (D) 330 to 350 ppm

308. CO_2 content of the atmosphere is likely to double by the year:
(A) 2001 (B) 2010 (C) 2020 (D) 2050

309. Gravity setting tanks/porous filters are used to remove particles larger than:
(A) 2 μm (B) 50 μm (C) 20 μm (D) 10 μm

310. Electronic precipitators are employed to remove particulate pollutants of the size of:
(A) 5-20 μm (B) 20-50 μm
(C) 50 μm and larger (D) 0.5-5.0 μm

311. B.O.D. is related to:
(A) Detergents (B) Putrescibility
(C) Inorganic pollutants (D) Organic pollutants

312. A persistent insecticide is:
(A) Organophosphates (B) Carbamates
(C) Thiocarbamates (D) Chlorinated hydrocarbons

313. An organoclorine insecticides is:
(A) DDT/aldrin (B) Malathion
(C) Tetraethyl pyrophosphate (D) Thiocarbamte

314. Domestic water is:
(A) Effluent (B) Biodegradable
(C) Nonbiodgradable (D) Air pollutants

315. Inadequate drainage in a soil will lead to:
(A) Floods (B) Soil erosion
(C) Aridity (D) Salination

316. Major reason of soil erosion in India is:
(A) Tilling of land
(B) Deforestation
(C) Overgrazing and deforestation
(D) Overgrazing

317. Jhuming occurs in India is:
(A) Hilly areas (B) Madhya Pradesh
(C) North – East (D) Malabar area

318. Ecological balance of a mining area is restored through:
(A) Conversion into agricultural area
(B) Revegetation of area
(C) Prevention of soil erosion
(D) Protection against grazing

319. Terrestrial life would become impossible with:
(A) Change in orbit of earth
(B) Change on atmosphere permitting all solar radiations to reach lithosphere
(C) Disappearance of moon
(D) Decrease in mean annual temperature by 10°C

320. Disease aggravated by pollution is:
(A) Scurvy (B) Rheumation
(C) Haemophilia (D) Bronchitis

321. Weather and climate are driven by what source(s) of energy?
(A) Earth's internal heat (B) Solar radiation
(C) Gravity (D) Both (B) and (C)

322. The difference between weather and climate is:
(A) Weather is a local phenomenon, climate is regional
(B) Weather is a day –to-day phenomenon, climate is a long term average
(C) Weather is a non-scientific term for climate
(D) There is no difference between the two

323. If Earth's albedo increased, then global mean temperature would probably:
(A) Increase (B) Decrease
(C) Saty the same (D) None of these

324. What change would increase the albedo of earth?
(A) Increase in cloudiness
(B) Increase in ice sheet size
(C) Increase in snow cover
(D) All of these

325. The green house effect is:
(A) Incoming shortwave radiation converted into long wave radiation, which is than trapped by atmospheric gasses
(B) Reflection of longwave radiation after converted shortwave radiation is trapped by atmospheric gasses
(C) Carbon dioxide buildup in the atmosphere
(D) Global warming due to ozone depletion

326. The uneven distribution of solar radiation, plus the pull of gravity on earth produces:
(A) Ocean currents and winds
(B) Earthquakes and volcanic eruptions
(C) Mass extinctions
(D) The greenhouse effect

327. The amount of energy a material needs to absorb to change its temperature defines its:
(A) Radiation (B) Heat capacity
(C) Conduction (D) Convection

328. The amount of energy a material needs to absorb to change its state from liquid to gas defines its:
(A) Radiation (B) Heat capacity
(C) Conduction (D) Latent heat of vaporization

329. Warm air in the tropics rises because:
(A) It has more energy (B) Of the Coriolis effect
(C) It is less dense (D) It is more dense

330. Cold air sinks because:
(A) It has less energy (B) Of the Coriolis effect
(C) It is less dense (D) It is more dense

331. Air masses always more horizontally from:
(A) High pressure to low pressure areas
(B) Low pressure to high pressure areas
(C) From oceans to continents
(D) From continents to oceans

332. Jet streams are:
(A) Flow west to east
(B) Are at high altitudes
(C) Are relatively narrow bonds
(D) All of these

333. Air around low pressure zones (cyclones) in the northern hemisphere:
(A) Spins clockwise
(B) Spins counterclockwise
(C) Spins sometimes clockwise, sometimes counterclockwise
(D) Moves directly towards the center of low pressure

334. Air around high pressure zones (anticyclones) in the northern hemisphere:
(A) Spins clockwise
(B) Spins counterclockwise
(C) Spins sometimes clockwise, sometimes counterclockwise
(D) Moves directly towards the center of low pressure

335. A rapidly moving cold front that encounters warm moist air may produce:
(A) A narrow band of intense thunderstorms
(B) A wide band of intense thunderstorms
(C) A wide band of intense rainfall without thunderstorms
(D) A wide band of moderate rainfall without thunderstorms

336. A slowly moving warm front that encounters cooler air may produce:
(A) A narrow band of intense thunderstorms
(B) A wide band of intense thunderstorms
(C) A wide band of intense rainfall without thunderstorms
(D) A wide band of moderate rainfall without thunder storms

337. Which of the following air movements are not cyclonic?
(A) Low pressure system (B) Tornado
(C) Hurricane (D) Jet stream

338. If you have an air mass with a certain relative humidity and you raise its temperature without adding or removing any water, the relative humidity:
(A) Will decrease (B) Will increase
(C) Will remain the same (D) Unknown

339. The latitude at which a rising air mass attains 100% relative humidity is called:
(A) At null buoyancy level
(B) The dew point
(C) A front
(D) Lifting condensation level

340. The Fujita scale quantifies tornadoes, based on:
(A) Length of time it is on the ground
(B) The extent of damages caused by wind
(C) Diameter of the funnel
(D) Speed at which the funnel moves across the ground

341. Thunderstorms most commonly occur:
(A) When warm moist air is heated
(B) On sunny days
(C) In early efternoon or evening
(D) All of these

342. Which of the following events are associated with thurnderstorms?
(A) Heavy rainfall and hail (B) Thunder and lighting
(C) Microbursta (D) Tornadoes
(E) All of these

343. Which of the following is not an effective greenhouse gas?
(A) Water vapour (B) Methane
(C) Oxygen (D) Carbon dioxide

344. Climate variations on the scale of decodes may be caused by:
(A) Solar radiation variation caused by orbital variations
(B) Changes in ocean currents
(C) Volcanic eruptions
(D) Plate tectonic cycles

345. A drought may be defined as:
(A) A desert region of low rainfall
(B) A famine condition
(C) A time of abnormally low rainfall
(D) A lack of water

346. Humans produce the largest amount of CO_2 by:
(A) Land clearing and burning
(B) Agriculture and land use changes
(C) Burning of fossil fuels
(D) Nuclear power generation

347. Some effects of global warming may be:
(A) Increase in droughts in some areas
(B) Increase in rainfall in some areas
(C) Shifts in forests and agriculture patterns
(D) Sea level rise
(D) All of these

348. Eargh's early atmosphere was rich in:
(A) Ozone (B) Oxygen
(C) Sulphur (D) Carbon dioxide

349. Earth's present atmosphere was rich in:
(A) Ozone (B) Oxygen
(C) Sulphur (D) Carbon dioxide

350. Carbon dioxide has been locked up over geological time as:
(A) Part of the mantle
(B) Limestone and fossils
(C) Material dissolved in the oceans
(D) Material lost into space

351. Earth's early green house climate resulted in a climate much like that of ——— today:
(A) Mercury (B) Venus
(C) Mars (D) Jupiter

352. A sedimentary rock that indicates a warm climate is:
(A) Till (B) Sandstone
(C) Limestone (D) Granite

353. The initiation of a glacial period requires:
(A) Polar landmasses
(B) Restricted ocean circulation
(C) Decrease green house gasses
(D) All of these

354. Climate cycles with periods of about 100000 years during the last 1 million years are probably caused by:
(A) Solar radiation variation caused by orbital variations
(B) Change in oceam currents
(C) Volcanic erutions
(D) Plate tectonic cycles

355. Hurrican Andrew in 1992 caused most of its damage by:
(A) Strong winds (B) Storm surge
(C) Flooding (D) Waves

356. Atmospheric pressure in the center of tropical cyclone is:
(A) Very high because of converging winds
(B) Very high because of sinking warm air
(C) Very low because of rising warm air
(D) Very low because of dense sinking air

357. The current rise in global sea level is about ——— feet per century:
(A) 1 (B) 3
(C) 10 (D) 30

358. Which ocean does not commonly experience tropical cyclones?
(A) North Atlantic (B) South Atlantic
(C) Western Pacific (D) Eastern Pacific

359. Common ocean waves are primarily caused by:
(A) Gravitational forces
(B) Tides
(C) Wind energy transferred to water
(D) Undersea earthquakes and volcanoes

360. The source of most beach sand is:
(A) Rivers (B) Submarine canyons
(C) Affshore bars (D) Headlands

361. North Atlantic hurricanes are most likely in:
(A) July (B) August
(C) September (D) October

362. What is the dew point when the dry bulbs temperature is 24°C and the wet bbulb temperature is 15°C?
(A) 8°C (B) -18°C
(C) 36°C (D) 4°C

363. In general, the probability of flooding decreases when there is an increase in the amount of:
(A) Precipitation (B) Infiltration
(C) Runoff (D) Snow melt

364. Liquid water can store more heat energy than an equal amount of any other naturally occurring substance because liquid water:
(A) Covers 71% of Earth's surface
(B) Has its greatest density at 4°C
(C) Has the higher specific heat
(D) Can be changed into a solid or a gas

365. Clouds usually from when:
(A) Air temperature reaches the dew point
(B) Evaporation has warmed the surrounding air
(C) Relative humidity is 0%
(D) Condensation nuclei have been removed from the air

366. The environment which has been modified by human activities is called:
(A) Anthropogenic environment
(B) Natural environment
(C) Manmade environment
(D) Both (A) and (C)

367. An eco-friendly slogan:
(A) 'Do and die'
(B) 'Birds of same feather fly together'
(C) 'Time and tide wait for none'
(D) 'Say no to crackers'

368. The depletion in the ozone layer is caused by:
(A) Nitrous oxide (B) Carbon dioxide
(C) Chlorofluorocarbons (D) Methane

369. 'Ozone –hole' means:
(A) Large sized hole in the ozone layer
(B) Thinning of the ozone layer
(C) Small holes scattered in the ozone layer
(D) Thickening of ozone in the ozone layer

370. Ozone layer is getting depleted because of:
(A) Excessive use of automobiles
(B) Excessive formation of industrial units
(C) Excessive use of man made compounds containing both fluorine and chlorine
(D) Excessive deforestation

371. Viruses, which cause hepatitis, are transmitted through:
(A) Air (B) Water
(C) Food (D) Personal contact

372. Vectors can be defined as:
(A) Animals carry the infecting agents from risk person to another healthy person
(B) Microorganisms which cause many diseases
(C) Infected person
(D) Diseased plants

373. Which of the following is a recently originated problem of environment?
(A) Ozone layer depletion (B) Green house effect
(C) Global warming (D) All of these

374. When we breathe in air, nitrogen also goes inside along with oxygen. What is the fate of this nitrogen?
(A) It moves alongwith oxygen into the cells
(B) It comes out with the CO_2 during exhalation
(C) It is absorbed only by the nasal cells
(D) Nitrogen concentration already move in the cells so it is not at all absorbed

375. Progress that meets the needs of the present without compromising the ability of future generations to meet their own needs is:
(A) The tragedy of commons
(B) Sustainable development
(C) Net primary productivity
(D) The impossibility theorem

376. Land and natural resources are considered:
(A) Capital accumulation
(B) Comon properly resources
(C) Non producible
(D) Output

377. Air pollution from automobile exhausts and water pollution of steel plants are example of:
(A) External economies (B) Negative externalities
(C) Internal spillover (D) Social distortion

378. The government levying taxes on polluters or changing a surcharge for pesticide use are:
(A) Examples of Coase's theorem
(B) Internalization of negative spillover effects
(C) Marginal abtement cost
(D) Example of a free rider

379. Which of the following is an example of tragedy of commons?
(A) Over fishing (B) Smoking is a public place
(C) Excessive rain (D) Common use of public toilets

380. Negative externality is also known as:
(A) External diseconomies (B) Marginal damage
(C) Public goods (D) Resource curse

381. Deforestation:
(A) Leads to localized flooding
(B) Reduces sustainable logging potential
(C) Reduces watershed stability
(D) Augments carbon restoration provides by forests

382. Irreversibility refers to:
(A) Natural resources that cannot be reproduced in the future if we fail to preserve them now
(B) Obtaining intellectual property rights for those developing and patenting new products
(C) Natural extinction of various species in DCs
(D) Industrialization replacing agriculture in LDCs

383. Many environmental resources are public goods, which are characterized by:
(A) Rivalsy and exclusion in consumption
(B) Non rivalsy and nnonexclusion in consumption
(C) Rivalsy but nonexclusion in production
(D) Nonrivalsy but exclusion in usage

384. Biodiversity:
(A) Includes genetic, species, ecosystem and functional diversities
(B) Refers to diversifying earth's nonrenewable resources
(C) Refers to recpnstruction of tropical rainforests
(D) Refers to biological effects on commercial plantation

385. The green house effects in thee phenomenon by which:
(A) Biological diversity is dominant in agricultural production
(B) The globe's water pollution affect plankton
(C) The earth's atmosphere traps infrared radiation
(D) Climate changes occur naturally in the forest

386. Which of the following is NOT a factor is creating climate?
(A) Latitude
(B) Interaction between land and water
(C) Atmospheric pressure systems
(D) Global wind patterns
(E) Daily wether variations

387. One of the following processes is not a step involved in the water cycle operating in nature:
(A) Evaporation (B) Temperature
(C) Precipitation (D) Photosynthesis

388. The term "water – pollution" can be defined in several ways, which of the following statements does not give the correct definitions?
(A) The addition of undesirable substances to water bodies
(B) The removal of desirable substances from water bodies
(C) A change in pressure of the water bodies
(D) A change in temperature of the water bodies

389. Which step is not involved in the carbon-cycle?
(A) Photosynthesis (B) Transpiration
(C) Respiration (D) Burning of fossil fuels

390. South-west monsoon contributes how much per cent of rainfall?
(A) 50 % (B) 75 % (C) 80 % (D) 95 %

391. Average rainfall in India is around:
(A) 2120 mm (B) 1120 mm
(C) 850 mm (D) 500 mm

392. A source of nonbiological oxygen is:
(A) Photodissociation
(B) Photolysis
(C) Formation of fats from carbohydrates
(D) Both (B) and (C)

393. A process that occurs without loss or gain of heat is:
(A) Coriolis effect (B) Eustatic effect
(C) Adiabatic (D) All of these

394. Adiabatic expansion resulting in cooling is called:
(A) Coriolis effect (B) Joule-Kelvin effect
(C) Le Chatelier Principle (D) Principle of Parsimony

395. Which area has four seasons:
(A) Tropics (B) Monsoonic
(C) Temperate (D) Arctic

396. Praying Mantis is green coloured predator insect that shows mimicry of the type:
(A) Concealing-protective (B) Warning-protective
(C) Alluring-aggressive (D) Concealing-aggressive

397. Male Atlas Moth has huge antennae that can detect pheromone released by female moth from a distance of:
(A) 1 km (B) 2 km (C) 5 km (D) 8 km

398. The biological amplication of DDT in the various trophic levels is known as:
(A) Green house effect (B) Biomagnification
(C) Entraphication (D) Pollution

399. Taj may be destroyed by:
(A) Flood in Yamuna
(B) Temperature mediated spoilage of marble
(C) Air pollutants from Mathura refinery
(D) All of these

400. Smog occurs in places of:
(A) Excess SO_2 (B) Low temperature
(C) High temperature (D) Excess NH_3

401. Acid rains is due to increase in atmospheric concentration of:
(A) Ozone and dust (B) CO_2 and CO
(C) SO_3 and CO (D) SO_2 and NO_2

402. Major pollution causing agents is:
(A) Hydrocarbons (B) Animals
(C) Human (D) Birds

403. Pollution is:
(A) Removal of top soil
(B) Release of toxic/undesirable materials in environment
(C) Conservation of energy (D) All of these

404. Effect of pollution is most marked on:
(A) Natural balanced of nature
(B) Natural geochemical cycles
(C) Natural flora of place (D) All of these

405. Chief sources of water and soil pollution is:
(A) Mining (B) Thermal power plant
(C) Agroindustry (D) All of these

406. Ultimate environment hazards to mankind is due to:
(A) Nuclear pollution (B) Water pollution
(C) Air pollution (d) Noise pollution

407. Which causes water pollution?
(A) Smoke/fly ash (B) Automobile exhausts
(C) Aeroplanes (D) 2, 4-D and pesticides

408. Air pollution is not caused by:
(A) Pollen grains (B) Hydroelectric power
(C) Industries (D) Automobiles

409. Pollution is controlled by:
(A) Banning atomic blasts
(B) Use of electrically operated automobiles
(C) Sewage treatment
(D) All of these

410. Carbon monoxide is harmful to human being as it is:
(A) Carcinogenic
(B) Antagonistic to CO_2
(C) With higher affinity for haemoglobin as compared to oxygen
(D) Destructive to O_3

411. Which one of the following statements about forest cover is wrong?
(A) North-eastern states and parts of Gujarat have a high percentage of un:classed forests
(B) J & K, AP, Kerala, Uttarakhand and WB have a high percentage of resourced forest
(C) Bihar, Punjab, Himachal Pradesh and Haryana have permanent forests estates
(D) Madhya Pradesh has the largest area under permanent forests

412. Which one of the following facts is wrong, about the role of community in conserving nature?
(A) Villagers in Rajasthan have encouraged the government officials to practice mining to saves their livelihood
(B) Each community has their own set of traditional methods of conserving nature
(C) Certain communities worship some particular plants and animals which they have preserved from time to time
(D) Tribal belief have preserved several virgin forests in pristine form called 'Sacred groves'

413. What measures is not clear from the dynamics of environment destruction and conservation from the following?
(A) Adopt strategies to save nature at any cost
(B) Accept only those economic activities that are people centric, environment-friendly and economically rewarding
(C) Local communities should not be involved in any resource management
(D) None of these

414. Which of these statements is not a valid reason for the depletion of flora and fauna?
(A) Agricultural expansion
(B) Grazing and fuel wood collection
(C) Large scale development projects
(D) Rapid industrialization and urbanization

415. Which factors does not cause depletion of biodiversity?
(A) Environment pollution
(B) Expansion on animals
(C) Judicious way of collecting products from forests
(D) Excessive hunting

416. Fog is a colloid of:
(A) Solid in gas (B) Liquid in gas
(C) Gas in liquid (D) Liquid in gas in solid

417. The ______________ zone is absent in ponds:
(A) Sublittoral (B) Limnetic
(C) Profundal (D) Gas in solids

418. Most stable ecosystem is:
(A) Desert (B) Mountain
(C) Ocean (D) Forest

419. The gas which causes depletion of ozone layer is the atmosphere is:
(A) Ethane (B) Methane
(C) Carbon dioxide (D) Chloro fluoro carbon

420. Disease caused by eating fish found in water contaminated with industrial waste having mercury is:
(A) Minamata disease (B) Bright's disease
(C) Hashimoto's disease (D) Osterosclerosis

421. Fish die in water bodies polluted by sewage due to:
(A) Pathogens (B) Clogging of gills by silt
(C) Reduction in oxygen (D) Foul smell

422. Lichens indicate pollution by:
(A) O_3 (B) SO_2 (C) NO_2 (D) CO

423. Drained sewage has B.O.D.:
(A) More than that of water (B) Less than that of water
(C) Equal to that of water (D) None of these

424. B.O.D. is connected with:
(A) Organic matter (B) Microbes
(C) Microbes and organic matter (D) None of these

425. Which one is the most important causative pollutants of soil?
(A) Glass junk (B) Iron junk
(C) Detergents (D) Plastics

426. BOD is:
(A) Biological oxygen deficit
(B) Biosphere oxygen demand
(C) Biological oxygen demand
(D) None of these

427. Pollution can bring about change in:
(A) Biogeochemical cycling (B) Abiotic environment
(C) Biotic environment (D) None of these

428. Which one is the major reason for pollution in big cities?
(A) Fossil fuel (B) Acid rain
(C) Heat dispersion (D) None of these

429. World environment day is:
(A) 28th Feb. (B) 5th June
(C) 7th Aug. (D) 10th April

430. Biodegradable pollutants is:
(A) Plastic (B) Asbestos
(C) Sewage (D) Mercury

431. Water pollution is due to:
(A) Sulphur dioxide (B) Carbon dioxide
(C) Oxygen (D) Industrial discharges

432. Major reason for soil erosion in India is:
(A) Tilling of land (B) Deforestation
(C) Overgrazing and deforestation
(D) Over grazing

433. SO_2 and NO_2 produce pollution by increasing:
(A) Alkalinity (B) Acidity
(C) Neutrality (D) Buffer action

434. Which one is found in higher concentration in the atmosphere?
(A) Ozone
(B) Nitrous oxides
(C) Carbon dioxide, carbon monoxide and sulphur dioxide
(D) All of these

435. Carbon dioxide is a major pollutant in:
(A) Air (B) Soil (C) Water (D) Noise

436. The nitrogen molecules present in air can be converted into nitrates and nitrites by:
(A) A biological process of nitrogen fixing bacteria present in soil
(B) A biological process of carbon fixing factor present in soil
(C) Any of the industries manufacturing nitrogenous compounds
(D) The plants used as cereal crops in field

437. Phosphate pollution is caused by:
(A) Weathering of phosphate rocks only
(B) Agricultural fertilizers
(C) Phosphate rocks and sewage
(D) Sewage and agricultural fertilizers

438. Which pigment protects plants from UV damage?
(A) Chlorophyll (B) Xanthophyll
(C) Carotenoids (D) Phycocyanin

439. Indescriminate use of fertilizers causes:
(A) Air pollution (B) Water pollution
(C) Land pollution (D) All of these

440. SO_2 pollution causes:
(A) Plasmolysis (B) Destruction of cell wall
(C) Destruction of Golgi bodies
(D) Destruction of chlorophyll

441. Depletion of ozone layer is due to:
(A) Oxides of nitrogen (B) Oxides of carbon
(C) Oxides of sulphur (D) None of these

442. Phosphate pollution is caused by:
(A) Agricultural fertilizers
(B) Weathering of phosphate rocks only
(C) Phosphate rocks ad sewage
(D) Sewage and agricultural fertilizers

443. Common indicator organism of water pollution is:
(A) *Escherichia coli* (B) *Cholera vibrio*
(C) *Salmonella typhi* (D) *Entamoeba histolytica*

444. Which pigment protects plants from UV damage?
(A) Chlorophyll (B) Xanthophyll
(C) Carotenoids (D) Phycocyanin

445. Component of smog is:
(A) O_3 (B) PAN
(C) PPN and PBN (D) Both (A) and (B)

446. The repeated movement of water between Earth,s surface and the atmosphere is called:
(A) Water cycle (B) Nitrogen cycle
(C) Carbon cycle (D) Phosphorus cycle

447. Most harmful type of environmental pollutants are:
(A) Human organic wastes
(B) Natural nutrients in excess
(C) Waste animal feed
(D) Nonbiodegradable chemicals

448. Which is not dangerous for life?
(A) Nuclear blast (B) Deforestation
(C) Ozone layer (D) Biopollutants

449. Spraying DDT on crops shall produce pollution of:
(A) Air and water (B) Air and soil
(C) Crops and air (D) Soil and water

450. Sewage water is purified by:
(A) Microorganisms (B) Light
(C) Fishes (D) Aquatic plants

451. Sewage treatment in which a portion of decomposer bacteria present in waste is recycled into the beginning of the process:
(A) Cyclic treatment
(B) Activated sludge treatment
(C) Primary treatment
(D) Tertiary treatment

452. Ozone hole refers to:
(A) Hole in ozone layers
(B) Reduction in thickness of ozone layer in stratosphere
(C) Reduction of thickness of ozone in troposphere
(D) Increase concentration of ozone

453. Photochemical smog is related to pollution of:
(A) Air (B) Water
(C) Soil (D) Nostoc

454. Among pollutants which has the most lasting effect:
(A) Carbon monoxide (B) Pesticides
(C) Sulphur dioxide (D) Smoke from chimneys

455. Pollution indicators plants:
(A) Are resistant to pollution
(B) Can purify the atmosphere
(C) Are very sensitive to pollutants
(D) Can indicate onset of rains

456. Depletion of ozone layer is due to:
(A) Oxides of nitrogen (B) Oxides of carbon
(C) Oxides of sulphur (D) None of these

457. Use of lichens in case of pollution is:
(A) Treatment of pollutant water
(B) Bioindicators of pollution
(C) Promote pollution
(D) Lichens have no relation with pollution

458. Peeling of ozone umbrella is due to:
(A) CFCs (B) PAN
(C) CO_2 (D) Coal burning

459. Ozone layer of stratosphere requires protection from indiscriminate use of:
(A) Fungicides, insecticides, bactericides and medicines
(B) Aerosols and high flying jets
(C) Atomic explosions and industrial wastes
(D) Baloons and turbo-prop aeroplanes

460. Environmental pollution affects:
(A) Biotic components
(B) Plants only
(C) Man only
(D) Biotic and abiotic components of environment

461. Anxiety and stress are caused by:
(A) Air pollution (B) Noise pollution
(C) Water pollution (D) Nuclear pollution

462. Water pollution is due to:
(A) Agricultural discharges
(B) Sewage and other wastes.
(C) Industrial effluents (D) All of these

463. Carbon monoxide kills because it destroys:
(A) Haemoglobin (B) Phytochrome
(C) Cytochrome (D) Both (A) and (B)

464. Water is often treated with chlorine to:
(A) Increase oxygen content
(B) Kill germs
(C) Remove hardness
(D) Remove suspended particles

465. Water blooms are formed by:
(A) Lemna (B) Hydrilla
(C) Water Hyacinth (D) Planktonic algae

466. DDT concentration recorded in human body has been:
(A) 0.3 – 0.7 ppm (B) 1.3 – 3.1 ppm
(C) 13 – 31 ppm (D) 30 - 70 ppm

467. Mottling of teeth is due to presence of an element in drinking water:
(A) Mercury (B) Fluorine
(C) Boron (D) Chlorine

468. Which one of the following is not pollutant?
(A) CO_2 (B) CO (C) SO_2 (D) NO_2

469. As it passes into food chain, the concentration of DDT:
(A) Remains same (B) Decreases
(C) Increases (D) Unpredictable

470. NEERI is:
(A) National Environmental Engineering Research Institute
(B) National Ecological and Environmental Research Institute
(C) National Ethological and Ecological Research Institute
(D) National Eugenics and Ecological Research Institute

471. A person has impaired nervous system and signs of madness due to continued intake of metal contaminated water. The metal is:
(A) Mercury (B) Calcium
(C) Manganese (D) Lead

472. Wildsterben is an effect of:
(A) Ozone (B) Nitrogen oxide and sulphur oxide
(C) Acid rain (D) All of these

473. Waldsterben is:
(A) Decline in forests
(B) Development of fog
(C) Harm to sculptures, status and buildings
(D) All of these

474. Ozone is:
(A) Pale blue (B) Highly reactive
(C) With penetrating odour (D) All of these

475. An ozone depleting chemical which is used in fire-extinguishers is:
(A) Nitrogen oxide (B) Methyl bromide
(C) Halon (D) Chlorofluorocarbon

476. Smoked foods are not favoured now because of the possible toxicity of:
(A) Benzpyrene (B) Nitrogen oxides
(C) Sulphur dioxide (D) All of these

477. Benzpyrene is an ingredient of:
(A) Automobile exhausts
(B) Wood smoke and tobacco smoke
(C) Coal tar
(D) All of these

478. Smoke or dust dispersed in air is called:
(A) Fog (B) Haze (C) Mist (D) Smog

479. Mist is fog with horizontal visibility of:
(A) 100 m (B) 200 m (C) 500 m (D) 1000 m

480. Fog is formed by:
(A) Smoke
(B) Condensation of water vapours
(C) Condensation of water vapour over dust particles
(D) Both (A) and (C)

481. The term green house effect was given by:
(A) Fourier (B) Tyndall
(C) Arrhenius (D) Linnaeus

482. Polychlorinated biphenyls are chemicals useful as:
(A) Fire – resistant (B) Fertilizers
(C) Pesticides (D) Germicides

483. Use of polychlorinated biphenyl is restricted because of their:
(A) Extreme toxicity (B) Persistent toxicity
(C) Biomagnification (D) Both (B) and (C)

484. Ecofriendly detergents do not contain:
(A) Phosphates and bleaches (B) Alcohols
(C) Water softners (D) Perborates

485. Sudden rattling of doors and windows is due to:
(A) Earthquake (B) Sonic boom
(C) Movements of crustal layers of earth
(D) Slipping of glaciers on the nearby mountain

486. Plantation of rows of trees and shrubs along roads, rails, around residential complexes, educational institutes, etc is called:
(A) Green muffler (B) Shelter belt
(C) Wind break (D) Both (B) and (C)

487. Exposure of UV radiations causes:
(A) Damage of eye sight (B) Sunburn
(C) Skin cancer (D) All of these

488. Third pollution is:
(A) Controlled tipping
(B) Land or landscape pollution
(C) Noise pollution
(D) UV pollution

489. Sewage is broken down in:
(A) Primary treatment (B) Secondary treatment
(C) Tertiary treatment (D) All of these

490. Tipping is a technique of:
(A) Sewage disposal (B) Waste disposal
(C) Smoke treatment (D) Effluent

491. Photocatalysis is useful in breakdown of:
(A) Acids and alkalies
(B) Chemicals containing heavy metals
(C) Toxic chemicals
(D) All of these

492. Chlorination of treated waste water required for:
(A) Killing all microbes
(B) Removing suspended materials
(C) Complete breakdown of organic matter
(D) All of these

493. Excess nitrate present in drinking water, vegetables and fruits shall result in:
(A) Methaemoglobinaemia (B) Mottling of teeth
(C) Minamata disease (D) Itai-itai

494. Bag filters are used for removal of pollutants from:
(A) Exhaust emissions (B) Effluent water
(C) Sewage water (D) Both (B) and (C)

495. Thermal plants have installed:
(A) Cyclone collector
(B) Porous filters
(C) Electrostatic precipitators
(D) Adsorbents

496. Substitute of chlorofluorocarbon is:
(A) Hydrofluorocarbon
(B) Hydrochlorofluorocarbon
(C) Fluorocarbon
(D) Both (A) and (B)

497. Smog that occurs due to low temperature and has reducing environment is:
(A) Photochemical smog (B) Los Angeles smog
(C) Nonsulphurous smog (D) London smog

498. Green audit is:
(A) Estimation of crop yield and crop area
(B) Estimation of forested area and pasture land
(C) Checking of impact of an establishment over environment
(D) Effect of environment over vegetation of an area

499. Runoff from fields is:
(A) Point source pollution (B) Line source pollution
(C) Area source pollution (D) Diffuse source pollution

500. Stick and leaf insects resembles the surroundings. They undertake mimicry of the type:
(A) Concealing-protective (B) Warning – protective
(C) Alluring-aggressive (D) Concealing-aggressive

501. Nitrogen oxides released in the stratosphere cause ozone depletion due to:
(A) Catalysing conversion of ozone to oxygen
(B) Binding ozone
(C) Reacting with ozone
(D) All of these

502. In foggy and hazy environment, birds are able to locate position of sun and their path by means of:
(A) Infra-red wavelength (B) Ultra-violet radiations
(C) Magnetic system (D) Pheromones

503. The amount of biogenic nutrients present in the abiotic environment is called:
(A) Standing state (B) Standing crop
(C) Detritus (D) Reserve nutrient

504. Carbon dioxide is being added to the cycling pool by:
(A) Respiration and combustion
(B) Sedimentation of carbonates
(C) Precipitation, skeletons and fossilization
(D) Both (B) and (C)

505. A source of nonbiological carbon dioxide is:
(A) Volcanic eruptions (B) Hot springs
(C) Weathering (D) All of these

506. A succession that does not follow the natural course due to human activity is:
(A) Subsere (B) Plagiosere
(C) Prisere (D) Asere

507. Which of the following disorder in mango is caused by low temperature?
(A) Black heart (B) Crotch injury
(C) Collor injury (D) All of these

508. Use of green plants and their associated microorganisms, soil amendments, and agronomic practices to remove, contain or render harmless environmental contaminants is known as:
(A) Phytoremediation (B) Phytostabilization
(C) Both (A) and (B) (D) None of these

509. Use of plants to sequester contaminants is cell wall lignins of root tissues, or to promote its binding to humus is known as:
(A) Phytoremediation (B) Phytostabilization
(C) Both (A) and (B) (D) None of these

510. The atmosphere of the earth is heated by radiations which are mainly:
(A) Radiated by the sun
(B) Re-radiated by land
(C) Re-radiated by water
(D) Re-radiated by land and water

511. If there were no atmosphere around the earth, the temperature of the earth will:
(A) Increase
(B) Go on decreasing
(C) Increase during day and decrease during night
(D) Be unaffected

512. What would happen, if all the oxygen present in the environment is converted to ozone :
(A) We will be protected more
(B) It will become poisonous and kill living forms
(C) Ozone is not stable, hence it will be toxic
(D) It will help harmful sun radiation to reach earth and damage many life forms

513. The two forms of oxygen found in the atmosphere are :
(A) Water and Ozone
(B) Water and Oxygen
(C) Ozone and Oxygen
(D) Water and Carbon dioxide

514. Rainfall pattern depend on:
(A) The underground water table
(B) The density pattern of human population in an area
(C) The number of water bodies in an area
(D) The prevailing season in an area

515. Among the given options, which one is not correct for the use of large amount of fertilizers and pesticides?
(A) They are eco-friendly
(B) There fields barren after some time
(C) The adversely affect the useful component from the soil
(D) They destroy the soil fertility

516. Suitable temperature for tuberization in potato is:
(A) 12-15°C (B) 18-20°C
(C) 25-27°C (D) 30-32°C

517. Which of the following ranges of temperature is ideal for wheat seed germination?
(A) 25-30°C (B) 20-25°C
(C) 15-20°C (D) 10-15°C

518. Which is not the cause of heavy rainfall?
(A) Flood (B) Destruction of crops
(C) Drought (D) Soil erosion

519. Desert regions receive the rainfall:
(A) More than 50 cm (B) Less than 50 cm
(C) 80-100 cm (D) None of these

520. High velocity wind whirls are called:
(A) Monsoon (B) Cyclones
(C) Earthquake (D) None of these

521. Which is the effect of anticyclone?
(A) Pleasant weather (B) Heavy rainfall
(C) Hail (D) Snow fall

522. The formation of upholds by horizontal compression is called:
(A) Anticyclones (B) Synclines
(C) Rift valley (D) None of these

523. For converting gifts of nature into a resource, man requires:
(A) Skill and technology (B) Machines
(C) Capital and labour (D) All of these

524. Which of the following is not a community resources?
(A) Fruit orchard (B) A village pond
(C) Burial (D) Play ground

525. Which one of the following is a natural resources:
(A) An apartment (B) A river
(C) A factory (D) A car

526. Rajasthan and Gujarat are the store houses of:
(A) Solar and wind energy (B) Coal and iron ore
(C) Tidal energy (D) Hydel energy

527. Which type of resource is solar energy?
(A) Exhaustive resource
(B) Soil resource
(C) Biotic resource
(D) Non-exhaustive resource

528. Which one of the following type of resource is iron ore?
(A) Renewable (B) Biotic
(C) Flow (D) Non renewable

529. Under which of the following type of resource, the tidal energy can be put?
(A) Exhaustible (B) Human-made
(C) Abiotic (D) Non-exhaustible

530. Which one of the following types of resources is forest resources?
(A) Biotic (B) Non-renewable
(C) Stock (D) Individual

531. Between 1951 and 1980, according to Forest Survey of India, how much of the forest land was brought under farming?
(A) 27200 sq. km (B) 22200 sq km
(C) 26200 sq km (D) 28300 sq km

532. Increased asthmatic attacks in certain seasons are related to:
(A) Inhalation of seasonal pollen
(B) Eating of seasonal vegetables
(C) Low temperature
(D) Wet and dry environment

533. When was the maximum damage of forest cover inflincted in India?
(A) After independence
(B) While developing commercial activities
(C) During mughal period
(D) During British period

534. Himalayan Yew is in trouble because of:
(A) Over- exploitation (B) Spread of diseases
(C) Fuel-wood collection (D) Grazing

535. When was the Indian Wildlife Protection Act implemented?
(A) 1962 (B) 1972 (C) 1982 (D) 1992

536. Which one of the following is not the aim of Indian Wildlife Protection Act?
(A) Encourage hunting
(B) Establish Wildlife sanctuaries
(C) Initiate projects for saving specific animals
(D) Legal protection to the habitats of animals

537. Which one of the following conservation strategies does not directly involve community participation?
(A) Joint Forest Management
(B) Chipko movement
(C) Beej Bachao Andolan
(D) Demarcation of wildlife sanctuaries

538. Ozone depletion in stratosphere shall results in:
(A) Forest fires
(B) Increased incidence of skin cancer
(C) Global warming
(D) None of these

539. Which one of the following dissolves more rapidly in blood haemoglobin than oxygen?
(A) Ozone (B) Nitrous oxide
(C) Sulphur dioxide (D) Carbon monoxide

540. Which of the following forms a toxic substance in blood by combining with haemoglobin?
(A) Carbon dioxide (B) Carbon monoxide
(C) Oxygen (D) Methane

541. DDT is:
(A) Green house gas
(B) Degradable pollutant
(C) Nondegradable pollutant
(D) None of these

542. Ultraviolet radiation from sunlight causes a reaction that produces:
(A) Fluorides (B) Carbon monoxide
(C) Sulphur dioxide (D) Ozone

543. Noise pollution is measured in:
(A) Hertz (B) Fathoms
(C) Nanometres (D) Decibels

544. Most hazardous metal pollutant of automobile exhausts is:
(A) Mercury (B) Cadmium
(C) Lead (D) Copper

545. Pollution is rising due to:
(A) Rains
(B) Research institutes
(C) Population explosion
(D) Automobiles and industries

546. Pollution can be checked through:
(A) Reducing discharge of waste
(B) Increasing industrial development
(C) Throwing wastes in open spaces
(D) All of these

547. Continuous sewage flow into a stream would lead to:
(A) Eutrophication (B) Algal bloom
(C) Depletion of oxygen (D) Increase in temperature

548. Sewage infection of water will be demonstrated by cysts of:
(A) Entamoeba (B) Escherichia coli
(C) Pseudomonas (D) Leishmania

549. Measurement of rate of oxygen utilization by a unit volume of water over a period of time is to measure:
(A) Fermentation
(B) Biogas generation
(C) Biosynthetic pathway
(D) Biological oxygen demand

550. Carbon dioxide is a major pollutant in:
(A) Air (B) Soil (C) Water (D) Noise

551. Increasing skin cancer and high mutation rate are due to:
(A) Acid rain (B) Ozone depletion
(C) CO pollution (D) CO_2 pollution

552. Sudden mass death of fish is more likely to occur in:
(A) Eutrophic lake (B) Mesotrophic lake
(C) Oligotrophic lake (D) Oxalotrophic lake

553. Which pollution is causes jaundice?
(A) Water (B) Air
(C) Land (D) Thermal

554. Which pollution causes leaf curling?
(A) CO (B) H_2S (C) O_3 (D) SO_2

555. Water pollution causes:
(A) Increased photosynthesis
(B) Increased deoxygenation and turbidity
(C) Decreased turbidity
(D) Increased oxygenation

556. Naural sink of stratospheric ozone is:
(A) Sulphuric dioxide (B) Sulphur fluxes of oceans
(C) Freons (D) HCFCs

557. Major source of methane in India is:
(A) Rice fields (B) Sugarcane plantation
(C) Wheat fields (D) Fruit orchards

558. Ozone hole is maximum over:
(A) Europe (B) Antarctica
(C) India (D) Africa

559. In coming years, skin related disorders will become more common due to:
(A) Excessive use of detergents
(B) Water pollution
(C) Air pollution
(D) Depletion of ozone layer

560. Chief air pollutant which is likely to deplete ozone layer is:
(A) Sulphur dioxide
(B) Carbon dioxide
(C) Carbon monoxide
(D) Nitrogen oxides and fluorocarbons

561. The idea that Earth has a number of geological plates that moves slowly across its surface is the theory of:
(A) Convection cells (B) Natural selection
(C) Evalution (D) Plate tectonics

562. ———— is the natural process of lower atmosphere heating that results from the trapping of incoming and reradiated solar energy by water vapoure, clouds and other green house gasses:
(A) The energy budget (B) Intertropical convergence
(C) Contenentality (D) Isolation

(E) Green house effect

563. Each of the following gases is responsible for the bulk of recent global warming except:
(A) Chlorofluorocarbon (B) Carnon monoxide
(C) Carbon dioxide (D) Methane

564. The term ——— is used ro describe inland climates with hot summers and cold, snowy winters such as those found in interior North America and Russia:
(A) Intertropical convergence
(B) Insolation
(C) Green house effect
(D) Continentality
(E) Steppe

565. Likely effects of global warming include:
(A) Shifts in major areas of agricultural production
(B) Flooding pf low-lying coastal areas, include entire countries
(C) Rising sea levels
(D) Melting of polar ice sheets and glaciers
(E) All of these

566. Which country contributes more to green house gas emissions those gasses responsible for global warming than any other country?
(A) United states (B) China
(C) Japan (D) India

567. Fresh water accounts for ——— per cent of the total global water supply. Of the fresh water supply, approximately ——— per cent is in the form of surface rivers and lakes:
(A) 3 and 1 (B) 15 and 70
(C) 30 and 30 (D) 97 and 1

568. The 1992 Earth Summit that formulated the first legal instrument to address global warming was held in:
(A) Kyoto, Japan
(B) San Francisco, California
(C) Frenkfart, Germany
(D) Rio de Janeiro, Brazil

569. An estimated ¾ of the population ——— will experience water shortage by the year 2025:
(A) South Asia (B) Africa
(C) Australia (D) Europe

570. Floods are responsible for approximately ——— of death from natural disasters:
(A) 40 per cent (B) 20 per cent
(C) 50 per cent (D) 10 per cent

571. As assemblage of local plants and animals hat is closely connected with a climate region is a(n):
(A) Ecology (B) Biome
(C) Flora and Fauna (D) Deciduous forest

572. Tropical deforestation is occurring at unprecendented rate because of:
(A) Globalization
(B) Inccreased demand for weed products
(C) Increased population
(D) Cattle ranching
(E) All of these

573. Desertification (the spread of desert like conditions) is:
(A) Decreased by more farming on marginal lands
(B) Slowing down because of salt buildup due to irrigation
(C) Currently threatening about 60% of the world's rangelands
(D) Not a threat in prairies and steppes

574. What do most current climate models predict as an increase in global temperature over the next century?
(A) 6 – 8°C (B) 2 – 4°C
(C) 0 – 2°C (D) No change

575. What are the main weaknesses of global climate models?
(A) Resolution, behavior of ice and human dimension
(B) Resolution, human dimension and scientific expertise
(C) Resolution, biological and chemical coupling, behavior of ice and human dimension
(D) Biological and chemical coupling

576. What is the residence time (in years) for an average carbon atoms in the ocean compared to the atmosphere?
(A) 400 : 5 (B) 400 : 40
(C) 40 : 500 (D) 5 : 400

577. In terms of the carbon cycle, where is the omissing carbon released into the atmosphere going?
(A) Dissolving in the ocean
(B) Terrestrial photosynthesis
(C) Both (A) and (B)
(D) Neither (A) nor (B)

578. Which international agreement decided to phase out the use of CFCs?
(A) Montreal Protocol of 1987
(B) Kyoto Protocol of 1997
(C) United Nations Framework Convention on climate change
(D) European Union energy policy

579. Which one of the following is NOT an accurate source of information about reconstructing past climate?
(A) Deposite of pollen in lakes and bogs
(B) Historical records
(C) Weather station records over the past ten years
(D) Growth of tree rings

580. Which of the following are not either positive or negative feed backs to accelerate and suppress in terms of environmental change?
(A) Replacement of coniferous forest by warmth-loving broadleaved forests and by agriculture will decrease planetary reflectance, causing cooling
(B) Increased transpiration in a warm world will lead to fewer clouds, warming the planet
(C) Warming will cause release of CO_2 from increased biomass decomposition, primarily in the forest regions of the world but also in the tundra, thus accelerating warming
(D) Warming will melt snow and ice, decreasing albedo and thus increasing warming, melting even more snow and ice

581. How many countries have ratified the Kyoto Protocol of 1997 to date?
(A) 175 (B) 275
(C) 150 (D) 75

582. During the KT boundary (65 million years ago), what percentage of (i) marine families, (ii) plant species and (iii) dinosaurs became extenct?
(A) 15, 75, 100 (B) 75, 15, 100
(C) 15, 100, 75 (D) 100, 15, 75

583. What fraction of global omissions of nitrous oxide are estimated to be atmosphere?
(A) One half (B) One third
(C) All (D) Two third

584. The compound used for artificial rain making/cloud seeding:
(A) Agl (B) Ni (C) Fe (D) Hg

585. Kyoto protocol came into force on:
(A) 16 February, 2004 (B) 14 February, 2006
(C) 14 February, 2004 (D) 16 February, 2006

586. Thicknes of ozone layer is measured in:
(A) Pascal (B) Danson units
(C) Cm/sec (D) Both (A) and (C)

587. Maximum oxygen cames from:
(A) Phytoplanktons (B) Grassland
(C) Forests (D) Herbs and shrubs

588. How many green house gases exist?
(A) Two (B) Three
(C) Four (D) Six

589. Annual world precipitation is:
(A) 7 x 10^{10}g (B) 7 x 10^{20}g
(C) 4.46 x 10^{20}g (D) 4.46 x 10^{20}g

590. ——— of stratosphere provides protection to our life:
(A) Ozone (B) Nitrogen
(C) Carbon (D) Hydrogen

591. The life supporting gases such as O_2, CO_2 and N_2 are chiefly concentrated in the:
(A) Troposphere (B) Stratosphere
(C) Mesosphere (D) Ionosphere

592. A natural phenomenon that becomes harmful due to pollution is:
(A) Global warming (B) Ecological balance
(C) Green house effect (D) Desertification

593. The atmospheric layer of ozone normally absorbs what proportion of UV radiation from sun?
(A) More than 99% (B) Around a quarter
(C) Less than 10% (D) About half

594. Acid rain is caused by:
(A) CO and CO_2 (B) SO_2 and O_2
(C) SO_2 and NO_2 (D) NO_2 and O_2

595. Ozone 'hole' in the stratosphere was discovered over the continent of:
(A) North America (B) Australia
(C) Green land (D) Antarctica

596. The optimum range of wind speeds for wind power generation is:
(A) 2 – 4 m/sec (B) 4 – 12 m/sec
(C) 12 – 14 m/sec (D) 2 -20 m/sec

597. The process of mixing two or more gasses is called:
(A) Condensation (B) Diffusion
(C) Sublimation (D) Solidification

598. Maximum density of water is at:
(A) 4°C (B) 0°C (C) 40°C (D) 100°C

599. Eco-labelling is governed by:
(A) ISO 14001 (B) ISO 14010
(C) ISO 14020 (D) ISO 14002

600. Precipitation and evaporation are important compounds of the:
(A) Nitrogen cycle (B) Carbon cycle
(C) Potassium cycle (D) Water cycle

601. Cyclone collector is used for minimising:
(A) Radioactive pollution (B) Air pollution
(C) Noise pollution (D) Water pollution

602. Which abiotic components of environment contributes in various ways to fulfillment of human needs?
(A) Air (B) Water (C) Soil (D) All of these

603. Environment consists of:
(A) Sum total of biotic factors
(B) Sum total of abiotic factors
(C) Both (A) and (B)
(D) Both (A) and (B) minus humans

604. Which constitutes environmental biotechnology?
(A) Waster treatment
(B) Waste disposal
(C) Technologies generating less waste
(D) Both (A) and (B)

605. Which constitutes 'end of pipe' technology?
(A) Waste treatment
(B) Waste disposal
(C) Technologies generating less waste
(D) Both (A) and (B)

606. Which constitutes 'front of pipe' technology?
(A) Waste treatment
(B) Waste disposal
(C) Technologies generating less waste
(D) Both (A) and (B)

607. Recalcitrants xenobiotic compounds are highly stable due to the presence of which substitution groups?
(A) Halogens and carbanyl groups
(B) Nitro- and sulphonate groups
(C) Amino- and methoxy groups
(D) All of these

608. Bioaccumulation/biomagnifications may occurs due to:
(A) Xenobiotic compounds are continuously taken up from the environment and accumulated in the lipid deposits of the body
(B) Consumption of these organisms by other organisms leads to further accumulation of xenobiotic compounds in the bodies of latter
(C) The xenobiotic compounds are recalcitrant and lipid soluble
(D) All of these

609. Which diseases spread through contaminated water/food?
(A) Amoebiosis and typhoid (B) Cholera and diarrhea
(C) Hepatitis (except B) (D) All of these

610. Which is not an example of physical pollutants?
(A) Dust and mechanical stress
(B) Sound pollution and electrical storms
(C) Heavy metals
(D) Radiation and fires

611. Presence of coliform bacteria in water is indicative of contamination by mammalian excreta and suggestive of the presence of which pathogenic bacteria?
(A) *Vibrio cholera* (B) *Salmonella*
(C) *Shigella* (D) All of these

612. Which pollutants pose immediate threat to humans and is included in the black list?
(A) Oils and hydro carbons from crude oils
(B) Heavy metals, metaloides and compounds containing them
(C) Organophosphates, cyanides and asbestos
(D) All of these

613. Various approaches to waster treatment may be grouped into which categories?
(A) Anaerobic digestion
(B) Biofilters and aerobic digestion
(C) Landfill and burning
(D) All of these

614. Landfill sites can be useful in which ways?
(A) Reclamation of derelict sites to develop landscaped gardens, etc
(B) As a source of biogas
(C) As a source of organic fertilizers
(D) Both (A) and (B)

615. Which microorganisms is the commonly found in aerobic digestion systems?
(A) Fungi (B) Bacteria
(C) Protozoa (D) Algae

616. Bioremediation of inorganic contaminants is primarily based on:
(A) Animal species (B) Microorganisms
(C) Plant species (D) All of these

617. The water (leachate) from landfills may contain :
(A) COD value of 1000-2000
(B) Total solids content upto 6000-7000 mg/l
(C) Sodium upto 800 mg/l
(D) All of these

618. Which statements is correct about landfill management?
(A) The risk due to methane can be removed by burning or tapping
(B) The risks of fires, offensive odours and increased vector populations may be circumvented by covering the waste with soil on daily basis
(C) The risk of damage from, landfill leachate may be avoided by lining the pit with an impermeable material
(D) All of these

619. The various parameters used to determine the level of pollution in water bodies, usually, determine:
(A) Amounts of organic matter present in the water
(B) Amounts of important inorganic nutrients
(C) Numbers of bacteria per mililitre of waste water
(D) Both (A) and (B)

620. Treatment of the water proceeds through five stages; select the correct order of different stages of water treatment:
(A) Sludge treatment → preliminary treatment → primary treatment → secondary treatment → tertiary treatment
(B) Preliminary treatment→ primary treatment → secondary treatment → tertiary treatment → Sludge treatment
(C) Preliminary treatment→ Sludge treatment → primary treatment → secondary treatment → tertiary treatment
(D) Preliminary treatment → secondary treatment→ tertiary treatment→ primary treatment → Sludge treatment

621. Which waste water treatments is an anaerobic process?
(A) Primary treatment (B) Secondary treatment
(C) Tertiary treatment (D) Sludge treatment

622. Preliminary treatment for waste water is necessary to prevent?
(A) Clogging of pipelines
(B) Damage to pump
(C) Interference with subsequent process
(D) All of these

623. Which is correct about aerobic digestion of waste water?
(A) Bacteria are the most common organisms
(B) Bacterial population may be more than 10^9 cells/ml
(C) Bacteria are responsible for the removal of about 80-90% of the BOD
(D) All of these

624. Practical application of microbes for xenobiotic degradation is facilitated by:
(A) Supply of sufficient nutrients or cometabolites
(B) Maintenance of the xenobiotic compounds at non-toxic levels
(C) Provision of microbial population or inoculums
(D) All of these

625. The quality of treated effluent from the digester depends mainly on :
(A) Sludge loading rate (B) Organic loading rate
(C) Residence time (D) All of these

626. NO_3^- is removed from water by the action of which bacteria?
(A) *Pseudomonas* (B) *Micrococcus*
(C) Alcaligenes (D) All of these

627. Sewage contains which pathogenic organisms?
(A) Bacteria causing typhoid, cholera, diarrhea
(B) Protozoa causing amoebic dysentery, giardiasis
(C) Viruses causing poliomyletis, hepatitis, gastroenteritis
(D) All of these

628. Protozoa are represented in waste waters by:
(A) Amoebal forms (B) Ciliates
(C) Flagellates (D) All of these

629. The most common type of protozoa found in waste waters:
(A) Amoebal forms (B) Ciliates
(C) Flagellates (D) All of these

630. Which form of protozoa is most diverse group?
(A) Amoebal forms (B) Ciliates
(C) Flagellates (D) All of these

631. Efficiency of anaerobic digestion can be enhanced by:
(A) By using solid support to retain biomass
(B) By recycling the active biomass
(C) By limiting the biomass loss to a rte compatible with high population density in the digester
(D) Any of these

632. Bioremediation of organic contaminants is primarily based on:
(A) Microorganisms naturally present at the site
(B) Microbial inoculants developed in the laboratory and introduced at the site
(C) Suitable plants planted at the site
(D) All of these

633. Oil spills cause severe damage to the ecosystems and pose which threats?
(A) Ground water pollution due to percolation
(B) Air pollution due to exoporation
(C) Fire
(D) All of these

634. Which is not correct?
(A) The US Environment Protection Agency and the Exon Campany used microorganisms to clean up Alaskan beaches contaminated by the Valdez oil spill
(B) In India, a consortium of bacterial species has been developed to combat oilspills and oily sludge; the inoculants is apply called 'oilzapper'
(C) Inoculation with Oilzapper reduced oilsludge contamination in soil to merely 0.5% in 360 days from the initial 13.41%
(D) Oilzapper has not been effective in field trials

635. The anaerobic digestion process involves a wide variety of organisms which digest the organic molecules, such as lipids, proteins, carbohydrates etc., into mainly:
(A) NO_3^- and PO_4^- (B) CH_4 and CO_2
(C) SO_4^- and CO (D) All of these

636. Which heterotrophic bacteria are responsible for oxidation of organic molecules?
(A) Pseudomonas and Escherichia
(B) Streptococcus and Salmonella
(C) Shigella and Aerobacter
(D) All of these

637. Which is not correct about halocarbons?
(A) They contain different numbers of halogen atoms in the place of He atoms
(B) They are used as solvents, in condenser units of cooling systems, etc
(C) The C_1-C_2 haloalkanes escape into the atmosphere where they destroys the protective ozone layer
(D) Pesticides leach into water bodies, where many of them are subjected to biomagnifications

638. Which is correct about alkylbenzyl sulphonates?
(A) These are surface-active detergents
(B) Sulphonate group present at one end resists microbial degradation
(C) At present, alkylbenzýl sulphonates are biodegraded by β–oxidation from their alkyl ends
(D) All of these

639. Phytoremediation may occur due to:
(A) Accumulation of the organic waste in plant tissues
(B) Translocation of the organic waste to leaf and volatilization from there
(C) The organic waste or in the rhizosphere itself
(D) Any of these

640. For phytoremediation of flyash, seedings of which of the following trees are established at abandoned flyash ponds/dams?
(A) Eucalyptus (B) Melia
(C) Poplar (D) All of these

641. Continued exposure of microorganisms to xenobiotic compounds can often lead to the evolution of metabolic processes needed to wholly or partly degrade the xenobiotic. These capabilities may arise due to:
(A) Mutation
(B) Transfer of plasmid –borne genes
(C) Transformation
(D) Both (A) and (B)

642. Continued exposure of microorganisms to xenobiotic compounds can often lead to the evolution of metabolic processes needed to wholly or partly degrade the xenobiotic. These can be expected to modify the active site of an enzyme so that it has an increased affinity for the xenobiotic compounds?
(A) Mutation
(B) Transfer of plasmid –borne genes
(C) Transformation
(D) Both (A) and (B)

643. Alicyclic hydrocarbons are present naturally in:
(A) Wax from plants (B) Microbial lipids
(C) Crude oils (D) All of these

644. Which is not related to green-house gas?
(A) CH_2 (B) CO_2
(C) O_2 (D) CFC

645. Fluoride pollution mainly affects:
(A) Heart (B) Kidney
(C) Brain (D) Teeth

646. Biochemical oxygen demand measures:
(A) Industrial pollution
(B) Air pollution
(C) Dissolved O_2 needed by microbes to decompose organic waste
(D) Oxygen need to support microorganism growth

647. Smog is the product of:
(A) Smoke (B) Fog
(C) Soot (D) Both (A) and (B)

648. Green house effect is related to:
(A) Global warming (B) Global afforestation
(C) Deforestation (D) Both (A) and (C)

649. Increasing skin cancer and rate of mutation are the result of:
(A) Acid rain (B) Ozone depletion
(C) SO_2 pollution (D) CO_2 pollution

650. Which diseases is caused by eating fish inhabitating mercury contaminated water?
(A) Minamata disease (B) Osteosclerosis
(C) Hashimoto disease (D) Brain fever

651. Depletion of ozone layer is caused by:
(A) CO_2 (B) SO_2
(C) CFC (D) Both (B) and (C)

652. The atmosphere of metro cities is polluted by:
(A) Pesticides (B) Automobile exhausts
(C) House-hold wastes (D) Radioactive fallout

653. Which associates with the blood haemoglobin more rapidaly than oxygen?
(A) CO (B) CO_2 (C) SO_4 (D) NO

654. Which decibles of sound becomes hazardous moise pollution?
(A) >20 (B) >80 (C) >120 (D) >140

655. Which plants can be used as indicators of SO2 pollution of air?
(A) Liverworts (B) Hornworts
(C) Epiphytic lichens (D) Fern

656. Attack of Asthama in certain persons may be due to:
(A) Inhalation of some airborne pollution
(B) Exposure to cold temperature
(C) Exposure to certain pollen grains
(D) All of these

657. Which gases contributes the maximum to the 'green – house effect'?
(A) CO_2 (B) CFC (C) CH_4 (D) N_2O

658. Which posses the greatest risk?
(A) Nuclear plants (B) Thermal power plants
(C) Biogas plants (D) Hydroelectric plants

659. Which is an example of nonrenewable resources?
(A) Petroleum (B) Biological agents
(C) Natural gas (D) Both (A) and (C)

660. Maximum deposition of DDT will occur in:
(A) Phytoplankton (B) Fish
(C) Crab (D) Sea gull

661. Which green-house gases is being generated by agricultural fields?
(A) Sulphur dioxide (B) SO_3
(C) Nitrous oxide (D) Ammonia

662. Ozone day is observed on:
(A) September 16 (B) June 5
(C) May 1 (D) December 25

663. Which is an example of a renewable resource?
(A) Plants (B) Animals
(C) Microorganisms (D) All of these

664. Ozone depletion in the stratosphere will cause:
(A) Gobal warming (B) Increased sea level
(C) Increased incidence of skin cancer
(D) All of these

665. Ozone layer of upper atmosphere is being destroyed by:
(A) Smog (B) Photochemcial oxidation
(C) CFC (D) Soot

666. Bioremediation of organic wastes is generally based on:
(A) Bacteria (B) Animal species
(C) Plant species (D) Both (B) and (C)

667. Which is most ecofriendly?
(A) Bacteria (B) Animal species
(C) Plant species (D) Both (B) and (C)

668. Ultra-violet light causes:
(A) Formation of pyrimidines
(B) Sticky metaphases
(C) Photodynamics action
(D) Destruction of hydrogen bonds between complementary DNA strands

669. Sulphur dioxide affects:
(A) Cell wall (B) Plasmodesmata
(C) All membrane systems (D) Nucleus

670. Addition of phosphates and nitrates/fertilizers into water leads to:
(A) Reduced algal growth
(B) Increased growth of decomposers
(C) Increased algal growth
(D) Nutrient enrichment/eutrophication

671. Fluoride pollution produces necrosis or chlorosis of:
(A) Leaf margin and tip (B) Stem
(C) Petiole but not lamina (D) Midrib

672. Air pollutants photochemical oxidants include:
(A) Nitrous oxide, nitric acid and nitric oxide
(B) Oxygen, chlorine and nitric acid
(C) Ozone, peroxyacetyl nitrate and aldehydes
(D) Carbon monoxide and sulphur dioxide

673. Ultraviolet radiations are lethal because of inactivation of:
(A) Proteins, nucleic acids and pigments
(B) Carbohydrates, fats and vitamins
(C) Minerals, air and water
(D) Water, carbon dioxide ad oxygen

674. Nonionising radiations with specific biological effects are:
(A) UV radiations (B) Beta-ray
(C) Gamma rays (D) X-rays

675. Carbon monoxide is pollutant as it:
(A) Inactibates nerves
(B) Inhibits glycolysis
(C) Combines with oxygen
(D) Combines with haemoglobin

676. Air pollutants that produce photochemical oxidants include:
(A) CO_2, CO and SO_2
(B) Nitrous oxide, nitric oxide and nitric acid
(C) Oxygen, clorine and nitric acid
(D) Ozone, chlorine and sulphur dioxide

677. Acid rains are produced by:
(A) Excess NO_2 and SO_2 from burning fossil fuels
(B) Excess production of NH_3 by industry and coal gas
(C) Excess release of carbon monoxide by incomplete combustion
(D) Excess formation of CO_2 by combustion and animal respiration

678. Spraying of DDT produces pollution of:
(A) Air (B) Air and water
(C) Air and soil (D) Air, water and soil

679. Most abundant water pollution is:
(A) Detergents (B) Pesticides
(C) Industrial waste (D) Ammonia

680. Atmospheric pollutants is:
(A) CO_2 (B) CO (C) O_2 (D) N_2

681. Burning of fossil fuels is the main source of pollution:
(A) Nitrogen oxide (B) Nitric oxide
(C) Nitrous oxide (D) Sulphur dioxide

682. Modes of controlling pollution in large cities include:
(A) Cleanliness and less use of insecticides
(B) Proper disposal of organic wastes, sewage and in dustrial effluents
(C) Broader roads and shifting of factories out of the residential areas
(D) All of these

683. Domestic waste constitutes:
(A) Nonbiodegradable pollution
(B) Biodegradable pollution
(C) Effluents
(D) Air pollution

684. Ozone layer of upper atmosphere is being destroy by:
(A) Chlorofluorocarbons
(B) SO_2
(C) Photochemical oxidants/O_2 and CO_2
(D) Smog

685. SO_2 pollution is indicated by:
(A) Deschampsia (grasses)
(B) Sphagnum (mosses)
(C) Usnea (lichens)
(D) Cucurbita (climbers)

686. The acidity of normal rainwater is due to :
(A) CO_2 (B) Cl_2 (C) NO_2 (D) SO_2

687. Percent of water on the world's surface representing fresh water is :
(A) 97 (B) 50
(C) 10 (D) 3

688. Concentration of CO_2 in present day atmosphere is :
(A) ~ 220 ppm (B) ~ 280 ppm
(C) ~ 360 ppm (D) ~ 390 ppm

689. In biological systems sulphur is largely bound in:
(A) Lipids (B) Proteins
(C) Nucleic acid (D) Proteins and nucelic acids

690. The source of thermal pollution are:
(A) Power plants (B) Cooling forests
(C) Industrial effluents (D) All of the above

691. Which one of the following is a waste recycling method of solid waste management?
(A) Pelletisation (B) Composting
(C) Incineration (D) Sanitary landfil

692. Natural gas is composed primarily of:
(A) Methane
(B) n-butane
(C) n-octane
(D) Mixture of n-butane and n-octane

693. The principal components of photochemical smog in urban areas are:
(A) SO_2 and NO_2 (B) SPM and CO
(C) SPM and NO_2 (D) Hydrocarbons and Ozone

694. Lightening in the atmosphere produces:
(A) NO (B) CO (C) CO_2 (D) NH_3

695. The Global Warming Potential (GWP) is the least for which of the following green house gases:
(A) CH_4 (B) CO2
(C) N_2O (D) SF_6

696. Scales of meterology are in the following order starting from the least:
(A) Macro-, Micro-, Meso-
(B) Macro-, Meso-, Micro-
(C) Meso-, Macro-, Micro-
(D) Micro-, Meso-, Macro-

697. The uppermost zone of atmosphere of earth, where short wave ultraviolet radiations are absorbed, is:
(A) Troposphere (B) Stratosphere
(C) Mesosphere (D) Thermosphere

698. In which part of the atmosphere the momentum, heat flux and moisture content are conserved?
(A) Surface boundary layer
(B) Mesosphere
(C) Troposphere
(D) Stratosphere

699. Perma frost represents:
(A) Permanently frozen sub surface soil
(B) Frozen leaves of oak trees
(C) Frozen needles of pine trees
(D) Temporarily frozen subsurface soil

700. Soil air contains CO_2 (%):
(A) 0.25 (B) 0.03
(C) 2.50 (D) 3.00

701. The CO_2 content of soil air is:
(A) 0.03% (B) 0.30%
(C) 21% (D) None of these

702. Atmosphere extends above mean sea level to a height of about:
(A) 160 km (B) 1600 km
(C) 1600 miles (D) 1600 m

703. Densest part of atmospheric strata is:
(A) Stratosphere (B) Troposphere
(C) Mesosphere (D) Thermosphere

704. Global warming is attributed to increase in concentration of green house gases like:
(A) CO_2 (B) CEC
(C) CH_4 (D) All of the above

705. Under seventh approximation soil classification, the number of soil order is:
(A) 8 (B) 11
(C) 14 (D) 15

706. The number of Agro-climatic zones of India in:
(A) 7 (B) 10
(C) 15 (D) 17

707. Solar constant is equal to (in cal/cm^2/min):
(A) 0.194 (B) 1.94
(C) 19.40 (D) 19.4

708. The grand period of rainfall in India is:
(A) South-west monsoon
(B) North-east monsoon
(C) Cold weather period
(D) Post monsoon

709. Horse latitude lies in between north and south latitudinal belt of equador:
(A) 10 to 15° (B) 0 to 5°
(C) 20 to 25° (D) 30 to 35°

710. Which of the following gases contributes maximum to global warming?
(A) CFC (B) CO_2
(C) Methane (D) N_2O

711. Dangerous gas for depletion of ozone layer is:
(A) Methane (B) CFC
(C) Ethane (D) Carbon-dioxide

712. The present level of carbon-dioxide in atmospheric is:
(A) 420-460 ppm (B) 490 ppm
(C) 295-300 ppm (D) 190 ppm

713. Impact of green house effect is:
(A) Sea formation (B) Ice melting
(C) Global warming (D) Sea level rise

714. Absolute zero is:
(A) 273°K (B) 273°F
(C) -273°C (D) 273°C

715. Continuous variation in atmospheric pressure is recorded by:
(A) Thermograph (B) Barograph
(C) Thermometer (D) Hydrograph

ANSWERSHEET: 2.AGRICULTURAL CLIMATOLOGY

1. (A); 2. (A); 3. (A); 4. (A); 5. (A); 6. (B); 7. (C); 8. (D); 9. (A); 10. (B); 11. (A); 12. (B); 13. (B); 14. (A); 15. (A); 16. (B); 17. (A); 18. (C); 19. (B); 20. (A); 21. (A); 22. (C); 23. (C); 24. (A); 25. (C); 26. (D); 27. (C); 28. (A); 29. (A); 30. (A); 31. (B); 32. (C); 33. (D); 34. (B); 35. (A); 36. (B); 37. (D); 38. (B); 39. (D); 40. (D); 41. (B); 42. (C); 43. (B); 44. (D); 45. (A); 46. (C); 47. (B); 48. (A); 49. (C); 50. (D); 51. (B); 52. (C); 53. (B); 54. (D); 55. (C); 56. (C); 57. (D); 58. (C); 59. (B); 60. (B); 61. (B); 62. (B); 63. (C); 64. (A); 65. (D); 66. (A); 67. (A); 68. (C); 69. (D); 70. (B); 71. (D); 72. (C); 73. (C); 74. (C); 75. (B); 76. (B); 77. (B); 78. (C); 79. (B); 80. (C); 81. (C); 82. (B); 83. (D); 84. (B); 85. (B); 86. (C); 87. (C); 88. (B); 89. (B); 90. (D); 91. (A); 92. (A); 93. (B); 94. (A); 95. (B); 96. (D); 97. (A); 98. (C); 99. (B); 100. (D); 101. (C); 102. (B); 103. (B); 104. (A); 105. (D); 106. (C); 107. (B); 108. (D); 109. (D); 110. (D); 111. (B); 112. (A); 113. (B); 114. (D); 115. (B); 116. (B); 117. (D); 118. (B); 119. (B); 120. (C); 121. (D); 122. (C); 123. (B); 124. (C); 125. (D); 126. (C); 127. (B); 128. (C); 129. (C); 130. (D); 131. (C); 132. (C); 133. (B); 134. (C); 135. (D); 136. (C); 137. (B); 138. (D); 139. (C); 140. (B); 141. (D); 142. (C); 143. (A); 144. (B); 145. (D); 146. (C); 147. (B); 148. (C); 149. (A); 150. (D); 151. (D); 152. (D); 153. (C); 154. (D); 155. (B); 156. (A); 157. (D); 158. (C); 159. (B); 160. (D); 161. (A); 162. (B); 163. (C); 164. (B); 165. (C); 166. (B); 167. (B); 168. (D); 169. (A); 170. (D); 171. (D); 172. (C); 173.(A); 174. (D); 175. (B); 176. (B); 177. (C); 178. (A); 179. (B); 180. (C); 181. (D); 182. (C); 183. (C); 184. (B); 185. (A); 186. (D); 187. (B); 188. (D); 189. (C); 190. (C); 191. (B); 192. (A); 193. (C); 194. (A); 195. (B); 196. (A); 197. (C); 198. (A); 199. (C); 200. (A); 201. (D); 202. (C); 203. (B); 204. (B); 205. (A); 206. (B); 207. (A); 208. (A); 209. (D); 210. (B); 211. (C); 212. (B); 213. (B); 214. (D); 215. (C); 216. (D); 217. (A); 218. (A); 219. (C); 220. (A); 221. (B); 222. (C); 223. (C); 224. (B); 225. (C); 226. (D); 227. (D); 228. (A); 229. (D); 230. (D); 231. (C); 232. (A); 233. (A); 234. (A); 235. (A); 236. (C); 237. (A); 238. (D); 239. (A); 240. (B); 241. (A); 242. (C); 243. (A); 244. (B); 245. (C); 246. (C); 247. (D); 248. (C); 249. (A); 250. (B); 251. (A); 252. (B); 253. (C); 254. (A); 255. (B); 256. (C); 257. (A); 258. (E); 259. (C); 260. (C); 261. (C); 262. (A); 263. (D); 264. (A); 265. (A); 266. (A); 267. (C);

268. (C); 269. (B); 270. (A); 271. (C); 272. (B); 273. (A); 274. (D), 275. (D); 276. (A); 277. (B); 278. (C); 279. (C); 280. (B); 281. (A); 282. (B); 283. (C); 284. (C); 285. (D); 286. (B); 287. (A); 288. (A); 289. (C); 290. (C); 291. (B); 292. (D); 293. (A); 294. (C); 295. (A); 296. (D); 297. (C); 298. (D); 299. (C); 300. (B); 301. (B); 302. (A); 303. (C); 304. (C); 305. (A); 306. (D); 307. (A); 308. (C); 309. (B); 310. (A); 311. (D); 312. (D); 313. (A); 314. (B); 315. (D); 316. (C); 317. (C); 318. (B); 319. (B); 320. (D); 321. (D); 322. (B); 323. (B); 324. (D); 325. (A), 326. (A); 327. (B); 328. (D); 329. (C); 330. (D); 331. (A); 332. (D); 333. (B); 334. (A); 335. (A); 336. (D); 337. (D); 338. (A); 339. (A); 340. (B); 341. (D); 342. (E); 343. (C); 344. (C); 345. (D); 346. (C); 347. (E); 348. (D); 349. (B); 350. (B); 351. (B); 352. (C); 353. (D); 354. (A); 355. (A); 356. (C); 357. (A); 358. (B); 359. (C); 360. (A); 361. (C); 362. (A); 363. (B); 364. (C); 365. (A); 366. (D); 367. (D); 368. (A); 369. (B); 370. (C); 371. (B); 372. (A); 373. (D); 374. (B); 375. (B); 376. (C); 377. (B); 378. (B); 379. (A); 380. (A); 381. (C); 382. (A); 383. (B); 384. (A); 385. (C); 386. (E); 387. (D); 388. (C); 389. (B); 390. (B); 391. (C); 392. (A); 393. (C); 394. (B); 395. (C); 396. (D); 397. (D); 398. (B); 399. (C); 400. (A); 401. (D); 402. (C); 403. (B); 404. (A); 405. (D); 406. (A); 407. (D); 408. (A); 409. (D); 410. (C); 411. (C); 412. (A); 413. (C); 414. (B); 415. (C); 416. (A); 417. (C); 418. (B); 419. (D); 420. (A); 421. (C); 422. (B); 423. (A); 424. (C); 425. (C); 426. (C); 427. (C); 428. (A); 429. (B); 430. (C); 431. (D); 432. (C); 433. (B); 434. (D); 435. (B); 436. (A); 437. (D); 438. (C); 439. (B); 440. (D); 441. (A); 442. (D); 443. (A); 444. (C); 445. (D); 446. (B); 447. (D); 448. (C); 449. (D); 450. (A); 451. (B); 452. (B); 453. (A); 454. (B); 455. (C); 456. (A); 457. (B); 458. (A); 459. (B); 460. (D); 461. (B); 462. (D); 463. (A); 464. (B); 465. (D); 466. (C); 467. (B); 468. (A); 469. (A); 470. (A); 471. (D); 472. (D); 473. (A); 474. (D); 475. (C); 476. (A); 477. (D); 478. (B); 479. (D); 480. (C); 481. (C); 482. (A); 483. (D); 484. (A); 485. (B); 486. (A); 487. (D); 488. (B); 489. (B); 490. (B); 491. (C); 492. (A); 493. (A); 494. (A); 495. (C); 496. (D); 497. (D); 498. (C); 499. (D); 500. (A); 501. (C); 502. (B); 503. (A); 504. (A); 505. (D); 506. (B); 507. (D); 508. (A); 509. (B); 510. (D); 511. (C); 512. (B); 513. (C); 514. (C); 515. (A); 516. (B); 517. (C); 518. (C); 519. (B); 520. (B); 521. (A); 522. (A); 523. (A); 524. (A); 525. (B); 526. (A); 527. (A); 528. (D); 529. (D); 530. (A); 531. (C); 532. (A); 533. (D); 534. (A); 535. (B); 536. (A); 537. (D); 538. (B); 539. (D); 540. (D); 541. (C); 542. (D); 543. (D); 544. (C); 545. (D); 546. (A); 547. (C); 548. (A); 549. (D); 550. (B); 551. (B); 552. (A); 553. (A); 554. (D); 555. (B); 556. (C); 557. (A); 558. (B); 559. (D); 560. (D); 561. (D); 562. (E); 563. (B); 564. (D); 565. (E); 566. (A); 567. (A); 568. (D); 569. (B); 570. (C); 571. (B); 572. (E); 573. (C); 574. (B); 575. (C); 576. (A); 577. (A); 578. (A); 579. (C); 580. (B); 581. (A); 582. (A); 583. (B); 584. (A); 585. (A); 586. (B); 587. (A); 588. (D); 589. (D); 590. (A); 591. (A); 592. (C); 593. (A); 594. (C); 595. (D); 596. (B); 597. (B); 598. (A); 599. (A); 600. (D); 601. (B); 602. (D); 603. (D); 604. (D); 605. (D); 606. (C); 607. (D); 608. (D); 609. (D); 610. (C); 611. (D); 612. (D); 613. (D); 614. (D); 615. (B); 616. (C); 617. (D); 618. (D); 619. (D); 620. (B); 621. (D); 622. (D); 623. (D); 624. (D); 625. (D); 626. (D); 627. (D); 628. (D); 629. (C); 630. (B); 631. (D); 632. (D); 633. (D); 634. (D); 635. (B); 636. (D); 637. (A); 638. (D); 639. (D); 640. (D); 641. (D); 642. (A); 643. (D); 644. (C); 645. (D); 646. (C); 647. (D); 648. (D); 649. (B); 650. (A); 651. (C); 652. (B); 653. (A); 654. (B); 655. (C); 656. (D); 657. (A); 658. (A); 659. (D); 660. (D); 661. (C); 662. (A); 663. (D); 664. (C); 665. (C); 666. (A); 667. (C); 668. (D); 669. (C); 670. (D); 671. (A); 672. (C); 673. (A); 674. (A); 675. (D); 676. (B); 677. (A); 678. (D); 679. (C); 680. (B); 681. (D); 682. (D); 683. (B); 684. (A); 685. (C). 686. (C); 687. (D); 688. (D); 689. (B); 690. (D); 691. (B); 692. (A); 693. (D); 694. (A); 695. (B); 696. (D); 697. (D). 698. (A); 699. (A); 700. (A); 701. (A); 702. (B); 703. (B); 704. (D); 705. (B); 706. (C); 707. (B); 708. (A); 709. (D). 710. (B); 711. (B); 712. (C); 713. (C); 714. (A); 715. (B).

Chapter 3

GENETICS AND PLANT BREEDING

1. The branch genetics started in
(A) 1857 (B) 1866 (C) 1900 (D) 1905

2. Sum total of characters which are transmitted from generation to generation constitute:
(A) Genetics (B) Phenocopy
(C) Heredity (D) Variation

3. Rediscovery of Mendel's work was made in:
(A) 1866 (B) 1900
(C) 1898 (D) None of these

4. The term Genetics was first used by:
(A) Johannsen (B) Punnett
(C) Bateson (D) Darlington

5. Genetics deals with:
(A) Genes (B) Heredity
(C) Variations (D) Heredity and Variations

6. The heredity variations are due to:
(A) Asexual reproduction (B) Sexual reproduction
(C) Vegetative reproduction (D) All of these

7. Concept of 'Pangenes' was given by:
(A) Wolff (B) Swammerdam
(C) Johannsen (D) Darwin

8. 'Genotype-phenotype concept' was given by:
(A) Johennsen (B) Charles Darwin
(C) Mendel (D) Bateson

9. When two genotype produce the same phenotype due to different environments one is called —— of other :
(A) Phenotype (B) Allelomorph
(C) Phenocopy (D) None of these

10. Mendel worked on:
(A) Pigeonpea (B) Garden pea
(C) Sweet pea (D) Chickpea

11. Mendel originally belonged to:
(A) Sweden (B) Austria
(C) Australia (D) Czechoslovakia

12. Mendel published the results of his experiments in:
(A) 1857 (B) 1866 (C) 1868 (D) 1900

13. Mendel studies how many pairs of contrasting characters?
(A) 1 (B) 2 (C) 7 (D) 10

14. Which of the following characters was not studied by Mendel?
(A) Plant height (B) Position of flower
(C) Colour of pod (D) Length of pod

15. The term 'Allele' was proposed by:
(A) Johennsen (B) Bateson
(C) Mendel (D) Punnett

16. Emasculation is achieved by:
(A) Removal of stigma
(B) Removal of anthers
(C) Removal of sepals and petals
(D) All of the these

17. Alleles are:
(A) Homologous chromosomes
(B) Linked genes
(C) Dominant genes
(D) Alternate forms of gene

18. If a plant heterozygous for tallness is selfed, the F_2 generation has both tall and dwarf plants. This proves the principle of:
(A) Dominance (B) Segregation
(C) Independent assortment (D) Incomplete dominance

19. In an experiment on pea, plants with yellow round seeds were crossed with plants with green wrinkled seeds; the phenotypic ratio of F_1 progeny is:
(A) 9 yellow round:3 round green : 3 wrinkled yellow : 1 green wrinkled
(B) All wrinkled green
(C) All yellow round
(D) 1 Yellow round :1 Green round : 1 Yellow wrinkled : 1 Green wrinkled

20. Red (RR) flower plant is crossed with white (WW), all the offsprings are pink, this indicates that R gene is:
(A) Hybrid (B) Recessive
(C) Incomplete dominant (D) Mutant

21. Cross between a heterozygous individual and homozygous recessive parent is called:
(A) Monohybrid cross (B) Dihybrid cross
(C) Back cross (D) Test cross

22. An organism with two identical alleles of gene in a cell is:
(A) Dominant (B) Homozygous
(C) Heterozygous (D) Recessive

23. In F_2 progeny of dihybrid cross, the expected genotypic proportions of individuals homozygous for both dominant characters is:
(A) 3/16 (B) 9/16 (C) 12/16 (D) 1/16

24. Mendel performed his work on pea plants during:
(A) 17th century (B) 18th century
(C) 19th century (D) 20th century

25. Who is known as "father of Genetics"?
(A) C. Darwin (B) G. Mendel
(C) Hugo de Vries (D) Weismann

26. When dominant epistasis is operative between two gene loci, the classical 9 : 3 : 3 : 1 ratio becomes modifies into:
(A) 9 : 3 : 4 ratio (B) 9 : 6 : 4 ratio
(C) 12 : 3 : 1 ratio (D) 15 : 1 ratio

27. One complete set of chromosomes inherited as a unit from one parent is known as:
(A) Genotype (B) Genome
(C) Gene pool (D) Karyotype

28. The geometrical device that helps in visualizing all the possible combinations of male and female gametes is known as:
(A) Bateson square (B) Morgan square
(C) Punnett square (D) Mendel square

29. The condition in which only one allele of a pair is present is called:
(A) Homozygous (B) Hemizygous
(C) Heterozygous (D) Incomplete dominance

30. A pure tall pea plant was reared in soil poor in the nutrition and reached the size of a pure dwarf pea plant, if this plant is selfed, then the genotype in the F_1 gen. is most likely to be:
(A) All tall (B) 50% tall : 50% dwarf
(C) All dwarf (D) Not possible to predict

31. A cross between a plant heterozygous for two factors and a plant recessive for both the factors, gives a phenotypic ratio of:
(A) 9 : 1 : 1 : 7 (B) 9 : 3 : 3 : 1
(C) 1 : 1 : 1 : 1 (D) 1 : 7 : 7 : 1

32. Which is back cross in a monohybrid cross?
(A) 9 : 3 : 3 : 1 (B) 1 : 1
(C) 1 : 2 : 1 (D) 3 : 1

33. Which term represents a pair of contrasting characters?
(A) Homozygous (B) Heterozygous
(C) Complementary genes (D) Allelomorphs

34. In monohybrid cross, the F_2 phenotypic ratio is:
(A) 1 : 2 (B) 2 : 1 (C) 3 : 1 (D) 1 : 3 : 1

35. The term heterosis represents:
(A) Hybrid incompatibility (B) Hybrid vigour
(C) Hybrid sterility (D) Structural hybridity

36. Genes which in homozygous condition result in non-viable progeny are:
(A) Duplicate genes (B) Linked genes
(C) Lethal genes (D) Epistatic genes

37. Which is called a 'pleitropic gene'?
(A) Control characters of human beings only
(B) Control character of plants
(C) Control one character in association with other
(D) Control more than one character

38. The genotype of a dominant parent is determined by crossing it with a hybrid; this cross is called:
(A) Back cross (B) Test cross
(C) Long cross (D) Red cross

39. Allosomes are:
(A) Bead like structures in chromosomes
(B) Node like structures on chromosomes
(C) Sex –chromosomes
(D) None of these

40. 'F_1' generation represents:
(A) First fertile generation
(B) First final generation
(C) First filial generation
(D) First federation generation

41. Independent assortment of genes ours due to the orientation of chromosomes at:
(A) Metaphase of mitosis
(B) Metaphase I of meiosis
(C) Metaphase II of meiosis
(D) Any phase of the cell division

42. Incomplete dominance was first reported in 4 o'clock plant by:
(A) Mendel (B) Morgan
(C) Batesen (D) Correns

43. Same genotypic and phenotypic ratios in F_2 generations are obtained in:
(A) Complete dominance (B) Blending inheritance
(C) Codominance (D) None of these

44. In which, incomplete dominance is seen?
(A) Wheat (B) Maize
(C) Mouse (D) Snapdragons

45. Lethal factors were reported in plants by:
(A) Cuenot (B) Baur
(C) Batesen (D) Morgan

46. The most common cause of the pleiotropic effect of a gene is due to:
(A) The same product of the given gene being involved in different metabolic pathways
(B) The gene making very different products in different cell types
(C) The DNA sequence of the gene getting changed in cell specific manner
(D) The gene not functioning in some cells

47. Gene interaction in which either of dominant gene is enough to give rise to same phenotypic effect is:
(A) Complementary gene (B) Duplicate gene
(C) Epistatic gene (D) Supplementary gene

48. When two pairs of genes interact to give 9 : 7 F_2 ratio, they are:
(A) Supplemetary factors (B) Complementary factors
(C) Epistatic factors (D) Duplicate factors

49. 12 : 3 : 1, F_2 ratio is obtained in interaction of .... Factors:
(A) Complementary (B) Supplementary
(C) Epistatic (D) Inhibitory

50. Recombination percentage in a diploid cannot exceed:
(A) 100 (B) 75 (C) 50 (D) 25

51. Exception to Mendel's law of independent assortment was first discovered by:
(A) Morgan (B) Sturtevant
(C) Bateson and Punnett (D) None of these

52. Phenomenon of linkage was first discovered in:
(A) Sweet pea (B) Garden pea
(C) Mirabilis (D) Snapdragon

53. Two linked genes a and b show 20% recombination. The individuals of a dihybrid cross between ++/++ x ab/ab shall show gametes:
(A) ++ 80 : ab : 20
(B) ++ 50 : ab : 50
(C) ++ 40 : ab 40 : + a10 : + b : 10
(D) ++ 30 : ab 30 : + a 20 : + b : 20

54. Coupling and repulsion phenomenon are concerned with:
(A) Crossing over (B) Linkage
(C) Mutation (D) None of these

55. Theory of linear arrangement of genes on chromosomes was formulated by:
(A) Sturtevant (B) Morgan
(C) Bateson (D) Mendel

56. When blue flower long pollen plant (BBLL) was crossed with red flowered round pollen plant (bbll) in Lathyrus and their F_1 was crossed with red round parent, then what is the ratio of blue long and red round obtained in this cross?
(A) 1/16 : 1/16 (B) 7/16 : 7/16
(C) 7/16 : 1/6 (D) 9/16 : 3/16

57. In a diploid organism, the genes A, B and C are present on the same chromosome in that order. The AB interval; is 10 map units and B interval is 20 map units. In an ABC/abc heterozygous individual, what will be the proportion of gametes that carry the genotype AbC?
(A) 1 % (B) 10 % (C) 20 % (D) 30 %

58. If 2n = 20 in a plant, then how many linkage groups are present in this?
(A) 20 (B) 10 (C) 5 (D) 5 or 10

59. Somatic crossing over was discovered in Drosophila by:
(A) Pontecorvo (B) Morgan
(C) Stern (D) Serebrousky

60. G. Pontecorvo observed non-germinal cover in:
(A) Maize (B) Drosophila
(C) Aspergillus (D) None of these

61. Most accepted theory of crossing over is:
(A) Copy Choice theory (B) Precosity theory
(C) Contact first theory (D) Breakage first theory

62. Phenotypic expression of the heterozygote being greater than that of either homozygote is called:
(A) Dominance (B) Midparent value
(C) Overdominance (D)Pseudodominance

63. Pseudoautosomal gene is present on:
(A) Both X-and Y- chromosomes
(B) Both X- and autosomes
(C) Both Y- and autosomes
(D) Only autosomes

64. Exchange of chromosomal segments between non-homologous chromosomes is:
(A) Crossing over (B) Inversion
(C) Translocation (D) None of these

65. Which of the following most accurately explains the cause for the abnormal numbers of chromosomes during human reproduction that an result in Down syndrome, Turner's syndrome, or Klinefelter syndrome?
(A) The occurrence of non-disjunction of homologous chromosomes during meiosis
(B) The duplicative production of extra chromosomes during DNA replication
(C) The abnormal pairing of non-homologous chromo somes during prophase of miosis I
(D) The selective loss of particular chromosomes from the sex cells after formation of the mature gamete

66. The process which involves reverse order of genes in chromosome is:
(A) Deficiency (B) Aneuploidy
(C) Inversion (D) Translocation

67. 'x' represents:
(A) Gametic number
(B) Basic chromosome number
(C) Haploid number
(D) All of the above

68. Presence of chromosome number, which is different from the multiple of basic chromosome no. is called?
(A) Hyperploidy (B) Heteroploidy
(C) Aneuploidy (D) Euploidy

69. 2n – 2 is a:
(A) Monosomic (B) Nullisomic
(C) Tetrasomic (D) Trisomic

70. If haploid chromosome no. in an organism is 12, what is the chromosome no. present in its trisomic?
(A) 11 (B) 13 (C) 25 (D) 23

71. In a three-factor cross among genetic markers A, B and C, the relative frequencies of recombinants were AB (1), AC (0.2) and BC (0.3). Which of the following is the most probable order of these markers on the genome?
(A) ABC
(B) BCA
(C) CAB
(D) The data do not reflect the order of genes

72. Reduction in size of vegetative and floral parts is found in:
(A) Autopolyploids (B) Allopolyploids
(C) Haploids (D) Diploids

73. 'Gigantism' is mostly found in:
(A) Autopolyploids (B) Allopolyploids
(C) Haploids (D) None of these

74. Polyploidy can be artificially induced by:
(A) 0.1 M NaCl solution (B) 0.25 M sucrose solution
(C) Zeatin (D) Colchicine

75. In Drosophila Y-chromosome is:
(A) Essential for male sex determination
(B) Required for development of secondary sex-characters
(C) Required for dosage compensation
(D) Not required for male sex determination

76. *Colchicine* was first obtained from *Colchicines autumnale* by:
(A) Dustin (B) Houde
(C) E.R. Sears (D) Blakeslee

77. Colchicine is a:
(A) Steroid (B) Alkaloid
(C) Cholesterol (D) None of these

78. Natural autopolyploidy is found in:
(A) Cynodon (B) Grapes
(C) Watermelon (D) None of these

79. One of the following statements is most appropriate about the Y chromosome in mammals:
(A) Essential for the development of embryo
(B) Non-essential for the development of the embryo but important for male sex determination
(C) Non-essential for sex determination by important for male fertility
(D) Essential for male sex determination and fertility

80. Complete linkage is observed in:
(A) Male Drosophila (B) Female Drosophila
(C) Birds (D) Maize

81. Percentage of crossing over will be more if:
(A) Linked genes are located far away from each other
(B) Linked genes are located close to each other
(C) Both (A) and (B)
(D) None of the above

82. What are the sexual phenotypes of the following genotypes in Drosophilla : XXY, XO?
(A) Male, female (B) Male, male
(C) Intersex, female (D) Female, male

83. Who is associated with Raphanobrassica?
(A) Muntzing (B) E.R. Sears
(C) Karpechenko (D) None of these

84. 'Triticale' is derived from cross between?
(A) Wheat and aegilops
(B) Wheat and rice
(C) Wheat and maize
(D) Wheat and rye

85. Why is the gene for pattern baldness in man exhibits dominance in man but acts recessively in women? Because:
(A) Baldness gene expression in regulated by hormones
(B) Baldness gene is autosomal
(C) Baldness gene in on X-chromosome
(D) Baldness gene in on Y-chromosome
(E) Mitotic failure in the early embryo

86. In common hexaploid bread wheat, the three diploid genomes are:
(A) A, B and C (B) A, C and D
(C) B, C and D (D) A, B and D

87. The three genomes of common hexaploid wheat show:
(A) Homologous relationship
(B) Non-homologous relationship
(C) Homoeologous relationship
(D) None of these

88. Each of the following chromosome abnormalities involves a 20 megabase region of the long arm of chromosome 5 (5q). Which abnormality is most likely to cause severe disease?
(A) Deletion of the region
(B) Duplication of the region
(C) A balanced translocation involving the region (i.e., in the translocation carrier)
(D) Pericentric inversion

89. Mutation was first discovered in:
(A) *Drosophila* (B) *Neurospora*
(C) *E. coli* (D) *Oenothera*

90. The term 'Sport' was given by:
(A) Hugo de Vries (B) Darwin
(C) Lamarck (D) None of these

91. The change due to structural changes in chromosomes:
(A) Can be reversed (B) Cannot be reverved
(C) Is not known (D) None of these

92. Which of the following will not cause triploidy:
(A) Fusion of an egg and polar body with subsequent fertilization by sperm cell
(B) Meiotic failure, producing diploid sperm or egg
(C) Dispermy
(D) All of these

93. Which of the following is called Drosophila of plant kingdom?
(A) *E. coli* (B) *Acetabularia*
(C) *Neurospora* (D) *Saccharomyces*

94. In mice, a mutation results in the production of yellow oat colour, as opposed to the normal gray oat. Crossing yellow mice with normal mice results in equal proportions of normal and yellow mice, whereas crossing yellow mice amongst themselves results in a ratio of 2 yellow : 1 normal progeny. The results suggest that:
(A) More than one gene is involved in conferring the yellow phenotype
(B) The yellow allele is lethal under homozygous condition
(C) The yellow allele exhibits cytoplasmic heredity
(D) The results cannot be explained by any genetic principles

95. A mutation in sheep called 'Ancon sheep' was recorded by:
(A) Hugo de Vries (B) Darwin
(C) Seth Wright (D) None of these

96. Mutagen is:
(A) Organism in which mutation has occurred
(B) Mutation inducing agent
(C) Both (A) and (B)
(D) None of the above

97. Frequency of mutation:
(A) Varies with character and organism
(B) Can be increased by X-rays
(C) Is greatly affected by environmental factors
(D) All of these

98. A cross between unlike organisms is called:
(A) Test cross (B) Heterosis
(C) Back cross (D) Hybrid

99. If a man who is colour-blind marries a woman who is homozygous for normal colour vision, the probability of their son being colour-blind is:
(A) 0 (B) 1 (C) 0.5 (D) 0.75

100. In nature, the rate of mutation is?
(A) 1 x 10^5 (B) 1 x 10^8
(C) 1 x 10^{-5} (D) 1 x 10^{12}

101. Mutations are
(A) Mostly useful (B) Always useful
(C) Rarely useful (D) Always harmful

102. Who used X-rays to cause mutation?
(A) Morgan (B) Muller
(C) Alterberg (D) Auerbach

103. Consider two linked loci, labeled A and B. Each locus has two alleles, labeled 1 and 2. The frequencies of the alleles in a population are: A1 = 0.6, A2 = 0.4, B1 = 0.7, B2 = 0.3. If there is linkage equilibrium between these two loci in the population, what is the expected frequency of chromosomes carrying a combination of the A1 and B2 alleles?
(A) 0.24 (B) 0.12 (C) 0.18 (D) 1.00

104. Changes due to mutation are:
(A) Small (B) Negligible
(C) Large (D) All of these

105. Pseudodominance is:
(A) There is incomplete dominance
(B) Dominance of recessive alleles in a single generation
(C) Deficiency in which a recessive allele will behave like a dominant allele
(D) There is no such phenomenon

106. Gene sanctuary refers to:
(A) An area of diversity protected against interference from man
(B) Gene collection
(C) Gene evolution
(D) Gene mutation

107. In a test cross of an individual heterozygous for each of three linked genes, the most frequent lasses of progeny were ABc/abc and abC/abc, and the least frequent lasses were ABC/abc and abc/abc. What is the order of the genes:
(A) A – C – B (B) B – A – C
(C) A – B – C (D) None of these

108. If an organism is having genotype AaBb, then what are the possible combinations of gametes after meiosis?
(A) AB, ab (B) Aa, Bb
(C) AB, aB, Ab, ab (D) Not possible to tell

109. Suppose DNA has base sequence as UAC, GAC, AGC, CGC, ACA, AAA and due to mutation only first base is deleted, what is the effect on coding?
(A) First amino acid will be different
(B) There will be one amino acid less
(C) Complete change in type and sequence of amino acids
(D) No change

110. Why are some autosomal dominant disorders (e.g., Marfan syndrome) seen more commonly in the offspring of older fathers?
(A) Replication errors accumulate as sperm-producing stem cells continue to divide
(B) Rate of nondisjunction increases in older males
(C) Recombination rates increase in older males
(D) None of these

111. Which of the following is non-ionizing radiation?
(A) X- rays
(B) β- rays
(C) γ rays
(D) UV-rays

112. Certain types of cancers can be correlated with specific changes in chromosome structure. In patients suffering from myelogenous leukemia, the abnormal chromosome detected was termed Philadelphia chromosome. Which of the following chromosomes is altered in this disease?
(A) Chromosome 10
(B) Chromosome 11
(C) Chromosome 20
(D) Chromosome 22

113. UV-rays as mutagenic agents were shown by:
(A) Auerbach (B) Stadler
(C) Altenberg (D) Muller

114. Chemicals as mutagenic agents were first shown by:
(A) Altenberg (B) Auerbach
(C) Stadler (D) None of these

115. People who carry an allele for normal hemoglobin and an allele for sickle cell hemoglobin are resistant to malaria. This phenomenon is called:
(A) Heterozygote advantage
(B) Polyploidy
(C) Mutation selection balance
(D) Recessive superiority

116. Hypophosphatemia is caused by an X-linked dominant gene in humans. If a hypophosphatemia man marries a normal woman, which of their children will have hypophosphatemia?
(A) All daughters (B) Half of their daughters
(C) All of the sons (D) Half of their sons

117. Which of the following statements are correct:
P. Down's syndrome/mongolism is due to extra 21st chromosome
Q. Down's syndrome is due to non-disjunction of chromosomes
R. Trisomy has chromosome complement of 2n + 1
S. Monosomics are 2n - 1
(A) P, Q (B) R, S
(C) P, Q, R (D) P, Q, R, S

118. Base substitution caused by 5-bromo uracil and BudR are called:
(A) Inversions
(B) Transitions
(C) Transversions
(D) Translocations

119. Which of the following statements are correct:
P. Cri-du-chat syndrome is due to chromosomal change involving deletion
Q. Formation of multivalents in meiosis is due to reciprocal translocation
R. Klinefelter's syndrome has 44 + XXY
S. In Down's syndrome of a male child, the sex complement is XY
(A) P, Q (B) R, S
(C) P, Q, R (D) P, Q, R, S

120. Proflavin and Acridine orange induce:
(A) Transitions (B) Transversions
(C) Inversions (D) Frameshift mutations

121. Which is not associate with mutations?
(A) Better progenies
(B) Death of organisms
(C) Change in genetic constitution
(D) Hybrid vigour

122. 'Mutations which are recessive are not lost by a gene pool' is called:
(A) Survival of dominance (B) Survival of recessive
(C) Hardy Weinberg law (D) None of these

123. Cytoplasmic inheritance is responsible for:
(A) Inheritance of kappa particles
(B) Iojap inheritance in corn
(C) Petite character in yeast
(D) All of these

124. A decentric chromosome is unstable because:
(A) It cannot resynthesize its telomeres during repli cation
(B) It pairs with non-homologous chromosomes in meiosis
(C) It pairs with non-homologous chromosomes in mitosis
(D) It is often simultaneously drawn to opposing spindle poles in mitosis

125. Who is credited to have differentiated between heritable variation and environmental variation?
(A) Darwin (B) Morgan
(C) Hugo de Vries (D) Bateson

126. The resistance developed by some insects against insecticides is due to:
(A) Hybrid vigour
(B) Genetic mutations
(C) Variations due to sexual reproduction
(D) Genetic recombinations

127. Who coined the term Heterosis?
(A) Mendel (B) G.H. Shull
(C) Schmidt (D) Schleiden

128. The force that can change the frequency of an allele in a population are:
(A) Forward mutation, gene conversion, neutral evolution and recombination
(B) Selection, mutation, migration, inbreeding and random genetic drift
(C) Dominance, family selection, fitness and diversification
(D) Gene interaction, gene transfer, gene mutation and out breeding

129. Which of the following crosses would result in 1 : 2 : 1 F_2 ratio?
(A) Yg x gg (B) Yg x yg
(C) Yg x yy (D) YY x gg

130. 'Gene Battery Model' was given by:
(A) Jacob and Monod
(B) McClinktock and Creigheten
(C) Bawden and Pierre
(D) Bridden and Davidson

131. The first 'hybrid' was produced by:
(A) Camerarius (B) Fairchild
(C) Lamarck (D) N. Grew

132. Sex determination was first of all studied in which plant?
(A) Coccinia (B) Melandrium
(C) Rumex (D) Humulus

133. Which of the following shows slowest cell division?
(A) Haploid (B) Diploid
(C) Polyploid (D) Same in all

134. 2n=x=7, represents a:
(A) Haploid (B) Diploid
(C) Monoploid (D) Polyploid

135. In wheat 2n=6x=42. No. of possible monosomics is:
(A) 42 (B) 21 (C) 7 (D) 6

136. Translocation was first reported in:
(A) Datura (B) Maize
(C) Oenothera (D) Doob grass

137. In snapdragon, when yellow leaved var. (Aurea) x Aurea, what is the ratio of yellow: green in progeny?
(A) 3 : 1 (B) 9 : 3
(C) 1 : 1 (D) 2 : 1

138. Mutations are mostly:
(A) Dominant (B) Incompletely dominant
(C) Codominant (D) Recesive

139. Suppose DNA has base sequence AAG GAG GAC CAA CCA, which one represents frameshift mutations?
(A) AAG GAG GAC CAA CCA
(B) AAG AAG GAC CAA CCA
(C) AAG GAG ACC AAC CAA
(D) AAG GAG GAC CAA CCA

140. No. of linkage groups is equal to:
(A) Diploid no. of chromosomes
(B) Haploid no. of chromosomes
(C) Tetraploid no. of chromosomes
(D) None of the above

141. Who discovered Mendel's principle in 1900 A.D.?
(A) Avery, McLeod, McCarty
(B) Sutton, Morgan, Bridges
(C) Bateson, Punnett and Bridges
(D) de Vries, Correns and Tschermak

142. Heterosis is also called:
(A) Hybrid vigour (B) Hybrid sterility
(C) Mutation (D) Variation

143. Why are haploids superior to diploids in study of mutations?
(A) They have shorter life time
(B) Smaller no. of chromosomes
(C) They allow expression of recessive mutation immediately
(D) Obtained in large number

144. The individuals having same maternal and paternal parents are called?
(A) Epistasis (B) Hypostasis
(C) Siblings (D) Gynander

145. When a pyrimidine is replaced by purine, it is called:
(A) Frameshift mutation (B) Transition
(C) Transversion (D) Reversible mutation

146. Picking up plants with superior phenotype for further propagation is:
(A) Mass selection (B) Pure line selection
(C) Hybrid vigour (D) Introduction

147. Picking up only homozygous plants for further cultivation is known as:
(A) Mass selection (B) Pure line selection
(C) Clone (D) Acclimitization

148. Homozygosity is obtained after inbreeding for:
(A) Two generations (B) Four generations
(C) Eight generations (D) Ten generations

149. PV-18, Kalyan sona and S-227 wheat varieties have developed through:
(A) Gamma ray irradiation (B) Multiple crossing
(C) Selection (D) Genetic engineering

150. Importing better varieties and plants from outside and acclimitising them to local environment is:
(A) Selection (B) Cloning
(C) Introduction (D) Heterosis

151. Heterosis is:
(A) Appearance of spontaneous mutations
(B) Induction of mutations
(C) Mixture of two or more traits
(D) Superiority of hybrids over their parents

152. Cutting of anthers from the intersexual flowers of female parents is:
(A) Emasculation
(B) Male sterile line
(C) Artificial pollination
(D) Selectivity or differential reproduction

153. Male sterile is due to:
(A) Cytoplasmic inheritance
(B) A recessive nuclear gene
(C) Interaction of nuclear gene and cytoplasmic factor
(D) All of these

154. Heterosis cannot be maintained in sexually reproducing plants as it disappears on:
(A) Outbreeding (B) Inbreeding
(C) Cross breeding (D) None of the above

155. Polyploidy gives:
(A) Hybrid vigour (B) Imrpoved varieties
(C) Gigas effect (D) All of these

156. Polyploidy involves:
(A) Increases in number of chromosomes
(B) Increases in number of genomes
(C) Increase in number of plasmons
(D) Deletion of part or whole of a chromosomes

157. Polypoidy is induced through:
(A) Colchicine (B) Irradiation
(C) Mutagenic chemicals (D) Ethylene

158. Best varieties of SugarBeet are:
(A) Tetraploids (B) Diploids
(C) Triploids (D) Allopolyploids

159. Most varieties of sugarcane are:
(A) Tetraploids (B) Triploids
(C) Hexaploids (D) Aneuploids

160. Main varieties of potato are:
(A) Tetraploids (B) Triploids
(C) Aneuploids (D) Diploids

161. Bread Wheat is:
(A) Autopolyploids (B) Allopolyploids
(C) Aneuploids (D) Trisomic

162. Seedless Water Melon is :
(A) Hexaploid (B) Tetraploid
(C) Triploid (D) Pentaploid

163. The first domesticated Wheat was:
(A) *Triticum boeoticum* (B) *T. dicocum*
(C) *T. dicoccoies* (D) *T. monococcum*

164. Induced mutation have produced a rice variety:
(A) Pelita-1 (B) Reimei
(C) Pelita-2 (D) All the above

165. Explant is:
(A) Plant collected after harvesting
(B) Exploited part of a plant
(C) Small part of the plant meant for tissue culture
(D) Uprooted for transplantation

166. A technique of micropropagation is:
(A) Multiple shoot production
(B) Multiple shoot production and somatic embryogenesis
(C) Growth of micro-organisms on culture medium
(D) Somatic embryogenesis

167. Pathogen free plants can be obtained through:
(A) Clean cultivation
(B) Embryoid cultivation
(C) Bud or shoot up propagation on culture medium
(D) Grafting

168. Embryoid is
(A) Nonzygotic embryo
(B) Nonfunctional embryo
(C) Parthenogenetic embryo
(D) An early stage in callus differentiation

169. Androgenic haploid culture was first performed by
(A) Guha and Maheswari
(B) Steward
(C) Skoog and Miller
(D) Halperrin and Wetherell

170. Name the tissue culture technique used to grow non-viable inter-specific hybrids
(A) Androgenic haploids (B) Shoot tip culture
(C) Somatic embruogenesis (D) Embryo rescue

171. Variations appearing in tissue culture are
(A) Culture variations (B) Auxotrophs
(C) Somaclonal variations (D) Pangenetic variations

172. Pollution tolerant variants can be produced through
(A) Large scale growth in polluted areas
(B) Tissue culture
(C) Protoplast technique
(D) Genetic engineering

173. Somatic hybridization is achieved through?
(A) Grafting
(B) Protoplast fusion
(C) Conjugation
(D) Recombinant DNA technology

174. Which one is required for protoplast fusion?
(A) Treatment with cellulose and pectinase
(B) Electrofusion or PEG treatment
(C) Both (A) and (B)
(D) Recombinant DNA technology

175. Restriction endonucleases are useful in
(A) Breaking DNA at particular places
(B) Producing sticky DNA ends
(C) Both (A) and (B)
(D) Crossing over

176. The growth kinetics that results from metabolizing one sugar before another is referred to as:
(A) Exponential growth (B) Diphasic growth
(C) Diauxic growth (D) Chemotaxis

177. In genetic engineering, for breaking DNA segments, one of the following is required
(A) RNA polymerase (B) DNA ligase
(C) Restriction endonuclease (D) Exonuclease

178. Colchicine brings about
(A) Gene mutations
(B) Chromosome aberrations
(C) Quick replication
(D) Duplication of chromosomes

179. Triticale has been developed through inter-generic hybridization between
(A) Wheat and Rye (B) Wheat and *Aegilops*
(C) Wheat and Rice (D) Rice and Maize

180. Mutations are generally induced by means of
(A) Bet rays (B) Alpha rays
(C) Gamma rays (D) UV rays

181. In crop improvement programme, haploids are important because they
(A) Require one half of nutrients
(B) Are helpful in study of meiosis
(C) Grow better under adverse conditions
(D) Form perfect homozygous individuals on diploidisation

182. What would be the effect of decreasing the number of tryptophane codons from 2 to 0 in the sequence of the tryptophan operon that codes for the leader peptide?
(A) The production of full – length m RNA would no longer be dependent upon the trp concentration
(B) When trp levels were low, higher amounts of full-length mRNA would be produced
(C) When trp levels were low, lower amounts of full-length mRNA would be produced
(D) Both A and B

183. Restriction endonucleases are used as
(A) Molecular scalpels for cutting DNA at specific sites
(B) Molecular cemet for combining DNA segments
(C) Molecular primer for building nucleotides in tendem
(D) Molecular degradation to DNA break up

184. Polyploidy is induced by
(A) Kinetin (B) GA_3
(C) Colchicine (D) IAA

185. A plant with multiple uses is
(A) Cotton (B) Coconut
(C) Jute (D) Sunflower

186. Which one is hexaploid wheat?
(A) *Triticum durum* (B) *T. monococcum*
(C) *T. tugidum* (D) *T. aestivum*

187. What is the role of the I gene of phage lambda?
(A) It encodes a protein that enables the phage to enter lysis
(B) Its product helps the prophage to excise from the host DNA
(C) It encodes the lambda repressor which represses the lytic cycle
(D) Its gene product helps the prophage to integrate into the host DNA

188. A clone is
(A) Homozygote produced by sexual reproduction
(B) Heterozygote produced by sexual method
(C) Homozygote obtained asexually
(D) Heterozygote obtained asexually

189. Heredity variations can be got with the help of
(A) X- rays (B) DDT
(C) Auxin (D) Gibberellin

190. Inbreeding is possible between two members of
(A) Order (B) Family
(C) Genus (D) Species

191. Improved wheat variety having genes for dwarfness and higher percentage of protein/lysine is
(A) Kalayan (B) Sharbati Sonora
(C) Sonalika (D) Lerma Rojo

192. Which of the following statements are true with respect to a special complex called 'dicot'?
P. It consists of deoxyribonuclease and DNA fragments
Q. It consists of ribonuclease and RNA fragments
R. It involved in gene silencing
S. It triggers apoptosis
(A) P, R (B) Q, R (C) P, S (D) Q, S

193. Majority of improved Indian rice varieties have been arisen from a cross between
(A) *Oryza sativa indica* X *Oryza nivara*
(B) *Oryza sativa indica* X *Oryza sativa japonica*
(C) *Oryza sativa japonica* X *Oryza nivara*
(D) Oryza rufipogon X *Oryza nivara*

194. Hybrid vigour is due to:
(A) Crossing over (B) Linkage
(C) Heterozygosity (D) Chiasma

195. Crop improvement is possible through:
(A) Judicious combination of selection, introduction and hybridization
(B) Selection
(C) Scientific improvement of cultivated plants
(D) Introduction

196. Which of the following is not directly related to gene regulation?
(A) Acetylation of histones
(B) Spicing
(C) Glycosylation in the endoplasmic reticulum
(D) Silencer

197. A breeder involving disease resistant variety will start with:
(A) Working out yield of different varieties
(B) Go through the subject in library
(C) Selection of parents
(D) Hybridisation

198. Emasculation of flowers is carried out through removal of:
(A) Stigma (B) Sepals and petals
(C) Anthers (D) Entire organs

199. Regulations governing movement of diseased plant material are called:
(A) Crop protection (B) Quarantine
(C) Plant regulation (D) Rotation

200. A tool in crop improvement involving identification of genes, transfer and integration is:
(A) Protoplast fusion and tissue culture
(B) Somaclonal hybridization
(C) Gene bank technology
(D) Genetic engineering

201. Aneuploidy is:
(A) Zygote having multiple of two gametic number
(B) Zygote having four times the number of gametic chromosomes
(C) Zygote having abnormal number of chromosomes
(D) Zygote having three times the number of gametic chromosomes

202. Intact duplex DNA is a substrate for:
(A) DNA pol I (B) DNA pol III
(C) RNA polymerase (D) DNA pol IV

203. Haploid plants are preferred over diploids for mutation study because in haploids
(A) Recessive mutations express immediately
(B) Induction of mutations is easier
(C) Culturing is easier
(D) Dominant mutations express immediately

204. Axenic culture is
(A) Culture without contamination
(B) Culture without a nutrient
(C) Culture of a gene
(D) Culture of a tissue

205. Plants can be made disease resistant through
(A) Colchicine treatment
(B) X-ray treatment
(C) Breeding with wild relatives
(D) Hormone treatment

206. Which of the following is not directly associated with regulation of eukaryotic gene expression?
(A) Acetylation of histones (B) Methylation of DNA
(C) Alternative splicing (D) Activation of caspases

207. Most of our crop plants are
(A) Autopolyploid in origin
(B) Allopolyploid in origin
(C) Mixed genotypic in origin
(D) Heterozygous in origin

208. In Rho-independent transcription termination, stem and loop structure is formed in:
(A) The transcript (B) The template DNA
(C) Coding strand (D) Both (A) and (B)

209. *Triticum aestivum* is:
(A) Hexaploid (B) Alloautopolyploid
(C) Heteropolyploid (D) Autopolyploid

210. Bombay green banana has developed from:
(A) Mass selection (B) Pure line selection
(C) Clonal selection (D) Hybridisation

211. Removal of anthers of some flowers during plant breeding as:
(A) Anthesis (B) Emasculation
(C) Pollination (D) For collection of pollen

212. Hybrid potato and tomato is:
(A) Tomapa (B) Pomato
(C) Potamato (D) None of the above

213. Androgenic haploids are used for:
(A) Raising stock for micropropagation
(B) Induction of mutations
(C) Knowing effect of all genes
(D) Preparation of homozygous diploids

214. The 'A' fragment of diphtheria toxin catalyzes the addition of an ADP-ribose group to the eukaryotic elongation factor EF-2. The source for the ADP-ribose group is:
(A) FAD (B) FMN
(C) NAD^+ (D) $NADP^{++}$

215. Crosses involving plants of the same variety are:
(A) Intravarietal (B) Interspecific
(C) Intervarietal (D) Inta-generic

216. An old breeding technique is:
(A) Introduction (B) Selection
(C) Mutation breeding (D) Hybridisation

217. There are important classes of pattern-regulating genes specify the basic features and functions during embryonic development are:
(A) Maternal effect genes (B) Segmentation genes
(C) Homeotic genes (D) All of these

218. Aim of plant breeding:
(A) Control pollution
(B) Keep soil fertile
(C) Produce improved varieties
(D) To maintain wild plants

219. Technique used in agriculture biotechnology is:
(A) Plant breeding (B) Transformation
(C) Dual replication (D) Tissue culture

220. Important use of agriculture biotechnology is to:
(A) Produce pest resistance
(B) Increase nitrogen content
(C) Increase plant weight
(D) Decrease seed number

221. Pure line breeds refer to:
(A) Homozygosity and independent assortment
(B) Homozygosity only
(C) Heteozygosity only
(D) Heterozygosity and linkage

222. The nucleotide sequence of DNA involved in binding to a transcription factor can be determined by:
(A) DNA footprinting
(B) S1 nuclease treatment
(C) DNA fingerprinting
(D) Northern hybridization

223. miRNA based silencing of genes is a type of:
(A) Transcription gene silencing
(B) Post transcription gene silencing
(C) Translation gene silencing
(D) Post translation gene silencing

224. Which will vanish in the absence of human help?
(A) Rice (B) Wheat
(C) Maize (D) Potato

225. *Gossypium hirsutum* is:
(A) New World tetraploid (B) Old World tetraploid
(C) New World diploid (D) Old World diploid

226. How many maternal effect genes are critical to the formation of the anterior-posterior axis?
(A) 3 (B) 4 (C) 5 (D) 6

227. New varieties of plants are produced by:
(A) Selection and hybridization
(B) Mutation and Selection
(C) Introduction and mutation
(D) Selection and introduction

228. Wheat revolution in green revolution has been due to introduction of disease resistant, fertilizer responsive, strong stemmed and dwarf varieties imparted from
(A) California (B) Beijing
(C) Mexico (D) Uzbekistan

229. Haploid plants are obtained by culturing:
(A) Root tip (B) Endosperm
(C) Young leaves (D) Pollen grain

230. Farmers connot use seeds of plants showing hybrid vigour in successive years because:
(A) Hybrid vigour is lost on inbreeding
(B) Dunkel's proposals
(C) They are not allowed to sow their own seeds
(D) It is cheaper to purchase fresh seeds

231. A major use of embryo culture is in:
(A) Induction of somaclonal variations
(B) Overcoming hybridization barriers
(C) Production of alkaloids
(D) Clonal propagation

232. The primer of the lagging strand during DNA replication is removed by:
(A) 3' to 5' exonuclease activity of DNA pol III
(B) DNA primase
(C) 5' to 3' exonuclease activity of DNA pol I
(D) 3' to 5' exonuclease activity of DNA pol I

233. 'Norin gene' of dwarfness in Wheat originated through spontaneous mutation in
(A) India (B) Japan
(C) Mexica (D) USSR

234. In postulating the theory of evolution by natural selection Darwin was greatly influenced by:
(A) Mutation theory of Hugo de Vries
(B) Lamarck's theory of acquired characters
(C) Malthus idea of population control
(D) Environmental factors

235. How many maternal genes involved in pattern formation have been identified?
(A) 15 (B) 20 (C) 25 (D) 30

236. The present giraffe has a long neck as compared to its ancestors. Lamarck believed it could be due to
(A) Natural selection
(B) Isolation
(C) Inheritance of acquired characters
(D) Speciation

237. A species inhabiting different geographical areas is known as
(A) Sympatric (B) Allopatric
(C) Sibling (D) Biosphere

238. Charles Darwin toured in a ship for 5 years
(A) Beagle (B) Vikrant
(C) Phillips (D) Alexander

239. Who gave the principle that population tends to multiply more rapidly than food supply?
(A) Darwin (B) Malthus
(C) Lamarck (D) Haldane

240. Role of isolation in evolution is
(A) Differentiation of species
(B) Maintenance of species
(C) Evolution of species
(D) Extermination of species

241. Four mutant strains of *E. coli* having the following characteristics were isolated:
Mutant strain 1: lac-repressor is not able to bind to operator because of a mutation in the operator region of the lac operon
Mutant strain 2: Allolactose is not able to bind to repressor because of a mutation in the repressor – encoding gene
Mutant strain 3: Catabolite activator protein (AP) is not able to bind to the promoter because of a mutation in the promoter region
Mutant strain 4: AMP is not able to bind to AP because of a mutation in the AP-encoding gene.
Lac genes are expressed in strain(s)
(A) 1 only (B) 1, 3 and 4 only
(C) 1 and 4 only
(D) 1 and 2 only, but in the presence of IPTG

242. The evolution of a species is based upon sum total of adaptive changes preserved by
(A) Natural selection (B) Isolation
(C) Speciation (D) Human conservation

243. The ultimate source of continuous variations is
(A) Natural selection (B) Sexual reproduction
(C) Hormonal action (D) Mutations

244. Which of the following concept is attributed to Charles Darwin?
(A) Use and disuse of organs is of great importance in evolution
(B) Every cell must come from a preexisting cell
(C) In the struggle for existence, the fittest would survive
(D) The gametes will carry only one of a pair of contrasting characters

245. Key point of Lamarck's view about organic evolution is that every offspring
(A) Is similar to its parents
(B) Inherits characters acquired by the parental generation
(C) Shows struggle for existence
(D) Repeats phylogeny in its ontogeny

246. The pair-rule gene proteins activate the transcription of the :
(A) Gap genes (B) Pair rule genes
(C) Segment-polarity genes (D) All of these

247. The fountain head of evolution is
(A) Mutations (B) Acquired characters
(C) Natural selection (D) Sexual reproduction

248. The idea of natural selection as the fundamental process for evolutionary change was reached
(A) Independently by Charles Darwin and Alfred Russel Wallace in 1900
(B) By Charles Darwin in 1866
(C) By Alfred Russel Wallace in 1901
(D) Independently by Charles Darwin and Alfred Russel Wallace in 1858

249. Match the terms in group I with their possible explanation in group 2:

Group 1	Group 2
P. Orthologs	1. A cell or an organism having foreign gene
Q. Paralogs	2. The complement of a protein expressed by a genome
R. Proteome	3. Genes from different species related to each other
S. Transgenic	4. Genes from same species related to each other

(A) P -2, Q – 4, R – 1, S - 3
(B) P -4, Q – 3, R – 2, S - 1
(C) P -3, Q – 4, R – 2, S - 1
(D) P -1, Q – 2, R – 3, S – 4

250. To be evolutionary successful, mutation must occur in:
(A) Somatic RNA
(B) Plasma protein
(C) Germplasm DNA
(D) Centrioles

251. Some organism resemble other organisms and thus escape from enemies. This phenomenon is known as:
(A) Homology (B) Variation
(C) Analogy (D) Mimicry

252. For food, light and space, the greatest competition is between two
(A) Closely related species of the same niche
(B) Closely related species of the different niche
(C) Unrelated species of the same niche
(D) Unrelated species of the different niche

253. The protein products of the gap genes interact with neighboring gap gene proteins to activate the transcription of the:
(A) Gap genes (B) Pair rule genes
(C) Segment-polarity genes (D) All of these

254. The first attempt to solve the problem of organic evolution was done by:
(A) Mendel (B) Haeckel
(C) Lamarck (D) Darwin

255. Lamarck's theory of evolution is known as
(A) Natural selection
(B) Inheritance of acquired characters
(C) Survival of fittest
(D) Mutation theory

256. Lamarck's theory of evolution fails to explain
(A) Indian women have been getting their ears pierced for thousands of years, but not a single baby with pierced ears is born
(B) Giraffes have long necks because they have to eat leaves from the tall trees on their habitat
(C) Deer have the capacity to escape from agile enemies because these can run very fast
(D) High altitude pines have branches which slope down so that snow fails to collect upon them in winter.

257. The greatest weakness in Darwin theory was his failure to explain:
(A) Over production (B) Struggle for existence
(C) Survival of the fittest (D) Variations

258. The ultimate source of all variations is:
(A) Sexual reproduction (B) Hormonal action
(C) Natural selection (D) Mutations

259. A culture of bacteria (all of one species) is grown on solid culture medium containing the antibiotic streptomycin. At first no growth is apparent, although bacteria on a similar medium without streptomycin grow rapidly. However, in a short time, two colonies of bacteria begin to grow on the streptomycin-containing medium. What is the most likely explanation?
(A) At least two bacteria adjusted to the streptomycin due to exposure to it. This streptomycin resistance was passed on to their offspring
(B) At least two bacteria were resistant to streptomycin before being exposed to it. These cells generated streptomycin-resistant colonies
(C) Streptomycin induced mutations in at least two bacterial cells that made them resistant. These mutated cells produced resistant colonies
(D) All bacteria will eventually grow in the presence of streptomycin if given long enough

260. Sudden heritable changes are called:
(A) Mutations (B) Variations
(C) Recombinations (D) Evolution

261. "Philosophic Zoologique" was written by:
(A) Linnaeus (B) Lamarck
(C) Wallace (D) Darwin

262. Continuity of germplasm was given by:
(A) Lamarck (B) Weismann
(C) Haeckel (D) Darwin

263. Genetic drift is on account of:
(A) Variations
(B) Mutation
(C) Increase in population
(D) Decrease in population

264. A species inhabiting different geographical areas is known as:
(A) Allopatric (B) Sympatric
(C) Sibling (D) Biospecies

265. Segmentation genes are expressed after:
(A) Ovulation (B) Fertilization
(C) Maturity (D) All of these

266. Lamarckism connot explain:
(A) Webbed toes in aquatic birds
(B) Weak muscles in the son of a wrestler
(C) Long narrow and limbless body of snakes
(D) Heterophylly

267. Darwinism connot explain:
(A) Mimicry (B) Adaptation
(C) Struggle for existence (D) Tusks of elephants

268. Zinc tolerant plant of Zwar mines is:
(A) Agrostis (B) Astragalus
(C) Haplopappus (D) Impatiens

269. According to Neo-Darwinism, natural selection operates through:
(A) Fighting between organisms
(B) Killing the weaker organism
(C) Differential reproduction
(D) Variations

270. The genes divide the embryo into broad regions and are so named because their mutations result in gap in the embryo's segmentation pattern is called:
(A) Gap genes
(B) Pair rule genes
(C) Segment-polarity genes
(D) All of these

271. Inspite of great disadvantageous nature of sickle cell anaemia it has a high frequency in some parts of Africa. It is because of the reason that:
(A) Sickle cell anaemics are resistant to malaria
(B) In South Africa oxygen concentration is very high
(C) It is a case of genetic drift
(D) All the above

272. Survival of fittest used by:
(A) Charles Darwin (B) Herbert Spencer
(C) Jean Baptiste Lamarck (D) Hugo de Vries

273. For its entry into gene pool a mutation must occur in:
(A) Centriole (B) Plasma proteins
(C) Somatic RNA (D) Germplasm DNA

274. Some organism escape detection from enemies by resembling other organisms. The phenomenon is:
(A) Homology (B) Mimicry
(C) Artificial selection (D) Natural selection

275. Who believed that an organ in disuse will become vestigial?
(A) Lamarck (B) De Vries
(C) Mendel (D) Darwin

276. The relation between the fraction of repetitive DNA and the time required for reassociation of its denatured strands in a specific genome is:
(A) Direct (B) Inverse
(C) Hyperbolic (D) None of these

277. A very important factor in evolution of a new species is:
(A) Extensive inbreeding (B) Extensive out breeding
(C) Reproductive isolation (D) Immigration

278. "Continuity of germplasm" theory was given by:
(A) De Vries (B) Weismann
(C) Darwin (D) Lamarck

279. The σ factor found in many bacteria is best described as a:
(A) Subunit of DNA pol that allows for synthesis in both 5' to 3' and 3' to 5' directions
(B) Subunit of RNA pol responsible for the specificity of the initiation of transcription of RNA from DNA
(C) Subunit of the 3S ribosome to which mRNA binds
(D) Subunit of the 50S ribosome that catalyzes peptide bond synthesis

280. A cross involving parents differing only in one trait is:
(A) Haploid (B) Diploid
(C) Monohybrid (D) Dihybrid

281. In mendelian crosses the genes responsible for different characters separate through:
(A) Fusion (B) Aggregation
(C) Segregation (D) Distribution

282. Galpagos Islands were visited by:
(A) Wallace (B) Darwin
(C) Lamarck (D) De Vries

283. The major objection against the theory of Natural selection is that it:
(A) Stresses on interspecific competition
(B) Explains natural calamities to take heavy toll of lives
(C) Explains adaptation
(D) Stresses on slow small variations

284. Density gradient centrifugation of genomic DNA of a specific organism revealed a lower density peak in addition to the main peak. This is due to:
(A) At rich satellite DNA
(B) GC rich satellite DNA
(C) Equal proportion of AT and GC rich satellite DNA
(D) Pyrimidine-rich satelite DNA

285. Most modern breeds of domestic Dog have evolved through:
(A) Sexual selection (B) Artificial selection
(C) Natural selection (D) Isolation

286. Most important theory of general biology was proposed by:
(A) Mendel and Morgan (B) Beadle and Tatum
(C) Watson and Crick (D) Darwin and Wallace

287. Hugo de Vries worked on the plant:
(A) Garden pea/*Pisum sativum*
(B) Sweet pea/*Lathyrus odoratus*
(C) *Primula sinensis*
(D)EveningPrimrose/*Oenothera lamarckiana*

288. Which one of the following is false for prokaryotic genes?
(A) Promoter of a gene in a genome cannot be present within the upstream open reading frame
(B) The ribosome binding site can be either upstream or downstream of the TATA box (Pribnow box)
(C) All prokaryotic promoters must have a – 10 and a – 35 region in order to allow transcription
(D) Sigma factors bind to the – 10 and -35 region of promoters

289. T.R. Malthus is famous for his book on:
(A) Population (B) Mathematics
(C) Geography (D) Genetics

290. Weismann cut off tails of mice generation after generation but tails neither disappeared nor shortened showing that
(A) Darwin as correct
(B) Tail is an essential organ
(C) Mutation theory is wrong
(D) Lamarckism was wrong in inheritance of acquired characters

291. C_0 t analysis provides an estimate of the:
(A) G + C content of the DNA
(B) T_m of the DNA
(C) Complexity of the genome
(D) Hyperchromic shift of the genome

292. Genetic drift is change of:
(A) Gene frequency in same generation
(B) Appearance of recessive genes
(C) Gene frequency from one generation to next
(D) None of the above

293. Sum total of genes in a population is:
(A) Genotype (B) Karyotype
(C) Genetic constitution (D) Gene pool

294. Which of the following best describes the function of the sigma subunit in the RNA polymerase of *E. coli*?
(A) It is essential for elongation of the RNA transcript
(B) It is essential for the recognition of and binding to the promoter sequences
(C) It increases RNA polymerase binding to any DNA template
(D) It is required for transcription termination

295. Improved race of Pigeon developed due to:
(A) Environmental selection (B) Natural selection
(C) Artificial selection (D) Protective selection

296. Evolution will not occur if:
(A) There is no environmental influence
(B) Genes are without effect
(C) Somatic variations are not heritable
(D) There is no genetic variations in individuals of a populations

297. Major defect of Darwinism was:
(A) Nondescription of survival of fittest
(B) Nondescription of inheritance of fittest
(C) Nondescription of overproduction of young ones
(D) Nondescription of reason for variations

298. Which book written by Charles Darwin?
(A) Origin of species (B) Species Plantarum
(C) Genera Plantarum (D) De Naturien Familien

299. Which one does not favour Lamarckian concept of inheritance of acquired characters?
(A) Lack of pigment in cave dwellers
(B) Absence of limbs in snakes
(C) Presence of webbed toes in aquatic birds
(D) Melanisation of Peppered Moth in industrial areas

300. Which one provides correct sequence of events in origin of new species according to Darwinism?
1. Natural selection 2. Variations and their in heritance
3. Survival of fittest 4. Struggle for existence
(A) 1,2,3,4 (B) 2,3,1,4
(C) 3,4,1,2 (D) 4,2,3,1

301. Which of the following species of denatured DNA will denature most rapidly in solution, under appropriate conditions of ionic strength, pH and temperature?
(A) Human live nuclear DNA
(B) Vaccinia virus DNA
(C) *E. coli* DNA
(D) Yeast nuclear DNA

302. Darwin is associated with
(A) Natural selection (B) Mutation
(C) Inorganic evolution (D) All of these

303. In which respect, Darwin's theory is wrong
(A) High efficiency of reproduction
(B) Origin of species
(C) Survival of fittest
(D) Arrival of fittest

304. Specificity of transcription by RNA polymerases holoenzyme in *E. coli* is altered by changing the:
(A) Alpha subunit (B) Sigma subunit
(C) Beta subunit (D) Beta prime subunit

305. Closely related, morphologically similar, reproductively isolated sympatric population are designated as:
(A) Clones (B) Clines
(C) Demes (D) Sibling species

306. Frequency of gene in a population will increase if the gene is:
(A) Lethal (B) Dominant
(C) Recessive (D) Favourable selected

307. Objection to natural selection theory of Darwin is:
(A) Vestigial and over-specialised organs
(B) Inheritance of acquired characters
(C) Struggle for existence
(D) Continuity of germplasm

308. Presence of vestigial organs supports:
(A) Natural selection
(B) Germplasm theory
(C) Evolution but not Lamarck's theory
(D) Synthetic theory

309. Which of the following is true of the relationship between genomes and physical complexity of organisms?
(A) Among mammals, genome sizes are roughly proportional to the mass of the adult animals
(B) The number of chromosomes is usually proportional to the complexity of an animal's adult form
(C) Species with very similar body plans often have genomes of very different sizes
(D) Larger genomes typically have fewer repeated sequences

310. Hexaploid Wheat developed through:
(A) Hybridomas
(B) Chromosom doubling
(C) Hybridisation
(D) Hybridisation and chromosome doubling

311. Lederberg's replica experiment is connected with:
(A) Adaptations (B) Evolution
(C) Genetic (D) Artificial selection

312. Which are immortal?
(A) Germ cells (B) Somatic cells
(C) Pituitary cells (D) Glomerular cells

313. Xeroderma pigmentosum is an autosomal recessive genetic disease that causes a variety of phenotypic changes in skin cells exposed to sunlight. This disease involves a defect in which of the following DNA repair mechanisms?
(A) Base excision repair
(B) Double-strand break repair
(C) Mismatch repair
(D) Nucleotide excision repair

314. Hybrid breakdown is failure of
(A) Hybrid zygote to develop into offspring
(B) Hybrid adult to produce fertile offspring
(C) Fusion failure of ova and sperms of two species
(D) All of these

315. Chief agent of evolution is:
(A) Natural selection (B) Mutations
(C) Acquired characters (D) Sexual reproduction

316. Genetic drift is found in:
(A) Small population with or without mutated genes
(B) Large population with random mating
(C) Animal population
(D) Plant population

317. The Alu element found in the human genome is:
(A) A 7SL RNA homologue
(B) A good RNA polymerase III template
(C) About 300 bp long repeat
(D) All of these

318. Who is not related to organic evolution?
(A) T. R. Malthus (B) Erasmus Darwin
(C) Darlington (D) Charles Darwin

319. Modern theory of organic evolution stresses on:
(A) Mutations (B) Isolation
(C) Population (D) All of these

320. Presence of different types of beaks in finches of Galapagos islands adapted to different feeding habits provides evidence for:
(A) Intraspecific variations
(B) Natural selection
(C) Intraspecific competition
(D) Interspecific competition

321. The SOS repair mechanism is activated by:
(A) 5 - bromouracil (B) 2 - aminopurine
(C) Hydroxylamine (D) Thymine dimmers

322. Theory of pangenesis was given by:
(A) Darwin (B) Lamarck
(C) Hugo de Vries (D) Oparin

323. Phenomenon of industrial melanism demonstrates:
(A) Geographical isolation (B) Natural selection
(C) Reproductive isolation (D) Induced mutation

324. Genetic drift:
(A) Is random change in gene frequency
(B) Has nothing in common with inbreeding
(C) Is an orderly change in gene frequency
(D) Produces greatest fluctuations in large population

325. Darwin wrote "Origin of Species" in:
(A) 1530 (B) 1975 (C) 1859 (D) 1895

326. Which is related to reproductive isolation?
(A) Genetic isolation (B) Behavioural isolation
(C) Temporal isolation (D) All of these

327. Gene pool of a population tends to remain stable if the population is large, without large scale mutations, without migration and with
(A) Random mating
(B) Moderate environmental changes
(C) Natural selection
(D) Reduction in predators

328. Bacterial cells that can be lysogenised by lambda phase, possess in their genome:
(A) 80S ribosomal genes (B) attB sites
(C) attP sites (D) attL sites

329. The theory that environment can bring about changes in individuals was proposed by:
(A) Mendel (B) Darwin
(C) Lamarck (D) Lyell

330. Darwin theory states that:
(A) Characters are acquired through inheritance
(B) Species change morphologically with time
(C) Nature selects organisms which can adapt
(D) Evolution is due to effect of environment

331. The animal that explains industrial melanism is:
(A) Lizard (B) Pepped Moth
(C) Monkey (D) Frog

332. During DNA replication, which of the following is not true at the replication fork?
(A) One core DNA polymerase III replicates the leading strand and another molecule replicates the tagging strand
(B) One DNA polymerase III molecule replicates the leading strand and one DNA polymerase I molecule replicates the tagging strand
(C) The sliding lamp protein is behind the DNA polymerase at the primer binding site
(D) A ring of DNA B helicase is placed between two polymerases

333. Chromosomes found in the salivary gland of Drosophila are:
(A) Lampbrush (B) Polytene
(C) Supernumerary (D) B-chromosomes

334. A giant chromosome with a number of chromonemata is:
(A) Lampbrush chromosome
(B) Heterochromosome
(C) Supernumerary chromosome
(D) Polytene chromosome

335. Lampbrush chromosomes occur in:
(A) Salivary gland (B)Lymph gland
(C) Cancer cells (D) Oocytes

336. Chromosomes ends are called:
(A) Satellites (B) Telomeres
(C) Centromeres (D) Kinetochores

337. The genome of a typical bacterium contains about 5 x 10^6 base pairs and an be replicated in about 41 minutes. The human genome is 600 x larger (3 x 10^9 base pairs) and at the rate of a bacterium would require 300 hours (~12 days) to be replicated; yet the entire human genome can be replicated with several hours. How is this possible?
(A) Eukaryotic DNA is simpler to replicate than prokaryotic DNA
(B) Human DNA polymerases work much faster than those of prokaryotes
(C) The nucleosomes of eukaryotic DNA allow for faster DNA replication
(D) Human DNA contains more origins of replication than prokaryotic DNA

338. A Chromosome with sub-terminal centromere is:
(A) Acentric (B) Acrocentric
(C) Metacentric (D) Telocentric

339. A chromosome with centromere near the middle is called:
(A) Metacentric (B) Submetacentric
(C) Acrocentric (D) Telocentric

340. Puffs or balbiani ring in salivary gland chromosomes are sites of:
(A) DNA replication (B) DNA duplication
(C) RNA synthesis (D) Protein synthesis

341. Chromatin fibres condense to form chromosomes during:
(A) G1-Phase (B) S-Phase
(C) Metaphase (D) Prophase

342. The stage of counting of chromosomes is:
(A) Early prophase (B) Mid prophase
(C) Late prophase (D) Metaphase

343. The genes which are present on the same chromosomes are called:
(A) Identical genes (B) Allelomorphs
(C) Linked genes (D) Dominant genes

344. Drosophila has four homologous pairs of chromosomes. What is the number of linkage groups in this animal?
(A) 4 (B) 2 (C) 8 (D) Uncertain

345. Genes for colour blindness in man is located on:
(A) X-chromosome only
(B) Y-chromosome only
(C) Either X or Y chromosome
(D) Both X and Y chromosomes

346. A colour blind daughter may be born if the:
(A) Father is normal and mother is colour blind
(B) Father is colour blind and mother is normal
(C) Father is normal and mother is carrier
(D) Father is colour blind and mother is a carrier

347. A somatic cell in human male contains:
(A) No gene on sex chromosomes
(B) Only one sex linked gene for each character
(C) Two genes for every sex linked character
(D) Genes only on sex chromosomes

348. Which of the following disease is sex linked?
(A) Colour blindness
(B) Malingnancy
(C) Hepatitis
(D) Leukemia

349. The genes for the eye colour and size of the wing in Drosophila are located on the same chromosome. They can be separated by:
(A) Non-disjunction (B) Crossing over
(C) Hybridization (D) Not be separated at any stage

350. The sex linked characters are those:
(A) Which are related to sexual physiology
(B) The genes of which are present on the sex chromosomes
(C) Which appear in male or in female
(D) Which are controlled by sex hormone

351. The chromosomes other than sex chromosomes are called:
(A) Autosomes (B) Heterosomes
(C) Karyosomes (D) None of the above

352. The points where the crossing over occurs between the homologous chromosomes of a pair are called:
(A) Centromeres (B) Chromomeres
(C) Chromocenters (D) Chiasmata

353. If the crossing over had occurred at two-strand stage in Neurospora, the ascospores would be arranged in:
(A) 1-1 position (B) 2-2 position
(C) 4-4 position (D) None of these

354. Complete linkage is found in:
(A) Birds (B) Snakes
(C) Female Drosophila (D) Male Drosophila

355. A linkage group is defined as:
(A) All the linked genes of a chromosomal pair
(B) Different groups of genes present on different chromosomes
(C) All the genes located on the same chromosome
(D) None of these

356. A phenomenon which works opposite to the linkage is:
(A) Independent assortment (B) Crossing over
(C) Segregation (D) Mutation

357. When two genes are situated very close to one another at a chromosome:
(A) The percentage of crossing over between them is very high
(B) Hardly any cross-overs are produced
(C) No crossing over can take place
(D) Only double cross-overs can occur between them

358. Greater is the distance between the two genes on a chromosome:
(A) Greater is the linkage strength
(B) Lesser is the linkage strength
(C) Linkage strength remains unchanged
(D) There is no relationship between the two

359. Crossing over occurs at:
(A) Single-strand stage of chromosomes during prophase
(B) Two-strand stage during zygotene
(C) Four-strand stage of pachytene
(D) Metaphase II of meiosis

360. The term coupling and repulsion signify the same phenomenon which is termed:
(A) Synapsis (B) Disjunction
(C) Linkage (D) Non-disjunction

361. The linked characters would always inherit together till they are:
(A) Delinked due to segregation
(B) Masked by dominance
(C) Mutated
(D) Separated due to crossing over

362. During meiosis, the crossing over which occurs between the homologous chromosomes is of:
(A) No genetic significance
(B) Importance to bring about recombination of characters
(C) Importance because it produces nutrition
(D) Importance because it causes reorganization of cy toplasm

363. A white eyed (recessive) and long winged male Drosophila was crossed to a vestigial winged (recessive) red eyed female. The two characters are linked in this animal. When a female F_1 was test crossed, the F_2 generation produced:
(A) Mostly white eyed with long wings
(B) Mostly red eyed with vestigial wings
(C) Mostly white eyed with long wings and red eyed with vestigial wings
(D) All white eyed and vestigial winged

364. *E coli* DNA ligase uses NAD^+ as a o factor. During the actual ligation reaction (joining of the 5' phosphate and 3' hydroxyl to form a phosphodiester bond), the leaving group is:
(A) NAD^+ (B) NMN
(C) ATP (D) AMP

365. The genes of different traits located on different loci on the same chromosome are:
(A) Alleles (B) Linked
(C) Pleiomorphic (D) Mutated

366. In Four O'clock plant, normal leaves (A) and variegated leaves (B) occur in different plants. If (B) male is crossed with (A) female, the hybrid has normal leaves, but when (B) female is crossed with (A) male, the hybrid has variegated leaves. It is a case of:
(A) Mutation
(B) Cytoplasmic inheritance
(C) Complementary genes
(D) Supplementary genes

367. Crossing over occurs between:
(A) Two non-sister chromatids of a homologous pair of chromosomes
(B) Two chromatids of any chromosome
(C) Two chromatids of same chromosome
(D) All the foregoing

368. Choose the correct match from a, b, c, and d:

Group I	Group II
P. 5' aping of mRNA	1. 5-methyl guanosine
Q. Ribozyme	2. Polyadenylate transferase
R. Promoter	3. Spliceosome
S. Poly-A-tail	4. 7-methyl guanosine
	5. RNA polymerase
	6. Catalytic RNA
	7. Polyadenylate polymerase

(A) P -1, Q – 6, R – 5, S - 2
(B) P -4, Q – 6, R – 5, S - 7
(C) P -1, Q – 5, R – 3, S - 7
(D) P -4, Q – 3, R – 6, S - 2

369. The frequency of crossing over would be higher if:
(A) Two genes are located closely
(B) Two genes are far apart on a chromosome
(C) Two genes are not located on the same chromosome
(D) None of these

370. Mendelian recombinations are due to:
(A) Mutation
(B) Linkage
(C) Crossing over
(D) Independent assortment of characters

371. The blue green algae and bacteria contain:
(A) One linked group (B) Two linked groups
(C) Three linked groups (D) None of these

372. Which one of the following character in man is controlled by a recessive gene?
(A) Colour blindness
(B) Woolly hair
(C) Brachydactyly
(D) Curly hairs

373. What is one conclusion that can be drawn from the observation that the genetic code is nearly identical in all cells on earth?
(A) Life on earth evolved only one
(B) Only four types of nitrogenous bases occurs in DNA
(C) Higher forms of life evolved from lower forms
(D) Natural selection

374. Crossing over may result into:
(A) Gene mutation (B) Genomatic mutation
(C) Genetic recombination (D) None of these

375. Maize has 10 pairs of chromosomes. How many linkage group does it have?
(A) 20 (B) 10 (C) 5 (D) 40

376. In flowering plants meiosis occurs at the time of:
(A) Germination of seed (B) Formation of endosperm
(C) Formation of embryo (D) Formation of pollens

377. Which of the following statement is correct concerning a homeobox?
(A) It is part of the promoter in eukaryotic genes that code for proteins involved in segmentation
(B) It is a conserved proteins structure found in glycotytic pathway enzymes
(C) It is a conserved protein structure found in tRNA-binding proteins
(D) It is a conserved DNA sequence found in genes that code for proteins that regulate development

378. In transversion:
(A) Purine is replaced by another purine
(B) Purine is replaced by pyrimidine or vice versa
(C) Some nitrogen bases are eliminated from a gene
(D) Some nitrogen bases are added in a gene

379. Colour blindness is more likely to occur in males than females because:
(A) Genes for this character are located on the X-chromosomes
(B) The trait is dominant in males and recessive in females
(C) Some males suffer from deficiency of vitamin A
(D) The Y-chromosome of males have the genes for distinguishing colours

380. Suppose that out of 1000 offspring in the F2 generation of a dihybrd cross. 748 are tall and hairy, 6 are tall and smooth, 4 are short and hairy, 242 are short and smooth. The reason is that the genes controlling these characters are:
(A) Lethal genes (B) Complementary genes
(C) Linked genes (D) Epistatic genes

381. An example of complete linkage is:
(A) Human beings (B) Neurospora
(C) Female Drosophila (D) Male Drosophila

382. In vitro translation of a polycistronic RNA generally results in formation of unequal amounts of protein products form the individual cistrons. The most likely reason is:
(A) Different strength of each promoter
(B) Rapid RNA degradation
(C) Rapid degradation of proteins
(D) Different strength of each of the ribosome binding sites

383. Most of the mutations remain unexpressed because they are:
(A) Too small
(B) Very significant
(C) Very few
(D) Mostly recessive

384. Ame's test is performed to determine:
(A) Carcinogenicity of chemicals
(B) Pathogenicity of a bacterium
(C) Viability of a bacterial culture
(D) Sporulation of bacterium

385. An exchange of segments between the two non-homologous chromosomes is called:
(A) Polyploidy (B) Chromosomal aberration
(C) Translocation (D) Inversion

386. If a part of a chromosome gets separated and reattached in reverse position to the same chromosome, the mutation is called:
(A) Inversion (B) Transversion
(C) Translocation (D) Gene mutation

387. The ribosome is involved in all of the following, except:
(A) Peptide bond formation
(B) Aminoacylation of tRNA
(C) Binding of protein factors during elon gation
(D) Binding of aminoacyl tRNA to mRNA

388. A sudden or spontaneous change in the structure and action of a particular gene is called:
(A) Linkage (B) Variation
(C) Mutation (D) Allelomorph

389. In case of an abnormal crossing over, the two homologous chromosomes happen to exchange unequal length of segments. The phenomenon is called:
(A) Transition (B) Duplication
(C) Inversion (D) Gene mutation

390. The first step of 'proof-reading' during protein synthesis is carried out by:
(A) Ribosomes (B) m-RNA
(C) t-RNA (D) Aminoacyl t-RNA synthetase

391. Mutation did not result in:
(A) Death of organism
(B) Better progenies
(C) Change in genetic constitution of the cell
(D) Hybrid vigour

392. Gene mutations are those which involve:
(A) The change in nature and sequence of base triplets of DNA
(B) The change in genome
(C) The change in all the genes
(D) The disappearance of certain part of chromosome

393. Ame's test is based on a hypothesis that if a substance is a mutagen, it will increase the rate at which Salmonella-auxotrophs revert to:
(A) Lysine synthesizers
(B) Arginine synthesizers
(C) Histidine synthesizers
(D) Tryptophan synthesizers

394. Aneuploidy is the term applied for the:
(A) Gene mutation
(B) Chromosomal mutation
(C) Chromosomal mutations involving the addition or loss of one or more chromosome
(D) Chromosomal mutation involving the addition of one or more complete set of chromosomes

395. What term is applied to the gene mutation where a base is replaced by another base?
(A) Duplication (B) Aneuploidy
(C) Euploidy (D) Substitution

396. A monosomic organism can best be represented as:
(A) 2n +1 (B) 2n +2
(C) n + 1 (D) 2n -1

397. Insulin is secreted as a dimmer consisting of two covalently linked polypeptides. These polypeptides are derived:
(A) By alternative splicing of the mRNA from a single gene
(B) By proteolytic processing of a single protein
(C) From two genes
(D) None of these

398. A flower was fertilised by pollen grains having the constitution (n-1). If the ovules are normal, which of the following is applicable to the seeds so produced?
(A) Allopolyploidy (B) Monosomy
(C) Nullisomy (D) Both (A) and (B)

399. Nullisomy is the term used for the condition when an organism has:
(A) An additional chromosome
(B) One chromosome less than normal
(C) A complete set of chromosomes except one homologous pair
(D) None of these

400. In genetics, suppression of a mutation refers to:
(A) Restoration of the original phenotype due to a second mutation
(B) Restoration of the original DNA sequence by mutation
(C) Prevention of expression of the mutant gene by metabolic regulation
(D) Appearance of the recessive phenotype in a heterozygous diploid

401. A classical example of allopolyploidy is:
(A) Brassica (B) Raphano-brassica
(C) Raphanus (D) All of these

402. The mutation which returns to the original state is called?
(A) Reversible mutation (B) Lethal mutation
(C) Backward mutation (D) Abnormal mutation

403. Mutations are:
(A) Always useful (B) Seldom useful
(C) Always harmful (D) May be useful or harmful

404. DNA polymerase contains a lysine residue that important for binding to DNA. Mutations were found that converted this lysine to either glutamate, glycine, valine or arginine. Which mutations would be practices to be the most and least harmful to the ability of the enzyme to bind DNA?

Most	Least
A. Glycine	Arginine
B. Arginine	Glycine
C. Glutamate	Valine
D. Glutamate	Arginine

405. Euploidy is the term applied for those mutations in which:
(A) There is multiplication of chromosomal number
(B) Involves addition of complete set of chromosomes
(C) There is loss of a few chromosomes
(D) None of these

406. Which is not applicable to mutations?
(A) They are inheritable genetic changes
(B) Many of them remain undetected
(C) They can be induced artificially
(D) They are possible only in diploid cells

407. Autopolyploids are those which have:
(A) More than one set of chromosomes
(B) Only one additional set of chromosomes
(C) The same basic set of chromosomes
(D) All of these

408. The drug chloramphenicol blocks:
(A) Cell-wall synthesis
(B) Translation-termination release factors
(C) Polypeptide chain elongation
(D) Polypeptide chain initiation

409. The frequency of mutations in nature is one in:
(A) 1×10^5 (B) 1×10^{12}
(C) 1×10^{-5} (D) 1×10^{-12}

410. When chromosome sets are present in multiple of 'n' the condition is termed:
(A) Diploidy (B) Haploidy
(C) Euploidy (D) Aneuploidy

411. Which of the following is not an aneuploidy?
(A) Monoploid (B) 2n -1
(C) Trisomic (D) 2n +2

412. The reason of genetic variability in sexually reproducing organisms is:
(A) Temperature
(B) Mutations
(C) Certain chemicals
(D) Evolution

413. Most of the high yielding varieties of wheat are:
(A) Imported
(B) Monosomic
(C) Trisomic
(D) Mutants

414. The smallest segment of a gene which can undergo mutation is called:
(A) Muton
(B) Recon
(C) Cistron
(D) Interferon

415. Which one of the following statements describes the ubiquitin-mediated degradation of proteins in the cytosol?
(A) One molecule of ubiquitin binds to the protein to be degraded
(B) The process is catalyzed by a single enzyme
(C) The process depends on edenosine triphosphate (ATP)
(D) The N-terminal residue of ubiquitin becomes covalently attached to the protein to be degraded

416. Mutation theory of evolution by Hugo de Vries was based on his studies on:
(A) *Pisum sativum* (B) *Oenothera lamarkiana*
(C) *Acetabularia* (D) *Lathyrus odoratus*

417. Most of the mutations are:
(A) Forward (B) Backward
(C) Reversible (D) Lethal

418. Mutagenes are:
(A) The genes which can mutate
(B) The organism which show mutation
(C) Genes that regulate mutations
(D) The agents which cause mutations

419. The mutations which occur in germ cells can be detected phenotypically in:
(A) Parents (B) Next generation
(C) Third generation (D) None of these

420. The frequency of mutation in a species can be increased by the use of:
(A) X-rays (B) UV - rays
(C) Nitrous acid (D) All of these

421. A pleiotropic mutation is the one which:
(A) Affects one character at a time
(B) Affects many characters at a time
(C) Is always lethal
(D) Easily gets preserved in natural conditions

422. Down's syndrome is a typical case of:
(A) Trisomy (B) Nullisomy
(C) Gene mutation (D) Monosomy

423. Which of the following statements can be used to support the endosymbiotic theory for the origin of mitochondria?
P. Mitochondrial DNA has a circular genome
Q. Mitochondria are about the same size as bacteria
R. Mitochondria; ribosomes are 70S
S. Mitochondrial ribisomes are inactivated by the antibiotic chloramphenicol
(A) P and Q (B) Q and R
(C) R and S (D) P, Q, R and S

424. Datura is a classical example of study of:
(A) Monosomics (B) Trisomics
(C) Triploids (D) Nullisomics

425. Which one is a hexaploid species of wheat?
(A) *Triticum monococcum* (B) *T. aestivum*
(C) *T. durum* (D) *T. turgidum*

426. When chromosome sets are present in multiple of 'n', the condition is called as:
(A) Diploidy (B) Aneuploidy
(C) Haploidy (D) Euploidy

427. Persons carrying the dominant gene(s) an taste a particular chemical and are called tastes. The others are nontasters. In a large random-mating human population at the Hardy-Weinberg equilibrium, there are 36% tasters and 64% non-tasters. The frequency of hetrozygotes in the population is:
(A) 6% (B) 8%
(C) 32% (D) 48%

428. X-rays cause mutations by:
(A) Breaking spindle
(B) Rupturing of nuclear membrane
(C) Changing the chromosome morphology
(D) Inducing karyokinesis

429. Colchicine interferes in:
(A) Spindle organization
(B) DNA replication
(C) Chromosome condensation
(D) Polyploidy

430. Gene mutation occur in:
(A) Chromosome (B) DNA
(C) Nucleus (D) RNA

431. It is thought that the Galapagos Islands were originally colonized by one species of finch. Today these Islands contain many species of finches that occupy diverse niches. This has most likely resulted from which of the following?
P. Biological magnification
Q. Convergent evolution
R. Adaptive radiation
(A) P and R
(B) Q and P
(C) R only
(D) Q and R

432. Haploids are used for researches because:
(A) They contain only one chromosome
(B) They contain two sets of chromosomes
(C) They contain three sets of chromosomes
(D) They contain only one set of chromosomes

433. Cytoplasmic inheritance differs from nuclear inheritance in the absence of:
(A) Similarity of reciprocal crosses
(B) Biparental contribution
(C) Effect on backcrossing
(D) All of these

434. Which of the following is characteristics of allelopatric speciation?
(A) Large populations
(B) Isolation through adaptation of alleles
(C) Geographic isolation
(D) Asexually reproducing populations

435. Cytoplasmic inheritance is also called maternal inheritance because it is:
(A) Without sexual reproduction
(B) Only female parent takes part in multiplication
(C) Most of the cytoplasm of the zygote is provided by ovum
(D) All of these

436. Number of chromatids present in a metaphasic chromosome are:
(A) 4 (B) 2 (C) 3 (D) 1

437. Number of chromatids per chromosome complex at metaphase is:
(A) One in mitosis and two in meiosis
(B) Two in mitosis and one in meiosis
(C) Two both in mitosis and meiosis
(D) Two in mitosis and four in meiosis

438. Choose the correct statement (s):
P. Similar traits resulting from similar selection pressures acting on similar gene pools is parallel evolution
Q. The evolution in one population in response to the evolution of another population, and vice versa is called co-evaluation
(A) P (B) Q
(C) P and Q (D) None of these

439. Chromosomes are formed by:
(A) DNA, RNA and proteins (B) DNA only
(C) RNA only (D) Proteins

440. Which of the following does not explain the term survival of the fittest?
(A) Fittest animals leave higher number of progeny than those which perish
(B) Fittest group of animals out populates its competitors
(C) Fittest animals are best predators
(D) Fittest group is rich in genetic variations

441. Centromere is connected with:
(A) Spindle formation
(B) Poleward movement of chromosomes
(C) Chromosome segregation
(D) DNA replication

442. Centromere is also called:
(A) Metamere (B) Satellite
(C) Chromocentre (D) Primary constriction

443. A chromosome with terminal centromere is:
(A) Telocentric (B) Metacentric
(C) Polycentric (D) Acrocentric

444. The terminal end of chromosome is:
(A) Chromomere (B) Centromere
(C) Telomere (D) Metamere

445. Which of the following statements is not consistent with the principle of totipotency?
(A) Cell specilization is based on position
(B) Plant cell can regenerate entire plants by mitosis
(C) Plant cell can differetiate into any cell type
(D) Cell specialization is based on gene content

446. An octamer of 4 histones complexed with DNA forms:
(A) Centromere (B) Nucleosome
(C) Mesosome (D) Endosome

447. A nucleosome is made of:
(A) Histone wrapped over octameric core of nucleic acid
(B) Histone
(C) DNA wrapped over octameric core of histone
(D) DNA

448. Choose the correct sequence of evolutionary events in one form of allopatric speciation, using the codes given below:
P. Geographical isolation
Q. Ecological isolation
R. Increased pre-mating reproductive isolation
S. Increased genetic divergence
T. Selection completed
(A) R – S - Q - T (B) P – S – R - T
(C) R – Q – S – T (D) P – Q – R - T

449. Polytene chromosomes were seen by:
(A) Ruckert (B) Balbiani
(C) Wilson (D) Heitz

450. Haploids are able to express both recessive and dominant alleles/mutations because there are:
(A) Many alleles for each gene
(B) Two alleles for each gene
(C) Only one allele for each gene in the individual
(D) Only one allele in a gene

451. Which of the following best describes pseudogenes?
(A) Two functional genes within an organism that arose from the duplication of one gene
(B) Genes that share the same ancestral sequence, but are found in different organisms
(C) Sequences of DNA that are very similar to functional genes, but do not produce a functional product
(D) Sequences of DNA that are very similar to a native genes, but do produce a functional product

452. The phenomenon of exchange of segments between paternal and maternal chromosome is called:
(A) Segregation (B) Recombination
(C) Crossing over (D) Linkage

453. Which one can have sex chromosomes?
(A) Unisexual flower (B) Bisexual flower
(C) Unisexual plant (D) Hermaphrodite plant

454. Crossing over produces:
(A) Recombination of linked genes
(B) Synapsis of linked genes
(C) Expression of recessive genes
(D) Linkage of dominant genes

455. The hypothesis that evolution occurs in spurts, with great amounts of evolutionary change followed by periods of stasis , is
(A) Puntuated equilibrium
(B) Allopatric speciation
(C) Gradualism
(D) Hardy – Weinberg equilibrium

456. Out of 8 ascospores formed in Neurospora, the arrangement is 2a : 4A : 2a showing:
(A) No crossing over
(B) Some meiosis
(C) Second generation division
(D) First generation division

457. Chromosomes connected with sex determination are:
(A) Autosomes (B) Heterosomes
(C) Oxysomes (D) B-chromosomes

458. An autosome is:
(A) Chromosome half
(B) Sex chromosome
(C) Chromosome other than sex
(D) None of these

459. Linkage in plants was first shown in:
(A) *Zea mays* (B) *Lathyrus odoratus*
(C) *Oenothera lamarkiana* (D) *Pisum sativum*

460. A DNA vaccine is:
(A) A DNA molecule that is recognized by an antibody
(B) A vaccine that works by stimulating the immune system to recognize pathogen DNA sequences
(C) A vaccine that is administered as DNA; the DNA is then expressed to produce a protein, which stimu lates an immune response
(D) All of these

461. Association of parental combination of characters in offspring in excess of dihybrid cross is due to:
(A) Polygeny (B) Linkage
(C) Pseudoalleles (D) Codominance

462. In a dihybrid cross, two recessive genes showed 10% recombinants. The distance between two genes are:
(A) 10 map units (B) 20 map units
(C) 30 map units (D) 40 map units

463. An individual with cd genes was crossed with wild type + +. On test crossing F_1, the progeny was + c 105, + d 115, cd 880 and ++ 900. Distance between cd genes is:
(A) 11 map units (B) 22 map units
(C) 44 map units (D) 88 map units

464. Which of the following may be part of the reason why thymine replaced uracil in DNA?
(A) Cytosine spontaneously deaminates to uracil
(B) Thymine is unstable
(C) Thymine synthetic pathway is easier
(D) Thymine has a stronger binding affinity for adenine than does uracil

465. Plant in which chromosomal basis of sex determination was discovered first is:
(A) Rumex (B) Melandium
(C) Coccinia (D) Sphaerocarpus

466. Sex linked characters are generally:
(A) Lethal (B) Recessive
(C) Dominant (D) Not inherited

467. Trisomy has chromosome complement of:
(A) 2n - 1 (B) 2n-1-1
(C) 2n+1+1 (D) 2n+1

468. The first successful gene therapy was done to which of the following enzyme:
(A) Homogentistic acid oxidase
(B) Adenosine decarboxylase
(C) Adenosine deaminase
(D) Adenyl cyclase

469. Mutations are commonly:
(A) Incomplete (B) Codominant
(C) Recessive (D) Dominant

470. Haploid chromosome number is 12. What is the number in a monosomic?
(A) 23 (B) 24 (C) 25 (D) 26

471. Gene mutation is due to:
(A) Change in base sequence
(B) Change in sequence of cistrons in DNA
(C) Linkage and crossing over
(D) Random segregation

472. Mitochondrial DNA is advantageous for evolutionary studies because:
(A) It is inherited only through the female parent and thus evolves in a way that allows trees of relation ship to be easily constructed
(B) It is inserted into the X chromosome
(C) It first appeared in humans and is not found in other animals
(D) It evolves more slowly than the genes in the nucleus

473. Name the mutagen:
(A) SO_2 (B) CO_2
(C) NO (D) HNO_2

474. Transversion is:
(A) Insertion of nitrogen base
(B) Replacement of a purine with a purine
(C) Replacement of a purine with pyrimidine
(D) Replacement of a pyrimidine with pyrimidine

475. Which of the following is not useful for phylogenetic analysis of human populations
(A) Ribosomal RNA genes (B) Microsatellites
(C) Mitochondrial DNA (D) Multiallelic genes

476. Recessive mutations are expressed in:
(A) Homozygous condition (B) Heterozygous condition
(C) Next generation (D) Same generation

477. Monosomic are:
(A) n- 1 - 1 (B) 2n+1
(C) 2n -2 - 1 (D) 2n -1

478. The plant on which de Vries worked in preparation of mutation theory was:
(A) Pansy (B) Evening Primerose
(C) Garden pea (D) Sweet pea

479. Sudden inheritable genetic change is:
(A) Mutation
(B) Inheritance of acquired trait
(C) Independent assortment
(D) Natural selection

480. Mutation is a change that is:
(A) Never inherited
(B) Inherited only in F2 generation
(C) Inherited
(D) Responsible for plant growth

481. An autosomal recessive disorder has a population frequency of 1:40000. The carrier frequency is therefore:
(A) 1/50 (B) 1/75 (C) 1/100 (D) 1/150

482. Colchicine interferes with:
(A) Chromosome replication
(B) Organisation of spindle
(C) Chromosome condensation
(D) Incorporation of nitrogen bases

483. Exact multiple of haploid number is:
(A) Euploid (B) Hyperploid
(C) Heteroploid (D) Aneuploid

484. Peptide nucleic acid (PNA):
(A) A polynucleotide analog in which the sugar phosphate backbone is replaced by amide bonds
(B) The use of molecular data to understand the genetic basis to variable phenotypes
(C) The use of molecular data to infer the evolutionary relationship between genomes
(D) The application of rigorous mathematical methods to analysis of variable characters

485. In pea, seed coat colour changes from grey to white. This is an example of:
(A) Induced mutation (B) Spontaneous mutation
(C) Pleiotropic mutation (D) Transformation

486. Polyploidy means occurrence of:
(A) Haploid sets of chromosomes
(B) Diploid sets of chromosomes
(C) More than diploid sets of chromosomes
(D) All of these

487. If the frequency of males affected with an x-linked recessive condition in a human population is 0.10 (one in ten), what will be the expected frequency of affected females?
(A) 0.00001 (B) 0.0001
(C) 0.001 (D) 0.01

488. Haploid number in 10. What shall be tetrasomic number?
(A) 22 (B) 25 (C) 30 (D) 40

489. A polyploidy with genomes derived from same original species is:
(A) Amphidiploid (B) Autopolyploid
(C) Autoallopolyploid (D) Allopolyploid

490. A polyploidy derived from F1 hybrid between two species is:
(A) Autopolyploid (B) Autodiploid
(C) Amphidiploid (D) Hexaploid

491. The likelihood of an individual in a population carrying two specific alleles of a human DNA marker, each of which has a frequency of 0.2, will be:
(A) 0.32 (B) 0.16 (C) 0.10 (D) 0.08

492. An auxotroph is:
(A) Plant capable of synthesizing own carbohydrates
(B) Plant showing quick bending response to sunlight
(C) A mutant having lost the ability to synthesize one or more nutrients
(D) An organism dependent on another for nutritional requirements

493. Ultimate source of organic variations is:
(A) Mutations (B) Natural selection
(C) Hormonal activity (D) Isolation

494. A trisomic individual possesses extra chromosomes:
(A) 1 (B) 2 (C) 3 (D) 4

495. If haploid chromosome number is 10. The monosomic number shall be:
(A) 9 (B) 12 (C) 18 (D) 19

496. Which of the following mutations are not hereditary:
(A) Gametic (B) Somatic
(C) Genetic (D) Germinal

497. Aneuploidy is zygotic chromosome number:
(A) Thrice of gametic number
(B) Twice of gametic number
(C) Quadruple of gametic number
(D) Abnormal

498. Both extra-nuclear as well as nuclear materials are involved in transmission of hereditary information as:
(A) Mitochondria and plastids are having DNA
(B) Both cytoplasm and nuclear material are equally involved in hereditary
(C) Each type of organisms have particular mitochondria or plastids
(D) Some cells are procaryotic

499. When present in tissue culture medium, gibberellins:
(A) Helps to break dormancy of buds and bulbs
(B) Promotes dormancy development in buds and bulbs
(C) Is regarded as plant growth inhibitor
(D) Prevent normal recognition of auxin molecule

500. An example of covergent evolution is:
(A) Australian marsupials and placental mammals
(B) The flippers in fish, penguins and dolphins
(C) The wings in birds, bats and insects
(D) All of these

501. Cytoplasmic male sterility is passed down:
(A) Biparentally (B) Paternally
(C) Maternally (D) None of these

502. Cytosterility in the pollen of maize is due to:
(A) Extra chromosomal inheritance
(B) Complementary genes
(C) Nuclear inheritance
(D) Hypostasis

503. Cytoplasmic male sterility in maize is manifestation of interaction between:
(A) Chloroplast and nuclear genes
(B) Mitochondrial and nuclear genes
(C) Chloroplast and mitochondrial genes
(D) Cytoplasmic factors and male sterile genes

504. Which is not an example of cytoplasmic inheritance?
(A) Female sterility in maize
(B) Plastic inheritance
(C) Kappa particle inheritance
(D) Stigma particle inheritance

505. When a certain character is inherited only through female parent, it probably represents:
(A) Multiple plastid inheritance
(B) Cytoplasmic inheritance
(C) Incomplete dominance
(D) Mendelian nuclear inheritance

506. The two eukaryotic organelles responsible for cytoplasmic inheritance are:
(A) Lysosomes and mitochondria
(B) Chloroplasts and lysosomes
(C) Mitochondria and chloroplasts
(D) Mitochondria and Golgi complex

507. Which crop variety is not due to induced mutations?
(A) Reimei of Rice (B) Prabhat of arhar
(C) Sharbati Sonora of wheat (D) Aruna of castor

508. Haploids are preferred over diploids for mutation studies because in haploids:
(A) Recessive mutations express immediately
(B) Dominant mutations express immediately
(C) Mutations are readily induced
(D) Tissue culture is easy

509. Vir genes are important for gene transfer to plants because of which of the following reason?
(A) They act on the origin of replication of the Ti plasmid
(B) They prevent oncogenes from working
(C) They are responsible for the mobilization of T-DNA
(D) They enable Agrobacterium to penetrate through intact plant surfaces

510. The cranial capacity of which one of the following prehistoric humans was almost the same as that of the modern man?
(A) Neanderthal man (B) Peking man
(C) Java ape man (D) Australopithecus man

511. Which of the following situations might create a population bottleneck?
(A) A small number of animals migrating from one geo graphic location to another, leaving behind the rest of the population
(B) Overhunting or overfishing of a particular species over a short period of time
(C) Natural disasters that leave behind a fraction of a population
(D) All of these statements describe situations that may lead to a population bottleneck

512. Orthologos
P. Are always located on the same chromosome
Q. Have similar amino acid sequence
R. Have identical promoters
S. Have similar functions
Which of the above statements are correct?
(A) P (B) R
(C) Q and R (D) Q and S

513. Alleles of different genes found on same chromosome may separate due to:
(A) Epistasis (B) Crossing over
(C) Continuous variations (D) Pleiotropy

514. Which of the following is suitable for experiment on linkage?
(A) aaBB x aaBB (B) AABB x aabb
(C) AaBb x AaBb (D) AAbb x AaBB

515. Consider the following
1. Tyrannosaurus (dinosaur)
2. Lobe-finned fishes
3. Great apes
4. Eohippus (Horse)
5. Trilobites
Which of the following represents the proper sequence, from the earliest in the fossil record to the latest, of the animals listed above?
(A) 1, 2, 5, 4, 3 (B) 2, 5, 1, 3, 4
(C) 2, 5, 1, 4, 3 (D) 5, 2, 1, 4, 3

516. Eucaryotic chromosome is made of:
(A) DNA (B) DNA + protein
(C) DNA + lipids (D) RNA + lipids

517. A strong mutagene is:
(A) Cold (B) Heat
(C) Ice (D) X-ray

518. The number of base substitutions possible in amino acid codons is:
(A) 461 (B) 494 (C) 525 (D) 549

519. Out of A-T, G-C pairing, bases of DNA may exist in alternate valency state owing to arrangement called:
(A) Anologue substitution
(B) Tautomerisational mutation
(C) Frameshift mutation
(D) Point mutation

520. Choose the false statements:
P. No circular RNA viral genome are known to exist (exception is Hepatitis-delta virus)
Q. No DNA viruses are known to have segmented genome
R. Although oxygenic photosynthesis evolved long aerobic respiration, the endosymbiotic theory has the evolution of mitochondria receding the evolution of chloroplasts
S. According to Carl Woese, eubacteria are likely to have evolved from an archaebacterial ancestor
(A) Q (B) S
(C) P and Q (D) Q and S

521. Linked genes separate due to:
(A) Mutation (B) Recombination
(C) Crossing over (D) None of these

522. Two genes situated very close on the chromosome show:
(A) High crossing over
(B) Hardly any crossing over is detected
(C) No crossing over
(D) Only double cross-over can occur

523. Genes located on Y-chromosome are:
(A) Autosomal genes (B) Mutant genes
(C) Sex linked genes (D) Holandric genes

524. Point at which polytene chromosomes appear attached is:
(A) Centromere (B) Chromocentre
(C) Chromomere (D) Centriole

525. Trisomy of 21st chromosones results in:
(A) Down's syndrome (B) Sickle cell anaemia
(C) Turner's syndrome (D) Klinefelter's syndrome

526. Match list – I with list – II and select the correct answer using the codes given below the lists:

List – I	List – II
P. Annual rhythm	1. The ability to time and repeat the functions at interval of 15 days
Q. Lunar rhythm	2. The ability to time and repeat the functions at interval of 24 hours
R. Ciradian rhythm	3. The ability to time and repeat the functions every

	year
S. Tidal rhythm	4. The ability to time and repeat the functions based on tidalphase

(A) P -1, Q – 2, R – 3, S - 4
(B) P -2, Q – 1, R – 3, S - 4
(C) P -3, Q – 2, R – 4, S - 1
(D) P -3, Q – 1, R – 2, S - 4

527. Which histone is not component of *nu* particle?
(A) H2 (B) H1 (C) H3 (D) H4

528. Complete haploid set of chromosomes of a species is:
(A) Genome (B) Allele
(C) Genetic code (D) Genotype

529. Circannual rhythm in an organism is:
P. Regulated by temperature
Q. Regulated by photoperiod
R. An innate character imprinted upon the organism by the environment
S. An acquired character developed by the animals in response to the frequent changes in the environment
Which of the above statements are correct?
(A) P and Q (B) Q and R
(C) Q and S (D) P and S

530. Which one is not related to chromosomal abnormality?
(A) Klinefelter's syndrome (B) Euploidy
(C) Aneuploidy (D) AIDS

531. Point mutation is:
(A) Loss of gene
(B) Change in a base of gene
(C) Addition of a gene
(D) Deletion of a segment of gene

532. Colchicine brings about:
(A) Polyploidy (B) Cell division
(C) Cell elongation (D) Cell differentiation

533. A man made allopolyploid is:
(A) Water melon (B) Gossypium
(C) Triticale (D) Triticum

534. Shoot organogenesis by culture results into:
(A) A bipolar structure that has no vascular connection with the explant
(B) A monopolar structure that has a strong connection with the preexisting vascular tissue of the explant
(C) A monopolar structure that has no vascular connection with the explant
(D) A bipolar structure that has a strong connection with the pre-existing vascular tissue of the explant

535. Genome comparisons have suggested that mouse DNA has mutated about twice as fast as human DNA. What is a possible explanation for his discrepancy?
(A) Mice are much smaller than humans
(B) Mice live in much less sanitary conditions than humans and are therefore exposed to wider range of mutation-causing substances
(C) Mice have a smaller genome size
(D) Mice have a much shorter generation time

536. Cis-trans expression of genes is an example of:
(A) Mutation
(B) Intergenic crossing over
(C) Intragenic crossing over
(D) Cytoplasmic inheritance

537. Match list – I with list – II and select the correct answer using the codes given below the lists:

List – I	List – II
P. Temperature affects size of animals	1. Gloger's rule
Q. Temperature affects colouration of animals	2. Jordan's rule
R. Temperature affects size of extremities of certain animals	3. Bergman's of rule
S. Temperature affects number of vertebrae in certain fish	4. Allen's rule

(A) P - 3, Q – 1, R – 4, S - 2 (B) P -3, Q – 4, R – 1, S - 2
(C) P -2, Q – 4, R – 1, S - 3 (D) P -2, Q – 1, R – 4, S - 3

538. The substance which causes a definite change in genes is called:
(A) Mutagene (B) Toxin
(C) Cytotoxin (D) Alkaloid

539. Recessive mutagen is not expressed in:
(A) Homozygous male (B) Heterozygous male
(C) Heterozygous female (D) Homozygous female

540. Which is an example of cytoplasmic inheritance?
(A) Eye color in Drosophila
(B) Flower colour in pea
(C) Height in pea
(D) Sterile pollen

541. Inclusive fitness:
P. Is the sum of the direct and indirect fitness effects of an individual's behaviour
Q. Provides an evolutionary explanation for altruistic behaviour among animal communication
(A) P (B) Q
(C) P and Q (D) None of these

542. In a chromosome the protein content is:
(A) Nil (B) Trace
(C) Half of DNA (D) Same as DNA

543. Number of linkage in a polynucleotide would be:
(A) Same as number of nucleotide
(B) Twice the number of nucleotides
(C) One less than the number of nucleotides
(D) One

544. What causes mutations?
(A) Colchicine (B) Cosmic rays
(C) Gamma rays (D) Crossing over

545. Match Group – I with Group – II and select the correct answer using the codes given below the lists:

Group – I	Group – II
P. Insect resistant cotton	1. Bt
Q. Golden rice	2. Round up
R. 'Flavr – Savr' tomato	3. 2, 4-D
S. Herbicide tolerant soyabean	4. Carotenoids
	5. Ferritin
	6. A -deaminase

(A) P -2, Q – 5, R – 1, S - 3
(B) P -1, Q – 4, R – 6, S - 2
(C) P -1, Q – 4, R – 6, S - 3
(D) P -2, Q – 4, R – 6, S – 1

546. Number of chromosomes can increase or decrease due to:
(A) Genetic repetition (B) Mutation
(C) Nondisjunction (D) All of these

547. Gene is formed of:
(A) Polynucleotide (B) Histone
(C) Hydrocarbon (D) Lipoprotein

548. Which one is a triploid?
(A) Guava (B) Apple
(C) Orange (D) Banana

549. RNA is supposed to be the first genetic material in the evolutionary process. It performs both catalytic and genetic functions. Which of the following is/are correct about RNA as catalyst (ribozyme)?
P. Display self splicing
Q. Catalyses peptide bond formation
R. RNase P is an example of ribozyme
S. Catalyzes formation of DNA
(A) P and Q (B) Q, R and S
(C) P, Q and R (D) P, Q, R and S

550. Mutation are responsible for:
(A) Extinction of organisms
(B) Variations in population
(C) Increase in population
(D) Decrease in population

551. Which of the following statements are true on DNA delivery methods during plant transformations?
P. Single stranded nicks are made in T-DNA border repeat by the VirD1, VirD2 and VirD3 protein complex
Q. VirA Gene products from the export apperaratus on the membrane for the transfer of T-DNA
R. Gold/Tungsten particles are used as microprojectiles in biolistic method
S Acceleration of DNA-coated microprojectiles is carried out with compressed CO_2
(A) P and S (B) R and S
(C) P and R (D) Q and S

552. Hexaploid or modern wheat developed through:
(A) Chromosome doubling
(B) Hybridomas
(C) Hybridisation
(D) Hybridisation and chromosome doubling

553. A fruitfly heterozygous for sex linked genes is mated with normal female fruitfly. Male specific chromosome will enter the egg cells in proportion of:
(A) 1 : 1 (B) 2 : 1 (C) 3 : 1 (D) 4 : 1

554. Genetic identity of human male is known by:
(A) Reverse mutation (B) Frameshift mutation
(C) Autosomes (D) Sex chromosomes

555. A mutation at one base of first codon of a gene forms a nonfunctional protein. It is:
(A) Nonsense mutation (B) Nucleolus
(C) Cell organelles (D) Mis-sense mutation

556. Which of the following comparison between RFLPs, RAPDs and AFLPa is/are correct?

	RFLPa	RAPDs	AFLPs
P. PCR based	No	Yes	Yes
Q. Polymorphism	Moderate	High	High
R. Codominant	Yes	No	No
S. Reproducibility	High	Medium	High

(A) P (B) Q
(C) P and S (D) P, Q, R and S

557. Process of altering code in DNA where message is reversed is:
(A) Deletion (B) Insertion
(C) Inversion (D) Movement

558. A change in chromosomal number is called:
(A) Polyploidy (B) Aneuploidy
(C) Chromosomal mutation (D) Somatic mutation

559. Which of the following statements are true about glyphosate tolerant transgenic plants?
P. Transgenic plants detoxify glyphosate
Q. Transgenic plants produce an altered enzyme that is not affected by glyphosate
R. Transgenic plants sequester glyphosate in vacuoles
S. Transgenic plants overcome the inhibition amino acid biosynthesis
(A) P and Q (B) P and R
(C) Q and S (D) R and S

560. Numerical change in chromosome number which is not the exact multiple of haploid genome is:
(A) Haploid (B) Autopolyploid
(C) Allopolyploid (D) Aneuploid

561. *Triticum aestivum* (Bread wheat) is:
(A) Tetraploid (B) Hexaploid
(C) Diploid (D) Triploid

562. Lampbrush chromosomes are also called diplotene chromosomes because they:
(A) Show chiasmata
(B) Resemble diplotene chromosomes
(C) Are the permanent diplotene stage
(D) All of these

563. Phytomediation processes includes:
P. Phytoextration: Uptake and concentration of sub stances from the environment into the plant biomass
Q. Phytostabilization: Reducing the mobility of substances in the environment
R. Phytotransformation: Chemical modification of environment substances as a direct result of plant metabolism
S. Phytostimulation: Enhancement of soil microbial activity for the degradation of contaminants
(A) P and Q (B) P, Q and R
(C) P, Q and S (D) P, Q, R and S

564. Mutation which are not inherited are:
(A) Somatic (B) Gametic
(C) Zygotic (D) Genetic

565. Which one is de novo mutation?
(A) TT x TT → TT (B) tt x tt →Tt
(C) Tt x tt → Tt (D) Tt x Tt → TT, Tt, tt

566. Lampbrush chromosomes were described by
(A) Balbiani (B) Benzer
(C) Dupraw (D) Ruckert

567. Basic structural unit of eucaryotic chromosome is:
(A) Chromatin fibre (B) DNA
(C) Chromonema (D) Nucleosome

568. Which of the following are commonly used as reporter genes?
P. NPT gene Q. Luciferase gene
R. CFTR gene S. GFP gene
(A) Q and S (B) P and R
(C) R and S (D) P and Q

569. A mutation results in change in:
(A) Sequence of amino acids in a protein
(B) tRNA sequence of nucleotides
(C) rRNA of ribosomes
(D) None of these

570. The two chromatids of a chromosome are attached in the region of:
(A) Telomere (B) Centromere
(C) Chromomere (D) None of these

571. Free mating is due to:
(A) Sex reversal by gene
(B) Environmental control of sex
(C) Hormonal control of sex
(D) Sex determination by chromosomes

572. Down's syndrome is due to:
(A) Autosome (B) Sex chromosome
(C) Sex linked disease (D) Duplication

573. A nucleosome is formed of:
(A) Histone and linker DNA
(B) Histone with wrapper DNA
(C) DNA
(D) RNA

574. Rearrangement of genes occurs due to:
(A) Translocation and duplication
(B) Translocation and deficiency
(C) Deletion and deficiency
(D) Translocation and inversion

575. Colchicine arrests cell division at:
(A) Prophase
(B) Metaphase
(C) Anaphase
(D) Telophase

576. DNA is associated with basic protein:
(A) Albumin (B) Nonhistone
(C) Histone (D) Both (A) and (B)

577. Hyperchromism is presence of:
(A) Some chromosome more than once
(B) Some type of chromosomes less than once
(C) Variable chromosomes in nucleus
(D) None of these

578. What are the assumptions of Hardy - Weinberg equilibrium?
(A) Small population size, random mating, no selection, no migration, no mutation
(B) Large population size, random mating, no selection, no migration, no mutation
(C) Large population size, random mating, heterozygote survive the best, no migration, no mutation
(D) Large population size, like individuals mate, no selection, no migration, no mutation

579. Euploidy is:
(A) One chromosome more than haploid set
(B) One chromosome more than diploid set
(C) One chromosome less than haploid set
(D) Exact multiple of haploid set of chromosomes

580. Chromosomes are made of:
(A) DNA + Pectin (B) RNA + DNA
(C) DNA + Histone (D) DNA only

581. Mutation refers to sudden change in:
(A) Genotype (B) Phenotype
(C) Metabolic rate (D) Genetic make up

582. In the endosymbiont hypothesis for the origin of eukaryotic cells, which of the following is proposed as the role of cyanobacteria?
(A) They become the chlorophyll of plant cells
(B) They are the ancestor of mitochondria
(C) They are the ancestor of the chloroplasts
(D) They close their DNA and become the vacuoles of plant cells

583. How many genomes are present in a typical green plant cell?
(A) 1 (B) 2 (C) 3 (D) 4

584. Formation of multivalents in meiosis is due to:
(A) Deletion (B) Inversion
(C) Monosomy (D) Reciprocal translocation

585. DNA parts which can switch their position are:
(A) Exons (B) Introns
(C) Cistron (D) Transposons

586. The concept of random fluctuations in gene frequency as an evolutionary force was originally proposed by:
(A) Muller (B) Hardy and Weinberg
(C) Sewall Wright (D) Wallace

587. Chromosome aberration occurs due to:
(A) Aneuploidy (B) Polyploidy
(C) Physical effects (D) All of these

588. Cytoplasmic inheritance is also called:
(A) Maternal inheritance (B) Cytoplasmic association
(C) Clonal inheritance (D) None of these

589. Linkage decreases frequency of:
(A) Recessive allele (B) Dominant allele
(C) Hybridisation (D) Both (B) and (C)

590. Which of the following statements defines anagenesis?
(A) Anagenesis describes the divergence between members of the same species into races
(B) Anagenesis is the process that occurs when one species gives rise to a new distinct species

(C) Anagenesis is the process whereby one species gives rise to two new species
(D) Anagenesis is the calculations used to develop a phylogenetic tree

591. Gene located on differential region of Y-chromosome are called:
(A) Mutant genes (B) Holandric genes
(C) Autosomal genes (D) XY linked genes

592. Exchange of one part of a chromosome with a part of the same or another chromosome is:
(A) Linkage (B) Crossing over
(C) Translocation (D) Inversion

593. A material which arrests cell division is obtained from:
(A) Colchicine (B) Chrysanthemum
(C) Cactus (D) Dalbergia

594. A reason for maternal inheritance is due to genes present in:
(A) Cytoplasm (B) Mitochondria
(C) Golgibody (D) Nucleolus

595. The genus *Rhagoletis* is a group of small, brightly coloured flies. Each species in the genus feeds, during its larval stage, on the fruit of just one plant family, and when the larvae hatch into adults, they court and mate on the same fruit. The species that feed on fruits from different plant families are reproductively isolated by:
(A) Ecological isolation (B) Temporal isolation
(C) Behavioral isolation (D) Mechanical isolation

596. Which is the main category of mutations?
(A) Genetic mutation (B) Zygotic mutation
(C) Somatic mutation (D) All of these

597. Diagram representing relative features and number of chromosomes is:
(A) Idiotype (B) Idiogram
(C) Karyogram (D) Karyotype

598. Number of linkage groups in Maize with ten pairs of chromosomes is:
(A) 5 (B) 10 (C) 15 (D) 20

599. Homology in antomical parts helps in determining evolutionary kinship because:
(A) Homologous body parts invariably perform similar functions
(B) Undergo similar genetic changes
(C) Display evolutionary adaptations
(D) Have common enbryological origin

600. Cross over frequency is proportional to:
(A) Phenotypic recombinant frequency
(B) Diploid number of chromosomes
(C) Haploid number of chromosomes
(D) Genotypic recombinant frequency

601. Chromosome complement 2n -1 is:
(A) Monosomy (B) Nullisomy
(C) Trisomy (D) Tetrasomy

602. The phenomenon of movement of DNA segment from one chromosome to another is:
(A) DNA replication (B) DNA transposition
(C) DNA recombination (D) DNA hybridization

603. Determination of percentage of crossing over between two linked genes is important in:
(A) Maintaining heterozygosity in population
(B) Indicating relative position of genes in chromosomes
(C) Fixation of heterosis in organism
(D) Explaining the phenomenon of coupling and repulsion

604. When two related populations that have been geographically isolated come back into contact, retain their reproductive isolation, and compete for resources, there may be a divergence in traits known as:
(A) Character convergence
(B) Sibling speciation
(C) Phyletic evolution
(D) Character displacement

605. Transition mutation is due to replacement of:
(A) GC by TA (B) AT by CG
(C) CG by GC (D) AT by GC

606. In Drosophila, XXY is female. In Humans it represents an abnormal male because:
(A) Y-chromosome induces male traits in humans
(B) Y-chromosome is essential for female sex in Drosophila
(C) Y-chromosome is not essential for male sex in human
(D) None of these

607. In Drosophila, during organ differentiation, one organ can be replaced by another like wings by legs. Genes responsible for it are:
(A) Plastid genes (B) Homeotic genes
(C) Complementary genes (D) Supplementary genes

608. Green fluorescent protein (GFP) cloned from jellyfish has now wide application in biological research. The fluorescence emitted by GPF is due to:
(A) Presence of two zinc ions in GPF molecule
(B) Heme, that serves as a prosthetic group in GPF molecule
(C) Three amino acids residues within GPF molecule
(D) Whole GPF molecule

609. Sex linked disorder is:
(A) Diphtheria (B) Colour blindness
(C) Tuberculosis (D) Leprosy

610. Extra chromosomal genetic material is:
(A) Prion (B) Cosmid
(C) Phasmid (D) Plasmid

611. A colour blind person cannot distinguish:
(A) Red and green (B) Green and blue
(C) Yellow and white (D) Black and yellow

612. Down's syndrome is due to extra chromosome:
(A) 20 (B) 21 (C) 24 (D) 26

613. The allele associated with sickle cell anemia apparently reached a high frequency in some human populations due to:
(A) Random mating
(B) Superior fitness of heterozygotes in areas where malaria was present
(C) Migration of individuals with the allele into other populations
(D) A high mutation rate at that specific gene

614. Place of gene on a chromosome is called:
(A) Locus (B) Chromomere
(C) Focus (D) Centrosome

615. One of the following is a random process:
(A) Variations (B) Adaptations
(C) Evolution (D) Mutations

616. Is it possible to say when a gene will mutate:
(A) Always (B) Sometimes
(C) Never (D) Rarely

617. What is the difference between a PAC and a BAC?
(A) One is a plasmid –based vector and the other is derived from a yeast chromosome
(B) One is derived from bacteriophage P1 can another is derived from bateriophage B1
(C) One is derived from the *E. coli* F-plasmid and other is derived from bacteriophage P1
(D) None of these

618. Turner's syndrome is due to:
(A) Trisomy of chromosome 21
(B) Trisomy of chromosome 18
(C) Autosomal recessive gene
(D) Absence of one sex chromosome

619. Genes located on mitochondrial DNA bring about:
(A) Paternal inheritance (B) Maternal inheritance
(C) Biparental inheritance (D) There is no inheritance

620. T-DNA is:
(A) DNA of plasmid origin which is transferred to the Agrobacterium chromosome
(B) DNA from the chromosome of Agrobacterium species which is transferred to the plant genome
(C) DNA of plasmid origin which is transferred to the plant genome
(D) Telomeric DNA of linear chromosome

621. Lyon's hypothesis is connected with:
(A) Number of barr bodies (B) Genetic compatibility
(C) Genetic incompatibility (D) Centromere position

622. Gene which increases frequency of mutations of other genes is:
(A) Mutagen (B) Complementary gene
(C) Mutator gene (D) Hypostatic gene

623. Frequency of recessive allele is 0.4. What is the frequency of heterozygous carriers in the population of 200?
(A) 96 (B) 100 (C) 116 (D) 126

624. The ratio of phenotypes in F_2 of a monohybrid cross is:
(A) 3 : 1 (B) 1 : 2 : 1
(C) 9 : 3 : 3 : 1 (D) 2 : 1 : 1

625. A pure tall pea was crossed with a pure dwarf pea. All the plants of F_1 were found to be tall. This is due to:
(A) Dominance
(B) Disappearance of factor for dwarfness in F_1 generation
(C) Coordination
(D) Segregation of factors

626. The monohybrid genotypic ratio 1 : 2 : 1 in F_2 generation indicates
(A) Segregation (B) Dominance
(C) Independent assortment (D) Incomplete dominance

627. Which of the following gene-therapy vectors preferentially infest nerve cells?
(A) Adeno-associated virus (B) Retrovirus
(C) Herpes virus (D) Adenovirus

628. The F_2 generation of a cross produced identical phenotypic and genotypic ratio. It is not an expected Mendelian result, and can be attributed to:
(A) Independent assortment (B) Linkage
(C) Incomplete dominance (D) Complete dominance

629. The law of independent assortment can be related to one of the following:
(A) Both homologous chromosomes of each pair are received by a single gamete
(B) The genes present on the same chromosome get randomly redistributed
(C) The non-homologous chromosomes show random distribution during anaphase – I of meiosis
(D) The exchange of segments between the non-homologous chromosomes

630. If organism is triploid, then Hardy-Weinberg theorem applicable will be:
(A) $(p + q)^2$ (B) $(p + q)$
(C) $(p + q + r)^2$ (D) $(p + q - r)^2$

631. The factors which represent the contrasting pairs of characters are called:
(A) Dominants (B) Alleles
(C) Homologous pairs (D) Dominant and recessive

632. Which of the following virus is not used as gene therapy vectors?
(A) Retroviruses (B) Adenoviruses
(C) Adeno-associated viruses (D) Polyomavirus

633. Chromosome walking is best described as:
(A) Aligning DNA sequences by computer to generate contigs
(B) Generating a map along a chromosome in a step-by-step manner
(C) Identifying clones whose inserts overlap to generate a library of clones that over a given segment of DNA
(D) Sequencing a genome at a time to ensure that no gaps are present at the end of the project

634. One of the following did not constitute the seven contrasting pairs of characters noticed by G.J. Mendel in pea:
(A) Colour of pod (B) Shape of the leaves
(C) Shape of pod (D) Height of the plants

635. In genetics the term test cross means:
(A) The crossing of F_1 individual with homozygous recessive
(B) Crossing an F_1 individual with either of the two parents
(C) Crossing F_1 individual with another F_1 individual
(D) Crossing F_1 individual with that of F_2

636. Which of the following is most likely to prevent extinction of a rare allele in which homozygosity represents a genetic lethal?
(A) New mutation (B) Genetic drift
(C) Founder effect (D) Balanced polymorphism

637. The hexokinase (Hk) locus in Indian sparrow has allelic frequencies of 0.8 and 0.2, respectively, for the alleles Hk-A and Hk-B. What is the probability that two heterozygous individuals will male, given that Hk is in Hardy-Weinberg equilibrium?
(A) 0.01 (B) 0.10 (C) 0.16 (D) 0.32

638. How many types of gametes are expected from the organism with genotype AABBCC?
(A) 1 (B) 2 (C) 4 (D) 6

639. Which of the following is not a feature of molecular polygenetics?
(A) The use of molecular data to reconstruct a phylogenetic tree
(B) The use of molecular data to understand the genetic basis to variable phenotypes
(C) The use of molecular data to infer the evolutionary relationship between genomes
(D) The application of rigorous mathematical methods to analysis of variable characters

640. The basis for development of leukemia in some children treated with a retrovirus – derived gene therapy vector is:
(A) Insertion of the viral genome adjacent to an oncogene
(B) Use of a virus that contains an oncogene
(C) Insertional mutation of a tumor suppressor gene
(D) Stimulation of immune response to cell surface antigens introduced by an inserted gene

641. A plant with a genotype Aa Bb is crossed with a plant having the genotype aa bb. The genotype of F_1 would be:
(A) AaBb, AABB (B) aabb, aaBb
(C) aaBB, AABb (D) AaBb, Aabb, aaBb, aabb

642. How many types of gametes are expected from a plant with genotype PpQq, provided there occurs independent assortment?
(A) 4 (B) 6 (C) 8 (D) 10

643. *Agrobacterium rhizogenes* can be used to genetically engineer plants because of which of the following statement?
(A) It contains a Ti plasmid
(B) Infected plants produce a good root system
(C) It infest all type of plants
(D) It causes gene transfer to plants

644. The lengths of the branches of a phylogenetic tree constructed from DNA sequence data indicate:
(A) The length of time since the organisms diverged
(B) The number of synonymous changes between the genes
(C) The degree of different between the genes represented by the nodes
(D) None of these

645. Which of the following crosses would produce a genotypic ratio of 1 : 2 : 1 in F_2?
(A) AB x AB
(B) ab x ab
(C) Ab x Ab
(D) ab x ab

646. In *Mirabilis jalapa* when two F_1 pink flowered plants were crossed with each other, the F_2 generation produced 40 red, 80 pink and 40 white flowering plants. This is a case of:
(A) Duplicate genes (B) Lethal genes
(C) Incomplete dominance (D) Epistasis

647. Genotype-phenotype concept was first proposed by:
(A) Bateson (B) Johannsen
(C) Sutton and Boveri (D) Punnet

648. Which of the following methods for studying loss of gene function does not involve any modification of the genome?
(A) RNA interference by injection of double stranded RNA
(B) Expression of an integrated antisense transgene
(C) Gene knockout by homologous recombination
(D) None of these

649. The phenomenon which defines the independent assortment is:
(A) Segregation (B) Crossing over
(C) Dominance (D) Linkage

650. The Biological clock in higher vertebrates is regulated by:
(A) The pituitary gland
(B) Cerebral cortex
(C) Supra-chiasmatic nucleus in hypothalamus
(D) Thymus

651. A pea with white flowers was crossed to another pea which is also white flower plant. When selfed the F_2 generation produced purple and white in the ratio 9 : 7. The reason for the results is that:
(A) It is typical monohybrid Mendelian ratio
(B) Purple flower colour is dominant over the white
(C) It is a complementary factor
(D) None of these

652. Possession by two or more species of a similar or identical character state that has not been derived by both species from their common ancestor is called:
(A) Homoplasy (B) Homology
(C) Homozygote (D) Homoerotic mutation

653. The law of segregation of characters is also called the law of purity of gametes because:
(A) Gametes have only one of the two alleles for each character
(B) Gametes cannot be contaminated
(C) Gametes are different type of cells
(D) It was just another name adopted accidentally

654. Identify the incorrect statement:
(A) 2, 4-dichlorophenoxyacetic acid is the most commonly used chemical analog of indole-3-acetic acid
(B) In somatic embryogenesis, embryo initiation needs a high concentration of 2, 4-D
(C) Crown-gall disease depends on the presence of Ti-plasmid in *Agrobacterium tumefaciens*
(D) *Agrobacterium tumefaciens* is responsible for hairy root formation in plants

655. The genes for same trait present on monohomologous chromosomes are:
(A) Alleles (B) Linked genes
(C) Multiple alleles (D) None of these

656. Recombinant live attenuated vain against hepatitis B was prepared from:
(A) Plasma of infected individual
(B) Recombinant yeast expressing hepatitis B surface antigen
(C) Recombinant vaccinia virus expressing hepatitis B surface antigen
(D) Transgenic plants expressing hepatitis B surface antigen

657. In a coral island in Atlantic Ocean, a natural calamity killed most of its population. The island is repropulated with the surviving individuals and their progeny. The present day population in the island shows high incidence of a rare recessive genetic disorder. What could be the most probably evolutionary force is responsible for this phenomenon?
(A) Geographic isolation
(B) Genetic drift
(C) Selective advantage of the recessive disorder
(D) Natural selection

658. Mendel formulated the law of heredity considering seven pairs of contrasting characters in the pea plant. If he had studied an eight pair, the law which would have been altered is:
(A) Law of segregation
(B) Law of dominance
(C) Law of independent assortment
(D) Law of purity of gametes

659. Medel was successful in formulating the law of inheritance whereas his predecessors were not because:
(A) He studied one clear –cut character at a time
(B) The characters studied by him were present on separate chromosomes
(C) Of the right choice of material
(D) He kept accurate records of his experiments

660. Appearance of hidden character in some progeny of F_2 population indicates:
(A) Law of purity of gametes
(B) Law of independent assortment
(C) Law of dominance
(D) Law of unit characters

661. Medel's law of segregation is based upon the F_2 ratio of:
(A) 2 : 1 : 2 (B) 9 : 3 : 3 : 1
(C) 1 : 2 : 1 (D) 3 : 1

662. The difference between transfection and transduction is:
(A) In transfection, the transgene is inserted in a plas mid, while in transduction, the transgene is inserted in a viral genome
(B) Transfection involves the transfer of naked DNA into the cell while transduction involves packaging the DNA into a virus particle, which then infest the cell
(C) There is no difference – the terms are synonymous
(D) None of these

663. From an evolutionary point of view, plant tendrils derived from leaves and those derived from stems are example of:
(A) Analogous structures resulting from convergence only
(B) Homologous structures resulting from convergence only
(C) Analogous structures resulting from divergence only
(D) Homologous structures resulting from divergence only

664. From an evolutionary point of view, plant tendrils derived from leaves and those derived from stems are example of:
(A) Analogous structures resulting from convergence only
(B) Homologous structures resulting from convergence only
(C) Analogous structures resulting from divergence only
(D) Homologous structures resulting from divergence only

665. A microarray is a large collection of specific DNA oligonucleotides spotted in a defined pattern on a microscope slide. What is the most useful experiment that can be done with such a tool?
(A) Predicting the presence of specific metabolites in a cell
(B) Comparing newly synthesized nuclear RNA with cytoplasmic RNA to locate introns
(C) Comparing RNA produced under two different physiological conditions to understand patterns of gene expression
(D) Comparing proteins produced under two different physiological conditions to understand their function

666. Genes do not occur in pairs in:
(A) Zygote
(B) Endosperm cells
(C) Somatic cell
(D) Gametes

667. If dwarf pea plant was treated with GA_3 it grew as tall as the pure tall pea plant. If this treated plant is crossed with a pure tall plant the phenotypic ratio F_1 is likely to be:
(A) All tall (B) All dwarf
(C) 50% tall and 50% dwarf (D) 75% tall and 25% dwarf

668. Sewall Wright's metaphor of evolution on an adaptive landscape describes a population as occupying a field of genotypes graded by contours delimiting gene combinations of equal fitness. The evolutionary mechanism most likely to move a population consistently "uphill" toward adaptive peaks on this landscape is:
(A) Chance
(B) Directed mutation
(C) Natural selection
(D) Gene flow between populations

669. Segregation of genes take place during:
(A) Metaphase
(B) Anaphase
(C) Prophase
(D) Embryo formation

670. A cross between hybrid and a parent is known as:
(A) Test cross (B) Back cross
(C) Monohybrid cross (D) Reciprocal cross

671. In an experiment on pea plant, pure plants with yellow round seeds were crossed with plants producing green wrinkled seeds. What will be phenotypic ratio of F_1 progeny?
(A) 9 yellow round : 3 round green : 3 wrinkled yellow : 1 green wrinkled
(B) All yellow round
(C) 1 yellow round : 1 round green : 1 wrinkled yellow : 1 green wrinkled
(D) All wrinkled green

672. Which one of the following are not the criteria for using ribosomal RNAs as evolutionary chronometers?
(A) The large size of ribosomal RNAs gives large number of possible sequence combinations
(B) Ribosomal RNAs are functionally invariant
(C) Ribosomal RNAs are universally distributed
(D) Ribosomal RNAs have enzyme activity

673. When a wheat variety of red kernels (homozygous for two nonallelic and independent dominant genes) is crossed with white kernelled wheat (homozygous for two recessive nonallelic independent genes), the phenotypic ratio in F_2 generation would be:
(A) 9 : 7 (B) 1 : 10 : 4 : 1
(C) 1 : 4 : 6 : 4 : 1 (D) 1 : 2 : 4 : 2 : 4 : 2 : 1

674. In multiple allele system a gamete possesses:
(A) One alleles (B) Two alleles
(C) Three allele (D) Several alleles

675. A pleiotropic gene is one which:
(A) Affects one character
(B) Affects more than one character
(C) Requires another gene for expression
(D) Supplements the effect of another gene

676. Sum total of all the genetic information in the breeding members of a population at a given time is known as:
(A) Gene pool (B) Gene frequency
(C) Genome (D) Genetic Drift

677. The spread of genes from one breeding population to another by migration which may result in changes in gene frequency is called:
(A) Genetic Drift (B) Gene flow
(C) Genetic clone (D) Gene frequency

678. A genetic clone is:
(A) Plants produced by asexual means
(B) Heterozygous plant produced by sexual means
(C) Homozygous plant produced by sexual means
(D) Hybrid produced by sexual means

679. Gaseous mixture used by Miller for synthesis of amino acids through heat and electric discharge included:
(A) Methane, ammonia, hydrogen and water vapours
(B) Methane, ammonia, nitrogen and water vapours
(C) Ammonia, carbon dioxide, hydrogen and water vapours
(D) Methane, carbon dioxide, nitrogen and water vapours

680. A pleitropic gene is:
(A) I^A (B) Hb^S
(C) Hb^A (D) I^B

681. Dominant gene for tallness is T and for yellow colour is Y. A plant heterozygous for both the traits is selfed, then the ratio of pure homozygous dwarf and green offspring would be:
(A) 4/16 (B) 3/16 (C) 2/16 (D) 1/16

682. A gene that shows its effect on more than one character is:
(A) Multiple gene (B) Pleiotropic gene
(C) Multi gene (D) Polygene

683. A dihybrid ratio of 1 : 4 : 6 : 4 : 1 is obtained instead of 9 : 3 : 3 : 1. This is an example of:
(A) Complementary genes (B) Pleiotropic genes
(C) Polygenic inheritance (D) Supplementary genes

684. A device for the detection of an analyte that combines a biological component with a physiochemical detector component is known as:
(A) Biosensor (B) Piezoelectric
(C) Biomimic (D) Bioremediation

685. Epistatic gene differs from dominant gene in:
(A) Epistatic gene is nonallelic
(B) Epistatic gene never express itself independently
(C) Epistatic and hypostatic genes are present at different loci
(D) All of these

686. Genotype is:
(A) Genetic composition of an organism
(B) Genetic composition of plastids
(C) Genetic composition of germ cells
(D) Genetic composition of an individual

687. The slope of the beaks of Darwin's flinches, industrial melanism, and the changes in horse teeth are all examples of:
(A) Artificial selection (B) Natural selection
(C) Covergent evolution (D) Homologous structure

688. Tall red flowered pea plant crossed to dwarf white flowered plant yields only tall red flowered plants. A test cross shall give a ratio of:
(A) 1 : 1 : 1 (B) 3 : 1 : 1
(C) 1 : 1 : 1 : 1 (D) 1 : 2 : 4 : 6 : 4 : 2 : 1

689. RR (red) *Antirhinum* is crossed with white (WW) one. Offspring RW are pink. This is an example of:
(A) Dominant – recessive (B) Incomplete dominance
(C) Hybrid (D) Supplementary genes

690. Dominant lethal gene is one which:
(A) Determines sex of offspring
(B) Allows the organism to survive but not reproduce
(C) Allows the organism to survive and reproduce
(D) Kills the organism

691. Which of the following is not true regarding adeno-associated virus?
(A) It contains double-stranded DNA
(B) It may integrate into the host chromosome
(C) It may exist in the ell as an episome
(D) It may be associated with insertional mutagenesis when used as a gene therapy vector

692. Which of the following is true of Mendelism but not of Darwism?
(A) It influenced human thought tremendously
(B) It was modified immediately
(C) It gave well defined principles even in early stage
(D) It was based on insufficient data

693. The three principles of Mendelism are:
(A) Dominance-recessiveness, segregation and independent assortment
(B) Linkage, segregation and independent assortment
(C) Linkage, dominance-recessiveness, and segregation
(D) Linkage, dominance-recesiveness and independent assortment

694. In a plant, the alleles of neither black flower colour nor white flower color are dominant. When both are present, the flower colour is grey. If two grey flowered plants are crossed the progeny will appear in the ratio of:
(A) All grey (B) Either black or white
(C) 2 grey : 1 black : 1 white (D) 1 black : 1 white

695. A pure tall pea plant is crossed with pure dwarf pea plant. The progeny is self pollinated. The ratio of true breeding tall pea plants to true breeding dwarf pea plants shall be:
(A) 2 : 1 (B) 1 : 1
(C) 3 : 1 (D) 1 : 2 : 1

696. Evalutionary changes in one species prompt corresponding changes in other species with which the former interacts ecologically. This process is known as:
(A) Coevaluation (B) Complementation
(C) Parallel evolution (D) Microevolution

697. Heterozygous tall (Tt) and dwarf (tt) pea plants are crossed. The ratio in the offspring shall be:
(A) All tall (B) All dwarf
(C) 3 tall : 1 dwarf (D) 1 tall : 1 dwarf

698. Alleles of a gene are found on:
(A) Same chromosome
(B) Homologous chromosomes
(C) Nonhomologous chromosomes
(D) Any chromosomes

699. Parents are heterozygous albino. What will be the first three children?
(A) Some normal, heterozygous and albino
(B) All normal
(C) No normal
(D) All heterozygous albino

700. Arabidopsis is advantageous for plant genetic research because:
(A) It is commercially important as a food crop
(B) It is an endangered species
(C) It is the closest to humans of any existing plant
(D) It is a small plant with a small genome size which can be raised inexpensively

701. Which is a dominant trait?
(A) Colour blindness
(B) Albinism
(C) Haemophilia
(D) Rh factor

702. TT mates with Tt. What will be characteristic of offspring?
(A) 25% recessive (B) 50% recessive
(C) 75% recessive (D) All dominant

703. The allele which is unable to express its effect in the presence of another is called:
(A) Codominant (B) Supplementary
(C) Complementary (D) Recessive

704. Lethal genes are:
(A) Causative for appearance of ancestral traits
(B) Always recessive
(C) Gene present on different chromosomes but influ encing single trait.
(D) Killer in homozygous state

705. The first vertebrates appeared in which one of the following periods of the Palaeozoic era?
(A) Ordovician (B) Silurian
(C) Devonian (D) Mississippian

706. The procedure adopted by Mendel which was different from his predecessors is:
(A) Keeping breeding records
(B) Employing many traits at one time
(C) Differentiating inherited traits
(D) Quantitative analysis of data

707. Mendel did not propose:
(A) Dominance (B) Incomplete dominance
(C) Segregation (D) Independent assortment

708. Genotype of F_1 individuals can be tested by:
(A) Backcross with homozygous recessive parent
(B) Reciprocal crossing
(C) Backcrossing with heterozygous parent
(D) Backcrossing with homozygous dominant parent

709. A dihybrid condition is:
(A) tt Rr (B) Tt rr (C) tt rr (D) Tt Rr

710. Which one of the following statements about circadian rhythm is correct?
(A) Circadian rhythm, dependent on endogenous clock, is restricted to birds and mammals
(B) An animal's circadian rhythm has a period of exact 24 hours
(C) The period of circadian rhythm is not exactly 24 hours
(D) Circadian rhythm is necessary because earth is titled on its axis

711. Which contributed to the success of Mendel?
(A) Qualitative analysis of data
(B) Observation of distinct inherited traits
(C) His knowledge of biology
(D) Consideration of one character at one time

712. Pleiotropic gene has:
(A) Single genotype (B) Multiple genotype
(C) Single phenotype (D) Multiple phenotype

713. An organism with two identical alleles is:
(A) Dominant (B) Hybrid
(C) Heterozygous (D) Homozygous

714. Hybrid of pure red flowered and white flowered pea is crossed to pure red flowered plant. The progeny shall be:
(A) White flowered
(B) Red flowered
(C) Both red and white flowered plants
(D) Mostly red flowered plants

715. A heterozygous round seeded plant is crossed to recessive wrinkled seeded plant. The progeny would be:
(A) 303 rounded : 301 wrinkled
(B) 301 rounded : 100 wrinkled
(C) 200 rounded : 100 wrinkled
(D) 99 rounded : 301 wrinkled

716. In a genetic cross having recessive epistasis, F_2 phenotypic ratio would be:
(A) 9 : 6 : 1 (B) 15 : 1 : 1
(C) 9 : 3 : 4 (D) 12 : 3 : 1

717. What is fate of most duplicated genes?
(A) Gene activation
(B) Gain of a novel function through subsequent mutation
(C) They are transferred to a new organism using lateral gene transfer
(D) They become orthologs

718. The contrasting pairs of factors in Mendelian crosses are called:
(A) Multiple alleles (B) Allelomorphs
(C) Alloloci (D) Paramorphs

719. In *Mirabilis/Antirrhinum,* hybrid between red and white flowered plants is pink flowered. This is:
(A) Heterosis (B) Segregation
(C) Incomplete dominance (D) Dominance

720. Mendel's laws apply only when:
(A) Characters are linked
(B) Parents are pure breeding
(C) F1 monohybrid ratio shows two types of individuals
(D) First pair of contrasting characters is dependent upon other pairs

721. An altruistic behavior is one in which an individual (the donor) performs an action that helps another animal (the recipient) with no apparent advantage to itself. This is known as:
(A) Kin selection
(B) Allelopathy
(C) Mutualism
(D) Competitive exclusion principle

722. Cross between hybrid and recessive parent is:
(A) Back cross (B) Test cross
(C) Monohybrid cross (D) Dihybrid cross

723. An exception to Mendel's principles/Mendel did not include in his discoveries:
(A) Dominance
(B) Purity of gametes/segregation
(C) Linkage
(D) Independent assortment

724. In a cross between AABB x aabb, the ratio of F_2 genotypes between AABB, AaBB, Aabb and aabb would be:
(A) 9 : 3 : 3 : 1 (B) 2 : 1 : 1 : 2
(C) 1 : 2 : 2 :1 (D) 7 : 5 : 3 : 1

725. Mendel formulated the law of purity of gametes on the basis of :
(A) Test cross (B) Back cross
(C) Monohybrid cross (D) Dihybrid cross

726. A cross between hybrid and its parent is:
(A) Back cross (B) Reciprocal cross
(C) Monohybrid cross (D) Dihybrid cross

727. Stabilizing selection favours:
(A) Both extreme forms of a trait
(B) Intermediate forms of a trait
(C) Environmental differences
(D) One extreme form over the other extreme form and over intermediate forms of a trait

728. Test cross is a cross between:
(A) Hybrid x Dominant parent
(B) Hybrid x Recessive parent
(C) Hybrid x Hybrid
(D) All of these

729. Multiple alleles control inheritance of:
(A) Phenylketonuria (B) Colour blindness
(C) Sickle cell anaemeia (D) Blood groups

730. A gene pair hides the effect of another. The phenomenon is:
(A) Epistasis (B) Dominance
(C) Mutation (D) None of these

731. Orthologous genes are which of the following?
(A) Genes that do not share a common ancestral origin
(B) Homologous genes that are present in the genomes of different organisms
(C) Homologous genes that are present in the same genomes
(D Nonhomologous genes that arose from convergent evolution

732. In a dihybrid cross, F_2 ratio of 15 : 1 is due to:
(A) Supplementary genes (B) Dominant epistasis
(C) Duplicate genes (D) Recessive epistasis

733. Polymorphism is mainly due to:
(A) Monogenic inheritance (B) Polygenic inheritance
(C) Both of the above (D) None of these

734. An allele is dominant if it is expressed in:
(A) Both homozygous and heterozygous states
(B) Second generation
(C) Heterozygous combination
(D) Homozygous combination

735. Percentage of heterozygous individuals obtained from selfing of Rr individuals is:
(A) 100 (B) 75 (C) 50 (D) 25

736. Hardy-Weinberg genetic equilibrium holds good for:
(A) Panmictic population
(B) Mendelian population
(C) Assortative mating population
(D) Dissortative mating population

737. Law of dominance-recessiveness is proved by:
(A) Back cross
(B) Incomplete dominance
(C) Monohybrid cross
(D) Dihybrid cross

738. In a test cross, the individual is crossed with:
(A) Homozygous dominant individual
(B) Heterozygous individual
(C) Homozygous recessive
(D) None of these

739. Which of the following is required to see a founder effect?
(A) High rate of new mutation
(B) Population bottleneck in which the population is reduced to a relatively small number of breeding individuals
(C) Lethality of homozygotes with a mutant alleles
(D) Selection if fevor of heterozygotes

740. A yellow seeded true breeding pea plant is crossed with green seeded true breeding plant. F_1 plants were yellow seeded showing that allele for yellow colour is:
(A) Heterozygus (B) Dominant
(C) Lethal (D) Epistatic

741. RrYy yields gamete types:
(A) One (B) Two (C) Three (D) Four

742. Heterozygous organism for two genes shall be:
(A) RRYY (B) RrYY
(C) RrYy (D) RRYy

743. Mendel's principle of segregation is based on separation of alleles during:
(A) Gamete formation (B) Seed formation
(C) Pollination (D) Embryonic development

744. Mendel's studied inheritance of seven pairs of traits in pea which can have 21 possible combinations. If you are told that in one of these combinations, independent assortment is not observed in later studies, your reaction will be:
(A) Independent assortment principle may be wrong
(B) Mendel might not have studied all the combinations
(C) It is impossible
(D) Later studies may be wrong

745. The Hardy-Weinberg law describes:
(A) Genotype frequencies of a population when evolutionary forces are not acting
(B) How sexual reproduction would change the relative gene frequencies in a population
(C) How mutations occur and balance each other
(D) Genotype frequencies of a population when evolutionary forces are acting

746. Hybrids are generally superior to parents due to:
(A) Heterosis
(B) Homozygosity
(C) Heterozygosity
(D) Parents are generally weak

747. Ratio 9 : 3 : 3 : 1 is due to:
(A) Duplicate genes (B) Lethal genes
(C) Monohybrid cross (D) Dihybrid cross

748. Ratio 9 : 7 is due to:
(A) Supplementary genes (B) Lethal genes
(C) Complementary genes (D) Epistatic genes

749. Monohybrid cross involves:
(A) Individuals of two different parents
(B) Individuals different in one traits
(C) Individuals different in two traits
(D) Individual with parent

750. Assuming Hardy-Weinberg equilibrium, the genotype frequency of heterozygotes, if the frequency of the two alleles at the gene being studied are 0.6 and 0.4, will be:
(A) 0.80 (B) 0.64 (C) 0.48 (D) 0.32

751. Mendel chose contrasting traits in pea:
(A) 4 (B) 5 (C) 6 (D) 7

752. How many contrasting traits of pea pod were chosen by Mendel?
(A) 1 (B) 2 (C) 3 (D) 3

753. Monohybrid ratio is:
(A) 3 : 1 (B) 9 : 7 (C) 1 : 2 (D) 9 : 3 : 3 : 1

754. Which of the following is not an assumption of the Hardy-Weinberg equilibrium?
(A) No mutation
(B) No more than two alleles at a locus
(C) Random mating with respect to genotype
(D) No selection on the basis of genotype

755. A cross between F_1 hybrid and a recessive parent (Tt x tt) gives a ratio of:
(A) 1 : 1 (B) 2 : 1 (C) 3 : 1 (D) 4 : 1

756. Different forms of a gene are called:
(A) Heterozygotes (B) Alleles
(C) Complementary genes (D) Supplementary genes

757. A homozygous red flowered pea plant is crossed with homozygous white flowered pea plant. The offspring would be:
(A) 50% red flowered (B) 50% pink flowered
(C) All red flowered (D) All pink flowered

758. Genetically identical progeny is produced when the individual:
(A) Performs self fertilization
(B) Produces identical gametes
(C) Practices inbreeding without meiosis
(D) Practices reproduction

759. If two populations are merged, each with different frequencies of an allele at a locus, and randomely mating occurs immediately, how long will it take to achieve a Hardy-Weinberg equilibrium in the new population?
(A) One generation
(B) Two generations
(C) Variable-depends on allele frequencies
(D) The population will never achieve equilibrium

760. Cross AABb x aaBb yields AaBB : AaBb:Aabb : aabb offspring in the ratio of:
(A) 0 : 3 : 1 : 1 (B) 1 : 2 : 1 : 0
(C) 1 : 1 : 1 : 1 (D) 1 : 2 : 1 : 1

761. The process of mating between closely related individuals is:
(A) Self breeding (B) Inbreeding
(C) Hybridisation (D) Cross breeding

762. A cross between pure tall pea plant with green pods and dwarf pea plant with yellow pods will produce short F_2 plants out of 16:
(A) 2 (B) 3 (C) 4 (D) 5

763. The physical similarity of body shape in dolphins, sharks and penguns results from:
(A) Parallel evolution
(B) Geographic isolation
(C) Convergent evolution
(D) A property of a common ancestor

764. A cross between homozygous recessive and heterozygous plant is:
(A) Monohybrid cross (B) Dihybrid cross
(C) Test cross (D) Back cross

765. In case of incomplete dominance, F_2 generation has:
(A) Genotypic ratio equal to phenotypic ratio
(B) Genotypic ratio is 3 : 1
(C) Phenotypic ratio is 3 : 1
(D None of these

766. Mendel's principles are related to:
(A) Variation (B) Reproduction
(C) Evolution (D) Heredity

767. Which of the following statements regarding Darwin and his theory of evolution is false?
(A) Darwin noticed that South American species of birds more closely resembled species found in other temperate regions that European species of birds
(B) Darwin proposed that species adapt over time
(C) Darwin proposed that natural selection produces changes in species over time
(D) Darwin was the original and sole developer of the theory of natural selection

768. Dihybrid cross is related to:
(A) Principle of dominance
(B) Principle of independent assortment
(C) Principle of segregation
(D) Principle of purity of gametes

769. Phenotypic dihybrid ratio is:
(A) 9 : 3 : 3 : 1 (B) 15 : 1 : 1
(C) 9 : 6 : 1 (D) 1 : 2 : 1

770. Mendel enunciated principles of inheritance :
(A) 2 (B) 3 (C) 4 (D) 5

771. In red-white flowered cross of *Mirabilis jalapa*, F_2 generation has red, pink and white flowered plants in the ratio of:
(A) 1 : 2 : 1 (B) 1 : 0 : 1
(C) 2 : 1 : 1 (D) 1 : 1 : 2

772. An increase in the inbreeding coefficient, F, is likely to result in:
(A) Reduced likelihood of heterozygotes being present in a population
(B) Higher proportion of genes that show linkage
(C) Higher proportion of genes with introns
(D) Lower level of difference between proteins in two daughter cells

773. Phenomenon of an allele of one gene suppressing the activity of allele of another gene is called:
(A) Dominance (B) Epistasis
(C) Suppression (D) Inactivation

774. Recombinations and variations in eucaryotic organisms are due to:
(A) Mitosis and meiosis
(B) Fertilization and mitosis
(C) Meiosis and fertilization
(D) Meiosis and amitosis

775. Mendel's law of independent assortment is based on F2 ratio of:
(A) 1 : 2 : 1 (B) 9 : 3 : 3 : 1
(C) 2 : 1 : 1 (D) 3 : 1 : 1

776. Pure red flowered and white flowered plants were crossed. It produced 120 offspring with :
(A) 90 white flowered and 30 red flowered
(B) 90 red flowered and 30 white flowered
(C) 60 red flowered and 60 white flowered
(D) All red flowered

777. A balanced polymorphonism occurs when:
(A) There is selection against the homozygous recessive individuals
(B) There is selection against heterozygotes
(C) There is selection against all genotypes
(D) There is selection against all homozygotes

778. Red seed colour shows an F_2 ratio of 1 : 4 : 6 : 4 : 1. It is due to:
(A) Two polygenes
(B) Multiple poly gene
(C) Different number of recessive genes
(D) Supplementary genes

779. Two dominant independently assorting genes react with each other. They are:
(A) Complementary (B) Supplementary
(C) Duplicate (D) Collaborative

780. Polygenes show:
(A) Different phenotypes (B) Different genotypes
(C) Both (A) and (B) (D) None of these

781. Epistasis in which dihybrid ratio of 9 : 3 : 3 : 1 between Aa Bb x AaBb is modified is:
(A) Dominance of one allele on another allele at the same locus
(B) Dominance of one allele on another allele at both its loci
(C) Interaction of two alleles at different loci
(D) Interaction between two alleles at same

782. Genes located on same locus but having different expressions are:
(A) Multiple alleles (B) Oncogenes
(C) Polygenes (D) Codominants

783. Which is true:
(A) Etiolation is genetic while albinism is physiological
(B) Etiolation is not reversible
(C) Etiolation is physiological but albinism is genetic
(D) Etiolation and albinism are synonyms

784. Allelomorphic pair implies:
(A) A pair of contrasting characters
(B) A pair of noncontrasting characters
(C) Any two characters
(D) Sex-linked characters

785. In a cross of red and white flowered Snapdragon, F_1 plants will have flowers:
(A) Red (B) White
(C) Both (A) and (B) (D) Pink

786. Cross between AaBB and aaBB will form:
(A) 1 AaBB : 1 aaBB (B) All AaBB
(C) 3 AaBB : 1 aaBB (D) 1 AaBB : 3 aaBB

787. Four daughter cells obtained from a single meiosis differ from one another because of:
(A) Crossing over
(B) Independent assortment
(C) Change in chromosome number
(D) Both (A) and (B)

788. Incomplete dominance produces in F_2 generation:
(A) Phenotypic ratio of 3 : 1
(B) Genotypic ratio of 3 : 1
(C) Phenotypic ratio similar to genotypic ratio
(D) No segregation occurs

789. Hybrid pea plant with yellow round seeds (YyRr) is self pollinated. Phenotypic ratio of next generation would be:
(A) 13 : 3 (B) 9 : 7
(C) 1 : 4 : 6 : 4 : 1 (D) 9 : 3 : 3 : 1

790. A cross between AaBb and aabb yields:
(A) AaBb (B) AaBb, Aabb, aaBb and aabb
(C) AaBb (D) Aabb

791. Heterosis is:
(A) Hybrid weakness
(B) Hybrid weakness and vigour
(C) Hybrid vigour
(D) Neither weakness nor vigour of hybrid

792. Mendel found certain traits not to assort independently. It is due to:
(A) Dominance (B) Linkage
(C) Crossing over (D) Amitosis

793. In *Lathyrus odoratus*, cross between two purple flowered plants gives a pink/white progeny. It is due to:
(A) Codominance (B) Incomplete dominance
(C) Epistatsis (D) Segregation

794. A pure red flowered pea plant is crossed with pure white flowered pea plant. The offspring will be:
(A) All red flowered (B) Half red flowered
(C) Half white flowered (D) Half pink flowered

795. How many type of gametes will be produced by individuals of AABbcc genotype?
(A) 2 (B) 4 (C) 6 (D) 9

796. If Mendel had studied 7 traits using a plant with 12 chromosomes instead of 14, he would have?
(A) Discovered incomplete dominance
(B) Not discovered the law of independent assortment
(C) Discovered sex linkage
(D) Mapped the chromosome

797. Crossing over in diploid organism is responsible for:
(A) Dominance of genes
(B) Linkage between genes
(C) Recombination of linked genes
(D) Segregation of alleles

798. Mendel's laws of heredity can be explained with the help of:
(A) Mitosis (B) Meiosis
(C) Cloning (D) Both (A) and (B)

799. Number of phenotypes possible from AaBbCc x AaBbCc is:
(A) 16 (B) 18 (C) 32 (D) 36

800. Phenotypic ratio of a monohybrid F_2 progeny of incomplete dominance is:
(A) 1 : 1 : 1 : 1 (B) 2 : 1 : 1
(C) 1 : 2 :1 (D) 3 : 1 : 1

801. Genotypic ratio of a dihybrid cross is:
(A) 12 : 3 : 1 (B) 15 :1 : 1
(C) 1 : 2 : 1 (D) 1 : 2 : 2 : 4 : 1 : 2 : 1 : 2 :1

802. Number of gamete types formed by AaBbCc individual is:
(A) 4 (B) 6 (C) 8 (D) 16

803. Which is connected with multiple phenotypes?
(A) Epistasis (B) Pleiotropy
(C) Polygenic inheritance (D) Mutations

804. In *Mirabilis jalapa*, crossing of red flowered and white flowered plants, yields pink flowered plants. This is due to:
(A) Crossing over (B) Complete dominance
(C) Incomplete dominance (D) Epistasis

805. Mendel did not observe linkage due to:
(A) Mutation (B) Synapsis
(C) Crossing over (D) Independent assortment

806. Organisms phenotypically similar but genetically different are:
(A) Heterozygus (B) Homozygous
(C) Monozygous (D) Multizygous

807. Alleles are:
(A) Genes found on allosomes
(B) Pair of genes governing a specific character like height
(C) Genes governing eye characters
(D) None of these

808. Incomplete dominance is found in:
(A) Neurospora crassa *(B) Lathyrus odoratus*
(C) Pisum sativum *(D) Antirrhinum*

809. Hybrid vigour is due to:
(A) Mixing of traits of male and female
(B) Homozygosity in gamete state
(C) Superiority of genes
(D) Heterozygosity

810. If both parents are albino, all the offspring shall be:
(A) Albino
(B) Some albino, some normal
(C) Homozygous normal
(D) Heterozygous normal

811. Two individuals with similar external appearance but different genetic make up have the similar:
(A) Genotype (B) Phenotype
(C) Heterozygote (D) Homozygote

812. Out of two alleles of the same gene, one finds morphological expression. The phenomenon is:
(A) Epistasis (B) Codominance
(C) Dominance (D) Incomplete dominance

813. In *Cucurbita*, W is epistatic over Y and y gene normally responsible for yellow and green colour fruits. It produces white flowers. What is the ratio of fruits in the progeny of cross WwYy x wwYy?
(A) 9 white : 7 yellow : 0 green
(B) 3 white : 4 yellow : 1 green
(C) 4 white : 3 yellow : 1 green
(D) 2 white : 1 yellow : 1 green

814. A recessive trait in Garden pea is:
(A) Wrinkled seeds (B) Tall stem
(C) Round seeds (D) Coloured seed coat

815. A cross involving two pairs of alleles is:
(A) Crossing over (B) Genetics
(C) Dihybrid cross (D) Monohybrid cross

816. Which is correct about traits chosen by Mendel?
(A) Terminal pod is dominant
(B) Constricted pod is dominant
(C) Green coloured pod is dominant
(D) Tall plants are recessive

817. Tallness (T-) is dominant over dwarfness (tt) while red flower colour (R-) is dominant over white colour (rr). A plant with genotype TtRr is crossed with plant of genotype ttrr. Percentage of progeny having tall plants with red flowers is:
(A) 25% (B) 50% (C) 75% (D) 100%

818. A cross between pure tall pea plant with green pods and dwarf pea plant with yellow pods will produce tall F_2 plants, out of 16:
(A) 15 (B) 13 (C) 12 (D) 7

819. "Gametes are never hybrid". It is a statement of law of:
(A) Dominance (B) Segregation
(C) Independent assortment (D) Random fertilization

820. Mendel could not find recombination and crossing over as:
(A) Traits he chose were either present on different chromosomes or were far apart
(B) Traits chosen by him were not influenced by genes
(C) He did not have a high power microscope
(D) He selected only pure types

821. Most favourite and ideal material for researches in genetics is:
(A) Housefly (B) Mosquito
(C) Frog (D) Fruitfly

822. The average life of a crop variety is:
(A) 25-30 years
(B) 5-15 years
(C) 15-25 years
(D) 1-5 years

823. Gene resistance to Brown Plant Hopper (Nilaparvata lugens) of modern rice has been obtained from:
(A) Wild rice (B) African rice
(C) Old rice varieties from Kerala (D) IR-8

824. A species restricted to a given area is:
(A) Endemic species (B) Allopatric species
(C) Sympatric species (D) Sibling species

825. Endangered plant species are conserved through:
(A) Herbarium (B) Gene library
(C) Gene bank (D) Reducing pollution

826. Which of the following are homologous organs?
(A) Wings of birds, flippers of whales and forearms of man
(B) Wings of butterfly, pterygium of flying squirrel and fins of flying fishes
(C) Tails of snake, scorpion and wall lizard
(D) Lungs of rabbit, trachea of cockroach and book of scorpion

827. Resistance to herbicide chlorsulfuron in plants is due to a change in:
(A) Glutamine synthetase (B) Threonine deaminase
(C) Acetolactate synthase (D) DNA polymerase

828. Which of the following may be part of the reason why replaced uracil in DNA?
(A) Cytosine spontaneously deaminates to uracil
(B) Thymine synthetic pathway is easier
(C) Thymine is unstable
(D) Thymine has a stronger binding affinity for adnine than does uracil

829. Consider the following statements:
P. Wings of insects and birds are homologous organs
Q. Wings of insects and bats are analogous organs
R. Wings of insects and birds are analogous organs
S. Wings of birds and bats are homologous organs
Which of the above statements are correct?
(A) P and Q (B) P and S
(C) Q, R and S (D) P, Q and S

830. The late Mesozoic era (the cretaceous period) was characterized by:
(A) Rise of dinosaurs, rise of toothed birds and rise of flowering plants
(B) Extreme specialization of reptiles, extinction of great reptiles and extinction of toothed birds
(C) Rise of archai mammals, rise of dinosaurs and appearance of first toothed birds
(D) Extinction of archai mammals, rise of birds based appearance of first dinosaurs

831. Gene with multiple effects is:
(A) Epistatic (B) Pleiotropic
(C) Codominant (D) Supplementary

832. Monohybrid test cross ratio is:
(A) 1 : 2 : 1 (B) 3 : 1
(C) 1 : 1 (D) 2 : 1

833. Intraallelic gene suppression is called:
(A) Epistatis (B) Hypostatis
(C) Dominance (D) None of these

834. Mendel may not be able to establish the law of independent assortment, if both characters were located on:
(A) Separate non-homologous chromosomes
(B) Same homologous chromosomes
(C) Separate chromosome but at different loci
(D) Sex chromosomes

835. Who coined the term linkage?
(A) Bateson and Punnett (B) T.H. Morgan
(C) Correns (D) de Vries

836. How many linkage groups are present in cyanophyceae?
(A) 0 (B) 1 (C) 2 (D) 4

837. Crossing over is more frequent in:
(A) Males (B) Females
(C) Equal in both (D) None of these

838. Bar eye character in Drosophila is due to:
(A) Defficiency (B) Duplication
(C) Translocation (D) Polyploidy

839. Which is called 'illegitimate crossing over'?
(A) Inversion (B) Translocation
(C) Transposons (D) None of these

840. Trisomic were first obtained in:
(A) Chinese spring wheat (B) Pigweed
(C) Jimson weed (D) Maize

841. Colchicine is used in what concentration for arresting and breaking the spindle?
(A) 0.001-0.05 % (B) 0.1-0.5 %
(C) 0.01-0.5 % (D) 1-5%

842. More the distance between the two genes on a chromosome:
(A) More is the linkage strength
(B) Lesser is the linkage strength
(C) Linkage strength remains unchanged
(D) None of these

843. Rephanobrassica is a classical example of:
(A) Autopolyploidy (B) Allopolyploidy
(C) Segmental polyploidy (D) None of these

844. Which is hexaploid sps. of wheat?
(A) *Triticum durum* (B) *T. monococcum*
(C) *T. aestivum* (D) *T. turgidum*

845. Mutation is:
(A) Change that is inherited
(B) Change in parents not inherited
(C) Plant growth controlling factor
(D) Change which affects the offsprings of F_2 gen.

846. Earliest record of point mutation goes back to:
(A) 1691 (B) 1791 (C) 1910 (D) 1940

847. Suppression of action of genes by genes not allelomorphic to those suppressed is called:
(A) Dominance (B) Episome
(C) Codominance (D) Epistasis

848. Which is point of mutation in a cell?
(A) Nucleus (B) Chromosome
(C) Centromere (D) Gene

849. Most of mutations are:
(A) Forward (B) Reverse directional
(C) Directionless (D) None of these

850. Cytoplasmic male sterility in maize is inherited:
(A) Paternally (B) Maternally
(C) Both (A) and (B) (D) None of these

851. Induced mutations are most common in:
(A) Coarse cereals (B) Pulses
(C) Higher plants (D) Cereals

852. Auxotrophs in *Neurospora* were produced by which radiation Beadle and Ttum?
(A) γ- rays (B) α-rays
(C) UV-rays (D) IR-rays

853. Which of the following mutagens can be best used in inducing mutation in micro-organisms?
(A) X- rays (B) β- rays
(C) UV-rays (D) γ- rays

854. Which is the most widely used chemical mutagen in plant mutation breeding nowadays?
(A) Mustard gas (B) EMS
(C) MMS (D) Nitrous acid

855. Who is associated with 'Green Revolution' in India?
(A) E.J. Butler (B) M.S. Swaminathan
(C) R.S. Paroda (D) B.P. Pal

856. Which of the following is an achievement of mutations?
(A) Todd's Mitcham var. (B) Aruna var
(C) Sharbati sonora var (D) All of these

857. Which type of mutations is more frequent?
(A) Transitions (B) Transversions
(C) Both equal (D) Not known

858. A dwarf pea plant was treated with GA_3, it grew as pure tall plant. If it is crossed with a pure tall plant, then phenotypic ratio of F_1 is likely to be:
(A) All dwarf (B) 50% tall, 50% dwarf
(C) 75% tall, 25% dwarf (D) All tall

859. Four inbred/pure lines of Maize are crossed. The cross is
(A) Double cross (B) Dihybrid cross
(C) Tetrahybrid cross (D) Tetraploid cross

860. In high yielding hybrid crop varieties, hybrid vigour can be exploited only when hybrid seeds are purchased a fresh every year because:
(A) Govt. of India has accepted Dunkel's proposals
(B) Hybrid vigour is lost due to inbreeding depression
(C) Farmers are not allowed to grow their own seeds
(D) It is cheaper to purchase fresh seed

861. Which of the following is not used for crop improvement?
(A) Inbreeding (B) Introduction
(C) Hybridization (D) Mutations

862. Heterosis is:
(A) Hybrid sterility (B) Hybrid vigour
(C) Hybrid incompatability (D) Hybrid inviability

863. Best method to increase crop yield is (e.g., Wheat):
(A) Using tractors
(B) Sowing seeds of improved varieties
(C) Eradication of weeds
(D) Reduce ration holders

864. Improved rice variety IR-8 has been introduced in India from:
(A) Taiwan (B) Japan
(C) Phillipines (D) Bangladesh

865. Method of selection in vegetatively propagated plants is:
(A) Pure line selection (B) Pedigree selection
(C) Mass selection (D) Clonal selection

866. Which variety is not due to induced mutations?
(A) Aruna of Castor (B) Prabhat of Arhar
(C) Reimi of Rice (D) Sharbati Sonora of wheat

867. The desired varieties of economically useful crops are raised by:
(A) Vernalisation (B) Mutations
(C) Natural selection (D) Hybridisation

868. The indica varieties of rice are crossed with japonica varieties as they are:
(A) High yielding (B) Cheaper
(C) Short life cycles (D) Resistant to diseases

869. A scientist wants to study the effect of virus on a plant. Which part he should exclude for inoculation:
(A) Shoot apex (B) Stem cortex
(C) Phloem (D) Pith

870. Mass selection is preferred in case of:
(A) Vegetatively propagated plants
(B) Cross pollinated plnats
(C) Self pollinated plants
(D) Both (B) and (C)

871. Plant with similar genotype are:
(A) Haploids (B) Autopolyploids
(C) Clone (D) All the above

872. Quarantine regulation is meant for:
(A) Preventing entry of diseases plants in the country
(B) Spraying diseases plants with insecticides
(C) Promoting dry farming
(D) Growing fruits trees in all the states

873. Hybrid vigour is best maintained in vegetatively reproducing plants because they are:
(A) Resistant to diseases
(B) Easily propagated
(C) With long life span
(D) Little liable to lose vigor due to absence of sexual reproduction

874. The first transgenic crop was:
(A) Pea (B) Tobacco
(C) Flax (D) Cotton

875. The use and disuse principles of evolution or theory of "inheritance of acquired characters" was proposed by:
(A) Hugo de Vries (B) Lamarck
(C) Weismann (D) Darwin

876. For evolutionary success, mutation must occur in:
(A) Somatic RNA (B) Somatic DNA
(C) Germplasm DNA (D) Germplasm RNA

877. A starfish with six arms may be a case of:
(A) Variation (B) Evolution
(C) Autotomy (D) Mutation

878. There is more competition for survival between:
(A) Different animals of same niche
(B) Same animals in same niche
(C) Different animals of different niche
(D) Same animals of different niche

879. Who wrote the "Origin of species"?
(A) Lamarck (B) Darwin
(C) De Vries (D) Mendel

880. Theory of natural selection was proposed by:
(A) Lamarck (B) Darwin
(C) De Vries (D) Haeckel

881. The evolution of species is based upon sum total of adaptive changes preserved by:
(A) Hybridization (B) Isolation
(C) Human conservation (D) Speciation

882. According to Neo-Darwinism variations which are the raw material of evolution occur:
(A) In an individual and remain restricted to him
(B) In gene pool of a population
(C) In all the membranes of one species
(D) In all the members of all species

883. Neo-Darwinism believes that new species develop through:
(A) Mutations with natural selection
(B) Continuous variations with natural selection
(C) Hybridisation
(D) Mutations

884. Theory of inheritance of acquired characters was given by:
(A) Wallace (B) Lamarck
(C) Darwin (D) De Vries

885. 'Hugo de Vries' contribution is:
(A) Theory of natural selection (B) Theory of mutations
(C) Law of Dominance (D) Law of segregation

886. Darwin was most influence by:
(A) Lamarck's theory of acquired characters
(B) Weismann's theory of germplasm
(C) Wallace's theory of origin of species
(D) Essay on population by Malthus

887. Theory of natural selection dwells on:
(A) Role of environment in evolution
(B) Natural selection acting on favourable variations
(C) Changes in gene complex resulting in heritable variations
(D) Inheritance of acquired characters

888. Frequency of a character increases when it is:
(A) Recessive (B) Dominant
(C) Inheritable (D) Adaptable

889. Natural selection theory of Darwin states that:
(A) Environment does not play any role in evolution
(B) Heritable variations arise through changes in gene complex of a species
(C) Natural selection acts on favourable variations which appear among individuals of a species
(D) Characters acquired during the life of an individual are inherited by its offspring

890. Light coloured Pepped Moth/*Biston betularia* gets changed to its darker carbonaria variety due to:
(A) Translocation of block of genes in response to heavy carbons
(B) Deletion of gene segment due to industrial pollution
(C) Mutation of single mendelian gene for survival in smoke laden industrial environment
(D) Industrial carbon deposited on wings

891. De Vries theory of mutation is:
(A) Opposed to natural selection theory
(B) Not opposed to natural selection theory
(C) Opposed to germplasm theory
(D) Opposed to Lamarck's theory

892. Gene frequency in a population remains stable unless and until there is:
(A) Random drift (B) Random mating
(C) Selection (D) Mutation

893. Darwin's Natural Selection is based on:
(A) Use and disuse
(B) Inheritance of acquired characters
(C) Variations
(D) Prodigality, struggle for existence and survival of fittest

894. Genetic drift operates in:
(A) Small population (B) Large population
(C) Island population (D) Mendelian population

895. Naturalist who sailed round the world in ship Beagle was:
(A) Charles Lyell (B) Charles Darwin
(C) Alfred Wallace (D) Lamarck

896. Lederberg's replica experiment explains:
(A) Lamarck's theory (B) Mutation theory
(C) Darwin's theory (D) None of these

897. Chromatid is:
(A) One half of chromosome
(B) Haploid chromosome
(C) Complete chromosome
(D) Duplicate chromosome

898. Mendel did not get any linkage in his experiments on pea. One of the reasons was that:
(A) He did not keep an exact record
(B) There is no linkage in pea
(C) He did not have means to detect linkage
(D) All the seven characters selected by him were present on different chromosomes or showed 50% cross-overs

899. *Neurospora crassa* is widely used in genetical studies because of all the following except one:
(A) It is a haploid plant and mutations can be easily detected
(B) It can be easily cultured
(C) Life cycle short
(D) Spores are not affected by mutagens

900. Plotting of specific genes on the chromosome is known as:
(A) Chromosome map (B) Linkage map
(C) Genetic map (D) All of these

901. In gene mutation, adenine is replaced by guanine. It is known as:
(A) Substitution (B) Point mutation
(C) Transition (D) Transversion

902. In Neurospora, both recessive and dominant alleles express their effect because:
(A) Two genes control each character
(B) There are two alleles for each character
(C) The organism contains only one allele for a gene
(D) Each gene only one allele

903. Genes located on X-chromosomes are called:
(A) Sex influenced (B) Sex limited
(C) Sex linked (D) Maternal genes

904. Mutation is:
(A) A change in phenotype of an organism caused by some environmental factor
(B) An inheritable change in genetic material
(C) A temporary change in structure of the nucleus
(D) Any one of these

905. 'Illegitimate crossing over' is another term for:
(A) Transition (B) Transversion
(C) Reciprocal translocation (D) None of these

906. Which of the following is a case of autoallopolyploidy?
(A) AA, AA, A (B) AA, AA, AB
(C) AA-1 (D) 2n -2

907. A mutation changes the original base sequence of DNA, GATAGCCG to a new sequence GGTAGCCG. What is the type of mutation?
(A) Transition (B) Transversion
(C) Translocation (D) Inversion

908. A genomatic mutation is the mutation involving:
(A) Change in gene
(B) Change in chromosomal structure
(C) Change in the number of chromosomes
(D) All of these

909. When a purine is replaced by a pyrimidine in a part of DNA, the mutation so produced is called:
(A) Transition (B) Transversion
(C) Deletion (D) Reversal

910. Mutations were first induced in Drosophila by the use of:
(A) X-ray (B) DDT
(C) High temperature (D) UV –rays

911. The mutations which prove fatal for the organism are called:
(A) Spontaneous (B) Induced
(C) Deleterious (D) Lethal

912. Cytoplasmic inheritance is due to:
(A) Plastids (B) Mitochondria
(C) Cytoplasmic particles (D) All of these

913. A trisomic individual has:
(A) One extra chromosome
(B) One less chromosome
(C) Two extra chromosomes
(D) One pair of extra chromosomes

914. It is a common observation that some insects have developed resistance to DDT. The reason may be attributed to:
(A) Poor quality of DDT (B) Genetic recombination
(C) Genetic mutation (D) Hybrid vigour

915. The frequency of mutations:
(A) Can be increased by recombination
(B) Varies in the character and the organism
(C) Is affected by environmental factors
(D) Can be increased by x-rays

916. Genome represents:
(A) Haploid chromosome set
(B) Complete chromosome set
(C) Diploid chromosome set
(D) All the genes present in the population

917. Centromere is part of:
(A) Cytoplasmic organelles (B) Ribosome
(C) Chromosome (D) Nucleolus

918. Allosomes are the name of:
(A) Sex chromosomes
(B) Swellings on the chromosomes
(C) Chromosomes other than the ones which determine sex
(D) Nucleolus organising regions of chromosomes

919. Down's syndrome is due to:
(A) Crossing over
(B) Linkage
(C) Sex-linked inheritance
(D) Nondisjunction of chromosomes

920. Crossing over takes place in:
(A) One strand stage (B) Two strand stage
(C) Three strand stage (D) Four strand stage

921. Two linked genes a and b show 20% recombination. The individuals of a dihybrid cross between ++/++ x ab/ab shall show gametes:
(A) ++ 80 : ab : 20
(B) ++ 50 : ab : 50
(C) ++ 40 : ab : 40 : + a10 : + b : 10
(D) ++ 30 : ab : 30 : + a 20 : + b : 20

922. Chromosome with median centromere and equal arms is:
(A) Acrocentric (B) Submetacentric
(C) Telocentric (D) Metacentric

923. Balbiani rings occur in:
(A) Polytene chromosomes
(B) Lampbrush chromosomes
(C) Polysomes
(D) Heterosomes

924. Which is a base analogue?
(A) 5-Bromouracil (B) Caffeine
(C) Colchicine (D) Nitrous acid

925. Mutations used in agriculture are commonly:
(A) Induced (B) Spontaneous
(C) Lethal (D) Recessive and lethal

926. The major source of variations are:
(A) Polyploidy
(B) Mutations
(C) Chromosome aberrations
(D) Segregation

927. The gene that controls the rate of mutation of another gene is:
(A) Regular gene (B) Inducer gene
(C) Mutable gene (D) Mutator gene

928. In Neurospora, 8 ascospores are formed instead of 4. This indicates:
(A) One meiosis
(B) Two meiosis
(C) Two mitosis
(D) Meiosis followed by mitosis

929. Which one can induced polyploidy?
(A) Colchicine (B) Acridines
(C) Ethylene (D) Maleic hydrazide

930. Harmful mutations do not get eliminated from gene pool because:
(A) They have survival value
(B) Genetic drift
(C) They are formed repeatedly
(D) They are recessive and carried by heterozygous individuals

931. Mutations can be brought about by:
(A) Aniline dye (B) X-rays
(C) Auxins (D) DDT

932. A normal green male maize is crossed with albino female. The progeny is albino because:
(A) Trait for albinism is dominant
(B) The albinos have biochemical to destroy plastids derived from green male
(C) Plastids are inherited from female parent
(D) Green plastids of male must have mutated

933. Which one of the following carries extra nuclear genetic material?
(A) Golgi apparatus (B) Ribosomes
(C) Chromosome (D) Plastids

934. Smallest segment of genetic material affected by mutation is:
(A) Recon (B) Cistron
(C) Muton (D) Exon

935. Chromosome pairing in salivary gland chromosomes resembles zygotene pairing in:
(A) It is followed by crossing over
(B) It can occur in nonhomologous segments
(C) Pairing occurs in homologous regions only
(D) It is followed by reductional separation

936. Two dominant nonallelic genes are 50 map units apart. The linkage is:
(A) Cis type (B) Trans type
(C) Complete (D) Absent/Incomplete

937. Which arrangement of ascospores in Neurospora does not represent second generation segregation?
(A) aaaaAAAA (B) AAaaAAaa
(C) AAaaaaAA (D) aaAAAAaa

938. Wheat plant is 6n=42. What will be the number of chromosomes in its monosomic, haploid number and monoploid?
(A) 43, 21 and 7 (B) 41, 21 and 7
(C) 15, 7 and 7 (D) 13, 7 and 7

939. In a cross between genotype AB and ++. 650 out of 1000 individuals were of parental type. The distance between A and B is:
(A) 35 map units (B) 45 map units
(C) 15 map units (D) 30 map units

940. Mutation can be induced in bacteria by:
(A) Growing different strains in same culture
(B) Starvation
(C) Providing growth substances
(D) High energy radiation

941. Mendel found that some traits do not assort independently. Later workers found it due to:
(A) Linkage
(B) Crossing over
(C) Amitosis
(D) Dominance of one trait over another

942. Number of genes on a chromosome is:
(A) 1 (B) 10 (C) 35 (D) Infinite

943. Different types of chromosomes can be recognised by the position of:
(A) Genes (B) Centromere
(C) Centrosome (D) Spindle

944. Exchange of segments between non-homologous chromosomes is:
(A) Translocation (B) Inversion
(C) Crossing over (D) Tetrasomy

945. In order to remain linked, the distance between two genes should not increase beyond:
(A) 10 map units (B) 20 map units
(C) 40 map units (D) 50 map units

946. Which pteridophyte has the maximum chromosome number?
(A) *Ophioglossum reticulatum*
(B) *Azolla pinnata*
(C) *Lycopodium cernuum*
(D) *Selaginella apus*

947. Lampbrush chromosomes are seen during :
(A) Meiotic metaphase (B) Mitotic metaphase
(C) Mitotic prophase (D) Meiotic prophase

948. Sudden change which breeds true is:
(A) Mutation
(B) Law of inheritance
(C) Inheritance of acquired characters
(D) Natural selection

949. Taylor demonstrated chromosome replication to be semiconservative in:
(A) *Vicia faba* (B) HeLa cell
(C) Pea (D) Mouse liver cell

950. *Drosophila melanogaster* possesses:
(A) 3 pairs autosomes + 1 pair sex chromosomes
(B) 2 pairs autosomes + 2 pairs sex chromosomes
(C) 1 pair autosomes + 3 pairs sex chromosomes
(D) 2 pairs autosomes + 1 pair sex chromosomes

951. In Down's syndrome of a male child, the sex complement is:
(A) XO (B) XY (C) XX (D) XXY

952. Mutation is:
(A) Change in gene frequency
(B) Genetic drift
(C) Change in base pairs in DNA molecule
(D) Environmental mechanism of evolution

953. A gene is:
(A) Three letter code in DNA
(B) Equivalent to chromosome
(C) Part of chromosome producing a set of enzymes (protein)
(D) Part of chromosome producing only one enzyme (protein)

954. Albinism and phenyl ketonuria are disorders due to:
(A) Recessive autosomal genes
(B) Dominant autosomal genes
(C) Dominant sex genes
(D) Recessive sex genes

955. If there is complete linkage in F_2 generation:
(A) Parental types and recombinations appear in equal ratio
(B) Recombinants are less than parental types
(C) Recombinants are more than parental types
(D) There will be only parental types

956. Lampbrush chromosomes occur inside:
(A) Nucleus of human cells
(B) Oocytes of frog
(C) Salivary glands of Drosophila
(D) Salivary gland of silkworm

957. Down's syndrome is due to trisomy of 21st chromosome caused by:
(A) Nondisjunction during egg formation
(B) Nondisjunction during sperm formation
(C) Addition of extra chromosome during mitosis of zygote
(D) Either (A) or (B)

958. Distance between two linked genes is measured in map units that depict:
(A) Ratio of crossing over between them
(B) Cross – over value
(C) Number of genes between them
(D) Both (B) and (C)

959. Extra 18th autosomal chromosome results in:
(A) Edward's syndrome (B) Patau's syndrome
(C) Down's syndrome (D) None of these

960. Queen Victoria of England was:
(A) Haemophiliac carrier (B) Colour blind
(C) AIDS patient (D) Deaf

961. Extra nuclear genes are present in:
(A) Cytoplasm
(B) E.R. and Cytoplasm
(C) Ribosome and Cytoplasm
(D) Mitochondria and cytoplasmic particles

962. The back cross is:
(A) A cross between F_1 individual and F_2 individual
(B) A cross between an F_1 individual with another F_1 individual
(C) Cross between F_1 and one of the two parents
(D) Cross between F_2 with one of the parents

963. A monohybrid cross is the one in which:
(A) Only a single plant is involved for the experiment
(B) A single pair of contrasting characters is considered for the genetic results
(C) A hybrid is crossed to a homozygous
(D) None of these

964. Even harmful mutations do not get eliminated from gene pool due to:
(A) Genetic drift
(B) Higher frequency due to dominant nature
(C) Being recessive and persisting in heterozygous condition
(D) Survival value

965. Turner's syndrome is represented by:
(A) XYY (B) XO (C) XXY (D) XXX

966. The law of segregation of characters postulated by Mendel can be related to:
(A) The presence of two genes for each character in a somatic cell
(B) A gamete receiving only one of the two homologous chromosomes during meiosis
(C) Presence of both genes on the same chromosome
(D) None of these

967. Mendel selected pea as material for his experiments because:
(A) It is an annual plant with comparatively short life cycle
(B) The flower are self pollinated
(C) The number of seeds produced is quite large
(D) All of these

968. At which stage is the fate of genetic constitution of gametes finally decided?
(A) Metaphase – I (B) Anaphase – I
(C) Anaphase – II (D) Interkinesis

969. How would you test a pea plant whether it is a pure or hybrid for tallness?
(A) Cross it with another tall pea plant of unknown genotype
(B) Cross it with a pure tall pea plant
(C) Cross with a homolozygous dwarf pea
(D) Cross it with any pea plant

970. According to the law of Independent assortment in a dihybrid cross:
(A) There are four genotypes in F_2
(B) F_2 contains 16 phenotypes
(C) There is a single individual which is homozygous recessive for both the characters
(D) It is not possible to forecast the different pheno types

971. The number of different type of gametes produced from a plant with genotype Aa Bb Cc is:
(A) 2 (B) 4 (C) 8 (D) 16

972. How many types of gametes are expected from the plants with genotypes AA, BB, aa, bb?
(A) 3 (B) 4 (C) 8 (D) 16

973. A haploid set of all the genes present in a gamete is called:
(A) Genotype (B) Phenotype
(C) Genome (D) Linkage group

974. The mendelian principle, which has always stood true is:
(A) The law of independent assortment
(B) The law of segregation
(C) The law of dominance
(D) All of these

975. A tobacco plant heterozygous for albinism (a recessive character) is self pollinated and 1200 seeds are subsequently germinated. How many seedlings would have the parental phenotype?
(A) 400 (B) 600 (C) 900 (D) 1200

976. Mendel is popular for postulating:
(A) Origin of species (B) Cell theory
(C) Linkage theory (D) Laws of inheritance

977. 'Like begets like' an important and universal phenomenon of life, is due to:
(A) Eugenics (B) Inheritance
(C) Dominance (D) Crossing over

978. Why pea plants more suitable than dogs for mendelian experiments?
(A) Pea plants can be self pollinated
(B) All pea plants are diploid
(C) There were no pedigree records of dogs
(D) Dogs have many genetic traits

979. Multiple alleles are present:
(A) At different loci in the same chromosome
(B) In different chromosomes
(C) At the same locus in one type of chromosomes
(D) None of these

980. A dwarf pea plant was treated with GA. The plant became tall. The treated plant was then crossed with a homozygous tall pea. The results in F_2 are expected to be:
(A) All tall
(B) Tall and dwarf in 3 : 1 ratio
(C) 50% tall
(D) All dwarf

981. Mendel observed red flowers in F_1 when he crossed red and white because of:
(A) Dominance
(B) Recessive gene
(C) Law of independent assortment
(D) Law of segregation

982. The genotypic ratio of a monohybrid cross will be:
(A) 3 : 1 : 1 (B) 1 : 1 : 1
(C) 1 : 2 : 1 (D) 2 : 1 : 1

983. A cross between unlike organisms is called:
(A) Test cross (B) Back cross
(C) Composite (D) Hybrid

984. Had Mendel decided to study those traits together which are determined by linked genes, he would not have found out:
(A) Principle of purity of gamete
(B) Principles of codominance
(C) Principle of segregation
(D) Principle of independent assortment

985. A pure tall plant is reared in a soil poor in nutrition and reached the size of dwarf plant. If this plant is selfed, the phenotype in the F1 generation is most likely to be:
(A) All tall plants (B) 50% tall and 50% dwarf
(C) All dwarf (D) Data insufficient

986. A modified dihybrid mendelian ratio of 9 : 3 : 4 indicates:
(A) Supplementary genes (B) Epistatic genes
(C) Lethal genes (D) Complementary genes

987. Genetics deals with:
(A) Heredity and Variation (B) Heredity
(C) Mutations (D) Variation

988. In a dihybrid cross, F2 phenotypic ratio is 13 : 3. It is case of:
(A) Complementary genes (B) Epistatic genes
(C) Multigenic inheritance (D) Incomplete dominance

989. Mendel conducted hybridization experiments on:
(A) Pigeonpea (B) Garden pea
(C) Wild pea (D) Sweet pea

990. F_2 generation of a cross between two white flowered strains of sweet pea yields 9 purple flowered plants: 7 white flowered plants. This is an example of:
(A) Epistasis (B) Complemetary genes
(C) Supplementary genes (D) Gene inhibition

991. The body feature are:
(A) Phenotype
(B) Genotype
(C) Sex of the individual
(D) Controlled by environment

992. Nonallelic genes having similar phenotypic effect, interact to produce a different trait and a ratio of 9 : 7 in F_2 generation are:
(A) Epistatic genes (B) Hypostatic genes
(C) Supplementary genes (D) Complementary genes

993. Mendel's second law is:
(A) Segregation
(B) Dominance
(C) Independent assortment
(D) Polygenic inheritance

994. When two hybrids rrTt and Rrtt are crossed, the phenotypic ratio of offspring shall be:
(A) 3 : 1 : 1 (B) 1 : 1 : 1 : 1
(C) 1 : 1 : 1 (D) 9 : 3 : 3 : 1

995. Gametes of AaBb individual can be:
(A) Aa, Bb (B) AB, ab
(C) AB, ab, aB (D) AB, Ab, aB, ab

996. Segregation of Mendelian factors (no linkage, no crossing over) occurs during:
(A) Anaphase I (B) Anaphase II
(C) Metaphase I (D) Metaphase II

997. Back cross is:
(A) F_1 x F_1
(B) F_1 x Recessive parent
(C) F_1 x Dominant parent
(D) F_1 x Any parent

998. Mendel's second law of independent assortment can be demonstrated by:
(A) Test cross (B) Back cross
(C) Monohybrid cross (D) Dihybrid cross

999. In a cross 45 tall and 14 dwarf plants were obtained. Genotype of parents was:
(A) TT x TT (B) TT x Tt
(C) Tt x Tt (D) TT x tt

1000. Which one is exception to Mendel's principle of dominance?
(A) Wild pea (B) Mirabilis
(C) Garden pea (D) Pigeonpea

1001. Genes controlling seven traits in pea studied by Mendel were actually located on:
(A) Two chromosomes (B) Three chromosomes
(C) Four chromosomes (D) Five chromosomes

1002. In a dihybrid cross AABB x aabb, F_2 progeny of AABB, AABb, AaBB and AaBb occurs in the ratio of:
(A) 1 : 1 : 1 : 1 (B) 9 : 3 : 3 : 1
(C) 1 : 2 : 2 :1 (D) 1 : 2 : 2 : 4

1003. F_1 generation is produced by:
(A) Crossing F_1 progeny with one of the parents
(B) Selfing the progeny of two individual parents
(C) Selfing the parents
(D) Recessive cross between individual parents

1004. Yellow colour is dominant over green colour. In a generation of 50% yellow and 50% green individuals, the parents are:
(A) Both homozygous
(B) Both heterozygous
(C) One of them is heterozygous
(D) None of these

1005. How many types of gametes are required for F_2 generation of a monohybrid cross?
(A) 4 (B) 8 (C) 16 (D) 32

1006. Two genetic loci produce identical phenotypes in *cis* as well as *trans* position. They are:
(A) Pseudoalleles (B) Multiple alleles
(C) Different genes (D) Parts of same gene

1007. The *Triticum aestivum* is a wheat:
(A) Diploid (B) Tetraploid
(C) Hexaploid (D) Monoploid

1008. The number of known species in genus Gossypium have been estimated at:
(A) 10 (B) 15 (C) 20 (D) 25

1009. The principle male sterile line introduced for hybrid pearlmillet production in India was:
(A) Tift-23 A (B) BJ-104
(C) MHB -118 (D) CO-2

1010. Self pollination is otherwise called as:
(A) Allogamy (B) Hamogamy
(C) Autogamy (D) Chasmogamy

1011. Cross pollination is otherwise called as:
(A) Allogamy (B) Hamogamy
(C) Autogamy (D) Chasmogamy

1012. The flower do not open and self pollination is the rule:
(A) Allogamy (B) Hamogamy
(C) Autogamy (D) Cleistogamy

1013. The differential maturation of male and female flowers is known as:
(A) Protandry (B) Protogyny
(C) Dichogamy (D) Dicliny

1014. The union of male and female gametes in sexual reproduction is known as:
(A) Fertilization (B) Pollination
(C) Infertility (D) Incompatibility

1015. The transfer of pollen to stigma of different flowers of same plant is called as:
(A) Chasmogamy (B) Cleistogamy
(C) Geitenogamy (D) Xenogamy

1016. The transfer of pollen to stigma of different flowers of two different plant is called as:
(A) Chasmogamy (B) Cleistogamy
(C) Geitenogamy (D) Xenogamy

1017. When mutation occur naturally that is called as:
(A) Somatic mutation (B) Germinal mutation
(C) Induced mutation (D) Spontaneous mutation

1018. Physical mutation is:
(A) EMS (B) Ethidium bromide
(C) X-rays (D) Nitrous acid

1019. Mutation agent is:
(A) X-rays (B) EMS
(C) β-rays (D) UV radiation

1020. Mutations occurring in bready cells, do not transmitted to next generation is called:
(A) Germinal mutation (B) Somatic mutation
(C) Induced mutation (D) Reverse mutation

1021. Development of a new mutant type from a wild type called as:
(A) Forward mutation (B) Reverse mutation
(C) Delinked mutation (D) Somatic mutation

1022. Gene mutation otherwise called as:
(A) Chromosomal aberration
(B) Chromosomal variation
(C) Point mutation
(D) Endogenous mutation

1023. The phenomenon of heterosis of the progeny over better parent is:
(A) Heterosis (B) Heterobeltiosis
(C) Relative heterosis (D) Standard heterosis

1024. Heterosis over mid parent value:
(A) Heterosis (B) Heterobeltiosis
(C) Relative heterosis (D) Standard heterosis

1025. Heterosis over the check parent:
(A) Heterosis (B) Heterobeltiosis
(C) Relative heterosis (D) Standard heterosis

1026. Which one of the following formula is used to calculate relative heterosis:
(A) $\frac{F - BP \times 100}{BP}$ (B) $\frac{F - MP \times 100}{MP}$
(C) $\frac{F - CV \times 100}{CV}$ (D) $\frac{F - CP \times 100}{CP}$

1027. Which one of the following formula is used to calculate relative heterobeltiosis:
(A) $\frac{F - BP \times 100}{BP}$ (B) $\frac{F - MP \times 100}{MP}$
(C) $\frac{F - CV \times 100}{CV}$ (D) $\frac{F - CP \times 100}{CP}$

1028. Which one of the following formula is used to calculate relative standard heterosis:
(A) $\frac{F - BP \times 100}{BP}$ (B) $\frac{F - MP \times 100}{MP}$
(C) $\frac{F - CV \times 100}{CV}$ (D) $\frac{F - CP \times 100}{CP}$

1029. The plant tissue or organ excised and used for invitro culture is known as:
(A) Transgenic plant (B) Bulb
(C) Explant (D) Callus

1030. Triticale has been developed by crossing wheat with:
(A) Barley (B) Phalaris
(C) Rye (D) Bajra

1031. Rice gene bank is situated at:
(A) India (B) Philippines
(C) China (D) Japan

1032. Different types of hybrids available in sorghum:
(A) Mono cross (B) Double cross
(C) Triple cross (D) Single cross

1033. A x B crossing in pearlmillet is done at the stage of:
(A) Certified seed (B) Breeder seed
(C) Foundation seed (D) Registered seed

1034. Top cross is otherwise known as:
(A) Inbred variety cross (B) Single cross
(C) Double cross (D) Triple cross

1035. In three way cross hybrid seed production, the pedigree involved in:
(A) A x B
(B) (A x B) x C
(C) (A x B) x open pollinated variety
(D) (A x B) x C (C x D)

1036. The first generation resulting from crossing of two inbreds is:
(A) Hybrid (B) Synthetics
(C) Composites (D) Variety

1037. The phenomenon of heterosis of the progeny over better parent is:
(A) Heterosis (B) Heterobeltiosis
(C) Relative heterosis (D) Standard heterosis

1038. The first generation resulting from the controlled crossing of an approved inbred line and open pollinated variety is:
(A) Single cross (B) Double cross
(C) Top cross (D) Four way cross

1039. A cross between parents differing in a single gene is:
(A) Double hybrid (B) Mono hybrid
(C) Single cross hybrid (D) Triple cross hybrid

1040. An alloautogamous oilseed is:
(A) Safflower (B) Groundnut
(C) Sesame (D) Mustard

1041. The average performance of a strain or genotype in a series of crosses or hybrid combinations and a measure of additive gene action is known as:
(A) General combining ability
(B) Specific combining ability
(C) Joint combining ability
(D) Directed combining ability

1042. The heterosis estimated over the superior parent is known as:
(A) Heterobeltiosis (B) Clonal degeneration
(C) Standard heterosis (D) Inbreeding depression

1043. The superiority of an F_1 hybrid over both its parents in terms of yield and some other characters is defined as:
(A) Inbreeding depression
(B) Heterosis
(C) Clonal degeneration
(D) Heterobeltiosis

1044. The progeny obtained by hybridization between two or more genetically dissimilar individuals, strains, pure lines etc., is called:
(A) Embryo (B) Explant
(C) Cybrid (D) Hybrid

1045. The homozygous line developed by continued inbreeding, usually selfing, accompanied by selection in cross pollinated crops is called:
(A) Hybrid derivative (B) Inbred line
(C) Ideotype (D)Cybrid lines

1046. The reduction or loss in vigour and fertility as a result of inbreeding is called:
(A) Inbreeding depression (B) Co-heritability
(C) Heterosis (D) Heritability

1047. Two lines, which are identical with respect to all other genes, except for the gene under investigation are known as:
(A) Multilines (B) Isogenic lines
(C) Hybrid lines (D) Transition lines

1048. The appearance of a plant with respect to particular character, such as plant height, flower colour etc., is known as:
(A) Genotype (B) Phenotype
(C) Pathotype (D) Biotype

1049. The observable variation present in a character in a population which includes genotypic and environmental components of variation is known as:
(A) Phenotypic variation (B) Genotypic variation
(C) Pathotype variation (D) Biotype variation

1050. National Biodiversity Board in India is situated at:
(A) Bangalore (B) New Delhi
(C) Mumbai (D) Chennai

1051. Progeny of a single homozygous plant of a self pollinated species is called:
(A) Dihybrid (B) Pure line
(C) Inbred (D) Monohybrid

1052. The variety produced by crossing in all combinations of a number of lines that combine well with each other is called:
(A) Varietal mixtures (B) Hybrid variety
(C) Composite variety (D) Synthetic variety

1053. The cross between a single cross hybrid (A x B) and inbred (C) is termed as:
(A) Three way cross (B) Double cross
(C) Single cross (D) Poly cross

1054. The male sterile parent is crossed with which one of the following line to produce hybrid seed:
(A) A line (B) B line
(C) R line (D) C line

1055. Hybrid rice seed production is carried out by:
(A) One line system (B) Multi line system
(C) Three line system (D) Four line system

1056. The first generation progeny of a cross between two single cross hybrids:
(A) Single cross hybrid (B) Double cross hybrid
(C) Back cross hybrid (D) Three way cross hybrid

1057. The dominance hybrid is first proposed by:
(A) Shull (B) Davenport
(C) Bruce (D) Jones

1058. A line became homozygous by repeated selfing upto:
(A) 3 generations (B) 5 generations
(C) 8 generations (D) 12 generations

1059. Over dominance hypothesis was independently proposed by:
(A) Shull (B) Collins
(C) Jones (D) East and Shull

1060. Single cross hybrid in maize is between:
(A) A x B (B) (C x D) x (C)
(C) (A x A) x (C x D) (D) (A x B) x C

1061. The breeding tool for cotton hybrid seed production:
(A) Self incompatibility
(B) Cytoplasmic male sterility
(C) Emasculation and dusting
(D) Genetic male sterility

1062. Presence of B line in A line of paddy hybrid seed production field is:
(A) Pollen shedder (B) Off-type
(C) Rogue (D) Weed

1063. Supplementary pollination is required for:
(A) Sorghum (B) Cotton
(C) Safflower (D) Rice

1064. Rope pulling is practiced for good seed set in hybrid seed production of:
(A) Castor (B) Sunflower
(C) Rice (D) Sorghum

1065. Genetic contaminants present in hybrid rice field:
(A) Shedding tassel (B) Self bolls
(C) Pollen shedder (D) Diseased plants

1066. The genetic contaminant present in hybrid maize field:
(A) Off-type plants (B) Weed plants
(C) Diseased plants (D) Shedding tassel

1067. Isolation distance for hybrid rice seed production:
(A) 200 m (B) 250 m
(C) 300 m (D) 400 m

1068. The breeding tool for pigeonpea hybrid seed production is:
(A) Cytoplasmic male sterility
(B) Genetic male sterility
(C) Emasculation and dusting
(D) Self incompatibility

1069. The seed replacement rate for hybrids:
(A) 25 % (B) 50 % (C) 75 % (D) 100 %

1070. Hand rubbing is practiced for good seed set in hybrid seed production of:
(A) Rice (B) Sorghum
(C) Sunflower (D) Castor

1071. Staggered sowing is practiced for:
(A) Easy processing
(B) Good field establishment
(C) Tolerance of pest and diseases
(D) Uniform flowering

1072. Detasseling is the technique used for hybrid seed production in:
(A) Maize (B) Tomato
(C) Castor (D) Cotton

1073. Presence of B line in A line in sunflower hybrid seed production plot is called:
(A) Off-type (B) Pollen shedder
(C) Rogue (D) Weed

1074. Isolation could be modified based on the number of border rows in:
(A) Maize (B) Rice
(C) Sunflower (D) Castor

1075. The technique used for hybrid seed production of sunflower is:
(A) CMS (B) GMS
(C) CGMS (D) Emasculation and dusting

1076. Which lines are used for production of cucumber:
(A) Monoecious (B) Gynoecious
(C) Gynomonoecious (D) Androgynoecious

1077. The crop in which the first Indian vegetable hybrid was released:
(A) Bottlegourd (B) Bittergourd
(C) Ridgegourd (D) Ashgourd

1078. In double cross hybrid seed production of maize, number of inbreds involved are:
(A) 2 (B) 3 (C) 4 (D) 5

1079. Chemical hybriding agent is:
(A) Sodium arsenate (B) $ZnSO_4$
(C) $MnSO_4$ (D) GA_3

1080. Chemicals used for killing the male gamete:
(A) Sodium Methyl arsenate
(B) Sodium chloride
(C) Sodium azide
(D) Sodium acetate

1081. R line in CMS hybrid seed production:
(A) Male fertile (B) Male sterile
(C) Female sterile (D) Female fertile

1082. Emasculation and dusting technique is practiced in:
(A) Maize (B) Paddy
(C) Sunflower (D) Cotton

1083. A line referred as:
(A) Male parent (B) Female parent
(C) F1 line (D) Fertile plant

1084. A type of male sterility that is conditioned by the cytoplasm rather than by the nuclear genes and transmitted only through the female parent is called:
(A) Genetic male sterility
(B) Cytoplasmic genetic male fertility
(C) Cytoplasmic male sterility
(D) Self incompatibility

1085. The male sterility that is dependent upon the action of genes carried in the nucleus with a particular cytoplasm is called:
(A) Genetic male sterility
(B) Cytoplasmic genetic male fertility
(C) Cytoplasmic male sterility
(D) Self incompatibility

1086. The chemicals, which are used to induce male sterility in crop plants is called:
(A) Nematicides (B) Herbicides
(C) Fungicides (D) Gametocides

1087. Normally, male sterility in plants can be used to develop:
(A) Purelines (B) Inbreds
(C) Hybrids (D) Synthetics

1088. The maintainer line in hybrid seed production denotes:
(A) A line (B) R line
(C) B line (D) C line

1089. The hybrid seed production technique adopted in rise is:
(A) CGMS (B) GMS
(C) Block method (D) CMS

1090. Cytoplasmic male sterile lines are also known as:
(A) R line (B) B line
(C) A line (D) Inbred

1091. Crossing over between non-sister chromatids of homologous chromosomes takes place during:
(A) Laptotene (B) Pachytene
(C) Diplotene (D) Zygotene

1092. The term heterosis was coined by:
(A) G.H. Shull (B) W. Bateson
(C) T.H. Morgan (D) E.M. East

1093. When a tranogenic plant is crossed with a non-transgenic, what would be zygosity status of the F1 plant?
(A) Homozygous (B) Heterozygous
(C) Hemizygous (D) Nullizygous

1094. Jacobsen Syndrome is due to:
(A) Deletion (B) Duplication
(C) Translocation (D) Inversion

1095. Pickout the biological function(s) of RNA silencing:
(A) Immunity against viruses or transporons
(B) Down-regulation of genes
(C) Upgradation of gene
(D) All of the above

1096. Polygenic traits in crops can be identified by:
(A) QTL mapping
(B) Cluster analysis
(C) Tandem analysis
(D) Gene mapping

1097. Among the following which is not an assumption of Hardy Weinberg rule?
(A) No mutation (B) Random mating
(C) No natural selection (D) Small population size

1098. The DNA segments may be mapped by locating the restriction sites through restriction enzymes called restriction mapping. When this is extended to complete chromosomes, it is called:
(A) Chromosome sketching
(B) Chromosome walking
(C) Chromosome reading
(D) Chromosome framing

1099. Acrosome of spermatozoon contains:
(A) Protease
(B) Hyaluronidase
(C) Acid phosphate
(D) All of the above

1100. The evolution of gene and the proteins mainly occur due to:
(A) Point mutation
(B) Chromosomal aberrations
(C) Gene recombinations
(D) Gene duplication and divergence

1101. The viral system used to delivers a therapeutic gene to a specified cell type in gene therapy:
(A) Retrovirus (B) Adenovirus
(C) Herpes simplex virus (D) All of the above

1102. What are arbuscules?
(A) The spore producing structures
(B) Propagation structures
(C) Site of nutrient exchange between the fungus and the host
(D) Structure formed for defensive purpose

1103. What is fuelgen reaction?
(A) DNA-leucobasic fuchsin reacts to form red product
(B) Sugar -leucobasic fuchsin reacts to form red product
(C) Lipid - leucobasic fuchsin reacts to form red product
(D)Prokin -leucobasic fuchsin reacts to form red product

1104. An evolutionary substitution of one base for another in an exon of a gene coding for a protein, such that the produced amino acid sequence is not modified is referred:
(A) Point mutation
(B) Gene duplication
(C) Synonymous mutations
(D) Non-synonymous mutations

1105. Sigma factor in RNA polymerase affects RNA synthesis by:
(A) Affecting elongation only
(B) Affecting initiation only
(C) Affecting both initiation and elongation
(D) Affecting termination only

1106. The sequence (AGGAGG), which is a ribosomal binding site in bacterial and archaeal messenger RNA, generally located around 8 bases upstream of the short codon AUG:
(A) Pribnow box (B) Hogness box
(C) Homeo box (D) Shine-Dalgarne box

1107. The gene whose phenotype is masked by the expression of an allele at a separate locus, in an epistasis event:
(A) Jumping gene
(B) Epistatic gene
(C) Supplementary gene
(D) Hypostatic gene

1108. The structure of DNA important for the critical biological processes of DNA recombination and repair mutations that occur in the cell:
(A) G-quadrets (B) Triplex DNA
(C) Hairpin DNA (D) Cruciform DNA

1109. Widely used computer programme used in bioinformatics for multiple sequence alignment with command line interface:
(A) BLAST (B) Target P
(C) CLUSTALW (D) Primer express

1110. Complex molecular machine that unwinds double stranded DNA into two single strands:
(A) Metasome (B) Helicsome
(C) Replisome (D) Spliceosome

1111. Genomically enceded non-coding RNAs that help regulate gene expression, particularly during development:
(A) miRNAs (B) siRNAs
(C) dsRNAs (D) ssRNAs

1112. Improved nutritional quality in crop plants by breeding is known as:
(A) Biomagnification (B) Biomining
(C) Biofortification (D) Bioremediation

1113. Genes transfer in rice plant to produce 'Golden rice' was obtained from:
(A) Carrot
(B) The plant Daffodil and a bacterium-erosion
(C) E. coli and Daffodil
(D) Sunflower and cotton

1114. Exon skipping is associated with .... :
(A) RNA processing mutations
(B) Nonsense mutations
(C) Regulatory mutations
(D) Silent mutations

1115. A nonsense mutation invorves ....:
(A) A regulatory sequence
(B) An AG splice acceptor site
(C) The creation of a different amino acid
(D) The creation of a stop codon

1116. The chromosome map unit is ....:
(A) Centimorgan
(B) Millimorgan
(C) Millimendel
(D) Centisutton

1117. The consensus sequence of Pribnow box is:
(A) TATATA
(B) TATAAT
(C) TTAATT
(D) TATTAT

1118. The topological form of DNA that does not occur in biological system is:
(A) A-DNA (B) B-DNA
(C) C-DNA (D) Z-DNA

1119. The enzymes that induce slight negative supercoiling in natural DNA is:
(A) DNA topoisomerase (B) DNA helicase
(C) DNA primase (D) DNA ligase

1120. Non-covalent chemical bonds that stabilize the DNA double helix by connecting the adjacent aromatic rings of base pairs is:
(A) Sigma bonds (B) Delta bonds
(C) Gamma bonds (D) Pi bonds

1121. In DNA triple helix, the third strand is attached to the double helix by:
(A) Watson & Crick pairing
(B) Hoogsteen pairing
(C) Agastic pairing
(D) Ionic pairing

1122. RNA polymerase binds to a gene on the 3' end of a DNA tempelate strand, termed :
(A) Repressor (B) Promoter
(C) Attenulator (D) Operator

1123. Intron homing is:
(A) Removal of introns from the exon sequences
(B) Splicing of introns in the hn RNA
(C) Insertion of introns to an intron less site
(D) Deletion of introns from a homologous site

1124. The snRNA are found in:
(A) Ribosome
(B) Lysosome
(C) Replisome
(D) Splicesome

1125. In humans, the gene with the greatest number of introns is for:
(A) Globin (B) Tifin
(C) Insulin (D) Cyclin

1126. Which one of the following terms describes the repeated selection on intervening population derived from interbreeding of selected individuals to accumulate the desirable genes for a particular quantitative character without loss of genetic variability?
(A) Master line selection
(B) Recurrent selection
(C) Mass selection
(D) Pureline selection

1127. Which among the following is not a nucleic acid database?
(A) EMBL (B) GenBank
(C) DDBJ (D) SWISS-PROT

1128. A cross in which the sexes of the parents are the reverse of another cross:
(A) Reciprocal cross (B) Back cross
(C) Test cross (D) Dihybrid cross

1129. BLAST is used:
(A) To find similarity between sequence
(B) To align sequences
(C) To design primers
(D) To amplify DNA

1130. Gene flow is a concept best used to describe an exchange between:
(A) Individuals (B) Species
(C) Chromosomes (D) Populations

1131. Shotgun cloning stands for:
(A) Cloning cDNAs
(B) Chromosome Walking
(C) Cloning a genome
(D) Restriction analysis of DNA

1132. Major cause of evolution of genes and protein is:
(A) Point mutation
(B) Chromosomal mutation
(C) Sexual reproduction
(D) Gene duplication and divergence

1133. Pureline breeding refers to the following:
(A) Homozygosity only
(B) Heterozygosity only
(C) Heterozygosity and linkage
(D) Homozygosity and self assortment

1134. What is the role of DMSO in plant biotechnology?
(A) Hormonal substitute
(B) Chelating agent
(C) Phenol absorbent
(D) Cryoprotectant

1135. Identify the characters of an offering developed as a result of inbreeding:
(A) Increase genetic disorders
(B) Higher infant mortality
(C) Depression on growth rate
(D) All of the above

1136. The plant part from which colchicins is isolated:
(A) Fruit and seed
(B) Stem and leaf
(C) Seeds and corms
(D) Fruit and bulks

1137. Linkage between genes affects:
(A) Vernalization (B) Fertilization
(C) Anaphase (D) Independent anortment

1138. DNA contains following number of nitrogenous bases repeated in various sequences:
(A) One (B) Two
(C) Four (D) Ten

1139. Animal and plant cells can be differentiated by:
(A) Conductivity
(B) Size
(C) Presence or absence of cell wall
(D) Shape

1140. The idea of particulate nature of inheritance was given by:
(A) Darwin (B) Mendel
(C) Jones (D) Bateson

1141. Crossing over during meiosis results in:
(A) Half mutation (B) Breaking linkage
(C) Promoting linkage (D) None of these above

1142. Alternate form of gene at the same locus are referred to as :
(A) Allel (B) Plantid
(C) Dominant (D) Chromosome

ANSWERSHEET: 3.GENETICS AND PLANT BREEDING

1. (C); 2. (C); 3. (B); 4. (C); 5. (D); 6. (B); 7. (D); 8. (A); 9. (C); 10. (B); 11. (B); 12. (B); 13. (C); 14. (D); 15. (B); 16. (B); 17. (D); 18. (B); 19. (C); 20. (C); 21. (D); 22. (D); 23. (D); 24. (C); 25. (B); 26. (C); 27. (B); 28. (C); 29. (B): 30. (A); 31. (C); 32. (B); 33. (D); 34. (C); 35. (B); 36. (C); 37. (C); 38. (B); 39. (C); 40. (C); 41. (B); 42. (D); 43. (B); 44. (D); 45. (A); 46. (A); 47. (B); 48. (B); 49. (C); 50. (C); 51. (C); 52. (A); 53. (C); 54. (B); 55. (A); 56. (B); 57. (A); 58. (B); 59. (C); 60. (C); 61. (B); 62. (C); 63. (A); 64. (A); 65. (A); 66. (C); 67. (B); 68. (C); 69. (B); 70. (C); 71. (D); 72. (C); 73. (A); 74. (D); 75. (D); 76. (B); 77. (B); 78. (B); 79. (D); 80. (A); 81. (A); 82. (D); 83. (C); 84. (D); 85. (A); 86. (D); 87. (C); 88. (A); 89. (A); 90. (B); 91. (B); 92. (D); 93. (C); 94. (B); 95. (C); 96. (B); 97. (D); 98. (D); 99. (A); 100. (C); 101. (C); 102. (B); 103. (C); 104. (D); 105. (C); 106. (A); 107. (A); 108. (C); 109. (C); 110. (A); 111. (D); 112. (D); 113. (C); 114. (B); 115. (A); 116. (A); 117. (D); 118. (B); 119. (D); 120. (D); 121. (D); 122. (C); 123. (D); 124. (D); 125. (C); 126. (B); 127. (B); 128. (B); 129. (D); 130. (D); 131. (B); 132. (D); 133. (C); 134. (C); 135. (B); 136. (C); 137. (D); 138. (D); 139. (C); 140. (B); 141. (D); 142. (A); 143. (C); 144. (C); 145. (C); 146. (A); 147. (B); 148. (D); 149. (C); 150. (C); 151. (D); 152. (A); 153. (D); 154. (B); 155. (C); 156. (B); 157. (A); 158. (C); 159. (D); 160. (A); 161. (B); 162. (C); 163. (D); 164. (D); 165. (C); 166. (B); 167. (C); 168. (A); 169. (A); 170. (D); 171. (C); 172. (B); 173.(B); 174. (C); 175. (C); 176. (C); 177. (C); 178. (D); 179. (A); 180. (C); 181. (D); 182. (A); 183. (A); 184. (C); 185. (B); 186. (D); 187. (C); 188. (C); 189. (A); 190. (D); 191. (B); 192. (B); 193. (B); 194. (C); 195. (A); 196. (C); 197. (C); 198. (C); 199. (B); 200. (D); 201. (C); 202. (C); 203. (A); 204. (A); 205. (B); 206. (D); 207. (A); 208. (A); 209. (A); 210. (C); 211. (B); 212. (B); 213. (D); 214. (C); 215. (A); 216. (D); 217. (D); 218. (C); 219. (D); 220. (A); 221. (A); 222. (A); 223. (B); 224. (C); 225. (A); 226. (B); 227. (A); 228. (C); 229. (D); 230. (A); 231. (B); 232. (C); 233. (B); 234. (C); 235. (D); 236. (C); 237. (B); 238. (A); 239. (B); 240. (C); 241. (A); 242. (B); 243. (B); 244. (C); 245. (B); 246. (C); 247. (A); 248. (D); 249. (C); 250. (C); 251. (D); 252. (A); 253. (B); 254. (C); 255. (B); 256. (A); 257. (D); 258. (D); 259. (B); 260. (A); 261. (B); 262. (B); 263. (D); 264. (A); 265. (B); 266. (B); 267. (D); 268. (D); 269. (C); 270. (A); 271. (A); 272. (B); 273. (D); 274. (B), 275. (A); 276. (B); 277. (C); 278. (B); 279. (B); 280. (C); 281. (C); 282. (B); 283. (C); 284. (A); 285. (B); 286. (D); 287. (D); 288. (B); 289. (A); 290. (D); 291. (C); 292. (C); 293. (D); 294. (B); 295. (B); 296. (D); 297. (D); 298. (A); 299. (A); 300. (D); 301. (A); 302. (A); 303. (D); 304. (B); 305. (D); 306. (D); 307. (A); 308. (C); 309. (C); 310. (D); 311. (A); 312. (A); 313. (D); 314. (B); 315. (B); 316. (A); 317. (D); 318. (C); 319. (D); 320. (B); 321. (D); 322. (A); 323. (B); 324. (A); 325. (C), 326. (D); 327. (A); 328. (B); 329. (C); 330. (C); 331. (B); 332. (B); 333. (B); 334. (D); 335. (D); 336. (B); 337. (D); 338. (B); 339. (B); 340. (C); 341. (D); 342. (D); 343. (C); 344. (A); 345. (A); 346. (D); 347. (B); 348. (A); 349. (B); 350. (B); 351. (A); 352. (D); 353. (C); 354. (D); 355. (A); 356. (B); 357. (B); 358. (B); 359. (C); 360. (C); 361. (D); 362. (B); 363. (C);

364. (D); 365. (B); 366. (B); 367. (A); 368. (B); 369. (B);
370. (D); 371. (A); 372. (A); 373. (A); 374. (C); 375. (B);
376. (D); 377. (D); 378. (B); 379. (A); 380. (C); 381. (D);
382. (D); 383. (D); 384. (A); 385. (C); 386. (A); 387. (B);
388. (C); 389. (B); 390. (D); 391. (D); 392. (A); 393. (C);
394. (C); 395. (D); 396. (D); 397. (B); 398. (B); 399. (C);
400. (A); 401. (B); 402. (A); 403. (D); 404. (D); 405. (B);
406. (D); 407. (C); 408. (C); 409. (A); 410. (C); 411. (A);
412. (B); 413. (D); 414. (A); 415. (C); 416. (B); 417. (A);
418. (D); 419. (B); 420. (D); 421. (B); 422. (A); 423. (D);
424. (B); 425. (B); 426. (D); 427. (D); 428. (C); 429. (A);
430. (B); 431. (C); 432. (D); 433. (D); 434. (C); 435. (C);
436. (B); 437. (D); 438. (C); 439. (A); 440. (C); 441. (B);
442. (D); 443. (A); 444. (C); 445. (D); 446. (B); 447. (C);
448. (D); 449. (B); 450. (C); 451. (C); 452. (C); 453. (C);
454. (A); 455. (A); 456. (C); 457. (B); 458. (C); 459. (B);
460. (C); 461. (B); 462. (A) 463. (A); 464. (A); 465. (B);
466. (B); 467. (D); 468. (C); 469. (C); 470. (A); 471. (A);
472. (A); 473. (D); 474. (C); 475. (A); 476. (A); 477. (D);
478. (B); 479. (A); 480. (C); 481. (C); 482. (B); 483. (A);
484. (A); 485. (B); 486. (C); 487. (D); 488. (A); 489. (B);
490. (C); 491. (D); 492. (C); 493. (A); 494. (A); 495. (D);
496. (B); 497. (D); 498. (A); 499. (A); 500. (D); 501. (C);
502. (A); 503. (D); 504. (A); 505. (B); 506. (C); 507. (D);
508. (A); 509. (C); 510. (A); 511. (D); 512. (D); 513. (B);
514. (B); 515. (D); 516. (B); 517. (D); 518. (D); 519. (B);
520. (C); 521. (C); 522. (B); 523. (D); 524. (B); 525. (A);
526. (D); 527. (B); 528. (A); 529. (C); 530. (D); 531. (B);
532. (A); 533. (C); 534. (B); 535. (D); 536. (B); 537. (A);
538. (A); 539. (C); 540. (D); 541. (C); 542. (D); 543. (D);
544. (C); 545. (C); 546. (C); 547. (A); 548. (D); 549. (C);
550. (B); 551. (C); 552. (D); 553. (A); 554. (D); 555. (D);
556. (D); 557. (C); 558. (C); 559. (C); 560. (D); 561. (B);
562. (D); 563. (D); 564. (A); 565. (B); 566. (D); 567. (D);
568. (A); 569. (A); 570. (B); 571. (C); 572. (A); 573. (B);
574. (D); 575. (B); 576. (C); 577. (A); 578. (B); 579. (D);
580. (C); 581. (D); 582. (C); 583. (B); 584. (D); 585. (D);
586. (C); 587. (D); 588. (A); 589. (C); 590. (B); 591. (B);
592. (C); 593. (A); 594. (B); 595. (A); 596. (A); 597. (B);
598. (B); 599. (D); 600. (A); 601. (A); 602. (B); 603. (B);
604. (D); 605. (D); 606. (A); 607. (B); 608. (C); 609. (B);
610. (D); 611. (A); 612. (B); 613. (B); 614. (A); 615. (D);
616. (C); 617. (C); 618. (D); 619. (B); 620. (C); 621. (A);
622. (C); 623. (A); 624. (A); 625. (A); 626. (A); 627. (C);
628. (C); 629. (C); 630. (C); 631. (B); 632. (D); 633. (C);
634. (B); 635. (A); 636. (A); 637. (B); 638. (A); 639. (B);
640. (A); 641. (D); 642. (A); 643. (D); 644. (A); 645. (C);
646. (C); 647. (B); 648. (A); 649. (D); 650. (C); 651. (C);
652. (A); 653. (A); 654. (D); 655. (D); 656. (B); 657. (B);
658. (C); 659. (A); 660. (A); 661. (D); 662. (B); 663. (A);
664. (A); 665. (C); 666. (D); 667. (A); 668. (C); 669. (B);
670. (B); 671. (B); 672. (A); 673. (C); 674. (A); 675. (B);
676. (A); 677. (B); 678. (A); 679. (A); 680. (B); 681. (D);
682. (B); 683. (C); 684. (A); 685. (D); 686. (D); 687. (B);
688. (C); 689. (B); 690. (D); 691. (A); 692. (C); 693. (A);
694. (C); 695. (B); 696. (A); 697. (D); 698. (B); 699. (A);
700. (D); 701. (D); 702. (B); 703. (D); 704. (D); 705. (A);
706. (D); 707. (B); 708. (A); 709. (D); 710. (A); 711. (D);
712. (D); 713. (D); 714. (B); 715. (A); 716. (C); 717. (A);
718. (B); 719. (C); 720. (B); 721. (A); 722. (B); 723. (C);
724. (C); 725. (C); 726. (A); 727. (B); 728. (B); 729. (D);
730. (A); 731. (B); 732. (C); 733. (B); 734. (A); 735. (C);
736. (A); 737. (C); 738. (C); 739. (B); 740. (B); 741. (D);
742. (C); 743. (A); 744. (B); 745. (A); 746. (A); 747. (D);
748. (C); 749. (B); 750. (C); 751. (D); 752. (D); 753. (D);
754. (B); 755. (A); 756. (B); 757. (C); 758. (B); 759. (A);
760. (B); 761. (B); 762. (B); 763. (C); 764. (C); 765. (A);
766. (D); 767. (D); 768. (B); 769. (A); 770. (B); 771. (A);
772. (A); 773. (B); 774. (C); 775. (B); 776. (D); 777. (D);
778. (A); 779. (A); 780. (A); 781. (B); 782. (A); 783. (C);
784. (A); 785. (D); 786. (A); 787. (D); 788. (D); 789. (D);
790. (B); 791. (C); 792. (B); 793. (D); 794. (A); 795. (A);
796. (B); 797. (C); 798. (B); 799. (A); 800. (C); 801. (D);
802. (C); 803. (B); 804. (C); 805. (D); 806. (A); 807. (B);
808. (D); 809. (C); 810. (A); 811. (B); 812. (C); 813. (C);
814. (A); 815. (C); 816. (C); 817. (B); 818. (C); 819. (B);
820. (A); 821. (D); 822. (B); 823. (C); 824. (A); 825. (C);
826. (A); 827. (C); 828. (A); 829. (C); 830. (C); 831. (B);
832. (C); 833. (C); 834. (B); 835. (B); 836. (B); 837. (B);
838. (B); 839. (B); 840. (C); 841. (C); 842. (B); 843. (B);
844. (C); 845. (A); 846. (B); 847. (D); 848. (D); 849. (A);
850. (B); 851. (D); 852. (C); 853. (C); 854. (B); 855. (B);
856. (D); 857. (A); 858. (D); 859. (C); 860. (B); 861. (A);
862. (B); 863. (B); 864. (C); 865. (D); 866. (B); 867. (D);

868. (A); 869. (A); 870. (C); 871. (C); 872. (A); 873. (C); 874. (B); 875. (B); 876. (C); 877. (A); 878. (B); 879. (B); 880. (B); 881. (B); 882. (B); 883. (A); 884. (B); 885. (B); 886. (D); 887. (B); 888. (D); 889. (C); 890. (C); 891. (B); 892. (C); 893. (D); 894. (A); 895. (B); 896. (C); 897. (A); 898. (D); 899. (D); 900. (D); 901. (C); 902. (C); 903. (C); 904. (B); 905. (C); 906. (B); 907. (B); 908. (C); 909. (B); 910. (A); 911. (D); 912. (D); 913. (A); 914. (C); 915. (D); 916. (A); 917. (C); 918. (A); 919. (C); 920. (D); 921. (C); 922. (D); 923. (A); 924. (A); 925. (A); 926. (B); 927. (D); 928. (D); 929. (A); 930. (D); 931. (B); 932. (C); 933. (D); 934. (C); 935. (C); 936. (D); 937. (A); 938. (B); 939. (A); 940. (D); 941. (A); 942. (D); 943. (B); 944. (A); 945. (C); 946. (A); 947. (D); 948. (A); 949. (A); 950. (A); 951. (B); 952. (C); 953. (D); 954. (A); 955. (D); 956. (B); 957. (D); 958. (B); 959. (A); 960. (A); 961. (D); 962. (C); 963. (B); 964. (C); 965. (B); 966. (B); 967. (D); 968. (C); 969. (C); 970. (C); 971. (D); 972. (B); 973. (C); 974. (B); 975. (B); 976. (D); 977. (B); 978. (A); 979. (C); 980. (B); 981. (A); 982. (C); 983. (D); 984. (D); 985. (A); 986. (A); 987. (A); 988. (B); 989. (B); 990. (B); 991. (A); 992. (D); 993. (C); 994. (B); 995. (D); 996. (B); 997. (D); 998. (D); 999. (C); 1000. (B); 1001. (C); 1002. (D); 1003. (B); 1004. (C); 1005. (C); 1006. (A); 1007. (C); 1008. (C); 1009. (A); 1010. (C); 1011. (A); 1012. (D); 1013. (C); 1014. (A); 1015. (C); 1016. (D); 1017. (D); 1018. (C); 1019. (B); 1020. (B); 1021. (A); 1022. (C); 1023. (B); 1024. (C); 1025. (D); 1026. (B); 1027. (A); 1028. (C); 1029. (C); 1030. (C); 1031. (B); 1032. (D); 1033. (C); 1034. (A); 1035. (B); 1036. (A); 1037. (B); 1038. (C); 1039. (B); 1040. (C); 1041. (A); 1042. (A); 1043. (B); 1044. (D); 1045. (B); 1046. (A); 1047. (B); 1048. (B); 1049. (A); 1050. (D); 1051. (B); 1052. (D); 1053. (A); 1054. (A); 1055. (C); 1056. (B); 1057. (B); 1058. (C); 1059. (D); 1060. (A); 1061. (C); 1062. (A); 1063. (D); 1064. (C); 1065. (C); 1066. (D); 1067. (A); 1068. (B); 1069. (D); 1070. (C); 1071. (D); 1072. (A); 1073. (B); 1074. (A); 1075. (A); 1076. (C); 1077. (A); 1078. (C); 1079. (A); 1080. (C); 1081. (A); 1082. (D); 1083. (B); 1084. (C); 1085. (B); 1086. (D); 1087. (C); 1088. (C); 1089. (A); 1090.(C); 1091. (B); 1092. (A); 1093. (C); 1094. (A); 1095.(D); 1096. (A); 1097. (D); 1098. (B); 1099. (D); 1100.(D); 1101. (D); 1102. (C); 1103. (A); 1104. (C); 1105. (B); 1106. (D); 1107. (D); 1108. (D); 1109. (C); 1110. (C); 1111. (A); 1112. (B); 1113. (B); 1114. (A); 1115. (D); 1116. (A); 1117. (B); 1118. (C); 1119. (A); 1120. (D); 1121. (B); 1122. (B); 1123. (C); 1124. (D); 1125. (B); 1126. (B); 1127. (D); 1128. (A); 1129. (A); 1130.(D); 1131. (A); 1132. (D); 1133. (A); 1134. (D); 1135.(D); 1136. (C); 1137. (D); 1138. (C); 1139. (C); 1140.(B); 1141. (B); 1142. (A).

Chapter 4

AGRICULTURAL BIOTECHNOLOGY

1. Biotechnology is:
(A) Technology used in biology
(B) Use of genetic engineering
(C) Development of technology for medical application
(D) Use of microorganisms in industrial processes

2. Who introduced the term Biochemistry?
(A) Marshall Nirenberg (B) Carl Neuberg
(C) John H.Northrop & Kunitz (D) Van Helmont

3. Biotechnology is a multidisciplinary field that encompasses inventions that are related to:
(A) Basic science (B) Applied science
(C) Both (A) and (B) (D) None of these

4. Transgenic plants containing genes for drought and salt tolerance, improved quality for value addition and increase shelf life are produced through:
(A) Genetic transformation (B) Genetic drift
(C) Genetic transduction (D) Genetic transfection

5. Agrobacterium – mediated transformation an indirect method of transformation that involves:
(A) *Agrobacterium tumefaciens*
(B) *E. coli*
(C) *S. cerevisiae*
(D) None of these

6. Lamda phage infects bacteria *E. coli* and replicates by a:
(A) Lysogenic pathway (B) Lytic pathway
(C) Both (A) and (B) (D) None of these

7. The vectors that are hybrids of lambda phages and plasmids, and their DNA can replicates in the cell like that of a plasmid or be packaged like that of a phage is called:
(A) Phagemids (B) Fosmids
(C) Plasmids (D) Cosmids

8. Plasmids capable of propagating and transferring (shuuling) genes between two different organism, one of which is typically a prokaryote (E. coli) and other a eukaryote (yeast) is called:
(A) Cloning vector insertion vector (B) Shuttle vector
(C) Viral vector (D) None of these

9. Cosmid DNA can be packaged into phage particles by using the:
(A) In vitro system (B) In vivo system
(C) Both (A) and (B) (D) None of these

10. Recombinants can therefore be identifies because the characteristics coded by the inactivated gene is no longer displayed by the host cells is called:
(A) Insertional inactivation (B) Transduction
(C) Transfection (D) None of these

11. Virus mediated gene transfer, in which the foreign DNA is packaged inside a virus particle is called:
(A) Insertional inactivation (B) Transduction
(C) Transfection (D) None of these

12. The use of chemical or physical methods to persuade cells to take up DNA from the culture medium or direct physical introduction of DNA into the cell is called:
(A) Insertional inactivation (B) Transduction
(C) Transfection (D) None of these

13. A fundamental tool in molecular genetics which takes advantage of the ability of individual single-stranded nucleic acid molecules to form double stranded molecules is called:
(A) Nucleic acid hybridization
(B) In situ hybridization
(C) Both (A) and (B)
(D) None of these

14. Nucleic acid hybridization can be used to detect homologous DNA or RNA sequences not only in cell extracts, but also in chromosomes or intact cells – a procedure is called:
(A) Nucleic acid hybridization
(B) In situ hybridization
(C) Both (A) and (B)
(D) None of these

15. The technique is used to detect the presence of a known piece of DNA in any given DNA preparation from an organism is known as:
(A) Southern blot (B) Northern blot
(C) Both (A) and (B) (D) None of these

16. A method of estimating the size and abundance of a particular transcript is known as:
(A) Southern blot (B) Northern blot
(C) Both (A) and (B) (D) None of these

17. The determine the 5'-end of transcript, the most reliable method is:
(A) Primer extension (B) Reporter assays
(C) Expression vector (D) None of these

18. To findout the expression patterns of a gene within the cell, its promoter can be fused to one of the several reporter genes and transfected inside living cells by:
(A) Primer extension (B) Reporter assays
(C) Expression vector (D) None of these

19. To maximize the expression of heterologous genes – these genes that are taken from different organisms and being expressed in bacterial, yeast, animal or plant cells - it is necessary to design a cloning vector, which allows optional transcription of the genes. Such vectors are also called:
(A) Primer extension (B) Reporter assays
(C) Expression vector (D) None of these

20. Recombinants are hybrid DNA molecules consisting of autonomously replicating DNA segment plus inserted elements. Such hybrid molecules are also called:
(A) Chimera (B) Ligation
(C) Transformation (D) Cloning

21. For plasmid purification the bacterial cells extract is deproteinized, the RNA removed, and the DNA concentrated by:
(A) Methane precipitation
(B) Ethanol precipitation
(C) CO_2 precipitation
(D) Ammonia precipitation

22. Autonomously replicating genetic element used to carry a fragment of genomic DNA into a host cell for the purpose of gene cloning:
(A) Cloning vector insertion vector
(B) Shuttle vector
(C) Viral vector
(D) None of these

23. The chromosome of the transgenic animal includes an endogenous coding sequence, preferably the c-myc gene, which is substantially the same as the:
(A) Oncogene sequence
(B) Coding sequence
(C) Viral sequence
(D) Homologous endogenous sequence

24. The transcription of the oncogene sequence is under the control of a promoter sequence different from the promoter sequence (or a synthetic promoter sequence) controlling transcription of the endogenous:
(A) Oncogene sequence
(B) Coding sequence
(C) Viral sequence
(D) Homologous endogenous sequence

25. The term oncogene encompasses the:
(A) Viral sequence
(B) Homologous endogenous sequence
(C) Both (A) and (B)
(D) Coding sequence

26. The promoter sequence controlling transcription of the oncogene sequence is:
(A) Oncogene sequence (B) Coding sequence
(C) Viral sequence (D) Inducible

27. Introduction of the oncogene sequence at the fertilized-oocytes stage ensures that the oncogene sequence will be present in transgenic animals is:
(A) Germ cell (B) Somatic cells
(C) Both (A) and (B) (D) None of these

28. Introduction of the oncogene sequence at a later embryonic stage might result in the oncogene's absence from some somatic cells of the founder animal, but the descendants of such an animal that inherit the gene will carry the activated oncogene in all of their:
(A) Germ cell (B) Somatic cells
(C) Both (A) and (B) (D) None of these

29. The process that promotes gene transfer between bacteria by direct cell-to-cell contact, comes closest to sex in higher organisms is called:
(A) Conjugation (B) Transduction
(C) Transformation (D) Transposons

30. The transfer of bacterial DNA from one cell (the donor) to another (the recipient) via the agency of bacteriophage is called:
(A) Conjugation (B) Transduction
(C) Transformation (D) Transposons

31. The mechanism of gene transfer involves uptake of free DNA from killed donor cells by a recipient cell is known as:
(A) Conjugation (B) Transduction
(C) Transformation (D) Transposons

32. The discrete genetic elements with the ability to move from one genomic position to other non homologous positions is known as:
(A) Conjugation (B) Transduction
(C) Transformation (D) Transposons

33. These single-stranded sequences (GGGCGGCGACCT/ AGGTCGCCGCCC) are complementary to each other and are called:
(A) COS (Cohesive sites) (B) Cis
(C) CS trans (D) Chi

34. Lambda is representative of a large group of bacteriophages that can propagate either through the standard lytic cycle or through a process called:
(A) Lysogeny (B) Caspid
(C) Both (A) and (B) (D) None of these

35. In mature virus particles, these genomes are packaged inside a protein coat called:
(A) Lysogeny (B) Caspid
(C) Both (A) and (B) (D) None of these

36. The represented phage genome in lysogenic cells is called:
(A) Prophage or provirus (B) Metaphage
(C) Transphage (D) Post phage

37. Who describes relative position of various genes on the chromosome(s) of an organism is?
(A) Genetic map (B) Synaptic map
(C) Physical map (D) None of these

38. Who is the first person to use the term molecular biology is?
(A) William Astbury, 1945 (B) John C Kendrew
(C) Hans Krebs in 1969 (D) K.C. Mehta

39. The *E. coli* chromosome is a circular duplex DNA of about:
(A) 4.6 x 10^5 base pairs (B) 4.6 x 10^6 base pairs
(C) 4.6 x 10^7 base pairs (D) 4.6 x 10^8 base pairs

40. The process of DNA replication in eukaryotes is ——— complex than that in prokaryotes:
(A) For more (B) Less
(C) Same (D) Unknown

41. Pure plasmid DNA was isolated from a bacterium. Restriction enzyme digestion of this plasmid with either Bam HI or Eo RI resulted in two DNA fragments. A double digestion of the same plasmid with both these enzymes resulted in three DNA fragments. From this we an conclude that the isolated plasmid DNA is:
(A) Double stranded and linear
(B) Double stranded and circular
(C) Single stranded and linear
(D) Single stranded and circular

42. The first ever vaccination was against which disease:
(A) Chicken pox (B) Rabies
(C) Small pox (D) Pneumonia

43. For cloning eukaryotic gene in prokaryotes, genes should be isolated from:
(A) cDNA library (B) Genomic library
(C) Eukaryotic host (D) None of these

44. In order to move from a molecular marker to the linked gene of interest, which of the following approaches can be used?
(A) Chromosome jumping
(B) Chromosome walking
(C) Colony hybridization
(D) Chromosome sorting

45. Essential constituents of a cloning vector are:
(A) Replicon, promoter and selectable marker
(B) Replicon, unique restriction site and selectable marker
(C) Promotor, operatop and restriction site
(D) Replicon, unique restriction site, selectable marker and promoter

46. Which is correct about heavy metal?
(A) Heavy metals refer to the metals which are present in the soil with a density greater than 5 or 6 g m^{-3}
(B) As, Cd, Cr, Hg, Cu, Ni, Pb are examples of heavy metals
(C) They are harmful for humans as well as plants
(D) All of these

47. Essential constituents of an expression vector are:
(A) Replicon, promoter and selectable marker
(B) Replicon, unique restriction site and selectable marker and promoter
(C) Promtor, operator and restriction site
(D) Replicon, unique restriction site and promoter

48. Algae have emerged as one of the most promising sources, especially for biodiesel production is:
(A) Algae an grow in places away from the farmlands and forests
(B) The yields of oil from algae are orders of magnitude higher than those from traditional oilseeds
(C) Algae can be grown in sewage(s) and next to power-plant smokestacks where they acid in pollution management
(D) All of these

49. Plant viral diseases can be controlled by
(A) Antisense RNA technology
(B) By using ribozymes
(C) RNAi technology
(D) Settellite RNA
(E) All of these

50. For patenting a plant variety it should be:
(A) DUS (B) DUD (C) DDS (D) SUS

51. Micro propagation can be done by:
(A) Through shoot bud proliferation
(B) Through adventitious buds
(C) Meristem tip culture
(D) All of these

52. Products will be applicable for patenting only if they are:
(A) Novel (B) Inventive (C) Useful (D) All of these

53. A CsCl gradient will separate DNA molecules by
(A) Adhesion (B) Adsorption
(C) Density (D) Absorption

54. IPR stands for:
(A) Intellectual Property Rights
(B) Indigenous Product Right
(C) Indian Property Rights
(D) International Product Rights

55. Which is correct about nutraceuticals:
(A) A combination of nutrition and pharmaceutical
(B) It refers to extracts of foods claimed to have a medicinal effect on human health
(C) It is usually contained in a medicinal format such as powder, tablet or capsule in a prescribed dose
(D) All of these

56. Virus free plants can be obtained by:
(A) Meristem culture (B) Pollen culture
(C) Ovary culture (D) Shoot tip culture

57. Selection of protoplast fusion hybrids can be done by:
(A) Visual selection (B) Fluorescent labels
(C) FACS (D) Any of these methods

58. Structural DNA nanotechnology is for:
(A) Nanoelectronics (B) Nanorobotics
(C) Smart material (D) All of these

59. In vitro techniques for germplasm storage are useful for:
(A) Clonally propagated crops
(B) Crops producing recalcitrant seeds
(C) Crops producing orthodox seeds
(D) Both (A) and (B)

60. Embryo therapy involves:
I. In vitro fertilization of egg
II. Production of several copies of the normal allele of defective gene
III. Introduction of the normal allele into zygote, or in the cells of developing embryo
IV. Integration of the normal allele (DNA), preferably in the place of the defective allele so that it may function normally
Options
(A) I, II, III, IV (B) I, II, III
(C) II, III, IV (D) I, III, IV

61. The copyright of a book may be held by:
(A) Author(s) (B) Editor(s)
(C) Publisher (D) Any of these

62. Which advantages is offered by in vitro technique of germplasm storage?
(A) The material can be stored over long periods
(B) They are ideal for germplasm exchange
(C) The risk of errors in labeling are minimized
(D) All of these

63. Which statement is correct?
(A) When available ESTs assembled in contiguous over lapping clusters are described as contigs
(B) When ESTs appear as singles and which can not be assembled in contigs are described as singletons
(C) Unigene set is a combined set of contigs and single-tons
(D) All of these

64. Which of the following enzyme is used in brewing?
(A) β amylase (B) Cellukase
(C) Dextranase (D) Pectinase

65. The combination of biology and IT is:
(A) Nanotechnology (B) Biotechnology
(C) Bio-informatics (D) Information technology

66. Bioinformatics can be used for :
I. Genome mapping
II. Identification of genes from gene sequences
III. Determination of protein structure and simulation of protein interactions
IV. Discovery of new therapeutic targets and design-ing of medicines aimed at specific target sites
Options
(A) I, II, III (B) II, III. IV
(C) I, II, III, IV (D) I, II

67. Technique for sequencing DNA and proteins is:
(A) NMR spectroscopy (B) X-ray crystallography
(C) Electron microscopy (D) All of these

68. Which is correct about biochemical oxygen demand (BOD)?
(A) The amount of O_2 required by aerobic microorgan-isms to decompose the organic matter present in a sample of water
(B) It is used to measure the degree of water pollution
(C) It is also called biological oxygen demand
(D) All of these

69. Which molecular techniques can be used to character-ized germplasm?
I. Techniques that simply describe DNA polymor-phism
II. Techniques based on DNA linked to a gene of inter-est
III. Techniques that characterized the gene of interest itself
Options
(A) I, II (B) I, III
(C) I, II, III (D) II, III

70. Immobilization of enzymes is done by:
(A) Physical adsorption (B) Cross linking
(C) Entrapment (D) All of these

71. Photorespiration occurs in
(A) Mitochondria (B) Peroxisomes
(C) Chloroplast (D) All of these

72. Important constituents of biogas:
(A) Methane (B) Ethane
(C) Carbon monoxide (D) SO_2

73. In USA, which systems are available for protection of IPRs is considered to be the most powerful and the most expansive in scope of coverage?
(A) The Plant Patents Act of 1930
(B) The Plant Variety Protection Act of 1970
(C) The Utility Patents Act of 1985
(D) All of these

74. Cosmid DNA an be packaged into phage particles by using the:
(A) *In vitro* system (B) Transduction
(C) Both (A) and (B) (D) None of these

75. DNA chip carries a large number of:
(A) Oligonucleotides (B) cDNAs
(C) DNA probes (D) All of these

76. Which is correct about bulbosum technique of haploid production?
I. The technique was developed with the cross *Hor deum bulbosum* x *H. vulgare*
II. It is based on the chromosome elimination during early embryo development
III. Generally embryo rescue is necessary
IV. This technique has been used with bread wheat
Options
(A) I, II, III, IV (B) I, II, III
(C) II, III, IV (D) I, III, IV

77. Biofertilizers can be used to achieve:
(A) To fix atmospheric nitrogen
(B) To solubilize soil nutrients like phosphate
(C) To stimulate plant growth
(D) All of these

78. Which plant is used for bio-diesel production?
(A) Jatropha (B) Soybean
(C) Hevea (D) Wolfia

79. DNA chip is a wafer of:
(A) Silicon (B) Al
(C) Carbon (D) Phosphorus

80. Biosensers are used as:
(A) Purified enzymes (B) Antibodies
(C) Whole microbial cells (D) All of these

81. In India, which of the following overseas the large-scale commercial use of GMOs and recombinant DNA prod-ucts:
(A) Department of Atomic Energy
(B) Council of Scientific and Industrial Research
(C) Ministry of Environment and Forest
(D) Department of Biotechnology

82. Burning of plastic bags, emits toxic gasses leading to:
(A) Bronchitis (B) Lung infection
(C) Respiratory problems (D) All of these

83. Antibiotic resistance is used as a selectable marker for:
(A) Selection of vector
(B) Selection of recombinant vectors
(C) Selection of transformed cells
(D) Selection of resistant bacteria

84. Which vector was used for development of hepatitis vaccine?
(A) pBluescript (B) PUC 18
(C) Cosmid (D) Yeast expression vector

85. The plants, which carry additional, stably integrated and expressed, foreign gene(s) from transpecies is known as:
(A) Genetic counseling (B) Genetic engineering
(C) Gene therapy (D) Transgenic plants

86. Genetic engineering of industrially useful microorganisms aims at developing organisms that have which of the following features?
I. Capacity to produce such compounds, which they are normally incapable of producing
II. Ability to produce substances in quantities beyond their natural capabilities
III. Elimination of their undesirable characteristics
Options
(A) I, II, III, IV (B) I, II, III
(C) II, III, IV (D) I, III, IV

87. AC-terminal KDEL motif will most often ensure:
(A) The protein to be folded by KSC70
(B) The protein to be degraded by the polyubiquitine pathway
(C) ER-retention of the protein
(D) Secretion of the protein

88. Which of the following vector can be used for cloning bigger DNA fragments?
(A) YACs (B) Cosmids
(C) Plamids (D) Phagemids

89. Type 2 diabetes mellitus occur due to
(A) Deficiency of insulin
(B) Inactivation of insulin recepters
(C) Both (A) and (B)
(D) None of these

90. Which is correct about HIV?
I. HIV uses receptors, viz., CD4 and CCR5, to infect T cells
II. CD4 is essential for T cells function, but CCR5 appears to be dispensable
III. Due to lack of CCR5 receptor in about 1% Caucasian population, such individuals naturally show immunity to AIDS
IV. Inhibition of CCR5 expression through appropriate siRNA molecules reduces the infection of T cells by HIV
Options
(A) II, III, IV (B) I, II
(C) I, IV (D) I, II, III, IV

91. Which plant is used to accumulate copper from factory waste?
(A) *Baccillus subtilis* (B) *Pseudomonas putida*
(C) *Zoogloea ramigera* (D) *E. coli*

92. Commonly used host in gene cloning is:
(A) *E. coli* (B) *Pseudomonas*
(C) Fungi (D) Bacillus

93. Which among the following microorganism is used as bio fertilizer?
(A) *Rhizobium* (B) *Bacillus*
(C) *Pseudomonas* (D) *E. coli*

94. Marker free selection can be done by:
(A) Co-transformation (B) Deletion
(C) Inversion (D) Insertion

95. Bacteria possessing restriction enzymes get not affected because:
(A) These restriction enzymes are in inactive form
(B) Their genome is methylated
(C) These enzymes are present in less concentration
(D) Their genome lack unique restriction sites for enzymes

96. DNA Plant Technology faced difficulties during release of Endless Summer due to the use of:
(A) Selectable marker *npt*II
(B) CaMV 35S peromoter and the ACC synthase gene
(C) Antisense RNA technology and *Agrobacterium tumerfaciens* mediated genetic transformation
(D) All of these

97. Indian has created a National Biodiversity Authority with headquarters at:
(A) New Delhi (B) Kolkata
(C) Chennai (D) Hyderabad

98. The copyright suffers from:
(A) It provides protection for a specific period and only from reproduction as such of the copyright material
(B) It does not prevent another person from using either the idea or he information contained in a copy right material
(C) In case of DNA sequence, one may get around this protection by designing an alternative sequence to encode the same protein
(D) All of these

99. Why have transgenic proteins present in crops often been tested in the presence of stomach acid before the plants were released?
(A) To see whether the protein will be mutated into a more harmful protein (e.g.prions)
(B) To see whether the protein can be digested to provide substantially equivalent nutrients.
(C) To see whether the protein will be digestible by an average consumer.
(D) To see whether the protein will persist long enough to cause an allergic reaction.

100. Molecular probes can be used:
I. Diagnosis of diseases and DNA fingerprinting
II. Screening of genomic and cDNA libraries
III. Location of genes or DNA sequences on chromosomes
IV. Identification of varieties of crop species
Options
(A) II, III, IV (B) I, II
(C) I, IV (D) I, II, III, IV

101. Transmission of chromosomal DNA from generation to generation is crucial to cell propagation id called:
(A) DNA conservation (B) Genetic engineering
(C) DNA replication (D) Transgenic plants

102. Carotenoids (carotene and xanthophylls) obtained from cyanobacteria are commercially used as :
(A) Food colouring
(B) Food additives to enhances flesh colour of salmon fish
(C) To enhance colour of egg yolk
(D) All of these

103. The specialized structures which cap the ends of eukaryotic chromosomes is known as:
(A) Centromere (B) Telomeres
(C) DNA replication (D) Transgenic plants

104. Which statement is correct?
(A) Convention on biological diversity (CBD) is con cerned with conservation of whole biodiversity while PGR is concerned with conservation of plant diversity
(B) SBBs stands for state biodiversity Boards
(C) BMCs stands for biodiversity management committees
(D) All of these

105. There is a group of additional histones, all loosely related to one another and collectively called:
(A) Linker histone (B) Linker nonhistone
(C) DNA replication (D) Chromatosomes

106. The PPVFR Authority (PPVFRA) began registration of plant varieties in which years?
(A) 2005 (B) 2006 (C) 2007 (D) 2008

107. Fusion protein also called:
(A) Hybrid protein (B) Chimeric protein
(C) Both (A) and (B) (D) None of these

108. Which is an example of geographical indications?
(A) Champagne wine (B) Scotch whiskey
(C) California wine (D) All of these

109. The most useful methods for studying protein-protein interactions are:
(A) Phage display (B) Yeast two hybrid system
(C) Both (A) and (B) (D) None of these

110. Consider the following statements about embryo culture?
I. It is routinely used for the recovery of hybrid plants from distant crosses
II. It is particularly useful where embryos fail to develop due to endosperm degeneration
III. It has been used extensively in the extraction of haploid barley plants from *H. bulbosum x H. vulgare* crosses
IV. Embryo culture is regularly used in orchid propagation and in breeding of those species that show dormancy , e.g., Iris
Options
(A) II, IV (B) I, II
(C) I, III (D) I, II, III, IV

111. Choose the right combination of components requires to set up a polymerase chain reaction from the following:
(A) Template DNA, two primers, dNTPs and DNA ligase
(B) Template DNA, two primers, NTPs and DNA ligase
(C) Template RNA, two primers, NTPs and DNA polymerase
(D) Template DNA, two primers, dNTPs and DNA polymerase

112. A certain purified DNA sample was cut with two restriction endonucleases E1 and E2. The following results were obtained from agarose gel electrophoresis:
Sample cut with E1 alone: two bands of size 35 kb and 15 kb
Sample cut with E2 alone: Two bands of size 40 kb and 10 kb
Sample cut simultaneously with E1 and E2: three bands of size 35 kb, 10 kb and 5 kb
(A) Two sites for E1 and one site for E2
(B) One site for E1 and two sites for E2
(C) One site each for E1 and E2
(D) Three sites for E1 and one site for E2

113. Which is an example of fortified food?
(A) Addition of iodine to table salt (iodized salt)
(B) Addition of calcium to fruit juice
(C) Addition of vitamin D to such foods as margarine
(D) All of these

114. People become tolerant to malaria if they are carriers for:
(A) Sickle cell anemia (B) Thalisemia
(C) Leishmenia (D) Kala azar

115. A nuclei aid sample is resistant to digestion with exonuclease. When heated it does not show typical melting curve of a linear double stranded DNA. On CsCl-ethidium bromide equilibrium density centrifugation it settles at the bottom of the centrifuge tube. The nuclei aid is:
(A) ccc pBR322 (B) Bacteriophage P22 DNA
(C) rRNA (D) RFIIM13DNA

116. Which technologies and/or raw materials were used to develop Endless Summer tomato by DNA Plant Technology?
(A) Selectable marker *npt*II
(B) CaMV 35S promoter and the ACC synthase gene
(C) Antisense RNA technology and *Agrobacterium tumerfaciens* mediated genetic transformation
(D) All of these

117. Geographical indications includes:
(A) Natural products (B) Agricultural goods
(C) Goods of handicraft (D) All of these

118. Which one of the following elements need not to present in an expression vector?
(A) Selection marker to select for host cells containing the vector
(B) Two different origins of replication
(C) Promoter sequence upstream of the
(D) Unique restriction enzyme sites for insertional cloning

119. The in-vitro strategies of germplasm storage are useful under:
I. When crop species are asexually propagated and/ or at least some of their genotypes do not produce seed
II. When crop species produce recalcitrant seeds
III. When asexually propagated species are highly heterozygous
IV. When materials developed by genetic transformation show instability
Options
(A) II, III, IV (B) I, II
(C) I, IV (D) I, II, III, IV

120. Microbes have a relatively ——— that allows the chemical conversions possible:
(A) High metabolic rate (B) Large surface area
(C) High multiplication rate (D) All of these

121. Chromosome walking involves:
I. Selection of a clone of interest from genomic library identified by a probe and subcloning of a small fragment from one end of the DNA insert from this clone
II. The subclones fragment of the selected clone is then hybridized with other clones in the library to identify a clone having DNA insert overlapping with that of the first one
III. The restriction map of each selected clone is prepared and compared to know the region of overlapping
IV. Identification of few overlapping restriction sites will amount to walking along the chromosome or along a long chromosome segment
Options
(A) II, III, IV (B) I, II
(C) I, IV (D) I, II, III, IV

122. Which can be produced from dead bodies of animals?
(A) Calmodulin and aprotinin
(B) Myelin and ubiquitin
(C) Gelatin and protein kinase
(D) All of these

123. Disorder, which occur due to single nucleotide change:
(A) Sickle cell anemia (B) Thalisemia
(C) Leishmenia (D) Anemia

124. Cybrids are produced:
(A) By *in vitro* fusion of gemetes
(B) By fusion of two cells and contain full nuclear genomes of the two parents
(C) By fusion of two cells and contain full nuclear genome of one and partial nuclear genome of the other parent
(D) By fusion of two cells and contain nuclear genome of one parent and cytoplasmic genome of the other parent

125. An otherwise excellent transgenic variety of which crops developed during 1990a could not be released due to IPR problems?
(A) Potato (B) Tomato
(C) Tobacco (D) All of these

126. PPVFR Act (2001) is comparable to UPOV Act (1991) with repect to:
(A) Requirement of novelty, distinctiveness uniformity and stability for registration of a variety
(B) Protection extended to commercial use of all the material of the protected variety
(C) Essentially-derived varieties being subjected to PBR protection granted to the concerned initial varieties
(D) Both (A) and (C)

127. Which one of the following statements is true?
(A) All microarrays are DNA microarrays
(B) Complete genome sequence should be known to make a microarray
(C) All the microarrays use radioisotopes
(D) Microarrays can be used to measure mRNA levels

128. Which of the following an be used for transferring the DNA into the host ells?
P. Transformation Q. Sonication
R. Transfection S. Electroporation
(A) Only P can be used
(B) Only Q and R can be used
(C) Only Q, R and S can be used
(D) Only P, R and S can be used

129. Cryopreservation is done to store cells/cultures etc. for a longer period at:
(A) High humidity (B) Very low temperature
(C) Low humidity (D) High temperature

130. Which of the following is most likely in the present in a BAC?
(A) cDNA (B) EST
(C) LB,RB (D) UTR

131. DNA is present in
(A) Nucleus (B) Mitochondria
(C) Chloroplast (D) All of these

132. DNA finger printing is useful for study of:
(A) Forensics (B) Polymorphism
(C) Identity and relationships (D) All of these

133. Match the column A and B carefully and select the correct option from those given below:

Column A	Column B
A. Proteomics	I. Single nucleotide polymorphism
B. DNA marker	II. mRNA content
C. Genomics	III. Proteins
D. Transcriptome	IV. DNA sequencing

Options:
(A) A - II, B – I, C – IV, D – III
(B) A - IV, B – III, C – II, D – I
(C) A - III, B – I, C – IV, D – II
(D) A - II, B – III, C – IV, D - I

134. Which of the following enzyme is used to overcome oxidative stress?
(A) SOD (B) PDH
(C) LDH (D) G6PDH

135. Which of the following pairs of species best demonstrates the c-value paradox?
(A) Arabidopsis and rice (B) Rice and maize
(C) Rice and poplar (D) Rice and wheat

136. Chromosme number of Arabidopsis is
(A) 4 (B) 5 (C) 6 (D) 7

137. Complete set of cloned DNA fragments of an organism is known as:
(A) Genomic library (B) cDNA library
(C) mRNA library (D) Any of these

138. During the process of reverse transcription:
(A) mRNA is synthesized from DNA
(B) DNA is synthesized from DNA
(C) Proteins are formed from mRNA
(D) DNA is synthesized from mRNA

139. Ribosomes are synthesized in
(A) Nucleus (B) Nucleolus
(C) Cytoplasm (D) All of these

140. Match the column A and B carefully and select the correct option from those given below:

Column A	Column B
A. Sexual hybridization	I. mRNA content
B. Protoplastfusion	II. Transgene organism
C. Recombinant DNA technology	III. Limited to closely related species
D. Transcriptome	IV. Difficult to predict outcome

Options:
(A) A - II, B – I, C – IV, D – III
(B) A - IV, B – III, C – II, D – I
(C) A - III, B – IV, C – II, D – I
(D) A - II, B – III, C – IV, D - I

141. Patent activity is the lowest in:
(A) USA (B) Japan (C) Europe (D) India

142. Which is not correct about geographical indications (GIs):
(A) Once enforced by legislation, GIs exclude others from using a GI as trademark
(B) The chief requirement for GI protection is that a given quality, reputation or some other characteristics of the product be essentially attributable to the locality of its origin
(C) The GIs do not cover manufactured products and even food products
(D) Indian has enacted the Geographical Indication Act (1999) to claim GI for a variety of goods, including 'Basmati rice'

143. Assertion (A) : Consumption of oils containing high amount of oleic acid is beneficial for human beings
Reason (R) : It has cholesterol lowering properties
Options:
(A) Both A and R are correct and R adequate explanation of A
(B) Both A and R are correct, but R does not explain A
(C) Only A is correct
(D) Only R is correct

144. Vaccine against bird flu have been developed by using complete sequence of – gene from an H5N1:
(A) Haemagglutinin (HA) (B) Protease gene
(C) Protease inhibitor gene (D) RNAase

145. Haploid production cannot be done through
(A) Pollen culture (B) Callus culture
(C) Anther culture (D) Hybridization

146. Which is correct about reverse transcriptase?
I. This enzyme was discovered by Temin and Baltimore
II. It is generally obtained from avian myeloblastosis virus
III. It is RNA dependent DNA polymerase
IV. This enzyme performs similar functions as DNA polymerase and has an absolute requirement for a primer with a free 3'-OH
Options
(A) II, III, IV (B) I, II
(C) I, IV (D) I, II, III, IV

147. Cloning is a means to:
(A) Replace original genotype
(B) Preserve genotype
(C) Production of HGH gene in Escherichia coli
(D) None of the above

148. Match the column A and B carefully and select the correct option from those given below:

Column A	Column B
A. Probe	I. Recombinant DNA technology
B. Transgene	II. Gene isolation
C. Embryo rescue	III. Cell fusion
D. Somatic hybridization	IV. Distant hybridization

Options:
(A) A - II, B – I, C – IV, D – III
(B) A - IV, B – III, C – II, D – I
(C) A - III, B – I, C – IV, D – II
(D) A - II, B – III, C – IV, D - I

149. The medium is added continuously to the fermentor tank to replace that, which has been fermented. This is true in case of:
(A) Batch technique (B) Continuous flow technique
(C) Both are true (D) None is true

150. Eukaryotic DNA replication is carried out by
(A) DNA Pol α (B) DNA Pol β
(C) DNA Pol μ (D) DNA Pol δ

151. Assertion (A) : The gene encoding thaumatin has been transferred and expressed in *E. coli* and yeast with a view to obtain thaumatin at commercially accepted prices.
Reason (R) : The cost of production remains unacceptably high due to rather low levels of expression
Options:
(A) Both A and R are correct and R adequately explains A
(B) Both A and R are correct, but R does not explain A
(C) Only A is correct
(D) Only R is correct

152. ABO blood groups is a good example for both——— & multiple alletism
(A) Co dominance (B) Dominance
(C) Recessive ness (D) Over dominance

153. Match the column A and B carefully and select the correct option from those given below:

Column A	Column B
A. Micropropagation	I. Haploid
B. Regeneration	II. Axillary bud break
C. Androgenesis	III. Interspecific hybrid
D. Embryo rescue	IV. Organogenesis and somatic embryogenesis

Options:
(A) A - II, B – IV, C – I, D – III
(B) A - IV, B – III, C – II, D – I
(C) A - III, B – I, C – IV, D – II
(D) A - II, B – III, C – IV, D - I

154. The ability to create transgenic organisms allows genetic characteristics to be transferred beyond:
(A) Genus (B) Species lines
(C) Across kingdom boundaries (D) All of these

155. Which advantage is offered by in vitro techniques of germplasm storage?
(A) Require relatively very small space
(B) Help in storage of germplasm free from diseases, insects etc.
(C) Materials are lean source of nucleus seed
(D) All of these

156. An enzyme, aminoglycoside-3-phosphotransferase II. [APH(3)II] is commonly called :
(A) β-glucuronidase
(B) Penicillinase
(C) Neomycin phosphotransferase II
(D) Streptomycin phosphotransferase

157. Why is the gene that was used by Sheller et al. to produce spider silk in transgenic plants called a "synthetic" gene?
(A) It produced silk fibre, which is a synthetic product in plants
(B) It is assembled from synthesized DNA oligomers.
(C) Some of the amino acids encoded do not occur in nature
(D) It is encodes synthetic components such as an epitope tag

158. Nucleic acid hybridization can be used to detect homologous DNA or RNA sequences not only in cell extracts, but also is chromosomes or intact cells- a procedure is called:
(A) *In situ* hybridization (B) *Ex situ* hybridization
(C) Both (A) and (B) (D) None of these

159. DNA synthesis can be specially measured by estimating the incorporation of radio-labelled:
(A) Thymidine (B) Deoxyribose sugar
(C) Uracil (D) Adnine

160. In 1970, Halmilton Smith and his co-worker first isolated a restriction enzyme from the bacterium of:
(A) Haemophilus influenzae strain Rd
(B) 200 base-pairs
(C) 256 base-pairs
(D) 4096 base-pairs

161. Undesirable pleiotropic effect of a transgene on the transgenic plants may be corrected by :
(A) Changing the genetic background of transgene through hybridization
(B) Targetting of the transgene product into an appropriate cellular compartment
(C) Mutagenesis in the transgene
(D) Both (A) and (B)

162. Which is a soluble dietary fibre products, and is used as a nutraceutical to prevent cancer?
(A) Psyllium seed husk (B) Broccoli (Sulgoraphane)
(C) Soy or clover products (D) Reseveratrol

163. Which life forms can not be patented in India under the provisions of Indian Patents Act (1970)
(A) Microorganisms (B) Animals
(C) Plants (D) All of these

164. Monitoring and tackling the IPR aspects of inventions has:
(A) Demand time
(B) Enhances cost
(C) May act as a distinctive for R & D efforts
(D) All of these

165. Which of the following attempts to increase the utility of fungal herbicides succeeded?
(A) Expression of a phytoxic protein at abnormally high levels in Fusarium.
(B) Expression of a phytotoxic protein from Fusarium in another fungal species.
(C) Expression of a phytoxin resistance protein from Fusarium in a crop plant.
(D) Expression of a phytotoxic protein from Fusarium in a weed.
(E) Expression of a phytotoxic protein from Fusarium in a crop plant.

166. Which is a soluble dietary fibre products, and is used reduce hypercholesteremia?
(A) Psyllium seed husk (B) Broccoli (Sulgoraphane)
(C) Soy or clover products (D) Reseveratrol

167. Assertion (A) : Deficiencies of micronutrients are common in the diet of developing countries.
Reason (R) : Deficiencies of micronutrients can be overcome by fortification/biofortification
Options:
(A) Both A and R are correct and R adequately over come the A
(B) Both A and R are correct, but R does not explain A
(C) Only A is correct
(D) Only R is correct

168. Which is a soluble dietary fibre products, and is used as antioxidant:
(A) Psyllium seed husk (B) Broccoli (Sulgoraphane)
(C) Soy or clover products (D) Reseveratrol

169. Who of the following scientists showed that *Saccharomyces cerevisiae* causes fermentation forming products such as beer and buttermilk?
(A) Louis Pasteur (B) Alexander Fleming
(C) Selman Waksman (D) Schatz

170. The hereditary material present in the bacterium *E. coli* is:
(A) Single standard DNA (B) Double standard DNA
(C) DNA (D) RNA

171. *Saccharomyces cerevisiae* is employed in production of
(A) Idli (B) Beer
(C) Bread (D) All of these

172. A compound produced by an organism that inhibits the growth of another micro-organisms is called:
(A) Antiseptic (B) Antibiotic
(C) Anticoagulent (D) Antibodies

173. The codons causing chain termination are:
(A) AGT, TAG, UGA (B) UAG, UGA, UAA
(C) TAG, TAA, TGA (D) GAT, AAT, AGT

174. A restriction endonuclease is a bacterial enzyme that cuts double-stranded DNA by leaving the phosphodiester bonds at specific nucleotide sequences known as:
(A) Restriction site (B) 200 base-pairs
(C) 256 base-pairs (D) 4096 base-pairs

175. Bacteria are present everywhere except in:
(A) Ice (B) Soil
(C) Sea water (D) Distilled water

176. Microbiology is a discipline of science dealing with the study of:
(A) Forms and structure of organism
(B) Malignant tissues
(C) Microbes
(D) Parasites

177. The plant virus was first isolated in crystalline form
(A) TMV (B) Bactriophages
(C) Canophages (D) None of these

178. Genetically engineered human insulin is manufactured by the use of:
(A) *E. coli* (B) *Rhizopus*
(C) *Pseudomonas* (D) *Ashbya gossypii*

179. Introduction of a tobacco mosaic virus transgene into plants in 1986 was used to demonstrate:
(A) Cross-expression (B) Co-inoculation
(C) Co-transformation (D) Cross-protection
(E) Cross –inoculation

180. Which of the following are not possible uses for fructans as presented in class?
(A) Natural insecticides (B) Fat substitute
(C) Paper production (D) Sweetners

181. Which of the following statements is true regarding the production of transgenic cassava?
(A) The release of cyanide was inhibited by immunomodulation.
(B) The release of cyanide was inhibited by antisense gene regulation.
(C) The release of cyanide was inhibited by immunolocalization.
(D) The production of cyanide was enhanced by over expression of an enzyme.
(E) A non-toxic cyanide derivative was produced by metabolic engineering.

182. Which method enhances vitamin C production?
(A) Polymerization of a series of simple sugars
(B) Partitioning of ascorbate into seed storage proteins
(C) Derivatization of tocopherols from the shikimate pathway
(D) Increased enzymatic activity in the beta-carotenoid pathway
(E) Enzymatic reduction of a component of pectin.

183. Which of the following plants produce endosperm following fertilization?
(A) Wheat (B) Canola
(C) Pine (D) Both (A) and (B)
(E) None of these

184. Alanine is found in:
(A) Skin (B) Muscles
(C) Brain (D) All of these

185. Which of the following statements concerning the RNA/DNA hybrid oligonucleotides is incorrect?
(A) They form a specific hairpin structure that is recognized by the host cell.
(B) They may be synthesized to contain a mutation that will be incorporated into the host genome.
(C) They provide a mechanism for targeted alteration of the plant genome.
(D) They are incorporated using a mechanism that involves host DNA repair machinery.
(E) They hybrid molecule must be introduced into the cell using a binary vector.

186. The molecular scissors used to cut DNA into specific genes of interest are called:
(A) Exonucleases (B) Restriction endonucleases
(C) Ligases (D) Polymerases

187. Enzyme TPA or tissue plasminogen activator is used for :
(A) Dissolving blood clots
(B) Maintaining plasma content
(C) Clearing turbidity of juices
(D) Stimulating thromboplastin production

188. Pyruvate is used to produce:
(A) Glucose (B) Fructose
(C) Sucrose (D) Maltose

189. What is a disadvantage of the current use of transgenic antibody fragments to induce resistance to the herbicide picloram?
(A) Each antibody molecular can neutralize only one herbicide molecule.
(B) Picloram is an auxinic herbicide, thus the antibodies will neutralize endogenous auxin.
(C) The antibodies induce morphological defects (e.g. epinasty)
(D) The antibodies are encoded by two separate polypeptide chains.
(E) The antibodies are quickly degraded by host enzymes, so they have little effect on picloram.

190. Which of the following is involved in seedless watermelon production?
(A) Apomixes (B) Interspecific crosses
(C) Tetraploidy (D) Both (A) and (B)

191. For cloning of the gene, which among following is not important?
(A) Vector (B) Host
(C) Restriction enzymes (D) DNase

192. Which of the following was an unexpected result commected to the transgenic production of the terpenoid linalool?
(A) Increased production of linalool resulted in lower levels of GA hormones
(B) Increased transgene expression was not strongly correlated with increased levels of linalool
(C) Transgene expression resulted in production of un expected enantiomers of linalool
(D) Linalool accumulation was localized to the nectary.
(E) Increased transgene expression resulted in higher levels of herbivory

193. Interferon is:
(A) Antibacterial (B) Antiviral
(C) Antianimal cell (D) Bacterium

194. The bond between sugar and nitrogenous base in case of DNA is known as:
(A) Glycosidic bond (B) Phosphodiester bond
(C) H- bond (D) Peptide bond

195. Transplastomics:
(A) Targets genes in the chloroplast
(B) provides exceptionally low yields of protein prod ucts
(C) Produces genes that are released in pollen
(D) Offers little opportunity for practical use

196. Somatic embryogenesis is:
(A) Germ line cells developing into embryos
(B) Non-germ line cells developing into embryos
(C) Embryos developing from zygotes
(D) Embryonic tissue becoming somatic

197. Manipulation of the genetic material towards a desired end in a directed and predetermined way is called:
(A) Genetic engineering
(B) Gene cloning
(C) Recombinant DNA technology
(D) All of these

198. Which of the following methods would give you the most precise and accurate information about where and when a given gene is expressed?
(A) in situ hybridization
(B) DNA microarray
(C) Protein microarray
(D) Reporter gene fusion including introns

199. Star Link is most closely associated with which of the following terms?
(A) Glufosinate (B) Event 176
(C) Cry_9 (D) Cry_1
(E) Glyphosate

200. Third generation vaccines are:
(A) Attenuated pathogen (B) Pathogen relative
(C) Monoclonal antigen (D) Synthetic antigen

201. Which of the following is present on a Ti plasmid, but not on any component of a binary vector system?
(A) Vir genes (B) GUS genes
(C) LB, RB (D) MCS
(E) Opine genes

202. Genetic modification of plants can be achieved by :
(A) Somatic hybridization (B) Genetic engineering
(C) Somaclonal variation (D) All of these

203. Why might you find two kanamycin resistance genes on different parts of a T-DNA plasmid?
(A) One of the genes could be used for positive selection, and one could be used for negative selection
(B) One of the genes could be used for selection in plants, and one could be used for selection in bacteria
(C) This arrangement facilities site-specific recombination for marker-free selection
(D) Each gene might have been introduced by a independent co-transformation event

204. RNA molecules involve in splicing are:
(A) Sn rRNA (B) 5s rRNA
(C) 16s RNA (D) tRNA

205. AIDS lead to destruction of immune system due to depletion of
(A) B-cells (B) T helper cells
(C) Macrophages (D) Plasma cells

206. Which of the following is true of glyphosate?
(A) It as an animal LD 50 of 0.4 g/hg
(B) It is more selective than 2,4-D
(C) It can be manufactures without a license from the patent holder
(D) It inhibits the synthesis of branch chain amino acids

207. Which peoples are considered under farmer's rights for sharing the profit earned from the development of high yielding varieties evolved from the use of raw materials/plant genetic resources?
(A) Tribal people
(B) Rural communities
(C) Traditional farming families
(D) All of these

208. Which countries has 30 years protection for inbred lines of maize and for clovers and a few grasses?
(A) France (B) Germany
(C) USA (D) UK

209. According to the assigned reading, in terms of total area of transgenic crops cultivated in 2001, the four major producers of transgenic crops (ranked highest to lowest) are:
(A) USA, Argentina, Canada, China
(B) USA, China, Canada, India
(C) USA, Canada, China, Argentina
(D) USA, China, India, Canada

210. Enzyme are immobilized through:
(A) Covalently attaching to solid support
(B) Cross-liking
(C) Entrapping in gel
(D) All of these

211. The quickest way to produce homozygous breeding lines from heterozygous parents is through:
(A) Half-seed technique (B) Doubled haploids
(C) Introgression (D) Aneuploidy

212. In a ——protocol, bacteria with engineered abilities to detoxify pollutants are intentionally released in an area.
(A) Microcosm establishment (B) Hybridization
(C) Bioremediation (D) Rhizosecretion

213. What does the CT-value (cutoff threshold) in real –time PCR experiments mean?
(A) The lower the value, the higher the transcript amount.
(B) The higher the value, the higher the transcript amount
(C) No transcript, if the value is larger than 30
(D) No transcript, if the value is lower than 30

214. Which of the following organelle lack membrane?
(A) Mitochondria (B) Ribosomes
(C) Microtubules (D) Peroxisomes

215. Enzyme immobilization is:
(A) Conversion of an active enzyme into inactive form
(B) Providing enzyme with protective covering
(C) Changing a soluble enzyme into insoluble state
(D) Changing pH so that enzyme is not able to carry out its function

216. Which organelles is/are semi autonomous in nature?
(A) Mitochondria (B) Chloroplast
(C) Nucleus (D) Both (A) and (B)

217. Prions are microorganism, which consist of ———:
(A) DNA and protein (B) DNA and RNA
(C) DNA, RNA and Proteins (D) Proteins only

218. Virions are ——:
(A) DNA viruses (B) RNA viruses
(C) Naked DNA viruses (D) Naked RNA viruses

219. Prokaryotic DNA Polymerase comprises five subunits, which are:
(A) αβμδε (B) $\alpha\ \beta_2\beta\sigma$
(C) σεδωω (D) αεδβθ

220. AFLP molecular markers are ——— in nature:
(A) Dominant (B) Co-dominant
(C) Recessive (D) Neutral

221. Which is an example of a prokaryote does not contain nucleus?
(A) Blue green algae (B) Sieve cell of phloem
(C) Fungi (D) Yeast

222. During splicing of proteins which part is removed?
(A) Exons (B) Introne
(C) Exteins (D) Inteins

223. Which of the following technique is used for amplification of DNA?
(A) HPLC (B) PCR
(C) TLC (D) ELISA

224. How many microorganism are generally present per gram of biofertilizers?
(A) $10^7 - 10^9$ (B) $10^2 - 10^3$
(C 10^5- 10^6 (D) $10^{12} - 10^{15}$

225. Taq polymerase used in PCR is
(A) Thermotolerant (B) Highly processive enzyme
(C) Osmotolerant (D) All of these

226. In Bt cotton, expression of crystal protein occurs in :
(A) Only in stem (B) In leaves
(C) Systemic expression (D) Both (A) and (B)

227. Most commonly used method for transformation of animals is:
(A) Agrobacterium mediated transformation
(B) Protoplast method
(C) Electrotransformation
(D) Micro injection

228. Which is an example of a prokaryote?
(A) Blue green algae (B) Sieve cell of phloem
(C) Fungi (D) Yeast

229. Franken food is the 'nick name' of:
(A) Fast food (B) Frozen food
(C) GM food (D) Fortified food

230. Colchicine arrest cells in Mitosis, because:
(A) It binds to monomeric tubulin and thus depolymerizes microtubules
(B) It prevents chromosome condensation by inhibiting histone-histone interactions
(C) It dephosphorylates tubulin monomers and thus depolymerises microtubules
(D) It binds to the + end of F-actin and prevents fur ther addition of G-actin to actin filaments

231. Transgenic expression study can be done by:
(A) Transcription profiling (B) RT-PCR
(C) Western hybridisation (D) All of these

232. Which is an example of a prokaryote may contain more than one nucleus per cell?
(A) Blue green algae (B) Sieve cell of phloem
(C) Fungi (D) Yeast

233. Hybridomas are employed for:
(A) Production of somatic hybrids
(B) Killing cancer cells
(C) Synthesis of antibiotics
(D) Synthesis of monoclonal antibodies

234. Which statements is correct?
(A) Earthworm (*Eisenia foetida*) has the ability to detoxify the toxic thermal power waste fly ash
(B) Uptake of high concentration of cadmium cause itai-itai disease if humans
(C) Excess of fluorine in drinking water causes dental and skeletal problems
(D) All of these

235. Which of the following are biological mutagens?
(A) UV radiation (B) Transposons
(C) Ionic radiation (D) Ethidium bromide

236. Which organisms is not used for preparation of biofertilizers?
(A) Azotobacter (B) Pseudomonas
(C) Azolla (D) Azospirillum

237. The first field tests were of what genetically altered organism?
(A) Bt Corn
(B) Vaccinia virus containing a gene from the rabies virus
(C) The "flavsavr" tomato
(D) Strawberry seedlings sprayed with "ice-minus" bac teria

238. Which is associated with transgenic cotton varieties producing coloured cottons?
(A) They would reduced pollution
(B) They are attacked by large number of pests due to their colour
(C) They require higher doses of inorganic fertilizers
(D) They would have enhanced fibre strength

239. When pyrimidine is substituted by purines and vice versa, this type of mutation is known as:
(A) Transition (B) Transversion
(C) Non sense mutation (D) Miss sense mutation

240. Usually which of the following RNA is used for production of cDNA?
(A) rRNA (B) mRNA
(C) tRNA (D) siRNA

241. Most commonly used method for transformation of plants is:
(A) Agrobacterium mediated transformation
(B) Protoplast method
(C) Electrotransformation
(D) Micro injection

242. Immobilization of enzyme is done by:
(A) Cross linking (B) Entrapment
(C) Physical adsorption (D) All of these

243. Bacterial genes lack:
(A) Exons (B) Introns
(C) Promoters (D) Operator

244. In Bt cotton expression of crystal protein occurs:
(A) On surface of leaves (B) Systemic expression
(C) Only in stem part (D) Stem and leaves

245. Which is an example of geographical indications?
(A) Champagne wine (B) Scotch whiskey
(C) California wine (D) All of these

246. Which is not correct about breeder's exemption?
(A) The use of a protected variety for the development of new varieties is exempted from protection
(B) The PBR for a new variety evolved from the initial variety will be of the breeder who developed the new variety and of the holder of PBR title of the initial variety as well
(C) Under the UPOV 1978 Act, all new varieties evolved using a protected variety were exempted from protection under this provision
(D) The UPOV 1991 Act has limited breeders exemption in that 'essentially-derived varieties' are covered by the PBR granted to the initially variety

247. Which is the main considerations for the development of PBR systems?
(A) It allows breeders to benefit from the varieties developed by them
(B) It encourages plant breeding activities
(C) Private sector is encouraged to invest in plant breeding and seed industry
(D) All of these

248. In phytoremediation plants are used to:
(A) Control microbial attack on plants
(B) Control microbial contamination of soil
(C) Accumulates pollutants from soil thus clears soil from pollutants
(D) All of these

249. Which of the following statements regarding manipulation of the flavonol bio-synthetic pathway in tomato as to true?
(A) Increased flavanol production led to decreased carotenoid production
(B) Increased flavanol production led to increased carotenoid production
(C) Increased flavanol production did not affect carotenoid production
(D) Decreased flavanol production did not increase carotenoid production
(E) Decreased flavanol production led to increased carotenoid production

250. The roots of a plant are converted into drug-producing structures in a process called:
(A) Microcosm establishment (B) Mibridization
(C) Bioremedition (D) Rhizosecretion

251. For clearing oil spills which microorganism is used?
(A) *Pseudomonas putida* (B) *Bacillus*
(C) *Rhizobium* (D) *E. coli*

252. Which statement is not correct?
(A) Fatty acid biosynthesis in plants is very similar to that in bacteria
(B) In case of plants, fatty acids are synthesized in peroxysomes
(C) High oleic acid oils are nutritionally more beneficial
(D) Transgenic variety Laurical has high lauric acid content

253. Bt cotton is resistant to bollworm, it is due to:
(A) Bt is repellent to bollworm
(B) Bt is contact poison for bollworm
(C) Bt is stomach poison; it changes the permeability of stomach due to generation of pores in alkaline environment of stomach
(D) Bt is causing death due to asphagsiation of insect

254. The PPVFR Authority (PPVFRA) began registration of plant varieties in which years?
(A) 2005 (B) 2006 (C) 2007 (D) 2008

255. Which is not correct about gamma-linolenic acid (GLA)?
(A) It is a non-essential fatty acid
(B) It is the first intermediate in the conversion of linoleic acid to arachidonic acid
(C) It is produced by desaturation of linoleic acid by the enzyme fatty acid desaturase
(D) It is beneficial in alleviation of many physiological and pathological conditions such as hypercholestereremia, cancer and diabetes

256. Which among the following is not used as vector for gene cloning?
(A) Cosmid (B) Phagemid (C) PAC (D) MAC

257. Which is not correct about protoplasts?
(A) They may be cultured on a suitable medium in a variety of ways
(B) They regenerate cell wall within 2-4 days of cul ture
(C) The first division ion protoplasts occurs generally after 2-7 days and multicellular colonies appears after 2-3 weeks
(D) Protoplasts are cultured in light

258. Which is correct about map based cloning?
(A) A critical prerequisite is that the target gene must be localized definitely with respect to a molecular markers
(B) It starts with an interesting phenotype observed at whole organism level and proceeds to clone the gene via genetic mapping and chromosome walking
(C) Non-availability of tightly linked flanking markers to the gene of interest in a problem
(D) All of these

259. GEM stands for:
(A) Genetically Engineered Metazoa
(B) Genetically Engineered Microbes
(C) Genetically Engineered Metabolism
(D) Genetically Engineered Mustard

260. During RNA isolation which chemical is used for breaking of plasma membrane?
(A) Phenol (B) Isopropanol
(C) CTAB/SDS (D) Ethanol

261. Bioreactor are used for :
(A) Production of enzymes
(B) Production of secondary metabolites
(C) Production of cell cultures
(D) All of these

262. Genetic improvement of crop plants through biotechnology needs conventional breeding due to:
(A) The elite cultivars developed by plant breeding will be parents of the next generation of trasgenics
(B) Field testing across locations or cropping systems and over years will be needed to select the best transgenic event
(C) Breeding may be required to minimize/eliminate pleitropic effects
(D) All of these

263. Which statements is correct about thaumatin?
(A) Gene encoding thaumatin is isolated from African shrubs, Thaumatococcus danielli
(B) It is a protein that is 100000 times sweeter than sugar on a molecular basis and has a flavour enhancing effect ay subsweetness levels
(C) The gene producing thaumatin has been transferred and expressed in *E. coli* and yeast
(D) All of these

264. Who said " Environment today is a question of survival. Our rivers are drying, cities are choking, underground water tables are depleting, the air is not fit for breathing. Only a strong people's movement can bring a change in our present situation. Unless we have a green vote bank, the attitude of political parties towards the environment will not change":
(A) Rajendra Prasad (B) M.C. Mehta
(C) R.T. Doshi (D) K. C. Mehta

265. Which acids is commonly used in foods?
(A) Fumaric and lactic acids
(B) Citric acid and vinegar
(C) Maleic acid and tartaric acid
(D) All of these

266. Synthetic seeds are:
(A) Protoplasts (B) Somaclones
(C) Cybris (D) Somatic embryos

267. DNA Plant Technology faced difficulties during release of Endless Summer due to the use of:
(A) Selectable marker *npt*II
(B) CaMV 35S peromoter and the ACC synthase gene
(C) Antisense RNA technology and *Agrobacterium tumerfaciens* mediated genetic transformation
(D) All of these

268. Geographical indications includes:
(A) Natural products (B) Agricultural goods
(C) Goods of handicraft (D) All of these

269. Cytoplasmic hybrids are known as:
(A) Hybrids (B) Cybrids
(C) Both (A) and (B) (D) None of the above

270. Protein purification can be done by
(A) Affinity chromatography
(B) Ion exchange chromatography
(C) Sephadex column
(D) All of these

271. Which phyto hormone is used for development of synthetic seeds?
(A) Auxin (B) GA (C) ABA (D) Ethylene

272. During stress which of the following amino acid is accumulated?
(A) Proline (B) Tryptophan
(C) Tryosine (D) Lysine

273. Transfer of genes (gene flow) in between different species is known as:
(A) Vertical gene flow (B) Horizontal gene flow
(C) Lateral gene flow (D) None of these

274. Which is not correct about Scarlet?
(A) It is a somaclonal variety of potato
(B) It was selected from shoot-tip culture derived plants
(C) This variety is comparable to the parent cultivar in yield and disease resistance
(D) It shows darker and more stable skin colour, which is a desirable quality trait

275. Horizontal gene flow can be reduced by expressing genes in:
(A) Nucleus (B) Chroloplast
(C) Both (A) and (B) (D) None of these

276. 'Kwashiorkar' is concerned with :
(A) Severe protein deficiency (B) Food poising
(C) Flourine toxicity (D) Nitrate toxicity

277. Most commonly used organism in SCP is:
(A) *Bacillus* (B) *E. coli*
(C) *Spirulina* (D) *Pseudomonas*

278. Double stranded DNA can be isolated from sDNA or RNA by using:
(A) Ion exchange column
(B) Sephadex column
(C) Hydroxyapatite column
(D) Affinity chromatography

279. Which sigma factor of RNA polymerase comes into play during stress?
(A) σ^{32} (B) σ^{70} (C) σ^{14} (D) σ^{60}

280. Which technologies and/or raw materials were used to develop Endless Summer tomato by DNA Plant Technology?
(A) Selectable marker *npt*II
(B) CaMV 35S promoter and the ACC synthase gene
(C) Antisense RNA technology and *Agrobacterium tumerfaciens* mediated genetic transformation
(D) All of these

281. Which is not correct about geographical indications (GIs):
(A) Once enforced by legislation, GIs exclude others from using a GI as trademark
(B) The chief requirement for GI protection is that a given quality, reputation or some other characteristics of the product be essentially attributable to the locality of its origin
(C) The GIs do not cover manufactured products and even food products
(D) Indian has enacted the Geographical Indication Act (1999) to claim GI for a variety of goods, including 'Basmati rice'

282. Development of high –density molecular linkage map of a particular mapping population can be used for:
(A) MAS (B) Map based cloning
(C) Species characterization (D) All of these

283. Which one of the following is used in genetic engineering?
(A) RNA polymerase (B) DNA polymerase
(C) Restriction endonuclease (D) Nuclease

284. Which is not correct about virus cross protection?
(A) The phenomenon of crosss protection was first observed by McKinney in 1929 in case of tobacco mosaic virus
(B) Plants infected with a mild strain of virus resist the development of symptoms upon infection by a more severe strain of the same virus
(C) A mild protecting strain effective for one region may be sévere for another region
(D) Selection of mild protecting strain is rather easy

285. Genetically engineered bacteria are become employed for production of:
(A) Human insulin (B) Thyroxine
(C) Cortisol (D) Epinephrine

286. During RNA isolation, which chemical is used to prevent RNAase contamination?
(A) DMSO (B) DEPC
(C) Glycerol (D) All of these

287. Which of the histone protein is known as linker protein?
(A) HI (B) H_2A (C) H_2B (D) H_3

288. During stress which of the following phytohormone is accumulated?
(A) Auxin (B) GA (C) ABA (D) Ethylene

289. Star activity of restriction enzyme means
(A) It cuts the restriction site more efficiently
(B) It cuts at only one site
(C) It does not cut at any site
(D) Due to change in ionic conc. or other components it cuts at more sites generating more no. of fragments

290. Plasmids are used in genetic engineering because they are:
(A) Easily available
(B) Able to replicate
(C) Able to integrate with host chromosomes
(D) Rarely available

291. Trade secret can not be applied for:
(A) Plant varieties (B) Equipment design
(C) Books (D) All of these

292. Government of India has now established the PPVFR Authority (PPVFRA) with head-quarters at:
(A) Chennai (B) Kolkata
(C) New Delhi (D) Mumbai

293. Giant mouse has been produced through:
(A) Tissue culture (B) Gene differentiation
(C) Gene manipulation (D) Gene interaction

294. Yeast two hybrid system is used to:
(A) Determine interaction between DNA and RNA
(B) Determine interaction between two domains of proteins (i.e., DNA binding and protein binding domains)
(C) Determine interaction between RNA and proteins
(D) Determine interaction between DNA and proteins

295. Patent activity is the lowest in:
(A) USA (B) Japan
(C) England (D) India

296. In tissue culture regeneration of shoot and root occurs by manipulating the balance of:
(A) Auxin and ABA (B) Auxin and Cytokinin
(C) Cytokinin and ABA (D) ABA and Ethylene

297. Which category of microbes are used in retting process of jute?
(A) Microbes tolerant to high temperature
(B) Which process high amount of hydrolytic enzyme activity
(C) Anaerobic microbes only
(D) Aerobic microbes only

298. Bacterium commonly used in plant genetic engineering is:
(A) *Agrobacterium* (B) *Corynebacterium*
(C) *Bacillus subtilis* (D) *Salmonella*

299. Which factor is responsible for detoxification of drugs in case of humans?
(A) p53 (B) P450
(C) p21 (D) All of the above

300. Minisatellite are also known as:
(A) VNTRs (B) RFLPs
(C) STRs (D) RAPDs

301. Which statement is not correct?
(A) Fatty acid biosynthesis in plants is very similar to that in bacteria
(B) In case of plants, fatty acids are synthesized in peroxysomes
(C) High oleic acid oils are nutritionally more beneficial
(D) Transgenic variety Laurical has high lauric acid content

302. Among the given macromolecules, which one is the richest source of energy?
(A) Carbohydrates (B) Proteins
(C) Lipids (D) Nucleic acids

303. Which amino acid provide twists in tertiary structure of proteins?
(A) Serine (B) Glycine
(C) Proline (D) Tryptophan

304. The techniques to clone the first gene were developed in:
(A) 1972 (B) 1974 (C) 1976 (D) 1977

305. Urokinase enzyme is produced by:
(A) *Baccillus fastidious* (B) *Baccillus subtilis*
(C) *Baccillus thuriengenesis* (D) *Pseudomonas putida*

306. Which among the following is released in photorespiration?
(A) Serine (B) Cysteine
(C) Proline (D) Tryptophan

307. Which life forms can not be patented in India under the provisions of Indian Patents Act (1970)
(A) Microorganisms (B) Animals
(C) Plants (D) All of these

308. Monitoring and tackling the IPR aspects of inventions has:
(A) Demand time
(B) Enhances cost
(C) May act as a distinctive for R & D efforts
(D) All of these

309. Which programmes converts the submitted amino acid sequence into nucleotide sequence and compares it with nucleotide sequence database?
(A) $BLAST_x$ (B) $BLAST_n$
(C) $tBLAST_n$ (D) $tBLAST_x$

310. Which enzyme plays important role in reverse transcription?
(A) DNA polymerase (B) RNA polymerase
(C) Reverse transcriptase (D) Ribonuclease

311. GAATIC is restriction sequence of CTTAAG:
(A) Eco RI (B) Bam HI
(C) Sau 3 (D) Nco I

312. In tissue culture disease resistance can be obtained by:
(A) Somaclonal variation (B) Meristem culture
(C) Somatic hybridization (D) Anther culture

313. In case of phytoremediation, plants are used for which of the following?
(A) To accumulate pollutants from soil
(B) To control microbial attack on plants
(C) To control microbial contamination of the water
(D) All of these

314. Which amino acid plays an important role in splicing of proteins?
(A) Serine (B) Glycine
(C) Proline (D) Tryptophan

315. Which is not correct about somaclonal varieties?
(A) A low neurotoxin somaclonal variety, Ratan, of khesari has been released for commercial cultivation in M.P.
(B) Pusa Jaikisan is a somaclonal variety of *B. juncea*
(C) A somaclonal variety Bio-13 of citronella java developed by NBRI, Lucknow
(D) Pusa Jaikisan was derived from Varuna

316. Bioreactors are used for:
(A) Production of ethanol
(B) Production of enzymes on large scale
(C) Production of cell culture on large scale
(D) All of these

317. Which of the following techniques is used to inactivate a gene by altering the DNA encoding it:
(A) Homologous recombination
(B) Antisense nucleic aid blocks
(C) Antibody microinjection
(D) Introduction of dominant inhibitory mutants

318. Mathanogenic bacteria are:
(A) Anaerobic in nature
(B) Aerobic in nature
(C) Micro aerophillic in nature
(D) Saprophitic in nature

319. First of all coloured cotton developed in India in the state of:
(A) Madhya Pradesh (B) Maharashtra
(C) Gujarat (D) Andhra Pradesh

320. First hybrid variety of sorghum was evolved in the year:
(A) 1946 (B) 1956 (C) 1964 (D) 1974

321. Which yeast strain is naturally used for ethanol production?
(A) *Saccharomyces cerevicae* (B) *S. diastaticus*
(C) Both (A) and (B) (D) None of these

322. Example of tandem gene cluster is
(A) rRNA genes (B) Histone gene cluster
(C) Both (A) and (B) (D) None of these

323. What are the components of PCR reaction?
(A) Buffer, Taq polymerase, dNTPs, primer and template
(B) Buffer, dNTPs, primer and template
(C) Buffer, Taq polymerase, primer an d template
(D) Buffer, Taq polymerase, dNTPs and template

324. The lineal use for the recombinant DNA technology is to have rapidly replicating bacteria produce large amounts of specific proteins (e.g., hormones). Expression of a eukaryotic gene in prokaryotes involves:
(A) A shine-Delgarno (SD) sequence in mRNA
(B) Absence of introns
(C) Regulatory elements upstream of the gene
(D) All of these

325. Stable transformation of genes in case of plants can be done by:
(A) Biolistic gun method
(B) Agrobacterium mediated transformation
(C) Protoplast fusion
(D) Microinjection

326. Artificial seed refers to:
(A) Seeds of a plant with desired characteristics synthesized with recombinant DNA technology
(B) Somatic embryos encapsulated in a gel
(C) Seeds of synthetic plants
(D) All of these

327. Which of the following is a dominant marker?
(A) RAPD (B) RFLP
(C) SSR (D) AFLP

328. Chaperon play an important role in:
(A) Protein folding (B) Protein denaturation
(C) Protein trafficking (D) Protein synthesis

329. Methods for which can not be patented?
(A) Treatment of disease (B) Diagnosis of disease
(C) Prevention of disease (D) All of these

330. PPVFR Act (2001) allows registration of:
(A) Farmer's varieties (B) Extant varieties
(C) New Varieties (D) All of these

331. First hybrid variety of bajra was evolved in the year:
(A) 1946 (B) 1956 (C) 1965 (D) 1974

332. What is the role of SDS in case of SDS PAGE?
(A) It increases speed of proteins in gel
(B) It protects proteins from denaturation
(C) It provides negative charge to all the proteins so that separation can be done only on the basis of molecular weight
(D) SDS is detergent so is used to emulsify contaminating fats if any in samples

333. In lac operon, allolactose act as:
(A) Inducer (B) Gratituous inducer
(C) Repressor (D) Substrate

334. High density molecular map can be developed by:
(A) RFLP (B) RAPD
(C) AFLP (D) SNPs

335. Translation of histone protein occur during:
(A) S phase of cell division only (B) G1 and S phase
(C) G2 phase (D) M phase only

336. Which is concerned with computer based analysis of large amounts of biological data sets?
(A) Molecular biology (B) Computer science
(C) Information technology (D) Bioinformatics

337. Which among the following contains only one type of nucleic acid?
(A) Bacteria (B) Virus
(C) Dead virus (D) Fungi

338. Histone octomer comprise of:
(A) HI, H2A, H2B and H3
(B) HI, H2A, H2B, H3 and H4
(C) HI, H2B, H3 and H4
(D) H2A, H2B, H3 and H4

339. Operon comprise of:
(A) Operator (B) Structural gene
(C) Promoter (D) Repressor
(E) All of the above

340. International Centre for Genetic Engineering and Biotechnology is located at :
(A) Trieste (B) New Delhi
(C) Bangalore (D) Both (A) and (B)

341. Histone proteins are —— in nature:
(A) Acidic (B) Basic
(C) Neutral (D) Any of these

342. You want to know whether the enzyme you have introduced as a transgene in poplar is being expressed in the intended cells at the intended development stage. Which technique would be most informative?
(A) Microarray (B) Immunolocalization
(C) RNA in situ hybridization (D) Promoter fusion
(E) QRT-PCR

343. Chaperone present in ER is:
(A) Hsp 60 (B) Hsp 43
(C) Hsp 80 (D) Bip protein

344. Which is correct about B cells?
(A) B cells develop and mature in bone marrow
(B) Mature B cells leave the bone marrow and migrate via blood stream to the secondary lymphoid organs or tissues, where they ultimately become plasma cells
(C) B cells play an important role in immune system of body
(D) All of these

345. What strategy can be adopted to overcome the chances of resistance development in case of insects against transgenic crops?
(A) Growing of refuge crops
(B) Tissue specific expression of genes
(C) Temporal gene expression
(D) Any of these

346. After transformation of plant cells through Agrobacterium, which antibiotic is used to carry out Agrobacterium free selection of cells/tissues?
(A) Ampicillin (B) Rifamycin
(C) Tetracyclin (D) Carbencillin/Kenamycin

347. Which of the following statements about basal promoters is true?
(A) Basal promoters can be located in the 3 UTR
(B) Basal promoters direct assembly of the core translation machinery
(C) Basal promoters are sufficient for expression in the shoot apical meristem
(D) Basal promoters are necessary for enhancer trapping
(E) Basal promoters are necessary for heterologous expression.

348. Removal of the topological strain by inducing the negative supercoilling is carried out by:
(A) Topoisomerases (B) DNA gyrases
(C) Primases (D) DNA polymerases

349. Which gene is used for development of salt tolerant crops?
(A) Heat shock proteins
(B) Genes involved in praline biosynthesis
(C) LEA genes
(D) All of these

350. In terms of base sequence of genomic DNA, we are more than 98% similar to:
(A) Mice (B) Chimpanzee
(C) Rabbit (D) Monkey

351. Chromosomal aberration in which two breaks occur in a chromosome and the intercalary segment reunites in a reverse order is categorized as:
(A) Duplication (B) Deletion
(C) Inversion (D) Translocation

352. In case of PPVFR Act (2001), the gene fund shall be used for:
(A) Paying compensation to communities for their con tributions to the development of a variety
(B) Conservation of genetic resources
(C) Sustainable use of genetic resources
(D) All of these

353. In the first instance, PPVFR began registration of varieties in how many of the crops?
(A) 6 (B) 12 (C) 18 (D) 20

354. Hybridomas are used for:
(A) Antibiotic synthesis
(B) Formation of monoclonal antibodies
(C) Killing cancer cells
(D) Formation of somatic hybrids

355. Assertion (A) : During preparation of embryo for culture , care should be taken to excise the young embryos with their suspensors intact
Reason (R) : Suspensor supplies GA_3 to young embryos
Options:
(A) Both A and R are correct and R adequately over come the A
(B) Both A and R are correct, but R does not explain A
(C) Only A is correct
(D) Only R is correct

356. The term transformation is used to?
(A) Conversion of a normal vertebrate cell into a tumorigenic cell
(B) Uptake of DNA by a cell
(C) Delivery of DNA into a bacterial cell by a virus
(D) Both (B) and (C)

357. Which of the following statements about astaxanthin is false?
(A) It is a carotenoid
(B) It is normally produced by salmon
(C) It is a red-orange pigment
(D) It is produced from terpenoids
(E) It has anti-oxidant properties

358. Which is called natural genetic engineer?
(A) *Agrobacterium tumefaciens*
(B) *Escherichia coli*
(C) *Neurospora crassa*
(D) *Pseudomonas putida*

359. Which non-infectious diseases is associated with SCP industries?
(A) Asthama (B) Allergy
(C) Dermatitis (D) All of these

360. Which is called stress hormone?
(A) GA3 (B) Ethylene
(C) ABA (D) Cytokinin

361. The unique properties of each amino acid are determined by its particular:
(A) R-groups (B) H-bonds
(C) Peptide (D) Amino groups

362. During catabolism, glucose is converted to G.6.P using A.T.P., while in anabolism:
(A) Glucose is formed from phosphate ester by hydrolysis.
(B) Glucose is formed from pyruvate.
(C) Pyruvate is produced from its enol phosphate by transphosphorylation to ADP
(D) None of these

363. Which ministries established Department of Biotechnology in 1986?
(A) Ministry of Human Resource Development
(B) Ministry of Environment
(C) Ministry of Science and Technology
(D) Ministry of Agriculture

364. Genetic engineering means:
(A) Meiotic division of cells
(B) Nucleotide transfer
(C) Deletion and repair mechanism
(D) None of these

365. When producing human serum albumin (HAS) in transgenic plants, what would be an advantage of incorporating the transgenic into the chloroplast genome?
(A) HAS forms inclusion bodies if expressed in cytoplasm
(B) HAS becomes allergenic if expressed in cytoplasm
(C) HAS is toxic to plants if expressed in cytoplasm
(D) HAS is degraded by plant enzymes if expressed in cytoplasm
(E) HAS diverts too many amino acids if expressed in cytoplasm

366. Which enzymes is used for the production of glucose from starch?
(A) Lipase (B) Glucoamylase
(C) α- amylase (D) Both (A) and (C)

367. Totipotency is the property of :
(A) Plant cells (B) Animal cells
(C) Cancer cells (D) Both (A) and (B)

368. Which of the following transgenic methods were used (in separate experiments) in an attempt to increase the prevalence of certain amino acids in seeds?
(A) Altered tRNA gene; overexpression of sunflower seed albumin (SSA).
(B) Altereed tRNA gene; overexpression of human serum albumin (HAS)
(C) Altered rRNA gene; overexpression of sunflower seed albumin (SSA)
(D) Altereed rRNA gene; overexpression of human serum albumin (HAS)
(E) Altereed tRNA gene; repressionof sunflower serum albumin (SAS) expression

369. Common food yeast is:
(A) Saccharomyces (B) Cryptococcus
(C) Torulopsis (D) Endomyces

370. Tissue plasminogen activator is:
(A) A vitamin
(B) An electric device
(C) An enzyme
(D) A chemical that stimulates tissue differentiation

371. How was luciferase (LUC) with an artificial stop codon used as a control in the experiments involving altered tRNA?
(A) Diminished LUC expression demonstrated that the altered tRNA could properly recognize stop codons.
(B) LUC expression demonstrated that the gene was being induced by Lys.
(C) LUC expression demonstrated that Lys was being substituted for the stop codon by the altered tRNA.
(D) Loss of LUC expression demonstrated that a trans version had occurred, in order to create a stop codon.
(E) The artificial LUC is an example of a synthetic gene.

372. Monoclonal antibodies are produced by:
(A) Sarcoma cells
(B) Germ cells
(C) Somatic cells
(D) Hybrid of carcinogen cell and bone cell

373. Transgenic animal has:
(A) Foreign DNA in all its cells
(B) Foreign RNA in all its cells
(C) Foreign DNA in some of the cells
(D) Both (B) and (C)

374. Which is not correct about abscisic acid?
(A) It is a naturally occurring growth inhibitor
(B) Precursor of ABA is maleic acid
(C) It is synthesized in leaves
(D) ABA counteracts the effects of GA and promotes dormancy in buds and seeds

375. Which methods is used for building of phylogenetic trees?
(A) Distance matrix methods
(B) Maximum parsimony methods
(C) Maximum likelihood methods
(D) All of these

376. In human beings, interferons are:
(A) Leucocytic (B) Immune
(C) Fibroblastic (D) All of these

377. The microbes play an important role in the catalytic complicated chemical reactions of:
(A) Wine production (B) Fermentation
(C) Dairy product (D) All of these

378. In the breakdown of glucose to pyruvic acid, how many enzymes catalyze these successive steps?
(A) 8 (B) 9 (C) 10 (D) 11

379. The gene fund shall be utilized for:
(A) Contribution
(B) Compensation
(C) Commercial production of varieties
(D) Collection of varieties

380. The technique of insertion of a desired gene into DNA of plasmid vector is:
(A) Gene drafting (B) Gene dressing
(C) Gene cloning (D) Gene splicing

381. Which one of the following makes use of RNA template to synthesize DNA?
(A) DNA polymerase
(B) RNA polymerase
(C) Reverse transcriptase
(D) DNA dependent RNA polymerase

382. In the area of biotechnology, which materials can be kept as trade secret?
(A) Cell lines
(B) Strains of microorganism
(C) Production processes
(D) All of these

383. Which biotechnological inventions can be patented?
(A) Microorganisms, cell lines, plant/animal lies obtained through biotechnological approaches
(B) DNA sequences/genes and the proteins encoded, if any, by them
(C) Biotechnological process/technologies for modification of the properties of various organisms
(D) All of these

384. *Bacillus thuringiensis* (Bt) strains have been used for designing novel:
(A) Biofertilizers
(B) Bio-metallurgical
(C) Bio –mineralization process
(D) Bio-insecticidal plants

385. What is the best method to identify the cellular location of a protein?
(A) Place a reporter gene next to the promoter of the gene encoding the portion, and identify the cellular location of the reporter protein
(B) Use of labeled antibody to identify the cellular location of the protein
(C) Separate the cellular compartments by centrifugation and screen the different compartments with an antibody
(D) Tag the protein with fluorescent amino acids and identify the cellular location of fluorescent microscopy

386. Optical isomerism arises from the presence of:
(A) An systematic carbon atom
(B) A centre of symmetry
(C) A line of symmetry
(D) None of these

387. Border sequences need to be incorporated into the design of plasmid vectors for agrobacterium mediated transformation to ensure:
(A) Greater promoter efficiency
(B) Oncogene deactivation
(C) Efficient replication of the plasmid
(D) Integration of the genes of interest into the host gene

388. What is a genomic library?
(A) A collection of recombinant molecules with inserts that contain all of the genes of an organism
(B) A collection of recombinant molecules with inserts that contain all of an organism's genome
(C) A collection of recombinant molecules that express all of the genes of an organism
(D) A collection of recombinant molecules that have been sequenced

389. Which of the following portions of a longer duplex DNA segment are likely to be recognition sequences of a restriction enzyme?

5'-AGTC -3' (A) 3'-TCAG-5'	5'-ATCG – 3' (B) 3'-TAGC -5'
5'-ACCT-3" (C) 3'-TGGA-5'	5'-ACGT-3' (D) 3'-TGCA-5'

390. Which non-infectious diseases is associated with SCP industries?
(A) Asthama (B) Allergy
(C) Dermatitis (D) All of these

391. Success in germplasm storage is determined by:
(A) Inherent longevity and physiological storage behaviour of the species
(B) Initial quality (e.g., moisture content) of the material stored
(C) Storage methods and conditions used
(D) All of these

392. Which is not correct about verminator technology?
(A) Zeneca, U.K. has developed what it calls improved plant germplasm invention
(B) Improved plant germplasm invention has been dubbed as verminator technology by RAFI, Canada
(C) This technology uses a gene from the fat tissues of a goat, the product of which interferes with normal plant growth
(D) If the transgenic crop is sprayed with a specific chemical, plants show normal growth

393. A restriction endonuclease has the recognition sequence G/AATTC, where "/" indicates the cut site. This sequence is found, on average, one every 'x' residues in a chromosome. 'x'=
(A) 146 base-pairs (B) 200 base-pairs
(C) 256 base-pairs (D) 4096 base-pairs

394. Lipids are source of :
(A) Vitamins
(B) Structural components of cell organization
(C) Energy
(D) All of these

395. Which is not correct about patent?
(A) It is the right of an individual or a group of individuals to gain profit from a particular invention or unique manufacturing process
(B) It is granted by the Government of a country to the applicant for an unlimited period of time
(C) When a patent is granted, the inventor becomes the owner of the patent like any other form of property
(D) Patents are granted for an invention (including a product), innovation/improvement in an invention, process/product of an invention and a concept

396. Which features is not concerned with the PPVFR Act (2001)?
(A) The Central Government is to constitute a National Gene Fund
(B) Citizens of convention countries will have the same rights as citizens of India under the Act
(C) Application for registration of a variety may be made in India within 2 years from the date of application for registration of the same plant variety made in the convention country
(D) The rights of PBR holder 'shall not be deemed in fringed by a farmer who was not aware of the exist ence of such right'

397. In case of medicine, biotechnology can be used for :
(A) Production of synthetic vaccines
(B) Diagnosis of disease
(C) Production of rare drugs
(D) All of these

398. Which is correct about food additives?
(A) These are substances added to food to preserve flavour or improve taste and appearance
(B) They may be natural or artificial
(C) To regulate these additives, and perform consumers, each additive is assigned a unique number, initially these were E numbers used in Europe for all approved additives
(D) All of these

399. Bacterial artificial chromosomes (BACs), cosmids, phages, plasmids, and yeast artificial chromosomes (YACs) are all commonly used cloning vectors that differ in their cloning capacities, with a range from approximately 100 bp to 3000 kb. Which of the following is the proper order for these vectors in terms of increasing cloning capacity?
(A) BAC, cosmid, phage, plasmid, YAC
(B) YAC, BAC, cosmid, phage, plasmid
(C) Plasmid, phage, cosmid, BAC, YAC
(D) Plasmid, cosmid, phage, BAC, YAC

400. Which statement is not correct about cryopreservation?
(A) It has proved to be the most reliable method for long term preservation of cell cultures
(B) It consists of choice of explants preculture and pretreatment, cryoprotection, freezing, storage, thawing and reculture
(C) DMSO is rarely used
(D) A general protocol applicable to all species and explants is not available

401. Using the strain of *E. coli* named A, a recombinant plasmid P was amplified. When plasmid DNA was isolated from strain A and was used to transform another E. coli strain B, no transformants could be obtained. Transformations using control DNA were fine. When the plasmid DNA from strain A was used to transform another strain, colonies were obtained. This is because:
(A) Strain B has an intact restriction-modification sys tem whereas A and do not
(B) Strain B is not capable or recombination
(C) A and C are exonuclease free whereas B is not
(D) Strain B is antibiotic sensitive

402. Many developing countries pay huge subsidies (e.g., USA pays – 50 billion dollars per year) to their farmers for not growing which of the following?
(A) Medicinal plants (B) Food crops
(C) Horticultural crops (D) Fodder crops

403. Which of the following is considered as primary physical containment?
(A) Containment equipment
(B) Special laboratory design
(C) Laboratory practices
(D) Both (A) and (C)

404. Which of the following organisms has the minimal genome?
(A) *E. coli*
(B) *Heliobacter pylori*
(C) *Methanococcus jannaschii*
(D) *Mycoplasma genitalium*

405. DNA replication occur in which stage of cell division?
(A) G1 phase (B) G2 phase
(C) S phase (D) M phase

406. Which of the following technique is used for DNA fingerprinting?
(A) HPLC (B) RFLD
(C) TLC (D) ELISA

407. Immobilisation protests an enzyme against:
(A) High temperature (B) Low temperature
(C) Amylases (D) Protease

408. cDNA synthetic DNA made from mRNA with the use of a special enzyme called:
(A) Reverse transcriptage (B) Expression vector
(C) Promoter gene (D) Reporter gene

409. Proceeded cheese is
(A) Soft cheese (B) Blended cheese
(C) Ripe cheese (D) Hard cheese

410. Gene producing a protein that can be detected and quantified using a simple assay is called:
(A) Structural gene (B) Reporter gene
(C) Promoter gene (D) Eucaryotic gene

411. Physical containment relates to:
(A) Laboratory design
(B) Containment equipment
(C) Special laboratory procedures
(D) All of these

412. A gene or DNA sequence having a known location on a chromosome and associated with a particular trait or gene is used as a:
(A) Genetic marker (B) Promoter gene
(C) Genetic map (D) Genetic engineering

413. Recombinants can therefore be identified because the characteristics coded by the inactivated gene is no longer displayed by the host cells is called:
(A) Transfection (B) Transduction
(C) Insertional inactivation (D) None of these

414. Immobilised yeast is:
(A) Preserved for future use
(B) Little active in fermentation
(C) 20 times more efficient
(D) Food yeast

415. Recombinants are hybrid DNA molecules consisting of autonomously replicating DNA segment plus inserted elements. Such hybrid molecules are also called:
(A) Chimera (B) Ligation
(C) Transformation (D) Cloning

416. Three terminal codons are:
(A) AUG, GUG, GUC (B) UAG, UAA, AGA
(C) UAA, UAC, AUG (D) UAA, UAG, UUA

417. A DNA microarray also known as:
(A) Gene chip (B) DNA chip
(C) Gene array (D) All of these

418. Which among the following is not unicellular?
(A) Bacteria (B) Yeast
(C) Fungi (D) None of these

419. Phage display method was first described by:
(A) George Smith (B) Ian Wilmut
(C) Flemming (D) Watson and Crick

420. Trade secret relates to:
(A) Certain materials, e.g., bacterial strains
(B) Processes
(C) Formulae
(D) Any one of these

421. Which biotechnology related maters can be patented?
(A) Method/process of production
(B) Products, etc
(C) Applications of the various processes/products'
(D) Any of these

422. Hybridoma cells are developed by fusion of:
(A) T cells with myeloma cells
(B) B cells with myeloma cells
(C) T cells with B cells
(D) B cells with APCs

423. The display of a heterologous target protein under study on phage particle is known as:
(A) Phage display (B) Chimeric protein
(C) Both (A) and (B) (D) None of these

424. Which bond in the DNA provides it the backbone?
(A) Glycosidic bond
(B) Phosphodiester bond
(C) H- bond
(D) Peptide bond

425. During meiosis actual reduction of chromosomes occur in which sub stage?
(A) Anaphase (B) Metaphase
(C) Telophase (D) Prophase

426. Which of the following amino acid participate in urea cycle?
(A) Arginine (B) Lysine
(C) Tryptophan (D) Histidine

427. Sequence tagged sites that are derived from DNA are called:
(A) Sequence tagged sites mapping
(B) Expressed sequence tags
(C) Both (A) and (B)
(D) None of these

428. Which factor is necessary for maintaining lysogeny in lemda phase?
(A) Cro (B) C1 repressor
(C) Both (A) and (B) (D) None of these

429. Shine Dalgarno sequence present on mRNA in case of prokaryotes plays as important role in protein synthesis. This sequence binds with —— during translation initiation:
(A) 16s RNA (B) tRNA
(C) 23S RNA (D) 5s RNA

430. Synthetic vaccines have been developed for:
(A) Typhoid and rabies
(B) Cancer (Feline leukaemia virus)
(C) Cholera and foot and mouth disease of cattle
(D) All of these

431. Which is used as a nutraceutical?
(A) Soluble dietary fibre products such as Psyllium seed husk, broccoli (sulforaphane), soy or clover (isoflavonoids)
(B) Reseveratrol from red grape products
(C) Alpha-linolenic acid from flax seeds
(D) All of these

432. Which sequence in case of eukaryotes is important for mRNA tailing?
(A) GAGAGA (B) GAATTC
(C) UACGAC (D) UACUAAC

433. The technique for correcting defective genes responsible for disease development is called:
(A) Gene therapy (B) Gene inhibition
(C) Gene cloning (D) All of these

434. The technique used to distinguish between individual of the same species using only samples of their DNA is called:
(A) DNA fingerprinting (B) DNA testing
(C) DNA typing/DNA profiling (D) All of these

435. Department of biotechnology promotes human resource development through:
(A) Master's programmes
(B) Post-doctoral programmes
(C) Certificate/Diploma programmes
(D) All of these

436. RAPD molecular markers are ——— in nature
(A) Dominant (B) Co-dominant
(C) Recessive (D) Dominant-Recessive

437. Which transgenic crops is being grown on commercial scale in India?
(A) Bt-cotton (B) Bt-brinjal
(C) Bt-rice (D) Bt-tobacco

438. Which is correct about dietary supplement?
(A) It is also called nutritional or food supplement
(B) It is a preparation intended to provide nutrients such as fatty acids, vitamins, minerals or amino acids that are missing or are not consumed in sufficient quantity in diet
(C) The hormones DHEA, pregnenolone and pineal hormone melatonin are marketed as dietary supplements in the USA
(D) All of these

439. Amniocentesis, a diagnosis technique that enables doctors to remove and culture fetal cells from pregnant women, has given doctors an important tool for the early diagnosis of fetal disorders is known as:
(A) Genetic counseling (B) Genetic engineering
(C) Gene therapy (D) Gene mapping

440. Plants cells tissue has the ability to develop into a complete plant. This property of plants is known as:
(A) Pluropotency (B) Totipotency
(C) Multipotency (D) None of these

441. The common reporter genes obtained from *E coli* is:
(A) Chloramphenicol acetyl transferase (CAT)
(B) β -galactosidase
(C) β-glucuronidase (GUS)
(D) All of these

442. In cell culture medium commonly used carbon source is:
(A) Fructose (B) Sucrose
(C) Glucose (D) Mannose

443. Sieve tubes of phloem tissues lack:
(A) Nucleus (B) Cell wall
(C) Mitochondria (D) Ribosomes

444. Which of the following DNA structure forms left hand helix?
(A) DNA A (B) DNA B
(C) DNA C (D) DNA Z

445. For molecular characterization of plant species which of the following molecular markers are used?
(A) RAPD (B) AFLP
(C) SSR/STMS (D) All of these

446. The whole process of transgenic plants development involving introduction, integration and expression of foreign gene(s) in the host is called:
(A) Genetic counseling (B) Genetic engineering
(C) Genetic transformation (D) Transgenic plants

447. Barr and Bertram observed a darkly staining body in interphase nerve cells of female acts that was absent in similar cells of males. This body described as:
(A) Sex chromatic body (B) Barr body
(C) Both (A) and (B) (D) None of these

448. Sterilisation of medium through autoclaving is done at:
(A) 10 pounds (B) 15 pounds
(C) 20 pounds (D) 25 pounds

449. For plasmid purification the bacterial cells extract is deproteinized, the RNA removed, and the DNA concentrated by:
(A) Ethanol precipitation
(B) Methane precipitation
(C) Carbon precipitation
(D) Ammonia precipitation

450. Problem (P): An inherent problem associated with the map-based cloning is the non-availability of tightly linked flanking markers to the gene interest
Solution (S): Transposon tagging provides solution for the isolation of useful genes
Options:
(A) Both P and S are correct and S adequately explains P
(B) Both P and S are correct, but S does not explain P
(C) Only P is correct
(D) Only S is correct

451. The vectors that are hybrids of lambda phages and plasmids, and their an replicate in the cell like that of a plasmid or be packaged like that of a phage is called:
(A) Cosmids (B) Fosmids
(C) Plasmids (D) Phagemids

452. Sterlization of antibiotics is done by:
(A) Autoclaving (B) Dry heat
(C) Use of filter (D) Any of the above methods

453. tRNA synthesis is done by:
(A) RNA Polymerase I (B) RNA Polymerase II
(C) RNA Polymerase III (D) All of these

454. Autonomously replicating genetic element used to carry a fragment of genomic DNA into a host cell for the purpose of gene cloning:
(A) Cloning vector (B) Insertion vector
(C) Shuttle vector (D) Viral vector

455. DNA is fit for making tools in case of nanotechnology. This is due to:
(A) Branching nature of DNA
(B) Flexibility in DNA conformation
(C) Small size of DNA
(D) All of these

456. The short tagged tracts of DNA sequence (200 to 500 base pairs) that is operationally unique and has a single occurrence in the genome is called:
(A) Sequence tagged sites mapping
(B) Expressed sequence tags
(C) Both (A) and (B)
(D) None of these

457. Lambda phage infest bacteria *E coli* and replicates by a:
(A) Lysogenic pathway (B) Lytic pathway
(C) Both (A) and (B) (D) None of these

458. A genetically engineered organism that carries one or more genes in its chromosomes that has been made inoperative is known as:
(A) DNA fingerprinting
(B) DNA testing
(C) DNA typing/DNA profiling
(D) Gene Knockout

459. DNA replication is:
(A) Conservation (B) Semi-conservative
(C) None of these (D) Both (A) and (B)

460. mRNA usually is being extracted using by:
(A) Poly T resin (B) RNase P
(C) a polyA resin (D) an rRNA trap

461. A set of disease resistance mechanisms that are not specific to a particular pathogen comes under?
(A) Adaptive immunity (B) Innate immunity
(C) Passive immunity (D) Active immunity

462. In USA, which systems are available for protection of IPRs related to plants?
(A) The Plant Patents Act (1930)
(B) The Plant Variety Protection Act of 1970
(C) The Utility Patents Act (1985)
(D) All of these

463. Which of these is/are the characteristic/s of the specific immunity?
(A) Antigenic specificity
(B) Diversity & self/nonself recognition
(C) Immuninologic memory
(D) All of these

464. Who coined the term Green revolution:
(A) Willianm Gadd (B) M.S. Swaminathan
(C) Norman E. Borlaoug (D) C. Subramanian

465. Diploid species of cotton are:
(A) Desi (B) American
(C) Both (A) and (B) (D) None of these

466. Tetraploid species of cotton are:
(A) Desi (B) American
(C) Both (A) and (B) (D) None of these

467. Silent valley has been preserved as it has:
(A) Recreational value (B) Valuable timber plants
(C) Natural forest (D) Rare plant and animals

468. Silent valley having rare plants and animals is located in:
(A) Kerala (B) Karnataka
(C) Jammu and Kashimir (D) Andhra Pradesh

469. Which one is connected with conservation of forests?
(A) Gir (B) Silent valley
(C) Kaziranga (D) Bharatpur

470. New approach to conservation is establishment of:
(A) Reserve forests (B) Santuries
(C) Biosphere reserves (D) National parks

471. Social forestry is:
(A) Management of forests by cooperative societies
(B) Growing and management of useful plants on common lands
(C) Growing one type of trees
(D) Growing different types of plants together

472. Decrease in species diversity in tropical countries is mainly due to:
(A) Urbanisation (B) Pollution
(C) Deforestation (D) Soil erosion

473. Adulteration of mustard oil with *Argemone mexicana* seed oil caused in the recent past:
(A) Tuberclosis (B) Diarrhoea
(C) Dropsy (D) None of these

474. Wheat is a:
(A) Monoecious plant (B) Diocious plant
(C) Hermaphrodite plant (D) None of these

475. A genetic process called 'gene conversion' is observed in:
(A) Yeast (B) Fungi
(C) Other eukaryotes (D) All of these

476. Which means that one of the allele is being converted from one type to other, possibly by DNA recombination during meiosis after fusion of the gametes is called:
(A) Gene conversion (B) Gene transcription
(C) Gene promoter (D) Gene disruption

477. The first step of gene expression when one strand of the double-stranded DNA encompassing the gene is copied to the functional RNA (synthesized from 5' to 3' direction) is called:
(A) Gene conversion (B) Gene transcription
(C) Gene promoter (D) Gene disruption

478. The part of the gene which controls its transcription is called ——— that is located upstream of the coding sequence is called:
(A) Gene conversion (B) Gene transcription
(C) Gene promoter (D) Gene disruption

479. The technique to introduce gene-specific mutations to chromosomes in intact organism is called:
(A) Gene conversion (B) Gene transcription
(C) Gene promoter (D) Gene disruption

480. Intervening the noncoding sequences are called:
(A) Exons (B) Introns
(C) Both (A) and (B) (D) None of these

481. Intervening the coding sequences are called:
(A) Exons (B) Introns
(C) Both (A) and (B) (D) None of these

482. Introns are more common for ——— coding genes:
(A) mRNA (B) sRNA (C) tRNA (D) DNA

483. The cloned piece of disrupted genomic DNA is called a:
(A) Gene conversion (B) Gene targeting DNA
(C) Gene promoter (D) Gene disruption

484. The set of Indian guidelines was also issued in 1990 by the Department of Biotechnology, Ministry of Science and Technology covering:
(A) Genetically engineered organisms
(B) Genetic transformation of plants and animals
(C) Containment measures at the laboratory level under different risk group
(D) Modalities of implementation

485. DNA measures are available for comparative determination of sizes of:
(A) Single stranded RNA molecules
(B) Gene targeting DNA
(C) Single stranded DNA molecules
(D) None of these

486. Mixtures of in-vitro transcribed RNA molecules are also available in the molecular weight ranges of:
(A) 0-500 bases
(B) 500-9000 bases
(C) 9000-10000 bases
(D) More than 10000 bases

487. Plasmids are extra chromosomal double stranded circular DNA molecules present in:
(A) Bacteria (B) Lower eukaryotes
(C) Both (A) and (B) (D) None of these

488. Ribosomal RNA constitutes 80-85% of total cellular RNA, transfer RNA and small nuclear RNA constitute about 10-15%, while messenger RNA constitutes only:
(A) 1-5% (B) 5-10%
(C) 10-15% (D) 15-20%

489. Which of the following organisms is used in the steroid synthesis?
(A) *E. coli*
(B) *Rhizopus stolonifer*
(C) *Ashbya gossypii*
(D) *Pseudomonas*

490. In the conversion of pyruvic acid to acetyl co-enzyme A, NAD^+ is
(A) Reduced (B) Oxidized
(C) Hydroxylated (D) None of these

491. The technique of insertion of a desired gene into DNA of plasmid vecter is:
(A) Gene splicing (B) Gene dressing
(C) Gene cloning (D) Gene drafting

492. One of the main technique in recombinant DNA technology is:
(A) DNA cloning (B) DNA molecules
(C) Both (A) and (B) (D) None of these

493. Which one is used in genetic engineering?
(A) RNA polymerase (B) DNA polymerase
(C) Restriction endonuclease (D) Nuclease

494. Pyruvate is used to produce:
(A) Glucose (B) Fructose
(C) Sucrose (D) Maltose

495. DNA fingerprinting is useful for study of:
(A) Forensics (B) Polymorphism
(C) Identify and relationships (D) All of these

496. Gene mouse has been produced through:
(A) Tissue culture (B) Gene differentiation
(C) Gene manipulation (D) All of these

497. Which is preferred for making artificial seeds?
(A) Potassium arginate (B) Sodium alginate
(C) Agar (D) Gel-Rite

498. A map of the locations of identifiable landmarks on chromosomes is known as:
(A) Structural gene (B) Genetic marker
(C) Genetic map (D) Eucaryotic gene

499. Plasmids capable of propagating and transferring (shuttling) genes between two different organism, one of which is typically a prokaryote (*E. coli*) and other a eukaryote (yeast) is called:
(A) Shuttle vector (B) Transduction
(C) Both (A) and (B) (D) None of these

500. The technique of obtaining large number of plantlets by tissue culture method is called:
(A) Plantlet culture (B) Organ culture
(C) Micropropagation (D) Macropropagation

501. Who is the father of plant tissue culture?
(A) P. Maheshwari (B) G. Haberlandt
(C) P.R. White (D) R.J. Gautheret

502. Transfection is a:
(A) Transfer of host genes from one cell to another by a virus
(B) Genetic transformation of animal cells
(C) Conversion of a normal animal cell into a cancerous cell
(D) A type of virus infection

503. In India, which of the following overseas That implements the research and development experiments utilizing GMOs and recombinant DNA:
(A) Department of Atomic Energy
(B) Council of Scientific and Industrial Research
(C) Ministry of Environment of Forest
(D) Department of Biotechnology

504. Strains of which of the following bacteria is used for the degradation of petroleum and management of oil spills?
(A) *Pseudomonas putida*
(B) *Clostridium acetobutylicum*
(C) *Methylophilus methylotrophus*
(D) *Pseudomonas fluorescence*

505. The ability to transfer or reprogram cultured cells with new genetic material (DNA and genes) has provided a major tool to molecular biologists wishing to study the cellular effects of the expression of these genes (new proteins) is called:
(A) Genetic counseling (B) Genetic engineering
(C) Gene therapy (D) All of these

506. A fundamental in molecular genetic which takes advantage of the ability of individual single – stranded nuleic acid molecules to form double-stranded molecules is called:
(A) Nuleic acid hybridization (B) ex situ hybridization
(C) In situ hybridization (D) None of these

507. Virus mediated gene transfer, in which the foreign DNA is packaged inside a virus particle is called:
(A) Transfection (B) Transduction
(C) Both (A) and (B) (D) None of these

508. Which drugs is produced through biotechnology?
(A) Erythropoeitin (B) Fibroblast interferon
(C) Insulin (D) All of these

509. Which form of RNA has a structure resembling clover leaf:
(A) rRNA (B) hnRNA
(C) mRNA (D) tRNA

510. Which organisms was discovered by Fleming to produce the antibiotic penicillin?
(A) *Aspergillus niger*
(B) *Penicillium notatum*
(C) *Penicillium griseofulvum*
(D) *Oenicillium chrysogenum*

511. In transgenics, expression of transgene in target tissue is determined by:
(A) Enhancer (B) Transgene
(C) Promoter (D) Reporter

512. During the World war II, technology for citric acid production was based on:
(A) *Aspergillus niger*
(B) *Aspergillus oryzae*
(C) *Rhizobium sp.*
(D) *Clostridium acetobutylium*

513. RNA interference is used to control different viral diseases in case of plants and animals. The active fragment involving in interference is known as:
(A) Si RNA (B) Micro RNA
(C) Both (A) and (B) (D) None of these

514. Which organisms is used for production of single cell protein:
(A) Pseudomonas (B) Azospirillum
(C) Spirulina (D) BGA

515. The use of chemical of physical methods to persuade cells to take up DNA from the culture medium or direct physical introduction of DNA into the cell is called:
(A) Transfection (B) Transduction
(C) Both (A) and (B) (D) None of these

516. Restriction site for cloning should be:
(A) Tendemly repeated (B) Palindromic in nature
(C) Repeated sequence (D) Hexanucleotide

517. In most of the cancer cases which gene is mutated?
(A) p53 (B) p21 (C) p16 (D) p15

518. Microsatellites are also known as:
(A) VNTRs (B) RFLPs
(C) STRs (D) RAPDs

519. Which one is the first to be used as a source of energy?
(A) Carbohydrates (B) Proteins
(C) Lipids (D) Fats

520. Golden rice is a transgenic crop of the future with the following improved trait:
(A) High insect resistance
(B) High lysine content
(C) High protein content
(D) High vitamin – A content

521. Genetic map is one that:
(A) Establishes site of the genes on a chromosomes
(B) Establishes the various stages in gene evolution
(C) Shows the stages during the cell division
(D) Shows the distribution of various species in a region

522. Three restriction endonucleases P, Q and R recognize 4bp, 6bp and 8 bp sequences respectively. The relative frequency of occurrence of these sequence on a bacterial genome is:
(A) P > Q > R (B) P > R < Q
(C) R < Q > P (D) Q < R > P

523. LCR using reverse transcriptase; specific for detection of RNA by:
(A) AG-LCR (B) Asymetrical PCR
(C) Both (A) and (B) (D) None of these

524. The two PCR primers used in a 100:1 ratio so that one primer is finished 10 cycles before the other; yields predominantly single-strand copies that can be used for DNA sequencing in:
(A) AG-LCR (B) Asymetrical PCR
(C) Both (A) and (B) (D) None of these

525. The application of computational techniques for the analysis of biological data (computational biology) that provides not just sequence analysis, but also structural bioinformatics and the stimulation of molecular interactions like metabolic pathways and gene networks is called:
(A) Bioinformatics (B) Bioethics
(C) Computer hardware (D) None of these

526. The information sciences, computer programming skills especially with scripting languages are useful:
(A) PERL (B) EMBL
(C) EST (D) GenBank

527. The important data bases in bioinformatics are used:
(A) SWISS-PROT (B) EMBL
(C) GenBank (D) Pfam
(E) All of these

528. The primary nucleotide sequence databases are:
(A) EMBL (B) GenBank
(C) DDBI (D) All of these

529. KRAF_HUMAN is the entry name for the raf-1 oncogene from:
(A) *Homo sapiens* (B) *Avena sativa*
(C) *Pisum sativum* (D) *Hoemophilus inflyence*

530. The database has sequences deposited normally after verification is:
(A) GenBank (B) DDBI
(C) EST (D) EMBL

531. The data base have cDNA single-pass sequences and contain a high degree of error is:
(A) GenBank (B) DDBI
(C) EST (D) EMBL

532. Deletions and insertions are collectively known as:
(A) Indels (B) Intels
(C) EST (D) None of these

533. Length of the DNA sequence affected by the:
(A) Mutational event (B) Point mutation
(C) Segmoidal mutation (D) KEGG database

534. To affect a single nucleotide are called:
(A) Mutational event (B) Point mutation
(C) Segmoidal mutation (D) KEGG database

535. The mutations that affect several adjacent nucleotides is:
(A) Mutational event (B) Point mutation
(C) Segmoidal mutation (D) KEGG database

536. The databases are those of metabolic pathways and resources is:
(A) Mutational event (B) Point mutation
(C) Segmoidal mutation (D) KEGG database

537. The KEGG database is an effort to computerize current knowledge of:
(A) Molecular (B) Cellular biology
(C) Both (A) and (B) (D) None of these

538. The protein sequence database are:
(A) SWISS-PROT (B) PIR
(C) Both (A) and (B) (D) None of these

539. The translated DNA databases are:
(A) PROTEIN (NCBI) (B) TREMBL
(C) Both (A) and (B) (D) None of these

540. The sequence derived databases are:
(A) PROTEIN (NCBI) (B) TREMBL
(C) Pfam (D) SWISS-PROT

541. The database for 3D structures of biological macromolecules determined by x-ray crystallography and NMR is:
(A) PDB (B) SCOP
(C) Pfam (D) PAM

542. Secondary data bases derived from PDB are the:
(A) PDB (B) SCOP
(C) Pfam (D) PAM

543. The dot plot was first described by:
(A) Gibbs and McIntyre (B) Margaret Dayhoff
(C) Senger etal (D) None of these

544. The steps in implementing the algorithm for alignment using dynamic programming is:
(A) Matrix initialization (B) Matrixfill
(C) Trace back (D) All of these

545. The programme is used to compare a protein sequence with a nucleotide (translated into all six frames) sequence database is:
(A) TBLASTN (B) TBLASTX
(C) PROSITE (D) PROMOT

546. The programme is used to compare a nucleotide (translated into all six frames) sequence with a nucleotide (translated into all ix frames) sequence database is:
(A) TBLASTN (B) TBLASTX
(C) PROSITE (D) PROMOT

547. A collection of functional sites and sequences patterns found in proteins, it also has search tools for matching patterns/motifs is:
(A) TBLASTN (B) TBLASTX
(C) PROSITE (D) PROMOT

548. The search tool can be used to match a sequence against the PROSITE database is:
(A) TBLASTN (B) TBLASTX
(C) PROSITE (D) PROMOT

549. The PAM matrix was developed by:
(A) Gibbs and McIntyre (B) Margaret Dayhoff
(C) Senger etal (D) None of these

550. The ancestral relationship of a set of species which determined by the comparison of specific features (characters) of a species with the natural assumption that similar species-with similar features are genetically close is:
(A) Phylogeny (B) Microarray
(C) LigBuilder (D) None of these

551. Phylogeny usually described:
(A) Plant (B) Tree
(C) Ferns (D) Algae

552. The function of genes identified using an ab initio method is identified by comparing the gene sequence against known proteins using:
(A) BLAST (B) TBLASTX
(C) PROSITE (D) PROMOT

553. The availability of the complete set of genes from an organism, with their sequences, has enabled the study of expression profiles using:
(A) PROSITE (B) BLAST
(C) Microarray (D) PROMOT

554. To introduce a mathematical formulation to determine the potential energy of a molecular called:
(A) Microarray
(B) Molecular force field
(C) Shot gun sequencing strategy
(D) Shot gun approach of sequence

555. Genome sequencing became possible on a large scale with the discovery of the:
(A) Microarray
(B) Molecular force field
(C) Shot gun sequencing strategy
(D) Shot gun approach of sequencing

556. The random selected clones are sequenced until all clones in the genomic library are analyzed, the sequences are aligned using computers by:
(A) Microarray
(B) Molecular force field
(C) Shot gun sequencing strategy
(D) Shot gun approach of sequencing

557. The partially digested genomic DNA is subjected to agarose gel electrophoresis or sucrose gradient configuration for separation from the mixture of fragments of appropriate size. The fragments are then inserted into a suitable vector for cloning by:
(A) Shot gun approach for gene cloning
(B) Molecular force field
(C) Shot gun sequencing strategy
(D) Shot gun approach of sequence

558. A de novo ligand design program which buildup ligands from a library of organic fragments taking into consideration the structural consist of the target protein is:
(A) LigBuilder (B) POCKET module
(C) GROW module (D) LINK module

559. LigBuilder progarmme basically consists of:
(A) POCKET (B) GROW
(C) LINK (D) None of these

560. The binding site of the target protein and derives the key interaction sites within this site is:
(A) LigBuilder (B) POCKET module
(C) GROW module (D) LINK module

561. Implements the sequential growing strategy of building up the molecular from a single starting block known as the seed by:
(A) LigBuilder (B) POCKET module
(C) GROW module (D) LINK module

562. Buildup the molecule from several seed structures which have been placed in the binding site so as to maximize interactions with the protein by:
(A) LigBuilder (B) POCKET module
(C) GROW module (D) LINK module

563. Which is concerned with determination of the function of a gene product?
(A) Functional proteomics (B)Structural proteomics
(C) Functional genomics (D) Structural genomics

564. The sequence of a human genome has how many nucleotide pairs?
(A) 4.2×10^6 (B) 4.6×10^9
(C) 3.2×10^9 (D) 3.8×10^9

565. One megabase pairs (Mb) is equal to how many base pairs?
(A) 10^5 bp (B) 10^6 bp
(C) 10^8 bp (D) 10^9 bp

566. Databases on DNA sequences contain which types of sequences?
(A) EST and GST sequences
(B) Genomic DNA and cDNA sequences
(C) Organellar DNA sequences and sequences of other molecules such as tRNA, small RNAs
(D) All of these

567. Which is an example of major nucleotide sequence database?
(A) DNA Databank of Japan
(B) Nucleotide Sequence Database maintained by EMBL
(C) Gene Bank held by NCBI, USA
(D) All of these

568. In which risk groups, agents are not associated with diseases in healthy humans?
(A) Risk group 3 (B) Risk group 1
(C) Risk group 2 (D) Risk group 4

569. Functional genomics includes answer to:
(A) How is a gene expressed?
(B) How is a gene product related in sequence and structure to products of other genes of the same organisms
(C) How does a gene interacts with the proteins?
(D) All of these

570. The α–amylase inhibitor from *Phaseolus vulgaris* (rajma) caused allergy when expressed in:
(A) Wheat (B) Maize
(C) Peas (D) Tomato

571. Which eukaryotic genome was sequenced in 1999?
(A) *Caenorhabditis elegans* (worm)
(B) *Drosophila melanogaster*
(C) *Saccharomyces cerevisiae*
(D) Both (A) and (B)

572. Non-infectious illnesses is associated with production of steroids:
(A) Increased body weight (B) Feminization of males
(C) Hypertension (D) All of these

573. The potential risks from novel DNA sequences used during transgenic development may be summed up as:
(A) Protein products of transgenes or products generated by them may be toxic
(B) The transgene product may be allergenic
(C) Antibiotic resistance genes may be taken up by human intestinal microflora, which, as a consequence, become resistant to these antibiotics
(D) All of these

574. Strategy of temporal and tissue-specific expression for preventing transgene transfer from transgenic varieties of crops is based on which techniques?
(A) Terminator technology
(B) Tapetum-specific promoters allow expression only when the suppressor is present
(C) Recoverable block function
(D) Inducible promoters allow expression only when the inducer is present

575. Which is not correct about printed oligonucleotide chips?
(A) These are produced by a light-directed printing technology
(B) These chips have double-stranded oligonucleotides of 20-25 bases
(C) Oligonucleotides are synthesized directly onto the chips
(D) Each sequence is represented by a set of 20 non-overlapping oligonucleotides to reduce false positive results

576. Lateral gene transfer is also called:
(A) Horizontal gene transfer (B) Vertical gene transfer
(C) Parallel gene transfer (D) Both (B) and (C)

577. Which statement is not correct about eukaryotic genomes?
(A) As the genome size increases, gene density decrease
(B) A high percentage of proteins encoded by human genome have orthologs in other eukaryotic genomes
(C) About 75% of repetitive DNA in human genome is due to LINEs and SINEs sequences
(D) Intron sequences are as variable as exon sequences

578. Which scientists proposed that amino acid sequences of proteins may be used to study evolutionary relationships among organisms?
(A) Zuckerland and Pauling (B) Sensen and Buchler
(C) Dayhoff and Rashidi (D) Tolin and Vidaver

579. The vector under host-vector I system includes:
(A) Nonconjugative plasmids
(B) Derivative of nonconjugative plasmids
(C) Variants of bacteriophages
(D) All of these

580. Most of the recombinant microorganism released in the environment of genes by:
(A) Transposon insertion
(B) Deletion of genes by molecular genetic procedure in vitro
(C) Genetic tagging of bacteria with specific
(D) Any of these

581. Insertional mutagenesis has been reported in:
(A) Mouse and Drosophila melanogaster
(B) Yeast and roundworm
(C) Zebrafish and Arabidopsis
(D) All of these

582. Which methods can be used for the identification/discovery of SNPs?
(A) DNA chips (B) Microarrays
(C) Analyzing the sequence data (D) All of these

583. Which is used for comparing the submitted nucleotide sequence with nucleotide database?
(A) BLASTn (B) BLASTx
(C) tBLASTx (D) tBLASTn

584. The rough draft of human genome was announced on:
(A) June 26, 2000 (B) June 26, 2001
(C) June 26, 1999 (D) June 26, 2003

585. The *Mycoplasma genitalium* complete genome was chemically synthesized and assembled in which year?
(A) 2005 (B) 2006 (C) 2007 (D) 2008

586. Which is not correct about homologous recombination?
(A) Exchange of identical segments occurs between two heterologous DNA molecules/chromosomes
(B) Toa chive transgene integration by homologous recombination, the marker gene is flanked by appropriate sequences from or around the target gene
(C) Mouse embryonic stem cells, may show upto 0.1% integration by homologous recombination
(D In prokaryotes, integration by homologous recombination may occur upto 90% of the time when suitable targeting vectors are used

587. Which is not correct about Gene trap vector?
(A) In this vector, the insertion element contains a reporter gene, e.g., lacZ or gusA, downstream of a splice acceptor site
(B) reporter gene will be expressed only when the vector inserts within the transcribed region of a gene
(C) It is used for detection and isolation of genes
(D) This vector was isolated from *Bacillus thuringiensis*

588. Which programmes can be used for the identifications/detection of genes from genome sequence of prokaryotes?
(A) Glimmer (B) GENMARK
(C) GeneFinder (D) Both (A) and (B)

589. India has not been involved in sequencing the genomes of which organisms as:
(A) Pigeon (B) Rat
(C) Pigeonpea (D) Humans

590. Which database is used by ENTREZ for bibliographic or citation search?
(A) SWISS-PROT (B) PubMed
(C) TrEMBL (D) PDB

591. Databases on DNA sequences contain which types of sequences?
(A) EST and GST sequences
(B) Genomic DNA and cDNA sequences
(C) Organellar DNA sequences and sequences of other molecular such as tRNA, small RNAs
(D) All of these

592. Effective dissemination of plant-associated microorganisms beyond the green house can be prevented by using:
(A) Organisms that have been genetically disabled to survive in nature
(B) Organisms whose natural mode of transmission requires injury to the host
(C) Organisms which require another organisms fro their survival
(D) Both (A) and (B)

593. The first recombinant microorganism (Tn5-containing strain of *Pseudomonas fluorescens*) was released in which countries?
(A) USA (B) Canada
(C) Netherlands (D) Germany

594. Which advantages is offered by Real-Time PCR?
(A) Multiplex PCR reactions can be used by employing molecular beacons/TagMan probes
(B) It eliminates the post – PCR processing of PCR products needed for quantitative measurements of PCR products
(C) Real-Time PCR offers a much wider dynamic range than conventional RT-PCR
(D) All of these

595. In case of experiments with animals, which procedures is followed at all biosafety levels?
(A) Animal carecass is disposed off to avoid it use as food
(B) A permanent record is maintained of the experimental use and disposal of each animal
(C) The animals are confined to security-fenced areas
(D) All of these

596. Which is not correct about NIH guidelines?
(A) A major revision of the guidelines was effected in 1982
(B) In the revision, the containment levels were made more stringent
(C) Experiments that were previously prohibited, were changed to category requiring review and approval by NIH
(D) In USA, the NIH guidelines are followed by all federal agencies that fund research on recombinant DNA

597. The short gun sequencing approach would be most suited for which genomes?
(A) *E. coli* (B) Yeast
(C) Humans (D) *Drosophila*

598. Which has operons similar to those of bacteria?
(A) *C. alegans* (B) *Drosophila*
(C) *Arabidopsis* (D) Mouse

599. Which is not correct about TaqMan probes?
(A) They are oligonucleotides longer than the primers
(B) They contain a fluorescent dye usually on the 5'-base, and a quenching dye typically on the 3'-base
(C) When Taq polymerase replicates a tempelate to which a TaqMan probe is paired, its 5'-exonuclease activity cleavage the TaqMan probe
(D) The fluorescence decreases proportionately to PCR amplification

600. Which is correct about genome annotation?
(A) GAME is a programme for describing experimental evidence to support annotation
(B) DAS is useful in indexing and visualization
(C) BioPerl 2001 is used for storing, manipulating and visualizing the genome annotation
(D) All of these

601. The type of regulation necessary during field trials should depend on:
(A) Ability of the modified plants to survive in nature
(B) Their ability for dispersal and reproduction
(C) Their ability to hybridize with crop and weed plants
(D) All of these

602. Which is not correct?
(A) RecA- strains lack recombination
(B) Transposon vectors lacking transposase gene are not able to transpose to other bacteria
(C) Auxotrophic mutants can survive in nature
(D) Transgenes integrated into plasmids may be transferred to other bacteria

603. Which terms is not used as synonyms of planned introduction?
(A) Field testing (B) Deliberate release
(C) Contained use (D) Deliberate introduction

604. The first comprehensive collection of amino acid sequences was compiled in the Atlas of Protein Sequence and Structure by:
(A) European Bioinformatics Institute
(B) National Biomedical Research Foundation
(C) National Centre for Bioinformatics Information
(D) European Molecular Biology Laboratory

605. Which is used for comparing the submitted nucleotide sequence with nucleotide database?
(A) BLASTn (B) BLASTx
(C) rBLASTx (D) tBLASTn

606. Transgenic plants could serve as vectors for transgene dissemination through :
(A) Seed
(B) DNA released from debris of transgenic plants
(C) Pollen
(D) All of these

607. For minimizing pollen-mediated gene flow from transgenic varieties of *Solanum tuberosum*, how many meters of isolation distance is recommended?
(A) 10 (B) 20 (C) 30 (D) 40

608. Which of the search tools is used for comparing the submitted nucleotide sequence with amino acid sequence database?
(A) BLASTn (B) BLASTx (C) rBLASTx (D) tBLASTn

609. Which is correct about transmission of transgenes from one crop species to another?
(A) In case of cross-pollinated species isolation by distance is rarely sufficient to prevent inter-population mating
(B) If male sterility is introduced into transgenic cross-pollinated crops, it can check the movement of transgenes through pollen
(C) It transgenes are integrated in the chloroplast genome, it would almost eliminated the transmission of transgene through pollen grains
(D) All of these

610. Which is not correct?
(A) European Molecular Biology laboratory endeavours to collect, organise and distribute nucleotide sequence data and related information
(B) National Centre of Bioinformatics was established in USA and serves as primary information databank and provider of information
(C) RNA data bank was established by Japan
(D) National Biomedical Research Foundation established the Protein Information System in 1984

611. Which of the search tools is used to compare submitted amino acid sequence with nucleotide sequence database?
(A) BLASTn (B) BLASTx
(C) rBLASTx (D) tBLASTn

612. Physical containment relates to:
(A) Containment equipment
(B) Laboratory design
(C) Special procedures
(D) All of these

613. Which is not correct?
(A) Oral exposure to transgenic tomato containing cry protein poses no additional risk to human and animal health
(B) In general, the quality of produce from cry gene carrying transgenic crops is regarded different from that from the nontransgenic plants of the same variety
(C) Cowpea trypsin inhibitors do show some adverse effects on rats in nutritional studies
(D) Significant hypertrophy of small intestine and altered enzyme activities in brush border occur due to wheat germ agglutinin

614. Which of the tools compares submitted amino acid sequence with amino acid database?
(A) BLASTn (B) BLASTx
(C) rBLASTx (D) tBLASTn

615. Single nucleotide polymorphisms (SNPs) may be located in:
(A) In the recognition sequence of an endonuclease
(B) Within the concerned gene
(C) Very close to the gene to show a tight linkage
(D) All of these

616. Which approaches can be used for gene identification?
(A) Exon trapping (B) cDNA capture
(C) cDNA selction (D) All of these

617. The biosafety guidelines are developed to ensure an adequate level of protection during which activities of modern biotechnology?
(A) Safe transfer (B) Handling
(C) Use of GMOs (D) All of these

618. In case of prokaryotes, which represents the frequency of unique ORFs?
(A) 50% (B) 100%
(C) 25% (D) 75%

619. The NIH, USA guidelines specify the practices for constructing and handling are:
(A) Recombinant DNA molecules
(B) Organisms and viruses containing recombinant DNA molecules
(C) Products obtained from B
(D) Both (A) and (B)

620. Which is not correct about TaqMan probes?
(A) They are oligonucleotides longer than the primers
(B) They contain a fluorescent dye usually on the 5'-base, and a quenching dye typically on the 3'-base
(C) When Taq polymerase replicates a template to which a TaqMan probe is paired, its 5'-exonuclease activity cleavage the TaqMan probe
(D) The fluorescence decreases proportionately to PCR amplification

621. Which is considered as primary physical containment?
(A) Laboratory design (B) Containment equipment
(C) Special procedures (D) Both (A) and (B)

622. It is feared that genetically engineered microorganisms may disturb the ecosystem and its processes, in which ways?
(A) They may rapidly multiply and out complete the native microbes
(B) They may transfer genes related to virulence or pathogenesis into the native bacterial populations
(C) Both (A) and (B)
(D) They may cause unexpected health hazards

623. DNA sequences/molecules used for preparing spotted DNA arrays may be obtained from:
(A) PCR amplification (B) Genomic DNA
(C) cDNA clones (D) All of these

624. Sequencing of the 24 human chromosomes was completed in which years?
(A) 2004 (B) 2005 (C) 2006 (D) 2007

625. Which eukaryotic genome was sequenced in 1999?
(A) *Caenohabditis elegans* (worm)
(B) *Drosophila melanogaster*
(C) *Saccharomyces cerevstae*
(D) Both (A) and (C)

626. Which is not correct about Cry proteins?
(A) Cry proteins are rapidly degraded by the stomach juices of vertebrates
(B) They could have harmful effects on non target insect species
(C) It has been reported that Cry proteins leak into soil from transgenic maize roots, where they persist for over 3 weeks
(D) The Cry proteins in the soil become associated with soil clay/humus and become prone to microbial degradation

627. In case of prokaryotes, which represents the frequency of unique ORFs?
(A) 25% (B) 50% (C) 75% (D) 100%

628. Which advantages is offered by Real-Time PCR
(A) Multiplex PCR reactions can be used by employing molecular beacons/TaqMan probes
(B) It eliminates the post-PCR processing of PCR products needed for quantitative measurements of PCR products
(C) Real-time PCR offers a much wider dynamic range than conventional RT-PCR
(D) All of these

629. The sequence of a human genome has how many nucleotide pairs?
(A) 4.2×10^6 (B) 4.6×10^9
(C) 3.2×10^9 (D) 3.8×10^9

630. Which is not correct?
(A) The public funded Human Genome Project employed the clone-by-clone Celera Genomics followed whole genome shot-gun approach
(B) Clone-by-clone approach is based on contigs
(C) In the whole genome shot-gun approach, genome DNA is sheared to obtain 2 to 10 kb fragments, which are then cloned and sequenced from both the ends
(D) Assigning biological functions to the DNA sequences using a variety of tools and techniques is described as structural genomics

631. Which is not correct about Real-time PCR?
(A) Real-time PCR can detect the size of the amplicon
(B) Design of primers, TaqMan probes and molecular beacons requires almost complete information of the target sequence
(C) With SYBR green method, primer dimer artifact is a problem
(D) The one-step RT_PCR can not be used for multiplex real-time RT-PCR

632. Which is not correct?
(A) In European Union, the maximum permissible limit for GMO content in conventional food products and food ingredients is 1% at the marketing stage
(B) All foods that contain more than 1% GMOs related to the different ingredients have to be labeled
(C) In case of organic farming, utilization and presence of GMOs is excluded
(D) Volunteers are not likely to pose problems in crops like rapeseed and mustard

633. Which is not correct about bromoxynil?
(A) It is a non-biodegradable herbicide
(B) It causes birth defects in laboratory animals
(C) It is toxic to fish and may cause cancer in humans
(D) Bromoxynil is absorbed through the skin

634. India has initiated genome sequencing project of:
(A) Rice (B) Tomato
(C) Pigeon (D) Human

635. The α–amylase inhibitor gene was transferred to modify which traits?
(A) Insect resistance (B) Amino acid balance
(C) Food digestibility (D) Both (A) and (C)

636. Which is not correct?
(A) Both TaqMan probes and molecule beacons allow multiplex reactions, and thereby, detection of multiple DNA sequences in a single PCR reaction
(B) In multiplexing, different probes/beacons are labelled with different reporter dyes
(C) Multiplexing does not allow amplification in a single PCR reaction of both target and endogeneous control sequences
(D) The use of probe/beacons avoids problems due to mispriming,primer-dimer artifacts, etc.

637. Which is not correct about genomics?
(A) In 1980, David Botstein and coworkers working at the Whitehead Institute of Biomedical Research, USA proposed that one could construct a genome map
(B) In 1995, the first bacterial genome to be completely sequenced was that of *Haemophilus influenza*
(C) The sequencing of whole genome of first multicellular organism, round worm (*Caenorhabdits elegans*), was published in 1999
(D) The sequencing of rice genome by International Rice Genome Sequencing Project was based on whole ge nome shot-gun approach

638. Which is considered as secondary physical containment?
(A) Laboratory design (B) Containment equipment
(C) Special procedures (D) Both (A) and (B)

639. Which statements is correct about comprehensive risk assessment?
(A) It is after the initial risk assessment
(B) The level of containment appropriate for an experiment is decided on the basis of comprehensive risk assessment
(C) The containment level for the experiment may, in the end, be comparable to that of the risk groups (RG) of the agent or it may be raised or lowered as a result of the comprehensive risk assessment
(D) All of these

640. Which is not correct about Gene trap vector?
(A) In this vector, the insertion element contains a reporter gene, e.g., *lacZ* or *gusA*, downstream of a splice acceptor site
(B) Reporter gene will be expressed only when the vector inserts within the transcribed region of a gene
(C) It is sued for detection and isolation of genes
(D) This vector was isolated from *Bacillus thuringiensis*

641. In which of the following is found the lowest density of transposable elements?
(A) *Arabidopsis* (B) *Drosophila*
(C) *C. elegans* (D) Humans

642. Which is not correct?
(A) The public funded Human Genome Project employed the clone-by-clone approach, while the private company Celera Genomics followed whole genome shot-gun approach
(B) Clone-by-clone approach is based on contigs
(C) In the whole genome shot-gun approach, genomic DNA is sheared to obtain 2 to 10 kb fragments, which are then cloned and sequenced from both the ends
(D) Assigning biological functions to the DNA sequences suing a variety of tools and techniques is described as structural genomics

643. Which is not correct?
(A) Transgene isolated from a food crop guarantees the food safety of the transgene product
(B) Acute toxic effect depends on the concentration of the compounds in question
(C) Psoralen is known to be toxic, carcinogenic and mutagenic
(D) Any outstanding insect resistance whose biochemical basis is not known should be considered as the prime candidate for toxicological evaluation

644. Which is not correct?
(A) In databases, base sequence of only one strand is listed
(B) Base seque... ...ns from the 3' to the 5'-directions
(C) In single letter codes, R denoted guanine or adenine, while Y denotes thymine or cytosine
(D) The nomenclature system adapted in bioinformatics is based on the recommendations of International Union of Pure and Applied Chemistry

645. One megabase pairs (Mb) is equal to how many base pairs?
(A) 10^4 bp (B) 10^6 bp
(C) 10^8 bp (D) 10^9 bp

646. It is feared that transgenic plants, may lead to:
(A) Transfer of their transgenic feature to weeds
(B) May themselves become weeds
(C) Degrade the environment
(D) All of these

647. Which is not correct about biosafety Guidelines in India?
(A) Initially, the Indian Recombinant DNA safety Guidelines and Regulation were prepared in 1990 b Recombinant DNA Advisory Committee, Department of Biotechnology, New Delhi
(B) The Indian Recombinant DNA safety Guide lines and Regulation were revised in 1994
(C) Field trials, in India using transgenic plants began in 1995
(D) Canada become the first country to begin commercial use of virus resistant transgenic tobacco and tomato in the early 1990

648. In case of *E. coli*, the host for any HV 1 system is:
(A) Strain K-12 (B) A derivative of strain K-12
(C) Either (A) or (B) (D) Strain chi 1776

649. In which of the following is found the highest density of transposable elements?
(A) *Arabidopsis* (B) *Drosophila*
(C) *C. elegans* (D) Humans

650. Which is not correct about genomics?
(A) Sequencing sequence assembly and annotation of the whole genome of an organism constitutes functional genomics
(B) The first plant genome to be completely sequenced is that of the weed, *Arabidopsis thaliana*
(C) *Arabidopsis thaliana* is a wild relative of mustards
(D) The Arabidopsis Genome Initiative followed the same method of sequencing as the Human Genome Project

651. Which is concerns with determination of the function of a gene product?
(A) Expression proteomics (B) Structural genomics
(C) Functional genomics (D) Structural proteomics

652. Which is correct about wheat germ agglutinin?
(A) It is effective against insects at concentration of 10 mg/kg body weight
(B) It produces cytotoxic effects in rats at a lower concentration (7 mg/kg)
(C) Safe food has it only at 0.3 mg/kg
(D) All of these

653. DNA sequences/molecules used for preparing spotted DNA arrays may be obtained from:
(A) PCR amplification (B) Genomic DNA
(C) cDNA clones (D) All of these

654. The *Mycoplasma genitalium* complete genome was chemically synthesized and assembled in which year?
(A) 2005 (B) 2006 (C) 2007 (D) 2008

655. Shot-gum sequencing was conceptualized by:
(A) Craig Venter (B) Michelmore
(C) Giovannoni (D) Young

656. Which phenomenon comes under post-translational modifications?
(A) Phosphorylation (B) Proteolysis
(C) Glycosylation (D) All of these

657. In the early 1970s which were considered as the possible potential health hazards of cloning recombinant DNA molecules?
(A) Molecular biologists may not be well-versed with the laboratory practices needed to handle serious human and animal pathogens
(B) 'Hybrid organisms' could be created with biological activities of an unpredictable nature
(C) 'Hybrid organisms' may escape from the laboratory with unpredictable consequences
(D) All of these

658. The chief objectives of risk assessment during laboratory research are to decide about which of the proposed research?
(A) Laboratory procedures (B) Biological containment
(C) Physical containment (D) All of these

659. Which approaches can be used for gene identification?
(A) Exon trapping (B) cDNA capture
(C) cDNA selection (D) All of these

660. Which is not correct about homologous recombination?
(A) Exchange of identical segments occurs between two heterologous DNA molecules/chromosomes
(B) To achieve transgene integration by homologous recombination, the marker gene is flanked by appropriate sequences from or around the target gene
(C) In prokaryotes, integration by homologous recombination may occur upto 90% of the time when suitable targeting vectors are used
(D) Mouse embryonic stem cells. May show upto 0.1% integration by homologous recombination

661. In the assessment of allergenicity, the source of transgene is the first consideration; the transgene may be obtained from:
(A) Commonly allergenic food
(B) Less commonly allergenic food
(C) No documented history of food allergenicity
(D) Any of these

662. Which software have been developed for the prediction/detection of genes from genome sequence of eukaryotes?
I. HMM Gene and GRAIL
II. GeneFinder and GENESCAN
III GENIE
IV. GENMARK
Options:
(A) I, II, III (B) II, III, IV
(C) I, II (D) I, II, III, IV

663. *E coli* genome has about how many operons?
(A) 200 (B) 600 (C) 1000 (D) 1200

664. Which is used to comparing the submitted nucleotide sequence with amino acid sequence data base??
(A) BLASTn (B) BLASTx
(C) tBLASTx (D) tBLASTn

665. A transgenic variety of which crops had to be withdrawn from cultivation following due to allergy?
(A) Maize (B) Soybean
(C) Tomato (D) *Brassica napus*

666. The assessment of risk during laboratory research is usually done in which steps?
(A) Initial risk assessment
(B) Comprehensive risk assessment
(C) Advanced risk assessment
(D) Both (A) and (B)

667. Which is used to compare submitted amino acid sequence with nucleotide sequence data base?
(A) BLASTn (B) BLASTx
(C) tBLASTx (D) tBLASTn

668. Which is used to compares submitted amino acid sequence with amino acid database?
(A) BLASTn (B) BLASTx
(C) tBLASTx (D) tBLASTn

669. Which is not correct about lectins?
(A) They are polysaccharides
(B) They are involved in defence mechanism of plants
(C) Several lectins have been identified as food allergens
(D) α-Amylase/tripsin inhibitor family proteins are potential allergens

670. Which PCR strategies monitors the fluorescence emitted during reaction to measure the amount of amplification product generated during the PCR cycles?
(A) Anchored PCR (B) Touchdown PCR
(C) Real-time PCR (D) Nested PCR

671. Which data base is used by ENTREZ for bibliographic or citation search?
(A) SWISS-PROT (B) PubMed
(C) TrEMBL (D) PDB

672. Which is not correct about NIH guidelines?
(A) The first NIH guidelines were prepared in 1975
(B) NIH guidelines were more liberal than the recommendations of the Asilomar Conference
(C) The guidelines were revised after two years
(D) In 1977, NIH prepared an Environmental Impact Statement

673. In risk assessment, a lack of scientific knowledge or scientific consensus should not necessarily be interpreted as indicating:
(A) A particular level of risk (B) Absence of risk
(C) Acceptable risk (D) Any of these

674. Which is not correct about genomics?
(A) Sequencing, sequence assembly and annotation of the whole genome of an organism constitutes functional genomics
(B) The first plant genome to be completely sequenced is that of the weed, *Arabidopsis thaliana*
(C) *Arabidopsis thaliana* is a wild relative of mustards
(D) The *Arabidopsis* Genome Initiatives followed the same methods of sequencing as the Human Genome Project

675. Which is correct about genome annotation?
(A) GAME is a programme for describing experimental evidence to support annotation
(B) DAS is useful in indexing and visualization
(C) BioPerl 2001 is used for storing, manipulating and visualizing the genome annotation
(D) All of these

676. The rate of gene-flow is affected by :
(A) Mode of seed dispersal (B) Mating system
(C) Mode of pollination (D) All of these

677. DNA sequences/molecules used for preparing spotted DNA arrays may be obtained from:
(A) PCR amplification (B) Genomic DNA
(C) cDNA clones (D) All of these

678. Insertional mutagenesis has been reported in:
(A) Mouse and *Drosophila melanogaster*
(B) Yeast and roundworm
(C) Zebrafish and Arabidopsis
(D) All of these

679. Which is not correct about prokaryotic genomes?
(A) The genome size varies considerably
(B) Pseudogenes do not occur in prokaryotic genomes
(C) The frequency of paralogues increases with the genome size
(D) The gene density is however, relatively constant at one gene per kilo base pairs

680. Which organism has the lowest density of transposable elements?
(A) Arabidopsis (B) Drosophila
(C) *C. elegans* (D) Humans

681. The amino acids were conventionally represented by three-letter symbols, while in bioinformatics, each of them is denoted by how many letters?
(A) Single letter (B) Two letter
(C) Three letters (D) Four letters

682. Which is not correct?
(A) In databases, base sequence of only one strand is listed
(B) Base sequence runs from the 3'- to the 5' – directions
(C) In single letter codes, R denotes guanine or adenine, while Y denotes thymine or cytosine
(D) The nomenclature system adapted in bioinformatics is based on the recommendations of International Union of Pure and Applied Chemistry

683. In case of which products, an empirical approach based on data obtained through experimentation about the source of potential hazards has proved satisfactory?
(A) Pesticides (B) Drugs
(C) Vaccines (D) All of these

684. India has initiated genome sequencing project of:
(A) Rice (B) Tomato
(C) Pigeonpea (D) Humans

685. Which progamme can be used for the identification/detection of genes from genome sequence of prokaryotes?
(A) Glimmer (B) GENMARK
(C) Gene-finder (D) Both (A) and (B)

686. Which is not correct about Real-time PCR?
(A) Real-Time PCR can detect the size of the amplicon
(B) Design of primers, TaqMan probes and molecular beacons requires almost complete information of the target sequence
(C) With SYBR green method, primer dimmer artifact is a problem
(D) The one-step RT-PCR can not be used for multiplex real-time RT-PCR

687. Which is found the highest density of transposable elements?
(A) Arabidopsis (B) Drosophila
(C) *C. elegans* (D) Humans

688. Which is concerned with the development and application of computer hardware and software for acquisition, storage, analysis and visualization of biological information?
(A) Computer science (B) Bio-infotechnology
(C) Bioinformatics (D) Information technology

689. Development of practices/procedures, assays/tests, biological agents, equipment and working design and structures to minimize the risks is called:
(A) Risk regulation (B) Risk management
(C) Risk perception (D) Risk determination

690. Task of regulators to review risk data and make regulatory decisions based on the data with a view to minimize risk. The process often is influenced by consideration of costs and benefits as well as by public perceptions:
(A) Risk regulation (B) Risk management
(C) Risk perception (D) Risk determination

691. It is the estimation of the overall risk posed by a GMO based on the evaluation of the likelihood and consequences of the identified adverse effects being realized is called:
(A) Risk regulation (B) Risk management
(C) Risk perception (D) Risk determination

692. Non experts intuitive judgment about risks, which often are not in agreement with the level of risk as judged by experts is known as:
(A) Risk regulation (B) Risk management
(C) Risk perception (D) Risk determination

693. Programme is used to convert protein sequence into nucleotide sequence and to compare it with a nucleotide sequence database is:
(A) tBLASTn (B) tBLASTx
(C) BLASTn (D) tBLAST

694. Programme is used to translate the submitted nucleotide sequence as well as the nucleotide sequence database into amino acid sequence and search for homology between the two is:
(A) tBLASTn (B) tBLASTx
(C) BLASTn (D) tBLAST

695. The programme is used to compare a nucleotide sequence with a nucleotide sequence database is:
(A) BLASTn (B) BLASTx
(C) BLASTp (D) tBLAST

696. It compare the submitted protein sequence against a protein database is:
(A) BLASTn (B) BLASTx
(C) BLASTp (D) tBLAST

697. The programme translated the submitted nucleotide sequence into amino acid sequence and compares the letter with a protein database is:
(A) BLASTn (B) BLASTx
(C) BLASTp (D) tBLAST

698. Sequencing of the 24 human chromosomes was completed in which year?
(A) 2004 (B) 2005 (C) 2006 (D) 2007

699. The term genomics was coined by:
(A) Thomas H. Roderick (B) H. Winkler
(C) A.H. Paterson (D) Benzer

700. According to biosafety guidelines of India, physical containment envisages to limit the spread of dangerous microorganisms by:
(A) Laboratory practices
(B) Safety equipment
(C) Laboratory design and facilities
(D) All of these

701. Non-infectious illness is associated with endotoxin production:
(A) Flu symptoms (B) Dermatitis
(C) Asthama (D) Cancer

702. Which procedures can be used for the identification/discovery of SNPs?
(A) DNA chips
(B) Microarrays
(C) Analyzing the sequence data
(D) All of these

703. In case of Real-time PCRm which methods can be used for the detection of amplification products?
(A) Molecular beacons (B) Scorpions
(C) TaqMan probes (D) All of these

704. Single nucleotide polymorphisms (SNPs) may be located in:
(A) In the recognition sequence of an endonuclease
(B) Within the concerned gene
(C) Very close to the gene to show a tight linkage
(D) All of these

705. Crop yields can be stabilized by introducing which genes in transgenic varieties?
(A) Genes for improvement nutritional quality
(B) Genes for better tolerance/resistance to biotic and/or abiotic stresses
(C) Genes for herbicide resistance
(D) All of these

706. Roundup Ready transgenic variety of soybean showed that on an average, how much yield decrease as compared to the non-transgenic parent varieties in USA, in a comparison basedon field trials at 8200 different locations?
(A) 15% (B) 10% (C) 5% (D) -5%

707. Which is not correct about prokaryotic genomes?
(A) The genome size varies considerably
(B) Pseudogenes do not occur in prokaryoptic genome
(C) The frequency of paralogues increases with the genome size
(D) The gene density is however, relatively constant at one per kilo base pairs

708. Which statement is not correct about eukaryotic genomes?
(A) As the genome size increases, gene density de creases
(B) A high percentage of proteins encoded by human genome have orthologs in other eukaryotic genomes
(C) About 75% of repetitive DNA in human genome is due to LINEs ad SINEs sequences
(D) Intron sequences are as variable as exon sequences

709. In case of Real-time PCR, which method can be used for the detection of amplification products?
(A) Molecular beacons (B) TaqMan probes
(C) Scorpions (D) All of these

710. Which features can be used to identify genes in bacterial genomes?
(A) Protein coding genes are composed of open reading frames
(B) Coding region begins with an initiation codon (usu ally ATG)
(C) Coding region ends with a termination codon (TAA, TAG or TGA)
(D) All of these

711. Which was the first prokaryotic genome to be sequenced?
(A) *E. coli*
(B) *Agrobacterium tumefaciens*
(C) *Haemophilus influenzae*
(D) *Proteus vulgaris*

712. Which has operons similar to those of bacteria?
(A) Arabidopsis (B) Drosophila
(C) C. elegans (D) Humans

713. Shot-gum sequencing was conceptualized by:
(A) Craig Venter (B) Michelmore
(C) Giovannoni (D) Young

714. Which is not correct about genomics?
(A) In 1980, David Botstein and coworkers working at the Whitehead Institute of Biomedical Research, USA proposed that one could construct a genome map
(B) In 1995, the first bacterial was that of *Haemophilis influenzae*
(C) The sequencing of whole genome of first multicel lular organism, round worm (*Caenorhabdits elegans*), was published in 1999
(D) The sequencing of rice genome by International Rice Genome Sequencing Project was based on whole ge nome shot-gun approach

715. The fragment connecting and sequential-grow approaches of building up a ligand and the whole building process is governed by a:
(A) Genetic algorithm (B) TAXONOMY BROWSER
(C) Phylogeny (D) None of these

716. To tool provides taxonomy information on various species by:
(A) Genetic algorithm (B) TAXONOMY BROWSER
(C) Phylogeny (D) None of these

717. The basic tasks of any decision, the following procedure must accomplish is:
(A) Characterization of the binding sites
(B) Positioning of the ligand into the binding site
(C) Evaluating the strength of interaction for a specific ligand receptor complex
(D) All of these

718. In 1997 Lipiski and coworkers analyzed 2245 compounds from the World Drug Index, and found that in a high percentage of compounds, poor absorption or permeation occurred when:
(A) There are more than five H-bond donors (expressed as the sum of OHs and NHs)
(B) The molecular weight is over 500
(C) The CLogP(calculated octanol/water partition coefficient) is over 5
(D) There are more than 10H-bond acceptors (expressed as the sum of Ns and Os)
(E) All of these

719. A network of protein-protein interaction generated from yeast two-hybrid data is:
(A) *Saccharomyces cereviseae* (B) *E. coli*
(C) Agrobacterium (D) None of these

720. The scoring matrix for protein sequences is:
(A) PAM matrix (B) BLOSUM matrix
(C) RNA sequences (D) Thure Etzold

721. The matrix for protein sequences is:
(A) PAM matrix (B) BLOSUM matrix
(C) RNA sequences (D) Thure Etzold

722. Dot plot are still used as quick and easy methods to identify regions of repeating sequences, where there are a series of parallel diagonal lines and potential hairpins is:
(A) PAM matrix (B) BLOSUM matrix
(C) RNA sequences (D) Thure Etzold

723. Sequence retrival system (SRS) developed by:
(A) PAM matrix (B) BLOSUM matrix
(C) RNA sequences (D) Thure Etzold

724. PCR based markers may be:
(A) Locus specific (SSR or SNP)
(B) Locus unspecific (RAPD or AFLP)
(C) Both (A) and (B)
(D) None of these

725. The strategy uses nucleic acid hybridization in colony hybridization or chromosome walking/jumping or is based on PCR is:
(A) Sequence – dependent screening
(B) Sequence – Tagged Microsatellites Sites (STMS)
(C) Sequence Tagged Site (STS)
(D) Single Locus Probe Analysis (SLP)

726. The microsatellite based markers are analysed by PCR amplification using a pair of primers specific to the unique sequences flanking a SSR sequence, alleles differ for the number of repeats of the short repeating unit is:
(A) Sequence – dependent screening
(B) Sequence – Tagged Microsatellites Sites (STMS)
(C) Sequence Tagged Site (STS)
(D) Single Locus Probe Analysis (SLP)

727. When pair of PCR primers amplifies only a single sequence within a haploid genome, the amplified region of the genome is called:
(A) Sequence – dependent screening
(B) Sequence – Tagged Microsatellites Sites (STMS)
(C) Sequence Tagged Site (STS)
(D) Single Locus Probe Analysis (SLP)

728. Use of a probe that identifies a single DNA fragment for fingerprinting is called:
(A) Sequence – dependent screening
(B) Sequence – Tagged Microsatellites Sites (STMS)
(C) Sequence Tagged Site (STS)
(D) Single Locus Probe Analysis (SLP)

729. Occuerence of different nucleotides at a specific site in the genome of different individuals of a species, human genome shows SNPs at about every 1000 bp is called:
(A) Single Nucleotide Polymorphism (SNP)
(B) Single Primer Amplification Reaction (SPAR)
(C) Single Strand Conformational Polymerase (SSCP)
(D) SSR-Anchored PCR (SSR-APCR)

730. DNA marker system that uses primers based on microsatellites and amplifies inter-SSR and DNA sequences, produces multiple fragments per reaction is called:
(A) Single Nucleotide Polymorphism (SNP)
(B) Single Primer Amplification Reaction (SPAR)
(C) Single Strand Conformational Polymerase (SSCP)
(D) SSR-Anchored PCR (SSR-APCR)

731. The marker relies on secondary and tertiary structure difference between denaturated and rapidly cooled amplified DNA fragments that differ slightly in their base sequence. Different SSCP alleles are resolved on non-denaturing acrylamide gel, usually at low temperature by:
(A) Single Nucleotide Polymorphism (SNP)
(B) Single Primer Amplification Reaction (SPAR)
(C) Single Strand Conformational Polymerase (SSCP)
(D) SSR-Anchored PCR (SSR-APCR)

732. This is essentially single primer amplification reaction, except that the primer is either anchored at the 3'-or 5' – terminus with two to four arbitrary nucleotides by:
(A) Single Nucleotide Polymorphism (SNP)
(B) Single Primer Amplification Reaction (SPAR)
(C) Single Strand Conformational Polymerase (SSCP)
(D) SSR-Anchored PCR (SSR-APCR)

733. The desired segment that is to be amplified by PCR is called:
(A) Target sequence (B) Telomeric sequence
(C) Recognition sequence (D) None of these

734. In yeast chromosomes, it is a 20-70 tandem repeat of the 6-base sequence 5'CCCCAA3' (its complementary sequence, 5'TTGGGG3', occurs in the other strand), there is a hairpin loop formation at the terminus, which makes the DNA duplex resistant to endonuclease action is called:
(A) Target sequence (B) Telomeric sequence
(C) Recognition sequence (D) None of these

735. The sequences recognized by restriction endonucleases for generating cleavage is called:
(A) Target sequence (B) Telomeric sequence
(C) Recognition sequence (D) None of these

736. Difference between individuals in terms of DNA regions either being or not being amplified in a polymerase hair reaction primed by random ~10 nucleotide long sequence by:
(A) Random Amplified Polymorphic DNA (RAPD)
(B) Pulsed-Field Gel Electrophoresis (PFGE)
(C) Restriction Fragment Length Polymorphism (RFLP)
(D) Reverse Transcription PCR (RT-PCR)

737. DNA is driven through the gel using alternating pulses of electrical current that are at different angles to each other; it can achieve a good separation of DNA fragments that vary from 50 kb to 10 mb is:
(A) Random Amplified Polymorphic DNA (RAPD)
(B) Pulsed-Field Gel Electrophoresis (PFGE)
(C) Restriction Fragment Length Polymorphism (RFLP)
(D) Reverse Transcription PCR (RT-PCR)

738. Difference between individuals in terms of the size of fragments generated by restriction enzyme digestion of genomic DNA and detected by hybridization with specific DNA sequences used as probe is:
(A) Random Amplified Polymorphic DNA (RAPD)
(B) Pulsed-Field Gel Electrophoresis (PFGE)
(C) Restriction Fragment Length Polymorphism (RFLP)
(D) Reverse Transcription PCR (RT-PCR)

739. First of all cDNA of the mRNA is produced using the enzyme reverse transcriptase; the cDNA single – strand is then used for amplification to obtain cDNA deplexes:
(A) Random Amplified Polymorphic DNA (RAPD)
(B) Pulsed-Field Gel Electrophoresis (PFGE)
(C) Restriction Fragment Length Polymorphism (RFLP)
(D) Reverse Transcription PCR (RT-PCR)

740. When PCR is used to achieve substractive cloning, this is called:
(A) Representational Difference Analysis
(B) Pulsed-Field Gel Electrophoresis (PFGE)
(C) Restriction Fragment Length Polymorphism (RFLP)
(D) Reverse Transcription PCR (RT-PCR)

741. Only one primer is used in combination with dideoxynucleotides; the primer is either radio or fluorescence labeled. The product of PCR reaction is subjected to PAGE for determining the base sequence of the template DNA is :
(A) Representational Difference Analysis
(B) Thermal Cycle Sequencing PCR
(C) Restriction Fragment Length Polymorphism (RFLP)
(D) Reverse Transcription PCR (RT-PCR)

742. A Ca 23 kb segment of Ti/Ri plasmid that has 24 bp direct repeat border sequence and it transferred into the host plant genome during pathogenesis; has genes for auxin, cytokinin and opion biosynthesis is called:
(A) T-DNA (B) Template DNA
(C) Z-DNA (D) Carrier DNA

743. DNA molecule used for synthesis of complementary sequences either in vivo or in PCR is called:
(A) T-DNA (B) Template DNA
(C) Z-DNA (D) Carrier DNA

744. The two strands of a DNA molecule are coiled together in a left-handed helix having diameter of 18Å, and the length of one turn of the helix is Ca 69Å; it contains 12 base pairs per turn in the helix is called:
(A) T-DNA (B) Template DNA
(C) Z-DNA (D) Carrier DNA

745. In genetic transformation; unrelated DNA added to the solution of DNA used for transformation, enhances transformation frequency, rearrangements and multicopy integration is called:
(A) T-DNA (B) Template DNA
(C) Z-DNA (D) Carrier DNA

746. Thiamine pyrophosphate (TPP) coenzyme is found in:
(A) Glutamin synthase
(B) Glyceraldehyde 3-phosphate dehydrogenase
(C) Pyruvate dehydrogenase complex
(D) DNA ligase

747. Dextran is employed in:
(A) Bleeding (B) Plasma extending
(C) Blood clotting (D) Blood preservation

748. Lyase class of enzyme does not contain :
(A) Decarboxylases (B) Dehydratases
(C) Aldolases (D) Proteases

749. Erythromycin and chloramphenicol are got from:
(A) *Penicillium species* (B) *Aspergillus species*
(C) *Streptomyces species* (D) *Bacillus species*

750. Which enzymes catalysis an irreversible reaction of glycolysis?
(A) Phosphofructokinase
(B) Glucose-6-isomerase
(C) Aldolase
(D) Phosphoglycerate mutase

751. All chemical reaction possess major properties:
(A) Kinetic (B) Thermodynamic
(C) Both (A) and (B) (D) Dynamic

752. Which diagnosed by anti-acetylcholine receptor?
(A) Polymyositis
(B) Scleroderma
(C) Systemic lupus erythematus
(D) Myasthenia gravis

753. Smith and Nathans are famous for discovery of:
(A) Gene therapy (B) Restriction enzyme
(C) Humulin (D) Second generation vaccine

754. Isomerases include:
(A) Intramolecular transferases (B) Epimerases
(C) Racemases (D) All of these

755. A broad spectrum antibiotic is effective:
(A) In low concentration
(B) On a variety of microorganisms
(C) Against all bacteria and viruses
(D) Against both pathogens and hosts

756. α-amylase is produced from:
(A) Malted barley (B) Bacillus
(C) Aspergillus (D) All of these

757. Antibiotics belong to the category of:
(A) Steroids (B) Toxins
(C) Medicines (D) Tonics

758. Flemming is famous for his discovery of:
(A) Actinomycetin (B) Penicillin
(C) Actinomycine (D) Neomycin

759. Streptomycin is prepared from:
(A) *Streptomyces antibioticus*
(B) *Streptomyces nodasus*
(C) *Streptomyces griscus*
(D) *Streptomyces rimosus*

760. Antibiotics are mostly got from:
(A) Fungi (B) Actinomycetes/Bacteria
(C) Both (A) and (B) (D) Cyanobacteria

761. *Streptomycin griscus* produces antibiotic:
(A) Terramycin (B) Chloramphenicol
(C) Neomycin (D) Streptomycin

762. Antibiotics destroy or inhibit:
(A) Bacteria, fungi and viruses
(B) Bacteria, algae and viruses
(C) Bacteria and fungi
(D) Bacteria and viruses

763. Proteases are used to digest away proteins present in:
(A) Blood stains (B) Dirt stains
(C) Grass stains (D) All of these

764. Who established the scientific basis of vaccination?
(A) Jenner (B) Pasteur
(C) Waksman (D) Kohler and Milstein

765. Which is not correct about aspartame?
(A) It is a dipeptide
(B) It is synthesized from an equal mixture of L-aspartic acid and methyl ester of L-phenylalanine
(C) It is 180 times sweeter than sucrose
(D) It is used as high coloric sweetner

766. Granulocyte colony stimulating factor can be used for the treatment of which diseases?
(A) AIDS (B) Cancer
(C) Bone marrow transplantation (D) All of these

767. An 'indel' defines a site in the genome that has ?
(A) Insertion of a sequence
(B) Deletion of a sequence
(C) Either (A) or (B)
(D) Inversion of a sequence

768. Most enzymes are used extra cellularily for which medical purposes?
(A) Removal of toxic substances
(B) Treatment of disorders within blood circulatory system
(C) Topical application
(D) All of these

769. Current hepatitis B vaccine is:
(A) An interferon
(B) First generation vaccine
(C) Second generation vaccine
(D) Third generation vaccine

770. Kohler and Milstein developed biotechnology for:
(A) Monoclonal antibiotics (B) Myclomas
(C) Immonilisation of enzymes (D) Steroid conversion

771. Synzymes must possess:
(A) Substrate binding site
(B) Catalytically effective site
(C) Allosteric site
(D) Both (A) and (B)

772. The term antibiotic was coined by:
(A) Flemming (B) Waksman
(C) Funk (D) Jenner

773. Which steps are involved in downstream processing?
(A) Separation of particles and disintegration of cells
(B) Extraction and concentration
(C) Purification and drying
(D) All of these

774. Which indicates the correct order of ascorbic acid production?
(A) Glucose→ Ascorbic acid→ D-sorbitol→ L-sorbose
(B) Glucose→ L-sorbose → D-sorbitol→ Ascorbic acid
(C) Glucose→D-sorbitol→ L-sorbose →Ascorbic acid
(D) D-sorbitol→ L-sorbose→ Glucose → Ascorbic acid

775. Which enzyme is isolated from hyperthermophiles?
(A) DNA polymerase (B) α- amylase
(C) α - Glucosidase (D) All of these

776. Waksman got nobel prize for his discovery of:
(A) Penicillin (B) Neomycin
(C) Chloromycetin (D) Streptomycin

777. Continuous flow reactors offer which advantages?
(A) Greater productivity per unit amount of enzyme
(B) Uniform product quality
(C) They can be used for substrates having low solubility
(D) All of these

778. Cells obtained from cancerous tissues are:
(A) Myelomas (B) Hybridomas
(C) Monoclonal (D) Lymphocytes

779. Hybridoma are used for:
(A) Killing cancer cells
(B) Formation of monoclonal antibiotics
(C) Antibiotic synthesis
(D) Formation of somatic hybrids

780. Biosensor for cocaine detection uses which biological agents?
(A) Whole cell (B) Antibody
(C) Enzyme (D) Either (A) or (C)

781. The first biotransformation to be commercialized was:
(A) Depxycorticosterone to contocosterone
(B) Ethanol to acetic acid
(C) Glucose to gluconic acid
(D) Fructose to mannitol

782. Biological component of biosensors perform which function?
(A) Recognise the analyte
(B) Interact with the analyte in such a manner, which produces some physical change detectable by the transducer
(C) Both (A) and (B)
(D) Inactivate toxic analytes

783. Ovulation may be stimulated by the administration of:
(A) Clomiphene
(B) Human chorionic gonadrotrophin (hCG)
(C) Human menopausal gonadotrophin (hMG)
(D) Any of these

784. Most of the eubacterial antibiotics are got from:
(A) Streptomyces (B) Pseudomonas
(C) Streptococcus (D) Bacillus

785. Restriction enzyme was discovered by:
(A) Smith and Nathans (B) Berger
(C) Walksman (D) Flemming

786. During 'invert syrup' production, small amounts of xylene are added to achieve:
(A) To prevent microbial contamination
(B) To facilitate product recovery
(C) Substrate activation
(D) To prevent product inhibition

787. The enzymology, abbrebiation EC stands for:
(A) Exotic collection (B) Enzyme commission
(C) Enzyme collection (D) Enzyme cofactor

788. *E coli* genome has how many operons?
(A) 200 (B) 600 (C) 1000 (D) 2000

789. The shot-gun sequencing approach would be the mosy suited for which genome?
(A) *E. coli* (B) Yeast
(C) Humans (D) Drosophila

790. Which is concerned with organisms which are associated with such human diseases, which are rarely serious and for which preventive or therapeutic intervention are often available?
(A) Risk group 4 (B) Risk group 3
(C) Risk group 2 (D) Risk group 1

791. Which hazards is associated with operation of bioreactor fermentation?
(A) Spillage
(B) Effluent contamination of gases
(C) Aerosols from reactor
(D) All of these

792. Lateral gene transfer is also called:
(A) Horizontal gene transfer
(B) Vertical gene transfer
(C) Parallel gene transfer
(D) Both (B) and (C)

793. Which is not correct about TaqMan probes?
(A) They are oligonucleotides longer than the primers
(B) They contain a fluorescent dye usually on the 5'-base , and a quenching dye typically on the 3' - base
(C) When Taq polymerase replicates a tempelate to which a TaqMan probe is paired, its 5'-exonuclease activity cleaves the TaqMan probe
(D) The fluorescence decreases proportionately to PCR amplification

794. In which years, a proposal for sequencing of human genome was prepared:
(A) 1984 (B) 1985 (C) 1986 (D) 1987

795. Which is not correct about printed oligonucleotide chips?
(A) These are produced by a light-directed printing technology
(B) These chips have double-stranded oligonucleotides of 20-25 bases
(C) Oligonucleotides are synthesized directly onto the chips
(D) Each sequence is represented by a set of ~20 non-overlapping oligonucleotides to reduce false positive results

796. Which is not correct?
(A) National Biomedical Research Foundation established the Protein Information System in 1984
(B) National centre for Bioinformatics Institute was established in USA and serves as primary information databank and provider of information
(C) RNA Data Bank was established by Japan
(D) European Molecular Biology Laboratory endeavours to collect, organize and distribute nucleotide sequence data and related information

797. Which was the first prokaryotic genome to be sequenced?
(A) *E. coli*
(B) *Agrobacterium tumefaciens*
(C) *Haemophilus influenza*
(D) *Proteus vulgaris*

798. Which phenomenon comes under post-translational modification?
(A) Phosphorylation (B) Proteolysis
(C) Glycosylation (D) All of these

799. Which proteins are responsible for the unwinding of the double – stranded DNA during replication?
(A) Ligages (B) Helicases
(C) Topoisomerases (D) Primases

800. Which is concerned with organisms, which cause serious or lethal disease for which preventive or therapeutic invention are not usually available?
(A) Risk group 1 (B) Risk group 2
(C) Risk group 3 (D) Risk group 4

801. Bacterium commonly used in plant genetic engineering is:
(A) Agrobacterium (B) Corynebacterium
(C) Bacillus subtilis (D) Salmonella typhisae

802. Antibiotics are drugs commonly used to cure disease of:
(A) Fungi (B) Viruses
(C) Protozoan (D) Bacteria

803. Genetically engineered bacteria are become employed for production of:
(A) Human insulin (B) Thyroxine
(C) Cortisol (D) Epinephrine

804. In 1928, a scientist discovered the first effective antibiotics. Scientist and antibiotics are:
(A) Flemmin – Streptomycin
(B) Flemming - Penicillin
(C) Waksman - Penicillin
(D) Waksman - Streptomycin

805. The nature of biomass obtained from energy crops is:
(A) Wood (B) Sugar and starch
(C) Hydrocarbon (D) All of these

806. Biotechnological production of antibiotics require:
(A) Search for antibiotic producing microorganisms
(B) Isolation of gene fro antibiotics production
(C) Join the gene for antibiotic with plasmid of Escherichea coli
(D) All of these

807. Which is not correct about the drug zidovudin (AZT)?
(A) The first successful drug used for the treatment of AIDS
(B) It is a nucleoside analogue and targeted at reverse transcriptase
(C) Incorporation of this base analogue into a cDNA chain enhances its extension
(D) Long term use produces several side effects and mutant viruses develop in treated patients

808. In case of AIDS, highly active antiretroviral therapy (HAART) has:
(A) It reduces viral load in blood below detection limits
(B) It lowers death rates due to AIDS
(C) This therapy is very expensive (upto US$ 15000 per day)
(D) All of these

809. Biohydrogen is produced by which routes?
(A) Bt photolysis of water
(B) During anaerobic fermentation
(C) An in vitro system created by coupling together the photosynthetic unit and hydrogenase
(D) All of these

810. The current processes of production of ethanol from cellulose are based on which of the following microorganisms?
(A) *Trichoderma reesei* (B) *Saccharomyces cerevisiae*
(C) Recombinant *E. coli* (D) All of these

811. The important reason for limiting the use of bioethanol as fuel?
(A) Ethanol run engines use about 10% more fuel than petrol
(B) Due to highly hydrophilic nature, ethanol can absorb moisture from atmosphere, which interferes with combustion and also causes corrosion
(C) The downstream processing for ethanol recovery is costly
(D) Ethanol produces engine starting problems when air is cool

812. Blood flows from the heart to the lugs in the pulmonary artery and returns from the lungs to the heart in the pulmonary vein. The blood in the pulmonary artery is:
(A) Higher in O_2 and lower in CO_2 content than the blood in the pulmonary vein
(B) Higher in both O_2 and CO_2 content than the blood in the pulmonary vein
(C) Lower in O_2 and higher in CO_2 content than the blood in the pulmonary vein
(D) Lower in both O_2 and CO_2 content than the blood in the pulmonary vein

813. Which is correct about conventional vaccines?
(A) Conventional vaccines consist of whole pathogenic organisms
(B) Pathogens may either be killed or live, but attenuated
(C) Conventional vaccines are produced in whole animals or cultured cells
(D) All of these

814. Bioethanole can be obtained from:
(A) Cereals (B) Tuber crops
(C) Root crops (D) All of these

815. Oxygen is carried in blood:
(A) Bound to haemoglobin
(B) Dissolved in the plasma
(C) Dissolved in the cytosol of erythrocytes
(D) All (A), (B) and (C)

816. Cardia output is the:
(A) Volume of blood pumped per minute by both ventricles
(B) Volume of blood flowing through the systemic circulation each minute
(C) Product of the number of heartbeats per minute and the volume pumped per beat
(D) Both (B) and (C)

817. Which statements are correct about hydrolytic and fermentative bacteria?
I. They include both obligate and facultative anaerobes
II. They may occur upto $10^8 - 10^9$ cells/ml of sewage sludge digesters
III. They remove O_2 and create anaerobic conditions
IV. Usually, only 50% of the polysaccharides present in the waster may be digested
Options:
(A) I, II (B) I, II, III, IV (C) I, II, IV (D) I, II, III

818. Which is organic acids?
(A) Long chain fatty acids (B) Formate and butyrate
(C) Acetate and propionate (D) All of these

819. Which produces hydrogen gas which can be used as a pollution free fuel?
(A) Chlamydomonas (B) Clostridium
(C) Both (A) and (B) (D) Green plants

820. The cellulose enzyme systems (produced by both bacteria and fungi) carry by:
(A) Binding to cellulose
(B) Hydrolysis at the end of a cellular chain
(C) Cleavage of cellulose chain at an internal site
(D) All of these

821. Consider the following with reference to the mechanism of action of a steroidal hormone on target cell:
P. Bonding of activated complex with specific genomic site
Q. Activation of transcriptional apparatus
R. Attachment with cytosolic receptor protein
S. Transportation of complex to the nucleus
T. Formation of new proteins in the cytoplasm
What is the correct sequence of the above in said mechanism?
(A) P, Q, R, S, T (B) S, P, R, Q,T
(C) Q, R, P, S, T (D) R, S, P, Q, T

822. During chemical transmission, the following events occur at synaptic junction:
P. Transmitters cross synaptic left and attach to receptors on the postsynaptic neurons
Q. Postsynaptic membrane is depolarized, initiating the impulse
R. Chemical excitation at the presynaptic terminal and release of neurotransmitters
S. Immediately after impulse is triggered, the recovery of pre and postsynaptic membranes follows and neurotransmitter is destroyed
Which of the statements given above is/are correct?
(A) P, R, S, Q (B) R, P, S, Q
(C) R, P, Q, S (D) P, R, Q, S

823. Therapeutic agents from nonrecombinant organisms ay originate from:
(A) Plant cell cultures (B) Animal cell cultures
(C) Microorganisms (D) All of these

824. Column A lists some tumour-assoiciated antigens, while column B lists different types of cancers in which antigen are found. Match both the Columns carefully and select the correct option from those given below them:

Column A	Column B
A.CEA	1. Some head, lung, etc. tumours
B.MUCI	2. Breast and ovarian tumours
C. HER 2	3. Breast cancer
D. EGFR	4. Colon cancer

Options:
(A) A – 3, B – 4, C – 1, D – 2
(B) A – 4, B – 3, C – 2, D – 1
(C) A – 3, B – 1, C – 4, D – 2
(D) A – 1, B – 3, C – 2, D – 4

825. Which diagnosed with anti-dsDDNA:
(A) Polymyositis
(B) Scleroderma
(C) Systemic lupus erythematus
(D) Myasthenia gravis

826. The recombinant *E. coli* used for pentose fermentation has transgenes encoding by which enzymes:
(A) Pyruvate decarboxylase
(B) Alcohol dehydrogenase
(C) Acetaldehyde dehydrogenase
(D) Both (A) and (C)

827. Methanogenic bacteria are:
(A) *Methanobacterium omelianskii*
(B) *Methanosarcina barkeri*
(C) Both (A) and (B)
(D) *Methylococcus capsulatus*

828. Monoclonal antibodies can be used for:
(A) Disease diagnosis (B) Disease treatment
(C) Immunopurification (D) All of these

829. Which enzyme is obtained commercially from Aspergillus?
(A) Protease and pectinase
(B) α-Amylase and glucoamylase
(C) Both (A) and (B)
(D) Papain

830. Hydrolytic and fermentative bacteria hydrolyse and ferent the organic materials and produce by:
(A) Organic acids (B) CO_2
(C) H_2 (D) All of these

831. Ethanol has which of the following advantages over petrol?
I. Ethanol has a lower latent heat of vaporization than petrol
II. Ethanol has a higher octane number (99) than petrol (80-100) as a result ; preignition does not occur when ethanol is used in engines set for petrol
III. Higher octane rating of ethanol allows the compression ratio of the engines to be increased; this results in increased production of power
IV. Ethanol is burnt more completely so that hydrocarbon emission is drastically lower as compared to that in case of petrol
Options:
(A) I, II (B) II, III, IV
(C) I, II, III, IV (D) I, II, III

832. Which is the examples of hydrophobic amino acids?
(A) Leucine and isoleucine
(B) Methionine and vaine
(C) Phenylalanine and often tyrosine
(D) All of these

833. Rennet enzyme is produced by:
(A) Mucor miehei (B) Calf stomach
(C) Calf liver (D) Both (A) and (B)

834. Anti-RNA polymerase I is used to diagnose which autoimmune diseases?
(A) Polymyositis
(B) Scleroderma
(C) Systemic lupus erythematus
(D) Myasthenia gravis

835. Enzyme 'diastase' was identified by:
(A) Dubrunfaut (B) Reamur
(C) Payen and Persoz (D) Hansen

836. In 1986, which of the scientists discovered the catalytic action of some RNA molecules?
(A) Baltimore (B) Smith
(C) Cech (D) Berg

837. The first antibiotic to be isolated was:
(A) Terramycin (B) Streptomycin
(C) Neomycin (D) Penicillin

838. Penicillin acylases use which of the following substrates?
(A) Penicillin G (B) Penicillin V
(C) Penicillins (D) Both (A) and (B)

839. Penicillin is got from:
(A) *Streptomyces griseus*
(B) *Penicillin chrysogenum*
(C) *Penicillum griseofulvum*
(D) *Aspergillus fumigatus*

840. Humilin is:
(A) Antibiotic (B) Human insulin
(C) Digestive enzyme (D) A form of chitin

841. Which yields highly purified enzymes in a single step?
(A) Gel filteration
(B) Affinity chromatography
(C) Ion exchange chromatography
(D) Both (B) and (C)

842. Which is a multienzyme complex?
(A) Urease (B) Tryptophan synthase from *E.coli*
(C) Aldoase (D) Both (B) and (C)

843. Which enzymes are used in the largest quantity?
(A) Textile industry (B) Alcohol production
(C) Detergents (D) Leather industry

844. Which is not an example of an energy crop that yields hydrocarbons?
(A) *Asclepia speciosa* (B) *Copaifera multijuga*
(C) *Euphorbia lathyrus* (D) *Eucalyptus globules*

845. Which metal ion is associated with urase enzyme?
(A) Fe (B) Ni (C) Mg (D) Cu

846. The principal purpose of transgenic plant containment is to avoid the unintentional transmission of a recombinant DNA containing plant genome including:
(A) Organellar hereditary material
(B) Nuclear hereditary material
(C) Release of recombinant DNA-derived organisms associated with plants
(D) Any of these

847. In which risk groups, is concerned with agents which are associated with such serious or lethal human diseases for which preventive or therapeutic interventions may be available?
(A) Risk group 3 (B) Risk group 2
(C) Risk group 1 (D) Risk group 4

848. Non-infectious illness like allergic asthama and dermatitis are associated with production of:
(A) Brewing (B) Single cell protein
(C) Antibiotic (D) All of these

849. The rough draft of human genome was announced on:
(A) June 26, 2000 (B) June 26, 2001
(C) June 26, 2002 (D) June 26, 2003

850. Which can be used to identify genes in bacterial genomes?
(A) Protein-coding genes are composed of open reading frames
(B) Coding region begins with an initiation codon (usually ATG)
(C) Coding region ends with a termination codon (TAA, TAG or TGA)
(D) All of these

851. Which scientists proposed that amino acid sequences of proteins may be used to study evolutionary relationships among organisms?
(A) Zuckerland and Pauling (B) Sensen and Buchler
(C) Dayhoff and Rashidi (D) Tolin and Vidaver

852. Shot-gun sequencing was conceptualized by:
(A) Craig Venter (B) Michelmore
(C) Giovannoni (D) Young

853. Which PCR strategies monitors the fluorescence emitted during reaction to measure the amount of amplification product generated during the PCR cycle?
(A) Anchored PCR (B) Touchdown PCR
(C) Real-Time PCR (D) Nested PCR

854. Which is not correct?
(A) Both TaqMan probes and molecular beacons allow multiplex reactions, and thereby, detection of multiple DNA sequences in a single PCR reaction
(B) In multiplexing, different probes/beacons are labeled with different reporter dyes
(C) Multiplexing does not allow amplification in a single PCR reaction of both target and endogeneous control sequences
(D) The use of probes/beacons avoids problems due to mispriming, primer dimmer artifacts, etc.

855. An increase in the number of copies of gene per genome of the organism in comparison to that naturally present is called:
(A) Gene amplification (B) Gene tagging
(C) Gene silencing (D) Gene expression

856. Use of Agrobacterium T-DNA, transposons etc., for identification and/or isolation of a gene by first integrating it into the concerned gene and isolating mutants so produced is known as:
(A) Gene amplification (B) Gene tagging
(C) Gene silencing (D) Gene expression

857. Suppression of transgene expression in later generations of transgenic plants; usually, due to transgene (including promoter) methylation; more common when the transgene is integrated in higher copy number is called:
(A) Gene amplification (B) Gene tagging
(C) Gene silencing (D) Gene expression

858. When a gene performs the specific function, which it is capable of, it is called:
(A) Gene amplification (B) Gene tagging
(C) Gene silencing (D) Gene expression

859. B cells marked by negative selection die by apoptosis is called :
(A) Clonal deletion (B) Clonal propagation
(C) Clonal selection (D) None of these

860. Multiplication of plants via asexual reproduction, e.g., via micropropagation, vegetative propagation, etc. is called:
(A) Clonal deletion (B) Clonal propagation
(C) Clonal selection (D) None of these

861. Phenomenon of selective prolifiration of lymphocyetes, e.g., B-cells, in response to their interaction with an antigen is called:
(A) Clonal deletion (B) Clonal propagation
(C) Clonal selection (D) None of these

862. A base sequence occurring in majority of the DNA segments serving the same function is called:
(A) Consensus sequence (B) Conserved sequence
(C) Both (A) and (B) (D) None of these

863. A base sequence present without any change in DNA sequences having the same function is called:
(A) Consensus sequence
(B) Conserved sequence
(C) Both (A) and (B)
(D) None of these

864. It is a method of breeding crops to increase their nutritional value; it can be achieved through conventional selective breeding or through genetic engineering is called:
(A) Consensus sequence (B) Conserved sequence
(C) Biofortification (D) None of these

865. A crop variety expressing a transgene, produced by genetic engineering is called:
(A) Genetically modified crops
(B) Genetically modified food
(C) Genetically modified organism
(D) Genetically modifies plants

866. Food and additives obtained from genetically modifies organisms is called:
(A) Genetically modified crops
(B) Genetically modified food
(C) Genetically modified organism
(D) Genetically modifies plants

867. An organism with a unique gene combination that does not occur naturally, and is produced either by using genetic engineering or protoplast fusion is called:
(A) Genetically modified crops
(B) Genetically modified food
(C) Genetically modified organism
(D) Genetically modifies plants

868. Plant with unique gene combinations that do not occur naturally and are produced either by using genetic engineering or protoplast fusion is called:
(A) Genetically modified crops
(B) Genetically modified food
(C) Genetically modified organism
(D) Genetically modifies plants

869. In vitro derived virus-free shoots are grafted onto healthy root-stocks seedlings grown in vitro to obtain virus free plants is called:
(A) Micrografting (B) Micropropagation
(C) Micro injection (D) None of these

870. In vitro production of a large number of small plantlets from cultured cells/tissues; through axillary branching, adventitious shoot bud regeneration or somatic embryogenesis is called:
(A) Micrografting (B) Micropropagation
(C) Micro injection (D) None of these

871. Delivery of DNA directly into the cytoplasm/nucleus of cells/male pronucleus of fertilized egg/young embryos; micro-manipulations are used to hold the eggs and microinject DNA is called:
(A) Micrografting (B) Micropropagation
(C) Micro injection (D) None of these

872. The plants with specific genes or traits obtained through genetic engineering or DNA recombinant technology is called:
(A) Transgenic plant (B) Transgenic animals
(C) Transgenic microbes (D) None of these

873. The animals having specific genes obtained from outside is known as:
(A) Transgenic plant (B) Transgenic animals
(C) Transgenic microbes (D) None of these

874. The microbes used in industry for producing different biochemical and various functions is called:
(A) Transgenic plant (B) Transgenic animals
(C) Transgenic microbes (D) None of these

875. Transgene may become deleted or modified by recombination, following their integration into the host genome is called:
(A) Transgene instability (B) Transgene silencing
(C) Transgene flow (D) Gene splicing

876. A decline in the level of transgene expression with the passage of generation of the transgenic line is called:
(A) Transgene instability (B) Transgene silencing
(C) Transgene flow (D) Gene splicing

877. Transfer of gene from a transgenic variety to non-transgenic varieties, wild species and wild relatives through hybridization is called:
(A) Transgene instability (B) Transgene silencing
(C) Transgene flow (D) Gene splicing

878. The enzymatic attachment of one gene, or part of a gene, to another is called:
(A) Transgene instability (B) Transgene silencing
(C) Transgene flow (D) Gene splicing

879. A transgenic rice that accumulates â-carotene in its seeds; expected to combat vitamin A deficiency areas where rice is the staple food is:
(A) Golden rice (B) Hybrid rice
(C) Boiled rice (D) Desi rice

880. A DNA sequence that appears at the end of a transcriptional unit and signals the end of transcription is called:
(A) Termination sequence (B) Terminal transferase
(C) Supercoiled DNA (D) Suppressor mutation

881. An enzyme that catalyzes the addition of nucleotide residues of a single kind to the 3' end of DNA chains is called:
(A) Termination sequence (B) Terminal transferase
(C) Supercoiled DNA (D) Suppressor mutation

882. DNA that twists upon itself because it is under-or over-wound (and thereby strained) relative to B-form DNA is called:
(A) Termination sequence (B) Terminal transferase
(C) Supercoiled DNA (D) Suppressor mutation

883. A mutation that totally or partially restores a function lost by a primary mutation; located at a site different from the site of the primary mutation is called:
(A) Termination sequence (B) Terminal transferase
(C) Supercoiled DNA (D) Suppressor mutation

884. A mutation is a gene that cause no detectable change in the biological characteristics of the gene product is called:
(A) Termination sequence (B) Terminal transferase
(C) Silent mutation (D) Suppressor mutation

885. An amino acid sequence, often at the amino terminus, that signals the cellular fate or destination of a newly synthesized protein is called:
(A) Termination sequence (B) Terminal transferase
(C) Signal sequence (D) Suppressor mutation

886. The process by which an extracellular signal (chemical, mechanical or electrical) is amplified and converted to a cellular response is called:
(A) Signal transduction (B) Terminal transferase
(C) Signal sequence (D) Suppressor mutation

887. Highly repeated, nontranslated segments of DNA in eukaryotic chromosomes; most often associated with the centromeric region is called:
(A) Satellite – DNA (B) Ribosomal RNA
(C) Small nuclear RNA (D) Transfer RNA

888. Which enzymes is obtained from bacteria?
(A) α- and β-amylases (B) Glucose isomerase
(C) Penicillin amidase (D) All of these

889. In case of glucose isomerase used for production of high fructose syrup, which is used as cofactor?
(A) Ca^{++} (B) Co^{++}
(C) Mg^{++} (D) Both (B) and (C)

890. Which statements is most correct about ribozymes?
(A) Enzyme activity associated with ribosomes
(B) Enzymes which catalyse both synthesis and degradation of small oligonucleotides
(C) Enzyme that use RNA as substrate
(D) RNA molecules having enzymatic properties

891. Streptomycin is useful for treatment of pathogens:
(A) Gram (+) (B) Gram (-)
(C) Both (A) and (B) (D) Gram neutral

892. Nuclease enzyme which begins its attack from free end of polynucleotide is:
(A) Endonuclease (B) Exonuclease
(C) Polymerase (D) Kinase

893. *Bacillus thuringiensis* can be used for the control of:
(A) Heliothis and other Lepidoptera
(B) Coleoptera pests
(C) Spodoptera litura
(D) Both (A) and (B)

894. Biotransformations are performed by microbes is far more efficient and economical due to:
(A) High metabolic rates of microorganisms
(B) Rapid microbial growth
(C) Both (A) and (B)
(D) More efficient enzymes

895. Which options lists the correct order of the mechanism of enzyme reaction?
(A) E + S → EP↔ ES ↔ E + P
(B) E + S ↔ ES↔ EP↔ E + P
(C) E + S → EP→ ES → E + P
(D) E + P → ES→ ES→ E + P

896. The bulk of enzymes, both in terms of quantity and variety, are produced from which of the following sources? The correct option lists these sources in describing order.
(A) Animals → Plants→ Microorganisms
(B) Microorganisms→ Animals→ Plants
(C) Microorganisms → Plants →Animals
(D) Plants → Microorganisms → Animals

897. Which of the following features is correct about asymmetric gap ligase chain reaction (AG-LCR)?
(A) It utilize reverse transcriptase
(B) It is used for detection of RNA
(C) It is used for detection of antibodies
(D) Both (A) and (B)

898. Erythropoietin (EPO) is used to stimulate red blood production in:
(A) Kidney dialysis (B) AIDS patients
(C) Cancer patients (D) Both (A) and (C)

899. Antibodies can be used as biological agents in which types of biosensors?
(A) Optical (B) Acoustic wave
(C) Potentiometric (D) Both (A) and (B)

900. Aspartate is an artificial sweetner, it is chemically is:
(A) Fatty acid (B) Carbohydrate
(C) Dipeptide (D) Oligopeptide

901. Modified antibiotics are got through:
(A) Vernalisation (B) Ultrafiltration
(C) Ultracentrifuge (D) Genetic engineering

902. Interferon is:
(A) Antibacterial (B) Antiviral
(C) Antianimal cell (D) Bacterium

903. Waksman and Woodruff (1941) is responsible for the isolation of:
(A) Antiseptic (B) Antibiotic
(C) Actinomycin (D) Antibodies

904. A packed bed reactor can take care of:
(A) Product inhibition (B) Substrate activation
(C) Reaction reversibility (D) All of these

905. Consider the following parts:
P. Reissner's membrane Q. *Tapetum lucidum*
R. *Fenestra rotunda* S. *Forea entrails*
Which of these are associated with auditory organs?
(A) P and Q (B) R and S
(C) P and R (D) Q and S

906. Which enzymes is not obtained commercially for malted barley?
(A) β - Glucanase (B) α- Amylase
(C) Glucoamylase (D) β – Amyase

907. Which techniques is used for obtaining information on three dimentional structure of enzymes?
(A) SDS-PAGE
(B) Nuclear magnetic resonance (NMR)
(C) X-ray differaction
(D) Both (B) and (C)

908. Enzymes are similar to catalysts in respects to:
(A) They lower the activation energy of reaction
(B) They do not participate in the reaction
(C) They only increase the reaction rate
(D) All of these

909. Consider the following statements:
P. Grey matter surrounds the white matter in the brain
Q. Grey matter surrounds the white matter in the spinal cord
R. White matter surrounds the grey matter in the spinal cord
S. White matter surrounds the grey matter in the brain brain
Of these statements:
(A) Q (B) S (C) P and S (D) P and R

910. Hydrolase class of enzymes does not include:
(A) Lipases (B) Esterases
(C) Oxygenases (D) Glycosidases

911. Which enzymes is not obtained commercially for Bacillus?
(A) Protease and glucose isomerase
(B) Penctinase
(C) α and β amylase
(D) *Penicillin amidase*

912. Which is false about antibiotics?
(A) Coined by Waksman (1942)
(B) Produced by microorganisms
(C) Produce allergy in some persons
(D) Cure all diseases

913. First hormone produced artificially by culturing bacteria is:
(A) Insulin (B) Thyroxine
(C) Testosterone (D) Adrenaline

914. Antiviral substance produced by vertebrates in case of viral infection is:
(A) Virion (B) Interferon
(C) Antigen (D) Antivirin

915. Which enzymes is not obtained from animals?
(A) Ficin (B) Lipase
(C) Trypsin (D) Catalase

916. Which enzymes is not obtained from plants?
(A) α - and β - amylase
(B) β - glucanase and papain
(C) Pectinase
(D) Ficin

917. Site- directed mutagenesis can be effected by which techniques?
(A) Oligonucleotide-directed mutageesis
(B) Casette mutagenesis
(C) Cut-paste mutagensis
(D) Both (A) and (B)

918. As a general rule of thumb, hydrophobic residues in a protein tend to be situated at:
(A) Extending out of the molecule
(B) Half exposed on the surface
(C) Interior of the molecule
(D) Exposed on the surface

919. Which is not affected by enzymes?
(A) Reaction rate (B) ΔG
(C) ΔG^{+}_{+} (D) Both (B) and (C)

920. Which biosensor is most sensitive?
(A) Potentiometric (B) Conductimetric
(C) Piezoelectric (D) Thermometric

921. Which statement is the most correct about enzymes?
(A) They are always proteins
(B) They are heat stable
(C) All the enzymes are highly specific for their substrates
(D) They can be inhibited by substrate analogues

922. Km is the substrate concentratuion at wgcug reaction rate if equal to:
(A) V_{max} (B) ½ V_{max}
(C) $2V_{max}$ (D) ¼ V_{max}

923. In an electrocardiogram, the QRS complex represents the:
(A) Depolarisation of the atria
(B) Repolarisation of the atria
(C) Depolarisation of the ventricles
(D) Repolarization of the ventricles

924. Column A lists of some enzymes, while column B lists their use. Match both the columns carefully and select the correct option from these listed below the.

Column A	Column B
A.Lysozyme	1. Gout
B. Uricase	2. Skin ulcer
C. Streptokinase	3. Blood clots
D. Rhodonase	4. Cyanide poising
	5. Antibiotic

Options:
(A) A – 3, B – 4, C – 1, D - 2
(B) A – 3, B – 1, C – 2, D – 4
(C) A – 5, B –1, C – 3, D – 4
(D) A – 4, B – 3, C – 1, D - 2

925. The glucose syrup composition is typically:
(A) 90-95% glucose, 10-15% maltose and 3-5% isomaltose
(B) 95-97% glucose, 1-2% maltose and 0.5-2.0% isomaltose
(C) 80-90% glucose, 5-10% maltose and 8-10% isomaltose
(D) 85-90% glucose, 5-10% maltose and 5-10% isomaltose

926. An enzyme used in detergents must have:
(A) Safe to use
(B) Cost-effective
(C) Able to function at pH between 8 to 10.5
(D) All of these

927. *Streptomyces venezuelae* produces antibiotics:
(A) Chloromycetin (B) Aureomycin
(C) Tetracycline (D) Streptomycin

928. Assertion (A): Immobilized cells are particularly vulnerable to contaminations
Reason (R): An antibiotic may have to be used in such cases
Options:
(A) Both A and R are correct and R is the correct remedy of A
(B) Both A and R correct, but R does not explain A
(C) Only A is correct
(D) Only R is correct

929. Na^{+}-K^{+} ATPase is related to:
(A) Active transport (B) Passive transport
(C) Group translocation (D) Both (A) and (C)

930. Unit of Km is:
(A) cm (B) moles/litre
(C) s-1 (D) It has no unt

931. A person has developed interferons in his body. He seems to carry infection of:
(A) Tetanus (B) Malaria
(C) Measles (D) Typhois

932. *Rhizobium leguminosarus* does not nodulate:
(A) Clover (*Trifolium spp*)
(B) Pea (*Pisum sativum*)
(C) Khesari (*Lathyrus sativus*)
(D) Lentil (*Lens culinaris*)

933. Consider the following statements:
Sympathetic nervous system is a characterized by
P. cetylcholine a neuro secretion
Q. Fight and flight activities
R. Longer preganglionic fibres
S. Non-medullated postganglionic fibres
T. Arising from thoracic-lumbar portion
Which of these are associated with auditory organs?
(A) Q and S (B) Q, S and T
(C) P, Q and R (D) P, R and T

934. β- oxidation of palmitic acid produces how many net ATPs?
(A) 96 (B) 146 (C) 130 (D) 123

935. At which of the following sites is the partial pressure of carbon dioxide highest?
(A) Exhaled gas
(B) Alveolar gas
(C) Systemic arterial blood
(D) Systemic venous blood

936. Which of the following growth factors is used for the treatment of osteoporosis?
(A) Fibroblast growth factor
(B) Thymosin α
(C) Insulin like growth factor
(D) Human growth hormone

937. Cerebellum of brain is concerned with:
(A) Static balance
(B) Initiation of muscular contraction
(C) Regulation of body posture and equilibrium
(D) Co-ordination of muscular movements

938. In case of covalent bonding, the most commonly employed matrix is:
(A) Cellulose (B) Polyacrylamides
(C) Agarose (D) All of these

939. In 1953, Sanger determined the amino acid sequence of which proteins?
(A) Casein (B) Insulin
(C) Haemoglobin (D) Heat shock protein

940. What is one of the main ethical questions involved with germ-lie therapy?
(A) Effects on future generations
(B) Effects on humans species
(C) Psychological effects
(D) Prejudice against those not able to conceive

941. Which of the following drugs for AIDS is not a protease inhibitor?
(A) Delavirdine (B) Indinavir
(C) Nelfinavir (D) Seqionavir

942. Due to which of the following features, enzymes are attractive for commercial application?
(A) They operate under mild conditions of temperature, pressure and pH, whih saves energy
(B) Undesirable by-products are not produced
(C) Certain stereochemical reactions are impractical with chemical methods
(D) All of these

943. Which is biological materials is most commonly used in biosensors?
(A) Antibodies (B) Nucleic acids
(C) Enzymes (D) Lectins

944. Assertion (A): Initially, plant and animal enzymes were preferred over microbial enzymes
Reason (R): Considerations of safety and the fear of contaminations by microorganisms or toxins limited the use of microbial sources
Options:
(A) Both A and R are correct and R roughly explains A
(B) Both A and R correct, but R does not explain A
(C) Only A is correct
(D) Only R is correct

945. Which is highly toxic?
(A) Carbodimides (B) Chloroformates
(C) Cyanogen bromide (D) Glutaraldehyde

946. Which enzymes is obtained commercially from animal pancreas?
(A) Trypsin (B) Rennet
(C) Lipase (D) All of these

947. Which of the following statements best describes the effect of increased carbon dioxide pressure in the blood?
(A) It affects oxygen affinity and oxygen carrying capacity
(B) It increases the oxygen affinity but has no effect on oxygen-carrying capacity
(C) It reduces the oxygen affinity and the oxygen carrying capacity
(D) It reduces oxygen affinity but has no effect on oxygen carrying capacity

948. Rhodopsin, · adrenergic receptor, and musarinic acetylcholine receptors share which of the following features?
(A) Each causes an inhibitory intracellular response
(B) Each activates a tyrosine kinase cascade
(C) Each is composed of an dimer
(D) Each functions through a heterotrimeric G-protein

949. Which is the basis of different VNTR loci?
(A) Sequence of repeating units
(B) Number of repeating units
(C) Length of repeating units
(D) Both (A) and (C)

950. Recombinant vaccines include:
(A) DNA vaccines
(B) Recombinant polypeptide vaccines
(C) Purified antigen vaccines
(D) Both (A) and (B)

951. Subtilisin inactivation by oxygen produced by bleaches is due to:
(A) Oxidation of the methionine residue at position 222
(B) Reduction of the methionine residue at position 222
(C) Oxidation of Ser residue in the active site
(D) All of these

952. Enzyme kinetics (i.e., activity) may be affected due to:
(A) Matrix used for immobilization
(B) Conformational changes in the enzyme molecules induced by the procedure of immobilization
(C) Both (A) and (B)
(D) Altered charge of enzyme molecules

953. Whole cells have been used to develop which type of biosensor devices?
(A) Potentiometric (B) Thermometric
(C) Amperomteric (D) All of these

954. Which is not is used for treatment of thrombosis?
(A) Interferon
(B) Granulocyte macrophage colony stimulating factor
(C) Streptokinase
(D) Didanosine

955. Stealth liposomes are made up of ?
(A) Lipid (B) Monoclonal antibodies
(C) PEG (D) All of these

956. The nature of interaction between analyte and the biological material used in bio-sensors may be of which types?
(A) The analyte may be converted into a new chemical molecule by enzymes
(B) The analyte may simply bind to the biological material present on biosensor
(C) The analyte may alter the biological agent used in the biosensor
(D) Both (A) and (B)

957. Packed bed enzyme reactors suffer from:
(A) Difficulty in temperature control
(B) Clogging by colloids
(C) Channel formation
(D) All of these

958. Enzyme ability in the presence of heavy metals can be improved by:
(A) Replacing Cys residue
(B) Replacing Met residue
(C) Removing surface carboxyl groups
(D) All of these

959. Which enzyme is regarded as catalytically perfect enzyme?
(A) Catalase
(B) Triphosphate isomerase
(C) Carbonic anhydrase
(D) All of these

960. Which of the following statement(s) concerning the rate of action potential propagation is/are true?
(A) It is faster in large-diameter axons than in small – diameter ones
(B) It is faster for a strong stimulus than for a weak one
(C) It s faster in myelinated nerve fibres than in non-myelinated ones
(D) Both (A) and (C)

961. Lowered pH results in:
(A) Increase in O_2 affinity of respiratory pigments
(B) Decrease in O_2 affinity of respiratory pigments
(C) Decrease in the number of red blood corpuscles
(D) Increase in the number of red blood corpuscles

962. At a VNTR locus, which of the following is the basis for different alleles?
(A) Sequence of repeating units
(B) Number of repeating units
(C) Length of repeating units
(D) Both (A) and (C)

963. Name of drug produced through biotechnology for cancer treatment:
(A) Insulin (B) Interferon
(C) HGH (D) TSH

964. Which of the following pairs are correctly matched?
P. Stenochaline Limited tolerance to change in the osmotic concentrations of external environment
Q. Euryhaline Wider range tolerance of osmotic concentrations
R. Osmoconformers Animals which are incapable of osmoregulation
S. Osmoregulators Animals which change in volume as the external osmotic concentration changes
Select the correct answer using the codes given below
(A) Q, R and S (B) P, Q and S
(C) P, R and S (D) P, Q and R

965. Which one of the following is not a neurotransmitter?
(A) Glutamine (B) Glutamate
(C) Glycine (D) Acetylcholine

966. The term vaccine is derived from:
(A) Vacca (B) Vacular
(C) Vaccination (D) Vascular

967. Penicillin inhibits bacterial multiplication because it:
(A) Checks RNA synthesis
(B) Checks DNA synthesis
(C) Destroys chromatin
(D) Inhibits cell wall formation

968. Which of the following diseases is diagnosed on the basis of probes?
(A) Kala-azar and filarial
(B) Malaria and sleeping sickness
(C) Herpex Virus type I, type 2
(D) All of these

969. Hybridoma cells are·
(A) Nerve cells
(B) Hybrid cells formed from myeloma cells
(C) Cells having oncogenes
(D) Product of spore formation in bacteria

970. Antibiotic cephalosporin is got from:
(A) Bacterium (B) Alga
(C) Fungus (D) Mycoplasma

971. In case of downstream processing, purification is achieved through :
(A) Crystallization
(B) Chromatographic methods
(C) Evaporation
(D) Both (A) and (B)

972. Which is correct about propteases?
(A) They help in removal of dirt
(B) Serine proteases are produced by Bacillus licheniformis and B. amyloliquefaciens
(C) They are packed inside dust-free granules for use in detergents
(D) All of these

973. In human beings, interferons are:
(A) Leucocytic (B) Immune
(C) Fibroblastic (D) All of these

974. Which is not antibiotic:
(A) Blastidin (B) Ceresin
(C) Penicillin (D) Cyclohexamide

975. Plasmids are used in genetic engineering because they are:
(A) Easily available
(B) Able to replicate
(C) Able to integrate with host chromosome
(D) Inert

976. The choice of reactor type depends on:
(A) Form (free or immobilized) of enzyme to be used
(B) Kinetics of reaction
(C) Scale of operation
(D) All of these

977. The chief limitations of membrane reactors are:
(A) Regular replacement of membrane adds to cost
(B) Need for substrate diffusion through the membrane
(C) Useful for production of small quantities of products
(D) All of these

978. Consider the following statements:
Constriction of the renal artery would result in an increase in mean arterial pressure due to
P. Increased rennin secretion
Q. Increased angitensin – II formation
R. Decreased plasma bradykinin levels
S. Increased aldosterone secretion
Which of the statements given above is/are correct?
(A) P and S (B) Q, R and S
(C) P, Q and R (D) P, Q, R and S

979. The sequence of flow of blood in the hepatic portal system of a mammal is:
(A) Caudal vein, Kidney, liver, posterior venacava
(B) Intestine, liver, hepatic vein, posterior venacava
(C) Kidney, liver, intestine, sinus, venosus
(D) Posterior venacava, intestine, liver, hepatic vein, sinus, venosus

980. Glucose isomerase enzyme is obtained commercially from:
(A) Bacillus (B) Streptomyces
(C) Aspergillus (D) Both (A) and (B)

981. Which is the most likely first event in colon cancer development?
(A) Loss of APC function (B) Loss of DCC function
(C) Activation of k-ras (D) Loss of p53 function

982. Which play the major role in hisyocompatibility?
(A) MHC loci (B) ABO blood group
(C) Rh blood group (D) MN blood group

983. Which is false about antibiotics?
(A) The term was coined by Waksman in 1942
(B) Antibiotics are capable of curing all diseases
(C) Some persons develop allergy by antibiotics
(D) Antibiotics are produced by microorganisms

984. K_m may be influenced by :
(A) Temperature (B) Ionic strength
(C) pH (D) All of these

985. Which is correct about BRCAI gene?
(A) It causes human breast cancer
(B) It is ·100 kb long
(C) It encodes a protein having over 1800 amino acids
(D) All of these

986. Typically,fermentation of 100 g glucose by selected strains of *Saccharomyces cerevisidae* and *Saccharpmyces carlsbergensis* yields how many grams of ethanol?
(A) 25-29 (B) 45-49 (C) 60-70 (D) 70-89

987. Monoclonal antibiotics are produced by:
(A) Sarcoma cells
(B) Germ cells
(C) Somatic cells
(D) Hybrid of carcinogen cell and bone cell

988. Homologues of BRCAI gene were not found in:
(A) Mice (B) Chickens
(C) Sheep (D) Pig

989. Which is not an example of immunosuppressive drugs?
(A) Methotrexate (B) Zidovudine
(C) Corticosteroids (D) Azathioprine

990. Which is not used in asymmetric gap-LCR?
(A) NAD
(B) DNA ligase
(C) DNA polymerase
(D) Reverse transcriptase

991. A class of RNA molecules serving as components of ribosomes is called:
(A) Satellite – DNA (B) Ribosomal RNA
(C) Small nuclear RNA (D) Transfer RNA

992. Any of several small RNA molecules in the nucleus; most have a role in the splicing reactions that remove introns from mRNA, TRNA and rRNA molecules is called:
(A) Satellite – DNA (B) Ribosomal RNA
(C) Small nuclear RNA (D) Transfer RNA

993. A class of RNA molecules, each of which combines covalently with a specific amino acid as the first step in protein synthesis is called:
(A) Satellite – DNA (B) Ribosomal RNA
(C) Small nuclear RNA (D) Transfer RNA

994. A mutation caused by insertion of gene or more extra bases, or a mutagen, between successive bases in DNA is called:
(A) Insertion mutation (B) Foot printing
(C) DNA chimera (D) DNA chip

995. A technique for identifying the nucleic acid sequence bound by a DNA or RNA-binding protein is called:
(A) Insertion mutation (B) Foot printing
(C) DNA chimera (D) DNA chip

996. DNA containing genetic information derived from two different species is called:
(A) Insertion mutation (B) Foot printing
(C) DNA chimera (D) DNA chip

997. An informal term for a DNA microarray, referring to the small size of typical microarrays is called:
(A) Insertion mutation (B) Foot printing
(C) DNA chimera (D) DNA chip

998. A collection of cloned DNA fragments is called:
(A) DNA library (B) DNA ligage
(C) DNA looping (D) DNA microarray

999. An enzyme that creates a phosphodiester bond between the 3'-end of one DNA segment and the 5' end of another is called:
(A) DNA library (B) DNA ligage
(C) DNA looping (D) DNA microarray

1000. In case of histocompatibility, which is recognized directly (without processing) by TH cells?
(A) Major histocompatibility complex antigens
(B) Minor histocompatibility antigens
(C) A and B antigens (ABO blood group)
(D) Both (A) and (B)

1001. Recombinant protein vaccines suffer from:
(A) Their cost is very high and often prohibitive
(B) They have to be stored at low temperature
(C) Storage and transportation, especially in developing countries are problematic
(D) All of these

1002. Candidate disorder for gene therapy are selected on the basis of which criteria?
I. Disease should be life-threatening
II. Gene responsible for disease should have been cloned
III. A precise regulation of the gene should be required
IV. A suitable delivery system should be available
Options:
(A) I, II, IV (B) II, III, IV
(C) I, II, III, IV (D) I, II, III

1003. Which is not used in gap-LCR?
(A) Reverse transcriptase (B) DNA ligase
(C) DNA polymerase (D) NAD

1004. Which statements is not correct about AIDS?
(A) In 1981, it was first reported in India
(B) About 30% of infants from women having AIDS get infected unless the mothers are treated with antiviral agents before delivery
(C) In general, antibodies against HIV proteins appear in the patients 3 months after infection
(D) The decrease in CD4+ T cells is the hallmark of AIDS

1005. The lead for development of the antihypertensive drug Captopril came from:
(A) South American pit viper (B) Banded Krait
(C) King Cobra (D) Rattle snake

1006. Protooncogenes (and their oncogenic counterparts) encode proteins that are involved in:
(A) Induction of cellular proliferation
(B) Regulation of programmed cell death
(C) Inhibition of cellular proliferation
(D) All of these

1007. Generally, the waste are of which types?
(A) Biological (B) Physical
(C) Chemical (D) All of these

1008. Which is correct about purifies antigen vaccines?
I. These vaccines are based on purified antigens of the concerned pathologens isolated from GMOs
II. They do not contain the organism so that the risk of pathogenicity is avoided
III. Successful examples of such vaccines are mostly from bacteria
IV. Cost of purified antigen vaccines is higher than that of conventional vaccines
Options:
(A) I, II, III (B) II, III, IV
(C) I, II, III, IV (D) I, II, IV

1009. Immuno-PCR detects the presence of:
(A) DNA (B) RNA
(C) Antigen (D) Immunogen

1010. SCID may result from a deficiency of:
(A) RAG-1/RAG-2 (B) ADA
(C) JAK-3 (D) All of these

1011. Which of the following is most likely to occur in colon cancer development?
(A) Loss of APC function (B) Loss of DCC function
(C) Loss of p53 function (D) Activation of k-ras

1012. Biogas produced by bacterial digestion of organic matter through which processes?
(A) Aerobic digestion
(B) Anaerobic digestion
(C) Both (A) and (B)
(D) Partially aerobic digestion

1013. Consider the following statements:
P. Chymotrypsinogen is converted into chymotrypsin by trypsin
Q. Cholecystokinin is secreted by stomach
R. Pepsin acts on peptide linkage in which tyrosine and phenylalanine are involved
Which of the statements given above is/are correct?
(A) P and Q (B) Q and R
(C) P and R (D) P, Q and R

1014. The term renal autoregulation refers in part to the fat that:
(A) The kidney does not require blood flow to sustain its active transport
(B) The kidney contains baroreceptors (pressure receptors that contribute to the regulation of ardia output
(C) Renal blood flow is relatively constant over a wide range of systemic arterial pressures
(D) Renal blood flow is not affected by activation of the sympathetic nerves that innervate the kidney

1015. Which of the following is not a function of the liver?
(A) Production of bile (B) Detoxication of drugs
(C) Storage of glucose (D) Storage of vitamin

1016. When antidiuretic hormone (ADH) is released by the pitutary gland it acts on the distal tubules and collecting ducts of the nephrons to increase their permeability to water and promote water retention by the body. This action of ADH on the kidney has which of the following effects on the body?
(A) Lowering blood osmolarity
(B) Raising blood pH
(C) Lowering blood pressure
(D) Raising blood glucose

1017. Which of the following diseases is not known to be caused by over-expression of trinucleotide repeat?
(A) Huntington's disease (B) Fragile X syndrome
(C) Alzheimer's disease (D) Kennedy's diseases

1018. Syntrophic H_2 producing bacteria are also called :
(A) Obligate H_2 producing bacteria
(B) Obligate proton educing bacteria
(C) Facultative O_2 reducing bacteria
(D) Both (A) and (B)

1019. Steroid biotransformation yields which adrenocortical hormones?
(A) Cortisone (B) Corticosterone
(C) Hydrocortisone (D) All of these

1020. Rhizobium can be used in:
(A) Pulses (B) Forage crops
(C) Cereals (D) Both (A) and (B)

1021. Bioethanol production on commercial scale is based on :
(A) Oil seed crops (B) Starch crops
(C) Sugar crops (D) Both (B) and (C)

1022. The correct sequence for sperm migration after its production in testes is:
(A) Seminiferous tubule → Epididymis→ Vas deferens→ Urethra
(B) Urethra → Vas deferens→ Epididymis→ Seminiferous tubule
(C) Epididymis→ Vas deferens→ Urethra → Seminiferous tubule
(D) Seminiferous tubule→ Vas deferens→ Epididymis→ Urethra

1023. Calcium is important in skeletal muscles contraction because it:
(A) Activates the myosin ATPase by binding to it
(B) Binds to troponin to remove a constant inhibition of cross-bridge attachment
(C) Prevents the formation of bonds between the myosin cross bridges and the actin filament
(D) Is required to detach the myosin head from the actin filament

1024. Salivary α–amylase becomes inactive in the stomach primarily due to:
(A) Inactivation by low pH
(B) Degradation by gastric pepsin
(C) Inhibition by chloride ion
(D) Inhibition by peptides

1025. Cerebrospinal fluid:
(A) Has a composition almost identical to that of the rain extracellular fluid
(B) Is produced by the choroid plexus
(C) Circulates from the subarachnoid space into the cerebral ventricles
(D) Is absorbed through the arachnoid villi

1026. If communication between the SA node and the AV node become blocked, which will most likely occur?
(A) The rate of ventricular contraction will decrease
(B) Afterload will increase
(C) The rate of atrial contraction will decrease
(D Stroke volume will increase to 5L/beat

1027. Biomass may be used in :
(A) Direct burning to generate heat
(B) Thermochemical conversion by pyrolysis/gasification
(C) Biological conversion
(D) All of these

1028. Rennet was purified by:
(A) Waksman (B) Flemming
(C) Hansen (D) Dubrunfaut

1029. Contraction of the abdominal muscles is important in:
(A) Normal (quiet) inspiration
(B) Forced (maximum) inspiration
(C) Normal (quiet) expiration
(D) Forced (maximum) expiration

1030. Biodiseal can be obtained from :
(A) Lipids accumulated in plants and algae
(B) Hydrocarbons produced by some plants and algae
(C) Both (A) and (B)
(D) Anaerobic digestion of domestic waste

1031. Which of the following statements is not true of the endocrine system?
(A) It is one of two major regulatory systems of the body
(B) It is composed of glands that secrete chemical messengers into the blood
(C) It is an important regulator of homeostatic mechanisms
(D) None of these

1032. Which is represents a gain of function?
(A) Loss of APC function (B) Loss of DCC function
(C) Activation of k-ras (D) Loss of p53 function

1033. In case of downstream processing, commonly used approach of drying is:
(A) Spray drying (B) Freeze drying
(C) Vacuum drying (D) All of these

1034. Tumour necrosis factor can be used for the treatment of:
I. Thrombolysis II. Sepsis
III. Tumours IV. Viral diseases
Options:
(A) I, II, III (B) III, IV
(C) I, II, III, IV (D) I, II, IV

1035. A Graffian follicle is:
(A) An immature developing follicle
(B) A mature follicle ready to ovulate
(C) A follicle undergoing apoptosis
(D) Ovulated follicle

1036. A shift in the oxygen-haemoglobin dissociation curve to the right occurs in:
(A) Hypothermia
(B) Carboxyhemoglobin
(C) Fetal hemoglobin S
(D) An increase in the 2-3-bisphosphoglycerate (2,3-BPG)

1037. High doses of antibiotics can destroy the bacterial flora of the large intestine. This can result in impaired:
(A) Absorption of protein (B) Blood coagulation
(C) Bone resorption (D) Respiratory control

1038. During the rising phase of an action potential:
(A) Voltage – gated Na^+ channels open
(B) Voltage – gated K^+ channels open
(C) Voltage-gated Na^+ channels lose
(D) Voltage-gated K^+ channels lose

1039. Uric acid is the chief end product of nitrogenous excretion in adult reptiles and birds. It is often excreted in concentrated crystalline form. This pattern of excretion is mainly an adaptation for:
(A) Conservation of urea producing enzymes
(B) Raising osmotic concentration of blood
(C) Conservation of water
(D) Storage of waste materials

1040. The function of the ileum is:
(A) Absorb nutrients
(B) Absorb vitamin B_{12} and bile salts
(C) To introduce bile and pancreatic juices
(D) Absorb alcohol and aspirin

1041. Which is correct about ideal vaccine?
(A) It should not contaminate the environment
(B) It should not be tumorigenic or toxic or pathogenic
(C) It should have very low levels of side-effects in normal individuals
(D) The technique of vaccination should be complex

1042. Polyesters of lactic acid and glycolic acid are used for achieving slow release of :
(A) Proteins (B) Oligonucleotides
(C) Polysaccharides (D) All of these

1043. Which statement are correct about cancer?
I. It refers specifically to malignant tumours
II. Cancers invade healthy tissues by a process called metastasis
III. They are classified as carcinomas, leukaemias/lymphomas and sarcomas according to the embryonic origin of the tissue from which the tumour is derived
IV. It is induced by a variety of agents, such as, chemical carcinogens, irradiation and certain viruses
Options:
(A) I, II, III (B) II, III, IV (C) I, II, III, IV (D) I, II, IV

1044. The resting membrane potential
(A) Occurs only in nerve and muscle cells
(B) Is the same in all cells
(C) Is oriented so that the cell's interior is positive with respect to the extracellular fluid
(D) None of these

1045. The rate of water reabsorption from the proximal tubule is determined primarily by the:
(A) Rate of dissolved particle (solute) reabsorption from the proximal tubule
(B) Concentration of ADH (antidiuretic hormone) in the blood
(C) Osmotic pressure developed by plasma proteins in the proximal tubule
(D) Active transport of water molecules by the proximal tubule cells

1046. An ECG would be useful for determining a patient's:
(A) Heart murmur
(B) Stroke volume
(C) Cardia output
(D) Blockage of condition of electrial signals between the atria and the ventricles

1047. The internal pacemaker that sets biological rhythms:
(A) Is located in the brain
(B) Is located in the heart
(C) Does not function in the absence of light or other environmental uses
(D) Both (B) and (C)

1048. Hormones:
(A) Are chemical regulators that are conveyed from one organ to another via the blood stream
(B) May be secreted by endocrine cells
(C) May be secreted by nerve cells
(D) All (A), (B) and (C)

1049. Piercidin has insecticidal activity; it is isolated from which microorganisms?
(A) *Streptomyces hygrosopicus*
(B) *Streptomyces sagamonensis*
(C) *Streptomyces monaraensis*
(D) *Streptomyces avermitilis*

1050. Which one of the following is not a heme protein?
(A) Catalase
(B) Cytochrome oxidase
(C) Microsomal P_{450} drug oxidizing system
(D) Fe-S protein complex

1051. Which of the following applies uniquely to secondary lamphoid organs?
(A) Presence of precursor B and T-cells
(B) Circulation of lymphocytes
(C) Terminal differentiation
(D) Cellular prolifiration

1052. For an action potential to occur:
(A) The stimulus must reach or exceed threshold
(B) Na^+ influx must exceed K^+ efflux
(C) The membrane must be out of the relative refractory period
(D) Both (A) and (B)

1053. As blood passes along the glomerular capillaries from the afferent to efferent arteriole, the net filtration pressure:
(A) Increases
(B) Decreases
(C) First decreases, reached a minimum about half way along the capillary, then increases
(D) First increases, reaches a maximum about half way along the capillary, then decreases

1054. Propagation of a regenerative action potential along an axon can be accelerated by which of the following?
(A) A decrease in the transmembrane resistance
(B) A decrease in the axoplasmic resistance
(C) Reduced myelin wrapping
(D) Shortened internodal lengths

1055. Commercial preparations of *B. thuringiensis* usually contain?
(A) Crystal proteins (B) Spores
(C) Inert carriers (D) All of these

1056. Which one of the following groups of enzymes is involved in the digestion of proteins?
(A) Pepsin, amylopsin and trypsin
(B) Amylopsin, trypsin and chymotrypsin
(C) Trypsin, chymotrypsin and steapsin
(D) Pepsin, trypsin and chymotrypsin

1057. A patient undergoes a total gastrectomy because of a proximally located gastric cancer. After the surgery, which of the following digestive enzymes will be produced in inadequate amounts?
(A) Amylase (B) Chymotrypsin
(C) Lipase (D) Pepsin

1058. Most of the CO_2 that is transported in blood:
(A) Is dissolved in the plasma
(B) Is bound to hemoglobin
(C) Is in carbonic acid
(D) Is in bicarbonate ion

1059. Which moiety of hemoglobin molecule binds 2,3-BPG at pH 7.0?
(A) Amino terminal (B) Carboxylic terminal
(C) Sulphadryl group (D) Alcoholic terminal

1060. Lipophillic hormones:
(A) Include peptide hormones such as EGF
(B) Are hormonally active in their original secreted form
(C) Must bind to membrane receptors to exert effects
(D) Pass through cell membranes into the cytoplasm

1061. All of the following are known to involve a calcium activated, vescile-mediated secretory event except:
(A) Synaptic transmissions
(B) Release of histamine from mast cells
(C) Sperm acrosomal reactions
(D) Constitutive secretion of collagens

1062. White matter of the brain is composed of:
(A) Glial cells
(B) Non-mylenated nerve fiber
(C) Mylenated nerve fiber
(D) Dendrites

1063. Which bacteria increases the solubility/availability of phosphate?
(A) *Glomus* (B) *Thiobacillus*
(C) *Bacillus* (D) Both (B) and (C)

1064. Which hydrolytic enzymes is produced in bulk quantity from plants?
(A) α- and β-amylase (B) Papain
(C) β-Glucanase (D) All of these

1065. When a person dives deep under water and resurfaces rapidly, which one of the following sequences of events causes decompression sickness?
(A) Increase in pressure - N_2 dissolves in blood - Rapid fall in presence - N_2 bubbles in blood
(B) Increase in pressure - Pressure on ear ossicles Loss of orientation - Imbalance
(C) Fall in pressure – Loss of O_2 from blood deoxygenation - Fatique
(D) Increased pressure – enhanced blood pressure – Headache – Loss of memory

1066. Complex media contain:
(A) Molasses (B) Cornsteep liquor
(C) Soybean meal (D) Any of these

1067. Biocontrol agents have which features?
(A) They should be specific to the target organism
(B) They should have a high multiplication rate
(C) They should not damage the environment (including human beings)
(D) All of these

1068. Some neurons in the vagus nerve terminate on sinoatrial (pacemaker) cells in the heart. These neurons secrete acetylcholine, which ultimately results in a decreased heart rate. this is an example of:
(A) Endorine control (B) Exorine control
(C) Hormonal control (D) Neural control

1069. In the statement "there are several ventricles in the brain and some of these ventricles are connected by foramen of Monro. This foraman connects X which are located in the Y of the brain", X and Y stand respectively for:
(A) Lateral ventricles and cerebral hemisphere
(B) Third and fourth ventricles in cerebral hemisphere and medulla oblongata
(C) Lateral ventricles and third ventricle in cerebral hemispheres and diencephalon
(D) Lateral ventricles and third ventricle in cerebral hemisphere and mesencephalon

1070. Which techniques are used to study of foetal cells for detection of genetic disorders?
I. Hybridization with oligonucleotide probes
II. RFLP analysis
III. Enzyme assay
IV. Karyotype analysis
Options:
(A) I, II, III (B) II, III, IV
(C) I, II, III, IV (D) I, II, IV

1071. Which one of the following events is not triggered by the hormone glucagon in liver?
(A) Activation of glycogen synthase
(B) Stimulation of cAMP production
(C) Activation of protein kinase
(D) Activation of phosphorylase kinase

1072. Oxidoreductase class of enzymes does not include:
(A) Aldolases (B) Dehydrogenases
(C) Oxidases (D) Peroxidases

1073. Which protective agents reduces the detrimental effects of lyophilization?
(A) Skimmed milk (B) Salt
(C) Sucrose (D) Both (A) and (C)

1074. What is the name and number of the nerve that transmits information from the basilar and tutorial membranes of the internal ear to the brain?
(A) Olfactory and I cranial nerve
(B) Optic and II cranial nerve
(C) Auditory and VIII cranial nerve
(D) Trigeminal and V nerve

1075. In which of the following combinations is the name of the hormone, its chemical type and its tissue of origin correctly matched?
(A) Aldosterone-peptide-pancrease
(B) Glucagon-peptide-adrenal cortex
(C) ACTH-polypeptide-adrenal cortex
(D) Vasopession – peptide-posterior pituitary

1076. During the cardiac cycle:
(A) The volume of blood leaving the left side of the heart is greater than that leaving the right side
(B) The pressure of blood leaving the right side of the heart is greater than that leaving the left side
(C) The duration of systole is greater than that of diastole
(D) The duration of diastole is greater than that of systole

1077. Trypsin differs from pepsin in that:
(A) Trypsin digests protein in an acidic medium while pepsin does so in an alkaline medium
(B) Trypsin digests protein in an alkaline medium while pepsin does so in an acidic medium
(C) Trypsin is secreted from the gastric glands while pepsin is erected from the pancreas
(D) Trypsin production is influenced by peptidergic neurohormones while pepsin is influenced by steroids

1078. Which of the following is the primary factor regulating normal coronary blood flow?
(A) Aortic diastolic pressure
(B) Coronary perfusion pressure
(C) Systolic wall tension
(D) Myocardial oxygen consumption

1079. The ability of a desert rodent's kidneys to form highly concentrated urine is largely attributable to:
(A) A high glomerular filteration rate
(B) The production of a very salty ultrafilterate in the glomerulus
(C) The ability of cells lining the proximal convoluted tubule to absorb water
(D) Long loops of Henle that pass through tissues of incresasing salt concentrations

1080. Chorionic gonadotropin hormone, synthesized in the placenta, plays an important role in the establishment and maintenance of pregnancy in:
(A) Cows (B) Pigs
(C) Rodents (D) Humans

1081. Which one of the following statements is correct?
(A) Chloride shift of Hamburger phenomenon occurs from the blood to liver in a fresh fish to maintain its osmotic balance
(B) Chloride shift takes place from stomach to the blood vascular system of man to produce HCl
(C) Chloride shift involves the passage of chloride ions from the plasma to RBC to balance the bicarbonate ions which have passed out from RBC to plasma
(D) Chloride shift occurs to balance chloride ions and sodium ions in a muscle cell during muscle contraction

1082. Which microbial culumns is used for studies on product isolation, identification and strain improvement?
(A) Assay culture (B) Development culture
(C) Research culture (D) Both (B) and (C)

1083. Which one of the following sequences is the correct route of circulation in the human heart?
(A) Right auricle – right ventricle – pulmonary artery – lungs – pulmonary veins – left auricle – left ven tricle - aorta
(B) Left auricle - Right auricle – right ventricle – pulmonary artery – lungs – pulmonary veins — left-ventricle - aorta
(C) Right auricle – left auricle - right ventricle – pul monary artery – lungs – pulmonary veins — left ventricle - aorta
(D) Right auricle – right ventricle – pulmonary veins– lungs — pulmonary artery - left auricle – left ventricle - aorta

1084. The rates of forward and reverse reactions depend on:
(A) Rate constants for the forward and backward reactions
(B) Concentrations of the products
(C) Concentrations of the reactants
(D) All of these

1085. Which is regarded as the third revolution in vaccines?
(A) Recombinant protein vaccines
(B) Recombinant polypeptide vaccines
(C) DNA vaccines
(D) All of these

1086. In human eyes, light exposure to the retinal photoreceptors:
(A) Causes its depolarization
(B) Causes its hyperpolarization
(C) Opens Na^+ channels of the photoreceptors
(D) Opens K^+ channels of the photoreceptors

1087. Which biosensor has the highest cost?
(A) Potentiometric (B) Conductimetric
(C) Piezoelectric (D) Thermometric

1088. In organisms with closed circulatory systems, fluid leaves the blood of capillary networks at the arterial end and returns to blood at the venous end for which of the following reasons?
(A) Osmotic pressure increases prior to dehydration
(B) Osmotic pressure difference dominates at the arterial end; hydrostatic pressure difference dominates at the venous end
(C) Hydrostatic pressure difference dominates at the arterial end; osmotic pressure difference dominates at the venous end
(D) Hydrostatic pressure drops during diastole

1089. Cutting the posterior root of a spinal nerve would:
(A) Impair motor control of skeletal muscle
(B) Interfere with the flow of sensory impulses
(C) Interfere with the ability of brain to transmit impulses
(D) Interfere with the circulation of CSF

1090. Insulin and thyroxine arrive at an organ at the same time. Thyroxine causes can effect on the organ but insulin does not. Why?
(A) Thyroxine is a lipid-soluble hormone and insulin is not
(B) The target cells in the organ are up-regulated by thyroxine
(C) The organ's cells have receptors for thyroxine but not for insulin
(D) Thyroxine is a local hormone and insulin is a circulating hormone

1091. Consider the following water conservation mechanisms:
P. Absence of sweat glands
Q. Dependence on metabolic water
R. Greater ADH level in the blood
S. Lying more on protein-rich diet
The kangaroo (*Dipodiomys*) living in deserts can survive without drinking water because of:
(A) P, Q and R (B) P, Q and S
(C) Q, R and S (D) P, R and S

1092. Release of which of the following peptide leads to an increase in the secretion of pancreatic-enzymes into the small intestine?
(A) Cholecystokinins (B) Gastrin
(C) Motilin (D) Secretin

1093. Drugs for AIDS are generally designed to block:
(A) Integrase (B) Reverse transcriptase
(C) Viral protease (D) All of these

1094. Compared with systemic arterial blood, pulmonary arterial blood has a higher:
(A) Oxygen content
(B) pH
(C) Bicarbonate ion concentration
(D) Hb concentration

1095. Consider the following statements:
For osmoregulation in aquatic medium
P. A freshwater fish has to drink water continuously
Q. A marine bony fish has to drink water continuously
R. A fresh water fish produces copious urine, which is hyposmotic to blood
S. A marine fish produces scanty urine, which is hyper osmotic to blood
Which of the statements given above is/are correct?
(A) P (B) S
(C) P and S (D) Q and R

1096. Polycythemia is a condition of blood in which the:
(A) Viscosity of blood is reduced due to deficiency in the blood count
(B) Surface tension of the blood is reduced due to the subnormal concentration of the haemoglobin in blood
(C) Viscosity of the blood is enhanced due to the solidification of the plasma
(D) Viscosity of the blood is increased due to the presence of excessive blood cells

1097. Which of the following are functions of sertolic cells?
1. Protection of developing spermatogenic cells
2. Nourishment of spermatocytes, spermatids and sperm
3. Phagocytosis of excess sperm cytoplasm as development proceeds
4. Mediation of the effects of testosterone and FSH
5. Control of movement of spermatogenic cells and release of sperm into the lumen of seminiferous tubules
(A) 1, 2, 4 and 5 (B) 1, 2, 3 and 5
(C) 2, 3, 4 and 5 (D) 1, 2, 3, 4 and 5

1098. Match list – I with list – II and select the correct answer using the codes given below the lists

List – I	List – II
P. Dehydration	1. Aldosterone
Q. Antidiuretic hormone	2. Osmoreceptor
R. Reabsorption of Na^+ in the kidney	3. Blood hyperosmotic urine
S. Water diuresis	4. Vasopressin
	5. Blood hyposmotic urine

(A) P –4, Q – 1, R – 2, S - 3
(B) P –4, Q – 4, R – 5, S - 1
(C) P –3, Q – 1, R – 2, S – 5
(D) P – 3, Q – 4, R – 1, S - 5

1099. Which of the following would cause a decrease in the binding affinity of haemoglobin for oxygen:
(A) Increased pH of the blood
(B) Increased temperature of the blood
(C) Decreased BPG levels in erythrocytes
(D) Both (A) and (B)

1100. Which of the following is not true about cortisol?
(A) Increases the circulating lamphoytes
(B) Increases the circulating eosinophils
(C) Decreases the production of prostaglandins
(D) Inhibits the production of fibroblasts

1101. Mesonephric kidney in which tubules extend posteriority throughout the length of coelom, such kidney is referred as:
(A) Pronephros (B) Opisthonephros
(C) Holonephros (D) Metanephros

1102. Which of the following statements accurately describes sex hormones?
(A) They bind specific membrane receptors
(B) They interact with DNA directly
(C) They cause release of a proteinaeous second messenger from the cell membrane
(D) They enhance transcription when bound to receptors

1103. Most of the enzymes are used extracellularily for which medical purposes?
(A) Removal of toxic substances
(B) Treatment of disorders within blood circulatory system
(C) Topical applications
(D) All of these

1104. Colorimetric test strips for glucose contain:
(A) Horse-radish peroxidase
(B) A weekly coloured chromogen
(C) Glucose oxidase
(D) All of these

1105. Which yields antibiotics?
(A) *Cladonia* (Lichens) (B) *Streptomyces*
(C) *Usnea* (D) All of these

1106. Which polymer matrices is commonly used for cell immobilization?
(A) Polyglycol oligomers (B) Polyacrylamide gel
(C) Calcium alginate (D) *K*-Carageenan

1107. Cell immobilization may be achieved in:
(A) Cells directly bound to water insoluble carriers
(B) They can be cross-linked to bi-or multifunctional reagents
(C) Polymer matrices may be used for entrapping
(D) All of these

1108. Sterilization can be achieved by:
(A) Heating (B) Filtration
(C) Radiation (D) All of these

1109. Glucose isomerase is obtained from:
(A) *Streptomyces sp.*
(B) *Actinoplanes missouriensis*
(C) *Bacillus coagulans*
(D) All of these

1110. Increased calcium influx across the membranes of presynaptic neurons has which of the following effects on synaptic transmission?
(A) Increases release of neurotransmitter
(B) Increases in voltage-dependent potassium efflux
(C) Decreases release of neurotransmitter
(D) Decrease in voltage-dependent potassium efflux

1111. Dissociation curve of hemoglobin shifts to the right when:
(A) CO_2 concentration decreases
(B) O_2 concentration decreases
(C) CO_2 concentration increases
(D) Chloride concentration increases

1112. Which is based on chemical interaction between the carrier and the product molecules?
(A) Covalent chromatography
(B) Gel Filtration
(C) Group specific affinity chromatography
(D) Both (A) and (C)

1113. In 1929, Fleming discovered which antibiotics?
(A) Streptomycin (B) Tetracyclin
(C) Penicillin (D) Gentamycin

1114. Which biosensor type has the lowest cost?
(A) Potentiometric (B) Conductimetric
(C) Piezoelectric (D) Thermometric

1115. Biotransformation of progesterone to 11 α – hydroxyprogesterone is achieved by which microorganisms?
(A) *Rhizopus arrhizus* (B) *Thiobacillus thiooxidans*
(C) *Rhizopus nigricans* (D) Both (A) and (C)

1116. Which biosensor has the fastest response?
(A) Potentiometric (B) Conductimetric
(C) Piezoelectric (D) Thermometric

1117. Luteinising hormone (LH) in female:
(A) Helps in the appearance of secondary sexual characters
(B) Stimulates ovary to secrete estradiol
(C) Helps in release of the ovum from the ovary
(D) Control the blood pressure

1118. Who is recognized as father of microbiology:
(A) Robert Koch (B) Edward Jenner
(C) Louis Pasteur (D) Aristotle

1119. N-butanol is used for making:
(A) Formaldehyde and urea (B) Petrol
(C) Resin (D) All of these

1120. Which acts as an agent of bioinformation?
(A) Resting and dried microbial cells
(B) Vegetative cells and spores of microbes
(C) Immobilized cells
(D) All of these

1121. Microbiology is widely consisted to have begun in 1857 when Pasteur reported his studies on :
(A) Lactic acid fermentation
(B) Microscopic features of the microorganisms
(C) A suitable medium for lactic acid fermentation
(D) All of these

1122. In industrial microorganisms, which one is commonly used?
(A) River mud (B) Soil
(C) Lake (D) All of these

1123. Interferon can be used for the treatment of:
(A) AIDS (B) Cancer
(C) Viral diseases (D) All of these

1124. Bestatin is produced by which organisms?
(A) *Trichoderma polysporum*
(B) *Rhizoctonia leguminicola*
(C) *Streptomyces olivoreticuli*
(D) *Streptomyces avermitilis*

1125. Gluconic acid is produced from:
(A) *Aspergillus niger*
(B) *Penicillum purpurogenous*
(C) *Rhizopus stolonifer*
(D) *Staphylococcus aureus*

1126. Which is not correct about Rhizobium?
(A) These are Gram-positive soil bacteria capable of forming root nodules in most leguminous plants
(B) *Azorhizobium caulinodans* nodulates the stems of *Sesbania rostrata*
(C) Root nodulating rhizobia are of common occurrence
(D) Rhizobium cells contain genes for nitrogen fixation (nif genes)

1127. Which products are obtained from yeast fermentation?
(A) Acetic acid, caproic acid and lactic acid
(B) Glycerol, ethanol, butanol and n-propanol
(C) Amyl alcohol and ethyl acetate
(D) All of these

1128. A germplasm bank of AMF (arbuscular mycorrhizal fungi) cultures was established at:
(A) National Bureau of Plant genetic Resources
(B) The Energy and Resources Institute, New Delhi
(C) Indian Institute Pulses Research, Kanpur
(D) Indian Agricultural Research Institute, New Delhi

1129. Correct sequence of hormone secretion from beginning of menstruation is
(A) FSH, progesterone, estrogen
(B) Estrogen, FSH, progesterone
(C) FSH, estrogen, progesterone
(D) Estrogen, progesterone, FSH

1130. Fungus *Ampelomyces quisqualis* is used for the control of which powdery mildew?
(A) Grapes (B) Tomato
(C) Straw berries (D) All of these

1131. Which is an example of broad spectrum antibiotics?
(A) Chloramphenicol (B) Tetracyclin
(C) Penicillin (D) Both (A) and (B)

1132. Which of the following is not true for adrenaline?
(A) Is synthesized by demethylation of noradrenaline
(B) Increases coronary blood flow
(C) Increases free fatty acids in the blood
(D) Mobilizes glycogen stores from the liver

1133. *Trichoderma parasitoid* can be used for the control of:
(A) Bollworms of cotton
(B) Stem borer (Sorghum)
(C) Sugarcane internode borer
(D) All of these

1134. Which bacteria induces leaching of copper as copper sulphate from ores?
(A) *Thiobacillus organoparus*
(B) *T. thiooxidans*
(C) *T. acidophilus*
(D) All of these

1135. Acid protease is produced by which microorganisms?
(A) *Mucor pusillus* (B) *Endothica parasitica*
(C) *Mucor miehei* (D) All of these

1136. Lactic acid is produced by activity of which microorganisms using molasses/starch as substrate?
(A) *Rhizopus*
(B) *Lactobacillus bulgaricus*
(C) *Streptococcus lactis*
(D) All of these

1137. In case of downstream processing separation of cells from liquid medium is achieved by:
(A) Centrifugation (B) Filtration
(C) Flocculation and floating (D) All of these

1138. Which is the commonly used techniques of filtration?
(A) Depth filtration
(B) Cross-flow filtration
(C) Rotary drum vacuum filtration
(D) Surface filtration

1139. Which diseases occurs due to a single gene mutation?
(A) Tay Sach's disease and sickle cell anaemia
(B) Cystic fibrosis and Huntington's chorea
(C) Haemophilia and Lesch-Nyhan syndrome
(D) All of these

1140. Bestatin shows which activities?
(A) Antitumour (B) Insecticidal
(C) Herbicide (D) Antifungal

1141. Disruption of microbial cells is usually difficult due to:
(A) Small cell size
(B) Strong cell wall
(C) High osmotic pressure inside cells
(D) All of these

1142. The high glucagons/insulin ratio seen in starvation:
(A) Promotes mobilization of fatty acids from adipose stores
(B) Stimulates beta oxidation by inhibiting the production of malonyl CoA
(C) Leads to increased concentrations of ketone bodies in the blood
(D) All of these

1143. Which of the following statement is true?
(A) All veins carry deoxygenated blood
(B) All arteries carry deoxygenated blood
(C) All veins except one carry oxygenated blood
(D) All arteries except one carry oxygenated blood

1144. Bioreactors are extensively used for:
(A) Fermentation (B) Food processing
(C) Waste management (D) All of these

1145. Azotobacter can be used in:
(A) Wheat (B) Vegetables
(C) Rice (D) All of these

1146. Which is correct about azathioprine?
(A) A mitotic inhibitor
(B) Acts on S phase cells
(C) Blocks synthesis of inosinic acid
(D) All of these

1147. The process involving *Thiobacillus ferrooxidans* is used to recover:
(A) Uranium (B) Gold
(C) Copper (D) All of these

1148. Typically, corn medium yields butanol, acetone and ethanol in which ratios?
(A) 3 : 3 : 1 (B) 6 : 3 : 1
(C) 3 : 6 : 1 (D) 1 : 3 : 6

1149. Which one of the following comparisons between oo-genesis and spermatogenesis in human is not correct?
(A) FSH promotes development in both eggs as well as sperms
(B) LH triggers ovulation in ovary and androgen production in testes
(C) Primary oocytes and primary spermatocytes follow similar pattern of development through meiosis
(D) An ovum is not produced until it fuses with the sperm whereas sperm is produced even in the absence of the ovum

1150. Fungus (*Verticillium lecanii*) can be used for the control of which insects?
(A) Aphids (B) Leaf hopper
(C) White flies (D) Both (A) and (C)

1151. Diuring penicillin G production which precursors is used?
(A) Phenylacetic acid (B) Acetadehyde
(C) Phenoxyacetic acid (D) Both (A) and (C)

1152. Which is used as antifoaming agent during antibiotic production?
(A) Mineral oil (B) Vegetable oil
(C) Silicones (D) Both (B) and (C)

1153. Aabomycin A shows which activities?
(A) Antifungal (B) Antibacterial
(C) Antiviral (D) Antiprotozoal

1154. Which indicates the correct order of steps involved in production of ethanol from starch substrate?
(A) Starch → Pyruvate→ Glucose → Acetaldehyde→ Ethanol
(B) Starch → Glucose→ Pyruvate → Acetaldehyde→ Ethanol
(C) Starch → Acetaldehyde→ Glucose→ Pyruvate→ Ethanol
(D) Starch → Pyruvate→ Acetaldehyde→ Glucose→ Ethanol

1155. Long chain immunosuppressive therapy increases the risk of:
(A) Hypertension (B) Cancer
(C) Metabolic bone disease (D) All of these

1156. Bioinformations are performed by:
(A) Plant cells (B) Animal cells
(C) Microbes (D) All of these

1157. Which is the commonly used to achieve cell disruption:
(A) Drying (B) Lysis
(C) Mechanical cell disruption (D) All of these

1158. Which is not correct about α–amylase?
(A) This enzyme is produced by bacillus amyloliquefaciens
(B) This is an exohydrolase
(C) It hydrolyses starch into components, which have 3 or more linear á -1, 4 glucan units
(D) The end products of starch hydrolysis are dextrins

1159. Recombinant insulin is produced in:
(A) *E. coli* (B) Continuous cell lines
(C) Yeast (D) Both (A) and (C)

1160. Which options indicates the correct order for bioinformation of penicillin G and penicillin V?
(A) Penicillin G/penicillin V→6-aminopenicillanic acid→ semisynthetic penicillins and cephalosporins
(B) Penicillin G/penicillin V→ semisynthetic penicillins and cephalosporins → 6-aminopenicillanic acid
(C) Penicillin G/penicillin V→ natural penicillin → semisynthetic penicillins
(D) Penicillin G/penicillin V→ semisynthetic penicillins and cephalosporins → synthetic antibiotics

1161. Pernicious anemia is due to:
(A) Blockage of vitamin B_{12} absorption
(B) Blockage of vitamin A absorption
(C) Deficiency of Vitamin C
(D) Deficiency of vitamin B_2

1162. Which bacteria induces leaching of copper as copper sulphate from ores?
(A) *Arthrobacter simplex* (B) *Thiobacillus ferroxidans*
(C) *Bacillus subtilis* (D) All of these

1163. Which vectors has been designated for targeted gene transfer on humans?
(A) Insertion vectors (B) Suicide vectors
(C) Replacement vectors (D) Both (A) and (C)

1164. Which options indicates the correct order of 11α – hydroxyprogesterone production?
(A) Diosgenin/stigmasterol→ progesterone → 11-α Hydroxyprogesterone
(B) Diosgenin →stigmasterol → 11 α – Hydroxyprogesterone
(C) Progesterone→diosgenic→11α–Hydroxyprogesterone
(D) Stigmasterol→ 11 α -Hydroxyprogesterone → Diosgenic

1165. In ruminants the correct order of the four chambers is:
(A) Rumen, reticulum, abomasums, omasum
(B) Rumen, reticulum, omasum, abomasums
(C) Omasum, abomasums, reticulum, rumen,
(D) Omasum, reticulum, rumen, abomasums

1166. Chemical fertilizers suffers from which limitations?
(A) They damage the environment
(B) They are costly
(C) They utilize petroleum (specially nitrogenous fertilizers)
(D) All of these

1167. Application of the process based on Thiobacillus for copper recovery has resulted in:
(A) Reduction in price of copper by about 2-3 fold
(B) Elimination of the release of SO_2 into the atmosphere
(C) Both (A) and (B)
(D) Increased purity of the product

1168. LCR can detect which molecules?
(A) DNA (B) Protein
(C) RNA (D) Both (A) and (C)

1169. Which bacteria can absorb metals on their cell surfaces?
(A) Proteus (B) Bacillus
(C) Streptococcus (D) All of these

1170. Which products generated by genetically engineered microbes is approved for marketing in India?
(A) Granulocyte macrophage colony stimulating factor and interferons
(B) Insulin and hepatitis-B surface antigen
(C) Granulocyte colony stimulating factor and insulin
(D) All of these

1171. A genetic marker that is detected as differential mobility of a protein/DNA \fragment or as differential amplification of a DNA sequence present in different individuals is:
(A) Molecular marker (B) Molecular biology
(C) Molecular breeding (D) Molecular farming

1172. A specific location on a chromosome that is defined by naked eye polymorphism, protein or DNA sequences; the inheritance of a genetic marker can be followed in a a mapping population is known as:
(A) Molecular marker (B) Genetic marker
(C) Molecular breeding (D) Molecular farming

1173. The plant breeding programme that are supported by the used of DNA-based markers and recombinant DNA technology is called:
(A) Molecular marker (B) Molecular biology
(C) Molecular breeding (D) Molecular farming

1174. Production of proteins encoded by transgenes in animal/plants; the protein is recovered from milk, urine, blood (in animals) seeds etc. (in plant) is called:
(A) Molecular marker (B) Molecular biology
(C) Molecular breeding (D) Molecular farming

1175. A branch of biology concerned with the study of molecular aspects of biology phenomenon is called:
(A) Molecular marker (B) Molecular biology
(C) Molecular breeding (D) Molecular farming

1176. It depicts the order of genetic markers and the relatives distances between them; measured by recombination frequency between the markers is called:
(A) Linkage map (B) Genetic map
(C) Genome map (D) Restriction map

1177. A schematic representation of various genes and molecular markers located in different chromosomes; different strategies are used to construct linkage maps, cytological or cytogenetic maps and physical maps is called:
(A) Linkage map (B) Genetic map
(C) Genome map (D) Restriction map

1178. A detailed schematic description of the structural and functional organization of the complete genome of an organism is called:
(A) Linkage map (B) Genetic map
(C) Genome map (D) Restriction map

1179. A linear map of a DNA molecule showing the relative positions of the recognition sites for a number of different restriction enzyme is known as:
(A) Linkage map (B) Genetic map
(C) Genome map (D) Restriction map

1180. A single protein species (interferons) produced by both leucocytes and fibroblasts in response to virus infection is:
(A) Interfron α (B) Interferon β
(C) Interferon γ (D) Interferon ω

1181. A single protein species produced by T-lymphocytes in response to antigen-activation is:
(A) Interfron α (B) Interferon β
(C) Interferon γ (D) Interferon ω

1182. A single protein species produced by virus infected cells is:
(A) Interfron α (B) Interferon β
(C) Interferon γ (D) Interferon ω

1183. It generally implies analysis of chromosome pairing relationship among genomes of different species is called:
(A) Genome analysis (B) Genome map
(C) Genome compaction (D) Genomic library

1184. Reduction in genome size achieved by reducing sequences not involved in genetic functions is called:
(A) Genome analysis (B) Genome map
(C) Genome compaction (D) Genomic library

1185. A collection of plasmid clones or phage lysates containing recombinant DNA molecule so that the sum total of DNA inserts in this collection, ideally, represents the entire genome of the concerned organisms is called:
(A) Genome analysis (B) Genome map
(C) Genome compaction (D) Genomic library

1186. Developoment of vectors permitting the synthesis of pure proteins encoded exclusively the DNA inserts is called:
(A) Transcriptional fusion
(B) Transcriptional silencing
(C) Transcription
(D) Transcription units

1187. Denotes a suppressions of transcription of the transgenes, it is the result of methylation of promoter region of transgene is called:
(A) Transcriptional fusion
(B) Transcriptional silencing
(C) Transcription
(D) Transcription units

1188. Production of complementary RNA copy of a DNA strand, catalyzed by RNA polymerase is called:
(A) Transcriptional fusion
(B) Transcriptional silencing
(C) Transcription
(D) Transcription units

1189. The segment of genome, which is transcribed into a single primary mRNA transcript; operon in case of prokaryotes, and one gene in eukaryotes is called:
(A) Transcriptional fusion
(B) Transcriptional silencing
(C) Transcription
(D) Transcription units

1190. An enhancer that makes the promoter activity stimulus specific is called:
(A) Stimulus-specific enhancer
(B) Stimulus-specific gene
(C) Stimulus-specific promoter
(D) None of these

1191. Gene expressed in response to only a specific stimulus is called:
(A) Stimulus-specific enhancer
(B) Stimulus-specific gene
(C) Stimulus-specific promoter
(D) None of these

1192. A promoter that becomes active/inactive only in response to a specific stimulus usually, due to enhancer/silencer sequences is called:
(A) Stimulus-specific enhancer
(B) Stimulus-specific gene
(C) Stimulus-specific promoter
(D) None of these

1193. Genes, which have major effect on the expression of the concerned traits is called:
(A) Oligogenes (B) Oncogene
(C) Luxury gene (D) Gene

1194. A gene involved in tumour development, in case of Agrobacterium, genes specifying auxin and cytokinin biosynthesis is called:
(A) Oligogenes (B) Oncogene
(C) Luxury gene (D) Gene

1195. Those genes, which are expressed either in specialized cells or in response to an specific stimuli is called:
(A) Oligogenes (B) Oncogene
(C) Luxury gene (D) Gene

1196. Molecular marker pattern of an individual obtained by using highly polymorphic markers; used for unequivocal identification of strains, individuals, criminals etc., is called:
(A) DNA fingerprint (B) DNA finger printing
(C) DNA footprinting (D) DNA sequencing

1197. Developing the DNA fingerprint of an individual is called:
(A) DNA fingerprint (B) DNA finger printing
(C) DNA footprinting (D) DNA sequencing

1198. It enables identification of the site on a DNA molecule to which a protein, such as RNA polymerase, binds is called:
(A) DNA fingerprint (B) DNA finger printing
(C) DNA footprinting (D) DNA sequencing

1199. Determination of the base sequence of a DNA fragment is called:
(A) DNA polymerase (B) DNA microarray
(C) DNA footprinting (D) DNA sequencing

1200. The enzyme that catalyses synthesis of DNA copies of template DNA, separate enzymes involved in DNA replication and DNA repair is called:
(A) DNA polymerase (B) DNA microarray
(C) DNA footprinting (D) DNA sequencing

1201. An assembly of a very large number of small does onto a solid support, each dot containing a short sequence of a different gene of the organism concerned is called:
(A) DNA polymerase (B) DNA microarray
(C) DNA footprinting (D) DNA sequencing

1202. A gene, which permits easy and rapid selection for the cell/organism in which it is present and expressed is:
(A) Molecular marker (B) Genetic marker
(C) Selectable marker (D) Scorable marker

1203. A gene that permits easy and rapid identification of the cell/organism in which it is present and expressed in:
(A) Molecular marker (B) Genetic marker
(C) Selectable marker (D) Scorable marker

1204. A gene that gives rise to a product involved in the regulation of the expression of another gene; for example, a gene coding for a repressor protein is called:
(A) Structural gene
(B) Regulatory gene
(C) Regulatory sequence
(D) Reverse transcriptase

1205. A gene coding for a protein or RNA molecule; as distinct from a regulatory gene is called:
(A) Structural gene
(B) Regulatory gene
(C) Regulatory sequence
(D) Reverse transcriptase

1206. A DNA sequence involved in regulating the expression of a gene, for example, a promoter or operator is called:
(A) Regulatory enzyme
(B) Regulatory gene
(C) Regulatory sequence
(D) Reverse transcriptase

1207. An RNA-directed DNA polymerase in retroviruses; capable of making DNA complementary to an RNA is called:
(A) RNA polymerase
(B) Regulatory gene
(C) Regulatory sequence
(D) Reverse transcriptase

1208. An enzyme that catalyzed the formation of RNA from ribonucleoside 5'-triphosphates, using a strand of DNA or RNA as a template is called:
(A) RNA polymerase
(B) Regulatory gene
(C) Regulatory sequence
(D) Reverse transcriptase

1209. Removal of introns and joining of exons in a primary transcript is called:
(A) RNA polymerase
(B) RNA splicing
(C) Regulatory sequence
(D) Reverse transcriptase

1210. The rate of sedimentation of a macromolecule can be enhanced many fold by applying:
(A) DNA polymerase I (B) DNA polymerase
(C) DNAlibrary (D) Centrifugal force

1211. Which varieties is the first variety in Indian developed by MAS?
(A) Pusa Basmati 1
(B) Pusa 1401
(C) Pusa 1121
(D) Improved Pusa Basmati 1

1212. In eukaryotic DNA replication:
(A) Only one replisome forms because there us a single origin of replication
(B) The Okazaki fragments are 1000 to 2000 nucleotides in length
(C) Helicase dissociates from DNA as soon as the initiation bubble forms
(D) FEN1 (flap endonuclease 1) is involved in removing the primer

1213. A strain of *E. coli* has the following genotype with respect to the lac operon: $I^-OZ^+Y^-A^+$. This strain will:
(A) Produce permease only in the presence of lactose
(B) Not produce beta galactosidase if lactose is present in the medium
(C) Not produce permease irrespective of the presence or absence of lactose in the medium
(D) Produce trans-acetylase only when induced

1214. What is incorrect about tRNA?
(A) It binds with an amino acid at its 3' end
(B) It has five double stranded regions
(C) It has a anticodon at one end which recognises codon of mRNA
(D) It looks like clover leaf in 3-dimentional structure

1215. Pusa 1406 has which of the following genes transferred through MAS?
(A) Xa21 (B) Xa13
(C) Both (A) and (B) (D) Xa14

1216. Which of the following statement about restriction endonucleases is not true?
(A) The enzymes are natural products of bacteria
(B) The enzymes cut DNA at defined sequences
(C) The enzymes cut only single-stranded DNA molecules
(D) The enzymes are named because they restrict the growth of bacteriophage

1217. SDS-PAGE does not involve:
(A) Separation of proteins based on their molecular weights
(B) protein with heat and chemicals
(C) Application of an electric field to proteins
(D) Creating a temperature gradient for denaturation of proteins

1218. Primer designing for sequence-tagged sites (STS) is based on:
(A) Random genomic DNA
(B) Ends of large genomic DNA fragments
(C) Sequences of cDNA (expressed sequence tags)
(D) All of these

1219. Improved Pusa Basmati 1 (Pusa 1406), the first variety in India developed through MAS, has been improved by transferring :
(A) Vitamin A (B) QTLs for grain size
(C) Disease resistance (D) Cooking quality

1220. Polynucleotide ligase joins two DNA molecules together by forming a covalent bond between :
(A) Two OH groups
(B) To phosphate groups
(C) Complementary nucleotides on opposite strands
(D) A 3'-OH group and a 5'-PO_4^{-2} group

1221. To test whether a new fertilizer is effective, rice is grown in some fields without fertilizer and in the other fields with fertilizer and the average yields from each of these fields is calculated in terms of its mean and variance. The statistical test to be used to find the effectiveness of the fertilizer would be:
(A) F – test (B) t -test
(C) Chi-square test (D) ANOVA

1222. Which markers systems has been used for developing Pusa 1460?
(A) RFLP (B) STMS
(C) ISSR (D) SCAR

1223. What would stop DNA replication in bacteria?
P. Addition of a DNA polymerase inhibitor
Q. Addition of dideoxyribonucleaotides (ddNTP)
R. Addition of lamin inhibitor
S. Addition of actin inhibitor
(A) P and Q (B) Q and R
(C) R and S (D) P and S

1224. Protein binding regions of DNA are identified by one of the following techniques:
(A) Finger printing (B) Footprinting
(C) Southern printing (D) Western printing

1225. Development of which of the following populations requires the greatest effort?
(A) F_2 (B) Backcross
(C) RILs (D) NILs

1226. Once an RAPD marker linked to a trait of interest is found, it can be converted into a more reproducible PCR type assay by use of :
(A) Allele-specific ligation
(B) Allele-specific PCR (AS-PCR)
(C) Sequence characterized amplified region (SCAR) assay
(D) All of these

1227. If 5-terminal of a linear DNA molecule with sticky ends are being modified in a way that the ends cannot be ligated with each other anymore (to form a circular molecule), what enzyme was used to modify the molecule?
(A) Methylase (B) Kinase
(C) Phosphatase (D) Dehydrogenase

1228. For a nucleic acid probe to hybridize, which of the following must be true?
(A) Both the probe and the sequence to which it hybridizes must be RNA
(B) Both the probe and the sequence to which it hybridizes must be DNA
(C) The sequence to which it hybridizes must be identical to that of the probe
(D) The sequence to which it hybridizes must be complementary to that of the probe

1229. Electrocardiogram (ECG):
(A) Measures electrical activity generated by the brain
(B) the hormonal activity of the heart over time
(C) Recording of the electrical activity of the heart over time
(D) None of these

1230. A major problem with the computational assembly of DNA sequences of complex eukaryotic genomes is the presence of:
(A) Multiple chromosomes (B) Mitochondrial DNA
(C) Introns within the genome (D) Repetitive sequence

1231. Positional cloning refers to:
(A) Using a selection procedure to lone a cDNA
(B) Cloning a portion of a gene using PCR
(C) Isolating a gene by PCR using primers from another species
(D) Mapping a gene to a chromosomal region and then identifying and cloning the genomic copy of the gene from the region

1232. Which step of translation does not consume high energy phosphate bond?
(A) Translocation
(B) Peptidyl transferase reaction
(C) Amino acid activation
(D) Aminoacyl tRNA binding to A-site

1233. The first case of linkage between a quantitative trait and a qualitative character was reported in *Phaseolus vulgaris* between:
(A) Seed size and seed colour
(B) Seed size and flower colour
(C) Seed size and pod colour
(D) Seed yield and seed colour

1234. All of the following statements about type I topoisomerase (Topo I) are true except:
(A) Topo I removes DNA supercoils in an ATP-dependent reaction
(B) Topo I from eukaryotic cells can remove both positive and negative supercoils
(C) Topo I from *E coli* an remove only negative superscoils
(D) Topo I is essential for viability in *E. coli*

1235. When a growing replication fork encounters an RNA polymerase complex engaged in transcription:
(A) Both DNA replication and transcription stops
(B) Replication stops but transcription continues
(C) Both replication and transcription continues but slowly
(D) Transcription pauses to allow replication fork to progress

1236. Spilicing of GU-AG intron:
(A) Involves transesterification reaction
(B) Occurs in prokaryotic mRNA
(C) Form Y-shaped branch structure
(D) Requires maturase activity

1237. The segregation ratios for dominant and codominant markers will differ in which of the following mapping populations:
(A) F_2 (B) Backcross
(C) RILs (D) NILs

1238. What electronic transitions are the most important in UV spectroscopy of organic molecules because they result in the absorbance of longer wavelength radiation outside the UV region?
(A) $\sigma \rightarrow \pi, n \rightarrow \pi$ (B) $\sigma \rightarrow \sigma^*, n \rightarrow \sigma^*$
(C) $\pi \rightarrow \pi^*, n \rightarrow n^*$ (D) $\pi \rightarrow \pi^*, n \rightarrow \pi^*$

1239. Which is not correct about molecular markers?
(A) The sequence tagged sites (STS) marker system has all the advantages of PCR technique
(B) RFLP is the first molecular marker that was developed in 1980s
(C) RFLP markers are highly reproducible
(D) A major limitation of RAPD markers in their nonreproducibility due to low annealing temperature

1240. What is the best method to identify the cellular location of a protein?
(A) Place a reporter gene next to the promoter of the gene encoding the protein, and identify the cellular location of the reporter protein
(B) Use of labeled antibody to identify the cellular location of the protein
(C) Separate the cellular compartments by centrifugation and screen the different compartments with an antibody
(D) Tag the protein with fluorescent amino acids and identify the cellular location of fluorescence microscopy

1241. To increase the stringency of a wash buffer so as to allow the detection of only completely matched hybrids you would?
(A) Lower the temperature and raise the salt concentration
(B) Lower the temperature and lower the salt concentration
(C) Raise the temperature and raise the salt concentration
(D) Raise the temperature and lower the salt concentration

1242. Transgenic plants are easier to produce the transgenic animals because:
(A) Plants can more easily be grown from single cultured cell into which foreign DNA has been introduced
(B) Plant DNA is easier to lone
(C) Plant cells can be transformed by bacterial infection
(D) DNA passes more readily through the plant cell wall than through the animal cell membrane

1243. Okazaki fragments are formed during DNA synthesis, because
P. DNA synthesis extends from 5' to 3' direction
Q, Their synthesis are opposite to the direction of replication fork
R. DNA ahead of replication fork is positively supercoiled
S. DNA synthesis is semi-conservative
(A) Q and R (B) P and Q
(C) R and S (D) S and P

1244. A recombination-dependent mechanism for repairing damaged DNA is:
(A) SOS repair
(B) Excision repair
(C) Mismatch repair
(D) Photo-reactivation repair

1245. Which mapping populations is used fro linkage mapping of molecular markers?
(A) Near-isogenic lines (NILs)
(B) Recombinant inbred lines (RILs)
(C) Backcross progeny
(D) Any of these

1246. A reporter gene is used to:
(A) Identity regulatory sequences from the upstream regions of other genes
(B) Determine if a protein binds to a given sequence element
(C) Determine if a gene contains introns
(D) Determine the stability of a protein

1247. Plasmid vectors for cloning:
(A) Can generally accommodate larger insects than phage vectors
(B) Grow within bacteria, and are present in bacterial colonies on an agar plate
(C) Can accommodate inserts of over 100 kilobases
(D) Include centromeres to allow propagation in yeast

1248. Eukaryotes differ from prokaryotes in DNA replication:
(A) Different enzyme for leading and lagging strands
(B) Use of DNA primer instead of RNA primer
(C) Discontinuous rather than semi-discontinous replication
(D) Unidirectional rather than bidirectional replication

1249. Which is not correct about recombinant inbred lines (RILs)?
(A) RILs are also called inbred-derived lines
(B) RILs are homozygous lines derived from a suitable cross
(C) RILs are created by single seed descent (SSD) through atleast five generations pf continued selfing
(D) At the end of SSD, seeds from each plant are harvested separately to obtain as many RILs as there are individual plants

1250. UV – B induced damage of DNA for formation of pyrimidine dimmers are repaired by photolyase on activation by:
(A) UV - C light (B) Green light
(C) Blue light (D) Infrared light

1251. Which of the following is an example of RNA editing?
(A) Removal of introns from an RNA transcript
(B) Degradation of an RNA molecule by nucleases
(C) Alternation of the nucleotide sequence of an RNA molecule
(D) Caping of the 5' end of an RNA transcript

1252. Why were the first genetically engineered human insulin genes split into separate A and B chain sequences and cloned in separate *E coli* strains?
(A) To control the proportions of A and B chains produced
(B) To prevent an *E. coli* strain producing an active insulin
(C) Because *E coli* cannot fold the resultant peptide with the chain present
(D) To prevent loss of insulin by secretion

1253. Sodium Dodecyl Sulphate (SDS) is used while separating proteins by PAGE because:
(A) It helps in solubilization of proteins thereby making it easier to separate
(B) It binds to proteins and confers uniform negative charge density thereby making move during electrophoresis
(C) Decreases the surface tension of the buffer used for electrophoresis
(D) Stabilizes the proteins

1254. Which of the following types of vector would be most suitable for introducing DNA into a human cell?
(A) Plasmid (B) Bacteriophage
(C) Cosmid (D) Adenovirus

1255. Which of the following activities is most loosely associated with alkaline phosphatase?
(A) Cutting DNA
(B) Preventing vector dimenrization
(C) Joining DNA fragments with cohesive ends
(D) Homopolymer tailing

1256. Group I introns are remarkable because:
(A) They are spliced by external RNA molecules with the help of protein
(B) They are spliced by protein molecules in the absence of external RNA molecules
(C) They are autocatalytic
(D) They are only present in mitochondrial and chloroplast genomes

1257. Which is not correct about the applications of molecular markers?
(A) The molecular markers that are uniformly distributed throughout the genome can be used to estimate the genetic contribution of each parent to each individual of a segregating population
(B) MAS can not be used to improve complex traits like yield
(C) Molecular markers would enable an effective selection for horizontal resistance even in the presence of vertical resistance genes
(D) Two rice varieties 'Angke' and 'Coonde', developed by MAS for bacterial leaf blight resistance, have been released in Indonesia

1258. Molecular breeding requires which:
(A) Genetic maps and molecular markers linked to agronomic traits
(B) Automated and high throughout diagnostic tech niques
(C) A modification if breeding practice that takes full advantages of the information provided by such diagnositic assays
(D) All of these

1259. Northern blotting is difficult from Southern blotting in that:
P. The target is DNA
Q. The probe should be selected taking introns into consideration
R. The label is not radioactive
S. The targets form hairpins
(A) Q (B) P and Q
(C) Q and S (D) P, Q and R

1260. Which of the following statements(s) concerning the Edman degradation method is/are true?
P. Phenyl isothiocyanate reacts with the amino terminal residue
Q. If a protein has a blocked amino-terminal residue (as does N-formyl methionine, for example), it can not react with phenyl isothiocyanate
(A) P (B) Q
(C) P and Q (D) None of these

1261. ATP-independent chromatin remodeling enzymes mediate:
(A) Nucleosome sliding
(B) Transfer of a histone octamer from one DNA strand to another
(C) Weakening of the interaction between DNA and histone octamer
(D) Replacement of a histone with a histone variant

1262. Which combination of the following statements with regard to gene expression is true?
P. Heteronuclear RNA represents contiguous segment of genomic DNA
Q. UTRs are part of heteronuclear RNA but not of mRNA
R. UTRs are part of exons
S. Translation start codon must be in the first exon
(A) P – True, Q – False, R – False, S - False
(B) P – True, Q – False, R – True, S - False
(C) P – True, Q – True, R – False, S - True
(D) P – False, Q – True, R – True, S - True

1263. Double chained DNA strand is made radioactive in both its chains. It is allowed to replicate twice in non-radioactive medium. The result would be:
(A) All strands have radioactivity
(B) Half the strands have radioactivity
(C) Three strands have radioactivity
(D) Radioactivity is absent in all strands

1264. In 1923, the theory of QTL mapping was described by:
(A) Stern (B) Sax
(C) Mather (D) Galton

1265. The polymerase chain reaction or PR is a technique that:
(A) Was used to demonstrate DNA as the genetic material
(B) Is used to determine the content of minerals in a soil sample
(C) DNA primers and a thermostable DNA polymerase to replicate specific DNA sequences in vitro
(D) Measures theriosome transfer rate during translation

1266. Which is not correct about application of molecular markers?
(A) Molecular markers do not permit the identification of putative resistant plants in the absence of disease of disease test
(B) In resistance breeding, molecular markers can be very useful in gene pyramidin
(C) Molecular markers may be useful in identifying heterotic combinations
(D) MAS can be successfully used to select for such traits whose phenotypic evaluation is either problematic or costly

1267. Why is the gene inactivation is a useful technique to determine the function of a gene?
(A) Gene activation provides information about the expression of a gene
(B) Gene inactivation provides information about the cellular location of a gene product
(C) Gene inactivation provides an opportunity to identify phenotypic changes associated with the loss of the functional gene
(D) Gene activation provides information about the structure of the gene product

1268. In an experiment designed to determine the number of phage particles in a sample 56 plaques were recorded when 100 l of a 10^{-4} dilution was plated. How many phage are there per ml in the original?
(A) 5.6 x 10^3 (B) 5.6 x 10^4
(C) 5.6 x 10^5 (D) 5.6 x 10^6

1269. The lineal use of recombinant DNA technology is to have rapidly replicating bacteria produce large amounts of specific proteins (e.g., hormone). Expression of a eukaryotic gene in prokaryotes involves:
(A) A shine-Dalgarno (SD) sequence in mRNA
(B) Introns
(C) Regulatory elements upstream of the gene
(D) All of these

1270. Begining with 600 template DNA molecules, after 25 cycles of PCR, how many amplicons will be produced?
(A) 2 X 10^{10} (B) 2^{25}
(C) 600 X 25 (D) 2 X 10^{25}

1271. Which is not correct about cleaved amplified polymorphic sequences (CAPS)?
(A) This marker captures some of the advantages of RFLP assay
(B) It avoids the disadvantages of Southern blot analysis
(C) CAPS was initially developed for use in pea
(D) CAPS detects DNA sequence variation in terms of the lengths of DNA fragments generated by the restriction digestion of PCR products

1272. DNA helicase in *E. coli*:
(A) Moves in the direction opposite of replication fork
(B) Binds with template of the leading strand
(C) Is a hexameric protein with ATPase activity
(D) Catalyzes formation of primer

1273. Hybridization between mature mRNA and DNA an identify introns in eukaryotic genes. When mature mRNA produced by the insulin gene is hybridized with denatured chromosomal DNA, which of the following will most likely be observed?
(A) No hybridization would occur under any conditions
(B) Hybridization of mRNA would occur with random sections of chromosomal DNA
(C) Hybridization of mRNA with DNA would occur in a continuous stretch that is equal to the length of the mRNA
(D) Hybridization of mRNA with DNA would occur but with many ingle – stranded loops of DNA

1274. Amplified fragment length polymorphism (AFLP)
P. A PCR – based tool
Q. Uses restriction enzymes to cut genomic DNA
R. Involves ligation of adaptors to the sticky ends of the restriction fragments
S. Involves amplification using primers complementary to the adaptor
(A) P and Q (B) R and S
(C) P, Q and S (D) P, Q, R and S

1275. Which of the following statements are correct?
P. The complement fixation test is a relatively simple procedure used in serological microbiology
Q. In an agglutination test, a visible lump of particles appears to the observer
R. In the precipitation test, cells and large particles enter the reaction and clump together
S. The double-diffusion Oudin test utilizes the precipitation reaction as its basis
(A) P and Q (B) Q and S
(C) P, Q and R (D) P, Q, R and S

1276. Degeneracy of genetic code implies that:
(A) The codons degenerate after the synthesis of polypeptide chain
(B) More than one codon can code for one amino acid
(C) Some codons degenerate as they are not involved in coding for any amino acid
(D) One codon can code for more than one amino acid

1277. DNA replication in prokaryotes requires:
(A) Primase to make small primers containing deoxyribonucleotides
(B) DNA polymerase III to join okazaki fragments
(C) DNA polymerase alpha to synthesize DNA
(D) DNA polymerase I to cut out primers and fill in gaps in the lagging strand

1278. Mapping of human chromosomes:
(A) Has been restricted to the sex chromosomes because of small family sizes
(B) Proceeded much more successfully as large number of DNA markers became available
(C) Has determined that the number of linkage groups is about twice the number of chromosomes
(D) Has demonstrated that almost all of the DNA is involved in coding for genes

1279. Two purified DNA samples A and B contain equal number of basepairs. Each of these DNA samples has one site each for EcoR1 and BamHI restriction enzymes. Complete digestion with both the enzymes yielded 3 DNA bands and 2 DNA bands respectively for A and B upon electrophoresis of the digestion products. Which one of the following explains the observation?
(A) A is circular DNA and B is linear
(B) B is circular DNA and A is linear
(C) A is circular DNA and B could be linear or circular
(D) B is circular DNA and A could be linear or circular

1280. Which of the following statements about DNA ligage is/are incorrect?
P. It forms a phosphodiester bond between a 5'-hydroxyl and a 3' –phosphate in duplex DNA
Q. It requires a cofactor, either NAD^+ or ATP, depending on the source of the enzyme, to provide the energy to form the phsophodiester bond
R. It catalyzes its reaction by a mechanism that involves the formation of a covalently lined enzyme adenylate
S. It catalyzes its reaction by a mechanism that involves the activation of a DNA phosphate through the formation of a phosphoanhydride bond with AMP
(A) P and S (B) P and R (C) S (D) P

1281. Which problems is associated with RAPD process?
(A) It produces difficulties in reproducing RAPD patterns in different laboratories
(B) RAPD patterns are not reproducible due to impurities in DNA preparation
(C) RAPD patterns are not reproducible due to the use of small amount of DNA
(D) Both (A) and (B)

1282. Which approaches can be used for the mapping of genes for monogenic traits in crop plants?
(A) Bulked-segregant analysis of F_2 population
(B) Use of near isogenic lines
(C) Bulked segregant analysis of RILs
(D) All of these

1283. What would happen if the concentration of dideoxynucleotides was too high in a chain termination sequencing reaction?
(A) The reactions would yield very long molecules and there would be little sequence data close to the primer
(B) The reactions would yield very short molecules
(C) The reactions would not proceed as the high con centrations of the dideoxynucleotides would inhibit the DNA polymerase
(D) The fluorescence of the sequencing product would be too high and difficult to read

1284. The function of the 3' to 5' exonuclease activity of a DNA polymerase is to:
(A) Remove the 5' end of the polynucleotide strand that is attached to the template strand that is being copied
(B) Remove damaged nucleotide from the template strand during DNA synthesis
(C) Remove nucleotides from the ends of DNA molecules to ensure the generation of blunt ends
(D) Remove incorrect nucleotide from the newly synthesized strand of DNA

1285. Which populations ,au contain released varieties as well?
(A) F_2 (B) Backcross
(C) NIls (D) RILs

1286. Mapping of disease resistance genes has been based on:
(A) ISSR (B) RAPD
(C) Microsatellite (D) All of these

1287. Which of the following statement about the 3' poly (A) tall of mRNA is false?
(A) It is transcribed for DNA sequence
(B) It is added to the primary transcript in the nucleus
(C) It is essential for protein synthesis
(D) It contributes to the stability and lifespan of the mRNA

1288. Telomerase :
P. Is a ribonucleoprotein
Q. Participates in replication of linear dsDNA
R. Is species specific
S. Is absent in somatic cells
(A) P, Q and R (B) P, R and S
(C) P, Q, R and S (D) P

1289. T4 polynucleotide kinase is used for:
(A) Remove 5'-phosphates from DNA and RNA
(B) Transfers terminal phosphate groups from ATP to 5'-OH groups
(C) Adds homopolymer tails to the 3'end of DNA
(D) Used in mapping studies

1290. A microorganism when viewed under a compound microscope with objective of 40X and eye piece of 10X magnification measured 400 μ in length. The same microorganism when viewed under a dissecting microscope with 10X lens, would measure:
(A) 100 μ (B) 10 μ
(C) 40 μ (D) 400 μ

1291. In bacterial RNA synthesis, which of the following is the function of factor rho?
(A) Allow proper initiation of transcription
(B) Eliminate the binding of RNA polymerase to the promoter
(C) Participate in the proper termination of transcription
(D) Increase the rate of RNA synthesis

1292. The tertiary structure of protein is detected by:
(A) X-ray diffraction/crystallography
(B) Spectrophotometry
(C) Electrophoresis
(D) Chromatography

1293. Which of the following sets of factors are needed for translation termination at an amber codon (UAG) and release of the fir··hed polypeptide chain?
(A) RF2, RF3, EF-G and RRF
(B) RF1, RF3, EF-Tu and RRF
(C) RF1, RF3, EF-G and RRF
(D) RF2, RF3, EF-G and RRF

1294. DNA sequencing by the Sanger method takes advantage of which property of DNA synthesis to generate a sequencing ladder?
(A) Addition of nucleotides requires a free OH group on the 3' of the DNA strand
(B) DNA has a free 5' phosphoryl group
(C) DNA polymerase has proofreading capabilities
(D) U is a base only found in RNA

1295. Which of the following technique is used to inactivate a gene by altering the DNA encoding it?
(A) Homologous recombination
(B) Antisense nucleic acid blocks
(C) Antibody microinjection
(D) Introduction of dominant inhibitory mutants

1296. Which of the following types of DNA replication or repair systems is dysfunctional in individuals with xeroderma pigmentosum?
(A) Mismatch repair (B) Base excision repair
(C) Nucleotide excision repair (D) DNA helicase

1297. Which of the following statements are correct? Very long DNA fragments (>100 kb) from eukaryotic genomes:

P. Can be efficiently packaged in a γ phage vector
Q. Can be propagated in yeast artificial chromosomes
R. Must have cohesive ends for cloning
S. Can be separated by pulsed field gel electrophoresis techniques

(A) P and Q (B) P and S
(C) Q and R (D) Q and S

1298. What is the advantage of probing a transcriptome with mRNAa bound to ribosomes?

(A) Eukaryotic mRNA molecules are difficult to isolate unless complexed to a ribosome
(B) mRNA molecules isolated from ribosomes represent those that are actively being translated into proteins
(C) mRNA molecules isolated from ribosomes are more stable than other mRNA molecules
(D) mRNA molecules that are not being translated by ribosome still contain their intron sequences

1299. In selecting a DNA sequence to be used as a template for construction of PCR primers, many investigators look for sequences rich in codons for tryptophan or methionine because:

(A) Tryptophan and methionine have only one codon each (i.e., there is no wobble)
(B) Tryptophan and methionine have three codon each
(C) Codon for tryptophan and methionine abundantly present in mRNA
(D) None of these

1300. In methyl-directed mismatch repair in *E. coli*, the daughter strand containing the mismatched base nicked by:

(A) MutH-endonuclease (B) UvrAB-endonuclease
(C) AP-endonuclease (D) 3' to 5' exonuclease

1301. Reverse transcriptase requires the following for the conversion of a single-strand RNA into a double-strand DNA:

P. An RNA template Q. A primer
R. A DNA template S. All four dNTPs

(A) P and Q (B) Q and R
(C) P and S (D) P, Q and S

1302. Among the following which statement is correct regarding the promoter of the gene:

(A) It is always located upstream of transcription start site
(B) It is always located downstream of translation start site
(C) Different promoter elements are recognized by different RNA polymerase
(D) Promoters are recognized due to their different secondary structures

1303. Humen DNA (3 x 10^6 Kb) is replicated in 5 hrs at a rate of 1 Kb/min. The number off origins of replication utilized are:

(A) 1 (B) 3
(C) 300 (D) 10000

1304. Hydroxyurea blocks DNA replication by:

(A) Inhibiting DNA polymerase
(B) Inhibiting ribonucleotide reductase
(C) Inhibiting lagging strand DNA synthesis
(D) Inhibiting replicative helicase

1305. Which of the following is true about RFLP?

(A) Acts as genetic marker
(B) RLFPs have more than two alleles
(C) Used in case of FISH
(D) Both (A) and (B)

1306. Two double stranded DNA samples that are identical with respect to the number of base pairs, but differ significantly in their GC content, can be separated by:

(A) Density gradient centrifugation
(B) electrophoresis
(C) Dialysis
(D) Oligo-dT column chromatography

1307. In which of the following splicing events, the intermediates do not form the lariat or branched structures?

(A) Trans splicing
(B) Group I intron splicing
(C) Group II intron splicing
(D) Spliceosome mediated splicing

1308. The product of gene X is a cytosolic protein. A repressor of gene X expression is overexpressed in a cell line using recombinant DNA technology. After the overexpression, the cell line is observed to be susceptible to UV lights and higher mortality rates are observed upon short exposures to the UV. What could be the function of the gene X related to?

(A) Photosynthesis
(B) DNA repair
(C) DNA replication
(D) X- chromosome assembly

1309. Which of the following is not part of the pathway for nuclear pre-mRNA splicing?

(A) Transesterification where the 5' OH of the downstream exon attacks the donor junction
(B) Transesterification where the 2' OH of the branched point attacks the donor junction
(C) Recognition of the donor junction by the UI snRNP
(D) Dissociation of U4 from U6

1310. The minimum components of an artificial yeast chromosomes include all of the following except:

(A) All the histones except H1
(B) An autonomously replicating sequence (ARS)
(C) A centromere (CEN) sequence
(D) A telomere (TEL) sequence

1311. A DNA sample when it is diluted by a factor of 100 gives an optical density reading at 260 nm of 0.369. Assuming 50 μg/ml of DNA gives an absorbance reading of 1, the concentration of the original DNA sample is:

(A) 36.9 μg/ml (B) 1845 μg/ml
(C) 36900 μg/ml (D) 184.5 μg/ml

1312. In a eukaryotic gene, one of the introns is mutated at the 3' end as shown:
Normal : 5'... TTTCCCACCCTTAG 3'
Normal : 5'.... TTTCCCACCCTTCG 3'
Which of the following processes is this mutation most likely to affect?
(A) Capping (B) Hybridization
(C) Polyadenylation (D) Splicing

1313. Which of the following is/are true?
P. RNA polymerase needs general transcription factors to elongate the message
Q. Tissue –specific transcription factors bind to TATA box
R. Generalized transcription factors bind to proximal upstream of the promoters
S. C-terminus of RNA polymerase II is an important control point
(A) P only (B) R only
(C) Q and R (D) R and S

1314. Sequence tagged sites have which of the following properties?
(A) They are present only one within a genome and posesses an RFLP site
(B) They are present only one within a genome and their sequence is known
(C) Their sequence is known and they must contain repetitive DNA sequences
(D) They must contain the sequences of a gene, and no-repetitive DNA sequences can be present

1315. An alpha-helical conformation of a globular protein in solution is best determined by which of the following?
(A) Ultraviolet-visible absorbance spectroscopy
(B) Fluorescence spectroscopy
(C) Electron microscopy
(D) Circular dichroism

1316. Which of the following reagents would be useful for labeling the oligodeoxyribo-nucleotide d(GGATATCC)?
P. [γ- ^{32}P] ATP
Q. Polynucleotide kinase
R. DNA-dependent RNA polymerase
S. DNA ligase
(A) P and Q (B) R and S
(C) P and R (D) Q and S

1317. What is the principal function of reporter molecules in a DNA hybridization assay?
(A) Enhances the stringency of the hybridization reaction
(B) Binds the target DNA to the solid support
(C) Aids in base pairing
(D) Aids in the detection of robe-target hybrids

1318. Standard deviation of a data is 6. When observation I increased by 1, then the standard deviation of new data is:
(A) 5 (B) 7 (C) 6 (D) 8

1319. Which of the following statements is incorrect?
(A) Most of the eukaryotic mRNAs have a 7-methylguanosine cap at their 5' end
(B) TATA binding protein is involved in the synthesis of mRNA, tRNA and rRNA
(C) Histones have no other function except in chromatin organization
(D) Eukaryotic RNA polymerase II consists of more than 4 subunits

1320. One of the significant discoveries which allowed the development of recombinant DNA technology was:
(A) The discovery of antibiotics used for selecting trans formed bacteria
(B) The identification and isolation of restriction endo nucleases permitting specific DNA cutting
(C) The discovery of DNA and RNA polymerase allowing workers to synthesize any DNA sequence
(D) The development of the polymerase chain reaction

1321. Choose the correct statement(s):
P. GrI intron show homing
Q. GrII intron show retrohoming
R. Many GrI and GrII introns have open reading frame
S. GrI intron acts as riboswitch
(A) P and Q (B) Only S
(C) P, Q and R (D) P and S

1322. UV-spectroscopy is not used to quantitative the following compounds based on absorbance:
(A) Nucleic acids have absorbance peak at 260 nm
(B) NAD(P)H have absorbance peak at 340 nm
(C) Aromatic amino acids have absorption maxima at about 280 nm
(D) Sulfur containing amino acids have absorbance at 270 nm

1323. A leader sequence in an mRNA of eukaryotes can be found:
(A) After the 'stop' codon
(B) Between transcriptional start site and translational start site
(C) Within the first exon
(D) In the exon-intron boundaries

1324. If you were designing a method to specifically inhibits prokaryotic transcription, but not eukaryotic transcription, interfering with which of the following would work best?
(A) DNase activity
(B) Recognition of the promoter region
(C) An interalating agent
(D) Ribosomal binding

1325. Khorana et al synthesized RNA with copolymer of UGUGUGUGUG ...... It produced a peptide with alternatives cysteine and valine. The codons for the two are:
(A) UGU and GUG (B) UUG and GGU
(C) GUG and UGU (D) UGG and GUU

1326. Electrophoretic separation of RNA molecules based on their molecular weight, requires a denaturant such as formaldehyde to be present in the gel system, because:
(A) RNAs are sensitive to nucleases
(B) RNAs are single stranded
(C) RNAs are usually present in high amounts
(D) RNAs are usually small

1327. Large quantities of useful products can be produced through genetic engineering involving:
(A) Bacteria containing recombinant plasmids
(B) Yeast carrying foreign genes
(C) Transgenic plants
(D) All of these

1328. Match items in group 1 with correct options from those given in group 2:

Group 1	Group 2
P. VNTR sequence	1. Gene regulation on the same chromosome
Q. Leader sequence	2. Ribosome binding site
R. SD sequence	3. DNA finger printing
S. Cis-acting sequence	4. Functions in at tenuation

(A) P – 3, Q – 1, R – 4, S – 2
(B) P - 2, Q – 3, R – 1, S - 4
(C) P - 3, Q – 4, R – 2, S – 1
(D) P – 3, Q – 1, R – 2, S – 4

1329. Four identical PCR reactions were carried out in tubes named I, II, III and IV. Besides the usual mix of dNTPs, each of the tubes respectively contained γ-^{32}P dATP, β-^{32}P dATP, α-^{32}P dATP and α -^{32}P rNTP. Which one of the tubes will have radiolabeled PCR product?
(A) Tube I (B) Tube II
(C) Tube III (D) Tube IV

1330. What is the minimum length of time required for the synthesis of a 110 KDa protein in *E. coli*?
(A) About 96 sec (B) About 55 sec
(C) About 5 sec (D) More than 200 sec

1331. With reference to protein synthesis, consider the following events:
P. Activation of amino acid
Q. Peptide bond formation
R. Formation of amino acid tRNA complex
S. Binding of aminoacyl tRNA complex with mRNA
What is the correct sequence of the above events in protein synthesis?
(A) P, R, Q, S (B) R, P, S, Q
(C) R, P, Q, S (D) P, R, S, D

1332. Some compounds called anti-inducers bind to repressors such as the *lac* repressor and inhibit the action of inducers that is, transcription is repressed and higher concentrations of inducer are required to induce transcription. Anti-inducers
(A) Compete for binding to the repressor
(B) Compete for binding to the operator
(C) Compete for binding to the RNA pol
(D) Compete for binding to the activator

1333. A viral genomic DNA that has two HindIII sites and two EcoR1 sites, was cut with both of the enzymes simultaneously for 5 hrs at 37^{0}C. The analysis of the restriction enzyme digestion was made of a 1% agarose gel. How many bands would one expect to see on the gel?
(A) 3 (B) 4 (C) 5 (D) 16

1334. What is genomic library?
(A) A collection of recombinant molecules with inserts that contain all of the genes of an organism
(B) A collection of recombinant of recombinant molecules with inserts that contain all an organism's genome
(C) A collection of recombinant molecules that express all of the genes of an organism
(D) A collection of recombinant molecules that have been sequenced

1335. An enhancer element shows all of the following properties except:
(A) Acts as a binding site for transcription factors
(B) Acts as a binding site for RNA polymerase
(C) Is functional when located upsteam or downstream of the transcription initiation site
(D) Is functional when located in an intron

1336. Choose the false statement
(A) Peptidyl-tRNA binds in the P-site
(B) Shine-Dalgarno is a polypurine sequence that helps in ribosome binding
(C) Eukaryotic mRNAs are mostly monocistronic
(D) Translocation of ribosome requires ATP hydrolysis

1337. Metaphase chromosomes were initially used for flourescent in situ hybridization, but the results were somewhat limiting because:
(A) Many regions of a chromosome are condensed and cannot hybridize to probes
(B) The probes will hybridize preferentially to repeated sequences present on multiple chromosome
(C) The chromosomes are not stable in the condensed stable, and the signal diffuses when the chromosome are relaxed
(D) Only low-resolution mapping is possible, as the chromosomes are condensed

1338. The coefficient of correlation r satisfies:
(A) 0 < r < 1 (B) I Γ I >1
(C) I Ã I =1 (D) - 1 < r < 0

1339. A restriction enzyme with a four-base recognition site would leave DNA with a statistical frequency of one every:
(A) 256 base pairs (B) 4 base pairs
(C) 4096 base pairs (D) 65.5 kb

1340. Simple tendem repeat polymorphisms in humans are most useful for:
(A) Solving criminal and paternity cases
(B) Reconstructing the relationships of humans and chimps
(C) Estimating relationships of humans and Neander thals
(D) Transferring disease resistance factors into bone marrow cells

1341. The DNA mixture containing unlabelled and N^{15} labelled DNA from phage T4 was denatured and allowed to reanneal. How many bands one would observe upon $CsCl_2$ density gradient centrifugation of the above mixture?
(A) 0 (B) 1 (C) 2 (D) 3

1342. Suppressor tRNA differ from normal tRNA in that:
(A) They are similar in size
(B) They do not carry an amino acid
(C) They have to be charged with a special aminoacyl-tRNA synthetase
(D) They do not obey regular genetic code

1343. How is the latiat structure formed during splicing of a GU-AG intron?
(A) After cleavage of the 5' splice site, a new phosphodiester bond is formed between the 5' nucleotide and the 2' carbon of the nucleotide at the 3' splice site
(B) After cleavage of the 5' splice site, a new phosphodiester bond is formed between the 5' nucle otide and the 2' carbon of an internal adenosine
(C) After cleavage of the 3' splice site, a new phosphodiester bond is formed between the 5' nucleotide and the 2' carbon of the nucleotide at the 5' splice site
(D) After cleavage of the 3' splice site, a new phosphodiester bond is formed between the 5' nucleotide and the 2' carbon of an internal adenosine

1344. Choose the mismatch:
(A) Pol I – primer removal and DNA repair
(B) Pol III- DNA replication
(C) Pol V – Translesion replication
(D) Pol γ – Nuclear DNA replication

1345. In *E. coli*, attenuation and antitermination utilize which structure?
(A) Stem loop structures in RNA
(B) Stem loop structures in DNA
(C) RNA/DNA hybrids
(D) Differential protein folding

1346. A mixture containing two proteins having similar molecular mass but different oligomeric properties can be separated by:
(A) SDS-PAGE analysis
(B) Native PAGE analysis
(C) Isoelectric focusing
(D) Both (B) and (C)

1347. During mismatch repair in *E. coli*, the parental strand is recognized by:
(A) Single-stranded breaks
(B) Glycosylated adenines
(C) Methylated adenines
(D) Methylation of the 6th position of guanine residues

1348. Match the item in group I with correct options in group II

Group I	Group II
P. DNA footprinting	1. Protein-protein interaction
Q. Yeast two-hybrid system	2. VNTR
R. DNA fingerprinting	3. DNA binding protein
S. SAGE	4. Transcriptome analysis

(A) P -1, Q – 2, R – 4, S – 3
(B) P -3, Q – 1, R – 2, S - 4
(C) P -3, Q – 4, R – 1, S – 2
(D) P -4, Q – 2, R – 1, S – 3

1349. Selenocystein:
(A) Only presents in prokaryotes
(B) Coded by UGA (stop codon)
(C) Derives from cysteine
(D) Forms post translationally

1350. Which of the following could not possibly give rise to a restriction fragment length polymorphism (RFLP)?
(A) A missense mutation within the protein coding region of a gene
(B) A silent mutation within the protein coding region of a gene
(C) A single base change within the intron sequence of a gene
(D) An error in RNA splicing that mistakenly removes an exon during RNA processing

1351. Histones have very high percentage of basic amino acids (15-30%). For this class of proteins which of the following reagents would be a suitable choice for generating peptide in determination of amino acid sequence of the protein:
(A) Cyanogen bromide
(B) Chymotrypsin
(C) Trypsin
(D) β – bromosuccinamide

1352. RNA interference works by which of the following methods?
(A) Using antisensce RNA molecules to block the translation of mRNA molecules
(B) Using RNA polymerase inhibitors to block the transcription of specific genes
(C) Using short, double-stranded RNA molecules that will cause the degradation of an mRNA molecule
(D) Using modified tRNA molecules to block the translation of mRNA molecules

1353. In Scanning Electron Microscope (SEM), to form an image of the specimen?
(A) Electron should pass through the specimen
(B) Electrons are scattered from the surface of the specimen
(C) A thin film of heavy metal is evaporated
(D) Specimens are stained

1354. siRNA interfere at:
(A) Transcriptional level
(B) Post-transcriptional level
(C) DNA replication level
(D) Translational level

1355. Which of the following techniques yields the most information about the positional and spatial characteristics of all atoms in a protein?
(A) Automated Edman analysis
(B) Circular dichroism spectroscopy
(C) Magnetic resonance spectroscopy
(D) Reversible unfolding/refolding

1356. Restriction endonucleases from two different organisms that recognize the same DNA sequence for cleavage are called:
(A) Isoschizomers (B)
(C) Concatemers (D) Palindromes

1357. Which of the following is a role of gRNA?
(A) RNA editing
(B) Self splicing
(C) Chemical modification of rRNA
(D) Polydenylation

1358. In a bacterial cell, a mutation in an aminoacyl-tRNA synthetase leads to charging of the entire $tRNA^{Ser}$ population with alanine. Which of the following describes the result of using these aminoacyl tRNAs for protein synthesis in the cell?
(A) Protein synthesized using the alanyl-$tRNA^{Ser}$ will contain neither alanine nor serine
(B) Protein synthesized using the alanyl-$tRNA^{Ser}$ will contain only serine where alanine would normally occur
(C) Protein synthesized using the alanyl-$tRNA^{Ser}$ will contain only alanine where alanine would normally occur
(D) Protein synthesized using the alanyl-$tRNA^{Ser}$ will randomly contain either alanine or serine where serine would normally occur

1359. Which is not correct about RILs?
(A) RILs are created by SSD through at least five generations of continued selfing
(B) Continued selfing yields a set of lines each of which contains a different combinations of linkage blocks from the original parents
(C) The different linkage blocks in RILs provide a basis from linkage analysis
(D) RILs are, theoretically, mortal

1360. PCR based DNA amplification is an essential feature of which of the following combination of molecular markers?
(A) RLFP, AFLP and SSR
(B) AFLP, SSR and RAPD
(C) RFLP, RAPD and SSR
(D) RAPD, RFLP and SSR

1361. Dominant markers supply as much information as codominant markers in :
(A) Doubled haploids
(B) One of the two backcross populatons in coupling phase
(C) RILs
(D) All of these

1362. Which of the following characteristics of eukaryotic genes is also commonly true of most prokaryotic genes?
(A) Genes are packaged into nucleosomes in chromatin
(B) Many genes contain introns
(C) Usually only one gene is contained on a given mRNA
(D) The initiation codon for protein synthesis is usually AUG

1363. An open reading frame is:
(A) All of the nucleotides of a gene that are transcribed into mRNA
(B) All nucleotides of a gene that make up the codons specifying amino acids
(C) The nucleotides of an mRNA molecule before the introns have been removed
(D) The amino acid sequence of a polypeptide

1364. What is the order of the following steps in Western blotting?
P. Protein denaturation
Q. Hydrogen peroxide reduction
R. Primary antibody binding
S. Transfer onto membrane
(A) P, Q, R, S (B) Q, P, R, S
(C) P, Q, S, R (D) P, S, R, Q

1365. Which of the following elements must be present in a plasmid cloning vector?
P. Origin of replication
Q. Selectable marker
R Unique restriction sites
S. Promoter element upstream of the unique restric tion site
(A) S and P (B) Q and R
(C) R and S (D) P, R and Q

1366. Which of the following statement are incorrect?
P. The frequency of recombination is a measure of linkage between genes on the same chromosome
Q. DNA polymerase I is the true DNA replicase in E. coli
R. The conserved element closest to the transcription initiation site is called the CAAT box
S. The introns in the nuclear pre-mRNAs are excised by spliceosomes
(A) P and Q (B) Q and R
(C) P and R (D) P and S

1367. RFLP analysis is a technique that:
(A) Uses hybridization to detect specific DNA restriction fragments in genomic DNA
(B) Is used to determine whether a gene is transcribed in specific cells
(C) Measures the transfer frequency of genes during conjugation
(D) Is used to detect genetic variation at the protein level

1368. Aminoacyl-tRNA synthetases must be capable of recognizing which of the following:
(A) A specific tRNA and a specific amino acid
(B) A specific rRNA and a specific amino acid
(C) A specific tRNA and the 40S ribosomal subunit
(D) A specific amino acid and the 40S ribosomal subunit

1369. In *E. coli*, which of the following codons are reorganized by the release factor RF1?
(A) UAG and UGA (B) UAA and UGG
(C) UAG and UAA (D) UAG and UUA

1370. Which of the following statements is/are correct about leucine zipper motif?
P. A DNA binding motif
Q. Homodimer or heterodimer
R. Present in transcription factors-Fos and Jun
S. Present in steroid recepters
(A) P only (B) P, Q and R
(C) Q, R and S and R (D) P and S

1371. Which one of the following transcription factor is involved in promoter recognition in eukaryotes?
(A) TFIIF (B) TFIIB
(C) TFIID (D) TFIIE

1372. Which of the following is not correct about freeze-fracture technique?
(A) It involves physical breaking of frozen biological samples
(B) Fixation in glutaraldehyde
(C) Involves vacuum sublimation of ice
(D) Cryoprotection with glycerol

1373. The chronological order in which the following landmark discoveries in molecular biology were made is:
(A) Recombinant DNA technology, DNA sequencing, polymerase chain reaction
(B) DNA sequencing, recombinant DNA technology, polymerase chain reaction
(C) Polymerase chain reaction, DNA sequencing, Recombinant DNA technology,
(D) Polymerase chain reaction, Recombinant DNA technology, DNA sequencing,

1374. When change in the phenotype or gene expression occur without changes in the underlying DNA sequence, the phenomenon is called:
(A) Mutation (B) Eugenics
(C) Epigenetics (D) Epistasis

1375. Klenow fragment is a key enzyme in molecular biology research. Which one of the following statements is not correct for the Klenow fragment isolated from *E. coli?*
(A) Klenow fragment is a constituent of DNA polymerase I
(B) Klenow fragment has DNA polymerase activity
(C) Klenow fragment has 3' to 5' exonuclease activity but it is devoid of the 5' to 3' exonulease activity
(D) While Klenow fragment is primarily associated with DNA repair function, DNA polymerase I is essential for DNA replication

1376. Edmann degradation is used for the:
(A) Determination of amino acid sequence from the N-terminus of a protein
(B) Nucleotide sequence of DNA
(C) Determination of amino acid sequence from the C-terminus of a protein
(D) Determination of RNA structure

1377. On average, how many fragments would a restriction enzyme which recognizes a specific 4 base sequence in DNA be expected to generate by leaving a double-stranded bacteriophag DNA with a genome size of 5000 bp into?
(A) About 2 (B) About 4
(C) About 20 (D) About 50

1378. Match the terms in group I with terms in group II

Group I	Group II
P. RNase-P	1. Lac operon
Q. Leucine zipper	2. rRNA gene transcription
R. RNA pol-I	3. tRNA gene transcription
S. Attenuation	4. Transcription factors
	5. Ribozymes
	6. trpoperon
	7. mRNA splicing

(A) P -7, Q – 5, R – 3, S - 1
(B) P -4, Q – 5, R – 2, S - 1
(C) P -5, Q – 4, R – 2, S - 6
(D) P -4, Q – 5, R – 3, S – 6

1379. The large (Klenow) fragment of *E. coli* DNA polymerase I contains which of the following activities?
(A) Reverse transcriptase activity and nick-translation activity
(B) Polymerase activity and 3'-5'- exonuclease activity
(C) Polymerase activity and nick-translation activity
(D) 5' – 3'- exonuclease activity

1380. The catabolite activating protein (CAP):
(A) Binds to allolactose
(B) Interacts with the â–subunits of RNA polymerase
(C) Binds the operator sequence of the la operon
(D) Causes a bending in the DNA near the start site of transcription

1381. Which of the following statements is false?
(A) β-mercapteothanol acts as reducing agent and used to break disulfide bonds
(B) Bromophenol blue used as tracking dye during electrophoresis
(C) Coomassie brilliant blue used for staining proteins
(D) Sodium dodecyl sulphate is a cationic detergent

1382. The yeast two-hybrid system is designated to identify which of the following?
(A) All of the components of a multiprotein complex
(B) Human proteins that are required for binding RNA polymerase
(C) Two proteins that directly interact with one another
(D) Two proteins that are involved in the same metabolic pathway

1383. What phenotype would you except for a mutation that eliminates the tetramerization domain of *lac* repressor:
(A) Recessive, inducible (B) Recessive, constitutive
(C) Dominant, inducible (D) Dominant, constitutive

1384. Wobble between a codon and an anticodon occurs between:
(A) The first nucleotide of the codon and the first nucleotide of the anticodon
(B) The first nucleotide of the codon and the third nucleotide of the anticodon
(C) The third nucleotide of the codon and the first nucleotide of the anticodon
(D) The third nucleotide of the codon and the third nucleotide of the anticodon

1385. A homogeneous protein of native molecular weight 100000 gave a single band of molecular weight 50000 on SDS-PAGE in presence of â– mercaptoethanol. N-terminal analysis gave two amino acids alanine and leucine in equal proportions. Hence the:
(A) Protein is a homodimer
(B) Protein is contaminated with another protein
(C) Protein has two polypeptides linked by disulfide bridges
(D) None of these

1386. Which of the following is characteristics of eukaryotic transcription but not prokaryotic transcriptions?
(A) Transcription and translation are coupled
(B) There is one form of RNA polymerase
(C) RNA polymerase II is responsible for synthesis of mRNA
(D) RNA can be polycistronic

1387. SDS-polyacrylamide gel electrophoresis and the isoelectric-focusing method for the separation of protein have which of the following characteristics in common? Both
(A) Separate native proteins
(B) Separate proteins according to their mass
(C) Required a pH gradient
(D) None of these

1388. Microsatellites are used more commonly than minisatellite as DNA markers because:
(A) Minisatellites are present in too many locations within the genomes
(B) Restriction enzymes can be used to type microsatellites but not minisatellite
(C) There are very few microsatellites in eukaryotic genome so they are easily identified and analyzed
(D) Microsatellites are present throughout eukaryotic genomes and are easily amplified using PCR

1389. In eukaryotic transcription by RNA polymerase II, formation of a pre-initiation complex:
(A) Begins with the binding of TBP to the TATA box of the promoter
(B) Involves the ordered addition of several transcription factors and the RNA polymerase
(C) Allows an ATP-dependent opening of the two strands of DNA
(D) All of these

1390. Antibiotic inhibiting interaction between tRNA and mRNA during protein synthesis in bacteria is:
(A) Tetrayline (B) Neomycin
(C) Erythromycin (D) Streptomycin

1391. If the *E coli* cells are grown in a medium containing both glucose and lactose, what is likely to happen?
(A) Both the sugars would be utilized simultaneously
(B) The culture will exhibit synchronous growth
(C) Lactose will be utilized first followed by glucose
(D) Glucose will be utilized first followed by lactose

1392. To join together two Okazaki fragments during DNA replication in *E. coli,* which of the following sets of enzymes are used?
(A) Pol I (5'→3' exonuclease), Pol I (polymerase), ligase
(B) Pol III (5'→3' exonuclease), Pol III (polymerase), ligase
(C) DNA gyrase, Pol III (polymerase), ligase
(D) SSB, Pol I, ligase

1393. Which of the following is true about poly (A) tail?
P. It is synthesized post-transcriptionally
Q. It is usually shorter than 300 nt
R. One of its functions is to protect the mRNA from cytoplasmic RNases
S. One of its functions is to stimulate translation
(A) P only (B) Q and R
(C) Q and S (D) P, Q, R and S

1394. In vertebrate genes, transcription regulatory regions that contain CpG islands are inactivated by which CpG modification?
(A) Methylation (B) Myristylation
(C) Phosphorylation (D) Acetylation

1395. Approximate molecular weight (KDa) of the product after translation of a 390 bases mRNA will be:
(A) 48 (B) 26 (C) 39 (D) 14

1396. A primer with the sequence ATCGGTAACGATTACATTC is to be used in a PCR reaction. Using the formula: $T_m = 2^0 (A + T) + 4^0 (G + C)$, what is the T_m of this oligonucleotide?
(A) 42^0 C (B) 52^0 C (C) 20^0 C (D) 62^0 C

1397. What is the purpose of performing a homology search with a DNA sequence
(A) To determine if any genes with similar sequences are present in the DNA databases
(B) if the sequence is already in the DNA database
(C) To search for consensus exon-intron boundaries
(D) To determine the codon bias for a specific gene

1398. A specific aminoacyl-tRNA always carries the same amino acid that is attached by a specific amioacyl-tRNA synthetase. Regions on tRNA that are required for the specificity of the enzyme is/are the:
(A) Anticodon and aceptor region
(B) D loop and acceptor region
(C) Anticodon alone
(D) The TØC loop and anticodon

1399. Which of the following is required for the initiation of translation in prokaryotes?
(A) The mRNA base pairing with the 3' end of 23S rRNA
(B) The Kozak sequence positioning
(C) Met –tRNA binding to the A site on the ribosome
(D) GTP bound to initiation fator 2 (IF 2)

1400. Gene conservation is observed when:
(A) During meiosis, one allele is replaced by another allele
(B) During meiosis, the expected 2:2 ratio of alleles is replaced with a 4:0 ratio
(C) During meiosis, the expected 2:2 ratio of alleles is replaced with a 3 : 1 ratio
(D) All of these

1401. Which of the following reagents would be useful for visualizing DNA restriction fragments that have been separated by electrophoresis in an agarose gel and remain in the wet gel?
(A) Ethidium bromide (B) $[\alpha\text{-}^{32}P]$ ATP
(C) Diphenylamine (D) $^{32}P_i$

1402. If a mutation occurred in the phosphorylation site in the CTD domain of RNA polymerase II, which of the following outcomes would be observed?
(A) Transcription of tRNA would occur normally
(B) Transcription of mRNA would not occur
(C) Transcription of precursor rRNA would not occur
(D) All of these

1403. How do eukaryotic ribosomes identify the start sites of proteins?
(A) Via their Shine-Dalgarno sequences
(B) By scanning from the 5' end of the mRNA for the first AUG
(C) The 'cap' structure of a eukaryotic mRNA overlaps with the first AUG
(D) The transcription start site of a eukaryotic mRNA overlaps with the translational start site

1404. After isolating and purifying to homogeneity a small enzyme (110 amino acids long) from a culture of bacteria, you are confused as to whether you grew wild-type bacteria or a mutant strain that produced the enzyme with a valine residue at position 66 instead of the glycine found in the wild-type strain. For quick determination of nature of protein you will use:
(A) Mass spectroscopy
(B) Ion exchange chromatography
(C) SDS-PAGE
(D) HPLC

1405. A mixture of proteins (W, X, Y, Z) elute from Sephadex G-200 column in the order W, X, Y, Z. The protein with maximum electrophorestic mobility on SDS-PAGE will be:
(A) W (B) (C) Y (D) Z

1406. Temperature probe is an essential component of:
(A) Centrifuge (B) Autoclave
(C) Vortex shaker (D) pH meter

1407. Which one of the following method cannot be used for quantitative assay of transgenes?
(A) RT-PCR (B) Southern
(C) PCR (D) Q-PCR

1408. Which of the following does not involve radio activity?
(A) AFLP (B) Southern
(C) PCR (D) Northern

1409. Individual loci controlling a complex trait is known as:
(A) Qualitative Trait Locus
(B) Quantitative Trait Locus
(C) Alleles
(D) Single locus

1410. Which one of the following is used for amplifying DNA samples?
(A) Gel documentation
(B) PCR
(C) Speed vac
(D) Thermo mixture comfort

1411. Role of macocaptoethanol in DNA finger printing technology is:
(A) To break cell wall
(B) To distrupt cell membrane
(C) To prevent phenolic oxidation
(D) To prevent cell membrane

1412. TE stands for:
(A) tri ethylene (B) Tris ethylene
(C) Tris EDTA (D) Triester

1413. For DNA quantification, the equipment used is:
(A) Nano drop (B) Thermo mixture
(C) Speed vac (D) PCR

1414. The extraction buffer used in analysis of seed protein by PAGE method is:
(A) Tris HCl (B) SDS
(C) Mercapto ethanol (D) tri ethylene

1415. For evaporating liquid from DNA samples, the following equipment is used:
(A) Speed vac (B) Water bath
(C) Microwave oven (D) Nano drop

1416. Which of the following characteristics is undesirable in cloning vectors?
(A) Readily isolated from cells
(B) Higher copy number
(C) Vulnerable at several sites to a restriction enzyme
(D) Small in size

1417. SmaI and XmaI are neoshizomers. A circular DNA has two SmaI and one XmaI sites. How many fragments will be generated by digestion with XmaI?
(A) 1 (B) 2 (C) 3 (D) 4

1418. Which is not correct about marker-assisted selection?
(A) Use of linked molecular markers would allow selection for desirable traits in early segregating generations at the seedling stage independent of environment influences
(B) Marker-assisted selection saves time and other resources needed for raising large segregating populations for several generations
(C) Indirect selection is expected to be less effective than the direct selection due to the absence of confounding effects of the environment in the case of latter
(D) Indirect selection allows pyramiding (accumulation) of genes for characters like disease resistance, which is difficult through the use of conventional methods of plant breeding

1419. A set of genes from *Bacillus subtillis* that encode the proteins required for sporulation have conserved DNA sequenes -35 and -10 nucleotides before the site of transcript initiation, although the sequence at -35 is different from that seen in most other genes from that species. Which of the following best explains this differences?
(A) A noval sigma factor is required for transcription initiation at these genes
(B) The -35 sequene is the binding site for a repressor of transcription
(C) The replication of these genes requires a specifically modified DNA polymerase
(D) Translation of the mRNAs transcribed from these genes requires specific ribosomes that recognize a modified shine-Dalgarno sequence

1420. It is the early 1960s. In an attempt to elucidate the genetic code you have performed a series of experiments similar to those of Khorana. You have prepared synthetic mRNA and used *E. coli* cell extracts to translate them into corresponding polypepetides. You have then characterized the amino acid composition of the protein. Your results are as follows:

mRNA	Amino acid components
5'UUUUUUUUU....3'	Phenylalanine
5'AUAUAUAUA....3'	Isoleucine, tyrosine
5'UAUUAUUA....3'	Asparagine, Isoleucine
5'AAAAAAAAA...3'	Lysine
5' GUAGUAGUA....3'	Serine, Valine
5'AUUAUUAUU....3'	Isoleucine, Leucine, Tyrosine

Based on the above information, the codon for isoleucine is
(A) AUA (B) UAA (C) AUU (D) UAU

1421. Enhancer elements are known to increase the rate of transcription when present at the upstream side of the promoter sequences. If the same enhancer element is placed in reverse orientation:
(A) The rate of transcription increases as compared to upstream side
(B) The rate of transcription decreases as compared to upstream side
(C) The rate of transcription remains same as compared to upstream side
(D) No transcription is observed

1422. How many loci can be detected from a single cross through AFLPs?
(A) >25 (B) > 400
(C) >70 (D) > 1000

1423. The methods which reveal genetic variability in protein composition are:
(A) Electrophoresis (B) HPLC and Electrophoresis
(C) Peroxidase test (D) Peroxidase test and HPLC

1424. The loss of DNA is referred as:
(A) Physical deterioration
(B) Physiological deterioration
(C) Biochemical deterioration
(D) Molecular deterioration

1425. Cryopreservation is used for storage of the following:
(A) Seed (B) Embryo
(C) Pods (D) Fruits

1426. Which activity of the cell is stopped during cryopreservation?
(A) Enzyme (B) Biochemical
(C) Metabolic (D) Respiration

1427. Cryopreservation is:
(A) Ethylene glycol (B) Calcium chloride
(C) Manganese sulphate (D) Manganese sulphide

1428. A place or storage in which seeds of rare plants or obsolute varieties are kept in order to prolong their viability is called:
(A) Plant bank (B) Seed bank
(C) Bureau bank (D) Cryo bank

1429. Kew garden is located at:
(A) USA (B) Mexico
(C) England (D) Germany

1430. Which institute is responsible for the germplasm collections of paddy:
(A) IRRI (B) CIMMYT (C) ITTA (D) AVRDC

1431. Which institute is responsible for the germplasm collections of sweet potato:
(A) IRRI (B) CIMMYT
(C) ITTA (D) AVRDC

1432. Thermodynamic predictions are based on:
(A) Stabilities of reactants
(B) Stabilities of products
(C) Both (A) and (B)
(D) Amounts of enzymes

1433. Which of the following techniques, based on migration of DNA fragments in a gel in the presence or absence of proteins, is used to identify proteins that bind to DNA?
(A) Nuclear magnetic resonance spectroscopy
(B) Gel retardation
(C) Nuclease protection
(D) DNA footprinting

1434. For DNA extraction, the suitable method is:
(A) Liquid nitrogen (B) cTAB method
(C) Minipep method (D) Nano drop

1435. Easy method for detection of GM crop is:
(A) ELISA method (B) Strip test method
(C) Finger printing method (D) Visual method

1436. ELISA plates are coated with:
(A) Enzymes (B) Lipids
(C) Antibody (D) Antigen

1437. Dongley key is an essential part of:
(A) Gel documentation unit (B) Fume hood
(C) Magnetic strirrer (D) Incubator

1438. The specific test recommended for varietal identification in sorghum is:
(A) KOH bleach test (B) Phenol test
(C) Fluorescent test (D) Vigour test

1439. Phenol test is recommended for varietal identification of:
(A) Paddy (B) Groundnut
(C) Blackgram (D) Brinjal

1440. The primary products of genes specific to each variety:
(A) Proteins (B) Enzymes
(C) Lipids (D) Carbohydrate

1441. The seed appendages are useful for:
(A) Variatal identification (B) Seed dispersal
(C) Seed development (D) Seed maturation

1442. In GM crops, many number of the transgenes can be analyzed by:
(A) Isoenzyme analysis (B) Urea Page
(C) Agarose gel (D) Q-PCR

1443. The function of cTAB is to:
(A) Distrupt the cell membrane
(B) Separate proteins
(C) Separate RNA
(D) Prevant phenolic oxidation

1444. Codon bias refers to which of the following?
(A) Some codons for an amino acids are more frequently used in all species
(B) Same amino acids are rarely used in the proteins of some organisms
(C) Some codons for an amino acids are more frequently used and the bias varies in different species
(D) Some codons code for rare amino acids such as salenocysteine in some species

1445. Which of the statement are correct? Spliceosomes
P. Are composed of RNA and protein
Q. Recognize RNA sequences that signal for removal of introns
R. Can produce different mRNA molecules by splicing at alternate site
(A) P and Q (B) P and R
(C) Q and R (D) P, Q and R

1446. The single primer amplification reaction (SPAR) is also called as:
(A) MP-PCR (B) RAMP
(C) SSLP (D) All of these

1447. All of the following cell types contain the enzyme telomerase which protects the length of telomeres at the end of chromosomes except:
(A) Germinal (B) Somatic
(C) Haemopoetic (D) Tumour

1448. The Shine-Dalgarno sequence is responsible for:
(A) Binding of RNA polymerase to gene during transcription
(B) Binding of DNA polymerases to origin of replication during DNA replication
(C) Binding of ribosomes to mRNA during initiation of translation
(D) Binding of snurps during splicing

1449. Which is not correct about sequence tagged microsatellite sites (STMS)?
(A) STMS markers are dominant
(B) These markers are extremely polymorphic and highly informative
(C) The most important uses for STMS markers are tagging genes for economic traits and for the study of gentic diversity
(D) STMS describes such microsatellite based markers that are analysed by PCR amplification of a short genomic region containing the repeated sequence

1450. A plasmid that will express a eukaryotic protein x in *E. coli* should contain:
P. Origin of replication
Q. Bacterial promoter
R. Mammalian promoter
S. Genomic sequence of gene
T. Genomic sequence of gene x
U. DNA sequence of gene x
(A) Q, S and T (B) P, R, S and T
(C) P, R, S and U (D) P, Q, S and U

1451. X-ray crystallography are:
P. Based on the elastic scattering of X-rays
Q. Non-destructive analytical techniques
R. Used in determining the arrangement of atoms within a crystal
S. Based on Bragg's law
(A) P and Q (B) R and S
(C) P, Q and R (D) P, Q, R and S

1452. Sequence tagged microsatellite sites (STMS) can be used as markers for:
(A) Identification of germplasm and analysis of gene pool varieties
(B) Cultivar identification and determination of hybrid purity
(C) Linkage mapping and diagnostic markers for traits of economic value
(D) All of these

1453. Which of the following event does not occur during pre-tRNA processing?
(A) 5' end cleavage by the RNaseP
(B) Addition of poly(A) site
(C) Addition of A sequence at 3' end
(D) Chemical modification of bases

1454. Proteins are separated in SDS-electrophoresis on the basis of their:
(A) Molecular size (B)Charge
(C) Amino acid composition (D) N-terminal sequence

1455. An RNA-dependent RNA polymerase is likely to be present in the virion of a:
(A) DNA virus that multiplies in the cytoplasm
(B) DNA virus that multiplies in the nucleus
(C) Minus-strand RNA virus
(D) Transforming virus

1456. Which is not correct about application of molecular markers?
(A) Molecular markers are very useful in unequivocal identification of plant varieties
(B) Molecular markers can not be used for elimination of undesirable genes from segregating population
(C) Molecular markers are used for mapping of QTLs
(D) They allow construction of high density linkage maps since the number of markers available per cross is very large

1457. The predictive value of molecular markers used in marker-assisted selection depends on :
(A) Their inherent repeatability
(B) Map position
(C) Linkage with economically important traits
(D) All of these

1458. During DNA replication significant proportions of newly synthesized DNA in the lagging strand exists as small Okazaki fragments. The size of these units in bacteria are approximately:
(A) 100 nucleotides (B) 1000 nucleotides
(C) 100 base pairs (D) 1000 base pairs

1459. What is the correct order of the following steps in protein synthesis?

1. A peptide bond is formed
2. The small ribosomal subunit is loaded with initiation factors, mRNA and initiation aminoacyl-tRNA
3. The intact ribosome slides forward three bases to read a new codon
4. The primed small ribosomal subunit binds with the large ribosomal subunit
5. Elongation factors deliver aminoacyl-tRNA to bind to the A site

(A) 1, 2, 5, 4, 3 (B) 2, 3, 4, 5, 1
(C) 4, 5, 1, 2, 3 (D) 2, 4, 1, 3, 5

1460. Three of the four eukaryotic rRNAs are synthesized from a single transcription unit consisting of the rDNA which one of the following does not belong to this group?
(A) 5.8S (B) 5S (C) 18S (D) 28S

1461. A solution contains DNA polymerase I, Mg^{++} salts of dATP, dGTP, dCTP and dTTP, and an appropriate buffer. Which of the following DNA molecules would serve as a template for DNA synthesis when added to this solution?
(A) A single –stranded closed circle
(B) A single-stranded closed circle base-paired to a shorter strand with a 3-terminal hydroxyl
(C) A single-stranded closed circle base-paired to a shorter strand with a 3-terminal phosphate
(D) A double-stranded closed circle

1462. The suitability of RFLP for elucidation of QTLs has been shown in :
(A) Tomato (B) Maize
(C) Potato (D) All of these

1463. Which of the most critical in marker-assisted selection (MAS)?
(A) Their inherent repeatability
(B) Map position
(C) Linkage with economically important traits
(D) All of these

1464. Which of the following is not a difference between a genomic library and a cDNA library?
(A) The presence or absence of introns in cloned genes
(B) The presence or absence of promoters in cloned genes
(C) The presence or absence of a prokaryotic origin of replication on the vector
(D) The presence or absence of DNA from intergenic regions

1465. A multiple cloning site:
(A) Contains many copies of a cloned gene
(B) Allows flexibility in the choice of restriction enzymes for cloning
(C) Allows flexibility in the choice of organism for cloning
(D) Contains many copies of the same restriction en zyme site

1466. Apart from the 20 naturally occurring amino acids, 700 unnatural amino acids have been reported in the literature. It all the 720 were to be accommodated in the genetic code, what is the minimum number of base needed in a codon:
(A) 4 (B) 5 (C) 6 (D) 7

1467. Codominant markers, such as RFLPs, are useful for:
(A) Marker-assisted selection (B) Evolutionary studies
(C) Linkage mapping (D) All of these

1468. Saturated molecular linkage maps have been developed for :
(A) Potato (B) Tomato
(C) Maize (D) All of these

1469. What of the following are required to be present in a plasmid, for it to be useful in a cloning experiment?
P. Multiple cloning site (polylinker)
Q. Antibiotic- resistance gene
R. Beta-galactosidase gene
S. Origin of replication
(A) P and Q (B) P, Q and R
(C) P, Q and S (D) P, Q, R and S

1470. Plasmid A and plasmid B were digested with BamHI and analysed by agarose gel electrophoresis. If plasmid A gave two fragments and plasmid B gave three fragments, then which of the following inferences are correct?
P. Plasmid A has three sites and is circular
Q. Plasmid B has three sites and is circular
R. Plasmid A has two sites and is linear
S. Plasmid B has two sites and is linear
(A) P and Q (B) Q and R
(C) P and S (D) Q and S

1471. The saturated molecular linkage maps initially developed in tomato. Potato and maize used which of the following molecular marker systems?
(A) AFLP (B) RFLP (C) SNP (D) DAF

1472. Which of the following is false about DNA pol V of *E. coli*?
(A) Belongs to Y-family of DNA polymerase
(B) Catalyzes translesion of DNA synthesis
(C) Has 5'→ 3' exonuclease activity
(D) Catalyzes error-prone replication

1473. The first strategy used for molecular mapping was based on:
(A) Recombinant inbred lines
(B) Bulked segregant analysis
(C) Near-isogenic lines
(D) F2 population

1474. A researcher is purifying a mixture of 60 kDa (P60) and 40 kDa (P40) polypeptides through a Sephadex G-100 column. What will be the fate of these two polypeptides when they pass through the column?
(A) P40 will run faster
(B) P60 will run faster
(C) Always a 50% mixture of p40 and P60 will come out of the column
(D) P40 and P60 will come out in a ratio of 3 : 2 at any given time.

1475. Development of which of the following populations requires the least amount of effort?
(A) F_2 (B) Backcross
(C) RILs (D) NILs

1476. A family of closely related proteins (interferons) produced by leucocytes in response to virus infection is:
(A) Interfron α (B) Interferon β
(C) Interferon γ (D) Interferon ω

1477. When DNA molecules of same density but different masses are to be separated by:
(A) Velocity sedimentation
(B) Equilibrium sedimentation
(C) Gel Electrophoretic separation
(D) Centrifugal force

1478. When DNA molecules of different density but same or different masses are to be separated by:
(A) Velocity sedimentation
(B) Equilibrium sedimentation
(C) Gel Electrophoretic separation
(D) Centrifugal force

1479. Separation of charged molecules under an electric field by:
(A) Velocity sedimentation
(B) Equilibrium sedimentation
(C) Gel Electrophoretic separation
(D) Centrifugal force

1480. The first cloning of a gene and its expression in a foreign organism were achieved in:
(A) 1971 (B) 1972 (C) 1973 (D) 1974

1481. The markers are available for comparative determination of sizes of single stranded RNA molecules is :
(A) RNA markers (B) DNA markers
(C) Genetic markers (D) DNA size markers

1482. Since many of the viral DNA genomes can easily be isolated in intact form, these molecules cleaved with a specific restriction endonuclease have been popular as:
(A) RNA markers (B) DNA markers
(C) Genetic markers (D) DNA size markers

1483. A specific location on a chromosome that is defined by naked eye polymorphism, protein or DNA sequence; the inheritance of a genetic marker can be followed in a mapping population is known as:
(A) RNA markers (B) DNA markers
(C) Genetic markers (D) DNA size markers

1484. Restriction endonuclease digest of plasmid DNA is also used as:
(A) RNA markers (B) DNA markers
(C) Genetic markers (D) DNA size markers

1485. The DNA fragments obtained by cleaving the genome of the bacterial virus lambda (λ) with the restriction endonuclease Hind III are very often used fro determining the sizes of DNA fragments between:
(A) 2 and 20 kb (B) 700 base to 8 kb
(C) 200 bases to 8 kb (D) None of these

1486. The same genome of the virus lambda (λ) when cleaved with the restriction endonuclease BstE II can be used as DNA markers between the size range of:
(A) 2 and 20 kb (B) 700 base to 8 kb
(C) 200 bases to 8 kb (D) None of these

1487. Fragments obtained by cleaving the genome of the bacteria; virus Ø x 174 with the restriction endonuclease Hae III are used as marker for DNA fragments between:
(A) 2 and 20 kb (B) 700 base to 8 kb
(C) 200 bases to 8 kb (D) None of these

1488. The first reported purification of chromosomal DNA was in:
(A) 1943 (B) 1944 (C) 1945 (D) 1946

1489. Plasmid DNA can be purified by ——— in a cesium chloride-ethidium bromide gradient is:
(A) Equilibrium centrifugation
(B) Centrifugal force
(C) Both (A) and (B)
(D) None of these

1490. There are many genetic engineering operations in which very pure plasmid DNA is:
(A) Required (B) Not required
(C) Unknown (D) None of these

1491. Once the cell lysate is obtained, RNA is isolated through steps similar to that of:
(A) DNA isolation (B) DNA purification
(C) DNA looping (D) DNA library

1492. Ribosomal RNA constitutes 80-85% of total cellular RNA, transfer RNA and small nuclear RNA (in case of eukaryotes) constitute about 10-15%, while messenger RNA constitutes only about:
(A) 1-5% (B) 5-10%
(C) 10-15% (D) 15-20%

1493. For most genetic engineering experiments which is most important:
(A) Messenger RNA (B) Messenger DNA
(C) D-RNA (D) cDNA

1494. Bacterial messenger RNA (mRNA) can be separated from ribosomal and transfer RNAs by :
(A) Velocity sucrose density-gradient centrifugation
(B) Gel electrophoresis
(C) Both (A) and (B)
(D) None of these

1495. Method for purification of eukaryotic mRNA was developed by Phillip Leder in:
(A) 1971 (B) 1972 (C) 1973 (D) 1974

1496. An important requirement in gene cloning is:
(A) Vecter (B) Vehicle
(C) Both (A) and (B) (D) None of these

1497. For gene cloning which following has been used :
(A) Plasmids (B) Viral DNA
(C) Both (A) and (B) (D) None of these

1498. The foreign DNA is then spliced to the DNA from the plasmid or virus by using the enzyme is:
(A) Plasmids (B) Viral DNA
(C) DNA ligage (D) Cloning vector

1499. The vecter containing the ligated foreign DNA is then introduced into the recipient organism by:
(A) Transformation (B) Transfection
(C) Both (A) and (B) (D) None of these

1500. Derivtives of the genome of bacteriophage lambda have been constricted to serve as:
(A) Plasmids (B) Viral DNA
(C) DNA ligage (D) Cloning vector

1501. Yeast one-hybrid system is used for cloning genes encoding a DNA-binding protein specific for a particular:
(A) DNA sequence
(B) DNA – binding Domain
(C) Yeast-two hybrid system
(D) Yeast three-hybrid system

1502. Yeast two hybrid system is based on the observation that most proteins involved in promoting transcription of a gene (transcription) factors) contain two distinct parts or:
(A) DNA sequence
(B) DNA – binding Domain
(C) Yeast-two hybrid system
(D) Yeast three-hybrid system

1503. A protein interacting with a specific DNA sequence, commonly known as DNA binding protein, can be studied and the DNA encoding it can be cloned using a method known as:
(A) Yeast one hybrid system
(B) DNA – binding Domain
(C) Yeast two- hybrid system
(D) Yeast three-hybrid system

1504. When two proteins interact with each other, their corresponding genes are known as interacting genes and can be cloned using a method known as the:
(A) DNA sequence
(B) DNA – binding Domain
(C) Yeast two- hybrid system
(D) Yeast three-hybrid system

1505. Interaction between protein and RNA molecules can be investigated using a method known as:
(A) DNA sequence
(B) DNA – binding Domain
(C) Yeast-two hybrid system
(D) Yeast three-hybrid system

1506. The genomic genes are often much larger that the:
(A) cDNA (B) mRNA
(C) rRNA (D) cRNA

1507. To identify particular locations in the genome by:
(A) Molecular markers (B) Genomes
(C) Selectable marker (D) Scorable marker

1508. Molecular markers rare:
(A) Based on nucleic acid hybridization (mostly DNA-DNA)
(B) Amplification based (PCR based)
(C) Both (A) and (B)
(D) None of these

1509. The human genome contains:
(A) 10000-20000 genes (B) 20000-30000 genes
(C) 30000-40000 genes (D) More than 40000 genes

1510. Each base in the genome I, therefore, a potential mutable unit. These alterations if present is less than 1% of the population are called:
(A) Mutations (B) Polymorphisms
(C) Gene (D) None of these

1511. Each base in the genome is, therefore, a potential mutable unit. These alterations if the frequency is higher, they are called:
(A) Mutations (B) Polymorphisms
(C) Gene (D) None of these

1512. Structural unit of inheritance in living organisms is:
(A) Mutations (B) Polymorphisms
(C) Gene (D) None of these

1513. The sequence of the ——— billion chemical base pairs that make up human DNA:
(A) 2 (B) 3 (C) 4 (D) 5

1514. Genes govern the functions of the cells, and thus of the entire organism:
(A) Structure (B) Metabolic
(C) Both (A) and (B) (D) None of these

1515. Molecular markers identify particular location in the:
(A) Genome (B) Gene
(C) DNA (D) All of these

1516. Molecular markers based on:
(A) Nucleic acid hybridization (mostly DNA-DNA)
(B) Amplification based (PCR based)
(C) Both (A) and (B)
(D) None of these

1517. The molecular markers are based on the hybridization of a cloned piece of DNA to total genomic:
(A) Genome (B) Gene
(C) DNA (D) All of these

1518. In prokaryotic gene regulation; a gene that encodes a:
(A) Regulator gene (B) Regulator protein
(C) Reporter gene (D) All of these

1519. A protein that either represses (= repressor protein) or activates (= activator protein) transcription of a transcription unit by binding to a DNA sequence in a highly specific manner is called:
(A) Regulator gene (B) Regulator protein
(C) Reporter gene (D) All of these

1520. A gene, which allows either easy selection (selectable markers) or identification (scorable markers) of the cells/organisms in which it is present and expressed is called:
(A) Regulator gene (B) Regulator protein
(C) Reporter gene (D) All of these

1521. Microbial culture used for the assay of biological activities of microbial products is:
(A) Assay culture (B) Batch culture
(C) Cell culture (D) Root culture

1522. Growth in such cultures shows a typical sigmoid curve, and they require subculture/termination after a specific period of time in:
(A) Assay culture (B) Batch culture
(C) Cell culture (D) Root culture

1523. In vitro culture of single or relatively small groups of cells, e.g., suspension cultures, this term is often used for callus culture as well is:
(A) Assay culture (B) Batch culture
(C) Cell culture (D) Root culture

1524. Culture of roots in vitro; initiated from piece of root containing root-tip; can be maintained indefinity is:
(A) Assay culture (B) Batch culture
(C) Cell culture (D) Root culture

1525. Use of jumping and linking libraries to determine the correct order in a chromosome of the DNA fragments generated by the restriction enzyme used to prepared these libraries is called:
(A) Chromosome jumping (B) Chromosome walking
(C) Cybridization (D) Hybridization

1526. Use of restriction maps to identify overlapping DNA fragments in a small region of a chromosome is called:
(A) Chromosome jumping
(B) Chromosome walking
(C) Cybridization
(D) Hybridization

1527. The process of cybrid production; usually, to transfer cytoplasm from one species in to other is called:
(A) Chromosome jumping
(B) Chromosome walking
(C) Cybridization
(D) Hybridization

1528. The process of producing, usually, by crossing; also by protoplast fusion. In case of nucleic acid, association of two complementary DNA sequence/RNA sequences or one RNA and one DNA sequence by complementary base pairing is:
(A) Chromosome jumping
(B) Chromosome walking
(C) Cybridization
(D) Hybridization

1529. A hybrid cell/individual derived from the fusion of a cytoplast from one parent and protoplast of the other patent, contains nucleus from one parent but cytoplasm from both the parents is:
(A) Cybrid (B) Hybrid
(C) Cybridization (D) Hybridization

1530. Individual obtained by crossing/mating individuals/strains/varieties of dissimilar genotypes; produced by sexual or protoplast fusion routes is:
(A) Cybrid (B) Hybrid
(C) Cybridization (D) Hybridization

1531. A science devoted broadly to the understanding of cellular and organism genomes is called:
(A) Hybrid (B) Genotype
(C) Genomes (D) Cybrid

1532. A genetic constitution of an organism, as distinct from its physical characteristics, or phenotype called:
(A) Genotype (B) Genome
(C) Cybrid (D) Hybrid

1533. "In glass"; that is, in the test tube is:
(A) In vitro (B) In vivo
(C) Either (A) or (B) (D) Neither (A) nor (B)

1534. "In life"; that is, in the living cell or organism is:
(A) In vitro (B) In vivo
(C) Either (A) or (B) (D) Neither (A) nor (B)

1535. For most genetic engineering experiments which RNA is important:
(A) mRNA (B) tRNA
(C) rRNA (D) nRNA

1536. Bacterial messenger RNA (mRNA) can be separated from ribosomal and transfer RNA by:
(A) Velocity sucrose density-gradient centrifugation
(B) Gel electrophoresis
(C) Both (A) and (B)
(D) None of these

1537. The PAC cloning system combines the features of the vectors derived from the bacteriophage P1 and the :
(A) BAC vectors (B) YAC vectors
(C) PAC vectors (D) None of these

1538. The process of reintroduction of a cloned DNA into a cell is known as:
(A) DNA transfection (B) Gene regulation
(C) Southern blot (D) Northern blot

1539. To study at the transcriptional level, that is how a given gene produces a transcript is called:
(A) DNA transfection (B) Gene regulation
(C) Southern blot (D) Northern blot

1540. To detect the presence of a known piece of DNA in any given DNA preparation from an organism by:
(A) DNA transfection (B) Gene regulation
(C) Southern blot (D) Northern blot

1541. The method of estimating the size and abundance of a particular transcript by:
(A) DNA transfection (B) Gene regulation
(C) Southern blot (D) Northern blot

1542. To determins the 5' end transcripts, the most reliable method is :
(A) Primer extension
(B) Gene tagging
(C) Gene therapy
(D) Chromosome engineering

1543. The method of inserting a recognizable DNA fragment within a gene such that the function of the gene is distrupted and the gene can be easily recognized by virtue of the inserted fragment is:
(A) Primer extension
(B) Gene tagging
(C) Gene therapy
(D) Chromosome engineering

1544. Correcting the illness by substituting an abnormal gene by a normal one is the basis of:
(A) Primer extension
(B) Gene tagging
(C) Gene therapy
(D) Chromosome engineering

1545. Manipulation of individual chromosomes or their parts is called:
(A) Primer extension
(B) Gene tagging
(C) Gene therapy
(D) Chromosome engineering

1546. Genes can be delivered during gene therapy in:
(A) in vivo (B) ex vivo
(C) Both (A) and (B) (D) None of these

1547. The procedures like electroporation, lipofection, etc, can all be used for gene therapy in the:
(A) in vivo (B) ex vivo
(C) Both (A) and (B) (D) None of these

1548. More precisely the regulation of gene expression can take place at different stages is called:
(A) Gene regulation (B) Gene silencing
(C) DNA (D) RNA levels

1549. When gene expression falls, resulting in:
(A) Gene regulation (B) Gene silencing
(C) DNA (D) RNA levels

1550. Transcriptional gene silencing takes place due to modification of the:
(A) Gene regulation (B) Gene silencing
(C) DNA (D) RNA levels

1551. Among the given supplements/additives which is an example of complex organic supplements/additive?
(A) Coconut milk and tomato juice
(B) Corn milk and malt extract
(C) Yeast extract and casein hydrolysate
(D) All of these

1552. Tobacco cultivar A.C. Change is resistant to:
(A) Leaf curl (B) Black rot
(C) Damping off (D) TMV

1553. Evergreen is a somaclone of:
(A) Blackberry (B) Potato
(C) Cucumber (D) Carrot

1554. Among the given surface which is best sterilizing agents is?
(A) Silver nitrate
(B) Cetrimide
(C) Calcium or sodium hypochlorite
(D) Bromine water

1555. Slow-growth shoot culture protocols have been developed for:
(A) Coconut and citrus
(B) Sweet potato and garlic
(C) Ginger and banana
(D) Both (B) and (C)

1556. Somaclone Scarlet has improvement in:
(A) Skin colour (B) Insect resistance
(C) Yield (D) Disease resistance

1557. Hyperhydracity of cultures may not be reduced by:
(A) Agar hydrolysate (B) Growth retardants
(C) Higher agar concentration (D) Gibberillic acid

1558. The cultivar A.C. Change and Delgold were released for commercial cultivation in:
(A) Japan (B) China
(C) Canada (D) Germany

1559. Which variety of rice was developed through haploid phase?
(A) IR 36 (B) Shin Shu
(C) Pragati (D) IR 8

1560. Andro is a somaclone of:
(A) Flax (B) Apple
(C) Maize (D) Cotton

1561. Somaclone Adro represents improvement in:
(A) Salt tolerance (B) Oil quality
(C) Disease resistance (D) Yield

1562. In protoplast isolation and culture, ionic osmoticum is not preferred because of:
(A) It reduced protoplast survival
(B) It reduces protoplast yield
(C) It reduces protoplast fusion
(D) It inhibits cell wall regeneration

1563. For long-term in vitro storage of plant germplasm, which is in practical use?
(A) Cryopreservation (B) Lyophilization
(C) Slow-growth cultures (D) Both (A) and (C)

1564. PEG-induced protoplast fusion is enhanced by:
(A) High pH PEG treatment medium
(B) High pH high Ca^{++} washing medium
(C) High Ca^{++} culture medium
(D) Low pH PEG treatment medium

1565. Distant crosses may fail due to:
(A) Lack of pollen germination
(B) Failure of pollen tube growth
(C) Degeneration of endosperm
(D) All of these

1566. Among the given wheat varieties, which one was developed through haploid phase?
(A) Jinghua 1 (B) Janak
(C) Arjun (D) Pb 543

1567. Somaclone Andro has improvement in which traits?
(A) Rust immunity
(B) Powdery mildew resistance
(C) Wilt resistance
(D) Leaf spot resistance

1568. Bio-13 is a somaclone of:
(A) *Mentha arvensis* (B) *B. juncea*
(C) *Zea mays* (D) *Cironella java*

1569. Intergeneric hybrids from which crosses have been recovered through embryo culture?
(A) *Horedeum vulgare x Triticum aestivum*
(B) *Triticum aestivum x Agropyron repens*
(C) *Hordeum vulgare x Secale cereal*
(D) All of these

1570. Anther culture ordinarily yields:
(A) Haploid plants (B) Disomic plants
(C) Polyploid plants (D) Doubled haploid plants

1571. Somaclonal variation arises due to:
(A) Change in chromosome number
(B) Deletion/duplication
(C) Gene mutation/gene amplification
(D) All of these

1572. Sugarcane somaclone Ono was isolated from:
(A) CO 1148 (B) Pindar
(C) BO 54 (D) COS 771

1573. 'New Alleles' of known genes and even 'new mutations' of genes not known so far have been recovered through:
(A) Embryo rescue (B) Somaclonal variation
(C) Somatic hybridization (D) Cybridization

1574. Which is not correct about micropropation?
(A) Vitrification may be a problem in some species
(B) Somaclonal variation may arise during in vitro culture
(C) Hugh production cost has limited the use of this technique to more valuable ornamental crops and some fruit trees
(D) Suitable techniques of micro-propagation are available for all valuable species

1575. Which is not correct about taxol?
(A) It acts on spindle like colchicines
(B) It promotes dissolution of microtubules into tubulin molecules
(C) It is used for treatment of breast and overian cancer
(D) It is obtained from *Trichosanthes sp.*

1576. Secondary metabolites include which compounds?
(A) Terpenoids (B) Alkaloids
(C) Phenylpronoids (D) All of these

1577. Biomass yields from cell cultures can generally be enhanced by which approaches?
(A) Use of larger inoculums size
(B) Modification of nutrient medium
(C) Selection of high producing cell clones
(D) Both (A) and (B)

1578. Biochemical production by cultural cells can be increased by using which approaches?
(A) Development of high producing clones
(B) Optimization of medium and culture conditions
(C) Use of elicitors
(D) Use of organ cultures and two –phase systems
Options:
(A) I, II, III (B) IIII, IV
(C) I, II, IV (D) I, II, III, IV

1579. Which is not an example of alkaloid?
(A) Morphine (B) Codeine
(C) Camphor (D) Berberine

1580. Somatic hybridization produces:
(A) Cybrids (B) Symmetric hybrids
(C) Asymmetric hybrids (D) All of these

1581. Somatic hybrids are also called:
(A) Parasexual hybrids (B) Cytoplasmic hybrids
(C) Vybrids (D) Both (A) and (B)

1582. Which is not correct about asymmetric hybrids?
(A) Asymmetric hybrids provide an opportunity for introgression of few desirable genes from one fusion partner into the other
(B) Degree of chromosome elimination does not appear to depend on the irradiation dose; it seems to depend mainly on the species combination used in the fusion
(C) Asymmetric hybrids can be produced by fusion of cytoplasts of a species with irradiated protoplasts of another species
(D) Several asymmetric hybrids have been produced in *Nicotiana, Brassica*, etc.

1583. Which is not correct about biotransformation?
(A) Biotransformation reactions are catalyzed by enzymes present in cells
(B) Products generated by biotransformation are more useful or valuable than the precursors used
(C) Low rate of biotransformation prevents commercially exploitation of the very large number of bio conversions known for plant cells
(D) Arbutin is produced commercially by biotransformation

1584. Which plant species, biotransforms hydroquinone into β –D-glucoside (Arbutin)?
(A) *Rauwolfia serpentine* (B) *Datura innoxia*
(C) *Catharanthus roseus* (D) All of these

1585. A spin-filter bioreactor is a god example of:
(A) Fluidized bed reactor (B) Continuous-flow bioreactor
(C) Air lift bioreactor (D) All of these

1586. Micropropagation of woody tress presents which problems?
(A) Difficulties in rooting (B) Browning in medium
(C) Poor growth in vitro (D) All of these

1587. Protoplasts can be used for genetic transformation by:
(A) PEF-mediated DNA uptake
(B) Liposome-mediated delivery
(C) Electroporation
(D) All of these

1588. Which limitations is associated with biochemical production from cultured plant cells?
(A) Often cultured plant cells do not produce compounds of high commercial value
(B) High production cost is the major disadvantage
(C) Due to lack of basic knowledge of the biosynthetic pathways, many systems are difficult to improve
(D) All of these

1589. Which is not a feature of somaclonal variation?
(A) A very effective selection can be practiced at the cell level for several traits
(B) Some new alleles or even 'new' mutations may be isolated, which were not available in the germplasm or through mutagenesis
(C) Somaclonal variation occurs in very low frequency
(D) Use of somaclonal variation may reduce the time required for the release of new variety as compared to mutation breeding

1590. For rapid development of homozygous lines haploids can be produced by which approaches?
(A) Anther culture (B) Ovary culture
(C) Interspecific hybridization (D) All of these

1591. Which elements is not essential for plant cell and tissue culture?
(A) Ni (B) K (C) Ca (D) P

1592. Which is not used as growth retardant?
(A) ABA and Chlomequat (CCC)
(B) TIBA and paclobutrazole
(C) Daminazide
(D) All of these

1593. In case of slow growth cultures, high osmotic concentrations as created by:
(A) Mannitol (B) Sorbitol
(C) Sucrose (D) All of these

1594. Which chemicals interferes with virus multiplication?
(A) Cycloheximide
(B) Actinomycin D
(C) Virazole (ribavirin)
(D) All of these

1595. Cryoprotectants reduce cryoinjury by:
(A) Reduction in cellular dehydration
(B) Reduced ice crystal formation
(C) Reduced dislocation of structural water
(D) All of these

1596. Cryoinjury results from:
(A) Ice crystal formation
(B) Toxic solution effect
(C) Dislocation of structural water
(D) All of these

1597. Cryoprotectants is done at which temperatures?
(A) 0°C (B) 5-10°C
(C) 10-15°C (D) 15-20°C

1598. Vitrification/encapsulation eliminated the need during cryopreservation:
(A) Controlled freezing (B) Cryoproetctant
(C) LN refrigerator (D) Both (A) and (B)

1599. Partial desiccation step can be combined with cryopreservation of:
(A) Somatic embryos
(B) Vitrified somatic embryos
(C) Encapsulated somatic embryos
(D) Both (A) and (C)

1600. Which is not correct about cryoprotectants?
(A) DMSO is highly permeable
(B) Sorbitol is nonpermeable
(C) Sucrose is slowly permeable
(D) Glycerol is highly permeable

1601. Which is not correct?
(A) Meristematic cells suffers from less cryoinjury than differentiated cells
(B) Dividing cells are more likely to survive cryopreservation than nondividing cells
(C) Cryoprotectants minimize is of structural water
(D) Partial desiccation of tissues reduces cryoinjury

1602. Which vitamins is essential for plant tissue cultures?
(A) Thiamine (B) Pyridoxine
(C) m-Inositol (D) Nicotinic acid

1603. Which was the first to be achieved?
(A) Embryo culture (B) Anther culture
(C) Somatic hybridization (D) Embryogenesis

1604. Which was the first to be applied to solve a practical problems?
(A) Embryo culture (B) Anther culture
(C) Somatic hybridization (D) Embryogenesis

1605. Which was the last to be achieved?
(A) Embryo culture (B) Anther culture
(C) Somatic hybridization (D) Embryogenesis

1606. A batch culture has following features:
(A) Typical sigmoidal growth curve
(B) Progressive nutrient depletion
(C) Limited culture duration
(D) All of these

1607. Which following techniques used for cell cloning are used for proplast culture as well?
(A) Microchamber technique
(B) Filter paper raft-nurse tissue
(C) Microdrop method
(D) Bergman's plating technique

1608. A high auxin concentration is required during:
(A) SE induction (B) SE development
(C) SE maturation (D) SE conversion

1609. GA3 has a beneficial effect during:
(A) SE induction (B) SE development
(C) SE maturation (D) SE conversion

1610. ABA is generally promotive during:
(A) SE induction (B) SE development
(C) SE maturation (D) SE conversion

1611. Rejuvenation may be brought about by:
(A) Repeated sprays of a cytokining
(B) Repeated grafting onto seedlings
(C) Repeated subculture on cytokinin containing media
(D) All of these

1612. GRs are not used during which stages of micropropagation?
(A) Preparation of mother plants
(B) Culture establishment
(C) Multiplication
(D) Rooting

1613. Auxin is used :
(A) Preparation of mother plants
(B) Culture establishment
(C) Multiplication
(D) Rooting

1614. In micropropagation, in vitro hardening is achieved by:
(A) PFG in medium
(B) Use of a Pseudomonas culture
(C) Cooling of culture vessels at bottoms
(D) All of these

1615. Which chemicals has been used to eliminate viruses from plants?
(A) Ribovirin (B) Actinomycin D
(C) Malachite green (D) All of these

1616. Virus free plants produced by various in vitro techniques can be used for:
(A) Breeding programmes
(B) For germplasm exchange
(C) Commercial cultivation
(D) Both (A) and (B)

1617. 'Genetic erosion' is the consequence of:
(A) Reduction in genetic variability among cultivated varieties
(B) Reduction in genetic variability within cultivated varieties
(C) Extinction of wild relatives of crop species
(D) All of these

1618. Which is a sweetening agent?
(A) Thaumatic (B) Shikonin
(C) Dioxgenin (D) Pyrethrin

1619. Which is a steroid?
(A) Thaumatic (B) Shikonin
(C) Dioxgenin (D) Pyrethrin

1620. Which has insecticidal activity?
(A) Thaumatic (B) Shikonin
(C) Dioxgenin (D) Pyrethrin

1621. Biochemical accumulation by cultured plant cells is affected by:
(A) Chemical composition of medium (B) Light quality
(C) Inoculum size (D) All of these

1622. Which can serve as elicitors?
(A) Chitosan (B) Cell wall pectins
(C) Some glycoproteins (D) All of these

1623. Certain biochemicals are produced by specialized cells present in organized structures. Which culture systems will be useful for production of such compounds?
(A) Immobilized cell cultures (B) Organ cultures
(C) Callus cultures (D) Suspension cultures

1624. Arbutin is commercially produced from hydroquinone by:
(A) Chemical reactions
(B) Biotransformation by Digitalis cells
(C) Biotransformation by Datura cells
(D) A combination of (B) and (A)

1625. In which of the following, dihaploids are used for breeding, while tetraploids are used as commercial varieties?
(A) Wheat (B) Potato
(C) Sugarcane (D) Tobacco

1626. In which plant species, YY males are used to produce all male populations?
(A) *Silene latifolia* (B) *Coccinia*
(C) *Asparagus* (D) *Trichosanthes*

1627. Which breeding procedures is applicable to only autopolyploid crop species?
(A) Hybrid sorting (B) Analytical breeding
(C) Somaclonal breeding (D) Both (A) and (B)

1628. Which involved in haploid production?
(A) Hybrid sorting (B) Analytical breeding
(C) Somaclonal breeding (D) Both (A) and (B)

1629. In most species, cells become committed to gametophytic development after which stages?
(A) Pollen mother cell (B) Uninucleate pollen
(C) Binuleate pollen (D) Trinucleate pollen

1630. Which have been shown to prevent precocious germination of cultured embryos?
(A) High sucrose levels (B) ABA
(C) Casein hydrolysate (D) All of these

1631. Asymmetric hybrids can be produced by:
(A) Spontaneous chromosome elimination
(B) Irradiation-incuded chromosome elimination
(C) Fusion of protoplasts of one species with microplats of the other species
(D) All of these

1632. Somatic hybridization can be resorted to only when which criteria are satisfied?
(A) Isolation of protoplasts in large quantities
(B) Totipotency of the isolated protoplasts
(C) Availability of effective strategy for selection of hybrid cells
(D) Both (A) and (B)

1633. It is desirable to preserve which of the following tissues to minimize the risk of genetic instability?
(A) Zygotic embryos (B) Shoot-tips
(C) Suspension cultures (D) Either (A) or (B)

1634. Which one is hybrid variety of mango:
(A) Neelam (B) Langra
(C) Sumarr Bahist (D) Chausa

1635. Amrapali is a variety of mango developed by cross of:
(A) Dashhari x Neelam (B) Neelam x Dashhari
(C) Chausa x Langra (D) Dashhari x Langra

1636. Mallika is a variety of mango developed by cross of:
(A) Dashhari x Neelam (B) Neelam x Dashhari
(C) Chausa x Langra (D) Dashhari x Langra

1637. Gaurav is a variety of mango developed by cross of:
(A) Dashhari x Totapari (B) Neelam x Dashhari
(C) Chausa x Langra (D) Dashhari x Langra

1638. Saurabh is a variety of mango developed by cross of:
(A) Dashhari x Fazali jafrani (B) Neelam x Dashhari
(C) Chausa x Langra (D) Dashhari x Langra

1639. Rajiv is a variety of mango developed by cross of:
(A) Dashhari x Romani (B) Neelam x Dashhari
(C) Chausa x Langra (D) Dashhari x Langra

1640. Which of the following practice is very much required for higher fruiting in bottle guard?
(A) Blanching (B) Chlorosis
(C) Staking (D) Digging

1641. Flower dropping of chilli can be checked by:
(A) By spraying of planofix
(B) By applying of nitrogen
(C) By spraying of clorine
(D) By spraying of urea solution

1642. Which is suitable method for seed production of cabbage:
(A) Head to seed method (B) Head intact method
(C) Stump method (D) All of these

1643. Alternate bearing of mango can be checked by:
(A) Deblossoming (B) Ringing
(C) Root prunnig (D) All of these

1644. Heading back pruning refers to:
(A) To remove extra and overlapping branches
(B) To remove terminal portion of all branches
(C) To remove the flowers only
(D) To remove flowers and top portion of main trunk

1645. Teonic in compatibility refers to:
(A) Scion and root stock are adversely affected by each other
(B) Scion and root stock are positively affected by each other
(C) No effect of scion and rootstock on each other
(D) None of these

1646. The regularity in mango bearing can be achieved by soil application of:
(A) Cultar @ 0.5 g/plant (B) 15 ppm NAA
(C) 50 ppm Ethylene (D) All of these

1647. Fruit thining and pre harvest of fruit can be controlled in mango and citrus by:
(A) 10-150 ppm NAA (B) 200 ppm NAA
(C) 10 ppm GA (D) 20 ppm IAA

1648. Which of the following is a evergreen temperate plant?
(A) Apricot (B) Chilgoza
(C) Plum (D) Almond

1649. Fruiting of apple occurs on:
(A) New fruiting spurs (B) Old branches
(C) Whole plant (D) All of these

1650. Fruit cracking in litchi can be controlled by:
(A) Spraying of water
(B) Spraying of hormones
(C) Spraying of 2,4-D
(D) Spraying of Bordeaux mixture

1651. Poor fruiting in loquat is mainly due to:
(A) Poor growth (B) Low fertility
(C) Self incompatibility (D) None of these

1652. Pusa Rasraj is a:
(A) A pureline variety of muskmelon
(B) A hybrid variety of muskmelon
(C) A Seedless variety of watermelon
(D) A hybrid variety of watermelon

1653. The primary objective of green house cultivation of vegetables is:
(A) To get early crop
(B) To get off season crop
(C) To protect the crop from disease
(D) To protect the crop from excessive rains

1654. Which of the following is a stem vegetable?
(A) Carrot (B) Sweet potato
(C) Knol khol (D) Radish

1655. Which of the following is a ripening hormone?
(A) Ethylene (B) Cytokinin
(C) Auxin (D) Gibberellin

1656. Which of the following root stock of grape is useful under high salt content and nematodes?
(A) Rumani (B) Olour
(C) Dogridge (D) Rangpur lime

1657. The bud dormancy in grapes can be broken by use of:
(A) Cultar (B) Hydrogen cynamide
(C) IAA (D) NAA

1658. Whipe tail is caused by deficiency of?
(A) Zn (B) P (C) Mo (D) B

1659. Buttoning is caused by:
(A) N deficiency (B) P deficiency
(C) K deficiency (D) Ca deficiency

1660. Browning is caused by:
(A) N deficiency (B) B deficiency
(C) K deficiency (D) Ca deficiency

1661. Which of the following factor is responsible for higher yield under protected cultivation?
(A) High photosynthetic efficiency
(B) High O_2 availability
(C) High temperature
(D) All of these

1662. Standard root stock of apple against woolly aphid is:
(A) Malling-Merton (B) DX-1
(C) PT-1 (D) None of these

1663. Which of the following is a very vigorous root stock is used for propagation of apple?
(A) Malling -12 (B) Malling-13
(C) Malling-16 (D) All of these

1664. Which of the following is a very dwarf root stock for propagation of apple?
(A) M-27 (B) M-12
(C) M-13 (D) None of these

1665. How much percentage of male plants should be planted in papaya orchard for proper pollination?
(A) 5% (B) 10% (C) 15% (D) 20%

1666. Stem of banana is a:
(A) Pseudomonas (B) Main stem
(C) Pseudostem (D) All of these

1667. Propping refers to:
(A) Staking of banana stem at the time of fruiting to avoid bending or breaking of stem due to heavy weight of fruits
(B) To remove extra suckers
(C) To provide shade to the fruits
(D) None of these

1668. Fruiting of guava during November–January is known as:
(A) Mrig bahar (B) Hast bahar
(C) Ambe bahar (D) Neno bahar

1669. Fruiting of guava during February-April is known as:
(A) Mrig bahar (B) Hast bahar
(C) Ambe bahar (D) Neno bahar

1670. Fruiting of guava during Rainy season is known as:
(A) Mrig bahar (B) Hast bahar
(C) Ambe bahar (D) Neno bahar

1671. Which of the root stock is used in citrus fruits?
(A) Karna Khatta (B) Jumberi
(C) Sour orange (D) All of these

1672. Frame working refers to:
(A) Change into old plant into high grade plants
(B) All side branches are grafted with high quality scions
(C) Only top part of plant is removed
(D) All of these

1673. Pruning is an essential cultural activity in fruit plants because:
(A) To provide proper shape
(B) To provide strong structure
(C) To beautify the plant
(D) All of these

1674. Notching of fruit plants refers to:
(A) To make a V shaped notch just below N above the bud in fruit plants
(B) It helps in vegetative growth
(C) It is common in Fig
(D) None of these

1675. Which of the following plant hormone is used to control pre harvest fruit drop?
(A) 2, 4-D (B) NAA
(C) 2, 4-D and NAA (D) None of these

1676. The cuttings are prepared by removing of cuttings from mother plant as:
(A) 3 cm below the bud at basal or lower part and 1.23 cm above the bud at upper part
(B) Removing of cuttings from upper part of the twig of mother plant
(C) Take centre part of the mother plant
(D) All of these

1677. Top working of fruit plants refers to:
(A) Modify the form of old plants
(B) Convert old plants into new plants
(C) Old plants are modified into new species
(D) All of these

1678. What is blanching?
(A) It is a method of fruit ripening
(B) It is a method of bunding
(C) It is a method of protecting fruits from light and hot
(D) It is a method of plant protecting from insects

1679. Blindness in cauliflower means:
(A) Full development of flower/fruit
(B) Full development of plant without flower
(C) Started plant growth
(D) All of these

1680. Training of fruit trees means:
(A) Remove few unwanted branches to make plant more beautiful and strong
(B) Remove few flowers
(C) Remove and cut few roots
(D) None of these

1681. Temperate fruit plants can be trained by:
(A) Leader method (B) Open centered method
(C) Modified leader system (D) All of these

1682. Which of the training is considered more suitable?
(A) Closed centered system (B) Open centered system
(C) Modified leader system(D) All of these

1683. In quincunx system, the total plants are nearly double than:
(A) Square method (B) Rectangular method
(C) Contour method (D) Hexagonal method

1684. In quincunx system, the filler plants are generally:
(A) Fast growing and early fruiting
(B) Slow growing but early fruiting
(C) Late fruiting
(D) All of these

1685. Loping of root stock refers to:
(A) To remove the upper part of root stock after setting of buds
(B) To remove the lower part of root stock after setting of buds
(C) To remove extra new branches from the root stock
(D) All of these

1686. Which of the following inhances rootings in cuttings?
(A) Auxin (B) Phenolic substances
(C) Auxin-Phenolic complex (D) All of these

1687. Which of the following chemical is used to enhances rootings in cuttings?
(A) IAA (B) IBA
(C) NAA (D) All of these

1688. Which of the following hormone is used for fast rootings in cuttings?
(A) Seredix-A (B) Vitamin A
(C) Vitamin E (D) None of these

1689. The rooting of cuttings can be induced by cuttings before planting in:
(A) IAA (B) GA
(C) ABA (D) Zn solution

1690. In above case, the concentration of IAA should be used:
(A) 100-500 ppm (B) 1000-5000 ppm
(C) 10-50 ppm (D) All of these

1691. IAA, belongs to which group of plant hormone:
(A) Auxin (B) Gibberellin
(C) Cytokinin (D) None of these

1692. Diploid variety of apple are?
(A) Self fruitful (B) Self unfruitful
(C) Both (A) and (B) (D) All of these

1693. Triploid varieties of apple are?
(A) Self fruitful (B) Self unfruitful
(C) Both (A) and (B) (D) All of these

1694. Which of the following is a diploid variety of apple?
(A) Jonathan (B) Rome beauty
(C) Newton wonder (D) All of these

1695. Which of the following is a triploid variety of apple?
(A) Baldvin (B) Blenhenn orange
(C) Cox's orange pippin (D) All of these

1696. Mahmud bahar is a cross of:
(A) Bombay green x Kallapady
(B) Neelam x Chausa
(C) Neelam x Langra
(D) Neelam x Dashhari

1697. Which of the following is a polyembryonic variety?
(A) Mazagaon (B) Chandra karan
(C) Neleswar (D) All of these

1698. Which of the root stock has dwarfing effect on Dashhari?
(A) Rumani (B) Olour
(C) Dogridge (D) Rangur lime

1699. Olour is a polyembryonic variety of:
(A) Guava (B) Citrus
(C) Mango (D) Strawberry

1700. Which of the following dwarfing root stock is used for mango?
(A) Olour (B) Totapari Red small
(C) Dashhari (D) Alfanso

1701. Dehorning refers to:
(A) Removal of branching from fruit trees
(B) Removal of terminal portion of branches
(C) Removal of flowers
(D) All of these

1702. Pinching refers to:
(A) To remove top portion of plant with few leaves
(B) It is like a head back pruning
(C) It is done in ornamental plants
(D) All of these

1703. Stripping refers to:
(A) Removal of barks in stripes of 1.25 – 2.5 cm length
(B Removal of extra flowers
(C) Removal of terminal portion of branches
(D) None of these

1704. Pollinizers are:
(A) Fruit plants that provide pollens to other plants
(B) Helps in fertilization and fruiting
(C) They are planted middle of fruiting plants
(D) All of these

1705. Fruiting can be enhanced by:
(A) Root pruning (B) Ringing
(C) Notching (D) All of these

1706. Fruiting can be induced by:
(A) NAA (B) 2,4-D
(C) IAA (D) All of these

1707. Spliting of fruits can be checked by spraying of:
(A) 1 ppm NAA (B) Orange rind oil
(C) Bordeaux mixture (D) All of these

1708. Periclinal chimeras in fruit plants can be produced by:
(A) Colchicine (B) X-rays
(C) High temperature (D) All of these

1709. Polygamous plants are those plants that produce:
(A) Male, female and hermaphrodite flowers
(B) Male flowers only
(C) Female flowers only
(D) Hermaphrodite flowers

1710. Which one is the hybrid variety of mango is:
(A) Alphanso (B) Dashahari
(C) Mallika (D) Langra

1711. Browning of cauliflower is caused by the deficiency of:
(A) Zinc (B) Manganese
(C) Iron (D) Boron

1712. The citrus diseases namely 'Reclamation Disease' and 'Die back' are caused by the deficiency of:
(A) Zinc (B) Iron
(C) Copper (D) Molybdenum

1713. Fastening of vines to a upright support of a tree called panagara (*Erithrina indica*) is known as:
(A) Kniffen system (B) Head system
(C) Single stake system (D) Cordon system

1714. Maximum genetic diversity of crop plants occurs where agriculture is:
(A) Run on commercial lines (B) Advanced
(C) Rainfed (D) Primitive

1715. Which is more suitable for germplasm collection:
(A) Centre of production (B) Homelands
(C) Tropical areas (D) Temperate areas

1716. Which one will be scarched for obtaining maximum genetic diversity:
(A) Recent varieties (B) Old varieties
(C) Wild relatives (D) All of these

1717. In a gene bank, genetic material is stored as:
(A) Genes
(B) Seeds
(C) Vegetative material/tissue culture
(D) Both (B) and (C)

1718. Genetic erosion is due to:
(A) Deforestation
(B) Shifting cultivation
(C) Adoption of genetically uniform varieties
(D) All of these

1719. Exchange of germplasm is carried mostly through shoot tips because it is:
(A) Cheap (B) Small and handy
(C) Virus free (D) Genetically stable

1720. The gynophore of groundnut is commonly referred to as the:
(A) Perinth (B) Peg
(C) Ovate (D) Monadelph

1721. The rice variety containing 'Dee-geo-woo-gen' gene is:
(A) Indrasan (B) Basmoti
(C) Tilak (D) IR 8

1722. Use of diploids in breeding of crops like potato offers:
(A) Much easier hybridization with diploid wild species, which offers many genes of great value
(B) Much simpler segregation ratios and as a result, the need for similar segregating populations
(C) A greater efficiency opf selection
(D) All of these

1723. Gametoclonal may occur due to :
(A) Residual mutation (B) Gene mutation
(C) Gene multiplication (D) All of these

1724. In 1922, Blakeslee and coworkers, for the first time, reported occurrence of haploids in Datura due to:
(A) Androgenesis (B) Gynogenesis
(C) Parthenogenesis (D) Mutagenesis

1725. During distant hybridization, problem of endosperm abortion can be overcome by:
(A) Embryo culture (B) Ovary culture
(C) Ovule culture (D) All of these

1726. In plant tissue culture, ABA may effect:
(A) Enhance shoot bud differentiation
(B) Promote somatic embryo maturation
(C) Induced somatic embryo conversion
(D) Both (A) and (B)

1727. Plants with gametic chromosome number are termed as:
(A) Diploid (B) Dihaploid
(C) Haploid (D) Monoploid

1728. Micropropagation can be achieved through:
(A) Enhanced axillary branching
(B) Somatic embryogenesis
(C) Production of adventitious shoot buds
(D) All of these

1729. Somatic embryo can be used for:
(A) Genetic transformation
(B) Production of uniform plant populations
(C) In vitro mutagenesis
(D) All of these

1730. Progress in haploid and dihaploid production is still slow due to:
(A) High levels of albinism, especially in monocots
(B) Low frequency of plantlet regeneration
(C) Genetic instability
(D) All of these

1731. Somaclonal variation occurs due to:
(A) Methylation and demethylation in the promoter region of genes
(B) Activation of silent genes in multigene families
(C) Non-reciprocal mitotic recombination and activation of transposable elements
(D) All of these

1732. Among the given hydrogels, which is most preferred?
(A) Agarose (B) Sodium alginate
(C) Agar (D) Gel-Rite

1733. GA3 is known to effect:
(A) Suppress callus formation
(B) Promote somatic embryo conversion
(C) Promote shoot regeneration in some species
(D) All of these

1734. Survival of encapsulated somatic embryo has often been reported to be lower than that of the non-encapsulated ones probably due to:
(A) Poor respiration under the almost anaerobic condition within the capsule
(B) Poor photosynthesis
(C) Poor storage of food material
(D) All of these

1735. Ethylene promotes:
(A) Activity of cell wall destroying enzymes
(B) Ageing of leaves
(C) Activity of chlorophyllase
(D) All of these

1736. Guha and Maheshwari produced haploids from anther culture of:
(A) Nicotiana (B) Datura
(C) Ginkgo (D) Hordeum

1737. Somaclonal variation generally serves the same purpose as:
(A) Analytical breeding (B) Mutation breeding
(C) Backcross breeding (D) Hybrid sorting

1738. Which is not correct about abscissic acid?
(A) It is a naturally occurring growth inhibitor
(B) It destroys the effect of GA and promotes dormancy in and seeds
(C) It is synthesized in leaves
(D) Precursor of ABA is malic acid

1739. How first reported interspecific hybridization through protoplast fusion in two species of Nicotiana?
(A) Powars and coworkers
(B) Carlson and coworkers
(C) Miller and coworkers
(D) Cocking and coworkers

1740. Which minimizes the risk of variation during in vitro conservation of germplasm?
(A) Shoot spices (B) Zygote embryos
(C) Somatic embryos (D) Both (A) and (B)

1741. Success in germplasm storage is determined by which factors?
(A) Inherent longevity and physiological storage behavior of the species
(B) Initial quality (e.g., moisture content) of material stored
(C) Storage methods and conditions used
(D) All of these

1742. Pusa Jaikisan (Bio 902) is a somaclonal variety of:
(A) Rapeseed (B) Indian mustard
(C) Groundnut (D) Sunflower

1743. Which material is conserved as gernplasm?
(A) Wild taxa related to crop species, weedy forms and land races
(B) Obsolate varieties and advanced cultivars
(C) Parents of hybrids, mutants and cytogenetic stocks
(D) All of these

1744. Which varieties developed through somatic hybridization followed by back cross?
(A) Delgold (B) Sigma
(C) A.C. Chang (D) Both (A) and (C)

1745. Which variety that was developed from somaclonal variants?
(A) Delgold (B) Sigma
(C) A.C. Chang (D) Both (A) and (C)

1746. Why conservation of genetic resources is so important?
(A) Due to piracy of genetic resource
(B) Due to regular use in plant breeding
(C) Due to genetic erosion and epidermics
(D) All of these

1747. Cryopreservation is based on:
(A) Liquid CO2 (B) Liquid helium
(C) Liquid nitrogen (D) Both (A) and (C)

1748. Which is thermolabile?
(A) Zeatin (B) Pectinase
(C) ABA (D) All of these

1749. Which scientists is regarded as father of plant tissue culture?
(A) Herbert E. Street (B) Gottlieb Haberlandt
(C) Jean P. Nitsch (D) Philip R. White

1750. The technique of embryo culture can be used to obtain:
(A) Interspecific hybrids
(B) Propagation of orchids
(C) Dormancy and to shorten breeding cycle
(D) All of these

1751. For log-term storage of germplasm, which is used?
(A) Carbon dioxide (B) Liquid oxygen
(C) Liquid nitrogen (D) Mineral oil

1752. Which enzymes is used for obtaining protoplasts?
(A) Macerozyme (B) Cellulase
(C) Pectinase (D) All of these

1753. In general, callus cultures are subcultured after:
(A) 4-6 days (B) 4-6 weeks
(C) 8-10 weeks (D) 2-3 months

1754. Somatic embryo maturation is not promoted by:
(A) Cytokinin (B) High sucrose
(C) Partial desiccation (D) Abscissic acid

1755. The technique of cryopreservation involves:
(A) Freezing and storage
(B) Thawing and reculture
(C) Freezing and reculture
(D) Both (A) and (B)

1756. The term somaclonal variation was coined by:
(A) Vasil and Vasil (B) Larkin and Scowcroft
(C) Skirvin and Karp (D) Evans and Sharp

1757. Protoplast fusion is induced by:
(A) High voltage electric pulse
(B) High pH (10.5) and high (50 m mol/l) Ca^{++} concentration
(C) Polyethylene glycol (PEG)
(D) All of these

1758. In general, suspension cultures are sub-cultured after which periods:
(A) 1-2 weeks (B) 10-15 days
(C) 3-10 days (D) 15-18 days

1759. Among the given auxins, which one is the most frequently used in plant tissue culture?
(A) 2, 4-D (B) IBA
(C) NAA (D) IAA

1760. Pusa Jai kisan (Bio 902) somaclone of Indian mustard was isolated from:
(A) Pusa Bold (B) Kranti
(C) Varuna (D) Vaibhav

1761. Delgold and A.C. Chang, developed through somatic hybridization, are cultivars of:
(A) *Nicotiana rustica*
(B) *Nicotiana tabacum*
(C) *Lycopersicon esculentums*
(D) *Brassica napus*

1762. Among the given compounds, which one is the most commonly used as cryoprotectant?
(A) Proline (B) Sorbitol
(C) DMSO (D) Glycerol

1763. In which tree species, embryo culture is used to overcome dormancy?
(A) Taxus (B) Iris
(C) Prunus (D) All of these

1764. Among the given variaties to tobacco, which one was developed through haploid phase?
(A) F 211 (B) DD 413
(C) GSH3 (D) All of these

1765. Sugarcane somaclone Ono, isolated from the parental variety Pindar, is resistant to a disease caused by:
(A) Fungus (B) Virus
(C) Mycoplasma (D) Bacterium

1766. Scarlet is somaclone of:
(A) Wheat (B) Sorghum
(C) Sweet potato (D) Potato

1767. Among the given agents, which one is used as cryoprotectants?
(A) DMSO, proline and valine
(B) DMSO, glycerol and proline
(C) DMSO, glycerol and acetic acid
(D) DMSO, arginine and acetic acid

1768. Delgold is a variety of:
(A) Cheroot tobacco (B) Flue-cured tobacco
(C) Chewing tobacco (D) Cigar and wrapper tobacco

1769. The first plant from a mature plant cell was regenerated by:
(A) Braun (B) Haberlandt
(C) Cocking (D) White

1770. During gynogenesis, haploid plants generally originated from:
(A) Synergids (B) Antipodal cells
(C) Egg cell (D) Polar nuclei

1771. Somatic hybridization has been the most successful in:
(A) Graminae (B) Leguminosae
(C) Malvaceae (D) Solanaceae and Crucifereae

1772. Post transcriptional gene silencing takes place at the:
(A) Gene regulation (B) Gene silencing
(C) DNA (D) RNA levels

1773. Cross breeding is also a very important method for getting useful:
(A) Hybrids (B) Conventional breeding
(C) Selection breeding (D) None of these

1774. The green revolution was a product of:
(A) Hybrids
(B) Conventional plant breeding
(C) Selection breeding
(D) None of these

1775. The process of choosing plants with desired characters followed by breeding is called:
(A) Hybrids (B) Conventional breeding
(C) Selection breeding (D) None of these

1776. Development of alleles from differentiated explants is termed as:
(A) Dedifferentiation phase (B) Cybrids
(C) In-vitro method (D) Ex-vitro method

1777. The variable cell resulting from the fusion of a cytoplast with a whole cell, thus creating a cytoplasmic hybrid is known as:
(A) Dedifferentiation phase (B) Cybrids
(C) In-vitro method (D) Ex-vitro method

1778. The material could be stored in the form of suspension cultures, meristem, callus cultures and embryoids by:
(A) Dedifferentiation phase (B) Cybrids
(C) In-vitro method (D) Ex-vitro method

1779. The method for preservation of genetic resources for cultivated as also the wild materials by:
(A) Dedifferentiation phase (B) Cybrids
(C) In-vitro method (D) Ex-situ method

1780. The method to preserve land races and wild relatives which could be used for future breeding programme by:
(A) In-situ method (B) Cybrids
(C) In-vitro method (D) Ex-vitro method

1781. A place where stores DNA for future analysis is:
(A) DNA bank (B) Gene bank
(C) Both (A) and (B) (D) None of these

1782. The Royal Botanic Garden (at Kew) DNA Bank contains over 13000 samples of plant genomic DNA (as at the beginning of 2002), all stored at:
(A) 0°C (B) -20°C (C) -40°C (D) -80°C

1783. The most common, straight forward and speedy method for the isolation of gene is:
(A) Restriction of gene
(B) Polymerase Chain Reaction (PCR)
(C) Recombinant DNA Technology
(D) None of these

1784. The speedy but little tricky method of gene isolation is:
(A) Restriction of gene
(B) Polymerase Chain Reaction (PCR)
(C) Recombinant DNA Technology
(D) None of these

1785. Which has been used extensively with microbial system, is also an important tool for the direct genetic manipulation of plant is called:
(A) Restriction of gene
(B) Polymerase Chain Reaction (PCR)
(C) Recombinant DNA Technology
(D) None of these

1786. *Agrobacterium tumefaciens* are capable of producing which disease in almost all dicotyledonous plants through wounded tissues:
(A) Crown galls (Plant tumors)
(B) Hairy root (extra root)
(C) Both (A) and (B)
(D) None of these

1787. *Agrobacterium rhizogens* are capable of producing which disease in almost all dicotyledonous plants through wounded tissues:
(A) Crown galls (Plant tumors)
(B) Hairy root (extra root)
(C) Both (A) and (B)
(D) None of these

1788. The T-DNA region includes important genes, which encode enzymes that synthesize the plant hormone auxin (IAA):
(A) iaaM (tms 1) (B) iaaH (tms2)
(C) Both (A) and (B) (D) None of these

1789. Vir genes are located on the Ti-plasmid itself beyond the region of:
(A) TDNA (B) RDNA
(C) CDNA (D) MDNA

1790. The interaction of proteins bound at distant sites on a DNA molecule so that the intervening DNA forms a loop is called:
(A) DNA library (B) DNA ligage
(C) DNA looping (D) DNA microarray

1791. A collection of DNA sequences immobilized on a solid surface, with individual sequence laid out in patterned arrays that can be probed by hybridization is called:
(A) DNA library (B) DNA ligage
(C) DNA looping (D) DNA microarray

1792. An enzyme that catalyzes template – dependent synthesis of DNA from its deoxyribonucleoside 5'-triphosphate precursors:
(A) DNA polymerase (B) DNA ligage
(C) DNA looping (D) DNA microarray

1793. The entire complex of enzymes and specializes proteins required in biological DNA replication is called:
(A) DNA replicase system (B) DNA supercoiling
(C) cDNA library (D) Complementary DNA

1794. The coiling of DNA upon itself, generally as a result of bending,, underwinding, or overwriting of the DNA helix is called:
(A) DNA replicase system (B) DNA supercoiling
(C) cDNA library (D) Complementary DNA

1795. DNA library consisting entirely of cloned cDNAs from a particular organism or cell type is called:
(A) DNA replicase system (B) DNA supercoiling
(C) cDNA library (D) Complementary DNA

1796. A DNA used in DNA cloning, usually made by reverse transcriptase, complementary to a given mRNA is called:
(A) DNA replicase system (B) DNA supercoiling
(C) cDNA library (D) Complementary DNA

1797. Proteins can be visualized by:
(A) Ethidium Bromide (B) Silver staining
(C) Acridine orange (D) Acetocarmine

1798. Who described the method of chemical sequencing in 1977 is:
(A) Maxam and Gilbert
(B) Sanger and coworkers
(C) Arthur Komberg and coworker
(D) Phillip Leder

1799. Who devised the enzymatic method of DNA base sequencing in 1977 is:
(A) Maxam and Gilbert
(B) Sanger and coworkers
(C) Arthur Komberg and coworker
(D) Phillip Leder

1800. Enzymatic synthesis in vitro of DNA complementary to a given DNA template is possible by using:
(A) DNA polymerase I (B) DNA polymerase
(C) DNA looping (D) DNA library

1801. Introduction of foreign genes for improving genotype is:
(A) Tissue culture (B) Immunisation
(C) Biotechnology (D) Genetic engineering

1802. Genetic engineering means:
(A) Meiotic division of cells
(B) Nucleotide transfer
(C) Deletion and repair mechanism
(D) None of these

1803. The set of techniques that enables DNA from different sources to be identified, isolated and recombined so that new characteristics can be introduced into an organism is called:
(A) Recombinant DNA technology
(B) Genetic engineering
(C) Molecular biology technology
(D) All of these

1804. The recombinant DNA technique was engineered by:
(A) Stanley Norman Cohen (B) Herbert Boyer
(C) Both of these (D) None of these

1805. Phosphatediester bond of DNA and RNA involves:
(A) 2'C and 1'C (B) 5'C and 3'C
(C) 5'C and 2'C (D) 3'C and 1'C

1806. Production of recombinant DNA and its transfer into a suitable host for multiplication/expression of the recombinant DNA. The recombinant DNA may or may not become integrated into the host genome is called:
(A) Genetic engineering (B) Genetic transformation
(C) Genetic material (D) Genetic resource

1807. The chemical that contains the genetic information, i.e., of which genes are composed; in most of the organisms, DNA is the genetic material; in only some viruses, RNA serves this purpose is called:
(A) Genetic engineering (B) Genetic transformation
(C) Genetic material (D) Genetic resource

1808. Sum total of all the genes and their alleles present in a species and its wild relatives is called:
(A) Genetic engineering (B) Genetic transformation
(C) Genetic material (D) Genetic resource

1809. Introduction of DNA or recombinant DNA in to a suitable cell/organism is called:
(A) Genetic engineering (B) Genetic transformation
(C) Genetic material (D) Genetic resource

1810. A DNA molecule created in the laboratory by ligation of two or more different pieces of DNA is called:
(A) Recombinant DNA (B) Recombinant clone
(C) Genetic counseling (D) Genetic diversity

1811. A bacterial clone containing a recombinant DNA molecules is called:
(A) Recombinant DNA (B) Recombinant clone
(C) Genetic counseling (D) Genetic diversity

1812. Educating perspective parents, either suffering from or suspected to be heterogygous for some genetic disease, on the risk of their children suffering from the same disease is called:
(A) Recombinant DNA (B) Recombinant clone
(C) Genetic counseling (D) Genetic diversity

1813. The variability present within a species, between individuals of a population or different population groups of the same species for one or a group of traits is called:
(A) Recombinant DNA (B) Recombinant clone
(C) Genetic counseling (D) Genetic diversity

1814. Genetic engineering was born on:
(A) 1972 (B) 1973 (C) 1974 (D) 1975

1815. Origin of genetic variation in a relatively high frequency, usually, much higher than that ascribable to spontaneous mutations is called:
(A) Genetic instability (B) Recombinant clone
(C) Genetic counseling (D) Genetic diversity

1816. A gene construct in which the gene is located in antisense orientation so that the natural sense strand of the gene is now transcribed to produce antisense RNA; used is antisense RNA Technology is called:
(A) Antisense gene construct
(B) Antisense RNA technology
(C) Both (A) and (B)
(D) None of these

1817. The process of an antisense construct of the target gene is integrated into the genome to suppress the concerned endogenous genes is called:
(A) Antisense gene construct
(B) Antisense RNA technology
(C) Both (A) and (B)
(D) None of these

1818. The antigenic peptides bind within the cleft of MHC I (Major Histrocompatibility Complex I) or MHC II molecules and are displayed on the surface of antigen processing cells (APCs) is called:
(A) Antigen presentation (B) Antisense orientation
(C) Antisense strand (D) None of these

1819. Reverse orientation of a gene in relation to its promoter; the promoter is now located at the 5'-end of the antisense strand so that the natural sense is now transcribed is called:
(A) Antigen presentation (B) Antisense orientation
(C) Antisense strand (D) None of these

1820. The strand of a native gene, which is used by RNA polymerase as template to genetrate its complementary copy of RNA; this RNA ultimately functions as messenger RNA; the promoter sequence is, as a rule, located at the 3'-end of the antisense strand is called:
(A) Antigen presentation (B) Antisense orientation
(C) Antisense strand (D) None of these

1821. Colonies from a master plates are either lifted or replica plates onto a nitrocellulose filter, cells are lysed, DNA is denatured and hybridized with radioactive probe to detect the colonies having DNA/RNA sequences complementary to the probe is called:
(A) Colony hybridization
(B) Northern hybridization
(C) Southern hybridization
(D) None of these

1822. RNA sample is subjected to gel electrophoresis, RNA transferred from the gel onto a nylon membrane, immobilized and hybridized with a single stranded radioactive probe; autoradiography revels the RNA band hybridized to the probe, used for detection of transcription of transgenes is called:
(A) Colony hybridization
(B) Northern hybridization
(C) Southern hybridization
(D) Somatic hybridization

1823. Restriction digested DNA is subjected to get electrophoresis, denatured, blotted onto a nitrocellulose/nylon membrane fixed, hybridized with a single-stranded DNA/RNA probe and hybridization of probe with DNA is detected by autoradiography; used for detection of transgene integration into host genome, RFLP detection, DNA fingerprinting, etc., is called:
(A) Colony hybridization
(B) Northern hybridization
(C) Southern hybridization
(D) None of these

1824. Production of hybrid plants via protoplast fussion; feasible in sexuality incompatible combinations, is sterile/non-flowering species, etc by:
(A) Colony hybridization
(B) Northern hybridization
(C) Southern hybridization
(D) Somatic hybridization

1825. A procedure for transferring denaturated RNA from an agarose gel onto a nitrocellulose filter to allow its hybridization with a probe by:
(A) Southern blotting (B) Northern blotting
(C) Both (A) and (B) (D) None of these

1826. A procedure includes electrophoresis of RNA sample and transfer of RNA from the gel onto a nylon membrane by:
(A) Southern blotting (B) Northern blotting
(C) Both (A) and (B) (D) None of these

1827. The characteristics two dimensional pattern (on paper or gel) formed by the separation of a mixture of peptides resulting from partial hydrolysis of a protein is known as:
(A) Genetic map (B) Peptide mapping
(C) Peptide map (D) None of these

1828. The DNA segment to be cloned is called:
(A) DNA ligase (B) DNA insert
(C) DNA array (D) DNA methylation

1829. An enzyme that seals nicks in DNA strands by catalyzing the formation of phosphodiester bond between the contiguous nucleotides is called:
(A) DNA ligase (B) DNA insert
(C) DNA array (D) DNA methylation

1830. A series of DNA sequence fixed as distinct spots on a suitable solid support, like glass chip; basically two types, spotted DNA array and printed oligonucleotide chips is called:
(A) DNA ligase (B) DNA insert
(C) DNA array (D) DNA methylation

1831. Addition of methyl (-CH3) residues, chiefly to C residues, usually, located in CG or CNG (N means any one of the four DNA bases) sequences, after they are incorporated in DNA is called:
(A) DNA ligase (B) DNA insert
(C) DNA array (D) DNA methylation

1832. The biotin-labelled mRNA from non-stressed cell is called:
(A) Driver mRNA (B) CDNA
(C) Both (A) and (B) (D) None of these

1833. The copy of DNA or complementary DNA produced by using RNA (Usually mRNA) as tempelate; produced by reverse transcriptase enzyme is called:
(A) Driver mRNA (B) CDNA
(C) Both (A) and (B) (D) None of these

1834. RNA produced by transcription of the antisense strand of a gene; in case of structural genes encoding proteins, it usually means mRNA is:
(A) Driver mRNA (B) CDNA
(C) Sense RNA (D) None of these

1835. A diagram showing the relative sequence and position of specific genes along a chromosomes is called:
(A) Genetic map (B) Genetic information
(C) Genetic code (D) Gene expression

1836. The hereditary information contained in a sequence of nucleotide bases in chromosomal DNA or RNA is known as:
(A) Genetic map (B) Genetic information
(C) Genetic code (D) Gene expression

1837. The set of triplet code words in DNA (or mRNA) coding for the amino acids of proteins is known as:
(A) Genetic map (B) Genetic information
(C) Genetic code (D) Gene expression

1838. Transciption, and in the case of proteins, translation, to yield the product of a gene, a gene is expressed when its biological product is present and active is known as:
(A) Genetic map (B) Genetic information
(C) Genetic code (D) Gene expression

1839. Recombination between two DNA molecules of similar sequence, occurring in all cells; occurs during meiosis and mitosis in eukaryotes is called:
(A) Homologous Genetic Recombination
(B) Recombinant DNA
(C) Recombinational DNA repair
(D) Recombination

1840. Peptide mapping is also known as:
(A) Peptide fingerprinting
(B) DNA fingerprinting
(C) DNA footprinting
(D) None of these

1841. The transfer of genetic information from one cell to another by means of a viral vector is called:
(A) Transmination (B) Transcription
(C) Transduction (D) Transformation

1842. Introductiohn of an exogenous DNA into a cell, causing the cell to require a new phenotype is called:
(A) Transmination (B) Transcription
(C) Transduction (D) Transformation

1843. The process in which the genetic information present in an mRNA molecule specifies the sequence of amino acids during protein synthesis is called:
(A) Translation (B) Transcription
(C) Transduction (D) Transformation

1844. The movement of gene or set of genes from one site in the genome to another is called:
(A) Transmination (B) Transposition
(C) Transduction (D) Transformation

1845. A DNA sequence at which RNA polymerase may bind, leading to initiation of transcription is called:
(A) Transmination (B) Transcription
(C) Promoter (D) Transformation

1846. DNA formed by the joining of genes in to new combinations is called:
(A) Recombinant DNA (B) Recombination
(C) Recombinational DNA repair (D) None of these

1847. Which of the variations of PCR aims to enhance specificity?
(A) Anchored PCR (B) Hotstart PCR
(C) Touchdown PCR (D) Both B and C

1848. Which *E. coli* vectors yields single-stranded copies of cloned DNA?
(A) pUC18/19 (B) λ - phage vectors
(C) Phage M13 vectors (D) Cosmid vectors

1849. Which bacteria has at least two circular chromosomes in its genome?
(A) *Bacillus* (B) *Pseudomonas*
(C) *Vibrio chlerae* (D) *Mycoplasma genitalium*

1850. Which is not correct about DNA?
(A) Double helix model of DNA was produced by Watson and Crix
(B) DNA molecules ordinarily consist of two polynucleotide chains/strands
(C) DNA double helix is stabilized by hydrogen bonds
(D) Two strands of a DNA molecule are oriented anti-parallel to each other

1851. Which is not correct?
(A) If the base sequence of one DNA strand is known, the base sequence of its other strand can be easily deduced
(B) The formation of hydrogen bonds between adenine and thymine, and between guanine and cytosine is known as complementary base pairing
(C) Usually, B-DNA molecule has right-handed coiling
(D) Hydrophobic interactions are important in tran scription

1852. Match the column A and B and select the correct option from those listed below them.

Column A	Column B
A. Genomics (functional)	1. Single nucleotide polymorphism
B. Marker	2. Proteins
C. Proteomics	3. Gene isolation
D. Probe	4.Function of all genes

Options:
(A) A – 1, B – 3, C – 2, D - 4
(B) A – 4, B – 1, C – 2, D - 3
(C) A – 2, B – 1, C – 3, D - 4
(D) A – 1, B – 4, C – 3, D - 2

1853. Multiple copies of a gene can be obtained by :
(A) Gene cloning
(B) Polymerase chain reaction
(C) Combination of PCCR with chemical synthesis
(D) All of these

1854. DNA polymerase III serves which type of functions?
(A) DNA repair
(B) DNA replication
(C) Initiation of DNA replication
(D) All of these

1855. Which is not correct about DNA replication?
(A) The presence of replication fork in E. coli chromosome was first demonstrated by Cairns in 1963
(B) DNA replication always proceeds in the 5'→ 3' direction
(C) Both strands of DNA replicate continuously
(D) DNA replication of DNA replication synthesis of a small RNA primer

1856. Which enzyme is called molecular scissor?
(A) DNA ligase (B) DNA gyrase
(C) Restriction enzyme (D) Topoisomerase

1857. A class of enzymes, which cleaves DNA only within or near those sites that have specific base sequences is called:
(A) Type I endonuclease
(B) Type III endonuclease
(C) Type II endonuclease
(D) Exonuclease

1858. Which scientists was awarded Nobel Prize in 1978 for the discovery of endonucleases?
(A) Smith (B) Nathans
(C) Arber (D) All of these

1859. Near-isogenic lines can be developed by:
(A) Single seed descent
(B) Mutagenesis
(C) Backcross method
(D) Continued inbreeding

1860. Which of the following nucleotide sequences is recognized by EcoRI?

G/ATTC (A) CTTAA/G	T/G A (B) AGC/T
A/AGCT T (C) TCGAG/A	A/GATCT (D) TTCGA/A

1861. IRGSP stand for:
(A) International Rice Genome Sequencing Project
(B) International Rice and Groundnut Sequencing Project
(C) International Research on Groundnut, Sorghum and Pigeonpea
(D) Indian Rice Genetic Resources Programme

1862. Most of the type II restriction endonucleases have recognition sites of how much length?
(A) 4 bp (B) 5 bp
(C) 6 bp (D) 4-6 bp

1863. Which restriction enzymes produces blunt ends?
(A) *Bam*H1 (B) *Alu*I
(C) *Pst*I (D) *Hind*III

1864. Which is an example of type I restriction endonuclease?
(A) *Eco*K (B) *Hinf*III
(C) *Hind*II (D) *Bgl*II

1865. Which one has the largest number of genes?
(A) *Arabidopsis* (B) Yeast
(C) *Caenorhabditis* (D) *Drosophila*

1866. Identification of protein using antibodies may be done by:
(A) Enzyme-linked immunosorbant assay (ELISA)
(B) Western blotting
(C) Precipitation and electrophoresis
(D) All of these

1867. Which scientists isolated ribosomal RNA genes in 1965?
(A) Maxam and Gilbert
(B) Wallace and Birnstiel
(C) Sanger and Southern
(D) Brown and Gilbert

1868. Which has the smallest genome size?
(A) *Mycoplasma genitalium*
(B) *Saccharomyces cerevisiae*
(C) *E. coli*
(D) *Caenorhabditis elegans*

1869. The lambda genome contains:
(A) Origin of replication
(B) Genes for head and tail proteins
(C) Proteins involved in lysogeny
(D) All of these

1870. Which population are used for linkage mapping?
(A) F_2 population
(B) Backcross population
(C) Recombinant inbred lines
(D) All of these

1871. Restrictive polyacrylamide gels are used for:
(A) Viroid test
(B) Analysis of point mutations
(C) DNA sequencing
(D) All of these

1872. During DNA isolation, if bacterial cell extract has a heavy load of proteins, it is treated with?
(A) Proteinase K (B) Trypsin
(C) Pronase (D) Both (A) and (C)

1873. Which is the least likely to accumulate property folded proteins encoded by an eukaryotic transgene?
(A) Animal cells (B) *Saccharomyces*
(C) *E. coli* (D) *Pichia*

1874. Which one has the largest genome size?
(A) *Arabidopsis* (B) Yeast
(C) *Caenorhabditis* (D) *Drosophila*

1875. In general, which is used for initial cloning?
(A) *B. subtilis* (B) *Agrobacterium*
(C) *E. coli* (D) *B. thuringiensis*

1876. Use of an antibiotic resistance and the *lacZ*a in a vector, eliminates the need for:
(A) Plating on the antibiotic containing
(B) Replica plating
(C) Plating on a second selection medium
(D) Both (B) and (C)

1877. AFLP may arise due to:
(A) A difference in restriction site
(B) Deletions and insertions within the amplified restriction fragments
(C) Mutations beyond the restriction sites that affect complementarity with the selection nucleotides of the AFLP primers
(D) All of these

1878. In case of linkage maps, the distance, between the markers are measured in terms of:
(A) Recombination frequency
(B) Mutation frequency
(C) Angstroms
(D) Nucleotide number

1879. Which of the following cleaves both the DNA strands at random sites?
(A) DNase I (B) S1 nuclease
(C) *Eco*R 1 (D) *Hind*III

1880. Which one has smallest genome size?
(A) *Arabidopsis* (B) Yeast
(C) *Caenorhabditis* (D) *Drosophila*

1881. Which of the following enzymes is used for converting protruding ends into blunt ends?
(A) Polynucleotide kinase
(B) S1 nuclease
(C) Klenow fragment of E. coli DNA polynuclease I
(D) Both (B) and (C)

1882. Which is the preferred strategy for converting protruding ends into blunts ends?
(A) Polynucleotide kinase
(B) S1 nuclease
(C) Klenow fragment of E. coli DNA polynuclease I
(D) Both (B) and (C)

1883. DNA fingerprinting is used for:
(A) Identification of varieties
(B) Resolving disputed parentages
(C) Identification of criminals
(D) All of these

1884. ARMS PCR achieves differential amplification of allele due to:
(A) Primer sequence
(B) Annealing temperature
(C) Ion concentration
(D) Specific PCR protocol

1885. In most PCR products, a single nucleotide overhang is of:
(A) A at the 3' - end (B) A at the 5'- end
(C) A at the 3' – end (D) T at the 5' - end

1886. The genome of eukaryotes was first to be sequenced:
(A) *Arabidopsis* (B) Yeast
(C) *Caenorhabditis* (D) *Drosophila*

1887. Which is not an example of insertion vector?
(A) λgt10 (B) λEMBL4
(C) λgt11 (D) λZAPII

1888. Identification of protein using antibodies may be done by:
(A) Enzyme-linked immunosorbant assay (ELISA)
(B) Precipitation and electrophoresis
(C) Western blotting
(D) All of these

1889. Replacement vector, λ ENBL4 can be used to clone DNA of upto which size?
(A) 10 kb (B) 20 kb
(C) 40 kb (D) 45 kb

1890. Bacterial cells are usually multiply in:
(A) Jensson medium
(B) Luria-Bertani (LB) medium
(C) YE medium
(D) Ganborg medium

1891. Pulsed-field gel electrophoresis is able to separate chromosomes of:
(A) Yeast (B) Bacteria
(C) Protozoa (D) All of these

1892. Who developed pBR322:
(A) Boliver Rodriguez (B) Berg Rosenberg
(C) Benfy Ris (D) Bonner Riggs

1893. Which vectors has been constructed for use in two different hosts?
(A) Shuttle vectors (B) Cosmids
(C) Phasmids (D) Phagemids

1894. DNA markers can be used for:
(A) Preparation of linkage map
(B) Map-based cloning of genes
(C) Mapping of quantitative trait loci (QTL)
(D) All of these

1895. Which is not correct about DNA replication?
(A) Mode of DNA replication is semi-conservative
(B) Mode of semi-conservative DNA replication was proposed by Watson and Crick
(C) Evidence for semi-conservative replication of DNA was first reported by Meselson and Stahl in 1958
(D) Semi-conservative mode of DNA replication was first demonstrated in T2 phage of *E. coli*

1896. Which is not correct about RNA?
(A) RNA molecules ordinarily consist of a single polyribonucleotide strand
(B) RNA molecules contain uracil in the place of cytosine
(C) Phosphodiester bonds are formed between ribonucleotides in the same manner as in the case of deoxyribonucleotides
(D) RNA was, most likely, the first genetic material, it was much later replaced by DNA

1897. Isozyme markers behave like:
(A) Recessive (B) Dominant
(C) Codominant (D) Overdominant

1898. During 2957, Frankel-Conrat and Singer proved that RNA is the genetic material in which organisms?
(A) Cauliflower mosaic virus (B) Tobacco mosaic virus
(C) Cucumber mosaic virus (D) Tomato mosaic virus

1899. Eukaryotic chromosomes consist of:
(A) Histone and non-histone proteins
(B) Nucleoproteins
(C) DNA
(D) Nucleic acids

1900. Lod score method was developed by:
(A) Paterson (B) Morton
(C) Peltier (D) Rigby

1901. Which is not correct?
(A) Double helix RNA generally follows the same rule of base-pairing as does DNA
(B) The smallest RNA molecule is tRNA
(C) Antisense RNA is also called micRNA
(D) Ribosomal RNA is processed from larger primary transcripts

1902. Direct repeat means:
(A) Repetition of certain length of nucleotides in reverse order
(B) Repetition of certain length of nucleotides in the same order
(C) Repetition of certain length of nucleotides in either the same or the reverse order
(D) Repetition of certain length of nucleotide to form a palindrome with rotational symmetry

1903. Which is not correct about *Taq* DNA polymerase?
(A) It is isolated from *Thermus aquaticus*
(B) It is stable at higher temperatures
(C) It has proof-reading property
(D) It is used in PCR for DNA amplification

1904. Among the following statements about B- and Z-DNA forms, which one is not common to both forms?
(A) Double helix
(B) Length of one full turn of helix is 34?
(C) Two strands of double helix are antiparallel
(D) Complementary base pairing

1905. Which is not correct about *E. coli*?
(A) *E coli* is rarely used for gene cloning
(B) Single generation of *E. coli* is completed within 20-30 minutes, under optimum conditions
(C) Chromosome length of *E. coli* is approximately 1 mm
(D) Rate of chromosome replication is 30-40 μ per minute under optimum conditions

1906. Biotin labeled probes are generally prepared by:
(A) Primer labeling (B) Nick translation
(C) End labeling (D) All of these

1907. Which is not correct?
(A) A mixture of fragments of genomic DNA is used to produce a genomic library
(B) Genomic library contains only those genes that are expressed in the concerned tissue/organism
(C) cDNA obtained from mRNA may be cloned to give rise to a cDNA library
(D) cDNA library contains DNA fragments that represent genes

1908. Who was appointed as the first Director of Human Genome Project:
(A) A. H. Paterson (B) P.M Gresshoff
(C) K.E. Davis (D) J.D. Watson

1909. A phagemid can be used as:
(A) A sequencing vector (B) An expression vector
(C) A cloning vector (D) All of these

1910. Who isolated ribosomal RNA genes in 1965?
(A) Maxam and Gilbert (B) Wallace and Birnstiel
(C) Sanger and Southern (D) Brown and Gilbert

1911. Enzyme procedure of DNA sequencing was developed by:
(A) Maxam and coworkers (B) Gilbert and coworkers
(C) Sanger and coworkers (D) Collins and coworkers

1912. Which is correct?
(A) Genome is the basic set of chromosomes, which is equivalent to the gametic chromosome complement number of a diploid species
(B) Genome map is the schematic depiction of structural and functional organization of all chromosomes in the genome of an organism
(C) Genome project was started to develop genome maps, genome sequence and its annotation in man and other experimental organisms
(D) All of these

1913. Which is used for non-radioactive labeling of probe/ nucleic acids?
(A) Digoxigenic (B) Fluorescent molecules
(C) Biotin (D) All of these

1914. Which techniques can be used for radioactive labeling of probes?
(A) Nick translation and direct labeling
(B) End labeling and primer extension
(C) Method based on RNA polymerases
(D) All of these

1915. Transgene becomes integrated chiefly by homologous recombination in:
(A) Yeast (B) Animals
(C) Plants (D) Both (B) and (C)

1916. Which is used in colony hybridization?
(A) Antibody (B) Enzyme
(C) Probe (D) Substrate

1917. Which enters *E. coli* cell in linear form?
(A) A cosmid vector (B) A λ phage vector
(C) pBR322 (D) Both (A) and (B)

1918. Which affect the level of recombinant protein accumulation?
(A) Promoter
(B) Context of AUG codon (initiator codon)
(C) Codon usage
(D) All of these

1919. RAPD is generated by using of:
(A) A single arbitrary primer
(B) Single-sequence specific primer
(C) Two arbitrary primers
(D) A pair of sequence-specific primers

1920. The term 'recombinant DNA' is specifically used to denote:
(A) A DNA molecules having sequences from two or more sources
(B) A vector into which a DNA insert is integrated
(C) A transgene having sequences from two different organisms
(D) A chromosome into which is transgene is integrated

1921. A cosmid vector can achieve which of the following that a plasmid vector can not achieve?
(A) Cloning of 40 kb DNA insert
(B) Packaging in empty λ phage heads
(C) Transfection of *E. coli* cells
(D) All of these

1922. Which restriction endonucleases is not used in gene cloning?
(A) Type I (B) Type II
(C) Type III (D) Both (A) and (C)

1923. Which markers are uses PCR as well as restriction enzymes?
(A) AFLP (B) RFLP
(C) RAPD (D) STMS (STR loci)

1924. Which sequences forms a palindrome with rotational summary?
(A) GAAAAG (B) GAAGAA
(C) GAATTC (D) Both A and B

1925. Which is a simple palindromic sequence?
(A) GAAAAG (B) GAAGAA
(C) GAATTC (D) Both (A) and (B)

1926. Which is a tandem direct repeat sequence?
(A) GAAAAG (B) GAAGAA
(C) GAATTC (D) Both (A) and (B)

1927. Which is a tendem direct repeat sequence?
(A) GAAAAG (B) GAAGAA
(C) GAATTC (D) Both (A) and (B)

1928. Which is not correct about amplified genomic library?
(A) Particular recombinant DNAs increase in frequency
(B) Particular recombinant DNAs decrease in frequency
(C) Some recombinant DNAs may be eliminated
(D) Specified recombinant DNAs are amplified

1929. Translational fusion offers:
(A) Enhanced translation efficiency
(B) Protection of transgene product
(C) May facilitate purification
(D) All of these

1930. Minisatellite variant PCR is an example of:
(A) Hotstart PCR (B) Touchdown PCR
(C) Nested PCR (D) ARMS PCR

1931. Which is not correct about minisatellite DNAs?
(A) They are usually 200-2000 bp (0.2-2.0 kb) long DNA fragments
(B) They consist of tandem repeats of 15-60 bp
(C) In human genome, they are concentrated in the protermnal regions of chromosomes
(D) They are a good marker system for human genome

1932. A phagemid vector has:
(A) Phage origin of replication
(B) Plasmid origin of replication
(C) Both (A) and (B)
(D) *cos* sequence from â phage

1933. ARMS PCR achieve:
(A) High efficiency amplification
(B) Highly specific amplification
(C) Differential amplification of alleles
(D) Both (A) and (B)

1934. Which aims at increasing the specificity of PCR without lowering the efficiency?
(A) Overlap extension PCR (B) Inverse PCR
(C) Touchdown PCR (D) Anchored PCR

1935. Which can be used when sequence of only one end of the desired segment is known?
(A) Overlap extension PCR (B) Inverse PCR
(C) Touchdown PCR (D) Anchored PCR

1936. Selection nucleotides are used in which of the marker systems?
(A) RAPD (B) RFLP
(C) AFLP (D) STMS (STR loci)

1937. A probe for colony hybridization may be obtained from:
(A) Sequence of the concerned gene
(B) Amino acid sequence of the concerned protein product
(C) Same gene of another organism
(D) All of these

1938. Which of the following observations provided evidence that Mendel's "units" were located on the chromosomes?
(A) Chromosomes occur as homologous pairs
(B) The number of linkage groups and chromosomes is always equal
(C) One member of each homologous pair goes to each gamete during meiosis
(D) All of these

1939. Technique of Southern blotting was originally based on:
(A) Capillary action (B) Electromotive force
(C) Vacuum (D) Both (A) and (C)

1940. Which is called molecular glue?
(A) DNA polymerase (B) DNA ligase
(C) Exonuclease (D) Endonuclease

1941. Theoretically, a recognition sequence of bases, e.g., 5'GAATTC3' (for EcoRI), is expected to occur at, on an average, a distance of how many nucleotides?
(A) 8188 bp (B) 1024 bp
(C) 4096 bp (D) 2048 bp

1942. Southern hybridization can be used for:
(A) Preparation of RLFP maps
(B) DNA fingerprinting
(C) Confirmation of transgene integration
(D) All of these

1943. At the end of n cycles of PCR, how many copies of the amplified segment (amplicon) are ideally expected?
(A) n^2 (B) 2^n (C) $n^2/2$ (D) 2n

1944. If 8 bases primer is used in the PCR, a sequence complementary to it may be expected to occur once every base pair of:
(A) 8^4 bp (B) 4^8 bp (C) 8^8 bp (D) 4^4 bp

1945. Complementation assay is based on:
(A) Protein function
(B) DNA-sequence homology
(C) Protein-protein interaction
(D) DNA-protein interaction

1946. Which of the following is the basis of Southern-Western assay?
(A) Protein function
(B) DNA-sequence homology
(C) Protein-protein interaction
(D) DNA-protein interaction

1947. In which of the PCR variants are used two pairs of primers?
(A) Asymmetric (B) Nested
(C) Anchored (D) Inverse

1948. Which of the PCR variants have flanking regions of the target gene are amplified?
(A) Asymmetric (B) Nested
(C) Anchored (D) Inverse

1949. When two selectable markers are used in a vector, where the DNA insert has to be integrated?
(A) In one of the selectable markers
(B) In both the selectable markers
(C) Anywhere in the vector
(D) Either (A) or (B)

1950. Which enzymes removes nucleotides from the 5'-ends?
(A) S1 nuclease (B) Exonuclease Bal31
(C) *E. coli* exonuclease III (D) λ exonuclease

1951. Which enzymes converts protruding ends into blunt ends?
(A) S1 nuclease (B) Exonuclease Bal31
(C) *E. coli* exonuclease III (D) λ exonuclease

1952. DNA microarrays are high throughout version of:
(A) Southern analysis (B) Differential screening
(C) Gene tagging (D) Colony hybridization

1953. The process of identification of genetic material began in 1928 with the experiments by:
(A) Griffith (B) Hershey and Chase
(C) Watson (D) Avery and coworkers

1954. Prokaryotic DNA polymerase I is known to function in:
(A) DNA replication (B) DNA repair
(C) Recombination (D) Both (A) and (B)

1955. Genome of which of the following organisms was the first to be sequenced?
(A) *Pseudomonas putida*
(B) *Mycoplasma genitalium*
(C) *Saccharomyces cerevisiae*
(D) *Haemophilus influenzae*

1956. Length of lambda genome is:
(A) 39 kb (B) 49 kb
(C) 59 kb (D) 29 kb

1957. A bacterial artificial chromosome vector can be used to clone DNA inserts of upto?
(A) 100 kb (B) 300 kb
(C) 800 kb (D) 600 kb

1958. Which enzymes is used in PCR?
(A) *Vent* polymerase (B) *Taq* polymerase
(C) *Pfu* polymerase (D) All of these

1959. PCR can be used for:
(A) Study of DNA polymorphism
(B) Detection of presence of transgene in transformed cells
(C) Production of cDNA
(D) All of these

1960. Meltinng temperature TM of DNA can be calculated using by:
(A) [2(G + C)] + [4(A + T)] (B) [4(G + C)] + [2(A + T)]
(C) [3(G + C)] + [2(A + T)] (D) (4G + C) + (2A + T)

1961. A plasmid vector can not achieve:
(A) Gene expression
(B) Propagation of recombinant DNA
(C) Propagation in two different hosts
(D) Packaging in empty λ particles

1962. Which endonucleases introduces single-strand nicks in DNA?
(A) DNase I (B) S1 nuclease
(C) *Eco*R 1 (D) *Hind*III

1963. Treatment with which of the following prevents vector circulation?
(A) Polynucleotide kinase (B) S1 nuclease
(C) Alkaline phosphatase (D) λ exonuclease

1964. Which vectors contains λ phage *cos* sequence?
(A) Cosmid (B) Phagemid
(C) Phasmid (D) Both (A) and (C)

1965. The λ exonuclease digests from:
(A) 3'- end having –OH (B) 5'- end having P residue
(C) 5'- end having –OH (D) Both (B) and (C)

1966. A good vector has at least how many marker genes?
(A) One (B) Two (C) Three (D) Four

1967. Which is a type II restriction endonuclease?
(A) *Eco*PI (B) *Hinf*III
(C) *Eco*P15 (D) *Pst*I

1968. Hershey and Chase used which organisms for their experiments to show that DNA is the genetic material?
(A) T_2 phage and *E. coli* (B) T_4 phage and *E. coli*
(C) λ phase and *E. coli* (D) T_7 phage and *E. coli*

1969. Germplasm is contained in:
(A) Somatic cells (B) Growing points
(C) Reproductive cells (D) Both (A) and (C)

1970. Which one will be best for germplasm collection:
(A) Britain (B) France
(C) Portugal (D) Peru

1971. A gram-negative soil bacterium that produces hairy-root disease of plants due to its Ri (Root inducing) plasmid; the natural genetic engineer is:
(A) *Agrobacterium rhizogenes* (B) *E. coli*
(C) *Agrobacterium tumifaciens* (D) None of these

1972. A gram negative soil bacterium that produces crown gall in dicots by transferring the T-DNA of its Ti (Tumour inducing) plasmid; the natural genetic engineer is:
(A) *Agrobacterium rhizogenes* (B) *E. coli*
(C) *Agrobacterium tumifaciens* (D) None of these

1973. Graft obtained from genetically different individuals of the same species is:
(A) Allograft (B) Pseudograft
(C) Nanograft (D) All of these

1974. A haploid obtained from an allo polyploidy is:
(A) Autohaploid (B) Allo haploid
(C) Androgenesis haploid (D) None of these

1975. Haploid obtained from the activity of male gamete is:
(A) Autohaploid (B) Allo haploid
(C) Androgenesis haploid (D) None of these

1976. Molecular maps have become increasingly useful in:
(A) Genetic analysis of animals and plants
(B) Location of functional genes within the genome
(C) Physical mapping of eukaryotic genomes
(D) All of these

1977. DNA ligase is used for:
(A) Joining of more than one fragment of DNA
(B) Synthesis of DNA
(C) Replication of DNA
(D) Cleavage of DNA

1978. Prokaryotic DNA polymerase I is known to function in:
(A) DNA replication (B) DNA repair
(C) Recombination (D) Both (A) and (B)

1979. Who proposed the concept of central dogma?
(A) Crick (B) Watson
(C) Meselson (D) Cairns

1980. Which have been involved in *Arabidopsis thaliana* genome evolution?
(A) Polyploidization
(B) Genome reduction
(C) Single cycle of (A) followed by (B)
(D) More than one cycle of (A) followed by (B)

1981. Who got Nobel Prize in 1980 for the development for recombinant DNA technology?
(A) W. Gilbert (B) F. Sanger
(C) D. Baltimore (D) P. Berg

1982. The S1 nuclease is isolated from:
(A) *Aspergillus oryzae* (B) *Escherichia coli*
(C) *Neisseria lactamica* (D) *Proteus vulgaris*

1983. Which is an example of virulence plasmid?
(A) T1 Plasmid (B) Col plasmid
(C) Ri plasmid (D) Both (A) and (C)

1984. The phenomenon of transformation was discovered by:
(A) N. Zinder (B) F. Griffith
(C) J. Leaderberg (D) E.L. Tatum

1985. Which is not correct about microsatellite DNAs?
(A) They are usually, larger than 2 kb long DNA fragments
(B) They consist of repeats of 2-3 bp sequences
(C) They are randomly and frequently distributed in all eukaryotic genomes, except yeast
(D) They are almost ideal markers for genome mapping

1986. Probes can be used for:
(A) To confirm the integration of a DNA insert into the host genome
(B) To prepare genome maps of eukaryotes
(C) To identify the desired recombinant clone
(D) All of these

1987. Which is an example of artificial chromosome vector?
(A) pBeloBAC11 (B) pBACe3.6
(C) pBAC108L (D) pBeloBAC1

1988. Which approaches is the most successful for chemical synthesis of DNA?
(A) Phosphodiester (B) Phosphite trimester
(C) Phosphotriester (D) Phosphite diester

1989. How many alleles are possible at an SNP locus?
(A) 1 (B) 2 (C) 4 (D) 3

1990. Which marker systems is corrently used for human DNA fingerprinting?
(A) RAPD (B) RFLP
(C) AFLP (D) STMS (STR loci)

1991. Synthetic oligonucleotides can be used to fulfill the:
(A) To produce a complete gene
(B) As linkers and adaptors for gene cloning
(C) For enzymatic DNA sequencing
(D) All of these

1992. Contigs facilitate:
(A) Rapid sequencing
(B) Sequence alignment
(C) Sequence annotation
(D) Sequence interpretation

1993. PCR amplification is sensitive to:
(A) Haem breakdown product (B) Humic acid
(C) Phenolics (D) All of these

1994. Which if not correct about PCR?
(A) Only intact genomic DNA can be used for PCR
(B) PCR does not use radioactivity
(C) Allele-specific primers can be used to detect specific alleles
(D) PCR requires nanogram quantities of DNA

1995. Which is not PCR based molecular marker?
(A) RAPD (B) RFLP (C) STS (D) AFLP

1996. In 1970, which scientists synthesized the gene for alanine-tRNA of yeast?
(A) Maxam and coworkers (B) Sanger and coworker
(C) Khorana and coworkers (D) Smith and coworkers

1997. The genome of which plants was the first to be sequenced?
(A) *Zea mays* (B) *Oryza sativa*
(C) *Hordeum vulgare* (D) *Arabidopsis thaliana*

1998. Which of the following is the basis for colony hybridization?
(A) Protein function
(B) DNA-sequence homology
(C) Protein-protein interaction
(D) DNA-protein interaction

1999. Which of the following is the basis of ELISA?
(A) Protein function
(B) DNA-sequence homology
(C) Protein-protein interaction
(D) DNA-protein interaction

2000. Initially, which marker systems was used for human DNA fingerprinting?
(A) RAPD (B) RFLP
(C) AFLP (D) STMS (STR loci)

2001. Transgene becomes integrated chiefly by homologous recombination in:
(A) Yeast (B) Animals
(C) Plants (D) Both B and C

2002. Which transgene integration generally occurs at random sites?
(A) Yeast (B) Animals
(C) Plants (D) Both (B) and (C)

2003. Which PCR variants generates single-stranded products?
(A) Asymmetric (B) Nested
(C) Anchored (D) Inverse

2004. Which plants has the smallest genome size?
(A) *Zea mays* (B) *Oryza sativa*
(C) *Hordeum vulgare* (D) *Arabidopsis thaliana*

2005. In which of the PCR variants is target DNA cleavage and ligation is involved?
(A) Asymmetric (B) Nested
(C) Anchored (D) Inverse

2006. In which of the PCR variants have the concentrations of one primer is much higher than that of the other?
(A) Asymmetric (B) Nested
(C) Anchored (D) Inverse

2007. Human genome is estimated to contain about how many genes?
(A) 35000 (B) 25000
(C) 50000 (D) 40000

2008. In which of the PCR variants are primers annealing to the 5'-ends of the target gene are used?
(A) Asymmetric (B) Nested
(C) Anchored (D) Inverse

2009. Which plants has the largest genome size?
(A) *Zea mays*
(B) *Oryza sativa*
(C) *Hordeum vulgare*
(D) *Arabidopsis thaliana*

2010. A cosmid vector has :
(A) A selectable marker (B) λ Phage *cos* site
(C) Terminase binding site (D) All of these

2011. Tailing uses which enzymes?
(A) Polynucleotide kinase
(B) Klenow fragments of *E. coli* DNA polymerase I
(C) Terminal deoxynucleotidyl transferase
(D) Both (B) and (C)

2012. Which enzymes removes nucleotide from 3'-ends?
(A) S1 nuclease (B) Exonuclease Bal31
(C) *E. coli* exonuclease III (D) λ exonuclease

2013. Which one has the smallest number of genes?
(A) *Arabidopsis* (B) Yeast
(C) *Caenorhabditis* (D) *Drosophila*

2014. Which enzymes is used to digest either the 3'- or the 5'-protruding ends from DNA fragments?
(A) S1 nuclease (B) Exonuclease Bal31
(C) *E. coli* exonuclease III (D) λ exonuclease

2015. Match the column A and B and select the correct option from those listed below them.

Column A	Column B
A. *Arabidopsis thaliana*	1. 20000
B. *Drosophila melanohaster*	2. 6000
C. *Caenorhabditis elegans*	3. 26000
D. *Saccharomyces cerevisiae*	4. 16000

Options:
(A) A – 1, B – 3, C – 2, D - 4
(B) A – 4, B – 1, C – 2, D - 3
(C) A – 3, B – 4, C – 1, D - 2
(D) A – 1, B – 4, C – 3, D - 2

2016. Which enzymes is used to shorten DNA fragments?
(A) S1 nuclease (B) Exonuclease Bal31
(C) *E. coli* exonuclease III (D) λ exonuclease

2017. Which is not correct about PCR?
(A) It does not use radioactivity
(B) It does not require specific skills
(C) DNA purity is not critical
(D) It always uses a pair of primers

2018. Which version of PCR is used for site-directed mutagenesis?
(A) Overlap extension PCR
(B) Inverse PCR
(C) Touchdown PCR
(D) Anchored PCR

2019. Most of the PCR products have which type of ends?
(A) Single nucleotide overhang (B) Blunt
(C) Cohesive (D) Either (B) or (C)

2020. Two-hybrid yeast system studies in:
(A) Protein function
(B) Heterosis
(C) Protein-protein interaction
(D) Protein expression

2021. An enzymatic process by which the linear arrangement of nucleic acid sequences in a chromosome is altered by cleavage and rejoining is called:
(A) Recombinant DNA (B) Recombination
(C) Recombinational DNA repair (D) None of these

2022. Recombinational processes that are directed at the repair of DNA strand breaks or cross-links, especially at inactivated replication forks is called:
(A) Recombinant DNA
(B) Recombination
(C) Recombinational DNA repair
(D) None of these

2023. To study the patterns of expression of all the different genes present in the genome of organisms that leads to the growth and development of the organism is called:
(A) Functional genome (B) Functional proteomics
(C) Gene regulation (D) DNA transfection

2024. Use of proteomics techniques to analyze the characteristics of molecular protein networked involved in a living cell is called:
(A) Functional genome (B) Functional proteomics
(C) Gene regulation (D) DNA transfection

2025. To study all the transcriptional level; that is how a given gene produces a transcript by:
(A) Functional genome (B) Functional proteomics
(C) Gene regulation (D) DNA transfection

2026. The process of reintroduction of a cloned DNA into a cell is known as:
(A) Functional genome (B) Functional proteomics
(C) Gene regulation (D) DNA transfection

2027. The construction of expression vectors is called:
(A) Vector engineering (B) Host engineering
(C) In vitro transcription (D) Codon bias

2028. Alanine has four codons: CGA, GCC, GCG and GCT. Humans use GCC four times more than GCG for codon alanine. This is known as:
(A) Vector engineering (B) Host engineering
(C) In vitro transcription (D) Codon bias

2029. To obtain optimal expression levels, such host factors have to be minimized by choosing an appropriate mutant of the host in which to express the heterologous gene. Such manipulation of the host for optimal expression of the introduced heterologous gene is known as:
(A) Vector engineering (B) Host engineering
(C) In vitro transcription (D) Codon bias

2030. In the cell, transcription takes place inside the nucleus or close to the nucleotide in bacteria by the action of the RNA polymerase on DNA. This process can also be mimicked outside the cell, that is closed DNA can be mixed with RNA polymerase and the four nucleotides in tube and under appropriate conditions, RNA transcripts can be formed at is does inside the cell. This is known as:
(A) Vector engineering (B) Host engineering
(C) In vitro transcription (D) Codon bias

2031. The *Agrobacterium tumefaciens* bacteria causes a cancerous growth in the infected plant due to the introduction of this DNA fragment derived from the plasmid. The plasmid is called the Ti plasmid and the introduced fragment of DNA is called the:
(A) T-DNA (B) Gene tagging
(C) Knockout organism (D) Gene therapy

2032. The method of insering a recognizable DNA fragment within a gene such that the function of the gene is distrupted and the gene can be easily recognized by virtue of the inserted fragment is known as:
(A) T-DNA (B) Gene tagging
(C) Knockout organism (D) Gene therapy

2033. Transgenic organisms can also be produced to inactivate a particular gene in which case it is known as:
(A) T-DNA (B) Gene tagging
(C) Knockout organism (D) Gene therapy

2034. Correcting the illness by substituting an abnormal gene by a normal one in the basis of:
(A) T-DNA (B) Gene tagging
(C) Knockout organism (D) Gene therapy

2035. Introgression of genes from improved varieties of crop plants into their wild relatives is called:
(A) Gene flow (B) Gene stacking
(C) Gene replacement (D) Gene correction

2036. Many transgenes are transferred into a single variety through a single transformation event is called:
(A) Gene flow (B) Gene stacking
(C) Gene replacement (D) Gene correction

2037. Depending upon the type of change one targets in the defective gene if the purpose is to increase its activity it is known as:
(A) Gene flow (B) Gene stacking
(C) Gene replacement (D) Gene correction

2038. The purpose is to replace a defective gene by a functional one is called:
(A) Gene flow (B) Gene stacking
(C) Gene replacement (D) Gene correction

2039. The method of determining the activity of the promoter is also known as:
(A) Reporter assay (B) Primer extension
(C) Expression vector (D) Codon bias

2040. A pBLUESCRIPT phagmid contains the genes for:
(A) Ampicillin resistance (B) Lac repressor (Lac I)
(C) β-galactosidase (Lac Z) (D) All of these

2041. Which polypeptide sequences are used as tags to facilitate recombinant protein purification?
(A) Glutathione-S-transfication
(B) Maltose-binding protein
(C) *myc* epitope
(D) All of these

2042. Which generates high level transient expression in some specific mouse cell lines?
(A) SV40 replicons (B) BPV replicons
(C) Murine-polymavirus (D) EBV replicons

2043. Agrobacterium proteins functions as an endonuclease?
(A) VirA (B) VirG
(C) VirD1 (D) VirD2

2044. Which Cry proteins is specific to nematodes?
(A) Cry II (B) Cry V
(C) Cry IV (D) Both (B) and (C)

2045. Which of the following is/are true?
P. Physical map unit is centimorgan
Q. Physical mapping is done using restriction enzymes
R. Physical maps help during genome sequencing projects
S. Physical maps are also called genetic maps
(A) P (B) R
(C) Q and R (D) P, R and S

2046. A mutant strain of Neurospora was crossed with the wild type. A total of 132 asci were analyzed:
First division segregation : 104
Second division segregation : 28
Findout the map distance (M) of the gene from the centromere.
(A) 5.3 (B) 8.3 (C) 10.6 (D) 21.2

2047. When a population of bacteria capable of conjugation transfers a specific chromosomal gene (say a gene coding for galactose metabolism) but no other genes, regardless of how long the bacteria are allowed to mate, these bacteria are said to be:
(A) F^+ (B) F^- (C) Hfr (D) F'

2048. Which leads to stable episomal transformation in various luman cell lines?
(A) SV40 replicons (B) BPV replicons
(C) Murine-polymavirus (D) EBV replicons

2049. Which statement is not correct?
(A) The enzyme thymidine kinase belongs to the salvage pathway
(B) The F-protein of Sendal Virus envelope mediates the fusion of viral envelope with cell plasma membrane
(C) Lipofection is the method of choice for transfection of mammalian cells invitro
(D) Virosomes contain viral genome

2050. The slow ripening transgenic tomato Flavr Savr was developed in USA using which approaches/technologies?
(A) Gene disruption
(B) Antisense RNA technology
(C) Ribozyme technology
(D) Cosupression approach

2051. Which insects shows 7 –fold (lowest) increase in resistance against Cry protein after 21 generations of laboratory selection?
(A) *Helicoverpa virescence* (B) *Cadra cutella*
(C) *Plodia interpunctella* (D) *Plutella xylostella*

2052. Drastic reduction in ethylene production can be achieved by:
(A) Expression of antisense constructs of ACC synthase or ACC oxidase
(B) Cosuppression of either of these genes
(C) Expression of such enzymes that metabolize S-adenosyl methionine (SAM), e.g., SAM hydrolase gene from bacteriophage
(D) All of these

2053. Production of recombinant human protein in insect larvae poses which problems?
(A) Protein folding (B) Downstream processing
(C) Both (A) and (B) (D) Glycosylation

2054. Tranforming principle was discovered by:
(A) Hershey and Chase
(B) Avery, MacLeod and McCarty
(C) Zinder and Lederberg
(D) Lederberg and Tatun

2055. Intein sequence facilitates which steps in recombinant protein production?
(A) Separation from other proteins
(B) Removal of tag
(C) Overexpression of the proteins
(D) Both (A) and (B)

2056. Anti-freeze protein promoter was isolated from:
(A) Frog (B) Fish
(C) Snake (D) Drosophila

2057. Which countries has the largest area under cultivation of transgenic crops?
(A) Canada (B) Brazil
(C) USA (D) Argentina

2058. Which Cry proteins is specific to diptera?
(A) Cry I (B) Cry II
(C) Cry IV (D) Cry V

2059. SV40 transducing vectors must have:
(A) SV40 origin including the surrounding region containing the transcriptional regulatory signals
(B) Genes encoding large-T, VP1, VP2 and VP3
(C) Total size 9including, that of the DNA insert) between 3900 bp and 5300 bp for packaging into virions
(D) All of these

2060. Which Agrobacterium proteins undergoes dimerization in response to acetosyringone?
(A) VirA (B) VirG
(C) VirD1 (D) VirD2

2061. Which is not correct?
(A) MAHYCO collaborated with Ciba Geigy, UA, for exploiting potential of Bt technology in cotton in India
(B) The Bt transgenic cotton showed significantly higher level of protection against *Helicoverpa armigera* with reduced number of insecticidal sprays
(C) Based on the extended biosafety analysis and agronomic trials three hybrids, viz., Bt MECH12, Bt MECH 162 and Bt MECH 184 were approved for release
(D) Cotton hybrids Bt MECH 12, Bt MECH 162 and Bt MECH 184 contain cry1Ac gene

2062. Which is correct?
(A) The sequence flanking the intein are exeins
(B) Intein sequence is derived from the VAM1 gene of Saccharomyces cerevisiae
(C) The chitin-binding domain (CBD) from *Bacillus circulans* has a high affinity for chitin column
(D) All of these

2063. A mechanism of genetic exchange that does not occur in a culture medium containing the enzyme DNase is:
(A) Specialized transduction
(B) Conjugation
(C) Transformation
(D) Generalized transduction

2064. A culture of tetracycline-sensitive bacteria was infected by a phage that is delivered from the lysis of a tetracycline-resistant bacterial strain. This results in developments of tetracycline-resistance in the original culture. What phenomenon has occurred?
(A) Conjugation (B) Recombination
(C) Transformation (D) Transduction

2065. Which is not correct about fish vectors?
(A) A typical fish vector is s plasmid
(B) These vectors usually contain a selectable marker
(C) The origin of replication (ori) from E. coli plasmid pBR322
(D) An enhancer/promoter sequence from CaMV

2066. Which techniques is used to obtain information on the relative levels of mRNA present in the tissues of different transgenic individuals?
(A) Reverse transcription PCR
(B) Nuclease protection assay
(C) In-situ hybridization
(D) All of these

2067. Which of the enzyme enables substitution of the carbon source in the medium?
(A) β-gluconidase
(B) Xylose isomerase
(C) Mannose-6-phosphate isomerase
(D) Both (B) and (C)

2068. Which constitutes gene silencing?
(A) Low level of transgene expression in TO
(B) Low level of transgene expression in TO
(C) High level of transgene expression in To, but reduced levels of expression in subsequent generation
(D) Both (A) and (B)

2069. Which Cry proteins is specific to coleoptera?
(A) Cry I (B) Cry II
(C) Cry IV (D) Cry V

2070. F-virosomes are prepared from Sendai Virus; they contain:
(A) Viral envelop (B) Viral F-protein
(C) Recombinant DNA (D) All of these

2071. Gene barnase is expressed in which tissues in order to produce male sterility?
(A) Pollen mother cell (B) Stamen
(C) Tapetum (D) Pollen grain

2072. Which insects shows resistance against *Bacillus thuringiensis* formulations?
(A) *Helicoverpa virescence* (B) *Cadra cutella*
(C) *Plodia interpunctella* (D) *Plutella xylostella*

2073. Which transgenes can be used for development of insect resistant transgenics?
(A) Lectin gene
(B) Cowpea trypsin inhibitor
(C) Serine protease inhibitor
(D) All of these

2074. Which is not correct about Knockout mice?
(A) In a knockout mouse, one gene is distrupted by the technique of site directed mutagenesis
(B) The name Knockout signifies the fact that a specific gene in the genomes of these animals has been made non-functional
(C) Production of Knockout mice employs the embryonic sten cell technology
(D) Knockout mice are invaluable for various kinds of basic and applied research

2075. Which Agrobacterium proteins guides the T-DNA into the plant cell nucleus?
(A) VirA (B) VirG
(C) VirD1 (D) VirD2

2076. RNAi suffers from which limitations?
(A) Exogenous siRNA must be delivered into the cells, which is not highly efficient
(B) RNAi based on exogenous siRNA is rather short-lived, e.g., for 4-6 days only
(C) Longer dsRNA molecules trigger antiviral; repose and a general suppression of gene expression
(D) All of these

2077. Which is the chief advantages of both late and early region replacement SV40 vectors?
(A) The recombinant DNA replicates to a high copy number
(B) The recombinant DNA is packaged into virions
(C) Both (A) and (B)
(D) The maximum size of DNA insert is -2.5 kb

2078. Virus mediated transfer of cellular genetic material from one bacterial cell to another is called:
(A) Transfection (B) Transformation
(C) Transposition (D) Transduction

2079. Which promoters specified anaerobic expression?
(A) Vicilin (B) Extensin
(C) 35S (D) Adh1

2080. Which is not correct about edible vaccines?
(A) They are edible parts of transgenic plants
(B) The transgenic encoded an antigen from plants
(C) They are consumed in large quantities
(D) They are in clinical trial stage

2081. The 'intein' sequence function in:
(A) RNA splicing (B) Protein splicing
(C) Both (A) and (B) (D) Recombination

2082. Which is not correct about *Agrobacterium tumefaciens?*
(A) It is sometimes called natural genetic engineer
(B) It is generally used for gene transfer in plants
(C) It is a Gram positive soil bacterium
(D) Optimum temperature for its growth is 28°C

2083. Human diseases caused by mutation in mitochondrial genome:
(A) Are inherited from both parents
(B) Are inherited from the father
(C) Are inherited from the mother
(D) Do not exist because the mutation is always complemented by the normal

2084. Which of the vector can be used for high level transient expression in insect larvae?
(A) Adeno-associated virus (B) Baculovirus
(C) Both (A) and (B) (D) Vaccinia virus

2085. Which poses problems in gene transfer in plants through genetic engineering?
(A) Low frequency of transformation
(B) Low level of transgene expression
(C) Gene silencing
(D) All of these

2086. In mouse cell lines, plasmid vectors having which replicons yield stable episomal transformation?
(A) SV40 replicons
(B) BPV replicons
(C) Murine-polymavirus
(D) EBV replicons

2087. Which is not obtained by Southern hybridization?
(A) Number of sites of transgene integration per genome
(B) Presence of mRNA/protein produced by transgene
(C) Both (A) and (B)
(D) Confirmation of integration of transgene

2088. How many type of gametes will be produced by individuals of AABb genotype?
(A) Two (B) Four (C) Six (D) Nine

2089. Genes encoding ultimately yields a substitute for cytokinin in the medium?
(A) β-glucuronidase
(B) Mannose-6-phosphate isomerase
(C) Xylose isomerase
(D) All of these

2090. Gene silencing may cause::
(A) Additional expenditure in evaluation
(B) Delayed release of transgenics as varieties
(C) Problems in registration of the transgenic varieties
(D) All of these

2091. In Mendel's experiments with Garden Pea, round seeds shape (RR) were dominant over wrinkled seeds (rr), yellow cotyledons (YY) were dominant over green cotyledons (yy). What are expected phenotypes if F2 generation RRYY x rryy?
(A) Only wrinkled seeds with green cotyledons
(B) Only wrinkled seeds with yellow cotyledons
(C) Only round seeds with green cotyledons
(D) Round seeds with yellow cotyledons and wrinkled seeds with yellow cotyledons

2092. The methotrexate (Mtx) resistant cell lines developed through cell selection show:
(A) Reduced Mtx uptake
(B) Overproduction of DHFR
(C) Reduced affinity of DHFR for Mtx
(D) All of these

2093. Which countries has the smallest area under cultivation of transgenic crops?
(A) Canada (B) Brazil (C) USA (D) Argentina

2094. Which Cry proteins is specific to Lepidoptera?
(A) Cry I (B) Cry II (C) Cry IV (D) Cry V

2095. Independent assortment of genes occurs due to the orientation of chromosomes at:
(A) Metaphase of mitosis
(B) Metaphase I of meiosis
(C) Metaphase II of meiosis
(D) Any phase of the cell division

2096. Which Agrobacterium proteins are phorphorylatedin response to acetosyringone?
(A) VirA (B) VirB11 (C) VirG (D) Both (A) and (C)

2097. Which is not correct about small interfering RNA (siRNA)?
(A) The two strands of an siRNA molecule are either synthesized chemically or they are produced invivo by transcription
(B) Chemically synthesized siRNA molecules have to be introduced into the cells, and they inhibit target gene expression for only 4-6 days
(C) There are two common mechanisms, viz., RNase H-dependent antisense mechanism, and siRNA mechanism by which short synthetic oligonucle otides can be used to modulate gene expression in mammalian cells
(D) The siRNA and RNase H-dependent mechanisms are not similar in any respect

2098. Sex linked inheritance:
(A) Occurs due to gene present on x-chromosomes only
(B) Occurs due to presence of gene in only one sex
(C) Occurs as a result of individuals sex hormone
(D) Not occurs in animals

2099. A women with normal vision but with colourblind father marries a colourblind man. The fourth child of the couple is a boy . This boy
(A) May or may not be colourblind
(B) Will be partially colourblind due to being heterozygous
(C) Must have normal vision
(D) Must be colourblind

2100. Tomato transgenic variety 'Endless summer' has which features?
(A) Delayed ripening
(B) Bacterial wilt resistance
(C) Delayed fruit softening
(D) Early ripening

2101. Which products is being produced in transgenic plants at commercial scale?
(A) Avidin
(B) Hirudin
(C) β-glucuronidase (GUS) enzyme
(D) All of these

2102. To linked genes a and b show 20% recombination. The individuals of a dihybrid cross between + + /+ + x ab/ab shall show gametes
(A) + + 80 : ab 20
(B) + + 50 : ab 50
(C) + + 40 : ab 40 : + a 10 : + b 10
(D) + + 30 : ab 30 : + a 20 : + b 20

2103. Which advantages is not offered by chloroplast transformation?
(A) Reduced risk of evolution of resistance in insect pests
(B) Ideal for eukaryotic gene expression
(C) Very high levels of transgene expression
(D) Elimination of the risk of transgene dispersal through pollen

2104. Which of the following is not correct?
(A) Huntington disease is autosomal dominant disease
(B) Myotonic dystrophy is X-linked dominant disease
(C) Gene involved in Kennedy disease contain trinucleotide repeat
(D) None of these

2105. Corn transgenic, 'KnockOut' has been developed by:
(A) Monsanto (B) Novartis
(C) DuPont (D) Anentis

2106. Which of the following is not used as a genetic marker?
(A) Single nucleotide polymorphisms (SNPs)
(B) RFLP
(C) SINE
(D) VNTR

2107. GFP fluoresces bright green when exposed to which?
(A) UV light (B) Blue light
(C) White light (D) Both (A) and (B)

2108. In case of fatty acid biosynthesis, which is sequentially added during chain elongation?
(A) Acetyl – CoA (B) Aceto-acetyl-CoA
(C) Malonyl-CoA (D) Butyryl-CoA

2109. Consider the following terms:
Protein, DNA, RNA, trait
Select the correct option that depicts their biological relationship from among those given below?

(A)	DNA	RNA	Protein	Trait
(B)	DNA	RNA	Protein	Trait
(C)	DNA	RNA	Protein	Trait
(D)	DNA	RNA	Protein	Trait

2110. An individual with d genes was crossed with would + +. On test crossing F_1, the progeny was + c105, + d 115, cd 880 and + + 900. Distance cd genes is:
(A) 11 maps units (B) 5.5 maps units
(C) 44 maps units (D) 88 maps units

2111. Which Agrobacterium proteins has ATPase activity?
(A) VirA (B) VirB11
(C) VirG (D) All of these

2112. Which is not correct?
(A) Useful vectors are created by joining together segments performing specific functions
(B) The uptake of foreign DNA by plant cells is called transfection
(C) When viral genome is introduced into plant cells with the help of Agrobacterium, it is called agroinfection
(D) A degenerate primer is a set of primers having the same sequence, except for one or more position where variation occurs

2113. Which of the following types of DNA modification results in silencing of a region of the genome in such a way that the silencing can be passed on to offspring?
(A) Acetylation (B) Methylation
(C) Phosphorylation (D) Ubiquitination

2114. Homologous repeats are involved in which of the following types of mutation?
(A) Frameshifts (B) Nonsense mutations
(C) Transition (D) Splicing mutations

2115. Linkage map of x-chromosome of fruitfly has 66 map units with yellow body gene (y) at one end and bobbed hair (b) at the other. The recombination frequency between Y and b gene would be:
(A) 66 % (B) >50 %
(C) =50 % (D) 100 %

2116. Genes encoding which can be used as a selectable marker when combined with a compound added to the medium?
(A) β-glucuronidase
(B) Mannose-6-phosphate isomerase
(C) Xylose isomerase
(D) All of these

2117. Gene silencing leads to:
(A) Reduced transgene expression
(B) Variable transgene expression
(C) Increased transgene expression
(D) Both (A) and (B)

2118. If a transgene has undesirable pleiotropic effect on the transgenic plants, it may be corrected by:
(A) Targetting the transgene product into the appropriate cellular compartment
(B) Changing the genetic background of transgene through hybridization
(C) Mutagenesis in the transgene
(D) Both (A) and (B)

2119. Which of the following would occur through specialized transduction?
(A) Acquisition of Hfr plasmid
(B) Transfer of genes for toxin production
(C) Transfer of genes for capsule formation
(D) Transfer of a plasmid with genes for degrading pesticides

2120. A recessive mutation was discovered in a small island population (1000 people). The mutation was observed in the heterozygous state, but never in the homozygous state. What is a possible explanation for this observation?
(A) Only individuals with opposite alleles at this locus reproduce
(B) The recessive allele is lethal in a homozygous state
(C) The heterozygous state is fixed in this population
(D) The population size is too small to give an explanation for this phenomenon

2121. Which produces high level transient expression in COS cell lines?
(A) SV40 replicons (B) BPV replicons
(C) Murine-polymavirus (D) EBV replicons

2122. Genetic use restriction technology (GURT) is also called:
(A) Terminator technology
(B) Control of plant gene expression
(C) Antifarmer technology
(D) All of these

2123. A disorder which is linked to the Y chromosome in humans (holandric) will:
(A) Only be expressed in male whose mothers were a carrier of the gene
(B) Never be passes from father to child
(C) Show a pattern of skipping generations in a family
(D) Be passed only from father to son

2124. Glyphosate resistance in plants can be generated by expressing in them transgenes that serve which purpose?
(A) Produce an enzyme that inactivates glycophosate
(B) Over-production of EPSPS enzyme
(C) Encode an EPSPS enzyme that is tolerant to glyphosate
(D) Any of these

2125. Which is not correct?
(A) Granule-bound starch synthase I (GBSS I) is the main enzyme of amylase biosynthesis
(B) Suppression of GBSS I activity in potato (by antisense RNA technology) yielded an amylase-free or waxy starch potato
(C) Production of waxy starch in potato is the first successful genetic modification of starch
(D) All of these

2126. Prophage, i.e., phase genome integrated into a bacterial genome, is selfish DNA because:
(A) It always propagates together with the genome of the lysogen
(B) It replicates independently of the genome of the lysogen
(C) It provides immunity against infection by further virions
(D) All of these

2127. Which Agrobacterium proteins are most likely involved in the tube formation between bacterial and plant cells?
(A) VirB (B) VirD4
(C) VirC (D) Both (A) and (B)

2128. Which is not correct about genetic transformation of chloroplasts?
(A) The first successful chloroplast transformation in higher plants was achieved in 1989, when spectinomycin resistance was transferred into tobacco
(B) pZS148 vector can be used fro plastid transformation
(C) Particle gun is widely used and effective method of transforming chloroplasts
(D) Transgenes integrated into the chloroplast genome generally show low levels of expression

2129. Enzyme aminoglycoside 3'-phosphotransferase II (APH (3') II) is commonly called:
(A) Luciferase
(B) Neomycin phosphotranferase II
(C) β-glucuronidase
(D) Acyltransferase

2130. A prophage is:
(A) An auxotrophic mutant
(B) A gene
(C) A phage DNA incorporated into the host genome
(D) Host DNA packed into viral heads

2131. Which is correct about SV40 vectors?
(A) It is a spherical virus with circular, double-stranded 5243 bp chromosome, which encodes 5 proteins
(B) Large-T is essential for viral replication
(C) VP1 (virion protein 1), VP2 and vP3 form viral capsid
(D) All of these

2132. Which of the following is not true about the human Y chromosome (in a male)?
(A) It is a single linear DNA molecule
(B) Its genes are known as holandric genes
(C) It is packaged in nucleosomes
(D) During mitosis, prior to metaphase, it contains only one chromatid

2133. The IV (intermediate vector) component of Ti plasmid cointegrate vector system does not have:
(A) Origin for replication in *E.coli*
(B) Origin for replication in Agrobacterium
(C) T-DNA sequence lacking borders
(D) A selectable marker

2134. A transition mutation:
(A) Occurs when a purine is substituted for a pyrimidine, or vie-versa
(B) Results from the insertion of one or two bases into the DNA chain
(C) Decreases in frequency in the presence of base analogs
(D) Results from substitution of one purine for another or of one pyrimidine for another

2135. Which is not correct about lectins?
(A) Lectin gene was isolated from snowdrop (*Galanthus nivalis*)
(B) Lectins are plant glycoproteins which act against piercing and sucking insects like aphids
(C) They are effective only when ingested in microgram quantities
(D) As a result of item 'C', it has strong insecticidal ef fectiveness in transgenic plants to be of practical usefulness

2136. Which is not correct?
(A) Introduction of a transgene encoding AFPase may increase the activity of this enzyme and starch yield
(B) Lauric acid occurs naturally in Brassica species
(C) The rate limiting step in fatty acid biosynthesis is the carboxylation of acetyl-CoA to malonyl-CoA
(D) Most successful example of fatty acid modification is the increased lauric acid content of *B. napus* transgenic, Laurical

2137. The karyotype designation 47, xx, +13 designates which of the following:
(A) Female with trisomy 13
(B) Female with monosom 13
(C) Female with 13 extra chromosomes
(D) Female with extra genetic material on chromosomes 13

2138. Which is used as processor for PHB synthesis?
(A) Acetyl – CoA (B) Aceto-acetyl-CoA
(C) Malonyl-CoA (D) Butyryl-CoA

2139. Choose the mismatch:
(A) Cri-du-chat syndrome – 5p^{-}
(B) Patan syndrome – 47, +13
(C) Retinoblstoma – 13q, 14
(D) Edwards syndrome – 47, +21

2140. What is the form of chromosomal aberration that leads to the fusion of two formerly separation non-homologous chromosomes into a single chromosome or to the separation of formerly joined sections of a single chromosome into two chromosomes?
(A) Reciprocal translocation
(B) Isochromosome
(C) Robertsonian translocation
(D) Centromeric fusion

2141. In bacteria, which enzyme binds single-stranded DNA, denatures double-stranded DNA and matches the single stranded DNA with complementary denatured DNA?
(A) RecA (B) RecBcD
(C) UvrABc (D) UvrD

2142. Which Agrobacterium proteins are DNA binding protein?
(A) VirG (B) VirD1
(C) VirE (D) All of these

2143. In many plants, certain traits related to quality can be improved by reducing or eliminating the expression of specific genes. The level of endogenous gene expression can be reduced by using which approaches?
(A) Antisense RNA
(B) Small interfering RNA (siRNA)
(C) Gene disruption
(D) Any of these

2144. A tryptophan auxotrophic Hfr strain of *E. coli* is mixed with F- histidine auxotroph and plated on to minimal medium to select for prototrophs. The prototrophs are formed due to:
(A) Conjugation (B) Transduction
(C) Transformation (D) Transfection

2145. Which transgenes shows protection from salt stress in tobacco?
(A) *mtl*1D (B) *bet*A
(C) P5CS (D) Any of these

2146. Which transgenes leads protection against heat stress in tobacco?
(A) *sac*B (B) *hsf*
(C) *bet*B (D) *ala*3

2147. Galatosemia is a recessive single gene genetic disorder, caused due to the mutation in any one of the three genes involved in galactose catabolism. A family consists of 10 normal children with both parents suffering from galatosemia. This is most likely because of:
(A) Epistatsis (B) Reversion
(C) Suppression (D) Complementation

2148. Which is not correct about Xenopus oocytes?
(A) Each mature oocyte has a large nucleus called germinal vesicle
(B) Nucleus is located in the darkly pigmented hemisphere of the oocytes
(C) Oocyte nucleus is arrested at first meiotic metaphase
(D) Oocytes serves as a valuable expression system for a wide variety of genes from both animals and plants

2149. The presence of transgene/DNA segment in plant cells can be detected by:
(A) DT blot assay (B) PCR amplification
(C) Southern hybridization (D) All of these

2150. Which of the following statements is not true for transposable element system?
(A) It consists of both autonomous and non-autonomous elements
(B) Transposases is transcribed by the central region of autonomous elements
(C) Dissociation elements are autonomous in nature
(D) Certain repeats in the genome remain fixed even after the element transposes out

2151. The genes, M, P and S are linked. crossover map distance determined by two-point crosses are:
P · C = 7, S – M = 10, C · M = 8, S · C = 2 and P – S = 5.
The relative positions of these four linked loci are:
(A) PSCM (B) SCPM
(C) PCSM (D) SCMP

2152. Freedom II is a transgenic variety of:
(A) Squash (B) Maize
(C) Cotton (D) Tobacco

2153. P element:
(A) Is a transposon
(B) Has a role in hybrid dysgenesis
(C) Occurs in *D. melanogaster*
(D) All of these

2154. Agroinfection achieves:
(A) Genetic transformation
(B) Introduction of viral genome into plant cells
(C) Crown gall formation
(D) Both (B) and (C)

2155. Temperature-sensitive mutant for a specific gene does not allow growth at the restrictive temperature by:
(A) Inhibiting the expression of the specific gene
(B) Expressing an unfolded form of the specific gene
(C) Degrading the protein as soon as it is expressed
(D) Inhibiting translation

2156. A mechanism that can cause a gene to move from one linkage group to another is:
(A) Translocation (B) Inversion
(C) Crossing over (D) Duplication

2157. Which transgenes generates protection against freeze stress in tobacco?
(A) *sac*B (B) *hsf*
(C) *bet*B (D) *ala*3

2158. A normal couple has seven children (2 daughters and 5 sons). Three of the sons suffer from a hereditary disorder but none of the daughter is affected. Which is the inheritance type?
(A) Sex-limited recessive (B) Autosomal dominant
(C) Sex-linked dominant (D) Sex-linked recessive

2159. According to classical genetic, which of the following statements is true?
(A) Recessive alleles are detected by the phenotype of the F_1 generation
(B) The closer two genes are, the more frequently they recombine
(C) Genes on different autosomes segregate independently
(D) Gene on sex chromosomes segregate with the same pattern as autosomal genes

2160. Which of the following is not correct about genomic imprinting?
(A) It is an epigenetic effect
(B) It occurs as a result of acetylation of histone protein
(C) DNA methylation is responsible for imprinting
(D) Paternal and maternal alleles different pattern of methylation

2161. In Drosophila sex is determined by:
(A) X and Y chromosomes
(B) Ratio of pairs of X-chromosomes to the pairs of autosomes
(C) Ratio of number of x-chromosomes to the sets of autosomes
(D) Whether the egg is fertilized or develops parthenogenetically

2162. The most common cause of the pleiotropic effect of a gene is due to:
(A) The same product of the given gene being involved in different metabolic pathways
(B) The gene making very different products in different cell types
(C) The DNA sequence of the gene getting changed in cell specific manner
(D) The gene not functioning in some cells

2163. Which transgenes shows protection against drought in tobacco?
(A) *sac*B (B) *hsf*
(C) *bet*B (D) *ala*3

2164. Which one the following statement is not correct?
(A) Almost all of the prokaryotes have a single circular chromosome
(B) Almost all of the prokaryotes have cell wall
(C) All prokaryotes have plasmids
(D) The genetic material in prokaryote is not enclosed in any structure surrounded by membrane

2165. In case of ribozyme approach of suppression of endogenous gene expression, the ribozyme encoding sequence is used as:
(A) Independent of the target gene
(B) Fused with a sequence of the target gene in sense orientation
(C) Fused with a sequence of the target gene in sense orientation
(D) Either (B) and (C)

2166. Soybean transgenic variety 'Roundup Ready' is resistant to:
(A) Phosphinothricin (B) Atrazine
(C) Yellow mosaic (D) Glyphosate

2167. The bar gene conferring resistance to phosphinothricin was isolated from:
(A) *Streptomyces sp.*
(B) *Azotobactor vinelandii*
(C) *Bacillus subtilis*
(D) *E. coli*

2168. The *barnase* and *barstar* genes were isolated from:
(A) *Bacillus thuringiensis*
(B) *Bacillus subtilis*
(C) *Agrobacterium rhizogenes*
(D) *Bacillus amyloliquefaciens*

2169. Which of the following is not related to fragile X syndrome?
(A) Trinucleotide repeat
(B) Inherited mental retardation
(C) Robertsonian translocation
(D) Defect in FMR1 gene

2170. Which is not correct about marker genes?
(A) Scorable markers provide valuable tools for identification of and tracking genetic modification in plants
(B) They are not useful in quantifying both transformation efficiency and gene expression in transformants
(C) They are particularly important where genetically modified plants can not be regenerated from single cells and direct selection is feasible or effective
(D) They identify and tag transformed cells

2171. Which presents problems in gene transfer in plants?
(A) Undesirable features of transonic plants
(B) Contamination by Agrobacterium
(C) Low frequency of regeneration from transformed cells
(D) All of these

2172. If a mutation in a gene makes the product of the gene defective in binding to its ligand, then the best what can be predicted about the gene and the mutation is:
(A) The gene codes for a receptor and the mutation is a gain – of - function mutation
(B) The gene codes for a kinase and the mutation is a gain – of – function mutation
(C) The gene codes for a receptor and the mutation is a loss – of – function mutation
(D) The gene codes for a kinase and the mutation is a loss – of – function mutation

2173. If Drosophila males of genotype AaBb are crossed with females of genotype AABB, what is the percentage of white-eyed wrinkled-winged offspring in the first generation (a-white eye, A-red eye; B- smooth wings, b-wrinkled wings)?
(A) 0 % (B) 10 % (C) 20 % (D) 25 %

2174. Which transgenes provides protection against osmotic stress in tobacco?
(A) *sac*B (B) *hsf*
(C) *bet*B (D) *ala*3

2175. In a dihybrid cross AABB x aabb F_2 progeny of AABB, AABb, AaBB and AaBb occurs in the ratio of:
(A) 1 : 1 : 1 : 1 (B) 9 : 3 : 3 : 1
(C) 1 : 2 : 2 : 1 (D) 1 : 2 : 2 : 4

2176. Which of the following feature is not associated with material inheritance?
(A) It is mostly uniparental
(B) It occurs due to nuclear genes
(C) Organelles genome is responsible for maternal in heritance
(D) Variegation in four O-clock plants is an example of maternal inheritance

2177. Which of the following does not participate during site specific recombination between genome of λ– phage and bacterial chromosome?
(A) λ integrase protein (λ Int)
(B) Integration host factor
(C) attB and attP
(D) RecBCD

2178. The high frequency of recombinant obtained from a cross between Hfr and F⁻ strains are because :
(A) The F for integrated into the chromosome mediates transfer of large regons of chromosomes into the recipient
(B) The strain bears high number of F pilli on its surface
(C) There are high number of copies of the F plasmid per cell
(D) It includes recombination at high rate

2179. In case of agrobacterium – mediated genetic transformation, 'wound response' ha a reference to:
(A) Release of a signal molecule from plant wounds
(B) Release of nutrients from plant wound sites
(C) Release of antibiotic compounds from plant wounds
(D) Both (A) and (B)

2180. Which is not correct about Golden Rice-2?
(A) During 2005, a team of researchers at biotechnology company Zeneca (now Syngenta) produced a variety of golden rice called Golden Rice – 2
(B) The team o researchers combined a phytene synthase gene from maize with a *lyc* and *cry*1 from the original golden rice
(C) Golden rice-2 produces 23 times more betacarotene (37 micrograms per gram) than golden rice (1.6 micrograms per gram)
(D) Consumption of golden rice, helps in removal of diseases associated with deficiencies of vitamin A and iron

2181. A silent mutation in a gene results in:
(A) No change in the nucleotide sequence of the mRNA encoded by the gene
(B) No change in the amino acid sequence of the encoded by the gene
(C) No expression of the protein encoded by the gene
(D) An amino acid substitution that has a significant effect on the functional activity of the protein encoded by the gene

2182. Male child with blood group AB is colorblind. His parents could be:
(A) Father normal vision with blood group A, mother colour bind with group O
(B) Father colour blind with group O, mother colour blind with blood group AB
(C) Father normal vision with blood group A, mother colour blind with blood group B
(D) Father colour blind with blood group O, mother normal vision with blood group O

2183. In a particle gun, acceleration of particles is achieved by:
(A) Blank cartridges
(B) Pressurised helium
(C) Electrostatic energy
(D) All of these

2184. A temperature phage differs from a virulant phage in that, temperature phage:
(A) Can exhibit only lytic cycle
(B) Can choose between lytic and lysogenic cycle
(C) Can exhibit only lysogenic cycle
(D) Engages in lysogenic cycle only at high temperature

2185. The cross between a mutant female mouse and a normal male mouse gives rise to offspring that are all mutant; the reciprocal cross gives rise to offspring that are all normal. This means that the mutation is likely to be:
(A) In the mitochondrial genome
(B) In the nuclear genome
(C) Dominant
(D) Recessive

2186. In a cross involving polygenic inheritance, only 2/125 of the offspring (F_2) were as extreme as one of the P_1 parents. How many gene pairs are involved?
(A) 3 (B) 1 (C) 4 (D) 2

2187. Which is the most successful approach for gene transfer in plants?
(A) Electroporation
(B) PEG induced DNA uptake
(C) Microinjection
(D) *Agrobacterium*-mediated transformation

2188. Match list-I with list II and select the correct answer using the codes given below the lists:

List –I (Character of man)	List – II (Example)
P. Sex-linked	1. Baldness
Q. Sex-influenced	2. Aquired immuno deficiency syndrome
R. Sex –limited	3. Klinefelter's syndrome
	4. Haemophilia
	5. Tuft of hairs (hyper trichosis) on pinna

(A) P – 4, Q – 1, R – 5
(B) P – 5, Q – 3, R - 2
(C) P – 5, Q – 1, R – 3
(D) P – 4, Q – 3, R - 2

2189. In Drosophila melanogaster, cinnabar eye (n) and vestigial wing (vg) are simple recessive traits. A female, heterozygous for both genes, was crossed with a male with cinnabar eyes and vestigial wings. The offspring resulting from this cross are listed below:

Class Phenotype	Number
I Wild type	445
II Cinnabar eye, wild type wing	51
III Wild type eye, vestigial wing	49
IV Cinnabar eye, vestigial wing	455

From the data shown, what is the distance between the n and vg genes and if n and vg were unlinked what percent of the offspring would be expected to have either cinnabar or vestigial wings or both?
(A) 500 map units and 100%
(B) 100 map units and 25 %
(C) 50 map units and 50%
(D) 10 map units and 75%

2190. The terminator technology uses:
(A) A lea promoter
(B) A recombine gene
(C) A ribosome inactivating protein gene
(D) All of these

2191. Which transgenic varieties could not be commercialized due to IPR problems?
(A) Endless Summer tomato
(B) Flavr Savr tomato
(C) Rainbow papaya
(D) Laurical B. napus

2192. Consider the following statements:
P. B-chromosomes are indispensable and their inheritance is non-Mendelian
Q. B-chromosomes may increase in number due to a drive involving non-disjunction during pollen mitosis
R. When B – chromosomes are present in high number, they suppress vigour and fertility
S. B-chromosome do not show homology with the basic A chromosomes
Which of the statements given above are correct?
(A) P, Q and R (B) Q, R and S
(C) P, R and S (D) P, Q and S

2193. *B. napus* transgenic variety 'Laurical' has a transgene that encoded:
(A) Acyl-ACP desaturase (B) Acyl-ACP thioesterase
(C) Acyl transferase (D) Both (A) and (C)

2194. A woman trying to get pregnant has a sister whose child has an autosomal recessive disease characteristic by dysfunction of mucus-secreting cells. As a result, this child has abnormally thick mucus that tends to precipitate into dense plugs that obstruct the pancreatic ducts, bronchic and bronchioles, as well as the bile ducts. Which of the following tests could be performed to determine if this woman and her husband are carriers of this disease?
(A) Northern blot (B) PR and sequencing
(C) Southern blot (D) Western blot

2195. Which of the following is not true?
P. RecBCD has both helicase and nuclease properties
Q. RecBCD activity is ATP dependent
R. RuVC resolves the Holiday junction
S. RecA protein play role in homologous recombination in bacteria
Which of the statements given above are correct?
(A) P and S (B) P R and S
(C) P, Q and S (D) P, Q, R and S

2196. 5-methylytosines are common sites for mutations because they:
(A) Can mispair with adenine
(B) Can determinate to thymidine
(C) Prevent discrimination between the daughter and parental strand
(D) Can deaminate to uracil

2197. Use of antibiotic resistance markers in production of transgenic plants may pose which risks?
(A) Toxicity
(B) Allergenicity
(C) Resistance in gut microflora
(D) All of these

2198. Choose the wrong statement(s):
P. QTLs (Qualitative trait loci) control polygonic traits
Q. RFLP is used as marker in QTL mapping
R. QTLs control a quantitative trait
S. QTLs are always present in single chromosome
(A) S (B) Q and R
(C) P and S (D) P and Q

2199. A suitable expression vector ordinarily contains which of the following?
(A) Promoter
(B) Transcription termination sequence
(C) Usually, a polyadenylation site
(D) All of these

2200. Which is required for the function of modified intein used in recombinant protein production?
(A) Low (4°C) temperature (B) β- mercaptoethanol
(C) High (40°C) temperature (D) Both (A) and (B)

2201. Which is based on homologous recombination?
(A) Targetted gene transfer (B) Gene disruption
(C) Gene replacement (D) All of these

2202. Genetic mosaics have
P. Highly methylated chromosome
Q. An extra chromosome
R. An inactivated chromosome
S. Nucleic containing Barr body
(A) Q and R (B) P, Q, R and S
(C) P, Q and S (D) P, R and S

2203. In a test cross of an individual heterozygous for each of the three lined genes, the most frequent classes of progeny were ABc/abc and abC/abc, and the least frequent classes were ABC/abc and abc/abc. What is the order of the genes?
(A) A – C – B (B) B – A – C
(C) A – B – C (D) None of these

2204. Which is the most commonly used vector for gene transfers in plants?
(A) Ri plasmid vectors
(B) Ti plasmid vectors
(C) 2 μ plasmid vectors
(D) Cauliflower mosaic virus

2205. Which type of chemical mutagens are incorporated into the genome by the DNA polymerase during genome replication?
(A) Alkylation agents (B) Base analogos
(C) Deaminating agents (D) Intercalating agents

2206. Match the following

Column – I	Column– II
A. Down's syndrome	P. An additional sex chromosome
B. Cri-dy-chat Syndrome	Q. Loss of a part is chromosome
C. Klinefelter's chromosome	R. Absence of an extra Syndrome
D. Turner's syndrome	S. Presence of an extra chromosome
	T. Presence of two extra chromosomes

(A) A – S, B – Q, C - P, D - R
(B) A – T, B – S, C - P, D - Q
(C) A – S, B – Q, C - Q, D – R
(D) A – S, B – Q, C - R, D – P

2207. Choose the correct statements:
P. Is element is a transposon
Q. Direct repeat acts as hallmark for transposable element
R. Transposons are only present in eukaryotes
S. Yeast Ty element is a retrotransposon
(A) Q and S (B) P and Q
(C) P, Q and S (D) P, Q, R and S

2208. In a particle gun, is the most commonly used:
(A) Blank cartridges (B) Pressurised helium
(C) Electrostatic energy (D) All of these

2209. Match the following:

Column –I (Type of interaction)	Column– II (F_2 phenotypic ratio)
P. Recessive epistasis	A. 12 : 3 : 1
Q. Dominant epistatsis	B. 13 : 3
R. Duplicate recessive epistasis	C. 9 : 6 : 1
S. Dominant and recessive epistasis	D. 9 : 3 : 4
	E 9 : 7
	F. 15 : 1

(A) P – B, Q – A, R – B, S – E
(B) P – D, Q – A, R – E, S - B
(C) P – F, Q – C, R – B, S – A
(D) P – A, Q – E, R – C, S - D

2210. Which of the following is true?
P. Transversional mutations are more harmful than transitional mutations
Q. Frameshift mutations may lead to premature termination
R. Missense mutations are more deleterious than nonsense mutations
S. GC mutation is a transversional mutation
Which of the statements given above are correct?
(A) P, Q, and R (B) P , Q, R and R
(C) P, Q and R (D) P, Q and S

2211. The controversial terminator technology seeks to provide :
(A) Transgenic varieties that can be grown by farmers generation after generations
(B) Protection from piracy of transgenes
(C) Seeds fit for consumption but not for use as seed
(D) Both (B) and (C)

2212. If only two segregating loci contribute to cob length and we represent the parental cross as AABB (9-in average cob length) x aabb (3-in average), the fraction of the F_2 expected to be 4.5 in is:
(A) 1/8 (B) 1/4
(C) 1/16 (D) 3/16

2213. Cells take in calcium phosphate-DNA precipitate particles by:
(A) Pincocytosis (B) Phagocytosis
(C) Both (A) and (B) (D) Lipofection

2214. Homologous recombination is exploited by:
(A) Specific gene function 'Knockout' mice
(B) Gene replacement (in gene therapy)
(C) Transgenic mice
(D) Both (A) and (C)

2215. A colour blind man and a phenotypically normal woman have four children (each has 46 chromosomes)- a boy with haemophilla but no other traits; a boy with colour blindness but no other traits; a colourblind girl but no other traits; and a phenotypically normal girl (all traits mentioned are inherited as x-linked recessives).What is the probability that the mother is heterozygous for colour blindness?
(A) ¼ (B) 1/2 (C) ¾ (D) 1

2216. Mendel could not find recombination and crossing over as:
(A) Traits he chosen were either present on different chromosomes or were far apart
(B) Traits chosen by him were not influenced by genes
(C) He did not have a high power microscope
(D) He selected only pure types

2217. A 12 year old boy develops a disorder also present in his father. No one else in the family is known to be affected. Which of the following modes of inheritance is least likely?
(A) Autosomal recessive (B) Autosomal dominant
(C) X-linked recessive (D) Y-linked

2218. Which is the most difficult problem in case of *Agrobacterium*-mediated transformation?
(A) Inability to transform apical meristems
(B) Inability to transform monocot cells/tissues
(C) High frequency of transient transformations
(D) Lack of 'wounds response' in certain dicots as well

2219. Among the given scorable markers, which one is generally used in plants?
(A) Nopaline synthase (nos)
(B) β-glucuronidase (gus)
(C) Luciferase (lux)
(D) Octopine synthase (ocs)

2220. In 1983, the first transgenic plant was produced in:
(A) Tobacco (B) Tomato
(C) Cotton (D) Maize

2221. Albinism is condition that results from the lack of normal pigmentation. Individuals possessing two recessive alleles (aa) are albino. Similarly, attached result from two recessive alleles (ee). In a family, the husband is an albino with attached earlobes and the wife if pigmented with non-attached earlobes. All their 20 children are pigmented with non-attached earlobes. What is the genotype of the mother?
(A) aaee (B) aaEE
(C) AaEe (D) AAEE

2222. The cryIIB protein is specific to:
(A) Coleoptera (B) Diptera
(C) Orthoptera (D) Lepidoptera

2223. The patent entitled 'Control of Plant Gene Expression' was granted in which country?
(A) USA (B) Japan
(C) European Union (D) India

2224. Consider the terms maternal inheritance and maternal effect and choose the correct statement below:
(A) There is no difference between the terms
(B) Maternal effect is the genetic information carried by the mRNAs deposited in the egg
(C) Maternal inheritance is the genetic information carried by the mRNAs deposited in the egg
(D) Chloroplasts only provide the genetic information that constitute the maternal inheritance

2225. Match list-I with list II and select the correct answer using the codes given below the lists:

List –I	List – II
P. Incomplete dominance	1. Human skin colour
Q. Codominance	2. Purple colour in maize due to anthrocyanin
R. Polygenes	3. Human being be longing to AB blood group
S. Compelementary genes	4. Pink flower in 4 o' clock plant

(A) P – 4, Q – 3, R – 1, S - 2
(B) P – 3, Q – 4, R – 2, S - 1
(C) P – 4, Q – 3, R – 2, S - 1
(D) P – 3, Q – 4, R – 1, S - 2

2226. A protein that is normally localized to nucleus has been mutated in such a way that its localization is no longer nuclear, but cytosolic instead. The function of the protein, however, is 99 of the wild type protein. Which of the followings could be said about the mutation?
(A) It is a frameshift mutation
(B) It is a nonsense mutation
(C) It is a silent mutation
(D) It is a small deletion

2227. The techniques for gene transfer and transgenic animal production were developed by using which animals as models?
(A) Sheep (B) Rabbits
(C) Mice (D) Pigs

2228. Which is correct about CAD (bacterial origin) protein:
(A) It is encoded by Cad gene
(B) It is a multifunctional enzyme and catalyzes the first three reaction of endogenous pyrimidine synthesis
(C) PALA-resistant cell lines developed through selection overproduce CAD protein due to a high amplification of the Cad gene
(D) All of these

2229. In a cross in Neurospora involving two alleles B and b, the following tetrad patterns were observed:

Tetrad pattern	Number
BBbb	36
bbBB	44
BbBb	4
bBbB	6
BbbB	3
bBBb	7

What will be the distance between the gene and the centromère
(A) 10 map unit (B) 15 map unit
(C) 40 map unit (D) None of these

2230. What could be the best term to describe effect of one gene on another in a way that one would hide the effect of another on a phenotype?
(A) Pleitropy (B) Homeostasis
(C) Epistasis (D) Hyperstasis

2231. Which is correct about development of insect resistance against Cry proteins?
(A) Resistance to Cry proteins may be due to reduced binding of the toxin to midgut epithelium (e.g., *P. xylostella*) or complete degradation of Cry proteins (e.g., *S. frugiperda*)
(B) The inheritance of resistance to Cry proteins in insect populations is usually recessive, and is governed by one or few major genes
(C) It appears that insects may evolve resistance to some groups of toxins which faster than no others
(D) All of these

2232. Phenylketonuria (PKU) is inherited disease that is characterized by:
(A) Elimination of genetisic acid in urine
(B) Increased occurrence of phenylalanine in blood and tissues
(C) Elimination of sugar in urine
(D) Decrease in phenylalanine in blood and tissues

2233. A recombination process occurring in bacteria always refers to the:
(A) Affect of a physical nature changing the nature of the genes
(B) Reversal of a microbial mutation
(C) Loss of genes from a cell
(D) Acquisition of DNA by an organism

2234. Which is an example of a dominant selectable marker?
(A) CAD protein
(B) DHFR and XGPRT
(C) Neomycin phosphotranferase
(D) All of these

2235. Which is not correct about methotexate (Mtx)?
(A) Normal nonmutant cultured cells are very sensitive to mtx
(B) It is a weak inhibitor of dihydrofolate reductase (DHFR)
(C) Enzyme DHFR is involved in endogenous synthesis of nucleotides
(D) The Chinese hamster ovary cell line 'A29' overproduces an altered DHFR so that it is extremely resistant to mtx

2236. Which of the following is not an example of allopolyploid?
(A) Rephanobrassia (B) Evening prime rose
(C) Triticale (D) *Tritihum aestivum*

2237. Which of the following feature is to related to inversion?
(A) Position effect
(B) Crossover suppression
(C) Formation of dicentric bridge
(D) Pseudodominance

2238. Consider the following pairs

P. Acridine compounds:	Frame-shift mutation
Q. Ultraviolet light:	Conversion of one base into another
R. Bromouracil:	Thymidine dimer

Which of the above is/are correctly matched?
(A) P (B) P and Q
(C) P and R (D) Q and R

2239. A gene can be obtained from:
(A) Genomic library (B) Chemical synthesis
(C) cDNA library (D) All of these

2240. Which transgenic varieties was the first to reach the market?
(A) Liberty Link (B) Endless Summer
(C) Flavr Savr (D) Bollgard

2241. In studying an *E. coli* strain containing a known frameshift mutation, you isolate a mutant in which this frameshift is suppressed. You therefore expect to find which of the following associated with the suppression? (Assume the frameshift has not been corrected in the coding for the peptide).
(A) Suppressor tRNAs encodes an amino acid for a stop signal
(B) Suppressor tRNAs encode a two-nucleotide anticodon
(C) Suppressor tRNAs encode a four-nucleotide anticodon
(D) Both (B) and (C)

2242. In tomato the following genes are located on chromosome 3:

+ tall plant	+ normal leaves	+ smooth fruit
d dwarf plant	m mottled leaves	p pubescent fruit

Results of the cross : +++//dmp X dmp/dmp were

+++	430
dmp	452
+mp	45
d++	38
++p	16
dm+	17
+m+	1
d+p	1

Which one of the following progeny groups represents double crossovers?
(A) +mp/d++ (B) ++p/dm+
(C) +m+/d+p (D) +++/dmp

2243. TA29 promoter has which features?
(A) Light-responsiveness
(B) Wound responsiveness
(C) Tissue specificity
(D) Development stage-specificity

2244. The term transformation is used to denote:
(A) Delivery of DNA into bacterial/plant cell by a virus
(B) Uptake of DNA by a cell
(C) Conversion of a normal vertebrate cells into a tumourigenic cells
(D) Both (B) and (C)

2245. The barnase gene was first used for producing male sterility in 1990 by:
(A) Thorpe and coworkers
(B) Paterson and coworkers
(C) Mariani and coworkers
(D) Scowcroft and coworkers

2246. Transformation through electroporation has been reported in:
(A) Rice (B) Tobacco
(C) Wheat (D) All of these

2247. An x-linked recessive gene produces red-green colour blindness in humans. A women with normal colour vision whose father was colour-blind marries a colour – blind man. What is the probability that their son will be colour-blind?
(A) 0 (B) 1/4 (C) ½ (D) ¾

2248. RecA is a protein involved in:
(A) Recombinational repair
(B) Mismatch repair
(C) Nucleotide excision repair
(D) Base excision repair

2249. Which has been the most often used promoter for gene transfers in plants?
(A) ShI early promoter (B) nos promoter
(C) 35S promoter (D) Adh1 promoter

2250. Which is not correct definition of transgene?
(A) A gene from a related species
(B) A gene from an unrelated species
(C) A gene transferred into an organism using recombinant DNA technology
(D) A synthetic gene

2251. The bacterial XCPRT gene serves as dominant selectable marker on a culture medium containing shih of the following?
(A) Xanthine and Adenine (B) Mycophenolic acid
(C) Aminopterin (D) All of these

2252. Which is correct about insertional mutagenesis?
(A) It often produces aberrant phenotypes
(B) It occurs by chance and can not be controlled
(C) Occurrence of insertional mutagenesis leads to accidental discovery of unsuspected genes and gene functions
(D) All of these

2253. 'Nature Gard' is a transgenic variety of:
(A) Brinjal (B) Potato
(C) Cotton (D) Corn

2254. DEAE-dextran brings about DNA uptake by :
(A) Pincocytosis (B) Endocytosis
(C) Active transport (D) Osmosis

2255. The concept of genetically engineered cross protection can be exploited using which approaches?
(A) Expression of satellite RNA and replicase sequences
(B) Expression of viral protease and movement proteins
(C) Use of satellite RNA and antisense RNA
(D) All of these

2256. A silent mutation is one that:
(A) Result in a truncated polypepetide
(B) Replaces an amino acid with an equivalent amino acid in a polypepetide
(C) Does not change the amino acid sequence of the polypeptide
(D) Changes the leading frame of the mRNA leading to an altered polypeptide

2257. Which of the following conditions is likely to interfere with the transfer of genetic material by conjugation in bacteria?
(A) Pre-treatment of the recipient cells with DNase
(B) Pre-treatment of the recipient cells by application of strong shearing forces
(C) Treatment of the recipient cells with cycloheximide
(D) Treatment of the mating cell pairs by application of strong shearing forces

2258. Neomycin phosphotransferase II (antII) inactivates by which phosphorylation:?
(A) Kanamycin (B) Neomycin
(C) G418 (D) All of these

2259. Which of the edible vaccines for humans?
(A) Honey (B) Banana
(C) Apple (D) All of these

2260. Which is not correct about SV40 plasmid vectors?
(A) Plasmid vectors are generally used for stable trans fection
(B) They integrate into the host genome with a frequency of 10-5 to 10-3
(C) Plasmid vectors have been mainly used to investigate sequences involved in transcriptional and post-transcriptional regulation
(D) pRSV series of vectors are derivatives of pSV vectors

2261. Among the given proteins, which has been evaluated for insect control?
(A) Lectin and aprotinin
(B) Cowpea trypsin inhibitor and serine protease inhibitor
(C) Cystein protease inhibitor and proteinase inhibitor II
(D) All of these

2262. The first transgenic plant was produced in 1983, when a tobacco line expressing which genes was produced?
(A) Ampicillin resistance
(B) Kanamycin resistance
(C) Hygromycin resistance
(D) Phosphinothricin resistance

2263. Which has been used in chrysanthemum and pelargonium to make these plants dwarf, delayed their leaf senescence and increased chlorophyll pigmentation:
(A) Suppression of endogenous genes of chalcone synthase
(B) Alteration of the cytokinins: auxin ratio by introducing and expressing specific transgenes
(C) Over expression of the gene encoding phytochrome A
(D) Gene disruption technology

2264. In Drosophila, XXY is female. In humans it represents an abnormal male beause:
(A) Y – chromosome induces male traits in humans
(B) Y – chromosome is essential for female sex in Drosophila
(C) Y – chromosome is not essential for male sex in humans
(D) None of these

2265. Crystal proteins from *B. thuringiensis* are solublized in the alkaline environment of insect midgut and then proteolytically processed to yield a toxic core fragment of which size?
(A) 50 kDa (B) 60 kDa
(C) 80 kDa (D) 70 kDa

2266. Stability/expression of transgenes in plants can be affected by:
(A) Gene silencing (B) Chromosomal rearrangements
(C) Pleiotropy (D) All of these

2267. *Petite* mutants in yeast stand for slow growing and small colony forming yeast cells. Genetic crosses between different haploid yeast strains show that the *Petite* mutations do not segregate with any known chromosome. The defect could be in the:
(A) Nucleus (B) Mitochondria
(C) Golgi body (D) ER

2268. The first animal vector was devised for :
(A) Simian Virus 40 (SV 40)
(B) Primate Papova Virus
(C) Papolloma Virus
(D) Both (A) and (B)

2269. Among the given tests which one is the most rapid and convenient?
(A) Immunoblotting
(B) ELISA
(C) Radioimmunoprecipitation
(D) Immunnohistochemical staining of tissue sections

2270. CryI proteins are insecticidal to:
(A) Coleoptera (B) Lepidoptera
(C) Orthoptera (D) Diptera

2271. Gene is in T-DNA determines the synthesis of:
(A) Cytokinin (B) Indole acetic acid
(C) Opine (D) Isopentenyl adenine

2272. How many different types of gametes could be produced by an individual with the arbitrary genotype of AAbbCCDdEe?
(A) Two (B) Four
(C) Six (D) Eight

2273. The TA29 promoter sequence was isolated from:
(A) Cotton (B) Tomato
(C) Tobacco (D) Cauliflower

2274. 'Control of Plant Gene Expression' was dubbed as 'Terminator Technology' by:
(A) Delta and Pine Land Company
(B) United States Department of Agriculture, USA
(C) Rural Advancement Foundation International, Canada
(D) Department of Biotechnology, New Delhi

2275. T-DNA is flanked on both the sides by direct repeat sequences of how many base-pairs?
(A) 24 bp (B) 44 bp
(C) 34 bp (D) 14 bp

2276. Modification of flower colour intensity in petunia, rose, carnation has been achieved by:
(A) Suppression of endogenous genes of chalcone syn thase
(B) Alteration of the cytokinins: auxin ratio by introducing and expressing specific transgenes
(C) Over expression of the gene encoding phytochrome A
(D) Gene disruption technology

2277. Who said that we need to have a National Biodiversity Act, a Plant Variety Act, Farmers Right's Act and a Geographic Appellation Act to deal with the problems created by terminator technology?
(A) M.S. Swaminathan (B) B.P. Pal
(C) R.S. Paro la (D) Panjab Singh

2278. SV 40 genome has been used to develop by:
I. Plasmid vectors
II. Transforming vectors
III. Tranducing vectors
IV. Early replacement vectors
(A) I, II, III, IV (B) I, II, III
(C) I, III, IV (D) II, III, IV

2279. Which is a more general approach for identification of transgenic animals?
(A) Dot blot techniques
(B) PC amplification
(C) Both (A) and (B)
(D) Southern hybridization

2280. Particle gun is suitable for transformation of:
(A) Embryogenic calli (B) Leaf discs
(C) Immature zygotic embryos (D) All of these

2281. Assume that two pairs of genes with two alleles each, Aa and Bb determine plant height additively in a population. The homozygote AABB is 50 m high and aabb 30 m. What is the height of the F_1 of a cross between them?
(A) 50 cm (B) 40 cm (C) 30 cm (D) 20 cm

2282. Stable integration of T-DNA has been reported in:
(A) Sunflower and Brassica (B) Cotton and tomato
(C) Petunia and tobacco (D) All of these

2283. Which sequences has been used to produce a virus resistant variety of squash?
(A) Defective virus genome (B) Ribozyme sequence
(C) cDNA of satellite RNA (D) Virus coat protein gene

2284. Which can be used to modify size and shape of ornamental plants?
(A) Suppression of endogenous genes of chalcone synthase
(B) Alteration of the cytokinins: auxin ratio by introducing and expressing specific transgenes
(C) Over expression of the gene encoding phytochrome A
(D) Gene disruption technology

2285. Potato transgenic variety 'New Leaf' is resistant to:
(A) Potato virus (B) Colorado potato beetle
(C) Late blight (D) Frost

2286. Teminism refers to:
(A) DNA-RNA hybridization (B) DNA denaturation
(C) Reverse transcription (D) Transcription

2287. Among the given gene transfer techniques, which is the most commonly used in case of mice embryos?
(A) Retroviral infection
(B) Microinjection
(C) Embryonic stem cell technology
(D) Electroporation

2288. Which of the following aims at transformation of cells in the developing inflorescence?
(A) Macroinjection (B) Electroporation
(C) Microinjection (D) Partcle gun

2289. If inducible gene expression is driven by metallothionein promoter, transcription is induced by which metal ions?
(A) Cadmium (B) Zinc
(C) Iron (D) Both (A) and (B)

2290. The term transfection is used to denote:
(A) Genetic transformation of bacteria via virus infection
(B) Genetic transformation of animal cells
(C) Delivery of recombinent DNA into animal cell via a virus
(D) All of these

2291. Liposime-mediated DNA uptake has been reported in protoplasts of:
(A) Petunia (B) Tobacco
(C) Carrot (D) All of these

2292. Which approaches can be used to increase the toxicity of Cry proteins?
(A) Increasing copy number of the cry genes
(B) Site directed mutagenesis within the toxin encoding region
(C) Production of hybrid cry genes by fusing selected toxin domains of two or more different cry genes
(D) Both (B) and (C)

2293. TA29 promoter is specific to:
(A) Tapetum (B) Pollen
(C) Ovule (D) Pollen mother cells

2294. If a chiasm forms between the loci of genes A and B in 20% of the tetrads of an individual of genotype AB/ab, the percentage of gametes expected to be Ab is:
(A) 40 (B) 20
(C) 10 (D) 5

2295. Transgenics have been produced in which animals?
(A) Cow and goat (B) Sheep and Rabbits
(C) Mice, fish and pigs (D) All of these

2296. Transgenic animal has:
(A) Foreign DNA in all its cells
(B) Foreign RNA in all its cells
(C) Foreign DNA in some of the cells
(D) Both (B) and (C)

2297. Which viral transduction expression vectors is used for in-vivo gene transfer in mammalian hosts?
(A) Adeno-associated virus (B) Baculovirus
(C) Both (A) and (B) (D) Vaccinia virus

2298. The chromatography column used to separate recombinant protein having polyhstidine tags has :
(A) Divalent nickel (B) Ca^{++}
(C) K^{+} (D) Both (A) and (B)

2299. Which is correct about toxic fragments of cry proteins?
(A) They have three domains, domain I functions in pore and iron channel formation, domain II is involved in binding of the receptor, and domain III is involved in receptor recognition
(B) They are -80000 times more toxic to the target insects than organophosphate insecticides
(C) They are highly selective in their action
(D) All of these

2300. Which states of India has the lowest area under cultivation of transgenics cotton varieties?
(A) Uttar Pradesh (B) Madhya Pradesh
(C) Maharashtra (D) Andhra Pradesh

2301. Which contributes to high level transgene expression?
(A) Inclusion of an intron in the expression cassette
(B) Removal of unnecessary untranslated sequences
(C) Codon optimization
(D) All of these

2302. Cloning is a mean to:
(A) Replace original genotype
(B) Preserve genotype
(C) Production of HGH gene in Escherichia coli
(D) None of these

2303. Which Agrobacterium proteins functions as a receptor?
(A) VirA (B) VirG
(C) VirD1 (D) VirD2

2304. 'Golden rice' has transgenes that provide which enzyme activities?
(A) Phytoene synthesase (B) Phytoene desaturase
(C) Lycopene β–cyclase (D) All of these

2305. Mammalian artificial chromosomes and human artificial chromosomes are linear vectors that contain :
(A) Centromeres and telomeres
(B) Origin of replication recognition sequences
(C) Transcriptionally active chromosomal domains
(D) All of these

2306. Which is not correct about promoter/enhancer?
(A) A promoter sequence is the site to which RNA polymerase first binds during the initiation of transcription
(B) Promotor must be located at the 5' –end of the antisense brand of the gene
(C) The promoter sequence determines the level of transcription of the gene and tissue or cell type where the gene will be expressed
(D) A suitable enhancer sequence may be used alongwith the promoter to enhance the promoter activity

2307. Which promoters has been the most wisely used in dicots?
(A) Vicilin (B) Extensin
(C) 35S (D) Adh1

2308. Which is correct about edible vaccines?
(A) They can be used as regular component of food
(B) Honey could be used as edible vaccine
(C) Their production cost is comparable to that of conventional vaccines
(D) They are useful only when the native antigen is orally inactive

2309. All the transfection techniques are applicable to cultured animal cells, but microinjection is ordinarily not used due to :
(A) Limited number of cells can be handled
(B) Tediousness of the technique
(C) Availability of less demanding alternative techniques
(D) All of these

2310. Which Agrobacterium proteins functions as transcription activator?
(A) VirA (B) VirG
(C) VirD1 (D) VirD2

2311. Which Cry proteins is specific to Lepidoptera and diptera?
(A) Cry II (B) Cry V
(C) Cry IV (D) Both B and C

2312. Maximize and yield gard are transgenic varieties of:
(A) Cotton (B) Maize (C) Cotton (D) Tobacco

2313. Early region replacement and late region replacement vectors of SV40 have:
(A) Transient expression
(B) Produce virions, which infect host cells
(C) Mammalian cells are hosts of SV40-derived vectors
(D) All of these

2314. Which statement is not correct?
(A) In case of electroporation, circular super-coiled DNA is far more efficient in transfection than linearized DNA
(B) The male pronucleus is chosen for microinjection because it is much larger than the female pronucleus
(C) The use of cationic lipids for construction of liposomes offers a distinct advantage
(D) Recombinant retroviruses produce virions

2315. In which insects, a >250- fold (highest) increase in the resistance against Cry proteins was observed after 36 generation of laboratory selection?
(A) *Helicoverpa virescence* (B) *Cadra cutella*
(C) *Plodia interpunctella* (D) *Plutella xylostella*

2316. Which states of India has the highest area under cultivation of transgenics cotton varieties?
(A) Uttar Pradesh (B) Madhya Pradesh
(C) Maharashtra (D) Andhra Pradesh

2317. The DNA of bacteriophage T4 contains approximately 200,000 nuclotide pairs. The a11 region of the T4 genome coupies about 1% of its total genetic length. Benzer has found that 300 sites are separable by recombination within the r11 region. What will be average number of nucleotides in each recon?
(A) About 20 (B) About 300
(C) About 7 (D) Cann't be determined

2318. Xenopus oocytes can be used for expression of :
(A) mRNAs obtained by in vitro transcription of eukaryotic genes using T7RNA polymerase
(B) Eukaryotic mRNAs isolated from various tissues
(C) Genomic eukaryotic DNA micro-injected into the nucleus
(D) All of these

2319. Which Agrobacterium proteins topoisomerase functions?
(A) VirA (B) VirG (C) VirD1 (D) VirD2

2320. Which is not correct?
(A) MAHYCO developed Bt cotton hybrids by transferring Bt (Cry 1Ac) gene from cotton transgenic variety Coker 312 into bred lines
(B) Cotton hybrids Bt ECH 12, bt MECH 162 and Bt MECH 184 were approved bt Genetic Engineering Approval Committee of the Ministry of Environment, Govt. of India, for commercial release
(C) Till March 2005, a total of 20 Bt-transgenic cotton hybrids has been released for commercial cultivation in India
(D) In 2004, of the 81.0 million hectares under transgenics, developing countries share was 53.4 mha 966%) as compared to 27.6 mha (34%) in the industrial countries

2321. Which is not correct about transgene integration in animal genomes?
(A) Usually, in a given cells, integration occurs at a single location
(B) In general, multiple copies are integrated when small amounts of DNA are used for transfection
(C) When multiple copies are integrated, they are mostly integrated at one site joined to each other head-to-tail
(D Transgene integration frequently leads to various forms of rearrangements

2322. A species of plant is discovered in which individual plants produce either white or purple flowers. True breeding plants that have short stems with white flowers are crossed with true breeding plants that have tall stems and purple flowers. The resulting offspring all have tall with purple flowers. When one of these tall, purple-flowered offspring is crossed with a short, white – flowered plant, plants with the following characteristics were obtained in equal proportions. White flowers on tall stems and purple flowers on tall stems. White flowers on short stems, purple flowers on short stems. Given the information above, which of the following is most likely true?
(A) The white-flower allele and the tall-stem allele occur at the same locus
(B) The genes for flower colour and stem length are loosely linked
(C) Extensive crossing-over between the purple-flower allele and the white-flower allele has occurred
(D) The genes for flower colour and stem length are not linked

2323. Which promoters is seed specific?
(A) Vicilin (B) Extensin
(C) 35S (D) Adh1

2324. Which can reduce transgene silencing?
(A) Same orientation of the transgenes in the recombinant DNA
(B) A.T composition of transgene simiar to that of the host chromosome
(C) Single copy integration of transgene
(D) All of these

2325. Papaya transgenic Rainberg is resistant to:
(A) Papaya Mosaic Virus
(B) Leaf Curl Virus
(C) Papaya Ring Spot Virus
(D) Distorted Ring Spot Virus

2326. An Hfr strain of *E. coli* contains:
(A) A bacterial chromosome with a human gene inserted
(B) A bacterial chromosome with the F factor inserted
(C) A human chromosome with a transposable element inserted
(D) A bacterial chromosome with a phage inserted

2327. The yeast artificial chromosomes have found regular use for cloning large genome DNA fragments in various genome sequencing projects is:
(A) BACs (B) PACs
(C) Both (A) and (B) (D) None of these

2328. For gene therapy, we need to have gene-sized human DNA fragments including their:
(A) Promoters (B) Control elements
(C) Both (A) and (B) (D) None of these

2329. The restriction maps provide the basic knowledge of the physical organization of a genomic region or a whole genome with respect to the cleavage sites of:
(A) Restriction endonuclease (B) Detritylation
(C) Coupling (D) Capping

2330. The chosen derivatized solid support, with the required 3' end nucleotide hooked onto it, is treated with trichloreacetic acid is called:
(A) Restriction endonuclease (B) Detritylation
(C) Coupling (D) Capping

2331. The nest nucleotide is chosen as the phosphoramide derivative whose 5' – OH group is also protected with dimethoxytrityl group is called:
(A) Restriction endonuclease (B) Detritylation
(C) Coupling (D) Capping

2332. There is always a possibility that some of the 3'-nucleotide (the first nucleotide from the 3' end in this case) may have escaped the coupling reaction and still have a reactive free 5'-OH group is known as:
(A) Restriction endonuclease (B) Detritylation
(C) Coupling (D) Capping

2333. The linkage between the first and the second nucleotide is then converted from the phosphate to the more stable phsphotriester by using iodine as the oxidizing agent and water as the oxygen donor is called:
(A) Oxidation (B) Detritylation
(C) Coupling (D) Capping

2334. The deletion of small phenolic compounds released by wounded plants resulting in autophosphorylation and then phohorylation of vir G by which vir gene?
(A) Vir A (B) Vir G
(C) Vir D1/D2 (D) Vir D2

2335. The activation of other vir gene by which gene:
(A) Vir A (B) Vir G
(C) Vir D1/D2 (D) Vir D2

2336. Recognize the 25 bp border sequence and produce a single-stranded endonucleallytic cleavage in the bottom strand of each border by which vir gene:
(A) Vir A (B) Vir G
(C) Vir D1/D2 (D) Vir D2

2337. Which vir gene is remains tightly associated with the 5' end of the T strand:
(A) Vir A (B) Vir G
(C) Vir D1/D2 (D) Vir D2

2338. The Ss binding protein rendering it resistant to nucleolytic degradation by which vir gene:
(A) Vir A (B) Vir G
(C) Vir E2 (D) Vir B

2339. Which vir gene provide energy for the translocation of the T-complex:
(A) Vir A (B) Vir G
(C) Vir E2 (D) Vir B

2340. All genetic transformations in monocot plants had been performed using direct:
(A) DNA delivery system (B) Cis-inactivation
(C) Cosuppression (D) Gene silencing

2341. When silencing takes place at the same place due to multiple insertion it can be referred to as:
(A) DNA delivery system (B) Cis-inactivation
(C) Cosuppression (D) Gene silencing

2342. The post-transcriptional gene silencing had been reported in multiple homologous transgenes and the endogeneous genes that share common coding sequences, the process referred to as:
(A) DNA delivery system (B) Cis-inactivation
(C) Cosuppression (D) Gene silencing

2343. Single copy transgene is also reported to be inactivated due to:
(A) DNA delivery system (B) Cis-inactivation
(C) Cosuppression (D) Gene silencing

2344. To demarcate those plants which does not bear a trans gene i.e. gene from other organism, rather it carries gene from itself but in a modified version is:
(A) Cis genic (B) Multi genic
(C) Uni genic (D) None of these

2345. The first genes available for genetic engineering crop plants for pest resistant were:
(A) Cis genic (B) Multi genic
(C) Uni genic (D) Cry genes

2346. The first genes available for genetic engineering crop plants for pest resistance were Cry genes a bacterium *Bacillus thuringiensis*, popularly known as:
(A) Bt genes (B) Multi genic
(C) Uni genic (D) Cry genes

2347. Bt cotton is resistant to a notorious insect pest of:
(A) *Helicoverpa armigera* (B) Termites
(C) Spodoptera (D) Aphids

2348. Bt cotton was adopted in India in:
(A) 2000 (B) 2001 (C) 2002 (D) 2003

2349. The transgenic technology provides an alternative and innovative method to improve pest-control management which is:
(A) Eco-friendly (B) Effective
(C) Sustainable and increase yield (D) All of these

2350. Bt cotton contains:
(A) Bt gene (B) Multi genic
(C) Cry IAC gene (D) Cry genes

2351. Genetic engineering can also be used to develop transgenic fruits that show:
(A) Delayed fruit softness (B) Improving nutrient
(C) Processing qualities (D) All of these

2352. Genetically engineered rice (golden rice), which is enriched in:
(A) Provitamin A (B) Vitamin B
(C) Mineral (D) Vitamin K

2353. The production of pathogen-free plants with uniform quality for agronomic cultured and also for germplasm conservation is:
(A) Micro propagation
(B) In-vitro culture
(C) Dedifferentiated cultures
(D) Differentiated cultures

2354. The production of secondary metabolites under controlled cultures conditions is:
(A) Micro propagation
(B) In-vitro culture
(C) Dedifferentiated cultures
(D) Differentiated cultures

2355. In vitro cultures are currently being used for secondary metabolite production of callus and suspension cultures is:
(A) Micro propagation
(B) In-vitro culture
(C) Dedifferentiated cultures
(D) Differentiated cultures

2356. In vitro cultures are currently being used for secondary metabolite production of transformed roots and shoots is:
(A) Micro propagation
(B) In-vitro culture
(C) Dedifferentiated cultures
(D) Differentiated cultures

2357. In 1958, Arthur Komberg and coworker purified an enzyme from the soluble fraction of the lysate of E. coli that could synthesize DNA in vitro and it was named:
(A) DNA polymerase I (B) DNA polymerase
(C) DNA looping (D) DNA library

2358. With a natural DNA tempelate, addition of all the four deoxyribonucleoside 5'-triphosphates was found essential for the poly-merization reaction catalyzed by the:
(A) DNA polymerase I (B) DNA polymerase
(C) DNA looping (D) DNA library

2359. A lambda cloning vecter can therefore be constructed by deletion (in-vivo or in-vitro by restriction deletion) of a part of the nonessential region such that the remainder is not less than 35 kb is:
(A) Insertion-type lambda vectors
(B) Substitution type lambda vectors
(C) Both (A) and (B)
(D) None of these

2360. The lambda vectors have two EcoR, sites or twp Bam HI sites in the nonessential; region is:
(A) Insertion-type lambda vectors
(B) Substitution type lambda vectors
(C) Both (A) and (B)
(D) None of these

2361. Cosmid vectors were first developed by Collins and coworkers in:
(A) 1977 (B) 1978 (C) 1979 (D) 1980

2362. Yeast artificial chromosomes was developed by D.T. Burke and C. F. Carle in:
(A) 1987 (B) 1988 (C) 1989 (D) 1990

2363. Bacterial artificial chromosomes was developed by Melsimon and coworkers in:
(A) 1990 (B) 1991 (C) 1992 (D) 1993

2364. A form of IPR protection, they cover such qualities, reputation or other characteristics of a product that can be essentially attributable to the geographical origin of the product, and that could be used to recognize that a product had originated from a particular terriority is called:
(A) Geographical Indications (B) Trademark
(C) Copyright (D) None of these

2365. It is an idea, a design, an invention, a manuscript etc., which can ultimately give rise to useful product/application is called:
(A) Intellectual Property
(B) Intellectual Property Right
(C) Both (A) and (B)
(D) None of these

2366. The right of an inventor to derive economic benefits from his invention, i.e., intellectual property, and to exclude others from doing so is called:
(A) Intellectual Property
(B) Intellectual Property Right
(C) Both (A) and (B)
(D) None of these

2367. The European Patent Convention (EPC) excludes from patent protection such inventions that are contrary to the public order of morality is under the act of:
(A) Article 53 (a) (B) Article 53 (b)
(C) Article 54 (a) (D) Article 54 (b)

2368. The European Patent Convention excludes from patent protection (1) plant and animal varieties and (2) essential biological processes for the production of plants and animals under the act of:
(A) Article 53 (a) (B) Article 53 (b)
(C) Article 54 (a) (D) Article 54 (b)

2369. A collection of right that are granted to an individual or groups of individuals for any work or creation that involved intellectual inputs is called:
(A) Intellectual Property
(B) Intellectual Property Right
(C) Both (A) and (B)
(D) None of these

2370. Intellectual Property is categories into:
(A) Industrial property
(B) Artistic and Literary property
(C) Both (A) and (B)
(D) None of these

2371. The protection to any kind of confidential information from the industry that has a commercial value or industrial significance is called:
(A) Trade secrete (B) Industrial design
(C) Geographical indications (D) Trade mark

2372. In plant-related inventions, the parent lines employed to produce a hybrid plant can be kept as a:
(A) Trade secrete (B) Industrial design
(C) Geographical indications (D) Trade mark

2373. Industrial property are:
(A) Trade secrete (B) Industrial design
(C) Geographical indications (D) Trade mark
(E) All of these

2374. Artistic and Literary property are:
(A) Copy right (B) Industrial design
(C) Geographical indications (D) Trade mark

2375. The deals with protection of the expression of an idea; however it does not protect the idea itself is called:
(A) Copy right (B) Industrial design
(C) Geographical indications (D) Trade mark

2376. The exclusive rights that are granted to the inventor to prevent third parties from making, using, selling, importing and offering for sale the invention, unless an authorization or license is obtained from the patentee is called:
(A) Patents (B) Industrial design
(C) Copy right (D) Trade mark

2377. In biotechnology, the significance of copyright would be:
(A) Possibility of copyrighting sequences
(B) Methods or manuals of exponential procedures
(C) Both (A) and (B)
(D) None of these

2378. The intellectual property that have protected by an alternative type of protection termed as:
(A) Sui Generic system (B) Trademark
(C) Copy right (D) None of these

2379. Sui Generic protection are:
(A) Plant variety protection
(B) Integrated computer circuits
(C) Data base protection
(D) All of these

2380. The TRIPs Agreement came is:
(A) 1994 (B) 1995 (C) 1996 (D) 1997

2381. The TRIPs Agreement came in to force in January 1, 1995 with immediate effect but with certain exceptions for which provisions were made which were through to came into force by:
(A) 1998 (B) 1999 (C) 2000 (D) 2001

2382. The patent Act, 1970 was amended first time in the year of:
(A) 1999 (B) 2002 (C) 2005 (D) 2006

2383. The patent Act, 1970 was amended second time in the year of:
(A) 1999 (B) 2002 (C) 2005 (D) 2006

2384. The patent Act, 1970 was amended third time in the year of:
(A) 1999 (B) 2002 (C) 2005 (D) 2006

2385. The term of protection shall not end before the expiration of a period of twenty years counted from the date of filing in:
(A) Article 33 (B) Article 34
(C) Article 35 (D) Article 36

2386. The time period accorded for protection of the invention has been set uniformly for all categories of inventions at:
(A) 10 years (B) 15 years
(C) 20 years (D) 25 years

2387. An 'invention' as 'a new product or process involving an inventive step and capable of industrial application' under:
(A) Section 2(1) (j) (B) Section 2(1) (ja)
(C) Section 3 (j) (D) Section 3 (ja)

2388. In 'inventive step' as 'inventive step means a feature of an invention that involves technical advance as compared to the existing knowledge of having economic significance or both and that makes the invention not obvious to a person skilled in the art' under:
(A) Section 2(1) (j) (B) Section 2(1) (ja)
(C) Section 3 (j) (D) Section 3 (ja)

2389. Essentially biological process falls under the exclusion of patentability under the section of the Indian patent act :
(A) Section 2(1) (j) (B) Section 2(1) (ja)
(C) Section 3 (j) (D) Section 3 (ja)

2390. Patentability requirements as per the Indian Patents Act are:
(A) It should be new
(B) It should involve an inventive step (nonobviousness of the invention)
(C) It should be capable of industrial application
(D) All of these

2391. Under Article 4 related to patents in TRIPs and ZEU Directives the following shall not be patentable:
(A) Plant and animal varieties
(B) Essentially biological processes of the production of plants and animals
(C) Both (A) and (B)
(D) None of these

2392. Essentially biological process falls under the exclusion of patentability under the section of the Indian Patent Act:
(A) Section 2(1) (j) (B) Section 2(1) (ja)
(C) Section 3 (j) (D) Section 3 (ja)

2393. The EU Directive Article 4.1 (b) appears to strongly exclude plant and animal varieties, but Article —— makes clear that – transgenic plants and animals are patentable, as they are produced by 'microbiological' or other technical process:
(A) Article 33 (B) Article 4.3
(C) Article 35 (D) Article 36

2394. The process by which used in making transgenic plants and animals are biological process, and so transgenic plants and animals should not be patentable:
(A) Transformation (B) Transfection
(C) Both (A) and (B) (D) None of these

2395. Whether or not a (non microbiological) process is to be considered as 'essentially biological' within the meaning of Article ——— EPC has to be judged on the basis of the essence of the invention, taking into account the totality of human intervention and its impact on the results to be achieved:
(A) Article 33 (B) Article 53(b)
(C) Article 35 (D) Article 36

2396. Gene splicing technique is one of the important techniques of:
(A) R-DNA technology (B) C-DNA technology
(C) m-DNA technology (D) D-DNA technology

2397. Patentable processes involving R-DNA technology are:
(A) Method of transferring cells or complete biological entity such as microbes, plants or animals
(B) Method of cloning and expression of desired gene in the same or different cell
(C) Method of optimizing gene expressions to produce useful therapeutic proteins
(D) Method of producing a transgenic plant or animal
(E) Method of serving and detection of specific DNA sequences
(F) Method of designing probes, primers, recombinant vectors
(G) Method of modifying genomic sequences
(H) All of these

2398. Patentable products involving R-DNA technology:
(A) Different cloning vehicle (vectors) such as plasmids, cosmids, phagemids, etc
(B) Probes, primers and oligonucleotides
(C) Vaccines and other useful proteins
(D) Transgenic/genetically modified organisms
(E) All of these

2399. For patenting of DNA/RNA sequences, ESTs, SNPs, the following steps should be undertaken:
(A) Identify novel genetic sequences
(B) Specify the resulting product from the sequence
(C) Specify how the product functions in nature (use)
(D) Enable one skilled in the field to use the sequence for its stated purpose
(E) All of these

2400. The genes are not patentable when they are in the:
(A) In-situ (B) Ex-situ
(C) Both (A) and (B) (D) None of these

2401. The major differences between the 1995 guidelines and the 2001 guidelines was the requirement that the claimed utility has to be:
(A) Substantial (B) Specific
(C) Credible (D) All of these

2402. The 2001 utility Examination Guidelines follow the 1999 Revised Interim Utility Guidelines in adopting the structure of:
(A) Specific, substantial and credible utility test
(B) Well established utility test
(C) Both (A) and (B)
(D) None of these

2403. Under both the 1999 and 2001 guidelines, a utility is 'specific' when it particularly concerns the:
(A) Subject matter claimed (B) Real world use
(C) DUS test (D) Novelty

2404. A European patent (EP 1417338) was granted in 2004 for genes and SNPs associated with eating disorders like:
(A) Anorexia nervosa (B) Bulimia
(C) Both (A) and (B) (D) None of these

2405. Under both 1999 and 2001 guidelines establish that a 'substantial utility' is one that defines a:
(A) Subject matter claimed (B) Real world use
(C) DUS test (D) Novelty

2406. A new variety shall be registered under PPVFR Act 2001 if it confirms to the criteria of novelty and fulfils the:
(A) Subject matter claimed (B) Real world use
(C) DUS test (D) Novelty

2407. The variety should not have been sold (offered for sale) or marketed earlier than one year for the country and four years for other countries where protection is desired :
(A) Subject matter claimed (B) Real world use
(C) DUS test (D) Novelty

2408. A US patent (US5,972,609) has been awarded in 1999 for a utrophin gene promoter that they may be used to control of:
(A) Transcription of heterologous sequences
(B) PPVFR Act 2001
(C) Distinctiveness
(D) Uniformity

2409. Indian chose to opt for a Sui Generis System for protection of plant and plant varieties by enacting the:
(A) Transcription of heterologous sequences
(B) PPVFR Act 2001
(C) Distinctiveness
(D) Uniformity

2410. The variety should be clearly distinguishable from any other variety of common knowledge is:
(A) Transcription of heterologous sequences
(B) PPVFR Act 2001
(C) Distinctiveness
(D) Uniformity

2411. The variety is uniform/homogeneous if, subject to the variation that may be expected from the particular features of it propagation, it is sufficiently uniform in it relevant characteristics is:
(A) Transcription of heterologous sequences
(B) PPVFR Act 2001
(C) Distinctiveness
(D) Uniformity

2412. The variety is stable if its relevant (essential) characteristics remain unchanged after repeated propagation is:
(A) Stability (B) Microorganisms
(C) Protection (D) None of these

2413. TRIPs allows members to exclude from patentability, plants and animals other than:
(A) Stability (B) Microorganisms
(C) Protection (D) None of these

2414. Plants found in the wild and other land races cannot be protected under an IPR system as they do not meet the requirements for:
(A) Stability (B) Microorganisms
(C) Protection (D) None of these

2415. All types of genes can be patented as long as an inventive step has been involved to:
(A) Identify
(B) Isolate
(C) Modify and/or transfer
(D) All of these

2416. There has been a decreasing intellectual property activity in the area of biotechnology since:
(A) 2000 (B) 2001 (C) 2002 (D) 2003

2417. Which Indian organisation is more active in the area of patenting:
(A) ICAR (B) CSIR (C) ICMR (D) BARC

2418. Some of the Indian companies active in the area of intellectual property right are:
(A) Ranbaxy (B) Dabur
(C) Ayur (D) All of these

2419. The institution of CSIR active in the area of intellectual property right are:
(A) NBRI (B) IICT
(C) CIMAP (D) All of these

2420. Patenting of stem cells may involved:
(A) Product patents (B) Process patents
(C) Both (A) and (B) (D) All of these

2421. There are different forms of protection for plant-related inventions are:
(A) Utility patents
(B) Plant Patent Act
(C) Plant Variety Protection Act
(D) All of these

2422. The utility patents and the plant protection patents are granted by:
(A) USPTO (B) PVTO
(C) Both (A) and (B) (D) None of these

2423. The plant variety protection act is granted by:
(A) Plant Variety Protection Officer
(B) Plant Breeder
(C) Plant Protection Officer
(D) All of these

2424. The patenting can also be done in:
(A) Transformations vectors
(B) Selectable marker genes
(C) Reporter genes
(D) All of these

2425. In India, animals in whole and any part thereof are not patentable under section of Patent Act:
(A) Section 2(1) (j) (B) Section 2(1) (ja)
(C) Section 3 (j) (D) Section 3 (ja)

2426. Which amendment of the Patent Act also also makes it mandatory to deposit the biological materials mentioned in the specification with IDA is:
(A) 2000 amendment (B) 2001 amendment
(C) 2002 amendment (D) 2003 amendment

2427. Plant varieties may be protected in India under:
(A) PVPFRA (B) PVRA
(C) Both (A) and (B) (D)None of these

2428. Under Section 3(j) of Indian Patents Act, 1970 plants are not considered patentable subject matter under section 3(h) of the Act are:
(A) Method of agriculture
(B) Method of horticulture
(C) Both (A) and (B)
(D) None of these

2429. The first plant patent was issued in 1931 to Henry Rosenberg of New Brunswick NJ for:
(A) Climbing rose (B) Trailing rose
(C) Either (A) or (B) (D) None of these

2430. The plant patent for a rose plant named —— granted to Ezio Bree in 2003 is:
(A) Breapum (B) Zembla
(C) Both (A) and (B) (D) None of these

2431. The patent for chrysanthemum plant named is:
(A) Breapum (B) Zembla
(C) Both (A) and (B) (D) None of these

2432. A patent for chrysanthemum plant named 'Zembla' was granted mainly for:
(A) Decorative inflorescence
(B) Excellent Past harvest longevity
(C) Both (A) and (B)
(D) None of these

2433. The invention for climbing roses bearing true:
(A) Blooming characters
(B) Vegetative characters
(C) Growth characters
(D) Transplanting characters

2434. In biotechnological components that are also patentable:
(A) Transformation vectors
(B) Selectable marker genes
(C) Reporter genes
(D) Promoter and enhancers
(E) All of these

2435. The act covers varieties of asexually propagated crops, e.g., ornamental and fruit trees by:
(A) The US Plant Patent Act (1930)
(B) The US Plant Variety Protection Act (1970)
(C) The Utility Patents Act (1985) of USA
(D) All of these

2436. The US version of the plant breeders right system followed by European Union and several other countries by:
(A) The US Plant Patent Act (1930)
(B) The US Plant Variety Protection Act (1970)
(C) The Utility Patents Act (1985) of USA
(D) All of these

2437. The act was originally meant to cover man-made industrial inventions and processes. Non-obviousness is the main criterion of utility patent; patent for plant varieties are now being granted under the provisions of this act is:
(A) The US Plant Patent Act (1930)
(B) The US Plant Variety Protection Act (1970)
(C) The Utility Patents Act (1985) of USA
(D) All of these

2438. The first patent of a life form was issued on:
(A) March 31, 1980 (B) March 31, 1981
(C) March 31, 1982 (D) March 31, 1983

2439. Trade secret relates to:
(A) Certain materials, e.g., bacterial strains
(B) Processes
(C) Formulae
(D) Any one of these

2440. Which is not correct about patent?
(A) It is the right of an individual or a group of indi viduals to gain profit from a particular invention or unique manufacturing process
(B) It is granted by the Government of a country to the applicant for an unlimited period of time
(C) When a patent is granted, the inventor becomes the owner of the patent like any other form of property
(D) Patents are granted for an invention (including a product), innovation/improvement in an invention, process/product of an invention and a concept

2441. In the area of biotechnology, which materials can be kept as trade secret?
(A) Cell lines
(B) Strains of microorganism
(C) Production processes
(D) All of these

2442. Trade secret can not be applied for:
(A) Plant varieties (B) Equipment design
(C) Books (D) All of these

2443. Which is not correct about breeder's exemption?
(A) The use of a protected variety for the development of new varieties is exempted from protection
(B) The PBR for a new variety evolved from the initial variety will be of the breeder who developed the new variety and of the holder of PBR title of the initial variety as well
(C) Under the UPOV 1978 Act, all new varieties evolved using a protected variety were exempted from protection under this provision
(D) The UPOV 1991 Act has limited breeders exemption in that 'essentially-derived varieties' are covered by the PBR granted to the initially variety

2444. Which peoples are considered under farmer's rights for sharing the profit earned from the development of high yielding varieties evolved from the use of raw materials/plant genetic resources?
(A) Tribal people
(B) Rural communities
(C) Traditional farming families
(D) All of these

2445. An otherwise excellent transgenic variety of which crops developed during 1990a could not be released due to IPR problems?
(A) Potato (B) Tomato
(C) Tobacco (D) All of these

2446. Indian has created a National Biodiversity Authority with headquarters at:
(A) New Delhi (B) Kolkata
(C) Chennai (D) Hyderabad

2447. The copyright of a book may be held by:
(A) Author(s) (B) Editor(s)
(C) Publisher (D) Any of these

2448. In USA, which systems are available for protection of IPRs related to plants?
(A) The Plant Patents Act (1930)
(B) The Plant Variety Protection Act of 1970
(C) The Utility Patents Act (1985)
(D) All of these

2449. In USA, which systems are available for protection of IPRs is considered to be the most powerful and the most expansive in scope of coverage?
(A) The Plant Patents Act (1930)
(B) The Plant Variety Protection Act of 1970
(C) The Utility Patents Act (1985)
(D) All of these

2450. The copyright suffers from:
(A) It provides protection for a specific period and only from reproduction as such of the copyright material
(B) It does not prevent another person from using either the idea or he information contained in a copyright material
(C) In case of DNA sequence, one may get around this protection by designing an alternative sequence to encode the same protein
(D) All of these

2451. PPVFR Act (2001) is comparable to UPOV Act (1991) with repect to:
(A) Requirement of novelty, distinctiveness uniformity and stability for registration of a variety
(B) Protection extended to commercial use of all the material of the protected variety
(C) Essentially-derived varieties being subjected to PBR protection granted to the concerned initial varieties
(D) Both (A) and (C)

2452. Which countries has 30 years protection for inbred lines of maize and for clovers and a few grasses?
(A) France (B) Germany
(C) USA (D) UK

2453. Which is the main considerations for the development of PBR systems?
(A) It allows breeders to benefit from the varieties developed by them
(B) It encourages plant breeding activities
(C) Private sector is encouraged to invest in plant breeding and seed industry
(D) All of these

2454. Government of India has now established the PPVFR Authority (PPVFRA) with head-quarters at:
(A) Chennai (B) Kolkata
(C) New Delhi (D) Hyderabad

2455. Which biotechnological inventions can be patented?
(A) Microorganisms, cell lines, plant/animal lies obtained through biotechnological approaches
(B) DNA sequences/genes and the proteins encoded, if any, by them
(C) Biotechnological process/technologies for modification of the properties of various organisms
(D) All of these

2456. Which features is not concerned with the PPVFR Act (2001)?
(A) The Central Government is to constitute a National Gene Fund
(B) Citizens of convention countries will have the same rights as citizens of India under the Act
(C) Application for registration of a variety may be made in India within 2 years from the date of application for registration of the same plant variety made in the convention country
(D) The rights of PBR holder 'shall not be deemed infringed by a farmer who was not aware of the existence of such right'

2457. Which biotechnology related maters can be patented?
(A) Method/process of production
(B) Products, etc
(C) Applications of the various processes/products'
(D) Any of these

2458. In case of PPVFR Act (2001), the gene fund shall be used fro:
(A) Paying compensation to communities for their contributions to the development of a variety
(B) Conservation of genetic resources
(C) Sustainable use of genetic resources
(D) All of these

2459. Methods for which can not be patented?
(A) Treatment of disease (B) Diagnosis of disease
(C) Prevention of disease (D) All of these

2460. Which life forms can not be patented in India under the provisions of Indian Patents Act (1970)
(A) Microorganisms (B) Animals
(C) Plants (D) All of these

2461. Patent activity is the lowest in:
(A) USA (B) Japan (C) Europe (D) India

2462. Which technologies and/or raw materials were used to develop Endless Summer tomato by DNA Plant Technology?
(A) Selectable marker *npt*II
(B) CaMV 35S promoter and the ACC synthase gene
(C) Antisense RNA technology and *Agrobacterium tumerfaciens* mediated genetic transformation
(D) All of these

2463. DNA Plant Technology faced difficulties during release of Endless Summer due to the use of:
(A) Selectable marker *npt*II
(B) CaMV 35S peromoter and the ACC synthase gene
(C) Antisense RNA technology and *Agrobacterium tumerfaciens* mediated genetic transformation
(D) All of these

2464. The PPVFR Authority (PPVFRA) began registration of plant varieties in which years?
(A) 2005 (B) 2006 (C) 2007 (D) 2008

2465. Which is an example of geographical indications?
(A) Champagne wine (B) Scotch whiskey
(C) California wine (D) All of these

2466. Geographical indications includes:
(A) Natural products (B) Agricultural goods
(C) Goods of handicraft (D) All of these

2467. Which is not correct about geographical indications (GIs):
(A) Once enforced by legislation, GIs exclude others from using a GI as trademark
(B) The chief requirement for GI protection is that a given quality, reputation or some other characteristics of the product be essentially attributable to the locality of its origin
(C) The GIs do not cover manufactured products and even food products
(D) Indian has enacted the Geographical Indication Act (1999) to claim GI for a variety of goods, including 'Basmati rice'

2468. Monitoring and tackling the IPR aspects of inventions has:
(A) Demand time
(B) Enhances cost
(C) May act as a distinctive for R & D efforts
(D) All of these

2469. PPVFR Act (2001) allows registration of:
(A) Farmer's varieties (B) Extant varieties
(C) New Varieties (D) All of these

2470. In the first instance, PPVFR began registration of varieties in how many of the crops?
(A) 6 (B) 12 (C) 2 (D) 20

2471. Physical containment relates to:
(A) Containment equipment (B) Laboratory design
(C) Special procedures (D) All of these

2472. Which is considered as primary physical containment?
(A) Laboratory design (B) Containment equipment
(C) Special procedures (D) Both (A) and (B)

2473. Which is considered as secondary physical containment?
(A) Laboratory design (B) Containment equipment
(C) Special procedures (D) Both (A) and (B)

2474. It is feared that transgenic plants, may lead to:
(A) Transfer of their transgenic feature to weeds
(B) May themselves become weeds
(C) Degrade the environment
(D) All of these

2475. In the early 1970s which were considered as the possible potential health hazards of cloning recombinant DNA molecules?
(A) Molecular biologists may not be well-versed with the laboratory practices needed to handle serious human and animal pathogens
(B) 'Hybrid organisms' could be created with biological activities of an unpredictable nature
(C) 'Hybrid organisms' may escape from the laboratory with unpredictable consequences
(D) All of these

2476. A transgenic variety of which crops had to be withdrawn from cultivation following due to allergy?
(A) Maize (B) Soybean
(C) Tomato (D) *Brassica napus*

2477. Which is not correct about NIH guidelines?
(A) The first NIH guidelines were prepared in 1975
(B) NIH guidelines were more liberal than the recommendations of the Asilomar Conference
(C) The guidelines were revised after two years
(D) In 1977, NIH prepared an Environmental Impact Statement

2478. Which is not correct about NIH guidelines?
(A) A major revision of the guidelines was effected in 1982
(B) In the revision, the containment levels were made more stringent
(C) Experiments that were previously prohibited, were changed to category requiring review and approval by NIH
(D) In USA, the NIH guidelines are followed by all federal agencies that fund research on recombinant DNA

2479. Which terms is not used as synonyms of planned introduction?
(A) Field testing (B) Deliberate release
(C) Contained use (D) Deliberate introduction

2480. The biosafety guidelines are developed to ensure an adequate level of protection during which activities of modern biotechnology?
(A) Safe transfer (B) Handling
(C) Use of GMOs (D) All of these

2481. The NIH, USA guidelines specify the practices for constructing and handling are:
(A) Recombinant DNA molecules
(B) Organisms and viruses containing recombinant DNA molecules
(C) Products obtained from B
(D) Both (A) and (B)

2482. In case of which products, an empirical approach based on data obtained through experimentation about the source of potential hazards has proved satisfactory?
(A) Pesticides (B) Drugs
(C) Vaccines (D) All of these

2483. In risk assessment, a lack of scientific knowledge or scientific consensus should not necessarily be interpreted as indicating:
(A) A particular level of risk (B) Absence of risk
(C) Acceptable risk (D) Any of these

2484. The assessment of risk during laboratory research is usually done in which steps?
(A) Initial risk assessment
(B) Comprehensive risk assessment
(C) Advanced risk assessment
(D) Both (A) and (B)

2485. The chief objectives of risk assessment during laboratory research are to decide about which of the proposed research?
(A) Laboratory procedures
(B) Biological containment
(C) Physical containment
(D) All of these

2486. Which is not correct about biosafety Guidelines in India?
(A) Initially, the Indian Recombinant DNA safety Guidelines and Regulation were prepared in 1990 by Recombinant DNA Advisory Committee, Department of Biotechnology, New Delhi
(B) The Indian Recombinant DNA safety Guide lines and Regulation were revised in 1994
(C) Field trials, in India using transgenic plants began in 1995
(D) Canada become the first country to begin commercial use of virus resistant transgenic tobacco and tomato in the early 1990

2487. Which statements is correct about comprehensive risk assessment?
(A) It is after the initial risk assessment
(B) The level of containment appropriate for an experiment is decided on the basis of comprehensive risk assessment
(C) The containment level for the experiment may, in the end, be comparable to that of the risk groups (RG) of the agent or it may be raised or lowered as a result of the comprehensive risk assessment
(D) All of these

2488. It is feared that genetically engineered microorganisms may disturb the ecosystem and its processes, in which ways?
(A) They may rapidly multiply and out complete the native microbes
(B) They may transfer genes related to virulence or pathogenesis into the native bacterial populations
(C) Both (A) and (B)
(D) They may cause unexpected health hazards

2489. Which is not correct?
(A) Oral exposure to transgenic tomato containing cry protein poses no additional risk to human and animal health
(B) In general, the quality of produce from cry gene carrying transgenic crops is regarded different from that from the nontransgenic plants of the same variety
(C) Cowpea trypsin inhibitors do show some adverse effects on rats in nutritional studies
(D) Significant hypertrophy of small intestine and altered enzyme activities in brush border occur due to wheat germ agglutinin

2490. In which risk groups, agents are not associated with diseases in healthy humans?
(A) Risk group 3 (B) Risk group 1
(C) Risk group 2 (D) Risk group 4

2491. In which risk groups, is concerned with agents which are associated with such serious or lethal human diseases for which preventive or therapeutic interventions may be available?
(A) Risk group 3 (B) Risk group 1
(C) Risk group 2 (D) Risk group 4

2492. Which is concerned with organisms which are associated with such human diseases, which are rarely serious and for which preventive or therapeutic intervention are often available?
(A) Risk group 3 (B) Risk group 1
(C) Risk group 2 (D) Risk group 4

2493. Which is concerned with organisms, which cause serious or lethal disease for which preventive or therapeutic invention are not usually available?
(A) Risk group 3 (B) Risk group 1
(C) Risk group 2 (D) Risk group 4

2494. The α–amylase inhibitor from *Phaseolus vulgaris* (rajma) caused allergy when expressed in:
(A) Soybean (B) Tomato
(C) Peas (D) Maize

2495. The α–amylase inhibitor gene was transferred to modify which traits?
(A) Insect resistance (B) Amino acid balance
(C) Food digestibility (D) Both (A) and (C)

2496. In case of *E. coli*, the host for any HV 1 system is:
(A) Strain K -12 (B) A derivative of strain K-12
(C) Either (A) or (B) (D) Strain chi 1776

2497. The vector under host-vector I system includes:
(A) Nonconjugative plasmids
(B) Derivative of nonconjugative plasmids
(C) Variants of bacteriophages
(D) All of these

2498. Effective dissemination of plant-associated microorganisms beyond the green house can be prevented by using:
(A) Organisms that have been genetically disabled to survive in nature
(B) Organisms whose natural mode of transmission requires injury to the host
(C) Organisms which require another organisms for their survival
(D) Both (A) and (B)

2499. The type of regulation necessary during field trials should depend on:
(A) Ability of the modified plants to survive in nature
(B) Their ability for dispersal and reproduction
(C) Their ability to hybridize with crop and weed plants
(D) All of these

2500. Which is correct about transmission of transgenes from one crop species to another?
(A) In case of cross-pollinated species isolation by distance is rarely sufficient to prevent inter-population mating
(B) If male sterility is introduced into transgenic cross pollinated crops, it can check the movement of transgenes through pollen
(C) It transgenes are integrated in the chloroplast genome, it would almost eliminated the transmission of transgene through pollen grains
(D) All of these

2501. Which is not correct about Cry proteins?
(A) Cry proteins are rapidly degraded by the stomach juices of vertebrates
(B) They could have harmful effects on non target insect species
(C) It has been reported that Cry proteins leak into soil from transgenic maize roots, where they persist for over 3 weeks
(D) The Cry proteins in the soil become associated with soil clay/humus and become prone to microbial degradation

2502. Which is not correct?
(A) RecA- strains lack recombination
(B) Transposon vectors lacking transposase gene are not able to transpose to other bacteria
(C) Auxotrophic mutants can survive in nature
(D) Transgenes integrated into plasmids may be transferred to other bacteria

2503. The first recombinant microorganism (Tn5-containing strain of Pseudomonas fluorescens) was released in which countries?
(A) USA (B) Canada
(C) Netherlands (D) Germany

2504. Most of the recombinant microorganism released in the environment of genes by:
(A) Transposon insertion
(B) Deletion of genes by molecular genetic procedure in-vitro
(C) Genetic tagging of bacteria with specific
(D) Any of these

2505. Non-infectious illnesses is associated with production of steroids:
(A) Increased body weight
(B) Feminization of males
(C) Hypertension
(D) All of these

2506. Non-infectious illness is associated with endotoxin production:
(A) Flu symptoms (B) Dermatitis
(C) Asthama (D) All of these

2507. Which hazards is associated with operation of bioreactor fermentation?
(A) Spillage
(B) Effluent contamination of gases
(C) Aerosols from reactor
(D) All of these

2508. Non-infectious illness like allergic asthama and dermatitis are associated with production of:
(A) Brewing (B) Single cell protein
(C) Antibiotic (D) All of these

2509. In case of experiments with animals, which procedures is followed at all biosafety levels?
(A) Animal carecass is disposed off to avoid it use as food
(B) A permanent record is maintained of the experimental use and disposal of each animal
(C) The animals are confined to security-fenced areas
(D) All of these

2510. According to biosafety guidelines of India, physical containment envisages to limit the spread of dangerous microorganisms by:
(A) Laboratory design and facilities
(B) Laboratory practices
(C) Safety equipment
(D) All of these

2511. Crop yields can be stabilized by introducing which genes in transgenic varieties?
(A) Genes for improvement nutritional quality
(B) Genes for better tolerance/resistance to biotic and/or abiotic stresses
(C) Genes for herbicide resistance
(D) All of these

2512. The potential risks from novel DNA sequences used during transgenic development may be summed up as:
(A) Protein products of transgenes or products generated by them may be toxic
(B) The transgene product may be allergenic
(C) Antibiotic resistance genes may be taken up by human intestinal microflora, which, as a consequence, become resistant to these antibiotics
(D) All of these

2513. Trangenic plants could serve as vectors for transgene dissemination through :
(A) Seed
(B) DNA released from debris of transgenic plants
(C) Pollen
(D) All of these

2514. Which is not correct?
(A) In European Union, the maximum permissible limit for GMO content in conventional food products and food ingredients is 1% at the marketing stage
(B) All foods that contain more than 1% GMOs related to the different ingredients have to be labeled
(C) In case of organic farming, utilization and presence of GMOs is excluded
(D) Volunteers are not likely to pose problems in crops like rapeseed and mustard

2515. The rate of gene-flow is affected by :
(A) Mode of seed dispersal
(B) Mating system
(C) Mode of pollination
(D) All of these

2516. Which is not correct about lectins?
(A) They are polysaccharides
(B) They are involved in defence mechanism of plants
(C) Several lectins have been identified as food allergens
(D) α-Amylase/tripsin inhibitor family proteins are potential allergens

2517. In the assessment of allergenicity, the source of transgene is the first consideration; the transgene may be obtained from:
(A) Commonly allergenic food
(B) Less commonly allergenic food
(C) No documented history of food allergenicity
(D) Any of these

2518. Which is correct about wheat germ agglutinin?
(A) It is effective against insects at concentration of 10 mg/kg body weight
(B) It produces cytotoxic effects in rats at a lower concentration (7 mg/kg)
(C) Safe food has it only at 0.3 mg/kg
(D) All of these

2519. Which is not correct?
(A) Transgene isolated from a food crop guarantees the food safety of the transgene product
(B) Acute toxic effect depends on the concentration of the compounds in question
(C) Psoralen is known to be toxic, carcinogenic and mutagenic
(D) Any outstanding insect resistance whose biochemical basis is not known should be considered as the prime candidate for toxicological evaluation

2520. Which is not correct about bromoxynil?
(A) It is a non-biodegradable herbicide
(B) It causes birth defects in laboratory animals
(C) It is toxic to fish and may cause cancer in humans
(D) Bromoxynil is absorbed through the skin

2521. For minimizing pollen-mediated gene flow from transgenic varieties of *Solanum tuberosum*, how many meters of isolation distance is recommended?
(A) 10 (B) 20 (C) 30 (D) 40

2522. Strategy of temporal and tissue-specific expression for preventing transgene transfer from transgenic varieties of crops is based on which techniques?
(A) Tapetum-specific promoters allow expression only when the suppressor is present
(B) Terminator technology
(C) Recoverable block function
(D) Inducible promoters allow expression only when the inducer is present

2523. Roundup Ready transgenic variety of soybean showed, on an average, how much yield decrease as compared to the non-transgenic parent varieties in USA, in a comparison basedon field trials at 8200 different locations?
(A) 15% (B) 10% (C) 5% (D) -5%

2524. The principal purpose of transgenic plant containment is to avoid the unintentional transmission of a recombinant DNA containing plant genome including:
(A) Organellar hereditary material
(B) Nuclear hereditary material
(C) Release of recombinant DNA-derived organisms associated with plants
(D) Any of these

2525. Which is often called molecular glue?
(A) DNA ligase (B) Synthetase
(C) DNA polymerase I (D) RNA polymerase

2526. Which assays involves antigen-antibody interaction?
(A) Southern hybridization
(B) Dot-blot assay
(C) ELISA
(D) Colony hybridization

2527. Organarily, the recognition sites of restriction enzymes are:
(A) Palindromes
(B) Nonpalindromic
(C) Asymmetrical
(D) Palindromes with rotational symmetry

2528. A RAPD fragment is generated when sequences complementary to the single arbitrary primer used in the PCR reaction are located at two sites in the target DNA, having:
(A) In the same direction
(B) In the opposite directions
(C) Within 5 kb distance
(D) Both (B) and (C)

2529. Totipotency is reflected by:
(A) Somatic embryogenesis
(B) Shoot regeneration
(C) Pollen embryogenesis
(D) All of these

2530. A vaccines contains :
(A) Inactivated pathogen
(B) Attenuated pathogen
(C) Purified antigen from pathogen
(D) Any of these

2531. Which vectors can be used to clone DNA inserts of about 40kb?
(A) Plasmid (B) Cosmid
(C) Phagemid (D) M13

2532. Which human chromosomes has the highest extent of segmental duplications?
(A) Autosome I (B) X chromosome
(C) Y chromosome (D) Autosome 13

2533. Which involves restriction mapping of DNA fragments?
(A) Chromosome jumping
(B) Primer walking
(C) Chromosome walking
(D) Both (B) and (C)

2534. Which uses two pairs of PCR primers?
(A) Nested PCR (B) Asymetrical PCR
(C) Inverse PCR (D) Anchored PCR

2535. Monoclonal antibodies are used for:
(A) Disease diagnosis (B) ELISA
(C) Immunopurification (D) All of these

2536. MHC molecules are present on the surface of:
(A) Helper T cells (B) Tc cells
(C) Antigen presenting cells (D) Pathogens

2537. Which markers behave as dominant marker?
(A) RFLP (B) RAPD
(C) AFLP (D) Both (B) and (C)

2538. Which molecular markers is expected to produce a single band?
(A) Sequence-tagged microsatellite marker
(B) RAPD
(C) Amplified fragment length polymorphism
(D) ISSR (Inter-simple sequence repeats)

2539. In case of single cell proteins, high content is problematic?
(A) Nucleic acids (B) Lipids
(C) Carbohydrates (D) Proteins

2540. Which molecules functions as a protein kinase?
(A) VirA (B) VirB
(C) VirD (D) VirG

2541. Which proteins is active against nematodes?
(A) CryII (B) CryIA
(C) CryIII (D) CryVI

2542. 'Laurical'variety of *B. napus* has which transgenes?
(A) Lauroyl-ACP desaturase
(B) Lauroyl-ACP thioesterase
(C) Lauroyl-ACP saturase
(D) Stearoyl-ACP desaturase

2543. Which research reagents is being produced from transgenic plants on a commercial scale?
(A) Avidin (B) β-glucuronidase
(C) Both (A) and (B) (D) Polygalacturonidase

2544. Tissue culture raised plantlets require hardening due to:
(A) Poor cuticle development
(B) Poor vascular bundle development
(C) Improper stomatal function
(D) All of these

2545. Which crop has heat shock induces androgenesis?
(A) Rice (B) *Brassica sp*
(C) Barley (D) Datura

2546. Somatic hybridization has been the most successful in:
(A) Solanaceae (B) Papilionaceae
(C) Graminae (D) Malvaceae

2547. Which is the least reliable strategy for virus indexing?
(A) Indicator plants (B) Serological tests
(C) Visible disease symptoms (D) Both (B) and (C)

2548. In cheese production, Rennet leads to curd formation by digesting:
(A) α-casein (B) β-casein
(C) κ-casein (D) All of these

2549. *Trichoderma viridae* is used for the control of:
(A) Fungal pathogen (B) Bacterial pathogen
(C) Viral pathogen (D) All of these

2550. Unfertilized ovaries/ovules are cultured for:
(A) Gynogenesis (B) In vitro pollination
(C) Both (A) and (B) (D) Embryo rescue

2551. Selection of hybrid cells based on which of the following is the most widely applicable?
(A) Visual markers
(B) Complementation
(C) Transgenic selectable markers
(D) Both (A) and (B)

2552. Which is the strongest auxin?
(A) IAA (B) 2,4-D (C) NAA (D) 2,4,5-T

2553. A test cross is made mainly to demonstrate:
(A) 1 : 1 ratio of gametes in monohybrids
(B) Random union of gametes
(C) Equal survival of zygotes
(D) Both (A) and (B)

2554. In case of species propagated by asexual reproduction, germplasm is represented by:
(A) Pollen grains (B) Seed
(C) Vegetative propagules (D) Zygotic embryos

2555. Which is a transgene?
(A) A human gene transferred into *E. coli*
(B) A human gene used for gene therapy
(C) Both (A) and (B)
(D) An *Aegilops sp.* gene transferred into *Triticum sp.* by backcrossing

2556. A biotechnological production process may use:
(A) A microbe (B) An animals/a plant cell
(C) An enzyme (D) Any of these

2557. S1 nuclease is used for:
(A) DNA footprinting
(B) Production of blunt ends from protruding ends
(C) DNA fingerprinting
(D) DNA sequencing

2558. Which of the following is used for positional cloning of genes?
(A) Chromosome walking (B) Chromosome jumping
(C) Chromosome sorting (D) Both (A) and (B)

2559. Orthologue means 'homologous' genes within:
(A) Different tissues of an organism
(B) Same tissue of an organism
(C) Individuals of different species
(D) Different individuals of the same species

2560. In case of retrovirus vectors, sequence is required for:
(A) Transcription
(B) Replication
(C) Reverse transcription
(D) Packaging into virions

2561. Restriction site recognition is affected by:
(A) Base substitution (B) Base deletion/addition
(C) Base methylation (D) All of these

2562. Which has enabled successful transformation of monocots using Agrobacterium?
(A) Exogenous acetosyringone
(B) Agrobacterium strain
(C) Explant tissue (D) Plant variety

2563. In case of mice, DNA is microinjected into:
(A) Male pronucleus (B) Female pronucleus
(C) Ovum cytoplasm (D) Both (A) and (C)

2564. Which marker systems would be preferred for testing the purity of hybrid seeds?
(A) RAPD (B) RFLP
(C) STMS (D) AFLP

2565. In which of the following types of cultures, a constant volume of spent medium is regularly replaced by equal volume of fresh medium?
(A) Perfusion cultures (B) Fed-batch cultures
(C) Agar cultures (D) Batch cultures

2566. In which of the following types of cultures, fresh medium is added gradually?
(A) Perfusion cultures (B) Fed-batch cultures
(C) Agar cultures (D) Batch cultures

2567. In which of the following types of cultures, there is an increase during the culture period?
(A) Perfusion cultures (B) Fed-batch cultures
(C) Agar cultures (D) Batch cultures

2568. Conversion of palm oil to cocoa butter is based on which enzymes?
(A) Lipase (B) Fatty acid synthase I
(B) Fatty acid desaturase (D) Fatty acid thioesterase

2569. Chitin synthase inhibition is used as screen for detection of microorganisms producing:
(A) Antibiotics (B) Antivirals
(C) Antifungals (D) Antitumour agents

2570. The biological component of biosensors is responsible for:
(A) Detection of analyte (B) Specificity of biosensor
(C) Sensitivity of biosensor (D) All of these

2571. Lipases are used in biphasic systems during the production of:
(A) Wax (B) Isomyl acetate
(C) Fatty acids (from lipid) (D) Both (A) and (C)

2572. Chromosome walking is used in case of eukaryotes to generate:
(A) Linkage map (B) Restriction map
(C) Physical map (D) Both (B) and (C)

2573. Anther culture is related to:
(A) Gynogenesis (B) Androgenesis
(C) Parthenogenesis (D) Embryogenesis

2574. Which is a Zn containing enzyme:
(A) Ureas (B) Alkaline phosphatase
(C) Hexokinase (D) Phosphoglycerate kinase

2575. Which techniques yields high purification of molecules in a single step?
(A) Affinity chromatography
(B) Gel filteration
(C) SDS-PAGE
(D) Column chromatography

2576. The terms 'single cell protein' refers to:
(A) Protein recovered from microbes
(B) Protein recovered from single cells
(C) Protein recovered from cultured plant and animal cells
(D) Microbial biomass used as food/feed supplement

2577. Alcohol dehydrogenase gene is induced by:
(A) Anaerobiosis (B) Wound
(C) Oxygen (D) Pathogen invasion

2578. RNA acts as the genetic material:
(A) Tobacco mosaic virus (B) Rhizobium
(C) Fungi (D) Archaebacterium

2579. Plant obtained by chromosome doubling of haploids are known as:
(A) Amphidiploids (B) Dihaploids
(C) Double haploids (D) Diploids

2580. Dot blot assay detects the presence of:
(A) Polypepetide (B) DNA sequence
(C) mRNA (D) Both (B) and (C)

2581. *Bacillus thuringiensis* strains are active against which group of insects?
(A) Diptera (B) Nematodes
(C) Coleoptera (D) All of these

2582. Genes whose products are necessary for the maintenance of all types of cells of a species are called:
(A) Smart genes (B) House-keeping genes
(C) Luxury genes (D) Both (A) and (B)

2583. Which uses as restriction enzymes?
(A) Chromosome jumping (B) Restriction mapping
(C) AFLP (D) All of these

2584. T-DNA is transferred into the plant cells most likely as:
(A) Single-stranded circular molecule
(B) Double-stranded circular molecule
(C) Single-stranded linear molecule
(D) Double-stranded linear molecule

2585. *Bacillus thuringiensis* Cry protein is inactivated by:
(A) UV light (B) Humidity
(C) Red light (D) Both (A) and (B)

2586. RuBISCO functions is:
(A) Photorespiration in C_3 plants
(B) Fixation of CO_2 in C_3 plants
(C) Respiration
(D) Both (A) and (B)

2587. In response to stress, which amino acids accumulates?
(A) Phenylalanine (B) Lysine
(C) Proline (D) Histidine

2588. Monoclonal antibodies are being used for diagnosis of which diseases?
(A) Hepatitis B (B) Cancer
(C) Venereal diseases (D) All of these

2589. Plant cell cultures are being used for the production of:
(A) Berrberine (B) Shikonin
(C) Ginseng (as biomass) (D) All of these

2590. The main features of light induced promoters of dicots relates to:
(A) An upstream AG or AGGA box
(B) TATA box
(C) Both TATA and AG/AGGA boxes
(D) CAAT box

2591. The action of chloranphenicol involves:
(A) Inhibition of DNA replication
(B) Inhibition of transcription
(C) Inhibition of peptidyl transferase
(D) Prevention of initiator tRNA from entering the P site

2592. The two complementary strands of a DNA molecule are held together chiefly by:
(A) Salt bridges (B) Ionic interactions
(C) Hydrogen bonds (D) Hydrophobic interactions

2593. Map based cloning requires:
(A) Saturated RFLP map (B) Genetic map
(C) Karyotype (D) Integrated (A) and (B)

2594. Molecular markers can be used for:
(A) Assessment of genetic variability
(B) Creation of genetic variability
(C) Selection of desirable genotype
(D) Both (A) and (C)

2595. Affinity chromatography techniques can be best used for:
(A) Enzyme identification
(B) Enzyme classification
(C) Enzyme purification
(D) Quantitative assays of enzyme activity

2596. The marker enzyme for mitochondria is:
(A) Cytochrome C (B) Catalase
(C) Amino acid permease (D) 5'-Nucleotidase

2597. A eukaryotic nuclear gene with a bacterial promoter can not be expressed in bacteria because of which differences between prokaryotes and eukaryotes?
(A) Translation (B) RNA processing
(C) DNA replication (D) Genetic code

2598. Rabies virus affects which organs?
(A) Thymus (B) Liver
(C) Kidney (D) Spinal cord

2599. Which is a prokaryotes:
(A) Yeast (B) Green algae
(C) Blue-green algae (D) Mushroom

2600. RNA is not involved in:
(A) Enzymatic activity
(B) Protein synthesis
(C) Chromosome structure (eukaryotes)
(D) Heredity

2601. Genetic material of tobacco mosaic virus will show:
(A) A # T (B) G = C
(C) A = U (D) A # U

2602. Which is considered as primary containment?
(A) Containment equipment
(B) Laboratory practices
(C) Both (A) and (B)
(D) Special laboratory design

2603. Which is considered as secondary containment?
(A) Containment equipment
(B) Laboratory practices
(C) Both (A) and (B)
(D) Special laboratory design

2604. Synthetic vaccines have been developed for:
(A) Typhoid and rabies
(B) Cancer (feline leukaemia virus)
(C) Cholera and foot and mouth disease of cattle
(D) All of these

2605. Which is the most suitable transfection technique for animal ova?
(A) Lipofection
(B) Calcium phosphate precipitation
(C) Microinjection
(D) Particle gun

2606. Technology of haploid production leads to:
(A) Genetic variation (B) Hybrid sorting
(C) Homozygous lines (D) All of these

2607. Which vitamins is commercially produced by microbes?
(A) Riboflavin (B) B12
(C) Carotene (D) All of these

2608. Which one became the first country to begin commercial use of virus resistant transgenic tobacco and tomato in early 1990s?
(A) India (B) USA
(C) China (D) Canada

2609. Which is responsible for the loss of biodiversity?
(A) Deforestation and overgrazing
(B) Biological invasion and hunting
(C) Destruction of habitats and overexploitation
(D) All of these

2610. Which viruses gas acquired a gene from its host genome?
(A) Adenoviruses
(B) Acute transforming retroviruses
(C) Epstein-Barr Virus
(D) Replication competent tumour retroviruses

2611. Which reason has already been responsible for extinction of biodiversity?
(A) Introduction of foreign species
(B) Habitat destruction
(C) Infrastructure development
(D) All of these

2612. Biological diversity, called biodiversity in short, refers to:
(A) Diversity among species
(B) Diversity among ecosystems
(C) Genetic diversity within a given species
(D) All of these

2613. Who said, "the cost of conserving biodiversity is far less than the penalty for allowing its degradation"
(A) M.S. Swaminathan (B) Edouard Sauma
(C) Suman Sahai (D) Panjab Singh

2614. The term 'biological diversity' was for the first time introduced by Norse and McManus in 1980. This term was shortened as 'biodiversity' in 1985 by which scientists?
(A) W.G. Rosen (B) M.S. Swaminathan
(C) N.I.Vavilov (D) J.R.Harlan

2615. Plant tissue culture is regarded to have began when Haberlandt attempted to culture in-vitro:
(A) Leaf discs (B) Isolated leaf cells
(C) Cambium tissue (D) Zygotic embryos

2616. Production of which of the following does not constitute biotechnology?
(A) Plastic (B) Antibiotic
(C) Idli (D) α-amylase

2617. Viruses from plants can be eliminated by:
(A) Thermotherapy (B) Cryotherapy
(C) Chemotherapy (D) Any of these

2618. A single-nucleotide polymorphic site has how many alleles?
(A) 1 (B) 4 (C) 2 (D) 3

2619. Nick translation uses which enzymes?
(A) *E. coli* DNA polymerase III
(B) *E. coli* Klenow fragment
(C) *E. coli* DNA polymerase II
(D) DNA polymerase I of E. coli

2620. Which marker is uses restriction of enzymes?
(A) RFLP
(B) Microsatellite markers
(C) AFLP
(D) Both (A) and (C)

2621. Trade secret can be used to protect:
(A) A formula (B) A design
(C) A manuscript (D) A transgene

2622. The unit of the DNA ladder is based on:
(A) Microgram (B) Base pair
(C) Micro litre (D) Voltage current

2623. Probe based markers are:
(A) Microsatellite (B) RAPD
(C) SSR (D) ISSR

2624. A hybridization based marker is:
(A) RFLP (B) ISSR
(C) SSR (D) SDS

2625. More number of targets can be analyzed simultaneously in:
(A) RT-PCR (B) Multiplex PCR
(C) Southern (D) ELISA

2626. Which of the following has the recognition sequence 5'GAATTC3'?
(A) DNase I (B) S1 nuclease
(C) *Eco*R 1 (D) *Hind*III

2627. Single cell proteins are obtained from:
(A) Bacteria (B) Fungi
(C) Algae (D) All of these

2628. 'Edible vaccines' belong to which category?
(A) Conventional vaccines
(B) Purified antigen vaccines
(C) Recombinant vaccines
(D) Both (A) and (B)

2629. The human Genome is estimated to contain how many genes?
(A) 25000 (B) 35000
(C) 45000 (D) 55000

2630. Embryo culture is routinely used for propagation of:
(A) Lily (B) Orchid
(C) Rose (D) Apple

2631. Negative cell selection allows which type of cells to divide?
(A) Wild type cells (B) Variant cells
(D) Both (A) and (B) (D) Neither (A) nor (B)

2632. Which contains nuclear genome from a single fusion parent?
(A) Asymetric hybrids (B) Cybrids
(C) Symmetric hybrids (D) Both (A) and 9B)

2633. Haploid plants can be obtained by:
(A) Anther culture
(B) Ovary culture
(C) Interspecific hybridization
(D) All of these

2634. Which Vir proteins functions as activator of vir operon?
(A) VirE (B) VirG
(C) VirH (D) VirD

2635. Which approaches for virus resistance has been the most successful?
(A) Coat protein gene
(B) cDNA of satellite RNA
(C) Antisense RNA approach
(D) Ribozyme approach

2636. Phylate is used for phosphate storage in seeds, and it interfere with absorption in human intestine:
(A) β-Carotene (B) Zinc
(C) Iron (D) Phosphate

2637. Which would reduce transgene silencing?
(A) Single copy integration in genome
(B) Transgene AT composition similar to host genome
(C) Each transgene having a different promoter and polydenylation signal
(D) All of these

2638. Affinity biosensors are based on:
(A) Enzyme (B) Antibody
(C) Nucleic acid (D) Either (B) or (C)

2639. Which enzymes is produced in largest quantity?
(A) β-Amylase (B) Lipase
(C) Lactase (D) α-Amylase

2640. *In vitro* regeneration potential is affected by:
(A) Oligogenes (B) Polygenes
(C) Plasma genes (D) All of these

2641. Which serves as cryoprotectant?
(A) DMSO (B) Proline
(C) Glycerol (D) All of these

2642. Frame-shift mutations may result from:
(A) Base substitution
(B) Trinucleotide expansion
(C) Tautomeric shift
(D) Transposition

2643. The term 'wound reponse' refers to the production of which of the following by the plant cells at wound sites?
(A) Protein VirA (B) Acetosyringone
(C) Protein VirG (D) Lentins

2644. Golgi bodies function in:
(A) Protein synthesis (B) Protein modification
(C) Protein packaging (D) Both (B) and (C)

2645. Which is the chief reason for the reduction in chromosome number from 2n to n as a result of meiosis?
(A) Chromosome pairing during MI
(B) Separation of chromosomes during AL
(C) The A above, followed by B
(D) Two nuclear divisions without an intervening S phase

2646. Which processes involves DNA-protein interaction?
(A) Replication initiation
(B) Transcription activation
(C) Transcription repression
(D) All of these

2647. North-western screening is based on:
(A) Antigen-antibody interaction
(B) RNA-protein interaction
(C) DNA: protein interaction
(D) RNA : DNA interaction

2648. Which has more than one chromosome in its genome?
(A) *E. coli* (B) *B. subtilis*
(C) *Agrobacterium* (D) *Vibrio cholera*

2649. Paralogues means 'homologous' genes within :
(A) Same cell (B) Same organism
(C) Same genus (D) Both (A) and (B)

2650. Which techniques does not use antigen-antibody interaction for screening of expression libraries?
(A) Complementation (B) ELISA
(C) FACS (D) Both (A) and (C)

2651. 'Nick translation' is used for:
(A) DNA labeling
(B) Recombinant protein production
(C) DNA replication
(D) Expression – based screening

2652. Beer production is estimated to have began about how many years ago:
(A) 4000 (B) 6000 (C) 400 (D) 600

2653. Which enzymes is used to prevent circulation of opened vectors during recombinant DNA production?
(A) Acid phosphatase
(B) Polynucleotide kinase
(C) Alkaline phosphatase
(D) Klenow fragment

2654. Which techniques identifies the DNA fragment to which a given protein binds?
(A) DNA footprinting (B) DNA fingerprinting
(C) Gel retardation assay (D) Restriction mapping

2655. Which techniques identifies the region of the given DNA fragment to which the concerned protein binds?
(A) DNA footprinting (B) DNA fingerprinting
(C) Gel retardation assay (D) Restriction mapping

2656. Which identifies is used for physical mapping?
(A) DNA footprinting (B) DNA fingerprinting
(C) Gel retardation assay (D) Restriction mapping

2657. Which techniques may used PCR?
(A) DNA footprinting (B) DNA fingerprinting
(C) Gel retardation assay (D) Restriction mapping

2658. Which of the following techniques identifies, DNase I digestion of DNA is involved?
(A) DNA footprinting (B) DNA fingerprinting
(C) Gel retardation assay (D) Restriction mapping

2659. Which generally produces transient transfection?
(A) DEAE-Dextran mediated transfection
(B) Calcium-phosphate precipitation
(C) Microinjection
(D) Retroviral infection

2660. Which proteins is specific to Diptera?
(A) CryI (B) CryII
(C) CryIII (D) CryIV

2661. Which *Agrobacterium* proteins has helicase activity?
(A) VirC (B) VirG
(C) $VirD_2$ (D) VirE

2662. Which of the proteins is single-strand binding protein?
(A) VirC (B) VirG
(C) $VirD_2$ (D) VirE

2663. Which genetic transformation techniques can be used in monocots?
(A) Particle gun (B) Agrobacterium-mediated
(C) Electroporation (D) All of these

2664. Change in medium pH is used as screen for detection of microbial strains producing:
(A) Protease (B) Urease
(C) Lipase (D) Amylase

2665. Which is used as the main nitrogen source for riboflavin production?
(A) Pancreatic digest of gelatin (B) Peanut granules
(C) Corn-steep liquor (D) Soybean meal

2666. Which is used as the nitrogen source for pencillin production?
(A) Pancreatic digest of gelatin (B) Peanut granules
(C) Corn-steep liquor (D) Soybean meal

2667. Lectins are used as ligands in affinity chromatography for the separation of:
(A) Antigens (B) Glycoproteins
(C) Polysaccharides (D) Both (B) and (C)

2668. Solid state fermentation is used for?
(A) Food processing (B) Enzyme production
(C) Organic acid production (D) All of these

2669. Which vectors is the most promosing for development of DNA vaccines?
(A) Vaccinia virus (B) Retrovirus
(C) Herpes virus (D) Adenovirus

2670. Glucose biosensor based on glucose oxidase detects which changes?
(A) Evolution of a gas
(B) Change in temperature
(C) Redox reaction
(D) Fluorescence

2671. Which can be used for in situ recovery of products?
(A) Vacuum (B) Two-phase systems
(C) Dialysis (D) All of these

2672. In plants, mitosis occurs during:
(A) Embryogenesis (B) Parthenogenesis
(C) Gametogenesis (D) All of these

2673. Nucleic acids are transformed from a gel onto a membrane filter in which techniques?
(A) Northern hybridization
(B) Dot-blot hybridization
(C) Calony hybridization
(D) Zoo blot hybridization

2674. Which is not a operation in a biotechnological process?
(A) Strain development
(B) Downstream processing
(C) Process engineering
(D) Safety evaluation

2675. Anzymes are:
(A) Antigen-antibody complexes
(B) Antigens with enzyme activity
(C) Antibodies with enzyme activity
(D) RNA molecules with enzyme activity

2676. The least used growth regulator in plant tissue cultures is:
(A) Kinetin (B) GA3
(C) BAP (D) 2,4-D

2677. Enzyme technology concerns with:
(A) Modification of the amino acid sequence of the enzymes
(B) Enzyme immobilization
(C) Both (A) and (B)
(D) Modification of glycosylation pattern of enzymes

2678. Gel filtration separates macromolecules on the basis of:
(A) Colour (B) Mass
(C) Charge (D) Specific binding

2679. Penicillin acylase (amidase) is being used industrially for production of:
(A) Detergents
(B) Semisynthetic penicillins
(C) Secondary metabolites
(D) Mitigation of penicillin allergy

2680. An endonuclease like DNase I is used in which techniques?
(A) Nick-translation (B) DNA foot-printing
(C) Both (A) and (B) (D) Chromosome walking

2681. An endonuclease like DNase II is used in which techniques?
(A) Nick-translation (B) DNA foot-printing
(C) Both (A) and (B) (D) Chromosome walking

2682. Immobilization of enzymes may involve:
(A) Covalent bonds (B) Cross-linking
(C) Ionic bonds (D) Any of these

2683. Nick-translation utilizes which enzymes?
(A) DNase I (B) DNA polymerase I
(C) Klenow fragment (D) Both (A) and (B)

2684. Stable integration of agene into plant genome can be conclusively shown by:
(A) PCR
(B) Northern hybridization
(C) Dot blot assay
(D) Southern hybridization

2685. Virus resistant transgenic plants can be produced by transferring into them:
(A) cDNA of the satellite RNA of the RNA virus
(B) Coat protein gene of the virus
(C) Defective genome of the virus
(D) Any of these

2686. Virus resistant transgenic plants has been exploited commercially:
(A) cDNA of the satellite RNA of the RNA virus
(B) Coat protein gene of the virus
(C) Defective genome of the virus
(D) Any of these

2687. A protein has isoelectric pH of 5.7. What will be the net charge of this protein at pH 7.0?
(A) Negative (B) Positive
(C) Neutral (D) Either (A) or (B)

2688. Introduction of purified phage DNA molecules into bacterial cells is called:
(A) Conjugation (B) Transfection
(C) Transduction (D) Transformation

2689. Which of the following shall be used when naked recombinant DNA is introduced into a bacterial cell?
(A) Conjugation (B) Transfection
(C) Transduction (D) Transformation

2690. Some introns perform which type of functions?
(A) Enhance promoter action in expression vectors
(B) Encode enzymes
(C) Perform self-splicing
(D) Any of these

2691. The recognition sites for restriction enzymes used in gene cloning are:
(A) Random sequences
(B) Palindromes with rotational symmetry
(C) Simple palindromes
(D) Nonpalindromic sequences

2692. The polysaccharides necessary for attachment of *A. tumefaciens* cells to plant cells is encoded by genes present in:
(A) vir region (B) bacterial chromosome
(C) T-DNA (D) Both (A) and (C)

2693. In a disarmed Ti plasmid, which genes are deleted?
(A) *iaa*M and *iaa*H (B) *ipt*
(C) *nos* (D) Both (A) and (B)

2694. Selection of toxin resistant cells for the development of disease resistant plant is the most successful in the case of pathogens producing is:
(A) Toxin (B) Specific toxin
(C) Nonspecific toxin (D) No toxin

2695. Isoenzyme are variants of a single enzyme differing in :
(A) Catalysis function
(B) Mobility during electrophoresis
(C) Substrate preference
(D) All of these

2696. Nitrate reductase consists of:
(A) Mo cofactor and one polypeptide
(B) One polypeptide
(C) Two polypeptides
(D) Three polyppetides

2697. Crystal formation in plant cells preserved in liquid nitrogen would take place if their temperature rises above:
(A) -130°C (B) -5°C (C) -190°C (D) -160°C

2698. Southern hybridization differs from northern hybridization in respect to:
(A) Membrane filter
(B) Need for denaturation
(C) Type of nucleic acid involved
(D) All of these

2699. In India, molecular markers can not be used to resolve issues related to:
(A) Biodiversity (B) Plant breeding rights
(C) Plant breeder's right (D) Farmer's rights

2700. An extremozyme is an enzyme that has which features?
(A) Extremely active
(B) Extremely long-lived
(C) Extremely mobile during electrophoresis
(D) Stable and active under extreme environments

2701. Chloramphenicol acetyltransferase gene has been used in gene transfer experiments as:
(A) Bacterial resistance gene (B) Selectable marker
(C) Scorbale marker (D) Both (B) and (C)

2702. A mutant of *Agrobacterium tumefaciens* defective in VirG shows which features?
(A) Can be used to transform dicots
(B) Can be used to transform monocots
(C) Can not be used for plant transformation
(D) Both (A) and (B)

2703. Acid phosphatase acts as a marker for:
(A) Mitochondria (B) Lysosome
(C) Golgi complex (D) Endoplamic reticulum

2704. For double-starnded DNA, which base ratios always equals one?
(A) (A+T)/(G+C) (B) (A+G)/(C+T)
(C) C/T (D) A/G

2705. Many of the *Bacillus species* are pathogenic to Lepidopteran pests due to:
(A) Endotoxin production (B) Exotoxin production
(C) Antibiotic production (D) Chitinase production

2706. AFLP utilize:
(A) Chromosome walking
(B) Southern hybridization
(C) PCR
(D) In situ hybridization

2707. Na+ · K+ ATPase is related to:
(A) Group translation (B) Active transport
(C) Passive transport (D) All of these

2708. Which has eukaryotic regulatory sequences?
(A) T-DNA (B) vir regulon
(C) ϕX174 (D) λphage

2709. In which organelles, phosphorylation is a common reaction?
(A) Chloroplast (B) Mitochondria
(C) Golgi bodies (D) Both (A) and (B)

2710. Which causes oncogenesis?
(A) DNA viruses (B) Retroviruses
(C) Mutation (D) All of these

2711. Which is known as Drosophilla of plant kingdom?
(A) *Nicotiana tabaccum*
(B) *Arabidopsis thaliana*
(C) *Daucus carota*
(D) *Capsicum annum*

2712. Which is correct about vinegar?
(A) Acetic acid is commonly called vinegar
(B) Used a souring agent, preservative and condiment
(C) It is 10-13% ripened acetic acid
(D) All of these

2713. Which strain of bacteria is used for the degradation of petroleum and management of oil spills?
(A) *Clostridium acetobutylicum*
(B) *Methylophilus methylotrophus*
(C) *Pseudomonas putida*
(D) *Pseudomonas fluorescens*

2714. In which transgene integration by homologous recombination is almost the rule?
(A) Yeast (B) Drosophila
(C) Mice (D) Humans

2715. It is feared that transgenic plants may pose which risks?
(A) Transfer their transgenic features to weeds
(B) Degrade the environment
(C) Theme selves become weeds
(D) All of these

2716. The gene encoding which proteins serves as a selectable marker for animal cells?
(A) Neomycin phosphotransferase
(B) Dihydrofolate reductase
(C) Thymidine kinase
(D) All of these

2717. In case of which vectors, selection for recombinant clones is based on replica plating technique?
(A) pBR322 (B) pUC18
(C) pUC8 (D) Both (B) and (C)

2718. Which of the marker genes leads to an amplification of the DNA insert linked to it?
(A) Neomycin phosphotransferase
(B) Dihydrofolate reductase
(C) Thymidine kinase
(D) All of these

2719. Risks assessment consists of:
(A) Hazard identification and hazard analysis
(B) Consequence analysis and determination
(C) Risks evaluation
(D) All of these

2720. Functional food provides the body with the required amount needed for healthy survival:
(A) Carbohydrates (B) Proteins and fats
(C) Vitamins (D) All of these

2721. Which features is correct about BABAC2 and TAC plant vectors?
(A) They can transfer multiple genes
(B) They can be used for positional cloning of genes identified by mutation
(C) Genomic libraries have been constructed using these vectors
(D) All of these

2722. Which can be used for induction of mutations in plants and also to serve as molecular tags fro the genes so mutated?
(A) T-DNA (B) Transposable elements
(C) EMS (D) Both (A) and (B)

2723. Which approaches are used for development of fungal disease resistant transgenic crops?
(A) Transgene encoding antifungal compounds
(B) Transgenes which encodes a product that leads to cell death upon pathogen infection
(C) The R gene may be transferred to initiate resistance response
(D) All of these

2724. Which features is correct about ribosome-inactivating proteins (RIPs)?
(A) An effective antifungal protein from barley seeds called RIP has been isolated
(B) It acts by inhibiting protein synthesis via N-glucosidase modification of fungal 28SrDNA
(C) Transgenic crop plants expressing RIP proteins are now being tested against many fungi
(D) All of these

2725. Molecular markers offer which advantages over morphological/other conventional markers?
(A) They show high degree of polymorphism
(B) They can be established at cellular and tissue levels
(C) They can be influenced by the environmental fluctuation
(D) All of these

2726. The ability to design and execute molecular strategies for generating male sterility has heen possible due to:
(A) Isolation, cloning and characterization
(B) Anther/pollen specific promoters
(C) Chemically inducible promoters
(D) All of these

2727. For the recovery of virus-free stocks of clonal crops, meristem culture is combined with:
(A) Thermotherapy (B) Chemotherapy
(C) Cryotherapy (D) Any of these

2728. Expressed sequence tags have been used for:
(A) As physical markers for genome maps
(B) Discovery of genes
(C) Differential – display PCR
(D) All of these

2729. Which is not correct about micropropagation?
(A) Mass clonal propagation of desired genotype
(B) Plantlets are usually pathogen free
(C) Shoot cultures of some species are maintained as 'slow growth cultured' for germplasm conservation
(D) There is no chance of somaclonal variation

2730. Which is correct about fluorescence in diacetate (FDA)?
(A) During testing of viability, cells are treated with 0.01% solution of FDA
(B) Live cells cleave FDA by esterase activity and produce fluorescencein
(C) When stained cells are exposed to UV rays, living cells appear green
(D) All of these

2731. Artificial seed is a bead of gel that contains :
(A) Pesticides and antibiotics
(B) Nutrients and growth regulators
(C) Shoot bud/somatic embryo
(D) All of these

2732. Which steps is necessary for the production of artificial seeds by desiccated system?
(A) Somatic embryo (SEs) are first hardened to withstand desiccation
(B) Hardened SEs are encapsulated in a suitable coating material
(C) SEs may be hardened neither by coating/treating mature SEs with a suitable polymer followed by drying during their maturation phase
(D) All of these

2733. Which is not correct about production of hydrated artificial seed?
(A) Somatic embryos are enclosed in gels, which are then desiccated
(B) Calcium alginate is the most suitable gel
(C) The artificial seed are sticky and difficult to handle
(D) The seeds are difficult to store for long periods

2734. Somatic embryos can be stored for prolonged periods through:
(A) Cryopreservation (B) Desiccation
(C) Cold storage (D) All of these

2735. The problem of endosperm abortion can be overcome by excising and culturing:
(A) Hybrid endosperms (B) Hybrid fruits
(C) Hybrid embryos (D) Hybrid zygote

2736. In cryopreservation, which favours storage of plant cells, tissues organs in liquid nitrogen at -196°C?
(A) All metabolic processes and growth are suppressed
(B) Occurrence of genetic, karyotypic morphological and biochemical changes are prevented
(C) It is the most reliable method for long term prevention of cell cultures
(D) All of these

2737. In case of plant variety is a chimera, which route of micropropagation can be used to conserve the chimaeric state?
(A) Somatic embryogenesis
(B) Adventitious shoot buds
(C) Shoot – tip culture
(D) Both (A) and (B)

2738. Which disrups chimaeras and produces nonchimeric progeny plants?
(A) Somatic embryogenesis
(B) Adventitious shoot buds
(C) Shoot – tip culture
(D) Both (A) and (B)

2739. Eukaryotic DNA segments are usually cloned initially in:
(A) Bacillus (B) Yeast
(C) E. coli (D) Both (A) and (B)

2740. Air – lift fermenters do not have provision for:
(A) Aeration (B) Stirring
(C) Temperature control (D) Sterilization

2741. Which vectors can not be used to efficiently clone DNA inserts longer than 5 kb?
(A) Cosmid (B) Plasmid
(C) Lambda phage (D) M13

2742. Which growth regulators suppresses callus formation?
(A) GA_3 (B) Auxin
(C) Cytokinin (D) Both (A) and (B)

2743. In which PCR procedures, target DNA is cut and circulated, then cut open to produce linear molecules?
(A) Anchored PCR (B) Asymetric PCR
(C) Inverse PCR (D) Nested PCR

2744. The melting temperature for a DNA sequence having 6G : C and 4 A : T base pairs will be:
(A) 32°C (B) 28°C (C) 36°C (D) 30°C

2745. In India, a transgenic variety can be protected by:
(A) Patent (B) Plant Breeder's Rights
(C) Both (A) and (B) (D) Trade Secret

2746. Which generated more number of different proteins from the same number of genes?
(A) Autosplicing (B) Trans-splicing
(C) Alternative splicing (D) RNA editing

2747. The term clonal selection signifies which in response to an antigen?
(A) B cell proliferation (B) T cell proliferation
(C) B/T cell maturation (D) Both (A) and (B)

2748. A helper Ti of binary vector system does have:
(A) vir region
(B) DNA insert
(C) Origin of replication for E. coli (D) T-DNA

2749. Which can be used to suppress the expression of endogenous genes?
(A) Consuppression (B) Antisense RNA
(C) RNAi (D) All of these

2750. In the case of barnase – barstar male sterility system, phosphinothricin resistance gene bar is used to achieve:
(A) Restoration of male fertility
(B) Reinforcing male sterility
(C) Enhancing Barstar action
(D) Eliminating male fertile plants

2751. The 'slow growth' cultures of tropical plants would suffer cold damage at which temperature?
(A) 4^0C (B) 8^0C
(C) Both (A) and (B) (D) 15^0C

2752. Androgenesis would yield a dihaploid:
(A) Rice (B) Potato
(C) Tobacco (D) Both (B) and (C)

2753. Somatic embryo maturation is promoted by:
(A) Auxin (B) ABA
(C) Cytokinin (D) GA3

2754. Disease resistant plants can be developed by cell selection using which as selection agents?
(A) Specific toxins (B) Nonspecific toxins
(C) Both (A) and (B) (D) Neither (A) and (B)

2755. A somatic hybrid plant may have chloroplasts from:
(A) Both the parents (B) First fusion parent
(C) Second fusion parent (D) Any of these

2756. Plant vitrification solutions are used for:
(A) Slow-growth cultures
(B) Cryopreservation
(C) Micropropagation
(D) Prevention of hyperhydration

2757. Which is the most commonly used for root regeneration?
(A) IBA (B) IAA
(C) 2,4-D (D) ABA

2758. In production of glucose syrup, -amylase brings about:
(A) Gelatinization (B) Liquefaction
(C) Saccharification (D) All of these

2759. Which solubilizes phosphate?
(A) *Thiobacillus sp.* (B) *Glomus sp.*
(C) Both (A) and (B) (D) *Azospirillum*

2760. In hypersensitive response, the avr gene of pathogen is responsible for generating:
(A) Toxin (B) Receptor
(C) Elicitor (D) Repressor

2761. Which is the most correct?
(A) Each normal chromosomes has, as a rule, a single centromere
(B) The cnetromeric activity is always confined to a single locus in the chromosomes
(C) In most species, the centromeric activity is confined to a single locus in the chromosomes
(D) Polycentric chromosomes represent structural aberrations

2762. Which is uses insertional mutagenesis?
(A) Gene tagging
(B) Exon trapping
(C) Differential screening
(D) Chromosomal jumping

2763. Klenow fragment is used for:
(A) Filling up of protruding ends
(B) Nick translation
(C) DNA sequencing
(D) Both (A) and (C)

2764. Which is the first recombinant protein to be approved for use in humans?
(A) Insulin (B) Interferon
(C) Hirudin (D) Erythroprotein

2765. A restriction enzyme has six-base recognition sequence. This sequence may be expected to occur at which intervals?
(A) 4^6bp (B) 6^4 bp
(C) 4 x 6 bp (D) $4^6/6$ bp

2766. Paralogues means 'homologous' genes within:
(A) Same cell (B) Same organism
(C) Same genus (D) Both (A) and (B)

2767. Which has more than one chromosome in its genome?
(A) *E. coli* (B) *B. subtilis*
(C) *Agrobacterium* (D) *Vibrio cholera*

2768. Improved Pusa Basmati I (Pusa 1460), the first variety developed in India through MAS, has been improved for which traits?
(A) VitA content
(B) Cooking quality
(C) Bacterial leaf blight resistance
(D) Blast resistance

2769. F-virosomes derived from Sendai virus do not contain:
(A) Viral envelope (B) Viral HN protein
(C) Recombinant DNA (D) Viral F protein

2770. The satellite RNA present in some RNA viruses may produce which effects on disease severity?
(A) Enhance disease severity
(B) Decrease disease severity
(C) No effect on disease severity
(D) All of these

2771. Enhanced metabolite production may be achieved by selection for:
(A) Auxotropihic mutants
(B) Analogue resistant mutants
(C) Revertant of a nonreproducing mutant
(D) All of these

2772. Which serves as inducer of pollulenase production?
(A) Maltose (B) Sucrose
(C) Dextrin (D) Starch

2773. In 1860 the first synthetic medium for microorganisms was used to study which fermentations?
(A) Lactic acid (B) Alcohol
(C) Acetic acid (D) Citric acid

2774. Which is used in gap –ligase chain reaction?
(A) Thermostable DNA ligase
(B) Thermostable DNA polymerase
(C) NAD
(D) All of these

2775. Which elements is not essential for plant cells?
(A) Ni (B) Na (C) K (D) P

2776. The Km of the substrate for any enzyme is increased by:
(A) Competitive inhibitor
(B) Noncompetitive inhibitor
(C) Substrate concentration (D) Both (A) and (C)

2777. Somaclonal variation is the least likely to occur in:
(A) Protoplast culture (B) Suspension culture
(C) Meristem culture (D) Callus culture

2778. Which statements is correct?
(A) All enzymes are heat stable
(B) Enzymes are always proteins
(C) Enzymes are always specific for a single substrate
(D) Enzymes can be inhibited by substrate analogues

2779. Antisen RNA technology may be used to achieve:
(A) Minimise the expression of a gene
(B) Enhance the expression of a gene
(C) Stabilize the expression of a gene
(D) Change the expression pattern of a gene

2780. Inactive precursors of enzymes are called:
(A) Isozymes (B) Zymogenes
(C) Allozymes (D) Pre-emnzymes

2781. Which techniques does not use radioactivity?
(A) RFLP (B) Western hybridization
(C) RAPD (D) Southern blotting

2782. In thermotherapy for virus elimination, the best results are obtained when high temperature treatment is applied to:
(A) Shoot-tip cultures
(B) Seeds
(C) Whole plants, followed by culture of their shoot-tips
(D) Whole plants

2783. Biological extraction of copper an duranium utilizes:
(A) *Thiobacillus thiooxidans* (B) *T. ferroxidans*
(C) Both (A) and (B) (D) *T. organoparus*

2784. Production of biochemicals by cultured plant cells is markedly improved by:
(A) Treatment with elicitors
(B) Selection of high producing clones
(C) Optimisation of culture conditions
(D) All of these

2785. Which agents is used for virus elimination?
(A) Ribavirin (B) Actinomycin D
(C) 2-Thiouracil (D) All of these

2786. Phenotypic variability having nongenetic basis is called:
(A) Epigenetic variation
(B) Soamclonal variation
(C) Gametoclonal variation
(D) Protoclonal variation

2787. The virus titre in shoot-tip merism cells is much lower than that in the differentiated cells; this may be due to:
(A) Absence of plasmodesmata
(B) Rapid cell division
(C) Inhibitory action of growth regulators
(D) All of these

2788. Lectin are proteins that have:
(A) DNA binding proteins
(B) Bind to glycoproteins
(C) Bind to lipoproteins
(D) Hormone binding proteins

2789. The most effective fusion treatment for plant protoplasts is:
(A) High pH and high Ca^{++}
(B) PEG
(C) PEG combined with high pH and high Ca^{++}
(D) $NaNO_3$

2790. Which *Agrobacterium tumefaciens* proteins is activated by acetosyringone?
(A) VirA (B) VirC
(C) VirG (D) VirD2

2791. Which is an approach of germplasm conservation?
(A) Slow-growth shoot cultures
(B) DNA banks
(C) Cryopreservation
(D) All of these

2792. ABA is related to:
(A) Flowering in plants (B) Dormancy
(C) Abscission of leaves (D) Both (B) and (C)

2793. A polyprotein is a protein that has:
(A) Several functions
(B) Homopolymer
(C) Heteropolymer
(D) Cleaved into two or more functional proteins

2794. WHO has launched Global Immunization Progarmme for which disease?
(A) Pertusis and Tetanus
(B) Diphtheria and Measles
(C) Polio and T.B.
(D) All of these

2795. A yeast minichromosome vector does not contain which molecules?
(A) ARS sequence (B) Telomere sequence
(C) CEN sequence (D) A functional yeast gene

2796. Cosmid contain which sequences from the ë phage genome?
(A) Terminase biding site
(B) cos site
(C) cos and terminase binding sites
(D) cos, terminase binding and cleavage sites

2797. A conditioned medium permits:
(A) Subculture at low cell density
(B) Elimination of lag phase
(C) Production of greater biomass
(D) Both (A) and (B)

2798. Hydrogen peroxide is degraded by the enzyme catalase in which organelles?
(A) Peroxisomes (B) Lysosomes
(C) Mitochondria (D) Glyoxysomes

2799. A transgene stably integrated into the plant genome shows:
(A) Cytoplasmic inheritance
(B) Segregation distortion
(C) Medelian inheritance
(D) All of these

2800. Restriction endonucleases are sometimes unable to digest the DNA even in the presence of the correct recognition sequence; this may be due to:
(A) Acetylation of DNA (B) Methylation of DNA
(C) Phosphorylation of DNA (D) All of these

2801. A circular plasmid DNA gives two bands of 6.5 kb and 5.5 kb with EcoRI and only one band of 12 kb with BamHI. How many bands will be expected when it is digested with both the enzymes?
(A) 2 (B) 4 (C) 3 (D) 5

2802. A double – stranded DNA molecule of 100000 bp length has about how many complete turns?
(A) 8000 (B) 10000
(C) 20000 (D) 12000

2803. Which plant is used as a model system for transgenic plant production?
(A) Potato (B) Arabidopsis
(C) Tobacco (D) Datura

2804. Which is gram positive bacterium?
(A) *Bacillus subtilis* (B) *Agrobacterium sp.*
(C) *Rhizobium sp.* (D) *E. coli*

2805. As a general rule of thumb, hydrophobic residues in a protein tend to be situated at:
(A) C-terminal
(B) N-terminal
(C) Interior of the molecules
(D) Surface of the molecules

2806. In which random transgene integration is far more frequent than that by homologous recombination?
(A) Animals (B) Plants
(C) Yeast (D) Both (A) and (B)

2807. Which can be used for DNA delivery into *E. coli* cells?
(A) $CaCl_2$ treatment (B) Microinjection
(C) Electroporation (D) Both (A) and (C)

2808. During gynogenesis, haploid plants generally originate from:
(A) Polar nuclei (B) Egg cell
(C) Antipodals (D) Synergids

2809. It is feared that transgenic plants may pose which risks?
(A) Transfer their transgenic features to weeds
(B) Degrade the environment
(C) Themselves become weeds
(D) All of these

2810. Which is not correct about insitu conservation of biodiversity?
(A) Plants are conserved in their natural habitats
(B) Only the wild relatives are candidates for insitu conservation
(C) Plants can continue to evolve
(D) Insitu conservation is rather simple

2811. Which is not correct about BIBAC2 and TAC plant vectors?
(A) Both have ampicilin resistance marker for selection in bacteria
(B) Both contain hygromycin resistance marker for se lection of transformed plant cells
(C) Both have the sacB marker (containing a polylinker fro DNA insert integration) for negative selection in bacteria
(D) Both have the Ri origin of maintenance in Rhizobium

2812. Nutraceticals differ from dietary supplements in respect to:
(A) They must not only supplement the diet, but should also aid in prevention or treatment of diseases or disorders
(B) They are represented for use as the conventional food or as the sole item of meal or diet
(C) They include some products from nonfood sources
(D) Both (A) and (B)

2813. The use of GM crops relates to:
(A) Biodiversity, ecology and biosafety
(B) Environmental and social concerns
(C) Ethical concerns
(D) All of these

2814. Which features is relevant to an ideal herbicide?
(A) No toxicity towards human, animals or soil organisms
(B) Kills selectively weeds
(C) Biodegradability
(D) All of these

2815. Selection of appropriate molecular marker for applied plant breeding would depend on:
(A) Trait(s) under consideration
(B) Crop in question
(C) Reproducibility of molecular marker
(D) All of these

2816. Which statement is correct about molecular markers?
(A) The ultimate goal of molecular marker research is to devlop plant varieties more efficiently and effectively
(B) Molecular markers tightly linked to male sterility gene can be used to identify male fertile sister plants at very early stage
(C) RAPD is cost effective and simple, but its reproducibility is not high
(D) All of these

2817. Terminator seed technology is seen to result in:
(A) Encourage monopoly of multinational seed companies
(B) Force the farmers to purchase seeds of even self-pollinated crops every year
(C) Jeoparadise food security
(D) All of these

2818. For toxicity or allergenicity is not reported:
(A) Cry protein
(B) Trypsin inhibitor
(C) Virus coat protein
(D) Both (A) and (C)

2819. Expressed sequence tags (ESTs) have been used for:
(A) Identification of genes in genomic clones
(B) Designing of PCR primers to generate large numbers of target sequences for cDNA microarrays
(C) Gene expression analysis
(D) All of these

2820. The protoplast fusion products recovered after a treatment with a fusion inducing agent consist of:
(A) Homokaryons
(B) Heterokaryons
(C) Unsafed protoplasts of the two species/strains
(D) All of these

2821. Which is not correct about meristem culture?
(A) Cultivation of axillary or apical shoot meristems
(B) The shoot apical meristem is the portion lying distal to the youngest leaf primordium and is 250ì m in length
(C) The shoot apex/shoot tip consists of the apical meristem plus 1-3 young leaf primordial
(D) Meristem culture usually involves the regeneration of a new shoot meristem

2822. Somatic embryos are used for micropropagation in:
(A) Coconut (B) Oil palm
(C) Orchids (D) Date palm

2823. Gene integration occurs generally by homologous recombination in:
(A) Yeast (B) Mammals
(C) Tobacco (D) Tomato

2824. 'Single cell protein' is represented by :
(A) Protein from microbes
(B) Protein from cell cultures
(C) Microbial biomass
(D) Both (A) and (B)

2825. The site where a protein binds a DNA molecule is determined by:
(A) DNA footprintig
(B) DNA fingerprinting
(C) Gel retardation assay
(D) FISH

2826. Somatic embryo maturation is promoted by:
(A) Auxin (B) ABA
(C) Cytokinin (D) GA_3

2827. GA_3 promotes shoot regeneration:
(A) Tobacco (B) Arabidopsis
(C) Carrot (D) Citrus

2828. Eukaryotic enzymes essentially functions as reverse transcriptase:
(A) DNA polymerase (B) Primase
(C) Telomerase (D) RNA polymerase

2829. Enhanced metabolite production may be achieved by selection for:
(A) Auxotrophic mutants
(B) Analogue resistant mutants
(C) Revertant of a nonproducing mutant
(D) All of these

2830. Enzyme involved in transcription:
(A) RNA polymerase
(B) DNA polymerase
(C) Kinase
(D) Endonuclease

ANSWERSHEET: 4.AGRICULTURAL BIOTECHNOLOGY

1. (D); 2. (B); 3. (C); 4. (A); 5. (A); 6. (C); 7. (D); 8. (B); 9. (A); 10. (A); 11. (B); 12. (C); 13. (A); 14. (B); 15. (A); 16. (B); 17. (A); 18. (B); 19. (C); 20. (A); 21. (B); 22. (A); 23. (A); 24. (B); 25. (C); 26. (D); 27. (C); 28. (C); 29. (A); 30. (B); 31. (C); 32. (D); 33. (A); 34. (A); 35. (B); 36. (A); 37. (A); 38. (A); 39. (B); 40. (A); 41. (A); 42. (B); 43. (A); 44. (A); 45. (B); 46. (D); 47. (B); 48. (D); 49. (E); 50. (A); 51. (D); 52. (D); 53. (C); 54. (A); 55. (D); 56. (A); 57. (D); 58. (D); 59. (D); 60. (A); 61. (D); 62. (D); 63. (D); 64. (A); 65. (C); 66. (C); 67. (D); 68. (D); 69. (C); 70. (D); 71. (D); 72. (A); 73. (C); 74. (A); 75. (D); 76. (A); 77. (D); 78. (A); 79. (A); 80. (D); 81. (C); 82. (D); 83. (C); 84. (D); 85. (D); 86. (B); 87. (C); 88. (A); 89. (B); 90. (D); 91. (C); 92. (A); 93. (A); 94. (A); 95. (B); 96. (D); 97. (C); 98. (D); 99. (D); 100. (D); 101. (C); 102. (D); 103. (B); 104. (D); 105. (A); 106. (C); 107. (C); 108. (D); 109. (C); 110. (D); 111. (D); 112. (C); 113. (D); 114. (A); 115. (C); 116. (D); 117. (D); 118. (B); 119. (D); 120. (D); 121. (D); 122. (D); 123. (A); 124. (D); 125. (B); 126. (D); 127. (D); 128. (C); 129. (B); 130. (D); 131. (D); 132. (D); 133. (C); 134. (A); 135. (D); 136. (B); 137. (A); 138. (D); 139. (B); 140. (C); 141. (D); 142. (C); 143. (A); 144. (A); 145. (B); 146. (D); 147. (B); 148. (A); 149. (B); 150. (D); 151. (A); 152. (A); 153. (A); 154. (D); 155. (D); 156. (C); 157. (B); 158. (A); 159. (A); 160. (A); 161. (D); 162. (B); 163. (D); 164. (D); 165. (B); 166. (C); 167. (A); 168. (A); 169. (A); 170. (B); 171. (D); 172. (B); 173.(B); 174. (A); 175. (D); 176. (C); 177. (A); 178. (A); 179. (D); 180. (A); 181. (E); 182. (E); 183. (C); 184. (B); 185. (E); 186. (D); 187. (A); 188. (A); 189. (A); 190. (A); 191. (D); 192. (B); 193. (B); 194. (A); 195. (A); 196. (B); 197. (D); 198. (D); 199. (C); 200. (D); 201. (E); 202. (D); 203. (B); 204. (A); 205. (B); 206. (C); 207. (D); 208. (A); 209. (A); 210. (D); 211. (B); 212. (C); 213. (A); 214. (B); 215. (B); 216. (D); 217. (D); 218. (D); 219. (B); 220. (B); 221. (B); 222. (D); 223. (B); 224. (A); 225. (A); 226. (C); 227. (D); 228. (A); 229. (C); 230. (A); 231. (D); 232. (C); 233. (D); 234. (D); 235. (B); 236. (B); 237. (C); 238. (A); 239. (B); 240. (B); 241. (A); 242. (D); 243. (B); 244. (B); 245. (D); 246. (B); 247. (D); 248. (C); 249. (C); 250. (D); 251. (A); 252. (B); 253. (C); 254. (C); 255. (A); 256. (D); 257. (D); 258. (D); 259. (B); 260. (C); 261. (D); 262. (D); 263. (D); 264. (B); 265. (D); 266. (D); 267. (D); 268. (D); 269. (B); 270. (D); 271. (C); 272. (A); 273. (B); 274. (A), 275. (B); 276. (A); 277. (C); 278. (C); 279. (A); 280. (D); 281. (C); 282. (D); 283. (C); 284. (D); 285. (A); 286. (B); 287. (A); 288. (C); 289. (D); 290. (B); 291. (D); 292. (C); 293. (C); 294. (B); 295. (D); 296. (B); 297. (B); 298. (A); 299. (B); 300. (A); 301. (B); 302. (C); 303. (C); 304. (A); 305. (A); 306. (A); 307. (D); 308. (D); 309. (C); 310. (C); 311. (A); 312. (A); 313. (A); 314. (A); 315. (C); 316. (D); 317. (A); 318. (A); 319. (B); 320. (C); 321. (A); 322. (C); 323. (A); 324. (D); 325. (B), 326. (B); 327. (A); 328. (A); 329. (D); 330. (D); 331. (C); 332. (C); 333. (B); 334. (D); 335. (A); 336. (D); 337. (B); 338. (D); 339. (B); 340. (D); 341. (B); 342. (B); 343. (D); 344. (D); 345. (D); 346. (D); 347. (D); 348. (B); 349. (D); 350. (B); 351. (C); 352. (D); 353. (B); 354. (B); 355. (A); 356. (D); 357. (B); 358. (A); 359. (D); 360. (C); 361. (A); 362. (A); 363. (C); 364. (C); 365. (D); 366. (D); 367. (A); 368. (A); 369. (C); 370. (C); 371. (C); 372. (D); 373. (A); 374. (B); 375. (D); 376. (D); 377. (D); 378. (D); 379. (A); 380. (B); 381. (C); 382. (D); 383. (D); 384. (D); 385. (A); 386. (A); 387. (D); 388. (B); 389. (D); 390. (D); 391. (D); 392. (C); 393. (D); 394. (D); 395. (B); 396. (C); 397. (D); 398. (D); 399. (C); 400. (C); 401. (A); 402. (B); 403. (D); 404. (D); 405. (C); 406. (B); 407. (D); 408. (A); 409. (B); 410. (B); 411. (D); 412. (A); 413. (C); 414. (C); 415. (A); 416. (B); 417. (D); 418. (C); 419. (A); 420. (D); 421. (D); 422. (B); 423. (A); 424. (B); 425. (A); 426. (A); 427. (B); 428. (B); 429. (A); 430. (D); 431. (D); 432. (D); 433. (A); 434. (D); 435. (D); 436. (A); 437. (A); 438. (D); 439. (A); 440. (B); 441. (D); 442. (B); 443. (A); 444. (D); 445. (D); 446. (C); 447. (C); 448. (B); 449. (A); 450. (A); 451. (A); 452. (C); 453. (C); 454. (A); 455. (D); 456. (A); 457. (C); 458. (D); 459. (B); 460. (A); 461. (B); 462. (D); 463. (D); 464. (A); 465. (A); 466. (B); 467. (D); 468. (A); 469. (B); 470. (C); 471. (B); 472. (C); 473. (C); 474. (A); 475. (D); 476. (A); 477. (B); 478. (C); 479. (D); 480. (B); 481. (A); 482. (A); 483. (B); 484. (C); 485. (A); 486. (B); 487. (C); 488. (A); 489. (B); 490. (A); 491. (B); 492. (A); 493. (C); 494. (A); 495. (D); 496. (C); 497. (B); 498. (C); 499. (A); 500. (C); 501. (B); 502. (B); 503. (D); 504. (A); 505. (B); 506. (A); 507. (B); 508. (D); 509. (D); 510. (B); 511. (B); 512. (A); 513. (C); 514. (C); 515. (A); 516. (B); 517. (A); 518. (C); 519. (A);

520. (D); 521. (A); 522. (A); 523. (A); 524. (B); 525. (A); 526. (A); 527. (E); 528. (D); 529. (A); 530. (A); 531. (C); 532. (A); 533. (A); 534. (B); 535. (C); 536. (D); 537. (C); 538. (D); 539. (C); 540. (C); 541. (A); 542. (B); 543. (A); 544. (D); 545. (A); 546. (B); 547. (C); 548. (D); 549. (B); 550. (A); 551. (B); 552. (A); 553. (C); 554. (B); 555. (C); 556. (D); 557. (A); 558. (A); 559. (A); 560. (B); 561. (C); 562. (D); 563. (C); 564. (C); 565. (B); 566. (D); 567. (D); 568. (B); 569. (D); 570. (C); 571. (D); 572. (D); 573. (D); 574. (D); 575. (B); 576. (A); 577. (D); 578. (A); 579. (D); 580. (B); 581. (D); 582. (D); 583. (A); 584. (B); 585. (D); 586. (A); 587. (D); 588. (D); 589. (D); 590. (B); 591. (D); 592. (D); 593. (A); 594. (D); 595. (D); 596. (B); 597. (A); 598. (A); 599. (D); 600. (D); 601. (D); 602. (C); 603. (C); 604. (B); 605. (A); 606. (D); 607. (B); 608. (C); 609. (D); 610. (C); 611. (D); 612. (D); 613. (B); 614. (B); 615. (D); 616. (D); 617. (D); 618. (B); 619. (D); 620. (D); 621. (D); 622. (C); 623. (D); 624. (C); 625. (D); 626. (D); 627. (D); 628. (D); 629. (C); 630. (D); 631. (A); 632. (D); 633. (A); 634. (A); 635. (A); 636. (C); 637. (D); 638. (C); 639. (D); 640. (D); 641. (B); 642. (D); 643. (A); 644. (B); 645. (B); 646. (D); 647. (D); 648. (C); 649. (D); 650. (A); 651. (C); 652. (D); 653. (D); 654. (D); 655. (A); 656. (D); 657. (D); 658. (D); 659. (D); 660. (A); 661. (D); 662. (A); 663. (B); 664. (C); 665. (B); 666. (D); 667. (D); 668. (B); 669. (A); 670. (C); 671. (B); 672. (B); 673. (D); 674. (A); 675. (D); 676. (D); 677. (D); 678. (D); 679. (B); 680. (B); 681. (A); 682. (B); 683. (D); 684. (C); 685. (D); 686. (A); 687. (D); 688. (C); 689. (A); 690. (B); 691. (D); 692. (C); 693. (A); 694. (B); 695. (A); 696. (C); 697. (B); 698. (C); 699. (A); 700. (D); 701. (A); 702. (D); 703. (D); 704. (D); 705. (B); 706. (C); 707. (B); 708. (D); 709. (D); 710. (D); 711. (C); 712. (C); 713. (A); 714. (D); 715. (A); 716. (B); 717. (D); 718. (E); 719. (A); 720. (A); 721. (B); 722. (C); 723. (D); 724. (C); 725. (A); 726. (B); 727. (C); 728. (D); 729. (A); 730. (B); 731. (C); 732. (D); 733. (A); 734. (B); 735. (C); 736. (A); 737. (B); 738. (C); 739. (D); 740. (A); 741. (B); 742. (A); 743. (B); 744. (C); 745. (D); 746. (C); 747. (B); 748. (D); 749. (C); 750. (A); 751. (C); 752. (D); 753. (B); 754. (D); 755. (B); 756. (D); 757. (C); 758. (B); 759. (C); 760. (C); 761. (D); 762. (A); 763. (D); 764. (B); 765. (D); 766. (D); 767. (C); 768. (D); 769. (C); 770. (A); 771. (D); 772. (B); 773. (D); 774. (C); 775. (D); 776. (D); 777. (D); 778. (A); 779. (B); 780. (D); 781. (A); 782. (C); 783. (D); 784. (D); 785. (A); 786. (A); 787. (B); 788. (B); 789. (A); 790. (C); 791. (D); 792. (A); 793. (D); 794. (C); 795. (B); 796. (C); 797. (C); 798. (D); 799. (A); 800. (D); 801. (A); 802. (D); 803. (A); 804. (B); 805. (D); 806. (D); 807. (C); 808. (D); 809. (D); 810. (D); 811. (C); 812. (C); 813. (D); 814. (D); 815. (D); 816. (D); 817. (C); 818. (D); 819. (C); 820. (D); 821. (D); 822. (C); 823. (D); 824. (B); 825. (C); 826. (D); 827. (D); 828. (D); 829. (C); 830. (D); 831. (B); 832. (D); 833. (D); 834. (B); 835. (C); 836. (C); 837. (D); 838. (D); 839. (B); 840. (B); 841. (B); 842. (B); 843. (C); 844. (D); 845. (B); 846. (D); 847. (A); 848. (B); 849. (B); 850. (D); 851. (A); 852. (A); 853. (C); 854. (C); 855. (A); 856. (B); 857. (C); 858. (D); 859. (A); 860. (B); 861. (C); 862. (A); 863. (B); 864. (C); 865. (A); 866. (B); 867. (C); 868. (D); 869. (A); 870. (B); 871. (C); 872. (A); 873. (B); 874. (C); 875. (A); 876. (B); 877. (C); 878. (D); 879. (A); 880. (A); 881. (B); 882. (C); 883. (D); 884. (C); 885. (C); 886. (A); 887. (A); 888. (D); 889. (D); 890. (D); 891. (C); 892. (B); 893. (D); 894. (C); 895. (B); 896. (C); 897. (D); 898. (D); 899. (D); 900. (C); 901. (D); 902. (B); 903. (C); 904. (D); 905. (C); 906. (C); 907. (D); 908. (D); 909. (D); 910. (C); 911. (B); 912. (D); 913. (A); 914. (B); 915. (A); 916. (C); 917. (D); 918. (C); 919. (B); 920. (D); 921. (D); 922. (B); 923. (C); 924. (C); 925. (B); 926. (D); 927. (A); 928. (A); 929. (A); 930. (B); 931. (C); 932. (A); 933. (B); 934. (C); 935. (A); 936. (C); 937. (D); 938. (D); 939. (B); 940. (A); 941. (A); 942. (D); 943. (C); 944. (A); 945. (C); 946. (D); 947. (D); 948. (D); 949. (D); 950. (D); 951. (A); 952. (C); 953. (D); 954. (C); 955. (D); 956. (D); 957. (D); 958. (D); 959. (D); 960. (D); 961. (B); 962. (B); 963. (B); 964. (D); 965. (A); 966. (A); 967. (D); 968. (D); 969. (B); 970. (D); 971. (D); 972. (D); 973. (D); 974. (D); 975. (B); 976. (D); 977. (D); 978. (D); 979. (B); 980. (D); 981. (B); 982. (A); 983. (B); 984. (D); 985. (D); 986. (B); 987. (D); 988. (B); 989. (B); 990. (C); 991. (B); 992. (C); 993. (D); 994. (A); 995. (B); 996. (C); 997. (D); 998. (A); 999. (B); 1000. (A); 1001. (D); 1002. (A); 1003. (A); 1004. (A); 1005. (A); 1006. (D); 1007. (D); 1008. (B); 1009. (C); 1010. (D); 1011. (C); 1012. (B); 1013. (C); 1014. (C); 1015. (D); 1016. (A); 1017. (C); 1018. (D); 1019. (D); 1020. (A); 1021. (D); 1022. (A); 1023. (B); 1024. (A); 1025. (B); 1026. (A); 1027. (D); 1028. (C); 1029. (D); 1030. (C); 1031. (D); 1032. (C); 1033. (C); 1034. (B); 1035. (B); 1036. (D); 1037. (B); 1038. (A); 1039. (C); 1040. (B); 1041. (D); 1042. (D); 1043. (C); 1044. (D); 1045. (A); 1046. (D); 1047. (B); 1048. (D); 1049. (C);

1050. (D); 1051. (C); 1052. (D); 1053. (B); 1054. (B);
1055. (D); 1056. (D); 1057. (D); 1058. (D); 1059. (A);
1060. (D); 1061. (D); 1062. (C); 1063. (D); 1064. (D);
1065. (B); 1066. (D); 1067. (D); 1068. (D); 1069. (C);
1070. (C); 1071. (A); 1072. (A); 1073. (D); 1074. (C);
1075. (B); 1076. (D); 1077. (B); 1078. (D); 1079. (D);
1080. (D); 1081. (C); 1082. (C); 1083. (A); 1084. (D);
1085. (C); 1086. (B); 1087. (D); 1088. (C); 1089. (B);
1090. (C); 1091. (C); 1092. (A); 1093. (D); 1094. (B);
1095. (D); 1096. (D); 1097. (D); 1098. (D); 1099. (B);
1100. (B); 1101. (B); 1102. (D); 1103. (D); 1104. (D);
1105. (D); 1106. (C); 1107. (D); 1108. (D); 1109. (D);
1110. (A); 1111. (C); 1112. (D); 1113. (C); 1114. (B);
1115. (D); 1116. (C); 1117. (C); 1118. (C); 1119. (D);
1120. (D); 1121. (D); 1122. (D); 1123. (D); 1124. (C);
1125. (B); 1126. (A); 1127. (D); 1128. (B); 1129. (C);
1130. (D); 1131. (D); 1132. (A); 1133. (D); 1134. (D);
1135. (D); 1136. (D); 1137. (D); 1138. (C); 1139. (D);
1140. (A); 1141. (D); 1142. (D); 1143. (D); 1144. (D);
1145. (D); 1146. (D); 1147. (D); 1148. (B); 1149. (C);
1150. (D); 1151. (D); 1152. (D); 1153. (C); 1154. (B);
1155. (D); 1156. (D); 1157. (A); 1158. (B); 1159. (D);
1160. (A); 1161. (A); 1162. (B); 1163. (D); 1164. (A);
1165. (B); 1166. (D); 1167. (C); 1168. (D); 1169. (D);
1170. (D); 1171. (A); 1172. (B); 1173. (C); 1174. (D);
1175. (B); 1176. (A); 1177. (B); 1178. (C); 1179. (D);
1180. (B); 1181. (C); 1182. (D); 1183. (A); 1184. (C);
1185. (D); 1186. (A); 1187. (B); 1188. (C); 1189. (D);
1190. (A); 1191. (B); 1192. (C); 1193. (A); 1194. (B);
1195. (C); 1196. (A); 1197. (B); 1198. (C); 1199. (D);
1200. (A); 1201. (B); 1202. (C); 1203. (D); 1204. (B);
1205. (A); 1206. (C); 1207. (D); 1208. (A); 1209. (B);
1210. (D); 1211. (D); 1212. (D); 1213. (C); 1214. (D);
1215. (C); 1216. (C); 1217. (D); 1218. (D); 1219. (C);
1220. (D); 1221. (C); 1222. (B); 1223. (A); 1224. (B);
1225. (D); 1226. (D); 1227. (C); 1228. (D); 1229. (C);
1230. (D); 1231. (D); 1232. (B); 1233. (A); 1234. (A);
1235. (D); 1236. (A); 1237. (A); 1238. (D); 1239. (B);
1240. (A); 1241. (D); 1242. (A); 1243. (B); 1244. (A);
1245. (D); 1246. (A); 1247. (B); 1248. (A); 1249. (A);
1250. (C); 1251. (C); 1252. (C); 1253. (B); 1254. (D);
1255. (B); 1256. (C); 1257. (B); 1258. (D); 1259. (C);
1260. (C); 1261. (D); 1262. (B); 1263. (B); 1264. (B);
1265. (C); 1266. (A); 1267. (C); 1268. (D); 1269. (D);
1270. (A); 1271. (C); 1272. (C); 1273. (D); 1274. (D);
1275. (B); 1276. (B); 1277. (D); 1278. (B); 1279. (B);
1280. (D); 1281. (D); 1282. (D); 1283. (B); 1284. (D);
1285. (C); 1286. (D); 1287. (A); 1288. (C); 1289. (B);
1290. (B); 1291. (C); 1292. (A); 1293. (C); 1294. (A);
1295. (A); 1296. (C); 1297. (D); 1298. (B); 1299. (A);
1300. (A); 1301. (D); 1302. (C); 1303. (D); 1304. (B);
1305. (A); 1306. (A); 1307. (B); 1308. (B); 1309. (A);
1310. (A); 1311. (D); 1312. (D); 1313. (D); 1314. (B);
1315. (D); 1316. (A); 1317. (D); 1318. (C); 1319. (C);
1320. (B); 1321. (C); 1322. (D); 1323. (B); 1324. (B);
1325. (A); 1326. (B); 1327. (D); 1328. (C); 1329. (C);
1330. (B); 1331. (D); 1332. (A); 1333. (C); 1334. (B);
1335. (B); 1336. (D); 1337. (D); 1338. (C); 1339. (A);
1340. (A); 1341. (D); 1342. (D); 1343. (B); 1344. (D);
1345. (A); 1346. (D); 1347. (C); 1348. (B); 1349. (B);
1350. (D); 1351. (C); 1352. (A); 1353. (B); 1354. (B);
1355. (C); 1356. (A); 1357. (A); 1358. (C); 1359. (D);
1360. (B); 1361. (D); 1362. (D); 1363. (B); 1364. (D);
1365. (D); 1366. (B); 1367. (A); 1368. (A); 1369. (C);
1370. (B); 1371. (C); 1372. (C); 1373. (A); 1374. (C);
1375. (D); 1376. (A); 1377. (C); 1378. (C); 1379. (B);
1380. (D); 1381. (D); 1382. (C); 1383. (D); 1384. (C);
1385. (A); 1386. (C); 1387. (D); 1388. (D); 1389. (D);
1390. (B); 1391. (D); 1392. (A); 1393. (D); 1394. (A);
1395. (D); 1396. (B); 1397. (A); 1398. (D); 1399. (D);
1400. (C); 1401. (A); 1402. (B); 1403. (B); 1404. (A);
1405. (D); 1406. (D); 1407. (C); 1408. (C); 1409. (B);
1410. (B); 1411. (C); 1412. (C); 1413. (C); 1414. (A);
1415. (A); 1416. (C); 1417. (C); 1418. (C); 1419. (A);
1420. (D); 1421. (C); 1422. (C); 1423. (B); 1424. (D);
1425. (B); 1426. (C); 1427. (A); 1428. (B); 1429. (C);
1430. (D); 1431. (D); 1432. (C); 1433. (B); 1434. (B);
1435. (B); 1436. (C); 1436. (B); 1438. (A); 1439. (A);
1440. (A); 1441. (A); 1442. (D); 1443. (A); 1444. (C);
1445. (D); 1446. (D); 1447. (B); 1448. (C); 1449. (A);
1450. (D); 1451. (D); 1452. (D); 1453. (B); 1454. (A);
1455. (C); 1456. (B); 1457. (D); 1458. (B); 1459. (D);
1460. (B); 1461. (B); 1462. (D); 1463. (C); 1464. (C);
1465. (B); 1466. (B); 1467. (D); 1468. (D); 1469. (C);
1470. (D); 1471. (B); 1472. (C); 1473. (C); 1474. (B);
1475. (A); 1476. (A); 1477. (A); 1478. (B); 1479. (C);
1480. (C); 1481. (A); 1482. (B); 1483. (C); 1484. (D);
1485. (A); 1486. (B); 1487. (C); 1488. (B); 1489. (A);
1490. (B); 1491. (A); 1492. (A); 1493. (A); 1494. (C);
1495. (B); 1496. (C); 1497. (C); 1498. (C); 1499. (C);

1500. (D); 1501. (A); 1502. (B); 1503. (A); 1504. (C);
1505. (D); 1506. (A); 1507. (A); 1508. (C); 1509. (C);
1510. (A); 1511. (B); 1512. (C); 1513. (B); 1514. (C);
1515. (A); 1516. (C); 1517. (C); 1518. (A); 1519. (B);
1520. (C); 1521. (A); 1522. (B); 1523. (C); 1524. (D);
1525. (A); 1526. (B); 1527. (C); 1528. (D); 1529. (A);
1530. (B); 1531. (C); 1532. (A); 1533. (A); 1534. (B);
1535. (A); 1536. (C); 1537. (A); 1538. (A); 1539. (B);
1540. (C); 1541. (D); 1542. (A); 1543. (B); 1544. (C);
1545. (D); 1546. (C); 1547. (C); 1548. (A); 1549. (B);
1550. (C); 1551. (D); 1552. (B); 1553. (A); 1554. (C);
1555. (D); 1556. (A); 1557. (D); 1558. (C); 1559. (B);
1560. (A); 1561. (A); 1562. (D); 1563. (D); 1564. (B);
1565. (D); 1566. (A); 1567. (A); 1568. (D); 1569. (D);
1570. (A); 1571. (D); 1572. (B); 1573. (B); 1574. (B);
1575. (D); 1576. (D); 1577. (D); 1578. (D); 1579. (C);
1580. (D); 1581. (A); 1582. (C); 1583. (D); 1584. (D);
1585. (B); 1586. (D); 1587. (D); 1588. (D); 1589. (C);
1590. (D); 1591. (A); 1592. (D); 1593. (D); 1594. (D);
1595. (D); 1596. (D); 1597. (A); 1598. (A); 1599. (D);
1600. (D); 1601. (B); 1602. (A); 1603. (A); 1604. (A);
1605. (C); 1606. (D); 1607. (C); 1608. (A); 1609. (A);
1610. (C); 1611. (D); 1612. (B); 1613. (D); 1614. (D);
1615. (D); 1616. (D); 1617. (D); 1618. (A); 1619. (C);
1620. (D); 1621. (D); 1622. (C); 1623. (B); 1624. (A);
1625.(B); 1626. (C); 1627. (B); 1628. (D); 1629. (C);
1630.(D); 1631. (D); 1632. (D); 1633. (D); 1634. (C);
1635. (A); 1636. (B); 1637. (A); 1638. (A); 1639. (A);
1640. (A); 1641. (A); 1642. (D); 1643. (D); 1644. (B);
1645. (A); 1646. (A); 1647. (A); 1648. (A); 1649. (A);
1650. (A); 1651. (C); 1652. (D); 1653. (B); 1654. (C);
1655. (A); 1656. (C); 1657. (B); 1658. (C); 1659. (A);
1660. (B); 1661. (D); 1662. (A); 1663. (D); 1664. (A);
1665. (B); 1666. (C); 1667. (A); 1668. (A); 1669. (B);
1670. (C); 1671. (D); 1672. (D); 1673. (D); 1674. (A);
1675. (C); 1676. (A); 1677. (D); 1678. (C); 1679. (B);
1680. (A); 1681. (D); 1682. (C); 1683. (A); 1684. (A);
1685. (A); 1686. (C); 1687. (D); 1688. (A); 1689. (B);
1690. (B); 1691. (A); 1692. (A); 1693. (B); 1694. (D);
1695. (D); 1696. (A); 1697. (D); 1698. (A); 1699. (C);
1700. (A); 1701. (A); 1702. (D); 1703. (A); 1704. (D);
1705. (D); 1706. (D); 1707. (D); 1708. (D); 1709. (A);
1710. (C); 1711. (D); 1712. (C); 1713. (C); 1714. (D);
1715. (B); 1716. (D); 1717. (D); 1718. (D); 1719. (C);
1720. (B); 1721. (D); 1722. (D); 1723. (D); 1724. (C);
1725. (D); 1726. (D); 1727. (C); 1728. (D); 1729. (D);
1730. (D); 1731. (D); 1732. (B); 1733. (D); 1734. (A);
1735. (D); 1736. (B); 1737. (B); 1738. (D); 1739. (B);
1740. (D); 1741. (D); 1742. (B); 1743. (D); 1744. (D);
1745. (B); 1746. (D); 1747. (C); 1748. (C); 1749. (B);
1750. (D); 1751. (C); 1752. (D); 1753. (B); 1754. (A);
1755. (D); 1756. (B); 1757. (D); 1758. (C); 1759. (A);
1760. (C); 1761. (B); 1762. (C); 1763. (D); 1764. (A);
1765. (B); 1766. (C); 1767. (B); 1768. (B); 1769. (A);
1770. (C); 1771. (D); 1772. (D); 1773. (A); 1774. (B);
1775. (C); 1776. (A); 1777. (B); 1778. (C); 1779. (D);
1780. (A); 1781. (A); 1782. (D); 1783. (A); 1784. (B);
1785. (C); 1786. (A); 1787. (B); 1788. (C); 1789. (A);
1790. (C); 1791. (D); 1792. (A); 1793. (A); 1794. (B);
1795. (C); 1796. (D); 1797. (B); 1798. (A); 1799. (B);
1800. (A); 1801. (D); 1802. (C); 1803. (D); 1804. (C);
1805. (B); 1806. (A); 1807. (C); 1808. (D); 1809. (B);
1810. (A); 1811. (B); 1812. (C); 1813. (D); 1814. (B);
1815. (A); 1816. (A); 1817. (B); 1818. (A); 1819. (B);
1820. (C); 1821. (A); 1822. (B); 1823. (C); 1824. (D);
1825. (A); 1826. (B); 1827. (B); 1828. (B); 1829. (A);
1830. (C); 1831. (D); 1832. (A); 1833. (B); 1834. (C);
1835. (A); 1836. (B); 1837. (C); 1838. (D); 1839. (A);
1840. (A); 1841. (C); 1842. (D); 1843. (A); 1844. (B);
1845. (C); 1846. (A); 1847. (B); 1848. (C); 1849. (C);
1850. (C); 1851. (D); 1852. (B); 1853. (D); 1854. (B);
1855. (C); 1856. (C); 1857. (C); 1858. (D); 1859. (C);
1860. (A); 1861. (A); 1862. (D); 1863. (B); 1864. (A);
1865. (A); 1866. (D); 1867. (B); 1868. (A); 1869. (D);
1870. (D); 1871. (D); 1872. (D); 1873. (C); 1874. (D);
1875. (C); 1876. (D); 1877. (D); 1878. (A); 1879. (A);
1880. (B); 1881. (D); 1882. (C); 1883. (D); 1884. (A);
1885. (A); 1886. (B); 1887. (B); 1888. (D); 1889. (B);
1890. (B); 1891. (D); 1892. (A); 1893. (A); 1894. (D);
1895. (D); 1896. (B); 1897. (C); 1898. (B); 1899. (B);
1900. (B); 1901. (B); 1902. (B); 1903. (C); 1904. (B);
1905. (A); 1906. (B); 1907. (B); 1908. (D); 1909. (D);
1910. (B); 1911. (C); 1912. (D); 1913. (D); 1914. (D);
1915. (A); 1916. (C); 1917. (D); 1918. (D); 1919. (A);
1920. (B); 1921. (D); 1922. (D); 1923. (A); 1924. (C);
1925. (A); 1926. (B); 1927. (A); 1928. (D); 1929. (D);
1930. (D); 1931. (D); 1932. (C); 1933. (C); 1934. (C);
1935. (D); 1936. (C); 1937. (D); 1938. (D); 1939. (A);
1940. (B); 1941. (C); 1942. (D); 1943. (B); 1944. (B);
1945. (A); 1946. (D); 1947. (B); 1948. (D); 1949. (A);

1950. (D); 1951. (A); 1952. (B); 1953. (A); 1954. (D);
1955. (D); 1956. (B); 1957. (B); 1958. (D); 1959. (D);
1960. (B); 1961. (D); 1962. (B); 1963. (C); 1964. (D);
1965. (B); 1966. (B); 1967. (D); 1968. (A); 1969. (C);
1970. (D); 1971. (A); 1972. (C); 1973. (A); 1974. (B);
1975. (C); 1976. (D); 1977. (A); 1978. (D); 1979. (A);
1980. (D); 1981. (D); 1982. (A); 1983. (D); 1984. (B);
1985. (A); 1986. (D); 1987. (C); 1988. (B); 1989. (C);
1990. (D); 1991. (D); 1992. (D); 1993. (D); 1994. (A);
1995. (B); 1996. (C); 1997. (D); 1998. (B); 1999. (C);
2000. (B); 2001. (A); 2002. (D); 2003. (A); 2004. (D);
2005. (D); 2006. (D); 2007. (B); 2008. (D); 2009. (C);
2010. (D); 2011. (C); 2012. (C); 2013. (B); 2014. (A);
2015. (C); 2016. (B); 2017. (D); 2018. (A); 2019. (A);
2020. (C); 2021. (B); 2022. (C); 2023. (A); 2024. (B);
2025. (C); 2026. (D); 2027. (A); 2028. (D); 2029. (B);
2030. (C); 2031. (A); 2032. (B); 2033. (C); 2034. (D);
2035. (A); 2036. (B); 2037. (C); 2038. (D); 2039. (A);
2040. (D); 2041. (D); 2042. (C); 2043. (D); 2044. (D);
2045. (C); 2046. (C); 2047. (D); 2048. (D); 2049. (D);
2050. (B); 2051. (B); 2052. (D); 2053. (D); 2054. (B);
2055. (B); 2056. (B); 2057. (C); 2058. (C); 2059. (D);
2060. (B); 2061. (A); 2062. (D); 2063. (C); 2064. (D);
2065. (D); 2066. (D); 2067. (D); 2068. (C); 2069. (B);
2070. (D); 2071. (C); 2072. (D); 2073. (D); 2074. (A);
2075. (D); 2076. (D); 2077. (C); 2078. (D); 2079. (D);
2080. (C); 2081. (B); 2082. (C); 2083. (C); 2084. (B);
2085. (D); 2086. (B); 2087. (B); 2088. (A); 2089. (A);
2090. (D); 2091. (D); 2092. (D); 2093. (B); 2094. (A);
2095. (B); 2096. (D); 2097. (D); 2098. (C); 2099. (A);
2100. (A); 2101. (D); 2102. (C); 2103. (B); 2104. (B);
2105. (B); 2106. (C); 2107. (D); 2108. (C); 2109. (A);
2110. (A); 2111. (B); 2112. (B); 2113. (B); 2114. (A);
2115. (C); 2116. (D); 2117. (D); 2118. (D); 2119. (C);
2120. (D); 2121. (A); 2122. (D); 2123. (D); 2124. (D);
2125. (D); 2126. (C); 2127. (D); 2128. (D); 2129. (B);
2130. (C); 2131. (D); 2132. (D); 2133. (B); 2134. (D);
2135. (D); 2136. (B); 2137. (A); 2138. (A); 2139. (D);
2140. (A); 2141. (A); 2142. (D); 2143. (D); 2144. (A);
2145. (D); 2146. (B); 2147. (D); 2148. (C); 2149. (D);
2150. (C); 2151. (A); 2152. (A); 2153. (D); 2154. (B);
2155. (A); 2156. (A); 2157. (D); 2158. (D); 2159. (C);
2160. (B); 2161. (C); 2162. (A); 2163. (A); 2164. (C);
2165. (C); 2166. (D); 2167. (A); 2168. (D); 2169. (C);
2170. (B); 2171. (D); 2172. (C); 2173. (A); 2174. (C);
2175. (D); 2176. (B); 2177. (D); 2178. (B); 2179. (A);
2180. (D); 2181. (B); 2182. (C); 2183. (D); 2184. (B);
2185. (A); 2186. (A); 2187. (D); 2188. (A); 2189. (D);
2190. (D); 2191. (A); 2192. (C); 2193. (B); 2194. (B);
2195. (D); 2196. (B); 2197. (D); 2198. (A); 2199. (D);
2200. (D); 2201. (D); 2202. (D); 2203. (A); 2204. (B);
2205. (B); 2206. (A); 2207. (C); 2208. (B); 2209. (B);
2210. (D); 2211. (D); 2212. (B); 2213. (B); 2214. (D);
2215. (D); 2216. (A); 2217. (C); 2218. (A); 2219. (B);
2220. (A); 2221. (D); 2222. (B); 2223. (A); 2224. (B);
2225. (A); 2226. (D); 2227. (C); 2228. (D); 2229. (A);
2230. (C); 2231. (D); 2232. (B); 2233. (D); 2234. (D);
2235. (B); 2236. (B); 2237. (D); 2238. (A); 2239. (D);
2240. (C); 2241. (D); 2242. (C); 2243. (C); 2244. (D);
2245. (C); 2246. (D); 2247. (C); 2248. (A); 2249. (C);
2250. (C); 2251. (D); 2252. (D); 2253. (D); 2254. (B);
2255. (D); 2256. (B); 2257. (D); 2258. (D); 2259. (D);
2260. (A); 2261. (D); 2262. (B); 2263. (C); 2264. (A);
2265. (B); 2266. (D); 2267. (B); 2268. (D); 2269. (B);
2270. (B); 2271. (C); 2272. (B); 2273. (C); 2274. (C);
2275. (A); 2276. (A); 2277. (A); 2278. (C); 2279. (C);
2280. (D); 2281. (B); 2282. (D); 2283. (D); 2284. (B);
2285. (B); 2286. (C); 2287. (B); 2288. (A); 2289. (D);
2290. (D); 2291. (D); 2292. (D); 2293. (A); 2294. (C);
2295. (D); 2296. (A); 2297. (C); 2298. (A); 2299. (D);
2300. (A); 2301. (D); 2302. (B); 2303. (A); 2304. (D);
2305. (D); 2306. (B); 2307. (C); 2308. (B); 2309. (D);
2310. (B); 2311. (A); 2312. (B); 2313. (D); 2314. (A);
2315. (C); 2316. (C); 2317. (C); 2318. (D); 2319. (C);
2320. (D); 2321. (B); 2322. (D); 2323. (A); 2324. (D);
2325. (C); 2326. (B); 2327. (C); 2328. (C); 2329. (A);
2330. (B); 2331. (C); 2332. (D); 2333. (A); 2334. (A);
2335. (B); 2336. (C); 2337. (D); 2338. (C); 2339. (D);
2340. (A); 2341. (B); 2342. (C); 2343. (D); 2344. (A);
2345. (D); 2346. (A); 2347. (A); 2348. (C); 2349. (D);
2350. (C); 2351. (D); 2352. (A); 2353. (A); 2354. (B);
2355. (C); 2356. (D); 2357. (B); 2358. (B); 2359. (A);
2360. (B); 2361. (B); 2362. (A); 2363. (A); 2364. (A);
2365. (A); 2366. (B); 2367. (A); 2368. (B); 2369. (B);
2370. (C); 2371. (A); 2372. (A); 2373. (E); 2374. (A);
2375. (A); 2376. (A); 2377. (C); 2378. (A); 2379. (D);
2380. (B); 2381. (C); 2382. (A); 2383. (B); 2384. (C);
2385. (A); 2386. (C); 2387. (A); 2388. (B); 2389. (C);
2390. (D); 2391. (C); 2392. (C); 2393. (B); 2394. (C);
2395. (B); 2396. (A); 2397. (H); 2398. (E); 2399. (E);

2400. (A); 2401. (D); 2402. (C); 2403. (A); 2404. (C);
2405. (B); 2406. (C); 2407. (D); 2408. (A); 2409. (B);
2410. (C); 2411. (D); 2412. (A); 2413. (B); 2414. (C);
2415. (D); 2416. (C); 2417. (B); 2418. (C); 2419. (D);
2420. (C); 2421. (D); 2422. (A); 2423. (A); 2424. (D);
2425. (C); 2426. (C); 2427. (A); 2428. (C); 2429. (C);
2430. (A); 2431. (B); 2432. (C); 2433. (A); 2434. (E);
2435. (A); 2436. (B); 2437. (C); 2438. (C); 2439. (D);
2440. (B); 2441. (D); 2442. (D); 2443. (B); 2444. (D);
2445. (B); 2446. (C); 2447. (D); 2448. (D); 2449. (C);
2450. (D); 2451. (D); 2452. (A); 2453. (D); 2454. (C);
2455. (D); 2456. (C); 2457. (D); 2458. (D); 2459. (D);
2460. (D); 2461. (D); 2462. (D); 2463. (D); 2464. (C);
2465. (D); 2466. (D); 2467. (C); 2468. (D); 2469. (D);
2470. (B); 2471. (D); 2472. (D); 2473. (C); 2474. (D);
2475. (D); 2476. (B); 2477. (B); 2478. (B); 2479. (C);
2480. (D); 2481. (D); 2482. (D); 2483. (D); 2484. (D);
2485. (D); 2486. (D); 2487. (D); 2488. (C); 2489. (B);
2490. (B); 2491. (A); 2492. (C); 2493. (D); 2494. (C);
2495. (A); 2496. (C); 2497. (D); 2498. (D); 2499. (D);
2500. (D); 2501. (D); 2502. (C); 2503. (A); 2504. (B);
2505. (D); 2506. (A); 2507. (D); 2508. (B); 2509. (D);
2510. (D); 2511. (B); 2512. (D); 2513. (D); 2514. (D);
2515. (D); 2516. (A); 2517. (D); 2518. (D); 2519. (A);
2520. (A); 2521. (B); 2522. (D); 2523. (C); 2524. (D);
2525. (A); 2526. (C); 2527. (D); 2528. (D); 2529. (D);
2530. (D); 2531. (B); 2532. (C); 2533. (C); 2534. (A);
2535. (D); 2536. (C); 2537. (D); 2538. (A); 2539. (A);
2540. (A); 2541. (C); 2542. (B); 2543. (C); 2544. (D);
2545. (B); 2546. (A); 2547. (C); 2548. (C); 2549. (A);
2550. (C); 2551. (C); 2552. (B); 2553. (A); 2554. (C);
2555. (C); 2556. (D); 2557. (B); 2558. (D); 2559. (C);
2560. (D); 2561. (D); 2562. (A); 2563. (A); 2564. (C);
2565. (A); 2566. (B); 2567. (B); 2568. (A); 2569. (C);
2570. (D); 2571. (D); 2572. (D); 2573. (B); 2574. (B);
2575. (A); 2576. (D); 2577. (C); 2578. (A); 2579. (C);
2580. (D); 2581. (D); 2582. (B); 2583. (D); 2584. (C);
2585. (D); 2586. (D); 2587. (C); 2588. (D); 2589. (D);
2590. (C); 2591. (B); 2592. (D); 2593. (D); 2594. (D);
2595. (C); 2596. (A); 2597. (B); 2598. (D); 2599. (C);
2600. (C); 2601. (D); 2602. (C); 2603. (D); 2604. (D);
2605. (C); 2606. (D); 2607. (D); 2608. (C); 2609. (D);
2610. (B); 2611. (D); 2612. (D); 2613. (B); 2614. (A);
2615. (B); 2616. (A); 2617. (D); 2618. (B); 2619. (D);
2620. (D); 2621. (A); 2622. (B); 2623. (B); 2624. (A);
2625. (B); 2626. (C); 2627. (D); 2628. (C); 2629. (A);
2630. (B); 2631. (A); 2632. (B); 2633. (D); 2634. (B);
2635. (A); 2636. (C); 2637. (D); 2638. (D); 2639. (D);
2640. (D); 2641. (D); 2642. (D); 2643. (B); 2644. (D);
2645. (D); 2646. (D); 2647. (B); 2648. (D); 2649. (D);
2650. (A); 2651. (A); 2652. (B); 2653. (C); 2654. (C);
2655. (A); 2656. (D); 2657. (B); 2658. (A); 2659. (A);
2660. (D); 2661. (A); 2662. (D); 2663. (D); 2664. (B);
2665. (A); 2666. (C); 2667. (D); 2668. (D); 2669. (A);
2670. (C); 2671. (D); 2672. (D); 2673. (A); 2674. (D);
2675. (C); 2676. (B); 2677. (C); 2678. (B); 2679. (B);
2680. (C); 2681. (D); 2682. (D); 2683. (D); 2684. (D);
2685. (D); 2686. (B); 2687. (A); 2688. (B); 2689. (D);
2690. (D); 2691. (B); 2692. (B); 2693. (D); 2694. (B);
2695. (B); 2696. (A); 2697. (A); 2698. (D); 2699. (D);
2700. (D); 2701. (C); 2702. (C); 2703. (B); 2704. (B);
2705. (A); 2706. (C); 2707. (A); 2708. (A); 2709. (D);
2710. (D); 2711. (B); 2712. (D); 2713. (C); 2714. (A);
2715. (D); 2716. (D); 2717. (A); 2718. (B); 2719. (D);
2720. (D); 2721. (D); 2722. (D); 2723. (D); 2724. (D);
2725. (D); 2726. (D); 2727. (D); 2728. (D); 2729. (D);
2730. (D); 2731. (D); 2732. (D); 2733. (A); 2734. (D);
2735. (C); 2736. (D); 2737. (C); 2738. (D); 2739. (C);
2740. (B); 2741. (D); 2742. (A); 2743. (C); 2744. (A);
2745. (B); 2746. (C); 2747. (D); 2748. (A); 2749. (D);
2750. (D); 2751. (C); 2752. (D); 2753. (B); 2754. (A);
2755. (D); 2756. (B); 2757. (A); 2758. (B); 2759. (C);
2760. (C); 2761. (C); 2762. (A); 2763. (D); 2764. (A);
2765. (A); 2766. (D); 2767. (D); 2768. (C); 2769. (C);
2770. (D); 2771. (D); 2772. (A); 2773. (B); 2774. (D);
2775. (A); 2776. (A); 2777. (C); 2778. (D); 2779. (A);
2780. (B); 2781. (C); 2782. (C); 2783. (C); 2784. (D);
2785. (D); 2786. (A); 2787. (D); 2788. (B); 2789. (C);
2790. (A); 2791. (D); 2792. (D); 2793. (D); 2794. (D);
2795. (B); 2796. (D); 2797. (D); 2798. (A); 2799. (C);
2800. (D); 2801. (C); 2802. (B); 2803. (C); 2804. (A);
2805. (C); 2806. (D); 2807. (D); 2808. (B); 2809. (D);
2810. (D); 2811. (A); 2812. (D); 2813. (D); 2814. (D);
2815. (D); 2816. (D); 2817. (D); 2818. (D); 2819. (D);
2820. (D); 2821. (D); 2822. (B); 2823. (A); 2824. (C);
2825. (A); 2826. (B); 2827. (B); 2828. (C); 2829. (D);
2830.(A).

Chapter 5

PLANT PHYSIOLOGY

1. The branch of science which deals with life activities of plants is called:
(A) Plant ecology (B) Plant physiology
(C) Genetics (D) Plant biochemistry

2. The branch of science which dealing with functional aspects of plant is called:
(A) Plant breeding (B) Microbilogy
(C) Plant physiology (D) Plant biochemistry

3. Plant physiology is derived by two words:
(A) Physis + logos (B) Phyton + logos
(C) Physis + logon (D) Phyton + logon

4. Who is known as father of plant physiology:
(A) Renner (B) Stephen Hales
(C) Dixon and Jelly (D) Hertig

5. When simple substances unite together to form a complex substances e.g. photosynthesis, lipid synthesis, protein synthesis is called:
(A) Annabolism (B) Catabolism
(C) Both (A) and (B) (D) None of these

6. When complex substances breaks down into a simple substances e.g. respiration, lipid breakdown, protein breakdown etc is called:
(A) Annabolism (B) Catabolism
(C) Both (A) and (B) (D) None of these

7. Anabolism is a:
(A) Constructive process (B) Synthetic process
(C) Destructive process (D) Both (A) and (B)

8. Catabolism is a :
(A) Destructive process (B) Catabolic process
(C) Both (A) and (B) (D) None of these

9. Movement of particles of a matter due to their own kinetic energy is known as:
(A) Diffusion (B) Osmosis
(C) Both (A) and (B) (D) None of these

10. When a solution is separated from its weaker solution by means of semipermeable membrane, than passes into stronger solution so as to equalize concentration on both side is known as:
(A) Diffusion (B) Osmosis
(C) Both (A) and (B) (D) None of these

11. It is a hypothetical term which describes the potential ability of a gas, liquid or solid to diffuse from its higher concentration to lower concentration is known as:
(A) Diffusion pressure (B) Osmotic pressure
(C) Both (A) and (B) (D) None of these

12. It is the pressure in atmosphere, which is required in opposite direction, so as to stop entry of water (solvent) molecules, when a semipermeable membrane separates concentrated solution from weaker solution is known as:
(A) Diffusion pressure (B) Osmotic pressure
(C) Both (A) and (B) (D) None of these

13. When concentration of outer solution (in which cell is placed) is equal to concentration of cell sap, it is called:
(A) Isotonic solution (B) Hypotonic solution
(C) Hypertonic solution (D) None of these

14. When concentration of outer solution (in which cell is placed) is lower than concentration of cell sap, it is called :
(A) Isotonic solution (B) Hypotonic solution
(C) Hypertonic solution (D) None of these

15. When concentration of outer solution (in which cell is placed) is higher than concentration of cell sap, it is called :
(A) Isotonic solution (B) Hypotonic solution
(C) Hypertonic solution (D) None of these

16. When a cell is placed in hypertonic solution then ——— ——— occurs:
(A) Exosmosis (B) Plasmolysis
(C) Endosmosis (D) Deplasmolysis

17. Due to exosmosis, the protoplasm shrinks and leaves the cell wall and thus cell becomes flaccid, which is called plasmolysed cell and this phenomenon is called:
(A) Exosmosis (B) Plasmolysis
(C) Endosmosis (D) Deplasmolysis

18. If the plasmolysed cell (flaccid cell) is placed in hypotonic solution then:
(A) Exosmosis (B) Plasmolysis
(C) Endosmosis (D) Deplasmolysis

19. If the plasmolysed cell (flaccid cell) is placed in hypotonic solution then endosmosis occurs, which makes the cell again turgid and this is known :
(A) Exosmosis (B) Plasmolysis
(C) Endosmosis (D) Deplasmolysis

20. The pressure that develops in a cell due to osmotic diffusion of water inside it is called:
(A) Turgor pressure (B) Wall Pressure
(C) Diffusion pressure (D) None of these

21. As cell wall is rigid, it will exert equal and opposite pressure to that of T.P., which is called :
(A) Turgor pressure (B) Wall Pressure
(C) Diffusion pressure (D) Suction pressure

22. The amount by which D.P. of solution is lower than that of its pure solvent, is called:
(A) Turgor pressure (B) Wall Pressure
(C) Diffusion pressure deficit (D) Suction pressure

23. In case of D.P.D., the solution will try to wipe off this difference in D.P. by sucking more water (solvent) molecules and hence D.P.D. is the measure or index of sucking power and hence it is called :
(A) Turgor pressure (B) Wall Pressure
(C) Diffusion pressure (D) Suction pressure

24. 80-90% of total water lost in transpiration is through specialized pores (stomata) on leaf surface is called:
(A) Stomatal transpiration
(B) Cuticular transpiration:
(C) Lenticular transpiration
(D) None of these

25. Transpiration directly through cuticle. 3-9% of total transpiration (in rare cases upto 20%) occurs through the cuticle. Thicker is the cuticle, lesser will be the transpiration is called:
(A) Stomatal transpiration
(B) Cuticular transpiration
(C) Lenticular transpiration
(D) None of these

26. Transpiration through lenticle or pores present in woody stem and in some fruits. It accounts for only 0.1% of total transpiration (rarely upto 1%) is called:
(A) Stomatal transpiration
(B) Cuticular transpiration:
(C) Lenticular transpiration
(D) None of these

27. The amount of water lost through transpiration per unit of dry matter produced during growing season of the plant:
(A) Transpiration ratio (B) Respiration ratio
(C) Both (A) and (B) (D) None of these

28. Which is wrong regarding guttation?
(A) Loss in liquid form (B) Through hydathodes
(C) Due to root pressure (D) Pure water

29. Which one of the following statements about sieve tube elements in plants is not correct?
(A) They are supported by companion cells
(B) They must die to become functional
(C) They translocate many organic nutrients
(D) They link end to end forming sieve tubes

30. Guttation occurs when:
(A) Warm days followed by cool nights
(B) Cool days followed by warm nights
(C) Plants growing in dry, warm soil
(D) Plants grown in humids

31. The highest loss of water through guttation takes place in:
(A) *Tropoeolum najus* (B) Oat
(C) Barley (D) *Colocasia antiquorum*

32. The stomatal movement is due to:
(A) Pressure of water vapour inside
(B) Turgor pressure of guard cells
(C) Genetic constitution
(D) None of the above

33. Transpiration efficiency refers to:
(A) Water absorption, transpiration ratio
(B) Transpiration in relation to ascent of sap
(C) Weight of dry matter produced per kg of water transpired
(D) None of the above

34. When relative humidity of atmosphere is cent per cent, then stomata will:
(A) Open (B) Closed
(C) Partially open (D) Partially closed

35. Which statement about the Mn centre of photosystem II is incorrect?
(A) The Mn centre has four possible oxidation states
(B) Electrons are transferred from the Mn centre to P 680
(C) A tyrosine residue on the D1 protein is an intermediate in electron transfer
(D) The O_2 released by the Mn centre comes from the oxidation of water

36. Blue green algae (cyanobacteria) form heterocysts when deprived of ammonia and nitrate. In this form, the algae lack nuleic and are attached to adjacent vegetative cells. Heterocysts have photosystem I activity but are entirely devoid of photosystem II activity. What is their role?
(A) N_2 fixation (B) Denitrification
(C) Proline synthesis (D) Spore formation

37. Guard cells differ from adjacent of epidermal cells is having:
(A) Mitochondria (B) Golgibodies
(C) Cell wall (D) Chloroplasts

38. Which of the following statement about the thylakoid membrane is incorrect?
(A) It contains photosystems I and ATP synthase in the unstaked regions
(B) It contains the cytochrome b6f complex in the unstaked regions only
(C) It contains photosystem II mostly in the stakd regions
(D) It facilitates communication between photosystem I and II by the circulation of plastoquinones and plasocyanines in the thylakoid space

39. The most important factors affecting transpiration is:
(A) Temperature (B) Atmospheric humidity
(C) Wind (D) Light

40. Match list - I with list – II and select the correct answer using the odes given below the lists:

List I	List II
P. Richmond Lang effect	1. Reversal of red-drop
Q. Emerson effect	2. Inhibition of glycolysis by oxygen
R.Pasteur effect	3. Inhibition of photosynthesis by oxygen
S. Warburg effect	4. Senshene retardation in detached leaves

(A) P -1, Q – 4, R – 2, S – 3
(B) P -4, Q – 1, R – 3, S - 2
(C) P -1, Q – 4, R – 3, S – 2
(D) P -4, Q – 1, R – 2, S - 3

41. The phenomenon of absorption of water that depends on water loss from aerial parts of plants (Transpiration) is:
(A) Active absorption
(B) Passive absorption
(C) Both passive and active absorption
(D) None of these

42. A nonliving continuum of cell walls and intercellular spaces is:
(A) Ground complex (B) Alburnum
(C) Apoplast (D) Desmotubule

43. Root hairs occur in the zone of:
(A) Cell division (B) Cell elongation
(C) Cell maturation (D) Mature cells

44. Water lightly held to soil particles is:
(A) Gravitational water (B) Capillary water
(C) Hygroscopic water (D) Free water

45. Which of the following is not correct about tarpene?
(A) It is the largest class of secondary metabolite
(B) Phenylalanine acts as a precursor
(C) It is synthesized by pathway mevalonic acid
(D) It plays defensive role

46. Which of the following statement is not true with respect to the photosystem II (PSII)?
(A) PSII is located in the grana
(B) Chrolophyll molecule in the reaction centre for the PSII is bound to a 110 Kd integral membrane protein
(C) A complex of cytochromes b and f transport electrons from PSII to PSI
(D) PSII reaction centre is bound to two Mn^{++} ions

47. Root pressure may be caused by:
(A) Osmotic flow of water into xylem of absorbing part of root
(B) Loss of water from xylem of the plant due to transpiration
(C) Low water potential of leaves
(D) High water potential of leaves

48. The osmotic theory of active water absorption was first given by:
(A) Kramer (1941) (B) Eaton (1943)
(C) Arkins (1916) (D) Priestley (1923)

49. Which of the following is not involved in maintaining oxygen homeostasis in nitrogen fixing nodules?
(A) Leghaemoglobin
(B) Cytochrome oxidase
(C) ATP dependent potassium pump
(D) Dinitrogenase reductase

50. Force for passive water absorption develops in:
(A) Xylem (B) Aerial parts
(C) Root (D) Root hairs

51. If radioactive carbondioxide ($^{14}CO_2$) is used in calvin cycle then which carbon atom of 3-phosphoglyerate will appear radioactive:
(A) Carbon – 3 (B) Carbon - 1
(C) Carbon – 2 (D) Carbon - 4

52. Force for active water absorption develops in:
(A) Xylem (B) Aerial parts
(C) Root (D) Root hairs

53. The lowest water potential is found in the xylem channels of:
(A) Stem
(B) Root
(C) Root in the root hairs zone
(D) Leaves

54. The two pathways required for the net synthesis of glucose from triglycerides in germinating groundnut seeds are:
(A) Hexose monophosphate shunt and gluoneogenesis
(B) Calvin cycle and glyoxalate cycle
(C) Glycolysis and cori cycle
(D) Glyoxalate cycle and gluoneogenesis

55. The initial stage of water absorption by root cell is by:
(A) Imbibition (B) Adsorption
(C) Osmosis (D) Desorption

56. The Pasteur effect refers to:
(A) An increase in hexokinase activity owing to increased oxidative phosphorylation
(B) An increase in ethanol formation from pyruvate on changing from aerobic to anaerobic metabolism
(C) An increase in glucose utilization on changing from anaerobic to aerobic metabolism
(D) A decrease in glucose utilization on changing from anaerobic to aerobic metabolism

57. Root cap has no role in water absorption because:
(A) It has no direct connection with the vascular system
(B) It has loosely arranged cells
(C) It has no cells containing chloroplasts
(D) It has no root hairs

58. The maximum absorption of water by roots occurs in the (region) zone of:
(A) Root cap (B) Cell division
(C) Cell elongation (D) Root hairs

59. Stomtal opening is measured by
(A) Planimeter (B) Potometer
(C) Porometer (D) Photometer

60. Transpiration cohesion tension theory operates in:
(A) Active absorption
(B) Passive absorption
(C) Both active and passive absorption
(D) None of these

61. The phenomenon of uptake of water at the expense of energy by the cell and usually against the osmotic gradient is known as:
(A) Osmosis (B) Active absorption
(C) Passive absorption (D) Imbibition

62. Rate limiting enzyme in ethylene biosynthesis is:
(A) α-aminolevulinic acid synthase
(B) Aminocyclopropane carboxylic acid synthase
(C) Polygaturanases
(D) Myo-inositol tetrakine phosphate

63. Which of the following statement is not correct:
(A) Plants absorb excess quantity of water
(B) Plants take in small quantity of mineral salts through the soil water
(C) Water and inorganic salts are taken in simultaneously by root hairs
(D) Plants absorb only one thing at a time, water or inorganic salts

64. Root pressure can be measured by means of:
(A) Porometer (B) Potometer
(C) Auxanometer (D) Manometer

65. How many ATO molecules can be derived from each molecule of acetyl-CoA that enters the Kreb's cycle (consider 1 NADH and $FADH_2$ give 3 and 2 ATP, Respectively)?
(A) 6 (B) 12 (C) 18 (D) 38

66. Rapid water absorption takes place by:
(A) Passive absorption (B) Active absorption
(C) Expenditure of energy (D) Osmosis

67. Microorganisms responsible for nitrification are:
(A) *Nitrosomonas* and *Nitrobacter*
(B) *Nostoc* and *Anabaena*
(C) *Rhizobium* and *Azatobacter*
(D) *Clostridium* and *Pseudomonas*

68. Which is correct about RNA polymerase III?
(A) It occurs in nucleoplasm
(B) It provides ~ 10% of total polymerase activity
(C) It transcribes tRNA and 5SRNA genes
(D) All of these

69. Assertion (A): In bulk of eukaryotic mRNAs, a poly – (A) tail is added at the 3'-end of primary transcripts or pre-mRNAs
Reason (R): Addition of a poly – (A) tail at the 3'-end of primary transcript is essential for the processing of pre-mRNAs
Options:
(A) Both A and R are correct and R adequately explains A
(B) Both A and R are correct, but R does not explain A
(C) Only A is correct
(D) Only R is correct

70. Which of the following statement about the mitochondrial protein gradient and membrane potential is correct?
(A) Either of them is sufficient to make ATP from ADP + Pi
(B) Both are required to make ATP
(C) Usually cancel one another since the system is at equilibrium
(D) Neither of them is required for ATP synthesis

71. Soil water is able to reach xylem elements of root because of:
(A) Imbibition
(B) Gradient of suction pressure
(C) Gradient of turgor pressure
(D) Ion concentration of water

72. Water potential found in root hair cells is generally:
(A) 1-2 atm (B) 1 -20 atm
(C) -1 to 4.0 atm (D) -1 to -4 atm

73. Path of water movement from soil to xylem is:
(A) Metaxylem → Protoxylem→ Cortex→ Soil→ Root hair
(B) Cortex → Root hair→ Endodermis→ Pericycle→ Protoxylem→ Metaxylem
(C) Soil→ Root hair→ Cortex→ Endodermis→ Pericycle→ Protoxylem→ Metaxylem
(D) Pericycle→ Soil → Root hair→ Cortex→ Endodermis → Protoxylem→ Metaxylem

74. The reactions of the kreb cycle:
(A) Take place in the cytosol of eukaryotic cells
(B) Generate ATP also by substrate phosphorylation
(C) Are important for the metabolism of carbohydrates but not other molecules
(D) Both (A) and (B)

75. In soil the water available for root absorption is:
(A) Gravitational water (B) Capillary water
(C) Hygroscopic water (D) Combined water

76. Rate of water absorption can be increased through:
(A) Decreased transpiration
(B) Decreased ion absorption
(C) Increased photosynthesis
(D) Increased transpiration

77. Which of the following statements completes the sentence more correctly? The rate of flow of electrons through the electron transport chain is most directly regulated by:
(A) The ATP : ADP ratio
(B) The concentration of acetyl-CoA
(C) The rate of oxidative phosphorylation
(D) Feedback inhibition by H_2O

78. The movement of water from one cell of cortex to the next in the root is due to:
(A) Water potential gradient
(B) Chemical potential gradient
(C) Accumulation of inorganic salts in the cells
(D) Accumulation of organic salts in the cells

79. Exudation of xylem sap on cutting of a shoot is due to:
(A) Guttation (B) Root pressure
(C) Transpiration (D) Osmosis

80. Water entering root due to diffusion is part of:
(A) Endocytosis (B) Guttation
(C) Passive absorption (D) Active absorption

81. Which of the following is not a substrate of ribulose 1, 5-bisphosphate carboxylase?
(A) CO_2
(B) Glyceraldehyde – 3- phosphate
(C) Ribulose 1, 5-bisphosphate
(D) O_2

82. Stump of well watered potted plant shows exudation of sap due to:
(A) Root pressure (B) Capillarity
(C) Cohesion force (D) Transpiration pull

83. A gene is alternatively spliced, and its exon 1 joined to any one or more of exons produces a functional protein. How many different functional proteins will be produced if this gene has 3 exons?
(A) 1 (B) 2 (C) 3 (D) 4

84. Assertion (A): Plant species with C_4 pathway generally have higher photosynthetic rate than C_3 plant species, specially under high light intensity
Reason (R): The main factor, which increase photosynthetic efficiency of C_4 plant species is the lack of measurable photorespiration in them
Options:
(A) Both A and R are correct and R adequately explains A
(B) Both A and R are correct, but R does not explain A
(C) Only A is correct
(D) Only R is correct

85. Water absorbed by root in order to meet the requirement of transpiration is due to:
(A) Transpiration pull (B) Osmosis
(C) Imbibition (D) Plasmolysis

86. Root hair absorbs water from soil through:
(A) Turgor pressure (B) Ion exchange
(C) Osmosis (D) D.P.D.

87. Which of the following inclusion bodies contain the enzymes responsible for carbon dioxide fixation in bacteria., blue green algae?
(A) Lysosomes (B) Peroxisomes
(C) Metachromatic granules (D) Carboxysome

88. Prolonged water-logging kills plants due to:
(A) Stoppage of root respiration
(B) Dilution of soil nutrients
(C) Dilution of plant cell sap
(D) Leaching of nutrients

89. Rate of water absorption is slow near freezing point because:
(A) Water absorption is a metabolic process
(B) Cell growth stops
(C) Transpiration is reduced
(D) Cell membranes become more viscous

90. Arrange root hair cell, inner cortical cell and mesophyll cell in ascending order of D.P.D.:
(A) Mesophyll cell, Root hair cell and Cortical cell
(B) Cortical cell, Mesophyll cell and Root hair cell
(C) Root hair cell, Cortical cell and Mesophyll cell
(D) Root hair cell, Mesophyll cell and Cortical cell

91. The upward movement of water, against the downward pull of gravity in plants depends on the process of:
(A) Absorption (B) Photosynthesis
(C) Respiration (D) Transpiration

92. Strasburger rejected the vital force theory on the ground that:
(A) Living cells are incapable of translocation
(B) Water rises in dead cells
(C) Respiration occurs in living cells
(D) Living cells are capable of growth

93. The enzyme nitrogenase that fixes atmospheric nitrogen in the root nodules of leguminous plants comprises two sub-units the composition of each respectively being:
(A) Fe-protein and Fe-Mo-protein
(B) Mo-protein and Fe-protein
(C) Mg-protein and Fe-Mo-protein
(D) Fe-protein and Fe-protein

94. Which of the following statement about the enzyme complexes of the electron transport system is correct?
(A) They are located in the mitochondrial matrix
(B) They cannot be isolated from one another in functional form
(C) They have very similar visible spectra
(D) They are integral membrane proteins located in the inner mitochondrial membrane

95. The pressure that develops in roots due to the metabolic activities of living cells is:
(A) Turgor pressure (B) Osmotic pressure
(C) Root pressure (D)Diffusion pressure

96. A gene is alternatively spliced, and its exon 1 joined to any one or more of exons produces a functional protein. How many different functional proteins will be produced if this gene has 4 exons?
(A) 4 (B) 6 (C) 7 (D) 8

97. Assertion (A): Enhancer sequences increase the activity of a promoter lying in their vicinity.
Reason (R): Enhancers most likely function by increasing the concentration of transcription factors in the vicinity of promoters
Options:
(A) Both A and R are correct and R adequately explains A
(B) Both A and R are correct, but R does not explain A
(C) Only A is correct
(D) Only R is correct

98. Which of the following statement is wrong in root-pressure concept of ascent of sap:
(A) Water can be raised to a height of 6' – 7'
(B) Water moves upward in the absence of root
(C) Water movement by this force is slow
(D) Root pressure operates in all the plants

99. Imbibitional theory was given by:
(A) Sachs (B) Boehm
(C) Scholander (D) Curtis

100. Which of the following statements about cyclic photophosphorylation is incorrect?
(A) It doesn't involve NADPH formation
(B) It is activated when ATP is limiting
(C) It does not generate O_2
(D) It involves a substrate-level phosphorylation

101. Cohesive force of water molecules is of the magnitude of (Dixon and Joly):
(A) 1-10 atm (B) 10-15 atm
(C) 45-200 atm (D) 15-45 atm

102. The following evidences rejected the atmospheric pressure theory. Which of them is incorrect:
(A) It can raise water to height of 33 feet only if complete vacuum is created
(B) Free surface of water is required for proper operation of atmospheric pressure which is readily available in plants
(C) Pressure falls below that of atmospheric pressure because of transpiration
(D) Water rises rapidly to compensate the water loss because of atmospheric pressure

103. Which of the following is not an example of N_2 fixation organism?
(A) *Frankia* (B) *Pisum sativum*
(C) *Rhizobium* (D) *Azospirillum*

104. Which of the following is not the least convincing in the imbibitional theory of ascent of sap:
(A) The imbibitional force is very high from 100-1000 atm
(B) The imbibitional force can raise the water to a height of 200' – 400' i.e. the tallest tree
(C) The movement of water is along the walls of the xylem vessels and not through the lumen (cavity) of the vessels
(D) The movement of water, no doubt is slow, but to some extent can keep pace with transpiration

105. The girding or ringing experiment is that:
(A) The metallic ring is tightly tied to a stem
(B) The ring tissue external to xylem is removed in a stem
(C) The area is simply marked in the form of ring in a stem
(D) None of the above

106. The girdling or ringing experiment to prove that the water moves through the xylem vessels was performed by:
(A) Stocking (B) Malpighi
(C) Molisch (D) Askenasy

107. Which of the following proteins in the photosynthetic electron-transport chain is not a transmembranes protein?
(A) ATP synthase (B) LHC
(C) PSII (D) Ferodoxin

108. Capillary theory was proposed by:
(A) Unger (B) Sachs
(C) Bohm (D) Mac Dougal

109. The main function of xylem vessel is:
(A) To provide mechanical support to the plant
(B) To conduct organic food from one part of the plant to another
(C) To conduct water and minerals from one part of the plant to another
(D) None of the above

110. Oxidative deamination is:
(A) The conversion of an amino acid to a keto acid plus ammonia
(B) The conversion of an amino acid to a carboxylic acid plus ammonia
(C) The transfer of an amino group from an amino acid to a keto acid
(D) The transfer of an amino group from an amino acid to a carboxylic acid

111. A girdled plant will eventually die because of:
(A) Absence of downward movement of water
(B) Absence of upward movement of water
(C) Absence of upward supply of organic nutrients
(D) Absence of downward movement of organic nutrients

112. A gene is alternatively spliced, and its exon 1 joined to any one or more of exons produces a functional protein. How many different functional proteins will be produced if this gene has 5 exons?
(A) 10 (B) 14 (C) 15 (D) 16

113. Which is not a constituents of protoplasm?
(A) Cell wall (B) Cell membrane
(C) Vacuole (D) Both (A) and (C)

114. The cut end of a shoot is dipped in eosine solution. What will happen?
(A) Ascent of sap does not occur
(B) Ascent of sap does not occur but the leaves remain fresh
(C) Leaves wilt but ascent of sap continues
(D) Ascent of sap occurs and the tracheary elements get stained

115. Which of the following conditions does not stimulate the activity of allosteric enzyme Rubisco?
(A) Increases in pH
(B) Increases in Mg^{++}
(C) Increases in NADPH concentration
(D) Increases in the concentration of K^+

116. Root pressure theory of ascent of sap is unacceptable because:
(A) Water can ascend without root or root pressure
(B) Root pressure cannot explain ascent of sap beyond 10 meters
(C) Root pressure is more during early morning than afternoon
(D) Root pressure does not occur in spring

117. Transpiration cohesion theory explains that the upwards pull of water is transmitted from top to bottom by cohesion of molecules caused by:
(A) Hydrophilic cell walls (B) Hydrogen bonds
(C) Oxygen bonds (D) Surface tension

118. Rising column of water does not break its connection with xylem walls despite negative pressure or tension due to:
(A) Cohesion amongst water molecules
(B) Strong transpiration pull
(C) Adhesion
(D) Surface tension

119. Ascent of sap is:
(A) Upward movement of water in the plant
(B) Downward movement of organic nutrients
(C) Upward and downward movement of water in the plant
(D) Redistribution of inorganic substances in the plant

120. The synthesis of ATP by hemiosmosis:
(A) Is endergonic and is coupled to exergonic electron transport
(B) Is exergonic and is coupled to endergonic electron transport
(C) Is due to electrons shuttled down the ET to the final electron acceptor, which NAD^+
(D) Produces less ATP than the Kreb's cycle and glycolysis combined

121. Instrument that can be used to demonstrate pull due to vaporisation of water is:
(A) Potometer (B) Atmometer
(C) Auxanometer (D) Anemometer

122. Further desaturation of 18:0 molecule occurs in association with?
(A) Cell wall (B) Mitochondria
(C) Endoplasmic reticulum (D) Peroxysomes

123. The transport of sap from root to top of the plant is:
(A) Ascent of sap (B) Conduction
(C) Transport (D) Translocation

124. If plants are grown for several days in an atmosphere containing $^{14}CO_2$ in place off $^{12}CO_2$, one would expect to find:
(A) Large amounts of radioactive oxygen released from the stomatas
(B) A large increase in ^{14}C in the starch
(C) A large decrease in the rate of carbon fixation in the guard cells
(D) An increase in the activity of RuBP carboxylase in the photosynthetic cells

125. In xylem, the ascent of sap takes place in:
(A) Tracheids with associated xylem parenchyma
(B) Walls of tracheary elements
(C) Xylem parenchyma
(D) Lumen of tracheary elements

126. The force with which water molecules are held together in xylem is:
(A) Imbibition pressure (B) Osmotic pressure
(C) Adhesion force (D) Cohesion force

127. Root pressure is unable to explain the ascent of sap because it is not found in:
(A) Bryophytes (B) All plants in all reasons
(C) Trees (D) Spring

128. The relay pump theory, in vital theories of ascent of sap, was given by:
(A) Strasburger (B) Dixon
(C) Godlewski (D) Unger

129. The apparatus that can measure root pressure is:
(A) Manometer (B) Potometer
(C) Auxanometer (D) Osmometer

130. Gibbrellic acid induced starch degradation in endosperm essentially requires gene expression, protein synthesis and protein transport. In an experiment to understand this, GFP tagged protein is seen to be moving from cytoplasm into the nucleus before alpha amylase could be detected. What do you think this GFP is tagged with?
(A) It is in fusion with a gene regulator
(B) It is in fusion with a signaling kinase
(C) It is in fusion with alpha-amylase
(D) It is tagged with a cell wall protein

131. In a gridled plant which dies first:
(A) Shoot (B) Root
(C) Both die simultaneously (D) Leaf

132. Guide RNA functions in:
(A) RNA processing (B) RNA editing
(C) RNA degradation (D) Transcription regulation

133. Most of the 18:0 molecules (stearic acid) are desaturated to 18:0 molecules (oleic acid) in which organelles?
(A) Plastids (B) Mitochondria
(C) Endoplasmic reticulum (D) Peroxysomes

134. A tree girdled upto xylem may survive for some time but ultimately dies because:
(A) Water does not move upwards
(B) Sugar does not move upwards
(C) Sugar does not move downwards
(D) Water does not move downwards

135. Water in plants is transported by or ascent of sap takes place through:
(A) Cambium (B) Phloem (C) Xylem (D) Epidermis

136. The most widely accepted theory for ascent of sap in trees is:
(A) Capillarity
(B) Role of atmospheric pressure
(C) Pulsating action of living cells
(D) Transpiration pull and cohesion theory of Dixon and Joly

137. Cell A has osmotic potential of -16 bars and pressure potential of 8 bars, whereas cell B has osmotic potential of -12 bars and pressure potential 2 bars. The direction of flow of water will be:
(A) From cell B to A (B) From cell A to B
(C) No flow (D) In both directions

138. Water rises in the stem due to:
(A) Cohesion and transpiration pull
(B) Turgor pressure
(C) Osmotic pressure
(D) None of these

139. Dixon and Joly are associated with:
(A) Light reaction and photosynthesis
(B) Anaerobic respiration
(C) Cohesion theory of ascent of sap
(D) Apical dominance

140. According to the vital force theory, ascent of sap is due to active pulsation of innermost layer of cortex. This theory was given by:
(A) J.C. Bose (B) Dixon
(C) Strasburger (D) Sachs

141. Root pressure is maximum when:
(A) Transpiration is high and absorption low
(B) Transpiration is very low and absorption high
(C) Transpiration is very high and absorption high
(D) Transpiration and absorption both are slow

142. Phototropism refers to how plants respond to sunlight by bending towards the light. This is the result of:
(A) Cell division on the side of the shoot apex closer to the light source
(B) Cell division on the side of the shoot apex away from the light source
(C) Cell differentiation on the side of the shoot apex away from the light source
(D) Cell elongation on the side of the shoot apex away from the light source

143. Xylem conducts sap from:
(A) Leaves to roots (B) Roots to leaves
(C) Roots to stems (D) Stems to roots

144. Which DNA polymerases functions in mitochndria?
(A) Pol β (B) Pol ε
(C) Pol γ (D) Pol δ

145. A plant will wilt when:
(A) A few roots are removed
(B) Phloem is blocked
(C) Xylem is blocked
(D) Both (B) and (C)

146. Cohesion-transpiration pull theory operates only in:
(A) Passive water absorption
(B) Active water absorption
(C) Conditions favouring transpiration
(D) Conditions restricting transpiration

147. Upward movement of water through xylem is best explained by:
(A) Cohesion theory (B) Pulsation theory
(C) Capillarity theory (D) Root pressure theory

148. Which contributes most to the transport of water from soil to the leaves of a tree:
(A) Root pressure
(B) Cohesion of water and transpiration pull
(C) Capillary rise of water inside xylem
(D) Hydrolysis of ATP

149. The principal pathway of water translocation in angiosperms is:
(A) Sieve cells (B) Sieve tube elements
(C) Xylem vessel system (D) Xylem and phloem

150. Cohesive force of water is due to:
(A) O-bonds (B) OH-bonds
(C) S-bonds (D) H-bonds

151. Water ascends in plants through:
(A) Xylem (B) Phloem
(C) Stele (D) Pith

152. Identify the correct statement:
(A) Cytokinins does not regulate cell division in plants
(B) Kinetin was discovered as a breakdown product of DNA
(C) Osmotic adjustment of cells does not help water balance in plants
(D) Cytokinins enhances leaf senescence

153. The force responsible for raising water in 100 feet tall plant is:
(A) Transpiration (B) Root pressure
(C) Air pressure (D) Capillary action

154. Water supply in the plant is due to:
(A) Osmosis (B) Guttation
(C) Cohesive force (D) Imbibition

155. The principle by which blotting paper absorbs water is:
(A) Capillary action (B) Transpiration pull
(C) Root pressure (D) Absorptive capacity

156. Which one of the following statements is correct with respect to the synthesis of starch and sucrose?
(A) Starch is synthesized in the mitochondria and sucrose in the cytosol
(B) Starch is synthesized in the cytosol and sucrose in the chloroplast
(C) Starch is synthesized in the chloroplast and sucrose in the mitochondria
(D) Starch is synthesized in the chloroplast and sucrose in the cytosol

157. In vascular plants, transport of sugar solutions is through tubes constructed of cells called:
(A) Vessel elements (B) Companion cells
(C) Tracheilds (D) Sieve-tube elements

158. If cohesion-tension transpiration pull theory is correct, a break in water column should:
(A) Increase water content of leaves
(B) Increase rate of photosynthesis
(C) Cause wilting of leaves
(D) No effect

159. Which is relevant for eukaryotes?
(A) Methylation of A in origin
(B) Methylation of C in origin
(C) Licensing factor
(D) DNA polymerase synthesis

160. The direction of water in Cycas leaflets is:
(A) Upward (B) Downward
(C) Lateral (D) Both (A) and (B)

161. Stomata open when guard cells are:
(A) Turgid (B) Flaccid
(C) Large (D) Small

162. Of the following four dorsiventral leaves which will show the maximum loss of weight:
(A) Smeared with Vaseline on both surfaces
(B) Smeared with Vaseline on the upper surface
(C) Smeared with Vaseline on the lower surface
(D) Unsmeared

163. Sir J.C. Bose proposed a theory of ascent of sap known as:
(A) Pulsation theory (B) Transpiration pull theory
(C) Relay pump theory (D) Capillary force theory

164. In alerone layer of graminaceous plants gibberellin promotes the synthesis of:
(A) Ribonuclease (B) α- amylase
(C) β-1, 3-glucanase (D) Tryptophan sunthase

165. Choose the correct set of words denoted by P, Q and R for the following statement. Pheophytin is a P molecule in which the central atom Q has been replaced by two atoms of R:
(A) P – Plastocyanin, Q – Copper, R - Hydrogen
(B) P - Plastocyanin, Q – Zinc, R - Oxygen
(C) P – Chlorophyll, Q – Magnesium, R - Oxygen
(D) P – Chlorophyll, Q – Magnesium, R - Hydrogen

166. Potometer are made on the principle that:
(A) The amount of water transpired is approximately equal to amount of water absorbed
(B) The amount of water transpired is more than the amount of water absorbed
(C) The amount od water transpired is less than the amount of water absorbed
(D) Humidity causes reduction in transpiration

167. Which mRNAs lacks poly-(A) tail?
(A) Histone mRNA (B) Globin mRNA
(C) Antibody mRNA (D) Vicilin mRNA

168 .In *E. coli*, DNA replication regulations is mostly likely based on:
(A) Methylation of A in origin
(B) Methylation of C in origin
(C) Licensing factor
(D) DNA polymerase synthesis

169. The process in which loss of water occurs in the form of water vapour is:
(A) Respiration (B) Guttation
(C) Transpiration (D) Exosmosis

170. Stomatal aperture is surrounded by guard cells and widens (opens) when guard cells are:
(A) Flaccid (B) Turgid
(C) Bean shaped (D) Dumb bell shaped

171. A mutant plant lacked plastocyanin. Upon illumination of this plant:
(A) Plastoquinol will remain in oxidized state
(B) Plastoquinol will be in reduced state
(C) Ferrodoxin will be in th ereduced state
(D) NAD will be in reduced state

172. The maximum loss of water in transpiration is from:
(A) Lenticels (B) Cuticle
(C) Stomata (D) Hydathades

173. The stomata are called sunken when:
(A) Guard cells are in line with epidermal cells
(B) Guard cells are situated below epidermal cells
(C) Guard cells are situated above epidermal cells
(D) Guard cells are situated in lower epidermis

174. Transpiration is high under:
(A) Dry environment
(B) Low atmospheric pressure
(C) High temperature
(D) All of these

175. Sunken stomata:
(A) Increase transpiration (B) Decrease transpiration
(C) Hinder transpiration (D) Stop transpiration

176. Stomatal frequency indicates:
(A) Number of stomata per unit area
(B) Rate of water loss
(C) Rate of gaseous exchange
(D) Width of stomatal aperture

177. DCMU (Dichlorophenyl dimethyl urea) is a herbicide which kills the plant by:
(A) Inhibition of respiration
(B) Destroying the chloroplasts
(C) Inhibiting flow of electrons from water to $NADP^+$
(D) Inhibiting PSI and photolysis of water

178. In dorsiventral leaf, the number of stomata per unit area are generally:
(A) Same on both the surface
(B) More on lower surface (epidermis)
(C) More on upper surface (Epidermis)
(D) Absent of upper surface (Epidermis)

179. In eukaryotic DNA replication. Promise action is due to:
(A) PCNA (B) Pol α
(C) Pol δ (D) MF 1

180. In isobilateral leaf, the number of stomata per unit area are:
(A) More on upper surface
(B) More on lower surface
(C) Approximately same on both the surfaces
(D) Absent on both the surfaces

181. In xerophytic leaf the stomata are situated:
(A) On both surfaces (B) On upper surface
(C) On lower surface (D) Absent from both surfaces

182. Assume a thylakoid is somehow punctured so that the interior of the thylakoid is no longer separated from the stroma. This damage will have the most direct effect on which of the following processes?
(A) The flow of electrons from photosystem II to photo system I
(B) The reduction of $NADP^+$
(C) The splitting of water
(D) The synthesis of ATP

183. The loss of water in the form of water drops is called:
(A) Transpiration (B) Respiration
(C) Guttation (D) Exosmosis

184. Guttation is from:
(A) Uninjured edges of leaves near vein ending
(B) Epidermal layers of leaf surfaces
(C) Injured edges of leaves
(D) None of the above

185. Transpiration is unavoidable evil because of:
(A) Structure of leaf and harmful effect
(B) Beneficial and harmful effect
(C) Maintenance of turgidity for growth
(D) Gaseous exchange for photosynthesis and respiration

186. Drooping of leaves due to loss of turgor at noon but recovery in the evening is referred to as:
(A) Temporary wilting (B) Incipient wilting
(C) Permanent wilting (D) Midday desiccation

187. F_1F_0-ATPase in chloroplasts is located on the:
(A) Inner chloroplast membrane with F1 facing the stroma
(B) Inner chloroplast membrane with F1 facing the intermembrane space
(C) Thylakoid membrane with F1 facing the stroma
(D) Thylakoid membrane with F1 facing the thylakoid lumen

188. Loss of water by cells without external sign of leaf drooping is called:
(A) Temporary wilting (B) Nascent wilting
(C) Incipient wilting (D) Permanent wilting

189. Which does not cause stomatal opening:
(A) Circadian rhythm (B) Influx of K^+ ions
(C) Light (D) High CO_2 concentration

190. Transpiration does not help in:
(A) Translocation of minerals (B) Excretion of sap
(C) Ascent of sap (D) Cooling

191. Plants with scotoactive stomata perform:
(A) C_4 photosynthesis (B) CAM photosynthesis
(C) C_3 photosynthesis (D) Anoxygenic photosynthesis

192. DNA replication does not is a primer?
(A) E. coli
(B) Vicia faba
(C) Production of replicative form of ϕX174 genome
(D) Replication of RF form of X174 genome

193. Day time loss of water in the vapour form from stomata is a trade off for intake of:
(A) Minerals (B) Oxygen
(C) Carbon di oxide (D) Energy

194. For keeping stomata open, besides K^+ ions the guard cells require a constant supply of:
(A) ABA (B) ATP
(C) Organic acids (D) Protons

195. Presence of stomata on the under surface of dorsiventral leaf is a mechanism of:
(A) Reduction in transpiration
(B) Protection from dust
(C) Proper regulation of transpiration
(D) Increase in transpiration

196. An antitranspirant is:
(A) Phenyl Mercuric Acetate (B) Abscisic acid
(C) Salicylic acid (D) All of these

197. Transpiration is a process related to:
(A) Osmosis (B) Diffusion
(C) Activated transport (D) Facilitated diffusion

198. Presence of hair on the leaf surface:
(A) Reduces transpiration
(B) Increases transpiration
(C) Helps in rapid exchange of gases
(D) Prevents guttation

199. Some plants possess modifications like phylloclades, cladodes, phyllodes, scale leaves etc for:
(A) Differentiation and evolution
(B) Decreasing transpiration
(C) Increasing transpiration
(D) Storage of absorbed water

200. Splicing yields is:
(A) rRNA (B) tRNA
(C) mRNA (D) All of these

201. Starch degradation activity of enzyme phosphorylase increases under:
(A) High pH (B) Low pH
(C) Neutral medium (D) Not connected with pH

202. According to Lloyd, the opening and closing of stomata is governed by:
(A) K^+ ion influx and efflux
(B) Change in pH
(C) Starch and sugar interconversion
(D) Photosynthetic activity of guard cell chloroplasts

203. Who proposed that opening and closing of stomata is connected with the change in pH of guard cells:
(A) Lyoyd (B) Von Mohl
(C) Sayre/Scarth (D) Levitt

204. Stomata remain open when relative humidity is:
(A) Above 70% (B) 50-70%
(C) 30-50% (D) Below 30%

205. Succinate dehydragenase is inhibited by malonate, oxalate and oxaloacetate. This is because they:
(A) Contain two anionic group and bind to a site other than the active site of the enzyme
(B) Are competitive inhibitors and bind to a site other than the active site of the enzyme
(C) Are non-competitive inhibitors and bind to the active site of the enzyme
(D) Contain two anionic groups and bind to the active site of the enzyme

206. Stomata close at relative humidity of:
(A) 60-70% (B) 50-60%
(C) 50-70% (D) Less than 50%

207. The class of seeds which must maintain a relatively high moisture content in order to remain viable are known as:
(A) Orthodox seeds (B) Recalcitrant seeds
(C) Labyrinth seeds (D) Ex-albuminous seeds

208. Which of the following hypostomatous leaves would dry up last:
(A) Both surface greased (B) Both surface ungreased
(C) Lower surface greased (D) Upper surface greased

209. In majority of plants, the guard cells are:
(A) Dumb bell shaped (B) Reniform
(C) Rounded (D) Polygonal

210. Spraying plant surfaces with phenyl mercuric acetate and silicon emulsion will result in:
(A) Increased photosynthesis
(B) Increased transpiration
(C) Decreased transpiration
(D) Exosmosis

211. Stomatal opening in a leaf:
(A) Involves losing of K^+ channels
(B) Occurs when there is an increase in turgor pressure in the guard cells
(C) Occurs in the dark
(D) Occurs in response to abscissic acid
(D) SD plant will flower and LD plant will remain vegetative

212. Sunken stomata are found in:
(A) Mesophytes (B) Hydrophytes
(C) Xerophytes (D) Epiphytes

213. Transpiration occurs from:
(A) Leaves (B) Stem
(C) All aerial parts (D) Roots

214. Transpiration efficiency/Transpiration ratio is:
(A) Water absorbed to water transpired
(B) Unit weight of drymatter synthesized in relation to units of water transpired by the plant
(C) Unit weight of drymatter in relation to water consumed
(D) Unit weight of water transpired to unit weight of drymatter synthesized

215. A mesophytic plant growing in well watered soil shows decreased transpiration in the afternoon due to:
(A) Closure of many stomata
(B) Contraction of cuticle
(C) Less water availability from soil
(D) High rate of photosynthesis

216. Guard cells differ from epidermal cells having:
(A) Mitochondria (B) Vacuoles
(C) Cell wall (D) Chloroplasts

217. When a short day plant (SD) and a long day plant (LD), both kept under favourable conditions for flowering, are exposed to a short flash of red light in the middle of the dark period:
(A) Both SD and LD plants will flower
(B) Both SD and LD plants will remain vegetative
(C) SD plant will remain vegetative and LD plant will flower
(D) None of these

218. Which of the following is most likely cause for wider opening of stomata?
(A) Atmosphere outside the stomata is becoming less saturated
(B) Secretion of salt molecules by adjacent guard cells is taking place
(C) Water molecules enter the guard cells
(D) The night temperature is going to fall

219. During high wind velocity, the stomata:
(A) Open more widely
(B) Close down
(C) Remain unaffected
(D) Remain unaffected but lose more water due to mass action

220. Guttation is the process of elimination of water from plants through:
(A) Stomata (B) Hydathodes
(C) Lenticels (D) Wounds

221. Water drops present on leaf margins of Tropaeolum. Balsam and grassy in early morning are due to:
(A) Guttation (B) Dew
(C) Osmosis (D) Transpiration

222. What is the action spectrum of transpiration?
(A) Green and ultraviolet (B) Orange and Red
(C) Blue and far red (D) Blue and red

223. Jasmonic acid:
(A) Is synthesized from linolenic acid
(B) Is growth promoter
(C) Delay leaf senescence
(D) Only present in mosses and ferns

224. Transpiration occurs when the outer atmosphere:
(A) Is wet
(B) Has more moisture content than the substomatal cavities
(C) Has less moisture than substomatal cavities
(D) Possesses the same moisture content as in substomatal cavities

225. Stomatal opening is under the control of:
(A) Epidermal cells (B) Palisade cells
(C) Spongy parenchyma cells (D) Guard cells

226. Dumb bell shaped guard cells are found in:
(A) Gymnosperms (B) Most dicot
(C) Cereals (D) Xerophytes

227. Transpiration rate of a dorsiventral leaf is:
(A) Higher on the lower surface
(B) Higher of the upper surface
(C) Equal on both the surface
(D) Dependent upon species

228. Growing plants in zero gravity on the space shuttle prevents:
(A) Phototropism because it is dark in space
(B) Photomorphogenesis because only near-red light filters into space
(C) Gravitropism because there is little gravity in space
(D) Aerotropism because there is no oxygen in space

229. Maximum transpiration takes place from:
(A) Stem (B) Leaves
(C) Roots (D) Flowers and Fruits

230. Stomata open during day time because the guard cells:
(A) Produce osmotically active sugars or organic acids
(B) Are thin walled (C) Are lean shaped
(D) Have to help in gaseous exchange

231. In which type, the stomata are present eclusively on the upper surface of the leaves:
(A) Potato type (B) Potamogeton type
(C) Barley type (D) Water lily type

232. Cobalt chloride is blue in dry state. In contact with moisture, it turns:
(A) Yellow (B) Pink (C) Red (D) Green

233. Stomata open at night and close during day time in:
(A) Xerophytes (B) Mesophytes
(C) Succulents (D) Hydrophytes

234. Percentage of water left in the soil when plant wilts is known as:
(A) Turgidity (B) Wilting coefficient
(C) Field capacity (D) Water retaining power of soil

235. Which of the following physiological effects is caused in plants by gibberellic acid?
(A) Shortening of genetically tall plants
(B) Elongation of genetically dwarf plants
(C) Rooting in stem cuttings
(D) Yellowing of young leaves

236. The conditions under which transpiration would be much rapid:
(A) High humidity
(B) Excess of water in soil
(C) Low humidity, high temperature, guard cells are turgid (open), moist soil
(D) Low velocity of wind

237. In which of the following plants would metabolism be hindered if the leaves are coated with wax on their upper surface:
(A) *Hydrilla* (B) Lotus (C) *Pistia* (D) *Vallisneria*

238. Wilting in plants occurs when:
(A) Phloem is blocked (B) Xylem is removed
(C) Pith is removed (D) A few leaves are removed

239. Number of stomata present per cm^2 of a common leaf is about:
(A) Less than 100 (B) More than 100000
(C) 1 Million (D) 1000

240. The following percentage of water absorbed by herbaceous plants is lost in transpiration:
(A) 80 (B) 90 (C) 99 (D) 100

241. Of the processes which occur in leaves the one which may lower their temperature (cooling effect) is:
(A) Respiration (B) Photosynthesis
(C) Transpiration (D) Hydrolysis

242. Which of the following statement provides a true example of both photomorphogenesis and phototropism?
(A) Phototropism is a growth towards blue light and photomorphogenesis is a growth towards red light
(B) Phototropism is a growth towards blue light and photomorphogenesis is a developmental process triggered by near –red light
(C) Phototropism is a growth towards blue light and photomorphogenesis is a germination triggered by blue light
(D) Phototropism is a movement towards blue light, that does not involve growth and photomorphogenesis is a movement towards red light that does involve growth

243. Transpiration from the two leaf surfaces are compared by dry paper strips preciously soaked in solution of:
(A) Sodium chloride (B) Calcium chloride
(C) Cobalt nitrate (D) Cobalt chloride

244. Transpiration is regulated by movements of:
(A) Guard cells (B) Subsidiary cells
(C) Epidermal cells (D) Mesophyll cells

245. Which of the following is the correct sequence of events in cytokinin signaling pathway?
(A) Receptor binding→ Activataion of His kinase→ Phos phorylation of an aspartate residues → Phosphory lation of AHP protein
(B) Receptor binding→ Phosphorylation of an aspartate residues → Activation of His kinase→ Phosphory lation of AHP protein
(C) Receptor binding→ Phosphorylation of AHP protein→ Activation of His kinase→ Phosphorylation of an aspartate residues
(D) Receptor binding→ Phosphorylation of AHP protein→ Phosphorylation of an aspartate residues→ Activation of His kinase

246. Transpiration rate is dependent upon:
(A) Stomatal frequency (B) Position of stomata
(C) State of stomata (D) All of these

247. Cytokinins receptor is:
(A) Similar to bacterial two component histidine kinase
(B) Acidic cytosolic protein
(C) GPR
(D) Tyrosine kinase

248. Transpiration differs from evaporation in:
(A) Rate of water loss
(B) Transpiration is a physiological process while evaporation is a physical process.
(C) Transpiration is a physical process while evaporation is a physiological process.
(D) Frequency of water loss

249. If a cell with a solute potential of -0.2 MPa and a pressure potential of 0.4 MPa is placed in a chamber filled with pure water that is pressurized with 0.5 MPa, what will happen?
(A) Water will flow out of the cell
(B) Water will flow into the cell
(C) The cell will be rushed
(D) The cell will explode

250. Transpiration increases in:
(A) Hot, damp and windy conditions
(B) Cool, damp and windy conditions
(C) Cool, dry and windy conditions
(D) Hot, dry and windy conditions

251. Common type of transpiration is:
(A) Stomatal (B) Cuticular
(C) Lenticular (D) Foliar

252. Match the terms in Group I with their definitions in Group II:

Group I	Group II
P. Ammonification	1. Conversion of atmo spheric nitrogen to ammonia
Q. Denitrification	2. Conversion of organic nitrogen into ammonia
R. Nitrification	3. Conversion of nitrite or nitrate into atmo spheric nitrogen
S. Nitrogen fixation	4. Conversion of ammonia into nitrite and nitrate

(A) P -2, Q – 3, R – 1, S - 4
(B) P -3, Q – 2, R – 4, S - 1
(C) P -3, Q – 2, R – 1, S - 4
(D) P -2, Q – 3, R – 4, S - 1

253. Hot weather wilting is caused by:
(A) Reduced water absorption
(B) Excessive transpiration
(C) Excessive transpiration and reduced absorption
(D) Excessive water absorption

254. The movement of water and minerals through xylem is explained by the:
(A) Pressure – flow theory
(B) Translocation theory
(C) Bulk-flow theory
(D) Cohesion – tension theory

255. Actual water vapour content of atmosphere is:
(A) Relative humidity (B) Differential humidity
(C) Absolute humidity (D) Positive humidity

256. Which one keeps its stomata open during night and closed during day:
(A) Cactus (B) Water lily
(C) Ivy (D) Hibiscus

257. How many ATP molecules are required for the conversion of one N_2 to $2NH_4$ during biological N_2 fixation?
(A) 8 ATP (B) 10 ATP
(C) 12 ATP (D) 16 ATP

258. Which wall of the guard cells is thick:
(A) Lateral (B) Inner
(C) Outer (D) Middle

259. What is the strongest reducing agent in photosynthetic electron-transfer reactions?
(A) Plastoquinone (B) P 680*
(C) P 700* (D) P 700

260. Higher CO_2 concentration around the leaves results in:
(A) Reduced photosynthesis
(B) Complete stomatal closure
(C) Wider opening of stomata
(D) Partial closure of stomata

261. Rate of transpiration is reduced with:
(A) Rise in temperature
(B) Decrease in light intensity
(C) Increase in wind velocity
(D) Increase in water uptake

262. Which one gives the most valid explanation for guard cells movements:
(A) Guard cells photosynthesis
(B) Starch hydrolysis theory
(C) Potassium influx and efflux
(D) Transpiration

263. Water movement against gravity is caused by:
(A) Imbibition (B) Transpiration
(C) Osmosis (D) Diffusion

264. When chrolophyll a is in its oxidized state:
(A) It fluorescence in the thylakoids
(B) O_2 is generated from sucrose
(C) It directly reduces $NADP^+$
(D) It results in protons being generated from H_2O

265. In thin leaved mesophytes, stomata open during the day and close during the night. They belong to:
(A) Barley type (B) Potato type
(C) Alfalfa type (D) Bean type

266. The large subunit of Rubisco is coded by:
(A) Chloroplast genome
(B) Nuclear genome
(C) Mitochondrial genome
(D) Both mitochondrial and nuclear genome

267. In terrestrial habitates, temperature and rainfall conditions are influenced by:
(A) Water transformations (B) Transpiration
(C) Thermoperiodism (D) Translocation

268. Low atmospheric pressure changes the rate of transpiration to:
(A) Decrease slowly (B) Increase
(C) Remain unchanged (D) Decrease rapidly

269. During wilting the sequence of events shall be:
(A) Endosmosis→ Plasmolysis→Temporary wilting→ Permanent wilting
(B) Exosmosis→Plasmolysis→Temporary wilting→ Permanent wilting
(C) Exosmosis→ deplasmolysis→ Plasmolysis→ Temporary wilting→Permanent wilting
(D) Exosmosis→ Plasmolysis→ Deplasmolysis→ Temporary wilting→ Permanent wilting

270. Escape of liquid from hydathodes (Uninjured leaf tips) is called:
(A) Evaporation (B) Transpiration
(C) Guttation (D) Osmosis

271. Dibromothymoquinone, an inhibitor of cytochrome bf complex of photosynthetic electron transport chain was added to the green alga chlamydomonas. After illumination it was found that in the alga:
(A) Plastiquinone was in reduced state
(B) P 700 was in reduced state
(C) Plastocyanin was in reduced state
(D) Both plastoquinone and plastocyanin were in oxidized state

272. Conversion of starch to organic acids is required for:
(A) Stomatal opening (B) Stomatal closing
(C) Stomatal formation (D) Stomatal activity

273. In which of the following organelles glycolate is oxidized to glyoxylate during photorespiration:
(A) Chloroplast (B) Peroxisome
(C) Mitochondria (D) Endoplasmic reticulum

274. Element involved in stomatal regulation is:
(A) Zinc (B) Magnesium (C) Potassium (D) Iron

275. In the pentose phosphate pathway:
(A) Only the C-1 carbon of glucose is oxidized to CO_2
(B) All the carbons of glucose is oxidized to CO_2
(C) No decarboxylation occurs
(D) C-4 and C-5 of glucose is oxidized to CO_2

276. In hot summer day, plant cooling occurs due to:
(A) Transport of water from root to all parts of the plant
(B) Loss of liquid water
(C) Water loss from entire plant
(D) Loss of water vapours from foliar surface

277. Stomatal opening in a leaf:
(A) Involves closing of K^+ channels
(B) Occurs when there is an increase in turgor pressure in the guard cells
(C) Occurs in the dark
(D) Occurs in response to abscissic acid

278. Stomata generally open during the day because the guard cells have:
(A) Outer thin wall (B) Chlorophyll
(C) Kidney –shaped (D) Larger nuclei

279. Tendrillar movement are examples of::
(A) Phototropism (B) Thigmotropism
(C) Geotropism (D) Hydrotropism

280. Transpiration is helpful to plants in:
(A) Cooling
(B) Loss of excess nutrients
(C) Upward conduction/ascent of sap
(D) Loss of excess water

281. A carotenoid –less mutant plant was grown under normal sunlight, it will experience:
(A) Increased photosynthetic rate
(B) Increased chlorophyll biosynthesis
(C) Reduced photorespiration
(D) Increased chlorophyll oxidation and necrosis

282. Temporary wilting is due to:
(A) Photosynthesis (B) Transpiration
(C) Respiration (D) Absorption of water

283. Water potential in leaf tissue is positive (near zero) during:
(A) Low transpiration (B) Excessive absorption
(C) Excessive transpiration (D) Guttation

284. Plastocyanin is a mobile electron carrier between:
(A) PS II and cytochrome bf complex
(B) Cytochrome bf complex and PS I
(C) Oxygen evolving complex and PS II
(D) PS I and ferrodoxin

285. A leafy twig of mesophytic plant dipped in water would demonstrate:
(A) Photosynthesis (B) Transpiration
(C) Respiration (D) Guttation

286. An adaptation for better gaseous exchange in plant leaves is:
(A) Hair on lower surface
(B) Multiple epidermis
(C) Waxy cuticle
(D) Stomata on lower surface away from direct sun rays

287. Which is not related to transpiration?
(A) Absorption and distribution of minerals
(B) Circulation of water
(C) Temperature
(D) Bleeding

288. At constant temperature, the rate of transpiration will be higher at:
(A) At sea level (B) 1 km below sea level
(C) 1 km above sea level (D) 1.5 km above sea level

289. Hydathodes occur on:
(A) Stem (B) Leaves
(C) Roots (D) All of these

290. The loss of water through cuticle reaches upto:
(A) 10% (B) 20% (C) 30% (D) 40%

291. Treatment of intact mitochondria with an uncoupler, such as 2, 4-dinitrophenol, in the presence of ADP, P succinate, and oxygen would have which of the following effects on the rates of electron transport and ATP synthesis? (+=Stimulation;) = No effect; - = inhibition)

	Rate of electron transport	Rate of ATP synthesis
(A)	+	+
(B)	-	-
(C)	+	-
(D)	-	+

292. A twig kept in water having some salt remains fresh for longer period due to:
(A) Decrease in bacterial degradation
(B) Exosmosis
(C) Decrease in transpiration rate
(D) Absorption of more water

293. When Darwin illuminated the coleoptile of a grass from one side, it bent towards the light source. This was due to:
(A) Diffusion of auxin from tip to the subterminal region
(B) Inreased cell division on dark side
(C) Diffusion of some chemicals from basal to the tip region
(D) Increased cell division on the illiminated side

294. Maximum transpiration occurs in:
(A) Algal cells (B) Xerophytic plants
(C) Hydrophytic plants (D) Mesophytic plants

295. Which of the following statement is not true with respect to the Photosystem II found in the chloroplasts?
(A) PS II is located in the grana
(B) Chl molecule in the reaction centre for the PS II is bound to a 110 Kd integral membrane protein
(C) A complex of cytochromes a and f transports electrons form PS II and PS I
(D) PS II reaction enter is bound to two Mn^{++} ions

296. Nonfunctional stomata can be seen in:
(A) Mango leaf (B) Pea leaves
(C) Hydrilla (D) Lotus

297. Phytohormone connected with closing of stomata is:
(A) ABA (B) Kinetin
(C) GA (D) IBA

298. Latest explanation for closure of stomata is:
(A) Starch glucose theory (B) Active K+ ions theory
(C) ABA theory (D) None of these

299. Choose the mismatch:

Inhibitor	Function
(A) Malnate	Prevents oxidation of succinate
(B) Cyanide	Inhibits cytochrome oxidase
(C) Oligomycin	Inhibits ATP synthase
(D) Rotenone	Blocks CoQH-cyt C oxidoreductase

300. In photorespiration, glycolate and gluoxylate are produced sequentially in the following organelles. Choose the correct sequence:
(A) Chloroplast and mitochondria
(B) Chloroplast and peroxisome
(C) Peroisome and mitochondria
(D) Peroxisome and chloroplast

301. When half the leaves are removed randomly transpiration will show:
(A) Higher magnitude but lower flux or rate per unit
(B) Lower magnitude but higher flux
(C) Both magnitude and flux increase
(D) Both magnitude and flux decrease

302. Which statement about seed germination in a diotuledon is false?
(A) The radicle elongates first and absorbs water needed for germination
(B) The hypocotyl elongates and carries the plumule into the light
(C) The plumule expands and functions in photosynthasis
(D) The shoot apical meristem is formed after germination

303. Who said that "Transpiration is a necessary evil"?
(A) Bose (B) Steward
(C) Anderson (D) Curtis

304. In C_4 plants, phosphoenol pyruvate carboxylase is located in:
(A) Cytosol (B) Chloroplast
(C) Peroxisome (D) Mitochondrion

305. Rate of transpiration is related to:
(A) Light and temperature
(B) Light, Temperature, atmospheric humidity and wind
(C) Light, temperature and wind
(D) Soil and temperature

306. Enzyme connected with opening and closing of stomata is:
(A) α amylase (B) Pyruvate kinase
(C) PEP carboxylase (D) RuDP carboxylase

307. In an attempt to create a nitrogen-fixing bioreactor, you set out to culture Rhizozbium from the root nodules of pea plants. After months of work, you are still having trouble getting the rhizobium to grow alone and fix nitrogen. What might explain this failure?
(A) Rhizobium only grows in association with plant root hairs
(B) Rhizobium from root nodules lacks a cell wall, so it will not grow in culture
(C) You aedentally put a fungicide in your growth medium that kills all of the Rhizobium
(D) Rhizobium from root nodules lack chromosomal DNA, so it will not grow in culture

308. In guard cells when sugar is converted into starch, the stomatal pore:
(A) Closes completely (B) Opens partially
(C) Opens fully (D) Remains unchanged

309. High CO_2 concentration in leaf interior will cause:
(A) Stomatal opening (B) Stomatal closure
(C) Photorespiration (D) Stomata are destroyed

310. Which one will reduce rate of transpiration?
(A) Rise in temperature
(B) Increase in water uptake
(C) Increase in wind velocity
(D) Decrease in light intensity

311. Stomata can open at nigh in:
(A) Xerophyte (B) Hydrophytes
(C) Gametophytes (D) Both (B) and (C)

312. Rate of transpiration is highest when:
(A) Soil is wet and air is dry
(B) Soil is wet and air is humid
(C) Soil is dry and air is humid
(D) Both soil and air is dry

313. In the following biosynthetic pathway X and Y represent:
X→ S-adenosyl methionine→ 1-aminocyclo propne-1-carboxylic acid → Y
(A)→ X Tryptophan Y→ IAA
(B)→ X Methionine Y→ Ethylene
(C)→ X Geranylgeranyl phosphate Y→ Gibberellin
(D) →X Tryptophan Y→ Ethylene

314. Red drop is:
(A) Drop in oxygen yield
(B) Drop in quantum yield
(C) Drop in organic yield
(D) Drop in photosynthetic yield

315. If plants are grown for several days in an atmosphere containing $^{14}CO_2$ in place of $^{12}CO_2$, one would except to find:
(A) Large amounts of radioactive oxygen released from the stomatas
(B) A large increase in 14 in the starch
(C) A large decrease in the rate of carbon fixation in the guard cells
(D) An increase in the activity of RuBP carboxylase in the photosynthetic cells

316. Light energy is converted into chemical energy through the formation of:
(A) $NADPH_2$ (B) ATP
(C) ATP and $NADPH_2$ (D) None of the above

317. Which one of the following pairs is correctly matched?
(A) Nutation : Movement due to injury
(B) Traumatotropism : Movement due to change in turgor pressure
(C) Thigmotropism : Movement of tendrils due to contact
(D) Seismonasty : Movement of stem tips of twining plants

318. Photophosphorylation consists of:
(A) Cyclic and non-cyclic phosphorylation
(B) Oxidative phosphorylation
(C) Substrate phosphorylation
(D) None of these

319. The by-product of photosynthesis is:
(A) Organic compounds (B) Oxygen
(C) Water (D) Energy

320. Plant shoot bend towards a light source as a result of:
(A) The increased amount of food synthesized by the leaves facing the light
(B) An unequal auxin distribution in their shoot apex
(C) The necessity of light for transcription
(D) Increased elasticity on the light-facing side

321. The first stable product in CAM/C_4 plants is:
(A) Starch (B) Oxalo-acetic acid
(C) Sugar (D) Malic acid

322. The co-operative photosynthesis is found in:
(A) C_4 plants (B) C_3 plants
(C) C_2 plants (D) Succulents

323. The sugar formed at photosynthetic site is:
(A) Utilised (B) Stored
(C) Transported (D) Assimilated

324. Stroma is the ground matrix of:
(A) Lysosomes (B) Oxysomes
(C) Ribosomes (D) Chloroplast

325. When chlorophyll absorbs light, it gets excited and emits (releases):
(A) Oxygen (B) Water
(C) Electrons (D) Energy rich compounds

326. During senescence of a plant:
(A) The plant displays sequential senescence of the leaves
(B) Metabolites like carbohydrates, protein etc. are transported from young leaves to the ageing leaves
(C) Metabolites like carbohydrates, protein etc. are transported out of the ageing leaves to the young leaves
(D) Secondary plant products accumulate in the tissues and organs during ageing

327. The excess energy of electron is used in the synthesis of:
(A) Organic compounds (B) ATP from ADP and iP
(C) $NADPH_2$ from NADP (D) None of these

328. The excess energy of electron is used in the synthesis of ATP and the process is called:
(A) Oxidative phosphorylation
(B) Substrate phosphorylation
(C) Photophosphorylation
(D) Phosphorylation

329. Match list - I with list – II and select the correct answer using the codes given below the lists:

List I	List II
P. Boron	1. Die-back
Q. Copper	2. Interveinal chlorosis
R. Iron	3. Little leaf disease
S. Zinc	4. Terminal leaf necrosis

(A) P -1, Q – 4, R – 3, S - 2
(B) P -1, Q – 4, R – 2, S - 3
(C) P -4, Q – 1, R – 3, S - 2
(D) P -4, Q – 1, R – 2, S - 3

330. Kranz anatomy is found in:
(A) Stems of C_4 plants (B) Stems of C_3 plants
(C) Leaves of C_4 plants (D) Leaves of C_3 plants

331. $C_{40}H_{56}$ is the empirical formula of:
(A) Chlorophyll (B) Carotene
(C) Xanthophyll (D) Anthocyanin

332. Identify the correct statements:
P. Vernalization is the process where flowering is promoted by heat shock
Q. The four different types of floral organs are initiated as separate whorls
R. The flowering stimulus is transported to the meristem via the xylem
S. Abscisic acid synthesis occurs via the acrotenoid biosynthetic pathway
(A) P and Q (B) Q and S (C) R and S (D) P and S

333. Calvin cycle operates in chloroplasts. Where does Hatch-Slack pathway occur:
(A) Mitochondria (B) Golgi bodies
(C) Chloroplasts (D) Cytoplasm

334. How many kreb cycle is needed for complete oxidation of one glucose molecule?
(A) 1 (B) 2 (C) 3 (D) 4

335. In case of eukaryoted, RNA processing is completed in:
(A) Cytoplasm (B) Nucleolus
(C) Nucleus (D) Vacuole

336. Emerson effect is related to:
(A) Decrease in photosynthesis in presence of high light intensity
(B) Decrease in photosynthesis when lights of two different wavelengths are provided together
(C) Increase in photosynthesis in presence of monochromatic light
(D) Increase in photosynthesis when lights of two different wavelengths are provided together

337. $C_{40}H_{56}O_2$ is an empirical formula of a :
(A) Xanthophyll (B) Carotene
(C) Anthocyanin (D) Chlorophyll

338. Which of the following equation sums up photosynthetic reaction most accurately:
(A) $CO + H_2O \rightarrow CH_2O + O_2$
(B) $6CO_2 + 6\ H_2O \rightarrow C_6H_{12}O_6 + 6O_2$
(C) $6CO_2 + 12\ H_2O \rightarrow C_6H_{12}O_6 + 6O_2 + H_2O$
(D) $6CO_2+12H_2O \xrightarrow[\text{Chlorophyll}]{\text{Light}} C_6H_{12}O_6+6O_2+6H_2O$

339. Which one is directly involved in light reaction of photosynthesis:
(A) Chlorophyll a (B) Chlorophyll b
(C) Carotenoids (D) All of these

340. What is the order of the following metabolic reactions in catabolism of a glucose molecule?

P. Pyruvate	Acetyl-CoA
Q. Glucose	Gluose-6-phosphate
R. Citrate	Isocitrate
S. ADP	ATP

(A) Q, S, P and R (B) Q, S, R and P
(C) P, Q, R and S (D) Q, R, P and S

341. Autumnal leaf colour is due to:
(A) Chlorophyll a (B) Carotenoids
(C) Anthrocyanin (D) Betacyanin

342. Grana are absent in the chloroplasts of:
(A) Shrubs (B) Herbs
(C) Algae (D) Trees

343. Photosynthetic pigments are located in the chloroplast in:
(A) Intrathylakoid
(B) Thylakoid membrane
(C) Intermembrane space
(D) Inner membrane of envelope

344. Phytochrome:
P. Is a blue protein pigment
Q. Show photoreversibility
R. Has a linear tetrapyrrole termed phytochromobilin
S. Regulates nytinastic movement of leaves
(A) P, Q and R (B) Q and R
(C) Q, R and S (D) P, Q, R and S

345. Chemosynthetic bacteria obtain energy from:
(A) Sun (B) Infra-red rays
(C) Organic substances (D) Inorganic chemicals

346. In cyclic photophosphorylation the electron released by reaction centre is ultimately accepted by:
(A) Ferredoxin (B) $NADP^+$
(C) Reaction centre (D) Plastocyanin

347. Which organelles mainly synthesize amino acids in the cells?
(A) Mitochondria (B) Ribosomes
(C) Chloroplasts (D) Both (A) and (B)

348. Which is correct about RNA polymerase I?
(A) It is present in nucleolus
(B) It constitutes 50-70% of total activity
(C) It transcribes genes for rRNA
(D) All of these

349. Plants do not store carbohydrate as glucose but do so as starch because glucose:
(A) Is unstable
(B) Attracts herbivores
(C) Will change nucleic acids
(D) Alters osmotic balance

350. Electron carriers involved in photophosphorylation are located in:
(A) Outer chloroplast membrane
(B) Inner chloroplasts membrane
(C) Stroma
(D) Thylakoid membranes

351. In noncyclic photophoshorylation the electron released by P 680 is replaced by:
(A) Plastoquinone (B) Water
(C) CO_2 (D) $NADP^+$

352. Chlorophyll b differs from chlorophyll a in that it does not:
(A) Become oxidized (B) Become reduced
(C) Become excited (D) Produce excited electrons

353. NADH is utilized during:
P. Glycolysis Q. Kreb's cycle
R. Electron transport chain S. Fermentation
(A) P and Q (B) Q and R
(C) R and S (D) P and S

354. NADPH is produced in photosynthesis during:
(A) Dark reaction
(B) Noncyclic photophosphorylation
(C) Pseudocyclic photophosphorylation
(D) Cyclic photophosphorylation

355. Which is correct about plastoquinones?
(A) They are special quinine present in plastids
(B) They carry two electrons from PS II tpward PS I
(C) They transport H+ from stroma into the thylakoid lumen
(D) All of these

356. The GC-AG rule does not apply to introns of:
(A) Chloroplast introns (B) Nuclear introns of yeast
(C) Mitochondrial introns (D) All of these

357. The process of photophosphorylation is similar to mitochondrial reaction called:
(A) Substrate-level phosphorylation
(B) Oxidative phosphorylation
(C) Carbohydrate phosphorylation
(D) Hydrolysis of phosphorylated compounds

358. Consider the following statements about alkaloids:
P. They are heterocyclic ring compounds containing nitrogen and have basic properties
Q. They occur rarely (barring a few exceptions) in cryptogams, gymnosperms or monocotyledons but they occur abundantly in dicotyledons
R. They are synthesized mostly in flowers
(A) P and Q (B) Q and R
(C) P and R (D) P, Q and R

359. Excited pigment molecule shifts and electron in an outer orbital having:
(A) Less energy (B) More energy
(C) Lesser spin (D) More spin

360. Source of protons within the chloroplast is:
(A) Water (B) Excited chlorophyll
(C) Carbon dioxide (D) Rubisco

361. Decline in quantumn yield was noticed by Emerson at a wavelength of:
(A) 400-450 nm (B) 600-650 nm
(C) 650-680 nm (D) more than 680 nm

362. Photosynthetic autotrophs get their energy requirement from:
(A) Heat (B) Light
(C) Inorganic chemicals (D) Organic chemicals

363. Energy required for ATP synthesis in PS II comes from:
(A) Proton gradient (B) Electron gradient
(C) Reduction of glucose (D) Oxidation of glucose

364. A heterotroph obtains its carbon supply from:
(A) Organic chemicals (B) CO_2
(C) Methane (D) Cyanide

365. Photosynthesis (nourishment in light) was discovered by:
(A) Stephen Hales (B) Priestley
(C) Ingenhousz (D) Van Mayer

366. Which statement is correct?
(A) PS I is located exclusively in stroma thylakoids and nonappressed region of grana that face the stroma
(B) The PS I core complex from barley contains 11 different polypeptides varying in size
(C) Out of the 11 polypeptide of PS I complex, six are coded for by nuclear genes
(D) All of these

367. Which is correct about RNA polymerase II?
(A) It is present in nucleoplasm
(B) It constitutes 20-40% of total activity
(C) It transcribes all the genes that produce mRNA
(D) All of these

368. Chrolophyll was first extracted by:
(A) Van Mayer
(B) Pelletier and Caventou
(C) Willstatter and Stoll
(D) Senebier

369. In 1772, Joseph Priestley demonstrated that:
(A) Plants foul the air
(B) Animals purify the air during day time
(C) Plants and animals restore air for each other
(D) Plants get killed in the air produced by animals

370. Chlorophyll consists of:
(A) A head of phytol and tail of four pyrrole rings
(B) A head of linked carbons and tail of linked pyrrole rings
(C) A head of four pyrrole rings and tail of linked nitrogens
(D) A head of four pyrrole rings and an alcoholic phytol tail

371. The source of hydrogen for carbon assimilation is:
(A) NADPH (B) $FADH_2$
(C) H_2O (D) RuBP

372. Which of the following pairs are correctly matched:

P. 2, 4-dinitrophenol	Uncoupling agent in oxidative phosphorylation
Q. Oligomycin	Inhibitor of ATP formation in oxidative phosphorylation
R. Valinomycin	Ionophore carriers potassium through the mitochondrial membrane
S. Iodoacetate	Separates the phosphorylating F1-ATPas from the inner mitochondrial membrane

(A) P, Q and R (B) P, R and S
(C) Q, R and S (D) P, Q and R

373. The first hypothesis that oxygen released during photosynthesis is derived from water, was proposed by:
(A) Engelmann (B) Priestley
(C) Van Niel (D) Blackman

374. Consider the following statements regarding the salt resistants plants:
P. Some salt resistant plants have salt glands at the surface of the leaves
Q. Accumulation of solute like sorbitol, glycine, betacine, proline
Which statement(s) is/are correct
(A) P only (B) Q only
(C) P and Q (D) None of these

375. The number of carbon atoms present in ribulose diphosphate is:
(A) 4 (B) 5 (C) 6 (D) 8

376. Photosynthesis is a process:
(A) Reductive and exergonic
(B) Reductive and catabolic
(C) Reductive, endergonic and catabolic
(D) Reductive, endergonic and anabolic

377. Dark reactions of carbon assimilation occur in:
(A) Cytoplasmic matrix (B) Mitochondria
(C) Leucoplasts (D) Chloroplasts

378. ATP molecules required for synthesis of a glucose molecule in Calvin-Benson cycle are:
(A) 38 (B) 36 (C) 32 (D) 18

379. Which is not correct about starch?
(A) It is stored in photosynthetically inactive organs like roots, modified shoots and seeds
(B) The constituents of starch grains are amylase and amylopectin
(C) Both amylase and amylopectin are made up of D-glucose units
(D) Starch molecules have a definite molecular weight and are assembled into starch grains

380. Mitochondria and plastids present in eukaryotic cells contain RNA polymerases that are analogous to:
(A) Eukaryotic polymerase (B) Phage polymerases
(C) Bacteria; polymerases (D) Both (B) and (C)

381. Who discovered that green plant parts and light are essential for photosynthesis?
(A) Englemann (B) Ingenhousz
(C) Sachs (D) Priestley

382. Lignin
P. Is a polysaccharides
Q. Is a second most abundant organic compound
R. Is a polymer of aromatic alcohols like coumaryl, sinapyl alcohol
S. Synthetic pathway is peroxidase dependent
(A) P and S (B) Q and R
(C) Q, R and S (D) P, Q, R and S

383. Electron donor to PS I is:
(A) Ferredoxin (B) FeS centre
(C) Plastoquinone (D) Plastocyanin

384. Wavelength of light least effective in photosynthesis is:
(A) Violet (B) Blue
(C) Green (D) Red

385. In Photosynthesis light forms:
(A) ATP and Sugar
(B) Hydrogen, Oxygen and Sugar
(C) ATP, Hydrogen donor and Oxygen
(D) ATP, Hydrogen and Oxygen donor

386. Which one of the following is limiting rate of photosynthesis in nature:
(A) Light (B) CO_2
(C) O_2 (D) Chlorophyll

387. Consider the following events involved in stomatal opening:
P. Turgor pressure of guard cells increases
Q. K^+ ions move into guard cells
R. pH of guard cells decreases
S. Water moves into guard cells
What is the correct sequence of these events leading to stomatal opening?
(A) Q, S, R, P (B) R, Q, S, P
(C) Q, R, S, P (D) R,P, Q, S

388. In PS I, the carrier that pick up electrons from P_{700} is:
(A) Fe protein (B) Fe-S protein
(C) Fe-Cu protein (D) Fe-Mg protein

389. Rate of photosynthesis is:
(A) Equal to that of respiration
(B) Less than that of respiration
(C) Depends upon chlorophyll content
(D) Faster than that of respiration

390. Chloroplasts are agranal in:
(A) Algae
(B) Bundle sheath cells of C_4 plants
(C) Mesophyll cells of C_4 plants
(D) Both (A) and (B)

391. Brassinosteroids:
P. Are steroid growth promoter
Q. Synthesized by short day plants only
R. Function is very similar to ABA
S. Are chemically similar to ecdysone hormone
(A) P and S (B) R and S
(C) P and Q (D) P and R

392. Illuminated isolated chloroplasts release O_2 in the absence of CO_2. Which will increase oxygen output?
(A) H_2O (B) $NADP^+$
(C) NAD^+ (D) Green light

393. Which helped in confirming that oxygen evolved in photosynthesis comes from water:
(A) $^{18}H_2O$ (B) $^{14}CO_2$
(C) $^{15}NO_3$ (D) $^{3}H_2O$

394. Consider the following compounds
P. Phosphoenol pyruvate Q. Malic acid
R. Pyruvic acid S. Oxaloacetic acid
What is the correct sequence of formation of these intermediate compounds in C_4 pathways?
(A) P, Q, R, S (B) Q, S, R, P
(C) S, P, Q, R (D) S, Q, R, P

395. In an experiment on O_2 evolution by photosynthesizing *Hydrilla* plant, a pinch of sodium bicarbonate was added to water. The rate of photosynthesis or O_2 evolution will:
(A) Increase (B) Decrease
(C) Stop (D) Not be affected

396. Properties of starch molecules are mainly determined by:
(A) Degree of branching (including branch length)
(B) Types of bonds linking the glucose units
(C) Chain length
(D) All of these

397. Which consensus sequences is recognized by more than one transcription factors?
(A) -35 sequence (B) CAAT box
(C) Octamer sequence (D) GC box

398. An illuminated plant is provided with 0.03% CO_2. The plant will:
(A) Die
(B) Just survive
(C) Show normal photosynthesis
(D) Have reduced respiration

399. Photosynthesis will be highest when the plant is exposed to:
(A) Continuous strong light
(B) Continuous weak light
(C) Alternate strong and weak light
(D) Intermittent light

400. Which is correct for photosynthetic organs:
(A) Light energy → Heat→ Chemical energy
(B) Chemical energy→ Light energy → Heat
(C) Heat→ Light energy→ Chemical energy
(D) Light energy→ Chemical energy → Heat

401. In photosynthesis:
(A) No ATP is formed
(B) ATP is formed in initial stages
(C) Water is not involved
(D) Oxygen comes from carbondioxide

402. Photolysis of water involves:
(A) Excitement of water
(B) Evolution of water
(C) Breakdown of water by light
(D) Splitting of water into ions, H^+ and OH^-

403. Which of the following statements are correct?
P. To remove seed dormancy by mechanical removal of seed-coat is called Scarification
Q. Vernallization is done at low temperature
R. The presence of auxins in a solution could be tested by *Avena sativa* stem tip test
S. The photoperiodic stimulus is perceived by leaves
(A) P and Q (B) P, Q and R
(C) Q and S (D) P, Q, R and S

404. The radiant energy absorbed by carotenoids is:
(A) Lost as heat
(B) Transferred to chlorophylls
(C) Be emitted as fluorescence or phosphorescence
(D) Used to perform photochemical act

405. High CO_2 – compensation point is found in:
(A) C_3 plants (B) C_4 plants
(C) CAM plants (D) Lichens

406. Consider the following statements regarding the plant growth substances:
P. Cytokinins make dormant buds active
Q. The richest concentration of cytokinins generally occur in endosperm
R. Ethylene is involved in the formation of aerenchyma in submerged roots and stems
S. Gibberellins are involved in the breaking of seed dormancy
Which of these statements are correct
(A) P and S (B) Q, R and S
(C) P, Q and S (D) P, Q, R and S

407. The substrate for photorespiration is:
(A) Malic acid (B) Oxaloacetic acid
(C) Gycolic acid (D) PGA

408. Which is correct?
(A) Flower senescence is promoted by ethylene
(B) Ethylene biosynthesis involves two enzymes, viz., ACC synthase and ACC oxidase
(C) ACC synthase enzyme is encoded by acs gene and ACC oxidase is encoded by aco gene
(D) All of these

409. Which is usually surrounded by a GC rich sequence?
(A) TATA box (B) Pribnow box
(C) GC box (D) CAAT box

410. During photosynthesis O_2 is liberated by oxidation of:
(A) H_2O (B) CO_2
(C) Phosphoglyceraldihyde (D) None of these

411. The first food substance plants make in photosynthesis is:
(A) Sugar (B) Starch
(C) Proteins (D) Fats

412. With reference to sucrose, consider the following statements::
P. Sucrose is a non-reducing sugar
Q. The linkage in it is an α_1- β_1 glycosidic bond
R. It is a disaccharide having glucose and β -fructose
Which of the statement(s) given below is/are correct
(A) P only (B) Q and R
(C) P and R (D) P, Q and R

413. Hypothesis for oxygen coming from water was put forward by:
(A) Hill (B) Warburg
(C) Blackman (D) Mendel

414. The most effective wavelength of light participating in photosynthesis is:
(A) 450-520 nm (B) 700-760 nm
(C) 660-700 nm (D) 520-560 nm

415. The first product of CO_2 fixation in C_4 plants is:
(A) Phosphoglyceric acid (PGA)
(B) Oxalo-acetate
(C) Malic acid
(D) Phosphoenol pyruvate

416. More free energy is released during the citric acid cycle than during glycolysis, but only 1 mole of ATP is produced for each mole of acetyl CoA that enters the cycle. what happens to most of the remaining free energy that is produced during the citric acid cycle?
(A) It is used to synthesize GTP
(B) It is used to reduce electron carriers
(C) It is lost as heat
(D) It is used to reduce pyruvate

417. The imperical formula for chlorophyll a is:
(A) $C_{28}H_{72}O_5N_4Mg$
(B) $C_{45}H_{72}O_6N_4Mg$
(C) $C_{55}H_{72}O_5N_4Mg$
(D) $C_{45}H_{72}O_6N_4Mg$

418. Gibberellin receptor:
P. Is membrane bound
Q. Interacts with G-protein
R. Increases the concentration of GMP when binds with gibberellin
S. Is cytosolic in nature
(A) P and Q (B) R and S
(C) P, Q and R (D) Q, R and S

419. The process of taking in CO_2 by plants and releasing O_2 is component of:
(A) Transpiration (B) Respiration
(C) End-osmosis (D) Photosynthesis

420. The bye product of photosynthesis is:
(A) CO_2 (B) H_2O
(C) Glucose (D) O_2

421. Which is not correct about transcription unit?
(A) It is that sequence of a DNA molecule that is transcribed into a single RNA molecule
(B) In case of all eukayotes, eachg gene is a distinct transcription unit
(C) In case of prokaryotes,each operon functions as a single transcription unit
(D) Enhancer sequences are present within promoters

422. Which occupies a fixed position given in the promoter at -25?
(A) TATA box (B) Pribnow box
(C) GC box (D) CAAT box

423. Our present day view regarding photosynthesis is that it:
(A) Converts light energy into chemical energy
(B) Creates useful energy
(C) Fixes CO_2 into carbohydrates
(D) Reverses the action of respiration

424. ABA:
P. Is a sesquiterpenoid
Q. Synthesized within chroloplast
R. Synthesized from carotenoid violaxanthin
S. Transport occurs through phloem only
(A) P and Q (B) P, Q and R
(C) P, R and S (D) P, Q, R and S

425. In which of the following process the light energy is converted into chemical energy:
(A) Digestive action (B) Respiration
(C) Fermentation (D) Photosynthesis

426. The oxygen in photosynthesis is released from:
(A) CO_2 (B) H_2O
(C) Carbohydrate (D) Chlorophyll

427. Photosynthesis is the fastest in:
(A) Blue light (B) Sun light
(C) Red light (D) Green light

428. Photophosphorylation is the process in which:
(A) CO_2 and O_2 unite
(B) Phosphoglyceric acid is produced
(C) Aspartic acid is formed
(D) Light energy is converted into chemical energy through production of ATP

429. The following 4 different solutions are prepared by mixing the components of electron transport chain. Which among them is expected to cause a net transfer of electron to cytochrome C?
(A) It is used to synthesize GTP
(B) It is used to reduce electron carriers
(C) It is lost as heat
(D) It is used to reduce pyruvate

430. The first step in photosynthesis is the:
(A) Joining of 3-carbon atom to form glucose
(B) Formation of ATP
(C) Ionization of water
(D) Excitement of an electron of chrolophyll by photon of light

431. Which of the following statements are true for auxin?
P. IAA is the principal auxin in plants
Q. Auxin shows polar transport
R. Distribution of auxin in cell is pH dependent
S. TIBA and NPA block polar transport of auxin
(A) P and Q (B) R and S
(C) P, Q and R (D) P, Q, R and S

432. Which is correct about transcription factors?
(A) These factors are required for transcription initiation at all the promoters of eukaryotes
(B) They join RNA polymerase II to form a complex around the startpoint
(C) They determine the site of transcription initiation
(D) All of these

433. ATP formation during photosynthesis is:
(A) Phosphorylation
(B) Photophosphorylation
(C) Oxidative phosphorylation
(D) None of these

434. Which of these statements about thioredoxin is correct?
(A) It contains a heme that cycles between two oxidation states
(B) Its oxidised form predominated while light absorption is taking place
(C) It activates some biosynthetic enzymes by reducing disulfide bridges
(D) It activates some degradative enzymes by reducing disulphide bridges

435. Chlorophyll is present:
(A) On the surface of chloroplast
(B) Dispersed throughout chloroplast
(C) In the stroma of chloroplasts
(D) In the grana of chloroplasts

436. Ethylene receptor is:
P. Similar to bacterial two component system
Q. Consists of histidine kinase
R. Amino terminal end contains ethylene binding site
S. Contains Cu^{++} ion
(A) P and Q (B) Q and S
(C) P Q and R (D) P, Q, R and S

437. The process in which water is split during photosynthesis is:
(A) Photolysis (B) Hydrolysis
(C) Plasmolysis (D) Hemolysis

438. Which functional sequence/module is not found in eukaryotic promoters?
(A) TATA box (B) Pribnow box
(C) GC box (D) CAAT box

439. Which of the following wave length of light is absorbed maximum for photosynthesis:
(A) 660 nm (B) 440 nm
(C) 680 nm (D) 700 nm

440. The number of protons translocated by the various vectorial localized on the inner mitochondrial membrane of an organism was cytochrome aa_3 complex: 4 and F_1F_0-ATPas: 3. One proton is also required for the transport of inorganic phosphate into the mitochondrial matrix:
I. The number of ATP molecules that can be synthesized by the oxidation of one NADH molecule is:
(A) 2 (B) 2.5
(C) 3 (D) 3.3
II. If the cytosolic NADH is transported to the matrix by the glycerol 3- phosphate shuttle, then the number of ATPs synthesized is:
(A) 1.5 (B) 2 (C) 2.5 (D) 3.3

441. The basic structure of chlorophyll comprises of:
(A) Porphyrin system (B) Cytochrome system
(C) Plastocyanin system (D) Flavoprotein only

442. The specific function of light energy in the process of photosynthesis is to:
(A) Activate chlorophyll (B) Split of water
(C) Reduced carbon di oxide (D) Synthesize glucose

443. In pigment system I, active chlorophyll is:
(A) P_{700} (B) P_{600}
(C) P_{608} (D) P_{720}

444. Discovery of Emerson effect has clearly shown the existence of:
(A) Two distinct photochemical reactions or processes
(B) Light and dark reactions in photosynthesis
(C) Photophosphorylation
(D) Photo-respiration

445. Suppose the interior of the thylakoids of isolated chloroplasts were made acidic and then transferred in the dark to a pH -8 solution. What would be likely to happen?
(A) The Calvin cycle will be activated
(B) The isolated chroloplasts will make ATP
(C) Cyclic photophosphorylation will occur
(D) Only a and b will occur

446. The site for light reaction is:
(A) Grana (B) Stroma (C) ER (D) Cytoplasm

447. When cell converts light energy into chemical energy, which of the following reaction would take place:
(A) ADP + iP = ATP (B) ATP + iP = ADP
(C) AMP + iP = ADP (D) GDP + iP = GTP

448. Which is often present in several copies in a single promoter?
(A) TATA box (B) Pribnow box
(C) GC box (D) CAAT box

449. Assimilatory power produced in Hill reaction and used in Blackman's reaction refers to:
(A) Generation of ATP and $NADPH_2$
(B) Reduction of CO_2
(C) Splitting of water
(D) Disintegration of plastids

450. Which of the following statement about photosystem II is incorrect?
(A) It is a multimolecular transmembrane assembly containing several polypeptides, several chrolophyll molecules, a special pair of chlorophyll (P680), pheophytin and plastoquinones
(B) It transfers electrons to photosystem I via the cytochrome bf complex
(C) It uses light energy to create a separation of charge whose potential energy can be used to oxidize H_2O and to produce a reductant, plastoquinol
(D) It uses an Fe^{++} - Cu^{+} centre as a charge accumulator to form O_2 without generating potentially harmful hydroxyl radicals, superoxide anions or H_2O_2

451. Water soluble pigment is:
(A) Chlorophyll (B) Carotene
(C) Anthocyanin (D) Xanthophyll

452. Photophosphorylation means synthesis of:
(A) ATP from ADP (B) NADP
(C) ADP from ATP (D) PGA

453. Which of the following statement regarding reduced glutathione is not true?
(A) It contains one λ–carboxyglutamate, one cysteine and one glycine residue
(B) It keeps the cysteine residues of proteins in their reduced states
(C) It is regenerated from oxidized glutathione by glutathione reductase
(D) It reacts with hydrogen peroxide and organic peroxides

454. In which wavelength of light, bacteria carryout photosynthesis:
(A) Utra violet (B) Blue
(C) Red (D) Far red

455. Pigment system is connected with:
(A) Flowering (B) Reduction of CO_2
(C) Photolysis of water (D) None of these

456. Besides water and light which is more essential as a raw material for food formation:
(A) CO_2 (B) NAD
(C) O_2 (D) Mineral salts

457. In pigment system II, active chlorophyll is:
(A) P_{680} (B) P_{683}
(C) P_{700} (D) P_{673}

458. Chloroplast contains maximum quantity of:
(A) RuDP-carboxylase (B) Hexokinase
(C) Pyruvic carboxylase (D) None of these

459. Which one is the precursor of chlorophyll:
(A) Bacteriochlorophyll
(B) Bacterio-viridin
(C) Tryptophan
(D) Chlorophyllide (proto chlorophyll)

460. Which is not correct about ribulose-1, 5-bisphosphate carboxylase?
(A) Makes up about 50 per cent of the chloroplast protein
(B) Is located in the same compartment as the oxygen-generating complex
(C) Is encoded by a combination of nuclear and chloroplast DNA
(D) Is located in the compartment where starch is stored

461. Dark reaction in photosynthesis is called so because:
(A) It can occur in dark alone
(B) It does not require direct light energy
(C) It cannot occur during day time
(D) It occur more rapidly at night

462. During ATP synthesis, electrons pass through:
(A) CO_2 (B) O_2
(C) H_2O (D) Cytochromes

463. While the number of ATP molecules generated by HMP pathway is less, microbes have this pathway to generate:
(A) Amino acid (B) Nuleic acid precursors
(C) Secondary metabolites (D) Fatty acid precursors

464. In case of C_4 pathway:
(A) CO_2 combines with PGA
(B) CO_2 combines with PEP
(C) CO_2 first combines with RuBP
(D) CO_2 combines with RMP

465. Where does the primary photo-chemical reaction occur in chloroplast:
(A) Stroma (B) Periplast cavity
(C) Quantasomes (D) Inner membrane of chloroplast

466. Ribulose biphosphate carboxylase enzyme catalyzes the reaction between:
(A) Oxaloacetic acid and acetyl CoA
(B) CO_2 and ribulose1,5,biphosphate
(C) Ribulose biphosphate and phosphoglyceraldehyde
(D) PGA and dihydroxy acetone phosphate

467. Which of the following statement is false? In citric acid cycle
(A) Four pairs of hydrogen atoms flow out of the cycle
(B) Two water molecules are formed
(C) One high energy compound (GTP) is produced
(D) Isocitrate is reduced to α–ketoglutarate and O_2

468. Carbon dioxide is formed in all the following except:
(A) Respiration in plants
(B) Burning of sugar
(C) Heating of lime stone
(D) Photosynthesis of plants

469. Which of the following enzyme is used for carboxylation of RuBP?
(A) Carboxydismutase (B) Hexosekinase
(C) Peroxidase (D) Phosphopentokinase

470. Grana refer to:
(A) Stacks of thylakoids in plastids of higher plants
(B) A constant inquantum equation
(C) Glycolysis of glucose
(D) By product of phorosynthesis

471. The reaction catalyed by phosphofructokinase – 1 (PFK -1)
(A) Is activated by high concentrations of ATP and citrate
(B) Uses fructose – 1- phosphate as substrate
(C) Is the rate-limiting reaction of the glycolytic pathway
(D) Is inhibited by fructose -2, 6-bisphosphate

472. During dark reaction of photosynthesis:
(A) Water is split
(B) CO_2 is reduced to organic compounds
(C) Chlorophyll is activated
(D) 6C sugar is broken down into 3C sugar

473. In C_3 plants first stable product of photosynthesis during dark reaction is:
(A) PGA (B) PGAI
(C) RuBP (D) RUDP

474. NADH→ Q→ X→ Cyt1→ Y → $Cyt_{(a+a3)}$→ O_2
Sequence of electron transfer in oxidative phosphorylation is given above. complete the missing sequence:
(A) X – Cyta and Y - Cytb (B) X – Cyta and Y - Cytc
(C) X – Cytb and Y - Cytc (D) X – Cytb and Y – Cytb1

475. CO_2 is necessary of photosynthesis. The chemical used to remove this gas most effectively from entering a control apparatus is :
(A) Potassium hydroxide solution
(B) Calcium oxide
(C) Sodium carbonate
(D) Distilled water

476. The isotope of carbon used extensively for studies in photosynthesis is:
(A) ^{12}C (B) ^{14}C (C) ^{13}C (D) ^{16}C

476. Glycolate accumulates in chloroplasts when there is:
(A) High CO_2 (B) Bright light
(C) Low temperature (D) Low CO_2

477. Better adapted individuals tend to produce more offspring than individuals within the same species less well adapted to that environment. This observation forms the principle of:
(A) Differential reproductive success
(B) Variation
(C) Heritability
(D) Use and disuse

478. Which of the following statements regarding Darwin and his theory of evolution is false?
(A) Darwin noticed that South American species of birds more loosely resembled species found in other temperate regions than European species of birds
(B) Darwin proposed that species adapt over time
(C) Darwin proposed that natural selection produces changes in species over time
(D) Darwin was the original and sole developer of the theory of natural selection

479. The C_4 plants are different from the C_3 plants with reference to the:
(A) The substance that accepts CO_2 in carbon assimilation
(B) Types of end product of photosynthesis
(C) The number of NADPH that are consumed in preparing sugar
(D) Types of pigments involved in photosynthesis

480. Red drop discovered by Emerson due to distruption of photochemical activity of:
(A) Carotenoids (B) PS I
(C) PS II (D) Both (B) and (C)

481. Which of the following is not an example of co-evolution?
(A) Adaptations of flowers and their exclusive pollinators
(B) Passion-flower vines and the butterfly *Heliconius*
(C) A parasite that is specific for one host
(D) Aposematic coloration of monarch butterflies and predators that learn not to eat them

482. The percentage of light energy utilized for photosynthesis by higher plants is:
(A) 100 (B) 50 (C) 10 (D) 1-2

483. NADP+ is hydrogen acceptor of:
(A) Krebs cycle
(B) Photolysis
(C) Hexose monophosphate shunt
(D) Both (B) and (C)

484. The first carbon dioxide acceptor in C_4 plants is:
(A) Phosphoenol –pyruvate
(B) Ribilose-1-5 diphosphate
(C) Oxalo-acetic acid
(D) Phosphoglyceric acid

485. A plant is kept in 300 ppm CO_2 concentration. What will happen to it:
(A) Plant will die soon
(B) Plant will not grow but will not die
(C) Plant will show normal photosynthesis
(D) Respiration will be greatly decreased

486. Sometime, speciation can be quite abrupt after long periods of stability. This is called:
(A) Founder effect (B) Crossing - over
(C) Macroevolution (D) Punctuated equilibrium

487. Which of the following have high CO_2 compensation point:
(A) C_2 plant (B) C_3 plant
(C) C_4 plant (D) Alpine herbs

488. The carbon dioxide acceptor in calvin cycle/C_3 plants is:
(A) Phosphoenol –pyruvate (PEP)
(B) Ribulose-1-5 diphosphate (RuBP)
(C) Phosphoglyceric acid (PGA)
(D) Ribulose-monophosphate (RMP)

489. In angiosperms, chlorophyll formation requires:
(A) Cytochrome (B) Carotenes
(C) Chloroplasts (D) Light

490. A prezygotic isolating mechanism prevents successful:
(A) Gamete production (B) Fertilization
(C) Zygote development (D) Reproduction of hybrids

491. Ninth-tenth of all photosynthesis of world is (85-90%) carried out by:
(A) Large trees with millions of branches and leaves
(B) Algae of the ocean
(C) Chlorophyll containing ferns of the forest
(D) Scientists in the laboratories

492. Which one of the following would not be limiting factors in photosynthesis?
(A) O_2 (B) CO_2
(C) Chlorophyll (D) Light

493. Which one directly transfer electrons in NADP+ during light reaction?
(A) PS I (B) PS II
(C) Cytochromes (D) Plastocyanin

494. Of the following species concepts, which one emphasizes reproductive isolation as the major cause for speciation:
(A) Biological species concept
(B) Evolutionary species concept
(C) Morphological species concept
(D) Phylogenetic species concept

495. PEP is CO_2 acceptor in:
(A) C_4 plants (B) C_3 plants
(C) C_2 plants (D) Both (A) and (B)

496. Maize, Sugarcane and some other tropical plants have high efficiency of CO_2 fixation because they operate:
(A) Calvin cycle (B) Hatch-Slack cycle
(C) TCA cycle (D) PP pathway

497. Genetic drift occurs when a few individuals of a species colonize an island. This particular phenomenon is known as:
(A) The bottleneck effect (B) The founder effect
(C) Assortative mating (D) Random mating

498. Compensation point is the value of a factors where there is:
(A) Beginning of photosynthesis
(B) Little photosynthesis
(C) Photosynthesis equal to rate of respiration
(D) Neither photosynthesis nor respiration

499. Which one of the following is the common storage product of photosynthesis:
(A) Protein (B) Fat
(C) Starch (D) Glucose/sucrose

500. When populations of a plant species are located adjacent to one another because of abrupt changes in soil conditions, they are called:
(A) Allopolyploid (B) Autopolyploid
(C) Sympatric (D) Parapatric

501. Chlorophyll takes part in:
(A) Absorption of light
(B) Absorption of CO_2
(C) Absorption of electrons
(D) Absorption of light and photochemical splitting of water

502. Starch is detected in food by its:
(A) White appearance
(B) Blue reaction with iodine solution
(C) Presence as an energy store
(D) Granular form even if cooked

503. An essential process connected with photosynthesis is:
(A) Photolysis of water (B) Synthesis of glucose
(C) Breakdown of glucose (D) Photophosphorylation

504. Which of the following is least effective in photosynthesis:
(A) Sunlight (B) Red light
(C) Blue light (D) Green light

505. Light energy is changed to chemical energy in the process of:
(A) Bioluminescence
(B) Photosynthesis
(C) Both (A) and (B)
(D) Photorespiration

506. An isolated population of humans, with approximately equal numbers of blue-eyed and brown-eyed individuals, was decimated by an earthquake. Only a few brown-eyed people remained to form the next generation. This kind of change in the gene pool is called a:
(A) Hardy-Weinberg equilibrium
(B) Blocked gene flow
(C) Bottleneck effect
(D) Founder effect

507. NAD is:
(A) Nicotinamide adenine dinucleotide
(B) Nicotinamide adenosine dinucleotide
(C) Nicotine adenosine dinucleotide
(D) Nicotinamide adenine diphosphate

508. The experimental material that has largely been responsible for making rapid advances in research on photosynthesis is:
(A) *Chlamydomonas* (B) *Chlorella*
(C) Spinach leaf (D) *Hydrilla*

509. Rate of photosynthesis is independent of:
(A) Intensity of light (B) Duration of light
(C) Quality of light (D) Temperature

510. The ingredient required for ATP synthesis is:
(A) Ca (B) S (C) P (D) Fe

511. Absorption of light brings about:
(A) Absorption of CO_2
(B) Liberation of O_2
(C) Oxidation of chlorophyll
(D) Reduction of chlorophyll

512. Energy changes in photosynthesis is:
(A) Electrical → Chemical
(B) Light → Chemical
(C) Chemical→ Light→ Electrical
(D) Light→ Electrical→ Chemical

513. Bacterial photosynthesis contains:
(A) PS I (B) PS II
(C) Both PS I and PS II (D) None of these

514. A taxonomic system based only on the traits that reflect the order in time in which branches arose in a phylogenetic tree is called:
(A) Phylogeny (B) Cladistics
(C) Classical evolutionary taxonomy (D) Phenetics

515. What is true for photosynthesis:
(A) Both carbon dioxide and water are oxidized
(B) Both carbon dioxide and water are reduced
(C) Carbon dioxide is oxidized and water reduced
(D) Carbon dioxide is reduced and water oxidized

516. Ina chlorophyll molecule, magnesium is present:
(A) At one end of porphyrin
(B) Centre of porphyrin
(C) Phytol unit
(D) Between porphyrin head and phytol tail

517. The best test of the relatedness of two species is in the similarly of their:
(A) Anatomy (B) DNA and proteins
(C) Development (D) Courtship behaviors

518. In aquaria green plant are grown for:
(A) CO_2 (B) Fresh food
(C) Oxygen (D) Both (A) and (B)

519. Leaves are green because they:
(A) Absorb green light
(B) Do not absorb but reflect green light
(C) Utilise green light
(D) Absorb and reflect green light

520. Algae often float on surface of water during day but sink down during night due to:
(A) Evolution and trapping of oxygen bubbles during the day in their photosynthesis
(B) Becoming light as they consume most of their food in the night
(C) Warming action of sun during the day
(D) Release of absorbed air by warming of water

521. North American organisms are different from south American organisms, because the two continents:
(A) Have very different climates
(B) Have never been connected
(C) Were physically separated for about 35 million years, until about 2 million years ago
(D) Have very different geographics

522. Carbon dioxide joins the photosynthetic pathways:
(A) PS I (B) PS II
(C) Light reaction (D) Dark reaction

523. According to Lamarck, a giraffe has a long neck because:
(A) A reactor designed it that way
(B) Catastrophes eliminated short-necked forms
(C) Its ancestors stretched their necks to get food
(D) Ancestral giraffes with slightly longer necks than others got more food and left more surviving offspring

524. The Darwinian fitness of an organism is a measure of:
(A) Its ability, relative to others in the population, to pass its genes to the next generation
(B) The number of offspring it produces
(C) Its lifespan
(D) Its physical vigor

525. Hill reaction occurs in:
(A) Dark (B) Light
(C) Both light and dark (D) All the time

526. Many bacteria are now resistant to penicillin, because:
(A) Penicillin causes gene mutations, some of which are beneficial
(B) Previously resistant forms survived and reproduced better than nonresistant forms
(C) The hospital environment inhibits competition among bacteria
(D) Penicillin triggers the synthesis of resistant proteins

527. The products of photochemical reaction are:
(A) O_2, ATP and NADPH
(B) O_2
(C) ATP and NADPH
(D) Organic compounds especially carbohydrates

528. When a short day plant (SD) and a long day plant (LD), both kept under favourable conditions for flowering, are exposed to a short flash of red light in the middle of the dark period:
(A) Both SD and LD plants will flower
(B) Both SD and LD plants will remain vegetative
(C) SD plant will remain vegetative and LD plant will flower
(D) SD plant will flower and LD plant will remain vegetative

529. Both respiration and photosynthesis require:
(A) Sunlight (B) Green cells
(C) Cytochrome (D) Organic substrate

530. Nobel prize was awarded to the Scientist for discovering the pathway of carbon assimilation:
(A) Watson (B) Krebs
(C) Calvin (D) Parnas

531. A photosynthetic organism which does not release oxygen is:
(A) Blue green algae (B) Green sulphur bacterium
(C) Green algae (D) Algal component of lichen

532. The sap of a plant cell has an osmotic potential of -10 bars and there is a wall pressure of 2 bars. When this cell is placed in a solution with an osmotic potential of -3 bars the force causing water to enter cell is:
(A) - 8 bar (B) - 7 bar
(C) - 5 bar (D) - 3 bar

533. PSII contains a nonchlorophyllous pigment in blue green algae:
(A) β-carotene (B) Zeaxanthin
(C) Phycocyanin (D) Cytochrome c

534. Which is not required for synthesis of carbohydrates:
(A) Chlorophyll (B) Water
(C) Nitrogen (D) Carbon dioxide

535. Isotopes employed to study photosynthesis are:
(A) ^{11}C and ^{32}P (B) ^{15}C and ^{32}P
(C) ^{16}C and ^{15}N (D) ^{14}C and ^{18}O

536. The evidence that during photosynthesis oxygen comes from water:
(A) Photosynthetic bacteria employ H_2S and CO_2 to form carbohydrates, water and sulphur
(B) Isolated illuminated chloroplasts release oxygen if provided with potassium ferrocyanide
(C) Isotopic ^{18}C provided as $H_2{}^{18}O$ appears as $^{18}O_2$ liber ated in photosynthesis
(D) All of these

537. Which one of the following pairs is correctly matched?
(A) Nutation: Movement due to injury
(B) Traumatotropism: Movement due to change in turgor pressure
(C) Thigmotropism: Movement of tendrils due to contact
(D) Seismonasty: Movement of stem tips of twining plants

538. Which one results in energy wastage:
(A) Photosynthesis (B) Photorespiration
(C) Dark respiration (D) Glycolysis

539. Which of the following constitutes Hill reaction:
(A) Photolysis by chloroplasts
(B) Photolysis of water by isolated chloroplasts forming ATP and NADPH
(C) Photolysis of water by isolated illuminated chloroplast causing reduction of some chemicals and liberation of oxygen
(D) Photolysis of water liberating oxygen and hydrogen by illuminated isolated chloroplasts

540. Thylakoids removed from chloroplasts were kept in illuminated culture having CO_2 and water. It did not produce sugar because of:
(A) Absence of enzymes
(B) Nonlinking of PS I and PS II
(C) Absence of light trapping
(D) Nonformation of assimilatory power

541. The substrate for photorespiration is:
(A) Phosphoglyceric acid (B) Glycolate
(C) Serine (D) Glycine

542. In C_4 plants, synthesis of sugar/final CO_2 fixation occurs in:
(A) Epidermal cells
(B) Spongy cells
(C) Undifferentiated mesophyll cells
(D) Bundle sheath cells

543. Which is produced during water stress that brings about stomatal closure?
(A) Abscisic acid (B) Ethylene
(C) Ferulic acid (D) Coumarin

544. As compared to chrolophyll a, bacteriochlorophyll has:
(A) One pyrrole ring with four hydrogen
(B) One pyrrole ring with three hydrogen
(C) One pyrrole ring with two hydrogen
(D) One pyrrole ring with one hydrogen

545. NADP$^+$ is reduced to NADPH in:
(A) PS I
(B) PS II
(C) Calvin cycle
(D) Noncyclic photophosphorylation

546. When the rate of translocation is slow, the rate of photosynthesis shall:
(A) Increase (B) Decrease
(C) Remain unchanged (D) Become zero

547. Active transport of ions across the membrane of a root hair cell can be assumed to be taking place if:
P. The cell produces more glutathione
Q. The cell has mitochondria
R. The uptake of ions stop when cyanide is added
S. The uptake of ions is against the concentration gradient
(A) P, R (B) R, S (C) Q, R (D) Q, S

548. Photosynthesis has two reaction complexes, one followed by the other. The second reaction complex:
(A) Traps light energy
(B) Synthesises starch
(C) Functions in the presence of light
(D) Fixes carbon dioxide

549. The chief source of nitrogen for green plants are:
(A) Atmospheric nitrogen
(B) Nitrates
(C) Ammonium salts
(D) Low molecular weight-organic nitrogenous compound

550. Quanta required for assimilation of one molecules of CO_2 or liberation of the molecule of oxygen are:
(A) 16 (B) 8 (C) 4 (D) 2

551. Ribulose biphosphate carboxylase oxygenase is located inside:
(A) Chloroplasts (B) Mitochondria
(C) Golgibodies (D) Peroxisomes

552. Chemosynthesis and photosynthesis are similar in being:
(A) Found in embryophytes
(B) Associated with heterotrophs
(C) Dependent on solar energy
(D) Autotrophic in nutrition

553. Phytochrome is:
(A) Monomer (B) Multimer
(C) Homodimer (D) Heterodimer

554. In case of prokaryotes, the net gain of ATP from complete oxidation of one molecule of glucose in respiration?
(A) 24 (B) 30 (C) 36 (D) 38

555. Gardeners pinch off the tips of plants to stimulate their side growth. Which one of the following is implicated in this phenomenon?
(A) Abscisic acid (B) Auxin
(C) Cytokinin (D) Gibberellin

556. Greatest producers of organic matter are:
(A) Crop plants
(B) Forests
(C) Plants of the land area
(D) Phytoplankton of oceans

557. Aerobic bacteria collect near illuminated phytoplankton due to:
(A) Manufactured food
(B) Light
(C) Oxygen
(D) Reduced CO_2 concentration

558. Which photosynthetic bacteria possess both PS I and PS II:
(A) Purple sulphur bacteria
(B) Cyanobacteria
(C) Purple nonsulphur bacteria
(D) Green sulphur bacteria

559. Which products of Hill reaction are used in Blackman's reaction:
(A) ATP, NADPH (B) ATP, NADH
(C) ADP, NAD (D) ATP, NAD

560. The light which stimulates flowering in an inducible plant:
(A) Is perceived by the apical meristem
(B) Is perceived throughout the plant
(C) Is perceived by the leaves
(D) None of the above are correct

561. Chlorophyll b is
(A) $C_{54}H_{70}O_6N_4Mg$ (B) $C_{55}H_{70}O_6N_4Mg$
(C) $C_{55}H_{72}O_5N_4Mg$ (D) $C_{45}H_{72}O_5N_4Mg$

562. The size of chlorophyll molecule is:
(A) Head 15 x 15 Å, tail 25 Å
(B) Head 20 x20 Å, tail 25 Å
(C) Head 15 x 15 Å, tail 20 Å
(D) Head 10 x 12 Å, tail 15 Å

563. Which statement about phloem transport is correct?
(A) Gravity influence phloem transport
(B) Phloem transport occurs unidirectionally
(C) Sugar transported in phloem as non-reducing sugar
(D) Ca^{2+} is most abundant action in phloem sap

564. Green light has a wavelength of:
(A) 400-450 nm (B) 500-580 nm
(C) 660-720 nm (D) 720-800 nm

565. During monsoon, the rice crop of eastern states of India shows lesser yield due to limiting factor of:
(A) CO_2 (B) Light
(C) Temperature (D) Water

566. Photosynthetic pigments found in the chloroplasts occur in:
(A) Thylakoid membranes (B) Plastoglobules
(C) Matrix (D) Chloroplast envelop

567. Dark reaction of photosynthesis occur in:
(A) Granal thylakoid
(B) Stromal lamella membranes
(C) Stroma outside photosynthetic lamellae
(D) Periplastidial space

568. Cytochrome b6-f complex consists of how many polypeptides?
(A) 1 (B) 2 (C) 3 (D) 4

569. Which blue green algae does not form heterocytes?
(A) *Anabaena* (B) *Nostoc*
(C) *Gleotrichia* (D) *Scytonema*

570. A Very efficient converter of solar energy with net productivity of 2-4 kg/m^2 or more in the crop of:
(A) Wheat (B) Sugarcane
(C) Rice (D) Bajra

571. Formation of ATP in photosynthesis and respiration is an oxidation process which utilizes the energy from:
(A) Cytochromes (B) Ferredoxin
(C) Electrons (D) Carbon dioxide

572. Cryptochrome is the pigment that absorbs:
(A) Red light (B) Green light
(C) Blue light (D) Yellow light

573. Algae employed by Calvin et al in experiments on photosynthesis belong to:
(A) Euglena (B) Chara
(C) Chlamydomonas (D) Chlorella

574. C_4 plants are also known as:
(A) Calvin type
(B) Calvin-Bassltam type
(C) Hatch and Slack type
(D) Emerson type

575. Path of dark reaction of photosynthesis was traced through the use of:
(A) ^{32}P (B) $^{14}CO_2$ (C) $^{18}O_2$ (D) X-rays

576. Conversion of PGA to PGAL involves:
(A) Reduction (B) Oxidation
(C) Hydrolysis (D) Photolysis

577. The plant growth regulator that retards senescence is:
(A) Cytokinin (B) Gibberellic acid
(C) Indoleacetic acid (D) Ethylene

578. When a plant is not reproducing, most of its cytokinins are produced in its:
(A) Leaves (B) Lateral roots
(C) Shoot apex (D) Roots

579. Cyclic photophosphorylation is carried out by:
(A) PS I (B) PS II
(C) Both (A) and (B) (D) Photolysis and PS II

580. In C_4 plants, Calvin cycle operates in:
(A) Stroma of bundle sheath chloroplasts
(B) Grana of bundle sheath chloroplasts
(C) Grana of mesophyll chloroplasts
(D) Stroma of mesophyll chloroplasts

581. Photosynthetic enhancement with flashing light was first observed by:
(A) Benson and Calvin (B) Hill and Calvin
(C) Hatch and Slack (D) Emerson and Arnold

582. Which is sensitive to longer wavelengths of light:
(A) PS II (B) PS I
(C) Phosphorylation (D) Photolysis

583. Maximum O_2 evolution occurs from:
(A) Forests (B) Marine phytoplankton
(C) Crops (D) Land mass

584. During cyclic photophosphorylation, which is electron donor of PS II?
(A) CO_2 (B) Water
(C) PS I (D) Cytochrome

585. Photosynthetic rate is the highest in which plants?
(A) C_3 plants (B) C_4 plants
(C) CAM plants (D) Both (B) and (C)

586. Which technique has helped in investigation of Calvin cycle?
(A) X-ray crystallography
(B) X-ray technique
(C) Radioactive isotope technique
(D) Intermittent light

587.

Group I (Hormone)	Group II (Functions)
P. Gibberellin	1. Phototropism and gravitrophism
Q. IAA	2. Stomatal movement
R. Cytokinin	3. Delay of senescence
S. Ethylene	4. Combating water deficit
	5. Seed germination
	6. Ripening of fruits

(A) P -6, Q – 4, R – 1, S - 2
(B) P -5, Q – 1, R – 3, S - 6
(C) P -3, Q – 4, R – 5, S - 1
(D) P -4, Q – 3, R – 1, S – 5

588. Downward movement of organic and inorganic solutes from leaves occurs through:
(A) Phloem by diffusion
(B) Phloem by protoplasmic streaming
(C) Parenchymatous cells by diffusion
(D) Phloem by mass flow

589. C_4 plants belongs to:
(A) Gramineae (B) Monocots
(C) Dicots (D) Both monocots and dicots

590. From the given statements identify the incorrect one:
(A) GA involves in flowering
(B) Ethylene is produced during ripening of the seeds
(C) Auxin helps in cell elongation and formation of root
(D) Cytokinin helps in embryo development and prevent leaf senescence

591. Kranz anatomy is typical of:
(A) C_4 plants (B) C_3 plants
(C) C_2 plants (D) CAM plants

592. Site of O_2 evolution and photophosphorylation is:
(A) Granal stacks
(B) Chloroplast surface
(C) Inner membrane of chloroplasts
(D) Stroma

593. Quantosomes occur in:
(A) Chloroplasts (B) Mitochondria
(C) Nucleus (D) Lysosomes

594. Sugarcane shows high efficiency of CO_2 fixation because it performs:
(A) Calvin cycle
(B) EMP pathway
(C) Hatch and Slack pathway
(D) TCA cycle

595. Which one of the following does not perform C_4 photosynthesis?
(A) *Saccharum*
(B) *Zea mays*
(C) *Triticum aestivum(=T. vulgare/Crotalaria*
(D) *Euphorbia milli (=E. splenedens)*

596. Photosystem II occurs in:
(A) Stroma (B) Cytochrome
(C) Grana (D) Mitochondrial surface

597. Which of the following acts as a precursor of IAA biosynthesis?
(A) Tryptophan
(B) Methionine
(C) Putrescine
(D) Geranyl geranyl pyrophosphate

598. The mitochondrial genome is the smallest in:
(A) Mammals (B) Yeasr
(C) Arabidopsis (D) Tomato

599. Kreb's cycle starts with entry of:
(A) Acetyl-CoA (B) Oxaloacetate
(C) Both (A) and (B) (D) Pyruvic acid

600. Through which of the following does an electron pass to $NADP^+$ in light reaction:
(A) Cytochrome (B) Ferredoxin
(C) Plastoquinone (D) Plastocyanin

601. Pigment system I performs independently:
(A) Noncyclic photophosphorylation
(B) Cyclic photophosphorylation
(C) Oxidative phosphorylation
(D) Photolysis

602. The mitochondrial genome is the smallest in:
(A) Mammals (B) Yeast
(C) Arabidopsis (D) Tomato

603. In case of eukaryotes the net gain of ATP from the complete oxidation of one molecule of glucose in respiration?
(A) 24 (B) 30 (C) 36 (D) 38

604. Ferredoxin is a constituent of:
(A) PS I (B) PS II
(C) Hill reaction (D) P_{680}

605. ATP is:
(A) Adenine triphosphate
(B) Adenosine tetraphosphate
(C) Adenosine triphosphate
(D) Adenine tetraphosphate

606. How many ATPs (gross ATP) are produced from complete oxidation of one molecule of glucose in respiration?
(A) 36 (B) 38 (C) 30 (D) 24

607. Which is due to cpDNA alone?
(A) Photosystem I
(B) Photosystem II
(C) Chlroplast RNA polymerase
(D) All of these

608. In case of eukaryotes, the net gain of ATP from complete oxidation of one molecule of glucose in respiration?
(A) 24 (B) 30 (C) 36 (D) 38

609. All type of plastids possess essentially the same structure because they:
(A) Perform the same function
(B) Store food materials like starch, fat and protein
(C) Occurs in aerial parts
(D) Can transform from one form to another

610. Chlorophyll is found in:
(A) Stroma of chloroplasts
(B) Grana of chloroplasts
(C) Covering of chloroplasts
(D) Dispersed from in the chloroplasts

611. For synthesis of a molecule of glucose, the requirement of ATP and NADPH is respectively:
(A) 15 and 10 (B) 33 and 22
(C) 12 and 8 (D) 18 and 12

612. Chlorophyll a occurs in:
(A) All photosynthetic autotophs
(B) In all higher plants
(C) All oxygen liberating autotrophs
(D) All parts except fungi

613. Which is due to cpDNA alone?
(A) Photosystem I
(B) Photosystem II
(C) Chloroplast RNA polymerase
(D) All of these

614. The enzyme that catalyses carbon dioxide fixation in C_4 plants is:
(A) RuBP carboxylase
(B) PEP carboxylase
(C) Carbonic anhydrase
(D) Carboxy dismutase

615. Photo-oxidation or protolysis of water (in photosynthesis) occurs in assimilation of:
(A) Cytochrome B (B) Plastocyanin
(C) PS II (D) PS I

616. The source of oxygen evolved during photosynthesis is:
(A) CO_2 (B) H_20
(C) Carbohydrate (D) Photosynthetic curve

617. Bacterial photosynthetic differs from that of higher plants in:
(A) Not liberating oxygen
(B) Nonrequirement of light
(C) Nonfixation of energy
(D) Requirement of host organism

618. Karntz type of leaf anatomy is found in:
(A) Maize (B) Amaranthus
(C) Sorghum (D) All of these

619. Chief functions of leaves are:
(A) Transpiration and photosynthesis
(B) Respiration and photosynthesis
(C) Respiration and digestion
(D) Respiration and transpiration

620. Food is transported to different parts of plant through:
(A) Xylem (B) Phloem
(C) Pith (D) Cortex

621. Oxidation in biological systems is accompanied by:
(A) Hydrogenation (B) Resonance stabilization
(C) Repulsion (D) Dehydrogenation

622. How many ATPs (gross ATP) are produced from complete oxidation of one molecule of glucose in respiration?
(A) 24 (B) 30 (C) 36 (D) 38

623. Presence of organic substances in the phloem can be demonstrated by means of:
(A) Ringing of stem (B) Grafting
(C) Root pressure (D) Defoliation

624. Emerson found photosynthesis to increase with light flashes. Blackman observed increase in photosynthesis with the increase in light and CO_2 till temperature became limiting. What of the following is true in light of above observation:
(A) Light intensity is temperature dependent
(B) Light harvesting occurs only in the presence of CO_2
(C) Photosynthesis has two phases, light and dark
(D) One factor is limiting photosynthesis at one time

625. Translocation of carbohydrate nutrients usually occurs in the form of:
(A) Glucose (B) Maltose
(C) Starch (D) Sucrose

626. Crassulacean acid metabolism (CAM) occurs in:
(A) Agave (B) Pineapple
(C) Opuntia (D) All of these

627. Which one die first when a ring of bark is removed:
(A) Shoot
(B) Root
(C) None
(D) Both shoot and root die simultaneously

628. At high oxygen concentration, the rate of photosynthesis decreases due to:
(A) Warburg effect
(B) Pasteur effect
(C) Emerson effect
(D) Richmond Lang effect

629. Kreb's cycle is also known as:
(A) Tricarboxylic acid cycle (TCA)
(B) Citric acid cycle
(C) Organic acid cycle
(D) All of these

630. Oxygen containing carotenoids are:
(A) Carotenes (B) Xanthophylls
(C) Phycobilins (D) Anthocyanins

631. In the two light reactions of photosynthesis:
(A) PS I produces strong oxidant while PS II a strong reductant
(B) PS I produces strong reductant NADPH while PS II a strong oxidant
(C) PS I produces ATP which is not formed by PS II
(D) PS I produces ATP which is not formed by PS II

632. Apical dominance is the result of:
(A) Auxin (B) Cytokinin
(C) Auxin and ABA (D) ABA

633. Which organelle is round and derives its name from the latin word for 'Kernel'?
(A) Nucleus (B) Lysosome
(C) Golgi complex (D) Mitochondria

634. When day light hours are increased, the rate of photosynthesis:
(A) Increases (B) Decreases
(C) Remains unchanged (D) None of the above

635. Photosynthesis is:
(A) Photochemical process (B) Chemical process
(C) Oxidation process (D) Reduction process

636. Unit membrane or plasma membrane is made of:
(A) Cellulose (B) Lipoprotein
(C) Phospho protein (D) Phospho lipids

637. Complete girdling ultimately kills the tree due to:
(A) Starvation of root
(B) Stoppage of photosynthesis
(C) Nonmovement of minerals
(D) Nonmovement of water

638. In CAM plants, formation of malic acid takes place during:
(A) Sunset (B) Morning
(C) Night (D) Afternoon

639. Photorespiration occurs in:
(A) Ribosomes (B) Mitochondria
(C) Peroxisomes (D) Lysosomes

640. How many ATPs are produced (if we start with Acetyl CoA) in Krebs cycle?
(A) 18 (B) 24 (C) 30 (D) 38

641. The real force causing entry of water into a cell from other cell is:
(A) Diffusion pressure deficit
(B) Turgor pressure
(C) Osmotic pressure
(D) Wall pressure

642. CAM occurs in:
(A) Thin green leaves with reticulate venation
(B) Thin green leaves with parallel venation
(C) Thin coloured leaves
(D) Fleshy green leaves

643. Minimum solar energy is trapped by:
(A) Planting trees (B) Cultivating crops
(C) Growing algae in tanks (D) Growing grasses

644. Excesive use of fertilizer causes death of plants due to:
(A) Endosmosis (B) Exosmosis
(C) Imbibition (D) Turdigity

645. Which of the following trees will die first:
(A) Deciduous (B) Pruned
(C) Hollowed trunk (D) Girdled

646. Two types of photosynthetic pigments are:
(A) Chlorophyll a and carotenoids
(B) Chlorophyll b and carotenoids
(C) Chlorophylls and carotenoids
(D) Chlorophyll a and chlorophyll b

647. PEP carboxylation takes place in:
(A) Mesophyll cells of C_4 plants
(B) Epidermal cells of C_4 plants
(C) Bundle sheath cells of C_4 plants
(D) Xylem cells of C_4 plants

648. Main difference between chlorophyll a and chlorophyll b is:
(A) $\cdot CH_3$ of chlorophyll a is replaced by $-CHO$ in chlo rophyll b
(B) Chlorophyll a is linear while chlorophyll b is branched
(C) Chlorophyll a has no Mg
(D) All of these

649. Light energy is used in photosynthesis for:
(A) Breakdown of H_2O
(B) Breakdown of CO_2
(C) Activation of chlorophyll
(D) Breakdown of $C_6H_{12}O_6$

650. The dry weight of a green leaf will be:
(A) Lesser in evening
(B) Greater in the afternoon
(C) Greater in the morning
(D) Almost same throughout the whole day

651. If plants stop photosynthesis which gas will disappear:
(A) CO_2 (B) N_2
(C) O_2 (D) NH_3

652. Glycolysis occurs in:
(A) Cytoplasm (B) Mitochondria
(C) Nucleus (D) Golgi bodies

653. Which plant shows chloroplast dimorphism:
(A) Wheat (B) Rice
(C) Sugar Beet (D) Sugar cane

654. Cytochromes are:
(A) O_2 acceptors (B) H_2 acceptors
(C) Electron acceptor (D) H_2O acceptors

655. Electron transport system take place in which part of mitochondria?
(A) Oxysome (B) Outer membrane
(C) Matrix (D) Both (A) and (C)

656. During dark reaction the three carbon atoms of 3-PGA are derived from:
(A) RuBP only (B) CO_2 only
(C) RuBP + CO_2 (D) RuBP + CO_2 + PEP

657. Pigment system is concerned with:
(A) Photolysis of water (B) Reduction of CO_2
(C) Flowering (D) None of these

658. First product of photorespiration is:
(A) Phosphoglycolate (B) Glycolate
(C) Glycine (D) None of the above

659. Assimilatory power formed in light reaction of photosynthesis is:
(A) ATP + $NADPH_2$ (B) ADP + $NADH_2$
(C) ATP (D) $NADPH_2$

660. Photoperiod is the most relevant for:
(A) Abscission (B) Flowering
(C) Shoot bud formation (D) Root growth

661. Which one is a C_4 plant:
(A) Maize (B) Sugarcane
(C) Sorghum (D) All the above

662. Which is correct about chlorophylls a and b leaves of higher plants:
(A) Both are present in equal proportion
(B) Chlorophyll a is more than chlorophyll b
(C) Chlorophyll a is less than chlorophyll b
(D) Chlorophyll b is ten times more than chlorophyll a

663. In C_4 plants, chloroplasts are also found in:
(A) Spongy parenchyma (B) Epidermal cells
(C) Guard cells (D) Bundle sheath cells

664. In C_4 plants, fixation of carbon dioxide occurs in:
(A) Transfusion tissue
(B Palisade tissue
(C) Cortex of stem
(D) Spongy mesophyll and bundle sheath cells

665. Liberation of oxygen when green cells in water are exposed to sunlight in presence of suitable acceptor is:
(A) Emerson effect (B) Blackman's reaction
(C) Hill's reaction (D) Arnon reaction

666. During cyclic photophosphorylation, which is electron donor of PS II?
(A) CO_2 (B) Water
(C) PS I (D) Cytochrome

667. Balance between CO_2 and O_2 is maintained by:
(A) Transpiration (B) Photosynthesis
(C) Photorespiration (D) C_4 pathway

668. Non-cyclic electron transport in photosynthesis is also known by the name of:
(A) S - scheme (B) Z- scheme
(C) Q – scheme (D) None of these

669. ATP synthesis during light reaction is:
(A) Oxidative phosphorylation
(B) Photolysis
(C) Photophosphorylation
(D) Phosphorylation

670. Pigment acting as a reaction centre during photosynthesis is:
(A) Carotene (B) Phytochrome
(C) P_{700} (D) Cytochrome

671. In response to stress, which amino acid is accumulated?
(A) Proline (B) Glycine
(C) Lysine (D) Phenylalanine

672. Photosystem has main light harvesting pigments:
(A) Chlorophyll a -660, chl a-670, chl a-680, chl a 690, chl a-700
(B) Chlorophyll a-660, chl b-, chl-700, chl-800 and carotenoid
(C) Chlorophyll a-680, chl-685, chl-695 and xanthophylls
(D) Chlorophyll a-700, chl-880, chl b, chl-685 and phytocobilins

673. Glycolysis occurs in:
(A) Cytoplasm (B) Mitochondria
(C) Nucleus (D) Golgi bodies

674. Wilting of leaves in hot weather is due to:
(A) Fe and Cu (B) Na and Cu
(C) Ca and K (D) Fe and Mg

675. C_4 cycle was discovered in:
(A) Groundnut (B) Sugarcane
(C) Chrysanthemum (D) Apple

676. Which one is the common product of both respiration and photosynthesis:
(A) ATP (B) Quinone
(C) Cytochrome (D) Chlorophyll

677. Chlorophylls absorb visible light of wavelengths:
(A) 400-500 nm only
(B) 300-400 nm only
(C) 600-800 nm only
(D) 400-500 nm and 600-700 nm

678. Chlorophyll a has:
(A) -CHO group (B) -COOH group
(C) -CH3 groups (D) No functional group

679. RuBISCO has which activities?
(A) Carboxylase (B) Reductase
(C) Oxygenase (D) Both (A) and (C)

680. Calvin cycle involves:
(A) Oxidative carboxylation
(B) Reductive carboxylation
(C) Phosphorylation
(D) Oxidative phosphorylation

681. Chlorophyll is soluble in:
(A) Water (B) Organic solvents
(C) Both (A) and (B) (D) None of these

682. Photorespiration is favoured by:
(A) High oxygen and low carbon dioxide
(B) High carbon dioxide and low oxygen
(C) High temperature and low oxygen
(D) High humidity and temperature

683. In which, electron transport leading to phosphorylation is a common reaction?
(A) Chloroplasts (B) Mitochondria
(C) Both (A) and (B) (D) Golgi bodies

684. Mitochondrial DNA does not code for:
(A) tRNA (B) Enzyme for DNA replication
(C) mRNA (D) rRNA

685. Photosynthesis is a process in which :
(A) ATP is generated
(B) NADH is reduced to NAD
(C) Oxidative phosphorylation occurs
(D) CO_2 is released

686. Carbon dioxide assimilation occurs through:
(A) Kreb cycle (B) Calvin cycle
(C) Glycolysis (D) Anaerobic respiration

687. Number of calvin cycles required to generate a molecule of hexose is:
(A) 2 (B) 4 (C) 6 (D) 8

688. Which one occurs both during cyclic and noncyclic modes of photophosphorylation:
(A) Involvement of both PS I and PS II
(B) Formation of ATP
(C) Release of O_2 (D) Formation of NADPH

689. Calvin cycle is:
(A) Dependent upon light (B) Independent of light
(C) Supported by light (D) Hindered by light

690. In C_4 plants, initial carbon dioxide fixation occurs in chloroplasts of:
(A) Palisade tissue (B) Spongy parenchyma
(C) Guard cells (D) Hypodermis and chlorenchyma

691. Phytochrome is a:
(A) Monomer (B) Multimer
(C) Homodimer (D) Heterodimer

692. Pyruvate kinase transfers a phosphate group from phosphoenolpyruvate to ADP, forming pyruvate and ATP. The reaction catalyzed by this enzyme is essentially irreversible. Which of the following is the best explanation for the irreversible nature of this reaction?
(A) The binding of pyruvate to the active site is very weak relative to the binding of phosphoenolpyruvate
(B) The reaction is coupled to the pyruvate dehydrogenase reaction
(C) The hydrolysis of ATP is highly favourble
(D) The change in free energy for the overall reaction is large and negative

693. Anaerobic respiration is also known as:
(A) Intra-molecular respiration
(B) Inter-molecula respiration
(C) Extra-molecular respiration
(D) Molecular respiration

694. Chlorophyll consists of:
(A) Porphyrin head and phytol tail
(B) Phosphate head and porphyrin tail
(C) Phytol head and porphyrin tail
(D) Porphyrin head and phosphate tail

695. Most of the reactions of glyoxylate cycle occurs in:
(A) Glyoxysomes (B) Cytosol
(C) Mitochondria (D) All of these

696. A free living anaerobic bacterium capable of nitrogen fixation in the soil is:
(A) *Azotobacter* (B) *Clostridium*
(C) *Rhizobium* (D) *Streptococcus*

697. Element is essential for photolysis of water is:
(A) Nitrogen (B) Oxygen
(C) Chlorine (D) Carbon

698. Enzymes PEP carboxylase and RuBP carboxylase are located in chloroplasts of:
(A) C_3 plants (B) CAM plants
(C) C_4 plants (D) Both (B) and (C)

699. Constituents of pigment system I are located on:
(A) Granal thylakoids
(B) Stomal thylakoids
(C) Outer surface of granal and stomatal thylakoids
(D) Stroma

700. Which are stress hormones?
(A) Ethylene and IAA (B) IAA and ABA
(C) ABA and Ethylene (D) ABA and IBA

701. Light energy is converted into chemical energy in the presence of:
(A) Chloroplasts (B) Pyrenoids
(C) Ribosomes (D) Mesosomes

702. Calvin cycle occurs in:
(A) Cytoplasm (B) Mitochondria
(C) Glyxosomes (D) Chloroplasts

703. Inclusion of a cell extract containing all enzymes of glycolysis with [ã32P] ATP and unlabeled inorganic phosphate, results in the formation of which of the following labeled compounds (assume that pyruvate kinase is inactivated)?
1. Glucose-6 - ^{32}phosphate
2. (3-^{32}P)-phosphoglycerate
3. (1-phospho – 3 – ^{32} phospho)-bisphosphoglycerate
4. (1-phospho – 6 – ^{32} phospho)- fructose bisphosphate
(A) 1 and 3 (B) 1, 2 and 3
(C) 2 and 4 (D) Only 4

704. Plants purify air during:
(A) Photosynthesis (B) Respiration
(C) Transpiration (D) Desiccation

705. Which organelles is involved in ATP production by photophosphorylation?
(A) Mitochondria (B) Chloroplasts
(C) Leucoplasts (D) Chromoplasts

706. In case of paints, cpDNA encodes which components of orgaeannellar ATPase?
(A) F1 ATPase
(B) F0 Factor
(C) Both (A) and (B)
(D) Some subunits of both (A) and (B)

707. Nitrite reductase contains which prosthetic groups?
(A) Siroheme ($Fe_4 – S_4$) cluster
(B) FMN (Fe-S) cluster
(C) Mo cofactor (Fe-S) cluster
(D) All of these

708. Photorespiration occurs in:
(A) Green plants (B) All living cells
(C) Mitochondria (D) Root

709. Thylakoids possess photosynthetic units called:
(A) Quantasomes (B) Glyoxysomes
(C) Polysomes (D) Ribosomes

710. Noncyclic photophosphorylation is performed by:
(A) Dark reaction (B) PS I
(C) PS II (D) Both (B) and (C)

711. Photorespiration is affected by:
(A) Temperature (B) Light intensity
(C) CO_2 and O_2 (D) All of these

712. A free living anaerobic bacterium capable of nitrogen fixation in the soil:
(A) Azotobacter (B) Clostridium
(C) Rhizobium (D) Streptococcus

713. At carbon atom III, chlorophyll a has:
(A) Methyl group (B) Aldehyde group
(C) Carboxyl group (D) Magnesium

714. One of the carbon atoms of a glucose molecule is [14]-labeled. If $^{14}CO_2$ is released during the conversion of pyruvate to acetyl coenzyme-A, which carbon atom of glucose was radio labeled?
(A) C3 but not C 4 (B) C 3 or C 4
(C) C 1 or C 6 (D) C 1 but not C 6

715. Photorespiration is characteristic of:
(A) C_3 plants (B) C_4 plants
(C) CAM plants (D) All of these

716. Which organelles is involved in citric acid cycle?
(A) Mitochondria (B) Chloroplasts
(C) Leucoplasts (D) Chromoplasts

717. In case of yeast mitochondria, mt – DNA encodes which components of mitochondrial ATPase?
(A) F1 ATPase
(B) F0 Factor
(C) Both (A) and (B)
(D) Some subunits of both (A) and (B)

718. C_4 plants differ from C_3 plants with respect to:
(A) First product
(B) Substrate which accepts carbon dioxide
(C) Number of ATP molecules consumed
(D) All of these

719. *nif*HDK genes are organized in which ways?
(A) They are contigous
(B) D and K are separated by 11 kb DNA
(C) Separated by 91 kb DNA located between genes D and K
(D) Separated by 55 kb DNA between genes D and K

720. Photooxidation is:
(A) Photorespiration
(B) Photolysis
(C) Light and oxygen induced breakdown
(D) All of these

721. Core metal of chlorophyll is:
(A) Cu (B) Mg
(C) Fe (D) Ni

722. Which statement is true of glucokinase?
(A) It catalyzes the phosphorylation of fructose
(B) It has a lower K_m for glucose than does hexokinase
(C) It is found in liver
(D) It is inhibited by glucose-6-phosphate

723. Which of the following statement about photosystem II is incorrect?
(A) It is multimolecular transmembrane assembly containing several polypeptides, several chrolophyll molecules, a special pair of chlorophyll (P680), pheophytin, and plastoquinones
(B) It transfers electrons to phtosystem I via the cytochrome bf complex
(C) It uses light energy to create a separation of charge whose potential I energy can be used to oxidize H_2O and to produce a reductant, plastoquinol
(D) It uses an Fe^{++} - Cu^{+} centre as a charge accumulator to form O_2 without generating potentially harmful hydroxyl radials, superoxide anions, or H_2O_2

724. RUBISCO content of chloroplast is:
(A) 20% (B) 5% (C) 11% (D) 16%

725. In case of plants, cpDNA encodes which components of organellar ATPase?
(A) F1 ATPase
(B) F0 factor
(C) Both (A) and (B)
(D) Some subunits of both (A) and (B)

726. Main substance involved in transfer of electrons in photosynthesis is:
(A) Photochrome (B) Cytochrome
(C) Isosomes (D) Both (A) and (B)

727. Which one is required for photosynthetic evolution of oxygen?
(A) F-enzyme (B) Z-enzyme
(C) Co-enzyme (D) M-enzyme

728. The electron transport system and oxidative phosphorylation are concerned with?
(A) Mitochondria (B) Chloroplasts
(C) Leucoplasts (D) Chromoplasts

729. Azolla contains ab ebdosymbiont which fix nitrogen; this oprganism belongs to which genera?
(A) *Aulosira* (B) *Azosprillum*
(C) *Nostoc* (D) *Anabaena*

730. C_4 plants have higher net photosynthetic rate as they have:
(A) No photorespiration
(B) PEP as CO_2 acceptor
(C) Kranz anatomy
(D) Photosynthesis even at low light intensity

731. Azolla contains an endosymbiont which fixes nitrogen; this organism belong to which genera?
(A) *Aulosira* (B) *Azospirillum*
(C) *Nostoc* (D) *Ananaena*

732. Accessory pigments transfer their energy to chlorophyll molecules through:
(A) Fluorescence (B) Resonance
(C) Radiation (D) Transduction

733. C_4 cycle is connected with:
(A) Respiration (B) Photosynthesis
(C) Transpiration (D) Grwoth and development

734. ATP formation in chloroplast and mitochondrian is explained by:
(A) Cholodny-Went model
(B) Chemi-osmotic theory of Mitchell
(C) Munch's mass flow theory
(D) Relay pump theory of Godlewski

735. Primary carboxylation occurs in C_3 and C_4 plants with the help of:
(A) PEP carboxylase and pyravate carboxylase
(B) PEP carboxylase and RuBP carboxylase
(C) RuBP carboxylase and PEPcarboxylase
(D) RuBP carboxylase and pyruvate carboxylase

736. Iodoacetamide irreversively inhibits few enzymes by reacting with the amino acid residue at the active site having the functional group:
(A) -SH (B) - COOH (C) - OH (D) - NH_2

737. In case of yeast mitochondria, mtDNA encodes which components of mitochondrial ATPase?
(A) F1 ATPase
(B) F0 factor
(C) Both (A) and (B)
(D) Some subunits of both (A) and (B)

738. A specific function of light in the process of photosynthesis is:
(A) Reduction of CO_2 (B) Activate other molecules
(C) Split water molecules (D) None of these

739. 2,4-dinitrophenol uncouples electron transport from oxidative phosphorylaton in mitochondria by:
(A) Causing dissipation of the proton gradient generated by the electron transport
(B) Allowing the proton translocating function of ATP synthase while inhibiting its ATP synthesizing activity
(C) Activating a second proton pump that sends back the protons into the mitochondrial matrix
(D) Neutralizing the proton gradient by absorbing the protons generated during the electron transport

740. C_4 plants differ from C_3 plants in:
(A) Number of ATP consumes per hexose molecule
(B) Type of end product
(C) Type of pigments
(D) Substrate for CO_2 in carbon assimilation

741. Which is an example of colourless plastid?
(A) Mitochondria (B) Chloroplasts
(C) Leucoplasts (D) Chromoplasts

742. nifHDK genes are organized in which ways?
(A) They are contiguous
(B) D and K are separated by 11 kb DNA
(C) Separated by 91 kb DNA located between genes D and K
(D) Separated by 55 kb DNA between genes D and K

743. Warburg effect is decrease in the rate of photosynthesis at:
(A) Low carbon dioxide concentration
(B) High oxygen content
(C) High carbon dioxide concentration
(D) Both (A) and (B)

744. Which one of the following contain agranal chloroplasts?
(A) C_3 plants (B) Succulants
(C) C_4 plants (D) Hydrophytes

745. Which of the plants bears intermediate features between C_3 and C_4 plants?
(A) Maize (B) *Moricandia arvensis*
(C) *Amaranthus* (D) Both (A) and (C)

746. Which is not an example of C_4 plant?
(A) Sugarcane (B) Sorghum
(C) Potulaca (D) Wheat

747. Cyanobacteria differ from purple and green photosynthetic bacteria because they:
(A) Have a membrane-enclosed nucleus
(B) Use H_2S as an electron donor
(C) Do not require light
(D) Produce oxygen during photosynthesis

748. Which element is required for photosynthetic evolution of oxygen?
(A) Cu (B) Fe (C) Zn (D) Mn

749. Thermogenin, the natural uncoupler in brown fat mitochondria, generates heat based on its ability to:
(A) Inhibits electron transport by binding to b-type cytochromes
(B) Allow protons to re-enter into the mitochondrial matrix
(C) Inhibit ATP production by binding to the ATP synthase
(D) Block electron transport to flavoproteins

750. The term carotenoid is related to:
(A) Carotene (B) Chlorophyll
(C) Xanthophyll (D) Both (A) and (C)

751. Nitrite reductase contains which prosthetic groups?
(A) Siroheme (Fe4-S4) cluster (B) FMN (Fe-S) cluster
(C) Mo cofactor (Fe-S) cluster (D) All of these

752. Protochlorophyll differs from chlorophyll in lacking:
(A) 4 hydrogen atoms in two pyrrole rings
(B) 2 hydrogen atoms in one of its pyrrole rings
(C) 2 hydrogen atoms in two pyrrole rings
(D) 1 hydrogen atoms in one pyrrole ring

753. Which is an example of C_4 plant?
(A) Maize (B) *Moricandia arvensis*
(C) *Amaranthus* (D) Both (A) and (C)

754. Which occurs during light reactions?
(A) Photolysis of water (B) Photophosphorylation
(C) Fixation of CO_2 (D) Both (A) and (B)

755. In photosynthesis, protolysis of water is used in:
(A) Reduction of $NADP^+$ (B) Oxidation of NADP
(C) Oxidation of FAD (D) None of these

756. Plant mitochondria were treated with antimycin A and sodium azide. This would lead to:
(A) Increased cyanide resistant respiration and reduced ATP synthesis
(B) Increased cyanide resistant respiration and in creased ATP synthesis
(C) Increased cytochrome oxidation
(D) Reduced nuclear –mitochondria interaction

757. Photosynthesis occurs in:
(A) Leaves (B) Leaf cells
(C) Plants (D) Chloroplasts

758. In green cell the enzyme catalase is located in:
(A) Peroxisomes (B) Cytoplasts
(C) Lysosomes (D) Vacuoles

759. The wavelength of light most absorbed during photosynthesis is:
(A) 700 nm (B) 600 nm
(C) 550 nm (D) 400 nm

760. Maximum photosynthesis occurs in:
(A) Blue light (B) Red light
(C) Green light (D) White light

761. Which is the major pigment of chloroplasts?
(A) Chlorophyll (B) Xanthophyll
(C) Carotene (D) All of these

762. Which are stress hormones?
(A) Ethylene and IAA (B) IAA and ABA
(C) ABA and ethylene (D) ABA and IBA

763. Special feature of C_4 plants is:
(A) Thin cuticle (B) Multi layeres epidermis
(C) Kranz anatomy (D) Both (A) and (B)

764. Factors which is not normally limiting in photosynthesis is:
(A) Chlorophyll (B) Light
(C) Water (D) CO_2

765. Which of the following contains copper?
(A) Quinone (B) Plastoquinone
(C) Plastocyanin (D) Cyt b_5

766. Radioactive tracer studies with ^{14}C have shown that:
(A) Phloem transports organic nutrients
(B) Phloem transports inorganic nutrients
(C) Xylem transports organic nutrients
(D) Xylem transports inorganic nutrients

767. Pigments of PS I occur in:
(A) Appressed part of granal thylakoids
(B) Stromal thylakoids and nonappressed parts of granal thylakoids
(C) Both appressed and nonappressed parts of granal thylakoids
(D) Stroma

768. Number of cell organelles involved in photorespiration is:
(A) One (B) Two
(C) Three (D) Four

769. Condition favouring cyclic photophosphorylation is:
(A) Anaerobic environment
(B) Aerobic and low light intensity
(C) Aerobic and optimum light
(D) Anaerobic and low light intensity

770. Which of the following statements about the enzyme complexes of the electron transport system is correct?
(A) They interact with one another via mobile electron carriers
(B) They are located in the mitochondria matrix
(C) They can not be separated from one another in a functional form
(D) They all have cytochromes

771. When sunlight is on the chloroplast, pH is lowest in the:
(A) Stroma
(B) Space enclosed by the inner and outer membranes
(C) Space enclosed by the thylakoid membranes
(D) Cytosol

772. Wave length of visible light is:
(A) 200-400 nm (B) 400-700 nm
(C) 700-900 nm (D) 100-200 nm

773. In photosynthesis synthesis of carbohydrates during CO_2 and:
(A) Water in the presence of sunlight with the help of chlorophyll
(B) Water in the presence of energy
(C) Water with the help of chlorophyll
(D) Presence of sunlight

774. Which is correct about C3 plants?
(A) During photosynthesis, RuBP carboxylase enzyme plays a key role in the fixation of CO_2
(B) 3- Phosphoglyceric acid (3-PGA) is the first stable products of photosynthesis
(C) Carbohydrates are formed through Calvin cycle
(D) All of these

775. During photosynthesis, dark reaction occurs in:
(A) Stroma (B) Thylakoids
(C) Oxysomes (D) Both (A) and (B)

776. Direction of conduction of food through phloem is:
(A) From below upwards
(B) From tip to bottom
(C) From leaves to roots
(D) Phloem never conducts food

777. Glycolysis uses, which substrate?
(A) Glucose (B) Sucrose
(C) Lipids (D) Fatty acids

778. RuBISCO has which activities?
(A) Carboxylase (B) Reductase
(C) Oxygenase (D) Both (A) and (C)

779. In photosynthesis oxygen is liberated during:
(A) Hydrolysis of carbohydrates
(B) Breakdown of proteins
(C) Reduction of CO_2
(D) Hydrolysis of water

780. DCMU inhibits photosynthesis at photosystem during:
(A) phe QA (B) QA Qb
(C) Cyt bfb P (D) Q8 ytbf6

781. The pigment present universally in all green oxygenic plants is:
(A) Chlorophyll a (B) Chlorophyll b
(C) Chlorophyll c (D) Chlorophyll m

782. Ferredoxin is:
(A) Phenol (B) Protein
(C) Fat (D) Carbohydrate

783. In which plants stomata are closed in day time and open during night?
(A) C_2 plants (B) C_4 plants
(C) CAM plants (D) All of these

784. Ethylene is mainly responsible for:
(A) Formation of flower (B) Formation of root hair
(C) Ripening of fruits (D) Formation of fruit

785. Green plants do not give out CO_2 during day time because they:
(A) Store the same
(B) Respire very slowly
(C) Do not respire
(D) Consume it in photosynthesis

786. The enzymes for the oxidation of amino acids, fatty acids and Kreb's cycle are present in:
(A) Lysosomes (B) Stroma
(C) Grana (D) Mitochondrial matrix

787. In response to stress, which amino acids is accumulated?
(A) Proline (B) Glycine
(C) Lysine (D) Phenylalanine

788. For fixing of one molecule of CO_2 in Calvin cycle, are required:
(A) 3 ATP + 1 $NADPH_2$ (B) 3 ATP + 2 $NADPH_2$
(C) 2 ATP + 3 $NADPH_2$ (D) 3 ATP + 3 $NADPH_2$

789. The rosette habit of cabbage can be changed drastically by the application of:
(A) GA (B) IAA
(C) ABA (D) CK

790. Enzyme required in early CO_2 fixation in C_4 cycle is:
(A) RuBP carboxylase (B) RuBP oxygenase
(C) PEP carboxylase (D) PGA dehydrogenase

791. During photosynthesis, light reaction occurs in:
(A) Stroma (B) Thylakoids
(C) Oxysomes (D) Both (A) and (B)

792. First reaction in photosynthesis is:
(A) Photolysis of water
(B) Excitation of chlorophyll molecules
(C) Formation of ATP
(D) Fixation of CO_2

793. Carbon dioxide is fixed in:
(A) Light reaction (B) Dark reaction
(C) Aerobic respiration (D) Anaerobic respiration

794. Three carbons of photophoglyceric acid formed during carbon fixation are derived from:
(A) PEP + CO_2 (B) RuBP
(C) CO_2 (D) RuBP + CO_2

795. Which of the following occurs in dark reaction of photosynthesis?
(A) Formation of ATP (B) Release of O_2
(C) Release of H_2 (D) Synthesis of PGAL

796. Photoperiod is the most relevant for:
(A) Abscission (B) Flowering
(C) Shoot bud formation (D) Root growth

797. Process of assimilation of CO_2 to produce carbohydrates is:
(A) Oxidative (B) Catabolic
(C) Reductive (D) All of these

798. Cytochrome b_6 and cytochrome *f* occur in:
(A) Ribosomes (B) Mitochondria
(C) Chloroplasts (D) Lysosomes

799. Light reaction of photosynthesis is also called:
(A) Calvin cycle (B) Hill reaction
(C) TCA cycle (D) All of these

800. Growth is a:
(A) Reversible increase in size
(B) Reversible increase in shape
(C) Irreversible increase in size
(D) None of these

801. Starch is stored in Potato tuber as:
(A) Sugar is synthesized in leaves
(B) Sugar is transported from leaves to tubers
(C) In the tuber sugar is changed to starch
(D) All of these

802. Which is correct about C_3 plants?
(A) During photosynthesis, PEP carboxylase plays a key role in the fixation of CO_2
(B) Malic is the first stable products of photosynthesis
(C) Carbohydrates are formed through Hatch Slack cycle
(D) All of these

803. Evidence for liberation of O_2 from water during photosynthesis comes from:
(A) Release of ^{18}O if water contains the same
(B) Isolated chloroplast supplied with reducing agent like potassium, ferrocyanide evolved O_2 even in absence of CO_2
(C) Photosynthetic bacteria do not liberate O_2 and they use H_2S for obtaining reducing power
(D) All of these

804. Bakanae disease is caused by:
(A) Fungus (B) Alga
(C) Bacterium (D) Virus

805. Which is present in stroma?
(A) DNA (B) RNA
(C) Ribosomes (D) All of these

806. Chemical which absorbs light energy and changes it to chemical energy is:
(A) Xanthophyll (B) Chlorophyll a
(C) Chlorophyll b (D) Chlorophyll c

807. In maize, mesophyll cells perform photosynthesis cycle:
(A) C_4 (B) C_3 (C) C_2 (D) C_1

808. The rate of photosynthesis, can be measured by:
(A) Amount of carbohydrate accumulated
(B) The volume of CO_2 consumed
(C) The volume of oxygen librated
(D) All of these

809. Carotene protect plants from:
(A) Photooxidation (B) Desiccation
(C) Photorespiration (D) Photosynthesis

810. An excised leaf does not turn yellow if it is induced to root. This is attributed to the synthesis in root of:
(A) Ethylene (B) Cytokinins
(C) Gibberellins (D) Auxins

811. Which have circular DNA?
(A) Chloroplasts (B) Mitochondria
(C) Bacteria (D) All of these

812. A natural growth regulator is:
(A) NAA (B) 2,4-D
(C) Benzaldehyde (D) Ethylene

813. The maximum growth phase is:
(A) Lag phase (B) Steady phase
(C) Log phase (D) Senescent phase

814. The main feature of light promoters of dicots is:
(A) TATA box
(B) An upstream AG or AGGA box
(C) Both TATA and AG/AGGA boxes
(D) CAAT box

815. The rate of growth is measured by:
(A) Auxanometer (B) Potometer
(C) Photometer (D) Hygrometer

816. During photosynthesis, which gases is evolved as byproduct?
(A) CO_2 (B) H_2
(C) O_2 (D) Both (B) and (C)

817. Anatomy of C4 plant leaf shows:
(A) Presence of perroxisomes
(B) Presence of bundle sheath cells
(C) Absence of mitochondria
(D) Absence of bundle sheath cells

818. The growth of plants differs from growth of animals in being:
(A) Localised (B) Diffused
(C) Localised and Diffused (D) None of these

819. Dark reaction is also called:
(A) Blackman reaction (B) Hill reaction
(C) Cyclic reaction (D) Both (A) and (B)

820. The term 'vernalization' was coined by:
(A) Garner and Allard (B) Darwin
(C) Geoffery (D) Lysenko

821. The alcohol dehydrogenase gene s induced by stress due to
(A) Heat (B) Anaerobiosis
(C) Wound (D) Pathogene invasion

822. Gibberellins were first extracted from:
(A) Coleoptile tip (B) Root tip
(C) Fungus (D) Bacterium

823. Hill used a dye for his famous hill reaction:
(A) Sulphur green (B) Eosine
(C) Methylene blue (D) Dichlorophenol indophenol

824. Etiolation in plants is caused when they are grown in:
(A) Red light (B) Blue light
(C) Dark light (D) Intense light

825. Translocation of carbohydrates mostly takes place in the form of :
(A) Maltose (B) Fructose
(C) Sucrose (D) Glucose

826. Substance which originated at the tip of stem and control growth elsewhere are:
(A) Enzymes (B) Hormones
(C) Vitamins (D) None of these

827. The short day plant is:
(A) Tomato (B) Sunflower
(C) Cotton (D) Potato

828. Which is not correct?
(A) Start point is typically a single base within the promoter
(B) The chief function of factor (of bacterial RNA polymerase) is to ensure that the holoenzyme binds stably at only promoter sequences
(C) Transcription factors are essential part of RNA polymerases
(D) Transcription factors can be divided into basal up stream and regulatory transcription factors

829. Which is correct about RNA polymerase I
(A) It is present in nucleolus
(B) It constitutes 50-70% of total activity
(C) It transcribes genes for rRNA
(D) All of these

830. In case of eukaryotes, RNA processing is completed in:
(A) Cytoplasm (B) Nucleolus
(C) Nucleus (D) Vacuole

831. Which is correct about RNA polymerase III?
(A) It occurs in nucleoplasm
(B) It provides -10% of total polymerase activity
(C) It transcribes tRNA and 5SRNA genes
(D) All of these

832. The long day plant is:
(A) Tomato (B) Potato
(C) Cotton (D) Spinach

833. The day neutral plant is:
(A) Tobacco (B) Tomato
(C) Wheat (D) Oat

834. Light reaction is also called:
(A) Blackman reaction (B) Hill reaction
(C) Cyclic reaction (D) Both (A) and (B)

835. The synthesis of amylase in certain cereals is stimulated by:
(A) GA (B) IAA (C) CK (D) ABA

836. Plant growth is:
(A) Irreversible (B) Increase in size
(C) Localised (D) All of the above

837. Plant response to environment is mainly through:
(A) Induction of dormancy (B) Abscission of parts
(C) Synthesis of pigments (D) Growth

838. Darwin and Darwin discovered that canary grass coleoptile bends towards light only when:
(A) It is 5 cm long
(B) The seedling is growing in blue light
(C) Coleoptile tip is intact and exposed to light
(D) Nights are cooler

839. In case of C_4 plants, RuBISCO exists only in:
(A) Mesophyll cells (B) Bundle sheath
(C) Guard cells (D) All of these

840. Etiolation is characterized by:
(A) Slender yellowish stems (B) Small yellowish leaves
(C) Subterminal hook (D) All of these

841. Fe protein of nitrogenase is:
(A) Homomeric dimer (B) Hetermeric dimer
(C) Homomeric tetramer (D) Hetermeric tetramer

842. How many ATP molecules are produced in anaerobic respiration of one glucose molecule?
(A) 2 (B) 4 (C) 6 (D) 8

843. On excess, CO_2:
(A) Accelerates growth (B) Does not influence growth
(C) Inhibits growth (D) Slow down growth

844. Which one regulates plant growth?
(A) Commensation (B) Climate
(C) Hormones (D) Both (B) and (C)

845. How many electrons are required for the reduction of one molecule of nitrogen to NH_4?
(A) 6 (B) 8 (C) 3 (D) 12

846. High C/N ratio produces:
(A) Solar tissues
(B) More mechanical tissues
(C) More growth hormones
(D) Growth inhibitors

847. Which is correct about C_3 plants?
(A) During photosynthesis, RuBP carboxylase enzyme plays a key role in the fixation of CO
(B) 3-Phosphogluceric acid (3-PGA) is the first stable product of photosynthesis
(C) Carbohydrates are formed through Calvin cycle
(D) All of these

848. HMP pathway is also known by:
(A) Pentose phosphate pathway
(B) Direct oxidase pathway
(C) EMP pathway
(D) Both (A) and (B)

849. Which one is not a growth promoter?
(A) ABA (B) GA
(C) IAA (D) CK

850. Which is not correct about transcription unit?
(A) It is that sequence of a DNA molecule that is transcribed into a single RNA molecule
(B) In case of all eukaryotes, each gene is a distinct transcription unit
(C) In case of prokaryotes, each operon functions as a single transcription unit
(D) Enhancer sequences are present within promoters

851. The GC-AG rule does not apply to introns of:
(A) Chloroplast introns (B) Nuclear introns of yeast
(C) Mitochondrial inrtons (D) All of these

852. Optimum growth occurs in :
(A) Blue light (B) Red light
(C) White light (D) Green light

853. DNP is:
(A) Cestrum (B) Cucumber
(C) Potato (D) Radish

854. Nitrogenase consists of:
(A) Mo protein (B) Fe-protein
(C) MoFe-protein (D) Both (B) and (C)

855. Major function of CK is to:
(A) Mobilise nutrients and prevent senescence
(B) Mobilise nutrients
(C) Prevent senescence
(D) Produce IAA

856. Growth promoter hormone are:
(A) IAA, ABA and CK (B) IAA, GA and ABA
(C) IAA, GA and CK (D) ABA, CK and GA

857. Glyoxysomes are involved in breakdown of:
(A) Sugar (B) Protein
(C) Fatty acids (D) All of these

858. Pyruvic acid is the end product of:
(A) Hydrolysis of starch (B) Citric acid cycle
(C) Photolysis (D) Glycolysis

859. Hormone involved in inhibition of growth is:
(A) ABA (B) Ethylene
(C) None of these (D) Both (A) and (B)

860. Which is correct about C_4 plants?
(A) During photosynthesis, PEP carboxylase enzyme plays a key role in the fixation of CO_2
(B) Malic acidis the first stable product of photosynthesis
(C) Carbohydrates are formed through Hatch-Slack cycle
(D) All of these

861. Cyanobacteria protect nitrogenase from oxygen by:
(A) Nodule formation (B) Heterocyst formation
(C) Phycocyanin (D) Leghaemoglobin

862. Cytokinins are mostly produced in:
(A) Shoot apex (B) Root apex
(C) Young leaves (D) Lateral buds

863. Nitrate reductase is found in:
(A) Vacuole (B) Mitochondria
(C) Cytoplasm (D) Chloroplasts

864. Cytokinins:
(A) Induce cell division and inhibit ageing
(B) Maintain dormancy
(C) Induced abscission
(D) Inhibit cell division

865. Guttation is due to:
(A) O.P (B) S.P.
(C) Root pressure (D) T.P.

866. Removal of apical buds makes the plant:
(A) Bushy (B) Grow rapidly
(C) Grow slowly (D) Dormant

867. In germinating cereals, amylase synthesis is stimulated by:
(A) Cytokinins (B) Abscisic acid
(C) Gibberellins (D) Auxin

868. In organs of plants there is a variation in rate of growth. It is first slow, then accelerated upto maximum and slows down to stand still. The period of vigorous growth is known as:
(A) Lag phase (B) Period of growth
(C) Grand growth period (D) Log phase

869. Cyanobacteria protect nitrogenase from oxygen by:
(A) Nodule formation (B) Heterocyst formation
(C) Phycocyanin (D) Leghaemoglobin

870. Bioassay for auxin is:
(A) Dwarf maize test (B) Avena curvature test
(C) Cell division test (D) Green leaf test

871. Bioassay for gibberellin is:
(A) Avena curvature test (B) Seed dormancy test
(C) Dwarf maize test (D) Green leaf test

872. Site of oxidative election transport in cells is:
(A) Golgi bodies (B) Mitochondria
(C) Nucleus (D) Chloroplasts

873. Stomata plays a key role in:
(A) Photoperiodism (B) Transpiration
(C) Respiration (D) Vernalisation

874. Internodal elongation is promoted by the action of:
(A) Auxin (B) Gibberellic acid
(C) Cytokinin (D) Ascisic acid

875. Acetylene reduction to ethylene is a measure of:
(A) N_2 fixation (B) Nitrate reduction
(C) Nitrite reduction (D) CO_2 fixation

876. A green plant bends towards the source of light when exposed to the light on only one side. Which of the following is the best explanation of the phenomenon?
(A) The apices of the stems are attracted by light
(B) The plant needs light for photosynthesis
(C) Auxin accumulates on the shaded side to induce great cell elongation on that side
(D) Light reduces length

877. Which is plants bears intermediate features between C_3 and C_4 plants?
(A) Maize (B) *Moricandia arvensis*
(C) Amarathus (D) Both (A) and (C)

878. Stomata are bound by a pair of fuard cells and open guard cells are:
(A) Flaccid (B) Large
(C) Turgid (D) Green

879. Delay in senescence is caused by the spray of:
(A) IBA (B) GA
(C) Cytokinin (D) ABA

880. Cytokinins are generally:
(A) Acids (B) Aminopurines
(C) Phenols (D) Glucosides

881. HMP pathway is also known by:
(A) Pentose phosphate pathway
(B) Direct oxidative pathway
(C) EMP pathway
(D) Both (A) and (B)

882. Apples are generally wrapped in waxed paper to:
(A) Prevent sunlight from changing its colour
(B) Prevent aerobic respiration by checking the entry of oxygen
(C) Prevent ethylene formation that hastens ripening
(D) Make the apples look costly

883. Seedless tomato are produced by:
(A) Supplying radioactive compounds
(B) Spraying hormones on flower
(C) Propagating tomato plants from cuttings
(D) Growing tomato plants in mineral solution

884. Which elements is related to cytochrome?
(A) P (B) Cu
(C) Fe (D) Zn

885. Avena coleoptile test is for which of the following hormone:
(A) GA (B) ABA
(C) IAA (D) NAA

886. Hormone involved in breaking dormancy of winter buds is:
(A) Ethylene (B) Gibberellin
(C) Auxin (D) Cytokinin

887. Which of the following is an essential part of photosynthesis?
(A) Glycolysis (B) Photorespiration
(C) Plasmolysis (D) Photolysis

888. Rapid growth of flowering axis in many plants is caused by:
(A) Lower level of absicisic acid
(B) Higher level of auxin
(C) Higher level of gibberellin
(D) Lower level of auxin

889. Which elements is related to cytochrome?
(A) P (B) Cu (C) Fe (D) Zn

890. In Plants, respiration includes:
(A) Glycolysis
(B) Kreb's cycle
(C) Electron transport system
(D) All of these

891. Differentiation of callus requires a specific ratio of two hormones:
(A) Auxin and gibberellin
(B) Auxin and abscisic acid
(C) Gibberellin and abscisic acid
(D) IAA and Cytokinin

892. Which is an example of C_4 plants?
(A) Maize (B) Moricandia arvensis
(C) Amarathus (D) Both (A) and C)

893. In case of C_4 plants, RuBISCO exists only in:
(A) Mesophyll cells (B) Bundle sheath
(C) Guard cells (D) All of these

894. Ferulic acid, coumarin and hesperidin are:
(A) Secondary plant products
(B) Hormones
(C) Enzymes
(D) Vitamins

895. Site of oxidative election transport in cells is:
(A) Golgi bodies (B) Mitochondria
(C) Nucleus (D) Chloroplasts

896. Gibberellins take part in:
(A) Bolting or rosette plants
(B) Replacing long day requirement
(C) Overcoming genetic dwarfism
(D) All of these

897. A decapitated plant exposed to unilateral light:
(A) Bends away source of illumination
(B) Bends towards the source of light
(C) Shows zigzag curvature
(D) Does not show bending movement

898. Correct sequence of different phases of growth:
(A) Cell division→ Cell differentiation→ Cell elongation
(B) Cell differentiation→ Cell division→ Cell elongation
(C) Cell elongation → Cell division → Cell differentiation
(D) Cell division→ Cell elongation → Cell differentiation

899. Glyoxosomes are involve in breakdown of:
(A) Sugar (B) Protein
(C) Fatty acids (D) All of these

900. Correct sequence of different stages of development is:
(A) Fruiting→ Flowering→ Juvenility → Germination
(B) Germination→ Juvenility→ Flowering→ Fruiting
(C) Flowering→ Fruiting→ Juvenility→ Germination
(D) Juvenility→ Flowering→ Fruiting→ Germination

901. Exponential growth is shown during:
(A) Lag phase (B) Log phase
(C) Maturation phase (D) Sebesence phase

902. During wall enlargement new wall materials are inserted amongst the older ones. The phenomenon is:
(A) Accretion (B) Integration
(C) Intussusception (D) Interpolation

903. Phosphorylation occurs in:
(A) Mitochondria (B) Glyoxysomes
(C) Chloroplasts (D) Both (A) and (C)

904. Temperature ranges for plant growth is:
(A) 25^0-35^0C (B) 10^0-35^0C
(C) 10^0-50^0C (D) 0^0-20^0C

905. Apical dominance is due to:
(A) C/N ratio (B) Photoperiodism
(C) Phototropism (D) Auxin cytokinin balance

906. Acetylene reduction to ethylene is a measure of:
(A) N_2 fixation (B) Nitrate reduction
(C) Nitrite reduction (D) CO_2 fixation

907. Which is contains DNA?
(A) Peroxysomes (B) Mitochondria
(C) Centrosomes (D) Prions

908. Pineapple can be made to flower with the help of:
(A) Kinetin
(B) Ethylene
(C) GA
(D) Exposure to low temperature

909. How many ATP molecules are produced in anaerobic respiration of one glucose molecule?
(A) 2 (B) 4 (C) 6 (D) 8

910. Growth of a shoot or a root involves:
(A) Cell enlargement (B) Cell division
(C) Cell differentiation (D) All of thesse

911. Agent Orange of Vietnam war was:
(A) 2,4,5-T (B) 2,4-D
(C) NAA (D) Mixture of (A) and (B)

912. In which plants stomata are closed in day time and open during night time?
(A) C_3 plants (B) C_4 plants
(C) CAM plants (D) All of these

913. During photosynthesis, oxygen is released from:
(A) CO_2 (B) H_2O (C) SO_2 (D) O_3

914. Hormone involved in the formation of winter buds is:
(A) GA (B) CK
(C) ABA (D) Ethylene

915. Leaf abscission is prevented by:
(A) Gibberellin (B) Ethylene
(C) Abscisic acid (D) Auxin

916. Leaf abscisin is promoted by:
(A) Ethylene (B) Abscisic acid
(C) Cytokinin (D) Tryptophan

917. The main function of leghaemoglobin in root nodule is concerned with:
(A) N_2 supply (B) O_2 supply
(C) Mineral supply (D) CO_2 supply

918. Leaf fall occurs when:
(A) Auxin content is more in lamina than in the stem
(B) Auxin content is low in the lamina as compared to stem
(C) Abscisic acid is formed
(D) Morphactins are produced

919. It will not be any advantage to produce seedless fruits in:
(A) Mangoes (B) Watermelon
(C) Pomegranate (D) Oranges

920. Which is correct about transcription factors?
(A) These factors are required for transcription initiation at all the promoters of eukaryotes
(B) They join RNA polymerase II to form a complex around the startpoint
(C) They determine the site of transcription initiation
(D) All of these

921. If auxin concentration is increased in plant roots, the water absorption will:
(A) Increase (B) Decrease
(C) Remain the same (D) None

922. Pyruvic acid is the end product of:
(A) Hydrolysis of starch (B) Citric acid cycle
(C) Photolysis (D) Glycolysis

923. Cell mitosis is induced by:
(A) Colchicine (B) Gibberellic acid
(C) Nitrosoguanidine (D) Kinetin

924. Nitrate reductase is found in:
(A) Vacuole (B) Mitochondria
(C) Cytoplasm (D) Golgi body

925. Vernalisation is related to:
(A) Cold treatment (B) Photoperiodism
(C) Development buffering (D) Male sterility

926. Chrysanthemum flower in winter because:
(A) They are short day plant
(B) They required low temperature treatment
(C) They are long day plants
(D) Both (A) and (B)

927. Gibberellic acid has been successfully employed to induce flowering:
(A) In short day plants under long day conditions.
(B) In long day plants under short day conditions
(C) In all plants
(D) None of these

928. Which enzymes is involved in conversion of ammonia to amino acid?
(A) Alanine transferase (B) Ntrate reductase
(C) Glutamine synthetase (D) Nitrite reductase

929. Plants which daily require shorter period of darkness for flowering are called:
(A) Short day plants (B) Long day plants
(C) Normal day plants (D) Ephemeral plants

930. Which have circular DNA?
(A) Chloroplasts (B) Mitochondria
(C) Bacteria (D) All of these

931. Which is not an example of C_4 plants?
(A) Sugarcane (B) Sorghum
(C) Portulaca (D) Wheat

932. Photoperiodism is associated with the formation of:
(A) Chlorophyll (B) Florigen
(C) Auxin (D) Gibberellin

933. Hormone is responsible for vernalization is:
(A) Florigen (B) Colchicine
(C) Abscisin (D) Vernalin

934. Stomata plays a key role in:
(A) Photoperiodism (B) Transpiration
(C) Respiration (D) Vernalization

935. Letuce seeds germinate only when:
(A) Red light is available
(B) Blue light is available
(C) Green light is available
(D) They are sown deep in the soil

936. Which of the following is required for the formation of flowers in the plants?
(A) Adequate oxygen (B) Definite photoperiod
(C) Adequate manure (D) Adequate water and mineral

937. Which is correct about RNA polymerase II?
(A) It is present in nucleoplasm
(B) It constitutes 20-40% of total activity
(C) It transcribes all the genes that produce mRNA
(D) All of these

938. Kinetin like properties are present in:
(A) Pineapple (B) Soybean
(C) Groundnut (D) Coconut milk

939. Low concentration of auxin inhibits growth in:
(A) Stem (B) Roots
(C) Leaves (D) Flowers

940. Apple growing on tree can have larger than normal size by:
(A) Thinning of blossoms (B) Defoliation
(C) Decapitation of stem apex (D) Applying hormones

941. Which enzyme are responsible for fixation of nitrogen?
(A) Nitrite reductase (B) Nitrate reductase
(C) Carboxylase (D) Nitrogenase

942. An apple tree can be made to bear larger sized fruits by:
(A) Decapitation (B) Defoliation
(C) Dehydration (D) Thinning of blossom

943. Dark reaction is also called?
(A) Blackman reaction (B) Hill reaction
(C) Cyclic reaction (D) Both (A) and (B)

944. Which occurs during light reaction:
(A) Photolysis of water (B) Photophosphorylation
(C) Fixation of CO_2 (D) Both (A) and (B)

945. Fruit drop occurs when fruit has:
(A) No auxin as in the stem
(B) Less auxin as compared to stem
(C) More auxin than the one present in the stem
(D) Auxin concentration equal to that of stem

946. How many electrons are required for the reduction of one molecules of nitrogen to NH_4^+?
(A) 6 (B) 8 (C) 10 (D) 20

947. During photosynthesis, water splits in:
(A) PS I (B) PS II
(C) Cytoplasm (D) Stroma

948. Auxin does not increase:
(A) The plasticity of cell walls
(B) The rate of photosynthesis
(C) The uptake of water by cells
(D) Rate of respiration

949. In plants, respiration includes:
(A) Glycolysis
(B) Kreb's cycle
(C) Electron transport system
(D) All of these

950. Induction of rooting in stem cutting would be most beneficial in:
(A) Marchantia (B) Wheat
(C) Cuscuta (D) Bougainvillea

951. Mitochondria and plastids present in eukaryotic cells contain RNA polymerases that are analogous to:
(A) Eukaryotic polymerase
(B) Phage polymerases
(C) Bacterial polymerases
(D) Both (B) and (C)

952. Weed killers have properties similar to:
(A) Hormones (B) Enzymes
(C) Insecticides (D) Vitamins

953. In Red (P_r) and Far-red (P_{fr}) (given alternatively) responses of plants, it is the:
(A) First treatment that is effective
(B) Middle treatment that is effective
(C) Last treatment that is important
(D) Sum total of all treatments that counts

954. During photosynthesis, dark reaction occurs in:
(A) Stroma (B) Thylakoids
(C) Oxysomes (D) Both (A) and (B)

955. Germination of some seeds is promoted by:
(A) Green light (B) Red light
(C) Far red light (D) Blue light

956. Flowering in some short day plants is promoted by:
(A) Auxins (B) Gibberellins
(C) Cytokinins/ABA (D) Ethylene

957. Mature root nodules mostly consists of:
(A) Diploid (B) Triploid
(C) Tetraploid (D) Pentaploid

958. Flowering in pineapple is promoted by:
(A) GA (B) IAA
(C) Ethylene (D) Both (B) and (C)

959. Reduction of nitrite to ammonium is catalyzed by nitrite reductase in:
(A) Chloroplasts (B) Mitochondria
(C) Nucleus (D) Golgi body

960. Mature root nodules mostly consist of which cells?
(A) Diploid (B) Triploid
(C) Tetraploid (D) Pentaploid

961. A decapitated plant does not bend towards light because:
(A) The root sends a message to shoot apex to do so
(B) There is accumulation of growth hormones at the cut end
(C) The hormone responsible for bending is not available
(D) Growth inhibitors are formed

962. During photosynthesis, light reaction occurs in:
(A) Stroma (B) Thylakoids
(C) Oxysomes (D) Both (A) and (B)

963. Low temperature treatment of some semigerminated seeds for some period often induces subsequently quicker flowering in plants. This process is called:
(A) Vernalization (B) Thermolysis
(C) Wintering (D) Devernalization

964. Which consensus sequences is recognized by more than one transcription factors?
(A) - 35 sequence (B) CAAT box
(C) Octamer sequence (D) GC box

965. Cytokinins have the ability to:
(A) Induce morphogenesis
(B) Substitute cold treatment
(C) Convert dwarf plants to tall plants
(D) None of these

966. Which contains DNA?
(A) Pereroxysomes (B) Mitochondria
(C) Centrosomes (D) Prions

967. Which one of the following is gaseous growth regulator?
(A) IAA (B) NAA
(C) IBA (D) Ethylene

968. Sprouting of potato can be prevented in storage by:
(A) Maleic hydrazide (B) Gibbrellins
(C) Indole acetic acid (D) Cytokinins

969. A response of length of day in plants is called:
(A) Phototropism (B) Photorespiration
(C) Photoperiodism (D) Photo-oxidation

970. Which enzymes is responsible for fixation of nitrogen?
(A) Nitrite reductase (B) Nitrate reductase
(C) Carboxylase (D) Nitrogenase

971. The substance which accelerates the growth in stem is:
(A) Vitamin (B) Etiolin
(C) Enzyme (D) Auxin

972. Which of the following effects of auxins on plants is the basis for commercial application?
(A) Callus formation
(B) Curvature of stem
(C) Induction of root formation in stem cuttings
(D) Stem growth

973. Leaves with yellow or colourless patches are termed as:
(A) Lecorine (B) Decolourized
(C) Etiolated (D) Variegated

974. Photomorphogenetic pigment in plants is:
(A) Phytochrome system (B) Chlorophyll
(C) Carotene (D) Chalcone pigment

975. The rate of photosynthesis can be measured by:
(A) Amount of carbohydrate accumulated
(B) The volume of CO_2 consumed
(C) The volume of oxygen librated
(D) All of these

976. Artificial ripening of fruits is accomplished by treatment with:
(A) Sodium chloride (B) IAA
(C) Ethylene gas (D) Kinetin

977. Plants bend towards light because:
(A) They need light for photosynthesis
(B) They need light for respiration
(C) Cells on the shaded side elongate more
(D) Lights attracts them

978. Which is usually surrounded by a GC rich sequence?
(A) TATA box (B) Pribnow box
(C) GC box (D) CAAT box

979. Which wavelength is more effective in photoperiodism:
(A) Green and yellow (B) Blue and red
(C) Blue and violet (D) Red and far-red

980. Oxygen evolved in photosynthesis is:
(A) Less than CO_2 consumed
(B) More than CO_2 consumed
(C) Equal to CO_2 consumed
(D) None of these

981. Abscisic acid (ABA) promotes:
(A) Cell division (B) Leaf fall
(C) Germination (D) Budding

982. Phytotron is a device by which:
(A) Electrons are bombarded
(B) Protons are liberated
(C) Plants are grown in controlled environment
(D) Mutation are produced in plants

983. During photosynthesis, which of the gas is evolved as byproduct?
(A) CO_2 (B) H_2
(C) O_2 (D) Both (A) and (B)

984. A green plant will turn yellow if placed in the dark. This is called:
(A) Phototropism (B) Dormancy
(C) Chlorosis (D) Etiolation

985. Which of the following structures controls the transport of the material into and out of living cells?
(A) Centrosome (B) Cell membrane
(C) Cell wall (D) Ribosome

986. Which of the following stimulates growth or cell enlargement?
(A) Gibberellin (B) Auxin
(C) Cytokinin (D) Ethylene

987. Which is occupies a fixed position?
(A) TATA box (B) Pribnow box
(C) GC box (D) CAAT box

988. Gibberellin was first extracted from:
(A) *Gibberella fujikuroi* (B) *Gelidium*
(C) *Aspergillus* (D) *Gracelaria*

989. Most of the reactions of glyxylate cycle occurs in:
(A) Glyoxysomes (B) Cytosol
(C) Mitochondria (D) All of these

990. Cytokinin:
(A) Is a hormone whose main function is the induction of cell division
(B) Is the process of cell division
(C) Refers to cell movements
(D) Causes dormancy

991. In autumn leaf fall occurs because:
(A) Formation of abscission layer at the bases
(B) Leaf becomes heavy
(C) Leaf does not remain green
(D) Of low temperature

992. Higher concentration of some synthetic auxins is used for:
(A) Weed control (B) Rooting of stem cuttings
(C) Parthenocarpy (D) Increasing fruit yield

993. Translation of carbohydrates mostly takes place in the form of:
(A) Maltose (B) Fructose
(C) Sucrose (D) Glucose

994. Ethylene gas:
(A) Is saturated hydrocarbon
(B) Slows down ripening of apples
(C) Retards ripening of tomatoes
(D) Speeds up of maturation of fruits

995. The main feature of light induced promoters of dicots is:
(A) TATA box
(B) An upstream AG or AGGA box
(C) Both TATA and AG/AGGA boxes
(D) CAAT box

996. Photochrome is involved in:
(A) Phototropism (B) Photorespiration
(C) Photoperiodism (D) Geotropism

997. Phosphorylation occurs in:
(A) Mitochondria (B) Glyoxysomes
(C) Chloroplasts (D) Both (A) and (C)

998. Evergreen trees remain green throughout the year on account of:
(A) Absence of leaf fall
(B) Leaves falling in small number at intervals
(C) Supply of moisture throughout the year
(D) Cold climate

999. Hormone involved in phototropism is:
(A) IAA (B) Gibbrellin
(C) Kinetin (D) 2,4-D

1000. Which organelles is involved in ATP production by photophosphorylation?
(A) Mitochondria (B) Chloroplasts
(C) Leucoplasts (D) Chromoplasts

1001. Phytochrome is found in:
(A) Algae (B) Fungi
(C) Vascular cryptogams (D) Flowering plants

1002. In many plants the change over from vegetative to reproductive phase takes place in response to:
(A) The length of the day
(B) The severity of temperature
(C) Mainly the food material available in the soil
(D) The oxygen present in the air

1003. Light reaction is also called?
(A) Blackman reaction (B) Hill reaction
(C) Cyclic reaction (D) Both (A) and (B)

1004. When the dark period of short day plants is interrupted by a brief exposure of light the plant:
(A) Will not flower at all
(B) Flowers immediately
(C) Gives more flowers
(D) Turns into a long day plant

1005. Which is present in stroma?
(A) DNA (B) RNA
(C) Ribosomes (D) All of these

1006. Which is not affected by light:
(A) Photosynthesis (B) Flowering
(C) Fertilization (D) Transpiration

1007. In unisexual plants sex can be changed by application of:
(A) Ethanol (B) Auxin
(C) Cytokinin (D) ABA

1008. The main function of leghaemoglobin in root nodules is concenrned with:
(A) N_2 supply (B) O_2 supply
(C) Mineral supply (D) CO_2 supply

1009. Which is used as a weedicide?
(A) Indole Acetic Acid (B) Napthalene Acetic Acid
(C) Indole Butyric Acid (D) 2,4-D

1010. Exogenous GA induces formation of male flowers on genetically female plants of:
(A) Cucurbits (B) Luffa
(C) Cucumis (D) Carica

1011. Citric acid cycle occurs in:
(A) Mitochondria (B) Chloroplasts
(C) Leucoplasts (D) Chromoplasts

1012. For plant tissue culture which among the following is required:
(A) Trypsin (B) Kinetin
(C) Caffeine (D) Counarin

1013. Gibberelins cause:
(A) Elongation of internodes
(B) Curvature of coleoptile
(C) Cell division
(D) Initiation of lateral roots

1014. Which functional sequence/module is not found in eukaryotic promoters?
(A) TATA box (B) Pribnow box
(C) GC box (D) CAAT box

1015. Substances formed at stem tip and controlling growth elsewhere are:
(A) Auxins (B) Vitamins
(C) Enzymes (D) Metabolic intermediates

1016. Ethylene is :
(A) A gaseous metabolite (B) A gaseous enzyme
(C) A gaseous hormone (D) A solid hormone

1017. The enzymes for the oxidation of amino acids, fatty acids and Kreb's cycle are present in:
(A) Lysosomes (B) Stroma
(C) Grana (D) Mitochondrial matrix

1018. Two climatic factors which affect growth are:
(A) Light and wind
(B) Light and temperature
(C) Rain and temperature
(D) Atmospheric humidity and temperature

1019. Herbicide DCMU kills plants by:
(A) Destroying chlorophyll
(B) Inhibiting PS II
(C) Inhibiting PS I
(D) Inhibiting oxidative phosphorylation

1020. Which one produces more ethylene?
(A) Green apple (B) Green Banana
(C) Ripening Banana (D) Fresh Potato tuber

1021. Which one shows red↔far-red interconversions?
(A) Carotenoids (B) Cytochromes
(C) Chlorophylls (D) Phytochrome

1022. The electron transport system and oxidative phosphorylation are concerned with:
(A) Mitochondria (B) Chloroplasts
(C) Leucoplasts (D) Chromoplasts

1023. Phytochrome is:
(A) Pigment that controls photomorphogenesis and development
(B) Regulatory protein that controls dark dependent development processes
(C) Hormone
(D) Photosynthetic hormone

1024. Plant material used in tissue culture is:
(A) *Cycas revolute* (B) *Pinus roxburghii*
(C) *Cocos nucifera* (D) *Borassus flabellifer*

1025. Natural plant hormones were isolated for the first time from:
(A) Human urine and Rice plant
(B) Human urine and Corn germ oil
(C) Rice plants, Spinach leaves and cotton fruits
(D) *Gibberella,* Spinach leaves and *Avena* coleoptile

1026. Glycolysis uses as:
(A) Glucose (B) Sucrose
(C) Lipids (D) Fatty acids

1027. IAA was first isolated from:
(A) Corn germ oil (B) *Gibberella*
(C) Human urine (D) *Rhizopus*

1028. Auxin suppresses growth of:
(A) Lateral axillary buds
(B) Apical buds
(C) Roots on stem cuttings
(D) Parthenocarpy

1029. Which is often present in several copies in a single promoter?
(A) TATA box (B) Pribnow box
(C) GC box (D) CAAT box

1030. Which one increases in the absence of light?
(A) Uptake of minerals (B) Uptake of water
(C) Elongation of internodes (D) Ascent of sap

1031. Bioassay of IAA is:
(A) Avena curvature test (B) Callus test
(C) Leaf disc test (D) α-amylase test

1032. Which is the colourless plastid?
(A) Mitochondria (B) Chloroplasts
(C) Leucoplasts (D) Chromoplasts

1033. The hormone capable of replacing the requirement of long photoperiods for flowering is:
(A) Ethylene (B) Auxin
(C) Gibberellin (D) Cytokinin

1034. Which is false about auxin?
(A) Promotes hyper elongation
(B) Promotes cell division
(C) Promotes stem elongation
(D) Inhibits lateral bud growth

1035. Which can replace the requirement of vernalization?
(A) Cytokinin (B) Ethylene
(C) Gibbrellins (D) Auxin

1036. Mowing grass lawn facilities better maintenance because:
(A) Wounding stimulates regeneration
(B) Removal of apical dominance and stimulation of intercalary meristem
(C) Removal of apical dominance
(D) Removal of apical dominance and promotion of lateral meristem

1037. The term carotenoid is related to:
(A) Carotene (B) Chlorophyll
(C) Xanthophyll (D) Both (A) and (C)

1038. Which of the following is used in root formation on stem cuttings?
(A) Kinetin (B) GA_3
(C) ABA (D) IBA

1039. Abscission layer is formed when the concentration of:
(A) Auxin increases (B) Auxin decreases
(C) Gibberellins decreases (D) Gibberellins increases

1040. Hormone primarily connected with cell division is:
(A) IAA (B) NAA
(C) Cytokinin/Zeatin (D) Gibberellic acid

1041. Physiologically active form of phytochrome is:
(A) P_{730} (B) P_{660} (C) P_{700} (D) P_{680}

1042. Which is the major pigment of chloroplasts?
(A) Chlorophyll (B) Xanthophyll
(C) Carotene (D) All of these

1043. Growth plotted against time gives a:
(A) Parabolic curve (B) Sigmoid curve
(C) Upright line (D) Horizontal line

1044. In plants, light is required for:
(A) Photosynthesis
(B) Controlling distribution of hormone
(C) Controlling growth and movement
(D) All of these

1045. Gibbrellins bring about:
(A) Yellowing of leaves
(B) Promotion of rooting
(C) Elongation of genetically dwarf plants
(D) Dwarfing of genetically tall plants

1046. Vernalization is:
(A) Growth curve related to light
(B) Effect of photoperiods on plant growth
(C) Speeding up ability to flower by low temperature treatment
(D) Diurnal photoperiodicity

1047. Phytochrome is involved in:
(A) Seed germination (B) Flowering
(C) Chloroplast orientation (D) All of these

1048. Flowering is influenced by:
(A) Soil water content (B) Soil acidity
(C) Photoperiod (D) Content of green pigment

1049. When seedling tip is cutoff, growth and light induced bending stop because it inhibits:
(A) Transpiration (B) Photorespiration
(C) Respiration (D) Perception of light stimulus

1050. Leaf fall occurs when content of:
(A) Abscisic acid decreases (B) Auxin decreases
(C) Auxin increases (D) Gibberellin decreases

1051. Gibbrellins promotes:
(A) Seed germination (B) Seed dormancy
(C) Leaf fall (D) Root elongation

1052. In case of prokaryotes, net gain of ATPs from complete oxidation of one molecule of glucose in respiration is:
(A) 40 (B) 38 (C) 36 (D) 30

1053. Apical dominance as found in higher plants is due to:
(A) Photoperiodism (B) Phototropism
(C) Hormones (D) Enzymes

1054. IAA precursor is:
(A) Tryptophan (B) Leucine
(C) Tyrosine (D) Phenylalanine

1055. Apical dominance is:
(A) Suppression of growth of apical bud by nearby lateral axillary buds
(B) Stimulation of growth of apical bud by removal of nearby axillary buds
(C) Suppression of growth of lateral axillary buds by removal of apical bud
(D) Suppression of growth of nearby lateral axillary buds by removal of apical bud

1056. Pigment taking part in photo-perception in flowering is:
(A) Carotene (B) Phytochrome
(C) Cytochrome (D) Lycopene

1057. Abscisic acid causes:
(A) Stomatal closure (B) Stem elongation
(C) Leaf expansion (D) Root elongation

1058. Abscisic acid controls:
(A) Cell division
(B) Leaf fall and dormancy
(C) Shoot elongation
(D) Cell elongation and wall formation

1059. Flowering dependent on low temperature exposure is:
(A) Vernalisation (B) Thermotrophy
(C) Cryoscopy (D) Cryostat

1060. Which is employed for artificial ripening of Banana fruits?
(A) Auxin (B) Coumarin
(C) Ethylene (D) Cytokinin

1061. Parthenocarpy can be achieved by:
(A) Zeatin (B) ABA
(C) Auxins (D) Kinetin

1062. Apical dominance is due to:
(A) Photoperiodism (B) Hormones
(C) Enzymes (D) Light demand

1063. Dwarfness can be controlled by treating the plant with:
(A) Cytokinin (B) Gibberellic acid
(C) Auxin (D) Antigibberellin

1064. Highest auxin concentration occurs:
(A) In growing tips
(B) In leaves
(C) At base of plant organs
(D) In xylem and phloem

1065. Cytokinins:
(A) Promote abscission
(B) Influence water movement
(C) Help retain chlorophyll
(D) Inhibit protoplasmic streaming

1066. Gibberellic acid induces flowering in:
(A) Some plants only
(B) Long day plants under short day conditions
(C) Short day plants under long day conditions
(D) Day neutral plants

1067. In short day plants, flowering is induced by:
(A) Photoperiod less than 12 hours
(B) Photoperiod below a critical length and unintruppted long night
(C) Long night
(D) Short photoperiod and interrupted long night

1068. Apical dominance is caused by:
(A) Abscisic acid in lateral bud
(B) Cytokinin in leaf tip
(C) Gibberellin in lateral buds
(D) Auxin in shoot tip

1069. How many cycles of Kreb's cycle is needed for complete oxidation of one glucose molecule?
(A) 1 (B) 3 (C) 2 (D) 4

1070. Which one is the test for gibbrellin?
(A) Bolting in cabbage
(B) Morphogenesis in tobacco callus
(C) Rapid division in carrot cells
(D) Elongation of oat coleoptile

1071. Which of the following hormone can replace vernalization?
(A) Auxin (B) Cytokinin
(C) Gibberellins (D Ethylene

1072. Cut or excised leaves remain green for long if induced to root or dipped in:
(A) Gibberellins (B) Cytokinins
(C) Auxins (D) Ethylene

1073. Bananas can be prevented from over ripining by:
(A) Maintaining them at room temperature
(B) Refrigeration
(C) Dipping in ascorbic acid solution
(D) Storing in a freezer

1074. Which is induced flowering in short day plants?
(A) Auxin (B) Cytokinin
(C) Gibbrellin (D) Propylene

1075. At the time of seed germination, digestive enzymes are induced to be formed under the influence of:
(A) Cytokinin (B) Gibberellins
(C) Ethylene (D) Auxin

1076. Foolish seedling disease/bakane of rice helped in discovery of harmone:
(A) Gibberellin (B) IAA
(C) Ethylene (D) Cytokinin

1077. Which one can induce flowering in long day plants?
(A) IBA (B) IAA
(C) GA_3 (D) NAA

1078. Photoperiodism is probably due to synthesis of:
(A) Cytokinins (B) Gibberellins
(C) Auxin (D) Florigen

1079. Gibberellin application to a plant brings about :
(A) Leaf fall
(B) Diversification of root system
(C) Elongation of shoot system
(D) Delay in senescence

1080. In root, the region of fastest growth is:
(A) Root hair zone (B) Behind the root tip
(C) Root tip (D) Before the root tip

1081. What is a stress hormone? or The hormone produced during adverse environmental conditions is:
(A) Benzyl aminopurine
(B) Dichlorophenoxy acetic acid
(C) Ethylene
(D) Abscisic acid

1082. Which organelles mainly synthesize amino acids in the cells?
(A) Mitochondria (B) Ribosomes
(C) Chloroplasts (D) Both (A) and (B)

1083. The regulator which retards ageing of plant parts is:
(A) Cytokinin (B) Auxin
(C) Gibberellin (D) Abscisic acid

1084. Richmond-Land effect is due to:
(A) Gibberellins (B) Auxin
(C) Ethylene (D) Kinetin

1085. Stomatal closure under stress conditions is due to:
(A) IAA (B) ABA
(C) IBA (D) 2, 4-D

1086. Phytochrome is associated with:
(A) P_r and P_{fr} reactions
(B) Absorption of red light of 660 nm
(C) Absorption of far-red light of 740 nm
(D) Absorption of blue light by leaves

1087. In tissue culture, differentiation of shoot and root is controlled by:
(A) Light intensity
(B) Temperature shocks
(C) Modifying auxin and cytokinin ratios
(D) Using tissue of right size

1088. Removal of apical buds results in:
(A) Formation of new apical buds
(B) Elongation of main stem
(C) Death of plant
(D) Formation of lateral branching

1089. Maximum of growth occurs in:
(A) Exponential phase (B) Decline phase
(C) Stationary phase (D) Lag phase

1090. Auxin synthesis occurs in :
(A) Root/shoot tips (B) Cortex
(C) Xylem (D) Phloem cells

1091. Phytotron is useful in:
(A) Protecting plants from excessive light
(B) Controlled plant growth
(C) Growth of drug plants
(D) Protection of plants

1092. Which one can absorb red and far red light?
(A) Carotene (B) Xanthophyll
(C) Chlorophyll (D) Phytochrome

1093. Which is correct about plastoquinones?
(A) They are special quinines present in plastids
(B) They carry two electrons from PS II towards PS I
(C) They transport H+ from stroma into th thylakoid lumen
(D) All of these

1094. Movement of auxin is:
(A) Centripetal (B) Basipetal
(C) Acropetal (D) Both (B) and (C)

1095. Growth is maximum in zone of:
(A) Cell elongation (B) Cell division
(C) Cell maturation (D) All of these

1096. Which does not take place in short day plants?
(A) An interrupted critical dark period
(B) Critical period is interrupted by light
(C) Dark period is interrupted by red light followed by far-red light
(D) Critical period is not interrupted by white or red light

1097. Which is not a phytohormone:
(A) Phytochrome (B) Florigen
(C) GA (D) IAA

1098. Florigen is produced in the region of:
(A) Leaves (B) Fruit
(C) Root (D) Trunk

1099. Genetically dwarf plant can be made to grow taller by application of:
(A) Ethylene (B) Gibberellin
(C) Auxin (D) Cytokinin

1100. Abscisic acid causes:
(A) Seed dormancy (B) Apical dominance
(C) Root initiation (D) Phototropism

1101. Induction in cell division and delay in senescence are due to:
(A) Gibberellin (B) Auxin
(C) Cytokinin (D) Ethylene

1102. 2, 4-D causes:
(A) Delay in senescence
(B) Bolting
(C) Increase in branches
(D) Increased growth of all parts

1103. Lateral axillary buds are not allowed to grow by:
(A) Auxin (B) Ethylene
(C) Gibberellin (D) Cytokinin

1104. Which blue green algae does not form heterocysts?
(A) *Anabaena* (B) *Nostac*
(C) *Gleotrichia* (D) *Scytonema*

1105. Which is not a natural plant hormone:
(A) GA_3 (B) GA_2
(C) IAA (D) 2, 4-D

1106. Cytokinins are mostly:
(A) Glucosides (B) Aminopurines
(C) Acidic (D) Phenolic

1107. Kinetin is:
(A) Indole butyric acid (B) Indole acetic acid
(C) Butyric acid (D) 6-furfuryl amino-purine

1108. One of the following is a growth retardant:
(A) Morphactin (B) Zeatin
(C) Adenine (D) Ascorbic acid

1109. Auxin takes part in:
(A) Promotion of cell division
(B) Promotion of cell enlargement
(C) Promotion of fruit ripening
(D) Inhibition of shoot growth

1110. Interfascicular cambium formation is induced by:
(A) Auxin (B) Cytokinin
(C) Gibberellin (D) Ethylene

1111. Hormone that has negative effect on apical dominance is:
(A) Cytokinin (B) Gibberellin
(C) Auxin (D) Both (B) and (C)

1112. Seasonality of plants is due to:
(A) Phototropism (B) Photosynthesis
(C) Photoperiodism (D) Photolysis

1113. Chemicals having profound effect on growth and development are:
(A) Enzymes (B) Phytohormones
(C) Catalytic agents (D) Manure

1114. Food products are kept in cold storage or refrigerator in order to:
(A) Make them tasty
(B) Use in off-season
(C) Maintenance of fressness, longevity and taste due to low respiration
(D) Keep them cool

1115. Which is not correct about starch?
(A) It is stored in photosynthetically inactive organs like roots, modified shoots and seeds
(B) The constituents of starch grains are amylase and amylopectin
(C) Both amylose and amylopectin are made up of D-glucose units
(D) Starch molecules have a definite molecular weight and are assembled into starch grains

1116. Food can be preserved by:
(A) Low temperature (B) High temperature
(C) Osmotic pressure (D) All of these

1117. Which one prevents photo-oxidation and pigment destruction?
(A) Phytochrome (B) Phytohormone
(C) Phycocyanin (D) Phyco-erythrin

1118. If a tree flowers thrice in a year in October, January and July in northern India, The plant is:
(A) Photo- and thermo sensitive
(B) Photo-and thermo insensitive
(C) Photo-sensitive and thermo-insensitive
(D) Thermo-sensitive but photo-insensitive

1119. Gibberellins induce:
(A) Cell division
(B) Leaf senescence
(C) Hydrolysing enzymes in germinating seeds
(D) Flowering

1120. Which is not caused by deficiency of mineral nutrition?
(A) Chlorosis (B) Internode shortening
(C) Necrosis (D) Etiolation

1121. In a plant subjected to continuous red light, phytochrome will show:
(A) Increased synthesis
(B) Decreased level
(C) Destruction
(D) Destruction and synthesis remain in balance

1122. The hormone which promotes flowering in long day conditions and controls sex expression is:
(A) Auxin (B) Cytokinins
(C) Gibberellins (D) Ethylene

1123. Properties of starch molecules are mainly determined by:
(A) Degree of branching (including branch length)
(B) Types of bonds linking the glucose units
(C) Chain length
(D) All of these

1124. Development of shoot and root is determined by:
(A) Cytokinin and auxin ratio (B) Enzymes
(C) Temperature (D) Plant nutrients

1125. Phytochrome absorbs visible light maximum in red region at:
(A) 600 nm (B) 650 nm
(C) 660 nm (D) 730 nm

1126. Hormone discovered through tissue culture technique is:
(A) Auxin (B) Cytokinin
(C) Gibberellin (D) Abscisic acid

1127. Sometimes potato sprouts during storage due to formation of:
(A) Ethylene (B) Gibberellin
(C) Cytoinin (D) Auxin

1128. Flash of light in dark inhibits flowering in:
(A) SDP (B) LDP
(C) LSDP (D) DNP

1129. Which one controls flowering in long day plants and sex expression in certain species?
(A) Antiauxins (B) Auxins
(B) Cytokinins (D) Gibberellins

1130. Agent orange is:
(A) Weedicide with dioxin
(B) Chemical used in luminous paint
(C) Biodegradable insecticide
(D) Colour used in fluorescent lamp

1131. Food is preserved at low temperature because:
(A) Bacterial attack is minimized
(B) For easy cooking
(C) For easy digestion
(D) All of these

1132. Photosynthetic rate is the highest in:
(A) C_3 plants (B) C_4 plants
(C) CAM (D) Both (B) and (C)

1133. Which one is used for defoliation of forest trees?
(A) Phosphon D (B) Maleic hydrazide
(C) AMO 1618 (D) 2, 4-D

1134. Which combination is used in fruit ripening?
(A) 80% CH_4 + 20% O_2 (B) 80% C_2H_4 + 20% CO_2
(C) 8% CO_2 + 20% CH_2 (D) 80% CH_4 + 20% CO_2

1135. Phytochrome is sensitive to:
(A) Blue light (B) Red light
(C) Red and far-red light (D) Green light

1136. Hormone produced during leaf fall is:
(A) Cytokinin (B) Florigen
(C) ABA (D) IAA

1137. Fruit drop is prevented by spraying:
(A) Auxin (B) Ethylene
(C) Gibberellins (D) Cytokinins

1138. Bakane disease of rice is due to:
(A) NAA (B) 2, 4-D
(C) IAA (D) GA

1139. Which of the following ion is pulled out in apoplast by auxin during growth?
(A) K^+ (B) H^+
(C) Na^+ (D) Mg^{2+}

1140. Substances that accumulate around a site of infection in plants is:
(A) Phytolexins (B) Proteinome
(C) Saponins (D) All of these

1141. Plant growth is:
(A) Limited (B) Unlimited
(C) Unlocalised (D) Diffused

1142. Photoperiodic stimulus is perceived by:
(A) Leaves (B) Flowers
(C) Buds (D) Meristem

1143. Kreb's cycle starts with entry of:
(A) Acetyl-CoA (B) Oxaloacetate
(C) Both (A) and (B) (D) Pyruvic acid

1144. Parthenocarpic fruits can be produced by application of auxin:
(A) IAA (B) IBA
(C) NAA (D) All of these

1145. Hormone found in liquid endosperm of coconut is:
(A) Gibberellin (B) Auxin
(C) Ethylene (D) Cytokinin

1146. From which was kinetin isolated:
(A) Coconut milk (B) Pineapple
(C) Soyabean (D) Groundnut

1147. Cardinal temperature is related to:
(A) Minimum temperature
(B) Optimum temperature
(C) Maximum temperature
(D) All of these

1148. High concentration of auxin is present in:
(A) Root apex (B) Shoot apex
(C) Petiole (D) Node

1149. Hormone responsible for plant and seed dormancy during drought is:
(A) IBA (B) NAA
(C) ABA (D) Zeatin

1150. Growth regulator ethylene is used for:
(A) Retarding ripening of tomatoes
(B) Speeding up ripening of fruits
(C) Slowing down ripening of apples
(D) Both (B) and (C)

1151. In short day plants, flowering is inhibited by:
(A) Interruption of dark by white and red light
(B) Dark interruption by far red light
(C) Dark interruption by red light followed by far red light
(D) Not possible

1152. First chemical identification of auxin, indole 3-acetic acid, was carried out by:
(A) Went (B) Thimann
(C) Darwin (D) Paal

1153. Plant hormone causing abscission of leaves, senescence and inhibition of cell division is:
(A) IAA (B) Ethylene
(C) Cytokinins (D) ABA

1154. During photosynthesis, oxygen is released from:
(A) CO_2 (B) H_2O
(C) SO_2 (D) O_3

1155. Reduction of nitrite to ammonium is catalyzed by nitrite reductase in:
(A) Chloroplasts (B) Mitochondria
(C) Nucleus (D) Golgi body

1156. In which electron transport leading to phosphorylation is a common reaction?
(A) Chloroplasts (B) Mitochondria
(C) Both (A) and (B) (D) Golgi bodies

1157. The alcohol dehydrogenase gene is induced by stress due to:
(A) Heat (B) Anaerobiosis
(C) Wound (D) Pathogene invasion

1158. Which is correct?
(A) Flower senescence is promoted by ethylene
(B) Ethylene biosynthesis involves two enzymes, viz, ACC synthase and ACC oxidase
(C) ACC synthase enzyme is encoded by acs gene and ACC oxidase is encoded by aco gene
(D) All of these

1159. The metabolic activities of roots cells are the main cause of water absorption in:
(A) Active absorption
(B) Passive absorption
(C) Both passive and active absorption
(D) None of these

1160. The movement of water is along:
(A) Turgor gradient (B) DPD gradient
(C) Diffusion gradient (D) Osmotic gradient

1161. The energy is required in absorption of water in:
(A) Active absorption
(B) Passive absorption
(C) Both passive and active absorption
(D) None of these

1162. The most dominant role, in the absorption of water, is that of:
(A) Active absorption
(B) Passive absorption
(C) Both passive and active absorption
(D) None of these

1163. No energy is required and roots play only the role of absorbing organs in:
(A) Active absorption
(B) Passive absorption
(C) Both passive and active absorption
(D) None of these

1164. Which is influenced by opening and closing of stomata:
(A) Active water absorption
(B) Passive water absorption
(C) Both types of water absorption
(D) Rate of growth

1165. Root pressure is maximum at the time of:
(A) Water absorption as well as transpiration are low
(B) Both water absorption and transpiration are high
(C) Absorption is low and transpiration is high
(D) Absorption is high and transpiration is low

1166. The glyoxylate cycle is found in plants and bacteria but not in animals. The lack of this cycle in animals results in the inability to:
(A) Synthesize oxaloacetate from isocitrate
(B) Synthesize glutamate from malate
(C) Perform gluconeogenesis from amino acids
(D) Perform gluconeogenesis from fatty acids

1167. Which of the following statements about ribulose 1, 5-bisphosphate carboxylase (Rubisco) is incorrect?
(A) It is present at low concentrations in the chloroplast
(B) It is activated by the addition of CO_2 to the å–amino group of a specific lysine to form a carbamate that then binds a divalent metal action
(C) It catalyzes, as one part of its reaction sequence, an extremely exergonic reaction, the cleavage of a six –carbon diol derivate of carabinitol to form 2,3-carbon compounds
(D) It catalyzes a reaction between ribulose 1, 5-bisphosphate and CO_2 that decreases the efficiency of photosynthesis

1168. The absorption of water is not affected when:
(A) Soil is water logged (poorly aerated)
(B) Soil temperature increases
(C) Soil solution is highly concentrated
(D) Soil water is available in maximum

1169. Guttation is loss of water:
(A) From root (B) Through stomata
(C) In liquid form (D) In vapour form

1170. For absorption of water, the root hairs act as:
(A) Sucking organs (B) Osmotic system
(C) Manometer (D) Anemometer

1171. Guttational drop comprises:
(A) Simple water
(B) Water and various salts dissolved in it
(C) Water and inorganic salts
(D) Water and organic salts

1172. A leafy shoot is enclosed air tightly in a flask. The flask becomes moist on the inner surface due to:
(A) Guttation (B) Evaporation
(C) Transpiration (D) Cooling effect

1173. Optimum temperature for water absorption is:
(A) 0^0-5^0C (B) 5^0-20^0C
(C) 20^0-35^0C (D) 40^0-50^0C

1174. Process of transpiration in plants helps in:
(A) Ascent of sap (B) Absorption of CO_2
(C) Opening of stomata (D) None of these

1175. At 0^0C soil temperature, the rate of water absorption:
(A) Increases (B) Not affected
(C) Decreases (D) Appreciable

1176. As absorbed water passes towards vascular cylinder, it must enter the cytoplasm of:
(A) Pericycle cells (B) Endodermal cells
(C) Cortical cells (D) Xylem parenchyma

1177. Which one of the following statement correctly describes the role of vernalisation?
(A) It reduces apical dominance and leads multiple branching
(B) It causes genetic modification that leads to early flowering
(C) It delays flowering and allows maximum vegetative growth leading to higher yield
(D) It causes epigenetic modification that promotes early flowering

1178. A living continuum of cells connected by plasmodesmata is:
(A) Dermal tissue (B) Ground complex
(C) Donnan free space (D) Symplast

1179. Q_{10} is:
(A) Temperature Quotient (B) Respiratory quotient
(C) Compensation point (D) Time factor

1180. Water is absorbed by root hair when the external medium is:
(A) Hypotonic (B) Hypertonic
(C) Isotonic (D) Viscous

1181. Transpiration is least in:
(A) Good soil moisture
(B) High wind velocity
(C) Dry environment
(D) High atmospheric humidity

1182. In light phase of photosynthesis there is formation of:
(A) ATP
(B) $NADPH_2$
(C) Both ATP and $NADPH_2$
(D) Carbohydrates

1183. Absorption of water by root is increased with the:
(A) Increase in transpiration
(B) Increase in rate of photosynthesis
(C) Decrease in transpiration
(D) Decrease in salt uptake

1184. Cryptochromes:
(A) A class of blue light photoreceptors
(B) Occurs in both plants and animals
(C) Belongs to a family of flavoproteins
(D) All of these

1185. Contribution of passive water absorption to total water absorption is:
(A) 50% (B) 70% (C) 80-90% (D) 96-100%

1186. Rate of water absorption generally follows closely the rate of:
(A) Transpiration (B) Photosynthesis
(C) Respiration (D) Growth

1187. Presently which view is considered best for turgor changes in guard cells:
(A) Photosynthesis of chloroplasts in guard cells
(B) Starch is converted into sugar in guard cells
(C) Starch is converted into glucose in guard cells
(D) Potassium is actively transported into guard cells

1188. Transpiration is reduced due to deposition of:
(A) Cutin (B) Pectin
(C) Mucilage (D) Cellulose

1189. Emersion effect indicates the existence of:
(A) Two pigment systems
(B) Two photosynthetic units
(C) Two photophosphorylations
(D) None of these

1190. The phenomenon related to active water absorption is:
(A) Transpiration (B) Root pressure
(C) Osmotic pressure (D) Translocation

1191. In succulent plants the stomata open in night and close by day. Which among following would be best hypothesis to explain the mechanism of stomatal action in night only:
(A) CO_2 accumulates, reduces pH, stimulates enzymes, resulting in accumulation of sugars.
(B) Increase in CO_2 concentration, conversion of organic acids into starch, resulting in the increased conversion into sugars, resulting in K^+ transport
(C) Low CO_2 concentration due to utilization, accumulation of organic acids resulting in the increased concentration of cell sap
(D) CO_2 used up, increases pH, results in accumulation of sugars

1192. Root pressure can be demonstrated by means of:
(A) Wilting (B) Guttation
(C) Transpiration (D) Exudation/bleeding

1193. The stomatal type of cereals which open only for a few hours during the day is:
(A) Barley type (B) Potato type
(C) Alfalfa type (D) Bean type

1194. Transpiring structure in plants is:
(A) Epidermis (B) Cortex
(C) Xylem (D) Phloem

1195. Dry weight of leaf is maximum during:
(A) Morning (B) Noon
(C) Afternoon (D) Night

1196. A soil is physiologically dry if it contains:
(A) No hygroscopic water
(B) No bound water
(C) No gravitational water
(D) High concentration of soil solution

1197. A water-logged soil is physiologically dry because of:
(A) Anaerobic conditions
(B) Nonmovement of water capillaries
(C) Increased viscosity of water
(D) Abundance of salts

1198. At field capacity the soil contains:
(A) Capillary and gravitational water
(B) Capillary and runaway water
(C) Capillary and hygroscopic water
(D) Capillary, hygroscopic and bound water

1199. Transpiration is high in:
(A) Rainy season/high humidity
(B) Winter season
(C) High temperature
(D) Low wind velocity

1200. Photosynthesis proceeds in sequence of:
(A) Dark phase and light phase
(B) Light phase alone
(C) Light phase and dark phase
(D) Dark phase alone

1201. The phenomenon which forces water upwards into tracheary elements of xylem in the root region is:
(A) Transpiration pull (B) Root pressure
(C) Turgor pressure (D) Imbibition pressure

1202. Root hairs absorb water when:
(A) They respire rapidly
(B) Soil solution is isotonic
(C) Salt concentration of cell sap is high
(D) Salt concentration of soil is high

1203. Transplanted seedlings often get killed because of:
(A) Damage to root hairs
(B) Damage to leaves
(C) Change in soil microorganisms
(D) Nonavailability of required minerals

1204. For process of photosynthesis all except one of the following items are essential point out the exception:
(A) Water, minerals
(B) Light, chlorophyll
(C) CO_2, Optimum temperature
(D) O_2, sucrose

1205. Rate of transpiration is inversely related to:
(A) Humidity (B) Light
(C) Temperature (D) Water

1206. Which is produced during water stress that brings about stomatal closure:
(A) Ethylene (B) Abscisic acid
(C) Ferulic acid (D) Coumarin

1207. Wilting appears due to excessive:
(A) Respiration (B) Photosynthesis
(C) Absorption (D) Transpiration

1208. Stomata open and close due to:
(A) Circadian rhythm
(B) Genetic clock
(C) Pressure of gases inside the leaves
(D) Turgor pressure of guard cells

1209. Which colour of light gives maximum absorption peak of chlorophyll a:
(A) Red (B) Blue
(C) Orange (D) Yellow

1210. PS II performs:
(A) Reduction of CO_2 (B) Photolysis of water
(C) Liberation of energy (D) Formation of water

1211. Stomata open when guard cells show:
(A) Increase in both osmotic and turgor pressure
(B) Decrease in both osmotic pressure and turgor pressure
(C) Increase in osmotic pressure but decrease in turgor pressure
(D) Decrease in osmotic pressure but increase in turgor pressure

1212. In guttation plants can:
(A) Excrete salts
(B) Compensate for loss of water
(C) Manufacture organic substances
(D) Get rid of excess water

1213. Light phase consists of:
(A) Two photochemical reactions
(B) Two photosynthetic units
(C) Two photorespiration reaction
(D) None of these

1214. An important factor in stomatal opening is:
(A) Hormone content of cells
(B) Protein content of cells
(C) Chlorophyll content of cells
(D) Shape of guard cells

1215. Which of the following shows guttation?
(A) *Pisum sativum* (B) *Ficus benghalensis*
(C) *Acacia nilotica* (D) *Tropaeolum*

1216. Scotoactive movement of stomata is that:
(A) Stomata open at night
(B) Stomata open during day
(C) Stomata close at night
(D) Stomata open both during day and night

1217. Which of the following does not affect light phase:
(A) Chlorophyll (B) Water
(C) Cytochrome (D) Temperature

1218. The most effective light for stomatal opening is:
(A) Yellow (B) Green
(C) Red (D) Blue

1219. The carbon di acceptor in CAM plants is:
(A) Malic acid (B) Oxalo-acetic acid
(C) Pyruvic acid (D) Phosphoenol pyruvic acid

1220. During photosynthesis the oxygen in glucose comes from:
(A) H_2O (B) CO_2
(C) Both from H_2O and CO_2 (D) Oxygen in air

1221. Kranz anatomy occurs in:
(A) Leaves (B) Stem
(C) Flower (D) Seed

1222. Guttation is mainly due to:
(A) Root pressure (B) Osmosis
(C) Transpiration (D) Imbibition

1223. A specialised multicellular structure in leaves which excretes water droplets is:
(A) Lenticel (B) Stomata
(C) Hydathode (D) Bordered pit

1224. Who discovered purification of air in light by green plants:
(A) Ingenhouse (B) Priestley
(C) De Saussure (D) Mayer

1225. Phenyl mercuric acetate (PMA) results in:
(A) Reduced photosynthesis
(B) Reduced transpiration
(C) Reduced respiration
(D) Killing of plants

1226. Red drop occurs in wavelength of:
(A) 492 nm (B) 535 nm
(C) 586 nm (D) 680 nm

1227. An internal factor in transpiration is:
(A) CO_2 (B) O_2
(C) N_2 (D) Stomata

1228. The two pigment system theory of photosynthesis was proposed by:
(A) Hill (B) Blackman
(C) Emerson (D) Arnon

1229. How much energy is utilized in the synthesis of one molecule of glucose:
(A) 673 kcal (B) 686 kcal
(C) 666 kcal (D) 696 kcal

1230. The energy change in photosynthesis is from:
(A) Light energy to electric energy
(B) Light energy to molecular energy
(C) Light energy to chemical energy
(D) Light energy to activation energy

1231. Action of spectrum of photosynthesis was first studied by:
(A) Blackmann (B) Van Mayer
(C) Engelmann (D) Boussingault

1232. Plants with high light compensation point are:
(A) C_3 plants (B) Sun plants
(C) Shade plants (D) Mesophytes

1233. C_4 plants are adapted to:
(A) Temperate humid climate
(B) Tropical humid climate
(C) Temperate dry climate
(D) Tropical dry climate

1234. Cyclic photophosphorylation produces:
(A) NADPH (B) ATP and NADPH
(C) ATP, NADPH and O_2 (D) ATP only

1235. Photosynthate is translocated through:
(A) Sieve tubes/Phloem cells (B) Trachids
(C) Xylem vessels (D) Latex ducts

1236. In C_4 plants, synthesis of sugars occurs in:
(A) Mesophyll cells
(B) Bundle sheath cells
(C) Spongy parenchyma cells
(D) Palisade parenchyma cells

1237. A photosynthesizing plant is releasing ^{18}O more than the normal. The plant must have been supplied with:
(A) O_2 (B) H_2O with ^{18}O
(C) CO_2 with ^{15}O (D) $C_6H_{12}O_6$ with ^{15}O

1238. Chemistry of photosynthetic pigments was first studied by:
(A) Willstatter and Stoll (B) Pelletier and Caventou
(C) Priestley (D) Engelmann

1239. Which one is a C_4 plant?
(A) Papaya (B) Pea
(C) Potato (D) Maize

1240. During day time the green plants show an ratio of CO_2 consumption to CO_2 production of:
(A) 1 : 1 (B) 5 : 1
(C) 10 : 1 (D) 20 : 1

1241. In which wave length of light photosynthesis is maximum:
(A) Red light (B) Green light
(C) Blue light (D) Ultra-violet light

1242. In C_4 plants Calvin cycle enzymes are:
(A) Mesophyll chloroplasts
(B) Bundle sheath chloroplasts
(C) Guard cell chloroplasts
(D) Epidermal chloroplasts

1243. Which one of the following pigments does not occur in chloroplast:
(A) Chlorophyll (B) Carotene
(C) Anthocyanin (D) Xanthophyll

1244. Chlorophyll absorbs:
(A) Green light (B) Red light only
(C) Blue light only (D) Blue as well as red light

1245. The number of chlorophyll molecules in a quantasome is:
(A) 50-100 (B) 200-250
(C) 300-400 (D) 500-600

1246. The law of limiting factors for photosynthesis was enunciated by:
(A) R-Hill (B) Calvin
(C) Krebs (D) Blackman

1247. Carbon cycle occurs in:
(A) Chloroplasts (B) Cytoplasm
(C) Mitochondria (D) Glyoxysomes

1248. In bacterial photosynthesis, the hydrogen donor is:
(A) H_2O (B) H_2SO_4 (C) NH_2 (D) H_2S

1249. In photosynthesis:
(A) ATP is formed in light reaction and glucose in dark reaction
(B) Both ATP and glucose are produced in dark reaction
(C) Both ATP and glucose are produced in light reaction
(D) Both ATP and glucose are formed in light and dark

1250. C_4 cycle was discovered by:
(A) Hatch and Slack (B) Calvin
(C) Hill (D) Arnone

1251. During photosynthesis when PGA is changed into phosphogylceraldehyde, which of the following reactions occurs?
(A) Oxidation (B) Reduction
(C) Electrolysis (D) Hydrolysis

1252. During day-light hours the rate of photosynthesis is higher than that is respiration and the ratio of O_2 produced to that of consumed is:
(A) 1: 1 (B) 10 : 1
(C) 50 : 1 (D) 5 : 1

1253. Which one is most efficient converter of sunlight?
(A) Sugarcane (B) Rice
(C) Wheat (D) Papaya

1254. Conversion of ATP to ADP produces:
(A) Electricity (B) Energy
(C) Hormone (D) Enzyme

1255. Hydrilla is used for demonstrating photosynthesis because it shows:
(A) Little respiration
(B) Little transpiration
(C) Rapid photosynthesis
(D) Evolution of O_2 bubbles which can be collected once water

1256. Dark reaction of photosynthesis occurs in:
(A) Grana (B) Stroma
(C) Matrix (D) Cytoplasm

1257. The loss of which will harm the tree most:
(A) Bark (B) Half the leaves
(C) Half of branches (D) All the leaves

1258. Mass flow hypothesis explains:
(A) Transpiration (B) Translation
(C) Transcription (D) Translocation

1259. Synthesis of ADP + iP→ATP in grana/photosynthesis is:
(A) Phosphorylation
(B) Photophosphorylation
(C) Oxidative phosphorylation
(D) Photolysis

1260. The number of light quanta required for evolution of one molecule of oxygen is called:
(A) Oxygen yield (B) Photosynthetic yield
(C) Quantum yield (D) Organic yield

1261. The photosynthetic unit having 250 chlorophyll molecules is called:
(A) Photon (B) Quantum
(C) Quantasome (D) Oxysomes

1262. Rate of transpiration is dependent upon:
(A) Negative turgor pressure
(B) Temperature
(C) O.P.
(D) Vapour pressure deficit

1263. Plants exhibit subepidermal evaporation of water during:
(A) Photosynthesis (B) Guttation
(C) Transpiration (D) Respiration

1264. Photophosphorylation is synthesis of:
(A) ADP from ATP
(B) Glucose 6-phosphate from glucose
(C) ATP from ADP
(D) $NADP^+$ from NAD^+

1265. Plant growing best in full sunlight are known as:
(A) Sciophytes (B) Therophytes
(C) Heliophytes (D) None of these

1266. Plant growing in sandy soil are called:
(A) Halophytes (B) Psammotphytes
(C) Lithophytes (D) None of these

1267. Generally growth of plants is best supported by:
(A) Sandy soil (B) Loam soil
(C) Silt soil (D) Clay soil

1268. Incoming solar radiation is known as:
(A) Insolation (B) Terrestrial radiation
(C) Convection (D) None of these

1269. Starch synthesis in leaves:
(A) Increases with a build up of phosphoglyceraldehyde in the stroma
(B) Is promoted in the light
(C) Occurs in the cytosol
(D) None of these

1270. C_4 plants are mostly abundant in:
(A) Tropical region with more dry condition
(B) Tropical region with more humid condition
(C) Temperate region with more humid condition
(D) Temperate region with dry condition

1271. The immediate precursor of chlorophyll a is:
(A) Protochlorophyll (B) Protoporphyrin IX
(C) Chlorophyllide a (D) None of these

1272. Carotine, the first caronoid to be named, was isolated from .... tissues by Wackenroder in 1831:
(A) Carrot root (B) Carrot leaf
(C) Carrot shoot (D) None of these

1273. The quantum yield in photosynthesis is:
(A) 8% (B) 12% (C) 25% (D) 50%

1274. During day light hours, the rate of photosynthesis is higher than that of respiration, and the ratio of oxygen produced to that of consumed is:
(A) 10 : 1 (B) 5 : 1 (C) 1 : 1 (D) 50 : 1

1275. Which of the following element is essential for photolysis of water?
(A) Chlorine (B) Nitrogen
(C) Oxygen (D) None of these

1276. Phosphorylation in a chloroplast is most similar to which of the under mentioned mitochondrial reactions?
(A) Substrate level phosphorylation
(B) Oxidative phosphorylation
(C) Oxidative decarboxylation
(D) None of these

1277. RuBP carboxylase was crystallized in 1971 from :
(A) Maize leaves (B) Rice leaves
(C) Chlorella (D) Tobacco leaves

1278. The research on photosynthesis began:
(A) In the 17th century (B) In the 18th century
(C) In the 19th century (D) In the early 20th century

1279. The bulk fixation of carbon through photosynthesis takes place in:
(A) Crop plants
(B) Tropical rain forest and crop plants
(C) Tropical rain forest
(D) Ocean

1280. The ultimate biological energy comes from:
(A) Sunlight (B) Mitochondria
(C) Glucose (D) ATP

1281. The percentage of light energy fixed in photosynthesis is generally around:
(A) 1 (B) 3 (C) 10 (D) 30

1282. The unit of photosynthesis is:
(A) Quantosome (B) Angstrom
(C) Nanometer (D) None of these

1283. Quantasome contains:
(A) 300 chlorophyll molecules
(B) 250 chlorophyll molecules
(C) 230 chlorophyll molecules
(D) 200 chlorophyll molecules

1284. In the process of photosynthesis, chlorophyll functions as:
(A) A hydrogen acceptor (B) A raw material
(C) An end-product (D) An energy converter

1285. Photo – oxidation of water in photosynthesis is in association of:
(A) Pigment system I (B) Pigment system II
(C) Plastocyanin (D) Cytochrome B6

1286. Rate of photosynthesis is independent of:
(A) Temperature (B) Quality of light
(C) Intensity of light (D) Duration of light

1287. Photosynthesis is an oxidation reduction process, the material that is oxidized is:
(A) CO_2 (B) NADP
(C) PGA (D) H_2O

1288. Plastocyanin is .... soluble copper protein:
(A) Water (B) lipid
(C) Both (A) and (B) (D) None of these

1289. Plastoquinone is .... soluble:
(A) Water (B) lipid
(C) Both (A) and (B) (D) None of these

1290. DCMU is an inhibitor of:
(A) PS II (B) PS I
(C) Both (A) and (B) (D) None of these

1291. DSMIB is an inhibitor of the:
(A) PS I (B) Cyt b6f complex
(C) Qa (D) Qb

1292. In photosynthesis, assimilatory power is generated during:
(A) Light reaction (B) Dark reaction
(C) Photolysis of water (D) None of these

1293. In photosynthesis, ultimate source of electron is:
(A) Water (B) Sunlight
(C) Phaeophytin (D) None of these

1294. The photosynthetic process is a complicated:
(A) Hydrogenetion process
(B) Oxidation-reduction process
(C) Only reduction process
(D) None of these

1295. Which of the following is not a component of electron transport chain in chloroplast?
(A) Water soluble Plastocyanin
(B) Lipid soluble Plastoquinone
(C) Ubiquinone
(D) None of these

1296. Which of the following does not form one electron redox system?
(A) Plastocyanin (B) Plastoquinone
(C) Cytochrome (D) None of these

1297. Ferredoxin (Fd) is a:
(A) Heme-iron protein
(B) Z-scheme
(C) Copper containing protein
(D) None of these

1298. First stable product of the dark reaction of photosynthesis is:
(A) 3PGALD (B) 3 PGA
(C) 1, 3-bis PGA (D) None of these

1299. How many enzymes are required in Calvin cycle?
(A) 12 (B) 13 (C) 14 (D) 15

1300. How many enzymes are involved in regeneration in calvin cycle?
(A) 10 (B) 11 (C) 12 (D) 13

1301. Photosynthetic green sulfur bacteria fix carbon via:
(A) Reductive carboxylic acid cycle
(B) Hydroxyl propionate Pathway
(C) Calvin cycle
(D) None of these

1302. Green non bacterium Chloroflexus fix carbon via:
(A) Hydroxy propionate pathway
(B) C_4 – cycle
(C) C_3 –cycle
(D) None of these

1303. Phycobillins are......... in water:
(A) Soluble (B) Insoluble
(C) Both (A) and (B) (D) None of these

1304. .... are also known as carotenols:
(A) Xanthophylls (B) Carotines
(C) Both (A) and (B) (D) None of these

1305. Chlorophyll a is distributed in all plant kingdom except:
(A) Bacteria (B) Algae
(C) Both (A) and (B) (D) None of these

1306. Chlorophyll b is distributed in:
(A) Higher plants (B) Green algae
(C) Both (A) and (B) (D) None of these

1307. Chlorophyll c is distributed in:
(A) Diatoms (B) Dinoflagellates
(C) Brown algae (D) All of these

1308. Chlorophyll e is distributed in:
(A) Tribonema (B) Zoospores of Vaucheria
(C) Both (A) and (B) (D) None of these

1309. Bacteriochlorophyll a is distributed in:
(A) Purple bacteria (B) Green bacteria
(C) Both (A) and (B) (D) None of these

1310. Bacteriochlorophyll g is distributed in:
(A) Heliobacteria (B) Lipophylic
(C) Both (A) and (B) (D) None of these

1311. Phytol tail of chlorophyll molecule is:
(A) Lipophilic (B) Hydrophilic
(C) Both (A) and (B) (D) None of these

1312. Which one is the most efficient converter of sunlight is:
(A) Rice (B) Wheat
(C) Sugarcane (D) None of these

1313. The stomatal pores open when:
(A) H+ ions are pumped into the guard cells
(B) K+ ions are pumped out of the guard cells
(C) Water moves into the guard cells in response to a decrease in water potential in the guard cells
(D) Water moves out of the guard cells in response to a decrease water potential of the guard cells

1314. Root pressure is due to:
(A) Increase in transpiration
(B) Active absorption
(C) Low osmotic potential in soil
(D) Passive absorption

1315. Decrease rate of photosynthesis at high concentration of oxygen is referred to as:
(A) Pasture effects (B) Emerson effects
(C) Warburg effects (D) None of these

1316. The rate limiting step in photosynthesis is:
(A) Light reaction (B) Dark reaction
(C) Oxygen evolution (D) CO_2

1317. In green sulfur bacteria, the reductant is:
(A) H_2O (B) H_2S
(C) Lactate (D) None of these

1318. In organisms such as plants, algae and cynobacteria, the reductant is:
(A) H_2O (B) H_2S
(C) Lactate (D) None of these

1319. In C_4 plants, phosphoenol pyruvate carboxylase is located in:
(A) Chloroplast (B) Peroxisome
(C) Mitochondria (D) Cytosol

1320. Mineral salts are absorbed from the soil solution in the form of:
(A) Cations ions (B) Anion
(C) Both (A) and (B) (D) None of these

1321. Which part of the root is involved in absorption in absorption of mineral salts in higher plants?
(A) Root cap (B) Meristematic region
(C) Zone of root hair (D) None of these

1322. First step in the absorption of mineral salts is the process of:
(A) Ion – exchange (B) Diffusion
(C) Imbibition (D) None of these

1323. Ion – exchange does not require metabolic energy but it generally facilitates mineral salt absorption. This statement is:
(A) True (B) False
(C) Unknown (D) None of these

1324. There is requirement of metabolic energy in:
(A) Active absorption of mineral salts
(B) Passive absorption of mineral salts
(C) Contact exchange of ions
(D) None of these

1325. When a plant is transferred from water to a salt solution, the rate of respiration increases. This is called:
(A) Ions respiration (B) Salt respiration
(C) Activated respiration (D) None of these

1326. Cytochrome pump theory for absorption of mineral salts was first proposed by:
(A) Lundergarth & Burstrom (B) Bennet-Clark
(C) Donnan (D) All of these

1327. Protein-lecithin theory for absorption of mineral salts was first proposed by:
(A) Lundergarth & Burstrom (B) Bennet-Clark
(C) Donnan (D) All of these

1328. According to Carrier Concept, the plasma membrane is .... to free ions:
(A) Impermeable (B) Permeable
(C) Both (A) and (B) (D) None of these

1329. Which of the following observation strongly supports the carrier concept of active absorption of mineral salts?
(A) Isotopic exchange (B) Saturation effect
(C) Specificity (D) All of these

1330. Protein associated with phosphatide is called:
(A) Lecithin (B) Proline
(C) Glycine betaine (D) None of these

1331. Which of the following enzyme has no role in Bennet-Clark's protein lecithin theory?
(A) Lecithinase (B) Choline esterase
(C) Choline acetylase (D) None of these

1332. Facilitated diffusion of solute through transport proteins across the membrane is:
(A) Active transport (B) Passive transport
(C) Both (A) and (B) (D) None of these

1333. The mineral salts in their ionic form move one cell to another by:
(A) Apoplastic pathways
(B) Trans membrane pathways
(C) Symplastic pathways
(D) All of these

1334. Anion channels are always:
(A) Outward rectifying (B) Inward rectifying
(C) Both (A) and (B) (D) None of these

1335. Ca^{2+} channels are always:
(A) Outward rectifying (B) Inward rectifying
(C) Both (A) and (B) (D) None of these

1336. K^+ channels are always:
(A) Outward rectifying (B) Inward rectifying
(C) Both (A) and (B) (D) None of these

1337. Among the following is an example of electroneutral pump?
(A) H+ - PPase (B) H+ - ATPase
(C) Ca^{2+} - ATPase (D) H^+/K^+ - ATPase

1338. Among the following, which ion can diffuse both outward and inward through channels across the plasmamembrane of plants?
(A) K+ (B) Ca^{2+}
(C) Both (A) and (B) (D) None of these

1339. In primary active transport, the required energy derived mainly from:
(A) Hydrolysis of GTP (B) Hydrolysis of ATP
(C) Both (A) and (B) (D) None of these

1340. In plant cell membrane the direction of electrogenic pump such as H+ - ATPase is:
(A) Inward (B) Outward
(C) Both (A) and (B) (D) None of these

1341. H^+ - ATPase pumps are also known as:
(A) P-type ATPase (B) L-type ATPase
(C) Q-type ATPase (D) None of these

1342. H^+ - ATPase of mitochondria and chloroplast are also known as:
(A) P-type ATPase (B) L-type ATPase
(C) F-type ATPase (D) None of these

1343. H^+-ATPase of plasma membrane and tonoplast are:
(A) Different (B) Same
(C) Not known (D) None of these

1344. H^+-ATPase of plasma membrane and tonoplast are inhibited by:
(A) Vabdate ions (B) Nitrate ion
(C) Cd ion (D) None of these

1345. H^+- PPase mainly found in:
(A) Tonoplast
(B) Plasmamembrane
(C) Chloroplast membrane
(D) None of these

1346. Secondary active transport is also known as:
(A) Co-transport mechanism
(B) Primary mechanism
(C) Both (A) and (B)
(D) None of these

1347. Photostationary equilibrium can be measured by:
(A) Spectrophotometer
(B) Difference spectroscopy
(C) Photoreversible equilibrium
(D) None of these

1348. Number of photons absorbed per unit surface area is called:
(A) Far-red treatment (B) White light treatment
(C) Fluence (D) None of these

1349. Based on the amount of light required or the fluence, the phytochrome mediated photoresponse can be grouped into how many categories?
(A) One (B) Two
(C) Three (D) Four

1350. Phytochrome was initially isolated from .... of etiolated turnip seedlings:
(A) Leaves (B) Cotyledons
(C) Shoots (D) Roots

1351. The carotenoid .... is blue light photoreceptor involved in stomatal opening:
(A) Riboflavin (B) Zeaxanthin
(C) FAD (D) None of these

1352. .... is blue light receptor that mediated chloroplast movement in plants:
(A) Zeaxanthin (B) Phototropin
(C) Chromophore (D) None of these

1353. The blue light stimulated stomatal opening can be reversed by:
(A) Red light (B) Green light
(C) Yellow light (D) None of these

1354. Photomorphogenetic response in plants are:
(A) Dependent on photosynthesis
(B) Independent of photosynthesis
(C) Found only in higher plants
(D) Found only in lower plants

1355. Which of the following is the most effective in inducing photomorphogenetic responses in plants?
(A) Blue and yellow light
(B) Blue and red light
(C) Orange and red light
(D) Red and far-red light

1356. Phytochrome is a:
(A) Flavoprotein (B) Glycoprotein
(C) Lipoprotein (D) Chromoprotein

1357. Which of the following is not a phytochrome mediated photoresponse?
(A) Germination of lettuce seeds
(B) De-etiolation of seedlings
(C) Phototropism
(D) Rotation of chloroplast in alga Mougeotia

1358. The effect of canopy shading can be described in terms of the ratio of:
(A) Red to far-red (B) Far-red to red
(C) Both (A) and (B) (D) None of these

1359. The movement of antherozoids towards the archegonium (egg) due to the secretion of some chemical substances is known as:
(A) Chemoreception (B) Chemotropism
(C) Chemotaxis (D) None of these

1360. Movement of hairs in *Drosera* is:
(A) Seismonastic (B) Thermonastic
(C) Photonastic (D) Thigmonastic

1361. Induce movement is also termed as:
(A) Spontaneous movement
(B) Paratonic movement
(C) Both (A) and (B)
(D) None of these

1362. Which of the following is not vital movement?
(A) Movement of locomotion
(B) Movement of curvature
(C) Hygroscopic movement
(D) All of these

1363. Movement of locomotion which occur in response to external stimulus are called:
(A) Tactic movements (B) Tropic movement
(C) Amoeboid movement (D) None of these

1364. In plants, the movement of curvature which occurs in response to external stimulus are called as:
(A) Paratonic (B) Autonomous
(C) Both (A) and (B) (D) None of these

1365. Haptonastic movements are found in the leaves of:
(A) *Dionaea and Drosera* (B) *Oxalis*
(C) *Hydrilla* (D) None of these

1366. Thermostatic movement is exhibited by;
(A) *Chlamydomonas* (B) *Desmodium gyrans*
(C) *Crocus* (D) None of these

1367. A movement in plant stimulated by oxygen is called:
(A) Thermotaxis (B) Sensitivity
(C) Nutation (D) Aerotaxis

1368. Tropic movements are basically due to:
(A) Cell elongation (B) Cell division
(C) Sudden water gain (D) Sudden water loss

1369. The minimum quantify of a stimulus required for bringing about a visible responses are called:
(A) Threshold value (B) Sensitivity
(C) Perceptibility (D) Summation

1370. The effects of stimulus subsides after a time called:
(A) Reaction time (B) Relaxation time
(C) Presentation time (D) Conduction time

1371. Time period from perception of stimulus and reaction is called:
(A) Restitution time (B) Presentation time
(C) Relaxation time (D) Reaction time

1372. The closure of lid of pitcher in pitcher plant is:
(A) A paratonic movement (B) A tropic movement
(C) A turgor movement (D) None of these

1373. Opening and closing of flowers represents a kind of:
(A) Autonomic movement (B) Nutation
(C) Tropic movement (D) Nastic movement

1374. Auxin –A is chemically:
(A) Auxentriolic acid (B) Auxenolonic acid
(C) Both (A) and (B) (D) None of these

1375. Auxin –B is chemically:
(A) Auxentriolic acid (B) Auxenolonic acid
(C) Both (A) and (B) (D) None of these

1376. Lateral branching of plant is controlled by the relative proportion of:
(A) Cytokinins and Auxins (B) ABA and Auxins
(C) Auxins and Ethylene (D) None of these

1377. Which one of the following is considered as true natural auxin in higher plants?
(A) Indole – 3 – acetic acid
(B) Indole – 3 – acetaldehyde
(C) Indole – 3 – pyruvic acid
(D) Indole – 3 – acetonitrile

1378. The primary physiological effect of auxin in plants is:
(A) Elongation of internodes (B) Cell division
(C) Elongation of cells in shoots (D) None of these

1379. Phytohormones are chemically:
(A) Amino (B) Purine
(C) Steroidal (D) Gaseous
(E) All of these

1380. Thimann isolated auxin from:
(A) Rhizopus (B) Mucor
(C) Penicillium (D) None of these

1381. The main site of auxin synthesis is:
(A) Roots (B) Meristem region of plant
(C) Shoots (D) None of these

1382. The most commonly used growth regulator to induce somatic embryogenesis is:
(A) BAP (B) IAA
(C) GA_3 (D) 2, 4-D

1383. Conjugated or bond auxin formed complex with:
(A) Carbohydrate (B) Protein or amino acid
(C) Inositol (D) All of these

1384. Which of the following amino acid is considered as precursor of auxin?
(A) Tryptophan (B) Tyrosine
(C) Proline (D) None of these

1385. Transport of auxin in plant is predominantly:
(A) Polar (B) Apolar
(C) Lateral (D) None of these

1386. Which of the following is an antiauxin?
(A) 2, 4-D (B) TIBA
(C) 2, 4-T (D) All of these

1387. Destruction of auxin in plants is mainly by:
(A) Conversion into bound auxin
(B) irradiance with x-rays and y-rays
(C) Oxidation of auxin
(D) None of these

1388. Auxin increases the .... of cell walls:
(A) Plasticity (B) Thickness
(C) Porosity (D) Rigidity

1389. Minimum lag time required for auxin induced growth is:
(A) 10 minutes (B) 15 minutes
(C) 20 minutes (D) 30 minutes

1390. The freeform of auxin are biologically .... forms:
(A) Active (B) Inactive
(C) Both (A) and (B) (D) None of these

1391. Gibberellins are derivatives of:
(A) Monoterpens (B) Sesquiterpenes
(C) Diterpens (D) Triterpenes

1392. Sprouting of potato is prevented by use of:
(A) Gibberellins (B) Auxin
(C) Cytokinin (D) None of these

1393. Auxin induces parthenocarpy in:
(A) Tomato (B) Banana
(C) Both (A) and (B) (D) None of these

1394. Auxin induces number of female flower in:
(A) Cereals (B) Cucurbit
(C) Pulses (D) None of these

1395. The primary precursor for formation of isoprenoid unit and synthesis of gibberellins is:
(A) Pyruvate (B) Acetate
(C) Oxalo acetate (D) None of these

1396. Antigibberellins such as cycocel (CCC) finds extensive use in:
(A) Food industry (B) Breweries
(C) Floriculture industry (D) None of these

1397. Which of the following is not an antigibberellins or growth retardant?
(A) Phosfon D (B) Cycocel
(C) Amo-1618 (D) NPA

1398. Production of large sized seedless grapes and malting of barley can be enhanced by treatment with:
(A) Auxins (B) Gibberellins
(C) Ethylene (D) Morphactins

1399. Gibberellins are more effective when applied to:
(A) Intact plant (B) Coleoptiles
(C) Internodes (D) Roots

1400. During seed germination stored food in seed is mobilized by:
(A) Ethylene (B) Cytokinin
(C) ABA (D) Gibberellin

1401. Antigibberellin compounds are growth:
(A) Retardant (B) Promoter
(C) Inhibitor (D) None of these

1402. Number of gibberellins present in a single plant organ can be:
(A) One (B) More than one
(C) Five in each cell (D) Ten in each organ

1403. The nature of all gibberellins is:
(A) Alkaline (B) Neutral
(C) Acidic (D) Buffer

1404. Phenolic inhibitors occur in plants as:
(A) Glycosides (B) Terpenoids
(C) Indoles (D) Phospholipids

1405. ABA occurs in plants predominantly in:
(A) Stems (B) Mature green leaves
(C) Roots (D) Flowers

1406. ABA is inactivated in plant cells by converting into:
(A) Phaseic acid (B) Dihydrophaseic acid
(C) ABA-glucosyl ester (D) All of these

1407. ABA is a:
(A) Sesquiterpene (B) Diterpene
(C) Triterpene (D) Tetraterpene

1408. Ethylene acceptors are located on:
(A) Plasmamembrane (B) Tonoplast
(C) Endoplasmic reticulum (D) All of these

1409. A potent inhibitor of ethylene action is:
(A) Silver ions (B) Cobalt ions
(C) Manganese ions (D) All of these

1410. Immediate precursor of ethylene biosynthesis in plant is:
(A) SAM (B) ACC
(C) Methionine (D) All of these

1411. The popular absorbent of ethylene is:
(A) Formalin (B) Potassium permanganate
(C) Potassium hydroxide (D) All of these

1412. Dark period is critical for:
(A) Long day plants (B) Short day plants
(C) Amphophotoperiodism (D) None of these

1413. In which case, flowering will be promoted in long day plants?
(A) Increasing the number of inductive cycles
(B) Continues inductive cycles
(C) Prolongation of light period
(D) All of these

1414. In plants, photoperiodic induction is perceived by:
(A) Stems (B) Leaves
(C) Roots (D) None of these

1415. The hormone produced as a result of photoperiodic induction is known as:
(A) Cytokinin (B) Vernalin
(C) Polyamines (D) Florigen

1416. Which of the followings is supported to be the precursor of Florigen?
(A) Auxin (B) Gibberellins
(C) Cytokinin (D) None of these

1417. Phytochrome is a soluble protein which is a:
(A) Heterodimer (B) Homodimer
(C) Both (A) and (B) (D) None of these

1418. In plants, richest source of phytochrome are:
(A) Seedling grown in light
(B) Etiolated seedlings
(C) Both (A) and (B)
(D) None of these

1419. Photochemical reaction occurs in:
(A) The plasma membrane of green plants
(B) The membranes of lysosomes
(C) The outer membrane of chloroplasts
(D) The thylakoid membranes

1420. Which of the following events involved in stomatal opening:
(A) Have CAM pathway
(B) True drought tolerant
(C) High water use efficiency
(D) True drought avoider

1421. Vernalization is a:
(A) Reversible process
(B) Low temperature treatment for seed germination
(C) Auxin dependent phenomenon
(D) Process in which genetic makeup of plants change

1422. What amount of water absorbed by plants is lost in transpiration?
(A) 85% (B) 90% (C) 95% (D) 100%

1423. Which statement is not correct regarding transpiration?
(A) Living cells are involved
(B) V.P., D.P., O.P. involved
(C) Causes dryness of surface
(D) Physiological process

1424. Stomatal transpiration accounts for what % of water loss?
(A) 75% (B) 80-90%
(C) 90% (D) 95%

1425. Which of the following is not correct ? CAM and C_4 plants as compared to C_3 plants are:
(A) Characteristics of hot arid areas
(B) More rapid photosynthesizers
(C) More economical in their use of water
(D) Characterized by reaching maximum photosynthetic rates at high radiation intensities

1426. Which of the following lacks stomata?
(A) Angiosperms (B) Bryophytes
(C) Algae (D) Pteridophytes

1427. Number of stomata/square cm of leaf is:
(A) 500-1000 (B) 1000-2000
(C) 1000-60000 (D) None of these

1428. Fully open stomata of leaf occupies what percent of leaf area?
(A) 0.05% (B) 0.5%
(C) 0.95% (D) 1-2%

1429. Barbell shaped guard cells are found in:
(A) Oilseeds (B) Cucurbits
(C) Grasses (D) Pulses

1430. Which of the following is false about glyoxylate cycle?
(A) It is a variant of citric acid cycle
(B) It results in the formation of succinate from acetyl-CoA
(C) It occurs in glyoxisome
(D) It occurs in all eukaryotic cells

1431. When stomata are equally distributed on both surfaces of leaf tissue, are:
(A) Potato type (B) Oat type
(C) Apple type (D) Lily type

1432. 'Photosynthetic production in guard cells' theory was put forwarded by:
(A) Scarth (B) Lloyd
(C) Von Mohl (D) Sayre

1433. Which ions are responsible for stomatal movement according to Levitt?
(A) Fe^{2+} (B) Zn^{2+}
(C) K^+ (D) Na^+

1434. Guttation occurs through:
(A) Lenticels (B) Hydathodes
(C) Stomata (D) Epidermis

1435. Scotoactive stomata are found in:
(A) Aloe (B) Salsola
(C) Sedium (D) All of these

1436. Maximum opening of stomata occurs at:
(A) 8.00 AM (B) 12.00 Noon
(C) 10.00 AM (D) 4.00 PM

1437. Cooperation of the two photosystems of the chloroplast is required for:
(A) ATP synthesis
(B) Reduction of $NADP^+$
(C) Oxidation of the reaction centre of photosystem I
(D) Generation of a proton – motive force

1438. Which of the following statement about ubiquinol is correct?
(A) It is the mobile electron carrier between cytochrome oxigoreductase and cytochrome oxidase
(B) It is an integral membrane protein
(C) Its oxidation involves the simultaneous transfer of two electrons to the Fe-S centre of cytochrome reductase
(D) It is a lipid –soluble molecule

1439. On per unit area basis, transpiration rate is more in:
(A) Smaller plants (B) Larger plants
(C) Equal in both (D) Not known

1440. Transpiration is maximum in:
(A) Blue light (B) Red light
(C) Far red light (D) Green light

1441. Transpiration is a 'necessary evil' called by:
(A) Kozlowski (B) Kramer
(C) Curtis (D) None of these

1442. The photosynthetic process is an oxidation-reduction reaction. Which of the following is an oxidation reaction in photosynthesis?
(A) Conversion of phosphoglyceric acid to glyceraldehyde phosphate
(B) Carboxylation of RuBP using Rubisco
(C) Splitting of H_2O to form oxygen
(D) Phosphorylation

1443. Which of the following is not an antitranspirants?
(A) ABA (B) Asprin
(C) PMA (D) Auxin

1444. Transpiration is measured by:
(A) Porometer (B) Potometer
(C) Photometer (D) Manometer

1445. Photosynthetically active radiation (PAR) is measured by:
(A) Thermosensor
(B) Albedometer
(C) Quantom sensor L1-190
(D) None of the above

1446. The amount of solar radiation continuously emited by sun is:
(A) 2.0 cal/cm^2/min (B) 2.0 cal/min
(C) 1.94 cal/cm^2/min (D) 1.94 cal/min

1447. The reflection of solar radiation reaching the earth surface back into the space is called:
(A) Lux (B) Albedo
(C) Net radiation (D) Phototaxis

1448. What is a C_4 plant?
(A) Rice (B) Wheat
(C) Soybean (D) Maize

1449. Photoperiodism is:
(A) Bending of shoot towards source of light
(B) Effect of light/dark durations on physiological processes
(C) Movement of chloroplast in cell in response to light
(D) Effect of light on chlorophyll synthesis

1450. The enzyme responsible for the fixation of CO_2 in mesophyll cells of C_4 plants is:
(A) Mallic enzyme
(B) RuBP carboxylase
(C) Phosphoenol pyruvate carboxylase
(D) Phosphoenol pyruvate carboxykinase

1451. Leaves wilt during the day but recover during night. It is:
(A) Incipient wilting (B) Temporary wilting
(C) Permanent wilting (D) None of the above

1452. Of the total water absorbed by the plant how much is lost in transpiration:
(A) 70% (B) 60%
(C) 99% (D) 40%

1453. Root pressure is high when:
(A) Transpiration is high and absorption of water is low
(B) Transpiration is low and abosorption of water is high
(C) Transpiration is low and abosorption of water is low
(D) Transpiration is high and absorption of water is high

1454. Wilting occurs when:
(A) Xylem is blocked (B) Phloem is blocked
(C) Combium is blocked (D) All correct

1455. The maximum loss of water in transpiration is from:
(A) Lenticels (B) Stomata
(C) Cuticle (D) Hydathodes

1456. Hydroponicks is also called as:
(A) Soil less cultivation of plant
(B) Parthenocarpy
(C) Vivipany
(D) Phytotron

1457. Flowering is influenced by:
(A) Soil water content (B) Soil acidity
(C) Photoperiod (D) Content of green pigment

1458. Photoperiodism is probably due to the synthesis of:
(A) Cytokinins
(B) Gibberellins
(C) Auxin
(D) Florigon

1459. Leghaemoglobin is a pink coloured pigment found in:
(A) Leg of dog
(B) Leg of human
(C) Roots of leguminous plants
(D) Brain of cat

1460. Leg-haemoglobin pigment is synthesized by:
(A) Cells of the host plant
(B) The bacterium
(C) Both (A) and (B)
(D) None of these

1461. Cells of root nodules are:
(A) Haploid (B) Diploid
(C) Polyploid (D) Without nucleus

1462. To obtain best growth of rice plants the following is added in the field:
(A) Chlorella (B) Nostoc
(C) Sodium chlorate (D) Thiobacillus denitrificans

1463. ADP + ip = ATP in grana is called:
(A) Phosphorylation
(B) Oxidative phosphorylation
(C) Photophosphorylation
(D) Photolysis

1464. The gas released in photosynthesis is:
(A) Pure CO_2 (B) Pure O_2
(C) Mainly CO_2 (D) Mainly O_2

1465. Photosyntheis is maximum during:
(A) Continuous weak light
(B) Continuous strong light
(C) Intermittent light
(D) None of the above

1466. Higher concentration (above 1%) of CO_2 have:
(A) No effect on photosynthesis
(B) Good effect on photosynthesis
(C) Inhibitory effect on photosynthesis
(D) None of the above

1467. Q10 means:
(A) Respiratory quotient
(B) Temperature quotient
(C) Both of the above
(D) Photosynthetic quotient

1468. Kranz anatomy is found in :
(A) C_2 plants (B) C_3 plants
(C) C_4 plants (D) Succulent plant

1469. Light reaction of photosynthesis form:
(A) Carbohydrates (B) ATP
(C) NADP + O_2 (D) $NADPH_2$, O_2 and ATP

1470. What is the first step in photosynthesis.
(A) ATP formation
(B) Ionization of water
(C) Excitation of electron of chlorophyll by photons of light
(D) Reaction of CO_2 with a pentose sugar

1471. 90% of total photosynthesis in the world is carried out by:
(A) Shrubs (B) Trees
(C) Grasses (D) Algal

1472. In photosynthesis PGAL is formed from PGA by a process of:
(A) Oxidation (B) Reduction
(C) Photolysis (D) Excitation

1473. Which element are essential for photolysis of water:
(A) Mg and P (B) Zn and Iodine
(C) Mn and Chlorine (D) Ca and Calcium

1474. Ratio of photosynthesis and respiration is approximately:
(A) 1 : 2 (B) 10 : 1 (C) 50 : 1 (D) 2 : 1

1475. Energy conversion efficiency of photosynthesis is nearly:
(A) 33% (B) 50%
(C) 66% (D) 100%

1476. Quantumn yield in photosynthesis is:
(A) 8 (B) 0.125
(C) 0.5 (D) 10

1477. Which plant is better photosynthesizer:
(A) Mango (B) Sugarcane
(C) Wheat (D) Rice

1478. Rate of photosynthesis is not affected by:
(A) Quality of light (B) Intensity of light
(C) Chlrophyll content (D) Size of leaf

1479. PCR cycle represents:
(A) C_3 cycle (B) C_4 cycle
(C) CAM cycle (D) TCA cycle

1480. PCR cycle occurs in :
(A) All photosynthetic eukaryotes
(B) Only photosynthetic bacteria
(C) Only chemosynthetic bacteria
(D) Only in CAM plants

1481. Heaviest pigment of chloroplast among the following is:
(A) Chlorophyll-a (B) Chlorophyll-b
(C) Carotenes (D) Xanthophylls

1482. Photosynthesis is a process:
(A) Reductive and excorgonic
(B) Reductive and catabolic
(C) Reductive, endergenic and catabolic
(D) Reductive, endergonic and anabolic

1483. Phaeophytin is a :
(A) Grey pigment (B) Green pigment
(C) Red pigment (D) Orange pigment

1484. Cybrids may arise through:
(A) Fusion of a normal protoplast with another nucleated protoplast of different origin
(B) Fusion between a normal protoplast and a protoplast containing viable nucleus
(C) Elimination of one of the nucleic from heterokaryon formed from two protoplasts of different origin
(D) Fusion of a normal cell with another nucleated cell

1485. The plant cells are considered totipotent. The phenomeon of a mature cell reverting to the meristematic state and farming undifferentiated callus tissue is called:
(A) Redifferentiation (B) Didifferentiation
(C) Organogenesis (D) Recalcitrancy

1486. Which one of the following ecosystem is represented by an inverted pyramid of numbers?
(A) Grassland (B) Pond
(C) Desert (D) Parasitic forest

1487. Raphides are deposits of:
(A) Calcium oxalate (B) Calcium carbonate
(C) Silica bodies (D) Protein bodies

1488. Which one of the following families shows bicollateral vascular bundle in the transverse section of the stem?
(A) Rutaceae (B) Asteraceae
(C) Cucurbitaceae (D) Malvaceae

1489. Arbuscular are highly branched structures, formed by:
(A) Ectomycorrhizae
(B) Endomycorrhizae
(C) Arbuteid mycorrhizae
(D) Monotropoid mycorrhizae

1490. Which one of the following combinations of polysaccharides is present in plant cell wall?
(A) Only cellulose, hemicellulose and fibrain
(B) Only cellulose, hemicellulose and lignin
(C) Only cellulose, hemcellulose and pectin
(D) Only cellulose, pectin and lignin

1491. Which one of the following statement is INCORRECT?
(A) Nitrogen fixation is a aerobic process
(B) Denitrogenase catalyzes reduction of nitrogen to ammonia
(C) Root nodules are found in Glycine max
(D) Nitrogen fixation is anaerobic process

1492. Which one of the following statement is INCORRECT with respect to photophosphorylation:
(A) ATP is generated without concomitant formation of NADPH
(B) Electron flows from Photosystem I to cytochrome of complex
(C) Photosystem II does not participate in cyclic photophosphorylation
(D) Cyclic photophosphorylation occurs when NADP/NADPH ratio is high

1493. Which of the following genera produces dimorphic seeds that help to broaden the time of germination in a viable habit?
(A) Xanthium
(B) Pisum
(C) Mangifera
(D) Linum

1494. The genes for micro RNA(miRNA) in plants are usually transcribed by:
(A) RNA polymerase I
(B) RNA polymerase II
(C) RNA polymerase III
(D) RNA polymerase IV

1495. Acid rain with a pH of 4.0 is more acidic than the rain with a pH of 6.0 by:
(A) 2 times (B) 10 times
(C) 100 times (D) 1000 times

1496. Which of the following plants produces Y lang-ylang oil?
(A) Cananga odorata
(B) Carcum copticum
(C) Pandanus odoratissimus
(D) Pimenta racemosa

1497. Which of the following stains is used to visualize callose under the microscope?
(A) Aleian blue (B) Aniline blue
(C) Totuidine blue (D) Thymol blue

1498. In the water oxidizing clock which metal element function as a pivotal component?
(A) Mg++ (B) Mn++
(C) Cu (D) Fe

1499. During the course of prolonged starvation and fasting glucose is synthesized from one of the following metabolic process:
(A) Glycogenesis (B) Glycolysis
(C) Gluconeogenesis (D) Fermentation

1500. What is meant by lichenometry?
(A) Study regarding the age of lichens
(B) Study of the age of exposed rock surface based on the size of lichen thalli
(C) Study regarding the calculation of the period in which lichen evolved
(D) Study regarding the succession of various forms of lichen

1501. The pH indicator in the litmus test in a day extracted from the lichen:
(A) Umbilicaria esculenta
(B) Parmelia saxatilis
(C) Roccella tinctoria
(D) Xanthoria parietina

1502. Identify the plant hormone known as anti senescence hormone:
(A) IAA (B) Gibberellin
(C) Cytokinin (D) ABA

1503. Which is the technique utilised to estimate the methylation of DNA?
(A) Ion mobility spectrometry
(B) Mass spectrometry
(C) Neutron triple-axis spectrometry
(D) Optical spectrometry

1504. Gamma diversity (y-diversity) is the total species diversity in a landscape. Who introduced this terminolgoy?
(A) Shannon (B) Simpson
(C) Odum (D) Whitlamer

1505. The bending of plants towards unilateral light is due to:
(A) Polar transport of auxin
(B) Auxin degradation in light
(C) Auxin synthesis in shaded area
(D) Lateral distribution of auxin towards shaded area

1506. Heavy charged particle, more toxic than other forms of radiation, net penetrating. These are the properties of which type of radiation?
(A) Alpha (B) Beta
(C) Gamma (D) All of the above

1507. Nitrogen fixation in weedy trees is accomplished through the microorganism:
(A) Rhizobium (B) Azotobacter
(C) Frankia (D) Azespirillum

1508. Of the different subunits of G protein, this is responsible for activation of Adenylate cyclase?
(A) Alpha subunit (B) Beta subunit
(C) Gamma subunit (D) Delta subunit

1509. Species selected for making conservation related decision, typically because protecting these species indirectly protects the many other species that make up the ecological community of its habitat:
(A) Indicator species (B) Keystone species
(C) Flagship species (D) Umbrella species

1510. A biome characterized by coniferous trees containing mostly of pines, species and laraches is referred as:
(A) Pine forests
(B) Temperate forests
(C) Temperate deciduous factors
(D) Boreal forests

1511. Phagotrophs are:
(A) Organisms that feed on dead organic matter
(B) Organisms that absorb dissolved organic matter
(C) Organisms that ingest other organisms or particular organic matter
(D) Organisms that synthesis food from inorganic substances

1512. Process in which organisms diversify rapidly from an ancestral species into a multitude of new forms is referred to as :
(A) Adaptive radiation
(B) Natural selection
(C) Convergent evolution
(D) Non-random evolution

1513. Which of the following terms best describes the character shared by a set of species but not present in their common ancestor?
(A) Cladogenesis (B) Paedomorphosis
(C) Allometric growth (D) Homoplasy

1514. Which of the following is not a greenhouse gas?
(A) Hydrofluoro carbons (B) Nitrogen oxides
(C) Sulphur hexafluoride (D) Sulphur dioxide

1515. Process by which organic material becomes a fossil through the replacement of the original material and the filling of the original pore species with minerals:
(A) Compression (B) Petrifactions
(C) Casts (D) Mineralization

1516. Distribution of plant species owing tropical regions of all continents is referred to as:
(A) Circumtropical (B) Pantropical
(C) Neo tropical (D) Palaeo tropical

1517. When did the convention on International Trade in Endangered Species of wild Fauna and Flora came into force:
(A) 1 July 1972 (B) 1 July 1971
(C) 1 July 1975 (D) 1 July 1974

1518. Cellulosic ethanol is:
(A) First generation biofuel
(B) Second generation biofuel
(C) Third generation biofuel
(D) Fourth generation biofuel

1519. The primitive mode of energy generating metabolic reaction is:
(A) Photosynthesis
(B) Reduction of nitrate
(C) Reduction of sulphate
(D) Oxidation of nitrate

1520. Name the biologist who coined the concept of ecosystem?
(A) Earnest Haeckel (B) A.G. Tansley
(C) Charles Elton (D) Charles Darwin

1521. In which year the UN convention on biological diversity was enacted?
(A) 1972 (B) 2004
(C) 2002 (D) 1992

1522. Which one of the following is the single largest air pollutant?
(A) Carbon dioxide (B) Nitrogen
(C) Carbon monooxide (D) PAN

1523. Which of the ecosystem has the lowest net primary productivity/square kilometer?
(A) Salt marsh (B) Open ocean
(C) Coral reef (D) Grassland

1524. Which of the following contribute to global warming? 1. Deforestation, 2. Fossile fuel burning, 3. Ground level Ozone
(A) 1, 2 only
(B) 1, 3 only
(C) None of the above
(D) All of the above

1525. Which one of the following is not an osmolyte?
(A) Proline (B) Sucrose
(C) Plastoquinone (D) Trehalose

1526. All of the following act to increase species diversity, except:
(A) Key stone species
(B) Competive exclusion
(C) Moderate disturbances
(D) Pachy environments

1527. Which is the inhibitor which inhibits the transfer of electrons from PS II to cytochrome complex?
(A) Paraquat (B) Methylviologen
(C) DCMU (D) DCBQ

1528. One of the following is an inhibitor of mitochondrial ETC:
(A) Rotenone (B) Chloramphenical
(C) Cyclohexamide (D) DCPIP

1529. One of the following is an inhibitor of electron transport in thylakoids:
(A) DCPIP (B) DCMV
(C) Rotenone (D) Zeaxanthin

1530. One of the following is not a stress protein :
(A) HSP (B) LEA
(C) Dehydrins (D) Phosphoprotein

1531. Inhibitor of GA:
(A) TIBA (B) Plaeobutrazol
(C) DAA (D) Cyclohexane

1532. An organisms that uses glucose as a source of both energy and carbon is called:
(A) Photoautotroph (B) Photoheterotroph
(C) Chemoautotroph (D) Chemoheterotroph

1533. The main mineral constituents of wood are salts of:
(A) Carbon, Magnesium and Potassium
(B) Calcium, Magnesium and Potassium
(C) Carbon, Nitrogen and Potassium
(D) Calcium, Potassium and Manganese

1534. Arctic animals maintain their body temperature because they have more:
(A) Transducing protein
(B) Uncoupling protein
(C) Myoglobin protein
(D) FOF1 ATPase

1535. Taxa which occupy mutually exclusive geographical areas are termed:
(A) Sympatric (B) Allopatric
(C) Pantropical (D) New world taxa

1536. Taxa shows distribution pattern which are interrupted by considerable areas from which the taxon is absent are termed:
(A) Vicariance (B) Disjunct
(C) Plesiomorphic (D) Alien

1537. The first herbarium in the world was founded in :
(A) Kew, UK (B) Padua, Italy
(C) Paris, France (D) Leningrad, Russia

1538. What are the ecological effects of Eutrophication?
(A) Decreased biodiversity
(B) New species invasion
(C) Algal blooms
(D) All of the above

1539. A group of one or more species derived from a sequential development pattern which involves continual and uniform changes from an extinct ancestral form on an evolutionary scale:
(A) Perinospecies (B) Chromospecies
(C) Ecospecies (D) Evospecies

1540. Succession beginning with sand is called:
(A) Lithosere (B) Sanosere
(C) Psammosere (D) Hydrosere

1541. Microevolution means :
(A) Evolution at geographical level
(B) Evolution in a small community
(C) Change in gene frequency within a population
(D) Spatial evolution

1542. Fruit drop is caused by:
(A) Accumulation of more auxin in fruit than in stem
(B) Accumulation of less auxin in fruit than in stem
(C) Absence of auxin in stem and roots
(D) Accumulation of auxin in roots

1543. National park is an example of:
(A) In vitro conservation
(B) Ex situ conservation
(C) In situ conservation
(D) All of the above

1544. Western Ghats passes through .... states in India:
(A) 6 (B) 4
(C) 7 (D) 5

1545. Respiration in plants is essentially a process related to the following:
(A) Evaporation (B) Transpiration
(C) Oxidation (D) None of the above

1546. Zinc is required for synthesis of:
(A) Fats (B) Proteins
(C) Tryptophan (D) Sugars

1547. Sulphur containing aminoacid:
(A) Lysine (B) Glutamine
(C) Methionine (D) Glycine

ANSWERSHEET: 5.PLANT PHYSIOLOGY

1. (D); 2. (B); 3. (A); 4. (C); 5. (C); 6. (B); 7. (B); 8. (C); 9. (C); 10. (D); 11. (C); 12. (D); 13. (B); 14. (C); 15. (C); 16. (D); 17. (C); 18. (B); 19. (D); 20. (A); 21. (B); 22. (C); 23. (C); 24. (D); 25. (B); 26. (B); 27. (C); 28. (D); 29. (D); 30. (A); 31. (D); 32. (B); 33. (C); 34. (A); 35. (A); 36. (A); 37. (D); 38. (B); 39. (B); 40. (D); 41. (B); 42. (C); 43. (C); 44. (C); 45. (B); 46. (D); 47. (A); 48. (C); 49. (C); 50. (B); 51. (B); 52. (C); 53. (D); 54. (D); 55. (A); 56. (D); 57. (D); 58. (D); 59. (C); 60. (B); 61. (B); 62. (B); 63. (C); 64. (D); 65. (B); 66. (A); 67. (A); 68. (D); 69. (A); 70. (A); 71. (B); 72. (D); 73. (C); 74. (B); 75. (B); 76. (D); 77. (A); 78. (A); 79. (B); 80. (C); 81. (B); 82. (A); 83. (C); 84. (A); 85. (B); 86. (D); 87. (D); 88. (A); 89. (D); 90. (C); 91. (D); 92. (B); 93. (A); 94. (D); 95. (C); 96. (C); 97. (A); 98. (D); 99. (A); 100. (D); 101. (C); 102. (B); 103. (B); 104. (C); 105. (B); 106. (B); 107. (D); 108. (C); 109. (C); 110. (A); 111. (D); 112. (B); 113. (D); 114. (D); 115. (D); 116. (A); 117. (B); 118. (C); 119. (A); 120. (A); 121. (B); 122. (C); 123. (A); 124. (B); 125. (D); 126. (D); 127. (B); 128. (C); 129. (A); 130. (B); 131. (B); 132. (B); 133. (A); 134. (C); 135. (C); 136. (D); 137. (B); 138. (A); 139. (C); 140. (A); 141. (B); 142. (D); 143. (B); 144. (C); 145. (C); 146. (A); 147. (A); 148. (B); 149. (C); 150. (D); 151. (A); 152. (B); 153. (A); 154. (C); 155. (A); 156. (B); 157. (D); 158. (C); 159. (C); 160. (C); 161. (A); 162. (D); 163. (A); 164. (B); 165. (D); 166. (A); 167. (A); 168. (A); 169. (C); 170. (B); 171. (B); 172. (C); 173.(B); 174. (D); 175. (B); 176. (A); 177. (C); 178. (B); 179. (B); 180. (C); 181. (C); 182. (D); 183. (C); 184. (A); 185. (D); 186. (A); 187. (C); 188. (C); 189. (D); 190. (B); 191. (B); 192. (D); 193. (C); 194. (B); 195. (A); 196. (D); 197. (B); 198. (A); 199. (B); 200. (C); 201. (A); 202. (C); 203. (C); 204. (A); 205. (D); 206. (D); 207. (B); 208. (A); 209. (B); 210. (C); 211. (B); 212. (C); 213. (C); 214. (D); 215. (A); 216. (D); 217. (C); 218. (C); 219. (B); 220. (B); 221. (A); 222. (D); 223. (A); 224. (C); 225. (D); 226. (C); 227. (A); 228. (C); 229. (B); 230. (A); 231. (D); 232. (B); 233. (C); 234. (B); 235. (B); 236. (C); 237. (B); 238. (B); 239. (D); 240. (C); 241. (C); 242. (B); 243. (D); 244. (A); 245. (A); 246. (D); 247. (B); 248. (B); 249. (B); 250. (D); 251. (A); 252. (D); 253. (C); 254. (D); 255. (C); 256. (C); 257. (D); 258. (B); 259. (C); 260. (D); 261. (B); 262. (C); 263. (B); 264. (D); 265. (C); 266. (A); 267. (B); 268. (B); 269. (B); 270. (C); 271. (A); 272. (A); 273. (B); 274. (C), 275. (A); 276. (D); 277. (B); 278. (B); 279. (B); 280. (C); 281. (D); 282. (B); 283. (C); 284. (B); 285. (B); 286. (D); 287. (D); 288. (D); 289. (C); 290. (D); 291. (C); 292. (C); 293. (B); 294. (D); 295. (D); 296. (C); 297. (A); 298. (C); 299. (D); 300. (B); 301. (B); 302. (D); 303. (D); 304. (A); 305. (B); 306. (C); 307. (B); 308. (A); 309. (B); 310. (D); 311. (A); 312. (A); 313. (B); 314. (B); 315. (B); 316. (C); 317. (C); 318. (A); 319. (B); 320. (B); 321. (D); 322. (A); 323. (C); 324. (D); 325. (C), 326. (C); 327. (B); 328. (D); 329. (D); 330. (C); 331. (B); 332. (B); 333. (C); 334. (B); 335. (C); 336. (D); 337. (A); 338. (D); 339. (A); 340. (A); 341. (B); 342. (C); 343. (B); 344. (D); 345. (D); 346. (C); 347. (D); 348. (D); 349. (D); 350. (D); 351. (B); 352. (A); 353. (C); 354. (B); 355. (D); 356. (D); 357. (A); 358. (D); 359. (B); 360. (A); 361. (D); 362. (B); 363. (A); 364. (A); 365. (C); 366. (D); 367. (D); 368. (B); 369. (C); 370. (D); 371. (A); 372. (D); 373. (C); 374. (C); 375. (B); 376. (D); 377. (D); 378. (D); 379. (D); 380. (D); 381. (B); 382. (C); 383. (D); 384. (C); 385. (C); 386. (D); 387. (B); 388. (B); 389. (D); 390. (D); 391. (A); 392. (B,, 393. (A); 394. (D); 395. (A); 396. (D); 397. (C); 398. (C); 399. (D);

400. (D); 401. (B); 402. (C); 403. (D); 404. (B); 405. (A); 406. (A); 407. (C); 408. (A); 409. (A); 410. (A); 411. (A); 412. (D); 413. (A); 414. (C); 415. (B); 416. (B); 417. (C); 418. (C); 419. (D); 420. (D); 421. (D); 422. (A); 423. (A); 424. (B); 425. (D); 426. (B); 427. (B); 428. (D); 429. (D); 430. (D); 431. (D); 432. (D); 433. (B); 434. (C); 435. (D); 436. (D); 437. (A); 438. (B); 439. (B); 440. i (B), ii (A) ; 441. (A); 442. (A); 443. (A); 444. (A); 445. (B); 446. (A); 447. (A); 448. (C); 449. (A); 450. (D); 451. (C); 452. (A); 453. (A); 454. (D); 455. (C); 456. (A); 457. (B); 458. (A); 459. (D); 460. (B); 461. (B); 462. (D); 463. (B); 464. (B); 465. (C); 466. (B); 467. (D); 468. (D); 469. (A); 470. (A); 471. (C); 472. (B); 473. (A); 474. (C); 475. (A); 476. (B); 477. (A); 478. (D); 479. (A); 480. (C); 481. (D); 482. (D); 483. (D); 484. (A); 485. (D); 486. (D); 487. (B); 488. (B); 489. (D); 490. (B); 491. (B); 492. (A); 493. (A); 494. (A); 495. (A); 496. (B); 497. (B); 498. (A); 499. (C); 500. (D); 501. (D); 502. (B); 503. (A); 504. (D); 505. (B); 506. (C); 507. (A); 508. (B); 509. (B); 510. (C); 511. (D); 512. (B); 513. (A); 514. (B); 515. (D); 516. (B); 517. (B); 518. (C); 519. (B); 520. (A); 521. (C); 522. (D); 523. (C); 524. (A); 525. (B); 526. (B); 527. (A); 528. (C); 529. (C); 530. (C); 531. (D); 532. (C); 533. (C); 534. (C); 535. (D); 536. (D); 537. (C); 538. (B); 539. (C); 540. (A); 541. (B); 542. (D); 543. (A); 544. (C); 545. (D); 546. (B); 547. (B); 548. (D); 549. (B); 550. (B); 551. (A); 552. (D); 553. (C); 554. (D); 555. (B); 556. (D); 557. (C); 558. (B); 559. (A); 560. (C); 561. (B); 562. (C); 563. (C); 564. (B); 565. (B); 566. (A); 567. (C); 568. (D); 569. (C); 570. (B); 571. (C); 572. (C); 573. (D); 574. (C); 575. (B); 576. (A); 577. (A); 578. (D); 579. (A); 580. (A); 581. (D); 582. (B); 583. (B); 584. (B); 585. (B); 586. (C); 587. (B); 588. (D); 589. (D); 590. (B); 591. (A); 592. (A); 593. (A); 594. (C); 595. (C); 596. (C); 597. (A); 598. (A); 599. (A); 600. (B); 601. (B); 602. (A); 603. (C); 604. (A); 605. (C); 606. (B); 607. (D); 608. (C); 609. (D); 610. (B); 611. (D); 612. (A); 613. (D); 614. (B); 615. (C); 616. (B); 617. (A); 618. (D); 619. (A); 620. (B); 621. (D); 622. (D); 623. (A); 624. (C); 625. (D); 626. (D); 627. (A); 628. (A); 629. (D); 630. (B); 631. (B); 632. (A); 633. (D); 634. (C); 635. (D); 636. (B); 637. (A); 638. (C); 639. (C); 640. (B); 641. (A); 642. (D); 643. (C); 644. (B); 645. (D); 646. (C); 647. (A); 648. (A); 649. (C); 650. (B); 651. (C); 652. (A); 653. (D); 654. (C); 655. (A); 656. (D); 657. (A); 658. (A); 659. (A); 660. (B); 661. (D); 662. (B); 663. (D); 664. (D); 665. (C); 666. (B); 667. (B); 668. (B); 669. (C); 670. (B); 671. (A); 672. (A); 673. (A); 674. (A); 675. (B); 676. (A); 677. (D); 678. (C); 679. (D); 680. (B); 681. (B); 682. (A); 683. (C); 684. (B); 685. (D); 686. (B); 687. (C); 688. (B); 689. (C); 690. (B); 691. (C); 692. (D); 693. (A); 694. (A); 695. (A); 696. (B); 697. (C); 698. (C); 699. (C); 700. (C); 701. (A); 702. (D); 703. (B); 704. (A); 705. (B); 706. (D); 707. (A); 708. (A); 709. (A); 710. (D); 711. (D); 712. (B); 713. (A); 714. (B); 715. (A); 716. (A); 717. (B); 718. (D); 719. (B); 720. (D); 721. (B); 722. (C); 723. (D); 724. (D); 725. (D); 726. (A); 727. (B); 728. (A); 729. (D); 730. (A); 731. (D); 732. (B); 733. (C); 734. (B); 735. (C); 736. (A); 737. (B); 738. (C); 739. (A); 740. (A); 741. (C); 742. (B); 743. (B); 744. (C); 745. (B); 746. (D); 747. (D); 748. (D); 749. (B); 750. (D); 751. (A); 752. (B); 753. (D); 754. (D); 755. (A); 756. (A); 757. (D); 758. (A); 759. (D); 760. (D); 761. (A); 762. (C); 763. (C); 764. (A); 765. (C); 766. (A); 767. (B); 768. (C); 769. (B); 770. (A); 771. (C); 772. (B); 773. (A); 774. (D); 775. (A); 776. (C); 777. (A); 778. (D); 779. (D); 780. (D); 781. (A); 782. (B); 783. (C); 784. (C); 785. (D); 786. (D); 787. (A); 788. (B); 789. (A); 790. (C); 791. (B); 792. (B); 793. (B); 794. (B); 795. (B); 796. (B); 797. (C); 798. (C); 799. (B); 800. (C); 801. (D); 802. (D); 803. (D); 804. (A); 805. (D); 806. (B); 807. (A); 808. (D); 809. (A); 810. (B); 811. (D); 812. (D); 813. (C); 814. (C); 815. (A); 816. (C); 817. (B); 818. (A); 819. (A); 820. (D); 821. (B); 822. (C); 823. (D); 824. (C); 825. (C); 826. (B); 827. (D); 828. (C); 829. (D); 830. (C); 831. (D); 832. (D); 833. (B); 834. (B); 835. (A); 836. (D); 837. (D); 838. (C); 839. (B); 840. (D); 841. (A); 842. (A); 843. (A); 844. (D); 845. (B); 846. (B); 847. (D); 848. (D); 849. (A); 850. (D); 851. (D); 852. (C); 853. (B); 854. (D); 855. (A); 856. (C); 857. (C); 858. (D); 859. (A); 860. (D); 861. (B); 862. (A); 863. (C); 864. (A); 865. (C); 866. (A); 867. (C); 868. (D); 869. (B); 870. (B); 871. (C); 872. (B); 873. (B); 874. (B); 875. (A); 876. (C); 877. (B); 878. (C);

879. (C); 880. (B); 881. (D); 882. (C); 883. (B); 884. (C);
885. (C); 886. (B); 887. (D); 888. (C); 889. (C); 890. (D);
891. (D); 892. (D); 893. (B); 894. (A); 895. (B); 896. (D);
897. (D); 898. (D); 899. (C); 900. (B); 901. (B); 902. (C);
903. (D); 904. (B); 905. (D); 906. (A); 907. (B); 908. (B);
909. (A); 910. (D); 911. (C); 912. (C); 913. (B); 914. (C);
915. (D); 916. (A); 917. (B); 918. (B); 919. (C); 920. (D);
921. (A); 922. (D); 923. (D); 924. (C); 925. (A); 926. (D);
927. (B); 928. (C); 929. (B); 930. (D); 931. (D); 932. (B);
933. (D); 934. (B); 935. (A); 936. (B); 937. (D); 938. (D);
939. (A); 940. (D); 941. (D); 942. (D); 943. (A); 944. (D);
945. (B); 946. (B); 947. (B); 948. (B); 949. (D); 950. (D);
951. (D); 952. (A); 953. (C); 954. (A); 955. (D); 956. (C);
957. (C); 958. (B); 959. (A); 960. (C); 961. (C); 962. (B);
963. (A); 964. (C); 965. (A); 966. (B); 967. (D); 968. (A);
969. (C); 970. (D); 971. (B); 972. (C); 973. (D); 974. (A);
975. (D); 976. (C); 977. (C); 978. (C); 979. (C); 980. (C);
981. (B); 982. (C); 983. (C); 984. (D); 985. (B); 986. (B);
987. (A); 988. (A); 989. (A); 990. (A); 991. (A); 992. (A);
993. (C); 994. (D); 995. (C); 996. (C); 997. (D); 998. (B);
999. (A); 1000. (B); 1001. (D); 1002. (A); 1003. (B);
1004. (A); 1005. (D); 1006. (C); 1007. (B); 1008. (B);
1009. (D); 1010. (C); 1011. (A); 1012. (B); 1013. (A);
1014. (B); 1015. (A); 1016. (C); 1017. (D); 1018. (B);
1019. (B); 1020. (C); 1021. (D); 1022. (A); 1023. (A);
1024. (C); 1025. (B); 1026. (A); 1027. (C); 1028. (A);
1029. (C); 1030. (C); 1031. (A); 1032. (C); 1033. (C);
1034. (A); 1035. (C); 1036. (B); 1037. (D); 1038. (D);
1039. (B); 1040. (C); 1041. (A); 1042. (A); 1043. (B);
1044. (D); 1045. (C); 1046. (C); 1047. (D); 1048. (C);
1049. (D); 1050. (C); 1051. (A); 1052. (B); 1053. (C);
1054. (A); 1055. (D); 1056. (B); 1057. (A); 1058. (B);
1059. (A); 1060. (C); 1061. (C); 1062. (B); 1063. (B);
1064. (A); 1065. (B); 1066. (B); 1067. (B); 1068. (D);
1069. (C); 1070. (A); 1071. (C); 1072. (B); 1073. (C);
1074. (B); 1075. (B); 1076. (A); 1077. (C); 1078. (D);
1079. (C); 1080. (B); 1081. (D); 1082. (D); 1083. (A);
1084. (D); 1085. (B); 1086. (A); 1087. (C); 1088. (D);
1089. (A); 1090. (A); 1091. (B); 1092. (D); 1093. (D);
1094. (B); 1095. (A); 1096. (B); 1097. (A); 1098. (A);
1099. (C); 1100. (A); 1101. (C); 1102. (D); 1103. (A);
1104. (C); 1105. (D); 1106. (B); 1107. (D); 1108. (A);
1109. (D); 1110. (B); 1111. (A); 1112. (C); 1113. (B);
1114. (C); 1115. (D); 1116. (A); 1117. (A); 1118. (B);
1119. (C); 1120. (D); 1121. (D); 1122. (C); 1123. (D);
1124. (A); 1125. (C); 1126. (B); 1127. (B); 1128. (A);
1129. (D); 1130. (A); 1131. (A); 1132. (B); 1133. (D);
1134. (B); 1135. (C); 1136. (C); 1137. (A); 1138. (D);
1139. (B); 1140. (A); 1141. (A); 1142. (A); 1143. (A);
1144. (D); 1145. (D); 1146. (A); 1147. (D); 1148. (B);
1149. (D); 1150. (B); 1151. (A); 1152. (B); 1153. (D);
1154. (B); 1155. (A); 1156. (C); 1157. (B); 1158. (A);
1159. (A); 1160. (B); 1161. (A); 1162. (B); 1163. (B);
1164. (B); 1165. (D); 1166. (B); 1167. (A); 1168. (D);
1169. (C); 1170. (B); 1171. (B); 1172. (C); 1173. (C);
1174. (A); 1175. (B); 1176. (B); 1177. (D); 1178. (D);
1179. (A); 1180. (A); 1181. (D); 1182. (C); 1183. (A);
1184. (D); 1185. (D); 1186. (A); 1187. (D); 1188. (A);
1189. (A); 1190. (B); 1191. (C); 1192. (D); 1193. (A);
1194. (A); 1195. (C); 1196. (D); 1197. (A); 1198. (D);
1199. (C); 1200. (C); 1201. (B); 1202. (C); 1203. (A);
1204. (D); 1205. (A); 1206. (B); 1207. (D); 1208. (D);
1209. (B); 1210. (B); 1211. (A); 1212. (D); 1213. (A);
1214. (D); 1215. (D); 1216. (A); 1217. (D); 1218. (D);
1219. (D); 1220. (B); 1221. (A); 1222. (A); 1223. (C);
1224. (B); 1225. (B); 1226. (D); 1227. (D); 1228. (C);
1229. (B); 1230. (C); 1231. (C); 1232. (B); 1233. (D);
1234. (D); 1235. (A); 1236. (B); 1237. (B); 1238. (A);
1239. (D); 1240. (C); 1241. (A); 1242. (A); 1243. (C);
1244. (D); 1245. (B); 1246. (D); 1247. (A); 1248. (D);
1249. (A); 1250. (A); 1251. (B); 1252. (B); 1253. (A);
1254. (B); 1255. (D); 1256. (B); 1257. (A); 1258. (D);
1259. (B); 1260. (C); 1261. (C); 1262. (D); 1263. (C);
1264. (C); 1265. (C); 1266. (B); 1267. (B); 1268. (A);
1269. (A); 1270. (A); 1271. (C); 1272. (A); 1273. (A);
1274. (A); 1275. (A); 1276. (B); 1277. (C); 1278. (B);
1279. (D); 1280. (A); 1281. (A); 1282. (A); 1283. (B);
1284. (D); 1285. (B); 1286. (D); 1287. (D); 1288. (A);
1289. (B); 1290. (A); 1291. (B); 1292. (A); 1293. (A);
1294. (B); 1295. (C); 1296. (B); 1297. (A); 1298. (B);

1299. (B); 1300. (A); 1301. (A); 1302. (A); 1303. (A);
1304. (A); 1305. (A); 1306. (C); 1307. (D); 1308. (C);
1309. (C); 1310. (A); 1311. (A); 1312. (C); 1313. (C);
1314. (B); 1315. (C); 1316. (D); 1317. (B); 1318. (A);
1319. (D); 1320. (C); 1321. (B); 1322. (A); 1323. (A);
1324. (A); 1325. (B); 1326. (A); 1327. (B); 1328. (A);
1329. (D); 1330. (A); 1331. (D); 1332. (B); 1333. (D);
1334. (A); 1335. (B); 1336. (C); 1337. (D); 1338. (A);
1339. (B); 1340. (B); 1341. (A); 1342. (C); 1343. (A);
1344. (A); 1345. (A); 1346. (A); 1347. (B); 1348. (C);
1349. (C); 1350. (B); 1351. (B); 1352. (B); 1353. (B);
1354. (B); 1355. (B); 1356. (D); 1357. (C); 1358. (A);
1359. (C); 1360. (D); 1361. (B); 1362. (C); 1363. (A);
1364. (A); 1365. (A); 1366. (A); 1367. (D); 1368. (A);
1369. (A); 1370. (B); 1371. (B); 1372. (A); 1373. (D);
1374. (A); 1375. (B); 1376. (A); 1377. (A); 1378. (C);
1379. (E); 1380. (A); 1381. (B); 1382. (D); 1383. (D);
1384. (A); 1385. (A); 1386. (D); 1387. (C); 1388. (A);
1389. (A); 1390. (A); 1391. (C); 1392. (B); 1393. (C);
1394. (B); 1395. (B); 1396. (C); 1397. (D); 1398. (B);
1399. (A); 1400. (D); 1401. (A); 1402. (B); 1403. (C);
1404. (A); 1405. (B); 1406. (D); 1407. (A); 1408. (C);
1409. (A); 1410. (B); 1411. (B); 1412. (B); 1413. (D);
1414. (B); 1415. (D); 1416. (B); 1417. (B); 1418. (B);
1419. (B); 1420. (B); 1421. (A); 1422. (C); 1423. (C);
1424. (B); 1425. (B); 1426. (C); 1427. (C); 1428. (D) ;
1429. (C); 1430. (D); 1431. (B); 1432. (C); 1433. (C);
1434. (B); 1435. (D); 1436. (C); 1436. (B); 1438. (D);
1439. (A); 1440. (B); 1441. (C); 1442. (C); 1443. (D);
1444. (B); 1445. (C); 1446. (C); 1447. (B); 1448. (D);
1449. (B); 1450. (C); 1451. (B); 1452. (C); 1453. (B);
1454. (A); 1455. (B); 1456. (A); 1457. (C); 1458. (D);
1459. (C); 1460. (A); 1461. (C); 1462. (B); 1463. (C);
1464. (B); 1465. (C); 1466. (C); 1467. (B); 1468. (C);
1469. (D); 1470. (C); 1471. (D); 1472. (B); 1473. (C);
1474. (B); 1475. (A); 1476. (B); 1477. (B); 1478. (D);
1479. (A); 1480. (A); 1481. (B); 1482. (D); 1483. (A);
1484. (C); 1485. (B); 1486. (D); 1487. (A); 1488. (C);
1489. (B); 1490. (B); 1491. (D); 1492. (D); 1493. (A);
1494. (B); 1495. (C); 1496. (A); 1497. (B); 1498. (B);
1499. (C); 1500. (B); 1501. (C); 1502. (C); 1503. (B);
1504. (B); 1505. (D); 1506. (A); 1507. (C); 1508. (A);
1509. (D); 1510. (D); 1511. (C); 1512. (A); 1513. (D);
1514. (D); 1515. (B); 1516. (B); 1517. (C); 1518. (B);
1519. (C); 1520. (B); 1521. (C); 1522. (C); 1523. (B);
1524. (D); 1525. (C); 1526. (B); 1527. (C); 1528. (A) ;
1529. (B); 1530. (D); 1531. (B); 1532. (D); 1533. (B);
1534. (B); 1535. (B); 1536. (B); 1537. (B); 1538. (D);
1539. (B); 1540. (C); 1541. (C); 1542. (B); 1543. (C);
1544. (A); 1545. (C); 1546. (C); 1547. (C).

Chapter 6

PLANT BIOCHEMISTRY

1. An antibody with enzymatic action, it is produced in response to stable analog of the transition state of the substrate for the given enzyme action is called:
(A) Abzyme (B) Enzyme
(C) Hormone (D) Activator

2. A phenolic compound produced by dicot plant cells at wound sites; act as signal molecule for the expression of vir regulation of *Agrobacterium tumefaciens* is:
(A) Acetosyringone (B) Abscisic acid
(C) Cytokinins (D) Auxin

3. The protein component of an enzyme that has a coenzyme/cofactor is:
(A) Abzyme (B) Enzyme
(C) Hormone (D) Apoenzyme

4. Enzyme that catalyze the transfer of amino groups from α – amino to α–keto acids is called:
(A) Transaminases (B) Amino transferases
(C) Both (A) and (B) (D) None of these

5. ATP-dependent enzymatic esterification of the carboxyl group of an amino acid to the 3'-hydroxyl group of its corresponding tRNA is called:
(A) Transaminases (B) Amino transferases
(C) Amino acid activation (D) None of these

6. An enzyme having a regulatory function through its capacity to undergo a change in catalytic activity by allosteric mechanisms or by covalent modification is called:
(A) Regulatory enzyme (B) Regulatory gene
(C) Regulatory sequence (D) Reverse transcriptase

7. Enzymatic transfer of an amino group from an α – keto acid is called:
(A) Transmination (B) Transcription
(C) Transduction (D) Transformation

8. The enzymatic process whereby the genetic information contained is one strand of DNA is used to specify a complementary sequence of bases in an mRNA chain is called:
(A) Transmination (B) Transcription
(C) Transduction (D) Transformation

9. The most common reporter proteins are the:
(A) Chloramphenicol acetyl transferase
(B) β-glucuronidase
(C) Luciferase
(D) Green Fluorescent protein
(E) All of these

10. Genes for chloramphenicol acetyl transferase, β-galactosidase and β-glucuronidase are obtained from:
(A) *E. coli*
(B) Firefly-*photinus pyralis*
(C) Jelly fish-*Aequaria victoria*
(D) None of these

11. Luciferase is obtained from:
(A) *E. coli*
(B) Firefly-*photinus pyralis*
(C) Jelly fish-*Aequaria victoria*
(D) None of these

12. Green fluorescent proteins is obtained from:
(A) *E. coli*
(B) Firefly-*photinus pyralis*
(C) Jelly fish-*Aequaria victoria*
(D) None of these

13. Proteins are clustered into families when they show close sequence homology, structural similarity and functional conservation is called:
(A) Protein family (B) Protein domain
(C) Protein fold (D) None of these

14. A functional unit of a protein in fold that is stable in solution can be defined as :
(A) Protein family (B) Protein domain
(C) Protein fold (D) None of these

15. The number of major secondary structural elements, their arrangement in three-dimensional space and their topological connectivity is called:
(A) Protein family (B) Protein domain
(C) Protein fold (D) None of these

16. There are covalently linked carbohydrate (PG) containing proteins is called:
(A) Glycoproteins (B) Lipoproteins
(C) Nulceoproteins (D) Phosphoproteins

17. There are covalently linked lipid (PG) containing proteins is called:
(A) Glycoproteins (B) Lipoproteins
(C) Nulceoproteins (D) Phosphoproteins

18. There are covalently linked nucleic acid (PG) containing proteins is called:
(A) Glycoproteins (B) Lipoproteins
(C) Nulceoproteins (D) Phosphoproteins

19. There are covalently linked phosphate (PG) containing proteins is called:
(A) Glycoproteins (B) Lipoproteins
(C) Nulceoproteins (D) Phosphoproteins

20. There are heme or iron porphyrin (PG) containing proteins is called:
(A) Hemeproteins (B) Phosphoproteins
(C) Hemoproteins (D) Glycoproteins

21. Which type of protein is found in casein of milk is called:
(A) Hemeproteins (B) Phosphoproteins
(C) Hemoproteins (D) Glycoproteins

22. Which type of protein is found in hemoglobin is:
(A) Hemeproteins (B) Phosphoproteins
(C) Hemoproteins (D) Glycoproteins

23. In some cases enzymes require additional organic and inorganic molecules is called:
(A) Cofactors (B) Enzyme inhibitors
(C) Both (A) and (B) (D) None of these

24. Many substances alter the activity of an enzyme and those which reduces the enzyme activity are called:
(A) Cofactors (B) Enzyme inhibitors
(C) Both (A) and (B) (D) None of these

25. The enzymatic rate constant (kcat/Km) of orotidine 5'-phosphate decarboxylase is $6 \times 10^{7} M^{-1}s^{-1}$ and the non-enzymatic rate constant (kn) is $3 \times 10^{-16}s^{-1}$. What is the value of the enzyme's catalytic proficiency?
(A) $2 \times 10^{23}M^{-1}$ (B) $5 \times 10^{-24}M$
(C) $12 \times 10^{-9}M^{-1}s^{-2}$ (D) 8.3×10^{7} Ms^{2}

26. Which one of the following is basic amino acid:
(A) Lysine (B) Threonine
(C) Valine (D) Glycine

27. Given the structure of DNA as described by Watson and Crick, all of the following must be true of DNA except:
(A) The 3' end of one of the strands must be aligned with the 5' end of the other strand
(B) The total number of purine bases must equal to the total number of pyrimidine bases
(C) The sequence of bases on one strand must be identical to those on the other strand
(D) The sequence of bases in the genome of one organism must be different from the sequence of bases in the genome of a different organism

28. The difference between genomic DNA and cDNA is:
(A) Genomic DNA is that which is expressed and cDNA is all the DNA
(B) Genomic DNA is found in the nucleus and cDNA is found in the cytoplasm
(C) Genomic DNA is all the DNA and cDNA is made only from genes that are expressed
(D) There is no difference between these two types of DNA

29. DNA polymerase III requires a/an ———— for synthesis of DNA to occur:
(A) DNA template
(B) RNA primer
(C) Free 3'-OH end of a DNA strand
(D) All of these

30. The subunit of DNA polymerase III is responsible for its ———— activity:
(A) 5'→ 3' polymerase (B) Sliding clamp
(C) 3'→ 5' exonuclease (proof reading)
(D) Ribosome assembly

31. Which of the following contain sulphur:
(A) Lipoic acid (B) Thiamin
(C) Riboflavin (D) Both (A) and (B)

32. The amino acids in polypeptide chains which contain sulphur (s) are:
(A) Cysteine, cystine and methionine (B) Cystine
(C) Methionine (D) Cysteine
(E) Cysteine and methionine

33. Which on of the following structure is a Zwitter ion:
(A) H N – CH (R) – COOH (B) H N + CH (R) – COO-
(C) H N – CH (R) – COO- (D) None of these

34. Acetylcholine on a smooth muscle cell is to phospholipase C as epinephrine on a liver cell is to?
(A) Protein kinase C (B) cAMP
(C) Adenylyl cyclase (D) GTP

35. A mixture of cytochrome - (MW-11.7 KD) and myoglobin (MW-17.2 KD) are to be separated by polyacrylamide gel electrophoresis. Their isoelectric pH (pI) values are 9.6 and 7.2 respectively. In which direction will each protein migrate at pH 8.5?
(A) Myoglobin will migrate to anode and cytochrome - C will migrate to cathode
(B) Myoglobin will migrate to cathode and cytochrome - C will migrate to anode
(C) Both will migrate to anode
(D) Both will migrate to cathode

36. *E. coli* DNA polymerase catalyzes polymerization in the ———— direction(s):
(A) 5'→ 3' (B) 3'→ 5'
(C) 2'→ 5' (D) Both 5'→3' and 3'→ 5'

37. Enzymes that remain bound to their nascent chains through many polymerization steps are said to be————.
(A) Constitutive (B) Distributive
(C) Conjugated (D) Processive

38. Which of the following is a monomeric unit of DNA:
(A) Nucleotide (B) 2-Deoxy ribofuranose
(C) Nucleoside (D) D-2-Deoxy-ribunucleotide

39. The pH at which the dipolar ion of amino acid do not migrate either to anode or cathode in an electric field is known as:
(A) Km Value (B) Vmax
(C) Turn over number (D) Isoelectric point

40. Fluorescent in-situ hybridization is used to:
(A) Locate sequences of DNA on a chromosome
(B) Identify mini-and microsatellite DNAs
(C) Identify homologous pairs of chromosomes
(D) Identify the centromere portion of chromosomes

41. It is difficult to determine either Km or Vmax from a graph of velocity vs. substrate concentration because:
(A) Too much substrate is required to determine them
(B) The graph is sigmoidal
(C) An asymptotic value must be determined from the graph
(D) The points on the graph are often not spread out on the hyperbola

42. Which of the following catalyzes the biosynthesis of DNA:
(A) Transcriptase (B) Nuclease
(C) Phosphorylase (D) DNA polymerase

43. The mechanism of action of lysozyme includes:
(A) Distortion of the substrat
(B) Acid catalysis
(C) Proximity effects
(D) Formation of a half – chair sugar form
(E) All of these

44. UV-spectrosopy is not used to quantitate the compounds based on absorbance:
(A) Nucleic acids have absorbance peak at 260 nm
(B) NAD (P) H have absorbance peak at 340 nm
(C) Aromatic amino acids have absorption maxima about 280 nm
(D) Sulphur containing amino acids have absorbance at 270 nm

45. Which of the following can not be digested by human:
(A) Glucose (B) Starch
(C) Cellulose (D) Dextrin

46. Serine and threonine are to As tyrosine is to receptor tyrosine kinases:
(A) Phosphatases (B) G proteins
(C) cAMP (D) Protein kinases (A) and (C)

47. When Keq of a reaction = 1, then
(A) The forward reaction is faster than the reverse reaction
(B) The reverse reaction is faster than the forward reaction
(C) The forward and reverse reaction rates are constants and equal
(D) More products are formed than reactants
(E) Fewer products are formed than reactants

48. The processively of E. coli DNA polymerase III holoenzyme accounts for all of the following except:
(A) The relatively small number of enzymes needed to replicate the entire chromosome
(B) The directionality of polymerization
(C) The rapid rate of polymerization
(D) The ability of a single holoenzyme to add many nucleotides to a growing DNA chain

49. The 3'→ 5' exonuclease activity of *E. coli* DNA polymerase III accounts for the ——of polymerization
(A) Low error rate (B) High speed
(C) Directionality (D) All of these

50. X-ray crystallographic examination of the active site of arginine kinase was possible because:
(A) It could be crystallized in the different structural forms
(B) Nitrate could be substituted for phosphate in the reaction
(C) It was readily purified (unlike most other enzymes)
(D) MgADP was involved in the reaction

51. Which of the following is a non-reducing sugar:
(A) Ribose (B) Galactose
(C) Sucrose (D) Fructose

52. Which of the following do not contain asymmetric carbon atom:
(A) Dihydroxy acetone (B) Fructose
(C) Glycine (D) Both (A) and (C)

53. Ultraviolet (UV) light can be used to estimate protein solution concentrations because:
(A) Phenylalanine absorbs at 260 nm
(B) All the amino acids absorb UV light
(C) Aromatic amino acids absorbs at 280 nm
(D) Tyrosine and tryptophan absorb at 280 nm
(E) All of these

54. The structure of different DNAs can vary in all the following ways except:
(A) The helix may be right-handed or left handed
(B) The sequence of nucleotide bases can be different
(C) The molecule can be fully hydrated as in the B form, or drier as in the A form
(D) The bases can be attached to the sugars or the phos phates of the backbone

55. Chymotrypsinogen in the native state and reduced unfolded states is loaded in wells A and B of a native polyacrylamide gel, respectively. After electrophoresis and staining, one of the following results is likely to be observed:
(A) Protein in well A would have moved longer distance than that in well B
(B) Protein in well A would have moved shorter distance than that in well B
(C) Protein in well A and B would have moved the same distance
(D) Protein in well A and B would have move at all

56. The dipeptide Lys-Glu on electrophoresis at pH 6.0 would:
(A) Move towards anode (+)
(B) Move towards cathode (-)
(C) Remain stationary
(D) Get degraded

57. Which of the following is not a reason that the actual number of human genes is still a mystery?
(A) The human genome is nowhere near completion
(B) It is not certain whether particular sequences of DNA actually contain genes
(C) Small genes with single exons can be missed by the analytical computers
(D) The genome's "dark matter" has yet to be accurately sequenced

58. Ryanodine and IP3 receptors are found on ——— membranes, whereas epinephrine and glucagons receptors are found on ——— membranes:
(A) Plasma, intracellular
(B) Intracellular, plasma
(C) Mitochondrial, endoplasmic reticulum
(D) Outer, inner

59. Concentrations of some proteins cannot estimated by UV spectrophotometry because they are:
(A) High in non-absorbing amino acids
(B) Low in tryptophan and tyrosine
(C) Low in protein
(D) High in aromatic amino acids

60. What is the biological function of lysozyme?
(A) Converts trypsinogen to trysin
(B) Is a ligase that accelerates the polymerization of glycogen
(C) It regulates vascular constriction in chickens
(D) Hydrolyzes polysaccharides of bacterial cell walls

61. Fluorescence microscopy is based on the ability of certain molecules to:
(A) Absorb light of a constant wavelength
(B) Absorb light of many different wavelength
(C) Absorb light of a given wavelength and then emit light of a longer wavelength
(D) Absorb light of a given wavelength and then emit light of a shorter wavelength

62. Polymerization of the lagging strand is catalyzed by:
(A) A second core complex on the same polymerase ho loenzyme that polymerizes the leading strand
(B) A completely separate replisome from that which polymerizes the leading strand
(C) DNA polymerase I holoenzyme
(D) The same polymerase holoenzyme that polymerized the leading strand, but with opposite directionality

63. The RNA primer at the beginning of each Okazaki fragment is removed by:
(A) DNA polymerase I
(B) DNA polymerase II
(C) DNA polymerase III
(D) 3' 5' exonuclease subunit of DNA polymerase III

64. Disulfide bridges acn form in proteins:
(A) Only between cysteine residues side-by-side in the protein sequence
(B) Between cysteine residues that are close in three-dimentional space, but not necessarily close in the primary structure
(C) Between two cystine residues in proteins
(D) Between any two methionines or cysteines

65. The process of RNA biosynthesis is known as:
(A) Translation (B) Transformation
(C) Replication (D) Transcription

66. Glyoside hydrolases such as bacterial cellulase have been shown to differ in mechanisms from that of enzyme. They:
(A) Are alkaline catalysts
(B) Form a boat of the sugar
(C) Form a covalent sugar-enzyme intermediate
(D) Do not distort the substrate

67. Which of the following is not a reason that the actual number of human genes is still a mystery?
(A) The human genome is nowhere near completion
(B) It is not certain whether particular sequences of DNA actually contain genes
(C) Small genes with single exons can be missed by the analytical computers
(D) The genome's "dark matter" has yet to be accurately sequenced

68. An inhibitors binds to a site other than the active site of the enzyme. Which statement below correlated with this observation?
(A) It must be a competitive inhibitor
(B) The inhibition must be irreversible
(C) It could be noncompetitive or uncompetitive inhibition
(D) It could be irreversible competitive, noncompetitive or uncompetitive. The data do not relate to the type of inhibition

69. The overall shape of a protein is greatly influenced by:
(A) Amino aid R group properties
(B) Charged amino acids
(C) Hydrophobic amino acids
(D) pH
(E) Hydrophilic amino acids

70. Which of the following is a heterpolysaccharide:
(A) Starch (B) Insulin
(C) Cellulose (D) Gum

71. The surplus glucose units in the liver of animals are stored in the from of:
(A) Sucrose (B) Cellulose
(C) Starch (D) Glycogen

72. Nick translation is a process for:
(A) Filling in and sealing the gap in DNA after removal of the RNA primer
(B) Synthesis of the protein for primase from mRNA
(C) Removal of the RNA primer by DNA polymerase I
(D) Removal of okazaki fragments

73. Cultured hamster cells treated with methotrexate become resistant to the drug by amplifying the gene for DHFR. This is an example of:
(A) Selective gene amplification
(B) Random gene duplication and selective survival
(C) Evolution (D) All of these

74. Okazaki fragments are:
(A) The smallest subunits of DNA polymerase III
(B) Short stretches of DNA formed on the lagging strand
(C) Short RNA primers needed for initiation of polymerization
(D) Fragments of DNA polymerase I that lack 5' 3' exonuclease activity

75. Time of action potential maximum for:
(A) Candiac muscle fibre (B) Nerve fibre
(C) Skeleton muscle fibre (D) Ganglionic fibre

76. Ramachandran determined the "allowed" values of the phi and psi angles primarily by considering:
(A) pKa values of the amino acids
(B) The hydropathy of amino acids
(C) Steric hindrance
(D) Hydrogen bonding effects

77. Induced fit studies by Koshland using the enzyme hexokinase showed that there are enzyme forms:
(A) Which react with the hydroxyl group of water
(B) Which react only with ATP and glucose present together
(C) Which are hydrophobic, excluding the competing hydroxyl group from water
(D) With and without glucose bound to each

78. One form of glycogen synthetase has a requirements for glucose-6-phosphate. What is the best explanation for this phenomenon?
(A) It is an example of negative feedback
(B) Glucose-6-phosphate is a substrate
(C) It is an example of allosteric activation
(D) It inhibits an enzyme that hydrolyses UDP-glucose

79. What is the function of the primosome?
(A) Removal of RNA from the lagging strand
(B) Synthesis of short RNA fragments
(C) Assembly of DNA polymerase III holoenzyme
(D) Detection of the origin of replication

80. Zymogens are inactive enzyme precursors which are made active by:
(A) Change in structure
(B) Selective proteolysis
(C) Secretion of new types of cells
(D) A, B and C

81. How many complete turns are there in an ideal α – helix that contains 15 amino aids and has a pith of 0.54 nm and a rise of 0.15 nm?
(A) 1 (B) 4 (C) 8 (D) 12

82. Insulin is entirely made up of:
(A) Fructose units
(B) Glucose and Fructose units
(C) Galactose units
(D) Glucose units

83. In eukaryotes, transposable elements arise by:
(A) Reverse transcription of RNA, followed by insertion into the genome
(B) Excision of a portion of the chromosome followed by insertion at a new place in the genome
(C) Duplication of transposable element's DNA followed by insertion into a new place in the genome
(D) All of these

84. Which is not a difference between B-DNA and A-DNA?
(A) A-DNA forms under more dehydrating conditions than B-DNA
(B) For polynucleotide strands containing the same number of nucleotide, the A-DNA strand will be shorter from end-to-end than the corresponding B-DNA
(C) Both are helical, but B-DNA is right –handed and A-DNA is left handed
(D) The planes of the nitrogenous bases are not perpendicular to the helix axis in A-DNA

85. What is a reasonable length for an RNA primer in *E.coli*?
(A) A few nucleotides (B) 50 nucleotides
(C) 100 nucleotides (D) 1000 nucleotide

86. The instant source of energy in human beings is:
(A) D-Glucose (B) Proteins
(C) Vitamins (D) Fats

87. In an enzyme reaction involving one enzyme and one substrate, the rate of the reaction depends on:
(A) Substrate concentration
(B) Enzyme concentration
(C) Both substrate and enzyme concentrations
(D) The enzyme concentration at first and the substrate concentration later on

88. Which one of the following bacteria contain T-DNA:
(A) *E. coli* (B) *Agrobacterium*
(C) *Rhizobium* (D) *Pneumococci*

89. At the beginning of an enzyme-catalyzed reaction the ——— is negligible:
(A) Formation of ES
(B) Formation of E + P
(C) Conversion of ES to E + S
(D) Disappearance of ES

90. Which of the following is not correct about fatty acid degradation and biosynthesis

	Degradation	Biosynthesis
(A) Location of pathway in the cell	Mitochondrion	Cytosol
(B) Electron acceptor or donor used	FAD, NAD^+	NADPH
(C) Participation of O_2	NO	Yes
(D) Carrier of intermediate acyl groups	AP-$(SH)_2$	CoA-SH

91. Fragments on the lagging strand are joined by the enzyme:
(A) DNA polymerase I (B) DNA synthase
(C) DNA primosome (D) DNA ligase

92. The concentrations of enzymes are determined by using pseudo first-order conditions in which:
(A) The concentration is kept the same and the rate is measured
(B) E concentration is kept the same and the rate is measured
(C) S concentration is constant, E concentration varies and the rate is measured
(D) E concentration is constant, S concentration varied and the rate is measured
(E) All of these

93. Which one of the following supports glycogen synthesis?
(A) High cyclic adenosine monophosphate (AMP) levels
(B) Inactive adenyl cyclase
(C) Active phosphorylase-alpha
(D) Epinephrine

94. Two strands of DNA double helix are held together tightly by:
(A) Electrostatic bonds
(B) Phosphodiester bonds
(C) Covalent bonds
(D) Hydrogen bonds

95. Which of the following is not a role that has been suggested for transposable elements in the process of evolution?
(A) Transposable elements bring previously unlinked segments together, possibly resulting in composite protein products
(B) Insertion of transposable elements into the genome results in gene duplication and the evolution of multigene families
(C) Transposable elements have no role in evolution; they are genetic parasites
(D) Natural selection always favours species with many transposable elements

96. Tachy candia means faster heart rate usually defined in adult person. It is ulases heart rate:
(A) By activating parasympathetic nervous system and blacking Na^+, Ca^{++} channels
(B) By activating sympathetic nervous system and blacking Na^+, Ca^{++} channels
(C) It deactivates sympathetic system and activates Na^+, Ca^{++} channels
(D) It activates sympathetic system, and generates action potential through the rapid entry of Na^+, Ca^{++} channels

97. When varying the substrate concentration at a fixed concentration of enzyme it is observed that at low concentrations of substrate the reaction is ———, while at high concentrations of substrate the reaction is—:
(A) Maximal, initial
(B) Initial, maximal
(C) Second order; first order
(D) First order; second order
(E) First order; zero order

98. Site – directed cross-linking is a technique that gives insight into what property of integral membrane proteins?
(A) Number of transmembrane domains
(B) On which side of the membrane the N and C termini are found
(C) Proximity of various membrane-spanning domains to each other
(D) None of these

99. Why does E. coli need both DNA polymerase III and DNA polymerase I?
(A) DNA polymerase III lacks the 5'→ 3' exonuclease activity needed to remove RNA primers
(B) Each polymerase is specific for only one strand of DNA. DNA polymerase III acts only
(C) DNA polymerase I is needed to join the Okazaki fragments
(D) The DNA replication is bidirectional; one polymerase is used for each direction
(E) Only DNA polymerase I has proof reading ability

100. In candiac muscle fibess, generation of action potential through:
(A) Rapid and slow entry of Na^+ and Ca^{++} ions, respectively
(B) Rapid entry of Na^+ only
(C) Slow entry of Ca^+ only
(D) Slow and rapid entry of Na^+ and Ca^{++} , respectively

101. Induced fit of enzyme activation is a result of:
(A) Contact (binding) of substrate
(B) Constant small and rapid motions of protein atoms
(C) Change to an active form
(D) All of (A), (B) and (C)

102. Active trypsin formation by the action of enteropeptidase can be viewed as the master activation step because:
(A) Enteropeptidase can activate its own zymogen
(B) It is allosterically controlled
(C) Trypsin activates other pancreatic zymogens
(D) All of these

103. Which one of the following correct about QRS complex in electrocardiogram:
(A) It is ventricular depolarisation having voltage smv
(B) It alrial depolarization having voltage (1-1.5) m.v.
(C) It is ventricular cystole and has a voltage of (1-1.5) m.v.
(D) It is ventricular depolarization, have a voltage of (1-1.5)

104. In cellulose, different D-Glucose units are linked by:
(A) a, D -1, 2-Glucosidic bonds
(B) b, D- 1, 4-Glucosidic bonds
(C) a, D- 1, 4- Glucosidic bonds
(D) b, D- 1, 4- Galactosidic bonds

105. The structure of two enantiomers (enantiomorps) of glucose are different only in the configuration of carbon atom at:
(A) C-1 (B) C- 2 (C) C-4 (D) C-5

106. Which might be a method used for distinguishing among several mechanistic possibilities in a multisubstrate reaction?
(A) Measure Km in the presence of one substrate at a time
(B) Measure the rate with respect to one substrate while varying the concentration of another substrate
(C) Measure the rate while adding inactivators for all but one substrate
(D) Any of these

107. Which of the following statements about glycogen storage are incorrect (more than one is correct):
P. Glycogen is stored in muscles and liver
Q. Glycogen is a major source of stored energy in brain
R. Glycogen reserves are less rapidly depleted than fat reserves during starvation
S. Glycogen storage ours in the form of dense granules in the cytoplasm of cells
(A) P and Q (B) P and R
(C) Q and R (D) Q and S

108. In inspired air which one of the following gas shows no change in its partial pressure
(A) O_2 (B) N_2
(C) CO_2 (D) H_2O vap

109. Proteins with SH2 domains are to ——— as adenylyl cyclase and phospholipase C are to G protein-coupled receptors:
(A) IRSs (B) Receptor tyrosine kinases
(C) Insulin (D) Autophosphorylation

110. When the concentration of a substrate inside a cell falls below the substrate concentration at half maximal velocity:
(A) ES is formed more easily
(B) ES dissociates to E + S
(C) ES is loser to the ground state than to the transition state
(D) E + S is in equilibrium with ES

111. The initial velocity of an enzyme reaction (v0) describes:
(A) The concentration of the enzyme at maximal velocity
(B) The concentration of substrate at maximal velocity
(C) The concentration of both at the start of the reaction
(D) The rate of reaction at when the substrate and enzyme are first mixed

112. Patent ducters antriosus generally occurs in case of:
(A) Tetralogy of fallot
(B) Blue baby syndrome
(C) When blood is purifies by lungs
(D) In newly born child when ducters does not lose properly

113. Which of the following is not true about ketogenesis:
(A) Acetyl – CoA acts as precursor
(B) HMG CoA synthase participates during reaction
(C) Occurs in the liver
(D) Acetone is the only product

114. Which of the following feature is not true about Z-DNA:
(A) Left handed helical structure
(B) Dinucleotide repeating unit
(C) 10.4 bp per helical turn
(D) Anti and syn conformation of glycosidic bond both

115. Ratio of end systolic volume stroke volume 2 end diasystolic volume:
(A) 1 : 1 : 2 (B) 3 : 7 : 9
(C) 1 : 1.5 : 3 (D) 2 : 6 : 8

116. Starting from a glucose residue in glycogen, how many net ATP molecules will be formed in the glycolysis of the residue to pyruvate?
(A) 1 (B) 2 (C) 3 (D) 4

117. Which mutant form of ras is likely to cause malignancy?
(A) ras that cannot hydrolyze GTP
(B) ras that cannot bind to GTP
(C) ras that cannot bind to Grb2 or Sos
(D) ras that cannot bind to Raf

118. Match the following:
a. Renett cell 1. Round worms
b. Solenocyte 2. Arthropode
c. Metarephridia 3. Molluscass
d. Coxal gland 4. Fasciola
(A) a-1, b-4, c-3, d-2 (B) a-4, b-3, c-1, d-2
(C) a-1, b-2, c-3, d-4 (D) a-1, b-4, c-2, d-3

119. A membrane is examined using freeze-fracture and freeze-etch analysis. The surfaces are shown to have almost no pits or bumps. Which of the following is the most likely source of the membrane?
(A) Mitochondria, inner membrane
(B) Myelin
(C) Red blood cell (D) All of these

120. In which of the following respect A from of DNA differs from B-form of DNA:
(A) Helic handedness
(B) Repeating unit
(C) Conformation of glycosidic bond
(D) Base pair per helical turn
(E) All of these

121. Which of the following is common features of both glycogen synthesis and glycogen breakdown?
(A) Both require UDO-glucose
(B) Both involve glucose-1-phosphate
(C) Both are driven in part by the hydrolysis of pyrophosphate
(D) Both use the same enzyme for branching and debranching

122. The MAP kinase cascade stimulates transcription via specific transcription factors, and stimulates translation via:
(A) SREs (B) MAPK
(C) PHAS-I and eIF4E (D) All of these

123. Some antibody molecules are enzymatic if they are formed against:
(A) Enzyme
(B) Enzyme bound to other proteins
(C) Transition state analogs bound to other proteins
(D) Transition state analogs

124. The substrate specificity of serine proteases is primarily due to:
(A) A specificity pocket in the protein
(B) The positions of specific side chains of serine, histidine and aspartate
(C) Distinct backbone conformations of the individual proteins
(D) All of (A), (B) and (C)

125. An update of Fisher's lock-and-key theory of enzyme specificity view the ——— as the lock and ——— as the key:
(A) Enzyme; substrate
(B) Substrate; enzyme
(C) Enzyme; transition state
(D) Transition state; enzyme

126. A helical wheel can be used to show:
(A) The pith of a helix structure
(B) The amphipathic nature of a helix
(C) DNA binding (D) Pleated sheets

127. The sequence of events that occurs after insulin binds to its receptor tyrosine kinase is:
(A) Binding of IRS → phosphorylation of proteins with SH2 domains → effect
(B) Binding of proteins with SH2 domains phosphory lation of IRS→ Effect
(C) Autophosphorylation and phosphorylation of IRS→ Phosphroylation of proteins with SH2 domains→ Effect
(D) Autophosphorylation and binding of IRS→ Phos phorylation of proteins with SH2 domains→ Effects

128. What second messenger has been implicated in blood pressure regulation?
(A) cAMP (B) NO (C) CO (D) IP3

129. Which is not true about β–sheets?
(A) The side-chains of all amino acids point to the same side of the sheet
(B) The polypeptide chains in the sheet are nearly fully extended
(C) The range of allowed phi and psi angles is broader than for those in the –helix
(D) In ant parallel sheets the hydrogen bonds between adjacent strands are nearly perpendicular to the backbones of the strands

130. After a meal, blood glucose enters cells and is stored as glycogen, particularly in the liver. Which of the following is the donor of new glucose molecules in glycogen?
(A) UDP-glucose-1-phosphate
(B) UDP-glucose
(C) UDP-glucose-6-phosphate
(D) Glucose-6-phosphate

131. Ab sandwich forms when:
(A) Two hydrophobic sides of b-sheets interact
(B) Two hydrophilic sides of b-sheets interact
(C) An a-helix separates two b-sheets
(D) Two amphipathic a-helices interact

132. Which one of the following organisms would you predict has the highest proportion of unsaturated fatty acids in their membranes?
(A) Antarctic fish (B) Cactus
(C) Bacteria from thermal hot springs (D) Humans

133. Which is not an activity for DNA polymerase I?
(A) 5'→ 3' polymerase activity
(B) Joining of Okazaki fragments
(C) 5'→ 3' exonuclease activity
(D) Nick translation

134. What is a Klenow fragment?
(A) A fragment of DNA polymerase I that lacks 5'→ 3' exonuclease activity
(B) A topoiromerase responsible for unwinding DNA at the replication fork
(C) A portion of the replisome responsible for RNA primer synthesis
(D) Short stretches of DNA formed during polymerization of the lagging strand

135. Super secondary structures that contain recognizable combinations of á–helices, β-strands and loops (e.g. the Greek key) are called:
(A) Domains (B) Folds
(C) Homologous regions (D) Motifs

136. The beta-oxidation of a molecule of palmitic acid:
(A) Yields 8 molecules of acetyl-CoA
(B) Yields 16 molecules of acetyl-CoA only
(C) Yields carbon dioxide and water only
(D) Does both involve oxygen

137. A mutation in the FAK gene would likely result in:
(A) Abnormal cell shape
(B) Abnormal would healing
(C) Abnormal cell growth
(D) All of these

138. The physiological roles of bile salts include which of the following:
P. They aid in the digestion of lipid
Q. They facilitate the absorption of sugars
R. They facilitate the absorption of lipid
S. They provide a means for excreting cholesterol
(A) P and Q (B) P and R
(C) Q and S (D) P, Q and S

139. How many monomers are there in an oligomeric protein to each other are:
(A) 2 (B) 3 (C) 4 (D) 5

140. Which of the following statements about the hormonal regulation of glycogen synthesis and degradation are correct?
P. Insulin increases the capacity of the liver to synthesize glycogen
Q. Insulin in secreted in response to low levels of blood glucose
R. Glucagon and epinephrine have opposing effects on glycogen metabolism
S. Glucagon stimulate the breakdown of glycogen, particularly in the liver
(A) P and Q (B) Q and R
(C) P and R (D) P and S

141. What would you expect about the formation of an α–helix for a segment of a protein chain that contains lysine approximately every fourth residue with all other residues being mostly hydrophobic?
(A) Helix formation would be favoured at low pH
(B) Helix formation would be favoured at high pH
(C) Helix formation would be favoured at neutral pH
(D) Helix formation would never our regardless of pH

142. NA primers are synthesized about once every:
(A) Millisecond (B) Second
(C) Minute (D) 40 minutes

143. Whch proteins are responsible for the unwinding of the double-stranded DNA during replication?
(A) Ligases (B) Helicases
(C) Toposiomerases (D) Primases

144. The sequence of genes and proteins used in the initiation and termination of DNA replication is:
(A) Tus→ ter→ origin→ DnaA
(B) dnaA→ DnaA → origin→ Tus→ ter
(C) Origin→ tus→ ter→ DnaA
(D) Ter→ Tus→ dnaA→ origin→ DnaA

145. The principle forces holding subunits of an oligomeric protein designated $\alpha\beta 2\gamma 2$?
(A) Peptide bonds
(B) Hydrophobic interactions
(C) Covalent bonds
(D) Disulfide bonds

146. A key role of the hydroxyl group at position 6 in the purine ring in the formation of a transition state by the enzyme adenosine deaminase is obtained by comparing a ——— and a ———:
(A) Competitive inhibitor; noncompetitive inhibitor
(B) Transition state analog; normal substrate
(C) Noncompetitive inhibitor; transition state analog
(D) Competitive inhibitor; transition state analog

147. The role of serine at the active site of serine proteases is to at as a (n) ——— catalyst, while the histidine residue serves as a(n) ——— catalyst:
(A) Strong; weak (B) Weak; strong
(C) Acid base; covalent (D) Covalent; acid base

148. A membrane-affiliated protein is isolated by detergent extraction, and found to be amphipathic and lacking any nonprotein components. To which class of membrane proteins does it belong?
(A) Integral proteins (B) Peripheral proteins
(C) Lipid-anchored proteins (D) None of these

149. Which of the following statements about the mobility of membrane components is true?
(A) Proteins and lipids move equally rapidly within the plane of the membrane
(B) Lipids generally move more rapidly than lipids within the plane of the membrane
(C) Proteins generally move more rapidly than lipids within the plane of the membrane
(D Lipids and proteins "flip-flop" between membrane leaflets more readily than they move within the plane of the membrane

150. Protein subunits in a multisubunit protein are held to each other primarily by:
(A) Covalent bonds
(B) Hydrophobic interactions exclusively
(C) Both strong and weak interactions
(D) Hydrophobic and other weak interactions

151. Which of the following is not about beta oxidation of fatty acid containing even number of carbons:
(A) End product – acetyl-CoA
(B) Cofactor required – NAD^+ and FAD
(C) Decreases during starvation
(D) Site – mitochondria

152. If phosphoglyceride A has a higher T_m than phosphoglyceride B, which of the following differences between A and B may exist? (In each case only one parameter-either chain length or double bonds-is compared)
P. A has shorter fatty acid chains than B
Q. A has longer fatty acid chains than B
R. A has more unsaturated fatty acid chains than B
S. A has more saturated fatty acid chains than B
(A) P and Q are correct (B) Only Q is correct
(C) Q and S are correct (D) R and S are correct

153. A tetramer protein contains:
(A) Four different subunits
(B) Four identical subunits
(C) Three subunits and one prosthetic group
(D) A or B only

154. Which of the following statements correctly describe what happens when acetyl CoA is abundant?
P. Pyruvate carboxylase is activated
Q. Phosphoenolpyruvate carboxykinase is activated
R. Phosphofrutokinase is activated
S. If ATP levels are high, oxaloacetate is diverted to gluconeogenesis
(A) P and Q (B) Q and R
(C) Q and S (D) P and S

155. Which of the following is mismatched?
(A) ter : Tus (B) dnaA : DnaA
(C) Ori : Primer (D) Tus : helicase inhibitor

156. The best enzyme to use in the Sanger method for DNA sequence determination is:
(A) DNA polymerase I
(B) Klenow fragment of DNA polymerase I
(C) DNA polymerase from a high temperature bacterium
(D) SSB protein

157. Which enzyme below is fastest?
(A) Kinase, $K_{cat} = 10^3$
(B) Papain, $K_{cat} = 10$
(C) Carboxypeptidase, $K_{cat} = 10^2$
(D) Catalase, $K_{cat} = 10^7$

158. Which of the following glycosidic bond is responsible for branching in amylopectin portion of starch:
(A) a, D-1, 4-Glucosidic bonds
(B) b, D-1, 4-Glucosidic bonds
(C) a, D-1, 6-Glucosidic bonds
(D) a, D-1, 2-Galactosidic bonds

159. The structure of two anomers of fructopyranose differ only in respect of orientation of –OH group at :
(A) C-2 (B) C-1
(C) C-5 (D) C-4

160. A hyperbolic binding curve differs from a sigmoidal binding curve in that the hyperbolic curve:
(A) Has a single equilibrium constant for oxygen binding
(B) Binds more oxygen after the initial proteins first bind oxygen
(C) Shows cooperativity
(D) Binds up to four molecules of oxygen

161. Transition state analogs should:
(A) Stabilize transition states
(B) Have a dissociation constant of 10^{-13} M or less
(C) Bind very tightly to the enzyme
(D) Both (A) and (C)

162. Which factors does not help to explain why many proteins exhibit quaternary structure?
(A) Oligomers are usually more stable than the free monomers
(B) Active sites can be formed when the protein chains associate
(C) The subunits always are able to maintain the same three-dimentional structure whether they are associated into an oligomer or not
(D) Increased efficiency by the sharing of the same sub unit with the same function among different proteins

163. The catalytic triad of hymotrypsin and other serine proteases consists of:
(A) Three subunits of the enzyme
(B) Three amino acid residues adjacent in the primary structure which at to make serine a strong nucleophile
(C) Three amino acid residues close enough in space to make serine a strong nuleophile
(D) Three enzymes with very similar structural features

164. Regulation of fatty acid biosynthesis occurs at the enzymatic step catalyzed by:
(A) Carnitine acyl transferase I
(B) Acetyl – coA carboxylase
(C) Pyruvate carboxylase
(D) Citrate-malate translocase

165. All of the following statements about acetyl-CoA carboxylase are correct except:
(A) It undergoes protomer-polymer interconversion during its physiological regulation
(B) It requires biotin
(C) It is inhibited by cAMP-mediated phosphorylation
(D) It is activated by both palmitoyl-CoA and cirate

166. Cooperative binding of oxygen by hemoglobin:
(A) Is induced by hemoglobin
(B) Is a result of different affinities for oxygen by each subunit protein
(C) Is induced by oxygenation
(D) Is a result of interaction with myoglobin

167. Which statement is false about the heme group?
(A) When oxygen binds to heme, the iron ion is oxidized from Fe^{2+} to Fe^{3+}
(B) If exposed to air, a free heme group (not associated with hemoglobin) is readily oxidized converting Fe^{2+} to Fe^{3+} and an no longer bind oxygen
(C) The heme group is tightly, but non-covalently, held in myoglobin molecule (D) The chemical structure of the heme groups in myoglobin and hemoglobin are identical

168. The Michaelis constant, Km, is equal to the:
(A) Maximum velocity that any given enzyme reaction can achieve
(B) Substrate concentration which gives the best enzyme assay for an enzyme reaction
(C) Substrate concentration when the rate is equal to half its maximal value
(D) Maximum velocity divided by two

169. Which one of the following set have a max different of pH?
(A) Saliva-uria (B) Bile-blood
(C) Bile-gastric juice (D) Saliva-gastric juice

170. The ratio of kcat/km for each of two different substrates present in equal concentrations in an enzyme reaction will measure enzyme affinity because:
(A) The rate of P formation are given by the values found for each substrate
(B) The rate of P formation are the same for each substrate
(C) It is an indication of the formation of ES for each substrate
(D) All of these

171. The main difference between prokaryotic and eukaryotic DNA replication mechanisms is:
(A) No Okazaki fragments are made in eukaryotes
(B) Both leading and lagging strands are contrinuous in eukaryotes
(C) No primer synthesis is needed in eukaryotes
(D) The replication fork moves slower in eukaryotes

172. Functions of DNA polymerases and neccessory proteins in eukaryotic DNA replication include:
(A) PCNA protein, a sliding clamp unity for DNA
(B) β-DNA polymerase, for repair of damaged DNA
(C) RPC protein, and equivalent of SSB proteins in bacteria
(D) RPC protein, used for mitochondrial DNA synthesis

173. In packaging the eukaryotic DNA into nucleosomes, it is though that the requirement for ———— and their synthesis is a cause of the slower rate of movement of the replication fork:
(A) Okazaki fragments
(B) Larger DNA molecules
(C) Increase number of accessory proteins
(D) Histone proteins

174. Membrane carbohydrates occur as:
(A) Either glycoproteins or glycolipids, but are not free in the membrane
(B) Either glycoproteins, or free in the membrane, but are not attached to lipids
(C) Either glycolipids or free in the membrane, but are not attached to proteins
(D) None of these

175. Substance X is a charged molecules of small molecular weight. It is found at higher concentrations outside of cells than inside. How might substance X enter cells?
(A) Diffusion through a channel
(B) Diffusion through the lipid bilayer
(C) Active transport
(D) None of these

176. An equimolar mixture of D-and L-alanine does not rotate the plane of polarized light is:
(A) True (B) False
(C) Unknown (D) Same

177. The proximity effect of speeding an enzyme-catalyzed reaction is explained by:
(A) A large loss of entropy when reactive groups are brought close to each other
(B) A large gain in entropy due to binding of the substrate
(C) A conversion of a reaction from endergonic to exergonic
(D) Increasing the flexibility of the substrate by increasing the degrees of freedom of rotation

178. The role of amino acid residues at the active site of enzymes can be determined by removing certain residues using the technique of:
(A) Specific hydrolysis
(B) Covalent binding
(C) Acylation of specific residues
(D) Site mutagenesis
(E) All of these

179. Which of the following disaccharide is made up of two D-Glucose units:
(A) Sucrose (B) Lactose
(C) Maltose (D) None of these

180. In the Lineweaver-Burk plot of an enzyme reaction, the Km is given by the:
(A) x-intercept
(B) y-intercept
(C) Negative reciprocal of the x-intercept
(D) Reciprocal of the y-intercept

181. The C-2 epimer of D-Glucose is:
(A) D-Mannose (B) D-Galactose
(C) D-Fructose (D) D-Altrose

182. Aspartate and lysine are in the active site of an enzyme. They are both known to participate directly in catalysis. The pKa's of the residues are found to be 3.2 and 9.6, respectively for aspartate and lysine. The optimum pH for the enzyme is 6.4. Which forms of these two residues will predominate when the enzyme is most active?
(A) Aspartate is protonated; lysine is deprotonated
(B) Both residues are protonated
(C) Aspartate is deprotonated; lysine is protonated
(D) Both residues are dprotonated

183. The Lineweaver-Burk plot and other linear transformation of the Michaelis-Menten curve of kinetics are valuable for:
(A) Determination of Km
(B) Determination of Vmax
(C) Determination of kcat
(D) Determination of types of enzyme inhibition
(E) All of these

184. The Km values for enzyme reactions such as $A + B \rightarrow C + D$
(A) Cannot be determined using the Lineweaver-Burk plot analysis
(B) Can be determined by holding one (A or B) at high concentration, while varying the concentration of the other substrate
(C) Can be determined for one substrate and not the other
(D) Do not indicate the efficiency of the enzyme

185. Gorter and Grendel's earliest evidence that membranes were formed from a lipid bilayer included:
(A) Microscopic examination of red blood cell membranes
(B) Measuring the surface area of lipids extracted from red blood cells, and comparing that to the surface area of the cells
(C) Measuring the mobility rates of membrane proteins

186. The distribution of K^+ across an artificial membrane was measured, and the concentrations were found to be equal on both sides. Which of the following statements is true about the distribution of K^+?
(A) K^+ must be at equilibrium across the cell membrane
(B) K^+ cannot be at equilibrium across the membrane
(C) There cannot be a membrane potential under these conditions
(D) More information is needed to determine whether K^+ is at equilibrium

187. Why is only one copy of DNA made per cell division cycle in eukaryotes despite the presence of multiple origins of replication?
(A) There is only one origin of replication per chromosome
(B) The rate of elongation proceeds at a rate that allows only one copy to be made by the time the cell divides
(C) The activity of an S-phase protein kinase controls the formation of the pre-replication complex
(D) Primase is activated by phosphorylation only at the onset of S-phase by S-phase protein kinase

188. The reaction catalyzed by DNA photolyase is an example of what type of repair?
(A) Recombination (B) Excision
(C) Mismatch (D) Direct

189. The assumptions made in calculating the Michaelis-Menten equation include:
(A) That the formation and decomposition of ES is the same for a period of time
(B) That the concentration of the substrate is much greater than the concentration of E
(C) That the value of K-2 can be ignored
(D) All of (A), (B) and (C)
(E) (A) and (B) only

190. Antibodies are:
(A) Globular proteins (B) Fibrous proteins
(C) Conjugated proteins (D) Metallo proteins

191. In which compound delocalisation is not possible?
(A) 2- butane (B) 1, 3-butadiene
(C) 1, 3, 5-hexatriene (D) Benzene

192. Stability of alkyl carbocations can be explained by:
(A) Induction effect (B) Hyperconjugation
(C) Both inductive (D) Electrometric effect

193. A mixture of three peptide A, B and C is loaded in a 18 reverse phase chromatography column. A mixture of water and acetonitrile is used as the solvent with increased concentration of the latter I a gradient mode. If the hydrophobicities of the peptides are in the order $A < B < C$, the order in which they will elute from the column would be:
(A) A followed by B followed by C
(B) B followed by B followed by A
(C) C followed by B followed by A
(D) A followed by C followed by B

194. Which of the following is not true of membrane phosphoglycerides?
(A) They include phosphatidylcholine, phosphatidyle thanolamine and cholesterol, among others
(B) They are amphipathic
(C) They have two fatty acids attached to a glycerol molecules, and a polar head group attached via a phosphate
(D) They have both saturated and unsaturated fatty acids

195. If a fresh water plant we put in sea water, its cells would:
(A) Undergo plasmolysis (B) Lyse
(C) Swell with turgor pressure (D) All of these

196. In photoreactivation, DNA damage caused by UV light is reversed by DNA photolyase which:
(A) Is activated by UV light to form a complex
(B) Removes nicks in the DNA backbone
(C) Forms a complex with thymine dimmers which absorbs visible light and reverses the dimerization
(D) Binds to the adenine bases opposite the thymine dimer
(E) All of these

197. Why is DNA damage in skin cells from exposure to excessive UV light not completely reversed by photoreactivation?
(A) DNA photolyase in humans is activated only under conditions of heat shock
(B) Humans lack a photoreactivation repair mechanism
(C) The presence of histones greatly slows photoreactivation, thus humans depends primarily on other forms of DNA repair
(D) Photoreactivation is specific for the removal of thym ine dimmers. Thymine dimmers do not form in human DNA

198. Which statement is true about the removal of reversible inhibitors from enzyme solutions?
(A) PAGE is the primary technique used for this purpose
(B) Dialysis and gel filtration are often used
(C) These inhibitors cannot be separated from the enzyme without treatment with Disopropylfluorophosphate
(D) Removal of reversible inhibitors is extremely difficult and can be achieved only after many purification steps

199. In a certain enzyme-catalyzed reaction the following steps our:
1. A phosphate group on substrate A is transferred to a side chain of an active residue of the enzyme
2. The dephosphorylated form of substrate A dissociates from the enzyme
3. Substrate B enters the active site and is phosphorylated with simultaneous regeneration of the enzyme in its original form

What kind of kinetic mechanism is described?
(A) Random (B) Sequential
(C) Ordered (D) Ping-pong

200. For determining the K_m values of enzyme catalyzed reaction, it is customary to plot the kinetic data in a variety of ways. The plot of V versus V/S yields a straight line and the intercepts on the Y-axis gives:
(A) K_m/V_{max} (B) V_{max}/K_m
(C) $1/V_{max}$ (D) Hill coefficient, n

201. What distinguishes reversible inhibitors from irreversible inhibitors?
(A) Reversible inhibitors are not covalently bound to enzymes but irreversible inhibitors are
(B) There is an equilibrium between bound and unbound reversible inhibitor. There usually is little back reaction for the binding of an irreversible inhibitor
(C) Reversible inhibitors are easier to purify from solutions of enzymes than irreversible inhibitors
(D) All of these

202. The reason to rewrite the Michaelis-Menten equation (such as the Lineweaver-Burk plot) is to:
(A) Visualize reactions better
(B) Form enzyme kinetic data as a hyperbolic curve
(C) Calculate catalytic proficiency
(D) Calculate Vmax and Km

203. Two substances are used to produced a certain biological product in an enzyme-catalyzed reaction. It is found that both substrates must bind to the enzyme, first one, then the other before the product is produced. This is an example of a/an:
(A) Linear reaction
(B) Ordered sequential reaction
(C) Random sequential reaction
(D) Ping-pong reaction

204. A thermodynamic pit occurs when:
(A) ES is not very stable
(B) ES forms faster than it dissociates
(C) ES is highly stable
(D) S is not bound tightly to an enzyme

205. Non classical competitive inhibition involves
(A) Binding of the substrate at both the active site and at the inhibitor site
(B) Binding of either the substrate to the active site or the inhibitors to its own binding site thus preventing the other from binding
(C) Binding of the inhibitor to the substrate followed by binding of this complex to the active site
(D) Chemical removal of the substrate form the active site of reaction with the inhibitor

206. Which among the given compound is most stable?
(A) 1- butane (B) 2, 3-dimethyl-2 butene
(C) cis-2-butene (D) trans-2-butene

207. A mixture containing protein – 1, -2, -3, -4 and -5 with molecular weights 5000, 10000, 25000, 65000 and 100000, respectively, were separated as Sephadex G-50 column. The order of elution of these proteins from the column will be:
(A) Protein -1, protein – 2, protein – 3, protein – 4, and protein - 5
(B) Protein -5, protein – 4, protein – 3, protein – 2, and protein - 1
(C) Protein -1, -2, and – 3 elute first, followed by protein – 5 and - 4
(D) Protein - 4, and – 5 elute first, followed by protein – 3, - 2, and - 1

208. The general excision repair pathway for DNA repair has the following order:
(A) Endonuclease cuts → helicase or endonuclease removes→ DNA polymerase→ DNA ligase
(B) DNA ligase→ Endonuclease cuts→ helicase or endonuclease removes→ DNA polymerase
(C) helicase or endonuclease removes→ Endonuclease cuts→ DNA polymerase→ DNA ligase
(D) DNA polymerase→ DNA ligase→ helicase or endonuclease removes→ Endonuclease cuts

209. In DNA repair, DNA glycosylases recognize deaminated bases within the DNA polymer and remove the deaminated:
(A) Nucleoside (B) Nucleotide
(C) Thymine dimmers (D) Bases

210. CH3 group of toluene is o, p-directing groups due to the:
(A) Hyperconjugation (B) Resonance
(C) Inductive effect (D) Electromeric effect

211. In a ping-pong reaction which does not occur?
(A) One product is released before a second substrate is bound
(B) The enzyme covalently binds a portion of the first substrate
(C) The enzyme is permanently converted to an altered form by the first substrate
(D) A group is transferred from one substrate to an other

212. Which of the following statements about gluconeogenesis are true?
P. It occurs actively in the muscle during periods of exercise
Q. It occurs actively in the liver during periods of exercise or fasting
R. It occurs actively in adipose tissue during feeding
S. It occurs actively in the kidney during periods of fasting
(A) P and Q (B) Q and R
(C) Q and S (D) P and S

213. Which statement is correct for inductive effect?
(A) It is permanent effect
(B) It is the property of single bond
(C) It is cause permanent polarization in the mole
(D) All of these

214. Which statement is correct for toluene:
(A) It has six vinylic carbons
(B) It has one benzyl carbon
(C) Benzyl hydrogen is acidic in character due to resonance
(D) All are correct

215. Match the transition state or chemical entity of each enzyme that is responsible for their catalytic function:

Column I	Column II
P. Ribonuclease	1. Oxyanion
Q. Lysozyme	2. Pentacovalent phosphorus
R. Chymotrypsin	3. Carbonium ion
S. Carboxypeptidase	4. Mixed anhydride

(A) P -3, Q – 2, R – 4, S - 1
(B) P -2, Q – 3, R – 1, S - 4
(C) P -2, Q – 1, R – 3, S – 4
(D) P -4, Q – 3, R – 2, S – 1

216. Which statement does not apply to transition states?
(A) Many have been detected experimentally
(B) Chemically bonds are in the process of being formed and broken
(C) Have lifetimes on the order of 10-14 to 10-13 seconds
(D) Differ in energy from the ground state by the activation energy

217. Two or more polypeptide chains present in the protein molecule are covalently cross linked by one or more:
(A) Diester bonds (B) Hydrogen bonds
(C) Disulphide bonds (D) Ionic bonds

218. Match the coefficients in group 1 with their corresponding downstream processing steps given in group 2:

Group 1	Group 2
P. Sedimentation coefficient	1. Aqueous two-phase extraction
Q. Partition coefficient	2. Ultrafiltration
R. Rejection coefficient	3. Dialysis
S. Activity coefficient	4. Centrifugation

(A) P -3, Q – 1, R – 4, S - 2
(B) P -2, Q – 1, R – 4, S 3
(C) P -4, Q – 3, R – 1, S – 2
(D) P -4, Q – 1, R – 2, S – 3

219. Protamins and histones form conjugated proteins in association with:
(A) Carbohydrates (B) Glycerides
(C) Nucleic acids (D) Vitamins

220. Most Km values of enzymes for their substrates are on the order of ———M:
(A) 10^{-2} (B) 10^{-3} (C) 10^{-4} (D) 10^{-5}

221. A histidine was determined to be the critical residue involved in an enzyme catalyzed reaction. If the Ka of the histidine is known to be 6.5 in the active site and the pH of maximum catalytic activity is 7.2. What is likely the primary role of histidine in the catalytic reaction?
(A) Act as a proton donor
(B) Forms a covalent bond with the substrate
(C) Stabilizes a charged intermediate
(D) Reduces the entropy of the substrate

222. The number of elelocalised ð–electrons in the given compound is:
(A) 4 (B) 6 (C) 8 (D) 10

223. Which statement is correct for CH_3- CH_2:
(A) It is paramagnetic is character
(B) It is neutral electrophile
(C) Formation take place hemolytic bond fission
(D) All of these

224. The T_m of phosphatidyl choline A is higher than T_m of phosphatidyl choline B because:
(A) A has shorter chain fatty acid and more unsaturated fatty acid than B
(B) A has longer chain fatty acid and more saturated fatty acid than B
(C) A has shorter chain fatty acid than B
(D) A has more is-unsaturated fatty acid than B

225. Deaminated bases must be removed from DNA because:
(A) They distort the molecule
(B) They form thymine dimmers
(C) They repair incorrectly during replication
(D) The mRNA formed may be irregular

226. In the holiday model describing recombination:
(A) Homologous sequences line up
(B) Strands of DNA are nicked
(C) DNA strands associate with the homologous strand ("strand invasion")
(D) Strands rotate then are cleaved at the crossover point
(E) All of these

227. In the nonpolar environment of most enzyme active sites, which statement applies to charge-charge interaction between the enzyme and the substrate?
(A) They are rare due to the non polar environment
(B) They are frequent, but not very strong in the non polar environment
(C) They are stronger in the non polar environment
(D) The polarity of the active site has no effect on the strength of the charge-charge interactions

228. The Bohr effect in hemoglobin refers to the:
(A) Effect of pH on hemoglobin and myoglobin
(B) Higher pH found in actively metabolizing tissues
(C) Increased affinity for O_2 at lower pH
(D) Reduced affinity for O_2 at lower pH

229. Primary structure of protein indicates:
(A) Sequence of amino acid present in it
(B) Helical and coiled shape of protein molecules
(C) Steric interactions of different amino acids present in it
(D) Both (A) and (C)

230. The test is used to detect amino acids, peptides and proteins is:
(A) Polarimeter test (B) Ninhydrin test
(C) Bardfored test (D) Biuret test

231. Which one of the following is iron containing protein:
(A) Cytochrome 'C' (B) Ascorbic acid anhydrase
(C) Alcohol dehydrogenase (D) Nitrogenase

232. In toluene methyl group has neither -ve charge nor lp even then it is o, p-directing group for arse reaction due to:
(A) Induction effect (B) Electrometric effect
(C) Hyperconjugation (D) Mesomeric effect

233. How does hemoglobin carry carbon dioxide generated in tissues back to the lungs:
(A) By coordination with heme
(B) By forming N-terminal carbamate
(C) By forming – terminal carbamate
(D) By linking to the epsilon –amino group of lysine

234. The enzyme has an active site which:
(A) Fits the substrate exactly
(B) Fits the transition state
(C) May contain hydrogen bonds which are covalent-like
(D) Both (B) and (C)

235. The role of ser-195 in hymotrypsin cleavage of a peptide bond is that of a(n):
(A) Acid catalyst (B) Proximity effector
(C) Strong nucleophile (D) Weak nucleophile

236. Which one of the following reagent is used in the determination of amino acid sequences of protein at the N-terminal of the protein chain:
(A) 2,4-Dinitrofluorobenzene
(B) Carboxypeptidase
(C) 2,4-Dinitropheny hydrazine
(D) Hydroxyl amine

237. The formation of the anlinotiazolinone derivative (ATZ-amino acid) in anhydrous trifluoroaetic acid, thereby liberating the N-terminal amino acid residue in a cyclic form, and leaving the remaining peptide truncated by one residue is called:
(A) Coupling (B) Cyclization
(C) Conversion (D) None of these

238. Which of the following fatty acids has the lowest melting point?
(A) Fatty acids with sites of unsaturation with is double bonds
(B) Fatty acids with sites of unsaturation with trans double bonds
(C) Fatty acids with no sites of unsaturation
(D) Fatty acids with longer hydrophobic tails

239. During recombination in E. coli ——— recognize (s) regions of sequence similarity and promote(s) strand invasion, while ——— bind(s) to a Holiday junction promoting strand migration:
(A) RuvA and RuvB; RecBCD
(B) RecBCD: RecA
(C) RecA; RecBCD
(D) RecA; RuvA and RuvC

240. The sequence of proteins used in bacterial recombination is:
(A) RuvC→ RuvB→RecA→ RecBCD
(B) RecBCD →RecA→ RuvAB→ RuvC
(C) RecA→RecBCD→RuvAB→ RuvC
(D) RuvAB→RuvC→RecBCD→ RecA

241. Casein found in milk is an example of:
(A) Phosphoprotein (B) Fibrous protein
(C) Glycoprotein (D) Flavoprotein

242. Experiments on the bacterial serine protease subtilisin show that even when all three residues of the catalytic triad are mutated, the catalytic rate of the enzyme is still 3000 times the uncatalyzed reaction. Which mode of catalysis is likely responsible for this remaining catalytic activity?
(A) Acid-base catalysis
(B) Covalent catalysis
(C) Transition-state stabilization
(D) Hydrophobic effects

243. Which amino acid is least likely to participate in acid-base catalysis?
(A) Lysine (B) Tyrosine
(C) Aspartate (D) Histidine

244. Active holoenzymes are formed from —— in the presence of—:
(A) Cofactors; proteins (B) Proteins; cofactors
(C) Apoenzymes; cofactors (D) Apoenzymes, proteins

245. Highly cooperative binding of a ligand to multiple binding sites on a macromolecule is best demonstrated by:
(A) Adair equation (B) Hill plot
(C) Lineweaver – Burk plot (D) Arrhenius plot

246. Because coenzymes are specific for the chemical groups that they accept and donate, they are referred to as:
(A) Cofactors (B) Reactive centres
(C) Activator ions (D) Group-transfer reagents

247. Ca^{++} or Mg^{++} are most likely to be part of ——, while Zn^{++} or Fe^{++} are present in:
(A) Metal-activated enzymes; metalloenzymes
(B) Metalloenzymes, metal-activated enzymes
(C) Cofactors; coenzymes
(D) Coenzyme; cofactors

248. Ethanol (CH_3CH_2OH) is the alcohol found in beverages. It is oxidized in the body to acetaldehyde by the enzyme alcohol dehydrogenase. Methanol (CH_3OH) also known as wood alcohol, is converted to formaldehyde by the same enzyme. Acetaldehyde is toxic, but formaldehyde is far more toxic to humans, which is why the ingestion of relatively small amounts of methanol can cause blindness or death. One treatment for mild methanol poisoning is the administration of ethanol. Why might a doctor choose this treatment?
(A) Ethanol must at as a competitive inhibitor for the alcohol dehydrogenase and therefore slows the formation of formaldehyde
(B) Ethanol likely irreversible binds to alcohol dehydrogenase which prevents the formation of formal dehyde
(C) The ethanol is likely an uncompetitive inhibitor and binds to a site other than the active site of the enzyme
(D) The doctor has given up on the patient and administers ethanol for sedation

249. A saturated characteristics common to lipids which allow them to function as good energy stores is:
(A) They are all hydrophobic
(B) They are all hydrophilic
(C) They have large numbers of carbon-phosphorus bonds
(D) They have large numbers of carbon-hydrogen bonds

250. Meselson and Stahl's experiments involved growing E. coli on a medium containing 15N (heavy nitrogen) which increased the density of the bacterial DNA. Cells were then grown on a medium containing only 14N. DNA samples were obtained for analysis after one, two and three rounds of replication (generations) and centrifuges on a density gradient. The double-helical DNA forms bands in the gradient depending on its density. How many bands did Meselson and Stahl see after two generation of cells were grown on the 14N medium?
(A) One with DNA containing both 15N and 14N
(B) Two, one with DNA containing only 15N and one with DNA containing only 14N
(C) Two, one with DNA containing both 15N and 14N and one with DNA containing only 14N
(D) Three, one with DNA containing only 15N, one with DNA containing both 15N and 14N and one with DNA containing only 14N

251. It takes about the same amount of time for eukaryotic chromosomes to be replicated in -vivo as it does for *E. coli*. Why is this true?
(A) Eukaryotes have multiple origins of replication, E. coli has one unique origin
(B) Most chromosomes are approximately the same size
(C) Eukaryotic replication occurs at a faster rate, enabling replication of the larger chromosomes in about the same amount of time
(D) The replication of DNA in eukaryotes is faster due to a quicker, more accurate repair of mismatches

252. The role of zinc in the mechanisms of carbonic anhydrase is to:
(A) Maintain the configuration of the holoenzyme
(B) Bind to three histidine residues
(C) Produce a nucleophilic attack on the substrate
(D) Promote ionization of bound water

253. Which of the following would produce a Hill plot with negative cooperativity:
(A) A protein has multiple subunits each with a single binding site. Binding a ligand to one site decreases affinity of other sites for ligand
(B) A protein is a single polypeptide chain with two ligand sites each with different ligand affinities
(C) A protein has a single binding site. But during purification the protein preparation is heterogeneous, hence there are different protein molecules with different ligand affinities
(D) A protein has multiple binding sites for different ligands. Binding of one type of ligand at a higher affinity site promotes binding of the other ligand at a lower affinity site

254. Site directed mutagenesis is used to study enzymes by:
(A) Changing the location of the active site
(B) Producing enzymes with different amino acid residues
(C) Producing enzymes with modified R groups
(D) Changing the pH of the environment

255. What is the role of the magnesium ion in kinases that require the magnesium-ATP complex to donate phosphoryl groups?
(A) Maintain the configuration of the holoenzyme
(B) Shield the charged phosphate groups of ATP
(C) Produce an electrophilic attack on the substrate
(D) Promote ionization of bound water

256. Which one of the following enzymes is used to selectively break the peptide bond adjacent to free C-terminal of polypeptide:
(A) Co-carboxylase (B) Carboxypeptidase
(C) Carbonic anhydrase (D) Catalase

257. SDS is used while separating proteins by polyacrylamide gel electrophoresis because:
(A) It helps in solubilization of proteins thereby making it easier to separate
(B) It binds to proteins and confers uniform negative charge density thereby making them move during electrophoresis
(C) Decreases the surface tension of the buffer used for electrophoresis
(D) Stabilizes the proteins

258. An ion commonly found in metalloenzymes and which can undergo reversible oxidation and reduction is:
(A) Ca^{++} (B) Mg^{++} (C) S^{-} (D) Fe^{++}

259. Replacement of the amino acid —— at or near an active site of an enzyme is more likely to change enzyme activity than the replacement of —— at or near the active site:
(A) Histidine; leucine (B) Leucine; histidine
(C) Leucine ; isoleucine (D) Histidine; aspartate

260. In an alpha-helix hydrogen bonds are:
(A) Within a single chain
(B) Between chains that run side by side
(C) Between polar amino acids and water
(D) Only between amino acids of opposite charge

261. Hemoglobin and myoglobin both have all of the following characteristics except:
(A) Consist of subunits designed to provide hydrogen bonds and nonpolar interaction with other subunits
(B) Highly alpha helical
(C) Bind one molecules of heme per globin chain
(D) Bind heme in a hydrophobic pocket

262. S-adenosylmethionine (SAM) is formed from the amino acid methionine so that:
(A) Methyl groups can be formed
(B) It can react with nucleophilic acceptors
(C) It can donate methyl groups in many biosynthetic reactions
(D) Both (B) and (C)

263. Which of the following reagent is used to break cross linkage of polypeptide chains of a complex protein:
(A) Performic acid (B) Periodic acid
(C) t-Butyl chromate (D) Potassium permanganate

264. A nucleotide sugar coenzyme involved in carbohydrate metabolism (UDP – glucose) is formed when UTP reacts with a glucose molecules. More UTP is made available for additional reactions by the transfer of a phosphate group from:
(A) Another carbohydrate (B) ATP
(C) ADP (D) UTP

265. NAD and NADP dependent dehydrogenases catalysis substrates by transferring ——— to C- 4 of NAD^{+} and $NADP^{+}$
(A) One electron (B) Two electron
(C) One electron and one proton
(D) Two electron and one proton

266. The following data were obtained in the presence and absence of inhibitor. What type of inhibition is shown?

Substrate concentration (millimolar)	Rate without inhibitor (mmol/min)	Rate with inhibitor (mmol/min)
0.100	2.5	1.6
0.200	4.2	2.9
0.500	6.6	5.1
0.750	7.4	6.2
1.000	9.9	9.8
2.000	10.0	10.1

(A) Competitive (B) Uncompetitive
(C) Noncompetitive (D) Irreversible

267. Which statement is false about regulatory enzymes that are controlled allosterically?
(A) Thy are always less active when a modulator is bound to them
(B) They are often larger than other enzymes
(C) They have more than one binding site
(D) They often catalyze the first step in a reaction pathway

268. In aldol addition reaction product is always:
(A) β-hydroxy aldehyde
(B) β -hydroxyl ketoner
(C) α- β –unsaturated aldehyde
(D) α- β -unsaturated ketenes

269. Which one of the following methods is used for protein purification:
(A) Dialysis (B) Osmosis
(C) HPLC (D) GLC

270. One dimensional polyacrylamide gel electrophoresis in the absence of SDS (native gel electrophoresis) resolves proteins based on their mass and:
(A) Charge (B) Hydrodynamic volume
(C) Secondary structure (D) Hydrophobicity

271. Insulin is a:
(A) Vitamin (B) Hormone
(C) Protein (D) Both (B) and (C)

272. Rank the following species in order of dareasing nucleophilicity in a polar solent:

1. $CH_3CH_2CH_2O^-$ 2. CH_3CH_2-CH_2S^- 3. $CH_3CH_2C(=O)$-O^-

(A) 3 > 1 > 2 (B) 2 > 3 >1
(C) 1 > 3 > 2 (D) 2 > 1 > 3

273. Which of the following statements is true:
(A) $CH_3CH_2S^-$ is both a stronger base and more than $CH_3CH_2O^-$
(B) $CH_3CH_2S^-$ is a both a stronger base but is less nucleophic than $CH_3CH_2O^-$
(C) $CH_3CH_2S^-$ is a weaker base and more nucleophilic than $CH_3CH_2O^-$
(D) $CH_3CH_2S^-$ is both a weaker base and less nucleophilic than $CH_3CH_2O^-$

274. Claisen-Shmidt reaction taken place between:
(A) Aldehyde & addehyde
(B) Aromatic aldehyde & kelone
(C) Kelone & kelon
(D) Aromatic aldehyde & compound having active methyl group

275. The bonds in protein structure that are not broken on denaturation:
(A) Hydrogen bonds (B) Peptide bonds
(C) Ionic bonds (D) Disulphide bonds

276. The amino acids are said to be ketogenic when the carbon skeleton is finally degraded to:
(A) Succinyl –CoA (B) Fumarate
(C) Acetyl-CoA (D) Pyruvate

277. Which does not return to its original form after each catalysis?
(A) Prosthetic groups (B) Cosubstrates
(C) Metalloenzyme (D) None of these

278. Which pair will give claisen condensation reaction?
(A) $C_6H_5COOC_9H_5$ and $HCOOC_6H_5$
(B) $C_6H_5COOC_9H_5$ and CH_3CHO
(C) $CH_3COOC_9H_5$ and $HCOOC_9H_5$
(D) $HCOOC_9H_5$ and $HCOOC_9H_5$

279. Clemmenson reduction is used for the reduction
(A) Carbonyl compound (B) Ester
(C) Nitro group (D) All of these

280. Which statement is false about allosteric regulation?
(A) It is usually the mode of regulation for the last step in reaction pathways since this step produces the final product
(B) Cellular response is faster with allosteric control than by controlling enzyme concentration in the cell
(C) The regulation usually is important to the conservation of energy and materials in cells
(D) Allosteric modulators bind non-covalently at sites other than the active site and induce conformational changes in the enzyme

281. A solution of DNA has an absorbance at 260 nm of 0.212. What is the concentration of the DNA solution expressed as millimolarity:
(A) 0.03 Mm (B) 05 Mm
(C) 07 Mm (D) 01 Mm

282. Among the various interactions that help in the sustenance of the tertiary structure of proteins, the weakest one is:
(A) Hydrogen bonding
(B) Van der Waals interactions
(C) S-S bonding
(D) Hydrophobic interactions

283. All of the following are correct about ornithine except:
(A) It may be formed from or converted to glutamic semialdehyde
(B) It can be converted to proline
(C) It plays a major role in the urea cycle
(D) It is chemically spermidine

284. An oligonucleotide with the sequence GAACTACGTTCGATCAAT is suspended in 750 µL of water. A 20 µL aliquot is diluted to a final volume of 1000 µL with water, and the absorbance at 260 nm is determined for the diluted sample. An OD of 0.242 is obtained. What is the concentration of the oligonucleotide stock solution in pmol/µL?
(A) 59.46 (B) 61.35
(C) 04.35 (D) 50.24

285. You have 5 mL of bacteriophage ë stock having a concentration of 2.5 x 1011 ë phage/mL. You will purify ë DNA using the entire stock. Assuming 80% recovery, how many micrograms of ë DNA will you obtain? Phage genome size = 48502 bp:
(A) 53.4 µg (B) 60.04 µg
(C) 12.65 µg (D) 25.40 µg

286. Purines and pyrimidines exist in two tautomeric forms (lactam and lactim); nucleosides exist in two conformations (syn and anti). The most common forms found in cells are:
(A) Lactum and anti (B) Lactum and syn
(C) Lactin and anti (D) Lactim and syn

287. In DNA, phosphodiester bonds form with hydroxyl groups of carbons number —— and —— of each deoxyribose sugar:
(A) 2'; 3' (B) 2'; 4'
(C) 3'; 4' (D) 3'; 5'

288. Cells fed 5-methyl uracil (radioactive in the methyl group) will form radioactive:
(A) RNA (B) DNA
(C) Ribosomes (D) tRNA

289. Which is a true statement concerning the transition state of an SN2 reatcion?
(A) Closely resembles a carbocation intermediate
(B) The electrophile is responsible for the reaction
(C) Lower is energy than the starting materials
(D) Involves both the nucleophic and electrophile

290. Increasing the concentration of an electrophile in a typical SN2 reaction by a factor of 3 and the concentration of the nucleophic by the a factor of 3 will change the reaction rate to:
(A) Increase by a factor of 6
(B) Increase by a factor of 9
(C) Decrease by a factor of 3
(D) Remain about the same

291. Consider the following statements:
1. Bridgehead halices are inert towards both SN^1 and SN^2, reactions (till of the ring size is eight member ring)
2. The first step in both SN^1 and E1 reactions is the same
3. SN^2 reactions proceeds with total reaction of configuration
4. E2 elimination are by the use of a solvent of low polarity and high concentration of a strong base

Which of the above statement are correct?
(A) 1, 2 and 4 (B) 1 and 3
(C) 2, 3 and 4 (D) 1, 2, 3 and 4

292. Deoxythymidine is a —— found mainly in DNA:
(A) Nucleotide (B) Deoxynucleotide
(C) Nucleotide (D) Deoxynucleoside

293. The difference between dGMP and GMP is:
(A) The type of phosphodiester linkage
(B) One hydroxyl group
(C) Syn versus anti conformation
(D) The presence of DNA or RNA

294. An enzyme is irreversibly inhibited by diisopropylfluorophosphate (DFP). What does this show?
(A) The enzyme has been denatured by DFP
(B) Serine is likely an important residue in the active site
(C) DFP is an allosteric modulator of the enzyme
(D) DFP is an analog of the enzyme's substrate

295. Adenylate and AMP each contain ———— phosphate groups:
(A) 0 (B) 1 (C) 2 (D) 3

296. One of the characteristics of a peptide bond is that:
(A) There is no restriction in rotation around peptide bond
(B) There is a restricted rotation around peptide bond
(C) Its bond energy is approximately 5 Kcal/mole
(D) It is longer than 0.2 nm

297. Which of the following statements regarding reduced glutathione is not true?
(A) It contains one γ-carboxyglutomate, one cysteine, and one glycine residue
(B) It keeps the cystein residues of proteins in their reduced states
(C) It is regerenerated from oxidized glutathione by glutathione reductase
(D) It reacts with hydrogen peroxide and organic peroxides

298. The abbreviation dGp indicates:
(A) 5' deoxyguanylate (B) 3' deoxyguanylate
(C) 3', 5' deoxyadenylate (D) 5'3' deoxyguanylate

299. The abbreviation pdAp indicates:
(A) 5', 3' deoxyadenylate (B) 3', 5' deoxyadenylate
(C) 5' deoxyadenylate (D) 3' deoxyadenylate

300. The abbreviation pppT indicates:
(A) TMP (B) TDP (C) TTP (D) 3'5'TP

301. According to Chargaffs observations of nucleotide composition of DN samples:
(A) % of (G + C) + % of (A + T) = 100%
(B) A=T
(C) G = C
(D) % A + % G + % T = 100%

302. Much of the stability of the double-stranded DNA structure is the result of:
(A) Hydrogen bonding between purines
(B) The phosphodiester backbone
(C) The angle of the lanes of the bases with respect to the helix axis
(D) The stacking interactions between base pair

303. The rise and pitch of B-DNA are 0.33 nm and 3.40 nm, respectively. About how many helical turns are there in a fragment 1 mm in length?
(A) 3030 (B) 294 (C) 330 (D) 0.0034

304. Which one of the following is responsible for pigment formation in skin and hair:
(A) Collagen (B) Elastin
(C) Fibroin (D) Melanin

305. Regions of DNA that are most easily unwound have:
(A) About half G and half C (B) Alternating A and G
(C) Greater G : C content (D) Greater A : T content

306. Which is not true of the different conformations of DNA?
(A) Z-DNA is a left-handed spiral
(B) A-DNA and B-DNA are right handed spirals
(C) A-DNA and Z-DNA segments are limited to short regions of DNA
(D) Both A-DNA and B-DNA are dehydrated

307. In addition to knowing the chemical structures of the nucleotide, Watson and Crick used ———— of Franklin and Wilkins and the chemical equivalencies of Chargaff in order to propose their model of DNA structure:
(A) Sequence information (B) UV spectra
(C) % (G + C) and % (A + T) (D) X-ray diffraction data

308. In proteins, amino acids are linked by peptide bonds; in poynucleotides, nucleotides are linked by:
(A) Phosphoanhydride bonds
(B) 3'-5' phosphodiester bonds
(C) 5'-3' phosphodiester bonds
(D) Both (B) and (C)

309. The tatranucleotide AGTC (in DNA) has a free hydroxyl group on:
(A) A (B) A G, T and C
(C) C (D) G, T and C

310. The enzyme containing non-protein group in addition to a-amino acids are known as:
(A) Apo-enzyme (B) Co enzyme
(C) Holoenzyme (D) Metal activator

311. Five prime to three prime description of a DNA strand refers to:
(A) The phosphorylated 5' carbon at one, the 3' at the other end
(B) The location of the hydroxyl groups
(C) The phosphodiester backbone of DNA
(D) The phosphorylated 5 carbon on one sugar and thc 3 carbon on the next sugar

312. Lysine is an amino acid with three ionizable groups. These are the á–COOH, 2á – amino and å – amino groups with pKa values of 2.2, 9.2 and 10.8, respectively. The isoelectric point (pI) for lysine is:
(A) 5.7 (B) 6.5 (C) 9.2 (D) 10.0

313. DNA polymerase contains a lysine residues that is important for binding to DNA. Mutations were found that converted this lysine to either glutamate, glyine, valine or arginine. Which mutations would be predicted to be the most and least harmful to the ability of the enzyme to bind DNA:

	Most	Least
(A)	Glycine	Arginine
(B)	Arginine	Glycine
(C)	Glutamate	Valine
(D)	Glutamate	Arginine

314. Which one of the following possesses enzyme activity:
(A) Coenzyme (B) Apoenzyme
(C) Cofactor (D) Holoenzyme

315. The nitrogenous base present in lecithin:
(A) Choline (B) Ethanolamine
(C) Inositol (D) Serine

316. As B-DNA is gradually heated, the absorbance at 260 nm:
(A) Increases
(B) Decreases
(C) Stays the same
(D) Is half way between that of poly (AT and poly, GC):

317. Adenine is a:
(A) Purine base (B) Pyrimidine base
(C) Pyridine base (D) Pteridene base

318. Which does not apply to most bacterial DNA?
(A) Circular
(B) Relaxed
(C) Not packed into nucleosomes
(D) Supercoiled

319. Which of the following is not true about alanine:
(A) At pH 1, the overall charge is +1
(B) At pH 1, it will move towards cathode
(C) An equimolar mixture of D and L alanine does not rotate the plane polarized light
(D) Contains branched side chain

320. Amino transferases:
(A) Usually require alpha-ketogluterate or glutamine as one of the reacting pair
(B) Catalyze reactions that results in a use or production of amino acids
(C) Catalyze irreversible reactions
(D) Require pyridoxal phosphate as an essential cofactor for the reaction

321. ———— plays and important role in storing the enrgy needed for the unwinding of DNA during replication, recombination, repair and transcription:
(A) A high percent of G : C base pairs
(B) The charge on basic side chains of histone proteins
(C) Negative supercoiling
(D) Overwinding of the helix

322. Scurvy disease is caused due to deficiency of:
(A) Retinol (B) Thiamine
(C) Nicotinic acid (D) Ascorbic acid

323. Which of the following is a molybdenum containing enzyme:
(A) Nitrogenase (B) Peroxidase
(C) Carbonic anhydrase (D) Catalase

324. Which type of RNA is the most abundant in living cells (by percent):
(A) Ribosomal (B) Messenger
(C) Small (D) Transfer

325. Nucleosides are condensation products of:
(A) Pentose sugar and phosphoric acid
(B) Nitrogen base and phosphoric acid
(C) Pentose sugar and nitrogen base
(D) Both (A) and (B)

326. Histidine is often found at the active-site of enzymes because:
(A) It has a cyclic group (B) It an form hydrogen bonds
(C) It has a pKa of 6.8 (D) It is an imino acid

327. Match the entries in column A against those in column B:

Column A	Column B
P. The amino acid with maximum number of codons	1. Cys
Q. The characteristic amino acid in lysozyme active site	2. Trp
R. The amino acid precursor in heme biosynthesis	3. Ser
	4. Gly
S. The amino acid contributing the most to protein absorbance at 280 nm	5. Glu

(A) P – 2, Q – 3, R – 1, S · 4
(B) P – 3, Q – 5, R – 4, S · 2
(C) P – 5, Q – 3, R – 2, S · 1
(D) P – 4, Q – 1, R – 5, S · 3

328. Which one of the following statements about prostaglandins is incorrect?
(A) Prostaglandins have a very short half-life
(B) Prostaglandins are synthesized only in the liver and the adrenal cortex
(C) Prostaglandins generally at locally on or near the tissue that produced them
(D) The common precursor of the prostaglandins us-arachidonic acid

329. Which type of RNA carries activated amino acids to the ribosome for assembly into proteins?
(A) rRNA (B) tRNA (C) mRNA (D) sRNA

330. Which is not an explanation of why site-directed mutagenesis is a more powerful tool than many other techniques for determining important residues in an enzyme?
(A) It is a more specific technique than irreversible inhibition experiments
(B) It can be used on enzymes for which no specific irreversible inhibitor is known
(C) It allows for testing of the specific function of amino acid side chains in an enzyme
(D) It is specially useful for enzymes whose sequence is not known

331. It is easier to melt DA richer in AT than GC because:
(A) It is more heat sensitive
(B) There is one less hydrogen bond
(C) The helix pitch is longer in AT rich regions
(D) All of these

332. What feature does a Ramachandran plot display?
(A) Allowed angles of phi and psi for a polypeptide back bone
(B) Preferred amino acids in an a-helix
(C) The hydropathy of amino acids
(D) The variation of pH versus volume of base added during titration to determine the pKa

333. A brown-eyed man has one allele for brown eyes and one for blue eyes. Half his sperm carry the allele for brown eyes, and half carry the allele for blue eyes. This is an illustration of:
(A) Mendel's law of independent segregation
(B) Mendel's law of independent assortment
(C) Mendel's law of segregation
(D) None of these

334. Both epinephrine and glucagons bind to G protein-coupled receptors and activate glycogen breakdown. Therefore epinephrine and glucagons must:
(A) Have very similar structures and bind to the same receptor
(B) Bind to receptors with different ligand-binding sites but similar functions
(C) Bind to different receptors and activate different second messengers
(D) Bind to the same receptors, one being intracellular and the other, extracellular

335. Which of the following is not a characteristic of DNA?
(A) It is less susceptible to alkaline hydrolysis than RNA
(B) It has a hydroxyl group attached to the 3' carbon at the 3' end
(C) It is stabilized by hydrogen bonding, van der Waals forces and covalent interaction
(D) It an exist in four or more than four different helical conformations

336. The amino acid with pK_r closest to physiological pH is:
(A) Serine (B) Histidine
(C) Threonine (D) Proline

337. What will be the molecular weight of poly (A) chain consisting of 100 residues, where the molecular weight of AMP is 300:
(A) 30000 (B) 29700
(C) 28218 (D) 29982

338. Which of the following enzyme catalyzes the hydrolysis of nucleic acid:
(A) Nuclease (B) Glycosidase
(C) Phosphorylase (D) Lipase

339. Glucagon binds to the glucagons receptor, which interacts with a membrane-bound G protein, whose a subunit releases GDP and binds GTP. The a subunit then dissociates from the G protein and diffuses to a membrane-bound adenylyl cyclase which produces cAMP. In this scenario, name the (1) transducer, (2) effector, (3) first messenger and (4) second messenger
(A) (1) Glucagon (2) cAMP (3) G Protein (4) GTP
(B) (1) G Protein (2) Adenylyl cyclase (3) Glucagon (4) cAMP
(C) (1) G protein (2) GTP (3) Glucagon (4) cAMP
(D) (1) Glucagon (2) G protein (3) adenylyl cyclase (4) cAMP

340. Which of the following contains magnesium:
(A) Nitrogenase (B) Chlorophyll
(C) b-carotene (D) Phytol

341. In RNA ribose is present in the form of:
(A) a, D-Ribofluranose (B) b, D-Ribofuranose
(C) b, D-Ribopyranose (D) a, D- Ribopyranose

342. Which is not a biological function of RNA?
(A) Participates in the modification of newly synthesized RNA
(B) It is the genome of some viruses
(C) Serves as a scaffold component for chromosomes
(D) Composes as integral part of the ribosome

343. Which one of the following statements about prostaglandin synthesis is incorrect?
(A) Prostaglandins synthesis is inhibited by glucocortioids and aspirin
(B) Prostagladins are derived from an essential fatty acid
(C) Prostaglandins are produced through the cation of lipoxygenase
(D) Arachidonic acid is a major precursor of prostaglandins

344. G proteins are said to be self-inactivating. What G – protein function accounts for this?
(A) GTPase activity of the a subunit
(B) Activation of the effector
(C) Binding to the receptor
(D) Dissociation of the subunits

345. Which one of the following amino acids has the largest number of rotatable bonds in the side chain:
(A) Tryptophan (B) Leucine
(C) Lysine (D) Cysteine

346. What is the number of hydrogen bonds in a double helical B-DNA structure of 100 base pairs with 20 adenines and 10 thymines in one of the two strands:
(A) 200 (B) 230 (C) 270 (D) 300

347. In nucleic acids, internucleotide linkage is known as:
(A) Glycosidic bond (B) 3'5' phosphodiester bonds
(C) Peptide bond (D) Simple ester bond

348. People who suffer from thyroid adenomas have constitutively activated G proteins in some cells of the thyroid gland. Tumor cells not only secrete copious quantities of thyroid hormones but also divide and grow excessively. Based on these observations, in what kind(s) of signal transduction pathways do the mutant G proteins participate?
(A) cAMP –regulated pathways
(B) MAP kinase pathways
(C) Both cAMP and MAP kinase pathways
(D) Neither cAMP anor MAP kinase pathways

349. Which of the following in not true about fatty acid synthesis
(A) Cofactors – NADPH (B) Precursor _ Two carbon unit
(C) Site – Cytosol (D) Carrier – carnitine

350. Folate derivatives are required for the synthesis of each deoxynucleotides?
(A) Adenylate and guanylate
(B) Cytidylate and thymidylate
(C) Adenylate, guanylate and thymidylate
(D) Adenylate and Cytidylate

351. Depending on the cell, increase in intracellular Ca^{2+} can do all of the following except:
(A) Stimulate the increase of intracellular calcium
(B) Stimulate the removal of intracellular calcium
(C) Stimulate the breakdown of cAMP
(D) Stimulate the breakdown of calmodulin

352. Which one of the following vitamin contains isollaxazine ring in its structure:
(A) Vitamin A (B) Vitamin B_2
(C) Vitamin B_1 (D) Vitamin B_{12}

353. Genetic material in plant viruses is mostly:
(A) DNA (B) RNA
(C) Protein (D) None of these

354. Epinephrine binds to the same type of receptors in liver, fat and smooth muscle cells. Yet in liver, glycogen breaks down; in act, trisylglycerols break down; and smooth muscle cells relax. How can the same hormone produce three such different responses
(A) The same receptors activate three different second messengers
(B) The same second messengers activate three different intracellular cascades
(C) The same enzyme that breaks down glycogen also breaks down fat and relaxes muscle
(D) None of these

355. Which one of the following is a carrier of oxygen to muscles in vertebrates:
(A) Muoglobin (B) Hemoglobin
(C) Cytochromes (D) Immunoglobulin

356. A cell responds to an external stimulus by increasing the intracellular Ca^{2+} concentration, releasing IP3 from the membrane, and producing salicylic acid. What kind of cell does this describe?
(A) Bacterial (B) Viral
(C) Plant (D) Animal

357. Which of the following favours Z-DNA conformation?
(A) 2'-OH steric hindrance
(B) Anti-G and 2'-endo sugar
(C) Syn-G and 3'-endo sugar
(D) G4-DNA quartet

358. The distance along a helix axis for one complete turn is called:
(A) Axial distance (B) Rise
(C) Screw length (D) Pitch

359. Which of the following contain cobalt in its structure:
(A) Vitamin D (B) Vitamin K
(C) Vitamin B_5 (D) Vitamin B_{12}

360. An ideal α –helix has a pith of 0.54 nm and a rise of 0.15 nm. What is the length along the helix axis for a segment of an ideal α –helix that contains 30 amino acids?
(A) 200 nm (B) 4.50 nm
(C) 16.20 nm (D) 55.5 nm

361. Which one of the following is smallest in size:
(A) r-RNA (B) m-RNA
(C) t-RNA (D) DNA

362. The reaction catalyzed by a certain phosphatase enzyme is found to follow ping-pong kinetics and involves the transfer of a phosphate group from substrate A to substrate B. Which mode of catalysis is likely for this reaction?
(A) Sequential catalysis (B) Acid-base catalysis
(C) Transfer catalysis (D) Covalent catalysis

363. The molecular weight of a bacterial DNA molecule is 2.64 x 10^9. The average molecular weight of a nucleotide pair is 660. Assume that the average protein is made up of a chain of 400 amino acid residues. What is the maximum number of proteins than can be coded by the bacterial DNA molecule?
(A) 20000 (B) 3333
(C) 6667 (D) None

364. Thiamine is a constituent of:
(A) Co enzyme A (B) Co enzyme Q
(C) NADP (D) Co-carboxylase coenzyme

365. An unwanted DNA and a supercoiled DNA with the same linking number are:
(A) Topologically and geometrically identical
(B) Topologically identical but geometrically different
(C) Topologically and geometrically different
(D) Topologically different by geometrically identical

366. Ribosomal RNA modification such as methylation or pseudouridylation are guided by which of the following RNA in eukaryotic cells:
(A) mi-RNA (B) si-RNA
(C) sno-RNA (D) sh-RNA

367. Which was a major scientific accomplishment by Linus Pauling?
(A) Determined the á-helix structure in keratin
(B) Discovered â – strands and described their assembly into sheets
(C) Determined the three-dimentional structure of hemoglobin and mycoglobin
(D) Determined the structure of penicillin

368. A protein has a molecular weight of 5600 Daltons; its subunits are about 1960 Daltons. There are ——— protein chains per oligomer:
(A) 4 (B) 3 (C) 2 (D) 1

369. Which of the following is not true about RNA
(A) RNA is chemically more reactive than DNA
(B) 2'-OH prevents RNA to adopt B-conformation
(C) RNA was discovered by Hoppe-Seyler
(D) Feulgen reaction is performed by both DNA and RNA

370. Which of the following is not true? When RNA undergo for alkali-catalyzed cleavage:
(A) Transesterification reaction occurs
(B) Cyclic 2'-3'-nucleoside monophosphate forms
(C) Cyclic 5'-3'-nucleoside monophosphate forms
(D) 2'-OH is converted into 2'-alkoxide ion

371. Metal cations may do all of the following except:
(A) Donate electron pairs to functional groups found in the primary structure of the enzyme protein
(B) Serve of Lewis acids in enzyme
(C) Participate in oxidation-reduction processes
(D) Stabilize the active conformation of an enzyme

372. Which of the following answers completes the sentence correctly? The first product of purine nucleotide biosynthesis that contains a complete purine ring (hypoxanthine) is:
(A) AMP (B) IMP
(C) GMP (D) Xynthylate

373. The steady – state hypothesis in enzyme kinetics states that:
(A) There is an equilibrium established between free enzyme (E) and substrate (S) versus the E·S complex
(B) There is an equilibrium established between the E·S complex and the products
(C) The rate of formation of the E·S complex in the same as the rate of breakdown of this complex either to products or back to reactants
(D) The rate of formation of products is the same as the rate of formation of E·S complex

374. The void volume of a gel filtration column, V0=3· mL. A monomeric protein with a known molecular weight of 25 kDa elutes at a volume of 45 mL. The protein that you are trying to purify elutes at a volume of 35 mL. Which of the following is a valid conclusion?
(A) Your protein is repelled by the gel filtration material
(B) Your protein Mr= 25 kDa
(C) Your protein Mr>25kDa
(D) Your protein Mr<25 kDa

375. Which is not an example of an oligomeric protein?
(A) Cytochrome (B) Potassium channel protein
(C) MS2 capsid protein (D) Hemoglobin

376. You have mixture of proteins with the following properties:
Protein 1 : Mr 12 kD pI = 10
Protein 2 : Mr 62 kD pI = 4
Protein 3 : Mr 28 kD pI = 8
Protein 4 : Mr 9 kD pI = 5
Predict the order of emergence of these proteins when a mixture of the four is chromatographed in DEAE-cellulose at pH 7.0
(A) 1, 3, 4, 2 (B) 2, 4, 3, 1
(C) Cannot be predicted (D) 5, 8, 4, 10

377. The main property of myoglobin and hemoglobin that makes them an efficient system for oxygen delivery from lungs to muscles is:
(A) Hydrophobicity
(B) Different binding affinities for oxygen
(C) Movement of the protein shapes
(D) Cooperativity

378. Which one of the following statements is correct concerning the Michaelis constant, K_m?
(A) Is measured in the same units as the rate of the enzyme-catalysed reaction
(B) Varies with the amount of enzyme used in measuring the reaction rate
(C) Can be considered to be a measure of the affinity an enzyme has for its substrate
(D) Has a low value to indicate a low affinity for substrate

379. Two proteins have the same molecular mass and have identical net charge at pH 7. The best way to separate them would be to use:
(A) SDS-polyacrylamide gel electrophoresis (SDS-PAGE)
(B) Native gel electrophoresis
(C) Cation-exchange chromatography
(D) Anion-exchange chromatography

380. The active site of a certain enzyme contains a serine residue. When the enzyme is incubated for a short time with its substrate, a form of the enzyme in which the active site serine is acetylated can be isolated and purified. In the native protein the serine is never found to be acetylated. This information supports:
(A) A covalent catalysis mode
(B) Intermediate state stabilization effects
(C) The acid base catalysis mode
(D) A polar group catalytic mechanism

381. Which statement is not true about enzyme inhibition?
(A) In competitive inhibition, the inhibitor binds to the active site of the enzyme
(B) In non-competitive inhibition, the inhibitor binds to the allosteric site of the substrate
(C) In irreversible inhibition, a poison binds to the enzyme so that it can never work again
(D) None of these

382. Three-dimensional images of the surface of the cells and tissues could be visualized through:
(A) Transmission electron microscope
(B) Scanning electron microscope
(C) Compound microscope
(D) Fluorescence microscope

383. From the data given below, identify the protein pair that would (P) give the least mobility band on a sodium dodecyl sulphate polyacrylamide gel electrophoresis (SDS-PAGE) experiment and (Q) elute last on an anion exchange e.g., DEAE column respectively:

Protein	pI	Subunit M.W.	Native M.W.
A	9	10000	20000
B	8	35000	35000
C	6	15000	90000
D	5	20000	80000
E	3	30000	30000

(A) (P) Protein C (Q) Protein A
(B) (P) Protein A (Q) Protein E
(C) (P) Protein B (Q) Protein E
(D) (P) Protein B (Q) Protein A

384. The relative proportion (by weight) of lipid and protein in a membrane:
(A) Is always about equal
(B) Varies depending upon the source of the membrane
(C) Is always characterized by more protein than lipid
(D) Is always characterized by more lipid than protein

385. The cell in number (188) above would be in an environment that was:
(A) Hypertonic (B) Hypotonic
(C) Isotonic (D) Hypo-osmotic

386. Glycogen is a branched polymer of glucose. It has:
(A) No reducing ends
(B) No non-reducing ends
(C) One reducing end and several non-reducing ends
(D) One non-reducing end and several reducing ends

387. Vitamin C is a vitamin but not a coenzyme because it:
(A) Does not bind to proteins
(B) Is a prosthetic group
(C) Is a reducing agent during hydroxylation of collagen
(D) Is not required human diets

388. Features of the fluid mosaic model of membrane structure include:
(A) A lipid bilayer
(B) Dynamic motion of both membrane lipids and proteins
(C) Proteins that may either rest on the surface or penetrate the membrane
(D) All of these

389. Which of the following is an example of an ion transport?
(A) Na + - K + ATPase (B) ABC transporter
(C) CFTR (D) All of these

390. Purified duplex DNA molecules cannot exist in which one of the following forms:
(A) Linear
(B) Circular and supercoiled
(C) Linear and supercoiled
(D) Circular and relaxed

391. Arrange the statement below in the proper sequence of events that occur in an action potential:
I. The resting potential is reestablished
II. A stimulus depolarizes the membrane to threshold
III. Na^+ gates open and Na^+ rushes into the cell
IV. K^+ gates open and K^+ rushes out of the cell
(A) II, III, IV, I (B) III, IV, II, I
(C) II, IV, III, I (D) I, II, IV, III

392. Which of the following answer complete the sentence incorrectly?
(A) Is a flavin nucleotide
(B) Is the major electron acceptor used in metabolism
(C) Contains a nicotinamide ring that accepts a hydride ion during reduction
(D) Loses a plus charge upon reduction

393. Coenzymes and prosthetic groups are found in all the key metabolic pathways. Which one of the following statements is accurate about these key molecules?
(A) Carbonic anhydrase contains a tightly bound zinc ion that is essential for its structure but not for its activity
(B) Some enzymes contain organic molecules called prosthetic groups that are essential for activity; prosthetic groups are vitamins and cannot be synthesized in the body
(C) Coenzymes are small polypepetides that are in volved in transfer reactions catalyzed by enzymes
(D) Coenzymes carry acyl or phosphate groups in many enzyme-catalyzed reactions that involve the transfer of these groups

394. Of the four classes of cellular macromolecules, which one is not a component of cell membrane?
(A) Proteins (B) Lipids
(C) Nucleic acids (D) Carbohydrates

395. A neurotransmitter that binds to a postsynaptic cell and opens K+ channels:
(A) Excites the postsynaptic cell
(B) Inhibits the postsynaptic cell
(C) Depolarizes the postsynaptic cell
(D) Must be acetylcholine

396. Prosthetic groups are:
(A) Required by all enzymes in the cell
(B) Loosely bound to enzymes via hydrogen bonds
(C) Sites on the enzyme molecule that permit allosteric modification of enzyme activity
(D) Tightly bound to enzymes and are required for their activity

397. The study of the energy changes during metabolic reactions is called:
(A) Bioinformatics (B) Metabodynamics
(C) Thermometrics (D) Bioenergetics
(E) Biological heat dynamics

398. An allosteric site on an enzyme is:
(A) The same as the active site
(B) Non-protein in nature
(C) Where ATP attaches and gives up its energy
(D) Often involved in feedback inhibition

399. How do membranes function as a locus of biochemical reactions?
(A) Membranes separate the reactants from the products of enzymatic reactions
(B) Membranes hold enzymes of sequential reactions in close proximity
(C) Membrane lipids catalize important cellular reactions
(D) Membrane increase the activation energy for catalysis

400. What region of polypeptide (backbone or side-chains) is involved in forming α –helices and β – sheets?
(A) The backbone (B) Side chain
(C) Both (A) and (B) (D) None of these

401. You put Protein X in methane, would you except Protein X to be:
1. You would expect that Protein X would be functional in methane
2. You would not expect that Protein X would be functional in methane
3. In a hydrophobic solution such as methane, it would be more energetically stable to have the hydrophobic amino acids on the outside of the structure and the polar amino acids on the inside of the structure.
4. The protein would be turned 'inside out' and no longer functional. Because some of these hydrophobic residues might also be involved in secondary structure, this level of protein structure would also be disrupted.

Correct answer is:
(A) 4 (B) 1 (C) 2 (D) 1, 3 and 4

402. What is the approximate diameter of a chromatin fiber?
(A) 3 cm (B) 30 nm (C) 300 nm (D) 30 mm

403. Much of what we know about the nicotinic acetylcholine receptor has come from studies of:
(A) The torpedo marmarota electric organ
(B) The human heart
(C) Simple, single-celled organisms
(D) *E. coli* bacteria

404. Which one of the following characteristics best apples to an allosteric effector?
(A) Completes with substrate for the catalytic site
(B) Binds to a site on the enzyme molecule distinct from the catalytic site
(C) Changes the nature of the product formed
(D) Changes the substrate specificity of the enzyme

405. A spontaneous chemical reaction always has a:
(A) Positive Gibb's free energy
(B) Negative Gibb's free energy
(C) Positive enthalpy (D) Negative enthalpy
(E) Positive entropy

406. In a chemical reaction, transition-state species have free energies:
(A) Lower than either the reactants or the products
(B) Higher than either the reactants or the products
(C) Lower than the reactants, but higher than the products
(D) Higher than the reactants, but lower than the products

407. More than 300 amino acids are present in cells but only 20 amino acid present in proteins called:
(A) Standard amino acids
(B) Nonstandard amino acids
(C) Both (A) and (B)
(D) None of these

408. Which is not a difference between RNA and DNA:
(A) The sugar ring of RNA is more oxidized than that in DNA
(B) RNA contains uracil; DNA usually does not
(C) RNA cannot form helices
(D) RNA is single-stranded; DNA is double-stranded

409. In a given reaction:

$$CH_3CH_2CHO + CH_3CHO \xrightarrow{\text{dil}} \text{NaOH}$$

(A) Three (B) Four
(C) Two (D) One

410. During the reduction of FAD
(A) A flavin group is transferred
(B) An equivalent of a hydride ion is transferred
(C) The isoalloxazine become charged
(D) Two hydrogen atoms are added to the isoallexazine ring

411. If you break liver cells by homogenization, separate the broken cell membranes (particulate) from the cytoplasm (soluble), and add epinephrine to the soluble portion, the result will be:
(A) The production of cAMP
(B) Binding of the epinephrine to its receptor
(C) Activation of adenylyl cyclase
(D) None of these

412. Why it is important that the enzymes in lysosomes are more active at acidic pH than at neutral pH?
(A) Since lysosomes are primarily found in the stomach acid of mammals, their pH dependence allows for maximum efficiency for the digestion of food stuffs
(B) It prevents their diffusion out of the lysosomes
(C) It maximizes the interaction with their substrates which are always bases
(D) It prevents them from accidentally degrading the macromolecules in the cytosol
(E) It allows for regulation of their uptake by the mitochondria

413. Match the function of following cofactor

Column I	Column II
P. Thiamine pyrophosphate	1. Acyl group transfer
Q. Coenzyme A	2. Transfer of one carbon component
R. Pyridoxal phosphate	3. Group transfer to/or from amino acid
S. Tetrahydrofolate	4. Aldehyde transfer

(A) P -4, Q – 3, R – 1, S - 2
(B) P -4, Q – 3, R – 2, S - 1
(C) P -4, Q – 1, R – 3, S – 2
(D) P -3, Q – 1, R – 4, S – 2

414. Chromatin was first observed and named by:
(A) Walter Flemming (B) Francis Crick
(C) Oswald Avery (D) Friedrich Miescher

415. Why were the traits that Mendel chose to study in the pea plant considered lucky choices?
(A) They were the only traits governed by genes
(B) None of the traits he chose acted as if they were part of the same linkage group
(C) All the traits he chose could be found on the same plant
(D) They all obeyed the law of segregation

416. Acid base catalysis is estimated to accelerate a typical enzymatic reaction by what factor?
(A) 1 to 2 fold increase (B) 10 to 100 fold increase
(C) 106 fold increase (D) 1023 fold increase

417. Which of the following is mismatched?
(A) eRNA : 80% of cellular RNA
(B) tRNA : Stable RNA carrying the coded information from DNA
(C) mRNA: stable RNA carrying the coded information from DNA
(D) Small RNA: catalytic with or without proteins

418. Heterotropic enzymes:
(A) Are different isoenzymes
(B) Are modified by their substrate concentration
(C) Are stimulated or inhibited by an effector or modulator molecule other than their substrate
(D) Have same V_{max} but different k_m values

419. Which of the following statements about ribizymes is false?
(A) Ribozymes are capable of self-replication, but can not catalyze other types of reactions
(B) Manfred Eigen found that RNA's could replicate themselves in solution without the aid of proteins
(C) In Tetrahymena thermophila, an intron was found that carries out its own excision and splicing
(D) A tRNA-processing enzyme containing RNA was found in which the RNA portion provided the catalysis

420. The observed optical rotation of an amino acid solution is directly proportional to the length of the light path through the polarimeter tube and the concentration of the amino acid solution is:
(A) True (B) False
(C) Unknown (D) Same

421. According to the RNA-world theory:
(A) RNA molecules were the first organic molecules formed on earth
(B) Life evolved an another planet called the 'RNA world'
(C) All RNA molecules in cells are 'rhizymes'
(D) Primitive RNA molecules evolved before protein and DNA

422. Stem-loop is a type of secondary structure not usually found in:
(A) rRNA (B) tRNA
(C) mRNA (D) Small RNA

423. You are an undergraduate researcher working in a lab that studies transport proteins. You have just identified a novel protein from human cells, Protein X, which you believe is involved in intracellular transport. After careful investigation, you determine the amino acid sequence of Protein X. Below is a portion of the sequence that contains the part of the site where Protein X binds the molecule that it transports:
Val-Ala-Phe-Gly-Gly-Leu-Asp-Asp-Gly-Glu-Ala-Trp-Tyr-Met-Leu-Lys-Leu. Bold represents amino acids in the binding site. Next, you find mutant versions of Protein X that have single amino acid substitutions.
A mutant in which one of the Asp residues is mutated to a Glu residue. Even with this mutation, Protein X is still functional. Why?
1. The two Asp amino acids are both in the binding site, so they are important for biding the molecule to be transported.
2. Asp and Glu negatively charged amino acids. Asp and Glu are also fairly similar in size.
3. Assuming that the negative charge is important for binding, Glu can substitute for Asp because the charge is conserved.
4. Insufficient data

Correct answers are:
(A) 4 (B) 2 (C) 1 and 2 (D) 1, 2 and 3

424. Considering previous question, A mutant in which the Val residue is mutated to a Phe residue. This mutation causes Protein X to be nonfunctional. Yes or NO?
1. Val in not in the binding site, so it is not directly involved in binding the transport molecule.
2. Val is changed to Phe. Both amino acids are hydro phobic or non polar
3. Phe, however, is a very large bulky amino acid com pared to Val. Thus, it is possible that the change to Phe causes some disruption in the tertiary structure that prevents Protein X from being functional.
4. Protein will be functional

Correct answer is:
(A) 4 (B) 2
(C) 1 and 2 (D) 1, 2 and 3

425. Continuing to previous question, what level of protein structure do you know is changed in all of these mutant proteins?
(A) Primary (B) Secondary
(C) Tertiary (D) Quaternary

426. Continuing to previous question, after further analysis of your protein, you find that α–helices and β–sheets are extremely important for maintaining the structure of the binding site. What level of structure do α- helices and β– sheets represent?
(A) Primary (B) Secondary
(C) Tertiary (D) Quaternary

427. Proline distinct among the 20 commonly found amino aids because:
(A) It is a ring compound
(B) It is hydrophilic and ionic
(C) The nitrogen of the amino group is in a ring
(D) The carbon of the carboxyl group is in a ring
(E) It has little effect on protein structure

428. Involves a nucleophilic attack of the N-terminal –amino group on the thionate carbon of phenylisothiocyanate (PITC), to form the phenylthioarbamyl-peptide derivative (PTC- peptide) is called:
(A) Coupling (B) Cyclization
(C) Conversion (D) Repulsion

429. An enzyme's active site contains and arginine residue and a glutamate residue with pKa's of 2.9 and 9,1, respectively. Both residues are actively involved in the catalytic mechanism and they are the only two ionizable residues in the active site. What you expect for the optimum pH of the enzyme?
(A) 6.0
(B) 7.0
(C) 7.4
(D) No pH can be determined since the information is irrivalent to the optimum pH

430. Supercoiling of pure circular DNA is a result of:
(A) Underwinding or overwinding of the helix
(B) Entanglements because of the long length of the strands
(C) The lack of associated histones
(D) The activity of exonucleases

431. The atoms of the pyrimidine ring are derived from:
(A) Glutamine
(B) Carbamyol phosphate and aspertate
(C) Glutamine and N10 formyl tetrahydrofolate
(D) Aspartate ands N 10 formyl tetrahydrofolate

432. Nitrogen fixation is a:
(A) Oxidative
(B) Reductive
(C) Oxidative followed by reductive
(D) Reductive followed by oxidative

433. Stem-loop structures in RNA resemble the ——— form of double-helical DNA:
(A) A (B) B
(C) Z (D) Supercoiled circular

434. Alleles are:
(A) Different forms of the gene for the same trait
(B) Genes for different traits
(C) Always identical
(D) Inherited in pairs, both members of which come from the same parent

435. In a given reaction:

$$CH_3CHO + C_6H_5CHO \xrightarrow{\text{dil}} NaOH$$

(A) Three (B) Four
(C) Two (D) One

436. Which type of organisms convert nitrogen gas into ammonia:
(A) Prokaryotic (B) Eukaryotic
(C) Both (A) and (B) (D) None of these

437. Superoxide dismutase enzyme catalysis is faster than the rate of diffusion because it:
(A) Is an acid base catalysis
(B) Is a two step reaction
(C) Has an electric field around the active site
(D) Occurs in very high quantities in cells

438. The biological process of nitrogen fixation is catalyzed by an enzyme complex is called:
(A) Nitrogenase complex (B) Dinitrogenase complex
(C) Both (A) and (B) (D) None of these

439. When statements is true about aeration with an equilibrium constant, Keq, equal to 1000?
(A) The forward rate constant is 1000 times greater than the reverse rate constant
(B) The forward rate constant is 3 times greater than reverse rate constant
(C) The forward rate constant is 1000 times smaller than the reverse rate constant
(D The forward rate constant is 3 times smaller than the reverse rate constant
(E) There is not enough information given to compare the forward and reverse rate constant

440. Tyrosine and tryptophan are less hydrophobic than phenylalanine because:
(A) Phenylalanine has an indole group
(B) Phenylalanine has no polar group in the side chain
(C) Phenylalanine is a phenol
(D) Tyrosine and tryptophan have smaller R groups
(E) All of these

441. Which of the following contributes nitrogen atoms to both purine and pyrimidine rings?
(A) Aspartate (B) Carbamoyl phosphate
(C) Carbon dioxide (D) Glutamine

442. A reaction that occurs with every collision between reactant molecules is called a/an:
(A) Saturation point
(B) Diffusion-controlled reaction
(C) Rate-determining step
(D) First order reaction

443. The rearrangement of the liberated ATZ derivative to the corresponding phenylthiohydantoin derivative (PTH-amino acid) by treatment with aqueous acid is called:
(A) Coupling (B) Cyclization
(C) Conversion (D) Repulsion

444. What is the function of DNA toposiomerases?
(A) Cleaving nucleotides from the ends of DNA
(B) Adding or removing supercoils
(C) Methylation of the nitrogenous bases
(D) Packaging of DNA into nucleosomes

445. Maize protein is deficient in:
(A) Betadine (B) Tryptophane and Lysine
(C) Lysine (D) Tryptophane

446. Zein is known as:
(A) Rice protein (B) Wheat protein
(C) Maize protein (D) Millet protein

447. Wheat protein is called as:
(A) Zein (B) Gluten
(C) Dhurin (D) Ricin

448. Essential oils are used in:
(A) Cosmetics and Toiletries
(B) Food beverage
(C) Paint industry (D) All of these

449. First Face Pack was developed in:
(A) France (B) Rome
(C) Greece (D) None of these

450. Oil content in fresh herbage in menthol mint is:
(A) 0.1-0.5 % (B) 0.5-0.8 %
(C) 1.0-1.5 % (D) 1.5 – 2.0 %

451. Major constituent of *Occimum cannum* is:
(A) Methyl chavical (B) Thymal
(C) Linalool (D) Engenol

452. Protein content in rajmash is:
(A) 10.9 % (B) 22.9 % (C) 25.9 % (D) 32.9 %

453. Carbohydrate content in rajmash is:
(A) 50.6 % (B) 60.6 % (C) 70.6 % (D) 80.6 %

454. Protein content in pea is:
(A) 10 % (B) 22 % (C) 25 % (D) 32 %

455. Carbohydrate content in potato is:
(A) 40.3 % (B) 70.3 % (C) 81.3 % (D) 91.3 %

456. In a signal transduction pathway, what is passed from the upstream member of the pathway to the downstream member?
(A) Electrons
(B) Protons
(C) Phosphates
(D) Information in the form of alterations in protein conformation

457. Which one of the following enzyme contains zinc:
(A) Peroxidase (B) Lipase
(C) Nitrogenase (D) Carbonix anhydrase

458. The protein molecules various amino acid residue are covalently linked in a head to tail arrangement by:
(A) Peptide bonds (B) Phosphodiester bonds
(C) Glycosidic bonds (D) Simple ester bonds

459. Given that alleles are located on chromosomes, and whole chromosomes are passed on to gametes, why don't offspring get all the alleles on a parental chromosome?
(A) Genetic recombination is a result of gene duplication during evolution
(B) Genetic recombination is a result of mutation
(C) Genetic recombination is a result of crossover between members of homologous pairs
(D) All traits on a given chromosomes are always found together in offspring

460. Cell signaling means all of the following except:
(A) Intercellular communication
(B) Export of cellular waste products
(C) Environmental monitoring
(D) Response to stimuli

461. Simple proteins on hydrolysis yield:
(A) L- Amino acids having – NH group on a- carbon
(B) D- Amino acids having – NH group on a- carbon
(C) L- Amino acids having – NH group on b- carbon
(D) D- Amino acids having – NH group on b- carbon

462. Rhodopsin is a protein complex the prosthetic group of which is:
(A) Retinol (B) Retinal
(C) Vitamin 'K' (D) Metal ion

463. The enzyme which exists in more than one structural forms in the same species is known as:
(A) Multienzyme complex (B) Zymogen
(C) Isozyme (D) Zymase

464. Changes in the topology and three-dimentional shapes of B-DNA can result from:
(A) Incubation with proteases (B) Supercoiling
(C) Isomerization (D) Hydration

465. Doing some more analysis of the function of Protein X reveals that in addition to binding the molecule it transports, it also binds ATP (at a different site). You have a transport assay that measures the activity of Protein X. You perform the transport assay under different conditions:

Transport Molecule	Nucleotide Added	Nucleotide Binding? Level	Transport Activity
Yes	ATP	+++	+++
Yes	ADP	+++	-
Yes	GTP	-	-
Yes	ATP-γS	+++	-

ATP –γS cannot substitute for ATP to transport the molecule because
(A) Impermeable to membrane
(B) Nonhydrolysable
(C) Both (A) and (B)
(D) It can substitute ATP

466. Continuing to question, GTP substitute for ATP to transport the molecule because:
1. This is most likely because ATP and GTP have different nitrogenous bases
2. Enzymes (and other proteins as well) have binding sites that are very specific. Due to the different structures of the nitrogenous bases, the ATP-binding site on Protein X cannot bind GTP.
3. Protein X could bind GTP; it probably sill could not substitute for ATP because GTP is not used as an energy source by the cell.
4. None

Correct answer is:
(A) 4 (B) 1 (C) 2 (D) 1, 2 and 3

467. Which one of the following enzymes contain copper as metal activator:
(A) Catalase (B) Carbonic anhydrase
(C) Cytochrome oxidase (D) Ascorbic acid oxidase

468. Watson and Crick used all of the following information in deducting the structure of the genetic material except:
(A) Avery, MacLeod and McCarty's identification of DNA as the "transforming principle"
(B) X-ray diffraction studies by Astbury, Franklin, Pauling and Wilkins
(C) Chargaff's rules of DNA base composition
(D) Microscopic examination of DNA structure

469. Individuals that share a haplotype are most likely to:
(A) Have all of the same genetic information
(B) Have the same SNPs in the region of the genome that contains that haplotype
(C) Be of different ethnic descent
(D) Be siblings

470. Which one of the following contain b-ionine ring in its structure:
(A) Chlorophyll (B) Purine
(C) Vitamin A (D) Inosine

471. In DNA structure propeller twist means:
(A) Two bases orient antiparallel to each other
(B) Two bases flip with respect to each other
(C) Two flat bases counter rotate relative to each other
(D) Two bases stack with respect to each other with 3, 4 Å space

472. Select out sulphur containing amino acid:
(A) Arginine (B) Isoleucine
(C) Phenyl alanine (D) Methionine

473. Which of the following is mismatched?
(A) B-DNA : dehydrated
(B) Z-DNA : left-handed helix
(C) A-DNA : dehydrated
(D) Z-DNA : no major or minor grooves

474. Which one of the following is an essential amino acid:
(A) Alanine (B) Serine
(C) Lysine (D) Proline

475. Triple stranded DNA:
(A) Is a characteristic of SV 40 genome
(B) Involves Hoogsteen base pair
(C) Involves only intermolecular interactions
(D) Is characteristics by in the presence of a string of alternating purine-pyrimidine bases

476. You work in a lab that studies bacterial membrane proteins. You have isolated a protein, pMem, that spans the membrane and you determined the sequence of that protein:
1 KKLLPCTAL* VMCAGMACAQ* AEERNDWHFN * IGAMYEIENV* EGYGED MDGL 50
51 AEPSVYFNAA* NGPWRIALAY* YQ EGP SA* GKRGTWFDRP* ELEVHY QFLE 100101 NDDFSFGLTG* GFRNYGYHYV* DEPGKDTA NM* QRWKIAPDWD* VKLTDDLRFN151 GWLSMYKFAN* DLNTTGYADT* RVETETGL QY* TFNETVALRV* NYYLERGFNM 200
201 DDSRNNGEFS*TQEIRAYLPL*T LGNHSVTPY* TRIGLDRWSN* WDWQDDIERE 250
251 GHDFNRVGLF* YGYDFQNGLS* VSLEYAFEWQ* DHDEGDSDKF* HYAGVGVNYS*F 301
Is the codon for phenyalanine (F) the last codon in the mRNA transcript?
(A) No. The last codon is always a stop codon
(B) It is F
(C) It can be any codon for amino acid
(D) None of these

477. Continue with above question, what is the best length (in nucleotides) of the mRNA that encodes for the above protein?
(A) 950 (B) 902 (C) 301 (D) 3303

478. Which one of the following yields two molecules of vitamin 'A' in intestine:
(A) a-carotene (B) Lycopene
(C) b-Carotene (D) g – Carotene

479. What factors are responsible for the specificity of G protein-coupled responses?
(A) Protein kinases can phosphorylate different proteins in different cells
(B) Receptors can exist as isoforms
(C) G-protein subunits can exist as isoforms
(D) All of these

480. Michaelis constants (Km) for enzyme – substrate pair of Catalase and H_2O_2 is:
(A) 25 mM (B) 0.4 mM
(C) 0.05 mM (D) None of these

481. Michaelis constants (Km) for enzyme – substrate pair of Hexokinase and ATP is:
(A) 25 mM (B) 0.4 mM
(C) 0.05 mM (D) None of these

482. Michaelis constants (Km) for enzyme – substrate pair of Hexokinase and D-Glucose is:
(A) 25 mM (B) 0.4 mM
(C) 0.05 mM (D) None of these

483. Michaelis constants (Km) for enzyme – substrate pair of Hexokinase and D-Fructose is:
(A) 25 mM (B) 0.4 mM
(C) 0.05 mM (D) 1.5mM

484. Michaelis constants (Km) for enzyme – substrate pair of Chymotrypsin and Glycyltyrosinglycine is:
(A) 25 mM (B) 9 mM
(C) 108 mM (D) None of these

485. Michaelis constants (Km) for enzyme – substrate pair of Chymotrypsin and N- Benzoyltyrosinamide is:
(A) 25 mM (B) 26 mM
(C) 111 mM (D) None of these

486. Michaelis constants (Km) for enzyme – substrate pair of Carbonic anhydrase and HCO3is:
(A) 7 mM (B) 8 mM
(C) 9 mM (D) None of these

487. The turnover number for Catalase is:
(A) $4x10^7$ (B) $9x10^7$
(C) $12x10^7$ (D) $17x10^7$

488. Enzymes are basically made up of:
(A) Proteins (B) Fat
(C) DNA (D) None of these

489. Heterotrophic enzymes are:
(A) Modified by their substrate concentration
(B) Different isoenzymes
(C) Stimulated and inhibited by an effectors or modulator molecule other than substrate
(D) All of these

490. Majority of enzymes belongs to the category of:
(A) Endoenzymes (B) Epoenzymes
(C) Both (A) and (B) (D) None of these

491. An excellent example of inducible enzyme in higher plant is:
(A) Nitrate reductase (B) α-amylase
(C) Both (A) and (B) (D) None of these

492. The substrate for ribozyme is:
(A) DNA (B) RNA
(C) Fat (D) None of these

493. The self splicing group I introns and RNAse P are the example of:
(A) Zymogens (B) Ribozymes
(C) Isozymes (D) None of these

494. The modern system of enzyme classification was introduced by International Union of Biochemistry (IUB) in:
(A) 1961 (B) 1962 (C) 1963 (D) 1964

495. All enzymes have a specific structure of:
(A) Three dimensional
(B) Two dimensional
(C) One dimensional
(D) No definite dimensional

496. .... Enzymes are thermolabile:
(A) All (B) Few
(C) Most of (D) None of

497. Km is not dependent on the concentration of:
(A) Enzyme (B) Substrate
(C) Both (A) and (B) (D) None of these

498. Allosteric enzymes have:
(A) Active sites (B) Allosteric sites
(C) Both (A) and (B) (D) None of these

499. The enzyme acts by:
(A) Lowering the energy of activation of reactions
(B) Increasing the energy of activation of reaction
(C) Conserving the energy of activation of reaction
(D) None of these

500. Which vitamin is necessary for coenzyme A synthesis?
(A) Pyridoxine (B) Pantothenic acid
(C) Biotin (D) None of these

501. Which of the following is the water soluble vitamin?
(A) Thiamines (B) Riboflavin
(C) Niacin (D) Biotin
(E) Panthothenic acid (F) Folic acid
(G) Cobalamin (H) Pyridoxin
(I) Ascorbic acid (J) All of these

502. Which of the following is the fat soluble vitamins?
(A) Vitamin A (B) Vitamin D
(C) Vitamin E (D) Vitamin K
(E) All of these

503. Catalytic antibodies function as enzymes on the principle of:
(A) Enzymatic conversion of antibodies
(B) Monoclonal antibodies with chemical capability
(C) Antigen-antibody affinity
(D) None of these

504. Which one of the following coenzyme pair correctly with the group transferred?
(A) NADP – phosphoryl
(B) Biotin – carbon dioxide
(C) Thiamin – pyrophosphate
(D) None of these

505. The Km value of an enzyme is:
(A) ½ of the Vmax
(B) The substrate concentration that gives ½ maximal velocity
(C) A dissociation constant
(D) None of these

506. Induced fit theory of enzyme action was proposed by:
(A) Koshland (B) Fischer
(C) Went (D) None of these

507. Among the following, which enzyme has the highest turnover number?
(A) Catalase (B) Lysozyme
(C) Lipase (D) None of these

508. Among the following, which enzyme has the lowest turnover number?
(A) Catalase (B) Lysozyme
(C) Lipase (D) None of these

509. The active site of enzymes is formed by some of its:
(A) Amino groups of amino acid
(B) Carboxyl groups of the amino acids
(C) Exposed sulfur bonds
(D) None of these

510. Abzymes are:
(A) Enzymes that hydrolyze antibodies
(B) Antibodies that have catalytic activity
(C) Both (A) and (B)
(D) None of these

511. Which one of the following vitamins does not act as a precursor for coenzymes?
(A) Thiamine (B) Biotin
(C) Folic acid (D) None of these

512. The enzymes where catalysis involve group transfer are named as:
(A) Transferases (B) Lyases
(C) Oxidoreductase (D) Isomerases

513. The amount of enzyme causing transformation of 1 molecule of substrate per minute at 25°C under optimal conditions of measurement is known as:
(A) Enzyme unit (B) The Katal (Kat)
(C) Rate constant (D) None of these

514. According to the new system of classification the enzymes are grouped into:
(A) Six major classes (B) Five major classes
(C) Four major classes (D) Three major classes

515. Which of the following is the carrier of sugar molecules such as glucose, galactose and their derivatives?
(A) UDP (B) PAPS
(C) TPP (D) None of these

516. The first ribozyme was discovered by:
(A) R. Cech (B) C. Woese
(C) F. Crick (D) L. Orgel

517. An allosteric modular influences enzyme activity by:
(A) Binding to a site on the enzyme molecule distinct from the catalytic site
(B) Competing for the catalytic site with the substrate
(C) Changing the specificity of the enzyme for its sub strate
(D) None of these

518. Which metabolic process in bacteria is directly inhibited by the antibiotic penicillin?
(A) ATP hydrolysis (B) Synthesis of fats
(C) Cellular respiration (D) None of these

519. The product of an enzyme-catalysed reaction can act as inhibitor of the reaction. This mechanism of control is known as:
(A) Feedback inhibition (B) Competitive inhibition
(C) Metabolic (D) None of these

520. Pepsinogen (inactive form of enzyme pepsin) is converted into its active form by:
(A) Hydrolysis (B) Isomerism
(C) Reduction (D) Oxidation

521. Most of the enzymes show optimum activity in:
(A) Alkaline medium (B) Neutral medium
(C) Acidic medium (D) Any of these

522. Some enzymes are synthesized in the cells in an inactive state and are called as:
(A) Zymogens (B) Proenzymes
(C) Ore-enzymes (D) All of these

523. How many forms of LDH have been discovered?
(A) Fours (B) Five
(C) Six (D) Seven

524. Lactate dehydrogenase (LDH) which catalyses pyruvate to lactate is an example of:
(A) Coenzyme (B) Isoenzyme
(C) Apoenzyme (D) None of these

525. The enzymes which act normally within cells are called:
(A) Ferment (B) Exoenzymes
(C) Endoenzymes (D) Apoenzymes

526. The lower value of Km means:
(A) Lower affinity of enzyme with substrate
(B) higher enzyme activity
(C) Higher substrate affinity of enzyme
(D) None of these

527. In a chemical reaction, transition state species have free energies:
(A) Higher than reactant, but lower than products
(B) Higher than products but lower than reactant
(C) Higher that either the reactant or the products
(D) None of these

528. In higher plants which of the following is the rich source of enzymes and are widely used in experimental work?
(A) Matured seeds (B) Germinating seeds
(C) Matured flowers
(D) Immature flowers (E) None of these

529. Which of the following is the common characteristic of any enzyme:
(A) Enzymes exhibit enormous catalytic power
(B) Enzymes are highly specific
(C) Most enzymes are proteins
(D) All of these

530. A competitive inhibitor:
(A) Decreases the Km of an enzyme
(B) Increases the Km of an enzyme
(C) Decrease the Km but increases the Vmax of an enzyme
(D) Increases both the Vmax and the Km of an enzyme

531. .... different types of peroxidase isoenzymes have been isolated from maize:
(A) 16 (B) 18 (C) 20 (D) 22

532. One of the first enzymes discovered in isoenzyme forms is:
(A) Lactate dehydrogenase (B) Phosphatase
(C) Lipase (D) None of these

533. The enzyme which are always present in the organism in constant amounts regardless of its metabolic state are called as:
(A) Constitutive enzymes (B) Inducible enzymes
(C) Isoenzymes (D) None of these

534. Which of the following statement is true about non-competitive inhibitors?
(A) Km decreases (B) Vmax increases
(C) Vmax decreases (D) Km increases

535. The enzyme glutathione peroxidase which catalyzes destruction of H_2O_2:
(A) Mo (B) Fe (C) Se (D) Zn

536. Which of the following organic compound acts as prosthetic groups in some enzyme reactions and as cofactors in some others?
(A) FMN (B) FAD
(C) Heme-iron (D) Biotin
(E) All of these

537. Among the following, which one is the accepted SI unit of enzyme activity?
(A) Katal (B) Talka
(C) Latka (D) None of these

538. Which of the following is the carrier of sulphate group?
(A) UDP (B) PAPS
(C) TPP (D) None of these

539. Transformation of 1 mol of substrate per second into product is known as:
(A) Enzyme unit (B) The Katal (Kat)
(C) Rate constant (D) None of these

540. Biotin is involved in which of the following type of reactions?
(A) Dehydrations (B) Carboxylations
(C) Hydroxylations (D) Decarboxylations

541. Km specifies affinity of enzyme with substrate when;
(A) $K_2 >>> K_1$ (B) $K_2 <<< K_1$
(C) $K_2 = K_1$ (D) None of these

542. Which of the following vitamin can act without phosphorylation:
(A) Thiamine (B) Lipoamide
(C) Niacin (D) Pyridoxine

543. When the velocity of the reaction is almost proportional to the substrate concentration, the rate of reaction is said to be:
(A) First order reaction (B) Second order reaction
(C) Zero order reaction (D) Third order reaction

544. If an enzyme substrate system has lower value of Km, then binding of substrate to its enzyme will be:
(A) More stronger (B) Less stronger
(C) Inhibited (D) None of these

545. Panthothenic acid is a constituent of the coenzyme involved in which of the following event?
(A) Acetylation (B) Reduction
(C) Decarboxylation (D) Dehydrogenation

546. Allosteric enzyme have:
(A) Allosteric sites (B) Active site
(C) Both (A) and (B) (D) None of these

547. Most of the digestive enzymes belong to:
(A) Oxidoreductase (B) Isomerases
(C) Lyases (D) Hydrolases

548. Which of the following vitamin is a precursor of coenzyme that is required in enzymatic reactions involving transfer of acyl groups?
(A) Biotin (B) Riboflavin
(C) Lipoic acid (D) p-aminobenzoic acid

549. An enzyme that catalyzes the conversion of an aldose suger to a ketose sugar would be classified as one of the:
(A) Isomerases (B) Transferase
(C) Hydrolases (D) Oxidoreductase

550. Proenzyme is the .... precursor of an enzyme:
(A) Active (B) Inactive
(C) Both (A) and (B) (D) None of these

551. Abzyme were first reported by:
(A) Lerner and Tramontano
(B) Northrop
(C) James B. Sumner
(D) None of these

552. Which of the following coenzyme is the carrier of hexose sugar molecules?
(A) TPP (B) PAPS
(C) UDP (D) None of these

553. Which of the following is considered as good source for studies of enzymes?
(A) Microorganism (B) Germinating seeds
(C) Both (A) and (B) (D) None of these

554. First enzyme to be isolated in pure crystalline form is:
(A) Invertase (B) Diastase
(C) Zymase (D) Urease

555. Which of the following enzyme was first isolated and purified in the form of crystal?
(A) Ribonuclease (B) Pepsin
(C) Urease (D) Amylase

556. Pepsinogen, trypsinogen, chymotrypsinogen and procarboxypeptidase are example of:
(A) Zymogens (B) Ribozymes
(C) Isozymes (D) None of these

557. Which of the following is not a zymogen?
(A) Trypsinogen (B) Pepsinogen
(C) Procarboxypeptidase (D) None of these

558. The non-protein part of enzyme is called:
(A) Prosthetic group (B) Apoenzyme
(C) Holoenzyme (D) None of these

559. An example of inductive or inducible enzyme in E.coli is:
(A) Nitrate reductase (B) α – amylase
(C) Both (A) and (B) (D) β – galactosidase

560. Multiple forms of enzymes with the same catalytic activity but different structures are known as:
(A) Isoenzymes (B) Isozymes
(C) Both (A) and (B) (D) None of these

561. Which of the following hydrolyzing enzyme does not hydrolyse ester bonds?
(A) Lipases (B) Phosphatase
(C) Both (A) and (B) (D) None of these

562. Which of the following coenzyme is the carrier of acyl group?
(A) Coenzyme II (B) Coenzyme A
(C) Coenzyme Q (D) None of these

563. The enzymes where calatysis involves transfer of electrons are named as:
(A) Transferases (B) Lyases
(C) Oxidoreductase (D) Isomerases

564. Sulfa drugs are effective antibodies which inhibit the formation of folic acid by acting as a:
(A) Non-competitive inhibitor
(B) Competitive inhibitor
(C) Uncompetitive inhibitor
(D) None of these

565. Which of the following is the example of ribozyme:
(A) Leadzyme (B) Hairpine
(C) Hammerhead (D) Hepatitis dalta virus
(E) Tetrahymena (F) All of these

566. Malonic acid inhibits the activity of the enzyme:
(A) Succinic dehydrogenase
(B) Threonine dehydrogenase
(C) Hexokinase
(D) None of these

567. An enzyme that catalyzes the conversion of an aldose sugar to a ketose sugar would be classified as one of the:
(A) Isomerases (B) Transferase
(C) Hydrolases (D) Oxidoreductase

568. Lactic dehydrogenase (LDH) is essential for lactic acid formation. This is an example of:
(A) Isoenzymes (B) Abzyme
(C) Apoenzyme (D) Chemienzyme

569. An uncompetitive inhibitor of enzyme binds to:
(A) Any other site and modifies part of an enzyme
(B) The active site of enzyme
(C) Site other than the active site
(D) None of these

570. Which of the following coenzyme acts as an electron sink to promote catalysis?
(A) Pyridoxal phosphate (B) Lipoamide
(C) CoA (D) Biotin

571. Induced fit model has sufficient experimental evidence from the:
(A) Tracer technique (B) X-ray diffraction
(C) Auto radiography (D) None of these

572. Most vitamins are:
(A) Covalently linked to enzyme
(B) Synthesized by humans
(C) Precursor of coenzymes
(D) None of these

573. Which of the following is a cofactor and not a coenzyme?
(A) Methylcobalamin (B) Copper
(C) Tetrahydrofolic acid (D) Biotin

574. The essential chemical compounds of many co-enzyme are:
(A) Vitamins (B) Carbohydrate
(C) Protein (D) None of these

575. Iodoacetate is an irreversible inhibitor of the enzyme:
(A) Papain
(B) Glyceraldehyde 3- phosphate dehydrogenase
(C) Both (A) and (B)
(D) None of these

576. E.C. number of the enzyme alcohol dehydrogenase is:
(A) 0.1.1.1 (B) 9.0.1.2
(C) 1.1.1.1 (D) None of these

577. Which of the following enzyme is known as meat tenderizer because of its continued action even at the high temperature of cooking?
(A) Papain (B) Oxaloacetic acid
(C) Mutases (D) None of these

578. A .... is obtained when velocity is plotted against the substrate concentration:
(A) Rectangular hyperbola (B) Rectangular parabola
(C) Liner (D) None of these

579. Carbohydrates are commonly referred as:
(A) Polyhydroxyaldehydes (B) Polyhydroxyketones
(C) Both (A) and (B) (D) None of these

580. Among the followings, which one is known as structural polysaccharide?
(A) Cellulose (B) Starch
(C) Inulin (D) More than one

581. Among the followings, which one is known as storage polysaccharide?
(A) Cellulose (B) Starch
(C) Inulin (D) Both (B) and (C)

582. Which of the following is the example of disaccharides?
(A) Sucrose (B) Maltose
(C) Lactose (D) All of these

583. Which of the following is the example of trisaccharides?
(A) Raffinose (B) Gentianose
(C) Both (A) and (B) (D) All of these

584. The sugar residue of amylase is:
(A) Fructose unit only (B) In β-1, 4 linkages
(C) Galactose units only (D) None of these

585. Sucrose is a:
(A) Non-reducing sugar (B) Reducing sugar
(C) Both (A) and (B) (D) None of these

586. Arabinose is also known as:
(A) L – Sugar (B) M – Sugar
(C) N –Sugar (D) None of these

587. D-Glucose is also known as:
(A) Grape sugar (B) Dextrose
(C) Both (A) and (B) (D) None of these

588. D-Fructose is also known as:
(A) Levulose (B) Fruit sugar
(C) Both (A) and (B) (D) None of these

589. Which of the following is a mucopolysaccharide?
(A) Pectin (B) Chitin
(C) Hemi-cellulose (D) None of these

590. The enzyme invertase hydrolyses sucrose into:
(A) Glucose phosphate and fructose
(B) Glucose and fructose
(C) Fructose
(B) None of these

591. Among the following, which one is known as most abundant organic polysaccharide on the earth?
(A) Cellulose (B) Chitin
(C) Starch (D) None of these

592. Raffinose consists of how many monosaccharide units?
(A) 3 (B) 4 (C) 5 (D) 6

593. Almost all the sugars in plants are:
(A) D-Sugars (B) L-Sugars
(C) Both (A) and (B) (D) None of these

594. In aldose sugars, aldehyde group or reducing centre always lies at carbon number:
(A) 1 (B) 2 (C) 3 (D) 4

595. In Ketose sugars, reducing centre always lies at carbon number:
(A) 1 (B) 2 (C) 3 (D) 4

596. Some organic compound follows the general formula of carbohydrate but these compounds are not carbohydrates. Example of such types of compounds are:
(A) Formaldehyde (B) Acetic acid
(C) Lactic acid (D) All of these

597. What are thee characteristics of monosaccharides?
(A) These cannot be hydrolysed
(B) These are the building units of complex carbohydrates
(C) These are simplest of carbohydrate and are known as sugars
(D) These are sweet – tasting, crystalline and soluble in water
(E) All of these

598. Which of the following is not a component of pectic substances?
(A) Arabinan (B) Galacturonan
(C) Liginin (D) None of these

599. Which of the following is not a reducing sugar?
(A) Sucrose (B) Maltose
(C) Ribose (D) Glucose

600. Which of the following is not an oligosaccharide?
(A) Raffinose (B) Sucrose
(C) Maltose (D) Inulin

601. Among the followings, which one is structural polysaccharide?
(A) Cellulose (B) Hemi-cellulose
(C) Pectic substances (D) Chitin
(E) Gum (F) Mucilage
(G) All of these

602. Which of the following stage is known as reactive form of monosaccharide?
(A) De-phosphorylated (B) Sulphonated
(C) Phosphorylated (D) All of these

603. Diastase enzyme consists of:
(A) R-enzyme (B) β-amylase and β-amylase
(C) Maltose (D) All of these

604. Six membered ring structure of aldohexoses is known as:
(A) Furanose form (B) Cyclic form
(C) Both (A) and (B) (D) Pyronose form

605. The highest number of asymmetric carbon atom is present in:
(A) Arabinose (B) Glucose
(C) Fructose (D) None of these

606. Which of the following character is correct regarding oligosaccharides?
(A) They are sweet in test (B) They are crystalline
(C) They are soluble sugar (D) All of these

607. Which of the following is the component of hemi-cellulose?
(A) Mannans and Glucomannans
(B) Xylans
(C) Galactons and arabogalactans
(D) All of these

608. Besides containing C, H and O, some carbohydrates also contain:
(A) Nitrogen (B) Sulfur
(C) Both (A) and (B) (D) None of these

609. In general, it is considered that plant metabolism is highly:
(A) Flexible (B) Rigid
(C) Both (A) and (B) (D) None of these

610. Which of the following is known as optically active and dextrorotatory?
(A) Glucose (B) Fructose
(C) Both (A) and (B) (D) None of these

611. What are the characteristics of polysaccharides?
(A) Amorphous in nature (B) Tasteless
(C) Insoluble in water (D) All of these

612. Sulfur activation is catalyzed by:
(A) ATP sulfurylase (B) ATP sulfotransferase
(C) ATP kinase (D) None of these

613. Sulfur containing amio acids which constitute proteins are:
(A) Cysteine and methionine
(B) Methionine and histidine
(C) Cysteine and serine
(D) None of these

614. Reduction of sulfate and it assimilation takes place in:
(A) Plastids of root (B) Plastids of shoot
(C) Plastids of leaves (D) Both (A) and (C)

615. In the process of sulfur uptake, reduction and transport in plants, sulfate is stored in:
(A) Vacuole (B) Chloroplast
(C) Mitochondria (D) None of these

616. The major protein content of wheat seed is:
(A) Albumin (B) Globulin
(C) Glutelin (D) Prolamin

617. Pyrimidins are:
(A) Cytosine (B) Thymine
(C) Adenine (D) Both (A) and (B)

618. Hydroxyl groups on sugar represent:
(A) Alcohol (B) Ethylene
(C) Starch (D) Fat

619. Fats are degraded by:
(A) Amylase (B) Protease
(C) Lipase (D) Endonuclease

620. Product of β-oxidation of fatty acids is:
(A) Acetyl COA (B) Mono glyceides
(C) Glycerol (D) Triglycerides

621. β-oxidation takes place in:
(A) Mitochondria (B) Cytiplasm
(C) Glyoxysome (D) Golgi apparatus

622. Key enzyme involved in glyoxylate of lipid break down during germination are:
(A) Phspho fructokinase
(B) Malate synthetase and isocitrate lyase
(C) Pyruvate kinase
(D) Pyruvate carboxylase

623. Which is the isomer of glucose:
(A) Fructose (B) Galactose
(C) Both (A) and (B) (D) Sucrose

624. Which one of the following is a disachharide:
(A) Fructose (B) Raffinose
(C) Starch (D) Sucrose

625. In polysaccharides, which of the following fact is no correct?
(A) Reducing
(B) Water insoluble
(C) Monosacharides joined by glucosidic bond
(D) Hydrolyzed by enzymes/acids

626. Major form of carbohydrate stored in seeds is:
(A) Sucrose (B) Starch
(C) Cellulose (D) Glucose

627. In rice, the predominant starch catabolic pathway is
(A) Hydrolytic (B) Phosphorylatic
(C) α-oxidation (D) β-oxidation

628. The major form of carbohydrate involved in translocation in plants:
(A) Glucose (B) Sucrose
(C) Starch (D) Hemi-cellulose

629. Paddy seeds lack in:
(A) Albumin (B) Zein
(C) Prolamin (D) Globulin

630. Glutelins are soluble in:
(A) Water (B) Acohols
(C) Dilute acids and alkali solution (D) Methane

631. Furanose is ring structure with:
(A) 5 member (B) 6 member
(C) 7 member (D) 8 member

632. Pyranose is ring structure with:
(A) 5 member (B) 6 member
(C) 7 member (D) 8 member

633. In D series, highest numbered asymmetric carbon has OH group directed towards:
(A) Left (B) Right
(C) Both (A) and (B) (D) None of these

634. In L series, highest numbered asymmetric carbon has OH group directed towards:
(A) Left (B) Right
(C) Both (A) and (B) (D) None of these

635. Most important monosaccharides in biological system is:
(A) Fructose (B) Glucose
(C) Galactose (D) Xylose

636. At room temperature, an aqueous solution of glucose can exist in ——— tautomeric form:
(A) One (B) Two
(C) Three (D) Four

637. Naturally occurring sugars are mostly:
(A) Pentoses (B) Hexoses
(C) Heptoses (D) Octoses

638. Staling of bread is related to:
(A) Gluten (B) Starch
(C) Albumin (D) None of these

639. Maltose is a disaccharide:
(A) Reducing (B) Non-reducing
(C) Both (A) and (B) (D) None of these

640. Fructo-oligosacchrides are commercially manufactured by which enzyme on sucrose:
(A) α-fructofuranosidase
(B) β-fructofuranosidase
(C) α-mannase
(D) β-mannase

641. Lactose can occur in a crystalline form of:
(A) One (B) Two
(C) Three (D) Four

642. Hydrolysis product of starch is:
(A) Glucose monomer (B) Glucose dimer
(C) Oligosaccharide (D) All of these

643. Enzyme used for starch hydrolysis is:
(A) Protease (B) Amylase
(C) Galactosidase (D) Cellulose

644. Isomerization of glucose to fructose is done with:
(A) Glucose isomerase (B) Galacto isomerase
(C) Mannose isomerase (D) None of these

645. Mannitol is commercially produced by catalytic hydrogenation of:
(A) D-mannose (B) D-glucose
(C) D-galactose (D) All of these

646. Starch granules are:
(A) Crystalline (B) Amylopectin
(C) Partly crystalline (D) None of these

647. Waxy starch contains:
(A) Amylose (B) Amylopectin
(C) Both (A) and (B) (D) None of these

648. Starch is composed of:
(A) Amylose (B) Amylopectin
(C) Both (A) and (B) (D) None of these

649. Stalling of bread occurs due to:
(A) Saccharification (B) Retrogradation
(C) Caramelization (D) Browning

650. Acid conversion of starch is carried out with:
(A) Hydrochloric acid (B) Sulfuric acid
(C) Both (A) and (B) (D) None of these

651. Example of substituted starch is:
(A) Starch acetate
(B) Starch monophosphatase
(C) Hydroxyl propyl starch
(D) All of these

652. Oxidation of starches leads to:
(A) Stabilization (B) Adhesion gelling
(C) Clarification (D) All of these

653. Which of the following is non starch polysaccharide:
(A) Hemicelluloses (B) Pentosans
(C) Galactomannan (D) All of these

654. Xylan consists of:
(A) Mannose (B) Xylose
(C) Both of (A) and (B) (D) None of these

655. Mannas consists of:
(A) Galactose (B) Mammose
(C) Both (A) and (B) (D) None of these

656. Cyclic polymers obtained by enzymatic treatment of starch is:
(A) Cyclodextrose (B) Cyclodextrin
(C) Cellulose (D) Hemicellulose

657. Cyclodextrin is used in:
(A) Enzyme preparation (B) Encapsulation
(C) Retrogradation (D) All of these

658. Polydextrose is a randomly bonded condensation polymer of:
(A) Fructose (B) Glucose
(C) Maltose (D) Starch

659. Polysaccharide that give highly viscous solution at low concentration:
(A) Pectin (B) Starch
(C) Gum (D) Cellulose

660. Polysaccharide with strong acid group is:
(A) Furcellaran (B) Carrageenan
(C) Both (A) and (B) (D) None of these

661. Guar gum is also known as:
(A) Cluster bean gum (B) Carob gum
(C) Gum elastic (D) None of these

662. Agar is extracted from:
(A) Algae (B) Fungi
(C) Bacteria (D) Crops

663. Agar generally contains:
(A) Agarose (B) Agaropectin
(C) Both (A) and (B) (D) None of these

664. Agar is obtained from:
(A) Red sea weed (B) Plants
(C) Both (A) and (B) (D) None of these

665. Sucrose is obtained from:
(A) Sugarcane (B) Sugarbeet
(C) Both (A) and (B) (D) None of these

666. Cellobiose is a:
(A) Disaccharide (B) Trisaccharide
(C) Tetrasaccharide
(D) Polysaccharide

667. Gentiobiose is a:
(A) Disaccharide (B) Trisaccharide
(C) Tetrasaccharide (D) Polysaccharide

668. Fucosidolactose is a:
(A) Disaccharide (B) Trisaccharide
(C) Tetrasaccharide (D) Polysaccharide

669. Isomaltose is a:
(A) Disaccharide (B) Trisaccharide
(C) Tetrasaccharide (D) Polysaccharide

670. Lactulose is a:
(A) Disaccharide (B) Trisaccharide
(C) Tetrasaccharide (D) Polysaccharide

671. Stachyose is a:
(A) Disaccharide (B) Trisaccharide
(C) Tetrasaccharide (D) Polysaccharide

672. Maltulose is a:
(A) Disaccharide (B) Trisaccharide
(C) Tetrasaccharide (D) Polysaccharide

673. Melibiose is a:
(A) Disaccharide (B) Trisaccharide
(C) Tetrasaccharide (D) Polysaccharide

674. Gentianose is a:
(A) Disaccharide (B) Trisaccharide
(C) Tetrasaccharide (D) Polysaccharide

675. Neotrehalose is a:
(A) Disaccharide (B) Trisaccharide
(C) Tetrasaccharide (D) Polysaccharide

676. Nierose is a:
(A) Disaccharide (B) Trisaccharide
(C) Tetrasaccharide (D) Polysaccharide

677. Isokestose is a:
(A) Disaccharide (B) Trisaccharide
(C) Tetrasaccharide (D) Polysaccharide

678. Kestose is a:
(A) Disaccharide (B) Trisaccharide
(C) Tetrasaccharide (D) Polysaccharide

679. Palatinose is a:
(A) Disaccharide (B) Trisaccharide
(C) Tetrasaccharide (D) Polysaccharide

680. Rutinose is a:
(A) Disaccharide (B) Trisaccharide
(C) Tetrasaccharide (D) Polysaccharide

681. Maltotriose is a:
(A) Disaccharide (B) Trisaccharide
(C) Tetrasaccharide (D) Polysaccharide

682. Maltotetraose is a:
(A) Disaccharide (B) Trisaccharide
(C) Tetrasaccharide (D) Polysaccharide

683. Trehalose is a:
(A) Disaccharide (B) Trisaccharide
(C) Tetrasaccharide (D) Polysaccharide

684. Panose is a:
(A) Disaccharide (B) Trisaccharide
(C) Tetrasaccharide (D) Polysaccharide

685. Raffinose is a:
(A) Disaccharide (B) Trisaccharide
(C) Tetrasaccharide (D) Polysaccharide

686. Umbelliferose is a:
(A) Disaccharide (B) Trisaccharide
(C) Tetrasaccharide (D) Polysaccharide

687. Sophorose is a:
(A) Disaccharide (B) Trisaccharide
(C) Tetrasaccharide (D) Polysaccharide

688. Limiting essential amino acid in wheat is:
(A) Leucine (B) Lysine
(C) Phenylalanine (D) Isoleucine

689. Oryzenin is a which type of protein:
(A) Globulin (B) Glutelin
(C) Prolamin (D) Histone

690. Leucosin is a which type of protein:
(A) Globulin (B) Glutelin
(C) Prolamin (D) Histone

691. Myosin is a which type of protein:
(A) Albumin (B) Glutelin
(C) Globulin (D) Histone

692. Glycinin is a which type of protein:
(A) Globulin (B) Glutelin
(C) Prolamin (D) Histone

693. Glutenin is a which type of protein:
(A) Globulin (B) Glutelin
(C) Prolamin (D) Histone

694. Zein is a which type of protein:
(A) Globulin (B) Glutelin
(C) Prolamin (D) Histone

695. Glidin is a which type of protein:
(A) Globulin (B) Glutelin
(C) Histone (D) Prolamin

696. Hordein is a which type of protein:
(A) Globulin (B) Glutelin
(C) Prolamin (D) Histone

697. Bread contains which type of vitamin:
(A) Thiamin (B) Riboflavin
(C) Niacin (D) All of these

698. Oryzenin is present in:
(A) Wheat (B) Rice
(C) Barley (D) Soybean

699. Glutenin is present in:
(A) Wheat (B) Rice
(C) Barley (D) Soybean

700. Gliadin is present in:
(A) Wheat (B) Rice
(C) Barley (D) Soybean

701. Zein is present in:
(A) Wheat (B) Rice
(C) Barley (D) Corn

702. Hordein is present in:
(A) Wheat (B) Rice
(C) Barley (D) Soybean

703. Glycinin is present in:
(A) Wheat (B) Rice
(C) Barley (D) Soybean

704. Actin is present in:
(A) Wheat (B) Rice
(C) Barley (D) Soybean

705. Whey protein is present in:
(A) Wheat (B) Rice
(C) Milk (D) Soybean

706. Casein protein is present in:
(A) Wheat (B) Rice
(C) Barley (D) Milk

707. Which of the following is considered as best quality protein:
(A) Cereal protein (B) Soya protein
(C) Egg protein (D) None of these

708. Denaturation of protein does not involve:
(A) Change in molecular structure
(B) Loss of biological quality
(C) Breaking of peptide bonds
(D) Change in solubility

709. Heat stability of casein is due to high content of:
(A) Proline (B) Hydroxyproline
(C) Both (A) and (B) (D) None of these

710. Histone protein are rich in:
(A) Lysine (B) Arginine
(C) Both (A) and (B) (D) None of these

711. Gluten proteins are rich in:
(A) Lysine (B) Methionine
(C) Glutamine (D) Tryptophan

712. Which of the following is non gluten forming protein:
(A) Albumin (B) Globulin
(C) Both (A) and (B) (D) None of these

713. Gluten protein have :
(A) Lysine (B) Methionine
(C) Tryptophan (D) All of these

714. Gliadin and glutenin are which type of protein in wheat:
(A) Storage (B) Gluten forming
(C) Both (A) and (B) (D) None of these

715. Protein are amphoteric due to presence of:
(A) Free amino group (B) Carboxyl group
(C) Both (A) and (B) (D) None of these

716. Which of the following is albumin protein:
(A) Leucosine (B) Legumeline
(C) Ovalbumin (D) All of these

717. Which of the following is globulin protein:
(A) Ovalbumin (B) Glycinine
(C) Legumine (D) All of these

718. Which of the following is prolamin protein:
(A) Glaidin (B) Zein
(C) Hordein (D) All of these

719. Which of the following is scleroprotein:
(A) Collagen (B) Elastin
(C) Keratin (D) All of these

720. Chromoproteins are the protein with:
(A) Carbohydrate (B) Lipid
(C) Pigment (D) Water

721. Which of the following is chromoprotein:
(A) Hemoglobin (B) Cytochrome
(C) Flavoprotein (D) All of these

722. Mucoproteins are the proteins which contain ——— as prosthetic group:
(A) Carbohydrate (B) Lipid
(C) Pigment (D) Water

723. Mucoproteins are also known as:
(A) Glycoproteins (B) Phosphoprotein
(C) Chromoprotein (D) All of these

724. Butyric acid is also known as:
(A) n-butanoic (B) n-hexonoic
(C) n-octanoic (D) n-decanoic

725. Caproic acid is also known as:
(A) n-butanoic (B) n-hexonoic
(C) n-octanoic (D) n-decanoic

726. Caprylic acid is also known as:
(A) n-butanoic (B) n-hexonoic
(C) n-octanoic (D) n-decanoic

727. Capric acid is also known as:
(A) n-butanoic (B) n-hexonoic
(C) n-octanoic (D) n-decanoic

728. Lauric acid is also known as:
(A) n-butanoic (B) n-hexonoic
(C) n-dodecanoic (D) n-decanoic

729. Myristic acid is also known as:
(A) n-tetradecanoic (B) n-hexonoic
(C) n-octanoic (D) n-decanoic

730. Palmitic acid is also known as:
(A) n-butanoic (B) n-hexadecanoic
(C) n-octanoic (D) n-decanoic

731. Stearic acid is also known as:
(A) n-butanoic (B) n-hexonoic
(C) n-octanoic (D) n-decanoic

732. Arachidic acid is also known as:
(A) n-butanoic (B) n-eicosanoic
(C) n-octanoic (D) n-decanoic

733. Valeric acid is also known as:
(A) n-pentanoic (B) n-hexonoic
(C) n-octanoic (D) n-decanoic

734. Which of the following is saturated fatty acid:
(A) Linolenic (B) Elaeostearic
(C) Arachidonic (D) None of these

735. Which of the following is unsaturated fatty acid:
(A) EPA (B) DHA
(C) Oleic (D) All of these

736. Myristoleic acid is also known as:
(A) Tetradec-9-enoic (B) Hexadec-9-enoic
(C) Octadec-6-enoic (D) Octadec-9-enoic

737. Palmitoleic acid is also known as:
(A) Tetradec-9-enoic (B) Hexadec-9-enoic
(C) Octadec-6-enoic (D) Octadec-9-enoic

738. Petroselinic acid is also known as:
(A) Tetradec-9-enoic (B) Hexadec-9-enoic
(C) Octadec-6-enoic (D) Octadec-9-enoic

739. Oleic acid is also known as:
(A) Tetradec-9-enoic (B) Hexadec-9-enoic
(C) Octadec-6-enoic (D) Octadec-9-enoic

740. Linoleic acid is also known as:
(A) Octa-9:12-dienoic
(B) Octadeca-9:12:15-trienoic
(C) Hexadec-9-enoic
(D) Octadec-9-enoic

741. Linolenic acid is also known as:
(A) Octa-9:12-dienoic
(B) Octadeca-9:12:15-trienoic
(C) Hexadec-9-enoic
(D) Octadec-9-enoic

742. Gadoleic acid is also known as:
(A) Octa-9:12-dienoic
(B) Octadeca-9:12:15-trienoic
(C) Hexadec-9-enoic
(D) Eicos-9-enoic

743. Major component of unsaponifiable fraction of most of the fat is:
(A) Sterol (B) Terepenic alcohols
(C) Aliphatic alcohols (D) All of these

744. DRI stand for:
(A) Daily Reference Intake
(B) Dietary Required Intake
(C) Dietary Reference Intake
(D) Daily Required Intake

745. Which of the following factor influence mineral bioavailabity of food:
(A) Solubility of mineral (B) Food ligand
(C) Reductant (D) All of these

746. Which of the following is a mineral bioavailability enhancer:
(A) Ascorbic acid (B) Citric acid
(C) Lactic acid (D) All of these

747. Which of the following is a mineral bioavailability inhibitor:
(A) Phytic acid (B) Polyphenol
(C) Tanin (D) All of these

748. Mineral composition of plant foods is affected by:
(A) Plant genetics (B) Soil fertility
(C) Environment (D) All of these

749. Food function of copper include:
(A) Catalyst (B) Colour modifier
(C) Texture stability (D) All of these

750. Chemical name of vitamin A is:
(A) Tocopherol (B) Calcipherol
(C) Retinol (D) None of these

751. Chemical name of vitamin C is:
(A) Tocopherol (B) Calcipherol
(C) Retinol (D) Ascorbic acid

752. Chemical name of vitamin D is:
(A) Tocopherol (B) Calcipherol
(C) Retinol (D) Ascorbic acid

753. Chemical name of vitamin E is:
(A) Tocopherol (B) Calcipherol
(C) Retinol (D) Ascorbic acid

754. Vitamin H is also known as:
(A) Thiamin (B) Niacin
(C) Biotin (D) Riboflavin

755. Precursor of Vitamin D_2 is:
(A) Cholecalciferol (B) Ergocalciferol
(C) Ergosterol (D) 7-dehydrocholesterol

756. Precursor of Vitamin D_3 is:
(A) Cholecalciferol (B) Ergocalciferol
(C) Ergosterol (D) 7-dehydrocholesterol

757. Vitamin B_6 is also known as:
(A) Niacin (B) Pyridoxine
(C) Riboflavin (D) Cyanocobalamine

758. Vitamin B_2 is also known as:
(A) Niacin (B) Pyridoxine
(C) Riboflavin (D) Cyanocobalamine

825. The extension approach used under IADP was:
(A) Proscriptive (B) Prescriptive
(C) Participatory (D) Commercial

826. The IAAP was launched in:
(A) 1962 (B) 1963
(C) 1964 (D) 1965

827. High Yielding Variety Programme was started in:
(A) 1962 (B) 1963
(C) 1964 (D) 1965

828. High yielding technologies are:
(A) Resource positive (B) Resource negative
(C) Resource neutral (D) None of these

829. Farm and Home unit of Akashvani was started in connection with:
(A) IAAP (B) HYVP
(C) IRDP (D) CDP

830. Radio Rural Forum was started in:
(A) 1963 (B) 1964
(C) 1965 (D) 1966

831. Broadcasting of rural programme in television started in:
(A) 1965 (B) 1966
(C) 1967 (D) 1968

832. Which of the following is not a key feature of T&V system:
(A) Professionalism
(B) Time bound work
(C) Single line of command
(D) Farming System Research

833. The base level extension functionary in T&V system:
(A) BDO (B) VLW
(C) AEO (D) SDAO

834. The first state in India to introduce Training and Visit system was:
(A) Punjab (B) Rajasthan
(C) Tamilnadu (D) Bihar

835. The ratio of extension agent to farmer in a typical T&V system is:
(A) 1-250 (B) 1-500
(C) 1-800 (D) 1-1500

836. T&V system of extension can be characterized as:
(A) Management system (B) Supervisory system
(C) Integrated system (D) Broad based system

837. Small Farmers Development Agency was set up in:
(A) 1969-70 (B) 1970-71
(C) 1971-72 (D) 1972-73

838. The basic approach followed by SFDA:
(A) Employment approach (B) Integrated approach
(C) Welfare approach (D) Target group approach

839. SFDA is meant for the farmers having land holding of:
(A) 1-2 hectare (B) 0-1 hectare
(C) 1-5 hectare (D) 2-5 hectare

840. SFDA-MFAL is associated with:
(A) Venkatpaiah committee
(B) Venkataraman committee
(C) Chelliah committee
(D) None of these

841. An agricultural labourer earns more than of his wage from agriculture:
(A) 33.33 per cent (B) 40.00 per cent
(C) 50.00 per cent (D) 66.66 per cent

842. A welfare programme for the rural poor:
(A) Krishi Vigyan Kendra (B) Lab to Land
(C) National Demostration (D) Minikit trials

843. The Ganga Kalyan Yojna is for:
(A) Consumption purpose of farmers
(B) Irrigation for small and marginal farmers
(C) Credit support scheme for rural women
(D) Credit support scheme for small farmers

844. Intensive Cattle Development Project was launched in:
(A) 1963-64 (B) 1964-65
(C) 1965-66 (D) 1966-67

845. The rural development programme launched on the basis of area development approach:
(A) DPAP (B) MWP
(C) RWP (D) IRDP

846. Desert Development Programme was launched on:
(A) 1977-78 (B) 1968-69
(C) 1971-72 (D) 1956-57

847. Tribal Area Development Programme was started in the year:
(A) 1970-71 (B) 1983-84
(C) 1984-85 (D) 1985-86

848. Integrated Wasteland Development Programme was launched on:
(A) 1989-90 (B) 1968-69
(C) 1971-72 (D) 1956-57

849. National Food Security Mission covers all except:
(A) Rice (B) Wheat
(C) Pulses (D) Sugarcane

850. Annual growth rate in agriculture sector expected during the eleventh plan:
(A) 8 per cent (B) 6 per cent
(C) 4 per cent (D) 3 per cent

851. In case of a mortised even repayment plan, the interest rate:
(A) Increases (B) Declines
(C) Remains constant (D) None of these

852. If liabilities>assets, then net worth is placed at:
(A) Liabilities side (B) Assets side
(C) Income side (D) None of these

853. Book value of durable refers to:
(A) Market price assets
(B) Purchase price of assets
(C) Production of assets
(D) Value of cost

582. During heat period, the fat per cent in milk will:
(A) Increase (B) Decrease
(C) No change (D) None of these

583. During heat period, the milk will:
(A) Increase (B) Decrease
(C) No change (D) Not known

584. How much energy in available from 1 gm milk?
(A) 8 calorie (B) 9 calorie
(C) 10 calorie (D) 12 calorie

585. Ca : P ratio in milk is:
(A) 1 : 1 (B) 1.43 : 1
(C) 1.5 : 1 (D) 2.11 : 1

586. What is the temperature range for survival of psychrophilic bacteria?
(A) 3-20^{0}C (B) 20-50^{0}C
(C) 50-70^{0}C (D) None of these

587. What is the temperature range for survival of mesophilic bacteria?
(A) 3-20^{0}C (B) 20-50^{0}C
(C) 50-70^{0}C (D) 75-80^{0}C

588. What is the temperature range for survival of thermophilic bacteria?
(A) 3-20^{0}C (B) 20-50^{0}C
(C) 50-70^{0}C (D) 75-80^{0}C

589. *Pseudomonas fragi* and *Pseudomonas fluorescens* belong to which of the following category?
(A) Psychrophilic bacteria (B) Mesophilic bacteria
(C) Thermophilic bacteria (D) All of these

590. *Streptococcus lactis* and *S. cremoris* belong to which of the following group?
(A) Psychrophilic bacteria (B) Mesophilic bacteria
(C) Thermophilic bacteria (D) Thermoduric bacteria

591. Which of the following is the thermophilic bacteria?
(A) *Micrococcus luteus*
(B) *Micrococcus condidus*
(C) *Streptococcus thermophilus*
(D) All of these

592. *Lactobacillus, thermophilus* and *Bacillus calidolactis* are the names of:
(A) Psychrophilic bacteria (B) Mesophilic bacteria
(C) Thermophilic bacteria (D) Thermoduric bacteria

593. Sweet curding of milk is caused by:
(A) *Streptococcus subtilis* (B) *Bacillus mycoides*
(C) *B. cereus* (D) All of these

594. The oiliness or cardboard flavour or sourness in milk is caused by:
(A) *Streptococcus liquefaciens*
(B) *Micrococcus pitiutoporus*
(C) *Pseudomonas flourescens*
(D) All of these

595. Oiliness of milk is caused mainly due to:
(A) Oxidation of fat (B) Reduction of fat
(C) Fermentation of lactose (D) Fermentation of fat

596. Sweet curdling is caused mainly due to:
(A) Fermentation of casein (B) Fermentation of fat
(C) Oxidation of ash (D) Fermentation of ash

597. Normal fermentation of milk is caused by:
(A) *Streptococcus lactis*
(B) *Bacillus subtilis*
(C) *Streptococcus liquefaciens*
(D) All of these

598. Rancid flavour in milk is caused due to:
(A) Fermentation of fat
(B) Fermentation of lactose
(C) Fermentation of protein
(D) Fermentation of ash

599. Sliminess or ropy fermentation is caused by:
(A) *Achromobactor aerogans*
(B) *Bacterium cloacae*
(C) *Alkaligenes viscosus*
(D) All of these

600. Sliminess is caused due to:
(A) Fermentation of milk cream
(B) Fermentation of lactose
(C) Fermentation of protein
(D) None of these

601. Frothiness in milk is caused due to:
(A) *Aerobacter aerogens* (B) *Torula cremoris*
(C) *Torula spherical* (D) All of these

602. Blue colour of milk is developed due to:
(A) *Pseudomonas syncyanea*
(B) *Streptococcus lutea*
(C) *Streptococcus marcescens*
(D) All of these

603. Red colour of milk is developed due to:
(A) *Pseudomonas syncyanea*
(B) *Streptococcus lutea*
(C) *Streptococcus marcescens*
(D) All of these

604. Yellow colour of milk is developed due to:
(A) *Pseudomonas syncyanea*
(B) *Streptococcus lutea*
(C) *Streptococcus marcescens*
(D) All of these

605. Which of the following yeast is found in milk?
(A) *Torula cremoris* (B) *Torula spherical*
(C) *Torula lactis* (D) All of these

606. Which of the following fungus involved in milk industry?
(A) *Penicillium glaucum* (B) *Penicillium rogueforti*
(C) *Oidium spp.* (D) All of these

607. Roquefort and Gorgonzola paneer are made with the help of:
(A) Bacteria (B) Yeast
(C) Fungi (D) All of these

608. Which of the following bacteria acts first on milk?
(A) *Lactobacillus* (B) *Streptococcus lactis*
(C) *Bacillus subtilis* (D) None of these

1059. (A); 1060. (A); 1061. (B); 1062. (D); 1063. (B); 1064. (C); 1065. (A); 1066. (B); 1067. (B); 1068. (A); 1069. (C); 1070. (A); 1071. (B); 1072. (B); 1073. (C); 1074. (A); 1075. (A); 1076. (A); 1077. (A); 1078. (B); 1079. (C); 1080. (A); 1081. (B); 1082. (C); 1083. (A); 1084. (D); 1085. (B); 1086. (C); 1087. (C); 1088. (D); 1089. (B); 1090. (A); 1091. (A); 1092. (D); 1093. (B); 1094. (A); 1095. (D); 1096. (B); 1097. (D); 1098. (B); 1099. (B); 1100. (B); 1101. (C); 1102. (D); 1103. (C); 1104. (C); 1105. (B); 1106. (B); 1107. (A) ; 1108. (B); 1109. (B); 1110. (D); 1111. (A); 1112. (C); 1113. (C); 1114. (A); 1115. (C); 1116. (A); 1117. (A); 1118. (D); 1119. (C); 1120. (D); 1121. (A); 1122. (B); 1123. (D); 1124. (C); 1125. (B); 1126. (A); 1127. (C); 1128. (D); 1129. (D); 1130. (C); 1131. (B); 1132. (A); 1133. (D) ; 1134. (C); 1135. (C); 1136. (A); 1137. (B); 1138. (C); 1139. (B); 1140. (C); 1141. (B); 1142. (B); 1143. (D); 1144. (C); 1145. (B); 1146. (A); 1147. (B); 1148. (C); 1149. (C); 1150. (C); 1151. (B); 1152. (B); 1153. (B); 1154. (B); 1155. (A); 1156. (C); 1157. (B); 1158. (A); 1159. (B); 1160. (C); 1161. (A); 1162. (B); 1163. (A); 1164. (B); 1165. (D); 1166. (B); 1167. (D); 1168. (A); 1169. (B); 1170. (A); 1171. (C); 1172. (B) ; 1173. (C); 1174. (B); 1175. (B); 1176. (B); 1177. (D); 1178. (A); 1179. (B); 1180. (C); 1181. (B); 1182. (B); 1183. (A); 1184. (D); 1185. (A); 1186. (B); 1187. (C); 1188. (B); 1189. (B); 1190. (A); 1191. (A); 1192. (A); 1193. (A); 1194. (C) ; 1195. (B); 1196. (A); 1197. (C); 1198. (D); 1199. (B); 1200. (B); 1201. (B); 1202. (B); 1203. (B); 1204. (D); 1205. (A); 1206. (A); 1207. (C); 1208. (C); 1209. (A); 1210. (B); 1211. (C); 1212. (C); 1213. (C); 1214. (B); 1215. (C); 1216. (C); 1217. (B); 1218. (A); 1219. (A); 1220. (C); 1221. (A); 1222. (C); 1223. (D); 1224. (A); 1225. (D); 1226. (A); 1227. (B); 1228. (A); 1229. (C); 1230. (A); 1231. (B); 1232. (D); 1233. (C); 1234. (A); 1235. (B); 1236. (B); 1237. (A); 1238. (C); 1239. (C); 1240. (B); 1241. (C); 1242. (B); 1243. (B); 1244. (A); 1245. (B); 1246. (C); 1247. (C); 1248. (C); 1249. (C); 1250. (C) ; 1251. (C); 1252. (D); 1253. (C); 1254. (C); 1255. (A); 1256. (C); 1257. (C); 1258. (C); 1259. (D); 1260. (D); 1261. (C); 1262. (B); 1263. (C); 1264. (A); 1265. (C); 1266. (D); 1267. (C); 1268. (A); 1269. (D); 1270. (C); 1271. (C); 1272. (C) ; 1273. (C); 1274. (B); 1275. (C); 1276. (C) ; 1277. (A);1278. (C); 1279. (A); 1280. (B); 1281. (B); 1282. (C); 1283. (A); 1284. (B); 1285. (A) ; 1286. (B); 1287. (C); 1288. (A); 1289. (B); 1290. (B); 1291. (A); 1292. (B); 1293. (D); 1294. (B); 1295. (A); 1296. (A); 1297. (C); 1298. (A); 1299. (A); 1300. (D); 1301. (C); 1302. (C); 1303. (D); 1304. (A); 1305. (B); 1306. (A); 1307. (B); 1308. (A); 1309. (D); 1310. (A); 1311. (C); 1312. (B) ; 1313. (B); 1314. (A); 1315. (A); 1316. (A); 1317. (D); 1318. (B); 1319. (B); 1320. (C); 1321. (C); 1322. (C); 1323. (B); 1324. (B); 1325. (B); 1326. (D); 1327. (C); 1328. (A); 1329. (A); 1330. (A); 1331. (A); 1332. (D); 1333. (B); 1334. (A); 1335. (B); 1336. (B); 1337. (A); 1338. (C); 1339. (C); 1340. (D); 1341. (D); 1342. (C); 1343. (C); 1344. (A); 1345. (D); 1346. (C); 1347. (D); 1348. (C); 1349. (C); 1350. (C); 1351. (C) ; 1352. (C); 1353. (B); 1354. (D) ; 1355. (B); 1356. (D); 1357. (A); 1358. (A); 1359. (B); 1360. (C); 1361. (A); 1362. (C); 1363. (B) ; 1364. (C); 1365. (A); 1366. (C); 1367. (D); 1368. (B); 1369. (A); 1370. (B); 1371. (B); 1372. (A); 1373. (B); 1374. (A); 1375. (A); 1376. (A); 1377. (D) ; 1378. (C); 1379. (B); 1380. (C); 1381. (B); 1382. (C); 1383. (B); 1384. (A); 1385. (B); 1386. (D); 1387. (D); 1388. (B); 1389. (B); 1390. (C); 1391. (B) 1392. (A); 1393. (A); 1394. (B); 1395. (B); 1396. (A); 1397. (B); 1398. (D); 1399. (B); 1400. (A); 1401. (B); 1402. (D); 1403. (A); 1404. (B); 1405. (C); 1406. (D); 1407. (A); 1408. (C); 1409. (D); 1410. (A); 1411. (A); 1412. (A); 1413. (D); 1414. (C); 1415. (A) ; 1416. (B); 1417. (B); 1418. (B); 1419. (B) ; 1420. (B); 1421. (C); 1422. (C); 1423. (D); 1424. (B); 1425. (A); 1426. (A); 1427. (B); 1428. (C); 1429. (D); 1430. (D); 1431. (A); 1432. (D); 1433. (D); 1434. (A); 1435. (B); 1436. (C); 1437. (C); 1438. (B); 1439. (A); 1440. (B); 1441. (C); 1442. (A); 1443. (B); 1444. (B); 1445. (C); 1446. (D); 1447. (C); 1448. (A); 1449. (B); 1450. (C); 1451. (B); 1452. (C); 1453. (A); 1454. (A); 1455. (A); 1456. (C) ; 1457. (A); 1458. (A); 1459. (C); 1460. (C); 1461. (A); 1462. (B); 1463. (A); 1464. (B); 1465. (A); 1466. (B); 1467. (C); 1468. (D); 1469. (C); 1470. (B); 1471. (C); 1472. (B); 1473. (B); 1474. (A); 1475. (A); 1476. (B); 1477. (C); 1478. (D); 1479. (A); 1480. (D); 1481. (C); 1482. (B); 1483. (C);1484. (D); 1485. (D); 1486. (C);1487. (D); 1488. (C); 1489. (C); 1490. (C); 1491. (D); 1492. (D) 1493. (D); 1494. (D); 1495. (D) ; 1496. (C); 1497. (D); 1498. (D); 1499. (B); 1500. (A); 1501. (A); 1502. (C); 1503. (B); 1504. (A); 1505. (D); 1506. (A); 1507. (B); 1508. (B) ; 1509. (A);

1510. (A); 1511. (C); 1512. (B); 1513. (B);1514. (B);
1515. (C); 1516. (A); 1517. (A); 1518. (B); 1519. (C);
1520. (C); 1521. (C);1522. (B); 1523. (A); 1524. (D);
1525. (A); 1526. (B);1527. (C); 1528. (D);1529. (D);
1530. (C); 1531. (B);1532. (A); 1533. (B);1534. (B);
1535. (A); 1536. (A); 1537. (B); 1538. (D); 1539. (C);
1540. (B); 1541. (B);1542. (C); 1543. (D); 1544. (B);
1545. (A); 1546. (A) ;1547. (D); 1548. (D); 1549. (B);
1550. (A); 1551. (C); 1552. (B); 1553. (D); 1554. (B);
1555. (A); 1556. (A); 1557. (C); 1558. (C); 1559. (A);
1560. (B); 1561. (A); 1562. (A); 1563. (A) ; 1564. (A);
1565. (C); 1566. (C); 1567. (C); 1568. (A); 1569. (B);
1570. (A); 1571. (B); 1572. (D); 1573. (C); 1574. (A);
1575. (A); 1576. (B) ; 1577. (C); 1578. (A); 1579. (B);
1580. (A); 1581. (A); 1582. (B); 1583. (B); 1584. (C);
1585. (C); 1586. (C); 1587. (A); 1588. (A); 1589. (C);
1590. (C); 1591. (C); 1592. (A); 1593. (B); 1594. (B);
1595. (B); 1596. (B); 1597. (B); 1598. (B) ; 1599. (C);
1600. (B); 1601. (C); 1602. (B);1603. (C); 1604. (B);
1605. (A); 1606. (A); 1607. (B); 1608. (C); 1609. (D);
1610. (A); 1611. (C); 1612. (B) ; 1613. (D); 1614. (B);
1615. (C); 1616. (A); 1617. (D); 1618. (A); 1619. (D);
1620. (B); 1621. (A); 1622. (A); 1623. (B); 1624. (C);
1625. (A) ; 1626. (C); 1627. (C); 1628. (D); 1629. (A);
1630. (B); 1631. (D); 1632. (B); 1633. (C); 1634. (A);
1635. (B); 1636. (A); 1637. (A); 1638. (A); 1639. (B);
1640. (C); 641. (D); 1642. (C); 1643. (D); 1644. (A);
1645. (B); 1646. (C); 1647. (B); 1648. (A); 1649. (A);
1650. (B); 1651. (A); 1652. (B); 1653. (C); 1654. (C);
1655. (D); 1656. (A); 1657. (D); 1658. (B); 1659. (C);
1660. (A); 1661. (D); 1662. (D); 1663. (B); 1664. (A);
1665. (B); 1666. (B); 1667. (A) ; 1668. (A); 1669. (A);
1670. (B); 1671. (D); 1672. (C); 1673. (B); 1674. (D);
1675. (A); 1676. (C); 1677. (D); 1678. (C); 1679. (A);
1680. (B); 1681. (C); 1682. (C); 1683. (D); 1684. (A);
1685. (A); 1686. (A); 1687. (D); 1688. (C); 1689. (C);
1690. (D) ; 1691. (A); 1692. (B); 1693. (A); 1694. (A);
1695. (A); 1696. (A); 1697. (D); 1698. (B); 1699. (C);
1700. (B); 1701. (A); 1702. (D); 1703. (C); 1704. (A);
1705. (B); 1706. (A); 1707. (B); 1708. (C); 1709. (B);
1710. (A); 1711. (C); 1712. (A); 1713. (B); 1714. (B);
1715. (A) ; 1716. (C); 1717. (D); 1718. (C); 1719. (D);
1720. (C); 1721. (B); 1722. (C); 1723. (C); 1724. (C);
1725. (D); 1726. (A); 1727. (B); 1728. (B); 1729. (B) ;
1730. (C); 1731. (B); 1732. (A); 1733. (D); 1734. (D);
1735. (B); 1736. (A); 1737. (D); 1738. (C); 1739. (C);
1740. (C); 1741. (D) ; 1742. (B) 1743. (B); 1744. (D);
1745. (C); 1746. (A); 1747. (B); 1748. (D); 1749. (C);
1750. (B); 1751. (A); 1752. (C); 1753. (A); 1754. (B);
1755. (A) ; 1756. (A); 1757. (C); 1758. (B); 1759. (D);
1760. (C); 1761. (B); 1762. (B); 1763. (B); 1764. (C);
1765. (B); 1766. (D); 1767. (A); 1768. (B); 1769. (D);
1770. (A); 1771. (A); 1772. (C); 1773. (B); 1774. (A);
1775. (D); 1776. (B); 1777. (B); 1778. (B); 1779. (B);
1780. (D); 1781. (C); 1782. (C); 1783. (B); 1784. (B);
1785. (D); 1786. (D); 1787. (A); 1788. (D); 1789. (C);
1790. (B); 1791. (C); 1792. (B); 1793. (A); 1794. (B);
1795. (C); 1796. (D); 1797. (D); 1798. (D); 1799. (D);
1800. (B); 1801. (C); 1802. (C) 1803. (C); 1804. (B);
1805. (B); 1806. (C); 1807. (D); 1808. (C); 1809. (B);
1810. (D); 1811. (B); 1812. (A); 1813. (B); 1814. (B);
1815. (C); 1816. (B); 1817. (C); 1818. (C); 1819. (A);
1820. (A); 1821. (A); 1822. (B); 1823. (C); 1824. (D);
1825. (B); 1826. (B); 1827. (D); 1828. (A); 1829. (B);
1830. (A); 1831. (C); 1832. (D); 1833. (B); 1834. (A);
1835. (B); 1836. (A); 1837. (B); 1838. (C); 1839. (D);
1840. (A); 1841. (A); 1842. (B); 1843. (C); 1844 (D);
1845. (A); 1846. (B); 1847. (C); 1848. (D); 1849. (A);
1850. (B); 1851. (C); 1852. (D); 1853. (A); 1854. (B);
1855. (C); 1856. (D); 1857. (B) 1858. (C); 1859. (D);
1860. (D) ; 1861.(A); 1862. (A); 1863. (B); 1864. (B);
1865. (B); 1866. (A); 1867. (A); 1868. (D); 1869. (B);
1870. (A); 1871. (D); 1872. (A); 1873. (D); 1874. (B);
1875. (B); 1876. (A); 1877. (B); 1878. (A); 1879. (A);
1880. (D); 1881. (D); 1882. (B); 1883. (B); 1884. (D);
1885. (A); 1886. (B) ; 1887. (B); 1888. (A); 1889. (A);
1890. (A); 1891. (A); 1892. (D); 1893. (A); 1894. (C);
1895. (C); 1896. (D); 1897. (C); 1898. (B);1899. (B);
1900. (C); 1901. (C); 1902. (B); 1903. (B); 1904. (A);
1905. (A); 1906. (D); 1907. (B); 1908. (C); 1909. (C);
1910. (D); 1911. (A); 1912. (B); 1913. (C); 1914. (C);
1914. (C); 1915. (A); 1916. (B); 1917. (C); 1918. (D);
1919. (A); 1920. (B); 1921. (C); 1922. (D); 1923. (A);
1924. (C); 1925. (D); 1926. (A); 1927. (B); 1928. (C);
1929. (D); 1930. (A); 1931. (C); 1932. (A) ; 1933. (B);
1934. (D); 1935. (D); 1936. (C); 1937. (D); 1938. (A);
1939.(B); 1940.(C); 1941. (D); 1942. (A); 1943. (C);
1944. (D); 1945. (A); 1946. (A); 1947. (A); 1948. (A);
1949. (A); 1950. (A) ; 1951. (B); 1952. (B); 1953. (D);
1954. (A); 1955. (B); 1956. (C); 1957. (D); 1958. (A);

1959. (C); 1960. (C); 1961. (D); 1962. (A); 1963. (D);
1964. (A); 1965. (C); 1966. (B); 1967. (A); 1968. (B);
1969. (D); 1970. (D); 1971. (A); 1972. (A); 1973. (B);
1974. (C); 1975. (D); 1976. (B); 1977. (A); 1978. (A);
1979. (A); 1980. (B); 1981. (B); 1982. (A); 1983. (B);
1984. (C); 1985. (C); 1986. (C); 1987. (C); 1988. (B);
1989. (B); 1990. (D); 1991. (A); 1992. (A); 1993. (C);
1994. (A); 1995. (B); 1996. (A); 1997. (C); 1998. (D);
1999. (A); 2000. (B); 2001. (A); 2002. (B); 2003. (C);
2004. (D); 2005. (A); 2006. (B); 2007. (C); 2008. (C);
2009. (D); 2010. (D) ; 2011. (D); 2012. (D); 2013. (C);
2014. (C) 2015. (C); 2016. (A); 2017. (B); 2018. (C);
2019. (D); 2020. (A); 2021. (B); 2022. (C); 2023. (D);
2024. (A); 2025. (A) ; 2026. (B); 2027. (C); 2028. (D);
2029. (A); 2030. (B); 2031. (C); 2032. (C); 2033. (D);
2034. (A); 2035. (A); 2036. (A); 2037. (B); 2038. (C); 2039. (A); 2040. (B); 2041. (C); 2042. (C); 2043. (D); 2044. (A);
2045. (A); 2046. (D); 2047. (B); 2048. (D); 2049. (C);
2050. (D); 2051. (D); 2052. (A); 2053. (D); 2054. (C);
2055. (C); 2056. (D); 2057. (A); 2058. (D); 2059. (B);
2060. (D); 2061. (B); 2062. (B); 2063. (B); 2064. (D);
2065. (B);2066. (C); 2067. (A); 2068. (A); 2069. (B);
2070. (A); 2071. (B); 2072. (A); 2073. (B); 2074. (C);
2075. (D); 2076. (D); 2077. (A); 2078. (B); 2079. (C);
2080. (D); 2081. (A); 2082. (C); 2083. (C); 2084. (C);
2085. (C); 2086. (C); 2087. (C); 2088. (A); 2089. (C);
2090. (C); 2091. (A); 2092. (A); 2093. (C); 2094. (D);
2095. (D); 2096. (A); 2097. (D); 2098. (B); 2099. (B);
2100. (D); 2101. (C); 2102. (B); 2103. (B); 2104. (A);
2105. (C); 2106. (D); 2107. (B); 2108. (B); 2109. (B);
2110. (A); 2111. (A); 2112. (C); 2113. (D); 2114. (B);
2115. (D); 2116. (B); 2117. (D) ; 2118. (B); 2119. (A);
2120. (A); 2121. (B); 2122. (A); 2123. (D); 2124. (B);
2125. (B); 2126. (A); 2127. (B); 2128. (B); 2129. (B);
2130. (C); 2131. (D); 2132. (B); 2133. (D); 2134. (B);
2135. (B); 2136. (B); 2137. (C); 2138. (C); 2139. (C);
2140. (A); 2141. (C); 2142. (A); 2143. (C) ; 2144. (A);
2145. (B); 2146. (A); 2147. (D); 2148. (A); 2149. (B);
2150. (D); 2151. (B); 2152. (C); 2153. (D); 2154. (C);
2155. (A); 2156. (C); 2157. (C); 2158. (C); 2159. (D);
2160. (B); 2161. (B); 2162. (A); 2163. (D); 2164. (A);
2165. (C); 2166. (D); 2167. (B); 2168. (C); 2169. (B);
2170. (B); 2171. (B); 2172. (D); 2173. (A); 2174. (C);
2175. (B); 2176. (C); 2177. (D); 2178. (B); 2179. (B);
2180. (B); 2181. (D); 2182. (C); 2183. (A); 2184. (A);
2185. (B); 2186. (B); 2187. (A); 2188. (D); 2189. (A);
2190. (A); 2191. (A); 2192. (A); 2193. (B); 2194. (C);
2195. (D) ; 2196. (D); 2197. (C); 2198. (A); 2199. (A);
2200. (B); 2201. (A); 2202. (A); 2203. (A); 2204. (A);
2205. (D); 2206. (A); 2207. (B); 2208. (D) ; 2209. (B);
2210. (B); 2211. (C); 2212. (C); 2213. (A); 2214. (C);
2215. (B); 2216. (C); 2217. (D); 2218. (D); 2219. (A);
2220. (A); 2221. (A) ; 2222. (A); 2223. (B); 2224. (C);
2225. (C); 2226. (A); 2227. (C); 2228. (A); 2229. (A);
2230. (A); 2231. (C); 2232. (C); 2233. (D); 2234. (C);
2235. (D); 2236. (D); 2237. (A); 2238. (B); 2239. (A);
2240. (C); 2241. (A); 2242. (A); 2243. (A); 2244. (A);
2245. (D); 2246. (A); 2247. (B) ; 2248. (A); 2249. (C);
2250. (C); 2251. (B); 2252. (C); 2253. (D); 2254. (D);
2255. (D); 2256. (A); 2257. (D); 2258. (B); 2259. (C);
2260. (B); 2261. (D) ; 2262. (C); 2263. (A); 2264. (B);
2265. (C); 2266. (C); 2267. (C); 2268. (C); 2269. (C);
2270. (B); 2271. (C); 2272. (C); 2273. (A); 2274. (A);
2275. (C); 2276. (D); 2277. (A); 2278. (A); 2279. (B);
2280. (C); 2281. (D); 2282. (D); 2283. (A); 2284. (B);
2285. (C); 2286. (D); 2287. (D); 2288. (A); 2289. (A);
2290. (A) ; 2291. (B); 2292. (B); 2293. (C); 2294. (C);
2295. (A); 2296. (B); 2297. (B); 2298. (A); 2299. (B) ;
2300. (C); 2301. (C); 2302. (A); 2303. (A); 2304. (B);
2305. (D); 2306. (A); 2307. (C); 2308. (B); 2309. (A);
2310. (C); 2311. (D); 2312. (D); 2313. (A); 2314. (B);
2315. (C); 2316. (A); 2317. (C); 2318. (A); 2319. (A);
2320. (D); 2321. (A); 2322. (B); 2323. (A); 2324. (A);
2325. (A); 2326. (B); 2327. (A); 2328. (C); 2329. (A);
2330. (C); 2331. (B); 2332. (C); 2333. (B); 2334. (B);
2335. (D); 2336. (A); 2337. (A); 2338. (A) ; 2339. (B);
2340. (A); 2341. (A); 2342. (A); 2343. (D); 2344. (C);
2345. (D); 2346. (B); 2347. (B); 2348. (C); 2349. (D);
2350. (D); 2351. (D); 2352. (E); 2353. (C); 2354. (B);
2355. (C); 2356. (C); 2357. (D); 2358. (D); 2359. (C);
2360. (C); 2361. (A); 2362. (D); 2363. (E); 2364. (D);
2365. (B); 2366. (C); 2367. (D); 2368. (C); 2369. (A);
2370. (C); 2371. (B); 2372. (C); 2373. (B); 2374. (C);
2375. (B); 2376. (B); 2377. (A); 2378. (B); 2379. (C);
2380. (A); 2381. (B); 2382. (A); 2383. (B); 2384. (C);
2385. (C); 2386. (A); 2387. (B); 2388. (B); 2389. (D);
2390. (A) ; 2391. (B); 2392. (C); 2393. (D); 2394. (A);
2395. (B); 2396. (C); 2397. (D); 2398. (B); 2399. (C);
2400. (D); 2401. (A); 2402. (A); 2403. (B); 2404. (C);
2405. (A); 2406. (D); 2407. (B); 2408. (B); 2409. (A);

2410. (B); 2411. (A); 2412. (C); 2413. (D); 2414. (A);
2415. (B); 2416. (C) ; 2417. (B); 2418. (B); 2419. (A);
2420. (A); 2421. (B); 2422. (C); 2423. (C); 2424. (A);
2425. (A); 2426. (B); 2427. (B); 2428. (A); 2429. (C);
2430. (B); 2431. (E); 2432. (A); 2433. (A); 2434. (B) ;
2435. (C); 2436. (D); 2437. (C); 2438. (B); 2439. (A);
2440. (A); 2441. (A); 2442. (B); 2443. (C); 2444. (A);
2445. (B); 2446. (C); 2447. (A); 2448. (B); 2449. (D);
2450. (A); 2451. (B); 2452. (C); 2453. (A); 2454. (B);
2455. (A); 2456. (B); 2457. (A); 2458. (B); 2459. (A);
2460. (C); 2461. (C); 2462. (A); 2463. (A); 2464. (B);
2465. (A) ;2466. (A); 2467. (D); 2468. (C); 2469. (C) ;
2470. (D); 2471. (A); 2472. (D); 2473. (B); 2474. (A);
2475. (C); 2476. (D); 2477. (A); 2478. (A); 2479. (B);
2480. (D); 2481. (A); 2482. (D); 2483. (A); 2484. (C);
2485. (A); 2486. (B); 2487. (C); 2488. (C); 2489. (D);
2490. (D); 2491. (A); 2492. (A); 2493. (B); 2494. (A);
2495. (C); 2496. (A); 2497. (A); 2498. (B); 2499. (C);
2500. (D); 2501. (D); 2502. (A); 2503. (B); 2504. (C);
2505. (D); 2506. (D); 2507. (B); 2508. (C); 2509. (A);
2510. (D); 2511. (A); 2512. (A); 2513. (A); 2514. (A);
2515. (B); 2516. (A); 2517. (B); 2518. (C); 2519. (D);
2520. (A); 2521. (B); 2522. (B); 2523. (B); 2524. (B);
2525. (B); 2526. (C); 2527. (C); 2528. (A); 2529. (C);
2530. (D); 2531. (B); 2532. (A); 2533. (C); 2534. (B);
2535. (C); 2536. (B); 2537. (A); 2538. (B); 2539. (A);
2540. (B); 2541. (A); 2542. (D); 2543. (B); 2544. (C);
2545. (C); 2546. (A); 2547. (B) ; 2548. (C); 2549. (D);
2550. (A); 2551. (C); 2552. (A); 2553. (B); 2554. (D);
2555. (A); 2556. (A); 2557. (B); 2558. (A); 2559. (A);
2560. (A) ; 2561. (A); 2562. (B); 2563. (C); 2564. (A);
2565. (B); 2566. (B); 2567. (C); 2568. (D); 2569. (A);
2570. (A); 2571. (D); 2572. (A); 2573. (D); 2574. (A);
2575. (D); 2576. (A); 2577. (A); 2578. (A); 2579. (B);
2580. (A); 2581. (B); 2582. (A); 2583. (B); 2584. (C);
2585. (D); 2586. (C); 2587. (D); 2588. (A); 2589. (A);
2590. (B); 2591. (C); 2592. (C); 2593. (A); 2594. (C);
2595. (A); 2596. (B); 2597. (C); 2598. (A); 2599. (B);
2600. (D); 2601. (D); 2602. (A); 2603. (B); 2604. (C);
2605. (D); 2606. (A); 2607. (B); 2608. (D); 2609. (C);
2610. (A); 2611. (C); 2612. (C); 2613. (D); 2614. (C);
2615. (B); 2616. (B); 2617. (C); 2618. (B); 2619. (C);
2620. (A); 2621. (C); 2622. (B); 2623. (A); 2624. (A);
2625. (C); 2626. (D); 2627. (D); 2628. (D); 2629. (D);
2630. (A); 2631. (D); 2632. (D); 2633. (A); 2634. (B);
2635. (A); 2636. (D); 2637. (C); 2638. (A) ; 2639. (B);
2640. (C); 2641. (A); 2642. (B); 2643. (B); 2644. (C);
2645. (A); 2646. (C); 2647. (B); 2648. (C); 2649. (D);
2650. (A); 2651. (D); 2652. (A); 2653. (B); 2654. (C);
2655. (D); 2656. (A); 2657. (A); 2658. (B); 2659. (C);
2660. (D); 2661. (A); 2662. (B); 2663. (D); 2664. (C)
2665. (A); 2666. (B); 2667. (C); 2668. (D); 2669. (A);
2670. (B); 2671. (C); 2672. (D); 2673. (C); 2674. (B);
2675. (C); 2676. (D); 2677. (A) ; 2678. (B); 2679. (A);
2680. (B); 2681. (C); 2682. (D); 2683. (A); 2684. (B);
2685. (C); 2686. (D); 2687. (A); 2688. (D); 2689. (A);
2690. (B); 2691. (B); 2692. (A); 2693. (C); 2694. (A);
2695. (C); 2696. (C); 2697. (C); 2698. (A); 2699. (D);
2700. (D); 2701. (C); 2702. (C); 2703. (A); 2704. (A);
2705. (D); 2706. (D); 2707. (A); 2708. (A); 2709. (A);
2710. (A); 2711. (B); 2712. (C); 2713. (A); 2714. (A);
2715. (C); 2716. (A); 2717. (C); 2718. (C); 2719. (A);
2720. (C); 2721. (C); 2722. (B); 2723. (A); 2724. (C);
2725. (D); 2726. (C); 2727. (B); 2728. (A); 2729. (A);
2730. (C); 2731. (B); 2732. (C); 2733. (C); 2734. (D);
2735. (C); 2736. (C); 2737. (B); 2738. (A); 2739. (C);
2740. (B); 2741. (C); 2742. (C) ; 2743. (D); 2744. (B);
2745. (B); 2746. (C); 2747. (C); 2748. (D); 2749. (A);
2750. (A); 2751. (C); 2752. (B); 2753. (D); 2754. (C);
2755. (A); 2756. (C); 2757. (A); 2758. (B); 2759. (B);
2760. (A); 2761. (D); 2762. (B); 2763. (C); 2764. (A);
2765. (C); 2766. (A); 2767. (A); 2768. (B); 2769. (B);
2770. (D); 2771. (C); 2772. (D); 2773. (A); 2774. (C);
2775. (C); 2776. (B); 2777. (D); 2778. (A); 2779. (D);
2780. (C); 2781. (D); 2782. (B); 2783. (D); 2784. (C);
2785. (A); 2786. (A); 2787. (A); 2788. (B); 2789. (D);
2790. (B); 2791. (C); 2792. (B); 2793. (C); 2794. (D);
2795. (D); 2796. (B); 2797. (D); 2798. (D); 2799. (B);
2800. (A); 2801. (C); 2802. (B); 2803. (C); 2804. (D);
2805. (D); 2806. (B); 2807. (C); 2808. (B); 2809. (B);
2810. (B); 2811. (C); 2812. (C); 2813. (B); 2814. (B).

Chapter 10

AGRICULTURAL ENGINEERING

1. The objective of tillage is:
(A) Seed bed preparation
(B) Weed control
(C) Soil and water conservation
(D) All of these

2. Which of the following is not a tillage conservation technique?
(A) Zero tillage (B) Reduced tillage
(C) Minimum tillage (D) All of these

3. Tillage operation which more soil to create desired soil configurations is called:
(A) Land forming (B) Land grading
(C) Land leveling (D) Land planning

4. Tillage operation that cuts and moves small layers of soil to provide a smooth and refined surface is called:
(A) Land forming (B) Land grading
(C) Land leveling (D) Land planning

5. Tillage operation employing rotary action to cut, break and mix soil is called:
(A) Ridge terrace (B) Ridge planting
(C) Rotary tillage (D) None of these

6. Tillage operations which constitute the initial major soil working operation is called:
(A) Primary tillage (B) Secondary tillage
(C) Preparatory tillage (D) Reduced tillage

7. The tillage operations following primary tillage to create a good seed bed for proper seeding/planting is called:
(A) Primary tillage (B) Secondary tillage
(C) Preparatory tillage (D) Reduced tillage

8. Various tillage operations carried out on the farm before the crops are sown to make a fine seedbed is called:
(A) Primary tillage (B) Secondary tillage
(C) Preparatory tillage (D) Reduced tillage

9. A tillage system in which the primary tillage operation is performed in conjunction with special planting procedure in order to reduce or eliminate secondary tillage operations is called:
(A) Primary tillage (B) Secondary tillage
(C) Preparatory tillage (D) Reduced tillage

10. Reduction in the specific volume of soil by means of mechanical manipulation is called:
(A) Primary tillage (B) Secondary tillage
(C) Preparatory tillage (D) Soil compaction

11. Mechanical manipulations of the soil that are used to provide necessary soil conditions favourable for the growth of crop is called:
(A) Tillage (B) Hoeing
(C) Sowing (D) All of these

12. Soil are tilled to create soil physical conditions which facilitates:
(A) Seed germination (B) Seedling emergence
(C) Development of rooting (D) All of these

13. Central Institute of Agricultural Engineering is situated at:
(A) Bhopal (B) Ludhiana
(C) New Delhi (D) Kolkata

14. The constituent of producer gas which occupies the highest percentage by volume and helps in increasing its overall calorific value is:
(A) CO (B) CO_2 (C) H_2 (D) CH_4

15. During field operation, the shank of a tractor drawn rigid tyne sweep type cultivator is mainly subjected to:
(A) Bending (B) Shear
(C) Torsion (D) Bending and torsion

16. Central Institute of Post Harvest Engineering and Technology is situated at:
(A) Bhopal (B) Ludhiana
(C) New Delhi (D) Kolkata

17. The power developed and the exhaust gas temperature of a diesel engine compared to a spark ignition engine of the same size and running at the same speed respectively, are:
(A) Higher and lower (B) Higher and higher
(C) Lower and higher (D) Lower and lower

18. In a semi-modular outlet, the discharge:
(A) Is independent of water levels in the distributary and the water course
(B) Depends upon the water levels of both distributary and water course
(C) Depends upon the water level in the distributary
(D) Depends upon the water level in the water course

19. The relationship between outflow Q in $m^3 s^{-1}$ and storage S in m^3 for an emergency spillway in a reservoir is Q = S/4000. Inflow, outflow and storage are assumed to be zero at time t = 0. If the inflow rate is 300 $m^3 s^{-1}$ at the end of t = 3 hours, the outflow rate in $m^3 s^{-1}$ is:
(A) 152.84 (B) 164.84
(C) 172.34 (D) 184.84

20. A trapezoidal grassed waterway is constructed along a longitudinal gradient of 4%. If the cross-sectional area of flow is 1.52 m^2, wetted perimeter is 12.5 m and Manning's n for the waterway is 0.04 $m^{-1/3}$ s, the flow through the waterway in m^3 s^{-1} is:
(A) 1.9 (B) 2.1 (C) 2.3 (D) 2.5

21. A single acting reciprocating pump discharges 3.5 litres of water per second at 40 rpm. The pump has a piston diameter of 150 mm and a stroke of 300 mm. The percentage slip is:
(A) 0.85 (B) 1.97 (C) 3.53 (D) 6.05

22. A pair of parallel glass panes, each of 3 mm thickness traps 2 mm layer of stagnant air. Thermal conductivities of glass and air are 0.5 and 0.02 W m^{-1} K^{-1}, respectively. If the film heat transfer coefficient of air is 10 W m^{-2} K^{-1}, then Biot Number is:
(A) 1.50 (B) 1.00 (C) 0.06 (D) 0.04

23. Two small parallel plane square surfaces, each measuring 4 mm × 4 mm are placed 0.5 m apart (centre to centre) with 30° angle between the radial distance and both the surface normals. The view factor between the two surfaces is:
(A) 1.53×10^{-5} (B) 1.76×10^{-5}
(C) 3.82×10^{-5} (D) 4.41×10^{-5}

24. Tomato catsup with 10 Paconsistency coefficient and 0.8 flow behaviour index is flowing in a pipe. Generalized coefficient of viscosity of catsup, in Pa is *nsns*:
(A) 2.66 (B) 6.93 (C) 15.91 (D) 23.87

25. A packed bed of 480 kg solid particles having particle size of 0.15 mm and density of 800 kg m^{-3} is fluidized using air at 25 °C and 1 atmospheric pressure. If the cross section of the empty bed is 0.45 m^2 and voidage at minimum fluidizing condition is 0.5, then the minimum height of the fluidized bed, in m is:
(A) 7.4 (B) 5.4 (C) 2.7 (D) 1.0

26. How much horse power (h.p.) is in one kW (Kilowatt) ?
(A)1.34 (B)1.90 (C)2.00 (D)3.80

27. A tractor power take-off (PTO) driven stationary peg tooth type wheat thresher operating at a cylinder speed of 540 rpm requires a torque of 250 Nm at PTO. The minimum net engine power required, in kW is:
(A) 12-14 (B) 15-17 (C) 17-19 (D) 22-22

28. A border strip of 8 × 250 m is being irrigated by a border stream of 50 lps. The infiltration capacity of the soil is 25 mm h^{-1} (assumed to be constant throughout the period of irrigation). The average depth of the advancing sheet of water over the land is 70 mm. The time required to irrigate the border strip, in minutes, will be:
(A) 15-18 (B) 23-26 (C) 52-57 (D) 65-69

29. Man power is considered equal to how much horse power (h.p.) ?
(A)0.1 (B)0.2 (C)0.3 (D)0.4

30. A tall silo having height to diameter ratio of 2 is holding 480 tons wheat of bulk density 960 kg m^{-3}. The angle of internal friction for wheat is 25° and for wheat and wall surface is 24°. Applying Airy formula, the maximum lateral pressure in kPa at the bottom of the bin section is:
(A) 40.24 (B) 41.79 (C) 42.83 (D) 42.92

31. One centimetre is equal to:
(A)5 mm (B)10 mm
(C)15 mm (D)None of these

32. The draft of 'Victory plough' is:
(A)80-100 kg (B)40-50 kg
(C)70-80 kg (D)10-20 kg

33. The inverse Laplace Transform of 23(3)*ss*- can be written as 32[2*teAtBtC* ++. The values of *A*, *B* and *C*, respectively are:
(A) 3, 5 and 7 (B) 2, 10 and 12
(C) 10, 12 and 4 (D) 9, 12 and 2

34. A two wheel drive tractor, weighing 15.84 kN with a wheel base of 2160 mm, has the static weight divided between the front and rear axles in the ratio of 30 : 70 on a horizontal level surface. The hitch point is at a height of 700 mm from the ground and at a horizontal distance of 120 mm to the rear side from the center of the rear axle. Pull acts at an angle of 12° downwards from the horizontal. The maximum pull in kN, when the front wheels would just start rising from the ground is:
(A) 1.48 (B) 14.46 (C) 39.04 (D) 85.54

35. A horizontal axis drag type wind mill with square blades and a horizontal axis lift type wind mill with airfoil section blades having same rotor size are installed at a height of 10 m above the ground. The average wind speed is 25 km h^{-1}. The maximum power coefficient for drag type and lift type wind mills is 0.148 and 0.593, respectively. If the maximum power extracted by drag type wind mill is 5 kW, the corresponding power extracted by lift type wind mill, in kW is:
(A) 8.43 (B) 12.63 (C) 18.03 (D) 20.03

36. The thresher of a wheat combine harvester has an optimal throughput capacity of 2400 kg (crop) per hour. The harvester has a forward velocity of 4.5 km h^{-1}. Sample tests have revealed that the yield of crop in the field is 3000 kg (grain) per ha. Grain to straw ratio is 60 : 40. If the above throughput is to be maintained, the width of cut of the harvester in m, neglecting turning losses, is:
(A) 0.71 (B) 1.07 (C) 1.78 (D) 2.96

37. In a disc clutch, the inside and outside radii of the clutch plate are 50 and 100 mm, respectively. If the axial force exerted on the disc is 4 kN, the maximum pressure experienced by the clutch plate in N mm-2 under uniform wear conditions is:
(A) 0.13 (B) 0.17 (C) 0.25 (D) 0.51

38. A regime channel carrying a discharge of 25 m^3 s^{-1} is designed using Lacey's regime theory. The side slope of the channel is ½ H : 1 V, and Lacey's silt factor is unity. The bottom width and depth of flow in the channel, in m, respectively are:
(A) 20.26, 1.38 (B) 20.26, 1.56
(C) 23.75, 1.56 (D) 32.78, 1.56

39. The length of 'Gunter chain' is:
(A)66 ft (B)100 ft (C)20 ft (D)90 ft

40. A trapezoidal canal, having a bottom width of 5.0 m and a side slope of 1 H : 1 V, is carrying a discharge of 20 $m^3 s^{-1}$. The critical depth, in m, is:
(A) 1.09 (B) 1.18 (C) 2.12 (D) 2.62

41. A 200 mm well fully penetrates a confined aquifer. After a long period of pumping at a rate of 1400 litres per minute, the drawdowns in the observation wells located at 25 m and 40 m from the pumping well are found to be 2.6 m and 1.9 m, respectively. The transmissivity of the aquifer in $m^2 day^{-1}$ is:
(A) 190 (B) 198 (C) 206 (D) 215

42. The area of an acre is:
(A)0.40 ha (B)0.80ha
(C)1.10 ha (D)None of these

43. It is proposed to construct bench terraces on a 10% hill slope. If the batter slope is ½ H : 1 V, the percentage area that will be lost for cultivation due to bench terracing is:
(A) 4.68 (B) 5.47 (C) 6.25 (D) 6.78

44. Air at 70 °C and 0.015 humidity ratio is cooled adiabatically by spraying water. The final temperature of the air is 55 °C. Specific heat capacities of dry air and water vapour are 1.005 and 1.88 kJ $kg^{-1} K^{-1}$, respectively and latent heat of vapourization of water at 0 °C is 2501.7 kJ kg^{-1}. The absolute humidity of the outlet air, in kg water vapour per kg dry air is:
(A) 0.017 (B) 0.019 (C) 0.021 (D) 0.023

45. Final mass flow rate of osmotically dehydrated cherries after finish drying from 18% dry basis moisture content to 11.5% wet basis moisture content is 5000 kg per hour. The dryer efficiency is 70%, latent heat of vapourization is 2345 kJ kg^{-1}, specific heat of air is 1.005 kJ $kg^{-1} K^{-1}$, drying temperature is 50 °C and the specific volume of ambient air at 25 °C is 0.866 $m^3 kg^{-1}$. The necessary air flow requirement for the drying system in $m^3 min^{-1}$ is:
(A) 477 (B) 587 (C) 625 (D) 702

46. A single effect vacuum evaporator has 100 tubes of 25 mm diameter. One thousand kg feed of milk per hour with 15% TS is concentrated to 20% TS in the evaporator. Film heat transfer coefficients on either sides of the tube are 5000 and 800 W $m^{-2} K^{-1}$. Thermal conductivity of 1.5 mm thick SS tubes is 15 W $m^{-1} K^{-1}$. Latent heat of vapourization under vacuum is 2309 kJ kg^{-1}. For 10 °C temperature difference across the tube wall, the height of each tube, in m is:
(A) 1.36 (B) 2.13 (C) 2.56 (D) 3.17

47. One thousand units of mixed fruit bar, each weighing 100 g with a surface area of 0.01 m^2, are frozen from 70 °C molten mass condition to -20 °C frozen storage condition within 3 hours. The specific heat capacity values of the bar are 3.6 kJ $kg^{-1} K^{-1}$ and 1.97 kJ $kg^{-1} K^{-1}$ before and after freezing point (0 °C) respectively. If the latent heat of crystallization is 250 kJ kg^{-1}, the cooling capacity of the refrigeration unit required in tons of refrigeration is:
(A) 0.77 (B) 1.43 (C) 1.66 (D) 4.32

Common Data for Questions 48 and 49:

48. A diesel engine running in dual fuel mode with diesel as pilot fuel and producer gas as primary fuel produces 3.5 kW at rated engine speed and is coupled directly to a generator for producing electricity. The amount of diesel and producer gas consumed per hour is 460 ml and 12.5 m^3, respectively. Assuming calorific value of diesel and producer gas as 35280 and 3.97 MJ m^{-3}, respectively, the brake thermal efficiency of the engine in percentage is
(A) 17.19 (B) 19.13 (C) 22.79 (D) 25.32

49. If generator efficiency is 90%, the maximum electricity produced, in kW is:
(A) 2.85 (B) 3.00 (C) 3.15 (D) 3.50

Common Data for Questions 50 and 51:
The hourly discharge observations at the mouth of a watershed due to 2 cm excess rainfall during 0 to 1 h and 3 cm excess rainfall during 1 to 2 h are given in the table below. Assume a constant base flow of 1 $m^3 s^{-1}$.

Time (h)	0	1	2	3	4	5	6
Discharge (m s^{-1})	1	7	26	37	27	13	1

50. The area of the watershed, in km^2:
(A) 7.56 (B) 8.24 (C) 8.35 (D) 8.86

51. The peak of 1 h unit hydrograph in $m^3 s^{-1}$ for the watershed and its time of occurrence in h, respectively are:
(A) 6, 1 (B) 7, 2 (C) 8, 2 (D) 9, 1

Statement for Linked Answer Questions 52 and 53:
Soybean is to be planted with a precision planter that meters 54 seeds per revolution of the metering disc powered from a ground wheel of diameter 490 mm. The desired plant population is 44800 per ha with a row to row spacing of 0.75 m. The germination percentage is 84. The planter is to be operated at 2.5 km h^{-1} with a 10% skid of ground wheel.

52. The angular speed of ground wheel in rpm is:
(A) 20.3 (B) 24.6 (C) 28.3 (D) 32.6

53. The angular speed ratio of metering disc to ground wheel for obtaining the desired plant population is:
(A) 0.125:1 (B) 0.150:1 (C) 0.225:1 (D) 0.250:1

Statement for Linked Answer Questions 54 and 55:
A 1 hp motor is used for running a dual cylinder reciprocating compressor of a refrigeration system based on R-134a refrigerant having 185 kJ kg^{-1} cooling capacity. COP of the system is 4.2 and overall efficiency of the compressor is 80%. Specific volume of the refrigerant vapour at suction temperature is 0.15 $m^3 kg^{-1}$. The compressor with bore diameters of 40 mm each runs at 1440 rpm.

54. The mass flow rate of the refrigerant in kg min^{-1}:
(A) 1.634 (B) 1.090 (C) 0.813 (D) 0.240

55. The compressor stroke length in mm is:
(A) 16.8 (B) 33.7 (C) 50.5 (D) 67.4

56. Poisson distribution having a mean of 5 will have v5 as:
(A) Median (B) Mode
(C) Standard deviation (D) Variance

57. During the testing of a spike-tooth type thresher for wheat crop, the throughput and the thresher capacity were found to be 750 kg h^{-1} and 445 kg h^{-1}, respectively. The grain-straw ratio (grain : straw) of the crop is 1.5 : 1. Material other than grain (MOG) collected at the main grain outlet is 0.5%. The total grain loss in percentage from all the sources will be:
(A) 0.33 (B) 0.50 (C) 1.11 (D) 1.61

58. The rear furrow wheel in a tractor mounted disc plough is provided to:
(A) Reduce the frictional power loss
(B) Maintain the uniform depth of cut
(C) Reduce the side draft
(D) Improve the penetration of the plough

59. The cumulative discharge of a tractor mounted hydraulic sprayer having 7 nozzles is 2.0 L min^{-1} when the tractor is operated at 4 km h^{-1}. If the nozzle spacing is 0.3 m, the discharge in L ha^{-1} is:
(A) 1000.02 (B) 285.72
(C) 166.67 (D) 142.86

60. A two-wheel drive tractor, while negotiating a terrain, indicates 100% slip of one of the rear wheels. Under such a condition, the use of differential lock causes:
(A) Equal speed and equal power distribution to both the drive wheels
(B) Equal speed and equal torque distribution to both the drive wheels
(C) Equal power and equal torque distribution to both the drive wheels
(D) Equal speed and unequal power distribution to both the drive wheels

61. A double acting hydraulic cylinder has a rod diameter equal to one-half the piston diameter. If the system pressure is maintained constant, the ratio of load carrying capacity of extension stroke to that of retraction stroke is:
(A) 0.75 (B) 1.00 (C) 1.33 (D) 4.00

62. Which of the following is best for driving machine from low power to slow speed?
(A) Belts & pulley (B)Spur gear
(C) Toothed wheel & chains (D) None of the above

63. The horse power of the motor running the compressor of a refrigerator having COP of 4.5 and extracting 200 kJ kg^{-1} of evaporating heat with 1.5 kg min^{-1} refrigerant flow rate is:
(A) 0.5 (B) 1.0 (C) 1.5 (D) 2.0

64. The gridiron pipe drainage system is more economical than the herringbone pipe drainage system because:
(A) It is adopted in the fields which do not require com plete drainage
(B) The number of main or sub-main lines is reduced
(C) The number of junctions and the double-drained area are reduced
(D) It has only main or sub-main lines

65. If the drainable porosity of a command area is 5% and the design rate of drop of the water table is 0.25 m day^{-1}, the drainage coefficient of the command area in mm day^{-1} will be:
(A) 250 (B) 12.5 (C) 1.25 (D) 0.0125

66. A soil has a void ratio of 0.75 and a specific gravity of 2.66. The value of critical hydraulic gradient at which quick sand condition will occur is:
(A) 0.95 (B) 1.05 (C) 2.09 (D) 6.64

67. The pressure that does not have any measurable influence on the void ratio or shearing resistance of the soil mass is:
(A) Pore water pressure
(B) Intergranular pressure
(C) Capillary pressure
(D) Surcharge pressure

68. The Rational method is used to estimate:
(A) Runoff volume (B) Peak runoff rate
(C) Runoff depth (D) Direct surface runoff

69. Pan evaporation data recorded at a certain location over a period of one week are 4.0, 4.3, 4.6, 4.9, 5.12, 5.18 and 6.21 mm. If irrigation scheduling based on ratio of irrigation water (IW) to cumulative pan evaporation (CPE) is practiced, the depth of irrigation at an interval of a week for IW/CPE = 0.9 is:
(A) 3.60 (B) 4.41 (C) 5.59 (D) 30.88

70. A 16 m high wind break is constructed to protect soil from wind erosion due to wind velocity of 18 m s^{-1} at 15 m height. The minimum wind velocity at 15 m height capable of moving the soil fraction is 8 m s^{-1}. The angle of deviation of prevailing wind direction from the perpendicular to the wind break is 30^0. The distance of full protection from the wind break in m is:
(A) 60.44 (B) 104.69
(C) 306.00 (D) 530.01

71. Animal driven cane planter is developed by—
(A) ICAR (B) CIAE (C) IISR (D) IARI

72. A 10 cm wide belt is running at a speed of 900 r.p.m. If the 15 H. P. is available, then how much plies of belt ?
(A) 3 (B) 10 (C) 6 (D) 4

73. How much width belt would be needed if 20 H. P. is to be transferred through a belt of 10 plies, where speed of belt is 1460 r.p.m.?
(A) 4 cm (B) 8 cm (C) 5 cm (D) 6 cm

74. A 10 cm wide belt, having the speed of 1350 meter per minute, would be appropriate to transfer the how much H. P.?
(A) 20 (B)15 (C) 25 (D) 40

75. A general farmer used deshi plough for the purpose of:
(A) Land ploughing (B) Collecting weeds
(C) Making soil powdery (D) Above all works

76. A two-wheel drive tractor pulls an implement that has a draft force of 11.5 kN. The total motion resistance of the tractor is 2.5 kN. Under these circumstances, the slip of the drive wheels is 20%. If the power loss in transmission is 20%, the percentage of power lost in converting engine power into drawbar power is:
(A)37.4-47.5 (B)47.4-47.5
(C)57.4-57.5 (D)67.4-67.5

77. An unconfined aquifer extends over an area of 1 km^2 and has hydraulic conductivity, total porosity and specific retention of 20 m per day, 30% and 10%, respectively. After pumping some groundwater from this aquifer, the water table dropped to a depth of 20 m from the ground level. If the water table was initially at 14.5 m below the ground level, the change in groundwater storage in million cubic meters would be:
(A)900000 (B) 1000000
(C)1100000 (D) 1200000

78. Two parallel canals 50 m apart fully penetrate a homogeneous unconfined aquifer resting on a horizontal impermeable layer. The aquifer has a hydraulic conductivity of 3 m day^{-1} and an effective porosity of 0.25. One-dimensional steady groundwater flow occurs from the upper canal to the lower canal with the height of water levels in the canals 10 m and 8.5 m from the aquifer bottom, respectively. If a sediment layer of 4 cm thick with hydraulic conductivity of 1.2×10^{-2} m day^{-1} is ultimately deposited on the inflow face, the groundwater discharge per 1000 m width between the two canals in m3 day^{-1}will be:
(A)590-595 (B)625-630
(C)650-655 (D)690-695

79. The overall heat transfer coefficient based on the outside surface area of a tubular heat exchanger decreased due to fouling during operation from 1000 W m^{-2} K^{-1} to 800 W m^{-2} K^{-1}. The fouling film coefficient of the heat exchanger in W m^{-2} K^{-1} is:
(A)400 (B)500 (C)600 (D)700

80. A high pressure dairy homogenizer operates under upstream and downstream pressures of 200 and 40 bar respectively homogenizing 30 L of whole milk per hour. Density and specific heat capacity of whole milk are 1030 kg m^{-3} and 3.8 kJ kg^{-1} K^{-1}, respectively. Assuming complete energy conservation, the temperature rise of whole milk in degree Celsius is:
(A)3-3.1 (B)4-4.2 (C)5-5.2 (D)6-6.2

81. A fish fillet of 5 mm thickness having 85% moisture (wet basis) is to be frozen using a plate freezer. The plates are at -35 °C and the heat transfer coefficient between the fillet and the freezer plates can be assumed to be 2.0 W m^{-2} K^{-1}. The initial freezing temperature of fish is -2.5 °C, latent heat of fusion is 330 kJ kg^{-1}, density of fish is 1100 kg m^{-3} and thermal conductivity of frozen fish is 1.5 W m^{-1} K^{-1}. The time required to freeze the fillet from the initial freezing temperature in hour(s) is:
(A)1.25-1.35 (B)2.25-2.35
(C)3.25-3.35 (D)4.25-4.35

82. Box 1 contains 15 balls out of which 3 are red. Box 2 contains 12 balls out of which 4 are red. If one ball is drawn at random from each box simultaneously, the probability of getting at least one red ball is:
(A) 0.07 (B) 0.47 (C) 0.53 (D) 0.75

83. A hemispherical vessel of 300 mm diameter is completely filled with oil and water. If the oil layer is 50 mm deep on the top, the volume of water in the vessel in litres is:
(A) 1.27 (B) 3.73 (C) 7.07 (D) 14.14

84. A tractor mounted off-set type reciprocating mower is driven by the PTO shaft. The maximum inertia force of 3.2 kN occurs along the pitman at 32^0 crank angle and 27^0 pitman angle with the horizontal plane. The knives of the cutterbar are riveted to the slider. If each of the allowable tensile and compressive stresses of the slider material is 50 MPa, the minimum cross-sectional area of the slider in mm^2 is:
(A) 29.05 (B) 33.91 (C) 54.27 (D) 57.02

85. The flywheel of a hand operated chaff cutter with two cutting knives is rotated at 30 rpm and is connected to a worm gear assembly for driving the feed rollers. The number of teeth of the worm gear is 24 and number of starts (threads) of the worm is 2. If diameter of each of the feed rollers is 15 cm, the chaff length in mm will be:
(A) 9.8 (B) 12.8 (C) 19.6 (D) 39.2

86. A piston pump is driven by a 5 m diameter horizontal axis wind turbine for supplying water from a borehole with a total pump head of 10 m. The mean velocity of air is 18 km h^{-1} and the density of air is 1.29 kg m^{-3}. The actual power coefficient of the wind turbine is 0.30 and the overall pump efficiency is 60%. Neglecting the transmission losses, the expected pump discharge in L s^{-1} will be:
(A) 2.90 (B) 5.80 (C) 28.50 (D) 32.27

87. A 4-cylinder, 4-stroke compression ignition engine has piston stroke of 10.5 cm and cylinder bore of 11 cm. At a mean piston speed of 7 m s^{-1}, the developed brake mean effective pressure is 650 kPa. The brake power in kW developed by the engine is:
(A) 39.40 (B) 43.24 (C) 86.48 (D) 172.96

88. A centrifugal pump having an overall efficiency of 75% requires 6 kW power at 1450 rpm to deliver water against a suction head of 5 m and a delivery head of 12 m. If the pump runs at 1650 rpm and frictional head losses are negligible, the total head developed by the pump in metres will be:
(A) 22. (B) 25.05 (C) 29.35 (D) 31.72

89. A 100 ha watershed received rainfall at a rate of 5 cm h^{-1} for 2 hours. If the runoff generated by the storm was at the rate of 1 m^3 s^{-1} for 10 hours, the runoff coefficient for the watershed would be:
(A) 3.6×10^{-3} (B) 6.0×10^{-2}
(C) 0.36 (D) 36

90. A 10 ha field has 1.2 m deep layer of sandy loam soil underlain by sandy soil up to a depth of 5 m. A pre-irrigation rainfall brings moisture content of the top 0.3 m layer to its field capacity. The moisture content of rest of the sandy loam layer remains at permanent wilting point. The volumetric moisture content at field capacity and permanent wilting point are 32 and 16%, respectively for the sandy loam soil. The field is irrigated with a stream size of 240 L s^{-1} for 24 hours. Considering the drainage from the sandy loam soil as deep percolation, application efficiency and deep percolation ratio in percent respectively are:
(A) 56.40 and 43.60 (B) 69.44 and 30.56
(C) 75.18 and 24.82 (D) 92.60 and 27.40

91. A watershed, with an area of 360 km², has a triangular shaped 4-h unit hydrograph with a base of 50 hours. The peak discharge of direct runoff hydrograph due to 3 cm of rainfall-excess in 4 hours from the watershed in $m^3 s^{-1}$ is:
(A) 13.33 (B) 40.00 (C) 120.00 (D) 160.00

92. A rotary dryer is used to dry 1200 kg h^{-1} of paddy containing 30% moisture (wet basis) to give a product containing 15% moisture (wet basis). Alternately, a portion of the dry product may be recycled and mixed with the fresh feed such that the mixed feed enters the dryer with moisture content of 20% (wet basis). The moisture evaporation rate without recycle and the paddy recycle rate in kg h^{-1} respectively in the dryer are:
(A) 211.76 and 2400 (B) 211.76 and 600
(C) 256.50 and 2400 (D) 256.50 and 600

93. The work index of a material is 6.25. If 80% of the feed and 80% of the product pass through IS Sieve No. 340 (3.25 mm opening) and IS Sieve No. 40 (0.42 mm opening), respectively, the power consumed in kW to crush 5000 kg h^{-1} of sorghum is:
(A) 9.77 (B) 20.49 (C) 26.29 (D) 32.29

94. In a screen separator, the mass flow rates of feed, overflow and underflow are 150, 140 and 10 kg h^{-1}, respectively. Mass fraction of material in the feed and overflow are 0.9 and 0.96, respectively. The effectiveness of separation in percentage is:
(A) 32 (B) 42 (C) 52 (D) 62

95. A milk fat globule of 2 im diameter is rising in whole milk of density 1030 kg m^{-3} and coefficient of viscosity 10^{-3} Poise. If the fat density is 950 kg m^{-3}, the time needed to rise 10 mm for this fat globule in min is:
(A) 0.57 (B) 34.57 (C) 35.57 (D) 95.57

Common Data for Questions 96 and 97:
A 9 × 20 cm fluted roller type seed drill is operated at a forward speed of 3 km h^{-1} in a field of size 120 m × 90 m. The effective ground wheel diameter of the seed drill is 0.5 m and the ratio of ground wheel rpm to the fluted roller rpm is 2. For one complete rotation of each fluted roller, 6.8 g seed is transferred from the seed box to the seed tube. The average time taken for each turn while operating length-wise is 50 s and the total time wasted in refilling the seed box for sowing the entire field is 40 min.

96. The seed rate in kg ha^{-1} will be:
(A) 108.32 (B) 122.55 (C) 136.99 (D) 240.71

97. The actual field capacity of the machine in ha h^{-1} is:
(A) 0.30 (B) 0.32 (C) 0.36 (D) 0.40

Common Data for Questions 98 and 99:
Specific heat capacity of dry air and water vapour are 1.005 and 1.88 kJ $kg^{-1} K^{-1}$, respectively. In an energy conserving system, 1 kg s^{-1} air at 30°C with constant absolute humidity of 0.02 kg water (kg dry air)$^{-1}$ is heated up to 65 °C. Water at the wet bulb temperature of air is then sprayed into the air so that the final temperature of the air-water vapour mixture is 40 °C. Latent heat of vapourization of water at 70°C and 40°C are 2334 and 2407 kJ kg^{-1}, respectively.

98. The thermal energy supplied per second during heating in kW is:
(A) 18.2 (B) 36.5 (C) 101.0 (D) 166.8

99. The absolute humidity of the exhaust air from the spray chamber in kg water (kg dry air)$^{-1}$ is:
(A) 0.027 (B) 0.029 (C) 0.031 (D) 0.033

Statement for Linked Answer Questions 100 and 101:
A two-wheel drive tractor weighing 21 kN has a wheel base of 2.1 m. The CG of the tractor is 0.7 m ahead of rear axle centre. The tractor is pulling a single-axle trailer with gross trailer weight of 50 kN on a leveled concrete road while maintaining the line of pull parallel to the ground surface. The tractor hitch point is 42 cm behind the rear axle centre and 52.5 cm above the ground surface. During operation, 20% of the gross trailer weight is transferred to the tractor hitch point. If the coefficient of rolling resistance for each of the tractor and trailer wheels is taken as 0.04, and their ground reactions are assumed to pass through their respective wheel centres,

100. The dynamic ground reaction against the tractor rear wheels in kN is:
(A) 14.0 (B) 24.0 (C) 24.5 (D) 26.4

101. The gross traction ratio developed by the tractor is:
(A) 0.061 (B) 0.082 (C) 0.108 (D) 0.123

Statement for Linked Answer Questions 102 and 103:
In an experimental setup, the discharge through a triangular notch is 0.0074 $m^3 s^{-1}$ at an operating head of 0.1 m. The coefficient of discharge for the notch is 0.7. If the required discharge is 0.1 $m^3 s^{-1}$,

102. the corresponding head in m is:
(A) 0.00015 (B) 0.035
(C) 0.283 (D) 67.13

103. the corresponding width of water surface in m is:
(A) 0.80 (B) 0.099
(C) 0.00043 (D) 0.00086

104. High compression petrol engines are used in some tractors and have high performance in:
(A) U.S.A. (B) Germany
(C) Japan (D) Holland

105. The distribution of fertilizers by aircraft is widely practised in:
(A) England (B) France
(C) Germany (D) New Zealand

106. A 2- row potato harvester, working in good conditions with 4 to 5 men on the machine, can do how much hectare per day ?
(A) 0.5-0.6 ha (B) 1.2-1.4 ha
(C) 2.5 ha (D) 1.8-1.9 ha

107. Mole drainage is practised extensively in country:
(A) New Zealand (B) Japan
(C) U.S.A. (D) China

108. Which is not included in the sources of energy, getting for agricultural work ?
(A) Diesel engine (B) Electric motor
(C) Bullocks (D) Cow

109. Which is included in dairy equipments ?
(A) Threshers (B) Lactometer
(C) Cane Planter (D) Winnowers

110. In our country, for the manufacturing of agricultural implements tools, mostly wood is used, because:
(A) Easy available at all the places
(B) Wood is cheaper
(C) Easy to repare tools made-up of wooden, compound to metal tools
(D) Above (A) to (C) are correct

111. Which of the following wood is not used for making handle of spade ?
(A) Sheesham (B) Babool
(C) Neem (D) Mango

112. Soil turning plough makes the furrow of which type (shape) ?
(A) 'V' shape (B) 'L' shape
(C) 'O' shape (D) No definite shape

113. Which of the following is one (single) handed soil turning plough ?
(A) Punjab plough (B) Praja plough
(C) Victory plough (D) U. P. No.1 plough

114. The purpose of tillage is/are:
(A) Soil clods breaking and suppressing in soil
(B) Eradication of weeds
(C) Leveling of soil
(D) Above (A), (B) and (C)

115. The main function of cultivator is:
(A) To turn the soil
(B) To make furrow in soil
(C) To pulverize the soil
(D) All above three functions

116. Bakhar is generally used in:
(A) U. P.
(B) Bundelkhand
(C) Vindhya Pradesh
(D) In all the States/parts of India (in All India)

117. Harrow is drawn by:
(A) Bullocks
(B) Tractor
(C) Diesel
(D) Bullocks and Tractor both
(E) It is automatic tool

118. Which of the following is not a secondary tillage implement ?
(A) Cultivators (B) Harrow
(C) Hoe (D) Meston plough

119. The term 'Olpad' in 'Olpad Thresher' is named on the name of:
(A) Scientist (B) Village
(C) Labourer (D) Farmer

120. Which of the following 'hoe' is bullock drawn ?
(A) Akola hoe (B) Sharma hoe
(C) Wheel hoe (D) Naini type hoe

121. The land leveling implement is:
(A) Patela (B) Roller
(C) Scrapper (D) All of these

122. *Patela* is used for:
(A) Sowing (B) Levelling
(C) Earthing (D) Weeding

123. Which of the material is not used for making rollers ?
(A) Wood (B) Stone
(C) Iron (D) Steel

124. In which condition, roller is used ?
(A) Where soil is wet
(B) Where clods are present in soil
(C) Where soil is friable
(D) None of these

125. Which one of the following work is not done by scrappers ?
(A) For leveling of soil
(B) For making irrigation channels
(C) For covering the sown-seeds by earth
(D) For making ridges

126. Dibbler is used for:
(A) Ploughing (B) Seed sowing
(C) Leveling of land (D) Interculture

127. The minimum expenditure is incurred by implements for sowing seeds among the following:
(A) Deshi plough (B) Cultivator
(C) Dibbler (D) Seed-drill

128. Which of the following tool is used for measuring the draft of agricultural implements?
(A) Dynamometer (B) Hydrometer
(C) Galvanometer (D) Barometer

129. Which of the following does not affect the draft of ploughs ?
(A) Width of furrow (B) Depth of furrow
(C) Length of furrow (D) Soil moisture

130. Swing-basket (Dhenkuli) is used for:
(A) Making furrow
(B) Lifting water from wells
(C) Destroying weeds
(D) Leveling of land

131. Which of the following is used maximum for lifting water from wells ?
(A) Persian wheel (Rahat) (B) Swing basket (Bedi)
(C) Dhenkuli (D) Don

132. Which of the following is used for lifting water from 8-10 m depth ?
(A) Washer Rahat (B) Hand pump
(C) Mayadas lift (D) None of these

133. In which implement. bullocks are not used for lifting water ?
(A) Buldev Balti (B) Charsa
(C) Egyptian screw (D) Rahat (Persian wheel)

134. The working efficiency per day of deshi plough is:
(A) 0.3 ha (B) 0.4 ha (C) 0.6 ha (D) 0.8 ha

135. Which of the method of ploughing is mostly practised ?
(A) Outside to inside ploughing
(B) Inside to outside ploughing
(C) Ploughing by putting furrow from one side of field
(D) Ploughing by halai making

136. The best method of ploughing through deshi plough is:
(A) Outside to inside ploughing
(B) Inside to outside ploughing
(C) Ploughing by making halai
(D) Ploughing by making furrow from one side of field

137. How much is the working efficiency of dibbler (ha per day) ?
(A) 0.15 (B) 0.25 (C) 0.35 (D) 0.45

138. Tillage includes:
(A) Ploughing of land
(B) Keep the land free from weeds
(C) Make the soil leveled
(D) All of the above operations

139. Function of the seed-drill is:
(A) Making furrow
(B) Dropping seeds
(C) Covering the seeds in furrow
(D) All of these

140. Wrought iron contains carbon (per cent) :
(A) 0.05-1% (B) 1-2% (C) 2-3% (D) 3-4%

141. The mould-board of a tractor drawn soil turning plough is the type of:
(A) General purpose (B) Stubble
(C) Sod (breaker) (D) High speed

142. Among the following, ridger is not used in crop:
(A) Maize (B) Gram
(C) Potato (D) Sugarcane

143. Belt mostly used is of which types:
(A) Rubber (B) Kirmich
(C) Leather (D) Cotton thread

144. The major defects of rubber belts are:
(A) Costly (B) Slips on wet
(C) Early rubbed (D) Increased due to heat

145. How many mm are in one foot length ?
(A) 304.8 (B) 404.8 (C) 204.8 (D) 104.8

146. The grooved pulleys are made-up of:
(A) Steel (B) Cast iron
(C) Wooden (D) Above (A) and (B) both

147. During a test, sound level was measured as 90 dB in the operator's cabin on a tractor. Taking reference sound pressure as 2×10^{-5} N m^{-2}, the measured RMS sound pressure in N m^{-2} is:
(A) 6.32 (B) 6.32×10^{-1}
(C) 1.8×10^{-3} (D) 6.32×10^{-10}

148. During land leveling of agricultural land for irrigation and drainage purposes, the acceptable deviation in elevation from the design value in metre is:
(A) 0.015 (B) 0.025
(C) 0.055 (D) 0.150

149. Reapers are used for:
(A) Crop cutting
(B) Threshing of harvested crop produce (lank)
(C) Seeds sowing
(D) Fodder cutting

150. Threshers (except Olpad threshers) are driven by:
(A) One pair of bullocks (B) Two pair of bullocks
(C) Diesel (D) All of three

151. Winnowing it called:
(A) Cutting a crop
(B) To separate straw etc. from threshed lank
(C) To thresh
(D) Cutting the fodder

152. Chaff-cutter is driven by:
(A) Hand (B) Bullocks
(C) Electric power (D) All of these

153. Sugarcane juice is extracted (%) from canes through bullock-drawn cane-crusher:
(A) 50-55 (B) 70-75
(C) 25-30 (D) 60-65

154. 'Olpad' thresher is used in:
(A) Oil extraction from mustard, toria etc.
(B) Extracting juice from cane
(C) Threshing of wheat, barley, pea etc.
(D) All (A), (B) and (C)

155. 'Seed dresser' is used for:
(A) Mixing/treating seeds with chemicals (pesticides/ fungicides etc.)
(B) Sowing seeds at proper distance
(C) Making seeds of high grade
(D) Keeping seeds effective upto longer period

156. 'Try square' is used by:
(A) Blacksmith (B) Carpenter
(C) Potter (D) Farmers

157. Which type of saw is not included ?
(A) Cross cut saw (B) Deshi saw
(C) Tenon saw (D) Teething saw

158. Which type of saw, used for cutting round shape in hole?
(A) Fret saw (B) Penal saw
(C) Tenon saw (D) Key hole saw

159. 'Draw-Knife' is used for:
(A) To fit galua in grinder
(B) Making round the corners of wood
(C) Smoothing the base of wood
(D) Sharping teeth of saw

160. The tool used for making deep pit in wood is:
(A) Adge (B) Saw
(C) Draw knife (D) Chisel

161. The main work of 'scrapper' is:
(A) For cutting the wood
(B) For scrapping the wood
(C) For smoothing the wood
(D) All of three

162. 'File' is used for:
(A) Scrapping the wood
(B) Cutting the wood
(C) Forcing the wood
(D) Making equal by scrapping the wood

163. The type of file is generally:
(A) Round (B) Triangular
(C) Flat (D) Above all types

164. Brace machine is used for:
(A) Making hole in wood (B) Scrapping the wood
(C) For smoothing the wood (D) None of these

165. Which one machine/tool is not used in making hole in wood ?
(A) Ordinary drill (B) Hand drill
(C) Pincer (D) Twist bit

166. 'Bar cramp' is a tool of:
(A) Wood cutting (B) Catching tightly wood
(C) Beating tool (D) Smoothing tool

167. Nail (Keel) pulling is done by:
(A) Claw-hammer (B) Pincer
(C) Plier (D) All of these

168. 'Forging' is said:
(A) Heating the iron
(B) Beating the hot iron
(C) Converting into desired shape by beating the iron
(D) Beating iron and making hole in it

169. 'Anvil' (Nihai) is used for:
(A) Beating the iron on keeping over this
(B) By beating iron through this
(C) Heating iron through handling this
(D) Cooling of hot iron

170. Fire is pulled-up through:
(A) Poker (B) Sewage block
(C) Showel (D) Anvil

171. 'Sledge' is used to:
(A) Lift heavy material
(B) Fire furnace
(C) Cut the iron in cold condition
(D) Cut the iron in hot condition

172. The roller of cane-crusher is made-up of which steel ?
(A) Cast iron (B) Gun metal
(C) High carbon steel (D) Wrought iron

173. The tool use for catching (handling) the claw and turning it hither and thither during beating is called:
(A) Sewage block (B) Chisel
(C) Plier (D) Tongs

174. Punches are of:
(A) Two types (B) Three types
(C) Four types (D) Several types

175. Agricultural tractors are generally having horse power (h.p.) :
(A) 20-50 (B) 50-70
(C)10-15 (D) 20-25

176. A three-tine furrow plough is ploughing at a speed of 2 km/hr and the each furrow size is 20 cm wide and 12 cm deep furrow. How much time, it would require to plough 5 hectare land ?
(A) 41 hours 40 minutes (B) 40 hours 55 minutes
(C) 20 hours 55 minutes (D) None of these

177. If a gear 'A', having 50 teeth and running at a speed of 200 r.p.m. is driving to another gear 'B' having 125 r.p.m. How many number of teeth in gear' B' ?
(A) 500 (B)400 (C) 80 (D)60

178. Milk enters into the heating section of a high temperature short time (HTST) pasteurization plant a ta temperature of 45°C and leaves at 72°C. Hot water at temperature of 95 °C enters counter currently into the heat exchanger and leaves at 77°C. The effectiveness of the heat exchanger is:
(A) 0.18 (B) 0.36 (C) 0.54 (D) 0.84

179. For an initial spore load equal to 25 spores per container inoculated with *Clostridium botulinum* having D_{121}= 0.25 min, the spoilage probability of the container subjected to F_{121} = 1.5 min is:
(A) 10^{-5} (B) 10^{-6} (C) 10^{-8} (D) 10^{-9}

180. A cold storage chamber is constructed with 10 mm mortar, 200 mm brick, 100 mm insulation and 5 mm wood-board having thermal conductivities of 0.8, 1.5, 0.025 and 0.2 W m^{-1} K^{-1}, respectively. The resistance of 4 K m^2 W^{-1} is offered by:
(A) Mortar (B) Brick
(C) Insulation (D) Wood-board

181. Soil erosion by wind brings about the serious damage in soil by changing the:
(A) Soil permeability (B) Soil structure
(C) Soil texture (D) Soil plasticity

182. A tractor driving pulley has its 25 cm diameter and revolving at a speed of 960 r.p.m. If on the shaft of a thresher, an attached pulley is revolving at a speed of 1600 r.p.m. what would be the diameter of this pulley ?
(A) 16 cm (B) 15 cm (C) 12 cm (D) 20 cm

183. A pulley of 21 cm diameter and revolving at 1600 r.p.m. Find out the speed of the belt running over it:
(A) 1120 m/minute (B) 1200 m/minute
(C) 1056 m/minute (D) None of these

184. If the distance between two pullies, having the diameter of 45 cm and 39 cm, is 3.2 metre, then what would be length of required flate belt ?
(A) 6.8 m (B) 7.72 m (C) 9.4 m (D) 7.2 m

185. Two pair of bullocks are ploughing a field by victory ploughs at a speed of 2 km/hr. If the average width of each furrow is 25 cm. How much time will be required to plough one hectare land? Where after each two hours working, 20 minutes rest is required to provide?
(A) 10 hours (B) 8 hours
(C) 11 hours 40 minutes (D) 8 hours 20 minutes

186. If the relative distance is 2.5 metre of centre points of two pullies, having SO cm and 25 cm diameters, then, how much length (metre) of 'V' belt would be required ?
(A) 6.18875 (B) 5.18775
(C) 6.28775 (D) 6.18775

187. A 'chain' contains how much metre ?
(A) 20.1168 (B) 21.1168
(C) 15.1168 (D) 18.1168

188. If the rate of electricity is rupee one per unit, then, a 20 h. p. electric motor, having approx. 100% efficiency will require how much cost (Rs.) on 100 hours running to irrigate the field?
(A) 500 (B) 750 (C) 1500 (D) 2000

189. 'Drip' irrigation is generally followed in the country:
(A) Israel (B) America
(C) Australia (D) Japan

190. In India, under canal net-work system of irrigation, generally, how much amount of water is allowed to run in small canal ?
(A)Less than 4 cusec (B)Less than 10 cusec
(C)Between 15 to 20 cusec (D)Between 25 to 30 cusec

191. Method of irrigation is:
(A)Surface irrigation (B)Sprinkler irrigation
(C)Drip irrigation (D)All of these

192. The Torque—a power to generate revolutions is expressed as (its unit is kg metre) :
(A)Length of arm (m) x Force (kg)
(B)Length of arm (cm) x Force (kg)
(C)Length of arm (m) x Force (g)
(D)Length of arm (rom) x Force (mg)

193. Work is expressed by the formula as:
(A)Work = Force (kg) x Distance (m)
(B)Work = Force (g) x Distance (m)
(C)Work = Force (mg) x Distance (m)
(D)Work = Force (kg) x Distance (cm)

194. The length of 'Engineer chain' is:
(A)100 ft (B)80 ft
(C)20 ft (D)None of these

195. A tractor operated single acting trailing type disc harrow has 8 discs on each gang. The gang angle of both the gangs is maintained at 35^0. The horizontal component of resultant soil reaction force on each disc is 600 N and it makes an angle of 30^0 with the gang axis. If the speed of operation is 6 km h^{-1}, the required drawbar power in kW to operate the harrow will be:
(A)14.4-14.6 (B)15.4-15.6
(C)16.4-16.6 (D)17.4-17.6

196. The part on the three-point coupling of a tractor used to set the cross angle of a hitched implement like a plough is called a:
(A) Top link (B) Levelling box
(C) Sensitivity element (D) None of these

197. Zero tillage system was first used successfully in 1950 in pasture removation in:
(A) Germany (B) Japan
(C) United Kingdom (D) USA

198. Though tilth is dynamic in nature, can be measured by:
(A) Aggregate analysis (B) Chemical analysis
(C) Biological analysis (D) None of these

199. According to CRIDA, Hyderabad, 15-20 cm deep ploughing is known as:
(A) Shallow (B) Medium
(C) Deep (D) None of these

200. Tilt angle of disc furrow is:
(A) 20^0 (B) 5^0 (C) 45^0 (D) 0^0

201. Loosing of the top soil hastens the development of a dry layer at the surface which reduces further loss of water from soil is known as:
(A) Shallow tillage (B) Traditional tillage
(C) Conventional tillage (D) Minimum tillage

202. The tillage practices is also associated with faster decomposition of soil organic matter, greater erosion risk and degradation of soil structure is known as:
(A) Shallow tillage (B) Traditional tillage
(C) Conventional tillage (D) Minimum tillage

203. The tillage envisages minimum disturbance of soil, residues are left in the field and weeds are killed with herbicides is called:
(A) Shallow tillage (B) Traditional tillage
(C) Conventional tillage (D) Minimum tillage

204. Tillage operations are considerably reduced to the minimum required for good seed bed preparation is called:
(A) Shallow tillage (B) Traditional tillage
(C) Conventional tillage (D) Minimum tillage

205. Ultimate stage of minimum tillage is called:
(A) Zero tillage (B) Traditional tillage
(C) Conventional tillage (D) Minimum tillage

206. Sowing of without any tillage operation is called:
(A) Shallow tillage (B) Zero tillage
(C) Conventional tillage (D) Minimum tillage

207. The tillage practices combined with crop residues, dry weeds and other types of vegetative material spread on the soil surface is called:
(A) Shallow tillage (B) Traditional tillage
(C) Mulch tillage (D) Minimum tillage

208. A tillage system in which only isolated bunds of soil are tilled is called:
(A) Strip tillage (B) Sub soiling
(C) Sub soil placement (D) Surface tillage

209. Breaking of compact sub soils without inverting them with a special narrow cultivator, showed or chisel at depth usually from 30-60 cm from the surface is called:
(A) Strip tillage (B) Sub soiling
(C) Sub soil placement (D) Surface tillage

210. Placement of fertilizers in the subsoil with the help of heavy power machinery usually followed in humid and sub humid region is known as:
(A) Strip tillage (B) Sub soiling
(C) Sub soil placement (D) Surface tillage

211. To till the surface without turning the tilled layers is called:
(A) Strip tillage (B) Sub soiling
(C) Sub soil placement (D) Surface tillage

212. To plough or cultivate soil for seeding is called:
(A) Strip tillage (B) Sub soiling
(C) Till (D) Surface tillage

213. The action of a tillage tool in executing a specific form of soil manipulation such as soil cutting, pulverizing or inversion is called:
(A) Tillage action (B) Tillage depth
(C) Tillability (D) Vertical mulching

214. Vertical distance from the initial soil surface to a specific point of penetrations of the tool is called:
(A) Tillage action (B) Tillage depth
(C) Tillability (D) Vertical mulching

215. The degree of ease with which a soil may be manipulated for a specific purpose is called:
(A) Tillage action (B) Tillage depth
(C) Tillability (D) Vertical mulching

216. Machine cuts the soil, pulverizes the soil to powder and spreads uniformly to level the field is:
(A) Rotavator
(B) Ridge and furrow makers
(C) Harrow
(D) None of these

217. A machine is used for making irrigation channel and boundaries of a field is:
(A) Rotavotor
(B) Ridge and furrow makers
(C) Harrow
(D) None of these

218. The speed of rotavotor is:
(A) 100-200 rpm (B) 200-300 rpm
(C) 300-400 rpm (D) 400-500 rpm

219. Harrows are tillage implement of:
(A) Primary tillage (B) Secondary tillage
(C) Both (A) and (B) (D) None of these

220. A ridger plough consists of:
(A) V – shaped (B) U – shaped
(C) L – shaped (D) M – shaped

221. An implement used for puddling the paddy field before transplanting is:
(A) Rotavotor
(B) Ridge and furrow makers
(C) Harrow
(D) Puddler

222. A two bladed flywheel type forage cutter rotates at 60 rpm. At a feed rate of 1·26 meters per minutes, theoretical length of cut of the forage will be:
(A) 21.0 mm (B) 42.0 mm
(C) 5.3 mm (D) 10.5 mm

223. The poisonous exhaust gas from a petrol engine is:
(A) Hydrochloric acid (B) Carbon monoxide
(C) Carbon dioxide (D) None of these

224. The oil that can be used in the braking system of a vehicle is:
(A) Of plant origin (B) Engine oil
(C) Mineral oil (D) None of these

225. The main advantage of 'PVC' pipes for drainage is the feasibility of:
(A) Mechanical laying
(B) Physical laying
(C) Chemical laying
(D) Physical, chemical laying both

226. 'ESCORT' tractor is manufactured at:
(A) Faridabad (B)Ghaziabad
(C) Kanpur (D) Chennai

227. The type of pump used in forced water cooling system of a tractor engine is:
(A) Piston (B) Centrifugal
(C) Gear (D) Vane

228. Which one of the following statements is NOT appropriate regarding cone index:
(A) It reflects strength of soil
(B) It is a composite parameter
(C) It is dimensionless
(D) It is measured at a constant penetration rate of 30 mm/s

229. The draft and total power requirement of a rotary cultivator operating in concurrent mode as compared to a spring tyne cultivator of equal cutting width under the same operating conditions, respectively are:
(A) Higher and higher (B) Lower and lower
(C) Lower and higher (D) Higher and lower

230. The soil erodibility factor needs to be determined for use in the universal soil loss equation. The length, in m and slope, in % of the experimental plot to be used for this purpose, respectively are:
(A) 19, 12 (B) 21, 11
(C) 22, 9 (D) 23, 8

231. The difference between Fore Bearing and Back Bearing of a traverse line is:
(A) Exactly 90^0 (B) Less than 180^0
(C) Exactly 180^0 (D) Greater than 180^0

232. A pumping device that combines the advantages of both centrifugal and reciprocating pumps is known as:
(A) Air lift pump (B) Hydraulic ram
(C) Jet pump (D) Rotary pump

233. If v is the kinematic viscosity of air – water vapour mixture and is D_{AB} the mass diffusivity of water vapour in air then the ratio V/D_{AB} is known as:
(A) Stanton number (B) Prandtl number
(C) Schmidt number (D) Sherwood number

234. Work index in size reduction can be obtained by multiplying Bond's energy constant with:
(A) 10 (B) v10 (C) 3v10 (D) 4v10

235. Approximate percentage of scores that fall within ±ó (standard deviation) of the mean in a normal distribution is:
(A) 34 (B) 68 (C) 95 (D) 99

236. Central Institute of Research on Jute and Allied Fibre Technology is situated at:
(A) Bhopal (B) Ludhiana
(C) New Delhi (D) Kolkata

237. Central Institute for Cotton technology is situated at:
(A) Bhopal (B) Ludhiana
(C) Mumbai (D) Kolkata

238. Indian Lac Research Institute is situated at:
(A) Ranchi (B) Ludhiana
(C) New Delhi (D) Kolkata

239. According to CRIDA, Hyderabad, less than 15 cm deep ploughing is known as:
(A) Shallow (B) Medium
(C) Deep (D) Very deep

240. According to CRIDA, Hyderabad, more than 15 cm deep ploughing is known as:
(A) Shallow (B) Medium
(C) Deep (D) Very deep

241. Rotavator mainly used for:
(A) Threshing (B) Ploughing
(C) Leveling (D) Ridge & furrow maker

242. Desi plough make the furrow of which shape?
(A) 'V' shapes (B) 'L' shapes
(C) 'O' shapes (D) No definite shapes

243. 'Mahindra' tractor is manufactured at:
(A) Hoshiarpur (B) Rudrapur & Jaipur
(C) Faridabad (D) Gurgaon

244. 'Massey furgusion' tractor is manufactured at:
(A) Hoshiarpur (B) Bangalaore
(C) Madurai & Alwar (D) Gurgaon

245. 'Eicher' tractor is manufactured at:
(A) Hoshiarpur (B) Bangalore
(C) Gurgaon (D) Faridabad & Bhopal

246. 'Ford' tractor is manufactured at:
(A) Hoshiarpur (B) Faridabad
(C) Guragaon (D) New Delhi

247. 'Swaraj' tractor is manufactured at:
(A) Hoshiarpur (B) Mohali
(C) Ludhiana (D) Gurgaon

248. 'Sonalika' tractor is manufactured at:
(A) Hoshiarpur (B) Ludhiana
(C) Faridabad (D) Gurgaon

249. Ferti cum seed drill is used for:
(A) Fertilizer application
(B) Seed sowing
(C) Both (A) and (B)
(D) None of these

250. Lesser land leveler is a machine used for:
(A) Ploughing (B) Leveling
(C) Harrowing (D) Lifting of water

251. Vikas and Vijay is a name of seed drill is manufactures at:
(A) Hoshiar pur (B) Meerut
(C) Chennai (D) Khorai

252. The draft of deshi plough is:
(A) 80-100 kg (B) 70-80 kg
(C) 50-60 kg (D) 30-35 kg

253. The draft of meston plough is:
(A) 80-100 kg (B) 70-80 kg
(C) 50-60 kg (D) 30-35 kg

254. Central Farm Machinery Training and Testing Institute us situated at:
(A) Bhopal, MP (B) Budni, MP
(C) Ludhiana, Pb (D) Jabalpur, MP

255. 'John Deere' tractor is manufactured at:
(A) Nabha, Pb (B) Faridabad
(C) Pune (D) Mohali

256. For a deep ploughing (more than 15 cm) which plough is used?
(A) Victory plough (B) Deshi plough
(C) Chisel plough (D) Rotavator

257. To break the hard pan of sub soil, which plough is used:
(A) Deshi plough (B) Chisel plough
(C) Rotavator (D) Victory plough

258. Chisel plough is required tractor of:
(A) 35 HP (B) 40 HP
(C) 45 HP (D) More than 50 HP

259. The draft of bullock drawn seed cum ferti drill plough is:
(A) 34-36 kg (B) 40-44 kg
(C) 50-54 kg (D) 60-64 kg

260. The draft of 'wheel hand hoe' is:
(A) 9-9.5 kg (B) 10-10.5 kg
(C) 11-11.5 kg (D) 12-12.5 kg

261. How much working efficiency of bullock drawn cum ferti drill (ha per hour):
(A) 0.30-0.40 (B) 0.50-0.80
(C) 0.80-0.90 (D) 1.0-1.2

262. How much working efficiency of wheel hand hoe (ha per hour):
(A) 0.10-0.12 (B) 0.20-0.25
(C) 0.30-0.40 (D) 0.40-0.50

263. The draft of 'Power drawn eight tined paddy transplanter' is:
(A) 250 kg (B) 275 kg (C) 325 kg (D) 350 kg

264. How much is the working efficiency of power drawn eight tined paddy transplanter (ha/day):
(A) 0.5 (B) 0.6 (C) 0.7 (D) 0.8

265. Rotavator works with the help of:
(A) Hydraulic pressure (B) PTO shaft
(C) Both (A) and (B) (D) None of these

266. Rotavator of 1.5 meter size can easily run by the tractor of:
(A) 40 HP (B) 45 HP (C) 50 HP (D) 55 HP

267. Rotavator of 1.75 meter size can easily run by the tractor of:
(A) 40 HP (B) 45 HP (C) 50 HP (D) 55 HP

268. Rotavator of 2.0 meter size can easily run by the tractor of:
(A) 40 HP (B) 45 HP (C) 50 HP (D) 55 HP

269. Rotavator of 2.25 meter size can easily run by the tractor of:
(A) 40 HP (B) 45 HP (C) 50 HP (D) 55 HP

270. The prime objective of conservation tillage is:
(A) Partial retention of plant residues
(B) Removal of plant residues
(C) Complete retention of plant residues
(D) All of these

271. In no tillage systems, the surface soil layers have:
(A) Lower bulk density
(B) Higher bulk density
(C) Lower water infiltration
(D) Higher total pore space

272. Conservation tillage tends to encourages:
(A) Higher microbial population
(B) Lower number of earthworms
(C) Residues soil fauna
(D) None of these

273. Winnowing of threshed crop can be done with the help of:
(A) Puddler (B) Thresher
(C) Winnowing fan (D) Reapers

274. Interculture operations can be done in the standing crop:
(A) Cultivator (B) Hand hoe
(C) Harrow (D) None of these

275. The physical condition of the soil resulting from tillage is known as:
(A) Tilth (B) Till
(C) Hoeing (D) None of these

276. Whom soil is brought to such a condition ideal for crop growth, it is called:
(A) Seed bed (B) Good tilth
(C) Both (A) and (B) (D) None of these

277. Findout the area sprayed by a full tank of a sprayer, if the volume of the sprayer tank is 225 litres and rate of spray application/ha is 159 litres:
(A) 1.25 ha (B) 1.50 ha (C) 1.75 ha (D) 2.00 ha

278. A soil clod having oven dry weight of 35 g ws resin/wax cooled and dipped into a measuring cylinder containing initial level of water 300 ml. The height of water level in the measuring cinder increased to 330 ml after dipping of clod. Findout the bulk density (g/cm^3) of clod:
(A) 1.40 (B) 1.50 (C) 1.17 (D) 1.08

279. The oven dry weight of soil per unit of its bulk volume is called:
(A) Bulk density (B) Particle density
(C) Both (A) and (B) (D) None of these

280. The mass per unit volume of soil particle is known as:
(A) Bulk density (B) Particle density
(C) Both (A) and (B) (D) None of these

281. Bulk density of soil is expressed as:
(A) g/cm^3 (B) g/cm^2 (C) g/cm (D) g/m

282. Particle density of soil is expressed as:
(A) g/cm^2 (B) g/cm^3 (C) g/cm (D) g/m

283. Soil having low bulk density exhibits:
(A) Favourable for good physical conditions
(B) Favourable for poor physical conditions
(C) No effect on physical conditions
(D) Unknown for physical conditions

284. Soil having high bulk density exhibits:
(A) Favourable for good physical conditions
(B) Favourable for poor physical conditions
(C) No effect on physical conditions
(D) Unknown for physical conditions

285. Bulk density of soil is also known as:
(A) Apparent specific density
(B) Particle density
(C) Either (A) or (B)
(D) Neither (A) nor (B)

286. How much is the working efficiency of 'Shallow Persian wheel' is?
(A) 0.43 ha (B) 0.33 ha (C) 0.23 ha (D) 0.20 ha

287. How much is the working efficiency of 'Deep Persian wheel' is?
(A) 0.15 ha (B) 0.24 ha (C) 0.30 ha (D) 0.35 ha

288. How much is the working efficiency of 'Single chain pump' is?
(A) 0.15 ha (B) 0.20 ha (C) 0.27 ha (D) 0.35 ha

289. How much is the working efficiency of 'Double chain pump' is?
(A) 0.15 ha (B) 0.20 ha (C) 0.30 ha (D) 0.40 ha

290. How much is the working efficiency of 'Washer Persian wheel' is?
(A) 0.23 ha (B) 0.20 ha (C) 0.15 ha (D) 0.10 ha

291. How much is the working efficiency of 'Egyptian screw' is?
(A) 0.40 ha (B) 0.30 ha (C) 0.20 ha (D) 0.15 ha

292. How much is the working efficiency of 'Swing basket' is?
(A) 0.08 ha (B) 0.12 ha (C) 0.18 ha (D) 0.28 ha

293. How much is the working efficiency of 'Charsa or mot' is?
(A) 0.28 ha (B) 0.22 ha (C) 0.18 ha (D) 0.08 ha

294. How much is the working efficiency of 'Jal chakra' is?
(A) 0.80 ha (B) 0.70 ha (C) 0.50 ha (D) 0.40 ha

295. How much is the working efficiency of 'Don' is?
(A) 0.10 ha (B) 0.18 ha (C) 0.28 ha (D) 0.38 ha

296. How much is the working efficiency of 'Mayadas lift (single)' is?
(A) 0.36 ha (B) 0.26 ha (C) 0.16 ha (D) 0.10 ha

297. How much is the working efficiency of 'Mayadas lift (double)' is?
(A) 0.20 ha (B) 0.25 ha (C) 0.30 ha (D) 0.40 ha

298. How much is the working efficiency of 'centrifugal pump (12.5 cm)' is?
(A) 1.13 ha (B) 1.0 ha (C) 0.90 ha (D) 0.80 ha

299. How much is the working efficiency of 'Hand pump' is?
(A) 0.40 ha (B) 0.30 ha
(C) 0.20 ha (D) 0.10 ha

300. 'New Holland' is manufactured at:
(A) Nabha, Pb (B) Pune, MH
(C) Buddi, HP (D) Mohali , HP

301 'Indo Farm' is manufactured at:
(A) Nabha, Pb (B) Pune, MH
(C) Baddi, HP (D) Mohali , HP

302. 'HMT' tractor is manufactured at:
(A) Pinjore (B) Panchkula
(C) Hyderabad (D) All of these

303. Which seeding and fertilizer application equipment is developed by CCSHAU, Hisar centre?
(A) Ridge seeder
(B) Two-row bullock drawn FESPO plough
(C) Tractor drawn planter
(D) Malviya seed cum fertilizer drill

304. Mechanisation of agriculture is required for:
(A) Better agricultural yield
(B) Multiple cropping
(C) Generation of employment
(D) All of these

305. How much farm power is available per hectare?
(A) 0.5 KW (B) 1.0 KW
(C) 1.5 KW (D) 2.0 KW

306. Level of farm mechanization is affected by:
(A) Size of holding
(B) Labour requirement
(C) Intensification of crop production
(D) All of these

307. Which of the following state is most mechanization in India?
(A) UP (B) Punjab
(C) Haryana (D) Kerala

ANSWERSHEET: 10.AGRICULTURAL ENGINEERING

1. (D); 2. (A); 3. (A); 4. (D); 5. (C); 6. (A); 7. (B); 8. (C); 9. (D); 10. (D); 11. (A); 12. (D); 13. (A); 14. (A); 15. (A); 16. (B); 17. (D); 18. (C); 19. (C); 20. (A); 21. (A); 22. (C); 23. (A); 24. (B); 25. (C); 26. (A); 27. (B); 28. (C); 29. (A); 30. (B); 31. (B); 32. (A); 33. (D); 34. (B); 35. (D); 36. (B); 37. (C); 38. (B); 39. (A); 40. (A); 41. (D); 42. (A); 43. (B); 44. (C); 45. (A); 46. (D); 47. (B); 48. (B); 49. (C); 50. (A); 51. (C); 52. (B); 53. (A); 54. (C); 55. (B); 56. (C); 57. (D); 58. (C); 59. (D); 60. (D); 61. (C); 62. (A); 63. (C); 64. (C); 65. (B); 66. (A); 67. (A); 68. (B); 69. (D); 70. (B); 71. (C); 72. (D); 73. (B); 74. (B); 75. (D); 76. (B); 77. (C); 78. (D); 79. (A); 80. (B); 81. (C); 82. (B); 83. (B); 84. (D); 85. (C); 86. (A); 87. (B); 88. (A); 89. (C); 90. (B); 91. (C); 92. (A); 93. (A); 94. (D); 95. (D); 96. (A); 97. (B); 98. (B); 99. (C); 100. (D); 101. (C); 102. (C); 103. (A); 104. (A); 105. (D); 106. (B); 107. (A); 108. (D); 109. (B); 110. (D); 111. (A); 112. (B); 113. (B); 114. (D); 115. (C); 116. (B); 117. (E); 118. (D); 119. (B); 120. (A); 121. (D); 122. (B) ; 123. (D); 124. (D); 125. (C); 126. (B); 127. (D); 128. (A) ; 129. (C); 130. (B); 131. (A); 132. (D); 133. (C); 134. (B); 135. (D); 136. (C); 137. (A); 138. (D); 139. (D); 140. (A); 141. (D); 142. (B); 143. (A); 144. (B); 145. (A); 146. (D); 147. (B); 148. (A); 149. (A); 150. (C); 151. (B); 152. (D); 153. (D); 154. (C); 155. (A); 156. (B); 157. (D); 158. (D); 159. (B); 160. (D); 161. (C); 162. (D); 163. (D); 164. (A); 165. (C); 166. (B); 167. (D); 168. (C); 169. (A); 170. (A); 171. (A); 172. (A); 173. (D); 174. (B); 175. (A); 176. (A); 177. (C); 178. (C); 179. (B); 180. (C); 181. (C); 182. (C); 183. (B); 184. (B); 185. (C); 186. (D); 187. (A); 188. (C); 189. (A); 190. (B); 191. (D); 192. (A); 193. (A); 194. (A); 195. (A); 196. (A); 197. (D); 198. (A); 199. (B); 200. (A); 201. (A); 202. (B); 203. (C); 204. (D); 205. (A); 206. (B); 207. (C); 208. (A); 209. (B); 210. (C); 211. (D); 212. (C); 213. (A); 214. (B); 215. (C); 216. (A); 217. (B); 218. (B); 219. (B); 220. (A); 221. (D); 222. (D); 223. (B); 224. (B); 225. (A); 226. (A); 227. (B); 228. (C); 229. (C); 230. (C); 231. (C); 232. (D); 233. (C); 234. (C); 235. (B); 236. (D); 237. (D); 238. (A); 239. (A); 240. (C); 241. (B); 242. (A); 243. (B); 244. (C); 245. (D); 246. (B); 247. (B); 248. (C); 249. (D); 250. (B); 251. (D); 252. (D); 253. (A); 254. (B); 255. (C); 256. (C); 257. (B); 258. (D); 259. (A); 260. (C); 261. (B); 262. (A); 263. (C); 264. (D); 265. (B); 266. (A) ; 267. (B); 268. (C); 269. (D); 270. (A); 271. (B); 272. (A); 273. (C); 274. (B); 275. (A); 276. (C); 277. (B); 278. (C); 279. (A); 280. (B); 281. (A); 282. (B); 283. (A); 284. (B); 285. (A); 286. (A); 287. (B); 288. (C); 289. (D); 290. (A); 291. (B); 292. (C); 293. (D); 294. (A); 295. (B); 296. (C); 297. (D); 298. (A); 299. (B); 300. (A); 301. (C); 302. (D); 303. (C); 304. (D); 305. (D); 306. (D); 307. (B).

Chapter 11

SOIL SCIENCE

1. Study of soil as a natural resource on the surface of the earth including fertilizer schedule soil formation, classification and mapping; physical, chemical and fertility properties of soils; and these properties in relation to the use and management of soils is called:
(A) Soil science (B) Padology
(C) Climatalogy (D) Microbiology

2. The different forms of earth on the surface of the rocks, formed by the breaking down or weathering of rocks is called:
(A) Soil science (B) Padology
(C) Climatalogy (D) Microbiology

3. To study of soils in their natural environment is called:
(A) Soil science (B) Padology
(C) Climatalogy (D) Microbiology

4. Padology word derived from:
(A) Romen (B) English
(C) Greek (D) Sanskrit

5. Component of soils are:
(A) Solid phase (B) Liquid phase
(C) Gasseous phase (D) All of these

6. Fertilier schedule includes:
(A) Dose only
(B) Source only
(C) Time of application
(D) Dose, source, time and method of application

7. The cheaper source of sulphur is:
(A) Single super phosphate
(B) Ammonium sulphate
(C) Elemental sulphur
(D) Gypsum

8. Nitrogen as starter solution is applied in:
(A) Paddy (B) Wheat
(C) Vegetable crops (D) Mustard

9. FYM is applied to the field:
(A) One day before sowing (B) 7 days before sowing
(C) 15 days before sowing (D) 30 days before sowing

10. The nutrient responsible for development of root is:
(A) Nitrogen (B) Sulphur
(C) Zinc (D) Phosphorus

11. Bio fertilizer supplied nitrogen to the extent of:
(A) 10 kg/ha (B) 20 kg/ha
(C) 40 kg/ha (D) 60 kg/ha

12. The soil pH for rice cultivation is:
(A) 5.5-6.5 (B) 8.5-9.5
(C) 4.5-5.5 (D) 8.0-8.5

13. The north-east alluvium of Uttar Pradesh:
(A) Light textured (B) Heavy textured
(C) High calcareous (D) Non calcareous

14. In alluvial soils of India, deficiency of the following nutrient is most wide spread:
(A) Iron (B) Zinc
(C) Copper (D) Manganese

15. Among major plant nutrients, the deficiency of the following nutrients is much common in soils of India:
(A) Nitrogen
(B) Potassium
(C) Nitrogen and Potassium both
(D) None of these

16. Clay fraction of Indo-Gangetic alluvium is dominated by the mineral:
(A) Vermiculite (B) Illite
(C) Kaolinite (D) Chlorite

17. Black cotton soils of India have originated from the rocks:
(A) Sand stones (B) Shales
(C) Lime stones (D) Deccan trap

18. The CEC SC mol (p+) kg^{-1} of black cotton soils of India ranges between:
(A) 15 to 25 (B) 5 to 20
(C) 35 to 50 (D) None of these

19. The clay minerals predominant in black cotton soils of India is/are:
(A) Montmorillonite (B) Beidellite
(C) Both (A) and (B) (D) None of these

20. Argillopedoturbation in black cotton soils of India is the result of:
(A) Slickensides (B) Shrinking and swelling of clays
(C) Gilgai (D) None of these

21. SiO_2/R_2O_2 of the clay fraction of black cotton soils usually lies between:
(A) 1.5 to 2.0 (B) 2.0 to 2.5
(C) 2.5 to 3.0 (D) 3.0 to 3.5

22. Black cotton soils of India are poor in:
(A) Nitrogen (B) Phosphorus
(C) Both (A) and (B) (D) None of these

23. Red soil of India occupy an area of nearly:
(A) 2000 km^2 (B) 3000 km^2
(C) 4000 km^2 (D) 5000 km^2

24. Red soils of India are:
(A) Calcareous (B) Rich in soluble salts
(C) Both (A) and (B) (D) None of these

25. The CEC SC mol (p+) kg^{-1} red soils is:
(A) < 20 (B) 20 to 30
(C) 30 to 40 (D) >40

26. Red soils have the dominance of the following clay mineral:
(A) Illite (B) Kaolinite
(C) Smectite (D) Vermiculite

27. The colour of the red soils is due to:
(A) High content of iron oxide
(B) Wide diffusion of iron
(C) Both (A) and (B)
(D) None of these

28 The laterites of India occupy nearly an area of:
(A) 50000 km^2 (B) 12500 km^2
(C) 250000 km^2 (D) 500000 km^2

29. Laterite soils have developed as:
(A) in situ formation
(B) Sedimentary formations
(C) Both (A) and (B)
(D) None of these

30. The SiO_2/R_2O_2 ratio of laterite clays varied between:
(A) 1.00 to 1.25 (B) 1.40 to 2.00
(C) 2.00 to 3.00 (D) > 3.00

31. Laterite soils are characterized by:
(A) Compact, honey combed/vesicular sub-soil mass
(B) Pallid zone/Plinthite above the water table
(C) Both (A) and (B)
(D) None of these

32. Laterites have a pH of:
(A) 4.5 to 6.0 (B) 6.0 to 7.0
(C) 7.0 to 8.0 (D) > 8.0

33. Laterite soils in India are commonly found in:
(A) Uttar Pradesh (B) Assam
(C) Both (A) and (B) (D) None of these

34. The essentiality of nitrogen for higher plants was first established in the year:
(A) 1801 (B) 1802 (C) 1803 (D) 1804

35. The number of aminoacids forming the building-blocks of plant proteins are:
(A) 19 (B) 21 (C) 23 (D) 25

36. Nitrogen is an essential constituent of the following compound in plants:
(A) Chlorophyll (B) Carotenes
(C) Xanthine (D) Xanthocyanin

37. The essential nature of nitrogen to higher plants was first established by:
(A) JustusVon Leibig (B) Sommer and Lipman
(C) Warrington (D) Theodore de Saussure

38. Nitrogen is an essential constituent of the plant growth substance:
(A) Gibberellins (B) Abscisin II
(C) Cytokinin (D) Calins

39. High N supply to plants influences shoot/root ratio in plants by:
(A) Increasing the ratio (B) Decreasing the ratio
(C) No alteration in the ratio (D) None of these

40. Excess N supply to plants leads to following changes in their carbohydrate accumulation:
(A) Higher accumulation
(B) Depressed accumulation
(C) No change in accumulation
(D) None of these

41. The susceptibility of crops to lodging alters with excessive N-supply to plants as:
(A) Lodging increases
(B) Lodging decreases
(C) Lodging remains unchanged
(D) None of these

42. Rice plants prefer absorption of fertilizer nitrogen in the from of:
(A) NO_3 (B) NH_4
(C) Amino acid (D) N_2

43. Deficiency symptoms of nitrogen in plants appear at first in:
(A) Lower older leaves (B) Middle leaves
(C) Top young leaves (D) All of these

44. Excessive N supply to plants promotes:
(A) Mildew and rusts (B) Root-rot
(C) Bacterial blight (D) Leaf mosaic

45. Under low N supply, tillering in plants:
(A) Decreases (B) Increases
(C) Does not alter (D) None of these

46. The critical value of N percentage for potato (in potato leaves 40-45 days old) is:
(A) 1.5 (B) 3.0 (C) 4.5 (D) 6.0

47. The critical value of N percentage for sugarcane (in 3rd to 6th sugarcane leaf from the top) is:
(A) 1.0 (B) 1.5 (C) 2.0 (D) 2.5

48. Good 'Indicator plant' for N deficiency is:
(A) Rice (B) Wheat
(C) Maize (D) Sunflower

49. The essential nature of phosphorus for higher plants was first established experimentally by:
(A) Arnon (B) E. Gris
(C) Ville (D) Johonson and Stout

50. The essentiality of phosphorus for higher plants was reported at first in the year:
(A) 1829 (B) 1839 (C) 1849 (D) 1859

51. The phosphorus in plant seeds is mainly stored as:
(A) Phospholipids (B) Nucleic acids
(C) Inorganic –P (D) Phytin

52. The phosphorus in cell membranes is mainly present as:
(A) Phospholipids (B) ATP
(C) NADP (D) Nucleoproteins

53. Phosphorus is a structural component of the cell organelles:
(A) Cell membranes
(B) Cell membranes and chloroplast
(C) Cell membrane, chloroplast and mitochondria
(D) None of these

54. As a constituent of coenzyme NAD and NADP, phosphorus plays an important role in:
(A) Energy transformation
(B) Hydrolytic reactions
(C) Decarboxylation
(D) Oxidation-reduction reactions

55. Phosphorus supply to plants promotes growth of:
(A) Fibrous roots
(B) Lateral roots
(C) Fibrous as well as lateral roots
(D) None of these

56. With an ample phosphorus supply to crops:
(A) Maturity period is delayed
(B) Maturity period is early
(C) Maturity period is unchanged
(D) None of these

57. Phosphorus supply to crops influences the fungal disease incidences by:
(A) Increasing disease incidence
(B) Decreasing disease incidence
(C) No change in disease incidence
(D) None of these

58. In legumes, phosphorus supply improves crop yield and protein content of crop through:
(A) Higher utilization of fertilizer – N
(B) Higher N – fixation
(C) Both (A) and (B)
(D) None of these

59. An ample supply of phosphorus to plants improves the availability of the following nutrients to plants:
(A) Molybdenum
(B) Zinc
(C) Both Molybdenum and zinc
(D) None of these

60. In plants, the dominant form of available inorganic phosphorus in soil below pH 6.7 exists in the inionic forms:
(A) $HPO_4^=$ (B) $H_2PO_4^-$
(C) $PO_4^=$ (D) None of these

61. In organic forms, plants can directly absorb organic phosphorus in the form:
(A) Phytin (B) Phospholipids
(C)Both (A) and (B) (D) None of these

62. The optimum soil pH range for phosphorus availability to crops is:
(A) 5.0 to 6.0 (B) 6.0 to 7.0
(C) 7.0 to 8.0 (D) None of these

63. Phosphorus deficiency in plants appears at first in:
(A) Older leaves (B) Younger leaves
(C) Both (A) and (B) (D) None of these

64. The critical value of P in phosphorus deficient leaves of maize plants before silking stage is:
(A) 0.25 per cent (B) 0.35 per cent
(C) 0.45 per cent (D) 0.55 per cent

65. Indicator plant for P deficiency indication is:
(A) Maize (B) Tomato
(C) Maize and tomato both (D) None of these

66. Potassium was experimentally established as an element essential to higher plants during the year:
(A) 1839 (B) 1876 (C) 1886 (D) 1896

67. The essentiality of potassium to higher plants was at first reported by:
(A) Russel (B) Sprengel
(C) Steinberg (D) Arnon

68. The potassium is almost entirely present as K^+ in the:
(A) Cytoplasm (B) Vacuole
(C) Both (A) and (B) (D) None of these

69. Within the cell, potassium helps to maintain:
(A) Permeability of cellular membranes
(B) Hydration of protoplasm
(C) Both (A) and (B)
(D) None of these

70. Potassium supply to plants has a favourable effect on:
(A) Protein synthesis (B) Respiration
(C) Both (A) and (B) (D) None of these

71. Potassium supply to plants decreases crop lodging through strengthening stalk by:
(A) Promoting cellulose and lignin deposition on walls
(B) Improving turgidity of cells
(C) Increasing parenchymatous tissues
(D) Maintaining electrical charge within the cell

72. Potassium deficiency leads to increase the severity of:
(A) Cotton wilt (B) Potato rot
(C) Both (A) and (B) (D) None of these

73. Plants utilize potassium in the form of:
(A) K_2 (B) K^+
(C) Both (A) and (B) (D) None of these

74. In general, potassium deficiency influences crop maturity by:
(A) Delaying maturity period
(B) Preceding maturity period
(C) Not changing maturity period
(D) None of these

75. Potassium deficiency symptoms in plants appear at first from:
(A) Lower most leaves
(B) Upper most leaves
(C) Middle leaves
(D) Scattered throughout the entire leaves

76. Deficiency of K in plants leads to an accumulation of:
(A) Methionine (B) Putrescine
(C) Ornithine (D) Isoleucine

77. In general, the shoot/root ratio of K-deficient crops is around:
(A) 1.0 (B) 2.0 (C) 3.0 (D) 4.0

78. Potassium is a structural component of:
(A) Nucleus (B) Mitochondria
(C) Chloroplast (D) None of these

79. Potassium deficiency in plants is characterized by:
(A) Leaf scorch (B) Tip burn
(C) Both (A) and (B) (D) None of these

80. The critical value of K percentage in petioles of deficient potato plants is:
(A) 4.0 (B) 5.0 (C) 6.0 (D) 7.0

81. The critical value for K deficiency in maize is:
(A) 0.5 per cent (B) 1.0 per cent
(C) 1.5 per cent (D) 2.0 per cent

82. Indicator plant for monitoring K deficiency is:
(A) Mustard (B) Pea
(C) Sunflower (D) Potato

83. Major part of dry weight of plant is derived from elements obtained from:
(A) Soil (B) Air
(C) Water (D) Decomposing organic matter

84. Major function of phosphorus is in the formation of:
(A) Cell membranes (B) Cell wall
(C) Enzymes (D) Carbohydrates

85. Minerals are absorbed by the plants through:
(A) Pressure flow (B) Diffusion
(C) Active transport (D) Osmosis

86. Magnesium is required for:
(A) DNA replication
(B) RNA synthesis
(C) Stomatal movements
(D) Synthesis of chlorophyll

87. Potassium is employed in:
(A) Chlorophyll synthesis
(B) Protein synthesis
(C) Cementing cells
(D) Opening and closing of stomata

88. Calcium is a component of:
(A) Middle lamella (B) Primary wall
(C) Secondary wall (D) Chlorophyll

89. Chlorosis results from deficiency of:
(A) Chlorine (B) Sodium
(C) Calcium (D) Magnesium

90. Both nitrogen and sulphur are required by plants for:
(A) Chlorophyll synthesis (B) Enzymes
(C) Cell wall (D) Stomatal movements

91. Micronutrients mainly function as:
(A) Osmotic constituents of cell sap
(B) Cofactors of enzymes
(C) Constituents of chlorophyll
(D) Components of important biochemicals

92. Critical elements are:
(A) Na, K and Ca (B) N, P and Mg
(C) N, P and K (D) Mn, B and Mo

93. Macronutrients are elements that:
(A) Play major role in plant nutrition
(B) Are require in large quantities in plants
(C) Form large molecules in plants
(D) None of these

94. Micronutrients/trace elements are those that:
(A) Can be removed from the plant without any effect on it
(B) Play a minor role in plant nutrition
(C) Form very small molecules in plants
(D) Are present/required in small quantities in plants

95. In plant mineral nutrition, elements are called macro or micro elements depending upon their:
(A) Relative presence in plant ash
(B) Relative importance in plant growth
(C) Relative amount required in plants
(D) Relative availability in soil

96. Minerals are absorbed in the form of:
(A) Molecules (B) Ions
(C) Compounds (D) Mixtures

97. Tracer elements are those that are:
(A) Required in small amount by the plant
(B) Radioactive and can be used to investigate a process in the plant
(C) First to be traced in the plant
(D) Present in traces in the soil solution

98. The absorption of minerals due to difference in the electropotential gradient without use of energy is:
(A) Active absorption (B) Passive absorption
(C) Osmotic absorption (D) None of these

99. Passive absorption of mineral salts is not dependent on:
(A) Diffusion (B) Osmosis
(C) Donnan equilibrium (D) Ionic exchange

100. Active absorption of minerals salts makes use of:
(A) Ionic exchange
(B) Carriers
(C) Phosphorylation of elements
(D) None of these

101. Which of the following does not operate in the carrier concept of active absorption of minerals:
(A) Electron transport
(B) Contractile protein
(C) Protein synthesis
(D) Ascending water stream

102. Absorbed minerals are transported to other parts of the plants by:
(A) Diffusion (B) Imbibition
(C) Ion exchange (D) Water stream

103. A primary deficiency is caused by insufficient absorption of:
(A) Magnesium (B) Manganese
(C) Calcium (D) Potassium

104. Down word passage of mineral, from upper soil layers to lower soil strata is called:
(A) Percolation (B) Leaching
(C) Weathering (D) Runway

105. A number of minerals like Ca, Mg and K are held over the surface of clay particles because the latter are:
(A) Negatively charged
(B) Positively charged
(C) Neutral
(D) Having both positive and negative residual valencies

106. Soil nitrate is more likely to leach than ammonium due to its:
(A) Small size (B) Negative charge
(C) Being useless to soil flora (D) Abundance

107. The largest reservoir of nitrogen is:
(A) Rocks (B) Sea
(C) Air (D) Soil

108. The enzymes required for nitrogen fixation is:
(A) Nitrogen deaminase (B) Nitrodioxidase
(C) Amino acid decarboxylase (D) Nitrogenase

109. Hydroponics is growing plants in:
(A) Laboratory (B) Phytotron
(C) Liquid culture medium (D) Solid culture medium

110. Aeroponics is growing plants in:
(A) Air
(B) Satellites
(C) Other planets and space
(D) Stands with liquid spray of their roots

111. Nodule formation of legume roots is reduced during deficiency
(A) Sulphur (B) Boron
(C) Nitrogen (D) Both (A) and (B)

112. Little leaf/leaf resetting is deficiency symptoms of:
(A) Zn (B) Mn
(C) Fe (D) B

113. Internal cork of apple/browning of cauliflower is deficiency symptoms of:
(A) Mo (B) B
(C) Cu (D) Zn

114. Exanthema is due to deficiency of:
(A) B (B) Mo
(C) Mn (D) Cu

115. Whiptail of crucifers is due to deficiency of:
(A) Mo (B) Cu
(C) Zn (D) P

116. A trace element is the one which is:
(A) Radioactive
(B) Heavy isotope
(C) Required in minute quantity
(D) Enters the protoplasm alongwith an other elements

117. Common free ion in the cell is:
(A) Iron (B) Potassium
(C) Sulphate (D) Phosphate

118. Which of the following are the most abundant elements in the living cells:
(A) C, H, S and P (B) C, O, N and P
(C) C, H, O and N (D) C, H, Mg and Ca

119. A fertilizer has a formula of three figures 19-9-9. They stand for percentage of:
(A) N, Ca and Mg (B) Mg, P and K
(C) Ca, N and Fe (D) N, P and K

120. The raw material most used by the plants is:
(A) O_2 (B) CO_2
(C) N_2 (D) K

121. Mineral salts which are absorbed by the root from the soil are in the form of:
(A) Very dilute solution
(B) Dilute solution
(C) Concentrated solution
(D) Very concentrated solution

122. Minerals absorbed by roots move to the leaf through:
(A) Xylem (B) Phloem
(C) Sieve tube (D) None of these

123. Plants absorb minerals salts from the soil solution through:
(A) Semipermeable membrane into the cytoplasm
(B) Perforation at the apex of root hairs only
(C) The cell wall which is semipermeable
(D) None of these

124. Which of the following is macronutrient:
(A) Ca (B) Mo
(C) Mn (D) Zn

125. Which of the following is a micronutrient:
(A) Mg (B) Zn
(C) Ca (D) P

126. Which of the following is not an essential micronutrient:
(A) Boron (B) Nickel
(C) Manganese (D) Molybdenum

127. Which one of the following elements is not required by plants for their healthy normal growth:
(A) Calcium (B) Magnesium
(C) Lead (D) Iron

128. Which one of the following essential elements, plants normally obtain directly from the air:
(A) Hydrogen (B) Carbon
(C) Nitrogen (D) Phosphorus

129. Which of the following elements is necessary for translocation of sugars in plants:
(A) Iron (B) Molybdenum
(C) Manganese (D) Boron

130. Which of the following is present in the core of chlorophyll molecule:
(A) Fe (B) Mg
(C) K (D) Mn

131. Which is not an essential element for plants:
(A) Iron (B) Potassium
(C) Zinc (D) Sodium

132. Which one is an inorganic nutrient:
(A) Cellulose (B) Vitamin
(C) Calcium (D) Protein

133. Chlorosis in plants occurs due to:
(A) High sunlight intensity
(B) Low sunlight intensity
(C) Absorption of yellow pigments from the soil
(D) Deficiency of Mg and Fe in the soil

134. Most of the plants obtain nitrogen from the soil in the form of:
(A) Free nitrogen gas (B) Nitric acid
(C) Nitrites (D) Nitrates

135. Mg and Fe are needed for plants in the:
(A) Energy transfer
(B) Synthesis of chlorophyll pigment in the leaves
(C) Stomatal opening
(D) Translocation of carbohydrates

136. Carbon becomes available to crop plant in the form of:
(A) Amino acids (B) Carbonates
(C) Carbon dioxide (D) Elemental carbon

137. On the basis of symptoms of chlorosis in leaves a student inferred that this Ws due to deficiency of nitrogen. This inference could be correct only if we assume that yellowing of leaves appeared first in:
(A) Young leaves
(B) Old leaves
(C) Young leaves followed by mature leaves
(D) Mature leaves followed by young leaves

138. Plants require two metallic compounds (minerals) for chlorophyll synthesis. They are:
(A) Fe and Ca (B) Fe and Mg
(C) Cu and Ca (D) Ca and K

139. Hydroponics is:
(A) Growing of aquatic plants
(B) Growing of floating aquatic plants
(C) Soil less cultivation of plants
(D) Growing of plants inside water

140. When chlorophyll is burnt which one is obtained:
(A) Ca (B) Na
(C) Mg (D) Mn

141. Importance of microelements was recognized late due to:
(A) Their toxicity
(B) Presence as contaminants in macro nutrients
(C) Their absence in plant ash
(D) Leakage from roots

142. Major nitrogen fixation is carried out by:
(A) Lighting (B) Chemical industries
(C) Symbiotic bacteria (D) Leaching

143. Which one does not contribute to nitrogen availability in the soil:
(A) Decay of organic matter (B) Electric storms
(C) *Azotobacter* (D) Liming

144. Which one plays an independent role in energy metabolisms:
(A) Calcium (B) Sodium
(C) Sulphur (D) Phosphorus

145. In an experiment a plant was dried, crushed and heated strongly in a crucible for long. The residue contained:
(A) Oxides and carbonates of about ten elements
(B) Oxides and carbonates of three elements
(C) Carbon only
(D) Starch and related compounds

146. Rotation of crops is useful for:
(A) Increasing mineral quantity of soil
(B) Increasing fertility of soil
(C) High protein content
(D) Denitrification of crops

147. Deficiency of Molybdenum causes:
(A) Poor development of vasculature
(B) Bending of leaf tips
(C) Yellowing of leaves
(D) Mottling and marginal necrosis of leaves

148. In plants a common symptom caused by deficiency of P, K, Ca and Mg is:
(A) Bending of leaf tips
(B) Formation of anthocynin
(C) Poor development of vasculature
(D) Appearance of dead necrotic cells

149. Active transport from outside to inside of molecule across a membrane requires:
(A) Cyclic AMP (B) Acetyl choline
(C) ATP (D) Phloroglucinol

150. Yellowing of leaves or chlorosis is the first deficiency symptoms of which element:
(A) Calcium (B) Nitrogen
(C) Phosphorus (D) Chlorine

151. Premature leaf fall is due to deficiency of:
(A) Phosphorus (B) Nitrogen
(C) Iron (D) Calcium

!52. Active uptake of minerals requires:
(A) Respiration (B) Expenditure of energy
(C) Root metabolism (D) Photophosphorylation

153. Ammonium sulphate is:
(A) Insecticide (B) Weedicide
(C) Fertilizer (D) Toxic chemical

154. Which element is needed in large amount:
(A) Nitrogen (B) Sulphur
(C) Phosphorus (D) Calcium

155. Phytotron is meant for:
(A) Controlled irradiation
(B) Induction of mutations
(C) Controlled humidity
(D) Growing plants under controlled environment

156. A nonessential element is:
(A) Calcium (B) Barium
(C) Potassium (D) Magnesium

157. Nonessential element taking part in plant growth is:
(A) Magnesium (B) Calcium
(C) Cobalt (D) Chlorine

158. Which is the role of molybdenum:
(A) Nitrogen fixation
(B) Flower induction
(C) Chromosome contraction
(D) Carbon assimilation

159. Which one of the following is required for nitrate reduction to nitrite:
(A) Fe (B) B (C) Zn (D) Mo

160. An essential element is that which:
(A) Is found in plant ash
(B) Is available in soil
(C) Improves health of plants
(D) Is irreplaceable and indispensable for growth of plants

161. Micronutrients are:
(A) Less important than macronutrients
(B) As important as macronutrients
(C) Having a minor role in plant nutrition
(D) Omitted from culture medium without any detrimental effect

162. A micronutrient is the one which is:
(A) More important than any major element
(B) Less important than major elements
(C) Needed in small quantity but is as important as a major element
(D) Found in small quantities in the soil

163. Framework elements are:
(A) C, H and O (B) Cu, Co and Fe
(C) Mg, Cu and Fe (D) Mn, Ca and N

164. Which of the following is a macronutrient:
(A) Mo (B) Mg
(C) Mn (D) Zn

165. Which one of the following is a micronutrient:
(A) Mg (B) Ca
(C) Mn/Mo (D) P

166. Plastocyanin contains:
(A) Mo (B) Fe
(C) Cu (D) Zn

167. Deficiency of Magnesium causes:
(A) Necrosis (B) Plasmolysis
(C) Hydrolysis (D) Chlorosis

168. A sulphur containing amino acid is:
(A) Methionine (B) Asparagine
(C) Serine (D) Proline

169. Plant ash indicates:
(A) Organic matter of plant
(B) Mineral salts absorbed by plants
(C) Both mineral salts and organic matter
(D) Silica absorbed by plants

170. Active ion uptake means:
(A) Ions moves passively
(B) Ions move freely
(C) There is expenditure of energy
(D) Ions are active

171. Iron deficiency results in:
(A) Leaf tip necrosis
(B) Small leaves
(C) Decreased protein synthesis
(D) Intervenial chlorosis appearing first in young leaves

172. A trace element, also required for plant growth, and in the form of radioactive isotope useful in cancer therapy is:
(A) Iron (B) Calcium
(C) Cobalt (D) Sodium

173. Theory suggesting that carbon dioxide produced in respiration helps in mineral absorption is called:
(A) Carbonic acid exchange theory
(B) Contact exchange theory
(C) Active mineral absorption
(D) Donnan equilibrium

174. Phosphorus and nitrogen ions generally depleted in soil because they usually occur as:
(A) Nautral ions
(B) Negatively charged ions
(C) Positively charged ions
(D) Both positively and negatively charged but disproportionate mixture

175. Passage of minerals from top soil to subsoil through seepage of water is known as:
(A) Leaching (B) Percolation
(C) Conduction (D) Transportation

176. NPK denotes:
(A) Nitrogen, Protein and Kinetin
(B) Nitrogen, Protein and Potassium
(C) Nitrogen, Potassium and Kinetin
(D) Nitrogen, Phosphorus and Potassium

177. Which one is a component of ferrodoxin:
(A) Zn (B) Mn
(C) Cu (D) Fe

178. Which one is an essential mineral, not constituents of any enzymes but stimulates the activity of many enzymes:
(A) Zn (B) Mn
(C) K (D) Mg

179. Potassium is involved in:
(A) Photosynthesis
(B) Promoting many enzymatic activities that regulate plant processes
(C) Providing reddish pigmentation to fruits
(D) Formation of vascular cambium

180. Which can function as carrier in active ion absorption:
(A) Cytochrome (B) Ferredoxin
(C) Lecithin (D) Plastoquinone

181. Stem and root tips die in deficiency of:
(A) Nitrogen (B) Phosphorus
(C) Calcium (D) Sulphur

182. Pungent principle, sinigrin of crucifers is a:
(A) Glycoside having sulphur
(B) Glycoside having cyanide
(C) Glycoside having special amino acids
(D) Tannin

183. Phosphorus brings about:
(A) Retardation of protein synthesis
(B) Healthy root growth
(C) Fruit ripening
(D) Retarded plant growth

184. Zinc is essential for:
(A) Stomatal closing
(B) Stomatal opening
(C) Biosynthesis of IAA
(D) Oxidation of carbohydrates

185. Brown heart disease is due to deficiency of:
(A) Iron (B) Boron
(C) Potassium (D) Molybdenum

186. Scientist credited with development of formula for balanced solution of mineral element is:
(A) Knopp (B) Went
(C) Hill (D) Emerson

187. A crop plant which can grow in nitrogen deficient soils without external supply of it is:
(A) *Allium sativum* (B) *Gossypium herbaccum*
(C) *Cajanus cajan* (D) *Helianthus annus*

188. Nitrogen fixing bacteria were discovered by:
(A) Koch (B) Winogradsky
(C) Pasteur (D) Lister

189. Ions are absorbed by the plants through:
(A) Molecular diffusion
(B) Carriers and pumps
(C) Difference in water potential
(D) Difference in DPD

190. Mineral absorption occurs:
(A) If soil solution is hypotonic
(B) If soil solution is hypertonic
(C) Independent of water potential
(D) Independent of water absorption

191. Whcih is not a micronutrient:
(A) P (B) Cu
(C) Zn (D) Mo

192. Essential macronutrients are:
(A) Manufactured during photosynthesis
(B) Produced by growth hormones
(C) Absorbed from soil
(D) Produced by enzymes

193. Active and passive transports across cell membrane differ in:
(A) Passive transport is nonselective
(B) Passive transport is along the concentration gradient while active transport is due to metabolic energy
(C) Active transport is more rapid
(D) Passive transport is confined to anions while active transport is confined to cations

194. Major role of phosphorus in plant metabolism is:
(A) O_2- evolution during photosynthesis
(B) CO_2 evolution during respiration
(C) Create anaerobic conditions
(D) Generate metabolic energy

195. Nitrogen fixing enzyme found in root nodules is:
(A) Nitrogen esterase (B) Nitrogenase
(C) Nitrase (D) Nitrosomonas

196. Sulphur is absorbed by plants as:
(A) SO_3 from soil (B) SO_2 from air
(C) SO_2 from soil (D) Both (B) and (C)

197. Mineral absorption is mostly:
(A) Physical process (B) Chemical process
(C) Active process (D) Passive process

198. Chlorosis is caused by deficiency of:
(A) iron (B) Calcium
(C) Copper (D) Boron

199. Which one group is of micro-elements:
(A) Mn, Cu, Ca (B) Na, K, B
(C) Na, Ni, Cl (D) Cu, Mo, Zn

200. Element playing important role in nitrogen fixation is:
(A) Mn (B) Mo
(C) Zn (D) Cu

201. Which one is not a micronutrient:
(A) Mo (B) B
(C) Mg (D) Zn

202. Culturing of plants in water/soil-less medium is:
(A) Hydroponics (B) Sand culture
(C) Aeroponics (D) Soil-less culture

203. Which is not a micronutrient:
(A) Cu (B) Mn
(C) B (D) Ca

204. Molybdenum is involved in plant metabolism in:
(A) Translocation of solutes
(B) Nitrate reduction
(C) Tryptophan synthesis
(D) Ascorbic acid synthesis

205. Element responsible for maintaining turgor is:
(A) K (B) Na
(C) Ca (D) Hg

206. Element required is least quantity is:
(A) Zn (B) Mn
(C) Mo (D) Cl

207. Role of inorganic nutrients in plant growth and metabolism was first discovered by:
(A) Knop (B) Woodland
(C) DeSassure (D) Steward

208. Organic compounds invariably contain:
(A) Phosphorus (B) Sulphur
(C) Magnesium (D) Carbon

209. Element which is a component of proteinaceous materials and protoplasm is:
(A) Potassium (B) Nitrogen
(C) Magnesium (D) Oxygen

210. Element essential for photolysis of water is:
(A) Nitrogen (B) Oxygen
(C) Chlorine (D) Carbon

211. Which is absent in plant ash:
(A) Mineral elements (B) Nitrogen
(C) Essential elements (D) Trace elements

212. Sink is related to:
(A) Phytochrome (B) Enzyme
(C) Stomata (D) Transport of minerals

213. Bidirectional translocation of mineral takes place through:
(A) Parenchyma (B) Cambium
(C) Xylem (D) Phloem

214. Fall of immature leaves is due to deficiency of:
(A) Potassium (B) Sodium
(C) Phosphorus (D) Zinc

215. Enzyme catalysed reactions can be inhibited by:
(A) Mg^{2+} (B) Cu^{2+}
(C) Hg^{2+} (D) Zn^{2+}

216. Permeability of protoplasm is accelerated by:
(A) Ca (B) Na
(C) P (D) K

217. Exothema is a result of deficiency of:
(A) Cu (B) Ca
(C) P (D) Na

218. Movement of minerals in plant roots through diffusion is:
(A) Endosmosis (B) Osmosis
(C) Active absorption (D) Passive absorption

219. Plants require Fe and Mg for:
(A) Synthesis of chlorophyll
(B) Opening and closing of stomata
(C) Energy transfer during photosynthesis and respi ration
(D) Translocation of carbohydrate

220. Nitrogen is an important constituent of:
(A) Polyphosphates (B) Carbohydrates
(C) Proteins (D) Lipids

221. Sand culture is a component of:
(A) Soil-less cultivation (B) Green house effect
(C) Photorespiration (D) None of the above

222. Inorganic nutrients are present in the soil as:
(A) Molecules (B) Atoms
(C) Electrically charged ions (D) Colloids

223. Clay and organic matter of the soil have negative charges. They attract positively charged ions like:
(A) Ca^{2+} (B) Mg^{2+}
(C) K^+ (D) All the above

224. Nitrogen absorbed by plants is:
(A) Converted to nitrate
(B) Reduced to ammonia
(C) Changed to nitrite
(D) Combined with oxygen

225. Micronutrient absorbed by foliage is:
(A) Zinc (B) Phosphorus
(C) Aluminium (D) Boron

226. Plants use zinc as:
(A) Zn (B) Zn^{2+}
(C) $ZnSO_4$ (D) $Zn(NO_3)_2$

227. Plants growing in urea sprayed but Mg deficient soil will show:
(A) Deep green foliage
(B) Early flowering
(C) Yellowing of leaves
(D) Loss of pigmentation in petals

228. Pedology is a study of:
(A) Locomotion of animals (B) Rocks
(C) Soil (D) Crop diseases

229. When is a man a part of biotic community:
(A) Always
(B) Only when he lives in big cities
(C) Only when he lives in forest
(D) Never

230. Laterite soils is rich in:
(A) Calcium (B) Magnesium
(C) Iron (D) Nitrogen

231. How much phosphorus content in casuarinas wood ash:
(A) Traces (B) 1.5%
(C) 5.9% (D) 1.0%

232. Non-expanding 1:1 clay mineral is:
(A) Chlorite (B) Illite
(C) Vermiculite (D) Kaolinite

233. One of the following sets of element has been termed as beneficial for plant growth:
(A) S, Na, Se, V (B) Cl, Si, Co, Al
(C) B, V, Si, Na (D) Si, Na, Se, Co

234. Harmful substances in urea is:
(A) Uric acid (B) Biuret
(C) Amide (D) Cyanide

235. DRIS system of fertility evaluation is:
(A) Diagnosis and Research Information System
(B) Development and Research Integrated system
(C) Diagnosis and Recommendation Integrated System
(D) Development and Research Information System

236. Which one is correctly matches?

	Plant nutrient	Common extractants
(A)	K+	KCl
(B)	$H_2PO_4^-$	NH_4OAC
(C)	NH_4^+	$NaHCO_3$
(D)	SO_4^-	$Ca(H_2PO_4)_2$

237. Which one is not correctly matched?

	Element	Deficiency disease
(A)	N	Khaira is paddy
(B)	Mn	Grey speck of oat
(C)	B	Brown heart in sugarcane
(D)	Mo	Whiptail of Brassica

238. Soil colours are generally due to:
(A) Iron oxides (B) Humus
(C) Clays (D) Iron oxide and humus

239. Most resistant mineral to weathering is:
(A) Quartz (B) Apatite
(C) Fluospar (D) Corundum

240. More weathering causing mineral is:
(A) Quartz (B) Apatite
(C) Fluospar (D) Corundum

241. *Axorhizobium candidas* forms nodules on the roots of:
(A) Legume (B) Non-leguminous trees
(C) Cereals (D) Sugarbeet

242. Complete oxidation of a gram of a dry organic matter gives rise to:
(A) 0.4 Kcal energy (B) 2.0 Kcal energy
(C) 4.0 Kcal energy (D) 6.0 Kcal energy

243. Group of salts formed in sodic soils are:
(A) Sodium carbonate and Sodium sulphate
(B) Sodium chloride and Sodium carbonate
(C) Sodium carbonate and Sodium bicarbonate
(D) Sodium chloride and Sodium sulphate

244. Read the following statements and mark your answer according to these code given below:
I. Alkali soils are structureless
II. Saline soils are poorly drained
III. Alkali soils are poorly drained
IV. Saline soils are devoid of organic matter
Answer code
(A) I and III are correct (B) II and III are correct
(C) I and IV are correct (D) III and IV are correct

245. Browning in cauliflower can be corrected by:
(A) Ammonium molybdate (B)Boric acid
(C) Zinc sulphate (D) Urea

246. Number of soil orders recognized in the taxonomy are:
(A) 10 (B) 11 (C) 12 (D) 13

247. Organic matter content of most Indian soils is:
(A) Less than 1.0% (B) Between 1.0 to 5.0%
(C) Between 6.0 to 10.0% (D) More than 10.0 %

248. Highest share of fertilizer consumption in India is by:
(A) Rice (B) Wheat
(C) Maize (D) Potato

249. Soil with less than 10 kg available P/ha are classified as:
(A) Low (B) Medium
(C) High (D) Extreme high

250. Soil with 10·24.6 kg available P/ha are classified as:
(A) Low (B) Medium
(C) High (D) Extreme high

251. Soil with more than 24.6 kg available P/ha are classified as:
(A) Low (B) Medium
(C) High (D) Extreme high

252. Soil with less than 280 kg available K/ha are classified as:
(A) Low (B) Medium
(C) High (D) Extreme high

253. Soil with 280-569 kg available K/ha are classified as:
(A) Low (B) Medium
(C) High (D) Extreme high

254. Soil with more than 560 kg available K/ha are classified as:
(A) Low (B) Medium
(C) High (D) Extreme high

255. Soil with less than 28 kg available N/ha are classified as:
(A) Low (B) Medium
(C) High (D) Extreme high

256. Soil with 28-56 kg available N/ha are classified as:
(A) Low (B) Medium
(C) High (D) Extreme high

257. Soil with more than 56 kg available N/ha are classified as:
(A) Low (B) Medium
(C) High (D) Extreme high

258. Soil with less than 0.5 % organic carbon in the soil are classified as:
(A) Low (B) Medium
(C) High (D) Extreme high

259. Soil with 0.5 to 0.75% organic carbon in the soil are classified as:
(A) Low (B) Medium
(C) High (D) Extreme high

260. Soil with more than 0.75% organic carbon is the soil are classified as:
(A) Low (B) Medium
(C) High (D) Extreme high

261. Soil with less than 18 kg available S/ha are classified as:
(A) Low (B) Medium
(C) High (D) Extreme high

262. Soil with 18 to 36 kg available S/ha are classified as:
(A) Low (B) Medium
(C) High (D) Extreme high

263 Soil with more than 36 kg available S/ha are classified as:
(A) Low (B) Medium
(C) High (D) Extreme high

264. General sufficiency or optimum range of N in plants is:
(A) 2.0-5.0% (B) 0.2-0.5%
(C) 1.0-5.0% (D) 0.1-1.0%

265. General sufficiency or optimum range of P in plants is:
(A) 2.0-5.0% (B) 0.2-0.5%
(C) 1.0-5.0% (D) 0.1-1.0%

266. General sufficiency or optimum range of K in plants is:
(A) 2.0-5.0% (B) 0.2-0.5%
(C) 1.0-5.0% (D) 0.1-1.0%

267. General sufficiency or optimum range of Ca in plants is:
(A) 2.0-5.0% (B) 0.2-0.5%
(C) 1.0-5.0% (D) 0.1-1.0%

268. General sufficiency or optimum range of S in plants is:
(A) 0.1-0.4% (B) 0.2-0.5%
(C) 1.0-5.0% (D) 0.1-1.0%

269. General sufficiency or optimum range of Mg in plants is:
(A) 2.0-5.0% (B) 0.1-0.3%
(C) 1.0-5.0% (D) 0.1-1.0%

270. General sufficiency or optimum range of Zn in plants is:
(A) 20-100 μg/g (B) 50-250 μg/g
(C) 20-300 μg/g (D) 5-20 μg/g

271. General sufficiency or optimum range of Fe in plants is:
(A) 20-100 μg/g (B) 50-250 μg/g
(C) 20-300 μg/g (D) 5-20 μg/g

272. General sufficiency or optimum range of Mn in plants is:
(A) 20-100 μg/g (B) 50-250 μg/g
(C) 20-300 μg/g (D) 5-20 μg/g

273. General sufficiency or optimum range of Cu in plants is:
(A) 20-100 μg/g (B) 50-250 μg/g
(C) 20-300 μg/g (D) 5-20 μg/g

274. General sufficiency or optimum range of B in plants is:
(A) 10-100 μg/g (B) 50-250 μg/g
(C) 20-300 μg/g (D) 5-20 μg/g

275. General sufficiency or optimum range of Mo in plants is:
(A) 20-100 μg/g (B) 0.1-0.5 μg/g
(C) 20-300 μg/g (D) 5-20 μg/g

276. General sufficiency or optimum range of Cl in plants is:
(A) 20-100 μg/g (B) 50-250 μg/g
(C) 2000-20000 μg/g (D) 5-20 μg/g

277. The clay minerals having CEC 200 meq/100 g belong to:
(A) Illite (B) Kaolinite
(C) Vermiculite (D) Montmorillonite

278. Amount of water available to the plants in the soil is called as:
(A) Holard (E+C) (B) Chresard
(C) Echard (D) None of these

279. Humus is defined as:
(A) Physical texture of soil
(B) Chemical composition of soil
(C) Decomposed organic matter in soil
(D) None of these

280. Almost decomposed organic matter in which original form is lost in:
(A) Duff (B) Litter
(C) Leaf mould (D) None of these

281. Edaphic factors are related to:
(A) Soil (B) Man
(C) Animals (D) Temperature

282. Which one of the following is not an edaphic factor:
(A) Mineral matter (B) Organic matter
(C) Rainfall (D) Soil water

283. The kind of soil water most useful to plant is:
(A) Hygroscopic water (B) Gravitational water
(C) Capillary water (D) None of these

284. The soil with poorest water holding capacity is:
(A) Clay (B) Loam
(C) Sandy (D) None of these

285. The organic matter of soil has the value:
(A) 25% (B) 50% (C) 40% (D) 10%

286. The size of clay particles is:
(A) Less than 0.002 mm (B) More than 0.002 mm
(C) 0.2 mm to 2 mm (D) None of these

287. Which of the following has smallest of soil particles:
(A) Loam soil (B) Sand
(C) Clay (D) Silt

288. Major constituent of soil is:
(A) Organic matter (B) Inorganic matter
(C) Soil air (D) Soil water

289. Formation of organic matter takes place by :
(A) Chemical weathering (B) Soil micro-organisms
(C) Herbivores (D) Soil water

290. Formation of soil takes place by:
(A) Weathering (B) Pedogenesis
(C) Melting (D) Both (A) and (B)

291. Soil erosion is more where:
(A) Rainfall is high (B) No rainfall
(C) Low rainfall (D) None of these

292. Solid matter from land washed annually into oceans is:
(A) 330 million tones (B) 3300 million tones
(C) 3.3 million tones (D) 33 million tones

293. Land mass occupies about:
(A) 14 million tones (B) 140 million tones
(C) 1400 million tones (D) 14000 million tones

294. Arable land constitutes:
(A) 36.6% (B) 30% (C) 22% (D) 11%

295. Soil forms the surface layer in percentage of land:
(A) 80% (B) 60% (C) 50% (D) 30%

296. Soil transported by air is:
(A) Alluvial (B) Colluvial
(C) Glacial (D) Eolian

297. River water deposits:
(A) Loamy soil (B) Alluvial soil
(C) Laterite soil (D) Sandy soil

298. Alluvial soils are found in:
(A) Northern India (B) Southern India
(C) Gangetic Jamuna plains (D) Eastern India

299. Ao layer of the soil is made of:
(A) Decaying litter
(B) Rocky matter
(C) Soil mixed with organic matter
(D) Soil rich in inorganic matter

300. Development of soil from rocky matter is:
(A) Pedogenesis (B) Pedology
(C) Edaphic factors (D) Edaphic climax

301. Soil particles arranged in a series of increasing size is:
(A) Silt → Sand → Clay
(B) Sand → Silt → Clay
(C) Clay → Sand → Silt
(D) Clay → Silt → Sand

302. Soil best suited for plant growth is:
(A) Clay (B) Loam
(C) Sandy (D) Gravel

303. Good soil is the one which:
(A) Allows water to percolate slowly
(B) Allows water to pass quickly
(C) Allows limited amount of water into it
(D) Holds whole of water entering it

304. Least porous soil is:
(A) Clay soil (B) Sandy soil
(C) Loam soil (D) Gravelly soil

305. Deep black soil is productive due to high proportion of:
(A) Sand and Zinc (B) Gravel and Calcium
(C) Clay and Humus (D) Silt and Earthworm

306. Humus is good for plant growth because:
(A) It improves physical condition of soil
(B) It makes the soil porous
(C) It increases water holding and aeration of soil
(D) All of these

307. Humus is important for plant growth as it is:
(A) Made of dead organic matter
(B) Derived from leaves
(C) Rich in nutrients and increases water holding capacity of soil
(D) Partially decomposed

308. Humus is example of:
(A) A fertilizer
(B) Component of soil structure
(C) Organic colloids
(D) Crystalloids

309. Soil fertility can be maintained without addition of nutrients due to:
(A) Floods
(B) Crop residue
(C) Activity of microorganisms
(D) Favourable temperature and water

310. Fertility of soil is measured by its ability to:
(A) Retain nutrients
(B) Hold organic materials
(C) Hold water
(D) Support life

311. A fertility agricultural soil appears deep coloured at the surface as compared to soil one metre down. The reason for colour of top soil is:
(A) More moisture
(B) Rich in organic matter
(C) Rich in Iron, calcium and magnesium
(D) Recent formation

312. Water holding capacity is highest in:
(A) Sandy soil (B) Silt soil
(C) Clay soil (D) Loam soil

313. Soil water available to root is:
(A) Surface water (B) Hygroscopic water
(C) Gravitational water (D) Capillary water

314. Water held tightly by soil particles as thin film over their surface is:
(A) Hygroscopic water (B) Capillary water
(C) Chemical water (D) Gravitational water

315. Which governs the solubility and availability of nutrients in the soil:
(A) Temperature (B) pH
(C) Aeration (D) Porosity

316. Water-logging commonly occurs in:
(A) Sandy soil (B) Gravelly soil
(C) Clayey soil (D) Loam soil

317. Soil conservation is:
(A) Conversion of sterile soil into fertile one
(B) Aeration of soil
(C) Erosion of soil
(D) Protecting against loss

318. Removal of top fertile soil by wind or water is:
(A) Siltation (B) Soil erosion
(C) Weathering of soil (D) Leaching

319. Soil erosion is prevented by:
(A) Afforestation
(B) Increased bird population
(C) overgrazing
(D) Removal of vegetation

320. Terracing helps in soil conservation in:
(A) Plains (B) Deserts
(C) Hilly areas (D) Wet areas

321. Sheet erosion is due to:
(A) Fast running rivers (B) Wind
(C) Heavy rains (D) Occasional rains

322. Maximum productivity (on global basis) is shown by:
(A) Forests (B) Aquatic ecosystem
(C) Deserts (D) Grasslands

323. Which one is biotic in origin:
(A) Texture and porosity of soil
(B) Rainfall
(C) Carbon dioxide content of soil
(D) Light intensity

324. Eroded soils are:
(A) Devoid of plant nutrients
(B) Richer in plant nutrients
(C) Unaltered in plant nutrients
(D) Fit for agriculture

325. Soil conservation is achieved through:
(A) Restricted human activities (B) Good plant cover
(C) Wind screen (D) Ploughing

326. Soil fertility is reduced by:
(A) Crop rotation
(B) Nitrogen fixing bacteria
(C) Decaying organic matter
(D) Intensive agriculture

327. Type of soil in Gangetic basin is:
(A) Eolian (B) Colluvial
(C) Residual (D) Alluvial

328. Soil particles determine its:
(A) Texture (B) Field capacity
(C) Water holding capacity (D) Soil flora

329. Soil showing permanently wilted plants has:
(A) Gravitational water (B) Capillary water
(C) Hygroscopic water (D) Ground water

330. Water is a resource:
(A) Nondegradable nonmaintainable
(B) Degradable maintainable
(C) Renewable
(D) Non renewable

331. Main cause of soil erosion is:
(A) Afforestation (B) Less rains
(C) Thinning of ozone layer (D) Deforestation

332. Largest amount of fresh water is found in:
(A) Lakes and streams (B) Underground
(C) Polar ice and glaciers (D) Rivers

333. Guano is the major source of:
(A) Nitrogen (B) Phosphorus
(C) Sulphur (D) Zinc

334. Sequence of humification and mineralization:
(A) Dead organic matter → Litter → Duff→ Humus →Minerals
(B) Humus → Minerals → Litter → Duff
(C) Mineral → Humus → Litter → Duff
(D) Dead organic matter → Duff→ Litter→ Minerals → Humus

335. Soil is composed of:
(A) Mineral + Water +Air
(B) Mineral+Organic matter + Water
(C) Mineral+Organic matter + Air + Water
(D) Organic matter + water

336. Common type of soil found in India is:
(A) Sandy
(B) Alluvial and black soils
(C) Red soil
(D) Both (B) and (C)

337. Plants die in water-logged soils due to:
(A) Nutrient leaching
(B) Stoppage of root respiration
(C) Dilution of soil nutrients
(D) Dilution of cell sap

338. In hilly areas, erosion is maintained by:
(A) Terracing (B) Manuring
(C) Ploughing (D) Mixed cropping

339. Edaphology is connected with:
(A) Plant and biosphere
(B) Soil and living microorganisms
(C) Animals and ecosystem
(D) Soil and biosphere

340. Top soil is darker and:
(A) Contains more Na and Mg
(B) Is drier than subsoil
(C) Contains more organic matter
(D) Is wetter than subsoil

341. Soil salinity is measured by:
(A) Porometer (B) Potometer
(C) Calorimeter (D) Conductivity meter

342. Wind velocity at 1 m height as compared to ground level is:
(A) Same (B) 2.5 times
(C) 10 times (D) 1.5 times

343. Soil covered by peaty fibrous layer is:
(A) Glei (B) Podsol
(C) Laterite (D) Chernozem

344. Reservoir, lakes, ponds, rivers and canals contain percentage of fresh water:
(A) 77.2% (B) 22.4%
(C) 0.36% (D) 30.3%

345. Red soils of India are:
(A) Laterite soils
(B) Deficient in lime, Mg, P and K
(C) Rich in iron and organic matter
(D) All of these

346. Macropores of soil take part in:
(A) Holding air
(B) Holding air and percolation of water
(C) Percolation of water
(D) Holding of capillary water

347. Soil strength for engineering purpose is determined by the characteristics of:
(A) Compressibility (B) Compatibility
(C) Bearing resistance (D) All of these

348. Type of water vapour movement occurs in soils is:
(A) Internal (B) External
(C) Both (A) and (B) (D) None of these

349. Capillary movement of water is complemented by:
(A) Stem elongation (B) Root extension
(C) Leaf orientation (D) None of these

350. When a plant is under soil moisture stress, the roots:
(A) Shrink in size (B) Expand in size
(C) Remain unaffected (D) Both (A) and (B)

351. The positive and high value of redox potential indicates:
(A) Reducing conditions
(B) Strong oxidizing conditions
(C) Equally reducing and oxidizing conditions
(D) None of these

352. The concept of soil as a natural body was first developed by:
(A) C.F. Marbut (B) K.D. Glinka
(C) V.V. Dakuchaev (D) E.W. Hillgard

353. In a waterlogged soil the concentration of which gas is high:
(A) Ethane (B) Methane
(C) Carbon dioxide (D) Carbon monoxide

354. A soil absorbs how much amount of incoming solar radiation:
(A) 5% (B) 10% (C) 15% (D) 20%

355. The global average of solar radiation reaching the earth is:
(A) 20% (B) 30% (C) 40% (D) 50%

356. Humus colloids are composed basically of:
(A) Carbon (B) Nitrogen
(C) Helium (D) Ozone

357. The most important silicate clays are known as:
(A) Photosilicates (B) Phyllosilicates
(C) Autosilicates (D) Simply silicates

358. Non-humic substances are less complex and are——— to microbial attack than those belonging to humic group:
(A) More resistant (B) Less resistant
(C) More susceptible (D) Less susceptible

359. Humus enhances the mineral breakdown and increases the availability of some:
(A) Anions (B) Cations
(C) Neutral ions (D) Nne of these

360. Which of the following is harmful to the higher plants:
(A) Phenol carboxylic acid (B) Vanillic acid
(C) Dihydroxystearic acid (D) All of these

361. Soil organic matter has many remarkable effects and one of them is given below, identify the same:
(A) Encouraged granulation
(B) Increased plasticity
(C) Enhanced cohesion
(D) Low cation exchange

362. The C:N ratio in the organic matter of the furrow slice (upper 15 cm) of arable soils commonly ranges from:
(A) 3:1 to 4:1 (B) 6:1 to 7:1
(C) 18:1 to 19:1 (D) 8:1 to 15:1

363. Which of the following rotations is likely to leave soil richer in organic matter:
(A) Continuous sorghum (B) Continuous maize
(C) Continuous oats (D) Maize-Oats-Clovers

364. Ammonium and potassium ions are apparently just the right size to fit into the cavities between are crystal units of:
(A) Kaolinite (B) Montmorillonite
(C) Vermiculite (D) None of these

365. Ammonia present in the soil can be lost in significant quantities from:
(A) Acid soils (B) Alkaline soils
(C) Both (A) and (B) (D) None of these

366. Etridiazol and nitrapyrin ——— oxidation of ammonia to nitrates:
(A) Accelerates (B) Maintains
(C) Inhibits (D) No effect

367. In soils where the negative charge is pH dependent, liming increases CEC which results in an increased potassium by the soil colloids and —— in solution:
(A) Increases (B) Decreases
(C) Little effect (D) None of these

368. The soils of which zones can supply adequate potassium for many years even under irrigations:
(A) Temperate (B) Humid
(C) Arid (D) Subtropical

369. The tropical soils high in hydrous oxides of iron and aluminium embody —— to torrential rains:
(A) Susceptibility (B) Resistance
(C) None of these (D) Both (A) and (B)

370. The prime objective of conservation tillage is:
(A) Removal of plant residues
(B) Complete retention of plant residues
(C) Partial retention of plant residues
(D) None of these

371. In no-tillage systems, the surface soil layers have:
(A) Higher bulk density
(B) Lower bulk density
(C) Higher total pore space
(D) Lower water infiltration

372. Conservation tillage tends to encourage:
(A) Higher microbial population
(B) Lower number of earthworms
(C) Reduces soil fauna
(D) None of these

373. Wind laden with soil particles embodied the abrasive action which is:
(A) Greatly increased (B) Greatly decreased
(C) Absolutely unaffected (D) None of these

374. The diameter of most erodible soil particles is:
(A) 0.2 mm (B) 0.4 mm
(C) 0.3 mm (D) 0.1 mm

375. Cost-sharing assistance for soil conservation has been in practice for many years in:
(A) India (B) England
(C) USA (D) China

376. Nitrogen fertilization in apples favours growth, but its excess:
(A) Hastens maturity (B) Delays maturity
(C) Improves colour (D) Improves quality

377. On the basis of productivity and fertility, the fertile and productive lakes have been classified as:
(A) Eutrotrophic (B) Oligotrophic
(C) Dystrophic (D) None of these

378. The idea of a map showing the use of land was mooted in 1919 by:
(A) Robert S. Plat (B) A.J. Hunt
(C) Charles K. Paul (D) Carl O. Sauer

379. Biogeochemical nutrient cycles have led to recognize that —— is very important:
(A) Balanced fertilization (B) Over fertilization
(C) Under fertilization (D) None of these

380. ————, nonessential elements in the ecosystem may accumulate at undesirable levels:
(A) Normally (B) Occasionally
(C) Mostly (D) Invariably

381. The soil without a well developed soil profile characteristics are known as azonal soil and include:
(A) Planosols (B) Latosols
(C) Lithosols (D) None of these

382. Geological distribution of clay minerals is of considerable importance, southern are predominantly found:
(A) Illite (B) Kaolinite
(C) Montmorillonite (D) None of these

383. Greasewood (vegetation) is found on desert soils where ——— problem prevails:
(A) Salinity (B) Acidity
(C) Alkalinity (D) Neutrality

384. The nodule in ——— plants tend to be large, pinkish, fleshy, singly or in clusters about the tap roots or the first formed lateral roots:
(A) Biennials (B) Perennials
(C) Annuals (D) All of these

385. The identification of haemoglobin in legume root nodules has been made by Kubo (1931) through its characteristic ——— band maxima of 575 mì and 540 mì and by comparing with haemoglobin of the horse:
(A) Reflection (B) Deflection
(C) Absorption (D) Adsorption

386. Nodule haemoglobin appeared to have the typical haemoglobin characteristic of taking up and giving off:
(A) Carbon dioxide (B) Hydrogen
(C) Nitrogen (gas) (D) Oxygen

387. The molecular nitrogen in a root nodule is reduced to ——— from which it changes rapidly to amino acids:
(A) Ammonium (B) Ammonia
(C) Nitrate (D) Nitrite

388. The humus myth was effectively deposed by ——— who presented a paper at a prominent scientific meeting:
(A) Justus Senebier (B) Jean Young
(C) Justus von Liebig (D) Justus Billright

389. Rothamsted Agricultural Experiment Station, England was founded by:
(A) J.B. Lawes
(B) J.H. Gilbert
(C) J.B. Lawes and J.H. Gilbert
(D) Lord Hastings

390. Some specialists have identified ——— factors that influence crop growth:
(A) 32 (B) 42 (C) 52 (D) 22

391. It has been suggested that the ——— of sunlight is generally most satisfactory for plant growth:
(A) Full spectrum
(B) Half spectrum
(C) Little more than full spectrum
(D) None of these

392. The maximum carbon dioxide concentration at normal atmospheric levels occur in relatively low light intensities of ——— foot candles for some plants:
(A) 2500-5000 (B) 6000-7500
(C) 1000-2000 (D) 2150-2450

393. Tomato plants grown on soils with low bulk densities of tended to be:
(A) High in sugar (B) Low in protein
(C) High in anthrocyanin (D) Showed no effect

394. ———in 1909 developed an equation that related growth to the supply of plant nutrients:
(A) V.J. Kilmer (B) W.J. Spillman
(C) J.W. Friedrich (D) E.A.Mitscherlich

395. The principle similar to that developed by E.A. Mitscherlich was also independently developed several years later by:
(A) F. Steenbjerg (B) W.J. Spillman
(C) M.E. Summer (D) J.G. Shannon

396. Based on dark metabolism, glucose consumption for protein production is about —— per cent when nitrogen is provided as nitrate rather than as ammonium:
(A) 20 (B) 30 (C) 40 (D) 50

397. The sequence in which amino acids are linked in proteins depends on:
(A) Climatic control (B) Agronomic control
(C) Genetic control (D) None of these

398. The tendency of the young upper leaves to remain as the lower leaves yellow or die is an indication of nitrogen ——— in the plant:
(A) Mobility (B) Immobility
(C) Stability (D) Degradability

399. Build up of soluble sugars and amino acids due to inadequate potassium, low molecular weight compounds encourage the activity of:
(A) Plant hormones (B) Plant Pathogens
(C) None of these (D) Both (A) and (B)

400. Blossem-end rot in tomato has been found to be caused by the deficiency of:
(A) Nitrogen (B) Phosphorus
(C) Calcium (D) Sulphur

401. Affinity of the enzyme RuDP carboxylase for carbon dioxide is enhanced by:
(A) Sulphur (B) Phosphorus
(C) Nitrogen (D) Magnesium

402. The sufficiency range of iron is normally between how much ppm:
(A) 10-40 ppm (B) 50-250 ppm
(C) 300-400 ppm (D) None of these

403. The presence of ectotrophic mycorrhia enhances the uptake of nutrients in general, but particularly of:
(A) Molybdenum (B) Calcium
(C) Magnesium (D) Phosphorus

404. The quantity of nutrients that can come in direct contact with the plant roots is the amount in a volume of soil ——— the volume of roots:
(A) Equal to (B) Less than
(C) More than (D) No relation with

405. ———ion effect represent the effect of absorbed ion on the release of another from he surface of a colloid:
(A) Supplementary (B) Complementary
(C) Coordinated (D) Integrated

406. The movement of nutrients, ions by mass flow is reduced at low temperatures because of:
(A) Low respiration (B) Low photosynthesis
(C) Low transpiration (D) None of these

406. Organic matter content under well drained conditions is higher under:
(A) Field crops (B) Fruit crops
(C) Grassland (D) Forest covers

407. The ——— oxidation of ammonia has been termed as nitrification:
(A) Physical (B) Chemical
(C) Biological (D) Physico-chemical

408. —— is considered to be most important of the soil organisms bringing about the conversion of NH_4^+ to NO_2:
(A) Nitrosomonas (B) Nitrobacter
(C) Azotobacter (D) Azospirillum

409. The rapid increase in denitrification at the elevated soil temperatures suggests that which microorganisms plays a major role in denitrification:
(A) Chemophilic (B) Photophilic
(C) Thermophilic (D) Hydrophilic

410. Application of urea, aqua-ammonia, anhydrous ammonia and diammonium phosphate form alkaline microsite at the point of application, and in rice —— encourage the taking up of nitrite:
(A) High NH_4^+ concentration (B) High pH
(C) Both (A) and (B) (D) None of these

411. ——of the NH_4^+ in a region of alkaline microsite helps to restore the conditions suitable for the normal conversion of NO_2^- to NO_3^-:
(A) Diffusion (B) Dilution
(C) Both (A) and (B) (D) None of these

412. The losses of ammonia usually become greater with:
(A) Increase in soil pH
(B) Decrease in soil pH
(C) Decrease in calcium
(D) Decrease in temperature

413. The highest percentage content of nitrogen among the commonly used nitrogenous fertilizers is embodied by:
(A) Urea (B) Urea sulphate
(C) Ammonium nitrate (D) Anhydrous ammonia

414. Permissible biuret level, in most of urea fertilizers is:
(A) 5 % (B) 4 % (C) 3 % (D) 2 %

415. In the fertilized soil with soil solution concentration of one ppm of phosphorus, mass flow contribution to phosphorus nutrition of crops could approach:
(A) 0 % (B) 40 % (C) 30 % (D) 20 %

416. The quantity of solid phase phosphorus that acts as reserve is often referred to as:
(A) Intensity factor (B) Quantity factor
(C) Quality factor (D) None of these

417. Phosphatase enzymes play a major role in —— of organic phosphates in soil:
(A) Mineralization (B) Immobilization
(C) Mass transportation (D) None of these

418. The mineralization of organic phosphorus has been studied in relation to — ratio:
(A) C/N/S (B) C/N/K
(C) C/N/P (D) C/N/Ca

419. The surfaces on which phosphate ions enter into sorption reaction are:
(A) Surfaces of constant charge
(B) Surfaces of variable charge
(C) Both (A) and (B)
(D) None of these

420. The mineral calcite develops a negative charge due to —— tendency for Ca^{++} ions than CO_3^{-}ions to go into solution:
(A) Lower (B) Greater
(C) Equal (D) None of these

421. Clays such as kaolinite will larger quantities of phosphorus with —— SiO_2/R_2O_3 ratio:
(A) Low (B) High
(C) Equal (D) None of these

422. Ammorphous aluminosilicate minerals like allophone have —— negative charge which is partly or entirely balanced by complex aluminium cations:
(A) Small (B) Zero
(C) Large (D) None of these

423. A minor portion of the phosphorus sorption capacity of soils originates from :
(A) Sodium bicarbonate (B) Sodium carbonate
(C) Calcium oxide (D) Calcium carbonate

424. The maximum phosphorus availability in most of the soils is in the pH range:
(A) 4.0-4.5 (B) 5.0-5.5
(C) 6.0-6.5 (D) 7.0-7.5

425. In a predominantly biological system, mineralization of phosphorus from organic matter or crop residues is dependent on soil —— activity:
(A) Chemical (B) Physical
(C) Physico-chemical (D) Biological

426. —— generally augments the utilization of phosphorus which is water soluble:
(A) Broadcast
(B) Broadcast and incorporation
(C) Band placement
(D) None of these

427. Rock phosphate is treated with —— to break the apatite bond and to render the contained phosphate more soluble:
(A) Acid alone (B) Alkali alone
(C) Heat alone (D) Either acid or heat

428. Various potassic fertilizers embody nearly an equal effectiveness in —— situations requiring its supplemental application:
(A) Climatic (B) Edaphic
(C) Cropping (D) Economic

429. The concentration of which ppm of SO_2^{-}ion in the soil or solution cultures have proved adequate for the growth of many plant species:
(A) 1-2 (B) 3-5
(C) 6-8 (D) 9-11

430. Proper —— nourishment for cabbage, potato and sugar beet etc., is essential since these are known to have it high requirement:
(A) Sulphur (B) Manganese
(C) Zinc (D) Calcium

431. The essentiality of copper for higher plants was first recognized in:
(A) 1911 (B) 1921
(C) 1931 (D) 1941

432. "Reclamation disease" due to copper deficiency occurs in plants grown on newly reclaimed acid:
(A) Histosols (B) Alfisols
(C) Aridosols (D) Ultisols

433. Among small grain crops which has an exceptional tolerance to low levels of soil copper supply:
(A) Barley (B) Wheat
(C) Rye (D) Oats

434. Repeated use of Bordeaux mixture, sewage sludge, pig and poultry manures and/or mine wastes may cause toxicity by:
(A) Iron (B) Aluminium
(C) Zinc (D) Copper

435. The accessibility and utilization of soil applied copper can be enhanced by:
(A) Reducing particle size
(B) Reducing number of granules
(C) Reducing droplets per unit soil valume
(D) None of these

436. Each unit increase in pH brings about a decrease in solubility of Fe^{++} of how many folds:
(A) 50 (B) 75 (C) 100 (D) 150

437. The total soluble iron concentration becomes sufficient to supply iron adequately to roots only at pH:
(A) 2 (B) 3 (C) 4 (D) 5

438. It has been observed that the iron chelated by:
(A) 25 % (B) 50 % (C) 75 % (D) 100 %

439. Low total iron content of soil is ——— the basis for plant deficiencies:
(A) Commonly (B) Often
(C) Seldom (D) Mostly

440. Iron deficiency usually occurs in soil in the pH range of:
(A) 5.4-6.5 (B) 7.4-8.5
(C) 9.0-9.5 (D) 9.6-10.1

441. There can actually be a build up of toxic concentrations of Fe^{++} in the soil solution of:
(A) Aridosol (B) Histosol
(C) Latosol (D) Vertisol

442. For each unit increase in pH, the concentration of Mn— decrease by how many folds:
(A) 50 (B) 100 (C) 150 (D) 200

443. Silicon is the ——— most abundant element in the earth's crust:
(A) Third (B) Fourth
(C) Sixth (D) Second

444. Properties of the silicon-deficient Hawaii soils include:
(A) High total aluminium content
(B) High base saturation
(C) High pH
(D) High total silicon

445. It is generally accepted that the Gramineae members contain how many times the silicon concentration normally found in legumes and other dicotyledonous plants:
(A) 5-10 (B) 10-20
(C) 20-30 (D) 30-40

446. Superphosphate containing ——— selenium may provide enough of it to the plants in selenium deficient areas:
(A) 10 ppm or more (B) 20 ppm or more
(C) 30 ppm or more (D) 40 ppm or more

447. In the room and pillar pattern of mining potassium ——— rooms are mined out and the pillars of ore are left for support:
(A) Hexagonal (B) Spherical
(C) Rectangular (D) Square

448. ——— is probably the most important growth limiting factor in acid soils especially those having pH below 5.0-5.5:
(A) Nitrogen (B) Calcium
(C) Potassium (D) Aluminium

449. At pH ——— H^+ ion toxicity is increased by the removal of ions:
(A) 4.5 or less (B) 5.5 or less
(C) 6.0 or less (D) 5.0 or less

450. If an acid soil is limed to the pH level ——— it enhances nitrification:
(A) 3.5-4.5 (B) 4.5-5.5
(C) 5.5-6.5 (D) 6.5-7.5

451. When the phosphorus is either band placed or broadcast, the relative efficiency is related to :
(A) Its soil status (B) Its applied quantum
(C) Both (A) and (B) (D) None of these

452. Crops during any season are generally able to recover only ——— of the added fertilizer phosphorus:
(A) Less than 25% (B) Less than 30%
(C) Less than 35% (D) None of these

453. Potassium salts are:
(A) More mobile than nitrates
(B) Less mobile than nitrates
(C) Less mobile than phosphorus
(D) As mobile as nitrates

454. Broadcast potassium, is ——— than banded:
(A) More efficient (B) As efficient as
(C) Less efficient (D) None of these

455. Winter application of phosphorus is followed in:
(A) Warm areas (B) Cool areas
(C) Both (A) and (B) (D) None of these

456. Potassium is utilized more efficiently if applied:
(A) Before/at planting (B) Side dressed
(C) Top dressed (D) None of these

457. Tillage of the soil ——— rate of disappearance of organic matter:
(A) Reduces (B) Increases
(C) Maintains (D) Has no effect

458. Deficiency of nitrogen in maize causes:
(A) Yellowing of leaves from tip to base in V shaped pattern
(B) Chlorotic spots at base of leaves near the margins
(C) Yellow streaking of leaf edges
(D) Whitish-yellow spots on leaves

459. If urdbean is being grown for the first time in a field, the seed should be inoculated with:
(A) *Azospirillum* (B) *Rhizobium*
(C) PSB (D) *Clostridium*

460. Who gave the criteria of essentiality of plant nutrients:
(A) Arnon and Stout (1939)
(B) Arnon and Stout (1941)
(C) Leibig (1949)
(D) None of these

461. When O.M. undergoes enzymatic oxidation, it librates:
(A) CO_2 (B) Water
(C) Energy (D) All of these

462. Humus is a:
(A) Amorphous (B) Colloidal
(C) Both (A) and (B) (D) None of these

463. What percent of humic group present in Humus?
(A) 60-80% (B) 50-60%
(C) 30-40% (D) 20-30%

464. Humic group contains:
(A) Polypherals (B) Polyquenones
(C) Both (A) and (B) (D) None of these

465. Which of the following is less resistant, light in colour, suitable in acid and alkali and less in molecular weight?
(A) Fulvic acid (B) Humic acid
(C) Humin acid (D) All of these

466. Humic acid is soluble in:
(A) Acid and alkali (B) Alkali
(C) Acid (D) None of these

467. Non humic group contains:
(A) Polysaccharide (B) Polyquenones
(C) Polyphenols (D) Both (B) and (C)

468. Bacteria responsible for nitrification process is:
(A) Nitrosomonas (B) Nitrobactor
(C) Rhizobium (D) Both (A) and (B)

469. Optimum temperature for nitrification process is:
(A) 15-20°C (B) 25-35°C
(C) 35-40°C (D) 40-45°C

470. Gaseous loss of N takes place in the form of:
(A) NO_3 (B) N_2
(C) N_2O (D) Both (B) and (C)

471. Cross inoculation group of clover is:
(A) *Rhizobium meliloti* (B) *Rhizobium trifoli*
(C) *Rhizobium lupini* (D) *Rhizobium phaseoli*

472. Cross inoculation group of alfalfa is:
(A) *Rhizobium meliloti* (B) *Rhizobium trifoli*
(C) *Rhizobium lupini* (D) *Rhizobium phaseoli*

473. Which is the slow release urea?
(A) Sulphur coated urea (B) Shellack coated urea
(C) Neem coated urea (D) All of these

474. Indian soils generally low in:
(A) Organic matter (B) Phosphorus
(C) Potassium (D) None of these

475. Under submerged soils, urea must be placed at:
(A) Oxidized zone (B) Reduced zone
(C) Root zone (D) All of these

476. Super granulated urea release slow N because of:
(A) Higher weight
(B) Big size
(C) More surface area in contact with soil
(D) Less surface area in contact with soil

477. Which phosphatic fertilizer contains phosphoric acid but not soluble in water or citric acid?
(A) SSP (B) Di calcium phosphate
(C) Rock phosphate (D) DAP

478. Application of MOP is not suitable for:
(A) Sugarcane (B) Sugarbeet
(C) Tobacco (D) All of these

479. The entry of inorganic materials into xylem occurs at:
(A) Leaves (B) Stem
(C) Root tips (D) None of these

480. Nitrate assimilation in plant is mainly carried out at:
(A) Toot (B) Stem
(C) Branches (D) Leaves

481. Conversion of organic form to inorganic form is known as:
(A) Mineralization (B) Immobilization
(C) Volatilization (D) None of these

482. Potassium is rich in:
(A) Albite (B) Biotite
(C) Silkinite (D) All the above

483. Alkali soil can be reclaimed by the application of:
(A) Gypsum (B) Basic slag
(C) Lime (D) Calcite

484. Acid soils can be reclaimed by:
(A) $CaCO_3$ (B) H_2SO_4
(C) $CaSO_4.2H_2O$ (D) HNO_3

485. Denitrification is more in:
(A) Waterlogged soils (B) Well drained soils
(C) Heavy soils (D) Light soils

486. Which of the following fertilizer has maximum plant nutrient?
(A) DAP (B) TSP
(C) MOP (D) Ammonium nitrate

487. The availability of phosphorus is optimum in the range of soil pH:
(A) <7 (B) 7-8
(C) 8-8.5 (D) 6-8

488. The species of rhizobium present in the root nodules of soybean:
(A) *Rhizobium japonicum* (B) *R. leguminosorum*
(C) *R. trifoli* (D) *R. phaseoli*

489. Blue green algae fix nitrogen mainly in:
(A) Rice field (B) Cereals field
(C) Sugarcane field (D) Legumes field

490. Organic carbon content in the organic matter is:
(A) 32 (B) 18 (C) 68 (D) 58

491. The acidity of H^+ in aqueous phase of soils is called as:
(A) Active acidity (B) Reserve acidity
(C) Permanent acidity (D) None of these

492. Soil is originated from word 'Solum'. The word 'Solum' is a:
(A) Latin word (B) Greek word
(C) Spanish word (D) English word

493. Edaphology refers the study of soil related to:
(A) Crop production (B) Origin of soil
(C) Classification of soil (D) None of these

494. Soil is a:
(A) Dynamic body (B) Static body
(C) Rock (D) None of these

495. Soil contains minerals approximately:
(A) 25 % (B) 45 % (C) 5 % (D) 10 %

496. Soil contains organic matter approximately:
(A) 2 % (B) 4 % (C) 6 % (D) 8 %

497. Soil contains porosity approximately:
(A) 25 % (B) 35 % (C) 50 % (D) 60 %

498. Soil contains water approximately:
(A) 75 % (B) 25 % (C) 35 % (D) 45 %

499. Sand particles contain lighter amount of:
(A) Quartz (B) Silica
(C) Iron (D) All of them

500. Fine particles of soil contain higher amount of:
(A) Quartz (B) Silica
(C) Iron (D) All of them

501. Different identified soil layers are known as:
(A) Horizon (B) Solum
(C) Profile (D) All of these

502. A horizon is known as:
(A) Organic (B) Elluvial
(C) Accumulation (D) All of these

503. O horizon is known as:
(A) Organic (B) Elluvial
(C) Accumulation (D) All of these

504. Illuvial horizon is a:
(A) O horizon (B) A horizon
(C) B horizon (D) None of these

505. The colour of different soil layer varies due to presence of:
(A) Minerals
(B) Organic matter
(C) Minerals and organic matter
(D) None of these

506. A and B horizons are collectively known as:
(A) Horizon (B) Solum
(C) Profile (D) All of these

507. Sial is a:
(A) Upper layer of earth's crust
(B) Lower of earth's crust
(C) Upper layer of atmosphere
(D) Lower layer of atmosphere

508. Sial is made of mainly by:
(A) Al silicate (B) Mg silicate
(C) Fe- silicate (D) All of them

509. Sima is situated just:
(A) Upper of Sial (B) Lower of Sial
(C) Middle of soil (D) None of these

510. Sima is made of:
(A) Al silicate (B) Mg silicate
(C) Fe- silicate (D) All of these

511. Oxy anions are generally:
(A) Lithophilic elements (B) Chaleophilic elements
(C) Biophilic elements (D) Siderophilic elements

512. N, O, He and Xe elements are grouped into:
(A) Lithophilic elements (B) Chaleophilic elements
(C) Biophilic elements (D) Atmosphilic elements

513. Chalophilic elements make covalent bonds with:
(A) Sulphide (B) Carbonate
(C) Oxide (D) Hydroxide

514. Igneous rocks are made of:
(A) Lava (B) Lime
(C) Marble (D) All of them

515. Silica percentage in acidic igneous rocks is:
(A) 55-60 % (B) 60-65 %
(C) 65-85 % (D) All of these

516. Silica percentage in basic rocks is:
(A) 55-60 % (B) 60-65 %
(C) 45-55 % (D) All of these

517. Granite is a:
(A) Acidic (B) Basic
(C) Saline (D) None of these

518. Basalt is a:
(A) Acidic (B) Basic
(C) Saline (D) None of these

519. Sedimetory rocks are:
(A) Layered rocks (B) Layer less rocks
(C) Metamorphic rocks (D) All of these

520. Gypsum is a example of:
(A) Igneous rock (B) Sedimentary rocks
(C) Metamorphic rocks (D) All of these

521. The list of rocks is given below:

Rocks	Metamorphic rocks
1. Sand stone	i. Quartzite
2. Shale	ii. Slate
3. Limestone	iii Marble
4. Granite	iv Gneiss

Which combinations is correct:
(A) 1. i 2. ii 3. iii 4. iv
(B) 1. i 2. iii 3. ii 4. iv
(C) 1. iv 2. ii 3. iii 4. i
(D) 1. i 2. iv 3. iii 4. ii

522. Regolith is known as:
(A) Elluvial + Illuvial layers
(B) Solum + Parent material
(C) O+A+B+C layers of soil profile
(D) All of them

523. Celluvial is the result of:
(A) Water (B) Air
(C) Gravity (D) None of these

524. Indo-Gangetic plains were developed by:
(A) Alluvial materials (B) Aeolian materials
(C) Colluvial materials (D) None of these

525. Sand dunes are the result of:
(A) Alluvial materials (B) Aeolian materials
(C) Colluvial materials (D) None of these

526. Peat and muck are the example of:
(A) Cumulose materials (B) Residual materials
(C) Transported materials (D) All of these

527. The plant roots are potent agent of weathering?
(A) Physical (B) Chemical
(C) Biological (D) None of these

528. Hematite is changed into limonite by the process of:
(A) Hydrolysis (B) Hydration
(C) Reduction (D) Carbonation

529. The colour of hematite is:
(A) Blue (B) Red
(C) Black (D) Yellow

530. Limonite is the source of:
(A) N (B) P
(C) Fe (D) Zn

531. Dolomite is a source of following elements?
(A) Ca (B) Mg
(C) Ca and Mg (D) K

532. Apatite is a source of following elements?
(A) Ca (B) Mg
(C) P (D) K

533. Which of the following mineral does not belong to primary mineral?
(A) Quartz (B) Orthoclase
(C) Biotite (D) Hematite

534. Size of particle of primary minerals is:
(A) = 2μ (B) < 2 μ
(C) = 2 μ (D) All of these

535. According to Mohr's scale, few minerals are arranged in order of its decreasing hardness. Which of the following is a correct order?
(A) Diamond > Corundum>Topaz>Talc
(B) Diamond > Topaz > Corundum>Talc
(C) Talc > Topaz > Corundum > Diamond
(D) None of these

536. Eluviation refers to:
(A) Percolation of minerals from upper layers to lower layers
(B) Deposition of percolated materials from upper to lower layers
(C) Erosion of soil
(D) None of these

537. Illuviation refers to:
(A) Percolation of minerals from upper layers to lower layers
(B) Deposition of percolated materials from upper to lower layers
(C) Erosion of soil
(D) None of these

538. Which of the following is the example of reduction process:
(A) $4FeO + O_2 \rightarrow 2Fe_2O_2$
(B) $2Fe_2O_3 + 2HOH \rightarrow 2Fe_2O_3.3H_2O$
(C) $2Fe_2O_3 \rightarrow O_2 + 4FeO$
(D) All of these

539. Humification helps in:
(A) Formation of Ao layer
(B) Decomposition of organic matter
(C) Synthesis of new organic materials
(D) All of these

540. Podzolization refers to:
(A) Decomposition of humus and sesquioxides in lower soil layers
(B) Leaching of humus and sesquioxides from upper soil layers
(C) This process occurs in humid temperate regions
(D) All of these

541. Laterization refers to:
(A) Fe and Al remain in upper layers
(B) Leaching of Si to lower layers
(C) This process occurs in humid tropical regions
(D) All of these

542. Classification refers to:
(A) This process in low rainfall areas
(B) It happens due to presence of higher amount of bases in parent materials
(C) $CaCO_3$ is accumulated in soil profile
(D) All of these

543. Solanization refers to:
(A) Accumulation of salts in soil profile
(B) High water table and poor drainage enhance this process
(C) The structure of such soils are very poor
(D) All of these

544. Which soil particle has highest surface areas?
(A) Sand (B) Silt
(C) Clay (D) All of these

545. Shape of clay particles is:
(A) Angular (B) Smooth
(C) Platy (D) All of these

546. Size of coarse sand particles is:
(A) 2 – 0.2 mm (B) 0.2 – 0.02 mm
(C) 0.02 – 0.002 mm (D) <0.002 mm

547. Size of fine sand particles is:
(A) 2 – 0.2 mm (B) 0.2 – 0.02 mm
(C) 0.02 – 0.002 mm (D) <0.002 mm

548. Size of silt particles is:
(A) 2 – 0.2 mm (B) 0.2 – 0.02 mm
(C) 0.02 – 0.002 mm (D) <0.002 mm

549. Size of clay particles is:
(A) 2 – 0.2 mm (B) 0.2 – 0.02 mm
(C) 0.02 – 0.002 mm (D) <0.002 mm

550. Whih is not a soil forming factor?
(A) Climate (B) Topography
(C) Lithosphere (D) All of these

551. Which is not a active soil forming factor?
(A) Climate (B) Topography
(C) Biosphere (D) All of these

552. The bulk density (g/cc) of clay soils is:
(A) 1.5-1.6 (B) 1.2-1.6
(C) 1.1-1.4 (D) 1.0-1.3

553. The bulk density (g/cc) of clay loam soils is:
(A) 1.5-1.6 (B) 1.2-1.6
(C) 1.1-1.4 (D) 1.0-1.3

554. The bulk density (g/cc) of loamy soils is:
(A) 1.5-1.6 (B) 1.2-1.6
(C) 1.1-1.4 (D) 1.0-1.4

555. The bulk density of organic matter is:
(A) 0.5 g/cc (B) 0.8 g/cc
(C) 1.0 g/cc (D) 1.2 g/cc

556. The particle density of organic matter is in the range of:
(A) 1.5-1.6 g/cc (B) 1.2-1.6 g/cc
(C) 1.7-1.9 g/cc (D) 1.0-1.4 g/cc

557. Clay particles are made of:
(A) Quartz (B) Feldspar
(C) Keolinite (D) All of these

558. Number of textural classes of soil as per American classification is:
(A) 4 (B) 8 (C) 10 (D) 15

559. Number of soil textural classes is:
(A) 3 (B) 4 (C) 6 (D) 8

560. Ca and Mg help in:
(A) Flocculation (B) Deflocculation
(C) Decomposition (D) All of these

561. Na helps in:
(A) Flocculation (B) Deflocculation
(C) Decomposition (D) All of these

562. The % pore space can be calculated by which of the under given formula:
(A) $\frac{\text{Bulk density}}{\text{Solid space}}$ (B) $\frac{\text{(100-Bulk density)}}{\text{Particle density}} \times 100$
(C) 110-Bulk density (D) All of these

563. The porosity of soil will decrease with increase of:
(A) Soil depth (B) Soil organic matter
(C) Soil texture (D) All of these

564. Which of the under given soil texture has highest porosity?
(A) Sandy soils (B) Loam soils
(C) Loamy clay soils (D) Clay loam soils

565. Which of the following soil structure has the highest porosity?
(A) Columnar (B) Compact
(C) Oblique (D) Granular

566. Munshel soil colour chart is used for measuring the:
(A) Soil colour (B) Soil water colour
(C) Light colour (D) None of these

567. Particle density is equal to:
(A) $\frac{\text{Mass of soil solids}}{\text{Volume of soil solids}}$ (B) $\frac{\text{Mass of soil solids}}{\text{Total soil volume}} \times 100$
(C) $\frac{\text{Mass of soil solids}}{\text{Volume of soil water}}$ (D) $\frac{\text{Mass of soil solids}}{\text{Volume of soil air}} \times 100$

568. Effective saturation is equal to:
(A) $\frac{\text{Volume of soil water}}{\text{Volume of soil air and water}}$ (B) $\frac{\text{Volume of soil water}}{\text{Volume of soil water}} \times 100$
(C) $\frac{\text{Volume of soil}}{\text{Volume of soil water}}$ (D) None of these

569. Void ratio is equal to:
(A) $\frac{\text{Volume of soil pores}}{\text{Volume of soil air}}$ (B) $\frac{\text{Volume of soil}}{\text{Volume of soil air}} \times 100$
(C) $\frac{\text{Volume of soil pores}}{\text{Volume of soil solids}}$ (D) All of these

570. Soil water potential (øw) can be expressed as:
(A) $\psi s = \psi g + \psi m + \psi o$ (B) $\psi s = \psi g - \psi m - \psi o$
(C) $\psi s = \psi g + \psi m - \psi o$ (D) $\psi s = \psi g + \psi m$

571. Loam soil has:
(A) Equal amount of sand and clay
(B) More sand and less clay
(C) Less sand and more clay
(D) None of these

572. The productivity level of different soils is given below on the basis of soil colour. Which of the following arrangement is correct?
(A) Black>Rust brown>Yellow>Grey
(B) Grey>Yellow>Rust brown>Black
(C) Black>Yellow>Rust brown>Grey
(D) All of these

573. Which of the soil has the highest infiltration rate?
(A) Loam soil (B) Clay soil
(C) Silt loam (D) Clay loam

574. Very slow infiltration rate will be called if the infiltration rate is:
(A) < 0.1 inch/hr (B) 0.1-0.5 inch/hr
(C) 0.5-1.0 inch/hr (D) > 1.0 inch/hr

575. The water is absorbed by soil due to the force of:
(A) Adhesion force (B) Cohesion force
(C) Suction force (D) None of these

576. The water particles are attracted to other water particle due to force of:
(A) Adhesion force (B) Cohesion force
(C) Suction force (D) All of these

577. pF is the unit of:
(A) Force (D) Distance
(C) Velocity (D) Weight

578. Hygroscopic coefficient is equal to:
(A) 1/3 bar (B) 5 bar
(C) 31 bar (D) None of these

579. Soil metrix is also known as:
(A) Soil solids (B) Soil water
(C) Soil air (D) Soil pores

580. The relationship between the soil water content on weight and volume basis can be expressed as:
(A) $Pv = \frac{Pw}{Bd}$ (B) Pv = Pw x Bd
(C) $Pv = \frac{Bd}{Pw}$ (D) Pw = Pv Bd

581. Which of the following gas is not found in reduced conditions?
(A) CH_4 (B) N_2
(C) H_2S (D) O_2

582. At which concentration of soil CO_2, the roots will die?
(A) 5 % (B) 10 % (C) 1 % (D) None of these

583. At which concentration of soil O_2, the root growth may be completely checked?
(A) > 15 % O_2 (B) 10 -15% O_2
(C) < 5 % O_2 (D) All of these

584. With increasing soil depth, concentration of CO_2 will be?
(A) Increased (B) Decreased
(C) Remained same (D) All of above

585. Which of the following is not correct statement?
(A) Soil has higher CO_2 concentration than atmospheric CO_2
(B) Soil has higher N_2 concentration than atmospheric N_2
(C) Soil has lower O_2 concentration than atmospheric O_2
(D) Soil has lower CO_2 concentration than atmospheric CO_2

586. 1 bar is equal to:
(A) pF of 3 (B) pF of 2
(C) pF of 4 (D) None of these

587. Osmotic potential is also known as:
(A) Solute potential (B) Pressure potential
(C) Metrix potential (D) None of these

588. Pascal is a unit of:
(A) Water potential (B) Work
(C) Distance (D) None of these

589. The curve indicating available water against soil water tension is called:
(A) Soil water curve
(B) Soil water tension curve
(C) Soil tension curve
(D) Water tension curve

590. Which of the following water is available to plant communities?
(A) Hygroscopic water (B) Gravitational water
(C) Capillary water (D) All of these

591. Moisture equivalent is equal to:
(A) 1/3 bar (B) 15 bar
(C) 0.3 - 1 bar (D) 1/3 – 31 bar

592. Which of the following is not responsible for soil water loss?
(A) Evaporation
(B) Transpiration
(C) Percolation
(D) Hydraulic conductivity

593. When all soil macrospores are water filled, the water movement in soil is called:
(A) Unsaturated flow (B) Saturated flow
(C) Run-off (D) None of these

594. How much heat is required for vaporization of 1 g water at 20°C?
(A) 500 calories (B) 540 calories
(C) 560 calories (D) None of these

595. Plant micronutrient is:
(A) Carbon (B) Boron
(C) Magnesium (D) Sulphur

596. Gypsum mainly found in:
(A) Bihar (B) Rajasthan
(C) Andhra Pradesh (D) Maharashtra

597. Pyrite mainly found in:
(A) Bihar (B) Rajasthan
(C) Andhra Pradesh (D) Maharashtra

598. Iron is an important component of:
(A) Siroheme (B) Ferrodoxin
(C) Cytochromoes (D) All of these

599. Which of the complex fertilizer?
(A) Ammonium sulphate nitrate
(B) Single super phosphate
(C) Ammonium phosphate
(D) Calcium ammonium nitrate

600. White colour of soil is due to the process of?
(A) Silica (B) Iron
(C) Manganese oxide (D) All of these

601. Which is not an organic matter in the following?
(A) Crude protein (B) Carbohydrate
(C) Vitamins (D) Hormones

602. At what stage are more nutrients essentially required in cereal crops?
(A) Growth stages of plants
(B) Formation of leaves
(C) Panicle initiation
(D) Maturity of crop

603. Saturated flow can be expressed by following law?
(A) Darcy's law (B) Newton's law
(C) William's law (D) None of these

604. Under Darcy law, water flow is inversely related to:
(A) Hydraulic conductivity (B) Hydraulic heat
(C) Flow length (D) Cross section area

605. Which of the following is not used for measuring soil water:
(A) Tensiometer (B) Zypsum block
(C) Pressure bomb (D) Neutron meter

606. Tensiometer can measure soil water up to soil moisture tension of:
(A) 0.0-0.85 bar (B) 0.0-0.50 bar
(C) -0.33-15.0 bar (D) All of these

607. Gypsum block can measure the soil water potential up to:
(A) 1-15 bar (B) 15-31 bar
(C) 31-1000 bar (D) All of these

608. Which of the following is the source of neutrons in neutron moisture meter?
(A) Radium (B) Thorium
(C) P^{32} (D) H^3

609. Which of the following can not prevent the soil water loss?
(A) Mulch (B) Hexadecanol
(C) Wind breakers (D) Deforestation

610. Which of the following gas has the highest concentration in atmosphere?
(A) CO_2 (B) O_2
(C) S_2 (D) N_2

611. Rhizobium bacteria is:
(A) Symbiotic bacteria (B) Nonsymbiotic bacteria
(C) Anaerobic bacteria (D) Autotropic bacteria

612. For calculating value of P, P_2O_5 is multiplied by:
(A) 2.29 (B) 0.437
(C) 1.20 (D) 3.00

613. For calculating value of P_2O_5, P is multiplied by:
(A) 2.29 (B) 0.437
(C) 1.20 (D) 3.00

614. If the soil moisture increases, the heat conductivity of soil will?
(A) Increase (B) Decrease
(C) Remain same (D) None of these

615. Which of the following can not diffuse across the semi permeable membrane?
(A) Crystalloid (B) Colloids
(C) Solid particles (D) None of these

616. The pF value at PWP is equal to:
(A) 4.2 (B) 5.2
(C) 3.2 (D) None of these

617. Which of the soil type has the lowest PWP?
(A) Sandy loam (B) Loam
(C) Clay loam (D) Clay

618. Water movement through soil columns is called:
(A) Permeability (B) Percolation
(C) Surface runoff (D) None of these

619. Water available in between hygroscopic coefficient and field capacity is known as:
(A) Capillary water (B) Gravitational water
(C) Hygroscopic water (D) None of these

620. Which of the soil type has the highest field capacity?
(A) Loam (B) Silt loam
(C) Clay loam (D) Clay

621. Sticky point refers to:
(A) Level of soil moisture at which water is completely absorbed by soil colloids
(B) Water status of soil
(C) Water level at which irrigation is applied
(D) None of these

622. Critical moisture point is equal to:
(A) Field capacity
(B) Permanent wilting point
(C) Sticky point
(D) None of these

623. Colloidal particles have:
(A) Positive charge (B) Negative charge
(C) Both (A) and (B) (D) None of these

624. Zeta potential refers to:
(A) Magnitude of electric charges of colloidal particles
(B) Size of colloidal particles
(C) Movement of colloidal particles
(D) None of these

625. Browning movement can be seen in:
(A) Colloidal particles (B) Crystalloid particles
(C) Urea (D) None of these

626. Colloid is a Greek word that is made of:
(A) Kolla + Eidos (B) Colla + Oid
(C) Co + Lloid (D) None of these

627. Which of the following is not a colloidal material?
(A) Starch (B) Zelatin
(C) Fog (D) Sugar

628. What is the diameter of colloidal particles:
(A) 0-4 mμ (B) 5-100 mμ
(C) 101-200 mì (D) None of these

629. Which of the following formula is used to calculate the soil moisture by volumetric method?
(A Pw = Pv x B.D. (B) Pv = Pw x B.D.
(C) Pv = Pv/B.D. (D) None of these

630. The uptake of which element is affected least under waterlogged conditions:
(A) N (B) P (C) K (D) Ca

631. The uptake of which element is affected most under waterlogged conditions:
(A) N (B) P (C) K (D) Ca

632. Which of the following harmful gas is produced under waterlogged conditions?
(A) Ethane (B) H_2S
(C) Methane (D) All of these

633. Which of the following disease incidence is increased under waterlogged conditions?
(A) Red rot of sugarcane (B) Boll shedding of cotton
(C) Wilt of indigo (D) All of these

634. Hexadecandole is a chemical used in soil particularly for:
(A) To increase evaporation
(B) To reduce evaporation
(C) To control runoff
(D) All of these

635. Higher salt concentration in soil solution increases osmotic pressure of solution but the available water will be:
(A) Increased (B) Decreased
(C) Remain same (D) None of these

636. What is the water tension of saturated flow of soil water?
(A) 1/3 bar (B) 15 bar
(C) 0 bar (D) None of these

637. Soil water flow is governed by:
(A) Conductivity of the medium
(B) Driving force of flow
(B) Proportionality factor
(D) All of these

638. Which of the following sail has the highest rate of saturated soil water flow?
(A) Sandy soil (B) Loamy soil
(C) Clay (D) None of these

639. Saturated soil water flow is inversely related to:
(A) Viscosity of liquid (B) Pressure difference
(C) Radius of tube (D) All of these

640. According to Darcy's law, flow velocity of any liquid is directly proportional to:
(A) Hydraulic conductivity
(B) Length of porous medium
(C) Viscosity of liquid
(D) All of these

641. Unsaturated water flow can be understood by:
(A) Darcy's law (B) Poiseuille law
(C) Capillary principle (D) All of these

642. Soil water is lost by:
(A) Evaporation (B) Transpiration
(C) Runoff (D) All of these

643. Which of the following elements become toxic to plants under waterlogged conditions?
(A) Fe and Mn (B) B and Mo
(C) N and P (D) N and K

644. Which of the following cation is not found under reduced conditions:
(A) NO_3 (B) Fe^{++} (C) Mn_{++} (D) NH_4^+

645. CO_2 concentration in the atmosphere is:
(A) 1 % (B) 2 % (C) 3 % (D) 0.03 %

646. Which of the flowing clay mineral absorbs less water?
(A) Kaolinite (B) Montmorillonite
(C) Vermiculite (D) None of these

647. Which of the following ion may cause more water absorption on soil colloids?
(A) Na^+ (B) Ca^{++}
(C) H^+ (D) All of these

648. Attraction force between two water molecules is known as:
(A) Cohesion (B) Adhesion
(C) Electromagnetic force (D) None of these

649. Attraction force between two different molecules is known as:
(A) Cohesion (B) Adhesion
(C) Electromagnetic force (D) None of these

650. Micelles or micro cells are:
(A) Micro clay colloids particles
(B) Micro cells
(C) Micro organisms
(D) None of these

651. The soil colloids of humid region have:
(A) Ca^{++} - H^+ complexes (B) Ca^{++} - Mg^{++} complexes
(C) Na^+ - Ca^{++} complexes (D) None of these

652. Tensiometer can measure soil water tension upto:
(A) - 0.85 bar (B) - 0.5 bar
(C) - 1.0 bar (D) None of these

653. Which of the following methods is used to measure soil moisture?
(A) Tensiometer
(B) Neutron moisture meter
(C) Zypsum block
(D) All of these

654. The source of fast neutrons in neutron moisture meter?
(A) H^{3+} (B) P^{32}
(C) Actinium + Baryllium (D) All of these

655. CO_2 concentration in soil is about:
(A) 0.03 % (B) 21.0 %
(C) 0.2-5.5 % (D) 78 %

656. Average O_2 concentration in soil is about:
(A) 0.03 % (B) 21.0 %
(C) 15.5 % (D) 78 %

657. Which of the following soil type has higher per cent of CO_2?
(A) Sandy soil (B) Loam soil
(C) Clay soil (D) None of these

658. With increasing soil depth, CO_2 concentration will be:
(A) Increased (B) Decreased
(C) No changed (D) Unknown

659. In comparison to fallow land, CO_2 concentration in cropped land will be:
(A) Less (B) More
(C) Same (D) Unknown

660. Which of the following is a kaolinite?
(A) $Al_4Si_4O_{10}(OH)_8$ (B) $Al_4Si_8O_{20}(OH)_4$
(C) $Al_2Si_4O_{10}(OH)_2K$ (D) None of these

661. Higher silica-sesquioxide ratio of soil indicates that soil has:
(A) Higher CEC (B) Higher WHC
(C) Higher soil fertility (D) All of these

662. Which of the following minerals is important for formation of silicate minerals?
(A) Feldspars (B) Mica
(C) Amphibols (D) All of these

663. Which of the following mineral has the highest anion exchange capacity?
(A) Kaolinite (B) Montmorillonite
(C) Illite (D) Chlorite

664. Which of the following mineral has the largest crystal size?
(A) Kaolinite (B) Montmorillonite
(C) Illite (D) Vermiculite

665. Which of the clay mineral has higher amount of K?
(A) Kaolinite (B) Montmorillonite
(C) Illite (D) Vermiculite

666. Which of the following mineral has the highest CEC?
(A) Kaolinite (B) Montmorillonite
(C) Illite (D) Vermiculite

667. Which of the following is a 2 : 2 type mineral?
(A) Chlorite (B) Illite
(C) Kaolinite (D) None of these

668. Which of the following is a 1 : 1 type clay mineral?
(A) Kaolinite (B) Halloysite
(C) Anauxite (D) All of these

669. Which of the following is a 2 : 1 type expanding mineral?
(A) Vermiculite (B) Montmorillonite
(C) Beidellite (D) All of these

670. Which of the following is a 2 : 1 type non expanding mineral?
(A) Vermiculite (B) Montmorillonite
(C) Illite (D) None of these

671. The soil colloids of tropical and subtropical regions have more:
(A) Ca-H complexes (B) Ca-Mg complexes
(C) Na-Ca complexes (D) None of these

672. The soil colloids of arid region have more:
(A) Ca-H complexes (B) Ca-Mg complexes
(C) Na-Ca complexes (D) None of these

673. The chemical composition of clay is:
(A) Fixed (B) Dynamic
(C) Loose (D) None of these

674. Which of the following element is found in tetrahedral layer of clay crystals?
(A) Se^{4+} (B) Al^{3+}
(C) Fe^{3+} (D) Zn^{2+}

675. Which of the following element is found in Octahedral layer of clay crystals?
(A) Se^{4+} (B) Al^{3+}
(C) Fe^{3+} (D) Zn^{2+}

676. Which of the following is a montmorillonite?
(A) $Al_4Si_4O_{10}(OH)_8$ (B) $Al_4Si_8O_{20}(OH)_4$
(C) $Al_2Si_4O_{10}(OH)_2K$ (D) None of these

677. List A has soil groups and B main clay minerals as below:

List A	List B
1. Alluvial soil	i. Illite
2. Black soil	ii. Montmorillonite
3. Red soil	iii. Kaolinite and Illite
4. Hilly soils	iv. Illite and chlorite

Which of the pair is correctly matched?
(A) 1. i. 2. ii. 3. iii. 4. iv.
(B) 1. i. 2. iv. 3. iii. 4. ii.
(C) 1. ii. 2. iii. 3. iv. 4. iii.
(D) 1. iv. 2. ii. 3. iii. 4. i.

678. List A has soil orders and B main clay minerals as below:

List A	List B
1. Mollisol	i. Montmorillonite & Illite
2. Vertisol	ii. Montmorillonite
3. Inceptisol	iii. Illite
4. Alfisol	iv. Kaolinite & Montmorillonite

Which of the pair is correctly matched?
(A) 1. i. 2. ii. 3. iii. 4. iv.
(B) 1. i. 2. iv. 3. iii. 4. ii.
(C) 1. ii. 2. iii. 3. iv. 4. iii.
(D) 1. iv. 2. ii. 3. iii. 4. i.

679. Base saturated soils are those which have fully saturated soil colloids with:
(A) Ca^{2+} (B) Mg^{2+} (C) H^+ (D) $Al(OH)^+_2$

680. Arid soils have higher soil pH because of:
(A) Higher cations on soil colloids
(B) Lower cations on soil colloids
(C) Higher anions on soil colloids
(D) All of these

681. If the base saturation of a soil is more the pH of soil will be:
(A) More (B) Low
(C) Normal (D) None of these

682. Which of the soil has higher CEC?
(A) Alluvial soil (B) Black soil
(C) Laterite soil (D) Red soil

683. In general, which of the following cation has higher concentration in soil?
(A) Ca^{2+} (B) Mg^{2+} (C) H^+ (D) Na^+

684. Which of the following anion has higher concentration in soil?
(A) SH^- (B) Cl^- (C) NO_3^- (D) $H_2PO_4^-$

685. Base saturation of soil can be calculated by which of the under given formula?

(A) $\frac{100 \times \text{total exchangeable cations}}{\text{CEC}}$

(B) $\frac{100 \times \text{total exchangeable anions}}{\text{CEC}}$

(C) $\frac{100 \times \text{CEC}}{\text{total exchangeable cation}}$

(D) All of these

686. Ion exchange is a:
(A) Static process
(B) Reversible process
(C) Only cations can be exchanged
(D) Only anions can be exchanged

687. Which of the soil has higher silica sesquioxide ratio?
(A) Black soil (B) Red soil
(C) Laterite soil (D) Sandy soil

688. Loss of NH_3 will be more at pH:
(A) <6.0 (B) <7.0
(C) <8.0 (D) None of these

689. Volatilization is a process in which:
(A) Loss of nitrogen in form of N_2 and N_2O
(B) Conversion of NH_4-N into NO_3-N
(C) Conversion of NO_3-N into NH_4-N
(D) None of these

690. In which form of nitrogen is lost at <6 soil pH:
(A) N_2 (B) N_2O
(C) NO (D) None of these

691. Denitifying bacterias are:
(A) Aerobic (B) Anaerobic
(C) Both of these (D) None of these

692. Denytrifying bacteria belongs to genera:
(A) Pseudomonas (B) Acromobactor
(C) Bacillus (D) All of these

693. Which of the following genera of bacteria is found mainly in soil?
(A) Pseudomonas and Achromobacter
(B) Thiobacillus
(C) Vibrio
(D) All of these

694. Denitrifying bacteria uses NO_3 as an electron acceptor in absence of:
(A) CO_2 (B) O_2 (C) H_2O (D) None of these

695. Optimum temperature for denitrification is:
(A) <25°C (B) >25°C
(C) 25-30 °C (D) >30°C

696. The name of rhizobia for lucerne group is:
(A) *Rhizobium meliloti* (B) *Rhizobium japonicum*
(C) *Rhizobium trifoli* (D) *Rhizobium phaseoli*

697. Loss of nitrogen under neutral soil pH occurs mainly in the form of:
(A) NO_2 (B) NO_3
(C) N_2O (D) NO

698. The red colour of legume nodules is due to the presence of:
(A) Haemoglobin (B) Leghaemoglobin
(C) RBC (D) None of these

699. Symbiotic N fixation governed by the group of bacteria is:
(A) Rhizobia (B) Azotobacter
(C) Azospirillum (D) All of these

700. Nitrogen fixation is a:
(A) Endothermic process (B) Exothermic process
(C) Hypothermic process (D) None of these

701. The main enzyme of non symbiotic N fixation is:
(A) Nitrogenase (B) Hydrogenase
(C) Isomerase (D) All of these

702. Azotobacter are active at soil pH of:
(A) >6.5-7.0 (B) 7.0-7.5
(C) 7.5-8.0 (D) 5.5-6.0

703. Average quantum of N fixed by non-symbiotic process is:
(A) 10-15 kg/ha (B) 25-30 kg/ha
(C) 50-60 kg/ha (D) None of these

704. The maximum decomposition shall be at soil WHC of:
(A) <30 % (B) 40-60 %
(C) 60-80 % (D) >80 %

705. The rate of decomposition is the lowest in:
(A) Sugars and proteins (B) Hemi cellulose
(C) Cellulose (D) Lignin

706. What is the end product of aerobic decomposition of sugars?
(A) CO_2 and H_2O (B) Pyruvic acid
(C) Glucose (D) None of these

707. What is the end product of anaerobic decomposition of sugars?
(A) Alcohols and acetones (B) CO_2 and H_2O
(C) Starch (D) None of these

708. Mineralization of nitrogen means:
(A) Conversion of organic N into inorganic N
(B) Conversion of inorganic N to organic N
(C) Conversion of inorganic N into N_2
(D) None of these

709. Conversion of proteins and organic N into amino acids is known as:
(A) Amonification (B) Amminization
(C) Nitrification (D) None of these

710. Conversion of amino acids into NH_3 and ammonium compounds are known as:
(A) Amonification (B) Amminization
(C) Nitrification (D) None of these

711. Ammonification rate will be higher at:
(A) 7-8 pH, 30-40 °C and 60-80 % WHC
(B) 4-5 pH, 30-40 °C and 60-80 % WHC
(C) 7-8 pH, 20-25 °C and >80 % WHC
(D) None of these

712. Nitrification is a:
(A) Aerobic process (B) Anaerobic process
(C) Both (A) and (B) (D) None of these

713. Which of the following statement is true?
(A) Nitrosomonas produces nitrite and nitrobacter produces nitrate
(B) Conversion rate of nirite to nitrate is faster than ammonium to nitrite
(C) Nitrosomonas oxidizes 35-40 mole NH_4, while nitrobacter 70-100 mole nitrite
(D) All of these

714. Nitrification rate is equal to:

(A) $\frac{\text{Quantity of } NO_3}{\text{Time}}$

(B) $\frac{\text{Quantity of } NO_2}{\text{Time}}$

(C) $\frac{\text{Quantity of } NO_2 \text{ \& } NO_3}{\text{Time}}$

(D) None of these

715. Optimum soil pH for nitrification is:
(A) 5.5-6.0 (B) 6.5-8.0
(C) 8.5-9.0 (D) >9.5

716. Which of the following is harmful for nitrification process?
(A) Hyposulphite (B) Bacteria
(C) Presence of Mo (D) None of these

717. Conversion of nitrate into nitrite and ammonia is called:
(A) Dinitrification (B) Assimilation
(C) Nitrification (D) None of these

718. Conversion of NO_3 into N_2 and N_2O is called:
(A) Denitrification (B) Nitrification
(C) Mineralization (D) None of these

719. What is the percentage of total exchangeable cations of CEC in normal fertile soils?
(A) 50-60 % (B) 60-70 %
(C) 70-80 % (D) 80-90 %

720. What is the percent of exchangeable H^+ in normal fertile soils?
(A) <10 % (B) <20 %
(C) <30 % (D) <40 %

721. Ca-clay has soil pH:
(A) 6.5-7.5 (B) 4.5-5.5
(C) 8.0-8.5 (D) None of these

722. Which of the clay mineral has highest anion exchange capacity?
(A) Kaolinite (B) Montmorillonite
(C) Vermiculite (D) All of these

723. In general, the range of soil pH is:
(A 5-7 (B) 2-5
(C) 4-10 (D) 7-12

724. Active acidity of soil is due to presence of:
(A) H^+ (B) Al^{3+} and H^+
(C) $Al(OH)^{+2}$ (D) All of these

725. Residual acidity of soil is due to presence of:
(A) H^+ (B) Al^{3+} and H^+
(C) $Al(OH)^{+2}$ (D) All of these

726. Which of the acidity is the highest in soils?
(A) Active (B) Exchangeable
(C) Residual (D) All of these

727. The availability of Ca and Mg to plants in humid region is reduced due to:
(A) Leaching of Ca and Mg
(B) Precipitation of Ca and Mg
(C) Increase in soil pH
(D) All of these

728. The availability of Ca and Mg to plants in arid region is reduced due to:
(A) Leaching of Ca and Mg
(B) Precipitation of Ca and Mg
(C) Increase in soil pH
(D) None of these

729. The highest availability of NO_3 –N to plants is at soil pH:
(A) 4.0-5.0 (B) 6.5-7.5
(C) 8.0-8.5 (D) None of these

730. The availability of P to plants is highest at soil pH:
(A) 4.0-5.0 (B) 6.0-7.0
(C) 7.5-8.5 (D) None of these

731. At which pH value, both ions H_2PO_4 and HPO_4^- are at equal concentration in soil solution?
(A) 5.0 (B) 6.0
(C) 7.0 (D) 8.0

732. S availability is highest at soil pH of:
(A) 5.0-6.0 (B) 7.0-8.0
(C) 6.0-8.0 (D) 8.0-9.0

733. Zinc availability is highest at soil pH of:
(A) 4.0 (B) 4.5
(C) 5.0 (D) 5.5

734. Exchangeable acidity of soil is due to presence of:
(A) H^+ (B) Al^{3+} and H^+
(C) $Al(OH)^{+2}$ (D) All of these

735. Which of the following has the highest capacity of N fixation?
(A) Rhizobia (B) Azotobacter
(C) Beijerinckia (D) Azospirillum

736. The rate of symbiotic N fixation is higher than non-symbiotic N fixation to the tune of:
(A) 2 times (B) 5/2 times
(C) 3 times (D) 4 times

737. Higher C:N ratio enhances the:
(A) N fixation (B) N reduction
(C) Denitrification (D) All of these

738. The optimum pH for rhizobium activity is:
(A) 4.5-5.5 (B) 5.5-6.0
(C) 5.5-7.0 (D) 7.5-8.5

739. P content in Phytin is:
(A) 16 % (B) 26 %
(C) 36 % (D) 46 %

740. Nucleoprotein contain P:
(A) 7-9 % (B) 10-16 %
(C) 1-5 % (D) 20-25 %

741. N contain P:
(A) 7-9 % (B) 13-14 %
(C) 16-21 % (D) 23-28 %

742. S reduction is mainly caused by:
(A) Desulphovibrio (B) Azotobacter
(C) Azospirillum (D) None of these

743. Mo is required for the activity of:
(A) Azotobacter (B) Legume bacteria
(C) Vibrio (D) None of these

744. B is required mainly by:
(A) Azotobacter (B) Azospirillum
(C) Legume bacteria (D) None of these

745. Humus is a:
(A) Part of organic matter
(B) Mixture of complex compounds
(C) Amorphous and colloidy
(D) All of these

746. Which of the following is correct for humus?
(A) It contains negative charge
(B) It is acidic
(C) Insoluble in water
(D) All of these

747. Which of the following pair is correct?
(A) Humus : C-55-60 % (B) Humus: N -3-6 %
(C) Humus: O_2 – 35-40 % (D) All of these

748. C : N ratio of humus is:
(A) 10 : 1 (B) 20 : 1
(C) 30 : 1 (D) 40 : 1

749. CEC of humus is:
(A) 150-300 C mol/kg (B) 50-100 C mol/kg
(C) 250-300 C mol/kg (D) All of these

750. Average C : N ratio of arable soil is:
(A) 10-12 : 1 (B) 20 : 1
(C) 80 : 1 (D) None of these

751. C : N ratio of arid soils in comparison to other arable soils is always:
(A) Higher (B) Lower
(C) Same (D) None of these

752. C : N ratio of hilly soils is:
(A) 9.2-13.7 : 1 (B) 6.4-11.4 : 1
(C) 20 : 1 (D) None of these

753. C : N ratio of legume fodder is:
(A) 10 : 1 (B) 25 : 1
(C) 80 : 1 (D) 225 : 1

754. C : N ratio of cereal straw is:
(A) 10 : 1 (B) 25 : 1
(C) 80 : 1 (D) 225 : 1

755. C : N ratio of saw dust is:
(A) 10 : 1 (B) 25 : 1
(C) 80 : 1 (D) 225 : 1

756. C : N ratio of FYM is:
(A) 20-30 : 1 (B) 10-12 : 1
(C) 80 : 1 (D) None of these

757. Acoding to modern classification, soils have been divided into:
(A) Order-Sub order-Great soil group-Sub group
(B) Order-soil group-family-series
(C) Order-family-series-types
(D) All of these

758. Soil orders have been divided into:
(A) Zonal – Interzonal - Azonal soils
(B) Zonal – Azonal – Series
(C) Zonal – Interzonal – Series – Azonal
(D) None of these

759. Which of the following soil type is not a interzonal soil?
(A) Hydromorphic soil (B) Halomorphic
(C) Calcimorphic soil (D) Alluvial soil

760. Calcimorphic soils have:
(A) Higher lime content (B) Higher Na content
(C) Low lime content (D) None of these

761. Azonal soils have three following great soil groups:
(A) Lithosol – Regosol – Alluvial
(B) Podzol – Planosol – Chernozems
(C) Latosol – Podzol – Lithosol
(D) None of these

762. Which is correct for Regosols?
(A) Soil profiles are not well developed
(B) Soils are made of marls, sand dunes and ash
(C) Parent materials affect the characters of soil
(D) All of these

763. Chernozems are like:
(A) Black cotton soils (B) Red soils
(C) Brown soils (D) Alluvial soils

764. Soil classification 7th approximation was developed in year:
(A) 1960 in America (B) 1950 in America
(C) 1960 in UK (D) 1960 in Russia

765. Epipedons are:
(A) Diagnostic surface horizons
(B) Diagnostic subsurface horizons
(C) Bed rocks
(D) None of these

766. Number of soil epipedons are:
(A) 6 (B) 10
(C) 16 (D) 21

767. As per new soil classification – 7th approximation, soil has been classified into following 6 categories
(A) Order – Suborder - Great soil group – Family - Series
(B) Order – Soil group – Family – Series – Type - Phase
(C) Order –Soil group – Family – Type – Series - Phase
(D) None of these

768. Number of soil orders are:
(A) 10 (B) 11
(C) 12 (D) 13

769. Which of the following soil order covers the highest area in the world?
(A) Vertisol (B) Entisol
(C) Aridisol (D) Alfisol

770. Usar was derived from 'Ustra word. Ustra is a word of:
(A) Latin (B) Greek
(C) Sanskrit (D) Bangla

771. Which of the following state has highest area under saline and alkaline soils:
(A) UP (B) Punjab
(C) Gujarat (D) Rajasthan

772. Finger and Toe disease in crucifer is caused in:
(A) Acidic soil (B) Neutral soil
(C) Alkaline soil (D) Saline soil

773. Total value of pH and POH is equal to:
(A) 7 (B) 10
(C) 14 (D) 21

774. pH word was first used by:
(A) Sorenson (B) Scofield
(C) Brady (D) Tisdale

775. The soil pH value of humid region is always:
(A) <7.0 (B) >7.0
(C) 7.0 (D) None of these

776. The fungi are highly active at soil pH of:
(A) Acidic (B) Neutral
(C) Alkaline (D) Saline

777. The highest activities of actinomycetes is found at soil pH of:
(A) Acidic (B) Neutral to slight alkaline
(C) Alkaline (D) All of these

778. Mo availability is higher under:
(A) Acidic condition (B) Neutral
(C) Alkaline (D) Saline

779. Fe availability is higher at soil pH of:
(A) 5.5 – 7.0 (B) 7.0 – 8.5
(C) 3.5 – 5.5 (D) None of these

780. B availability is highest at soil pH of:
(A) 3.5-5.5 (B) 5.5-8.0
(C) 8.0-8.5 (D) 9.0-9.5

781. Which of the following word is used for saline and alkaline soils?
(A) Rakar (B) Usar
(C) Kaller (D) All of these

782. Saline soil is also known as:
(A) White alkali soil (B) Black alkali soil
(C) Solonetz soil (D) All of these

783. Alkali soil is also known as:
(A) White alkali soil (B) Black alkali soil
(C) Usar soil (D) All of these

784. Which of the following is the characteristics of saline soils?
(A) EC>4 ds/m, ESP<15, pH<8.5
(B) EC>4 ds/m, ESP>15, pH<8.5
(C) EC<4 ds/m, ESP>15, pH>8.5
(D) None of these

785. Which of the following is the characteristics of alkali soils?
(A) EC>4 ds/m, ESP<15, pH<8.5
(B) EC>4 ds/m, ESP>15, pH<8.5
(C) EC>4 ds/m, ESP>15, pH>8.5
(D) None of these

786. Which of the following is the characteristics of saline and alkali soils?
(A) EC>4 ds/m, ESP<15, pH<8.5
(B) EC>4 ds/m, ESP>15, pH<8.5
(C) EC<4 ds/m, ESP<15, pH>8.5
(D) None of these

787. Which of the anion is not found in saline soils?
(A) CO_3^- (B) Cl^-
(C) SO_4^- (D) HCO_3^-

788. Saline and alkaline soils are mainly found in:
(A) Arid and semiarid region
(B) Humid region
(C) Temperate region
(D) All of these

789. Which of the following chemical is used for reclamation of saline and alkaline soils?
(A) Zypsum (B) Phosphozypsum
(C) El.Sulphur (D) All of these

790. Empirical formula of $1 - \frac{\text{Bulk density}}{\text{Particle density}}$ is:
(A) Bulk density (B) Percent pore space
(C) Pore space (D) Particle density

791. Saline and alkaline soils were classified by:
(A) D. Sigmoid (B) Arnon
(C) Wilcox (D) None of these

792. Which of the following process takes place in the development of saline and alkaline soils?
(A) Salanization (B) Alkalization
(C) Salodization (D) All of these

793. At which salt concentration in soil solution, the water absorption by plants is reduced?
(A) 0.5 % (B) 1.0 %
(C) 1.5 % (D) 2.0 %

794. The water absorption by plants is completely checked at salt concentration in soil solution:
(A) 10 % (B) 20 %
(C) 30 % (D) 40 %

795. Which of the following is more toxic to roots?
(A) Nacl (B) Na_2SO_4
(C) Na_2CO_3 (D) $NaHCO_3$

796. The critical limit of boron in soil is:
(A) 0.7 ppm (B) 1.0 ppm
(C) 1.5 ppm (D) 2.0 ppm

797. Which of the following method is used for pH determination?
(A) Hydrogen electrode method
(B) Glass electrode method
(C) Calorimetric method
(D) All of these

798. Which of the following type of soil has the largest area?
(A) Alluvial soil
(B) Black cotton soil
(C) Red soil
(D) Laterite and lateritic soil

799. Old alluvial plains are known as:
(A) Bangar (B) Khadar
(C) Regur (D) None of these

800. New alluvial plains are known as:
(A) Bangar (B) Khadar
(C) Regur (D) None of these

801. Alluvial soils are generally deficient in:
(A) N (B) Humus
(C) P (D) All of these

802. Black cotton soils are black due to presence of:
(A) Titaniferrous magnetite (B) $MgCO_3$
(C) $CaCO_3$ (D) All of these

803. Karail is a:
(A) Type of black soils
(B) Black soils of lower lands of both banks of Ganga in UP
(C) These soils have high lime and Mg
(D) All of these

804. Which of the following is correct for black cotton soils?
(A) Excess in lime and K_2O
(B) Deficient in N, P_2O_5 and organic matter
(C) Excess in Fe, $CaCO_3$ and $MgCO_3$
(D) All of these

805. Which of the following is correct for red soils?
(A) Made of gneiss, schist and granite rocks
(B) Deficient in N, humus, P and lime
(D) Excess of Fe and Al oxides
(D) All of these

806. Which of the following is correct for laterite soils?
(A) Lower layers are made of Al and Fe oxides
(B) Deficient in P, K and lime
(C) Generally rich in organic matter
(D) All of these

807. Rakar is a:
(A) Red soil (B) Black soil
(C) Alluvial soil (D) None of these

808. Which of the following is correct for Tarai soils?
(A) Deficient in Zn (B) High organic matter
(C) High soil fertility (D) All of these

809. Arid and desert soils are characterized by:
(A) High evaporation (B) Low organic matter
(C) High Ca CO_3 (D) All of these

810. Efflorescent deposits are found in:
(A) Saline and alkaline soils (B) Alluvial soils
(C) Black soils (D) None of these

811. Which of the following is correct for peat and marshy soils?
(A) Low pH (B) High organic matter
(C) Blue colour due to Fe^{++} (D) All of these

812. Dhakar soils are characterized as:
(A) Good for rice soils (B) Found in eastern UP
(C) Light acidic soils (D) All of these

813. Saline and alkali soils are characterized as:
(A) Excess of Na_2CO_3 and NaCl
(B) Absence of $CaSO_4$
(C) Hard rock at subsoil
(D) All of these

814. In general, Indian soils have been classified into:
(A) 4 types (B) 6 type
(C) 8 types (D) None of these

815. Which of the following order of Indian soil has highest area:
(A) Alfisols (B) Entisols
(C) Inceptisols (D) Vertisols

816. Who classified Indian soils into 25 great soil groups?
(A) Govindarajan (1965) and Dutta Vishwash (1968)
(B) FAO (1965)
(C) ICAR (1937)
(D) Brady (1972)

817. Black soils are classified in order:
(A) Vertisols (B) Alfisols
(C) Spodosol (D) Histosols

818. Red soils are classified in order:
(A) Vertisols (B) Alfisols
(C) Spodosol (D) Histosols

819. Which of the following epipedon is present in mollisols?
(A) Mollic (B) Umbric
(C) Ochric (D) Histic

820. Which of the following is correct for histosols?
(A) They are bog and half bog soils
(B) They are developed under waterlogged conditions
(C) They contain 20-30% organic carbon
(D) All of these

821. Soil monoliths are:
(A) Permanent soil profile records
(B) Temporary soil profile records
(C) Types of soil classification
(D) None of these

822. Soil is divided on the basis of land capability into:
(A) 4 classes (B) 6 classes
(C) 8 classes (D) 10 classes

823. Land capability classification is based on:
(A) Utility of land (B) Soil erosion
(C) Techniques to protect soil (D) All of these

824. Which of the class is unfit for agricultural uses:
(A) Class I (B) Class IV
(C) Class VIII (D) Class VI

825. List 'A' and 'B' represent the land capability classes and colour on soil maps, respectively below:

A	B
1. Class I	i. Light green
2. Class II	ii. Yellow
3. Class III	iii. Red
4. Class IV	iv. Blue

Which of the pair is matched correctly:
(A) 1. i 2. ii 3. iii 4. iv
(B) 1. iii 2. ii 3. iv 4. i
(C) 1. ii 2. i 3. iii 4. iv
(D) 1. iv 2. ii 3. i 4. iii

826. Na per cent in list A and irrigation water quality groups in B are given below:

A	B
1. <20 %. Na	i. Excellent
2. 20-40 % Na	ii. Good
3. 40-60 % Na	iii. Permissible
4. >60 % Na	iv. Unsuitable

Which of the pair is matched correctly:
(A) 1. i 2. ii 3. iii 4. iv
(B) 1. iii 2. ii 3. iv 4. i
(C) 1. ii 2. i 3. iii 4. iv
(D) 1. iv 2. ii 3. i 4. iii

827. List 'A' has SAR values and 'B' the quality of irrigation water below:

A	B
1. 0-10	i. Low Na
2. 10-18	ii. Medium Na
3. 18-26	iii. High Na
4. >26	iv. Very high Na

Which of the pair is matched correctly:
(A) 1. i 2. ii 3. iii 4. iv
(B) 1. iii 2. ii 3. iv 4. i
(C) 1. ii 2. i 3. iii 4. iv
(D) 1. iv 2. ii 3. i 4. iii

828. Which of the following is matched correct:

A	B
1. 0-250 micro mhos/cm	i. Low saline water
2. 250-500 micro mhos/cm	ii. Moderate saline water
3. 250-750 micro mhos/cm	iii. Moderate to high saline water
4. 750-2250 micro mhos/cm	iv. High saline water

Which of the pair is matched correctly:
(A) 1. i 2. ii 3. iii 4. iv
(B) 1. iii 2. ii 3. iv 4. i
(C) 1. ii 2. i 3. iii 4. iv
(D) 1. iv 2. ii 3. i 4. iii

829. Which of the biological agent is used for amelioration of saline and alkali soils?
(A) FYM (B) Molasses
(C) *Argemone mexicana* (D) All of these

830. Which of the following grass is tolerant to alkali soils?
(A) Nandi grass (B) Rhodes grass
(C) Dinanth grass (D) All of these

831. Irrigation water is classified on the basis of Na per cent by:
(A) Wilcox (B) Staff Richards
(C) Thorne and Peterson (D) Kanwar

832. Gypsum contains Ca and S:
(A) 23 % Ca and 18 % S (B) 12 % Ca and 12 % S
(C) 40 % Ca and 12 % S (D) 40 % Ca and 20 % S

833. Pyrite contains S:
(A) 22-24 % (B 12-18 %
(C) 28-30 % (D) 30-32 %

834. 1 ds/m EC is equal to:
(A) 640 ppm salt concentration
(B) 64 ppm salt concentration
(C) 6.4 gm/litre
(D) None of these

835. Sodium percentage is equal to:
(A) $\frac{Na^+ \times 100}{Ca^{++} + Mg^{++} + K^+ + Na^+}$ (B) $\frac{Na^+ \times 100}{Ca^{++} + Mg^{++}}$
(C) $\frac{\sqrt{Na}+ 100}{Ca^{++} + Mg^{++}}$ (D) None of these

836. SAR is equal to:
(A) $\frac{Na^+}{\frac{\sqrt{Ca^{++} + Mg^{++}}}{2}}$ (B) $\frac{Na^+}{\sqrt{Ca^{++} + Mg^{++}}}$
(C) $\frac{Na^+}{\sqrt{Ca^{++} + Mg^{++} + K^+ + Na^+}}$ (D) None of these

837. Lime requirement for acid soils was worked out by:
(A) Shoemaker etal (1961) (B) Kanwar (1972)
(C) Arnon (1961) (D) Biswas (1965

838. Over liming may cause:
(A) P availability reduced
(B) Fe, Mn, Cu and Zn availability reduced
(C) Adverse effect on soil particles
(D) All of these

839. Which of the following can be used for improvement of acid soils?
(A) NH_4^+ (B) NO_3^-
(C) Ca^{++} (D) Mg^{++}

840. Generally acid soils have pH range of:
(A) <4.0 (B) 4.0-6.8
(C) 6.5-7.5 (D) >7.5

841. Acid soils generally found in:
(A) Humid region (B) Arid region
(C) Temperate region (D) Tundra region

842. Which of the crop is sensitive to boron concentration in irrigation water?
(A) Grape (B) Apple
(C) Lemon (D) Sugarbeet

843. The safe limit of K^+ ions in irrigation water is:
(A) <5 me/l (B) 5-10 me/l
(C) 10-20 me/l (D) >20 me/l

844. The safe limit of NO_3^- ions in irrigation water is:
(A) 2.5-5.0 me/l (B) 5-10 me/l
(C) 10-15 me/l (D) >15 me/l

845. If residual sodium carbonate (RSC) is more than 2.50 me/l, the quality of irrigation water shall be:
(A) Safe (B) Marginal
(C) Unsafe (D) All of these

846. The good quality of irrigation water should have HCO_3^- quantity in water:
(A) Less than Ca^{++} and Mg^{++} quantity
(B) More than Ca^{++} and Mg^{++} quantity
(C) More than Ca^{++} quantity
(D) More than Mg^{++} quantity

847. Which of the following can be used for improvement of acid soils?
(A) Marl (B) Chalk
(C) Quick lime (D) All of these

848. Average P in soils is:
(A) 0.02 – 0.5 % (B) 0.5 – 1.5 %
(C) 1.6 – 2.2 % (D) 2.0-2.5 %

849. Plant contains N:
(A) 1-4 % of total dry weight
(B) 0.5-0.8% of total dry weight
(C) 5-7 % of total dry weight
(D) None of these

850. Excess N causes:
(A) Soft stem (B) High growth
(C) Late maturity (D) All of these

851. N deficiency causes:
(A) Chlorosis (B) Stunted growth
(C) Reduced tillering (D) All of these

852. Average N per cent in arable soils is:
(A) 0.02 – 0.10 % (B) 0.2 – 0.3 %
(C) 0.5 – 0.7 % (D) 1.0 - 1.5 %

853. Na is essential for:
(A) Green algae (B) Sugar beet
(C) Diatoms (D) Sugarcane

854. Va is essential for:
(A) Green algae (B) Sugar beet
(C) Rice (D) Wheat

855. Role of Mo is:
(A) Sugar synthesis
(B) NH_3 and NO_3 reduction
(C) N fixation by free living bacteria
(D) All of these

856. Role of Cu is:
(A) Synthesis of Indole acetic acid
(B) Control respiration
(C) Synthesis of chlorophyll
(D) All of these

857. Role of Zn is:
(A) Control disintegration of RNA
(B) Water absorption
(C) Sugar translocation
(D) All of these

858. Role of B is:
(A) Translocation of sugars (B) Pollen formation
(C) N absorption (D) All of these

859. Which of the following element was included recently in the list of essential plant nutrients?
(A) Co (B) Na
(C) Ni (D) Se

860. Plant requires micronutrients:
(A) 1 ppm (B) < 1ppm
(C) > 1 ppm (D) None of these

861. Auxin synthesis is catalysed by element:
(A) N (B) P
(C) Mo (D) S

862. Which of the following element helps in fat and lipid metabolism?
(A) N (B) P
(C) K (D) Zn

863. Drought resistance is increased by element:
(A) N (B) P
(C) K (D) Mo

864. Which of the following element is required for oil synthesis?
(A) N (B) P
(C) K (D) S

865. Alkali soils are deficient in:
(A) Ca (B) N
(C) N and Ca (D) None of these

866. Soil fertility refers to:
(A) Capacity of soil to provide all essential elements in balanced amounts to plants
(B) Availability of N, P and K elements
(C) High crop productivity
(D) None of these

867. Soil productivity refers to:
(A) Capacity of soil to produce high economic yield
(B) Capacity of soil to provide essential plant nutrients to plant
(C) Capacity of soil to produce high soil organisms
(D) All of these

868. Lag phase of growth indicates:
(A) Slow growth rate
(B) It is observed at early plant growth stage
(C) Decreasing growth rate with time
(D) All of these

869. Lag plant growth phase refers to:
(A) Maximum growth rate
(B) Increasing growth rate with time
(C) Maximum dry matter accumulation
(D) All of these

870. Who established the relationship between crop yield and growth factor?
(A) Wilcox (B) Mitscherlich
(C) Brown (D) None of these

871. The Baule unit refers to:
(A) Unit of any growth factor that provides 50% yield of expected yield
(B) It was proposed by German Scientist, Baule
(C) It varies with inputs
(B) All of these

872. Presently total essential plant nutrients are:
(A) 15 (B) 16
(C) 17 (D) 18

873. Plant nutrient contributes the total plant dry matter approximately:
(A) 2-5 % (B) 5-10 %
(C) 12-16 % (D) 17-19 %

874. Inverse nitrogen yield concept was given by:
(A) O.W. Wilcox (B) Mitscherlish
(C) Spillman (D) Tisdale

875. The optimum temperature range for most of the crops in:
(A) 10-25 °C (B) 15-40 °C
(C) 35-45 °C (D) > 45 °C

876. Mitscherlich's equation is:
(A) dy/dx=(A-Y)C (B) Y = K/n
(C) Y=CX – A (D) None of these

877. The baule units in lbs/acre for N, P_2O_5 and K_2O, respectively are:
(A) 218, 45, 70 (B) 223, 45, 76
(C) 255, 50, 75 (D) 280, 55, 80

878. Which of the following crop is not tolerant to acid soils?
(A) Berseem (B) Potato
(C) Tea (D) Rice

879. Lime requirement of sandy soil in comparison to clay soil at a specific soil pH will be:
(A) More (B) Less
(C) Equal (D) None of these

880. Among soil organisms, earthworms account nearly:
(A) 50-75 % weight of total soil organisms
(B) 75-85 % weight of total soil micro organisms
(C) > 85 % weight of total soil micro organisms
(D) None of these

881. How much dry soil is produced by earthworms annually per acre?
(A) 10 tonnes (B) 15 tonnes
(C) 20 tonnes (D) None of these

882. What will be the effect of earthworms on C/N ratio of soil?
(A) Increase (B) Decrease
(C) Same (D) All of these

883. Earthworms increase the highest availability of:
(A) K (B) Mg
(C) N (D) P

884. Number of earthworms may be present per m² area of arable land is:
(A) 30-300 (B) 500-800
(C) 1000-12000 (D) None of these

885. At which soil pH, the activity of earthworms is highest?
(A) <4.5 (B) 4.5-5.5
(C) 5.5-6.6 (D) 7.0-8.5

886. Which of the following reduces the population of earthworms:
(A) Drought (B) Frost
(C) Cold (D) All of these

887. Pasture lands and non-ploughed soils have earthworms population in comparison to ploughed soils:
(A) 6.5-11 % higher (B) 12-15 % higher
(C) 16-18 % higher (D) > 20 % higher

888. Which of the following is not a soil microorganisms?
(A) Nematodes (B) Protozoans
(C) Rotifers (D) Sludge and snails

889. Rotifers are:
(A) Parasite (B) Saprophytes
(C) Predators (D) None of these

890. Higher population of rotifers are available in:
(A) Dry region (B) Temperate region
(C) Humid region (D) None of these

891. Mycorrhiza is a
(A) Association of fungi with roots of higher plants
(B) Association of fungi and bacteria
(C) Association of bacteria and higher plants
(D) None of these

892. Maximum decomposition shall occur at temperature:
(A) 20-25°C (B) 25-30°C
(C) 30-40° C (D) > 40° C

893. If the C/N ratio is high in organic matter, the rate of decomposition will be:
(A) Increased (B) Reduced
(C) Unchanged (D) None of these

894. Which of the following is found highest in plant body?
(A) Sugar and Starch (B) Hemicellulose
(C) Cellulose (D) Lignin

895. Peat soils may contains organic matter:
(A) 20-95% (B) 0.1-5 %
(C) 5-10 % (D) >10 %

896. Mineral soils may contains organic matter:
(A) 1-5 % (B) 5-10 %
(C) 10-15 % (D) >20 %

897. Ammonia is converted into nitrite with the help of:
(A) Nitrosomonas (B) Nitrobacter
(C) Desulfovibrio (D) Bacillus

898. Which of the following bacteria helps in conversion of nitrite to nitrate:
(A) Nitrosomonas (B) Nitrobacter
(C) Desulfovibrio (D) Bacillus

899. Optimum soil pH for nitrogen fixing bacteria is:
(A) 4.5-5.5 (B) 5.5-6.5
(C) 6.5-8.0 (D) >8.0

900. Mycorrhiza is generally found in:
(A) N and P deficient soils (B) P and K deficient soils
(C) Ca deficient soils (D) All of these

901. Highest activity of actinomycetes is found at soil pH:
(A) 4.5-5.5 (B) 5.5-6.5
(C) 6.5-8.0 (D) >8.0

902. Which of the following is not a fungi?
(A) Nocardia (B) Mucor
(C) Rhizopus (D) Yeast

903. Which of the following is not a actinomycetes?
(A) Nocardia (B) Mucor
(C) Micromonospora (D) Streptomycetes

904. Potato scab disease is caused by:
(A) Bacteria (B) Fungi
(C) Actinomycetes (D) Algae

905. Sweet potato pox is caused by:
(A) Bacteria (B) Fungi
(C) Actinomycetes (D) Algae

906. Which of the following is present highest in soil?
(A) Bacteria (B) Fungi
(C) Actinomycetes (D) Algae

907. Which of the following type of bacteria is the largest in size?
(A) Bacillus (B) Coccus
(C) Spirillum (D) None of these

908. Average P in plants is:
(A) 0.05-1.0 % (B) 1 – 3 %
(C) 3 – 5 % (D) 4 – 5 %

909. Highest P is found in:
(A) Leaf (B) Stem
(C) Seed (D) Root

910. Organic P is available in the form of:
(A) Phytin (B) Nucleoproteins
(C) Phospholipids (D) None of these

911. Which of the following P ion is available in acidic conditions?
(A) $H_2PO_4^-$ (B) HPO_4^-
(C) PO_4^- (D) PO_3^-

912. Which of the following P ion is absorbed by plants?
(A) $H_2PO_4^-$ (B) HPO_4^-
(C) PO_4^- (D) PO_3^-

913. The lower leaves of maize plant develop violet green colour due to deficiency of:
(A) N (B) P
(C) K (D) Zn

914. K % in plant is:
(A) Nearly 2% of total dry weight
(B) <1 % of total dry weight
(C) >2% of total dry weight
(D) None of these

915. Highest K content is found in:
(A) Leaf (D) Stem
(C) Seed (D) Root

916. Yellow white mottling of cotton is caused by:
(A) K deficiency (D) N deficiency
(C) S deficiency (D) Zn deficiency

917. Necrosis on lower leaf margin is caused by:
(A) K deficiency (B) N deficiency
(C) S deficiency (D) Zn deficiency

918. Tea yellow is caused by:
(A) K deficiency (B) N deficiency
(C) S deficiency (D) P deficiency

919. Interveinal chlorosis on lower leaves is caused by:
(A) Mg (B) Ca
(C) P (D) Cl

920. Average Mg content in plant is:
(A) 0.5-1.5 % (B) 1.7-1.9 %
(C) 2.0-2.2 % (D) 2.5-2.7 %

921. Average S content in plant is:
(A) 0.01-0.1 % (B) 1.7-1.9 %
(C) 2.0-2.2 % (D) 2.5-2.7 %

922. Papery white is an important deficiency symptoms of:
(A) Fe (B) N
(C) Cu (D) Mo

923. Marsh spots in pea is caused by the deficiency of:
(A) N (B) Mg
(C) Mn (D) Cu

924. If the EC of soil is >4 ds m^{-1}, the soil is rated as:
(A) Normal (B) Injurious
(C) Sensitive (D) None of these

925. If the soil nutrient index is <1.67, the soil is rated as:
(A) Low fertile (B) Medium fertile
(C) High fertile (D) None of these

926. If the available P in soil test value is >24.6 kg P/ha, the soil is rated as:
(A) Low in available P (B) Medium in available P
(C) High in available P (D) None of these

927. If the available P in soil test value is 10.0-24.6 kg P/ha, the soil is rated as:
(A) Low in available P (B) Medium in available P
(C) High in available P (D) None of these

928. If the available P in soil test value is <10.0 kg P/ha, the soil is rated as:
(A) Low in available P (B) Medium in available P
(C) High in available P (D) None of these

929. If the available N in soil test value is >280 kg N/ha, the soil is rated as:
(A) Low in available N
(B) Medium in available N
(C) High in available N
(D) None of these

930. If the available N in soil test value is 280-560 kg N/ha, the soil is rated as:
(A) Low in available N
(B) Medium in available N
(C) High in available N
(D) None of these

931. If the available N in soil test value is >560 kg N/ha, the soil is rated as:
(A) Low in available N
(B) Medium in available N
(C) High in available N
(D) None of these

932. Metabolites play a significant role in ion uptake and process is known as:
(A) Passive uptake (B) Active uptake
(C) Mass flow (D) All of these

933. Which of the following cakes is not edible?
(A) Castor cake (B) Mustard cake
(C) Til cake (D) Groundnut cake

934. Ion uptake caused by electro-chemical potential is known as:
(A) Passive uptake (B) Active uptake
(C) Mass flow (D) All of these

935. Which of the following is not responsible for transport of nutrient ions from soil solution to root surface?
(A) Root inception (B) Mass flow
(C) Diffusion (D) Soil solution

936. Which of the following is a B rich mineral in soil?
(A) Tourmaline (B) Tenorite
(C) Powellite (D) None of these

937. Which of the following is a chelating material?
(A) EDTA (B) DPTA
(C) NTA (D) All of these

938. Which of the following is correct for chelating agent?
(A) It is a organic material
(B) It combines or attracts the cations
(C) It increases the availability of chelated nutrients
(D) All of these

939. Khaira disease of rice is caused by:
(A) K deficiency (B) N deficiency
(C) Zn deficiency (D) P deficiency

940. Little leaf or mottle leaf is caused by:
(A) K deficiency (B) Zn deficiency
(C) S deficiency (D) P deficiency

941. White bud of maize is the deficiency symptoms of:
(A) Zn deficiency (B) N deficiency
(C) S deficiency (D) Mg deficiency

942. Yellow spot of citrus is caused by:
(A) K deficiency (B) N deficiency
(C) Mo deficiency (D) P deficiency

943. Which of the following diseases are caused by B deficiency:
(A) Heart rot of sugar beet (B) Yellow tip of alfalfa
(C) Brown heart of turnip (D) All of these

944. Average values of N, P and K in tank silt is:
(A) 3, 0.3 and 0.3 % (B) 0.1, 0.2 and 0.3 %
(C) 1.2, 0.5 and 0.5 % (D) None of these

945. Which of the following is a concentrated organic manure?
(A) Dried bone meal (B) Oilseed cakes
(C) Guano (D) All of these

946. Optimum N : P : K ratio for cereal crop production is:
(A) 4 : 2 : 1 (B) 2 : 1 : 1
(C) 2 : 1 : 2 (D) 2 : 2 : 2

947. Optimum N : P : K ratio for pulses is:
(A) 4 : 2 : 1 (B) 2 : 1 : 1
(C) 2 : 1 : 2 (D) 2 : 2 : 2

948. Optimum N : P : K ratio for root crops is:
(A) 4 : 2 : 1 (B) 2 : 1 : 1
(C) 2 : 1 : 2 (D) 2 : 2 : 2

949. Muriate of potash is known as:
(A) KCl (B) K_2SO_4
(C) K_2CO_3 (D) None of these

950. S content in single super phosphate is:
(A) 12-16 % (B) 8-12 %
(C) 16-20 % (D) 11-12 %

951. Ca content in single super phosphate is:
(A) 18-21 % (B) 15-18 %
(C) 12-15 % (D) 11-12 %

952. Which of the following formula is correct?
(A) $P_2O_5 = P \times 2.29$ (B) $P_2O_5 = P \times 2.09$
(C) $P_2O_5 = P \times 1.29$ (D) $P_2O_5 = P \times 0.43$

953. Rhanania phosphate contains P_2O_5:
(A) 24-25 % (B) 16-17 %
(C) 32-34 % (D) 34-36 %

954. Pelo phos phosphate contains P_2O_5:
(A) 16-17 % (B) 16-17 %
(C) 21-23 % (D) 24-26 %

955. P_2O_5 in dicalcium phosphate is:
(A) 14-18 % (B) 21-25 %
(C) 34-39 % (D) 41-42 %

956. P_2O_5 in basic slage is:
(A) 14-18 % (B) 21-25 %
(C) 34-39 % (D) 41-42 %

957. P_2O_5 in triple super phosphate is:
(A) 16 % (B) 32 %
(C) 18 % (D) 46 %

958. P_2O_5 in double super phosphate is:
(A) 16 % (B) 32 %
(C) 18 % (D) 46 %

959. P_2O_5 in single super phosphate is:
(A) 16 % (B) 32 %
(C) 18 % (D) 46 %

960. Ammonium Nitrate contains N:
(A) 21 % (B) 33 %
(C) 40 % (D) 46 %

961. Liquid ammonia contains N:
(A) 21 % (B) 46 %
(C) 58 % (D) 80%

962. Calcium cyanamide contains N:
(A) 21 % (B) 32 %
(C) 38 % (D) 46 %

963. Sulpho nitrate is known as:
(A) Ammonium sulphate
(B) Calcium Ammonium Nitrate
(C) Ammonium Nitrate
(D) Ammonium Sulphate Nitrate

964. Which of the following nutrient should be applied by placement method?
(A) N (B) P
(C) K (D) P and K

965. Incomplete manure provide:
(A) Only one nutrient
(B) Only one or two nutrients
(C) Most of the plant nutrients
(D) None of these

966. Complete manure provide:
(A) Only one nutrient
(B) Only one or two nutrients
(C) Most of the plant nutrients
(D) None of these

967. Sodium nitrate is a:
(A) Alkaline fertilizer (B) Acidic fertilizer
(C) Neutral fertilizer (D) None of these

968. Ammonium sulphate is a:
(A) Alkaline fertilizer (B) Acidic fertilizer
(C) Neutral fertilizer (D) None of these

969. Which of the following element is not found in stimulative and prophylactic manures?
(A) N (B) B
(C) Mn (D) Mo

970. In foliar spray, the solution should not contain the biurate concentration?
(A) < 1 % (B) > 1 %
(C) 3 % (D) 4 %

971. Sona is a trade name of:
(A) Urea
(B) Calcium Ammonium Nitrate
(C) Ammonium Chloride
(D) Sulphate of Potash

972. Flame photometer is used for the estimation of:
(A) Available K (B) Available P
(C) Available N (D) Available Ca

973. Walkley and Black method is used for estimation of:
(A) N (B) P
(C) K (D) S

974. A value technique is used for estimation of:
(A) Available K in soil (B) Available P in soil
(C) Available N in soil (D) Available Ca in soil

975. Cunninghmella plague method is used for testing the soil value of:
(A) N (B) P
(C) K (D) S

976. Which of the following nutrient is mobile in soil?
(A) N (B) P
(C) K (D) S

977. List A and B indicate the nutrient and indictor plants that shows deficiency of respective nutrient respectively below:

List A	List B
1. N	i. Cauliflower
2. P	ii. Toria
3. K	iii. Potato
4. B	iv. Sunflower

Which of the pair is correctly matched
(A) 1. i 2. ii 3. iii 4. iv
(B) 1. iii 2. ii 3. iv 4. i
(C) 1. ii 2. i 3. iii 4. iv
(D) 1. iv 2. ii 3. i 4. iii

978. Phosphatic fertilizers and its P_2O_5 content are given below:

Phosphatic fertilizers	P_2O_5 content (%)
1. Steamed bonemeal	i. 22
2. Rock phosphate	ii. 20-40
3. Udaipur rock phosphate	iii. 30-32
4. Massori rock phospahte	iv. 10-20

Which of the pair is correctly matched
(A) 1. i 2. ii 3. iii 4. iv
(B) 1. iii 2. ii 3. iv 4. i
(C) 1. ii 2. i 3. iii 4. iv
(D) 1. iv 2. ii 3. i 4. iii

979. Which of the following is a nitrification inhibitor?
(A) N serve (B) AM
(C) Thiourea (D) All of these

980. N-lignin contains N %:
(A) 18 (B) 20
(C) 22 (D) 25

981. Which of the following is a slow release N fertilizer?
(A) IBDU (B) CDU
(C) N lignin (D) All of these

982. Which of the following is correct for Ankur?
(A) Liquid N fertilizer (B) Contains 32 % N
(C) Useful for rice crop (D) All of these

983. Aqua ammonium contains ammonia:
(A) 18 % (B) 20 %
(C) 22.4 % (D) 25 %

984. Anhydrous ammonia contain ammonia:
(A) >90 % (B) <90 %
(C) >99% (D) None of these

985. Which of the following is a source of Mo?
(A) Sodium molybdate
(B) Ammonium molybdate
(C) Molybdenum sulphate
(D) All of these

986. Copper sulphate contains Cu:
(A) 15 % (B) 20 %
(C) 25 % (D) 30 %

987. Nitrophosphate contains N, P and K respectively:
(A) 20 : 20 : 0 (B) 20 : 20 : 20
(C) 20 : 45 : 0 (D) 20 : 40 : 40

988. Nitrophosphate potash grade I contains N, P and K respectively:
(A) 18 : 18 : 9 (B) 18 : 18 : 0
(C) 17 : 17 : 17 (D) 20 : 20 : 20

989. Nitrophosphate potash grade II contains N, P and K respectively:
(A) 18 : 18 : 9 (B) 18 : 18 : 18
(C) 15 : 15 : 15 (D) 20 : 20 : 20

990. Zinc sulphate contains S:
(A) 15 % (B) 20 %
(C) 14.6 % (D) 17.8 %

991. Borax contains Boron:
(A) 5 % (B) 10 %
(C) 11 % (D) 13 %

992. Boric acid contain Boron:
(A) 15 % (B) 17 %
(C) 18 % (D) 19 %

993. Cu deficiency may be corrected by application of:
(A) Copper sulphate (B) Cupperous oxide
(C) Cupric oxide (D) All of these

994. Wood ash is generally:
(A) Acidic (B) Basic
(C) Neutral (D) None of these

995. KCl is not suitable for:
(A) Wheat (B) Rice
(C) Gram (D) Tobacco

996. K_2O in KCl is:
(A) 60 % (B) 42%
(C) 50 % (D) 51 %

997. K_2O in potassium carbonate is:
(A) 21 % (B) 46 %
(C) 65 % (D) 54 %

998. Which of the following is known as salt petre of nitre?
(A) KCl (B) KNO_3
(C) K_2SO_4 (D) None of these

999. KNO_3 contains:
(A) 12-13 % N and 44 % K_2O
(B) 12-13 % N and 36 % K_2O
(C) 18 % N and 48 % K_2O
(D) 12-13 % N and 48 % K_2O

1000. Which of the following is a basic fertilizer?
(A) KCl (B) Single super phosphate
(C) KNO_3 (D) None of these

1001. Which of the following element is found maximum in ash?
(A) N (B) P
(C) K (D) All of these

1002. K_2O in ash is approximately?
(A) 0.5-12 % (B) 17-18 %
(C) 25-28 % (D) 28-30%

1003. S in K_2SO_4 is:
(A) 12 % (B) 21%
(C) 17.6 % (D) 20.6 %

1004. Complete complex fertilizers contains:
(A) Only N (B) Only P
(C) Only K (D) N, P and K

1005. Complex fertilizers generally contain total nutrients:
(A) >20 % (B) <20 %
(C) 10-15 % (D) 5-10 %

1006. P in nitrophosphate is:
(A) Water soluble (B) Citrate soluble
(C) Water and Citrate soluble (D) None of these

1007. Which of the following is a nitro phosphate fertilizer?
(A) Suphla (B) Ammonium chloride
(C) Super phosphate (D) Urea

1008. Which of the following is a complex fertilizer?
(A) Urea
(B) Muriate of potash
(C) Mono Ammonium Phosphate
(D) None of these

1009. Which of the following element is mainly found in guano?
(A) N (B) P
(C) N and P (D) K

1010. N content in Guano is found in the form of:
(A) NO_3^- (B) NH_4^+
(C) N_2O (D) N_2

1011. Guano is a:
(A) Fish by product (B) Beat of sea birds
(C) Mixture of cakes (D) Inorganic fertilizer

1012. Average N, P and K content of poultry manure are:
(A) 3.8, 3.5 and 1.7 % (B) 1.2, 0.5 and 0.7 %
(C) 12, 7 and 3% (D) 1.2, 5 and 7.0 %

1013. N content in fishmeal is:
(A) 4-10 % (B) 1-2 %
(C) 0.1-1.5 % (D) >4-1.0 %

1014. Pressmud contains mainly:
(A) Lime and P (B) N and P
(C) P and K (D) N

1015. Which of the crop is used for green manuring in saline-alkaline soils?
(A) Berseem (B) Dhaincha
(C) Sunnhemp (D) Mung

1016. Phosphate absorber bio fertilizer is:
(A) VAM (B) BGA
(C) FYM (D) Azolla

1017. Which of the following is the phosphate mobilizing bio-fertilizers?
(A) *Bacillus polymyza* (B) *Pseudomonas strata*
(C) *Aspergillus niger* (D) All of these

1018. Average C : N ratio of digested gobar gas slurry is:
(A) 10-12 (B) 15-16
(C) 20-23 (D) 25-30

1019. The average values of different parameters of vermicompost are given below:

1. N	i. 0.56 %
2. P	ii. 1.48 %
3. K	iii. 0.36 %
4. C :N ratio	iv. 34-35

Which of the pair is correctly matched

(A) 1. i	2. ii	3. iii	4. iv
(B) 1. iii	2. ii	3. iv	4. i
(C) 1. ii	2. i	3. iii	4. iv
(D) 1. iv	2. ii	3. i	4. iii

1020. The values of different parameters of compost are given below:

Parameters	Average values
1. N	i. 2.0-2.4 %
2. P	ii. 1.0-1.2 %
3. K	iii. 3.5-3.6 %
4. C :N ratio	iv. 15-17

Which of the pair is correctly matched
(A) 1. i 2. ii 3. iii 4. iv
(B) 1. iii 2. ii 3. iv 4. i
(C) 1. ii 2. i 3. iii 4. iv
(D) None of these

1021. Average values (%) of N, P_2O_5 and K_2O in sludge are:
(A) 3.5, 2.5 and 0.5 (B) 0.5, 0.25 and 0.75
(C) 2.5, 0.5 and 2.5 (D) None of these

1022. Which of the following crop is used for green manuring?
(A) Sunnhemp (B) Dhaincha
(C) Indigo (D) All of these

1023. Which of the following bacteria fixes nitrogen in non leguminous crops?
(A) Azotobacter (B) Azospirillum
(C) Beijerinckia (D) All of these

1024. Dicyandiamide and S.T. (2-Silphathiazole) are:
(A) Nitrification inhibitors
(B) N slow release fertilizers
(C) High grade N fertilizer
(D) None of these

1025. Low grade fertilizers have total percentage of N, P and K:
(A) 14 (B) >14
(C) <14 (D) None of these

1026. Medium grade fertilizers have total percentage of N, P and K:
(A) 15-25 (B) >35
(C) <15 (D) None of these

1027. High grade fertilizers have total percentage of N, P and K:
(A) 15-25 (B) >35
(C) <15 (D) None of these

1028. Concentrated fertilizers have total percentage of N, P and K:
(A) 15-25 (B) >35
(C) <15 (D) None of these

1029. Trench method of preparing FYM was developed by:
(A) Dr. C.N. Acharya (B) Dr. Panjab Singh
(C) B.N. Ghildyal (D) None of these

1030. Which of the following micronutrient is found highest in FYM:
(A) Cu (B) Zn
(C) Mn (D) B

1031. Average C : N ratio of FYM is:
(A) 0.5 : 1 (B) 2 : 1
(C) 10-12 : 1 (D) None of these

1032. Average N content in FYM nearly on dry weight basis:
(A) 0.2 % (B) 0.6 %
(C) 0.9 % (D) 1.0 %

1033. Organic carbon content in FYM is :
(A) 5 % (B) 8 %
(C) 11 % (D) 15 %

1034. Digested gobar gas slurry contains N % on dry weight basis nearly:
(A) 1.8-1.9 % (B) 1.51-1.55 %
(C) 2.11-2.24 % (D) None of these

1035. Conditioners are mixed with mixed fertilizers because:
(A) To improve physical condition
(B) Avoid clouding
(C) To avoid formation of volatile compounds
(D) All of these

1036. Fillers are mixed in fertilizers just to:
(A) Improve physical condition of fertilizers
(B) Increase weight of fertilizers
(C) Improve chemical composition of fertilizers
(D) None of these

1037. Fertilizer control order was announced in year:
(A) 1947 (B) 1957
(C) 1965 (D) 1970

1038. Which of the following is not a bulky organic manure?
(A) FYM (B) Compost
(C) Green manure (D) Fish manure

1039. India's share in soil organic carbon stock of the world is only:
(A) 3 % (B) 0.3 %
(C) 0.03 % (D) 0.003 %

1040. Sizeable amount of nitrogen could be escaped by volatilization is:
(A) Ammonia (B) Nitrate
(C) Nitrite (D) Amide

1041. Sizeable amount of nitrogen could be escaped by denitrification is:
(A) Nitrous oxide or chemical nitrogen
(B) Nitrate
(C) Nitrite
(D) Amide

1042. Sizeable amount of nitrogen could be escaped by underground water is:
(A) Ammonia (B) Nitrate
(C) Nitrite (D) Amide

1043. NPK removal ratio of crops is:
(A) 4 : 1 : 4 (B) 1 : 1 : 5
(C) 5 : 1 : 5 (D) 5 : 5 : 1

1044. During 1998-99, consumption of N : P_2O_5 : K_2O ratio was:
(A) 1.33: 4.10 : 11.32 (B) 4.10: 11.32: 1.33
(C) 11.32 : 4.10 : 11.33 (D) 11.32 : 4.10 : 1.33

1045. Since last 25 years the Indian soils are experiencing on an average net negative balance of:
(A) 8-10 mt/annum (B) 10-12 mt/annum
(C) 12-14 mt/annum (D) 14-16 mt/annum

1046. Arrangement of soil sparticle is called:
(A) Soil texture (B) Soil structure
(C) Bulk density (D) Particle density

1047. With increase in organic matter, the density will:
(A) Decrease (B) Increase
(C) Not change (D) Unknown

1048. The weight per unit volume of the solid particle is called:
(A) Bulk density (B) Particle density
(C) Soil porosity (D) Soil Consistency

1049. The soil consisting 40% or more of clay particles are classes as:
(A) Silt loam (B) Silt
(C) Clay loam (D) Clay

1050. Red soil develops from the parent material of:
(A) Granite (B) Schist
(C) Both (A) and (B) (D) None of these

1051. Quartz, feldspar and mica are:
(A) Primary mineral (B) Secondary mineral
(C) Tertiary mineral (D) None of these

1052. Dolomite, gypsum, apatite and hematite are:
(A) Primary mineral (B) Secondary mineral
(C) Tertiary mineral (D) None of these

1053. Calcium, magnesium and sulphur plant nutrients are called:
(A) Primary (B) Secondary
(C) Micro (D) Trace

1054. Non essential but useful plant nutrients are:
(A) Cobalt & Iodine (B) Strontium & Flourine
(C) Vanadium & Bromine (D) All of these

1055. Nitrogen mainly absorbed in the form of:
(A) Ammonia (B) Nitrate
(C) Both (A) and (B) (D) None of these

1056. Under humid condition potassium fixation more in silicate clay of:
(A) Illite (B) Monmorillonite
(C) Kaolinite (D) None of these

1057. First stage in the evaluation of saline soils is:
(A) Solonchek (B) Solonetz
(C) Solodi (D) None of these

1058. SAR is equal to:

(A) $Na^{+}\sqrt{Ca^{+2}+Mg^{+2}}$

(B) $Na^{+}\sqrt{Ca^{+2}+Mg^{+2/2}}$

(C) $Na^{+}\sqrt{CEC}$

(D) $Ca^{+2}\sqrt{Na^{+}+Mg^{+2/2}}$

1059. Facultative free living 'N' fixing bacterium:
(A) Rhizobium (B) Azotobacter
(C) Klebsiella (D) None of these

1060. Cyst formation is characteristics feature of :
(A) *Acetobacter* (B) *Arthrobacter*
(C) *Azotobacter* (D) None of these

1061. Bacteria having especially high rate of respiration:
(A) *Rhizobium* (B) *Azotobacter*
(C) *E. coli* (D) *Acetobacter*

1062. "Ray Fungi" is the name given to:
(A) Rhizopus (B) Yeast
(C) Actinomycetes (D) Mycoplasma

1063. Actinomycetes are:
(A) Gram +ve, aerobic (B) Gram +ve, anaerobic
(C) Gram –ve, aerobic (D) Gram –ve, anaerobic

1064. Example for microaerophillic N_2 fixer:
(A) *Rhizobium* (B) *Azotobacter*
(C) *E. coli* (D) *Frankia*

1065. Phosphorus solublizing microorganism:
(A) VAM (B) Mucor
(C) Rhizopus (D) *E. coli*

1066. Liquid N_2 has a temperature:
(A) -196°C (B) -186°C
(C) -176°C (D) -172°C

1067. Low pH of media inhibits the growth of:
(A) Bacteria (B) Moulds
(C) Both (A) and (B) (D) None of these

1068. Which of the following genera is most likely to contain organisms capable of surviving high temperature?
(A) Clostridium (B) Penicillium
(C) Torula (D) Pseudomonas

1069. Which of the following is/are Bio-insecticides?
(A) *Bacillus thuringiensis* (B) *Bacilllus popilliae*
(C) Both (A) and (B) (D) None of these

1070. Funaric acid is produced by:
(A) *Aspergillus terrus* (B) *Rhizopus nigricans*
(C) *Aspergillus niger* (D) *Asergillus fumigatus*

1071. ATP needed for N_2 fixation:
(A) 12 (B) 16
(C) 28 (D) 32

1072. Factor which does not affect legume root nodulation?
(A) Temperature (B) Light
(C) Combined N_2 (D) H^+ ion conc.

1073. Microaerophillic N_2 fixing bacteria:
(A) *Azotobacter* (B) *Azospirillium*
(C) *Clostridium* (D) *Enterobacter*

1074. Bacteria which can ferment sugar faster than yeast is:
(A) *Clostridium* (B) *Zygomonas*
(C) *Thiobacillus* (D) *Bacillus*

1075. In leghaemoglobin the heme portion is specified by:
(A) Plant genes (B) Bacterial genes
(C) Both (A) and (B) (D) None of these

1076. Leghaemoglobin is structurally and functionally similar to:
(A) Haemoglobin (B) Actin
(C) Myosin (D) Myogolobin

1077. Anaerobic free living 'N' fixing bacterium:
(A) Azotobacter (B) Acetobacter
(C) Kelbseilla (D) Clostridium

1078. Microorganism which are associated with non legumes:
(A) Anabaena (B) Klebseilla
(C) Both (A) and (B) (D) None of these

1079. Bacteria are commonly seen in soils such as:
(A) Acidic (B) Neutral
(C) Alkaline (D) Saline

1080. Nitrat reductase enzyme present in:
(A) Cytoplasm (B) Chloroplast
(C) Mitochondria (D) Nucleus

1081. Microbes involved in fibre retting:
(A) *Micrococcus sp.* (B) *Bacillus subtilis*
(C) *Clostridium butrycium* (D) *Lactobacillus sp.*

1082. Microbe involved in tobacco curing:
(A) *Micrococcus sp.* (B) *Bacillus subtilis*
(C) *Lactobacillus sp.* (D) *Aspergillus niger*

1083. Urea degrading bacteria:
(A) *Bacillus pasturi*
(B) *Clostridium pasteurianum*
(C) *A. niger*
(D) *Micrococcus sp.*

1084. 'P' solubilizing bacteria:
(A) *Bacillus megatherium*
(B) *Clostridium pasteurianum*
(C) *A. niger*
(D) *Micrococcus sp.*

1085. The size of microbe as:
(A) Viriod<Virus<Bacteria<Fungi
(B) Virus<Viriod<Bacteria<Fungi
(C) Viriod<Virus<Fungi<Bacteria
(D) Virus<Viriod<Fungi<Bacteria

1086. Type of relationship between *Acetobacter diazotrophicus* and sugarcane is:
(A) Symbiotic (B) Associative
(C) Endophytic (D) Free living

1087. Example of VAM fungi:
(A) Pisoletecis (B) Sclerotenia
(C) Glomus (D) Trichoderma

1088. For calculating value of K, K_2O is multiplied by:
(A) 2.29 (B) 0.437
(C) 1.20 (D) 3.00

1089. For calculating value of S, SO_4 is multiplied by:
(A) 2.29 (B) 0.437
(C) 1.20 (D) 3.00

1090. For calculating value of Ca, CaO is multiplied by:
(A) 1.40 (B) 3.00
(C) 1.286 (D) 4.43

1091. For calculating value of Ca, $Ca(OH)_2$ is multiplied by:
(A) 1.40 (B) 3.00
(C) 1.286 (D) 4.43

1092. For calculating value of N, NH_4 is multiplied by:
(A) 1.40 (B) 3.00
(C) 1.286 (D) 4.43

1093. For calculating value of N, NO_3 is multiplied by:
(A) 1.40 (B) 3.00
(C) 1.286 (D) 4.43

1094. For calculating value of organic carbon, organic matter is multiplied by:
(A) 1.724 (B) 3.00
(C) 1.286 (D) 4.43

1095. Which of the following crops cause maximum reduction in soil alkalinity:
(A) Cowpea (B) Maize
(C) Paddy (D) Wheat

1096. Which one of the following is not a micronutrient?
(A) Cu (B) B
(C) S (D) Mn

1097. Choose the micronutrient:
(A) Magnesium (B) Zinc
(C) Phosphorus (D) Potassium

1098. How much urea contains nitrogen?
(A) 16% (B) 26% (C) 36% (D) 46%

1099. How much calcium ammonium nitrate contains nitrogen?
(A) 16% (B) 26% (C) 25% (D) 46%

1100. How much ammonium liquid contains nitrogen?
(A) 16% (B) 20-25% (C) 36% (D) 46%

1101. How much ammonium nitrate contains nitrogen?
(A) 16% (B) 26% (C) 33-34% (D) 46%

1102. How much sodium nitrate contains nitrogen?
(A) 16% (B) 26% (C) 36% (D) 46%

1103. How much ammonium chloride contains nitrogen?
(A) 24-26% (B) 26% (C) 36% (D) 46%

1104. How much Ammonium sulphate nitrate contains nitrogen?
(A) 16% (B) 20-21% (C) 36% (D) 46%

1105. How much Calcium cynamide contains nitrogen?
(A) 16% (B) 21% (C) 36% (D) 46%

1106. How much Ammonium anhydrous contains nitrogen?
(A) 16% (B) 26% (C) 36% (D) 82%

1107. How much oxamide contains nitrogen?
(A) 16% (B) 26% (C) 31.8% (D) 46%

1108. How much ammonium phosphate contains nitrogen?
(A) 16-20% (B) 26% (C) 36% (D) 46%

1109. How much ammonium sulphate nitrate contains nitrogen?
(A) 16% (B) 26% (C) 36% (D) 46%

1110. How much ammonium phosphate contains phosphorus?
(A) 16% (B) 20% (C) 36% (D) 46%

1111. How much single super phosphate contains phosphorus?
(A) 16-18% (B) 26% (C) 36% (D) 48%

1112. How much dicalcium phosphate contains phosphorus?
(A) 16% (B) 26% (C) 32-34% (D) 46%

1113. How much triple super phosphate contains phosphorus?
(A) 16% (B) 26% (C) 36% (D) 46-48%

1114. How much rhenania phosphate contains phosphorus?
(A) 16% (B) 25% (C) 36% (D) 46%

1115. How much pelophos contains phosphorus?
(A) 17% (B) 26% (C) 36% (D) 46%

1116. How much rock phosphate (imported) contains phosphorus?
(A) 16% (B) 26% (C) 30-40% (D) 46%

1117. How much mussorie rock phosphate contains phosphorus?
(A) 16% (B) 23-24% (C) 36% (D) 46%

1118. How much Jhabua rock phosphate contains phosphorus?
(A) 16% (B) 26% (C) 31-38% (D) 46%

1119. How much basic slage contains phosphorus?
(A) 14-18% (B) 26% (C) 31-38% (D) 46%

1120. How much bone meal contains phosphorus?
(A) 16% (B) 20-25% (C) 36% (D) 46%

1121. How much stream bone meal contains phosphorus?
(A) 16% (B) 22% (C) 36% (D) 46%

1122. How much muriate of potash contains potash?
(A) 16% (B) 26% (C) 36% (D) 60%

1123. How much sulphate of potash contains potash?
(A) 16% (B) 26% (C) 48-52% (D) 46%

1124. How much potassium nitrate contains potash?
(A) 16% (B) 44% (C) 36% (D) 46%

1125. How much potassium magnesium sulphate contains potassium?
(A) 16% (B) 25-30% (C) 36% (D) 46%

1126. How much potassium magnesium contains magnesium?
(A) 12% (B) 26% (C) 36% (D) 46%

1127. How much bone meal contains nitrogen?
(A) 1-2% (B) 3-4% (C) 4-5% (D) 5-6%

1128. How much nitrogen content in coal ash?
(A) 0.7% (B) 0.4% (C) 0.5% (D) 1.0%

1129. How much phosphorus content in coal ash?
(A) 0.7% (B) 0.4% (C) 0.5% (D) 1.0%

1130. How much potassium content in coal ash?
(A) 0.7% (B) 0.4% (C) 0.5% (D) 1.0%

1131. How much nitrogen content in house hold ash?
(A) 0.5-1.9% (B) 0.4% (C) 0.5% (D) 1.0%

1132. How much phosphorus content in house hold ash?
(A) 0.7% (B) 1.6-4.2% (C) 0.5% (D) 1.0%

1133. How much potassium content in house hold ash?
(A) 0.7% (B) 0.4% (C) 2.3-12.0% (D) 1.0%

1134. How much nitrogen content in cane trash ash?
(A) 0.1-0.2% (B) 0.4% (C) 0.5% (D) 1.0%

1135. How much phosphorus content in cane trash ash?
(A) 0.7% (B) 0.8-1.3% (C) 0.5% (D) 1.0%

1136. How much potassium content in cane trash ash?
(A) 0.7% (B) 0.4% (C) 1.5-3.1% (D) 1.0%

1137. How much nitrogen content in babool wood ash?
(A) 0.1-0.2% (B) 0.4% (C) 0.5% (D) 1.0%

1138. How much phosphorus content in babool wood ash?
(A) 0.7% (B) 2.5-3.0% (C) 0.5% (D) 1.0%

1139. How much potassium content in babool wood ash?
(A) 0.7% (B) 0.4% (C) 3.5-4.5% (D) 1.0%

1140. How much nitrogen content in casuarinas wood ash?
(A) Traces (B) 1.5% (C) 0.5% (D) 1.0%

1141. How much nitrogen content in Eucalyptus wood ash?
(A) Traces (B) 1.5% (C) 0.5% (D) 1.0%

1142. How much nitrogen content in Tobacco wood ash?
(A) Traces (B) 1.5% (C) 0.5% (D) 1.0%

1143. How much phosphorus content in Eucalyptus wood ash?
(A) Traces (B) 1.5% (C) 5.9% (D) 1.0%

1144. How much phosphorus content in tobacco wood ash?
(A) 0.7% (B) 2.6% (C) 5.9% (D) 1.0%

1145. How much potassium content in casuarina wood ash?
(A) 14.0% (B) 23.8% (C) 36.0% (D) 40.0%

1146. How much potassium content in Eucalyptus wood ash?
(A) 14.0% (B) 23.8% (C) 36.0% (D) 40.0%

1147. How much potassium content in tobacco wood ash?
(A) 0.7% (B) 23.8% (C) 36.0% (D) 40.0%

1148. How much nitrogen contains content in dung of cow and bull?
(A) 0.40% (B) 1.00% (C) 0.75% (D) 1.35%

1149. How much nitrogen contains content in urine of cow and bull?
(A) 0.40% (B) 1.00% (C) 0.75% (D) 1.35%

1150. How much nitrogen contains content in dung of sheep and goat?
(A) 0.40% (B) 1.00% (C) 0.75% (D) 1.35%

1151. How much nitrogen contains content in urine of sheep and goat?
(A) 0.40% (B) 1.00% (C) 0.75% (D) 1.35%

1152. How much nitrogen contains content in dung of Horse?
(A) 0.40% (B) 0.55% (C) 0.75% (D) 1.35%

1153. How much nitrogen contains content in urine of Horse?
(A) 0.40% (B) 0.55% (C) 0.75% (D) 1.35%

1154. How much nitrogen contains content in dung of pigs?
(A) 0.40% (B) 0.55% (C) 0.75% (D) 1.35%

1155. How much nitrogen contains content in urine of pigs?
(A) 0.40% (B) 0.35% (C) 0.75% (D) 1.35%

1056. How much phosphorus contains content in dung of cow and bull?
(A) 0.20% (B) Trace (C) 0.30% (D) 0.50%

1157. How much phosphorus contains content in urine of cow and bull?
(A) 0.20% (B) Trace (C) 0.30% (D) 0.50%

1158. How much phosphorus contains content in dung of sheep and goat?
(A) 0.20% (B) Trace (C) 0.30% (D) 0.50%

1159. How much phosphorus contains content in urine of sheep and goat?
(A) 0.20% (B) Trace (C) 0.30% (D) 0.50%

1160. How much phosphorus contains content in dung of Horse?
(A) 0.20% (B) Trace (C) 0.30% (D) 0.50%

1161. How much phosphorus contains content in urine of Horse?
(A) 0.20% (B) Trace (C) 0.30% (D) 0.50%

1162. How much phosphorus contains content in dung of pigs?
(A) 0.20% (B) Trace (C) 0.30% (D) 0.50%

1163. How much phosphorus contains content in urine of pigs?
(A) 0.10% (B) Trace (C) 0.30% (D) 0.50%

1164. How much potassium contains content in dung of cow and bull?
(A) 0.10% (B) 0.45 % (C) 1.35% (D) 2.10%

1165. How much potassium contains content in urine of cow and bull?
(A) 0.10% (B) 0.45 % (C) 1.35% (D) 2.10%

1166. How much potassium contains content in dung of sheep and goat?
(A) 0.10% (B) 0.45 % (C) 1.35% (D) 2.10%

1167. How much potassium contains content in urine of sheep and goat?
(A) 0.10% (B) 0.45 % (C) 1.25% (D) 2.10%

1168. How much potassium contains content in dung of Horse?
(A) 0.40% (B) 0.45 % (C) 1.25% (D) 2.10%

1169. How much potassium contains content in urine of Horse?
(A) 0.10% (B) 0.45 % (C) 1.25% (D) 2.10%

1170. How much potassium contains content in dung of pigs?
(A) 0.40% (B) 0.45 % (C) 1.25% (D) 2.10%

1171. How much potassium contains content in urine of pigs?
(A) 0.10% (B) 0.45 % (C) 1.25% (D) 2.10%

1172. How much nitrogen is fixed in the soil by *Crotolaria juncea*?
(A) 75-85 kg/ha (B) 70-78 kg/ha
(C) 55-65 kg/ha (D) 45-52 kg/ha

1173. How much nitrogen is fixed in the soil by *Sesbania aculeata*?
(A) 75-85 kg/ha (B) 70-78 kg/ha
(C) 55-65 kg/ha (D) 45-52 kg/ha

1174. How much nitrogen is fixed in the soil by *Phaseolus tribolus*?
(A) 75-85 kg/ha (B) 70-78 kg/ha
(C) 55-65 kg/ha (D) 45-52 kg/ha

1175. How much nitrogen is fixed in the soil by *Vigna radiata*?
(A) 75-85 kg/ha (B) 70-78 kg/ha
(C) 55-65 kg/ha (D) 45-52 kg/ha

1176. How much nitrogen is fixed in the soil by *Vigna mungo*?
(A) 42-53 kg/ha (B) 55-58 kg/ha
(C) 60-67 kg/ha (D) 120-125 kg/ha

1177. How much nitrogen is fixed in the soil by *Vigna sinensis*?
(A) 42-53 kg/ha (B) 55-58 kg/ha
(C) 60-67 kg/ha (D) 120-125 kg/ha

1178. How much nitrogen is fixed in the soil by *Cyamopsis tetragonoloba*?
(A) 42-53 kg/ha (B) 55-58 kg/ha
(C) 60-67 kg/ha (D) 120-125 kg/ha

1179. How much nitrogen is fixed in the soil by *Melilotus alba*?
(A) 42-53 kg/ha (B) 55-58 kg/ha
(C) 60-67 kg/ha (D) 120-125 kg/ha

1180. How much nitrogen is fixed in the soil by *Trifolium alexandrinum*?
(A) 42-53 kg/ha (B) 58-64 kg/ha
(C) 62-75 kg/ha (D) 120-125 kg/ha

1181. How much nitrogen is fixed in the soil by *Lathyrus sativus*?
(A) 42-53 kg/ha (B) 58-64 kg/ha
(C) 62-75 kg/ha (D) 120-125 kg/ha

1182. How much nitrogen contains in castor cake?
(A) 4.3% (B) 2.5% (C) 5.2% (D) 3.0%

1183. How much nitrogen contains in mahua cake?
(A) 4.3% (B) 2.5% (C) 5.2% (D) 3.0%

1184. How much nitrogen contains in neem cake?
(A) 4.3% (B) 2.5% (C) 5.2% (D) 3.0%

1185. How much nitrogen contains in coconut cake?
(A) 4.3% (B) 2.5% (C) 5.2% (D) 3.0%

1186. How much nitrogen contains in cotton seed cake?
(A) 6.4% (B) 7.3% (C) 4.9% (D) 5.2%

1187. How much nitrogen contains in groundnut cake?
(A) 6.4% (B) 7.3% (C) 4.9% (D) 5.2%

1188. How much nitrogen contains in linseed cake?
(A) 6.4% (B) 7.3% (C) 4.9% (D) 5.2%

1189. How much nitrogen contains in mustard cake?
(A) 6.4% (B) 7.3% (C) 4.9% (D) 5.2%

1190. How much nitrogen contains in safflower cake?
(A) 6.4% (B) 7.9% (C) 4.9% (D) 5.2%

1191. How much nitrogen contains in sesame cake?
(A) 6.4% (B) 7.3% (C) 6.2% (D) 5.2%

1192. How much potassium contains in castor cake?
(A) 1.3% (B) 1.4% (C) 1.8% (D) 2.2%

1193. How much potassium contains in mahua cake?
(A) 1.3% (B) 1.4% (C) 1.8% (D) 2.2%

1194. How much potassium contains in neem cake?
(A) 1.3% (B) 1.4% (C) 1.8% (D) 2.2%

1195. How much potassium contains in coconut cake?
(A) 1.3% (B) 1.4% (C) 1.8% (D) 2.2%

1196. How much potassium contains in cotton seed cake?
(A) 1.3% (B) 1.4% (C) 1.8% (D) 2.2%

1197. How much potassium contains in groundnut cake?
(A) 1.3% (B) 1.4% (C) 1.8% (D) 2.2%

1198. How much potassium contains in linseed cake?
(A) 1.2% (B) 1.3% (C) 1.9% (D) 2.0%

1199. How much potassium contains in mustard cake?
(A) 1.2% (B) 1.3% (C) 1.9% (D) 2.0%

1200. How much potassium contains in safflower cake?
(A) 1.2% (B) 1.3% (C) 1.9% (D) 2.0%

1201. How much potassium contains in sesame cake?
(A) 1.2% (B) 1.3% (C) 1.9% (D) 2.0%

1202. How much phosphorus contains in castor cake?
(A) 0.8% (B) 1.0% (C) 1.8% (D) 1.9%

1203. How much phosphorus contains in mahua cake?
(A) 0.8% (B) 1.0% (C) 1.8% (D) 1.9%

1204. How much phosphorus contains in neem cake?
(A) 0.8% (B) 1.0% (C) 1.8% (D) 1.9%

1205. How much phosphorus contains in coconut cake?
(A) 0.8% (B) 1.0% (C) 1.8% (D) 1.9%

1206. How much phosphorus contains in cotton seed cake?
(A) 1.4% (B) 1.5% (C) 1.8% (D) 2.9%

1207. How much phosphorus contains in groundnut cake?
(A) 1.4% (B) 1.5% (C) 1.8% (D) 2.2%

1208. How much phosphorus contains in linseed cake?
(A) 1.4% (B) 1.5% (C) 1.8% (D) 2.2%

1209. How much phosphorus contains in mustard cake?
(A) 1.4% (B) 1.5% (C) 1.8% (D) 2.2%

1210. How much phosphorus contains in safflower cake?
(A) 1.4% (B) 1.5% (C) 1.8% (D) 2.2%

1211. How much phosphorus contains in sesame cake?
(A) 1.4% (B) 1.5% (C) 1.8% (D) 2.0%

1212. How much nitrogen contains in blood meal?
(A) 10-12% (B) 4-10% (C) 13% (D) 15.2%

1213. How much nitrogen contains in fish meal?
(A) 10-12% (B) 4-10% (C) 13% (D) 15.2%

1214. How much nitrogen contains in horn and hoof meal?
(A) 10-12% (B) 4-10% (C) 13% (D) 15.2%

1215. How much phosphorus contains in blood meal?
(A) 1-2% (B) 3-9% (C) 10% (D) 15.2%

1216. How much phosphorus contains in fish meal?
(A) 1-2% (B) 3-9% (C) 10% (D) 15.2%

1217. How much potassium contains in fish meal?
(A) 1-2% (B) 3-9% (C) 0.3-1.5% (D) 15.2%

1218. How much nitrogen contains in farm yard manure?
(A) 0.5-0.6% (B) 0.50% (C) 1.0-1.5% (D) 1.5%

1219. How much nitrogen contains in rural compost?
(A) 0.5-0.6% (B) 0.50% (C) 1.0-1.5% (D) 1.5%

1220. How much nitrogen contains in urban compost?
(A) 0.5-0.6% (B) 0.50% (C) 1.0-1.5% (D) 1.5%

1221. How much nitrogen contains in sewage and sludge?
(A) 0.5-0.6% (B) 0.50% (C) 1.0-1.5% (D) 1.5%

1222. How much nitrogen contains in Human urine?
(A) 0.60% (B) 1.1-1.6% (C) 1.50% (D) 3.0%

1223. How much nitrogen contains in bird beat?
(A) 0.60% (B) 1.1-1.6% (C) 1.50% (D) 3.0%

1224. How much nitrogen contains in molasses?
(A) 0.60% (B) 1.1-1.6% (C) 1.50% (D) 3.0%

1225. How much nitrogen contains in vermicompost?
(A) 0.60% (B) 1.1-1.6% (C) 1.50% (D) 3.0%

1226. How much phosphorus contains in farm yard manure?
(A) 0.25% (B) 0.40% (C) 1.0% (D) 1.0-1.25%

1227. How much phosphorus contains in rural compost?
(A) 0.25% (B) 0.40% (C) 1.0% (D) 1.0-1.25%

1228. How much phosphorus contains in urban compost?
(A) 0.25% (B) 0.40% (C) 1.0% (D) 1.0-1.25%

1229. How much phosphorus contains in sewage and sludge?
(A) 0.25% (B) 0.40% (C) 1.0% (D) 1.0-1.25%

1230. How much phosphorus contains in Human urine?
(A) 0.1% (B) 0.40% (C) 1.0% (D) 1.0-1.25%

1231. How much phosphorus contains in bird beat?
(A) 0.25% (B) 0.40% (C) 1.0% (D) 1.0-1.20%

1232. How much phosphorus contains in molasses?
(A) 0.25% (B) 0.40% (C) 1.0% (D) 3.0%

1233. How much phosphorus contains in vermicompost?
(A) 0.25% (B) 0.40% (C) 1.0% (D) 1.25%

1234. How much potassium contains in farm yard manure?
(A) 0.5% (B) 0.50-1.0% (C) 0.8% (D) 1.5%

1235. How much potassium contains in rural compost?
(A) 0.5% (B) 0.50-1.0% (C) 0.8% (D) 1.5%

1236. How much potassium contains in urban compost?
(A) 0.5% (B) 0.50-1.0% (C) 0.8% (D) 1.5%

1237. How much potassium contains in sewage and sludge?
(A) 0.5% (B) 0.50-1.0% (C) 0.8% (D) 1.5%

1238. How much potassium contains in Human urine?
(A) 0.80% (B) 1.1-1.6% (C) 1.50% (D) 3.0%

1239. How much potassium contains in bird beat?
(A) 0.5% (B) 0.50-1.0% (C) 0.8% (D) 6.1-8.5%

1240. How much potassium contains in vermicompost?
(A) 0.60% (B) 1.1-1.6% (C) 1.50% (D) 3.0%

1241. How much sulphur contains in gypsum (Ag grade)?
(A) 12% (B) 13% (C) 15% (D) 16%

1242. How much sulphur contains in phosphogypsum?
(A) 12% (B) 13% (C) 15% (D) 16%

1243. How much sulphur contains in pyrite (Ag grade)?
(A) 12% (B) 13% (C) 15% (D) 22%

1244. How much sulphur contains in copper sulphate?
(A) 12% (B) 13% (C) 15% (D) 16%

1245. How much sulphur contains in ferrus sulphate?
(A) 12% (B) 13% (C) 15% (D) 16%

1246. How much sulphur contains in magnesium sulphate?
(A) 12% (B) 13% (C) 15% (D) 16%

1247. How much sulphur contains in zinc sulphate?
(A) 12% (B) 13% (C) 15% (D) 11·16%

1248. How much calcium contains in gypsum (Ag grade)?
(A) 21% (B) 13% (C) 15% (D) 16·19%

1249. How much calcium contains in phopshogypsum?
(A) 21% (B) 13% (C) 15% (D) 16·19%

1250. How much magnesium contains in magnesium sulphate?
(A) 12% (B) 13% (C) 15% (D) 16%

1251. How much boron contains in borax?
(A) 10.5% (B) 13% (C) 15% (D) 16%

1252. How much boron contains in solabar?
(A) 10% (B) 13% (C) 19% (D) 26%

1253. How much copper contains in copper sulphate?
(A) 20% (B) 22% (C) 24% (D) 26%

1254. How much iron contains in ferrus sulphate?
(A) 12% (B) 14% (C) 14% (D) 16%

1255 How much mangnese contains in magnesium sulphate?
(A) 20% (B) 30.5% (C) 35% (D) 40%

1256. How much zinc contains in zinc sulphate?
(A) 21 or 33% (B) 22% (C) 24% (D) 26%

1257. How much molybdenum contains in sodium molybdate?
(A) 17·39% (B) 30·45%
(C) 24·39% (D) 26·50%

1258. How much zinc contains in zinc chelate?
(A) 10% (B) 12% (C) 14% (D) 16%

1259. How much molybdenum contains in ammonium molybdate?
(A) 39% (B) 45% (C) 54% (D) 60%

1260. The nitrogen is removed by grain of paddy:
(A) 66 kg/ha (B) 39 kg/ha
(C) 180 kg/ha (D) 75 kg/ha

1261. The nitrogen is removed by straw of paddy:
(A) 66 kg/ha (B) 39 kg/ha
(C) 180 kg/ha (D) 75 kg/ha

1262. The nitrogen is removed by grain of wheat:
(A) 66 kg/ha (B) 39 kg/ha
(C) 180 kg/ha (D) 75 kg/ha

1263. The nitrogen is removed by straw of wheat:
(A) 66 kg/ha (B) 39 kg/ha
(C) 180 kg/ha (D) 75 kg/ha

1264. The nitrogen is removed by grain of maize:
(A) 45 kg/ha (B) 70kg/ha
(C) 80 kg/ha (D) 95 kg/ha

1265. The nitrogen is removed by straw of maize:
(A) 45 kg/ha (B) 70kg/ha
(C) 80 kg/ha (D) 95 kg/ha

1266. The nitrogen is removed by grain of sorghum:
(A) 45 kg/ha (B) 70kg/ha
(C) 80 kg/ha (D) 95 kg/ha

1267. The nitrogen is removed by straw of sorghum:
(A) 45 kg/ha (B) 70kg/ha
(C) 80 kg/ha (D) 95 kg/ha

1268. The nitrogen is removed by grain of barley:
(A) 45 kg/ha (B) 70kg/ha
(C) 80 kg/ha (D) 95 kg/ha

1269. The nitrogen is removed by straw of barley:
(A) 45 kg/ha (B) 70kg/ha
(C) 80 kg/ha (D) 95 kg/ha

1270. The nitrogen is removed by mustard:
(A) 45 kg/ha (B) 70kg/ha
(C) 95 kg/ha (D) 150 kg/ha

1271. The nitrogen is removed by groundnut:
(A) 45 kg/ha (B) 70kg/ha
(C) 95 kg/ha (D) 150 kg/ha

1272. The nitrogen is removed by sugarcane:
(A) 45 kg/ha (B) 70kg/ha
(C) 95 kg/ha (D) 150 kg/ha

1273. The nitrogen is removed by tobacco in light soils:
(A) 35 kg/ha (B) 70kg/ha
(C) 95 kg/ha (D) 150 kg/ha

1274. The nitrogen is removed by tobacco in heavy soils:
(A) 48 kg/ha (B) 70kg/ha
(C) 95 kg/ha (D) 150 kg/ha

1275. The nitrogen is removed by potato:
(A) 45 kg/ha (B) 70kg/ha
(C) 95 kg/ha (D) 150 kg/ha

1276. The nitrogen is removed by cotton seed:
(A) 45 kg/ha (B) 70kg/ha
(C) 80 kg/ha (D) 150 kg/ha

1277. The nitrogen is removed by jute:
(A) 45 kg/ha (B) 70kg/ha
(C) 95 kg/ha (D) 150 kg/ha

1278. The nitrogen is removed by grain of soybean:
(A) 45 kg/ha (B) 70kg/ha
(C) 95 kg/ha (D) 210 kg/ha

1279. The phosphorus is removed by grain of paddy:
(A) 15 kg/ha (B) 24 kg/ha
(C) 35 kg/ha (D) 70 kg/ha

1280. The phosphorus is removed by straw of paddy:
(A) 15 kg/ha (B) 24 kg/ha
(C) 35 kg/ha (D) 70 kg/ha

1281. The phosphorus is removed by grain of wheat:
(A) 15 kg/ha (B) 24 kg/ha
(C) 35 kg/ha (D) 70 kg/ha

1282. The phosphorus is removed by straw of wheat:
(A) 15 kg/ha (B) 24 kg/ha
(C) 35 kg/ha (D) 70 kg/ha

1283. The phosphorus is removed by grain of maize:
(A) 27 kg/ha (B) 30 kg/ha
(C) 41 kg/ha (D) 48 kg/ha

1284. The phosphorus is removed by straw of maize:
(A) 27 kg/ha (B) 30 kg/ha
(C) 41 kg/ha (D) 48 kg/ha

1285. The phosphorus is removed by grain of sorghum:
(A) 27 kg/ha (B) 30 kg/ha
(C) 41 kg/ha (D) 48 kg/ha

1286. The phosphorus is removed by straw of sorghum:
(A) 27 kg/ha (B) 30kg/ha
(C) 41 kg/ha (D) 48 kg/ha

1287. The phosphorus is removed by grain of barley:
(A) 22 kg/ha (B) 25 kg/ha
(C) 38 kg/ha (D) 48 kg/ha

1288. The phosphorus is removed by straw of barley:
(A) 22 kg/ha (B) 25 kg/ha
(C) 38 kg/ha (D) 48 kg/ha

1289. The phosphorus is removed by mustard:
(A) 22 kg/ha (B) 25kg/ha
(C) 38 kg/ha (D) 48 kg/ha

1290. The phosphorus is removed by groundnut:
(A) 22 kg/ha (B) 25 kg/ha
(C) 38 kg/ha (D) 48 kg/ha

1291. The phosphorus is removed by sugarcane:
(A) 45 kg/ha (B) 58 kg/ha
(C) 65 kg/ha (D) 190 kg/ha

1292. The phosphorus is removed by tobacco in light soils:
(A) 45 kg/ha (B) 58 kg/ha
(C) 65 kg/ha (D) 190 kg/ha

1293. The phosphorus is removed by tobacco in heavy soils:
(A) 45 kg/ha (B) 58 kg/ha
(C) 65 kg/ha (D) 190 kg/ha

1294. The phosphorus is removed by potato:
(A) 45 kg/ha (B) 58 kg/ha
(C) 65 kg/ha (D) 190 kg/ha

1295. The phosphorus is removed by cotton seed:
(A) 30 kg/ha (B) 35 kg/ha
(C) 60 kg/ha (D) 150 kg/ha

1296. The phosphorus is removed by jute:
(A) 30 kg/ha (B) 35 kg/ha
(C) 60 kg/ha (D) 150 kg/ha

1297. The phosphorus is removed by grain of soybean:
(A) 30 kg/ha (B) 35 kg/ha
(C) 60 kg/ha (D) 150 kg/ha

1298. The potassium is removed by grain of paddy:
(A) 18 kg/ha (B) 52 kg/ha
(C) 96 kg/ha (D) 98 kg/ha

1299. The potassium is removed by straw of paddy:
(A) 18 kg/ha (B) 52 kg/ha
(C) 96 kg/ha (D) 98 kg/ha

1300. The potassium is removed by grain of wheat:
(A) 18 kg/ha (B) 52 kg/ha
(C) 96 kg/ha (D) 98 kg/ha

1301. The potassium is removed by straw of wheat:
(A) 18 kg/ha (B) 52 kg/ha
(C) 96 kg/ha (D) 98 kg/ha

1302. The potassium is removed by grain of maize:
(A) 20 kg/ha (B) 35 kg/ha
(C) 90 kg/ha (D) 105 kg/ha

1303. The potassium is removed by straw of maize:
(A) 20 kg/ha (B) 35 kg/ha
(C) 90 kg/ha (D) 105 kg/ha

1304. The potassium is removed by grain of sorghum:
(A) 20 kg/ha (B) 35 kg/ha
(C) 90 kg/ha (D) 105 kg/ha

1305. The potassium is removed by straw of sorghum:
(A) 20 kg/ha (B) 35 kg/ha
(C) 90 kg/ha (D) 105 kg/ha

1306. The potassium is removed by grain of barley:
(A) 30 kg/ha (B) 75 kg/ha
(C) 89 kg/ha (D) 110 kg/ha

1307. The potassium is removed by straw of barley:
(A) 30 kg/ha (B) 75 kg/ha
(C) 89 kg/ha (D) 110 kg/ha

1308. The potassium is removed by mustard:
(A) 30 kg/ha (B) 75 kg/ha
(C) 89 kg/ha (D) 110 kg/ha

1309. The potassium is removed by groundnut:
(A) 30 kg/ha (B) 75 kg/ha
(C) 89 kg/ha (D) 110 kg/ha

1310. The potassium is removed by sugarcane:
(A) 60 kg/ha (B) 70 kg/ha
(C) 210 kg/ha (D) 265 kg/ha

1311. The potassium is removed by tobacco in light soils:
(A) 60 kg/ha (B) 70 kg/ha
(C) 210 kg/ha (D) 265 kg/ha

1312. The potassium is removed by tobacco in heavy soils:
(A) 60 kg/ha (B) 70 kg/ha
(C) 210 kg/ha (D) 265 kg/ha

1313. The potassium is removed by potato:
(A) 60 kg/ha (B) 70 kg/ha
(C) 210 kg/ha (D) 265 kg/ha

1314. The potassium is removed by cotton seed:
(A) 22 kg/ha (B) 30 kg/ha
(C) 85 kg/ha (D) 150 kg/ha

1315. The potassium is removed by jute:
(A) 22 kg/ha (B) 30 kg/ha
(C) 85 kg/ha (D) 150 kg/ha

1316. The potassium is removed by grain of soybean:
(A) 22 kg/ha (B) 30 kg/ha
(C) 85 kg/ha (D) 150 kg/ha

1317. Which rhizobium species is suitable for pea and lentil?
(A) *Rhizobium leguminosorum*
(B) *Rhizobium phaseoli*
(C) *Rhizobium meliloti*
(D) *Rhizobium japonicum*

1318. Which rhizobium species is suitable for arhar, mung, urd and lobia?
(A) *Rhizobium leguminosorum*
(B) *Rhizobium phaseoli*
(C) *Rhizobium meliloti*
(D) *Rhizobium japonicum*

1319. Which rhizobium species is suitable for lucern?
(A) *Rhizobium leguminosorum*
(B) *Rhizobium phaseoli*
(C) *Rhizobium meliloti*
(D) *Rhizobium japonicum*

1320. Which rhizobium species is suitable for soyabean?
(A) *Rhizobium leguminosorum*
(B) *Rhizobium phaseoli*
(C) *Rhizobium meliloti*
(D) *Rhizobium japonicum*

1321. Which rhizobium species is suitable for lupin?
(A) *Rhizobium lupine*
(B) *Rhizobium trifoli*
(C) *Rhizobium meliloti*
(D) *Rhizobium japonicum*

1322. Which rhizobium species is suitable for berseem?
(A) *Rhizobium lupine*
(B) *Rhizobium trifoli*
(C) *Rhizobium meliloti*
(D) *Rhizobium japonicum*

1323. Hilly soils are formed by:
(A) Granite and Quartzite rocks
(B) Schist and Biotite rocks
(C) Sand stone
(D) Lime stone

1324. Nitrate formation in soil is carriedout by the microbe named?
(A) Azatobactor (B) Azospirillum
(C) Nitrobactor (D) Nitrosomonas

1325. Organic matter in soil is an index of:
(A) Available nitrogen (B) Available phosphorus
(C) Available potassium (D) Available zinc

1326. $CaSO_4 + 2H_2O \rightarrow CaSO_4.2H_2O$ is an example of the weathering process is known as:
(A) Hydration (B) Hydrolysis
(C) Oxidation (D) Reduction

1327. Rock phosphate can be safely used in:
(A) Saline soil (B) Alkali soil
(C) Calcareous soil (D) Acid soil

1328. Nitrate inhibitors inhibit the formation of:
(A) NH_4^+ (B) NO_2^-
(C) NO_3^- (D) NH_3

1329. Silt and clay content can be determined with the help of:
(A) Tensiometer (B) Thermometer
(C) Hydrometer (D) Osmometer

1330. Organism whose population in soil is maximum:
(A) Bacteria (B) Virus
(C) Earthworm (D) Fungi

1331. Fraction of humus solubilized by acid or alkali is known as:
(A) Humic acid (B) Fulvic acid
(C) Humin (D) None of these

1332 Easily oxidized rocks contain:
(A) Fe (B) K
(C) Mg (D) $CaCO_3$

1333. Soil organisms obtaining energy from the oxidation of mineral constituents are known as:
(A) Autotrophic (B) Heterotrophic
(C) Facultative (D) Aerobic

1334. Soils with high silica/sesquioxide ratio are called:
(A) Siliceous (B) Sesquioxide
(C) Both (A) and (B) (D) None of these

1335. Soils with low silica/sesquioxide ratio are called:
(A) Siliceous (B) Sesquioxide
(C) Both (A) and (B) (D) None of these

1336. Due to presence of iron and aluminium, the colour of soil is:
(A) Red to yellow (B) Yellow to red
(C) Red to Brown (D) Brown to red

1337. Mycorrhizae refers to:
(A) Root environment (B) Soil-root interaction
(C) Fungus – root association (D) Mycology

1338. Optimum temperature for nitrification is:
(A) 18°C (B) 22°C
(C) 30°C (D) 35°C

1339. Aggregates are more stable in soils dominated by:
(A) Kaolinite (B) Montmorillonite
(C) Vermiculite (D) Chlorite

1340. Heavy soils refer to:
(A) High density soils
(B) Soils dominated by high density particles
(C) Soils which are difficult to work
(D) Soils rich in nutrients

1341. Bulk density of $1m^3$ soil weighing 1.4 mg is:
(A) 1.33 mg/m_3 (B) 1.44 mg/m_3
(C) 1.50 mg/m_3 (D) 1.40 mg/m_3

1342. Number of soil orders are:
(A) 10 (B) 11 (C) 12 (D) 13

1343. Number of master horizons are:
(A) 3 (B) 4 (C) 5 (D) 6

1344. Exchange of O_2 and CO_2 between soil pores and atmospheric air refers to:
(A) Soil-air capacity (B) Soil-oxygen diffusion
(C) Soil-air permeability (D) Soil aeration

1345. Most porous soil structure is:
(A) Granular (B) Crumb
(C) Prismatic (D) Platy

1346. Liming of soil is done to reclaim:
(A) Saline soil (B) Sodic soil
(C) Acidic soil (D) Saline-sodic soil

1347. Regolith refers to:
(A) A horizon (B) A+B horizon
(C) A+B+C horizon (D) Solum

1348. Carbon exchange capacity is found maximum in:
(A) Montmorillonite (B) Kaolinite
(C) Vermiculite (D) Humus

1349. Di-calcium phosphate is written as:
(A) $Ca(H_2PO_4)_2$ (B) $CaHPO_4$
(C) $Ca_3(PO_4)_2$ (D) None of these

1350. Hydraulic conductivity of soil is directly proportional to:
(A) Radius of capillary pores (B) Capillary porosity
(C) Total porosity (D) Water holding capacity

1351. Application of urea super granules (USG) is beneficial in:
(A) Water logged soil (B) Saline soils
(C) Both (A) and (B) (D) None of these

1352. The 'white bud' disease of maize is caused by:
(A) Deficiency of B (B) Deficiency of Fe
(C) Deficiency of S (D) Deficiency of Zn

1353. Azolla may fix atmospheric nitrogen in the soil upto:
(A) 251 kg/ha (B) 400 kg/ha
(C) 451 kg/ha (D) None of these

1354. Azolla is nitrogen fixing:
(A) Bacteria (B) Fern
(C) Fungi (D) Algae

1355. Nitrogen deficiency symptoms in plants first appears on:
(A) Lower leaves
(B) Middle leaves
(C) Upper leaves
(D) All of the above at the same time

1356. The process of loosening for separating fibre from plant is known as:
(A) Curing (B) Retting
(C) Wetting (D) Netting

1357. Azolla is used for nitrogen fixation in:
(A) Wheat crop (B) Gram crop
(C) Paddy crop (D) Arhar crop

1358. Tick the correct group of plant nutrients which acts as energy exchange from the following:
(A) K, Na, Ca and Mg
(B) C, N, P and S
(C) N and O
(D) Fe, Mn, Mo, Cu, B and Zn

1359. Which gas is released from paddy fields:
(A) CH_4 (B) H_2S (C) CO_2 (D) NH_3

1360. Peat soil are generally characterized as:
(A) Black, heavy and alkaline soil
(B) Red, light and acidic soil
(C) Black, light and alkaline soil
(D) Black, heavy and neutral soil

1361. Solubility of rock phosphate can be improved by:
(A) Nitrosomonas (B) Nitrobactor
(C) Azotobactor (D) *Bacillus polymexa*

1362. The principal way in which P and K ions more from the soil to the root of field crops:
(A) Root interception (B) Mass flow
(C) Diffusion (D) None of these

1363. Biuret content in urea should not exceed in foliar spray on wheat crop more than the following:
(A) 0.5% (B) 1.0% (C) 1.5% (D) 2.0%

1364. Which element causes dispersion of soil particles:
(A) Na (B) K (C) Ca (D) Mg

1365. VA micorrhiza promotes mainly the plant uptake of:
(A) Nitrogen absorption
(B) Phosphorus absorption
(C) Potassium absorption
(D) Molybdenum absorption

1366. Immobilization of sulphur takes place when the 'S content' of organic matter is less than:
(A) 0.60% (B) 0.45%
(C) 0.30% (D) 0.15%

1367. Read the following statements and mark your answer according to the code given below:
I. Soil acidity increases with exchangeable Al^{++}
II. Soil acidity increases with exchangeable H
III. Soil alkalinity increases with exchangeable calcium
IV. Soil alkalinity increases with exchangeable aluminium
Answer code
(A) I and III are correct (B) I and II are correct
(C) II and III are correct (D) III and IV are correct

1368. Consider the following statement:
Assertion (A): Alkali soils are poor in nitrogen
Reason (R): Nitrogen is leached from alkali soil
Select the correct answer from the code given below:
(A) A is true but R is wrong (B) R is true but A is false
(C) Both A and R are correct (D) Both A and R are false

1369. Carbon dioxide content in soil air is:
(A) 0.03% (B) 0.003%
(C) 0.25% (D) 0..50%

1370. How much is the porosity percentage if the particle density of soil is 2.6 and bulk density is 1.3:
(A) 35% (B) 40% (C) 50% (D) 55%

1371. Read the following statements and mark your answer according to the code given below:
I. Organic matter improves the structure of clayey soil
II. Organic matter improves the structure of sandy soil
III. Sandy soils are heavy in texture
IV. Clayey soils are lighter in texture
Answer code
(A) I and II are correct (B) I and III are correct
(C) I and IV are correct (D) III and IV are correct

1372. Read the following statements and mark your answer according to the code given below:
I. Alfisols are lighter in texture
II. Vertisols are higher in texture
III. Entisols have low potassium content
IV. Ultisols have low phosphorus fixation
Answer code
(A) I and II are correct (B) I and III are correct
(C) I and IV are correct (D) III and IV are correct

1373. Potassium content is maximum in:
(A) Feldspar (B) Pyroxine
(C) Amphibole (D) Olivine

1374. Which element has greater significance in biological nitrogen fixation:
(A) Nickel (B) Sulphur
(C) Potassium (D) Molybdenum

1375. Chemical formula of single super phosphate is:
(A) $Ca\ (H_2PO_4)_2$
(B) $CaHPO_4$
(C) $Ca_3(PO_4)_2$
(D) $Ca(H_2PO_4)_2.CaSO_4.2H_2O$

1376. The hilly soils of Uttar Pradesh are formed by:
(A) Granite and quartzite rocks
(B) Schist and biotite rocks
(C) Sand stone
(D) Lime stone

1377. Iron containing enzyme responsible for the biological nitrogen fixation is:
(A) Nitrate reductase (B) Nitrite reductase
(C) Nitrogenase (D) Peroxidase

1378. Decomposition of organic matter in submerged soil is carried out by:
(A) Bacteria (B) Actinomycetes
(C) Fungi (D) Earthworm

1379. The highest nitrogen content is found in:
(A) Farm Yard Manure (B) Vermicompost
(C) Poultry manure (D) Biogas slurry

1380. 'Azolla' (a nitrogen fixing organism) is in symbiotic association with:
(A) Fern and blue green algae
(B) Bryophyta and blue green algae
(C) Fungi and blue green algae
(D) Nostoc and blue green algae

1381. Urea hydrolysis in the soil produces:
(A) Uric acid (B) Ammonium carbonate
(C) Ammonium carbonate (D) Amide

1382. Sodic soils are characterized by:
(A) High soluble salt concentration and low pH
(B) High soluble salt concentration
(C) Low pH and high exchangeable sodium
(D) High exchangeable sodium and high pH

1383. Which of the following is not correctly matched:
(A) Magnesium — Macro nutrient
(B) Sandy loam — Soil texture
(C) Rhizobium — Symbiotic N_2-fixation
(D) Tensiometer — Soil porosity

1384. Iron deficiency is prominent in:
(A) Neutral soil (B) Calcareous soil
(C) Acidic soil (D) Waterlogged soil

1385. Direct application of rock phosphate as phosphorus source for crops is recommended in:
(A) Acid soils (B) Neutral soils
(C) Alkaline soils (D) Saline soils

1386. The availability of phosphorus is higher in:
(A) Between 4-5 pH (B) Between 6-7 pH
(C) Between 7.5-8.5 pH (D) Between 9.0-9.5 pH

1387. Organic matter content in organic soils in the range of:
(A) 10-15% (B) 16-20%
(C) 20-25% (D) Above 25%

1388. Consider the following statements:
Assertion (A): Nitrification inhibitors are the chemicals used to control nitrogen transformation in the soils
Reason (R): They promote the activities of nitrifying bacteria
Select the correct answer using the code given below:
(A) A is true but R is false
(B) A is false but R is true
(C) Both A and R are true and R is the correct explanation of A
(D) Both A and R are true but R is not the correct explanation of A

1389. Integrated Plant Nutrients Management pertains with:
(A) Use of organic manures
(B) Use of chemical fertilizers
(C) Use of biofertilizers
(D) All of these

1390. For foliar spray, zinc sulphate fertilizer is dissolved in water at the rate of:
(A) 0.25% (B) 0.50% (C) 0.75% (D) 1.0%

1391. White bud of maize is caused by the deficiency of:
(A) Fe (B) Mn (C) Zn (D) B

1392. Which organism is responsible for phosphorus availability to plants?
(A) Azotobacter (B) Nitrosomonas
(C) Mycorrhiza (D) Earthworms

1393. With increasing soil pH, base saturation % is:
(A) Decrease
(B) Increase
(C) Increases upto pH 7 and then decline
(D) Decrease upto pH 7 and then decline

1394. Zinc sulphate should not be mixed with:
(A) Urea (B) Muriate of potash
(C) DAP (D) FYM

1395. Which clay mineral is rich in potash?
(A) Montmorillonite (B) Kaolinite
(C) Illite (D) Chlorite

1396. Decomposition of organic matter in submerged soil is carried out by:
(A) Bacteria (B) Actinomycetes
(C) Fungi (D) Earthworm

1397. Which soil has the highest cation exchange capacity?
(A) Loam (B) Loamy sand
(C) Sandy loam (D) Clay loam

1398. The carbohydrates produced in leaves cannot be translocated to different growing parts of plants in the absence of:
(A) Manganese (B) Boron
(C) Zinc (D) Iron

1399. Application of potash increases:
(A) Disease resistance in plants
(B) Resistance for water logging
(C) Frost resistance in plants
(D) None of these

1400. Ammonia is converted into nitrate by:
(A) Fungus (B) Algae
(C) Earthworms (D) Bacteria

1401. Electrical conductivity of saturation extract of saline-sodic soils in mmhos per cm is:
(A) 2.0 (B) Between 2.0 to 3.0
(C) 3.5 (D) More than 4.0

1402. When organic matter is added to the soil:
(A) Bulk density is increased
(B) Bulk density is decreased
(C) Bulk density does not change
(D) None of these

1403. The availability of molybdenum is more at:
(A) pH 5.0 (B) pH 7.0
(C) pH 7.5 (D) pH 9.0

1404. Azotobacter fixes atmospheric nitrogen:
(A) Symbiotically
(B) Non-symbiotically
(C) Both symbiotically and non-symbiotically
(D) None of these

1405. The soil is well flocculated if it has more of:
(A) Sodium (B) Zinc
(C) Potassium (D) Calcium

1406. Rock phosphate is used in:
(A) Neutral soil (B) Acidic soil
(C) Alkaline soil (D) Saline soil

1407. Montmorillonite clay has silica and alumina layers respectively in the following proportion:
(A) 2 : 1 (B) 1 : 2 (C) 3 : 2 (D) 2 : 3

1408. Which one of the following microorganisms is involved in nitrification in the soil:
(A) Pseudomonas (B) Micro-coccus
(C) Thio-bacillus (D) Nitrosomonas

1409. Bactria responsible for converting nitrite into nitrate is called:
(A) Pseudomonas (B) Nitrobacter
(C) Heterosomonas (D) Rhizobium

1410. Sulphur is needed for which group of the following amino acids:
(A) Lysine and glutamine
(B) Methionine and cystine
(C) Tryptophane and cystine
(D) None of these

1411. Energy metabolism in plants is performed by the element:
(A) Potassium (B) Magnesium
(C) Phosphorus (D) Iron

1412. Brayer and others discovered which of the essential elements?
(A) Copper (B) Boron
(C) Chlorine (D) Molybdenum

1413. Potassic fertilizer containing maximum quantity of potash is:
(A) MOP (B) Sulphate of potash
(C) Potassium nitrate (D) Potassium ammonium nitrate

1414. Which one of the following pairs is commonly matched:
(A) Peat soil : Calcareous
(B) Alkali soil : pH 9
(C) Black soil : Rich in kaolinite
(D) Laterite soil : pH 9

1415. Nitrogen is a major nutrient because:
(A) It is available in soil more
(B) It is constituent of protein
(C) It is available in atmosphere
(D) It is required by plants more

1416. Peat soils are generally:
(A) Black, heavy and acidic
(B) Red, light and acidic
(C) Black, light and basic
(D) Black, heavy and neutral

1417. A branch of science that deals with the study of origin of the soil, its classification and description is known as:
(A) Entomology (B) Ecology
(C) Pedology (D) Geology

1418. Soil amendments used to reclaim usar (barren) land is:
(A) Calcium sulphate (B) Sodium sulphate
(C) Potassium sulphate (D) Calcium carbonate

1419. The salt affected soils are divided in how many classes depending upon electrical conductivity and sodium contents of their solution extract:
(A) Two (B) Three
(C) Four (D) Five

1420. Which microorganisms is present in large quantities in the soil?
(A) Algae (B) Bacteria
(C) Fungi (D) Protozoa

1421. Propounding of the essentiality of the following elements for plants was made in different years:
(A) Nitrogen (B) Manganese
(C) Zinc (D) Copper
Which of the following chronological order is correct in respect of year of propounding:
(A) 2, 3, 1, 4 (B) 3, 2, 4, 1
(C) 1, 3, 2, 4 (D) 1, 2, 3, 4

1422. Scorching or burning on margins of bottom leaves manifests the deficiency of:
(A) Nitrogen (B) Zinc
(C) Copper (D) Potassium

1423. Browning in cauliflower can be corrected by use of:
(A) Ammonium sulphate (B) Zinc sulphate
(C) Boric acid (D) Urea

1424. Non – essential but beneficial element to plants is:
(A) Calcium (B) Sodium
(C) Potassium (D) Magnesiumm

1425. Carbon is an essential element for plant growth. Plant take it from:
(A) Soil (B) Water
(C) Air (D) Light

1426. Hydrogen is an essential element for plant growth. Plant take it from:
(A) Soil (B) Water
(C) Air (D) Light

1427. Oxygen is an essential element for plant growth. Plant take it from:
(A) Soil (B) Water
(C) Air (D) Light

1428. Near the canal banks soils become:
(A) Acidic (B) Neutral
(C) Alkaline (D) Well drained

1429. Azotobacter is useful in fixing of nitrogen form air in:
(A) Non-leguminnous crops like rice and wheat
(B) Leguminous crops
(C) Soybean
(D) None of these

1430. Most of the plants obtained nitrogen from the soil in the form of:
(A) Nitrite (B) Nitrate
(C) Free nitrogen gas (D) Nitric acid

1431. Which of the following is micronutrients:
(A) Calcium (B) Zinc
(C) Nitrogen (D) Phosphorus

1432. Deficiency of which of the following element causes general yellowing and poor growth of the plant?
(A) Phosphorus (B) Nitrogen
(C) Calcium (D) boron

1433. Most effective phosphatic fertilizer for crops in acidic soil is:
(A) DAP (B) Rock phosphate
(C) SSP (D) None of these

1434. The availability of zinc is reduced when zinc sulphate is mixed with:
(A) Nitrogenous fertilizers (B) Phosphatic fertilizers
(C) Potassium fertilizers (D) None of these

1435. Kadar and Bangar are the soil types of belonging to:
(A) Black soil (B) Laterite soil
(C) Alluvial soil (D) Arid soil

1436. Laterite soil is formed by the presence of:
(A) Transportation (B) Weathering
(C) Leaching (D) Deposition

1437. This soil is silt and loamy in valley areas and coarse grained in upper slopes. It is called:
(A) Alluvial soil (B) Khadar soil
(C) Forest soil (D) Arid soil

1438. Select the most appropriate combination to indicates the proportion of important relief features in India:
(A) Plains 43%, Plateau 25%, Mountains 32%
(B) Plains 30%, Plateau 42%, Mountains 28%
(C) Plains 43%, Plateau 27%, Mountains 30%
(D) None of these

1439. The general structural formula of montmorrilonite is:
(A) $X_{0.8}(Al_{0.3}Si_{7.7})Al_{2.6}Fe^{3+}{}_{0.9}Mg_{0.5})O_{20}(OH)_4nH_2O$
(B) $K_2Al_2Si_6Al_4O_{20}(OH)_4$
(C) $K_2Al_2Si_6(Fe^{2+},Mg)_6O_{20}(OH)_4$
(D) $Si_4Al_4O_{10}OH_8$

1440. The structural formula of muscovite is:
(A) $X_{0.8}(Al_{0.3}Si_{7.7})Al_{2.6}Fe^{3+}{}_{0.9}Mg_{0.5})O_{20}(OH)_4nH_2O$
(B) $K_2Al_2Si_6Al_4O_{20}(OH)_4$
(C) $K_2Al_2Si_6(Fe^{2+},Mg)_6O_{20}(OH)_4$
(D) $Si_4Al_4O_{10}OH_8$

1441. The structural formula of biotite is:
(A) $X_{0.8}(Al_{0.3}Si_{7.7})Al_{2.6}Fe^{3+}{}_{0.9}Mg_{0.5})O_{20}(OH)_4nH_2O$
(B) $K_2Al_2Si_6Al_4O_{20}(OH)_4$
(C) $K_2Al_2Si_6(Fe^{2+},Mg)_6O_{20}(OH)_4$
(D) $Si_4Al_4O_{10}OH_8$

1442. The structural formula of kaolinite is:
(A) $X_{0.8}(Al_{0.3}Si_{7.7})Al_{2.6}Fe^{3+}{}_{0.9}Mg_{0.5})O_{20}(OH)_4nH_2O$
(B) $K_2Al_2Si_6Al_4O_{20}(OH)_4$
(C) $K_2Al_2Si_6(Fe^{2+},Mg)_6O_{20}(OH)_4$
(D) $Si_4Al_4O_{10}OH_8$

1443. The minerals of montmorillonite or smectic isomorphism series are freely expandable layer silicate are:
(A) Beidellite (B) Monmorillonite
(C) Saponite (D) All of these

1444. Quartz and feldspar are:
(A) Primary minerals (B) Secondary minerals
(C) Tertiary minerals (D) Accessory minerals

1445. Silicate minerls, phyllo-silicates or clay minerals are:
(A) Primary minerals (B) Secondary minerals
(C) Tertiary minerals (D) Accessory minerals

1446. Apatite, Pyroxins, Tourmaline and Magnetite are:
(A) Primary minerals (B) Secondary minerals
(C) Tertiary minerals (D) Accessory minerals

1447. Feldspar are alluminosilicates of:
(A) Potassium (B) Sodium
(C) Calcium (D) All of these

1448. As the inter-layer potassium in subjected to depletion through chemical equilibrium with the soil solution, the exchangeable ions and inter-layer water is gained and leads to another mineral according to following relation:
(A) Micas → Illite → Vermiculite → Montmorillonite
(B) Micas → Vermiculite → Illite → Montmorillonite
(C) Montmorillonite→Illite→Vermiculite→Micas
(D) Vermiculite → Micas → Illite → Montmorillonite

1449. Maximum number of bacteria are available in:
(A) Horizon A (B) Solum
(C) Horizon B (D) Horizon C

1450. Guano is not seen in the excreta of:
(A) Scals
(B) Bats
(C) Sea birds
(D) Whales

1451. Nitrogen bases do not contain:
(A) Phosphorus
(B) Hydrogen
(C) Nitrogen
(D) Carbon

1452. Nitrogen fixation in blue green algae is done by the specialized structures known as:
(A) Akinete
(B) Hormogonium
(C) Nanocyst
(D) Heterocys

1453. Vermicomposting is a type of composting process involving:
(A) Earthworms
(B) Wooden ash
(C) Plant material
(D) Animal manure

1454. N, P, K proportion in balanced fertilization is:
(A) 4 : 1 : 1
(B) 2 : 1 : 1
(C) 6 : 2 : 1
(D) 6 : 3 : 1

1455. Name of the trace element is:
(A) Sulphur
(B) Zinc
(C) Calcium
(D) Potassium

1456. Soil sickness is caused due to:
(A) Accumulation of autotoxins
(B) N-deficiency
(C) Accumulation of phosphates
(D) Zinc deficiency

1457. Nitrogen use efficiency is lower in:
(A) Wheat
(B) Maize
(C) Paddy
(D) Vegetable

1458. Phosphorus use efficiency in most of the crops is:
(A) 10-15%
(B) 20-25%
(C) 30-35%
(D) 40-45%

1459. Availability of which element is decreased under waterlogged conditions?
(A) Fe
(B) Zn
(C) P
(D) Mn

1460. Saline-alkali soils are generally found in:
(A) Arid region
(B) Humid region
(C) Temperate region
(D) Flood region

1461. Acidic soils are generally found in:
(A) Arid region
(B) Humid region
(C) Temperate region
(D) Flood region

1462. Phosphate Solubilizing Bacteria (PSB) improves soil fertility by:
(A) Fixing atmospheric nitrogen
(B) Fixing atmospheric phosphorus
(C) Releasing fixed phosphorus in the soil
(D) All of these

1463. Soils most suitable for agriculture are those whose pH value in between:
(A) 5 to 6
(B) 6 to 7
(C) 7 to 8
(D) 8 to 9

1464. At present the number of elements essential for plant growth:
(A) 15 (B) 16 (C) 17 (D) 18

1465. Which element is now essential for plant growth:
(A) Na
(B) Si
(C) Ni
(D) None of these

1466. How much percentage of soluble salts are present in alkali soils?
(A) 1-2 %
(B) 0.1-0.2 %
(C) 1.5-2.5 %
(D) 2-3 %

1467. The categories of soil classification and its number are given in 'A' and 'B' list below:

A	B
1. Orders	i. 54
2. Suborders	ii. 11
3. Greatgroups	iii. 254
4. Subgroups	iv. 1200

Which of the pair is matched correctly:

(A) 1. i	2. ii	3. iii	4.iv
(B) 1. iii	2. ii	3. iv	4.i
(C) 1. ii	2. i	3. iii	4.iv
(D) 1. iv	2. ii	3. i	4.iii

1468. Phosphorus is called macronutrient because it is:
(A) Available in soil in relatively large amount
(B) Available in plant in relatively in large amount
(C) Used by plants in relatively large amount
(D) Found in grains in relatively large amount

1469. Seed inoculation in pulses is done with:
(A) Azotobacter
(B) Rhizobium
(C) Azospirillum
(D) Phosphotica

1470. Which elements are useful in energy storage, transfer and bonding?
(A) NPK
(B) NSP
(C) NKS
(D) None of these

1471. Which of the following elements play an important role in nitrogen metabolism?
(A) Magnesium
(B) Manganese
(C) Molybdenum
(D) Iron

1472. The leaching loss of nitrogen (N) is more in the form of:
(A) Ammonia
(B) Nitrate
(C) Nitrogen
(D) Water soluble N

1473. Which is not a natural factor affecting soil fertility?
(A) Topography
(B) Soil age
(C) Air
(D) Parent material

1474. Empirical formula of $100 - \left[\frac{\text{Bulk density}}{\text{Particle density}} \times 100\right]$ is:
(A) Bulk density
(B) Percent pore space
(C) Pore space
(D) Particle density

1475. Granite is a .... rock:
(A) Metamorphic
(B) Igneous
(C) Sedimentary
(D) None of these

1476. Sulphur coated urea contains N:
(A) 21 % (B) 33 % (C) 40 % (D) 26 %

1477. What culture should be given priority in groundnut cultivation?
(A) Myccorhiza
(B) Rhizobium
(C) Phosphobacteria
(D) Azospirillum

1478. Which group of plant nutrients involves in N_2 fixation?
(A) P, S, Mo (B) P, S, Co
(C) P, Mo, Co (D) All of these

1479. Which one of the following is not a primary nutrient?
(A) S (B) N
(C) K (D) P

1480. Percentage of P_2O_5 in Pelophos fertilizer is:
(A) 12 % (B) 16 %
(C) 11 % (D) 18 %

1481. Soil fertility is reduced due to:
(A) Over irrigation
(B) Poor drainage
(C) Imbalanced use of fertilizers
(D) Continuous cropping

1482. Chemical formula of pyrite is:
(A) FeS (B) CuS
(C) MnS (D) FeS_2

1483. For waterlogged rice cultivation, the fertilizer that may be available is:
(A) Sodium nitrate (B) Urea
(C) Ammonium chloride (D) None of these

1484. Biofertilizers are:
(A) Organic manure
(B) Green manure
(C) Culture of microorganism
(D) None of these

1485. Which of the following micronutrients are most widely deficient in Indian soils?
(A) Zinc and boron (B) Zinc and Iron
(C) Zinc and Menganese (D) Zinc and copper

1486. Acid soils, as per USDA, are those which in 1 : 1 soil water suspension, have pH less than:
(A) 5.5 (B) 6.0 (C) 6.5 (D) 7.0

1487. The equivalent acidity of $(NH_4)_2SO_4$ fertilizer is:
(A) 100 (B) 110 (C) 115 (D) 118

1488. The pH of 0.1 N NaOH solution would theoretically be :
(A) 14 (B) 13 (C) 12 (D) 11

1489. The microorganism, constituting the largest share in the total microbial population in soil is:
(A) Nematodes (B) Fungi
(C) Bacteria (D) Actinomycetes

1490. One Baule unit of P (lb P_2O_5/acre) corresponds to:
(A) 50 (B) 55 (C) 45 (D) 40

1491. Irrigation water is not suitable for irrigation if it contains boron (ppm) greater than:
(A) 1.00 (B) 1.25 (C) 1.50 (D) 2.50

1492. The value of `K' (lb/acre) used in the Wilcox's inverse yield N^2 law is:
(A) 218 (B) 318 (C) 418 (D) 518

1493. An irrigation water with EC 1.0 d^s/m would have total soluble salt concentration (mg/litre or ppm):
(A) 32 (B) 64 (C) 320 (D) 640

1494. The indicator plant for B deficiency in soil is:
(A) Soybean (B) Rice (C) Mustard (D) Sunflower

1495. The diameter (in mm) of the soil particles which are carried away due to saltation mode of soil erosion is:
(A) 0.02-0.2 (B) 0.01-0.1
(C) 0.1-0.5 (D) 0.5-1.0

1496. A highly beneficial and essential element for rice along with ohter macro and micro nutrients is :
(A) CO (B) V
(C) Ni (D) Si–

1497. The most common growth limiting factor in acid soil is:
(A) Sodium (B) Calcium
(C) Aluminium (D) Potassium

1498. Well decomposed compost have C : N ratio of :
(A) 10-15 : 1 (B) 20-25 : 1
(C) 30-35 : 1 (D) 40-45 : 1

1499. Rock phosphates are generaly used in soils which are:
(A) Strongly acidic (B) Strongly alkaline
(C) Neutral soils (D) Saline soils

1500. Browning of curd of cauliflower is due to the deficiency of:
(A) Iron (B) Manganese
(C) Boron (D) Copper

1501. Percent calcium content in single superphosphate is:
(A) 12 (B) 16 (C) 20 (D) 24

1502. Biuret content of urea should more approximately by:
(A) <0.5% (B) <1.0% (C) <2.0% (D) <3.0%

1503. Movement of P and K from soil to the root surface take place due to:
(A) Root interception (B) Diffusion
(C) Mass flow (D) Osmosis

1504. What is the chemical formula of slaked lime?
(A) $CaCO_3$ (B) CaO
(C) Ca $Mg(CO_3)_2$ (D) $Ca(OH)_2$

1505. Biosuper is:
(A) Rock phosphate + S-oxidizing bacteria
(B) Rock phosphate
(C) Rock phosphate + S + Zn
(D) Rock phosphate + S + Thiobacillus

1506. The optimum temperature for nitrification is:
(A) 10-20°C (B) 20-25°C
(C) 25-30°C (D) 30-25°C

1507. Which one of the following is a vertisol?
(A) Black cotton soil
(B) Red sandy loam soil
(C) Sandy loam sodic soil
(D) Sub mountaine (tarai) soil

1508. Which of the following is commonly referred to as muriate of potash?
(A) Potassium nitrate (B) Potassium chloride
(C) Potassium sulphate (D) Potassium silicate

1509. For applying 100 kg of nitrogen, how much urea would one use?
(A) 45 kg (B) 111 kg
(C) 222 kg (D) 333 kg

1510. The crop highly sensitive to Mo deficiency :
(A) Cauliflower (B) Tomato
(C) Carrot (D) Wheat

1511. Which of the following is not a chelate compound?
(A) EDTA (B) CDMA
(C) DTPA (D) EDDHA

1512. In the process of nitrogen mineralization, the first step is called :
(A) Immobilization (B) Ammonification
(C) Amminization (D) Nitrification

1513. Which species of Rhizobium is suitable for fenugreek ?
(A) Meliloti (B) Leguminosarum
(C) Phaseoli (D) Trifoli

1514. Which of the following is Na liking crop?
(A) Chickpea (B) Maize
(C) Soybean (D) Sugarbeet

1515. Bacterial **Thiobacillus thioxidans** is involved in:
(A) Reduction of nitrogen to ammonia
(B) Phosphorus solubilization
(C) Oxidation of S in soil
(D) Conversion of urea to ammonia

1516. In cotton, Zn deficiency is known as :
(A) White bud (B) Little leaf
(C) Mottle leaf (D) Fern leaf

1517. Which of the following nutrients is involved in sugar and starch formation, lipid metabolism and N-fixation?
(A) Nitrogen (B) Phosphorus
(C) Potassium (D) Sulphur

1518. Van Bemmel factor for computing soil organic matter from soil organic carbon is:
(A) 6.250 (B) 2.292
(C) 1.724 (D) 0.895

1519. The half life of ^{32}P is:
(A) 13.8 days (B) 14.3 days
(C) 14.8 days (D) 27.6 days

1520. Which of the following is an essential nutrient for higher plants?
(A) Co (B) Na (C) pH (D) Ni

1521. Akochi disease in paddy is caused by :
(A) Reduction of sulphate to sulphide
(B) Oxidation of S to SO_4
(C) Transformation of H_2S into SO_2
(D) Transformation of SO_2 into SO_4

1522. Which group of microorganisms is most active at the terminal stage of composting?
(A) Actinomycetes (B) Bacteria
(C) Fungi (D) Protozoa

1523. At the end point of titration for determination of organic carbon, colour changes from:
(A) Blue to red
(B) Red to green
(C) Green to blue
(D) Blue to green

1524. The first micronutrient reported deficient in Indian soil is:
(A) Zn (B) Mn
(C) Fe (D) Mo

1525. Who is known as father of agricultural chemistry?
(A) Kathrotull (B) Justis Von Liebig
(C) J.H. Gilbert (D) J.B. Lawes

1526. In which crop, responses to Si application has been established widely?
(A) Onion (B) Rice
(C) Potato (D) Chickpea

1527. As per land capability classification, the classes of soils suitable for cultivation of crops are:
(A) Classes I-IV (B) Classes V-VI
(C) Classes VII-VIII (D) All of the above

1528. Two molecules that have the same chemical formula but are mirror images of each other are called:
(A) Enantiomers (B) Isomers
(C) Isotopes (D) Mirrormers

1529. The concept of nutrient mobility in soils was first proposed by:
(A) R.H. Bray (B) M.L. Jackson
(C) E. Mitscherlich (D) E.W. Russell

1530. Which micro nutrient is usually deficient in calcarious soils?
(A) Mn (B) Mo (C) Cu (D) B

1531. The C-axis spacing of smectite group of silicate clays is:
(A) 0.66 nM (B) 0.72 nM
(C) 1.01 nM (D) 1.40 nM

1532. The saturation extract of a soil has EC 2.0 dS/m. What would be its osmotic potential (in bars)?
(A) -0.18 (B) -0.36
(C) -0.72 (D) -1.08

1533. By increasing the electrolytic concentration, the thickness of diffuse double layer:
(A) Increases (B) Decreases
(C) Remain unaffected (D) Depends on pH

1534. If the nitrate formation in a soil were independent of the amount of ammoniacal-N, in that case the reaction would be considered as:
(A) Zero order
(B) First order
(C) Second order
(D) Either first or second order

1535. The first N fertilizer manufactured in India was:
(A) Ammonium chloride (B) Ammonium sulphate
(C) Urea (D) CAN

1536. Which one is a major component of organic P in soils P:
(A) Nucleic acids (B) Phospholipids
(C) Inositol phosphate (D) ATF

1537. Which position at the layer silicate has the maximum specificity for K^+ ion?
(A) e-position (B) p-position
(C) i-position (D) Both e- and p- position

1538. Which micronutrient is fixed on the same sites of the soil colloids as is phosphate?
(A) Boron (B) Copper
(C) Zinc (D) Molybdenum

1539. Which micronutrient is most strongly bound to organic matter?
(A) Fe (B) Cu (C) Zn (D) B

1540. The molarity of concentrated H_2SO_4 (AR grade) is :
(A) 10 M (B) 12 M (C) 16 M (D) 18 M

1541. Which one is not a pedogenic process?
(A) Humification (B) Calcification
(C) Laterization (D) Ferruginization

1542. Division of the factors of soil formation into active and passive groups was proposed by:
(A) H. Jenny (B) C.F. Marbut
(C) J.S. Joffe (D) V.V. Dokuchaiev

1543. A subsoil layer of high bulk density which is brittle when moist and very hard when dry is known as:
(A) Fragipan (B) Duripan
(C) Durinodes (D) Plinthite

1544. Ultisols are comparable with alfisols except for having:
(A) High base saturation of the exchange complex
(B) Low base saturation of the exchange complex
(C) Uniform soil profile
(D) High water table

1545. Total number of agro-ecological sub-regions (AESR) in India delineated by NBSS & LUP is:
(A) 21 (B) 42 (C) 60 (D) 78

1546. Dehydrogenase activity is a good index of biological activity of soil because it plays role in:
(A) Carbon metabolism
(B) Respiration
(C) Synthesis of cell macro molecules
(D) Cell division

1547. Association of two microorganisms in which one is benefitted while other remains unaffected is known as:
(A) Commensalism (B) Protocooperation
(C) Ammensalism (D) Competition

1548. The presence of `Hartig net' is a characteristic of :
(A) Endomycorrhizae (B) Ectomycorrhize
(C) Arbuscular micrrohizae (D) Ericold mycorrhizae

1549. In soil taxonomy, great groups are defined largely by :
(A) Soil foundation process
(B) Presence or absence of dignostic horizon
(C) Thermal regione
(D) Moisture regione

1550. Rhizosphere is dominated by:
(A) Gran positive spore forming bacteria
(B) Gram negative spore forming bacteria
(C) Gram negative non-spore forming bacteria
(D) Gram positive non-spore forming bacteria

1551. The toxicity of oxygen for strict anaerobic is due to the production of:
(A) Hydroxyl radical (B) Superoxide radical
(C) Hydrogen peroxide (D) Cyamide

1552. The specificity between legume host and rhizobium spp. is govern by:
(A) Flavonoids (B) Tryptophan
(C) Polysaccharides (D) Indole acitic acid

1553. Some microorganisms produce organic ligands to chelete iron in the soil and make it available to the plants. These organic ligands are known as:
(A) Porphyrines (B) Siderophores
(C) Chromophores (D) Ferrodoxins

1554. Soil-test based fertilizer recommendation for targeted yield of crops was first introduced in India by:
(A) J.S. Kanwar (B) N.P. Dutta
(C) B. Ramamoorthy (D) R.V. Tamhane

1555. TDR is a method of monitoring :
(A) Soil wetness (B) Varpour pressure
(C) Salt concentration (D) Solar radiation

1556. Which one is not a process of chemical weathering?
(A) Hydration (B) Hydrolysis
(C) precipitation (D) Carbonation

1557. In the Jenny's factor of soil formation, factor `b' refers to:
(A) Biological factor (B) Biochemical factor
(C) Base exchange (D) Biosphere

1558. Sediments deposited in fresh water lakes are called:
(A) Delta (B) Aeolian
(C) Lacustrine (D) Outwash

1559. The number of diagnostic surface horizons (epideons) included in soil taxonomy is:
(A) 6 (B) 9 (C) 12 (D) 15

1560. Argillie horizon is rich in:
(A) Fe and Al-oxides (B) Sodium
(C) $CaCO_3$ (D) Silicate clay

1561. Phosphate fixation is minimum at soil pH between:
(A) 8 and 9 (B) 7 and 8
(C) 6 and 7 (D) 5 and 6

1562. Chlorite are the minerals belonging to:
(A) Tectosilicates (B) Cyclosilicates
(C) Indosilicates (D) Phyllosilicates

1563. Intensity of blue colour in phosphorus determination is measured at a wavelength of:
(A) 420 nm (B) 470 nm
(C) 540 nm (D) 660 nm

1564. Montmorillomite minerals have external surface area (m^2/g) of:
(A) 80-150 (B) 150-180
(C) 180-250 (D) 250-300

1565. Which of the following is not a slow-release fertilizer?
(A) N-serve (B) LCU
(C) SCU (D) NCCU

1566. Capillary porosity is given by the formula
(A) 1-Field capacity
(B) Field capacity × Bulk desnity
(C) Bulk density × Particle density
(D) $\left[1 - \frac{\text{Bulk density}}{\text{FieldCapacity}} \times 100\right]$

1567. Azomethine -H is used in determination of:
(A) Molybdenum (B) Sulphur
(C) Boron (D) Calcium

1568. The first compound formed on urea hydrolysis is :
(A) Ammonium carbonate
(B) Ammonium carbamate
(C) Anhydrous ammonia
(D) Ammonium hydroxide

1569. In the universal soil loss equation, factor `s' strands for:
(A) Slope length
(B) Slope-gradient
(C) Soil texture
(D) Soil erodibility

1570. The process of bringing bulk soil sample to the desired quality is called:
(A) Reduction (B) Minimization
(C) Quartering (D) Partitioning

1571. Double rings infilterometer is used for:
(A) Maximizing the vertical and horizontal flows
(B) Reducing the vertical flow
(C) Reducing the upward flow
(D) Minimizing the horizontal flow

1572. Which is not a resource conservation technology (RCT)?
(A) Zero tillage (B) Puddling
(C) Laser levelling (D) Bed planting

1573. Aspergillus awamori is a :
(A) K-mobilizer
(B) Organic matter decomposer
(C) S-oxidizer
(D) P-solubilizer

1574. In India, the work on `biogas' was first carried out by:
(A) J.K. Mukherjee (B) S.V. Desai
(C) N.P. Datta (D) N.N. Goswami

1575. cfu is a unit for measuring:
(A) Soil pollution (B) Metal contamination
(C) Soil quality (D) Microbial population

1576. Acid sulphate soils are formed mainly due to:
(A) Reduction of sulphates
(B) Oxidation of sulphides
(C) Evolution of hydrogen sulphide
(D) Rice cultivation using ammonium sulphate

1577. Findout the most appropriate order of organic soil nitrogen in order of their ease of mineralization:
(A) Urea > amino acids > proteins > nucleic acids > amino sugars > humified N
(B) Urea > amino sugars > proteins > amino acids > humified N> nucleic acid
(C) Urea > nucleic acids > proteins > amino acids > humified N > amino sugars
(D) Urea > amino acids > nucleic acids > humified N > proteins > amino sugars

1578. Number of recognised land capability classes given by soil conservation service of USA is :
(A) IV (B) VI
(C) VIII (D) X

1579. If total soil water potential at a point A in soil is -15 DCM and at point B is -200 cm, then
(A) Water will flow from point A to B
(B) Water will flow from point B to A
(C) Water will be no flow of water between points of A and B
(D) More information is needed to draw conclusion on direction of flow of water

1580. Rhizobium japonicum fixes N is symbiotic relationship with :
(A) Pea group (B) Lupin group
(C) Soybean group (D) Phaseolus group

1581. If 2 ML of 1 N $K_2Cr_2O_4$ solution was consumed for oxidation of organic C present in 1 gram of a soil, the organic matter percentage of the soil is:
(A) 0.3 (B) 0.6 (C) 1.0 (D) 1.5

1582. Eriochrome Black-T is used for determination of :
(A) Lime requirement (B) Gypsum requirement
(C) Organic-C (D) Boron determinations

1583. Anion exchange capacity is highest in:
(A) Vermiculite (B) Montomorillonite
(C) Kaolinite (D) Illite

1584. DRIS approach was given by
(A) Bray (B) Beaufils
(C) Jackson (D) Nemeth

1585. Amount of micronutrients required by various crops can be arranged in an order:
(A) Fe > Mn>Zn>Cu>Mo (B) Mn>Zn>Fe>Cu>Mo
(C) Fe>Zn>Mn>Cu>Mo (D) Mo>Cu>Zn>Mn>Fe

1586. Mitscherlich related the potential yield of a crop to the amount of a given nutrient with the following equation :
Log (A - Y) = log A - Cb
In the equation `b' denotes
(A) Proportionality constant
(B) Potential yield
(C) Amount of given form of nutrient present in the soil
(D) Growth factor specific to a crop when all other factors are not limiting

1587. Which one of the following organisms is an obligate chemoautotraph:
(A) Frankia (B) Nitrosomonas
(C) Rhizobium (D) Anabaena

1588. Which one of the following minerals would weather most rapidly:
(A) Muscovite (B) Albite
(C) Quatz (D) Microline

1589. `BIOLOG' is used for determining
(A) Genetic fingerprints of soil microorganisms
(B) Taxonomic changes of soil microorganism
(C) Metabolic fingerprints of soil bacteria
(D) All of the above

1590. Carbon - 14 (14c) is used for C dating. What is the half life of 14c?
(A) 8370 years (B) 5730 years
(C) 1620 years (D) 13 years

1591. Prokaryotes in soil include the following groups of organisms
(A) Fungi, Actinomycetes, Bacteria, Algae
(B) Bacteria, Archaea, Actinomycetes, BGA
(C) Protozoa, Bacteria, Fungi, Archaea
(D) Algae, Protozoa, Fungi, Bacteria

1592. There are number of fundamental assumptions upon which the validity of Stokes' law is based. Which one of the assumptions given below is not correct?
(A) The particles must be large in comparison to liquid molecules
(B) Practice must not be rigid and smooth
(C) There must be no slipping between the particle and the liquids
(D) The velocity of the fall must not exceed a certain critical value so that the viscosity of the liquid remains the only resistance to the fall of the particles.

1593. Suppose that 100 kg of green manure containing 40% C and 2% N is decomposed by bacteria with C to N ratio of 5 : 1 and C assimilation efficiency of 5%. What will be the quantity of N mineralized or immobilized from the addition of green manuring?
(A) 0.4 Kg N immobilized (B) 1.6 Kg N Mineralized
(C) 2 Kg N Immobilized (D) 2 Kg N Mineralized

1594. Which of the following processes explains the formation of secondary crystalline colloids from the mica and feldspars present in the parent rock?
(A) Hydrolysis (B) Solution
(C) Coagulation (D) Carbonation

1595. When a mixture of rhizobium, Azotobacter and Vibrio was applied to rhizosphere, fixation of atmospheric nitrogen was increased. it was due to the activity of:
(A) Azospirillium and Vibrio
(B) Azotobacter and Vibrio
(C) Azotobacter and Azospirillum
(D) All of the above

1596. An element in soil will be considered as a trace element if its concentration is:
(A) 1·2 wt % (B) <0.1 wt %
(C) >2 wt % (D) <1.0 wt %

1597. Which one of the following can be estimated by Flame Photometer?
(A) Sodium and Potassium (B) Cadmium and Cobalt
(C) Chlorine and Bramine (D) Mercury and Arsenic

1598. Walkley and Black rapid titration method is used for the determination of ?
(A) Organic carbon content of soil
(B) Nitrate content of soil
(C) Phosphorus content of soil
(D) Fluoride content of soil

1599. In biological systems sulphur is largely bound in:
(A) Lipids (B) Proteins
(C) Nucleic acid (D) Proteins and nucleic acid

1600. pH of 0.01 M HNO_3 is :
(A) 0.1 (B) 1 (C) 2 (D) 10

1601. Lightening in the atmosphere produces :
(A) NO (B) CO_2 (C) CO (D) NH_3

1602. p OH of 0.001 M solution of Hcl is:
(A) 0.1 (B) 1 (C) 10 (D) 11

1603. Which one of the following category of earth worms is most suitable for vermicomposting?
(A) Epigeic (B) Anecic
(C) Endogenic (D) All of the above

1604. Carbon dioxide evolved from soil mainly comes from:
(A) Microbial respiration
(B) Root respiration
(C) Small animal respiration
(D) All of the above

1605. Which one of the following in the most predominent in a majority of igneous rocks?
(A) Al (B) Fe (C) O (D) Si

1606. Which one of the following is used to determine total organic matter by Walkley and Black method?
(A) KOH and H_2SO_4 (B) $Na_2S_2O_3$ and H_2SO_4
(C) $K_2Cr_2O_7$ and H_2SO_4 (D) HNO_3 and H_2SO_4

1607. The POH of a 0.001 M solution of HCl is:
(A) 12 (B) 10 (C) 11 (D) 13

1608. Under anaerobic conditions denitrifying Pseudomonas changes,
(A) Nitrate to molecular nitrogen
(B) Nitrae to ammonia
(C) Nitrate to Nitrite
(D) Nitrite to Nitrate

1609. Aerobic bacteria is most active in:
(A) Moist and saturated soil
(B) Moist and unsaturated soil
(C) Alluvial soil
(D) Permafrost

1610. Andisols are formed by:
(A) Biological activity (B) Erosion by wind
(C) Erosion by waves (D) Volcanoes

1611. Hydraulic conductivity is a function of:
(A) Medium alone
(B) Fluid alone
(C) Either medium or fluid
(D) Both fluid and medium

1612. What is OH– ion concentration of HCl whose pH is 3?
(A) ·3 (B) 3 (C) 10^{-3} (D) 10^{-11}

1613. Particles which have maximum ability to attract and hold K^+, Ca^{++} and NH_4^+ ions on their surface are:
(A) Clay (B) Sand
(C) Loam (D) Loamy sand

1614. Laterite soil contains more of :
(A) Iron and Aluminium
(B) Magnesium and Boron
(C) Manganese and Silicate
(D) Potassium and Lead

1615. Peaty soil is found more in :
(A) Kerala (B) Uttar Pradesh
(C) Maharashtra (D) Gujarat

1616. Brown forest soil is also known as:
(A) Entisols (B) Altisols
(C) Spedosole (D) Mollisols

1617. The soil type which is good for agriculture is:
(A) Podzols (B) Latosols
(C) Serpent soil (D) Solonachak

1618. Laterite represents:
(A) Regolith soil (B) Glacial soil
(C) Black cotton soil (D) Red soil

1619. The mineral, most resistant to chemical weathering:
(A) Olivine (B) Quartz
(C) K-feldspar (D) Biotic

1620. Removal of top fertile soil by water is called:
(A) Leaching (B) Siltation
(C) Weathering of soil (D) Soil erosion

1621. Indian monsoon period is from
(A) October to November (B) December to February
(C) June to September (D) March to May

1622. Normality of 0.25 M phosphoric acid is:
(A) 0.25 (B) 0.50 (C) 0.75 (D) 2.50

1623. **Azolla pinnata** is a :
(A) Blue green algae (B) Green algae
(C) Red algae (D) Ferns

1624. Which of the following material has the highest hydraulic conductivity?
(A) Clay (B) Sand stone
(C) Lime stone (D) Quartzite

1625. Which of the following mixture of gases is called biogas?
(A) CO_2, CH_4, NH_3, H_2S, H_2O (Vapour)
(B) CO, CH_4, NH_3, H_2S, H_2O (Vapour)
(C) CO_2, CH_4, N_2O, NH_3, H_2O (Vapour)
(D) CO_2, NO_4, H_2O, CH_4

1626. The chemical used in a fermenter with molasses as a substrate is:
(A) Diammonium sulphate
(B) Diammonium phosphate
(C) Diammonium nitrate
(D) Diammonium chloride

1627. Permanent wilting coefficient represents ..... form of water:
(A) Hygroscopic (B) Runaway
(C) Capillary (D) Chemical combined

1628. "Law of restitution" was proposed by:
(A) Olsen (B) Das
(C) Liebig (D) Bray

1629. Stem nodulating Rhizobium spp. is:
(A) Sino-rhizobium (B) Bradyrhizobium
(C) Azo-rhizobium (D) All of the above

1630. Who is considered as father of soil testing?
(A) Colwell (B) Troug
(C) Parker (D) Puri

1631. The value of Von Bemmelen factor is:
(A) 1.724 (B) 1.200
(C) 2.290 (D) 2.000

1632. Who developed the DTPA tests for available soil micronutrients
(A) Kurtz LT (B) Lindsay WL
(C) Larsen (D) Jackson ML

1633. The tension of the water at the upper plastic limit is equivalent to pF value:
(A) 1.5 (B) 0.5 (C) 2.0 (D) 2.9-3.3

1634. Thermal conductivity follows the order
(A) Sand > loam > clay > peat
(B) Sand > Peat > loam > clay
(C) Loam > sand > peat > clay
(D) Peat > sand > loam > clay

1635. Nutrient index between 1.5 and 2.5 is categorised as :
(A) Low (B) Medium
(C) High (D) None of these

1636. Hartig net in soil produce by :
(A) Rhizobium (B) Ectomycorrhiza
(C) Endomycorrhiza (D) None of the above

1637. Soil order characterized by excess of weathering with dominant iron and aluminium oxide clays, common in tropics is:
(A) Entisols (B) Mollisols
(C) Alfisols (D) Oxisols

1638. Water stable aggregates in soil increase by addition of:
(A) N-fertilizer (B) P-fertilizer
(C) K-fertilizer (D) Micronutrients

1639. Fertilizer gradient approach was given by:
(A) B.V. Mehta (B) B. Ramamoorthy
(C) A.B. Ghosh (D) B. Murthy

1640. Which of the following nutrients is lost in the greatest amount by allowing exposure of manure pile to vain:
(A) N (B) P (C) Zn (D) Ca

1641. Nitrogen deficiency in plants leads to :
(A) Chlorosis (B) Excessive growth
(C) Profuse flowering (D) Dark green colour

1642. Soil productivity takes into accounts the following:
(A) Soil structure
(B) Soil moisture
(C) Soil fertility
(D) Soil fertility and productivity

1643. Ammonium sulphate is:
(A) Straight fertilizer (B) Binary fertilizer
(C) Tertiary fertilizer (D) Mixed fertilizer

1644. Gypsum equivalent in iron sulphate is:
(A) 1.24 (B) 1.78 (C) 1.62 (D) 0.77

1645. Peat soils are:
(A) Acidic (B) Saline
(C) Alkaline (D) Saline-alkaline

1646. Essentiality of which element was established by "Arnon and Stout"?
(A) Nitrogen (B) Magnesium
(C) Chlorine (D) Molybdenum

1647. Tailor fertilizer for legume is:
(A) MOP (B) DAP
(C) SSP (D) NPK

1648. In average fertilizer use ratio of NPK in India is about?
(A) 4.6 : 2 : 1 (B) 7.3 : 3 : 1
(C) 4 : 2 : 1 (D) 6 : 3 : 1

1649. The permanent property of soil is:
(A) Texture (B) Structure
(C) Aggregation (D) pH

1650. Government of India started subsidy on fertilizers with effect from:
(A) Nov. 1, 1966 (B) Nov. 1, 1969
(C) Nov. 1, 1977 (D) Oct. 2, 1988

1651. The capacity of a soil to resist appreciable change in pH value is called?
(A) Percent base saturation
(B) CEC
(C) Buffering capacity
(D) Anion exchange capacity

1652. Mechanical analysis of soil makes use of :
(A) Darcy's law (B) Stoke's law
(C) Schofield's law (D) Ohm's law

1653. The process of replacement of one atom by another atom of similar size, in a crystal lattice of a soil clay without disrupting or changing the crystal structure of the mineral is termed as :
(A) Ion exchange (B) Isomerism
(C) Isomorphic substitution (D) Polymorphism

1654. The element which does not enter into permanent organic combination in plants is:
(A) Nitrogen (B) Phosphorus
(C) Magnesium (D) Potassium

1655. The plateau with little growth rate and continuing accumulation of nutrient element in the plants can be defined as:
(A) Severe deficiency range (B) Toxic range
(C) Moderate deficiency range (D) Sufficiency range

1656. The major organic cementing agent in soil aggregate formation is:
(A) Lipids
(B) Protein and protein derivatives
(C) Polysaccharides
(D) Organic acid

1657. Biuret content in urea should not exceed .... according to fertilizer control order, 1957:
(A) 1.0% (B) 1.5% (C) 2.0% (D) 2.5%

1658. For which type of fertilizer in India is fully dependent on imports:
(A) N-fertilizer (B) P-fertilizer
(C) K-fertilizer (D) None of the above

1659. Remote sensing helps in studying:
(A) Cropped area (B) Soil characters
(C) Underground water (D) All of the above

1660. Amongst oil cakes, the highest nitrogen content is in:
(A) Castor cake (B) Neem cake
(C) Groundnut cake (D) Coconut cake

1661. Maximum residual acidity is associated with the continuous soil application of:
(A) Urea (B) Ammonium nitrate
(C) Ammonium sulphate (D) CAN

1662. The most important potential contaminant of food produced on sewage sludge ammended soils:
(A) Chromium (B) Lead
(C) Zinc (D) Cadmium

1663. Which of the following phytohormone is involved in nutrient mobilization?
(A) Auxin (B) Gibberellin
(C) ABX (D) Cytokinin

1664. Heavy application of potassium fertilizer often lead to reduced absorption of:
(A) Mg and Ca (B) B and Mn
(C) N and P (D) Zn and Cu

1665. If a plant shows extensive interveinal chlorosis of the older leaves and late this reaches the younger leaves, the symptoms is due to lack of :
(A) Phosphorus (B) Magnesium
(C) Copper (D) Calcium

1666. The concept of Q/1 relationship was developed by:
(A) Mitscherlich (B) Beckett
(C) Schofield (D) Martin

1667. The organic residues with wider C : N ratios as compared to narrow C : N ratios are decomposed at
(A) Faster rate
(B) Slow rate
(C) Equal rate
(D) Faster rate followed by slow rate

1668. Application of organic materials with C:N ratio wider than 20 to 30:1 leads to ...... of soil N:
(A) Nitrification (B) Fixation
(C) Immobilization (D) Mineralization

1669. `A' value concept was given by:
(A) Sorenson (B) Beckett
(C) Schofield (D) Fried and Dean

1670. The process of formation of nitrogen and nitrous oxides gasses from ammonical fertilizers in soil is known as:
(A) Ammonification (B) Nitrification
(C) Denitrification (D) Mineralization

ANSWERSHEET: 11.SOIL SCIENCE

1. (A); 2. (A); 3. (B); 4. (A); 5. (D); 6. (D); 7. (D); 8. (C); 9. (D); 10. (D); 11. (B); 12. (A); 13. (D); 14. (B); 15. (A); 16. (B); 17. (D); 18. (C); 19. (C); 20. (B); 21. (D); 22. (D); 23. (A); 24. (D); 25. (A); 26. (B); 27. (B); 28. (B); 29. (C); 30. (A); 31. (C); 32. (A); 33. (B); 34. (D); 35. (B); 36. (A); 37. (D); 38. (C); 39. (A); 40. (B); 41. (A); 42. (B); 43. (A); 44. (A); 45. (A); 46. (C); 47. (A); 48. (C); 49. (C); 50. (B); 51. (D); 52. (A); 53. (C); 54. (D); 55. (C); 56. (B); 57. (B); 58. (B); 59. (A); 60. (A); 61. (A); 62. (B); 63. (A); 64. (A); 65. (C); 66. (A); 67. (B); 68. (C); 69. (C); 70. (A); 71. (A); 72. (C); 73. (B); 74. (B); 75. (A); 76. (B); 77. (A); 78. (D) ;79. (C); 80. (D); 81. (B); 82. (D); 83. (B); 84. (A); 85. (C); 86. (D); 87. (D); 88. (A); 89. (D); 90. (B); 91. (B); 92. (C); 93. (B); 94. (D); 95. (C); 96. (B); 97. (B); 98. (B); 99. (B); 100. (B); 101. (D); 102. (D); 103. (D); 104. (B); 105. (A); 106. (B); 107. (C); 108. (D); 109. (C); 110. (D); 111. (D); 112. (A); 113. (D); 114. (D); 115. (A); 116. (B); 117. (C) ; 118. (C); 119. (D); 120. (B); 121. (A); 122. (A); 123. (A); 124. (A); 125. (B); 126. (B); 127. (C); 128. (B); 129. (D); 130. (B) ; 131. (D); 132. (C); 133. (D); 134. (D); 135. (B); 136. (C); 137. (B); 138. (B); 139. (C); 140. (C); 141. (B); 142. (C); 143. (C) ; 144. (D); 145. (A); 146. (B); 147. (D); 148. (D); 149. (C); 150. (B); 151. (B); 152. (B); 153. (C); 154. (A); 155. (D); 156. (B) ; 157. (C); 158. (A); 159. (D); 160. (D); 161. (B); 162. (C); 163. (A); 164. (B); 165. (C); 166. (C); 167. (D); 168. (A); 169. (B); 170. (C); 171. (D); 172. (C); 173. (A); 174. (B); 175. (A); 176. (D); 177. (B); 178. (C); 179. (B); 180. (C); 181. (C); 182. (A) ; 183. (B); 184. (C); 185. (B); 186. (A); 187. (C); 188. (B); 189. (B); 190. (D); 191. (A); 192. (C); 193. (B); 194. (D); 195. (B); 196. (D); 197. (C); 198. (A); 199. (D); 200. (B); 201. (C); 202. (A); 203. (D); 204. (B); 205. (A); 206. (C); 207. (A); 208. (D); 209. (B); 210. (C); 211. (B); 212. (D); 213. (D); 214. (C); 215. (C); 216. (D); 217. (A); 218. (D); 219. (A); 220. (C); 221. (A); 222. (C); 223. (D); 224. (B); 225. (A); 226. (B); 227. (C); 228. (C); 229. (A); 230. (C); 231. (B); 232. (D); 233. (C); 234. (B); 235. (C); 236. (A); 237. (A); 238. (D); 239. (A); 240. (C); 241. (B); 242. (B); 243. (C); 244. (C); 245. (B); 246. (A); 247. (A); 248. (B); 249. (A); 250. (B); 251. (C); 252. (A); 253. (B); 254. (C); 255. (A);

256. (B); 257. (C); 258. (A); 259. (B); 260. (C) ; 261. (A); 262. (B); 263. (C); 264. (A); 265. (B); 266. (C) ; 267. (D); 268. (A); 269. (B); 270. (A); 271. (B); 272. (C); 273. (D) ; 274. (A); 275. (B); 276. (C); 277. (A); 278. (B); 279. (C); 280. (C); 281. (A); 282. (C); 283. (C); 284. (C); 285. (D); 286. (A); 287. (C); 288. (B); 289. (B); 290. (D); 291. (A); 292. (B); 293. (D); 294. (D); 295. (A); 296. (D); 297. (B); 298. (C); 299. (A); 300. (A); 301. (D); 302. (B); 303. (A); 304. (A); 305. (C); 306. (D); 307. (C); 308. (C); 309. (C); 310. (D); 311. (B); 312. (C); 313. (D); 314. (A); 315. (B); 316. (C); 317. (D); 318. (B); 319. (A); 320. (C); 321. (C); 322. (B); 323. (C); 324. (A); 325. (B); 326. (D); 327. (D); 328. (A); 329. (C); 330. (B); 331. (D); 332. (C); 333. (B); 334. (A); 335. (C); 336. (D); 337. (B); 338. (A); 339. (B); 340. (C); 341. (D); 342. (B); 343. (D); 344. (C); 345. (D); 346. (B); 347. (D); 348. (C); 349. (B); 350. (A); 351. (B); 352. (C); 353. (B); 354. (B); 355. (D); 356. (A); 357. (B); 358. (B); 359. (B); 360. (C); 361. (A); 362. (D); 363. (D); 364. (C) ; 365. (B); 366. (C); 367. (B); 368. (C); 369. (B); 370. (C); 371. (A); 372. (A); 373. (A); 374. (D); 375. (C); 376. (B); 377. (A) ; 378. (A); 379. (A); 380. (A); 381. (C); 382. (B); 383. (C); 384. (C); 385. (C); 386. (D); 387. (B); 388. (C); 389. (C); 390. (C); 391. (A); 392. (C); 393. (C); 394. (D); 395. (B); 396. (D); 397. (C); 398. (A); 399. (B); 400. (C); 401. (D); 402. (B); 403. (D) ; 404. (A); 405. (B); 406. (C); 407. (C); 408. (A); 409. (C); 410. (C); 411. (C); 412. (A); 413. (D); 414. (D); 415. (D); 416. (B); 417. (A); 418. (C); 419. (C); 420. (B); 421. (A); 422. (C); 423. (D); 424. (C); 425. (D); 426. (C); 427. (D); 428. (D); 429. (B) ; 430. (D); 431. (C); 432. (A); 433. (C); 434. (D); 435. (A); 436. (C); 437. (B); 438. (D); 439. (D); 440. (B); 441. (C); 442. (B) ; 443. (D); 444. (A); 445. (B); 446. (B); 447. (C); 448. (D); 449. (A); 450. (C); 451. (C); 452. (A); 453. (B); 454. (C); 455. (A); 456. (A); 457. (B); 458. (A); 459. (B); 460. (A); 461. (D); 462. (C); 463. (A); 464. (C); 465. (A); 466. (D); 467. (A); 468. (D); 469. (B); 470. (D); 471. (B); 472. (A); 473. (D); 474. (A); 475. (B); 476. (D); 477. (C); 478. (C); 479. (C); 480. (D); 481. (A); 482. (C); 483. (C); 484. (A); 485. (C); 486. (A); 487. (B); 488. (A); 489. (A); 490. (D); 491. (A); 492. (A); 493. (A); 494. (A); 495. (B);

496. (B); 497. (C); 498. (B); 499. (A); 500. (B); 501. (A); 502. (B); 503. (A); 504. (C); 505. (C); 506. (B); 507. (A); 508. (A); 509. (B); 510. (B); 511. (A); 512. (D); 513. (A); 514. (A); 515. (C); 516. (C); 517. (A); 518. (B); 519. (A); 520. (B); 521. (A); 522. (C); 523. (C); 524. (A); 525. (B); 526. (A); 527. (C); 528. (B); 529. (B); 530. (C); 531. (C); 532. (C); 533. (D); 534. (A) ; 535. (A); 536. (A); 537. (B); 538. (C); 539. (D); 540. (D); 541. (D); 542. (D); 543. (D); 544. (C); 545. (C); 546. (A); 547. (B) ; 548. (C); 549. (D); 550. (C); 551. (B); 552. (D); 553. (C); 554. (B); 555. (A); 556. (C); 557. (C); 558. (C); 559. (C); 560. (A) ; 561. (B); 562. (B); 563. (A); 564. (D); 565. (D); 566. (A); 567. (A); 568. (A); 569. (C); 570. (A); 571. (A); 572. (A); 573. (A) ; 574. (A); 575. (A); 576. (B); 577. (A); 578. (C); 579. (A); 580. (B); 581. (D); 582. (C); 583. (C); 584. (B); 585. (D); 586. (A) ; 587. (A); 588. (A); 589. (B); 590. (C); 591. (C); 592. (D); 593. (B); 594. (B); 595. (B); 596. (B); 597. (C); 598. (D); 599. (C) ; 600. (C); 601. (D); 602. (A); 603. (A); 604. (C); 605. (C); 606. (A); 607. (A); 608. (A); 609. (D); 610. (D); 611. (A); 612. (A); 613. (B); 614. (A); 615. (A); 616. (A); 617. (A); 618. (B); 619. (A); 620. (D); 621. (A); 622. (B); 623. (B); 624. (A); 625. (A); 626. (A); 627. (D); 628. (B); 629. (B); 630. (B); 631. (C); 632. (D); 633. (D); 634. (B); 635. (B); 636. (C); 637. (D); 638. (A); 639. (A); 640. (A); 641. (C); 642. (D); 643. (A); 644. (A); 645. (D); 646. (A); 647. (A); 648. (A); 649. (B); 650. (A); 651. (A) ; 652. (A); 653. (D); 654. (D); 655. (C); 656. (C); 657. (C); 658. (A); 659. (B); 660. (A); 661. (D); 662. (D); 663. (A); 664. (A) ; 665. (C); 666. (D); 667. (A); 668. (D); 669. (D); 670. (C); 671. (B); 672. (C); 673. (B); 674. (A); 675. (B); 676. (B); 677. (A) ; 678. (A); 679. (A); 680. (A); 681. (A); 682. (B); 683. (A); 684. (A); 685. (A); 686. (B); 687. (A); 688. (C); 689. (A); 690. (C) ; 691. (B); 692. (D); 693. (A); 694. (B); 695. (C); 696. (A); 697. (C); 698. (B); 699. (A); 700. (A); 701. (A); 702. (B); 703. (B); 704. (C); 705. (D); 706. (A); 707. (A); 708. (A); 709. (B); 710. (A); 711. (A); 712. (A); 713. (D); 714. (A); 715. (B); 716. (A) ; 717. (B); 718. (A); 719. (D); 720. (B); 721. (A); 722. (A); 723. (C); 724. (A); 725. (C); 726. (C); 727. (A); 728. (B); 729. (B); 730. (B); 731. (C); 732. (C); 733. (D); 734. (B); 735. (A); 736. (B); 737. (A); 738. (C); 739. (B); 740. (A); 741. (B); 742. (A) ; 743. (A); 744. (C); 745. (D); 746. (D); 747. (D); 748. (A); 749. (A); 750. (A); 751. (B); 752. (B); 753. (B); 754. (C); 755. (D) ; 756. (A); 757. (A); 758. (A); 759. (D); 760. (A); 761. (A); 762. (D); 763. (A); 764. (A); 765. (A); 766. (A); 767. (A); 768. (B) ; 769. (C); 770. (C); 771. (A); 772. (A); 773. (C); 774. (A); 775. (A); 776. (A); 777. (B); 778. (C); 779. (A); 780. (B); 781. (D) ; 782. (A); 783. (B); 784. (A); 785. (C); 786. (B); 787. (A); 788. (A); 789. (D); 790. (C); 791. (A); 792. (D); 793. (A); 794. (C); 795. (C); 796. (D); 797. (D); 798. (A); 799. (A); 800. (B); 801. (D); 802. (A); 803. (D); 804. (D); 805. (D); 806. (D); 807. (A); 808. (D); 809. (D); 810. (A); 811. (D); 812. (D); 813. (D); 814. (C); 815. (C); 816. (A); 817. (A); 818. (B); 819. (A); 820. (D) ; 821. (A); 822. (C); 823. (D); 824. (C); 825. (A); 826. (A); 827. (A); 828. (A); 829. (D); 830. (B); 831. (A); 832. (A); 833. (A) ; 834. (A); 835. (A); 836. (A); 837. (A); 838. (D); 839. (A); 840. (B); 841. (A); 842. (A); 843. (C); 844. (A); 845. (C); 846. (A); 847. (D); 848. (A); 849. (A); 850. (D); 851. (D); 852. (A); 853. (B); 854. (A); 855. (D); 856. (D); 857. (D); 858. (D); 859. (B); 860. (B); 861. (A); 862. (B); 863. (C); 864. (D); 865. (C); 866. (A); 867. (A); 868. (D); 869. (D); 870. (B); 871. (B); 872. (C); 873. (B); 874. (A); 875. (B); 876. (A); 877. (B); 878. (A); 879. (B); 880. (A); 881. (B); 882. (B) ; 883. (A); 884. (A); 885. (C); 886. (D); 887. (A); 888. (A); 889. (A); 890. (C); 891. (A); 892. (C); 893. (B); 894. (C); 895. (A) ; 896. (A); 897. (A); 898. (B); 899. (C); 900. (A); 901. (C); 902. (A); 903. (B); 904. (C); 905. (C); 906. (A); 907. (C); 908. (A); 909. (C); 910. (D); 911. (B); 912. (A) ; 913. (B); 914. (A); 915. (A);916. (A); 917. (A); 918. (C); 919. (A); 920. (A); 921. (A); 922. (A); 923. (C); 924. (A); 925. (A); 926. (C); 927. (B); 928. (A); 929. (A); 930. (B); 931. (C); 932. (B); 933. (A); 934. (A); 935. (D); 936. (A); 937. (D) ; 938. (D); 939. (C); 940. (D); 941. (A); 942. (C); 943. (D); 944. (A); 945. (D); 946. (A); 947. (B); 948. (C); 949. (A); 950. (D); 951. (A); 952. (A); 953. (A); 954. (A); 955. (C); 956. (A); 957. (D); 958. (B); 959. (A); 960. (B); 961. (D); 962. (A); 963. (D); 964. (D); 965. (B); 966. (C); 967. (A); 968. (B); 969. (A); 970. (B); 971. (B); 972. (A); 973. (A); 974. (B); 975. (B); 976. (A); 977. (A); 978. (A) ; 979. (D); 980. (A); 981. (D); 982. (D); 983. (C); 984. (C); 985. (D); 986. (C); 987. (A); 988. (A); 989. (C); 990. (D); 991. (C) ; 992. (B); 993. (D); 994. (B); 995. (D); 996. (A); 997. (C); 998. (B);

999. (A); 1000. (C); 1001. (B); 1002. (A); 1003. (C); 1004. (D); 1005. (A); 1006. (C); 1007. (A); 1008. (C); 1009. (C); 1010. (B); 1011. (B); 1012. (A); 1013. (A); 1014. (A); 1015. (B); 1016. (A); 1017. (D); 1018. (C); 1019. (A); 1020. (A); 1021. (A); 1022. (D); 1023. (D); 1024. (A); 1025. (C); 1026. (A); 1027. (B); 1028. (B); 1029. (A); 1030. (C); 1031. (C); 1032. (C); 1033. (C); 1034. (A); 1035. (D); 1036. (B); 1037. (B); 1038. (D); 1039. (A); 1040. (A); 1041. (A); 1042. (B) ; 1043. (C); 1044. (D); 1045. (A); 1046. (B); 1047. (A); 1048. (B); 1049. (D); 1050. (C); 1051. (A); 1052. (B); 1053. (B); 1054. (D); 1055. (B) ; 1056. (A); 1057. (A); 1058. (B); 1059. (C); 1060. (C); 1061. (B); 1062. (C); 1063. (A); 1064. (D); 1065. (A); 1066. (A); 1067. (A); 1068. (A) ; 1069. (C); 1070. (B); 1071. (B); 1072. (B); 1073. (B); 1074. (B); 1075. (B); 1076. (A); 1077. (D); 1078. (C); 1079. (B); 1080. (A); 1081. (C) ; 1082. (A); 1083. (A); 1084. (A); 1085. (A); 1086. (C); 1087. (C); 1088. (C); 1089. (D); 1090. (A); 1091. (B); 1092. (C); 1093. (D); 1094. (A) ; 1095. (C); 1096. (C); 1097. (B); 1098. (D); 1099. (C); 1100. (B); 1101. (C); 1102. (A); 1103. (A); 1104. (B); 1105. (B); 1106. (D); 1107. (C) ; 1108. (A); 1109. (B); 1110. (B); 1111. (A); 1112. (C); 1113. (D); 1114. (B); 1115. (A); 1116. (C); 1117. (B); 1118. (C); 1119. (A); 1120. (B); 1121. (B); 1122. (D); 1123. (C); 1124. (B); 1125. (B); 1126. (A); 1127. (B); 1128. (A); 1129. (B); 1130. (C); 1131. (A); 1132. (B); 1133. (C); 1134. (A); 1135. (B); 1136. (C); 1137. (A); 1138. (B); 1139. (A); 1140. (A); 1141. (A); 1142. (A); 1143. (C); 1144. (B); 1145. (A); 1146. (B) ; 1147. (C); 1148. (A); 1149. (B); 1150. (C); 1151. (D); 1152. (B); 1153. (D); 1154. (B); 1155. (A); 1156. (A); 1157. (B); 1158. (D); 1159. (B); 1160. (C); 1161. (B); 1162. (D); 1163. (A); 1164. (A); 1165. (C); 1166. (B); 1167. (D); 1168. (A); 1169. (C); 1170. (A); 1171. (B); 1172. (A) ; 1173. (B); 1174. (C); 1175. (D); 1176. (A); 1177. (B); 1178. (C); 1179. (D); 1180. (C); 1181. (B); 1182. (A); 1183. (B); 1184. (C); 1185. (D); 1186. (A); 1187. (B); 1188. (C); 1189. (D); 1190. (B); 1191. (C); 1192. (A); 1193. (C); 1194. (B); 1195. (C); 1196. (D); 1197. (A); 1198. (B); 1199. (A); 1200. (C); 1201. (A); 1202. (C); 1203. (A); 1204. (B); 1205. (D); 1206. (D); 1207. (B); 1208. (A); 1209. (C); 1210. (D); 1211. (D) ; 1212. (A); 1213. (B); 1214. (C); 1215. (A); 1216. (B); 1217. (C); 1218. (A); 1219. (B); 1220. (C); 1221. (D); 1222. (A); 1223. (B); 1224. (D); 1225. (C); 1226. (A); 1227. (B); 1228. (C); 1229. (C); 1230. (A); 1231. (D); 1232. (D); 1233. (D); 1234. (B); 1235. (C); 1236. (D) ; 1237. (A) ; 1238. (A); 1239. (D); 1240. (C); 1241. (B); 1242. (D); 1243. (D); 1244. (B); 1245. (A); 1246. (C); 1247. (D); 1248. (D); 1249. (A) ; 1250. (D) ; 1251. (A); 1252. (C); 1253. (C); 1254. (A); 1255. (B); 1256. (A); 1257. (A); 1258. (B); 1259. (C); 1260. (A); 1261. (B); 1262. (C); 1263. (D) ; 1264. (D); 1265. (A); 1266. (C); 1267. (A); 1268. (C); 1269. (A); 1270. (B); 1271. (C); 1272. (D); 1273. (A); 1274. (A); 1275. (D); 1276. (C) ; 1277. (B); 1278. (D); 1279. (B); 1280. (A); 1281. (D); 1282. (C); 1283. (C); 1284. (B); 1285. (D); 1286. (A); 1287. (C); 1288. (A); 1289. (D) ; 1290. (B); 1291. (D); 1292. (A); 1293. (B); 1294. (C); 1295. (B); 1296. (A); 1297. (C); 1298. (A); 1299. (C); 1300. (B); 1301. (D); 1302. (C) ; 1303. (A); 1304. (B); 1305. (D); 1306. (A); 1307. (B); 1308. (C); 1309. (D); 1310. (D); 1311. (A); 1312. (B); 1313. (C); 1314. (B); 1315. (A) ; 1316. (C); 1317. (A); 1318. (B); 1319. (C); 1320. (D); 1321. (A); 1322. (B); 1323. (B); 1324. (C); 1325. (A); 1326. (A); 1327. (D); 1328. (B); 1329. (C); 1330. (A); 1331. (C); 1332. (B); 1333. (B); 1334. (A); 1335. (B); 1336. (A); 1337. (C); 1338. (C); 1339. (D); 1340. (B); 1341. (D) ; 1342. (C); 1343. (C); 1344. (D); 1345. (B); 1346. (C);1347. (C); 1348. (D); 1349. (B); 1350. (A); 1351. (A); 1352. (D); 1353. (D); 1354. (B); 1355. (A); 1356. (B); 1357. (C); 1358. (B); 1359. (A); 1360. (A); 1361. (D); 1362. (C); 1363. (A); 1364. (A); 1365. (B); 1366. (C); 1367. (B) ; 1368. (A); 1369. (C); 1370. (C); 1371. (D); 1372. (A); 1373. (A); 1374. (D); 1375. (A); 1376. (B); 1377. (C); 1378. (A); 1379. (C); 1380. (A) ; 1381. (B); 1382. (D); 1383. (D); 1384. (B); 1385. (A); 1386. (B); 1387. (A); 1388. (D); 1389. (D); 1390. (A); 1391. (B); 1392. (C); 1393. (B) ; 1394. (A); 1395. (C); 1396. (B); 1397. (D); 1398. (B); 1399. (A); 1400. (D); 1401. (D); 1402. (B); 1403. (D); 1404. (B); 1405. (D); 1406. (B) ; 1407. (A); 1408. (D); 1409. (B); 1410. (C); 1411. (C); 1412. (C); 1413. (A); 1414. (B); 1415. (D); 1416. (A); 1417. (C); 1418. (A); 1419. (B) ; 1420.

(B); 1421. (D); 1422. (D); 1423. (C); 1424. (B); 1425. (C); 1426. (B); 1427. (C); 1428. (C); 1429. (A); 1430. (B); 1431. (B); 1432. (B) ; 1433. (B); 1434. (B); 1435. (C); 1436. (C); 1437. (C); 1438. (C); 1439. (A); 1440. (B); 1441. (C); 1442. (D); 1443. (D); 1444. (A); 1445. (B); 1446. (D); 1447. (D); 1448. (A); 1449. (B); 1450. (D); 1451. (A); 1452. (D); 1453. (A); 1454. (B); 1455. (B); 1456. (A); 1457. (C); 1458. (A) ; 1459. (C); 1460. (A); 1461. (B); 1462. (C); 1463. (B); 1464. (C); 1465. (A); 1466. (B); 1467. (C); 1468. (C); 1469. (B) ; 1470. (C); 1471. (C); 1472. (B); 1473. (C); 1474. (B); 1475. (C); 1476. (B); 1477. (C); 1478. (A); 1479. (A); 1480. (C); 1481. (C); 1482. (D) ; 1483. (D); 1484. (C); 1485. (B). 1486. (A); 1487. (B) ; 1488. (B); 1489. (C); 1490. (C). 1491. (D); 1492. (C) ; 1493. (D); 1494. (D); 1495. (C). 1496. (D); 1497. (C) ; 1498. (A); 1499. (A); 1500. (C). 1501. (A); 1502. (A); 1503. (C); 1504. (D); 1505. (D); 1506. (D) ; 1507. (A); 1508. (B); 1509. (C); 1510. (A); 1511. (B); 1512. (B); 1513. (A); 1514. (D); 1515. (C); 1516. (B); 1517. (C); 1518. (C); 1519. (B) ; 1520. (D); 1521. (C); 1522. (B); 1523. (B); 1524. (A); 1525. (B) 1526. (B); 1527. (A); 1528. (B); 1529. (A); 1530. (D); 1531. (A); 1532. (C) ; 1533. (B); 1534. (A); 1535. (B); 1536. (C); 1537. (C); 1538. (D); 1539. (B); 1540. (D); 1541. (B); 1542. (C); 1543. (A); 1544. (B); 1545. (C); 1546. (B); 1547. (A); 1548. (B); 1549. (B); 1550. (B); 1551. (B); 1552. (A); 1553. (B); 1554. (C); 1555. (A); 1556. (C); 1557. (D); 1558. (C) ; 1559. (B); 1560. (D); 1561. (C); 1562. (D); 1563. (D); 1564. (A); 1565. (A); 1566. (D); 1567. (C); 1568. (B); 1569. (B) ; 1570. (A); 1571. (D); 1572. (B); 1573. (D); 1574. (B); 1575. (D); 1576. (B); 1577. (B); 1578. (C); 1579. (A); 1580. (C); 1581. (A); 1582. (B) ; 1583. (C); 1584. (B); 1585. (A). 1586. (D); 1587. (B) ; 1588. (C); 1589. (D); 1590. (B). 1591. (B); 1592. (B) ; 1593. (B); 1594. (A); 1595. (C). 1596. (B); 1597. (A) ; 1598. (A); 1599. (B); 1600. (C) 1601. (A); 1602. (D); 1603. (A); 1604. (A); 1605. (C); 1606. (C) ; 1607. (C); 1608. (A); 1609. (B); 1610. (D); 1611. (D); 1612. (D); 1613. (A); 1614. (A); 1615. (A); 1616. (A); 1617. (B); 1618. (A); 1619. (B) ; 1620. (D); 1621. (C); 1622. (C); 1623. (D); 1624. (B); 1625. (A) 1626. (B); 1627. (A); 1628. (C); 1629. (C); 1630. (B); 1631. (A); 1632. (B) ; 1633. (B); 1634. (A); 1635. (B); 1636. (B); 1637. (D); 1638. (B); 1639. (B); 1640. (A); 1641. (A); 1642. (D); 1643. (A); 1644. (C); 1645. (A); 1646. (D); 1647. (D); 1648. (B); 1649. (A); 1650. (C); 1651. (C); 1652. (B); 1653. (C); 1654. (D); 1655. (D); 1656. (C); 1657. (B); 1658. (C) ; 1659. (D); 1660. (C); 1661. (C); 1662. (D); 1663. (A); 1664. (D); 1665. (B); 1666. (B); 1667. (B); 1668. (C); 1669. (D) ; 1670. (C);

Chapter 12

ENTOMOLOGY

1. Pest is:
(A) Organism feeding on and damaging others
(B) Fungal pathogen attacking plants and animals
(C) Bacterial pathogen attacking plants and animals
(D) Viral pathogen attacking organisms

2. DDT and 2,4-D were introduced:
(A) Towards end of World War II
(B) Towards beginning of World War II
(C) During World War I
(D) Just after the World War I

3. Bordeaux mixture is named after:
(A) Scientist (B) Chemical composition
(C) Country (D) University

4. Insecticides kill:
(A) Harmful insects
(B) Both harmful and useful insects
(C) Specific insects
(D) Only plant pests

5. Which is organochlorine:
(A) Malathion (B) Parathion
(C) Aldicarb (D) DDT

6. Which one is an organophosphate:
(A) BHC (B) Aldrin
(C) Fenitrothion (D) Endosulphan

7. A carbamate pesticide is:
(A) Propoxur
(B) Simazine
(C) Atrazine
(D) Tetraethyl pyrophosphate

8. Urea derivative pesticides belong to the category of:
(A) Organochlorines (B) Organophosphates
(C) Carbamates (D) Triazines

9. Carbamates employed as insecticide are:
(A) Phenylcarbamates (B) Thiocarbamates
(C) Propoxur (D) Dithiocarbamates

10. Carbamates used as fungicides are:
(A) Thiocarbamates
(B) Dithiocarbamates
(C) Carbofuron
(D) Carbofuron and Aldicarb

11. The pesticides having very low biodegradation and strong affinity for fatty tissues are:
(A) Organochlorines (B) Organophosphates
(C) Triazines (D) Pyrethroids

12. Pesticide ingredient methyl isocyanate which killed thousands in Bhopal gas tragedy is intermediate of:
(A) Mustard gas (B) Nitrous acid
(C) Carbaryl (D) Carbon tetrachloride

13. The most common pesticide used in India is:
(A) DDT (B) Endrin
(C) BHC (D) Baygon

14. Pesticides comprise:
(A) Herbicides and insecticides
(B) Insecticides only
(C) Insecticides, nematicides and rodenticides
(D) Fungicides, herbicides, insecticides, nematicides and rotendicides

15. Herbicide carbamates are:
(A) Dithiocarbamates (B) Thiocarbamates
(C) Phenyl carbamates (D) Both (B) and (C)

16. Household pesticides consist of:
(A) Finit, Baygon, Mosquitorepellent cakes and coils
(B) Finit and Baygon
(C) Finit/Flit
(D) DDT

17. DDT is:
(A) Carbamate (B) Organophosphate
(C) Organochlorine (D) Triazine

18. First commercial pesticides was:
(A) Burgandy mixture (B) Bordeaux mixture
(C) DDT (D) 2,4-D

19. Mode of pesticides action is similar in:
(A) DDT and parathion
(B) Organochlorine and organophosphates
(C) Organochlorine and carbamates
(D) Organophosphates and carbamates

20. The wheat aphid is:
(A) *Lipaphis erysimi* (B) *Myzus persicae*
(C) *Macrosiphum miscanthi* (D) *Aphis neri*

21. The wheat seeds are treated with chlorpyriphos 20 EC against termite @:
(A) 3 to 4 ml/kg (B) 15-20 ml/kg
(C) 25 to 30 ml/kg (D) 35 to 40 ml/kg

22. Eggs of *Atherigona naqvi* are laid in the:
(A) Buds
(B) Flowers
(C) Inside the leaves
(D) At the base of stem and under side of leaves

23. The field rates are controlled by:
(A) Zinc chloride (B) Zinc phosphide
(C) Zinc oxide (D) Copper oxide

24. Which one of the following caste of termite is sterile:
(A) King (B) Queen
(C) Soldiers (D) Winged termite

25. White grubs damage:
(A) Roots (B) Stems
(C) Leaves (D) Flowers

26. Which one of the following is 'June Beetle':
(A) *Holotrichia consanguinea*
(B) *Xanthopimia sp.*
(C) *Rodolia sp.*
(D) *Coccinella sp.*

27. Sugarcane shoot borer damages:
(A) Roots (B) Stem
(C) Leaves (D) All of these

28. 'Bunchy top' in sugarcane is caused by:
(A) Root borer (B) Pyrilla
(C) White fly (D) Scale insects

29. 'Anal-tufts' are found in the:
(A) Adult pyrilla (B) Nymphs of pyrilla
(C) Nymphs and adults both (D) None of these

30. *Beauvaria brassiana* is effective against control of:
(A) Termite (B) Pod borer
(C) Grasshopper (D) Grubs

31. There is a pupal stage in:
(A) Pyrilla (B) Termite
(C) White fly (D) Jassids

32. The female of which insect has a tuft of hairs at its anal end?
(A) Sugarcane top borer
(B) Sugarcane leaf hopper
(C) Sugarcane root borer
(D) Sugarcane white grubs

33. *Epiricanea melanoleuca* is a parasite of pyrilla:
(A) Nymphs only (B) Adults only
(C) Nymphs and adult both (D) Eggs

34. *Amrasca biguttula* is:
(A) Jassid (B) Leaf roller
(C) Whitefly (D) Stem weevil

35. 'Hibernacula' is found in:
(A) Flower beetle (B) Stem weevil
(C) Spotted boll worm (D) Pink boll worm

36. Trycho card is very effective against control of:
(A) Termite (B) Pod borer
(C) Grasshopper (D) Cutworm

37. Spotted boll worm of cotton pupaes is:
(A) Inside the boll
(B) Out of the boll but naked
(C) Out of the boll in a coccon
(D) Inside the seeds

38. Which one of the following is cotton leaf roller?
(A) *Oxycarenus laetus*
(B) *Sylepta derogata*
(C) *Dysdercus cingulatus*
(D) *Pempherulus affinis*

39. The larvae of *Sylepta derogata* are:
(A) Greenish bluish (B) Pinkish
(C) Blackish (D) Yelowish

40. Nymphs of red cotton bug also damage:
(A) Wheat (B) Sugracane
(C) Gram (D) Lady's finger

41. Which one of the following insects come from store to infest crop in field?
(A) Red cotton bug (B) Pink boll worm
(C) Leaf roller (D) Spotted boll worm

42. In which of the following insects a greenish strip is found in forewings?
(A) *Earia vitella* (B) *Earias insulana*
(C) *Oxycarenus leatus* (D) *Pectinophora gossypii*

43. Which one of the following lays stalked eggs?
(A) *Bemisia tabaci* (B) *Aphis gossypii*
(C) *Myllocerms sp* (D) *Pempherulus affinis*

44. Larvae of *Helicoverpa armigera* damage:
(A) Pods (B) Roots
(C) Stem (D) Buds

45. Which one of the following is a cut worm:
(A) Plume moth (B) Pod borer
(C) *Agrotis sp* (D) *Heliothis sp.*

46. Seed treatment of *Beauvaria brassiana* @ 10 g/kg is effective against control of:
(A) Termite (B) Pod borer
(C) Grasshopper (D) Cutworm

47. Against *Helicoverpa armigera*, NPV is applied @:
(A) 250 LE/ha (B) 100 LE/ha
(C) 50 LE/ha (D) 10 LE/ha

48. Cutworms pupate:
(A) Inside the pods
(B) Inside the stem
(C) On upper surface of soil
(D) In an earthen shell inside the soil

49. The plume moth belongs to the family:
(A) Noctudae (B) Hespiridae
(C) Pterophoridae (D) Gelehiidae

50. *Agrotis flammatra* larvae cut the:
(A) Leaves (B) Buds
(C) Flowers (D) Stems

51. A dark smoky spot is present on the centre of fore wings is:
(A) *Helicoverpa armigera* (B) *Agrotis ipsilon*
(C) *Etiella zinckenella* (D) *Exelastris atomosa*

52. Which one of the following is better for the control of gram pod borer?
(A) Parathion (B) Phosphomidon
(C) Chlorpyriphos (D) Malathion

53. Pheromone trap is used for capturing:
(A) Larvae (B) Pupae
(C) Adult males (D) Adult females

54. Agricultural chemicals denote:
(A) Pesticides
(B) Pesticides and fertilizers
(C) Fertilizers and growth regulators
(D) Pesticides, fertilizers and growth regulators

55. Growth regulators produce:
(A) Stimulation of growth
(B) Retardation of growth
(C) Some cause stimulation while others bring about retardation
(D) Shedding of leaves

56. Loss of agricultural productivity due to pests and diseases is:
(A) 1% (B) 30% (C) 5% (D) 15%

57. Household pesticide sprays are:
(A) Finit (Flit)
(B) Finit and Baygon
(C) Finit, Baygon, Mosquito coils and Çakes
(D) DDT

58. Cabamates include:
(A) Fungicides, herbicides and insecticides
(B) Insecticides only
(C) Herbicides, insecticides and nematicides
(D) Insecticides, nematicides and rotendicides

59. The first organic pesticide to be used commercially was:
(A) Bordeaux mixture (B) Burgandy mixture
(C) DDT (D) 2,4-D

60. Bordeaux mixture is:
(A) Organochlorine (B) Antibiotic
(C) Organophosphate (D) Inorganic fungicide

61. The most famous pesticide of world is:
(A) BHC (B) DDT
(C) Baygon (D) Flit

62. In India, the most commonly used pesticide BHC is:
(A) Carbamate (B) Triazine
(C) Organochlorine (D) Antibiotic

63. The pesticide used in foundations of buildings for preventing termite attack is:
(A) DDT (B) BHC
(C) Endrin (D) Aldrin

64. The chemical present in Temik is:
(A) Malathion (B) Propoxur
(C) Carbofuran (D) Aldicarb

65. The chemical present in flit (finit) is:
(A) Malathion (B) DDT
(C) BHC (D) Aldicarb

66. The number of organophosphates synthesized is:
(A) 1000 (B) 100
(C) 10000 (D) 100000

67. The mode of action of insecticides is through the attack on:
(A) Respiratory system (B) Nervous system
(C) Muscular system (D) Both (A) and (B)

68. One of earliest pesticides employed by humans was:
(A) Nicotine
(B) Margosa (Neem) leaves
(C) Pyrethrum
(D) DDT

69. An example of pesticide treadmill is:
(A) Forty times increase in requirement for cotton
(B) Development of mosquito resistance to DDT, BHC, malathion and fenitrothion
(C) Decreasing use of pesticides due to introduction of biological control
(D) Decreasing use of pesticides due to decrease in the population of pests

70. DDT concentration increases in the human body because it is soluble in:
(A) Water (B) Vitamins
(C) Fat (D) Protein

71. White grub is an important insect of:
(A) Sugarcane (B) Potato
(C) Ground nut (D) Mustard

72. Army worm is a serious pest of:
(A) Wheat (B) Chickpea
(C) Rice (D) Groundnut

73. At which stage, Gandhi bug damages the rice crop:
(A) Tillering stage (B) Panicle initiation stage
(C) Milking stage (D) Maturity stage

74. Yellow stem borer of rice is:
(A) *Scirpophaga innotata*
(B) *Scirpophaga incertulas*
(C) *Orseolia oryzae*
(D) *Dicladispa armigera*

75. 'Hopper burn' is caused by:
(A) *Leptocorisa varicornis* (B) *Bemisia tabaci*
(C) *Spodoptera litura* (D) *Nilaparvata lugens*

76. 'Dead heart' is caused by:
(A) Paddy stem borer (B) Termite
(C) Rice gandhi bug (D) Paddy stem weevil

77. 'Silver shoot' is caused by:
(A) Grass hopper (B) Stem borer
(C) White fly (D) Gall midge

78. 'Chaffy appearance' in the paddy is caused on which part of plant:
(A) Roots (B) Tillers
(C) Flowers (D) Ears

79. Rice Gandhi bug attacks appear on the plant at:
(A) Seedling stage (B) Flowering stage
(C) Milky stage (D) Tillering stage

80. Female of paddy hispa lays eggs:
(A) Inside the roots (B) Inside the stem
(C) Inside the leaves (D) Inside the flowers

81. Systemic granules are applied @ 1.5 kg a.i.:
(A) Per hectare (B) Per acre
(C) Per 100 m^2 (D) Per 1000 m^2

82. *Odontotermis obesus* is controlled by the application of:
(A) Phosphamidon (B) Aluminium chloride
(C) Chlorpyriphos (D) Brestanol

83. Nuptial flight is found in:
(A) Termite (B) Grasshopper
(C) White fly (D) Gujhia weevil

84. Which one of the following is called as 'Army worm'?
(A) Leptocorisa (B) Mythimna
(C) Spodoptera (D) Tanymecus

85. Wheat shoot fly attacks the plants at:
(A) Seedling stage (B) Flowering stage
(C) Milky stage (D) Maturity stage

86. *Atherigona naqvii* belongs to family:
(A) Curculionidae (B) Aphididae
(C) Anthonomidae (D) Pyralidae

87. Which one of the following is Gujhia weevil?
(A) *Tanymcuns* (B) *Mylabris pustulatus*
(C) *Proceras indicus* (D) *Tryporyza incertulas*

88. BHC is:
(A) Benzene hexachloride (B) Benzine heptachloride
(C) Benzene hydrochloride (D) Bromide herbicide

89. *Drasicha mangiferae* is known as:
(A) Mango mealy bug (B) Pink boll worm
(C) Cutworm (D) None of these

90. Mango mealy bug is harmful to:
(A) Mango (B) Guava
(C) Pipal (D) Ber

91. Herbicides prevent:
(A) Transpiration (B) Water absorption
(C) Translocation of salts (D) Photolysis of water

92. Pesticides generally attacks:
(A) Muscular system (B) Respiratory system
(C) Nervous system (D) Circulatory system

93. Pests resistant to insecticides have:
(A) Enzymes for metabolizing toxins
(B) Ability to store toxin in fat
(C) Less permeable cuticles
(D) All of these

94. Most effective pesticides are:
(A) Carbamates (B) Organochlorine
(C) Organophosphates (D) All of these

95. DDT content in Indian population has touched:
(A) 1·12 ppm (B) 11·30 ppm
(C) 25·50 ppm (D) 48·98 ppm

96. Which one is not matched:
(A) Organochlorine – DDT
(B) Pyrethroid – Atrazine
(C) Organophosphate – Malathion
(D) Carbamate - Carbofuran

97. Pesticides with very low biodegradation but strong affinity for fatty tissues are:
(A) Organochlorines (B) Organophosphates
(C) Pyrethrines (D) Triazines

98. Bordeaux mixture was formulated by:
(A) Leeuwehoek (B) Tillet
(C) De Bary (D) Millardet

99. Baygon contains:
(A) Malathion (B) Carbofuran
(C) Propoxur (D) Aldicarb

100. Which ingredient killed hundreds of people in Bhopal gas tragedy:
(A) Nitrous acid (B) Mustard gas
(C) Carbon tetrachloride (D) Methyl isocyanate

101. DDT is:
(A) Biodegradable (B) Non-biodegradable
(C) Antibiotic (D) Growth regulator

102. Which is a fungicide:
(A) 2,4-D (B) DDT
(C) Bordeaux mixture (D) BHC

103. Zinc phosphide is a pesticide used for:
(A) Spraying (B) Fumigation
(C) Killing of weevils (D) Killing of rodents

104. Rotenone is:
(A) A bioherbicide (B) A natural hormone
(C) An insect hormone (D) A natural insecticide

105. 2,4-D is an effective:
(A) Weedicide (B) Pesticide
(C) Fungicide (D) Rodenticide

106. Organophosphate inhibit:
(A) Cholinesterases (B) Enterokinase
(C) Urease (D) Carbohydrases

107. Fungicide Bordeaux mixture is:
(A) Magnesium hydroxide
(B) Copper sulphate + Sodium hydroxide
(C) Copper sulphate + Calcium hydroxide
(D) Magnesium sulphate + Calcium hydroxide

108. Which of the following is cholinesterase inhibitor:
(A) Endosulphan (B) Aldrin
(C) BHC (D) Malathion

109. Which is major component of Bordeaux mixture:
(A) Copper sulphate (B) Calcium chloride
(C) Sodium chloride (D) Magnesium sulphate

110. Malathion, parathion and fumitrothion belong to group of pesticides:
(A) Triazines (B) Carbamates
(C) Pyrethroids (D) Organophosphates

111. Ethylene dibromide is:
(A) Stomach poison (B) Contact poison
(C) Fumigant (D) Biological control

112. A widely but periodically occupying plant disease is:
(A) Andemic (B) Epidemic
(C) Sporodic (D) Epiphytotic

113. Insecticides generally attack:
(A) Respiratory system (B) Nervous system
(C) Muscular system (D) Circulatory system

114. Carbamates used as fungicides are:
(A) Thiocarbamates
(B) Dithiocarbamates
(C) Carbofuran
(D) Carbofuran and Aldicarb

115. Biopesticides include:
(A) Bioherbicides
(B) Bioinsecticides
(C) Bioherbicides and bioinsecticides
(D) Bioherbicides, bioinsecticdes and biofertilizers

116. A transgenic crop that is resistant to hornworm larvae is:
(A) Sunflower (B) Tomato
(C) Tobacco (D) Sugarcane

117. A major pest eradicated through release of sterile males is:
(A) Screw worm (B) Aphids
(C) Ladybug (D) Praying mantis

118. Pheromone are chemicals that are used in animals for:
(A) Sex attractants
(B) Sending alarm signals
(C) Marking trails
(D) All of these

119. Rotenone is a:
(A) Natural herbicide (B) Bioherbicides
(C) Natural insecticide (D) Insect hormone

120. A natural insecticide obtained from plants is:
(A) Azadirachtin
(B) Rotenone
(C) Nicotine, pyrethrum and cinerin
(D) All of these

121. Cochineal insect has been used in checking the wild growth of:
(A) Opuntia (B) Eichhornia
(C) Aphids (D) Screw worm

122. Pheromone traps have:
(A) Sterile males (B) Male pheromones
(C) Female pheromones (D) Sterile females

123. In confusion technique, the pheromone coated paper strips are thrown over an area in order to:
(A) Attract insects and kill them
(B) Repel insects from an area
(C) Confuse males so that they are unable to locate females
(D) Confuse females so that they are unable to locate males

124. Juvenile hormone causes:
(A) Early sexual maturity
(B) Late sexual maturity
(C) Completion of larval stages
(D) Development of giant larva without ability to undergo maturation

125. Pyrethrin is a common ingredient of:
(A) Mosquito coils (B) Fly sprays
(C) Mosquito mats (D) All of these

126. Thurioside is:
(A) Insecticide (B) Fungicide
(C) Antibiotic (D) Weedicide

127. Antifeedant property occurs in:
(A) Nicotine (B) Azadirachtin
(C) Rotenone (D) Cinerin

128. Ladybug is a predator of:
(A) Moths (B) Beetles
(C) Bacteria (D) Aphids

129. Which one is an effective plant insecticide:
(A) Pyrethrin (B) Cinerin
(C) Nicotine (D) All of these

130. Pyrethrin is extracted from:
(A) *Azadiracta indica*
(B) *Helianthus annus*
(C) *Poa indica*
(D) *Chrysanthemum cinerarifolium*

131. The earliest pesticide was:
(A) Margosa (B) Pyrethrum
(C) Nicotine (D) DDT

132. First bioinsecticide developed on commercial scale was:
(A) Quinone (B) DDT
(C) Organophosphate (D) Sporeine

133. Sporoneine was developed in:
(A) Germany (B) France
(C) USA (D) Russia

134. The phenomenon of using a predator for controlling a pest is:
(A) Biological control (B) Genetic engineering
(C) Artificial control (D) Confusion technique

135. Which is a microbial insecticide:
(A) *Bacillus thuringensis* (B) *B. subtilis*
(C) *B. polymixa* (D) *B. brexis*

136. IPM (Integrated Pest Management) involves :
(A) Biological control (B) Pesticides
(C) Confusion technique (D) Biofertilizers

137. *Corpora allata* is removed from a nymph. It will:
(A) Remain nymph for life
(B) Become adult
(C) Change to next nymph and remain in that state forever
(D) Die after some time

138. To insects ecdysone takes part in:
(A) Growth and development
(B) Maturation into adult
(C) Moulting till pupa is formed
(D) Secretion of cuticle

139. Thurioside is proteinaceous toxin obtained from:
(A) Biofertilier (B) Green manure
(C) Bacterial origin (D) Farmyard manure

140. Latest trend in plant disease control is:
(A) Chemical control
(B) Biological control
(C) Good manure and fertilizer
(D) Breeding for disease resistance

141. A third generation pesticide is:
(A) Pheromone
(B) Weedicide
(C) Juvenile hormone analogue
(D) Sterilogastrone

142. Genus *Radopholus* of nematode was first described in 1893 on the basis of a male specimen obtained from necrotic roots of ——— growing in Fiji Islands:
(A) Sugarbeet (B) Sugarcane
(C) Banana (D) Coconut

143. The nematode *Radopholus similes* has shown the maximum numbers at a soil depth of:
(A) 5-15 m (B) 20-30 m
(C) 32-42 m (D) 1-4 m

144. Stem borer can be controlled by:
(A) Putting mixture of chloroform and carbon bisulphide into the holes
(B) Spray of monocrotophos
(C) Insect may be killed with the help of wire
(D) All of these

145. Red spider can be controlled by:
(A) Spray of sulphur (B) Spray of zinc
(C) Spray of iron (D) Spray of 2,4-D

146. *Paratylenshus projectus* nematode has been found to cause considerable damage in red clover yield at —— nematodes/kg of soil:
(A) 2-7 (B) 8-13
(C) 14-19 (D) 20-25

147. *Xiphinema cobbi* (nematode) has been described after 1980 from:
(A) Brazil (B) South Africa
(C) India (D) Italy

148. It has been estimated that insects had existed on this earth for at least:
(A) 100 million (B) 150 million
(C) 200 million (D) 250 million

149. Stored grain pests can be killed by —— hours exposure to 51.5-54.5 ^{0}C:
(A) 2 (B) 3 (C) 5 (D) 7

150. Most of the insects cannot survive the temperatures above:
(A) 29.9 ^{0}C (B) 39.9 ^{0}C
(C) 49.9 ^{0}C (D) 59.9 ^{0}C

151. Ultra – Low Volume (ULV) or Low Volume Concentration (LVC) refers to the use of spray volume (liquid) less than :
(A) 5 litres (B) 7 litres
(C) 9 litres (D) 11 litres

152. Malathion is a insecticide:
(A) Slow fumigant (B) Contact plus systemic
(C) Contact plus stomach (D) Systemic

153. Metasystox – R is a ——— insecticide:
(A) Sytemic (B) Fumigant
(C) Contact (D) Contact plus stomach

154. Phorate is a ——— insecticide:
(A) Systemic (B) Contact plus systemic
(C) Contact (D) Contact plus stomach

155. Phosdrin is a ———insecticides:
(A) Slow fumigant (B) Systemic
(C) Systemic plus contact (D) Contact

156. A minimum period of —— years is to be given before assessing the effectiveness of an important biocontrol agent:
(A) 2 (B) 3 (D) 4 (D) 5

157. The queen termite produces ——— eggs per second:
(A) 40 (B) 50 (C) 60 (D) 70

158. Honey bee queen produces ——— thousand eggs per day:
(A) 0.5-1.5 (B) 2-3
(C) 4.0-4.5 (D) 5.0-5.5

159. An average locust swarm spreads over:
(A) 5 sq km (B) 10 sq km
(C) 15 sq km (D) 20 sq km

160. One kilometer of locust swarm weights as much as —— tones of locusts:
(A) 100 (B) 200
(C) 300 (D) 400

161. The duration of locust plague period lasts for:
(A) 2-3 years (B) 3-4 years
(C) 5-10 years (D) 11-16 years

162. A locust swarm as large as ——— square kilometers is on record:
(A) 100 (B) 200
(C) 300 (D) 400

163. Dead heart in rice is the symptom of insect damage:
(A) Stem borer (B) Leaf roller
(C) Grass hopper (D) Gandhi bug

164. Pod borer of pulses can be controlled by:
(A) Spray of Endosulphan 35 EC @ 1.5 l/ha
(B) Spray of Monocrotophos @ 0.75 l/ha
(C) Spray of Quinolphos 25 EC @ 1.25 l/ha
(D) All of these

165. Grasshoppers lay eggs:
(A) On soil surface (B) Below soil surface
(C) In nest (D) All of these

166. The colour of insects is:
(A) Red (B) Green and yellow
(C) Voilet (D) Pink

167. Suprannal plate is also called as:
(A) 11th tergite (B) Podical plate
(C) Hook (D) All of these

168. Holopneustic respiration system refers to:
(A) All spiracles open and work effectively
(B) Only first one pair spiracles open and work
(C) Only last one pair spiracles open and work
(D) Only first and last pair of spiracles open and work

169. Propneustic respiration system refers to:
(A) All spiracles open and work effectively
(B) Only first one pair spiracles open and work
(C) Only last one pair spiracles open and work
(D) Only first and last pair of spiracles open and work

170. Metapneustic respiration system refers to:
(A) All spiracles open and work effectively
(B) Only first one pair spiracles open and work
(C) Only last one pair spiracles open and work
(D) Only first and last pair of spiracles open and work

171. Johnson's organ is a:
(A) Same organ (B) Respiratory organ
(C) Circulatory organ (D) All of these

172. Johnson's organ is found at:
(A) 2nd segment of antennal (B) Tympanum
(C) Mouth cavity (D) Overyduct

173. Stadium is a:
(A) Stage of insects during moulting
(B) Time between two moulting
(C) Play ground for insects
(D) None of these

174. Apneustic respiration means:
(A) All spiracles remain closed
(B) Only few spiracles closed
(C) All spiracles open
(D) Only few spiracles open

175. Sound producing and receiving organs are called :
(A) Scolophorous organs (B) Olfactory organs
(C) Gustatory organs (D) None of these

176. A metomorphosis or Ameta bola is a process in which:
(A) No moulting occurs
(B) No metamorphosis occurs
(C) Only growth in size and colour may occur
(D) All of these

177. Which of the following is the most important character of diptera order?
(A) Only 2 wings are found
(B) Larvae are called maggot
(C) Mendibles absent
(D) All of these

178. How many moulting stages are found in grasshopper from egg laying to adult?
(A) 2 – 3 (B) 3 – 4
(C) 5 – 6 (D) 6 – 7

179. Levelling of insect should indicate the information:
(A) Name of insect and its host plant
(B) Date and place of collection
(C) Name of collector
(D) All of these

180. Potato crop is mainly damaged by:
(A) Crow (B) Wild boar
(C) Parrot (D) None of these

181. Which of the following insect is predator?
(A) Coccinellids (B) Chrysopa sp.
(C) Mantis (D) All of these

182. Blue triable on insecticide container indicates:
(A) Highly poisonous (B) Poisonous
(C) Medium poisonous (D) Less poisonous

183. The dark yellow triangle on insecticide container indicates:
(A) Highly poisonous (B) Poisonous
(C) Medium poisonous (D) Less poisonous

184. Dark green triable on insecticide container indicates:
(A) Highly poisonous (B) Poisonous
(C) Medium poisonous (D) Less poisonous

185. Which of the following poison is used in killing bottle?
(A) Potassium cynide (B) Hydrocynic acid
(C) H_2S (D) None of these

186. Snails and sludge are:
(A) Phytophagous (B) Polyphagous
(C) Saprophytes (D) None of these

187. Yellowing of flour is caused by:
(A) Rice weevil (B) Red rust flour beetle
(C) Khapra beetle (D) All of these

188. Which stage of pulse weevil is harmful?
(A) Grub only (B) Adult only
(C) Grub and adult (D) None of these

189. Khapra beetle is also known as:
(A) Pai (B) Grain moth
(C) Red surahi (D) All of these

190. Bihar hairy caterpillar can be controlled by:
(A) Chlorpyriphos (B) Endosulphan
(C) Nuvan (D) All of these

191. Honey dew secretion is found in:
(A) Grasshopper (B) Pyrilla
(C) White fly (D) None of these

192. Sugar cane shoot borer completes its life cycle in:
(A) 34-44 days (B) 24-26 days
(C) 14-16 days (D) 44-54 days

193. Pyrilla can be controlled by spraying of:
(A) Endosulphan (B) Rogor
(C) Phosphemidon (D) All of these

194. Red pumpkin beetle is a harmful insects of:
(A) Tomato (B) Cucurbits
(C) Rice (D) Potato

195. Gandhi bug can be controlled by:
(A) Thiodan 4% dust @ 20 kg/ha
(B) Malathion 5% dust @ 30 kg/ha
(C) Carbaryl 10% dust @ 20 kg/ha
(D) All of these

196. Life cycle of fruit fly are:
(A) Egg – Larvae – Adult
(B) Egg – Nymph – Adult
(C) Egg – Maggot – Pupa – Adult
(D) All of these

197. Life cycle of Gandhi bug are:
(A) Egg – Nymph – Adult
(B) Egg – Larvae – Pupa – Adult
(C) Egg – Maggot – Adult
(D) All of these

198. Life cycle of red pumpkin beetle are:
(A) Egg – Grub – Pupa – Adult
(B) Egg – Larvae – Pupa – Adult
(C) Egg – Maggot – Pupa – Adult
(D) Egg – Nymph – Adult

199. Ovoviviparous insects can:
(A) Adult can yield nymph either directly or from eggs
(B) Adult produces only larvae
(C) Adult can produce nymphs for eggs only
(D) All of these

200. Ovoviviparous insect is:
(A) Grasshopper (B) Mustard aphid
(C) Fruit fly (D) Shoot borer

201. Which of the following is a natural enemy of mustard aphid as:
(A) *Cocciniella septumpunctata*
(B) *Chrysopa spp.*
(C) *Syrphid maggot*
(D) All of these

202. Gram caterpillar is harmful to:
(A) Chickpea (B) Sunflower
(C) Tomato (D) All of these

203. Scientific name of gram caterpiller is:
(A) *Heliothis armigera* (B) *Athalia spp*
(C) *Nezara viridula* (D) None of these

204. Scientific name of mustard saw fly is:
(A) *Athalia proxima* (B) *Bagrada cruciferaum*
(C) *Trogogerma granarium* (D) None of these

205. Which of the following stage of mealy bug is harmful?
(A) Larvae only
(B) Female adult only
(C) Male adult only
(D) Larvae and female adult

206. Caterpillar of rice weevil is called as:
(A) Nymph (B) Larvae
(C) Maggot (D) Grub

207. Most harmful stage of khapra beetle is:
(A) Grub (B) Adult
(C) Pupa (D) All of these

208. Which of the following stage of lesser grain moth is harmful to grains?
(A) Grub only (B) Adult only
(C) Grub and adult (D) Nymph only

209. Maximum harm occurs due to presence of pulse weevil during:
(A) July – September
(B) September – December
(C) January – February
(D) None of these

210. Red rust flour remain most active during:
(A) Winter season (B) Summer season
(C) Rainy season (D) All season

211. Grain moth is also known as:
(A) Grain and Flour caterpillar
(B) Potato tuber moth
(C) Rice weevil
(D) Cut worm

212. Which of the life stage of grain moth is harmful?
(A) Egg and grub (B) Caterpillar
(C) Nymph and adult (D) Adult

213. Khpara beetle may be controlled at:
(A) <16 % O_2 (B) >16 O_2
(C) <21 % O_2 (D) >21 % O_2

214. The life cycle of pulse weevil is:
(A) 15-20 days (B) 25-28 days
(C) 30-35 days (D) 40-42 days

215. Net like structures are formed in flour due to presence of:
(A) Khapra beetle (B) Red rust flour beetle
(C) Rice weevil (D) None of these

216. Following insecticides may be used to control of mealy bug is:
(A) Phosphomedon 100 EC (B) Malathion 50 EC
(C) Thiodan 35 EC (D) All of these

217. Mango leaf hopper is most active in the month of:
(A) February – November
(B) December – January
(C) September – October
(D) All of these

218. *Sitotroga cerealella* is a scientific name of:
(A) Grain moth (B) Pulse beetle
(C) Rice weevil (D) None of these

219. Which of the following part of grains is damaged by khapra beetle?
(A) Seed (B) Endosperm
(C) Embryo (D) Cotyledons

220. Azospirillum biofertilizer induce resistance in sorghum against pest of:
(A) Shoot fly (B) Head borer
(C) Cut worm (D) None of these

221. Which of the preventive measure should be adopted for safe storage?
(A) Fumigation of godowns with E.D.C.T.
(B) Bags should be treated with 0.1 % malathion solution
(C) Put E.D.B. ampules 3.0 ml/q grain
(D) All of these

222. Heat spots in stored grains are:
(A) Favourable to insect multiplication due to optimum temperature
(B) Harmful to insects
(C) Spots of low temperature and high humidity
(D) None of these

223. Which of the storage pest may survive at $\leq$ 10 % moisture in stored grains?
(A) Rice weevil (B) Khapra beetle
(C) Pulse beetle (D) All of these

224. The process of which insect has many life stage, is called as:
(A) Monocious (B) Diocious
(C) Polymorphism (D) None of these

225. Mustard aphid can be controlled by spray of:
(A) Phosphomidon 100 EC @ 250 ml/ha
(B) Rogor 30 EC @ 1000 ml/ha
(C) Thiodan 35 EC @ 1250 ml/ha
(D) All of these

226. Painted mustard bug can be controlled by spraying of:
(A) Malathion @ 1.25 l/ha (B) Dimecron @ 250 ml/ha
(C) Rogor @ 1.00 l/ha (D) All of these

227. Cut worm generally attack on plants at:
(A) Day only (B) Day and night
(C) Night only (D) None of these

228. Highest production of mustard aphid is found in the month of:
(A) November – December (B) January – February
(C) February – March (D) March – April

229. Which of the following is a parasite insect of Gurdaspur borer?
(A) *Trichogramma spp.* (B) *Apanteles flavipes*
(C) *Sturmiopsis interens* (D) All of these

230. Dead heart in sugarcane is caused by:
(A) Top borer (B) Pyrilla
(C) Gandhi bug (D) Shoot borer

231. Sugarcane crop is damaged severely by shoot borer in the month of:
(A) April – June (B) July – September
(C) October – December (D) January –March

232. Termite control in house building is possible by using:
(A) 25% Tri chlorobenzene in fuel oil
(B) 5% penta chlorophenol
(C) 1 % Chlordane
(D) All of these

233. Cactus can be controlled by which insect:
(A) *Cactoblastic cactorium* (B) *Coccinellids*
(C) *Apanteles* (D) None of these

234. The virus of Bunchy top disease of banana is transmitted by:
(A) Fruit fly (B) Pentalonia
(C) Thrips (D) None of these

235. Plant tolerance refers to:
(A) Plant kills insects
(B) Plant dies after insect attack
(C) Plant does not show any side effect of insect incidence
(D) None of these

236. Antibiosis refers to:
(A) Ability of plant to suppress the reproduction potential of the insect
(B) Production of antibodies by insect
(C) Insecticide to kill insects
(D) None of these

237. Phytoecdysone is a:
(A) Plant origin hormone similar to moulting hormone of insect
(B) Insect moulting hormone
(C) Plant steroids to kill insects
(D) None of these

238. The pheromone of honey bee that suppresses sex differentiation among workers is known as:
(A) 9- Keto-trans-2 decenoic acid
(B) Gyplure
(C) Dolphene
(D) None of these

239. Which of the following is a repellent for mosquitoes?
(A) Dolphene (B) Citronella oil
(C) Camphor oil (D) All of these

240. Baygon acts as antifedant against:
(A) Cotton ball weevil (B) Moth worm
(C) Cut worm (D) None of these

241. Which of the following is a protoplasmic poison?
(A) Nitrophenols (B) Lead arsinate
(C) Sodium tetra borate (D) All of these

242. Which of the following is a fumigants?
(A) Methyl bromide (B) HCN
(C) EDCT (D) All of these

243. Which of the following is a plant origin?
(A) Nicotine sulphate (B) Pyrethrins
(C) Ryania (D) All of these

244. Which of the following is a stomach poison?
(A) Arsenic salt
(B) Flourine compounds
(C) Chlorinated hydrocarbons
(D) All of these

245. Rogor is also known as:
(A) Metasystox (B) Dimethoate
(C) Phosdrin (D) None of these

246. Endosulphan is also known as:
(A) Lindane (B) Thiodan
(C) Aldrin (D) BHC

247. Sterility method as a synergist if it is mixed with?
(A) Pyrethrum (B) Lindane
(C) Endosulphan (D) None of these

248. The chemicals that make insects sterile are known as:
(A) Systemic poison (B) Chemosterilants
(C) Contact poison (D) Pheromones

249. DDVP is known as:
(A) Nuvan (B) Malathoate
(C) Thiodan (D) Sulfec

250. DDT is a:
(A) Fumigant (B) Contact poison
(C) Stomach poison (D) All of these

251. Polyhedrosis virus can control:
(A) *Heliothis armigera* (B) *Heliothis virescence*
(C) *Alfalfa caterpillar* (D) All of these

252. Which of the radio active material was used for γ rays in sterile male technique?
(A) Ra ^{232}
(B) P ^{32}
(C) Co^{60}
(D) Ba^{232}

253. Which of the insect was first of all controlled by knipling techniques?
(A) Screw worm fly
(B) Mealy bug
(C) Mango hopper
(D) Mustard aphid

254. Cannibalism is a process in which insects:
(A) Fight and kill each other
(B) Protect each other
(C) Attract for copulation
(D) None of these

255. Stored grains can be saved from insect damage if the grain moisture content is:
(A) < 10%
(B) > 10 %
(C) 10 %
(D) None of these

256. The grasshopper always move towards:
(A) South-east direction
(B) South-west direction
(C) North-east direction
(D) North-west direction

257. Which of the following is the optimum temperature for insect growth and development?
(A) 5-10°C
(B) 13-40°C
(C) 42-55°C
(D) 15-20°C

258. Which of the following is the antidote of insecticide poisoning?
(A) Atropine sulphate
(B) Diacetyl monohexane
(C) Pyridine aldoxymet
(D) All of these

259. Apposition refers to:
(A) Mosaic vision
(B) It is found in diurnal insects
(C) Separate image is made in each omatidium
(D) All of these

260. On pH scale, insect blood is a:
(A) Acidic
(B) Basic
(C) Neutral
(D) None of these

261. Typical or primitive type of mouth parts are found in:
(A) Grubs
(B) Honey bee
(C) Bugs
(D) Mosquitoes

262. Trachea and tracheoles are found in:
(A) RBC
(B) WBC
(C) Blood
(D) All of these

263. Insects can live in:
(A) Water only
(B) Land only
(C) Air only
(D) All of these

264. Exoskeleton of insects is made of:
(A) Hemi-cellulose
(B) Cellulose
(C) Chitin
(D) None of these

265. Which is the most important pest of cauliflower is?
(A) Sawfly
(B) Aphids
(C) Cabbage borer
(D) Flea beetle

266. Which is the most important pest of onion is?
(A) Thrips
(B) Maggot
(C) Tobacco caterpiller
(D) Flea beetle

267. Which is the most important pest of bitter guard is:
(A) Red Pumpkin beetle
(B) Aphid
(C) Cutworm
(D) Fruitfly

268. Fruit fly is pest of bottle guard and damaged fruits so, fruit fly can be cotrolled by:
(A) By staking
(B) BY seed treatment
(C) Bt spray of Aldrin dust
(D) By spray of 0.2 % Sevin solution at flowering stage

269. Which is the most important pest of carrot is:
(A) Cut worm
(B) Wire worm
(C) White grub
(D) All of these

270. Sweet potato aphinx attacks mainly on:
(A) Leaves
(B) Stem
(C) Roots
(D) All of these

271. Insects hide at low temperature is known as:
(A) Aestivation
(B) Hibernation
(C) Natural safety
(D) None of these

272. Insects hide at high temperature is known as:
(A) Aestivation
(B) Hibernation
(C) Natural safety
(D) None of these

273. Which of the following rat species is found in agricultural farms only:
(A) *Bandicota bengalensis*
(B) *Nesokia indica*
(C) *Tetera indica*
(D) All of these

274. Red triable on insecticide container indicates:
(A) Highly poisonous
(B) Poisonous
(C) Medium poisonous
(D) Less poisonous

275. Relaxing insects of insect means:
(A) To collect dead insects in a special bottle for establishing its bodies
(B) To kill the insects
(C) To mount the insects
(D) None of these

276. Which of the following insects are insect parasites?
(A) *Trichogramma spp.*
(B) *Microbacon spp.*
(C) *Sarcophaga spp.*
(D) All of these

277. Which of the following is a nocturnal insect?
(A) Butterfly
(B) Beetle
(C) Moth
(D) Honey bee

278. Dragon fly belongs to order:
(A) Odonota
(B) Orthoptera
(C) Isoptera
(D) Hemipetra

279. Termites belong to order:
(A) Odonota
(B) Hemiptera
(C) Isoptera
(D) None of these

280. Simple metamorphosis refers to:
(A) Simple change in size from Nymph to adult. Nymph are almost similar to adult,
(B Egg – Nymph – Adult stages are found
(C) It is found only in exopterygota insects
(D) All of these

281. Which of the following is known as incomplete metamorphosis?
(A) A metamorphosis
(B) Complete metamorphosis
(C) Simple metamorphosis
(D) None of these

282. Mciller's organ is a:
(A) Sense organ
(B) It is a group of small many sense organs
(C) It is found in Tympanum
(D) All of these

283. Complete metamorphosis is found in:
(A) Exopterygota (B) Endopterygota
(C) Apterygota (D) None of these

284. Mustard aphid has a mouth organs:
(A) Sucking type
(B) Piercing and sucking type
(C) Siphoning type
(D) Chewing type

285. Which of the following stage of mustard aphid is harmful to crop plants?
(A) Nymph (B) Egg
(C) Adult (D) Nymph and Adult

286. Which stage of mustard saw fly is harmful to crop plants?
(A) Egg (B) Grubs
(C) Adult (D) Grub and Adult

287. Harmful stage of painted mustard bug is:
(A) Nymph (B) Egg
(C) Adult (D) Nymph and Adult

288. Harmful stage of red pumpkin beetle is:
(A) Nymph (B) grubs
(C) Adult (D) Grubs and Adult

289. Which of the following is a natural enemy of mustard aphid is:
(A) *Cocciniella septumpunctata*
(B) *Chrysopa spp.*
(C) *Syrphid maggot*
(D) All of these

290. Harmful stage of mango leaf hopper is:
(A) Nymph (B) Egg
(C) Adult (D) Nymph and Adult

291. Mango leaf roller commonly known as:
(A) Mango chanpa (B) Lassi
(C) Tella (D) All of these

292. Mango leaf hopper has mouth parts:
(A) Piercing type (B) Sucking type
(C) Piercing and sucking type (D) Chewing type

293. Scientific name of mango hopper is:
(A) *Amritodus atkinsoni* (B) *Athalia spp*
(C) *Heliothis spp* (D) None of these

294. Stingless honey bee is:
(A) *Melipona spp.* (B) *Apis dorsata*
(C) *Apis florae* (D) None of these

295. Sting honey bee is found only in:
(A) Queen (B) Drone
(C) Worker (D) All of these

296. Queen honey bee completes its life cycle in:
(A) 15-16 days (B) 18-20 days
(C) 23-24 days (D) 25-26 days

297. Drone honey bee completes its life cycle in:
(A) 15-16 days (B) 18-20 days
(C) 23-24 days (D) 25-26 days

298. Workers completes its life cycle in:
(A) 15-16 days (B) 18-20 days
(C) 23-24 days (D) 25-26 days

299. One honey bee hive contains nearly total honey bees:
(A) 100-500 (B) 1000-5000
(C) 30000-50000 (D) More than 50000

300. One honey bee hive has drones:
(A) 1 - 5 (B) 10 - 20
(C) 200 - 300 (D) More than 300

301. One honey bee hive contains nearly queen honey bees:
(A) 1 (B) 2 (C) 3 (D) 4

302. Which of the following honey bee is found in India?
(A) *Apis dorsata* (B) *Apis florae*
(C) *Apis indica* (D) All of these

303. Which of the following honey bee smallest?
(A) *Apis dorsata* (B) *Apis florae*
(C) *Apis indica* (D) *Melipona spp.*

304. Which of the following honey bee produce the sweetest honey?
(A) *Apis dorsata* (B) *Apis florae*
(C) *Apis indica* (D) *Melipona spp.*

305. Which of the following honey bee is largest one?
(A) *Apis dorsata* (B) *Apis florae*
(C) *Apis indica* (D) *Melipona spp.*

306. Air conditioning system is found only in:
(A) Honey bee (B) Silkworm
(C) Butterfly (D) None of these

307. Which of the following is most common silk insects in India?
(A) *Bombyx mori* (B) *Antheraca puphia*
(C) *Antheraea ossima* (D) *Phlocania ricini*

308. Which of the following is a moonga silk insects?
(A) *Bombyx mori* (B) *Antheraca puphia*
(C) *Antheraea ossima* (D) *Phlocania ricini*

309. Silk worm completes its life cycle in:
(A) 20-30 days (B) 40-50 days
(C) 50-60 days (D) 60-0 days

310. Silk production and its related aspects are studied under:
(A) Sericulture (B) Apiculture
(C) Floricultre (D) Pisciculture

311. Stifiling is a process in which:
(A) Cocoon are boiled to kill the insect
(B) Cocoons are stored at 4^0C for 15 days
(C) Adults are allowed to remove from cocoons
(D) None of these

312. Lac is a:
(A) Resinous matter (B) Solid matter
(C) Liquid (D) Gaseous matter

313. Lac is mainly used in:
(A) Gramophone industry (B) Leather industry
(C) Printing industry (D) Paper industry

314. Which of the following state is the largest producer of lac in India?
(A) Bihar (B) UP
(C) MP (D) AP

315. Lac is generally produced by:
(A) Female Lac insect
(B) Male Lac insect
(C) Both male and female Lac insects
(D) None of these

316. Which of the following is a raw lac?
(A) Eric lac (B) Stick lac
(C) Seed lac (D) Shell lac

317. Lac insects may be reared on following trees?
(A) *Butea monosperme* (B) *Zizyphus spp.*
(C) *Schleichera spp* (D) All of these

318. Scientific name of lac insects is:
(A) *Kerria lacca* (B) *Apis dorsata*
(C) *Bombyx mori* (D) *Phlocania ricini*

319. Which of the following is the predatory insect of lac insect?
(A) *Crysopa spp.* (B) *Eublema amabilis*
(C) *Holcocera pulvarea* (D) All of these

320. European honey bee species is:
(A) *Apis mellifera* (B) *Apis dorsata*
(C) *Melipona spp* (D) *Apis florae*

321. Honey robber is also known as:
(A) *Acherontia styx* (B) Wax moth
(C) *Pebrine* (D) None of these

322. Honey contains sugar:
(A) 55 % (B) 78 % (C) 91 % (D) 99 %

323. Polymorphism is found in:
(A) Mustard aphid (B) Lemon butterfly
(C) Cotton leaf roller (D) None of these

324. Mustard aphid is a serious pest of:
(A) Rape seed and mustard (B) Chickpea
(C) Pigeonpea (D) Sesame

325. Scientific name of mustard aphid is:
(A) *Lipaphis erysimi* (B) *Lipophis cruciferaum*
(C) *Athalia proxima* (D) None of these

326. Which of the following is a natural enemies of honey bees?
(A) Wax moth (B) *Acherontia styx*
(C) *Vespa cincta* (D) All of these

327. Which is the origin place of silk?
(A) China (B) India
(C) Egypt (D) Malaysia

328. Mustard aphid mainly attacks on:
(A) Stem (B) Leaf
(C) Flower (D) All of these

329. Which of the following silk race is found in our country?
(A) Univoltine race (B) Multi-voltine race
(C) Bi-voltine race (D) None of these

330. Pink boll worm is mainly harmful to:
(A) Stem (B) Flower
(C) Bolls (D) All of these

331. The species of silk worm that complete only one life cycle in a year, is called:
(A) Univoltine race (B) Multi-voltine race
(C) Bi-voltine race (D) None of these

332. Central Silk Research Board is situated at:
(A) Bahrampur (WB)
(B) Cuttack (Orissa)
(C) Dehradune (Uttarakhand)
(D) Lucknow (UP)

333. Which of the country is the highest producer of lac?
(A) India (B) China
(C) Nepal (D) Bangladesh

334. How many times the female lac insect is larger than male lac insects?
(A) Two times (B) Three times
(C) Four times (D) Five times

335. Which of the following organ is found in female lac insects?
(A) Eyes (B) Legs
(C) Wings (D) Antennae

336. Adult lac male insect survive only for:
(A) 1-2 days (B) 3-4 days
(C) 5-7 days (D) More than 7 days

337. Which of the instar does not have eyes, legs and antennae?
(A) Ist (B) 2nd
(C) 3rd (D) 4rth

338. The colour of the lac insect is:
(A) Pink and Red (B) Green – red
(C) Blue – red (D) Black

339. Brood lac is a:
(A) The branch of a plant that contains lac cells
(B) Caterpiller of lac insect
(C) Pupa of lac insect
(D) None of these

340. Which of the following is a rat species?
(A) *Rattus rattus* (B) *Rattus norvegicus*
(C) *Mus musculus* (D) All of these

341. Brown or Norvey rat is:
(A) *Rattus rattus* (B) *Rattus norvegicus*
(C) *Mus musculus* (D) None of these

342. Mites have legs?
(A) 4 pairs (B) 3 pairs
(C) 2 pairs (D) Do not have legs

343. The mouth organ of mites is:
(A) Cutting, piercing and sucking type
(B) Chewing and lapping type
(C) Sponging type
(D) None of these

344. The middle stage between egg and adult of mites is called as:
(A) Nymph (B) Larvae
(C) Maggot (D) None of these

345. Miticides is used to control of:
(A) Mites (B) Rats
(C) Insects (D) Nematodes

346. Which of the following poison is used to control of rats?
(A) Zinc phosphide (B) Barium carbonate
(C) Arsenium oxide (D) All of these

347. Which of the fumigants can be used against control of rats?
(A) Sulphos (B) Cynogas dust
(C) Hydrocynic acid (D) All of these

348. Flying poxes are:
(A) Bats (B) Porcupine
(C) Parrot (D) None of these

349. Larvae are preserved in:
(A) 70% alcohol solution
(B) 5 % salt solution
(C) 80 % sugar solution
(D) 10% sulphuric acid solution

350. Which of the following is not a ingredient of universal antidote for insecticide poisoning?
(A) Charcoal (B) Tannic acid
(C) Milk of magnesia (D) All of these

351. Insect requires relative humidity for its better growth and development?
(A) 30-40 % (B) 40-50%
(C) 50-60 % (D) 60-80 %

352. Coccinellids are predator on:
(A) Coccinellids (B) Chrysopa spp
(C) Mantis (D) All of these

353. Most of the insects become inactive at temperature?
(A) <40°F (B) >40°F
(C) 40°F (D) 50°F

354. Most of the insects become active at temperature?
(A) <140°F (B) >140°F
(C) 80°F (D) 100°F

355. Insects may be killed by using mainly:
(A) Low frequency waves (B) High frequency waves
(C) Ultrasound waves (D) None of these

356. *Rodolia cardinalis* (lady bird beetle) is a predator insect of:
(A) *Icerya purchasi* (B) *Pseudococcus gahani*
(C) *Pseudococcus nipoe* (D) None of these

357. *Pseudoapliyeus utilis* is a parasite insect of:
(A) Mealy bug (B) Sugarcane weevil
(C) Cottony cushion scale (D) None of these

358. *Cytrohinus mundulus* is a predator insect of:
(A) Mealy bug (B) Sugarcane weevil
(C) Cottony cushion scale (D) None of these

359. *Aphelinus mali* is predator insects of:
(A) Wooly aphid (B) Potato tuber moth
(C) Fruit fly (D) None of these

360. Milky disease in Japani beetle is caused by:
(A) Protozoa (B) Bacteria
(C) Fungi (D) Virus

361. Insects can be trapped by using the light?
(A) Infrared light (B) Ultra violet light
(C) Far infrared light (D) None of these

362. Insects may be attracted maximum at light intensity of:
(A) 3650A (B) 4000A
(C) 5000 A (D) More than 6000 A

363. Fruit fly can be biologically controlled by:
(A) *Opius vandenboschi* (B) *Rodolia cardinalis*
(C) *Aphelinus mali* (D) None of these

364. Insects were controlled by parasitic viruses first of all in:
(A) USA (B) Canada
(C) China (D) India

365. *Cocobacillis acridiorum* bacteria can be used against:
(A) Locust (B) Housefly
(C) Grasshopper (D) Mites

366. Japanese beetle is controlled by bacteria:
(A) *Bacillus popilliae* (B) *Bacillis lentimarbus*
(C) Both (A) and (B) (D) None of these

367. *Bacillus thuringiensis* is used against:
(A) Larvae of lapidoptera (B) Beetle
(C) Cockroach (D) Grasshopper

368. Empusa fungus can be used against:
(A) Housefly (B) Housefly and Locust
(C) Moth (D) Grubs

369. *Icerya purchasi* is a lemon pest that came in India from which country:
(A) Australia (B) USA
(C) UK (D) Canada

370. *Aspidiotus prenicious* is a lemon pest that came in India from which country:
(A) China (B) Australia
(C) UK (D) Canada

371. *Erriosoma lanigerum* is a lemon pest that came in India from which country:
(A) China (B) USA
(C) Europe (D) Srilanka

372. Acaricides are used against:
(A) Beetles (B) Mites
(C) Moths (D) Grubs

373. Sesamin acts as a synergist if it is mixed with:
(A) Pyrethrum (B) Lindane
(C) Endosulphan (D) Monocrotophos

374. Which of the following is a chemosterilants?
(A) Tepa (B) Matepa
(C) Colchison (D) All of these

375. Which of the following is a systemic poison?
(A) Lindane (B) Thiodan
(C) Aldrin (D) All of these

376. Dimethoate is very effective against:
(A) Housefly (B) Fruit fly
(C) Aphids (D) All of these

377. Which of the following is a respiratory poison?
(A) Hydrogen sulphide (B) Carbon monoxide
(C) Phosphin (D) All of these

378. Which of the following form of BHC is most toxic?
(A) α BHC (B) α BHC
(C) γ BHC (D) δ BHC

379. Which of the following is a physical repellent?
(A) Water barriers
(B) Tar
(C) Oil bands around tree trunks
(D) All of these

380. Which of the following is not attractant?
(A) Pheromones (B) Sound
(C) Light (D) Chemosterilants

381. Nerve poisons are:
(A) Chlorinated hydrocarbons
(B) Organo phosphates
(C) Plant origin insecticides
(D) All of these

382. Which of the following is a contact poison?
(A) Chlorinated hydrocarbons
(B) Organo phosphate
(C) Carbamates
(D) All of these

383. NPV refers to:
(A) Nuclear Plyhedrosis Virus
(B) Non Polyhedrosis Virus
(C) Nuclear Polyhedrosis Viroids
(D) None of these

384. DDT refers to:
(A) Dichloro-diphenyl-trichloro ethane
(B) Dichloro-dinitro-tetra chloride
(C) Diphenyl-dichloride-trichloro ethane
(D) None of these

385. Pheromone is released by:
(A) Male insects (B) Female insects
(C) Male and female insects (D) None of these

386. Antifeedants are substances that:
(A) Kill the insects
(B) Repell the insects
(C) Repell insects without killing them
(D) None of these

387. Which of the following is a predator insects?
(A) Trichogramma (B) Coccinellids
(C) Apanteles (D) None of these

388. Which of the following is a predator insects?
(A) Trichogramma (B) Coccinellids
(C) Apanteles (D) None of these

389. Termites are very active during:
(A) Day (B) Night
(C) Day and Night (D) None of these

390. Six segmented head is found in:
(A) Myriapods (B) Arthopoda
(C) Annelida (D) None of these

391. Homeostasis refers to:
(A) Stability of population in a ecosystem
(B) Change in insect population
(C) Increase of insect population
(D) None of these

392. Gyplure is a:
(A) Synthetic compound related to natural compounds of gypsy moth
(B) Natural compound of gypsy moth
(C) Natural sex hormone of fruit fly
(D) None of these

393. Which of the following is not a locust species?
(A) *Locusta migraioria* (B) *Pantanga succinata*
(C) *Schistocerca gregarta* (D) *Cactoblastic spp.*

394. Which stage of pink boll worm is harmful to plants?
(A) Larvae only (B) Larvae and Adult
(C) Adult only (D) None of these

395. The only plant that remained untouched by *Schistocerca gregarta* (Desert locust):
(A) Mango (B) Ber
(C) Neem (D) None

396. Lantana can be controlled by:
(A) Lantana bug (B) Scale insect
(C) Fruit fly (D) None of these

397. Sexone is a:
(A) Insecticide (B) Weedicide
(C) Hormone mimics (D) Pesticide

398. Which of the following is a repellent for mosquitoes?
(A) Dolphene (B) Citronella oil
(C) Camphore oil (D) All of these

399. Monophagy refers to:
(A) Insect feeding on simple plant species
(B) Insect feeding on large variety of plants
(C) Insect feeding on few plants
(D) None of these

400. Oligophagy refers to:
(A) Insect feeding on simple plant spp.
(B) Insect feeding on large number of plants
(C) Insect feeding on few plants
(D) None of these

401. Which of the following stage of termite is harmful to plants?
(A) Nymph (B) Adult
(C) Nymph and adult (D) Pupae

402. Termite caste refers to:
(A) Sexual differentiation
(B) Physical differentiation
(C) Difference in mouth organs
(D) Difference in colour

403. Which of the following is the sterile caste of termite?
(A) Queen (B) Workers
(C) Soldiers (D) Workers and soldiers

404. Termite can be controlled by use of:
(A) BHC (B) Aldrin
(C) Chlordane (D) All of these

405. Gandhi bud is active in south India during:
(A) June – October (B) November – December
(C) February – March (D) May - June

406. Gandhi bud is active in north India during:
(A) June – October (B) November – December
(C) February – March (D) May – June

407. Order of fruit fly is a:
(A) Orthoptera (B) Diptera
(C) Isoptera (D) Hemiptera

408. Order of red pumpkin beetle is a:
(A) Orthoptera (B) Coleoptera
(C) Isoptera (D) Hemiptera

409. Order of gandhi bug is a:
(A) Orthoptera (B) Diptera
(C) Isoptera (D) Hemiptera

410. Pyrilla is a severe insect of:
(A) Rice (B) Sugarcane
(C) Maize (D) Wheat

411. Gurdaspur borer is a pest of:
(A) Rice (B) Sugarcane
(C) Maize (D) Wheat

412. Fruit fly lays eggs:
(A) On leaves (B) Inside fruits
(C) In soil (D) All of these

413. Which of the stage of pyrilla is more harmful:
(A) Adult (B) Nymph
(C) Pupa (D) Adult and Nymph

414. Which of the stage of fruit fly is more harmful:
(A) Adult (B) Nymph
(C) Maggot (D) Adult and Nymph

415. Gandhi bug has mouth organs:
(A) Piercing and sucking type
(B) Sponging type
(C) Siphoning type
(D) All of these

416. Which of the polyphagous insect?
(A) Termite (B) Bihar hairy caterpillar
(C) Locust (D) All of these

417. Fruit fly is harmful to:
(A) Papaya (B) Guava
(C) Cucurbits (D) All of these

418. Mouth parts of red pumpkin beetle is:
(A) Siphoning type
(B) Sponging type
(C) Cutting and chewing type
(D) None of these

ANSWERSHEET: 12.ENTOMOLOGY

1. (A); 2. (A); 3. (D); 4. (B); 5. (D); 6. (C); 7. (A); 8. (D); 9. (C); 10. (B); 11. (A); 12. (C); 13. (C) ; 14. (D); 15. (D); 16. (A); 17. (C); 18. (B); 19. (D); 20. (C); 21. (A); 22. (D); 23. (B); 24. (C); 25. (A) ;26. (A); 27. (B); 28. (A); 29. (B); 30. (A); 31. (C); 32. (A); 33. (C); 34. (A); 35. (D); 36. (B); 37. (C); 38. (B) ; 39. (A); 40. (D); 41. (B); 42. (B); 43. (A); 44. (A); 45. (B); 46. (A); 47. (A); 48. (D); 49. (C); 50. (A); 51. (A) ; 52. (D); 53. (C); 54. (D); 55. (C); 56. (D); 57. (B); 58. (A); 59. (C); 60. (D); 61. (B); 62. (C); 63. (D); 64. (D) ;65. (A); 66. (D); 67. (D); 68. (B); 69. (A); 70. (C); 71. (C); 72. (C); 73. (C); 74. (B); 75. (D); 76. (A); 77. (D); 78. (D) ;79. (C); 80. (C); 81. (A); 82. (C); 83. (A); 84. (B); 85. (A); 86. (C); 87. (A); 88. (A); 89. (A); 90. (A); 91. (D) ;92. (C); 93. (D); 94. (C); 95. (B); 96. (B); 97. (A); 98. (D); 99. (C); 100. (D); 101. (B); 102. (C); 103. (D); 104. (D); 105. (A); 106. (A); 107. (C); 108. (D); 109. (A); 110. (D); 111. (C); 112. (D); 113. (A); 114. (B); 115. (C); 116. (B); 117. (A); 118. (D); 119. (C); 120. (D); 121. (A); 122. (C); 123. (C); 124. (D); 125. (D); 126. (A); 127. (B); 128. (D); 129. (D); 130. (D); 131. (A); 132. (D); 133. (A); 134. (A); 135. (A); 136. (A); 137. (A); 138. (C); 139. (C); 140. (B); 141. (C); 142. (C); 143. (B); 144. (D); 145. (A); 146. (A); 147. (C); 148. (C); 149. (B); 150. (D); 151. (A); 152. (C); 153. (C); 154. (B); 155. (A); 156. (B); 157. (C); 158. (B); 159. (B); 160. (C); 161. (C); 162. (C); 163. (A); 164. (A); 165. (B); 166. (B); 167. (A); 168. (A); 169. (B); 170. (C); 171. (A); 172. (A); 173. (A); 174. (A); 175. (A); 176. (D); 177. (D); 178. (C); 179. (D); 180. (B); 181. (D); 182. (C) ; 183. (B); 184. (D); 185. (A); 186. (A); 187. (B); 188. (C); 189. (A); 190. (D); 191. (B); 192. (A); 193. (D); 194. (B); 195. (D) ; 196. (C); 197. (A); 198. (A); 199. (A); 200. (B); 201. (D); 202. (D); 203. (A); 204. (A); 205. (D); 206. (D); 207. (A); 208. (C) ; 209. (A); 210. (C); 211. (A); 212. (B); 213. (A); 214. (B); 215. (B); 216. (D); 217. (A); 218. (A); 219. (B); 220. (A); 221. (D) ; 222. (A); 223. (B); 224. (C); 225. (D); 226. (D); 227. (C); 228. (B); 229. (D); 230. (A); 231. (A); 232. (D); 233. (A); 234. (B); 235. (C); 236. (A); 237. (A); 238. (A); 239. (D); 240. (A); 241. (D); 242. (D); 243. (D); 244. (D); 245. (B); 246. (B); 247. (A) ; 248. (B); 249. (A); 250. (D); 251. (D); 252. (C); 253. (A); 254. (A); 255. (A); 256. (A); 257. (B); 258. (D); 259. (D); 260. (A) ; 261. (A); 262. (C); 263. (D); 264. (B); 265. (B); 266. (A); 267. (A); 268. (D); 269. (A); 270. (A); 271. (B); 272. (A); 273. (D) ; 274. (A); 275. (A); 276. (D); 277. (C); 278. (A); 279. (C); 280. (D); 281. (C); 282. (D); 283. (B); 284. (B); 285. (D); 286. (B); 287. (D); 288. (D); 289. (D); 290. (D); 291. (D); 292. (C); 293. (A); 294. (A); 295. (C); 296. (A); 297. (C); 298. (B); 299. (C) ; 300. (C); 301. (A); 302. (D); 303. (D); 304. (B); 305. (A); 306. (A); 307. (A); 308. (B); 309. (C); 310. (A); 311. (A); 312. (A); 313. (A); 314. (A); 315. (A); 316. (A); 317. (D); 318. (A); 319. (D); 320. (A); 321. (A); 322. (B); 323. (A); 324. (A); 325. (A) ; 326. (D); 327. (A); 328. (D); 329. (B); 330. (C); 331. (A); 332. (A); 333. (A); 334. (B); 335. (D); 336. (B); 337. (B); 338. (A) ; 339. (A); 340. (D); 341. (B); 342. (A); 343. (A); 344. (A); 345. (A); 346. (D); 347. (D); 348. (A); 349. (A); 350. (D); 351. (D) ; 352. (D); 353. (A); 354. (B); 355. (B); 356. (A); 357. (A); 358. (B); 359. (A); 360. (B); 361. (B); 362. (A); 363. (A); 364. (B) ; 365. (A); 366. (C); 367. (A); 368. (B); 369. (A); 370. (A); 371. (C); 372. (B); 373. (A); 374. (D); 375. (D); 376. (D); 377. (D) ; 378. (C); 379. (D); 380. (D); 381. (D); 382. (D); 383. (A); 384. (A); 385. (B); 386. (C); 387. (B); 388. (A); 389. (B); 390. (A); 391. (A); 392. (A); 393. (D); 394. (A); 395. (C); 396. (A); 397. (C); 398. (D); 399. (A); 400. (C); 401. (C); 402. (A); 403. (D) ; 404. (D); 405. (A); 406. (B); 407. (B); 408. (B); 409. (D); 410. (B); 411. (B); 412. (B); 413. (B); 414. (C); 415. (A); 416. (D) ; 417. (D); 418. (C).

Chapter 13

PLANT PATHOLOGY

1. Studies of plant diseases is called:
(A) Virology (B) Plant Pathology
(C) Mycology (D) None of these

2. Who studies the plant diseases:
(A) Virologist (B) Pathologist
(C) Plant Pathologist (D) Mycologist

3. Plant Pathology word derived from three Greek words. Which of the following is not related to it?
(A) Phyton (B) Pathos
(C) Logus (D) Logy

4. Who is known as father of plant pathology?
(A) Anton de Bery (B) Luis Pasteur
(C) R.L. Tulasne (D) Julious Coon

5. Who laid the foundation of experimental plant pathology?
(A) Anton de Bery (B) Luis Pasteur
(C) Morren (D) Tillet

6. Downey mildew fungus was discovered by:
(A) Anton de Bery (B) Luis Pasteur
(C) Morren (D) Tillet

7. Virus is found in cells in the form of:
(A) Crystalline groups (B) Liquid form
(C) Suspension form (D) None of these

8. Indian Phytopathological Society was established by:
(A) E.T. Butler (B) S.N. Dasgupta
(C) R. Verma (D) B.B. Mundkur

9. Who is the father of Indian plant pathology?
(A) E.T. Butler (B) K.C. Mehta
(C) R.S. Singh (D) R. Verma

10. Who told, 'Virus is a contagium vivum fluidum'?
(A) Beijerinck (B) Ivanowski
(C) Meyor (D) R. Kotch

11. Who is associated with the study of wheat rust in India:
(A) B.B. Mundkur (B) E.J. Butler
(C) K.C. Mehta (D) J.F. Dastur

12. Which type of rust does not occur on wheat:
(A) White rust (B) Black rust
(C) Brown rust (D) Yellow rust

13. On germination, the Teliospore produces:
(A) Germ tube (B) Germ sporangium
(C) Promycelium (D) None of these

14. In a rust fungus, how many types of spores are produced:
(A) One (B) Two
(C) Four (D) Five

15. Which one is considered as the perfect spore:
(A) Pycniospore (B) Aeciospore
(C) Teliospore (D) Uredospore

16. Heteroecious rust completes its life cycle on:
(A) One host (B) Two unrelated hosts
(C) Three hosts (D) None of these

17. Pustules are often arranged in rows in:
(A) Black rust (B) Brown rust
(C) Yellow rust (D) None of these

18. *Puccinia recondita* is the causal organism of:
(A) Black rust (B) Brown rust
(C) Yellow rust (D) White rust

19. Appearance of wheat rusts in India takes place through the dissemination of:
(A) Pycniospores (B) Aeciospores
(C) Uredospores (D) Teliospores

20. *Erysiphe graminis tritici* is:
(A) Obligate saprophyte (B) Obligate parasite
(C) Facultative parasite (D)Facultative saprophyte

21. Ear cockle of wheat is a :
(A) Fungal disease (B) Bacterial disease
(C) Viral disease (D) Nematode disease

22. The pathogen of loose smut of wheat survives:
(A) With the seeds (B) On the seeds
(C) Within the seeds (D) None of these

23. Which one is the abiotic disease of rice:
(A) Blast (B) False smut
(C) Khaira (D) Brown spot

24. Only the naked rachis remains in the last in case of:
(A) Covered smut of barley (B) Loose smut of wheat
(C) Grain smut of Jowar (D) Bajra smut

25. Solar energy treatment against loose smut of wheat was evolved by:
(A) Mehta (B) Mundkur
(C) Butler (D) Luthra and Sattar

26. The causal organism of ear cockle of wheat is:
(A) *Ustilago tritici* (B) *Puccinia triticina*
(C) *Anguina tritici* (D) *Alternaria triticina*

27. Whip smut occurs in:
(A) Maize (B) Sorghum
(C) Bajra (D) Sugarcane

28. Karnal bunt is the name because:
(A) Symptoms appear in Kernal
(B) Reported from Karnal
(C) Reported by Karnal
(D) None of these

29. False smut is the name because its causal organism does not belong to:
(A) Masticomycotina (B) Zygomycotina
(C) Ascomycotina (D) Basidiomycotina

30. Disease development is rapid in Karnal bunt due to:
(A) Frequent light rains
(B) Cloudy weather
(C) Low temperature
(D) Combination of above three factors

31. Which disease has caused the famous 'Bengal famine':
(A) Rust of wheat
(B) Red rot of sugarcane
(C) Tikka disease of groundnut
(D) Brown spot of paddy

32. The toxin 'Cochliobolin' is produced by the causative fungus of:
(A) Chickpea grey mould (B) Sugarcane wilt
(C) Rice brown spot (D) Rice blast

33. In blast of rice, the symptoms appear on:
(A) Leaves only (B) Leaf-sheath only
(C) Panicle only (D) On all above parts

34. A disease which occurs widely but periodically is:
(A) Epidemic disease (B) Endemic disease
(C) Sporadic disease (D) Pandemic disease

35. A disease which occurs at a short interval and its severity is more called:
(A) Epidemic disease (B) Endemic disease
(C) Sporadic disease (D) Pandemic disease

36. The pathogen of brown spot of rice is:
(A) Soil borne (B) Seed borne
(C) Soil and seed borne (D) None of these

37. Smuts, in general are considered as:
(A) Obligate parasites
(B) Obligate saprophytes
(C) Facultative parasites
(D) Facultative saprophytes

38. The conidia of *Pyricularia oryzae* are:
(A) One celled (B) Two celled
(C) Three celled (D) Many celled

39. Mycellium of powdery mildew pathogen is:
(A) Ectophytic (B) Endophytic
(C) Intercellular (D) Intracellular

40. In red rot of sugarcane, the reddening is found in:
(A) Parenchyma (B) Pith
(C) Phloem (D) Xylem

41. Setae are produced in:
(A) *Fusarium* (B) *Pythium*
(C) C*olletotrichum* (D) *Cercospora*

42. In *Pyricularia oryzae*, the spots on leaves are mostly:
(A) Reniform (B) Pyriform
(C) Doliform (D) Spherical

43. In Ascochyta blight, the spots on leaves are mostly:
(A) Angular (B) Circular
(C) Oblong (D) Irregular

44. Fusaric acid plays an important role in pathogenesis of:
(A) Wilt (B) Blight
(C) Rust (D) Smut

45. The perfect size of *Colletotrichum falcatum* is:
(A) *Pyrenophora graminae*
(B) *Mycosphaerella arachidicola*
(C) *Glumerellatu cumanensis*
(D) *Cochliobolus miyabeanus*

46. In red rot of sugarcane, symptoms also appear on:
(A) Tip of the leaves (B) Margin of the leaf
(C) Midrib of the leaves (D) None of these

47. In *Colletrotrichum falcatum*, the conidia are produced in:
(A) Pycnidia (B) Acervuli
(C) Synnemata (D) Sporodochia

48. Karathane is basically used for the control of:
(A) Downey mildew (B) Powdery mildew
(C) Wilt (D) Blast

49. Bordeaux mixture is a:
(A) Copper fungicide (B) Sulphur fungicide
(C) Dithiocarbamate (D) Systemic fungicide

50. Bacterial blight of cotton is also called as:
(A) Tar leaf spot (B) Angular leaf spot
(C) Zonate leaf spot (D) Red leaf spot

51. Bacterial blight of cotton does not produce the symptoms on:
(A) Seedlings (B) Stems
(C) Bolls (D) Roots

52. The disease is a typical vascular wilt in rice:
(A) Blast (B) Brown leaf spot
(C) Bacterial leaf streak (D) Bacterial leaf blight

53. The main source of primary inoculum of bacterial blight of cotton is:
(A) Seed (B) Soil
(C) Air (D) Water

54. Kresek phase is found in the rice disease:
(A) Bacterial leaf streak (B) Bacterial blight
(C) Blast (D) Sheath blight

55. Which factor does not favour the rapid spread of bacterial blight of rice:
(A) Rainy weather
(B) Strong wind
(C) High humidity
(D) Temperature below 20^{0}C

56. The symptoms only appear on leaf in case of:
(A) Angular leaf spot of cotton
(B) Bacterial leaf streak of rice
(C) Blast of rice
(D) Brown spot of rice

57. In the genus *Fusarium*, conidia are produced in:
(A) Pycnidia (B) Acervuli
(C) Sporodocia (D) Synnemata

58. Wilt disease can be controlled by eliminating soil borne infection through:
(A) Crop rotation (B) Seed treatment
(C) Spraying of fungicide (D) None of these

59. Wilt pathogen is mainly:
(A) Soil borne (B) Seed borne
(C) Air borne (D) None of these

60. The size of pesticide molecules, if all other conditions are constant, greatly influences it is:
(A) Absorption (B) Solubilization
(C) Adsorption (D) Volatilization

61. Some pesticides with positively charged groups are also absorbed by:
(A) Illite clays (B) Kaolinite clays
(C) Monmorillonite clays (D) Silicate clays

62. In general, herbicides are —— than either fungicides or insecticides:
(A) More mobile (B) Less mobile
(C) Equally mobile (D) More immobile

63. Some degradation products of pesticides are even —— than the original pesticides:
(A) Less toxic (B) More toxic
(C) Non-toxic (D) Equally toxic

64. The total number of bacteria frequently —— after fumigation:
(A) Increases (B) Decreases
(C) Remains constant (D) None of these

65. The process of ammonification is often —— by the use of pesticides:
(A) Harmful (B) Benefitted
(C) Not affected (D) None of these

66. The principle that in modern times "the distribution of plant species on the earth is not uniform" was established by:
(A) Charles Darwin (B) De Candolle
(C) Nikolai Vavilov (D) Carl Von Linnaeus

67. Vavilov recognized primary centres of probable origin of cultivated plants numbering:
(A) 4 (B) 5 (C) 8 (D) 10

68. Species and subdivisions in plants were designated respectively at Linneons and Jordanons by:
(A) W.F. Libby (B) L.P. Lotsy
(C) R.E. Reeves (D) M.Y. Nuttonson

69. The settled village community farming based on small grain and animal domestication occurred in Near East about ——— years ago:
(A) 5000 (B) 7500
(C) 10000 (D) 10500

70. The causal organism of black rust in wheat is:
(A) *Pucinia recondita tritici*
(B) *Puccinia streiformis*
(C) *Puccinia graminis tritici*
(D) *Ustilago nuda tritici*

71. Karnal bunt or partial bunt was first observed in experimental seed plots at:
(A) Kernal (B) Pantnagar
(C) New Delhi (D) Hisar

72. Brown stripe downy mildew of maize can be controlled by use of:
(A) Bordeaux mixture (4:450)
(B) Bavistin
(C) Karathane-45
(D) Dithane M-45

73. The causal agent of anthracnose disease in sorghum is:
(A) *Puccinia purpurea*
(B) *Helminthosporium turcicum*
(C) *Colletotrichum graminicola*
(D) *Gleourcospora sorghi*

74. The chickpea wilt is caused by:
(A) *Fusarium oxysporum*
(B) *Ascochyta rabi*
(C) *Bortrylis cineraria*
(D) *Scelerotinia sclerotiorum*

75. Chickpea pod borer can be controlled by the spray of:
(A) Monocrotophos @ 1 ml/litre of water
(B) Aldrin @ 15 kg/ha
(C) Seed treatment by captan (0.25%) @ 4g/ha of seed
(D) BHC @ 10 kg/ha

76. The seed of safflower should be treated with Agrosan GN at the rate of:
(A) 30 g/kg seed (B) 30mg/kg seed
(C) 30 decigram/kg seed (D) 100 decigram/kg seed

77. Safflower rust can be controlled by:
(A) Spray of Zineb
(B) Seed reatment with thiram
(C) Destroying the infected material
(D) Following the 3 to 4 years crop rotation

78. Treatment of cotton seed with organo-mercurial compound (s) helps to minimize the disease:
(A) Soil borne (B) Air borne
(C) Water borne (D) Seed borne

79. The most destructive disease of sugar cane is:
(A) Red rot (B) Wilt
(C) Smut (D) Albino

80. To avoid rotting in early planting of potato, seeds should be treated with:
(A) 0.5 per cent Dithane M-45
(B) 2.0 per cent Dithane M-45
(C) 1.5 per cent Aretan
(D) 2.0 per cent Tafasan

81. Treat the sugarcane seed setts with——— against fungal diseases:
(A) Diathane M-45 (B) Zineb
(C) Thiram (D) Aretan

82. The ——— botanist Micheli (1679-1737) was the first scientist who in 1729 studied fungi and show their spores:
(A) German (B) Dutch
(C) Italian (D) Austrian

83. In ———, the French botanist Tillet published a paper on bunt or stinking smut of wheat:
(A) 1555 (B) 1655 (C) 1755 (D) 1855

84. By 1857, Berkley and Schacht discovered root knot nematodes and cyst nematode in———:
(A) Sweet sorghum (B) Sugarcane
(C) Potato (D) Sugarbeet

85. The discovery of——— mosaic by Meyer in 1886 was the beginning of studies on viruses as incidants of diseases in plants:
(A) Tomato (B) Mungbean
(C) Tobacco (D) Soybean

86. ———, before 1910 initiated exhaustive study of Indian fungi and diseases caused by them:
(A) E.J. Butler (B) J.G. Moseman
(C) R.E. Nelson (D) A. Livney

87. Dr. K. C. Mehta of Agra College made an outstanding contribution to plant pathology by discovering the disease cycle of cereal in India:
(A) Smuts (B) Rusts
(C) Wilts (D) Blights

88. In bacterial brown rot and wilt of potato the amount of ——— produced by pathogens are proportional to severity of symptoms:
(A) Gibberellins (B) Lipids
(C) Polysaccharides (D) Proteins

89. Some ———varieties of apple exude toxic wax on leaf surface to prevent germination of conidia of *Podosphaera leucotricha*:
(A) Scab resistant
(B) Leaf blight resistant
(C) Powdery mildew resistant
(D) Rust resistant

90. Red scales of ———- (red varieties) contain protocatechuic acid and catechol which may exude in drops to resist the attack of *Colletrotrichum circinans*:
(A) Garlic (B) Onion
(C) Leek (D) Celery

91. Some varieties of ——— are resistant to wilt due to the presence of hydrocyanides (HCN) in root exudates:
(A) Sunflower (B) Safflower
(C) Castor (D) Linseed (flax)

92. Oil from American elm seed contains 55 per cent capric acid toxic to *Ceratocystis ulmi,* the pathogen causing the famous disease:
(A) English elm (B) Dutch elm
(C) American elm (D) French elm

93. The epidermal cells of——— plant become filled with toxic terpenoid aldehyde after they are few days old and these cells are rarely penetrated by pathogens:
(A) Jute (B) Flax
(C) Cotton (D) Okra

94. In *P. coffeae* (nematode), a ——— species, the optimum temperature for multiplication is reported to be 29.5 °C on rough lemon:
(A) Temperate (B) Tropical
(C) Humid (D) Sub-humid

95. Strawberry plants infected with P. penetrans (nematode) have been successfully treated with exposure in hot water at:
(A) 49.4 °C for 17.5 min. (B) 52.8 °C for 6 min.
(C) 54.4 °C for 4 min. (D) 56 °C for 2 min.

96. Leaf curl is a serious disease of:
(A) Moong (B) Pea
(C) Gram (D) Lentil

97. Ear cockle disease is found in:
(A) Sugarcane (B) Wheat
(C) Cotton (D) Rice

98. Blight in pulses can be controlled by:
(A) Spray of manganese carbamate @ 2.0 kg /ha
(B) Ziram 80% @ 2.0 kg/ha before flowering
(C) Seed treatment with Benlate and Thiram
(D) All of these

99. Downey mildew in pulses can be controlled by spraying of:
(A) Mencozeb @ 0.2 %
(B) Thiofinate methyl 70% @ 0.15 %
(C) Grow resistant variety
(D) All of these

100. Powdery mildew in pea can be controlled by spraying of:
(A) Dusting of Pendimethalin @ 2.0-3.0 l/ha
(B) Karathan @ 0.06% spray
(C) Stomp @ 2.0 l/ha
(D) None of these

101. Khaira disease is caused due to deficiency of:
(A) Iron (B) Zinc
(C) Sulphur (D) Nitrogen

102. Wilt is an important disease of:
(A) Pea (B) Arhar
(C) Mustard (D) Wheat

103. The wilt resistant variety of chickpea is:
(A) K-850 (B) KWR-108
(C) Sadabahar (D) Radhey

104. Seed treatment of sugarcane is done by:
(A) Agalol (B) Thiram
(C) Vitavax (D) Zinc sulphate

105. Which one the important disease of groundnut is:
(A) Phyllody (B) Wilt
(C) Tikka (D) Cercospora leaf spot

106. Which one the important disease of sesame is:
(A) Phyllody (B) Wilt
(C) Tikka (D) Cercospora leaf spot

107. Sterility mosaic is a disease related to:
(A) Fungal (B) Bacterial
(C) Viral (D) Physiological

108. Black tip is a common disease of:
(A) Banana (B) Grape
(C) Orange (D) Mango

109. Which wheat rust normally appears earlier in plains of Uttar Pradesh:
(A) Black rust (B) Yellow rust
(C) Brown rust (D) None of these

110. Bacterial disease are controlled by:
(A) Antibiotics (B) Viricide
(C) Fungicide (D) Kethane

111. Seed treatment of pulses is done with:
(A) Azotobactor (B) Rhizobium
(C) Azospirillum (D) Phosphotica

112. Dr. K.C. Mehta was associated with:
(A) Cereal rust in India (B) Smut research
(C) Bacterial research (D) Virus research

113. Viroid is a particle with:
(A) Naked nucleic acid (B) Covered nucleic acid
(C) Protein capsid (D) None of these

114. Potato spindle tuber disease is caused by:
(A) Virus (B) Viroid
(C) Mycoplasma (D) Fungus

115. All plant pathogenic bacteria are:
(A) Gram positive
(B) Gram negative
(C) Gram positive and gram negative
(D) None of these

116. Multiplication of bacteria is caused by:
(A) Mitosis (B) Meiosis
(C) Binary fussion (D) All of these

117. The cell wall of oomycetes is made of:
(A) Chitin (B) Cellulose
(C) Both chitin and cellulose (D) None of these

118. Cell wall of fungi is made of:
(A) Chitin (B) Cellulose
(C) Both chitin and cellulose (D) All of these

119. Infectious disease is caused by:
(A) Parasites (B) Non parasites
(C) Virus (D) All of these

120. Non infectious disease is caused by:
(A) Nutrient deficiency (B) Mineral deficiency
(C) Pollutants (D) All of these

121. Saprophytic pathogens feed on:
(A) Only dead matter (B) Only living organisms
(C) Both (A) and (B) (D) None of these

122. Parasites feed on:
(A) Only dead matter (B) Only living organisms
(C) Both (A) and (B) (D) None of these

123. Which of the following has its own metabolic activity?
(A) Fungus (B) Virus
(C) Viroid (D) Plasmids

124. Litchen is a example of:
(A) Mutalistic association (B) Symbiotic association
(C) Parasitism (D) All of these

125. A plant disease spreads on a larger area with great loss is called:
(A) Endemic disease (B) Epidemic disease
(C) Sporadic disease (D) All of these

126. A plant disease spreads after irregular intervals at different locations in big way, is called:
(A) Endemic disease (B) Epidemic disease
(C) Sporadic disease (D) All of these

127. A plant disease spreads on a regular intervals is called:
(A) Endemic disease (B) Epidemic disease
(C) Sporadic disease (D) All of these

128. An obligate pathogen can:
(A) Grow only on living organism
(B) Grow on dead matters
(C) Grow on both living and non living matters
(D) None of these

129. Fungi is devoid of:
(A) Chlorophyll (B) Cell wall
(C) Nucleus capsid (D) None of these

130. Yellow mosaic in soybean spread by:
(A) Seeds (B) Soils
(C) Air (D) White fly

131. Loose sumt of wheat in controlled by:
(A) Seed treatment (B) Soil solarization
(C) Crop rotation (D) Thiram spray on crop

132. Homothallic fungus is:
(A) Dioecious (B) Monoecious
(C) Both (A) and (B) (D) None of these

133. The fungi grown on dung of herbivorous animals are called as:
(A) Coprophilous fungi (B) Keratinaceous fungi
(C) Amphibious fungi (D) All of these

134. The fungi grown on dead bodies are called as:
(A) Coprophilous fungi (B) Keratinaceous fungi
(C) Amphibious fungi (D) All of these

135. The root like structure of fungi is called as:
(A) Rhizoid (B) Stroma
(C) Rhizomorph (D) Sclerotium

136. Karnal bunt is a disease of:
(A) Viral (B) Nematodes
(C) Fungal (D) Bacterial

137. The pigeonpea wilt is caused by:
(A) *Fusarium oxysporum* (B) *Fusarium udum*
(C) *Ascochyta rabi* (D) *Bortrytis cineraria*

138. Bakanae disease is caused by:
(A) Fungus (B) Alga
(C) Bacterium (D) Virus

139. *Chenopodium album* is a weed which mostly grown in the crop:
(A) Kharif (B) Rabi
(C) Zaid (D) None of these

140. The main objective of Integrated Pest management is:
(A) To reduce pest status
(B) To kill pests
(C) To prevent pest entry
(D) To reduce use of insecticides

141. Aflatoxin is produced by:
(A) *Aspergillus flavus*
(B) *Phytophthora infestans*
(C) *Albugo candida*
(D) *Puccinia graminis tritici*

142. Under integrated system of insect control we integrate:
(A) Cultural and biological methods
(B) Biological and legal methods
(C) Physical and chemical methods
(D) All methods

143. The least toxic insecticides to human being belong to the following group:
(A) Organo chlorines (B) Organo phosphates
(C) Carbamates (D) Synthetic pyrethroids

144. Which of the following is a facultative parasite?
(A) Fusarium (B) Rhizoctonia
(C) Alternaria (D) All of these

145. Which of the following is a facultative saprophyte?
(A) Smut (B) Helminthosporium
(C) Leaf curl fungus (D) All of these

146. In which of the following group of fungi has similar sporangiophore to its vegetative hyphae?
(A) Pythium (B) Albugo
(C) Permospora (D) Sclerospora

147. Aflatoxin poison is produced by:
(A) *Aspergillus flavus* (B) *Claviceps purpurea*
(C) *Pencillium spp* (D) All of these

148. Gibberellic acid is formed by:
(A) *Gibberella fujikuroi* (B) *G. mosshata*
(C) *Rhizopus niger* (D) *Aspergillus flavus*

149. The member of oomycetes are:
(A) Uniflagellate (B) Biflagellate
(C) Triflagellate (D) None of these

150. Amphigynous antheridium is found in:
(A) Pythium (B) Phytophthora
(C) Both (A) and (B) (D) None of these

151. Which of the following is a obligate parasite?
(A) Puccinia (B) Pernospora
(C) Albugo (D) All of these

152. Which of the following is not saprophyte?
(A) Erysiphe (B) Aspergillus
(C) Mucor (D) Penicillium

153. Indole acetic acid is obtained from:
(A) *Gibberella spp.* (B) *Rhizopus niger*
(C) *Aspergillus flavus* (D) *Claviceps purpurea*

154. Alkaloid can be obtained from:
(A) *Gibberella spp.* (B) *Rhizopus niger*
(C) *Aspergillus flavus* (D) *Claviceps purpurea*

155. Pin moulds are also known as:
(A) Mucorales (B) Algal fungi
(C) Yeast fungi (D) None of these

156. Seed treatment with vitavex is the mainly control :
(A) Loose smut (B) Rust
(C) Downy mildew (D) Powdery mildew

157. Loose smut of wheat can be controlled by:
(A) Seed treatment by vitavex before sowing
(B) Seed treatment with hot water
(C) Use of disease resistant varieties
(D) All of these

158. The important symptoms of black rot in cauliflower is:
(A) Chlorosis
(B) Drying of leaves
(C) Burning of leaves
(D) Black spots on petioles and veins

159. Purple black of onion is caused by:
(A) Fungus (B) Bacteria
(C) Virus (D) None of these

160. Onion smut can be avoided by:
(A) Spray of Dithane M-45
(B) Seed treatment by captan/Thiram
(C) Weed control
(D) Irrigation

161. Carrot yellow is caused by:
(A) Fungus (B) Algae
(C) Bacteria (D) Virus

162. Bacterial soft rot is the result of:
(A) High temperature (B) Very low temperature
(C) Waterlogging (D) None of these

163. Black rot is main disease of sweet potato and it can be controlled by:
(A) Selection of healthy cuttings
(B) Proper drainage facilities
(C) Cuttings should be treated with 6.25% Areton solution for 8-10 minutes
(D) All of these

164. Wilt is the main disease of nursery of sweet potato and it is caused by:
(A) Bacteria (B) Fungus
(C) Virus (D) Nematodes

165. Why waterlogging area is not suitable for chilli cultivation?
(A) Damping off (B) Bacterial spot
(C) Leaf curl (D) All of these

166. Leaf curl of chilli is mainly spread by:
(A) Air (B) White fly
(C) Water (D) Nematode

167. Mango black tip is caused by:
(A) Bacteria (B) Virus
(C) Smoke of Brick kline (D) Fungi

168. Which of the following disease is the most harmful to mango orchard?
(A) Mango wither tip (B) Anthracnose
(C) Bunchy top (D) Black tip

169. Which of the following disease is not found in papaya orchard?
(A) Anthracnose (B) Collar rot
(C) Papaya mosaic (D) Black tip

170. Panama disease of banana is caused by:
(A) *Fusarium oxysporum* (B) *Bacillus spp.*
(C) Actinomycetes (D) Virus

171. Bunchy top of banana is caused by:
(A) Bacteria (B) Fungi
(C) Virus (D) Actinomycetes

172. Which of the following disease is most harmful to citrus plants?
(A) Citrus canker (B) Gummosis
(C) Anthracnose (D) All of these

173. Citrus canker is caused by:
(A) *Phytophthora infestans*
(B) *Xanthomonas citri*
(C) *Colletotrichum spp.*
(D) *Bacillus moniliformis*

174. Citrus canker can be controlled by :
(A) Blitox-50 @ 0.25% solution
(B) Endosulphan @ 0.2% solution
(C) 2,4-D @ 1.5 kg a.i./ha
(D) None of these

175. Powdery mildew of apple can be controlled by:
(A) Spray of Krathane @ 0.1% solution
(B) Spray of Endosulphan @ 0.2% solution
(C) Spray of Diathane -78 @ 0.1% solution
(D) All of these

176. Which of the following is an important disease of peach?
(A) Blight (B) Leaf curl
(C) Gummosis (D) All of these

177. Suitable bioagents for controlling of wilt disease of chickpea, lentil and pigeonpea is:
(A) *Trychoderma harzianum*
(B) *Trichoderma viridi*
(C) Both (A) and (B)
(D) None of these

178. Sexual reproduction in yeast occurs under:
(A) Favourable condition
(B) Unfavourable condition
(C) Both (A) and (B)
(D) None of these

179. Asperigillosis is caused by:
(A) Aspergillus fungus (B) Bacteria
(C) Virus (D) None of these

180. Wood is deteriorated by the species of Panicillium:
(A) *Penicillium divaricatum*
(B) *Penicillium expensum*
(C) *Penicillium glacuum*
(D) None of these

181. Rust fungi mainly a:
(A) Saprophyte (B) Parasite
(C) Facultative parasite (D) All of these

182. Which of the alkaloid is produced by ergot fungi?
(A) Ergotin (B) Penicillin
(C) Cutrinin (D) All of these

183. The initial stage of ergot infection is known as:
(A) Honey dew stage (B) Scleotial stage
(C) Spore formation stage (D) All of these

184. The sexual fruiting body of ergot fungi is:
(A) Apothecium (B) Cleistothecium
(C) Synnemata (D) Perithecium

185. Which of the nucleic acid is found in the nucleus of cauliflower mosaic virus?
(A) RNA (B) DNA
(C) Both RNA and DNA (D) All of these

186. England famine in 1846 was caused mainly by:
(A) Late blight of potato
(B) Helminthosporium leaf spot of rice
(C) Rust of wheat
(D) Leaf curl virus in tomato

187. One of the most important reason of Bengal famine in 1943 was:
(A) Late blight of potato
(B) Helminthosporium leaf spot of rice
(C) Rust of wheat
(D) Leaf curl virus in tomato

188. Smut fungi are:
(A) Obligate parasite (B) Obligate saprophyte
(C) Facultative saprophyte (D) Facultative parasite

189. Rust fungi are:
(A) Polymorphic (B) Homothallic
(C) Monoecious (D) All of these

190. Smut means:
(A) Sooty powder (B) Wet powder
(C) Grey powder (D) All of these

191. Cuscuta is a:
(A) Complete stem parasite
(B) Complete root parasite
(C) Semi stem parasite
(D) Semi root parasite

192. Loranthus is a:
(A) Complete stem parasite
(B) Complete root parasite
(C) Semi stem parasite
(D) Semi root parasite

193. Striga is a:
(A) Complete stem parasite
(B) Complete root parasite
(C) Semi stem parasite
(D) Semi root parasite

194. Orobanche is a:
(A) Complete stem parasite
(B) Complete root parasite
(C) Semi stem parasite
(D) Semi root parasite

195. Eu-form rust are also known as:
(A) Long cycled rust (B) Short cycled rust
(C) Demi cycled rust (D) None of these

196. Which is not an example of long cycled rust?
(A) Wheat rust (B) Sunflower rust
(C) Lucerne rust (D) White rust of Brassica

197. The life cycle of rust fungi, in which uredinial stage is not found is known as:
(A) Macro cycle (B) Demi cycle
(C) Micro cycle (D) None of these

198. The life cycle of rust fungi, in which all five stages are not found is known as:
(A) Macro cycle or long cycle (B) Demi cycle
(C) Micro cycle (D) None of these

199. The life cycle of rust fungi where only dicoryotic telial stage is not found is known as:
(A) Macro cycle (B) Demi cycle
(C) Micro cycle (D) None of these

200. Autocious fungi complete their life cycle on:
(A) Single host (B) Two host
(C) Dead matter (D) None of these

201. Heterocious fungi complete their life cycle on:
(A) Single host (B) More than one host
(C) Dead matter (D) None of these

202. Leaf curl of tomato is caused by:
(A) Tobacco virus 16 (B) Tomato virus 16
(C) Virus No. 10 (D) Virus No.16

203. Most appropriate time of black rust attack in Northern Plains is:
(A) Late February – March (B) December- January
(C) October-November (D) All of these

204. Most appropriate time of black rust attack in southern Plains is:
(A) Late February – March (B) December- February
(C) October-November (D) All of these

205. Optimum temperature for black rust spread is:
(A) 17-20^{0}C (B) 25-30^{0}C
(C) 30-35^{0}C (D) 35-40^{0}C

206. Which is the alternate host of yellow rust?
(A) *Tholictrum spp* (B) *Berberis spp*
(C) Wild wheat (D) Not known

207. Which of the following is the collateral host of yellow rust?
(A) *Agropyron semicostatum*
(B) *Bromus catharticus*
(C) *Bromus japonicus*
(D) All of these

208. Which of the following urediniospores of rust can require the highest temperature for its germination?
(A) Black rust (B) Yellow rust
(C) Brown rust (D) All of these

209. Covered smut of barley is a disease of:
(A) Externally seed borne (B) Internally seed borne
(C) Air borne (D) None of these

210. In melampsora, the teliospores are:
(A) Sessile and unicellular
(B) Strigmated and unicellular
(C) Sessile and multinucleate
(D) None of these

211. Which of the following stages is the alternate host of black rust?
(A) 0 and I stage (B) II and III stages
(C) IV and 0 stages (D) All of these

212. Which of the following is the collateral host of stem rust?
(A) *Birza minor* (B) *Bromes patulus*
(C) *Avena fatua* (D) All of these

213. Mature teliospore has a:
(A) Heploid nucleus (B) Diploid nucleus
(C) Dikaryotic (D) None of these

214. Which of the following sequence of rust occurrence in UP:
(A) Leaf rust > Stripe rust > Stem rust
(B) Stripe rust > Leaf rust > Black rust
(C) Black rust > Brown rust > Yellow rust
(D) Yellow rust > Leaf rust > Black rust

215. The causal organism of powdery mildew of wheat is:
(A) *Erysiphe graminis var tritici*
(B) *Erysiphe polygoni*
(C) *Erysiphe aestivum*
(D) None of these

216. Pin moulds are also known as sugar fungi because of:
(A) They use simple carbohydrates
(B) They do not use carbohydrate
(C) They use polysaccharides
(D) All of these

217. The pathogen that can grow only under controlled environment is called:
(A) Ecologically obligate
(B) Bio-chemically obligate
(C) Facultative parasite
(D) All of these

218. Bajra smut is a:
(A) Air borne (B) Seed borne
(C) Soil borne (D) Soil and seed borne

219. The colour of bajra smut spores is:
(A) Black (B) Yellow
(C) Brownish green (D) Red

220. The sorus of bajra smut is formed usually on:
(A) Leaf (B) Stem
(C) Ovary (D) All aerial parts

221. The main symptoms of early blight of potato is:
(A) Formation of concentric ringes on infected spots
(B) Formation of cottony growth
(C) Formation of chlorosis
(D) All of these

222. Optimum temperature for development of early blight of potato is:
(A) 12-19^{0}C (B) 25-26^{0}C
(C) 30-36^{0}C (D) 38-40^{0}C

223. Which of the following spores are not formed in Fusarium wilt?
(A) Micro conidia (B) Macro conidia
(C) Chlamydospores (D) Onidia

224. The symptoms of false smut of rice appears on:
(A) Leaf (B) Stem
(C) Panicle (D) Root

225. The main symptoms of false smut of rice is:
(A) Apearance of round bluish spots on leaves
(B) Conversion of few seeds into scleratial bodies covering by green colour layer
(C) Formation of black powder in whole panicle
(D) All of these

226. The main symptoms of rice blast is:
(A) Round black spots on leaves
(B) Oval shaped round spots on leaves
(C) Boat shaped spots with grey red borders and grey - ash colour centre
(D) None of these

227. Rice blast is also known as neck rot because of:
(A) Neck of panicle becomes shriveled due to severity of disease
(B) Neck becomes stronger as fungus produces pyricularin toxin
(C) It produces hyaline spores
(D) None of these

228. The cell wall of bacteria is made of:
(A) Pectin (B) Chitin
(C) Mycopeptide (D) None of these

229. Fission fungi is also known as:
(A) Bacteria (B) Virus
(C) Nematodes (D) None of these

230. The coma (,) shaped bacteria are called :
(A) Bacillus (B) Vibrio
(C) Streptococcus (D) None of these

231. Bacteria without flagella is known as:
(A) Atrichous (B) Monotrichous
(C) Lophotrichous (D) Petritrichous

232. Which type of bacteria can be killed by Penicillin?
(A) Gram +ve (B) Gram – ve
(C) Both (A) and (B) (D) None of these

233. The pathogenic bacteria are generally:
(A) Gram +ve (B) Gram –ve
(C) Both (A) and (B) (D) None of these

234. The main function of mesosomes in bacterial cell is:
(A) To help in cell division
(B) To help in respiration
(C) To assist in photosynthesis
(D) All of these

235. Bacteria having flagella on its whole body is known as:
(A) Monotrichous (B) Amphitrichous
(C) Petritrichous (D) None of these

236. Bacteria having a bunch of flagella at one end is known as:
(A) Monotrichous (B) Lophotrichous
(C) Petritrichous (D) None of these

237. Which of the following disease is caused by bacteria?
(A) Brown rot of potato
(B) Angular leaf spot of cotton
(C) Fire blight of apple and pear
(D) All of these

238. Which of the following disease is caused by bacteria?
(A) Red stripe of sugarcane (B) Soft rot of cotton
(C) Leaf roll of potato (D) Stalk rot of maize

239. Which of the following is a autotrophic bacteria?
(A) *Chlorobium* (B) *Thiospirillum*
(C) *Nitrobacter* (D) All of these

240. Which of the following is a heterotrophic bacteria?
(A) *Chlorobium* (B) *Pseudomonas*
(C) *Xantthomonas* (D) All of these

241. Fungi have been originated from:
(A) Polygeneric line (B) Phytogeneric line
(C) Polygendric line (D) None of these

242. The carrier of leaf curl tomato virus is:
(A) White fly (B) Beetle
(C) Bacteria (D) Fungus

243. Cyst formation is very common in which of the following bacteria?
(A) Azotobacter (B) Myxococcus
(C) Chondromyces (D) All of these

244. Nucleus of bacteria cell is called as:
(A) Nucleoplasm (B) Nucleoid
(C) Both (A) and (B) (D) None of these

245. Which of the following feature of gram positive bacteria?
(A) Homogenous cell wall that contains >85% mucopeptides and its polymers
(B) Teichoic acids, ribitol, glutamic acid and alanine are found in more quantity
(C) Violet or blue colour after staining with solution of crystal violet and iodine
(D) All of these

246. Which of the following feature of gram negative bacteria?
(A) Cell wall is thin and made of mainly lipoprotein and its polymers
(B) They react with counter stain safranine and become red to pink in colour
(C) Teichoic acid is not formed
(D) All of these

247. Generally plant virus posses a:
(A) Single stranded RNA (B) Double stranded RNA
(C) Doubled stranded DNA (D) None of these

248. Viroid is a:
(A) Naked RNA particles
(B) Naked DNA particles
(C) Naked RNA and DNA particles
(D) None of these

249. Prions are:
(A) Infectious protein particles
(B) Infectious DNA particles
(C) Both (A) and (B)
(D) None of these

250. The virus enzymes are collectively known as:
(A) Transcriptase (B) Polymerase
(C) Peplomers (D) None of these

251. Virus can be classified under the Phylum of:
(A) Procaryota (B) Vira
(C) Eucaryota (D) Virophyta

252. Viruses are resistant to:
(A) Acids (B) Bases
(C) Salts (D) All of these

253. Virus is a:
(A) Obligate parasite
(B) Obligate saprophyte
(C) Facultative parasite
(D) Facultative saprophyte

254. Chemically virus is a:
(A) Nucleoprotein (B) Protein
(C) Nucleic acid (D) All of these

255. Virus has mainly two parts, one is nucleic acid core and second part is:
(A) Cell wall (B) Nucleic membrane
(C) Protein core (D) None of these

256. Which of the following disease is caused by virus?
(A) Small pox (B) Measles
(C) Common cold (D) All of these

257. Ultra microscopic particle capable of infecting live cells is called:
(A) Plasmic (B) Nucleiod
(C) Virion (D) None of these

258. The size of virus particle is:
(A) 10-300 μ (B) 10-20 μ
(C) 50-80 μ (D) 25-35 μ

259. Virus can be seen only with:
(A) Naked eyes (B) Microscope
(C) Electron microscope (D) All of these

260. Yellow Jack is a:
(A) Yellow fever (B) Small pox
(C) Swine fever (D) None of these

261. Cuscuta is a:
(A) Holoparasite on stem (B) Semi-parasite on stem
(C) Holoparasite on root (D) Semi-parasite on root

262. Loranthus is a:
(A) Stem Holoparasite (B) Stem semi-parasite
(C) Root Holoparasite (D) Root semi-parasite

263. Orobanche is a:
(A) Stem Holoparasite (B) Stem semi-parasite
(C) Root Holoparasite (D) Root semi-parasite

264. Striga is a:
(A) Stem Holoparasite (B) Stem semi-parasite
(C) Root Holoparasite (D) Root semi-parasite

265. Broom rape is also known as:
(A) Orobanche (B) Striga
(C) Loranthus (D) Cuscuta

266. Witch weed is also known as:
(A) Orobanche (B) Striga
(C) Loranthus (D) Cuscuta

267. Vrikshabhaksha is known as:
(A) Loranthus (B) Banda
(C) Giant mistletoe (D) All of these

268. Nematodes are:
(A) Saprophyte (B) Predator
(C) Pathogenic (D) All of these

269. Nematodes may cause following symptoms?
(A) Cellular hypertrophy/hyperplasia
(B) Suppression of mitosis
(C) Cell necrosis
(D) All of these

270. Plant nematodes always have:
(A) Stylet (B) Respiratory organs
(C) Circulatory system (D) None of these

271. Which of the following system is not present in plant nematodes?
(A) Respiratory system (B) Circulatory system
(C) Reproductive system (D) None of these

272. Which of the following nematicides are used to control of nematodes?
(A) Nemagon (B) Nemaphos
(C) Hexanema (D) All of these

273. Which stage of root knot nematode attacks to plant?
(A) First larvae (B) Second larvae
(C) Third larvae (D) Fourth Larvae

274. Male nematodes differ with female nematodes in:
(A) Male nematodes are cylindrical and filliform while female are bigger in size with round and oval shaped
(B) Male nematodes have respiratory system
(C) Females do not have stylets
(D) None of these

275. Main symptoms of root knot of tomato is:
(A) Knot formation in roots
(B) Yellowing of leaves
(C) Stunting growth
(D) All of these

276. The nematodes take their food from host with the help of:
(A) Mouth (B) Stylet
(C) Haustoria (D) None of these

277. The important feature of plant parasitic nematodes is:
(A) Body cavity absent
(B) Transparent segmentless body
(C) Respiratory and circulation systems absent
(D) All of these

278. Which of the following nematodes acts as carrier of viruses?
(A) *Xiphineiva spp* (B) *Longidorus spp*
(C) *Trix chodorous spp* (D) All of these

279. Cyst forming nematodes spp is:
(A) *Meloidogyne spp* (B) *Heterodera spp*
(C) *Aphelenchoides spp* (D) *Pratylenchus spp*

280. Leaf nematodes spp is:
(A) *Meloidogyne spp* (B) *Heterodera spp*
(C) *Aphelenchoides spp* (D) *Pratylenchus spp*

281. Root injury creating nematodes spp is:
(A) *Meloidogyne spp* (B) *Heterodera spp*
(C) *Aphelenchoides spp* (D) *Pratylenchus spp*

282. Which of the following nematode species is the causal agent for root knot disease in plants?
(A) *Meloidogyne spp* (B) *Heterodera spp*
(C) *Aphelenchoides spp* (D) *Pratylenchus spp*

283. Reverse transcription is found only in:
(A) Bacteria (B) Virus
(C) Protozoa (D) Fungus

284. Which of the following disease is caused by MLO's?
(A) Papaya bunchy top
(B) Sugarcane grassy shoot
(C) Witch's broom of groundnut
(D) All of these

285. Stinking smut is also known as:
(A) Hill bunt of wheat (B) Karnal bunt
(C) Covered smut of jowar (D) None of these

286. Which of the following is used for disinfestations of ware houses?
(A) $CuSO_4$
(B) Formaldehyde
(C) Fumigation with chloropicrin tear gas
(D) All of these

287. Exclusion includes:
(A) Compulsory quarantines and inspection
(B) Embargoes
(C) Certification
(D) All of these

288. Which of the following gas is used for sterilization?
(A) Chlorine (B) Sulphur dioxide
(C) Formadehyde (D) All of these

289. Tyndalization is a:
(A) Fractional sterilization (B) Complete sterilization
(C) Incomplete sterilization (D) None of these

290. Tyndalization is used for:
(A) Sterilization of medium
(B) Sterilization of inoculation materials
(C) Sterilization of inoculating pathogens
(D) None of these

291. Vericides are used to control:
(A) Bacteria (B) Virus
(C) Nematodes (D) Fungus

292. Damping off of tobacco is caused by:
(A) *Pythium aphanidermatum*
(B) *Pythium infestans*
(C) *Phytophthora infestans*
(D) None of these

293. White Blister disease is caused by:
(A) *Pythiaceae* (B) *Albuginaceae*
(C) *Perouosporaceae* (D) None of these

294. Downy mildew disease is caused by:
(A) *Pythiaceae* (B) *Albuginaceae*
(C) *Perouosporaceae* (D) None of these

295. Late blight of potato is caused by:
(A) *Phytophthora infestans* (B) *Albugo candida*
(C) *Peronospora pisi* (D) None of these

296. Downy mildew of pea disease is caused by:
(A) *Phytophthora pisi* (B) *Peronospora pisi*
(C) *Albigo candida* (D) None of these

297. Green ear disease is caused by:
(A) *Albugo candida*
(B) *Sclerospora graminicola*
(C) *Erysiphe graminis*
(D) None of these

298. White blister of crucifers disease is caused by:
(A) *Albugo candida* (B) *Erysiphe graminis*
(C) *Peronospora pisi* (D) None of these

299. Downy mildew of brassica is caused by:
(A) *Albugo candida* (B) *Peronospora brassicae*
(C) *Phytophthora infestans* (D) None of these

300. Downey mildew of maize is caused by:
(A) *Sclerospora sacchari*
(B) *Sclerospora rayssiae var zeae*
(C) *Sclerospora philippinensis*
(D) All of these

301. Stem gall of coriander is caused by:
(A) *Phycomycetes* (B) *Ascomycetes*
(C) *Basidiomycetes* (D) All of these

302. Causal organism of stem gall of coriander is:
(A) *Protomyces macrosporus*
(B) *Taphrina deformans*
(C) *Raphrina maculans*
(D) *Penicillium expensum*

303. Spot rot of apple is caused by:
(A) *Protomycese macrosporus*
(B) *Penicillium expensum*
(C) *Raphrina mauclans*
(D) *Taphrina deformans*

304. Leaf spot of turmeric is caused by:
(A) *Taphrina maculans*
(B) *Erasiphe polygoni*
(C) *Erysiphe graminis*
(D) All of these

305. Peach leaf curl is caused by:
(A) *Taphrina maculans* (B) *Erasiphe polygoni*
(C) *Erysiphe graminis* (D) All of these

306. Powdery mildew of pea is caused by:
(A) *Erysiphe graminis* (B) *Erysiphi polygoni*
(C) *Uncinula necator* (D) All of these

307. Powdery mildew of cereals is caused by:
(A) *Erysiphe graminis* (B) *Erysiphi polygoni*
(C) *Taphrina deformans* (D) All of these

308. Powdery mildew of apple is caused by:
(A) *Uncinula necator*
(B) *Podosphaera leuxotricha*
(C) *Sclerotium oryzae*
(D) All of these

309. Stem rot of paddy is caused by:
(A) *Sclerotium oryzae* (B) *Claviceps purpurea*
(C) *Ustilaginoidea virens* (D) All of these

310. False smut of paddy is caused by:
(A) *Ustilaginoidea virens* (B) *Claviceps purpurea*
(C) *Erysiphe graminis* (D) None of these

311. Ergot of bajra is caused by:
(A) *Claviceps microcephala* (B) *Claviceps purpurea*
(C) *Sclerotium oryzae* (D) None of these

312. Black rust of wheat is caused by:
(A) *Puccinia graminis var tritici*
(B) *Puccinia recondita*
(C) *Puccinia striformis*
(D) *Melampsora lini*

313. Leaf rust of wheat is caused by:
(A) *Puccinia graminis* (B) *Puccinia recondita*
(C) *Puccinia striformis* (D) *Melampsora lini*

314. Stripe rust of wheat is caused by:
(A) *Puccinia graminis* (B) *Puccinia recondita*
(C) *Puccinia striformis* (D) *Melampsora lini*

315. Stem rust of wheat is also known as:
(A) Black rust (B) Brown rust
(C) Yellow rust (D) White rust

316. Leaf rust of wheat is also known as:
(A) Black rust (B) Brown rust
(C) Yellow rust (D) White rust

317. Stripe rust of wheat is also known as:
(A) Black rust (B) Brown rust
(C) Yellow rust (D) White rust

318. Causal organism of pea rust is:
(A) *Uromyces pisi* (B) *Uromyces aestivum*
(C) *Puccinia recondita* (D) None of these

319. Causal organism of linseed rust is:
(A) *Uromycese pisi* (B) *Uromyces fabae*
(C) *Melampsora lini* (D) None of these

320. Causal organism of loose smut of wheat is:
(A) *Ustilago nuda*
(B) *Ustilago nuda var tritici*
(C) *Ustilago hordei*
(D) *Tolyposporium penicillariae*

321. Causal organism of grain smut of bajra is:
(A) *Sphacelotheca sorghi*
(B) *Sphacelotheca hordi*
(C) *Tolyposporium penicillariae*
(D) None of these

322. Causal organism of smut of bajra is:
(A) *Fusarium oxysporum*
(B) *Tolyposorium penicillariae*
(C) *Ustilago nuda*
(D) None of these

323. Causal organism of early blight of potato is:
(A) *Alternaria solani* (B) *Alternaria infestans*
(A) *Pyricularia oryzae* (D) *Cercospora personata*

324. Causal organism of Alternaria leaf blight of wheat is:
(A) *Alternaria tritici* (B) *Alternaria aestivum*
(C) *Pyricularia triticina* (D) None of these

325. Causal organism of lentil wilt is:
(A) *Fusarium oxysporum* (B) *Fusarium udum*
(C) *Fusarium solani* (D) *Fusarium pisi*

326. Causal organism of arhar wilt is:
(A) *Fusarium oxysporum* (B) *Fusarium udum*
(C) *Fusarium solani* (D) *Fusarium pisi*

327. Causal organism of linseed wilt is:
(A) *Fusarium oxysporum* (B) *Fusarium udum*
(C) *Fusarium solani* (D) *Fusarium pisi*

328. Causal organism of tikka disease of groundnut is:
(A) *Cercospora personata*
(B) *Cercospora arachidicola*
(C) Both (A) and (B)
(D) None of these

329. Which of the following causal organism of tikka disease of groundnut is more fatal?
(A) *Cercospora personata*
(B) *Cercospora arachidicola*
(C) Both (A) and (B)
(D) None of these

330. Which of the following disease is caused by mycoplasma like organisms (MLO)?
(A) Rice yellow dwarf
(B) Sugarcane grassy shoot
(C) Potato witch's broom
(D) All of these

331. Which of the following disease is caused by Rickettsia like microorganisms (RLO's)?
(A) Ratoon stunting (B) Aster yellow
(C) Khaira disease (D) All of these

332. Which of the following characteristics is not associated with MLO's?
(A) Unicellular (B) Usually immobile
(C) Cell wall absent (D) Cell wall present

333. Khaira disease is a example of:
(A) Endemic disease (B) Epidemic disease
(C) Non infectious disease (D) Infectious disease

334. Green ear disease is a example of:
(A) Endemic disease (B) Epidemic disease
(C) Sporodic disease (D) Infectious disease

335. Late blight of potato disease is a example of:
(A) Endemic disease (B) Epidemic disease
(C) Sporodic disease (D) Infectious disease

336. Necrotrophs are pathogens that:
(A) Kill host plants
(B) Do not kill host plant
(C) Provide nutrition to host plants
(D) None of these

337. Which of the following is a example of necrotroph pathogen?
(A) Pythium (B) Rhizoctonia
(C) Soft rot bacteria (D) All of these

338. Facultative parasites are:
(A) Primarily saprophytes but can attack on living organisms under favourable conditions
(B) Primarily saprophytes and do not attack on living organisms
(C) Primarily obligate parasites
(D) All of these

339. Main symptoms of karnal bunt is:
(A) Formation of black powder below the pericarp of seed
(B) Produces rotten fish odour
(C) Bad odour is due to Trimethylamine
(D) All of these

340. Hill bunt can be controlled by:
(A) Seed treatment with 0.2% vitavex
(B) Use of resistant varieties
(C) Clean the field
(D) All of these

341. Which of the following culture is a natural medium for isolation of pathogen?
(A) Potato dextrose agar medium
(B) Milk
(C) Egg
(D) All of these

342. Trychoderma can be used as:
(A) Soil (B) Seed
(C) Both (A) and (B) (D) None of these

343. Root knot disease of tomato is caused by:
(A) *Meloidogyne javanica* (B) *Heterodera avenaei*
(C) *Anguina tritici* (D) None of these

344. The spores of grain smut of jowar are attached with long solid structure known as:
(A) Columella (B) Pith
(C) Paraphysis (D) None of these

345. The collateral host of linseed rust is:
(A) *Linum usitatissimum* (B) *Linum mysorense*
(C) *Linum pisi* (D) All of these

346. LSD is a:
(A) Stimulant (B) Pain killer
(C) Antibiotic (D) None of these

347. LSD is abbreviated as:
(A) Least Stimulant Dose
(B) d-Lysergic Acid diethylamide
(C) Least Statistical Deviation
(D) None of these

348. Irish blight is also known as:
(A) Late blight of potato
(B) White blight of mustard
(C) Yellow rust
(D) Loose smut of wheat

349. Polymorphism refers to the process having:
(A) > 2 life cycle stages (B) Only one life cycle
(C) < 2 life cycle stages (D) All of these

350. Which of the fruiting body is related to sexual reproduction?
(A) Cleistothecia (B) Perithecia
(C) Apothecia (D) All of these

351. Which of the following disease is caused by virus?
(A) Fiji disease of sugarcane
(B) Leaf curl of papaya
(C) Ratoon stunting of sugarcane
(D) None of these

352. Which of the crop is affected mostly by striga?
(A) Sugarcane (B) Maize
(C) Millets (D) All of these

353. Which of the following gas can stimulate the seeds of striga?
(A) Ethylene (B) Nitrous oxide
(C) Carbon mono oxide (D) Oxygen

354. Strigol is used to control:
(A) Striga (B) Cuscuta
(C) Orobanche (D) Loranthus

355. Which of the following is a synthetic culture medium?
(A) Oat agar medium (B) Plant extracts
(C) Rechard's solution (D) None of these

356. Hill bunt of wheat is caused by:
(A) *Tiletia foetida* (B) *Neovossia indica*
(C) *Uromystis tritici* (D) None of these

357. Which of the following is used for tree surgery?
(A) Bordeaux mixture (B) Chaubatia paste
(C) Santar A (D) All of these

358. Penicillin antibiotic is prepared by:
(A) *Penicillium chrysogenum*
(B) *Penicillium notatum*
(C) *Penicillium albidum*
(D) *Penicillium citrinum*

359. Citrinin is prepared by:
(A) *Penicillium citrinum* (B) *Penicillium flavus*
(C) *Penicillium albidum* (D) *Penicillium citrinum*

360. Penicillin B is prepared by:
(A) *Penicillium flavum* (B) *Penicillium notatum*
(C) *Penicillium albidum* (D) *Penicillium citrinum*

361. The fruit like apple, pear and grapes are spoiled by:
(A) *Penicillium expensum* (B) *Penicillium notatum*
(C) *Penicillium flavum* (D) *Penicillium citrinum*

362. Which of the *penicillium spp* is harmful for food products?
(A) *Penicillium expensum* (B) *Penicillium glaccusm*
(C) *Penicillium digitatum* (D) All of these

363. Aspergillin antibiotic was prepared by which fungus?
(A) *Aspergillus niger* (B) *Aspergillus fumigatus*
(C) *Aspergillus flavus* (D) All of these

364. The causal organism of powdery mildew of grape vine is:
(A) Unicellular nector
(B) Podosphaera leucotricha
(C) Sclerotium oryzae
(D) All of these

365. Most important symptoms of ergot disease is:
(A) Transformation of kernels into black structure
(B) Dispersion of spores
(C) Both (A) and (B)
(D) None of these

366. The symptoms of powdery mildew of pea appear first of all on:
(A) Leaves (B) Stem
(C) Whole plant (D) Roots

367. Ghost light is a result of:
(A) Bacterial luminescence (B) Phosphorescent glow
(C) Bioluminescence (D) All of these

368. The main symptoms of black leg or soft rot of potato is:
(A) Brown – black or jet black colour of stem at soil level
(B) Black leaf coloration
(C) Black crust on tubers
(D) None of these

369. Virus can be passed through:
(A) Brekfeld and chamberlain filters
(B) Only brekfeld filter
(C) Only chamberlain filter
(D) None of these

370. Virus can be precipitated with the help of:
(A) Acetone (B) Safranin
(C) Ethyl alcohol (D) All of these

371. If the virus multiplies in host bacterium which can undergoes complete lysis to release the daughter phages, this type of virus is called:
(A) Lytic bacterio phages
(B) Lysogenic bacteriophages
(C) Transformation
(D) None of these

372. Root knot nematodes have higher growth at :
(A) 15-30^{0}C (B) 20-25^{0}C
(C) 25-28^{0}C (D) 30-35^{0}C

373. Which of the following disease is caused by MLO's?
(A) Little leaf of egg plant (B) Citrus greening
(C) Purple top roll of potato (D) All of these

374. Silica content in leaf helps rice plant to resist against:
(A) Blast (B) False smut
(C) Bacterial blight (D) None of these

375. Which of the following asexual method is very common in bacteria?
(A) Conidia formation (B) Endospore formation
(C) Binary fission (D) None of these

376. Which of the following stages are found in rust fungi?
(A) Pycnial, Aerial, Uredinal, Telial and Basidial stages
(B) Pycnial, Aerial, Apothecial, Telial and Basidial stages
(C) Conidial, Aerial, Uredinal, Telial and Basidial stages
(D) None of these

377. Conidia are produced in basidiomycetes on:
(A) Primary mycelium (B) Secondary mycelium
(C) Both (A) and (B) (D) None of these

378. Aseptic basidium is known as:
(A) Probasidium (B) Phragmobasidium
(C) Holobasidium (D) None of these

379. Telial stage of rust fungi is always found on:
(A) Primary host (B) Secondary host
(C) Dead matter (D) All of these

380. How many physiological strains of black rust are found?
(A) 100 (B) 200
(C) 250 (D) 300

381. Which of the following is not the collateral host of yellow rust?
(A) *Agropyron semicostatum*
(B) *Bromes cathartics*
(C) *Puccinia striiformis*
(D) All of these

382. Which of the following is the alternate host of black rust?
(A) Wheat (B) Barbary
(C) Calotropis (D) *Phalaris minor*

383. Basidiospores of smut fungi are called as:
(A) Sporidia (B) Pycniospores
(C) Oidiospores (D) None of these

384. Basidiospores of stem rust of wheat attack on:
(A) Wheat plant (B) Barbary plant
(D) *Phalaris minor* (D) *Chenopodium album*

385. Which of the following plant is the natural alternate host of *Puccinia recondita*?
(A) *Berberis vulgaris* (B) *Isopyron fumarioides*
(C) *Avena fatua* (D) *Triticum aestivum*

386. Which of the following rust requires the lowest temperature for its over summering?
(A) Black rust (B) Leaf rust
(C) Strip rust (D) All of these

387. Linseed rust is a:
(A) Obligate parasite
(B) Obligate saprophyte
(C) Facultative parasite
(D) Facultative saprophyte

388. Linseed rust fungus is a:
(A) Autocious (B) Heterocious
(C) Both (A) and (B) (D) None of these

389. In A*lternaria solani*, the conidia are formed at:
(A) Top end of the condiophores
(B) Intercalary formation
(C) Both (A) and (B)
(D) None of these

390. The infected seed of bajra smut are converted into:
(A) Sori (B) Nod
(C) Mycelium (D) None of these

391. Grain smut of jowar may cause following disease in humen beings:
(A) Hey fever (B) Cholera
(C) Cold (D) None of these

392. Loose smut is a:
(A) Internally seed borne disease
(B) Air borne disease
(C) Soil borne disease
(D) Externally seed borne disease

393. The spores of covered smut of barley are covered by:
(A) Yellow membrane
(B) Black membrane
(C) Shining silvery membrane
(D) None of these

394. Mechanism of wilting can be understood by:
(A) Plugging of vessels by fungus hyphae
(B) Toxin theory-production of toxins like Fusaric acids
(C) Enzyme theory-Production of enzymes like Pectolytic enzyme
(D) All of these

395. *Alternaria solani* is a:
(A) Soil borne disease (B) Seed borne disease
(C) Air borne disease (D) None of these

396. Which of the perfect stage of red rot of sugarcane:
(A) *Glomerella tucumansis*
(B) *Physalospora tucumansis*
(C) Both (A) and (B)
(D) None of these

397. The perfect stage of tikka disease of groundnut is:
(A) *Mycosphaerella berkeleyii*
(B) *Mycosphaerella arachidicola*
(C) Both (A) and (B)
(D) None of these

398. Rotten neck disease is known as:
(A) Blast of rice (B) Red rot of sugarcane
(C) Leaf rust of wheat (D) None of these

399. What is the main symptoms of red rot of sugarcane?
(A) Dull red pith (B) Stunted growth
(C) Shorter internodes (D) None of these

400. Which of the main symptoms of leaf blight of wheat?
(A) Small, oval, discolored, irregular lesions scattered on leaves
(B) Leaf starts drying from tip
(C) Black powdery conidia of the fungus cover the lesion under moist conditions
(D) All of these

401. The bacteria arranged like grapes bunches are known as:
(A) Streptococcus (B) Staphylococcus
(C) Sarcinae (D) None of these

402. Which of the following fruiting body is formed in red rot of sugarcane?
(A) Acervuli (B) Pycnidia
(C) Apothecia (D) All of these

403. Photosynthetic bacteria have photosynthetic pigments made of:
(A) Lamellae and chromatophores
(B) Ribosomes
(C) Endoplasmic reticulum (D) All of these

404. Which of the following plant disease is caused by MLO's?
(A) Witch brooms of potato (B) Mulbery dwarf
(C) Aster yellows (D) All of these

405. Which of the following feature is related to MLO.s?
(A) Can be grown on cell free media
(B) They are resistant to Penicillin and susceptible to tetracycline
(C) They can pass through anti bacterial filters
(D) All of these

406. Which of the following genus comes under moniliales order of imperiti fungi?
(A) Alternaria (B) Helminthosporium
(C) Cercospora (D) All of these

407. Which of the following class of deuteromycetes has false mycelium and reproduction by budding?
(A) Blastomycetes (B) Hypomycetes
(C) Coelomycetes (D) All of these

408. Infected hyphae of grain smut of fungi is a:
(A) Uninucleate (B) Dikaryotic
(C) Diploid (D) None of these

409. The optimum temperature for development of leaf blight of wheat is:
(A) 15°C (B) 20°C
(C) 25°C (D) 30°C

410. Literary meaning of virus is:
(A) Morbid poison (B) Acute poison
(C) Poison (D) All of these

411. Antibiotics are prepared mainly from:
(A) Actinomycetes (B) Rhizobium
(C) Pseudomonas (D) Azotobacter

412. Grain smut disease of jowar is a disease of:
(A) Internally seed borne (B) Externally seed borne
(C) Air borne (D) Soil borne

413. *Escherichia coli* is a:
(A) Monomorphic (B) Dimorphic
(C) Both (A) and (B) (D) None of these

414. Which of the following does not posses mitochondria?
(A) Fungi (B) Bacteria
(C) Virus (D) MLO's

415. The virus that attacks on bacteria is called:
(A) Bacteria parasite (B) Bacteria phages
(C) Virophages (D) None of these

416. Viroids have only:
(A) Naked nucleic acid particles
(B) Nucleic acid protein core
(C) Only protein core
(D) None of these

417. What is optimum temperature for spread of loose smut disease?
(A) 18°C (B) 23°C
(C) 28°C (D) 33°C

418. Which of the following does not possess metabolic ability?
(A) Fungi (B) Bacteria
(C) Virus (D) All of these

419. Protein core of virion is also known as:
(A) Capsid (B) Capsomer
(C) Structural unit (D) All of these

420. Ear cockle of wheat or Sehun disease is caused by:
(A) *Anguina tritici* (B) *Heterodera avenae*
(C) *Meloidogyne javanica* (D) All of these

421. Molya disease of barley is caused by:
(A) Bacteria (B) Virus
(C) Nematodes (D) None of these

422. The causal organism of molya diseases is:
(A) *Heterodera avenae* (B) *Melodogne spp.*
(C) *Anguina spp.* (D) None of these

423. Molya disease of barley first reported at:
(A) Gujarat (B) UP
(C) Punjab (D) Rajasthan

424. What is optimum temperature for germination of Urediniospores?
(A) 3-5°C (B) 15-16°C
(C) 30-35°C (D) 40-45°C

425. Which of the following crops are affected by orobanche?
(A) Tobacco (B) Brinjal
(C) Tomato (D) All of these

426. Ear cockle of wheat is often associated with:
(A) Yellow ear rot or Tundu disease
(B) Loose smut
(C) Rust
(D) None of these

427. Which of the following striga species are found in India?
(A) *Striga lutea* (B) *Striga densiflora*
(C) *Striga euphrasiodes* (D) All of these

428. Loranthus is parasite of:
(A) Mangoes (B) Guava
(C) Pipal (D) All of these

429. Prophylaxis includes:
(A) Exclusion (B) Eradication
(C) Direct protection (D) All of these

430. The medium or culture is generally sterilized by:
(A) Autoclave (B) Hot air oven
(C) Light (D) None of these

431. The different spore stages of rust fungi are known as:
(A) 0, I, II, III and IV stages
(B) 1, 2, 3, 4 and 5 stages
(C) 0, I, ii, iii, iv and iv stages
(D) None of these

432. What is causal organism of yellow rust of wheat?
(A) *Puccinia striiformis* (B) *Puccinia recondita*
(C) *Puccinia tritici* (D) All of these

433. Mycoses disease is caused by:
(A) Fungi (B) Bacteria
(C) Virus (D) Fly

434. Heterothallic fungus is:
(A) Dioecious (B) Monoecious
(C) Both (A) and (B) (D) None of these

435. The conjugation of oogonium and antheridium is known as:
(A) Oogamy (B) Karyogamy
(C) Plasmogamy (D) All of these

436. Zygospore is formed after conjugation of:
(A) Two similar gametangia only
(B) Two gametangia
(C) Oogonium and anthesidium
(D) None of these

437. Oospore is formed after conjugation of:
(A) Two similar gametangia
(B) Two similar spores
(C) Oogonium and anthesidium
(D) Two dissimilar gametangia

438. The sucking part of fungi, that draws their neourishment is known as:
(A) Hypha (B) Haustoria
(C) Septa (D) Spore

439. Biotrophs are pathogens that:
(A) Complete their life cycle on living cells
(B) Produce haustoria
(C) Rarely kill to host plants
(D) All of these

440. Vegatative life cycle in yeast mainly occurs by:
(A) Budding (B) Fission
(C) Spermatization (D) Conjugation

441. Vegetative life cycle in yeast occurs under:
(A) Favourable condition
(B) Unfavourable condition
(C) Both (A) and (B)
(D) None of these

442. Asexual reproduction in Aspergillus fungi occurs through:
(A) Conidia (B) Aplanospores
(C) Oospores (D) Ascospores

443. Ascus are naked in which of the following class:
(A) Hemiascomycetes (B) Plectosmycetes
(C) Pyrenomycetes (D) Discomycetes

444. Which viruses is likely to be commercialized in India for control of *Helicoverpa (Heliothis) armigera*?
(A) Nucleopolyhedrosis viruses (NPV)
(B) Granulosis viruses (GV)
(C) Advenoviruses
(D) All of these

445. Pusa 1406 has resistance for :
(A) Bacterial leaf blight (B) Blast
(C) Tungro (D) Grassy stunt

446. Annual loss due to pests, diseases and weeds annually is estimated nearly:
(A) >50000 crores (B) <50000 crores
(C) > 1.0 crores (D) < 1.0 crores

447. India is using biopesticides presently:
(A) 1% of total pesticide (B) 5 % to total pesticide
(C) 10 % of total pesticide (D) 15 % of total pesticide

448. Average biopesticide use globally is:
(A) 5 % of total pesticide (B) 10 % of total pesticide
(C) 12 % of total pesticide (D) 15 % of total pesticide

449. Total pesticide consumption in India during 2006-07 is:
(A) 35.65 million tonnes (B) 37.95 million tonnes
(C) 39.26 million tonnes (D) 42.56 million tonnes

450. Directorate of Plant Protection Quarantine and Storage was started in:
(A) 1907 (B) 1914
(C) 1921 (D) 1946

451. Destructive Insects and Pest Act was launched in India in:
(A) 1907 (B) 1914
(C) 1921 (D) 1946

452. Which of the airport does not have plant quarantine station?
(A) Lucknow (B) Kolkata
(C) New Delhi (D) Madras

453. Browning of cauliflower is caused by the deficiency of:
(A) Zinc (B) Manganese
(C) Iron (D) Boron

454. The citrus diseases namely 'Reclamation Disease' and 'Die back' are caused by the deficiency of:
(A) Zinc (B) Iron
(C) Copper (D) Molybdenum

455. Papaya transgenic Rainberg is resistant to:
(A) Papaya Mosaic Virus
(B) Leaf Curl Virus
(C) Papaya Ring Spot Virus
(D) Distorted Ring Spot Virus

456. Tobacco cultivar A.C. Change is resistant to:
(A) Leaf curl (B) Black rot
(C) Damping off (D) TMV

457. Horizontal resistance is:
(A) Resistance against one race of pathogen
(B) Resistance against all the races of pathogen
(C) Resistance against lodging
(D) None of these

458. What is causal organism of yellow rust of wheat?
(A) *Puccinia striiformis* (B) *Puccinia recondita*
(C) *Puccinia tritici* (D) All of these

459. What is optimum temperature for germination of Urediniospores?
(A) 3-5^{0}C (B) 15-16^{0}C
(C) 30-35^{0}C (D) 40-45^{0}C

460. What is optimum temperature for spread of loose smut disease?
(A) 18^{0}C (B) 23^{0}C
(C) 28^{0}C (D) 33^{0}C

461. Grain smut disease of jowar is a disease of:
(A) Internally seed borne (B) Externally seed borne
(C) Air borne (D) Soil borne

462. Which of the following genus comes under moniliales order of imperiti fungi?
(A) Alternaria (B) Helminthosporium
(C) Cercospora (D) All of these

463. Which of the following class of deuteromycetes has false mycelium and reproduction by budding?
(A) Blastomycetes (B) Hypomycetes
(C) Coelomycetes (D) All of these

464. Infected hyphae of grain smut of fungi is a:
(A) Uninucleate (B) Dikaryotic
(C) Diploid (D) None of these

465. The optimum temperature for development of leaf blight of wheat is:
(A) 15^{0}C (B) 20^{0}C
(C) 25^{0}C (D) 30^{0}C

466. Which of the following fruiting body is formed in red rot of sugarcane?
(A) Acervuli (B) Pycnidia
(C) Apothecia (D) All of these

467. Mechanism of wilting can be understood by:
(A) Plugging of vessels by fungus hyphae
(B) Toxin theory-production of toxins like Fusaric acids
(C) Enzyme theory-Production of enzymes like Pectolytic enzyme
(D) All of these

468. Silica content in leaf helps rice plant to resist against:
(A) Blast (B) False smut
(C) Bacterial blight (D) None of these

469. Mycoplasma was first discovered from:
(A) Pleuropneumonia disease of sheep
(B) *Bacillis vibrio*
(C) Anthrax of cattle
(D) None of these

470. The sedimentation coefficient of bacterial ribosome is:
(A) 60 s (B) 70 s
(C) 80 s (D) 90 s

471. The pilli or Fimbria are hair like structures found on bacterial cell and its main function is related to:
(A) Conjugation (B) Movement
(C) Cell division (D) None of these

472. The unit of sedimentation is:
(A) Pascal (B) Svedberg
(C) Virus (D) MLO's

473. Which of the following asexual method is very common in bacteria?
(A) Conidia formation (B) Endospore formation
(C) Binary fission (D) None of these

474. The method through which the genetic material of bacteria is transferred from one cell to other by bacteriophages, is known as:
(A) Conjugation (B) Transformation
(C) Transduction (D) None of these

475. Ghost light is a result of:
(A) Bacterial luminescence (B) Phosphorescent glow
(C) Bioluminescence (D) All of these

476. The main symptoms of black leg or soft rot of potato is:
(A) Brown – black or jet black colour of stem at soil level
(B) Black leaf coloration
(C) Black crust on tubers
(D) None of these

477. Virus can be passed through:
(A) Brekfeld and chamberlain filters
(B) Only brekfeld filter
(C) Only chamberlain filter
(D) None of these

478. Plasmid are found in:
(A) Bacterial cells (B) Viral cells
(C) Fungal cells (D) None of these

479. Plasmid are mainly:
(A) Extra chromosomal particles of DNA
(B) Particles of proteins
(C) Micro organisms
(D) None of these

480. Virus can be precipitated with the help of:
(A) Acetone (B) Safranin
(C) Ethyl alcohol (D) All of these

481. If the virus multiplies in host bacterium which can undergoes complete lysis to release the daughter phages, this type of virus is called:
(A) Lytic bacterio phages
(B) Lysogenic bacteriophages
(C) Transformation
(D) None of these

482. Root knot nematodes have higher growth at :
(A) 15-30°C (B) 20-25°C
(C) 25-28°C (D) 30-35°C

483. Which of the following disease is caused by MLO's?
(A) Little leaf of egg plant (B) Citrus greening
(C) Purple top roll of potato (D) All of these

484. Which of the following disease is caused by virus?
(A) Fiji disease of sugarcane
(B) Leaf curl of papaya
(C) Ratoon stunting of sugarcane
(D) None of these

485. The process in which bacterial genome is multiplied with bacteriophages is called?
(A) Transformation (B) Transduction
(C) Reproduction (D) None of these

486. Which of the crop is affected mostly by striga?
(A) Sugarcane (B) Maize
(C) Millets (D) All of these

487. Which of the following gas can stimulate the seeds of striga?
(A) Ethylene (B) Nitrous oxide
(C) Carbon mono oxide (D) Oxygen

488. Strigol is used to control:
(A) Striga (B) Cuscuta
(C) Orobanche (D) Loranthus

489. Which of the following is a synthetic culture medium?
(A) Oat agar medium (B) Plant extracts
(C) Rechard's solution (D) None of these

490. Hill bunt of wheat is caused by:
(A) *Tiletia foetida* (B) *Neovossia indica*
(C) *Uromystis tritici* (D) None of these

491. Which of the following is used for tree surgery?
(A) Bordeaux mixture (B) Chaubatia paste
(C) Santar A (D) All of these

492. Main symptoms of karnal bunt is:
(A) Formation of black powder below the pericarp of seed
(B) Produces rotten fish odour
(C) Bad odour is due to Trimethylamine
(D) All of these

493. Hill bunt can be controlled by:
(A) Seed treatment with 0.2% vitavex
(B) Use of resistant varieties
(C) Clean the field
(D) All of these

494. Which of the following culture is a natural medium for isolation of pathogen?
(A) Potato dextrose agar medium (B) Milk
(C) Egg (D) All of these

495. Trychoderma can be used as:
(A) Soil (B) Seed
(C) Both (A) and (B) (D) None of these

496. Which of the filters are used to sterilize the medium?
(A) Brekfeld filter (B) Seitz filter
(C) Chamber lain filter (D) None of these

497. Root knot disease of tomato is caused by:
(A) *Meloidogyne javanica* (B) *Heterodera avenaei*
(C) *Anguina tritici* (D) None of these

498. The spores of grain smut of jowar are attached with long solid structure known as:
(A) Columella (B) Pith
(C) Paraphysis (D) None of these

499. Coelomycetes class of imperfect fungi is characterized by:
(A) Well developed mycelium
(B) Budding absence
(C) Pycnidium and acervulus fruiting bodies present
(D) All of these

500. The collateral host of linseed rust is:
(A) *Linum usitatissimum* (B) *Linum mysorense*
(C) *Linum pisi* (D) All of these

501. Covered smut of barley is a disease of:
(A) Externally seed borne (B) Internally seed borne
(C) Air borne (D) None of these

502. In melampsora, the teliospores are:
(A) Sessile and unicellular
(B) Strigmated and unicellular
(C) Sessile and multinucleate
(D) None of these

503. The causal organism of powdery mildew of wheat is:
(A) *Erysiphe graminis var tritici*
(B) *Erysiphe polygoni*
(C) *Erysiphe aestivum*
(D) None of these

504. Pin moulds are also known as sugar fungi because of:
(A) They use simple carbohydrates
(B) They do not use carbohydrate
(C) They use polysaccharides
(D) All of these

505. Ascogonium produces one tubular structure that receives the nuclei of antheridium. This structure is known as:
(A) Trichogyne (B) Crosier
(C) Fertilization tube (D) All of these

506. How many ascospores are produced by an ascus?
(A) Two (B) Four
(C) Eight (D) Twelve

507. Some filaments protrude out through ostiole of perethecium are known as:
(A) Paraphyses (B) Periphyses
(C) Appendages (D) All of these

508. The spores and spore bearing hyphae are some times grouped into small or large masses or clusters, known as:
(A) Synnemata (B) Sori
(C) Pycnidia (D) None of these

509. Disc like, saucer shaped or cup like fruiting body raised on long or short stalk is known as:
(A) Aecium (B) Sporodochium
(C) Apothecium (D) All of these

510. Aeciospores are formed in:
(A) Pycnidia (B) Synnemata
(C) Acervuli (D) Aecium

511. Setae is present in which of the following fruiting body:
(A) Pycnidia (B) Acervuli
(C) Aecium (D) All of these

512. The thallus that is converted fully into a reproductive structure is known as:
(A) Holocarpic (B) Eucarpic
(C) Monoecious (D) Dioecious

513. In gametangial contact, ascogonia contact with antheridia with the help of a special structure known as:
(A) Trichogyne (B) Dikaryon hyphae
(C) Stroma (D) All of these

514. Which of the following is motyle spore?
(A) Chlamydospore (B) Conidia
(C) Odia (D) Zoospore

515. Which of the following is a biotroph pathogen?
(A) Viruses
(B) Rust fungi
(C) Downy and Powdery mildew
(D) All of these

516. The pathogen that can grow only under controlled environment is called:
(A) Ecologically obligate
(B) Bio-chemically obligate
(C) Facultative parasite
(D) All of these

517. Apisomes are also known as:
(A) Plasmids (B) RNA particles
(C) Golgi bodies (D) All of these

518. Fungi have been originated from:
(A) Polygeneric line (B) Phytogeneric line
(C) Polygendric line (D) None of these

519. The carrier of leaf curl tomato virus is:
(A) White fly (B) Beetle
(C) Bacteria (D) Fungus

520. Leaf curl of tomato is caused by:
(A) Tobacco virus 16 (B) Tomato virus 16
(C) Virus No. 10 (D) Virus No.16

521. Which of the following sexual reproduction method is found in bacteria?
(A) Conjugation (B) Transformation
(C) Transduction (D) All of these

522. Hemaglutination is the process in which bacteria cells can:
(A) Attach with RBC cells
(B) Attach with water surface
(C) Attach with virons
(D) None of these

523. Bacterial nucleus is characterized by:
(A) Undeveloped nucleus
(B) Not covered by nucleus membrane
(C) Multiplied by binary fission
(D) All of these

524. Which of the nucleic acid is found in the nucleus of cauliflower mosaic virus?
(A) RNA (B) DNA
(C) Both RNA and DNA (D) All of these

525. Who did classify plant diseases for the first time?
(A) Zollinger (1973)
(B) J.F. Dastur (1886-1910)
(C) Anton de Bary (1831-1888)
(D) Robert Koch (1843-1910)

526. Plant pathology also called as:
(A) Plankton pathology (B) Pathogenesis
(C) Phyto pathology (D) Economic pathology

527. Who was the first Indian plant pathologists who studied the fungi and plant diseases caused by them?
(A) Zollinger (1973)
(B) J.F. Dastur (1886-1910)
(C) Anton de Bary (1831-1888)
(D) Robert Koch (1843-1910)

528. Any thing causes plant disease is called:
(A) Plant pathogen (B) Plant disease
(C) Pathogenicity (D) None of these

529. Any deviation from the nrmal conditions of the plant may be called:
(A) Plant pathogen (B) Plant disease
(C) Pathogenicity (D) None of these

530. The ultra microscopic entities, which are obligate in parasitism, functioning as living organisms inside the host and as non-living molecules outside the host is called:
(A) Viruses (B) Viroids
(C) Both (A) and (B) (D) None of these

531. The naked strands of nucleic acids and are non-encapsulated in protein coat like viruses is called:
(A) Viruses (B) Viroids
(C) Both (A) and (B) (D) None of these

532. A cell, spore, hyphae or mycelium may be called as:
(A) Heterokaryotic (B) Homokaryotic
(C) Both (A) and (B) (D) None of these

533. If all the nuclei are identical, the terms applied are:
(A) Heterokaryotic (B) Homokaryotic
(C) Homokaryon (D) Both (A) and (B)

534. A cell, spore, hyphae or mycelium are possesses nuclei of more than one genetic type. Such nuclear condition is called:
(A) Heterokaryotic (B) Homokaryotic
(C) Heterokaryon (D) Homokaryon

535. When two genetically different nuclei fuse to complete the life cycle, the condition is called:
(A) Heterothallic (B) Amphithallic
(C) Pseudo-homothallic (D) None of these

536. Heterothallic fungi in which some spores contain two or more nuclei of only one mating type while others contain two or more nuclei with compatible mating type genes compared with pseudo-homothallic mating system is called:
(A) Heterothallic
(B) Amphithallic
(C) Pseudo-homothallic
(D) None of these

537. Heterothallic mating systems in which all or nearly all spores of the fungus contain two or more nuclei with compatibl mating type genes is called:
(A) Heterothallic (B) Amphithallic
(C) Pseudo-homothallic (D) None of these

538. If a disease is characterized by th necrosis, hypertrophy and wilting, all these symptoms will collectively be called as:
(A) Disease syndrome (B) Disease diagnosis
(C) Both (A) and (B) (D) None of these

539. The process in which facts about the cause and pathogenesis of the disease are assimilated and determines is called:
(A) Disease syndrome (B) Disease diagnosis
(C) Both (A) and (B) (D) None of these

540. The pathogen exhibits its ill influence on whole of the plant is known as:
(A) Incubation period (B) Colonization
(C) Systemic infection (D) None of these

541. After infection the pathogen develops within the host tissues and reproduces therein. This condition is called as;
(A) Invasion (B) Colonization
(C) Both (A) and (B) (D) None of these

542. The disease causing uniy that incites diseases for the first time is called as:
(A) Primary inoculums (B) Secondary inoculums
(C) Tertiary inoculums (D) None of these

543. The disease infection from inoculums produced during the same growing season is called as:
(A) Primary inoculums (B) Secondary inoculum
(C) Tertiary inoculums (D) None of these

544. Any form or part of a pathogen that can initiate disease development is termed as:
(A) Pathogenecity (B) Host
(C) Inoculum (D) None of these

545. A disease causing agent which may be living or nonliving in nature is called:
(A) Pathogen (B) Parasitism
(C) Both (A) and (B) (D) None of these

546. An organism that lives upon the other living organism and desire food material from it for its survival and existence is called:
(A) Pathogen (B) Parasitism
(C) Both (A) and (B) (D) None of these

547. The disease are those which occur at very irregular intervals and locations and in relatively few instances is called:
(A) Endemic disease (B) Epidemic disease
(C) Sporodic disease (D) Pandemic disease

548. When disease is observed to be prevalent through out the country, continent or the world wide and cause mass mortality to a crop, such disease is called as:
(A) Endemic disease
(B) Epidemic disease
(C) Sporodic disease
(D) Pandemic disease

549. Green ear disease of pearlmillet is an example of:
(A) Endemic disease (B) Epidemic disease
(C) Sporodic disease (D) Pandemic disease

550. Late blight of potato in Iris famine was an example of:
(A) Endemic disease (B) Epidemic disease
(C) Sporodic disease (D) Pandemic disease

551. Endemic disease is also known as:
(A) Epiphytotic disease (B) Endemic disease
(C) Sporodic disease (D) Pandemic disease

552. Late blight of potato occurred in 1913 in Rangpur now in Pakistan was an example of:
(A) Endemic disease (B) Epiphytotic disease
(C) Sporodic disease (D) Pandemic disease

553. The toxic effect induced by the salivary components secreted by the arthropods specially while feeding is called:
(A) Phytoxaemia (B) Etiolation
(C) Chlorosis (D) All of these

554. Chlorosis is usually seen is disease incited by:
(A) Mineral deficiencies (B) Viruses
(C) Both (A) and (B) (D) None of these

555. Bordeaux mixture is a:
(A) Systemic fungicide
(B) Fungicide and bactericide
(C) Tropical inorganic fungicide
(D) Tropical organic fungicide

556. Nimbin and Nimbidin are extracted from:
(A) Neem (B) Ashwagandha
(C) Amla (D) Corainder

557. Aflatoxin are produced by:
(A) Bacteria (B) Algae
(C) Virus (D) Fungi

558. Bacillus thuringiensis is commercially useful as:
(A) Bio-fungicide (B) Bio-weedicide
(C) Bio-insecticide (D) Bio-fertilizer

559. Udbatta disease is associated with which of the following crops?
(A) Sorghum (B) Pearlmillet
(C) Rice-false urgot (D) Cotton

560. The 'Akiochi' disease of rice is due to:
(A) B deficiency (B) Al toxicity
(C) Fe toxicity (D) H_2S poisoning

561. White tip of maize in caused due to deficiency of:
(A) Cu (B) Mo (C) Zn (D) Fe

562. Ergot disease is caused by which pathogen on which host?
(A) *Claviceps purpurea* on rye
(B) *Puccinia recondita* on wheat
(C) *Drechlera sorokiniana* on wheat
(D) *Albugo candida* on mustard

563. The devasting impact of plant diseases on human suffering and survival was first realized by epidemic of:
(A) Brown spot of rice in Bengal
(B) Late blight of potato in USA
(C) Late blight of potato in Europe
(D) Rust of wheat in India

564. Bengal famine was caused by:
(A) *Helminthosporium oryzae*
(B) *Phytophthora infestans*
(C) *Puccinia graminis*
(D) *Erysiphae graminicola*

565. Which of these viruses is used as a pesticide for crops infested by insects susceptible to contraction?
(A) Cauliflower mosiac virus
(B) Cucumber mosaic virus
(C) Rice tungro virus
(D) Nuclear polyhedrosis virus

566. Coffee rust is caused by:
(A) Puccinia graminis
(B) Cephaleuros coffeae
(C) Helminthosporium coffeae
(D) Hemileia vastatrix

567. Specialized cell typical of many fungal plant pathogens that is used to infect host plants:
(A) Hyphae (B) Vesicles
(C) Appressorium (D) Germ tube

568. Pickout an endemic disease from the following?
(A) Wheat stem rust *(Puccinia graminis tritici)*
(B) Late blight of potato *(Phytophthora infestans)*
(C) Red rot of sugarcane *(Collectotrichum falcatum)*
(D) Citrus canker *(Xanthomonas axonopodis pv. citri)*

569. Partial or complete loss of virulence in pathogens is called:
(A) Abortive parasitism (B) Altenuation
(C) Neutralism (D) Susceptibility

570. Karnal bunt is caused by:
(A) Nevosia indica (B) Albugo candida
(C) Phytophthora infestans (D) None of the above

571. The crop having the highest pesticide use is:
(A) Oilseeds (B) Rice
(C) Vegetables (D) Cotton

572. Dose of NPV is:
(A) 250 L.E. (B) 500 L.E.
(C) 750 L.E. (D) 1000 L.E.

573. Jassid is a vector responsible for which disease of paddy?
(A) Blast (B) Tungro
(C) Brown spot (D) Bacterial blight

574. Who advanced the gene for gene concept of disease resistance and susceptibility in 1946?
(A) Flor (B) Vander plank
(C) Gaumann (D) Muller

575. A bacteria which have a large number of flagella all over the cell is categorized in:
(A) Atrichous (B) Amphitrichous
(C) Lophotrichos (D) Peritrichous

576. In bacteria, variability is cause by:
(A) Conjugation (B) Transformation
(C) Transduction (D) All of the above

577. Citrus greening disease mainly confined to:
(A) North India (B) South India
(C) East India (D) West India

578. Select the organism which can not synthesize protein by own enzymes:
(A) Bacteria (B) Mycoplasms
(C) RLO (D) Virus

579. The adjacent cells of the same fungal hypha brought into communication with each other by means of loops is known as:
(A) Transduction (B) Transformation
(C) Clamp connection (D) Conjugation

580. The sexual spores of fungi is:
(A) Chlamydospores (B) Sporangiospores
(C) Zoospores (D) Zygospores

581. The bacterial colony is known as:
(A) Spore (B) Mycelium
(C) Ooze (D) Hypha

582. *Albugo candida* casues white blisters or white rust of crucifers in a/an :
(A) Obligate parasite
(B) Obligate saprophite
(C) Facultative parasites
(D) Facultative saprophite

583. *Albugo candida* produces :
(A) Basidiospores (B) Ascospores
(C) Zoospores (D) Oospores

584. The downey mildew disease of pearlmillet is primarily a :
(A) Seed borne (B) Air borne
(C) Soil borne (D) Water borne

585. VAM is :
(A) Bacteria (B) Fungi
(C) Virus (D) Algae

586. Micrografting is used to produce plants free from?
(A) Virus (B) RLOs
(C) MLOs (D) Bacteria

587. Hot water treatment of seed is useful for control of :
(A) Loose smut (B) Covered smut
(C) Rust (D) Powdery mildew

588. RSD is caused by:
(A) Gram (-) fastidious bacteria
(B) Gram (+) fastidious bacteria
(C) Clavibacter xyli
(D) Both (B) and (C)

589. MLO disease transmitted by:
(A) Leaf hopper (B) Aphid
(C) Whitefly (D) Animals

590. Sterility mosaic disease of pigeonpea spread by :
(A) Virus (B) Aphids
(C) Whitefly (D) Mites

591. Virus caprid is made up of :
(A) Protein (B) Carbohydrate
(C) Lipid (D) Nucleic acid

592. Phyllody disease of sesamum spread by:
(A) Leaf hopper (B) Jassid
(C) Aphid (D) Whitefly

593. In north India and central India, the black rust inoculum cause from:
(A) South (B) Hilly area
(C) USA (D) Nepal

594. Alternate host of black rust is:
(A) Barberi (B) Bajra
(C) Jowar (D) Wheat

595. Viruses contain:
(A) RNA
(B) DNA
(C) Both RNA and DNA
(D) Either RNA or DNA, never both

596. Suicidal germination take place in:
(A) Dodder (B) Striga
(C) Loranthus (D) Dendrophthoe falcata

597. The nature particle of a plant virus is generally known as:
(A) Caprid (B) Vivoid
(C) Virion (D) Nucleocapsid

598. Decomposition of organic matter in submerged soil is carried out by:
(A) Bacteria (B) Actinomycetes
(C) Fungi (D) Earthworm

599. Application of potash increases :
(A) Resistance to waterlogging
(B) Frost resistance in plants
(C) Disease resistance in plants
(D) None of these

600. For quick and accurate detection of viruses can be done by :
(A) ELISA (B) IEM
(C) HADAS (D) All of the above

601. The sexual, thick walled and resting spores of the rust fungi is known as :
(A) Basidiospores (B) Urediospores
(C) Teliospores (D) Aeciospores

602. When plant showed the partial resistance against all the races of pathogen then it is a type of?
(A) Horizontal resistance (B) Vertical resistance
(C) Induced resistance (D) No-host resistance

603. Which one of the following cannot be detected by ELISA technique?
(A) Virus (B) Bacteria
(C) Viroid (D) Fungus

604. Which is most recent widely used fungicide for the control of powdery mildew?
(A) Calaxin (B) Sulphur dust
(C) Dithan M-45 (D) Apron SD-35

605. Which fungicide is used against downey mildew?
(A) Matalaxyl (B) Karathene
(C) Oxathin (D) Streptomycin

606. The major storage fungi that effects the foodgrains is:
(A) Rhizobium
(B) Mucor
(C) Cercospora
(D) Aspergillus

607. A bacteria which have a large number of flagella all over the cell is categorized in:
(A) Atrichos (B) Amphitrichous
(C) Lophotrichous (D) Peritrichous

608. In bacteria, variability is caused by:
(A) Conjugation (B) Transformation
(C) Trandsuction (D) All of the above

609. Early blight of potato produces:
(A) Conidia (B) Telia
(C) Uredia (D) Acacia

610. The rust fungi completing their life cycle on one host are called:
(A) Polymorphic (B) Atoecious
(C) Heteroecious (D) None of the above

611. Promycelium in rust fungi bears:
(A) Basidiospores
(B) Aeciospores
(C) Uredia
(D) Telia

612. Loose smut of barley and wheat is/are:
(A) Internally seed borne
(B) Externally seed borne
(C) Soil borne
(D) All of the above

613. Karnal bunt of wheat first time reported by Mitra in Karnal:
(A) 1929 (B) 1931
(C) 1941 (D) 1951

ANSWERSHEET: 13.PLANT PATHOLOGY

1. (B); 2. (C); 3. (D); 4. (A); 5. (A); 6. (A); 7. (A); 8. (D); 9. (A); 10. (A); 11. (C); 12. (A); 13. (C) ; 14. (D); 15. (C); 16. (B); 17. (C); 18. (B); 19. (C); 20. (B); 21. (D); 22. (C); 23. (C); 24. (B); 25. (D) ;26. (C); 27. (D); 28. (B); 29. (D); 30. (D); 31. (D); 32. (C); 33. (D); 34. (A); 35. (B); 36. (C); 37. (D); 38. (C) ;39. (A); 40. (B); 41. (C); 42. (B); 43. (B); 44. (A); 45. (C); 46. (C); 47. (B); 48. (B); 49. (A); 50. (B); 51. (D) ;52. (D); 53. (A); 54. (B); 55. (D); 56. (B); 57. (C); 58. (A); 59. (A); 60. (C); 61. (D); 62. (A); 63. (B); 64. (A) ;65. (B); 66. (C); 67. (C); 68. (B); 69. (D); 70. (A); 71. (A); 72. (A); 73. (C); 74. (A); 75. (A); 76. (C); 77. (B); 78. (D) ;79. (A); 80. (A); 81. (D); 82. (C); 83. (C); 84. (D); 85. (C); 86. (A); 87. (C); 88. (C); 89. (C); 90. (B); 91. (D) ;92. (B); 93. (C); 94. (B); 95. (A); 96. (A); 97. (B); 98. (D); 99. (D); 100. (B); 101. (B); 102. (B); 103. (B); 104. (A); 105. (C); 106. (A); 107. (C); 108. (D); 109. (B); 110. (A); 111. (B); 112. (A); 113. (A); 114. (B); 115. (C); 116. (C); 117. (B); 118. (D); 119. (A); 120. (D); 121. (A); 122. (B); 123. (A); 124. (A); 125. (B); 126. (C); 127. (A); 128. (A); 129. (A); 130. (D); 131. (A); 132. (B); 133. (A); 134. (B); 135. (A); 136. (C); 137. (B); 138. (A); 139. (B); 140. (D); 141. (A); 142. (D); 143. (D); 144. (D); 145. (D); 146. (A); 147. (A); 148. (A); 149. (B); 150. (B); 151. (A); 152. (A); 153. (B); 154. (D); 155. (A); 156. (A); 157. (D); 158. (D); 159. (A); 160. (B); 161. (D); 162. (C); 163. (D); 164. (B); 165. (A); 166. (B); 167. (C); 168. (C); 169. (D); 170. (A); 171. (C); 172. (A); 173. (B); 174. (A); 175. (A); 176. (D); 177. (C); 178. (B); 179. (A); 180. (A); 181. (B); 182. (A) ; 183. (A); 184. (D); 185. (B); 186. (A); 187. (B); 188. (C); 189. (A); 190. (A); 191. (A); 192. (C); 193. (D); 194. (B); 195. (A) ; 196. (D); 197. (B); 198. (A); 199. (C); 200. (A); 201. (B); 202. (A); 203. (A); 204. (C); 205. (A); 206. (D); 207. (D); 208. (B) ; 209. (A); 210. (A); 211. (A); 212. (D); 213. (B); 214. (A); 215. (A); 216. (A); 217. (A); 218. (A); 219. (C); 220. (C); 221. (A) ; 222. (B); 223. (D); 224. (C); 225. (B); 226. (C); 227. (D); 228. (C); 229. (A); 230. (B); 231. (A); 232. (A); 233. (B); 234. (A) ; 235. (C); 236. (B); 237. (D); 238. (C); 239. (D); 240. (D); 241. (B); 242. (A); 243. (D); 244. (A); 245. (D); 246. (D); 247. (A); 248. (A); 249. (A); 250. (A); 251. (B); 252. (D); 253. (A); 254. (A); 255. (C); 256. (D); 257. (C); 258. (A); 259. (C); 260. (A) ; 261. (A); 262. (B); 263. (C); 264. (D); 265. (A); 266. (B); 267. (D); 268. (D); 269. (D); 270. (A); 271. (A); 272. (D); 273. (B) ; 274. (A); 275. (D); 276. (B); 277. (D); 278. (D); 279. (B); 280. (C); 281. (D); 282. (A); 283. (B); 284. (D); 285. (A); 286. (D) ; 287. (D); 288. (D); 289. (A); 290. (A); 291. (B); 292. (A); 293. (B); 294. (C); 295. (A); 296. (B); 297. (B); 298. (A); 299. (B); 300. (B); 301. (B); 302. (A); 303. (B); 304. (A); 305. (A); 306. (B); 307. (A); 308. (B); 309. (A); 310. (A); 311. (A); 312. (A) ; 313. (B); 314. (C); 315. (A); 316. (B); 317. (C); 318. (A); 319. (C); 320. (B); 321. (A); 322. (B); 323. (A); 324. (A); 325. (A); 326. (B); 327. (A); 328. (C); 329. (A); 330. (D); 331. (A); 332. (D); 333. (C); 334. (C); 335. (B); 336. (A); 337. (D); 338. (A) ; 339. (D); 340. (D); 341. (D); 342. (C); 343. (A); 344. (A); 345. (B); 346. (A); 347. (B); 348. (A); 349. (A); 350. (D); 351. (D) ; 352. (D); 353. (A); 354. (A); 355. (C); 356. (A); 357. (D); 358. (A); 359. (A); 360. (B); 361. (A); 362. (D); 363. (A);

364. (A) ; 365. (A); 366. (A); 367. (D); 368. (A); 369. (A); 370. (D); 371. (A); 372. (C); 373. (D); 374. (A); 375. (C); 376. (A); 377. (C) ; 378. (C); 379. (A); 380. (C); 381. (C); 382. (B); 383. (A); 384. (B); 385. (B); 386. (C); 387. (A); 388. (A); 389. (A); 390. (A) ; 391. (A); 392. (A); 393. (C); 394. (D); 395. (A); 396. (C); 397. (C); 398. (A); 399. (A); 400. (D); 401. (B); 402. (A); 403. (A) ; 404. (D); 405. (D); 406. (D); 407. (A); 408. (B); 409. (C); 410. (A); 411. (A); 412. (B); 413. (B); 414. (A); 415. (B); 416. (A); 417. (B); 418. (C); 419. (D); 420. (A); 421. (C); 422. (A); 423. (D); 424. (B); 425. (D); 426. (A); 427. (D); 428. (D); 429. (D); 430. (A); 431. (A); 432. (A); 433. (A); 434. (A); 435. (A); 436. (B); 437. (C); 438. (B); 439. (D); 440. (A); 441. (A); 442. (A) ; 443. (A); 444. (A); 445. (A); 446. (A); 447. (A); 448. (C); 449. (B); 450. (D); 451. (B); 452. (A); 453. (D); 454. (C); 455. (C); 456. (B); 457. (B); 458. (A); 459. (B); 460. (B); 461. (B); 462. (D); 463. (A); 464. (B); 465. (C); 466. (A); 467. (D); 468. (A); 469. (A); 470. (B); 471. (A); 472. (B); 473. (C); 474. (C); 475. (D); 476. (A); 477. (A); 478. (A); 479. (A); 480. (D); 481. (A); 482. (C); 483. (D); 484. (D); 485. (B); 486. (D); 487. (A); 488. (A); 489. (C); 490. (A); 491. (D); 492. (D); 493. (D); 494. (D); 495. (C); 496. (D); 497. (A); 498. (A); 499. (D); 500. (B); 501. (A); 502. (A); 503. (A); 504. (A); 505. (A); 506. (C); 507. (B); 508. (B); 509. (C); 510. (D); 511. (D); 512. (A); 513. (A); 514. (D); 515. (D); 516. (A); 517. (A); 518. (B); 519. (A); 520. (A); 521. (D); 522. (A); 523. (D); 524. (B); 525. (A); 526. (C); 527. (B); 528. (A); 529. (B); 530. (A); 531. (B); 532. (A); 533. (D); 534. (C) ; 535. (A); 536. (B); 537. (C); 538. (A); 539. (B); 540. (C); 541. (C); 542. (A); 543. (B); 544. (C); 545. (A); 546. (B); 547. (C) ; 548. (D); 549. (C) 550. (D); 551. (A); 552. (B); 553. (A); 554. (C); 555. (C); 556. (A); 557. (A); 558. (C); 559. (C); 560. (D); 561. (C); 562. (A); 563. (C); 564. (D); 565. (D); 566. (D); 567. (C); 568. (---); 569. (B); 570. (A); 571. (D) ; 572. (A); 573. (B) 574. (A); 575. (D); 576. (D); 577. (A); 578. (D); 579. (C); 580. (D); 581. (C); 582. (A); 583. (D); 584. (C); 585. (B); 586. (A); 587. (A); 588. (D); 589. (A); 590. (D); 591. (A); 592. (A); 593. (A); 594. (A); 595. (D); 596. (B); 597. (C); 598. (A); 599. (C); 600. (D); 601. (C); 602. (A); 603. (C); 604. (A); 605. (A); 606. (D); 607. (D); 608. (D); 609. (A); 610. (B); 611. (A); 612. (A); 613. (B).

Chapter 14

HORTICULTURE

1. Growing of flowers, fruits, ornamental plants and vegetables in small grdens or farm is called:
(A) Horticulture (B) Agriculture
(C) Sericulture (D) Apiculture

2. Growing of flowers and ornamental plants is called:
(A) Pomology (B) Olericulture
(C) Floriculture (D) Both (B) and (C)

3. Growing of fuits in garden and farm is called:
(A) Pomology (B) Olericulture
(C) Floriculture (D) Vegetable Science

4. Growing of vegetables in farm is called:
(A) Pomology (B) Olericulture
(C) Floriculture (D) Vegetable Science

5. Which type of garden are of geometric design and symmetrical shape are called:
(A) Formal garden (B) Informal garden
(C) Free style garden (D) Natural garden

6. The style reflects naturalistic effect of total view and represents natural beauty is called:
(A) Formal garden (B) Informal garden
(C) Free style garden (D) Natural garden

7. The combination of both formal and informal garden is called:
(A) Formal garden (B) Informal garden
(C) Free style garden (D) Natural garden

8. Rose garden of Ludhiana is an example of:
(A) Formal garden (B) Informal garden
(C) Free style gardem (D) Natural garden

9. Persian and mughal garden is an example of:
(A) Formal garden (B) Informal garden
(C) Free style garden (D) Natural garden

10. Japanese garden is an example of:
(A) Formal garden (B) Informal garden
(C) Free style garden (D) Natural garden

11. Decarnation of a tract of land with plants and other garden materials so as to produce a picturesque and naturalistic effect in a limited space is called:
(A) Land scape gardening (B) Informal gardening
(C) Free style gardening (D) Natural gardening

12. Plants have special structures to climbs and support is called:
(A) Climbers (B) Creepers
(C) Both (A) and (B) (D) None of these

13. Important annual climbers are:
(A) Morning glory (B) *Thunbergia gradiflora*
(C) *Clerodendrom splendens* (D) All of these

14. Ideal climbers for pergola are:
(A) *Quisqualis* (B) *Poiveria densiflorum*
(C) *Banisyeria taurifolia* (D) All of these

15. Mughal garden is famous for:
(A) Rose (B) Chrysanthemum
(C) Dahelia (D) Merigold

16. Brandavan garden is situated in:
(A) Mysore (B) Mathura
(C) New Delhi (D) Agra

17. Queen of winter is also called:
(A) Rose (B) Merigold
(C) Chrysanthemum (D) Dahalia

18. National flower of India is:
(A) Rose (B) Lotus
(C) Chrysanthemum (D) Merigold

19. Growing of plants in a small containers is called:
(A) Herbaceous border (B) Shrubary border
(C) Nursery (D) Bonsai

20. Valley of flowers is found in:
(A) Nilgiri Hills
(B) North Eastern Himalayas
(C) North Western Himalayas
(D) Satpura Hills

21. Cucumber (kheera) was known in France in ninth century and was common in England in:
(A) 927 (B) 1127
(C) 1227 (D) 1327

22. Bottle guard (*Legenaria siceraria*) was being cultivated in India even before B.C.:
(A) 500 (B) 1000
(C) 1500 (D) 2000

23. Which is the cultivar of bottle guard:
(A) Pusa Meghdoot (B) CO-1
(C) Autralian Green (D) Straight Eight

24. The ——is probably a native of South Asia and perhaps also of Africa, Australia and Polynesia:
(A) Smooth sponge gourd (B) Bottle guard
(C) Pointed guard (D) Bitter guard

25. The number of seeds of bitter guard per 100 g are about:
(A) 350 (B) 450
(C) 550 (D) 650

26. About 3 kg of seed or ——root tubers of pointed guard are enough to plant an area of one hectare:
(A) 20000 (B) 30000
(C) 40000 (D) 50000

27. Turnip (*Brassica campestris*) was introduced into England probably in ——, while is 1606 in USA:
(A) 1590 (B) 1490
(C) 1390 (D) 1290

28. One gramme of turnip seed holds how many seeds:
(A) 460-500 (B) 360-400
(C) 560-660 (D) 260-360

29. Radish seems to have reached England in about 1548 but, introduced in America in :
(A) 1429 (B) 1529 (C) 1629 (D) 1729

30. One gramme radish contains how many seeds:
(A) 700-1000 (B) 500-600
(C) 600-675 (D) 1100-1150

31. One gramme carrot contains how many seeds:
(A) 550 (B) 650 (C) 750 (D) 850

32. Carrot seed will not germinate below———°C:
(A) 7.9 (B) 6.9 (C) 5.9 (D) 3.9

33. The length of sweet potato cutting should normally be ——— with 4-5 sound (healthy) buds:
(A) 40 cm (B) 30 cm
(C) 20 cm (D) 10 cm

34. On weight basis ——— kg of cutting of sweet potato are required per hectare for planting:
(A) 300 (B) 400 (C) 500 (D) 600

35. Potato tuber production is the maximum at ———°C:
(A) 15 (B) 20 (C) 25 (D) 30

36. The young potato plant grows the best at ———°C:
(A) 16 (B) 20 (C) 24 (D) 28

37. In general, potato needs —— mm of irrigation water for the optimum production:
(A) 500 (B) 600 (C) 700 (D) 800

38. Elephant yam corns (buds) weighing 100-200 g are used for planting of the —— years:
(A) First (B) Second
(C) Third (D) Fourth

39. Elephant yam corns at the end of——— year weight 6.8-9.0 kg:
(A) Second (B) Third
(C) Fourth (D) Fifth

40. One gramme of onion, as estimated, generally contains how many seeds:
(A) 180 (B) 240 (C) 300 (D) 360

41. Garlic freezes at the average temperature of——— °C:
(A) -5 (B) -4 (C) -3 (D) -2

42. The best storage temperature for garlic is 0 to 2.2 °C with ——— per cent of relative humidity:
(A) 60 (B) 50 (C) 40 (D) 30

43. Garden pea was an important crop in eleventh century in:
(A) Holland (B) Poland
(C) England (D) New Zealand

44. Garden pea was probably distinguished from field pea in 1536, but their common use began after:
(A) 1586 (B) 1600 (C) 1675 (D) 1700

45. One gramme of celery seed holds about how many seeds:
(A) 1570 (B) 2070 (C) 2270 (D) 2470

46. Celery seed takes nearly ——— to germinate:
(A) 10 days (B) 20 days
(C) 30 days (D) 40 days

47. Pomegranate, fig, olive and grape have been placed in —— salt tolerant group of fruit trees:
(A) High (B) Low
(C) Medium (D) Very high

48. Browning of cauliflower is caused by the deficiency of:
(A) Zinc (B) Manganese
(C) Iron (D) Boron

49. The citrus diseases namely 'Reclamation Disease' and 'Die back' are caused by the deficiency of:
(A) Zinc (B) Iron
(C) Copper (D) Molybdenum

50. Fastening of vines to a upright support of a tree called panagara (*Erithrina indica*) is known as:
(A) Kniffen system (B) Head system
(C) Single stake system (D) Cordon system

51. In whip tongue grafting the length of the cut should be about how many times the diameter of the plants in question:
(A) 3 (B) 4 (C) 5 (D) 6

52. It has been observed that the Malta sweet orange budded on rough lemon produced trees how many times larger than those budded on citrus (*C. medica*):
(A) Two (B) Three
(C) Four (D) Five

53. A fruit that develops from a hollow pear shaped fleshy receptable (inflorscence: hypanthodium) which encloses a number of minute male and female flowers is known as:
(A) Aggregate fruit (B) Hesperidium
(C) Pome fruit (D) Syconus fruit

54. The budded plants commence bearing in how many years:
(A) Second (B) Third
(C) Fourth (D) Fifth

55. Mandarins reached Erope in:
(A) 1605 (B) 1700 (C) 1790 (D) 1805

56. For hand pollination in date-palm 10 male trees are sufficient for ——— female plants:
(A) 50 (B) 75 (C) 100 (D) 125

57. Hillawi is very popular variety of ——— throughout the world due to adaptability to high humidity and rains:
(A) Coconut (B) Date-palm
(C) Cashewnut (D) Arecanut

58. Fig in India is grown in:
(A) Madhya Pradesh (B) Punjab
(C) Kerala (D) Gujarat

59. Guava is the native of:
(A) Tropical South-east Asia (B) Tropical Africa
(C) Tropical America (D) Sub-tropical India

60. Allahabadi safeda is the variety of:
(A) Mango (B) Guava
(C) Loquat (D) Grapes

61. ——— is grownins a shade tree in coffee, areca and cardamom plantations:
(A) Jack fruit (B) Coconut
(C) Mango (D) Guava

62. The stone, bark, leaves etc. of ——— are used as an effective medicine against diabetes:
(A) Litchi (B) Mango
(C) Indian blackberry (jamun) (D) Jujube (ber)

63. The first school of Ikebana called Ikenobo meaning "the arrangement of flowers of the hermitage along a pond" was started by Budhist Monk Semmu around ——— A.D.:
(A) 421 (B) 521 (C) 621 (D) 721

64. The Moribana meaning "Piled flowers" denoted the arrangement of flowers in :
(A) Shallow containers (B) Deep containers
(C) Paper wrappings (D) Paper cones

65. Rose cut flower must be cut at tight bud stage at least ——— cm stem length for export:
(A) 20 (B) 40 (C) 60 (D) 80

66. The Mandor Garden, Jodhpur (Rajasthan) was established by Raja ——————— (1724-1749)
(A) Abhai Singh (B) Ajay Singh
(C) Hari Singh (D) Man Singh

67. *Lantana camara* syn. *L. aculeate* belong to family ——— and is a native of Jamaica:
(A) Lythraceae (B) Rosaceae
(C) Acanthaceae (D) Verbenaceae

68. Aristolochia, family Aristolochiaceae, represent a group of:
(A) Herbaceous annuals (B) Perennial shrubs
(C) Annual flowering plant (D) Climbers

69. Golden shower (*Cassia fistua*) may take ——— years to flower:
(A) 5 (B) 10 (C) 15 (D) 18

70. In *Casia nodosa* ———flowers appear in bunches from the nodes all along the branches:
(A) Yellow (B) Red
(C) Pink (D) Purple

71. In potato tree (*Solanum macranthum* syn. S. maroniense) flowers 2.5 cm across and resemble ———flowers:
(A) Tomato (B) Egg plant
(C) Potato (D) Sweet potato

72. Madonna lily is a:
(A) *Lillium candidum* (B) *L. pardalimum*
(C) *L. tigrinum* (D) *L. auratum*

73. Fruit setting is most adversely affected by:
(A) Moist air (B) Moist and cold air
(C) Dry and cold air (D) Dry and hot air

74. Which of the following is a natural colouring agent of chilli?
(A) Oleoresin (B) Turmeric
(C) Resinocine (D) None of these

75. The cutting are planted by:
(A) Putting 2/3 part in soil and 1/3 part above the soil surface
(B) Half in soil and half above the soil
(C) 2/3 part in air and 1/3 part in soil
(D) All of these

76. Scion is a:
(A) Fruit bearing part of the grafted plant
(B) Root bearing part of the grafted plant
(C) Both (A) and (B)
(D) None of these

77. Angoorlata is a variety of::
(A) Tomato (B) Potato
(C) Pomato (D) Grape

78. Pear is mainly propagated by:
(A) Air layering (B) Stool layering
(C) Surpentine layering (D) Simple layering

79. Litchi is mainly propagated by:
(A) Air layering (B) Grafting
(C) Gootee (D) Simple layering

80. Mango is mainly propagated by:
(A) Grafting (B) Gootee
(C) Surpentine layering (D) Seed

81. Guava is mainly propagated by:
(A) Seed (B) Root cutting
(C) Surpentine layering (D) Grafting

82. Rose is mainly propagated by:
(A) Budding (B) Grafting
(C) Seed (D) Simple layering

83. Which of the fruit plant is propagated by root cuttings?
(A) Apple (B) Litchi
(C) Guava (D) All of these

84. The soft cutting are generally planted during:
(A) Spring season (B) Summer season
(C) Winter season (D) Rainy season

85. Suitable season for planting of root cutting is:
(A) Summer season (B) Spring season
(C) Rainy season (D) Winter season

86. Cuttings of deciduous plants are planted in:
(A) December-January (B) February-March
(C) May –June (D) Any season

87. Leaf cutting is used for the propagation of:
(A) Bigonia (B) Mango
(C) Fig (D) Plum

88. Herbaceous cutting is used for propagation of:
(A) Guava (B) Colius
(C) Geranium (D) All of these

89. Cuttings of ever green plants are planted in:
(A) December-January (B) February-March
(C) May –June (D) June-July

90. Equilateral triangular system of orchard layout is also known as:
(A) Square (B) Hexagonal
(C) Rectangular (D) Quincunx

91. The suitable method of orchard layout near the cities is:
(A) Square (B) Rectangular
(C) Quincunx (D) Hexagonal

92. In hills, the suitable method of orchard layout is:
(A) Contour (B) Square
(C) Hexagonal (D) Rectangular

93. Which of the very common method of orchard layout in India is:
(A) Square (B) Rectangular
(C) Hexagonal (D) Quincunx

94. In hexagonal system how much plants can be accommodated than square system?
(A) 10% more (B) 15% more
(C) 20% more (D) 25% more

95. Most suitable planting time of evergreen plants is:
(A) Winter season (B) Spring season
(C) Rainy season (D) Summer season

96. Most suitable planting time for deciduous fruit plants is:
(A) Winter season (B) Spring season
(C) Rainy season (D) Summer season

97. Most suitable time for root pruning of mango tree is:
(A) May-June (B) July-August
(C) Novemeber-December (D) February-March

98. Dehorning refers to:
(A) Removal of branching from fruit trees
(B) Removal of terminal portion of branches
(C) Removal of flowers
(D) All of these

99. Pinching refers to:
(A) To remove top portion of plant with few leaves
(B) It is like a head back pruning
(C) It is done in ornamental plants
(D) All of these

100. Stripping refers to:
(A) Removal of barks in stripes of 1.25 – 2.5 cm length
(B Removal of extra flowers
(C) Removal of terminal portion of branches
(D) None of these

101. Pollinizers are:
(A) Fruit plants that provide pollens to other plants
(B) Helps in fertilization and fruiting
(C) They are planted middle of fruiting plants
(D) All of these

102. Fruiting can be enhanced by:
(A) Root pruning (B) Ringing
(C) Notching (D) All of these

103. Fruiting can be induced by:
(A) NAA (B) 2,4-D
(C) IAA (D) All of these

104. Spliting of fruits can be checked by spraying of:
(A) 1 ppm NAA (B) Orange rind oil
(C) Bordeaux mixture (D) All of these

105. Jelly is a:
(A) Transparent (B) Opaque
(C) Semi-transparent (D) All of these

106. Periclinal chimeras in fruit plants can be produced by:
(A) Colchicine (B) X-rays
(C) High temperature (D) All of these

107. Brix percentage of sauce is:
(A) 15 (B) 20 (C) 25 (D) 30

108. Brix percentage of ketchup is:
(A) 15-16 (B) 20-22 (C) 25-26 (D) 28-30

109. Which of the following is not a fruit preservation method?
(A) Drying (B) Canning
(C) Freezing (D) Grafting

110. Which of the following fruit is used for Jam preparation?
(A) Apple (B) Grape
(C) Papaya (D) All of these

111. Commercial method of mango propagation is:
(A) Inarching (B) Root grafting
(C) Vaneer grafting (D) All of these

112. Papaya is propagated by:
(A) Cutting (B) Layering
(C) Seed (D) Grafting

113. Flowering in mangoes occurs on:
(A) One year old terminal buds
(B) Two years old terminal buds
(C) Whole plants
(D) Old branches

114. Flowering of mango occurs during:
(A) June-July (B) September-October
(C) November-February (D) March-June

115. Polygamous plants are those plants that produce:
(A) Male, female and hermaphrodite flowers
(B) Male flowers only
(C) Female flowers only
(D) Hermaphrodite flowers

116. Optimum planting time of mango in north India is:
(A) July-August (B) September-October
(C) November-December (D) February-March

117. Optimum planting time of mango in south India is:
(A) July-August (B) September-October
(C) November-December (D) February-March

118. Optimum planting time of papaya is:
(A) July-August (B) September-October
(C) February-March (D) Whole year

119. Optimum planting time of banana is:
(A) June-July (B) September-October
(C) February-March (D) Whole year

120. Optimum planting time of banana in Kerala is:
(A) July-August (B) September-October
(C) February-March (D) Whole year

121. Optimum planting time of guava is:
(A) July-August (B) September-October
(C) February-March (D) November-December

122. Pit size of mango planting should be:
(A) 60 x 60 x 60 cm (B) 90 x 90 x 90 cm
(C) 100 x 100 x 100 cm (D) 90 x 90 x 60 cm

123. Optimum plant spacing of papaya is:
(A) 2.5-3.0 m (B) 5.0-6.0 m
(C) 6.0-6.5 m (D) 7.0-7.5 m

124. Optimum planting distance of banana is:
(A) 2.0-3.0 m (B) 1.0-2.0 m
(C) 3.0-4.0 m (D) 4.0-5.0 m

125. Optimum planting distance of guava is:
(A) 5.0 m (B) 6.0 m
(C) 7.0 m (D) 8.0 m

126. Optimum planting distance of sweet orange is:
(A) 4.0 m (B) 5.0 m
(C) 6.0 m (D) 7.0 m

127. Optimum planting distance of litchi is:
(A) 5 x 5 m (B) 8 x 8 m
(C) 10 x 10 m (D) 12 x 12 m

128. Optimum planting distance of loquat is:
(A) 5 x 5 m (B) 8 x 8 m
(C) 10 x 10 m (D) 12 x 12 m

129. Optimum planting distance of peach is:
(A) 4 x 4 m (B) 6 x 6 m
(C) 7 x 7 m (D) 9 x 9 m

130. Optimum planting distance of straw berry is:
(A) 45 x 45 cm (B) 1 x 1 m
(C) 2 x 2 m (D) 3 x 3m

131. Optimum planting distance of ber is:
(A) 5.5 – 5.5 m (B) 6.5 x 6.5 m
(C) 7.5 x 7.5 m (D) 8.5 x 8.5

132. Optimum planting distance of pineapple is:
(A) 60 x 30 cm (B) 45 x 20 cm
(C) 90 x 60 cm (D) 100 x 100 cm

133. Optimum planting distance of gladiolus is:
(A) 15 x 15 cm (B) 30 x 15 cm
(C) 45 x 30 cm (D) 60 x 80 cm

134. Which of the root stock has dwarfing effect on Dashhari?
(A) Rumani (B) Olour
(C) Dogridge (D) Rangur lime

135. Olour is a poly embryonic variety of:
(A) Guava (B) Citrus
(C) Mango (D) Strawberry

136. Which of the following dwarfing root stock is used for mango?
(A) Olour (B) Totapari Red small
(C) Dashhari (D) Alfanso

137. Optimum planting distance of rose is:
(A) 30 x 30 cm (B) 60 x 60 cm
(C) 1 x 1 m (D) 2 x 2 m

138. Pruning of peach should be done in the month of:
(A) September (B) November
(C) December (D) January

139. Pruning of rose should be done in the month of:
(A) September (B) October
(C) December (D) January

140. Optimum planting time of plum:
(A) June-July (B) September-October
(C) January-February (D) March-April

141. Optimum planting time of pear:
(A) June-July (B) September-October
(C) February- March (D) March-April

142. Optimum planting time of straw berry in hills is:
(A) July-August (B) September-October
(C) January-February (D) March-April

143. Optimum planting time of Straw berry in plains is:
(A) June-July (B) September-October
(C) January-February (D) March-April

144. Optimum planting time of gladiolus in north India is:
(A) June-July (B) September-October
(C) January-February (D) March-April

145. Which of the following is sub-tropical fruit?
(A) Grape (B) Guava
(C) Litchi (D) All of these

146. Which of the following is tropical fruit?
(A) Mango (B) Banana
(C) Pine apple (D) All of these

147. Which of the following is temperate fruit?
(A) Apple (B) Pear
(C) Cherry (D) All of these

148. Fruit growing at 1200 msl is:
(A) Peach (B) Plum
(C) Straw berry (D) All of these

149. Fruit growing at 1500-2000 msl is:
(A) Apple (B) Pear
(C) Cherry (D) All of these

150. Which of the following is a early maturing variety of mango?
(A) Bombay green (B) Alfanso
(C) Gopal Bhog (D) All of these

151. Which of the following is a medium maturing variety of mango?
(A) Langra (B) Dashhari
(C) Safeda (D) All of these

152. Which of the following is a late maturing variety of mango?
(A) Fazli (B) Neelam
(C) Kanchan (D) All of these

153. Mahmud bahar is a cross of:
(A) Bombay green x Kallapady
(B) Neelam x Chausa
(C) Neelam x Langra
(D) Neelam x Dashhari

154. Which of the following is a polyembryonic variety?
(A) Mazagaon (B) Chandra karan
(C) Neleswar (D) All of these

155. Which of the following is a variety of sweet orange?
(A) Wheetaker (B) Blood Red
(C) Pine apple (D) All of these

156. Which of the following variety of sweet orange is cultivated in south India?
(A) Sathgudi (B) Batavian
(C) Mosambi (D) All of these

157. Which of the following is a early maturing variety of litchi?
(A) Early large red (B) Early seed less
(C) Pickling (D) All of these

158. Which of the following is a medium maturing variety of litchi?
(A) Dehradun (B) Rose scented
(C) Pyazi (D) All of these

159. Which of the following is a late maturing variety of litchi?
(A) Kalkattia (B) Late seed less
(C) Large seed red (D) All of these

160. Which of the following is a early maturing variety of loquat?
(A) Golden yellow
(B) Improved golden yellow
(C) Thames pride
(D) All of these

161. Which of the following is a medium maturing variety of loquat?
(A) Pale yellow (B) Large Agra
(C) Fire ball (D) All of these

162. Which of the following variety of grape is suitable for North India?
(A) Parlett (B) Black prince
(C) Bedana (D) All of these

163. Which of the following variety of grape is suitable for kismish preparation is?
(A) Sultana (B) Thomson seed less
(C) Munakka (D) All of these

164. Which of the following variety of grape is suitable for vine preparation?
(A) Muscat (B) Black prince
(C) Bangalare blue (D) All of these

165. Which of the following is a imported grape variety?
(A) Muscat hamburg (B) Foster's seedless
(C) Black Prince (D) All of these

166. Diploid variety of apple are?
(A) Self fruitful (B) Self unfruitful
(C) Both (A) and (B) (D) All of these

167. Triploid varieties of apple are?
(A) Self fruitful (B) Self unfruitful
(C) Both (A) and (B) (D) All of these

168. Which of the following is a diploid variety of apple?
(A) Jonathan (B) Rome beauty
(C) Newton wonder (D) All of these

169. Which of the following is a triploid variety of apple?
(A) Baldvin (B) Blenhenn orange
(C) Cox's orange pippin (D) All of these

170. Which of the following is a early maturing variety of apple?
(A) Summer golden (B) Pippin
(C) James grieve (D) All of these

171. Which of the following is a mid season apple variety is?
(A) Orange pippin (B) Jonathan
(C) King of pippin (D) All of these

172. Which of the following is a variety of apple is grown in Uttrakhand hills?
(A) Blenhein orange pippin (B) Delicious
(C) Rome Beuty (D) All of these

173. Which of the following is a variety of apple mainly grown in Kashmir valleys?
(A) Ambari (B) Blood red
(C) Baldwin (D) All of these

174. Which of the following is a variety of apple mainly grown in kullu valleys?
(A) Red astrachan (B) Red delicious
(C) Golden delicious (D) All of these

175. Which of the following apple variety should be planted with rymer variety for better fruiting?
(A) Golden delicious (B) Buckinghum
(C) Early shan burry (D) Red delicious

176. Peach variety J.H. Hale is a:
(A) Self fruitful (B) Self unfruitful
(C) Both (A) and (B) (D) None of these

177. Which of the following variety of peach can be grown in plains?
(A) Florida Sun (B) Sharvati
(C) Sun Red (D) All of these

178. Which of the following variety of peach is grown at high hills?
(A) Alexander (B) Elberta
(C) Noblesse (D) All of these

179. Which of the following variety of straw berry?
(A) Kavaliar (B) Elberton
(C) Red Ford (D) All of these

180. Which of the following variety of straw berry is grown in Uttaranchal hills?
(A) Chaubatia Arun (B) Chaubatia Manjul
(C) Chaubatia Abhichal (D) All of these

181. Which of the following variety of pine apple?
(A) Kew (B) Queen
(C) Mauritius (D) All of these

182. Ratna variety of mango is a cross of:
(A) Neelam x Dashhari (B) Dashhari x Neelam
(C) Neelam x Alphonso (D) None of these

183 Garden Tea is a:
(A) Type of rose (B) Tea grown in garden
(C) Variety of tea (D) None of these

184. High density planting of fruit plants is useful because of:
(A) Achieve high productivity
(B) More efficient use of inputs
(C) Easy harvest distance
(D) All of these

185. High density planting in Dashhari variety of mango can be achieved by using of:
(A) Closer spacing (B) Canopy management
(C) Use of cultar (D) All of these

186. Rose oil is extracted from:
(A) *Rosa damascene* (B) *Rosa borboniana*
(C) *Rosa moschata* (D) *Rosa chinensis*

187. Gulkand is prepared from:
(A) *Rosa damascene* (B) *Rosa borboniana*
(C) *Rosa moschata* (D) *Rosa chinensis*

188. Gladiolus is propagated by:
(A) Runners (B) Suckers
(C) Corms (D) S. Runners

189. Pineapple is propagated by:
(A) Rhizomes (B) Suckers
(C) Runners (D) Budding

190. Aonla is propagated by:
(A) Air layering (B) Cuttings
(C) Inarching (D) Budding

191. Ber is propagated by:
(A) Seed (B) Shield budding
(C) Ring budding (D) All of these

192. Straw berry is propagated by:
(A) Runners (B) Suckers
(C) Air layering (D) Grafting

193. Apple is propagated by:
(A) Shield budding (B) Bench grafting
(C) Tongue grafting (D) All of these

194. Old apple's orchards are rejuvenated mainly by:
(A) Top grafting (B) Air layering
(C) Bench grafting (D) All of these

195. Grape can be propagated by:
(A) Cutting (B) Air layering
(C) Shield budding (D) All of these

196. Grape is propagated in India by:
(A) Hard wood cuttings (B) Soft wood cuttings
(C) Gootee (D) Grafting

197. Jackfruit is propagated by:
(A) Seed (B) Cuttings
(C) Layering (D) Grafting

198. Loquat can be propagated by:
(A) Seed (B) Gootee
(C) Shield budding (D) All of these

199. Main method of litchi propagation in India is:
(A) Cutting (B) Air layering
(C) T budding (D) Inarching

200. Mandarin is propagated mainly by:
(A) T budding (B) Seed
(C) Cutting (D) None of these

201. *Citrus aurandifolia* is propagated by:
(A) Seed (B) Cutting
(C) Gootee (D) All of these

202. Sweet orange is propagated mainly by:
A) Cutting (B) Seed
(C) Shield budding (D) Grafting

203. Guava is mainly propagated by:
(A) Seed (B) Cutting
(C) Inarching (D) Shield

204. Banana is propagated by:
(A) Layering (B) Inarching
(C) Suckers (D) Cuttings

205. Training of fruit trees means:
(A) Remove few unwanted branches to make plant more beautiful and strong
(B) Remove few flowers
(C) Remove and cut few roots
(D) None of these

206. Temperate fruit plants can be trained by:
(A) Leader method
(B) Open centered method
(C) Modified leader system
(D) All of these

207. Which of the training is considered more suitable?
(A) Closed centered system
(B) Open centered system
(C) Modified leader system
(D) All of these

208. In quincunx system, the total plants are nearly double than:
(A) Square method (B) Rectangular method
(C) Contour method (D) Hexagonal method

209. In quincunx system, the filler plants are generally:
(A) Fast growing and early fruiting
(B) Slow growing but early fruiting
(C) Late fruiting
(D) All of these

210. Loping of root stock refers to:
(A) To remove the upper part of root stock after setting of buds
(B) To remove the lower part of root stock after setting of buds
(C) To remove extra new branches from the root stock
(D) All of these

211. Which of the following inhances rootings in cuttings?
(A) Auxin (B) Phenolic substances
(C) Auxin-Phenolic complex (D) All of these

212. Which of the following chemical is used to enhances rootings in cuttings?
(A) IAA (B) IBA
(C) NAA (D) All of these

213. Which of the following hormone is used for fast rootings in cuttings?
(A) Seredix-A (B) Vitamin A
(C) Vitamin E (D) None of these

214. The rooting of cuttings can be induced by cuttings before planting in:
(A) IAA (B) GA
(C) ABA (D) Zn solution

215. In above case, the concentration of IAA should be used:
(A) 100-500 ppm (B) 1000-5000 ppm
(C) 10-50 ppm (D) All of these

216. The optimum size of cuttings is:
(A) 15-18 cm (B) 20-25 cm
(C) 25-30 cm (D) 30-35 cm

217. Which of the following method is used for temporary preservation of vegetables?
(A) Drying
(B) Use of low concentration of antiseptic
(C) Pasteurization
(D) All of these

218. Which of the following method is used for permanent preservation of vegetables?
(A) Cold storage (B) Bottling
(C) Drying (D) All of these

219. Which of the following vegetables can be stored in salt solution?
(A) Potato (B) Cauliflower
(C) Carrot (D) All of these

220. IAA, belongs to which group of plant hormone:
(A) Auxin (B) Gibberellin
(C) Cytokinin (D) None of these

221. Yellow coloured fruits and vegetables are rich source of:
(A) Vitamin A (B) Vitamin B
(C) Vitamin C (D) Vitamin E

222. Pant Lohit is a variety of:
(A) Onion (B) Garlic
(C) Fenugreek (D) Fennel

223. Pant Bahar is a variety of:
(A) Onion (B) Garlic
(C) Tomato (D) Fennel

224. Pant Uphar is a variety of:
(A) Onion (B) Pea
(C) Tomato (D) Potato

225. Pant Anupma is a variety of:
(A) French bean (B) Carrot
(C) Garlic (D) Fennel

226. Pant Ragini is a variety of:
(A) Onion (B) Garlic
(C) Fenugreek (D) Fennel

227. Pant Haritma is a variety of:
(A) Onion (B) Garlic
(C) Fenugreek (D) Coriander

228. Pant C-1 is a variety of:
(A) Onion (B) Garlic
(C) Chilli (D) Fennel

229. Pant Ruchika is a variety of:
(A) Onion (B) Ajowan
(C) Chilli (D) Fennel

230. Pant Krishna is a variety of:
(A) Black cumin (B) Garlic
(C) Fenugreek (D) Fennel

231. What is main ingredient of pungency in chilli?
(A) Alkaloid (B) Acids
(C) Capsicin (D) Sulphur

232. Which is not a Turnip variety?
(A) Purple top white globe (B) Golden ball
(C) Pusa Kanchan (D) Arka suryamukhi

233. Which is a not variety of carrot?
(A) Nantes (B) Chentne
(C) Pusa Kesar (D) CO2

234. What is the scientific name of sponge guard?
(A) *Luffa acutangula* (B) *Lufa cylindrica*
(C) *Lagenaria siceraria* (D) None of these

235. What is the scientific name of ridge guard?
(A) *Luffa acutangula* (B) *Lufa cylindrical*
(C) *Lagenaria siceraria* (D) None of these

236. Pusa nasdar is a variety of:
(A) Ridge guard (B) Snake guard
(C) Sponge guard (D) None of these

237. What is the scientific name of musk melon?
(A) *Cucumis melo* (B) *Cucumis sativus*
(C) *Cucurbita moschata* (D) None of these

238. What is the scientific name of pumpkin?
(A) *Cucumis melo* (B) *Cucumis sativus*
(C) *Cucurbita moschata* (D) None of these

239. Which is not a variety of muskmelon?
(A) Pusa Sarvati (B) Pusa Madhurus
(C) Green Madhu (D) Pusa Alankar

240. The main gradient of garlic oil is:
(A) Alicin
(B) Alile propel disulphide
(C) Alicin and Alicin propel disulphide
(D) Acetelin

241. Which of the garlic varieties are more pungent:
(A) White varieties (B) Green varieties
(C) Red varieties (D) All of these

242. Bitter guard is very useful for patients of:
(A) AIDS (B) Rickets
(C) Diabetes (D) Malaria

243. Garlic is useful medicine for patients having problems of:
(A) High blood pressure (B) Low blood pressure
(C) AIDS (D) All of these

244. Which is the main ingredients of special smell in onion?
(A) Sulphur dioxide (B) Methane
(C) Allyl propile disulphide (D) All of these

245. Onion is very useful against:
(A) Hot air and cholera (B) Malaria
(C) Diabetes (D) All of these

246. Scientific name of cabbage is:
(A) *Brassica oleracea var oleracea*
(B) *Brassica oleracea var capitata*
(C) *Brassica juncea*
(D) *Brassica sativum*

247. Scientific name of cauliflower is:
(A) *Brassica oleracea var oleracea*
(B) *Brassica oleracea var capitata*
(C) *Brassica juncea*
(D) *Brassica sativum*

248. Scientific name of knol khol is:
(A) *Brassica oleracea var oleracea*
(B) *Brassica oleracea var caulorapa*
(C) *Brassica juncea*
(D) *Brassica sativum*

249. What is blanching?
(A) It is a method of fruit ripening
(B) It is a method of bunding
(C) It is a method of protecting fruits from light and hot
(D) It is a method of plant protecting from insects

250. Blindness in cauliflower means:
(A) Full development of flower/fruit
(B) Full development of plant without flower
(C) Started plant growth
(D) All of these

251. Cabbage can be successfully grown under:
(A) Cold and moist (B) Cold and hot
(C) Cold and dry (D) All of these

252. Which of the following is a early varieties of cabbage?
(A) Pride of India (B) Red cabbage
(C) Pusa drum head (D) All of these

253. Which of the following is a late varieties of cabbage?
(A) Pride of India (B) Coperhegan market
(C) Pusa drum head (D) All of these

254. White Viana is a variety of knoll khol:
(A) Early maturing (B) Middle maturing
(C) Late maturing (D) Very late maturing

255. Yellow Viana is a variety of knoll khol:
(A) Early maturing (B) Middle maturing
(C) Late maturing (D) Very late maturing

256. Which group belongs to late sown varieties of cauliflower?
(A) Japanese Improved, Hisar-1, Pusa synthetic
(B) Pusa snowball-1, and 2, Snowball -16
(C) Pusa Depali, Pant Shubhra, Pusa Kartic
(D) 2345, Snowball-16, Pusa Depali

257. Which type of onion varieties have good storage capacity?
(A) Red skin varieties (B) White skin varieties
(C) Yellow skin varieties (D) All of these

258. White skin varieties of onion is used for following purpose?
(A) For allopathic medicines
(B) For ayurvedic medicines
(C) For cosmetics
(D) All of these

259. Which garlic variety has larger sized cloves?
(A) Type 56-4 (B) G-4
(C) Solan (D) All of these

260. Which group of varieties belongs to bitter guard?
(A) Arka Chandan, Arka Suryamukhi CO-1
(B) Pusa Rasmi, Pusa Chetki, Japanese Safed
(C) Pusa Red, Pusa White, EC 7193
(D) Pusa do Mousmi, Arka green, Coimbatore Long

261. Scientific name of bitter guard is:
(A) *Mamordica cherantia* (B) *Trichosanthes dioica*
(C) *Coccinia indica* (D) *Trichosanthes anguina*

262. Scientific name of Bottle guard is:
(A) *Mamordica cherantia* (B) *Trichosanthes dioica*
(C) *Lagenaria siceraria* (D) *Trichosanthes anguina*

263. Hybrid variety of bottle guard is:
(A) Arka Bahar (B) Pusa Meghdoot
(C) Pusa summer prolific long (D) All of these

264. Which variety of bottle guard has round fruits?
(A) Kalyanpur long (B) Pusa majri
(C) Pusa meghdoot (D) NDBG-1

265. Which is not a European radish:
(A) Rapid red white tipped (B) White Isicle
(C) Pusa Himani (D) Japanese white

266. Which variety of sweet potato contains higher amount of carotene?
(A) Pusa Lal (B) Pusa Safed
(C) Pusa Sunhari (D) EC 7193

267. Which group belongs to chilli varieties?
(A) NP 46 A, Pant C-1, Pusa Jwala, Andhra Jyoti
(B) NP 46, MSFH-8, Pant Kanchan, Pusa Jwala
(C) Pant C-1, Pusa long, Nantes, Pusa Mukti
(D) All of these

268. Which of the following is a variety of brinjal?
(A) Pant Smart (B) Pant Rituraj
(C) Pant Bagain – 4 (D) All of these

269. The bitterness in colocasia corms is due to the presence of:
(A) Calcium oxalate (B) Calcium carbonate
(C) Calcium chloride (D) Potassium oxalate

270. Which of the onion variety is suitable for kharif season:
(A) Pusa Madhvi (B) Agri Found dark red
(C) Punjab red round (D) None of these

271. Which country is largest consumer of flowers and plants in the world is:
(A) Norway (B) Germany
(C) Switzerland (D) Denmark

272. Which of the following country is largest producer of cut flowers:
(A) Netherlands (B) USA
(C) Germany (D) India

273. Windbreaks protect the garden:
(A) Up to 2 times distance of its height
(B) Up to 3 times distance of its height
(C) Up to 4 times distance of its height
(D) Up to 6 times distance of its height

274. The cuttings are prepared by removing of cuttings from mother plant as:
(A) 3 cm below the bud at basal or lower part and 1.23 cm above the bud at upper part
(B) Removing of cuttings from upper part of the twig of mother plant
(C) Take centre part of the mother plant
(D) All of these

275. Which direction is most sensitive to sun burn effects in plants:
(A) East – South (B) South – West
(C) West – North (D) North – East

276. Top working of fruit plants refers to:
(A) Modify the form of old plants
(B) Convert old plants into new plants
(C) Old plants are modified into new species
(D) All of these

277. Semi-hard wood cuttings should be:
(A) One year old twigs (B) One season old twigs
(C) New growth (D) All of these

278. Semi-hard wood cuttings is used for propagation is:
(A) Fig (B) Rose
(C) Pear (D) All of these

279. The length of soft wood cuttings should be:
(A) 5-10 cm (B) 10-15 cm
(C) 15-25 cm (D) 25-30 cm

280. Which fruit has the highest frost resistance?
(A) Orange (B) Malta
(C) Sweet lime (D) Lemon

281. Frost is harmful to:
(A) Mango (B) Litchi
(C) Papaya (D) All of these

282. Frame working refers to:
(A) Change into old plant into high grade plants
(B) All side branches are grafted with high quality scions
(C) Only top part of plant is removed
(D) All of these

283. Pruning is an essential cultural activity in fruit plants because:
(A) To provide proper shape
(B) To provide strong structure
(C) To beautify the plant
(D) All of these

284. Notching of fruit plants refers to:
(A) To make a V shaped notch just below N above the bud in fruit plants
(B) It helps in vegetative growth
(C) It is common in Fig
(D) None of these

285. Which of the following preservative is used in the preparation of squash?
(A) Formaldehyde (B) NaCl
(C) Potassium-meta bi-sulphide (D) Na benzoate

286. Which of the following plant hormone is used to control pre harvest fruit drop?
(A) 2, 4-D (B) NAA
(C) 2, 4-D and NAA (D) None of these

287. Optimum temperature for growth and development of mango tree is:
(A) 40-50°F (B) 75-80°F
(C) 100-105°F (D) >105°F

288. How much percentage of male plants should be planted in papaya orchard for proper pollination?
(A) 5% (B) 10% (C) 15% (D) 20%

289. Papain is obtained from papaya tree :
(A) Mango (B) Papaya
(C) Banana (D) Guava

290. Stem of banana is a:
(A) Pseudomonas (B) Main stem
(C) Pseudostem (D) All of these

291. Propping refers to:
(A) Staking of banana stem at the time of fruiting to avoid bending or breaking of stem due to heavy weight of fruits
(B) To remove extra suckers
(C) To provide shade to the fruits
(D) None of these

292. Fruiting of guava during November –January is known as:
(A) Mrig bahar (B) Hast bahar
(C) Ambe bahar (D) Neno bahar

293. Fruiting of guava during February-April is known as:
(A) Mrig bahar (B) Hast bahar
(C) Ambe bahar (D) Neno bahar

294. Fruiting of guava during Rainy season is known as:
(A) Mrig bahar (B) Hast bahar
(C) Ambe bahar (D) Neno bahar

295. Which of the root stock is used in citrus fruits?
(A) Karna Khatta (B) Jumberi
(C) Sour orange (D) All of these

296. Which of the following variety belongs to lemon?
(A) Eureka (B) Genova
(C) Lisbon (D) All of these

297. Which of the following is required for better growth of litchi plant?
(A) Frost free climate
(B) High humidity
(C) High organic matter content in the soil
(D) All of these

298. Fruiting and flowering in litchi occur on:
(A) New growth (B) Old growth
(C) One year old branches (D) Old branches

299. Flowering of litchi occurs in the month of:
(A) March-April (B) September-October
(C) February-March (D) June –July

300. Fruiting and flowering in loquat occur on:
(A) New growth
(B) Old growth
(C) One year old branches
(D) One month old branches

301. Flowering of loquat occurs in the month of:
(A) March-April (B) September-October
(C) February-March (D) June –July

302. Flowering of jackfruit occurs in the month of:
(A) March-April (B) September-October
(C) January -February (D) June –July

303. The optimum time of pruning in grape in north India is:
(A) January-February (B) August-September
(C) June-July (D) October-November

304. Fruiting in grape occurs only on:
(A) Fruiting spurs (B) 3 months old branches
(C) One month old branches (D) One year old branches

305. In spur system of pruning of grape is:
(A) The branches are removed from base leaving only 2-3 buds
(B) Removal of new branches from base
(C) One year old branches are removed from base
(D) All of these

306. In cane system of pruning in grape:
(A) Half portion of one year old branches is removed
(B) Only upper portion of last season fruiting spurs are removed
(C) Both (A) and (B)
(D) None of these

307. Which of the following part has heavy fruiting in grape?
(A) Foundation spur (B) Fruiting spur
(C) Spur (D) All of these

308. Which of the following is a very vigorous root stock is used for propagation of apple?
(A) Malling -12 (B) Malling-13
(C) Malling-16 (D) All of these

309. Which of the following is a very dwarf root stock for propagation of apple?
(A) M-27 (B) M-12
(C) M-13 (D) None of these

310. Standard root stock of apple against woolly aphid is:
(A) Malling-Merton (B) DX-1
(C) PT-1 (D) None of these

311. Peach is mainly propagated by:
(A) Shield budding (B) Air layering
(C) Cutting (D) None of these

312. Fruiting in peach occurs on:
(A) New growth (B) One year old branch
(C) Whole stem (D) Whole plant

313. Flowering in peach should be done in the month of:
(A) February-March (B) October-November
(C) November-December (D) April-June

314. Which of the following is a variety of pear?
(A) Conference (B) Bartlett
(C) Winter Nalis (D) All of these

315. Which of the following is a variety of plum?
(A) Beauty (B) Early Transparent
(C) Santa Rosa (D) All of these

316. Which of the following is a variety of Ber?
(A) Banarasi Karaka (B) Umran
(C) Narma Thornless (D) All of these

317. Flowering occurs in Aonla in the month of:
(A) January-February (B) March-April
(C) July-August (D) September-October

318. Which of the following root stock of grape is useful under high salt content and nematodes?
(A) Rumani (B) Olour
(C) Dogridge (D) Rangpur lime

319. The bud dormancy in grapes can be broken by use of:
(A) Cultar (B) Hydrogen cynamide
(C) IAA (D) NAA

320. Cultar is a name of:
(A) Paclobutrazol (B) Methyl oxide
(C) Calcium carbonate (D) None of these

321. Sodium benzoate is added for the preservation of fruit juice at the rate of:
(A) 0.01-0.5% (B) 0.06-0.10%
(C) 1.10-1.5% (D) 1.6-2.0%

322. Whipe tail is caused by deficiency of?
(A) Zn (B) P (C) Mo (D) B

323. Buttoning is caused by:
(A) N deficiency (B) P deficiency
(C) K deficiency (D) Ca deficiency

324. Browning is caused by:
(A) N deficiency (B) B deficiency
(C) K deficiency (D) Ca deficiency

325. How much seed of cauliflower is required for sowing of one hectare area?
(A) 500-600 g (B) 800-1000 g
(C) 200-300 g (D) 1000-1500 g

326. Which is the optimum spacing for late sown cauliflower?
(A) 45 x 45 cm (B) 60 x 45 cm
(C) 30 x 30 cm (D) 60 x 60 cm

327. Fertilizer requirement of cabbage is:
(A) 60 N : 60 P_2O_5 : 40 K_2O kg/ha
(B) 100 N : 60 P_2O_5 : 40 K_2O kg/ha
(C) 120 N : 80 P_2O_5 : 40 K_2O kg/ha
(D) 150 N : 80 P_2O_5 : 60 K_2O kg/ha

328. How much seed of cauliflower is required for sowing of one hectare area?
(A) 500-600 g (B) 800-1000 g
(C) 1000-1250 g (D) 1500-2000 g

329. Which is the best time for nursery sowing of knoll khol?
(A) July-August (B) September-October
(C) November-December (D) January-February

330. What is optimum sowing time of knoll khol in hills:
(A) May-June (B) July-August
(C) November-December (D) January-February

331. Optimum planting spacing of knoll khol is:
(A) 45 x 30 cm (B) 45 x 20 cm
(C) 60 x 45 cm (D) 60 x 60 cm

332. Fertilizer requirement of onion is:
(A) 60 N : 60 P_2O_5 : 40 K_2O kg/ha
(B) 100 N : 60 P_2O_5 : 80 K_2O kg/ha
(C) 120 N : 80 P_2O_5 : 40 K_2O kg/ha
(D) 150 N : 80 P_2O_5 : 60 K_2O kg/ha

333. Optimum planting spacing of onion is:
(A) 20 x 10 cm (B) 25 x 10 cm
(C) 30 x 15 cm (D) 20 x 20 cm

334. How much seed of onion is required for sowing of one hectare area?
(A) 5-6 kg (B) 6-8 kg
(C) 8-10 kg (D) 10-12 kg

335. Fertilizer requirement of seed production of onion is:
(A) 60 N : 60 P_2O_5 : 40 K_2O kg/ha
(B) 75 N : 60 P_2O_5 : 60 K_2O kg/ha
(C) 120 N : 80 P_2O_5 : 40 K_2O kg/ha
(D) 150 N : 80 P_2O_5 : 60 K_2O kg/ha

336. Optimum planting spacing of garlic is:
(A) 15 x 7-8 cm (B) 25 x 10 cm
(C) 30 x 15 cm (D) 20 x 20 cm

337. How much seed of garlic is required for sowing of one hectare area?
(A) 500-800 kg cloves (B) 800-1000 kg cloves
(C) 1000-1200 kg cloves (D) 1200-1500 kg cloves

338. Fertilizer requirement of garlic is:
(A) 60 N : 60 P_2O_5 : 40 K_2O kg/ha
(B) 75 N : 60 P_2O_5 : 60 K_2O kg/ha
(C) 120 N : 80 P_2O_5 : 40 K_2O kg/ha
(D) 150 N : 60 P_2O_5 : 100 K_2O kg/ha

339. Optimum sowing time of garlic is:
(A) August-September (B) September-October
(C) October-November (D) December-January

340. Optimum seed rate of bitter gourd is:
(A) 5-6 kg/ha (B) 6-7 kg/ha
(C) 7-8 kg/ha (D) 8-10 kg/ha

341. Optimum seed rate of summer crop of bottle gourd is:
(A) 5-6 kg/ha (B) 6-7 kg/ha
(C) 4-5 kg/ha (D) 8-10 kg/ha

342. Fertilizer requirement of bottle gourd is:
(A) 60 N : 60 P_2O_5 : 40 K_2O kg/ha
(B) 75 N : 60 P_2O_5 : 60 K_2O kg/ha
(C) 75 N : 60 P_2O_5 : 40 K_2O kg/ha
(D) 60 N : 80 P_2O_5 : 60 K_2O kg/ha

343. Optimum sowing time of muskmelon is:
(A) December-January (B) January-February
(C) February-March (D) June-July

344. Optimum seedrate of muskmelon is:
(A) 2-3 kg/ha (B) 3-4 kg/ha
(C) 4-5 kg/ha (D) 5-6 kg/ha

345. Fertilizer requirement of carrot is:
(A) 60 N : 50 P_2O_5 : 80 K_2O kg/ha
(B) 75 N : 60 P_2O_5 : 60 K_2O kg/ha
(C) 100 N : 80 P_2O_5 : 40 K_2O kg/ha
(D) 100 N : 60 P_2O_5 : 60 K_2O kg/ha

346. Optimum planting spacing of summer pumpkin is:
(A) 60 x 60 cm (B) 75 x 75 cm
(C) 100 x 75 cm (D) 150 x 60 cm

347. Optimum planting spacing of radish is:
(A) 30 x 10 cm (B) 25 x 10 cm
(C) 15 x 15 cm (D) 20 x 20 cm

348. Optimum seedrate of summer pumpkin is:
(A) 3-4 kg/ha (B) 4-5 kg/ha
(C) 7-8 kg/ha (D) 8-10 kg/ha

349. Optimum seedrate of radish is:
(A) 3-4 kg/ha (B) 4-5 kg/ha
(C) 7-8 kg/ha (D) 8-10 kg/ha

350. Pusa Deepali is a variety of:
(A) Cabbage (B) Cauliflower
(C) Tomato (D) Brinjal

351. Optimum seedrate of European carrot is:
(A) 3-4 kg/ha (B) 5-6 kg/ha
(C) 7-8 kg/ha (D) 8-10 kg/ha

352. Optimum seedrate of desi carrot is:
(A) 3-4 kg/ha (B) 4-5 kg/ha
(C) 7-8 kg/ha (D) 8-10 kg/ha

353. Optimum planting spacing of carrot is:
(A) 20 x 10 cm (B) 30 x 20 cm
(C) 40 x 8 cm (D) 40 x 10 cm

354. Fertilizer requirement of sweet potato is:
(A) 60 N : 50 P_2O_5 : 80 K_2O kg/ha
(B) 75 N : 60 P_2O_5 : 60 K_2O kg/ha
(C) 100 N : 60 P_2O_5 : 80 K_2O kg/ha
(D) 100 N : 60 P_2O_5 : 60 K_2O kg/ha

355. What is the main practice of sweet potato sowing:
(A) Planting of cuttings (B) Sowing of bulbs
(C) Bulbs and cuttings both (D) All of these

356. How much cuttings are required for one hectare sowing of sweet potato is:
(A) 30000 cuttings (B) 35000 cuttings
(C) 40000 cuttings (D) 45000 cuttings

357. How much weight of cuttings for one hectare sowing of sweet potato is:
(A) 5-7 q (B) 10-12 q
(C) 14-16 q (D) 18-20 q

358. Optimum size of cutting of sweet potato is:
(A) 10-15 cm (B) 20-25 cm
(C) 40-45 cm (D) 50-75 cm

359. Optimum planting time of sweet potato is:
(A) June-July (B) August-September
(C) October-November (D) February-March

360. Optimum row to row spacing of sweet potato is:
(A) 40 cm (B) 50 cm
(C) 60 cm (D) 70 cm

361. Fertilizer requirement of turnip is:
(A) 60 N : 50 P_2O_5 : 50 K_2O kg/ha
(B) 75 N : 60 P_2O_5 : 60 K_2O kg/ha
(C) 100 N : 60 P_2O_5 : 80 K_2O kg/ha
(D) 100 N : 60 P_2O_5 : 60 K_2O kg/ha

362. Optimum planting time of turnip is:
(A) June-July (B) August-September
(C) October-November (D) February-March

363. Optimum row to row spacing of turnip is:
(A) 25 x 10 cm (B) 30 x 10 cm
(C) 40 x 10 cm (D) 40 x 15 cm

364. Optimum seedrate of turnip is:
(A) 3-4 kg/ha (B) 4-5 kg/ha
(C) 7-8 kg/ha (D) 8-10 kg/ha

365. Optimum planting time of chilli in frost affected areas is:
(A) June-July (B) August-September
(C) May-June (D) February-March

366. Optimum row to row spacing of chilli is:
(A) 25 x 30 cm (B) 30 x 30 cm
(C) 40 x 30 cm (D) 45 x 30 cm

367. Optimum seedrate of chilli is:
(A) 1-1.5 kg/ha (B) 2-3 kg/ha
(C) 4-5 kg/ha (D) 5-6 kg/ha

368. Optimum planting time of shimla mirch during spring season is:
(A) June-July (B) August-September
(C) November-December (D)February-March

369. Which of the following factor is responsible for higher yield under protected cultivation?
(A) High photosynthetic efficiency
(B) High O_2 availability
(C) High temperature
(D) All of these

370. Which of the following fruits are indigenous of India:
(A) Mango, Banana (B) Bel, Jackfruit
(C) Aonla, Phalsa (D) All of these

371. Which of the following is a ripening hormone?
(A) Ethylene (B) Cytokinin
(C) Auxin (D) Gibberellin

372. Which of the following floricultural crops are important for cut flowers trade?
(A) Rose (B) Chrysanthemum
(C) Carnation (D) All of these

373. Which of the following chemical is used to control the moisture in the storage plants?
(A) $CaCl_2$ (B) Na Cl
(C) $CaSO_4$ (D) All of these

374. Pusa Rasraj is a:
(A) A pureline variety of muskmelon
(B) A hybrid variety of muskmelon
(C) A Seedless variety of watermelon
(D) A hybrid variety of watermelon

375. The primary objective of green house cultivation of vegetables is:
(A) To get early crop
(B) To get off season crop
(C) To protect the crop from disease
(D) To protect the crop from excessive rains

376. Which of the following is a stem vegetable?
(A) Carrot (B) Sweet potato
(C) Knoll khol (D) Radish

377. Fruits are the source of mainly:
(A) Minerals (B) Vitamins
(C) Carbohydrates (D) All of these

378. Vitamin C is found in:
(A) Grape (B) Straw berry
(C) Papaya (D) All of these

379. Vitamin B_2 is found in:
(A) Papaya (B) Banana
(C) Litchi (D) All of these

380. Vitamin B_1 is found in:
(A) Ber (B) Apple
(C) Cherry (D) All of these

381. Folic acid is found in:
(A) Grape (B) Banana
(C) Lemon (D) All of these

382. Fruits are not the source of:
(A) Vitamins (B) Fat
(C) Minerals (D) Carbohydrates

383. Banana cultivation requires:
(A) High RH and high temperature
(B) High RH and low temperature
(C) Low RH and high temperature
(D) Low RH and low temperature

384. T budding is also known as:
(A) Shield budding (B) Patch budding
(C) Ring budding (D) Air layering

385. Which of the ornamental plant is most suitable for wind breaks?
(A) Silver oak (B) Mahua
(C) Ber (D) Bel

386. Which of the filler plant is used in the orchards of litchi and mango in North-West plains?
(A) Strawberry (B) Peach
(C) Papaya (D) Phalsa

387. Root pruning is performed to:
(A) Increase the C:N ratio in the plant
(B) Increase the fruiting
(C) Remove some smaller roots
(D) All of these

388. Optimum soil pH for gardening is:
(A) 4.5-5.5 (B) 5.5-6.5
(C) 6.5-7.5 (D) 7.5-8.5

389. Optimum temperature for mango growth is:
(A) 23.4 – 26.6°C (B) 18.5-20.5°C
(C) 35-40°C (D) >40°C

390. Optimum planting distance of jackfruit is:
(A) 5 x 5 m (B) 6 x 6 m
(C) 8 x 8 m (D) 10 x 10 m

391. Optimum planting distance of peach is:
(A) 5 x 5 m (B) 6 x 6 m
(C) 7 x 7 m (D) 8 x 8 m

392. For temporary fruit preservation the fruits should be pasteurized at:
(A) 160-180°F for 20-30 minutes
(B) 120-130°F for 20-30 minutes
(C) 160-180°F for 5 minutes
(D) 140-150°F for 15-20 minutes

393. For temporary preservation, fruits and vegetables can be stored at temperature:
(A) 10-20°C (B) 15-20°C
(C) 30-35°C (D) 39-45°C

394. Fruits and vegetables should be sterilized for permanent preservation at:
(A) 160°F for 10-15 minutes
(B) 180°F for 12-35 minutes
(C) 212°F for 12-35 minutes
(D) 225°F for 30 minutes

395. Which of the following hormone is used for blossom thinning?
(A) 250-500 ppm NAA (B) 1000 ppm NAA
(C) 1500 ppm NAA (D) All of these

396. Which of the following plant hormone is used for ripening of apples?
(A) 4 ppm 2,4,5-T (B) 10 ppm 2,4,5-T
(C) 15 ppm 2,4,5 –T (D) 20 ppm 2,4,5-T

397. Which of the following variety belongs to papaya?
(A) Honey dew (B) Washington
(C) Gujrati (D) All of these

398. Mango is a rich source of:
(A) Vitamin A (B) Vitamin C
(C) Vitamin A and Vitamin C (D) Vitamin E

399. Scientific name of papaya is:
(A) *Carica papaya* (B) *Mangifera indica*
(C) *Musa paradisiaca* (D) None of these

400. Which of the following species of papaya can be grown in hills?
(A) *Carica papaya* (B) *Carica candamarcensis*
(C) *Carica monoica* (D) *Carica paradisiaca*

401. Fruits can be obtained from papaya tree after planting of:
(A) 5-6 months (B) 8-10 months
(C) 10-14 months (D) 16-18 months

402. Which of the suckers are used for multiplication of banana?
(A) Tall suckers (B) Dwarf suckers
(C) Sword suckers (D) Water suckers

403. Which of the following variety of banana is used for vegetables only:
(A) Nendran (B) Hill banana
(C) Lal Kali (D) Safed velchi

404. Which of the following is a variety of banana?
(A) Poovan (B) Basraidwarf
(C) Rasthali (D) All of these

405. Optimum temperature for ripening of banana fruits is:
(A) 50°F (B) 60°F
(C) 70°F (D) 80°F

406. Which of the following is a variety of Guava?
(A) L-49 (B) Safeda
(C) Kerala (D) Lalit

407. Flowering and fruiting in guava always occur on:
(A) One year old branches (B) New growth
(C) Two years old branches (D) All of these

408. Optimum time of flowering of malta is:
(A) February-March (B) June-July
(C) September-October (D) December-January

409. Litchi is a rich source of:
(A) Fe and P (B) Vitamin C
(C) Vitamin D (D) All of these

410. Fruit cracking in litchi can be controlled by:
(A) Spraying of water
(B) Spraying of hormones
(C) Spraying of 2,4-D
(D) Spraying of Bordeaux mixture

411. Poor fruiting in loquat is mainly due to:
(A) Poor growth (B) Low fertility
(C) Self incompatibility (D) None of these

412. Loquat fruits mature in the month is:
(A) March-April (B) June-July
(C) September-October (D) December-January

413. Which of the training system of grape is considered most important:
(A) Head system (B) Trellis system
(C) Pergola system (D) Single stake system

414. Pergola system of grape training is very popular in:
(A) Tamilnadu (B) Punjab
(C) Uttar Pradesh (D) Maharashtra

415. Head system of grape training is very popular in:
(A) Tamilnadu (B) Punjab
(C) Uttar Pradesh (D) Maharashtra

416. Which of the following variety belongs to loquat?
(A) Tanaka (B) California advance
(C) Thames pride (D) All of these

417. Which of the following is a variety of jackfruit is:
(A) Rudrakshi (B) Singapur
(C) Khaja (D) All of these

418. Sugar content in grape is:
(A) 5% (B) 10% (C) 20% (D) 25%

419. Optimum time of apple planting is:
(A) January-February (B) March-April
(C) June-July (D) September-October

420. Optimum time of flowering in apples is:
(A) September-October (B) November-December
(C) March-April (D) June-July

421. Fruiting of apple occurs on:
(A) New fruiting spurs (B) Old branches
(C) Whole plant (D) All of these

422. Most suitable pruning method in apple is:
(A) Open centre form (B) Leader system
(C) Modified leader system (D) All of these

423. Scientific name of aonla is:
(A) *Ziziphus meritiana* (B) *Phyllanthus emblica*
(C) *Prunus armenica* (D) *Carissa carandas*

424. Scientific name of pineapple is:
(A) *Ananas comosus* (B) *Crewin asiatica*
(C) *Artocarpus heterophyllus* (D) *Malus sylvestris*

425. The regularity in mango bearing can be achieved by soil application of:
(A) Cultar @ 0.5 g/plant (B) 15 ppm NAA
(C) 50 ppm Ethylene (D) All of these

426. Fruit thining and pre harvest of fruit can be controlled in mango and citrus by:
(A) 10-150 ppm NAA (B) 200 ppm NAA
(C) 10 ppm GA (D) 20 ppm IAA

427. Vapour heat treatment is required for disinfestations of:
(A) Fresh fruits (B) Dry fruits
(C) Grains and pulses (D) Preserved fruits

428. Which of the following is a evergreen temperate plant?
(A) Apricot (B) Chilgoza
(C) Plum (D) Almond

429. Scientific name of pear is:
(A) *Pyrus communis* (B) *Prunus persica*
(C) *Prunus demestica* (D) *Carissa carandes*

430. Scientific name of jackfruit is:
(A) *Artocarpus heterophyllus*
(B) *Acacia tortilis*
(C) *Prunus paradisi*
(D) *Pisum sativum*

431. Scientific name of grape is:
(A) *Vitis vinifera* (B) *Prunus sylvestris*
(C) *Prunus paradisi* (D) *Mangifera indica*

432. Scientific name of loquat is:
(A) *Malus sylvestris* (B) *Mangifera indica*
(C) *Erioborya japonica* (D) *Prunus persica*

433. Banana is a plant of:
(A) Tropical climate (B) Temperate climate
(C) Subtropical climate (D) All of these

434. The best climate of banana is available in India at:
(A) Thana Bombay (B) Pune
(C) West Bengal (D) Andhra Pradesh

435. Optimum climate for banana cultivation is:
(A) 95°F temperature and 80-95% RH
(B) 110°F temperature and 95-99% RH
(C) 60°F temperature and 70-75% RH
(D) 80°F temperature and 40-45% RH

436. Alternate bearing of mango can be checked by:
(A) Deblossoming (B) Ringing
(C) Root prunnig (D) All of these

437. Marmalade can be prepared generally from:
(A) Apple (B) Mango
(C) Citrus fruits (D) Banana

438. Jelly can be prepared from the fruits that has high amount of:
(A) Sugar (B) Water
(C) Pectin (D) Carbohydrate

439. Which of the following solution can be used for preservation of fruits and their products?
(A) 6-8% salt (B) 42-50% sugar
(C) 0.5-1.5% acetic acid (D) All of these

440. Heading back pruning refers to:
(A) To remove extra and overlapping branches
(B) To remove terminal portion of all branches
(C) To remove the flowers only
(D) To remove flowers and top portion of main trunk

441. Teonic in compatibility refers to:
(A) Scion and root stock are adversely affected by each other
(B) Scion and root stock are positively affected by each other
(C) No effect of scion and rootstock on each other
(D) None of these

442. Which of the following plant is suitable for hedges?
(A) *Thevetia neriefolia* (B) *Citrus karma*
(C) *Carissa carandas* (D) All of these

443. If the potato is to be stored for 30 weeks, the optimum temperature an [illegible]tive humidity of storage should be:
(A) 2-3°C and 85-90 % (B) 4-5°C and 85-90 %
(C) 2-3°C and 75-80 % (D) 2-3°C and 65-70 %

444. Which of the following antiseptic solutions can be used for the preservation of vegetables?
(A) 2% Acetic acid (B) 2-3 % salt
(C) 60-70 % sugar (D) All of these

445. The most of the vegetables can be stored safely at temperature is:
(A) < 10°C (B) 15°C
(C) 20°C (D) 25°C

446. Wax coating treatment enhances the self life of fruits because it blocks:
(A) Transpiration (B) Respiration
(C) Ripening (D) Photosynthesis

447. Flower dropping of chilli can be checked by:
(A) By spraying of planofix
(B) By applying of nitrogen
(C) By spraying of clorine
(D) By spraying of urea solution

448. Most critical stages of chilli for irrigation:
(A) Seedling stage (B) Flowering stage
(C) Pod development stage (D) All of these

449. Optimum dose of nitrogen for chilli crop is:
(A) 60 kg/ha (B) 75 kg/ha
(C) 80 kg/ha (D) 100 kg/ha

450. Chilli can be grown upto height of:
(A) 500 m (B) 1000 m
(C) 1500 m (D) 2000 m

451. The European carrot varieties should be harvested at the stage that has root diameter equal to:
(A) 2-3 cm (B) 4-5 cm
(C) 3-4 cm (D) 5-6 cm

452. Optimum temperature for better colour root development of carrot is:
(A) 15-16°C (B) 17-18°C
(C) 19-20°C (D) 20-22°C

453. Optimum temperature for radish cultivation is:
(A) 10-15°C (B) 17-18°C
(C) 19-20°C (D) 20-22°C

454. Which of the following practice is very much required for higher fruiting in bottle guard?
(A) Blanching (B) Chlorosis
(C) Staking (D) Digging

455. The sweetness in muskmelon shall be higher at:
(A) Low temperature and high rainfall
(B) High temperature and high rainfall
(C) High temperature and dry climate
(D) Low temperature and dry climate

456. Scientific name of onion is:
(A) *Allium cepa* (B) *Allium sativum*
(C) *Allium capitata* (D) *Allium garlic*

457. Scientific name of garlic is:
(A) *Allium cepa* (B) *Allium sativum*
(C) *Allium capitata* (D) *Allium garlic*

458. What is the maturity period of garlic:
(A) 100-120 days (B) 150-160 days
(C) 170-180 days (D) 180-190 days

459. Average yield of garlic is:
(A) 50-60 q/ha (B) 60-70q/ha
(C) 70-80 q/ha (D) 80-100 q/ha

460. Optimum temperature for cold storage of garlic is:
(A) 0-2°C (B) 2-5°C
(C) 5-8°C (D) 8-10°C

461. Cabbage head can be stored long time at:
(A) 0°C and 90-95% RH
(B) 5-6°C and 90-95 % RH
(C) 10°C and 90-95 % RH
(D) 10°C and 70-75 % RH

462. Which is suitable method for seed production of cabbage:
(A) Head to seed method (B) Head intact method
(C) Stump method (D) All of these

463. Averge yield of early cabbage is:
(A) 250-300 q/ha (B) 300-350 q/ha
(C) 400-450 q/ha (D) 450-500 q/ha

464. Cabbage is a rich source of:
(A) Vitamin B (B) Vitamin C
(C) Vitamin D (D) Vitamin E

465. Which vegetable does not belong to the leguminaceae family:
(A) *Vicia faba* (B) *Vigna sinensis*
(C) *Amaranthus oleracea* (D) *Dolichos lablab*

466. Which vegetable does not belong to cole crops:
(A) Cauliflower (B) Knoll khol
(C) Cabbage (D) Lady's finger

467. Which is not a root vegetable?
(A) *Daucus carota*
(B) *Brassica rapa*
(C) *Ipomoea batatas*
(D) *Dioscorea alata var purpurea*

468. What is the optimum size of bulbs of onion for seed production is:
(A) 2.5-3.0 cm diameter (B) 3.0-3.5 cm diameter
(C) 3.5-4.0 cm diameter (D) 4.0-4.5 cm diameter

469. What is optimum sowing time of bulbs of onion for seed production is:
(A) First week of October
(B) Second week of October
(C) Third week of October
(D) Forth week of October

470. What is optimum temperature for proper growth and development of garlic is:
(A) 0-10°C (B) 13-14°C
(C) 18-23°C (D) >23°C

471. The root stock used in the propagation of ber is:
(A) Jherberi (B) M-8
(C) X-7 (D) All of these

472. Which of the following fruit is not suitable for jam making:
(A) Mango (B) Banana
(C) Aonla (D) Lemon

473. Which one is hybrid variety of mango:
(A) Neelam (B) Langra
(C) Sumarr Bahist (D) Chausa

474. Pusa Rubi is a variety of:
(A) Carrot (B) Tomato
(C) Onion (D) Raddish

475. Which one is the early variety of tomato:
(A) Kalyanpur Kuber (B) Kalyanpur T 1
(C) HS 101 (D) HS 102

476. Which one is the late variety of tomato:
(A) HS102 (B) Kalyanpur angoorlata
(C) Best of all (D) Kalyanpur T1

477. The suitable time of planting for medium variety of tomato is:
(A) July – Aug. (B) Sep. - Oct.
(C) Aug.- Sep. (D) Oct. – Nov

478. Which variety of tomato is more suitable for growing as a summer crop:
(A) Kalyanpur Kuber (B) Kalyanpur T 1
(C) Kalyanpur Angoorlata (D) S 152

479. Mango belongs to the family:
(A) Myrtaceae (B) Anacariaceae
(C) Musaceae (D) None of these

480. Which one is the hybrid variety of mango is:
(A) Alphanso (B) Dashahari
(C) Mallika (D) Langra

481. The suitable time for planting of mango is:
(A) July-Sept. (B) Sept.-Nov.
(C) Dec.-Feb. (D) None of these

482. Spinach produces seed stalks in about ——— and seed can be harvested in about 150 days:
(A) 45 days (B) 55 days
(C) 65 days (D) 75 days

483. Early snowball is a variety of:
(A) Cauliflower (B) Cabbage
(C) Potato (D) Carrot

484. Pride of India is a variety of:
(A) Brinjal (B) Cabbage
(C) Tomato (D) Caulifower

485. Pusa Sawani is a variety of:
(A) Bhindi (B) Cucumber
(C) Cabbage (D) Potato

486. Arkil is a variety of :
(A) Garden Pea (B) Field Pea
(C) Both (A) and (B) (D) None of these

487. Agri Found dark and light red are the varieties of:
(A) Garlic (B) Onion
(C) Tomato (D) Potato

488. Which variety of *Brassica perkinensis* is a useful leaf vegetable for plains and hills of N. India:
(A) Chinese (B) America
(C) Canadian (D) Japanese

489. Orach or chakwat (*Artriplex hortensis* L.) or mountain spinach is mainly grown on small scale during winters in:
(A) Maharashtra and Karnataka
(B) Gujarat and Madhya Pradesh
(C) Tamilnadu and Kerala
(D) Himachal Pradesh and J &K

490. Sev is a variety of:
(A) Apple (B) Guava
(C) Ber (D) Mango

491. Lalit is a varety of:
(A) Apple (B) Guava
(C) Aonla (D) Mango

492. Winter banana is a variety of:
(A) Banana (B) Apple
(C) Guava (D) None of these

493. Sugar baby is a variety of:
(A) Water melon (B) Musk melon
(C) Both (A) and (B) (D) None of these

494. Beauty bath is a tropical variety of :
(A) Apple (B) Pine apple
(C) Peach (D) Plum

495. Propping and wrapping is a important cultural practice of:
(A) Banana (B) Grape
(C) Mango (D) Citrus

496. Dwarf cavandish is a high yielding variety of:
(A) Mango (B) Apple
(C) Banana (D) Guava

497. Amrapali is a variety of mango developed by cross of:
(A) Dashhari x Neelam (B) Neelam x Dashhari
(C) Chausa x Langra (D) Dashhari x Langra

498. Mallika is a variety of mango developed by cross of:
(A) Dashhari x Neelam (B) Neelam x Dashhari
(C) Chausa x Langra (D) Dashhari x Langra

499. Gaurav is a variety of mango developed by cross of:
(A) Dashhari x Totapari (B) Neelam x Dashhari
(C) Chausa x Langra (D) Dashhari x Langra

500. Saurabh is a variety of mango developed by cross of:
(A) Dashhari x Fazali jafrani
(B) Neelam x Dashhari
(C) Chausa x Langra
(D) Dashhari x Langra

501. Rajiv is a variety of mango developed by cross of:
(A) Dashhari x Romani (B) Neelam x Dashhari
(C) Chausa x Langra (D) Dashhari x Langra

502. Allahabadi safeda is a variety of:
(A) Guava (B) Mango
(C) Apple (D) Ber

503. Poovan is a variety of:
(A) Banana (B) Ber
(C) Aonla (D) Guava

504. Malbhog is a variety of:
(A) Banana (B) Ber
(C) Amla (D) Guava

505. Queen Alizabeth is a variety of:
(A) Rose (B) Chrysanthemum
(C) Dahalia (D) Merigold

506. First lady is a variety of:
(A) Merigold (B) Rose
(C) Dahalia (D) Chrysanthemum

507. First love is a variety of:
(A) Rose (B) Merigold
(C) Chrysanthemum (D) Dahalia

508. The optimum temperature for brinjal-germination is ——^{0}C:
(A) 5 (B) 10 (C) 15 (D) 20

509. An isolation distance of ——between varieties of water melon is considered enough for pure seed production:
(A) 100-200 m (B) 200-300 m
(C) 300-375 m (D) 400-500 m

510. Squash melon (*Citullus vulgaris*) seed should be sown when the temperature (air) warms upto ——0C:
(A) 15.6 (B) 20.6
(C) 24.6 (D) 26.6

511. In squash melon 100 g of seed accounts for about —— seeds:
(A) 100 (B) 200
(C) 350 (D) 400

512. The cold storage temperature for melons is 0^0C, the relative humidity (%) will be:
(A) 50-60 (B) 60-70
(C) 40-50 (D) 80-90

513. The seed rate (kg/ha) of long melon is:
(A) 1-2 (B) 2-3
(C) 3-4 (D) 4-5

514. Cucumber (*Cucumis sativus* var kheera) was domesticated by the ancient Greeks and Romans in about ——— B.C.:
(A) 100 (B) 200
(C) 300 (D) 400

515. Flowering in pumpkin (*Cucurbita moschata*) starts:
(A) 30-40 DAS (B) 50-60 DAS
(C) 60-80 DAS (D) 85-95 DAS

516. Akbar is a hybrid of:
(A) Grape (B) Mango
(C) Apple (D) Papaya

517. Queen is the variety of:
(A) Grape (B) Guava
(C) Pineapple (D) Mango

518. Commercial method of propagation of bael is:
(A) Grafting (B) Budding
(C) Cutting (D) Seed

519. Mango hybridization work was started in:
(A) 1911 (B) 1921
(C) 1931 (D) 1941

520. Average yield (t/ha) of mango is:
(A) 5 (B) 6 (C) 7 (D) 8

521. Cultivated strawberry is:
(A) Diploid (B) Triploid
(C) Tetraploid (D) Octaploid

522. India's share in global production of fruits is:
(A) 5 % (B) 10 % (C) 15 % (D) 20 %

523. The chemical by which seedless grapes can be produced is called:
(A) Gibberellins (B) Ethanol
(C) Auxins (D) Catalyses

524. Which citrus species is mono embryonic unlike other species which:
(A) *Citrus limon* (B) *Citrus reticulate*
(C) *Citrus grandis* (D) *Citrus aurantifolia*

525. In citrus, which growth regulator acts as a good fruit thinner:
(A) GA (B) NAA
(C) IBA (D) ABA

526. Guava belongs to which family:
(A) Myrtaceae (B) Caricaceae
(C) Rosaceae (D) Musaceae

527. Papaya belongs to which family:
(A) Myrtaceae (B) Caricaceae
(C) Rosaceae (D) Musaceae

528. Apple belongs to which family:
(A) Myrtaceae (B) Caricaceae
(C) Rosaceae (D) Musaceae

528. Custard apple belongs to which family:
(A) Myrtaceae (B) Caricaceae
(C) Rosaceae (D) Annonaceae

530. What is the moisture content to maintain seed viability for jack seed?
(A) 39 % (B) 20 % (C) 13 % (D) 10 %

531. Mango seed germination is improved by:
(A) Seed husking
(B) Soaking in GA_3
(C) Soaking in IAA
(D) Seed husking and soaking in GA_3

532. In which papaya variety vivipary occurrence is more?
(A) CO 1 (B) CO 2
(C) CO 6 (D) CO 7

533. Fruit type of mangosteen is:
(A) Drupe (B) Berry
(C) Capsule (D) Synconium

534. Fruit type of litchi is:
(A) Berry (B) Capsule
(C) Nut (D) Drupe

535. Litchi is propagated through:
(A) Seed
(B) Vegetative propagules
(C) Both seed and vegetative propagules
(D) Tissue culture

536. Fig is propagated through:
(A) Seed
(B) Vegetative propagules
(C) Both seed and vegetative propagules
(D) Tissue culture

537. Mangosteen is propagated through:
(A) Seed
(B) Vegetative propagules
(C) Both seed and vegetative propagules
(D) Tissue culture

538. Minimum seed moisture content for storage of mangosteen is:
(A) 7-8 % (B) 8-10 %
(C) 10-15 % (D) 53-54 %

539. Inflorescence of anthurium is:
(A) Spathe (B) Spadix
(C) Receme (D) Solitary head

540. An example for protogyny flower is:
(A) Anthurium (B) Orchid
(C) China aster (D) Chrysanthemum

541. Anthesis time of carnation is:
(A) 6-7 am (B) 7-8 am
(C) 8-10 am (D) 10-12 am

542. Type of fruit found in anthurium is:
(A) Berry (B) Capsule
(C) Achene (D) Drupe

543. Example for the protendrous type of cut flower is:
(A) Gerbera (B) Orchid
(C) Gladiolus (D) Rose

544. Type of fruit found in chrysanthemum is:
(A) Berry (B) Capsule
(C) Achene (D) Drupe

545. Pollination of orchid is carried out by:
(A) Insects (B) Beetles
(C) Wind (D) Water

546. Fruit of rose is called:
(A) Pod (B) Hip
(C) Berry (D) Achene

547. Carnation belongs to the family of:
(A) Araceae (B)Caryophyllaceae
(C) Asteraceae (D) Alstromeraceae

548. Anthurium seed attains physiological maturity at:
(A) 30 DAA (B) 40 DAA
(C) 50 DAA (D) 60 DAA

549. Petunia can be propagated by means of:
(A) Cutting (B) Budding
(C) Layering (D) Seeds

550. Gladiolus seed attains physiological maturity at:
(A) 30-35 DAA (B) 40-45 DAA
(C) 50-60 DAA (D) 60-65 DAA

551. Carnation seed attains physiological maturity at:
(A) 30-35 DAA (B) 40-45 DAA
(C) 50-55 DAA (D) 60-65 DAA

552. Branched cluster type of inflorescence can be seen in:
(A) Anthurium (B) Carnation
(C) China aster (D) Gladiolus

553. Dormancy in orchid can extend up to:
(A) One month (B) Two months
(C) Three months (D) Four months

554. Exogenous and endogenous dormancy is exhibited in which of the following crop:
(A) Rose (B) Orchid
(C) Dahlia (D) Petunia

555. The commercial propagation of gladiolus is:
(A) Corms (B) Crowns
(C) Slips (D) Suckers

556. Which of the following crop seeds has a tendency of sweating?
(A) Rose (B) Orchid
(C) Dahlia (D) Petunia

557. Celosia belongs to the family of:
(A) Oleaceae (B) Amaranthaceae
(C) Apocynaceae (D) Ericaceae

558. The seed colour of bougainvillea is:
(A) Red (B) White
(C) Black (D) Yellow

559. Number of seeds produced by crossandra per bloom is about:
(A) One (B) Two
(C) Three (D) Four

560. The inflorescence type of cosmos is:
(A) Solitary (B) Spike
(C) Raceme (D) Spadix

561. Fruit type of cactus and succulents is:
(A) Achenes (B) Capsule
(C) Pod (D) Drupe

562. Number of seeds present per capsule in Azalea is about:
(A) 200 (B) 300
(C) 400 (D) 500

563. Drooping panicle type inflorescence can be observed in:
(A) *Heliconia* (B) *Hemerocallis*
(C) *Ghomphrena* (D) Lotus

564. Zinnia attains seed maturity at:
(A) 20 DAA (B) 30 DAA
(C) 40 DAA (D) 50 DAA

565. Celosia can be pollinated by:
(A) Self (B) Insects
(C) Bees (D) Wind

566. Lotus can be pollinated by:
(A) Birds (B) Insects
(C) Bees (D) Wind

567. Marigold attains seed maturity at:
(A) 20-25 DAA (B) 30-40 DAA
(C) 40-50 DAA (D) 50-60 DAA

568. Rose attains seed maturity at:
(A) 100 DAA (B) 125 DAA
(C) 150 DAA (D) 175 DAA

569. Chrysanthemum is a:
(A) Self pollinated crop
(B) Often cross pollinated crop
(C) Cross pollinated crop
(D) Self incompatible crop

570. Seed of which annual germinate only in dark:
(A) Zinnia (B) Nigella
(C) Marigold (D) Aster

571. RRL-85-1 is a important high yielding varieties of:
(A) Celery (B) Plum
(C) Pear (D) Chicory

572. Which species that produce suckers for vegetative reproduction:
(A) Citrus (B) Mango
(C) Sapota (D) Banana

573. The vegetative propagation method that is commonly practiced for apple is:
(A) Leaf cuttings (B) Stem cuttings
(C) Grafting (D) Suckers

574. Grapes are propagated through:
(A) Stem cuttings (B) Root cuttings
(C) Grafting (D) Layering

575. The best example for propagation through stem cutting is:
(A) Coffee (B) Peach
(C) Tea (D) Pear

576. The grafting method is commonly practiced:
(A) Apple (B) Peach
(C) Plum (D) Poplar

577. The plant material chosen for tissue culture purpose is called:
(A) Tissue (B) Explant
(C) Plantlet (D) Stock plant

578. Who is the pioneer in micro propagation?
(A) Skoog Miltan
(B) Toshiba Murashige
(C) Fredrick Campion Steward
(D) Cornell Redaunbaugh

579. In which year onwards the plant tissue culture is practiced for micro propagation?
(A) Late 1940s and early 1950s
(B) Late 1950s and early 1960s
(C) Late 1960s and early 1970s
(D) Late 1970s and early 1980s

580. In micro propagation, the explants are surface sterilized by:
(A) Potassium permengnate
(B) Sodium thiosulphate
(C) Calcium chloride
(D) Sodium hypochlorite

581. Fruits are generally deficient in:
(A) Proteins (B) Carbohydrate
(C) Vitamins (D) Water

582. The prominent organic acid present in grapes is:
(A) Malic acid (B) Citric acid
(C) Tarteric acid (D) Stearic acid

583. Yellow fruits contains:
(A) Chlorophyll (B) Carotenoids
(C) Anthrocyanins (D) None of these

584. Among the following , highest amount of calcium is found in:
(A) Amla (B) Carambola
(C) Dates (D) Apricot

585. Among the following , highest amount of fat is found in:
(A) Apple (B) Avocado
(C) Dates (D) Cherries

586. Among the following , highest amount of minerals is found in:
(A) Apple (B) Apricot
(C) Dates (D) Figs

587. Among the following , highest amount of phosphorus is found in:
(A) Avicado (B) Banana
(C) Dates (D) Grape

588. Mango is a rich source of:
(A) Vitamin C (B) Carotene
(C) Calcium (D) Phosphorus

589. Wood apple is a rich source of:
(A) Calcium (B) Vitamin C
(C) Water (D) None of these

590. The highest protein content is present in:
(A) Wood apple (B) Pear
(C) Prunes (D) Sapota

591. The enzyme responsible for enzymatic browning in cut fruits is:
(A) PPO (B) Catalase
(C) Peroxidase (D) Lipase

592. Enzymatic browning can be prevented by:
(A) Sugar (B) Lemon juice
(C) Both (A) and (B) (D) None of these

593. Upon ripening the acid content of fruit:
(A) Increases (B) Decreases
(C) Remains constant (D) None of these

594. Upon ripening the protein content of fruit:
(A) Increases (B) Decreases
(C) Remains constant (D) None of these

595. Upon ripening the mineral content of fruit:
(A) Increases (B) Decreases
(C) Remains constant (D) None of these

596. Upon ripening the fat content of fruit:
(A) Increases (B) Decreases
(C) Remains constant (D) None of these

597. Upon ripening the vitamin C content of fruit:
(A) Increases (B) Decreases
(C) Remains constant (D) None of these

598. The respiration of fruits after harvest is:
(A) Continues (B) Ceases
(C) Increase manifold (D) None of these

599. The metabolic activities of fruit at low temperature:
(A) Increases (B) Decreases
(C) Remains constant (D) None of these

600. Unripe fruit has:
(A) High starch, low sugar
(B) Low starch, high sugar
(C) Equal levels of starch & sugar
(D) Low starch & low sugar

601. Upon harvesting, the water insoluble pectin substances in fruit:
(A) Increases (B) Decreases
(C) Remains constant (D) None of these

602. Upon harvesting, the water soluble pectin substance in fruit:
(A) Increases (B) Decreases
(C) Remains constant (D) None of these

603. The softening of fruits during ripening and storage is due to:
(A) Pectin substances (B) Pectin
(C) Protopectin (D) None of these

604. The organic acid contents of fruits at ripening is:
(A) Increases (B) Decreases
(C) Remains constant (D) None of these

605. Jaffa, Valencies and Washington navel are varieties of:
(A) Apple (B) Sweet orange
(C) Lemon (D) Guava

606. Tangerine is also known as:
(A) Lemon (B) Mandarin
(C) Sweet orange (D) Lime

607. Santra, Kamala and Kinnow are varieties of:
(A) Lemon (B) Sweet orange
(C) Mandarin (D) Lime

608. A grape fruit is a hybrid between:
(A) Pummelo and Sweet orange
(B) Lime and Lemon
(C) Sour and Sweet orange
(D) Tangerine and Lime

609. Apricots, cherries, peaches and plums belong to genus:
(A) Citrus (B) Prunus
(C) Rubus (D) None of these

610. The high alcohol content liquor obtained from sweet cherry is known as:
(A) Kirschwasser (B) Lemonado
(C) Sake (D) None of these

611. Free stones and cling stones are examples of:
(A) Apricot (B) Cherry
(C) Plums (D) Peach

612. The oldest cultivated plant among the following is:
(A) Cherry (B) Peach
(C) Grape (D) Plum

613. The cultivation of grapes is called:
(A) Pisciculture (B) Misciculture
(C) Viticulture (D) Sericulture

614. Bangalore blue, Bhokri, Gulabi and Thompson seedless are example of:
(A) Apricot (B) Plum
(C) Mango (D) Grape

615. The ripe grapes contains:
(A) Sucrose (B) Glucose and Fructose
(C) Both (A) and (B) (D) None of these

616. The principal acid of grapes is:
(A) Citric and malic acid
(B) Tarteric and citric acid
(C) Tarteric and malic acid
(D) None of these

617. The green colour of grapes is due to:
(A) Carotenoids (B) Chlorophyll
(C) Anthrocyanin (D) All of these

618. The red, blue and black colour of grapes is due to:
(A) Carotenoids (B) Chlorophyll
(C) Anthrocyanin (D) All of these

619. Musk, Cantaloupe and Honey dew are varieties of:
(A) Grapes (B) Apple
(C) Melon (D) Pears

620. Ambri, Kashmiri, Golden Delicious and Baldwin are varieties of:
(A) Pear (B) Apple
(C) Plum (D) Grapes

621. Principal sugar of apple is:
(A) Sucrose (B) Fructose
(C) Glucose (D) All of these

622. Cidar, a wine is prepared from:
(A) Grapes (B) Apple
(C) Pears (D) Plums

623. Indian gooseberry is also known as:
(A) Grape (B) Amla
(C) Peach (D) Lime

624. Nira, a sweetened drink is obtained from:
(A) Grapes (B) Dates
(C) Pears (D) Mango

625. Kapa and Barka are varieties of:
(A) Mango (B) Papaya
(C) Jamuns (D) Jackfruit

626. Wood apple is also known as:
(A) Elephant apple (B) Monkey fruit
(C) Curd fruit (D) All of these

627. The seed of the fruit are actually matured:
(A) Ovary (B) Ovules
(C) Stigma (D) Calyx

628. The strong flavor of ginger and onion is due to which volatile compound:
(A) Phosphorus (B) Sulphur
(C) Iron (D) Zinc

629. Agathi is the richest source of:
(A) Calcium (B) Iron
(C) Phosphorus (D) Riboflavin

630. Agathi, curry leaves and drum stick are particularly rich in:
(A) Water (B) Protein
(C) Phosphorus (D) Iron

631. Sauerkraut is prepared from:
(A) Cauliflower (B) Cabbage
(C) Spinach (D) Onion

632. The principal pigment present in red cabbage is:
(A) Carotenoids (B) Chlorophyll
(C) Anthrocyanin (D) Cryptoxanthin

633. The Chinese cabbage is known as:
(A) Pak-choi (B) Tak-thoi
(C) Cak-phoi (D) None of these

634. Brinjal is also known as:
(A) Hen plant (B) Egg plant
(C) Cock plant (D) Duck plant

635. Ridged, bottle, bitter, snake and sponge are varieties of :
(A) Potato (B) Radish
(C) Gourd (D) Carrot

636. Vegetables are canned in:
(A) Sugar syrup (B) Salt brine
(C) Water (D) None of these

637. Fruits are canned in:
(A) Sugar syrup (B) Salt brine
(C) Water (D) None of these

638. Maturity and ripening are similar:
(A) True (B) False
(C) Cannot be determines (D) None of these

639. Maturity comes ———— than ripeness:
(A) Sooner (B) Later
(C) Before (D) Equal

640. The TSS of fruit juice is measured by:
(A) Refractometer (B) Hydrometer
(C) Thermometer (D) Both (A) and (B)

641. Fruit juices are rich in:
(A) Vitamin C (B) Vitamin E
(C) Vitamin B (D) All of these

642. Blanching of fruits and vegetable is used to:
(A) Inactivate enzymes
(B) Increase sweetness
(C) Improve the vitamin content
(D) None of these

643. The target organism for canning is:
(A) *Bacillus cereus* (B) *E. coli*
(C) *Pediococcus botulinum* (D) All of these

644. The sweetness of corn is reduced due to conversion of:
(A) Starch to sugar (B) Sugar to starch
(C) Sugar to organic acid (D) None of these

645. Which of the following spice is used for pungency:
(A) Cinnamon (B) Clove
(C) Pepper (D) Saffron

646. Which of the following spice is not used for pungency:
(A) Ginger (B) Chillies
(C) Mustard (D) Cinnamon

647. Which of the following spice is an aromatic spice:
(A) Clove (B) Turmeric
(C) Mustard (D) Fenugreek

648. Which of the following spice is not an aromatic spice:
(A) Cardamom (B) Nutmeg
(C) Mac (D) Paprika

649. Clove consists of:
(A) Eugenol (B) Allicin
(C) Menthol (D) None of these

650. Which of the following spice is used for colour:
(A) Paprika (B) Pepper
(C) Cassia (D) Anise

651. Major and minor are two distinct varieties of:
(A) Pepper (B) Cardamom
(C) Cinnamon (D) Clove

652. The queen of spices is:
(A) Pepper (B) Cardamom
(C) Cinnamon (D) Clove

653. Ginger is modified:
(A) Root (B) Underground stem
(C) Fruit (D) Bud

654. The pungent constituents of ginger are:
(A) Oxymethyl phenols (B) Terpinyl acetate
(C) Ethyl alcohol (D) None of these

655. Tabasco sauce is prepared from:
(A) Bird chilli (B) Hen chilli
(C) Both (A) and (B) (D) None of these

656. Turmeric is an:
(A) antidepressant (B) Antipyretic
(C) Antioxidant (D) All of these

657. White, yellow and red are varieties of:
(A) Nutmeg (B) Mace
(C) Onion (D) Asafetida

658. Good source of protein in:
(A) Almond (B) Cashewnut
(C) Groundnut (D) All of these

659. The coconut water is present in:
(A) Exocarp (B) Endocarp
(C) Endosperm (D) None of these

660. Copra is obtained from:
(A) Cashewnut (B) Walnut
(C) Coconut (D) Peanut

661. Copra is obtained from:
(A) Ripe coconut (B) Unripe coconut
(C) Coconut leaves (D) None of these

662. The coconut sap is also known as:
(A) Peera (B) Jeera
(C) Neera (D) Beera

663. Brazil nut is similar to:
(A) Walnut (B) Hazelnut
(C) Almond (D) None of these

664. Tea is obtained from:
(A) Root (B) Bud
(C) Leaves (D) Stem

665. Which of the following is a type of tea:
(A) Black (B) Green
(C) Oolong (D) All of these

666. Which of the following is a fermented tea:
(A) Black (B) Green
(C) Oolong (D) All of these

667. Which of the following is a non fermented tea:
(A) Black (B) Green
(C) Oolong (D) All of these

668. The flavor of the tea is due to:
(A) Caffeine (B) Polyphenols
(C) Essential oils (D) All of these

669. Enzyme involved in fermentation of tea is:
(A) PPO (B) Lipase
(C) Protease (D) None of these

670. Which of the following is an artificial sweetener:
(A) Aspartame (B) Saccharin
(C) Acesulfame potassium (D) All of these

671. Which of the following is a class of soft drink:
(A) Gin (B) Wine
(C) Beer (D) None of these

672. The acid mostly used in soft drinks is:
(A) Malic acid (B) Citric acid
(C) Tarteric acid (D) Succinic acid

673. Which of the following is an important constituent of soft drinks:
(A) CO (B) O_2
(C) N_2 (D) CO_2

674. Fruit juice which has been concentrated by removal of water is called:
(A) Fruit juice concentrate (B) Fruit punch
(C) Squash (D) Cordial

675. Which of the following is distilled liquor:
(A) Beer (B) Whisky
(C) Tea (D) Wine

676. Which of the following is not a distilled liquor:
(A) Beer (B) Gin
(C) Vodka (D) None of these

677. Champagne is also called:
(A) Tingling wine (B) Sparkling wine
(C) Sprinkling wine (D) None of these

678. Port and slurry are types of:
(A) Brandy (B) Beer
(C) Whisky (D) Fortified wine

679. In preparation of red wine:
(A) Both Juice and skin are used
(B) Only juice
(C) Only skin
(D) None of these

680. Dry wine is:
(A) Sweet (B) Not sweet
(C) Dehydrated wine (D) None of these

681. Wines containing less than 14% alcohol are known as:
(A) Table wine (B) Dessert wine
(C) Fortified wine (D) None of these

682. Wines containing more than 14% alcohol are known as:
(A) Table wine (B) Dessert wine
(C) Fortified wine (D) Both (B) and (C)

683. Fermentation of grapes for wine is done with:
(A) Bacteria (B) Yeast
(C) Virus (D) None of these

684. Beer is chiefly prepared from:
(A) Sorghum (B) Barley
(C) Wheat (D) Maize

685. Ale, stout lager and porter are types of:
(A) Wine (B) Fortified wine
(C) Beer (D) Distilled liquor

686. Saki is prepared from:
(A) Barley (B) Wheat
(C) Rice (D) Maize

687. Toddy is prepared from:
(A) Barley (B) Maize
(C) Rice (D) Coconut

688. Rum is prepared from:
(A) Barley (B) Wheat
(C) Molasses (D) Maize

689. NHB stands for:
(A) National Horticulture Board
(B) Non-Horticulture Board
(C) Nutritional Health Board
(D) None of these

690. IGPB stands for:
(A) Indian Grape Producing Board
(B) Indian Grape Processing Board
(C) Indian Grape Production Bureau
(D) None of these

691. IICPT stands for:
(A) Indian Institute of Crop Processing Technology
(B) Indian Institution of Crop Processing Technology
(C) Indian Institute of Crop Production Technology
(D) None of these

692. Kimchi is a fermented product of:
(A) Mixed vegetable in Chinese cabbage
(B) Cucumber
(C) Vegetables and milk
(D) Vegetable and rice

693. Sajur asin is a fermented product of:
(A) Mixed vegetable in Chinese cabbage
(B) Cucumber
(C) Vegetables and milk
(D) Vegetable and rice

694. Tarhana is a fermented product of:
(A) Mixed vegetable in Chinese cabbage
(B) Cucumber
(C) Vegetables and milk
(D) Vegetable and rice

695. Kish K is a fermented product of:
(A) Rice (B) Vanilla beans
(C) Dough and milk (D) Lettuce

696. Poi is a fermented product of:
(A) Taro (B) Cabbage
(C) Milk (D) Rice

697. Sonti is a fermented product of:
(A) Malt (B) Fruit
(C) Molasses (D) Rice

698. Pulque is a fermented product of:
(A) Guava (B) Agave
(C) Molasses (D) Malt

699. Pulque is a fermented product of:
(A) LAB (B) Yeast
(C) Both (A) and (B) (D) Mould

700. The tenderness of pea is measured by:
(A) Tenderometer (B) Succulometer
(C) Penetrometer (D) All of these

701. Pitting is observed in:
(A) Apples (B) Melons
(C) Berries (D) Banana

702. Which of the following fruits have the lowest pH
(A) Mango (B) Pumpkin
(C) Pear (D) Prunes

703. Mustard seeds and oil contains following toxic substances:
(A) Elaidic acid (B) Erucic acid
(C) DHA & EPA (D) Muric acid

704. The dipping of fruit in solution of $CaCl_2$:
(A) Reduce browning
(B) Decrease flesh softening
(C) Increase ripening
(D) Both (A) and (B)

705. Which of the following is hazardous ripening agent:
(A) Ethylene (B) Ethephon
(C) Calcium carbide (D) All of these

706 The transparency of apple juice is measured in:
(A) NTU (B) BTU
(C) CTU (D) GTU

707. Jam is produced from:
(A) Fruit juice (B) Pulp
(C) Clear extracts (D) Fruit slices

708. Jelly is produced from:
(A) Fruit juice (B) Pulp
(C) Clear extracts (D) Fruit slices

709. Marmalade is produced from:
(A) Fruit juice (B) Pulp
(C) Clear extracts (D) Both (A) and (B)

710. Jams are solid gel which involves:
(A) Pulp (B) Sugar
(C) Pectin (D) All of these

711. The pectin mainly present in the:
(A) Pulp (B) Seeds
(C) Skin (D) Juice

712. The acid added to jam to attain desired acidity is:
(A) Citric (B) Malic
(C) Tartaric (D) All of these

713. For setting of a gel structure, important factor is:
(A) Sugar (B) Acid
(C) Pectin (D) All of these

714. Which of the following foods is not exempted from food labeling?
(A) Coffee beans
(B) Dehydrated vegetable condiments
(C) Instant tea
(D) Unpopped popcorn

715. Solanine is present in:
(A) Tomato (B) Coffee
(C) Potato (D) Tea

716. Pressure canners used in canning is known as:
(A) Steam blanchers (B) Retorts
(C) Plate exchangers (D) Sublimators

717. Fruits and vegetables discolor when cut due to:
(A) Caramelization (B) Sulfating
(C) Dehydration (D) Enzymatic browning

718. Kalonji is the common name for:
(A) Turmeric (B) Cumin black
(C) Fennel (D) Fenugreek

719. Jaepatri is the common name for:
(A) Fennel (B) Fenugreek
(C) Nutmeg (D) Mace

720. Jam should have minimum per cent of soluble solids:
(A) 28 (B) 48 (C) 68 (D) 88

721. Marmalade is obtained by ——— of fruit juice
(A) Boiling (B) Fermentation
(C) Homogenization (D) Irradiation

722. Bangalore, blue, Gulabi, Perlette and Sonaka are varieties of:
(A) Melon (B) Banana
(C) Grapes (D) Lemon

723. Kesar, Rajapuri, Pairi and Malda are varieties of:
(A) Melon (B) Banana
(C) Mango (D) Lemon

724. Which of the following is a high pectin fruit:
(A) Gooseberries (B) Peaches
(C) Pears (D) Figs

725. Pectin is a polymer of:
(A) Polygalacturonic acid (B) Maltose
(C) Glucose (D) None of these

726. Pst harvest losses can be minimized by:
(A) Increasing ethylene
(B) Decreasing temperature
(C) Increasing temperature
(D) Increasing oxygen

727. Fruit Product Order came in:
(A) 1954 (B) 1955 (C) 1956 (D) 1957

728. Self decomposition of fruits and vegetables products can be prevented by:
(A) Inactivating food enzymes
(B) By keeping them at room temperature
(C) By ripening
(D) None of these

729. Which of the following is not as method of preservation:
(A) Pickle making
(B) Dehydration
(C) Modified atmospheric packaging
(D) None of these

730. Canning involves:
(A) Low heating
(B) Low temperature processing
(C) Sterilization
(D) Pasteurization

731. Curing, pickling and chutney preparation are examples of:
(A) Preservation (B) Deterioration
(C) Fermentation (D) None of these

732. Fruit juice consists of:
(A) Juice and seeds (B) Juice and skin
(C) Juice and peel (D) Juice

733. Commonly added preservative in fruit juice is:
(A) Potassium metasulphite (B) Sodium benzoate
(C) Both (A) and (B) (D) None of these

734. The bitterness of citrus fruit juice can be reduced by:
(A) Keeping in open air
(B) By dipping in 2% boiling NaOH
(C) By exposure to high temperature
(D) By keeping at low temperature

735. Apple and potatoes are:
(A) Non-perishable (B) Semi-perishable
(C) Perishable (D) Dry

736. Controlled removal of water from fruits and vegetables is called:
(A) Dehydration (B) Sundrying
(C) Canning (D) None of these

737. The presence of salt for preservation causes:
(A) Precipitation of enzyme protein
(B) Reduction in sulfur linkage in enzyme protein
(C) Reaction with aldehyde
(D) None of these

738. An antioxidant:
(A) Prevents oxidation (B) Allows oxidation
(C) Does not get consumed (D) All of these

739. The tenderization of vegetables is done by:
(A) Cellulase (B) Protease
(C) Glucosidase (D) Pectinase

740. Safrole is a toxicant found in:
(A) Beans (B) Spices
(C) Almonds (D) Spinach

741. Prussic acid is a toxicant found in:
(A) Beans (B) Spices
(C) Almonds (D) Spinach

742. Oxalic acid is found in:
(A) Beans (B) Spices
(C) Almonds (D) Spinach

743. Which of the following has longest storage life:
(A) Fruits (B) Leafy vegetables
(C) Root crops (D) Dried fruits

744. Canned tomato juice is heated by:
(A) Convection (B) Conduction
(C) Radiation (D) Both (A) and (B)

745. When the water in the intercellular space of fruits and vegetables freezes rapidly form:
(A) No ice crystal (B) Small ice crystal
(C) Large ice crystal (D) None of these

746. During blanching:
(A) No nutrient loss
(B) Water soluble components are lost
(C) Minerals lost
(D) Both (B) and (C)

747. The spoilage of canned foods occur due to:
(A) Underprocessing
(B) Inadequate cooling
(C) Infection from leakage in cans
(D) All of these

748. When ends of the canned food remains flat and contents become sour, it is called:
(A) Flipper (B) Flat sour
(C) TA spoilage (D) Sulfur stinker

749. Jam, jellies and fruit juices are spoiled by:
(A) Yeast (B) Mold
(C) Bacteria (D) All of these

750. The surface of the fruit juice may be contaminated by:
(A) Yeast (B) Mold
(C) Bacteria (D) All of these

751. Fruits and vegetables concentrated are mostly spoiled by:
(A) Yeast (B) Mold
(C) Bacteria (D) All of these

752. Jams, jellies and preserves are mainly spoiled by:
(A) Osmophilic yeast (B) Mold
(C) Bacteria (D) All of these

753. Soft kraut and black kraut are:
(A) Varieties of sauerkraut
(B) Spoilage
(C) Sauerkraut with different acid fermentation
(D) None of these

754. Soft pickles is due to:
(A) *Penicillium* (B) *Fusarium*
(C) Both (A) and (B) (D) None of these

755. Black pickles is due to:
(A) Yeast
(B) Mold
(C) Hydrogen sulfide produced by bacteria
(D) All of these

756. Slippery pickles results by action of:
(A) Yeast (B) Mold
(C) Bacteria (D) All of these

757. The type of storage for equalization of moisture to a desired level in dried fruits is:
(A) Sweating (B) Chilling
(C) Freezing (D) None of these

758. Perry is wine from:
(A) Peaches (B) Plums
(C) Pears (D) Cherries

759. Mead is wine from:
(A) Peaches (B) Honey
(C) Pears (D) Cherries

760. Most of the vegetables require ———— heat treatment than fruits:
(A) More (B) Less
(C) Equal (D) None of these

761. The common class I preservative is:
(A) Sugar (B) Salt
(C) Both (A) and (B) (D) None of these

762. Brine is:
(A) Salt solution
(B) Sugar solution
(C) Solution of chemical preservatives
(D) All of these

763. An orange ripens on tree:
(A) Sugar increase, acid decrease
(B) Sugar decrease, acid increase
(C) Sugar and acid levels are equal
(D) None of these

764. Peels of citrus fruits is particularly rich in:
(A) Bitter oils (B) Citric acid
(C) Iron (D) Calcium

765. Anthrocyanin are found in:
(A) Capsicum (B) Carrots
(C) Blueberries (D) Banana

766. Richest source of carotene among the following fruits is:
(A) Amla (B) Apricot
(C) Banana (D) Figs

767. When a cut apple is exposed to air browning occurs due to presence of ——— substances in fruit:
(A) Phenolic (B) Protein
(C) Fat (D) Watery

768. In immature fruits, pectin is present as:
(A) Myopectin (B) Protopectin
(C) Propectin (D) Pectin

769. On ripening of the fruit, protopectin changes to:
(A) Pectin (B) Pectic acid
(C) Pectinic acid (D) Both (A) and (B)

770. Pectins are polymer of:
(A) Sugar and protein
(B) Protein and fat
(C) Sugar and acid derivatives
(D) Acid and acid derivatives

771. The natural enzyme present in fruit that hydrolyze pectin is:
(A) Pectin methyl esterase
(B) Pectin ethyl esterase
(C) Pectin methyl peroxidase
(D) Pectin ethyl peroxidase

772. The pigments and their precursors found in fruits and vegetables occurs in:
(A) Chloroplast (B) Chromoplast
(C) Both (A) and (B) (D) None of these

773. Long berries are hybrid of:
(A) Blue and cran berries (B) Black and blue berries
(C) Black and raspberries (D) None of these

774. Papaya is principle source of enzyme:
(A) Esterase (B) Papain
(C) Cathepsins (D) All of these

775. Papain is commercially used for:
(A) Tenderization of meat
(B) Fixation of red colour of meat
(C) Removal of fat from meat
(D) All of these

776. The fat content of vegetables is:
(A) High (B) Low
(C) Very high (D) None of these

777. Cabbage, Kale and cauliflower belong to which group of crops:
(A) Root (B) Cole
(C) Fruit (D) Cucurbits

778. "Eyes" are found in:
(A) Pursley (B) Carrot
(C) Potato (D) Agathi

779. Potato and tapioca are particularly rich in:
(A) Protein (B) Starch
(C) Minerals (D) Fat

780. Onion, garlic and leek belong to which group:
(A) Tuber (B) Cole
(C) Bulb (D) Bud

781. Edible mushrooms are:
(A) Poisonous (B) Non-poisonous
(C) Both (A) and (B) (D) None of these

782. The skin of vegetables are removed by:
(A) Hot dipping in alkali solution
(B) Iye solution
(C) Sun drying
(D) Both (A) and (B)

783. The firmness, softness, juiciness, chewiness of fruits and vegetables are factors of:
(A) Appearance (B) Flavour
(C) Textural (D) Colour

784. Wholeness, damage, gloass, colour, transparency of fruit and vegetables are factors of:
(A) Appearance (B) Flavour
(C) Textural (D) Colour

785. Sweet, sour, salty, bitter and aromas of fruits and vegetable are factors of:
(A) Appearance (B) Flavour
(C) Textural (D) Colour

786. Flavour of a food product is a combination of:
(A) Taste and smell (B) Taste and appearance
(C) Smell and appearance (D) None of these

787. The self life of a good is the time for which the food is in:
(A) Fresh (B) Mature
(C) Acceptable (D) Ripe

788. The major causes of deterioration of fruit and vegetables is:
(A) Bacteria (B) Yeast
(C) Molds (D) All of these

789. Canning of fruits and vegetables is a process of:
(A) Heat (B) Cold
(C) Irradiation (D) Microwave

790. The storage life of fruits and vegetables in cold storage:
(A) Decrease (B) Increase
(C) No effect (D) None of these

White and black are principal types of:
(A) Mustard (B) Garlic
(C) Fennel (D) Fenugreek

792. Distilled 'kaju' is made from:
(A) Feni (B) Wine
(C) Brandy (D) Whisky

793. Cobnut is another name of:
(A) Brazil nut (B) Haze nut
(C) Walnut (D) Chestnut

794. Pecan nuts are:
(A) Edible (B) non-edible
(C) Toxic (D) None of these

795. Which is a type of:
(A) Green (B) White
(C) Yellow (D) All of these

796. Trained juice containing moderate quantity of fruit pulp to which sugar is added is called:
(A) Fruit juice (B) Fruit drink
(C) Squash (D) Cordial

797. Fruit squash from which all suspended material is removed and is clear is called:
(A) Fruit juice (B) Fruit drink
(C) Squash (D) Cordial

798. Cooling drink of sweeted and diluted fruit juice is called:
(A) Fruit punch (B) Sherbet
(C) Cordial (D) Fruit juice

799. Natural juice pressed out of fruit is known as:
(A) Fruit juice (B) Fruit drink
(C) Sherbet (D) Fruit punch

800. The chemical preservatives added to fruit juices are:
(A) Sodium benzoate (B) Sodium bisulfate
(C) Both (A) and (B) (D) Citric acid

801. Wine is chiefly prepared from:
(A) Grapes (B) Apple
(C) Pineapple (D) Sapota

802. Scotch whiskey is prepared from:
(A) Sorghum (B) Barley malt
(C) Rice (D) Wheat

803. Brandy is distilled liquor from:
(A) Grape (B) Apple
(C) Sapota (D) Coconut

804. Which type of water is used for beer making:
(A) Hard (B) Soft
(C) Both (A) and (B) (D) None of these

805. Poi is a fermented product of:
(A) Cabbage (B) Milk
(C) Taro (D) Rice

806. Which of the following is bacteria fermented product:
(A) Dill pickle (B) Sauerkraut
(C) Tarhana (D) All of these

807. Which fruit can ripen after harvest:
(A) Mango (B) Lemon
(C) Pineapple (D) Grape fuit

808. Which of the following is an example of climacteric fruit:
(A) Pear (B) Kiwi fruit
(C) Sapota (D) Pomegranate

809. Which of the following is an example of nonclimacteric fruit:
(A) Papaya (B) Passion fruit
(C) Sapota (D) Both (A) and (B)

810. Upon maturation, the acidity of fruit:
(A) Increases (B) Decreases
(C) Remains same (D) None of these

811. The oil content of fruits with the onset of maturity is:
(A) Increases (B) Decreases
(C) Remains same (D) None of these

812. The juice content of fruits with the onset of maturity is:
(A) Increases (B) Decreases
(C) Remains same (D) None of these

813 Blanching is a heat treatment for:
(A) Mild (B) High
(C) Not known (D) Unknown

814. The adequacy of blanching is indicated by:
(A) Peroxidase (B) Catalase
(C) Lipoxygenase (D) Both (A) and (B)

815. PPC refers to:
(A) Para Processing Contamination
(B) Post Processing Contamination
(C) Pre Processing Contamination
(D) None of these

816. The enzyme is responsible for cut fruits and vegetables browning:
(A) PPO (B) Lipase
(C) Phosphorylase (D) Catalase

817. Fruits and vegetables discolor when cut due to:
(A) Proteins (B) Water
(C) Carbohydrate (D) Fat

818. For canning, fruits are used:
(A) Whole (B) Cut
(C) Peeled and cut (D) All of these

819. The blanching time is influenced by:
(A) Type of vegetable (B) Size of vegetable
(C) Temperature (D) All of these

820. Enzyme which causes the loss of eating and nutritional quality in fruits and vegetables:
(A) Lipoxygenase (B) PPO
(C) Polygalacturonase (D) All of these

821. Function of blanching is:
(A) Reduce number of contaminating micro organisms
(B) Inactive enzymes
(C) Softens vegetable tissue
(D) All of these

822. Coating of fruits and vegetables:
(A) Improve appearances (B) Enhance flavours
(C) Modify texture (D) All of these

823. Major acid is apple is:
(A) Malic acid (B) Tartaric acid
(C) Quinic acid (D) None of these

824. Major acid is grapes is:
(A) Malic acid (B) Tartaric acid
(C) Quinic acid (D) None of these

825. Major acid is lemon is:
(A) Malic acid (B) Tartaric acid
(C) Quinic acid (D) Citric acid

826. Highest oxalic acid content is present in:
(A) Amaranth leaves (B) Spinach
(C) Amla (D) Beet root

827. Dehydrofreezing of fruits and vegetables involves:
(A) Only freezing
(B) Dehydration before freezing
(C) Dehydration after freezing
(D) Dehydration

828. Jams, jelly and honey fall into which category of product:
(A) High moisture (B) Low moisture
(C) Intermediate moisture (D) None of these

829. Preservative is used in jam, jellies and pickles:
(A) Benzoates (B) Sorbates
(C) Acetate (D) Both (A) and (B)

830. Scalding of vegetables is a:
(A) Freezing treatment (B) Heat treatment
(C) Irradiation (D) None of these

831. Lloyd Botanical Garden, Darjeeling (West Bengal) was laid out by ——— on a piece of land donated by William Lloyd:
(A) Sir Robert Llyod (B) Sir George King
(C) Sir Peter Hunt (D) Sir Holdgate Baird

832. The highest per capita consumption of flower in the world is in:
(A) The USA (B) India
(C) Switzerland (D) The Netherlands

833. Which organic acid, often used as a preservative, occurs naturally in cranberries, prunes, cinnamon and cloves?
(A) Citric acid (B) Benzoic acid
(C) Tarteric acid (D) Lactic acid

834. Kemin yellow oil from Kemin Industries, USA is the product of:
(A) Rose (B) Carnation
(C) Merigold (D) Calendula

835. A popular lawn grass species Zoysia japonica is commonly known as:
(A) Blue-green grass
(B) Doob grass
(C) Japanese or Korean grass
(D) Nut grass

836. Which of the following catci is almost spineles?
(A) Zygocactus (B) Opuntia
(C) Hawarthia (D) Echinopsis

837. Zen garden are also known as:
(A) Mughal garden (B) English garden
(C) Japanese garden (D) Italian garden

838. Which of the following Indian rose varieties is believed to have contributed chocolate brown colour to the present day roses?
(A) Pusa muskan (B) Mohini
(C) Mrinalini (D) Manmatha

839. The earliest example of verticle garden is in :
(A) Ryoan-ji garden in Japan
(B) Haging garden of Babylon
(C) Villa Medici at Fiesole in Italy
(D) Vaux-le-Vicomte in France

840. Which of the following indoor plants is poisonous?
(A) Cardyline (B) Chlorophytum
(C) Dieffenbachia (D) Draceana

841. Major constituent of rose oil:
(A) Geraniol (B) Citronellol
(C) Citral (D) Rosal

842. Fruit of rose is known as:
(A) Bulbs (B) Hips
(C) Caropsis (D) Berry

843. A series of arches known as:
(A) Pergola (B) Arch
(C) Terrace (D) Topiary

844. To avoid continuous shade of certain portion of a multispan green house, the most ideal orientation is:
(A) East-West (B) North-South
(C) South-East (D) North-East

845. Terrarium is an art of growing plants:
(A) In Terraces (B) In water
(C) In rocks (D) In glass container

846. International Registration Authority of Bougainvillea is located at:
(A) The Netherlands (B) New Delhi
(C) China (D) Thailand

847. Dried rose petals are known as:
(A) Gul Roghan (B) Gulkand
(C) Pankhuri (D) Hips

848. Wazid Ali Shah is a variety of :
(A) Barleria (B) Rose
(C) Bougainvillea (D) Chrysanthomum

849. Fruits of tomato is known as:
(A) Bulbs (B) Hips
(C) Caropsis (D) Berry

850. The quickest method of laying a lawn is:
(A) Seedling (B) Turfing
(C) Dibbling (D) Dung plastering

851. The Rashtrapati Garden is example of which type of gardens?
(A) Mughal (B) Japanese
(C) English (D) Italian

852. Canna is also known as:
(A) Spanish day (B) African lily
(C) Indian shot (D) Century plant

791. Art of training plants into shape of different statues, birds or animals known as:
(A) Edge (B) Topiary
(C) Hedge (D) Pergole

854. Gladiolus in hill as planting during:
(A) February-March (B) March-April
(C) May-June (D) September - October

855. Stenting is practiced in
(A) Dahlia (B) Ficus
(C) Rose (D) Ranuncules

856. Which is national flower of South Africa?
(A) Chrysanthemum (B) Bird of Paradise
(C) Protea (D) Gladiolus

857. Gladiolus fruit is a:
(A) Berry (B) Capsule
(C) Pod (D) Drupe

858. Epinasty in painsettis is due to:
(A) Ethylene (B) Gibbrellin
(C) Maleic hydrazide (D) Aluminium sulphate

859. The first botanical garden in the world was made during 16th Century at :
(A) Pauda (Italy) (B) Kyato (Japan)
(C) Singapore (D) San Francisco (USA)

860. Lipstick is a variety of :
(A) Carnation
(B) Red Ginger
(C) Gerbera
(D) Heliconia pstitacorum

861. Which is the common method of propagation in Dahlia for taking large sized flowers?
(A) Cuttings (B) Seeds
(C) Budding (D) Grafting

862. Hanging garden, a seven wonder located at:
(A) Vatican City (B) New Delhi
(C) Babylonia (D) Moscow

863. Work on Hibiscus improvement and development of varieties was extensively carried at:
(A) IARI, New Delhi (B) NBRI, Lucknow
(C) IIHR, Bengaluru (D) YSRUHF, Solan

864. Bangkok golden apple is variety of:
(A) Guava (B) Pineapple
(C) Avocado (D) Pone granate

865. Rajat Rekhi is a mutant of :
(A) Rose (B) Carnation
(C) Tube rose (D) Chrysanthemum

866. Star Gazer is a popular variety of:
(A) Tiger lily (B) Asiatic lily
(C) Oriental lily (D) Torch lily

867. Deepika is a variety of:
(A) Zinnia (B) Balsam
(C) Holly hock (D) Rose

868. Rosa wichuriana is a :
(A) Tetraploid (B) Diploid
(C) Triploid (D) Aneuploid

869. Repetition of same object at equidistance in landscape is called:
(A) Design (B) Rhythm
(C) Focal point (D) Layout

870. First Islamic garden in India was developed by:
(A) Ferozshah (B) Jahangir
(C) Akbar (D) Babar

871. Contour system of orchard planting is generally followed in :
(A) Punjab (B) Hills
(C) Saline soils (D) UP

872. "T" budding is also referred to as:
(A) Shield budding (B) Patch budding
(C) Annular budding (D) Ring budding

873. To check the variability in rootstocks one should opt for:
(A) Clonal rootstock (B) Seedling rootstock
(C) Seasonal rootstock (D) Any of the above

874. Modified leader system of training fruit tree is:
(A) Unscientific (B) discarded
(C) Good for plains only (D) Most acceptable

875. Most of the fruit tree varieties are highly heterozygous, therefore, for commercial plantation, their propagation is recommended by:
(A) True seeds (B) A Sexual propagation
(C) Mutation (D) All of the above

876. The following crop does not belong to cole group:
(A) Kale (B) Turnip
(C) Cabbage (D) Cauliflower

877. Pineapple originated in :
(A) Brazil (B) New Zealand
(C) India (D) China

878. Commercial propogation of banana is done through:
(A) Budding (B) Layering
(C) Seeds (D) Cutting

879. In apple dwarfing, rootstock is good for:
(A) Low chilling (B) Scab resistance
(C) Fruit colour (D) high density planting

880. "June drop" in apple is mainly due to:
(A) Lock of pollination (B) Alternate bearing
(C) Moisture stress (D) Calcium deficiency

881. Spongy tissue of mango is a :
(A) Bacterial disease (B) Fungal disease
(C) Viral disease (D) Physiological disorder

882. Semi hard wood cutting is mostly used to propagate the following:
(A) Evergreen fruit plants (B) Apple
(C) Vegetable crops (D) Peach

883. Cryo preservation is done in liquid nitrogen at a temperature of:
(A) 0°C (B) 4°C
(C) 200°C (D) -196°C

884. Most of the varieties of mango grown in South India are:
(A) Disease free (B) Seedless
(C) Mono embryonic (D) Polyembryonic

885. Growth regulator CCC is used in grapes for :
(A) Increase vegetative growth
(B) Increases TSS
(C) Increase fruitfullness
(D) None of the above

886. McIntosh is a variety of :
(A) Apple (B) Plum
(C) Peach (D) Pear

887. India contributes following percentage of Mango production in the world:
(A) 20% (B) 10% (C) 56% (D) 80%

888. In India, banana contributes following percentage of total fruit production:
(A) 80% (B) 60% (C) 31% (D) 10%

889. Pear is adopted to:
(A) Asiatic climate (B) Tropical climate
(C) Temperate climate (D) European climate

890. In Nilgiries the main crop of mandorin orange is harvested in:
(A) July (B) August-October
(C) December-January (D) March

891. Stooling method of propagation is generally used in:
(A) Grapes (B) Walnut
(C) Guava (D) Apple

892. More than 50 percent of the area under grapes in India is under the variety:
(A) Thompson seedless (B) Dikhush
(C) Palette (D) Anab-e-shahi

893. Rains during ripening of grapes causes:
(A) Low TSS (B) Fruit drop
(C) Berry cracking (D) Higher marketable yield

894. Seedless varieties of guava bear following types of fruits:
(A) All seeded (B) All seedless
(C) (A) and (B) both (D) None of the above

895. Kniffin system of training is used in:
(A) Litchi (B) Papaya
(C) Mango (D) Grapes

896. Which kind of variety in tomato needs stalking?
(A) Dwarf (B) Determinate
(C) Indeterminate (D) None of the above

897. Which of the following variety of radish is round in shape:
(A) Pusa Himani (B) Japanese white
(C) RRWT (D) Chinese pink

898. Brocoli is a :
(A) Cole crop (B) Root crop
(C) Fruit crop (D) Legume crop

899. Among vegetable crops, leaving aside potatoes the maximum production in the world:
(A) Cucumber (B) Brinjal
(C) Onion (D) Tomato

900. Seed production of early cauliflower can be done :
(A) In Europe only (B) In plains
(C) In hills only (D) None of these

901. Pusa Mukta variety of cabbage is resistant to:
(A) Virus (B) Bolting
(C) Cabbage caterpillar (D) Black rot

902. Nectarine is fuzzyness mutant of:
(A) Plum (B) Almond
(C) Peach (D) Apricot

903. Best time of pruning peach is:
(A) Spring (B) Mid winter
(C) Autumn (D) Mid summer

904. Sweet orange is generally trained to:
(A) Single branches (B) Two branches
(C) Multiple branches (D) None of these

905. Frontier is a mid season variety of :
(A) Apricot (B) Grape
(C) Plum (D) Peach

906. Most peach varieties bear flowers of following colour:
(A) Blue (B) Pink
(C) Yellow (D) White

907. Pride of India is a variety of :
(A) Radish (B) Turnip
(C) Cabbage (D) Cauliflower

908. In general, fruits and vegetables are not a rich source of?
(A) Vitamin C (B) Vitamin B1
(C) Vitamin A (D) Minerals

909. Most varieties of onion grown in India are :
(A) Chilling type (B) Day neutral
(C) Short day types (D) Long day types

910. Pinching is generally done in quality production of:
(A) Gladiolus (B) Carnation
(C) Dahlia (D) Aster

911. Pusa Narangi is a variety of :
(A) Marigold (B) Jasmine
(C) Carnation (D) Rose

912. First Red King is generally referred to in:
(A) Marigold (B) Rose
(C) Vinta (D) Larkspur

913. Chrysanthemum is a :
(A) Day neutral (B) Long day plant
(C) Short day plant (D) None of the above

914. Cymbidium is a kind of:
(A) Orchid (B) Rose
(C) Vegetables (D) Fig

915. Gladiolus is propagated through:
(A) Suckers (B) Budding
(C) Corms (D) Cuttings

916. Quincunx system of planting is used in:
(A) Fruit trees (B) Flowers
(C) Potato (D) Vegetable crops

917. Ornamental value of baugainvilia lies in:
(A) Green foliage (B) Seeds
(C) Colourful bracts (D) Fruits

918. Magnolia is best propagated through:
(A) Air layering (B) Cutting
(C) Seedling (D) None of these

919. Gynomoecious means:
(A) Having female and bisexual flowers
(B) Having all female flowers
(C) Having male flowers only
(D) None of these

920. Betelvine is a :
(A) Hermaphrodite (B) Dioecious plant
(C) Andromonoecious (D) Monoecious plant

921. Mango and banara are:
(A) Non-climacteric fruit
(B) Climacteric fruit
(C) Nature of ripening depends upon varieties
(D) None of these

922. Aurocaria is a tree suited to following regions:
(A) Cold desert (B) Marshy land
(C) Sub-tropical (D) None of these

923. Increase in ethylene production with ripening is characteristics of:
(A) Citrus fruits (B) Grapes
(C) Climacteric fruits (D) Non-climacteric fruits

924. Modified Atmospheric Packaging (MAP) of fruits and vegetables percent the build up of following:
(A) CO_2 and O_2 (B) CO_2 and C_4H_4
(C) Sugars (D) Proteins

925. Sulphites and SO_2 are added to preserve fruits and vegetables products to check:
(A) Reduce acidity (B) Microbial activity
(C) Ethylene production (D) Methane production

926. Which vegetable crop is known as vegetable of Twentieth century?
(A) Winged bean (B) Clusterbean
(C) Chekkurmanis (D) Amaranthus

927. Large cardamom is commercially propagated through:
(A) Suckers (B) Seeds
(C) Layers (D) Cuttings

928. The part used as vegetable in gerusalem artichoke is:
(A) Flower bud (B) Root tuber
(C) Stem (D) Leaf stalk

929. Bitter principle in bittergourd is:
(A) Capsorubin (B) Allyl propyl disulphide
(C) Cucurbitacin (D) Phenol

930. Pickout the method which is not available for extraction of floral perfumes:
(A) Maceration (B) Solvent extraction
(C) Enfleurage (D) Distillation

931. Which is one of the following is a well-known cardiotonic:
(A) Cinchona (B) Periwinkle
(C) Eucalyptus (D) Digitalis

932. When a seed in stored at freezing temperature, it is known as:
(A) Ex-situ conservation
(B) Orthodox seed
(C) In-situ conservation
(D) Recalcitrant seed

933. The variety of mango seriously affected by spongy tissue is:
(A) Neelam (B) Dashahari
(C) Amrapali (D) Alphanso

934. The plant bioregular that acts as antigibberellin is:
(A) Paclobutazole (B) Indole acetonitrite
(C) Zeatin (D) Morphactin

935. The pear fruit tree is trained in:
(A) Open center system
(B) Head system
(C) Modified central leader system
(D) None of these

936. Following is the dwarfing rootstock of apple:
(A) M25 (B) M15
(C) M9 (D) M1

937. Which of the following is not a cytokinin?
(A) Diphenylurea (B) Ethrel
(C) Zeatin (D) IBA

938. The class of plant bioregulator used commercially in grape production is:
(A) Cytokinin (B) Gibberellin
(C) Ethylene (D) Auxin

939. The plant bioregulator used as substitute for chilling is:
(A) GA (B) IAA
(C) BA (D) MH

940. Which of the following is rich source of fat?
(A) Aonla (B) Avocado
(C) Bael (D) Pomegrane

941. The origin place of the fruit guava is:
(A) China (B) Australia
(C) India (D) Tropical America

942. Kinnow cultivator of citrus is a hybrid between:
(A) King × Willow leaf (B) King × Ovari
(C) King × Cleopatra (D) King × Dancy

943. The term Ikebana is related to:
(A) Flower arrangement (B) Topiary
(C) Carpet bedding (D) Bonsai culture

944. Triploidy has been put to commerical use in following vegetables
(A) Watermelon (B) Brinjal
(C) Tomato (D) Cauliflower

945. For complete preservation the percentage of sugar in fruit jam should be nearly:
(A) 90% (B) 70% (C) 50% (D) 30%

946. Which one of the following is not a food preservatives?
(A) Glacial acetic acid (B) Benlate
(C) KMS (D) Sodium benzoate

947. Which of the following medicinal plants is called Indian Gooseberry?
(A) Arjuna (B) Aonla
(C) Bohera (D) Harar

948. Following is recommended for maintaining freshness in cut leafy vegetables?
(A) Cytokinin (B) ABA
(C) Auxin (D) Gibberellin

949. "Heroine" one of the major chemicals responsible for "drug abuse" is obtained from:
(A) Podophylum sp. (B) Datura sp.
(C) Poppy (D) Perriwinkle

950. Mango tree bears fruits on:
(A) Last year growth
(B) Current season growth
(C) Both (A) and (B)
(D) None of these

951. Seedlessness in most grape varieties is due to:
(A) Stenospermocarpy (B) Parthemocarpy
(C) Polyembryony (D) Apomixis

952. The ultra dwarf rootstock of apple is:
(A) M27 (B) MM106
(C) M9 (D) MM104

953. The dwarfin rootstock of guava is:
(A) Brazilian guava (B) Cattleya guava
(C) Pineapple guava (D) Chinese guava

954. "Protective food" includes:
(A) Fruis and Vegetables (B) Cereals
(C) Fruits (D) Vegetables

955. In peaches the traditional system of training is:
(A) Open center
(B) Multiple stem
(C) Center leader
(D) Modified central leader

956. Marmalade is prepared from the fruits of:
(A) Citrus (B) Apple
(C) Mango (D) Guava

957. Citrus canker is caused by:
(A) Virus (B) Fungus
(C) Bacteria (D) Insects

958. The heat treatment given to inactivate enzymes is called:
(A) Blanching (B) Cooling
(C) Dehydration (D) Oxidation

959. Waxing of fruits is done mainly to reduce:
(A) Respiration
(B) Transpiration
(C) Ripening
(D) Respiration & Transpiration

960. Most of the banana species are:
(A) Triploid (B) Diploid
(C) Haploid (D) Hexaploid

961. Avenue term is mostly related to:
(A) Shrubs only (B) Annulas only
(C) Trees only (D) None of these

962. Sweetest sugar in fruit is:
(A) Glucose (B) Fructose
(C) Sucrose (D) Galactose

963. Lye peeling is done by:
(A) 10% hot acid solution (B) 10% cool acid solution
(C) 1% cool alkali solution (D) 1% hot alkali solution

964. Chrysanthemum is a :
(A) Long day plant (B) Short day plant
(C) Day neutral plant (D) None of these

965. Mango has bearing habit which is:
(A) Axillary (B) Terminal
(C) Alternate (D) None of these

966. Flower which is white and scented?
(A) Marigold (B) Jasmine
(C) Mangolia (D) Both (B) and (C)

967. Dwarf mutant of papaya is:
(A) Pusa nanha (B) Pusa giant
(C) Pusa dwarf (D) None of these

968. The permissible limit of colours to be added in preservative:
(A) 100 ppm (B) 250 ppm
(C) 300 ppm (D) 200 ppm

969. Multistem system is observed in which of the following?
(A) Pomegranate (B) Mango
(C) Aonla (D) Jackfruit

970. Cidar is prepared from:
(A) Apple (B) Cashew
(C) Datepalm (D) Pear

971. Large numbers of hybrids have been developed for which of the following vegetables?
(A) Tomato (B) Onion
(C) Brinjal (D) Potato

972. Best candy can be prepared from:
(A) Aonla (B) Jamun
(C) Ash gourd (D) Bael

973. Most ancient type of garden is:
(A) Truck garden (B) Kitchen garden
(C) Market garden (D) None of these

974. Cauliflower is:
(A) Self pollinated (B) Cross pollinated
(C) Often cross pollinated (D) None of the above

975. Transgenic are mostly made these days for which of the following fruit crops?
(A) Mango (B) Grapes
(C) Guava (D) Banana

976. Which of the following variety of apple has comparatively longer self life?
(A) Ambari (B) Red delicious
(C) Golden delicious (D) Ambrich

977. Which of the following is dioecious?
(A) Bitter gourd (B) Pointed gourd
(C) Snake gourd (D) Bottle gourd

978. Red colour of carrot is due to presence of:
(A) Anthocyanin (B) Carotene
(C) Xanthophyll (D) None of the above

979. Which of the following garden known as nature in miniatures?
(A) Mughal (B) Japanese
(C) English (D) Persian

980. Which is known as queen of spices?
(A) Cardamon (B) Black pepper
(C) Turmeric (D) Ginger

981. Yellow coloured fruits are rich in:
(A) Vitamin A (B) Vitamin B
(C) Vitamin C (D) Vitamin D

982. Collocasin contains:
(A) Calcium corbonate (B) Calcium oxalates
(C) Calcium arsenate (D) None of these

983. Charbagh is developed in:
(A) Japanese style (B) Persian style
(C) French style (D) None of these

984. In preserve the portion of fruit juice is:
(A) 25% (B) 35% (C) 45% (D) 55%

985. Generally seedless watermelon is:
(A) Diploid (B) Triploid
(C) Tetraploid (D) Hexaploid

986. Contour system of orchard planting is generally followed in:
(A) Punjab (B) Hills
(C) Saline soils (D) UP

987. Which fruit is known as "Miracle fruit of China"?
(A) Kiwi fruit (B) Mango
(C) Litchi (D) Loquat

988. "Hillocks are made to create a natural mountaing scenery" are the characteristic feature of (informal style)
(A) Persian garden
(B) Mughal garden
(C) Japanese garden
(D) Rose garden of Ludhiana

989. Garden lanterns, garden pagoda, garden bridges, tea garden and flat garden are the unique feature of:
(A) Japanese garden
(B) Mughal garden
(C) English garden
(D) Rose garden

990. Running water, high protecting wall, terminal budding and Baradari are the important features of:
(A) Persian garden (B) Mughal garden
(C) English garden (D) Rose garden of Ludhiana

991. Art of training plants into different shapes of birds, domes, animals, umbrellas etc is known as:
(A) Topiary (B) Edges
(C) Standard Shrubs (D) Baradari

ANSWERSHEET: 14.HORTICULTURE

1. (A); 2. (D); 3. (A); 4. (D); 5. (A); 6. (B); 7. (C); 8. (C); 9. (A); 10. (B); 11. (A); 12. (A); 13. (D) ; 14. (D); 15. (A); 16. (A); 17. (C); 18. (B); 19. (D); 20. (C); 21. (D); 22. (D); 23. (A); 24. (A); 25. (D) ; 26. (D); 27. (A); 28. (A); 29. (C); 30. (A); 31. (D); 32. (D); 33. (B); 34. (D); 35. (B); 36. (C); 37. (C); 38. (B) ;39. (C); 40. (B); 41. (B); 42. (A); 43. (A); 44. (D); 45. (D); 46. (C); 47. (C); 48. (D); 49. (C); 50. (C); 51. (C) ; 52. (C); 53. (D); 54. (D); 55. (D); 56. (C); 57. (B); 58. (D); 59. (C); 60. (B); 61. (A); 62. (C); 63. (C); 64. (A) ;65. (C); 66. (A); 67. (D); 68. (D); 69. (B); 70. (C); 71. (C); 72. (A); 73. (D); 74. (A); 75. (A); 76. (A); 77. (A); 78. (B) ;79. (A); 80. (A); 81. (B); 82. (A); 83. (D); 84. (D); 85. (B); 86. (B); 87. (A); 88. (D); 89. (D); 90. (B); 91. (C) ;92. (A); 93. (A); 94. (B); 95. (C); 96. (B); 97. (C); 98. (A); 99. (D); 100. (A); 101. (D); 102. (D); 103. (D); 104. (D) ; 105. (A); 106. (D); 107. (C); 108. (D); 109. (D); 110. (D); 111. (A); 112. (C); 113. (A); 114. (C); 115. (A); 116. (A); 117. (C) ; 118. (D); 119. (A); 120. (B); 121. (A); 122. (B); 123. (A); 124. (A); 125. (B); 126. (B); 127. (C); 128. (C); 129. (C); 130. (A); 131. (C); 132. (A); 133. (B); 134. (A); 135. (C); 136. (A); 137. (C); 138. (C); 139. (B); 140. (C); 141. (C); 142. (A); 143. (B) ; 144. (B); 145. (D); 146. (D); 147. (D); 148. (D); 149. (D); 150. (D); 151. (D); 152. (D); 153. (A); 154. (D); 155. (D); 156. (D) ; 157. (D); 158. (D); 159. (D); 160. (D); 161. (D); 162. (D); 163. (D); 164. (D); 165. (D); 166. (A); 167. (B); 168. (D); 169. (D); 170. (D); 171. (D); 172. (D); 173. (D); 174. (D); 175. (B); 176. (B); 177. (D); 178. (D); 179. (D); 180. (D); 181. (D); 182. (C) ; 183. (C); 184. (D); 185. (D); 186. (A); 187. (B); 188. (C); 189. (B); 190. (C); 191. (D); 192. (A); 193. (D); 194. (A); 195. (D) ; 196. (A); 197. (A); 198. (D); 199. (B); 200. (A); 201. (D); 202. (C); 203. (C); 204. (C); 205. (A); 206. (D); 207. (C); 208. (A) ; 209. (A); 210. (A); 211. (C); 212. (D); 213. (A); 214. (A); 215. (B); 216. (B); 217. (D); 218. (D); 219. (D); 220. (A); 221. (A) ; 222. (B); 223. (C); 224. (B); 225. (A); 226. (C); 227. (D); 228. (C); 229. (B); 230. (A); 231. (C); 232. (D); 233. (D); 234. (B) ; 235. (A); 236. (A); 237. (A); 238. (C); 239. (D); 240. (C); 241. (C); 242. (C); 243. (A); 244. (C); 245. (A); 246. (B); 247. (A); 248. (B); 249. (C); 250. (B); 251. (A); 252. (A); 253. (C); 254. (A); 255. (C); 256. (B); 257. (A); 258. (B); 259. (C); 260. (D) ; 261. (A); 262. (C); 263. (B); 264. (B); 265. (D); 266. (C) ; 267. (A);

268. (D); 269. (A); 270. (B); 271. (A); 272. (A); 273. (D) ; 274. (A); 275. (A); 276. (D); 277. (B); 278. (D); 279. (B); 280. (D); 281. (D); 282. (D); 283. (D); 284. (A); 285. (C); 286. (C) ; 287. (B); 288. (B); 289. (B); 290. (C); 291. (A); 292. (A); 293. (B); 294. (C); 295. (D); 296. (D); 297. (D); 298. (A); 299. (C) ; 300. (A); 301. (A); 302. (C); 303. (A); 304. (A); 305. (A); 306. (C); 307. (B); 308. (D); 309. (A); 310. (A); 311. (A); 312. (B) ; 313. (A); 314. (D); 315. (D); 316. (D); 317. (B); 318. (C); 319. (B); 320. (A); 321. (B); 322. (C); 323. (A); 324. (B); 325. (A) ; 326. (B); 327. (C); 328. (C); 329. (B); 330. (A); 331. (A); 332. (B); 333. (A); 334. (C); 335. (B); 336. (A); 337. (A); 338. (D); 339. (B); 340. (C); 341. (C); 342. (D); 343. (A); 344. (B); 345. (C); 346. (D); 347. (C); 348. (B); 349. (D); 350. (B); 351. (B); 352. (C); 353. (C); 354. (C); 355. (A); 356. (B); 357. (B) ;358. (D); 359. (A); 360. (C); 361. (A); 362. (B); 363. (A); 364. (A) ; 365. (C); 366. (D); 367. (A); 368. (C);369. (D); 370. (C); 371. (A); 372. (D); 373. (C); 374. (D); 375. (B); 376. (C); 377. (D) ; 378. (D); 379. (D); 380. (D); 381. (D); 382. (B); 383. (A); 384. (A); 385. (A); 386. (B); 387. (D); 388. (C); 389. (A); 390. (D); 391. (B); 392. (A); 393. (D); 394. (C); 395. (A); 396. (A); 397. (D); 398. (C); 399. (A); 400. (B); 401. (C); 402. (C); 403. (A) ; 404. (D); 405. (B); 406. (B); 407. (B); 408. (A); 409. (D); 410. (A); 411. (C); 412. (A); 413. (B); 414. (A); 415. (B); 416. (D); 417. (D); 418. (D); 419. (A); 420. (C); 421. (A); 422. (C); 423. (B); 424. (A); 425. (A); 426. (A); 427. (A); 428. (A); 429. (A); 430. (A); 431. (A); 432. (C); 433. (A); 434. (A); 435. (B); 436. (D); 437. (D); 438. (C); 439. (D); 440. (B); 441. (A); 442. (D) ; 443. (A); 444. (D); 445. (A); 446. (A); 447. (A); 448. (D); 449. (C); 450. (C); 451. (C); 452. (B); 453. (A); 454. (A); 455. (C); 456. (A) ; 457. (B); 458. (C); 459. (D); 460. (A); 461. (A); 462. (D); 463. (A); 464. (B); 465. (C); 466. (D); 467. (B); 468. (A); 469. (A) ; 470. (B); 471. (A); 472. (D); 473. (C); 474. (B); 475. (A); 476. (C); 477. (B); 478. (C); 479. (A); 480. (C); 481. (A); 482. (C); 483. (A); 484. (B); 485. (A); 486. (A); 487. (B); 488. (D); 489. (A); 490. (C); 491. (C); 492. (B); 493. (A); 494. (A); 495. (A) ; 496. (C); 497. (A); 498. (B); 499. (A); 500. (A); 501. (A); 502. (A); 503. (A); 504. (A); 505. (A); 506. (A); 507. (A); 508. (B); 509. (D); 510. (D); 511. (C); 512. (D); 513. (B); 514. (C); 515. (C); 516. (C); 517. (C); 518. (B); 519. (A); 520. (D); 521. (D) ; 522. (B); 523. (A); 524. (C); 525. (B); 526. (A); 527. (B); 528. (C); 529. (D); 530. (A); 531. (D); 532. (B); 533. (B); 534. (C) ; 535. (B); 536. (C); 537. (A); 538. (D); 539. (A); 540. (A); 541. (D); 542. (A); 543. (C); 544. (C); 545. (A); 546. (B); 547. (B) ; 548. (D); 549. (D); 550. (C); 551. (C); 552. (B); 553. (C); 554. (A); 555. (A); 556. (B); 557. (B); 558. (C); 559. (A); 560. (A); 561. (A); 562. (C); 563. (A); 564. (C); 565. (C); 566. (A); 567. (B); 568. (C); 569. (C); 570. (B); 571. (A); 572. (D); 573. (C) ; 574. (A); 575. (C); 576. (B); 577. (B); 578. (C); 579. (B); 580. (D); 581. (A); 582. (C); 583. (B); 584. (C); 585. (B); 586. (C) ; 587. (A); 588. (B); 589. (A); 590. (A); 591. (A); 592. (C); 593. (B); 594. (C); 595. (C); 596. (C); 597. (A); 598. (A); 599. (B) ; 600. (A); 601. (B); 602. (A); 603. (B); 604. (B); 605. (B); 606. (B); 607. (C); 608. (A); 609. (B); 610. (A); 611. (D); 612. (C); 613. (C); 614. (D); 615. (B); 616. (C); 617. (B); 618. (D); 619. (C); 620. (B); 621. (B); 622. (B); 623. (B); 624. (B); 625. (D) ; 626. (D); 627. (B); 628. (B); 629. (A); 630. (B); 631. (B); 632. (C); 633. (A); 634. (B); 635. (C); 636. (B); 637. (A); 638. (B); 639. (A); 640. (D); 641. (D); 642. (A); 643. (C); 644. (B); 645. (C); 646. (D); 647. (D); 648. (D); 649. (A); 650. (A); 651. (B) ; 652. (B); 653. (B); 654. (A); 655. (A); 656. (B); 657. (C); 658. (D); 659. (B); 660. (C); 661. (A); 662. (C); 663. (D); 664. (C); 665. (D); 666. (A); 667. (C); 668. (D); 669. (A); 670. (D); 671. (D); 672. (B); 673. (D); 674. (A); 675. (B); 676. (D); 677. (B); 678. (C); 679. (A); 680. (B); 681. (A); 682. (D); 683. (B); 684. (B); 685. (C); 686. (C); 687. (D); 688. (C); 689. (A); 690. (A) ; 691. (A); 692. (A); 693. (D); 694. (C); 695. (C); 696. (A); 697. (D); 698. (B); 699. (B); 700. (D); 701. (B); 702. (D); 703. (B) ; 704. (D); 705. (C); 706. (A); 707. (B); 708. (C); 709. (D); 710. (D); 711. (C); 712. (D); 713. (D); 714. (D); 715. (C); 716. (B); 717. (D); 718. (B); 719. (D); 720. (C); 721. (B); 722. (C); 723. (C); 724. (A); 725. (A); 726. (B); 727. (B); 728. (A); 729. (D) ; 730. (C); 731. (A); 732. (D); 733. (C); 734. (B); 735. (B);

736. (A); 737. (A); 738. (A); 739. (A); 740. (B); 741. (C); 742. (D) ; 743. (D); 744. (D); 745. (B); 746. (D); 747. (D); 748. (B); 749. (A); 750. (B); 751. (A); 752. (A); 753. (B); 754. (C); 755. (C); 756. (A); 757. (A); 758. (C); 759. (B); 760. (A); 761. (C); 762. (A); 763. (A); 764. (A); 765. (C); 766. (B); 767. (B); 768. (B) ; 769. (D); 770. (C); 771. (A); 772. (C); 773. (C); 774. (B); 775. (A); 776. (B); 777. (B); 778. (C); 779. (B); 780. (C); 781. (B); 782. (D); 783. (C); 784. (A); 785. (B); 786. (A); 787. (C); 788. (D); 789. (C); 790. (B); 791. (A); 792. (A); 793. (B); 794. (A) ; 795. (D); 796. (C); 797. (D); 798. (B); 799. (A); 800. (C); 801. (A); 802. (B); 803. (A); 804. (A); 805. (C); 806. (D); 807. (A); 808. (D); 809. (D); 810. (B); 811. (A); 812. (A); 813. (A); 814. (D); 815. (B); 816. (A); 817. (B); 818. (D); 819. (D); 820. (D); 821. (D); 822. (D); 823. (A); 824. (B); 825. (D); 826. (A); 827. (B); 828. (C); 829. (D); 830. (B); 831. (B); 832. (D); 833. (B); 834. (A); 835. (C); 836. (A); 837. (C); 838. (B); 839. (B); 840. (C); 841. (B); 842. (B); 843. (B); 844. (B); 845. (D); 846. (B); 847. (C); 848. (C); 849. (D); 850. (B); 851. (A); 852. (C); 853. (B); 854. (B); 855. (C); 856. (C); 857. (B); 858. (A); 859. (A); 860. (D); 861. (A); 862. (C); 863. (C); 864. (A); 865. (C); 866. (C); 867. (C);

868. (A); 869. (B); 870. (D); 871. (B); 872. (A); 873. (A); 874. (D); 875. (B); 876. (B); 877. (A); 878. (C); 879. (D); 880. (C); 881. (D); 882. (A); 883. (D); 884. (D); 885. (C); 886. (A); 887. (C); 888. (C); 889. (C); 890. (B); 891. (C); 892. (A); 893. (C); 894. (C); 895. (D); 896. (C); 897. (C); 898. (A); 899. (D); 900. (B); 901. (D); 902. (C); 903. (B); 904. (A); 905. (C); 906. (B); 907. (C); 908. (A); 909. (C); 910. (C); 911. (A); 912. (B); 913. (C);914. (A); 915. (C); 916. (A); 917. (C); 918. (A); 919. (A); 920. (B); 921. (B); 922. (C); 923. (A); 924. (B); 925. (B); 926. (A); 927. (A); 928. (B); 929. (C); 930. (C); 931. (B); 932. (D); 933. (D); 934. (A); 935. (C); 936. (C); 937. (B); 938. (B); 939. (A); 940. (B); 941. (D); 942. (A); 943. (A); 944. (A); 945. (B); 946. (B); 947. (B); 948. (A); 949. (C); 950. (B); 951. (A); 952. (A); 953. (B); 954. (A); 955. (C); 956. (A); 957. (C); 958. (A); 959. (D); 960. (A); 961. (C); 962. (B); 963. (D); 964. (B); 965. (B); 966. (D); 967. (A); 968. (D); 969. (A); 970. (A); 971. (A); 972. (C); 973. (B); 974. (C); 975. (D); 976. (A); 977. (B); 978. (A); 979. (B); 980. (A); 981. (A); 982. (B); 983. (B); 984. (C); 985. (B); 986. (A); 987. (A); 988. (A); 989. (A); 990. (B); 991. (A).

Chapter 15

AGRICULTURAL EXTENSION

1. The word Extension is derived from:
(A) Latin (B) Greek
(C) Both (A) and (B) (D) None of these

2. The word "tentio" in extension is derived from:
(A) Greek word (B) Latin word
(C) English word (D) None of these

3. Extension literary means:
(A) Stretching out (B) Talking with
(C) Deciding for (D) Thinking about

4. The latin roots for the word 'extension':
(A) *ex* and *tensio* (B) *ex* and *tension*
(C) *ex* and *luceo* (D) *ex* and *termo*

5. Extension is a/an:
(A) Informal education
(B) Out of school systems of education
(C) Both (A) and (B)
(D) Formal education

6. Extension is:
(A) Education for all village people
(B) Teaching through learning by doing and seeing is believing
(C) Two way channel
(D) All of these

7. The purpose of introducing agricultural universities in India is to improve the standard of agriculture:
(A) Extension (B) Teaching
(C) Research (D) All of these

8. Which extension method is evolved in result demonstration, posters, movies, slides and film strips?
(A) Spoken (B) Visual
(C) Written (D) None of these

9. An extension worker, while extending, scientific knowledge to the farmers, starts with:
(A) Lectures
(B) Practical demonstration
(C) Distribution of leaflet
(D) All of these

10. There are how many degrees of evaluation in Extension:
(A) 4 (B) 5 (C) 3 (D) 2

11. The extension activities was started first time in:
(A) USA (B) Russia
(C) Germany (D) India

12. Extension Education Institute (EETs) impart training in the area of:
(A) Subject matter
(B) Training & communication methodology
(C) Subject matter, training and communication methodology
(D) None of these

13. The basic guide, philosopher and friend in rural development is:
(A) BDO (B) VDO
(C) Pradhan (D) Pramukh

14. The term 'University Extension' was first used in:
(A) 1866 (B) 1972 (C) 1901 (D) 1921

15. The person considered as father of extension:
(A) James Stewatt (B) Seeman A Knapp
(C) Edgar Dale (D) Paul Leagans

16. The term 'Extension Education' was first coined in:
(A) USA (B) UK
(C) Netherlands (D) India

17. The person considered as father of extension in India:
(A) Adivi Reddy (B) O P Dahama
(C) K N Singh (D) J B Chitambar

18. 'Vulgarisation' in French is equivalent to the word:
(A) Social taboos (B) Education
(C) Extension (D) Social change

19. Extension when aims at empowering or uplifting the poor:
(A) Emancipatory extension (B) Formative extension
(C) Informative extension (D) Persuasive extension

20. Extension as a science is classified under:
(A) Hard Science (B) Soft Science
(C) Pure Science (D) Physical Science

21. Extension is not:
(A) Voluntary (B) Family centered
(C) Conclusion oriented (D) Flexible

22. Extension can be considered as a:
(A) Service (B) Profession
(C) Discipline (D) All of these

23. Extension basically is a two way process between:
(A) Farmer and researcher
(B) Researcher and planner
(C) Teacher and planner
(D) Researcher and teacher

24. Extension that emphasize on human resource development:
(A) Emancipatory extension (B) Formative extension
(C) Informative extension (D) Persuasive extension

25. Extension when used as a policy instrument for achieving societal objective:
(A) Emancipatory extension (B) Formative extension
(C) Informative extension (D) Persuasive extension

26. Extension that moves beyond the traditional crop production technologies to include all land based activities:
(A) Universal extension (B) Cyber-extension
(C) Emancipatory extension (D) Broad based extension

27. Extension as an instrument for helping people to make well-considered choices:
(A) Emancipatory extension (B) Formative extension
(C) Informative extension (D) Persuasive extension

28. The instrument used by extension to bring change:
(A) Communication (B) Teaching
(C) Learning (D) Adoption

29. Extension can be effective only through:
(A) Voluntary change (B) Democratic change
(C) Autocratic change (D) Persuasive change

30. The main mode of extension in the farmer first model:
(A) Farmer to farmer (B) Agent to farmer
(C) Researcher to agent (D) Researcher to farmer

31. The satellite model concept of extension was formulated by:
(A) M S Swaminathan (B) G Perumal
(C) Curt Lewin (D) Anil Gupta

32. The fundamental objective of extension work is:
(A) Development of farmer
(B) Development of individual
(C) Development of culture
(D) Development of society

33. The ultimate objective of extension is:
(A) Increase productivity
(B) Increase income
(C) Rural development
(D) Realize ones fullest potential

34. Agricultural extension deals with:
(A) Farmers
(B) Change agents
(C) Farmers and change agents
(D) None of these

35. An extension worker help farmers to attain desirable changes in his:
(A) Temperament (B) Behaviour
(C) Health (D) Wealth

36. Leadership development in extension is:
(A) A means (B) An end
(C) Both means and end (D) None of these

37. Most useful leader in agricultural extension is:
(A) Voluntary leader (B) Bureaucratic leader
(C) Autocratic leader (D) Democratic leader

38. An extension worker is a:
(A) Professional leader (B) Lay leader
(C) Local leader (D) Voluntary leader

39. An extension agent is considered to be a:
(A) Personal cosmopolite
(B) Impersonal cosmopolite
(C) Personal localite
(D) Impersonal localite

40. The full form of A.T.M.A. is:
(A) Agriculture Technology Management Association
(B) Agriculture Technology Management Agency
(C) Agriculture Technology Mission Agency
(D) None of above

41. T & V system is a good example of:
(A) Extension approach
(B) Training approach
(C) Co-operative self help approach
(D) Integrated development approach

42. To get the intelligence quotient (I.Q.) of the individual, the formula will be:

(A) $I.Q. = \frac{\text{Chronological age}}{\text{Mental age}} \times 100$

(B) $I.Q. = \frac{\text{Mental age}}{\text{Chronological age}} \times 100$

(C) $I.Q. = \frac{\text{Chronological age}}{\text{Mental age}}$

(D) $I.Q. = \frac{\text{Mental age}}{\text{Chronological age}}$

43. Which one of the step is not included in programme planning?
(A) Collect facts (B) Analyze situation
(C) Identify problems (D) Execute plan

44. ________ is the primary institution of society:
(A) Family (B) Village
(C) Individual (D) Block

45. Which of the following extension approaches has got the highest intensity of influence?
(A) Mass approach (B) Group approach
(C) Community approach (D) Individual approach

46. Which of the following scheme is exclusively meant for self – employment for rural youths?
(A) NREP (B) TRYSEM
(C) IRDP (D) DPAP

47. First Agricultural University in India was established at:
(A) New Delhi (B) Hyderabad
(C) Bikaner (D) Pantnagar

48. Which of the following is not the form of social interaction?
(A) Competition (B) Cooperation
(C) Coordination (D) Conflict

49. Which of the following pair is not matched:
(A) Communication : Berlo
(B) Rural weakers : Small farmers
(C) Mass media : Demonstration
(D) Helping poorest of the poor first : IRD programme

50. General meeting are broadly, the meeting of:
(A) Heterogeneous participation
(B) Homogeneous participation
(C) Community participation
(D) Society participation

51. Campaign is an:
(A) Intensive teaching (B) Extensive teaching
(C) Invasive teaching (D) None of these

52. Programme planning is a procedure of:
(A) Working by people (B) Working for the people
(C) Working with the people (D) None of these

53. A series of still pictures on one roll is called:
(A) Channel (B) Film strip
(C) Mock-up (D) Black board

54. A bulletin should contain:
(A) 4-12 pages (B) 12-24 pages
(C) 24-48 pages (D) None

55. In India, slightly less than half the villagers are inhabited ———— persons:
(A) 500 (B) 800
(C) 1000 (D) 1500

56. About ¾ Indian villages have a human population of less than:
(A) 600 (B) 700
(C) 800 (D) 1000

57. According to CACP, sugarcane prices announced is:
(A) MSP (B) Procurement prices
(C) Statuary (D) Flat prices

58. B.R.Mehta team stimulated an active consideration of through democratic bodies in India:
(A) Decentralization (B) Stabilization
(C) Centralization (D) Mobilization

59. Most people retain ———— % of what they read:
(A) 5-10 (B) 10-15
(C) 20-25 (D) 30-35

60. Most people kept in mind ————% of what they seen:
(A) 10-15 (B) 20-25
(C) 30-35 (D) 50-55

61. Up to ———— % of what is taught is kept in mind by majority of people, if they participate actively, and all the senses (read, hear, and seen) are involved:
(A) 50 (B) 65
(C) 80 (D) 90

62. Symposium is a short series of lectures, usually by speaker:
(A) 1-2 (B) 2-5
(C) 5-7 (D) 7-10

63. Phillips 66 format of hurdle system is related with:
(A) Panel (B) Form
(C) Buzz section (D) Work shop

64. Small group interaction designed to encourage the free introduction of ideas on an unrestricted basis without any limitations to feasibility is known as:
(A) Conference (B) Seminar
(C) Buzz section (D) Brain storming

65. ———— essentially follow the seminar method and the focus is on any particular subject or problem:
(A) Work shop (B) Syndicate studies
(C) Institute (D) None of these

66. News paper articles are included in which method of extension:
(A) Individual (B) Group
(C) Mass (D) None of these

67. "Stay at home" type of people can be contacted by which method of extension:
(A) Individual (B) Group
(C) Mass (D) None of these

68. ———— is essentially a recognizable imitation or replica of the original, whether workable or not:
(A) Model (B) Specimens
(C) Mock-up (D) Objects

69. A working model is known as:
(A) Objects (B) Model
(C) Mock-up (D) Poster

70. "The communication process in Rural development" is written by:
(A) Leagans (B) Berlo
(C) A.A. Reddy (D) Ray

71. ———— are expressions of the ends towards which our efforts are directed:
(A) Goal (B) Objectives
(C) Needs (D) Message

72. ———— is defined as the distance in any given direction one expects to go during a given period of time:
(A) Goal (B) Objectives
(C) Needs (D) Message

73. The ultimate objective of extension work is the full development of ————:
(A) Individual (B) Group
(C) Mass (D) Village

74. The gap between the situation and objective is the area of ————:
(A) Goal (B) Needs
(C) Interest (D) None

75. In good administration the offices of a ———— are arranged in a pyramid of authority and responsibility:
(A) Autocracy (B) Monarchy
(C) Bureaucracy (D) Democracy

76. ————believes that if you leave workers alone the work will be done:
(A) Autocratic leader (B) Democratic leader
(C) Laissez-faire leader (D) None of these

77. ———— is one who has received specialized training in the field in which he work as full time and paid for his work:
(A) Operational leader
(B) Nominal leader
(C) Professional leader
(D) Prominent talent leader

78. Best method for selection of leader is
(A) Sociometry (B) Election
(C) Discussion method (D) Key informants

79. __________ are those things which person ought or ought not to do in society
(A) Tabbo (B) Mores
(C) Folkways (D) Ritual

80. Which one of the following is a farmer agency?
(A) FCI (B) IFFCO
(C) Agro service centers (D) Shree Ram fertilizers

81. The basic unit of development of IRDP is:
(A) Community Development Programme
(B) Village
(C) District
(D) Family

82. Krishi Vigyan Kendra (KVKs) can only be under control of:
(A) Agricultural University
(B) Voluntary organisation
(C) Govt. and Voluntary organization
(D) Agricultural University, ICAR and Voluntary Organisation

83. "The Lab to Land" programme also included demonstration but they were different from early demonstrations in terms of:
(A) Target groups
(B) Type of demonstration
(C) Size of demonstration
(D) Place of demonstration

84. The first Krishi Vigyan Kendra in India was established at:
(A) Bihar (B) Rajasthan
(C) Punjab (D) Pondichery

85. Which of the following is not a group approach device:
(A) Symposium (B) Film
(C) Demonstration (D) Workshop

86. The subject extension is a:
(A) Means (B) Discipline
(C) End (D) Branch of mean

87. The extension system used to provide precise and reliable scientific information in simple language about a simple practice is called:
(A) Leaflet (B) Newspaper
(C) Literature (D) Circular letter

88. The benefit of first hand knowledge good will and creating confidence can be developed by:
(A) Telephone
(B) Visiting home and farms
(C) Organising village fairs
(D) None of these

89. IVLP stands for:
(A) Indian Village Linkage Programme
(B) Indian Village Linkage Plan
(C) Institution Village Linkage Programme
(D) Integrated Village Lab to Land Programme

90. Audio-visual aids is:
(A) Radio (B) Wireless
(C) Telephone (D) Television

91. Which of the following country was first use the term 'extension' to imply extension knowledge from agriculture research institution to the farmers through the process of informal education?
(A) U.S.A. (B) U.S.S.R.
(C) India (D) Japan

92. Which of the following will give the clear idea of proportion?
(A) Model (B) Pie chart
(C) Radio (D) TV

93. For providing the precise and reliable scientific information in simple language the extension system used is:
(A) Charts (B) Demonstration
(C) Leaflets (D) None of these

94. CONA stands for:
(A) Comparative Oriented Need Assessment
(B) Community Oriented Need Assessment
(C) Commercial Oriented Natural Assignment
(D) None of these

95. Which of the following visual aids are commonly used by agricultural extension workers in India?
(A) Posters, photographs, flash card and fenal graphs
(B) Slides, film-strips, models and puppets
(C) Bulletin boards, cultural programmes and black boards
(D) All of these

96. Which of the following is not true?
(A) Programme planning is a spontaneous process
(B) Programme planning is an educational process
(C) Programme planning is a continuous process
(D) Programme planning is a co-ordinating process

97. Which of the following is a function of the RRB's?
(A) To develop rural economy
(B) To provide credit for agriculture and allied activities
(C) To encourage rural industries
(D) All of these

98. Evaluation done by the agency is known as:
(A) Self evaluation (B) Internal evaluation
(C) External evaluation (D) None of these

99. The all – round development of Indian rural people can be achieved by means of:
(A) Extension education (B) Adult education
(C) More money distribution (D) All of these

100. Integrated Rural Development Programme (IRDP) has been conceived:
(A) Essentially to development anti hoarding programme
(B) Essentially to development adult education facilities in the rural areas
(C) Essentially as an anti-poverty programme
(D) Essentially to develop fisheries in the area

101. Coding of the data is done to assign:
(A) Numerical to answer (B) Symbol to answer
(C) Both (A) and (B) (D) None of these

102. Slide projector gives the image:
(A) Direct (B) Indirect
(C) Both (a) and (b) (D) Reflected

103. What is the specific objectives of the GATT?
(A) Globilization
(B) Increase world production
(C) Complete use of resources
(D) All of these

104. High yielding Variety programme has given emphasis on:
(A) Land reform
(B) Package and practice with dwarf varieties
(C) Increase production through dry farming
(D) None of these

105. Which of the following is the first step in developing of programme?
(A) Selection of site
(B) Collection of information
(C) Both (A) and (B)
(D) Find felt need of people

106. Kisan Mitra is an employee of:
(A) Central Govt (B) State Govt.
(C) Corporation (D) None of these

107. Unit of the IRDP:
(A) Village (B) Block
(C) District (D) State

108. Puppet show in village as a method for transfer of technology is an example of:
(A) Mass media (B) Group media
(C) Individual media (D) Tribal media

109. The main purpose of T and V programme is to:
(A) Identify research problems
(B) Bring about co-ordination in the administration of agricultural extension
(C) Update the technical knowledge and skills of agricultural extension workers
(D) Help farmers in the adoption of latest innovations in agriculture

110. Consider the following statements:
In interpersonal channels of communication
(1) Message flow tends to be one way
(2) Quantum of feedback available is very high
(3) Communication is face to face
(4) Speed of reaching a large audience is relatively slow of these statements
(A) 1, 2, 3 and 4 are correct (B) 1 and 4 are correct
(C) 2, 3 and 4 are correct (D) 2 and 3 are correct

111. Read the following statements and mark your answer according to the code given below:
(1) Newspaper is a popular medium of communication in rural areas.
(2) Newspaper is written form of communication
(3) Film show is audio-visual form of communication
(4) Film show is popular medium of communication in rural areas.
Answer code
(A) 1 and 2 are correct (B) 2 and 3 are correct
(C) 3 and 4 are correct (D) 1 and 4 are correct

112. Which one of the following is not correctly matched?

List-I	List-II
(A) Mass	Exhibition
(B) Individual	Farm visit
(C) Group	Mela
(D) Mass	Newspaper

113. Which of the following is the group contact teaching method?
(A) Farm and home visit
(B) Bulletins
(C) Result demonstration
(D) Method demonstration

114. Match list – I and list – II and select the correct answer using the code given below:

List – I (Aids)	List-II (Method of contact)
(A) Model	(1) Group contact
(B) Slide	(2) Mass contact
(C) Telephone call	(3) Personal contact
(D) Poster	(4) Mass contact

Code:
(A) A-2, B-3, C-4, D-1 (B) A-1, B-2, C-3, D-4
(C) A-3, B-2, C-2, D-1 (D) A-3, B-2, C-4, D-1

115. Read the following statements and mark your answer according to the code given below:
(1) C.D. is a programme (2) C.D. is a cultural event
(3) C.D. is a social event (4) C.D. is a project
Answer code:
(A) 1 and 2 are correct (B) 1 and 4 are correct
(C) 2 and 3 are correct (D) 2 and 4 are correct

116. The benefit of first hand knowledge, good will and create confidence can be developed by:
(A) Visiting homes and farm (B) Telephone
(C) Village fair (D) All of the above

117. To create interest in a wider range of people and to influence people to adopt better practices the extension method generally used is:
(A) Exhibition (B) Tours
(C) Fair (D) Competition

118. The extension system used to provide precise and reliable scientific information in simple language about a simple practice is called:
(A) Leaflet (B) Literature
(C) Newspaper (D) Circular letter

119. Extension of recent agricultural knowledge to rural peoples through trained persons of the same locality is called the principle of:
(A) Panchayati Raj (B) Centralized power
(C) Local leadership (D) None of the above

120. Effective extension service is a process of:
(A) Helping the people which helps themselves
(B) Profiting the local leaders
(C) Helping the B.D.O.
(D) All of the above

121. Krishi Gyan Kendra's are functioning through the financial assistance of:
(A) State Govt. (B) UNICEF
(C) ICAR (D) IARI

122. While working as an extension specialist in a village one finds that the package of practice are not being followed despite the best communication methods one had employed for the purpose one finds the following reasons for non-adaptivity:
(1) 60% are laggards
(2) 20% are innovative but orthodox
(3) 20% are old and orthodox
The best remedial measure would be to:
(A) Adopt 1 and 3
(B) Adopt 2 and above
(C) Adopt 3
(D) Adopt 2 and make then team leaders

123. The concept of demonstration was given by:
(A) Lewis Jones (B) O.P. Dahama
(C) Rudramurthy (D) Dr. Seeman A. Knapp

124. The functions of pre-extension in the development block was:
(A) To survey block area
(B) To prepare ground for intensive stage
(C) To conduct agricultural demonstrations
(D) All of these

125. Block samiti is guided by:
(A) B.D.O. and his extension officers
(B) V.L.W.
(C) Zila Parishad
(D) Panchayat Samiti

126. The development approach of National Extension Service Block was based mainly on:
(A) Self help (B) Govt. help
(C) State Govt. help (D) Help by village Panchayat

127. Which of the following statement is not correct?
(A) Village Panchayat is a non-official body
(B) Village Panchayat is headed by V.L.W.
(C) The head of a block is B.D.O.
(D) All of these

128. The responsibility of development work at village level is of:
(A) Collector (B) B.D.O.
(C) V.L.W. (D) Gram Pradhan

129. The number of V.L.W's is an I.A.D.P. block is:
(A) 5 (B) 10
(C) 15 (D) 20

130. National demonstrations are first line demonstrations which is conducted on the farmer's field by:
(A) Extension workers (B) V.L.W.
(C) Research Scientists (D) B.D.O.

131. To create interest in a wider range of people to adopt better practices, the extension method generally used is:
(A) Field trials (B) Exhibitions
(C) Competition (D) Fair

132. The elements of teaching learning process are:
(A) Teaching subject matter and learner
(B) Teaching equipments and physical facilities
(C) Both (A) and (B)
(D) None of these

133. "Cone of experience" was devised by:
(A) Rogers (B) O.P. Bhatnagar
(C) A.A. Reddy (D) Edger Dale

134. The programme for transfer of technology started by ICAR at the golden jubilee year occasion was:
(A) Operational Research Project
(B) Lab to Land Programme
(C) National Demonstration
(D) I.R.D.P.

135. Extension workers are:
(A) Professional leaders (B) Autocratic leader
(C) Lay leaders (D) None of these

136. Match the following

List-I (Aids)	List-II (Method of contact)
(a) Model	(1) Group contact
(b) Slide	(2) Mass contact
(c) Telephone call	(3) Personal contact
(d) Poster	(4) Mass contact

Code	a	b	c	d
(A)	1	2	3	4
(B)	2	3	4	1
(C)	3	4	2	1
(D)	4	1	3	2

137.

List-I (Approach)	List-II (Technique)
(a) Mass	(1) Exhibition
(b) Individual	(2) Farm visit
(c) Group	(3) Model
(d) Mass	(4) Newspaper

Code	a	b	c	d
(A)	2	3	4	1
(B)	1	2	3	4
(C)	4	3	2	1
(D)	2	1	3	4

138. VLW (Multi Purpose Worker) concept emerged in which project?
(A) T & V Project (B) NEP
(C) KVK (D) Etawah Pilot Project

139. Which of the extension programmes in India advocated single line of command and strong research-extension linkages?
(A) KVKs (B) T&V
(C) NATP (D) ORP

140. SREP is:
(A) Scientific Research Extension Programmes
(B) Strategic Research Extension Plan
(C) Specific Research Extension Plan
(D) Special Resource and Enterprise Focus Programme

141. ATMA is a:
(A) Registered society at district level for decentralized planning
(B) Non-Govt. organization working very near to the poor and down trodden people
(C) Technical package of information dissemination
(D) None of these

142. Which of the following programmes is being implemented by SAUs/Institutes to strengthen institute-village linkages?
(A) ATIC (B) NATP
(C) IVLP (D) UPDASP

143. "Farmers Interest Group" concept in agricultural extension is:
(A) A group of farmers meant for cultural activities
(B) A group of farmers in the villages created to watch the interests of the farmers
(C) A group of farmers coming together voluntarily to take up some activities
(D) A group of farmers interested in employment generation for rural youths

144. The internal state which stimulate the individual to carry out certain activities is:
(A) Perception (B) Motivation
(C) Intuition (D) Learning

145. The motivational type most important for extension is:
(A) Extrinsic motivation (B) Intrinsic motivation
(C) Situational motivation (D) Classical motivation

146. "If I have to get that gold medal, I should work hard at studies" reflects:
(A) Determination (B) Intrinsic motivation
(C) Industriousness (D) Extrinsic motivation

147. Which of the following is not an approach to education:
(A) Conventional (B) Prevocational
(C) Progressive (D) Liberating

148. Effective teaching and learning is a matter of:
(A) Chance and luck
(B) Design and effect
(C) Material and non material
(D) Formal and informal

149. Education is shaped by:
(A) Society (B) Teacher
(C) Student (D) Knowledge

150. Cone of experience classify the instructional methods according to:
(A) Experiential correctness
(B) Difficulty in understanding
(C) Ease in use and application
(D) Level of abstraction in thinking

151. The cone of experience was developed by:
(A) Egder Dale (B) Berlo C.K.
(C) Kuldeep Nair (D) S C Parmar

152. The type of experience a farmer have when he saws wheat in lines under the supervision of extension worker:
(A) Symbolic experience (B) Iconic experience
(C) Supervisory experience (D) Enactive experience

153. When a person involved in iconic experiences, learning occurs through:
(A) Active participation (B) Observation
(C) Abstract thinking (D) Doing

154. Which among the following is an example of direct purposeful experience:
(A) Ploughing (B) Model of a plough
(C) Blow up of a plough (D) Diagram of a plough

155. The extension worker tells about the varieties characters to a farmer, the farmer learns by:
(A) Abstraction (B) Learning
(C) Intelligence (D) Questioning

156. Classification of extension methods into individual, group and mass is based on:
(A) Foam (B) Use
(C) Number of people contacted (D) Periodicity

157. To get first hand knowledge on a farm situation we use:
(A) Village fairs (B) Farm & home visit
(C) Telephone (D) Questionnaire

158. The ratio of 'takes' to 'acceptance' is the highest in:
(A) Farm and Home visit
(B) Method demonstration
(C) Result demonstration
(D) Campaign

159. Media forum is classified under:
(A) Individual contact method
(B) Group contact method
(C) Mass contact method
(D) All of these

160. When the farmer meets the extension worker at his work place:
(A) Farmer's call (B) Extension campaign
(C) Result demonstration (D) Office visit

161. The extension method that act as a safety valve in reducing tension:
(A) Group discussion
(B) Method demonstration
(C) Result demonstration
(D) Campaign

162. The best extension method suited to all situation:
(A) Demonstration (B) Campaign
(C) Farm publication (D) None of these

163. Method demonstration aims more at improving ones:
(A) Cognitive skills (B) Perceptual skills
(C) Motor skills (D) Analytical skills

164. Seeman A Knapp conducted result demonstration in:
(A) Cotton (B) Maize
(C) Barley (D) Wheat

165. A method which build the confidence of the researcher, extension worker and the farmer:
(A) Minikit trial (B) Result demonstration
(C) Method demonstration (D) All of these

166. A method of motivating the people to adopt a new practice by showing what has actually been achieved by applying the practice under field conditions:
(A) Campaign (B) Result demonstration
(C) Field day (D) Farm bulletin

167. Maintenance of record is absolutely necessary in case of:
(A) Method demonstration (B) Minikit trial
(C) Field day (D) Field trip

168. The method used when the farmer is not there in the field while the extension workers make a visit:
(A) Farm and home visit
(B) Method demonstration
(C) Result demonstration
(D) Flag method

169. The normal group boundaries get obliterated in:
(A) Method demonstration (B) Farm and home visit
(C) Office calls (D) Campaign

170. A body of general principles or laws of a field of knowledge is called:
(A) Television (B) Philosophy
(C) Mandate (D) Principle

171. The basic philosophy of extension is to teach people:
(A) How to do (B) What to do
(C) When to do (D) Where to do

172. A person, who believes in equal rights and opportunities for women and men:
(A) Communalist (B) Feminist
(C) Secularist (D) Patriachal

173. The fundamental unit of civilization according to Horton is:
(A) Home (B) Land
(C) Individual (D) Capital

174. The difference between what is and what ought to be:
(A) Need (B) Aim
(C) Goal (D) Objective

175. The condition which people after study decided needs change is:
(A) Problem (B) Programme
(C) Project (D) Solution

176. A blue print for action is:
(A) Plan (B) Programme
(C) Objective (D) Project

177. The distance in any direction one expects to go is:
(A) Aim (B) Goal
(C) Objective (D) Principle

178. Process or procedure of attaining an objective:
(A) Programme (B) Method
(C) Plan (D) Media

179. The level of objective operates at field extension work:
(A) Fundamental objective (B) Working objective
(C) General objective (D) None of these

180. Which of the following is not a level of objective:
(A) Basic objective
(B) Fundamental objective
(C) Working objective
(D) General objective

181. Extension work grew out of:
(A) Problem (B) Situation
(C) Condition (D) Participation

182. The need which are not realized as important are:
(A) Felt need (B) Unfelt need
(C) Ignored need (D) Unrealized need

183. Which of the following is the most effective method in identifying real rural needs:
(A) Election method (B) Sociometric method
(C) Workshop method (D) Selection method

184. Measuring performance against predetermined goals is called:
(A) Management (B) Evaluation
(C) Measurement (D) Supervision

185. The evaluation of any programme should be based on:
(A) Financial outlay
(B) Number of participation
(C) Objectives of the programme
(D) None of these

186. The element that take time, effort and resource to move from one event to another:
(A Event (B) Activity
(C) Critical path (D) Slack

187. The largest possible time to perform an activity assuming everything goes well is:
(A) Optimizing time (B) Pessimistic time
(C) Most likely time (D) Expected time

188. IVLP uses the concept of:
(A) Agro-system analysis
(B) Agro-ecosystem analysis
(C) Eco-system analysis
(D) Socio-ecosystem analysis

189. ATMA is an important component under:
(A) NARRS (B) NATP
(C) NAEP (D) KVK

190. ATMA operated at:
(A) Block level (B) District level
(C) State level (D) National level

191. The number of ATMA developed under NATP:
(A) 22 (B) 24
(C) 25 (D) 28

192. The ATMA under NATP is a:
(A) Statutory body (B) Registered society
(C) Cooperative society (D) Krishi Vigyan Kendra

193. The apex training institute at state level to give training support to ATMA:
(A) MANAGE (B) EEI
(C) CAPART (D) SAMETI

194. SREP is an important component in the functioning of:
(A) NARS (B) ATMA
(C) ATIC (D) KVK

195. Indian Council of Agricultural Research is a:
(A) Educational organization
(B) Research institution
(C) Registered cooperative
(D) Autonomous body

196. Extension teaching at the under graduate level was started by:
(A) Madras university (B) Meerut university
(C) Calcutta university (D) Mumbai university

197. The first line system of extension in India is controlled by:
(A) ICAR institutes/SAUs
(B) Ministry of Agriculture
(C) Non Government Organisations
(D) Private Extension Agencies

198. The demonstration in case of AICPND was done by:
(A) Farmer (B) Media personal
(C) Research scientist (D) Extension person

199. The relatively non-mobile among the TOT programmes of ICAR:
(A) LLP (B) ORP
(C) AICPND (D) KVK

200. The funding of KVK is done by:
(A) ICAR (B) SAU
(C) State government (D) Central government

201. Farm Science Centre is a synonym of:
(A) FTC (B) FDG
(C) KVK (D) Teleclub

202. The extension component of NATP:
(A) Discipline oriented research
(B) Organizational and management reforms
(C) Agro-ecosystem research
(D) Innovation in technology dissemination

203. The extension service in USA is called:
(A) National extension service
(B) US extension work
(C) Cooperative extension service
(D) Rural extension service

204. The actual physical product of source-encounter is:
(A) Receiver (B) Message
(C) Channel (D) Feedback

205. The three important dimensions of message include:
(A) Code, content, treatment
(B) Source, self, receiver
(C) Encode, decode, feedback
(D) Simplicity and clarity

206. The way in which message is handled before placing in the channel:
(A) Decoding (B) Treatment
(C) Distortion (D) Feedback

207. The link that connects the source with the receiver:
(A) Message (B) Communication effect
(C) Channel (D) Treatment

208. Television is primarily a:
(A) Media mix (B) Source
(C) Medium (D) Channel

209. Trustworthiness and competency are elements of:
(A) Credibility (B) Fidelity
(C) Empathy (D) None of these

210. Acceptance and rejection are the one of the kinds of:
(A) Failure of channel (B) Feedback
(C) Treatment of message (D) Empathetic feelings

211. The term class media refers to:
(A) Mass media (B) Group media
(C) Interpersonal media (D) Intra personal media

212. While communicating, the communicator can write facing the audience in:
(A) Black board (B) Flash card
(C) Slide (D) OHP

213. Which of the following statement about audio-visual aids is true:
(A) Replaces a teacher
(B) Decorates learning situation
(C) Emphasis non-verbal experience
(D) Includes text book

214. The term electronic chalkboard refers to:
(A) LCD projector (B) OHP projector
(C) VHS projector (D) Beta chrom projector

215. The film projector commonly used in extension:
(A) 8 mm (B) 16 mm
(C) 32 mm (D) 70 mm

216. The object for projection in a slide projector is kept between:
(A) Reflector and condenser
(B) Condenser and objective
(C) Bulb and condenser
(D) Objective and screen

217. The principle made used in slide-cum-film projector:
(A) Direct projection (B) Indirect projection
(C) Reflected projection (D) Diffused projection

218. Which of the following instructional aids essentially involved two sense organs:
(A) Television (B) Radio
(C) Poster (D) All of these

219. Reflected projector are used in:
(A) OHP (B) Slide projector
(C) Lantern projector (D) Epidiascope

220. Primary colours used in extension are:
(A) Red, Blue, Yellow (B) Red, Blue, Green
(C) Red, Blue, Purple (D) Red, yellow, green

221. Black and white colours are:
(A) Chromatic colours (B) Achromatic colours
(C) Diligent colours (D) Symbolic colours

222. The number of pages a leaflet usually will contain is:
(A) Two (B) Four
(C) Six (D) Eight

223. Non-photographic reconstruction of reality is:
(A) Illustration (B) Pictures
(C) Models (D) Diagrams

224. Exact visual recording of things is called:
(A) Illustration (B) Pictures
(C) Models (D) Photographs

225. The extension system used to provide precise and reliable scientific information in simple language about a simple practice is called:
(A) Leaflet
(B) Literature
(C) News paper
(D) Circular letter

226. The coverage of agricultural news in newspaper is:
(A) 3 to 5 %
(B) 20%
(C) 1 to 2 %
(D) 50%

227. Radio is an example for:
(A) Hot media
(B) Cold media
(C) Optical media
(D) Fiber media

228. The effective rate of delivery for a radio talk is:
(A) 80-100 words/minute
(B) 120-140 words/minute
(C) 160-180 words/minute
(D) 200-220 words/minute

229. The best style of presentation for radio talks with the farmers:
(A) Lecture
(B) Group discussion
(C) Interview
(D) Songs

230. The knowledge base for a technology usually derives from:
(A) Basic research
(B) Applied research
(C) Technology testing
(D) Technology development

231. Technology development process starts with:
(A) Technology testing
(B) Technology production
(C) Technology generation
(D) Technology dissemination

232. Technology generation mainly happens at:
(A) Research station
(B) Farmers fields
(C) Training centres
(D) Educational institutes

233. The function of extension starts from:
(A) Technology testing
(B) Technology production
(C) Technology integration
(D) Technology dissemination

234. Fine testing and packaging of technologies into specific recommendations is called:
(A) Technology testing
(B) Technology production
(C) Technology integration
(D) Technology dissemination

235. Linkage is the technological term used to denote communication that is:
(A) Inter-organizational
(B) Intra-organizational
(C) Inter – personal
(D) Intra-personal

236. The technology flow in the farmer first model is basically from:
(A) Agent to farmer
(B) Researcher to agent
(C) Farmer to farmer
(D) Researcher to farmer

237. The process of delivery of technology to farmers:
(A) Technology testing
(B) Technology production
(C) Technology integration
(D) Technology dissemination

238. Apart from introduction of technology, to make use of it, extension should concentrate:
(A) Adoption of indigenous ideas
(B) Mobilization of farmers
(C) Diffusion of technology
(D) Large farm demonstration

239. Evaluation of technology means:
(A) Value judgment
(B) Valere
(C) Positive and negative
(D) None of these

240. A technology that represents a major improvement over the existing one embodies:
(A) Compatibility
(B) Visibility
(C) Utility
(D) Complexity

241. The first stage of on-farm research:
(A) Technology evaluation
(B) Technology refinement
(C) Technology verification
(D) Technology dissemination

242. The belief that technology causes changes in the society:
(A) Technological fundamentalism
(B) Technological materialism
(C) Technological structuralism
(D) Technological determinism

243. "If people utilize all the presently available knowledge, there will not be any hunger":
(A) Transfer of technology paradigm
(B) Emancipatory extension paradigm
(C) Informative extension paradigm
(D) Persuasive extension paradigm

244. Transfer of technology is related with:
(A) Local idea
(B) Progressive idea
(C) Foreign idea
(D) None of these

245. Farmers interact with:
(A) Many sources
(B) Few sources
(C) Intra personal sources
(D) Information sources

246. The collectivity of units which are functionally differentiated and engaged in joint problem solving:
(A) Social norms
(B) Social system
(C) Social organisation
(D) Social unit

247. Established behaviour patterns for the members of the social system:
(A) Norms
(B) Values
(C) Attitude
(D) Beliefs

248. The patternized social relationship among the members of a social system:
(A) Social structure
(B) Social system
(C) Social organization
(D) Social unit

249. The alterations in the structure and functions of the social system:
(A) Social change
(B) Social movement
(C) Social alteration
(D) Social interaction

250. The social change where in the source of the change is from within the system:
(A) Local change
(B) Immanent change
(C) Localized change
(D) Direct change

251. Microanalysis approach in change analysis refers to:
(A) Change at individual level
(B) Change at family level
(C) Change iat group level
(D) Change at social system level

252. Extension agent need to necessitate change initially on:
(A) Adoption only
(B) Knowledge and adoption only
(C) Belief and value system
(D) Infrastructural facilities

253. Voluntary agencies are characterized by:
(A) Less acceptability (B) Greater continuity
(C) Less commitment (D) Less intensity

254. Essential characteristics of a primary group is:
(A) Face to face contact
(B) Personal and emotional relationship
(C) Interpersonal proximity
(D) Unity in diversity

255. A person who actually initiates the action in the community is:
(A) Professional leader (B) Assured leader
(C) Operational leader (D) Popular leader

256. The leader who is well trained and paid:
(A) Autocratic leader (B) Democratic leader
(C) Lay leader (D) Professional leader

257. Contact farmer can be considered as:
(A) Opinion leader (B) Professional leader
(C) Official leader (D) Lay leader

258. Which among the following find the greatest application in desire stage of extension teaching:
(A) Farm publication (B) Radio
(C) Exhibition (D) Mass media

259. Poster is most effective at which stage:
(A) Awareness (B) Evaluation
(C) Affection (D) Persuasion

260. The terminology that does to relates with the most innovative member of a social system:
(A) Drones (B) Ultra adopters
(C) Progressists (D) Light houses

261. The chairman of District Rural Development Agency:
(A) District Collector
(B) District Planning Officer
(C) District Panchayat President
(D) Block Development Officer

262. The authority responsible for planning and implementation of rural development programme at state level:
(A) Planning commission
(B) State Development Committee
(C) Ministry of Rural Development
(D) Chief Minister

263. The chairman of State Development Committee:
(A) Governor
(B) Chief Minister
(C) Chief Secretary
(D) Minister of Rural Development

264. The Gandhian approach of rural development have its basic emphasis on:
(A) Village industries
(B) Development of inner man
(C) Local self government
(D) Production improvement

265. CDP differs from IADP as the former have:
(A) Area based approach
(B) Decentralized administration
(C) Degree of investment
(D) Multi purpose approach

266. The extension approach used under IADP was:
(A) Prescriptive (B) Perscriptive
(C) Participatory (D) Commercial

267. The base level extension functionary in T & V system:
(A) BDO (B) VLW
(C) AEO (D) SDAO

268. The new name of VLW is:
(A) VDO (B) VEO
(C) CDO (D) VDW

269. The ratio of extension agent to farmer in a typical T&V system is:
(A) 1·250 (B) 1·500
(C) 1·800 (D) 1·1500

270. Single line of command is one of the important feature of:
(A) T&V (B) CDP
(C) IAAP (D) IRDP

271. ABC of posters means:
(A) Attractive, Brief and Complete
(B) Attractive, Brief and Clear
(C) Attractive and Creative
(D) None of these

272. A blue print for action is:
(A) Plan (B) Programme
(C) Project (D) None of these

273. A codified system of mores is generally called:
(A) Norms (B) Laws
(C) Customs (D) Values

274. A critical activity must have:
(A) Positive float (B) Negative float
(C) Zero float (D) None of these

275. A five minute radio talk has:
(A) 400 words (B) 300 words
(C) 600 words (D) 800 words

276. A new story technology can be published in:
(A) Extension bulletin (B) Folder
(C) Newspaper (D) Sports magazine

277. A poster has letter height of 22 inches. It can be clearly viewed up to the distance of:
(A) 10 feet (B) 20 feet
(C) 60 feet (D) 80 feet

278. A predetermined course of action is called:
(A) Objective (B) Project
(C) Plan (D) Management

279. A recognizable imitation or replica of an object whether workable or not is called:
(A) Chart (B) Specimen
(C) Mock-up (D) Model

280. A three dimentional recognizable imitation of an object is:
(A) Illustration (B) Model
(C) Mock up (D) Specimen

281. A working model is known as:
(A) Objects (B) Model
(C) Mockup (D) Poster

282. A visual aid has caption whose letter height is 0.5 inch. It can be visible clearly from:
(A) 10 Feet (B) 20 Feet
(C) 40 Feet (D) 60 Feet

283. Affective domain relates to:
(A) Knowledge (B) Speech
(C) Attitude (D) None of these

284. After a suitable extension teaching, the villagers attain desirable changes in his:
(A) Behaviour (B) Health
(C) Habits (D) All of these

285. Action research programme is considered as an example for:
(A) Formal training (B) Non formal training
(C) Direct training (D) Distant training

286. Action-reaction interdependence in communication is called:
(A) Credibility (B) Feed back
(C) Message treatment (D) Fidelity

287. Age, sex and caste indicates:
(A) Social status (B) Class status
(C) Ascribed status (D) Achieved status

288. An extension worker while extending scientific knowledge to the farmers starts with:
(A) Practical demonstration (B) Lectures
(C) Leaflets (D) All of these

289. An innovation has relatively high degree of complexity, its adoption will be:
(A) Quick
(B) Fast
(C) Slow
(D) Cannot be said anything

290. Aristotle model of communication contains how many ingredients:
(A) Five (B) Four
(C) Three (D) Two

291. Any decision and/or action to be realistic has to be based on:
(A) Facts (B) Opinion
(C) Value judgement (D) Moral grounds

292. Andragogy deals with:
(A) Change (B) Adult learning
(C) Society (D) Ancient civilization

293. An individual letter written by the extension worker to a farmer in connection with extension work is called:
(A) Circular (B) News letter
(C) Personal letter (D) None of these

294. At which level, communication is fact and problem oriented:
(A) Explorative level (B) Participative level
(C) Convenient level (D) None of these

295. At which of following level of farmers information and advisory centres under ATMA operates:
(A) District level (B) Block level
(C) Tehsil level (D) Village level

296. At which of following level Jawahar Gram Samridhi Yojna is implemented:
(A) District panchayat
(B) Intermediate panchayat
(C) Village panchayat
(D) Nager panchayat

297. ATMA Governing board is a/an:
(A) Advisory body (B) Statutory body
(C) Autonomous body (D) Financial body

298. ATMA management committee consists of:
(A) 6 members (B) 12 members
(C) 18 members (D) 24 members

299. ATMA management meets:
(A) Twice in a month (B) Once in a month
(C) Once in a crop season (D) Once in a year

300. Behaviour of an individual comprises of:
(A) Knowledge and attitude
(B) Attitude and Skill
(C) Skill and knowledge
(D) Knowledge, attitude and Skill

301. Behavioral changes in human beings in extension education include:
(A) Changes in knowledge (B) Changes in attitude
(C) Changes in skill (D) All of these

302. Campaign is an:
(A) Intensive teaching (B) Extensive teaching
(C) Invasive teaching (D) None of these

303. Communication behaviour essentially consists of:
(A) Information seeking
(B) Information processing
(C) Information dissemination
(D) All of these

304. Folkways which deals with moral aspects are:
(A) Taboos (B) Laws
(C) Mores (D) Norms

305. Folkways have:
(A) Moral validity (B) Conventional validity
(C) Legal validity (D) All of these

306. In National Programme of Doordarshan, Krishi Darshan is broadcasted at:
(A) 7:00 p.m. to 7:30 p.m. (B) 7:00 a.m. to 7:30 a.m.
(C) 7:30 a.m. to 8:00 a.m. (D) 6:30 p.m. to 7:00 p.m.

307. Systems approach is also known as :
(A) Contingency theory (B) Integrated theory
(C) Modern theory (D) Ecological theory

308. System approach to training has been developed by:
(A) S. Tyson (B) Alfred York
(C) B. Richard (D) Sameul Paul

309. The ability to use the most effective means to reach a given goal is:
(A) Rationality (B) Objectivity
(C) Competency (D) Relativity

310. The accuracy or precision of a measuring instrument is:
(A) Validity (B) Measurement
(C) Reliability (D) Readability

311. The amount of information that can be transmitted per unit of time is called:
(A) Channel capacity (B) Rate of transmission
(C) Transmission power (D) Transmission ratio

312. The approach to leadership which assumes that " leaders are born and not made" is:
(A) Innate approach (B) Trait approach
(C) Contingency approach (D) Behavioral approach

313. The communication of innovations:
(A) Involves a time factor
(B) Does not involve time factor
(C) May or may not involve time factor
(D) Has no relevance

314. The component of message is/are:
(A) Code (B) Content
(C) Treatment (D) All of these

315. The concept of feedback in communication is generally used to reflect:
(A) Resource orientation (B) Receiver orientation
(C) Process orientation (D) All of these

316. The essence of the scalar principle is:
(A) Unity of command (B) Span of control
(C) Communication (D) Coordination

317. The first line extension system mainly comprises of:
(A) ICAR and SAUs
(B) NGO
(C) Ministry of Agriculture and Cooperation
(D) Ministry of Rural Development

318. The shortest possible time to perform an activity, assuming that everything goes well is:
(A) Optimistic time (B) Pessimistic time
(C) Most likely time (D) None of these

319. The social network model ideally suited or imparting knowledge change:
(A) Star (B) Wheel
(C) Circle (D) Slash

320. There are how many basic social processes:
(A) 5 (B) 4 (C) 3 (D) 2

321. DASP stands for:
(A) Development Agriculture Service Project
(B) Diversified Agriculture Support Project
(C) Diversified Agriculture Service Project
(D) Development Agriculture Support Project

322. Farmers and parliament is published from:
(A) Pantnagar (B) New Delhi
(C) Hisar (D) Lucknow

323. Integrated crop management is a combination of:
(1) Integrated nutrient management
(2) Integrated water management
(3) Integrated Weed management
(4) Integrated pest management
Correct answer is:
(A) 1 and 2 (B) 2 ad 3
(C) 1, 3 and 4 (D) 1, 2, 3, and 4

324. Haryali yojna is related to:
(A) Food grain production (B) Fertilizer production
(C) Irrigation (D) Grass management

325. New name of ATMA is:
(A) National Agricultural Extension and Technology Mission
(B) National Food Security Mission
(C) Agriculture Technology Management Agency
(D) Integrated Scheme for Oilseeds, pulses, palmoil and maize

326. Which one of the following is the correct match?
(A) Etawah pilot project - Albert Mayer (1948)
(B) Grow more food campaign - J.L. Nehru (1947)
(C) Co-operative movement - Nicholas (1904)
(D) Sewagram project - M.K. Gandhi (1927)

327. The first KVK was established in 1974 at:
(A) Nagpur (B) Nilokheri
(B) Ludhiana (D) Pondichery

328. Which of the following rural development project was launched before independence?
(A) Etawah pilot project
B) Co-operative movement
(C) Community development
(D) National Extension service

329. Lab to Land programme was stared by:
(A) NBSSLUP (B) Govt. of U.P.
(C) Smt. Indira Gandhi (D) ICAR

330. National Academy of Agriculture Research Management (NAARM) is located at:
(A) New Delhi (B) Hyderabad
(C) Bangalore (D) Cuttack

331. Community development project was started in India in:
(A) 1947 (B) 1949
(C) 1952 (D) 1953

332. T and V system of extension was started by:
(A) D. Benor (B) M. Jackson
(C) M. Anderson (D) O. P. Dhama

333. High yielding variety programme was started in the year:
(A) 1960 (B) 1963
(C) 1966 (D) 1968

334. "Kisan Bharti" periodical is published from:
(A) Delhi (B) Lucknow
(C) Hisar (D) Pant Nagar

335. Gurgaon project was organized by:
(A) Mahatma Gandhi (B) F.L. Bryne
(C) Hatch (D) Albert Mayer

336. Sevagram attempt was started under the supervision of:
(A) F.L. Bryne (B) R.N. Tagore
(C) J.L. Nehru (D) M.K. Gandhi

337. Shriniketan attempt was started by R.N. Tagore in collaboration with Elmhirst in:
(A) Uttar Pradesh (B) Bengal
(C) Travancore (D) Etawah

338. Marthandum attempt was started by:
(A) Howard (B) Hatch
(C) Albert Mayer (D) F.L. Bryne

339. Which one of the following pairs is not correct:

Attempt	Year of start
(A) Gurgaon	1920
(B) Sevagram	1920
(C) Marthandum	1921
(D) Etawah	1950

340. NREP was started in the year:
(A) 1977 (B) 1978
(C) 1979 (D) 1982

341. The government sponsored Firka development scheme of Madras state was launched in:
(A) 1956 (B) 1946-47
(C) 1976 (D) 1980

342. The first KVK, Pondicherry was established in 1974 related with:
(A) RAU, Pusa (B) TAU, Coimbatore
(C) AAU, Hyderabad (D) ICAR

343. Lab to land programme was launched by the ICAR as a part of its———jubilee celebration in 1979:
(A) Centenary (B) Platinum
(C) Golden (D) Silver

344. The Rural System Research idea was motivated by M.S.Swaminathan in:
(A) 1978 (B) 1988
(C) 1998 (D) 2004

345. The government of India set up Planning Commission is:
(A) 1948 (B) 1950
(C) 1952 (D) 1954

346. The Chairman of Planning Commission is:
(A) President (B) Prime Minister
(C) Finance Minister (D) Speaker of Lok Sabha

347. Operational Research Projects (ORP) were initiated in:
(A) 1960 – 61 (B) 1966 – 67
(C) 1970 – 71 (D) 1975 – 76

348. State Agricultural Universities in India were setup on the pattern of Land Grant College of:
(A) U.K. (B) France
(C) Italy (D) U.S.A.

349. The key village scheme was initiated in August 1952 comprising first systematic attempt to improve the:
(A) Quality and productivity of buffaloes and cattle in India
(B) Quality and productivity of sheep and goat in India
(C) Quality of pigs
(D) All of these

350. Farm demonstration work began in ——— in USA by Dr. Seeman:
(A) 1903 (B) 1908
(C) 1955 (D) 1960

351. Extension work in Japan was first started by:
(A) Government organization
(B) Farmer's organization
(C) Labour's organization
(D) Business's organization

352. Which of the following is not correct:
(A) Department of revenue, agriculture and commerce 1871
(B) The famine commission 1880
(C) The famine commission 1901
(D) None of these

353. The Royal commission's report came in:
(A) 1920 (B) 1924 (C) 1928 (D) 1931

354. "Grow more food campaign" was started in:
(A) 1944 (B) 1942
(C) 1952 (D) 1953

355. Intensive Agricultural Development Programme (IADP) popularly also known as package programme started originally in 7 districts in:
(A) 1959 (B) 1960-61
(C) 1964 (D) 1970

356. Intensive Agriculture Area Programme came into operation in:
(A) March, 1964 (B) March, 1969
(C) October, 1969 (D) None

357. Which Five year plan aimed at attaining self sufficiency in food?
(A) 3rd (B) 4th (C) 5th (D) 6th

358. For social justice government started new programme:
(A) Small Farmers Development Agency (SFDA)
(B) Marginal Farmers and Agriculture Labours Scheme (MFALS)
(C) Drought Prone Area Programme
(D) All of these

359. The leader of the Boodan movement was:
(A) Acharya Vinoba Bhave (B) Spencer Hatch
(C) F.L. Bryne (D) T. Prakasam

360. Young Farmers Association was formed in India under the guidance of Dr. P.S. Deshmukh in:
(A) April, 1956 (B) April, 1966
(C) April, 1976 (D) April, 1980

361. National commission on farmers was set up in India in:
(A) March, 1999 (B) March, 2002
(C) February, 2004 (D) June, 2004

362. Who was the first chairman of National Commission on Farmers:
(A) M.S. Swaminathan (B) Sompal
(C) Balram Jakhar (D) V.L. Chopra

363. National Agricultural Science Museum was inaugurated on 3rd Nov, 2004 by:
(A) Dr. Man Mohan Singh
(B) Dr. A.P.J. Abdul Kalam
(C) A.B. Vajpayee (D) Sonia Gandhi

364. National Agricultural Science Museum located at:
(A) New Delhi (B) Bangalore
(C) Kolkata (D) Chennai

365. National Food for work programme was launched on:
(A) November14, 2000 (B) November 14, 2004
(C) October 2, 2000 (D) October 2, 2004

366. National Food for work programme implemented as a:
(A) 100% centrally sponsored scheme
(B) 80:20 center state sponsored scheme
(C) 50:50 centre state sponsored scheme
(D) None of these

367. Swaranjayanti Gram Swarozgar Yojana (SGSY) was launched in:
(A) April, 1999 (B) October, 1999
(C) January, 2000 (D) January, 2002

368. Which one of the following pairs not correct:
(A) Sompoorna Grameen Rozgar Yojana - 2001
(B) Pradhan Mantri Gramodaya Yoajna - 2000-01
(C) Rural Employment Generation Programme 1995
(D) Pradhan Mantri Gram Sadak Yojana - 1999

369. The government sponsored Firka development scheme was launched under the guidance of
(A) T. Prakasam (B) Jayachandran
(C) Spencer Hatch (D) Murlidharan

370. In India ——— of the total population lives in villages
(A) One third (B) Half
(C) Two third (D) Three fourth

371. National Agriculture Technology Project was started in:
(A) 1995-96 (B) 1998-99
(C) 2000-01 (D) 2002-03

372. NATP was funded from
(A) USA (B) World Bank
(C) GOI (D) ICAR

373. National Institute of Agriculture Marketing (NIAM) is located at:
(A) New Delhi (B) Jaipur
(C) Hisar (D) Meerut

374. Under NRDP, the state government has its share:
(A) 20% (B) 75%
(C) 80% (D) 50%

375. National Demonstration was started in the year of:
(A) 1964 (B) 1965
(C) 1970 (D) 1975

376. In India, Community Development Projects were launched in:
(A) 1950 (B) 1951
(C) 1952 (D) 1962

377. The High Yielding Variety Programme started in the year of:
(A) 1966 (B) 1964
(C) 1971 (D) 1976

378. I.R.D.P. was launched in:
(A) 1980 (B) 1960
(C) 1970 (D) 1978

379. Which of the following state has no panchayati raj:
(A) Rajasthan (B) Meghalaya
(C) Bihar (D) Orissa

380. The first Agriculture University of India:
(A) P.A.U., Ludhiana (Punjab)
(B) H.A.U., Hissar (Haryana)
(C) G.B.P.U.A. & T., Pantnagar (U.A.)
(D) A.P.A.U., Hyderabad (A.P.)

381. What does the term DPAP denotes?
(A) Dry Project Area Programme
(B) Drought Prone Area Programme
(C) Development Project Area Programme
(D) None of these

382. Leads Bank scheme was started in the year:
(A) 1965 (B) 1969 (C) 1975 (D) 1964

383. The scheme of Regional Rural Banks was launched in India on:
(A) 1st Oct.1975 (B) 15th Aug. 1978
(C) 2nd Oct. 1975 (D) 1st Jan. 1970

384. Fourth five year plan was started in the year:
(A) 1966 (B) 1969
(C) 1967 (D) 1960

385. National Agricultural Extension Project was launched by:
(A) ICAR (B) IARI
(C) Govt. of India (D) State Govt.

386. There are_________ degrees of evaluation:
(A) Four (B) Three
(C) Five (D) Two

387. The S.F.D.A. and M.F.A.L. Programme were started in:
(A) 1970 (B) 1965
(C) 1967 (D) 1969

388. Command Area Development Programme was started in:
(A) 1962 (B) 1965
(C) 1978 (D) 1974

389. Who is the chairman of S.F.D.A.?
(A) Agriculture Minister (B) Governor
(C) Chief Minister (D) Collector

390. The ninth five year plan was started in:
(A) 1995-96 (B) 1996-97
(C) 1997-98 (D) 1998-99

391. The Taccavi loan is given by:
(A) The bank (B) The government
(C) The co-operative society (D) None of these

392. The principle "one village one society" was suggested by:
(A) Ashok Mehta Committee
(B) Mec Lagen Committee
(C) Wanchoo Committee
(D) Rajamannan Committee

393. T and V system of extension was developed by:
(A) M. Jackson (B) D. Benor
(C) M. Anderson (D) O.P. Dhama

394. Chairman of planning commission is:
(A) President (B) Governor
(C) Finance Minister (D) Prime Minister

395. The major 14 banks were nationalized in the year:
(A) 1969 (B) 1970
(C) 1973 (D) 1975

396. I.R.D.P. was started in the year:
(A) 1965-66 (B) 1978-79
(C) 1964-65 (D) 1980-81

397. Which of the following guides the Block samiti:
(A) Block Pramukh
(B) V.D.O.
(C) B.D.O. and other extension officers
(D) None of these

398. Desert Development programme was started in:
(A) 1977-78 (B) 1980-81
(C) 1981-82 (D) 1983-84

399. Which of the following method is the best for selection of a leader?
(A) Statistic method (B) Psycho-metric method
(C) Socio- metric method (D) Bio-metric method

400. Albert Mayer of USA started a project in September, 1948 known as:
(A) Gurgaon experiment (B) Marthandam Project
(C) Etawah pilot project (D) None of these

401. The Gurgaon experiment was started by:
(A) R.N. Tagore (B) Vinoba Bhave
(C) F.L. Brayne (D) Mahatama Gandhi

402. The effective media of self-dependent of village of Rajasthan is:
(A) Formulation of village sponsored economic plans
(B) Extension of cities
(C) Extension of village education
(D) Service to rural unemployment in cities

403. First KVK was established in Rajasthan was:
(A) Badgaon (1952) (B) Fathehpur (1976)
(C) Beechwal (1972) (D) Banswara (1975)

404. National Bank for Agriculture and Rural Development came into existence on:
(A) July 22, 1984 (B) July 12, 1982
(C) October 15, 1986 (D) August 10, 1983

405. The T & V system of Agricultural extension was first introduced in the year:
(A) 1964 (B) 1954
(C) 1984 (D) 1974

406. The programme which have been merged under IRDP were:
(A) SFDA, MFAL, DPAP and CADP
(B) MFAL, SFDA, IAAP and IADP
(C) SFDA, MFAL and Lab to Land
(D) DRDA, MFAL and TV

407. The first five year plan was started in:
(A) 1947-48 (B) 1950-51
(C) 1951-52 (D) 1952-53

408. In India the use of television as a medium of communication for rural development was started in:
(A) 1964 (B) 1965
(C) 1966 (D) 1967

409. National extension service was started in the year:
(A) 1951 (B) 1952
(C) 1953 (D) 1954

410. In India major commercial banks were nationalised in:
(A) 1965 (B) 1967
(C) 1969 (D) 1971

411. Kisan Mandal System under T and V system was initiated in the year of:
(A) 1996 (B) 1995
(C) 1994 (D) 1993

412. Fore runner of C.D. programme is:
(A) Etawah project (B) Mrthundum project
(C) Gurgaon project (D) Sevagram

413. Which of the following is the special scheme launched by the ICAR for transfer of technology to the farmers?
(A) All India Co-ordinated Research Project
(B) Operational Research Project
(C) Lab to Land Programme
(D) National Demonstration Programme

414. Which of the following rural development project was launched before independence:
(A) Etawah Pilot Project
(B) Co-operative Project
(C) Community Development
(D) National Extension Service

415. The new name of community development programme is:
(A) Intensive Agriculture Programme
(B) C.A.D.P.
(C) Rural Development Programme
(D) None of the above

416. For demonstration the value of a new extension programme to rural peoples of the locality, the programme should be sponsored by group of:
(A) Rural people of locality
(B) Urban people
(C) Local leaders
(D) Gram Pradhan and V.L.W.s

417. The Etawah pilot project for rural development was started in:
(A) 1971 (B) 1978 (C) 1952 (D) 1948

418. The national rural development programme was replaced by food and work programme during:
(A) 1982 (B) 1980 (C) 1979 (D) 1976

419. In India, the use of television as a medium of communication for rural development was started in:
(A) 1960 (B) 1965 (C) 1966 (D) 1967

420. National demonstration was started in:
(A) 1960 (B) 1962 (C) 1950 (D) 1965

421.

List-I (Programme)	List-II (Starting year)
(a) Gurgaon project	(1) 1920
(b) Marthandom Project	(2) 1948
(c) Rural development	(3) 1935
(d) F.A.O.	(4) 1928

Code	a	b	c	d
(A)	2	3	4	1
(B)	3	4	2	1
(C)	1	4	3	2
(D)	1	2	3	4

422.

List-I (Programme)	List-II (Worker)
(a) Indian Village Service	(1) S.K. Dubey
(b) Etawah Pilot Project	(2) Spencer Hatch
(c) Nilokheri Project	(3) W.H. Wisher
(d) Marthandom Project	(4) Albert Mayer

Code	a	b	c	d
(A)	3	4	1	2
(B)	3	2	1	4
(C)	4	3	2	1
(D)	1	2	4	3

423.

List-I (Programme)	List-II (Founder)
(a) Planning Commission	(1) Nicholos
(b) Bhoodan Movement	(2) Madras Govt.
(c) Firka development Scheme	(3) Viinoba Bhave
(d) Co-operative Movement	(4) Govt. of India

Code	a	b	c	d
(A)	4	3	2	1
(C)	3	4	2	1
(C)	3	2	1	4
(D)	1	2	3	4

424.

List-I (Programme)	List-II (Starting year)
(a) Multiple cropping scheme	(1) 1950
(b) Rural development	(2) 1908
(c) Sriniketan	(3) 1960
(d) Sarvodaya Programme	(4) 1935

Code	a	b	c	d
(A)	2	3	1	4
(B)	3	4	2	1
(C)	4	3	2	1
(D)	1	2	3	4

425.

List-I	List-II
(a) IIHR	(1) Syria
(b) CTRL	(2) Bangalore
(c) IRRI	(3) Matunga
(d) ICARDA	(4) Philippines

Code	a	b	c	d
(A)	1	4	3	2
(B)	4	3	2	1
(C)	3	2	1	4
(D)	2	3	4	1

426.

List-I	List-II
(a) NABARD	(1) 1935
(b) RBI	(2) 1982
(c) CACO	(3) 1975
(d) RRB	(4) 1965

Code	a	b	c	d
(A)	4	3	2	1
(B)	3	4	2	1
(C)	2	1	4	3
(D)	1	2	3	4

427.

List-I	List-II
(a) AGMARK	(1) 1953
(b) ICAR	(2) 1973
(c) NES	(3) 1937
(d) T and V system	(4) 1929

Code	a	b	c	d
(A)	1	4	3	2
(B)	4	3	2	1
(C)	3	4	1	2
(D)	2	3	4	1

428.

List-I	List-II
(a) First KVK of India	(1) Sultanpur
(b) Total KVK in India	(2) Pondichery
(c) First KVK in Rajasthan	(3) 261
(d) First KVK in U.P.	(4) Fatehpur

Code	a	b	c	d
(A)	2	3	4	1
(B)	3	2	1	4
(C)	3	1	2	4
(D)	1	2	3	4

429.

List-I	List-II
(a) CPCRI	(1) Manila
(b) CTCRI	(2) Jodhpur
(c) IRRI	(3) Trivendrum
(d) CAZRI	(4) Kesargod

Code	a	b	c	d
(A)	1	2	4	3
(B)	4	3	1	2
(C)	3	2	4	1
(D)	2	3	4	1

430. Name of the project in which 'Gaon Sathi' worked:
(A) Etawah Pilot Project
(B) Sriniketan Project
(C) Yamuna par Punarnirvan Project
(D) Marthandam Project

431. Who aspired for 'Villagers to be self-reliant and self – sufficient'?
(A) Indira Gandhi (B) M.K. Gandhi
(C) Lal Bahadur Shastri (D) Jawahar Lal Nehru

432. Balwant Rai Mehta Committee recommended:
(A) Three Tier Panchayati Raj System
(B) T & V System
(C) Small Farmers Development Agency
(D) None of these

433. Community Development Project started with the agreement between?
(A) Indo-Japan Govt.
(B) Indo-US Govt.
(C) Indo-British Govt.
(D) Indo-Netherland Govt.

434. "Agricultural Administrative Committee" was constituted in 1958 to review the CD programmes under whose chairmanship?
(A) R.N. Tagore
(B) Balwant Rai Mehta
(C) Raja Sunder Singh Nalagarh
(D) F.L. Brayne

435. Ford Foundation team was invited to help in:
(A) The study of agricultural scenario of India
(B) To support grow more food campaign
(C) To support food crises by way of ensuring supplies
(D) None of these

436. Ford Foundation Committee recommended:
(A) Efforts in specific areas and specific crops
(B) Research in major crops
(C) Opening of institutions for agricultural research and education
(D) Major emphasis on extension

437. Post graduate programme in extension was started in:
(A) 1950 (B) 1952
(C) 1955 (D) 1960

438. The first extension education institute in India is established in:
(A) Nilokheri (B) Anand
(C) Hyderabad (D) Jorhat

439. NATP was launched in the year:
(A) 1998 (B) 1999
(C) 2000 (D) 2001

440. NATP is funded by:
(A) ICAR (B) FAO
(C) World Bank (D) Reserve Bank

441. Agriculture was made a state subject through:
(A) Government of India Act (1919)
(B) Royal Commission Report (1928)
(C) Ford Foundation Report (1959)
(D) GMF Enquiry Report (1955)

442. Bharti Balika was a component of:
(A) Gurgaon experiment (B) Sevagram
(C) Sriniketan (D) Etawah project

443. The place where Gandhiji started his sewagram project:
(A) Wardha (B) Pohri
(C) Gwalior (D) Rajkot

444. The concept of village Guide was introduced by:
(A) R N Tagore (B) B Pant
(C) M K Gandhi (D) F L Brayne

445. Bombay plan was formulated in:
(A) 1943 (B) 1947
(C) 1950 (D) 1951

446. People's plan was envisaged by:
(A) Jawaharlal Nehru (B) Viswaswrayya
(C) M N Roy (D) Mahalanobis

447. Indian Village Service was started in:
(A) Wardha (B) Pohri
(C) Gwalior (D) Lucknow

448. The person actively involved with Indian Village Service:
(A) Albert Mayer (B) F L Brayne
(C) S K Dey (D) A T Mosher

449. The concept of multipurpose village level worker was evolved from:
(A) Etawah Pilot Project
(B) Community Development Project
(C) National Extension Service
(D) Gurgaon Project

450. Etawah Pilot Project received financial assistance from:
(A) US Point 4 programme
(B) US Point 7 programme
(C) US Point 20 programme
(D) World Bank

451. The project started to rehabilitate the refugees from Pakistan during partition:
(A) Gurgaon project (B) Nilokheri project
(C) Sriniketan project (D) Sewagram project

452. 'Betting on the strong horse' is a slogan associated with:
(A) IRDP (B) HYVP
(C) IADP (D) SFMA

453. The IADP was popularly known as:
(A) Package programme
(B) Training programme
(C) Community programme
(D) Agriculture programme

454. Linking farm plan with credit is a characteristics feature of:
(A) IRDP (B) IADP
(C) CDP (D) Etawah project

455. The first state in India to introduce Training and Visit system was:
(A) Punjab (B) Rajasthan
(C) Tamilnadu (D) Bihar

456. National Extension Service (NES) in India was initiated on:
(A) October 2, 1952 (B) October 2, 1953
(C) January 26, 1950 (D) August 15, 1947

457. Ethopia is the origin home of:
(A) Coffee (B) Maize
(C) Rice (D) Cabbage

458. Cabbage originated in:
(A) Mexico and Central America
(B) Mediterranean
(C) Peruvian Andes
(D) South west Asia

459. Maize evolved in:
(A) Brazil
(B) Mexico and Central America
(C) USA
(D) Asia minor and Afganistan

460. The centre of origin of Almond and Apple is:
(A) South West Asia
(B) South East Asia
(C) China
(D) Asia minor and Afganistan

461. The origin home of rice is:
(A) South West Asia (B) South East Asia
(B) Mediterranean (D) China

462. Peruvian Andes are believed to be centre of origin of:
(A) Tomato
(B) Potato
(C) Tomato and Potato
(D) Tomato, Potato and Chillies

463. The origin of sunflower is believed to be in:
(A) Mexico and Central America
(B) Brazil
(C) USA
(D) Peruvian Andes

464. South East Asia is thought to be centre for the origin of:
(A) Rice
(B) Rice and Sugarcane
(C) Rice, Sugarcane and Mango
(D) Rice, Sugarcane, Mango and Banana

465. The famous cultivated plant which developed in China is:
(A) Tea (B) Coffee
(C) Cotton (D) Orange

466. Production centre of Hevea Rubber is:
(A) Malayasia (B) Brazil
(C) Peruvian Andes (D) South West Asia

467. The centre of origin of Wheat is:
(A) South West Asia (B) South East Asia
(C) Asia minor and Afganistan (D) Mexico

468. Which is not a new world crop:
(A) Tapioca (B) Peanut
(C) Cashewnut (D) Poppy

469. Which one belong to old world:
(A) Squash (B) Pumpkin
(C) Water Melon (D) Guava

470. Select a crop which belongs to new world:
(A) Jackfruit (B) Papaya
(C) Spinach (D) Pear

471. Which crop belong to new world:
(A) Red Pepper (B) Black pepper
(C) Onion (D) Garlic

472. Which is the crop which originated in Peru and Brazil but is now mainly grown in India is:
(A) Groundnut (B) Potato
(C) Maize (D) None of these

473. The crop which originated in Peru but is most extensively grown in East Europe is:
(A) Coffee (B) Potato
(C) Rubber (D) Maize

474. Name the crop which had its birthplace in Tropical America but the centre of production is now Midwest USA:
(A) Pineapple (B) Oil palm
(C) Cocoa (D) Maize

475. Christopher Columbus landed in island of Bahmas (Carribean sea) in:
(A) 1692 (B) 1592
(C) 1492 (D) 1792

476. Tropical forests are being lost at the annual rate of:
(A) 11 million ha/yr (B) 110 million ha/yr
(C) 1 million ha/yr (D) 1100 million ha/yr

477. Post Vavilov megadiversity centre is:
(A) Australia (B) Siberia
(C) Andaman (D) Tibet

478. The number of centres of origin of crop plants recognized by Vavilo is:
(A) 8 (B) 11 (C) 7 (D) 21

479. Guarana is:
(A) Fodder plant (B) Oil plant
(C) Beverage plant (D) Food plant

480. The Central Marketing Department was established in India in:
(A) 1930 (B) 1947
(C) 1935 (D) 1955

481. The Food Corporation in India was born in:
(A) 1952 (B) 1960
(C) 1965 (D) 1971

482. Cotton Corporation of India is a ideal agency of:
(A) Cereal (B) Pulses and Oilseeds
(C) Cotton (D) Jute

483. Central Warehousing Corporation was set up in:
(A) 1930 (B) 1957
(C) 1966 (D) 1970

484. Indian Standard Institute was set up in:
(A) 1935 (B) 1942
(C) 1947 (D) 1960

485. International Organisation for Standardization (I.S.O.) came into existence in:
(A) 1947 (B) 1950
(C) 1958 (D) 1965

486. The Commission or Agricultural Costs and Prices was setup in:
(A) 1950 (B) 1960
(C) 1965 (D) 1970

487. Council of State Agricultural Marketing Board was established in:
(A) 1970 (B) 1975
(C) 1988 (D) 1990

488. National Agricultural Co-operative Marketing Federation (NAFED) was established in:
(A) 1955 (B) 1958
(C) 1965 (D) 1968

489. Cultivation of hybrid rice is most popular in :
(A) China (B) India
(C) Phillipines (D) Canada

490. Central Upland Rice Research Station is located in:
(A) Masodha (B) Orissa
(C) Hyderabad (D) Bihar

491. Green revolution in wheat basically attributed to:
(A) Greater use of seeds of high yielding varieties
(B) Higher use of N, P and K fertilized
(C) Greater availability of irrigation
(D) Greater use of plant protection chemicals

492. The land/man ratio in India at present is:
(A) 0.10 ha (B) 0.15 ha
(C) 0.20 ha (D) 0.25 ha

493. A severe drought occurred during the last decade is:
(A) 1985 (B) 1986
(C) 1987 (D) 1988

494. Total flood prone area in India is approximately:
(A) 30 mha (B) 35 mha
(C) 40 mha (D) 45 mha

495. Area sown more than once is maximum in the state of:
(A) UP (B) MP
(C) Rajasthan (D) Gujarat

496. National Cooperative Development Corporation (NCDC) was established in:
(A) 1960 (B) 1961
(C) 1962 (D) 1963

497. When the quantitative restriction (QR) was over in W.T.O.?
(A) 1951 (B) 1994
(C) 1999 (D) 2001

498. National Agricultural Policy was declared in:
(A) 1951 (B) 1961
(C) 1971 (D) 1981

499. National Seed Corporation (NSC) was established in:
(A) 1960 (B) 1963
(C) 1965 (D) 1968

500. Directorate of Marketing and Inspection is situated in:
(A) Kanpur (UP) (B) Faridabad (Haryana)
(C) Kolkata (WB) (D) Hyderabad (AP)

501. Out of the total cropped area of India, forest occupy only:
(A) 26% (B) 33%
(C) 35% (D) 19.5%

502. The first Director General of Indian Council of Agricultural Research was:
(A) Dr. M.S. Swaminathan (B) Dr N.S. Randhawa
(C) Dr. B.P Pal (D) Dr. O.P. Gautam

503. Indian Institute of Subtropical Horticulture is situated in:
(A) Bangalore (B) Lucknow
(C) Mysore (D) New Delhi

504. The agro-climatic zones in India is divided into:
(A) Ten (B) Twelve
(C) Fifteen (D) Twenty

505. The thrust of the agricultural programmes during the Eight Plan is on:
(A) Increasing production of rice
(B) Increasing production of wheat
(C) Increasing production of oilseeds, fruits, vegetables and pulses
(D) Increasing production of cotton

506. On October 2, 1952, was launched:
(A) National Extension Service
(B) Integrated Rural Development Programme
(C) Training and Visit System
(D) Community Development Programme

507. Cooperative movement in India was started in:
(A) 1914 (B) 1947
(C) 1904 (D) 1935

508. When was the 'Minimum Wages Act' for agricultural laboures passed?
(A) 1947 (B) 1948
(C) 1951 (D) 1956

509. Training and Visit system of extension started in UP in:
(A) 1970 (B) 1980
(C) 1985 (D) 1990

510. Which one is the latest programme in India?
(A) KVK (B) NARP
(C) NAEP (D) NATP

511. Which one is the latest programme in India?
(A) ATMA (B) NATP
(C) RKVY (D) ISOPALM

512. National Extension Services in India was established in:
(A) 1952 (B) 1953
(C) 1990 (D) 1995

513. Discouraging the application of chemical fertilizers and advocating organic farming:
(A) Principle of transfer
(B) Principle of set
(C) Principle of disassociation
(D) Principle of association

514. The extension scientific knowledge to villagers by specialists of the same area:
(A) Principle of local leadership
(B) Principle of Panchayati Raj
(C) Principle of grass root organisation
(D) Principle of self activity

515. Learning to use a sprayer correctly requires repeated trying of the instrument:
(A) Principle of motivation
(B) Principle of attitude
(C) Principle of practice
(D) Principle of satisfaction

516. If a farmer has learned the technique of water management in a particular crop, they should be able to use this method in other crops as well:
(A) Principle of achievement
(B) Principle of transfer
(C) Principle of disassociation
(D) Principle of association

517. Issue of advertisement regarding public health and sanitation at the on set of an epidemic outbreak is a classical example of:
(A) Principle of set (B) Principle of time
(C) Principle of association (D) Principle of transfer

518. When planting a crop in lines gives better yield, the farmers may be advised not to practice broadcasting:
(A) Principle of achievement
(B) Principle of transfer
(C) Principle of disassociation
(D) Principle of association

519. When farmers are ready to cooperate, with good guidance, they may be able to form a cooperative society:
(A) Principle of motivation
(B) Principle of attitude
(C) Principle of readiness
(D) Principle of satisfaction

520. When a farmer develop a favourable set towards scientific method of cultivation, they shall learn more on the topic:
(A) Principle of motivation
(B) Principle of attitude
(C) Principle of readiness
(D) Principle of satisfaction

521. The attainment of perfection demands that undesirable and useless movements are replaced by desirable and useful movements:
(A) Principle of motivation
(B) Principle of attitude
(C) Principle of readiness
(D) Principle of satisfaction

522. Favourable experience of planting trees motivate tribal farmers to collect saplings from the forest nursery:
(A) Principle of motivation
(B) Principle of attitude
(C) Principle of readiness
(D) Principle of satisfaction

523. When insect have appeared or are likely to appear in crops, farmers shall learn about plant protection:
(A) Principle of timing
(B) Principle of adaptation
(C) Principle of maximum response
(D) Principle of satisfaction

524. When villagers assist the extension worker at different levels in implementing a new programme:
(A) Principle of cooperation
(B) Principle of satisfaction
(C) Principle of implementation
(D) Principle of interest

525. When farmers use their crop loan for the very purpose for which the loan is sanctioned:
(A) Principle of achievement
(B) Principle of completion
(C) Principle of clarity of objectives
(D) Principle of satisfaction

526. The principle under which custom religion is findout?
(A) Principle of social change
(B) Principle of cultural differences
(C) Principle of adaptability
(D) Principle of whole family

527. The principle under which children and family are included:
(A) Principle of social change
(B) Principle of cultural differences
(C) Principle of adaptability
(D) Principle of whole family

528. The basis of any extension approach is:
(A) Individual is the end
(B) Individual is a means to an end
(C) Rural development
(D) Transfer of technology

529. Extension approach in India generally follows the norms of:
(A) Cooperation (B) Participation
(C) Prescription (D) Mobilization

530. The most preferred approach in extension work is:
(A) Autocratic (B) Democratic
(C) Lessiz faire (D) None of the above

531. The interactive socio-political process motivated by desire to meet an individuals needs through collective actions on mutual benefit basis:
(A) Participation (B) Cooperation
(C) Extension (D) Collaboration

532. People participate by taking initiatives independent of external institutions to change systems:
(A) Self-mobilization
(B) Interactive participation
(C) Functional participation
(D) Passive participation

533. People participation by being told what is going to happen or has already happened:
(A) Self-mobilization
(B) Interactive participation
(C) Functional participation
(D) Passive participation

534. The participatory approach of extension recognizes the importance of:
(A) Exploitation of local resources
(B) Farm planning
(C) Democratic way of problem solving
(D) Learning from farmers

535. Participatory development approach in otherwards is called:
(A) Bottom up approach (B) Top down approach
(C) Circle centered approach (D) Centralized approach

536. The extension approach wherein short-run change is the measure of success:
(A) Project approach
(B) Commodity specialized approach
(C) Cost sharing approach
(D) General agricultural extension approach

537. The extension approaches wherein success is measured in terms of farm people's willingness and ability to share some of the cost:
(A) Project approach
(B) Commodity specialized approach
(C) Cost sharing approach
(D) General agricultural extension approach

538. The privatization approach wherein government grants private entity authority to provide a particular service within a specific geographical area:
(A) Contracting (B) Franchise
(C) Vouchers (D) Self help

539. The privatization approach where in the public pays and private delivers:
(A) Service within a specific geographical area
(B) Contracting
(C) Franchise
(D) Vouchers
(E) Self help

540. The privatization approach most commonly used in sugarcane production sector:
(A) Contracting (B) Franchise
(C) Vouchers (D) Self help

541. The extension approaches wherein measure of success is the farming people's attendance at and participation in the school's agricultural extension activities:
(A) Project approach
(B) Commodity specialized approach
(C) Educational institution approach
(D) General agricultural extension approach

542. Extension is fairly centralized and is oriented towards one commodity or crop and the agent has many functions:
(A) Project approach
(B) Commodity specialized approach
(C) Educational institution approach
(D) General agricultural extension approach

543. The extension approaches wherein the success is measured in terms of production increases of the particular crops covered by the programme:
(A) Training and visit approach
(B) Commodity specialized approach
(C) Educational institution approach
(D) General agricultural extension approach

544. The extension approaches wherein success is measured by the extent to which farming people adopt the technologies developed by the programme and continue using them over time:
(A) Commodity specialized approach
(B) General agricultural extension approach
(C) Cost sharing approach
(D) Farming system development approach

545. The farming system development approach essentially aims at:
(A) Development of technology which fits the need
(B) Development of farmer's innovation
(C) Increasing farm production and productivity
(D) Integrated development of the farm

546. "How do I help people achieve what they themselves want to achieve":
(A) "DO TO" strategy
(B) "DO FOR" strategy
(C) "DO WITH" strategy
(D) "DO AGAINST" strategy

547. The communication strategy embedded in the transfer of technology paradigm:
(A) Persuasive + Paternalistic
(B) Persuasive + Participatory
(C) Educational + Paternalistic
(D) Educational + Participatory

548. The strategy that often lead to the so called progressive farmer approach in agricultural extension:
(A) "DO TO" strategy
(B) "DO FOR" strategy
(C) "DO WITH" strategy
(D) "DO AGAINST" strategy

549. "How do I develop an offering which my target clients want" relates with:
(A) "DO TO" strategy
(B) "DO FOR" strategy
(C) "DO WITH" strategy
(D) "DO AGAINST" strategy

550. The most important factor in FSR/E is:
(A) Farmer (B) Researcher
(C) Extensionist (D) All of these

551. Which of the followings is not a factor contributing to the limited application is FSR/E:
(A) Action based approach
(B) Multidisciplinary approach
(C) Resource poor clientale
(D) Limited research output

552. The extension approach wherein success is measured in terms of the rate of take-up of the recommendations, and increases in national production:
(A) Project approach
(B) Commodity specialized approach
(C) Cost sharing approach
(D) General agricultural extension approach

553. Under subsistent farming situation agricultural development is characterized by:
(A) Commercial orientation
(B) Subsidy orientation
(C) Information orientation
(D) Companion-mission orientation

554. Which among the following is a bottom up approach:
(A) Farming system approach
(B) Commodity specialized approach
(C) Training and visit approach
(D) Cost sharing approach

555. ——— is a statement of situation, objectives, problems and solutions:
(A) A plan of work (B) Extension programme
(C) A calendar of work (D) A project

556. ——— is a plan of work arranged chronologically:
(A) A project (B) A calendar of work
(C) A plan of work (D) Programme planning

557. A sound extension programme building based on:
(A) Analysis of the facts in the situation
(B) Select problems based on needs
(C) Determine objective and solutions
(D) All of these

558. First step for making a programme planning includes:
(A) Evaluation
(B) Analysis of the situation and determining problems
(C) Deciding on objectives
(D) Teaching

559. The main objective of evaluation in extension is:
(A) To facilitate effective decision making directly goes to conclusion
(B) To facilitate effective decision making without jumping to conclusion
(C) To see weak points in programme
(D) None of these

560. When an extension programme is successfully completed and the farmer and his family members are satisfied by solving the problem and getting a new idea, it is called:
(A) The principle of achievement
(B) The principles of completion
(C) The principles of maximum response
(D) The principles of satisfaction

561. The strategy for supporting the rapidly expanding agricultural programme requires:
(A) That cultivators are educated in favour of modern farming
(B) That efforts are made to manufacture more fertilizers
(C) That efforts are made to provide more irrigation facilities
(D) All of these

562. A good extension programme should be:
(A) Rigid (B) Flexible
(C) Both (A) and (B) (D) None of these

563. Determination of extension programme consists of:
(A) Definition and finding solutions
(B) Preparing plan of work and carry out the plan of work
(C) Collection and analysis of data's determination of needs and objectives
(D) Both (A) and (B)

564. ATIC has been initiated with:
(A) Single window concept of technology dissemination
(B) Diversification of agricultural technology
(C) Intensification of technologies
(D) Intensification and diversification of technologies

565. PARA extension workers are considered to be:
(A) The workers working with NGOs
(B) Rural youths trained to take up small extension works on payment basis
(C) Extension workers having specialization in rural areas and enterprises
(D) Workers with knowledge of all the enterprises attached to farm house hold

566. Which are the main reasons for the limited success of T & V in India:
(A) Heavy infrastructure investment
(B) Assumptions at different levels
(C) Top to down approach
(D) All of these

567. Participating extension management envisages:
(A) Participation of all the stakeholders in the project
(B) Participation of farmers in the project
(C) Technology transfer by extension agencies
(D) None of these

568. What is "Cyber extension":
(A) Extension through group approach
(B) Extension through use of television and other mass media
(C) Extension through computers
(D) Extension through NGOs

569. Gender development in extension programmes is related to:
(A) Empowering farm women
(B) Development of both genders on equal basis
(C) Development of households with emphasis on gender
(D) None of these

570. People's participate in an extension programme is significant when:
(A) Local leaders participate
(B) Literate section of villagers participate
(C) Gram panchayat members participate
(D) Majority of villagers participate

571. ....is that of working with rural people through out of school education along with their current interest and needs for getting overall development of rural families:
(A) Extension education (B) Extension process
(C) Extension jobs (D) Extension service

572. The extension programme is a statement of:
(A) Situation and objectives (B) Solution and problems
(C) Problems and situation (D) All of these

573. Who, what, when and where are part of:
(A) Plan of work (B) Calendar of work
(C) Programme evaluation (D) Decision making

574. The plan of work when arranged chronologically is:
(A) Calendar of work (B) Calendar of operation
(C) Objective (D) Programme

575. A sound programme must have:
(A) Permanence with rigidity
(B) Permanence with flexibility
(C) Temporarity with rigidity
(D) Temporarity with flexibility

576. Extension programme planning is a:
(A) Educational process
(B) Administrative process
(C) Effective process
(D) Uniform process

577. The first step in extension programme planning at local level is:
(A) Determination of objectives
(B) Analysis of existing situation
(C) Decision of extension teaching method
(D) Determination of target

578. The participation of people in extension programme is:
(A) Compulsory (B) Voluntary
(C) Involuntary (D) None of these

579. The pilot study in extension programme represent:
(A) Post evaluation (B) Concurrent evaluation
(C) Pre evaluation (D) Monitoring

580. Last stage of programme planning process is:
(A) Evaluation (B) Auditing
(C) Supervision (D) Reconsideration

581. Under T & V system is mainly responsible to make regular contact with farmers:
(A Deputy Director (Extension)
(B) Kisan Sahayak/Sewak
(C) SDAEO
(D) BDO

582. The extension work must be based on needs and interests of:
(A) People (B) Leader
(C) Officers (D) Government

583. Need based and skill oriented vocational training to practicing farmers and employed extension personnel is provided through:
(A) Training and visit
(B) Krishi Vigyan Kendra
(C) National Demonstration
(D) Lab to Land programme

584. Which of the following scheme have been merged under Jawahar Rozgar Yojna:
(1) NREP (2) RLEGP
(3) IRDP (4) TRYSEM
(A) 1 & 2 (B) 1 & 3
(C) 3 & 4 (D) All of these

585. Operational flood programme relate to:
(A) Conserving soil (B) Controlling floods
(C) Boosting milk supply (D) Funds for flood victims

586. The first step in extension programme planning is:
(A) Analysis of situation (B) Fixing objectives
(C) Collection of facts (D) Determination of goals

587. The best method to convince people is:
(A) Video (B) Demonstration
(C) Classroom teaching (D) Group discussion

588. I.B.R.D. is synonymous to:
(A) World Bank
(B) Asian Development Bank
(C) International Monetary Fund
(D) World Trade Organisation

589. Lab to Land Programme was started by:
(A) Government of India
(B) Government of UP
(C) Smt. Indira Gandhi
(D) ICAR

590. Read the following statement and mark your answer according to code given below:
Principle of extension is to:
I. Plan for Gram Panchayat members
II. Plan for whole village
III. Plan for youth, women and farmers
IV. Plan for village leaders
Answer code:
(A) I and II are correct (B) II and III are correct
(C) III and IV are correct (D) IV and IV are correct

591. The components of programme planning are:
(A) Simple (B) Complex
(C) Easy (D) None of these

592. Components of programme planning is/are:
(A) People their needs (B) Useful technology
(C) The educational process (D)Analysing situation
(E) All of these

593. Principles of programme planning is/are:
(A) It is based on an analysis f the facts of the situation
(B) Based on the needs and interests of the local people
(C) It determines priorities, define objectives and solution which differ satisfaction
(D) All of these

594. A plan of economic development of India is a:
(A) The Bombay plan (B) Peoples plan
(C) Visvesvaraya plan (D) Gandhian plan

595. Bombay plan was prepared by eight business man of Bombay in:
(A) 1944 (B) 1945
(C) 1946 (D) 1947

596. The main objective of Bombay plan was:
(A) To double up the per capita income within the period of 5 years
(B) To double up the per capita income within the period of 10 years
(C) To double up the per capita income within the period of 15 years
(D) To double up the per capita income within the period of 20 years

597. How many stages of Bombay plan:
(A) The people plan
(B) The Visvesvaraya plan
(C) The Gandhian plan
(D) All of these

598. People's plan was prepared by:
(A) M.N. Roy (B) M. Visvesvaraya
(C) S.N Agarwal (D) None of these

599. Visvesvaraya plan was prepared by:
(A) M.N. Roy (B) M. Visvesvaraya
(C) S.N Agarwal (D) None of these

600. Gandhian plan was prepared by:
(A) M.N. Roy (B) M. Visvesvaraya
(C) S.N Agarwal (D) None of these

601. Basic principle of people's plan was:
(A) Production for uses distinct from production for exchange
(B) To raise the country's economic efficiency to a level at which the average citizen would be able to find employment and to earn a livelihood
(C) To raise the material as well as the cultural level of the masses to a basic standard of living within a period of 10 years.
(D) All of these

602. Basic principle of Visvesvaraya plan was:
(A) Production for use s distinct for production for exchange
(B) To raise the country's economic efficiency to a level at which the average citizen would be able to find employment and to earn a livelihood
(C) To raise the material as well as the cultural level of the masses to a basic standard of living within a period of 10 years.
(D) All of these

603. Basic principle of Ganghian plan was:
(A) Production for use s distinct for production for ex change
(B) To raise the country's economic efficiency to a level at which the average citizen would be able to find employment and to earn a livelihood
(C) To raise the material as well as the cultural level of the masses to a basic standard of living within a period of 10 years.
(D) All of these

604. The concept of T and V system was evolved in:
(A) 1971 (B) 1972
(C) 1973 (D) 1974

605. The concept of T and V system was evolved in 1973 by:
(A) Daniel Benor (B) Albert Mayer
(C) F. L. Bryne (D) Arthur T.Mosher

606. T and V system firstly evolved in:
(A) France (B) Turkey
(C) Nigeria (D) USA

607. Government sponsored rural development prograamme is:
(A) Shantineketan programme
(B) Sarvodaya programme
(C) T and V
(D) None of these

608. T and V system introduced in India in:
(A) 1973 (B) 1974
(C) 1075 (D) 1976

609. T and V system is introduced in Indian in 1974 in:
(A) Rajasthan canal area
(B) Chambal command area in MP
(C) Both (A) and (B)
(D) None of these

610. The basic objective of T and V system is/are:
(A) To improve the effectiveness of agricultural extension
(B) Transfer of technical literacy increases agricultural production
(C) Transfer of technical literacy increases agricultural production
(D) To provide training to youth for agricultural and home development

611. Training and visit programme also called:
(A) Intensive Agricultural Extension
(B) Sarvodaya programme
(C) Indian Village Scheme
(D) Shatineketan programme

612. The credit for the success of KVK is goes to:
(A) Dr. MS. Swaminathan
(B) Dr. Vergees Kurien
(C) Dr. Chandrika Prasad
(D) Dr. B.P. Pal

613. In the panchayati raj system, village panchayat is the:
(A) Lowest unit (B) Middle unit
(C) Higher unit (D) Highest unit

614. General body of panchayat is:
(A) Zila sabha (B) Tehsil sabha
(C) Block sabha (D) Gram sabha

615. The basic idea and concept of community development propgramme was borrowed from:
(A) England (B) USA
(C) Canada (D) China

616. 4H-Club was established in:
(A) 1900 (B) 1910
(C) 1920 (D) 1930

617. 4 H – Club was established in:
(A) Russia (B) Canada
(C) America (D) Japan

618. 4H-Club means Heart, Head, Hand and Health were explained by:
(A) Dr. MS. Swaminathan (B) Dr. Vergees Kurien
(C) Dr. Chandrika Prasad (D) A.B. Graham

619. Panchayati raj was first started at:
(A) Nagaur (Rajasthan) (B) Andhra Pradesh
(C) Both (A) and (B) (D) None of these

620. Panchayati raj was started in:
(A) 1958 (B) 1959
(C) 1960 (D) 1961

621. The main objective of KVK is:
(A) To improve training at the kendras or out of the centre
(B) Transfer of technical literacy increases agricultural production
(C) To provide training to youth for agricultural and home development
(D) To improve the effectiveness of agricultural extension

622. First KVK was started at Pandicherry in 1974 with the aim of:
(A) To improve training at the kendras or out of the centre
(B) Transfer of technical literacy increases agricultural production
(C) To provide training to youth for agricultural and home development
(D) To improve the effectiveness of agricultural extension

623. The goal of Extension Education is to promote:
(A) Social status of the farmers
(B) Area of the new crop
(C) Scientific outlook
(D) Production of the crops

624. The basic element of the core of Extension Education is:
(A) Men himself (B) Man and environment
(C) Man created devices (D) All of these

625. Principle of extension education is:
(A) Learning by doing (B) Learning by seeing
(C) Learning by reading (D) Learning by hearing

626. Goal of extension education is:
(A) To promote income of farmers
(B) To promote production of the crops
(C) To promote new crops
(D) To promote area of crops

627. Purpose of extension evaluation is:
(A) To identify the weak points
(B) To identify the strong points
(C) To identify the gaps and errors
(D) To identify all the above

628. The fundamental objective of extension education is:
(A) To provide the farmer knowledge and help that will enable him to increase his income
(B) To encourage the farmer to grow own food, eat well and live well
(C) The development of the people
(D) None of these

629. Extension education is:
(A) A discipline
(B) A profession
(C) Discipline and profession both
(D) None

630. The objectives of education are decided by:
(A) Social sciences (B) Physical sciences
(C) Normative sciences (D) Biological sciences

631. Less intensive and less effective method of extension is:
(A) Individual (B) Group
(C) Mass (D) Both (A) and (B)

632. Who coined the term Green revolution?
(A) Willianm Gadd (B) M.S. Swaminathan
(C) Norman e. Borlaoug (D) C. Subramanian

633. Extension education is different from formal education in a sense that:
(A) We start from practical and end up with theory
(B) We start from theory and end up with practical
(C) We teach practical only
(D) We teach theory only

634. Extension education aims at:
(A) Increasing agricultural production
(B) Increasing Milk production
(C) Changing knowledge, skills and attitudes
(D) Changing administrative system

635. In extension education communication is a:
(A) Process (B) Method
(C) Idea (D) Expression of thoughts

636. To induce emotional participation of the local community and to create a conductive psychological climate for adoption of a new practice, the method of extension generally applied is:
(A) Competition (B) Exhibitions
(C) Fair (D) Field trails

637. The method of extension education for effectively impressing the people about a particular recommended practice suitable for a particular area is:
(A) Radio (B) Tours
(C) Charts (D) Demonstrations

638. The objective of education are decided by:
(A) Biological science (B) Physical science
(C) Normative science (D) Social science

639. The extension of scientific knowledge to villagers by specialists of the same area is known as the principle of:
(A) Panchayat Raj (B) Local leadership
(C) Modernization (D) None of them

640. In extension we prefer:
(A) Laiser fair approach (B) Democrative approach
(C) Automatic approach (D) None of these

641. The two important principles of extension education are:
(A) Participation and leadership
(B) Participation and class room teaching
(C) Leadership and classroom teaching
(D) Leadership and subject matter specialist

642. Which of the following extension programmes linked farm plan with credit for farmers?
(A) Integrated Rural Development Programme
(B) Community Development Programme
(C) Intensive Agricultural Development Programme
(D) Etawah Pilot Project

643. Which one of the following categories the discipline of extension education belong?
(A) Pure science
(B) Applied biological science
(C) Applied behaviour science
(D) Art

644. An extension worker practically explains Japanese method of paddy cultivation to a group of farmers and also show its superiority over the traditional method. In the process, which of the following communication methods have been used by them?
(1) Method demonstration (2) Result demonstration
(3) Field visit (4) Study tour
Select the correct answer using the codes given below:
(A) 1, 2, 3 and 4 (B) 1 and 3
(C) 2 and 3 (D) 1, 2 and 4

645. Consider the following statements:
Extension in the long run means change in:
(A) Knowledge (2) Skill
(3) Work (4) Attitude
Of these statements:
(A) 1, 2 and 3 are correct (B) 1, 3 and 4 are correct
(C) 1, 2 and 4 are correct (D) 2, 3 and 4 are correct

646. Extension workers should be well versed with economic principles:
(A) Never
(B) Partially
(C) Both (A) and (B)
(D) Without it they can't give effective advice from business point of view

647. Extension education focuses upon development of:
(A) An individual (B) A family
(C) A social group (D) A community

648. Aim of extension work is to bring about the changes:
(A) Toward betterment (B) In the heart of peoples
(C) In rural areas (D) None of these

649. The objective of TRYSEM is to provide technical skills to the Rural youth from families below the poverty line. The target group comprises rural youth between the age of:
(A) 20 and 25 (B) 18 and 35
(C) 14 and 35 (D) 14 and 3

650. An extension education job is:
(A) To educate the people
(B) To advice the Urban people
(C) To advice the Rural people
(D) All of these

651. Extension work must be based on:
(A) The interest of Gram Pradhan
(B) The interest of BDO
(C) The needs and interest of the people
(D) None of these

652. In India, the development of poorers from all the sides is possible only by:
(A) More money supply (B) Making strong rules
(C) Extension education (D) All of these

653. In India the development of poorer sections from all the sides possible only by:
(A) Extension education (B) Making strong rules
(C) More money supply (D) All of these

654. The basic element at the core of extension education is:
(A) Man itself (B) Man and Environment
(C) Institutions (D) Farmers

655. Extension education is:
(A) Formal education (B) Informal education
(C) Non formal education (D) All of these

656. The basic objective of extension education is to create opportunities for:
(A) Greater awareness (B) Effective learning
(C) Promoting knowledge (D) Feeling aloof

657. The basic logic applicable in extension education:
(A) Horizontal logic (B) Deductive logic
(C) Inductive logic (D) Lateral logic

658. Education is a process of:
(A) Interaction (B) Learning
(C) Socialization (D) Institutionalization

659. Subject and target group specificity are characteristics feature of:
(A) Formal education (B) Informal education
(C) Non formal education (D) Elementary education

660. The objectives of education are decided by:
(A) Biological science (B) Physical science
(C) Normative science (D) Social science

661. The last stage in the extension education process is:
(A) Evaluation (B) Reconsideration
(C) Adoption (D) Teaching

662. Education is essentially considered as production of desirable change in:
(A) Knowledge (B) Attitude
(C) Practice (D) Behaviour

663. The two important principles of extension education are:
(A) Participation and leadership
(B) Participation and classroom teaching
(C) Leadership and classroom teaching
(D) Leadership and subject knowledge

664. The method of extension education for effectively impressing the people about a particular recommended practice suitable for a particular area is:
(A) Radio (B) Demonstration
(C) Tours (D) None of these

665. To prove the worth of a new practice suited to a particular area, the best method of extension education is:
(A) Method demonstration (B) Result demonstration
(C) Field trials (D) Farm visits

666. The steps in extension teaching was put forward by:
(A) Paul Leagans (B) Curl Lewin
(C) Ensminger (D) Wilson & Gallup

667. For effective Extension teaching, one must use:
(A) Combination of appropriate aids
(B) Only visual aids
(C) Only audio aids
(D) Printed literature

668. Correct sequence of steps in extension teaching is:
(A) Attention – Interest – Desire – Conviction – Action - Satisfaction
(B) Desire – Interest – Attention – Conviction – Action - Satisfaction
(C) Interest – Attention – Desire – Conviction – Action - Satisfaction
(D) Action – Attention – Interest – Conviction – Desire - Satisfaction

669. Which methods of extension teaching is a way of showing people the value of an improved practice:
(A) Result demonstration
(B) Method demonstration
(C) Newspaper
(D) Circular letter

670. Leaflet is a single sheet of paper folded to make a ———— page piece of printed matter:
(A) 4 (B) 6 (C) 8 (D) 10

671. For one talk, how many flash cards should be used?
(A) 10 – 12 (B) 12 – 14
(C) 14 – 18 (D) 20 – 22

672. Steps of extension teaching are:
(A) Source, Message, Channel, Recommendation
(B) Awareness, Interest, Trial, Evaluation
(C) Attention, Interest, Desire, Conviction, Action and Satisfaction
(D) None of these

673. The process of arranging situation in which the things to be learned are called to the attention of the learners, their interest developed, desire aroused and action promoted:
(A) Communication (B) Training
(C) Teaching (D) Learning

674. Steps in teaching is shortly called as:
(A) AIETA (B) AIDCAS
(C) KPDC (D) AIDACS

675. In extension teaching the teacher should first know:
(A) Subject matter (B) Teaching aids
(C) Place of teaching (D) Attitude of the learner

676. Effective teaching is:
(A) For informing people
(B) For transforming people
(C) Like filling a bucket
(D) Done by trial and error

677. When an extension worker approaches large number of people, spread over large areas for disseminating new improved production technology, it is called:
(A) Mass Communication
(B) Undirectional communication
(C) Public communication
(D) Group communication

678. Written form of extension teaching methods does not include:
(A) Bulletins (B) News articles
(C) Circular letters (D) None of these

679. To show the relative worth of a new practice, the extension method best suited is:
(A) Result demonstration
(B) Method demonstration
(C) Campaign
(D) Group discussion

680. Result demonstration is conducted by:
(A) Researcher (B) Extension worker
(C) Farmer (D) All of these

681. Method demonstration works on the principle of:
(A) Learning is believing (B) Learning by teaching
(C) Seeing and doing (D) Learning by doing

682. Result demonstration is done in:
(A) Research station (B) Farmer's field
(C) District farm (D) None of these

683. Result demonstration works on the principle of:
(A) Learning is believing (B) Learning by teaching
(C) Seeing is believing (D) Learning by doing

684. An extension method that can also be used as a group teaching method:
(A) Campaign (B) Method demonstration
(C) Office calls (D) Farm and home visits

685. Education cause:
(A) Change in knowledge (B) Change in attitude
(C) Change in skills (D) All of these

686. Which extension method is evolved in result demonstration, posters and movies?
(A) Spoken (B) Visual
(C) Written (D) None of these

687. The term extension was first used by:
(A) USA (B) England
(C) India (D) China

688. The basic unit of I.R.D.P. programme is:
(A) A village
(B) A community development block
(C) A family
(D) A district

689. Extension education is:
(A) A discipline
(B) A profession
(C) Discipline and profession both
(D) None o these

690. Agriculture became state subject in year:
(A) 1910 (B) 1919
(C) 1947 (D) 1952

691. When did ARS service start?
(A) 1947
(B) 1952
(C) 1973
(D) 1982

692. Extension education aims at:
(A) Increasing agricultural production
(B) Increasing milk production
(C) Changing knowledge, skills and attitudes of the participants
(D) Changing administrative system

693. Sh. Rabindra Nath Tagore established a rural reconstruction institute in Shantineketan in:
(A) 1921
(B) 1922
(C) 1923
(D) 1924

694. The main principles of extension education is:
(A) Principles of interest and need
(B) Principles of cultural change
(C) Principles of learning by doing
(D) Principles of leadership
(E) All of these

695. Rural welfare movement on a mass scale was:
(A) Work at Shantineketan
(B) Gurgaon experiment
(C) Marthandam Programme
(D) Indian Village Service

696. Self help with internate expert counsel is the technique of:
(A) Work at Shantineketan
(B) Gurgaon experiment
(C) Marthandam Programme
(D) Indian Village Service

697. The aim of Etawah Pilot Project was:
(A) Introducing intensive work on the rural recontraction front
(B) Promotion of cottage industries and spread of literacy
(C) The development of village panchayat
(D) None of these

698. Who is the founder of Indian Village Service (IVS):
(A) Arthor T. Mosher
(B) B.N. Gupta
(C) Both (A) and (B)
(D) None of these

699. The main features of Sarvodaya Programme were:
(A) Simplicity and non violence
(B) Holiness of laboures
(C) Reconstruction of human value
(D) All of these

700. The main aims of Sarvodaya Programme were:
(A) Raising the standard of living
(B) Scientific development of agriculture
(C) Promotion of cottage industries and spread of literacy
(D) Spread of medical and health facilities
(E) The development of village panchayat
(F) All of these

701. The intensive Agricultural District Programme was popularly known as:
(A) Package programme
(B) Knowledge programme
(C) Innovative Programme
(D) None of these

702. Government sponsored and aimed adoptment of the Gandhian ideas of Gram Swaraj is:
(A) Work at Shantineketan
(B) Gurgaon experiment
(C) Firka Development Scheme
(D) Indian Village Service

703. The main aims of Firka development scheme of Madras were:
(A) Educational and economic improvements
(B) Sanitary improvement
(C) Making the people self confident
(D) All of these

704. The package programme name was given of I.A.D.P. due to:
(A) Collective application of all improved practices
(B) Simultaneous application of all improved practices
(C) Collective and simultaneous application of all improved practices
(D) None of these

705. In which scheme, the Collector, who was primarily responsible for the successful working of the scheme in the district:
(A) Work at Shantineketan
(B) Gurgaon experiment
(C) Firka Development Scheme
(D) Indian Village Service

706. IADP programme was started in 1960 in the month of:
(A) July
(B) August
(C) September
(D) October

707. Etawah Pilot Project was started in 1948 in which month:
(A) July
(B) August
(C) September
(D) October

708. Extension education centres around:
(A) Farmers or rural peoples
(B) Innovation or inventions
(C) Extension workers
(D) All of these

709. The technology development of education can be related to various field of extension like:
(A) Agriculture
(B) Home Science
(C) Agriculture and Home Science
(D) None of these

710. Agricultural extension is concerned with agricultural education assisting farmers to bring about continuous improvement in their:
(A) Physical
(B) Economic
(C) Social
(D) Physical, economic and social

711. Agricultural extension makes available to the farming community to solve the problems of agriculture including:
(A) Animal husbandry
(B) Gardening
(C) Agricultural engineering
(D) All of these

712. Learning is motivated by:
(A) Self
(B) Group
(C) Environment
(D) All of these

713. Learning is an internal process mainly controlled by:
(A) The learner (B) The teacher
(C) The subject matter (D) The learning situation

714. Learning the rope refers to:
(A) Training (B) Socialization
(C) Both (A) and (B) (D) None of these

715. NAARM, Hyderabad was established to train scientists in which year?
(A) 1949 (B) 1973
(C) 1976 (D) 1986

716. Learning is a process which continues:
(A) Upto college education (B) Upto employment
(C) Life long (D) Occasionally

717. Which is not a principle of learning:
(A) Learning is the growth like and continuous
(B) Learning should be meaningful
(C) Learning should be challenging and satisfying
(D) Learning should not develop functional understanding of learners

718. Which one of the following is not an element of good learning situation:
(A) Physical facilities (B) Extension worker
(C) Subject matter (D) Transfer process

719. The significant aspect of non formal learners:
(A) Undirectional flow of knowledge
(B) Flexible curriculum
(C) Hierarchically structure system
(D) Homogenous learners

720. Learning is a internal process mainly controlled by:
(A) Teacher (B) Learner
(C) Subject matter (D) Teaching materials

721. The process of acquiring or imparting the ability to perform a behavioral pattern through experience and practice:
(A) Communication (B) Learning
(C) Perception (D) Adoption

722. Learning become the most rapid by maximizing concentration on:
(A) One sense at a time
(B) Two sense at a time
(C) Three sense at a time
(D) All the sense simultaneously

723. "People learn more rapidly and permanently when the learning experience is pleasant or enjoyable" relate to:
(A) Law of readiness (B) Law of exercise
(C) Law of effect (D) Law of belonging

724. To Do, By Doing and For Doing are the fundamental objectives of:
(A) Learning (B) Teaching
(C) Training (D) Perception

725. The learning curve in teaching follows a:
(A) S-Shape (B) U-Shape
(C) V-Shape (D) n - Shape

726. The experimental technique of learning emphasis on:
(A) Use of psychoanalysis
(B) Use of simulation games
(C) Use of psychiatry
(D) Use of learner groups

727. Progressive behaviour adaptation of an individual to a stimulus is called:
(A) Learning (B) Teaching
(C) Perceiving (D) Listening

728. Learning that suddenly appears when a reward or incentives for performance is given:
(A) Discovery learning
(B) Individualized learning
(C) Rote learning
(D) Latent learning

729. The general approach in progressive mode of learning is:
(A) Authoritarian (B) Paternalistic
(C) Humanitarian (D) Democratic

730. New learning must be related to the old:
(A) Principle of readiness
(B) Principle of association
(C) Principle of practice
(D) Principle of timing

731. Life long process of learning happening through daily experience and exposure:
(A) Informal education
(B) Non formal education
(C) Formal education
(D) Professional education

732. Learning must result in functional understanding:
(A) Principle of readiness
(B) Principle of association
(C) Principle of practice
(D) Principle of timing

733. Learning must make sense to the learner:
(A) Principle of readiness (B) Principle of clarity
(C) Principle of practice (D) Principle of timing

734. A relatively enduring change in overt and covert response as a result of perceived stimulus is called:
(A) Endurance (B) Attribute
(C) Learning (D) Perception

735. Adult learning is:
(A) Problem centered (B) Subject centered
(C) Instructor centered (D) Institute centered

736. Adult literacy programme can be considered:
(A) Formal education (B) Informal education
(C) Non-formal education (D) All of these

737. Which of the following statements represents true concepts of extension evaluation?
(A) Extension evaluation begins before the start of an extension programme
(B) Extension evaluation begins at the mid step of the programme
(C) Extension evaluation begins at the last step of the programme
(D) Extension evaluation begins inbuilt in each step of programme

738. Important element of evolution process is:
(A) Observations, criteria to observations and making some decisions
(B) Observations and criteria to observations
(C) Observations and collection of some information and final decision
(D) All of these

739. The science that deals with adult learning is:
(A) Androgogy (B) Pedagogy
(C) Pedology (D) Edaohology

740. The emphasis of androgogical method in comparison with pedogological method is on:
(A) Content and Process (B) Content
(C) Process (D) None of these

741. Androgogy literally means:
(A) Adult leading (B) Adult learning
(C) Child leading (D) Child learning

742. Adult learning is:
(A) Problem centered (B) Subject centered
(C) Instructor centered (D) Institute centered

743. The central problem in adult learning is:
(A) Age (B) Literacy
(C) Sex difference (D) Motivation

744. Adult literacy programme can be considered as:
(A) Formal education (B) Informal education
(C) Non-formal education (D) All of the above

745. "Learning the rope" refers to:
(A) Training (B) Socialization
(C) Both (A) and (B) (D) None of the above

746. Learning is facilitated mostly by:
(A) Taste (B) Hearing
(C) Touch (D) Sight

747. Communication is more meaningly:
(A) Extension approach
(B) Subject of broadcasting
(C) Exchange of ideas, facts, feelings etc
(D) Transference of meaning

748. Communication is sharing together and is for:
(A) Apathy (B) Commonness
(C) Differences (D) Popularity

749. Communicator is the characteristic element in the model of:
(A) Leagans (B) Lezwell
(C) Berlo (D) Rogers

750. Leagans model of communication contains:
(A) 6 elements (B) 5 elements
(C) 4 elements (D) 3 elements

751. Leagans model of communication process is:
(A) Communication- Message- Channel- Treatment- Audience- Audience response
(B) Sender- Encoding- Channel- Decoding- Receiver
(C) Communication source- Encoder- Message- Channel- Decoder-Communication receiver
(D) None of these

752. Linear model of communication was given by:
(A) Barlo (B) Weaver
(C) Sanane (D) Leagans

753. Real object taken out of their natural settings is called as:
(A) Objects (B) Specimens
(C) Model (D) Mock-up

754. Sound can be recorded in ———— ways:
(A) 1 (B) 2 (C) 3 (D) 4

755. The disk recording is made by the ———— process:
(A) Mechanical (B) Magnetic
(C) Optical (D) None of these

756. Tape and wire recording are made by ———— process:
(A) Magnetic (B) Mechanical
(C) Optical (D) None of these

757. The movie film recording is made by———— process:
(A) Magnetic (B) Mechanical
(C) Optical (D) None of these

758. Which one of the following is not a example of still pictures:
(A) Drawing (B) Painting
(C) Etching (D) None of these

759. Flash cards are a series of illustrated cards when flashed or presented before a group in proper sequence tell a:
(A) Part of story (B) Complete story
(C) Incomplete story (D) None of these

760. A B C of posters means:
(A) Attractive, Brief and Complete
(B) Attractive, Brief and Clear
(C) Attractive, Brief and Creative
(D) None of these

761. ———— consist of a series of individual charts which are bound together and hung on a supporting stand:
(A) Bar charts (B) Job charts
(C) Flip charts (D) Tree charts

762. Who coined the word Green revolution:
(A) Willianm Gadd (B) M.S. Swaminathan
(C) Norman E. Borlaoug (D) C. Subramanian

763. Which of the following scientist has given the theory of information of communication?
(A) Shannon and Weaver (B) Leagon
(C) Berlo (D) Aristotle

764. Which one is not the part of communication module?
(A) Communicator (B) Message
(C) Advice (D) Audience

765. The number of elements in the communication model of Berlo:
(A) 4 elements (B) 5 elements
(C) 6 elements (D) 7 elements

766. The smallest meaningful unit of speech is called:
(A) Heuristic (B) Productive
(C) Morphemes (D) Phonemes

767. The communication model better applicable for a person reading newspaper:
(A) Shannon-Weaver model
(B) Leagan's model
(C) Berlo's model
(D) Barker Wiseman model

768. The basic elements in Aristotle's model of communication is:
(A) Speaker (B) Speech
(C) Listener or Audience (D) All of these

769. The S-M-C-R model of communication is given by:
(A) Berlo (B) Laswell
(C) Rogers (D) Leagans

770. The influence of mass media according to hypodermic model of communication is:
(A) Direct (B) Indirect
(C) Secondary (D) No effect

771. Communication IQ was studied by:
(A) Bhanja (B) Paul Vegans
(C) Hovland (D) None of these

772. Communication competency concept was stressed by:
(A) Wiesmann and Backlund (B) Hymes
(C) Sannon and Weaver (D) Thayer

773. Source-transmitter-signal-receiver-destination is the model of communication given by:
(A) Aristotle (B) Shannon and Weaver
(C) Laswel (D) Westley and Machean

774. The first basic persuasive communication model:
(A) Aristotle (B) Shannon and Weaver
(C) Laswell (D) Berlo

775. The communication model that matches well with political symbolism:
(A) Aristotle (B) Shannon and Weaver
(C) Laswell (D) Berlo

776. Mathematical theory of communication was put forwarded by:
(A) Aristotle (B) Shannon and Weaver
(C) Laswell (D) Berlo

777. The amount of information that can be transmitted per unit of time:
(A) Redundancy (B) Channel capacity
(C) Coupling (D) Rhetoric

778. A communication model developed based on behavioral research:
(A) EM Rogers (B) Lazerfield
(C) D Berlo (D) Stevens

779. Which of the following extension 4 sources of communication do you consider most credible for non progressive/ remote village situation:
(A) Radio (B) Demonstration
(C) Print media (D) Exhibition

780. Which one is not the mode of radio broadcast?
(A) Straight talks (B) Interviews
(C) Songs (D) Lecture

781. Which is the traditional media of communication?
(A) Print (B) Tape recorder
(C) Radio (D) Ram leela

782. Which of the following is a traditional and simple aid of communication?
(A) Model (B) Puppet
(C) Radio (D) TV

783. Attitude of communication source is important from the point of view of:
(A) Code, content, treatment
(B) Source, self, receiver
(C) Encode, decode, feedback
(D) Self, audience, content

784. The degree to which individuals involved in a communication act differ in certain characteristics:
(A) Homophily (B) Credibility
(C) Empathy (D) Heterophily

785. The communication skills that get emphasized on listening and reading:
(A) Attitude (B) Effect
(C) Encoding (D) Decoding

786. Speaking and writing involve which communication skills:
(A) Encoding (B) Decoding
(C) Feedback (D) Knowledge

787. The process of selecting information that is relevant to a problem while ignoring distractions:
(A) Selective encoding (B) Selective decoding
(C) Selective comparison (D) None of these

788. Communication behaviour essentially consist of:
(A) Information seeking
(B) Information processing
(C) Information disseminating
(D) All of these

789. Communication barriers include:
(A) Social (B) Semantic
(C) Technical (D) All of these

790. The communication barrier wherein people attach different meaning to things:
(A) Semantic barrier (B) Personal barrier
(C) Technical barrier (D) None of these

791. Problem of homogeneity is the communication barrier related to:
(A) Source (B) Message
(C) Channel (D) Receiver

792. Providing too much information is a communication barrier related to:
(A) Communicator (B) Message
(C) Treatment (D) Receiver

793. Which among the following calls for effective communication:
(A) Listening (B) Learning
(C) Training (D) Looking

794. Indirect and inferential feedback is available in:
(A) Group media (B) Mass media
(C) Audio media (D) Visual media

795. The communication type best apply to largely illiterate audience:
(A) Written communication
(B) Informal communication
(C) Oral communication
(D) Formal communication

796. Reading newspaper is:
(A) Intra personal communication
(B) Inter personal communication
(C) Mass communication
(D) Dyadic communication

797. Constant feedback between the sender and the receiver is the characteristic feature of:
(A) Inter personal communication
(B) Mass communication
(C) Written communication
(D) Vertical communication

798. The main problem with mass communication is:
(A) Costly (B) No local dialect
(C) Same information (D) Cumbersome

799. Mass communication is generally identified with modern mass media like:
(A) Television and radio
(B) Cinema and news paper
(C) Farm and home visit
(D) All of these

800. An informal communication type involved in the spread of runner:
(A) Horizontal communication
(B) Vertical communication
(C) Diagonal communication
(D) Grapevine communication

801. Cybernetics is the science of:
(A) Decision making (B) Organizations
(C) Ecology (D) Communication

802. Communication breakdown is a common feature of:
(A) Ulterior transaction (B) Cross transaction
(C) Straight transaction (D) Vertical transaction

803. The art of communication is:
(A) Learning (B) Listening
(C) Feeling (D) None of these

804. Which of the following are the elements of communication process?
(A) Source – Message – Channel – Receiver
(B) Production process – Farmers – Farm – Farm business
(C) Research – Education – Extension – Training
(D) Information – Interest – Evaluation – Persuasion

805. ____________ is a unit of two or more people in reciprocal communication and interaction with each other:
(A) Family (B) Group
(C) Society (D) Social stratification

806. The term community development appears to have originated from in England in 1946:
(A) London (B) France
(C) Cambridge (D) Belfast

807. Among the community, voluntary agencies are characterized by:
(A) Less intensity (B) Greater continuity
(C) Greater commitment (D) Less accountability

808. ____________ is the process by which two or more people exchange ideas, facts, feelings or impressions.
(A) Adoption (B) Teaching
(C) Learning (D) Communication

809. Leagans model of communication process is
(A) Communication-message-channel-treatment-Audience-Audience response
(B) Sender-encoding-channel-decoding-receiver
(C) Communicationsource-Encoding-Message-Channel-Decoder-Communication receiver
(D) None of these

810. The term 'intercommunication' means:
(A) Communication on any issue
(B) Communication between two persons
(C) Communication between various groups
(D) All of these

811. The term 'human communication' means:
(A) Exchange of service
(B) Exchange of liabilities
(C) Exchange of ideas
(D) Exchange of ideas with commodity

812. Which one of the sources of the communication is most credible for non-progressive/remote village situation?
(A) Farm radio (B) Demonstration
(C) Printed material (D) Exhibition

813. In extension education exhibition is helpful for:
(A) Individual learning (B) Group communication
(C) Mass communication (D) None of these

814. In extension education communication is a:
(A) Process (B) Method
(C) Idea (D) Expression of thoughts

815. The authority of PACS is known as:
(A) President (B) Secretary
(C) General body (D) Executive body

816. Action-reaction inter-dependable communication is termed as:
(A) Barrier (B) Feed back
(C) Fidelity (D) Treatment

817.

List-I	List-II
(a) Demonstration means	(1) Short type demonstration
(b) Method demonstration is	(2) Showing by doing
(c) Result demonstration is	(3) Long type demonstration
(d) National Demonstration is	(4) Front line extension project

Code	a	b	c	d
(A)	2	1	3	4
(B)	3	4	2	1
(C)	4	3	2	1
(D)	1	2	3	4

818. Under limited resource of a manpower, time and money, we should select:
(A) Individual contact methods
(B) Group contact methods
(C) Mass contact methods
(D) All of these

819. An extension method that can be considered both spoken and visual:
(A) Posters (B) Radio
(C) Campaign (D) Agricultural clinics

820. Flag method is classified under:
(A) Spoken method (B) Written method
(C) Group method (D) Mass method

821. Agricultural clinic is classified under:
(A) Group contact method
(B) Individual contact method
(C) Mass contact method
(D) All of these

822. Educational television is classified under:
(A) Broadcast media (B) Screen media
(C) Printed media (D) None of these

823. Communication is the process by which message are transferred from source to:
(A) Channel (B) Message
(C) Receiver (D) Effect

824. Which of the following is not a method of group communication?
(A) Sabarmati (B) Sevagram
(C) Shriniketan (D) Bhatinda

825. Which of the following is not a method of group communication?
(A) Conference (B) Field trips
(C) National demonstration (D) Farm and home visits

826. The most basic function of communication:
(A) Information (B) Instruction
(C) Integration (D) Influence

827. Communication is more meaningfully:
(A) Extension approach
(B) Subset of broadcasting
(C) Exchange of facts and feeling
(D) None of these

828. Level of communication are:
(A) Conventional (B) Exploratory
(C) Participative (D) All of these

829. On which level, communication is fact and problem oriented:
(A) Explorative level (B) Participative level
(C) Conventional level (D) None of these

830. The essence of communication is:
(A) Sharing information
(B) Imparting knowledge
(C) Sharing understanding
(D) Transmitting information

831. The art of communication is:
(A) Learning (B) Listening
(C) Feeling (D) None of these

832. The most important element in the communication process:
(A) Source (B) Receiver
(C) Channel (D) Message

833. Action-reaction inter-dependence in communication process is:
(A) Feedback (B) Homophily
(C) Competence (D) Fidelity

834. Which of the following is not communication process:
(A) Source (B) Encoding
(C) Transmission (D) Meetings

835. Communication process is stabilized by:
(A) Feedback (B) Credibility
(C) Channel (D) Empathy

836. The type of audience suited for communication process is called:
(A) Active audience (B) Potential audience
(C) Available audience (D) None of these

837. The communication channels most active in diffusion process:
(A) Mass media channel (B) Interpersonal channel
(C) Both (A) and (B) (D) None of these

838. The communication channel most suited to inform the audience about the existence of an innovation:
(A) Mass media channel (B) Interpersonal channel
(C) Both (A) and (B) (D) None of these

839. Mass communication is intended to be consumed immediately hence it is called:
(A) Rapid (B) Transient
(C) General (D) None of these

840. Feedback is inferential in case of communication made through:
(A) Village Level Worker
(B) Group meeting
(C) Radio
(D) Method Demonstration

841. Mass media had direct, immediate and powerful effects on a mass audience:
(A) Two step flow model
(B) One step flow model
(C) Hypodermic needle model
(D) None of these

842. The controlling of the flow of information through a communication channel:
(A) Gate keeping (B) Gate locking
(C) Gate security (D) Gate guard

843. A system of decision centres interconnected by patterned flows of information:
(A) Communication techniques
(B) Communication network
(C) Communication level
(D) Communication channel

844. The number of communication network links possible in a social system of 100 members:
(A) 1000 (B) 10000
(C) 9876 (D) 4950

845. Two linked individuals in a network having personal communication networks that overlap:
(A) Communication proximity
(B) Communication preference
(C) Communication acceptance
(D) None of these

846. The communication network ideally suited for imparting knowledge change:
(A) Star network (B) Wheel network
(C) Circle network (D) Slash network

847. Which of the following is not a decentralized communication networks:
(A) Star (B) Chain
(C) Circle (D) Both (A) and (B)

848. ATIC under NATP functions under:
(A) State Department of Agriculture
(B) Ministry of Agriculture, GOI
(C) Ministry of RD, GOI
(D) ICAR

849. AKIS expands to:
(A) Agricultural Knowledge and Information Syndrome
(B) Agricultural Knowledge and Information Service
(C) Agricultural Knowledge and Information Science
(D) Agricultural Knowledge and Information System

850. Agricultural and rural Industries Information System Network to strengthen Small and Micro Enterprises (SMEs):
(A) AFPINET (B) ARISNET
(C) AGRISNET (D) VISTARNET

851. Land Information System network linking all institutions involved in land and water management for agricultural productivity and production systems:
(A) LISNET (B) ARINET
(C) AGRISNET (D) VISTARNET

852. Agricultural Research Information System Network:
(A) HORTNET (B) ARISNET
(C) AGRISNET (D) VISTARNET

853. Agricultural Extension Information System Network:
(A) COOPNET (B) FERTNET
(C) AGRISNET (D) VISTARNET

854. A network that connect agricultural produce wholesale markets and rural markets:
(A) AFPINET (B) AGMARKNET
(C) HORTNET (D) VISTARNET

855. An infrastructure network upto block level agricultural offices facilitating agricultural extension services and agribusiness activities to user in rural prosperity:
(A) AGRISNET (B) ARISNET
(C) HORTNET (D) VISTARNET

856. The toll free telephone number on kisan call centre is:
(A) 1551 (B) 1331
(C) 5115 (D) 3113

857. When started kisan call centre:
(A) 1970 (B) 1971
(C) 1972 (D) 1973

858. Warna village network project was started in the state of:
(A) Uttar Pradesh (B) Maharastra
(C) Tamilnadu (D) Madhya Pradesh

859. The e-choupal concept in extension delivery is promoted by:
(A) ITC (B) MSSRF
(C) KINFRA (D) CAPART

860. "Rice Doctor" an interactive pest diagnostic expert system is developed by:
(A) ICAR (B) IRRI
(C) ICRISAT (D) INIBAP

861. Krishi Darshan programme in India is telecasted from:
(A) Mumbai (B) Kolkata
(C) New Delhi (D) Bangalore

862. Prasar Bharti came into existence on:
(A) 15th October, 1995 (B) 18th January, 1996
(C) 20th December, 1996 (D) 23rd January, 1997

863. The 4-H club colours are:
(1) Green (2) Yellow
(3) White (4) Red
(5) Blue
Select the right answer using the codes given below:
(A) 1 and 2 (B) 3 and 4
(C) 1 and 3 (D) 1 and 5

864. The 4-H club motto is:
(A) To make the better best
(B) To make the best better
(C) To make the good better
(D) To make the good best

865. The 4-H clubs are functioning in USA, 4 –H implies:
(A) Head, Hand, Heart, Health
(B) Head, Home, Health, Hands
(C) Hand, Health, Hope, Heart
(D) Home, Heart, Health, Head

866. The 4 H of communication are:
(A) Informal, Instruction, Influence and Integration
(B) Intermediation, Information, Interest and Instruction
(C) Information, Instruction, Influence and Integration
(D) Information, Integration, Influence and Instruction

867. The abbreviation COIK in COIK fallacy means:
(A) Clear Only If Known
(B) Clarity Of Informational Knowledge
(C) Capacity Of Indigenous Know-how
(D) Concept of Imaging Knowledge

868. The COIK fallacy is associated with which of the following:
(A) Communication channels
(B) Audio visual aids
(C) Communicator
(D) Receiver

869. The 4-H club was forwarded by:
(A) O.H. Benson (B) A.B. Graham
(C) Seaman A. Knapp (D) A.B. Ross

870. The motto of 4 H club is to make:
(A) The poorer rich (B) Better best
(C) Adult motivated (D) Child educated

871. Which one is not the part of communication module:
(A) Communicator (B) Message
(C) Advice (D) Audience

872. Round table discussion is called:
(A) Panel (B) Symposium
(C) Forum (D) Workshop

873. Intelligence Quotient determines:
(A) Capacity age (B) School going age
(C) Chronological age (D) Mental age

874. The extension work must be based on needs and interests of:
(A) People (B) Leader
(C) Worker/Official (D) Government

875. Which of the following is not the element of communication process?
(A) Message (B) Feedback
(C) Channel (D) Fidelity

876. A person receiving favourable response from other is called:
(A) Authoritative (B) Psychological support
(C) Physical support (D) Social support

877. The most conspicuous development in the globalization context has been the:
(A) Information technology influx
(B) Communication technology influx
(C) Information and communication technology influx
(D) None of these

878. Developing world is gradually catching up with the technological advancement to solve its:
(A) Sociological problem (B) Economical problem
(C) Both (A) and (B) (D) None of these

879. The set of activities that facilitates, by electronic means the capturing, storage, processing, transmission and display of information is:
(A) Communication
(B) Information and communication technology
(C) Agricultural extension
(D) Telecommunication technology

880. The objective of Information and Communication Technology is/are:
(A) To elucidate the role of some of the major ICT initiatives in rural development
(B) To study the impact of ICT efforts on agriculture, health, women empowerment and panchayat raj
(C) To print out the impediments for the effective implementation of ICT based project
(D) All of these

881. ICT tackle key problems in mainly:
(A) Agriculture and cooperation sector
(B) Health and education
(C) Panchayat raj and women empowerment
(D) All of these

882. Which programme is designed to assist their development and develop their information, education and communication base of rural people is:
(A) PRA (B) ICT
(C) AV AIDS (D) None of these

883. ICT programme came into existence in India is:
(A) 1982 (B) 1983
(C) 1984 (D) 1985

884. ICT programme was given a vent in 1984 by the:
(A) Congress Govt. (B) BJP Govt
(C) Janta Party Govt. (D) BSP Govt

885. ICT programme was given a vent in 1984 by Rajiv Gandhi and the technologist:
(A) R. S. Paroda (B) Sam Pitroda
(C) M. S. Swaminathan (D) B.P. Pal

886. A high power national task force on information and software development was set up in:
(A) 1996 (B) 1997
(C) 1998 (D) 1999

887. A high power national task force of information technology and software development was set up in 1998 by the:
(A) Congress Govt. (B) BJP Govt
(C) Janta Party Govt. (D) BSP Govt

888. The national task force in information technology prepared the blue print for making the adoption of IT or a national movement by establishing a wide network of empowered task force at:
(A) Government level (B) Nongovernment level
(C) Both (A) and (B) (D) None of these

889. The new telecommunication policy was enacted in :
(A) 1999 (B) 2000
(C) 2001 (D) 2002

890. The new telecommunication Policy 1999 was stipulated targets in terms of establishing a telecom network with a view to achieving teledensity by 2005:
(A) 5 per cent (B) 7 per cent
(C) 10 per cent (D) 15 per cent

891. The new telecommunication Policy 1999 was stipulated targets in terms of establishing a telecom network with a view to achieving teledensity by 2010:
(A) 5 per cent (B) 7 per cent
(C) 10 per cent (D) 15 per cent

892. The concept of knowledge society means that it will incorporate new knowledge will take over the role of capital and labour in the economy further leading to:
(A) Structural changes (B) Functional changes
(C) Both (A) and B) (D) None of these

893. Sponsoring agency of Cyber Extension Project is:
(A) MANAGE, Hyderabad (B) ICAR, New Delhi
(C) NAARM, Hyderabad (D) CSAUA&T, Kanpur

894. Sponsoring agency of Agriculture Technology Information Centre is:
(A) MANAGE, Hyderabad (B) ICAR, New Delhi
(C) NAARM, Hyderabad (D) CSAUA&T, Kanpur

895. Sponsoring agency of Agricultural Gateway of India is:
(A) MANAGE, Hyderabad (B) ICAR, New Delhi
(C) NAARM, Hyderabad (D) CSAUA&T, Kanpur

896. Sponsoring agency of Helpline Service is:
(A) MANAGE, Hyderabad (B) ICAR, New Delhi
(C) NAARM, Hyderabad (D) CSAUA&T, Kanpur

897. Sponsoring agency of Warna Wired Village Project is:
(A) NIC
(B) Warna Cooperative Complex, Warn district Kolhapur (MH)
(C) Both (A) and (B)
(D) CSAUA&T, Kanpur

898. Sponsoring agency of Gyandoot Project is:
(A) MANAGE, Hyderabad
(B) ICAR, New Delhi
(C) CSAUA&T, Kanpur
(D) Gram Panchyat Dhar Dist. MP

899. Sponsoring agency of Interactive Touch Screen Information , KIOSK is:
(A) Dept. of International Development, Rajiv Gandhi College of Animal Sciences (Pondicherry)
(B) University of Reading (UK)
(C) Both (A) and (B)
(D) CSAUA&T, Kanpur

900. Sponsoring agency of Lokmitra Project is:
(A) National Information Centre
(B) Gram Panchayat Hamirpur dist HP
(C) Both (A) and (B)
(D) CSAUA&T, Kanpur

901. Sponsoring agency of E. choupal is:
(A) Imperial Tobacco Company
(B) EID Parry (Tamilnadu)
(C) Nagarjuna Group
(D) MS Swaminathan Research Foundation, Chennai

902. Sponsoring agency of Rural Information Kiosks is:
(A) Imperial Tobacco Company
(B) EID Parry (Tamilnadu)
(C) Nagarjuna Group
(D) MS Swaminathan Research Foundation, Chennai

903. Sponsoring agency of I-Kisan Portal is:
(A) Imperial Tobacco Company
(B) EID Parry (Tamilnadu)
(C) Nagarjuna Group
(D) MS Swaminathan Research Foundation, Chennai

904. Sponsoring agency of Village Knowledge Centre is:
(A) Imperial Tobacco Company
(B) EID Parry (Tamilnadu)
(C) Nagarjuna Group
(D) MS Swaminathan Research Foundation, Chennai

905. Sponsoring agency of Computer of Wheels is:
(A) Imperial Tobacco Company
(B) Pingali Rajeswari Enterpreneur, A.P.
(C) Nagarjuna Group
(D) MS Swaminathan Research Foundation, Chennai

906. Sponsoring agency of www.doctoranywhere.com is:
(A) Imperial Tobacco Company
(B) EID Parry (Tamilnadu)
(C) Tata Council for Community Initiatives (TCCI), Pune (MS)
(D) MS Swaminathan Research Foundation, Chennai

907. Important features of Cyber Extension are:
(A) Village Information Centres
(B) Institutional support to other ICT projects
(C) Both (A) and (B)
(D) None of these

908. Important features of Agriculture Technology Information Centre is:
(A) Call centre
(B) Expert system
(C) Researchers and farmers linkage
(D) All of these

909. Important features of Agricultural Gateway of India are:
(A) Helpline service through telephones
(B) Researchers-farmers linkage
(C) Both (A) and (B)
(D) None of these

910. Important features of Helpline Service are:
(A) Helpline service through telephones
(B) Researchers-farmers linkage
(C) Both (A) and (B)
(D) None of these

911. Important features of Warna Wired Village Project are:
(A) Wide Area Network (B) Computer booths
(C) Both (A and (B) (D) None of these

912. Important features of Gyandoot Project are:
(A) Soochanalayas
(B) Portal partial recovery
(C) Panchayat community partnership
(D) All of these

913. Important features of Interactive Touch Screen Information Kiosk are:
(A) Information Kiosks text
(B) Picture with sound backup
(C) Both (A) and (B)
(D) None of these

914. Important features of Lokmitra Project are:
(A) Lokmitra Soochana Kendra in 25 panchayat
(B) Panchayat Community partnership
(C) Researcher-farmers linkage
(D) None of these

915. Important features of E-choupal are:
(A) e-choupal (B) Crop specific internet
(C) Both (A) and (B) (D) None of these

916. Important features of Rural Information Kiosks are:
(A) Framers managed Kiosks portal
(B) Public private partnership in applying ICT improvations and content development
(C) Both (A) and (B)
(D) None of these

917. Important features of I-kisan portal are:
(A) e-choupal
(B) Portal in regional language
(C) Both (A) and (B)
(D) None of these

918. Important features of Village Knowledge Centre are:
(A) Knowledge centre
(B) Pro-poor
(C) Pronature prowomen
(D) Community ownership
(E) All of these

919. Important features of computer on wheels are:
(A) Internet on motor bikes (B) Portals
(C) Remote area (D) Resource poor
(E) All of these

920. Important features of www.doctoranywhere.com are:
(A) Telemedicine Service through websites
(B) Telemedicine Service through computer
(C) Both (A) and (B)
(D) None of these

921. Major thrust or areas or level of cyber extension project are:
(A) Post harvest processing
(B) Women and child welfare
(C) Both (A) and (B)
(D) None of these

922. Major thrust or areas or level of Agricultural Technology Information Centre project are:
(A) Agronomic practices (B) Technologies inputs
(C) Both (A) and (B) (D) None of these

923. Major thrust or areas or level of Agricultural Gateway of India are:
(A) Cultivation practices (B) Plant protection
(C) New technologies (D) All of these

924. Major thrust or areas or level of Helpline service are:
(A) Cultivation practices (B) Plant protection
(C) New technologies (D) All of these

925. Major thrust or areas or level of Warna Wired Village Project are:
(A) Cultivation practices (B) Pest and diseases
(C) Marketing (D) All of these

926. Major thrust or areas or level of Gyandoot Project are:
(A) Agricultural produce (B) Market intelligence
(C) Auction rates (D) Land records
(E) All of these

927. Major thrust or areas or level of Interactive Touch Screen Information Kiosk are:
(A) Information about important cattle disease
(B) Management of cattle
(C) Both (A) and (B)
(D) None of these

928. Major thrust or areas or level of Lokmitra Project are:
(A) Classified components (B) Job vacancies
(C) Market rates (D) Tenders
(E) All of these

929. Major thrust or areas or level of E-choupal are:
(A) Market price (B) Cultivation practices
(C) Weather data (D) All of these

930. Major thrust or areas or level of Rural Information kiosks are:
(A) Farm practices (B) Farm advisory
(C) Market intelligence (D) Weather
(E) All of these

931. Major thrust or areas or level of I - Kisan portal are:
(A) Agricultural practices (B) Plant protection
(C) Animal husbandry (D) Weather
(E) All of these

932. Major thrust or areas or level of Computer on wheels are:
(A) Market (B) Weather
(C) Plant protection (D) Animal health
(E) All of these

933. Major thrust or areas or level of www.doctoranwhere.com are:
(A) Medical council
(B) Medical treatment in PHC
(C) Both (A) and (B)
(D) None of these

934. Which ICT project is related to cattle diseases and management of cattle is:
(A) Interactive Touch Screen Information kiosks
(B) E - choupal
(C) Lokmitra project
(D) Gyandoot project

935. Which ICT project is related to medical treatment in PHC is:
(A) Interactive Touch Screen Information kiosks
(B) www.doctoranywhere.com
(C) Lokmitra project
(D) Gyandoot project

936. E-choupal launched by ITC in:
(A) June, 1998 (B) June, 1999
(C) June, 2000 (D) June, 2001

937. ITC's E - choupal is a unique example of using ICT for:
(A) Agricultural development
(B) Animal management
(C) Human health improvement
(D) All of these

938. Which ITC's has already become the largest initiatives among all internet based interventions in rural India is:
(A) E-choupal (B) I-kisan portal
(C) Lokmitra project (D) Gyandoot project

939. E - choupal services render in which crops:
(A) Soybean and Coffee (B) Wheat and rice
(C) Pulses (D) All of these

940. E - choupal covers across the:
(A) MP and Karnataka (B) AP and UP
(C) Maharashtra and Rajasthan (D) All of these

941. Disseminating various farming information to the farmers at various places through ICT centres by:
(A) I- Kisan portal (B) E – choupal
(C) Gyandoot project (D) Lokmitra project

842. Which ICT's provide strong effective coverage and dissemination of livestock out reach information to the farming community, e – mail conferencing system are:
(A) goat-net-cirg.nic.in
(B) livestock-net@cirg.nic.in
(C) fishnet@cirg.nic.in
(D) All of these

943. Under the animal health project funded by DFID Rajiv Gandhi College of Veterinary and Animal Science, Pondicherry in collaboration with University of Reading, UK has installed a:
(A) Interactive Touch Screen Information, Kiosk
(B) Agricultural Technology Information System
(C) Warna Wired Village Project
(D) Agricultural Gateway of India

944. Specific features of Interactive Touch Screen Information, Kiosk are:
(A) Cattle diseases
(B) Management of cattle
(C) Method of acquiring information
(D) All of these

945. Apollo hospital started telemedicine project on:
(A) 1999 (B) 2000
(C) 2001 (D) 2002

946. The ISRO and EDCU has initiated several telemedicine project for human health in:
(A) Remote areas (B) Rural areas
(D) Urban areas (D) Tribal areas

947. Sankara Nethralaya is a super-specility eye hospital in Chennai launched a telemedicine project through satellite connectivity is:
(A) goat-net-cirg.nic.in
(B) livestock-net@cirg.nic.in
(C) fishnet@cirg.nic.in
(D) Tele-ophthalmology

948. www.doctoranywhere.com is a telemedicine service project with aim that headquarters connected with the PHC's with atleast how many doctors always present at headquarter to give advice and medicine to the rural people:
(A) Only one doctor (B) Team of two doctors
(C) Team of 5-6 doctors (D) Team of many doctors

949. Lokmitra project launched in:
(A) May, 2000 (B) May, 2001
(C) May, 2002 (D) May, 2003

950. For Panchayat raj ICT project are:
(A) Lokmitra (B) Gyandoot and Saytu
(C) E-Gram Suvidha (D) All of these

951. Lokmitra Soochana Kendras were initially set up in how many panchayats:
(A) 25 (B) 30 (C) 35 (D) 40

952. Who provide the Lokmitra software and necessary training to the soochaks (operators):
(A) National Information Centre (NIC)
(B) Agricultural Technological Information Centre (ATIC)
(B) Both (A) and (B)
(D) None of these

953. ICT for panchayat raj initiated at Dhar district of MP is:
(A) Lokmitra (B) E-choupal
(C) Gayandoot (D) None of these

954. ICT for panchayat raj was successfully implemented in Maharashra is:
(A) SAYTU (B) E-Gram Suvidha
(C) E-choupal (D) I-Kisan portal

955. ICT is Geographical Information System (GIS) based on management information system is:
(A) SAYTU (B) E-Gram Suvidha
(C) E-choupal (D) I-Kisan portal

956. E-gram suvidha is currently operated in five district of MP:
(A) MP (B) AP
(C) WB (D) UP

957. Name of website of E-Gram Suvidha is:
(A) goat-net-cirg.nic.in
(B) livestock-net@cirg.nic.in
(C) fishnet@cirg.nic.in
(D) www.nic.in

958. ICT programmee is related to women's empowerment is:
(A) Lokmitra (B) Village Knowledge Centre
(C) Gayandoot (D) E-choupal

959. Village Knowledge Centre provided:
(A) Proper training to the women
(B) Imparting skills to the women
(C) Proper training and Imparting skills to the women
(D) None of these

960. The ICT for co-operative development is:
(A) Dairy cooperative
(B) Cooperative societies
(C) Household good cooperative
(D) None of these

961. ICT project 'Dairy cooperative' developed by:
(A) AMUL (B) NDDB
(C) PARAG (D) None of these

962. Anand Milk Union Limited also introduced a ICT programme is:
(A) Dairy cooperative
(B) Cooperative societies
(C) Household good cooperative
(D) Amul cooperative

963. Amul cooperative are:
(A) Milk procurement only
(B) Milk distribution only
(C) Milk procurement and distribution both
(D) None of these

964. Important features of dairy cooperative are:
(A) Measure butter fat content in milk
(B) Test the quality of the milk
(C) Help to make payment to the farmers
(D) All of these

965. Warna Wired Village Project also introduced a ICT programme is:
(A) Dairy cooperative (B) Cooperative societies
(C) Sugar cooperative (D) Amul cooperative

966. ICT programme sugar cooperative on project of Warna Wired Village Project covering 70 villages of:
(A) Kolhapur district of Maharashra
(B) Sangli district of Maharashtra
(C) Kolhapur and Sangli districts of Maharashtra
(D) None of these

967. Village Information Centre of sugar cooperative has providing indepth knowledge and information about:
(A) Sugarcane cultivation
(B) Marketing to the farmers
(C) Both (A) and (B)
(D) None of these

968. ICT programmes have significant inputs in the development paradigm are:
(A) Reduction in transition cost
(B) Time saving
(C) Awareness and utilization by entitlement
(D) All of these

969. Through ICT enabled services, a farmer could save this travelling time and transaction costs by more than:
(A) 25-30% (B) 40-50%
(C) 60-70% (D) 80-90%

970. By using e-choupal (soya choupal), the farmers could save transaction costs by:
(A) 25-30% (B) 40-50%
(C) 67-68% (D) 80-90%

971. Soya choupal saves charges towards freight to providing factory by:
(A) 35-36% (B) 40-50%
(C) 60-70% (D) 80-90%

972. Farmers increases in the sale value for their Soyabean produce by ——— due to the premium price offered by soyachoupal for their quality farm produced:
(A) 1% (B) 2% (C) 3% (D)4%

973. Soyachoupal is more beneficial mainly for:
(A) Small holding farmers
(B) Medium holding farmer
(C) Large holding farmers
(D) None of these

974. Farmers could place their requirement of inputs to the ———, who place bulk demand with the ICT:
(A) Sanchalak (B) Scientist
(C) Leader (D) Soochanalayas

975. Gyandoot kiosks, farmers could get the land records in village itself and save transaction costs by:
(A) 25-30% (B) 40-50%
(C) 66-67% (D) 80-90%

976. Through gyandoot, the farmer particularly getting higher returns on the basis of the timely access to market information in:
(A) Small holding farmers
(B) Medium holding farmer
(C) Large holding farmers
(D) None of these

977. E-choupal has:
(A) LAN connectivity (B) WAN connectivity
(C) VSAT connectivity (D) WLL connectivity

978. COW has:
(A) LAN connectivity (B) WAN connectivity
(C) VSAT connectivity (D) GPRS connectivity

979. KIOSKs is internet connectivity through technological device such as:
(A) LAN connectivity (B) WAN connectivity
(C) VSAT connectivity (D) WLL connectivity
(E) All of these

980. It is widely expected that ICT's are playing role in the:
(A) Developmental process (B) Management process
(C) Transactional process (D) All of these

981. KIOSKs is internet connectivity for small towns and villages, that is just similar to the:
(A) PCO booths (B) STD booths
(C) Both (A) and (B) (D) None of these

982. Farmer-owned Amul Cooperative in Anand has shown that since the:
(A) 1950s (B) 1960s
(C) 1970s (D) 1980s

983. Which ICT project become a model for most dairy development project in India:
(A) AMUL (B) PARAG
(C) VISAKHA (D) Namestey India

984. India is a developing nation that is making remarkable progress toward development through the:
(A) Information strategy
(B) Communication strategy
(C) Both (A) and (B)
(D None of these

985. The aims of the task force was to formulate the draft of a national policy on Information to enable India to emerge as in Information Technology super power within the next:
(A) 1 year (B) 2 years
(C) 5 years (D) 10 years

986. Indigenous communication channels are available in many forms such as:
(A) Folk media
(B) Social gathering
(C) Deliberate instruction
(D) Record and unstructured channels
(E) All of these

987. Traditional communication channels such as:
(A) Puppet show and folk drama
(B) Community tea house and open market
(C) Story telling and village meeting
(D) All of these

988. Traditional communication media are deeply rooted in the socio-cultural moves of the society are:
(A) Folk songs (B) Folk drama
(C) Both (A) and (B) (D) None of these

989. Folk media may constitute a part of the:
(A) Indigenous communication system
(B) Traditional communication system
(C) Both (A) and (B)
(D) None of these

990. In India, the puppetry styles are:
(A) Sutradharika
(B) Rod puppets shadow puppets
(C) Hand puppets
(D) All of these

991. Traditonal media, folk theatre forms in Maharashtra are:
(A) Tamasha (B) Powada or Pawada
(C) Dashavatar (D) All of these

992. Traditional media, Folk theatre forms in Gujarat are:
(A) Bhavai (B) Dangi tamasha
(C) Both (A) and (B) (D) None of these

993. Traditional media, Folk theatre forms in UP are:
(A) Burrakatha (B) Yakshagana
(C) Ramleela (D) Noutanki

994. Traditional media, Folk theatre forms in North India are:
(A) Burrakatha (B) Yakshagana
(C) Ramleela (D) Noutanki

995. Traditional media, Folk theatre forms in Karmataka are:
(A) Burrakatha (B) Yakshagana
(C) Ramleela (D) Noutanki

996. Traditional media, Folk theatre forms in AP are:
(A) Burrakatha (B) Yakshagana
(C) Ramleela (D) Noutanki

997. Traditional media, Folk theatre forms in Bihar are:
(A) Bidesia (B) Jaljatin
(C) Both (A) and (B) (D) Noutanki

998. Traditional media, Folk theatre forms in Tamilnadu are:
(A) Therukoothu (B) Yakshagana
(C) Burrakatha (D) Noutanki

999. Traditional media, Folk theatre forms in Bengal and Orissa are:
(A) Burrakatha (B) Yakshagana
(C) Ramleela (D) Jatra

1000. Traditional media, keertana in Maharashtra are:
(A) The Naraditya Harikatha
(B) The Warkari Haarikatha
(C) The National Harikatha
(D) All of these

1001. Traditional media, folksongs in UP are:
(A) Chaiti (B) Kaijri
(C) Both (A) and (B) (D) None of these

1002. Traditional media, folksongs/folk verse/ballads/folk saying and riddles in Kerala is:
(A) Kolkali pattus (B) Bihu
(C) Mnad and Panihari (D) Rauf and Chhakri

1003. Traditional media, folksongs/folk verse/ballads/folk saying and riddles in Assam is:
(A) Kolkali pattus (B) Bihu
(C) Mnad and Panihari (D) Rauf and Chhakri

1004. Traditional media, folksongs/folk verse/ballads/folk saying and riddles in Rajasthan is:
(A) Kolkali pattus (B) Bihu
(C) Mnad and Panihari (D) Rauf and Chhakri

1005. Traditional media, folksongs/folk verse/ballads/folk saying and riddles in Kashmir is:
(A) Kolkali pattus (B) Bihu
(C) Mnad and Panihari (D) Rauf and Chhakri

1006. Traditional media, folksongs/folk verse/ballads/folk saying and riddles in Madhya Pradesh is:
(A) Sua and Dadaria (B) Mando and Dhalo
(C) Boli (D) Lavani

1007. Traditional media, folksongs/folk verse/ballads/folk saying and riddles in Goa is:
(A) Sua and Dadaria (B) Mando and Dhalo
(C) Boli (D) Lavani

1008. Traditional media, folksongs/folk verse/ballads/folk saying and riddles in Eastern Punjab is:
(A) Sua and Dadaria (B) Mando and Dhalo
(C) Boli (D) Lavani

1009. Traditional media, folksongs/folk verse/ballads/folk saying and riddles in Maharashtra is:
(A) Sua and Dadaria (B) Mando and Dhalo
(C) Boli (D) Lavani

1010. Indigenous communication, folk media are:
(A) Story telling (B) Poetry
(C) Plays and Puppet shows (D) All of these

1011. Indigenous communication, Indigenous organization and social gatherings are:
(A) Religious groups (B) Village meetings
(C) Irrigation associations (D Mothers clubs
(E) Loan associations (F) All of these

1012. Indigenous communication, deliberate instruction are:
(A) Parents teach children
(B) Raft people instruct apprentices
(C) Elder guide young people
(D) Adolescents undergo initiation rites
(E) All of these

1013. Indigenous communication, many societies keep records in:
(A) Written (B) Carved
(C) Painted (D) Memorized forms
(E) All of these

1014. Unstrumented channels of communication media are:
(A) Talk at home and at the well
(B) Talk in the field and on the road
(C) Talk in tea house and coffee house
(D) Talk in chief's house and at the market
(E) All of these

1015. Communication does not have to be intentional is:
(A) Unstructured channels (B) Structured channels
(C) Both (A) and (B) (D) None of these

1016. The traditional media are the closest to the:
(A) Hearts (B) Minds
(C) Both (A) and (B) (D) None of these

1017. The traditional media appeals is at a:
(A) Personal (B) Intimate levels
(C) Both (A) and (B) (D) None of these

1018. The colloquial dialects used make for clarity in communication are:
(A) Familiar format (B) Content
(C) Both (A) and (B) (D) None of these

1019. The traditional media are:
(A) Credible (B) Culturally compatible
(C) Both (A) and (B) (D) None of these

1020. The traditional media are:
(A) Inexpensive (B) Informal
(C) Both (A) and (B) (D) None of these

1021. The greatest advantage of the traditional media over the electronic media in accommodating new themes is:
(A) Flexibility (B) Rigidness
(C) Unaffected (D None of these

1022. The traditional media have:
(A) Entertainment (B) Informative value
(C) Both (A) and (B) (D) None of these

1023. Indian government established a Song and Drama Division in:
(A) 1954 (B) 1955
(C) 1956 (D) 1957

1024. All India Radio also used folk media in its rural broadcasting in the form of a :
(A) Daily programme (B) Weekly programme
(C) Monthly programme (D) None of these

1025. Traditional media are important channels of:
(A) Innovation (B) Communication
(C) PRA (D) Learning

1026. Several traditional methods can be used in secondary school for teaching:
(A) Extension (B) Biotechnology
(C) Agriculture (D) Home Science

1027. Among different folk media, story telling, folk songs and dances are outstanding to augment in imparting education in an:
(A) Interesting manner
(B) Effective manner
(C) Interesting and effective manner both
(D) None of these

1028. Folk song can be both effective and efficient as a communication medium or a society which has a:
(A) Folk theatre forms (B) Strong oral tradition
(C) Both (A) and (B) (D) None of these

1029. The pioneering institution to draw upon folk music is:
(A) All India Radio (B) Doordarshan
(C) Akashwani (D) Star Plus

1030. The radio, doordarshan, drama and theatre have integrated various:
(A) Folk theatre forms (B) Strong oral tradition
(C) Both (A) and (B) (D) None of these

1031. Traditional media have remarkable impact on the rural society because of their:
(A) Acceptable idioms
(B) Functional significance
(C) Emotional touch
(D) Entertainment component
(E) All of these

1032. Traditional media can more effective in rural areas because:
(A) The techniques are simple and easy
(B) The techniques are tough and complex
(C) The techniques are no understandable
(D) The techniques are politically accepted

1033. Traditional media can more effectively stir and felling and affect the attitude of:
(A) Literate masses (B) Illiterate masses
(C) Leaders (D) Servicemen

1034. Folk arts laying a meaningful role in rural areas in educating the:
(A) Urban people (B) Rural people
(C) Tribal people (D) Student

1035. Traditional communication forum such as the market place, tea or coffee house pubs, transportation system and village meetings should be used to reach:
(A) Urban audience (B) Tribal audience
(C) Rural audience (D) None of these

1036. Midwife is the best communicator in many parts of:
(A) Asia (B) Africa
(C) America (D) Russia

1037. Which type of market plays an important role in the life of the tribal people is:
(A) Daily market (B) Seasonal market
(C) Weekly market (D) Monthly market

1038. The tribal market brings together people from different ethnic group for not only economic in the tribal region but also:
(A) Secular activities
(B) Religious activities
(C) Secular and religious activities
(D) None of these

1039. The market is the most powerful channel of communication in the:
(A) Tribal region (B) Urban region
(C) Rural region (D) None of these

1040. The impact of weekly market on their traditional life has shown an attitude of accepting:
(A) Diffusion (B) Communication
(C) Innovation (D) All of these

1041. Africa has a number of unique traditional communication technique, among them are the:
(A) Taking Drums (B) Taking horns
(C) Both (A) and (B) (D) None of these

1042. The Comila project in Bangladesh was built into traditional communication approaches such as:
(A) Village meeting (B) Block meeting
(C) Taluka meeting (D) District meeting

1043. The important source of information to village women is:
(A) Barber's wife (B) Dai
(C) Washer women (D) Bangle sellers
(D) All of these

1044. Traditional media may compute for attention will modern media such as:
(A) Radio (B) Television
(C) Cassettes or video (D) All of these

1045. Which bank used puppetry show in Pantnagar Farmer's fair to arouse the interest of the visitors in bank savings and loans?
(A) Punjab Bank (B) State Bank of India
(C) Central Bank (D) Allahabad Bank

1046. Traditional channel of communication (Munadi) is more effective than ——— by KVK worker and newspaper in disseminating message of local importance in the tribal villages:
(A) Interpersonal contact (B) Intra personal contact
(C) Official contact (D) None of these

1047. The process of effecting an interchange of understanding between two or more people is called:
(A) Communication (B) Innovation
(C) Diffusion (D) None of these

1048. The process of communication is:
(A) Teaching (B) Learning
(C) Teaching and learning (D) None of these

1049. Teaching act as:
(A) Analyst (B) Catalyst
(C) Analyst and catalyst (D) All of these

1050. Communication is a:
(A) One way process (B) Two way process
(C) Three way process (D) Multi way process

1051. Communication involves:
(A) Source and his message
(B) Receiver and his response
(C) Both (A) and (B)
(D) None of these

1052. The chief emphasis in communication on:
(A) Target audience (B) Target communicator
(C) Target receiver (D) Target channel

1053. In which communication system of agencies works:
(A) Knowledge, Generating and System
(B) Knowledge, Disseminating and System
(C) Knowledge, Consuming and System
(D) None of these

1054. In which communication system of mass media works:
(A) Knowledge, Generating and System
(B) Knowledge, Disseminating and System
(C) Knowledge, Consuming and System
(D) None of these

1055. In which communication system of farmers works:
(A) Knowledge, Generating and System
(B) Knowledge, Disseminating and System
(C) Knowledge, Consuming and System
(D) None of these

1056. Individual approaches in communication and extensions are:
(A) Personal visits (B) Personal letters
(C) Both (A) and (B) (D) None of these

1057. Group approaches in communication and extension are:
(A) Demonstration (B) Training the leaders
(C) Discussion meetings (D) Field days and tours
(E) All of these

1058. Mass approaches in communication and extensions are:
(A) Film and slides
(B) Flash cards, printed material and Newspaper
(C) Radio, television and recorded talks
(D) Charts, diagrams and circular letters
(E) Campaign
(F) All of these

1059. Projected audio-visual aids are:
(A) Film projector (B) Slide projector
(C) Overhead projector (D) Opaque projector
(E) All of these

1060. Non-projected audio-visual aids are:
(A) Flash cards and groups (B) Charts and leaflets
(C) Folders and black boards (D) All of these

1061. Audio aids are:
(A) Radio (B) Tape recorder
(C) Both (A) and (B) (D) None of these

1062. Visual aids are:
(A) Flash cards (B) Overhead projector
(C) Slides (D) All of these

1063. Audio-visual aids are:
(A) Puppet (B) Drama
(C) Television (D) Cinema
(E) All of these

1064. The basic principles of demonstration is learning by:
(A) Seeing only (B) Doing only
(C) Seeing and doing both (D) All of these

1065. Discussion meeting approach of communication are:
(A) Group discussion and Panel discussion
(B) Symposium, workshop and Forum
(C) Group interview and dialogue or public conversation
(D) All of these

1066. The single practice demonstrations is:
(A) Method demonstration
(B) Result demonstration
(C) Frontline demonstration
(D) Single line demonstration

1067. Father of method demonstration is:
(A) Mr Arthur T. Mosher (B) F. L. Bryne
(C) Dr. Seeman A Knapp (D) Mr Albert Mayer

1068. In which demonstration process, extension worker may take help of audio-visual aids:
(A) Method demonstration
(B) Result demonstration
(C) Frontline demonstration
(D) Single line demonstration

1069. In which demonstration practices are shown in comparison to existing practices in:
(A) Method demonstration
(B) Result demonstration
(C) Frontline demonstration
(D) Single line demonstration

1070. In which demonstration cannot compare between old and new techniques or skill:
(A) Method demonstration
(B) Result demonstration
(C) Frontline demonstration
(D) Single line demonstration

1071. Which is correct:

(A) Short type demonstration	Result demonstration
(B) Comparison to existing practices can be done	Result demonstration
(C) Comparison between old and new techniques cannot be done	Result demonstration
(D) May take help of audio-visual aids	Method demonstration

1072. Which is correct:
(1) Method demonstration is the short type demonstration
(2) In method demonstration cannot compare between old and new techniques
(3) The result demonstration practices are shown in comparison to existing practices
(4) In result demonstration process, the extension worker may take the help of audio-visual aids
Options
(A) 1, 2 and 4 (B) 1, 3 and 4
(C) 1, 3 and 4 (D) 1, 2, 3 and 4

1073. The important points of result demonstration is/are:
(A) Never try to discover new truths
(B) Do not repeat demonstration repeatedly
(C) Use audio-visual aids to support the result demonstration
(D) Better to conduct more than one demonstration in the same village on the same subject
(E) All of these

1074. There are methods of recording the sound:
(A) Mechanical process (B) Magnetic process
(C) Optical process (D) All of these

1075. Disc recording is done by:
(A) Mechanical process (B) Magnetic process
(C) Optical process (D) All of these

1076. Tape and wire recording is done by:
(A) Mechanical process (B) Magnetic process
(C) Optical process (D) All of these

1077. Recording of sound on movie film by the:
(A) Mechanical process (B) Magnetic process
(C) Optical process (D) All of these

1078. In India and England, projector is also called as Epidiascope:
(A) Opaque projector (B) Slide projector
(C) Overhead projector (D) Movie projector

1079. The projector are based on principle of direct projection:
(A) Opaque projector (B) Slide projector
(C) Overhead projector (D) Movie projector

1080. The projector are based on principle of indirect projection:
(A) Opaque projector (B) Slide projector
(C) Overhead projector (D) Movie projector

1081. Using the principle of magnetic process recording can be done on tapes made of:
(A) Paper (B) Plastic
(C) Metal (D) All of these

1082. The disc or gramophone records, the disc is made of:
(A) Lac (B) Plastic
(C) Both (A) and (B) (D) None of these

1083. Leaflets contain number of pages:
(A) 4-12 (B) 12-24
(C) 24-48 (D) Unlimited

1084. Pamphlets contain number of pages:
(A) 4-12 (B) 12-24
(C) 24-48 (D) Unlimited

1085. Bulletin contain number of pages:
(A) 4-12 (B) 12-24
(C) 24-48 (D) Unlimited

1086. Quick and cheap method of written communication and proves very effective if done well is:
(A) Leaflet (B) Pamphlet
(C) Bulletin (D) Circular letter

1087. Which written method of communication can teach villagers easily:
(A) Circular letters (B) Leaf lets
(C) Pamphlets (D) Bulletin

1088. Brainstorming is basically a:
(A) Creative technique (B) Motivation technique
(C) Evaluation technique (D) Reasoning technique

1089. Which type of communication is called as public relations:
(A) Across communication
(B) Interpersonal communication
(C) Internal communication
(D) External communication

1090. An official communication either oral or written is called:
(A) Informal communication
(B) Formal communication
(C) Both (A) and (B)
(D) None of these

1091. A formal communication may be:
(A) Mandatory (B) Indicative
(C) Both (A) and (B) (D) None of these

1092. A mandatory communication is mostly:
(A) Vertical (B) Horizontal
(C) Both (A) and (B) (D) None of these

1093. The indicative or explanatory communication may exist:
(A) Diagonal (B) Horizontal
(C) Vertical (D) All of these

1094. The communication type which provides instructions, informal and clarification is called:
(A) Informal communication
(B) Formal communication
(C) Administrative communication
(D) Both (A) and (B)

1095. The communication that take place between two or more functionaries at the same level under the same supervisors:
(A) Downward communication
(B) Upward communication
(C) Horizontal communication
(D) Grapevine communication

1096. Which one of the following is not an element of communication:
(A) Consistency (B) Uniformity
(C) Rigidity (D) Adequacy

1097. A large number of communication models exist because of:
(A) Contextual variation
(B) Behavioural variation
(C) Sequential variation
(D) Geographical variation

1098. 'Communicator' is the characteristic element in the model of:
(A) Leagans (B) Lazwell
(C) Berlo (D) Rogers

1099. Flag method is classified under:
(A) Written (B) Spoken
(C) Visual (D) Audio-visual

1100. Flash cards, pull charts, slides and film strips are the type of aids:
(A) Audio (B) Visual
(C) Audio-visual (D) None of these

1101. Flash cards are a series of illustrated cards when flashed or presented before a group in proper sequence tell a:
(A) Part of story (B) Complete story
(C) Incomplete story (D) None of these

1102. Television broadcast for rural development in India, started in:
(A) 1947 (B) 1957 (C) 1967 (D) 1977

1103. Radio mass medium is characterized by:
(A) One way with immediate feedback
(B) One way without instant audience response
(C) One way and colourful
(D) One way and timeless

1104. Which one of the following is not correctly matched?

Approach	Technique
(A) Mass	Exhibition
(B) Individual	Farm visit
(C) Group	Mela
(D) Mass	Newspaper

1105. Read the following statements and mark your answers according to the code given below:
(a) Newspaper is popular medium of communication in rural areas.
(b) Newspaper is written form of communication.
(c) Film show is audio visual form of communication.
(d) Film show is popular medium of communication in rural areas.
Answer code:
(A) A and C are correct (B) B and C are correct
(C) C and D are correct (D) A and D are correct

1106. Consider the following statements:
Assertion (A): Film is not effective medium in rural areas.
Reason (R): Film can not be projected with low voltage power.
Select the correct answer from the code given below:
(A) A is true but R is false
(B) R is true but A is false
(C) Both A and R are correct and R is correct explanation of A
(D) Both A and R are correct but R is not the correct explanation

1107. Match list I with list II and select the correct answer using the code given below.

List I (Form)	List II (Technique)
(a) Spoken	1. Folder
(b) Written	2. Tape recorder
(c) Projected	3. Film
(d) Non projected	4. Puppet

Code:	a	b	c	d
(A)	2	1	3	4
(B)	1	2	3	4
(C)	2	3	1	4
(D)	4	3	2	1

1108. Match list I with list II and select the correct answer using the code given below:

List I (Aids)	List II (Methods of contact)
(a) Model	1. Group contact
(b) Slide	2. Mass contact
(c) Telephone	3. Personal contact
(d) Poster	4. Mass contact

Code:	a	b	c	d
(A)	2	3	4	1
(B)	1	2	3	4
(C)	3	4	2	1
(D)	3	2	4	1

1109. Which of the following is not a method of group communication?
(A) Demonstration (B) Symposium
(C) Flannel graph (D) Circular letter

1110. Which one of the following is not correctly matched?

Approach	Technique
(A) Mass	Television
(B) Mass	Film
(C) Mass	Tape recorder
(D) Mass	Demonstration

1111. Which source of interview is the best in a rural society?
(A) Key informant (B) Group
(C) Focused group (D) None of these

1112. Method of mass communication is:
(A) Demonstration (B) Farm visit
(C) Group discussion (D) News paper

1113. Which one is not the part of communication module?
(A) Communicator (B) Message
(C) Advice (D) Audience

1114. Which one of the following is not a method of mass contacts:
(A) Circular letters (B) Television
(C) Leaflets (D) Result demonstration

1115. Which one of the following is a method of contact group:
(A) Farms and home visits (B) Office calls
(C) Personal letters (D) Tours

1116. Which one of the following is an audio – visual method of extension teaching:
(A) Drama (B) Recordings
(C) Flash board (D) Tape recorders

1117. Which one is a non – projected aid:
(A) Cinema (B) Slide
(C) Over head projector (D) Pictures

1118. Puppet shows are included in:
(A) Audio aids (B) Visual aids
(C) Audio – visual aids (D) None of these

1119. Flash cards, pull charts, slides and films strips are the type of aids:
(A) Audio (B) Visual
(C) Audio – visual aids (D) None of these

1120. Visual teaching aid flannel graph is also known as:
(A) Chalk board (B) Bulletin board
(C) Khadder graph (D) None of these

1121. Television, Fairs, Radios and Exhibitions are media of:
(A) Group communication
(B) Individual communication
(C) Community communication
(D) Mass communication

1122. If we want to impress the people about a particular recommended practice suitable for a specific area, then effective extension method is:
(A) TV (B) Radio
(C) Demonstration (D) Pumphlets

1123. Radio is the medium of mass communication in which, use is made up of:
(A) Audio device (B) Visual device
(C) Both (A) and (B) (D) None of these

1124. Bulletins, leaflets, circular letters, posters, fairs, exhibitions, radio and television are the media of:
(A) Public communication
(B) Official communication
(C) Group communication
(D) Mass communication

1125. Puppet, Drama, Television and Cinema are example of:
(A) Audio-aids (B) Visual – aids
(C) Audio-visual aids (D) None of these

1126. Radio and tape recorder is the examples of:
(A) Audio-aids (B) Visual – aids
(C) Audio-visual aids (D) All of these

1127.

List-I	List-II
(a) Mass contact	(1) Bulletin
(b) Group contact	(2) Method demonstration
(c) Individual contact	(3) Symposium
(d) Individual contact	(4) Result demonstration

Code	a	b	c	d
(A)	1	3	4	2
(B)	2	3	4	1
(C)	4	3	2	1
(D)	1	2	3	4

1128.

List-I	List-II
(a) Personal contact	(1) Mass method
(b) Public address equipment	(2) Mass method
(c) Circular letter	(3) Individual method
(d) Television	(4) Group method

Code	a	b	c	d
(A)	3	4	1	2
(B)	4	3	2	1
(C)	1	2	3	4
(D)	2	1	3	4

1129. The fastest method of delivering of message in the village is:
(A) Newspaper (B) Radio
(C) Television (D) Group discussion

1130. Television, radio, fairs and exhibitions are the media of:
(A) Group communication
(B) Individual communication
(C) Community communication
(D) Mass communication

1131. Which one of the following is not correctly matched?

Approach	Technique
(A) Mass	Television
(B) Mass	Demonstration
(C) Mass	Tape recorder
(D) Mass	Film

1132. Which one is related to mass approach method?
(A) Film, Demonstration and Radio
(B) Radio, TV and Lecture
(C) Phone call, Radio and TV
(D) Exhibition, Newspaper and Radio

1133. The best method for showing the value of new project is:
(A) Result demonstration (B) Exhibition
(C) Farmers fair (D) Meeting

1134. Which one of the sources of communication is most credible for non-progressive/remote village situation?
(A) Farm radio (B) Demonstration
(C) Printed material (D) Exhibition

1135. Method of mass communication is:
(A) Demonstration (B) Farm visit
(C) Group discussion (D) Newspaper

1136. Best method for showing the value of new practice is:
(A) Result demonstration (B) Exhibition
(C) Farmer's fair (D) Meeting

1137. In a good chart facts are presented:
(A) Only in words
(B) Only in figures
(C) Only in photographs
(D) Both in figures and photographs

1138. Non-institutional communication method is:
(A) Exhibition (B) Family member
(C) Extension literature (D) Radio

1139. Which one of the following aids involves two sensory organs simultaneously?
(A) Poster (B) Model
(C) Radio (D) Television

1140. Identify the projected audio-visual aid:
(A) Puppet (B) Film
(C) Drama (D) Folk song

1141. Which one is not the example of nonverbal communication?
(A) Ram Leela (B) Movements of arms
(C) Facial expression (D) Gestures

1142. We learn maximum through:
(A) Taste (B) Touch
(C) Hearing (D) Sight

1143. The extension service is called " Cooperative Extension Service" in:
(A) USA (B) India
(C) Mexico (D) China

1144. The main functions of an agricultural university in India are:
(A) Technology generation, technology transfer and technology adoption
(B) Drawing attention, creating interest and taking action
(C) Teaching, research and extension
(D) Preparation, deliberation and action

1145. General meeting are broadly the meeting of:
(A) Heterogeneous participation
(B) Homogeneous participation
(C) Community participation
(D) Social participation

1146. A photograph is called:
(A) A diagnostic representation of facts or ideas
(B) A meaningful exchange of ideas
(C) An exact visual recording of a thing
(D) An act of influencing others

1147. Identify one which is not correctly matched in the following pairs:
(A) Method demonstration : Group approach
(B) Circular letter : Mass approach
(C) Radio : Individual approach
(D) Campaign : Mass approach

1148. Non projected teaching method is:
(A) Cinema (B) Slides
(C) Puppets (D) 3-D film

1149. Which one is not the group approach device?
(A) Demonstration (B) Symposium
(C) Workshop (D) Film

1150. Television, fairs, radios and exhibition are the media of:
(A) Group communication
(B) Individual communication
(C) Commodity communication
(D) Mass communication

1151. Powerful audio-visual media for group communication is:
(A) Television (B) Film
(C) Puppet (D) None of these

1152. Puppet can be made out of easily available and cheap material in the community like:
(A) Paper match and cloth (B) Cardboard and mud
(C) Saw dust and hay (D) All of these

1153. Puppetry in an 'INDIAN' medium of communication which attracts people of:
(A) Young (B) Old
(C) Children (D) All age group

1154. Video tapes can be used for promoting audio-visual literacy for:
(A) Motivation and attitudinal change
(B) Behaviour reinforcement
(C) Community participation and entertainment
(D) All of these

1155. Indian folk theatre is full of exploitative stories of kings and priests tuned into Gods, picked up from different Hindu epics like:
(A) Ramayana
(B) Mahabharat
(C) Ramayana and Mahabharat
(D) None of these

1156. Like video, similar results even with more clarity and immediacy can be now used through:
(A) Computer system (B) Television system
(C) Gramophone system (D) All of these

1157. These are how many degree of evaluation:
(A) 4 (B) 3 (C) 5 (D) 2

1158. Diffusion is the subset of:
(A) Communication (B) Teaching
(C) Learning (D) Motivation

1159. Diffusion has:
(A) 3 elements (B) 4 elements
(C) 6 elements (D) 8 elements

1160. Diffusion and adoption of a new technology are respectively decided by:
(A) Society and Society
(B) Individual and Individual
(C) Individual and Society
(D) Society and Individual

1161. First step for making a programme planning includes:
(A) Evaluation
(B) Analysis of the situation and determining problems
(C) Deciding on objectives
(D) Teaching

1162. At which of the stage of adoption group discussion is most effective:
(A) Awareness (B) Evaluation
(C) Testing (D) Persuasion stage

1163. The correct sequence of innovation – decision process or adoption process is:
(A) Interest – awareness – evaluation – trial – adoption
(B) Evaluation – awareness — interest - trial – adoption
(C) Awareness -interest — evaluation – adoption - trial
(D) Awareness – interest – evaluation – trial – adoption

1164. ———— are the first farmers to adopt a new idea:
(A) Innovators (B) Early adopters
(C) Early majority (D) Laggards

1165. ———— are the last to adopt new idea:
(A) Innovators (B) Early adopters
(C) Laggards (D) Early majority

1166. Which one of the following pair is not correctly matched:

Adopter Category	Percentage
(A) Innovators	2.5
(B) Early adopters	13.5
(C) Early majority	40
(D) Laggards	16

1167. Normal rate of adoption requires how many years from the introduction of the innovation to its adoption throughout the community:
(A) 2 – 6 (B) 6 – 10
(C) 10 – 14 (D) 14 – 18

1168. In case of diffusion of innovation, the overt behavior means:
(A) Adoption or rejection
(B) Remembering or forgetting
(C) Fooding or lodging
(D) Mental weighing or field trial

1169. At the adoption stage, information sources is:
(A) Village level change agent and to some extent mass media
(B) Extension agency and other farmers
(C) Informal personal sources including family members
(D) Self-experience gained at the trial stage

1170. Adopter category which respect and regarded by many others in the social system as a role model are:
(A) Innovators (B) Early adopters
(C) Late majority (D) Laggards

1171. The spread of technology in a social system:
(A) Diffusion (B) Teaching
(C) Learning (D) Adoption

1172. Diffusion is essentially concerned with messages that are:
(A) Profitable (B) New ideas
(C) Vital (D) Widely known

1173. The elements of diffusion includes all except:
(A) Innovation (B) Social system
(C) Adoption (D) Time

1174. The oldest research tradition in diffusion research is from:
(A) Communication (B) Sociology
(C) Education (D) Anthropology

1175. The major drawback of past diffusion studies was their inability to take up:
(A) Source research (B) Channel research
(C) Message research (D) Audience research

1176. Largest number of diffusion studies are conducted in the field of:
(A) Communication (B) Sociology
(C) Education (D) Anthropology

1177. The model of diffusion as proposed by Rogers assumes:
(A) The homogeneity of population
(B) The progressiveness of farmers
(C) Availability of infrastructural facilities
(D) The differential farming situation

1178. Diffusion and adoption of a new technology are respectively decided by:
(A) Society and Society
(B) Individual and Individual
(C) Individual and Society
(D) Society and Individual

1179. Diffusion can be considered as a subject of:
(A) Communication (B) Teaching
(C) Learning (D) Motivation

1180. The degree to which the results of an innovation may be diffused to others:
(A) Credibility (B) Compatibility
(C) Communicability (D) Relative advantage

1181. Homophilous diffusion pattern with in a system helps in the spread of ideas:
(A) Horizontally (B) Vertically
(C) Diagonally (D) None of these

1182. Which of the following is the first in motivating the farmers:
(A) Food (B) Money
(C) Farm literature (D) All of these

1183. CONA expands to:
(A) Career Oriented Need Assessment
(B) Community Oriented Need Assessment
(C) Creative Organizational Need Assessment
(D) Career Oriented Need Analysis

1184. The idea which is perceived as new is termed as:
(A) Information (B) Innovation
(C) Perception (D) Invention

1185. An innovation may have in general:
(A) One component (B) Two component
(C) Three component (D) Four component

1186. The material aspect of an innovation is:
(A) Idea component (B) Object component
(C) Product component (D) Relative component

1187. Symbolic adoption take place in case of innovation with:
(A) Idea component (B) Object component
(C) Product component (D) Relative component

1188. To be called as an innovation, an idea, practice or object must be:
(A) Objectively new (B) Subjectively new
(C) All of these (D) None of these

1189. The characteristic of an innovation important from extension point of view:
(A) Intrinsic (B) As people see them
(C) Extraordinary (D) Typological

1190. An innovation can be considered to have originated from:
(A) Farmer research (B) Extension programme
(C) Research trials (D) All of these

1191. An innovation include all except:
(A) Material products
(B) Ideological belief
(C) Social movement
(D) Traditional technologies

1192. An innovation is liked or disliked based on:
(A) Innovation effects (B) Innovation decision
(C) Perceived attributes (D) Traditional makeup

1193. The perceived newness of the innovation can be expressed in terms of:
(A) Knowledge (B) Decision to adopt
(C) Persuasion (D) All of these

1194. The perceived attribute of an innovation that reduces the risk component:
(A) Complexability (B) Compatibility
(C) Trialability (D) Absorbability

1195. Providing subsidy to farmers on introduction of farm machinery is an example for:
(A) Relative advantage (B) Complexity
(C) Compatibility (D) Observability

1196. An innovation with low relative advantage may have:
(A) Slow rate of adoption
(B) High rate of discontinuance
(C) Low return on investment
(D) All of these

1197. Compatibility is the degree to which an innovation is perceived consistent with:
(A) Clients needs for innovation
(B) Previously introduced needs
(C) Socio-cultural values and benefits
(D) All of these

1198. The luxuriant growth of the plant on the application of urea relates:
(A) Compatibility (B) Trialability
(C) Observability (D) Complexity

1199. 'Preventive innovations' have less adoption rate mainly due to:
(A) Low relative advantage (B) High complexity
(C) Low triability (D) Low observability

1200. Providing free seed samples to the farmers is closely associated with:
(A) Relative advantage (B) Complexity
(C) Trialability (D) Observability

1201. The perceived attribute of an innovation that shows a negative correlation with adoption:
(A) Complexability (B) Compatibility
(C) Trialability (D) Observability

1202. A compactable innovation offers the adopter:
(A) Greater risk (B) Less security
(C) All of these (D) None of these

1203. The attributes of the innovation as matters by whom is most important:
(A) Research scholars (B) Manufactures
(C) Potential adopters (D) Change agents

1204. Time is very important consideration in the process of:
(A) Communication (D) Adoption
(C) Diffusion (D) Innovation

1205. Individual adopters in a social system is described in terms of his:
(A) Economic conditions (B) Time of adoption
(C) Frequency of adoption (D) Concurrency ability

1206. Frequency of adoptors when plotted over time follows a:
(A) Normal curve (B) Skewed curve
(C) Linear curve (D) Curvilinear shape

1207. The rate of adoption of an innovation in a social system over time follows:
(A) Normal curve (B) Skewed curve
(C) Linear curve (D) S- shaped curve

1208. The S shape adoption curve increases at an increasing rate for:
(A) Innovators (B) Early adopters
(C) Early majority (D) All of these

1209. The length of time required to pass through the innovation decision process:
(A) Innovation decision matrix
(B) Innovation decision period
(C) Innovation decision index
(D) Innovation decision time

1210. Innovation decision period is actually a:
(A) Gestation period (B) Production period
(C) Mental period (D) None of these

1211. The innovation decision process can be shortly called as:
(A) KPDIC (B) AIDTCA
(C) NAIDTEA (D) AIETA

1213. The chief concern of farmer at the persuasion stage is:
(A) How to information (B) How do information
(C) How much information (D) None of these

1214. The individual become more psychologically involved with the innovation at:
(A) Knowledge stage (B) Persuasion stage
(C) Decision stage (D) Confirmation stage

1215. The outcome of persuasion leads to subsequent change in:
(A) Covert behaviour (B) Overt behaviour
(C) Significant behaviour (D) None of these

1216. Persuasion function of innovation decision process is mainly:
(A) Cognitive (B) Affective
(C) Behavioural (D) Normative

1217. The reinforcement of innovation decision making occurs at:
(A) Persuasion stage (B) Knowledge stage
(C) Confirmation stage (D) Decision stage

1218. Conflicting information should be strictly avoided during:
(A) Readiness stage (B) Knowledge stage
(C) Decision stage (D) Confirmation stage

1219. The decision to adopt or reject made after the prior innovation decision is called:
(A) Optimal decision (B) Contingent decision
(C) Collective decision (D) Authority decision

1220. The basic unit of decision making in the optimal innovation decision model is:
(A) Change agent (B) Social system
(C) Researcher (D) Individual

1221. The type of innovation decisions that resulted in the emergence of classical diffusion model:
(A) Optimal innovation decisions
(B) Collective innovation decisions
(C) Authority innovation decisions
(D) None of these

1222. Generally the fastest rate of adoption of innovation results from:
(A) Optimal innovation decisions
(B) Collective innovation decisions
(C) Authority innovation decisions
(D) Contingent innovation decisions

1223. The unit of decision making and unit of adoption are different in case of:
(A) Optimal decision (B) Collective decision
(C) Authority decision (D) Mental decision

1224. In collective decision making who screens the idea for approving or puts sanction on it:
(A) Stimulators (B) Initiators
(C) Actors (D) Legitimizers

1225. Which of the following is not an element in the model of innovation decision process:
(A) Antecedents (B) Variability
(C) Consequences (D) Processes

1226. Dissonance does not occur at which of the stages of innovation decision process:
(A) Knowledge (B) Persuasion
(C) Decision (D) Confirmation

1227. The degree to which an innovation is changed or modified by a user in the process of its adoption:
(A) Re-invention (B) Re-engineering
(C) Re-modeling (D) Re-making

1228. An agricultural innovator is a:
(A) Traditional leader (B) Opinion leader
(C) Functional leader (D) Formal leader

1229. The decision to make use of a innovation as best course of action:
(A) Adoption (B) Diffusion
(C) Innovation (D) Technology

1230. Adoption is basically a:
(A) Mental process (B) Social process
(C) Psychological process (D) Physical process

1231. The scientist first to recognize adoption of innovation occurs in stages:
(A) Wilkening
(B) Rogers and Shoemaker
(C) Johnson and Haver
(D) Ryan and Gross

1232. The stages of adoption are dynamic and not static according to:
(A) Ryan and Gross (B) Wilkening
(C) Johnson (D) Y P Singh

1233. Who for the first time reported that adoption process involves decision:
(A) Pederson (B) Wilkening
(C) Holmberg (D) Mason

1234. The first stage in the process of adoption according to Rogers is:
(A) Attention (B) Interest
(C) Desire (D) Satisfaction

1235. The stage wherein the individual makes full use of the innovation:
(A) Evaluation (B) Adoption
(C) Awareness (D) Trial

1236. The stage wherein the individual decides to continue the full use of the innovation:
(A) Evaluation (B) Adoption
(C) Awareness (D) Trial

1237. The stage wherein an individual is exposed to the innovation but lacks complete information about it:
(A) Attention (B) Interest
(C) Desire (D) Satisfaction

1238. The most important source of information at the attention stage of adoption:
(A) Mass media (B) Group media
(C) Extension worker (D) Personal localities

1239. The stage of adoption wherein group discussion is most effective:
(A) Awareness (B) Evaluation
(C) Testing (D) Persuasion

1240. Unfreezing existing behaviour and motivating people to change is expressed at:
(A) Desire (B) Interest
(C) Conviction (D) Attention

1241. The extension activity that need to be emphasized in the interest stages of adoption:
(A) Limited trial is good
(B) Rapport building is needed
(C) Group training is needed
(D) Specialists are needed

1242. Symbolic adoption is an example for:
(A) Covert behaviour (B) Overt behaviour
(C) Affection (D) Perception

1243. Adopter categories are the classification of the members of the social system on the basis of:
(A) Innovativeness
(B) Adoption process
(C) Diffusion of innovation
(D) Innovation decision period

1244. The time of adoption of an innovation by all the members in a social system:
(A) Always same (B) Always vary
(C) May vary (D) All of these

1245. The classification of adopters into various categories is based on:
(A) Mean and variance
(B) Mean and standard deviation
(C) Mean and median
(D) Median and standard deviation

1246. The adopter category considered as deviant by other members of the society:
(A) Innovators (B) Early majority
(C) Early adopters (D) Laggards

1247. The adopter category considered as friends and neighbours by other members in the social system:
(A) Innovators (B) Early majority
(C) Early adopters (D) Laggards

1248. The adopter category considered by many as the men to "check with":
(A) Innovators (B) Early majority
(C) Early adopters (D) Laggards

1249. Opinion leaders belong to the category of:
(A) Innovators (B) Early majority
(C) Early adopters (D) Late majority

1250. The point of reference for laggards is:
(A) Past (B) Present
(C) Future (D) All of these

1251. The adopter categories who is also considered as the custodians of indigenous knowledge:
(A) Innovators (B) Early majority
(C) Early adopters (D) Laggards

1252. The person who adopt an innovation before every average member of the society does it:
(A) Innovators (B) Early majority
(C) Early adopters (D) Late adopters

1253. Agricultural innovators are generally viewed as:
(A) Opinion leader (B) Social deviants
(C) Skeptical (D) Traditional

1254. 'Legitimization' is done by:
(A) Innovators (B) Early adopters
(C) Early majority (D) Late majority

1255. Who among the following is more concerned about triability of the innovation:
(A) Early adopters (B) Early majority
(C) Innovators (D) Laggards

1256. "Be not the last to lay the old aside nor be the first by which the new is tried" is the motto of:
(A) Innovators (B) Early majority
(C) Early adopters (D) Laggards

1257. The percentage of population constituted by the innovators:
(A) 2.5 per cent (B) 7.5 per cent
(C) 16.5 per cent (D) 33 per cent

1258. The character best represent an innovator:
(A) Venturesome (B) Skeptical
(C) Traditional (D) Respectfulness

1259. The adoption of innovation rather vigourously when experts feel that it should be stopped is called:
(A) Excessive adoption (B) Over adoption
(C) Vigorous adoption (D) Overt adoption

1260. A type of rejection which consists of considering adoption of the innovation but then deciding not to adopt it:
(A) Active rejection (B) Passive rejection
(C) Pro-active rejection (D) None of these

1261. Who incorporates the innovations into a specific plan of action adopted to needs of the system:
(A) Stimulators (B) Initiators
(C) Actors (D) Legitimizers

1262. 'Complementary adoption' is usually associated with:
(A) Agricultural adoption
(B) Technological adoption
(C) Industrial adoption
(D) Both (B) and (C)

1263. The phenomena wherein adoption of one innovation requires the previous adoption of another innovation:
(A) Interaction effect (B) Diffusion effect
(C) Technology effect (D) Adoption effect

1264. The individuals in the same adopter category have:
(A) Varied degree of innovativeness
(B) Multitude of innovativeness
(C) Similar degree of innovativeness
(D) None of these

1265. Consequences are the changes that occur is a social system as a result of:
(A) Adoption of innovation
(B) Rejection of innovation
(C) Both (A) and (B)
(D) None of these

1266. Discontinuance occurs only after an innovation has been:
(A) Fully adopted (B) Partially adopted
(C) Not at all adopted (D) None of these

1267. A decision to reject an idea as a result of dissatisfaction with it's performance:
(A) Disenchantment discontinuance
(B) Replacement discontinuance
(C) Betterment discontinuance
(D) Distinguishable discontinuance

1268. The decision to reject an idea in order to adopt a better idea:
(A) Disenchantment discontinuance
(B) Replacement discontinuance
(C) Betterment discontinuance
(D) Distinguishable discontinuance

1269. The dysfunctional effect of an innovation to an individual or to a social system:
(A) Direct consequence
(B) Indirect consequence
(C) Desirable consequence
(D) Undesirable consequence

1270. The change to an individual or a social system that occur in immediate response to an innovation:
(A) Direct consequence
(B) Indirect consequence
(C) Desirable consequence
(D) Undesirable consequence

1271. The changes from an innovation that is neither intended nor recognized by the members of a social system:
(A) Direct consequence
(B) Indirect consequence
(C) Desirable consequence
(D) Unanticipated consequence

1272. Acceptance and rejection are the one of the kinds of:
(A) Failure of channel (B) Feed back
(C) Treatment of message (D) Empathetic feelings

1273. Adoption is a/an:
(A) Social process (B) Mental process
(C) Psychological process (D) Individual process

1274. Serialise the following stages of adoption process:
(1) Interest (2) Evaluation
(3) Trial (4) Awareness
Pick out the answer from the following
(A) 1, 2, 3, 4 (B) 3, 1, 2, 4
(C) 4, 1, 3, 2 (D) 3, 2, 1, 4

1275. Purpose of extension evaluation is:
(A) To identify the weak points
(B) To identify the strong points
(C) To identify the gaps and errors
(D) To identify all of these

1276. Read the following statements and mark your answer according to the code given below:
I. Extension evaluation is a planned process
II. Extension evaluation is spontaneous
III. Extension evaluation is indivisible portion of teaching
IV. Extension evaluation is beginning process
Answer code:
(A) I and III are correct (B) II and III are correct
(C) I and IV are correct (D) II and IV are correct

1277. In case of diffusion of innovation, the overt behavior means:
(A) Adoption or rejection
(B) Remembering or forgetting
(C) Fooding or lodging
(D) Mental weighing or field trial

1278. The collection of data in PRA is done by:
(A) Local Extension agents
(B) Progressive farmers
(C) Research assistants
(D) A multidisciplinary team

1279. For a study of farming system, the best PRA exercise is:
(A) Transect walk (B) Resource mapping
(C) Time line (D) Chapatti diagram

1280. PRA is known for:
(A) People Resource Appraisal
(B) Public Resource Appraisal
(C) Participatory Rural Appraisal
(D) People Rural Agency

1281. The concept of PRA was given by:
(A) Henry Ford (B) Robert Chambers
(C) Neils Roling (D) Nila Mukerjee

1282. The person considered as father of PRA:
(A) Neils Rolling (B) Robert Chambers
(C) Van den Ban (D) Paul Leagans

1283. The term PRA was first used in:
(A) Kenya (B) USA
(C) UK (D) Israel

1284. PRA was introduced in India by:
(A) Government departments (B) ICAR
(C) NGO's (D) Ford foundation

1285. PRA had it's origin during late:
(A) Sixties (B) Seventies
(C) Eighties (D) Nineties

1286. The main innovation in PRA is:
(A) Methods (B) Behaviour
(C) Process (D) Actors

1287. The role of extension personnel's in PRA is:
(A) Elective (B) Extractive
(C) Participatory (D) None of these

1288. Which of the following is not the characteristic of PRA:
(A) Quick (B) Rigid
(C) Interactive (D) Collective

1289. The collection of data in PRA is based on:
(A) Extension workers
(B) Multi disciplinary team
(C) Research scientists
(D) Progressive farmers

1290. The type of PRA used to understand broad patterns and trends in a rural society:
(A) Exploratory (B) Topical
(C) Participatory (D) Monitoring

1291. Which among the following is not a principle of PRA:
(A) Optimal ignorance (B) Setting uniformity
(C) Offsetting bias (D) Triangulation

1292. The PRA technique that shows cause effect relationship:
(A) Venn diagrams (B) Flow diagram
(C) Social mapping (D) Livelihood analysis

1293. The PRA technique wherein rural elders narrates life histories:
(A) Time line (B) Village transits
(C) History sheets (D) Ranking

1294. Social calendar is a tool most commonly used in:
(A) PRA (B) RRA
(C) LFA (D) FSR

1295. Which of the following is not a tool for conducting PRA:
(A) Transect walk (B) Social map
(C) Time line analysis (D) Role play

1296. The PRA technique that represent the degree of importance of individuals and institutions in decision making:
(A) Social mapping
(B) Venn diagram
(C) Semi structured interview
(D) None of these

1297. People's participation in an extension programme is significant when:
(A) Literate section of villagers participate
(B) Gram panchayat members participate
(C) Majority of villagers participate
(D) Local leader participate

1298. Consider the following statements:
Assertion (A): Farmers posses keen interest in training programme at block headquarter
Reason (R): Farmers enjoy free meal during training period.
Select the correct answer from the code given below
(A) A is true but R is false (B) R is true but A is false
(C) Both A and R are correct and R is the correct explanation of A
(D) Both A and R are true but R is not the correct explanation of A

1299. Participatory rural appraisal is directly related to:
(A) Soil science (B) Entomology
(C) Agricultural economics (D) Agricultural extension

1300. Systematic process of involvement of people for the analysis of their own situations and finding out solution for improving their present condition is called:
(A) Participatory research (B) Explorative research
(C) Amiliorative research (D) None of these

1301. An effort to introduce innovation not known to social system is called:
(A) Participatory research (B) Explorative research
(C) Amiliorative research (D) None of these

1302. It emphasizes in the improvement of what people are doing (at present). It is an effort to make the good better (Compatible) is:
(A) Participatory research (B) Explorative research
(C) Amiliorative research (D) None of these

1303. The characteristics of participatory research is/are:
(A) An activity carried out by group of people to understand their own situation for their own benefit
(B) It is a collective effort to bring desirable results taking into consideration the various aspects of problem in a collective manner
(C) It is an analysis that lead to action
(D) Understanding of personal realities and larger social realities it will facilitate awareness raising about the connections between micro-aspects of their life and the real macro-aspects of society at large
(E) All of these

1304. Participatory research is a:
(A) Cyclic process (B) Noncyclic process
(C) Both (A) and (B) (D) None of these

1305. The limitations of the cyclic process of participatory research to be the:
(A) More participation of practioners
(B) Least participation of practioners
(C) Either (A) or (B)
(D) None of these

1306. Steps in conducting a participatory research are:
(A) To know the problem
(B) Solution is needed or not
(C) To know the present status and additional information needed for the problem to be tackled
(D) All of these

1307. A well known method of data collection is to carry out a:
(A) Big story (B) Medium story
(C) Small story (D) None of these

1308. A small number of people (10-25) come together to start experiences and decide for any action through:
(A) Individual discussion (B) Group discussion
(C) Individual interview (D) Group interview

1309. A well known method of data collection is to carry out a:
(A) Small survey (B) Big survey
(C) Reconscence survey (D) None of these

1310. A small number of people (10-25) come together to store experiences and decide for any action through:
(A) Individual discussion (B) Group discussion
(C) Individual interview (D) Group interview

1311. Sometimes the group engaged in participatory research may find itself uncomfortable with:
(A) Written method
(B) Oral method
(C) Written and oral method both
(D) None of these

1312. Women can be engaged to draw their perceptions of the problem on:
(A) Blackboard
(B) Paper or cloth
(C) Black and white photographs
(D) Cutting from old magazines
(E) All of these

1313. A group discussion or individual interview can be taped and then can be:
(A) Replayed for analysis (B) Data collection
(C) Questionnaire (D) None of these

1314. A group of people can investigate a particular problem by using ———— to known the opinions of a larger number of people:
(A) Discussion (B) Survey
(C) Questionnaire (D) None of these

1315. The questionnaire should be single have only questions?
(A) 4-6 (B) 6-8
(C) 8-10 (D) 10-12

1316. Each question of questionnaire should be:
(A) Short, complex and straight forward
(B) Long, simple and straight forward
(C) Short, complex and straight forward
(D) Long, complex and straight forward

1317. A questionnaire of self survey should possess:
(A) Should have few (8-10) questions
(B) Should be short, simple and straight forward
(C) Should have easy in conversation
(D) All of these

1318. Group discussion is one of the most popular and easy method of combining:
(A) Data collection (B) Analysis
(C) Action planning (D) All of these

1319. The important consideration of group discussion is to insure the:
(A) Uncomfortable atmosphere
(B) Comfortable atmosphere
(C) Positive atmosphere
(D) None of these

1320. Which method can easily be combined with other methods is/are:
(A) Individual discussion (B) Group discussion
(C) Individual interview (D) Group interview

1321. The important characteristics of peoples camp is/are:
(A) May be one or two days programme
(B) Common lodging and boarding arrangement
(C) Peoples comes from longer distance
(D) All of these

1322. Which method can provide an informal and free atmosphere:
(A) People's camp (B) Exhibition
(C) Campaign (D) Seminar

1323. In a people's camp how many peoples can participate easy:
(A) 30-40 (B) 50-100
(C) 100-200 (D) Unlimited

1324. It is always beneficial to visit an area where a similar problem has been solved for which the group is interested is:
(A) Demonstrations (B) Group discussion
(C) Exhibition (D) Campaign

1325. Indian folk theatre has a important role to play in the process of:
(A) Social changes
(B) Political changes
(C) Economic changes
(D) Socio-economic changes both

1326. Folk theatre media are:
(A) Low cost
(B) Indigenous
(C) May provide better for focusing and solving rural problems
(D) All of these

1327. Rural poor can communicate better and convince people by making their:
(A) Own theatre (B) Own house
(C) Leased house (D) None of these

1328. Any inanimate thing animated on stage to communicate any subject is called:
(A) Puppets (B) Drama
(C) Kisan mela (D) All of these

1329. All the instincts basic to human learning are stimulated and get satisfied through puppet show:
(A) Communicative dramatic
(B) Artistic musical
(C) Inquisitive constructive
(D) All of these

1330. Groups of local people analyzing their own conditions and choosing their own means of improving them is:
(A) Participatory Rural Appraisal
(B) Rapid Rural Appraisal
(C) Both (A) and (B)
(D) None of these

1331. The technique developed for analyzing the needs of rural community quickly, cost effectively and with little destruction to every day life is:
(A) Participatory Rural Appraisal
(B) Rapid Rural Appraisal
(C) Both (A) and (B)
(D) None of these

1332. The techniques are a range of survey methods to be used principally in the extraction of qualitative data is:
(A) Participatory Rural Appraisal
(B) Rapid Rural Appraisal
(C) Both (A) and (B)
(D) None of these

1333. Rapid Rural Appraisal (RRA) evolved in the :
(A) 1979 (B) 1980
(C) 1981 (D) 1982

1334. PRA which began in the late:
(A) 1970s (B) 1975s
(C) 1980s (D) 1990s

1335. Which method emphasized multi-disciplinary teams, careful observation, semi-structured interviewing and focus group in:
(A) Participatory Rural Appraisal
(B) Rapid Rural Appraisal
(C) Both (A) and (B)
(D) None of these

1336. The method which involve group rather than individuals and visual representations rather than solely verbal communication in:
(A) Participatory Rural Appraisal
(B) Rapid Rural Appraisal
(C) Both (A) and (B)
(D) None of these

1337. Key issue of Participatory Rural Appraisal is:
(A) To highlight the different PRA techniques available for understanding the access strengths and constraints of the rural poor
(B) To describe the best condition for use of PRA techniques
(C) To highlights the constraints of using PRA techniques
(D) All of these

1338. Assumptions of PRA are:
(A) The environmental problem is a common concern, the solution of which requires the active and responsible involvement of the entire community
(B) The collective action is possible only when a stakeholders of a community develop a clear common understanding about the issue
(C) Both (A) and (B)
(D) None of these

1339. Rapid Rural Appraisal is a:
(A) Extractive in nature
(B) Elicited
(C) Information owned by outsiders
(D) All of these

1340. Participatory Rural Appraisal is a:
(A) Learning with local people
(B) Information owned by rural people but shared with outsiders
(C) Both (A) and (B)
(D) None of these

1341. Principles of PRA is/are:
(A) Using optimal ignorance
(B) Offsetting biases
(C) Triangulation
(D) Learning from and with rural people
(E) Learning rapidly and progressively
(F) All of these

1342. Good features of PRA is/are:
(A) Iterative (B) Innovative
(C) Interactive (D) Informal
(E) In the community (F) All of these

1343. The menu of PRA is/are:
(A) Secondary data review
(B) Semi structured interviews
(C) Analytical games
(D) Stories ad portraits
(E) Diagrams
(F) Workshop
(G) All of these

1344. Helpers of PRA is/are:
(A) What and Why (B) Who and How
(C) Where and When (D) All of these

1345. Important tips of PRA is/are:
(A) Look (B) Listen
(C) Hear (D) All of these

1346. Participatory approaches are not substitutes for , but are rather a integral part of:
(A) Long term dialogue (B) Sustained interaction
(C) Both (A) and (B) (D) None of these

1347. A single, brief participatory exercise with a group of local people will not lead to:
(A) Positive and lasting change
(B) Negative and lasting change
(C) Positive and integral change
(D) Negative and integral change

1348. PRA are not a panacea of:
(A) Qualitative surveying
(B) Quantitative surveying
(C) Either (A) or (B)
(D) Neither (A) nor (B)

1349. PRA work must effectively where they are carried out over a sufficient:
(A) Length of time (B) Money
(C) Research (D) None of these

1350. Primary importance of PRA is:
(A) Behaviour (B) Attitudes
(C) Longevity (D) All of these

1351. A method of linking different survey methods is order to cross-check the information collected from each method is:
(A) Triangulation (B) Pilot experiment
(C) Action research (D) Survey design

1352. Triangulation is of particular use when employing:
(A) Mapping (B) Ranking and scoring
(C) Flow and venn diagrams (D) All of these

1353. One of the strengths of PRA is that many of the methods are visual and therefore, accessible to a:
(A) Longer group of people (B) Small group of peoples
(C) Peoples are not required (D) None of these

1354. Participatory methods have contributed to adjusting the interview to make it more conversational, while still:
(A) Controlled
(B) Structured
(C) Controlled and structured both
(D) None of these

1355. What are the characteristics of informal conversational interview:
(A) Questions emerge from the immediate context and are asked in the natural course of things
(B) There is no predetermination of question topics or wording
(C) Either (A) or (B)
(D) Both (A) and (B)

1356. What are the strengths of informal conversational interview:
(A) Increases the silence and relevance of questions
(B) Interview the built on and emerge from observations
(C) The interview can be matched to individuals and circumstances
(D) All of these

1357. Key interview methods in PRA system are:
(A) Informal conversational interview
(B) Interview guide approach
(C) Standardised open-ended interview
(D) Closed quantitative interviews
(E) All of these

1358. What are the weaknesses of informal conversational interview are:
(A) Different information collected from different people with different questions
(B) Less systematic and comprehensive
(C) Data organization and analysis can be quite difficult
(D) Requires maximum attention by interviewer
(E) All of these

1359. What are important characteristics of interview guide approach:
(A) Topics and issues to be covered are specified in advance, in out line form
(B) Interviewer decide sequence and working of questions in the course of the interview
(C) Both (A) and (B)
(D) None of these

1360. What are the strengths of Interview guide approach:
(A) The outline increases the comprehensiveness of the data
(B) Makes data collection systematic for each respondent
(C) Interviews remain fairly conversational and situational
(D) All of these

1361. What are the weaknesses of interview guide approach:
(A) Important and salient topics may be inadventently omitted
(B) Interviewer flexibility in working questions can result in incomparatively of responses
(C) Both (A) and (B)
(D) None of these

1362. Important characteristics of Standardised open-ended interview:
(A) The exact wording and sequence of questions are determined in advance
(B) All interviewees are asked the same basic questions in the same order
(C) Both (A) and (B)
(D) None of these

1363. Important strengths of standardized open-ended interview:
(A) Respondents answer the same questions, thus increasing comparability of response
(B) Reduces interviewer bias when several interviewers are used
(C) Facilitates organization and analyses of data
(D) All of these

1364. What are the weakness of standardized open - ended interview:
(A) Little flexibility in rilchig the interview to particular individuals and circumstances
(B) Requires maximum attention by interviewer
(C) Data organization and analysis can be quite difficult
(D) All of these

1365. Important characteristics of closed quantitative interviews:
(A) Questions and response categories are determined in advance
(B) Responses are fixed
(C) Respondent chooses from among these fixed responses
(D) All of these

1366. Important strengths of closed quantitative interviews:
(A) Data analysis is simple
(B) Responses can be directly compared and easily aggregate
(C) Man questions can be asked in a short time
(D) All of these

1367. Important weaknesses of closed quantitative interviews:
(A) Respondents must fit their experience and fallings into the researchers categories
(B) May be perceived as impersonal, irrelevant and mechanistic
(C) Can distort what respondents really mean or experienced
(D) All of these

1368. Participatory methods tend to make use if the:
(A) Informal conversational interview
(B) Inter view guide approach
(C) Both (A) and (B)
(D) None of these

1369. Informal conversational interview and interview guide approach are:
(A) Semi-structured interviews
(B) Long-structured interviews
(C) Short-structured interviews
(D) All of these

1370. Ranking and scoring have long been used to assess:
(A) People's expectations (B) Beliefs and attitudes
(C) Preferences and opinions (D) All of these

1371. Which is useful tool to be used in generating basic information which helps to focus further questioning is:
(A) Ranking (B) Scoring
(C) Ranking and scoring (D) None of these

1372. What are the important categories of ranking:
(A) Preference ranking (B) Wealth ranking
(C) Matrix ranking (D) All of these

1373. The way of quickly identifying main preferences as experienced by individual villager or village groups:
(A) Preference ranking (B) Matrix ranking
(C) Wealth ranking (D) None of these

1374. Ranking can be used to express other issues, where a two dimensional comparison is useful:
(A) Preference ranking (B) Matrix ranking
(C) Wealth ranking (D) None of these

1375. The most common technique for ranking is:
(A) Preference ranking (B) Matrix ranking
(C) Wealth ranking (D) Card sorting

1376. Closed scoring can be done in:
(A) Each 'box' (element for that criteria) can be scored between eg. 0-5, this is repeated over the whole matrix
(B) A fixed number of 'points' can be awarded for each criteria (row or column), and these are distributed between the elements in that two or column in relation to their perceived importance
(C) A fixed number of points is allowed for the entire matrix, and there must be distributed between both elements and criteria as participants decide
(D) All of these

1377. Ranking enables villagers to divide households in the community according to economic and other well being categories is:
(A) Preference ranking (B) Matrix ranking
(C) Wealth ranking (D) None of these

1378. Maps can be used to identify the comparative location and importance of different resources within an area is called:
(A) Participatory mapping (B) Social mapping
(C) Contour mapping (D) Flow diagram

1379. Diagrams are used for the systematic analysis of a wide range of issues whereby a whole stories of cause and effect relationships are examined is known as:
(A) Participatory mapping (B) Social mapping
(C) Contour mapping (D) Flow diagram

1380. Maps can be used to locate houses services and infrastructure within an area is called:
(A) Participatory mapping (B) Social mapping
(C) Contour mapping (D) Flow diagram

1381. Flow diagram can act as a basis for discussing the relationship between:
(A) Different groups
(B) Individuals or issues
(C) Can demonstrate potential multiplier effects
(D) All of these

1382. The diagrams are used to depict key institutions, organizations and individuals, and their interaction with the local community is called:
(A) Venn diagram (B) Flow diagram
(C) Both (A) and (B) (D) None of these

1383. In the rural transport context, venn diagram can be used to demonstrate the interaction between:
(A) Local villagers (B) Transport operators
(C) Local government (D) All of these

1384. The process of making the community inventory is sometimes called:
(A) Semi-structured interviewing
(B) High-structured interviewing
(C) Both (A) and (B)
(D) None of these

1385. The techniques widely used in recent years in most of the developmental programme:
(A) Participatory Rural Appraisal
(B) Rapid Rural Appraisal
(C) Participatory Resource Appraisal
(D) All of these

1386. A process that empowers local people to change their own condition and situation:
(A) Participatory Rural Appraisal
(B) Rapid Rural Appraisal
(C) Participatory Resource Appraisal
(D) All of these

1387. A way of learning for local community about their natural resources is called:
(A) Participatory Rural Appraisal
(B) Rapid Rural Appraisal
(C) Participatory Resource Appraisal
(D) Participatory Appraisal of Natural Resources

1388. The important principles which PRA is governed are:
(A) Listening and learning
(B) Offsetting biases
(C) Proper utilization of precious community time
(D) Seeking diversity
(E) Cross checking
(F) All of these

1389. Components of PRA are:
(A) Attitude and Behaviour
(B) Process of participation
(C) Methods
(D) Sharing
(E) Facilitation
(F) All of these

1390. PRA is a:
(A) Unidimensional approach
(B) Multidimensional approach
(C) Straight dimensional approach
(D) All of these

1391. Sharing of information and knowledge is integral in the process of :
(A) Participation (B) Communication
(C) Learning (D Adoption

1392. Attitude and behavior of the agent constitute the base for approaching and establishing links with:
(A) Local communities (B) Outside communities
(C) Urban communities (D) None of these

1393. PRA is an essential component of:
(A) Rural developmental projects
(B) Rural developmental schemes
(C) Both (A) and (B)
(D) None of these

1394. Advantages of PRA is/are:
(A) Empowering the poor and weak
(B) Diversity
(C) Research priority
(D) Organizational changes
(E) All of these

1395. PRA has extensive range of application in research and planning for:
(A) Land tenure
(B) Urban planning
(C) Natural resource management and domestic violence
(D) All of these

1396. Sharing of information and knowledge has magic effort in the process of :
(A) Participation (B) Communication
(C) Learning (D Adoption

1397. Successful training requires encouraging new ways of learning:
(A) Within oragnisation
(B) Outside the organization
(C) Both (A) and (B)
(D) None of these

1398. Training courses are always only part of the:
(A) Questions (B) Answer
(C) Questionnaire (D) None of these

1399. PRA, a field based methodology which has evolved and spread in the:
(A) 1980s (B) 1985s
(C) 1990s (D) 2000s

1400. The primary focus of PANR is an local natural resources from:
(A) Local community (B) Urban community
(C) Rural community (D) None of these

1401. The conventional techniques help the experts to arrive at numerical value for appraising:
(A) Manmade resources (B) Artificial resources
(C) Natural resources (D) None of these

1402. PRA is a technique for involving people to work for their:
(A) Other development (B) Own development
(C) Rural development (D) Urban development

1403. All principles of PRA are framed on the:
(A) NGO involvement
(B) Government involvement
(C) People's involvement
(D) None of these

1404. Basic aim of 4H-Club is:
(A) To improve training at the kendras or out of the centre
(B) Transfer of technical literacy increases agricultural production
(C) To provide training to youth for agricultural and home development
(D) To improve the effectiveness of agricultural extension

1405. The work at Shantineketan with the objectives of:
(A) To create a real interest in people for rural improvement
(B) To study rural problems
(C) To improve village sanitation
(D) All of these

1406. Major thrust or areas or level of Village Knowledge Centre are:
(A) Agronomic practices (B) Cattle feed
(C) Weather (D) All of these

1407.

List-I	List-II
(a) Audio	(1) Flash cards
(b) Visual	(2) Radio
(c) Audio	(3) Drama
(d) Audio-visual	(4) Tape recorder

Code	a	b	c	d
(A)	2	1	4	3
(B)	3	2	1	4
(C)	4	3	2	1
(D)	1	2	3	4

1408. Toll free number of Kisan Call Centtre is:
(A) 1800-180-5151 (B) 1800-180-1561
(C) 1800-180-1551 (D) 1800-180-5511

1409. Early majority adopter category in a social system :
(A) 2.5% (B) 13.5% (C) 34% (D) 16%

1410. A statement of situation, objectives, problems and solution is known as:
(A) Extension programme (B) Evaluation
(C) Plan of Work (D) Calender of Work

1411. In which type of organizational communication, messages flow from superior to subordinates:
(A) Upward (B) Downward
(C) Horizontal (D) External

1412. Technically web site address is known as:
(A) FTP (B) URL
(C) WWW (D) Gopher

1413. Edger Dale is recognized with :
(A) Demonstration (B) Cone of experience
(C) Field day (D) Exhibition

1414. A short series of lectures by 5-7 experts on subtopics of a main topic is called?
(A) Seminar (B) Workshop
(C) Symposium (D) Conference

1415. The concept of demonstration was popularized by
(A) Robert Chambers (B) Paul Leagans
(C) E.M. Rogors (D) Seaman Knapp

1416. Measurement in equal appearing internal seal is at:
(A) Interval level (B) Ratio level
(C) Ordinal level (D) Nominal level

1417. The degree to which an innovations results are visible is known as:
(A) Compexity (B) Observability
(C) Compatibility (D) Triability

1418. The concept of MBO was first put forth by:
(A) Peter Drucker (B) F.W. Taylor
(C) Henry Fayol (D) Luthor Gullick

1419. McClelland's theory of achievement motivation emphasized that enterpreneur should have need for:
(A) Employment (B) Achievement
(C) Affiliation (D) Money

1420. Organization develoment does not include :
(A) Change managment
(B) Team building
(C) Quality of work life
(D) Centralized management

1421. Treatment referes to the way of handling of :
(A) Message (B) Reciever
(C) Channel (D) Communicator

1422. Nature of learners in extension education is :
(A) Homogenous (B) Heterogenous
(C) Cosmopolite (D) Localite

1423. m-Kisan is the initiative of:
(A) ITC (B) GOI
(C) Tata (D) IIIT

1424. The first step in extension programme planning is:
(A) Setting objectives
(B) Collection of facts
(C) Analysis of situation
(D) Identification of problems

1425. Foundation stones of an extension program are:
(A) Problems (B) Facts
(C) Solutions (D) Objectives

1426. Fidelity in communication is associated with:
(A) Noise (B) Distortion
(C) Effectiveness (D) Barrier

1427. Which of the following program is responsible for green revolution?
(A) HYVP (B) IAAD
(C) WDP (D) IRDP

1428. The social relationships are direct, face to face and informal in:
(A) Primary group (B) Secondary group
(C) Voluntary group (D) Congregate group

1429. One of the following is not a PRA techniques
(A) Transect walk (B) Resource mapping
(C) Venn diagram (D) Mind mapping

1430. SGSY is a programme for promotion of?
(A) Infrastructure community
(B) Wage employment
(C) Community development
(D) Self employment

1431. Goal directed behaviour is known as:
(A) Performance
(B) Motivation
(C) Achievement
(D) Encouragement

1432. Extension method best suited to tackle a burning problems at an opportune time is :
(A) Field trip (B) Campaign
(C) Group discussion (D) General meeting

1433. NATP was launched in:
(A) 2001 (B) 1998
(C) 2006 (D) 1992

1434. The last step in teaching-learning process is
(A) Desire (B) Satisfaction
(C) Action (D) Conviction

1435. Poster is a :
(A) Projected visual aid
(B) Audio-visual aid
(C) Audio aid
(D) Non-projected visual aid

1436. Ability of an instrument to provide consistent results is referred as:
(A) Validity (B) Practicticability
(C) Reliability (D) Applicability

1437. In maxi min con principle this is to be maximised :
(A) Experimental variance (B) Total variance
(C) Error variance (D) Random error

1438. Person who influences innovation decisions informally in a social system is termed as:
(A) Facilitator (B) Opinion leader
(C) Change agent (D) Key communicator

1439. The people having appreciation for innovation and perceptiveness are grouped under one of the following types of interpersonal styles:
(A) Challenges (B) Casuals
(C) Visionaries (D) Stabilizers

1440. The process of division of society into social group is termed as:
(A) Social stratification (B) Social institution
(C) Social control (D) Social change

1441. The group of Swarozgaris SHG in SGSY programme get the subsidy to the extent of:
(A) 50% (B) 25% (C) 33% (D) 10%

1442. The technique of dividing a large group into much smaller ones, in which a topic is discussed within a limited period of time?
(A) Buzz session (B) In basket exercise
(C) Fish bowl exercise (D) Syndicate method

1443. Round table discussion is called:
(A) Symposium
(B) Forum
(C) Panel
(D) Workshop

1444. National demonstration is the example of:
(A) Frontline Demonstration
(B) Method demonstration
(C) Composite demonstration
(D) Result demonstration

1445. Extension teaching method is focussed to:
(A) Subject matter specialist
(B) Learner
(C) Subject
(D) None of these

1446. Which one of the following body is responsible for development work at village level?
(A) Gram Panchayat
(B) Block
(C) Tehsil
(D) District

1447. Which one is most effective method of communication?
(A) Radio
(B) Television
(C) Kisan Gosthi
(D) Personal visit

1448. Most effective extension service is:
(A) Profiting Local Leaders
(B) Helping Block Development Officer
(C) Helping Gram Pradhan
(D) Helping the farmers to help themselves

1449. For one talk, how many flash cards should be used ?
(A) 10-12
(B) 12-14
(C) 14-18
(D) 20-22

1450. Full name of NITI Ayog :
(A) National Institution for Transforming India
(B) National Insitution for Technologies India
(C) National Indian for Technology Institute
(D) National Indian forTransformation Institution

1451. NITI Ayog fermed in :
(A) 1 January, 2014
(B) 1 January, 2015
(C) 1 January, 2016
(D) 1 January, 2017

1452. NITI Ayog in the premier policy of :
(A) Think Tank of GoI
(B) Providing both directional and policy input
(C) Both (A) and (B)
(D) None of the above

ANSWERSHEET: 15.AGRICULTURAL EXTENSION

1. (A); 2. (B); 3. (A); 4. (A); 5. (C); 6. (D); 7. (D); 8. (C); 9. (B); 10. (D); 11. (A); 12. (B); 13. (B) ;14. (A); 15. (A); 16. (B); 17. (C); 18. (C); 19. (A); 20. (B); 21. (C); 22. (D); 23. (A); 24. (B); 25. (D) ;26. (D); 27. (C); 28. (A); 29. (A); 30. (A); 31. (B); 32. (B); 33. (D); 34. (D); 35. (B); 36. (A); 37. (A); 38. (A) ;39. (A); 40. (B); 41. (B); 42. (B); 43. (D); 44. (A); 45. (D); 46. (B); 47. (D); 48. (C); 49. (C); 50. (A); 51. (A) ; 52. (C); 53. (B); 54. (C); 55. (A); 56. (D); 57. (C); 58. (A); 59. (B); 60. (C); 61. (D); 62. (B); 63. (C); 64. (D) ; 65. (B); 66. (C); 67. (A); 68. (A); 69. (C); 70. (A); 71. (B); 72. (A); 73. (A); 74. (B) ;75. (C); 76. (C); 77. (C); 78. (A) ;79. (B); 80. (B); 81. (D); 82. (D); 83. (C); 84. (D); 85. (B); 86. (A); 87. (A); 88. (B); 89. (C); 90. (D); 91. (A) ;92. (B); 93. (C); 94. (B); 95. (D); 96. (A); 97. (A); 98. (C); 99. (D); 100. (C); 101. (A); 102. (A); 103. (D); 104. (B) ; 105. (D); 106. (B); 107. (A); 108. (B); 109. (C); 110. (C); 111. (B); 112. (C); 113. (D); 114. (B); 115. (B); 116. (A); 117. (A) ; 118. (A); 119. (C); 120. (A); 121. (A); 122. (D); 123. (D); 124. (D); 125. (A); 126. (A); 127. (B); 128. (B); 129. (C); 130. (C) ; 131. (B); 132. (C); 133. (D); 134. (B); 135. (A); 136. (A); 137. (B); 138. (D); 139. (D); 140. (B); 141. (A); 142. (C); 143. (C) ; 144. (B); 145. (B); 146. (D); 147. (B); 148. (B); 149. (A); 150. (A); 151. (A); 152. (D); 153. (B); 154. (A); 155. (A); 156. (B) ; 157. (B); 158. (A); 159. (B); 160. (A); 161. (A); 162. (D); 163. (C); 164. (A); 165. (A); 166. (C); 167. (B); 168. (D); 169. (D); 170. (B); 171. (A); 172. (B); 173. (A); 174. (A); 175. (A); 176. (A); 177. (B); 178. (B); 179. (B); 180. (A); 181. (B); 182. (B) ; 183. (B); 184. (B); 185. (C); 186. (B); 187. (B); 188. (B); 189. (B); 190. (B); 191. (D); 192. (B); 193. (D); 194. (B); 195. (D) ; 196. (C); 197. (A); 198. (C); 199. (D); 200. (A); 201. (C); 202. (D); 203. (C); 204. (B); 205. (A); 206. (B); 207. (C); 208. (C) ; 209. (A); 210. (B); 211. (B); 212. (D); 213. (C); 214. (B); 215. (B); 216. (B); 217. (A); 218. (A); 219. (D); 220. (A); 221. (B) ; 222. (B); 223. (A); 224. (D); 225. (A); 226. (A); 227. (A); 228. (B); 229. (B); 230. (A); 231. (C); 232. (A); 233. (A); 234. (C) ; 235. (A); 236. (C); 237. (D); 238. (B); 239. (A); 240. (C); 241. (C); 242. (D); 243. (A);

244. (B); 245. (B); 246. (B); 247. (A); 248. (A); 249. (A);
250. (B); 251. (A); 252. (C); 253. (B); 254. (B); 255. (C);
256. (D); 257. (A); 258. (A); 259. (A); 260. (A) ; 261. (A);
262. (B); 263. (B); 264. (A); 265. (D); 266. (B) ; 267. (B);
268. (A); 269. (C); 270. (A); 271. (B); 272. (A); 273. (B) ;
274. (C); 275. (C); 276. (C); 277. (C); 278. (C); 279. (D);
280. (C); 281. (C); 282. (A); 283. (C); 284. (A); 285. (B);
286. (B); 287. (C); 288. (B); 289. (C); 290. (C) ; 291. (A);
292. (B); 293. (C); 294. (A); 295. (A); 296. (C); 297. (A);
298. (B); 299. (B); 300. (D); 301. (D); 302. (A); 303. (D);
304. (C); 305. (B); 306. (D); 307. (C); 308. (B); 309. (A);
310. (C); 311. (A); 312. (B) ; 313. (A); 314. (D); 315. (A);
316. (A); 317. (A); 318. (A); 319. (A); 320. (A); 321. (B);
322. (B); 323. (D); 324. (C); 325. (A); 326. (A); 327. (D);
328. (B); 329. (D); 330. (B); 331. (C); 332. (A); 333. (C);
334. (D); 335. (B); 336. (D); 337. (B); 338. (B); 339. (D);
340. (A); 341. (B); 342. (B); 343. (C); 344. (B); 345. (B);
346. (B); 347. (C); 348. (D); 349. (A); 350. (A); 351. (B);
352. (C); 353. (C); 354. (B); 355. (B); 356. (A); 357. (B);
358. (D); 359. (A); 360. (A); 361. (C); 362. (B); 363. (B);
364. (A); 365. (B); 366. (A); 367. (A); 368. (D); 369. (A);
370. (D); 371. (B); 372. (B); 373. (B); 374. (D); 375. (B);
376. (C); 377. (A); 378. (D); 379. (B); 380. (C); 381. (B);
382. (B); 383. (C); 384. (B); 385. (A); 386. (C); 387. (A);
388. (D); 389. (D); 390. (C) ; 391. (B); 392. (B); 393. (B);
394. (D); 395. (A); 396. (A); 397. (C); 398. (A); 399. (C);
400. (C); 401. (C); 402. (A); 403. (B) ; 404. (B); 405. (D);
406. (A); 407. (C); 408. (D); 409. (C); 410. (C); 411. (D);
412. (A); 413. (C); 414. (C); 415. (C); 416. (A); 417. (D);
418. (B); 419. (D); 420. (D); 421. (D); 422. (A); 423. (A);
424. (B); 425. (C); 426. (C); 427. (C); 428. (A); 429. (A) ;
430. (C); 431. (B); 432. (A); 433. (B); 434. (C); 435. (A);
436. (A); 437. (C); 438. (A); 439. (A); 440. (C); 441. (A);
442. (C) ; 443. (A); 444. (D); 445. (A); 446. (C); 447. (D);
448. (D); 449. (A); 450. (A); 451. (B); 452. (C); 453. (A);
454. (B); 455. (B); 456. (B) ; 457. (A); 458. (B); 459. (B);
460. (B); 461. (B); 462. (D); 463. (C); 464. (D); 465. (A);
466. (A); 467. (A); 468. (D); 469. (C) ; 470. (B); 471. (B);
472. (B); 473. (B); 474. (D); 475. (C); 476. (A); 477. (A);
478. (B); 479. (C); 480. (C); 481. (C); 482. (C) ; 483. (B);
484. (C); 485. (D); 486. (C); 487. (C); 488. (B); 489. (A);
490. (A); 491. (A); 492. (B); 493. (C); 494. (C); 495. (A) ;
496. (D); 497. (D); 498. (A); 499. (B); 500. (B); 501. (D);
502. (C); 503. (B); 504. (C); 505. (C); 506. (D); 507. (C);
508. (B) ; 509. (C); 510. (D); 511. (C); 512. (B); 513. (C);
514. (A); 515. (C); 516. (B); 517. (B); 518. (C); 519. (C);
520. (B); 521. (C); 522. (A); 523. (A); 524. (A); 525. (C);
526. (A); 527. (D); 528. (B); 529. (C); 530. (B); 531. (A);
532. (A); 533. (D); 534. (D) ; 535. (A); 536. (A); 537. (C);
538. (B); 539. (C); 540. (A); 541. (C); 542. (B); 543. (A);
544. (D); 545. (A); 546. (A); 547. (A) ; 548. (A); 549. (B);
550. (A); 551. (C); 552. (D); 553. (D); 554. (A); 555. (B);
556. (B); 557. (D); 558. (B); 559. (B); 560. (D) ; 561. (D);
562. (B); 563. (C); 564. (A); 565. (B); 566. (D); 567. (A);
568. (C); 569. (B); 570. (D); 571. (B); 572. (D); 573. (A) ;
574. (A); 575. (B); 576. (A); 577. (B); 578. (B); 579. (C);
580. (D); 581. (B); 582. (A); 583. (B); 584. (A); 585. (C);
586. (C) ; 587. (B); 588. (A); 589. (D); 590. (B); 591. (B);
592. (E); 593. (D); 594. (A); 595. (B); 596. (C); 597. (D);
598. (A); 599. (B) ; 600. (C); 601. (A); 602. (B); 603. (C);
604. (C); 605. (A); 606. (B); 607. (C); 608. (B); 609. (C);
610. (A); 611. (A); 612. (D) ; 613. (A); 614. (D); 615. (B);
616. (A); 617. (C); 618. (D); 619. (C); 620. (B); 621. (A);
622. (B); 623. (C); 624. (A); 625. (A) ; 626. (A); 627. (D);
628. (C); 629. (C); 630. (C); 631. (C); 632. (A); 633. (A);
634. (C); 635. (D); 636. (B); 637. (D); 638. (C) ; 639. (B);
640. (B); 641. (A); 642. (A); 643. (C); 644. (C); 645. (C);
646. (D); 647. (B); 648. (A); 649. (B); 650. (A); 651. (C) ;
652. (C); 653. (A); 654. (A); 655. (C); 656. (B); 657. (C);
658. (C); 659. (A); 660. (C); 661. (B); 662. (D); 663. (A);
664. (B) ; 665. (B); 666. (D); 667. (A); 668. (A); 669. (A);
670. (A); 671. (A); 672. (C); 673. (C); 674. (B); 675. (D);
676. (B); 677. (A) ; 678. (A); 679. (A); 680. (D); 681. (D);
682. (B); 683. (C); 684. (B); 685. (D); 686. (C); 687. (B);
688. (C); 689. (C); 690. (B) ; 691. (C); 692. (C); 693. (A);
694. (E); 695. (B); 696. (C); 697. (A); 698. (C); 699. (D);
700. (F); 701. (A); 702. (C); 703. (D) ; 704. (C); 705. (C);
706. (A); 707. (B); 708. (D); 709. (C); 710. (D); 711. (D);
712. (A); 713. (A); 714. (B); 715. (C); 716. (C) ; 717. (D);
718. (D); 719. (B); 720. (B);721. (B); 722. (A); 723. (C);
724. (A); 725. (A); 726. (B); 727. (A); 728. (D); 729. (B) ;
730. (B); 731. (A); 732. (C); 733. (B); 734. (C); 735. (A);
736. (C); 737. (B); 738. (A); 739. (A); 740. (C); 741. (A);
742. (A) ; 743. (D); 744. (C); 745. (B); 746. (D); 747. (C);

748. (B); 749. (A); 750. (A); 751. (A); 752. (D); 753. (B);
754. (C); 755. (A); 756. (A); 757. (C); 758. (D); 759. (B);
760. (B); 761. (C); 762. (A); 763. (A); 764. (C); 765. (C);
766. (C); 767. (D); 768. (D); 769. (A); 770. (A); 771. (A);
772. (C); 773. (A); 774. (C); 775. (C); 776. (B); 777. (A);
778. (C); 779. (A); 780. (D); 781. (D) ; 782. (B); 783. (D);
784. (D); 785. (D); 786. (A); 787. (B); 788. (D); 789. (D);
790. (A); 791. (D); 792. (A); 793. (B); 794. (B) ; 795. (C);
796. (A); 797. (A); 798. (B); 799. (A); 800. (D); 801. (D);
802. (B); 803. (B); 804. (A); 805. (B); 806. (C); 807. (C) ;
808. (D); 809. (A); 810. (D); 811. (C); 812. (B); 813. (C);
814. (D); 815. (B); 816. (B); 817. (B); 818. (C); 819. (C);
820. (A) ; 821. (B); 822. (A); 823. (C); 824. (C); 825. (A);
826. (D); 827. (C); 828. (D); 829. (A); 830. (C); 831. (B);
832. (B); 833. (A) ; 834. (D); 835. (A); 836. (A); 837. (C);
838. (A); 839. (B); 840. (C); 841. (C); 842. (A); 843. (B);
844. (D); 845. (A); 846. (A); 847. (C) ; 848. (D); 849. (D);
850. (B); 851. (A); 852. (B); 853. (D); 854. (B); 855. (A);
856. (A); 857. (C); 858. (B); 859. (A); 860. (B) ; 861. (C);
862. (D); 863. (C); 864. (B); 865. (A); 866. (C); 867. (A);
868. (B); 869. (A); 870. (B); 871. (C); 872. (A); 873. (B) ;
874. (A); 875. (D); 876. (C); 877. (C); 878. (C); 879. (B);
880. (D); 881. (D); 882. (B) ; 883. (C); 884. (A); 885. (B);
886. (C) ; 887. (B); 888. (C); 889. (A); 890. (B); 891. (D);
892. (C); 893. (A); 894. (B); 895. (C) ; 896. (D); 897. (C);
898. (D); 899. (C); 900. (C); 901. (A); 902. (B); 903. (C);
904. (D); 905. (B); 906. (C); 907. (C); 908. (D); 909. (C);
910. (C); 911. (C); 912. (D); 913. (C); 914. (A); 915. (C);
916. (C); 917. (C); 918. (E); 919. (E); 920. (C); 921. (C);
922. (C); 923. (D); 924. (D) ; 925. (D); 926. (E); 927. (C);
928. (E); 929. (D); 930. (E); 931. (E); 932. (E); 933. (C);
934. (A); 935. (B); 936. (C); 937. (A) ; 938. (A); 939. (D);
940. (D); 941. (A); 942. (D); 943. (A); 944. (D); 945. (B);
946. (A); 947. (D); 948. (C); 949. (B); 950. (D); 951. (A);
952. (A); 953. (C); 954. (A); 955. (B); 956. (A) ; 957. (D);
958. (B); 959. (C); 960. (A); 961. (B); 962. (D); 963. (C);
964. (D); 965. (C); 966. (C); 967. (C); 968. (D); 969. (D);
970. (C); 971. (A); 972. (B); 973. (A); 974. (A); 975. (C);
976. (A); 977. (C); 978. (D); 979. (E); 980. (A); 981. (C);
982. (B); 983. (A); 984. (A); 985. (D); 986. (E); 987. (D);
988. (C); 989. (A); 990. (D) ; 991. (D) ; 992. (C); 993. (C);
994. (D); 995. (B); 996. (A); 997. (C); 998. (A); 999. (D);
1000. (D); 1001. (C); 1002. (A); 1003. (B); 1004. (C);
1005. (D); 1006. (A); 1007. (B); 1008. (C); 1009. (D);
1010. (D); 1011. (F); 1012. (E); 1013. (E); 1014. (E);
1015. (A); 1016. (C) ; 1017. (C); 1018. (C); 1019. (C);
1020. (C); 1021. (A); 1022. (C); 1023. (A); 1024. (A);
1025. (B) ; 1026. (D) ; 1027. (C); 1028. (B); 1029. (B);
1030. (A); 1031. (E); 1032. (A); 1033. (B); 1034. (B);
1035. (C); 1036. (A); 1037. (C); 1038. (C) ; 1039. (A);
1040. (C); 1041. (C); 1042. (A); 1043. (E); 1044. (D);
1045. (B); 1046. (A); 1047. (A); 1048. (C); 1049. (C);
1050. (B); 1051. (C) ; 1052. (A); 1053. (A); 1054. (B);
1055. (C) ; 1056. (C); 1057. (E); 1058. (F); 1059. (E);
1060. (D); 1061. (C); 1062. (D); 1063. (E); 1064. (C);
1065. (D); 1066. (A); 1067. (C); 1068. (B) ; 1069. (B);
1070. (A); 1071. (B); 1072. (D); 1073. (E); 1074. (A);
1075. (D); 1076. (B); 1077. (C); 1078. (A) ; 1079. (B);
1080. (C); 1081. (D); 1082. (C); 1083. (A); 1084. (B);
1085. (C); 1086. (D); 1087. (A); 1088. (A); 1089. (D);
1090. (B); 1091. (C) ; 1092. (C); 1093. (B); 1094. (C);
1095. (C); 1096. (C); 1097. (A); 1098. (A); 1099. (B);
1100. (C); 1101. (B); 1102. (C); 1103. (B) ; 1104. (C);
1105. (B); 1106. (B); 1107. (A) ; 1108. (B;); 1109. (D);
1110. (D); 1111. (C); 1112. (D); 1113. (C); 1114. (D);
1115. (D); 1116. (A) ; 1117. (D) ; 1118. (C); 1119. (B);
1120. (C) ; 1121. (D); 1122. (C); 1123. (A); 1124. (D);
1125. (C); 1126. (A); 1127. (D); 1128. (A); 1129. (B);
1130. (D) ; 1131. (B); 1132. (D); 1133. (A) ; 1134. (A);
1135. (D); 1136. (A); 1137. (D); 1138. (D); 1139. (D);
1140. (B); 1141. (A); 1142. (D) ; 1143. (A) ; 1144. (C);
1145. (B); 1146. (C) ; 1147. (C); 1148. (C); 1149. (D);
1150. (D); 1151. (C); 1152. (D); 1153. (D); 1154. (D);
1155. (C) ; 1156. (A) ; 1157. (C); 1158. (A); 1159. (C) ;
1160. (D); 1161. (B); 1162. (B); 1163. (D); 1164. (A);
1165. (C); 1166. (C); 1167. (B); 1168. (A) ; 1169. (D) ;
1170. (B); 1171. (A); 1172. (B); 1173. (C); 1174. (D);
1175. (A); 1176. (B); 1177. (A); 1178. (D); 1179. (A);
1180. (C); 1181. (A) ; 1182. (D) ; 1183. (B); 1184. (B);
1185. (B) ; 1186. (B); 1187. (A); 1188. (A); 1189. (B);
1190. (D); 1191. (B); 1192. (C); 1193. (D); 1194. (C) ;
1195. (A) ; 1196. (D); 1197. (D); 1198. (C) ; 1199. (D);
1200. (C); 1201. (A); 1202. (D); 1203. (C); 1204. (C);
1205. (B); 1206. (A); 1207. (D); 1208. (D); 1209. (B);

1210. (A); 1211. (A); 1212. (A); 1213. (A); 1214. (B); 1215. (B); 1216. (B); 1217. (C); 1218. (D); 1219. (B); 1220. (B) ; 1221. (A); 1222. (C); 1223. (C); 1224. (D); 1225. (B); 1226. (B); 1227. (A); 1228. (C); 1229. (A); 1230. (A); 1231. (D); 1232. (D); 1233. (B) ; 1234. (A) ; 1235. (D); 1236. (B); 1237. (A); 1238. (A); 1239. (B); 1240. (A); 1241. (B); 1242. (A); 1243. (A); 1244. (B); 1245. (B); 1246. (A) ; 1247. (B); 1248. (C); 1249. (C); 1250. (A); 1251. (D); 1252. (B); 1253. (B); 1254. (B); 1255. (A); 1256. (B); 1257. (A); 1258. (A); 1259. (B); 1260. (A); 1261. (B); 1262. (A); 1263. (A); 1264. (C); 1265. (C); 1266. (A); 1267. (A); 1268. (B); 1269. (D); 1270. (A); 1271. (D); 1272. (B); 1273. (B); 1274. (C); 1275. (D); 1276. (C); 1277. (A); 1278. (D); 1279. (B); 1280. (C); 1281. (A); 1282. (A); 1283. (A); 1284. (A); 1285. (C) ; 1286. (B); 1287. (C); 1288. (B); 1289. (B) ; 1290. (C) ; 1291. (B); 1292. (B); 1293. (A); 1294. (A); 1295. (D); 1296. (B); 1297. (A); 1298. (D) ; 1299. (D); 1300. (A); 1301. (B); 1302. (C); 1303. (E); 1304. (A); 1305. (B); 1306. (D); 1307. (C); 1308. (B); 1309. (A); 1310. (B); 1311. (A) ; 1312. (E); 1313. (A); 1314. (C); 1315. (C) ; 1316. (C); 1317. (D); 1318. (C); 1319. (B); 1320. (B); 1321. (D); 1322. (A); 1323. (B); 1324. (A) ; 1325. (A); 1326. (D); 1327. (A); 1328. (A) ; 1329. (D); 1330. (A); 1331. (B); 1332. (A); 1333. (B); 1334. (C); 1335. (B); 1336. (A); 1337. (D); 1338. (C); 1339. (D); 1340. (C); 1341. (F) ; 1342. (F); 1343. (G); 1344. (D); 1345. (D); 1346. (C); 1347. (A); 1348. (A); 1349. (A); 1350. (D); 1351. (A); 1352. (D); 1353. (A); 1354. (C); 1355. (D); 1356. (D); 1357. (E); 1358. (E); 1359. (C); 1360. (D); 1361. (C); 1362. (C); 1363. (D); 1364. (D) ; 1365. (D); 1366. (D); 1367. (D); 1368. (C); 1369. (A); 1370. (D); 1371. (C); 1372. (D); 1373. (A); 1374. (B); 1375. (D); 1376. (D) ; 1377. (C); 1378. (A); 1379. (D); 1380. (B) ; 1381. (D); 1382. (A); 1383. (D); 1384. (A); 1385. (D); 1386. (A); 1387. (D); 1388. (F); 1389. (F) ; 1390. (B) ; 1391. (C); 1392. (A); 1393. (C); 1394. (E); 1395. (D); 1396. (A); 1397. (A); 1398. (B); 1399. (C); 1400. (A); 1401. (C); 1402. (B) ; 1403. (C); 1404. (B); 1405. (D); 1406. (D) ; 1407. (A); 1408. (C); 1409. (C); 1410. (A); 1411. (B) ; 1412. (B); 1413. (B); 1414. (C); 1415. (D); 1416. (A) ; 1417. (B); 1418. (A); 1419. (B); 1420. (D); 1421. (A) ; 1422. (B); 1423. (B); 1424. (B); 1425. (B); 1426. (C) ; 1427. (A); 1428. (A); 1429. (D); 1430. (D); 1431. (B) ; 1432. (B); 1433. (B); 1434. (B); 1435. (D); 1436. (C) ; 1437. (A); 1438. (B); 1439. (C); 1440. (A); 1441. (A) ; 1442. (A); 1443. (C); 1444. (C); 1445. (B); 1446. (A) ; 1447. (D); 1448. (D); 1449. (A); 1450. (A); 1451. (B) ; 1452. (C).

Chapter 16

AGRICULTURAL ECONOMICS

1. Science which deals the principles and methods of economics are applied to the special conditions of agricultural industry" is called:
(A) Agricultural economics (B) Economics
(C) Industrial economics (D) Pure economics

2. The scope of agricultural economics is exactly covered by the subject of agricultural economics. It covers mainly three things:
(A) Subject matter of agricultural economics
(B) Nature of agricultural economics
(C) Limitations
(D) All of these

3. Subject matter of agricultural economics may be divided into:
(A) Consumptions (B) Production
(C) Exchange (D) Distribution
(E) Public finance (F) All of these

4. Agricultural economics puts facts before us.
(A) Positive science (B) Normative science
(C) Applied science (D) None of these

5. Agricultural economics establishes certain norms or ideals.
(A) Positive science (B) Normative science
(C) Applied science (D) None of these

6. Agricultural economics is also an applied science because the principles and practices of economics solves the different agricultural problems of farmers.
(A) Positive science (B) Normative science
(C) Applied science (D) None of these

7. Indian agriculture is characterized by:
(A) Low productivity (B) Subsistence farming
(C) Labour dominance (D) All of these

8. The cost of production can be minimized by using the principle of:
(A) Law of return
(B) Law of opportunity cost
(C) Law of least-cost combination
(D) Cost principle

9. Who defined Economics is a wealth of science:
(A) Smith (B) Marshall
(C) Robins (D) Pigue

10. Who defined Economics as the study of human behaviour in the daily business of life?
(A) Smith (B) Marshall
(C) Robins (D) Pigue

11. Who defined economics as the study of human welfare?
(A) Smith (B) Marshall
(C) Robins (D) Pigue

12. Who defined economics as the study of choice making behaviour of human being?
(A) Smith (B) Marshall
(C) Robins (D) Pigue

13. At optimum level of production or output, the marginal return should be:
(A) Positive (B) Negative
(C) Zero (D) Equal to average return

14. Which of the following is a subject matter of economics?
(A) Consumption (B) Production and distribution
(C) Public finance (D) All of these

15. Which is not a factor of production?
(A) Land (B) Labour
(C) Capital (D) Crops

16. Which of the following function is related to an economic system?
(A) What goods and services are to be produced?
(B) How to best organize and combine recourse of production?
(C) Allocation or distribution of income to owners of resources
(D) All of these

17. With the globalization of trade, the importance of the following has increased:
(A) Food processing (B) Steel industries
(C) Machines and Tools (D) Chemicals

18. The first five year plant in India was started on:
(A) Ist April 1958 (B) Ist April 1951
(C) Ist April 1952 (D) Ist April 1953

19. The chairman of the Planning Commission in India is:
(A) Minister of Planning (B) Minister of Industries
(C) Prime Minister (D) President of India

20. A marginal farmer owns cultivated land of:
(A) 1 to 2 hectares (B) 2 to 3 hectares
(C) Below 1 hectare (D) 3 hectares and above

21. A large farmer owns cultivated land of:
(A) 1 to 2 hectares (B) 2 to 3 hectares
(C) Below 1 hectare (D) Above 2 hectares

22. Which of the following organization governs international trade:
(A) U.N.O. (B) F.A.O.
(C) I.M.F. (D) W.T.O.

23. The fourth Five Year Plan gave first priority to:
(A) Industries (B) Transport
(C) Health and Education (D) Agriculture

24. The Agricultural Prices are recommended by:
(A) I.C.A.R. (B) C.S.I.R.
(C) C.A.C.P. (D) R.B.I.

25. The Agricultural Prices are recommended on the basis of:
(A) Average cost of production
(B) Minimum cost of production
(C) Bulk line cost
(D) Maximum cost of production

26. Which of the following organization refinance agriculture:
(A) I.D.B.I. (B) I.C.I.C.
(C) N.A.B.A.R.D. (D) S.B.I.

27. Regional Rural Banks provide credit to:
(A) Big farmers
(B) Weaker section of urban population
(C) Weaker section of rural population
(D) None of these

28. Loan for purchase of tractor is known as:
(A) Short term loan (B) Medium term loan
(C) Long term loan (D) Cop plan

29. First priority to industry was given during:
(A) First Five Year Plan (B) Second Five Year Plan
(C) Third Five Year Plan (D) Fourth Five Year Plan

30. Agricultural Produce and Marketing Act was passed in:
(A) 1937 (B) 1935
(C) 1947 (D) 1949

31. The growth rate of food production in India after 1980 has been:
(A) Increasing (B) Decreasing
(C) Constant (D) Negative

32. Increase in Indian population is due to:
(A) High birth rate
(B) Low birth rate
(C) High birth rate and low death rate
(D) Defective Family Planning

33. The export of Agricultural commodities since Nineties is:
(A) Increasing (B) Decreasing
(C) Stand still (D) Negative

34. I.R.D.P. Programme is meant for:
(A) Poverty alleviation
(B) Employment generation
(C) Disposal of agricultural produce
(D) Agro-processing

35. The Zamidari Abolition Act was passed in UP in:
(A) 1950 (B) 1952
(C) 1954 (D) 1956

36. Payment made to labourers for their services is called:
(A) Rent (B) Wages
(C) Interest (D) None of these

37. The agro-climatic Regional Planning approach was started in:
(A) Sixth Five Year Plan
(B) Seventh Five Year Plan
(C) Eighth Five Year Plan
(D) Ninth Five Year Plan

38. In India land reforms have been equated with total:
(A) Urban development
(B) Agrarian development
(C) Industrial development
(D) Educational development

39. The term land reforms should be used in broadest sense to serve as:
(A) Inscription (B) Prescription
(C) Description (D) Formulation

40. CAPART expands as:
(A) Council for Action and people Advancement in Rural Technology
(B) Council for Advancement of People Action and Rural Technology
(C) Commission for Action and People Advancement in Rural Technology
(D) Committee for Advancement of People Action and Rural Technology

41. CAPART is an autonomous wing of:
(A) Ministry of Agriculture and Cooperation
(B) Ministry of Rural Development
(C) ICAR
(D) Ministry of HRD

42. Khadi and Village Industries Commission was established on:
(A) 1957 (B) 1959
(C) 1976 (D) 1982

43. The International Institute of Rural Reconstruction is located at:
(A) Philipines (B) Hyderabad
(C) Bombay (D) New Delhi

44. A monthly newsletter published by Ministry of Rural Development:
(A) Grameen Bharat (B) Grama Sabha
(C) Panchayat Sabdh (D) Rural Panchayat

45. The Right to Information Act passed in the year:
(A) 2005 (B) 2006
(C) 2007 (D) 2008

46. Mission 2007 refers to:
(A) Medicinal Plants (B) Micro irrigation
(C) Plantation sector (D) Knowledge centres

47. The recent National Commission on Agriculture was headed by:
(A) M.S. Swaminathan (B) Chidambaram
(C) Sharad Pawar (D) Panjab Singh

48. The millennium development goal of reducing world poverty by half is aimed by:
(A) 2020 (B) 2015
(C) 2010 (D) 2012

49. Zamidari system was introduced by:
(A) Mughals (B) British
(C) Marathas (D) All of these

50. The land revenue under Zamidari system was charged:
(A) Through village panchayat
(B) Through landlords
(C) Directly from the farmers
(D) All of these

51. 'Zamidari' system was abolished in India on:
(A) 1952 (B) 1954
(C) 1968 (D) 1960

52. In Mahalwari system land revenue was charged:
(A) Through landlords
(B) Directly from the farmers
(C) Through village panchayat
(D) All of these

53. The land revenue in Ryotwari system was charged:
(A) Through landlords
(B) Directly from the farmers
(C) Through village panchayat
(D) All of these

54. The department of agriculture, revenue and commerce was set up in:
(A) 1861 (B) 1871
(C) 1891 (D) 1899

55. The election of panchayat is normally held after a period of:
(A) 3 years (B) 4 years
(C) 5 years (D) 6 years

56. The number of tier for Panchayats in a state with population below 20 lakhs:
(A) Two (B) Three
(C) Four (D) One

57. The proportion of seat in a panchayat to be filled from women through election:
(A) Two-third (B) One-third
(C) One-fifth (D) One-fourth

58. Which of the following states have no Panchayati Raj:
(A) Karnataka (B) Bihar
(C) Orissa (D) Meghalaya

59. The 73rd amendment act does not apply to the stages of:
(A) Nagaland (B) Mizoram
(C) Meghalaya (D) All of these

60. Panchayati Raj system was adopted first in:
(A) Rajasthan and Madhya Pradesh
(B) Andhra Pradesh and West Bengal
(C) Rajasthan and Andhra Pradesh
(D) Tamilnadu and Rajasthan

61. The basic unit of Panchayati Raj institutions:
(A) Zila Parishad (B) Grama Sabha
(C) Panchayat Samiti (D) Block Panchayat

62. The 'Year of Grama Sabha' was observed in:
(A) 1999-2000 (B) 2001-2002
(C) 2004-2005 (D) 2006-2007

63. The total number of panchayats in the country:
(A) 2.31 lakhs (B) 2.23 lakhs
(C) 2.47 lakhs (D) 2.79 lakhs

64. The intermediate level in Panchayati Raj institutions:
(A) Zila Parishad (B) Grama Sabha
(C) Panchayat Samiti (D) Block panchayat

65. The total number of panchayat at intermediate level:
(A) 6000 (B) 7000
(C) 8000 (D) 9000

66. The three tier of Panchayati Raj is:
(A) Zilla Parishad (B) Grama Sabha
(C) Panchayat Samiti (D) Block Panchayat

67. The responsibility of village panchayat include:
(A) Agricultural production
(B) Rural Industrial development
(C) Maternity and child welfare
(D) All of these

68. The need for constitutional sanction of village panchayats was first felt by:
(A) Balwantrai Mehta Committee
(B) Ashoka Mehta Committee
(C) GVK Rao Committee
(D) L M Singhvi Committee

69. The planning commission was setup in:
(A) 1949 (B) 1950
(C) 1951 (D) 1952

70. The first chairman of planning commission:
(A) Dr. Rajendra Prasad (B) Dr S Radhakrishnan
(C) Pt Jawahrlal Nehru (D) Dr. B. R. Ambedkar

71. Planning commission is a:
(A) Statutory body (B) Advisory body
(C) Autonomous body (D) Elected body

72. Which one of the following is not an objective of India's economic planning:
(A) Population growth (B) Industrial growth
(C) Self reliance (D) Poverty alleviation

73. The long term objective of development in India has been:
(A) High economic growth (B) Economic self reliance
(C) Social justice (D) All of these

74. The National Development Council gets its administrative support from:
(A) Parliament (B) Planning commission
(C) Finance commission (D) Union cabinet

75. The planning commission of India is a:
(A) Constitutional body (B) Statutory body
(C) Non-statutory body (D) Autonomous body

76. Who is not member in the National Development Council:
(A) Prime minister (B) President
(C) Chief minister (D) Cabinet ministers

77. Who was the first deputy chairman of planning commission of India:
(A) Jawaharlal Nehru (B) Rajendra Prasad
(C) V T Krishnamachari (D) C Rajagopalachari

78. The main objective of National Development Council is:
(A) To promote common economic policies in all vital spears
(B) To mobilize the efforts and resources of the nation in support of plan
(C) To ensure balanced and rapid development of all parts of the country
(D) To secure cooperation of states in the execution of the plan

79. The book entitled ' Planned Economy for India' was the attempt of:
(A) Mahalanobis (B) Viswesvarayya
(C) Jawaharlal Nehru (D) M N Roy

80. The highest policy making body in maters relating to national planning is the:
(A) Planning commission
(B) Finance commission
(C) Interstate council
(D) National Development Council

81. The National Development Council was set up on:
(A) 1950 (B) 1951
(C) 1952 (D) 1953

82. Reviewing the implementation of national plans from time to time is the function of:
(A) Research Programmes committee
(B) Union Council of Ministers
(C) Planning commission
(D) National Development Council

83. The programme Evaluation Organization is an integral part of the:
(A) National Development Council
(B) Central Secretariat
(C) Department of Public Enterprises
(D) Planning Commission

84. The plan implementation and its impact evaluation is done by:
(A) NDC (B) Union Cabinet
(C) PEO (D) Planning commission

85. Which of the following is not a function of planning commission in India:
(A) Formulation of a plan
(B) Execution of development plans
(C) Appraisal of plan progress
(D) Policy and administrative recommendation

86. The rolling plan concept in national planning was introduced by:
(A) Congress Government
(B) UDF Government
(C) Janata Party Government
(D) NDA Government

87. The credit for coining and applying the concept of indicative planning goes to:
(A) USSR (B) USA
(C) Britain (D) France

88. Economic planning is a subject included under the:
(A) Union list (B) State list
(C) Concurrent list (D) Unspecified list

89. In 1938 the National Planning Committee was set up under the chairmanship of:
(A) Jawaharlal Nehru
(B) Vallabhbhai Patel
(C) Moulana Abdul Kalam Azad
(D) Mohammad Ali Jinnah

90. Encyclopedia of social work in India is published by:
(A) NIRD (B) Planning commission
(C) ICSSR (D) UNESCO

91. The first five year plan was submitted to the parliament in:
(A) 1950 (B) 1951
(C) 1952 (D) 1953

92. Harrod-Domar model was adopted in:
(A) First five year plan (B) Second five year plan
(C) Fifth five year plan (D) Sixth five year plan

93. Removal of poverty was the foremost objective of which five year plan:
(A) Fourth (B) Fifth
(C) Sixth (D) Seventh

94. Which decade of planning has been the best for agricultural growth:
(A) Fifties (B) Sixties
(C) Seventies (D) Eighties

95. Till which plan employment strategy has been growth linked:
(A) Third plan (B) Fourth plan
(C) Fifth plan (D) Sixth plan

96. Command Area Development Programme was launched during:
(A) Fifth five year plan (B) Sixth five year plan
(C) Seventh five year plan (D) Eight five year plan

97. The opening of Indian economy and signing of WTO happened during:
(A) Seventh five year plan (B) Eight five year plan
(C) Ninth five year plan (D) Tenth five year plan

98. Perspective plan is drawn out for a period of:
(A) One year (B) 2 to 3 years
(C) 5 years (D) 15 to 20 years

99. The term "Hindu rate of growth" was coined by:
(A) J N Bhagawathi (B) K N Raj
(C) Raj Krishna (D) Mahalanobis

100. Attainment of economic self reliance and removal of poverty were the main objectives of:
(A) First five year plan (B) Fourth five year plan
(C) Fifth five year plan (D) Sixth five year plan

101. Which one of the following did the Mahalanobis model lay greater emphasis:
(A) Agriculture
(B) Heavy industry
(C) Power generation
(D) Science and technology

102. India achieved self reliance in food production during:
(A) Fourth five year plan (B) Fifth five year plan
(C) Sixth five year plan (D) Seventh five year plan

103. Micro level planning is an example for:
(A) Imperative planning (B) Five year planning
(C) Regional planning (D) Target group planning

104. Planning in India as it is at present can be termed as:
(A) State planning (B) Centralized planning
(C) Decentralized planning (D) Indicative planning

105. Indicative planning started from:
(A) Eight five year plan (B) Seventh five year plan
(C) Sixth five year plan (D) Fifth five year plan

106. Five year plan is a:
(A) Perspective plan (B) Medium term plan
(C) Adhoc plan (D) Contingent plan

107. Why was plan holiday necessary in the national planning:
(A) First Indo-Pakistan war
(B) Second Indo-Pakistan war
(C) Third Indo-Pakistan war
(D) Indo-China war

108. The highest priority of the first five year plan was given for:
(A) Agriculture and irrigation
(B) Medium and large industry
(C) Unemployment and poverty
(D) Communication and education

109. The goal of socialistic pattern of society was presented before the nation in the:
(A) First plan (B) Second plan
(C) Third plan (D) Fourth plan

110. The concept of mixed economy in planning was introduced during:
(A) Second plan (B) Fifth plan
(C) Seventh plan (D) Ninth plan

111. A rolling plan is a plan for:
(A) One year (B) Two year
(C) Three year (D) Year to year basis

112. The idea of family planning as a concept of population control was widely popularized during:
(A) Fifth plan (B) Sixth plan
(C) Seventh plan (D) Eight plan

113. The current five year plan is:
(A) Ninth plan (B) Eleventh plan
(C) Tenth plan (D) Twelfth plan

114. The eleventh five year plan ended on:
(A) 2011 (B) 2008
(C) 2009 (D) 2010

115. The GDP growth envisaged in the tenth plan is:
(A) 8.10 per cent (B) 8.50 per cent
(C) 9 per cent (D) 10 per cent

116. The percentage of reduction in poverty envisaged at the end of tenth plan is:
(A) 2 per cent (B) 5 per cent
(C) 10 per cent (D) 15 per cent

117. The approximate number of new employment opportunities envisaged in the tenth plan is:
(A) 50 million (B) 40 million
(C) 60 million (D) 30 million

118. The allocation of funds for rural development programme in the tenth plan is:
(A) 77,474 crores (B) 42,276 crores
(C) 85,123 crores (D) 94 million

119. The Panchayati Raj institutions are intended to strengthen by:
(A) Structural planning (B) Indicative planning
(C) Multilevel planning (D) Functional planning

120. The origin of the word 'Democracy' is derived from:
(A) Roman (B) Sanskrit
(C) Greek (D) Persian

121. The word 'Democracy' means:
(A) Peoples participation (B) Peoples rules
(C) Peoples government (D) People judgement

122. The term panchayati raj was christened by:
(A) M.K. Gandhi (B) J. L. Nehru
(C) B. R. Mehta (D) M. S. Mehta

123. Panchayati raj was introduced in India in:
(A) 1950 (B) 1952
(C) 1958 (D) 1967

124. Panchayati raj institution is formed based on:
(A) Ashok Mehta Committee
(B) Hanumantha Rai Committee
(C) Dantiwala Committee
(D) Balwant rai Mehta Committee

125. Panchayat raj is a system of:
(A) Local government (B) Local administration
(C) Local self government (D) None of these

126. Annual growth rate in agriculture sector expected during the eleventh plan:
(A) 8 per cent (B) 6 per cent
(C) 4 per cent (D) 3 per cent

127. The increase in the production of rice expected at the end of eleventh plan:
(A) 8 million tones (B) 10 million tones
(C) 12 million tones (D) 15 million tones

128. The rural development programmme launched on the basis of area development approach:
(A) DPAP (B) MWP
(C) RWP (D) IRDP

129. Small farmers Agri-business Consortium was set up in:
(A) 1992 (B) 1994
(C) 1971 (D) 1986

130. A farmer cooperative society among the following is:
(A) IFFCO (B) FCI
(C) ICAR (D) NIRD

131. A welfare programme for the rural poor:
(A) KVK (B) Lab to Land
(C) National Demonstration (D) Minikit Trial

132. The Ganga Kalyan Yojna is for:
(A) Consumption purpose of farmers
(B) Irrigation for small and marginal farmers
(C) Credit support scheme for rural women
(D) Credit support scheme for small farmers

133. The Food Corporation of India was set up in:
(A) 1960 (B) 1962
(C) 1965 (D) 1967

134. An Agricultural labourer earns more than of his wage from agriculture:
(A) 33.33 per cent (B) 40.00 per cent
(C) 50.00 per cent (D) 66.66 per cent

135. "India's food crises and steps to meet it" is a:
(A) Book on food situation in India
(B) Report of ford foundation team
(C) Formula to solve food problem
(D) A slogan for campaign

136. The third in chronological order among the following report is:
(A) Royal Commission Report
(B) Agricultural Administration Report
(C) GMF Enquiry Committee Report
(D) Nalagarh Committee Report

137. Bhoodan movement was initiated on:
(A) 1947 (B) 1948
(C) 1951 (D) 1963

138. Bhoodan movement was started by:
(A) Acharya Vinoba Bhave (B) Mahatma Gandhi
(C) Jai Prakash Narayan (D) Jawaharlal Nehru

139. Bhodan movement was started at:
(A) Santhinietan (B) Wardha
(C) Panchampalli (D) Pavnar

140. The land collected in bhoodan is distributed among:
(A) Marginal farmers (B) Landlless peasants
(C) Medium farmers (D) Krishi pandits

141. The report on All India Rural Credit Enquiry Survey was submitted on:
(A) 1947 (B) 1954
(C) 1960 (D) 1961

142. Which of the following network connect agricultural produce, wholesale markets and rural markets:
(A) AFPINET (B) AGMARKNET
(C) HORTNET (D) VISTARNET

143. Which one is the land Information System Network linking all institutions involved in land and water management for agricultural productivity and production system:
(A) LISNET (B) ARINET
(C) AGRISNET (D) VISTARNET

144. Which one is the most effective method is identifying real rural needs:
(A) Election method (B) Selection method
(C) Workshop method (D) Socio-metric method

145. Which one is the purpose of social control:
(A) Social order (B) Stability
(C) Custom (D) None of these

146. Which one of the following is not an objective of India's economic planning:
(A) Population growth (B) Industrial growth
(C) Self reliance (D) Poverty alleviation

147. What is five year plan:
(A) Perspective plan (B) Medium term plan
(C) Adhoc plan (D) Contingent plan

148. Which of the following is co-operation among family members:
(A) Non formal (B) Voluntary
(C) Formal (D) Informal

149. A tool for optimum utilization of the available resources in the maximization of the attainable income of the farmers is known as:
(A) Farm planning (B) Home planning
(C) Programme planning (D) Extension planning

150. Allocation of funds for different sectors of programme:
(A) Controlling (B) Budgeting
(C) Financing (D) Recruiting

151. Basically GATT is aimed at:
(A) Facilitating accords on the most favoured nation status to the trading partners
(B) Reciprocity and transparency in global trade
(C) Tariff reduction through trade negotiations
(D) All of above

152. By whom the Rural infrastructural fund set up:
(A) RBI (B) NABARD
(C) IDRB (D) SBI

153. Central co-operative banks and Cooperative societies were established under Sir Daniel Hemilton Scheme in:
(A) 1920 (B) 1922
(C) 1924 (D) 1926

154. First census of population in India was conducted in the year:
(A) 1761 (B) 1771
(C) 1781 (D) 1791

155. First Five year Plan of India adopted:
(A) Harrod Domar Model (B) Lewis-Fei Model
(C) Maholonobis model (D) Keynesian Model

156. Food Grain Price Committee was appointed by the Government of India in:
(A) 1954 (B) 1958
(C) 1962 (D) 1964

157. Food grain production includes the production of:
(A) Cereals only
(B) Pulses only
(C) Cereals and Pulses
(D) Cereals, pulses and Oilseeds

158. For which of following a cooperative society is an example:
(A) Formal group (B) Informal group
(C) Primary group (D) Territory group

159. For which of following logical framework technique is greatly used:
(A) Project planning (B) Input utilization
(C) Work scheduling (D) Policy formulation

160. For which of the following Kasturba Gandhi Balika Vidyalaya are meant for promoting Education:
(A) Gild children (B) Senior citizen
(C) Rural youth (D) Agricultural labourers

161. Land Development Banks in India are owned by whom:
(A) Govt. of India (B) State Governments
(C) Commercial banks (D) Cooperative banks

162. Land holding of either 10 or more hectares has been classified as which of following in India:
(A) Semi-medium (B) Medium
(C) Semi large (D) Large

163. Land reforms have been equated with total:
(A) Urban development
(B) Agrarian development
(C) Industrial development
(D) Educational development

164. Lead bank scheme was introduced in the year:
(A) 1965 (B) 1969
(C) 1976 (D) 1964

165. Major commercial banks were nationalized in:
(A) 1965 (B) 1967
(C) 1969 (D) 1971

166. Mandi ordinance of agricultural production was enforced in:
(A) 1949 (B) 1950
(C) 1955 (D) 1959

167. Mention the type of correlation that operates between rurality and population density:
(A) Positive (B) Negative
(C) Accidental (D) Scattered

168. National demonstration was started in the year:
(A) 1975 (B) 1974
(C) 1963 (D) 1964

169. National demonstration is an example for:
(A) Firstline demonstration
(B) Frontline demonstration
(C) Futureline demonstration
(D) All of these

170. Planning at districts and block levels is generally considered as:
(A) Micro planning (B) Meso planning
(C) Macro planning (D) All of these

171. Planning in India as it is at present can be termed as which of the following:
(A) State planning (B) Centralized planning
(C) Decentralized planning (D) Indicative planning

172. Providing subsidy to farmers on introduction of farm machinery is an example for which of the following:
(A) Relative advantage (B) Complexity
(C) Compatibility (D) Observability

173. The election to panchayat is normally held after a period of how many years:
(A) 3 years (B) 4 years
(C) 5 years (D) 6 years

174. The family in an agrarian society is:
(A) Patriarchal (B) Matriarchal
(C) Matrilocal (D) Nuclear

175. The Panchayati Raj institutions are intended to strengthen by which:
(A) Structural planning (B) Indicative planning
(C) Multilevel planning (D) Functional planning

176. The Planning Commission was set in India in accordance with the directive principles in:
(A) Article 38 (B) Article 39
(C) Article 42 (D) Article 51

177. The scheme of Regional Rural Banks was launched in India on:
(A) Ist Oct. 1975 (B) 15th Aug. 1978
(C) 2nd Oct. 1975 (D) 1st Jan. 1970

178. The tool for optimum utilization of the available resources in the maximization of the attainable income of the farmers is:
(A) Farm planning (B) Resource allocation
(C) Planned change (D) Farm layout

179. The ——— year plan sought to maintain a clear cut distinction between the minimum support price and the procurement price:
(A) Second (B) Third
(C) Fifth (D) Sixth

180. Following the support price policy, the Government of India in 1984-85 raised the price of rice (paddy) by:
(A) 32.2% (B) 44.2%
(C) 36.8% (D) 26.7%

181. The Food Corporation of India was set up in:
(A) 1960 (B) 1962
(C) 1965 (D) 1967

182. Land holding of either 10 or more hectares have been classified as ——— in India:
(A) Semi-medium (B) Medium
(C) Semi-large (D) Large

183. Land reforms have been equated with total ——— development:
(A) Urban (B) Agrarian
(C) Industrial (D) Educational

184. Profits are defined as $\partial = R\text{-}(FC + VC)$, where, FC = total fixed cost and VC = total variable cost and ——:
(A) R=Revenue (B) R = Reduction
(C) R = Regression (D) R = Co-efficient

185. According to Hung-Chao Tai land reforms aim at aiming just relationship among ——— population:
(A) Urban (B) Industrial
(C) Traders (D) Agricultural

186. The term "land reform" should be used in broadest sense to serve as ——— for action:
(A) Inscription (B) Prescription
(C) Description (D) Formulation

187. In Zamidari system was based on ———— to support British Imperialism in India:
(A) Exploitation (B) Elimination
(C) Regulation (D) Augmentation

188. The main objective of land reform measures undertaken after independence was removal of institutional and ———— obstacles:
(A) Emotional (B) Motivational
(C) Conventional (D) National

189. In Zamidari abolition, the elimination of intermediaries led to excessive burden of:
(A) Revenue (B) Reservation
(C) Supplementation (D) Compensation

190. The reform of tenancy laid down that the rent should not exceed ———— per cent of gross produce:
(A) 20-25 (B) 10-15
(C) 25-30 (D) 35-40

191. According to the Directorate of Marketing and Inspection ———— per cent may be taken as the average marketing margin for all surplus food grains in India:
(A) 15 (B) 20
(C) 25 (D) 30

192. NABARD Act was passed in————, 1979:
(A) January (B) February
(C) March (D) April

193. NABARD came into existence on July 1, ————:
(A) 1980 (B) 1981
(C) 1982 (D) 1983

194. According to the 1951-52 survey, the proportions of indebtness was about 60 per cent for all the families, of which ———— per cent were cultivators:
(A) 70 (B) 60
(C) 80 (D) 90

195. The average debt for an indebted cultivator's family during 1951-52 and 1961-62 rose by ———— per cent:
(A) 60 (B) 45
(C) 35 (D) 25

196. The spurt in Zero Marginal Productivity type of exercise was given by ———— in 1954:
(A) W. Arthur Lewis
(B) W. David Hopper
(C) W. Nelson Michael
(D) W. Benzene Shoemaker

197. Balwantrai Mehta Team stimulated an active consideration of ———— through democratic bodies:
(A) Centralization (B) Mobilization
(C) Stabilization (D) Decentralization

198. Maharashtra Government appointed a committee in ———— under the chairmanship of Vasantrao Naik to examine the question of democratic decentralization in Maharashtra:
(A) 1961 (B) 1959
(C) 1957 (D) 1955

199. Maharashtra Zila Parishads and Panchayat Samiti Act was inacted in:
(A) 1967 (B) 1965
(C) 1963 (D) 1961

200. In ———— village panchayats were established under the Bombay Village Panchayat Act, 1958:
(A) Madhya Pradesh (B) Gujarat
(C) Rajasthan (D) Maharashtra

201. According to 1981 census, ———— of working force was still engaged in the primary sector:
(A) 61 (B) 67
(C) 71 (D) 77

202. One of the significant models of economic development of a labour surplus economy has been advanced by:
(A) Hung-Chao Tai (B) Huag P. Lewis
(C) Chan K. Heady (D) Foe C. Gray

203. In an expanding industrial sector, technological development in industry tend to ———— labour:
(A) Displace (B) Replace
(C) Engage (D) No change

204. In countries of ———— labour problem has become all the more serious owing to demographic explosion:
(A) America (B) Europe
(C) Asia (D) Whole world

205. The problem of allocation of capital is essentially a farm management problem aiming at ———— utilization of the available production resources:
(A) Maximizing (B) Optimizing
(C) Stabilizing (D) Minimizing

206. ———— production function is expressed as Y = marx, where Y = total production, x = quantity of variable factor, m = maximum output attained from fixed technical unit, r = ratio of addition of increments to production and a = maximum output added by the particular variable factor:
(A) Douglas (B) Bruce
(C) Spillman (D) Hillman

207. The constant elasticity of substitution function (CES) was used by J.W. Swan and R.M. Solow in:
(A) 1926 (B) 1936
(C) 1946 (D) 1956

208. Input-output production function is based on the general equilibrium model of ———— (1874):
(A) Leon Walras (B) Leon Michael
(C) Wilson Arrow (D) David Solow

209. According to law of ———— proportions, "If equal increments of one input are added, the inputs of other production services being held constant, beyond a certain point, the resulting increments of the product will decrease, that is, marginal product will diminish"
(A) Constant (B) Linear
(C) Variable (D) Inverse

210. ———— rightly remarked, "Let divisibility enter through door, the law of variable proportions rushes out through the window"
(A) M.M. Bober (B) C.A. Watson
(C) T. Sutovsky (D) R.M. Solow

211. When fertilizer input is increased from one to two units, total output increases from 103 to 174, that is, the output has increased only by ——— per cent:
(A) 90 (B) 85
(C) 61.9 (D) 71.9

212. ——— was the first economist to give a clear formulation of price elasticity (Ep) which is the ratio of a relative change in quality of relative change in price (Ep = relative change in quantity or relative change in price):
(A) Alfred Marshall (B) A.M. Khusro
(C) E.J. Long (D) P. Dorner

213. Food Grain Price Commission was constituted in India by:
(A) 1951 (B) 1953
(C) 1947 (D) 1950

214. The Food Grain Price Commission was appointed in ——— by the Govt. of India:
(A) 1954 (B) 1958
(C) 1970 (D) 1973

215. The Agricultural Price Commission (APC) was set up in ——— to evolve a balanced and integrated price structure:
(A) 1955 (B) 1965
(C) 1970 (D) 1973

216. Rural population is about ——— of the world population and lives mainly in developing countries:
(A) One-third (B) One-fourth
(C) One-fifth (D) Two-third

217. In India ——— of the population lives in villages:
(A) Half (B) One-third
(C) Three-fourth (D) Two-third

218. Efforts to revive panchayats in India were made in ——— by introducing Village Panchayat Act in several states:
(A) 1945 (B) 1935
(C) 1925 (D) 1915

219. The Village Panchayati Act came into existence in:
(A) 1956 (B) 1958
(C) 1960 (D) 1954

220. ——— concentrates on a rational structure with universal application in social institutions:
(A) Bureaucracy (B) Autocracy
(C) Anarchy (D) Monarchy

221. Formation of clique ——— an organization:
(A) Damages (B) Improves
(C) Stabilizes (D) None of these

222. Micro economics is concerned with:
(A) Analyzing the economic behaviour of the individual farm or business
(B) Analyzing the total economy of nation
(C) Only deals with scarcity of resources
(D) None of these

223. Macro economics deals with:
(A) Analyzing the economic behaviour of the individual farm or business
(B) Analyzing the total economy of nation
(C) Scarcity only
(D) Net profit only

224. US economy is an example of:
(A) Mixed economy (B) Closed economy
(C) General economy (D) Capitalistic economy

225. Jouzier definition of agricultural economics is related to:
(A) Prosperity to the enterprise
(B) Net profit
(C) Humen behaviour
(D) Wealth activities

226. Capitalistic economy refers to:
(A) Private ownership of resources
(B) Public ownership of resources
(C) Public and private ownership of resources
(D) All of these

227. The law of diminishing returns states that:
(A) The output increases with increasing input levels continuously
(B) The output increases with input levels irregularily
(C) The output increases with increasing input levels upto certain level and than declines
(D) All of these

228. In Ist irrational stage of production function, marginal return is:
(A) Equal to average return
(B) Equal to total output
(C) Either lower or equal to average return
(D) All of these

229. In IIrd irrational stage of production function, the marginal return is always:
(A) Negative (B) Positive
(C) Equal to average (D) All of these

230. Agriculture differs with industry in:
(A) Nature plays an important role in agricultural production
(B) Land is the most important factor for agricultural production
(C) Elasticity of price and supply of agricultural products are lower than industrial products
(D) All of these

231. Which of the following function is related to production function analysis?
(A) Total physical product
(B) Average physical product
(C) Marginal physical product
(D) All of these

232. Principle of least-cost combination explains:
(A) Output is the function of two or more than two variable inputs
(B) Use of two variable inputs in such a way to get maximum profit
(C) Use of two variable inputs in such a way to get same level of output at comparatively lowest cost
(D) All of these

233. Law of equi-marginal return states that:
(A) Each unit of input is invested in different enterprises in such a way so that a farmer can get maximum marginal return
(B) Each unit of input is invested to get maximum return
(C) Each unit of input invested to get maximum average return
(D) None of these

234. Principal of equi-marginal return states that:
(A) Maximum satisfaction is obtained from a given level of expenditure on two goods when the last penny spent on either good gives the same amount of satisfaction
(B) Equal marginal return is obtained by substitution of one good by other
(C) Both (A) and (B)
(D) None of these

235. Factor-product relationship is also known as:
(A) Input-output relationship
(B) Output-output relationship
(C) Demand-supply relationship
(D) Output-Input relationship

236. Law of equi-marginal return also known as:
(A) Law of opportunity cost
(B) Alternative cost principle
(C) Both (A) and (B)
(D) Law of equi-marginal return

237. Law of variable proportions is also known as:
(A) Law of returns
(B) Law of factor-product relationship
(C) Law of input-output relationship
(D) All of these

238. Law of variable proportions deals with:
(A) Input-output relationship where the output is the function of input
(B) Cost of output and input
(C) Supply of output
(D) All of these

239. Maximum net return from one variable input will occur when:
(A) Additional return from an additional unit of output from employing an additional unit of input is equal to the additional cost of that unit of input
(B) Value of marginal return is equal to cost of the input
(C) Marginal return x selling price of product = marginal input cost
(D) All of these

240. The production possibilities curve shows:
(A) Alternative combination of two outputs that may be produced with a given amount of resource during a period of time
(B) Combination of inputs that can produce equal level of production
(C) Combination of outputs that can be produced with same amount of input
(D) All of these

241. The percentage share of agriculture in National economy is:
(A) 50 (B) 40
(C) 45 (D) 25

242. The first chairman of Agriculture Price Commission was:
(A) Sri. L.L. Jha (B) Sri. M.L. Dantewala
(C) Sri Ashok Mehta (D) Sri Dharam Narayan

243. Loan for marriage known as:
(A) Direct productive loan (B) Unproductive loan
(C) Indirect productive loan (D) None of these

244. Economy of a developing country is marked by:
(A) Predominance of agriculture
(B) Predominance of industry
(C) Predominance of transportation
(D) None of these

245. The chairman of S.F.D.A. is:
(A) Prime Minister (B) Chief Minister
(C) Finance Minister (D) District Magistrate

246. The owner of loan in state farming is:
(A) Farmer (B) Capitalist
(C) Government (D) None of these

247. A farmer who has land holding between 1 and 2 hectares will be called as:
(A) Marginal farmer (B) Small farmer
(C) Medium farmer (D) Large farmer

248. The Regional Rural Bank was established through:
(A) RRB Act 1974 (B) RRB Act 1975
(C) RRB Act 1976 (D) None of these

249. India's working population engaged in agriculture is about:
(A) 50-52% (B) 67-69%
(C) 75-80% (D) 90-95%

250. The population of agricultural goods manufactured with agricultural contents contributes to India's export by about:
(A) 30% (B) 50%
(C) 70% (D) 90%

251. The percentage share of agricultural export in total export of India is about:
(A) 10% (B) 18%
(C) 25% (D) 30%

252. The growth of Indian economy can be sustained only if agriculture and allied activities grow at an annual growth rate of:
(A) 4% (B) 6%
(C) 8% (D) 10%

253. The scheme of agriculture in the National income has been continuously:
(A) Increasing (B) Decreasing
(C) Constant (D) None of these

254. As compared to other developing countries, the share of agriculture in National income of India is:
(A) Much lower (B) Much higher
(C) Slightly higher (D) Slightly lower

255. The working population engaged in agriculture is very high only in:
(A) Developing countries (B) Developed countries
(C) European countries (D) USA

256. The sugar industries in India depend for their raw materials on:
(A) Industrial product (B) Mines and minerals
(C) Agriculture (D) Crafts

257. At point of inflection in production function curve, MR is:
(A) $>$AR (B) $<$AR
(C) MR=AR (D) All of these

258. Marginal cost is equal to:

(A) $\frac{\text{Change in total cost}}{\text{Additional output}}$ (B) $\frac{\text{Change in total revenue}}{\text{Additional output}}$

(C) $\frac{\text{Change in total value product}}{\text{Additional input}}$

(D) All of these

259. Marginal rate of substitution is equal to:

(A) $\frac{\text{Change in number of y units}}{\text{Change in number of x units}}$

(B) $\frac{\text{Change in price of x commodity}}{\text{Change in price of y commodity}}$

(C) $\frac{\text{Price of x commodity}}{\text{Price of y commodity}}$

(D) None of these

260. The main purpose of farmers is:
(A) High production
(B) Minimum cost of cultivation
(C) High living standard
(D) High net profit

261. If the increase or decrease one product increases or decreases the other product, the relationship between both products is known as:
(A) Complementary (B) Supplementary
(C) Competitive (D) Symbiotic

262. If the increase or decrease in the production level of one product does not affect the production level of the other product, the relationship is called as:
(A) Complementary (B) Supplementary
(C) Competitive (D) Symbiotic

263. If the increase in the production level of one product necessarily decreases the production level of the other product, the relationship is called as:
(A) Complementary (B) Supplementary
(C) Competitive (D) Symbiotic

264. Production possibility curve is also called as:
(A) Iso-revenue curve (B) Opportunity curve
(C) Iso-return curve (D) All of these

265. Kisan Diwas in Uttar Pradesh is organized in every month of:
(A) 3rd Wednesday (B) 3rd Thursday
(C) 3rd Friday (D) 3rd Saturday

266. Types of farming refers to:
(A) Types of enterprises are grown or taken
(B) Ownership of farm and method or practices of production
(C) Both (A) and (B)
(D) None of these

267. Law of demand indicates:
(A) Inverse relationship between demand and price
(B) Linear relationship between demand and price
(C) Inverse relationship between market demand and marketable surplus
(D) Linear relationship between supply and demand

268. Price line or budget line indicates:
(A) A capacity of a consumer to purchase within a given budget
(B) A capacity of a consumer to provide marketable surplus
(C) The graph between demand and satisfaction
(D) None of these

269. Price elasticity of demand is always:
(A) Positive (B) Zero
(C) Negative (D) None of these

270. The demand of any commodity is affected by:
(A) Price of the commodity
(B) Taste and liking of the consumers
(C) Income of the consumers
(D) All of these

271. Which of the following item has the highest income elasticity?
(A) Bread (B) Butter
(C) Milk (D) Wine

272. Income elasticity is generally:
(A) Zero (B) Positive
(C) Negative (D) Infinite

273. The supply elasticity of perishable commodities is:
(A) Elastic (B) Inelastic
(C) Zero (D) Equal to one

274. The supply elasticity of bulky products is normally:
(A) Less elastic (B) High elastic
(C) Complete elastic (D) All of these

275. Supply of a commodity is governed by:
(A) Price of the commodity
(B) Price of production factors
(C) State of technology
(D) All of these

276. The price elasticity of demand for agricultural commodities is always:
(A) <1 (B) >1 (C) Zero (D) Infinite

277. The price elasticity of luxury items is always:
(A) <1
(B) >1
(C) Zero
(D) Infinite

278. If the price elasticity is >1, the total revenue will be increased at:
(A) Decrease in price
(B) Increase in price
(C) No change in price
(D) Unknown

279. If the price elasticity is <1, the total revenue will be increased at:
(A) Decrease in price
(B) Increase in price
(C) No change in price
(D) Unknown

280. The classical function curve exhibits following production stage:
(A) Stage I – Irrational production stage
(B) Stage II- Rational production stage
(C) Stage III- Irrational production stage
(D) All of these

281. Marginal return is equal to:
(A) $\frac{\Delta Y}{\Delta X}$
(B) $\frac{Y}{X}$
(C) $\frac{Px}{Px}$
(D) $\frac{Px}{Py}$

282. In rational stage of production function, both marginal and average returns are:
(A) At decreasing rate
(B) At increasing rate
(C) Both are equal
(D) None of these

283. An isoquant line shows:
(A) Combination of two inputs that may produce a given level of output
(B) Combination of two inputs that may produce maximum level of output
(C) Different marginal returns
(D) None of these

284. An isocost line shows:
(A) Combination of two variable inputs that cost different amount
(B) Combination of two variable inputs that cost the same amount
(C) Combination of two variable inputs that produce the same level of production
(D) All of these

285. What is the problem of Indian agriculture?
(A) Low irrigation facilities
(B) Low and erratic rainfull
(C) Indebtness of Indian farmers
(D) All of these

286. Which of the following principle is employed in production when resources are scare?
(A) Equi-marginal principle
(B) Comparative advantage
(C) Law of returns
(D) None of these

287. Production possibility curve indicates:
(A) Various combinations of the products y_1 and y_2 obtained with fixed level of resources combinations
(B) Maximum production with unlimited resources
(C) Marginal return with limited resources
(D) None of these

288. Operating ratio is equal to:
(A) $\frac{\text{Total expenses}}{\text{Gross income}}$
(B) $\frac{\text{Fixed expenses}}{\text{Gross income}}$
(C) $\frac{\text{Operating expenses}}{\text{Gross income}}$
(D) None of these

289. Gross ratio is equal to:
(A) $\frac{\text{Total expenses}}{\text{Gross income}}$
(B) $\frac{\text{Fixed expenses}}{\text{Gross income}}$
(C) $\frac{\text{Operating expenses}}{\text{Gross income}}$
(D) None of these

290. Fixed ratio is equal to:
(A) $\frac{\text{Total expenses}}{\text{Gross income}}$
(B) $\frac{\text{Fixed expenses}}{\text{Gross income}}$
(C) $\frac{\text{Operating expenses}}{\text{Gross income}}$
(D) None of these

291. Fixed cost is also known as:
(A) Over head cost
(B) Prime cost
(C) Alternate cost
(D) All of these

292. Variable cost is also known as:
(A) Overhead cost
(B) Prime cost
(C) Alternate cost
(D) All of these

293. Opportunity cost is also known as:
(A) Overhead cost
(B) Prime cost
(C) Alternate cost
(D) All of these

294. Indifference curve shows a:
(A) Relative willingness to consume one good rather than another
(B) Relationship of demand and supply
(C) Relationship between demand and price
(D) Relationship between supply and price

295. If price elasticity is -1, it indicates that:
(A) An equal change in demand and price
(B) Percentage change on quantity in smaller than the percentage price change
(C) Percentage quantity change is greater than the percentage price change
(D) None of these

296. If the price elasticity of demand is more than one, the elasticity is known as:
(A) Elastic
(B) Inelastic
(C) Infinity
(D) None of these

297. If the price elasticity of demand is less than one, the elasticity is known as:
(A) Elastic (B) Inelastic
(C) Infinity (D) None of these

298. Changes in supply elasticity occurs mainly during:
(A) Short period (B) Long period
(C) Winter season (D) Rainy or summer season

299. The elasticity of the agricultural commodity in very long run will be always:
(A) Elastic (B) Inelastic
(C) Both (A) and (B) (D) None of these

300. The supply slope is normally:
(A) Negative (B) Positive
(C) Both (A) and (B) (D) None of these

301. The luxury good has income elasticity:
(A) 1 (B) <1
(C) >1 (D) Between o to 1

302. Monopoly implies to:
(A) Single firm
(B) Special or unique product
(C) Firm is price maker
(D) All of these

303. Monopsony refers to:
(A) Single buyer (B) Few buyers
(C) Many buyers (D) No buyers

304. Market price of a commodity means:
(A) Balance between demand and supply of a commodity for a specific day
(B) It changes very frequently
(C) It increases or decreases and also remains nearly normal price line
(D) All of these

305. Normal price of a commodity means:
(A) It is almost fixed price of a commodity
(B) The consumption and production rates of a commodity remains same at its normal price
(C) Price plays a great role in normalizing the price of a commodity
(D) All of these

306. Agricultural commodities have greater price fluctuations mainly because of:
(A) Inelastic demand and supply
(B) Elastic demand and supply
(C) More control of government
(D) All of these

307. Which of the following is a government measure to deal with uncertainty in agriculture?
(A) Guaranteed agricultural prices
(B) Buffer stock schemes
(C) Crop insurance
(D) All of these

308. Duopoly refers to:
(A) Two either firms (sellers) or two buyers in the com petition
(B) Few firms or buyers
(C) Many firms or buyers
(D) None of these

309. Uncertainity in agriculture refers to:
(A) Yield uncertainty (B) Price uncertainty
(C) Tenurial uncertainty (D) All of these

310. Uncertainty can be reduced mainly at farmers level by:
(A) Diversification (B) Flexibility
(C) Capital rationing (D) All of these

311. What is the important measure to stabilize the agricultural prices and incomes?
(A) Buffer stock scheme
(B) Crop insurance scheme
(C) Control over agricultural production
(D) All of these

312. Instability in agriculture is caused by:
(A) Fluctuations in price
(B) Fluctuations in income
(C) Fluctuations in outcome
(D) All of these

313. Farm management should answer the following question:
(A) How to organize a farm
(B) How to operate it
(C) How to achieve the maximum net profit
(D) All of these

314. Farm management is:
(A) An art (B) A science
(C) A business (D) All of these

315. Farm layout means:
(A) Physical arrangement of farm in the form of plots, building, roads, channels etc.
(B) Planning of farm to maximize the profit
(C) Production of crops
(D) None of these

316. Farm layout is required because of:
(A) Newly acquired land
(B) Change in power and enterprises
(C) Change in technologies
(D) All of these

317. Farm profit is equal to:
(A) Gross revenue – cost
(B) Gross revenue – net returns
(C) Gross income – gross revenue
(D) All of these

318. Farm profit is affected by:
(A) Total produce
(B) Market and farm prices
(C) Input cost
(D) All of these

319. Farm budget refers to:
(A) A tentative estimation of expenses and returns of a farm
(B) A blueprint of production and cost
(C) Loan and its interest
(D) All of these

320. Complete farm budget refers to:
(A) Farm budget for a entire farm as a single unit
(B) Farm budget separately for each enterprise
(C) It is a budget for a Gram Panchayat as a single unit
(D) It is a budget for whole area of the region

321. Partial farm budget refers to:
(A) Farm budget for a entire farm as a single unit
(B) Farm budget separately for each enterprise
(C) It is a budget for a Gram Panchayat as a single unit
(D) It is a budget for whole area of the region

322. Farm inventory refers to:
(A) List of items possessed by a farm on a specific date
(B) Land under cultivation
(C) Inventory of new farm
(D) All of these

323. Farm production records indicate:
(A) Financial position of farm
(B) Production in quantity
(C) Production in quality
(D) All of these

324. Which of the following is a farm production record?
(A) Livestock record
(B) Crop cultivation record
(C) Crop production record
(D) All of these

325. The motto of land reform in India is:
(A) Land to tiller (B) Land to landlord
(C) Land to Talukedar (D) All of these

326. The land revenue was charged in cash first time during the reign of:
(A) Ashoka (B) Chandra Gupta I
(C) Tamur (D) Akbar

327. The land revenue during Kautilya era was:
(A) 1/6th of total produce (B) 1/5th of total produce
(C) 1/4th of total produce (D) 1/3rd of total produce

328. The land revenue during Sher Shah Suri reign was:
(A) 1/6th of total produce (B) 1/5th of total produce
(C) 1/4th of total produce (D) 1/3rd of total produce

329. The land revenue during Akbar regime was:
(A) 1/6th of total produce (B) 1/5th of total produce
(C) 1/4th of total produce (D) 1/3rd of total produce

330. Agricultural holding refers to:
(A) Total area of land which is held for cultivation as a single unit
(B) Total area of land used as a single unit
(C) Total area owned by a cultivator
(D) All of these

331. Operational holding refers to:
(A) Total area of land which is held for cultivation as a single unit
(B) Total area including fallow and barren used as a single unit
(C) Total area owned by a owner
(D) All of these

332. Fragmentation of holding refers to:
(A) It is a breaking up of a single holdings into a number of small pieces situated at a distance from each other
(B) Consolidating the small holdings
(C) Breaking of holdings into economic holdings
(D) None of these

333. An optimum holding refers to the size that:
(A) Should not be more than 3 times of economic holding
(B) Should be > 3 times of economic holding
(C) Should be equal to operational holding
(D) Should be <2 ha land

334. The cooperative movement was started first in:
(A) England (B) Germany
(C) USA (D) India

335. Which was the first cooperative society setup in the world?
(A) Rashdel Poineers Equitable Society
(B) Primary Cooperative Credit Society
(C) National Dairy Development Board
(D) Anand Milk Union Limited

336. Which of the following is a first cooperative credit society opened in the world?
(A) Reiffeisen Society, Germany
(B) Milk Cooperative Society, Denmark
(C) 4 H Club, USA
(D) None of these

337. Which of the following is a first cooperative bank opened in the world?
(A) Cooperative Bank of China
(B) Bank of London
(C) Shultz Delite Bank of Germany
(D) IFCI Bank

338. Which of the non-institutional agency of that provides agricultural loan?
(A) Traders and money lenders
(B) Commercial banks
(C) Land Development Bank
(D) Regional Rural Banks

339. Which of the following credit agency that provides institutional loan to farmers?
(A) Cooperative Societies
(B) Commercial Banks
(C) Land Development Bank
(D) All of these

340. Which is the most important source of loan in India?
(A) Money lenders
(B) Land Development Banks
(C) Government
(D) Cooperative Societies

341. Which of the following credit agency provides long term loan to farmers?
(A) State Land Development Bank
(B) Primary Land Development Bank
(C) Money lenders
(D) All of these

342. Primary Land Development Bank is established at:
(A) State level (B) District level
(C) Tehsil level (D) Village level

343. Central Land Development Bank is established at:
(A) State level (B) District level
(C) Tehsil level (D) Village level

344. Regional rural banks provide loan to:
(A) Small and marginal farmers
(B) Landless labourers
(C) Rural artisans
(D) All of these

345. Which type of loan is supplied agricultural service societies?
(A) Short term (B) Medium term
(C) Long term (D) All of these

346. National Agricultural Credit Council was constituted in year:
(A) 1952 (B) 1962
(C) 1969 (D) 1972

347. The main function of National Agricultural Credit Council is:
(A) To help in distribution of agricultural credit card estimation of total credit demand
(B) To provide agricultural loan to farmers
(C) To control over RBI activities
(D) To check the inflation rate

348. The main objective of Agricultural Finance Corporation is:
(A) To provide credit to agricultural schemes
(B) To encourage the role of commercial banks in agricultural development
(C) Both (A) and (B)
(D) None of these

349. Which type of loan is given by ARDC?
(A) Short term
(B) Medium term
(C) Long term
(D) Medium and long term

350. Marketing involves the activities from:
(A) Production to end consumer
(B) Production only
(C) Production to middlemen
(D) Production to storage

351. Monopoly market refers to:
(A) Single producer or seller in market
(B) Single buyer in market
(C) More number of producers and single buyers
(D) None of these

352. Monopsony market refers to:
(A) Market having only one buyer or institution
(B) Buyer is free to decide the quantity to be purchased
(C) Buyer is free to change the price policy
(D) All of these

353. Duopsony refers to:
(A) Market having only two buyers
(B) Market having only two sellers
(C) Market having only two buyers and sellers each
(D) None of these

354. Oligopsony refers to:
(A) Few buyers in a market
(B) Only one buyer in a market
(C) Only one seller in a market
(D) None of these

355. Oligopoly refers to:
(A) More than sellers in market
(B) Both product homogeneity and differentiation may be found
(C) Product substitution is necessary
(D) All of these

356. Marketable surplus refers to:
(A) Actual amount that is marketed
(B) Excess amount than requirement for consumption
(C) Both (A) and (B)
(D) None of these

357. Marketed surplus refers to:
(A) Actual amount that is marketed
(B) Excess amount than requirement for consumption
(C) Both (A) and (B)
(D) None of these

358. Normally marketed surplus is higher than marketable surplus because of:
(A) To repay the loan
(B) Poor facilities of storage at farms
(C) Inability to estimate the total family consumption
(D) All of these

359. Which of the following is a market for agricultural commodities?
(A) Whole sale market (B) Retail market
(C) Fairs (D) All of these

360. Demand of a particular commodity can be estimated by:
(A) Yardstick approach
(B) Regression technique
(C) Simultaneous equation model
(D) All of these

361. Marketing middlemen refers to:
(A) Institutions or persons that work between the transfer process of a commodity from producers to consumers
(B) Institutions or persons that purchase the produce from producers
(C) Institutions or persons that sell the produce to consumers
(D) None of these

362. Kachcha commission agent is related to:
(A) Collection of agricultural produce
(B) Distribution of agricultural produce
(C) Both (A) and (B)
(D) None of these

363. Packka commission agent is related to:
(A) Collection of agricultural produce
(B) Distribution of agricultural produce
(C) Both (A) and (B)
(D) None of these

364. Market structure refers to:
(A) Organizational structure that affects the competition and price in market
(B) Behavioral pattern of buyers and sellers through which prices are affected
(C) It indicates about the role of commission agents in market only
(D) All of these

365. Market conduct refers to:
(A) Organizational structure that affects the competition and price of a commodity
(B) Behavioral pattern of market that affect the commodity price
(C) Involvement of brokers in marketing
(D) All of these

366. Which of the following is a important element of market structure?
(A) Degree of seller's concentration
(B) Degree of buyer's concentration
(C) Degree of product differentiation
(D) All of these

367. Which of the following is a important element of market conduct?
(A) Rules and regulations of production firms
(B) Product policy
(C) Product promotion policy
(D) All of these

368. Market clearing refers to:
(A) Equilibrium state of market price
(B) Stage of marketing at which brokers mediates in deciding the price of a commodity
(C) Change in price with increase supply
(D) None of these

369. Price discovery refers to:
(A) Process through which buyer and sellers decide a price of a commodity at a given place
(B) Stage of deciding equilibrium market price
(C) Search of market price
(D) All of these

370. Which of the following is a component of marketing cost?
(A) Labour (B) Transportation
(C) Packing and storage (D) All of these

371. Price spread is also known as:
(A) Marketing cost (B) Marketing margin
(C) Farm price (D) None of these

372. Which of the statement is correct for monopoly market?
(A) MC = MR (B) MC > AC
(C) MC < AC (D) None of these

373. New name of Indian Standard Institution is:
(A) Bureau of Indian Standards
(B) Institute of Indian Standards
(C) Standard Institute of India
(D) None of these

374. The Bureau of Indian Standard is situated at:
(A) Lucknow (B) New Delhi
(C) Hyderabad (D) Mumbai

375. The price which necessarily covers the variable cost and also the fixed price is:
(A) The long run price (B) The short run price
(C) The market price (D) The equilibrium price

376. Marginal Physical Product of a variable input is negative:
(A) Only in I stage (B) Only in II stage
(C) Only in III stage (D) In I and II stage

377. Average cost is equal to:
(A) Total cost + opportunity cost (B) $\frac{\text{Total cost}}{\text{Output}}$
(C) $\frac{\text{Average variable cost}}{\text{Output}}$ (D) Cost B + Cost C

378. Fixed cost is equal to:
(A) $\frac{\text{Total fixed cost}}{\text{Output}}$ (B) $\frac{\text{Total cost}}{\text{Output}}$
(C) $\frac{\text{Total variable cost}}{\text{Output}}$ (D) None of these

379. Variable cost is equal to:
(A) $\frac{\text{Total cost}}{\text{Output}}$ (B) $\frac{\text{Total variable cost}}{\text{Unit of output}}$
(C) Cost A + Cost B + Cost C (D) All of these

380. Marginal cost is equal to:
(A) $\frac{\text{Total cost}}{\text{Output}}$ (B) $\frac{\text{Increase in variable cost}}{\text{Increase in output}}$
(C) Cost C + Rent paid to leased rent
(D) $\frac{\text{Total variable cost}}{\text{Increase in output}}$

381. Demand is a:
(A) Desire and purchase of a commodity at a specified time
(B) Desire for a commodity but not able to purchase it
(C) Ability to purchase a commodity but not purchase it
(D) All of these

382. If the price elasticity of barley is -0.5, it means:
(A) Barley demand will be decreased by 0.5%, if the price increased by 1%
(B) Barley demand will be increased by 0.5% if the price is increased by 1%
(C) If the price of barley is increased by 0.5%, the demand will also be increased by 0.5%
(D) Price and demand both either increased or de creased by 0.5%

383. If the petrol price fall from Rs. 10 to 8 for 1 litre and as a result the demand of petrol is increased from 5 units to 9 units. What will be the price elasticity of demand?
(A) -1 (B) -2 (C) -3 (D) -4

384. Cross elasticity of demand refers to:

(A) $$\frac{\text{\% change in quantity demanded of X}}{\text{\% change in income}}$$

(B) $$\frac{\text{\% change in quantity demanded of X}}{\text{\% change in price}}$$

(C) $$\frac{\text{\% change in income}}{\text{\% change in quantity demanded}}$$

(D) None of these

385. Income elasticity of demand refers to:

(A) $$\frac{\text{\% change in quantity demanded}}{\text{\% change in income}}$$

(B) $$\frac{\text{\% change in quantity demanded}}{\text{\% change in price}}$$

(C) $$\frac{\text{\% change in income}}{\text{\% change in quantity demanded}}$$

(D) None of these

386. Price elasticity of supply is equal to:

(A) $$\frac{\text{\% change in quantity supplied of a commodity A}}{\text{\% change in price of A}}$$

(B) $$\frac{\text{\% change in demand of commodity A}}{\text{\% change in price of A}}$$

(C) $$\frac{\text{\% change in income}}{\text{\% change in price}}$$

(D) All of these

387. The production or supply of a product will be closed by a firm if:
(A) Production cost < cost of marginal product
(B) Production cost > cost of marginal product
(C) Variable cost > production cost
(D) All of these

388. Which of the following condition is required for perfect competition in the market?
(A) More number of firms in the market
(B) Homogeneous product
(C) Free entry or exit in the market
(D) All of these

389. What will happen if the market price of a commodity is higher than normal price?
(A) Demand will be reduced and supply will be increased in the market
(B) Demand will be increased
(C) Supply will be reduced
(D) None of these

390. What will happen if the market price of a commodity is lower from its normal price?
(A) Demand of a commodity will be increased
(B) Supply will be decreased
(C) Market price will increase towards the normal price line
(D) All of these

391. Income elasticity of inferior goods is:
(A) Positive (B) Negative
(C) Between zero to one (D) Infinite

392. An income elasticity of necessary is:
(A) 1 (B) < 1
(C) >1 (D) Between zero to one

393. Which of the following is a feature of monopolistic competition?
(A) Many firms
(B) Firm is price maker
(C) Products/services differentiated through promotion, quality and design
(D) All of these

394. Cyclic fluctuations in prices of agricultural commodities means:
(A) Ups and downs in the prices after certain intervals
(B) Increase in prices after certain intervals
(C) Decrease in prices after certain intervals
(D) None of these

395. The fluctuations in agricultural prices due to operation of Cobweb theorem shows:
(A) Cyclic fluctuations
(B) Long term fluctuations
(C) Seasonal fluctuations
(D) All of these

396. Long term fluctuations in agricultural prices and incomes are caused mainly due to:
(A) Change in population
(B) Opening of new markets
(C) Development of new technology
(D) All of these

397. Monopoly is a perfect stage of:
(A) Imperfect competition (B) Pure competition
(C) Monopolistic competition (D) None of these

398. The condition of market where few buyers are present is called as:
(A) Monopsony (B) Oligopsony
(C) Oligopoly (D) Monopoly

399. Mechanisation of agriculture refers to:
(A) Use of power driven machines for various farm jobs
(B) Ploughing with tractor operated plough only
(C) Crushing of sugarcane with crusher only
(D) All of these

400. At present, the farming system of India has become:
(A) Total fixed cost/100
(B) Total fixed cost/total variable cost
(C) Total fixed cost/output
(D) Total fixed cost/input

401. Average product is equal to marginal product when:
(A) Average product is maximum
(B) Average product is minimum
(C) Marginal product is zero
(D) None of these

402. The wage is not paid to:
(A) Permanent labour (B) Contract labour
(C) Family labour (D) Casual labour

403. Break even points is determined by:
(A) Marginal cost = Marginal returns
(B) Marginal cost = Average variable cost
(C) Marginal cost = Average cost = Average returns

404. Farm planning refers to:
(A) Reorganization and/or alteration of present farm organization and operation in order to get maximum return from limited resources
(B) Activities for crop production on farm
(C) Physical arrangement of farm resources
(D) All of these

405. The basic objective of farm planning is:
(A) To guide a farmer to move in the right direction
(B) To indicate the destination
(C) Proper use and allocation of input
(D) All of these

406. Farm price means to:
(A) The price that farmer receives for his produce
(B) Price prevailing in the market
(C) Price declared by Govt.
(D) All of these

407. Farm plan refers to:
(A) Systematic pre-thinking of future activities related to farm business
(B) It is a tentative future plan for farm business
(C) It is related to maximize the farm profit
(D) All of these

408. Farm production efficiency can be measured with following overall standard?
(A) Input per unit of output (B) Total production index
(C) Output per hectare (D) All of these

409. The main objective of land reform in India is:
(A) Abolition of intermediates
(B) Confirment of ownership right to tenants
(C) Imposition of ceiling on agricultural land holdings
(D) All of these

410. The maximum limit is land holding under best soil is:
(A) 10-18 acres (B) > 27 acres
(C) > 34 acres (D) None of these

411. The maximum limit of land holding if the soil is poor is:
(A) 20-22 acres (B) > 37 acres
(C) > 54 acres (D) None of these

412. What is the main provision of land consolidation act?
(A) Fixation of minimum standard area for regulating transfers
(B) Prohibition of fragmentation below standard area
(C) Filling of objectives, appeals and penalties
(D) All of these

413. UP consolidation of Holding Act was passed in years:
(A) 1935 (B) 1943
(C) 1953 (D) 1958

414. UP consolidation of Holding (Amendment) Act was passed in years:
(A) 1938 (B) 1943
(C) 1953 (D) 1958

415. Royal commission on agriculture was appointed in year:
(A) 1919 (B) 1921
(C) 1927 (D) 1932

416. Farmers can get long term loan from:
(A) Money lenders
(B) Land Development Banks
(C) Government
(D) All of these

417. Long term loan is given mainly for:
(A) Purchase of seeds, fertilizers and weedicide
(B) Purchase of agricultural implements
(C) Purchase of land and its development
(D) All of these

418. Farmers can get short term loan from:
(A) Money lenders
(B) Land Development Banks
(C) Government
(D) All of these

419. Short term loan is given mainly for:
(A) Purchase of seeds, fertilizers and weedicide
(B) Purchase of agricultural implements
(C) Purchase of land and its development
(D) All of these

420. What is the main principle of good system of credit to agriculture?
(A) Repayment capacity
(B) Risk bearing capacity
(C) Estimation of returns to credit
(D) All of these

421. The lead Bank Scheme was started in 1969 on the recommendation of:
(A) Nariman Committee
(B) All India Rural Credit Survey Committee
(C) Committee on Cooperative Credit
(D) Expert group of RBI on credit scheme

422. Abolition of bounded labour was recommended by:
(A) Dhebar Commission
(B) National Agriculture Commission
(C) Planning Commission
(D) Labour Commission

423. The bounded labour practiced was abolished in India by the notification made on:
(A) January 26, 1975 (B) August 15, 1975
(C) October 2, 1975 (D) October 25, 1975

424. Agricultural marketing involves the operations:
(A) Movement of food and raw material from field to final consumer
(B) Handling of produce at farm
(C) Processing, grading and packing
(D) All of these

425. The main objective of agricultural marketing is:
(A) To make available agricultural produce to consumer at optimum prices
(B) To help farmers to get maximum profit from their produce
(C) To maintain optimum market prices of agricultural commodities
(D) All of these

426. Functional approach of agricultural marketing deals with:
(A) Different functions (B) Different institutions
(C) Different commodities (D) None of these

427. Which of the following is a facilitating function of agricultural marketing:
(A) Standardization (B) Financing
(C) Risk bearing (D) All of these

428. Which of the following is physical function of agricultural marketing?
(A) Buyers and sellers (B) Transportation
(C) Processing and storage (D) All of these

429. Which of the following is known as a middlemen?
(A) Whole sellers (B) Retailers
(C) Commission agents (D) All of these

430. Specialised market refers to:
(A) Sell-buy of specialized kind of a commodity
(B) Sell-buy of different kind of a commodity
(C) Communication, transport and storage play an important role
(D) All of these

431. Terminal market are found at:
(A) Large cities and sea ports
(B) Village levels
(C) Town levels
(D) District levels

432. Local market generally found at:
(A) Large cities and sea ports
(B) Village level
(C) Town levels
(D) District level

433. Which of the marketing system was prevalent in the ancient time?
(A) Barter system (B) Modern system
(C) European system (D) None of these

434. Computer based market refers:
(A) Prices are decided on interest or phone
(B) Exchange of commodities are based on computer information
(C) Payment can also be made through computer
(D) All of these

435. Complete competitive market should have:
(A) Large number of sellers and buyers
(B) No product differentiation
(C) Dynamic factors of production
(D) All of these

436. The market that is organized under some rules and regulations made by trade body federation etc. is known as:
(A) Regulated market (B) Organized market
(C) Economic market (D) Unorganized market

437. The market that is under control of government rules and regulations is called as:
(A) Regulated market (B) Organized market
(C) Economic market (D) Unorganized market

438. Agricultural products are characterized as:
(A) Bulky raw material (B) Perishable
(C) Essential for life (D) All of these

439. Agricultural production is characterized as:
(A) Annual variability
(B) Seasonal variation
(C) Varying costs of production
(D) All of these

440. Which of the following statement is correct?
(A) Marketable surplus > marketed surplus
(B) Marketable surplus < Marketed surplus
(C) Marketable surplus = marketed surplus
(D) All of these

441. Marketable surplus is affected by:
(A) Prices of agricultural produce
(B) Support prices
(C) Availability of resources and its prices
(D) All of these

442. Marketed surplus is affected by:
(A) Family size (B) Production level
(C) Sale prices (D) All of these

443. Which of the following government agency is responsible for marketing and distribution of agricultural produce in the country?
(A) FCI
(B) State Trading Corporation Ltd
(C) Agricultural Price Commission
(D) All of these

444. Which of the following government agency is responsible for marketing and distribution of tobacco produce in the country?
(A) FCI
(B) State Trading Corporation Ltd
(C) Tobacco Board
(D) Agricultural Price Commission

445. Which of the following government agency is responsible for international trading of agricultural produce in the country?
(A) FCI
(B) State Trading Corporation Ltd
(C) Agricultural Price Commission
(D) All of these

446. Crop area production and productivity data are available with:
(A) Directorate of Economics and Statistics
(B) Forward Market Commission
(C) Central Storage Corporation
(D) All of these

447. Marketing margin can be defined as:
(A) Part of total cost that goes to marketing firms
(B) It is a difference between producer's selling cost and consumer's purchasing cost
(C) It is the produce of total services of marketing firms
(D) All of these

448. Absolute margin equal to:
(A) Sale price – Purchase price
(B) Sale rice + Purchase price
(C) Sale price – Total marketing cost
(D) None of these

449. Net margin is equal to:
(A) Absolute margin – absolute marketing expenditure
(B) Sale price – Purchase price
(C) Sale price – total cost
(D) Absolute margin/sale price x 100

450. Marketing margin can be calculated by:
(A) Concurrent margin method
(B) Time lag method
(C) Both (A) and (B)
(D) None of these

451. Which of the following is a first regulated market opened in India?
(A) Karanjia cotton market
(B) Brar cotton market
(C) Cashew market of Kerala
(D) None of these

452. The new name of Agricultural Price Commission is:
(A) Commission on Agricultural Cost and Prices (CACP)
(B) Commission on agricultural Prices
(C) Agricultural Prices and Cost Commission
(D) None of these

453. The CACP came into existence in year:
(A) 1982 (B) 1985
(C) 1987 (D) 1990

454. Which of the following basis is adopted in determination of prices of agricultural commodities in our country?
(A) Production cost (B) Parity cost
(C) Prevalent market price (D) All of these

455. Which of the following aspect is related to macro economics?
(A) Agricultural employment
(B) Instability of agriculture
(C) Tenurial system
(D) All of these

456. The principle of least-cost combination of two factors indicates that profit can be maximized at:
(A) $\frac{\Delta X2}{\Delta X2} = \frac{Px1}{Px2}$ (B) $\frac{\Delta X2}{\Delta X1} = \frac{Px2}{Px1}$
(C) $\frac{\Delta X1}{X2} = \frac{Px1}{Px2}$ (D) None of These

457. In the production possibility curve, the maximum profit will be at:
(A) $\frac{\Delta Y2}{\Delta Y1} = \frac{\cdot Py1}{Py2}$ (B) Py2,Δ Y2=Py1,Δ Y1
(C) Both (A) and (B) (D) None of these

458. Which of the following is representing Cobb-Douglas production function?
(A) $Q = KL^{\alpha} C^{\beta}$ (B) $Q = L^{\alpha} C^{\beta} M^{\gamma}$
(C) $Q = KC^{\alpha} M^{\beta}$ (D) All of these

459. Elasticity of production is equal to:
(A) $\frac{\Delta Y1}{\Delta X} \times \frac{X}{Y}$ (B) $\frac{MP}{AP}$
(C) $\frac{\text{Present change in output}}{\text{Present change in input}}$
(D) All of these

460. If MP, AP and EP represent marginal product, average product and elasticity of production, which of the following inference is correct?
(A) When MP = AP, Ep=1 (B) When MP>AP, Ep>1
(C) When MP<AP, Ep<1 (D) All of these

461. The least cost combination of inputs will be at:
(A) $\frac{\Delta X1}{\Delta X2} = \frac{Px2}{Px1}$ (B) $\frac{\Delta X1}{\Delta X2} > \frac{Px2}{Px1}$
(C) $\frac{\Delta Px1}{\Delta Px2} < \frac{Py2}{Py1}$ (D) None of these

462. The size of holding that gives full employment to the farm family and reasonable living standard is known as:
(A) Family holding (B) Optimum holding
(C) Economic holding (D) Non-economic holding

463. Operating holding refers to:
(A) Land area owned by a farmers
(B) Cropped area cultivated by a farmer
(C) Net land area cultivated by a farmer
(D) Land area leased in by a farmer

464. The contribution of livestock to gross farm income in mixed farming is:
(A) At least 10% (B) At least 30%
(C) At least 40 % (D) More than 50%

465. Under diversified farming, the income from the single enterprise should not be:
(A) < 50% of total farm income
(B) > 50% of total farm income
(C) > 33% of total farm income
(D) None of these

466. Under specialized farming, the income from the single enterprise should not be:
(A) < 50% of total farm income
(B) > 50% of total farm income
(C) > 75% of total farm income
(D) None of these

467. Ranching refers to:
(A) Growing of crops only
(B) Growing of livestock only
(C) Growing of crops and livestock
(D) All of these

468. Price elasticity of demand is equal to:
(A) $\frac{\text{Percentage change in quantity demanded}}{\text{Percentage change in price}}$
(B) $\frac{\text{Percentage change in supply of quantity}}{\text{Percentage change in price}}$
(C) $\frac{\text{Percentage change in quantity demanded}}{\text{Percentage change in quantity supplied}}$
(D) All of these

469. The income elasticity is generally:
(A) 1% increase income increase only 0.2% demand
(B) 1% increase income decreases 0.2 % demand
(C) Both (A) and (B)
(D) None of these

470. International trade is more difficult in arrange than inter regional trade within the country because:
(A) Factors of production are less mobile
(B) Each country uses a different currency
(C) Government interferes with foreign trade for political reasons
(D) All of these

471. Risk and uncertainty differs in:
(A) Risk has known probability but uncertainty has no probability
(B) Risk is temporary by uncertainty is sure
(C) Risk is more fatal than uncertainty in agriculture
(D) None of these

472. The difference of average price and minimum support price of a particular commodity is paid to each farmer if the average price is lower than minimum support price, under the scheme of:
(A) Crop Insurance (B) Forward price
(C) Deficiency payment (D) All of these

473. Labour efficiency can be increased by:
(A) Improving farm layout
(B) Farm work simplification
(C) Training of labourers
(D) All of these

474. Farm management refers to:
(A) Managing the farm
(B) The art of managing the farm
(C) The art of managing the farm successfully as measured by the test of profitability
(D) All of these

475. Farm building is important because of:
(A) It saves feed, both in quantity and quality
(B) It increases safety in farm work
(C) It helps to increase production
(D) All of these

476. Input-output relationship is called:
(A) Cost function (B) Production function
(C) Price function (D) None of these

477. Which of the following is a partial standard of measuring farm production efficiency?
(A) Crop yield index
(B) Production per unit livestock
(C) Yield per hectare of the most important crop
(D) All of these

478. Farm survey refers to:
(A) An enquiry in the composition activities and living conditions of farmers and farm
(B) Collection of information related expenses and profit of the farms
(C) Visit of farms
(D) All of these

479. Which of the following step (s) are involved in farm survey?
(A) Planning of survey and collection of data
(B) Analysis of data
(C) Presentation and interpretation of the findings
(D) All of these

480. Which type of sampling method is used for collection of data if the population heterogeneous?
(A) Random sampling (B) Stratified sampling
(C) Purposive sampling (D Sequential sampling

481. Sequential sampling method is widely used in industries for:
(A) Quantity control (B) Quality control
(C) Efficiency measures (D) None of these

482. Cluster sampling method is used for:
(A) Quality control (B) Geological survey
(C) Efficiency measures (D) All of these

483. The system of accounting refers to:
(A) Posting of entry or transactions in books of account
(B) Entry of items in register
(C) Listing of production records
(D) Physical verification of records

484. Toy holding refers to:
(A) Very small size of holding, generally in sq. m and sq. yards
(B) Very large size of holding generally > 10 acre of land
(C) Both (A) and (B)
(D) None of these

485. Who proposed the concept of agricultural banks?
(A) Sir Willian Wedderburn and Justice Ranade
(B) Dr. Dadgil
(C) H. Dupernix
(D) Henry Wolff

486. Which of the following credit agency provides short and medium credit to farmers?
(A) State Cooperative Banks
(B) District Cooperative Bank
(C) Cooperative Credit Society
(D) All of these

487. Which of the following is not accepted as mortgage against long term loan by land development bank?
(A) Land (B) House
(C) Waterlogged soil (D) All of these

488. Regional Rural banks were established on:
(A) 26 September, 1975 (B) 19 July, 1969
(C) Ist July, 1965 (D) 2 October, 1969

489. Market refers to:
(A) A place where producers and sellers are assembled for selling and purchasing of goods
(B) Sellers and producers have free but competitive relationship
(C) Prices of commodities remain at normal prices
(D) All of these

490. Market process refers to:
(A) Activities through which a commodity is reached from producer to final consumer
(B) Grading and storage of a commodity
(C) Middlemen through which a commodity is either sold or purchase
(D) None of these

491. Which of the following approach useful and simple to study the agricultural marketing?
(A) Functional approach (B) Institutional approach
(C) Commodity approach (D) Behavioral approach

492. What is difference between Dalal and Commission agent?
(A) Commission agents have more rights than Dalal
(B) Dalal does not have physical control of produce
(C) Commission agents can sell the produce and charge the commission
(D) All of these

493. Which of the following statement is correct?
(A) All regulated markets should be organized markets
(B) All organized markets should be regulated markets
(C) All regulated markets may not be organized markets
(D) All unorganized markets may not be regulated markets

494. Which of the following middlemen accept the ownership of produce during the marketing process?
(A) Itinerant traders (B) Village traders
(C) FCI (D) All of these

495. Which of the following agency is responsible for creating awareness about quantity control of agricultural products?
(A) Export Inspection Standard Council of India
(B) Indian Standards Institute
(C) Indian Society for Quality control
(D) All of these

496. Which of the following is related to market development?
(A) Advertisement (B) Quality control
(C) Packing (D) All of these

497. Which of the following is a trade practices of marketing?
(A) Personal selling
(B) Advertisement based sale
(C) Sales by inspection
(D) All of these

498. Percentage marketing margin is equal to:
(A) Sale price – purchase price
(B) Absolute margin/sale price x 100
(C) Sale price – total marketing cost
(D) None of these

499. Agricultural Production Mandi Act was passed in UP in year:
(A) 1962 (B) 1964
(C) 1966 (D) 1968

500. Which of the following is a feature of a regulated market?
(A) Proper marketing system
(B) Proper communication of marketing information
(C) Availability of optimum grading, storage and packing facilities
(D) All of these

501. What is the objective of keeping farm records?
(A) To know annual performance of farm
(B) To know the present worth
(C) To compare the performance over a period of years
(D) All of these

502. Programme related to elimination of rural poverty is:
(A) JRY (B) RLEGP
(C) IRDP (D) All of these

503. Which of the following balances is needed in the development of underdeveloped countries?
(A) Crosswise balance (B) Backward balance
(C) Monetary balance (D) All of these

504. RBI was nationalized in:
(A) 1945 (B) 1947
(C) 1949 (D) 1939

505. Short term credit needs of a farmer is assessed by taking into account:
(A) Area under different crops
(B) Existing level of inputs used in crop production
(C) Recommended doses of inputs needed in crop production
(D) Recommended doses of inputs needed – (minus) existing level of inputs used in crop production

506. India's largest nationalised commercial bank is:
(A) Reserved Bank of India (B) Central Bank of India
(C) State Bank of India (D) Bank of India

507. Which of the following is the most effective way to overcome the defects of agricultural marketing?
(A) Cooperative marketing (B) Forward marketing
(C) Regulated marketing (D) Monopoly

508. A farm plan can be prepared by using:
(A) Partial budgeting (B) Complete budgeting
(C) Input-output model (D) Break-even point

509. The cost of production of the crop can be minimized by using economic principle:
(A) Law of diminishing return
(B) Law of opportunity cost
(C) Cost principle
(D) Principle of least cost combination

510. Principle of equi-marginal returns is applied when:
(A) The resources are unlimited
(B) The resource prices are high
(C) The prices of resources are varying
(D) The resources are limited

511. Do the T.C. and V.C. increases with increase in the output?
(A) Never (B) Yes
(C) Partially (D) With decrease in output

512. When a new variety of crop is introduced:
(A) Partial budgeting is required
(B) Full budgeting is a must
(C) No budgeting is needed
(D) Only planning is needed

513. The objective of preparation of calendar of operations in:
(A) To prepare a cropping scheme
(B) To calculate the cost of production
(C) To assess the need of human and bullock power
(D) To adjust farm enterprises

514. The farm plan would be based on the availability of efficiency of various factors of production:
(A) True (B) False
(C) Partially true (D) Incomplete statement

515. Farm budget is a tool of farm management analysis:
(A) True (B) False
(C) Never include cost (D) It is a futile exercise

516. Mixed farming is when:
(A) Crop production is combined with livestock enterprise and 50% : 50% income is from both
(B) Entire farm land is sown under mixed crops
(C) Only milch cattle are raised on the farm
(D) Income from milch animal is 95%

517. When marginal product is increasing at increasing rate than total product increases at a decreasing rate:
(A) Total product increases at a decreasing rate
(B) Total product increases at a increasing rate
(C) Average product decreases
(D) None of these

518. Price elasticity of supply of food products is generally:
(A) Infinite (B) Less than one
(C) More than one (D) Zero

519. One of the important fixed cost on a farm:
(A) Seed cost (B) Land rent
(C) Plant protection cost (D) Fertilizer cost

520. When marginal product is more than marginal cost, further investment on production would be:
(A) Profitable (B) Harmful
(C) Equal (D) No effect

521. The record of farm produce is entered in:
(A) Log book (B) Inventory book
(C) Farm stock register (C) Cash book

522. Price theory is a part of:
(A) Micro economics (B) Macro economics
(C) Farm management (D) Marketing

523. When marginal product is declining but remains more than zero, the total product will/will be:
(A) Zero
(B) Increase with increasing rate
(C) Increase with decreasing rate
(D) Initially decreasing than increasing

524. Which one of the following is not included in the factors of production:
(A) Enterprise (B) Organisation
(C) Land (D) Time

525. Which of the following is not a Consumer Right?
(A) Right to Freedom of Religion
(B) Right to Information
(C) Right to safety
(D) Right to Equality

526. Which of the following is not a feature of money?
(A) It includes currency – paper notes and coins and demand deposits
(B) It is without any use of its own
(C) It is authorized by the government of the country
(D) Gold has high liquidity so it is also included in the money

527. The modern currency is without any use of its own but still it is accepted as a medium of exchange because:
(A) The currency is authorized by the World Bank
(B) The currency is authorized by the government of the country
(C) The currency is authorized by RBI
(D) There is no other thing which can be used as a medium of exchange

528. Which of the following issues currency notes in India?
(A) RBI (B) SBI
(C) ICICI (D) President

529. When a farm is managed by Government officials it is called a:
(A) Peasant farm (B) Capitalistic farm
(C) Collective farm (D) State farm

530. When the depreciation of an asset is measured by divide the first cost by its estimated life, it is called:
(A) Diminishing value method
(B) Sum of the years digit method
(C) Straight line method
(D) Revaluation methods

531. When $\frac{\Delta x1}{\Delta x2} = \frac{\Delta x2}{\Delta x1}$ then we can :
(A) Employ any combination of x1 and x2
(B) Do nothing
(C) Go on substituting x1 ad x2
(D) Employ x2 only

532. Level of optimum production is available in:
(A) I stage of production
(B) II stage of production
(C) III stage of production
(D) None of these

533. Which one of the following kind of budgeting would you advise to a new farm planner:
(A) Enterprise budgeting (B) Complete budgeting
(C) Partial budgeting (D) None of these

534. Which of the following is the most effective way to overcome the defects of agricultural marketing?
(A) Cooperative marketing (B) Forward marketing
(C) Regulated marketing (D) Monopoly

535. A farm plan can be prepared by using:
(A) Partial budgeting (B) Complete budgeting
(C) Input-output model (D) Break even point

536. When marginal product is increasing at increasing rate then:
(A) Total product increases at a decreasing rate
(B) Total product increases at a increasing rate
(C) Average product decreases
(D) None of these

537. Farming system approach ensures the best use of:
(A) Land
(B) Labour
(C) Capital
(D) Land + Labour + Capital + Management

538. The price which necessarily covers the variable cost and also the fixed cost is:
(A) The long run price (B) The short run price
(C) The market price (D) The equilibrium price

539. Read the following statements and mark your answer according to code given below:
I. For variable input; APP = MPP in the Ist stage
II. For variable input; APP = MPP at the end of Ist stage
III. For variable input; APP = MPP at the start of Ist stage
IV. For variable input; APP = MPP in the IInd stage
Answer code
(A) I and III are correct (B) II and III are correct
(C) I and IV are correct (D) II and IV are correct

540. MPP (Marginal Physic Product) of a variable input is negative:
(A) Only in I stage (B) Only in II stage
(C) Only in III stage (D) In II and III stage

541. Match list I with list II and select the correct answer using the code give below:

List I (Farm Income)	List II (Measure)
a. Family labour income	1. G.I. – cost A1
b. Farm business income (owner farmer)	2. G.I. – cost A2
c. Net income	3. G.I. – cost B
d. Farm business income (tenant farmer)	4. G.I. cost C

Code:

	a	b	c	d
(A)	2	1	3	4
(B)	1	2	4	3
(C)	1	2	3	4
(D)	3	1	4	2

542. Marketing surplus is:
(A) Total production – Total family requirement
(B) Total consumption – Marginal consumption
(C) Total cost – marginal cost
(D) Total produce – Total cost

543. Line passing through the least cost points in the isoquants map is called:
(A) Isoline (B) Ridge line
(C) Expansion path (D) Isoquant curve

544. The demand of the commodity is different mainly by its:
(A) Quality (B) Price
(C) Storage (D) Supply

545. Formula which indicates the least-cost combination point is:
(A) MR = MC (B) $\Delta Y/\Delta x1$
(C) MRSY1, Y2 (D) MRSx1 x2 = PX2/PX1

546. The chief function of money is:
(A) Reserve base of credit expansion
(B) Providing liquidity
(C) Medium of exchange
(D) None of these

547. Support prices of agricultural commodities are worked out by:
(A) Finance ministry
(B) Agricultural Price Commission
(C) Food and Agricultural ministry
(D) None of these

548. According to the law of Diminishing Return the optimum profit will be at the point where:
(A) MP = AP (B) MP > AP
(C) AP > MP (D) MC = MP

549. The best level of output is obtained at point where:
(A) MR < MC (B) MR = MC
(C) MR > MC (D) MR = AC

550. Most prevalent "system of farming" in India is:
(A) Cooperative farming (B) Collective farming
(C) Peasant farming (D) State farming

551. When the depreciation of an asset is measured by dividing the first cost by its estimated life, it is called:
(A) Diminishing value method
(B) Sum of years digit method
(C) Straight line method
(D) Revaluation method

552. The commission for agricultural cost and prices is located in:
(A) Lucknow (B) Hyderabad
(C) New Delhi (D) Faridabad

553. The record of farm produce is entered in:
(A) Log book (B) Cash book
(C) Farm stock register (D Inventory book

554. The marginal cost –curve intersects the average cost-curve when average cost is:
(A) Maximum (B) Minimum
(C) Rising (D) Falling

555. The overhead cost is also known as:
(A) Variable cost (B) Fixed cost
(C) Total cost (D) Marginal cost

556. The subsidiary service of a marketing system is:
(A) Welfare (B) Assembling
(C) Dispersion (D) Processing

557. According to law of diminishing return, the optimum profit will be at the point where:
(A) MP = AP (B) MP > AP
(C) AP > MP (D) MC = MR

558. The vertical demand curve for a commodity shows that the demand is:
(A) Perfectly inelastic (B) Highly elastic
(C) Unit elastic (D) Perfectly elastic

559. The demand of agricultural product is mainly dependent on:
(A) Increase in agricultural holding
(B) Increase in income
(C) Increase in population
(D Increase in cost

560. Type of farming is based on:
(A) Law of comparative advantage
(B) Law of equi-marginal return
(C) Law of diminishing return
(D) Law of substitution

561. The ultimate aim of farm management is to:
(A) Increase gross income
(B) Optimize factor utilization
(C) Reduce total cost
(D) None of these

562. Which one of the following is a gift of nature?
(A) Land (B) Labour
(C) Capital (D) None of these

563. The efficiency of marketing channel is determined mainly on the basis of:
(A) Price spread
(B) Difference between retail and wholesale prices
(C) Regulated market
(D) Producer's share in consumer's rupees

564. Which one is not the function of a production manager?
(A) Planning (B) Organisation
(C) Marketing (D) Monitoring

565. Which system of farming does not involve farmer?
(A) Collective farming (B) Dry farming
(C) State farming (D) Diversified farming

566. Livestock insurance is included to:
(A) Fixed cost (B) Special cost
(C) Variable cost (D) Marginal cost

567. Farm management is a:
(A) Intra-farm science
(B) Inter-farm science
(C) Inter-marginal farm science
(D) International farm science

568. Which one is not a tool of farm management?
(A) Farm planning
(B) Farm budgeting
(C) Production and cost functions
(D) Farm stock keeping

569. One of the important fixed cost on a farm is:
(A) Seed cost (B) Land cost
(C) Irrigation cost (D) Fertilizer cost

570. When marginal product is at the maximum, the marginal cost is at the:
(A) Higher point (B) Lowest point
(C) Parallel (D) No change

571. If the marginal product is more than marginal cost than further investment in production is:
(A) Profitable (B) Harmful
(C) No effect (D) Equal

572. If the substitution ratio is less than the price ratio, than the cost:
(A) Remains the same (B) Increases
(C) Decreases (D) None of these

573. Principle of equi-marginal return is applied when:
(A) The resources are unlimited
(B) The price of resources is high
(C) The prices of resources vary
(D) The resources are limited

574. Which of the following is also referred to as "Opportunity cost"?
(A) Money cost (B) Alternate cost
(C) Social cost (D) Fixed cost

575. The law of diminishing return refers to an eventual fall in:
(A) Marginal product of the variable factor
(B) Average income of the farm
(C) Total production of all factors
(D) Total earning of the farm

576. Which one of the following is not correctly matched?

Type of market	Basis of classification
(A) Unorganised market	Organization
(B) Regulated market	Control
(C) Perfect market	Competition
(D) Terminal market	Area

577. Which of the following cost is subtracted from gross income (GI) to get the value of family labour income?
(A) Cost A1 (B) Cost A2
(C) Cost B (D) Cost C

578. Marketing surplus is:
(A) Total production Total family consumption
(B) Total consumption Marginal consumption
(C) Total cost Marginal cost
(D) Total production Total cost

579. Read the following statements and mark your answer according to code given below:
I. For variable input, APP = MPP in the I stage
II. For variable input, APP = MPP at the end of I stage
III. For variable input, APP = MPP at the start of II stage
IV. For variable input, APP = MPP in the II stage
Answer code
(A) I and III are correct (B) II and III are correct
(C) I and IV are correct (D) II and IV are correct

580. The concept of farming is best explained by:
(A) Secretarial planning approach
(B) Mixed farming approach
(C) Holistic farming approach
(D) Diversified farming approach

581. When MPP is zero, TPP is at its maximum, because:
(A) Further addition to MPP is negative
(B) Further addition to MPP is positive
(C) MC = MR
(D) Elasticity of productive is more than one

582. Crop farming and milk production is an example of:
(A) Competitive enterprise
(B) Complementary enterprise
(C) Supplementary enterprise
(D) None of these

582. When marginal product increases at increasing rate, then:
(A) Total product increases at a decreasing rate
(B) Total product increases at a increasing rate
(C) Average product decreases
(D) None of these

583. When increase or decrease in the level of one product does not affect the production level of the other product, it is known as:
(A) Supplementary relationship
(B) Complementary relationship
(C) Competitive relationship
(D) None f these

584. According to A. Marshall the " law of diminishing return" is applicable to:
(A) Industry (B) Agriculture
(C) Service sector (D) All of these

585. Why people prefers to receive payments in money?
(i) Because a person holding money can easily exchange it for any commodity or service that he/she might want
(ii) Because money raises the status of a person
(A) (i) is true but (ii) is false
(B) (i) is false but (ii) is true
(C) Both (i) and (ii) are true
(D) Both (i) and (ii) are false

586. What is barter system?
(i) A system under which goods are directly exchanged without the use of money
(ii) A system under which different currencies are exchanged
(iii) A system under which banks work as mediator between buyer and seller
(A) Only (i) is true
(B) (iii) is true
(C) Both (i) and (ii) are true
(D) Both (i) and (ii) are false

587. What is double coincidence of wants?
(A) A situation under which seller is ready to sell the product
(B) A situation under which buyer is ready to buy the product
(C) A situation under which both buyers and sellers are ready to exchange the product
(D) A situation under which neither the buyer nor the seller is ready to exchange the product

588. Which of the following is most important medium of exchange?
(A) Money (B) Gold
(C) Silver (D) Copper

589. Which of the following are part of medium forms of money?
(i) Paper notes and coins (ii) Demand deposits
(ii) ATM card
(A) Only (i) (B) Only (i) and (ii)
(C) All of these (D) None of these

590. Which of the following is not true with reference to Indian currency?
(A) It is issued by the Reserve Bank of India
(B) As per Indian law, no other individual or organization is allowed to issue currency
(C) No individual in India can legally refuse a payment made in rupees
(D) All the currency notes are signed by the President of the country

591. The rupees is widely accepted as a medium of exchange in India which of the following is/are appropriate reason for the same?
(i) It is issued by the RBI on the behalf of the central government
(ii) The law legalizes the use of rupee as a medium of payment
(iii) It is the only currency available in India
(A) Only (i) and (iii) (B) Only (ii) and (iii)
(C) Only (i) and (ii) (D) All of these

592. Why the deposits in the bank accounts are known as demand deposits?
(i) Because these can be withdrawn on demand
(ii) Because these creates demand of currency in the market
(iii) Because the banks pay an interest rate on the deposits
(A) Only (i)
(B) Only (i) and (ii)
(C) Only (i) and (iii)
(D) All of these

593. Which of the following facility is provided by the demand deposits
(i) Credit card
(ii) Cheque
(A) Only (i)
(B) Only (ii)
(C) Only (i) and (ii)
(D) All of these

594. What is a cheque?
(A) It is a paper instructing the bank to pay money
(B) It is a paper instructing the bank to deposit money
(C) It is a paper issued by bank against a loan
(D) It is a rate of interest

595. For which purpose the bank's use the major portion of their deposits?
(A) To pay the depositors who come to withdraw money
(B) They invest in share market
(C) To extend loans
(D) They deposit the excess money to RBI

596. Which of the following work as a mediator between those who have surplus funds and those who are in need of these funds?
(A) RBI
(B) Commercial banks
(C) Landlords
(D) Government of India

597. What is the major source of revenue for the commercial banks?
(A) The difference between what is charged from borrower and what is paid to depositors
(B) Revenue from the share market
(C) Fixed deposits
(D) None of these

598. Read the given statement and choose the right word from the given below:
An agreement in which the lender supplies the borrower with money, goods or services in return for the promise of future payment.
(A) Credit
(B) Demand deposit
(C) Currency
(D) Draft

599. In rural areas, the main demand for the credit is:
(A) For industralisation
(B) For crop production
(C) For marriages
(D) For education

600. ....is an asset that the borrower owns and uses this as a guarantee to the lender until the loan is repaid:
(A) Will
(B) Collatoral
(C) Fixed deposit
(D) Education

601. Interest rate, collateral and documentation requirement, and the mode of repayment together comprise. What is called the:
(A) Terms of credit
(B) Procedure of credit
(C) Method of credit
(D) Terms of trade

602. Which of the following is a formal source of credit?
(A) Money-lender
(B) Relatives
(C) Trader
(D) Commercial Banks

603. Which of the following is a informal source of credit?
(i) Money-lender
(ii) Relatives
(iii) Trader
(iv) Commercial Banks
(A) Only (i) and (ii)
(B) Only (ii), (iii) and (iv)
(C) Only (iii) and (iv)
(D) All of these

604. Which of the following agency supervises the formal sources of credit?
(A) RBI
(B) SBI
(C) CBI
(D) Police

605. Which of the following is the main source of credit for the rural households in India:
(A) Commercial Banks
(B) Co-operative societies
(C) Land-lords
(D) Money lenders

606. Which of the following charges higher rate of interest from the borrower?
(A) Informal sources of credit
(B) Formal sources of credit
(C) Both formal as well as informal sources of credit
(D) Government of India

607. Which of the following is the main source of credit for the rich house holds?
(A) Informal
(B) Formal
(C) Both formal and informal
(D) Neither formal nor informal

608. With reference to source of credit find the odd one out:
(A) Money lenders
(B) Traders
(C) Friends
(D) Commercial banks

609. Poor households are still dependent on formal sources of credit. Which of the following is/are reasons for the same?
(i) Banks are not present everywhere in rural India
(ii) Banks charges higher rate of interest than other sources
(iii) Bank loans require proper documents and collateral but poor don't have these
(A) Only (i)
(B) Only (ii)
(C) Only (i) and (iii)
(D) Only (ii) and (iii)

610. Which of the following promotes for house hold to take loan from money lender?
(i) In most of the cases there is no need for any collatoral
(ii) They charges low rate of interest
(iii) The borrower can, if necessary approach the money lender even without repaying their earlier loans
(A) Only (i)
(B) Only (ii)
(C) Only (i) and (iii)
(D) Only (i) and (ii)

611. What is SHG?
(A) Self Help Groups
(B) Self Handicraft Groups
(C) Social Help Groups
(D) Society Help Groups

612. A typical Self Help Group has how members:
(A) 100-150
(B) 150-200
(C) 15-20
(D) 5-6

613. People prefer to take loan from SHG because:
(A) No rate of interest is charges
(B) Rate of interest is less than what the information sources charges
(C) Rate of interest is less than what the formal sources charges
(D) Most of numbers are women

614. Which of the following is correctly matched:
(i) Banks : Formal source of credit
(ii) SHG : Informal source of credit
(iii) Land : Collatoral
(A) Only (i) (B) Only (ii)
(C) Only (iii) (D) Only (i) and (ii)

615. Who takes most of the decisions regarding in the SHG's?
(A) Banks (B) Members
(C) NGO's (D) RBI

616. What is MNC?
(A) Multi Nation Corporations
(B) Multinational Corporations
(C) Moving National Corporations
(D) Mid National Cooperatives

617. A company that owns or controls production in more than one nation. Choose the correct word:
(A) MNC (B) MVC
(C) KMC (D) VMC

618. Which of the following is not a feature of MNC's?
(i) A MNC owns or controls production in more than one nation
(ii) The goods and services are produced globally
(iii) Production is organized in a complex way
(iv) Most of the MNCs are investing in developed countries
(A) Only (i) and (ii) (B) Only (ii) and (iii)
(C) Only (iii) and (iv) (D) Only (iv)

619. Which of the following country provides cheap manufacturing location?
(A) China (B) Mexico
(C) USA (D) UK

620. Which of the following is not an important locational factor for a MNC?
(A) Cheap manufacturing location
(B) Availability of cheap labour
(C) Market
(D) Type of government i.e., democratic or non democratic

621. MNC are investing in India which of the following is/are appropriate reason for the same?
(i) Availability of cheap skilled as well as unskilled labour
(ii) Market
(iii) Good infrastructure
(iv) Efficient Government
(A) Only (i) and (ii) (B) Only (ii) and (iii)
(C) Only (iii) and (iv) (D) All of these

622. What is investment?
(A) Money that is spent to buy assets such as land, building, machines and other equipment
(B) Money that is deposited in banks
(C) Money that is spent on imports
(D) Profit of the businessman

623. Investment made by MNC is called:
(A) Internal investment (B) External investment
(C) Foreign investment (D) Multi investment

624. Which of the following methods is/are used by MNC's to spread/increase their production?
(A) They buy up local companies
(B) They place orders to the local producers
(C) They produce jointly with local producers
(iv) They sell their products at cheap rates
(A) Only (i) and (ii) (B) Only (i), (ii) and (iii)
(C) Only (ii) and (iii) (D) Only (iii) and (iv)

625. What is globalisation? Select the most appropriate answer:
(A) It is the process of rapid integration or interconnection between countries
(B) It is increase in foreign trade
(C) It is the process under which developing countries are using modern technology
(D) It is the process of movement of people from one region to another

626. Which of the following is not a feature of foreign trade?
(i) It connects different markets of the world
(ii) Choice of goods in the market rises
(iii) Prices of similar goods in the two markets tend to become equal
(iv) Developing countries gain more than the developed countries
(A) Only (i) and (ii) (B) Only (iv)
(C) Only (iii) and (iv) (D) All of these

627. .... is the process of rapid integration or interconnection between countries:
(A) Privatisation (B) Liberalisation
(C) Globalisation (D) None of these

628. Which of the following has been the major factor that has stimulated the globalization process:
(A) Technology (B) Ship industry
(C) Democracy (D) Socialism

629. Which of the following is a trade barrier?
(i) Technology (B) Tax on imports
(iii) Tax on exports
(A) Only (i) (B) Only (i) and (ii)
(C) Only (ii) and (iii) (D) Only (iii)

630. Why the countries use trade barriers?
(i) Countries put up trade barriers to increase or decrease foreign trade
(ii) Countries put up trade barriers to decide what kind of goods and how much of each, should come into the country
(iii) To create social and political equality
(A) Only (i) (B) Only (i) and (ii)
(C) Only (iii) (D) Only (i) and (ii)

631. The Indian government, after independence has put trade barriers to foreign trade and foreign investments:
(A) To protect the producers within the country from foreign competition
(B) To promote Swadeshi Movement
(C) Because foreign goods were very expensive
(D) Because foreign goods were not liked by the Indian people

632. When was India's foreign trade liberalized?
(A) 1960s (B) 1970s (C) 1980s (D) 1990s

633. Removing barriers or restrictions setup by the government is known as:
(A) Privatisation (B) Globalisation
(C) Liberalisation (D) Industrialisation

634. Which of the following is/are result of liberalization?
(i) Businesses are allowed to make decisions freely
(ii) Government removes unwanted restrictions from trade
(iii) Government provides political and fundamental rights to all
(iv) Poverty is removed
(A) Only (i) and (ii) (B) Only (i), (ii) and (iii)
(C) Only (iii) and (iv) (D) Only (iv)

635. Indian government wants to check the free flow of cheap Chinese electronic goods into Indian market. Which of the following method can government use?
(A) Trade barrier
(B) Complaint to WTO
(C) Government can increase export tax
(D) Government can complaint to World Bank

636. What is WTO?
(A) World Tower Organisation
(B) World Trade Organisation
(C) World Tax Organisation
(D) World traders Organisation

637. What is the basic aim of WTO?
(A) To check the free flow of goods and services across the world
(B) To bridge gap between rich and poor countries
(C) To liberalise international trade
(D) To support MNC

638. How many countries are the members of WTO?
(A) 95 (B) 120 (C) 149 (D) 200

639. Which of the following organization/s liberalises international trade?
(i) WTO (ii) MNC
(iii) Commercial bank
(A) Only (i) (B) Only (i) and (ii)
(C) Only (ii) and (iii) (D) Only (iii)

640. Which of the following has benefitted the most from the globalization?
(A) Small producers (B) Farmers
(C) Poor (D) Consumers

641. Which of the following is/are Indian MNC's?
(i) Samsung (ii) Nokia
(iii) Tata motors (iv) Asian Paints
(A) Only (i) and (ii) (B) Only (ii) ad (iii)
(C) Only (iii) and (iv) (D) Only (i) and (iv)

642. Which of the following has benefitted the most from the process of globalisation?
(A) IT
(B) Agro based industries
(C) Textile industries
(D) Iron and steel industry

643. Which of the following have suffered the most because of the globalisation?
(A) IT
(B) Small manufectures of vegetable oil, capacitors, plastic etc.
(C) MNC's
(D) Automobile industries

644. Which of the following has benefitted the most from the globalization?
(A) People with education, skill and wealth
(B) People living below poverty line
(C) Farmers
(D) Local traders

645. Which of the following method can be used by the government for a fair globalization?
(A) It can impose trade barriers
(ii) It can negotiate at the WTO for fairer rules
(iii) It can align with other developing countries
(iv) It can close its market for foreign trade
(A) Only (i) and (ii) (B) Only (ii) and (iv)
(C) Only (i), (ii) and (iii) (D) All of these

646. Who is a consumer?
(A) Anyone who purchase goods and services from the market and pays for these
(B) Anyone who sells goods and services in the market
(C) Anyone who manufactures goods
(D) Anyone who invests in the market

647. Anyone who produce goods and services and sell these in the market is known as:
(A) Producer (B) Consumer
(C) Buyer (D) Entrepreneur

648. Under which of the following condition the market do not work in a fair manner?
(i) When producers are few and powerful whereas consumers purchase in small amounts and are scattered
(ii) When consumers are few and producers are in a large number
(A) Only (i) and (ii) (B) Only (i)
(C) Only (iii) (D) Only (ii)

649. Which of the following is an appropriate reason for the Consumer Movement?
(A) It arose out dissatisfaction of the consumers as many unfair practices were being indulged in by the sellers
(B) It arose because of spread of democracy
(C) It arose because of cheap goods and services
(D) It arose because of MNC's

650. Which of the following factor's gave birth to the Consumer Movement in India?
(i) Food shortage
(ii) Hoarding
(iii) Black marketing and adulteration
(iv) Government of India
(A) Only (i) and (ii)
(B) Only (i), (ii) and (iii)
(C) Only (iii)
(D) Only (i) and (ii)

651. The Consumer Movement in an organized form was started in:
(A) 1950s
(B) 1960s
(C) 1970s
(D) 1980s

652. Why is the year 1986 known as a turning point in the history of Consumer Movement?
(A) COPRA was enacted
(B) CHOPRA was enacted
(B) Policy of liberalization was followed in this year
(C) WTO was established

653. What is COPRA?
(A) Consumer Preservation Act
(B) Consumer Protection Act
(C) Consumer Protection Activities
(D) None of these

654. "We do find bad quality products in the market". Which of the following is an appropriate reason for the same?
(i) Production is being done by MNC's
(ii) Consumer movement is still very weak
(iii) Consumer are not aware of their rights and duties
(iv) Prevailing laws to safe guard the consumers are very weak
(A) Only (i)
(B) Only (ii), (iii) and (iv)
(C) Only (iii) and (iv)
(D) Only (i) and (ii)

655. What is MRP?
(A) Minimum Retail Price
(B) Maximum Retail Price
(C) Marginal Retail Price
(D) Marginal Revenue Price

656. If a manufacture has sold you a pressure cooker with improper safety value, which of the following right can help you to pull up the manufacturer?
(A) Rights to safety
(B) Right to Choice
(C) Right to Consumer Education
(D) Right to Representation

657. Under which right the producer is suppose to provide details about the products, its ingredient, price, batch number, date of manufacturing, etc?
(A) Right to Informed
(B) Right to Information
(C) Right to Choose
(D) Right to Represent

658. What is RTI?
(A) Right to Informed
(B) Right to Information
(C) Right to Choose
(D) Right to Represent

659. Which of the following right you will use in case you want to have information about the functioning of a government department?
(A) Right to Informed
(B) Right to Information
(C) Right to Choose
(D) Right to Seek Redressal

660. You have joined an education institution and had paid the fee for the whole year but you found that the institution is not providing what it promised. Now you want to change the institution but the management refused to refund your fee. Which of your right is being violated?
(A) Right to Informed
(B) Right to Information
(C) Right to Choose
(D) Right to Represent

661. Suppose you want to buy toothpaste, and the shop owner says that he/she can sell the toothpaste only if you buy a toothbrush. Which of your right is being violated by the shopkeeper?
(A) Right to Informed
(B) Right to Information
(C) Right to Choose
(D) Right to Seek redressal

662. Which of the following guides consumers regarding how to file cases in the Consumer Court?
(A) PRO
(B) Consumer Forums
(C) Police
(D) A lawyer

663. Under which of the following right a consumer can represent his/her own case by himself?
(A) Right to Informed
(B) Right to Information
(C) Right to Choose
(D) Right to Represent

664. Under which of the following rights the government gives advertisements in the newspapers about the consumer's rights and print slogans like "Jago Grahak Jago":
(A) Right to Informed
(B) Right to Information
(C) Right to Education
(D) Right to Represent

665. Under COPRA a .... tier judicial machinery has been setup to provide justice to the exploited consumers:
(A) 2
(B) 3
(C) 4
(D) 5

666. The district level Consumer courts deals with the cases involving claims up to Rs.:
(A) 20 lakh
(B) 40 lakh
(C) 60 lakh
(D) 80 lakh

667. The state level Consumer courts deals with the cases involving claims up to Rs.:
(A) 20 lakh - 1 crore
(B) 40 lakh- 2 crore
(C) 60 lakh - 3 crore
(D) 80 lakh – 4 core

668. The national level Consumer courts deals with the cases involving claims up to Rs.:
(A) 20 lakh
(B) 40 lakh
(C) 60 lakh
(D) 1 crore

669. While buying a product from a market which of the following logo assures you of quality?
(A) ISI or Agmark
(B) LTTE
(C) KCF
(D) BJP

670. Which day is celebrated as the National Consumer Day in India?
(A) 24th January (B) 24th December
(C) 24th October (D) 26th January

671. Why is 24th December celebrated as the National Consumer Day?
(A) It was the day when the Indian Parliament enacted the Consumer Protection Act 1986
(B) It was the day when the Indian Parliament celebrates Gandhi Jayanti
(C) On this day the first Consumer Court was set up in Delhi
(D) On this day the Supreme Court was setup

672. A chemist sold you a medicine of expiry date. Under which consumer right you can approach the Consumer court?
(A) Right to Informed (B) Right to Information
(C) Right to safety (D) Right to Equality

673. To which court you will approach if your claim is more than Rs 40 lakh and less than Rs one crore?
(A) State Consumer Court
(B) National Consumer Court
(C) District Consumer Court
(D) The Supreme Court

674. Which of the following is not a Consumer Right?
(A) Right to Informed (B) Right to Information
(C) Right to safety (D) Right to Equality

675. Find the odd one out:
(A) Right to Informed (B) Right to Information
(C) Right to Education (D) Right to Equality

676. Success in rural development project depends upon:
(A) Regular training of workers
(B) Regular contacts of workers
(C) Amount of subsidy
(D) Participation of beneficiaries

677. Role of different agencies for village development is included in:
(A) Resource mapping (B) Time line
(C) Transect walk (D) Chapati diagram

678. Rural development depends on:
(A) Research
(B) Research – Extension
(C) Research – Teaching – Extension
(D) None of these

679. The basic unit of development under IRD program is:
(A) A village
(B) A community development block
(C) A family
(D) A district

680. From among the following identify one which is three – tier system of the local self – government:
(A) Gram panchayat – Khand samiti – Zila parishad
(B) Gram panchayat –Village school– Village cooperative
(C) IRD programme – TRYSEM – NREP
(D) Lab to land – KVK – KGK

681. The University Education Commission was headed by Dr. S. Radha Krishanan recommended the establishment of:
(A) Urban Universities (B) Rural Universities
(C) Both (A) and (B) (D) None of these

682. There are———— development blocks in India:
(A) 5000 (B) 6000
(C) 7000 (D) 8000

683. ————is a dominant social institution permeating social and economic relations
(A) Society (B) Creed
(C) Caste (D) Customs

684. Rural community differ from urban community in relation to
(A) Population density (B) Culture
(C) Environment (D) All of these

685. Which of the following statements about "joint family system of rural areas" is correct?
(A) Joint families are always a cause
(B) Joint families are both an impediment and a factor or rural development
(C) There are no joint families in the rural areas
(D) Joint families are the only factors of rural development

686. Rural development programme at district level is being looked after by:
(A) DRDA
(B) MP of that constituency
(C) Additional S.P.
(D) None of these

687. Family is included under:
(A) Secondary group
(B) Primary group
(C) Tertiary group
(D) Depends upon the situation

688. Integrated Rural Development Programme was started in:
(A) 1978 (B) 1974 (C) 1975 (D) 1979

689. Co-operative society is basic institutions for:
(A) Political growth of the villagers
(B) Social growth of the villagers
(C) Cultural growth of the villagers
(D) Socio-economic growth of the villagers

690. According to Balwantrai Mehta committee, Panchayati Raj should be a
(A) Two tier structure of local self government bodies from village to district
(B) Three tier structure of local self government bodies from village to district
(C) Two tier structure of local self government bodies from village to block
(D) Three tier structure of local self government bodies from village to block

691. Panchayati Raj was first time started on 2nd October, 1959 in:
(A) Bombay (Maharastra)
(B) Nagaur (Rajasthan)
(C) Shimla (Himachal Pradesh)
(D) Patna (Bihar)

692. The village Panchayati Act came into existence in:
(A) 1958 (B) 1959
(C) 1960 (D) 1961

693. Panchayati Raj came after:
(A) Pre-determined programme
(B) Self determined programme
(C) Fact determined programme
(D) None of these

694. Panchayati Raj in India was first recommended by:
(A) Balwant Rai Mehta Committee
(B) Mahatma Gandhi
(C) Kriplani committee
(D) Pandit Jawaharlal Nehru

695. Panchayati Raj was first time introduced in which of the following state:
(A) Rajasthan (B) Punjab
(C) Andhra Pradesh (D) Orissa

696. Panchayati Raj was first started at:
(A) Nagpur (B) Nagaur
(C) Nasik (D) Narendrea Nagar

697. Kishak Bharti magazine is published from:
(A) Kanpur (B) Pantnagar
(C) New Delhi (D) Hisar

698. List-I / List-II
(a) Theory of rent — (1) Edward H.
(b) Quasi rent — (2) Adam Smith
(c) Monopolistic competition — (3) Alfred Marshall
(d) Theory of Wealth of Nation — (4) Ricardio

Code	a	b	c	d
(A)	3	4	2	1
(B)	4	3	1	2
(C)	4	2	1	3
(D)	2	3	4	1

699. The establishment of Panchayati Raj bodies at various level fulfils:
(A) Principle of cooperation
(B) Principle of leadership
(C) Principle of participation
(D) Principle of grass root

700. RRB were set up under the recommendation of:
(A) Nariman Committee
(B) Narasimhan committee
(C) Sen Committee
(D) CRAFICARD Committee

701. The new Panchayat Raj Act (73rd Constitution amendment) was enacted all over the country from:
(A) 24th April, 1993 (B) 24th April, 1994
(C) 24th April, 1995 (D) None of these

702. Which of the following rural development project was launched before independence:
(A) Etawah Pilot Project
(B) Cooperative movement
(C) Community development
(D) National Extension Service

703. Regional Rural Bank was set up in:
(A) 1972 (B) 1973
(C) 1974 (D) 1975

704. Success of rural projects depend upon:
(A) Regular training of staff
(B) Constant visit of staff
(C) Quantum of subsidies
(D) Partnership of beneficiaries

705. Programme related to elimination of rural poverty is:
(A) JRY (B) RLEGP
(C) IRDP (D) All of these

706. Name the 'Poverty Eradication Programme' in which bank play a dominant role:
(A) MNP (B) IRDP
(C) RBY (D) JRY

707. When National Development Council was formed?
(A) August 1947 (B) August 1950
(C) August 1952 (D) August 1960

708. Which one of the following scheme is exclusively meant for self-employment of rural youth?
(A) NREP (B) TRYSEM
(C) IRDP (D) DPAP

709. Which one of the following is not the example of social system?
(A) Village (B) Communicator
(C) Block (D) District

710. The first cooperative societies established in India were:
(A) Credit societies (B) Marketing societies
(C) Agricultural societies (D) Housing societies

711. In distinguished unemployment is:
(A) Marginal physical productivity becomes zero
(B) Average physical productivity becomes zero
(C) Total physical productivity becomes zero
(D) None of these

712. Social changes refer to changes in:
(A) Social structure of the society
(B) Culture of the society
(C) Adoption behavior of the people
(D) Communication network

713. Capital is the result of savings and its formation depends upon:
(A) Capacity to save
(B) Desire to save
(C) Facility to save
(D) All of these

714. Selecting 'poorest among poor first' is the approach under:
(A) Jawahar Rojgar Yojna
(B) Training and Visit
(C) Integrated Rural Development Programme
(D) National Agricultural Technology Programme

715. Which one of the following does directly affect the division of labour?
(A) Capital (B) Climate
(C) Demand of labour (D) Nature of work

716. The basic unit of development under IRD programme is:
(A) Village Community (B) Development block
(C) Family (D) District

717. An effective facilitator in the development of any society is:
(A) Communication technology
(B) Information and communication technology
(C) Information technology
(D) Rural Development

718. Rural India reflects the very essence of:
(A) Indian culture (B) Tradition
(C) Both (A) and (B) (D) None of these

719. The most serious economic problem of India are:
(A) Poverty and unemployment
(B) Unemployment and stagnation
(C) Underdevelopment and poverty
(D) Corruption and unemployment

720. The percentage of people below poverty line in India is:
(A) 25.1 (B) 27.8
(C) 29.3 (D) 21.5

721. The National Rural Employment Guarantee Act was enacted initially to cover:
(A) 175 districts (B) 200 districts
(C) 257 districts (D) 304 districts

722. "Attacking Rural Poverty" is a book written by:
(A) Paul leagans (B) A.T. Mosher
(C) Coomb and Ahamed (D) Henry Parker

723. The important philosophy behind community development is:
(A) Work based on felt needs
(B) Peoples participation
(C) Philosophy of self help
(D) Emancipatory training

724. The number of unemployed people living in India is around:
(A) 13.10 million (B) 9.02 million
(C) 7.98 million (D) 10.51 milion

725. Community development project was started on the basis of:
(A) Nalagarh Commission Report
(B) Royal Commission Report
(C) Ford Foundation Report
(D) GMF Enquiry Report

726. The population of India according to 2001 census is:
(A) 102.21 crores (B) 104.56 crores
(C) 107.56 crores (D) 102.70 crores

727. The sex ratio of India according to 2001 census is:
(A) 927 (B) 933
(C) 938 (D) 942

728. India's population projection for 2026 is:
(A) 1.21 billion (B) 1.27 billion
(C) 23.50 billion (D) 1.40 billion

729. The literacy rate of India according to 2001 census is:
(A) 75.85 (B) 54.16
(C) 62.67 (D) 65.38

730. The concept of multipurpose village level worker was evolved from:
(A) Etawah Pilot Project
(B) Community Development Project
(C) National Extension Service
(D) Gurgaon Project

731. The first bank to launch Kisan Credit Card:
(A) Punjab National Bank (B) State Bank of India
(C) Bank of Baroda (D) Allahabad Bank

732. Etawah district is in the state of:
(A) Madhya Pradesh (B) Uttar Pradesh
(C) Haryana (D) Bihar

733. Nilokheri project was also called:
(A) Mazdoor Manzil (B) Kisan Manzil
(C) Mazdoor Madad (D) Kisan Madad

734. The term community development have originated in:
(A) Oxford (B) Cambridge
(C) Ohio (D) San Francisco

735. Firka development scheme was started by:
(A) Bihar government
(B) Bengal government
(C) Maharashtra government
(D)Madras government

736. DRDA was established in:
(A) 1919 (B) 1980
(C) 1967 (D) 1947

737. The chairman of District Rural Development Agency is:
(A) Block Development Officer
(B) District Planning Officer
(C) District Panchayat President
(D) District Collector

738. The authority responsible for planning and implementation of rural development programme at state level is:
(A) Planning Commission
(B) State Development Committee
(C) Ministry of Rural Development
(D) Chief Minister

739. The chairman of State Development Committee is:
(A) Governor
(B) Chief Minister
(C) State Agriculture Minister
(D) Minister of Rural Development

740. The Gandhian approach of rural development have its basic emphasis on:
(A) Village industries
(B) Development of inner man
(C) Local self government
(D) Production improvement

741. The annual outlay for the ministry of rural development for the year 2006-07 was:
(A) 36000 crores (B) 31000 crores
(C) 21000 crores (D) 14000 crores

742. India's position in the UNDP's Human Development Report for 2006 was:
(A) 212 (B) 214 (C) 126 (D) 198

743. The National Commission of Women was set up is:
(A) 31st January, 1991 (B) 31st January, 1992
(C) 31st January, 1993 (D) 31st January, 1994

744. The chief proponent of the concept of "Empowerment" is:
(A) Alfred York (B) David Werner
(C) Paulo Fiere (D) Osgood

745. Which among the following is a correct equation:
(A) Motivation + Empowerment = Enabling
(B) Motivation + Enabling = Empowerment
(C) Enabling + Empowerment = Motivation
(D) Motivation + Empowerment + Enabling = 0

746. The transfer of authority from central government to state government implies:
(A) Decentralization (B) Delegation
(C) Devolution (D) Deconcentration

747. In India land reforms have been equated with total:
(A) Urban development
(B) Agrarian development
(C) Industrial development
(D) Educational development

748. The term land reforms should be used in broadest sense to serve as:
(A) Inscription (B) Prescription
(C) Description (D) Formulation

749. Ministry of Panchayat Raj was established on:
(A) 27th May, 2004 (B) 15th August, 2002
(C) 2nd October, 2001 (D) 15th August, 2006

750. The backward class commission was headed by:
(A) B.N. Roy (B) Mahatma Gandhi
(C) Kaka Kalelkar (D) None of these

751. NIRD under Ministry of Rural Development is located at:
(A) New Delhi (B) Hyderabad
(C) Lucknow (D) Chennai

752. Journal of Rural Development is published from:
(A) MANAGE
(B) Ministry of Rural Development
(C) NIRD
(D) IARI

753. Which among the following is a rural development journal:
(A) Kurukshetra (B) Yojana
(C) Gram vikash (D) Gramin Soochna

754. The number of languages in which Yojana is published:
(A) 10 (B) 11 (C) 12 (D) 13

755. International Women's day is celebrated on:
(A) 8th March (B) 8th October
(C) 21st November (D) 2nd December

756. World Habitat day is celebrated on the first Monday of:
(A) October (B) December
(C) May (D) July

757. Tehsil Divas in Uttar Pradesh is organized on every month of:
(A) 1st and 2nd Monday (B) 1st and 3rd Monday
(C) 1st and 3rd Tuesday (D) 2nd and 4rth Tuesday

758. The Lakhina experiment is associated with:
(A) District Administration
(B) Panchayat Raj Adminsitration
(C) Rural Development Administration
(D) Tribal Development Administration

759. Panchayati Raj system was adopted first in:
(A) Rajasthan and Madhya Pradesh
(B) Andhra Pradesh and West Bengal
(C) Rajasthan and Andhra Pradesh
(D) Tamilnadu and Rajasthan

760. The basic unit of Panchayati Raj institutions:
(A) Zila Parishad (B) Grama Sabha
(C) Panchayat Samiti (D) Block Panchayat

761. The 'Year of Grama Sabha' was observed in:
(A) 1999-2000 (B) 2001-2002
(C) 2004-2005 (D) 2006-2007

762. The total number of panchayats in the country:
(A) 2.31 lakhs (B) 2.23 lakhs
(C) 2.47 lakhs (D) 2.79 lakhs

763. The intermediate level in Panchayati Raj institutions:
(A) Zila Parishad (B) Grama Sabha
(C) Panchayat Samiti (D) Block panchayat

764. The total number of panchayat at intermediate level:
(A) 6000 (B) 7000
(C) 8000 (D) 9000

765. The third tier of Panchayati Raj is:
(A) Zilla Parishad (B) Grama Sabha
(C) Panchayat Samiti (D) Block Panchayat

766. The responsibility of village panchayat include:
(A) Agricultural production
(B) Rural industrial development
(C) Maternity and child welfare
(D) All of these

767. The need for constitutional sanction of village panchayats was first felt by:
(A) Balwantrai Mehta Committee
(B) Ashoka Mehta Committee
(C) GVK Rao Committee
(D) L M Singhvi Committee

768. Vikas Volunteer Vahini is sponsored by:
(A) DRDA (B) RRB's
(C) NGO's (D) NABARD

769. Rural infrastructural fund was set up by:
(A) RBI (B) NABARD
(C) IDRB (D) SBI

770. The Department of Rural Development under Government of India was started on:
(A) 1919 (B) 1928
(C) 1931 (D) 1947

771. People's plan was envisaged by:
(A) Jawaharlal Nehru (B) Visweswrayya
(C) M.N. Roy (D) Mahalanoibis

772. The report on All India Rural Credit Enquiry Survey was submitted on:
(A) 1947 (B) 1954
(C) 1960 (D) 1961

773. Community Development projects were inaugurated in India on:
(A) 26th January, 1952 (B) 2nd October, 1952
(C) 25th August, 1954 (D) 8th September, 1953

774. Grow More Food enquiry committee was constituted in:
(A) 1952 (B) 1953
(C) 1955 (D) 1956

775. Grow More Food enquiry committee was headed by:
(A) V.T. Krishnamachari (B) M.S. Swaminathan
(C) M.K. Gandhi (D) F.L. Brayne

776. Sarvodaya Scheme was started by:
(A) R.N. Ragore (B) Vinoba Bhave
(C) M.K. Gandhi (D) F.L. Brayne

777. Sarvodaya Scheme was started in:
(A) 1945 (B) 1946
(C) 1947 (D) 1948

778. Key Village Scheme was started in:
(A) 1949 (B) 1950
(C) 1951 (D) 1952

779. Rural Reconstruction Institute at Sundarban district was started by:
(A) R.N. Tagore (B) Daniel Hamilton
(C) M.K. Gandhi (D) F.L. Brayne

780. The structure of CD blocks were taken from:
(A) Etawah Project (B) Firka Project
(C) GMF Campaign (D) Nilokheri Experiment

781. The technical cooperation agreement of CDP was signed with:
(A) Africa (B) France
(C) USA (D) Netherland

782. CD is sponsored by:
(A) Government agencies (B) Private agencies
(C) Voluntary agencies (D) Both (A) and (B)

783. The community development programme was launched in India as:
(A) Development organisation
(B) Administrative setup
(C) As a means to an end
(D) Local self government

784. "A Guide to Community Development" was written by:
(A) M. Horton (B) D. Hamilton
(C) T. Moscher (D) Ensminger

785. The CDP was planned and conceived as peoples:
(A) Planned programme (B) Local programme
(C) Self help programme (D) Organize programme

786. The number of community development projects implemented in India:
(A) 54 (B) 55 (C) 56 (D) 57

787. The operational area of each community development project includes:
(A) 100 villages (B) 200 villages
(C) 300 villages (D) 400 villages

788. Under community development project each village level worker is appointed for:
(A) 1-5 villages (B) 5-10 villages
(C) 10-15 villages (D) 15-20 villages

789. "Extension Education in Community Development" is published by:
(A) Ministry of Planning
(B) Ministry of Food and Agriculture
(C) Ministry of Rural Development
(D) Ministry of Information and Broadcasting

790. National extension service was started in:
(A) 1952 (B) 1953
(C) 1954 (D) 1955

791. NES was considered basically as a:
(A) Method (B) Agency
(C) Project (D) Plan

792. Community Development Blocks was started in:
(A) 1952 (B) 1953
(C) 1954 (D) 1955

793. A community development project have an operational area of:
(A) 400-500 square miles (B) 600-700 square miles
(C) 700-800 square miles (D) 800-1000 square miles

794. The entire nation was covered under CD programme by the year:
(A) 1957 (B) 1958
(C) 1959 (D) 1960

795. The committee constituted to review CD and NES projects was headed by:
(A) Ashok Mehta (B) Hanumantha Rao
(C) O.P. Dahama (D) Balwant Ray Mehta

796. Panchayat Raj was introduced in India in:
(A) 1951 (B) 1952
(C) 1953 (D) 1954

797. Panchayat Raj institutions is formed based on:
(A) Ashok Mehta Committee
(B) Hanumantha Rao Committee
(C) Dantwala Committee
(D) Balwant Ray Mehta committee

798. Panchayat Raj is a system of:
(A) Local government (B) Local administration
(C) Local self-government (D) None of these

799. The term Panchayat Raj was christened by:
(A) M.K. Gandhi (B) J.L. Nehru
(C) B.R. Mehta (D) M.S. Mehta

800. The Panchayat Raj institutions are intended to strengthen:
(A) Structural planning (B) Indicative planning
(C) Multilevel planning (D) Functional planning

801. The origin of the word democracy is from:
(A) Roman (B) Sanskrit
(C) Greek (D) Persian

802. The word "democracy" means:
(A) Peoples participation (B) Peoples rule
(C) Peoples government (D) Peoples judgement

803. The process of dispersing decision making closer to the point of service or action:
(A) Decentralization (B) Delegation
(C) Devotion (D) Deconcentration

804. The constitutional amendment act was passed in:
(A) 1996 (B) 2001
(C) 1992 (D) 1997

805. The election to panchayat is normally held after a period of:
(A) 3 years (B) 4 years
(C) 5 years (D) 6 years

806. The number of tiers for Panchayat in a state with population below 20 lakhs:
(A) Two (B) Three
(C) Four (D) Five

807. The proportion of seat in a panchayat to be filled from women through election:
(A) Two-third (B) One-third
(C) One-fifth (D) One-fourth

808. Net return per hectare is:
(A) Overall efficiency measure
(B) Partial efficiency measure
(C) Specific efficiency measure
(D) None of these

809. The 73rd amendment act does not apply to the states of:
(A) Nagaland (B) Mizoram
(C) Meghalaya (D) All of these

810. The total number of district in India:
(A) 256 (B) 594
(C) 488 (D) 604

811. Market demand curved is faced by:
(A) Monopolist (B) Competitive firm
(C) Joint firm (D) Oligopolistic firsm

812. The pioneer state in India to bring the whole of it under democratic decentralization:
(A) Uttar Pradesh (B) Bihar
(C) Rajasthan (D) Madhya Pradesh

813. The constitutional 74th amendment act relates to:
(A) Urban local bodies
(B) Election to Gram Sabha
(C) Free primary education
(D) Reservation in panchayat

814. Which of the following is a committee on block level planning:
(A) Hanumantha Rao Committee
(B) Dantewada Committee
(C) G.V.K. Rao Committee
(D) L.M. Singhvi Committee

815. Which of the following is a committee on district planning:
(A) Hanumantha Rao Committee
(B) Diwakar Committee
(C) Dantewada Committee
(D) Santhanam Committee

816. Ashok Mehta Committee recommended the establishment of:
(A) Mandal Panchayat (B) Nagar Panchayat
(C) Panchayat Samiti (D) Gram Panchayat

817. The need for constitutional sanction of village panchayat was first felt by:
(A) Balwantray Mehta Committee
(B) Ashok Mehta Committee
(C) G.V.K. Rao Committee
(D) L.M. Singhvi Committee

818. Ford foundation was invited to India on:
(A) 1958 (B) 1959
(C) 1960 (D) 1961

819. "India's food crisis and step to meet it" is a:
(A) Book on food situation in India
(B) Report of ford foundation team
(C) Formula to solve food problems
(D) A slogan for campaign

820. The programme started based on the report of ford foundation:
(A) CDP (B) IADP
(C) IRDP (D) Etawah project

821. The IADP was launched in:
(A) 1960 kharif (B) 1960 rabi
(C) 1961 kharif (D 1961 rabi

822. "Betting on the strong horse" is a slogan associated with:
(A) IRDP (B) HYVP
(C) IADP (D) SFMA

823. Linking farm plan with credit is a characteristic feature of:
(A) IRDP (B) IADP
(C) CDP (D) Etawah Project

824. CDP differs from IADP as the former have:
(A) Area based approach
(B) Decentralized administration
(C) Degree of investment
(D) Multi purpose approach

825. The extension approach used under IADP was:
(A) Proscriptive (B) Prescriptive
(C) Participatory (D) Commercial

826. The IAAP was launched in:
(A) 1962 (B) 1963
(C) 1964 (D) 1965

827. High Yielding Variety Programme was started in:
(A) 1962 (B) 1963
(C) 1964 (D) 1965

828. High yielding technologies are:
(A) Resource positive (B) Resource negative
(C) Resource neutral (D) None of these

829. Farm and Home unit of Akashvani was started in connection with:
(A) IAAP (B) HYVP
(C) IRDP (D) CDP

830. Radio Rural Forum was started in:
(A) 1963 (B) 1964
(C) 1965 (D) 1966

831. Broadcasting of rural programme in television started in:
(A) 1965 (B) 1966
(C) 1967 (D) 1968

832. Which of the following is not a key feature of T&V system:
(A) Professionalism
(B) Time bound work
(C) Single line of command
(D) Farming System Research

833. The base level extension functionary in T&V system:
(A) BDO (B) VLW
(C) AEO (D) SDAO

834. The first state in India to introduce Training and Visit system was:
(A) Punjab (B) Rajasthan
(C) Tamilnadu (D) Bihar

835. The ratio of extension agent to farmer in a typical T&V system is:
(A) 1·250 (B) 1·500
(C) 1·800 (D) 1·1500

836. T&V system of extension can be characterized as:
(A) Management system (B) Supervisory system
(C) Integrated system (D) Broad based system

837. Small Farmers Development Agency was set up in:
(A) 1969-70 (B) 1970-71
(C) 1971-72 (D) 1972-73

838. The basic approach followed by SFDA:
(A) Employment approach (B) Integrated approach
(C) Welfare approach (D) Target group approach

839. SFDA is meant for the farmers having land holding of:
(A) 1·2 hectare (B) 0·1 hectare
(C) 1·5 hectare (D) 2·5 hectare

840. SFDA-MFAL is associated with:
(A) Venkatpaiah committee
(B) Venkataraman committee
(C) Chelliah committee
(D) None of these

841. An agricultural labourer earns more than of his wage from agriculture:
(A) 33.33 per cent (B) 40.00 per cent
(C) 50.00 per cent (D) 66.66 per cent

842. A welfare programme for the rural poor:
(A) Krishi Vigyan Kendra (B) Lab to Land
(C) National Demostration (D) Minikit trials

843. The Ganga Kalyan Yojna is for:
(A) Consumption purpose of farmers
(B) Irrigation for small and marginal farmers
(C) Credit support scheme for rural women
(D) Credit support scheme for small farmers

844. Intensive Cattle Development Project was launched in:
(A) 1963-64 (B) 1964-65
(C) 1965-66 (D) 1966-67

845. The rural development programme launched on the basis of area development approach:
(A) DPAP (B) MWP
(C) RWP (D) IRDP

846. Desert Development Programme was launched on:
(A) 1977-78 (B) 1968-69
(C) 1971-72 (D) 1956-57

847. Tribal Area Development Programme was started in the year:
(A) 1970-71 (B) 1983-84
(C) 1984-85 (D) 1985-86

848. Integrated Wasteland Development Programme was launched on:
(A) 1989-90 (B) 1968-69
(C) 1971-72 (D) 1956-57

849. National Food Security Mission covers all except:
(A) Rice (B) Wheat
(C) Pulses (D) Sugarcane

850. Annual growth rate in agriculture sector expected during the eleventh plan:
(A) 8 per cent (B) 6 per cent
(C) 4 per cent (D) 3 per cent

851. In case of a mortised even repayment plan, the interest rate:
(A) Increases (B) Declines
(C) Remains constant (D) None of these

852. If liabilities>assets, then net worth is placed at:
(A) Liabilities side (B) Assets side
(C) Income side (D) None of these

853. Book value of durable refers to:
(A) Market price assets
(B) Purchase price of assets
(C) Production of assets
(D) Value of cost

854. Marginal value products of scarce resources is known bby:
(A) Linear programming (B) Production function
(C) Planning (D) Both (A) and (B)

855. Tangible benefit of project is:
(A) Income distribution (B) Standard of living
(C) National prosperity (D) Highei yield level

856. Shadow prices of resources are used in:
(A) Economic analysis (B) Financial analysis
(C) Econometric analysis (D) Technical analysis

857. Benefit cost ratio is used to evaluate mostly:
(A) Construction of dam
(B) Construction of building
(C) Construction of house
(D) Construction of cattleshed

858. Elasticity cannot be explained by:
(A) Regression (B) GLS
(C) OLS (D) Correlation

859. Marginal utility of money cannot be:
(A) Positive (B) Negative
(C) Zero (D) Both (A) and (B)

860. Indifference curves for perfect substitutes will be:
(A) Convex to origin (B) Concave to origin
(C) Perpendicular to origin (D) Straight line

861. If income consumption curve is bending towards Y axis than commodity X is:
(A) Inferior (B) Luxury
(C) Normal (D) None of these

862. For Giffen goods price consumption curve will be:
(A) Backward going (B) Horizontal
(C) Downward sloping (D) Upward sloping

863. Quantity purchased will vary inversely with price when income effect is for normal goods:
(A) Zero (B) Positive
(C) Negative (D) None of these

864. When with a change in price the total outlay on commodity remains constant, it is case of:
(A) Perfect elasticity (B) Unit elasticity
(C) Perfect inelasticity (D) Zero elasticity

865. Change in quantity demand refers to:
(A) Upward shift of demand (B) Downshift of demand
(C) Movement on same curve(D) None of these

866. For inelastic goods, fall in the price leads to:
(A) Increase in total elasticity
(B) No change in total elasticity
(C) Decrease in total elasticity
(D) None of these

867. Cross elasticity of demand between petrol and automobiles is:
(A) Negative (B) Zero
(C) High (D) Infinite

868. Marginal rate of product substitution on the isoclines will be:
(A) Positive (B) Zero
(C) Negative (D) Constant

869. Marginal cost = marginal returns = average cost = average returns shows the long run, equilibrium positive of the:
(A) Monopolist firm (B) Oligopolistic firm
(C) Competitive firm (D) None of these

870. The least cost principle is use in:
(A) What to produce (B) How much to produce
(C) How to produce (D) When to produce

871. When total produce means constant, the marginal product will be:
(A) Zero (B) Negative
(C) Positive (D) Constant

872. When average product < marginal product, average product is:
(A) Remains constant (B) At maximum
(C) Increasing (D) Decreasing

873. A production function will unitary elasticity indicates:
(A) Increasing returns (B) Decreasing returns
(C) Nothing (D) Constant returns

874. In the rational region of production function, choice of indicator is:
(A) Factor-factor price ratio
(B) Factor-product prices
(C) Product-product price ratio
(D) Factor-product quality

875. The relationship between total product and total cost is:
(A) Direct proportional (B) No relationship
(C) Inversely related (D) None of these

876. When marginal cost is at lowest, marginal product will be:
(A) Decreasing (B) Maximum
(C) Increasing (D) Minimum

877. In balanced zone:
(A) Marginal cost = Average cost
(B) Marginal cost > Average cost
(C) Marginal cost < Average cost
(D) None of these

878. If marginal rate of product substitution between two products is negative, then two goods are:
(A) Substituting (B) Complementary
(C) Supplementing (D) No relationship

879. In case of diminishing substitution the curve will be:
(A) Convex to origin (B) Straight line
(C) Concave to origin (D) Vertical or Horizontal

880. Optimum choice of enterprises or combination of enterprises is made based on:
(A) Principle of least cost
(B) Principle of variable proportions
(C) Principle of opportunity cost
(D) Principle of cost

881. In L.P., the availability and input-output coefficients as well as prices are known with certainty then assumption is:
(A) Linearity (B) Additively
(C) Finiteness (D) Single value expectations

882. In maximum returns interaction, the outgoing column will be located with the help of:
(A) Highest positive value
(B) Lowest positive value
(C) Lowest negative value
(D) Highest negative value

883. Marketable surplus is given by:
(A) Marketable surplus = Total production + Total requirement
(B) Marketable surplus = Total production + Total requirement/2
(C) Marketable surplus = Total production – Total requirement
(D) Marketable surplus = Total requirement – Total production

884. First Land mortgage bank in 1920 established at:
(A) Karnataka (B) Bihar
(C) West Bengal (D) Punjab

885. Capital formation in agriculture over the years:
(A) Decreasing (B) Increasing
(C) Both (A) and (B) (D) Constant

886. Private investment is to public investment is agriculture:
(A) Less than (B) More than
(C) Equal (D) Can't say

887. Nature of economic law is:
(A) Normative (B) Destructive
(C) Exactness (D) Positive

888. Law of substitution is given by:
(A) Marshall (B) Smith
(C) Gossen (D) Pigon

889. When marginal physical product is zero:
(A) Marginal cost is maximum
(B) Marginal cost become vertical
(C) Marginal cost is minimum
(D) Marginal cost become horizontal

890. When the demand curve is relatively flatter, then demand is:
(A) Relatively inelastic (B) Relatively elastic
(C) Unitary elastic (D) All of these

891. If the demand is elastic, for monopoly:
(A) Higher prices, more profit
(B) Lower prices, more profit
(C) Higher price, less profit
(D) Lower price, lower profit

892. Benefit cost ratio is used to evaluate:
(A) Social project (B) Public project
(C) Both (A) and (B) (D) None of these

893. Perfect elasticity is shown by:
(A) Vertical straight line
(B) Horizontal straight line
(C) Slopping curve from left to right
(D) Slopping curve from right to left

894. Production is a function of:
(A) Profit (B) Factors
(C) Price (D) Cost

895. Price theory is a part of:
(A) Micro economics (B) Macro economics
(C) Farm management (D) Marketing

896. The overhead cost is also known as:
(A) Variable cost (B) Total cost
(C) Marginal cost (D) Fixed cost

897. Livestock insurance is included in:
(A) Variable cost (B) Special cost
(C) Fixed cost (D) Marginal cost

898. Farm management is an:
(A) Intra farm science
(B) Inter farm science
(C) Inter regional farm science
(D) International farm science

899. Principle of equi marginal return is applied when:
(A) The resources are unlimited
(B) The price of resources is high
(C) The price of resources vary
(D) The resources are limited

900. Regulated market ensures:
(A) Procurement price (B) Support price
(C) Remunerative price (D) Fair price

901. Farming system approach ensures the best use of:
(A) Land
(B) Land + labour
(C) Land + labour + capital
(D) Land + labour + capital + management

902. IRDP was introduced in India on:
(A) 1977-78 (B) 1978-79
(C) 1979-80 (D) 1980-81

903. The basic approach followed in IRDP is:
(A) Individual based (B) Group based
(C) Community based (D) Family based

904. IRDP was conceived essentially as a:
(A) Adult education programme
(B) Anti-poverty programme
(C) Anti-boarding programme
(D) Fisheries programme

905. The basic unit of development under IRDP is:
(A) District (B) Village
(C) CD blocks (D) Family

906. Which of the following is not a characteristic of IRDP:
(A) Unidimensional in nature
(B) Adopt a holistic approach
(C) Calls for higher coordination
(D) Poverty alleviation programme

907. The total credit disbursed in agricultural sector comes around:
(A) 1800 billion (B) 2000 billion
(C) 2200 billion (D) 2300 billion

908. The share of cooperatives in the total credit disbursed in agricultural sector is:
(A) 70 per cent (B) 60 per cent
(C) 50 per cent (D) 40 per cent

909. Short term credit to farmers with a limit of 3 lakhs is available at an interest of:
(A) 7 per cent (B) 8 per cent
(C) 9 per cent (D) 10 per cent

910. The total number of kisan credit distributed in India:
(A) 642 lakhs (B) 556 lakhs
(C) 612 lakhs (D) 621 lakhs

911. Total number of bank linked SHG in India:
(A) 1.25 lakhs (B) 2.25 lakhs
(C) 3.25 lakhs (D) 4.25 lakhs

912. The National Agricultural Insurance Scheme was launched on:
(A) 1999-2000 Rabi (B) 1999-2000 Kharif
(C) 2000-01 Rabi (D) 2001-01 Kharif

913. The recently launched weather insurance scheme of government of India:
(A) Ganga Beema (B) Varsha Beema
(C) Dhan Phuhar Beema (D) Barsat Beema

914. The agricultural and Processed Food Products Export Development Authority (APEDA) came into existence in:
(A) 1986 (B) 1987
(C) 1988 (D) 1989

915. The GDP growth recorded in agriculture sector for the year 2006-07:
(A) 2.70 (B) 2.80
(C) 3.12 (D) 3.71

916. The special relief package for distress farmers recently announced does not cover the state of:
(A) Kerala (B) Tamilnadu
(C) Andhra Pradesh (D) Karnataka

917. The IRDP beneficiaries should include SC/ST at the rate of:
(A) 20 per cent (B) 25 per cent
(C) 30 per cent (D) 35per cent

918. The SFDA has been merged into:
(A) MFAL (B) SFDA
(C) SISI (D) CAPART

919. The agency constituted for implementing IRDP:
(A) DRDA (B) SFDA
(C) NREP (D) TRUSEM

920. DRDA expands to:
(A) District Rural Development Authority
(B) District Rural Development Agency
(C) District Rural Development Act
(D) District Rural Development Association

921. DRDA is located at:
(A) Block level (B) District level
(C) Tehsil level (D) Village level

922. The meetings of District Development Committee is presided by:
(A) BDO (B) DDO
(C) CDO (D) District Collector

923. Minimum need programme was launched during:
(A) Third five year plan (B) Fourth five year plan
(C) Fifth five year plan (D) Seventh five year plan

924. TRYSEM was introduced in the year:
(A) 1978 (B) 1979
(C) 1979 (D) 1980

925. Training under TRYSEM is mainly given for:
(A) Young scientists (B) Rural Youth
(C) Young students (D) Young women

926. DWCRA was implemented in:
(A) 1982 (B) 1979
(C) 1978 (D) 1977

927. The sill requirement for the members of DWCRA was met through:
(A) SFDA (B) IRDP
(C) NREP (D) TRYSEM

928. The programme aimed at creating income generating assets through a mix of bank credit and government subsidies:
(A) SGSY (B) NREP
(C) SITRA (D) JGSY

929. The Swarnajayanti Gram Swarozgar Yojna (SGSY) was launched on:
(A) April, 1999 (B) October, 1999
(C) November, 1999 (D) March, 1999

930. SGSY was formed by integrating all of the below programmes except:
(A) SITRA (B) PMRY
(C) MWS (D) GKY

931. The beneficiaries under SGSY is:
(A) Projapathis (B) Rural youth
(C) Swarozgaris (D) Farm family

932. The share of centre and state in the funding of SGSY is:
(A) 75:25 (B) 60:40
(C) 50:50 (D) 25:75

933. The minimum net monthly earning envisaged for a beneficiary of SGSY is:
(A) 1500 (B) 2000
(C) 2500 (D) 3000

934. The largest self employment programme for rural poor:
(A) SGSY (B) NREP
(C) SJSRY (D) JGSY

935. SGSY adopts:
(A) Cluster approach (B) Project approach
(C) Target group approach (D) All of these

936. The BPL families in a SHG formed under SGSY should be:
(A) 80 per cent (B) 83.33 per cent
(C) 90 per cent (D) 91.33 per cent

937. Which of the following is not a type of production function?
(A) Linear (B) Quadratic
(C) Square root (D) Rectangular

938. Maximum profit can be achieved in agri-business when marginal return (MR) is equal to:
(A) Marginal cost (B) Average output
(C) Total output (D) Total cost

939. Production function illustrates the relationship between:
(A) Input levels and corresponding output
(B) Input levels and corresponding marginal input
(C) Demand and supply pattern
(D) Demand and competition

940. The number of people living under the below poverty line in India:
(A) 9.56 crores (B) 9.52 crores
(C) 10.72 crores (D) 11.21 crores

941. The reservation for women as beneficiary under SGSY is:
(A) 50.00 per cent (B) 60.00 per cent
(C) 40.00 per cent (D) 33.33 per cent

942. The reservation for SC/ST as beneficiary under SGSY is:
(A) 50.00 per cent (B) 60.00 per cent
(C) 40.00 per cent (D) 33.33 per cent

943. The maximum subsidy under SGSY for individuals
(A) 7000 (B) 7500
(C) 8000 (D) 9000

944. The maximum subsidy under SGSY for SHG's:
(A) 1.00 lakh (B) 1.25 lakhs
(C) 1.50 lakhs (D) 2.00 lakhs

945. Bank credit under SGSY is available on group corpus up to:
(A) 3 times (B) 4 times
(C) 5 times (D) 6 times

946. Swarna Jayanti Shahari Rozgar Yojna (SGSRY) was launched in the country from:
(A) 1st December, 1997 (B) 2nd October, 1998
(C) 14th November, 1998 (D) 2nd October, 1999

947. The programme exclusively for urban poor:
(A) SJSRY (B) PMRY
(C) NSAP (D) SGSY

948. The programme developed to cover one crore poorest of the poor:
(A) SFDA (B) TRYSEM
(C) NAEP (D) AAY

949. Which of the following programme have clubbed together to form SJSYR:
(A) NRY (B) UBSP
(C) PMIUPEP (D) All of these

950. Antyadaya Anna Yojna was launched in:
(A) 1999 (B) 2000
(C) 2001 (D) 2002

951. The amount of rice and wheat together supplied per family in a month under AAY:
(A) 20 kg (B) 25 kg (C) 30 kg (D) 35 kg

952. The per kg rice and wheat under AAY is:
(A) Rs 2 and Rs 3 (B) Rs 3 and Rs 3
(C) Rs 2 and Rs 2 (D) Rs 3 and Rs 2

953. The number of children covered under the mid day meal scheme:
(A) 12 crores (B) 15 crores
(C) 20 crores (D) 21 crores

954. Mid Day Meal Scheme covers children studying at:
(A) Primary level (B) Secondary level
(C) High School level (D) Collegiate level

955. The Mid Day Meal Scheme was started in:
(A) 1995 (B) 2000
(C) 2005 (D) 2010

956. The nutrition content recommended in the mid Day Meal Scheme:
(A) 450 calories and 12 gram protein
(B) 500 calories and 30 gram protein
(C) 350 calories and 24 gram protein
(D) 300 calories and 12 gram protein

957. A scheme aimed at ensuring food security of senior citizens:
(A) SJSRY (B) PMRY
(C) NSAP (D) Annapurma Scheme

958. The Annapurna Scheme was started in:
(A) 1998 (B) 1999
(C) 2000 (D) 2001

959. The amount of food grains supplied to senior citizens not covered under National Old Age Pension Scheme under Annapurna Scheme:
(A) 5 kg/month (B) 10 kg/month
(C) 15 kg/month (D) 25 kg/month

960. The on-going National Social Assistance Programme (NSAP) provides benefits under:
(A) National Old Age Pension Scheme
(B) National Family Benefit Scheme
(C) National Maternity Benefit Scheme
(D) All of these

961. Day of elderly was observe on:
(A) December 1 (B) October 1
(C) September 1 (D) May 1

962. Life expectancy at birth in India according to the 2001-2005 estimates:
(A) 60 years (B) 62 years
(C) 63 years (D) 65 years

963. The scheme to provide safeguard against starvation during the period of natural calamity or during lean season:
(A) NREP (B) Village Gramin Bank
(C) IRDP (D) RLEGP

964. Which of the following is not a poverty removal programme:
(A) NREP (B) HYVP
(C) IRDP (D) RLEGP

965. Food for work was started in:
(A) 1977 (B) 1978 (C) 1979 (D) 1980

966. The food for work programme has been restructured as:
(A) IRDP (B) MFAL
(C) NREP (D) DPAP

967. NREP was started in:
(A) 1979
(B) 1980
(C) 1981
(D) 1982

968. NREP was launched to check:
(A) Unemployment
(B) Population
(C) Brain drain
(D) Drought

969. The programmes designed to create community assets on the one hand and providing wage employment on the other hand:
(A) NREP
(B) IRDP
(C) JRY
(D) DWCRA

970. Which of the following has been started to supplement NREP:
(A) DPAP
(B) RLEGP
(C) IRDP
(D) NATP

971. The National Rural Employment Programme dependent on:
(A) Rural employment
(B) Self employment
(C) Wage employment
(D) Revolving employment

972. The programme emphasizing on employment generation among the following is:
(A) CADP
(B) JRY
(C) KADP
(D) CDP

973. JRY started in:
(A) 1988 (B) 1989 (C) 1990 (D) 1991

974. JRY is combination of following programmes:
(A) FFW and NAEP
(B) PIREP and RLEGP
(C) SFDA and MFAL
(D) NREP and RLEGP

975. Jawahar Rozgar Yojana was started by:
(A) Jawaharlal Nehru
(B) Rajiv Gandhi
(C) Indira Gandhi
(D) Sanjay Gandhi

976. The employment Assurance Scheme was launched on:
(A) 1992
(B) 1993
(C) 1994
(D) 1995

977. Jawahar Gram Samridhi Yojana (JGSY) was launched in:
(A) April, 1999
(B) October, 1999
(C) November, 1999
(D) March, 1999

978. Jawahar Gram Samridhi Yojana was formed modifying:
(A) IRDP
(B) PMRY
(C) MWS
(D) JRY

979. The programme aimed at creation of demand driven village infrastructure including durable assets at the village level to enable the rural poor to increase the opportunities for sustained employment:
(A) SGSY
(B) PMRY
(C) JGSY
(D) DWACRA

980. Jawahar Gram Samridhi Yojana is implemented at the level of:
(A) District panchayat
(B) Intermediate panchayat
(C) Village panchayat
(D) Nagar panchayat

981. The most recently started programme for providing employment to rural people is:
(A) JGSY
(B) NREP
(C) JRY
(D) SGRY

982. JGSY and EAS were merged to form:
(A) Sampoorna Grameen Rozgar Yojana
(B) National Social Assistance Programme
(C) Pradhan Mantri Gram Sadak Yojana
(D) SJSRY

983. Sampoorna Gramen Rozgar Yojana (SGRY) was started on:
(A) 1999
(B) 2000
(C) 2001
(D) 2002

984. A scheme exclusively implemented by the Panchayat Raj Institutions:
(A) SGRY
(B) NSAP
(C) PMGSY
(D) SJSRY

985. Under which scheme center supply food grains to state government free of cost every year:
(A) SGRY
(B) NSAP
(C) PMGSY
(D) SJSRY

986. The National Rural Employment Guarantee Act was passed in the parliament on:
(A) 2002
(B) 2003
(C) 2004
(D) 2005

987. The minimum wage prescribed for unskilled labours in India:
(A) 45
(B) 50
(C) 55
(D) 60

988. The minimum number of days of employment a year guaranteed under the NREG Act:
(A) 90 days
(B) 100 days
(C) 120 days
(D) 150 days

989. Jobs ubder NREG Act has to provided to registered job card holders within:
(A) 5 days
(B) 15 days
(C) 20 days
(D) 30 days

990. Prime Minister's Rozgar Yojana (PMRY) was started in:
(A) 1993
(B) 1994
(C) 1995
(D) 1996

991. The age group to get subsidy under PMRY:
(A) 18-35
(B) 20-40
(C) 15-35
(D) 25-50

992. The maximum amount of subsidy eligible per entrepreneur under PMRY:
(A) 10000
(B) 12500
(C) 15000
(D) 20000

993. The maximum income limit to get subsidy under PMRY:
(A) 75000
(B) 80000
(C) 90000
(D) 100000

994. The maximum finance one can get under PMRY for starting joint venture projects:
(A) 1 lakh
(B) 2 lakhs
(C) 5 lakhs
(D) 10 lakhs

995. The minimum level of education to make eligibility under PMRY:
(A) IV pass (B) VIII pass
(C) X pass (D) XII pass

996. Which of the following schemes does not come under Nehru Rozgar Yojana:
(A) TRYSEM (B) SUME
(C) SHASU (D) SUKE

997. Pradhan Mantry Gram Sadak Yojana was launched during:
(A) December, 2000 (B) August, 2002
(C) October, 2001 (D) August, 2003

998. The sharing of cost between center and state in the Pradhan Mantri Gram Sadak Yojana is:
(A) 100:0 (B) 75:25
(C) 60:40 (D) 50:50

999. The Pradhan Mantri Gramodaya Yojana is a programme related to:
(A) Rural sanitation (B) Rural water supply
(C) Rural housing (D) Primary education

1000. The programme launched to meet the housing needs of the poor:
(A) IAY (B) JRY
(C) MWS (D) NREP

1001. The programme launched to meet the housing needs of the poor in Uttar Pradesh:
(A) IAY (B) JRY
(C) MWS (D) KAY

1002. IAY was launched as a sub scheme of:
(A) RLEGP (B) JRY
(C) MWS (D) NREP

1003. IAY was de-linked from JRY and become an independent scheme by:
(A) 1994 (B) 1995
(C) 1996 (D) 1997

1004. The ceiling on construction assistance under the IAY is:
(A) Rs. 15000/- per unit (B) Rs. 25000/- per unit
(C) Rs. 35000/- per unit (D) Rs. 45000/- per unit

1005. Which of the following is the function of JRY:
(A) Land development and social forestry
(B) Soil and water conservation
(C) Flood protection and drainage
(D) All of these

1006. The primary objective of JRY is:
(A) Creation of employment
(B) Creation of social assets
(C) Assistaing rural people
(D) All of these

1007. A time bound plan for rural infrastructure development by the government of India:
(A) Bharat Nirman (B) Bharat Parivartan
(C) Naya Bharat (D) Hamara Bharat

1008. Prime Minister's Rozgar Yojana strategy is embedded with:
(A) Self employment (B) Wage distribution
(C) Credit disbursement (D) Capital creation

1009. The concept of Bharat Nirman covers all of the following areas except:
(A) Rural connectivity (B) Rural housing
(C) Rural water supply (D) Primary education

1010. The number of un-electrified villages taken for electrification under Bharat Nirman:
(A) 0.50 lakhs (B) 1.25 lakhs
(C) 2.25 lakhs (D) 3.75 lakhs

1011. The additional houses to be constructed for rural poor under Bharat Nirman:
(A) 60 lakhs (B) 62 lakhs
(C) 70 lakhs (D) 75 lakhs

1012. The number of habitats not covered with pure drinking water in India:
(A) 55000 (B) 75000
(C) 100000 (D) 125000

1013. All weather roads under Bharat Nirman connects habitats with population above:
(A) 5000 (B) 2000
(C) 1000 (D) 500

1014. The targets aimed under Bharat Nirman is to be achieved by:
(A) 2008 (B) 2009
(C) 2010 (D) 2011

1015. The decadal growth in population for the period 1991-2001 is:
(A) 13.86 per cent (B) 21.34 per cent
(C) 29.86 per cent (D) 38.54 per cent

1016. The land availability per person as per 1991 census is:
(A) 0.42 hectares (B) 0.61 hectares
(C) 0.64 hectares (D) 0.75 hectares

1017. The density of population of India as per 2001 census:
(A) 297 per km^2 (B) 324 per km^2
(C) 366 per km^2 (D) 378 per km^2

1018. The proportion of our population in the working age group of 15-64 years:
(A) 62.9 per cent (B) 67 per cent
(C) 60 per cent (D) 55 per cent

1019. Accelerated Rural Water Supply Programme was launched in:
(A) 1972-73 (B) 1973-74
(C) 1974-75 (D) 1975-76

1020. National Drinking Water Mission was launched in:
(A) 1985 (B) 1986
(C) 1987 (D) 1988

1021. Rajiv Gandhi National Drinking Water Mission was launched on:
(A) 1991 (B) 1992
(C) 1993 (D) 1994

1022. The recently launched Community-based rural drinking water supply programme:
(A) Swajaldhara (B) Jaldhara
(C) Neerdhara (D) Samritdhara

1023. Farmers functional literacy programme was started in the year:
(A) 1968 (B) 1958 (C) 1978 (D) 1948

1024. National adult education programme was launched in:
(A) 1968 (B) 1978 (C) 1988 (D) 1984

1025. National Literacy Mission was launched in the year:
(A) 1990 (B) 1991
(C) 1992 (D) 1993

1026. National Literacy Mission aims at imparting functional literacy to non-literates in the age-group of:
(A) 5-15 (B) 10-25
(C) 15-35 (D) 35-50

1027. Operation black board aims to improve:
(A) Primary education
(B) Adult education
(C) Informal education
(D) Professional education

1028. The constitutional amendment made to ensure free and compulsory education to all children:
(A) 86th Amendment (B) 74th Amendment
(C) 73rd Amendment (D) 72nd Amendment

1029. Free and compulsory education to all children in the age group of 6-14 years is envisaged in:
(A) Sarva Shiksha Abhiyan
(B) Kasturba Gandhi Balika Vidyalaya
(C) National Literacy Mission
(D) National Adult Education Programme

1030. All children complete eight years of elementary schooling under Sarva Shiksha Abhiyan by:
(A) 2008 (B) 2009 (C) 2010 (D) 2011

1031. Sarva Siksha Abhiyan was started in:
(A) 1994 (B) 1995 (C) 1996 (D) 1997

1032. National Service Scheme was started in:
(A) 1967 (B) 1968 (C) 1969 (D) 1970

1033. Youth is classified to belong under the age group of:
(A) 5-15 (B) 10-25
(C) 15-35 (D) 35-50

1034. Nehru Yuva Kendras was started in:
(A) 1962 (B) 1972 (C) 1982 (D) 1992

1035. Nehru Yuva Kendra was started for:
(A) Working youth (B) Non student youth
(C) Student youth (D) School dropout

1036. Kasturba Gandhi Balika Vidyalaya are meant for promoting education of:
(A) Girl children (B) Senior citizen
(C) Rural youth (D) Agricultural labourers

1037. The organization under Department of Science and Technology established with the main goal of providing institutional support in protecting grassroots green innovations:
(A) MANAGE
(B) NISTAT
(C) National Innovation Foundation
(D) National Science foundation

1038. "Files to field programme" has been introduced in:
(A) Lucknow (B) Ahmedabad
(C) Ahmednagar (D) New Delhi

1039. National Rural Health Mission was launched in:
(A) 2003 (B) 2004
(C) 2005 (D) 2006

1040. The comprehensive housing scheme launched recently with a view to ensuring integrated provision of shelter, sanitation and drinking water:
(A) RLEGP (B) JRY
(C) IAY (D) Samagra Awas Yojana

1041. The centrally sponsored Rural Sanitation Programme was launched in:
(A) 1985 (B) 1986
(C) 1987 (D) 1988

1042. The most important agency responsible for distribution of agricultural credit is:
(A) Cooperative banks (B) Commercial banks
(C) Money lenders (D) NABARD

1043. "Each for all and all for each" forms the basic principle of:
(A) Cooperative banks (B) ICAR
(C) Planning Commission (D) NABARD

1044. National Agricultural Cooperative Marketing Federation of India Ltd has its headquarters at:
(A) Chennai (B) Lucknow
(C) New Delhi (D) Hyderabad

1045. The first cooperative society was started in:
(A) 1904 (B) 1905
(C) 1906 (D) 1907

1046. The minimum members required to register a co-operative society:
(A) 10 (B) 15 (C) 20 (D) 25

1047. Nationalization of commercial banks happened in:
(A) 1967 (B) 1968
(C) 1969 (D) 1970

1048. Lead bank scheme was introduced in:
(A) 1968 (B) 1969
(C) 1970 (D) 1971

1049. Principle of equi-marginal return is also known as:
(A) Product-product relationship
(B) Output-output relationship
(C) Substitution between output/products
(D) All of these

1050 Law of diminishing marginal utility states that:
(A) Utility varies with time and space
(B) A commodity generates different utilities for different people and for the same person at different times
(C) Utility of additional unit of a commodity to any consumer commodity to already consumption increases
(D) All of these

1051. Principle of least –cost combination is also known as:
(A) Substitution between inputs
(B) Factor-factor relationship
(C) Rate of technical substitution
(D) All of these

1052. Welfare definition of economics was given by:
(A) Adam Smith (B) Alfred Marshall
(C) Rebbins (D) Keynes

1053. Which one of the following is not correct about Isoquant curve?
(A) It is used in factor-factor relationship
(B) They are not intersecting
(C) Concave to origin
(D) Placed above the another represent higher input

1054. "Profit maximization" is the objective of which farming system?
(A) Peasant farming (B) Capitalistic farming
(C) State farming (D) Cooperative farming

1055. Which of the following is not a type of farming?
(A) Specialized farming (B) Collective farming
(C) Diversified farming (D) Ranching farming

1056. Contribution of Central Government in the capital share of regional rural bank is:
(A) 50% (B) 35%
(C) 15% (D) 25%

1057. A farmer having 2.0 hectare of land will be grouped under:
(A) Small (B) Medium
(C) Semi-medium (D) Large

1058. Total number of principles of farm management is:
(A) 4 (B) 7
(C) 5 (D) 8

1059. National Agriculture Insurance Scheme started in:
(A) 1998 (B) 1999
(C) 1996 (D) 1997

1060. Levy price applies to:
(A) Farmers (B) Consumers
(C) Middle men (D) Traders

1061. National Institute of Agricultural Marketing is located at:
(A) Delhi (B) Mumbai
(C) Bhopal (D) Jaipur

1062. In India, mechanization index in highest for which crop?
(A) Paddy (B) Wheat
(C) Maize (D) Sugarcane

1063. Collective farming system has originated from:
(A) Poland (B) France
(C) Bulgaria (D) Soviet Union

1064. A market guided by rules and regulations is called:
(A) Seasonal market (B) Regulated market
(C) Terminal market (D) Perfect market

ANSWERSHEET: 16.AGRICULTURAL ECONOMICS

1. (A); 2. (D); 3. (F); 4. (A); 5. (B); 6. (C); 7. (D); 8. (C); 9. (A); 10. (B); 11. (D); 12. (C); 13. (C) ; 14. (D); 15. (D); 16. (D); 17. (A); 18. (C); 19. (A); 20. (C); 21. (D); 22. (D); 23. (D); 24. (C); 25. (A) ;26. (C); 27. (C); 28. (C); 29. (B); 30. (A); 31. (B); 32. (C); 33. (A); 34. (A); 35. (B); 36. (B); 37. (D); 38. (B) ;39. (B); 40. (B); 41. (B); 42. (A); 43. (A); 44. (A); 45. (A); 46. (D); 47. (A); 48. (B); 49. (A); 50. (B); 51. (A) ;52. (C); 53. (B); 54. (B); 55. (C); 56. (A); 57. (B); 58. (D); 59. (D); 60. (C); 61. (B); 62. (A); 63. (A); 64. (C) ; 65. (A); 66. (A); 67. (D); 68. (D); 69. (B); 70. (C); 71. (B); 72. (A); 73. (D); 74. (B); 75. (A); 76. (B); 77. (C); 78. (D); 79. (B); 80. (B); 81. (C); 82. (D); 83. (D); 84. (C); 85. (B); 86. (C); 87. (D); 88. (C); 89. (A); 90. (B); 91. (B) ;92. (A); 93. (B); 94. (B); 95. (C); 96. (A); 97. (B); 98. (D); 99. (C); 100. (C); 101. (B); 102. (A); 103. (C); 104. (D) ; 105. (A); 106. (B); 107. (A); 108. (A); 109. (B); 110. (A); 111. (D); 112. (B); 113. (B); 114. (A); 115. (A); 116. (B); 117. (A) ; 118. (A); 119. (C); 120. (C); 121. (B); 122. (B); 123. (C); 124. (D); 125. (C); 126. (C); 127. (B); 128. (A); 129. (B); 130. (A) ; 131. (B); 132. (B); 133. (C); 134. (C); 135. (B); 136. (D); 137. (C); 138. (A); 139. (C); 140. (B); 141. (B); 142. (B); 143. (A) ; 144. (D); 145. (B); 146. (A); 147. (B); 148. (D); 149. (B); 150. (B); 151. (D); 152. (B); 153. (C); 154. (A); 155. (A); 156. (D) ; 157. (D); 158. (A); 159. (A); 160. (A); 161. (D); 162. (D); 163. (B); 164. (B); 165. (C); 166. (A); 167. (B); 168. (D); 169. (A) ; 170. (B); 171. (D); 172. (A); 173. (C); 174. (A); 175. (C); 176. (B); 177. (B); 178. (A); 179. (C); 180. (B); 181. (C); 182. (D); 183. (B); 184. (A); 185. (D); 186. (B); 187. (A); 188. (B); 189. (D); 190. (A); 191. (C); 192. (C); 193. (A); 194. (C); 195. (A) ; 196. (B); 197. (A); 198. (D); 199. (D); 200. (C); 201. (B); 202. (D); 203. (C); 204. (B); 205. (C); 206. (D); 207. (A); 208. (C); 209. (A); 210. (D); 211. (A); 212. (C); 213. (D); 214. (B); 215. (A); 216. (C); 217. (B); 218. (B); 219. (B); 220. (A); 221. (C) ; 222. (A); 223. (B); 224. (A); 225. (A); 226. (A); 227. (C); 228. (C); 229. (A); 230. (D); 231. (D); 232. (C); 233. (A); 234. (C) ; 235. (A); 236. (C); 237. (D); 238. (A); 239. (D); 240. (A); 241. (D); 242. (A); 243. (B);

244. (A); 245. (D); 246. (C); 247. (B) ; 248. (B); 249. (B);
250. (C); 251. (B); 252. (A); 253. (B); 254. (B); 255. (A);
256. (C); 257. (C); 258. (A); 259. (A); 260. (D); 261. (A);
262. (B); 263. (C); 264. (D); 265. (A); 266. (B) ; 267. (A);
268. (A); 269. (C); 270. (D); 271. (D); 272. (B); 273. (B) ;
274. (A); 275. (D); 276. (A); 277. (B); 278. (A); 279. (B);
280. (D); 281. (A); 282. (A); 283. (A); 284. (B); 285. (D);
286. (B) ; 287. (A); 288. (C); 289. (A); 290. (B) ; 291. (A);
292. (B); 293. (C); 294. (A); 295. (A); 296. (A); 297. (B);
298. (A); 299. (A) ; 300. (B); 301. (C); 302. (D); 303. (A);
304. (D); 305. (D); 306. (A); 307. (D); 308. (A); 309. (D);
310. (A); 311. (D); 312. (D) ; 313. (D); 314. (D); 315. (A);
316. (D); 317. (A); 318. (D); 319. (A); 320. (A); 321. (B);
322. (A); 323. (B); 324. (D); 325. (A); 326. (C); 327. (C);
328. (D); 329. (D); 330. (A); 331. (B); 332. (A); 333. (A);
334. (A); 335. (A); 336. (A); 337. (C); 338. (A); 339. (D);
340. (A); 341. (D); 342. (C); 343. (A); 344. (D); 345. (D);
346. (C); 347. (A); 348. (C); 349. (D); 350. (A); 351. (A) ;
352. (D); 353. (A); 354. (A); 355. (D); 356. (B); 357. (A);
358. (D); 359. (D); 360. (D); 361. (A); 362. (A); 363. (B);
364. (A) ; 365. (B); 366. (D); 367. (D); 368. (A); 369. (A);
370. (D); 371. (A); 372. (A); 373. (A); 374. (B); 375. (B);
376. (C); 377. (B) ; 378. (A); 379. (B); 380. (B); 381. (A);
382. (B); 383. (D); 384. (A); 385. (A); 386. (A); 387. (B);
388. (D); 389. (A); 390. (D) ; 391. (B); 392. (D); 393. (D);
394. (A); 395. (A); 396. (D); 397. (A); 398. (B); 399. (A);
400. (C); 401. (A); 402. (C); 403. (D) ; 404. (A); 405. (D);
406. (D); 407. (D); 408. (D); 409. (D); 410. (A); 411. (C);
412. (D); 413. (C); 414. (D); 415. (C); 416. (D); 417. (C);
418. (D); 419. (A); 420. (D); 421. (A); 422. (A); 423. (D);
424. (D); 425. (D); 426. (A); 427. (D); 428. (D); 429. (D) ;
430. (A); 431. (A); 432. (B); 433. (A); 434. (D); 435. (D);
436. (B); 437. (A); 438. (D); 439. (D); 440. (B); 441. (D);
442. (D) ; 443. (A); 444. (C); 445. (B); 446. (A); 447. (D);
448. (A); 449. (A); 450. (C); 451. (A); 452. (A); 453. (C);
454. (D); 455. (D); 456. (A) ; 457. (C); 458. (A); 459. (D);
460. (D); 461. (A); 462. (C); 463. (C); 464. (A); 465. (A);
466. (B); 467. (B); 468. (A); 469. (A) ; 470. (D); 471. (A);
472. (C); 473. (D); 474. (D); 475. (D); 476. (B); 477. (D);
478. (A); 479. (D); 480. (B); 481. (B); 482. (B) ; 483. (A);
484. (A); 485. (A); 486. (D); 487. (C); 488. (A); 489. (D);
490. (A); 491. (A); 492. (D); 493. (A); 494. (D); 495. (D) ;
496. (D); 497. (D); 498. (B); 499. (B); 500. (D); 501. (D);
502. (D); 503. (D); 504. (C); 505. (A); 506. (C); 507. (C);
508. (C); 509. (D); 510. (D); 511. (B); 512. (A); 513. (A);
514. (A); 515. (A); 516. (A); 517. (B); 518. (B); 519. (B);
520. (A); 521. (B); 522. (A); 523. (B); 524. (D); 525. (A);
526. (D); 527. (B); 528. (A); 529. (D); 530. (C); 531. (B);
532. (B); 533. (B); 534. (C) ; 535. (C); 536. (B); 537. (D);
538. (A); 539. (B); 540. (C); 541. (D); 542. (A); 543. (C);
544. (D); 545. (A); 546. (B); 547. (B) ; 548. (C); 549. (C);
550. (C); 551. (C); 552. (C); 553. (D); 554. (B); 555. (B);
556. (D); 557. (A); 558. (A); 559. (C); 560. (A) ; 561. (B);
562. (A); 563. (C); 564. (D); 565. (C); 566. (A); 567. (A);
568. (A); 569. (B); 570. (B); 571. (A); 572. (C); 573. (D) ;
574. (B); 575. (A); 576. (D); 577. (C); 578. (A); 579. (B);
580. (D); 581. (D); 582. (B); 583. (A); 584. (B); 585. (A);
586. (A); 587. (C); 588. (A); 589. (B); 590. (D); 591. (B);
592. (A); 593. (A); 594. (A); 595. (C); 596. (B); 597. (A);
598. (A); 599. (B); 600. (B); 601. (A); 602. (D); 603. (B);
604. (A); 605. (D); 606. (A); 607. (B); 608. (D); 609. (C);
610. (C); 611. (A); 612. (C); 613. (B); 614. (D); 615. (B);
616. (B); 617. (A); 618. (D); 619. (A); 620. (D); 621. (A);
622. (A); 623. (C); 624. (B); 625. (A) ; 626. (B); 627. (C);
628. (A); 629. (C); 630. (B); 631. (A); 632. (D); 633. (C);
634. (A); 635. (A); 636. (B); 637. (C); 638. (C) ; 639. (B);
640. (D); 641. (C); 642. (A); 643. (B); 644. (A); 645. (C);
646. (A); 647. (A); 648. (B); 649. (A); 650. (B); 651. (B);
652. (A); 653. (B); 654. (B); 655. (B); 656. (A); 657. (A);
658. (B); 659. (B); 660. (C); 661. (D); 662. (B); 663. (D);
664. (C) ; 665. (B); 666. (A); 667. (A); 668. (D); 669. (A);
670. (B); 671. (A); 672. (C); 673. (A); 674. (D); 675. (C);
676. (D); 677. (D); 678. (C); 679. (A); 680. (A); 681. (B);
682. (A); 683. (C); 684. (D); 685. (B); 686. (A); 687. (B);
688. (A); 689. (D); 690. (B) ; 691. (B); 692. (A); 693. (C);
694. (A); 695. (A); 696. (B); 697. (A); 698. (B); 699. (D);
700. (B); 701. (A); 702. (B); 703. (D) ; 704. (D); 705. (D);
706. (B); 707. (C); 708. (B); 709. (B); 710. (A); 711. (B);
712. (A); 713. (D); 714. (C); 715. (D); 716. (C) ; 717. (B);
718. (C); 719. (A); 720. (B); 721. (B); 722. (C); 723. (C);
724. (A); 725. (D); 726. (D); 727. (B); 728. (D); 729. (D) ;
730. (A); 731. (B); 732. (B); 733. (A); 734. (B); 735. (D);
736. (B); 737. (D); 738. (B); 739. (B); 740. (A); 741. (A);
742. (C); 743. (B); 744. (B); 745. (B); 746. (C); 747. (B);

748. (B); 749. (A); 750. (C); 751. (B); 752. (C); 753. (A); 754. (D); 755. (A) ; 756. (A); 757. (C); 758. (B); 759. (C); 760. (B); 761. (A); 762. (A); 763. (C); 764. (A); 765. (A); 766. (D); 767. (D); 768. (D); 769. (B); 770. (C); 771. (C); 772. (B); 773. (B); 774. (A); 775. (A); 776. (B); 777. (D); 778. (D); 779. (B); 780. (A); 781. (C) ; 782. (D); 783. (C); 784. (D); 785. (C); 786. (B); 787. (C); 788. (B); 789. (B); 790. (B); 791. (B); 792. (C); 793. (A); 794. (B) ; 795. (D); 796. (B); 797. (D); 798. (C); 799. (B); 800. (C); 801. (C); 802. (B); 803. (A); 804. (C); 805. (C); 806. (A); 807. (B) ; 808. (A); 809. (D); 810. (D); 811. (B); 812. (C); 813. (A); 814. (B); 815. (A); 816. (A); 817. (D); 818. (B); 819. (B); 820. (B) ; 821. (A); 822. (C); 823. (B); 824. (D); 825. (B); 826. (C); 827. (C); 828. (A); 829. (B); 830. (D); 831. (C); 832. (D); 833. (B); 834. (B); 835. (C); 836. (A); 837. (B); 838. (D); 839. (A); 840. (A); 841. (C); 842. (B); 843. (B); 844. (B); 845. (A); 846. (A); 847. (A) ; 848. (A); 849. (D); 850. (C); 851. (C); 852. (B); 853. (D); 854. (D); 854A); 868. (D); 869. (C); 870. (C); 871. (A); 872. (C); 873. (D); 874. (B); 875. (C); 876. (B); 877. (B); 878. (A); 879. (A); 880. (C); 881. (D); 882. (D); 883. (C); 884. (D); 885. (A); 886. (B) ; 887. (D); 888. (A); 889. (D); 890. (B); 891. (B); 892. (C); 893. (B); 894. (B); 895. (A) ; 896. (D); 897. (C); 898. (A); 899. (D); 900. (D); 901. (D); 902. (A); 903. (D); 904. (B); 905. (C); 906. (A); 907. (A); 908. (A); 909. (A); 910. (A); 911. (B); 912. (A); 913. (B); 914. (A); 915. (A); 916. (B); 917. (C); 918. (B); 919. (A); 920. (B); 921. (B); 922. (D); 923. (C); 924. (B); 925. (B); 926. (A); 927. (D); 928. (A); 929. (A); 930. (B); 931. (C); 932. (A); 933. (B); 934. (A); 935. (D); 936. (A); 937. (D) ; 938. (A); 939. (A); 940. (A); 941. (C); 942. (A); 943. (B); 944. (B); 945. (B); 946. (A); 947. (A); 948. (D); 949. (D); 950. (B) ; 951. (D); 952. (D); 953. (A); 954. (A); 955. (A); 956. (D); 957. (D); 958. (C); 959. (B); 960. (D); 961. (B); 962. (C); 963. (B) ; 964. (B) ; 965. (A); 966. (C); 967. (B); 968. (B); 969. (A); 970. (B); 971. (C); 972. (B); 973. (B); 974. (D); 975. (B); 976. (B); 977. (A); 978. (D); 979. (C); 980. (C); 981. (D); 982. (A); 983. (C); 984. (A); 985. (A); 986. (D); 987. (A); 988. (B); 989. (B); 990. (A); 991. (A) ; 992. (A); 993. (D); 994. (D); 995. (B); 996. (A); 997. (A); 998. (A); 999. (C); 1000. (A); 1001. (D); 1002. (A); 1003. (A) ; 1004. (B); 1005. (A); 1006. (D); 1007. (A); 1008. (A); 1009. (D) ; 1010. (B); 1011. (A); 1012. (A); 1013. (C); 1014. (B); 1015. (B); 1016. (A) ; 1017. (B); 1018. (A); 1019. (A); 1020. (B); 1021. (A); 1022. (A); 1023. (A); 1024. (C); 1025. (C) ; 1026. (A) ; 1027. (A); 1028. (A); 1029. (A); 1030. (C); 1031. (A); 1032. (C); 1033. (C); 1034. (B); 1035. (B); 1036. (A); 1037. (C); 1038. (C); 1039. (C); 1040. (D); 1041. (B); 1042. (A); 1043. (A); 1044. (C); 1045. (A); 1046. (A); 1047. (C); 1048. (B); 1049. (D); 1050. (C); 1051. (D). 1052. (B); 1053. (C); 1054. (B); 1055. (B); 1056. (A). 1057. (A); 1058. (B); 1059. (B); 1060. (D); 1061. (D); 1062. (B). 1063. (D); 1064. (B).

Chapter 17

ANIMAL HUSBANDRY AND DAIRYING

1. The branch of agriculture which concerned with the care and breeding of domestic animals such as cattle, goats, sheep, hogs, and horses are called:
(A) Plant Husbandry (B) Animal Husbandry
(C) Dairying (D) None of these

2. The first domesticated livestock animal, which was tamed around 9000 B.C. in North Iraq:
(A) Sheep (B) Goat
(C) Cattle (D) Pig

3. Around 6500 B.C., which domestic animal were kept in the same region:
(A) Sheep (B) Goat
(C) Cattle (D) Pig

4. About 6000 B.C, which domestic animal was domesticated in Iraq:
(A) Sheep (B) Goat
(C) Cattle (D) Pig

5. By 5900 B.C. (and perhaps 3,000 years earlier), which animal was domesticated in Chad:
(A) Sheep (B) Goat
(C) Cattle (D) Pig

6. While independently about 5500 B.C., which animal was domesticated in South-west Iran:
(A) Sheep (B) Goat
(C) Cattle (D) Pig

7. Around 3500 B.C., which animal was domesticated on the Eurasian steppes:
(A) Horse (B) Goat
(C) Cattle (D) Pig

8. How many cattle breeds have been recognized in India:
(A) 7 (B) 17 (C) 26 (D) 30

9. How many buffalo breeds have been recognized in India:
(A) 7 (B) 17 (C) 26 (D) 30

10. In India, at present 94.3 per cent of the milch cows yield
(A) Less than 1 kg milk per day
(B) More than 2 kg milk per day
(C) No milk producer
(D) None of these

11. In India only 0.5 per cent of the milch cows yield:
(A) Less than 1 kg milk per day
(B) More than 2 kg milk per day
(C) No milk producer
(D) None of these

12. Which domesticated animal are better milk producers:
(A) Buffaloes (B) Cow
(C) Goat (D) Sheep

13. About 18-20 per cent of the milch buffaloes yield:
(A) Less than 1 kg milk per day
(B) More than 2 kg milk per day
(C) No milk producer (D) None of these

14. About 19 per cent of the milch buffaloes yield :
(A) Less than 1 kg milk per day
(B) More than 2 kg milk per day
(C) No milk producer
(D) None of these

15. Mating system in which the males and females mated to beget progeny are more closely related than the average of the population from which they come is called:
(A) Inbreeding (B) Cross breeding
(C) Out breeding (D) Line breeding

16. Which type of breeding makes more pairs of genes in the population homozygous.
(A) Inbreeding (B) Cross breeding
(C) Out breeding (D) Line breeding

17. Which type of breeding can be used to farm distinct lines of families within a breed:
(A) Line breeding (B) Cross breeding
(C) Out breeding (D) Inbreeding

18. In laboratory animals, highly inbred lines are very useful in many experiments and tests:
(A) Inbreeding (B) Cross breeding
(C) Out breeding (D) Line breeding

19. The objective is not to increase homozygosity but to maintain a high relationship to an outstanding ancestor with inbreeding of :
(A) Inbreeding (B) Cross breeding
(C) Out breeding (D) Line breeding

20. This is a breeding system in which individuals, less related than the average of the population to which they belong, are mated is called:
(A) Inbreeding (B) Cross breeding
(C) Out breeding (D) Line breeding

21. Which type of breeding makes more pairs of genes in the population homozygous:
(A) Inbreeding (B) Out breeding
(C) Cross breeding (D) Line breeding

22. Mating of unrelated animals within the same pure breed is termed as:
(A) Inbreeding (B) Cross breeding
(C) Out breeding (D) Line breeding

23. Mating of animals of two or more different breeds is known as:
(A) Inbreeding (B) Cross breeding
(C) Out breeding (D) Line breeding

24. Which breeding method is mainly used for commercial production:
(A) Inbreeding (B) Cross breeding
(C) Out breeding (D) Line breeding

25. Crossing of individuals from two species is referred to as:
(A) Species hybridization (B) Breeding management
(C) Animal management (D) None of these

26. Which management practices are often the root cause of lowered fertility (as distinct from sterility) in livestock is called:
(A) Species hybridization
(B) Breeding management
(C) Animal management
(D) None of these

27. Earliest animal to the domesticated by primitive men was:
(A) Goat (B) Dog
(C) Horse (D) Cat

28. The most used domesticated animal by Eskimo is:
(A) Cow (B) Sheep
(C) Goat (D) Husky

29. Huskies are:
(A) Yaks (B) Thick coated dog
(C) Donkeys (D) Water buffaloes

30. Best source for dietary protein for a vegetarian is:
(A) Soybean (B) Gram
(C) Groundnut (D) Milk

31. Zebu cattle is:
(A) Water Buffalo (B) Indian Buffalo
(C) Cow (D) Sheep

32. Water buffalo is:
(A) European breed of buffalo that prefers living in water for most of the day
(B) Buffalo-like animal living in rivers
(C) Llama
(D) Buffalo

33. Exotic breeds is:
(A) Require specific environment
(B) Hardy
(C) Sturdy
(D) Take less food

34. Exotic breeds are:
(A) Used for cross breeding
(B) Allowed to multiply and replace local breeds
(C) Cheaper
(D) Resistant to local pests and pathogens

35. Number of breeds of Zebu cattle is:
(A) 16 (B) 26 (C) 6 (D) 13

36. Most economic meat making animal is:
(A) Yak (B) Pig
(C) Goat (D) Sheep

37. Polo game played with the help of:
(A) Donkey (B) Horse
(C) Camel (D) Cock

38. Bhutia is a breed of:
(A) Chicken (B) Goat
(C) Sheep (D) Horse

39. Ship of desert is:
(A) Elephant (B) Camel
(C) Sheep (D) Goat

40. The animal most useful on difficult terrains is:
(A) Mule (B) Yak (C) Camel (D) Elephant

41. Bacterian camel is characterized by:
(A) Two humps and long neck
(B) Two humps and long limbs
(C) Two humps and thick coat
(D) Single hump and thick coat

42. A breast of burden which needs little care is:
(A) Pig (B) Donkey
(C) Mule (D) Yak

43. The largest land animal is:
(A) Camel (B) Elephant
(C) Rhino (D) Python

44. Elephants are slought after for:
(A) Skin (B) Hair
(C) Meat (D) Ivory

45. Silk industry is related to:
(A) Sericulture (B) Apiculture
(C) Pisciculture (D) Horticulture

46. A queen honey bee lays eggs of:
(A) One type from which all castes develop
(B) Two types, one forming queen and workers and second forming drones
(C) Three types forming queen, drone and workers
(D) Unfertilised die while fertilized ones form all castes

47. Lac is a product of:
(A) Faecal matter
(B) Secretion from body
(C) Excretion from body
(D) Excess food oozing out of the body

48. Silk is produced by:
(A) Larva (B) Adult moth
(C) Both (A) and (B) (D) Cocoon/Pupa

49. Which among the following is real product of Honey Bee:
(A) Honey (B) Pollen
(C) Bee wax (D) Propolis

50. Super ovulation and embryo transplantation are meant for improving:
(A) Human race (B) Livestock
(C) Poultry (D) Plants

51. Worker bees are:
(A) Fertile males (B) Fertile females
(C) Sterile females (D) Sterile males

52. The young chicken raised specifically for meat are called:
(A) Broilers (B) Cockerels
(C) Pellets (D) Hen

53. Honey is collected by:
(A) Wasp (B) Housefly
(C) Butterfly (D) Bee

54. Ranikhet disease is connected with:
(A) Honey Bee (B) Hens
(C) Fishes (D) Pigs

55. Induced breeding is carried out in case of:
(A) Pisciculture (B) Apiculture
(C) Sericulture (D) Lac culture

56. Silk is the product of:
(A) Cuticle of larva (B) Cuticle of adult
(C) Salivary gland of larva (D) Salivary gland of adult

57. Rearing of bees is:
(A) Horticulture (B) Apiary
(C) Apiculture (D) Poultry

58. Which is a product of insect body:
(A) Honey, wax and silk
(B) Silk, lac and wax
(C) Royal jelly, wax and lac
(D) All of these

59. Lac is produced by:
(A) Males
(B) Females
(C) More by females than males
(D) More by males than females

60. Silkworm larva spins silk from:
(A) Inside to outside
(B) Outside to inside
(C) Random fashion
(D) Anterior to posterior side

61. Three carp fishes, *Catla labeo* and *Cirrhina*, can be grown together in the same pond more economically as they have:
(A) Positive interactions
(B) Commensalism
(C) Symbiosis
(D) No competition for food

62. Fish introduced in India by foreigners is:
(A) *Labeo rohita* (B) *Mysnis singhala*
(C) *Pomphret* (D) *Clasius batrachits*

63. Best aquarium is located at:
(A) Vishakhapatnam (B) Chennai
(C) Tarapur, Mumbai (D) Z.S.I., Calcutta

64. To increase milk yield, cow is given:
(A) Sorbitol (B) Stilbesterol
(C) Prolactin (D) Gonadotropin

65. In silkworm, if juvenile hormone is absent at the time of larval moulting, the worm will:
(A) Die (B) Moult into larval stage
(C) Moult into pupa (D) Moult into adult

66. Name of sheep cloned for the first time is:
(A) Dolly (B) Polly
(C) Molly (D) Holly

67. Nagpuri buffalo is:
(A) Milker (B) Draught cattle
(C) Dual purpose (D) Grazer

68. Which amongst the following is used in raising super milk cows:
(A) Artificial insemination with pedigree bull
(B) Embryo transplantation
(C) Superovulation of high yielding cow
(D) All of these

69. Inland fisheries are:
(A) Deep sea fishes
(B) Capturing fishes from sea coast
(C) Raising and capturing fishes in fresh water
(D) Oil extraction from fish

70. Silkworm is a:
(A) Beetle (B) Worm
(C) Fly (D) Moth

71. Which of the following is not a plant product:
(A) Hemp (B) Silk
(C) Cotton (D) Flax

72. Pisciculture is rearing and production of:
(A) Fishes (B) Birds
(C) Reptiles (D) Wool yielding animals

73. Which is not a plant fibre:
(A) Coir (B) Silk
(C) Sunnhemp (D) Manila Hemp

74. Hormone used for inducing lactation in sterile cows is:
(A) Esterogen (B) Progesterone
(C) Relaxin (D) Stilbestrol

75. The bulls of which bred of buffalo are extensively used for upgrading inferior stock:
(A) Ravi (B) Bhadawari
(C) Murrah (D) Jaffarabadi

76. There are about ——— breeds of cattle in India besides hybrids:
(A) 14 (B) 20 (C) 26 (D) 32

77. There are —— breeds of buffaloes in India, in addition to a large number of non-descripts:
(A) 4 (B) 6 (C) 10 (D) 12

78. An examination of the rainfall map of India will how that it is invariably in the higher elevations, and in humid areas that the cattle are:
(A) Large and non-descript
(B) Large and well-descript
(C) Small and well descript
(D) Small and non-descript

79. ——— is a bred of cattle:
(A) Bhadawari (B) Tharparkar
(C) Kashmiri (D) Landrace

80. ——— is a breed of buffalo:
(A) Berkshire (B) Beetal
(C) Jaffarabadi (D) Sindhi

81. ——— is a breed of goat:
(A) Berbari (B) Large White Yorkshire
(C) Sahiwal (D) Nili

82. ——— is a breed of swine:
(A) Malwi (B) Ravi
(C) Gaddi (D) Chester white

83. Hereford is a breed of:
(A) Cattle (B) Goat
(C) Swine (D) Buffalo

84. Durock is a breed of:
(A) Swine (B) Cattle
(C) Buffalo (D) Goat

85. Tamworth and middle White Yorkshire are breeds of:
(A) Buffalo (B) Goat
(C) Cattle (D) Swine

86. Jamanapari and Barabari are breeds of:
(A) Cattle (B) Buffalo
(C) Swine (D) Goat

87. Surti and Beetal are breeds of:
(A) Swine (B) Goat
(C) Cattle (D) Buffalo

88. The discovery of living cells would have been difficult, if most impossible, before the compound microscope was invented by Zachaias Jansen of——— in 1590:
(A) England (B) Poland
(C) France (D) Holland

89. ——— of England applied the term "cell" to the cavities he saw in sections of cork:
(A) Robert Hooke (B) William Winter
(C) Ralph Turner (D) Kleiner Baird

90. Marcello Malpighi in ——— published "Anatomy of Plants", the first systematic study of cell structure:
(A) 1575 (B) 1675
(C) 1775 (D) 1875

91. In 1839, ———, an animal anatomist formulated the cell theory which set forth the concept that "elementary parts of all tissues are formed of cells though highly diversified in manner":
(A) Robert Hooke (B) Mercello Malphighi
(C) Jheodor Schwann (D) Alred White

92. The nucleus contains——— per cent of the RNA in the cell:
(A) 4-10 (B) 6-12
(C) 7-17 (D) 8-20

93. The ribosome is composed of two sub-units and has an approximate size of :
(A) 200 Å (B) 210 Å
(C) 200 Å (D) 195 Å

94. Jersey calves suffering from ——— have broad skull, large eye-sockets, short and broad nasal bones and forehead broader than normal:
(A) Aguatha
(B) Cerebral hernia
(C) Umblical hernia
(D) Bulldog head (prognathism)

95. Sex linkage has been indicated for more than 200 traits in:
(A) Cattle (B) Sheep
(C) Horse (D) Humans

96. The sac of the scrotium in bulk is thermoregulatory in function, that is, it provides protection to the tests and maintains them at a temperature lower by ——— ^{0}F than the abdomen which is necessary for active spermatogenesis:
(A) 2-4 (B) 5-7
(C) 8-10 (D) 11-13

97. In the case of bulls the seminal vesicles are about ——— cm long 5 cm wide:
(A) 13.5-16.0 (B) 10.0-12.5
(C) 14.0-15.5 (D) 13.0-14.5

98. In each ovary (of farm animals) there may be as many ——— follicles:
(A) 25000 (B) 50000
(C) 75000 (D) 100000

99. In case of———, the ovaries are enclosed by membranous covering which is continuous with the wall of the tube to ensure that the discharged ova passes into the tubes and is not lost into the body:
(A) Dog (B) Cow
(C) Goat (D) Rabbit

100. In case of cow, the vagina is about ——— cm long:
(A) 7.5-10.0 (B) 12.5-15.0
(C) 16.5-19.0 (D) 20.0-25.0

101. The process of spermatogenesis in the male and oogenesis in the female finally gives rise to ——— daughter cells:
(A) 2 (B) 4 (C) 6 (D) 8

102. The egg after the formation of fertilization membrane decreases in diameter by——— per cent due to the secretion of some fluid:
(A) 4 (B) 8 (C) 11 (D) 14

103. The parathyroid glands were discovered by sandstorm in:
(A) 1840 (B) 1860
(C) 1880 (D) 1900

104. ——— compounds have been isolated from the adrenal cortex which are all sterol derivatives:
(A) Twenty-eight (B) Twenty-six
(C) Twenty-four (D) Twenty-three

105. The critical hormones control electrolytes as well as water metabolism which helps in retention of——— in the body:
(A) Iodine (B) Sulphur
(C) Sodium (D) Phosphorus

106. The ovary can be examined upto ——— months of pregnancy after which it sinks deeper into abdominal cavity:
(A) One (B) Two
(C) Three (D) Four

107. Dystokia, among the domesticate species is mostly met with in :
(A) Dogs (B) Cats
(C) Monkeys (D) Dairy cattle

108. The amount of ——— needed by an adult cow for efficient reproduction is about 10-12 g per day:
(A) Potassium (B) Phosphorus
(C) Calcium (D) Sodium

109. The calcium to phosphorus ratio for the maximum reproduction efficiency in cattle should between:
(A) 1:2 and 2:1 (B) 1:1 and 3:1
(C) 1:3 and 3:1 (D) 1:4 and 4:1

110. Vitamins ——— are synthesized by the ruminants in their body:
(A) A and D (B) B and K
(C) E and D (D) C and D

111. ——— per cent of animals contracting bang's diseases become either temporarily or permanently sterile:
(A) 10-15 (B) 15-20
(C) 18-23 (D) 25-30

112. The milk exudes from ——— separate gland tubes that open at the base of the "mammary hairs":
(A) 50-95 (B) 100-150
(C) 40-80 (D) 155-195

113. The udder usually increases about ——— in size between milking, depending upon the quality of the udder and the amount of milk secreted:
(A) One-fourth (B) Half
(C) One-third (D) Two-thirds

114. In Holstein cow, between two months and the time of puberty (7-10 months), the mammary gland grows three and half times faster than the body weight and reaches the peak pubertal development at ——— months of age:
(A) 6-8 (B) 10-12
(C) 11-13 (D) 14-16

115. It is estimated that ——— volumes of blood flow through the mammary gland for each volume of milk synthesized:
(A) 600 (B) 500 (C) 400 (D) 300

116. An acceptable record is 70 per cent non-returns upto 60 days after breeding or 60 per cent non-returns upto ——— days after breeding:
(A) 70 (B) 80 (C) 90 (D) 100

117. A fertile herd takes less than ——— months between calving:
(A) 13 (B) 14 (C) 15 (D) 16

118. A cow during her productive life must remain pregnant only for ——— months in every year:
(A) 6 (B) 7 (C) 8 (D) 9

119. In unipara as the cow, mare etc., usually single ovum is released, in sheep one or two and occasionally three, whereas in swine several ova upto ——— are released in a single ovulation:
(A) 15 (B) 20 (C) 25 (D) 30

120. According to legend artificial insemination had its origin in ——— at which time and Arab Chieftain used it to impregnate a prized mare with semen stealthily collected by night from the sheath of an enemy's stallion:
(A) 1310 (B) 1315 (C) 1319 (D) 1322

121. The first scientific research on artificial insemination of domestic animals was conducted with ——— by the Italian psychologist, Lazarro Spallanzani in 1780:
(A) Dogs (B) Sheep
(C) Goats (D) Cows

122. The discovery of spermatozoa dates back to ——— when the compound microscope was invented by Leewenhock, the Dutch Naturalist:
(A) 1627 (B) 1677 (C) 1697 (D) 1707

123. Mikhail Fyalorovich Ivanoy, a Russian Physiologist commenced his investigation on artificial insemination in farm animals as early as :
(A) 1879 (B) 1889 (C) 1897 (D) 1899

124. In Denmark, Edward Sorensen and Jens Gylling Ham organized the first Cooperative Artificial Breeding Association in:
(A) 1916 (B) 1926
(C) 1936 (D) 1946

125. Dr. Sampath Kumaron employed artificial insemination method for the first time in India in ——— at the Palace Dairy Farm, Mysore:
(A) August, 1939 (B) September, 1939
(C) August, 1938 (D) September, 1938

126. "Broad helix" of "Spiral body" or mitochondrial sheath is ——— rich structure believed to be derived from mitochondria:
(A) Carbohydratte (B) Lipid
(C) Protein (D) Minerals

127. The human prostatic secretion contains fibrinolysin that is so powerful that 2 cc of the same can liquidify 100 cc of clotted blood in ——— hours at 37°C:
(A) 18 (B) 16 (C) 14 (D) 12

128. The prostatic secretion represents the main source of citric acid and acid phosphate for whom ———semen:
(A) Stallion (B) Bull
(C) Human (D) Dog

129. The identification of the seminal sugars as ——— by Mann (1956) opened up the way for detailed studies on its generation capacity of the accessory tissues:
(A) Glucose (B) Fructose
(C) Sucrose (D Ribulose

130. In species viz., ——— semen after ejaculation remains liquid unlike in others wherein it tends to gelate or coagulate shortly after ejaculation:
(A) Horse and Ass (B) Sheep and Goat
(C) Bull and Dog (D) Cat and Camel

131. Haemacytometer contains in the centre 16 big squares divided into ——— small squares:
(A) 128 (B) 256 (C) 384 (D) 512

132. The percentage of abnormal semen in a normal bull should not exceed:
(A) 21-26 (B) 5-12 (C) 15-20 (D) 26-30

133. Roughages are feeds embodying more than ——— per cent of crude fibre:
(A) 24 (B) 22 (C) 20 (D) 18

134. In practice, the food value of ether extract has been taken as yielding ——— k cal of gross energy per gramme of the product while carbohydrates and proteins yield about five K cal per gramme:
(A) 6.35 (B) 7.35 (C) 8.35 (D) 9.35

135. The term protein (from word Protrus) was suggested in 1840 by:
(A) Mulder (B) Shudder
(C) Coleure (D) Waternail

136. There may be about ——— different enzymes in a single cell:
(A) 250 (B) 500 (C) 750 (D) 1000

137. Wool is rich in cystine and contains ——— per cent of sulphur:
(A) 1.5 (B) 2.0 (C) 3.0 (D) 4.0

138. About ——— per cent of the plasma copper is firmly bound to an alpha-2 globulin as ceruloplasmin, the form in which it is transported in the body:
(A) 96 (B) 94 (C) 92 (D) 90

139. As early as ———, a Swiss physician, J. Francois Coindet, first recommended iodine as a remedy for goiter:
(A) 1800 (B) 1820 (C) 1840 (D) 1860

140. By late —— synthetic vitamin A became available on the commercial market:
(A) 1925 (B) 1930 (C) 1935 (D) 1940

141. Vitamin K was identified by ——— in 1935 as a factor present in green leaves:
(A) Dam (B) Dixon
(C) Dukes (D) Diggins

142. Jansen and Donath in ——— succeeded in crystaling Vitamin B_1 in pure form:
(A) 1906 (B) 1916 (C) 1926 (D) 1936

143. The intestine represents a long tube approximately ——— feet long in average dairy cattle:
(A) 180 (B) 160 (C) 200 (D) 220

144. Rumination is a long process and occupies about —— hours out of 24 hours:
(A) 6 (B) 7 (C) 8 (D) 9

145. The large intestine of ——— is the shortest and simplest of all the domestic animals:
(A) Cat (B) Dog (C) Cow (D) Horse

146. The rate of gas production in the rumen is most rapid immediately after meal and in the cow may exceed ——— per hour:
(A) 15 (B) 20 (C) 25 (D) 30

147. It has been estimated that the human stomach wall is replete with ———glands:
(A) 100000000 (B) 150000000
(C) 300000000 (D) 350000000

148. In bovine — per cent of the total rumen bacteria have a proteolytic activity:
(A) 32 (B) 34 (C) 36 (D) 38

149. In tail to tail system of barn arrangement ——— per cent of time is spent behind the cows:
(A) 40 (B) 50 (C) 60 (D) 70

150. The central walk (of a cow stall) should have a width of 5-6 feet exclusive of gutters when cows face out, and ——— feet when they face in:
(A) 2-3 (B) 3-4 (C) 4-5 (D) 1.5-2

151. The manure gutter should have a gradient of one inch for every —— feet length:
(A) 15 (B) 14 (C) 12 (D) 10

152. A calving box should have an area of about ——— square feet with ample of soft bedding:
(A) 40-50 (B) 75-95
(C) 60-75 (D) 45-55

153. An area of 100 square feet per head for an young stock of ——— calves and an increase of 50 square feet for every additional calf makes a good paddock:
(A) 12 (B) 10 (C) 8 (D) 6

154. A system of grazing known as Henbenhein system was developed in:
(A) Holland (B) Poland
(C) England (D) Germany

155. Bone meal upto ——— q/ha is supposed to be the best fertilizer for pastures:
(A) 7-8 (B) 5-6 (C) 3-4 (D) 2-3

156. The digestibility of colostrums increases when it is given at a temperature between:
(A) 89-92 °F (B) 93-98 °F
(C) 97-98 °F (D) 99-102 °F

157. In electrical dehorning the rod embodies an automatic control to maintain the temperature at about ——— °F and its application to horn button for 10 seconds destroys the horn cells:
(A) 1000 (B) 900 (C) 800 (D) 700

158. The best time for castration of calves, lambs and pigs is between ———, 2 and 1 weeks, respectively:
(A) 2-4 (B) 5-7 (C) 8-10 (D) 11-13

159. On attainment of maturity, between first and second year a bull can serve ——— cows:
(A) 18 (B) 20 (C) 23 (D) 25

160. Ordinarily, lambs are not used for servicing until ——— months of age:
(A) 20 (B) 18 (C) 16 (D) 14

161. Boars reach sexual maturity between :
(A) 60-90 days (B) 100-180 days
(C) 190-200 days (D) 210-220 days

162. Cow come in heat regularly at intervals of 16-24 days and heat lasts :
(A) 8-12 hours (B) 13-15 hours
(C) 16-20 hours (D) 21-24 hours

163. In most of the cows ovulation occurs ——— hours after the end of heat:
(A) 5-9 (B) 10-14 (C) 15-17 (D) 18-20

164. Ewes remain in heat for about ——— hours and ovulate shortly before the end of heat:
(A) 20 (B) 25 (C) 30 (D) 33

165. A lose of 10 per cent of water causes restlessness, trembling and weakness, and loss of ——— per cent of water results in death through thirst:
(A) 14-16 (B) 17-19
(C) 20-22 (D) 23-25

166. Dairy cows and buffaloes require ———— gallons of water per diem and an additional amount of about one kilogramme need be allowed for each pound (nearly half a kilogramme) of milk produced:
(A) 4-5 (B) 5-6 (C) 7-8 (D) 9-10

167. Barned variety of Plymoth Rock was developed in about ——— from the cross of the Dominique and Black Cochin breeds:
(A) 1835 (B) 1845 (C) 1855 (D) 1865

168. The poultry breed New Hampshire was developed both for meat and egg production in New Hampshire from single comb Rhode Island Red and Red Stock during:
(A) 1910 (B) 1920 (C) 1930 (D) 1940

169. The poultry breed New Hampshire developed in —— — about 200 years ago, primarily as a table bird:
(A) Poland (B) Holland
(C) New Zealand (D) England

170. The commercial egg farms in India maintain Austro White, a cross between:
(A) Australorp male x White Leghorn female
(B) Australorp male x leg horn female
(C) Australorp male x New Hampshire female
(D) Australorp mole x Wyandotte female

171. Artificial hatching was practiced some ——— years ago in China and Egypt:
(A) 2500 (B) 3000
(C) 3500 (D) 4000

172. In 1770, an Englishman named ———, hatched chicks by passing hot air through the room where the eggs were located:
(A) John Watchman (B) John William
(C) John Champion (D) John Laggard

173. Bonneman, a French physician, in ———— hatched chicks by heating a hatching oven with circulated hot water:
(A) 1770 (B) 1777 (C) 1779 (D) 1787

174. The most convenient folding unit to house poultry is the one made for ——— hens:
(A) 15 (B) 20 (C) 25 (D) 30

175. It takes about 2.5 hours for feed to go from mouth to cloaca in the laying hen, and —— hours:
(A) 4 (B) 6 (C) 8 (D) 10

176. By water-glass and lime-water methods eggs can be preserved for about ———— months
(A) 2 (B) 4 (C) 6 (D) 8

177. Dipping of eggs in hot water at a temperature of 135-145 ^{0}F for about ———minutes kill the germ of the egg:
(A) 10 (B) 15 (C) 20 (D) 25

178. Provision of ——— hours of light artificial or natural maximizes egg production:
(A) 10 (B) 12 (C) 14 (D) 16

179. Exposure of hens to brilliant light say —— watt –bulb flashed on for 15 seconds effectively stimulates winter egg production:
(A) 750 (B) 4000 (C) 1250 (D) 1500

180. In goats oestrus is usually 24-48 hours, but in certain breeds like Indian Beetal may be as short as ——— hours on an average:
(A) 18 (B) 16 (C) 14 (D) 12

181. Breeding of goats at the age from ——— months yields good results:
(A) 8-12 (B) 14-18 (C) 10-13 (D) 19-22

182. The average life of a goat is usually ——— years and should give milk for about ten lactations:
(A) 8 (B) 10 (C) 12 (D) 14

183. A well fed buck keeps fit for breeding till ——— years of age:
(A) 5-7 (B) 8-10 (C) 11-13 (D) 14-16

184. The ——— is the most economical animal to keep:
(A) Cow (B) Buffalo
(C) Swine (D) Goat

185. ———— are excellent feed convertors:
(A) Goats (B) Sheep
(C) Pigs (D) Hens

186. One kilogramme of work can be produced on———— kilogrammes of feed:
(A) 1.5-2.0 (B) 2.5-3.5
(C) 4.0-4.5 (D) 5.0-5.5

187. The test single measures of the ability of a sow to produce is the weight of her litter at ——— day, the usual weaning time:
(A) 60 (B 58 (C) 56 (D) 54

188. The diameter of mature follicles is about :
(A) 4-5 mm (B) 7-8 mm
(C) 9-10 mm (D) 11-12 mm

189. The diameter of sow-corpora lutea is about 15-18 cm, while the body of the uterus is about ——— cm long:
(A) 3 (B) 4 (C) 5 (D) 6

190. Most of the gilts reach puberty between ——— months:
(A) 4-5 (B) 6-8 (C) 9-10 (D) 11-12

191. The mammary gland in sow are ——— number, are arranged along the belly in two rows:
(A) 14-16 (B) 10-12 (C) 6-8 (D) 17-19

192. In sow fertilization occurs only ——— hours after mating:
(A) 2-5 (B) 6-10 (C) 11-13 (D) 14-15

193. A value of ———— per cent of calcium in the diet is probably adequate in both male and female breeding swine:
(A) 0.45 (B) 0.55 (C) 0.65 (D) 0.75

194. The role of zinc in preventing skin lesions (parakeratosis in pigs) was first reported by Tucker and Salmon in:
(A) 1935 (B) 1945 (C) 1955 (D) 1965

195. The onset of puberty in a female (swine) normally occurs at around —— days of age and 90 kg of live weight:
(A) 160 (B) 170 (C) 180 (D) 190

196. Most of the growth of the unborn young pig is during last ——— days of pregnancy:
(A) 20 (B) 30 (C) 40 (D) 50

197. The energy value of swine milk is ——— MJ/R and total solid contents of 20 g/100g of fresh milk:
(A) 2.5 (B) 3.5 (C) 4.5 (D) 5.5

198. Pig milk contains ——— per cent fat, 5 per cent milk sugar and 6 per cent protein:
(A) 10 (B) 8 (C) 6 (D) 5

199. The lactation pig can produce ——— g of protein in her milk every day:
(A) 50-100 (B) 110-220
(C) 350-375 (D) 400-500

200. The quantitative production of pig response to a change in feed dose lies at the sore of a computer model developed at the university of ——— in 1970:
(A) Michigan (B) Kansas
(C) Edinburgh (D) California

201. It has been estimated that half a dozen average-size pigs will throw off the energy at the rate of about 1 kwh which should be actively dispersed to maintain normal body temperature at:
(A) 39 °C (B) 38 °C (C) 37 °C (D) 40 °C

202. Gestation period in pregnant ewes last:
(A) 135 days (B) 140 day
(C) 147 days (D) 170 days

203. The number of lambs born per ewe, on an average rise until the age of ——— years:
(A) 6 (B) 5 (C) 4 (D) 3

204. For most of the farm animals the ratio is of the order of ——— parts by weight of water to every one part of dry matter eaten:
(A) 1-2 (B) 3-4 (C) 5-6 (D) 7-8

205. ——— poisoning is common in areas of USA where it is known as alkali disease or blind staggers:
(A) Cobalt (B) Chlorine
(C) Magnesium (D) Selenium

206. The finest woolled sheep of the Duwn breeds and the South down (of Britain), has a count of ——— microns (1000000 microns = 1 meter):
(A) 27-28 (B) 25-26
(C) 23-24 (D) 29-30

207. Wool yields vary from about 1.3 kg for a Welsh ewe to 2.6 kg for a Suffolk and ——— kg for an English Leicester:
(A) 3 (B) 4 (C) 5 (D) 6

208. Match the somatic chromosome number with the animal:

Somatic chromosome number	Animal species
(a) 78	(i) Sheep
(b) 54	(ii) Cattle
(c) 60	(iii) Human
(d) 46	(iv) Hens

Answer:
(A) (a) i (b) ii (c) iii (d) iv
(B) (a) i (b) iii (c) ii (d) iv
(C) (a)iv (b) i (c) ii (d) iii
(D) (a) iv (b) ii (c) iii (d) i

209. Match the sex chromosome number with the animals embodying it:

Chromosomes number	Animal
(a) 23 pairs	(i) Cattle
(b) 30 pairs	(ii) Sheep
(c) 27 pairs	(iii) Wheat
(d) 21 pairs	(iv) Human

Answer:
(A) (a) i (b) ii (c) iii (d) iv
(B) (a) i (b) iii (c) ii (d) iv
(C) (a) iv (b) i (c) ii (d) iii
(D) (a) iv (b) ii (c) iii (d) i

210. Match the heat period with the animal:

Heat period	Animal
(a) 12-24 hours	(i) Ewe
(b) 1-2 days	(ii) Mare
(c) 4-5 days	(iii) Sow
(d) 2-3 days	(iv) Cow

Answer:
(A) (a) i (b) ii (c) iii (d) iv
(B) (a) iv (b) i (c) ii (d) iii
(C) (a) i (b) iii (c) ii (d) iv
(D) (a) iv (b) ii (c) iii (d) i

211. Match the term used in pasteurization with the animal:

Pasteurization	Animal
(a) Foaling	(i) Ewe
(b) Lambing	(ii) Sow
(c) Farrowing	(iii) Bitch
(d) Whelping	(iv) Mare

Answer:
(A) (a) i (b) ii (c) iii (d) iv
(B) (a) iv (b) i (c) ii (d) iii
(C) (a) i (b) iii (c) ii (d) iv
(D) (a) iv (b) ii (c) iii (d) i

212. Match the disease with the symptom(s):

Disease	Symptoms
(a) Salpingitis	(i) Uterus inflammation
(b) Piosalpingitis	(ii) Pus accumulation in uterus
(c) Metritis	(iii) Fallopian tube-in flammation
(d) Pyometra	(iv) Pus-filled fallopian tube

Answer:
(A) (a) i (b) ii (c) iii (d) iv
(B) (a) iv (b) i (c) ii (d) iii
(C) (a) i (b) iii (c) ii (d) iv
(D) (a) iii (b) iv (c) i (d) ii

213. Match the percentage with the component as estimated in colostrums:

Value (per cent)	Colostrum
(a) 14.0	(i) Sugar (lactose)
(b) 2.7	(ii) Fat
(c) 6.7	(iii) Total solids
(d) 23.9	(iv) Protein

Answer:
(A) (a) iv (b) i (c) ii (d) iii
(B) (a) i (b) ii (c) iii (d) iv
(C) (a) i (b) iii (c) ii (d) iv
(D) (a) iv (b) ii (c) iii (d) i

214. Match the per cent value with the component as estimated in Indian cow milk:

Per cent value — Component
(a) 13.52 — (i) Fat
(b) 2.78 — (ii) Total solids
(c) 4.56 — (iii) Lactose
(d) 3.67 — (iv) Protein

Answer:
(A) (a) iv (b) i (c) ii (d) iii
(B) (a) i (b) ii (c) iii (d) iv
(C) (a) i (b) iii (c) ii (d) iv
(D) (a) ii (b) iv (c) iii (d) i

215. Match the reproductive rate (s) with the animal(s):

Reproductive rate — Animal
(a) One offspring/year — (i) Sheep and goat
(b) One to three progeny/year — (ii) Swine
(c) Ten progeny twice/year — (iii) Poultry
(d) 100 or more progeny/year — (iv) Cattle, buffalo,horse

Answer:
(A) (a) iv (b) i (c) ii (d) iii
(B) (a) i (b) ii (c) iii (d) iv
(C) (a) i (b) iii (c) ii (d) iv
(D) (a) iv (b) ii (c) iii (d) i

216. Match the animals breed with the parents:

Animal breed — Parents
(a) Hisar dale sheep — (i) Lincoln rams x Rambouillet ewes
(b) Columbia — (ii) Tamworth x landrace
(c) Panama — (iii) Bikaneri owes x Merino rams
(d) Minnesoda No 1 swine — (iv) Rambouilet rams x Lincoln ewes

Answer:
(A) (a) iv (b) i (c) ii (d) iii
(B) (a) iii (b) i (c) iv (d) ii
(C) (a) i (b) iii (c) ii (d) iv
(D) (a) iv (b) ii (c) iii (d) i

217. Match the protein content (per cent) with the concentrate (feed):

Feeds — Protein content
(a) Wheat bran — (i) 20.0
(b) Rice bran — (ii) 25.0
(c) Chickpea chuni — (iii) 12.0
(d) Linseed cake — (iv) 46.0
(e) Groundnut cake — (v) 10-15

Answer:
(A) (a) iv (b) i (c) ii (d) iii (e) v
(B) (a) v (b) iii (c) i (d) ii (e) iv
(C) (a) i (b) iii (c) ii (d) iv (e) v
(D) (a) iv (b) ii (c) v (d) i (e) ii

218. Match the feeding stuff with digestibility coefficient:

Feeding stuff — Digestibility coefficient
(a) Egyptian clover — (i) 61
(b) Guinea grass — (ii) 45
(c) Maize (feed) — (iii) 81
(d) Mixed grass hay — (iv) 0
(e) Rice/wheat straw — (v) 58

Answer:
(A) (a) iv (b) i (c) ii (d) iii (e) v
(B) (a) iii (b) v (c) i (d) ii (e) iv
(C) (a) i (b) v (c) ii (d) iv (e) iii
(D) (a) v (b) ii (c) iii (d) I (e) iv

219. Match the feeding stuff with digestibility coefficient:

Constituent (grain/seed) — Digestibility coefficient
(a) Barley and Oats — (i) 68
(b) Maize — (ii) 78
(c) Chickpea — (iii) 70
(d) Cotton seed — (iv) 74
(e) Cowpea hay — (v) 66

Answer:
(A) (a) iv (b) i (c) ii (d) iii (e) v
(B) (a) ii (b) iv (c) v (d) iii (e) i
(C) (a) i (b) v (c) ii (d) iv (e) iii
(D) (a) v (b) ii (c) iii (d) I (e) iv

220. Match the feeding stuff with digestibility coefficient:

Constituent — Digestibility coefficient
(a) Mustard cake — (i) 92
(b) Rape cake — (ii) 84
(c) Sesame cake — (iii) 90
(d) Groundnut cake — (iv) 55
(e) Maize silage — (v) 85

Answer:
(A) (a) iv (b) i (c) ii (d) iii (e) v
(B) (a) ii (b) v (c) i (d) iii (e) iv
(C) (a) i (b) v (c) ii (d) iv (e) iii
(D) (a) v (b) ii (c) iii (d) I (e) iv

221. Match the feeding stuff with crude protein content:

Feeding stuff — Crude protein content
(a) Fish meal — (i) 14.0
(b) Rice bran — (ii) 17.4
(c) Wheat bran — (iii) 51.8
(d) Berseem (feed) — (iv) 70.1
(e) Groundnut cake — (v) 11.4

Answer:
(A) (a) iv (b) i (c) ii (d) iii (e) v
(B) (a) iv (b) i (c) v (d) ii (e) iii
(C) (a) i (b) v (c) ii (d) iv (e) iii
(D) (a) v (b) ii (c) iii (d) I (e) iv

222. Match the feeding stuff with crude protein content:

Feeding stuff — Crude protein content
(a) Sesame cake — (i) 18.1
(b) Rape cake — (ii) 36.1
(c) Mustard cake — (iii) 46.3
(d) Cotton seed — (iv) 36.4
(e) Chickpea grain — (v) 18.0

Answer:
(A) (a) iv (b) i (c) ii (d) iii (e) v
(B) (a) iii (b) iv (c) ii (d) v (e) i
(C) (a) i (b) v (c) ii (d) iv (e) iii
(D) (a) v (b) ii (c) iii (d) I (e) iv

223. Match the feeding stuff with crude protein content:

Feeding stuff — Crude protein content
(a) Oat grain — (i) 9.5
(b) Barley grain — (ii) 3.3
(c) Rice straw — (iii) 10.1
(d) Wheat straw — (iv) 14.0
(e) Rice bran — (v) 3.8

Answer:
(A) (a) iv (b) i (c) ii (d) iii (e) v
(B) (a) iii (b) i (c) v (d) ii (e) iv
(C) (a) i (b) v (c) ii (d) iv (e) iii
(D) (a) v (b) ii (c) iii (d) I (e) iv

224. Match the feeding stuff with crude protein content:

Feeding stuff	Crude protein content
(a) Cowpea hay	(i) 7.9
(b) Mixed grass hay	(ii) 17.4
(c) Maize silage	(iii) 7.6
(d) Maize hay	(iv) 15.3
(e) Berseem silage	(v) 5.1

Answer:

(A) (a) iv	(b) i	(c) ii	(d) iii	(e) v
(B) (a) iv	(b) v	(c) i	(d) iii	(e) ii
(C) (a) i	(b) v	(c) ii	(d) iv	(e) iii
(D) (a) v	(b) ii	(c) iii	(d) I	(e) iv

225. Match the feeding stuff with nutritive ratio:

Feeding stuff	Nutritive ratio
(a) Wheat bran	(i) 0.2
(b) Rice bran	(ii) 22.0
(c) Fish meal	(iii) 7.1
(d) Mixed grass hay	(iv) 9.0
(e) Oats grain	(v) 7.3

Answer:

(A) (a) iv	(b) i	(c) ii	(d) iii	(e) v
(B) (a) iii	(b) v	(c) i	(d) ii	(e) iv
(C) (a) i	(b) v	(c) ii	(d) iv	(e) iii
(D) (a) v	(b) ii	(c) iii	(d) I	(e) iv

226. Match the feeding stuff with nutritive ratio:

Feeding stuff	Nutritive ratio
(a) Berseem feed	(i) 10.5
(b) Guinea grass feed	(ii) 16.8
(c) Maize feed	(iii) 22.0
(d) Maize silage	(iv) 3.6
(e) Mixed grass hay	(v) 13.2

Answer:

(A) (a) iv	(b) i	(c) ii	(d) iii	(e) v
(B) (a) iv	(b) i	(c) iv	(d) i	(e) iii
(C) (a) i	(b) v	(c) ii	(d) iv	(e) iii
(D) (a) v	(b) ii	(c) iii	(d) I	(e) iv

227. Match the feeding stuff with nutritive ratio:

Feeding stuff	Nutritive ratio
(a) Cow pea hay	(i) 9.0
(b) Barley grain	(ii) 3.9
(c) Oats grain	(iii) 10.5
(d) Maize grain	(iv) 6.0
(e) Chickea grain	(v) 10.7

Answer:

(A) (a) iv	(b) i	(c) ii	(d) iii	(e) v
(B) (a) ii	(b) v	(c) i	(d) iii	(e) iv
(C) (a) i	(b) v	(c) ii	(d) iv	(e) iii
(D) (a) v	(b) ii	(c) iii	(d) I	(e) iv

228. Match the feeding stuff with nutritive ratio:

Feeding stuff	Nutritive ratio
(a) Cotton seed	(i) 1.7
(b) Mustard cake	(ii) 1.0
(c) Rape cake	(iii) 0.7
(d) Sesame cake	(iv) 1.8
(e) Groundnut cake	(v) 6.1

Answer:

(A) (a) iv	(b) i	(c) ii	(d) iii	(e) v
(B) (a) v	(b) i	(c) iv	(d) ii	(e) iii
(C) (a) i	(b) v	(c) ii	(d) iv	(e) iii
(D) (a) v	(b) ii	(c) iii	(d) I	(e) iv

229. Which of the following animal is not a ruminant?
(A) Cow (B) Goat
(C) Pig (D) Sheep

230. The stomach of ruminant is divided into:
(A) 2 parts (B) 3 parts
(C) 4 parts (D) 5 parts

231. Which of the following is not a compartment of ruminant stomach?
(A) Rumen (B) Reticulum and Omasum
(C) Abomasum (D) Abdomen

232. Which of the part of ruminant stomach is the largest one?
(A) Rumen (B) Reticulum
(C) Abomasum (D) Omasum

233. Regurgitation takes place in:
(A) Rumen (B) Reticulum
(C) Abomasum (D) Omasum

234. Which of the following has not a simple digestive tract?
(A) Swine (B) Poultry
(C) Goat (D) Men

235. The main function of proventriculus in poultry is:
(A) Digestive juices are secreted and mixed with feed
(B) Grinding of feed into small prices
(C) Chewing of feed
(D) All of these

236. The main function of Gizzard is:
(A) Grind and crush the coarse food
(B) Digestive juices are secreted and mixed with food
(C) Chewing of feed
(D) All of these

237. Which of the following part is not found in the digestive systems of poultry?
(A) Crop (B) Proventriculus
(C) Gizzard (D) Abomasum

238. The main function of abomasums is:
(A) Storage of feed
(B) Squeezing of water from feed
(C) Cheaning of feeds
(D) Secretion of gastric juice

239. The true stomach in ruminants is known as:
(A) Rumen (B) Reticulum
(C) Abomasum (D) Omasum

240. The function of reticulum is:
(A) To hold foreign harmful bodies
(B) To furnish additional storage space
(C) Both (A) and (B)
(D) Squeezing out the water from feed

241. What is true for omasum?
(A) Third compartment of ruminant stomach
(B) It has strong muscular walls
(C) It helps in squeezing out water from feed before it enters to abomasums
(D) All of these

242. American buffalo is known as:
(A) Bison (B) Toda
(C) Mehsana (D) Jaffrabadi

243. Which of the following breed has highest fat in milk:
(A) Surti (B) Mehsana
(C) Bhadawari (D) Murrah

244. Which of the breed of buffalo gives highest milk production?
(A) Bhadawari (B) Murrah
(C) Jaffrabadi (D) Surti

245. How many breeds of cattle are found in India?
(A) 10 (B) 15
(C) 21 (D) 27

246. Deoni is a:
(A) Draft breed (B) Dairy breed
(C) Dual purpose breed (D) All of these

247. Which of the following is not a dairy breed of cattle?
(A) Sahiwal (B) Sindhi
(C) Nagore (D) Gir

248. Santa Gertrudis is a cross of:
(A) Bull of Indian breed x Shorthorm cow of USA
(B) Jersey x Local breed
(C) Kankry x Dargi
(D) None of these

249. Which of the Indian cattle breed was exported to Brazil?
(A) Sahiwal (B) Jersey
(C) Kankrej (D) All of these

250. Which of the following cattle breed in the smallest one?
(A) Jersey (B) Ayrshire
(C) Brown swiss (D) Holstein – Friesians

251. Which of the following breed considered the most beautiful?
(A) Red deoni (B) Ayreshire
(C) Jersey (D) Tharparkar

252. Which of the following is considered best draft breed of buffalo?
(A) Surti (B) Nagpuri
(C) Nili (D) Murrah

253. Which of the following exotic cattle breed is known as dual purpose type?
(A) Jersey (B) Red Dane
(C) Brown swiss (D) Holstein - Friesian

254. Which of the following is not a dual purpose cattle breed?
(A) Kankrej (B) Haryana
(C) Gir (D) Ongole

255. Cow Jill is a cross of:
(A) Ayreshire x Haryana
(B) Ayrshire x Holstein Friesian
(C) Jersey x Sindhi
(D) Sahiwal x Sindhi

256. Hisarsale is a cross of:
(A) Bikaneri ewes x Merino rams
(B) Merino ewes x Bikaneri rams
(C) Rambouillet ewes x Merino rams
(D) Merino ewes x Rambonillet rams

257. Whch of the following is a oldest cross bred wool breed of sheep?
(A) Corriedale (B) Hissardale
(C) Bharat merino (D) None of these

258. Which of the following is a sheep breed of southern region of India?
(A) Deccani (B) Nellore
(C) Mandy (D) All of these

259. Which of the following sheep breed is found in arid western region?
(A) Kathiawari (B) Marwari
(C) Bikaneri (D) All of these

260. Which of the following sheep breed is most popular for its fine wool?
(A) Rambouillet (B) Corriedale
(C) Merino (D) Gurej

261. Which of the following sheep breed of temperate region?
(A) Gurej (B) Bhakarwal
(C) Gaddi (D) Lohi

262. What is correct for Bharat merino sheep:
(A) Fine wool sheep breed
(B) Indigenously developed cross breed
(C) Cross of Rambouillet and Merino with indigenous breeds
(D) All of these

263. Indian sheep produce mostly:
(A) Carpet wool (B) Fur wool
(C) Fine wool (D) Long wool

264. How many recognized breeds of goat is found in India?
(A) 15 (B) 18
(C) 21 (D) 23

265. Which of the following goat breed is the highest milk producer?
(A) Barbari (B) Bengil black
(C) Jamanapari (D) Assam local

266. The most striking feature of Jamanapari goat is:
(A) Romen nose coupled with a shortened upper lip
(B) High milk production
(C) High reproductive rate
(D) All of these

267. Which of the following is a horse breed?
(A) Kathiawari (B) Marwari
(C) Bhutia (D) All of these

268. Which of the following horse breed is found in Pakistan and India both?
(A) Baluchi (B) Hizzai
(C) Unmol (D) All of these

269. Which of the following horse breed is generally known as a race horse?
(A) Arabi (B) Thoroughbred
(C) Spiti (D) None of these

270. Which of the following horse breed is popular in hilly areas of India?
(A) Manipuri (B) Spiti
(C) Bhutia (D) All of these

271. Which of the following horse breed is in great demand for polo and racing?
(A) Manipuri (B) Spiti
(C) Kathiawari (D) Bhutia

272. Which of the following breed of donkey is the smallest one?
(A) Indian ass (B) Poitou
(C) Egyptian white (D) Arab

273. Which of the following breed of donkey is the largest one in the world?
(A) Poitou (B) White Egyptian
(C) Arab (D) Persian

274. Which of the following donkey breed is found in France?
(A) Poitou (B) White Egyptian
(C) Arab (D) Persian

275. Which of the following is a breed of a donkey?
(A) White Egyptian (B) Damascus
(C) Persian (D) All of these

276. Which of the following is considered as a parent of a present day donkey?
(A) Nubian ass (B) Ponies
(C) Mare (D) Mules

277. Mules are most suitable for:
(A) Hilly tracts (B) Military operations
(C) Stress climate (D) All of these

278. Which of the following is a baggage type of camel breed in India?
(A) Hill camel (B) Reverine camel
(C) Both (A) and (B) (D) None of these

279. Which of the camel breed has two humps?
(A) Bacterial camel (B) Arabian camel
(C) Both A and B (D) None of these

280. Which of the following breed of camel is found in India?
(A) Riverine camel (B) Desert camel
(C) Hill camel (D) All of these

281. Which of the following camel breed is used for riding purposes?
(A) Hill camel (B) Riverine camel
(C) Desert camel (D) All of these

282. Which of the following camel breed is not found in India?
(A) Bacterian (B) Arabian
(C) Both (A) and (B) (D) None of these

283. Riverine camel is mostly found in:
(A) UP and Punjab (B) Rajasthan and TN
(C) Assam and Uttaranchal (D) Kerala and AP

284. Which of the following pig breed produces high quality pork?
(A) Middle white x Yorkshire
(B) Berkshire
(C) Danish Landrace
(D) All of these

285. Which of the following is a exotic pig breed?
(A) Hampshire (B) Poland China
(C) Tamworth (D) All of these

286. Which of the following is a exotic pig breed?
(A) Chester white (B) Duroc
(C) Danish Landrace (D) All of these

287. Which of the following pig breed is known as bacon hog?
(A) Wessex Saddleback (B) Hampshire
(C) Duroc (D) Bershire

288. Which of the pair is correct?
(A) Tamworth – Red colour
(B) Hampshire – Black but white belt encircling
(C) Berkshire – Black
(D) All of these

289. Which of the following is a American fowl?
(A) Plymoth Rock (B) Rhode Island Red
(C) New Hampshire (D) All of these

290. Which of the following poultry breed is suitable for drier areas?
(A) White Leghorn (B) Australorp
(C) White Rock (D) White Cornish

291. Which of the following poultry breed is suitable for wet and heavy rainfall regions?
(A) Australorp (B) White leghorn
(C) White Rock (D) White Cornish

292. Which of the following is a most popular egg breed of fowl in all over world?
(A) White Leghorn (B) Plymoth Rock
(C) New Hampshire (D) White Rock

293. Which of the following breed is known as best round breed of all poultry?
(A) White Leghorn (B) Plymoth Rock
(C) Rhode Island Red (D) White Rock

294. Which of the following is the largest breed of poultry?
(A) Black Minorca (B) White Leghorn
(C) Light Sussex (D) New Hampshire

295. Which of the following belongs to duck breeds?
(A) Sylhet met (B) Nageswari
(C) Indian Runner (D) All of these

296. Khaki Campbell is a breed of:
(A) Hen (B) Pigeon
(C) Duck (D) Pea cock

297. Which of the duck breed produces the largest size of egg?
(A) Khaki Campbell (B) Sylhet Mete
(C) Nagaswari (D) Indian runner

298. Most popular duck breed in India is:
(A) Khaki Campbell (B) Sylhet Mete
(C) Nagaswari (D) Indian runner

299. Which of the following is a geese breed?
(A) Indian goose (B) Farn chinese
(C) Emden (D) All of these

300. The heaviest breed of geese is:
(A) Emden (B) English white
(C) Romen (D) White Chinese

301. The lightest breed of geese is:
(A) Emden (B) Indian goose
(C) English grey (D) Toulouse

302. Which of the following is a breed of Turkey?
(A) Beltville white (B) Breasted Bronze
(C) Norfolks (D) All of these

303. Which of the following a not a breed of Turkey?
(A) Cambridge (B) British white
(C) New Hampshire (D) Norfolks

304. Which of the following breed of Turkey is the most popular and heaviest one?
(A) Cambridge
(B) Broad Breasted Bronze
(C) Western white
(D) Beltsville white

305. Indubrazil is a cattle breed:
(A) Cross between Indian breed x Local of Brazil cow
(B) New cross breed of Brazil
(C) Suitable for beef and milk
(D) All of these

306. Which of the following is not a draft cattle breed?
(A) Nagore (B) Amritmahal
(C) Gir (D) Halikar

307. Which of the following pair is not matched correctly?
(A) Amritmahal – Mysore (B) Malvi – Malva tract
(C) Nagore – Nagore (D) Gir – South India

308. What is correct to Tharparkar breed of Cattle?
(A) Dual purpose breed
(B) Originated from Tharparkar district of Western Pakistan
(C) Main breed of Jaisalmer and Jodhpur in Rajasthan
(D) All of these

309. Which of the cattle breed of south India?
(A) Rath (B) Mewati
(C) Tharparkar (D) Ongole

310. Which of the following is not matched correctly?
(A) Toda - Nilgiri hills (B) Bhadawari – UP
(C) Parlakmedi – Orissa (D) Surti – Punjab

311. Which of the following group belongs to buffalo breeds?
(A) Murrah, Jaffarabadi, Surti and Mehsana
(B) Gir, Nagore, Rath and Knakrej
(C) Pashmina, Beetal, Barbari and Malabari
(D) Lohi, Merino, Rambouillet and Corriedale

312. Which of the following group belongs to sheep breeds?
(A) Murrah, Jaffarabadi, Surti and Mehsana
(B) Gir, Nagore, Rath and Knakrej
(C) Pashmina, Beetal, Barbari and Malabari
(D) Lohi, Merino, Rambouillet and Corriedale

313. Which of the following group belongs to goat breeds?
(A) Murrah, Jaffarabadi, Surti and Mehsana
(B) Gir, Nagore, Rath and Knakrej
(C) Pashmina, Beetal, Barbari and Malabari
(D) Lohi, Merino, Rambouillet and Corriedale

314. The highest milk producer cattle breed is:
(A) Holstein Friesian
(B) The Trademarks sable Fashion
(C) Jersey
(D) Brown swiss

315. Which of the following pair is correct:
(A) Holstein Friesian – Holland
(B) Ayrshire – Scotland
(C) Brown swiss – Switzerland
(D) Gurney – USA

316. The name of breed and fat % are given below:

Breed	Fat %
1. Brown swiss	i. 4 %
2. Guensey	ii. 5 %
3. Holstein Friesian	iii. 3.67 %
4. Jersey	iv. 5 %

Which of the pair is correctly matched
(A) 1. i 2. ii 3. iii 4. iv
(B) 1. iii 2. ii 3. iv 4. i
(C) 1. ii 2. i 3. iii 4. iv
(D) 1. iv 2. ii 3. i 4. iii

317. Which of the following criteria is used for selecting the dairy milking animals?
(A) Breed and pedigree
(B) Production and records
(C) Physical appearance
(D) All of these

318. Which of the most important criteria is used for selecting the dairy animals?
(A) General appearance
(B) Dairy character
(C) Body appearance
(D) Mammary appearance

319. Which of the following criteria is used for selecting the dairy bull?
(A) Dairy character (B) Body capacity
(C) Physical appearance (D) All of these

320. Which of the following method is used to dry up cows:
(A) Intermittent milking (B) Incomplete milking
(C) Abrupt milking (D) All of these

321. Which of the following method of drying off is most suitable for high yielder animals?
(A) Intermittent milking (B) Incomplete milking
(C) Abrupt milking (D) All of these

322. Strip cup is used for:
(A) Removing two –three streams of milk from each quarter before milking
(B) Checking the normalcy of milk
(C) Cleaning of treat canals
(D) All of these

323. Milk replacer refers to:
(A) Readymade dry feed mixture
(B) Substitute of milk
(C) It is also fortified with vitamin A, D and antibiotics
(D) All of these

324. Milk replacer should be given to young calves from:
(A) 5 days after birth (B) 10 days after birth
(C) Just after birth (D) 15 days after birth

325. Milk replacer can be given to young calves upto?
(A) 1 month old calves (B) 2 months old calves
(C) 3 months old calves (D) 4 months old calves

326. The bull matures at a age of:
(A) 2 years (B) 3 years
(C) 4 years (D) 5 years

327. The cattle matures, in general, at the age is about:
(A) 2 years (B) 3 years
(C) 4 years (D) 5 years

328. Most common system for housing dairy cattle is:
(A) Stanchion barn
(B) Open barn
(C) Half open and half closed system
(D) All of these

329. Housing space for a breeding bull is:
(A) 6 sq. meter (B) 8 sq. meter
(C) 9 sq. meter (D) 10 sq. meter

330. Calves start eating green fodder after:
(A) 5 days after birth (B) 10 days after birth
(C) 30 days after birth (D) 15 days after birth

331. Silage should be given to calves after:
(A) 5 days after birth (B) 10 days after birth
(C) 30 days after birth (D) 15 days after birth

332. The size of stanchion type barn for each adult cow should be:
(A) 5-8 cubic meter (B) 11-14 cubic meter
(C) 17-19 cubic meter (D) 21-23 cubic meter

333. The size of milk house for a cow should be:
(A) 4 sq. meter (B) 6 sq. meter
(C) 10.5-17.5 sq. meter (D) None of these

334. Which of the following method is used for market of calves:
(A) Ear notches (B) Tattooing
(C) Metal ear tags (D) All of these

335. Which of the following marketing method of calves is considered the best one?
(A) Ear notches (B) Tattooing inside
(C) Metal ear tags (D) All of these

336. Adult cattle are generally marked successfully with:
(A) Ear notches
(B) Hot iron branding on hind flanks
(C) Neck chains
(D) All of these

337. Which of the following marking method is used for marketing of buffaloes:
(A) Horn branding
(B) Hot iron branding
(C) Totooning the inside of the root of the tail
(D) All of these

338. Urea in the cattle ration should not be:
(A) 1% of total dry matter
(B) 3 % of the concentrate mixture
(C) Both (A) and (B)
(D) None of these

339. Under the feeding rule, the roughage should provide the total dry matter to dairy animal is:
(A) 1/3 (B) 2/3 (C) >2/3 (D) ½

340. Which of the following statement is correct?
(A) If roughage contains no legumes, the protein in concentrate ration should be 17-20 %
(B) If cow is receiving all high quality legume hay or legume silage, concentrate ration should have 9-10 % protein
(C) If animal is fed on green fodder, there is no need to provide concentrate ration
(D) All of these

341. The oestrus cycle of cow occurs at interval of:
(A) 21 days (B) 32 days
(C) 45 days (D) 51 days

342. The heat period of cows lasts on an average in:
(A) 10 hrs (B) 15 hrs
(C) 18 hrs (D) 21 hrs

343. The heat period of heifers lasts on an average in:
(A) 10 hrs (B) 15 hrs
(C) 18 hrs (D) 21 hrs

344. She buffalo comes into heat after every:
(A) 15-17 days (B) 21-23 days
(C) 27-29 days (D) 31-33 days

345. The heat period in sheep occurs at interval of:
(A) 17-19 days (B) 12-14 days
(C) 21-13 days (D) 26-28 days

346. The duration of heat period in sheep is:
(A) 13-15 hrs (B) 21-39 hrs
(C) 42-48 hrs (D) 55-65 hrs

347. The duration of heat period in sheep ranges in between:
(A) 1-3 days (B) 4-6 days
(C) 7-9 days (D) 10-12 days

348. The breeding season of sheep is:
(A) March-April (B) June-July
(C) October-November (D) All of these

349. Breeding male in sheep is known as:
(A) Ewe (B) Lamb
(C) Ram (D) Male sheep

350. Which of the breeding season considered as best mating season of sheep?
(A) March-April (B) June-July
(C) October-November (D) All of these

351. The lowest mating season in sheep occurs during the breeding season of:
(A) March-April (B) June-July
(C) October-November (D) All of these

352. The rams can be allowed for breeding after attaining the age of:
(A) 12 months (B) 18 months
(C) 24 months (D) 30 months

353. The breeding period of a normal rams is in between:
(A) 1-3 years (B) 1.5 to 3.5 years
(C) 2.5 to 5 years (D) 3-7 years

354. What is the optimum ratio of Ewe-Ram in a flock?
(A) 30 : 1 (B) 40 : 1
(C) 50 : 1 (D) 60 : 1

355. The ewes attain maturity at the age of:
(A) 1 year (B) 1.5 year
(C) 2 year (D) 2.5 year

356. Average productive life of ewes is:
(A) 2-3 years (B) 2-5 years
(C) 2-7 years (D) 2-10 years

357. The normal breeding season of goat is:
(A) September-February (B) November-June
(C) June – January (D) All of these

358. The heat period of goat repeats after:
(A) 15 days (B) 18-21 days
(C) 30-32 days (D) 35-38 days

359. The duration of heat period of a goat is:
(A) 2-3 days (D) 4-5 days
(C) 1-2 days (D) 6-7 days

360. Female goat is known as:
(A) Buck (B) Doe
(C) Wether (D) Ram

361. Breeding age of a doe is:
(A) 1 year (B) 6 months
(C) 1.5 year (D) 2 year

362. Male goat is known as:
(A) Buck (B) Doe
(C) Wether (D) Ram

363. Indian goats tend to come in heat during:
(A) April-May (B) June-July
(C) November-December (D) All of these

364. Which of the following goat breeds births twice in a year?
(A) Jamanapari (B) Barbari
(C) Beetal (D) Saanen

365. The doe should be allowed for mating at the body weight of:
(A) 25 kg (B) 35 kg
(C) 45 kg (D) 55 kg

366. The buck should be allowed for mating at a age of:
(A) 2 years (B) 3 years
(C) 4 years (D) 5 years

367. How many goats can be served by a single buck in a year?
(A) 20-30 (B) 50-60
(C) 75-100 (D) 110-120

368. The male horse used for breeding purposes is known as:
(A) Filly (B) Stallion
(C) Foal (D) Mare

369. Young female horse is known as:
(A) Filly (B) Stallion
(C) Foal (D) Mare

370. The stallion attains maturity at the age of:
(A) 2 years (B) 4.5-5 years
(C) 6 years (D) 7 years

371. The female horse (Filly) should be bred at the age of:
(A) 1 year (B) 2 years
(C) 3 years (D) 3.5 years

372. The average age of puberty in a filly is:
(A) 10-12 months (B) 15-18 months
(C) 36-38 months (D) 48-50 months

373. How many filly could be allowed to stallion?
(A) 10-20 (B) 30-40
(C) 60-70 (D) 80-100

374. The oestrus cycle of mare averages:
(A) 21 days (B) 27 days
(C) 32 days (D) 48 days

375. The duration of heat period of a mare is:
(A) 3-4 days (B) 12-24 days
(C) 7-9 days (D) 11-13 days

376. The donkey can be bred at age of:
(A) 2.5-3.0 years (B) 1.0-1.5 years
(C) 3.5-4.0 years (D) 5 years

377. The average oestrus cycle of donkey is:
(A) 21-28 days (B) 31-33 days
(C) 17-18 days (D) 37-42 days

378. The heat period of donkey lasts in:
(A) 1-2 days (B) 3-7 days
(C) 8-11 days (D) 13-15 days

379. Rut or Musth in camel refers to:
(A) Sexual instinct (B) Castrated camel
(C) Diseased camel (D) None of these

380. The breeding season of camel comes during:
(A) November-February/March
(B) June-July
(C) February – March
(D) September-October

381. What is main symptoms of musth camel?
(A) Constantly grind the teeth
(B) Foams at the mouth
(C) Involuntarily betches out
(D) All of these

382. The male camel should not be allowed to serve before attaining the age of:
(A) 4 years (B) 5 years
(C) 6 years (D) 8 years

383. Bull camel attains full sexual maturity at the age of:
(A) 5 years (B) 6 years
(C) 8 years (D) 10 years

384. How many female camels can be bred with a bull camel in a breeding season?
(A) 40 (B) 50 (C) 70 (D) 100

385. A female camel is sexually mature at the age of:
(A) 2 years (B) 3 years
(C) 4 years (D) 5 years

386. The breeding age of female camel is:
(A) 10 years (B) 15 years
(C) 20 years (D) 25 years

387. The male pig is known as:
(A) Hog (B) Swine
(C) Sow (D) Piglet

388. The female pig is known as:
(A) Hog (B) Swine
(C) Sow (D) Piglet

389. The breeding pig is known as:
(A) Hog (B) Boar
(C) Sow (D) Piglet

390. Which of the following pair is correct:
(A) Cattle – Calving (B) Horse – Faoling
(C) Pig – Farrowing (D) All of these

391. The young female pig is called:
(A) Piglet (B) Gilt
(C) Sow (D) Hog

392. The gilts should be bred at the age of:
(A) 5-6 months (B) 10-12 months
(C) 11-13 months (D) 15-16 months

393. The gilts in general show their first puberty at age of:
(A) 6 months (B) 8 months
(C) 9 months (D) 12 months

394. The heat period of sows usually lasts in:
(A) 15-20 hrs (B) 40-65 hrs
(C) 3-4 days (D) 5-6 days

395. The heat period in pigs repeats at the interval of:
(A) 1 weeks (B) 2 weeks
(C) 3 weeks (D) 4 weeks

396. Normally one mature boar can be mated yearly with:
(A) 10-15 gilts (B) 20-25 gilts
(C) 50-60 gilts (D) 80-100 gilts

397. Breeding age of a boar is:
(A) 1 year (B) 2 years
(C) 5 years (D) 6 years

398. At which age, the young boar should be allowed for mating?
(A) 8 month (B) 1 year
(C) 1.5 year (D) 2 years

399. Which of the chemical is used for blocking the heat cycle in sows?
(A) ICI-33828 (B) IC – 35
(C) IAA (D) All of these

400. How many sows can be bred artificially by a single ejaculation of boar?
(A) 10 (B) 15 (C) 25 (D) 50

401. Male fowl is known as:
(A) Cock (B) Layer
(C) Hen (D) All of these

402. Female fowl is known as:
(A) Cock (B) Layer
(C) Hen (D) All of these

403. Layers refers to the:
(A) 0-6 weeks old fowl
(B) 6-14 weeks old fowl
(C) 14-20 weeks old fowl
(D) >20 weeks old fowl

404. Broiler refers to:
(A) Young chickens about 8-10 weeks old
(B) Raised mainly for meat purposes
(C) Broiler may weight 1-1.5 kg weight
(D) All of these

405. Pullet refers to:
(A) 0-6 weeks old fowl (B) 6-14 weeks old fowl
(C) 14-20 weeks old fowl (D) >20 weeks old fowl

406. The pullet start laying eggs at the age of:
(A) 4 months (B) 5 to 5 ½ months
(C) 7 months (D) 1 year

407. Male duck is called as:
(A) Drake (B) Duck
(C) Cock (D) Geese

408. How many female ducks can be allowed to breed with one male duck?
(A) 5-6 (B) 8-10
(C) 12-15 (D) 20-22

409. The average egg production of turkey per annum is:
(A) 60 eggs (B) 75 eggs
(C) 105 eggs (D) 120 eggs

410. Turkey matures at the age of:
(A) 1.5 years (B) 2 years
(C) 3 years (D) 4 years

411. The average hatching period of turkey eggs is:
(A) 21 days (B) 25 days
(C) 28 days (D) 33 days

412. What is optimum time for breeding of milking cows?
(A) 30 days after calving (B) 60 days after calving
(C) 70 days after calving (D) 85 days after calving

413. Average gestation period for a cow is:
(A) 210 days (B) 250 days
(C) 280 days (D) 310 days

414. Average gestation period for a buffalo is:
(A) 210 days (B) 250 days
(C) 280 days (D) 310 days

415. Average gestation period for a sheep is:
(A) 125-130 days (B) 143-151 days
(C) 160-165 days (D) 170-180 days

416. Average gestation period for a goat is:
(A) 125-130 days (B) 135-142 days
(C) 110-118 days (D) 145-152 days

417. The average gestation period for a mare is:
(A) 145 days (B) 250 days
(C) 340 days (D) 365 days

418. The gestation period of donkey is:
(A) 11.5-12.0 months (B) 2-2.5 months
(C) 5 months (D) 7 months

419. The average gestation period for a female camel is:
(A) 145 days (B) 389 days
(C) 265 days (D) 421 days

420. The average gestation period of sow is:
(A) 114 days (B) 145 days
(C) 221 days (D) 345 days

421. How much space is required for a cow shed?
(A) 5 sq. meter (B) 6 sq. meter
(C) 10 sq. meter (D) 15 sq. meter

422. Shearing in sheep is done during:
(A) Winter season
(B) Summer season
(C) After winter and rainy season
(D) All of these

423. The space requirement for a sheep is:
(A) 1-2 sq. meter (B) 3-4 sq. meter
(C) 4-5 sq. meter (D) 8 sq. meter

424. The height of fencing or boundary wall of a sheep ward should be:
(A) 2 m high (B) 5 m high
(C) 7 m high (D) 8 m high

425. The space requirement for a goat under standard feeding is:
(A) 0.75 m x 1.2 m (B) 1.0 m x 1.2 m
(C) 1.2 m x 1.2 m (D) 1.5 m x 1.5 m

426. The space requirement for milking of a goat should be:
(A) 1.0 m length x 0.55 m wide space and 0.5-0.75 m high
(B) 1.0 m length x 1.0 m wide space and 0.5 m high
(C) 1.0 m length x 1.0 m wide space and 1.0 m high
(D) None of these

427. The goats are reared under the housing system of:
(A) Loose housing system (B) Closed housing system
(C) Stanchion system (D) All of these

428. Which of the housing system is considered the best one for goats?
(A) Loose housing system (B) Closed housing system
(C) Stanchion system (D) All of these

429. The space requirement for a buck is:
(A) 1.2 sq. meter (B) 2.0 sq. meter
(C) 3.4 sq. meter (D) 5.2 sq. meter

430. The space requirement for a kid is:
(A) 1.2 sq. meter (B) 1.0 sq. meter
(C) 0.8 sq. meter (D) 0.4 sq. meter

431. The goat wards (house) could be disinfected with the help of:
(A) 1-2 % solution of caustic soda (NaOH)
(B) 4 % solution of Na_2CO_3
(C) 30% solution of bleaching powder
(D) All of these

432. 10% solution of ammonia is used mainly for:
(A) To kill the coccidice spores on goat floors
(B) To kill all type of bacteria and viruses
(C) To kill protozoa
(D) None of these

433. Kerosol is used for:
(A) To disinfect the goat floor by killing most the bacteria and its spores
(B) To kill the external parasites of goats
(C) To kill internal parasites of goats
(D) None of these

434. Which of the following is a common type of goat keeping systems?
(A) Subsistence (B) Extensive
(C) Intensive (D) All of these

435. The housing place of a horse is known as:
(A) Ward (B) Stable
(C) Brood (D) None of these

436. The space requirement of a layer bird is:
(A) 0.20 sq. meter (B) 0.23 -0.2 sq. meter
(C) 0.50 sq. meter (D) 0.72 sq. meter

437. A house of 6 m x 12 m may accommodate:
(A) 200 hens of heavy breeds
(B) 200 hens of light breeds
(C) 300 hens of light breeds or 260 hens of heavy breeds
(D) 260 hens of light breeds or 300 hens of heavy breeds

438. Calves should be dehorned at the age of:
(A) 3-10 days old (B) 15-30 days old
(C) 45-60 days old (D) 85-90 days old

439. Which of the following method is used for dehorning?
(A) Chemical (B) Hot Iron
(C) Clippers and saws (D) All of these

440. Which chemical is used for dehorning of calves?
(A) Caustic potash (B) Potassium chloride
(C) Benzene (D) All of these

441. The temperature of electric dehorner should be about:
(A) 100°C (B) 250°C
(C) 310°C (D) 537°C

442. The old cattle may be dehorned with the help of:
(A) Electric dehorner (B) Chemicals
(C) Clippers and saws (D) All of these

443. The age of cattle can be determined with the help of:
(A) Tail (B) Teeth
(C) Health (D) Pregnancy

444. Castration means:
(A) Feed properly to cattle
(B) Removal of male parts from male calves
(C) Complete milking of cow
(D) None of these

445. Age of male calves for castration is:
(A) 6 months (B) 12 months
(C) 15 months (D) 20 months

446. Best season for castration is:
(A) Rainy season (B) Spring season
(C) Winter season (D) All of these

447. Which of the following method is recommended for castration the male calves?
(A) Removal of testicles by making incision
(B) Crush and destroy spermatic cord and blood vessels with the help of Burdizzo castrator
(C) Removal of whole male organs
(D) All of these

448. The lambs should be castrated at the age of:
(A) 7-14 days (B) 15-18 days
(C) 21-28 days (D) > 30 days

449. Male goats should be castrated at a age of:
(A) Just after birth (B) 5 days after birth
(C) 10 days to 3 weeks of age (D) All of these

450. Castrated male goat is called as:
(A) Ram (B) Buck
(C) Wether (D) Chhanga

451. Disbudding of goats can be done at the age of:
(A) 5 days (B) 10 days
(C) 30 days (D) 45 days

452. Which of the following method is used for identification of goats:
(A) Tattooing (B) Ear notches
(C) Leather neck bands (D) All of these

453. Average lactation period of a goat is:
(A) 5 months (B) 6 months
(C) 8 months (D) 10 months

454. How much a dairy animal needs rest between successive lactation period?
(A) 30 days (B) 45-60 days
(C) 120 days (D) 104 days

455. Which of the following fur wool is most valuable?
(A) Persian lamb (B) Broadtail
(C) Karakul (D) All of these

456. Persian lamb fur is collected from:
(A) <10 days old lambs (B) Newly born lambs
(C) >2 weeks old lambs (D) > 3 weeks old lambs

457. The sheep of southern region are suited mainly for:
(A) Fine wool (B) Carpet wool
(C) Meat (D) Fine wool and meat

458. Which of the following pair is matched correctly?
(A) Merino – Spain
(B) Rambouillet – France
(C) Corriedale – New Zealand (D) All of these

459. Which of the following pair is matched correctly?
(A) Cheviot – Scotland (B) Leicester – England
(C) Karakul – USA (D) All of these

460. Wethers refer to:
(A) Male lambs castrated at an early age before they reach sexual maturity
(B) Male lambs ready for slaughter
(C) Lambs bearing a full year's growth of wool
(D) Mating male sheep

461. Shearer lambs refers to:
(A) One year full growth of wool
(B) Fattened male ready for slaughter
(C) Mating male sheep
(D) Castrated male sheep

462. Feeder sheep refers to:
(A) Over one year old fattened sheep for marketing purposes
(B) < 1 year old fattened sheep
(C) >2 years old fattened sheep
(D) All of these

463. Flushing refers to:
(A) Good feeding to sheep before mating
(B) Good feeding to sheep before marketing
(C) Good feeding to sheep before shearing
(D) All of these

464. Flushing is useful for:
(A) It brings ewes into heat early
(B) Higher conception rate
(C) Possibility of more twins
(D) All of these

465. Grafting in case of sheep refers to:
(A) To put dead skin of a lamb over the lamb of another
(B) To remove fleece
(C) To remove tails
(D) None of these

466. Docking refers to:
(A) Removal of testicles (B) Removal of tails
(C) Removal of wool (D) Removal of fleece

467. Docking can be performed with the help of:
(A) Knife and Emasculator (B) Hot iron
(C) Rubber band (D) All of these

468. Which of the following method is used for marketing of flock/ewes/rams?
(A) Paint brands (B) Metal tags
(C) Ear notches (D) All of these

469. Tagging in sheep refers to:
(A) Removal of wool from around the dock before breeding season starts
(B) Putting tags on necks for their identification
(C) Making colour marks for their identification
(D) All of these

470. Eyeing in sheep refers to:
(A) Clipping of wool from around the eyes
(B) Eye operation in case of eye problems
(C) Tearing of eyes
(D) All of these

471. Shearing in sheep refers to:
(A) Clipping of wool from whole body of rams and ewes before breeding season
(B) Removal of wool from eyes only
(C) Removal of wool from dock only
(D) None of these

472. Ringing refers to:
(A) Removal of wool from the neck and around the penis of rams before breeding season
(B) Making of colour rings in necks fur identification
(C) Putting belt around the neck of sheep
(D) All of these

473. The diameter of fine sheep wool ranges between:
(A) 0.50-0.75 μ (B) 1-5 μ
(C) 8-12 μ (D) 15-25 μ

474. Which of the following is a recognizes quality of sheep wool for export?
(A) Joria (B) Harnai
(C) Rajpootana (D) All of these

475. The sheep wool is divided on the basis of length into:
(A) Carpel and apparel wool
(B) Course and rough wool
(C) Klemp and fine wool
(D) All of these

476. Souring of wool refers to:
(A) Removal of grease dirt and vegetable matters from wool
(B) Removal of thick and short hairs from fine wool
(C) Removal of clotted fleece from the wool
(D) All of these

477. Klemp is a:
(A) Fine hair (B) Course hair
(C) Very fine hairs (D) All of these

478. Shrinkage of sheep wool refers to:
(A) Weight of grease, dirt etc expressed as a percentage of the grease fleece weight
(B) Shrinking of hairs
(B) Loss of wool weight
(D) None of these

479. Who referred goat as a poor man's cow?
(A) M.K. Gandhi
(B) J.L. Nehru
(C) Indira Gandhi
(D) Sardar Ballabh Bhai Patel

480. Why does goat rearing is beneficial profession?
(A) Quick onset of reproduction and high fecundity
(B) High survival rate under adverse conditions
(C) Production of meat, milk, skins and fibre
(D) All of these

481. The meat of goat is known as:
(A) Chevan (B) Mutton
(C) Beef (D) Pork

482. The meat of sheep is known as:
(A) Chevan (B) Mutton
(C) Beef (D) Pork

483. The meat of cow is known as:
(A) Chevan (B) Mutton
(C) Beef (D) Pork

484. The meat of pig is known as:
(A) Chevan (B) Mutton
(C) Beef (D) Pork

485. The meat of poultry is known as:
(A) Chevan (B) Mutton
(C) Chicken (D) Pork

486. Which of the following animal is known as ship of the desert?
(A) Donkey (B) Horse
(C) Camel (D) Zebra

487. Foal-heat refers to:
(A) Mares come into heat just after 4-11 days of foaling
(B) Mating of mares with stallion
(C) First heat of a mare
(D) None of these

488. The usual working age of donkey is:
(A) 5 years (B) 8 years
(C) 12 years (D) 16-17 years

489. The usual working age of mule is:
(A) 5 years (B) 8 years
(C) 12 years (D) 18-20 years

490. Mule is a cross of:
(A) She ass x Stallion pony
(B) Male donkey x Mare
(C) Male horse x Female donkey
(D) Zebra x Horse

491. Mules are:
(A) Fertile (B) Infertile
(C) Both (A) and (B) (D) None of these

492. Mule become ready to work at the age of:
(A) 1 years (B) 2 years
(C) 3 years (D) 5 years

493. The average speed of desert camel is:
(A) 4 km/hr (B) 6 km/hr
(C) 10 km/hr (D) 20 m/hr

494. The average speed of Reverine camel is:
(A) 4 km/hr (B) 6 km/hr
(C) 10 km/hr (D) 20 m/hr

495. The average speed of a riding camel is:
(A) 4-5 km/hr (B) 8-11 km/hr
(C) 16-22 km/hr (D) 25-27 km/hr

496. At what age, the camel should be allowed for heavy and strong work?
(A) 4 years (B) 5 years
(C) 6 years (D) 7 years

497. The best age to break a camel for riding is:
(A) 2 years (B) 3 years
(C) 4 years (D) 5 years

498. How much distance can be covered by a riding camel in a day?
(A) 20-25 km (B) 30-35 km
(C) 40-50 km (D) 60-80 km

499. An adult camel can be loaded with:
(A) 1-2 q (B) 2-3 q
(C) 3-4 q (D) 4-5 q

500. A well feed camel can work continuously for:
(A) 3-4 days (B) 5-6 days
(C) 7-9 days (D) 10-15 days

501. Average life of a camel is:
(A) 20 years (B) 25 years
(C) 30 years (D) >40 years

502. How much wool can be obtained by a camel in a single clipping?
(A) 1-1.5 kg (B) 1.5-2.0 kg
(C) 2.5-3.0 kg (D) 3.0-3.5 kg

503. How many needle teeth are found in newly born piglets?
(A) 2 in each jaw (B) 4 in each jaw
(C) 5 in each jaw (D) 6 in each jaw

504. Which of the following method is commonly used in pigs identification?
(A) Iron branding (B) Ear notching
(C) Neck strips (D) All of these

505. Pigs should be castrated at the age of:
(A) 5 days (B) 1 weeks
(C) 2-3 weeks (D) 6 weeks

506. Weaning of piglets should be started at the age of:
(A) 1 week (B) 2 weeks
(C) 3-5 weeks (D) 6 weeks

507. Ringing pigs refer to:
(A) To put ring or iron ring in the nose of pig
(B) To make a ring by removing hair from the neck
(C) To brand the hogs with round hot iron ring
(D) All of these

508. The light requirement of hens for good egg production is at least:
(A) 10 hrs (B) 14 hrs
(C) 16 hrs (D) 18 hrs

509. 60 watt bulb in the layer house is sufficient for the floor space of:
(A) 8 sq. meter (B) 10 sq. meter
(C) 19 sq. meter (D) 21 sq. meter

510. The light should be paced at the height of:
(A) 1 meter (B) 2 meter
(C) 3 meter (D) 4 meter

511. Cannibalism in hens refers to:
(A) Picking of each other by hens
(B) Laying of eggs
(C) Breaking of eggs by other hens
(D) None of these

512. Brooder house refers to:
(A) House for chicks (B) House of pullets
(C) House of layers (D) House of cocks

513. The space requirement of each chick for below one month is:
(A) 0.045 sq meter (B) 0.09 sq. meter
(C) 0.18 sq. meter (D) 0.36 sq. meter

514. The space requirement of each chick for 2-3 month is:
(A) 0.045 sq. meter (B) 0.09 sq. meter
(C) 0.18 sq. meter (D) 0.36 sq. meter

515. Chicks start to roost at the age of:
(A) 4-5 weeks (B) 6-7 weeks
(C) 1-2 weeks (D) 2-3 weeks

516. The broiler are marketed at the age of:
(A) 2 weeks (B) 6 weeks
(C) 10 weeks (D) 12 weeks

517. Caponette refers to:
(A) Female chicks produced after implanting of synthetic female hormones in young cockerels
(B) Chicks without male organs
(C) Both (A) and (B)
(D) None of these

518. Capon refers to:
(A) Female chicks produced after implanting of synthetic female hormones in young cockerels
(B) Chicks without male organs
(C) Both (A) and (B)
(D) None of these

519. Which one of the hardiest breed of poultry?
(A) Rhode Island Red (B) White leghorn
(C) Australorp (D) New Hempshire

520. Gummar refers to:
(A) Sheep having no teeth in her mouth
(B) Sheep that has broken teeth
(C) Sheep with only 2 teeth
(D) None of these

521. The condition, Broken mouth in case of sheep, refers to:
(A) All teeth disappeared
(B) Few teeth broken
(C) All teeth have become red and brown
(D) All of these

522. Feeder lambs refers to:
(A) Under one year age used for slaughter
(B) Under 2 year age used for slaughter
(C) Lambs are fed properly and make ready for mating
(D) Castrated lambs

523. Lamb is a:
(A) Newly born sheep kid (B) Upto 1 year old sheep
(C) Upto 2 years old sheep (D) All of these

524. Cashgora refers to:
(A) Goat breed
(B) It is a cross between Changthangi x Angora
(C) Both (A) and (B)
(D) None of these

525. Which of the following is the highest producer of milk?
(A) Alpine (B) Anglo-nubian
(C) Saanen (D) Toggenburg

526. Most beautiful goat breed is:
(A Barbari (B) Jamanapari
(C) Beetal (D) Surti

527. Pashmina is a:
(A) It is a goat breed
(B) It is found in Ladakh, Lahaul and Spiti valleys
(C) It produces fine and silky hair
(D) All of these

528. Which of the following is a dual purpose goat breed?
(A) Barbari (B) Jamanapari
(C) Mehsana (D) All of these

529. Which of the following goat breeds are suitable for meat prurpose?
(A) Marwari (B) Bengal goat
(C) Barbari (D) All of these

530. Which of the following goat breed is found in Peninsular India?
(A) Osmunabadi (B) Sangamneri
(C) Malabari (D) All of these

531. Which of the following breed is found in Himalayan region of India?
(A) Angora (B) Kashmiri
(C) Guddi and Chamba (D) All of these

532. Barbari goat is more useful because of:
(A) Good milk and meat producer
(B) More prolific as it may kid twice with two kids
(C) Higher butter fat (5%)
(D) All of these

533. Billy refers to:
(A) A goat breed
(B) It is found in Barren Island in Andamans
(C) It survive on sea water
(D) All of these

534. Which is correct for Angora breed of goat?
(A) It is a goat breed and its hair are known as Mohair
(B) It is found in the Himalayan mountain region
(C) It originated in Asia-minor and Turkey
(D) All of these

535. The minimum standards of milk and milk product quality is maintained by the rules/Acts of:
(A) Prevention of Food Adulteration Rules, 1955 and amended in 1976
(B) Food Adulteration Act 1954
(C) Indian Dairy Corporation
(D) None of these

536. The standard of buffalo milk should contain:
(A) 6.0% Fat and 9% SNF
(B) 5% Fat and 9.5% SNF
(C) 3.5% Fat and 8.5% SNF
(D) 4.0% Fat and 5% SNF

537. The standard of cow milk should contain:
(A) 5.5% Fat and 9% SNF
(B) 5% Fat and 9.5% SNF
(C) 3.5% Fat and 8.5% SNF
(D) 3.0% Fat and 5% SNF

538. The standard of raw mixed milk should contain:
(A) 5.5% Fat and 9% SNF
(B) 5% Fat and 9.5% SNF
(C) 4.5% Fat and 8.5% SNF
(D) 4.0% Fat and 5% SNF

539. The standard of toned milk should contain:
(A) 5.0% Fat and 9% SNF
(B) 4% Fat and 9.5% SNF
(C) 3.0% Fat and 8.5% SNF
(D) 3.0% Fat and 8% SNF

540. The standard of double toned milk should contain:
(A) 4.0% Fat and 9% SNF
(B) 1.5% Fat and 9.0% SNF
(C) 3.5% Fat and 8.5% SNF
(D) 2.0% Fat and 5% SNF

541. The standard of skimmed milk should contain:
(A) 1.0% Fat and 9% SNF
(B) 0.75% Fat and 9.5% SNF
(C) 0.5% Fat and 8.7% SNF
(D) 0.25% Fat and 5% SNF

542. Goat milk is easily digested because of:
(A) Small size fat globules and soft curd
(B) Low fat per cent
(C) Fat and protein per cent are equal to human milk
(D) All of these

543. The average milk productin of goat per lactation is:
(A) 50 litres (B) 60 litres
(C) 70 litres (D) 80 litres

544. The average milk production of per lactation of Barbari goat is:
(A) 80 litres (B) 90 litres
(C) 100 litres (D) 110 litres

545. The average fat and protein content in milk of mare is:
(A) 1.14 and 2.05 %, resp. (B) 2.0 and 2.5 %, resp.
(C) 2.5 and 3.5 %, resp. (D) 3.0 and 4.0%, resp.

546. The average protein and fat content in milk of donkey is:
(A) 1.85 and 1.37 %, resp. (B) 2.0 and 2.0 %, resp.
(C) 2.5 and 3.0 %, resp. (D) 3.0 and 3.5%, resp.

547. The average fat and protein content in milk of camel is:
(A) 1.50 and 2.05 %, resp. (B) 2.0 and 3.0 %, resp.
(C) 2.5 and 3.5 %, resp. (D) 4.0 and 3.07%, resp.

548. The normal milk should contain:
(A) 3.25% Fat and 8.5% SNF
(B) 4.5% Fat and 7.5% SNF
(C) 4.8% Fat and 7.2% SNF
(D) 5.0 % Fat and 7.7 % SNF

549. In general Indian cows have average fat per cent in the milk:
(A) 4.2 % (B) 4.5 %
(C) 4.9 % (D) 5.1 %

550. What is the average protein in the milk of Indian cows:
(A) 3.2 % (B) 3.7 %
(C) 3.9 % (D) 4.1 %

551. What is the average lactose in the milk of Indian cows:
(A) 4.2 % (B) 4.6 %
(C) 5.2 % (D) 5.5 %

552. What is the average total solids in the milk of Indian cows:
(A) 12.5 % (B) 13.5 %
(C) 13.9 % (D) 14.1 %

553. What is the average ash in the milk of Indian cows:
(A) 0.5 % (B) 0.6 %
(C) 0.7 % (D) 0.8 %

554. What is the average fat in the milk of Indian buffaloes:
(A) 6.2 % (B) 7.2 %
(C) 8.2 % (D) 9.2 %

555. What is the average protein content in the milk of Indian buffaloes:
(A) 3.5 % (B) 3.6 %
(C) 3.7 % (D) 3.8 %

556. What is the average lactose per cent in the milk of Indian buffaloes:
(A) 4.4 % (B) 4.6 %
(C) 4.8 % (D) 5.0 %

557. What is the average ash per cent in the milk of Indian buffaloes:
(A) 0.52 % (B) 0.65 %
(C) 0.76 % (D) 0.82 %

558. What is the average S.N.F. per cent in the milk of Indian buffaloes:
(A) 9.3 % (B) 9.8 %
(C) 10.2 % (D) 10.7 %

559. What is the total solids per cent in the milk of Indian buffaloes:
(A) 16.5 % (B) 16.9 %
(C) 17.2 % (D) 17.5 %

560. Which of the following cow has the highest fat per cent in the milk?
(A) Sindhi (B) Haryana
(C) Tharparkar (D) Shahiwal

561. The colour of milk is:
(A) Colourless (B) White
(C) Grey (D) None of these

562. The main reason of milk fat rancidity is:
(A) Conversion of fat into methyl ketones
(B) Conversion of fat into butyric acid
(C) Conversion of fat into hydroperoxide
(D) All of these

563. Which of the following protein is available in curd?
(A) Lactose (B) Casein
(C) Glycerol (D) All of these

564. Which of the following casein is highest available in milk?
(A) α Casein (B) β Casein
(C) γ Casein (D) All of these

565. Lacto globulin is known mainly for:
(A) It provides immunity to newly born babies
(B) It saves people from heat strokes
(C) It increases body weight
(D) All of these

566. Lactose in milk is a:
(A) Carbohydrate (B) Fat
(C) Protein (D) All of these

567. Which of the following amylase is present in milk?
(A) α amylase (B) β amylase
(C) γ amylase (D) All of these

568. What is casein per cent in milk?
(A) 2.0-3.5 (B) 4.5-5.5
(C) 5.5-6.5 (D) None of these

569. What is the ratio of lactose : protein : ash in cow milk is?
(A) 15 : 9 : 2 (B) 10 : 9 : 2
(C) 15 : 8 : 2 (D) None of these

570. In comparison to cow and goat milk has fat globules of:
(A) Larger size (B) Smaller size
(C) Same (D) All of these

571. Which of the following is considered TB free milk?
(A) Goat milk (B) Cow milk
(C) Sheep milk (D) Camel milk

572. Lactometer is used to measure the:
(A) Relative density of milk
(B) Relative density of water
(C) Relative density of fat
(D) Relative density of protein

573. What will the effect of mixing of SNF in milk:
(A) Relative density will increase
(B) Relative density will decrease
(C) Relative density will not change
(D) All of these

574. What is the average acidity of milk?
(A) 0.11 – 0.16 % (B) 0.18 – 0.20 %
(C) 0.25 – 0.30 % (D) 0.35-0.40 %

575. What is the average viscosity of milk?
(A) 0.7 – 1.0 poise (B) 1.5 – 2.0 poise
(C) 2.8- 3.0 poise (D) 3.1-3.5 poise

576. What is the ratio of lactose : protein : ash in buffalo milk is?
(A) 5 : 9 : 2 (B) 10 : 9 : 2
(C) 12 : 10 : 2 (D) 15 : 1 : 2

577. Which of the following will increase with advancement of lactation period?
(A) Ca and P (B) Cl
(C) Fat (D) All of these

578. What will be the impact of high temperature on milk fat?
(A) Increase (B) Decrease
(C) Same (D) All of these

579. If the interval of milking increases, the fat per cent in milk will:
(A) Increase (B) Decrease
(C) No change (D) Not known

580. The fat per cent in evening milk is found higher because of:
(A) Less interval in milking
(B) Cow produces more concentrated milk in evening than morning
(C) Cow eats more feed in day and gives higher fat per-ent in evening
(D) None of these

581. Which of the following milk has higher per cent of fat?
(A) Fore milk (B) Middle milk
(C) Stripped milk (D) All of these

582. During heat period, the fat per cent in milk will:
(A) Increase (B) Decrease
(C) No change (D) None of these

583. During heat period, the milk will:
(A) Increase (B) Decrease
(C) No change (D) Not known

584. How much energy in available from 1 gm milk?
(A) 8 calorie (B) 9 calorie
(C) 10 calorie (D) 12 calorie

585. Ca : P ratio in milk is:
(A) 1 : 1 (B) 1.43 : 1
(C) 1.5 : 1 (D) 2.11 : 1

586. What is the temperature range for survival of psychrophilic bacteria?
(A) 3-20°C (B) 20-50°C
(C) 50-70°C (D) None of these

587. What is the temperature range for survival of mesophilic bacteria?
(A) 3-20°C (B) 20-50°C
(C) 50-70°C (D) 75-80°C

588. What is the temperature range for survival of thermophilic bacteria?
(A) 3-20°C (B) 20-50°C
(C) 50-70°C (D) 75-80°C

589. *Pseudomonas fragi* and *Pseudomonas fluorescens* belong to which of the following category?
(A) Psychrophilic bacteria (B) Mesophilic bacteria
(C) Thermophilic bacteria (D) All of these

590. *Streptococcus lactis* and *S. cremoris* belong to which of the following group?
(A) Psychrophilic bacteria (B) Mesophilic bacteria
(C) Thermophilic bacteria (D) Thermoduric bacteria

591. Which of the following is the thermophilic bacteria?
(A) *Micrococcus luteus*
(B) *Micrococcus condidus*
(C) *Streptococcus thermophilus*
(D) All of these

592. *Lactobacillus, thermophilus* and *Bacillus calidolactis* are the names of:
(A) Psychrophilic bacteria (B) Mesophilic bacteria
(C) Thermophilic bacteria (D) Thermoduric bacteria

593. Sweet curding of milk is caused by:
(A) *Streptococcus subtilis* (B) *Bacillus mycoides*
(C) *B. cereus* (D) All of these

594. The oiliness or cardboard flavour or sourness in milk is caused by:
(A) *Streptococcus liquefaciens*
(B) *Micrococcus pitiutoporus*
(C) *Pseudomonas flourescens*
(D) All of these

595. Oiliness of milk is caused mainly due to:
(A) Oxidation of fat (B) Reduction of fat
(C) Fermentation of lactose (D) Fermentation of fat

596. Sweet curdling is caused mainly due to:
(A) Fermentation of casein (B) Fermentation of fat
(C) Oxidation of ash (D) Fermentation of ash

597. Normal fermentation of milk is caused by:
(A) *Streptococcus lactis*
(B) *Bacillus subtilis*
(C) *Streptococcus liquefaciens*
(D) All of these

598. Rancid flavour in milk is caused due to:
(A) Fermentation of fat
(B) Fermentation of lactose
(C) Fermentation of protein
(D) Fermentation of ash

599. Sliminess or ropy fermentation is caused by:
(A) *Achromobactor aerogans*
(B) *Bacterium cloacae*
(C) *Alkaligenes viscosus*
(D) All of these

600. Sliminess is caused due to:
(A) Fermentation of milk cream
(B) Fermentation of lactose
(C) Fermentation of protein
(D) None of these

601. Frothiness in milk is caused due to:
(A) *Aerobacter aerogens* (B) *Torula cremoris*
(C) *Torula spherical* (D) All of these

602. Blue colour of milk is developed due to:
(A) *Pseudomonas syncyanea*
(B) *Streptococcus lutea*
(C) *Streptococcus marcescens*
(D) All of these

603. Red colour of milk is developed due to:
(A) *Pseudomonas syncyanea*
(B) *Streptococcus lutea*
(C) *Streptococcus marcescens*
(D) All of these

604. Yellow colour of milk is developed due to:
(A) *Pseudomonas syncyanea*
(B) *Streptococcus lutea*
(C) *Streptococcus marcescens*
(D) All of these

605. Which of the following yeast is found in milk?
(A) *Torula cremoris* (B) *Torula spherical*
(C) *Torula lactis* (D) All of these

606. Which of the following fungus involved in milk industry?
(A) *Penicillium glaucum* (B) *Penicillium rogueforti*
(C) *Oidium spp.* (D) All of these

607. Roquefort and Gorgonzola paneer are made with the help of:
(A) Bacteria (B) Yeast
(C) Fungi (D) All of these

608. Which of the following bacteria acts first on milk?
(A) *Lactobacillus* (B) *Streptococcus lactis*
(C) *Bacillus subtilis* (D) None of these

609. Germicidal property of milk is due to presence of:
(A) Lactenin (B) Fatty acids
(C) Amino acids (D) All of these

610. Food poisoning is caused by:
(A) *Staphylococcus aureus* (B) *Bacillus spp.*
(C) *Streptococcus lactis* (D) All of these

611. Milking pot may be cleaned with:
(A) Clorine water (B) Hot water
(C) Soda water (D) All of these

612. Milking pot may be sterilized by:
(A) Keep pot in water at 82^{0}C for 15 minutes
(B) Keep pot in water at 101.6^{0}C for 15 minutes
(C) Keep pot in clorine water for 15 minutes
(D) None of these

613. The pathogenic and non-pathogenic bacteria may be killed if the milk is heated at the temperature:
(A) 62.8^{0}C for 30 minutes (B) 60^{0}C for 15 minutes
(C) 62.8^{0}C for 15 minutes (D) 80^{0}C for 45 minutes

614. At which temperature the milk is heated for pasteurization at temperature long time (L.T.L.T.) method:
(A) 62.8^{0}C for 30 minutes (B) 60^{0}C for 15 minutes
(C) 62.8^{0}C for 15 minutes (D) 70^{0}C for 30 minutes

615. Batch pasteurization method refers to:
(A) Milk is heated at 62.8 to 65.6^{0}C for 30 minutes
(B) Milk is heated with the help of either stream or hot water
(C) Milk is filled in big containers
(D) All of these

616. Bottle pasteurization refers to:
(A) Hot milk is filled in hot bottles before pasteurization
(B) The bottle are then heated in hot water at 62.8 to 65.6^{0}C for 30 minutes
(C) The filled bottles may be heated in hot air chamber at 64^{0}C for 30 minutes
(D) All of these

617. What is the benefit of bottle pasteurization:
(A) High keeping quality
(B) Free from pathogenic bacteria
(C) Free from *Bacillus coli*
(D) All of these

618. High temperature short time (HTST) method refers to:
(A) Continuous pasteurization
(B) Flash method
(C) Heating milk at 71.7^{0}C for 15 minutes
(D) All of these

619. Vacuum pasteurization refers to:
(A) Milk is heated at 72.85^{0}C for few seconds
(B) Milk is cooled at 43.5^{0}C
(C) It is a continuous process
(D) All of these

620. Stassanization refers to:
(A) Discovered by Dr. Stassano for milk pasteurization
(B) Milk is heated at 74^{0}C for 15-16 seconds
(C) Milk is cooled at 15^{0}C temperature
(D) All of these

621. Uperization of milk means:
(A) Milk is heated at 148.8^{0}C for 1 second
(B) Milk is depressed to remove bad flavour
(C) Milk is aerated just to avoid loss of vitamins and Vitamin C
(D) All of these

622. What is the benefit of uperization of milk?
(A) Milk is completely sterilized
(B) No effect on quality of milk
(C) Bad flavour is removed
(D) All of these

623. Bactofugation refers to:
(A) A method developed by Prof. S. Monart to remove bacteria from milk instead of killing bacteria
(B) Milk is heated at 75^{0}C
(C) Milk is run under speed centrifuge section to remove 99.99% bacteria
(D) All of these

624. What is the benefit of bactofugation of milk?
(A) Bacteria killed up to 99.99%
(B) Milk flavour remains like fresh milk
(C) Higher keeping quality than pasteurization milk
(D) All of these

625. Which of the thermoduric bacteria is found in pasteurized milk?
(A) *Micrococcus epidermidis*
(B) *Microbacterium liquefaciens*
(C) *Streptococcus faecalis*
(D) All of these

626. Which of the thermophilic bacteria is found in pasteurized milk?
(A) *Micrococcus epidermidis*
(B) *Lactobacillus thermophilus*
(C) *Streptococcus lactis*
(D) All of these

627. Pasteurized milk may be infected with:
(A) Mesophilic bacteria
(B) Psychrophilic bacteria
(C) Thermophilic bacteria
(D) Thermoduric bacteria

628. Sterilization of milk refers:
(A) Milk is heated at 93.3 to 99.4^{0}C for 30 minutes
(B) Milk is heated at 80^{0}C for 20 minutes
(C) Milk is heated at 90^{0}C for 20 minutes
(D) All of these

629. Bottle milk is sterilized at:
(A) 100-110^0 C for one hour (B) 98^{0}C for 40 minutes
(C) 100^{0}C for 20 minutes (D) All of these

630. Which of the following is the main benefit of sterilization of milk?
(A) Increase keeping quality (B) Free from all bacteria
(C) Increase digestibility (D) All of these

631. Homogenization of milk refers to:
(A) To break fat globules and mix uniformly in the milk
(B) To avoid deposition of creamy layer at surface
(C) To break fat globules and allow to deposit at surface
(D) All of these

632. Homogenization of milk requires:
(A) 60-65.5°C temperature and 2000-2500 lbs per square inch pressure
(B) 40°C temperature and 2000 lbs per square inch pressure
(C) 60-65.5°C temperature and 1500 lbs per square inch pressure
(D) All of these

633. At what temperature milk is chilled at chilling centres?
(A) 4.4°C (B) 18.2°C
(C) 17.4°C (D) 20.1°C

634. What are the cooling agent is used for cooling the milk?
(A) Salt solution (B) Ammonia solution
(C) Ice (D) All of these

635. Cream is a:
(A) Part of milk which has higher fat per cent than milk
(B) Generally it has >25% fat
(C) It is collected from milk through gravity or centrifugal forces
(D) All of these

636. Cream has more dark yellow colour than milk is due to:
(A) Lower carotene and xanthophylls in milk
(B) Higher carotene and xanthophylls in cream
(C) Cream has higher density than milk
(D) Milk has higher density than cream

637. What is percentage of fat in table cream?
(A) 20-25 (B) 30-40
(C) 65-85 (D) 70-75

638. What is percentage of fat in whipping cream?
(A) 20-25 (B) 30-40
(C) 50-60 (D) 70-75

639. What is percentage of fat in plastic cream?
(A) 20-30 (B) 40-50
(C) 65-85 (D) 70-75

640. What is the relative density of cream?
(A) 0.93 (B) 0.95
(C) 1.03 (D) 1.30

641. The separator slime is consisted of:
(A) Dust particles
(B) Fodder and feed particles
(C) Bacteria and animal body parts
(D) All of these

642. What is the average fat per cent in cream that is separated by centrifugal method?
(A) 18-45 (B) 25-50
(C) 35-55 (D) 45-65

643. What is the average fat per cent in cream that is separated by gravity method?
(A) 10-25 (B) 25-50
(C) 30-50 (D) 45-65

644. Sterilized canned cream has fat per cent :
(A) 20 (B) 25 (C) 30 (D) 35

645. Frozen canned cream has fat per cent :
(A) 10 (B) 20 (C) 30 (D) 40

646. Fruit canned cream has fat per cent :
(A) 20-25 (B) 25-30
(C) 30-35 (D) 35-40

647. Which of the following is responsible for better flavour in cream?
(A) Diacetyl
(B) Acidic and Propionic acid
(C) Acetyl methyl carbinol
(D) All of these

648. What is the optimum amount of diacetyl in cream for better flavour?
(A) 0.2-0.6 ppm (B) 0.7-1.5 ppm
(C) 1.0-2.0 ppm (D) 1.2-2.5 ppm

649. Separata refers to:
(A) Milk without cream
(B) Milk after separated butter
(C) Milk with cream
(D) All of these

650. What is the fat per cent in separated milk?
(A) 3.2-3.5 (B) 1.5-2.0
(C) 1.2-1.8 (D) 0.01-0.8

651. What is the Solid Not Fat (SNF) per cent in separated milk?
(A) 4.5-5.3 (B) 9.52-10.15
(C) 15.5-16.5 (D) 20.1-25.0

652. Butter milk refers to:
(A) The part of milk and cream that left after separating butter
(B) Milk without cream
(C) Part of milk left after separating butter
(D) All of these

653. Butter milk contain fat per cent:
(A) 0.35-0.60 (B) 0.9-1.0
(C) 1.0-1.8 (D) 1.9-2.2

654. SNF per cent in butter milk is:
(A) 0.65-0.73 (B) 1.0-1.5
(C) 1.6-2.0 (D) 2.5-3.2

655. Whey refers to:
(A) Part of milk left after separating paneer
(B) Part of milk left after separating cream
(C) Part of milk after separating butter
(D) None of these

656. What is the casein per cent in whey?
(A) 0.02 (B) 0.05
(C) 0.10 (D) 0.40

657. What is the average fat per cent in whey?
(A) 0.70 (B) 0.60
(C) 0.50 (D) 0.30

658. What is the average water per cent in whey?
(A) 93.2 (B) 91.3
(C) 90.5 (D) 85.5

659. Which of the following has higher SNF?
(A) Whey (B) Milk
(C) Butter milk (D) Separata

660. Which of the following has higher mineral matters?
(A) Ghee (B) Separata
(C) Butter milk (D) Milk

661. What is the suitable temperature for cream pasteurization?
(A) 44°C for 30 minutes (B) 58°C for 30 minutes
(C) 62°C for 30 minutes (D) 70°C for 20 minutes

662. Why cream pasteurization is required?
(A) To get higher fat per cent in butter
(B) To improve taste, texture and flavour of butter
(C) To get homogenous butter
(D) All of these

663. What is the breaking stage of cream churning?
(A) Separation of fat globules from lassi/butter milk/separata
(B) Formation of fat globules like sorghum seed size
(C) Churning glass become clear
(D) All of these

664. What is the suitable cream churning temperature during summer season?
(A) 9-11°C (B) 11-13°C
(C) 13-15°C (D) 15-18°C

665. What is the suitable cream churning temperature during winter season?
(A) 9-11°C (B) 11-13°C
(C) 13-15°C (D) 15-18°C

666. Which of the following is used for colour development in butter?
(A) Annatto (B) Yellow
(C) Blue (D) Red

667. What per cent of milk is used for ice cream preparation?
(A) 0.8 (B) 1.8
(C) 2.8 (D) 3.8

668. Ice cream should have minimum fat per cent:
(A) 10 (B) 15
(C) 18 (D) 21

669. Ice cream should have minimum acidity per cent?
(A) 0.25 (B) 0.35
(C) 0.50 (D) 0.75

670. Maximum sugar per cent in ice cream:
(A) 10 (B) 12
(C) 15 (D) 18

671. Ice cream should have minimum total solids:
(A) 36 (B) 37
(C) 42 (D) 50

672. What is the fat per cent in ice cream?
(A) 2-4 (B) 6-8
(C) 10-12 (D) 15-18

673. Pearson method is used for:
(A) Standardization of fat in milk and cream
(B) Milk adulteration
(C) Judging the quality of cheese
(D) All of these

674. Red colour formation after mixing milk and resorcinol indicates the adulteration of milk with:
(A) Water (B) Sucrose
(C) Starch (D) All of these

675. Boudovin test is used for detection of:
(A) Adulteration of sesame oil in ghee
(B) Adulteration of tallow in ghee
(C) Adulteration of cotton oilin ghee
(D) All of these

676. Nitric acid is used for detection of:
(A) Adulteration of wax, vegetable ghee and animal fat in ghee
(B) Adulteration of vegetable oil in ghee
(C) Adulteration of cotton oil in ghee
(D) All of these

677. Alezarine test is useful for testing the:
(A) Milk pH (B) Milk fat
(C) Milk flavour (D) All of these

678. Soda ash test is used to detect:
(A) Adulteration of vegetable fat and animal fat in ghee
(B) Adulteration of preservatives in ghee
(C) Adulteration of potato in ghee
(D) All of these

679. Milk pasteurization can be detected by:
(A) Phosphate test (B) Boudovine test
(C) Nickel test (D) All of these

680. Milk fat can be tested with the help of:
(A) Butyrometer (B) Lactometer
(C) Thermometer (D) All of these

681. Which of the following are tested under organolaptic tests:
(A) Milk appearance (B) Milk taste
(C) Milk flavour (D) All of these

682. Hansa test is famous for:
(A) Detection of buffalo milk in mixture
(B) Detection of *Lactobacillus lactis* in cream
(C) Detection of antiserum in cow milk
(D) None of these

683. Halphen test is used to detect:
(A) Adulteration of vegetable oil in ghee
(B) Adulteration of animal fat in ghee
(C) Adulteration of cotton oil in ghee
(D) All of these

684. Iodine test is used to detect the milk adulterated with:
(A) Starch (B) Sucrose
(C) Water (D) All of these

685. Valenta test is used to detect:
(A) Adulteration of vegetable oil in pure ghee
(B) Adulteration of butter in pure ghee
(C) Adulteration of animal fat in ghee
(D) None of these

686. Nitric acid test detects the milk adulteration with:
(A) Starch (B) Separata
(C) Milk powder (D) All of these

687. Bomer's phytosterol test is used to detect:
(A) Adulteration of vegetable oil in pure ghee
(B) Adulteration of butter in pure ghee
(C) Adulteration of tallow in ghee
(D) None of these

688. Which of the following is used as milk preservative?
(A) Formalin (B) Borex and Boric acid
(C) Salicylic acid (D) All of these

689. Nitrate test is used for detection of:
(A) Nitrate in milk (B) Formalin in milk
(C) Borex in milk (D) None of these

690. Brine solution is used as:
(A) Cleaning agent (B) Refrigerant
(C) Heating agent (D) None of these

691. Which of the following can be mixed in ghee?
(A) Vegetable ghee (B) Lard and tallow
(C) Safflower oil (D) All of these

692. Pure ghee should have Reichert Meissal value:
(A) >28 (B) <24
(C) >15 (D) None of these

693. Vieth ratio may be used for:
(A) Adulteration of milk (B) Urea in milk
(C) Fat estimation in milk (D) None of these

694. Which of the following is a vieth ratio?
(A) 3 : 9 : 2 – Lactose : Protein :Mineral matter
(B) 3 : 9 : 2 - Protein : Lactose : Mineral matter
(C) 3 : 9 : 2 – Fat : Protein : Lactose
(D) All of these

695. Milk powder may be prepared from:
(A) Milk (B) Separata
(C) Cream (D) All of these

696. Condensed milk may be prepared from:
(A) Normal milk (B) Separata milk
(C) Both milk and separata (D) All of these

697. Condensed milk should be stored properly at:
(A) 4.5°C (B) 5.5°C
(C) 6.5°C (D) 7.5°C

698. Evaporated milk may be made of:
(A) Sugar free milk (B) Sweet milk
(C) Both type (D) None of these

699. Bitterness in cheese is caused by:
(A) High acidity (B) Poor quality of milk
(C) Low salt per cent (D) All of these

700. Rancidity in cheese is caused by:
(A) Lipolytic organisms (B) High acidity
(C) Low salt per cent (D) None of these

701. Pasty in cheese is caused by:
(A) High moisture and acidity
(B) Low moisture and acidity
(C) High salt per cent
(D) All of these

702. If the humidity is less than 70%, the cheese may get:
(A) Dry (B) Wet
(C) Fungus growth (D) All of these

703. If the humidity is more than 70%, the cheese may get:
(A) Dry (B) Wet
(C) Fungus growth (D) All of these

704. Which of the following cheese has the highest protein per cent:
(A) Cheddar (B) Swiss
(C) Limberger (D) Camembert

705. Evaporated milk means:
(A) Sugar free milk is condensed in vaccum
(B) Sweet milk is condensed in open pan
(C) Milk is condensed at 40°C
(D) None of these

706. Evaporated milk is homogenized at:
(A) 45-50°C (B) 55-60°C
(C) 60-65°C (D) 80-85°C

707. Paraffining of cheese means:
(A) To cover cheese with thin layer of wax
(B) To cover cheese with thin layer of salt
(C) To cover cheese with thin layer of colour
(D) All of these

708. The paraffining of cheese is done at temperature:
(A) 106-115°C (B) 100-105°C
(C) 95-98°C (D) 80-85°C

709. Unripened paneer is known as:
(A) Cheddar (B) Swiss
(C) Cottage (D) All of these

710. Which of the following is a soft paneer?
(A) Limberger (B) Camembert
(C) Cottahe (D) All of these

711. Which of the following paneer is without gas holes?
(A) Cheddar (B) Swiss
(C) Limberger (D) All of these

712. Which of the following paneer is with gas holes?
(A) Cheddar (B) Swiss
(C) Limberger (D) All of these

713. Paneer is mainly source of:
(A) Milk fat (B) Milk protein
(C) Milk lactose (D) All of these

714. Which type of paneer is most popular in the world?
(A) Swiss (B) Cheddar
(C) Limberger (D) Roquefort

715. Cheddar paneer is prepared by which milk:
(A) Buffalo milk (B) Cow milk
(C) Both milk (D) None of these

716. How much per cent of khoa can be prepared from cow milk?
(A) 15.5 (B) 18.3
(C) 20.6 (D) 22.5

717. How much per cent of khoa can be prepared from buffalo milk?
(A) 15.5 (B) 18.3 (C) 21.6 (D) 22.5

718. Which of the following is considered the most important parameter for judging the quality of khoa?
(A) Flavour (B) Body and texture
(C) Colour (D) Packing

719. Which of the following has higher protein per cent?
(A) Milk (B) Ice cream
(C) Cream (D) Butter

720. Rancidity of icecream is caused by:
(A) High acidity (B) Hydrolysis of fat
(C) High over run (D) None of these

721. Foamy of icecream is caused by:
(A) High acidity (B) Hydrolysis of fat
(C) High over run (D) None of these

722. Curdy of icecream is due to:
(A) High acidity (B) Hydrolysis of fat
(C) High over run (D) None of these

723. Yellow level ghee is known as:
(A) Cow ghee (B) Buffalo ghee
(C) Normal ghee (D) Special ghee

724. Blue level ghee is known as:
(A) Cow ghee (B) Buffalo ghee
(C) Normal ghee (D) Special ghee

725. Red level ghee is known as:
(A) Cow ghee (B) Buffalo ghee
(C) Normal ghee (D) Special ghee

726. Grey level ghee is known as:
(A) Cow ghee (B) Buffalo ghee
(C) Normal ghee (D) Special ghee

727. Tollowy flavour in ghee is caused by:
(A) Oxidation (B) Reduction
(C) Hydrolysis (D) Protolysis

728. What is the melting point of ghee?
(A) 30-41°C (B) 35-40°C
(C) 45-50°C (D) 51-55°C

729. The iodine value of ghee is:
(A) 26-35 (B) 40-50
(C) 55-60 (D) 80-90

730. What is the refractive index of ghee?
(A) 1.4538-1.4578 (B) 1.4696- 1.4718
(C) 1.4580-1.4620 (D) 1.4596- 1.4649

731. Which of the following milk has higher amount of lower fatty acid (i.e. Capric, Caprylic and Caproic acids)?
(A) Cow milk (B) Goat milk
(C) Buffalo milk (D) Camel milk

732. What is fat per cent of horse milk?
(A) 1.67 (B) 2.17
(C) 2.57 (D) 2.87

733. What is fat per cent in elephant milk?
(A) 17.57 (B) 19.57
(C) 15.57 (D) 20.57

734. What is fat per cent in camel milk?
(A) 4.07 (B) 3.07
(C) 5.07 (D) 6.07

735. The fat per cent of diarrhea infected animal will be:
(A) Low (B) More
(C) Same (D) None of these

736. The fat per cent of mastitis infected animal will be:
(A) Low (B) More
(C) Same (D) None of these

737. The milk pH of mastitis infected animal will be:
(A) Increase (B) Decrease
(C) Same (D) Unknown

738. Which of the protein is not found in milk?
(A) Casein (B) Albumin
(C) Globulin (D) All of these

739. Who reported first time the germicidal property in goat milk?
(A) Fokker (B) R. Hook
(C) Mc Millon (D) Michael

740. Milk pasteurization refers to:
(A) To kill bacteria before its use
(B) To heat milk at 60°C for 20 minutes
(C) It does not affect the quality of milk and cream layer
(D) All of these

741. What is benefit of pasteurization of milk?
(A) No effect on Ca, P and Vit. A, D and B complex
(B) No or little effect on flavour
(C) No effect on cream layer
(D) All of these

742. What are the benefit of homogenization of milk?
(A) No creamy layer at milk surface
(B) High digestibility
(C) Better colour and flavour
(D) All of these

743. What should be the milk temperature at cream separation?
(A) 20-25°C (B) 30-37°C
(C) 40-42°C (D) 45-47°C

744. Separator slime is a:
(A) Concentrated and sticky product of milk that is found in the inner walls of bowl hood
(B) The colour is grey white to ash
(C) It is a mixture of impurities in milk
(D) All of these

745. Which country used centrifugal force first time to separate the cream from milk?
(A) Germany (B) Sweden
(C) Denmark (D) Pakistan

746. Which country used centrifugal force first time to separate the cream from milk?
(A) India (B) Sweden
(C) Denmark (D) Germany

747. Butter refers to:
(A) It is manufactured from either cream or milk
(B) It should not have <80% fat and >2% other ingredients
(C) It should not have >16% water
(D) All of these

748. At what temperature the butter should be removed from curd?
(A) >27°C (B) <27°C
(C) 12-13°C (D) 35°C

749. What is used for cream standardization?
(A) Water (B) Milk
(C) Separata (D) Whey

750. At which temperature the cream should be stored?
(A) 15-21°C (B) 25-30°C
(C) 32-37°C (D) 35-40°C

751. How much starter is used for ripening of cream?
(A) 3-4% of cream (B) 5-6% of cream
(C) 1-2%of cream (D) 7-8% of cream

752. What is most suitable temperature for cream ripening?
(A) 18-21°C (B) 25-27°C
(C) 30-32°C (D) 35-40°C

753. Cream standardization is used only when cream contains fat per cent:
(A) >35% (B) <35%
(C) 25% (D) 50%

754. How are fat per cent and cream acidity related for cream churning?
(A) Positive (B) Linear
(C) Inverse (D) Negative

755. Fishy flavour in butter is developed due to:
(A) Hydrolysis of lecithin
(B) Oxidation of butter due to high acidity
(C) Oxidation of Fe and Cu
(D) All of these

756. Salty butter contains salt per cent:
(A) 0.5-1.0 (B) 1.5-2.0
(C) 2.5-3.0 (D) 3.0-3.5

757. What should be curd per cent in butter:
(A) 0.5 (B) 1.0
(C) 1.5 (D) 2.0

758. What should be lactose per cent in butter:
(A) 0.5 (B) 0.6
(C) 0.7 (D) 0.8

759. What should be water per cent in butter:
(A) 14 (B) 16
(C) 18 (D) 20

760. Mottled butter is caused by:
(A) Improper washing of butter
(B) Improper salt mixing
(C) Improper working of butter
(D) All of these

761. Rancidity of butter is caused by:
(A) Hydrolysis of fat in cream by lipase enzyme
(B) High acidity of cream
(C) Storage at low temperature
(D) All of these

762. Dahi refers to:
(A) Fermented milk products
(B) Lactose is converted into lactic acid
(C) Starter is mixed at 21°C milk temperature
(D) All of these

763. Which of the following bacteria is responsible for good flavour in dahi?
(A) *Streptococcus lactis*
(B) *Streptococcus diacetyl aromaticus*
(C) *Lactonostoc citrovorum*
(D) All of these

764. What are the following bacteria present in starter?
(A) *Lactobacillus acidophilus*
(B) *Streptococcus lactis*
(C) *L. bulgaricus*
(D) All of these

765. What are the incubation period of Dahi?
(A) 5-6 hours (B) 6-7 hours
(C) 8-10 hours (D) 12-15 hours

766. Dahi has more nutritive value than milk because of:
(A) Dahi has more Ca and P
(B) Dahi has bacteria that helps to prepare Vit. B and C
(C) Dahi casein is more digestible than milk casein
(D) All of these

767. Dahi like product refers to:
(A) Dahi made of *Lactobacillus acidophilus*
(B) Dahi made of *Streptococcus lactis*
(C) Dahi made of artificial milk
(D) All of these

768. In general, ghee has fat per cent :
(A) 99-99.5 (B) 80-85
(C) 75-80 (D) 70-75

769. What is optimum temperature for ghee storage?
(A) 0°C (B) 15-20°C
(C) 25-30°C (D) 30-35°C

770. What is the melting point of ghee?
(A) 30-41°C (B) 36-45°C
(C) 46-55°C (D) 56-65°C

771. What is the per cent of volatile fatty acid in ghee?
(A) 5-6 (B) 8-10
(C) 12-15 (D) 16-20

772. What is the per cent of nonvolatile fatty acid in ghee?
(A) 70-80 (B) 80-90
(C) 90-92 (D) 98-99

773. What is the main component of ghee adulteration?
(A) Vegetable fat (B) Pig fat
(C) Edible oil (D) None of these

774. Fat per cent in ice cream is important for:
(A) Softness (B) Smoothness
(C) Flavour (D) All of these

775. Ice cream may be pasteurized at:
(A) 60°C for 30 minutes (B) 68°C at 30 minutes
(C) 71°C for 30 minutes (D) All of these

776. Ice cream should be stored at:
(A) -2 to -3°C (B) -5 to -10°C
(C) 2-3 °C (D) None of these

777. Egg solids in ice cream increases:
(A) Whipping capacity (B) Body and texture
(C) Flavour (D) All of these

778. Ice cream contains fat per cent in comparison to milk:
(A) 4-5 times more (B) 7-8 times more
(C) 10-12 times more (D) 14-15 times more

779. Ice cream is hardened at:
(A) -31 to -41°C (B) -5 to -10°C
(C) -42 to -50°C (D) -15 to -20°C

780. Colour of cow khoa is:
(A) White (B) Light yellow
(C) Yellow (D) Colourless

781. Colour of buffalo khoa is:
(A) White (B) Light yellow
(C) Yellow (D) Colourless

782. In general , khoa contains total solids:
(A) 70-75% (B) 80-85%
(C) 60-65% (D) 20-25%

783. Chhana refers to:
(A) It is a milk product
(B) It is prepared after curding of milk
(C) The solid particles are separated from paneer water
(D) All of these

784. How much citric acid or lactic acid is used for curding of milk?
(A) 1-2% (B) 2-3%
(C) 3-4% (D) 4-5%

785. How much per cent of chhana can be prepared from cow milk?
(A) 13.8 (B) 20.8
(C) 30.3 (D) 33.3

786. How much per cent of chhana can be prepared from buffalo milk?
(A) 13.9 (B) 20.9
(C) 30.9 (D) 33.9

787. Better quality chhana has:
(A) Rose flavour (B) Rancid flavour
(C) Nutty flavour (D) None of these

788. Pasty chhana can be prepared by:
(A) Cow milk (B) Sheep milk
(C) Colostrum (D) Goat milk

789. Export quality cheddar paneer contains water per cent:
(A) <36 % (B) >36%
(C) <16 % (D) <26%

790. The fat per cent in milk should be in preparation of export quality of cheddar paneer:
(A) >3.5-4.0 (B) >5.0-6.0
(C) >7.0-8.0 (D) All of these

791. How much lactic acid bacteria culture is added in milk for paneer preparation?
(A) 0.2 - 1.0 % (B) 1.5 – 2.0 %
(C) 2.0 – 2.5 % (D) 2.5 - 3.0 %

792. The paneer curd is ripened at:
(A) 25-30°C (B) 36-40°C
(C) 45-50°C (D) 45-50°C

793. The prepared paneer curd should be dried at:
(A) 5-10°C (B) 10-12°C
(C) 15-20°C (D) 20-25°C

794. Which of the following is main reason of paraffining of cheese?
(A) The holes of cheese are closed
(B) The cheese moisture is not changed
(C) The cheese is protected from fungal growth
(D) All of these

795. When should colour be added in milk in the process of cheddar preparation?
(A) After adding of lactic acid or before rennet adding
(B) After rennet adding or before lactic acid adding
(C) Before adding of lactic acid or rennet
(D) All of these

796. Evaporated milk means:
(A) Sugar free milk is condensed in vaccum
(B) Sweet milk is condensed in open pan
(C) Milk is condensed at 40°C
(D) None of these

798. Milk powder means:
(A) Evaporated milk in powder form with <3% moisture
(B) Milk powder has 30% water
(C) It is like a condensed milk
(D) None of these

799. Condensed milk means:
(A) Milk with 30% water
(B) Milk with 30% total solids
(C) Milk with 9% fat
(D) All of these

800. What is the optimum sugar per cent in condensed milk?
(A) 10-15 (B) 15-20
(C) 25-30 (D) 30-35

801. What should be minimum fat per cent in milk powder prepared by normal milk?
(A) 15 (B) 20
(C) 25 (D) 30

802. What should be minimum fat per cent in milk powder prepared by cream?
(A) 45 (B) 55
(C) 65 (D) 75

803. If lactometer reading of a milk sample is 32, what will be the relative density?
(A) 1.032 (B) 1.034
(C) 1.036 (D) 1.038

804. Relative density of milk can be calculated with the help of following formula:
(A) $\frac{\text{Lactometer reading} + 1}{1000}$ (B) $\frac{1 - \text{Lactometer reading}}{1000}$
(C) $\frac{\text{Lactometer reading} - 1}{100}$ (D) None of these

805. What is the water content in the milk of cows?
(A) Nearly 86 % (B) Nearly 87 %
(C) Nearly 88 % (D) Nearly 89 %

806. Iodine value of milk is:
(A) 22-24 (B) 26-35
(C) 35-40 (D) 45-50

807. Polenske value of cow milk is:
(A) 0.7-3.5 (B) 4.2-4.5
(C) 4.5-6.5 (D) 6.6-7.0

808. Polenske value of buffalo milk is:
(A) 0.7-3.5 (B) 0.8-5.0
(C) 4.5-6.5 (D) 6.6-7.0

809. Pure ghee has a special odour due to presence of the following:
(A) Lactose (B) Diacetyle
(C) Vit. B complex (D) Minerals

810. Colestrol is a:
(A) Milk fat sterole (B) Phytosterole
(C) Phosphonized fat (D) All of these

811. What is sponification value of milk?
(A) 210-235 (B) 250-260
(C) 261-265 (D) 266-270

812. What is surface tension of normal milk?
(A) 40-60 dynes/cm (B) 80-100 dynes/cm
(C) 100-120 dynes/cm (D) 125-150 dynes/cm

813. On an average, goat milk has:
(A) 4.5 % fat and 9.0 % SNF
(B) 3.5 % fat and 8.5 % SNF
(C) 3.0 % fat and 8.5 % SNF
(D) 3.0 % fat and 8.0 % SNF

814. Which of the following two elements are not found in proper amount in normal milk?
(A) Fe and Cu (B) Ca and Mg
(C) K and Ca (D) All of these

815. Which of the following essential amino acid is not found in optimum dose in normal milk?
(A) Riboflavin (B) Nicotinic acid
(C) Thymine (D) None of these

816. The optimum acidity of butter making cream is:
(A) 0.2-0.4% (B) 0.5-0.6%
(C) 0.8-1.0% (D) 1.0-1.2%

817. What is the disadvantages of deshi method of butter preparation?
(A) Butter may have cooked flavour
(B) Loss of Vitamin B_2 and C due to more heating of milk
(C) More water in butter
(D) All of these

818. What will happen if cream of more acidity is churned for butter?
(A) To reduce fat recovery
(B) Bad odour developed
(C) Reduced keeping quality of butter
(D) All of these

819. What should be average salt per cent in butter?
(A) 1.0 (B) 1.2
(C) 1.4 (D) 1.6

820. Cheesy butter is caused due to:
(A) Breaking of casein by proteolytic bacteria
(B) Low salt in butter
(C) Pasteurization at high temperature
(D) All of these

821. What is the reason for leaky butter?
(A) Churning of cream at higher temperature
(B) Low working
(C) Storage at high temperature
(D) All of these

822. The butter should be stored at:
(A) 0°C (B) <5°C
(C) >5°C (D) 10°C

823. What is the optimum acidity of dahi?
(A) 0.40-0.50% (B) 0.65-0.75%
(C) 0.80-0.90% (D) 1.00-1.20%

824. What should be the acidity of milk for dahi preparation?
(A) 0.21% (B) 0.19%
(C) 0.17% (D) 0.15%

825. Keeping quality of dahi in comparison to milk is:
(A) More (B) Less
(C) Same (D) Unknown

826. Which of the following affects the quality of dahi?
(A) Flavour (B) Texture and body
(C) Acidity (D) All of these

827. Moisture per cent in ghee should not be:
(A) >1% (B) <1%
(C) >1.5% (D) <1.5

828. Which of the following has highest marks in judging the quality of ghee?
(A) Flavour (B) Body and texture
(C) Colour (D) Acidity

829. B.T. test is used for identification of:
(A) Vegetable oil in ghee (B) Cotton seed oil in ghee
(C) Pig fat in ghee (D) None of these

830. Sandiness of ice cream is due to:
(A) High fat % (B) Low lactose %
(C) Lactose crystallization (D) None of these

831. Sandiness of ice cream can be reduced by:
(A) Increasing fat % (B) Increasing lactose %
(C) Reducing fat % (D) None of these

832. Ice cream mixture should be stored for ageing at:
(A) 0-4.4°C (B) 4.5-5.0°C
(C) 6-8°C (D) 9-10°C

833. What is the role of ice cream ageing?
(A) It solidifies the ice cream
(B) It increases whipping ability
(C) It also improves flavour of ice cream
(D) All of these

834. 100 g ice cream is equal to:
(A) 206 calories (B) 218 calories
(C) 306 calories (D) 350 calories

835. What is main difference between cow and buffalo khoa?
(A) Cow khoa contains higher per cent of water, protein and lactose
(B) Buffalo khoa contains higher per cent of fat and ash
(C) Cow khoa contains higher Fe than buffalo khoa
(D) All of these

836. At what temperature khoa should be stored?
(A) 5-10°C (B) 15-20°C
(C) 25-30°C (D) 35-40°C

837. At what temperature, citric acid should be mixed in milk for curding?
(A) 50°C (B) 60°C
(C) 70°C (D) 82°C

838. The quality parameters of judging chhana are given below in decreasing importance. Which of the following order is correct?
(A) Flavour>Texture>Colour and body>Modified product
(B) Texture>Flavour>Colour and body>Modified product
(C) Flavour>Colour and body>Texture>Modified product
(D) All of these

839. What is colouring agent is mixed in milk for paneer preparation?
(A) 0.003-0.012% (B) 0.04-0.05%
(C) 0.5-0.7 (D) 1.0-1.2%

840. Which of the following parameter is most important for judging quality of cheese?
(A) Flavour (B) Body
(C) Texture (D) Colour

841. What is the freezing point of normal milk?
(A) 0.544°C (B) -0.125°C
(C) -0.025°C (D) None of these

842. Krischner value of pure cow ghee should be in the range of:
(A) 20-25 (B) 25-30
(C) 30-35 (D) 35-40

843. Salt is added in paneer curd at the rate of:
(A) 1.2-1.5% (B) 2.0-2.5%
(C) 3.0-3.5 (D) 3.5-4.0%

844. One bull is sufficient for a bar of cows having:
(A) 20 cows (B) 30 cows
(C) 50 cows (D) 60 cows

845. Nurse cow refers to:
(A) A cow that feeds the calves of another cow
(B) A feeding cow of her own calves
(C) A diseased cow
(D) None of these

846. Which of the following method is considered best, feeding system of calves?
(A) Liberal milk feeding
(B) Limited milk feeding plus a dry calf starter
(C) Using milk replacers
(D) None of these

847. Which of the following is a calf feeding system?
(A) Liberal milk feeding
(B) Using milk replacers
(C) Limited milk feeding plus a dry calf starter
(D) All of these

848. Which of the following is a external parasite of cattle?
(A) Screw worm (B) Live fluke
(C) Protozoa (D) All of these

849. Which of the following is a infectious disease of cattle?
(A) Rinderpest (B) FMD
(C) Brucellosis (D) All of these

850. Which of the following disease is not a infectious disease?
(A) FMD
(B) Black quarter
(C) Haemorrhagic Septicaemia
(D) Mastitis

851. High temperature and bloody discharge from natural opening are the major symptoms of:
(A) FMD (B) Mastitis
(C) Anthrex (D) Black quarter

852. Which of the following is a serious threat to young calves?
(A) Pneumonia (B) Milk fever
(C) Warts (D) Calf scours

853. Milk fever is caused by:
(A) Ca deficiency in blood (B) P deficiency
(C) Virus (D) Bacteria

854. Milk fever disease is found in:
(A) Cows (B) High producing cows
(C) Calves (D) Camel

855. Which of the following is a internal parasite of cattle?
(A) Screw worm (B) Ticks
(C) Liver fluke (D) Ring worm

856. Which of the following disease of cattle is not caused by virus?
(A) Rinderpest (B) Cowpox
(C) Black quarter (D) FMD

857. Tympanitis is a:
(A) Disorder caused by over feeding resulting more gas production known as a bloat
(B) Disease of cattle caused by bacteria
(C) Instrument used for castrating male animals
(D) It is highly infectious disease of cattle

858. What is correct for mastitis?
(A) Caused by bacteria
(B) It is a disease of lactating dairy cattle
(C) Milk will be stringy and sometimes bloody
(D) All of these

859. What is correct for ketosis?
(A) It is a disease caused by shortage of sugar in body
(B) High producing cows are most affected
(C) Affected animals are nervous and loss their appetite
(D) All of these

860. Which of the following is known as Bang's disease?
(A) Brucellosts (B) FMD
(C) Rinderpest (D) Black quarter

861. Which of the following method is used for making of flock/ewes/rams?
(A) Paint brands (B) Metal tags
(C) Ear notches (D) All of these

862. Which of the external parasite of sheep?
(A) Blow fly (B) Lice
(C) Ticks (D) All of these

863. Bloody severe diarrhea is a main symptoms of:
(A) Coccidiosis (B) Black quarter
(C) Rinderpest (D) Mastitis

864. Shooting diarrhea with fetid smell and high temperature are the symptoms appear in sheep due to the disease of:
(A) Sheep pox (B) Rinderpest
(C) Black quarter (D) Anthrax

865. Sheep scab is caused by:
(A) Mites (B) Lice
(C) Fungus (D) Bacteria

866. The continuous sniffing and nose run in sheep are caused by:
(A) Liver fluke (B) Sheep bots
(C) Ticks (D) Lice

867. Which of the bacterial disease of goat?
(A) Anthrax (B) Brucellosis
(C) Vibriosis (D) All of these

868. Which of the following goat's disease is caused by virus?
(A) Goat pox (B) Rinderpest
(C) Pleuropnemonia (D) All of these

869. Dystocia is a:
(A) Condition in which goat has difficulty in kidding due to malposition of kids
(B) Disease caused by internal parasites
(C) Bacterial disease
(D) Viral disease

870. Trichomoniasis is:
(A) Contagious protozoan veneral disease of both sexes
(B) Infectious bacterial disease of goats
(C) Infectious viral disease of goats
(D) None of these

871. Ketosis is a serious disease of goat herd caused by:
(A) Metabolic disorder of carbohydrates and fats
(B) Bacteria
(C) Virus
(D) External parasites

872. Which of the following parasite causes disease in goat herds?
(A) Flukes (B) Tapeworms
(C) Protozoan (D) All of these

873. Florosis in goats is caused by:
(A) Excess florine (B) Excess P
(C) P deficiency (D) None of these

874. Mycotic dermetitis is caused by:
(A) Fungus (B) Bacteria
(C) Roundworm (D) Liver flukes

875. Which of the following disease is not found in equines?
(A) Epizoatic Lymphangitis (B) Tetanus
(C) Colic (D) FMD

876. Which of the most important disease of camel?
(A) Camel pox (B) Rabies
(C) Surra (D) All of these

877. Surra disease is transmitted with the help of:
(A) Water
(B) Housefly
(C) Biting and blood sucking flies
(D) Mosquitoes

878. The important symptoms of surra is:
(A) High body temperature (B) Anemia
(C) Emaciation (D) All of these

879. Rabies is transmitted to camels by:
(A) Bites of rabid dogs (B) Jackal
(C) Wolves (D) All of these

880. Which of the following disease is found in camel?
(A) Surra (B) Camel pox
(C) Anthrax (D) All of these

881. The main symptoms of rabies in camel is:
(A) Aggressiveness
(B) Glassy look
(C) Dribbling of saliva from mouth
(D) All of these

882. The main symptoms of Anthrax of camel is:
(A) High body temperature
(B) Cessation of rumination
(C) Rapid breathing
(D) All of these

883. Camel pox is generally found in:
(A) Young camel (B) Adult camel
(C) Old camel (D) None of these

884. Camel pox is a:
(A) Contagious disease
(B) Non contagious disease
(C) Infectious disease
(D) Non-infectious disease

885. The common symptoms of camel pox is:
(A) Lesion on skin of upper and lower lips
(B) Lesions on other body parts too at acute form
(C) High temperature
(D) All of these

886. Which of the following disease of pigs is caused by virus?
(A) FMD (B) Hog Cholera
(C) Swine pox (D) All of these

887. Diamond skin disease of pigs is known as:
(A) Anthrax (B) Swine crysipelas
(C) Swine pox (D) FMD

888. Which of the following disease is found in newly born pigs?
(A) Dysentery (B) Piglet influenza
(C) Novel ill (D) None of these

889. Which of the following internal parasite causes damage to pigs?
(A) Roundworm (B) Mange
(C) Lice (D) None of these

890. Mange disease of pigs is caused by:
(A) Lice (B) Mites
(C) Roundworm (D) Virus

891. The bio control of all external parasites of pigs is:
(A) 0.5% DDT solution
(B) 1.0% BHC solution
(C) Pasting of ash of maize cobs
(D) None of these

892. New castle disease (Ranikhet) in chicks refers to:
(A) Viral disease first discovered in Ranikhet
(B) Highly infections and caused death from 25 to 100%
(C) Chicks sneeze, gasp and have difficult in breathing
(D) All of these

893. Coccidiosis disease is very serious in poultry because of:
(A) Usually occurs in 3-9 weeks old chickens and caused by protozons
(B) Blood diarrhea leads to quick death
(C) Outbreaks occur most often during warm and humid region
(D) All of these

894. Fowl pox (chicken pox or Sore-Head) refers to:
(A) It is a viral infections poultry disease
(B) It has two forms either affects skin and lambs or mucous membranes
(C) Small blister like spots on face, comb and wattles are major symptoms
(D) All of these

895. Larygotracheitis refers to:
(A) Highly communicable and fetal viral poultry disease
(B) Labored breathing, gasping, rattling with head extended and beak open are main symptoms
(C) Vaccinate all birds at the age of 6 weeks to avoid this disease
(D) All of these

896. Which of the following external parasite is a serious threat to poultry?
(A) Round worms (B) Tapeworms
(C) Ticks (D) All of these

897. Infectious Bovine Rhinotrachitis (IBR) is a:
(A) Disease of bovines caused by virus mainly excretes through semen
(B) Bacterial disease
(C) Protozoan disease
(D) Nutritional disorder

898. Infectious coryza is a:
(A) Serious disease of poultry in India
(B) Caused by *Hemophilus gallinarum* bacterium
(C) Nasal discharge and sneezing are common symptoms
(D) All of these

899. Bronchitis in poultry refers to:
(A) Viral disease but death loss is highest among young
(B) Raspy cough, gasp and discharge from nose and eyes
(C) Egg production suddenly drops
(D) All of these

900. C- ELISA kit is useful for:
(A) Diagnosis and treatment of rinderpest disease
(B) It is a vaccine against T.B.
(C) It is a technique to test milk fat in a adulterated milk
(D) It is a machine used for quick milking

901. How much total digestible crude protein and TDM are required by a draft bullock weighing 400 kg for normal work?
(A) 0.45 kg crude protein and 4.0 kg TDM
(B) 0.56 kg crude protein and 4.9 kg TDM
(C) 0.57 kg crude protein and 4.9 kg TDM
(D) 0.71 kg crude protein and 6.4 kg TDM

902. The ration of young dairy stock should contain the following(s)?
(A) Vit. A, D and B complex (B) Protein
(C) Antibiotics (D) All of these

903. How much milk is to be fed upto first to 3-4 weeks old calves?
(A) 1/8 to 1/10th of body weight
(B) 1/5to1/10th of body weight
(C) 1/3 to 1/4th of body weight
(D) All of these

904. How much does a 300 kg weight heifer requires digestible crude protein, TDM, Ca and P, respectively for proper growth and development?
(A) 0.47 kg crude protein, 4.0 kg TDM, 13 g Ca and 12 g P
(B) 0.40 kg crude protein, 3.0 kg TDM, 13 g Ca and 12 g P
(C) 0.48 kg crude protein, 5.0 kg TDM, 12 g Ca and 12 g P
(D) 0.28 kg crude protein, 2.0 kg TDM, 13 g Ca and 11 g P

905. A calf starter refers to:
(A) High energy ration for calves
(B) A kind or liquid milk fed to calves
(C) A kind of dried milk to calves
(D) All of these

906. Generally calf starter contains:
(A) Ground grain of cereals
(B) Iodized salt and aureomycin
(C) Steamed bone meal
(D) All of these

907. How much aureomycin hydrochloride (Aureomycin) can be given to calf per day?
(A) 50 mg (B) 80 mg
(C) 100 mg (D) 120 mg

908. PETA refers to:
(A) People for Ethical Treatment of Animals
(B) People for Exotic Treated Animals
(C) Problems in Exotic Treated Animals
(D) None of these

909. Feed requirement of pregnant milking goat is:
(A) 5.0-6.0 % DCP + 50-60 % TDN of body weight
(B) 3.5-4.0 % DCP + 50-60 % TDN of body weight
(C) 2.0-2.5 % DCP + 30-40 % TDN of body weight
(D) 4-5 % DCP + 30-40 % TDN of body weight

910. Which of the following feed stuff is most suitable for horses?
(A) Oat (B) Gram
(C) Barley (D) Ragi

911. How much water is required daily for a horse?
(A) 20-25 litres (B) 30-35 litres
(C) 45-55 litres (D) 60-65 litres

912. The main function of camel hump is:
(A) Acts as an energy food store in form of fatty acids
(B) It stores water
(C) It helps in riding
(D) All of these

913. Camel can tolerate the water loss upto:
(A) 20 % of body weight (B) 30% of body weight
(C) 40% of body weight (D) 50% of body weight

914. How much water a camel can drink at a time?
(A) 100 litres (B) 150 litres
(C) 165 lires (D) 175 litres

915. Generally a daily requirement of water to a camel is:
(A) 10-12 litres (B) 15-16 litres
(C) 18-36 litres (D) 45-50 litres

916. The camel can sustain without water for longer period because of:
(A) It has higher water in the blood plasma
(B) It releases out meager water in urine and faeces
(C) Both (A) and (B)
(D) None of these

917. Flushing refers to:
(A) Proper feeding of sows/gilts before two weeks of breeding
(B) Measuring of fat % in pig's carcasses
(C) Removing extra teats of boar
(D) None of these

918. The daily requirement of feed to 45 kg boar is:
(A) 1.0 – 1.25 kg (B) 2.0 - 2.5 kg
(C) 2.5 – 3.0 kg (D) 3.0 – 3.5 kg

919. The daily requirement of a mature sow is:
(A) 1.5 – 2.0 % of its body weight
(B) 0.5 – 1.0 % of its body weight
(C) 2.5 – 3.0 % of its body weight
(D) 2.0 – 2.5 % of its body weight

920. How much supplement ration should be given to a breeding sows?
(A) 100 gm (B) 150 gm
(C) 230-450 gm (D) 500-550 gm

921. The starter feed of chicken should contain the protein percentage of:
(A) 15-17 % (B) 20-22 %
(C) 28-30 % (D) 38-40 %

922. Milk and milk product order, 1992 is related to:
(A) Registration of dairy man if he handles>10000 litres or 500 MT milk solids/year
(B) Production of clean milk
(C) Production of butter
(D) Production of ghee

923. The weight of A grade extra large size egg is:
(A) >60 gm (B) 53-59 gm
(C) 45-52 gm (D) 38-44 gm

924. The weight of large size egg is:
(A) >60 gm (B) 53-59 gm
(C) 45-52 gm (D) 38-44 gm

925. The weight of medium size egg is:
(A) >60 gm (B) 53-59 gm
(C) 45-52 gm (D) 38-44 gm

926. The weight of small size egg is:
(A) >60 gm (B) 53-59 gm
(C) 45-52 gm (D) 38-44 gm

927. The average egg weight of geese is:
(A) 65 gm (B) 75 gm
(C) 85 gm (D) 128 gm

928. Fryer-raoster turkey refers to:
(A) 16 weeks old young turkeys of either sex having tender soft meat
(B) 5-7 months old young female turkeys with soft and tender meat
(C) 5-7 month old young male (Tom) turkeys
(D) All of these

929. Young hen turkey refers to:
(A) 16 weeks old young turkeys of either sex having tender soft meat
(B) 5-7 months old young female turkeys with soft and tender meat
(C) 5-7 month old young male (Tom) turkeys
(D) All of these

930. Young tom turkey refers to:
(A) 16 weeks old young turkeys of either sex having tender soft meat
(B) 5-7 months old young female turkeys with soft and tender meat
(C) 5-7 month old young male (Tom) turkeys
(D) All of these

931. The average weight of turkey's egg is:
(A) 70 gm (B) 75 gm
(C) 85 gm (D) 90 gm

932. Turkey production is not popular in India because of:
(A) High cost
(B) Low growth
(C) Poor laying capacity in summer
(D) All of these

933. The true stomach in ruminants is known as:
(A) Rumen (B) Abomasum
(C) Reticulum (D) Omasum

934. The main function of abomasums is:
(A) Storage of feed
(B) Squeezing of water from feed
(C) Cheating of feeds
(D) Secretion of gastric juice

935. Which of the following animal has not a simple digestive tract?
(A) Swine (B) Poultry
(C) Goat (D) Men

936. How much amino acids are essential for poultry ration?
(A) 7 (B) 10
(C) 11 (D) 15

937. How much amino acids are essential for simple stomach animals?
(A) 7 (B) 10
(C) 11 (D) 15

938. Which of the following antibiotic is added in the animal feed?
(A) Aureomycin (B) Terramycin
(C) Penicillin (D) All of these

939. Which of the following is a major mineral required for animal's growth and development?
(A) Salt (B) P
(C) Ca (D) All of these

940. Stilbestrel refers to:
(A) An artificial hormone
(B) Resembles with natural female hormone estrogen
(C) Improves growth rate of animals
(D) All of these

941. Transquilizers have:
(A) Calming effect upon livestock
(B) Growth stimulating effect
(C) They also reduces shrink in cattle when given prior to shipping
(D) All of these

942. Iodinated casein is a:
(A) Similar to thryroxine hormone
(B) It induces greater milk production
(C) It is more useful to brood sows
(D) All of these

943. Main function of Gizzard is:
(A) Grind and crush the coarse feed
(B) Digestive juices are secreted and mixed with feed
(C) Chewing of feed
(D) All of these

944. The function of reticulum is:
(A) To hold foreign harmful bodies
(B) To furnish additional storage space
(C) Both (A) and (B)
(D) Squeezing out the water from feed

945. Roughages contain:
(A) Large amount of fibre
(B) Large amount of non digestible materials
(C) Low digestibility
(D) All of these

946. Normal body temperature of cattle is:
(A) 38.3°C (B) 37.3°C
(C) 39.3°C (D) 40.5°C

947. Normal body temperature of buffalo is:
(A) 37.2-38.0°C (B) 39.2-39.9°C
(C) 40-41°C (D) 42-44°C

948. Respiration rate of cattle is:
(A) 15-20 per minute (B) 25-30 per minute
(C) 40-60 per minute (D) 70-80 per minute

949. Normal body temperature of sheep is:
(A) 38.2°C (B) 38.8°C
(C) 39.4°C (D) 40.1°C

950. Pulse rate of sheep is:
(A) 50-55 per minute (B) 60-65 per minute
(C) 68-90 per minute (D) 95-100 per minute

951. Respiration rate of sheep is:
(A) 15-20 per minute (B) 20-30 per minute
(C) 40-50 per minute (D) 45-55 per minute

952. Normal body temperature of goat is:
(A) 38.5°C (B) 39.4°C
(C) 37.9°C (D) 40.1°C

953. Pulse rate of goat is:
(A) 50-55 per minute (B) 60-65 per minute
(C) 68-90 per minute (D) 95-100 per minute

954. Respiration rate of goat is:
(A) 15-20 per minute (B) 20-30 per minute
(C) 40-50 per minute (D) 45-55 per minute

955. Normal body temperature of horse is:
(A) 36.8°C (B) 37.8°C
(C) 38.8°C (D) 39.8°C

956. Pulse rate of horse is:
(A) 36-40 per minute (B) 42-45 per minute
(C) 46-52 per minute (D) 55-60 per minute

957. Respiration rate of horse is:
(A) 8-16 per minute (B) 18-20 per minute
(C) 25-28 per minute (D) 30-35 per minute

958. Normal body temperature of donkey is:
(A) 37.5-38.0°C (B) 38.5-39.5°C
(C) 39.5-40.0°C (D) 40.2-40.8°C

959. Pulse rate of donkey is:
(A) 35 per minute (B) 38 per minute
(C) 40 per minute (D) 42 per minute

960. Respiration rate of donkey is:
(A) 12 per minute (B) 15 per minute
(C) 18 per minute (D) 21 per minute

961. Normal body temperature of camel is:
(A) 35.7°C (B) 36.1°C
(C) 37.2°C (D) 38.5°C

962. Pulse rate of camel is:
(A) 32-44 per minute (B) 40-46 per minute
(C) 50-52 per minute (D) 55-57 per minute

963. Normal body temperature of pig is:
(A) 38.2°C (B) 38.9°C
(C) 39.1°C (D) 39.6°C

964. The causal organism of T.B. is:
(A) *Mycobacterium tuberculosis*
(B) *Mycobacterium typhii*
(C) *Streptococcus typhii*
(D) *Bacillus tuberculosis*

965. Causal organism of undulant fever is:
(A) *Brucella melitensis* (B) *B. abortus*
(C) *Salmonella typhosa* (D) All of these

966. Scalet fever is caused by:
(A) *S. scartaline* (B) *S. lactis*
(C) *B. dysenteriae* (D) *Salmonelle sp.*

967. Typhoid fever is caused by:
(A) *S. lactis* (B) *Brucella abortus*
(C) *Salmonella typhosa* (D) All of these

968. Diphtheria is caused by:
(A) *S. lactis* (B) *Diplitheria bacillis*
(C) *B. dysenteriae* (D) All of these

969. Abortion is caused in cattle mainly by
(A) *Brucella abortus* (B) *Brucella melitensis*
(C) *Salmonella spp* (D) All of these

970. Paratyphoid fever is caused by:
(A) *Salmonella typhosa* (B) *Salmonda paratyphi*
(C) *Brucella abortus* (D) None of these

971. The colour of colostrums is yellow due to presence of:
(A) Carotene (B) Blood particles
(C) Immune globulin (D) All of these

972. Which of the following element has lower content in colostrums than milk:
(A) Clorine (B) Iron
(C) Potassium (D) All of these

973. Colostrum is more concentrated than milk due to presence of:
(A) Total solids (B) Clorine ions
(C) Globulin (D) All of these

974. Vitamin A is stored in:
(A) Liver (B) Pancrease
(C) Gall bladder (D) Lungs

975. Which of the following is rich in coline?
(A) Ghee (B) Egg
(C) Liver (D) All of these

976. The fat per cent in egg is:
(A) 5.6% (B) 8.9%
(C) 11.8% (D) 12.8%

977. The protein per cent is egg is:
(A) 5.6% (B) 8.9%
(C) 11.8% (D) 12.8%

978. The dry fodder can be treated with urea at the rate of:
(A) 1% solution (B) 2% solution
(C) 4-5% solution (D) 5% solution

979. What should be the ratio of urea and molasses for treating the dry fodder?
(A) 1 : 1 (B) 1 : 5
(C) 1 : 9 (D) 1 : 11

980. The nutritive value of a fodder can be evaluated by which method:
(A) Digestibility trials
(B) Metabolic experiments
(C) Calorimetric measurement
(D) All of these

981. The crude protein content in succulent roughages is:
(A) 15-20% (B) 25-28%
(C) 35-47% (D) 41-43%

982. The water content in succulent roughages is:
(A) 25% (B) 48%
(C) 75% (D) 85%

983. Pulse rate of pig is:
(A) 25-30 per minute (B) 45-50 per minute
(C) 55-75 per minute (D) 80-85 per minute

984. Respiration rate of pig is:
(A) 10-15 per minute (B) 20-30 per minute
(C) 35-40 per minute (D) 45-50 per minute

985. Gastro-enteritis is caused mainly by:
(A) *Staphylococcus aureus* (B) *Proteus vulgaris*
(C) *Clostridium welchii* (D) All of these

986. Septic sore throat is caused by:
(A) *Vibrio sp.* (B) *S. scarlatine*
(C) *Diphtheria bacilli* (D) *S. epidermicus*

987. Undulant fever is very common in:
(A) Goat (B) Cow
(C) Sheep (D) All of these

988. Which among the following has the highest milk protein content:
(A) Cow (B) Buffalo
(C) Human (D) Goat

989. The principal sugar/carbohydrate in milk is:
(A) Glucose (B) Sucrose
(C) Lactose (D) Maltose

990. Milk is deficient in vitamin:
(A) Vitamin A (B) Vitamin B
(C) Vitamin D (D) Vitamin C

991. The yellow colour of milk products like ghee is due to:
(A) Cholesterol (B) Riboflavin
(C) Carotene (D) Xanthophyll

992. The greenish yellow colour of whey is due to:
(A) Riboflavin (B) Carotene
(C) Xanthophylls (D) Casein

993. Natural milk is in nature of:
(A) Basic (B) Acidic
(C) Highly acidic (D) Neutral

994. Sandiness in ice cream is due to:
(A) Casein (B) Lactose
(C) Phospholipids (D) Calcium

995. The milk pigment riboflavin is:
(A) Fat soluble (B) Water soluble
(C) Both (A) and (B) (D) None of these

996. Higher pH values of milk is due to:
(A) Casein (B) Mastitis
(C) Bacterial action (D) Light exposure

997. Lower pH values of milk is due to:
(A) Casein (B) Mastitis
(C) Bacterial action (D) Light exposure

998. Which among the following has highest specific gravity:
(A) Whole milk (B) Buffalo milk
(C) Skim milk (D) Standardized milk

999. Which among the following has highest specific gravity:
(A) Fat (B) Protein
(C) Lactose (D) Salt

1000. Addition of water to milk , the specific gravity:
(A) Decreases (B) Increases
(C) Do not affect (D) None of these

1001. The specific gravity of milk is increased upon addition of:
(A) Fat (B) Water
(C) Both (A) and (B) (D) Skim milk

1002. Milk freezes at a temperature:
(A) Similar to water (B) Less than water
(C) Greater than water (D) Milk does not freeze

1003. The heat stability of milk is judged by:
(A) MBRT (B) COB
(C) Lactometer (D) Freezing point

1004. Bactofugation is a process to remove bacteria by:
(A) Sedimentation (B) Centrifugation
(C) Precipitation (D) Homogenization

1005. Pearson's square method is used to:
(A) Detect adulteration in milk
(B) Bacteria removal
(C) Form a standardized milk
(D) None of these

1006. Homogenization of milk is carried out at:
(A) 2500 psi (B) 2000 psi
(C) 1500 psi (D) 1000 psi

1007. Acidophillus milk microorganisms is:
(A) *Lactobaccilus acidophilus*
(B) *Lactococcus ecidophilus*
(C) *Streptococcus acidophilus*
(D) *Streptococus lactis*

1008. Yoghurt microorganism is:
(A) *Streptococcus thermophillus*
(B) *Lactobacillus bulgaricus*
(C) Both (A) and (B)
(D) None of these

1009. In milk, whey proteins are present in:
(A) True solution (B) Suspended particle
(C) Coloidal solution (D) None of these

1010. Which of the following has highest particle size in milk:
(A) Fat globules (B) Casein
(C) Whey protein (D) Lactose

1011. Milk is a type of emulsion:
(A) Oil in water (B) Water in oil
(C) Not emulsion (D) Both (A) and (B)

1012. Most abundant milk proteins are:
(A) Whey proteins
(B) Albumin
(C) Casein
(D) Fat globule membrane proteins

1013. Which of the following is not a type of milk casein:
(A) Alpha (B) Beta
(C) Gama (D) Omega

1014. Milk serum proteins are also called:
(A) Whey protein (B) Casein
(C) Both (A) and (B) (D) None of these

1015. Which of the following enzyme is not found in milk system:
(A) Peroxidase (B) Catalase
(C) Phosphatase (D) Diastage

1016. The sunlight flavor of milk is due to oxidation of:
(A) Fat (B) Carbohydrate
(C) Protein (D) Vitamin

1017. Milk fat mainly consists of:
(A) Free fatty acids (B) TGA
(C) Mono and diglycerides (D) Phospholipid

1018. Fresh milk is:
(A) Slightly basic (B) Extremely acidic
(C) Slightly acidic (D) Extremely basic

1019. Developed acidity in milk is due to the accumulation of:
(A) Salts (B) Lactic acid
(C) Calcium ions (D) Phosphatas

1020. Bitter taste in milk is due to enzymatic breakdown of:
(A) Fats (B) Carbohydrates
(C) Proteins (D) Calcium

1021. Cowy flavour of milk is primarily due to:
(A) Acetanemia (B) Mastitis
(C) Ketosis (D) Both (A) and (B)

1022. Rancid flavor in milk is primarily due to:
(A) Caproic acid (B) Butyric acid
(C) Palmitic acid (D) Stearic acid

1023. Plastic cream consists of per cent of milk fat:
(A) 20-25 (B) 30-40
(C) 65-85 (D) More than 90

1024. The per cent of fat is ideal for whipping of cream:
(A) 25-30 (B) 30-35
(C) 35-40 (D) 40-45

1025. High milk fat per cent in cream is:
(A) Sour (B) Plastic
(C) Butter (D) Sweet

1026. Overworking of butter causes:
(A) Rancidity (B) Yeasty flavor
(C) Greasy texture (D) Gummy texture

1027. Gummy texture in butter arises due to:
(A) Over working (B) Under working
(C) High melting point fats (D) Both (A) and (B)

1028. Crumby texture in butter arises due to:
(A) Over working (B) Under working
(C) High melting point fats (D) Both (A) and (B)

1029. Butter is an emulsion of:
(A) Oil in water (B) Water in oil
(C) Not an emulsion (D) None of these

1030. Higher churning temperature results in the body of butter is:
(A) Strong (B) Weak
(C) Both (A) and (B) (D) None of these

1031. The optimum temperature for churning is:
(A) 9-11°C (B) 18-20°C
(C) 20-30°C (D) 30-35°C

1032. Greater the viscosity of cream during the churning period:
(A) Longer (B) Smaller
(C) Do not affect (D) None of these

1033. Size of fat globules during the churning period is:
(A) Smaller (B) Longer
(C) Do not affect (D) None of these

1034. Milk fat is a which type of fat:
(A) True (B) Mixed
(C) Both (A) and (B) (D) None of these

1035. The yellow colour of butter is due to:
(A) Alizarin (B) Sunset yellow
(C) Annatto (D) None of these

1036. The flavor of butter is due to:
(A) Ethanol (B) Esters
(C) Diacetyl (D) Propionates

1037. Normal butter has a curd content of:
(A) 0.6-1.0 % (B) 1-6 %
(C) 0.1-0.6 % (D) None of these

1038. Greater the size of fat globules, the fat loss in butter milk:
(A) Higher (B) Lower
(C) No effect (D) None of these

1039. Increased proportion of soft fats, the churning period will be:
(A) Longer (B) Shorter
(C) No effect (D) None of these

1040. Which of the following is not a property of annatto:
(A) Non toxic (B) Have flavor
(C) Water soluble (D) Oil soluble

1041. Riboflavin is also called:
(A) Vitamin C (B) Vitamin D
(C) Vitamin G (D) Vitamin K

1042. Milk SNF does not include:
(A) Protein (B) Sugar
(C) Mineral (D) Fat

1043. Corn syrup is a mixture of:
(A) Dextrose + maltose (B) Dextrose + lactose
(C) Dextrose + galactose (D) Maltose + maltose

1044. The principal flavouring material in vanilla is:
(A) Villin (B) Vanillin
(C) Vaniella (D) Vanole

1045. Continuous freezing take place, time than batch freezing:
(A) Longer (B) Shorter
(C) Equal (D) None of these

1046. Overrun in ice cream is due to:
(A) Addition of air (B) Addition of stabilizer
(C) Both (A) and (B) (D) None of these

1047. Dry ice is also known as:
(A) Solid O_2 (B) Solid N_2
(C) Solid CO_2 (D) Freon

1048. Curdy melt down in ice cream is due to:
(A) Addition of curd (B) High acidity of mix
(C) Low acidity of mix (D) None of these

1049. Sogginess in ice cream may be due to:
(A) High overrun (B) Low sugar content
(C) Excessive stabilizer (D) None of these

1050. Fluffy texture in ice cream is due to:
(A) Low overrun (B) Low sugar content
(C) High solid content (D) Low milk solids

1051. Icy texture of ice cream is due to:
(A) Fast hardening (B) Fast freezing
(C) High stabilizer (D) Low solid content

1052. Packaging of ice cream is done prior to freezing:
(A) True (B) False
(C) Cannot be determined (D) None of these

1053. Hardening of ice cream is done after packaging:
(A) True (B) False
(C) Cannot be determined (D) None of these

1054. Sandiness defect is mainly due to:
(A) Sucrose (B) Sucralose
(C) Galactose (D) Lactose

1055. Lactose is a sweeter than sucrose
(A) More (B) Less
(C) Equal (D) None of these

1056. In ice cream manufacturing ageing is done:
(A) Post freezing of mix (B) At the time of freezing
(C) Before freezing (D) After hardening

1057. Milk fat is delivered in the form of:
(A) Liquid fat (B) AMF
(C) Milk (D) None of these

1058. Which of the following is example of hard cheese:
(A) Cammenbert (B) Blue
(C) Roquefort (D) Swiss

1059. Rennet enzyme is composed of:
(A) Rennin and trypsin (B) Rennin and pepsin
(C) Trypsin and pepsin (D) None of these

1060. The bitterness in cheese is due to:
(A) Pepsin (B) Trypsin
(C) Rennin (D) None of these

1061. Function of salting in cheese is:
(A) Removal of whey
(B) Hardening of curd
(C) Retards formation of lactic acid
(D) All of these

1062. Parafinning is done to:
(A) Improve the flavor of cheese
(B) Improve the nutrient quality
(C) Reduce loss of moisture
(D) None of these

1063. Species of camel are:
(A) Bactrianus (B) Dromedatius
(C) Both (A) and (B) (D) None of these

1064. Species of dunkey is:
(A) Asinus (B) Arise
(C) Hircus (D) Cuballus

1065. The pH of the duodena juice from pigs ranges from:
(A) 4.5-5.6 (B) 6.2-7.9
(C) 8.4-8.9 (D) 8.0-8.2

1066. Zebu is a name of Indian breeds of cattle in:
(A) Europe (B) South America
(C) Both (A) and (B) (D) None of these

1067. Which of the following is not a exotic breed of cattle?
(A) Jersey (B) Hallikar
(C) Holstein –Friesian (D) Brown swiss

1068. Syneresis of cheese curd is called:
(A) Expulsion of whey (B) Contraction of curd
(C) Weeping of curd (D) All of these

1069. Pizza cheese is:
(A) Limburger (B) Mozarella
(C) Ricotta (D) Cottage

1070. Cottage cheese is:
(A) Unripened (B) Mold ripened
(C) Surface ripened (D) None of these

1071. Cottage cheese is:
(A) Semi hard (B) Semi soft
(C) Soft (D) Hard

1072. The curing of cheese is also known as:
(A) Ripening (B) Souring
(C) Maturing (D) All of these

1073. Green cheese refers to:
(A) Green colour cheese (B) Unripened cheese
(C) Not fully ripened (D) Both (B) and (C)

1074. Example of soft cheese is:
(A) Cheddar (B) Gouda
(C) Swiss (D) Camembert

1075. Eye cheese is:
(A) Cheddar (B) Gouda
(C) Swiss (D) Camembert

1076. Cheddaring is done to:
(A) Improve texture (B) Improve taste
(C) To improve shelf life (D) None of these

1077. Milling of cheese is done for:
(A) Size reduction (B) Texture development
(C) Improve shelf life (D) None of these

1078. Cottage cheese is prepared from:
(A) Whole milk (B) Skim milk
(C) Tonned milk (D) Double toned milk

1079. Pin hole defect in cheese occurs due to:
(A) Addition of pins
(B) Excess fat
(C) Contamination by microorganisms
(D) High acid

1080. Swiss holes is due to:
(A) Propionic bacteria
(B) Lactic acid bacteria
(C) Yeast
(D) Mold

1081. Cracked paraffin leads to:
(A) Appearance defect
(B) Increase in moisture content of cheese
(C) Increase contamination by microorganisms
(D) All of these

1082. The brown pigment causing discoloration of condensed milk is:
(A) Alizarin (B) Melanoidin
(C) Melano (D) Riboflavin

1083. Alcohol-alizarin test is used for:
(A) Heat stability of milk
(B) Fat determination
(C) Amino acid determination
(D) Mineral determination

1084. The crystal defect in condensed milk is chiefly due to:
(A) Sucrose (B) Lactose
(C) Maltose (D) Glucose

1085. The introduction of lactose in very fine powder form during cooling process to provide nuclei for crystallization is known as:
(A) Feeding (B) Seeding
(C) Heeding (D) Beading

1086. Evaporated milk differs from condensed milk in:
(A) Having more sugar (B) No added sugar
(C) No water (D) None of these

1087. Additive step in manufacture of evaporated milk is:
(A) Homogenization (B) Pasteurization
(C) Sterilization (D) Cooling

1088. Bloat defect in condensed milk is due to:
(A) Bacteria (B) Yeast
(C) Virus (D) Mold

1089. WMP stands for:
(A) White Milk Powder (B) Whole Milk Powder
(C) White Milk Paste (D) Whole Milk Paste

1090. SMP stands for:
(A) Skim Milk Powder (B) Small Milk Powder
(C) Skim Milk Paste (D) Small Milk Powder

1091. NFDM stands for:
(A) Non Fat Dairy Milk (B) New Fat Dry Milk
(C) Non Fat Dry Milk (D) Non Fat Dried Milk

1092. WPN stands for:
(A) Whey Protein Nitrogen (B) Whey Protein Nitrate
(C) White Powder Nitrogen (D) None of these

1093. Permitted antioxidant added to milk powder is:
(A) Phospholipids (B) BHA
(C) Magnesium chloride (D) None of these

1094. Khoa is similar to:
(A) Condensed milk (B) Evaporated milk
(C) Clotted cream (D) None of these

1095. Kheer is similar to:
(A) Condensed milk (B) Evaporated milk
(C) Clotted cream (D) None of these

1096. Rabri is similar to:
(A) Condensed milk (B) Evaporated milk
(C) Clotted cream (D) None of these

1097. Kulfi is similar to:
(A) Condensed milk (B) Evaporated milk
(C) Clotted cream (D) Ice cream

1098. Dahi is similar to:
(A) Condensed milk (B) Evaporated milk
(C) Yoghurt (D) Butter milk

1099. Panner is similar to:
(A) Cheese (B) Evaporated milk
(C) Clotted cream (D) None of these

1100. Chhana is similar to:
(A) Cheese (B) Evaporated milk
(C) Clotted cream (D) Green cheese

1101. Makkhan is similar to:
(A) Yoghurt (B) Butter
(C) Clotted cream (D) None of these

1102. Ghee is similar to:
(A) Yoghurt (B) Evaporated milk
(C) Butter oil (D) Butter milk

1103. Lassi is similar to:
(A) Condensed milk (B) Evaporated milk
(C) Clotted cream (D) Butter milk

1104. Which is a fermented milk product:
(A) Kheer (B) Khoa
(C) Chhana (D) Chakka

1105. Which is a fermented milk product:
(A) Lassi (B) Khoa
(C) Chhana (D) Chakka

1106. Which is a fermented milk product:
(A) Kheer (B) Khoa
(C) Shrikhand (D) Chakka

1107. Which of the following is a milk concentrate:
(A) Kheer (B) Khoa
(C) Mawa (D) All of these

1108. Which of the following is a milk concentrate:
(A) Butter (B) Kulfi
(C) Chhana (D) Paneer

1109. Kheer is also known as:
(A) Shrikhand (B) Rabri
(C) Chhana (D) Basundi

1110. Khoa is also known as:
(A) Mawa (B) Rabri
(C) Chhana (D) Basundi

1111. Pindi type khoa is used for preparation of:
(A) Pista burfi (B) Burfi
(C) Peda (D) All of these

1112. Dhap type khoa is used for preparation of:
(A) Pista burfi (B) Burfi
(C) Peda (D) Gulab jamun

1113. Danedaar type khoa is used for preparation of:
(A) Pista burfi (B) Kalakand
(C) Peda (D) All of these

1114. Color of cow milk khoa is:
(A) Pale yellow (B) Whitish
(C) Both (A) and (B) (D) None of these

1115. Color of buffallo milk khoa is:
(A) Pale yellow (B) Whitish
(C) Both (A) and (B) (D) None of these

1116. Body of cow milk khoa is:
(A) Slightly hard (B) Hard
(C) Soft (D) Very soft

1117. Body of buffalo milk khoa is:
(A) Slightly hard (B) Hard
(C) Soft (D) Very soft

1118. Texture of cow milk khoa is:
(A) Smooth (B) Granular
(C) Both (A) and (B) (D) Slightly sandy

1119. Texture of baffalo milk khoa is:
(A) Smooth (B) Granular
(C) Both (A) and (B) (D) None of these

1120. Which of the following is a characteristic of cow milk khoa is:
(A) Slightly salty (B) Slightly sweet
(C) Both (A) and (B) (D) None of these

1121. Which of the following is a characteristic of buffalo milk khoa is:
(A) Slightly salty (B) Slightly sweet
(C) Both (A) and (B) (D) None of these

1122. Which of the following occurs during khoa making from milk is:
(A) Change in state
(B) Change in colour intensity
(C) Homogenization of milk fat
(D) All of these

1123. Which of the following khoa is highly suitable for sweet making:
(A) Cow milk khoa (B) Buffalo milk khoa
(C) Camel milk khoa (D) Goat milk khoa

...ge is due to action of
(A) Am...
(B) Protease
(C) Lipase
(D) All of these

1125. Grity texture of khoa is due to:
(A) Sand like particles
(B) Lactose crystals
(C) Both (A) and (B)
(D) None of these

1126. Moldy surface of khoa may be due to :
(A) Excessive high humidity of storage room
(B) Excessive high moisture content of khoa
(C) Air leaky packaging of khoa
(D) All of these

1127. Coarse texture of khoa is due to:
(A) High acidity of milk
(B) Low fat content of milk
(C) High temperature of dehydration
(D) All of these

1128. Khurchan is a , whole milk product:
(A) Concentrated
(B) Sweetened
(C) Both (A) and (B)
(D) None of these

1129. Which of the following culture is used in sweet dahi preparation:
(A) *Streptococcus lactis*
(B) *S. diacetilactic*
(C) *S. cremoris*
(D) All of these

1130. Which of the following culture is used in sour dahi making:
(A) *Lactobacillus bulgaricus*
(B) *Streptococcus thermophilus*
(C) Both (A) and (B)
(D) None of these

1131. Which of the following is a type of sweetened dahi:
(A) Misti dahi
(B) Lal dahi
(C) Payodhi dahi
(D) All of these

1132. Which of the following is a characteristic of misti dahi:
(A) Cooked flavor
(B) Caramelized flavor
(C) Firm body
(D) All of these

1133. Dahi can be used for the preparation of:
(A) Chakka
(B) Shrikhand
(C) Makkhan
(D) All of these

1134. Shrikhand is generally prepared from:
(A) Skim milk
(B) Whole milk
(C) Both (A) and (B)
(D) None of these

1135. Basic ingredient for shrikhand is:
(A) Chakka
(B) Makkhan
(C) Ghee
(D) All of these

1136. Paneer is associated with:
(A) Rennet coagulation
(B) Soft cheese
(C) Both (A) and (B)
(D) None of these

1137. Which of the following affect the body and texture of chhana:
(A) Fat content of milk
(B) Acidity of milk
(C) Adulteration of milk
(D) All of these

1138. Chhana is used in the preparation of:
(A) Rossogulla
(B) Sandesh
(C) Pantooa
(D) All of these

1139. Which of the following is a traditional Indian dairy product:
(A) Ghee
(B) Makkhan
(C) Both (A) and (B)
(D) None of these

1140. Makkhan is also known as:
(A) Ghee
(B) Desi ghee
(C) Desi butter
(D) All of these

1141. Which of the following is the richest source of milk fat:
(A) Ice cream
(B) Butter
(C) Ghee
(D) Buttermilk

1142. Which of the following is associated with ghee making:
(A) Milk
(B) Curd making
(C) Butter
(D) All of these

1143. Which of the following is a method of ghee making
(A) Desi method
(B) Creamy butter method
(C) Direct cream method
(D) All of these

1144. Which of the following is a small scale method of ghee making
(A) Desi method
(B) Creamy butter method
(C) Direct cream method
(D) All of these

1145. Which of the following is a by-product of desi method of ghee making:
(A) Butter milk
(B) Ghee residue
(C) Both (A) and (B)
(D) None of these

1146. Which of the following is a by-product of creamy butter method of ghee making:
(A) Butter milk
(B) Ghee residue
(C) Skim milk
(D) All of these

1147. Which of the following is a by-product of direct cream method of ghee making:
(A) Butter milk
(B) Ghee residue
(C) Skim milk
(D) All of these

1148. Which of the following method give strong aroma of ghee:
(A) Pre-stratification method
(B) Direct cream method
(C) Continuous method
(D) Desi method

1149. Large grains in ghee is produced by:
(A) Desi method
(B) Creamery butter method
(C) Direct cream method
(D) All of these

1150. Granular form of ghee is developed during:
(A) Clarification
(B) Cooling
(C) Both (A) and (B)
(D) None of these

1151. Melting point (°C) of buffalo ghee is:
(A) 25
(B) 29
(C) 31
(D) 37

1152. Melting point (°C) of cow ghee is:
(A) 25
(B) 29
(C) 31
(D) 37

1153. In desi method of ghee making, item of fat loss is:
(A) Lassi (B) Ghee residue
(C) Handling (D) All of these

1154. In creamery butter method of ghee making, item of fat loss is:
(A) Butter milk (B) Ghee residue
(C) Skim milk (D) All of these

1155. Ag in AGMARK represents:
(A) Agronomy (B) Animal
(C) Agriculture (D) Dairy

1156. When high grade ghee is melted, it should be:
(A) Clear (B) Transparent
(C) Sediment free (D) All of these

1157. Ghee is used in:
(A) Cooking (B) Confectionery
(C) Direct consumption (D) All of these

1158. Lassi is also known as:
(A) Chhachh (B) Matha
(C) Both (A) and (B) (D) None of these

1159. Skim milk can be utilized for the preparation of:
(A) Dried skim milk (B) Cottage cheese
(C) Casein (D) All of these

1160. Buttermilk can be utilized for the preparation of:
(A) Condensed buttermilk (B) Dried buttermilk
(C) Soft cheese (D) All of these

1161. Whey can be utilized for the preparation of:
(A) Whey beverage (B) Ricotta cheese
(C) Primost cheese (D) All of these

1162. Ghee residue can be utilized for the preparation of:
(A) Ghee toffee (B) Sweet meat
(C) Both (A) and (B) (D) None of these

1163. Yoghurt is similar to:
(A) Dahi (B) Paneer
(C) Condensed milk (D) Lassi

1164. Which of the following is used as a sweetner in yoghurt:
(A) Sucrose (B) Glucose
(C) Maltose (D) All of these

1165. Which of the following is used as stabilizer in yoghurt:
(A) Starch (B) Alginate
(C) Carrageenan (D) All of these

1166. Kishk is made from:
(A) Low fat yoghurt (B) Salt
(C) Burghol (D) All of these

1167. Burghol is:
(A) Parboiled wheat (B) Parboiled rice
(C) Parboiled pearlmillet (D) None of these

1168. Quarg is generally made from:
(A) Skim milk (B) Whole milk
(C) Both (A) and (B) (D) None of these

1169. Which of the following is a property of cultured cream:
(A) Slight acidic (B) Mild cheese flavor
(C) Viscosity (D) All of these

1170. pH of synthetic milk is:
(A) 5.8 (B) 6.8
(C) 8.8 (D) 10.2

1171. Taste of synthetic milk is:
(A) Palatable (B) Bitter
(C) Sour (D) Sweet

1172. Synthetic milk become ——— on heating:
(A) White (B) Light green
(C) Yellowish (D) Purple

1173. Synthetic milk become ——— on storage:
(A) White (B) Light green
(C) Yellowish (D) Purple

1174. Which of the following gives positive test of urea:
(A) Natural milk (B) Synthetic milk
(C) Both (A) and (B) (D) None of these

1175. Synthetic milk gives test of urea:
(A) Strongly positive (B) Weakly positive
(C) Strongly negative (D) Weakly negative

1176. Natural milk gives test of urea:
(A) Strongly positive (B) Weakly positive
(C) Strongly negative (D) Weakly negative

1177. Which of the following is not a part of milk plasma:
(A) Fat globules (B) Casein micelles
(C) Globular protein (D) Lipoprotein

1178. Which of the following is not a part of milk serum:
(A) Fat globules (B) Casein micelles
(C) Globular protein (D) Lipoprotein

1179. Which of the following enzyme is present in milk:
(A) Lactoperoxidase (B) Superoxidismutase
(C) Lipoprotein lipase (D) All of these

1180. Which of the following enzyme is present in plasma portion of milk:
(A) Sulfhydryl oxidase (B) Superoxidismutase
(C) Both (A) and (B) (D) None of these

1181. Which of the following enzyme is present in casein micelles of milk:
(A) Lipoprotein (B) Plasmin
(C) Both (A) and (B) (D) None of these

1182. Which of the following enzyme is present in serum portion of milk:
(A) Lactoperoxidase (B) Ribonuclease
(C) Both (A) and (B) (D) None of these

1183. Which of the following decrease the water activity of milk:
(A) Concentration (B) Evaporation
(C) Both (A) and (B) (D) None of these

1184. Which of the following is associated with milk processing:
(A) Thermalization (B) Separation
(C) Homogenization (D) All of these

1185. Consistency of butter depends on:
(A) Temperature (B) Storage condition
(C) Working (D) All of these

1249. The foaming properties of egg is due to:
(A) Egg yolk (B) Egg white
(C) Both (A) and (B) (D) None of these

1250. The major portion of egg yolk solids is:
(A) Lipovitellins (B) Phosvitin
(C) Lipoprotein (D) All of these

1251. The colour of egg yolk is due to:
(A) Carotenoids (B) Xanthophylls
(C) Both (A) and (B) (D) Anthocyanin

1252. Pigments are present in:
(A) Egg white (B) Egg yolk
(C) Egg shell (D) None of these

1253. Most abundant mineral present in egg is:
(A) Phosphorus (B) Calcium
(C) Magnesium (D) Iron

1254. Most abundant mineral present in egg yolk is:
(A) Phosphorus (B) Calcium
(C) Magnesium (D) Iron

1255. The size of hen egg depends in:
(A) Season of laying (B) Stage of laying
(C) Body size of hen (D) All of these

1256. Smaller sized eggs contains ——— proportion of egg yolk:
(A) Higher (B) Lower
(C) Similar (D) None of these

1257. Yolk index is:
(A) Height of yolk
(B) Diameter/height of yolk
(C) Height/diameter of yolk
(D) Diameter of yolk

1258. Albumen index is:
(A) Height of thick albumen
(B) Diameter of thick albumen
(C) Height/diameter of albumen
(D) None of these

1259. An egg is graded as large if its weight (g) is:
(A) 60-65 (B) 53-59
(C) 45-52 (D) 38-44

1260. An egg is graded as medium if its weight (g) is:
(A) 60-65 (B) 53-59
(C) 45-52 (D) 38-44

1261. An egg is graded as small if its weight (g) is:
(A) 60-65 (B) 53-59
(C) 45-52 (D) 38-44

1262. As the alkalinity increases the egg white becomes:
(A) Thicker (B) Thinner
(C) No effect (D) None of these

1263. The most common contaminant of hen egg is:
(A) Salmonella (B) Pediococcus
(C) Bacillus (D) Candida

1264. Most of the egg production takes place in:
(A) Winter (B) Summer
(C) Rainy (D) Spring

1265. Thermostabilization is a method for prolonging storage life of egg. It involves heating egg at:
(A) 54°C/15 minutes
(B) Boiling water/5 seconds
(C) 68°C/30 minutes
(D) Both (A) and (B)

1266. When the egg is frozen, its shell:
(A) Cracks (B) Contracts
(C) No effect (D) None of these

1267. The yolk coagulates at a temperature ——— than egg white:
(A) Lower (B) Higher
(C) Similar (D) None of these

1268. The presence of salt, the coagulation temperature:
(A) Increase (B) Decrease
(C) No effect (D) None of these

1269. Soft and hard cooked eggs are those prepared:
(A) In shell (B) Outside shell
(C) Frying egg (D) None of these

1270. Designer eggs are those which have:
(A) Unique shapes
(B) Different vitamin content
(C) Lower cholesterol
(D) Both (A) and (B)

1271. The yolk is formed in ——— of hen:
(A) Ovaries (B) Oviduct
(C) Lower oviduct (D) None of these

1272. The albumen is formed in ——— of hen:
(A) Ovaries (B) Oviduct
(C) Lower oviduct (D) None of these

1273. Poultry refers to fowl reared for their:
(A) Flesh (B) Egg
(C) Feathers (D) All of these

1274. Chicken of 8-10 weeks with tender meat is called:
(A) Broiler (B) Rooster
(C) Stag (D) None of these

1275. Chicken of 3-5 months of age is called:
(A) Broiler (B) Rooster
(C) Stag (D) None of these

1276. A mature chicken cock under 10 months of age is called:
(A) Broiler (B) Rooster
(C) Stag (D) None of these

1277. A mature chicken more than 10 months of age is called:
(A) Stewing chicken (B) Rooster
(C) Stag (D) Both (A) and (B)

1278. A mature male chicken more than 10 months of age is called:
(A) Cock (B) Fowl
(C) Stag (D) Both (A) and (B)

1279. Which of the following is a hot water treatment:
(A) Roasting (B) Scalding
(C) Frying (D) None of these

1280. Scalding is done for:
(A) Cooking (B) Defeathering
(C) Bleeding (D) All of these

1281. Highest protein is present in which meat:
(A) Goose (B) Duck
(C) Turkey (D) Fowl

1282. Lowest protein is present in which meat:
(A) Goose (B) Duck
(C) Turkey (D) Fowl

1283. Highest fat is present in which meat:
(A) Goose (B) Duck
(C) Turkey (D) Fowl

1284. Lowest fat is present in which meat:
(A) Goose (B) Duck
(C) Turkey (D) Fowl

1285. Tandoor chicken is:
(A) Barbecued (B) Boiled
(C) Fried (D) Poached

1286. FPC stands for:
(A) Fat Protein Concentrate
(B) Fish Protein Concentrate
(C) Fish Putrid Concentrate
(D) None of these

1287. AMD stands for:
(A) Age Mediated Disease
(B) Age related Macular degeneration
(C) Age Mediated Deficiency
(D) None of these

1288. PCB stands for:
(A) Poly Chlorinated Biphenyls
(B) Poly Chlorine Biphenyls
(C) Poultry Chicken and Birds
(D) None of these

1289. Muslim method of slaughter is:
(A) Schechita (B) Kosher
(C) Halal (D) Terepha

1290. In Jhatka method of slaughter is damaged:
(A) Medulla oblongata (B) Back bone
(C) Stomach (D) Leg

1291. At rigor mortis:
(A) Meat is tender
(B) Meat is tough
(C) Meat water absorption capacity is low
(D) Both (B) and (C)

1292. Jewish slaughter is called:
(A) Schechita (B) Kosher
(C) Schochet (D) Terepha

1293. PSE stands for:
(A) Pale Soft and Exudate (B) Poly Soft and Exudate
(C) Protein Soft and Exudate (D) None of these

1294. DFD stands for:
(A) Dark Firm and Dry (B) Dry Firm and Dark
(C) Dry Fat and Dark (D) Dark Fat and Dry

1295. Bulk of the meat is formed of:
(A) Skeletal muscles (B) Cardiac muscles
(C) Unstriated muscles (D) None of these

1296. Female sheep which has not yet borne a lamb is called:
(A) Ether (B) Gimmer
(C) Both (A) and (B) (D) None of these

1297. Ideal age for slaughter of buffalo calves for meat is:
(A) 6-7 months (B) 18-24 months
(C) 3-4 years (D) 6-7 years

1298. Pigs are slaughtered at the age of:
(A) 6-7 months (B) 12-25 months
(C) 3-4 years (D) 6-7 years

1299. Male sheep castrated early in life are called:
(A) Ether (B) Gimmer
(C) Both (A) and (B) (D) None of these

1300. Meat from sheep of more than 12 months of age is called:
(A) Lamb (B) Chevan
(C) Mutton (D) Veal

1301. Meat from sheep of less than 12 months of age is called:
(A) Lamb (B) Chevan
(C) Mutton (D) Veal

1302. Nellore, madras red, trichy black are breeds of:
(A) Goat (B) Sheep
(C) Buffalo (D) Cow

1303. Mehsan, surti, gaddi and Marwari are breeds of:
(A) Goat (B) Sheep
(C) Buffalo (D) Cow

1304. A milch breed of cow is:
(A) Haryana (B) Malvi
(C) Kankrej (D) Sahiwal

1305. A draught breed of cattle is:
(A) Red Sindhi (B) Gir
(C) Malvi (D) Haryana

1306. An exotic breed of cow is:
(A) Ongole (B) Friesian
(C) Halliker (D) Deoni

1307. Hybrid breed of cattle is:
(A) Sunandini (B) Holstein
(C) Brown swiss (D) Kankrej

1308. Gestation period for buffalo is:
(A) 9 months (B) 14 months
(C) 10 moths (D) 21-22 months

1309. Cows and buffaloes remain in heat for:
(A) 24-36 hours (B) 24-36 days
(C) 7-10 days (D) 15-20 days

1310. Infertility of local breeds of cattle can be overcome by use of:
(A) Cross breeding with exotic breeds
(B) Good nourishment
(C) Stilbesterol
(D) Gonadotropin

1311. Stilbesterol is used for:
(A) Induction of lactation (B) Artificial insemination
(C) Super-ovulation (D) Cryopreservation

1312. Surrogate mother is:
(A) Mother without lactation
(B) Future mother with embryo implanted from another
(C) Carrying several embryos
(D) Artificially inseminated female

1313. Artificial insemination is better than natural insemination in cattle because:
(A) Semen of good bulls can be provided everywhere
(B) There is no likelihood of contagious diseases
(C) It is economical
(D) All of these

1314. Cryopreservation is:
(A) Preservation of living being in chemicals
(B) Preservation at very low temperature
(C) Preservation through exposure to irradiation
(D) Preservation through use of gases

1315. Cryopreservation is useful for:
(A) Preservation of semen (B) Very young fetuses
(C) Living cells (D) All of these

1316. Which one is most recent domestication:
(A) Buffalo (B) Sheep
(C) Turkey (D) Silkworm

1317. Sheep are reared mainly for:
(A) Meat (B) Mutton
(C) Wool (D) Milk

1318. Pashmina is obtained from a variety of:
(A) Sheep (B) Goat
(C) Rabbit (D) Yak

1319. Famous angora wool is got from a:
(A) Sheep (B) Rabbit
(C) Goat (D) Yak

1320. Lohi is:
(A) A breed of sheep
(B) A breed of goat
(C) A breed of Domestic Fowl
(D) A breed of Geese

1321. Merino sheep has been imported for:
(A) Increasing wool output through merino farms
(B) Increasing mutton availability
(C) Improving indigenous breeds
(D) Developing hybrid between sheep and goat

1322. Which one is poor men's cow:
(A) Sheep (B) Yak
(C) Goat (D) Llama

1323. Goats feed on:
(A) Low herbs (B) Grass
(C) Low shrubs (D) Shrubs and low trees

1324. Flesh of goat is:
(A) Pork (B) Meat
(C) Mutton (D) Beef

1325. Flesh of cow is
(A) Pork (B) Meat
(C) Mutton (D) Beef

1326. Flesh of pig is
(A) Pork (B) Meat
(C) Mutton (D) Beef

1327. Flesh of hen is:
(A) Pork (B) Chicken
(C) Mutton (D) Beef

1328. Domesticated livestock of high mountains is:
(A) Yak (B) Goat
(C) Sheep (D) Horse

1329. Genus of Cattle is:
(A) Bos (B) Bubalis
(C) Ovis (D) Capra

1330. Genus of buffalo is:
(A) Bos (B) Bubalis
(C) Ovis (D) Capra

1331. Genus of sheep is:
(A) Bos (B) Bubalis
(C) Ovis (D) Capra

1332. Genus of goat is:
(A) Bos (B) Bubalis
(C) Ovis (D) Capra

1333. Genus of swine is:
(A) Sus (B) Camelus
(C) Equus (D) Homo

1334. Genus of cammel is:
(A) Sus (B) Camelus
(C) Equus (D) Homo

1335. Genus of dunkey and horse is:
(A) Sus (B) Camelus
(C) Equus (D) Homo

1336. Genus of human is:
(A) Sus (B) Camelus
(C) Equus (D) Homo

1337. Species of cattles are:
(A) Taurus (B) Indicus
(C) Both (A) and (B) (D) None of these

1338. Species of buffalo is:
(A) Bubalis (B) Arise
(C) Hircus (D) Cuballus

1339. Species of sheep is:
(A) Bubalis (B) Arise
(C) Hircus (D) Cuballus

1340. Species of goat is:
(A) Bubalis (B) Arise
(C) Hircus (D) Cuballus

1341. Species of horse is:
(A) Bubalis (B) Arise
(C) Hircus (D) Cuballus

1342. Species of swine are:
(A) Scrofa (B) Indicus
(C) Domestic (D) All of these

ANSWERSHEET: 17.ANIMAL HUSBANDRY AND DAIRYING

1. (B); 2. (A); 3. (B); 4. (D); 5. (C); 6. (C); 7. (A); 8. (C); 9. (A); 10. (A); 11. (B); 12. (A); 13. (B) ; 14. (A);15. (A); 16. (A); 17. (D); 18. (C); 19. (D); 20. (C); 21. (B); 22. (C); 23. (B); 24. (B); 25. (A) ; 26. (B); 27. (B); 28. (D); 29. (B); 30. (D); 31. (C); 32. (D); 33. (A); 34.(A); 35. (B); 36. (B); 37. (B); 38. (D) ; 39. (B); 40. (A); 41. (C); 42. (B); 43. (B); 44. (D); 45. (A); 46. (B); 47. (B);48. (D); 49. (C); 50. (B); 51. (C) ; 52. (A); 53. (D); 54. (B); 55. (A); 56. (C); 57. (C); 58. (B); 59. (C); 60. (B); 61. (D); 62. (C); 63. (C); 64. (B); 65. (C); 66. (A); 67. (A); 68. (D); 69. (C); 70. (D); 71. (B); 72. (A); 73. (B); 74. (D); 75. (C); 76. (C); 77. (C); 78. (D); 79. (B); 80. (C); 81. (A); 82. (D); 83. (C); 84. (A); 85. (D); 86. (D); 87. (B); 88. (D); 89. (A); 90. (B); 91. (C) ; 92. (D); 93. (A); 94. (D); 95. (D); 96. (B); 97. (B); 98. (C); 99. (A); 100. (D);101. (B); 102. (D); 103. (C); 104. (A) ; 105.(C); 106. (C); 107. (D); 108. (D); 109. (A); 110. (B); 111. (D); 112. (B); 113. (C); 114. (B); 115. (B); 116. (C);117. (A); 118. (D); 119. (C); 120. (D); 121. (A); 122. (B); 123. (D); 124. (C); 125. (A); 126. (B); 127. (A); 128.(C); 129. (B); 130. (C); 131. (B); 132. (C); 133. (D); 134. (D); 135. (A); 136. (D); 137. (D); 138. (A); 139. (B); 140. (D); 141. (A); 142. (C); 143. (A) ; 144. (C); 145. (B); 146. (D); 147. (D); 148. (D); 149. (C); 150. (C); 151. (D); 152. (A); 153. (B); 154. (D); 155. (B); 156.(D); 157. (A); 158. (C); 159. (D); 160. (D); 161. (B); 162.(C); 163. (B); 164. (C); 165. (C); 166. (C); 167. (D); 168. (C); 169. (D) ; 170. (A); 171. (D); 172. (C); 173. (B); 174. (C); 175. (D); 176. (C); 177. (B); 178. (C); 179. (D); 180. (A); 181. (B); 182. (C); 183. (B); 184.(D); 185. (C); 186. (B); 187. (C); 188. (B); 189. (C); 190. (B); 191. (B); 192. (B); 193. (D); 194. (C); 195. (C); 196. (B); 197. (D); 198. (B); 199. (D); 200. (C); 201. (A); 202.(C); 203. (D); 204. (B); 205. (D); 206. (A); 207. (C); 208. (C) ; 209. (C); 210. (B); 211. (B); 212. (D); 213. (A); 214.(D); 215. (A); 216. (B); 217. (B); 218. (B); 219. (B); 220.(B); 221. (B); 222. (B); 223. (B); 224. (B); 225. (B); 226. (B); 227. (B); 228. (B); 229. (C); 230. (C); 231. (D); 232. (A);

233. (A); 234. (C);235. (A); 236. (A); 237. (D); 238. (D); 239. (C); 240. (C); 241. (D); 242. (A); 243. (C); 244. (B); 245. (C); 246. (B);247. (C) ; 248. (A); 249. (A); 250. (A); 251. (B); 252. (B); 253. (B); 254. (C);255. (A); 256. (A); 257. (A); 258. (D); 259. (D); 260. (C);261. (D); 262. (D); 263. (A); 264. (C); 265. (C); 266. (D);267. (D); 268. (D); 269. (B);270. (D); 271. (A); 272. (A); 273. (A);274.(A); 275. (D); 276. (A); 277. (D); 278. (C);279. (A); 280. (D); 281. (C); 282. (C); 283. (A); 284. (A); 285. (D); 286. (D); 287. (A); 288. (D); 289. (D); 290. (A); 291. (A);292. (A); 293. (C); 294. (A); 295. (D); 296. (C); 297. (A); 298. (A); 299. (D);300. (A); 301. (B); 302. (D); 303. (C);304. (B); 305. (D); 306. (C); 307. (D); 308. (D); 309. (D); 310. (D); 311. (A); 312. (D); 313.(C);314. (A); 315. (D); 316. (A); 317. (D); 318. (A); 319. (C); 320. (D); 321. (C); 322. (D); 323. (D); 324. (A); 325. (B);326. (C); 327. (B); 328. (A); 329. (C); 330. (D); 331. (C); 332. (B); 333. (C); 334. (C); 335. (B); 336. (B); 337. (C); 338. (C); 339. (C); 340. (D); 341. (A); 342. (C); 343. (B); 344. (B); 345. (A); 346. (B); 347. (A); 348. (D); 349. (C); 350. (B); 351. (C) ; 352. (B); 353. (C); 354. (B); 355. (C); 356. (C); 357. (B); 358. (B); 359. (A); 360. (B); 361. (A); 362. (A); 363. (A); 364. (B); 365. (B); 366. (A); 367. (C); 368. (B); 369. (A); 370. (B); 371. (C); 372. (B); 373. (B); 374. (A); 375. (C); 376. (A); 377. (A) ;378. (A); 379. (A); 380. (A); 381. (D); 382. (C); 383. (C); 384. (C); 385. (C); 386. (C); 387. (A); 388. (C); 389. (B);390. (D); 391. (B); 392. (C); 393. (C); 394. (B); 395. (C); 396. (C); 397. (B); 398. (B); 399. (A); 400. (C); 401. (A); 402. (C); 403. (D); 404. (D); 405. (C); 406. (B); 407. (A); 408. (A); 409. (B); 410. (C); 411. (C); 412. (B); 413. (C); 414. (D); 415. (B); 416. (D); 417. (B);418. (A); 419. (B); 420. (A); 421. (B); 422. (C); 423. (A); 424. (A); 425. (A); 426. (A); 427. (D); 428. (A); 429. (C) ;430. (D); 431. (D); 432. (A); 433. (A); 434. (D); 435. (B); 436. (B); 437. (C); 438. (A); 439. (D); 440. (A); 441. (D); 442. (C) ; 443. (B); 444. (B); 445. (B); 446. (C); 447. (B); 448. (A); 449. (C); 450. (C); 451. (A); 452. (D); 453. (A); 454. (B);

, 456. (A); 457. (C); 458. (D); 459. (D); 460. (A);
A); 462. (A); 463. (A); 464. (D); 465. (A); 466. (B);
7. (D); 468. (D); 469. (A); 470. (A); 471. (A);
472. (A); 473. (D); 474. (D); 475. (A); 476. (A);
477. (B); 478. (A); 479. (A); 480. (D); 481. (A); 482. (B);
483. (C); 484. (D); 485. (C); 486. (C); 487. (A); 488. (D);
489. (D); 490. (B); 491. (B); 492. (C); 493. (C); 494. (A);
495. (B); 496. (C); 497. (B); 498. (D); 499. (B); 500. (D);
501. (D); 502. (A); 503. (B); 504. (B); 505. (C); 506. (C);
507. (A); 508. (B) ; 509. (C); 510. (B); 511. (A); 512. (A);
513. (A); 514. (B); 515. (B); 516. (C); 517. (A); 518. (B);
519. (A); 520. (A); 521. (A) ; 522. (A); 523. (B); 524. (C);
525. (D); 526. (B); 527. (D); 528. (D); 529. (D); 530. (D);
531. (D); 532. (D); 533. (D); 534. (D); 535. (A); 536. (A);
537. (C); 538. (C); 539. (C); 540. (B); 541. (C); 542. (D);
543. (B); 544. (C); 545. (A); 546. (A); 547. (D) ; 548. (A);
549. (C); 550. (B); 551. (B); 552. (B); 553. (D); 554. (B);
555. (D); 556. (C); 557. (C); 558. (A); 559. (A); 560. (A);
561. (A); 562. (D); 563. (B); 564. (A); 565. (A); 566. (A);
567. (A); 568. (A); 569. (A); 570. (B); 571. (A); 572. (B);
573. (A); 574. (A); 575. (B); 576. (C); 577. (D); 578. (A);
579. (B); 580. (A); 581. (C); 582. (A); 583. (B); 584. (B);
585. (B); 586. (A) ; 587. (B); 588. (C); 589. (A); 590. (B);
591. (D); 592. (C); 593. (D); 594. (D); 595. (A); 596. (A);
597. (A); 598. (A); 599. (D) ; 600.(A); 601. (D); 602. (A);
603. (C); 604. (B); 605. (D); 606. (D); 607. (C); 608.(B);
609. (A); 610. (A); 611. (D); 612. (A); 613. (A); 614. (A);
615. (D); 616. (D); 617. (D); 618. (D); 619. (D); 620.(D);
621. (D); 622. (D); 623. (D); 624. (D); 625. (D) ; 626.(B);
627. (C); 628. (A); 629. (A); 630. (D); 631. (A); 632. (A);
633. (A); 634. (D); 635. (D); 636. (B); 637. (A); 638.(B);
639. (C); 640. (A); 641. (D); 642. (A); 643. (A); 644. (A);
645. (D); 646. (A); 647. (D); 648. (B); 649. (A); 650.(D);
651. (B); 652. (A); 653. (A); 654. (A); 655. (A); 656.(C);
657. (D); 658. (A); 659. (D); 660. (B); 661. (C); 662.(D);
663. (A); 664. (C); 665. (B); 666. (A); 667. (A);
668.(A); 669. (A); 670. (C); 671. (A); 672. (A); 673. (A);
674. (B); 675. (A); 676. (D); 677. (A) ; 678. (D); 679.(A);
680. (A); 681. (D); 682. (A); 683. (C); 684. (A); 685. (C);
686. (C); 687. (A); 688. (D); 689. (A); 690. (B); 691. (B);
692. (A); 693. (A); 694. (A); 695. (D); 696. (D); 697. (A);
698. (A); 699. (D); 700. (A); 701. (A); 702. (A);703. (C) ;
704. (B); 705. (A); 706. (A); 707. (A); 708. (A); 709. (C);
710. (D); 711. (A); 712. (B); 713. (B); 714. (B); 715. (C);
716. (B); 717. (C); 718. (A); 719. (D); 720. (B); 721. (C);
722. (A); 723. (A); 724. (B); 725. (D); 726. (C); 727. (A);
728. (A); 729. (A); 730. (A); 731. (B); 732. (A); 733. (B);
734. (B); 735. (B); 736. (A); 737. (A); 738. (B); 739. (A);
740. (D); 741. (D); 742. (D); 743. (B); 744. (D);745. (A);
746. (D); 747. (D); 748. (A); 749. (A); 750. (A); 751. (A);
752. (A); 753. (A); 754. (C); 755. (D); 756. (B);757. (B);
758. (B); 759. (B); 760. (D); 761. (D); 762. (D); 763. (D);
764. (D); 765. (C); 766. (D); 767. (A); 768. (A) ;769. (A);
770. (A); 771. (B); 772. (C); 773. (A); 774. (D);775. (D);
776. (A); 777. (D); 778. (A); 779. (A); 780. (B); 781. (A);
782. (A); 783. (D); 784. (A); 785. (A); 786. (B);787. (C);
788. (C); 789. (A); 790. (A); 791. (A); 792. (B); 793. (B);
794. (D); 795. (A); 796. (A); 797. (A); 798. (A); 799. (D);
800. (C); 801. (C); 802. (C); 803. (A); 804. (A);805. (A);
806. (B); 807. (A); 808. (B); 809. (B); 810. (A); 811. (A);
812. (A); 813. (A); 814. (A); 815. (B); 816. (A); 817. (D);
818. (D); 819. (C); 820. (A) ; 821. (D); 822. (B);823. (B);
824. (C); 825. (A); 826. (D); 827. (A); 828. (A); 829. (A);
830. (C); 831. (A); 832. (A); 833. (D); 834. (A); 835. (D);
836. (A); 837. (D); 838. (A); 839. (A); 840. (A); 841. (A);
842. (A); 843. (B); 844. (D); 845. (A); 846. (A); 847. (D);
848. (A); 849. (D); 850. (D); 851. (C); 852. (A); 853. (A);
854. (B); 855. (C); 856. (C); 857. (A); 858. (D); 859. (D);
860. (A); 861. (D); 862. (D); 863. (A); 864. (B);865. (A);
866. (B); 867. (D); 868. (D); 869. (A); 870. (A); 871. (A);
872. (D); 873. (A); 874. (A); 875. (D); 876. (C); 877. (C);
878. (D); 879. (D); 880. (D); 881. (D); 882. (D); 883. (A);
884. (A); 885. (D); 886. (D) ; 887. (B); 888. (C);889. (A);
890. (B); 891. (C); 892. (D); 893. (D); 894. (D); 895. (D) ;
896. (D); 897. (A); 898. (D); 899. (D); 900. (A);901. (A);
902. (D); 903. (A); 904. (A); 905. (A); 906. (D); 907. (B);
908. (A); 909. (A); 910. (B); 911. (C); 912. (A) ; 913. (B);
914. (C); 915. (C); 916. (C); 917. (A); 918. (A); 919. (A);
920. (C); 921. (B); 922. (A); 923. (A); 924. (B) ; 925. (C);
926. (D); 927. (D); 928. (A); 929. (B); 930. (C); 931. (C);
932. (D); 933. (B); 934. (D); 935. (C); 936. (C); 937. (B) ;
938. (D); 939. (D); 940. (D); 941. (D); 942. (D);943. (A);
944. (C); 945. (D); 946. (A); 947. (A); 948. (C); 949. (C);
950. (C) ; 951. (B); 952. (B); 953. (C); 954. (B); 955. (B);

956. (A); 957. (A); 958. (A); 959. (B); 960. (B); 961. (B); 962. (A); 963. (A) ; 964. (A) ; 965. (A); 966.(A);967. (C); 968. (B); 969. (A); 970. (B); 971. (A); 972. (A); 973. (C); 974. (A); 975. (D); 976. (C); 977. (D) ; 978. (C); 979. (C); 980. (D); 981. (C); 982. (C); 983. (C); 984. (C); 985. (D); 986. (D); 987. (A); 988. (B); 989. (C); 990. (D) ; 991. (C) ; 992. (A); 993. (B); 994. (B); 995. (B); 996. (B);997. (C); 998. (C); 999. (D); 1000. (A); 1001. (D); 1002. (B); 1003. (B) ; 1004. (B); 1005. (C); 1006. (A); 1007. (A); 1008. (C); 1009. (C) ; 1010. (A); 1011. (A); 1012. (C); 1013. (D) ; 1014. (A); 1015. (D); 1016. (A); 1017. (B); 1018. (C); 1019. (B); 1020. (C); 1021. (D); 1022. (B); 1023. (C); 1024. (B); 1025. (B) ; 1026. (C) ; 1027. (B); 1028. (C); 1029. (B) ; 1030. (B); 1031. (A); 1032. (A); 1033. (B); 1034. (B); 1035. (C); 1036. (B); 1037. (A); 1038. (B) ; 1039. (B); 1040. (C); 1041. (C); 1042. (D) ; 1043. (A); 1044. (B); 1045. (B); 1046. (C); 1047. (C); 1048. (B); 1049. (C); 1050. (D); 1051. (D); 1052. (B); 1053. (A); 1054. (D); 1055. (B) ; 1056. (C); 1057. (B); 1058. (D); 1059. (B); 1060. (A); 1061. (D); 1062. (C); 1063. (C); 1064. (A) ; 1065. (C); 1066. (C); 1067. (B); 1068. (D); 1069. (B); 1070. (A); 1071. (C); 1072. (D); 1073. (D); 1074. (D); 1075. (C); 1076. (A); 1077. (A); 1078. (B); 1079. (C); 1080. (A); 1081. (D) ; 1082. (B); 1083. (A); 1084. (B); 1085. (B); 1086. (B); 1087. (C); 1088. (B); 1089. (B); 1090. (A); 1091. (C); 1092. (A); 1093. (B); 1094. (B); 1095. (A); 1096. (C); 1097. (D); 1098. (C); 1099. (A); 1100. (D); 1101. (B); 1102. (C); 1103. (D); 1104. (); 1105. (A); 1106. (C); 1107. (D) ; 1108. (B); 1109. (D); 1110. (A); 1111. (D); 1112. (D); 1113. (B); 1114. (A); 1115. (B); 1116. (A) ; 1117. (C) ; 1118. (D); 1119. (C); 1120. (A) ; 1121. (B); 1122. (D); 1123. (B); 1124. (C); 1125. (C); 1126. (D); 1127. (D); 1128. (C); 1129. (D) ; 1130. (C) ; 1131. (D); 1132. (D); 1133. (D); 1134. (B); 1135. (A); 1136. (C); 1137. (D); 1138. (D); 1139. (C); 1140. (C); 1141. (C); 1142. (D) ; 1143. (D) ; 1144. (A); 1145. (C); 1146. (D) ; 1147. (B); 1148. (D); 1149. (A); 1150. (B); 1151. (B); 1152. (C); 1153. (D); 1154. (D); 1155. (C) ; 1156. (D) ; 1157. (D); 1158. (C); 1159. (D) ; 1160. (D); 1161. (D); 1162. (C); 1163. (A); 1164. (D); 1165. (D); 1166. (D); 1167. (A); 1168. (A) ; 1169. (D); 1170. (D); 1171. (B); 1172. (C) ;

1173. (C); 1174. (C); 1175. (A); 1176. (B); 11
1178. (B); 1179. (D); 1180. (C); 1181. (C) ; 1182.
1183. (C); 1184. (D); 1185. (D) ; 1186. (B); 1187.
1188. (D); 1189. (D); 1190. (C); 1191. (A); 1192. (B); 1193. (C); 1194. (A) ; 1195. (B) ; 1196. (C); 1197. (D); 1198. (D) ; 1199. (C); 1200. (A); 1201. (B); 1202. (B); 1203. (A); 1204. (D); 1205. (A); 1206. (D); 1207. (C); 1208. (D); 1209. (A); 1210. (B); 1211. (C) ; 1212. (B); 1213. (D); 1214. (C); 1215. (A); 1216. (A); 1217. (A); 1218. (B); 1219. (D); 1220. (A) ; 1221. (A); 1222. (A); 1223. (B); 1224. (A) ; 1225. (B); 1226. (B); 1227. (D); 1228. (C); 1229. (B); 1230. (A); 1231. (B); 1232. (A); 1233. (D) ; 1234. (C); 1235. (B); 1236. (A); 1237. (A); 1238. (C); 1239. (C); 1240. (C); 1241. (B); 1242. (A); 1243. (A); 1244. (B); 1245. (B); 1246. (A); 1247. (A) ; 1248. (A); 1249. (B) ; 1250. (D); 1251. (C); 1252. (B); 1253. (B); 1254. (A); 1255. (C); 1256. (A); 1257. (C); 1258. (A); 1259. (B); 1260. (C); 1261. (D); 1262. (B) ; 1263. (A) ; 1264. (D); 1265. (D); 1266. (A) ; 1267. (B); 1268. (B); 1269. (A); 1270. (D); 1271. (A); 1272. (B); 1273. (D); 1274. (A); 1275. (B); 1276. (C) ; 1277. (D); 1278. (A); 1279. (B); 1280. (B); 1281. (D); 1282. (B); 1283. (B); 1284. (C); 1285. (A) ; 1286. (B); 1287. (B); 1288. (A); 1289. (C) ; 1290. (A) ; 1291. (D); 1292. (A); 1293. (A); 1294. (A); 1295. (A); 1296. (B); 1297. (B); 1298. (A) ; 1299. (A) ; 1300. (C); 1301. (A); 1302. (B) ; 1303. (A); 1304. (D); 1305. (C); 1306. (B); 1307. (A); 1308. (C); 1309. (A); 1310. (D); 1311. (A); 1312. (B) ; 1313. (D); 1314. (B); 1315. (D); 1316. (C); 1317. (C); 1318. (B); 1319. (B); 1320. (A); 1321. (C); 1322. (C); 1323. (D); 1324. (C); 1325. (D); 1326. (A); 1327. (B); 1328. (A) ; 1329. (A); 1330. (B); 1331. (C); 1332. (D); 1333. (A); 1334. (B); 1335. (C); 1336. (D); 1337. (C); 1338. (A); 1339. (B); 1340. (C); 1341. (D); 1342. (D);

455. (A)

461. (A

467

ICULTURAL STATISTICS

1. Who is considered as father of statistics:
(A) R.A. Fisher (B) Pearson
(C) A.L. Bowley (D) Boddington

2. When statistics is used in plural sense it means:
(A) Statistical methods (B) Collect statistics
(C) Both (A) and (B) (D) None of these

3. Which measure of central tendency is applied for calculation of regression and correlation:
(A) Arithmetic mean (B) Mode
(C) Median (D) Geometric mean

4. The problem of effect of the observer is minimized by:
(A) Concealing the behavior
(B) Concealing the observer
(C) Concealing the observed
(D) None of these

5. Coefficient of variation cannot find a place in interval level of measurement because of the problem in:
(A) Standard Deviation (B) Mean
(C) Both (A) and (B) (D) None of these

6. Select the two dimension diagrams of data it means:
(A) Squares (B) Rectangular bodies
(C) Cubes (D) Pictures

7. A hypothesis when tested true becomes a :
(A) Principle (B) Theory
(C) Philosophy (D) Law

8. When a test measure what it is intended to measure, we calls it:
(A) Reliable (B) Valid
(C) True (D) None of these

9. Linkert scale for measurement of attitude falls under:
(A) Nominal (B) Ordinal
(C) Interval (D) Ratio

10. In a symmetrical distribution of data which pair is correct
(A) Mean=Median=Mode (B) Mean>Median>Mode
(C) Mean<Median<Mode (D) Mean=Median>Mode

11. The first psychological test was constructed:
(A) Binet-Simon intelligence scale
(B) Army Alpha test
(C) Thorndike Handwritting scale
(D) Woodworths scale

12. Reliability essentially is a property of:
(A) Test score (B) Test as such
(C) Norms of test (D) Sample of test

13. Which measure of central tendency requires data arrangement in ascending or descending order for its estimation:
(A) Arithmetic mean (B) Median
(C) Mode (D) Harmonic mean

14. If any items of the series is zero, which mean of central tendency become zero:
(A) Harmonic mean
(B) Geometric mean
(C) Arithmetic mean
(D) Weighted harmonic mean

15. The term ex post facto was first coined by:
(A) Chapin and Greenwood (B) Festinger and Katz
(C) Goode and Hatt (D) Campbell and Katona

16. To calculate the average speed, which item of central tendency is most suited:
(A) Mode (B) Median
(C) Weighted HM (C) GM

17. Freedom from variable errors is known as:
(A) Reliability (B) Validity
(C) Credibility (D) Sustainability

18. Which is the simplest measure of dispersion:
(A) Mean deviation (B) Standard deviation
(C) Range (D) Variance

19. The first central moment (μ_1) is always:
(A) 0 (B) 2
(C) 4 (D) 6

20. A pilot study to gin more insight into the research problem form the characteristic feature of:
(A) Explorative design (B) Diagnostic design
(C) Descriptive design (D) Experimental design

21. The curves which are very highly peaked and have the value of $\beta_2 > 3$ are called:
(A) Mesokurtic (B) Leptokurtic
(C) Platykurtic (D) All of these

22. The mean of the binomial distribution is:
(A) np (B) npq
(C) $\sqrt{npq}$ (D) None of these

23. A hypothesis is a must in:
(A) Explorative design (B) Diagnostic design
(C) Descriptive design (D) Experimental design

24. The first and third moments about the mean of normal distribution is
(A) 0 (B) δ^2
(C) $3\delta^4$ (D) None of these

25. To study the effect of independent variable:
(A) Factorial design (B) CRD
(C) LSD (D) RBD

26. β_1 and β_2 constants of the normal distribution are ...., respectively:
(A) 0, 3 (B) 3, 0
(C) 0, 0 (D) 3, 3

27. Experimental and non-experimental research shares in common which of the following characteristics:
(A) Control (B) Manipulation
(C) Measurement (D) All of these

28. Student's t test is used when sample size is .... and population standard deviation is not known:
(A) Large (B) Small
(C) Both (A) and (B) (D) None of these

29. Opinion polls falls under:
(A) Explorative design (B) Diagnostic design
(C) Descriptive design (D) Experimental design

30. For comparison of two means from independent sample which test is applicable:
(A) t-test (B) x^2-test
(C) F-test (D) z-test

31. Experimenter has the highest control in manipulation of variable in:
(A) Laboratory method (B) Library method
(C) Simulation method (D) Field experiment

32. x^2 test is applied for:
(A) Testing of independence of attributes
(B) Genetic problem and detection of linkages
(C) Testing the expectation of a ratio
(D) All of these

33. The following description " I will bet this would happen if I did this" best represent to:
(A) Explorative design
(B) Confirmatory experiment
(C) Crucial experiment
(D) Pilot experiment

34. The range of multiple correlation coefficient lies between:
(A) 0 to +1 (B) -1to +1
(C) 0 to ∞ (D) - ∞ to + ∞

35. A research design without hypothesis among the following is:
(A) Explorative design (B) Diagnostic design
(C) Descriptive design (D) Experimental design

36. The principle of making use of greater homogeneity in groups of experimental unit to reduce the experimental error is referred as:
(A) Replication (B) Randomization
(C) Local control (D) None of these

37. Which of the following is not a characteristic feature of a good research design:
(A) High external validity (B) Low external validity
(C) High internal validity (D) None of these

38. The completely randomized design is appropriate when the experimental material is
(A) Limited and homogenous
(B) Unlimited and heterogeneous
(C) Limited and heterogeneous
(D) Unlimited and homogenous

39. A sample of 12 specimen taken from a normal population is expected to be 50 kg. To test Ho: μ= 50Vs: Hi not μ= 50, the test used is:
(A) t- test (B) X^2- test
(C) Z test (D) F-test

40. When the fertility gradient of the field goes in two directions, which design is most appropriate:
(A) Augmented design (B) CRD
(C) LSD (D) SPD

41. Equality of two regression coefficient can be tested with:
(A) 't' test (B) 'X^2' test
(C) Z test (D) ANOVA

42. In randomized block design, how many treatments can be adopted without any loss of efficiency:
(A) 10 (B) 15
(C) 18 (D) 20

43. When the experimental material is very less and number of genotypes are very large, which design is most suited:
(A) Split Plot Design (B) LSD
(C) RBD (D) Augmented Design

44. Analysis of variance utilizes:
(A) t- test (B) X^2- test
(C) Z test (D) F-test

45. Hypothesis that the population variance has a specified value can be tested by:
(A) t- test (B) X^2- test
(C) Z test (D) F-test

46. The critical difference (CD) is a form of t-test and its formula is given by:
(A) CD=t.SE(d) (B) CD=t+SE
(C) CD=t/SE(d) (D) CD=F.SE(d)

47. The ANOVA is a tool by which total variation may be spit up into several physically assignable components was defined by:
(A) Karl Pearson (B) RA Fisher
(C) Horace Secrist (D) AL Bowley

48. The degree of freedom for statistics for paired t test of n pairs of observations is:
(A) 2 (n-1) (B) n – 1
(C) 2n – 1 (D) None of these

49. The value of the sum of deviations from mean is:
(A) Zero (B) Positive
(C) Negative (D) Different from zero

50. The correlation coefficient remains unaffected by change of:
(A) Origin (B) Scale
(C) Both (A) and (B) (D) None of these

51. The manipulated variable in an experiment is:
(A) Independent variable (B) Dependent variable
(C) Active variable (D) Both (A) and (C)

52. The variance that reflects the systematic differences between groups of measures:
(A) Experimental variance (B) Extraneous variance
(C) Error variance (D) None of these

53. Assignment of values to alternatives is called:
(A) Observation (B) Judgment
(C) Measurement (D) Evaluation

54. The basic focus of Thematic Apperception Test is on:
(A) Product (B) Process
(C) Input (D) Both (A) and (B)

55. The regression coefficient are independent of change of
(A) Origin (B) Scale
(C) Both (A) and (B) (D) None of these

56. Questionnaire involves presentation of:
(A) Oral-verbal stimuli (B) Written-verbal stimuli
(C) Oral-written stimuli (D) None of these

57. On the spot reliability method among the following is:
(A) Test re-test method (B) Odd even method
(C) Parallel foams method (D) Comparable method

58. When the value of correlation coefficient is 1 then two variable are:
(A) Uncorrelated (B) Partially correlated
(C) Perfectly correlated (D) Highly correlated

59. Index of discrimination is otherwise known as:
(A) Item validity index
(B) Item of reliability index
(C) Item difficulty index
(D) Item credibility index

60. Cluster sample are more effective if the population taken is:
(A) Cluster of small size
(B) Clusters with largest within variation
(C) Clusters with least within variation
(D) None of these

61. The measure of association or relationship between two or more variables:
(A) Dispersion (B) Correlation
(C) Central tendency (D) Regression

62. To test the hypothesis about proportions of items in a class, we use:
(A) t- test (B) X^2- test
(C) Z test (D) F-test

63. Which transformation is most appropriate for percentages:
(A) Square root (B) Arc sine
(C) Logarithmic (D) All of these

64. If a frequency distribution is positively skewed, the mean of the distribution is:
(A) Greater than the mode (B) Less than the mode
(C) Equal to the mode (D) None of these

65. The correlation coefficient is the geometric mean between two:
(A) Median (B) Standard deviation
(C) Regression coefficient (D) Mode

66. Usually the explanatory value of a regression equation of y on x is measured using:
(A) Correlation coefficient
(B) Coefficient of determination
(C) Coefficient of explanation
(D) Coefficient of variance

67. If there are 5 treatments with 4 replication to each, the error degree of freedom for CRD will be:
(A) 15 (B) 20
(C) 12 (D) 9

68. If correlation =1, the angle between two lines of regression is:
(A) 0 (B) 45
(C) 60 (D) 90

69. A variance shared by two or more variables:
(A) Covariance
(B) Correlation
(C) Common factor variance
(D) None of these

70. Which design provides maximum number of degrees of freedom for the estimation of error as compared with other design for the given number of experimental units:
(A) RBD (B) LSD
(C) Augmented design (D) CRD

71. For testing the independence of two attributes, we use:
(A) t- test (B) X^2- test
(C) Z test (D) F-test

72. If 4 levels of sowing dates are laid out in main plot, 5 levels of nitrogen applied in sub plot and they are replicated three times, then what will be the degree of freedom for error (b):
(A) 32 (B) 6
(C) 20 (D) 60

73. The level of measurement associated with Mann-Whitney U test:
(A) Nominal (B) Ordinal
(C) Interval (D) Ratio

74. For any two events A and B, P(A-B) is equal to:
(A) P(A) – P(B) (B) P(B)-P(A)
(C) P(B)-P(AB) (D) P(A)-P(AB)

75. Which design follow the number of rows = number of column= Number of treatments:
(A) SPD (B) Strip plot design
(C) LSD (D) RBD

76. The right hand tail of a frequency distribution is found to the mirror image of the left-hand tail. The distribution is:
(A) Positively skewed (B) Negatively skewed
(C) Asymmetric (D) Symmetric

77. When coefficient of skewedness is negative:
(A) $Q_2 + Q_3 = 2Q_1$ (B) $Q_3 + Q_1 < 2Q_2$
(C) $Q_3 + Q_1 > 2Q_2$ (D) $Q_3 + Q_2 < 2Q_1$

78. When we want to conduct a experiment on a long strip of land, which design will be preferred:
(A) RBD (B) CRD
(C) SPD (D) LSD

79. Select the most appropriate design when all factors are not of equal importance in experimentation:
(A) Augmented (B) Strip plot design
(C) LSD (D) Split plot design

80. Indian Agricultural Statistical Research Institute is located at:
(A) Lucknow (B) Jaipur
(C) New Delhi (D) Kolkata

81. Attitude summarizes:
(A) Present experience (B) Past experience
(C) Future planning (D) None of these

82. At present time how many number of forecasts issued by Government of India on 34 important crops:
(A) 50 (B) 60
(C) 70 (D) 80

83. Pie diagram is a type of:
(A) One dimensional diagram
(B) Two dimensional diagram
(C) Three dimensional diagram
(D) None of these

84. Sum of squares of deviations taken from A is minimum when A is:
(A) Zero (B) Mean
(C) Mode (D) Median

85. Comparison of several independent samples is tested by:
(A) Quade test
(B) Friedmann test
(C) Wilcoxon Signed –rank test
(D) Kruskal-Wallis H test

86. A test that rates a persons potential for learning skills required by various occupations is called:
(A) Vocational interest test (B) Interest test
(C) Attitude test (D) Aptitude test

87. Attitude scales are used to measure:
(A) Opinion (B) Judgment
(C) Fact (D) None of these

88. When both the variables are not normally distributed, in such case we may use:
(A) Partial correlation (B) Multiple correlation
(C) Rank correlation (D) Both (A) and (B)

89. Role play forms an excellent example for:
(A) Expressive technique (B) Association technique
(C) Constructive technique (D) Completion technique

90. A method in which the individual is asked to state his or her willingness to admit members of a particular group to various levels of social closeness:
(A) Social distance scale (B) Interview
(C) Open ended interview (D) Attitude scale

91. A good item is characterized by:
(A) High ambiguity
(B) High discrepancy power
(C) High discriminatory items
(D) None of these

92. An aptitude test of higher validity coefficient will make a better:
(A) Prediction (B) Result
(C) Method (D) None of these

93. Ranking of raw data in numerical order is called:
(A) Series (B) Scores
(C) Frequency distribution (D) None of these

94. Unrestricted probability sampling method:
(A) Area sampling
(B) Simple random sampling
(C) Quota sampling
(D) Systematic sampling

95. Rank correlation is tested by:
(A) Quade test (B) Friedmann test
(C) t-test (D) F-test

96. If four levels of depth of ploughing is applied to vertical strip, five levels of moisture regimes is applied to horizontal strip and replicated three times, then what will be the degree of freedom for error (c).[(r-1),(a-1)(b-1):
(A) 8 (B) 6
(C) 10 (D) 24

97. The relation between standard deviation and mean deviation is:
(A) 3MD = 2SD (B) 5MD = 4SD
(C) 6MD = 5SD (D) MD = SD

98. Which of the following is correct:
(A) $r_{12.34} = r_{13.24}$ (B) $r_{12.3} = r_{21.3}$
(C) $r_{13} = r_{23}$ (D) $r_{13.2} = r_{13.4}$

99. The square of the standard deviation is known as the:
(A) Variance (B) Standard deviation
(C) Mean deviation (D) Coefficient of variance

100. A subset or portion of a population is called:
(A) Population (B) Sample
(C) Representative (D) All of these

101. Rating scales are used to measure:
(A) Actual behavior (B) Perceived behavior
(C) Both (A) and (B) (D) None of these

102. When 4 levels of sowing dates is applied to vertical strip, five levels of irrigation is applied to horizontal strip, four levels of nitrogen randomly allocated within the intersection plots of vertical and horizontal strip and replicated three times, then what will be the degree of freedom for error (d).[ab(r-1).(c1)] (Strip split plot design):
(A) 72 (B) 120
(C) 24 (D) 8

103. A set of different measures of an individual or group, each of which is expressed in the same unit of measure:
(A) Profile (B) Biodata
(C) Set (D) Character

104. A measure of the peakedness or convexity of a curve is known as:
(A) Skewness (B) Kurtosis
(C) Ogive (D) Histogram

105. With the help of ogive one can determine:
(A) Mean (B) Median
(C) Mode (D) Geometric mean

106. Greater importance to bigger items of a given series is given by:
(A) Median (B) Mode
(A) Arithmetic mean (D) None of these

107. The basic unit containing the elements of the population to be sampled:
(A) Sampling unit (B) Sampling frame
(C) Sampling element (D) None of these

108. Which one of the following is not a measure of central tendency?
(A) Mean (B) Median
(C) Mode (D) Variance

109. An example for two way non-parametric analysis of variance among the following is:
(A) F test (B) Kvuskal Wallis test
(C) Friedman test (D) Z test

110. Geometric mean is better than other mean when:
(A) Data are positive as well as negative
(B) Data are in ratios or percentage
(C) Data are binary
(D) Data are on interval scale

111. Mean of ten observations is 20. If a value 11 is removed, what is the mean now?
(A) 19 (B) 20
(C) 21 (D) 22

112. Greater the distance of Lorenz curve from the line of equal distribution:
(A) More is inequality (B) Less in equality
(C) Nothing can be inferred (D) None of these

113. Mean of eight observations is 5. When one value x is added, mean remains same. What is the value of x?
(A) 3 (B) 5
(C) 8 (D) 13

114. The technique employed to find the degree of correlation is:
(A) Experimental (B) Observational
(C) Statistical (D) None of these

115. Which one of the following is not a dependent variable:
(A) Chronological age (B) Rate of response
(C) REM rebound (D) Reaction time

116. Mean of a frequency distribution is m. If every frequency is doubled . What is new mean?
(A) m (B) m/2
(C) 2m (D) m+2

117. The logical process that operates when a researcher moves from hypothesis testing to making of prediction:
(A) Vertical logic (B) Horizontal logic
(C) Inductive logic (D) Deductive logic

118. Standard error of number of successes is given by:
(A) $\sqrt{pq/n}$ (B) $\sqrt{npq}$
(C) npq (D) n^2p^2

119. Nominal level of measurement can categories objects:
(A) Qualitatively and Quantitatively
(B) Only Qualitatively
(C) Only Quantitatively
(D) None of these

120. Validity of a measuring instrument in simple terms is the freedom from:
(A) Variable errors (B) Systematic errors
(C) Random errors (D) Non-systematic errors

121. The correlation type in which an increase in one measure is matched by an increase in the other:
(A) Positive (B) Negative
(C) Both (A) and (B) (D) None of these

122. If for a distribution, coefficient of kurtosis $\gamma 2<0$, the frequency curve is:
(A) Leptokurtosis (B) Platykurtosis
(C) Mesokurtosis (D) None of these

123. Range of the values 15, 12, 27, 6, 9, 18, 21 is:
(A) 6 (B) 21
(C) 4.5 (D) 3

124. From the cumulative frequency diagram which one of the following types of average can be estimated?
(A) Geometric mean (B) Harmonic mean
(C) Median (D) Mode

125. While testing the significance of the difference of two sample means in case of small samples, the degrees of freedom is calculated by:
(A) $v = n_1 + n_2 - 2$ (B) $v = n_1 + n_2 - 1$
(C) $v = n_1 - n_2 - 2$ (D) $v = n_1 + n_2 + 2$

126. — error are made when we accept a hypothesis which is not true:
(A) Type I (B) Type II
(C) Both (A) and (B) (D) None of these

127. The standard distribution of Poisson's distribution is: (m=mean)
(A) m (B) $\sqrt{m}$
(C) m^2 (D) $1/\sqrt{m}$

128. Correct formula for variance is:
(A) $1/n\ \Sigma X_i^2 - X^{-2}$ (B) $1/n\ \Sigma X^2 - nX^{-2}$
(C) $1/n\ \Sigma X_i^2 - X^{-2}/2$ (D) None of these

129. Theoretically the best average in the construction of index number is:
(A) Median (B) Geometric mean
(C) Mode (D) Arithmetic mean

130. Variance of twenty observations is 8 if every observation is multiplied by 2, what is the value of variance now?
(A) 4 (B) 8
(C) 16 (D) 32

131. The coefficient of correlation is independent of change of:
(A) Origin (B) Scale
(C) Both (A) and (B) (D) None of these

132. During one year the ratio of prices of two commodities is 3, whereas during the next year it is 2. For averaging these ratios:
(A) Arithmetic mean is most appropriate
(B) Harmonic mean is most appropriate
(C) Geometric mean is most appropriate
(D) All the three, arithmetic, harmonic and geometric means are equally appropriate

133. Total number of possible out comes in a cricket match is:
(A) 1 (B) 2
(C) 3 (D) 11

134. Measurement in social science is concerned with:
(A) Objects
(B) Properties of objects
(C) Phenomena
(D) Properties of phenomena

135. Consistency and reproductibility of test score are indicated by:
(A) Reliability (B) Validity
(C) Credibility (D) Sustainability

136. Reliability is found to increase when we go for:
(A) Less number of items
(B) Homogeneous items
(C) High discriminatory items
(D) Less difficult items

137. Attitude can be considered as a mixture of:
(A) Belief and emotion (B) Emotion and feeling
(C) Thought and emotion (D) Belief and thought

138. If A contains B, what can be said about the value of r = P (A)/ P (B):
(A) r < 1 (B) r = 0
(C) r = 1 (D) 0 < r < 1

139. Each item in the attitude scale is given a scale value in:
(A) Guttman scale
(B) Thurstone and Clave scale
(C) Adorne scale
(D) Likert scale

140. The primary elements of the universe of possible outcome is called:
(A) Sample point (B) Sample space
(C) Sample event (D) None of these

141. Sampling involves:
(A) How to select the elements
(B) Hoe to estimate the population characteristics from the sampling units
(C) Both (A) and (B)
(D) None of these

142. Correlation coefficient is considered as that of two regression coefficients:
(A) Mean (B) Geometric mean
(C) Square (D) None of these

143. If standard deviation is greater than variance, the value of standard deviation is:
(A) 1
(B) Less than 1
(C) Greater than 1
(D) Nothing definite can be said

144. The tendency to act in ways that influence experimental results so that we obtain results that are consistent with our expectations:
(A) Self fulfilling prophecy (B) Cock tail party effect
(C) Sapir-Whorf hypothesis (D) None of these

145. If A and B are mutually exclusive events. P(A) = ¼ and P (A U B) = ¾, what is the value of P (B)?
(A) 0 (B) ½
(C) 3/8 (D) 3/16

146. "After this, therefore caused by this" is usually denoted to indicate:
(A) Experimental design (B) Survey design
(C) Ex Post facto design (D) Explorative design

147. Which among the following is the testable foam in a scientific research:
(A) Research problem (B) Research hypothesis
(C) Research variable (D) All of these

148. To discriminate groups from one another on the basis of sets of measures, we use:
(A) Profile analysis (B) Factor analysis
(C) Discriminate analysis (D) Disaster analysis

149. When compared with reality, measurement procedure and number system are:
(A) Isomorphic (B) Polymorphic
(C) Monomorphic (D) Oligomorphic

150. A die is rolled once. What is the probability of obtaining two or four points:
(A) ½ (B) 1/3
(C) ¼ (D) 1/6

151. For statistical stability and reliability the ideal number of cards needed in Q distribution is:
(A) 1-30 (B) 31-60
(C) 61-90 (D) 91-120

152. Rorschach projection tests foams an example for:
(A) Association test
(B) Construction technique
(C) Completion technique
(D) Ordering technique

153. The measurement of temperature falls under:
(A) Nominal level (B) Ordinal level
(C) Interval level (D) Ratio level

154. In order to study fertility fluctuations, which one of the following is used?
(A) Randomization (B) Replication
(C) Local control (D) Uniformity trials

155. Information gained through direct observation and measurement is called:
(A) Empirical evidence
(B) Experimental variance
(C) Observational evidence
(D) None of these

156. The first step in summarizing the data is:
(A) Tabulation (B) Classification
(C) Interpretation (D) Observation

157. The hypothesis type stated for the purpose of rejection:
(A) Null hypothesis (B) Alternate hypothesis
(C) Parallel hypothesis (D) Both (B) and (C)

158. The underlying philosophy of scientific method followed in different disciplines of science are:
(A) Exactly the same (B) Almost same
(C) Some what the same (D) Entirely different

159. In a 2 x 2 contingency table, what is the degree of freedom in x^2 – test?
(A) 0 (B) 1
(C) 2 (D) 4

...is the main purpose of design of experiments?
(A) To compare means only
(B) To compare variances only
(C) To compare means and variances both
(D) To compare neither means nor variances

161. If there are 8 blocks and 6 treatments in RBD, what are the degrees of freedom corresponding to total sum of squares?
(A) 5 (B) 7
(C) 3 (D) 47

162. What is the geometric mean of 1, 4 and 16?
(A) 2 (B) 4
(C) 6 (D) 8

163. Age is classified under:
(A) Discontinuous variable (B) Discrete variable
(C) Continuous variable (D) Intervening variable

164. A test used to compare the variance of two independent samples:
(A) t- test (B) X^2- test
(C) Z test (D) F-test

165. If the correlation coefficient between X and Y is r, the correlation between x^2 and y^2 is:
(A) r (B) r^2
(C) 0 (D) None of these

166. Any categorization scheme must be on:
(A) Four level of discourse
(B) Three level of discourse
(C) Two level of discourse
(D) One level of discourse

167. If correlation between two variables (x and y) is negative, the regression coefficient of y on x is:
(A) Positive (B) Negative
(C) Not certain (D) Zero

168. Which one of the following is not a fundamental principle of experimental design?
(A) Randomization (B) Replication
(C) Validation (D) Local control

169. If r is the correlation coefficient, r^2 is:
(A) Coefficient of determination
(B) Coefficient of non determination
(C) Coefficient of alienation
(D) None of these

170 If all the variate values are the same (e.g. 8, 8, 8), the standard deviation will be:
(A) 4 (B) 8
(C) 16 (D) 0

171. What are the degrees of freedom in testing the difference between means by t-test?
(A) $n_1 + n_2 - 1$ (B) $n_1 + n_2 + 1$
(C) $n_1 + n_2 - 2$ (D) $n_1 + n_2 + 2$

172. The ratio of regression sum of squares to the total sum of squares is:
(A) Error means square
(B) Coefficient of determination
(C) Coefficient of non determination
(D) Both (A) and (B)

173. If correlation coefficient is 0.7, then coefficient of determination is:
(A) 0.7 (B) 0.07
(C) 0.49 (D) None of these

174. To study average rate of change in population, we use:
(A) Median (B) Mode
(C) Arithmetic mean (D) Geometric mean

175. Which measure of dispersion is least affected by extreme values:
(A) Range (B) Quartile deviation
(C) Standard deviation (D) Mean deviation

176. The simplest way of visualizing a correlation is to construct a:
(A) Graph (B) Table
(C) Scatter diagram (D) None of these

177. Local control is a device to maintain homogeneity:
(A) Among blocks (B) Within Blocks
(C)Both (A) and (B) (D) None of these

178. To test the hypothesis about proportions of items in a class, we use:
(A) t- test (B) X^2- test
(C) Z test (D) F-test

179. Ranking of raw data in numerical order is called:
(A) Series (B) Scores
(C) Frequency distribution (D) None of these

180. Index numbers used in social science measure are having:
(A) Arbitrary zero point (B) Absolute zero point
(C) Natural zero point (D) Both (A) and (C)

181. In a series of values if one value is zero, G.M. is:
(A) Positive (B) Zero
(C) Indeterminate (D) None of these

182. Probable error is:
(A) 0.6745 S.E. (B) 0.6457 S.E.
(C) 0.6753 S.E. (D) 0.7645 S.E.

183. The first step in analysis of data is:
(A) Manipulation (B) Categorization
(C) Interpretation (D) Summarization

184. If two events A and B are dependent, the conditional probability of A given B, i.e., P(A/B) is calculated as:
(A) P (A) + P (B) (B) P (A) x P (B)
(C) P (A) – P (B) (D) P (AB) / P (B)

185. If a constant 50 is subtracted from each of the value of X and Y, the regression coefficient is:
(A) Reduced by 50 (B) 1/50th of original
(C) Increased by 50 (D) Not changed

186. If all the variate values are negative, the standard deviation will be:
(A) Negative (B) Positive
(C) No sign (D) Cannot calculate

187. The level of measurement attached to categorical variable is:
(A) Nominal (B) Ordinal
(C) Interval (D) Ratio

188. When neither variable is the cause of the other, the relationship is called:
(A) Symmetrical (B) Reciprocal
(C) Asymmetrical (D) None of these

189. An interval in which scores are grouped is called:
(A) Point (B) Value
(C) Class interval (D) None of these

190. If there is an increase in a series at constant rate, the graph will be:
(A) Convex curve (B) Concave curve
(C) Parabola (D) Straight line

191. If both the regression coefficients are negative, the correlation coefficient would be——:
(A) Negative (B) Positive
(C) Zero (D) None of these

192. A sub set or portion of a population is called:
(A) Population (B) Sample
(C) Representative (D) All of these

193. The primary elements of the universe of possible outcome is called:
(A) Sample point (B) Sample space
(C) Sample event (D) None of these

194. The under root of two——— coefficient gives us the value of correlation coefficient:
(A) Regression (B) Standard Error
(C) Standard Deviation (D) None of these

195. Survey method would most typically used in:
(A) Clinical assessment (B) Opinion poll
(C) Laboratory experiment (D) Psychoanalysis

196. Sampling frame is a must in the case of:
(A) Random sampling (B) Systematic sampling
(C) Stratified sampling (D) Cluster sampling

197. The number of possible samples of size two from a population of four units in a simple random sample without replacement is:
(A) 2 (B) 4
(C) 6 (D) 8

198. A constant value of 50 is subtracted from each observation of a set, the mean of the set is:
(A) Increased by 50 (B) Decreases by 50
(C) Not affected (D) Zero

199. Take the odd one out:
(A) Area sampling (B) Snowball sampling
(C) Quota sampling (D) Incidental sampling

200. Cross section of population is represented less in:
(A) Simple random sampling
(B) Multistage sampling
(C) Cluster sampling
(D) Stratified random sampling

201. If sum of the product of deviations of x and y series from their means, i.e. Ó xy is zero, the coefficient of correlation shall be:
(A) +1 (B) 0
(C) -1 (D) None of these

202. There will be only one regression line variables if:
(A) $r = 0$ (B) $r = \pm 1$
(C) $r - 0.5$ (D) None of these

203. To have an accurate picture of how large segments the population feel about current issues:
(A) Case study (B) Lab experiment
(C) Field survey (D) None of these

204. Unrestricted probability sampling method:
(A) Area sampling
(B) Simple random sampling
(C) Quota sampling
(D) Systematic sampling

205. The sample type wherein similarity between immediate neighbours and difference between distinct neighbours are both taken care of:
(A) Random sample (B) Cluster sample
(C) Systematic sample (D) Stratified sample

206. If each observation of a set is multiplied by 10, the mean of the new set of observation:
(A) Remains the same
(B) 10 times the original mean
(C) 1/10th of the original mean
(D) Increased by 10

207. When information is to be collected from a geographically dispersed population we go for:
(A) Stratified sampling
(B) Systematic sampling
(C) Cluster sampling
(D) Multipurpose sampling

208. Regression coefficient is independent of change of:
(A) Origin but not of scale
(B) Both origin and scale
(C) Scale but not of origin
(D) None of these

209. The best sample size from a perfectly homogenous population in respect of a character:
(A) A large sample (B) A small sample
(C) A single item (D) No item

210. The regression lines cut each other at the point of:
(A) Average of x and y (B) Average of x only
(C) Average of y only (D) None of these

211. Sum of measure divided by the number of the measure:
(A) Mean (B) Mode
(C) Median (D) None of these

212. A constant of 5 is subtracted from each observation of a set, the variance is:
(A) Reduced by 5 (B) Reduced by 25
(C) Unaltered (D) Increased by 25

213. Greater importance to bigger items of a given series is given by:
(A) Median (B) Mode
(C) Arithmetic mean (D) None of these

160. Wha e regression coefficients cannot be:
(A) ss than one
(B) Greater than one
(F Equal to one
(D) None of these

.5. Normal distributions are ———— around their mean:
(A) Symmetric
(B) Asymmetric
(C) Both (A) and (B)
(D) None of these

216. To study the average of quantity prices, we use:
(A) Median
(B) Harmonic mean
(C) Arithmetic mean
(D) Geometric mean

217. The mean, median and mode of a normal distribution are:
(A) 1
(B) -1
(C) Equal
(D) Unequal

218. To study the average height of plants, we use:
(A) Mode
(B) Harmonic mean
(C) Arithmetic mean
(D) Geometric mean

219. Mention the correct answers. The ranks according to two attributes in a sample are given below:

R1	1	2	3	4	5
R2	5	4	3	2	1

The rank correlation between them is:
(A) O
(B) +1
(C) -1
(D) None of these

220. The AM of the age of 20 students in a class is 15 years. When the teachers age also is added, the mean increase by one year. The age of the teacher is:
(A) 16 years
(B) 21 years
(C) 36 years
(D) 25 years

221. The area under the normal curve is equal to:
(A) -1
(B) 1
(C) 0
(D) 2

222. Formula for calculating coefficient of variation is:
(A) SD/Mean x 100
(B) Mean/SD x 100
(C) Mean x SD/100
(D) 100/Mean x SD

223. The method of assembling correspondence without experiment control over them is called:
(A) Dispersion
(B) Correlation
(C) Central tendency
(D) None of these

224. Mean deviation is minimum when taken about:
(A) Mean
(B) Median
(C) Mode
(D) Zero

225. The relation between arithmetic mean, geometric mean and harmonic mean is:
(A) AM = GM = HM
(B) AM>GM>HM
(C) AM<GM<HM
(D) AM<HM<GM

226. The most frequent score in a given distribution:
(A) Mean
(B) Mode
(C) Median
(D) None of these

227. In case of positively skewed distribution the relation between mean,, median and mode that hold is:
(A) Median > Mean > Mode
(B) Mean > Median > Mode
(C) Mean = Median = Mode
(D) None of these

228. Which mean of dispersion is most affected by extreme values:
(A) Range
(B) Quartile deviation
(C) Standard deviation
(D) Mean deviation

229. Experimental method of data collection among the following is:
(A) Field study
(B) Sample survey
(C) Library method
(D) Field experiment

230. Interview is the widely used technique in:
(A) Local study
(B) Public studies
(C) Empirical studies
(D) Non-scientific studies

231. The measure of dispersion which ignores signs of deviation from the central value:
(A) Range
(B) Quartile deviation
(C) Standard deviation
(D) Mean deviation

232. Which measure of dispersion is least affected by extreme values:
(A) Range
(B) Quartile deviation
(C) Standard deviation
(D) Mean deviation

233. The range of simple correlation coefficient is:
(A) 0 to ∞
(B) - ∞ to + ∞
(C) -1 to +1
(D) 0 to 1

234. Range is not a reliable measure of:
(A) Mean
(B) Dispersion
(C) Median
(D) Mode

235. Empirical relation between quartile deviation and standard deviation is.
(A) 3QD = 2 SD
(B) 4 QD = 3 SD
(C) 6 QD = 5SD
(D) 5 QD = 4 SD

236. Correlation coefficient is independent of:
(A) Scale
(B) Origin
(C) Both (A) and (B)
(D) Neither (A) nor (B)

237. Who first discovered the normal distribution?
(A) Fisher
(B) K.F.Gauss
(C) Cochran
(D) de Moivre

238. The average of sum of squares of deviation about the mean is called:
(A) Absolute deviation
(B) Variance
(C) Standard deviation
(D) Mean deviation

239. Which of the following is the unit less measure of dispersion:
(A) Standard deviation
(B) Mean deviation
(C) Coefficient of variance
(D) Range

240. The most commonly used method in a social survey is:
(A) Observation
(B) Interview
(C) Questionnaire
(D) Case study

241. The value of co-efficient of concordance ranges between:
(A) 0 to + 1
(B) - ∞ to + ∞
(C) -1 to +1
(D) -1 to 0

242. The degree of linear relationship between variables are expressed through:
(A) Coefficient of correlation
(B) Standard deviation
(C) Arithmetic mean
(D) None of these

243. A constant of 5 is subtracted from each observation of a set, the variance is:
(A) Reduced by 5
(B) Reduced by 25
(C) Unaltered
(D) Increased by 25

244. Survey research employs which of the following techniques:
(A) Panel technique
(B) Personal interview
(C) Controlled observation
(D) All of these

245. If each of a set is divided by 10, the standard deviation of a new observation is:
(A) 1/10th of SD of originals
(B) 1/100th of SD of originals
(C) Not altered
(D) 10 times of SD of originals

246. The best control for practice effects in an experiment is:
(A) Random selection
(B) Counter balancing
(C) Equivalent group
(D) Split half presentation

247. The unit of correlation coefficient is:
(A) Kg/cc
(B) per cent
(C) No unit
(D) None of these

248. In a normal distribution the points of inflexion occur at:
(A) $X + \sigma$
(B) $x + 2\sigma$
(C) $X + 3\sigma$
(D) None of these

249. If an event cannot take place, probability will be:
(A) +1
(B) -1
(C) 0
(D) None of these

250. The circular test is satisfied when:
(A) $P_{12} \times P_{23} \times P_{31} = 0$
(B) $P_{12} \times P_{32} \times P_{31} = 1$
(C) $P_{12} \times P_{23} \times P_{31} = 1$
(D) $P_{12} \times P_{23} \times P_{13} = 1$

251. If β xy and β yx are two regression coefficient, they have:
(A) Same sign
(B) Opposite sign
(C) Either same or opposite sign
(D) Nothing can be said

252. An example for non parametric test from among the following is:
(A) t- test
(B) X^2- test
(C) Z test
(D) F-test

253. When degree of freedom of X^2 test is more than 100, t test is approximated to:
(A) t- test
(B) X^2- test
(C) Z test
(D) F-test

254. The normal distribution is the most important and most widely used distribution in statistics. It is sometimes called the:
(A) Bell curve
(B) Gaussian curve
(C) Both (A) and (B)
(D) None of these

255. Degree of freedom from X^2 in case of 4 x 3 contingency table is:
(A) 12
(B) 9
(C) 6
(D) 8

256. The measure of association or relationship between two or more variables:
(A) Dispersion
(B) Correlation
(C) Central tendency
(D) Regression

257. Time reversal test is satisfied when:
(A) $P_{01} \times P_{23} = 0$
(B) $P_{01} \times P_{10} = 1$
(C) $P_{02} \times P_{10} > 1$
(D) $P_{01} \times P_{10} < 1$

258. Addition theorem states that if two events A and B are mutually exclusive the probability of occurrence of either A or B is given by:
(A) P (A) + P (B)
(B) P (A) x P (B)
(C) P (A) – P (B)
(D) None of these

259. The normal distribution with x = 0 and á = 1 is known as:
(A) Standard normal distribution
(B) Poisson's distribution
(C) Both (A) and (B)
(D) None of these

260. Which among the correlation coefficient represent the strongest relationship:
(A) -0.86
(B) +0.86
(C) +0.10
(D) +0.09

261. The most important factors causing seasonal variations are:
(A) Growth of population
(B) Technological improvement
(C) Weather and social customs
(D) Change in fashions

262 If correlation =1, the angle between two lines of regression is:
(A) 0
(B) 45
(C) 60
(D) 90

263. Which one of the following average can be computed from a cumulative frequency curve:
(A) Geometric mean
(B) Harmonic mean
(C) Mode
(D) Median

264. —— asserts that there is no true difference in the sample and the population in the particular matter under consideration:
(A) Null hypothesis
(B) Alternate hypothesis
(C) Both (A) and (B)
(D) None of these

265. Coefficient of determination is defined as:
(A) r^3
(B) $1- r^2$
(C) $1 + r^2$
(D) r^2

266. If X and Y are independent, the value of regression coefficient β xy is equal to:
(A) 0
(B) 1
(C) a
(D) Any positive value

267. The farther the two regression lins cut each other:
(A) The greater will be degree of correlation
(B) The lesser will be degree of correlation
(C) Does not really matter
(D) None of these

268. Historically the first index was constructed in:
(A) 1764 (B) 1844
(C) 1850 (D) 1744

269. If the correlation coefficient between X and Y is r, the correlation between x^2 and y^2 is:
(A) r (B) r^2
(C) 0 (D) None of these

270. If two events A and B are independent, the probability that they will both occur is given by:
(A) P (A) + P (B) (B) P (A) x P (B)
(C) P (A) – P (B) (D) P(A)+P(B)-P(AB)

271. — error is committed when the hypothesis is true but out test rejects it:
(A) Type I (B) Type II
(C) Both (A) and (B) (D) None of these

272. If the standard deviation of a variable x is s and if y = a + bs, where a and b are constants, the standard deviation of y would be:
(A) a + bs (B) bs
(C) vb.s (D) b^2s

273. If correlation between two variables (x and y) is negative, the regression coefficient of y on x is:
(A) Positive (B) Negative
(C) Not certain (D) Zero

274. "Hit and run" style of interview is adopted in:
(A) Invasion method (B) Immersion method
(C) Case method (D) None of these

275. The most ubiquitous kin of variance in research is:
(A) Experimental variance (B) Extraneous variance
(C) Error variance (D) None of these

276. In an experiment measuring the effect of temperature on dilation of blood vessels, the level of temperature is the:
(A) Control variable (B) Antecedent variable
(C) Dependent variable (D) Independent variable

277. Scientific data must be:
(A) Objective (B) Measurable
(C) Predictable (D) All of these

278. The value of correlation coefficient lies in between:
(A) O (B) -1 to +1
(C) 1 to 2 (D) 0 to -1

279. Which among the following accounts for existing data, predict new observation and guide further research:
(A) Hypothesis (B) Theory
(C) Definition (D) Method

280. Research design is not essentially a:
(A) Plan (B) Product
(C) Structure (D) Strategy

281. If correlation coefficient is 0.7, then coefficient of determination is:
(A) 0.7 (B) 0.07
(C) 0.49 (D) None of these

282. The classification of variable into independent and dependent is based on:
(A) Uses (B) Kinds
(C) Foam (D) Both (A) and (C)

283. The word hypothesis literally means:
(A) Less than generally held view
(B) More than generally held view
(C) Same as generally held view
(D) None of these

284. To penetrate deep into the inner sanctum of the group we go for:
(A) Case study
(B) Participatory observation
(C) Formal discussion
(D) Interview

285. If r is the correlation coefficient, r^2 is:
(A) Coefficient of determination
(B) Coefficient of non determination
(C) Coefficient of alienation
(D) None of these

286. The most widely used method of measuring seasonal variations is:
(A) Ratio-to-moving average method
(B) Ratio-to trend method
(C) Link relative method
(D) None of these

287. The ratio of regression sum of squares to the total sum of squares is:
(A) Error means square
(B) Coefficient of determination
(C) Coefficient of non determinant
(D) Both a and b

288. To study the average size of shoes sold in the market, we use:
(A) Median (B) Mode
(C) Arithmetic mean (D) Geometric mean

289. Correlation coefficient is considered as that of two regression coefficients:
(A) Mean (B) Geometric mean
(C) Square (D) None of these

290. When we want average of rates of change or ratios or index number, which measure of central tendency is suitable:
(A) Geometric mean (B) Mode
(C) Median (D) Arithmetic mean

291. Under non-parametric the comparison of several related samples is tested by:
(A) Quade test (B) Friedmann test
(C) F-test (D) Both (A) and (B)

292. The correlation type wherein more than one dependent and independent variables are included in the regression model:
(A) Illusory correlation
(B) Spurious correlation
(C) Cannonical correlation
(D) Hallsinative correlation

293. In Confounding, generally which order of interaction is confounded:
(A) Lower order (B) First order
(C) Higher order (D) All of these

294. In a experiment, a treatment requires large area and b treatment requires smaller area (different precision), which design is most appropriate:
(A) Split Plot Design (B) LSD
(C) RBD (D) Strip Plot Design

295. The technique employed to find the degree of correlation is:
(A) Experimental (B) Observational
(C) Statistical (D) None of these

296. Local control is not applied for the design:
(A) RBD (B) CRD
(C) LSD (D) SPD

297. Which test is used to test of significance of the difference between two means:
(A) t-test (B) F-test
(C) z-test (D) x^2-test

298. Normal distribution is due to the work of
(A) James Bernoulli (B) SD Poisson
(C) Laplace and Gauss (D) Demoivre

299. Mean and variance of a standard normal distribution are respectively:
(A) 0 and 1 (B) 1 and 0
(C) Both (A) and (B) (D) None of these

300. A normal variate takes values in the range of:
(A) 0 to 1 (B) 0 to ∞
(C) - ∞ to + ∞ (D) -11.96 to +1.96

301. To study the average height of plants, we use:
(A) Median (B) Mode
(C) Arithmetic mean (D) Geometric mean

302. The population for sampling is defined in terms of:
(A) Time (B) Element
(C) Extent (D) All of these

303. "Items on a mathematical ability are examined to see whether they seem to be related to this ability" is an example for:
(A) Face validity (B) Content validity
(C) Concurrent validity (D) Predictive validity

304. Correlation coefficient is:
(A) More than +1
(B) Less than -1
(C) Lies between -1 and +1
(D) Lies between +1 and -1

305. The correlation type in which an increase in one measure is matched by an increase in the other:
(A) Positive (B) Negative
(C) Both (A) and (B) (D) None of these

306. Quartile deviation is:
(A) $\frac{4}{5}\sigma$ (B) $\frac{3}{2}\sigma$
(C) $\frac{2}{3}\sigma$ (D) $\frac{5}{4}\sigma$

307. Standard error can be:
(A) 0 (B) 1
(C) >1 (D) All of these

308. A relationship in correlation coefficient that does not form a straight line is:
(A) Linear relationship (B) Negative relationship
(C) Curvilinear relationship (D) Positive relationship

309. The value of x^2 always lies between:
(A) 0 to 1 (B) 0 to ∞
(C) -1 to +1 (D) - ∞ to + ∞

310. Pick the odd man out:
(A) F test (B) Kvuskal Wallis test
(C) Friedman test (D) Z test

311. .... is known as the 'Ideal' formula for constructing index numbers:
(A) Fisher's Ideal Index (B) Laspeyre's Index
(C) Paasche's Index (D) None of these

312. Mean ± 3σ covers
(A) 99.37 % items (B) 90 % items
(C) 99.9 % items (D) 99.73 % items

313. The degrees of freedom for statistics for paired t test of n pairs of observations is:
(A) 2 (n-1) (B) n-1
(C) 2n-1 (D) none of these

314. Ogive for more than type and less than type distribution intersect at:
(A) Mean (B) Median
(C) Mode (D) None of these

315. Measures of dispersion includes the range and:
(A) Standard deviation (B) Mode
(C) Median (D) None of these

316. Unit of variance of a data in cm is:
(A) No unit (B) Cm^2
(C) cm (D) cm^3

317. The range of multiple correlation coefficient R is:
(A) 0 to 1 (B) 0 to ∞
(C) -1 to +1 (D) - ∞ to + ∞

318. Coefficient of quartile deviation is calculated by the formula:
(A) $\frac{Q_3 + Q_1}{4}$ (B) $\frac{Q_3 + Q_1}{2}$
(C) $\frac{Q_3 - Q_1}{Q_3 + Q_1}$ (D) $\frac{Q_3 + Q_1}{Q_3 - Q_1}$

319. Which of the following is correct:
(A) $r_{12.34} = r_{13.24}$ (B) $r_{12.3} - r_{21.3}$
(C) $r_{13} = r_{23}$ (D) $r_{13.2} = r_{13.4}$

320. If the correlation coefficient between x and y is r, the correlation coefficient of y and x is:
(A) -r (B) r^2
(C) r (D) 1/r

321. For testing the significance of difference between two means, we use:
(A) t- test (B) X^2- test
(C) Z test (D) F-test

322. Usually the explanatory value of a regression equation of y on x is measured using:
(A) Correlation coefficient
(B) Coefficient of determination
(C) Coefficient of explanation
(D) Coefficient of variance

323. The difference of two means in case of small samples is tested by the formula:

(A) $T = \frac{X1 - X2}{S}$

(B) $T = \frac{X1 + X2}{S} \quad \frac{\sqrt{n1 + n2}}{n1 - n2}$

(C) $T = \frac{X1 + X2}{S} \quad \frac{\sqrt{n1\ n2}}{n1 + n2}$

(D) None of these

324. Size of critical region is also known as:
(A) Power of the test
(B) Size of type II error
(C) Critical value of the test statistics
(D) Size of the test

325. Equality of two regression coefficient can be tested with:
(A) t-test (B) x^2 test
(C) Z test (D) ANOVA

326. To test the hypothesis about proportions of items in a class, we use:
(A) F test (B) x^2 test
(C) Z test (D) t test

327. For testing the significance of correlation coefficient, we use:
(A) F test (B) ANOVA
(C) x^2 test (D) t test

328. Standard deviation of sample mean is given by:
(A) σ (B) n
(C) σ/root n (D) None of these

329. Who is considered as father of statistics:
(A) Ronald Fisher (B) Emile Dukeman
(C) Francis Galton (D) Carl Pearson

330. The non-parametric substitute for the parametric t test:
(A) x^2 test
(B) Kruskal Wallis test H test
(C) Friedman test
(D) Mann-Whitney U test

331. To study the average of quantity prices, we use:
(A) Median (B) Harmonic mean
(C) Arithmetic mean (D) Geometric mean

332. To get a short hand summary of large number of observation we go for:
(A) Pure statistics (B) Descriptive statistics
(C) Numerical statistics (D) None of these

333. Statistics serves to provide a short hand description of large amounts of:
(A) Scores (B) Data
(C) Points (D) None of these

334. Guttman scale is basically a:
(A) Summated rating scale
(B) Equal appearing interval scale
(C) Cumulative scale
(D) None of these

335. Which among the concepts is mathematically considered the biggest:
(A) Reliability (B) Validity
(C) Both (A) and (B) (D) None of these

336. The internal and external validity for laboratory experiment are respectively:
(A) High and High (B) High and Low
(C) Low and Low (D) Low and High

337. Test scores can be used for scientific purposes if they are:
(A) True and effective (B) Reliable and true
(C) Reliable and valid (D) None of these

338. An example for two way non-parametric analysis of variance among the following is:
(A) X^2 test
(B) Kruskal Wallis test H test
(C) Friedman test
(D) Z test

339. Opinion level of difficulty of a test corresponds to:
(A) Maximum item inter-correlation
(B) Maximum item total correlation
(C) Higher reliability of test scores
(D) All of these

340. To discriminate groups from one another on the basis of sets of measures, we use:
(A) Profile analysis (B) Factor analysis
(C) Discriminate analysis (D) Distracter analysis

341. Harmonic mean gives weightage to:
(A) Small values (B) Positive values
(C) Large values (D) Negative values

342. An example for one way non-parametric analysis of variance among the following is:
(A) X^2 test
(B) Kruskal Wallis test H test
(C) Friedman test
(D) Z test

343. The level of measurement associated with Mann-Whitney U test:
(A) Nominal (B) Ordinal
(C) Interval (D) Ratio

344. The value of coefficient of kurtosis can be:
(A) Less than 3 (B) Greater than 3
(C) Equal to 3 (D) All of these

345. If there are ten treatments and three replications in a CRD experiment, the error degrees of freedom is:
(A) 18 (B) 20
(C) 27 (D) 29

346. A family of parametric distribution with mean is equal to variance is:
(A) Binomial (B) Normal
(C) Poisson (D) Negative binomial

347. Sum of square divided by the number of the measure:
(A) Mean (B) Mode
(C) Median (D) None of these

348. When information is to be collected from a geographically dispersed population we go for:
(A) Stratified sampling
(B) Systematic sampling
(C) Cluster sampling
(D) Multipurpose sampling

349. If a<b and b<c, then a<c denotes:
(A) Transitive Property of Inaquality
(B) Addition Property of Equality
(C) Commutative Property of Addition
(D) Multiplication Property of Equality

350. In an RBD with 10 treatment and 4 replications, what will be the error degree of freedom if two observations are lost:
(A) 25 (B) 2
(C) 27 (D) 28

351. The test used in selection process to measure a persons interests, existing skills, and potential for acquiring skills:
(A) Intelligence tests (B) Aptitude tests
(C) Vocation tests (D) Personality test

352. Which of the following is not an ability test:
(A) Aptitude test
(B) Social intelligence test
(C) Gottshehold test
(D) General intelligence test

353. If a=b, then a+C = b+C denotes:
(A) Transitive Property of Inequality
(B) Addition Property of Equality
(C) Commutative Property of Addition
(D) Multiplication Property of Equality

354. Distribution wherein mean is always greater than variance is:
(A) Binomial (B) Poisson
(C) Normal (D) Both (A) and (B)

355. The units within and between each stratum in a stratified random sampling are respectively:
(A) Homogenous and homogenous
(B) Homogenous and Heterogeneous
(C) Heterogeneous and Heterogeneous
(D) Heterogeneous and Homogenous

356. Degree of freedom of X^2 test for 4 x 3 contingency table is:
(A) 12 (B) 6
(C) 1 (D) 7

357. Harmonic mean is better than other means if data are for:
(A) Speeds or rates (B) Heights and lengths
(C) Binary values 0 and 1 (D) Ratio and proportion

358. The variance of a binomial distribution with n and p as parameters is:
(A) n (B) n(1-p)
(C) np (D) np (1-p)

359. The sampling procedure most commonly used in order to ensure a representative sample:
(A) Random sampling (B) Systematic sampling
(C) Cluster sampling (D) Quota sampling

360. A normal curve is all but not:
(A) Symmetrical (B) Continuous
(C) Unimodal (D) Asymbiotic

361. If X is binomial variant with parameters n and p. If n=1, the distribution is:
(A) Poison (B) Normal
(C) Bernoulli (D) Binomial

362. Theoretical value of test statistics for five per cent when compared with one per cent is:
(A) Same (B) Greater
(C) Lesser (D) None of these

363. The value of probability ranges from:
(A) -1 to +1 (B) 0 to ∞
(C) 0 to 1 (D) None of these

364. Probability of two persons being born on the same day is:
(A) 1/49 (B) 1/365
(C) 1/7 (D) None of these

365. Mean Deviation is minimum when taken about:
(A) Median (B) Mode
(C) Arithmetic mean (D) Zero

366. Average performance of a representative sample on a given test is called:
(A) Norm (B) Validity
(C) Reliability (D) Objectivity

367. Sampling size is influenced by:
(A) Heterogeneity of population
(B) Sampling method
(C) Degree of accuracy
(D) All of these

368. Sampling frame is a must in the case of:
(A) Random sampling (B) Systematic sampling
(C) Stratified sampling (D) Quota sampling

369. A hypothesis which specifies the population distribution completely is called:
(A) Single hypothesis (B) Composite hypothesis
(C) Null hypothesis (D) Alternate hypothesis

...bility of a leap year will have 53 Sundays is:
(A) 1/7 (B) 2/7
(C) 2/53 (D) 52/53

371. The limiting relative frequency approach of probability is known as:
(A) Statistical probability
(B) Classical probability
(C) Mathematical probability
(D) Asymmetrical probability

372. The most frequent score in a given distribution:
(A) Median (B) Mode
(C) Arithmetic mean (D) Geometric mean

373. If it is known an event A has already occurred, the probability of an event E given A is:
(A) Statistical probability
(B) Classical probability
(C) Mathematical probability
(D) Asymmetrical probability

374. Accepting a hypothesis when it is false:
(A) Size of Type I error (B) Size of Type II error
(C) Both of the these (D) None of these

375. The probability that the throw of two dice yields a total of 5 or 8 is:
(A) ¼ (B) 1/5
(C) 2/5 (D) None of these

376. The method used to identify the units or dimensions behind the measures:
(A) Multiple regression analysis
(B) Discriminate analysis
(C) Profile analysis
(D) Factor analysis

377. For testing the significance of correlation coefficient, we use:
(A) X^2 test (B) F test
(C) Z test (D) t test

378. A structural observations is used:
(A) For testing a causal hypothesis
(B) In exploratory analysis
(C) In formulative studies
(D) All of these

379. According to which process the cone of experience classify the instructural methods:
(A) Experimental correctness
(B) Difficulty in understanding
(C) Ease in use and application
(D) Level of abstraction in thinking

380. After this therefore caused by this is usually denoted to indicate:
(A) Experimental design (B) Survey design
(C) Ex post facto design (D) Exploratory design

381. Age is classified under:
(A) Discontinous variable (B) Discrete variable
(C) Continuous variable (D) Intervening variable

382. An exact test for testing the independence of attributes is a contingency table of order (2 x 2) was given by:
(A) Karl Pearson (B) Pascal
(C) De moivre (D) R A Fisher

383. If a card is drawn from a pack of cards the probability of getting either a king or a queen is:
(A) 2/13 (B) 5/15
(C) ¼ (D) None of these

384. In 1933, the theory of testing of hypothesis was propounded by:
(A) R A Fisher (B) J Nijman
(C) E L Lehman (D) Karl Pearson

385. In case of a positively skewed distribution:
(A) Median > mean > mode
(B) Mean > median > mode
(C) Mean = median = mode
(D) None of these

386. In case of conditional probability (A and B) = ?
(A) P (B) x P (A/B) (B) P (B) x P (A)
(C) P (A) + P (B) (D) None of these

387. In Sequential Probability Ratio Test (SPRT), the sample size is:
(A) Fixed (B) Fixed but small
(C) Fixed but large (D) A random variable

388. Mention the type of correlation that operated between rurality and population density:
(A) Positive (B) Negative
(C) Accidental (D) Scattered

389. Out of two errors, the most severe is:
(A) Type I error (B) Type II error
(C) Both (A) and (B) (D) None of these

390. Student's t test is applicable in case of:
(A) Small sample (B) For sample size 5 – 30
(C) Large samples (D) None of these

391. Student's t test was invented by:
(A) R A Fisher (B) G W Snesecor
(C) W S Gosset (D) W.G. Cochran

392. Student's t test is applicable only when:
(A) Variate values are independent
(B) The variate is distributed normally
(C) The sample is not large
(D) All of these

393. The coefficient of reliability at a perfectly homogenous experimental condition will be:
(A) 0 (B) 1
(C) 0.5 (D) 2

394. The measure of dispersion which ignores signs of deviation from the central value is:
(A) Range (B) Quartile deviation
(C) Standard deviation (D) Mean deviation

395. The most ubiquitous kind of variance in research is:
(A) Experimental variance (B) Sample variance
(C) Population variance (D) Error variance

396. The probability of an impossible event is:
(A) 0 (B) 1
(C) 8 (D) None of these

397. The number of possible samples of size two from a population of four units in a simple random sample without replacement is:
(A) 2 (B) 4
(C) 6 (D) 8

398. The range of multiple correlation coefficient R is:
(A) 0 to 1 (B) 0 to ∞
(C) -1 to +1 (D) -∞ to +∞

399. The ratio of difference between sample variance and with in sample variance follows:
(A) F- distribution (B) X^2 – distribution
(C) Z – distribution (D) t- distribution

400. The suitable measure to find the average speed when time for each speed is fixed would be:
(A) Arithmetic mean (B) Geometric mean
(C) Harmonic mean (D) None of these

401. To control variance is the major purpose of:
(A) Statistical test
(B) Research design
(C) Ex post facto method of research
(D) Interview schedule

402. Any categorization scheme must be on:
(A) One level of discourse
(B) Three level of discourse
(C) Two level of discourse
(D) Four level of discourse

403. Take the odd one out:
(A) Area sampling (B) Snowball sampling
(C) Quota sampling (D) Incidental sampling

404. To findout the average height of plants we should use:
(A) Arithmetic mean (B) Median
(C) Mode (D) Harmonic mean

405. To find the average of quality prices, the measure of central tendency to be used is:
(A) Harmonic mean (B) Geometric mean
(C) Arithmetic mean (D) None of these

406. To findout the average size of shoes sold in the market, we should be used:
(A) Arithmetic mean (B) Geometric mean
(C) Median (D) Mode

407. To test the hypothesis about proportions of items in a class, the usual test is:
(A) T test (B) Z test
(C) F test (D) All of these

408. When information is to be collected from a geographically dispersed population we go for:
(A) Stratified sampling
(B) Systematic sampling
(C) Cluster sampling
(D) Multipurpose sampling

409. When degree of freedom for x^2 is 100 or mo[illegible] proximated to:
(A) t-distribution (B) F distribution
(C) Z -distribution (D) None of these

410. Whether a test is one sided of two sided depends on:
(A) Alternate hypothesis (B) Composite hypothesis
(C) Null hypothesis (D) Simple hypothesis

411. Classification of research data should follow which among the following rules:
(A) Categories should be exhaustive
(B) Mutually exclusive
(C) Independent
(D) All of these

412. Which of the following is the unit less measure of dispersion:
(A) Standard deviation (B) Mean deviation
(C) Coefficient of variation (D) Range

413. Which of the following statistics are used to divide a normal adoptor distribution into categories:
(A) Mean and Variance
(B) Mean and Standard deviation
(C) Mean and Regression
(D) Correlation and Standard deviation

414. Which among the following accounts for existing data, predict new observations and guide further research:
(A) Hypothesis (B) Theory
(C) Definition (D) Method

415. The process of drawing conclusions or inferences after a careful analysis of the collected data:
(A) Interpretation (B) Tabulation
(C) Classification (D) Evaluation

416. The basic research difference between classical research and participatory research lies in their:
(A) Control (B) Content
(C) Power (D) Substance

417. The summary and conclusion chapter in a research thesis is otherwise known as:
(A) Epilogue (B) Prologue
(C) Alphalogue (D) Symatologue

418. Opinion polls falls under:
(A) Descriptive design (B) Explorative design
(C) Diagnostic design (D) Experimental design

419. If all values of a variable are multiplied by a given number, which one of the following will NOT be affected?
(A) Mean (B) Median
(C) Range (D) Coefficient of variation

420. A symbol to which numerals or values are assigned:
(A) Mean (B) Mode
(C) Variable (D) Median

421. Variables are essentially a:
(A) Property (B) Symbol
(C) Constructs (D) All of these

370. Prob... antity which can vary from one individual to an-...r individual is called:
(A) ... Mean (B) Mode
(C) Variable (D) Median

423. Assertion (A): Variance is always greater than the standard deviation
Reason (R): Variance is the square of the standard deviation
(A) Both A and R are individually true and R is the correct explanation of A
(B) Both A and R are individually true but R is not the correct explanation of A
(C) A is rue but R is false
(D) A is false but R is true

424. The condition under investigation that act as potential cause of some change in behaviour:
(A) Dependent variable (B) Independent variable
(C) Intervening variable (D) None of these

425. Independent variables are all but not:
(A) Presumed cause (B) Predicted to
(C) Manipulated (D) Antecedent

426. The variable that reflects changes in the independent variable is:
(A) Dependent variable (B) Independent variable
(C) Extraneous variable (D) None of these

427. The manipulate variable in an experiment is:
(A) Independent variable (B) Dependent variable
(C) Active variable (D) Both (A) and (C)

428. All variables that can be measured in the smallest degree of finness is called:
(A) Continuous variable (B) Discrete variable
(C) Qualitative variable (D) Moderator variable

429. Age is classified under:
(A) Dicotonimous variable (B) Discrete variable
(C) Continuous variable (D) Intervening variable

430. Sex falls under which of the following type of variable:
(A) Dichotomies variable (B) Polytomies variable
(C) Continuous variable (D) Qualitative variable

431. Response variable are a function of:
(A) Orgasmic variable (B) Stimulus variable
(C) Both (A) and (B) (D) None of these

432. If r is the correlation coefficients between the two variables, then:
(A) $0 = r = 1$ (B) $-1 = r = 0$
(C) $-1 = r = 1$ (D) $1 = r = 2$

433. The price index that uses base year quantities as weights is called:
(A) Passche's index (B) Laspeyere's index
(C) Fisher's ideal index (D) Wholesale price index

434. An example for quantitative variable from among the following is:
(A) Sex (B) Age (C) Race (D) Religion

435. An example for discrete variable from among the following is:
(A) Age (B) Sex
(C) Intelligence (D) Height

436. The word hypothesis literary means:
(A) Less than generally held view
(B) More than generally held view
(C) Same as generally held view
(D) None of these

437. A hypothesis is essentially a:
(A) Interrogative statement (B) Empirical statement
(C) Relational proposition (D) Generally held view

438. Which one of the following statistical measures is based on all observation?
(A) Quartile deviation (B) Harmonic mean
(C) Range (D) Mode

439. Which among the following is the testable form in a scientific research:
(A) Research problem (B) Research hypothesis
(C) Research variable (D) All of these

440. The hypothesis type stated for the purpose of rejection:
(A) Null hypothesis (B) Alternate hypothesis
(C) Parallel hypothesis (D) Both (A) and (C)

441. A hypothesis when tested true become:
(A) Principle (B) Theory
(C) Philosophy (D) Law

442. The accuracy of a measurement signifies the closeness with which a estimate approaches the:
(A) True value (B) Average value
(C) Both (A) and (B) (D) None of these

443. The collection of data in research falls under:
(A) Statistical design (B) Observational design
(C) Operational design (D) Sampling design

444. In the field experimentation, local control is used for:
(A) Validity of estimate of error
(B) Validity of treatment effects
(C) Diminution of error
(D) All of these

445. Experimental method of data collection among the following is:
(A) Field study (B) Sample survey
(C) Library method (D) Field experiment

446. Survey method would typically used in:
(A) Clinical assessment (B) Opinion poll
(C) Laboratory experiment (D) Psychoanalysis

447. Formula of linear correlation coefficient (r_{yx}) is:

(A) $r_{yx} = \dfrac{B_{yx} + b_{xy}}{2}$ (B) $ryx = \dfrac{\sqrt{byx.\ bxy}}{}$

(C) $r_{yx} = \dfrac{2}{\left(\dfrac{1}{Byx} + \dfrac{1}{bxy}\right)}$

(D) $r_{yx} = \dfrac{W1.b_{yx} + W2.b_{xy}}{W_1 + W_2}$

448. Survey research employs which of the following techniques:
(A) Panel technique (B) Personal interview
(C) Controlled observation (D) All of these

449. The best control for practice effects in an experiment is:
(A) Random selection (B) Counter balancing
(C) Equivalent group (D) Split half presentation

450. The R rank correlation was suggested by:
(A) James Bernoulli (B) SD Poisson
(C) C Spearman (D) Laplace and Gauss

451. If some of the counts are very small or zero, the most appropriate transformed variate will be:
(A) √x+1/2 (B) √x
(C) √x+1/2 (D) √x+1

452. The most commonly used method in a social survey is:
(A) Observation (B) Interview
(C) Questionnaire (D) Case study

453. Interview is widely used technique in:
(A) Local studies (B) Public studies
(C) Empirical studies (D) Non-scientific studies

454. Assignment of values to alternatives is called:
(A) Observation (B) Judgement
(C) Measurement (D) Evaluation

455. Questionnaire involves presentation of:
(A) Oral verbal stimuli (B) Written-verbal stimuli
(C) Oral written stimuli (D) None of these

456. When β2 < 3 the distribution is:
(A) Leptokurtic (B) Platykurtic
(C) Mesokurtic (D) None of these

457. In a normal distribution:
(A) Mean and median are equal
(B) Median and mode are equal
(C) All the three are equal
(D) None of these

458. If fertility variation in the field is in two direction at right angles, which of the experimental design is suitable:
(A) CRBD (B) RBD
(C) Latin square (D) CRD

459. As a general rule, the number of classes in the frequency distribution should be about:
(A) 30 (B) 15
(C) 6 (D) 20

460. A hypothesis which does not specify the population distribution completely:
(A) Simple hypothesis (B) Composite hypothesis
(C) Null hypothesis (D) Alternate hypothesis

461. Range of distribution is:
(A) The difference between largest and smallest observations
(B) Averages of all observations
(C) Middle of the series
(D) All of these

462. The second moment about mean is:
(A) $\frac{\Sigma (x - x)^2}{N}$ (B) $\frac{(x - x)^2}{N}$
(C) $\frac{(x - x)^2}{N}$ (D) None of these

463. In the regression line Y = α+ βx, β is called:
(A) Intercept (B) Slope of the line
(C) Mean of the X values (D Mean of the Y values

464. Which mean of dispersion is most affected by extreme values:
(A) Range (B) Quartile deviation
(C) Standard deviation (D) Mean deviation

465. If β_{xy} and β_{yx} are two regression coefficients, they have:
(A) Same sign
(B) Opposite sign
(C) Either same or opposite sign
(D) Nothing can be said

466. In an experimental plot, 7 rows were sown with 50 cm row spacing and 5 m length, while harvesting one row on either side of the plots were discarded. The net plot size will be:
(A) 2.0 x 5.0 m (B) 2.5 x 5.0 m
(C) 3.5 x 5.0 m (D) 4.0 x 5.0 m

467. In a field experiment, the net plot size was 2.5 x 4.0 m and the yields were recorded in kg. What is the conversion factor for q/ha:
(A) 10 (B) 100
(C) 50 (D) 70

468. In a fertilizer trial, 40 kg N/ha has to be applied to the plot of 10.0 sq.m area. What quantity of urea is needed:
(A) 95 g (B) 89 g
(C) 115 g (D) 100 g

469. If the mode frequency distribution is Mo=16 and its mean = 16 then median is:
(A) Zero (B) 16
(C) 32 (D) <16

470. Biometry deals with:
(A) Observation with living things
(B) Collection of observation
(C) Interpretation of data
(D) Analysis of data

471. Arithmetic mean is most commonly used because:
(A) Based on all the observations
(B) Largest and smallest observations
(C) Located in the middle of the series
(D) Based on largest data only

472. Standard deviation is:
(A) Expressed in the same unit of measurement
(B) Free from unit of measurement
(C) It is an percentage
(D) It is in weight

473. The test of significance of differences between two means in small samples is known as:
(A) T-test
(B) F-test
(C) On the basis of normal deviate
(D) x^2 –test

474. When two variables change together in such a way that an increase in one variable is accompanied by an increase in the other. The variables are said to be:
(A) Independent (B) Positively correlated
(C) Negatively correlation (D) None of these

475. During experimentation, we lost some information and we want to get idea about these values, which technique will be useful:
(A) Field plot technique (B) Missing plot technique
(C) Seed plot technique (D) Uniformity trial

476. F-test can be used for testing the significance of:
(A) Several differences
(B) A single difference
(C) Not used for testing the difference
(D) Two differences

477. When the fertility gradient of the field is in one direction, which design is appropriate:
(A) CRD (B) LSD
(C) RBD (D) SPD

478. X^2 test is applied to:
(A) Quantitative observations
(B) Qualitative observations
(C) In both the cases
(D) In none of the cases

479. The probability of committing type I error is known as:
(A Test of significance (B) Level of significance
(C) Composite hypothesis (D) None of these

480. Which type of average would be suitable to calculate the size of agricultural holding:
(A) Mode (B) Median
(C) Weighted HM (D) Weighted AM

481. When a frequency distribution is not symmetrical about the mean it is said to be:
(A) Kurtosis (B) Moment
(C) Skewed (D) None of these

482. The randomization of treatments in the field experiments is essential:
(A) The statistical procedures hold good only when randomization is done
(B) To average out the effects of environmental differences
(C) For comparison of treatment effects
(D) For comparison of plot wise differences

483. Which test is applied for the test of significance between the two variances:
(A) t-test (B) z-test
(C) x^2-test (D) F-test

484. Local control helps in:
(A) Reducing the experimental error
(B) Homogeneity of experimental units
(C) Both (A) and (B)
(D) None of these

485. Which of following is not a measure of dispersion?
(A) Range (B) Coefficient of variation
(C) Mean deviation (D) Standard deviation

486. What is the correct formula of coefficient of variation?
(A) $\frac{SD}{AM}$ x 100 (B) $\frac{AM}{SD}$ x 100
(C) $\frac{SD}{AM}$ (D) $\frac{AM}{SD}$

487. For testing the independence of two attributes, the test used is:
(A) F-test (B) T-test
(C) x^2-test (D) Z-test

488. For testing the significance of correlation coefficient, the test used is:
(A) F-test (B) T-test
(C) x^2-test (D) Z-test

489. Latin square design is suitable for comparing:
(A) 2 to 4 treatment (B) 3 treatment
(C) 15 treatment (D) 5-12 treatment

490. The reasoning process wherein the researcher collect observation and formulate hypothesis Based upon it:
(A) Inductive reasoning (B) Deductive reasoning
(C) Both (A) and (B) (D) None of these

491. For the field experimentation the accurate idea of the fertility variation can be had by the help of:
(A) Experimental field (B) Unformity trial
(C) Coordinated trials (D) All of these

492. An abstract formed from generalization from particular is called:
(A) Principle (B) Hypothesis
(C) Theory (D) Concept

493. Concepts are powerful tools that make us to think:
(A) Absurdly (B) Abstractly
(C) Positively (D) Negatively

494. Do such and such things in so and so manner is the characteristic feature of:
(A) Operative definition
(B) Constitutive definition
(C) Conceptual definition
(D) Observational definition

495. Defining intelligence in terms of verbal reasoning analytical ability, comprehension and numerical capacity forms the characteristic feature of:
(A) Operative definition
(B) Constitutive definition
(C) Conceptual definition
(D) Observational definition

496. The variation due to uncontrolled factors is spoken as
(A) Standard error (B) Experimental error
(C) Treatment effects (D) All of these

497. Pure research differs from applied research in the sense the former:
(A) Illuminates theory
(B) Uses highly technical language
(C)Shares multipurpose view
(D) Concentrate on the "Why" aspect

498. The basic difference between classical research and participatory research lies in their:
(A) Control (B) Content
(C) Power (D) Substance

499. The basic function of a good design is to:
(A) Decrease experimental variance
(B) Maximize extraneous variance
(C) Minimize error variance
(D) Explain metaphysical observation

500. Poisson distribution deals with:
(A) Discrete variables (B) Continuous variables
(C) Both (A) and (B) (D)None of these

501. A symbol to which numerals or values are assigned:
(A) Mean (B) Mode
(C) Variable (D) Median

502. Variables are essentially a:
(A) Property (B) Symbol
(C) Constructs (D) All of these

503. A quantity which can vary from one individual to another individual is called:
(A) Mean (B) Mode
(C) Variable (D) Median

504. Correlation does not demonstrate:
(A) Causation (B) Effect
(C) Reason (D) None of these

505. If two factors are involved in experimentation and both the factors requires long plot sizes, the design suited for this will be:
(A) Split plot (B) Strip plot
(C) CRD (D) RBD

506. Median of seven observations 9, 3, 5, 4, 2, 3, 2 is:
(A) 2 (B) 3
(C) 4 (D) 5

507. The condition under investigation that act as potential cause of some change in behavior:
(A) Dependent variable (B) Independent variable
(C) Intervening variable (D) None of these

508. Two unbiased coins are tossed once What is the probability of getting two heads?
(A) 0 (B) ¼
(C) ½ (D) 1

509. If a<b and b<c, then a<c denotes:
(A) Transitive Property of Inequality
(B) Addition Property of Equality
(C) Cumulative Property of Addition
(D) Multiplication Property of Equality

510. Independent variables are all but not:
(A) Presumed cause (B) Predicted to
(C) Manipulated (D) Antecedent

511. Harmonic mean gives weightage to:
(A) Small values (B) Positive values
(C) Large values (D) Negative values

512. A constant value of 50 is subtracted from each observation of a set, the mean of the set is:
(A) Increased by 50 (B) Decreased by 50
(C) Not affected (D) Zero

513. A relationship in correlation coefficient that does not form a straight line is:
(A) Linear relationship (B) Negative relationship
(C) Curvilinear relationship (D) Positive relationship

514. The variable that reflects changes in the independent variable is:
(A) Dependent variable (B) Independent variable
(C) Extraneous variable (D) None of these

515. The rule of correspondent between a dependent variable and an independent variable:
(A) Function (B) Equation
(C) Argument (D) Premise

516. All variables that can be measured in the smallest degree of fitness is called:
(A) Continuous variable (B) Discrete variable
(C) Qualitative variable (D) Moderator variable

517. The most important measure of dispersion used in industry for quality control:
(A) Standard deviation (B) Mean deviation
(C) Coefficient of variation (D) Range

518. What is the mean of 5, 13, 0, 6?
(A) 3 (B) 4
(C) 6 (D) 8

519. The x^2 (chi-square) test for goodness of fit and independence of attributes come under:
(A) Parametric tests (B) Non-parametric test
(C) Both (A) and (B) (D) Not classified

520. When the same interaction is confounded in all the replications, it is known as:
(A) Partial confounding (B) Complete confounding
(C) Both (A) and (B) (D) None of these

521. A variable selected to prevent alteration of the relationship between primary independent variable with the dependent variable:
(A) Active variable (B) Intervening variable
(C) Attribute variable *(D) Moderator variable*

522. The values of regression coefficient lies between:
(A) - 8 to + 8 (B) 0 to 1
(C) -1to +1 (D) 0 to 8

523. How many number of treatments can be adopted in LSD:
(A) 5 to 8 (B) 5 to 12
(C) Both (A) and (B) (D) Not estimated

For x^2 test no theoretical cell frequency should be small, it should be at least:
(A) 2 (B) 5
(C) 8 (D) 12

525. Sex falls under which of the following type of variable:
(A) Dichotomies variable (B) Polytomies variable
(C) Continuous variable (D) Qualitative variables

526. When the distribution of data is continuous, which distribution is applicable:
(A) Normal distribution (B) Binomial
(C) Poisson (D) None of these

527. Degree of freedom of x^2 test for 4 x 3 contingent table is:
(A) 12 (B) 6
(C) 1 (D) 7

528. If all the variate values are negative, the standard deviation will be:
(A) Negative (B) Positive
(C) No sign (D) Cannot calculate

529. A true dicotominous variable among the following is:
(A) Male-female (B) Literate-illiterate
(C) Intelligent-unintelligent (D) Introvert-extrovert

530. Take the odd man out:
(A) Independent variable (B) Active variable
(C) Stimulus variable (D) Orgasmic variable

531. Response variable are a function of:
(A) Orgasmic variable (B) Stimulus variable
(C) Both (A) and (B) (D) None of these

532. A variance produced in the dependent variable by the manipulation of the active variable by the experimenter:
(A) Experimental variance (B) Extraneous variance
(C) Error variance (D) None of these

533. Which among the following is not a character of eroor variance:
(A) Self compensation (B) Unpredictability
(C) Generalizability (D) Skewedness

534. A hypothesis is essentially a:
(A) Interrogative statement (B) Empirical statement
(C) Relational proposition (D) Generally held view

535. Hypothesis that states the negation of the assertion is commonly called as:
(A) Null hypothesis (B) Parallel hypothesis
(C) Alternate hypothesis (D) Both (B) and (C)

536. The bridge between theory and empirical enquiry is connected through:
(A) Research principle (B) Research hypothesis
(C) Research problem (D) Research output

537. Finding the cause of behavior from number of observation is called;
(A) Inductive reasoning (B) Horizontal reasoning
(C) Vertical reasoning (D) Deductive reasoning

538. When we want average of rates of change or ratios or index number, which measure of central tendency is suitable:
(A) Geometric mean (B) Mode
(C) Median (D) Arithmetic mean

539. Under non-parametric the comparison of several related samples is tested by:
(A) Quade test (B) Friedmann test
(C) F-test (D) Both (a) and (b)

540. In Confounding, generally which order of interaction is confounded:
(A) Lower order (B) First order
(C) Higher order (D) All of these

541. A tentative explanation of an event or observation:
(A) Variable (B) Hypothesis
(C) Argument (D) Principle

542. A sampling technique used when "Natural" but relatively homogeneous groupings are evident us a statistical population is called:
(A) Stratified sampling (B) Cluster sampling
(C) Both (A) and (B) (D) None of these

543. Whether a test is one tailed or two tailed depends on:
(A) Alternate hypothesis (B) Null hypothesis
(C) Composite hypothesis (D) Simple hypothesis

544. The research method which focus independently on different aspects of a single subject:
(A) Field studies (B) Case studies
(C) Experimental studies (D) None of these

545. Local control is not applied for the design:
(A) RBD (B) CRD
(C) LSD (D) SPD

546. Which test is used to test of significance of the difference between two means:
(A) t-test (B) F-test
(C) z-test (D) x^2-test

547. Normal distribution is due to the work of
(A) James Bernoulli (B) SD Poisson
(C) Laplace and Gauss (D) Demoivre

548. Which among the following can be thought as natural clinical tests of the effects of unusual variables:
(A) Observational variables (B) Empirical studies
(C) Experimental studies (D) Case studies

549. Which among the following is the main disadvantage of mailed questionnaire:
(A) Dishonest response
(B) Low return rate
(C) Time consuming and costly
(D) Postal lose

550. A sampling technique often used in marketing research is:
(A) Stratified sampling (B) Cluster sampling
(C) Both (A) and (B) (D) None of these

551. The method of assembling correspondence without experimental control over then is called:
(A) Dispersion (B) Correlation
(C) Central tendency (D) None of these

552. The intelligency, ability or efficiency can be measured by use of:
(A) Median (B) Mode
(C) Weighted HM (D) Geometric mean

553. When there are several factors with different levels to be experimented simultaneously with the same level of precision, which design is most appropriate:
(A) CRD (B) LSD
(C) RBD (D) Factorial design

554. The collection of data in research falls under:
(A) Statistical design (B) Observational design
(C) Operational design (D) Sampling design

555. The interview type where in conversation are informal and the interviewee determines the subjects for discussion:
(A) Direct observation (B) Projective tes
(C) Unstructured interview (D) Structured interview

556. What is/are the basic principles of field experimentation:
(A) Randomization (B) Replication
(C) Local control (D) All of these

557. For paired observation, which test is used for testing the significance of a mean difference:
(A) Paired t-test (B) t-test
(C) F-test (D) x^2-test

558. Binomial distribution is a very useful distribution for dealing with ______ variates:
(A) Continuous (B) Discrete
(C) Both (A) and (B) (D) None of these

559. The idea of randomization was put forward by:
(A) Chapman and Greenwood (B) Kerlinger
(C) Ronald A Fisher (D) Sukathmae

560. The assignment of numbers to certain characteristics of objects according to rules:
(A) Measurement (B) Evaluation
(C) Observation (D) Survey

561. The most frequent occurred value of data or whose frequency is maximum is known as:
(A) Arithmetic mean (B) Mode
(C) Median (D) GM

562. The degree of linear relationship between variables are expressed through:
(A) Coefficient of correlation (B) Standard deviation
(C) Arithmetic mean (D) None of these

563. If X is multiplied by 10 and Y by 20, b_{xy}, the regression coefficient is:
(A) Same as b_{xy} (B) Half by b_{xy}
(C) 4 times of b_{xy} (D) Double of b_{xy}

564. The exact procedures used to represent a concept:
(A) Operational definition
(B) Constitutive definition
(C) Conceptual definition
(D) Observational definition

565. A functional variate for establishing a para[...] called:
(A) An estimate (B) An estimator
(C) A frame (D) A statistic

566. The most eliminatory level of measurement is:
(A) Nominal (B) Ordinal
(C) Interval (D) Ratio

567. Stratified sampling is:
(A) Always more efficient than simple random sampling
(B) Atleast equally efficient as simple random sampling
(C) Less efficient than simple random sampling
(D) None of these

568. Transitivity postulate find its application right from:
(A) Nominal level (B) Ordinal level
(C) Interval level (D) Ratio level

569. The lowest level of measurement having an arbitrary zero is:
(A) Nominal level (B) Ordinal level
(C) Interval level (D) Ratio level

570. The level of measurement that goes smooth with non-parametric statistics is:
(A) Nominal-Ordinal (B) Ordinal – Interval
(C) Nominal - Interval (D) Ordinal – Ratio

571. Intelligence, aptitude and personality measures strictly speaking falls under:
(A) Nominal level (B) Ordinal level
(C) Interval level (D) Ratio level

572. If the measure of kurtosis is greater than 3, then the curve is:
(A) Platykutosis (B) Mesokurtosis
(C) Leptokurtosis (D) Normal

573. The experimental design wherein the heterogeneity in the experimental units are in two directions:
(A) CRD (B) RBD
(C) LSD (D) BIBD

574. Q methodology was put forward by:
(A) Fisher (B) Osgood
(C) Mereno (D) William Stephenson

575. Reflective, transitive and symmetrical relationship is the characteristic feature of:
(A) Nominal level (B) Ordinal level
(C) Interval level (D) Ratio level

576. To test the agreement between observed frequencies and expected frequencies, we use:
(A) F test (B) ANOVA
(C) X^2 test (D) t test

577. Height and weight are obvious examples for:
(A) Nominal level (B) Ordinal level
(C) Interval level (D) Ratio level

578. The error degree of freedom. for a RBD experiment with 5 treatments and 3 replications is:
(A) 15 (B) 12
(C) 8 (D) 14

579. Paper pencil test is otherwise called as:
(A) Verbal test (B) Non verbal test
(C) Performance test (D) Non-language test

580. To judge the capacity of individuals in synthesis of ideas and critical thinking:
(A) New type test (B) Paper pencil test
(C) Free answer test (D) Selection type test

581. The item type best suited for measuring individual achievement is:
(A) Matching item (B) Multiple choice item
(C) Two alternative item (D) None of these

582. Degree of freedom is related to the number of:
(A) People involved
(B) Hypothesis under set
(C) Independent observations
(D) None of these

583. The chance for guessing from a two alternative item is:
(A) 100 per cent (B) 75 per cent
(C) 50 per cent (D) 25 per cent

584. Thematic Apperception Test is used to study:
(A) Leadership behavior
(B) Individual creativity
(C) Conflict management
(D) Achievement motivation

585. When the experimental units are homogenous, the design used is:
(A) CRD (B) RBD
(C) LSD (D) None of these

586. An example for constructive techniques from among the following is:
(A) Rorschach Projection Test
(B) Thematic Apperception Test
(C) Sentence Completion Test
(D) Word Association Test

587. Projetcile test are:
(A) Objective test (B) Subjective test
(C) Attitude test (D) Both (B) and (C)

588. Which design is most appropriate for the laboratory experiments:
(A) CRBD (B) CRD
(C) LSD (D) Factorial RBD

589. The stimulus used in a projectile test must be:
(A) Structured ambiguous
(B) Unstructured ambiguous
(C) Structured unambiguous
(D) Unstructured unambiguous

590. For testing the significance of difference between two means, we use:
(A) F test (B) Z test
(C) X^2 test (D) t test

591. The lowest level of measurement that facilitate us to go for addition and subtraction is:
(A) Nominal level (B) Ordinal level
(C) Interval level (D) Ratio level

592. Expressive test gives basic emphasis on:
(A) Product (B) Process
(C) Input (D) Both (A) and (B)

593. The principle of experimental design include:
(A) Randomization (B) Local control
(C) Both (A) and (B) (D) None of these

594. The method used to observe and measure the psychological meaning of concept:
(A) Somatic differentiation (B) Projectile technique
(C) Content analysis (D) Q methodology

595. The method of analyzing communication in a systematic, objective and quantitative manner to measure variable is called:
(A) Content analysis (B) Process analysis
(C) Hypothesis analysis (D) Variable analysis

596. Correlation coefficient lies between:
(A) - ¥ < r < 0 (B) 0 £ r £ 1
(C) -1 < r < + 1 (D) -¥ £ r £ ¥

597. Which measure of central tendency is appropriate for index numbers:
(A) Harmonic mean (B) Arithmetic mean
(C) Mode (D) Geometric mean

598. If there are 5 varieties and 4 replications to each, the error degree of freedom for RBD will be:
(A) 20 (B) 19
(C) 12 (D) 9

599. Ogive is a method of:
(A) Diagrammatic representation
(B) Graphic representation
(C) Both (A) and (B)
(D) None of these

600. Find the modal value in 9, 13, 18, 14, 21, 18 and 13:
(A) 9 (B) 13
(C) 18 (D) 21

601. Who has developed the technique of analysis of variance?
(A) Cox (B) Fisher
(C) Kempthorne (D) Rao

602. To study the average size shoes sold in the market, we use:
(A) Mode (B) Harmonic mean
(C) Arithmetic mean (D) Geometric mean

603. The smallest unit of content analysis is:
(A) Theme (B) Words
(C) Characters (D) Items

604. The t test tends to be Z test when sample size becomes more than:
(A) 100 (B) 50
(C) 30 (D) 5

605. The computer programme used for content analysis:
(A) General inquirer (B) Dream weaver
(C) Mind mapper (D) Coral draw

606. To quantify readability in content analysis we go for:
(A) Fog index (B) Log index
(C) Content index (D) Variable index

607. Who is the author of the book Sampling Techniques':
(A) W G Cochran (B) A L Bowley
(C) R L Ackoff (D) Wison M.

608. Which one of the following is generally used for testing the equality of two variances?
(A) t - test (B) Z- test
(C) x^2 – test (D) F – test

609. Sum of absolute deviations about median is:
(A) Least (B) Greatest
(C) Zero (D) Equal

610. Naturalistic observation only provides description of:
(A) Experience (B) Feeling
(C) Behaviour (D) None of these

611. The non-parametric test used for two independent samples is:
(A) t-test (B) z-test
(C) F-test (D) Mann-Whitney –test

612. Which design gives precision high enough to reduce the standard error to less than 1%:
(A) SPD (B) Strip plot design
(C) RBD (D) LSD

613. In confounding, the precision on the main effects and certain interactions of lower order:
(A) Increases (B) Decreases
(C) Remain same (D) None of these

614. For the comparison of two correlated samples under non-parametric alternative to paired F-test is:
(A) z-test
(B) F-test
(C) Wilcoxon Signed –rank test
(D) Kruskal-Wallis H test

615. The three R's in the SQ3R strategy of comprehension does not include:
(A) Read (B) Recite
(C) Review (D) Reward

616. When 5 is added to every observation, which of the one of the following does:
(A) Mean (B) Median
(C) Mode (D) Variance

617. A method in which people are asked to respond to questions by expressing their personal evaluations:
(A) Attitude scale (B) Interview
(C) Self report method (D) All of these

618. Mode can be calculated by the formula:
(A) Mode = 4 median – 3 mean
(B) Mode = 3 median – 2 mean
(C) Mode = median –mean
(D) None of these

619. For testing HO : m = 0 in a normal distribution N(m, 1), which one of the following test is used?
(A) t – test (B) Normal – test
(C) x^2 – test (D) F – test

620. Statistical conclusion are always:
(A) Absolutely true (B) Absolutely wrong
(C) Either (A) or (B) (D) True on average

621. The observational procedure in explorative studies is more likely:
(A) Structured (B) Unstructured
(C) Partially structured (D) None of these

622. A low error variance results in a:
(A) High validity (B) Low validity
(C) High reliability (D) Both (A) and (C)

623. The alpha coefficient of variability indicates:
(A) Temporal stability (B) Internal consistency
(C) Both (A) and (B) (D) None of these

624. F test is used to test:
(A) Ratio of two variances (B) Ratio of two means
(C) Square of normal variate (D) None of these

625. Equality of several variance is done by:
(A) F (B) X^2
(C) Z (D) None of these

626. The concept of semantic differentiation was discovered by:
(A) Fisher (B) Osgood
(C) Mereno (D) William Stephenson

627. Which of the following is not a unit of content analysis:
(A) Themes (B) Readability
(C) Characters (D) Items

628. The lowest common denominator of a test is otherwise called:
(A) Stem (B) Item
(C) Root (D) Base

629. A high inter-correlation among different items of a test indicates:
(A) High validity (B) Low validity
(C) Low reliability (D) High reliability

630. The arithmetic mean of the Poisson distribution is:
(A) $3m^2+m$ (B) $\sqrt{m}$
(C) m (D) 1/m

631. For test of goodness of fit, which statistical test is applied:
(A) x^2-test (B) F-test
(C) z-test (D) t-test

632. The allocation of the treatments to the different experimental units in a random manner is known as:
(A) Randomization (B) Replication
(C) Local control (D) Sampling

633. The hypothesis of no difference is known as:
(A) Simple hypothesis
(B) Null hypothesis
(C) Composite hypothesis
(D) Alternative hypothesis

634. The external validity of the findings are the lowest in:
(A) Experimental design (B) Explorative design
(C) Diagnostic design (D) Survey design

635. The reliability of test with a high error score will be:
(A) Constant (B) High
(C) Low (D) Zero

636. The type of validity attached to an instrument when it appears to measure what it sets out to measure:
(A) Face validity (B) Content validity
(C) Criteria related validity (D) Construct validity

637. Hypothesis that stated the negation of the assertion is commonly called as:
(A) Null hypothesis (B) Parallel hypothesis
(C) Alternate hypothesis (D) Both (A) and (C)

638. Coefficient of variation is useful for comparing:
(A) Comparison between different populations
(B) Same population
(C) Both (A) and (B)
(D) None of these

639. Attitude scale giving emphasis for intensity of altitude expression:
(A) Cumulative scale
(B) Equal appearing interval scale
(C) Summated rating scale
(D) None of these

640. The research method which focus independently on different aspects of a single subject:
(A) Field studies (B) Case studies
(C) Experimental studies (D) None of these

641. The type of validity that assume most significance when some set standards are compared with the measuring score:
(A) Face validity (B) Content validity
(C) Criteria related validity (D) Construct validity

642. The coefficient of reliability at a perfectly homogenous experimental condition will be:
(A) Zero (B) One
(C) 0.5 (D) Two

643. The parent population in analysis of variance is assumed to follow:
(A) Normal distribution (B) Binomial distribution
(C) Poisson distribution (D) None of these

644. When degree of freedom of x^2 test is more than 100, t test is approximated to:
(A) F test (B) X^2 test
(C) Z test (D) None of these

645. The statement "All that is reliable is valid but all those is valid is not reliable" is:
(A) Completely false (B) Partially true
(C) Completely true (D) Partially false

646. The range of correlation coefficient lies between
(A) -1to+1 (B) 0 to 1
(C) 0 to ∞ (D) 0 to 100

647. Arithmetic mean and variance are always equal in:
(A) Normal distribution (B) Poisson distribution
(C) Binomial (D) All of these

648. The property of a test that gives us the same score when corrected by different people:
(A) Validity (B) Reliability
(C) Objectivity (D) Credibility

649. Compared to laboratory experiment, the internal and external validity of field experiment are respectively:
(A) High-Low (B) Low-High
(C) Low-Low (D) High-High

650. The value of x^2 ranges from .... and is always positive:
(A) -1 to +1 (B) 0 to 2
(C) 0 to ∞ (D) None of these

651. External validity is the highest for:
(A) Field experiment (B) Field study
(C) Sample survey (D) Case study

652. A variance shared by two or more variables:
(A) Covariance (B) Correlation
(C) Common factor variance (D) None of these

653. Jury opinion is used to establish:
(A) Reliability (B) Validity
(C) Creativity (D) Adjustability

654. A sample of 12 specimens taken from a normal population is expected to be 50 kg. To test H0:μ = 50 Vs Hi: ì not = 50, the test used is:
(A) Z-test (B) t-test
(C) x^2 test (D) F-test

655. An example for summated rating scale from among the following:
(A) Guttman scale (B) Thurstone scale
(C) Adorno scale (D) Likert scale

656. A test used to compare the variance of two independent samples:
(A) F test (B) X^2 test
(C) Z test (D) None of these

657. The area under standard normal curve beyond Z= + 1.96 is:
(A) 95 per cent (B) 99 per cent
(C) 5 per cent (D) 1 per cent

658. Reduction in the size of the test results in:
(A) Decrease in its power (B) Increase in its power
(C) No change in power (D) All of these

659. To test the hypothesis about proportions of items in a class, we use:
(A) F test (B) X^2 test
(C) Z test (D) None of these

650. Analysis of variance utilizes:
(A) F test (B) x^2 test
(C) Z test (D) t test

661. The ratio of difference between sample variance and with in sample variance follows:
(A) Z distribution (B) t distribution
(C) F distribution (D) X^2 distribution

662. Measures of dispersion includes the range and:
(A) Standard deviation (B) Mode
(C) Median (D) None of these

663. An observable phenomena substituted for an less observable phenomena is called:
(A) Index (B) Ratio
(C) Linkage (D) Concept

664. An example for summated rating scale from among the following:
(A) Guttman scale (B) Thurstone scale
(C) Adorno scale (D) Likert scale

665. A tentative explanation of an even or observation:
(A) Variable (B) Hypothesis
(C) Argument (D) Principle

666. The word 'scale' is correctly used in which of the following cases:
(A) Achievement scales (B) Attitude scales
(C) Intelligent scales (D) Both (A) and (B)

667. The basic emphasis while placing item in the cumulative scale is on:
(A) Scalability of set of items
(B) Items and their placement
(C) Scale position of individuals
(D) Both (A) and (C)

668. For calculation of x^2, N should be at least:
(A) 20 (B) 30
(C) 40 (D) 50

669. Attitude scale giving emphasis for intensity of attitude expression:
(A) Cumulative scale
(B) Equal appearing interval scale
(C) Summated rating scale
(D) None of these

670. Guttman scale is basically a:
(A) Summated rating scale
(B) Equal appearing interval scale
(C) Cumulative scale
(D) None of these

671. The difficulty indices of items producing maximum variance among examinees lies between:
(A) 0.2 to 0.4 (B) 0.4 to 0.6
(C) 0.6 to 0.8 (D) 0.8 to 1.0

672. The probability of an impossible event is:
(A) 0 (B) 1
(C) 8 (D) None of these

673. Analysis of variance utilizes:
(A) F test (B) X^2 test
(C) Z test (D) T test

674. A lower scale value in Thurstone and Clave scale indicates:
(A) Positive attitude (B) Negative attitude
(C) Neutral attitude (D) Zero attitude

675. Tests designed to measure individual potential and to predict future achievements:
(A) Attitude tests (B) Achievement tests
(C) Aptitude tests (D) Personality tests

676. Kruskall Wallis analysis of data is meant for:
(A) No classification (B) One way classification
(C) Two way classification (D) None of these

677. The first step in summarizing the data is:
(A) Tabulation (B) Classification
(C) Interpretation (D) Observation

678. The decision criterian in SPRT depends in the function of:
(A) Type I error (B) Type II error
(C) Type I and II error (D) None of these

679. An exact test for testing the independence of attributes in a contingency table of order (2 x 2) is based on the calculation of:
(A) The value of x^2
(B) Probability of configuration
(C) Value of Z
(D) None of these

680. Area of critical region depends upon:
(A) Size of type I error
(B) Size of type II error
(C) Value of statistics
(D) Number of observation

681. For any two evens A and B, P (A-B) is equal to:
(A) P (A) – P (B) (B) P (B) – P (A)
(C) P (B) – P (AB) (D) P (A) – P (AB)

682. The measure of central tendency to be used to study average rate of change in the population is:
(A) Arithmetic mean (B) Mode
(C) Geometric mean (D) Median

683. An interval in which scores are grouped is called:
(A) Point (B) Value
(C) Class interval (D) None of these

684. The first step in analysis of data is:
(A) Manipulation (B) Categorization
(C) Interpretation (D) Summarization

685. Equality of several normal population means can be tested by:
(A) F test (B) X^2 test
(C) Z test (D) t test

686. The t test is used to test the significance of:
(A) Two means of large sample
(B) Two means of small sample
(C) Correlation coefficient
(D) Both (B) and (C)

687. The reliability method that take into consideration both temporal stability and internal consistency dimensions:
(A) Test re-test method
(B) Odd even method
(C) Parallel foams method
(D) Split half method

688. If there is an increase in a series at constant rate, the graph will be:
(A) Convex curve (B) Concave curve
(C) Parabola (D) Straight line

689. Assessment of numbers of the football players of a team is an example of:
(A) Ratio measurement
(B) Internal measurement
(C) Ordinal measurement
(D) Nominal measurement

690. The hypothesis under test is:
(A) Simple hypothesis (B) Alternate hypothesis
(C) Null hypothesis (D) None of these

691. The key of accurate sampling is:
(A) Randomness (B) Opinion
(C) Prediction (D) All of these

692. Cross section of population is represented less in:
(A) Simple random sampling
(B) Multistage sampling
(C) Cluster sampling
(D) Stratified random sampling

693. The sample type wherein similarity between immediate neighbours and difference between distinct neighbours are both taken care of:
(A) Stratified sampling
(B) Systematic sampling
(C) Cluster sampling
(D) Multipurpose sampling

694. If A and B are two events which have no points in common, the events A and B are:
(A) Mutually exclusive
(B) Independent
(C) Complimentary to each other
(D) None of these

695. The best sample size from a perfectly homogenous population in respect of a character:
(A) A large sample (B) A small sample
(C) A single item (D) No item

696. Mean and variance of a standard normal distributing respectively:
(A) 0 and 1 (B) 1 and 0
(C) Both (A) and (B) (D) None of these

697. If A is an event the probability of A given A is equal to:
(A) Zero (B) One
(C) Infinite (D) Indeterminate

698. When r = +1, the two regression lines are:
(A) Coincident
(B) Parallel
(C) Perpendicular to each other
(D) Perpendicular to X axis

699. For testing the independence of two attributes, we use:
(A) F test (B) Z test
(C) x^2 test (D) t test

700. The sampling type that had developed from the principles of sociometry:
(A) Area sampling (B) Snowball sampling
(C) Quota sampling (D) Incidental sampling

701. Square of a standard normal variate is:
(A) F (B) X^2
(C) Z (D) None of these

702. Pick the odd man out:
(A) x^2 test
(B) Kruskal Wallis test H test
(C) Friedman test
(D) Z test

703. Statistical conclusions are always:
(A) Absolutely true (B) Absolutely wrong
(C) Either (A) or (B) (D) True on average

704. Which of the following is not a discrete distribution:
(A) Binomial (B) Poisson
(C) Normal (D) Both (A) and (B)

705. Area of critical region depends upon:
(A) Size of Type I error (B) Size of Type II error
(C) Both (A) and (B) (D) None of these

706. Mean deviation is minimum when deviants are taken from:
(A) Mean (B) Median
(C) Mode (D) None of these

707. Who is the author of the book 'Sampling Techniques':
(A) W.G. Cochran (B) A.L.Bowley
(C) R.L. Ackoff (D) Wilson M

708. The AM of the age of 20 students in a class is 15 years. When the teachers age also added, the mean increase by one year. The age of the teacher is:
(A) 16 years (B) 26 years
(C) 36 years (D) 46 years

709. The sampling type that had developed from the principles of sociometry:
(A) Area sampling (B) Snowball sampling
(C) Quota sampling (D) Incidental sampling

710. Sampling size is influenced by:
(A) Heterogeneity of population
(B) Sampling method
(C) Degree of accuracy
(D) All of these

711. The measure of dispersion which ignores signs of deviation from the central value:
(A) Range (B) Quartile deviation
(C) Standard deviation (D) Mean deviation

712. Range is not a reliable measure of:
(A) Mean (B) Dispersion
(C) Median (D) Mode

713. The key of accurate sampling is:
(A) Randomness (B) Opinion
(C) Prediction (D) All of these

714. The most important measure of dispersion used in industry for quality control:
(A) Standard deviation (B) Mean deviation
(C) Coefficient of variation (D) Range

715. The population for sampling is defined in terms of:
(A) Time (B) Element
(C) Extent (D) All of these

716. The best measure of dispersion among the following is:
(A) Standard deviation (B) Mean deviation
(C) Coefficient of variation (D) Range

717. The range of variable ratio F is:
(A) -1 to +1 (B) $-\infty$ to $+\infty$
(C) 0 to ∞ (D) 0 to 1

718. Empirical relation between quartile deviation and standard deviation is:
(A) 3QD = 2SD (B) 4QD = 4SD
(C) 6QD = 5SD (D) 5QD = 4SD

719. The average of sum of squares of deviation about the mean is called:
(A) Absolute deviation (B) Variance
(C) Standard deviation (D) Mean deviation

720. The sampling procedure most commonly used in order to ensure a representative sample:
(A) Random sampling (B) Systematic sampling
(C) Cluster sampling (D) Quota sampling

721. Which of the following is the unit less measure of dispersion:
(A) Standard deviation (B) Mean deviation
(C) Coefficient of variation (D) Range

722. An abstract formed generalization for particular is called:
(A) Principle (B) Hypothesis
(C) Theory (D) Concept

723. To find the average speed of the vehicle, when the same distance is covered with different speeds, the suitable measure is:
(A) AM (B) HM
(C) GM (D) None of these

724. The most frequent score in a given distribution:
(A) Mean (B) Mode
(C) Median (D) None of these

725. The first step is summarizing the data is:
(A) Tabulation (B) Classification
(C) Graphical representation(D) Collection

726. The value of coefficient of concordance ranges between:
(A) 0 + 1 (B) $-\infty$ to $+\infty$
(C) - 1 to +1 (D) -1 to 0

727. The unit of correlation coefficient is:
(A) Kg/cc (B) Per cent
(C) No unit (D) None of these

728. Level of significance is the probability of:
(A) Type I error (B) Type of II error
(C) Non committing error (D) All of these

729. Classification of research data should follow which among the following rules:
(A) Categories should be exhaustive
(B) Mutually exclusive
(C) Independent
(D) All of these

730. Which among the correlation coefficient represent the strongest relationship:
(A) - 0.86 (B) +0.86
(C) +0.10 (D) +0.09

731. The variance is the square of
(A) Mean deviation (B) Standard deviation
(C) Range (D) Correlation

732. Regression coefficient is independent of:
(A) Origin
(B) Scale
(C) Both origin and scale
(D) Neither origin nor scale

733. When is the normality of populations is used in analysis of experiments?
(A) At the time of planning of experiments
(B) At the time of computing various sums of squares
(C) At the time of constructing ANOVA table
(D) At the time of testing of hypothesis

734. The basic unit containing the elements of the population to be sampled:
(A) Sampling unit (B) Sampling frame
(C) Sampling element (D) None of these

735. The process of drawing of conclusions or inferences after a careful analysis of the collected data:
(A) Interpretation (B) Tabulation
(C) Classification (D) Evaluation

736. What is the main purpose of local control?
(A) To reduce the number of plots
(B) To reduce the number of replications
(C) To reduce the experimental error
(D) To reduce the degrees of freedom

737. Regression coefficient takes values:
(A) $-\infty$ to 1 (B) 1 to ∞
(C) $-\infty$ to $+\infty$ (D) -1 to +1

738. Mean deviation is the least when calculated about (minimal property)
(A) Arithmetic mean (B) Median
(C) Mode (D) Geometric mean

739. If X and Y are independent, the value of regression coefficient β_{xy} is equal to:
(A) 0 (B) 1
(C) a (D) Any positive value

740. Which mean of dispersion is most affected by extreme values:
(A) Range (B) Quartile deviation
(C) Standard deviation (D) Mean deviation

741. An example for quantitative variable from among the following is:
(A) Sex (B) Age
(C) Race (D) Religion

742. The word statistics is used as:
(A) Singular (B) Plural
(C) Singular and plural both (D) None of these

743. What is the probability that the sum of points in a toss of two dice is 14?
(A) 0 (B) 1/6
(C) 1/36 (D) 1

744. In CRD, number of treatments are 5 and number of replications are 6. What is the error degrees of freedom:
(A) 20 (B) 25
(C) 29 (D) 4

745. The critical value for a t-distribution with 120 degrees of freedom at 5% level of significance in a two tailed test is:
(A) 1.24 (B) 1.98
(C) 2.54 (D) 3.13

746. If p is the probability of any event, which one of the following is true?
(A) $-x < p < 0$ (B) $0 = p = 1$
(C) $1 < p < x$ (D) $-x = p$ $= x$

747. The class interval of this grouped data 10-19, 20-29, 30-39, 40-49 is:
(A) 9 (B) 10
(C) 14.5 (D) 39

748. An example of discrete variable from among the following is:
(A) Age (B) Sex
(C) Intelligence (D) Height

749. Scientific data must be:
(A) Objective (B) Measurable
(C) Predictable (D) All of these

750. If the reliability coefficient of a test is one, the standard error will be:
(A) +1 (B) -1
(C) 0.5 (D) 0

751. All the three principles of randomization, replication and local control are used in:
(A) CRD only (B) RBD only
(C) CRD and RBD both (D) None of these

752. If X is multiplied by 10 and Y by 20, b_{xy}, the regression coefficient is:
(A) Same as b_{xy} (B) Half by b_{xy}
(C) 4 times of b_{xy} (D) Double of b_{xy}

753. Mean and variance of a standard normal distribution are respectively:
(A) 0 and 1 (B) 1 and 0
(C) Both (A) and (B) (D) None of these

754. The second central moment (about the arithmetic mean) is always equal to:
(A) Variance (B) Mean deviation
(C) Standard deviation (D) None of these

755. A normal variate takes values in the range of:
(A) 0 to 1 (B) 0 to ∞
(C) $-\infty$ to $+\infty$ (D) -11.96 to +1.96

756. Correlation does not demonstrate:
(A) Causation (B) Effect
(C) Reason (D) None of these

757. Unit less measure of dispersion is
(A) Standard deviation (B) Mean deviation
(C) Range (D) CV

758. Standard Error can be:
(A) 0 (B) 1
(C) >0 (D) All of these

759. The range of multiple correlation coefficient R is:
(A) 0 to 1 (B) 0 to ∞
(C) -1 to +1 (D) $-\infty$ to $+\infty$

760. The coefficient of variation can be calculated by using the formula:
(A) Standard deviation/arithmetic mean x 100
(B) Arithmetic mean/standard deviation x 100
(C) Square root of standard deviation/arithmetic mean x 100
(D) Standard deviation/geometric mean x 100

761. When r = +1, the two regression lines are:
(A) Coincident
(B) Parallel
(C) Perpendicular to each other
(D) Perpendicular to X axis

762. .If the correlation coefficient between x and y is r, the correlation coefficient of y and x is:
(A) -r (B) r^2
(C) r (D) 1/r

763. The value of X^2 always lies between:
(A) 0 to 1 (B) 0 to ∞
(C) -1 to +1 (D) $-\infty$ to $+\infty$

764. For testing the significance of correlation coefficient, we use:
(A) t- test (B) X^2- test
(C) Z test (D) F-test

765. To test the agreement between observed frequencies and expected frequencies, we use:
(A) t- test (B) X^2- test
(C) Z test (D) F-test

766. The t test tends to be Z test when sample size becomes more than:
(A) 100 (B) 50
(C) 30 (D) 5

767. The parent population in analysis of variance is assumed to follow:
(A) Normal distribution (B) Binomial distribution
(C) Poisson distribution (D) None of these

768. Kruskall Wallis analysis of data is meant for:
(A) No classification (B) One way classification
(C) Two way classification (D) None of these

769. Ogives for more than type and less than type distributions intersect at:
(A) Mean (B) Median
(C) Mode (D) None of these

770. Hypothesis that the population variance has a specified value can be tested by:
(A) F test (B) Z test
(C) x^2 test (D) t test

771. The simplest way of visualizing a correlation is to construct a:
(A) Graph (B) Table
(C) Scatter diagram (D) None of these

772. Equality of several normal population means can be tested by:
(A) t- test (B) X^2- test
(C) Z test (D) F-test

773. Formula for calculating coefficient of variance is:
(A) SD/Mean x 100 (B) Mean/SD x 100
(C) Mean x SD/100 (D) 100/Mean x SD

774. Regression coefficient is independent of:
(A) Origin
(B) Scale
(C) Both origin and scale
(D) Neither origin nor scale

775. In the regression line Y = α + β x, β is called:
(A) Intercept (B) Slope of the line
(C) Mean of the X values (D) Mean of the Y values

776. The 't' test is used to test the significance of:
(A) Two means for large sample
(B) Two means for small sample
(C) Correlation coefficient
(D) Both (B) and (C)

777. Degree of freedom from x^2 in case of 4 x 3 contingency table is:
(A) 12 (B) 9
(C) 6 (D) 8

778. Greater the distance of Lorenz curve from the line of equal distribution:
(A) More is inequality (B) Less is inequality
(C) Nothing can be inferred (D) None of these

779. If for a distribution, coefficient of kurtosis ā2 <0, the frequency curve is:
(A) Leptokurtosis (B) Platykurtosis
(C) Mesokurtosis (D) None of these

780. F – test is used to test:
(A) Ratio of two variance (B) Ratio of two means
(C) Square of normal variate (D) None of these

781. Square of a standard normal variate is:
(A) F (B) X^2
(C) Z (D) None of these

782. If P (AEB) = 0.8 and P(A)) = P(B) = 0.3, what is the value of P(A c B)?
(A) 0.1 (B) 0.2
(C) 0.3 (D) 0.5

783. 'Annawari' procedure is a:
(A) Traditional method for estimation of crop yield
(B) Method for weather forecasting
(C) Method for rainfall forecasting
(D) All of these

784. For the expression of standard error of mean which formula is applicable:
(A $S/\sqrt{n}$ (B) $2S/\sqrt{n}$
(C) $\sqrt{S/n}$ (D) $n/\sqrt{S}$

785. The statistical procedure for estimating whether the difference under study is significant or non significant is known as:
(A) Test of significance (B) Standard error
(C) Level of significance (D) Sampling

786. For application of t – test, it is assumed that population is:
(A) Negatively skewed (B) Normal
(C) Positively skewed (D) None of these

787. Equality of several variance is done by:
(A) F (B) X^2
(C) Z (D) None of these

788. Standard deviation of sample mean is given by:
(A) σ (B) n
(C) σ/root n (D) None of these

789. The ratio of difference between sample variance and with in sample variance follows:
(A) Z distribution (B) t distribution
(C) F distribution (D) X^2 distribution

790. The area under standard normal curve beyond Z =+ 1.96 is:
(A) 95 per sent (B) 99 per cent
(C) 5 per cent (D) 1 per cent

791. A sampling technique where we select a group of subject (a sample) for study from a larger group (a population) is known as:
(A) Random sampling (B) Quata sampling
(C) Cluster sampling (D) Stratified sampling

792. Statistics serves to provide a short hand description of large amounts of:
(A) Scores (B) Data
(C) Points (D) None of these

793. The non parametric substitute for the parametric 't' test:
(A) X^2 test (B) Kvuskal Wallis test
(C) Friedman test (D) Mann-Whitney U test

794. An example for one way non-parametric analysis of variance among the following is:
(A) F test (B) Kvuskal Wallis test
(C) Friedman test (D) Z test

795. The value of coefficient of kurtosis can be:
(A) Less than 3 (B) Greater than 3
(C) Equal to 3 (D) All of these

796. If the measure of kurtosis is greater than 3, then the curve is:
(A) Platykurtic (B) Mesokurtic
(C) Leptokurtic (D) Normal

797. A population is divided into clusters and it has been found that all items within a cluster are a like, which of the following sampling procedure would you adopt?
(A) Simple random sampling
(B) Cluster sampling
(C) Systematic sampling
(D) Stratified sampling

798. A functional variate for establishing a parameter is called:
(A) An estimate (B) An estimator
(C) A frame (D) A statistic

799. The idea of randomization was put forward by:
(A) Chapman and Greenwood
(B) Kerlinger
(C) Ronald A Fisher
(D) Sukathamae

800. The principle of experimental design include:
(A) Randomization (B) Local control
(C) Both (A) and B) (D) None of these

801. When the experimental units are homogeneous, the design used is:
(A) CRD (B) RBD
(C) LSD (D) None of these

802. Degree of freedom is related to the number of:
(A) People involved
(B) Hypothesis under test
(C) Independent observations
(D) None of these

803. The experimental design wherein the heterogeneity in the experimental units are in two directions:
(A) CRD (B) RBD
(C) LSD (D) BIBD

804. The error d.f. for a RBD experiment with 5 treatments and 3 replications is:
(A) 15 (B) 12
(C) 8 (D) 14

805. If there are ten treatments and three replications in a CRD experiment, the error degrees of freedom is:
(A) 18 (B) 20
(C) 27 (D) 29

806. If ten treatment are there, how many blocks are needed to have a minimum error degree of freedom of 12 in RBD:
(A) 1 (B) 2
(C) 3 (D) 4

807. In an RBD with 10 treatment and 4 replications, what will be the error degree of freedom if two observations are lost:
(A) 25 (B) 26
(C) 27 (D) 28

808. A family of parametric distribution with mean is equal to variate is:
(A) Binomial (B) Normal
(C) Poisson (D) Negative binomial

809. The variance of a binomial distribution with n and p as parameters is:
(A) n (B) n (1-p)
(C) np (B) np(1-p)

810. Which of the following is not a discrete distribution:
(A) Binomial (B) Poisson
(C) Normal (D) Both (A) and (B)

811. Distribution wherein mean is always greater than variance is:
(A) Binomial (B) Poisson
(C) Normal (D) Both (A) and (B)

812. A normal curve is all but not:
(A) Symmetrical (B) Continuous
(C) Unimodal (D) Asymbiotic

813. If X is binomial variant with parameters n and p. If n=1, the distribution is:
(A) Binomial (B) Poisson
(C) Normal (D) Bernoulli

814. Theoretical value of test statistics for five per cent when compared with one per cent is:
(A) Same (B) Greater
(C) Lesser (D) None of these

815. The value of probability ranges from:
(A) -1 to +1 (B) 0 to ∞
(C) 0 to 1 (D) None of these

816. Probability of two persons being born on the same day is:
(A) 1/49 (B) 1/365
(C) 1/7 (D) None of these

817. Probability of a leap year will have 53 Sundays is:
(A) 1/7 (B) 2/7
(C) 2/53 (D) 52/53

818. If A and B are two events which have no points in common, the events A and B are:
(A) Mutually exclusive
(B) Independent
(C) Complementary to each other
(D) None of these

819. The probability of an impossible event is:
(A) 0 (B) 1
(C) 2 (D) None of these

820. The limiting relative frequency approach of probability is known as:
(A) Statistical probability
(B) Classical probability
(C) Mathematical probability
(D) Asymmetric probability

821. If it is know that an event A has already occurred, the probability of an event E given A is:
(A) Statistical probability
(B) Conditional probability
(C) Mathematical probability
(D) Asymmetric probability

822. If A is an event the probability of A given A is equal to:
(A) Zero (B) One
(C) Infinite (D) Indeterminate

823. The data reveals that 10 patient die in a particular type of operation. A doctor performed 9 operations and all of them survived. What is the chance of the 10th patient die on being operated:
(A) 100% (B) >10% but <100%
(C) <10% (D) 10%

824. The basis of sampling theory is:
(A) Probability (B) Randomization
(C) Local control (D) Relative frequency

825. List of sampling units is:
(A) Sample population (B) Sampling frame
(C) Sample element (D) Sampling unit

826. The nature of relationship between sample size and sapling error is:
(A) Positive (B) Negative
(C) Zero (D) Constant

827. Difference between estimated value and actual value is measured as:
(A) Non sampling error (B) Bias
(C) Sampling error (D) None of these

828. Level of significance is the probability of:
(A) Type I error (B) Type II error
(C) Not committing error (D) Any one of these

829. Rejecting a hypothesis when it is true:
(A) Type I error (B) Type II error
(C) Not committing error (D) Any one of these

830. Area beyond Z = ±1.96 in a standard normal distribution:
(A) 95 (B) 5
(C) 10 (D) 1

831. Rejecting a hypothesis when it is true:
(A) Type I error (B) Type II error
(C) Not committing error (D) Any one of the above

832. Area of critical region depends upon:
(A) Size of Type I error (B) Size of Type II error
(C) Both (A) and (B) (D) None of these

833. Accepting a hypothesis when it is false:
(A) Type I error (B) Type II error
(C) Both (A) and (B) (D) None of these

834. The method used to identify the units or dimensions behind the measures:
(A) Multiple regression analysis
(B) Discriminant analysis
(C) Profile analysis
(D) Factor analysis

835. Which of the following method consists in the collection of data concerning the living, the working conditions of people in a given area with a view to formulate practical social measures for their betterment and welfare?
(A) Ideal types (B) Social survey
(C) Sociometry (D) Statistics

836. Circular systematic sampling was first used by:
(A) W.G. Cochran (B) M.H. Hanson
(C) D.B. Lahiri (D) P.C. Mahalanobis

837. Consider the following statements:
1.Median is not useful while dealing with skewed distribution such as income distribution or price distribution
2.Median can be determined graphically by drawing only one give
3.Mode can be used to describe qualitative phenomenon
Which of the above statements are correct?
(A) 1 and 2 (B) 1 and 3
(C) 2 and 3 (D) 1, 2 and 3

838. The following equation indicates the relationship among average, median and mode:
(A) Mode = 3 Median – 2 Mean
(B) Mean = 2 Mode – 3 Median
(C) Median = 3 Mean – 2 Mode
(D) None of these

839. What is the median of:
36, 72, 49, 45, 60, 33, 61, 72, 52?
(A) 33 (B) 52
(C) 61 (D) 60

840. The data reveals that 10 patient die in a particular type of operation. A doctor performed 9 operations an all of them survived. What is the chance of the 10th patient die on being operated:
(A) 100% (B) >10% but <100%
(C) <10% (D) 10%

841. In a moderately asymmetrical distribution:
(A) AM>GM>HM (B) AM<GM<HM
(C) AM<GN>HM (D) None of these

842. The geometric mean of 8, 64 and 512 is:
(A) 194 (B) 64
(C) 16 (D) None of these

843. If the reliability coefficient of a test is one, the standard error will be:
(A) +1 (B) -1
(C) 0.5 (D) 0

844. A statistical method involving the selection of elements from an ordred sampling frame is called:
(A) Systematic sampling (B) Cluster sampling
(C) Quata sampling (D) Stratified sampling

845. The range of simple correlation coefficient is:
(A) 0 to ∞ (B) - ∞ to + ∞
(C) -1 to +1 (D) 0 to 1

846. The best measure of dispersion among the following is:
(A) Standard deviation (B) Mean deviation
(C) Coefficient of variation (D) Range

847. Cluster sample are more effective if the population taken is:
(A) Cluster of small size
(B) Clusters with largest within variation
(C) Clusters with least within variation
(D) None of these

848. The points of intersection of the 'less than' and the 'more than' gives correspond to:
(A) Mean (B) Median
(C) Geometric mean (D) None of these

849. In CRD, which one of the following is the alternative hypothesis?
(A) All the means are unequal
(B) At least one mean is different
(C) At least half of the means are unequal
(D) All the means are equal

850. A distribution is said to be symmetrical if the values are uniformly distributed around the:
(A) Mean (B) Mode
(C) Median (D) Mean deviation

851. The validity of regression coefficient is tested by:
(A) z –test (B) t-test
(C) x2-test (D) Both a and b

852. An example for non parametric test from among the following is:
(A) Z test (B) t test
(C) F test (D) x^2 test

853. The method of moments is fairly simple and yields consistent estimators (under very weak assumptions), though these estimator are often:
(A) Biased (B) Unbiased
(C) Unknonwn (D) None of these

854. Histogram is useful to determine graphically the value of:
(A) Arithmetic mean (B) Median
(C) Mode (D) None of these

855. A statistical measure that indicates the level of peakedness of a probability distribution is called:
(A) Moment (B) Platykurtosis
(C) Leptokurtosis (D) None of these

856. Coefficient of variation is calculated by the formula:
(A) $\frac{X}{\Sigma} \times 100$ (B) $\frac{\sigma}{x} \times 100$
(C) $\frac{X}{\Sigma}$ (D) $\frac{\sigma}{X}$

857. If a=b, then a+c=b+c denotes:
(A) Transitive Property of Inequality
(B) Addition Property of Equality
(C) Commutative Property of Addition
(D) Multiplication Property of Equality

858. Mean of numbers 0, 1, 2, 3, ...10:
(A) 10 (B) 11
(C) 5.5 (D) 5

859. The sum of squared deviations is minimum when taken from:
(A) Mean (B) Median
(C) Mode (D) None of these

860. The validity of correlation coefficient is tested by:
(A) z-test (B) t-test
(C) x^2-test (D) F-test

861. Which measure of central tendency is especially suited in the field of business:
(A) Median (B) Arithmetic mean
(C) Harmonic mean (D) Mode

862. Karl Pearson's coefficient of skewednes is ...... Bowley's coefficient of skewed for any skewed distribution:
(A) Equal to (B) Less than
(C) Greater than (D) Not related to

863. Geometric mean of a given series is always less than its:
(A) Harmonic mean
(B) Weighted mean
(C) Arithmetic mean
(D) Weighted harmonic mean

864. x^2 is calculated with the help of formula:
(A) $x^2 = \Sigma(O-E)^2/E$ (B) $x^2 = \Sigma(O-E)/(E)^2$
(C) $x^2 = \Sigma(O-E)^2/\Sigma O$ (D) $x^2 = \Sigma(O-E)^2/\Sigma(O)^2$

865. Coefficient of determination is a square of
(A) Correlation coefficient (B) Regression coefficient
(C) Both (A) and (B) (D) None of these

866. If there are 6 levels of moisture regime and 6 replications each, the error degree of freedom for LSD will be:
(A) 25 (B) 35
(C) 36 (D) 20

867. Harmonic mean of two observations 4 and 6 is:
(A) 12/5 (B) 12/10
(C) 24/5 (D) 28/10

868. To get a short hand summary of large number of observation we go for:
(A) Pure statistics (B) Descriptive statistics
(C) Numerical statistics (D) None of these

869. The measure of variation that is least affected by extreme observation is:
(A) Range (B) Mean deviation
(C) Standard deviation (D) Quartile deviation

870. Difference between estimated value and actual value is measured as:
(A) Non sampling error (B) Bias
(C) Sampling error (D) None of these

871. When coefficient of skewed is zero the distribution is:
(A) J-shaped (B) U-shaped
(C) Symmetrical (D) L-shaped

872. In a negatively skewed distribution is:
(A) Mode>Median>Mean (B) Median>Mode>Mean
(C) Mode<Median<Mean (D) None of these

873. The nature of relationship between sample size and sampling error is:
(A) Positive (B) Negative
(C) Zero (D) Constant

874. The coefficient of correlation:
(A) Has no limits (B) Can be less than one
(C) Can be more than one (D) Varies between ± 1

875. List of sampling units is:
(A) Sample population (B) Sampling frame
(C) Sample element (D) Sampling unit

876. In the context of normal distribution N (m, 1), which one of the following is a simple hypothesis?
(A) H : m <0 (B) H : m >0
(C) H : m = 0 (D) H : -1 < m < 1

877. The relationship between independent and dependent variables is computed by:
(A) Regression coefficient (B) Correlation coefficient
(C) Both (A) and (B) (D) None of these

878. The significance of multiple correlation (R) is tested by:
(A) t-test (B) F-test
(C) z-test (D) Paired t-test

879. Crop cutting survey on sugarcane recommended the plot size for estimation of crop yield is:
(A) 20 m x 10 m (B) 10 m x 10 m
(C) 10 m x 5 m (D) 5 m x 5 m

880. The minimum error degree of freedom should be at least:
(A) 6 (B) 12
(C) 18 (D) 20

881. The repetition of the treatment under investigation is known as:
(A) Randomization (B) Replication
(C) Local control (D) None of these

882. The optimum plot size in the experiment is/are determined by:
(A) Maximum curvative method
(B) Fair field Smith's variance law
(C) Both (A) and (B)
(D) Uniformity trial

883. Which one of the following is true for n (degrees of freedom) in large sample tests?
(A) n <5 (B) 5 £ n <20
(C) n^2 30 (D) no restrictions

884. When one regression coefficient is negative the other would be:
(A) Negative (B) Positive
(C) Zero (D) None of these

885. Where r is zero the regression lines but each other making an angle of:
(A) 45^0 (B) 60^0
(C) 90^0 (D) None of these

886. If with a rise of 10% in prices the wages are increased by 20%, the real wage increased is by:
(A) 10% (B) More than 10%
(C) Less than 10% (D) 20%

887. Fisher's ideal index is:
(A) Arithmetic mean of Laspeyre's and Paasche's index
(B) Median of Laspeyre's and Paasche's index
(C) Geometric mean of Laspeyre's and Paasche;s index
(D) None of these

888. Laspeyre's index is based on:
(A) Base year quantities
(B) Current year quantities
(C) Average of current and base year
(D) None of these

889. A good index number is one that satisfies:
(A) Unit test
(B) Time test
(C) Time Reversal test and Factor Reversal test
(D) None of these

890. Cyclical fluctuations are caused by:
(A) Wars (B) Earth quakes
(C) Floods (D) None of these

891. Indian Livestock Census is conducted after every:
(A) 2 years (B) 5 years
(C) 10 years (D) 20 years

892. In India the available acreage statistics for the total geographical area is:
(A) 51% (B) 61%
(C) 71% (D) 100%

893. If E(X) is the mathematical expectation of the variable X and C is constant then E(CX) =?
(A) E(C) + E(X) (B) E(C) x E(X)
(C) CE(X) (D) None of these

894. E (x + y) = ?
(A) E(x) + E(y) (B) E(x) x E(y)
(C) E(xy) (D) None of these

895. Which of the following is not a characteristic feature of a good research design:
(A) High external validity (B) Low external validity
(C) High internal validity (D) None of these

896. If A and B are mutually exclusive events, P(AB) = ?
(A) Zero (B) P (A) + P (B)
(C) P (A) (D) None of these

897. A research design without hypothesis among the following is:
(A) Explorative design (B) Diagnostic design
(C) Descriptive design (D) Experimental design

898. If two events A and B are dependent, the conditional probability of B given A, i.e. P(A/B) is calculated as:
(A) P (A) + P (B) (B) P (A) x P (B)
(C) P (A) – P (B) (D) P (AB) / P (A)

899. To pilot study to gain more insight into the research problem form the characteristic feature of:
(A) Explorative design (B) Diagnostic design
(C) Descriptive design (D) Experimental design

900. ${}^5C_2$ is equal to:
(A) 20 (B) 10
(C) 30 (D) 100

901. Two cards are drawn from a well – shuffled pack of 52 cards. Find the probability that they are both aces if the first is replaced:
(A) $\frac{1}{169}$ (B) $\frac{2}{52}$
(C) $\frac{1}{221}$ (D) None of these

902. The standard deviation of Binomial distribution is:
(A) $\sqrt{npq}$ (B) npq
(C) $n^2p^2q^2$ (D) np

903. The basic concept of a good research design is to:
(A) Decrease experimental variance
(B) Maximize extraneous variance
(C) Minimize error variance
(D) Explain metaphysical observation

904. In case of normal distribution x + 2σ covers:
(A) 95.45% (B) 95.54%
(C) 94.45% (D) 99.73%

905. If in Binomial distribution mean is 10 and standard deviation 2, q will be:
(A) 0 (B) 1
(C) 0.4 (D) 0.8

906. After this, therefore caused by this" is usually denoted to indicate:
(A) Experimental design (B) Survey design
(C) *Ex post facto* design (D) Exploratory design

907. The normal distribution is an approximation to:
(A) Binomial distribution (B) Poisson's distribution
(C) Both (A) and (B) (D) None of these

908. The mean plus minus 1.96 standard deviation includes —— per cent of the items of normal distribution:
(A) 75 (B) 47.5
(C) 74.5 (D) None of these

909. The standard deviation of sampling distribution is called:
(A) Standard error (B) Probable error
(C) Both (A) and (B) (D) None of these

910. Research design is not essentially a:
(A) Plan (B) Product
(C) Structure (D) Strategy

911. A hypothesis is a must in:
(A) Experimental design (B) Explorative design
(C) Descriptive design (D) Both (A) and (B)

912. Dispersions are NOT calculated about:
(A) Mean (B) Median
(C) Mode (D) Geometric mean

913. The following description " I will bet this would happen if I did this" best represent to:
(A) Explorative design
(B) Confirmatory experiment
(C) Crucial experiment
(D) Pilot experiment

914. To study the effect of independent variable on a single experiment we go for:
(A) Factorial design (B) CRD
(C) LSD (D) RBD

915. Large sample theory is applicable when:
(A) N>30 (B) N<30
(C) N=30 (D) None of these

916. If the mode of frequency distribution is Mo = 16 and its mean = 16 then median is:
(A) Zero (B) 16
(C) 32 (D) <16

917. To study average rate of change in population, we use:
(A) Median (B) Mode
(C) Arithmetic mean (D) Geometric mean

918. Student's 't' Distribution was discovered by:
(A) Karl Pearson (B) Laplace
(C) Fisher (D) Gosset

919. Experimenter has the highest control in manipulation of variable in:
(A) Laboratory method (B) Library method
(C) Simulation method (D) Field experiment

920. Consider the following measures:
1. Correlation Coefficient
2. Covariance
3. Coefficient of variation
4. Index number
Which of these are unit-free?
(A) 1 and 2 (B) 1 and 3
(C) 2 and 4 (D) 1, 3 and 4

921. The units within and between each stratum in a stratified random sampling are respectively:
(A) Homogenous and homogenous
(B) Homogenous and Heterogenous
(C) Heterogenous and Homogenous
(D) Heterogenous and Heterogenous

922. Harmonic mean is better than other means if data are for:
(A) Speeds or rates (B) Heights and lengths
(C) Binary values 0 and 1 (D) Ratio and proportion

923. Regression coefficient takes values:
(A) $-\infty$ to 1 (B) 1 to ∞
(C) $-\infty$ to $+\infty$ (D) -1 to +1

924. If each observation of a set is multiplied by 10, the mean of the new set of observation:
(A) Remains the same
(B) 10 times the original mean
(C) 1/10th of the original mean
(D) Increased by 10

925. Standard deviation : Variability : : Mean : ————.
(A) Central tendency (B) Discrimination
(C) Level of difficult (D) Correlation

926. Which out of the following is correct about median?
(A) It is a measure of variability
(B) It is the 50th percentile
(C) It is most stable measure of central tendency
(D) It is significantly affected by extreme scores

927. Correlation coefficient `r' lies between:
(A) 0 and 1 (B) -1 and 1
(C) -1 and 0 (D) 0 and ∞

928. The relationship between Arithmatic mean (A), Harmonic mean (H) and Geometric mean (G) is
(A) $G^2 = AH$ (B) $G = \sqrt{A+H}$
(C) $H^2 = AH$ (D) $A^2 = GH$

929. .... is the square root of variance :
(A) Standard error
(B) Standard deviation
(C) Coefficient of variation
(D) Correlation coefficient

930. The coefficient of variation (CV) is calculated by the formula:
(A) (Mean/SD) x 100 (B) (SD/Mean) x 100
(C) SD/Mean (D) Mean/SD

931. All research process begins with:
(A) Observation
(B) Pilot study
(C) Hypothesis
(D) Reviewing the data

932. This refers to the process of selecting units from a whole population of interest :
(A) Survey (B) Census
(C) Sampling (D) Exclusion

933. The primary aim of .... is to give structure to data in the form of categories :
(A) Coding (B) Pilot study
(C) Statistical analysis (D) Testing

934. Which of the following is not a source of data which is appropriate for quantitative study?
(A) Experiments (B) Historical data
(C) Observations (D) Case study reports

935. Deciding what data is best for your research analysis depends upon which of the following?
(A) Individual preference of the researcher
(B) The research problem and the samples
(C) The tool used for the study
(D) All the above

936. A meta analysis would allow the researcher to ·
(A) Explore the variations or circumstances in the outcomes of lots of studies:
(B) Identify the antecedents of a behaviour
(C) Assess the reability of a study
(D) Helps to frame the hypothesis in a more convincing manner

937. A .... is done when the study variables involves observation and analysis of more than one statistical outcome variable at a time :
(A) Regression (B) Chi square
(C) Correlation (D) Multivariate analysis

938. What sort of data is `Height'?
(A) Independent (B) Qualitative
(C) Dependent (D) Quantitative

939. Which of the following assumption/s are normally TRUE for doing a two way ANOVA:
(A) The group must have not equal number of samples
(B) The sample must not be independent
(C) The population variances must not be equal
(D) The population must be close to a normal distribution

940. If the standard deviation of a population is 20 and the standard error of mean is 4, then the sample size is:
(A) 25 (B) 80
(C) 5 (D) 100

941. In a multiple regression model, the explained variance per d.f. is 50 and unexplained variance per d.f. is 10 :
(A) 2.5 (B) 5
(C) 25 (D) 0.2

942. Which one of the following is not a random sampling method?
(A) Stratified sampling (B) Cluster sampling
(C) Systematic sampling (D) Judgement sampling

943. Identify the random sampling method among the following:
(A) Judgment sampling (B) Quota sampling
(C) Convenience sampling (D) Stratified sampling

944. For 5 degree of freedom, the variance of x^2 distribution is:
(A) 10 (B) 5 (C) 16 (D) 4

945. If the standard deviation of a population is 20, the population and sample means are 35 and 33, respectively and the t-statistics at 95% confidence level is 2.5, the sample size is:
(A) 100 (B) 125 (C) 250 (D) 625

946. In a multiple regression model, the F-ratio is used to test the :
(A) Variance of the data
(B) Standard error of mean
(C) R^2 value
(D) Overall goodness of fit of the model

947. The third moment about the mean of a distribution of a set of observation is:
(A) Mode (B) Skewness
(C) Kurtosis (D) Variance

948. If the mean of a sample is 20, the standard error of mean is 1 and the t-statistic for 95%, level of confidence is 2.5, the population mean will be in the range :
(A) 17.5 to 22.5 (B) 15 to 25
(C) 20 to 25 (D) 15 to 20

949. The following statistical test is used to ascertain whether there is significant difference between the variance of two sets of observations:
(A) t-test (B) F-test
(C) Chi-square test (D) Regression

950. Wind rose represents statistical distribution is :
(A) Wind velocity in vector form
(B) Wind speeds in scaler form
(C) Wind speeds in π diagram
(D) Wind speeds in the form of histograms

951. In most of the studies, a large sample size in anticipated to :
(A) Get a low level of precision
(B) Maximizing the sample error
(C) Get a high level of precision
(D) Maximize the standard deviation

952. The geometric mean of 4, 8 and 16 is :
(A) 9.3 (B) 8.0
(C) 4.8 (D) 10.2

953. The mean and standard deviation of a Binomial distribution are 9 and 1, respectively. The first moment of the distribution is:
(A) 9 (B) 3 (C) 1 (D) 0

954. The covariance between two data of N observations each represents by variables X and Y is given by :

(A) $\frac{\Sigma(X-\overline{X})(Y-\overline{Y})}{N^2}$ (B) $\frac{\Sigma(X-\overline{X})(Y-\overline{Y})^2}{N}$

(C) $\frac{\Sigma(X-\overline{X})^2(Y-\overline{Y})^2}{N^2}$ (D) $\frac{\Sigma(X-\overline{X})(Y-\overline{Y})}{N}$

955. A population, from where samples are drawn is called:
(A) Total population
(B) Target population
(C) Accessible population
(D) Universal population

956. The rate of variations of population (N) with time (t) represented by equation dN/dt = y N, follows :
(A) J-shaped curve (B) S-shaped curve
(C) Z-shaped curve (D) Parabolic curve

957. The standard deviation of weights of contain 1 kg packets of milk is 10 grams. A random sample of 20 packets showed a standard deviation of 15 grams. The value of X^2 statistic for the sample is:
(A) 30 (B) 45 (C) 1.5 (D) 0.66

958. For degrees of freedom (df) >1, the mean (μ) of the t-distribution is :
(A) Zero (B) 1
(C) Depends on df (D) 2

959. Which one of the following conditions would indicate that the dataset is not bell shaped?
(A) The mean is much smaller than median
(B) The range is equal to five times the standard deviation
(C) The range is larger than interquartile range
(D) The range is twice the standard deviation

960. Assertion (A) : X^2 distribution is a non-parametric distribution.
Reason (R) : X^2 is a sample statistic having no corresponding population parameter
Codes:
(A) Both (A) and (R) are correct and (R) is the correct explanation of (A)
(B) Both (A) and (R) are correct and (R) is not the correct explanation of (A)
(C) (A) is true, but (R) is false
(D) Both (A) and (R) are true

961. In a sample regression analysis of y and x, the standard error or estimate of y on x, Syx = 5, number of observations N is 30, and ? y2 = 2000. The unexplained variance is
(A) 1500 (B) 750
(C) 500 (D) 250

962. Two normal population have variance 6_1^2 =10 and 6_2^2 =20 Two random samples of sizes 25 and 20, independently selectd from these populations have variance of S_1^2 = 8 and S_2^2 = 15, respectively. What is the F(24, 19) statistic ?
(A) 1 (B) 2 (C) 2.81 (D) 3.6

963. The Pearson Linear correlation coefficient (r) for the following paired data
(x,y) : (2, 1.4), (4, 1.8), (8, 2.1), (8, 2.3), (9, 2.6) is
(A) 0.623 (B) -0.572
(C) 0.957 (D) 0.824

964. A random sample of size 26 has a mean of 20. The sum of squares of the deviations taken from the mean is 200. If the population mean is 18, what is the value of t-statistics?
(A) ~ 0.9 (B) ~ 2.1
(C) ~ 3.6 (D) ~ 5.2

965. If there are 10 treatments replicated thrice in RBD design, then error degree of freedom will be:
(A) 30 (B) 18
(C) 16 (D) 7

966. Which design has equal number of replication and treatments?
(A) CRD (B) RBD
(C) LSD (D) SPD

ANSWERSHEET: 18.AGRICULTURAL STATISTICS

1. (A); 2. (B); 3. (A); 4. (B); 5. (B); 6. (A); 7. (B); 8. (B); 9. (C); 10. (A); 11. (A); 12. (A); 13. (B) ; 14. (B); 15. (A); 16. (C); 17. (A); 18. (C); 19. (A); 20. (A); 21. (B); 22. (A); 23. (D); 24. (A); 25. (A); 26. (A); 27. (C); 28. (B); 29. (A); 30. (A); 31. (A); 32. (D); 33. (B); 34. (A); 35. (A); 36. (B); 37. (B); 38. (A) ; 39. (A); 40. (C); 41. (D); 42. (D); 43. (D); 44. (D); 45. (B); 46. (A); 47. (B); 48. (B); 49. (A); 50. (C); 51. (D); 52. (A); 53. (B); 54. (A); 55. (A); 56. (B); 57. (B); 58. (C); 59. (A); 60. (B); 61. (D); 62. (C); 63. (B); 64. (A); 65. (C); 66. (B); 67. (C); 68. (A); 69. (C); 70. (C); 71. (B); 72. (A); 73. (B); 74. (D); 75. (C); 76. (B); 77. (A); 78. (D); 79. (D); 80. (C); 81. (B); 82. (C); 83. (B); 84. (B); 85. (D); 86. (D); 87. (A); 88. (C); 89. (A); 90. (A); 91. (B) ; 92. (A); 93. (C); 94. (B); 95. (C); 96. (D); 97. (B); 98. (B); 99. (A); 100. (B); 101. (B); 102. (B); 103. (A); 104. (B); 105. (B); 106. (A); 107. (A); 108. (D); 109. (C); 110. (B); 111. (C); 112. (A); 113. (B); 114. (C); 115. (A); 116. (A); 117. (D); 118. (B); 119. (B); 120. (B); 121. (A); 122. (B); 123. (B); 124. (C); 125. (A); 126. (B); 127. (B); 128. (A); 129. (B); 130. (D); 131. (C); 132. (C); 133. (C); 134. (B); 135. (A); 136. (C); 137. (A); 138. (C); 139. (B); 140. (A); 141. (C); 142. (B); 143. (B); 144. (C); 145. (A); 146. (C); 147. (B); 148. (C); 149. (A); 150. (B); 151. (C); 152. (A); 153. (C); 154. (D); 155. (A); 156. (B); 157. (A); 158. (A); 159. (B); 160. (C); 161. (D); 162. (B); 163. (C); 164. (D); 165. (B); 166. (D); 167. (B); 168. (C); 169. (A); 170. (D); 171. (C);

172. (C); 173. (B); 174. (D); 175. (B); 176. (B); 177. (B); 178. (C); 179. (C); 180. (A); 181. (B); 182. (A); 183. (B); 184. (D); 185. (D); 186. (B); 187. (A); 188. (A); 189. (B); 190. (D); 191. (A); 192. (B); 193. (A); 194. (A); 195. (A); 196. (A); 197. (C); 198. (B); 199. (A); 200. (C); 201. (B); 202. (B); 203. (C); 204. (B); 205. (C); 206. (B); 207. (A); 208. (A) ; 209. (C); 210. (A); 211. (A); 212. (C); 213. (B); 214. (B); 215. (A); 216. (B); 217. (C); 218. (C); 219. (C); 220. (C); 221. (B); 222. (A); 223. (B); 224. (B); 225. (B); 226. (B); 227. (C); 228. (B); 229. (A); 230. (D); 231. (D); 232. (B); 233. (C); 234. (B) ; 235. (A); 236. (C); 237. (D); 238. (B); 239. (C); 240. (B); 241. (A); 242. (A); 243. (C); 244. (D); 245. (A); 246. (B); 247. (C) ; 248. (A); 249. (C); 250. (C); 251. (A); 252. (B); 253. (C); 254. (C); 255. (C); 256. (D); 257. (B); 258. (A); 259. (A); 260. (A) ; 261. (C); 262. (A); 263. (D); 264. (A); 265. (D); 266. (A); 267. (B); 268. (A); 269. (B); 270. (B); 271. (A); 272. (B); 273. (B) ; 274. (A); 275. (C); 276. (D); 277. (D); 278. (B); 279. (B); 280. (B); 281. (C); 282. (A); 283. (A); 284. (B); 285. (A); 286. (A); 287. (B); 288. (C); 289. (B); 290. (B); 291. (D); 292. (C); 293. (C); 294. (A); 295. (C); 296. (B); 297. (C); 298. (D); 299. (A); 300. (C); 301. (C); 302. (B); 303. (A); 304. (C); 305. (A); 306. (C); 307. (D); 308. (C); 309. (B); 310. (D); 311. (A); 312. (D); 313. (B); 314. (B); 315. (A); 316. (B); 317. (A); 318. (C); 319. (B) ; 320. (C); 321. (A); 322. (B); 323. (C); 324. (D); 325. (C); 326. (C); 327. (D); 328. (C); 329. (A); 330. (D); 331. (B); 332. (B); 333. (B); 334. (C); 335. (A); 336. (B); 337. (C); 338. (C); 339. (D); 340. (C); 341. (A); 342. (B); 343. (B); 344. (D); 345. (B); 346. (C); 347. (A); 348. (A); 349. (A); 350. (A); 351. (D); 352. (A); 353. (B); 354. (A); 355. (B); 356. (B); 357. (A); 358. (D); 359. (D); 360. (A); 361. (C); 362. (C); 363. (C); 364. (C); 365. (A); 366. (A); 367. (D); 368. (A); 369. (A); 370. (B); 371. (A); 372. (B); 373. (B); 374. (B); 375. (A); 376. (D); 377. (C); 378. (A); 379. (A); 380. (C); 381. (C); 382. (D); 383. (A); 384. (D); 385. (B); 386. (A); 387. (D); 388. (B); 389. (B); 390. (B); 391. (C); 392. (D); 393. (A); 394. (D); 395. (D); 396. (A); 397. (A); 398. (A); 399. (A); 400. (A); 401. (B); 402. (A); 403. (A); 404. (A); 405. (A); 406. (D); 407. (B); 408. (A); 409. (C); 410. (A); 411. (D); 412. (C); 413. (B); 414. (B); 415. (A); 416. (A); 417. (A); 418. (A); 419. (D); 420. (C); 421. (D); 422. (C); 423. (A); 424. (B); 425. (B); 426. (A); 427. (D); 428. (A); 429. (C); 430. (A); 431. (C); 432. (C); 433. (B); 434. (B); 435. (B); 436. (A); 437. (C); 438. (B); 439. (B); 440. (A); 441. (A); 442. (A); 443. (B); 444. (C); 445. (D); 446. (B); 447. (B); 448. (D); 449. (B); 450. (C); 451. (A); 452. (B); 453. (C); 454. (B); 455. (B); 456. (B); 457. (C); 458. (C); 459. (A); 460. (B); 461. (A); 462. (A); 463. (B); 464. (A); 465. (A); 466. (B); 467. (A); 468. (B); 469. (B); 470. (A); 471. (A); 472. (A); 473. (A); 474. (B); 475. (B); 476. (A); 477. (C); 478. (B); 479. (B); 480. (A); 481. (C); 482. (A); 483. (D); 484. (C); 485. (D); 486. (A); 487. (C); 488. (B); 489. (D); 490. (A); 491. (B); 492. (D); 493. (B); 494. (A); 495. (B); 496. (B); 497. (A); 498. (A); 499. (C); 500. (A); 501. (C); 502. (D); 503. (C); 504. (A); 505. (B); 506. (B); 507. (C); 508. (B); 509. (A); 510. (B); 511. (A); 512. (B); 513. (C); 514. (A); 515. (A); 516. (A); 517. (D); 518. (C); 519. (B); 520. (B); 521. (D) ; 522. (A); 523. (C); 524. (B); 525. (A); 526. (A); 527. (B); 528. (B); 529. (A); 530. (D); 531. (C); 532. (A); 533. (D); 534. (C); 535. (A); 536. (B); 537. (A); 538. (A); 539. (D); 540. (C); 541. (B); 542. (B); 543. (A); 544. (B); 545. (B); 546. (C); 547. (D) ; 548. (D); 549. (B); 550. (B); 551. (B); 552. (A); 553. (D); 554. (B); 555. (C); 556. (D); 557. (A); 558. (B); 559. (C); 560. (A) ; 561. (B); 562. (A); 563. (B); 564. (A); 565. (B); 566. (A); 567. (A); 568. (B); 569. (C); 570. (A); 571. (B); 572. (C); 573. (C) ; 574. (D); 575. (A); 576. (C); 577. (D); 578. (C); 579. (A); 580. (C); 581. (A); 582. (C); 583. (C); 584. (D); 585. (A); 586. (B) ; 587. (D); 588. (B); 589. (B); 590. (D); 591. (C); 592. (B); 593. (C); 594. (A); 595. (A); 596. (C); 597. (D); 598. (C); 599. (C) ; 600. (B); 601. (B); 602. (A); 603. (B); 604. (C); 605. (A); 606. (A); 607. (A); 608. (D); 609. (A); 610. (C); 611. (D); 612. (D); 613. (A); 614. (D); 615. (D); 616. (D); 617. (C); 618. (B); 619. (A); 620. (D); 621. (B); 622. (D); 623. (B); 624. (A); 625. (B) ; 626. (B); 627. (B); 628. (B); 629. (D); 630. (C); 631. (A); 632. (A); 633. (B); 634. (A); 635. (C); 636. (A); 637. (A); 638. (A) ; 639. (C); 640. (B); 641. (C); 642. (A); 643. (A); 644. (C); 645. (C); 646. (A); 647. (B); 648. (C); 649. (B); 650. (C); 651. (C) ; 652. (C); 653. (B); 654. (B); 655. (D); 656. (A); 657. (C); 658. (A); 659. (C); 660. (A); 661. (C); 662. (A); 663. (A); 664. (D) ; 665. (B); 666. (B); 667. (D); 668. (D); 669. (C); 670. (C); 671. (B); 672. (A); 673. (A); 674. (A); 675. (B);

676. (B); 677. (A); 678. (C); 679. (B); 680. (A); 681. (D); 682. (C); 683. (C); 684. (B); 685. (A); 686. (D); 687. (C); 688. (D); 689. (D); 690. (C) ; 691. (A); 692. (C); 693. (B); 694. (A); 695. (C); 696. (A); 697. (B); 698. (A); 699. (C); 700. (B); 701. (B); 702. (D); 703. (D); 704. (C); 705. (A); 706. (B); 707. (A); 708. (C); 709. (B); 710. (D); 711. (D); 712. (B); 713. (A); 714. (D); 715. (D); 716. (A); 717. (C); 718. (A); 719. (B); 720. (A); 721. (C); 722. (D); 723. (B); 724. (B); 725. (B); 726. (A); 727. (C); 728. (A); 729. (D); 730. (A); 731. (B); 732. (A); 733. (C); 734. (A); 735. (A); 736. (C); 737. (C); 738. (B); 739. (A); 740. (A); 741. (B); 742. (C) ; 743. (A); 744. (A); 745. (B); 746. (B); 747. (B); 748. (B); 749. (D); 750. (D); 751. (B); 752. (B); 753. (A); 754. (A); 755. (C) ; 756. (A); 757. (D); 758. (D); 759. (A); 760. (A); 761. (A); 762. (C); 763. (B); 764. (A); 765. (B); 766. (C); 767. (A); 768. (B) ; 769. (B); 770. (C); 771. (C); 772. (D); 773. (A); 774. (A); 775. (B); 776. (D); 777. (C); 778. (A); 779. (B); 780. (A); 781. (B) ; 782. (B); 783. (A); 784. (A); 785. (A); 786. (B); 787. (B); 788. (C); 789. (C); 790. (C); 791. (A); 792. (B); 793. (D); 794. (B); 795. (D); 796. (C); 797. (B); 798. (B); 799. (C); 800. (C); 801. (A); 802. (C); 803. (C); 804. (C); 805. (B); 806. (C); 807. (A) ; 808. (C); 809. (D); 810. (C); 811. (A); 812. (D); 813. (D); 814. (C); 815. (C); 816. (C); 817. (B); 818. (A); 819. (A); 820. (A) ; 821. (B); 822. (B); 823. (D); 824. (B); 825. (B); 826. (B); 827. (C); 828. (A); 829. (A); 830. (B); 831. (A); 832. (A); 833. (B); 834. (D); 835. (B); 836. (C); 837. (B); 838. (A); 839. (D); 840. (D); 841. (A); 842. (B); 843. (D); 844. (A); 845. (C); 846. (A); 847. (B); 848. (B); 849. (B); 850. (A); 851. (D); 852. (D); 853. (A); 854. (C); 855. (B); 856. (B); 857. (B); 858. (D); 859. (A); 860. (B) ; 861. (D); 862. (D); 863. (C); 864. (A); 865. (A); 866. (D); 867. (C); 868. (B); 869. (B); 870. (C); 871. (C); 872. (A); 873. (B); 874. (D); 875. (B); 876. (C); 877. (A); 878. (B); 879. (D); 880. (B); 881. (B); 882. (C); 883. (C); 884. (A); 885. (D); 886. (C); 887. (C); 888. (A); 889. (C); 890. (D); 891. (B); 892. (C); 893. (C); 894. (A); 895. (B); 896. (A); 897. (A); 898. (D); 899. (A); 900. (B); 901. (A); 902. (A); 903. (C); 904. (A); 905. (C); 906. (C); 907. (A); 908. (B); 909. (A); 910. (B); 911. (A); 912. (D); 913. (B); 914. (A); 915. (A); 916. (B); 917. (D); 918. (D); 919. (A); 920. (D); 921. (B); 922. (A); 923. (C); 924. (B); 925. (A); 926. (B); 927. (B); 928. (C); 929. (B); 930. (B); 931. (A); 932. (C); 933. (A); 934. (A); 935. (D); 936. (A); 937. (D); 938. (D); 939. (D); 940. (A); 941. (B); 942. (B); 943. (A); 944. (A); 945. (D); 946. (D); 947. (B); 948. (A); 949. (B); 950. (A); 951. (C); 952. (B); 953. (D); 954. (A); 955. (C); 956. (A); 957. (C); 958. (A); 959. (A); 960. (A); 961. (B); 962. (A); 963. (C); 964. (C); 965. (B); 966. (C).

Chapter 19

RESEARCH METHODOLOGY

1. The process of designing and maintaining an environment in which individuals working together in groups effectively accomplish selected aims:
(A) Administration (B) Management
(C) Policy making (D) Planning

2. The term "management" was derived from:
(A) Latin (B) French
(C) Greek (D) Sanskrit

3. The literal meaning of the word "Management" is:
(A) House keeping
(B) Personal administration
(C) Store keeping
(D) Staff placement

4. Management is called a process because it is:
(A) Applicable to manufacturing process
(B) Relevant for social organization
(C) Involves a series of action
(D) None of these

5. Social units, or human grouping, deliberately considered to seek specific goals:
(A) House keeping
(B) Personal administration
(C) Institution
(D) Organization

6. Output-input ratio within a time period with due consideration for quality:
(A) Productivity (B) Profitability
(C) Quality control (D) Sustainability

7. 'POSCORB' as management function was developed by:
(A) Gullick & Urwick (B) Urwick
(C) L. Gulick (D) H. Fayol

8. The acronym 'POSCORB' symbolizes:
(A) The activities of government
(B) The principles of administration
(C) The activities of an administrator
(D) The objectives of organization

9. Management interest or practice followed for a period or time with exaggerated zeal:
(A) Management fad (B) Management zeal
(C) Management outlook (D) Management culture

10. Work tends to expand to fill the time available:
(A) Parkison's law
(B) Marconi law
(C) Management by objective
(D) Commitment principle

11. The management skill most needed for people working at top position:
(A) Technical Skills (B) Conceptual Skills
(C) Clerical Skills (D) Leadership Skills

12. Middle management is:
(A) Consists of departmental managers
(B) Transmit orders down
(C) Motivate lower management
(D) All of these

13. Selecting missions and objectives and the actions to achieve them:
(A) Planning (B) Organizing
(C) Leading (D) Coordinating

14. General programs of action and development or resources to attain comprehensive objectives:
(A) Plan (B) Strategy
(C) Rule (D) Policies

15. A statement of expected results expressed in numerical terms:
(A) Budget (B) Programme
(C) Plan (D) Strategy

16. The inputs for formulating policies are otherwise called as:
(A) Objectives (B) Procedures
(C) Programmes (D) Projects

17. Plans should cover a period of time that is long enough to fulfill commitments involved in decisions made today:
(A) Commitment principle
(B) Principles of planning
(C) Management strategy
(D) Management by objective

18. A comprehensive management system that integrates many key management activities in a systematic manner and that is consciously directed towards efficient and effective accomplishment of organizational and individual objectives:
(A) Key Performance Method
(B) Critical Programme Method
(C) Management Information System
(D) Management by Objective

19. The concept of Management by Objective was first put forward by:
(A) Edward Weidner (B) Montgomery
(C) Peter F Druner (D) F.W. Riggs

20. A modern tool for analysis of the situation:
(A) Management by objective
(B) Win-loose game
(C) 'TOWS' matrix
(D) Brainstorming

21. When a company plans to liquidate or retrench, the business strategy is more of:
(A) 'SO' strategy (B) 'WO' strategy
(C) 'ST' strategy (D) 'WT' strategy

22. Planning Research and Action Institute is located at:
(A) New Delhi (B) Hyderabad
(C) Lucknow (D) Bhopal

23. Selection of a course of action from among the alternatives:
(A) Planning (B) Decision making
(C) Controlling (D) Leading

24. The core of planning is:
(A) Span of control (B) Coordination
(C) Hierarchy (D) Decision making

25. Any decision or action to be realistic has to be based on:
(A) Facts (B) Opinion
(C) Value judgement (D) Moral grounds

26. Ability to make prompt and good decisions is known as:
(A) Decisiveness (B) Initiative
(C) Cohesiveness (D) Social insight

27. The decisions concerned with structured or routine problems:
(A) Programmed decisions
(B) Non-programmed decisions
(C) Natural decisions
(D) Illogical decisions

28. Strategic decisions in general are:
(A) Programmed decisions
(B) Non-programmed decisions
(C) Natural decisions
(D) Illogical decisions

29. Which among the following is a quantitative tool used for decision making:
(A) Break even analysis
(B) Finite element analysis
(C) Factor analysis
(D) None of these

30. 'With or without principle' in decision making approach related to:
(A) Coast benefit analysis (B) Factor analysis
(C) Differential analysis (D) Marginal analysis

31. The character that distinguish a non programmed decision from programmed decision:
(A) Routine nature (B) Subjectivity
(C) Structured (D) Precedence

32. The process of establishing an internal structure of roles for people to fill in an organization:
(A) Supervision (B) Controlling
(C) Managing (D) Organizing

33. The process of grouping of activities into units for the purpose of administration is characteristic of:
(A) Span of control (B) Hierarchy
(C) Unity of command (D) Dapartmentation

34. The cases of organizations working in various shifts is a clear example of departmentation based on:
(A) Territory (B) Function
(C) Product (D) Time

35. The span of control depends on:
(A) Personality of the supervisors
(B) Age of the organization
(C) Element of space
(D) All of these

36. In which of the following cases span of control is smaller:
(A) Well planned work schedule
(B) Utilizing the staff assistance
(C) Effective communication system
(D) Heterogeneous work environment

37. Which of the following can widen the span of control:
(A) Delegation (B) Simple task
(C) Ineffective meetings (D) Both (A) and (B)

38. The informal organization is characterized by:
(A) Emotional (B) Customary
(C) Personal (D) All of these

39. The ability of an individual to induce or influence the beliefs or action of other persons of group:
(A) Power (B) Leadership
(C) Authority (D) Control

40. The right in a position to exercise discretion in making decisions affecting others;
(A) Responsibility (B) Hierarchy
(C) Authority (D) Unity of command

41. The concept of Zone of Indifferent is associated with:
(A) Decision making (B) Leadership
(C) Authority (D) Motivation

42. Which of the follo~~g is legitimate power:
(A) Intimidation (B) Authority
(C) Manipulation (D) Suggestion

43. The most logical criterion to distinguish a line function from staff is:
(A) Authority relationship
(B) Functional relationship
(C) Grouping of functions
(D) Departmentalization

44. Taking decision and acting upon them are the main responsibility of:
(A) Line authority (B) Staff authority
(C) Super authority (D) Functional authority

45. Which one of the following is not a function of staff agencies:
(A) Planning (B) Advising
(C) Consultation (D) Achieving goals

46. Which of the following is not true of line activity:
(A) It has operational ability
(B) It stands for though and facilitation
(C) It decides the organizational goal
(D) It has authority

47. Which of the following factors could place limitations on the authority of a superior:
(A) Group behaviour (B) Social relationships
(C) Organizational norms (D) All of these

48. The chief reason for delegation of authority from headquarters to field officers is:
(A) Inculcate skill in the later
(B) Facilitate local decision making
(C) Decongest the headquarter
(D) Control the field officier

49. The concept of unity of command is comlimentary to the principle of:
(A) Supervision (B) Scalar chain
(C) Span of control (D) Administration

50. When the delegation goes by a climate of understanding between the superior and subordinate, it is known as:
(A) General delegation (B) Specific delegation
(C) Formal delegation (D) Informal delegation

51. For field observation method preferred is:
(A) Conference (B) Primary source
(C) Correspondence (D) Interview

52. Sociology are primarily interested in:
(A) Similarities between human behaviours and the behaviour of other animals
(B) Way in which human beings mutually influence each other
(C) Individual since he is convinced that there really is no such thing as group
(D) Anything which is related in a casual way to the behaviour of mankind

53. Correct chronology of study method suggested by A. Comte is:
(A) History, Observation, Experiment and Comparison
(B) Experiment, Observation, History and Comparison
(C) Observation, Experiment, History and Comparison
(D) Experiment, Observation, Comparison and History

54. Procedures which cannot prove useful for ensuring objective fact finding is:
(A) Senior people should be sent to the field
(B) Some counter sample-checking should be done
(C) Trained people should be set to the field
(D) Respondents should be given monetary incentives

55. Hinderance in objective finding is:
(A) Investigator is unbiased
(B) Respondent is aware of the magnitude of the problem
(C) Respondent is unaware of the magnitude of the problem
(D) Investigator is serious in investigating a problem

56. Dependable method for obtaining objectivity is:
(A) Questionnaire method (B) Interview method
(C) Field survey (D) Group discussion

57. Primary methodology of functionalists is:
(A) Library work (B) Positivism
(C) Conjectural history (D) Field work

58. Objectivity in investigating social problem can be possible only when investigator has:
(A) Close mind (B) Pre-conceived notions
(C) Scientific mind (D) Pre-determined ideas

59. Experiments are:
(A) Quite common in sociology
(B) Rare but possible in sociology
(C) Impossible in sociology
(D) Easiest in sociology

60. Pareto termed the rational explanations for non-logical action as:
(A) Conceits (B) Archi-types
(C) Derivations (D) Pseudo-judgements

61. Social research is:
(A) Collection of facts on a given subject
(B) The relevant facts put together
(C) Collection of the required data, maps, charts, observations etc.
(D) Logical arrangement of facts which mutually explain each other

62. For collecting information about Indian society method preferred is:
(A) Participant observation method
(B) Social survey method
(C) Questionnaire method
(D) Interview method

63. Basic difference between E. Durkheim and M. Weber in terms of methodology is:
(A) Durkheim believed in subjective approach and Weber in the objective approach
(B) Weber believed in objective approach and the Durkheim in the subjective approach
(C) Both adopted each other method
(D) None of these

64. Error regarding an abstraction as a real phenomenon is called:
(A) Fallacy of abstraction (B) Fallacy of confusion
(C) Fallacy of ratification (D) Fallacy of verification

65. Objectivity in social investigation can be possible when the data:
(A) Has been manipulated
(B) Has been collected in particular manner
(C) Is sufficient
(D) Is unrelated to the problem

66. While investigating a social problem objective finding is:
(A) Very much possible
(B) Very difficult
(C) Possible without great difficulty
(D) None of these

67. By objectivity we generally mean:
(A) Impersonal attitude
(B) Unbiassed collection of data
(C) Unconditional motion of tangible and intangible facts
(D) All of these

68. According to Weber, values influence our judgement about the ...... of a specific event:
(A) Viability (B) Cause
(C) Repercussions (D) Effect

69. Quality for an investigator of social problem is that he should:
(A) Have a closed mid
(B) Have capacity to twist the facts
(C) Be capable of observing facts impartially
(D) Be guided by emotions

70. Which of the following methods is usually not used for objective fact finding?
(A) Historical method (B) Library method
(C) Questionnaire method (D) All of these

71. Individuals with a combination of low-ascribed and high-achieved statuses often act "pushy" in their attempts to emphasize their achieved status. This conceptualized as:
(A) Mediocracy (B) Double jeopardy
(C) Status inconsistency (D) Relative mobility

72. Holistic approach in sociology is associated with:
(A) Evolutionary approach
(B) Neo diffusionist approach
(C) Diffusionist approach
(D) Structural-functional approach

73. In which of the following methods of data collection where maximum danger of losing one's sharpeness of observation is possible because of too much familiarity?
(A) Interview
(B) Questionnaire
(C) Observation
(D) Participant observation

74. An observation, uninfluenced by personal bias is called:
(A) Functional (B) Critical
(C) Objective (D) Subjective

75. By which of the following means objective facts cannot be secured or established by the researcher?
(A) By observation
(B) By research
(C) By objective analysis of data
(D) From material collected on the basis of oath and affirmation

76. Objectivity cannot be introduced at the time of:
(A) Data collection
(B) Data interpretation
(C) Field survey
(D) Meeting non-respondents

77. Which method is very frequently used by the sociologist?
(A) Ideal type method (B) Sample method
(C) Sociometry (D) Field research method

78. Which of the following is define as a method of measuring individuals social behaviour?
(A) Sociometry (B) Sample method
(C) Verstehen method (D) Interview method

79. First hand study of a small community is called:
(A) Ethnology (B) Ethnography
(C) Ethics (D) Aesthetics

80. Participant observation has a clear advantage over other methods of data collection in the sense that it:
(A) More uniform
(B) More gratifying
(C) Yields in-depth information
(D) More precise

81. Sometimes mailed questionnaire is preferable to interview method because:
(A) Face-to-face contact is possible
(B) People immediately return it
(C) Illiterates can also be covered
(D) It works out cheaper and easy

82. Which of the following is a social fact?
(A) Tangible fact (B) Intangible fact
(C) Social custom (D) All of these

83. The scientific study of animal behaviour is:
(A) Ethymology (B) Ethology
(C) Pathology (D) Socio-biology

84. Second stage of scientific or experimental method is:
(A) Classification (B) Recording
(C) Observation (D) Prediction

85. According to Durkheim, biology has three principal division, viz.:
(A) Social morphology, social biology and general sociology
(B) Social morphology, social psychology and general sociology
(C) Social statistics, social dynamics and general sociology
(D) None of these

86. Correct sequence of the processes in the scientific method is:
(A) Hypothesis, observation, recording, classification, prediction and verification
(B) Observation, recording, classification, hypothesis, verification and prediction
(C) Recording, observation, classification, verification, hypothesis and prediction
(D) None of these

87. Recording data is called:
(A) Concept (B) Fact
(C) Hypothesis (D) Label

88. A man who studies effects of the social structure of a society on its form of government is most likely a/an:
(A) A sociology
(B) A physical anthropology
(C) An economist
(D) A historian

89. Which of the following method is a form of qualitative analysis involving the very useful and complete observation of a person, a situation or an institution?
(A) Inverse deductive (B) Case study
(C) Social survey (D) Historical method

90. A study of relations between organism and their environment is called:
(A) Topography (B) Geography
(C) Ecology (D) Biology

91. Social research is:
(A) Logical arrangement of facts which mutually explain each other
(C) Collection of the required data-maps, charts, observations, etc
(C) Collection of facts on a given subject
(D) All the relevant facts put together

92. In the social sciences:
(A) Statistical methods are not applicable
(B) Social survey method is the only reliable method of data collection
(C) Prediction is possible
(D) None of these

93. Which of the following theories refutes the assumption that richness I\in itself bring down birth rate and in poverty the rate is bound to go up?
(A) Optimum theory of population
(B) Theory of increasing prosperity
(C) Theory of rationalism
(D) Theory of voluntarism

94. Sociology cannot make use of:
(A) Historical method
(B) Ideal types method
(C) Scientific method
(D) Social survey method

95. Method of participant observation is associated with:
(A) Psychology
(B) Social Anthropology
(C) Archaeology
(D) Biology

96. Technique more suitable for study in depth is:
(A) Mailed questionnaire
(B) Interview guide
(C) Interview schedule
(D) Structured interview

97. Maximum subjectivity in social investigation can come when:
(A) Problem is being identified
(B) Problem has been identified
(C) Questions to be asked are being finalised
(D) Report is being finalized

98. Study of Human society involves the study of:
(A) Man
(B) Group
(C) Culture
(D) family

99. Field study is a method of research in which the observations are taken in the:
(A) Natural setup
(B) Observed set up
(C) Laboratories
(D) field

100. The comparative method was used by:
(A) Positivists
(B) Neo-positivist
(C) Diffusionist
(D) Evolutionist

101. The role of statistical methods in the study:
(A) Objectivity
(B) Complexity
(C) Subjectivity
(D) None of these

102. An interviewer can collect good material when the relationship between interviewer and the interviewee is:
(A) Indifferent
(B) Uninterested
(C) Cordial
(D) Ethical

103. The nature of objectivity is:
(A) Always based on the mental cognition of the person concerned
(B) To come upon objectivity consistent thinking with logical approach is to be maintained
(C) Degree of objectivity differs from person to person in different or same situations
(D) All of these

104. Social survey studies:
(A) Economic problem
(B) Social problem
(C) Historical problems
(D) Political problems

105. Which of the following statements carry the correct meaning of the objectivity approach?
(A) Selection of facts should not be subject to one's like or dislikes
(B) Analysis should not be influenced by subjective approach
(C) Approach should be impersonal
(D) All of these

106. Objectivity exists in:
(A) Partially interpreting data
(B) Scientific attitude
(C) Highlighting one's own view point
(D) Creating ambiguities

107. Objective finding is hindered because:
(A) Investigators are educated
(B) Respondents are usually biased
(C) Investigators are illiterate
(D) Respondents are illiterate

108. Objectivity is:
(A) Seeks scientific temperature
(B) Is biased on cause-effect relationship
(C) In collecting facts, it fulfils the manifestation of problems in time and space relationship
(D) All of these

109. Objectivity is:
(A) Requires a detached impersonal view
(B) Requires a person to give secondary importance to his beliefs
(C) Requires a person to discard his own beliefs and moral convictions
(D) None of these

110. Objective facts refer to facts when they have been:
(A) Theoretically established
(B) Rationally determined
(C) Logically derived
(D) Empirically verifies

111. An important criterion of a good sample is its:
(A) Small size
(B) Large size
(C) Representativeness
(D) Ambiguities

112. Attitude is ——— in nature and character:
(A) Vague
(B) Clear
(C) Subjective
(D) Objective

113. Main goal of sociological study is:
(A) Social reform
(B) Social construction
(C) Social reconstruction
(D) Study society in its existing form

114. The interview guide will be used during interview by:
(A) Interviewee
(B) Interviewer
(C) Different people
(D) Both interviewer and interviewee

115. Most widely used method of data collection among sociologists is:
(A) Participant observation
(B) Interview method
(C) Mailed questionnaire
(D) Simple observation

116. Sociology is a science because it is:
(A) Objective (B) Quantitative
(C) Methodical (D) Qualitative

117. Which one of the following is a thought or, an idea that sample for a phenomenon or, a group of facts under a common lable?
(A) Concept (B) Definition
(C) Sign (D) Theory

118. Comparative method was generally used by:
(A) Positivists (B) Anti-positivists
(C) Evolutionists (D) Empericist

119. Which is the following means a looking at things?
(A) Observation (B) Recording
(C) Prediction (D) Verification

120. Crucial feature of scientific data is its:
(A) Reliability
(B) Quality
(C) Quantifiability
(D) Universal applicability

121. Most frequently used method of research in sociology is/are:
(A) Experiment method
(B) Explorative method
(C) Descriptive method
(D) Explorative and descriptive methods

122. Social scientists deal with ideas:
(A) So that they can reinforce them
(B) To ascertain their origin
(C) To ascertain whether they are or can be turned into knowledge
(D) To prove them false

123. Objectivity requires a:
(A) Person to give secondary importance to his own biases and preferences
(B) Person to discard his own belief and moral conviction
(C) Person's detached and an empersonal view
(D) None of these

124. While conducting a social survey, the questionnaire technique is not applicable if:
(A) There is widespread literacy
(B) Area to be covered is vast
(C) There is male literacy
(D) There is over-population

125. A social researcher can use interview than a questionnaire to collect good and more material in getting information in a short time when he meets persons of:
(A) A particular area (B) Different areas
(C) A bigger areas (D) A wider areas

126. A set of techniques used to measure attractions and repulsions during interpersonal relations in quantitative and diagrammatic terms is called:
(A) Sociometry (B) Social survey
(C) Quantitative method (D) Social statistics

127. Emotions, sensations, perceiving, desiring, etc come under the category of:
(A) Integrative events (B) Freudian events
(C) Objective events (D) Subjective events

128. The reaction of a child when scolded can be studied under controlled conditions by applying the technique of:
(A) Observation (B) Interview
(C) Questionnaire (D) Field survey

129. Case study aims to:
(A) Establish statistical correlation
(B) Bring out the structure of the units as a whole
(C) Expose persons danger to society
(D) Treats cases requiring self-help

130. A tentative proposition, the authenticity of which is yet to be ascertained is:
(A) Index (B) Hypothesis
(C) Introduction (D) Conclusion

131. Observation means to see things:
(A) With a purpose
(B) With many purposes
(C) Without a purpose
(D) For important discussion

132. Which of the following method consists in the collection of data concerning the living, the working conditions of people in a given area with a view to formulate practical social measures for their betterment and welfare?
(A) Ideal types (B) Social survey
(C) Sociometry (D) Statistics

133. If wishes or value do not consciously or unconsciously distort reporter's description of the situation, then report is:
(A) Subjective (B) Objective
(C) Absolute (D) Original

134. Objectivity cannot be introduced at the time of:
(A) Selection of respondents
(B) Selecting field
(C) At the time of giving publicity through media
(D) Report writing

135. According to Weber, the subject-matter we select for study determine:
(A) Material benefits
(B) Long-term perspective
(C) Prestige associated with the work
(D) Values

136. Objective finding is hindered because:
(A) Investigators are educated
(B) Respondents are usually biased
(C) Investigators are illiterate
(D) Respondents are illiterate

137. Objective fact finding in social research means:
(A) Some specific object in view
(B) Demands of a particular organization
(C) Impartially
(D) Some pre-conceived objectives

138. Which of the following is a social fact?
(A) Tangible fact (B) Unknown fact
(C) Intangible fact (D) All of these

139. Things external to the mind, material objects and overt bodily behaviour are:
(A) Physical events (B) External events
(C) Objective events (D) Subjective events

140. Schedule is a list of questions which will be answered in an interview by a/an:
(A) Interviewer (B) Respondent
(C) Surveyor (D) Researcher

141. Questionnaire method is not applicable if:
(A) There is overpopulation
(B) There is wide spread literacy
(C) There is illiteracy
(D) Area is vast

142. Use of an interview method by a social researcher is expensive because it covers:
(A) More group of an area
(B) A few group of an areas
(C) More group of different areas
(D) A group of different areas

143. Social research can be conducted to study the problems related to:
(A) An individual (B) A group
(C) All groups (D) Some groups

144. Case study involves:
(A) Very careful observation of a person
(B) Complete observation of a person
(C) Careful observation of a person
(D) Very careful and complete observation of a person

145. Sociologist in study of human behaviour:
(A) Treats values as data which he studies
(B) Determines which values are of greatest importance
(C) Leaves values completely along since they have no place in human behaviour
(D) Attempts to justify the values of his society if he is a good sociologist

146. In an experimental group dynamic situation and new social norms arise, when:
(A) The experimenter gives an instruction
(B) Individuals are faced with ambiguous situations
(C) Instrumental roles are performed
(D) Expressive roles are performed

147. Thurston scale is a type of:
(A) Interview schedule (B) Attitude scale
(C) Opinion poll (D) Questionnaire

148. Training Performance Index does not includes:
(A) Training Effectiveness Index
(B) Training Utility Index
(C) Training Cost Index
(D) Training Participation Index

149. A science is to be judged by three criteria, reliability of its knowledge, its organization and:
(A) Its method
(B) Its capacity to expose universal laws
(C) Its quantifiability
(D) Bulk of qualitative data it can produce

150. Respondent is the term used in:
(A) Formal sociology
(B) Sociology of knowledge
(C) Kinship studies
(D) Social survey research

151. Which of the following statements carry the correct meaning of the objective approach?
(A) Selection of facts should not be subject to one's likes and dislike
(B) Analysis should not get influenced by subjective approach
(C) Approach should be impersonal and unbiased
(D) All of these

152. Participant observation refers to:
(A) Working with a group
(B) Studying a people by participating in their activities
(C) Observing a people's activities closely
(D) Observation of the participants

153. Objectivity in sociology requires a person:
(A) To negate his own culture entirely
(B) To discard his beliefs and values
(C) To give secondary importance to his own biases and preferences
(D) To take a detached, impersonal view

154. Set a techniques to measure attractions and repulsions in inter-personal relations in quantitative and diagrammatic terms is:
(A) Social statistics (B) Sociometry
(C) Social survey (D) Quantitative method

155. The prime concern of a sociologist with human being and their relations to one another is a/an:
(A) Disciplined inquirer (B) Advocate
(C) Preacher (D) Reformer

156. When area of research will be large, the appropriate research method is:
(A) Participant observation method
(B) Sample survey method
(C) Questionnaire method
(D) Mailing questionnaire

157. Objectivity means:
(A) Absence of personal values
(B) Lack of personal values and bias in one's work
(C) Using scientific method of study
(D) Full of personal value

158. Which of the following is important for the study of sociology?
(A) It helps in solving the social problems
(B It enriches our knowledge
(C) It helps in raising the standards of the people
(D) All of these

159. Correct chronology steps involved in sociological investigation are:
(A) Formulation of problem, planning research design, field work, analyzing data, drawing the conclusion
(B) Analysing data, planning research design, field work, formulation of problem, drawing conclusion
(C) Formulation of problem, analyzing data, planning research design, drawing conclusion
(D) Field work, planning research design, formulation of problem, field work and drawing conclusion

160. The method which consists of observation, recording, classification, hypothesis, verification and prediction is:
(A) Historical method
(B) Scientific method
(C) Inverse-deductive method
(D) Ideal-type method

161. Which of the following fall under the scope of sociology?
(A) Study of society
(B) Study of social relationship
(C) Study of social action
(D) All of these

162. Which of the following method involves comparisons of various kinds or groups of people in order to find out the differences as well as similarities in their ways of life and thus to find out clues to man's social behaviour?
(A) Historical method
(B) Comparative method
(C) Experimental method
(D) Inverse-deductive method

163. Experimentation is:
(A) Impossible is sociology
(B) Most important method in sociology
(C) Though it is possible not preferred in sociology
(D) Rarely possible in sociology

164. According to which method, an ideal in constructed from concrete cases and then a particular case is evaluated according to the degree of its approximation to the ideals?
(A) Inverse deductive (B) Comparative
(C) Ideal type (D) Experimental

165. Which method has limitation that cannot be used among illiterates?
(A) Participant observation
(B) Non-participant observation
(C) Questionnaire
(D) Interview

166. A man who studies effects of the social structure of a society on its form of government is most likely a/an:
(A) Historian (B) Sociologist
(C) Economist (D) Anthropologist

167. The field of sociology is distinguished from other disciplines by:
(A) Its method
(B) Types of questions it asks and the particular perspectives it uses
(C) Its social phenomenon measurement scales
(D) The special interest in explaining human behaviour

168. An important feature of a good sample is:
(A) Smaller quantity
(B) Big quantity
(C) Correct representativeness
(D) None of these

169. Case study involves:
(A) Careful observation of a person
(B Complete observation of a person
(C) Very careful observation of a person
(D) Very careful and complete observation of a person

170. Sometimes mailed questionnaire is preferable to interview because:
(A) It works out cheaper
(B) Face –to-face contact is possible
(C) People immediately return it
(D) None of these

171. To obtain entirely reliable knowledge of the exact value of a parameter, observations must be made on:
(A) All the elements in a single stratum of the stratified population
(B) A stratified sample of the population
(C) A complete census of the population
(D) A simple random sampling of the population

172. Which of the following method is less expensive?
(A) Case study (B) Mailing questionnaire
(C) Interview (D) Questionnaire

173. In participant observation method, observes can collect material from a group to a large extent if he acts:
(A) Emotionally (B) Passively
(C) Objectively (D) Subjectively

174. Correct chronology of Comte's method is:
(A) Experiment, observation, History and Sociology
(B) Observation, Experiment, Comparison and History
(C) History, Political Science, Biology and Sociology
(D) History, Physics, Biology and Sociology

175. There is maximum danger of loosing one's sharpness while gathering data due to excessive rapport in:
(A) Interview (B) Questionnaire
(C) Participant observation (D) Mailing questionnaire

176. Which of the following is most appropriate?
(A) All social sciences study society
(B) Only sociology and social anthropology study society
(C) Only sociology, social anthropology and history study society
(D) Only sociology studies society

177. Filling and keeping filled, the positions in the organization structure:
(A) Planning (B) Organizing
(C) Staffing (D) Leading

178. Discussion papers based on in-depth study and research is presented and discussed under the guidance of experts:
(A) Colloquium (B) Seminar
(C) Buzz session (D) Workshop

179. The activities of staffing include:
(A) Recruitment (B) Promotion
(C) Separation (D) All of these

180. Detailed biographical information about a job applicant is referred to as:
(A) Questionnaire (B) Bio-data
(C) Interview (D) Discussion

181. The last stage in the process of recruitment is:
(A) Appointment (B) Orientation
(C) Placement (D) Promotion

182. Managers tends to be promoted to the level of their incompetence is said in:
(A) Parkinson's law
(B) Marconi law
(C) Peter principle
(D) Commitment principle

183. The essence of promotion is:
(A) Change of title
(B) Change of pay
(C) Change of assignment
(D) Change of duties and punishment

184. The basic objective of an interview is to assess the candidate's:
(A) Intellectual capacity (B) General knowledge
(C) Personal qualities (D) Power of expression

185. The technique that gives managers a clear idea of their path of development:
(A) Managerial appraisal (B) Assignment centres
(C) Planned progression (D) None of these

186. Purpose of time study is:
(A) To remove wastage of time
(B) To make workers punctual
(C) To determine fair day's work
(D) To give timely assistance

187. To identify exactly what workers do and what skills or knowledge they need to succeed in a job we go for:
(A) Job analysis (B) Item analysis
(C) Factor analysis (D) None of these

188. The situation that an employee must be able to cope with if he is going to succeed in a particular job:
(A) Critical incidents (B) Work behaviour
(C) Work motivation (D) Work environment

189. A lower job satisfaction will lead to:
(A) High absenteeism (B) Low moral
(C) High employee turn over (D) Both (A) and (B)

190. An expression of discontent or dissatisfaction with any aspect of organization is:
(A) Stress (B) Conflict
(C) Grievance (D) Competition

191. Which of the following is a source of stress in official situation:
(A) Overload of work (B) Lack of adjustment
(C) Lack of autonomy (D) All of these

192. Organizational conflict does not occur when there exist:
(A) Autocratic supervision (B) Incompatible goals
(C) Unlimited resources (D) Heterogeneous group

193. Which of the following is an informal means of coordination:
(A) Planning (B) Leadership
(C) Conferences (D) Budgeting

194. Formation of cliques in an organization ultimately creates:
(A) Damaging effect (B) Performing effect
(C) Stabilizing effect (D) No effect

195. Measuring and correcting of activities of subordinates to ensure the events confirm to plans:
(A) Supervision (B) Controlling
(C) Managing (D) Leading

196. The principle of supervision is inherent in the principle of:
(A) Span of control (B) Coordination
(C) Hierarchy (D) Unity of command

197. The technique of supervision include:
(A) The promulgation of service standard
(B) Budgetary limitations upon the operations
(C) Reporting system on work progress
(D) All of these

198. The aspect of supervision does not include:
(A) Institutional aspect (B) Personal aspect
(C) Political aspect (D) Technical aspect

199. Each individual in an organization must receive his or her instructions for a particular operation from only one person is known as:
(A) Span of control (B) Unity of command
(C) Hierarchy (D) Power

200. The essence of the scaler principle is:
(A) Unity of command (B) Span of control
(C) Communication (D) Coordination

201. "If it is violated, authority is under mind, discipline is in jeopardy, order disturbed and stability threatened" refers mainly to the principle of:
(A) Hierarchy (B) Span of control
(C) Coordination (D) Unity of control

202. Discharging of duties according to the letter and spirit of the command signifies the principles of:
(A) Control (B) Authority
(C) Accountability (D) Responsibility

203. The span of control depends on:
(A) Personality of superior
(B) Nature of work to be supervised
(C) Age of the organization
(D) All of these

204. The principle of unity of command is complimentary to the principle of:
(A) Span of control
(B) Authority
(C) Scaler chain
(D) Supervision

205. The principles that is directly concerned with preventing overlap, conflict and friction in an organization:
(A) Direction
(B) Supervision
(C) Coordination
(D) Span of control

206. Which of the following is the limiting factor in the development of coordination:
(A) Time
(B) Size
(C) Informal organization
(D) Both (A) and (B)

207. The most important means of coordination is:
(A) Hierarchy
(B) Consultation
(C) Planning
(D) Centralization

208. Allocation of funds for different sectors of programme:
(A) Controlling
(B) Budgeting
(C) Financing
(D) Supervising

209. Dividing a large group into much smaller ones, in which a topic is discussed within a limited period of time:
(A) Colloquium
(B) Seminar
(C) Buzz session
(D) Symposium

210. Which of the following is not an empirical method of assessing training needs:
(A) Job analysis
(B) Performance evaluation
(C) Buzz session
(D) Questionnaire method

211. A cooperative gathering of individuals who discuss, learn and apply practical skills under expert supervision:
(A) Colloquium
(B) Seminar
(C) Buzz session
(D) Workshop

212. The characteristic feature of the fourth generation of evaluation is:
(A) Measurement oriented
(B) Description oriented
(C) Judgement oriented
(D) Collaboration oriented

213. The element in the evaluation framework that covers the trainees opinion about the programme:
(A) Reaction
(B) Learning
(C) Behaviour
(D) Result

214. The extent of change in the behvaiour interms of job performance as a result of training is determined by:
(A) Process evaluation
(B) Impact evaluation
(C) Plan evaluation
(D) Terminal evaluation

215. Objective-training evaluation model was given by:
(A) Ralph W Tyler
(B) Daniel Stufflebeam
(C) Robert Stake
(D) Michael Scriven

216. In a situation of abundant supply and low demand the village industry should thrive on:
(A) Export promotion
(B) Value addition
(C) Distress sale
(D) Consumer education

217. The evaluation model that uncovers the utility and overall impact of the training programme without any set criteria:
(A) Objective model
(B) Goal free model
(C) Expert model
(D) Naturalistic model

218. A systematic, integrated and planned approach for improving enterprise effectiveness:
(A) Organizational development
(B) Futures organization
(C) Management grid
(D) Management development

219. The highest quality organization is:
(A) Transcendental
(B) Mutualistic
(C) Parasitic
(D) Accidental

220. The relationship wherein one gains from an association at the expense of the other:
(A) Transcendental
(B) Mutualistic
(C) Parasitic
(D) Accidental

221. The term "entrepreneur" is coined by:
(A) Richard Cantinlon
(B) Tony Pears
(C) Curt Lewin
(D) Gilford Pinchot

222. The most important for an entrepreneur to have is:
(A) High need for achievement
(B) High knowledge about the business
(C) High operational skills
(D) High financial resources

223. Technology recommended for an enterpreneurship must be:
(A) Appropriate
(B) Innovative
(C) Risky
(D) Costly

224. Enterpreneur are people who take:
(A) High risk
(B) Low risk
(C) No risk
(D) Moderate risk

225. Successful entrepreneurs are:
(A) Risky people
(B) Rational people
(C) Power hungry
(D) Empire builders

226. National Institute of Entrepreneurship and Small Business Development is located at:
(A) New Delhi
(B) Jaipur
(C) Kolkata
(D) Bangalore

227. An entrepreneur who worked within the organizational environment:
(A) Entrepreneur
(B) Intrapreneur
(C) Intrepreneur
(D) Entrapreneur

228. One of the basic elements in the leadership relationship is:
(A) Task
(B) Faith
(C) Motivation
(D) Education

229. The Path-Goal approach of leadership theory was formulated by:
(A) Robert House
(B) Fred E Fiedler
(C) B F Skinner
(D) L W Porter

230. Take the odd man out from the following:
(A) Country club leadership
(B) Middle road leadership
(C) Lassez-faire leadership
(D) Team leadership

231. Leader are born and not made:
(A) Behavioural theory of leadership
(B) Great man theory of leadership
(C) Situational theory of leadership
(D) None of these

232. A basically lazy approach that avoids as much work as possible:
(A) Impoverished management
(B) Country Club management
(C) Authority-compliance
(D) Team management

233. People define roles for themselves and others based on social learning and reading:
(A) Role theory of leadership
(B) Great man theory of leadership
(C) Situational theory of leadership
(D) None of these

234. Successful leadership is based in definable, learnable behavior:
(A) Behavioural theory (B) Situational theory
(C) Trait theory (D) Contingency theory

235. People are more committed to action where they have involved in the relevant decision making:
(A) Transactional leadership
(B) Situational leadership
(C) Participative leadership
(D) Contingency leadership

236. Contingency model of leadership was formulated by:
(A) Robert House (B) Fred E. Fiedler
(C) B F Skinner (D) L W Porter

237. The leader has a low concern for people and uses such methods as threats and other fear-based methods to achieve conformance:
(A) Exploitive authoritative
(B) Benevolent authoritative
(C) Consultative leadership
(D) Participative leadership

238. The best approach when the following are expert and their advice is both needed and they expect to be able to give it:
(A) Supportive leadership
(B) Directive leadership
(C) Participative leadership
(D) Achievement-oriented leadership

239. When people have agreed to do a job, a part of the deal is that they cede all authority to their manager:
(A) Transactional leadership
(B) Situational leadership
(C) Transformational leadership
(D) Contingency leadership

240. A leadership style in which the leader himself develops a plan of action and achieves the challenging goal is called:
(A) Instrumental (B) Supportive
(C) Participative (D) Achievement-oriented

241. According to which theory, leadership is multidimentional:
(A) Behavioural theory (B) Situational theory
(C) Trait theory (D) None of these

242. Benevolent autocrat un Linkert's four system of management comes under:
(A) System I (B) System 2
(C) System 3 (D) System 4

243. Transformational leadership theory was formulated by:
(A) Robert House (B) Bass, B.M.
(C) B F Skinner (D) L W Porter

244. The term "Motivation" means:
(A) To move or incite into action
(B) To bring forward
(C) To energize inner self
(D) To act according to wishes

245. Motivation is a general term which includes:
(A) Drives and desire (B) Desire and needs
(C) Emotions and feelings (D) Both (A) and (B)

246. Which of the following is true regarding Maslow's theory:
(A) Based on psychoanalysis
(B) First theory of motivation
(C) Published in 1942
(D) Based on prepotency of needs

247. The behavioral type embedded in the basic concept of motivation:
(A) Risk taking behavior
(B) Goal satisfying behavior
(C) Decision making behavior
(D) None of these

248. The process theory of motivation does not includes the models of:
(A) Victor Vroom (B) Herald Kelly
(C) Mc Gregor (D) Porter Lawler

249. The middle level need in Maslow's need hierarchy is:
(A) Social needs (B) Security needs
(C) Esteem needs (D) Self actualization needs

250. The need for power, prestige, status and self-confidence belongs to:
(A) Esteem needs (B) Security needs
(C) Social needs (D) Self actualization needs

251. Motivation in Herzberg's need classification belong to which level of Maslow's need hierarchy:
(A) Social needs
(B) Security needs
(C) Physiological needs
(D) Self actualization needs

252. Two factor theory of motivation was proposed by:
(A) F. Herzberg (B) V. H. Vroom
(C) J. S. Adams (D) L. W. Porter

253. The satisfiers according to Herzberg does not include:
(A) Achievement (B) Recognition
(C) Challenging work (D) Work condition

254. The strength of a person's motivation is a function of:
(A) Valence (B) Expectancy
(C) Both (A) and (B) (D) None of these

255. According to Maslow:
(A) Human needs are always conscious
(B) Man is perpetually wanting animals
(C) A satisfied need does not motivate
(D) Both (A) and (C)

256. Latest theory of motivation among the following is:
(A) Expectancy theory
(B) ERG theory
(C) Theory X and Y
(D) Maturity immaturity theory

257. Valence expectancy theory of motivation was proposed by:
(A) C. Argyris (B) V.H. Vroom
(C) A.H. Maslow (D) Mc Gregor

258. ERG theory of motivation was put forward by:
(A) F. Herzberd (B) V. H. Vroom
(C) Clayton Alderfer (D) L.W. Porter

259. The term self actualization was first coined by:
(A) Abraham Maslow (B) Kurt Goldstein
(C) Mc Gregor (D) Herzberg

260. The first major exponent of the classical theory of motivation:
(A) Maslow (B) Fayol
(C) Taylor (D) Mc Gregor

261. The most important hygiene factor according to Herzberg is:
(A) Policy and administration
(B) Achievement
(C) Salary
(D) Working condition

262. Which of the following are correct regarding Herzberg hygiene factors:
(A) They are external to the job
(B) They reflect the Adam nature of man
(C) They are related to the job content
(D) Both (A) and (B)

263. The book " An introduction to Motivation" is written by:
(A) J. W. Atkinson (B) D C Mc Clelland
(C) R A Clark (D) E L Lowell

264. Which of the following is not included under the hygiene factors of motivation by Herzberg:
(A) Pay and job security
(B) Working conditions
(C) Advancement and growth
(D) Interpersonal relations

265. The need theory of motivation was developed by:
(A) David Mc Clelland (B) A H Maslow
(C) Douglas Mc Gregor (D) None of these

266. Management grid was developed by:
(A) Samuel David (B) B F Skinner
(C) Blake and Mouton (D) Peter and Lawter

267. The most important of all human needs:
(A) Physiological need
(B) Safety needs
(C) Esteem needs
(D) Self actualization need

268. Carrot and stick approach of motivation can be effectively linked with:
(A) Natural learning methods
(B) Behavior control methods
(C) Controlled exposure methods
(D) Expository methods

269. Reinforcement theory of motivation was put forward by:
(A) F Herzberg (B) V H Vroom
(C) B F Skinner (D) L W Porter

270. The equity theory of motivation was put forward by:
(A) F Herzberg (B) V H Vroom
(C) J S Adams (D) L W Porter

271. The term "Theory X" was coined by:
(A) Mc Gregor (B) Marks
(C) Abrams (D) Jewel

272. Workers enjoy autonomy and are willing to accept responsibility is in assumption of:
(A) Theory X (B) Theory Y
(C) Theory Z (D) Balance theory

273. Which one of the following is not a proposition in Theory X:
(A) Man wished to avoid responsibility
(B) Man has inherent liking for work
(C) Man must be coerced for working
(D) Man must be threatened with punishment

274. Workers must be goaded or guided into being productive is the assumption of:
(A) Theory X (B) Theory Y
(C) Theory Z (D) Balance theory

275. The central principle which derive from Mc Gregor Theory Y is:
(A) Direction of control (B) Coordination
(C) Scaler principle (D) Integration

276. According to path goal theory the effectiveness of leadership style is contingent upon:
(A) Evolutional characters
(B) Characteristics of subordinates
(C) Work environment
(D) Both (B) and (C)

277. Administration is:
(A) Policy making
(B) Execution of policy
(C) Giving direction to things
(D) None of these

278. The term "Development Administration" was coined by:
(A) Edward Weidner (B) Montegomery
(C) George Gant (D) E W Riggs

279. The hallmark of public administration is:
(A) Consistency of treatment
(B) Public responsibility
(C) Community service
(D) External financial control

280. The word watch of public administration is:
(A) Public accountability
(B) Uniformity of treatment
(C) Service orientation
(D) Political direction

281. The gangplank idea is found in the works of:
(A) Fayol (B) Follett
(C) Barnard (D) Simon

282. To whom administration and authority are intimately linked:
(A) Fayol (B) Barnard
(C) Weber (D) Simon

283. Which one of the following is, according to Max Webber, the defining characteristic of authority:
(A) Power (B) Coercion
(C) Tradition (D) Legitimacy

284. Who among the following authors attached more importance to structure rather than persons in organizations:
(A) Luther Gulick (B) Elton Mayo
(C) Chester Barnard (D) Mary Porter Follett

285. The influence of socio-psychological factors on worker's output was first proved by:
(A) M P Follett (B) F W Raylor
(C) E Mayo (D) R Linkert

286. Who among the following does not belong to the new human relations approach:
(A) C Argyris (B) E Mayo
(C) R Linkert (D) W Bennis

287. The theory of bureaucracy was mainly developed by:
(A) Psychologists (B) Sociologists
(C) Political scientists (D) Economists

288. The emphasis of scientific management theory is on:
(A) Scholarly view (B) Explaining behavior
(C) Improving practice (D) None of these

289. The classical theory of organization was developed during the period:
(A) 1900-1950 (B) 1800-1850
(C) 1700-1750 (D) 1600-1650

290. The classical theory is chiefly concerned with:
(A) Formal organizational structure
(B) Efficiency at the operational level
(C) Basic management process
(D) Both (A) and (C)

291. The most important exponent of the classical theory is:
(A) Luther Gulick (B) Henry Fayol
(C) JD Mooney (D) Lyndall Urwick

292. The Hawthorne studies were conducted by:
(A) Elton Mayo (B) Hugo
(C) FW Taylor (D) Fayol

293. According to human relations theory, the productivity of employees mainly depends on:
(A) Physical condition at work
(B) Relationship of superiors and colleagues
(C) Amount of salary received
(D) None of these

294. According to Marx, class membership was determined by ones:
(A) Family
(B) Relationship to the state
(C) Education
(D) Relation to the means of production

295. The greatest resistance to scientific management came from:
(A) Human relationists (B) Workers
(C) Managers (D) Trade unions

296. According to human relations theory, an organization consist of which of the following elements:
(A) Informal groups (B) Formal groups
(C) Intergroup relations (D) All of these

297. Which of the following is a characteristic of classical theory:
(A) Specialization (B) Impersonality
(C) Hierarchy (D) All of these

298. Scientific management means:
(A) One best way of doing things
(B) Scientific division of work
(C) Functional foremanship
(D) Mental revolution

299. The idea of scientific management was introduced by:
(A) Gilbreth (B) Mayo
(C) Lewin (D) Taylor

300. Scientific management uses:
(A) Time and motion studies (B) Task analysis
(C) Job specialization (D) All of these

301. Which of the following theories views leadership as a manipulative function:
(A) Human relations theory
(B) Scientific management theories
(C) Classical theory
(D) Decision making theory

302. Which of the following have advocated positional theory of authority:
(A) Weber and Barnard (B) Fayol and Simon
(C) Barnard and Simon (D) Fayol and Weber

303. The office of district collector of India is based on the principle of:
(A) Delegation (B) Decetralization
(C) Devolution (D) Deconcentration

304. The Indian Institute of Public Administration is located at:
(A) Mumbai (B) New Delhi
(C) Kolkata (D) Kanpur

305. The scientific management stressed upon:
(A) Rationality (B) Specialization
(C) Predictability (D) All of these

306. The experiment conducted by Taylor was intended to understand the:
(A) Human behavior in organization
(B) Process of mental revolution
(C) Motivation of workers
(D) Best way to carry out specific tasks

307. The human relations theory lays emphasis on:
(A) People (B) Motivation
(C) Informal group (D) All of these

308. A hierarchical arrangement among the different parts of an organization with the divisibility of functions and authority is an important feature of:
(A) Span of control
(B) Bureaucracy
(C) Management by objective
(D) Organizational climate

309. Which one the following is the major drawback of hierarchical organization:
(A) Defined structure
(B) Red tapism
(C) Rule orientation
(D) Fixed decision making levels

310. Taylor's concept of functional foremanship leads to:
(A) Multiple command
(B) Specialization
(C) Negation unity of command
(D) All of these

311. Weber's analysis of bureaucracy was from:
(A) Political perspective
(B) Sociological perspective
(C) Economic perspective
(D) Cultural perspective

312. The behavioural characteristics of bureaucracy are:
(A) Rationality and role specificity
(B) Role specificity and impersonality
(C) Impersonality and role orientation
(D) Rationality and system of rules

313. Theory X, or scientific management, is concerned primarily with improving:
(A) Output (B) Productivity
(C) Work efficiency (D) Work environment

314. The structural characteristics of bureaucracy includes:
(A) Hierarchy (B) Rule orientation
(C) Role specificity (D) All of these

315. According to Max Webber, bureaucracy is superior to any other foam of organization in:
(A) Precision (B) Stability
(C) Discipline (D) All of these

316. Which of the following is not a behavioral characteristics of bureaucracy:
(A) Impersonality (B) Role specificity
(C) Rationality (D) Rule orientation

317. The essential focus of the human relations theory was on:
(A) Physiological factors (B) Social factors
(C) Economic factors (D) Both (A) and (B)

318. The democratic and participatory style of leadership become popular during the era of:
(A) Behavioural theory
(B) Classical theory
(C) Systems theory
(D) Human relations theory

319. Plans and actions should be evaluated by their consequences:
(A) Utilitarian theory (B) Theory of rights
(C) Theory of justice (D) Parkinson's law

320. Administration will differ from one field to another according to the subject matter:
(A) Administrative view (B) Wider view
(C) Integral view (D) Narrower view

321. Which of the following is described as anti-theorist, anti-positivist and anti-hierarchical:
(A) Development communication
(B) New public administration
(C) Comparative communication
(D) Administrative development

322. The principal disadvantage of hierarchy is:
(A) Rigidity in organization
(B) Creates superiority or inferiority feeling
(C) Delay in disposal of case
(D) Decentralization decision making

323. Which of the following is not a feature of system approach:
(A) Adaptive (B) Deterministic
(C) Multi-variate (D) Descriptive

324. Behavioural approach in management is not concerned with:
(A) Prescription (B) Quantification
(C) Mathematization (D) Theorization

325. Which of the following approaches gave rise to contingency management:
(A) Behavioural approach
(B) Classical approach
(C) Systems approach
(D) Human Relations approach

326. Systems approach in management is also known as:
(A) Contingency theory (B) Integral theory
(C) Ecological theory (D) Modern theory

327. Training is basically is:
(A) Product (B) Process
(C) Activity (D) Programme

328. Training, looking in its true sense:
(A) Stimulate and create potential
(B) Stimulate potential
(C) Create potential
(D) None of these

329. Training basically is not a:
(A) Overt process (B) Covert process
(C) Communication process (D) Learning process

330. The focus of training and learning are respectively on:
(A) Events and self (B) Events and outcome
(C) Outcome and self (D) Outcome and Process

331. Training is a function of:
(A) Knowledge only (B) Attitude only
(C) Skill only (D) All of these

332. The most emphasized behaviour dimension in training is:
(A) Knowledge (B) Attitude
(C) Skill (D) None of these

333. The behavior domain that gets the highest emphasis in training and education are:
(A) Cognitive and Connotative
(B) Cognitive and Effective
(C) Affective and Cognitive
(D) Affective and Psychomotive

334. Training differs from development in the sense the former is:
(A) Long term (B) Short term
(C) Equal term (D) None of these

335. Training is related to learning in the same way:
(A) Sowing to seed (B) Soil to seed
(C) Seed to soil (D) None of these

336. The learning curve in training follows a:
(A) S-shape (B) U-shape
(C) V-shape (D) C-shape

337. The place in which a training is conducted is technically called:
(A) Training institute (B) Training organization
(C) Training association (D) Training group

338. The training skill that need to be emphasized for top level management:
(A) Conceptual (B) Technical
(C) Design (D) Both (A) and (B)

339. Training programme for VLW's should essentially focus on:
(A) Technical skill (B) Conceptual skill
(C) Design skill (D) Leadership skill

340. Extension education institutes focus their training for functionaries working under:
(A) Grass root level (B) Middle level
(C) Top level (D) Both (A) and (B)

341. Institution share in the training course is the highest for:
(A) ICAR (B) SAUs
(C) ACTs (D) CGO

342. Foundation training is otherwise known as:
(A) Portal (B) Vestibule
(C) Inductive (D) All of these

343. The training type that prepares an employee to take a new assignment is:
(A) Re-training (B) Refresher training
(C) Inductive training (D) Orientation training

344. Training and visit system essentially make use of:
(A) Refresher training (B) Inductive training
(C) Re training (D) Development training

345. The extension worker trains the farmer at their farm situation is termed as:
(A) Pipeline training (B) Vestibule training
(C) Peripatatic training (D) Inservice training

346. Action research programme is considered as an example for:
(A) Formal training (B) Non formal training
(C) Direct training (D) Distant training

347. The training type that can also be grouped under pre service training:
(A) Inductive training (B) Development training
(C) Orientation training (D) Overseas training

348. The NAARM training for freshly recruited ARS trainees falls under:
(A) Pre-training (B) Induction training
(C) Orientation training (D) Refresher training

349. Summer institute is a good example for:
(A) Induction training
(B) Developmental training
(C) Orientation training
(D) Refresher training

350. Learning by doing is called:
(A) Formal training (B) Informal training
(C) In-service training (D) Post-entry training

351. Training principle that best explain the use of analogue is:
(A) Association (B) Readiness
(C) Capacity (D) Intensity

352. Academic orientation is the characteristic feature of:
(A) Process approach
(B) System approach
(C) Traditional approach
(D) Experimental approach

353. The training approach developed from engineering concept:
(A) Discrete approach (B) System approach
(C) Process approach (D) Performance approach

354. The concept of training is considered very loose and less precise in:
(A) Discrete approach (B) Process approach
(C) Experimental approach (D) System approach

355. Training model widely used by policy makers and managers:
(A) Decision model (B) Expert model
(C) Experimental model (D) Objective model

356. Cyclic model of training was developed by:
(A) Lynton and Pareek (B) R D Miller
(C) A York (D) DG Reay

357. System concept of training was developed by:
(A) A York (B) R D Miller
(C) DG Reay (D) D Kolb

358. The number of steps in the discrete approach developed by RD Miller is:
(A) Three (B) Four
(C) Five (D) Six

359. System approach of training is basically a:
(A) Diagnostic approach
(B) Productive approach
(C) Descriptive approach
(D) Experimental approach

360. Training model that shares an empathetic attitude towards training institution:
(A) Elaborated model (B) Discrete model
(C) Simple model (D) Cyclic model

361. The training stage that does not form a component of cyclic model:
(A) Preparatory training (B) Pre-training
(C) Training (D) Post training

362. Moving phase of Curt Lewin matches with:
(A) Preparatory training phase
(B) Pre training phase
(C) Training phase
(D) Post training phase

363. The training phase that is otherwise known in the adjustment phase:
(A) Preparatory phase (B) Pre training phase
(C) Training phase (D) Post training phase

364. The motivational phase in training is otherwise called:
(A) Preparatory phase (B) Pre-training phase
(C) Training phase (D) Post training phase

365. Receptability of organization plays a crucial role in:
(A) Preparatory phase (B) Pre-training phase
(C) Training phase (D) Post training phase

366. The probable outcome at the end of post training phase is:
(A) Enhanced motivation
(B) Internalization of new knowledge
(C) Expectancy building
(D) Institutionalization of change

367. The rational training need analysis method among the following is:
(A) Observation (B) Check list
(C) Critical incident method (D) Card sort method

368. The first step in SATNA need assessment method is:
(A) Job description (B) Problem identification
(C) People identification (D) KAS identification

369. "A need is something whose absence or deficiency proves harmful" represent:
(A) Democratic view (B) Autocratic view
(C) Diagnotic view (D) Discrepancy view

370. Logical analysis is the characteristic feature of:
(A) Rational method (B) Empirical method
(C) Discovery method (D) Encounter method

371. Internal training strategy covers everything except:
(A) Individual receptability
(B) Training scheduling
(C) Logistical arrangement
(D) Curriculum development

372. Training methods differ from training modality in the sense:
(A) Method is broader (B) Modality is broader
(C) Both are same (D) None of the same

373. The analogue of heart and blood vessels in training best represent to:
(A) Methods and Curriculum
(B) Strategy and Objective
(C) Curriculum and Methods
(D) Need and Methods

374. The level in the training need where a person can perform a job effectively but takes more time:
(A) Precision level (B) Perfection level
(C) Sufficiency level (D) Deficiency level

375. Phenomena which shows continuous change over time is termed as:
(A) Systematic (B) Process
(C) Erratic (D) None of these

376. "Training for Organizational Transformation" is a book written by:
(A) Lynton and Pareek
(B) Berger and Berger
(C) PN Singh
(D) Pareek, Rao and Parimjee

377. Curriculum literacy means:
(A) Keep on running (B) Content in training
(C) Theory in training (D) Printed material

378. The S in the SMART objective refers to:
(A) Simple (B) Specific
(C) Sincere (D) Secure

379. What is the first step in SQ3R method:
(A) Study (B) Select
(C) Speculate (D) Survey

380. Training objective does not contain:
(A) Criteria (B) Behaviour
(C) Condition (D) Goal

381. The latest priority component in the concentric approach in curriculum development is:
(A) Could know (B) Should know
(C) Must know (D) Would know

382. Outer and inner core of human brains are respectively the seats of:
(A) Emotional behavior
(B) Logical behavior
(C) Emotional and logical behavior
(D) Logical and Emotional behavior

383. The curriculum development approach that equate with the principle of thermostat:
(A) Reductionlist approach
(B) Constructionalist approach
(C) Cybernetic approach
(D) Goal management approach

384. Kolb is associated with:
(A) Programmed instruction (B) Learning style
(C) Role play (D) ELC cycle

385. The learning style wherein people like hand on experience is:
(A) Accommodates (B) Diverges
(C) Assimilators (D) Convergers

386. People with high imaginative ability and sensitivity falls under:
(A) Accommodates (B) Diverges
(C) Assimilators (D) Convergers

387. Integration of ideas and formulation of theories is the characteristic feature of:
(A) Accommodates (B) Diverges
(C) Assimilators (D) Convergers

388. The learning style with more emphasis on pragmatism rather than conceptualization is:
(A) Accommodates (B) Diverges
(C) Assimilators (D) Convergers

389. The stage in learning cycle where the trainees become aware about the changes around them and develop interest about the advances in the subject:
(A) Unconscious incompetence
(B) Conscious incompetence
(C) Conscious competence
(D) Unconscious competence

390. Transfer of learning from concrete experience to reflective observation needs which mode of learning:
(A) Accommodative (B) Divergent
(C) Assimilative (D) Convergent

391. The first stage in ELC cycle is:
(A) Experiencing (B) Publishing
(C) Generalizing (D) Applying

392. The fulcrum of ELC cycle is:
(A) Experiencing (B) Processing
(C) Generalizing (D) Applying

393. Theory can be integrated in the experiential learning cycle at:
(A) Experiencing (B) Processing
(C) Generalizing (D) Applying

394. The last phase in experiential learning cycle is:
(A) Application (B) Generalization
(C) Publication (D) Experience

395. The first and last step in the seven step model of ELC are:
(A) Climate setting and closure
(B) Experience and closure
(C) Experiencing and Application
(D) Objective clarification and closure

396. The role of trainer in experiential approach is:
(A) Facilitator (B) Contributor
(C) Controller (D) Enabler

397. Sensitivity training normally needs a time period of:
(A) 15-20 minutes (B) 1-2 hours
(C) 1-2 days (D) 5-10 days

398. "He is passive in laboratory culture, take it in with no struggle or fight and back home will put quickly to the old shell":
(A) Tourist (B) Expatriate
(C) Missionary (D) Easy convert

399. The best bet for transfer of learning for a training programme is:
(A) Easy convert (B) Self-mystic
(C) Learner critic (D) Expatriate

400. The basic training orientation that gives equal importance to content and concept:
(A) Laboratory (B) Academic
(C) Activity (D) Person development

401. Going from specific to general in a training programme make use of:
(A) Inductive reasoning (B) Deductive reasoning
(C) Horizontal reasoning (D) Vertical reasoning

402. Relating class room learning with real life experience takes place in:
(A) Experiencing (B) Processing
(C) Generalizing (D) Applying

403. Transfer of learning to work situation does not take place in:
(A) Tourist model (B) Missionary model
(C) Self mystic model (D) Learner critic model

404. The amount of learning that takes place through non-verbal communication is:
(A) 10 per cent (B) 20 per cent
(C) 30 per cent (D) 40 per cent

405. Participants role in field training is:
(A) Outsider – learner
(B) Player – Observer
(C) Initiator – Contributor
(D) Contributor – Reporter

406. The focus of trainer in a training sessions is essentially on:
(A) Learner (B) Subject matter
(C) Learning climate (D) Learning outcome

407. The responsibility for the outcome of training rest solely on:
(A) Learner (B) Trainer
(C) Institution (D) Both (A) and (B)

408. The training needs of individual employees in an organization has to be assessed by:
(A) Employee (B) Colleagues
(C) Employer (D) Training organization

409. The process of transforming the experiences into knowledge as a basic for action is called:
(A) Continued learning
(B) Experimental learning
(C) Topical learning
(D) Experimental learning

410. Human beings continues to learn till:
(A) Adulthood (B) Childhood
(C) College days (D) Life end

411. Institutional aids used in a training session can:
(A) Provide verbal experience
(B) Replace trainer
(C) Includes worksheet
(D) Decorates learning situation

412. Instructional aids used in a training for:
(A) Supplement learning (B) Supplant learning
(C) Both (A) and (B) (D) None of these

413. The phase of interactive demonstration wherein the participants make attempt to intellectualize the skill:
(A) Cognitive phase (B) Fixation phase
(C) Autonomous phase (D) Association phase

414. The behavior domain that acquires the most prominence in simulation method:
(A) Knowledge (B) Skill
(C) Attitude (D) All of these

415. Compared to education training show variation in all except:
(A) Vocational component (B) Facilitator role
(C) Inductive process (D) Delayed usage

416. Training by feeling is:
(A) Concrete experience
(B) Reflective observation
(C) Abstract conceptualization
(D) Active experimentation

417. The number of trainees attending an open learning programme will be:
(A) Very small (B) Small
(C) Medium (D) Large

418. Which type of training centre is Krishi Vigyan Kendra:
(A) Trainers Training Centre
(B) Extension Workers Training Centre
(C) Vocational Training Centre
(D) Leadership Training Centre

419. The apex body to offer training for extension functionaries:
(A) MANAGE (B) NAARM
(C) NCAP (D) NIRD

420. The normal pace of delivery in the lecture method is:
(A) 120 words per minutes (B) 160 words per minutes
(C) 200 words per minutes (D) 240 words per minutes

421. The behavior domain that assumes significance in lecture presentation:
(A) Cognitive (B) Effective
(C) Connative (D) All of these

422. Printed or types sheet of paper used for quick communication before lecture:
(A) Handout (B) Lead letter
(C) Circular letter (D) Notice

423. Question following inductive reasoning:
(A) Divergent question (B) Rhetorical question
(C) Lead question (D) Convergent question

424. The stage in group development characterized by uncertainty:
(A) Forming (B) Storming
(C) Norming (D) Performing

425. Use of psychological tools in training is otherwise known as:
(A) Instrumental exercise (B) Simulation exercise
(C) Sensitivity exercise (D) Labortaory exercise

426. The training method best suites to all situation:
(A) Simulation method (B) Lecture method
(C) Sensitive method (D) None of these

427. Simulation method does not include:
(A) In basket exercise (B) Role play
(C) Business game (D) Brainstorming

428. Win loose game is staged to demonstrate:
(A) Intra-team collaboration
(B) Inter-team collaboration
(C) Inter-personal collaboration
(D) Intra-personal collaboration

429. Microlab exercise in a training programme is conducted at its:
(A) Beginning (B) Middle
(C) End (D) Any time

430. The purpose of microb exercise in a training programme is:
(A) Concept building (B) Rapport building
(C) Strategy building (D) Experience building

431. Win-loose game when made to play with the same group for a second time will gives:
(A) Better results (B) Average result
(C) Best result (D) No result

432. T-group training is grouped under:
(A) Transactional method (B) Sensitivity method
(C) Experiential method (D) Laboratory method

433. The training type that standardized data from real life for systematic step by step learning of certain type:
(A) Encounter method
(B) Inspiratory method
(C) Expository method
(D) Individualized method

434. Generation of data during training from real life situation is not the characteristic feature of:
(A) Individualized method (B) Simulation method
(C) Laboratory method (D) Field method

435. The co-trainer concept of training is very much related to:
(A) T group (B) L group
(C) O group (D) Q group

436. The training method that is liked most by people with divergent learning style:
(A) Role play (B) Lecture
(C) Clinical study (D) Library review

437. "Handbook of Structural Experience" – a collection of training methods is published by:
(A) Pfeiffer and Jones (B) David G Reay
(C) D Kolb (D) Lynton and Pareek

438. The training method that gives more importance to cognitive domain:
(A) Lecture (B) Role play
(C) Brainstorming (D) Workshop

439. The stage of brainstorming in which free flow of ideas takes place:
(A) Red stage (B) Green stage
(C) Yellow stage (D) Orange stage

440. Evaluation of ideas in brainstorming takes place in:
(A) Red stage (B) Green stage
(C) Yellow stage (D) Orange stage

441. Brainstorming method was developed by:
(A) Osgood (B) Gorden
(C) Andrews (D) Kolb

442. The training method best suits to large audience:
(A) Discussion method (B) Simulation method
(C) Sensitivity method (D) Transactional method

443. A programmed instruction takes participants:
(A) Outside social context
(B) Nearer to the participants
(C) Nearer to the trainer
(D) Both (A) and (B)

444. Computer based training is classified under:
(A) Expository method
(B) Encounter method
(C) Discovery method
(D) Individualized method

445. The training method that work on the principle of empathy:
(A) Role play (B) Case study
(C) Transactional analysis (D) Sensitivity method

446. Transactional analysis is classified under:
(A) Laboratory method
(B) Simulation method
(C) Discussion method
(D) Individualized method

447. Sensitivity training includes:
(A) T group (B) L group
(C) Process work (D) All of these

448. Sensitivity training is a:
(A) Self-disclosure method
(B) Trainer dominated method
(C) Process work method
(D) Both (A) and (C)

449. The essential difference between games and exercise in a training session rest on:
(A) Games can be repeated
(B) Exercise can be repeated
(C) Games do not have rule
(D) None of these

450. "Showing method" among the following is:
(A) Lecture (B) Seminar
(C) Brainstorming (D) Demonstration

451. The number of participants is minimum in:
(A) Group discussion (B) Huddle method
(C) Buzz session (D) Symposium method

452. Training method that samples real life is:
(A) Incidence and Case
(B) Role play and Case
(C) Incidence and Craftmanship
(D) Study tours and Case

453. Counseling as a training method comes under:
(A) Simulation method (B) Laboratory method
(C) Individualized method (D) Discussion method

454. The training method that emphasis on problem diagnosis:
(A) Case method (B) Role play
(C) Group discussion (D) Project method

455. The focus of sensitivity training essentially is on:
(A) Blind self (B) Unknown self
(C) Open self (D) Hidden self

456. The training method that gets its name from a particular behavior of honey bee:
(A) Brainstorming (B) Huddle method
(C) Buzz method (D) Role play

457. The training method that does not represent on the job training:
(A) Craftmanship (B) Process work
(C) Apprenticeship (D) Both (A) and (C)

458. Syndicate is essentially a:
(A) Telling method (B) Doing method
(C) Showing method (D) Reading method

459. Reciprocal communication between trainees is most effective in:
(A) Lecture (B) Role play
(C) Group discussion (D) Seminar

460. Role play wherein an individual plays his own role is called:
(A) Socio-drama (B) Culturo-drama
(C) Psycho-drama (D) Arto-drama

461. T group training with trainers from different discipline:
(A) Stranger group (B) Family group
(C) Cousin group (D) Informal group

462. Buzz session is normally conducted for:
(A) 5-10 minutes (B) 15-20 minutes
(C) 25-30 minutes (D) 55-60 minutes

463. Computer based training make use of:
(A) Digital image (B) Analogue image
(C) Iconic image (D) Enactive image

464. The loss of productive days is on a minimum for:
(A) Laboratory training (B) Field training
(C) Simulated training (D) Library training

465. Trainer role in brainstorming session is:
(A) Stimulator (B) Facilitator
(C) Concepualiser (D) Controller

466. Data in role play method is:
(A) Self generated (B) Brought from outside
(C) Standardized (D) General

467. Behaviour conditioning in interactive demonstration takes place in:
(A) Cognitive phase (B) Fixation phase
(C) Autonomous phase (D) Association phase

468. 'Covedale' methodology of training is the characteristic methodology followed in:
(A) IARI (B) MANAGE
(C) IIM (D) SIETI

469. To make people to reach precision level the type of training method best suits is:
(A) Operational training (B) Behavioural training
(C) Sensitivity training (D) Applied training

470. Data is very abstract and general in:
(A) Lecture (B) Business game
(C) Case (D) Brainstorming

471. Trainer-trainee ratio is maximum in:
(A) Syndicate (B) Brainstorming
(C) Case (D) Lecture

472. Trainer-trainee ratio is minimum in:
(A) Syndicate (B) Brainstorming
(C) Role play (D) Lecture

473. The training setup under Training and Visit system can be considered as:
(A) Short arm long finger (B) Long arm short finger
(C) Short arm short finger (D) Long arm long finger

474. Training programme for VLWs in training and visit system is organized on a:
(A) Weekly basis (B) Fortnightly basis
(C) Monthly basis (D) Half yearly basis

475. Interaction and involvement of participants is in the highest degree in:
(A) Field method (B) Case method
(C) Laboratory method (D) Lecture method

476. Feedback from the participants is the highest in:
(A) Instrumental method (B) Discussion method
(C) Simulation method (D) Field method

477. Very intensive learning is a characteristic feature of:
(A) Field method (B) Simulation method
(C) Laboratory method (D) Instructional method

478. Cost of training is the highest for:
(A) Field training (B) Laboratory training
(C) Simulation training (D) Instructional training

479. Role play can be considered as a:
(A) Projective method (B) Simulation method
(C) Encounter method (D) All of these

480. "Doing" method among the following is:
(A) Seminar (B) Brainstorming
(C) Movie (D) Role play

481. The Lynton and Pareek classification of training method is basically on:
(A) Involvement of trainee (B) Cognitive emphasis
(C) Sensory capacity (D) Involvement of trainer

482. The "Showing, Doing, Telling' classification of training methods was done by:
(A) Lynton and Pareek (B) D Kolb
(C) D G Reay (D) Pfeiffer and Jones

483. The tower building exercise was developed by:
(A) Alfred York (B) Danieel Stufflebeam
(C) Robert Stake (D) Michael Scriven

484. The concept of microlab was first used in:
(A) USA (B) Canada
(C) Japan (D) India

485. When trainers for the T-group are selected from the same department:
(A) Cousin group (B) Father group
(C) Family group (D) Stranger group

486. When trainers in the T-group are selected from only social scientists:
(A) Cousin group (B) Father group
(C) Family group (D) Stranger group

487. A role-play wherein a person directs a blind fold man to perform best represents:
(A) Role rotation (B) Role reversal
(C) Role multiplication (D) Role simplification

488. The three dimentional classification of training method by Pareek and Rao is not based on:
(A) Cognitive emphasis (B) Learner activity
(C) Sensory perception (D) Use of experience

489. To remove the afternoon sleep from the participants, we use:
(A) Ice breaks (B) Microlab
(C) Climate setting (D) All of these

490. Experiential games and exercise are conducted during:
(A) Beginning of the session (B) Middle of the session
(C) End of the session (D) Both (A) and (C)

491. Hypnosis as a training method is classified under:
(A) Expository method
(B) Individualized method
(C) Controlled exposure method
(D) Behaviour control method

492. Entrepreneurship development training does not contain:
(A) Stimulus phase (B) Supplementary phase
(C) Sustenance phase (D) Support phase

493. Low learner activity and high cognitive emphasis is the characteristic feature of:
(A) Encounter method (B) Discovery method
(C) Expository method (D) Enactive method

494. Use of experience, learner activity and cognitive emphasis is high in:
(A) Discovery method (B) Enactive method
(C) Encounter method (D) Inspiratory method

495. The learner activity and use of experimentation in encounter method are:
(A) High-Low (B) Low-High
(C) High-High (D) Low-Low

496. For large group we can adopt:
(A) Buzz session
(D) Brainstorming
(C) Debate
(D) Symposium

497. The adult ego state in transitional analysis best represent:
(A) Though concept of life
(B) Taught concept of life
(C) Felt concept of life
(D) Both (A) and (B)

498. The concept of transitional analysis was first proposed by:
(A) Eric Berne
(B) David G Reay
(C) Pfeiffer and Ballow
(D) Lynton and Pareek

499. The number of human birth in transitional analysis is:
(A) One
(B) Two
(C) Three
(D) Four

500. The child ego state in transitional analysis best represent:
(A) Thought concept of life
(B) Taught concept of life
(C) Felt concept of life
(D) Both (A) and (B)

501. A child who holds his cry when his mother beats is a:
(A) Normal child
(B) Adaptive child
(C) Creative child
(D) Dominant child

502. The part of personality of a trainee that is known only to him:
(A) Blind self
(B) Unknown self
(C) Open self
(D) Hidden self

503. The basic life position that denotes over dominance:
(A) I am OK You are OK
(B) I am OK You are not OK
(C) I and You are not OK
(D) I am not OK but You are OK

504. Collection of bad and negative stamps in transactional analysis is otherwise called:
(A) Gold stamps
(B) Silver stamps
(C) Black stamps
(D) Grey stamps

505. The basic life scripts are recorded before as age of:
(A) One
(B) three
(C) Five
(D) Ten

506. The unspoken psychological flow of communication that runs in parallel:
(A) Transactions
(B) Perception
(C) Empathy
(D) Opinion

507. A state in which people behave, feel and think in response to what is going on in the 'here-and-now', using all of their resources as an adult human being with many years of life experience to guide them:
(A) Neopsychic
(B) Exteropsychic
(C) Archaepsychic
(D) None of these

508. The modified form of panel discussion in which three or more resource persons discuss a specific topic is called:
(A) Symposium
(B) Colloquium
(C) Workshop
(D) Brainstorming

509. Pooling of experience and opinion among group of people who have special qualifications in the area is called:
(A) Symposium
(B) Conference
(C) Workshop
(D) Brainstorming

510. A discussion method, usually more than one day in duration and giving emphasis on lectures, individual conferences with emphasis on small working groups:
(A) Conference
(B) Panel discussion
(C) Symposium
(D) Workshop

511. In which of the following training methods do the trainees expose their idiosyncrasies to their fellow-trainees who rub them off through mutual interaction:
(A) Syndicate method
(B) Case study method
(C) Lecture method
(D) T-Group method

512. The process through which a trainer communicate to a group of participants through the spoken words supplemented by audio-visual aids and discussion is:
(A) Group discussion
(B) Lecture
(C) Brainstorming
(D) Workshop

513. The training type wherein speakers covers sub-divisions of a specific topic for a definite period of time:
(A) Seminar
(B) Symposium
(C) Panel discussion
(D) Lecture

514. The method of verbal presentation on a topic by a speaker to a group of audience:
(A) Lecture
(B) Seminar
(C) Panel discussion
(D) Workshop

515. A meeting in which a small group of resource persons present short prepared papers on a given topic:
(A) Colloquium
(B) Seminar
(C) Buzz session
(D) Symposium

516. Which of the following directly measures competence?
(A) Personality tests
(B) Paper pencil tests
(C) Standardized tests
(D) Performance tests

517. Motivation is meant to:
(A) Create interest in the task
(B) Reduce anxiety
(C) Enhance task related anxiety
(D) Develop personality

518. Fill in the blank:
In asking higher order questions, teacher must ask —————— questions.
(A) Closed ended
(B) Fact based
(C) Concept based
(D) Convergent type

519. Pickout the odd one:
(A) Development proceeds from centre to periphery
(B) Development proceeds from head to tail
(C) Childhood reflects the future adult's personality
(D) Development of mathematical concepts starts with number value

520. Which practice in questioning brings about more classroom interaction?
(A) Asking questions before calling on a student
(B) Focussing on divergent questions
(C) Focussing on convergent questions
(D) Asking questions after calling on a student

521. Greatest weakness of intelligence tests is that they:
(A) Are not reliable
(B) Are not fair to persons of different background
(C) Do not provide motivation to different task takers
(D) Measure intelligence rather than ability

ANSWERSHEET: 19.RESEARCH METHODOLOGY

1. (B); 2. (B); 3. (A); 4. (B); 5. (D); 6. (A); 7. (C); 8. (C); 9. (A); 10. (A) ; 11. (B); 12. (D) ; 13. (A) ; 14. (B); 15. (A); 16. (A); 17. (A); 18. (D); 19. (C); 20. (C); 21. (D); 22. (C); 23. (B); 24. (D); 25. (A) ; 26. (D); 27. (A); 28. (B); 29. (A); 30. (B); 31. (B); 32. (D); 33. (D); 34. (D); 35. (D); 36. (D); 37. (D); 38. (D) ; 39. (A); 40. (D); 41. (C); 42. (B); 43. (A); 44. (A); 45. (D); 46. (B); 47. (D); 48. (C); 49. (B); 50. (D); 51. (D); 52. (B); 53. (C); 54. (D); 55. (B); 56. (B); 57. (D); 58. (C); 59. (B); 60. (C); 61. (D); 62. (D); 63. (B); 64. (A) ; 65. (C); 66. (B); 67. (D); 68. (B); 69. (C); 70. (A); 71. (C); 72. (D); 73. (D); 74. (C); 75. (D); 76. (D); 77. (B) ;78. (A) ; 79. (B); 80. (B); 81. (D); 82. (D); 83. (B); 84. (B); 85. (B); 86. (A); 87. (B); 88. (A); 89. (B); 90. (C) ; 91. (A) ; 92. (B); 93. (A); 94. (B); 95. (B); 96. (D); 97. (D); 98. (A); 99. (A); 100. (A); 101. (A) ; 102. (C); 103. (D); 104. (B); 105. (D); 106. (B); 107. (B); 108. (D); 109. (A); 110. (D); 111. (C); 112. (C); 113. (D); 114. (B); 115. (B); 116. (C); 117. (D); 118. (C); 119. (A); 120. (A); 121. (D); 122. (C); 123. (C); 124. (C); 125. (A); 126. (A); 127. (D); 128. (A); 129. (B); 130. (B); 131. (B); 132. (B); 133. (B); 134. (C); 135. (D); 136. (B); 137. (C); 138. (D); 139. (C); 140. (B); 141. (C); 142. (B); 143. (C); 144. (D); 145. (A); 146. (B); 147. (B); 148. (C); 149. (A); 150. (D); 151. (D); 152. (B); 153. (D); 154. (B); 155. (A); 156. (B); 157. (B); 158. (A); 159. (A); 160. (B); 161. (D); 162. (B); 163. (D); 164. (C); 165. (C); 166. (B); 167. (B); 168. (B); 169. (D); 170. (A); 171. (A); 172. (B); 173. (C); 174. (B); 175. (C); 176. (A); 177. (C); 178. (B); 179. (D); 180. (B); 181. (B) 182. (C) 183. (C) ; 184. (C); 185. (C); 186. (C); 187. (A); 188. (A); 189. (D); 190. (C); 191. (D); 192. (C); 193. (B); 194. (A); 195. (B); 196. (C); 197. (D); 198. (C); 199. (B); 200. (A); 201. (D); 202. (C); 203. (D); 204. (C); 205. (C); 206. (D); 207. (C); 208. (B); 209. (C); 210. (C); 211. (D); 212. (D); 213. (A); 214. (B); 215. (A); 216. (B); 217. (B); 218. (A); 219. (A); 220. (C); 221. (A); 222. (A); 223. (A); 224. (D); 225. (D); 226. (A); 227. (B); 228. (B); 229. (A); 230. (C); 231. (B); 232. (A); 233. (A); 234. (A); 235. (C); 236. (B); 237. (A); 238. (C); 239. (A); 240. (D); 241. (B); 242. (B); 243. (B); 244. (A); 245. (D); 246. (A); 247. (B); 248. (C); 249. (A); 250. (A); 251. (D); 252. (A); 253. (D); 254. (C); 255. (D); 256. (B); 257. (B); 258. (C); 259. (B); 260. (C); 261. (C); 262. (D); 263. (A); 264. (C); 265. (B); 266. (C) ; 267. (A); 268. (B); 269. (C); 270. (C); 271. (A); 272. (B); 273. (B) ; 274. (A); 275. (D); 276. (D); 277. (A); 278. (C); 279. (B); 280. (B); 281. (A); 282. (C); 283. (A); 284. (A); 285. (C); 286. (B); 287. (B); 288. (C); 289. (A); 290. (D) ; 291. (B); 292. (A); 293. (B); 294. (D); 295. (D); 296. (D); 297. (D); 298. (D); 299. (D); 300. (D); 301. (A); 302. (D); 303. (D); 304. (B); 305. (D); 306. (D); 307. (D); 308. (B); 309. (B); 310. (D); 311. (B); 312. (C); 313. (C); 314. (D); 315. (D); 316. (B); 317. (D); 318. (D); 319. (A); 320. (C); 321. (B); 322. (C); 323. (B); 324. (A); 325. (C); 326. (D); 327. (B); 328. (B); 329. (B); 330. (A); 331. (D); 332. (B); 333. (C); 334. (B); 335. (A); 336. (A); 337. (A); 338. (D); 339. (A); 340. (B); 341. (C); 342. (D); 343. (A); 344. (A); 345. (C); 346. (B); 347. (C); 348. (B); 349. (D); 350. (B); 351. (A); 352. (C); 353. (C); 354. (A); 355. (A); 356. (A); 357. (A); 358. (C); 359. (D); 360. (B); 361. (A); 362. (C); 363. (D); 364. (C); 365. (D); 366. (D); 367. (A); 368. (B); 369. (C); 370. (A); 371. (A); 372. (B); 373. (C); 374. (A); 375. (B); 376. (A); 377. (A); 378. (A); 379. (D); 380. (D); 381. (D); 382. (C); 383. (C); 384. (B); 385. (A); 386. (B); 387. (C); 388. (D); 389. (D); 390. (B); 391. (A); 392. (B); 393. (C); 394. (A); 395. (A); 396. (A); 397. (A); 398. (D); 399. (C); 400. (B); 401. (A); 402. (C); 403. (A); 404. (D); 405. (A); 406. (A); 407. (A); 408. (C); 409. (D); 410. (D); 411. (C); 412. (A); 413. (A); 414. (C); 415. (D); 416. (A); 417. (D); 418. (B); 419. (A); 420. (A); 421. (A);422. (A); 423. (A); 424. (A); 425. (A); 426. (D); 427. (D); 428. (B); 429. (A); 430. (B); 431. (D); 432. (B); 433. (D); 434. (A); 435. (A); 436. (A); 437. (A); 438. (A); 439. (A); 440. (B); 441. (A); 442. (A); 443. (A); 444. (D); 445. (A); 446. (A); 447. (D); 448. (D); 449. (B); 450. (D); 451. (C); 452. (A); 453. (C); 454. (A); 455. (B); 456. (C); 457. (B); 458. (B); 459. (B); 460. (C); 461. (A); 462. (A); 463. (A); 464. (B); 465. (B); 466. (A); 467. (B); 468. (B); 469. (D); 470. (A); 471. (D); 472. (B); 473. (B); 474. (B); 475. (C); 476. (A); 477. (C); 478. (B); 479. (D); 480. (D); 481. (A); 482. (C); 483. (A); 484. (A); 485. (C); 486. (A); 487. (A); 488. (C); 489. (A); 490. (A); 491. (D); 492. (B); 493. (C); 494. (A); 495. (C); 496. (A); 497. (A); 498. (A); 499. (D); 500. (C); 501. (A); 502. (D); 503. (B); 504. (D); 505. (C); 506. (A); 507. (A); 508. (B); 509. (B); 510. (D); 511. (D); 512. (B); 513. (B); 514. (A); 515. (D); 516. (D); 517. (A); 518. (C); 519. (C); 520. (B); 521. (D).

Chapter 20

AGROFORESTRY

1. Growing trees along with agricultural crops is known as:
(A) Silviculture (B) Horticulture
(C) Social forestry (D) Agroforestry

2. The sustainable land use systems involving trees combined with crops and/or animals on the same unit of land is known as:
(A) Agroforestry (B) Silviculture
(C) Social forestry (D) Horticulture

3. The practice of forestry in and around the farms/village land is called:
(A) Social forestry (B) Community forestry
(C) Farm forestry (D) All of these

4. Alley cropping means:
(A) Growing of pastures in between two widely spaced rows of fast growing trees
(B) Growing of field crops in between two widely spaced rows of fast growing trees
(C) Growing of only short-duration crops in between two widely spaced rows of fast growing trees
(D) Growing of only fodder crops in between two widely spaced rows of fast growing trees

5. Alley cropping is also known as:
(A) Hedge row intercropping
(B) Avenue cropping
(C) Both (A) and (B)
(D) Woody hedge

6. The type of forests in which Teak plants grow is:
(A) Alpine (B) Deciduous
(C) Evergreen (D) Scrub jungle

7. The process of cultivating wild species to suit human needs is called:
(A) Hybridization (B) Acclimatization
(C) Selection (D) Domestication

8. The main criteria for good agroforestry design are:
(A) Productivity (B) Sustainability
(C) Adaptability (D) All of these

9. Based on the nature of components, Nair (1985) classified the agroforestry system into how many groups:
(A) Four (B) Five
(C) Six (D) Seven

10. The practice of intensive exploitation of land capability by regular harvesting of such trees which produce numerous shoots from the stump after cutting is called:
(A) Coppice farming (B) Cover crop
(C) Crop competition (D) Crop canopy

11. A concept involving the farming of fast growing plants/trees for the purpose of providing biomass that can be used directly as fuel or converted in to other forms of fuel or energy products is called:
(A) Energy farming (B) Organic farming
(C) Diversified farming (D) Energy productivity ratio

12. A type of agro-forestry system in which trees are planted with food crops which serve as a nursery for the young tree seedlings is called:
(A) Taungya cultivation (B) Shifting cultivation
(C) Both (A) and (B) (D) None of these

13. The art of growing and reproducing a forest is called:
(A) Horticulture (B) Silviculture
(C) Alley cropping (D) Agroforestry

14. The land management system in which agriculture and forest crops are grown in combined form on same unit of the land is called:
(A) Conservation farming system
(B) Ley farming system
(C) Alley cropping system
(D) Agroforestry system

15. The scientific basis for the practice of silviculture is:
(A) Silvics (B) Forest management
(C) Forest biology (D) None of these

16. In India, common types of forest is:
(A) Tropical thorn forests
(B) Sal and Teak forests
(C) Tropical most deciduous forest
(D) Tropical dry deciduous forest

17. Rotation of arable crops with two or more years of sown pasture is called as:
(A) Mixed farming (B) Mixed cropping
(C) Land use systems (D) Lay farming

18. Integration of trees with grasses is known as:
(A) Aqua forestry (B) Silvipasture
(C) Apiculture (D) Sericulture

19. The practice of managing rows of closely planted woody plants with annual crops planted in alleys in between hedges is called:
(A) Hedge row intercropping
(B) Woody hedgerows
(C) Conservation hedges
(D) Multipurpose tree gardens

20. The deliberated use of land for the production of crops including trees and fruit tree is called:
(A) Agri silvohorticulture (B) Agripasture
(C) Agrisilvipasture (D) Agrisilviculture

21. Integration of trees with pasture grasses and crops is called:
(A) Agri horticulture (B) Agripasture
(C) Agrisilvipasture (D) Agrisilviculture

22. Integration of horticultural tree with different types of crops is called:
(A) Agri horticulture (B) Agripasture
(C) Agrisilvipasture (D) Agrisilviculture

23. Integration of pasture grasses with crops is called:
(A) Agri horticulture (B) Agripasture
(C) Agrisilvipasture (D) Agrisilviculture

24. The practice of forestry for the simultaneous use of a forest area for two or more purpose is called:
(A) Multiple use forestry
(B) Mixed use forestry
(C) Multipurpose trees and shrubs
(D) Multispecies tree garden

25. The practice of forestry with the object of obtaining maximum gain per unit area is called:
(A) Mixed forestry (B) Multiple use forestry
(C) Intensive forestry (D) All of these

26. Home gardens found in tropics are one of the classic example of:
(A) Agrisilvicultural systems
(B) Silvopastoral systems
(C) Agrosilvopastoral systems
(D) Apiculture with trees

27. Hedge row intercropping is:
(A) Agrisilvicultural systems
(B) Silvopastoral systems
(C) Agrosilvopastoral systems
(D) None of these

28. Most common agroforestry system practiced in arid regions are:
(A) Agrisilvicultural systems
(B) Silvopastoral systems
(C) Agrisilvopastoral systems
(D) Aquaforestry

29. Which forestry practice is not included in the scope of social forestry:
(A) Farm forestry (B) Extension forestry
(C) Recreational forestry (D) Protection forestry

30. Which social forestry programme is purely managed by private efforts:
(A) Farm forestry (B) Community forestry
(C) Urban forestry (D) Silvopastoral

31. Shifting cultivation is most prevalent in:
(A) North-east India (B) Eastern part of India
(C) Northern part of India (D) Western part of India

32. In north-eastern hill region, shifting cultivation is called:
(A) Jhum (B) Podu
(C) Kumari (D) Slash and burn

33. Jhuming is the other name for:
(A) Shifting cultivation (B) Taungya
(C) Both (A) and (B) (D) None of these

34. Cutting and burning of forests or woodlands to create field for agriculture or pasture for livestock, or for a variety of other purposes is called:
(A) Industrial agriculture
(B) Slash and Burn agriculture
(C) Bio-dynamic agriculture
(D) Community supported agriculture

35. The scientific management of forests for the continued output of products is called:
(A) Dendrology (B) Forestry
(C) Ecology (D) Forest mensuration

36. Hoe, das and digging sticks are used in which of the following farming types?
(A) Subsistence farming
(B) Intensive subsistence farming
(C) Shifting cultivation
(D) Commercial farming

37. Practice of raising forest or fruit trees on the society land to reduce atmospheric pollution is known as:
(A) Farm forestry (B) Silvipasture
(C) Agroforestry (D) Social forestry

38. Lay farming refers:
(A) Growing of field crops in rotation with grasses and fodder legumes
(B) Growing trees in rotation with grasses and fodder legumes
(C) Growing of pastures in rotation with grasses and fodder legumes
(D) Growing of grasses and fodder legumes in forest areas

39. Extensive fallow systems are:
(A) Shifting cultivation system
(B) Bush fallow or grass fallow systems
(C) Fallow system
(D) None of these

40. Intensive fallow systems are:
(A) Shifting cultivation system
(B) Bush fallow or grass fallow systems
(C) Fallow system
(D) None of these

41. A rotation of arable crops requiring annual cultivation and artificial pastures occupying field for two years or longer is called:
(A) Ley farming (B) Alley cropping
(C) Silvi pastoral system (D) None of these

42. The practice of cultivating clearings scattered in the reservoir of natural vegetation and of abandoning them as soon as the soil is exhausted; and this includes the practice of shifting homesteads in order to fallow the cultivator's search for new fertile land is called:
(A) Row intercropping (B) Shifting cultivation
(C) Strip cropping (D) None of these

43. A type of agro-forestry system in which trees are planted with food crops which serve as a nursery for the young tree seedlings is called:
(A) Taungya cultivation (B) Shifting cultivation
(C) Both (A) and (B) (D) None of these

44. The practice of forestry with the object of raising flowering trees is called:
(A) Recreation forestry (B) Urban forestry
(C) Social forestry (D) Farm forestry

45. Plantations grow as linear strips of lands on the sides of public roads, canals and railway lines are called:
(A) Linear strip plantations (B) Commercial forestry
(C) Avenue plantations (D) Urban forestry

46. In which agroforestry system, agricultural crops is of secondary importance?
(A) Silvipasture (B) Agrisilviculture
(C) Silviagriculture (D) Hortisilviculture

47. Tree planted in silvi-agriculture should be oriented in the:
(A) East-west direction (B) North-south direction
(C) East-south direction (D) West-south direction

48. Silvipasture means:
(A) Growing trees with grass
(B) Growing trees with annual crops
(C) Growing trees with fruit crops
(D) Growing trees with crops and horticultural crops

49. Agri-silviculture means:
(A) Growing trees with grass
(B) Growing trees with annual crops
(C) Growing trees with fruit crops
(D) Growing trees with crops and horticultural crops

50. Silvi-horticulture means:
(A) Growing trees with grass
(B) Growing trees with annual crops
(C) Growing trees with fruit crops
(D) Growing trees with crops and horticultural crops

51. Agri-silvi-horticulture means:
(A) Growing trees with grass
(B) Growing trees with annual crops
(C) Growing trees with fruit crops
(D) Growing trees with crops and horticultural crops

52. Agri -silvipasture means:
(A) Growing trees with grass and crops
(B) Growing trees with annual crops
(C) Growing trees with fruit crops
(D) Growing trees with crops and horticultural crops

53. The most prevalent agroforestry systems in high rainfall areas:
(A) Silvipasture (B) Silviagriculture
(C) Horti-silvi-pasture (D) Home gardens

54. The most prevalent agroforestry systems in low rainfall (arid) regions is:
(A) Agrisilvicultural system (B) Silvopastoral system
(C) Agrosilvopastoral system(D) All of these

55. In which agroforestry system, agricultural crops is of secondary importance?
(A) Silvipasture (B) Agrisilviculture
(C) Silviagriculture (D)Hortisilviculture

56. Identification of trees according to species is called:
(A) Forestry (B) Silviculture
(C) Ecology (D) Dendrology

57. Central Board of Forestry was constituted in the year:
(A) 1947 (B) 1948
(C) 1952 (D) 1988

58. Who initiated Van Mahotsava?
(A) K.M. Munshi (B) Mahatma Gandhi
(C) J.L. Nehru (D) A.K. Benejee

59. Van Mahotsava is a festival of:
(A) Wild life (B) Tree plantation
(C) Timbers (D) All of these

60. Van Mahotsava was started in:
(A) 1847 (B) 1850
(C) 1947 (D) 1950

61. National Wasteland Development Board came into existence during the period:
(A) 1970 (B) 1975
(C) 1980 (D) 1985

62. The India, plantation forestry was started in:
(A) 1842 (B) 1944
(C) 1947 (D) 1950

63. Joint forest management evolved in early:
(A) 1950 (B) 1960
(C) 1970 (D) 1980

64. ICAR had launch All India Co-ordinated Research Project on Agroforestry in the country in:
(A) 1974 (B) 1979
(C) 1981 (D) 1983

65. First Indian Forest Act was drafted in the year:
(A) 1865 (B) 1848
(C) 1894 (D) 1927

66. The last forest policy was formulated during the year:
(A) 1998 (B) 1972
(C) 1988 (D) 1952

67. Forest conservation act was enacted during:
(A) 1880 (B) 1980
(C) 1988 (D) 1998

68. The first independent National Forest Policy was enunciated during:
(A) 1950 (B) 1952
(C) 1954 (D) 1956

69. The first Forest Policy was enunciated during:
(A) 1964 (B) 1874
(C) 1884 (D) 1894

70. Declaration of national parks is covered in which section of wildlife protection act, 1972?
(A) Section 25 (B) Section 35
(C) Section 36 (D) Section 37

71. Conservation of ecological fragile ecosystem and preservation of biological diversity in terms of flora and fauna was emphasized in:
(A) Forest policy, 1894 (B) Forest policy, 1952
(C) Forest policy, 1988 (D) None of these

72. Wildlife protection act was enacted during:
(A) 1975 (B) 1980
(C) 1972 (D) 1995

73. The project tiger was launched in the year:
(A) 1972 (B) 1995
(C) 1973 (D) 1980

74. India's first Forest Policy was enunciated during:
(A) 1894 (B) 1927
(C) 1952 (D) 1988

75. Wildlife protection act of 1972 was amended during the year:
(A) 1988 (B) 1990
(C) 1991 (D) 1980

76. In Wildlife Protection Act, 1972, NilgiriTahr comes under:
(A) Schedule I Part II (B) Schedule I Part I
(C) Schedule II Part II (D) Schedule III

77. World Forest Day is celebrated on:
(A) 15th June (B) 1st July
(C) 1st October (D) 21st March

78. The Environment (Protection) Act is attributed to the year:
(A) 1972 (B) 1980
(C) 1986 (D) 1988

79. The Indian Board of wildlife was constituted in the year:
(A) 1952 (B) 1972
(C) 1966 (D) 1994

80. The forests are brought on concurrent list by 42nd amendment to the constitution in the year:
(A) 1972 (B) 1976
(C) 1980 (D) 1988

81. Forest law is:
(A) A general law (B) A special law
(C) An international law (D) A constitutional law

82. Man And Biosphere (MAB) programmes was started during:
(A) 1971 (B) 1980
(C) 1982 (D) 1990

83. Sub chapter 6.2 of the Indian Forest Act, 1927 deals with the:
(A) Protected forests
(B) Penalities and procedures
(C) Duty on timber and other forest produce
(D) Reserved forests

84. National Forest Policy, 1988 envisages that ——— % of the total geographical area of India should be under forest:
(A) 22 (B) 33
(C) 40 (D) 40

85. Forest and wildlife were brought under concurrent list as per:
(A) 40th amendment (B) 42nd amendment
(C) 44th amendment (D) 48th amendment

86. Indian Grassland and Fodder Research Institute is located at:
(A) Jhansi (B) Kanpur
(C) Jaipur (D) Bhopal

87. National Research Centre for Agroforestry is located at:
(A) Jhansi (B) Kanpur
(C) Jaipur (D) Bhopal

88. ICRAF stands for:
(A) International Council for Research in Agro Forestry
(B) Indian Council for Research in Agro Forestry
(C) Indian Centre for Rainfall Assessment in Forest
(D) International Centre for Rainfall Assessment in Forest

89. Centre for International Forestry Research (CIFOR) is in:
(A) Nairobi, Kenya (B) Vienna, Austria
(C) Bogor, Indonesia (D) Ibadan, Nigeria

90. International Centre for Research on Agro Forestry is located at:
(A) Nairobi, Kenya (B) Vienna, Austria
(C) Bogor, Indonesia (D) Ibadan, Nigeria

91. Tropical Forest Research Institute is located at:
(A) Kanpur (B) Jabalpur
(C) Jhansi (D) Jodhpur

92. Forest Research Institute is located at:
(A) Jodhpur (B) Jhansi
(C) Dehradun (D) New Delhi

93. ICFRE is located at:
(A) New Delhi (B) Dehradun
(C) Jhansi (D) Ibanan

94. The product of form factor and total height of tree:
(A) Form - quotient (B) Form – class
(C) Tree - height (D) Form - height

95. Institute of Tropical Forestry is situated at:
(A) Jodhpur (B) Dehradun
(C) Puerto Rica (D) Kenya

96. National Research Institute for Arid Horticulture is situated at:
(A) New Delhi (B) Jodhpur
(C) Jhansi (D) Dehradun

97. Central Arid Zone Research Institute is located at:
(A) Jodhpur (B) New Delhi
(C) Jhansi (D) Dehradun

98. National Institute on Subtropical Horticulture is located at:
(A) New Delhi (B) Jodhpur
(C) Jhansi (D) Lucknow

99. Deforestation brings about:
(A) Soil erosion (B) Weed control
(C) Decreases drought (D) Increased sunlight

100. Deforestation is caused due to:
(A) Silviculture (B) Conservation of roads
(C) Rainfall (D) Plantation of trees

101. Deforestation is the major causal agent of:
(A) Depletion of natural resources
(B) Environmental pollution
(C) Desertification of habitat
(D) Genetic erosion

102. Deforestation will decrease:
(A) Soil erosion (B) Land slides
(C) Soil fertility (D) Rainfall

103. When the use of land is directed towards satisfying basic needs and is managed by the owner is said to be:
(A) Commercial Agroforestry systems
(B) Intermediate Agroforestry systems
(C) Subsistence Agroforestry systems
(D) All of these

104. The practice of forestry in areas devoid of tree growth and other vegetation outsides the conventional forest areas is:
(A) Social forestry (B) Extension forestry
(C) Farm forestry (D) Afforestation

105. Homegarden agroforestry system is practices in:
(A) Humid tropical climates (B) Semi arid lands
(C) High lands (D) Sub humid low lands

106. A forest which has been reproduced through sprouting is called:
(A) Epicormic (B) Stocked
(C) Coppice (D) All of these

107. The forest products can be broadly classified into:
(A) Major Timber Products
(B) Non Timber Forest Products
(C) Both (A) and (B)
(D) Secondary Timber Products

108. Food, fuel, fodder and fibre obtained from the forests is its:
(A) Protective role (B) Productive role
(C) Ameliorative role (D) Developmental role

109. The closeness of the tree tops is called:
(A) Closed canopy (B) Open canopy
(C) Crown density (D) Carrying capacity

110. A method of harvesting that removes all of the trees in the stand is called:
(A) Clearing (B) Clear cutting
(C) Total cutting (D) Total removal

111. A cutting which removes trees of poor quality, undesirable species or those in poor condition is called:
(A) Improvement cutting (B) Upgrading
(C) Thinning (D) Defect clearing

112. *Darbergia sisso* is:
(A) Frost hardy
(B) Moderately frost hardy
(C) Frost tender
(D) None of these

113. *Shorea robusta* is:
(A) Susceptible to drought
(B) Drought hardy
(C) Moderately drought hardy
(D) None of these

114. Deciduous trees are:
(A) *Melia azedarach* (B) *Acacia catechu*
(C) *Ailanthus excelsa* (D) All of these

115. Evergreen trees are:
(A) *Abies pindrow* (B) *Michelia chamoaca*
(C) *Hopea parviflora* (D) All of these

116. Light demanders trees for moist temperate forest are:
(A) *Pinus wallichiana* (B) *Populus ciliate*
(C) Both (A) and (B) (D) None of these

117. Shade demander trees for tropical moist deciduous forest are:
(A) *Syzygium cumini* (B) *Schleichera oleosa*
(C) *Mallotus philinensis* (D) All of these

118. Shade bearer trees for tropical wet evergreen forest are:
(A) *Artocarpus spp.* (B) *Dipterocarpus* spp.
(C) *Calophyllum* (D) *Mesua ferrea*

119. Frost tender tree species are:
(A) *Azadiracta indica* (B) *Tectona grandis*
(C) *Terminalia arjuna* (D) All of these

120. Frost hardy tree species are:
(A) *Acacia catechu* (B) *Anogeissus pendula*
(C) *Toona ciliate* (D) All of these

121. Drought hardy tree species are:
(A) *Acacia arabica*
(B) *Diospyros melanoxylon*
(C) *Syzygium cumini*
(D) All of these

122. The commercial bidi leaves are plucked from:
(A) *Diospyros melanoxylon* (B) *Shorea robusta*
(C) *Pterosperma* spp (D) *Gmelina arborea*

123. Indigenous populous tree species is/are:
(A) *Populus ciliate* (B) *Populus robusta*
(C) *Populus deltoides* (D) *Populus rubrapoiret*

124. Agar wood tree is/are:
(A) *Aquilaria malaccensis* (B) *Teactona grandis*
(C) *Dalbergia sissoo* (D) *Acacia catechu*

125. The silvicultural stem in which regeneration fellings are distributed over the whole felling series and the crop is always irregular is known as:
(A) Systems of diffused regeneration
(B) Accessory system
(C) System of concentrated regeneration
(D) Coppice system

126. Most suitable silvicultural system of deodar in hilly terrain for control of soil erosion is:
(A) Selection system (B) Clear felling
(C) Coppice system (D) All of these

127. Extensive planting of trees to increase forest cover is called:
(A) Deforestation (B) Agroforestry
(C) Social forestry (D) Afforestation

128. Forests control drought by:
(A) Preventing soil erosion
(B) Increasing oxygen
(C) Increasing humidity and rainfall
(D) Preventing floods

129. The silvicultural system in which regeneration fellings instead of being done uniformly all over the compartment are carried out in scattered group is defined as:
(A) Uniform systems (B) Group systems
(D) Selection systems (D) Shelter wood strip systems

130. Silvicultural systems in which clear felling are done in the form of strips which progress successively in one direction is known as:
(A) Alternate strip system
(B) Shelterwood strip system
(C) Clear strip system
(D) Clear-felling with standards

131. Which of the silvicultural system is used for the regeneration and management of high forests:
(A) Clear felling (B) Shelter wood
(C) Selection felling (D) All of these

132. The silvicultural system followed for black wattle (*Acacia mearnsii*) in the Nilgiris is:
(A) Clear felling followed by natural regeneration
(B) Clear felling followed by artificial regeneration
(C) Uniform system
(D) Simple coppice system

133. Silvicultural systems prevalent during early part of 20th century was:
(A) Shelter wood system (B) Clear felling systems
(C) Selection systems (D) Coppice systems

134. Shading effect of trees can be overcome by:
(A) Canopy management (B) Root management
(C) Water (D) Fertilizer

135. Multistory arrangement reduces:
(A) Productivity (B) Nutrition
(C) Competition (D) All of these

136. The oldest known agroforestry systems is:
(A) Alley cropping (B) Taungya systems
(C) Home gardens (D) Shifting cultivation

137. Social forestry programme was launched in the:
(A) Fifth five year plan (B) Sixth five year plan
(C) Seventh five year plan (D) Eight five year plan

138. Carrying capacity is more related to:
(A) Agrisilviculture
(B) Improvement of grasslands
(C) Improvement of nutritive value of grasslands
(D) None of these

139. Social forestry was first coined by:
(A) Singh & Sharma (B) Nair
(C) Brandis (D) Westoby

140. Social forestry was first adopted successfully in:
(A) Gujarat (B) Karnataka
(C) Uttar Pradesh (D) Rajasthan

141. Neem and mango are preferred for:
(A) Railway line plantation (B) Canal bank plantation
(C) Tank bed plantation (D) Avenue plantation

142. TCIP stands:
(A) Tree Cultivation Inducement Programme
(B) Tree Cultivation Intensive Programme
(C) Tree Cultivation Incentive Programme
(D) Tree Cultivation Income Programme

143. Component of social forestry is:
(A) Farm forestry (B) Extension forestry
(C) Recreational forestry (D) All of these

144. *Prosopis cineraria* and *Cenchrus ciliaris* is a good fodder tree with grass combination for:
(A) Semi-arid climate with red soil
(B) Temperate climate
(C) Arid climate with sandy soil
(D) Semi-arid black cotton soil

145. Tree competes with the neighbouring crops mainly for:
(A) Light (B) Moisture
(C) Nutrients (D) All of these

146. The fundamental attributes of all agroforestry systems are:
(A) Recover depleted soil nutrients
(B) Increase crop yields
(C) Production and protection
(D) Stabilize the soil structure

147. Planting of nectar producing trees on the boundary of the agricultural field is called:
(A) Aquaforestry
(B) Apiculture with sericulture
(C) Multipurpose wood lots
(D) Woody hedgerows

148. Agroforestry D&D covers entire ecological zone within a country is:
(A) Macro D&D (B) Micro D&D
(C) Both (A) and (B) (D) None of these

149. Agroforestry D&D covers within the chosen ecological zone within a country is:
(A) Macro D&D (B) Micro D&D
(C) Both (A) and (B) (D) None of these

150. Which is false about shelter belt:
(A) Single tree species
(B) Multiple tree species
(C) Used to arrest wind
(D) Prevent sand dune formation

151. The ratio of height and width in shelter belt is:
(A) 1:10 (B) 1:15
(C) 1:20 (D) 1:25

152. The cross sectional shape of any shelter belt is:
(A) Parallel (B) Neloid
(C) Cylinder (D) Pyramid

153. Taungya was first introduced by:
(A) Nair (B) Brandis
(C) Khanna (D) Clements

154. Taungya system was originated in:
(A) North Bengal (B) West Bengal
(C) Myanmar (D) China

155. Taungya is a word derived from:
(A) Nepali (B) Chinese
(C) Burmese (D) Bengali

156. In southern India, the taungya system is also called:
(A) Jhum (B) Kumari
(C) Podu (D) None of these

157. First taungya plantation was established in:
(A) West Bengal (B) North Bengal
(C) Rajasthan (D) Uttar Pradesh

158. The main aim of raising crops in departmental taungya is to:
(A) Reduce the weed growth
(B) To provide additional income
(C) Maximum utilization of site
(D) All of these

159. The queen of timber is:
(A) Shisham (B) Vagai
(C) Teak (D) Eucalyptus

160. Loyal timber of the poor is:
(A) Acacia (B) Prosopis
(C) Bamboo (D) Teak

161. Shisham is also called:
(A) Rose wood (B) Soft wood
(C) Hard wood (D) None of these

162. Energy plantation is established for:
(A) Pulp wood production (B) Fuel wood production
(C) Soil conservation (D) Wind break

163. Based on temporal arrangement of agroforestry system, 'home garden' is the example of:
(A) Coincident (B) Concurrent
(C) Interpoleted (D) Alley cropping

164. Home gardening is highly suitable for:
(A) Humid/sub humid region
(B) Arid/sub arid region
(C) Tropical region
(D) Temperate region

165. The home garden is characterized by:
(A) 2-3 vertical strata (B) 3-4 vertical strata
(C) 3-5 vertical strata (D) 5-6 vertical strata

166. In the manufacture of Braile paper which tree species is used:
(A) *Eucalyptus globules*
(B) *Eucalyptus tereticornis*
(C) *Eucalyptus camandulensis*
(D) *Eucalyptus urophylla*

167. *Indigofera tinctoria* is used for:
(A) Rasin industry (B) Dye industry
(C) Tannin industry (D) Gum industry

168. The major tree species are most suitable for planting on tank bed is:
(A) *Acacia leucophloea* (B) *Prosopis juliflora*
(C) *Acacia nilotica* (D) *Eucalyptus* spp.

169. Heliotropic type of branching in trees should be preferred for:
(A) Roadside planting (B) Canal planting
(C) Railway planting (D) Wasteland planting

170. Which type of trees are suitable for noise abatement?
(A) Deciduous trees (B) Evergreen trees
(C) Semi evergreen trees (D) All of these

171. Cutting of large trees which throw shade on the young trees is called:
(A) Thinning (B) Selection cutting
(C) Shade tree cutting (D) Release cutting

172. Removing trees damaged by fire, insects or disease is called:
(A) Salvage cut (B) Clean cutting
(C) Clear cut (D) Sanitation cut

173. Most suitable tree species for reclamation of saline-sodic soils is:
(A) *Prosopis juliflora*
(B) *Albizia lebbeck*
(C) *Azadiracta indica*
(D) *Casuarina equisetifolia*

174. Most suitable tree species for reclamation of sandy soils is:
(A) *Acacia species* (B) *Casuarina equisetifolia*
(C) *Melia dubia* (D) *Azadiracta indica*

175. "Block wood" is obtained from:
(A) Acacia (B) Pongamia
(C) Dalbergia (D) Albizia

176. As the ability to regenerate by shoot or root suckers, or a forest so established is called:
(A) Coppice farming (B) Cover crop
(C) Crop competition (D) Crop canopy

177. Generally shelterbelt assumes the shape of:
(A) Quadrangle (B) Rectangle
(C) Triangle (D) trapezoidal

178. Characteristics of multipurpose tree suitable for agroforestry is:
(A) High adaptability (B) Nitrogen fixing
(C) Coppicing ability (D) All of these

179. Basic set of elements or components of agroforestry are:
(A) Tree (B) Herbs
(C) Animals (D) All of these

180. The National Afforestation programme is being implemented through a which tier structure:
(A) 2 (B) 3
(C) 4 (D) 5

181. Wind breaks are:
(A) Strips of trees (B) Line of trees
(C) Both (A) and (B) (D) None of these

182. The average size of tree patta land could be:
(A) 5.0 (B) 4.0
(C) 3.0 (D) 2.0

183. Which is false about classification of agroforestry based on arrangement of components in respect of time:
(A) Coincident (B) Mixed dense
(C) Concomitant (D) Interploted

184. The tree species having xerophytic characteristics are suitable for:
(A) Humid Bengal Assam Basin
(B) Arid zones
(C) Sub humid areas
(D) Humid to semi arid areas

185. In North India agroforestry the combination of which crop is common:
(A) Rice with poplar (B) Rice with eucalyptus
(C Rice with melia (D) Rice with shisham

186. The combination of wheat with eucalyptus was common in:
(A) Uttar Pradesh (B) Punjab
(C) Maharashtra (D) Bihar

187. The important tree species for shape in tea estates is:
(A) *Grevillea robusta* (B) *Eucalyptus* spp.
(C) *Melia dubia* (D) All of these

188. Which tree species is preferred for water logged areas:
(A) Coconut (B) Eucalyptus
(C) Neem (D) Casuarina spp.

189. Agroforestry practice of growing *Albizia lebbek* with sorghum is called:
(A) Xeromorphic system (B) Hydromorphic system
(C) Mesophytic system (D) None of these

190. Which is not correct about an advantages of agroforestry system?
(A) Maximum production
(B) Improve soil fertility
(C) Maximize adverse impact on climatic factors
(D) Aid industrial growth

191. In order to get maximum net return from agroforestry system, we have to:
(A) Choose suitable tree species
(B) Choose high yielding agricultural crops
(C) Plant clonally propagated tree seedlings
(D) Adopt suitable tree crop combination

192. Which is false about shelterbelt:
(A) Single species
(B) Multiple tree species
(C) Used to arrest wind
(D) Prevalent sand dune formation

193. Which is false about planting of forest trees in agroforestry system:
(A) Reduces temperature
(B) Reduces evaporation
(C) Increase irrigation requirement
(D) Increase humidity

194. The important practices in silvicultural operation in agroforestry is:
(A) Pruning (B) Thinning
(C) Weeding (D) Climber cutting

195. Protein bank is a sub system of:
(A) Agri -silviculture (B) Agri - pastoral
(C) Agri -horticulture (D) Agri-silvi-pastoral

196. Which tree species is called green gold is?
(A) Sal (B) Sandal
(C) Bamboo (D) Teak

197. The system in which woody component is combined with agricultural crop is known as:
(A) Agri -silviculture (B) Agri - pastoral
(C) Agri -horticulture (D) Agri-silvi-pastoral

198. Essential attributes of agroforestry systems are:
(A) Adaptability (B) Productivity
(C) Sustainability (D) All of these

199. Shelterbelt should posses a typical:
(A) Triangular shape (B) Linear belt
(C) Square shape (D) Rectangular shape

200. A common example of the zonal pattern is:
(A) Hedgerow intercropping
(B) Home gardens
(C) Silvopastoral systems
(D) Agrosilvopastoral system

201. Commercial agroforestry system is used:
(A) Production of perennial cash crops
(B) Satisfying basic needs
(C) Sale of the production of output
(D) All of these

202. Shading effect of trees can be overcome by:
(A) Canopy management (B) Root management
(C) Water management (D) Nutrient management

203. Which classification of agroforestry is more important and commonly adopted?
(A) Socioeconomic (B) Structural
(C) Floristic (D) Ecological

204. The broad and well known classification of agroforestry systems was given by:
(A) Chundawat (B) Gautam
(C) Nair (D) Sharma

205. The term social forestry first time used by:
(A) Shah (B) Brandis
(C) Nair (D) Jack Westoby

206. Social forestry was first recommended by:
(A) National Commission on Agriculture
(B) Van Mahotsava
(C) Ninth Commonwealth Forestry Congress
(D) All of these

207. The first state in India who created a separate social forestry wing in the forest department was:
(A) Gujarat (B) Maharashtra
(C) Tamilnadu (D) Kerala

208. Social Forestry Project Phase – I was launched in:
(A) 1981 (B) 1985
(C) 1988 (D) 1992

209. Social Forestry Project Phase – II was launched in:
(A) 1981 (B) 1985
(C) 1988 (D) 1992

210. Major thrust to social forestry concept was given by:
(A) NCA report
(B) Forest Conservation Report
(C) National Forest Policy
(D) Indian Forest Act

211. Which is not a factor in which the agroforestry systems are classified?
(A) Floristic (B) Function
(C) Agro climatic zones (D) None of these

212. The Diagnosis and Design (D&D) survey in agroforestry was given by:
(A) P.K.R. Nair (B) Dwivedi
(C) Sharma (D) Singh

213. Forest products contribution to world gross domestic product (GDP) is:
(A) 1.0 % (B) 1.7 %
(C) 2.0 % (D) 2.7 %

214. When an immediate supply of a single necessary factor falls below the combined demand of the plant is called:
(A) Allelopathy (B) Allelospoly
(C) Both (A) and (B) (D) None of these

215. The direct or indirect of one plant upon another through the production of chemical inhibitors is called:
(A) Allelopathy (B) Allelospoly
(C) Both (A) and (B) (D) None of these

216. Allelopathy denotes:
(A) Positive interaction (B) Negative interaction
(C) Both (A) and (B) (D) None of these

217. Allelopathy is the best example for:
(A) Association (B) Ammensalism
(C) Symbiosis (D) Photo cooperation

218. Allelochemicals produced by plants are:
(A) Primary plant metabolites
(B) Secondary plant metabolites
(C) Both (A) and (B)
(D) None of these

219. Allelochemicals are released by means of:
(A) Volatilization (B) Leaching
(C) Exudation (D) All of these

220. Examples of allelochemicals:
(A) Gallic acid (B) Vanilic acid
(C) Syringic acid (D) All of these

221. Strips of trees planted to protect fields, homes, canal from wind is referred to:
(A) Shelter belts (B) Wind breaks
(C) Alley cropping (D) All of these

222. Characteristics of multipurpose tree (MPTs) suitable for agroforestry is:
(A) High adaptability (B) Nitrogen fixing
(C) Coppicing ability (D) All of these

223. Suitable tree species for MPTs are:
(A) Casuarina (B) Acacia
(C) Gliricidia (D) All of these

224. Suitable tree species for nitrogen fixing tree (NFTs) is/are:
(A) Teak (B) Leucaena
(C) Neem (D) All of these

225. Suitable herbs for live fencing is:
(A) Euphorbia (B) Gliricidia
(C) Acacia (D) Sesbania

226. Nitrogen fixing ability of Casuarina is attributed to:
(A) Azotobacter (B) Frankia
(C) Rhizobium (D) Azospirilum

227. Benefit sharing concept is a core objective in
(A) Joint Forest Management
(B) Sustainable Forest Management
(C) Community Forestry
(D) Social Forestry

228. Joint Forest Management concept was first implemented at:
(A) Ahmedabad (B) Arabari
(C) Mumbai (D) Pune

229. In Joint Forest Management programme, the unit of management is:
(A) Village (B) Block
(C) Taluka (D) District

230. In Joint Forest Management programme, the objectives are to be decided at:
(A) Division level (B) District level
(C) Taluka level (D) Village level

231. Protection and reforestation of degraded forests with the involvement of local people is:
(A) Joint Forest Management
(B) Sustainable Forest Management
(C) Community Forestry
(D) Social Forestry

232. Silviculture systems followed for Bamboo is:
(A) Clear felling systems (B) Shelter wood systems
(C) Group selection systems (D) Culm selection system

233. Silviculture system for road side plantation is:
(A) Simple coppice systems (B) Shelter wood systems
(C) Clear strip systems (D) None of these

234. Chipko movement is an example of forest conservation through:
(A) Tehri-Garhwal district
(B) Uttarakhand
(C) Public awareness and participation
(D) Political issue

235. Chipko movement is related to:
(A) Forest conservation (B) Soil conservation
(C) Water conservation (D) Wetland conservation

236. Chipko movement was launched for protection of:
(A) Forests (B) Grasslands
(C) Wetlands (D) Live stocks

237. Conservation of species in its natural habitat is:
(A) In-situ (B) Ex-situ
(C) In-vitro (D) Both (B) and (C)

238. Development of botanical garden is what type of conservation:
(A) Ex-situ (B) A common
(C) An easy (D) In-situ

239. An agroforestry system suitable for small orchards prone to wind damage is:
(A) Taungya (B) Home garden
(C) Windbreak (D) Shifting cultivation

240. A well designed windbreak can offer protection upto how many times of its height towards leeward side:
(A) 10 (B) 20
(C) 30 (D) 40

241. Swidden is a synonym for one of the following agroforestry system. Identify the correct one:
(A) Taungya (B) Home garden
(C) Perkland system (D) Shifting cultivation

242. The suitable intercrop in *Populus deltoids* based agrisilvicultural system is:
(A) Wheat (B) Rice
(C) Maize (D) Ginger

243. Indian lilac is the common name for:
(A) *Azadiracta indica* (B) *Tectonia grandis*
(C) *Casuarina equisetifolia* (D) *Acacia nilotica*

244. In Kerala the most important agroforestry systems is:
(A) Alley cropping (B) Home garden
(C) Lay farming (D) Shifting cultivation

245. Kangeyam region in Tamilnadu, the famous silvipastoral system is:
(A) *Cenchrus ciliaris* and *Acacia nilotica*
(B) *Cenchrus ciliaris* and *Acacia leucophlea*
(C) *Eulaliopsis binata* and *Acacia nilotica*
(D) *Eulaliopsis binata* and *Acacia leucophlea*

246. "The Journal of Agroforestry Today" is published from:
(A) Kenya (B) Denmark
(C) Japan (D) Canada

247. "Rangeland and Agroforestry" published from:
(A) New Delhi (B) Jhansi
(C) Jodhpur (D) Dehradun

248. Which among tree species are best suited for alley cropping?
(A) *Eucalyptus tereticornis*
(B) *Casuarina equisetifolia*
(C) *Cassia siamea*
(D) *Ailanthus excels*

249. Which among these factors has degraded the Indian grasslands the most?
(A) Adverse site factors (B) Poor management
(C) Poor productivity (D) Over grazing

250. The literary meaning of agroforestry is:
(A) Agri + Forest (B) Agri + Horti
(C) Agri + Fish (D) Agri + Honey bee

251. The literary meaning of agri-horticulture is:
(A) Agri + Forest (B) Agri + Horti
(C) Agri + Fish (D) Agri + Honey bee

252. The literary meaning of agri-apiculture is:
(A) Agri + Forest (B) Agri + Horti
(C) Agri + Fish (D) Agri + Honey bee

253. The literary meaning of agri -pissiculture is:
(A) Crops + Forest (B) Crops + Horti
(C) Crops + Fish (D) Crops + Honey bee

254. The literary meaning of silvi-pastoral is:
(A) Forest + Pasture (B) Forest + Horti
(C) Forest + fish (D) Forest + Honey bee

255. The literary meaning of silvi- horticulture is:
(A) Forest + Pasture (B) Forest + Horti
(C) Forest + fish (D) Forest + Honey bee

256. The literary meaning of silvi- pissi culture is:
(A) Forest + Pasture (B) Forest + Horti
(C) Forest + fish (D) Forest + Honey bee

257. The literary meaning of silvi-apiculture is:
(A) Forest + Pasture (B) Forest + Horti
(C) Forest + fish (D) Forest + Honey bee

258. The literary meaning of agri-silvi-pissiculture system is:
(A) Crops + Forest + Pasture
(B) Crops + Forest + Horti
(C) Crops + Forest + fish
(D) Crops + Forest + Honey bee

259. The literary meaning of agri-silvi-pastoral system is:
(A) Crops + Forest + Pasture
(B) Crops + Forest + Horti
(C) Crops + Forest + fish
(D) Crops + Forest + Honey bee

260. The literary meaning of agri-silvi-horticulture system is:
(A) Crops + Forest + Pasture
(B) Crops + Forest + Horti
(C) Crops + Forest + fish
(D) Crops + Forest + Honey bee

261. The literary meaning of agri-silvi-apiculture system is:
(A) Crops + Forest + Pasture
(B) Crops + Forest + Horti
(C) Crops + Forest + fish
(D) Crops + Forest + Honey bee

262. The literary meaning of alley cropping system is:
(A) Crops + Herb tree
(B) Trees around house + Vegetables
(C) Crops + Legume grasses + Grasses
(D) Crops + Forest + Honey bee

263. The literary meaning of home gardening system is:
(A) Crops + Herb tree
(B) Trees around house + Vegetables
(C) Crops + Legume grasses + Grasses
(D) Crops + Forest + Honey bee

264. The literary meaning of lay farming system is:
(A) Crops + Herb tree
(B) Trees around house + Vegetables
(C) Crops + Legume grasses + Grasses
(D) Crops + Forest + Honey bee

265. The literary meaning of homestead agroforestry system is:
(A) Crops + Herb tree
(B) Trees around house + Vegetables
(C) Crops + Legume grasses + Grasses
(D) Crops + Forest + Honey bee

266. The literary meaning of timfib system is:
(A) Crops + Herb tree
(B) Trees around house + Vegetables
(C) Crops + Legume grasses + Grasses
(D) Timber tree + fibre tree

267. The branch of forestry which deals with the determination of dimension, form and age and increment of single trees, stands or whole woods either standing or after feeling is called:
(A) Dendrology (B) Forest management
(C) Dendrometry (D) Forest protection

268. Management of forests for multiple uses involves which of the following uses:
(A) Timber (B) Recreation
(C) Water (D) All of these

269. The practical application of scientific, technical and economical principles of forestry is called:
(A) Forest management (B) Forest mensuration
(C) Forest economics (D) Forest legislation

270. Silviculture rotation is applicable to:
(A) Aesthetic (B) Recreational purpose
(C) Both (A) and (B) (D) None of these

271. Which type of rotation is useful for aesthetic and recreational purpose?
(A) Silvicultural rotation
(B) Physical rotation
(C) Technical rotation
(D) Rotation of maximum volume production

272. Which rotations controlling the financial returns?
(A) Silvicultural (B) Physical rotation
(C) Both (A) and (B) (D) None of these

273. Which type of rotation is most favourable for natural reproduction?
(A) Silvicultural rotation (B) Physical rotation
(C) Technical rotation (D) Financial rotation

274. Which social forestry practice is not included in the scope of social forestry?
(A) Farm forestry (B) Extension forestry
(C) Recreational forestry (D) Protection forestry

275. Social forestry include:
(A) Farm forestry (B) Extension forestry
(C) Recreational forestry (D) All of these

276. Which is false about type of rotation?
(A) Regular (B) Silvicultural
(C) Technical (D) Financial

277. The rotation in which a species retains satisfactory vigour of growth and reproduction on a given site is:
(A) Silvicultural rotation
(B) Financial rotation
(C) Technical rotation
(D) Rotation with maximum volume production

278. The rotation followed for sandal wood in south India is:
(A) Silvicultural rotation (B) Financial rotation
(C) Technical rotation (D) Physical rotation

279. Rotation of roadside and avenue trees is:
(A) Silvicultural rotation
(B) Financial rotation
(C) Physical rotation
(D) Rotation with maximum volume production

280. Tropical dry evergreen forests are found in:
(A) Carnatic coast
(B) Andaman and Nicobar island
(C) Coastal areas of Kerala
(D) North east India

281. Most suitable silvicultural system of deodar in hilly terrain for control of soil erosion is:
(A) Selection system (B) Clear felling
(C) Coppice system (D) All of these

282. The silviculture of India trees was written by:
(A) L.S. Khanna (B) M.L. Chaudhary
(C) R.S. Singh (D) Benerjee

283. Mimocin is present in:
(A) Teak (B) Leucaena
(C) Neem (D) Acacia

284. Eucalyptus is a:
(A) Bad coppicer
(B) Moderately good coppicer
(C) Good coppicer
(D) Very bad coppicer

285. Non leguminous nitrogen fixing tree is:
(A) *Alunus nepalensis*
(B) *Leucaena leucocephala*
(C) *Acacia nilotica*
(D) *Calliandra callothyrsus*

286. A silviculture prescription:
(A) Changes in forest density
(B) Is a type of forest inventory
(C) Plans for pesticide application to control insects
(D) Usually contains cultural operations relating to objectives

287. A very complex community of associated trees, shrubs, other plants and animals is a/an:
(A) Arboretum (B) Forest
(C) Green house (D) Park

288. Removal of lower branches of a tree is called:
(A) Thinning (B) Cutting
(C) Prunning (D) Pinching

289. The prunned materials of sesbania in alley cropping can be utilized as:
(A) For feed
(B) For moisture conservation
(C) For organic matter
(D) All of these

290. Khejari tree in Rajasthan is used for:
(A) For feed to animals (B) For vegetable
(C) Both (A) and (B) (D) None of these

291. *Prosopis julifloa, Acacia nilotica, Casuarina equisetifolia* and *Tamerix articulate* can be successfully grown at pH of:
(A) Less than 7.0 (B) 8.2 – 9.0
(C) 9.1 – 10.0 (D) More than 10.0

292. *Terminalia arjuna, Albizzia lebbeck, Pongamia pinnata* and *Sesbania sesban* can be successfully grown at pH of:
(A) Less than 7.0 (B) 8.2 – 9.0
(C) 9.1 – 10.0 (D) More than 10.0

293. *Eucalyptus tereticornis, Parkinsonia auleata, Salvidora Persia* and *Capparis decidua* can be successfully grown at pH of:
(A) Less than 7.0 (B) 8.2 – 9.0
(C) 9.1 – 10.0 (D) More than 10.0

294. *Butea monosperma, Kijellea pinnata, Acacia auriculiformis* and *Moringa olifera* can be successfully grown at pH of:
(A) Less than 7.0 (B) 8.2 – 9.0
(C) 9.1 – 10.0 (D) More than 10.0

295. *Grevillia robusta, Azadiracta indica, Populus deltoids* and *Melia azadirach* can be successfully grown at pH of:
(A) Less than 7.0 (B) 8.2 – 9.0
(C) 9.1 – 10.0 (D) More than 10.0

296. *Acacia albida* is used for:
(A) Timber (B) Nitrogen fixation
(C) Fodder (D) All of these

297. *Albizia amara, A, lebbeck, A. Procera* are used for:
(A) Shade (B) Nitrogen fixation
(C) Fodder (D) All of these

298. *Annona squamosa* and *Annona muricata* are used for:
(A) Fruits (B) Nitrogen fixation
(C) Fodder (D) All of these

299. *Areca catchu* is used for:
(A) Fuits (B) Mulching
(C) Both (A) and (B) (D) None of these

300. *Casuarina spp* is used for:
(A) Timber (B) Nitrogen fixation
(C) Energy (D) All of these

301. *Bombax insignis* is used for:
(A) Timber (B) Fodder
(C) Both (A) and (B) (D) None of these

302. *Cassia siamea* is used for:
(A) Mulching (B) Nitrogen fixation
(C) Fodder (D) Soil conservation
(E) All of these

303. *Dalbergia sissoo* and *D. latifolia* are used for:
(A) Timber (B) Nitrogen fixation
(C) Fodder (D) All of these

304. *Gliricidia sepium* is a MPTs tree mainly used for:
(A) Timber (B) Nitrogen fixation
(C) Fodder (D) Mulching
(E) Fencing (F) All of these

305. *Genelina arborea* is used for:
(A) Timber (B) Soil amendment
(C) Fodder (D) All of these

306. *Moringa oleifera* is used for:
(A) Vegetables (B) Medicine
(C) Fodder (D) All of these

307. *Psidium guajava* is used for:
(A) Fruits (B) Nitrogen fixation
(C) Fodder (D) All of these

308. *Sesbania grandiflora* is used for:
(A) Timber (B) Nitrogen fixation
(C) Vegetables (D) All of these

309. *Mallotus phillipnensis* is used for:
(A) Timber (B) Nitrogen fixation
(C) Fodder (D) All of these

310. *Grewia optiva* a..u *C. oppositifolia* are used for:
(A) Timber (B) Soil amendment
(C) Fodder (D) All of these

311. *Leucaena leucocephala* is used for:
(A) Mulching (B) Nitrogen fixation
(C) Fodder (D) Soil conservation
(E) All of these

ANSWERSHEET: 20.AGROFORESTRY

1. (D); 2. (A); 3. (D); 4. (C); 5. (C); 6. (B); 7. (D); 8. (D); 9. (A); 10. (A); 11. (A); 12. (A); 13. (B); 14. (D); 15. (A); 16. (D); 17. (D); 18. (B); 19. (A); 20. (A); 21. (C); 22. (A); 23. (B); 24. (A); 25. (C) ; 26. (C); 27. (A); 28. (A); 29. (D); 30. (A); 31. (A); 32. (A); 33. (A); 34. (B);35. (B); 36. (C); 37. (D); 38. (A) ; 39. (A); 40. (B); 41. (A); 42.(B); 43. (A); 44. (A); 45. (A); 46. (C);47. (A); 48. (A); 49. (B); 50. (C); 51. (D) ; 52. (A); 53. (D); 54. (A); 55. (C); 56.(D); 57. (B); 58. (A); 59. (B); 60. (D); 61. (D); 62. (A); 63.(C); 64. (D) ; 65. (A); 66. (C); 67. (B); 68. (B); 69. (D); 70. (B); 71. (C); 72. (C); 73. (C); 74. (A); 75. (C); 76. (B); 77. (D); 78. (C) ; 79. (A); 80. (B); 81. (D); 82. (A); 83. (D); 84. (B); 85. (B); 86. (A); 87. (A); 88. (B); 89. (C); 90. (A); 91. (B) ; 92. (C); 93. (B); 94. (D); 95. (C); 96. (B); 97. (A); 98. (D); 99. (A); 100. (B); 101. (C); 102. (D); 103. (A); 104. (B) ; 105. (A); 106. (C); 107. (C); 108. (B); 109. (C); 110. (A); 111. (A); 112. (A); 113. (A); 114. (D); 115. (D); 116. (C); 117. (D); 118. (A); 119. (D); 120. (D); 121. (D); 122. (A); 123. (A); 124. (A); 125. (A); 126. (A); 127. (D); 128. (C); 129. (B); 130. (C) ; 131. (D); 132. (A); 133. (B); 134. (A); 135. (C); 136. (D); 137. (A); 138. (B); 139. (D); 140. (A); 141. (D); 142. (C); 143. (D); 144. (C); 145. (D); 146. (C); 147. (B); 148. (A); 149. (B); 150. (B); 151. (A); 152. (B); 153. (B); 154. (C); 155. (C); 156. (B); 157. (B); 158. (D); 159. (C); 160. (B); 161.(A); 162. (B); 163. (C); 164. (A); 165. (A); 166. (A); 167. (B); 168. (C); 169. (A) ; 170. (B); 171. (D); 172. (A); 173. (A); 174. (B); 175. (A); 176. (A); 177. (C); 178. (D); 179. (D); 180. (A); 181. (A); 182. (D) ; 183. (B); 184. (B); 185. (A); 186. (B); 187. (A); 188. (B); 189. (A); 190. (C); 191. (D); 192. (A); 193. (C); 194. (A); 195. (B) ; 196. (C); 197. (A); 198. (D); 199. (A); 200. (A); 201. (C); 202. (A); 203. (B); 204. (C); 205. (D); 206. (A); 207. (A); 208. (A) ; 209. (C); 210. (A); 211. (C); 212. (A); 213. (B); 214. (B); 215. (A); 216. (C); 217. (C); 218. (B); 219. (D); 220. (D); 221. (B) ; 222. (D); 223. (D); 224. (B); 225. (A); 226. (B); 227.(A); 228. (B); 229. (A); 230. (B); 231. (A); 232. (D); 233. (C); 234. (A) ; 235. (A); 236. (A); 237. (A); 238. (A); 239.(C); 240. (B); 241. (D); 242. (A); 243. (A); 244. (B); 245. (B); 246. (A); 247. (B) ; 248. (C); 249. (D); 250. (A); 251. (B); 252. (D); 253. (C); 254. (A); 255. (B); 256.(C); 257. (D); 258. (C); 259. (A); 260. (B) ; 261. (D); 262.(A); 263. (B); 264. (C); 265. (B); 266. (D); 267. (C); 268.(D); 269. (A); 270. (C); 271. (A); 272. (C); 273. (A) ; 274.(D); 275. (D); 276. (A); 277. (A); 278. (D); 279. (C); 280.(A); 281. (A); 282. (D); 283. (B); 284. (C); 285. (A); 286. (D); 287. (B); 288. (C); 289. (D); 290. (C); 291.(D); 292. (C); 293. (C); 294. (B); 295. (B); 296. (D); 297.(D); 298. (A); 299. (C) ; 300. (D); 301. (C); 302. (E); 303.(D); 304. (F); 305. (D); 306. (D); 307. (A); 308. (D); 309.(B); 310. (D); 311. (E).

Chapter 21

FISH CULTURE/PISICULTURE

1. Rearing of fish is called:
(A) Pisciculture (B) Sericulture
(C) Agriculture (D) Aquaculture

2. In 1937, a Russian scientist succeeded in inducing a significant number of sturgeons, Acipenser stellatus.
(A) Gerbiskii
(B) Hussain
(C) Hiralal Choudhuri
(D) Ramaswamy and Sunderaraj

3. India is the country in the world to make the technique an integral part of its piscicultural programme:
(A) First (B) Second
(C) Third (D) Fourth

4. The first attempt at hypophysation in India was made by in 1937:
(A) Hamid Khan
(B) Hussain
(C) Hiralal Choudhuri
(D) Ramaswamy and Sunderaraj

5. Hamid Khan in 1937 when he tried to induce spawning by the injection of mammalian pituitary gland:
(A) *Labeo rohita* (B) *Cirrhinus mrigala*
(C) Both (A) and (B) (D) None of these

6. The next attempt was made by Hussain in 1945 with certain hormones like 80-120 RV Prolan and Antuitrin-S into female of:
(A) *Labeo rohita* (B) *Cirrhinus mrigala*
(C) Both (A) and (B) (D) None of these

7. In 1955, Hiralal Choudhuri succeeded in inducing spawning in Esomus danricus by intraperitonal injection of pituitary extract of:
(A) *Labeo rohita* (B) *Cirrhinus mrigala*
(C) *Catla catla* (D) None of these

8. Hiralal Choudhuri also succeeded in the breeding of :
(A) *Labeo rohita*
(B) *Cirrhinus mrigala*
(C) *Catla catla*
(D) *Pseudotropius atherinoide*

9. Ramaswamy and Sunderaraj succeeded breeding in 1955 and 1956 respectively in:
(A) *Heteropneustes fossilis*
(B) *Clarias batrachus*
(C) Both (A) and (B)
(D) None of these

10. The first success in induced breeding of Indian major carps through hypophysation was achieved in 1957 by:
(A) Hamid Khan
(B) Hussain
(C) Hiralal Choudhuri and Alikunhi
(D) Ramaswamy and Sunderaraj

11. CIFE stands for:
(A) Central Institute of Food Education
(B) Central Institute of Fisheries Education
(C) Central Institute of Fashion Education
(D) None of these

12. CIFE is located at:
(A) Mumbai (B) Kolkata
(C) Bhuvneshwar (D) Cochin

13. CIFRI stands for:
(A) Central Institute of Fashion Research in India
(B) Central Inland Fashion Research Institute
(C) Central Inland Fisheries Research Institute
(D) None of these

14. CIFRI is located at:
(A) Mumbai (B) Barrackpur
(C) Bhuvneshwar (D) Cochin

15. CIFA is stands for:
(A) Central Institute of Freshwater Aquaculture
(B) Central Institute of Fashion Aquarium
(C) Central Institute of Fisheries Aquarium
(D) Central Inland Fisheries Aquaculture

16. Central Institute of Freshwater Aquaculture is located at:
(A) Mumbai (B) Barrackpur
(C) Bhuvneshwar (D) Cochin

17. CIFT is stands for:
(A) Central Inland Fisheries Technology
(B) Central Institute of Fisheries Technology
(C) Central Inland Fisheries Testing laboratory
(D) Central Institute of Fisheries Testing

18. Central Institute of Fisheries Technology is located at:
(A) Mumbai (B) Barrackpur
(C) Bhuvneshwar (D) Cochin

19. NBFGR is stands for:
(A) National Bureau of Fish Genetic Resources
(B) National Bureau of Fashion Genetic Resources
(C) National Board of Fish Genetic Resources
(D) National Board of Fashion Generation Research

20. National Bureau of Fish Genetic Resources is located at:
(A) Cochin (B) Mumbai
(C) Lucknow (D) Buvneshwar

21. Meat Products Orders came in:
(A) 1972 (B) 1973
(C) 1974 (D) 1975

22. National Meat and Poultry Processing Board is located at:
(A) Bhuvneshwar (B) Kolkata
(C) Mumbai (D) Hyderabad

23. National Meat and Poultry Processing Board was constituted in:
(A) 2007 (B) 2008
(C) 2009 (D) 2010

24. Preservation of Food Adulteration Act came in:
(A) 1953 (B) 1954
(C) 1955 (D) 1956

25. Food Safety and Standards Act came in:
(A) 2003 (B) 2004
(C) 2005 (D) 2006

26. The system works well in paddy fields fed copiously by rivers or lakes reared along with fishes. Such type of fish farming system is known as:
(A) Paddy cum fish farming
(B) Integrated fish cum poultry farming
(C) Integrated fish cum horticulture farming
(D) Azolla aquacultre

27. India has a traditional system of paddy - cum - fish culture largely practiced in the coastal states of :
(A) Kerala (B) West Bengal
(C) Both (A) and (B) (D) None of these

28. Major or minor carps are cultured in paddy fields. Such type of culture system is known as:
(A) Paddy cum air breathing fishculture
(B) Paddy cum carp culture
(C) Integrated fish cum horticulture farming
(D) Azolla aquacultre

29. Air breathing cat fish like singhi and magur are cultured in paddy fields in most rice grown areas. Such type of fish farming system is called:
(A) Paddy cum air breathing fishculture
(B) Paddy cum carp culture
(C) Integrated fish cum horticulture farming
(D) Azolla aquacultre

30. Aqua-farming which could be considerably improved through judicious use of dykes for production of vegetables and fish feed. Such type of farming system is known as:
(A) Integrated fish cum duck farming
(B) Integrated fish cum poultry farming
(C) Integrated fish cum horticulture farming
(D) Azolla aquacultre

31. Rearing of fishes alongwith biofertilization through application of heterocystous blue-green algae and related members. Such type of fish farming system is known as:
(A) Integrated fish cum duck farming
(B) Integrated fish cum poultry farming
(C) Integrated fish cum horticulture farming
(D) Azolla aquacultre

32. Rearing of fish with poultry. Such type of fish farming system is called:
(A) Integrated fish cum duck farming
(B) Integrated fish cum poultry farming
(C) Integrated fish cum horticulture farming
(D) Azolla aquacultre

33. Integrated fish-cum-duck farming is the most common practice in :
(A) China (B) America
(C) Africa (D) Australia

34. Integrated fish-cum-duck farming is the most common practice in Indian states like:
(A)West Bengal & Assam
(B) Tamilnadu & Andhra Pradesh
(C) Kerala & Bihar
(D) All of these

35. As ducks use both land and water as a habitat, their integration with the fish is to utilise the mutual benefits of a biological relationship. Such type of fish farming is known as:
(A) Integrated fish cum duck farming
(B) Integrated fish cum poultry farming
(C) Integrated fish cum horticulture farming
(D) Azolla aquacultre

36. The raising of pigs with fish by constructing pig - sties on the pond embankment or near the pond so that the pig wastes are directly drained into the pond or lifted from the pig house and applied to the pond. Such type of farming is known as:
(A) Integrated fish farming web
(B) Fish cum cattle farming
(C) Fish cum prawn farming
(D) Fish cum pig farming

37. Through a lot of work has been done on composite fish culture incorporating Indian major carps and exotic carps having different feeding habits, and a considerable production achieved, no large scale polyculture of prawns and fish has been attempted. Such type farming is known as:
(A) Integrated fish farming web
(B) Fish cum cattle farming
(C) Fish cum prawn farming
(D) Fish cum pig farming

38. Fish farming by using cattle manure has long been practiced in our country. Such type of farming is known as:
(A) Integrated fish farming web
(B) Fish cum cattle farming
(C) Fish cum prawn farming
(D) Fish cum pig farming

39. Various types of combinations of aquaculture, agriculture, animal husbandry and horticulture can constitute the:
(A) Integrated fish farming web
(B) Fish cum cattle farming
(C) Fish cum prawn farming
(D) Fish cum pig farming

40. Scientific name of big or stripped murrel or snake head fish is:
(A) *Channa straitus* (B) *Channa punctatus*
(C) *Channa marulius* (D) *Clarias batrachus*

41. Scientific name of Spotted murrel fish is:
(A) *Channa straitus* (B) *Channa punctatus*
(C) *Channa marulius* (D) *Clarias batrachus*

42. Scientific name of Giant murrel fish is:
(A) *Channa straitus* (B) *Channa punctatus*
(C) *Channa marulius* (D) *Clarias batrachus*

43. Scientific name of Magur fish is:
(A) *Channa straitus* (B) *Channa punctatus*
(C) *Channa marulius* (D) *Clarias batrachus*

44. Scientific name of Singhi fish is:
(A) *Heteropneustes fossilis* (B) *Anabas testudineus*
(C) *Channa marulius* (D) *Clarias batrachus*

45. Scientific name of Koi or climbing perch fish is:
(A) *Heteropneustes fossilis* (B) *Anabas testudineus*
(C) *Channa marulius* (D) *Clarias batrachus*

46. Which type of fish has highest demand in the markets and is also commands a higher price:
(A) *Heteropneustes fossilis* (B) *Anabas testudineus*
(C) *Channa striatus* (D) *Clarias batrachus*

47. Trout is either grown as a food fish or sport fish, are released into natural waters for sport fishermen is called as:
(A) Trout culture (B) Pen culture
(C) Cage culture (D) None of these

48. Fish possessing accessory respiratory organ that enables them to take atmospheric oxygen when required is called:
(A) Air breathing fish (B) Water breathing fish
(C) Both (A) and (B) (D) None of these

49. The rearing of aquatic organisms under controlled or semi-controlled conditions is called:
(A) Horticulture (B) Aqua culture
(C) Fish culture (D) Poly culture

50. The intermediate type of aquatic environment where freshwater mingles with marine water is called:
(A) Brackish water (B) Reverine water
(C) Fresh water (D) None of these

51. Particular area of the body of water where breeding of a fish species takes place is known as:
(A) Breeding ground (B) Spawning ground
(C) Both (A) and (B) (D)None of these

52. Part of the year when a fish species is sexually active is known as:
(A) Breeding season (B) Spawning season
(C) Both (A) and (B) (D) None of these

53. Groups of mature male and female fishes reared for breeding purpose is known as:
(A) Breeding stock (B) Rearing stock
(C) Both (A) and (B) (D) None of these

54. Sexually mature fish, ready to spawn is known as:
(A) Adult fish (B) Brood fish
(C) Both (A) and (B) (D) None of these

55. A fish larger than a fry but not of marketable size. Though not rigidly defined, fmgerling fishes are generally between 4 and 10 cm long is called:
(A) Fingerline fishes (B) Fry fishes
(C) Adult fishes (D) None of these

56. A sequence of organisms, each of which provides food for the next, from primary producers to ultimate consumers, or top carnivores is called:
(A) Breeding chain (B) Rearing chain
(C) Water chain (D) Food chain

57. The reciprocal of food conversion ratio times 100, expressed as a percentage is known as:
(A) Food conservation efficiency
(B) Food conversion ratio
(C) Both (A) and (B)
(D) None of these

58. In aquaculture, the amount of food fed divided by weight gain is known as:
(A) Food conservation efficiency
(B) Food conversion ratio
(C) Both (A) and (B)
(D) None of these

59. The lower the FCR, the ——— efficient the animal is at converting feed into new tissue is known as:
(A) More (B) Less
(C) Very less (D) Limited

60. A stage of fish next to spawn stage when yolk sac has already been absorbed and active feeding commenced. It has the external characteristics of the adult but are smaller than fingerlings is called:
(A) Fingerline fishes (B) Fry fishes
(C) Adult fishes (D) None of these

61. Number of fish alive after a specific period of time expressed as percentage of the initial number of hatchlings is called:
(A) Food conservation efficiency
(B) Food conversion ratio
(C) Survival rate
(D) None of these

62. Integated fish and animal husbandry systems evolved recently are the:
(A) Fish-cum-duck culture
(B) Fish-cum-poultry culture
(C) Fish-cum-pig culture
(D) All of these

63. The main aim of fish culture is to achieve the highest possible fish production from ponds and water resources is called:
(A) Mono fish culture (B) Composite fish culture
(C) Both (A) and (B) (D) None of these

64. The culture of a many species of fish in a pond is called:
(A) Mono fish culture (B) Composite fish culture
(C) Both (A) and (B) (D) None of these

65. The fish pond is a ecosystem:
(A) Simple (B) Complex
(C) Both (A) and (B) (D) None of these

66. The culture of a single species of fish in a pond is called:
(A) Mono fish culture (B) Composite fish culture
(C) Both (A) and (B) (D) None of these

67. If only one species is introduced into a pond, due to the same dietary habits, all the fish congregate at one place is called:
(A) Mono fish culture (B) Composite fish culture
(C) Both (A) and (B) (D) None of these

68. Composite fish culture is undoubtedly more superior over:
(A) Mono fish culture (B) Dual fish culture
(C) Both (A) and (B) (D) None of these

69. The Chinese carps are cultured together. These are silver carp - *Hypophthalamichthys molitrix*, grass carp - *Ctenopharyngodon idella* and common carp - *Cyprinus carpio* is known as:
(A) Chinese system (B) Indian system
(C) European system (D) None of these

70. These are also called as exotic fishes in India.
(A) Chinese system (B) Indian system
(C) European system (D) None of these

71. The Indian carps are cultured together and are also cultured with Chinese carps is known as:
(A) Chinese system (B) Indian system
(C) European system (D) None of these

72. These carps are rohu – *Labeo rohita*, catla - *Catla catla* and mrigal - *Crirrhina mrigala* are cultured as:
(A) Chinese system (B) Indian system
(C) European system (D) None of these

73. The main species cultured is the common carp - *Cyprinus carpio* is known as:
(A) Chinese system (B) Indian system
(C) European system (D) None of these

74. Other Chinese carps used for fish culture are : bighead carp - *Aristichthys nobilis,* mud carp - *Cirrhinus molitorella* and black carp - *Mylopharyngodon piceus*:
(A) Mono fish culture (B) Dual fish culture
(C) Brackish water culture (D) Composite fish culture

75. The predatory catfish and murrels can also be incorporated in which type of fish culture system.
(A) Mono fish culture
(B) Composite fish culture
(C) Dual fish culture
(D) Brackish water culture

76. The fringe-lipped carp and the milk fish are commonly cultured in the composite fish culture in which type of water culture system.
(A) Fresh water culture (B) Saline water culture
(C) Brackish water culture (D) Composite fish culture

77. The airbreathing fishes like murrels, catfishes and koi are also cultured together in the freshwater culture system.
(A) Fresh water culture (B) Saline water culture
(C) Brackish water culture (D) Composite fish culture

78. In India and China, polyculture is more popular unlike in the:
(A) Arab countries (B) European countries
(A) African countries (D) American countries

79. Indian major carps are more riverine in nature and these do not ordinarily breed in:
(A) Confined waters (B) Brackish water
(C) Pond water (D) Well water

80. Generally fish production increases with the increase in the number of fish stocked per unit area to a maximum and then starts:
(A) Increasing (B) Decreasing
(C) No change (D) Unknown

81. The percentage of stocking of the Catla and silver carp fishes is:
(A) 45 % (B) 30 - 35 %
(C) 15-20 % (D) 5-10 %

82. The percentage of stocking of the rohu fishes is:
(A) 45 % (B) 30 - 35 %
(C) 15-20 % (D) 5-10 %

83. The percentage of stocking of the grass carp fishes is:
(A) 45 % (B) 30 - 35 %
(C) 15-20 % (D) 5-10 %

84. The percentage of stocking of the mrigal and common carp fishes is:
(A) 45 % (B) 30 - 35 %
(C) 15-20 % (D) 5-10 %

85. How many species combination excluding grass carp, the optimal stocking ratios are catla 6(30%) : rohu 3(15%) : mrigal 5(25%) : common carp 4(20%) : silver carp 2(10%) of:
(A) 8 species (B) 5 species
(C) 4 species (D) 3 species

86. How many species combination excluding silver carp and grass carp, the optimal stocking ratios are - catla 6(30%) : rohu 3 (15%) : mrigal 6(30%) : common carp 5(25%) of:
(A) 8 species (B) 5 species
(C) 4 species (D) 3 species

87. How many species combination excluding exotic carps, the optimal ratios are - catla 4 (40%) : rohu 3 (30%) : mrigal 3 (30%) of:
(A) 8 species (B) 5 species
(C) 4 species (D) 3 species

88. How many species combination is also possible for composite fish culture, where milk fish and fringe-lipped carps are included in the culture system along with Indian and Chinese major carps of:
(A) 8 species (B) 5 species
(C) 4 species (D) 3 species

89. The milk fish is a:
(A) Reverine water fish (B) Brackish water fish
(C) Well water fish (D) Pond water fish

90. Usually the stocking ratio is catla: rohu : mrigal: common carp : silver carp : grass carp : fringe-lipped carp : milk fish are:
(A) 2 : 2 : 4 : 3 : 5 : 2 : 1 : 1 (B) 4 : 3 : 2 : 3 : 5 : 2 : 1 : 1
(C) 3 : 4 : 4 : 3 : 5 : 2 : 4 : 1 (D) 3 : 3 : 4 : 3 : 4 : 2 : 1 : 4

91. All over the world, the major cultivable fishes, especially for polyculture belong to the family of:
(A) Mrigal (B) Carp
(C) Milk fish (D) None of these

92. Harvesting of fish is generally advocated after how many years of rearing:
(A) 6 months (B) 1 year
(C) 2 years (D) 3 years

93. Shorter rearing periods may also be resorted to depending on the pond conditions and size preference in the local markets. An individual fish grows to the size of 12 months:
(A) 0.8-1kg (B) 2 kg
(C) 3kg (D) 4 kg

94. Grass carp have a faster growth rate and attains a size of weight in a year:
(A) 0.8-1kg (B) 2 kg
(C) 3kg (D)4 kg

95. The systems are the oldest and its farming is the use of the environment as the fish farm:
(A) Open system (B) Semi closed system
(C) Closed system (D) None of these

96. Water is taken from natural sources or ground water and is directed into specially designed ponds and race ways. Such type of culture is referred to as:
(A) Open system (B) Semi closed system
(C) Closed system (D) None of these

97. No water is exchanged and the water is subjected to extensive treatment. Extremely high densities of organisms may be raised under these conditions. Farmer has complete control over growing conditions of fishes. Such type of culture is referred to as:
(A) Open system (B) Semi closed system
(C) Closed system (D) None of these

98. Natural resources can be used as culture systems and organisms to be cultured are stocked in the water body:
(A) Open system (B) Semi closed system
(C) Closed system (D) None of these

99. Capital expenses are low for the open culture systems. There is less management than in the other systems:
(A) Open system (B) Semi closed system
(C) Closed system (D) None of these

100. The conditions are more natural and uncrowded in the culture environment, less time is required in monitoring the condition of the culture organisms in:
(A) Open system (B) Semi closed system
(C) Closed system (D) None of these

101. Open system of fish rearing are:
(A) Cages system (B) Long lines system
(C) Floats system (D) Rafts system
(E) Trays system (F) Clam beds system
(G) All of these

102. Semi closed system of fish rearing are:
(A) Pond Culture (B) Raceway culture
(C) Both (A) and (B) (D) None of these

103. It is the culture of fish or other organisms in a river, lake or bays by holding them in cages. The system of fish culture is:
(A) Cage culture (B) Pen culture
(C) Raft culture (D) Rack culture

104. Pens are the specially designed nylon or bamboo made enclosures constructed in a water body into which fish are released for culture. Such type of culture is referred to as:
(A) Cage culture (B) Pen culture
(C) Raft culture (D) Rack culture

105. Rafts are generally made of bamboo poles or metal rods with buoys at the top for floating in the water. These are used in the culture:
(A) Cage culture (B) Pen culture
(C) Raft culture (D) Rack culture

106. Racks are constructed in brackish water areas and inshore areas for rearing oysters, mussels, seaweeds, etc. Such types of culture is referred to as:
(A) Cage culture (B) Pen culture
(C) Raft culture (D) Rack culture

107. The majority of aquaculture throughout the world is conducted in ponds. Such types of culture is referred to as:
(A) Pond Culture (B) Raceway culture
(C) Both (A) and (B) (D) None of these

108. Earthen ponds or reinforced concrete ponds are used for culturing the fish, shrimp, prawn, etc. in:
(A) Freshwater (B) Brackish waters
(C) Both (A) and (B) (D) None of these

109. A series of earthen or cement tanks are constructed along the course of a river or stream and are used for fish culture.
(A) Pond Culture (B) Raceway culture
(C) Both (A) and (B) (D) None of these

110. The system where culture chamber that is generally long and narrow. Water enters at one end and leaves through the other end in most cases.
(A) Pond Culture (B) Raceway culture
(C) Both (A) and (B) (D) None of these

111. In India largely center round the three species of Indian major carps are:
(A) *Catla catla* (B) *Labeo rohita*
(C) *Cirrhinus mrigala* (D) All of these

112. The exotic species of fishes are:
(A) Tilapia (B) Common carp
(C) Silver carp (D) Grass carp
(E) All of these

113. No water is exchanged and the water is subjected to extensive treatment. Fish or prawn culture in water recirculation systems is good example for :
(A) Open system (B) Semi closed system
(C) Closed system (D) None of these

114. The fish seed is obtained from the sources of:
(A) Riverine (B) Hatcheries
(C) Bundhs (D) All of these

115. The collection of seed from riverine source was an age old practice. This method is strenuous and we get the mixture of wanted and unwanted fish seed. Such method of fish collection is called:
(A) Riverine (B) Hatcheries
(C) Bundhs (D) All of these

116. In recent time the best way of getting fish seed. Such method of fish collection is called:
(A) Riverine (B) Hatcheries
(C) Bundhs (D) All of these

117. To collect the fish seed by creating a natural habitat. Such method of fish collection is called:
(A) Riverine (B) Hatcheries
(C) Bundhs (D) All of these

118. The northern rivers are perennial and support commercial fisheries:
(A) Rich (B) Poor
(C) Very poor (D) Unknown

119. The peninsular rivers is both in the upper and middle reaches are support of commercial fisheries:
(A) Rich (B) Poor
(C) Very poor (D) Unknown

120. Generally shooting nets are used to collect the seed of fishes in the rivers is:
(A) Shooting net (B) Gamcha
(C) Benchi jal (D) Midnapur net

121. At the tail end of the net, there is a stitched - inring of split bamboo or cane, and to this is attached, during the operation, a receptacle, termed for collection of fish seed is:
(A) Shooting net (B) Gamcha
(C) Benchi jal (D) Midnapur net

122. To collect the seed of fishes in Bengal is used:
(A) Shooting net (B) Gamcha
(C) Benchi jal (D) Midnapur net

123. In Bengal, especially in the south-western parts, to collect the seed of fishes is used:
(A) Shooting net (B) Gamcha
(C) Benchi jal (D) Midnapur net

124. The collection of seed of fishes from which source was an age old practice:
(A) Riverine (B) Hatcheries
(C) Bundhs (D) All of these

125. The Brazilians, thus, were the first to use the technique of fish breeding successfully through
(A) Hypophysation (B) Induced breeding
(C) In breeding (D) Cross breeding

126. Different types of carp hatcheries are:
(A) Earthen hatching pits
(B) Earthen pot hatcheries
(C) Cement hatching pits
(D) Hatching hapas
(E) All of these

127. This is the oldest method adopted for hatching. The percentage of hatching is about 40%:
(A) Earthen hatching pits
(B) Earthen pot hatcheries
(C) Cement hatching pits
(D) Hatching hapas

128. This is the method of hatching of fishes. The percentage of hatching is 30-50%:
(A) Earthen hatching pits
(B) Earthen pot hatcheries
(C) Cement hatching pits
(D) Hatching hapas

129. The method of hatchery is most extensively used. The hapa is fixed in the water with the help of bamboo poles in shallow waters:
(A) Earthen hatching pits
(B) Earthen pot hatcheries
(C) Cement hatching pits
(D) Double Hatching hapas

130. The percentage of survival rate of which hatchery system is about 30-40%:
(A) Earthen hatching pits
(B) Earthen pot hatcheries
(C) Cement hatching pits
(D) Hatching hapas

131. The percentage of survival rate of which hatchery system is about 40%:
(A) Earthen hatching pits
(B) Earthen pot hatcheries
(C) Cement hatching pits
(D) Hatching hapas

132. The percentage of survival rate of which hatchery system is about 30-50%:
(A) Earthen hatching pits
(B) Earthen pot hatcheries
(C) Cement hatching pits
(D) Hatching hapas

133. The percentage of survival rate of which hatchery system is about 50-60%.:
(A) Earthen hatching pits
(B) Earthen pot hatcheries
(C) Cement hatching pits
(D) Hatching hapas

134. The percentage of survival rate of which hatchery system is about 50-70%:
(A) Floating hapas
(B) Tub hatchery
(C) Cemented cisternae hatchery
(D) Vertical jar hatchery

135. Floating hapas hatchery was introduced in:
(A) Madhya Pradesh (B) West Bengal
(C) Orissa (D) Andhra Pradesh

136. The percentage of survival rate of this hatchery system is about 50-70%:
(A) Floating hapas
(B) Tub hatchery
(C) Cemented cisternae hatchery
(D) Vertical jar hatchery

137. The percentage of survival rate of which hatchery system is about 50-70%:
(A) Floating hapas
(B) Tub hatchery
(C) Cemented cisternae hatchery
(D) Vertical jar hatchery

138. The percentage of survival rate of which hatchery system is about 90 per cent:
(A) Floating hapas
(B) Tub hatchery
(C) Cemented cisternae hatchery
(D) Vertical jar hatchery

139. The percentage of survival rate of which hatchery system is about 70-80%:
(A) Plastic bin hatchery
(B) Plastic bucket hatchery
(C) Chinese hatchery
(D) D-variety Hatcheries

140. The percentage of survival rate of which hatchery system is about 90%:
(A) Plastic bin hatchery
(B) Plastic bucket hatchery
(C) Chinese hatchery
(D) D-variety Hatcheries

141. The survival rate of which hatchery system is 90 per cent:
(A) Plastic bin hatchery
(B) Plastic bucket hatchery
(C) Chinese hatchery
(D) D-variety Hatcheries

142. The survival rate of which hatchery system is 92-95%:
(A) Plastic bin hatchery
(B) Plastic bucket hatchery
(C) Chinese hatchery
(D) D-variety Hatcheries

143. Particularly the spawn from dry bundhs as this source yields of pure spawn:
(A) 100% (B) 90%
(C) 80 % (D) 70 %

144. Growing fish is a very ancient practice in:
(A) Reservoirs (B) Ponds
(C) Canals (D) Wells

145. Fish were cultured as long ago as ——————— in China:
(A) 2698 B.C. (B) 2600 B.C.
(C) 2500 B.C. (D) 2400 B.C.

146. Fish culture was done in ancient Egypt and in China, which has had a continuous civilization for over:
(A) 2000 years (B) 3000 years
(C) 3500 years (D) 4000 years

147. The fish farms in which fishes are bred to raise the :
(A) Fry (B) Fingerlings
(C) Both (A) and (B) (D) None of these

148. Water ponds are supplied by ground water, either through natural springs at their bottom or through others lying adjacent to them. The spring water is good for fish culture because it is clean and has no unwanted fish or fish eggs in it are:
(A) Spring water ponds
(B) Rain water ponds
(C) Well water ponds
(D) Flood plain ox-bow ponds

149. These are filled with rain water and the extent of their filling depends upon the amount of the rainfall:
(A) Spring water ponds
(B) Rain water ponds
(C) Well water ponds
(D) Flood plain ox-bow ponds

150. These are filled with well-water and considered very good for fish culture. They may be adequately supplied with water which has no contaminants are:
(A) Spring water ponds
(B) Rain water ponds
(C) Well water ponds
(D) Flood plain ox-bow ponds

151. Water for these ponds is supplied by the stream. These are highly productive due to the accumulation of organic materials and periodic flooding are:
(A) Spring water ponds
(B) Rain water ponds
(C) Well water ponds
(D) Flood plain ox-bow ponds

152. These ponds are usually filled by rainfall or by spring water. A spring, for example, sends water flowing through a small valley or down a slope into a low place. Or, a spring bubbles from the ground into a natural depression. The pond is formed by collecting water at the base of the valley and in the low places are:
(A) Barrage ponds (B) Diversion ponds
(C) Well water ponds (D) Head ponds

153. These ponds are made by diverting water from another source like a stream or river. Channels are dug to carry the water from the water source to the pond are:
(A) Barrage ponds (B) Diversion ponds
(C) Well water ponds (D) Head ponds

154. These ponds are built one after another in a string. All the ponds drain into each other and must be managed as if they were one pond. If the first pond in the series with a water inlet is full of predators which must be poisoned, all the other ponds have to be harvested and drained before the first pond can be poisoned are:
(A) Parallel system (B) Rosary system
(C) Both (A) and (B) (D) None of these

155. Each pond has its own inlet and outlet. Therefore, each pond can be managed as a separate pond:
(A) Parallel system (B) Rosary system
(C) Both (A) and (B) (D) None of these

156. This pond is usually constructed near a perennial source of water. The main purpose of the pond is to meet the water requirements of the entire farm, taking into consideration the losses through seepage, evaporation etc. called:
(A) Barrage ponds (B) Diversion ponds
(C) Well water ponds (D) Head ponds

157. These ponds are also called as spawning ponds:
(A) Hatching ponds (B) Nursery ponds
(C) Rain water ponds (D) None of these

158. These ponds are also called as sky ponds:
(A) Hatching ponds (B) Nursery ponds
(C) Rain water ponds (D) None of these

159. These are small and mostly in the form of small tanks or plastic pools, made near the spawn collection centres. Hapas are fixed in these ponds. The eggs are collected and kept in the hapas for hatching. Similar ponds are also constructed in the fish farm are:
(A) Hatching ponds (B) Nursery ponds
(C) Rearing ponds (D) Stocking ponds

160. These are also called transplantation ponds:
(A) Hatching ponds (B) Nursery ponds
(C) Rain water ponds (D) None of these

161. These are seasonal ponds and are constructed near the spawning and rearing ponds. The main object is to create a suitable condition of food availability and growth of fry because at this stage they are most susceptible to hazards like the wave action and predators called:
(A) Hatching ponds (B) Nursery ponds
(C) Rearing ponds (D) Stocking ponds

162. These should be slightly larger but not proportionally deep. These should be located near the nursery pond and their number may vary depending upon culture called :
(A) Hatching ponds (B) Nursery ponds
(C) Rearing ponds (D) Stocking ponds

163. These are the largest ponds and are more deep, with a depth of about 2-2.5 m. The size of the pond may vary from 0.2-2.0 ha., but these should preferably be 0.4-0.5 ha in size. These are rectangular in shape. The fingerlings and advance fingerlings are reared upto marketable size for about 6 months. One year old fishes may grow upto kg. or more in weight called:
(A) Hatching ponds (B) Nursery ponds
(C) Rearing ponds (D) Stocking ponds

164. Management of ponds is one of the most important aspect for successful fish culture practices. The hatchlings or spawn are reared to fry stages in small ponds called:
(A) Hatching ponds (B) Nursery ponds
(C) Rearing ponds (D) Stocking ponds

165. The predatory fishes are which cause maximum harm to spawn, and use them as food. These are:
(A) *Channa* sp. (B) *Wallago attu*
(C) *Heteropneustes fossilis* (D) *Clarias batrachus*
(E) *Anabas testudineus* (F) All of these

166. Weed fishes are small sized and uneconomic fishes, which prey on carp spawn. These are:
(A) *Salmostoma sp.*
(B) *Amblypharyngodon mola*
(C) *Barbus sp.*
(D) *Esomus danricus*
(E) All of these

167. Endrin at 0.01 ppm, dieldrin at 0.01 ppm, aldrin 0.2 ppm and nuvan at 30 ppm are useful to eradicate the forage fishes and all other fish enemies for:
(A) 1-2 months (B) One week
(C) 6 Months (D) I Year

168. Derris root powder (4 ppm) is good to eradicate forage fish from nursery pond and it is effective for:
(A) 1-2 months (B) One week
(C) 6 Months (D) I Year

169. Mahua oil cake *(Madhuca latifolia)* at 250 ppm is lethal to forage fish. It should be applied:
(A) A fortnight before stocking
(B) Two fortnight before stocking
(C) Three fortnight before stocking
(D) Four fortnight before stocking

170. Sugarcane jaggery at concentration is also lethal to the fish and its active poison is saponin:
(A) 1% (B) 2%
(C) 3 % (D) 4 %

171. Tea seed cake is lethal to fish seed at the rate of :
(A) 500 kg/ha (B) 600 kg/ha
(C) 700 kg/ha (D) 800 kg/ha

172. Application of powdered seed kernel of *Croton tiglium* is also effective against forage and predator fishes:
(A) 3-5 ppm (B) 2-6 ppm
(C) 20 ppm (D) 12 ppm

173. Powdered root of *Milletia pachycarpa* is also effective against forage and predator fishes:
(A) 3-5 ppm (B) 2-6 ppm
(C) 20 ppm (D) 12 ppm

174. Powdered seed of *Barringtonia accutangula* is also effective against forage and predator fishes:
(A) 3-5 ppm (B) 2-6 ppm
(C) 20 ppm (D) 12 ppm

175. Powdered unripe *Randia dumetorum* is also effective against forage and predator fishes:
(A) 3-5 ppm (B) 2-6 ppm
(C) 20 ppm (D) 12 ppm

176. Powdered bark of *Walsula piscidia* is also effective against forage and predator fishes:
(A) 10 ppm (B) 2-6 ppm
(C) 20 ppm (D) 12 ppm

177. Liming is most essential to maintain the:
(A) pH of water (B) Electrical conductivity
(C) SAR ratio (D) None of these

178. The water should be slightly alakaline as it is useful for the eradication of microorganisms in the pond and also to help maintain the hygienic condition of water:
(A) Lime (B) Gypsum (C) Pyrite (D) Iron

179. Which is useful to neutralise the acidic condition which will result while manuring:
(A) Lime (B) Gypsum
(C) Pyrite (D) Iron

180. When water is saline in nature the lime is applied at the rate of :
(A) 250 kg/ha (B) 500 kg/ha
(C) 750 kg/ha (D) 1000 kg/ha

181. When water is highly saline in nature the lime is applied at the rate of :
(A) 250 kg/ha (B) 500 kg/ha
(C) 750 kg/ha (D) 1000 kg/ha

182. The nursery pond has to be filled with water upto a depth of:
(A) One metre (B) Two metres
(C) Three metres (D) Four metres

183. It is essential to be done after filling the pond with water:
(A) Manuring (B) Fertilization
(C) Immunization (D) None of these

184. The main objective of manuring is production of adequate quantities of plankton, which is useful for carp seed:
(A) Medicine (B) Natural food
(C) Breeding material (D) Rearing of eggs

185. Several types of manures are available to increase the productivity of the pond. The most common , best and cheap of all the manures is:
(A) Well rotten FYM (B) Gobar gas slurry
(C) Raw cattle dung (RCD) (D) Nadep compost

186. Raw cattle dung at the rate of 10,000 kg/ha produces a good bloom of zooplankton in 10 days.
(A) 1000 kg/ha (B) 5000 kg/ha
(C) 10000 kg/ha (D) 15000 kg/ha

187. The application of 5,000 kg/ha of poultry manure also produces good amount of plankton in pond. However, it is better to find a suitable manure which produces plankton within:
(A) 3 days (B) 3-4 days
(C) 5-6 days (D) More than one week

188. A mixture of 5,000 kg/ha raw cattle dung, 250 kg/ha of single super phosphate and 250 kg/ha groundnut oil cake (GNO) has been found to yield plankton in about:
(A) 3 days (B) 3-4 days
(C) 5-6 days (D) More than one week

189. Manures are useful to fertilize the soil instead of water:
(A) Inorganic (B) Organic
(C) Both (A) and (B) (D) None of these

190. The elemental ratio of N:P is required for phytoplankton growth:
(A) 5 : 1 (B) 10:1
(C) 5 : 2 (D) 10 : 2

191. Inorganic fertilizers are usually applied in equal monthly instalments at the rate of 100-150 kg/ha/yr:
(A) 5 (B) 10 (C) 15 (D) 20

192. Notonecta, Ranatra, Cybister, Lethoceros, Nepa, Hydrometra and Belostoma are highly destructive to the:
(A) Carp seed (B) Adult fishes
(C) Both (A) and (B) (D) None of these

193. The insects can be eradicated by using:
(A) Watering (B) Manuring
(C) Hatchery (D) Oil emulsions

194. After manuring the nurseries, they should be treated with:
(A) Watering (B) Oil emulsions
(C) Hatchery (D) None of these

195. The spraying of oil emulsion is before stocking the spawn in nursery pond so as to eradicate the insects:
(A) 5-10 hours (B) 12-24 hours
(C) 2-3 days (D) 5-6 days

196. The oil emulsion are sufficient to treat one hectare of water:
(A) 60 kg of oil and 20 kg soap
(B) 80 kg of oil and 20 kg soap
(C) 60 kg of oil and 40 kg soap
(D) 80 kg of oil and 40 kg soap

197. An emulsion of which oil and 560 ml of Teepol is also useful to treat one hectare of water:
(A) 40 kg mustard oil (B) 56 kg of mustard oil
(C) 40 kg mahua oil (D) 56 kg of mahua oil

198. An emulsion can also be prepared with ________ oil and any detergent:
(A) Petrol (B) Mobil oil
(C) Diesel boiler (D) Bitumin

199. Since soap has become very costly, one effective method is to use 50 cc of Hyoxyde-10 mixed in 5 litres of water with ________ high speed diesel oil for a hectare of water:
(A) 20 litres (B) 30 litres
(C) 40 litres (D) 50 litres

200. The mixture of Herter W.P (0.6-1.0 ppm) and oil extracted from plant *Calophyllum inophyllum* is effective to insects as well as prawns like *Paleamon lamenii,* which is usually found in nurseries. A mixture of gamma isomer of benzene hexachloride and ethyl alcohol is also highly toxic to insects:
(A) 10 ppm (B) 5 ppm
(C) 2.5 ppm (D) 0.01 ppm

201. After rearing the fish seed upto fingerlings in rearing ponds, these fingerlings are reared to marketable size in:
(A) Nursery ponds (B) Stocking ponds
(C) Rearing pods (D) Hatchery

202. This stocking material is fry stage, which is reared up to fingerling stage for about 3 months. The stocking density of fry is:
(A) 0.2-0.3 millions/ha (B) 0.3-0.4 millions/ha
(C) 0.4-0.5 million/ha (D) 0.5-0.6 million/ha

203. In 15 days of nursery rearing, the spawn grows to 20-30 mm size fry. At this stage, these fry could be transferred to:
(A) Nursery ponds (B) Stocking ponds
(C) Rearing pods (D) Hatchery

204. Predatory and trash fishes of the pond are:
(A) *Notopterus* (B) *Mastacembelus*
(C) *Rosbora* (D) All of these

205. Predatory and trash fishes of the pond are:
(A) *Esomus*
(B) *Amblypharyngodon mola*
(C) *Salmostoma bacaila*
(D) All of these

206. Predatory and trash fishes of the pond are:
(A) *Ambasis* (B) *Puntius ticto*
(C) *P. conchoneus* (D) All of these

207. The fish utilize submerged weeds like in that order of preference.
(A) *Hydrilla* (B) *Najas*
(C) *Ceratophyllum* (D) *Ottelia*
(E) *Nechamandra* (F) *Vallisnaria*
(G) All of these

208. The young fish prefer smaller floating plants like:
(A) *Wolffia* (B) *Lemna*
(C) *Azolla* (D) *Spirodela.*
(E) All of these

209. The other herbivorous fish which utilize aquatic weeds are:
(A) *Pulchelluspulchellu* (B) *Oreochromis*
(C) *Etroplus* (D) All of these

210. Though an omnivore, *Cyprinus carpio* feeds well on filamentous algae like:
(A) *Pithophora* (B) *Cladophora*
(C) Both (A) and (B) (D) All of these

211. Biological control of weeds through stocking the water with weed-eating fish, such as
(A) Grass carp (B) Silver carp
(C) Common carp (D) None of these

212. Common carp, *Cyprinus carpio* and Katti, *Acrossocheilus hexagonalepsis* and ducks are also used for biological control of:
(A) Annual weeds (B) Marshy weeds
(C) Aquatic weeds (D) Perennial weeds

213. Application of biodegradable organophosphates like Fumadol, Sumithion, Baytex, Dipterex, etc. are useful to kill the insects:
(A) 0.25 to 3 ppm (B) 5-6 ppm
(C) 8-10 ppm (D) More than 100 ppm

214. Carps (Indian and exotic) are the main species cultured in:
(A) Ponds (B) Swamp
(C) Both (A) and (B) (D) Flowing river

215. Catfishes, murrels, freshwater prawns and mollusks for pearl culture are also being cultured in:
(A) Ponds (B) Swamp
(C) Both (A) and (B) (D) Flowing river

216. Magur, singhi, koi, murrel, chital and gourami are also being cultured in:
(A) Ponds (B) Swamp
(C) Both (A) and (B) (D) Flowing river

217. Air breathing fishes are:
(A) Magur (*Clarias batrachus*)
(B) Singhi (*Heteropneustes fossilis*)
(C) Koi (*Anabas testudineus*)
(D) Murrel (*Ophiocephalus (=Channa)* spp.)
(E) Chital (*Notopterus* spp).
(F) All of these

218. An exotic fishes is/are:
(A) Magur (*Clarias batrachus*)
(B) Singhi (*Heteropneustes fossilis*)
(C) Gourami (*Osphronemus gorami*)
(D) Murrel (*Ophiocephalus* spp.)

219. It is the most preferred food on which the fish will thrive best is called:
(A) Main food (B) Occasional food
(C) Eniergency food (D) Basic food

220. It has relatively high nutritive values and is liked and consumed by fish whenever the opportunity presents itself is called:
(A) Main food (B) Occasional food
(C) Eniergency food (D) Basic food

221. It is fed upon accepted when other food material is not available is called:
(A) Main food (B) Occasional food
(C) Eniergency food (D) Basic food

222. It is normally eaten by the fish and comprises most of the gut contents is called:
(A) Main food (B) Occasional food
(C) Eniergency food (D) Basic food

223. It is frequently consumed in smaller quantities by the fishesis known as:
(A) Secondary food (B) Incidental food
(C) Obligatory food (D) Basic food

224. It is consumed rarely by fishes is known as:
(A) Secondary food (B) Incidental food
(C) Obligatory food (D) Basic food

225. The fish consumes this food in the absence of basic food is called:
(A) Secondary food (B) Incidental food
(C) Obligatory food (D) Basic food

226. They feed on plant material, which forms more than 75% of gut contents these fishes known as:
(A) Herbivorous fishes (B) Omnivorous fishes
(C) Carnivorous fishes (D) None of these

227. They consume both plant and animal food these fishes are called:
(A) Herbivorous fishes (B) Omnivorous fishes
(C) Carnivorous fishes (D) None of these

228. They feed on animal food, which comprises of more than 80% of the diet such type of fishes is called:
(A) Herbivorous fishes (B) Omnivorous fishes
(C) Carnivorous fishes (D) None of these

229. They consume only phyto- and zooplankton such type fishes are called:
(A) Planktophagous fishes (B) Detritophagous fishes
(C) Both (A) and (B) (D) None of these

230. They feed on detritus:
(A) Planktophagous fishes (B) Detritophagous fishes
(C) Both (A) and (B) (D) None of these

231. Omnivores fishes can also be grouped into :
(A) Herbi-omnivores (B) Cami-omnivores
(C) Both (A) and (B) (D) None of these

232. Fishes feed more on plant material than animal food such type of fishes are called:
(A) Herbi-omnivores (B) Cami-omnivores
(C) Both (A) and (B) (D) None of these

233. Fishes feed more on animal food than plant material such type of fishes are called:
(A) Herbi-omnivores (B) Cami-omnivores
(C) Both (A) and (B) (D) None of these

234. Carnivores fishes are also classified into insectivores which feed:
(A) Feed on insects (B) Feed on crustaceans
(C) Feed on molluscans (D) Feed on other fishes

235. Carcinophagous fishes are also classified into insectivores which feed:
(A) Feed on insects (B) Feed on crustaceans
(C) Feed on molluscans (D) Feed on other fishes

236. Malacophagous fishes are also classified into insectivores which feed:
(A) Feed on insects (B) Feed on crustaceans
(C) Feed on molluscans (D) Feed on other fishes

237. Piscivorous fishes are also classified into insectivores which feed:
(A) Feed on insects (B) Feed on crustaceans
(C) Feed on molluscans (D) Feed on other fishes

238. Larvivorous fishes are also classified into insectivores which feed:
(A) Feed on larva (B) Feed on crustaceans
(C) Feed on molluscans (D) Feed on other fishes

239. Some fishes are:
(A) Cannibalistic (B) Omnibalistic
(C) Larvibalistic (D) None of these

240. Rohu, common carp, silver carp and grass carp feed preferably:
(A) Phytoplankton (B) Zooplankters
(C) Both (A) and (B) (D) None of these

241. Rohu shows preference to chlorophyceae, as also the crustacean preferably:
(A) Phytoplankton (B) Zooplankters
(C) Both (A) and (B) (D) None of these

242. Mrigal feeds on bottom dwellers like:
(A) Plankton (B) Benthos
(C) Detritus (neuston) (D) All of these

243. Shrimp feeds upon:
(A) Zoo phytoplankton (B) Phytoplankton
(C) Both (A) and (B) (D) None of these

244. This technique is frequently used in ponds where shrimps can not be easily removed or concentrated and where the ponds are undrainable is called:
(A) Pond treatment (B) Bath treatment
(C) Dip treatment (D) Flush treatment

245. The method of treatment is effective only in small water bodies, aquaria, cisterns and pools. Moreover, only low concentrations of chernotherapeut can be used as they must be dispersed by natural processes is called:
(A) Pond treatment (B) Bath treatment
(C) Dip treatment (D) Flush treatment

246. However, this is a very effective method of prophylactic treatment of shrimps for external parasites is called:
(A) Pond treatment (B) Bath treatment
(C) Dip treatment (D) Flush treatment

247. This method is useful in culture facilities having sions for rapid flow of water. Alternatively, aquaria, sized plastic or aluminium vessels may also be useful is called:
(A) Pond treatment (B) Bath treatment
(C) Dip treatment (D) Flush treatment

248. In this method the shrimps are placed in a hand net arid dipped into a concentrated solution of chemotherapeutant for one minute or less is known as:
(A) Pond treatment (B) Bath treatment
(C) Dip treatment (D) Flush treatment

249. This method has been found highly effective in treatment of acute diseases, but it may cause additional stress on the affected shrimps is called:
(A) Pond treatment (B) Bath treatment
(C) Dip treatment (D) Flush treatment

250. In this technique, the entire doses of the chemical is added at the inlet and allowed to pass through the flow of water into the pond is called:
(A) Pond treatment (B) Bath treatment
(C) Dip treatment (D) Flush treatment

251. This method is useful where the water supply is contaminated is known as:
(A) Constant flow treatment (B) Feed treatment
(C) Paranteral treatment (D) Tropical treatment

252. The shrimps are constantly exposed to pathogens under these conditions and constant presence of drugs may be necessary to prevent outbreaks is called:
(A) Constant flow treatment (B) Feed treatment
(C) Paranteral treatment (D) Tropical treatment

253. This technique is highly popular to administer drugs to shrimps for systemic infection is called:
(A) Constant flow treatment (B) Feed treatment
(C) Paranteral treatment (D) Tropical treatment

254. Paranteral injections are applicable only in the case of valuable broodstock, berried spawners, etc. is called:
(A) Constant flow treatment (B) Feed treatment
(C) Paranteral treatment (D) Tropical treatment

255. Shrimps and fishes suffer from many external parasite, fungal and bacterial infections, which respond to topical application of drugs is known as:
(A) Constant flow treatment (B) Feed treatment
(C) Paranteral treatment (D) Tropical treatment

256. Fresh iced fish are generally spoiled by:
(A) Bacteria (B) Yeast
(C) Mold (D) None of these

257. Salted and dried fishes are generally spoiled by:
(A) Bacteria (B) Fungi
(C) Both (A) and (B) (D) None of these

258. Which of the following decreases during fish spoilage:
(A) Trimethylamine oxide (B) Creatine
(C) Taurine (D) All of these

259. Which of the following is produced by fish spoilers:
(A) Propanol (B) Ethanol
(C) Isopropanol (D) All of these

260. In detection of microbial fish spoilage, TMAO is reduced into:
(A) TMV (B) TMN
(C) TMA (D) None of these

261. Which of the following is not found in fresh fish:
(A) TMAO (B) TMA
(C) Both (A) and (B) (D) None of these

262. Which of the following is used as fish spoilage indicator:
(A) Histamine (B) Dismine
(C) Total volatiles (D) All of these

263. Which of the following is included in total volatile compound in fish:
(A) TVB (B) TVA
(C) TVN (D) All of these

264. Total volatile acids in fish includes:
(A) Acetic (B) Propionic
(C) Both (A) and (B) (D) None of these

265. Finfish and shell fish are:
(A) Inedible (B) Edible
(C) Both (A) and (B) (D) None

266. Molluse and crustaceans are examples of:
(A) Finfish (B) Shellfish
(C) Both (A) and (B) (D) None of these

267. Oysters, clams, scallops are examples of:
(A) Finfish (B) Mollusks
(C) Crustaceans (D) None of these

268. Lobsters, crabs, shrimp are examples of:
(A) Finfish (B) Mollusks
(C) Crustaceans (D) None of these

269. Examples of fresh water fishes are:
(A) Sardine (B) Mullet
(C) Catfish (D) All of these

270. Lean fishes have fat per cent:
(A) <2 (B) 2·5
(C) 5 (D) >5

271. Medium fishes have fat per cent:
(A) <2 (B) 2·5
(C) 5 (D) >5

272. Fat fishes have fat per cent:
(A) <2 (B) 2·5
(C) 5 (D) >5

273. Example of lean fish are:
(A) Bombay duck (B) Salmon
(C) Sardine (D) Tuna

274. Which of the following is not a lean fish:
(A) Halibut (B) Cod
(C) Perch (D) Salmon

275. Examples of fat fish is:
(A) Sardine (B) Salmon
(C) Tuna (D) All of these

276. Majority of the fish fat is located in:
(A) Liver (B) Pyrolic caecum
(C) Mesenteria (D) All of these

277. Fish proteins are:
(A) Highly digestible (B) Poorly digestible
(C) Non digestible (D) None of these

278. Fish found in the surface and middle water layer of sea is called:
(A) Pelagic (B) Demersal
(C) Both (A) and (B) (D) None of these

279. Fish found in the bottom water layer of sea is called:
(A) Pelagic (B) Demersal
(C) Both (A) and (B) (D) None of these

280. The removal of internal organs of fish is known as:
(A) Gutting (B) Putting
(C) Mutting (D) Cutting

281. The length wise cuts of fin fish is called:
(A) Sticks (B) Steaks
(C) Fillets (D) None of these

282. Cross cut section of fin fish is called:
(A) Fillets (B) Steaks
(C) Sticks (D) None of these

283. Fishes are preserved by:
(A) Drying (B) Salting
(C) Canning (D) All of these

284. The fishes are mainly categorized based on their:
(A) Fat (B) Carbohydrates
(C) Protein (D) Minerals

285. Mass of the egg of fish is called:
(A) Mutt (B) Roe
(C) Caecum (D) None of these

286. The principal component of fish oils is:
(A) Triglycerides (B) Monoglycerides
(C) Diglycerides (D) None of these

287. Majority of fatty acids in the fish are:
(A) Saturated (B) Unsaturated
(C) Hydragenated (D) None of these

288. Fish fat from fish eggs is high in:
(A) Cholesterol (B) Waxes
(C) Lecithin (D) None of these

289. Fish is a ———— source of protein:
(A) Rich (B) Poor
(C) Deficient (D) None of these

290. The protein content of fin fish is:
(A) 10 % (B) 20 %
(C) 30 % (D) 40 %

291. The protein content of shel fish is ———— than fin fish:
(A) Higher (B) Similar
(C) Lower (D) None of thes

292. Sat water fishes have ——— iron than fresh water fishes:
(A) More (B) Less
(C) Similar (D) Unknown

293. The iodine content of ———— fish is exceptionally high:
(A) Fresh water (B) Salt water
(C) Both (A) and (B) (D) None of these

294. The iodine content of salt water fish is exceptionally —— than fresh water fish:
(A) Higher (B) Lower
(C) Similar (D) None of these

295. Omega 3 oils found in fish helps in prevention of:
(A) AMD (B) Scurvy
(C) Goiter (D) Pellagra

296. Shrimp, clams and oysters contain what level of mercury:
(A) High (B) Low
(C) Extremely high (D) None of these

297. Higher level of mercury is found in:
(A) Tuna (B) Lobster
(C) Grouper (D) Cod

298. Fishes are ———— perishable:
(A) Highly (B) Semi
(C) Lesser (D) None of these

299. Shell fish are ———— perishable than finfish:
(A) Less (B) More
(C) Equally (D) None of these

300. Rigor mortis in fish is relatively ——— as compared to other mammals:
(A) Shorter (B) Longer
(C) Similar (D) None of these

301. The fish odour is due to:
(A) Trimethyl amine (B) Oxidation of fat
(C) Both (A) and (B) (D) None of these

302. The iodine content of marine fish per 100 g of fish meat is:
(A) 2-3 mg (B) 1-2 mg
(C) 0.5-1.0 mg (D) 0.01-0.5 mg

303. Iodine is necessary for prevention of:
(A) Scurby (B) Goitre
(C) Rickets (D) Pellagre

304. Fiesh oil is richest source of vitamin:
(A) K (B) D
(C) A (D) Both (A) and (B)

305. Shell fish are ———— in vitamin B than fin fish:
(A) Rich (B) Poor
(C) Deficient (D) None of these

306. Demersal fish are ——— in fat:
(A) Poorly (B) Highly
(C) Either (A) or (B) (D) Neither (A) nor (B)

307. Pelagic fish are ——— in fat:
(A) Poorly (B) Highly
(C) Either (A) or (B) (D) Neither (A) nor (B)

308. Fish flour contain how much per cent protein:
(A) 50-60 (B) 60-70
(C) 70-80 (D) 85-90

309. Fish meal has ——— per cent of moisture:
(A) 6-12 (B) 15-20
(C) 20-25 (D) 25-30

310. Fish meal has ——— per cent of minerals:
(A) 25-30 (B) 20-25
(C) 15-29 (D) 10-12

311. Fish meal has ——— per cent of fat:
(A) 2-5 (B) 5-8
(C) 9-12 (D) 12-15

312. Fish meal has ——— per cent of proteins:
(A) 30-40 (B) 40-50
(C) 50-70 (D) 70-80

313. Broiling, baking and frying are examples of ——— heat treatment:
(A) Moist (B) Dry
(C) Smoking (D) None of these

314. Physchrotrophic bacteria occur in fish during freezing:
(A) *Pseudomonas* (B) *Acinetobacterium*
(C) *Moraxella* (D) All of these

315. Physchrotrophic bacteria occur in fish during freezing:
(A) *Alcaligenes* (B) *Flavobacterium*
(C) Both (A) and (B) (D) None of these

316. Microbial spoilage in fish can be reduced by:
(A) Dry salting (B) Heating
(C) Freezing (D) All of these

317. Maximum limit (g/kg) of sodium ascorbate in grozen fish fillets is:
(A) 1.0 (B) 2.0
(C) 3.0 (D) 4.0

318. Maximum limit (ppm) of calcium sorbate in salted fish is:
(A) 100 (B) 200
(C) 300 (D) 400

319. Maximum limit (ppm) of potassium sorbate in salted fish is:
(A) 100 (B) 200
(C) 300 (D) 400

320. Maximum limit (g/kg) of pectin in canned finfish is:
(A) 2.5 (B) 5.0
(C) 10.0 (D) 20.0

321. Maximum limit (g/kg) of xanthium gum in canned finfish is:
(A) 2.5 (B) 5.0
(C) 10.0 (D) 20.0

322. Maximum limit (g/kg) of sodium alginate in canned finfish is:
(A) 2.5 (B) 5.0
(C) 10.0 (D) 20.0

323. Maximum limit (g/kg) of distarch glycerol in canned finfish is:
(A) 40 (B) 50
(C) 60 (D) 70

324. Maximum limit (g/kg) of distarch phosphate in canned finfish is:
(A) 40 (B) 50
(C) 60 (D) 70

325. Maximum limit (g/kg) of natural flavor in canned finfish is:
(A) 40 (B) 50
(C) 60 (D) GMP

326. A phenomenon marked by the dominance of phytoplankton in a fish pond system due to overgrazing of zooplankton particularly by catla and other such zooplanktivores as bighead carp *(Aristichthys nobilis)* and Thai magur *(Clariasgariepinus)* is known as:
(A) *Ichthyo-eutrophication* (B) *Cyclomorphosis*
(C) Both (A) and (B) (C) None of these

327. A phenomenon where the same organism exhibits a number of morphological characteristics as the seasons change for example, *Daph-nia,* in summer, exhibits a sharppointing head and in winter, a round one is known as:
(A) *Ichthyo-eutrophication* (B) *Cyclomorphosis*
(C) Both (A) and (B) (C) None of these

328. These are found on the body surface, fins and gills eg. *Argulus, Lernaea, Ergasilus,* leaches is called:
(A) Ectoparasites (B) Endoparasites
(C) Both (A) and (B) (C) None of these

329. These are found inside the body of fishes is called:
(A) Ectoparasites
(B) Endoparasites
(C) Both (A) and (B)
(D) None of these

330. These parasites are found in the cells eg. *Microsporidia, Glugia* of fishes is called:
(A) Cytozoic parasites (B) Histozoic parasites
(C) Coenozoic parasites (D) None of these

331. These are found in the tissues of fishes is called:
(A) Cytozoic parasites (B) Histozoic parasites
(C) Coenozoic parasites (D) None of these

332. These are found in the body cavity or inside the alimentary canal of fishes eg. *Diphyllobothrium,* nematodes is known as:
(A) Cytozoic parasites (B) Histozoic parasites
(C) Coenozoic parasites (D) None of these

333. The disease is caused by *Chondroccus columnaris and Cytophaga columnaris* in many freshwater aquarium fish is called:
(A) Columnaris (B) Tail or fin rot
(C) Vibriosis (D) Dropsy

334. The disease of fishes is caused by *Aeromonas salmonicid and A.liquefaciens* is known as:
(A) Columnaris (B) Tail or fin rot
(C) Vibriosis (D) Dropsy

335. Vibrio bacteria are the causative agents of disease in salmon and many other fishes is called:
(A) Columnaris (B) Tail or fin rot
(C) Vibriosis (D) Dropsy

336. *Pseudomonas punctata* is the causative agent of this disease is called:
(A) Columnaris (B) Tail or fin rot
(C) Vibriosis (D) Dropsy

337. The filamentous bacteria, *Flexibacteria* is the causative agent of this disease is called:
(A) Cotton mouth disease
(B) Bacterial gill disease
(C) Saproligniasis disease
(D) Branchiomycosis disease

338. This disease is caused by Myxobacteria in salmon fish:
(A) Cotton mouth disease
(B) Bacterial gill disease
(C) Saproligniasis disease
(D) Branchiomycosis disease

339. This disease is also called as cotton wool or water mould disease:
(A) Cotton mouth disease
(B) Bacterial gill disease
(C) Saproligniasis disease
(D) Branchiomycosis disease

340. This disease is caused by *Saprolignia parasitica.* It is the most common fungus affecting fishes, especially major carps is called:
(A) Cotton mouth disease
(B) Bacterial gill disease
(C) Saproligniasis disease
(D)Branchiomycosis disease

341. This is also called as gill rot:
(A) Cotton mouth disease
(B) Bacterial gill disease
(C) Saprolligniasis disease
(D) Branchiomycosis disease

342. This disease is caused by *Byanchiomyces demigrans* and *B.sanguinis* is called:
(A) Cotton mouth disease
(B) Bacterial gill disease
(C) Saprolligniasis disease
(D) Branchiomycosis disease

343. It is also known as reeling disease:
(A) *Ichthyophonosi* (B) Whirling disease
(C) *Costiasis* (D) *Ichthyophthiriasis*

344. It is characterized by swinging movement of the infected fish. It is caused by *Ichthyophonus hoferi* is known as:
(A) *Ichthyophonosi* (B) Whirling disease
(C) *Costiasis* (D) *Ichthyophthiriasis*

345. This disease is caused by a myxosporidian protozoan, *Myxosoma cerebralis* only in salmon fishes is known as:
(A) *Ichthyophonosi* (B) Whirling disease
(C) *Costiasis* (D) *Ichthyophthiriasis*

346. This is caused by a mastigophore, *Costia necatrix* in culture fishes is called:
(A) *Ichthyophonosi* (B) Whirling disease
(C) *Costiasis* (D) *Ichthyophthiriasis*

347. This disease of fishes is caused by a ciliate, *Ichthyophthirius multifilis*:
(A) *Ichthyophonosi* (B) Whirling disease
(C) *Costiasis* (D) *Ichthyophthiriasis*

348. This disease of fishes is also called as ich or white spot disease:
(A) *Ichthyophonosi* (B) Whirling disease
(C) *Costiasis* (D) *Ichthyophthiriasis*

349. The monogenic trematode, *Dactylogyrus* is reported to cause serious infection in fishes. *D. exitensis, D. vastator and D. lamellatus* are found in carps. These are found on the body, fins and gills is is called as:
(A) *Dactylogyrosis* (B) *Gyrodactylosis*
(C) *Ergasillus* (D) S*almincola*

350. Another monotreme trematode, *Gyrodactylus* also causes disease in culture ponds. This also lives on fins and on the body of the fish is called:
(A) *Dactylogyrosis* (B) *Gyrodactylosis*
(C) *Ergasillus* (D) S*almincola*

351. Leeches like *Piscicola, Myzobdella and Hemiclepsis* hold the skin of the fish and suck fish blood is called:
(A) *Dactylogyrosis* (B) *Gyrodactylosis*
(C) *Leeches disease* (D) S*almincola*

352. These two parasites are responsible for huge mortality of fishes in the culture systems is called as:
(A) *Ergasillus* (B) S*almincola*
(C) Both (A) and (B) (D) None of these

353. Cyanophyceae member, *Oscillatoria* is responsible for fish mortality. It is found on gills and fish body in large numbers and produce toxic substances is known as:
(A) *Ergasillus* (B) S*almincola*
(C) Both (A) and (B) (D) Algal disease

354. Bacteria are responsible for many fatal diseases in fishes like:
(A) Furunculosis (B) Columnaris
(C) Fin or tail rot (D) Vibriosis
(E) Dropsy (F) Cotton mouth
(F) Tuberculosis (G) All of these

355. The fungal diseases in fishes are :
(A) Saproligniasis (B) Branchiomycosis
(C) Ichthyophonosis (D) All of these

356. The protozoan diseases in fishes are :
(A) Whirling disease (B) Costiasis
(C) Ichthyophthiriasis (D) All of these

357. The helminthic diseases in fishes are:
(A) Dactylogyrus (B) Gyrodactylosis.
(C) Both (A) and (B) (D) None of these

358. The crustacean diseases in fishes are :
(A) Argulosis (B) Lernaeasis
(C) Ergasillus (D) Salmincola
(E) All of these

ANSWERSHEET: 21.FISH CULTURE/PISICULTURE

1. (A); 2. (A); 3. (C); 4. (A); 5. (B); 6. (C); 7. (C); 8. (D); 9. (C); 10. (C); 11. (B); 12. (A); 13. (C) ; 14. (B); 15.(A); 16. (C); 17. (B); 18. (D); 19. (A); 20. (C); 21. (B); 22. (B); 23. (C);24. (B); 25. (D) ; 26. (A); 27. (C); 28.(B); 29. (A); 30. (C); 31. (D); 32. (B); 33. (A); 34. (D); 35. (A); 36. (D); 37. (C); 38. (B) ; 39. (A); 40. (A); 41.(B); 42. (C); 43. (D); 44. (A); 45. (B); 46. (C); 47.(A); 48. (A); 49. (B); 50. (A); 51. (C) ; 52. (C); 53. (A); 54.(B); 55. (A); 56. (D); 57. (A); 58. (B); 59. (A); 60. (B); 61.(C); 62. (D); 63. (B); 64. (B); 65. (B); 66. (A); 67. (A); 68.(A); 69. (A); 70. (A); 71. (B); 72. (B); 73. (C); 74. (D); 75. (B); 76. (C); 77. (A); 78. (B) ; 79. (A); 80. (B); 81.(B); 82. (C); 83. (D); 84. (A); 85. (B); 86. (C); 87. (D); 88.(A); 89. (B); 90. (A); 91. (B) ; 92. (B); 93. (A); 94.(C); 95. (A); 96. (B); 97. (C); 98. (A); 99. (A); 100. (A); 101.(G); 102. (C); 103. (A); 104. (B) ; 105. (C); 106. (D); 107. (A); 108. (C); 109. (B); 110. (B); 111. (D); 112. (E); 113. (C); 114. (D); 115. (A); 116. (B); 117.(C) ; 118. (A); 119. (B); 120. (A); 121. (B); 122. (C); 123.(D); 124. (A); 125. (A); 126. (E); 127. (B); 128. (C); 129. (D); 130. (A) ; 131. (B); 132. (C); 133. (D); 134. (A); 135.(A); 136. (B); 137.(C); 138. (D); 139. (A); 140. (B); 141. (C); 142. (D); 143. (A) ; 144. (B); 145. (A); 146. (D); 147.(C); 148. (A); 149. (B); 150. (C); 151. (D); 152. (A); 153. (B); 154. (B); 155. (A); 156. (D); 157. (A); 158. (C); 159. (A); 160. (B); 161. (B); 162. (C); 163. (D); 164. (B); 165. (F); 166. (E); 167. (A); 168. (B); 169. (A) ; 170. (A); 171. (B); 172. (A); 173. (B); 174. (C); 175. (D); 176. (A); 177. (A); 178. (A); 179. (A); 180. (A); 181. (D); 182. (A) ; 183. (A); 184. (B); 185. (C); 186. (C); 187. (B); 188.(A); 189. (A); 190. (B); 191. (B); 192. (A); 193. (D); 194.(B); 195. (B); 196. (A); 197. (B); 198. (C); 199. (D); 200.(D); 201. (B); 202. (A); 203. (C); 204. (D); 205. (D); 206.(D); 207. (G); 208. (E) ; 209. (D); 210. (C); 211. (A); 212.(C); 213. (A); 214. (A); 215. (A); 216. (B); 217. (F); 218. (C); 219. (A); 220. (B); 221. (C) ; 222. (D); 223. (A); 224.(B); 225. (C); 226. (A); 227. (B); 228. (C); 229. (A); 230. (B); 231. (C); 232. (A); 233. (B); 234. (A); 235. (B); 236. (C); 237. (D); 238. (A); 239. (A); 240. (A); 241. (B); 242. (D); 243. (C); 244. (A); 245. (A); 246. (A); 247. (B) ; 248. (C); 249. (C); 250. (D); 251. (A); 252. (A); 253. (B); 254. (C); 255. (D); 256. (A); 257. (B); 258. (D); 259.(D); 260. (C) ; 261. (B); 262. (D); 263. (D); 264. (C); 265.(B); 266. (B) ; 267. (B); 268. (C); 269. (D); 270. (A); 271.(B); 272. (D); 273. (A) ; 274. (D); 275. (D); 276. (D); 277.(A); 278. (A); 279. (B); 280. (A); 281. (C); 282. (B); 283. (D); 284. (A); 285. (B); 286. (A) ; 287. (B); 288. (C); 289. (A); 290. (B) ; 291. (C); 292. (A); 293. (B); 294.(A); 295. (A); 296. (B); 297. (C); 298. (A); 299. (B) ; 300.(A); 301. (B); 302. (D); 303. (B); 304. (C); 305. (A); 306. (A); 307. (B); 308. (D); 309. (A); 310. (D); 311. (A); 312. (C) ; 313. (B); 314. (D); 315. (C); 316. (D); 317.(A); 318.(B); 319. (B); 320. (A); 321. (D); 322. (D); 323.(C); 324. (C); 325. (D) ; 326. (A); 327. (B); 328. (A); 329.(B); 330. (A); 331. (B); 332. (C); 333. (A); 334. (B); 335.(C); 336. (D); 337. (A); 338. (B) ; 339. (C); 340. (C); 341.(D); 342.(D); 343. (A); 344. (A); 345. (B); 346. (C); 347.(D); 348. (D); 349. (A); 350. (B); 351. (C) ; 352. (C); 353.(D); 354.(G); 355. (D); 356. (D); 357. (C); 358. (E).

APPENDIX

ICAR RESEARCH INSTITUTES

AGRICULTURAL ENGINEERING

1. CIAE	Central Institute of Agricultural Engineering, Nabi Bagh, Berasia road, Bhopal- 462038 (MP)
2. CIPHET	Central Institute of Post Harvest Engineering & Technology, Ludhiana- 141004 (Punjab)
3. CICT	Central Institute of Cotton Technology, P.B. No. 16640, Mumbai – 400019 (Maharashtra)
4. ILRI	Indian Lac Research Institute, Ranchi- 834010 (Jharkhand)
5. NIRJAFT	National Institute of Research on Jute & Allied Fibre Technology, Ranikuthi, Kolkata -700120 (W.B.)

AGRICULTURE EXTENSION

1. NRCWA	National Research Centre for Women in Agriculture, 1199 Jagmara, Bhubneshwar-751030 (Orrisa)
2.DIPA	Directorate of Information and Publication of Agriculture, KAB-I, Pusa, New Delhi-110012

ANIMAL SCIENCES

1. CARI	Central Avian Research Institute, Izzatnagar , Bareili-243122(U.P)
2. CIRB	Central Institute for Research on Buffaloes, Sirsa Road, Hisar- 125001 (Haryana)
3. CIRG	Central Institute for Research on Goats, Makhdoom, Mathura-281122 (U.P)
4. IVRI	Indian Veterinary Research Institute, Izzatnagar, Bareili-243122(U.P)
5. NAINP	National Institute of Animal Nutrition & Physiology, Adugodi, Bangalore-560030 (Karnataka)
6. CSWRI	Central Sheep and Wool Research Institute, Avikanagar- 304501 (Rajasthan)
7. NBAGR	National Bureau of Animal Genetic Resources, P.B. No. 129, Kernal-132001 (Haryana)
8. NDRI	National Dairy Research Institute, Karnal-132001(Haryana)
9. NRCC	National Research Centre on Camel, Jobber, P.B. No. 07, Bikhaner-334001 (Rajasthan)
10. NRCM	National Research Centre for Mithun, Jharnapani, Medziphema, Nagaland-500059 (A.P.)
11. NRCM	National Research Centre on Meat, Saidabad, CRIDA campus, Hyderabad- 500059 (A.P.)
12. NRCY	National Research Centre on Yak, Dirang, West Kameng-790101 (Arunachal Pradesh)
13. NRCP	National Research centre on Pigs, Panjabari Road, 6th mile, Guwahati-785037 (Assam)
14. PDADMS	Project Directorate on Animal Disease Monitoring and Surv`llance, Hebbal, Bangalore-560024 (Karnataka)
15. PDC	Project Directorate on Cattle, Grass farm Road, P.B. No. 17, Meerut Cant, Meerut- 250001 (U.P.)
16. PDFMD	Project Directorate on Foot and Mouth Disease, IVRI, Mukteshwar, Kumaon- 263138 (Uttaranchal)
17. PDP	Project Directorate on Poultry, Hyderabad-500030(A.P.)
18. NRCE	National Research Centre for Equines, Hisar - 125001

CROP SCIENCES

1. CICR	Central Institute for Cotton Research, P.B. No. 2, Nagpur-440010 (Maharashtra)
2. CRIJAF	Central Research Institute for Jute & Allied Fiber, Nilganj, Barrackpore-700120 (W.B.)
3. CRRI	Central Rice Research Institute, Cuttak, 753006 (Orrisa)
4. CTRI	Central Tobacco Research Institute, Rajahmundry-533105(A.P.)
5. DMR	Directorate of Maize Research, IARI Pusa Campus, New Delhi -110012
6. DOR	Directorate of Oilseed Research, Rajendranagar, Hyderabad-500030 (A.P.)
7. DRR	Directorate of Rice Research, Hyderabad-500030 (A.P.)
8. DWR	Directorate of Wheat Research, PB No.158, Karnal -132001 (Haryana)
9. IARI	Indian Agricultural Research Institute, New Delhi-110012
10. IGFRI	Indian Grassland and Fodder Research Institute, Jhansi-284003 (U.P.)
11. IIPR	Indian Institute of Pulses Research, Kanpur-208024
12. IISR	Indian Institute of Sugarcane Research, Lucknow-226602
13. NBAIMO	National Bureau of Agriculturally Important Microorganisms, Kusmaur, Mau Nath Bhanjan-275101 (U.P.)
14. NBPGR	National Bureau of Plant Genetic Resources, New Delhi-110012
15. NCIPM	National Centre for Integrated Pest Management, LBS Building, Pusa campus, New Delhi -110012
16. NRPB	National Research Centre on Plant Biotechnology, Pusa campus, New Delhi-110012
17. DGR	Directorate of Groundnut Research, PB No. 5, Ivnagar Road, Junagarh-362001 (Gujarat)
18. DRMR	Directorate of Rapeseed and Mustard Research, Bharatpur - 321303 (Rajasthan)
19. DSR	Directorate of Sorghum Research, Rajendranagar, Hyderabad-500030 (A.P.)
20. DSR	Directorate of Soybean Research, Indore-452017 (M.P.)
21. NABII	National Bureau of Agriculturally Important Insect, PB.No. 2491, Hebbal, Banglore-560024 (Karnataka)
22. VPKAS	Vivekanand Parvatiya Krishi Anusandhan Sansthan, Almora-263601 (Uttaranchal)

23.DSR Directorate of Seed Research, P.B. No. 11, Village Kushmour, P.O. Kaithuli, Mau – 275901
24. SBI Sugarcane Breeding Institute, Coimbatore – 641007

FISHERIES

1. CIBA Central Institute for Brackish Water Aquaculture, 75 Santhome High Road, RA Puram, Chennai-600028 (T.N.)
2. CIFA Central Institute for Fresh Water Aquaculture, Kausalyaganga, Bhubneshwar-751002 (Orissa)
3. CIFA Central Institute for Fisheries Education, Mumbai-400061 (Maharashtra)
4. CIFRI Central Inland Fisheries Research Institute, Barrackpore- 700120 (W.B)
5. CIFT Central Institute of Fisheries Technology, Cochin-682029 (Kerala)
6. CMFRI Central Marine Fisheries Research Institute, PB.No. 1603, Tatpuram P.O. Kochi-682018 (Kerala)
7. NBFGR National Bureau of Fish Genetic Resources, P.O. Dilkusha, Lucknow-226002 (U.P)
8. DCFR Directorate of Coldwater Fisheries Research, Saurabh Cottage, Bhimtal, Nainital-263136 (Uttaranchal)

HORTICULTURE

1. CARI Central Agricultural Research Institute, PB No. 181, Port Blair- 744101(A&N Island)
2. CIAH Central Institute of Arid Horticulture, Sri Ganganagar Highway, Bikaner-334006(Rajasthan)
3. CISH Central Institute for Subtropical Horticulture, Rehmankheda, PO Kakori, Lucknow-227107 (U.P)
4. CITH Central Institute of Temperate Horticulture, Old Air Field, Rangreth-190007 (J&K)
5. CPCRI Central Plantation Crop Research Institute, Kasargod-671124 (Kerela)
6. CPRI Central Potato Research Institute, Shimla-171001 (H.P.)
7. CTCRI Central Tuber Crops Research Institute, Sreekariam, Thiruvanthapuram-695017 (Kerela)
8. IIHR Indian Institute of Horticultural Research, Hassarghatta Lake post, Bangalore-560089 (KN)
9. IISR Indian Institute of Spices Research, PB.No. 1701, Marikunnu PO, Kozhikode-673012 (Kerela)
10. IIVR Indian Institute of Vegetable Research, PB No. 01, PO.Jakini, Sahansapuram, Varanasi- 221305 (U.P.)
11. NRCB National Research Centre for Banana, Thogamalai Main Road, Thayar PO Thiruchirapalli -620102 (TN)
12. DCR Directorate of Cashew Research, Puttur-574202 Dakshina Kannada (KN)
13. NRCC National Research Centre for Citrus, PB No. 464 Shankarnagar PO, Nagpur-440010
14. NRCSS National Research Centre for Seed Spices, Tabiji, Ajmer-305206 (Rajasthan)
15. NRCG National Research Centre for Grapes, P B No. 3, Manjri Farm Post, Solapur Road, Pune-412307 (MH)
16. NRCL National Research Centre for Litchi, Manchi House, Club Road, Muzaffarpur-842002 (Bihar)
17. NRCM National Research Centre for Mushroom, Chambaghat, Solan-173213 (H.P)
18. DOR Directorate of Oil Palm Research, Near Jawahar Navodaya Vidyalaya Pedavegi-534450, West Godavari dist. (A.P)
19. NRCO National Research Centre for Orchids, Pakyang, Sikkim- 737106
20. NRCOG National Research Centre for Onion & Garlic, Ragurunagar-410505, Pune (MH)

MANAGEMENT AND STATISTICS

1. IASRI Indian Agricultural Statistics Research Institute, Library Avenue, Pusa, New Delhi -110012
2. NAARM National Academy of Agricultural Research and Management, Hyderabad-500030 (A.P.)
3. NCAEPR National Centre for Agricultural Economics & Policy Research, Library Avenue, Pusa, New Delhi-110012

NATURAL RESOURCES MANAGEMENT

1. CAZRI Central Arid Zone Research Institute, Jodhpur-342003 (Rajasthan)
2. CRIDA Central Research Institute for Dryland Agriculture, Santhoshnagar, Hyderabad-500059 (A.P.)
3. CSSRI Central Soil Salinity Research Institute, Kernal- 132001 (Haryana)
4. CSWCRTI Central Soil & Water Conservation Research & Training Institute, 218 Kaulagarh Road, Dehradun -248195 (Uttaranchal)
5. ICARRCE ICAR Research Complex of Eastern Region, Walmi Complex, PO Phulwari Sharif, Patna-801505 (Bihar)
6. ICARRCG ICAR Research Complex for Goa, Ela , Old Goa-403402
7. ICARRCNEH ICAR Research Complex for NEH Region, Barapani-793103 (Meghalaya)
8. IISS Indian Institute of Soil Science, Nabi bagh, Berasia Road, Bhopal-462038 (M.P.)
9. NBSSLUP National Bureau of Soil Survey and Land Use Planning, Nagpur-440010 (MH)
10. NRCAF National Research Centre for Agroforestry, Near Pahujdam, Jhansi-284003 (UP)
11. DWRS Directorate of Weed Science Research, Maharajpur, Adhartal, Jabalpur-482004 (M.P.)
12. PDFSR Project Directorate on Farming Systems Research, Meerut-250110

13. DWM — Directorate of Water management, Chandra Sekharpur, PO SE Railway, Project Complex, Bhubneswar-751023 (Orissa)
14. NIASM — National Institute of Abiotic Stress Management, Malegaon, Baramati – 413115 (Pune- Maharashtra)

OTHER IMPORTANT NATIONAL ORGANISATIONS

1. APEDA — Agricultural and Processed Food Export Development Authority, NCU 1 Building, 3 Siri Industrial Area , August Kranti marg, Hauzkhas, New Delhi-110016
2. BARC — Bhabha Atomic Research Centre, Trombay, Mumbai
3. CFTRI — Central Food Technological Research Institute, Mysore
4. CIMAP — Central Medicinal and Aromatic Plants, Lucknow
5. CPPTI — Central Plant Protection Training Institute, Hyderabad
6. CSIR — Council of Scientific and Industrial Research, New Delhi
7. CSMCRI — Central Salt and Marine Chemical Research Institute, Bhavnagar
8. FRI — Forest Research Institute, Dehradun
9. IAS — Indian Academy of Science, Pune
10. ICFRI — Indian Council of Forestry Research and Education, Dehradun
11. ICM — Institute of Cooperative Management, Ahmedabad
12. IFGTB — Institute of Forest Genetics and Tree Breeding, Coimbatore
13. IGNOU — Indira Gandhi National Open University, New Delhi
14. IIM — Indian Institute of Management, Ahmedabad
15. INSA — Indian National Science Academy, Bahadur Shah Zafar Marg, NewDelhi
16. IWST — Institute of Wood Science & Technology, Banglore
17. JARI — Jute Agricultural Research Institute, Barrackpore
18. JTRI — Jute Technology Research Institute, Kolkata
19. MANAGE — National Institute of Agricultural Extension Management, Hyderabad
20. — MS Swaminathan Research Foundation, 3rd Cross Street, Taramani Institutional Area, Chennai-600133(TN)
21. NABARD — National Bank of Agriculture and Rural Development
22. NAFED — National Agricultural Cooperative Marketing Federation of India Ltd., NAFED House, Siddartha Enclave, Asharm Chowk , New Delhi
23. NBRI — National Botanical Research Institute, Lucknow
24. NCDC — National Cooperative Development Corporation, New Delhi
25. NCERT — National Council of Education Research & Training, New Delhi
26. NCL — National Chemical Laboratory, Pune
27. NDDB — National Dairy Development Board, Safadarjung Enclave, New Delhi
28. NIAG — National Institute of Animal Genetics, Kernal (Harayana)
29. NIN — National Institute on Nutrition, Hyderabad
30. NIRD — National Institute for Rural Development, Hyderabad
31. NOVDB — National Oil Seed & Vegetable Development Board, New Delhi
32. NPL — National Physical Laboratory, New Delhi
33. NRL — Nuclear Research Laboratory, IARI, New Delhi
34. NSC — National Seed Corporation, Bee Bhawan, Pusa, New Delhi
35. NSI — National Sugar Institute, Kalyanpur, Kanpur
36. PPII — Patash Research Institute of India, Gurgaon (Haryana)
37. RSAC — Remote Sensing Application Centre, Lucknow
38. UGC — University Grant Commission, Bahadur Shah Zafar Marg, New Delhi
39. WLMTRI — Water and Land Management Training Research Institute, Hyderabad
40. WTC — Water Technology Centre, IARI, New Delhi

INTERNATIONAL ORGANISATIONS

1. AIT — Asian Institute of Technology, P.O. Box 2734, Bankok, Thailand-10501
2. AVRDC — Asian Vegetable Research and Development Centre, P.O.Box 42 , Shanhua, Taiwan 74199, Taiwan
3. BARC — Bangladesh Agricultural Research Council, Dhaka
4. CGIAR — Consultative Group on International Agriculture Research, Washington (USA)
5. CIAT — Centro International de Agricultural Tropical (International Centre for Tropical Agriculture, Coli, Columbia, South Africa
6. CIDA — Canadian International Development Agency, Cuba
7. CIMMYT — Centro International deMajormiento de Maizy Trigo (International Centre for Improvement of Maize, Wheat and Barley),
8. CIP — International Potato Centre (Centro Internationally de la Papa), Peru, Mexico
9. CSIRO — Commonwealth Scientific and Industrial Research Organization, Near Perth, Australia
10. FAO — Food and Agriculture Organization, Rome, Italy

11. IBPGR	International Plant and Genetic Resources, Italy
12. IBSNAT	International Benchmark Soil Network for Agrotechnology Transfer, Hawaii
13. ICARDA	International Centre for Agricultural Research for Dryland Areas, Syrea
14. ICRAF	International Centre for Research in Agroforestry, Nairobi, Kenya
15. ICRISAT	International Crop Research Institute for the semi Arid Tropics, Patancheru, Hyderabad
16. IDRC	International Development Research Centre, Canada
17. IFAD	International Fund for Agriculture Development, Rome
18. IFDC	International Fertilizer Development Centre, Muscale Shoala, Alobama (USA)
19. IFPRI	International Food Policy Research Institute, Washington DC (USA)
20. IIASA	International Institute for Applied Systems Analysis, Luxemberg, Vienna, Austria
21. IITA	International Institute for Tropical Agriculture, Ibadan, Nigeria
22. IIASA	International Institute for Applied Systems Analysis, Luxemberg, Vienna, Austria
23. INSFFER	International Network on Soil Fertility and Fertilizer Evaluation of Rice, Manila, Philippines
24. INSNAR	International Service for National Agricultural Research, The Hague, Netherland
25. IPGRI	International Plant Genetic Research Institute
26. IRRI	International Rice Research Institute, Manila Philippines
27. ISRIC	International Soil Reference and Information Centre, Netherland
28. IWMI	International Water Management Institute
29. LURI	Land Use Research Institute, Aberdeen, Scotland
30. NIAS	National Institute on Agricultural Science, Japan
31. PPIC	Potash and Phosphate Institute, Canada
32. TPRI	Tropical Pesticides Research Institute, Arushe, Tanzania
33. WARDA	West Africa Rice Development Association, Manrovia, Liberia
34. WMO	World Meteorological Organization, Geneva, Switzerland

STATE AGRICULTURAL UNIVERSITIES

1. AAU	Anand Agricultural University, Anand (Gujarat)
2. AAU	Assam Agricultural University, Jorhat (Assam)
3. ANGRAU	Acharya NG Ranga Agricultural University, Rajendranagar, Hyderabad (A.P.)
4. BAU	Birsa Agricultural University, Kanke, Ranchi (Jharkhand)
5. BCKV	Bidhan Chandra Krishi Vishwavidhalaya, Mohanpur, Nadia (WB)
6. BSKKV	Dr. Babasaheb Sawant Kokan Krishi Vidyapeeth, Dapoli, Ratnagiri, (Maharashtra)
7. CAU	Central Agricultural University, P O Box No. 23, Imphal, (Manipur)
8. CCSHAU	Chaudhary Charan Singh Haryana Agricultural University, Hisar (Haryana)
9. CSAUA&T	Chandra Shekhar Azad University of Agriculture & Technology, Kanpur, (U.P.)
10. CSKKV	Ch. Sarwan kumar Krishi Vishwavidyalaya, Palampur (HP)
11. CVV&GAS	U.P. Pandit Deen Dayal Upadhyaya Pashu Chiktsha Vigyan Vishwavidhayala evam Go Anushandhan Sansthan, Mathura, (U.P.)
12. GBPUA&T	Gobind Ballabh Pant University of Agriculture & Technology, Pantnagar- 263145, Udam Singhnagar (Uttaranchal)
13. IGKV	Indira Gandhi Krishi Vishwavidyalaya, Krishak negar, Raipur-492012 (Chhattisgarh)
14.JAU	Junagarh Agricultural University, Junagarh-362001 (Gujarat)
15. JNKVV	Jawaharlal Nehru Krishi Vishwa Vidyalaya, Jabalpur – 482004 (MP)
16.KAU	Kerala Agricultural University, Vellanikara, Trichur-680656 (Kerala)
17. MAS&FU	Maharastra Animal Science & Fishery University, Seminary Hillls, Nagpur-440006 (Maharashtra)
18.MAU	Marathwada Agricultural University, Parbhani-431 402 (Maharastra)
19.MPKV	Mahatma Phule Krishi Vidyapeeth, Rahuri-413 402 (Maharastra)
20. MPUA&T	Maharana Pratap University of Agriculture & Technology, Udaipur-3133001 (Rajasthan)
21. NAU	Navsari Agricultural University, Navsari (Gujarat)
22.NDAU&T	Narendra Deo University of Agriculture & Technology, Kumar Ganj, Faiza bad-224 229 (UP)
23.OUA&T	Orissa University of Agriculture & Technology, Bhubneshwar-751003 (Orissa)
24.PAU	Punjab Agriculture University, Ludhiana -141004 (Punjab)
25.PDKV	Dr. Panjabrao Deshmukh Krishi Vidyapeeth, Krishinagar, Akola-444 104 (Maharashtra)
26.RAU	Rajasthan Agriculture University, Bikaner-33400 (Rajasthan)
27. RAU	Rajendra Agriculture University, Pusa, Samastipur-848 125 (Bihar)
28. SBBPUA&T	Sardar Ballabh Bhai Patel University of Agriculture & Technology, Modipuram, Meerut-250 110 (UP)
29. SKDAU	Sardar Krushinagar Dantiwada Agricultural University, Sardar Krushinagar, Banskantha-385506 (Gujarat)
30. SKUAC&T	Sher-E-Kashmir University of Agricultural Sciences & Technology, Jammu -188812 (J&K)
31. SKUAC&T	Sher-E-Kashmir University of Agricultural Sciences & Technology, Kashmir -191121 (J&K)
32.TAU	Tamilnadu Agricultural University, Coimbattor-641003 (Tamilnadu)
33. TV&ASU	Tamilnadu Veterinary & Animal Sciences University, Chennai-600051 (Tamilnadu)

34. UAS	University of Agricultural Sciences GKVK Bangalore-560065 (Karnataka)
35. UAS	University of Agricultural Sciences, Dharwad-580005 (Karnataka)
36. UBKV	Uttar Banga Krishi Vishwavidyalaya, Pundibari, Cooch behar – 736 165 (WB)
37. WUA&FS	West Bengal University of Animal & Fishery Sciences, 68 K Sahani, Kolkata-700037 (WB)
38. YSPUH&F	Dr. Yashwant Singh Parmar University of Horticulture & Forestry, Solan, Nauni-173 230 (HP)
39. UHF	University of Horticulture and Forestry, Pauri, Uttarakhand
40. MKJUAT	Manyawar Kansiram Ji University of Agriculture and Technology, Banda (UP)
41. SVVU	Sri Venkateswara Veterinary University, Trupati – 517502 (AP)
42. APUH	Andhra Pradesh University of Horticulture, Venkataramanagudem, West Godawari – 534001 (AP)
43. RVSKV	Rajmata Vijayaraje Scindhia Krishi Vishwavidyalaya, Gwalior – 474002 (MP)
44. MARFSU	Maharashtra Animal & Fisheries Sciences University, Nagpur – 4400006
45.GADVASU	Guru Angad Dev Veterinary And Animal Science University, Ludhiana – 141004
46.SHIATS	Sam Higginbottom Institute of Agriculture Technology and Sciences, Naini, Allahabad – 211007
47. KUFOS	Kerala University of Fisheries and Ocean Studies, Pananagad, Kochi – 682506
48. KV&ASU	Kerala Veterinary and Animal Sciences University, Pookot, Wayanand – 673576
49.LLRUVAS	Lala Lajpat Rai University of Veterinary & Animal Sciences, Hisar (Haryana)
50.RUV&AS	Rajasthan University og Veterinary and Animal Sciences, Bijay Bhawan Palace Complex, Near Pt. Deendayal Circle, Bikaner – 334006 (Rajasthan)
51. BAU	Bihar Agriculture University, Sabour, Bhagalpur – 813210 (Bihar)
52. MPPCVV	Madhya Pradesh Pashu Chikitsa Vigyan Vishwavidhyalaya, Jabalpur – 482004 (MP)
53.UHS	University of Horticultural Sciences, Sector-60, Navanagar, Bagalkot – 587102
54. UAS	University of Agriculture Science, Raichur – 584102 (Karnataka)

ICAR DEEMED AGRICULTURAL UNIVERSITY

IARI	Indian Agricultural Research Institute, New Delhi – 110012
IVRI	Indian Veterinary Research Institute, Izatnagar, Berrreili - 243122 (UP)
NDRI	National Dairy Research Institute, Kernal – 132001 (Haryana)
CIFE	Centrral Institute of Fisheries Education, Mumbai – 400061

Central Universities related to Agriculture discipline

AMU	Aligarh Muslim University, Aligarh
BHU	Banaras Hindu University, Varanasi
VISWA	Viswa Bharti (Palli Siksha Bhawan), P.O. Sriniketan – 731236
NU	Nagaland University (School of Agricultural Sciences and Rural Development), Modziphema - 797106

ABBREVIATIONS/ACRONYMS

A3P	Accelerated Pulse Production Programme
ACIAR	Australian Council on International Agriculture Research
AGRIS	International Information System for Agricultural Science and Technology
AICRP	All India Coordinated Research Project
AOAC	Association of Official Analytical Chemists
AP Cess Fund	Agricultural Product Cess Fund
APAS	Andhra Pradesh Academy of Science
APEDA	Agriculture & Processed Products Export Authority
APWSS	Asian Pacific Weed Science Society
ARIC	Agricultural Research Information Centre
ARS	Agricultural Research Services
ASA	American Society of Agronomy
ASRB	Agricultural Scientist Recruitment Board
ASSOD	Assessment of the Status of Human –included Soil Degradation
ATIC	Agricultural Technology Information Centre
ATMA	Agriculture Technology Management Agency
BHC	Benzene Hexa chloride
CAB	Commonwealth Agricultural Bureau
CCIS	Comprehensive Crop Insurance Scheme
CDD	Community Development Directorate
CEP	Central Exchange Programme
CFB	Corrugated Fibre Board
CIDA	Canadian International Development Agency
CSC	Central Seed Committee
CSSA	Crop Science Society of America
CVRC	Central Variety Release Committee

DAP	Di Ammonium Phosphate
DARE	Department of Agricultural Research and Education
DBT	Department of Bio Technology
DDP	Desert Development Programme
DDT	Dichloro Diphenyle Trichloro Ethane
DPAP	Drought Prone Area Programme
DPD	Directorate of Pulses Development
DST	Department of Science and Technology
EAS	Employment Assurance Scheme
EDB	Ethylene Di Bromide
EDCIL	Educational Development Consultant India Ltd
ESCAP	Economic and Social Commission for Asia and the Pacific
FAI	Fertilizer Association of India
FAO	Food and Agriculture Organisation
FCI	Food Corporation of India
FDA	Food and Drug Adulteration Act
FPC	Fertilizer Promotion Campaign
FPTC	Food Processing and Training Centre
FRC	Food Research Council
FSC	Farm Science Centre
FYM	Farm Yard Manure
GLASOD	Global Assessment of the Status of Human-included Soil Degradation
GOI	Government of India
GSFC	Gujarat State Fertilizer Corporation Ltd
HCN	Hydro Cyanic Acid
HETA	Horticulture Experiment & Training Centre
HIL	Hindustan Insecticide Limited
HSDC	Haryana Seed Development Corporation
HYVP	High Yielding Variety Programme
IAAP	Intensive Agriculture Area Programme
IADP	Intensive Agriculture District programme
IARC	International Agricultural Research Centre
IBRD	International Bank for Reconstruction and Development
ICDS	Integrated Child Development Scheme
ICID	International Commission on Irrigation and Drainage
IEL	Indian Explosive Limited
IFFCO	Indian Farmers Fertilizers Corporation Ltd
IFPC	Intensive Fertilizer Promotion Campaign
IFWA	International Federation for Woman Development
IJO	International Jute Organisation
ILCA	International Livestock Centre for Africa
IRDC	International Research and Development Council
IRDP	Integrated Rural Development Programme
IRTP	International Rice Testing Programme
ISA	Indian Society of Agronomy
ISOPOM	Integrated Scheme for Oilseeds, Pulses, Oilpalm and Maize
ISRIC	International Soil Reference and Information Centre
ISSS	Indian Society of Soil Science
ISSS	International Soil Science Society
ISTS	Indian Society of Tobacco Science
ITEC	Indian Technical and Economic Corporation
IUFRO	International Union of Forest Research Organisation
IWSS	Indian Weed Science Society
KGK	Krishi Gyan Kendra
KRIBHCO	Krishak Bharti Corporation Limited
KVK	Krishi Vigyan Kendra
LCU	Lac Coated Urea
LEISA	Low External Input and Sustainable Agriculture
LLP	Lab to Land Programme
LTPC	Long Term Programme of Cooperation
MCFC	Maharastra Cooperative Fertilizer and Chemical Limited
MEPDA	Marine Product Export Development Authority

MFAL	Marginal Farmers and Agricultural Labour Programme
MGREGS	Mahatma Gandhi Rural Employment Guarantee Scheme
MIS	Management Information System
NAAS	National Academy of Agriculture Science
NACA	Network of Aquaculture Centre in Asia –Pacific
NAFED	National Agriculture Cooperative Marketing Federation of India Limited
NAIP	National Agriculture Innovative Project
NARP	National Agricultural Research Project
NARS	National Agricultural Research System
NATP	National Agricultural Technology Project
NCA	National Commission on Agriculture
NCAER	National Council of Applied Economic Research
NCDC	National Cooperative Development Corporation
NCIPM	National Centre for Integrated Pest Management
NCL	National Chemical Laboratory
NCU	Neem coated Urea
NDDB	National Dairy Development Board
NDP	National Development Programme
NET	National Eligible Test
NFL	National Fertilizer Limited
NFSM	National Food Security Mission
NGO	Non Government Organisation
NIPMA	North Indian Pesticide Manufacture Association
NICRA	National Initiative on Climate Resilient Agriculture
NOVDB	National Oilseed and Vegetable Development Board
NPC	National Productivity Council
NPL	National Physical Laboratory
NRDC	National Research Development Corporation
NREGA	National Rural Employment Guarantee Agency
NREP	National Rural Employment Programme
NSKS	Neem Kernel Seed Extract
NWDPRA	National Watershed Development Project for Rainfed Area
OFC	Orissa Fertilizer and Chemical Limited
ORP	Oilseeds Research Project
PARC	Pakistan Agriculture Research Council
PCARR	Philippines Council of Agriculture and Resources Research
PCARRD	Philippines Council for Agriculture Forestry and Natural Resources Research and Development
PCNB	Penta Chloro Nitro Benzene
PIRCOM	Project for Intensification of Regional Research on Cotton, Oilseeds and Millets
PPCL	Pyrites Phosphate and Chemical Limited
PRA	Participatory Rural Appraisal
PRAI	Planning Research and Action Institute
PSSC	Punjab State Seed Corporation
PTD	Participatory Technology Development
RAPI	Rural Agro Processing Industries
RKVY	Rastriya Krishi Vikas Yojna
RLEGP	Rural Landless Employment Guarantee Programme
RNE	Renewable Natural Energy
RRA	Rapid Rural Appraisal
SAARC	South Asian Association for Regional Cooperation
SADCC	Southern African Development Coordination Conference
SAMETI	State Agricultural Management and Extension Training Institute
SAU	State Agricultural Universities
SFCI	State Farms Corporation of India
SFDA	Small Farmers Development Agency
SIDA	Swedish International Development Agency
SISSTA	South Indian Sugarcane and Sugar Technology Association
SITE	Satellite Instructional Television Experiment
SOTER	Soils and Terrain Digital Database
SRI	System of Rice Intensification
SSDC	State Seed Development Corporation
SSP	Single Super Phosphate

SURE	Society to Uplift Rural Economy
SVRC	State Variety Release Committee
TAS	Time Allocation Study
TDC	Terai Development Corporation
TMO	Technology Mission on Oilseeds
TRIMS	Trade Related Investment Measures
TRIPS	Trade Related Intellectual Property Rights
TRYSEM	Training of Rural Youth Self Employment Programme
UGC	University Grand Commission
UNDP	United Nations Development Programme
UNEP	United Nations Environment Programme
UNESCO	United Nations Educational Scientific and Cultural Organisation
UNICEF	United Nations International Children Emergency Fund
UPCAR	Uttar Pradesh Council of Agricultural Research
USAID	United States Agency for International Development
USDA	United State Department of Agriculture
USIF	United State Indian Fund
WAPCCS	Water and Power Consultative Service India Limited
WMO	World Meteorological Organisation